Lecture Notes in Artificial Intelligence 4682

Edited by J. G. Carbonell and J. Siekmann

Subseries of Lecture Notes in Computer Science

De-Shuang Huang Laurent Heutte
Marco Loog (Eds.)

# Advanced Intelligent Computing Theories and Applications

## With Aspects of Artificial Intelligence

Third International Conference on
Intelligent Computing, ICIC 2007
Qingdao, China, August 21-24, 2007
Proceedings

Springer

Series Editors

Jaime G. Carbonell, Carnegie Mellon University, Pittsburgh, PA, USA
Jörg Siekmann, University of Saarland, Saarbrücken, Germany

Volume Editors

De-Shuang Huang
Chinese Academy of Sciences
Institute of Intelligent Machines, China
E-mail: dshuang@iim.ac.cn

Laurent Heutte
Université de Rouen
Laboratoire LITIS
76800 Saint Etienne du Rouvray, France
E-mail: Laurent.Heutte@univ-rouen.fr

Marco Loog
University of Copenhagen
Datalogical Institute
2100 Copenhagen Ø, Denmark
E-mail: loog@diku.dk

Library of Congress Control Number: 2007932602

CR Subject Classification (1998): I.2.3, I.2, F.4.1, F.1, I.5, F.2, G.2, I.4

LNCS Sublibrary: SL 7 – Artificial Intelligence

ISSN 0302-9743
ISBN-10 3-540-74201-8 Springer Berlin Heidelberg New York
ISBN-13 978-3-540-74201-2 Springer Berlin Heidelberg New York



Springer is a part of Springer Science+Business Media

springer.com


Printed in Germany

Typesetting: Camera-ready by author, data conversion by Scientific Publishing Services, Chennai, India
Printed on acid-free paper SPIN: 12107902 06/3180 5 4 3 2 1 0

# Preface

The International Conference on Intelligent Computing (ICIC) was formed to provide an annual forum dedicated to the emerging and challenging topics in artificial intelligence, machine learning, bioinformatics, and computational biology, etc. It aims to bring together researchers and practitioners from both academia and industry to share ideas, problems and solutions related to the multifaceted aspects of intelligent computing.

ICIC 2007, held in Qingdao, China, August 21-24, 2007, constituted the Third International Conference on Intelligent Computing. It built upon the success of ICIC 2006 and ICIC 2005 held in Kunming and Hefei, China, 2006 and 2005, respectively.

This year, the conference concentrated mainly on the theories and methodologies as well as the emerging applications of intelligent computing. Its aim was to unify the picture of contemporary intelligent computing techniques as an integral concept that highlights the trends in advanced computational intelligence and bridges theoretical research with applications. Therefore, the theme for this conference was "**Advanced Intelligent Computing Technology and Applications**". Papers focusing on this theme were solicited, addressing theories, methodologies, and applications in science and technology.

ICIC 2007 received 2875 submissions from 39 countries and regions. All papers went through a rigorous peer review procedure and each paper received at least three review reports. Based on the review reports, the Program Committee finally selected 496 high-quality papers for presentation at ICIC 2007, of which 430 papers have been included in three volumes of proceedings published by Springer: one volume of *Lecture Notes in Computer Science* (LNCS), one volume of *Lecture Notes in Artificial Intelligence* (LNAI), and one volume of *Communications in Computer and Information Science* (CCIS). The other 66 papers will be included in four international journals.

This volume of *Lecture Notes in Artificial Intelligence* (LNAI) includes 139 papers.

The organizers of ICIC 2007, including the Ocean University of China and the Institute of Intelligent Machines of the Chinese Academy of Science, made an enormous effort to ensure the success of ICIC 2007. We hereby would like to thank the members of the ICIC 2007 Advisory Committee for their guidance and advice, the members of the Program Committee and the referees for their collective effort in reviewing and soliciting the papers. We would like to thank Alfred Hofmann, executive editor from Springer, for his frank and helpful advice and guidance throughout and for his support in publishing the proceedings. In particular, we would like to thank all the authors for contributing their papers. Without the high-quality submissions from the authors, the success of the conference would not have been possible. Finally, we are especially grateful to the IEEE Computational Intelligence Society, the International Neural Network Society and the National Science Foundation of China for their sponsorship.

June 2007

De-Shuang Huang
Laurent Heutte
Marco Loog

# ICIC 2007 Organization

**General Co-chairs**

De-Shuang Huang, China
Luonan Chen, Japan

**International Advisory Committee**

Moonis Ali, USA
Shun-Ichi Amari, Japan
Zheng Bao, China
John L. Casti, USA
Guoliang Chen, China
Diane J. Cook, USA
Ruwei Dai, China
John O Gray, UK
Aike Guo, China
Fuchu He, China
Xingui He, China
Tom Heskes, Netherlands
Mustafa Khammash, USA
Okyay Knynak, Turkey
Yanda Li, China
Marios M. Polycarpou, USA
Songde Ma, China
Erke Mao, China
Michael R. Lyu, Hong Kong
Yunyu Shi, China
Harold Szu, USA
Stephen Thompson, UK
Mathukumalli Vidyasagar, India
Shoujue Wang, China
Paul Werbos, USA
George W. Irwin, UK
DeLiang Wang, USA
Youshou Wu, China
Xin Yao, UK
Nanning Zheng, China
Yixin Zhong, China
Mengchu Zhou, USA
Qingshi Zhu, China
Xiang-Sun Zhang, China

**Steering Committee Co-chairs**

Sheng Chen, UK
Xiao-Ping Zhang, Canada
Kang Li, UK

**Program Committee Chair**

Laurent Heutte, France

**Organizing Committee Co-chairs**

Guo Chen, China
Ming Lv, China
Guangrong Ji, China
Ji-Xiang Du, China

**Publication Chair**

Marco Loog, Denmark

**Special Session Chair**

Wanquan Liu, Australia

**International Liaison Chair**

Prashan Premaratne, Australia

**Tutorial Chair**

Robert Hsieh, Germany

## Publicity Co-chairs

Liyanage C. De Silva , New Zealand
Vitoantonio Bevilacqua, Italy
Kang-Hyun Jo, Korea
Jun Zhang, China

## Exhibition Chair

Bing Wang, China

## International Program Committee

Andrea Francesco Abate, Italy
Waleed H. Abdulla, New Zealand
Shafayat Abrar, Pakistan
Parag Gopal Kulkarni, UK
Vasily Aristarkhov, Russian Federation
Masahiro Takatsuka, Australia
Costin Badica, Romania
Soumya Banerjee, India
Laxmidhar Behera, India
Vitoantonio Bevilacqua, Italy
Salim Bouzerdoum, Australia
David B. Bracewell, Japan
Toon Calders, Belgium
Vincent C S Lee, Australia
Gianluca Cena, Italy
Pei-Chann Chang, Taiwan
Wen-Sheng Chen, China
Hong-Qiang Wang, Hong Kong
Rong-Chang Chen, Taiwan
Geoffrey Macintyre, Australia
Weidong Chen, China
Chi-Cheng Cheng, China
Ziping Chiang, Taiwan
Min-Sen Chiu, Singapore
Tommy Chow, Hong Kong
Mo-Yuen Chow, USA
Rasoul Mohammadi Milasi, Canada
Alexandru Paul Condurache, Germany
Sonya Coleman, UK
Pedro Melo-Pinto, Portugal
Roman Neruda, Czech Republic
Gabriella Dellino, Italy
Grigorios Dimitriadis, UK
Mariagrazia Dotoli, Italy
Minh Nhut Nguyen, Singapore
Hazem Elbakry, Japan
Karim Faez, Iran
Jianbo Fan, China
Minrui Fei, China
Mario Koeppen, Japan
Uwe Kruger, UK
Fausto Acernese, Italy
Qing-Wei Gao, China
Takashi Kuremoto, Japan
Richard Lathrop, USA
Agostino Lecci, Italy
Marco Loog, Denmark
Choong Ho Lee, Korea
Jinde Cao, China
Kang Li, UK
Peihua Li, China
Jin Li, UK
Xiaoli Li, UK
Chunmei Liu, USA
Paolo Lino, Italy
Ju Liu, China
Van-Tsai Liu, Taiwan
Wanquan Liu, Australia
Brian C. Lovell, Australia
Hongtao Lu, China
Mathias Lux, Austria
Sheng Chen, UK
Jinwen Ma, China
Yongjun Ma, China
Guido Maione, Italy
Vishnu Makkapati, India
Filippo Menolascina, Italy
Damien Coyle, UK
Cheolhong Moon, Korea
Angelo Ciaramella, Italy
Tark Veli Mumcu, Turkey
Michele Nappi, Italy
Kevin Curran, UK
Giuseppe Nicosia, Italy
Kenji Doya, Japan
Ahmet Onat, Turkey
Ali Ozen, Turkey
Sulin Pang, China
Antonino Staiano, Italy
David G. Stork, USA
Fuchun Sun, China
Zhan-Li Sun, Hong Kong
Maolin Tang, Australia
John Thompson, UK
Amir Atiya, Egypt
Anna Tramontano, Italy
Jose-Luis Verdegay, Spain
Sergio Vitulano, Italy
Anhua Wan, China
Chengxiang Wang, UK
Bing Wang, China
Kongqiao Wang, China
Zhi Wang, China
Hong Wang, China
Hong Wei, UK
Xiyuan Chen, China
Chao-Xue Wang, China
Yong Wang, Japan
Xue Wang, China
Mike Watts, New Zealand
Ling-Yun Wu, China

## Reviewers

Elham A. Boroujeni, Khalid Aamir, Ajith Abraham, Fabrizio Abrate, Giuseppe M.C. Acciani, Ali Adam, Bilal Al Momani, Ibrahim Aliskan, Roberto Amato, Claudio Amorese, Senjian An, Nestor Arana Arexolaleiba, Sebastien Ardon, Khaled Assaleh, Amir Atiya, Mutlu Avci, Pedro Ayrosa, Eric Bae, Meng Bai, Amar Balla, Zaochao Bao, Péter Baranyi, Nicola Barbarini, Edurne Barrenechea, Marc Bartels, Edimilson Batista dos Santos, Devon Baxter, Yasar Becerikli, Ammar Belatreche, Domenico Bellomo, Christian Benar, Vitoantonio Bevilacqua, Daowei Bi, Ida Bifulco, Abbas Bigdeli, Hendrik Blockeel, Leonardo Bocchi, Gennaro Boggia, David Bracewell, Janez Branj, Nicolas Brodu, Cyril Brom, Dariusz Burak, Adrian Burian, Jose M. Cadenas, Zhiyuan Cai, David Camacho, Heloisa Camargo, Maria Angelica Camargo-Brunetto, Francesco Camastra, Ricardo Campello, Galip Cansever, Bin Cao, Dong

Dong Cao, Alessandra Carbotti, Jesus Ariel Carrasco-Ochoa, Deborah Carvalho, Roberto Catanuto, Xiujuan Chai, Kap Luk Chan, Chien-Lung Chan, Ram Chandragupta, Hong Chang, Hsueh-Sheng Chang, Clément Chatelain, Dongsheng Che, Chun Chen, Chung-Cheng Chen, Hsin-Yuan Chen, Tzung-Shi Chen, Xiaohan Chen, Y.M. Chen, Ying Chen, Ben Chen, Yu-Te Chen, Wei-Neng Chen, Chuyao Chen, Jian-Bo Chen, Fang Chen, Peng Chen, Shih-Hsin Chen, Shiaw-Wu Chen, Baisheng Chen, Zhimin Chen, Chun-Hsiung Chen, Mei-Ching Chen, Xiang Chen, Tung-Shou Chen, Xinyu Chen, Yuehui Chen, Xiang Cheng, Mu-Huo Cheng, Long Cheng, Jian Cheng, Qiming Cheng, Ziping Chiang, Han-Min Chien, Min-Sen Chiu, Chi Yuk Chiu, Chungho Cho, Sang-Bock Cho, Soo-Mi Choi, Yoo-Joo Choi, Wen-Shou Chou, T Chow, Xuezheng Chu, Min Gyo Chung, Michele Ciavotta, Ivan Cibrario Bertolotti, Davide Ciucci, Sonya Coleman, Simona Colucci, Patrick Connally, David Corne, Damien Coyle, Cuco Cristi, Carlos Cruz Corona, Lili Cui, Fabrizio Dabbene, Weidi Dai, Thouraya Daouas, Cristina Darolti, Marleen De Bruijne, Leandro De Castro, Chaminda De Silva, Lara De Vinco, Carmine Del Mondo, Gabriella Dellino, Patrick Dempster, Da Deng, Yue Deng, Haibo Deng, Scott Dexter, Nele Dexters, Bi Dexue, Wan Dingsheng, Banu Diri, Angelo Doglioni, Yajie Dong, Liuhuan Dong, Jun Du, Wei-Chang Du, Chen Duo, Peter Eisert, Mehdi El Gueddari, Elia El-Darzi, Mehmet Engin, Zeki Erdem, Nuh Erdogan, Kadir Erkan, Osman Kaan Erol, Ali Esmaili, Alexandre Evsukoff, Marco Falagario, Shu-Kai Fan, Chin-Yuan Fan, Chun-I Fan, Lixin Fan, Jianbo Fan, Bin Fang, Yikai Fang, Rashid Faruqui, Markus Fauster, Guiyu Feng, Zhiyong Feng, Rui Feng, Chen Feng, Yong Feng, Chieh-Chuan Feng, Francisco Fernandez Periche, James Ferryman, Mauricio Figueiredo, Vítor Filipe, Celine Fiot, Alessandra Flammini, Girolamo Fornarelli, Katrin Franke, Kechang Fu, Tiaoping Fu, Hong Fu, Chaojin Fu, Xinwen Fu, Jie Fu, John Fulcher, Wai-keung Fung, Zhang G. Z., Sebastian Galvao, Junying Gan, Zhaohui Gan, Maria Ganzha, Xiao-Zhi Gao, Xin Gao, Liang Gao, Xuejin Gao, Xinwen Gao, Ma Socorro Garcia, Ignacio Garcia-del-Amo, Lalit Garg, Shuzi Sam Ge, Fei Ge, Xin Geng, David Geronimo, Reza Ghorbani, Paulo Gil, Gustavo Giménez-Lugo, Tomasz Gingold, Lara Giordano, Cornelius Glackin, Brendan Glackin, Juan Ramón González González, Jose-Joel Gonzalez-Barbosa, Padhraig Gormley, Alfredo Grieco, Giorgio Grisetti, Hanyu Gu, Xiucui Guan, Jie Gui, Aaron Gulliver, Feng-Biao Guo, Ge Guo, Tian-Tai Guo, Song Guo, Lingzhong Guo, Yue-Fei Guo, P Guo, Shwu-Ping Guo, Shengbo Guo, Shou Guofa, David Gustavsson, Jong-Eun Ha, Risheng Han, Aili Han, Fengling Han, Hisashi Handa, Koji Harada, James Harkin, Saadah Hassan, Aboul Ella Hassanien, Jean-Bernard Hayet, Hanlin He, Qingyan He, Wangli He, Haibo He, Guoguang He, Pilian He, Yanxiang He, Pawel Herman, Francisco Herrera, Jan Hidders, Grant Hill, John Ho, Xuemin Hong, Tzung-Pei Hong, Kunjin Hong, Shi-Jinn Horng, Lin Hou, Eduardo Hruschka, Shang-Lin Hseih, Chen-Chiung Hsieh, Sun-Yuan Hsieh, Jih-Chang Hsieh, Chun-Fei Hsu, Honglin Hu, Junhao Hu, Qinglei Hu, Xiaomin Hu, Xiaolin Hu, Chen Huahua, Xia Huang, Jian Huang, Xiaojing Huang, Gan Huang, Weitong Huang, Jing Huang, Weimin Huang, Yufei Huang, Zhao Hui, Sajjad Hussain, Thong-Shing Hwang, Giorgio Iacobellis, Francesco Iorio, Mohammad Reza Jamali, Horn-Yong Jan, Dar-Yin Jan, Jong-Hann Jean, Euna Jeong, Mun-Ho Jeong, Youngseon Jeong, Zhen Ji, Qing-Shan Jia, Wei Jia, Fan Jian, Jigui Jian, Peilin Jiang, Dongxiang Jiang, Minghui Jiang, Ping Jiang, Xiubao Jiang, Xiaowei Jiang, Hou Jiangrong, Jing Jie, Zhang Jihong, Fernando Jimenez, Guangxu Jin, Kang-Hyun Jo,

Guillaume Jourjon, Jih-Gau Juang, Carme Julià, Zhou Jun, Dong-Joong Kang, Hee-Jun Kang, Hyun Deok Kang, Hung-Yu Kao, Indrani Kar, Cihan Karakuzu, Bekir Karlik, Wolfgang Kastner, John Keeney, Hrvoje Keko, Dermot Kerr, Gita Khalili Moghaddam, Muhammad Khurram Khan, Kavi Umar Khedo, Christian Kier, GwangHyun Kim, Dae-Nyeon Kim, Dongwon Kim, Taeho Kim, Tai-hoon Kim, Paris Kitsos, Kunikazu Kobayashi, Sarath Kodagoda, Mario Koeppen, Nagahisa Kogawa, Paul Kogeda, Xiangzhen Kong, Hyung Yun Kong, Insoo Koo, Marcin Korze, Ibrahim Kucukdemiral, Petra Kudova, Matjaz Kukar, Parag Kulkarni, Saravana Kumar, Wen-Chung Kuo, Takashi Kuremoto, Janset Kuvulmaz, Jin Kwak, Lam-For Kwok, Taekyoung Kwon, Marcelo Ladeira, K. Robert Lai, Darong Lai, Chi Sung Laih, Senthil Kumar Lakshmanan, Dipak Lal Shrestha, Yuk Hei Lam, M. Teresa Lamata, Oliver Lampl, Peng Lan, Vuokko Lantz, Ana Lilia Laureano-Cruces, Yulia Ledeneva, Vincent C S Lee, Narn-Yih Lee, Malrye Lee, Chien-Cheng Lee, Dong Hoon Lee, Won S Lee, Young Jae Lee, Kyu-Won Lee, San-Nan Lee, Gang Leng, Agustin Leon Barranco, Chi Sing Leung, Cuifeng Li, Fuhai Li, Chengqing Li, Guo-Zheng Li, Hongbin Li, Bin Li, Liberol Li, Bo Li, Chuandong Li, Erguo Li, Fangmin Li, Juntao Li, Jinshan Li, Lei Li, Ming Li, Xin Li, Xiaoou Li, Xue li, Yuan Li, Lisa Li, Yuancheng Li, Kang Li, Jun Li, Jung-Shian Li, Shijian Li, Zhihua Li, Zhijun Li, Zhenping Li, Shutao Li, Xin Li, Anglica Li, Wanqing Li, Jian Li, Shaoming Li, Xiaohua Li, Xiao-Dong Li, Xiaoli Li, Yuhua Li, Yun-Chia Liang, Wei Liang, Wuxing Liang, Jinling Liang, Wen-Yuan Liao, Wudai Liao, Zaiyi Liao, Shizhong Liao, Vicente Liern, Wen-Yang Lin, Zhong Lin, Chih-Min Lin, Chun-Liang Lin, Xi Lin, Yu Chen Lin, Jun-Lin Lin, Ke Lin, Kui Lin, Ming-Yen Lin, Hsin-Chih Lin, Yu Ling, Erika Lino, Erika Lino, Paolo Lino, Erika Lino, Shiang Chun Liou, Ten-Yuang Liu, Bin Liu, Jianfeng Liu, Jianwei Liu, Juan Liu, Xiangyang Liu, Yadong Liu, Yubao Liu, Honghai Liu, Kun-Hong Liu, Kang-Yuan Liu, Shaohui Liu, Qingshan Liu, Chen-Hao Liu, Zhiping Liu, Yinyin Liu, Yaqiu Liu, Van-Tsai Liu, Emmanuel Lochin, Marco Loog, Andrew Loppingen, Xiwen Lou, Yingli Lu, Yao Lu, Wen-Hsiang Lu, Wei Lu, Hong Lu, Huijuan Lu, Junguo Lu, Shangmin Luan, Jiliang Luo, Xuyao Luo, Tuan Trung Luong, Mathias Lux, Jun Lv, Chengguo Lv, Bo Ma, Jia Ma, Guang-Ying Ma, Dazhong Ma, Mi-Chia Ma, Junjie Ma, Xin Ma, Diego Magro, Liam Maguire, Aneeq Mahmood, Waleed Mahmoud, Bruno Maione, Agostino Marcello Mangini, Weihua Mao, Kezhi Mao, Antonio Maratea, Bogdan Florin Marin, Mario Marinelli, Urszula Markowska-Kaczmar, Isaac Martin, Francesco Martinelli, Jose Fco. Martínez-Trinidad, Antonio David Masegosa Arredondo, Louis Massey, Emilio Mastriani, Marco Mastrovito, Kerstin Maximini, Radoslaw Mazur, Daniele Mazzocchi, Malachy McElholm, Gerard McKee, Colin McMillen, Jian Mei, Belen Melian, Carlo Meloni, Pedro Melo-Pinto, Corrado Mencar, Luis Mesquita, Jianxun Mi, Pauli Miettinen, Claudia Milaré, Rasoul Milasi, Orazio Mirabella, Nazeeruddin Mohammad, Eduard Montseny, Inhyuk Moon, Hyeonjoon Moon, Raul Morais, J. Marcos Moreno, José Andrés Moreno, Philip Morrow, Santo Motta, Mikhal Mozerov, Francesco Napolitano, David Naso, Wang Nengqiang, Mario Neugebauer, Yew Seng Ng, Wee Keong Ng, Tam Nguyen, Quang Nguyen, Thang Nguyen, Rui Nian, James Niblock, Iaobing Nie, Eindert Niemeijer, Julio Cesar Nievola, Haijing Niu, Qun Niu, Changyong Niu, Asanao Obayashi, Kei Ohnishi, Takeshi Okamoto, Jose Angel Olivas, Stanley Oliveira, Kok-Leong Ong, Chen-Sen Ouyang, Pavel Paclik, Tinglong Pan, Sanjib Kumar Panda, Tsang-Long Pao, Emerson Paraiso, Daniel Paraschiv, Giuseppe

Patanè, Kaustubh Patil, Mykola Pechenizkiy, Carlos Pedroso, Zheng Pei, Shun Pei, Chang Pei-Chann, David Pelta, Jian-Xun Peng, Sheng-Lung Peng, Marzio Pennisi, Cathryn Peoples, Eranga Perera, Alessandro Perfetto, Patrick Peursum, Minh-Tri Pham, Phuong-Trinh Pham-Ngoc, Lifton Phua, Son Lam Phung, Alfredo Pironti, Giacomo Piscitellei, Elvira Popescu, Girijesh Prasad, Prashan Premaratne, Alfredo Pulvirenti, Lin Qi, HangHang Qi, Yu Qiao, Xiaoyan Qiao, Lixu Qin, Kai Qin, Jianlong Qiu, Ying-Qiang Qiu, Zhonghua Quan, Thanh-Tho Quan, Chedy Raïssi, Jochen Radmer, Milo Radovanovi, Bogdan Raducanu, Humera Rafique, Thierry Rakotoarivelo, Nini Rao, Ramesh Rayudu, Arif Li Rehman, Dehua Ren, Wei Ren, Xinmin Ren, Fengli Ren, Orion Reyes, Napoleon Reyes, Carlos Alberto Reyes-Garcia, Alessandro Rizzo, Giuseppe Romanazzi, Marta Rosatelli, Heung-Gyoon Ryu, Hichem Sahbi, Ying Sai, Paulo Salgado, Luigi Salvatore, Nadia Salvatore, Saeid Sanei, Jose Santos, Angel Sappa, Heather Sayers, Klaus Schöffmann, Bryan Scotney, Carla Seatzu, Hermes Senger, Murat Sensoy, Carlos M.J.A. Serodio, Lin Shang, Li Shang, XiaoJian Shao, Andrew Shaw, Sheng Yuan Shen, Yanxia Shen, Yehu Shen, Linlin Shen, Yi Shen, Jinn-Jong Sheu, Mingguang Shi, Chaojian Shi, Dongfeng Shi, June-Horng Shiesh, Yen Shi-Jim, Zhang Shuhong, Li Shundong, Nanshupo Shupo, Oliver Sinnen, Sukree Sinthupinyo, Silvia Siri, Ernest Sithole, Nicolas Sklavos, Stanislav Slusny, Pilar Sobrevilla, Ignacio Solis, Anthony Solon, Andy Song, Liu Song, Qiankun Song, Zheng Song, Yinglei Song, Nuanwan Soonthornphisaj, Aureli Soria-Frisc, Jon Sporring, Kim Steenstrup Pedersen, Domenico Striccoli, Juhng Perng Su, Shanmugalingam Suganthan, P. N. Suganthan, Youngsoo Suh, Yonghui Sun, Xinghua Sun, Ning Sun, Fuchun Sun, Lily Sun, Jianyong Sun, Jiande Sun, Worasait Suwannik, Roberto T. Alves, Tele Tan, Taizhe Tan, Xuan Tan, Xiaojun Tan, Hong Zhou Tan, Feiselia Tan, Hong Tang, Chunming Tang, David Taniar, Michele Taragna, David M.J. Tax, Ziya Telatar, Zhi Teng, John Thompson, Bin Tian, Ching-Jung Ting, Fok Hing Chi Tivive, Alexander Topchy, Juan Carlos Torres, Ximo Torres, Joaquin Torres-Sospedra, Hoang Hon Trinh, Chia-Sheng Tsai, Chieh-Yuan Tsai, Huan-Liang Tsai, Wang-Dauh Tseng, Yuan-Jye Tseng, Yifeng Tu, Biagio Turchiano, Cigdem Turhan, Anna Ukovich, Muhammad Muneeb Ullah, Nurettin Umurkan, Mustafa Unel, Daniela Ushizima, Adriano Valenzano, Pablo A. Valle, Bram Van Ginneken, Christian Veenhuis, Roel Vercammen, Enriqueta Vercher, Silvano Vergura, Brijesh Verma, Raul Vicente Garcia, Boris X. Vintimilla Burgos, Gareth Vio, Stefano Vitturi, Aristeidis Vlamenkoff, John Wade, Manolis Wallace, Li Wan, Shijun Wang, Xiaodong Wang, Xue Wang, Zhi Wang, Bing Wang, Chih-Hung Wang, Chao Wang, Da Wang, Jianying Wang, Le Wang, Min Wang, Rui-Sheng Wang, Sheng Wang, Jiahai Wang, Guanjun Wang, Linshan Wang, Yanyan Wang, Xuan Wang, Xiao-Feng Wang, Yong Wang, Zidong Wang, Zhongsheng Wang, Zhengyou Wang, Yen-Wen Wang, Shiuh-Jeng Wang, Shouqi Wang, Ling Wang, Xiang Wang, Lina Wang, Qing-Guo Wang, Yebin Wang, Dingcheng Wang, Dianhui Wang, Meng Wang, Yi Wang, Bao-Yun Wang, Xiaomin Wang, Huazhong Wang, Jeen-Shing Wang, Haili Wang, Haijing Wang, Jian Wang, Yoshikazu Washizawa, Yuji Watanabe, Wiwat Watanawood, Michael Watts, Richard Weber, Lisheng Wei, Zhi Wei, Yutao Wei, Hong Wei, Li Weigang, Dawid Weiss, Hou Weiyan, Guo-Zhu Wen, Brendon Woodford, Derek Woods, Lifang Wu, Zikai Wu, Ke Wu, Xinan Wu, Hsien-Chu Wu, QingXiang Wu, Shiqian Wu, Lihchyau Wuu, Jun-Feng Xia, Li Xia, Xiao Lei Xia, Zhiyu Xiang, Kui Xiang, LiGuo Xiang, Tao Xiang, Jing Xiao, Min Xiao, Liu

Xiaodong, Zhao Xiaoguang, Xiangpeng Xie, Zhijun Xie, Shaohua Xie, Jiang Xie, Hong Xie, Rui Xing, Li Xinyu, Wei Xiong, Huan Xu, Jiangfeng Xu, Jianhua Xu, Yongjun Xu, Jun Xu, Hongji Xu, Bingji Xu, Yu Xue, Yun Xue, Mehmet Yakut, Xing Yan, Jiajun Yan, Hua Yan, Yan Yang, Hsin-Chang Yang, Tao Yang, Chengfu Yang, Banghua Yang, Ruoyu Yang, Zhen Yang, Zhichun Yang, Wu-Chuan Yang, Ming Yang, Cheng-Zen yang, Shouyi Yang, Ming-Jong Yao, Kim-Hui Yap, Hao Ye, Chia-Hsuan Yeh, James Yeh, Jun-Heng Yeh, Shwu-Huey Yen, Sang-Soo Yeo, Yang Yi, Tulay Yildirim, PeiPei Yin, Junsong Yin, Lin Ying, Ling Ying-Biao, Yang Yongqing, Kaori Yoshida, Tomohiro Yoshikawa, Qi Yu, Wen Yu, Wen-Shyong Yu, Kun Yuan, Kang Yuanyuan, Chen Yuepeng, Li Yun, Kun Zan, Chuanzhi Zang, Ramon Zatarain-Cabada, Faiz ul Haque Zeya, Zhihui Zhan, Changshui Zhang, Yongping Zhang, Jie Zhang, Jun Zhang, Yunchu Zhang, Zanchao Zhang, Yifeng Zhang, Shihua Zhang, Ningbo Zhang, Junhua Zhang, Jun Zhang, Shanwen Zhang, Hengdao Zhang, Wen-sheng Zhang, Haoshui Zhang, Ping Zhang, Huaizhong Zhang, Dong Zhang, Hua Zhang, Byoung-Tak Zhang, Guohui Zhang, Li-Bao Zhang, Junping Zhang, Junpeng Zhang, Jiye Zhang, Junying Zhang, JingRu Zhang, Jian Zhang, Duanjin Zhang, Xin Zhang, Huaguang Zhang, Guo Zhanjie, Jizhen Zhao, Zhong-Qiu Zhao, Li Zhao, Ming Zhao, Yinggang Zhao, Ruijie Zhao, Guangzhou Zhao, Liu Zhaolei, Fang Zheng, Ying Zheng, Chunhou Zheng, Cong Zheng, Guibin Zheng, Qinghua Zheng, Wen-Liang Zhong, Jinghui Zhong, Jiayin Zhou, Jie Zhou, Xiaocong Zhou, Fengfeng Zhou, Chi Zhou, Sue Zhou, Mian Zhou, Zongtan Zhou, Lijian Zhou, Zhongjie Zhu, Xinjian Zhuo, Xiaolan Zhuo, Yanyang Zi, Ernesto Zimmermann, Claudio Zunino, Haibo Deng, Wei Liu.

# Table of Contents

## Neural Networks

## Independent Component Analysis and Blind Source Separation

## Combinatorial and Numerical Optimization



## Neural Computing and Optimization

## Knowledge Discovery and Data Mining

## Artificial Life and Artificial Immune Systems



## Ensemble Methods



## Manifold Learning Theory



## Evolutionary Computing and Genetic Algorithms

## Fuzzy Systems and Soft Computing

## Particle Swarm Optimization and Niche Technology



## Swarm Intelligence and Optimization

## Kernel Methods and Support Vector Machines



## Fine Feature Extraction Methods

## Intelligent Fault Diagnosis

## Fuzzy Control



## Intelligent Human-Computer Interactions for Multi-modal and Autonomous Environment



## Computational Systems Biology



## Intelligent Robot Systems Based on Vision Technology



## Intelligent Computing for Motion Picture Processing

## Particle Swarm Optimization: Theories and Applications



## Recent Advances of Intelligent Computing with Applications in the Multimedia Systems



## Computational Intelligence in Chemoinformatics



## Strategy Design and Optimization of Complex Engineering Problems

## Traffic Optimization



## Intelligent Mobile and Wireless Sensor Networks



## Intelligent Prediction and Time Series Analysis

## Intelligent Computing in Neuroinformatics

# A New Watermarking Approach Based on Neural Network in Wavelet Domain

Xue-Quan Xu[1], Xian-Bin Wen[1], Yue-Qing Li[2], and Jin-Juan Quan[1]

[1] School of Computer Science and Technology, Tianjin University of Technology
300191 Tianjin, P.R. China
[2] Beijing Polytechnic College 100042 Beijing, P.R. China
xb_wen@yahoo.com.cn

**Abstract.** A new digital watermarking algorithm based on BPN neural network is proposed. Watermark embed processing is carried out by transforming the host image in wavelet domain. Watermark bits are added to the selected coefficients blocks. Because of the learning and adaptive capabilities of neural networks, the trained neural networks can recover the watermark from the watermarked images. The experimental results show that this watermarking algorithm has a good preferment.

## 1 Introduction

With the development of modern society, multimedia becomes more and more important in people's daily life. However, the more use the more illegal duplications of multimedia products can be readily spread through internet. It is a crucial time for us to take some measures to protect the copyright of media. Toward this aim, many techniques have been proposed in the literatures in the last few years, in which digital watermarking is quite efficient and promising. A significant merit of digital watermarking is that multimedia data can still be utilized by users although they are embedded with an invisible digital watermark. These watermarks cannot be removed by unauthorized persons and they can be extracted by legal author.

In recent years, a number of invisible watermarking techniques for digital images have been reported. Generally speaking, there exits two typical watermarking techniques including: spatial domain methods and transform domain methods [1]-[4]. Between these two methods, embedding the watermark into the transform domain can increase the security, imperceptibility and robustness of watermark, and is widely adopted in many digital watermark methods.

In this paper, a new blind watermarking scheme based on neural networks in the wavelets domain is proposed. To ensure the watermark safety and imperceptibly, embedding the watermark bits into the edges and textures of the image we make use of the statistical properties of the DWT and of the human visual system (HVS). Due to neural network [5]-[6] possessing the learning capability from given training patterns, our methods can recover the watermark from the watermarked images without the original images. The watermarked images are tested for different type of attacks and the results prove the validity of the proposed approach.

D.-S. Huang, L. Heutte, and M. Loog (Eds.): ICIC 2007, LNAI 4682, pp. 1–6, 2007.

## 2 The Proposed Watermarking Method

The proposed method embeds watermark by decomposing the host image. Dividing these coefficients into small blocks, calculating the standard deviation of these blocks, deciding whether this block can be use for embedding watermark. The watermark bits are added to the selected coefficient blocks without any perceptual degradation for host image. The watermark used for embedding is a binary logo image, which is very small compared with the size of the host image. During the watermark recovery, the trained neural network is employed to extract the watermark.

### 2.1 Watermark Embedding Algorithm

The algorithm for embedding a binary watermark is formulated as follows:

**Step 1:** Decompose the host image by L-levels using DWT. The watermark is operated by Arnold transform (the size of watermark is $N \times N$ ).

**Step 2:** Splitting the wavelet coefficients (mainly in HL and LH) into many non-overlapping small blocks, each block size is 3×3. Calculating the standard deviation of each block use Eq. (2), arranging these values on ascending, then can decide the threshold $T_1$ for the watermark embedding. Choose the ( $N \times (N+1)$ ) standard deviation as $T_1$ value. The fontal $N \times N$ blocks are selected for watermarking.

The formula of the block average value as follow:

$$ave = \sum_{m=-1}^{1} \sum_{n=-1}^{1} I(i+m, j+n)/9 . \qquad (1)$$

Then the formula for calculating standard deviation as follow:

$$stdev = (\sum_{m=-1}^{1} \sum_{n=-1}^{1} (I(i+m, j+n) - ave)^2 /8)^{1/2} . \qquad (2)$$

where $I(i+m, j+n)$ is the coefficient of each small block, $I(i, j)$ is the central coefficient of selected block, variable m, n stand for the surrounding coefficients in the same block corresponding.

**Step 3:** The largest value of block standard deviation is evaluated to $T$ . Then the strength of watermark for each block can be calculated by the ratio of the block standard deviation to $T$ , this ratio is evaluated to $\alpha$ .

**Step 4:** Adding watermark bits to these blocks central items using Eq (3).

$$I'(i, j) = I(i, j) + \alpha(2w(k) - 1) . \qquad (3)$$

where $I(i, j)$ is the central coefficient of selected block, $\alpha$ is the watermark embedding strength, $w(k)$ is the watermark bit. The value of $\alpha$ is alterable, so the imperceptible and robust of the watermark is very well.

**Step 5:** After embedding watermark bits, L-level inverse wavelet transform of the image, get the watermarked image.

### 2.2 The BPN Neural Network

The BPN is one type of supervised learning neural networks. It is a very popular modal in neural networks. The principle behind BPN [7]-[8] involves using the steepest gradient descent method to reach small approximation. The general modal has architecture as follow describing. There are three layers, including input layer, hidden layer, and output layer. Each layer has one or more neurons and each neuron is fully connected to its adjacent layers. Two neurons of each adjacent layer are directly connected to one another, which is called a link. Each link has a weighted value, representing the relational degree between two neurons. These weighted values are determined by the training algorithm described by the following equations:

$$net_j(t) = \sum_i \alpha_{i,j} o_i(t) - \theta_j . \tag{4}$$

$$o_j(t+1) = f_{act}(net_j(t)) . \tag{5}$$

Where $net_j(t)$ is the activation of the neuron $j$ in iteration $t$, $o_j(t+1)$ is output of the neuron $j$ in iteration $t+1$, $f_{act}(x)$ is the activation function of a neuron, which is usually a sigmoid function in hidden layer and a pureline function in output layer. Generally all initial weight values $\alpha_{i,j}$ are assigned using random values. In each iteration process, all $\alpha_{i,j}$ are modified using the delta rule according the learning sample. After training, the BPN can act as an approximating function.

### 2.3 Watermark Extracting

Here, the BPN neural network is used as the method of watermark extraction, which can transform the watermarked coefficients into the watermark data. Firstly, the watermarked image is L-level wavelet decomposed. Then divide these coefficients into small blocks size in 3×3, calculating the standard deviation of each block. If the result is not large than $T_2$ ($T_2 > T_1$), then this block can be used for extracting.

According to our method, we construct three layers BPN neural network with 8, 4 and 1 neurons in the input, hidden and output layer respectively. The input signals are the neighbors of watermarked coefficients and the output signals are the watermark data. Aiming at the correction of watermark bit extraction, we should training the neural network firstly. Some watermarked coefficients neighbors and the watermark data are used to train the BPN neural network, which are destroyed by the attack software. For example: for a selected coefficient block, the central item of the block is $\bar{I}(i,j)$ . The network is trained with its 3×3 neighbors, i.e., let $\{\bar{I}(i-1,j-1), \bar{I}(i-1,j), \bar{I}(i-1,j+1), \bar{I}(i,j-1), \bar{I}(i,j+1), \bar{I}(i+1,j-1), \bar{I}(i+1,j),$ $\bar{I}(i+1,j+1)\}$ as input vector and the value $\tilde{I}(i,j)$ as output value. After the training, the BPN neural network has become a robust digital watermark extraction network, which can easily and correctly extract the watermark data from the watermarked image. The extract watermark bits can be described as follow:

$$w'(k) = \begin{cases} 1 & \text{if } \bar{I}(i,j) \geq \tilde{I}(i,j) \\ 0 & \text{otherwise} \end{cases} \quad k = 1, \cdots, N*N \tag{6}$$

### 2.4 Watermark Detecting

Peak Signal to Noise Ration (PSNR) [3] is used to measure quality of watermarked image while Normalized Cross Correlation (NC) [4] is used to measure quality of watermark after recovery.

$$PSNR = 10\log_{10}\frac{255^2}{MSE} \tag{7}$$

where $MSE$ is the mean-square error between a watermarked or an attacked watermarked image and its original image.

$$NC = \frac{\sum_k w(k)w'(k)}{\sum_k |w(k)|^2} \tag{8}$$

If $NC$ >0.7, we can draw the conclusion that the watermark extracted is the same as the original watermark, otherwise it is not the same watermark which is embedded into the original image and the extraction is false.

## 3 Experiment and Results

In our experiments, we take the "TJUT logo" as the watermark $W$ and the logo is a binary image with size 64×64. The 3-level wavelet decomposition bior 5.5 filter coefficients are used. Here the results are presented for grayscale 8-bit Lena image of size 512×512. Original Lena and logo images are shown is Fig. 1(a) and (b) respectively. Watermarked Lena image having PSNR value of 37.9 is shown in Fig. 1(c). If the original and the watermarked Lena images are observed we cannot find any perceptual degradation. Extracted logo from the watermarked image is shown in Fig. 1(d).

To prove the robustness of the new type of scheme, we investigate the effect of common signal distortions on the watermarked images. Such as AWGN with the SNR=11.4db, median filtering, Gaussian filtering, cropping, adding salt and pepper noise randomly. After these operations, the images are greatly degraded and lots of data are lost, but the extracted logos are still recognizable. These results are shown in Fig. (2). The watermarked Lena image is also tested for JPEG 2000 compression operate. Fig. (3) show the extracted watermarks from the JPEG compressed version of the watermarked images with various compression qualities.

To confirm the validity of our method, we compare the correlation of original watermark and extracted watermark by our method and the method proposed by reference [8]. We calculate the value of $NC$, $NC_1$ stand for our method, $NC_2$ indicate the method proposed by reference [8], the results are shown in table 1. From the table we

**Fig. 1.** (a) Original Lena image, (b) Original logo image, (c) Watermarked Lena image, (d) extracted logo image ( $NC=1$)

**Fig. 2.** Logo extracted after (a) AWGN, (b) median filtering, (c) pepper and salt noise, (d) cropping, (e) Gaussian filtering

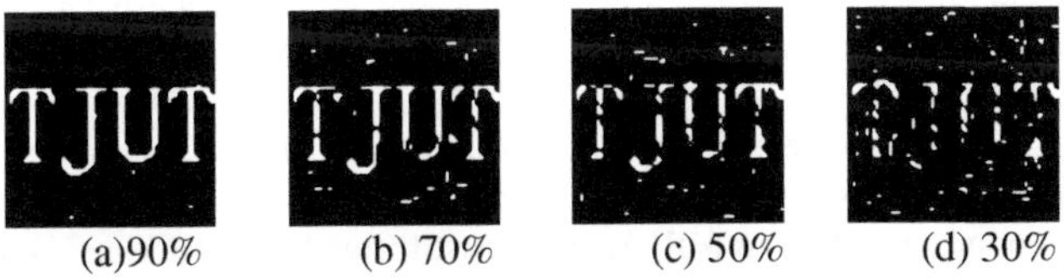


**Fig. 3.** Robustness to JPEG compression

can see, the watermarked image gone through a variety of attacks, including AWGN,media filtering, cropping, Gaussian filtering and JPEG compression, then the watermark data are extracted by our method and the method in reference [8] respectively. The $NC$ values of our method are 0.969, 0.945, 0.901, 0.965, 0.875. Whereas the $NC$ values of the method in reference [8] are 0.906, 0.927, 0.877, 0.914, 0.751. These data prove that our method have a better performance than the method in reference [8].

**Table 1.** Compare the correlation of original watermark and extracted watermark by this method and reference [8]

| Operation | AWGN | Media filtering | cropping | Gaussian filtering | JPEG(30%) |
|---|---|---|---|---|---|
| $NC_1$ | 0.969 | 0.945 | 0.901 | 0.965 | 0.875 |
| $NC_2$ | 0.906 | 0.927 | 0.877 | 0.914 | 0.715 |

## 4 Conclusion

This paper presents a blind digital watermarking algorithm based on BPN neural network. The host image is decomposed into wavelet domain, then watermark bits embedded in the selected coefficients blocks. In watermark extraction, the original watermark is retrieved by neural network.

## Acknowledgements

This work is supported in part by the National Natural Science Foundation of China (No. 60375003), the Aeronautics and Astronautics Basal Science Foundation of China (No. 03I53059), the Science and Technology Development Foundation of Tianjin Higher-learning (2006BA15).

## References

1. Chen, Y.H., Su, J.M., Fu, H.C., Huang ,H.C.,. Pao, H.T.: Adaptive Watermarking Using Relationships Between Wavelet Coefficients, IEEE International Symposium on Circuits and Systems, 5 (2005) 4979-4982
2. Khelifi, F., Bouridane, A., Kurugollu, F., Thompson, A.I.: An Improve Wavelet-based Image Watermarking Technique, IEEE Conference on Advanced Video and Signal Based Surveillance, (2005) 588-592
3. Nafornita, C.: Improved Detection for Robust Image Watermarking, International Symposium on Signals, Circuits and Systems, 2 (2005) 473-476
4. Temi, C., Choomchuay, S., Lasakul, A.: A Robust Image Watermarking Using Multiresolution Analysis of Wavelet, IEEE International Symposium on Communications and Information Technology, 1 (2005) 623-626
5. Wang, Z.F., Wang N.C., Shi, B.C.: A Novel Blind Watermarking Scheme Based on Neural Network in Wavelet Domain, The Sixth World Congress on Intelligent Control and Automation, 1 (2006) 3024-3027
6. Zhang ,X.H., Zhang,F.: A Blind Watermarking Algorithm Based on Neural Network, International Conference on Neural Networks and Brain, 2 (2005) 1073-1076
7. Chang ,C.Y., Su, S.J.: A Neural Network Based Robust Watermarking Scheme, IEEE International Conference on Systems, Man and Cybernetics, 3 (2005) 2482-2478
8. Zhang, J., Wang, N.C., Xiong, F: A Novel Watermarking for Images Using Neural Networks, International Conference on Machine Learning and Cybernetics, 3 (2002)1405-1408

# Analysis of Global Convergence and Learning Parameters of the Back-Propagation Algorithm for Quadratic Functions

Zhigang Zeng

School of Automation,
Wuhan University of Technology, Wuhan, Hubei, 430070, China
zhigangzeng@163.com

**Abstract.** This paper analyzes global convergence and learning parameters of the back-propagation algorithm for quadratic functions. Some global convergence conditions of the steepest descent algorithm are obtained by directly analyzing the exact momentum equations for quadratic cost functions. In addition, in order to guarantee the convergence for a given learning task, the method is obtained to choose the proper learning parameters. The results presented in this paper are the improvement and extension of the existed ones in some existing works.

## 1 Introduction

Back-propagation (BP) is one of the most widely used algorithms for training feedforward neural networks [1]. However, it is seen from simulations that it takes a long time to converge. Consequently, many variants of BP have been suggested. One of the most well-known variants is the back-propagation with momentum terms (BPM) [2], in which the weight change is a combination of the new steepest descent step and the previous weight change. The purpose of using momentum is to smooth the weight trajectory and speed the convergence of the algorithm [3]. It is also sometimes credited with avoiding local minima in the error surface. BP can be shown to be a straightforward gradient descent on the least squares error, and it has been shown recently that BP converges to a local minimum of the error. While it is observed that the BPM algorithm shows a much higher rate of convergence than the BP algorithm.

Although squared error functions are only quadratic for linear networks, they are approximately quadratic for any smooth error functions in the neighborhood of a local minimum. (This can be shown by performing a Taylor series expansion of the error function about the minimum point [3].) Phansalkar and Sastry [1] analyze the behavior of the BPM algorithm and show that all local minima of the least squares error are the only locally asymptotically stable poits of the algorithm. Hagiwara and Sato [5], [6] show that the momentum mechanism can be derived from a modified cost function, in which the squared errors are exponentially weighted in time. They also derive a qualitative relationship between themomentumterm, the learning rate and the speed of convergence. Qian [7]

D.-S. Huang, L. Heutte, and M. Loog (Eds.): ICIC 2007, LNAI 4682, pp. 7–13, 2007.

demonstrates an analogy between the convergence of the momentum algorithm and the movement of Newtonian particles in a viscous medium. By utilizing a discrete approximation to this continuous system, Qian also derives the conditions for stability of the algorithm. Torii and Hagan [4] analyze the effect of momentum when minimizing quadratic error functions, and provide necessary and sufficient conditions for stability of the algorithm and present a theoretically optimal setting for the momentum parameter to produce fastest convergence.

In this paper, some global convergence conditions of the steepest descent algorithm are obtained by directly analyzing the exact momentum equations for quadratic cost functions. Those conditions can be directly derived from the parameters (different from eigenvalues that are used in [4]) of the Hessian matrix. The results presented in this paper are the improvement and extension of the existed ones in [8]. In addition, in order to guarantee the convergence for a given learning task, the method is obtained to choose the proper learning parameters.

## 2 Problem Description

Our objective is to determine a set of network weights that minimize a quadratic error function. The quadratic function can be represented by

$$F(\mathbf{x}) = \frac{1}{2}\mathbf{x}^T\mathbf{H}\mathbf{x} + \mathbf{d}^T\mathbf{x} + c, \tag{1}$$

where $\mathbf{H}$ is a symmetric Hessian matrix with nonnegative eigenvalues (since the error function must be positive semidefinite). The standard steepest descent algorithm is

$$\Delta\mathbf{x}(k) = -\alpha\nabla F(\mathbf{x}(k)). \tag{2}$$

This algorithm is stable if $\alpha$ times the largest eigenvalue of the matrix $\mathbf{H}$ is less than 2 [1]. If we add momentum, the steepest descent algorithm becomes

$$\Delta\mathbf{x}(k) = \gamma\Delta\mathbf{x}(k-1) - (1-\gamma)\alpha\nabla F(\mathbf{x}(k)). \tag{3}$$

where the momentum parameter $\gamma$ will be in the range $0 < \gamma < 1$.

Some global convergence conditions for (2) and (3) are obtained to in [8]. In fact, (2) can be regarded as a special case of the following algorithm:

$$\Delta\mathbf{x}(k) = -\text{diag}\{\alpha_1, \alpha_2, \cdots, \alpha_n\}\nabla F(\mathbf{x}(k)), \tag{4}$$

where $\alpha_i$ $(i = 1, 2, \cdots, n)$ are learning parameters, $n$ is the dimension number of the matrix $\mathbf{H}$.

In addition, (3) can be regarded as a special case of the following algorithm:

$$\begin{aligned}\Delta\mathbf{x}(k) = {} & \text{diag}\{\gamma_1, \gamma_2, \cdots, \gamma_n\}\Delta\mathbf{x}(k-1) \\ & -\text{diag}\{(1-\gamma_1)\alpha_1, (1-\gamma_2)\alpha_2, \cdots, (1-\gamma_n)\alpha_n\}\nabla F(\mathbf{x}(k)).\end{aligned} \tag{5}$$

where the momentum parameters $\gamma_i$ $(i = 1, 2, \cdots, n)$ will be in the range $0 < \gamma_i < 1$.

The gradient of the quadratic function is

$$\nabla F(\mathbf{x}) = \mathbf{H}\mathbf{x} + \mathbf{d}, \tag{6}$$

where the matrix $\mathbf{H} = (h_{ij})_{n\times n}$.

# 3 Steepest Descent Without Momentum

## 3.1 Analysis of Global Convergence

Let

$$\bar{h}_{ij} = \begin{cases} \alpha_i h_{ii}, & i = j, \\ -\alpha_i |h_{ij}|, & i \neq j, \end{cases} \qquad \tilde{h}_{ij} = \begin{cases} 2 - \alpha_i h_{ii}, & i = j, \\ -\alpha_i |h_{ij}|, & i \neq j. \end{cases}$$

Denote matrices $\mathbf{H}_1 = (\bar{h}_{ij})_{n\times n}, \mathbf{H}_2 = (\tilde{h}_{ij})_{n\times n}$.

*Theorem 1.* If $\mathrm{rank}(\mathbf{H}) = \mathrm{rank}(\mathbf{H}, \mathbf{d})$, and when $\alpha_i h_{ii} \in (0,1], i \in \{1, 2, \cdots, n\}$, $\mathbf{H}_1$ is a nonsingular $M$-matrix; when $\alpha_i h_{ii} \in [1,2), i \in \{1, 2, \cdots, n\}$, $\mathbf{H}_2$ is a nonsingular $M$-matrix, then the algorithm (4) is globally convergent.

Let $N_1 \bigcup N_2 = \{1, 2, \cdots, n\}, N_1 \bigcap N_2$ is empty.

*Theorem 2.* If $\mathrm{rank}(\mathbf{H}) = \mathrm{rank}(\mathbf{H}, \mathbf{d})$, and when $i \in N_1, \alpha_i h_{ii} \in (0,1), \alpha_i h_{ii} - \sum_{j=1, j\neq i}^{n} \alpha_i |h_{ij}| > 0$; when $l \in N_2, \alpha_l h_{ll} \in [1,2), (2 - \alpha_l h_{ll}) - \sum_{j=1, j\neq l}^{n} \alpha_l |h_{lj}| > 0$, then the algorithm (4) is globally convergent.

## 3.2 Analysis of Learning Parameters

Let

$$|H| = \begin{cases} h_{ii}, & i = j, \\ -|h_{ij}|, & i \neq j, \end{cases}$$

*Corollary 1.* If $|H|$ is a nonsingular $M$-matrix, $\alpha_i h_{ii} = 1$, then the algorithm (4) is globally convergent.

*Remark 1.* When $|H|$ is a nonsingular $M$-matrix, there exist positive constants $\gamma_1, \gamma_2, \cdots, \gamma_n$ such that

$$\gamma_i h_{ii} - \sum_{j=1, j\neq i}^{n} \gamma_j |h_{ij}| > 0.$$

According to the proof of Theorem 1, $\forall i \in \{1, 2, \cdots, n\}$,

$$|x_i(t) - x_i^*| \leq \max_{1\leq i\leq n} \{|x_i(0) - x_i^*|\} \Big\{ \sum_{j=1, j\neq i}^{n} \frac{\gamma_j |h_{ij}|}{\gamma_i h_{ii}} \Big\}^t,$$

where $t$ is natural number.

*Corollary 2.* If $h_{ii} - \sum_{j=1, j\neq i}^{n} |h_{ij}| > 0$, then the algorithm (4) is globally convergent with the estimation

$$|x_i(t) - x_i^*| \leq \max_{1\leq i\leq n} \{|x_i(0) - x_i^*|\} \max_{1\leq i\leq n} \Big\{ \sum_{j=1, j\neq i}^{n} \frac{|h_{ij}|}{h_{ii}} \Big\}^t,$$

where $t$ is natural number, $x^* = (x_1^*, x_2^*, \cdots, x_n^*)^T$ is a convergent point of the algorithm (4).

*Remark 2.* If $h_{ii}-\sum_{j=1,j\neq i}^{n}|h_{ij}|>0$, by choosing the algorithm (2), according to the results in [8], we can obtain $|x_i(t)-x_i^*|\leq \max_{1\leq i\leq n}\{|x_i(0)-x_i^*|\}\max_{1\leq i\leq n}\Big\{(1-\alpha h_{ii})+\alpha\sum_{j=1,j\neq i}^{n}|h_{ij}|\Big\}^{t}$, where $t$ is natural number, $x^*=(x_1^*,x_2^*,\cdots,x_n^*)^T$ is a convergent point of the algorithm (2).

We will compare the algorithm (2) with the algorithm (4) by an example.

## 4 Steepest Descent with Momentum

Let

$$\hat{h}_{ij}=\begin{cases}(1-\gamma_i)\alpha_i h_{ii}-2\gamma_i, & i=j,\\ -(1-\gamma_i)\alpha_i|h_{ij}|, & i\neq j,\end{cases}$$

$$\check{h}_{ij}=\begin{cases}2-(1-\gamma_i)\alpha_i h_{ii}-2\gamma_i, & i=j,\\ -(1-\gamma_i)\alpha_i|h_{ij}|, & i\neq j.\end{cases}$$

Denote matrices $\mathbf{H}_3=(\hat{h}_{ij})_{n\times n},\mathbf{H}_4=(\check{h}_{ij})_{n\times n}$.

*Theorem 3.* If $\text{rank}(\mathbf{H})=\text{rank}(\mathbf{H},\mathbf{d})$, and when $(1-\gamma_i)\alpha_i h_{ii}-\gamma_i\in(0,1],i\in\{1,2,\cdots,n\}$, $\mathbf{H}_3$ is a nonsingular $M$-matrix; when $(1-\gamma_i)\alpha_i h_{ii}-\gamma_i\in[1,2),i\in\{1,2,\cdots,n\}$, $\mathbf{H}_4$ is a nonsingular $M$-matrix, then the algorithm (5) is globally convergent.

## 5 Example

Consider a quadratic function represented by

$$F(\mathbf{x})=\frac{1}{2}\mathbf{x}^T\mathbf{H}\mathbf{x}+c, \tag{7}$$

where

$$\mathbf{H}=\begin{pmatrix}2, 1\\ 2, 4\end{pmatrix}.$$

By choosing the algorithm (2),

$$(1-\alpha h_{11})+\alpha\sum_{j=1,j\neq 1}^{2}|h_{ij}|=1-\alpha,$$

$$(1-\alpha h_{22})+\alpha\sum_{j=1,j\neq 2}^{2}|h_{ij}|=1-2\alpha.$$

In addition, $\alpha h_{11}\leq 1,\alpha h_{22}\leq 1$. Hence, by choosing $\alpha=0.25$,

$$\max_{1\leq i\leq 2}\Big\{(1-\alpha h_{ii})+\alpha\sum_{j=1,j\neq i}^{2}|h_{ij}|\Big\}=\max_{1\leq i\leq 2}\{1-\alpha,1-2\alpha\}=0.75.$$

According to the results in [8], we can obtain $|x_i(t)| \le y_1(t) = \max_{1\le i\le 2}\{|x_i(0)|\}$ $(0.75)^t$, where $t$ is natural number.

By choosing the algorithm (4),

$$(1-\alpha_1 h_{11}) + \alpha \sum_{j=1, j\neq 1}^{2} |h_{ij}| = 1-\alpha_1,$$

$$(1-\alpha_2 h_{22}) + \alpha_2 \sum_{j=1, j\neq 2}^{2} |h_{ij}| = 1-2\alpha_2.$$

In addition, $\alpha_1 h_{11} \le 1, \alpha_2 h_{22} \le 1$. Hence, by choosing $\alpha_1 = 0.5, \alpha_2 = 0.25$,

$$\max_{1\le i\le 2}\left\{(1-\alpha_i h_{ii}) + \alpha_i \sum_{j=1, j\neq i}^{2} |h_{ij}|\right\} = \max_{1\le i\le 2}\{1-\alpha_1, 1-2\alpha_2\} = 0.5.$$

According to Corollary 2, we can obtain $|x_i(t)| \le y_2(t) = \max_{1\le i\le 2}\{|x_i(0)|\}$ $(0.5)^t$, where $t$ is natural number.

Hence, the algorithm (4) does more accurately than the algorithm (2).

| Times | The algorithm (2) | | | The algorithm (4) | | |
|---|---|---|---|---|---|---|
| | Actual value of $x_1$ | Actual value of $x_2$ | Estimate value $y_1$ | Actual value of $x_1$ | Actual value of $x_2$ | Estimate value $y_2$ |
| 1 | -1.0000 | 0.5000 | 1.5000 | -1.0000 | 0.5000 | 1.0000 |
| 3 | -0.4375 | 0.3125 | 0.8438 | -0.2500 | 0.1250 | 0.2500 |
| 6 | -0.1387 | 0.1016 | 0.3560 | -0.0156 | 0.0313 | 0.0313 |
| 9 | -0.0442 | 0.0323 | 0.1502 | -0.0039 | 0.0020 | 0.0039 |
| 12 | -0.0141 | 0.0103 | 0.0634 | -0.0002 | 0.0005 | 0.0005 |
| 15 | -0.0045 | 0.0033 | 0.0267 | -0.0001 | 0.0000 | 0.0001 |

The actual values and estimate values of $x_i$ in the algorithms (2) and (4) with the initial string $(-1, 2)^T$.

| Times | The algorithm (2) | | | The algorithm (4) | | |
|---|---|---|---|---|---|---|
| | Actual value of $x_1$ | Actual value of $x_2$ | Estimate value $y_1$ | Actual value of $x_1$ | Actual value of $x_2$ | Estimate value $y_2$ |
| 1 | -0.7500 | 1.0000 | 0.5000 | 0.5000 | 1.0000 | 1.5000 |
| 3 | -0.4063 | 0.3125 | 0.1250 | 0.1250 | 0.2500 | 0.8438 |
| 6 | -0.1309 | 0.0957 | -0.0313 | -0.0313 | -0.0156 | 0.3560 |
| 9 | -0.0417 | 0.0305 | 0.0020 | 0.0020 | 0.0039 | 0.1502 |
| 12 | -0.0133 | 0.0097 | -0.0005 | -0.0005 | -0.0002 | 0.0634 |
| 15 | -0.0042 | 0.0031 | 0.0000 | 0.0000 | 0.0001 | 0.0267 |

The actual values and estimate values of $x_i$ in the algorithms (2) and (4) with the initial string $(-2, -1)^T$.

| Times | The algorithm (2) | | | The algorithm (4) | | |
|---|---|---|---|---|---|---|
| | Actual value of $x_1$ | Actual value of $x_2$ | Estimate value $y_1$ | Actual value of $x_1$ | Actual value of $x_2$ | Estimate value $y_2$ |
| 1 | 0.0000 | -0.5000 | 1.5000 | -1.0000 | -0.5000 | 1.0000 |
| 3 | 0.0625 | -0.0625 | 0.8438 | -0.2500 | -0.1250 | 0.2500 |
| 6 | 0.0215 | -0.0156 | 0.3560 | 0.0156 | 0.0313 | 0.0313 |
| 9 | 0.0068 | -0.0050 | 0.1502 | -0.0039 | -0.0020 | 0.0039 |
| 12 | 0.0022 | -0.0016 | 0.0634 | 0.0002 | 0.0005 | 0.0005 |
| 15 | 0.0007 | -0.0005 | 0.0267 | -0.0001 | 0.0000 | 0.0001 |

The actual values and estimate values of $x_i$ in the algorithms (2) and (4) with the initial string $(1,2)^T$.

| Times | The algorithm (2) | | | The algorithm (4) | | |
|---|---|---|---|---|---|---|
| | Actual value of $x_1$ | Actual value of $x_2$ | Estimate value $y_1$ | Actual value of $x_1$ | Actual value of $x_2$ | Estimate value $y_2$ |
| 1 | -1.2500 | 1.0000 | 1.5000 | -1.2500 | 1.0000 | 1.0000 |
| 3 | -0.5938 | 0.4375 | 0.8438 | -0.5938 | 0.4375 | 0.2500 |
| 6 | -0.1895 | 0.1387 | 0.3560 | -0.1895 | 0.1387 | 0.0313 |
| 9 | -0.0604 | 0.0442 | 0.1502 | -0.0604 | 0.0442 | 0.0039 |
| 12 | -0.0192 | 0.0141 | 0.0634 | -0.0192 | 0.0141 | 0.0005 |
| 15 | -0.0061 | 0.0045 | 0.0267 | -0.0061 | 0.0045 | 0.0001 |

The actual values and estimate values of $x_i$ in the algorithms (2) and (4) with the initial string $(-2,1)^T$.

## 6 Conclusion

In this paper, we analyze global convergence and learning parameters of the back-propagation algorithm for quadratic functions, present some theoretical results on global convergence conditions of the steepest descent algorithm with momentum (and without momentum) by directly analyzing the exact momentum equations for quadratic cost functions. In addition, in order to guarantee the convergence for a given learning task, the method is obtained to choose the proper learning parameters. The results presented in this paper are the improvement and extension of the existed ones in some existing works.

## Acknowledgement

This work was supported by the Natural Science Foundation of China under Grant 60405002 and Program for the New Century Excellent Talents in University of China under Grant NCET-06-0658.

## References

1. Phansalkar, V.V., and Sastry, P.S.: Analysis of the Back-propagation Algorithm with Momentum. IEEE Trans. Neural Networks, **5** (1994) 505-506
2. Rumelhart, D.E., Hinton, G.E., and Williams, R.J.: Learning Representations by Back-propagating Errors. Nature, **323** (1986) 533-536

3. Hagan, M.T., Demuth, H.B., and Beale, M.H.: Neural Network Design. Boston, MA: PWS, (1996)
4. Torii, M., and Hagan, M.T.: Stability of Steepest Descent with Momentum for Quadratic Functions. IEEE Trans. Neural Networks, **13** (2002) 752-756
5. Hagiwara, M., and Sato, A.: Analysis of Momentum Term in Back-propagation. IEICE Trans. Inform. Syst., **8** (1995) 1-6
6. Sato, A.: Analytical Study of the Momentum Term in A Backpropagation Algorithm. Proc. ICANN91, (1991) 617-622
7. Qian, N.: On the Momentum Term in Gradient Descent Learning Algorithms. Neural Networks, **12** (1999) 145-151
8. Zeng, Z.G., Huang, D.S., and Wang, Z.F.: Global Convergence of Steepest Descent for Quadratic Functions. In: Yang, Z.R. et al. (eds.): Intelligent Data Engineering and Automated Learning C IDEAL 2004. Lecture Notes in Computer Science, Vol. 3177. Springer-Verlag, Berlin Heidelberg New York (2004) 672-677

# Application Server Aging Prediction Model Based on Wavelet Network with Adaptive Particle Swarm Optimization Algorithm

Meng Hai Ning[1], Qi Yong[1], Hou Di[1], Pei Lu Xia[1], and Chen Ying[2]

[1] School of Electronics and Information Engineering, Xi'an Jiaotong University, 710049 Xi'an, China
[2] IBM China Research Laboratory, 100094 Beijing, China
mengning2001ji@gmail.com

**Abstract.** According to the characteristic of performance parameters of application sever, a new software aging prediction model based on wavelet network is proposed. The dimensionality of input variables is reduced by principal component analysis, and the parameters of wavelet network are optimized with adaptive particle swarm optimization (PSO) algorithm. The objective is to observe and model the existing systematic parameter data series of application server to predict accurately future unknown data values. By the model, we can get the aging threshold before application server fails and rejuvenate the application server in autonomic ways before observed systematic parameter value reaches the threshold. The experiments are carried out to validate the efficiency of the proposed model and show that the aging prediction model based on wavelet network with adaptive PSO algorithm is effective and more accurate than wavelet network model with Genetic algorithm (GA).

**Keywords:** Application server, software aging, Particle swarm optimization, wavelet network, time series prediction, software reliability.

## 1 Introduction

Recent studies have reported the phenomenon of software aging [1, 2] in which the state of system performance degrades with time. The primary symptoms of this degradation include exhaustion of system resources, data corruption and instantaneous error accumulation. This may eventually lead to performance degradation, crash/hang failure, or other unexpected effects. Aging has not only been observed in software used on a mass scale but also in specialized software used in high-availability and safety-critical applications. In order to enhance system reliability and performance and prevent degradation or crash, such a preventive technique called software rejuvenation was introduced [1]. This involves occasionally stopping the running software, cleaning its internal state and then restart. For optimizing the timing of such a preventive maintenance, it is important to detect software aging and predict the time when the resource exhaustion reaches the critical level. Our final objective is

D.-S. Huang, L. Heutte, and M. Loog (Eds.): ICIC 2007, LNAI 4682, pp. 14–25, 2007.

to predict software aging of application server and then take preventive maintenance technique such as software rejuvenation to improve the reliability and availability of application server, thus it leads to lower maintenance cost and more reliable application server that are under the effect of software aging.

Most of the previous measurement techniques for dependability evaluation were based on data from failure events [3, 4]. Estimation of the failure rate and mean time to failure of widely distributed software was presented in [3]. The approach for failure prediction was described in [5], which based on an increase in observed error rate, an error number threshold, a CPU utilization threshold or a combination of the above factors. For the reason that software aging cannot be detected or estimated via collecting data at failure events only, by contrast, periodically monitoring and recording of the activity parameters of software in operation is adopted in our works. The data relative to system parameters are extracted from application server at regular intervals, therefore the extracted data can be considered as the time series of system parameters. So far, many kinds of methods for time series prediction have been proposed, such as neural network [6], principal component analysis [7], wavelet network [8-10], Bayesian theory [11] and support vector machine [12]. Neural networks [8] are powerful tools for fitting nonlinear time series. However, the implementation of neural networks has disadvantages in prediction precision, network convergence rate, determining the parameters of neurons and constructing network topology, and the training processes often settle in undesirable local minimal of the error surface. Wavelet networks [8] can make up for the deficiencies of both wavelet and neural network and construct network topology efficiently. The key problem is to design an algorithm to determine the network structure and train the network to adjust the parameters to minimize the cost function. The procedure for designing wavelet network structure essentially involves selecting the input layer, hidden layer, and output layer. According to Occam's razor principle, the fewer weights in the network, the greater confidence that over-training has not result in noise being fitted. The selection of input layer mainly depends on the consideration of which input variables are necessary for prediction the output ones. From the complexity viewpoint, it would be required to reduce the number of input nodes to an absolute minimum of essential nodes. In this regard, principal component analysis [7] is used here to reduce the number of impact factors and keep the accuracy of prediction model. On the other hand, Genetic algorithm is used as optimization method in our research before [10, 16]. PSO algorithm [13-15] can be used to train neural network as genetic algorithms. However, PSO has not complex encoding, crossover and mutation as genetic algorithm. The particles in a PSO system have their own positions representing the current solutions and velocities reflecting the changing rate of the solution in each generation. PSO algorithm need not adjust lots of parameters and has characteristics of rapid convergence. Thus, PSO algorithm is adopted to help search the optimum parameters of wavelet network.

In this paper, wavelet network method with adaptive PSO algorithm is proposed to predict resource usage for the purposes of detecting aging in application sever. Firstly, principal component analysis (PCA) is introduced to preprocess the original multi-objective variables, and the principal components of original variables are considered as the input of Wavelet network, which cuts down the dimensions of input, thus improves the convergence rate and stability of wavelet network and simplifies the wavelet network structure. Then the parameters of wavelet network are optimized by adaptive PSO algorithm. The experiment results are demonstrated to validate the efficiency of the proposed method, and show that the aging prediction model based on wavelet network with PSO algorithm is superior to the wavelet networks model with GA [10] in the aspects of convergence rate and prediction precision.

## 2 Software Aging in Application Server

Application server is a complex software system on which enterprise applications are deployed and executed. Due to Application server presents high-level abstractions that simplify the development of enterprise applications, programmers are shielded from handling issues such as transactions, database interactions, concurrency, and memory. An application server may have more than a hundred parameters that relate to software aging. The parameters include the size of multiple thread pools, queues and cache, session bean count, response time, throughput, JVM heap memory usage and JVM free heap memory.

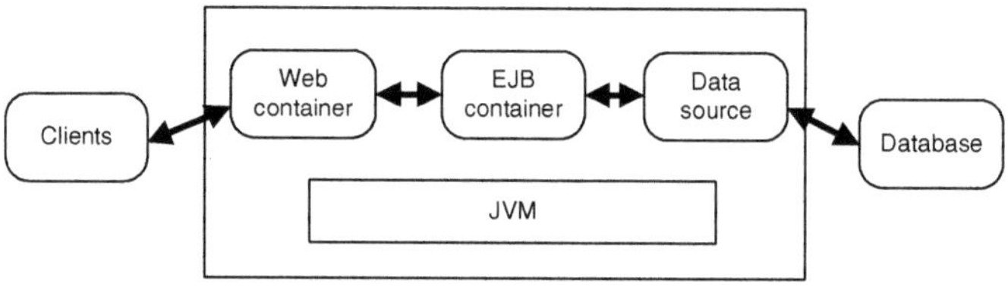


**Fig. 1.** Application server architecture

Fig.1 shows the architecture of a J2EE application server and the components with which interacts. An application server can be thought of consisting of three components: a web container, the component corresponding to the presentation layer, where JSPs, static HTML pages, and servlets execute, an EJB Container, the component corresponding to the business logic layer, where Enterprise Java Beans (EJBs) execute, and the Data Source layer, an abstraction of a database or other back-end, where transactions and interactions to persistent data stores are handled. Clients request service to application server. Requests flow from Web Containers to EJB containers to Data Sources and to a database.

## 3 Application Server Aging Prediction Model

### 3.1 Preprocess Based on Primary Component Analysis

In application sever, the stateful session bean count, stateless session bean count, container management persistence count and bean management persistence count increase with run time of application sever, which increases the JVM heap memory usage and enhances the happening probability of aging of application server. And the increase of JVM heap memory usage directly results in response time and throughout decrease. The relationship among them can be expressed as such a particular mathematical function analytic formula:

$$y = f(x_1, x_2, x_3, \cdots, x_n) \ . \tag{1}$$

where y denotes the amount of JVM memory usage and $x_1$, $x_2$,..., $x_n$ is the impact factors of aging of application sever.

Nevertheless, multiple factors usually reduce the efficiency of prediction, the principal component analysis is used here to reduce the number of impact factors and keep the accuracy of prediction model. The samples are represented as $X=(X_1, X_2,\ldots, X_n)^T$. The steps arc as follows:

Step1. The samples $X$ (factors) are normalized to remove dimension affection.

Step2. Calculate the relative matrix $P$ and covariance matrix $S$ of the sample data and the characteristic roots and vectors of matrix $S$.

Step3. Calculate the contribution rate of each component respectively. If the accumulation contribution rate of the first m components is more than 85 percent, the first $m$ factors $x_1,x_2,x_3,\ldots,x_m$ are principal components.

After dealing with principal component analysis, response time and throughput are selected as impact factors of software aging. Thus, formula (1) can be reduced to formula as follow:

$$y = f(x_1, x_2) \tag{2}$$

where $y$ denotes the amount of JVM heap memory usage, $x_1$ is response time and $x_2$ is throughput amount.

### 3.2 Wavelet Network (WN) Aging Prediction Model

The primary components of aging factors in application server are predicted using wavelet network. Fig.2 illustrates the basic design schema of wavelet network.

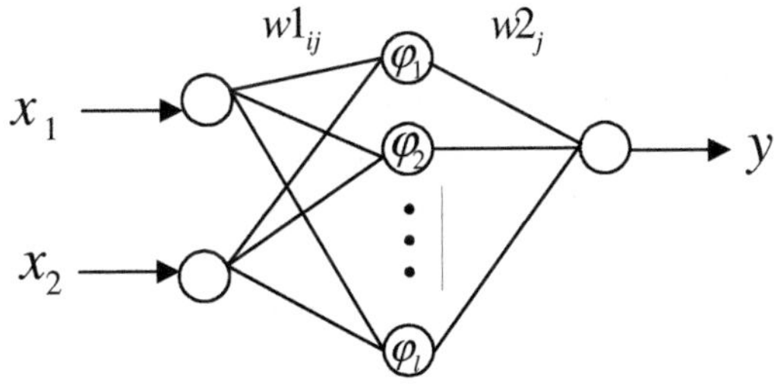


$i = 1, 2; \quad j = 1, 2, \cdots, l;$

**Fig. 2.** Wavelet network aging prediction model

The wavelet network includes three layers:

Layer 1 includes input variables $x_1, x_2$;

Layer 2 consists of wavelet function substituting for activation function. Weight $w1$ links the input nodes and the hidden nodes. Wavelet function is expressed as follows:

$$\varphi_j(x) = s_j^{-\frac{1}{2}} \psi\left(\frac{x - t_j}{s_j}\right) \qquad (j = 1, 2, \cdots, l) \tag{3}$$

where $s_j$, $t_j$ are dilation and translation factors of mother wavelet $\psi$. $\varphi$ is a set of daughter wavelets generated by dilation s and translation t from a mother wavelet $\psi$. In this paper, Morlet wavelet is chosen as a mother wavelet expressed as follows:

$$\psi(x) = \cos(1.75x) e^{-\frac{x^2}{2}} \tag{4}$$

By substitution (3) to (4), the following formula is drawn.

$$\varphi_j(x) = s_j^{-\frac{1}{2}} \cos\left(1.75 \cdot \left(\frac{x - t_j}{s_j}\right)\right) e^{-\frac{\left(\frac{x-t_j}{s_j}\right)^2}{2}} \qquad (j = 1, 2, \cdots, l) \tag{5}$$

Layer 3 is an output layer that sums the production of output value of the hidden nodes and the output connection weight $w2$ between the hidden nodes and the output nodes. The output formula of wavelet network is expressed as follows:

$$y = \sum_{j=1}^{l} w2_j \varphi_j \tag{6}$$

From the theory above, the wavelet network formula can be deduced as follows:

$$y(x) = \sum_{j=1}^{l} w2_j s_j^{-\frac{1}{2}} \psi\left(\frac{\sum_{i=1}^{n} w1_{ij} x_i - t_j}{s_j}\right) \tag{7}$$

Two key problems in designing of WNN are how to determine WNN architecture, what learning algorithm can be effectively used for training the WNN, and how to find the proper orthogonal or nonorthogonal wavelet basis.

### 3.3 Iterative Gradient Descent-Based Method with Additive Momentum

Put the input and actual value of $p$ samples into the wavelet network and calculate the output values and the system error of the network. The training is base on the minimization of the following cost function:

$$E = \frac{1}{2}\sum_{l=1}^{p}(d_l - y_l)^2 \tag{8}$$

where $y_l$ is the computing value of the $l$ th sample on the output node in the wavelet network. And $d_l$ is the actual value of the output node.

The minimization is performed by iterative gradient descent-based method with additive momentum. The partial derivative of the cost function with respect to $\theta = [w1\ w2\ t\ s]$ is as follows:

$$\frac{\partial E}{\partial \theta} = \sum_{l=1}^{p}(d_l - y_l)\frac{\partial y_l}{\partial \theta} \tag{9}$$

$$\text{Weight } w_2: \quad \frac{\partial y}{\partial w2_j} = \varphi_j \tag{10}$$

$$\text{Weight } w_1: \quad \frac{\partial y}{\partial w1_{ij}} = w2_j \frac{\partial \varphi_j}{\partial w1_{ij}} \tag{11}$$

$$\text{Dilation } s_i: \quad \frac{\partial y}{\partial s_j} = w2_j \frac{\partial \varphi_j}{\partial s_j} \tag{12}$$

$$\text{Translation } t_i: \quad \frac{\partial y}{\partial t_j} = w2_j \frac{\partial \varphi_j}{\partial t_j} \tag{13}$$

$(j = 1, 2, \cdots, l,\ i = 1, 2)$.

The parameters $\theta = [w1\ w2\ t\ s]$ are adjusted according to the formula as follows:

$$\theta_{k+1} = \theta_k + \Delta\theta_k \tag{14}$$

$$\Delta\theta_k = -(1-\alpha)\eta\frac{\partial E}{\partial \theta} + \alpha\Delta\theta_{k-1} \tag{15}$$

where $\eta$ is learning rate parameter, $0 < \eta < 1$, $\alpha$ is the momentum constant, and $0 < \alpha < 1$.

### 3.4 PSO Algorithm for Training WN Aging Prediction Model

It is difficult to decide the best parameters of the wavelet network. The learning algorithm of wavelet network often settles in undesirable local minima and converges slowly. PSO algorithm is adopted here to help search the optimum number of hidden nodes and parameters of wavelet network such as the connection weights. $w1_{ij}$, $w2_j$, the wavelet translation factor $t_i$, and dilation factor $s_i$.

### 3.4.1 Principle of PSO Algorithm

PSO is an evolutionary computation technique developed by Kennedy and Eberhart in 1995 [13].

Given an optimization function $f(X)$, where $X$ is a $n$-dimensional random vector. The PSO initializes a swarm of particles. Each particle $i$ has its velocity $V_i=(v_{i1},v_{i2},\ldots,v_{ij})$ and position $X_i=(x_{i1},x_{i2},\ldots,x_{ij})$, $i=1,2,\ldots,q$, $j=1,2,\ldots,n$, where $q$ is the swarm size. Particle $i$ is a candidate solution to the optimization function and it flies through the problem space to search the global optimum resolution. In each generation, particles $i$ adjusts its velocity $V_i$ and position $X_i$ according to the following formula:

$$v_i(k+1) = wv_i(k) + c_1 \times \varphi_1 \times (p_i(k) - x_i(k)) + c_2 \times \varphi_2 \times (p_g(k) - x_i(k)) \tag{16}$$

$$x_i(k+1) = x_i(k) + v_i(k) \tag{17}$$

where $c_1$ and $c_2$, termed as cognition and social components respectively, are the acceleration constants which change the velocity of a particle. $\varphi_1$ and $\varphi_2$ are uniform random functions (i.e. rand()) in the range of [0, 1], $v_i$ is particle's current velocity, $x_i$ is particle's current position. $p_i(k)$ is the position at which the particle has achieved its best fitness, $p_g(k)$ is the position at which the best global fitness has been achieved. $w$ is generation weight. If $w$ is bigger, then the algorithms has strong global search ability, otherwise, the algorithms tends to local search. $w$ can be adjusted as follows:

$$w(k) = w_{\max} - \frac{w_{\max} - w_{\min}}{k_{\max}} \times k \tag{18}$$

where $w_{max}$ is the initial weight, $w_{min}$ is the final weight, $k$ is the current generation number and $k_{max}$ is the maximum generation number.

In general, the velocity formula of a PSO particle in equation (16) comprises three parts. The first is the momentum part, which prevents abrupt velocity change. The second is the "cognitive" part, which represents learning achieved from its own search experience. The third is the "social" part that represents the cooperation among particles that learn from the group best's search experience. The generation weight $w$ controls the balance of global and local search ability.

### 3.4.2 Fitness Evaluation

The least-squared error function $e$ is used to represent the unfitness value of the PSO wavelet network associated with one particle. Thus, the fitness function $f$ is defined as follows:

$$f = \frac{1}{e} = \frac{1}{\frac{1}{2}\sum_{l=1}^{p}(d_l - y_l)^2} \tag{19}$$

where $y_l$ is the computing value of the $l$ th sample on the output node in the wavelet network, $p$ enumerates the points of training data set, and $d_l$ is the actual value of the output node accordingly.

### 3.4.3 Adaptive PSO Algorithm for Training WN

According to the general principle of PSO algorithm. The main steps of training wavelet network with adaptive PSO algorithm can be summarized as the following algorithm 1.

**Algorithm 1.** Adaptive PSO algorithm for training wavelet network

1. Input data and generate initial swarm $G(0)$ at random, and set $i$=0;
2. Encode the candidate solution as $x=\{w1_{ij}, w2_j, t_i, s_i\}$, $i$=1,2, $j$=1,2,...$l$, where $t_i$, $s_i$ are dilation and translation factors of wavelet, $w1_{ij}$ denotes the connection weights between input nodes and hidden nodes, $w2_j$ denotes the connection weights between hidden nodes and output nodes, and $l$ is the number of hidden nodes. Thus, $x$ is a $6l$ dimension vector;
3. Initialize the position and velocity of each particle from the domain (-1,1) using random generator;
4. Initialize best fitness $pbest_i$ of each particle and the global best fitness $gbest$;
5. REPEAT
   a) Use iterative gradient descent-based method training wavelet network parameters, evaluate the fitness value of each particle in the swarm according to the training results;
   b) Compare each particle's current fitness value with particle's $pbest_i$. If current fitness value is better than $pbest_i$ value, then set $pbest_i$ value equal to the current value. Compare particle's current fitness value with Global best *gbest*. If current value is better than *gbest*, then set *gbest* value equal to the current value.
   c) Update the velocity and position of the particle according to the equation (16) and (17) respectively.
   d) Set $i=i+1$;
6. UNTIL termination criterion is satisfied or generation number reaches the given maximum generation number.

## 4 Experimental Results and Discussions

### 4.1 Experimental Setup Schema

The experimental platform simulates a monitoring and recording system for application server. The experimental environment consists of a J2EE application server, clients and database server. In the client, the load generator is used to generate requests to the application server through standards-based HTTP or SOAP protocols. The application server connects and queries database server, and then returns results to clients. By load generator model and resource monitor model, the dynamic parameters in clients and application server are periodically monitored and recorded in a certain format separately. The experimental setup schema is presented in Fig.3.

In the experiment, all the machines involved are 2.0 GHz Pentium IV system running Windows XP, with 2.0GB of memory. The application server is IBM Websphere application Server 5.1 with maximum JVM heap memory 256M. The

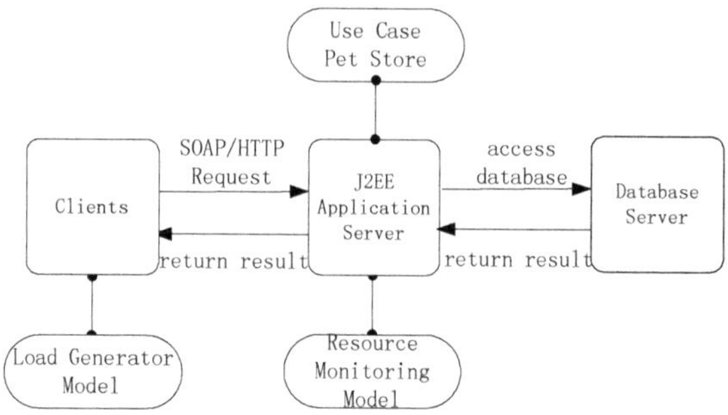


**Fig. 3.** Experimental setup schema

database server is CloudScape 4.0 that is integrated into Websphere application server. The machines are connected on a same local area network with 100Mbps Ethernet. Tivoli Performance Viewer is used to monitor the parameter data of Websphere application server. Pet Store is used as deployment application. The sampling interval is ten minutes. System dynamic parameters for about eleven days are extracted from recorded file to predict aging of Websphere application server.

### 4.2 Prediction Results and Analysis

Fig.4 shows the amount of used JVM heap memory of application server was increasing over time until the maximum JVM memory of 256M is fully occupied. Fig.5 shows how the free heap memory of JVM exhibits with the application server running. We can see the free heap memory of JVM changes steadily. Thus we predict the forward value of the JVM heap memory usage to grasp the application server aging threshold. Normalized mean square error (NMSE) is adopted as indicator of performance evaluation for aging prediction. NMSE is defined as follow:

$$NMSE = \frac{1}{\sigma^2 n} \sum_{k=1}^{n} (x(k) - \hat{x}(k))^2 \tag{20}$$

where $x(k)$ is the actual value of the time series, $\hat{x}(k)$ is the prediction value, $n$ enumerates the points of training data set, and $\sigma^2$ is the variance of the actual value of time series over the prediction period.

The wavelet function is taken as Morlet wavelet. The number of hidden nodes $l$ is selected as 20. Each wavelet network has 2 input nodes, 20 hidden layer nodes and 1 output node in the double WN model. Population size of the PSO-based training algorithms is 50. The Number of particles is 100. $w_{max} = 0.9$, $w_{min} = 0.4$, $w$ is adjusted adaptively according to formula (18). $c_1 = c_2 = 2$. The connection weight changes between [-1,1]. The maximum generation is determined as 600. Fig.6 displays the prediction data for one-step forward prediction model of JVM heap memory usage and the error between original data and prediction data. We can see the proposed model can predict application server aging with lower error and application server performance decreases with time. Table 1 presents approximation performance based on wavelet network with adaptive PSO algorithm model compared with wavelet network with genetic algorithm. The table is shown that prediction precision of wavelet network with adaptive PSO algorithm model is superior to wavelet network with GA algorithm. For the aging prediction model based wavelet network with PSO, when generation number reaches to 297, the NMSE value convergences to 0.0212 and the maximum fitness has been achieved. For the aging prediction model based wavelet network with GA, when generation number reaches to 482, the NMSE value convergences to 0.0267.

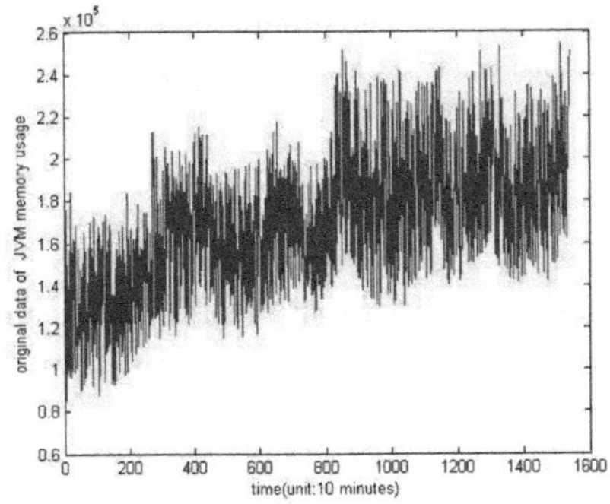


**Fig. 4.** JVM heap memory usage

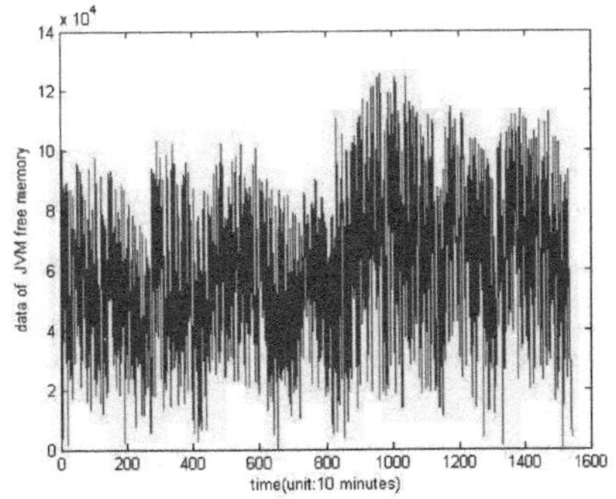


**Fig. 5.** JVM free heap memory

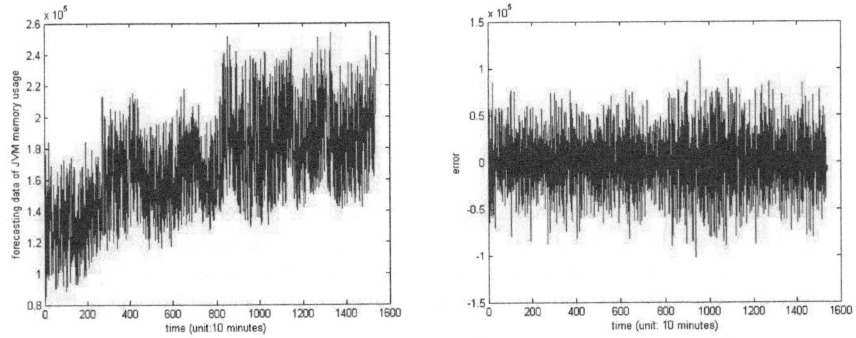


(a) (b)

**Fig. 6.** One-step forward prediction JVM heap memory usage (a) Prediction data, (b) Error between original data and prediction data

**Table 1.** Comparison of approximation performance

| Models | Generation Number | NMSE |
|---|---|---|
| Wavelet network with adaptive PSO algorithm | 297 | 0.0212 |
| Wavelet network with GA algorithm in ref [10] | 482 | 0.0267 |

## 5 Conclusions

The effectiveness of wavelet network with adaptive PSO algorithm for aging prediction has been investigated. The original time series is preprocessed by primary component analysis. Then the primary components are predicted by means of wavelet network and an algorithm of back-propagation based on adaptive iterative gradient descent method with adaptive PSO algorithm is proposed for wavelet network learning. PSO algorithm can optimize the parameters of wavelet network in the same BP training process. Thus, the local minimum problem in the training process will be overcome efficiently. Compared with previous work on wavelet network with GA, the method proposed in this paper has superiority in aspects of convergence rate and prediction precision.

It is important to predict the critical resource usage such as memory usage for application server. Software aging can be detected and the aging threshold before server crashed can be evaluated using the prediction model. Future work includes the aging prediction model considering more causations of resource.


## Acknowledgements

The author would like to thank the sponsors of the National Natural Science Foundation of China under Grant No. 60473098 and IBM China Research Laboratory Joint Project.


## References


1. Garg, S., Puliafito, A., Telek, M., Trivedi, K.S.: A Methodology for Detection and Estimation of Software Aging. Int. Symp. On Software Reliability Engineering, ISSRE (1998)
2. Huang, Y., Kintala, C., Kolettis, N., Fulton. N: Software Rejuvenation: Analysis, Module and Applications. IEEE Int. Symposium on Fault Tolerant Computing, FTCS 25 (1995).
3. Chillarege, R., Biyani, S., Rosenthal, J.: Measurement of failure rate in widely distributed software. In Proc. of 25th IEEE Intl. Symposium on Fault-Tolerant Computing, Pasadena, CA (1995) 424–433
4. Tang, D., Iyer, R. K.: Dependability Measurement Modeling of a Multicomputer System. IEEE Transactions on Computers, 31 (1993)
5. Lin, T.T., Siewiorek, D. P.: Error Log Analysis: Statistical Modeling and Heuristic Trend Analysis. IEEE Transactions on Reliability, 39 (1990) 419–432
6. Amir, B., Geva: Scalenet-Multiscale Neural network Architecture For Time Series Prediction. IEEE Transactions on Neural Networks, 9 (1998) 1471–1482
7. Rattan, Sanjay S.P., Hsieh, William,W: Complex-valued Neural Networks for Nonlinear Complex Principal Component Analysis, Neural Networks, 18 (1) (2005) 61–69
8. Zhang, Q., Benvenise, A.: Wavelet Network. IEEE Transactions on Neural Network, 3 (1992) 889–898
9. Bashir, Z., El-Hawary, M.E.: Short Term Load Forecasting By Using Wavelet Neural Networks. The IEEE Conference on Electrical and Computer Engineering, Canadian (2000) 163 –166

10. Meng, H. N., Qi, Y., Hou, D. (ed.): Study on Application Server Aging Prediction based on Wavelet Network with Hybrid Genetic Algorithm. International Symposium on Parallel and Distributed Processing and Applications, Sorrento, Italy (2006) 573–583.
11. Chris, C., Holmes, B., Mallick, K.: Bayesian Wavelet Networks for Nonparametric Regression. IEEE transactions on neural networks, 11 (2000)
12. Zhang, X. (ed.): Robust Multiwavelets Support Vector Regression Network. International Con-ference on Control and Automation, Budapest, Hungary (2005) 27–29
13. Kennedy, J, Eberhart, R. C.: Particle swarm optimization, Proceedings of IEEE International Conference on Neural Networks, Perth, Australia (1995) 1942–1948
14. Zhang, C., Shao,H., Li, Y.: Particle swarm optimization for evolving artificial neural network. Proceedings of the IEEE International Conference on Systems, Man, And Cybernetics, 4 (2000) 2487–2490
15. Settles, M., Ryiander, B.: Neural Network Learning using Particle Swarm Optimizers. Advances in Information Science and Soft Computing, (2002) 224–226
16. Meng, H. N., Qi, Y., Hou, D. (ed.): Software Aging Prediction Model based on Fuzzy Wavelet Network with Adaptive Genetic Algorithm. 18th IEEE International Conference on Tools with Artificial Intelligence (2006)

# Edge Detection Based on Spiking Neural Network Model

QingXiang Wu, Martin McGinnity, Liam Maguire, Ammar Belatreche, and Brendan Glackin

School of Computing and Intelligent Systems, University of Ulster at Magee Campus Derry, BT48 7JL, Northern Ireland, UK
{q.wu,tm.mcginnity,lp.maguire,a.belatreche,b.glackin}@ulster.ac.uk

**Abstract.** Inspired by the behaviour of biological receptive fields and the human visual system, a network model based on spiking neurons is proposed to detect edges in a visual image. The structure and the properties of the network are detailed in this paper. Simulation results show that the network based on spiking neurons is able to perform edge detection within a time interval of 100 ms. This processing time is consistent with the human visual system. A firing rate map recorded in the simulation is comparable to Sobel and Canny edge graphics. In addition, the network can separate different edges using synapse plasticity, and the network provides an attention mechanism in which edges in an attention area can be enhanced.

**Keywords:** Edge detection, spiking neural networks, receptive field, attention, visual system.

## 1 Introduction

The visual cortex has a highly ordered structure [1-2], and it has attracted considerable attention from theoretical neurobiologists and computer scientists. For example, various network models for the visual cortex have been simulated using spiking neurons since the Hodgkin and Huxley equations [3] were regarded as a basic spiking neuron model [4]. Retinal ganglion cells convey the visual image from the eye to the brain [1-2]. Neurobiologists have found that various receptive fields exist in the visual cortex [1-2]. However, an accurate representation of the neuron circuits for the visual cortex is still not very clear. Various neural network models have been proposed to explain how the visual system is able to process on image efficiently. Knoblauch and Palm have proposed a network [5-6] consisted of three areas (retina, primary visual cortex, and central visual area). Each area is composed of several neuron populations and reciprocally connected. The network has been applied to scene segmentation by means of spike synchronization. A dynamically coupled neural oscillator network is proposed to segment image in [7]. By means of attention-guided object selection and novelty detection, an oscillatory model is proposed to recognise objects by combining consecutive selection of objects and discrimination between new and familiar objects [8].

D.-S. Huang, L. Heutte, and M. Loog (Eds.): ICIC 2007, LNAI 4682, pp. 26–34, 2007.

A model of self-organizing maps of spiking neurons has been applied in computational modelling of the pattern interaction and orientation maps in the primary visual cortex [9-11]. Spiking neurons with leaky integrator synapses have been used to model image segmentation and binding by synchronization and desynchronization of neuronal group activity. The model, which is called RF-SLISSOM, integrates the spiking leaky integrator model with the RF-LISSOM structure, modelling self-organization and functional dynamics of the visual cortex at a more accurate level than earlier models.

These neural network models can be applied to explain some of the behaviours of the visual system in the human brain. The spike synchronization network in [5-6] can be applied to explain why the visual system can perform high-level visual processing tasks in a limited time of 100-150 ms. This model is based on a firing order encoding scheme which is called spike wave, in which neurons are allowed to fire only once during a period. The model can explain how this information embedded in the first wave of spikes generated in the retina can be decoded by post-synaptic neurons, and how it can propagate in a feed-forward way through a simple hierarchical model of the visual system, to implement fast and reliable object recognition. Although to date there has been no experimental observation to directly confirm the model, there is also no direct experimental evidence of the contrary. The literature shows that many experimental results tend to favour the hypothesis in the model. Actually, many neuron models and receptive field models have been described in neuroscience [2]. In this paper, different receptive field models [2] are used to construct a spiking neural network which is used to simulate the visual cortex for edge detecting. Firstly, a network model based on integrate-and-fire neurons is detailed in Section 2. The receptive fields of spiking neurons play a crucial role for edge detecting in the network. The behaviours of the neurons with the receptive fields are analyzed in Section 3. Simulation results for edge detecting and comparison with other edge detecting algorithms are shown in Section 4. Discussions about the network are presented in Section 5.

## 2 Spiking Neural Network Model for Edge Detection

The human visual system performs edge detection very efficiently. Neuroscientists have found that there are various receptive fields from simple cells in the striate cortex to those of the retina and lateral geniculate nucleus (see page 236-248 in [2]) and the neurons can be simulated by the Hodgkin and Huxley neuron model. Based on these receptive fields and the neuron model, a network model is proposed to detect edges in a visual image in this paper. The structure of the network is shown in Fig. 1. Suppose that the first layer represents photonic receptors. Each pixel corresponds to a receptor. The intermediate layer is composed of four types of neurons corresponding to four different receptive fields respectively. 'X' in the synapse connections represents an excitatory synapse. 'Δ' represents an inhibitory synapse. Each neuron in the output layer integrates four corresponding outputs from intermediate neurons. The firing rate map of the output layer forms an edge graphic corresponding to the input image.

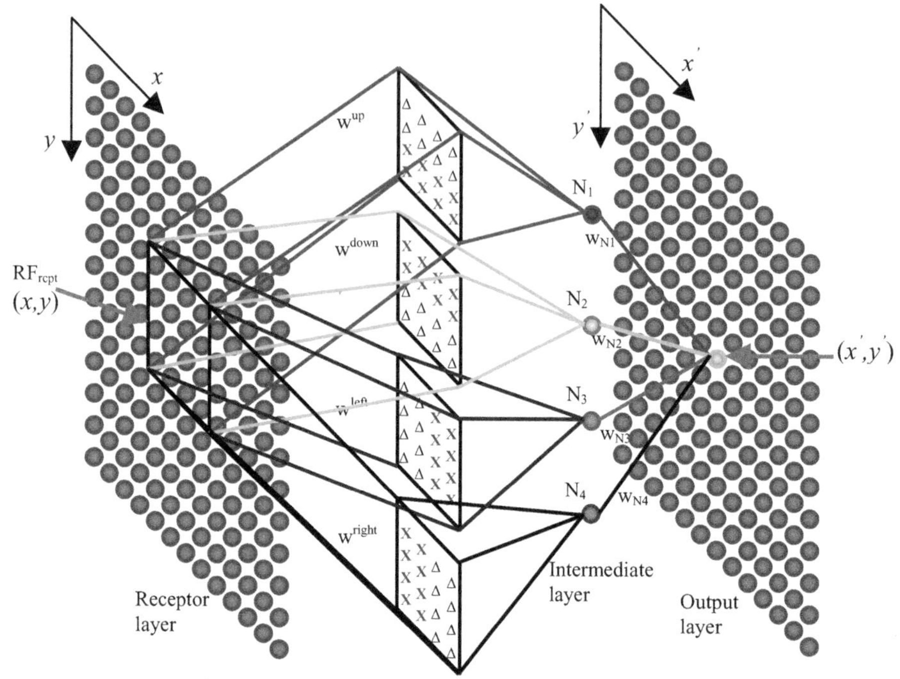


**Fig. 1.** Spiking Neural Network Model for Edge Detecting

There are four parallel arrays of neurons in the intermediate layer each of the same dimension as the Receptor layer. These arrays are flagged as N1, N2, N3 and N4 and only one neuron in each array is shown in Figure 1 for simplicity. Each of these layers perform the processing for up, down, left and right edges respectively and are connected to the receptor layer by differing weight matrices. These weight matrices can be of varying sizes to represent the width of the receptive field under consideration. For example in Figure 1 neuron N1 connects to receptive field $RF_{rcpt}$ in the receptor layer through synapse strength distribution matrix $w^{up}$, and responds to an up-edge within the field. If a uniform image within $RF_{rcpt}$ makes a uniform output, the outputs through synapses in $w^{up}$ reach neuron N1. Connections through the upper-half of the weight matrix represent inhibitory synapses which depress the membrane potential of Neuron N1 while connections through the lower-half excitatory synapses potentiate the membrane potential of Neuron N1. Therefore the membrane potential of Neuron N1 has not been changed, and no spikes are generated by Neuron N1. However, if an edge image within the $RF_{rcpt}$ is incident on lower-half receptors with a strong signal and the upper-half receptors with a very weak signal, then the strong signal will potentiate (due to the excitatory synapses) neuron N1, but the weak signal will not depress the membrane potential significantly. The membrane potential of Neuron N1 rise up fast and generates spikes frequently to respond to an up-edge within its receptive field. The synapse distribution matrix $w^{up}$ plays a role as a filter for up-edge within the receptive field. By analogy, neuron N2 with synapse strength distribution $w^{down}$ can best respond to a down-edge within the receptive field; neuron N3 with synapse strength distribution $w^{left}$ can best respond to a left-edge; and neuron N4 with synapse strength distribution $w^{right}$ can best respond to a right-edge. Neuron ($x'$, $y'$) in the output layer integrates the outputs from these four neurons from the neuron arrays in the intermediate layer, and can respond to any direction edge within receptive field $RF_{rcpt}$. The network model is presented in following sections.

## 3 Spiking Neuron Model and Receptive Fields

Simulation results show that the conductance based integrate-and-fire model is very close to the Hodgkin and Huxley neuron model [11-16]. The conductance based integrate-and-fire model is applied to the aforementioned network model. Let $G_{x,y}$ represent gray scale at $(x,y)\in RF_{rcpt}$, $q^{ex}_{x,y}$ represent peak conductance caused by excitatory current from a receptor at $(x,y)$, and $q^{ih}_{x,y}$ represent peak conductance caused to inhibitory current from a receptor at $(x,y)$. For simplicity, suppose that each receptor can transform a value of gray scale to peak conductance by the following expressions.

$$q^{ex}_{x,y} = \alpha G_{x,y}; \qquad q^{ih}_{x,y} = \beta G_{x,y} \tag{1}$$

where $\alpha$ and $\beta$ are constants. According to the conductance based integrate-and-fire model [15-16], neuron N1 is governed by the following equations.

$$\frac{g^{ex}_{x,y}(t)}{dt} = -\frac{1}{\tau_{ex}} g^{ex}_{x,y}(t) + \alpha G_{x,y} \tag{2}$$

$$\frac{g^{ih}_{x,y}(t)}{dt} = -\frac{1}{\tau_{ih}} g^{ih}_{x,y}(t) + \beta G_{x,y} \tag{3}$$

$$\begin{aligned} c_m \frac{dv_{N1}(t)}{dt} = g_l(E_l - v_{N1}(t)) + \sum_{(x,y)\in RF_{rcpt}} \frac{w^{up\_ex}_{x,y} g^{ex}_{x,y}(t)}{A_{ex}}(E_{ex} - v_{N1}(t)) \\ + \sum_{(x,y)\in RF_{rcpt}} \frac{w^{up\_ih}_{x,y} g^{ih}_{x,y}(t)}{A_{ih}}(E_{ih} - v_{N1}(t)) \end{aligned} \tag{4}$$

where $g^{ex}_{x,y}(t)$ and $g^{ih}_{x,y}(t)$ are the conductance for excitatory and inhibitory synapses respectively, $\tau_{ex}$ and $\tau_{ih}$ are the time constants for excitatory and inhibitory synapses respectively, $v_{N1}(t)$ is the membrane potential of neuron *N1*, $E_{ex}$ and $E_{ih}$ are the reverse potential for excitatory and inhibitory synapses respectively, $c_m$ represents a capacitance of the membrane, $g_l$ represents the conductance of membrane, *ex* is short for excitatory and *ih* for inhibitory, $w^{up\_ex}_{x,y}$ represents the strength of excitatory synapses, $w^{up\_ih}_{x,y}$ represents the strength of inhibitory synapses, $A_{ex}$ is the membrane surface area connected to a excitatory synapse, and $A_{ih}$ is the membrane surface area connected to a inhibitory synapse. According to the description of biological receptive fields [2], values for $w_{x,y}^{up\_ex}$ and $w_{x,y}^{up\_ih}$ are expressed as follows.

$$w_{x,y}^{up\_ex} = \begin{cases} 0 & if\ (y-y_c) \le 0 \\ w_{e\max} e^{-\frac{(x-x_c)^2}{\delta_x^2} - \frac{(y-y_c)^2}{\delta_y^2}} & if\ (y-y_c) > 0 \end{cases} \tag{5}$$

$$w_{x,y}^{up\_ih} = \begin{cases} 0 & if\ (y-y_c) > 0 \\ w_{i\max} e^{-\frac{(x-x_c)^2}{\delta_x} - \frac{(y-y_c)^2}{\delta_y}} & if\ (y-y_c) \le 0 \end{cases} \tag{6}$$

where $(x_c, y_c)$ is the centre of receptive field $RF_{rcpt}$, $(x,y) \in RF_{rcpt}$, $\delta_x$ and $\delta_y$ are constants, $w_{emax}$ and $w_{imax}$ are the maximal weights for excitatory synapses and inhibitory synapses respectively.

By analogy, Neuron N2, N3, and N4 are governed by a set of equations similar to that for neuron N1. When the membrane potential reaches a threshold $v_{th}$ the neuron generates a spike, and then it enters a refractory state. After period $\tau_{ref}$ the neuron can integrate inputs to generate another spike. Let $S_{N1}(t)$ represent a spike train which is generated by neuron N1.

$$S_{N1}(t) = \begin{cases} 1 & if\ neuron\ N1\ fires\ at\ time\ t. \\ 0 & if\ neuron\ N1\ does\ not\ fire\ at\ time\ t. \end{cases} \tag{7}$$

By analogy, let $S_{N2}(t)$, $S_{N3}(t)$ and $S_{N4}(t)$ represent spike trains for neurons N2, N3 and N4 respectively. Neuron $N_{x',y'}$ in the output layer is governed by the following equations.

$$\frac{g_{x',y'}^{ex}(t)}{dt} = -\frac{1}{\tau_{ex}} g_{x'y'}^{ex}(t) + (w_{N1}S_{N1}(t) + w_{N2}S_{N2}(t) + w_{N3}S_{N3}(t) + w_{N4}S_{N4}(t)) \tag{8}$$

$$c_m \frac{dv_{x',y'}(t)}{dt} = g_l(E_l - v_{x',y'}(t)) + \frac{g_{x',y'}^{ex}(t)}{A_{ex}}(E_{ex} - v_{x',y'}(t)) \tag{9}$$

Note that Neuron $N_{x',y'}$ is connected to intermediate neurons only by excitatory synapses. Let $S_{x',y'}(t)$ represent spike a train generated by Neuron $N_{x',y'}$ in output layer. The firing rate for Neuron $N_{x',y'}$ is calculated by the following expression.

$$r_{x',y'} = \frac{1}{T} \sum_{t}^{t+T} S_{x',y'}(t) \tag{10}$$

By plotting this firing rate as an image with a colour bar an edge graphic for the input image is obtained.

## 4 Simulation Results

The network model was implemented in Matlab using a set of parameters for the network: $v_{th}$ = -60 mv. $v_{reset}$ = -70 mv. $E_{ex}$= 0 mv. $E_{ih}$= -75 mv. $E_l$= -70 mv. $g_l$=1.0 $\mu$s/mm$^2$. $c_m$=10 nF/mm$^2$. $\tau_{ex}$=4 ms. $\tau_{ih}$=10 ms. $\tau_{ref}$ =6 ms. $A_{ih}$=0.028953 mm$^2$. $A_{ex}$=0.014103 mm$^2$. These parameters are consistent with biological neurons [3].

Synapse strengths are controlled by $w_{emax}$ and $w_{imax}$. The proportion between $w_{emax}$ and $w_{imax}$ can be adjusted to ensure that the neuron does not fire in response to a uniform image within its receptive field. Contrasting the maximal weights $w_{emax}$ provided in [15] , $w_{emax}$ is set to 0.7093 for excitatory synapses, and $w_{imax}$ is set to 0.3455 for inhibitory synapses. Image gray scale values are normalized in a real number in the range of 0 to 1. Therefore, $\alpha$ and $\beta$ are set to 1/max_value_in_image. The size of $RF_{rcpt}$ may be set in the range 2×2 to 6×6. The parameters $\delta_x$ and $\delta_y$ can be applied to control sensitiveness to edges. Experiments for different values of $\delta_x$ , $\delta_y$ and size of $RF_{rcpt}$ have been done. The results show that the larger of $\delta_x$ , $\delta_y$ and size of $RF_{rcpt}$, the lower is the detector's sensitivity to noise. On the other hand, the larger of $\delta_x$ , $\delta_y$ and size of $RF_{rcpt}$, the edge become more vague. There is a tradeoff for selection of the values. For the synapse strength distribution matrix $w^{up}$ and $w^{down}$, $\delta_x$ should be set to $\delta_x > \delta_y$ to get a horizontal shape that will be consistent with the receptive field in biological system [2]. In the results presented, $\delta_x$ =6, $\delta_y$=2, and the size of $RF_{rcpt}$ is set to the 5×5. For example, the 5×5 receptive field matrices for $w^{up\_ex}$ and $w^{up\_ih}$ , which are calculated according to (5) and (6), are shown as follows.

$$w^{up\_ex} = \begin{bmatrix} 0 & 0 & 0 & 0 & 0 \\ 0 & 0 & 0 & 0 & 0 \\ 0 & 0 & 0 & 0 & 0 \\ .31 & .34 & .35 & .34 & .31 \\ .11 & .12 & .13 & .12 & .11 \end{bmatrix} \quad w^{up\_ih} = \begin{bmatrix} .11 & .12 & .13 & .12 & .11 \\ .31 & .34 & .35 & .34 & .31 \\ 0 & 0 & 0 & 0 & 0 \\ 0 & 0 & 0 & 0 & 0 \\ 0 & 0 & 0 & 0 & 0 \end{bmatrix}$$

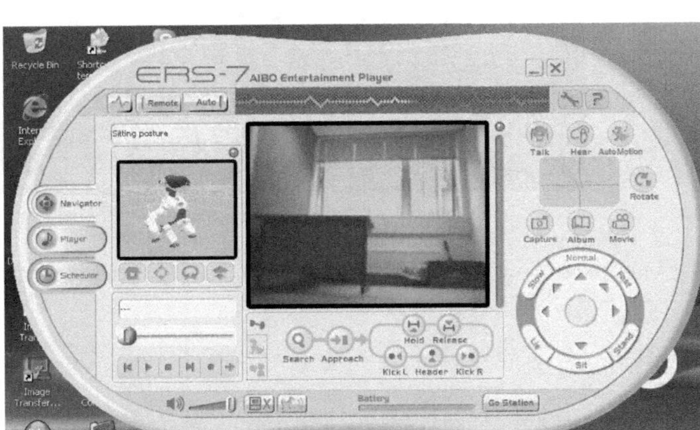


**Fig. 2.** Screen shot image from AIBO robot control system

If a screen shot, which is shown in Fig. 2, is presented to the network, the firing rate map on the output layer is obtained as shown in Fig. 3 reflecting the edges for the input image. Bright lines show that the corresponding neurons fires with a high frequency and indicate the edges with high contrast. Dark lines show that the corresponding neurons fires with a low frequency and indicate the edges with low contrast. Using the firing rates, different contrast edges can be separated.

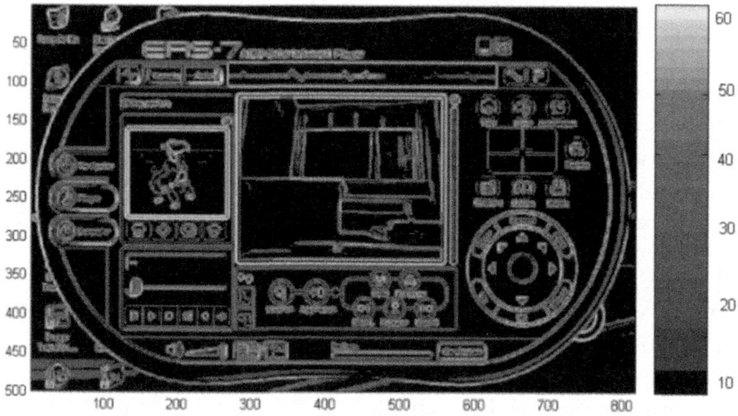


**Fig. 3.** Firing rate map from output layer

In order to compare with Sobel and Canny edge detection methods, the results for benchmark image Lena photo are shown in Fig. 4.

Lena photo | Sobel edges | Canny edges | Neuron firing rate map

**Fig. 4.** Comparison of neuron firing rate map with other edge detecting methods

## 5 Discussion

Spiking neural networks are constructed by a hierarchical structure that is composed of spiking neurons with various receptive fields and plasticity synapses. The spiking neuron models provide powerful functionality for integration of inputs and generation of spikes. Synapses are able to perform different computations, filters, adaptation and dynamic properties [17]. Various receptive fields and hierarchical structures of spiking neurons enable a spiking neural network to perform very complicated computations, learning tasks and intelligent behaviours in the human brain. This paper demonstrated how a spiking neural network can detect edges in an image. Although the neuron circuits in the brain for edge detection are not very clear, the proposed network model is a possible solution based on spiking neurons. In the simulation, the neuron firing rate map for edges can be obtained with a virtual biological time interval 100 ms. This time interval is consistent with the biological visual system. If the model is simulated by Matlab program in a PC with CPU 1.2G, it takes about 50 seconds to get a firing-rate map for an image with 500x800 pixels. If the network model is implemented in parallel on hardware, the edge detection can be achieved within 100 ms. Therefore, this model can be applied to artificial intelligent systems.

If synaptic plasticity is considered, different scales of firing rate map for edges can be obtained. For example, the human visual system can focus attention on a selected area and enhance resolution and contrast. Based on this model, an attention area can be enhanced by simply strengthening $w_{emax}$ and $w_{imax}$. Fig. 5 shows that an attention area around point (650,350) is enhanced. Within the attention area, $w_{emax}$=0.7093 and $w_{imax}$=0.3455. Outside of the attention area, $w'_{emax}= w_{emax}/4$ and $w'_{imax}= w_{imax}/4$.

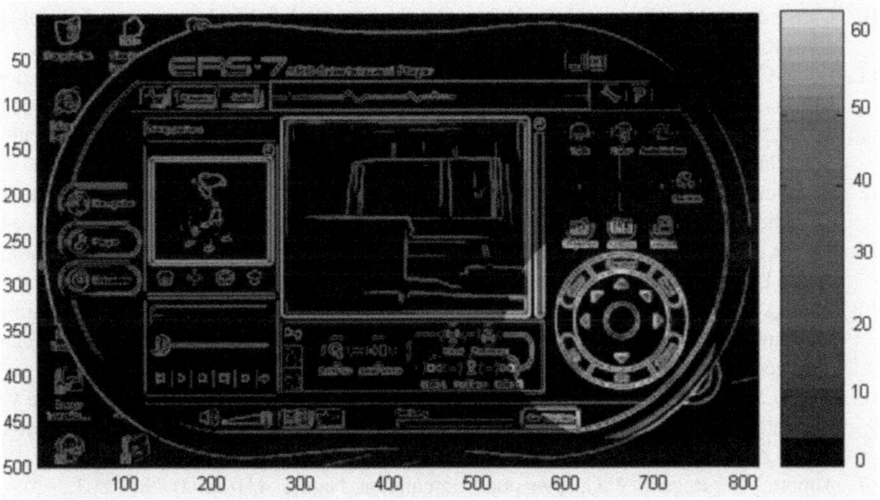


**Fig. 5.** Attention area around (650,350)

By adjusting neuron thresholds in the intermediate layer and output layer, the resolution and contrast in the attention area can also be enhanced. This paper has only investigated edge detection based on spiking neurons. Future work will consider different approaches to further improve the network and investigate the use of lateral connections within the intermediate layers or output layer.

## References

1. Hosoya, T., Baccus, S.A., Meister, M.: Dynamic Predictive Coding by The Retina. Nature, 436 (2005) 71 - 77
2. Kandel, E.R., Shwartz, J.H.: Principles of Neural Science. Edward Amold (Publishers) Ltd. (1981)
3. Hodgkin, A., Huxley, A.: A Quantitative Description of Membrane Current and Its Application to Conduction and Excitation in Nerve. Journal of Physiology. (London). 117 (1952) 500-544
4. Neuron Software download website: http://neuron.duke.edu/
5. Knoblauch, A., Palm, G.: Scene Segmentation by Spike Synchronization in Reciprocally Connected Visual Areas. I. Local Effects of Cortical Feedback, Biol Cybern. 87(2002) 151-67
6. Knoblauch, A., Palm, G.: Scene Segmentation by Spike Synchronization in Reciprocally Connected Visual Areas. II. Global Assemblies and Synchronization on Larger Space and Time Scales. Biol Cybern. 87 (2002) 168-84
7. Chen, K., Wang, D.L.: A Dynamically Coupled Neural Oscillator Network for Image Segmentation. Neural Networks. 15(3) (2002) 423-439
8. Purushothaman, G., Patel, S.S., Bedell, H.E., Ogmen, H.: Moving Ahead Through Differential Visual Latency, Nature. 396 (1998) 424-424.
9. Choe, Y., Miikkulainen, R.: Contour Integration and Segmentation in A Self-organizing Map of Spiking Neurons. Biological Cybernetics. 90(2) (2004) 75-88
10. Borisyuk, R.M., Kazanovich, Y.B.: Oscillatory Model of Attention-guided Object Selection and Novelty Detection. Neural Networks. 17(7) (2004) 899-915
11. Koch, C.: Biophysics of Computation: Information Processing in Single Neurons. Oxford University Press. (1999)
12. Dayan, P., Abbott, L.F.: Theoretical Neuroscience: Computational and Mathematical Modeling of Neural Systems. The MIT Press, Cambridge, Massachusetts. (2001)
13. Gerstner, W., Kistler, W.: Spiking Neuron Models: Single Neurons, pulations, Plasticity. Cambridge University Press. (2002)
14. Müller, E.: Simulation of High-Conductance States in Cortical Neural Networks, Masters thesis, University of Heidelberg, HD-KIP-03-22. (2003)
15. Wu, Q.X., McGinnity, T.M., Maguire, L.P., Glackin, B., Belatreche, A.: Learning Mechanism in Networks of spiking Neurons. Studies in Computational Intelligence, Springer-Verlag. 35 (2006) 171–197
16. Wu, Q.X., McGinnity, T.M., Maguire, L.P., Belatreche, A., Glackin, B.: Adaptive Co-Ordinate Transformation Based on Spike Timing-Dependent Plasticity Learning Paradigm, LNCS, Springer. 3610 (2005) 420-429
17. Abbott, L.F., Regehr, W.G.: Synaptic Computation. Nature. 431(2004) 796 – 803

# Gait Parameters Optimization and Real-Time Trajectory Planning for Humanoid Robots

Shouwen Fan and Min Sun

School of Mechatronics Engineering
University of Electronic Science and Technology of China
ChengDu SiChuan, P.R. China
Shouwenfan@263.net

**Abstract.** Trajectory planning of humanoid robots not only is required to satisfy kinematic constraints, but also other criteria such as staying balance, having desirable upper and lower postures, having smooth movement etc, is needed to maintain certain properties. In this paper, calculation formulas of driving torque for each joint of humanoid robot are derived based on dynamics equation, mathematic models for gait parameters optimization are established via introducing energy consumption indexes. gait parameters are optimized utilizing genetic algorithm. A new approach for real-time trajectory planning of humanoid robots is proposed based on fuzzy neural network (FNN), Zero Moment Point (ZMP) criteria, B-spline interpolation and inverse displacement analysis model. The minimum energy consumption gait, which similar with human motion, are used to train FNN, b-spline curves are utilized to fit dispersive Center of Gravity (COG) position and body posture datas, based on above models and inverse displacement model, trajectory of COG and desired body posture can be mapped into trajectory of joint space conveniently. Simulation results demonstrate feasibility and effectiveness of above real-time trajectory planning method. Numeric examples are given for illustration.

**Keywords:** Humanoid Robot, Trajectory Planning, Gait Optimization, Energy Consumption Index, Fuzzy Neural network.

## 1 Introduction

Research on humanoid robots is currently one of the most exciting topics in the field of robotics and there are many ongoing projects[1-9]. Development of humanoid robots with natural and efficient movements presents many challenging problems to the humanoid robot researchers. For all humanoid robots, trajectory generation is the core problem that mainly contributes to the quality of their movements. Humanoid robot trajectory generation is generally more complicated than that of the conventional industrial robots. This is due to influences of the impact force, the balance constraint condition, variation of kinematics and dynamics models in different phases throughout walking cycles; that is, a single supporting phase and an instantaneous double supporting phase. Due to the high DOF in humanoid robot mechanisms, complex computational requirements in task planning and trajectory generation are

D.-S. Huang, L. Heutte, and M. Loog (Eds.): ICIC 2007, LNAI 4682, pp. 35–46, 2007.

expected. Furthermore, to allow adaptability and flexibility in generating movement, humanoid robot trajectory generation is required to carry out in real time.

Generally, humanoid robot trajectory planning can be categorized into three main approaches: ①ZMP- based Trajectory Planning[9].②Trajectory is directly resolved through a dominant dynamics of the robot[5,6].③Trajectory planning as an optimization problem[4,7]. Trajectory planned in the first approach is limited by the preplanned lower-body movement and the ZMP trajectory. The second approach may suffer from stability due to an inadequate model used, this approach needs to rely heavily on the quality of feedback signal. The third approach may require extensive computational burden.

In general, trajectory generation in humanoid robots is not only required to satisfy the given task constraints such as footprint locations and obstacle locations, which are typically expressed in terms of leg trajectories but also other criteria needed to maintain certain properties. These required properties are such as staying balance, having desirable upper and lower postures, minimizing energy consumption, and having smooth movement etc.

For realizing such style of movement as walking, the gait pattern should he planned in real-time. In order to generate a humanoid robot gait parameters in a short enough time for real time applications, we utilize a FNN to generate gait parameters of humanoid robot on-line. In order to gain smooth motion for humanoid robot, we utilize b-spline curve to fit dispersive COG position and body posture datas. We also establish forward and inverse displacement model for humanoid robot, using above models we can map trajectory of COG and desired body posture into trajectory of joint space conveniently.

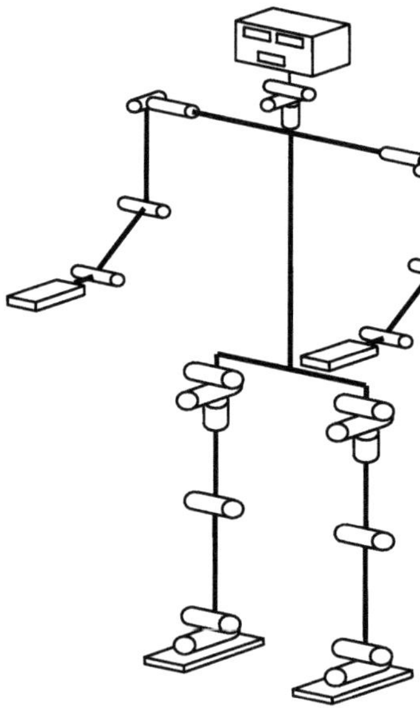

**Fig. 1.** Structure Scheme of humanoid robot

## 2 Structure Scheme of Humanoid Robot

In this paper, we study a virtual humanoid robot, which is composed of six segments: head, body, arm, upper leg, lower leg and the foot. Virtual humanoid robot has 6 DOFs (degree of freedom) on each leg, 5 DOFs on each arm, and 3 DOFs on head, with the result that the virtual humanoid robot has 25 DOFs. The structure of humanoid robot and the DOFs are presented in Fig. 1. To actualize virtual humanoid robot a transmission mechanism is employed, all joints are driven by DC motors, almost all joints have harmonic drive gears and pulleys for gaining a drive torque.

## 3 Biped Model and ZMP Calcuation

During walking, the arms of the humanoid robot will be fixed on the chest. Therefore, it can be considered as a five-link biped robot in the saggital plane, as shown in Fig. 2.

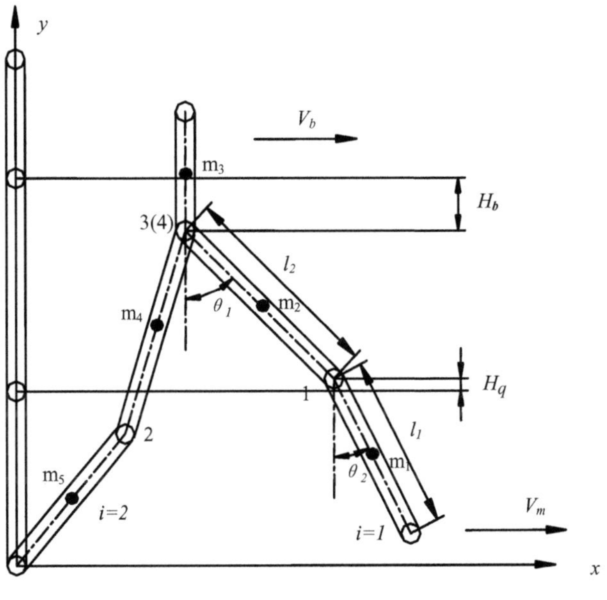


**Fig. 2.** Simplified five-link model

Nowadays, theory of Zero Moment Point(ZMP) is employed widely in humanoid robot balance control, ZMP is mostly used as standard evaluation of stability of humanoid robot and firstly introduced by Vukobratovic[1]. ZMP is defined as the point on the floor at which the moment $T:(T_x,T_y,T_z)$ generated by the reaction force and reaction torque satisfies $T_x=0$ and $T_y=0$. If ZMP is in convex hull of the foot-support area then humanoids robot can stand or walk without falling down.

The motion of the biped robot is considered to be composed from a single support phase, an instantaneous double support phase. The friction force between the robot's feet and the ground is considered to be great enough to prevent sliding. ZMP position can be calculated using following formulas[4]

$$\overline{X}_{ZMP} = \frac{\sum_{i=1}^{5} m_i(\ddot{\overline{z}}_i + \ddot{z}_w + g_Z)\overline{x}_i - \sum_{i=1}^{5} m_i(\ddot{\overline{x}}_i + \ddot{x}_w)(\overline{z}_i + z_w)}{\sum_{i=1}^{5} m_i(\ddot{\overline{z}}_i + \ddot{z}_w + g_Z)} \tag{1}$$

where $m_i$ is the mass of the particle $i$, $x_w$ and $z_w$ are the coordinates of the waist with respect to the coordinate system at the ankle joint of supporting leg, $\overline{x}_i$ and $\overline{z}_i$ are the coordinates of the mass particle $i$ with respect to the $O_1X_1Z_1$ coordinate system. $\ddot{\overline{x}}_i$ and $\ddot{\overline{z}}_i$ are the acceleration of the mass particle $i$ with respect to the $O_1X_1Z_1$ coordinate system.

# 4 Gait Parameters Optimization

## 4.1 Object Function

Firstly, gait related parameters are defined as follows, $l_1$ length of upper leg, $l_2$ length of lower leg, $D$ step length, $H_q$ height of knee rise, $H_b$ height of sciatic rise, $T_b$ walking period of humanoid robot, $V_b$ walking velocity of humanoid robot, $V_b = D/T_b$, $T_q$ walking period of knee joint, $V_q$ walking velocity of knee joint, $V_q = D/T_q$.

During walking, humanoid robots adopt smooth wave gait. Three energy consumption indexes are introduced as follows

1) Mean power

Suppose j is number of joint (each leg has two joints j=1, 2), i is the number of leg (humanoid robot has two legs i=1, 2). The power of mechanism is defined as product of driving torque of motor and angle velocity of joint, so, the mean power can be calculated using following equation

$$P_{av} = \frac{1}{T}\sum_{i=1}^{2}\sum_{j=1}^{2}\int_0^T \left|\tau_{i,j}(t)\cdot\dot{\theta}_{i,j}(t)\right|dt \tag{2}$$

Where $\tau$ is driving torque of motor, $\dot{\theta}$ is angle velocity of joint.

2) Mean power derivation

Although mean power is an important index for optimization, but it may occurs during motion process of humanoid robot that instantaneous power of mechanism system reach infinite, under this circumstance, the mean power may be a small value, but instantaneous peak may do great harm to humanoid robot system. So it is necessary to establish another optimization object describing distribution of instantaneous power around mean power.

$$P_i(t)=\sum_{i=1}^{2}\sum_{j=1}^{2}\tau_{i,j}(t)\cdot\dot{\theta}_{i,j}(t) \tag{3}$$

$$D_{av}=\sqrt{\frac{1}{T}\int_0^T(P_i(t)-P_{av})^2\,dt} \tag{4}$$

Where $P_i$ is instantaneous power of mechanism system.

3) Mean torque consumption

Mean torque consumption can be calculated using following equation

$$P_L=\frac{1}{T}\sum_{i=1}^{2}\sum_{j=1}^{2}\int_0^T(\tau_{i,j}(t))^2\,dt \tag{5}$$

Overall object function can be constructed as follows

$$F_{\min}=P_{av_{\min}}+D_{av_{\min}}+P_{L_{\min}} \tag{6}$$

Constrain equations of humanoid robot system can be defined as

$$\begin{cases}0\le l_1+l_2\le 1.5\\ 0\le D\le 1\\ H_b\le H_q,\ \ H_b\le l_2,\ \ H_q\le l_1\end{cases} \tag{7}$$

## 4.2 Derivation of Equations for Joint Driving Torque

Suppose $m$ is mass of upper leg, lower leg has the same mass as upper leg, $m_0$ is mass of humanoid robot's body (namely, overall mass removing the mass of upper leg and lower leg and foot).

According to dynamic principle, particle, which located at upper leg, away from coordinate orgin $r_1$, with mass of $dm$, its kinetic energy can be expressed as

$$dK_1=\frac{1}{2}(dm)\cdot(r_1\cdot\theta_1)^2=\frac{1}{2}\cdot\frac{m}{l_1}\cdot\theta_1^2\cdot r_1^2\cdot dr_1 \tag{8}$$

Its potential energy can be expressed as

$$dP_1=-(dm)\cdot g\cdot r_1\cdot\cos\theta_1=-\frac{m}{l_1}\cdot g\cdot\cos\theta_1\cdot r_1\cdot dr_1 \tag{9}$$

Integrating both side of above two equation in scope of $0\le r\le l_1$, we can obtain

$$\begin{aligned}K_1&=\frac{1}{6}\cdot m\cdot l_1^{\ 2}\cdot\theta_1^{\ 2}\\ P_1&=-\frac{1}{2}m\cdot g\cdot l_1\cdot\cos\theta_1\end{aligned} \tag{10}$$

Using the same method, kinetic energy and Potential energy of lower leg can be expressed as

$$K_2 = \frac{1}{2} m \cdot l_2{}^2 \left[ \theta_1^2 + \frac{1}{3}\theta_2^2 + \theta_1 \cdot \theta_2 \cdot \cos(\theta_1 - \theta_2) \right]$$
$$P_2 = -m \cdot g \cdot l_2 \left( \cos\theta_1 + \frac{1}{2}\cos\theta_2 \right) \tag{11}$$

Suppose the length ratio of upper leg and lower leg of humanoid robot is the same as that of real human being, namely, $l_2 = l_1 = l$ . Based on lagrange dynamics equation, we can derive following dynamics formulas

$$\begin{cases} \tau_1 = \frac{1}{2} m \cdot l^2 [\frac{480}{\pi}\theta_1 + \frac{180}{\pi}\theta_2 \cdot \cos(\theta_1 - \theta_2) \\ \quad + \theta_2^2 \cdot \sin(\theta_1 - \theta_2)] + \frac{3}{2} m \cdot g \cdot l \cdot \sin\theta_1 \\ \tau_2 = \frac{1}{5} m \cdot l^2 [\frac{540}{\pi}\theta_1 \cdot \cos(\theta_1 - \theta_2) \\ \quad - 3\theta_1^2 \cdot \sin(\theta_1 - \theta_2) + 2\theta_2^2] + \frac{3}{5} m \cdot g \cdot l \cdot \sin\theta_2 \end{cases} \tag{12}$$

Foot of supporting leg is pressed by reaction force of the ground, two driving torques of supporting leg can be expressed as

$$\begin{cases} \tau_1' = \tau_1 - R_x (l \cdot \cos\theta_1 + l \cdot \cos\theta_2) \\ \quad - R_y \cdot (l \cdot \sin\theta_1 + l \cdot \sin\theta_2) \\ \tau_2' = \tau_2 - R_x \cdot l \cdot \cos\theta_2 - R_y \cdot \sin\theta_2 \end{cases} \tag{13}$$

Where $R_x, R_y$ are two component of reaction force.

Considering the case of single support phase, calculation formulas for $R_x, R_y$ can be derived as

$$\begin{cases} R_x = \frac{1}{2} m \cdot l \cdot (\dot{\theta}_1 \cdot \cos\theta_1 - \theta_1^2 \cdot \sin\theta_1 + \dot{\theta}_2 \cdot \cos\theta_2 \\ \quad - \theta_2^2 \cdot \sin\theta_2) \\ R_y = -\frac{1}{2} m \cdot l \cdot (\dot{\theta}_1 \cdot \sin\theta_1 + \theta_1^2 \cdot \cos\theta_1 + \dot{\theta}_2 \cdot \sin\theta_2 \\ \quad + \theta_2^2 \cdot \cos\theta_2) + \frac{1}{3}(m_0 + m)g \end{cases} \tag{14}$$

### 4.3 Optimization Result

Suppose parameters of virtual humanoid robot are presented in Table 1.

**Table 1.** Parameters of humanoid robot

| | Body | Lower leg | Upper leg | Lower leg+foot |
|---|---|---|---|---|
| Mass[kg] | 12.000 | 2.930 | 3.890 | 4.090 |
| Inertia[kg.m2] | 0.190 | 0.014 | 0.002 | 0.017 |
| Length[m] | 0.600 | 0.400 | 0.400 | 0.568 |

Using above optimization model, programming and calculating using genetic algorithm, a set of gait parameter optimization solutions are derived and listed in Table 2.

**Table 2.** Result of gait parameters optimization

| Gait parameter | $D$ | $H_b$ | $H_q$ | $V_b$ | $V_q$ |
|---|---|---|---|---|---|
| Optimization solution | 0.43m | 0.11m | 0.08m | 0.16m/s | 0.21m/s |

## 5 Displacement Analysis Model

Among all displacement analysis model for robotic manipulators, Denavit-Hartenberg model is most widely used. Here, we utilize Denavit-Hartenberg model for displacement analysis of humanoid robot. Coordinate system representation for humanoid robot is shown as Fig. 3.

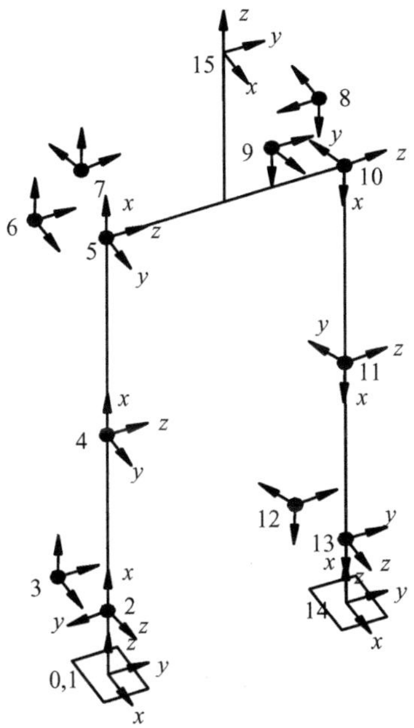

**Fig. 3.** Coordinate system representation of humanoid robot

### 5.1 Forward Displacement Model

The difference between humanoid robot and industrial robot lies on transformation matrix ${}^0_1T$, because coordinate system 1 is fixed on right foot of humanoid robot, during walking, right foot of humanoid robot may have different position and orientation relative to ground.

COG position and body posture of humanoid robot respect to base coordinate system can be expressed as

$$ {}^{0}_{15}T = {}^{0}_{1}T\,{}^{1}_{2}T\,{}^{2}_{3}T\,{}^{3}_{4}T\,{}^{4}_{5}T\,{}^{5}_{6}T\,{}^{6}_{7}T\,{}^{7}_{15}T \tag{15} $$

Where ${}^{0}_{1}T = \begin{bmatrix} r_{11} & r_{12} & r_{13} & l_x \\ r_{21} & r_{22} & r_{23} & l_y \\ r_{31} & r_{32} & r_{33} & l_z \\ 0 & 0 & 0 & 1 \end{bmatrix}$

$r_{11} = \cos\alpha_1 \cos\beta_1 \quad r_{12} = \cos\alpha_1 \sin\beta_1 \sin\gamma_1 - \sin\alpha_1 \cos\gamma_1$

$r_{13} = \cos\alpha_1 \sin\beta_1 \cos\gamma_1 + \sin\alpha_1 \sin\gamma_1 \quad r_{21} = \sin\alpha_1 \cos\beta_1$

$r_{22} = \sin\alpha_1 \sin\beta_1 \sin\gamma_1 + \cos\alpha_1 \cos\gamma_1$

$r_{23} = \sin\alpha_1 \sin\beta_1 \cos\gamma_1 - \cos\alpha_1 \sin\gamma_1$

$r_{31} = -\sin\beta_1 \quad r_{32} = \cos\beta_1 \sin\gamma_1 \quad r_{33} = \cos\beta_1 \cos\gamma_1$

$\alpha_1$、$\beta_1$、$\gamma_1$ are three rotation angle of coordinate system 1 relative to coordinate system 0 around three coordinate axis $z$、$y$、$x$.

$$ {}^{i}_{i+1}T = \begin{bmatrix} \cos\theta_i & -\sin\theta_i \cos\alpha_i & \sin\theta_i \sin\alpha_i & a_i \cos\theta_i \\ \sin\theta_i & \cos\theta_i \cos\alpha_i & -\cos\theta_i \sin\alpha_i & a_i \sin\theta_i \\ 0 & \sin\alpha_i & \cos\alpha_i & d_i \\ 0 & 0 & 0 & 1 \end{bmatrix} $$

Similarly, position and orientation of humanoid robot's left foot respect to base coordinate can be expressed as

$$ {}^{0}_{14}T = {}^{0}_{7}T\,{}^{7}_{8}T\,{}^{8}_{9}T\,{}^{9}_{10}T\,{}^{10}_{11}T\,{}^{11}_{12}T\,{}^{12}_{13}T\,{}^{13}_{14}T \tag{16} $$

## 5.2 Inverse Displacement Model

Because humanoid robot possesses high degree of freedom, if we use the position and orientation of left foot to solve each joint angle, the derivation of inverse displacement analysis equations may be too complex. Fortunately, in case of inverse displacement analysis, usually, not only the position and posture of left foot relative to ground but also the COG position and body posture of humanoid robot relative to ground are known. So kinematic chain from right foot to left foot can be divided into two subchain, namely, left leg subchain and right leg subchain. In this way, the calculation burden of inverse displacement analysis can be reduced greatly, at the same time, phenomena, such as inverse displacement analysis equation cannot be solved, can be avoided effectively.

*1) Inverse displacement equations for right leg*

Transformation matrix of humanoid robot's body relative to base coordinate system 0 can be expressed as

$$ {}^{0}_{15}T = {}^{0}_{1}T\,{}^{1}_{2}T\,{}^{2}_{3}T\,{}^{3}_{4}T\,{}^{4}_{5}T\,{}^{5}_{6}T\,{}^{6}_{7}T\,{}^{7}_{15}T \tag{17} $$

Multiple both side of above equation using matrix ${}^{0}_{1}T^{-1}$, we can obtain

$$ {}^{0}_{1}T^{-1}\,{}^{0}_{15}T = \begin{bmatrix} k_{11} & k_{12} & k_{13} & k_{14} \\ k_{21} & k_{22} & k_{23} & k_{24} \\ k_{31} & k_{32} & k_{33} & k_{34} \\ 0 & 0 & 0 & 1 \end{bmatrix} \tag{18} $$

Utilizing relationship of corresponding items of two side of above equation equal, we can derive inverse displacement equations for right leg[10].

*2) Inverse displacement equations for left leg*

Transformation matrix of humanoid robot's left foot relative to coordinate system 15 can be expressed as

$$ {}^{15}_{14}T = {}^{15}_{8}T\,{}^{8}_{9}T\,{}^{9}_{10}T\,{}^{10}_{11}T\,{}^{11}_{12}T\,{}^{12}_{13}T\,{}^{13}_{14}T \tag{19} $$

Multiple both side of above equation using matrix ${}^{15}_{8}T^{-1}$, we can derive

$$ {}^{15}_{8}T^{-1}\,{}^{15}_{14}T = \begin{bmatrix} h_{11} & h_{12} & h_{13} & h_{14} \\ h_{21} & h_{22} & h_{23} & h_{24} \\ h_{31} & h_{32} & h_{33} & h_{34} \\ 0 & 0 & 0 & 1 \end{bmatrix} \tag{20} $$

Utilizing relationship of corresponding items of two side of above equation equal, we can derive inverse displacement equations for left leg[10].

In order to plan trajectory of humanoid robot, we can preset the COG position and body posture of humanoid robot relative to ground in advance, then we can utilize above inverse displacement analysis model to solve real-time trajectory of humanoid robot in joint space. In the meantime, we can utilize the redundant DOFs, which derives from different position and orientation of humanoid robot's foot relative to ground, to optimize humanoid robot's ZMP position. In this way, not only the desired COG position and body posture of humanoid robot can be obtained but also stability of humanoid robot system can be guaranteed.

## 6 Real-Time Trajectory Planning for Humanoid

In order to generate a humanoid robot gait in a short enough time for real time applications, we utilize a FNN, which gives good results for the approximation problems. A FNN model provides high accuracy, fast training (identification), and is computationally and algorithmically simple. In many applications, the FNN approximation has superior accuracy and training time compared to multilayered perceptron networks. To teach the Neural Network (NN), the optimized gait parameters are used as training datas. One of the advantages of FNN is that the FNN can be used to approximate any gait within the range of pre-computed optimal gaits. After training, the FNN can quickly generate the minimum energy consumption gait on-line.

Real-time trajectory generation steps for humanoid robot can be described as follows

Step 1: Plan COG velocity trajectory of humanoid robot using smooth "S-shape" curve
Step 2: Preset body posture of humanoid robot in dispersive positions of COG trajectory
Step 3: Generate minium energy consumption gait parameters of humanoid robot based on FNN
Step 4: Interpolate COG position and body posture datas of humanoid robot using b-spline curve
Step 5: Optimize redundant DOFs based on ZMP criteria
Step 6: Inverse displacement solution for humanoid robot
Step 7: verify joint angle trajectories, joint torque trajectories to guarantee not violating the allowed limits, and verify ZMP position to guarantee stability
Step 8: If joint angle trajectories, joint torque trajectories violate the allowed limits or stability criteria doesnot meet, then modify gait parameters or body posture, go to Step 1
Step 9: Output trajectories plan results (including joint angle trajectories, joint torque trajectories, ZMP trajectories etc)

## 7 Numerical Examples and Simulation Results

The joint angle trajectories and joint torque trajectories of humanoid robot are shown in Fig. 4. and Fig. 5. respectively. The humanoid robot gait generated are very similar with the results presented in [3]. The robot posture is straighter, like the human walking. It can be seen that torque value is low and the torques change smoothly during simulation procedure, which ensure a minimum consumed energy. The ZMP trajeories for humanoid robot are presented in Fig. 6. The ZMP is all the time inside the sole, which ensure a stable walking motion. Kinematic simulation resultes of humanoid robot are shown in Fig.7.

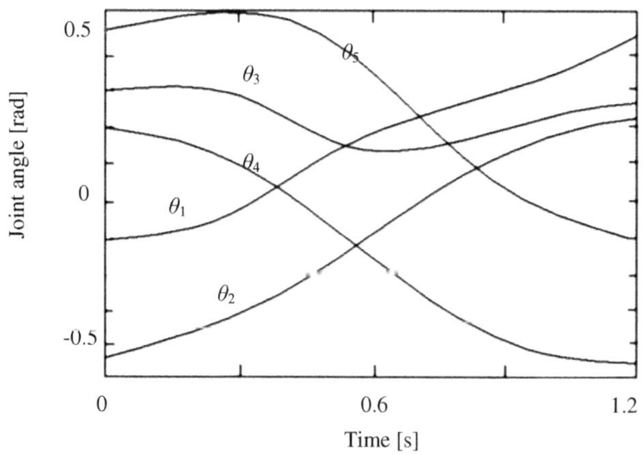


**Fig. 4.** Joint angle trajectories of humanoid robot

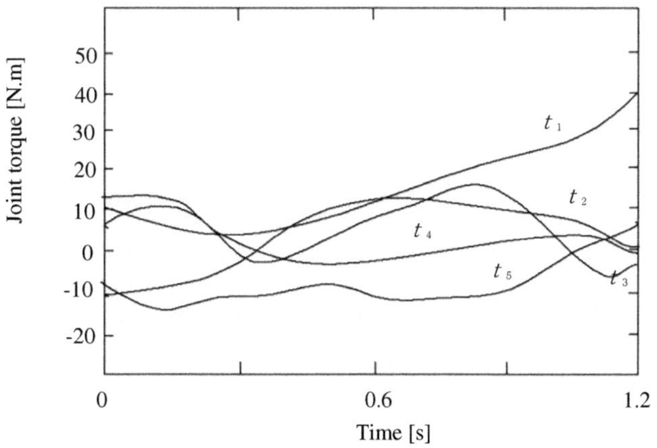


**Fig. 5.** Joint torque trajectories of humanoid robot

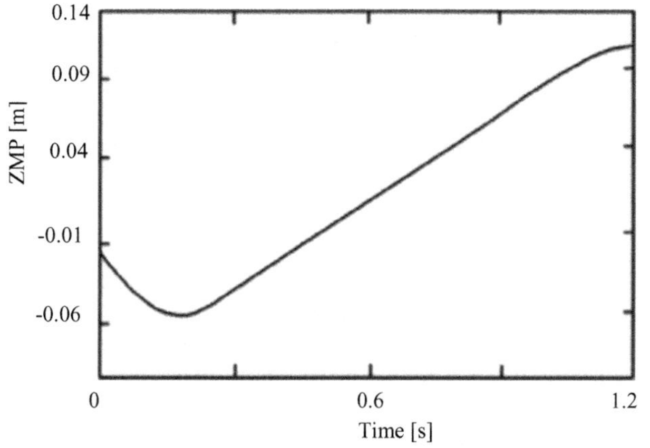


**Fig. 6.** ZMP trajectories of humanoid robot

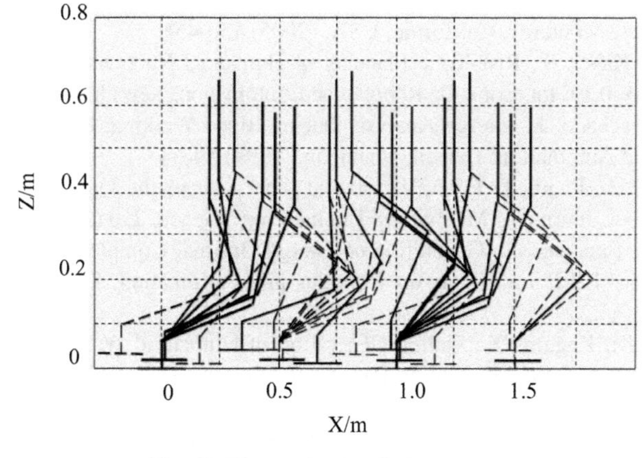


**Fig. 7.** Kinematic simulation graph

Based on the simulation results, it can be seen that minimum consumed energy gait of humanoid robot is similar to that of real human walking.

## 8 Conclusions

This paper presented a new method for real-time trajectory planning of humanoid robot. The final goal of this research is to create an autonomous humanoid robot, able to operate in unknown environments and generate the real-time optimal gait on-line. The performance evaluation is carried out by simulations utilizing the virtual humanoid robot. Based on the simulation results, we conclude

- The time needed by FNN to generate the gait parameters is very short;
- The optimal gait generated by FNN is stable and the impact of the foot with the ground is small.
- The motion of humanoid robot is smooth and continuous.
- The gait of humanoid is similar to that of of real human being, with minium energy consumption.

Above research works provide theorical base for dynamics, stability analysis, precise motion control of humanoid robot. The implementation of the proposed model and method to real application of humanoid robot is considered to be the future works of our research.

## References


1. Vukobratovic, M., Borovac, B., Surla, D., Stokic, D.: Biped Locomotion, Dynamics, Stability, Control and Application. Springer, Berlin (1990)
2. Capi, G. , Nasu, Y.: Application of Genetic Algorithms for Biped Robot Gait Synthesis Optimization During Walking and Going Up-stairs. Advanced Robotics Journal, 15 (2001) 675–695
3. Capi, G. , Nasu, Y.: Real Time Gait Generation for Autonomous Humanoid Robots:A Case Study for Walking. Robotics and Autonomous Systems, 42 (2003) 107–116
4. Capi, G., Yokota, M.: A New Humanoid Robot Gait Generation based on Multiobjective Optimization. Proceedings of IEEE/ASME Int. Conf. On Advanced Intelligent Mechatronics. Monterey, California, USA, (2005) 450-454
5. Harada, K., Kajita, S.: Real-Time Planning of Humanoid Robot's Gait for Force Controlled Manipulation. IEEE Int. Conf. On Robotics and Automation. New Orleans, LA, (2004) 616-622
6. Silva, F., Machado, J.: Energy Analysis During Biped Walking. Proc. IEEE Int. Conf. On Robotics and Automation, Detroit, Michigan, (1999) 59-64
7. Channon, P.H., Pham, D.T.: A Variational Approach to the Optimization of Gait for a Bipedal Robot. Journal of Mechanical Engineering Science, 210 (1996) 177-186
8. Roussel, L., Canudas, C.: Generation of Energy Optimal Complete Gait Cycles for Biped Robots. Proc. IEEE Int. Conf. on Robotics and Automation, Leuven, Belgium, (1998) 2036-2041
9. Nishiwaki, K., Kagami, S.: Online Generation of Humanoid Walking Motion based on a Fast Generation Method of Motion Pattern that Follows Desired ZMP. Proc. of IEEE/RSJ Int. Conf. on Intelligent Robots and Systems, Zurich, (2002) 2684-2689
10. Yang, D.C., Liu, L.: Kinematic Analysis of Humanoid Robot. Chinese J. of Mechanical Engineering, 39 (2003) 70–74

# Global Asymptotic Stability of Cohen-Grossberg Neural Networks with Multiple Discrete Delays

Anhua Wan[1,2], Weihua Mao[3,4,⋆], Hong Qiao[2], and Bo Zhang[5]

[1] School of Mathematics and Computational Science, Sun Yat-Sen University, 510275 Guangzhou, China
anhuawan@163.com, anhua.wan@ia.ac.cn
[2] Institute of Automation, Chinese Academy of Sciences, 100080 Beijing, China
[3] Department of Applied Mathematics, College of Science, South China Agricultural University, 510642 Guangzhou, China
[4] College of Automation Science and Engineering, South China University of Technology, 510641 Guangzhou, China
mao.weihua@mail.scut.edu.cn
[5] Institute of Applied Mathematics, Chinese Academy of Sciences, 100080 Beijing, China

**Abstract.** The asymptotic stability is analyzed for Cohen-Grossberg neural networks with multiple discrete delays. The boundedness, differentiability or monotonicity condition is not assumed on the activation functions. The generalized Dahlquist constant approach is employed to examine the existence and uniqueness of equilibrium of the neural networks, and a novel Lyapunov functional is constructed to investigate the stability of the delayed neural networks. New general sufficient conditions are derived for the global asymptotic stability of the neural networks with multiple delays.

## 1 Introduction

Cohen-Grossberg neural networks model is an important recurrently connected neural networks model[2]. The model includes many significant models from neurobiology, population biology and evolutionary theory([6]), among which is Hopfield-type neural networks model([15]) and Volterra-Lotka biological population model. Meanwhile, the model has extensive applications in many important areas such as signal processing, image processing, pattern classification and optimization([6]). Therefore, the study of Cohen-Grossberg neural networks has been the focus of interest(see, e.g., [1],[7],[18],[19],[20],[21],[23]).

Due to the finite switching speed of neurons and amplifiers, time delays inevitably exist in biological and artificial neural networks([1],[10],[12],[17]). In this paper, we consider Cohen-Grossberg neural networks with multiple discrete delays. The model is described by the following functional differential equations

⋆ Corresponding author.

D.-S. Huang, L. Heutte, and M. Loog (Eds.): ICIC 2007, LNAI 4682, pp. 47–58, 2007.

$$\frac{du_i(t)}{dt} = -a_i(u_i(t))\Big[b_i(u_i(t)) - \sum_{k=0}^{K}\sum_{j=1}^{n} w_{ij}^{(k)} f_j^{(k)}(u_j(t-\tau_{ij}^{(k)})) + J_i\Big], \quad i = 1,2,\ldots,n, \tag{1}$$

where $n \geq 2$ is the number of neurons in the networks, $u_i(t)$ denotes the neuron state vector, $a_i$ denotes an amplification function, $b_i$ denotes a self-signal function, $W^{(k)} = (w_{ij}^{(k)})_{n\times n}$ denotes the connection weight matrix, $f_j^{(k)}$ denotes an activation function, $\tau_{ij}^{(0)} \equiv 0$, $\tau_{ij}^{(k)} \geq 0\,(k = 1,2,\ldots,n)$ are discrete delays caused during the switching and transmission processes, and $J_i$ represents the constant external input.

The initial conditions associated with system (1) are of the following form

$$u_i(s) = \phi_i(s) \in C\left([t_0-\tau, t_0], R\right),\; s \in [t_0-\tau, t_0],\; i = 1,2,\ldots,n, \tag{2}$$

where $\tau = \max\{\tau_{ij}^{(k)} : 1 \leq i,j \leq n, 1 \leq k \leq K\} \in [0,+\infty)$ and $C([t_0-\tau,t_0],R)$ denotes the space of all real-valued continuous functions defined on $[t_0-\tau,t_0]$. Denote $\phi(s) = (\phi_1(s),\phi_2(s),\ldots,\phi_n(s))^T$.

The special cases of system (1) consist of system with pure delays([20])

$$\frac{du_i(t)}{dt} = -a_i(u_i(t))\Big[b_i(u_i(t)) - \sum_{j=1}^{n} w_{ij} f_j(u_j(t-\tau_{ij})) + J_i\Big],\; i = 1,2,\ldots,n, \tag{3}$$

where $\tau_{ij} \geq 0$ are delays caused during the switching and transmission processes, $W = (w_{ij})_{n\times n}$ is the delayed connection weight matrix; hybrid system with discrete delays([7])

$$\frac{du_i(t)}{dt} = -a_i(u_i(t))\Big[b_i(u_i(t)) - \sum_{j=1}^{n} w_{ij} f_j(u_j(t)) - \sum_{j=1}^{n} w_{ij}^{\tau} f_j(u_j(t-\tau_{ij})) + J_i\Big], \quad i = 1,2,\ldots,n, \tag{4}$$

where $\tau_{ij} \geq 0$ are delays caused during the switching and transmission processes, $W = (w_{ij})_{n\times n}$ and $W^{\tau} = (w_{ij}^{\tau})_{n\times n}$ respectively denote the normal and the delayed connection weight matrix; and system with multiple delays([21],[23])

$$\frac{du_i(t)}{dt} = -a_i(u_i(t))\Big[b_i(u_i(t)) - \sum_{k=0}^{K}\sum_{j=1}^{n} w_{ij}^{(k)} f_j(u_j(t-\tau_k))\Big], i = 1,2,\ldots,n, \tag{5}$$

where the delays $\tau_k \geq 0$ are arranged such that $0 = \tau_0 < \tau_1 < \cdots < \tau_K$.

In addition, system (1) includes many other popular models as special cases, for example, Hopfield-type neural networks with discrete delays([5],[22])

$$C_i\frac{du_i(t)}{dt} = -\frac{u_i}{R_i} + \sum_{j=1}^{n} w_{ij} f_j(u_j(t-\tau_{ij})) + J_i, \quad i = 1,2,\ldots,n. \tag{6}$$

The stability of neural networks is of crucial importance for the designing and successful applications of neural networks([14]). Time delays are often the sources of oscillation and even instability in neural networks, and thus will dramatically change the dynamic behavior of neural networks. Hence, it is necessary and significant to examine the stability of delayed neural networks. This paper aims to present new general sufficient conditions for the asymptotic stability of multiple-delayed neural networks (1).

In this paper, we only make the following assumptions:

($A_1$) Each $a_i(\cdot)$ is continuous, and there exists a positive constant $\grave{\alpha}_i$ such that $\grave{\alpha}_i \leq a_i(r)$, $\forall r \in R$.

($A_2$) Each $b_i(\cdot)$ is continuous, and there exists a constant $\lambda_i > 0$ such that for any $r_1, r_2 \in R$, $(r_1 - r_2)\big[b_i(r_1) - b_i(r_2)\big] \geq \lambda_i (r_1 - r_2)^2$.

($A_3$) Each $f_j^{(k)}(\cdot)$ is Lipschitz continuous. Denote $m_j^{(k)}$ the minimal Lipschitz constant of $f_j^{(k)}$, i.e., $m_j^{(k)} = \sup\limits_{s_1, s_2 \in R, s_1 \neq s_2} \frac{|f_j^{(k)}(s_1) - f_j^{(k)}(s_2)|}{|s_1 - s_2|}$.

Since the monotonicity or boundedness assumption on activation functions makes the results inapplicable to some important engineering problems([3],[11]), we make neither boundedness nor monotonicity or differentiability assumption on $f_j^{(k)}$. Meanwhile, we do not impose any restriction on the matrix $W^{(k)}$. Thus, a much broader connection topology for the networks is allowed.

## 2 Preliminaries

In this section, we will present some preliminary concepts which will be used in the next section.

**Definition 1.** ([13]) *Suppose that $\Omega$ is an open subset of Banach space $X$ and $F : \Omega \to X$ is an operator. The constant*

$$\alpha_\Omega(F) = \sup_{x, y \in \Omega, x \neq y} \frac{1}{\|x - y\|} \lim_{r \to +\infty} \big[\|(F + rI)x - (F + rI)y\| - r\|x - y\|\big] \quad (7)$$

*is called the generalized Dahlquist constant of $F$ on $\Omega$.*

**Lemma 1.** ([16]) *If $\alpha_\Omega(F) < 0$, then $F$ is a one-to-one mapping on $\Omega$. If in addition $\Omega = X$, then $F$ is a homeomorphism of $X$ onto $X$.*

## 3 Global Asymptotic Stability of Neural Networks (1)

Let $R^n$ be the $n$-dimensional real vector space. In this paper, we will always use the $l^p$-norm, that is, for each vector $x = (x_1, x_2, \ldots, x_n)^T \in R^n$, $\|x\|_p = \left(\sum_{i=1}^n |x_i|^p\right)^{1/p}$, $p \in [1, +\infty)$. For any two operators $F$ and $G$, $FG$ denotes the composition of operators, that is, $FG(x) = F(G(x))$, $\forall x \in D(G)$, where $D(\cdot)$ denotes the domain of an operator. Let sign$(r)$ denote the sign function of $r \in R$, i.e., sign$(r) = \{1, r > 0;\ 0, r = 0;\ -1, r < 0\}$.

We first present the following result for the existence and uniqueness of an equilibrium point for the delayed neural networks (1).

**Theorem 1.** *Suppose that* $(A_1)$-$(A_3)$ *hold. Then for each set of external input* $J_i$, *the delayed neural networks* (1) *has a unique equilibrium point* $u^*$, *if there exist a real number* $p \in [1, +\infty)$ *and four sets of real numbers* $d_i > 0$, $c_i > 0$, $r_{ij}^{(k)} > 0$, $s_{ij}^{(k)}$ *such that*

$$\sum_{k=0}^{K} \Big\{ m_i^{(k)} \sum_{j=1}^{n} \Big( \frac{d_j}{d_i} (r_{ji}^{(k)})^{p-1} |w_{ji}^{(k)}|^{2-p+(p-1)s_{ji}^{(k)}} \Big) + (p-1) \sum_{j=1}^{n} \Big( \frac{m_j^{(k)} d_i c_j}{d_j c_i} (r_{ij}^{(k)})^{-1} |w_{ij}^{(k)}|^{2-s_{ij}^{(k)}} \Big) \Big\} < p\lambda_i, \quad i = 1, 2, \ldots, n. \tag{8}$$

*Proof.* Define an operator $G : R^n \to R^n$ by $G(x) = \big(G_1(x), G_2(x), \ldots, G_n(x)\big)^T$ and

$$G_i(x) = -\Big[ b_i(x_i) - \sum_{k=0}^{K} \sum_{j=1}^{n} w_{ij}^{(k)} f_j^{(k)}(x_j) + J_i \Big], \quad i = 1, 2, \ldots, n.$$

Then, $u^*$ is an equilibrium point of (1) if and only if $G(u^*) = 0$.

Let $Q = \text{diag}(d_1, d_2, \ldots, d_n)$ and $P = \text{diag}(c_1, c_2, \ldots, c_n)$. Below we will show that $\alpha_{R^n}(QGQ^{-1}P) < 0$ in the sense of the $l^p$-norm.

It is easy to verify that in the sense of the $l^p$-norm,

$$\alpha_{R^n}(QGQ^{-1}P) = \sup_{y,z \in R^n, y \neq z} \frac{\sum_{i=1}^{n} \Big\{ \big[ (QGQ^{-1}P)_i(y) - (QGQ^{-1}P)_i(z) \big] |y_i - z_i|^{p-1} \text{sign}(y_i - z_i) \Big\}}{\|y - z\|_p^p}.$$

For all $y, z \in R^n$, we have

$$\begin{aligned}
&\sum_{i=1}^{n} \Big\{ \big[ (QGQ^{-1}P)_i y - (QGQ^{-1}P)_i z \big] |y_i - z_i|^{p-1} \text{sign}(y_i - z_i) \Big\} \\
&= \sum_{i=1}^{n} \Big\{ - d_i \{ b_i(d_i^{-1} c_i y_i) - b_i(d_i^{-1} c_i z_i) \\
&\qquad - \sum_{k=0}^{K} \sum_{j=1}^{n} w_{ij}^{(k)} \big[ f_j^{(k)}(d_j^{-1} c_j y_j) - f_j^{(k)}(d_j^{-1} c_j z_j) \big] \} |y_i - z_i|^{p-1} \text{sign}(y_i - z_i) \Big\} \\
&\leq \sum_{i=1}^{n} \Big\{ - d_i |b_i(d_i^{-1} c_i y_i) - b_i(d_i^{-1} c_i z_i)| |y_i - z_i|^{p-1} \\
&\qquad + d_i \sum_{k=0}^{K} \sum_{j=1}^{n} |w_{ij}^{(k)}| |f_j^{(k)}(d_j^{-1} c_j y_j) - f_j^{(k)}(d_j^{-1} c_j z_j)| |y_i - z_i|^{p-1} \Big\} \\
&\leq \sum_{i=1}^{n} \Big\{ - \lambda_i c_i |y_i - z_i|^p + d_i \sum_{k=0}^{K} \sum_{j=1}^{n} \big( m_j^{(k)} d_j^{-1} c_j |w_{ij}^{(k)}| |y_j - z_j| |y_i - z_i|^{p-1} \big) \Big\} \\
&= - \sum_{i=1}^{n} \lambda_i c_i |y_i - z_i|^p \\
&\quad + \sum_{i=1}^{n} \sum_{j=1}^{n} \sum_{k=0}^{K} \Big\{ d_i m_j^{(k)} d_j^{-1} c_j \big( (r_{ij}^{(k)})^{\frac{p-1}{p}} |w_{ij}^{(k)}|^{\frac{2-p+(p-1)s_{ij}^{(k)}}{p}} |y_j - z_j| \big) \times \\
&\qquad \big( (r_{ij}^{(k)})^{-\frac{p-1}{p}} |w_{ij}^{(k)}|^{\frac{(p-1)(2-s_{ij}^{(k)})}{p}} |y_i - z_i|^{p-1} \big) \Big\}
\end{aligned}$$

$$\begin{aligned}
&\le -\sum_{i=1}^{n} \lambda_i c_i |y_i - z_i|^p \\
&\quad + \sum_{i=1}^{n}\sum_{j=1}^{n}\sum_{k=0}^{K} \Big\{ d_i m_j^{(k)} d_j^{-1} c_j \cdot \frac{1}{p}\big[(r_{ij}^{(k)})^{p-1} |w_{ij}^{(k)}|^{2-p+(p-1)s_{ij}^{(k)}} |y_j - z_j|^p \\
&\qquad\qquad + (p-1)(r_{ij}^{(k)})^{-1} |w_{ij}^{(k)}|^{2-s_{ij}^{(k)}} |y_i - z_i|^p\big]\Big\} \\
&= -\sum_{i=1}^{n} \lambda_i c_i |y_i - z_i|^p \\
&\quad + \frac{1}{p}\sum_{j=1}^{n}\sum_{k=0}^{K}\sum_{i=1}^{n} \{ d_i m_j^{(k)} d_j^{-1} c_j (r_{ij}^{(k)})^{p-1} |w_{ij}^{(k)}|^{2-p+(p-1)s_{ij}^{(k)}} |y_j - z_j|^p \} \\
&\quad + \frac{p-1}{p}\sum_{i=1}^{n}\sum_{k=0}^{K}\sum_{j=1}^{n} \{ d_i m_j^{(k)} d_j^{-1} c_j (r_{ij}^{(k)})^{-1} |w_{ij}^{(k)}|^{2-s_{ij}^{(k)}} |y_i - z_i|^p \} \\
&= -\sum_{i=1}^{n} \Big\{ \lambda_i c_i - \frac{1}{p}\sum_{k=0}^{K} m_i^{(k)} d_i^{-1} c_i \sum_{j=1}^{n} \{ d_j (r_{ji}^{(k)})^{p-1} |w_{ji}^{(k)}|^{2-p+(p-1)s_{ji}^{(k)}} \} \\
&\qquad - \frac{p-1}{p} d_i \sum_{k=0}^{K}\sum_{j=1}^{n} \{ m_j^{(k)} d_j^{-1} c_j (r_{ij}^{(k)})^{-1} |w_{ij}^{(k)}|^{2-s_{ij}^{(k)}} \}\Big\} |y_i - z_i|^p \\
&= -\sum_{i=1}^{n} \frac{c_i}{p} \Big\{ p\lambda_i - \sum_{k=0}^{K} m_i^{(k)} \sum_{j=1}^{n} \{ \tfrac{d_j}{d_i} (r_{ji}^{(k)})^{p-1} |w_{ji}^{(k)}|^{2-p+(p-1)s_{ji}^{(k)}} \} \\
&\qquad - (p-1)\sum_{k=0}^{K}\sum_{j=1}^{n} \{ \tfrac{m_j^{(k)} d_i c_j}{d_j c_i} (r_{ij}^{(k)})^{-1} |w_{ij}^{(k)}|^{2-s_{ij}^{(k)}} \}\Big\} |y_i - z_i|^p.
\end{aligned}$$

Therefore, it follows by (8) that

$$\begin{aligned}
&\alpha_{R^n}(QGQ^{-1}P) \\
&\le -\min_{1\le i\le n} \Big\{ \lambda_i c_i - \frac{1}{p}\sum_{k=0}^{K}\sum_{j=1}^{n} m_i^{(k)} d_i^{-1} c_i d_j (r_{ji}^{(k)})^{p-1} |w_{ji}^{(k)}|^{2-p+(p-1)s_{ji}^{(k)}} \\
&\qquad - \frac{p-1}{p}\sum_{k=0}^{K}\sum_{j=1}^{n} m_j^{(k)} d_i d_j^{-1} c_j (r_{ij}^{(k)})^{-1} |w_{ij}^{(k)}|^{2-s_{ij}^{(k)}} \Big\} < 0.
\end{aligned}$$

By virtue of Lemma 1, we conclude that $QGQ^{-1}P$ is a homeomorphism of $R^n$. Since $Q$ and $P$ are invertible, we confirm that $G(u) = 0$ has and only has one solution $u^*$. Thus, system (1) has a unique equilibrium $u^*$.

*Remark 1.* Theorem 1 presents general and relaxed sufficient conditions for the existence and uniqueness of an equilibrium of the multiple-delayed neural networks model (1). The incorporation of a positive number $p \in [1, +\infty)$ and the four sets of adjustable parameters $d_i > 0, c_i > 0, r_{ij}^{(k)} > 0, s_{ij}^{(k)}$ into condition (8) endows the criteria with much flexibility and generality. Through specific choice of the parameters $p, d_i, c_i, r_{ij}^{(k)}, s_{ij}^{(k)}$ in (8), a number of new criteria for the existence and uniqueness of an equilibrium of the multiple-delayed neural networks (1) can be directly deduced.

Now we investigate the global asymptotic stability of the delayed neural networks (1).

**Theorem 2.** *Suppose that* ($A_1$)-($A_3$) *and* (8) *hold. Then for each set of external input* $J_i$*, the delayed neural networks* (1) *has a unique equilibrium* $u^*$*, which is globally asymptotically stable and independent of the multiple delays.*

*Proof.* It follows from Theorem 1 that system (1) has a unique equilibrium $u^* = (u_1^*, u_2^*, \ldots, u_n^*)^T$.

Let $x_i(t) = \frac{d_i}{c_i^{\frac{p-1}{p}}}(u_i(t) - u_i^*)$, $i = 1, 2, \ldots, n$ and $x(t) = (x_1(t), x_2(t), \ldots, x_n(t))^T$. Substitution of $u_i(t) = \frac{c_i^{\frac{p-1}{p}}}{d_i} x_i(t) + u_i^*$ into (1) leads to

$$\begin{aligned}\frac{\mathrm{d}x_i(t)}{\mathrm{d}t} = & -\frac{d_i}{c_i^{\frac{p-1}{p}}} a_i\Big(\frac{c_i^{\frac{p-1}{p}}}{d_i} x_i(t) + u_i^*\Big)\Big\{b_i\Big(\frac{c_i^{\frac{p-1}{p}}}{d_i} x_i(t) + u_i^*\Big) - b_i(u_i^*) \\ & -\sum_{k=0}^{K}\sum_{j=1}^{n} w_{ij}^{(k)}\Big[f_j^{(k)}\Big(\frac{c_j^{\frac{p-1}{p}}}{d_j} x_j(t-\tau_{ij}^{(k)}) + u_j^*\Big) - f_j^{(k)}(u_j^*)\Big]\Big\}, \\ & i = 1, 2, \ldots, n.\end{aligned} \tag{9}$$

Let $p_i(x_i(t)) = a_i\big(\frac{c_i^{\frac{p-1}{p}}}{d_i} x_i(t) + u_i^*\big)$, $q_i(x_i(t)) = b_i\big(\frac{c_i^{\frac{p-1}{p}}}{d_i} x_i(t) + u_i^*\big) - b_i(u_i^*)$, and $s_j^{(k)}(x_j(t-\tau_{ij}^{(k)})) = f_j^{(k)}\big(\frac{c_j^{\frac{p-1}{p}}}{d_j} x_j(t-\tau_{ij}^{(k)}) + u_j^*\big) - f_j^{(k)}(u_j^*)$. Then (9) reduces to

$$\begin{aligned}\frac{\mathrm{d}x_i(t)}{\mathrm{d}t} = & -\frac{d_i}{c_i^{\frac{p-1}{p}}} p_i(x_i(t))\Big\{q_i(x_i(t)) - \sum_{k=0}^{K}\sum_{j=1}^{n} w_{ij}^{(k)} s_j^{(k)}(x_j(t-\tau_{ij}^{(k)}))\Big\}, \\ & i = 1, 2, \ldots, n.\end{aligned} \tag{10}$$

It is clear that 0 is the unique equilibrium of (10).

We define the following novel Lyapunov functional

$$\begin{aligned}V(x(t)) = \sum_{i=1}^{n}\Big\{&\int_0^{x_i(t)} \frac{p|s|^{p-1}}{p_i(s)}\mathrm{sign}(s)\mathrm{d}s \\ &+ \sum_{k=0}^{K}\sum_{j=1}^{n} m_j^{(k)}\frac{d_i}{d_j}(r_{ij}^{(k)})^{p-1}|w_{ij}^{(k)}|^{2-p+(p-1)s_{ij}^{(k)}}\int_{t-\tau_{ij}^{(k)}}^{t}|x_j(s)|^p\mathrm{d}s\Big\}.\end{aligned} \tag{11}$$

Estimating the derivative of $V$ along the solution trajectory $x(t)$ of (10), we deduce

$$\begin{aligned}&\frac{\mathrm{d}V(x(t))}{\mathrm{d}t} \\ &= -\sum_{i=1}^{n} p|x_i(t)|^{p-1}\mathrm{sign}(x_i(t))\frac{d_i}{c_i^{\frac{p-1}{p}}}\Big\{q_i(x_i(t)) - \sum_{k=0}^{K}\sum_{j=1}^{n} w_{ij}^{(k)} s_j^{(k)}(x_j(t-\tau_{ij}^{(k)}))\Big\} \\ &\quad + \sum_{i=1}^{n}\sum_{k=0}^{K}\sum_{j=1}^{n} m_j^{(k)}\frac{d_i}{d_j}(r_{ij}^{(k)})^{p-1}|w_{ij}^{(k)}|^{2-p+(p-1)s_{ij}^{(k)}}|x_j(t)|^p \\ &\quad - \sum_{i=1}^{n}\sum_{k=0}^{K}\sum_{j=1}^{n} m_j^{(k)}\frac{d_i}{d_j}(r_{ij}^{(k)})^{p-1}|w_{ij}^{(k)}|^{2-p+(p-1)s_{ij}^{(k)}}|x_j(t-\tau_{ij}^{(k)})|^p\end{aligned}$$

$$
\begin{aligned}
&\leq -p\sum_{i=1}^{n}|x_i(t)|^{p-1}\lambda_i|x_i(t)| \\
&\quad +p\sum_{i=1}^{n}\sum_{k=0}^{K}\sum_{j=1}^{n}|w_{ij}^{(k)}||x_i(t)|^{p-1}\frac{d_i}{c_i^{\frac{p-1}{p}}}m_j^{(k)}\frac{c_j^{\frac{p-1}{p}}}{d_j}|x_j(t-\tau_{ij}^{(k)})| \\
&\quad +\sum_{i=1}^{n}\sum_{k=0}^{K}\sum_{j=1}^{n}m_j^{(k)}\tfrac{d_i}{d_j}(r_{ij}^{(k)})^{p-1}|w_{ij}^{(k)}|^{2-p+(p-1)s_{ij}^{(k)}}|x_j(t)|^p \\
&\quad -\sum_{i=1}^{n}\sum_{k=0}^{K}\sum_{j=1}^{n}m_j^{(k)}\tfrac{d_i}{d_j}(r_{ij}^{(k)})^{p-1}|w_{ij}^{(k)}|^{2-p+(p-1)s_{ij}^{(k)}}|x_j(t-\tau_{ij}^{(k)})|^p \\
&= -p\sum_{i=1}^{n}\lambda_i|x_i(t)|^p \\
&\quad +p\sum_{i=1}^{n}\sum_{j=1}^{n}\sum_{k=0}^{K}\Big\{m_j^{(k)}\big((\tfrac{d_i}{d_j})^{\frac{1}{p}}(r_{ij}^{(k)})^{\frac{p-1}{p}}|w_{ij}^{(k)}|^{\frac{2-p+(p-1)s_{ij}^{(k)}}{p}}|x_j(t-\tau_{ij}^{(k)})|\big)\times \\
&\qquad\qquad \big((\tfrac{d_ic_j}{d_jc_i})^{\frac{1}{p}}(r_{ij}^{(k)})^{-\frac{1}{p}}|w_{ij}^{(k)}|^{\frac{2-s_{ij}^{(k)}}{p}}|x_i(t)|\big)^{p-1}\Big\} \\
&\quad +\sum_{i=1}^{n}\sum_{k=0}^{K}\sum_{j=1}^{n}m_j^{(k)}\tfrac{d_i}{d_j}(r_{ij}^{(k)})^{p-1}|w_{ij}^{(k)}|^{2-p+(p-1)s_{ij}^{(k)}}|x_j(t)|^p \\
&\quad -\sum_{i=1}^{n}\sum_{k=0}^{K}\sum_{j=1}^{n}m_j^{(k)}\tfrac{d_i}{d_j}(r_{ij}^{(k)})^{p-1}|w_{ij}^{(k)}|^{2-p+(p-1)s_{ij}^{(k)}}|x_j(t-\tau_{ij}^{(k)})|^p \\
&\leq -p\sum_{i=1}^{n}\lambda_i|x_i(t)|^p \\
&\quad +p\sum_{i=1}^{n}\sum_{j=1}^{n}\sum_{k=0}^{K}\Big\{m_j^{(k)}\cdot\frac{1}{p}\big[\tfrac{d_i}{d_j}(r_{ij}^{(k)})^{p-1}|w_{ij}^{(k)}|^{2-p+(p-1)s_{ij}^{(k)}}|x_j(t-\tau_{ij}^{(k)})|^p \\
&\qquad\qquad +(p-1)\tfrac{d_ic_j}{d_jc_i}(r_{ij}^{(k)})^{-1}|w_{ij}^{(k)}|^{2-s_{ij}^{(k)}}|x_i(t)|^p\big]\Big\} \\
&\quad +\sum_{i=1}^{n}\sum_{k=0}^{K}\sum_{j=1}^{n}m_j^{(k)}\tfrac{d_i}{d_j}(r_{ij}^{(k)})^{p-1}|w_{ij}^{(k)}|^{2-p+(p-1)s_{ij}^{(k)}}|x_j(t)|^p \\
&\quad -\sum_{i=1}^{n}\sum_{k=0}^{K}\sum_{j=1}^{n}m_j^{(k)}\tfrac{d_i}{d_j}(r_{ij}^{(k)})^{p-1}|w_{ij}^{(k)}|^{2-p+(p-1)s_{ij}^{(k)}}|x_j(t-\tau_{ij}^{(k)})|^p \\
&= -\sum_{i=1}^{n}\Big\{p\lambda_i-\sum_{k=0}^{K}\Big[m_i^{(k)}\sum_{j=1}^{n}\tfrac{d_j}{d_i}(r_{ji}^{(k)})^{p-1}|w_{ji}^{(k)}|^{2-p+(p-1)s_{ji}^{(k)}} \\
&\qquad\qquad +(p-1)\sum_{j=1}^{n}m_j^{(k)}\tfrac{d_ic_j}{d_jc_i}(r_{ij}^{(k)})^{-1}|w_{ij}^{(k)}|^{2-s_{ij}^{(k)}}\Big]\Big\}|x_i(t)|^p \\
&\leq -\mu\|x(t)\|_p^p<0,
\end{aligned}
$$

where

$$
\begin{aligned}
\mu=\min_{1\leq i\leq n}\Big\{p\lambda_i-\sum_{k=0}^{K}\Big[m_i^{(k)}\sum_{j=1}^{n}\frac{d_j}{d_i}(r_{ji}^{(k)})^{p-1}|w_{ji}^{(k)}|^{2-p+(p-1)s_{ji}^{(k)}} \\
+(p-1)\sum_{j=1}^{n}m_j^{(k)}\frac{d_ic_j}{d_jc_i}(r_{ij}^{(k)})^{-1}|w_{ij}^{(k)}|^{2-s_{ij}^{(k)}}\Big]\Big\}.
\end{aligned}
$$

We deduce

$$
V(x(t))+\mu\int_{t_0}^{t}\|x(s)\|_p^p\mathrm{d}s\leq V(x(t_0)).
$$

On the other hand,

$$
\begin{aligned}
&V(x(t_0))\\
&= \sum_{i=1}^{n} \Big\{ \int_0^{x_i(t_0)} \tfrac{p|s|^{p-1}}{p_i(s)} \mathrm{sign}(s) \mathrm{d}s\\
&\quad + \sum_{k=0}^{K}\sum_{j=1}^{n} m_j^{(k)} \tfrac{d_i}{d_j} (r_{ij}^{(k)})^{p-1} |w_{ij}^{(k)}|^{2-p+(p-1)s_{ij}^{(k)}} \int_{t_0-\tau_{ij}^{(k)}}^{t_0} |x_j(s)|^p \mathrm{d}s \Big\}\\
&\le \sum_{i=1}^{n} \Big\{ \tfrac{1}{\grave{\alpha}_i} |x_i(t_0)|^p\\
&\quad + \sum_{k=0}^{K}\sum_{j=1}^{n} m_j^{(k)} \tfrac{d_i}{d_j} (r_{ij}^{(k)})^{p-1} |w_{ij}^{(k)}|^{2-p+(p-1)s_{ij}^{(k)}} \int_{t_0-\tau_{ij}^{(k)}}^{t_0} \Big| \tfrac{d_j}{c_j^{\frac{p-1}{p}}} (u_j(s) - u_j^*) \Big|^p \mathrm{d}s \Big\}\\
&= \sum_{i=1}^{n} \Big\{ \tfrac{1}{\grave{\alpha}_i} \Big| \tfrac{d_i}{c_i^{\frac{p-1}{p}}} (u_i(t_0) - u_i^*) \Big|^p\\
&\quad + \sum_{k=0}^{K}\sum_{j=1}^{n} m_j^{(k)} d_i \tfrac{d_j^{p-1}}{c_j^{p-1}} (r_{ij}^{(k)})^{p-1} |w_{ij}^{(k)}|^{2-p+(p-1)s_{ij}^{(k)}} \tau_{ij}^{(k)} \sup_{s\in[t_0-\tau,t_0]} |u_j(s) - u_j^*|^p \Big\}\\
&\le \max_{1\le i\le n} \Big\{ \tfrac{1}{\grave{\alpha}_i} \big( \tfrac{d_i}{c_i^{\frac{p-1}{p}}} \big)^p + \sum_{k=0}^{K}\sum_{j=1}^{n} m_j^{(k)} d_i \big(\tfrac{d_j}{c_j}\big)^{p-1} (r_{ij}^{(k)})^{p-1} |w_{ij}^{(k)}|^{2-p+(p-1)s_{ij}^{(k)}} \tau_{ij}^{(k)} \Big\} \times\\
&\qquad\qquad \sup_{s\in[t_0-\tau,t_0]} \sum_{i=1}^{n} |\phi_i(s) - u_i^*|^p < +\infty,
\end{aligned}
$$

which implies $\|x(t)\|_p^p \in L^1(t_0, +\infty)$. By [4, Lemma 1.2.2], we deduce that $\|x(t)\|_p^p \to 0$ as $t \to +\infty$, i.e., $u_i(t) \to u_i^*$ as $t \to +\infty$, $i = 1, 2, \ldots, n$, and therefore, the equilibrium $u^*$ is globally asymptotically stable for system (1).

*Remark 2.* Ye et al. [23] proved the global asymptotic stability of a special case of system (1) when in particular $f_j^{(k)} = f_j$, but they additionally required that the matrix $\sum_{k=0}^{K} W^{(k)}$ is symmetric and each $f_j \in C^1(R,R)$ is a sigmoidal function. Liao et al. [8] analyzed the global asymptotic stability of a special case of system (1) when in particular for $i = 1, 2, \ldots, n$, $a_i(u_i) = 1$, $b_i(u_i)$ is linear and $f_i^{(k)} = f_i$ is monotonically nondecreasing.

Particularly, if we take respectively $p = 1$ and $p = 2$ in Theorem 2, then we can derive the following two corollaries.

**Corollary 1.** *Suppose that* $(A_1)$-$(A_3)$ *hold and there exist a set of real numbers* $d_i > 0$ *such that*

$$
\sum_{k=0}^{K} \Big\{ m_i^{(k)} \sum_{j=1}^{n} \frac{d_j}{d_i} |w_{ji}^{(k)}| \Big\} < \lambda_i, \quad i = 1, 2, \ldots, n. \tag{12}
$$

*Then for each set of external input* $J_i$*, the delayed neural networks (1) has a unique equilibrium* $u^*$*, which is globally asymptotically stable and independent of the delays.*

**Corollary 2.** *Suppose that* $(A_1)$-$(A_3)$ *hold and there exist a* $p \in [1, +\infty)$ *and four sets of real numbers* $d_i > 0$, $c_i > 0$, $r_{ij}^{(k)} > 0$, $s_{ij}^{(k)}$ *such that*

$$\sum_{k=0}^{K} \left\{ m_i^{(k)} \sum_{j=1}^{n} \left( \frac{d_j}{d_i} r_{ji}^{(k)} |w_{ji}^{(k)}|^{s_{ji}^{(k)}} \right) + \sum_{j=1}^{n} \left( \frac{m_j^{(k)} d_i c_j}{d_j c_i} (r_{ij}^{(k)})^{-1} |w_{ij}^{(k)}|^{2-s_{ij}^{(k)}} \right) \right\} < 2\lambda_i, \quad i = 1, 2, \ldots, n. \tag{13}$$

*Then for each set of external input* $J_i$, *the delayed neural networks* (1) *has a unique equilibrium* $u^*$, *which is globally asymptotically stable and independent of the delays.*

*Remark 3.* Corollary 2 improves the criteria in [21]. (i) Wang et al. [21, Theorem 1] deduced the global asymptotic stability of system (5), but they additionally required that each $b_i$ be differentiable, each $f_i$ be bounded, each $a_i$ be bounded from above (i.e., there exists a positive constant $\acute{\alpha}_i$ such that $a_i(r) \leq \acute{\alpha}_i$, $\forall r \in R$) and

$$\min_{1 \leq i \leq n} \left\{ \frac{\grave{\alpha}_i}{\acute{\alpha}_i} \lambda_i - m_i \sum_{k=0}^{K} \sum_{j=1}^{n} |w_{ji}^{(k)}| \right\} > 0. \tag{14}$$

Clearly, condition (14) is more restrictive than the special case of (12) when $m_i^{(k)} = m_i$ and $d_i = 1$.

(ii) Wang et al. [21, Theorem 2] deduced the global asymptotic stability of system (5), but they additionally required that each $b_i$ be differentiable, each $f_i$ be bounded, $a_i(r) \leq \acute{\alpha}_i$ ($\forall r \in R$) and

$$\min_{1 \leq i \leq n} \left\{ 2 \frac{\grave{\alpha}_i}{\acute{\alpha}_i} \lambda_i - \sum_{k=0}^{K} \sum_{j=1}^{n} \left( m_i^2 |w_{ji}^{(k)}| + |w_{ij}^{(k)}| \right) \right\} > 0. \tag{15}$$

Clearly, condition (15) is stronger than the special case of (13) when $m_i^{(k)} = m_i$, $d_i = c_i = s_{ij}^{(k)} = 1$ and $r_{ij}^{(k)} = m_j$.

Denote $(A_3)'$: Each $f_j(\cdot)$ is Lipschitz continuous with the minimal Lipschitz constant $m_j = \sup_{s_1, s_2 \in R, s_1 \neq s_2} \frac{|f_j(s_1) - f_j(s_2)|}{|s_1 - s_2|}$.

Since system (4) is a special case of system (1), we can obtain the following result for the global asymptotic stability of system (4).

**Corollary 3.** *Suppose that* $(A_1), (A_2), (A_3)'$ *hold. If there exist a* $p \in [1, \infty)$ *and six sets of real numbers* $d_i > 0$, $l_i > 0$, $r_{ij} > 0$, $\tilde{r}_{ij} > 0$, $s_{ij}$, $\tilde{s}_{ij}$ *such that*

$$\sum_{j=1}^{n} \left\{ \frac{d_j}{d_i} r_{ji}^{p-1} |w_{ji}|^{2-p+(p-1)s_{ji}} + (p-1) \frac{l_j}{l_i} r_{ij}^{-1} |w_{ij}|^{2-s_{ij}} + \frac{d_j}{d_i} \tilde{r}_{ji}^{p-1} |w_{ji}^{\tau}|^{2-p+(p-1)\tilde{s}_{ji}} + (p-1) \frac{l_j}{l_i} \tilde{r}_{ij}^{-1} |w_{ij}^{\tau}|^{2-\tilde{s}_{ij}} \right\} < p \frac{\lambda_i}{m_i}, \quad i = 1, 2, \ldots, n, \tag{16}$$

*then for each set of external input* $J_i$, *system* (4) *has a unique equilibrium* $u^*$, *which is globally asymptotically stable and independent of the delays.*

*Remark 4.* Lu [9, Theorems 2 and 3] derived the global asymptotic stability of a special case of system (4) when $a_i(u_i) = 1$ and $\tau_{ij} = \tau_j$, but [9] additionally required that each $b_i$ be differentiable, and the derived sufficient conditions are special cases of condition (16) when $p = 2$, $r_{ij} = \tilde{r}_{ij} = 1, s_{ji} = \tilde{s}_{ij} = 1$ and $l_i$ with several fixed values.

From Corollary 3, we can deduce the following result for the global asymptotic stability of system (3).

**Corollary 4.** *Suppose that* $(A_1), (A_2), (A_3)'$ *hold and there exist a* $p \in [1, \infty)$ *and four sets of real numbers* $d_i > 0$, $l_i > 0$, $r_{ij} > 0$, $s_{ij}$ *such that*

$$\sum_{j=1}^{n} \left( \frac{d_j}{d_i} r_{ji}^{p-1} |w_{ji}|^{2-p+(p-1)s_{ji}} + (p-1)\frac{l_j}{l_i} r_{ij}^{-1} |w_{ij}|^{2-s_{ij}} \right) < p\frac{\lambda_i}{m_i}, \; i = 1, 2, \ldots, n. \tag{17}$$

*Then for each set of external input* $J_i$*, system* (3) *has a unique equilibrium* $u^*$*, which is globally asymptotically stable and independent of the delays.*

Respectively letting $p = 1$ and $p = 2$ in Corollary 4, we can derive the following two corollaries.

**Corollary 5.** *Suppose that* $(A_1), (A_2), (A_3)'$ *hold and there exists a set of real numbers* $d_i > 0$ *such that*

$$m_i \sum_{j=1}^{n} \frac{d_j}{d_i} |w_{ji}| < \lambda_i, \quad i = 1, 2, \ldots, n. \tag{18}$$

*Then for each set of external input* $J_i$*, system* (3) *has a unique equilibrium* $u^*$*, which is globally asymptotically stable and independent of the delays.*

**Corollary 6.** *Suppose that* $(A_1), (A_2), (A_3)'$ *hold and there exist four sets of real numbers* $d_i > 0$, $l_i > 0$, $r_{ij} > 0$ *and* $s_{ij}$ *such that*

$$\sum_{j=1}^{n} \left\{ \frac{d_j}{d_i} r_{ji} |w_{ji}|^{s_{ji}} + \frac{l_j}{l_i} r_{ij}^{-1} |w_{ij}|^{2-s_{ij}} \right\} < \frac{2\lambda_i}{m_i}, \quad i = 1, 2, \ldots, n. \tag{19}$$

*Then for each set of external input* $J_i$*, system* (3) *has a unique equilibrium* $u^*$*, which is globally asymptotically stable and independent of the delays.*

As for the asymptotic stability of Hopfield-type neural networks with discrete delays (6), we have the following corollary.

**Corollary 7.** *Suppose that* $(A_3)'$ *holds and there exist a set of real numbers* $d_i > 0$ *such that*

$$\max_{1 \le i \le n} \left\{ m_i R_i \sum_{j=1}^{n} \frac{d_i}{d_j} |w_{ji}| \right\} < 1. \tag{20}$$

*Then for each set of external input* $J_i$*, system* (6) *has a unique equilibrium point* $u^*$*, which is globally asymptotically stable and independent of the delays.*

*Proof.* Clearly, system (6) is the special case of system (3) when $a_i(u_i) = 1/C_i$ and $b_i(u_i) = u_i/R_i \, (i = 1, 2, \ldots, n)$. It is easily seen that conditions $(A_1), (A_2)$ are naturally satisfied and $\lambda_i = 1/R_i$. Condition (20) implies (18) holds, and thus this corollary directly follows from Corollary 5.

*Remark 5.* Zhang [24, Corollary 3.1] is a special case of Corollary 7 when $d_i = 1 \, (i = 1, 2, \ldots, n)$. Driessche et al. [3, Theorem 2.1] derived the same result as [24, Corollary 3.1], however, they additionally required that each $f_i$ be bounded.

## 4 Conclusions

This paper is concerned with the asymptotic stability of Cohen-Grossberg neural networks model with multiple discrete delays. Only assuming the activation functions to be globally Lipschitz continuous, we derive new sufficient conditions for the global asymptotic stability of the discrete-delayed neural networks (1), which are very general and improves many existing results.

## Acknowledgements

The authors gratefully acknowledge the support of China Postdoctoral Science Foundation under Grant No. 20060400117, K. C. Wong Education Foundation, Hong Kong, the National Natural Science Foundation of China under Grant No. 60675039, and the National High Technology Research and Development Program of China under Grant No. 2006AA04Z217.

## References

1. Chen, T.P., Rong, L.B.: Delay-independent Stability Analysis of Cohen-Grossberg Neural Networks. Physics Letters A **317** (2003) 436–449
2. Cohen, M.A., Grossberg, S.: Absolute Stability and Global Pattern Formation and Partial Memory Storage by Competitive Neural Networks. IEEE Transactions on Systems, Man and Cybernetics **SMC-13** (1983) 815–826
3. van den Driessche, P., Zou, X.: Global Attractivity in Delayed Hopfield Neural Network Models. SIAM J. Appl. Math. **58** (1998) 1878–1890
4. Gopalsamy, K.: Stability and Oscillations in Delay Differential Equations of Population Dynamics. Dordrecht: Kluwer, 1992
5. Gopalsamy, K., He, X.Z.: Stability in Asymmetric Hopfield Nets with Transmission Delays. Physica D **76** (1994) 344–358
6. Grossberg, S.: Nonlinear Neural Networks: Principles, Mechanisms, and Architectures. Neural Networks **1** (1988) 17–61
7. Liao, X.F., Li, C.G., Wong, K.W.: Criteria for Exponential Stability of Cohen-Grossberg Neural Networks. Neural Networks **17** (2004) 1401–1414
8. Liao, X.F., Li, C.D.: An LMI Approach to Asymptotical Stability of Multi-delayed Neural Networks. Physica D **200** (2005) 139–155
9. Lu, H.T.: On Stability of Nonlinear Continuous-time Neural Networks with Delays. Neural Networks **13**(10) (2000) 1135–1143

10. Marcus, C., Westervelt, R.: Stability of Analog Neural Networks with Delay. Physics Review A **39** (1989) 347–359
11. Morita, M.: Associative Memory with Non-monotone Dynamics. Neural Networks **6**(1) (1993) 115–126
12. Peng, J.G., Qiao, H., Xu, Z.B.: A New Approach to Stability of Neural Networks with Time-varying Delays. Neural Networks **15** (2002) 95–103
13. Peng, J.G., Xu, Z.B.: On Asymptotic Behaviours of Nonlinear Semigroup of Lipschitz Operators with Applications. Acta Mathematica Sinica **45**(6) (2002) 1099–1106
14. Qiao, H., Peng, J.G., Xu, Z.B.: Nonlinear Measures: A New Approach to Exponential Stability Analysis for Hopfield-type Neural Networks. IEEE Transactions on Neural Networks **12**(2) (2001) 360–370
15. Tank, D.W., Hopfield, J. J.: Simple "Neural" Optimization Networks: An A/D Converter, Signal Decision Circuit, and a Linear Programming Circuit. IEEE Transactions on Circuits and Systems **33**(5) (1986) 533–541
16. Wan, A.H., Mao, W.H., Zhao, C.: A Novel Approach to Exponential Stability Analysis of Cohen-Grossberg Neural Networks. International Symposium on Neural Networks, Advances in Neural Networks-ISNN 2004 **1** (2004) 90–95
17. Wan, A.H., Peng, J.G., Wang, M.S.: Exponential Stability of a Class of Generalized Neural Networks with Time-varying Delays. Neurocomputing **69**(7-9) (2006) 959–963
18. Wan, A.H., Qiao, H., Peng, J.G., Wang, M.S., Delay-independent Criteria for Exponential Stability of Generalized Cohen-Grossberg Neural Networks with Discrete Delays. Physics Letters A **353**(2-3) (2006) 151–157
19. Wan, A.H., Wang, M.S., Peng, J.G., Qiao, H., Exponential Stability of Cohen-Grossberg Neural Networks with a General Class of Activation Functions. Physics Letters A **350**(1-2) (2006) 96–102
20. Wang, L., Zou, X.F.: Exponential Stability of Cohen-Grossberg Neural Networks. Neural Networks **15** (2002) 415–422
21. Wang, L., Zou, X.F.: Harmless Delays in Cohen-Grossberg Neural Network. Physica D **170**(2) (2002) 162–173
22. Wang, L.S., Xu, D.Y.: Stability of Hopfield Neural Networks with Time Delays. Journal of Vibration and Control **8** (2002) 13–18
23. Ye, H., Michel, A.N., Wang, K.: Qualitative Analysis of Cohen-Grossberg Neural Networks with Multiple Delays. Physics Review E **51** (1995) 2611–2618
24. Zhang, J.Y.: Global Stability Analysis in Delayed Cellular Neural Networks. Computers and Mathematics with Applications **45** (2003) 1707–1720

# Global Exponential Stability of Cohen-Grossberg Neural Networks with Reaction-Diffusion and Dirichlet Boundary Conditions

Chaojin Fu[1,2] and Chongjun Zhu[1]

[1] Department of Mathematics, Hubei Normal University, Huangshi, Hubei, 435002, China
chaojinfu@126.com
[2] Hubei Province Key Laboratory of Bioanalytical Technique, Hubei Normal University, Huangshi, Hubei, 435002, China

**Abstract.** In this paper, global exponential stability of Cohen-Grossberg neural networks with reaction-diffusion and Dirichlet boundary conditions is considered by using an approach based on the delay differential inequality and the fixed-point theorem. Some sufficient conditions are obtained to guarantee that the reaction-diffusion Cohen-Grossberg neural networks are globally exponentially stable. The results presented in this paper are the improvement and extension of the existed ones in some existing works.

## 1 Introduction

Recurrent neural networks (RNNs) have been found useful in areas of signal processing, image processing, associative memories, pattern classification. As dynamic systems, RNNs frequently need to be analyzed for stability.

The buds of some recurrent neural network models may be traced back to the nonlinear difference-differential equations in learning theory or prediction theory [1]. The global exponential stability for such systems was analyzed. In particular, a general neural network, which is called the Cohen-Grossberg neural network (CGNN) and can function as stable associative memory, was developed and studied [2].

As a special case of the Cohen-Grossberg neural network, the continuous-time Hopfield neural network (HNN) [3] was proposed and applied for optimization, associative memories, pattern classification, image processing, etc. In parallel, cellular neural networks (CNNs) [4] were developed and have attracted much attention due to their great perspective of applications. CNNs and delayed cellular neural networks (DCNNs) have been applied to signal processing, image processing, and pattern recognition. The stability criteria of equilibrium points are established in a series of papers; e.g., [5]-[12].

Moreover, both the biological neural networks and the artificial neural networks, strictly speaking, diffusion effects cannot be avoided when electrons are moving in asymmetric electromagnetic fields. So we must consider that the activations vary in space as well as in time. The stability of the neural networks

D.-S. Huang, L. Heutte, and M. Loog (Eds.): ICIC 2007, LNAI 4682, pp. 59–65, 2007.

with diffusion terms has been considered in [13] and [14], which are expressed by partial differential equations. The boundary conditions of the investigated reaction-diffusion neural networks in [13] and [14] are all the Neumann boundary conditions.

Motivated by the above discussions, our aim in this paper is to consider the global exponential stability of Cohen-Grossberg neural networks with reaction-diffusion and Dirichlet boundary conditions.

This paper consists of the following sections. Section 2 describes some preliminaries. The main results are stated in Sections 3. Finally, concluding remarks are made in Section 4.

## 2 Preliminaries

Throughout of this paper, let $C\big([-\tau,0]\times\Re^m,\Re^n\big)$ be the Banach space of continuous functions which map $[-\tau,0]\times\Re^m$ into $\Re^n$ with the topology of uniform converge, where $\tau$ is a constant. Let $\Omega=\big\{(x_1,x_2,\cdots,x_m)^T\big|\ \ |x_i|<l_i, i=1,2,\cdots,m\big\}$ be an open bounded domain in $\Re^m$ with smooth boundary $\partial\Omega$. Denote $mes\Omega>0$ as the measure of $\Omega$. $L^2(\Omega)$ is the space of real functions on $\Omega$ which are $L^2$ in the Lebesgue measure. It is a Banach space for the norm

$$||u(t)||_2=\left[\sum_{i=1}^{n}||u_i(t)||_2^r\right]^{1/r},$$

where $u(t)=\big(u_1(t),\cdots,u_n(t)\big)^T$, $||u_i(t)||_2=\big(\int_\Omega|u_i(t,x)|^2dx\big)^{1/2}$, and $r\geq 1$. Consider the following reaction-diffusion delayed recurrent neural networks with the Dirichlet boundary conditions:

$$\begin{cases}\frac{\partial u_i(t,x)}{\partial t}=\sum_{k=1}^{m}\frac{\partial}{\partial x_k}\big(a_{ik}\frac{\partial u_i}{\partial x_k}\big)-\alpha_i(u_i(t,x))\big[\beta_i(u_i(t,x))\\ \quad -\sum_{j=1}^{n}c_{ij}f_j(u_j(t,x))\\ \quad -\sum_{j=1}^{n}d_{ij}g_j(u_j(t-\tau_j(t),x))-I_i\big], & (x,t)\in\Omega\times[0,+\infty),\\ u_i(t,x)=0, & (x,t)\in\partial\Omega\times[-\tau,+\infty),\\ u_i(t,x)=\phi_i(t,x), & (x,t)\in\partial\Omega\times[-\tau,0],\end{cases} \quad (1)$$

where $i=1,2,\cdots,n$, $n$ is the number of neurons in the networks; $x=(x_1,x_2,\cdots,x_m)^T\in\Omega\subset\Re^m$, $u(t,x)=(u_1(t,x),u_2(t,x),\cdots,u_t(t,x))^T\in\Re^n$ and $u_i(t,x)$ is the state of the $i$-th neurons at time $t$ and in point $x$, smooth function $a_{ik}>0$ represents the transmission diffusion operator along the $i$-th unit, $b_i>0$ represents the rate with which the $i$-th the unit will reset its potential to the resting state in isolation when disconnected from the networks and external inputs, $c_{ij}$ denotes the strength of the $j$-th unit on the $i$-th unit at time $t$ and in point $x$, $d_{ij}$ denotes the strength of the $j$-th unit on the $i$-th unit at time $t-\tau_j(t)$ and in point $x$, $\tau_j(t)$ corresponds to time-varying transmission delay along the axon of the $j$-th unit and satisfies $0\leq\tau_j(t)\leq\tau$, $f_j(u_j(t,x))$ denotes the activation function of the $j$-th unit at time $t$ and in point $x$, $g_j(u_j(t-\tau_j(t),x))$ denotes the activation function of the $j$-th unit at time $t-\tau_j(t)$ and in point $x$, $\phi(t,x)=(\phi_1(t,x),\phi_2(t,x),\cdots,\phi_n(t,x))^T$ and $\phi_i(t,x)$ are continuous functions.

For any $\varphi(t,x) \in C([-\tau,0] \times \Omega, \Re^n)$, we define

$$||\varphi||_2 = \left[\sum_{i=1}^{n} ||\varphi_i||_2^r\right]^{1/r},$$

where $\varphi(t,x) = (\varphi_1(t,x), \cdots, \varphi_n(t,x))^T, ||\varphi_i||_2 = \left(\int_\Omega |\varphi_i(x)|_\tau^2 dx\right)^{1/2}, |\varphi_i(x)|_\tau$ $= \sup_{-\tau \le s \le 0} |\varphi_i(s,x)|$, $|\varphi(t,x)|^{(\tau)} = \max_{1\le i\le n} |\varphi_i(x)|_\tau$.

In this paper, we always assume that for $i = 1, 2, \cdots, n$,

$A_1$: there exist constants $\bar{\alpha}_i > 0, \underline{\alpha}_i > 0$ such that $0 < \underline{\alpha}_i \le \alpha_i(u_i(t,x)) \le \bar{\alpha}_i$, for all $u_i(t,x) \in \Omega$;

$A_2$: $\beta_i(0) = 0$, and there exist constants $\bar{b}_i > 0, \underline{b}_i > 0$ such that $0 < \underline{b}_i \le \frac{\beta_i(u_i(t,x)) - \beta_i(v_i(t,x))}{u_i(t,x) - v_i(t,x)} \le \bar{b}_i$, for all $u_i(t,x) \in \Omega, v_i(t,x) \in \Omega, u_i(t,x) \ne v_i(t,x)$.

$A_3$: that the activation functions $f_j$ and $g_j$ $(j = 1, 2 \ldots, n)$ are globally Lipschitz continuous; i.e., $\forall j \in \{1, 2, \cdots, n\}, \forall r_1, r_2, r_3, r_4 \in \Re$, there exist real number $\ell_j$ and $\mu_j$ such that

$$|f_j(r_1) - f_j(r_2)| \le \ell_j |r_1 - r_2|, \quad |g_j(r_3) - g_j(r_4)| \le \mu_j |r_3 - r_4|.$$

It is easy to find that $f_j(\theta) = (1 - e^{\lambda\theta})/(1 + e^{\lambda\theta}), 1/(1 + e^{\lambda\theta})(\lambda > 0)$, $\arctan(\theta), \max(0,\theta), (|\theta+1| - |\theta-1|)/2$ are all globally Lipschitz continuous.

**Definition 1:** An equilibrium point $u^* = (u_1^*, u_2^*, \cdots, u_n^*)^T$ of the recurrent neural network (1) is said to be globally exponentially stable, if there exist constant $\varepsilon > 0$ and $\Upsilon \ge 1$ such that for any initial value $\phi$ and $t \ge 0$,

$$||u(t,x) - u^*||_2 \le \Upsilon ||\phi - u^*||_2 e^{-\varepsilon t}.$$

**Definition 2:** Let $f : \Re \to \Re$ be a continuous function. The upper right Dini-derivative $D^+ f$ is defined as

$$D^+ f(t) = \lim_{h \to 0^+} \sup \frac{f(t+h) - f(t)}{h}.$$

**Lemma 1:** Let $h(x)$ be a real-valued function belonging to $C^1(\Omega)$ which vanish on the boundary $\partial\Omega$ of $\Omega$; i.e., $h(x)|_{\partial\Omega} = 0$. Then

$$\int_\Omega h^2(x) dx \le l_i^2 \int_\Omega |\frac{\partial h}{\partial x_i}|^2 dx \tag{2}$$

**Proof:** If $x \in \Omega$, then

$$h(x) = \int_{-l_i}^{x_i} \frac{\partial}{\partial x_i} h(x_1, \cdots, x_m) dx_i, \tag{3}$$

$$h(x) = -\int_{x_i}^{l_i} \frac{\partial}{\partial x_i} h(x_1, \cdots, x_m) dx_i. \tag{4}$$

From (3) and (4), we can obtain

$$2|h(x)| \leq \int_{-l_i}^{l_i} |\frac{\partial}{\partial x_i} h(x_1, \cdots, x_m)| dx_i. \tag{5}$$

From (5) and the Schwarz's inequality,

$$|h(x)|^2 \leq \frac{l_i}{2} \int_{-l_i}^{l_i} |\frac{\partial}{\partial x_i} h(x_1, \cdots, x_m)| dx_i. \tag{6}$$

Integrating both sides of (6) with respect to $x_1, x_2, \cdots, x_m$, we get

$$\int_{\Omega} h^2(x) dx \leq l_i^2 \int_{\Omega} |\frac{\partial h}{\partial x_i}|^2 dx. \tag{7}$$

## 3 Main Results

Denote $\bar{A} = diag\{\sum_{k=1}^{m} \frac{a_{1k}}{l_k^2} + \underline{\alpha}_1 \underline{b}_1, \sum_{k=1}^{m} \frac{a_{2k}}{l_k^2} + \underline{\alpha}_2 \underline{b}_2, \cdots, \sum_{k=1}^{m} \frac{a_{nk}}{l_k^2} + \underline{\alpha}_n \underline{b}_n\}$, $|C| = (|c_{ij}|)_{n\times n}$, $|D| = (|d_{ij}|)_{n\times n}$, $\bar{\alpha} = diag\{\bar{\alpha}_1, \bar{\alpha}_2, \cdots, \bar{\alpha}_n\}$, $\ell = diag\{\ell_1, \ell_2, \cdots, \ell_n\}$, $\mu = diag\{\mu_1, \mu_2, \cdots, \mu_n\}$.

Based on assumptions $A_1$-$A_3$, it is well known that the equilibrium points of the neural network (1) exist if $\bar{A} - |C|\ell\bar{\alpha} - |D|\mu\bar{\alpha}$ is a nonsingular $M$ matrix. Let $u^* = (u_1^*, u_2^*, \cdots, u_n^*)^T$ be an equilibrium point of the neural network (1).

**Theorem 1:** If $\bar{A} - |C|\ell\bar{\alpha} - |D|\mu\bar{\alpha}$ is a nonsingular $M$ matrix, then the neural network (1) is globally exponentially stable.

**Proof:** Suppose $u(t,x)$ is an arbitrary solution of the neural network (1) with initial conditions $\varphi_u(t,x) \in C([-\tau, 0] \times \Omega, \Re^n)$. Let $z(t,x) = (u(t,x) - u^*)/T$, $\varphi_z(t,x) = (\varphi_u(t,x) - u^*)/T$, where the constant $T \neq 0$. Then from (1), for $i = 1, 2, \cdots, n$,

$$\begin{aligned} \frac{\partial z_i(t,x)}{\partial t} &= \sum_{k=1}^{m} \frac{\partial}{\partial x_k}\left(a_{ik}\frac{\partial z_i(t,x)}{\partial x_k}\right) - \alpha_i(z_i(t,x) + u_i^*)\big[\beta_i^*(z_i(t,x)) \\ &-\frac{1}{T}\sum_{j=1}^{n} c_{ij}(f_j(u_j(t,x)) - f_j(u_j^*)) \\ &-\frac{1}{T}\sum_{j=1}^{n} d_{ij}(g_j(u_j(t-\tau_j(t),x)) - g_j(u_j^*))\big], \end{aligned} \tag{8}$$

where for $i = 1, 2, \cdots, n$,

$$\beta_i^*(z_i(t,x)) := \beta_i(z_i(t,x) + u_i^*) - \beta_i(u_i^*).$$

Multiplying both sides of the above equation (8) by $z_i(t,x)$ and integrating with respect to $x$ over the domain $\Omega$, for $i=1,2,\cdots,n$,

$$\begin{aligned}\frac{1}{2}\frac{d}{dt}\int_\Omega \left(z_j(t,x)\right)^2 dx = & \sum_{k=1}^{m}\int_\Omega z_i(t,x)\frac{\partial}{\partial x_k}\left(a_{ik}\frac{\partial z_i(t,x)}{\partial x_k}\right)dx \\ & -\int_\Omega \alpha_i(z_i(t,x)+u_i^*)\big[z_i(t,x)\beta_i^*(z_i(t,x)) \\ & -\frac{1}{T}\sum_{j=1}^{n} c_{ij}z_i(t,x)(f_j(u_j(t,x))-f_j(v_j(t,x)))dx \\ & -\frac{1}{T}\sum_{j=1}^{n} d_{ij}z_i(t,x)(g_j(u_j(t-\tau_j(t),x)) \\ & -g_j(v_j(t-\tau_j(t),x)))\big]dx. \end{aligned} \tag{9}$$

From the Green's formula and the Dirichlet boundary condition, we have

$$\sum_{k=1}^{m}\int_\Omega z_i(t,x)\frac{\partial}{\partial x_k}\left(a_{ik}\frac{\partial z_i(t,x)}{\partial x_k}\right)dx = -\sum_{k=1}^{m}\int_\Omega a_{ik}\left(\frac{\partial z_i(t,x)}{\partial x_k}\right)^2 dx \tag{10}$$

Furthermore, from Lemma 1,

$$\begin{aligned}\sum_{k=1}^{m}\int_\Omega z_i(t,x)\frac{\partial}{\partial x_k}\left(a_{ik}\frac{\partial z_i(t,x)}{\partial x_k}\right)dx = & -\sum_{k=1}^{m}\int_\Omega a_{ik}\left(\frac{\partial z_i(t,x)}{\partial x_k}\right)^2 dx \\ \le & -\sum_{k=1}^{m}\int_\Omega \frac{a_{ik}}{l_k^2}\left(z_i(t,x)\right)^2 dx. \end{aligned} \tag{11}$$

From (9), (11), and the Holder inequality, we have:

$$\begin{aligned}\frac{d}{dt}||z_i(t,x)||_2^2 \le & -\sum_{k=1}^{m}\frac{2a_{ik}}{l_k^2}||z_i(t,x)||_2^2 - 2\underline{\alpha}_i\underline{b}_i||z_i(t,x)||_2^2 \\ & +2\bar{\alpha}_i\sum_{j=1}^{n}|c_{ij}|\ell_j||z_i(t,x)||_2||z_j(t,x)||_2 \\ & +2\bar{\alpha}_i\sum_{j=1}^{n}|d_{ij}|\mu_j||z_i(t,x)||_2||z_j(t-\tau_j(t),x)||_2; \end{aligned} \tag{12}$$

i.e.,

$$\begin{aligned}\frac{d||z_i(t,x)||_2}{dt} \le & -\left(\sum_{k=1}^{m}\frac{a_{ik}}{l_k^2}+\underline{\alpha}_i\underline{b}_i\right)||z_i(t,x)||_2 + \sum_{j=1}^{n}|c_{ij}|\bar{\alpha}_i\ell_j||z_j(t,x)||_2 \\ & +\sum_{j=1}^{n}|d_{ij}|\bar{\alpha}_i\mu_j||z_j(t-\tau_j(t),x)||_2. \end{aligned} \tag{13}$$

Since $\bar{A} - |C|\ell\bar{\alpha} - |D|\mu\bar{\alpha}$ is a nonsingular $M$ matrix, there exist positive numbers $\gamma_1, \cdots, \gamma_n$ such that

$$\gamma_i\Big(\sum_{k=1}^{m}\frac{a_{ik}}{l_k^2} + \underline{\alpha}_i\underline{b}_i\Big) - \sum_{j=1}^{n}\gamma_j(|c_{ij}|\bar{\alpha}_i\ell_j + |d_{ij}|\bar{\alpha}_i\mu_j) > 0. \tag{14}$$

Let $y_i(t,x) = ||z_i(t,x)||_2/\gamma_i$. From (12),

$$D^+y_i(t,x) \le -\Big(\sum_{k=1}^{m}\frac{a_{ik}}{l_k^2} + \underline{\alpha}_i\underline{b}_i\Big)y_i(t,x) + (\sum_{j=1}^{n}\gamma_j|c_{ij}|\ell_j\bar{\alpha}_i y_j(t,x)$$
$$+\sum_{j=1}^{n}\gamma_j|d_{ij}|\mu_j\bar{\alpha}_i y_j(t-\tau_j(t),x))/\gamma_i. \tag{15}$$

From (13) there exists a constant $\theta > 0$ such that

$$\gamma_i\Big(\sum_{k=1}^{m}\frac{a_{ik}}{l_k^2} + \underline{\alpha}_i\underline{b}_i\Big) - \sum_{j=1}^{n}\gamma_j\bar{\alpha}_i(|c_{ij}|\ell_j + |d_{ij}|\mu_j e^{\theta\tau}) \ge 0. \tag{16}$$

Let $\nu(0,x) = \max_{1\le i\le n}\{\sup_{-\tau\le s\le 0}\{y_i(s,x)\}\}$. Then $\forall t \ge 0$,

$$||y(t,x)|| \le \nu(0,x)\exp\{-\theta t\}. \tag{17}$$

Otherwise, there exist $t_2 > t_1 > 0$, $q \in \{1,2,\cdots,n\}$ and sufficiently small $\varepsilon > 0$ such that $\forall s \in [-\tau, t_1]$, (16) holds, and

$$y_i(s,x) \le \nu(0,x)\exp\{-\theta s\} + \varepsilon, \;\; s \in (t_1,t_2], i \in \{1,2,\cdots,n\}, \tag{18}$$

$$D^+y_q(t_2,x) + \theta\nu(0,x)\exp\{-\theta t_2\} > 0. \tag{19}$$

But from (14), (15) and (17),

$$D^+y_q(t_2,x) + \theta\nu(0,x)\exp\{-\theta t_2\} \le 0. \tag{20}$$

Hence, from this conclusion of absurdity, it shows that (16) holds.

If $a_{ik} \equiv 0$, consider recurrent neural networks with time-varying delays

$$\frac{\partial u_i(t,x)}{\partial t} = -\alpha_i(u_i(t,x))\big[\beta_i(u_i(t,x)) - \sum_{j=1}^{n}c_{ij}f_j(u_j(t,x))$$
$$-\sum_{j=1}^{n}d_{ij}g_j(u_j(t-\tau_j(t),x)) - I_i\big], \tag{21}$$

where $i = 1,\cdots,n$.

Denote $B = diag\{\underline{\alpha}_1\underline{b}_1, \underline{\alpha}_2\underline{b}_2, \cdots, \underline{\alpha}_n\underline{b}_n\}$.

**Corollary 1:** If $B - |C|\ell\bar{\alpha} - |D|\mu\bar{\alpha}$ is a nonsingular $M$ matrix, then the neural network (21) is globally exponentially stable.

## 4 Concluding Remarks

In this paper, using the delay differential inequality, we have obtained some sufficient conditions to guarantee that the Cohen-Grossberg neural networks with reaction-diffusion and Dirichlet boundary conditions is globally exponentially stable. The results presented in this paper are the improvement and extension of the existed ones in some existing works.

**Acknowledgement.** This work was supported by the Key Project of Hubei Provincial Education Department of China Under Grant B20052201.

## References

1. Grossberg, S.: Nonlinear Difference-differential Equations in Prediction and Learning Theory. Proceedings of the National Academy of Sciences, **58** (1967) 1329-1334
2. Cohen, M.A., Grossberg, S.: Absolute Stability of Global Pattern Formation and Parallel Memory Storage by Competitive Neural Networks. IEEE Transactions on Systems, Man, and Cybernetics, SMC, **13** (1983) 815-826
3. Hopfield, J.J.: Neurons with Graded Response Have Collective Computational Pproperties like Those of Two-state Neurons. Proc. Natl. Academy Sci., **81** (1984) 3088-3092
4. Chua, L.O., and Yang, L.: Cellular Neural Networks: Theory. IEEE Trans. Circuits Syst., **35** (1988) 1257-1272
5. Forti, M., Tesi, A.: New Conditions for Global Stability of Neural Networks with Application to Linear and Quadratic Programming Problems. IEEE Trans. Circ. Syst. I, **42** (1995) 354-366
6. Yi, Z., Heng, A., Leung, K.S.: Convergence Analysis of Cellular Neural Networks with Unbounded Delay. IEEE Trans. Circuits Syst. I, **48** (2001) 680-687
7. Yuan, K., Cao, J.D., Li, H.X.: Robust Stability of Switched Cohen-Grossberg Neural Networks with Mixed Time-varying Delays. IEEE Transactions on Systems Man and Cybernetics B-Cybernetics, **36** (2006) 1356-1363
8. Yang, Z.C., Xu, D.Y.: Impulsive Effects on Stability of Cohen-Grossberg Neural Networks with Variable Delays. Applied Mathematics and Computation, **177** (2006) 63-78
9. Wang, Z.D., Liu, Y.R., Li, M.Z., and Liu, X.H.: Stability Analysis for Stochastic Cohen-Grossberg Neural Networks with Mixed Time Delays. IEEE Transactions on Neural Networks, **17** (2006) 814-820
10. Liao, X.F., Li, C.D.: Global Attractivity of Cohen-Grossberg Model with Finite and Infinite Delays. Journal of Mathematical Analysis and Applications, **315** (2006) 244-262
11. Cao, J.D., Li, X.L.: Stability in Delayed Cohen-Grossberg Neural Networks: LMI Optimization Approach. Physica D-Nonlinear Phenomena, **212** (2005) 54-65
12. Chen, T.P., Rong, L.B.: Robust Global Exponential Stability of Cohen-Grossberg Neural Networks with Time-delays. IEEE Transactions on Neural Networks, **15** (2004) 203-206
13. Song, Q.K., Cao, J.D.: Global Exponential Stability and Existence of Periodic Solutions in BAM Networks with Delays and Reaction Diffusion Terms. Chaos, Solitons & Fractals, **23** (2005) 421-430
14. Song, Q.K., Cao, J.D., Zhao, Z.J.: Periodic Solutions and Its Exponential Stability of Reaction-Diffusion Recurrent Neural Networks with Continuously Distributed Delays. Nonlinear Analysis: Real World Applications, **7** (2006) 65-80

# Global Exponential Stability of Fuzzy Cohen-Grossberg Neural Networks with Variable Delays and Distributed Delays

Jiye Zhang, Dianbo Ren, and Weihua Zhang

Traction Power State Key Laboratory, Southwest Jiaotong University, Chengdu 610031, China
jyzhang@home.swjtu.edu.cn

**Abstract.** In this paper, we extend the Cohen–Grossberg neural networks from classical to fuzzy sets, and propose the fuzzy Cohen–Grossberg neural networks (FCGNN). The global exponential stability of FCGNN with variable delays and distributed delays is studied. Based on the properties of M-matrix, by constructing vector Liapunov functions and applying differential inequalities, the sufficient conditions ensuring existence, uniqueness, and global exponential stability of the equilibrium point of fuzzy Cohen–Grossberg neural networks with variable delays and distributed delays are obtained.

**Keywords:** Neural networks; global exponential stability; fuzzy; time delay.

## 1 Introduction

Since Cohen and Grossberg proposed a class of neural networks in 1983 [1], this model have attracted the attention of the scientific community due to their promising potential for tasks of classification, associative memory, and parallel computation and their ability to solve difficult optimization problems. In applications to parallel computation and signal processing involving solution of optimization problems, it is required that the neural network should have a unique equilibrium point that is globally asymptotically stable. Thus, the qualitative analysis of dynamic behaviors is a prerequisite step for the practical design and application of neural networks [2-14]. The stability of Cohen–Grossberg neural networks with delays has been investigated in [8-14]. Yang extended the cellular neural networks (CNNs) from classical to fuzzy sets, and proposed the fuzzy cellular neural networks (FCNNs), and applied it to the image processing [15,16]. Some conditions ensuring the global exponential stability of FCNNs with variable time delays were given in [17-19].

In the paper, we extend the Cohen–Grossberg neural networks from classical form to fuzzy sets, and propose the fuzzy Cohen–Grossberg neural networks (FCGNN), which contains both variable delays and distributed delays. By constructing proper nonlinear integro-differential inequalities involving variable delays and distributed delays, applying the idea of vector Liapunov method, we obtain the sufficient conditions of global exponential stability of FCGNN.

D.-S. Huang, L. Heutte, and M. Loog (Eds.): ICIC 2007, LNAI 4682, pp. 66–74, 2007.

## 2 Notation and Preliminaries

For convenience, we introduce some notations. $x^{\mathrm{T}}$ and $A^{\mathrm{T}}$ denote the transpose of a vector $x$ and a matrix $A$, where $x\in R^n$ and $A\in R^{n\times n}$. $[A]^s$ is defined as $[A]^s=[A^{\mathrm{T}}+A]/2$. $|x|$ denotes the absolute-value vector given by $|x|=(|x_1|,|x_2|,\cdots|x_n|)^{\mathrm{T}}$ and $|A|$ denotes the absolute-value matrix given by $|A|=(|a_{ij}|)_{n\times n}$. $\|x\|$ denotes the vector norm defined by $\|x\|=(x_1^2+\cdots+x_n^2)^{1/2}$ and $\|A\|$ denotes the matrix norm defined by $\|A\|=(\max\{\lambda:\lambda$ is an eigenvalue of $A^{\mathrm{T}}A\})^{1/2}$. $\bigwedge$ and $\bigvee$ denote the fuzzy AND and fuzzy OR operation, respectively.

The dynamical behavior of FCGNNs with indefinite time delays can be described by the nonlinear differential equations as follows

$$\dot{x}_i=\theta_i(x)[-c_i(x_i(t))+\sum_{j=1}^{n}a_{ij}f_j(x_j(t))+\sum_{j=1}^{n}b_{ij}f_j(x_j(t-\tau_{ij}(t)))$$

$$+\bigwedge_{j=1}^{n}\alpha_{ij}\int_{-\infty}^{t}k_{ij}(t-s)f_j(x_j(s))\mathrm{d}s+\bigvee_{j=1}^{n}\beta_{ij}\int_{-\infty}^{t}k_{ij}(t-s)f_j(x_j(s))\mathrm{d}s+J_i],\ (i=1,2,\cdots n),\quad(1)$$

where $x_i$ is the state of neuron $i$, $i=1,2,\cdots,n$, and $n$ is the number of neurons; $J_i$ denotes bias of the $i$th neuron, respectively; $\theta_i(x)$ is an amplification function; $f_i$ is the activation function of the $i$th neuron; $a_{ij}$ are elements of feedback template; $\alpha_{ij}$ and $\beta_{ij}$ are elements of fuzzy feedback MIN template, fuzzy feedback MAX template, respectively.

The initial conditions associated with equation (1) are of the form $x_i(s)=\phi_i(s)$, $s\le 0$, where it is assumed that $\phi_i\in C((-\infty,0],R)$, $i=1,2,\cdots,n$. Time delays $\tau_{ij}(t)\in[0,\tau]$ for all $t\ge 0$, where $\tau$ is a constant, $i,j=1,2,\cdots,n$.

Let $A=(a_{ij})_{n\times n}$, $B=(b_{ij})_{n\times n}$ $\alpha=(\alpha_{ij})_{n\times n}$, $\beta=(\beta_{ij})_{n\times n}$, $J=(J_1,J_2,...,J_n)^{\mathrm{T}}$, $f(x)=(f_1(x_1),...,f_n(x_n))^{\mathrm{T}}$.

*Assumption 1.* For each $i\in\{1,2,...,n\}$, $f_i:R\to R$ is globally Lipschitz with constants $L_i>0$, i.e.

$$|f_i(u)-f_i(v)|\le L_i|u-v|$$

for all $u,v$. Let $L=\mathrm{diag}(L_1,\cdots,L_n)>0$.

*Assumption 2.* For each $i\in\{1,2,\cdots,n\}$, $c_i:R\to R$ is strictly monotone increasing, i.e., there exists constant $d_i>0$ such that, $[c_i(u)-c_i(v)]/(u-v)\ge d_i$ for all $u$, $v$ $(u\ne v)$. Let $D=\mathrm{diag}(d_1,d_2,\cdots,d_n)$.

*Assumption 3.* For each $i\in\{1,2,\cdots,n\}$, $\theta_i:R^n\to R$ is a continuous function and satisfies $0<\sigma_i\le\theta_i$, where $\sigma_i$ is a constant, $i$=1,2,...,$n$.

*Assumption 4.* The kernel functions $k_{ij}:[0,+\infty)\to[0,+\infty)$ ($i,j=1,2,\cdots,n$) are piecewise continuous on $[0,+\infty)$ and satisfy

$$\int_0^{+\infty} \mathrm{e}^{\beta s} k_{ij}(s)\mathrm{d}s = p_{ij}(\beta),\ i,j=1,2,\cdots,n\ ,$$

where $p_{ij}(\beta)$ are continuous functions in $[0,\delta)$, $\delta>0$, and $p_{ij}(0)=1$.

If the delay-kernels in (1) are taken to be of the type:

$$k_{ij}(s)=\left(\frac{1}{\gamma_{ij}}\right)^{m+1}\frac{s^m e^{-s/\gamma_{ij}}}{m!},\ \gamma_{ij}\in(0,\infty),\ m=0,1,2,\ldots;\ i,j=1,2,\ldots,n,$$

then

$$\int_0^{+\infty} \mathrm{e}^{\beta s} k_{ij}(s)\mathrm{d}s = \left(\frac{1}{1-\gamma_{ij}\beta}\right)^{m+1}.$$

So the delay-kernels satisfy the Assumption 4. In this paper, in order to study the exponential stability of neural networks (1), conveniently, we adopt the Assumption 4 for the kernel functions.

*Note.* In papers [8-12], the boundedness of function $\theta_i$ was assumed. However, in this paper, the Assumption 3 is only needed. It is obvious that the function $\theta_i$ satisfied Assumption 3 maybe an unbounded one.

*Definition 1.* The equilibrium point $x^*$ of (1) is said to be globally exponentially stable, if there exist constants $\lambda>0$ and $M>0$ such that

$$|x_i(t)-x_i^*|\le M\,\|\phi-x^*\|\,\mathrm{e}^{-\lambda t}$$

for all $t\ge 0$, where $\|\phi-x^*\|=\max_{1\le i\le n}\{\sup_{s\in(-\infty,0]}|\phi_i(s)-x_i^*|\}$.

*Lemma 1.* [14] Let $A=(a_{ij})_{n\times n}$ be a matrix with non-positive off-diagonal elements. Then the following statements are equivalent:

(i) $A$ is an M-matrix;
(ii) There exists a vector $\xi>0$, such that $\xi^{\mathrm{T}}A>0$;
(iii) $A$ is nonsingular and all elements of $A^{-1}$ are nonnegative;
(iv) There exists a positive definite $n\times n$ diagonal matrix $Q$ such that matrix $AQ+QA^{\mathrm{T}}$ is positive definite.

*Lemma 2.* [3] If $H(x)\in C^0$ is injective on $R^n$, and $\|H(x)\|\to\infty$ as $\|x\|\to\infty$, then $H(x)$ is a homeomorphism of $R^n$.

*Lemma 3.* [15] Suppose $x$ and $y$ are two states of system (1), then

$$|\bigwedge_{j=1}^{n}\alpha_{ij}f_j(x_j)-\bigwedge_{j=1}^{n}\alpha_{ij}f_j(y_j)|\le\sum_{j=1}^{n}|\alpha_{ij}||f_j(x_j)-f_j(y_j)|,\ (i=1,2,\cdots n)$$

$$|\bigvee_{j=1}^{n}\beta_{ij}f_j(x_j)-\bigvee_{j=1}^{n}\beta_{ij}f_j(y_j)|\le\sum_{j=1}^{n}|\beta_{ij}||f_j(x_j)-f_j(y_j)|,\ (i=1,2,\cdots n).$$

## 3 Existence and Uniqueness of the Equilibrium Point

In the section, we study the existence and uniqueness of the equilibrium point of (1). We firstly study the nonlinear map associated with (1) as follows:

$$H_i(x)=-c_i(x_i)+\sum_{j=1}^{n}(a_{ij}+b_{ij})f_j(x_j)+\bigwedge_{j=1}^{n}\alpha_{ij}f_j(x_j)+\bigvee_{j=1}^{n}\beta_{ij}f_j(x_j)+J_i\ . \tag{2}$$

Let $H(x)=(H_1(x),H_2(x),...,H_n(x))^{\mathrm{T}}$. If map $H(x)$ is a homeomorphism on $R^n$, then there exists a unique point $x^*$ such that $H(x^*)=0$. We have

$$\begin{aligned}\dot{x}_i&=\theta_i(x^*)[-c_i(x_i^*)+\sum_{j=1}^{n}(a_{ij}+b_{ij})f_j(x_j^*)+\bigwedge_{j=1}^{n}\alpha_{ij}\int_{-\infty}^{t}k_{ij}(t-s)f_j(x_j^*)\mathrm{d}s\\&\quad+\bigvee_{j=1}^{n}\beta_{ij}\int_{-\infty}^{t}k_{ij}(t-s)f_j(x_j^*)\mathrm{d}s+J_i]\\&=\theta_i(x^*)[-c_i(x_i^*)+\sum_{j=1}^{n}(a_{ij}+b_{ij})f_j(x_j^*)+\bigwedge_{j=1}^{n}\alpha_{ij}f_j(x_j^*)\mathrm{d}s+\bigvee_{j=1}^{n}\beta_{ij}f_j(x_j^*)+J_i]\\&=\theta_i(x^*)H_i(x^*)\,.\end{aligned}$$

So the solution of $H(x)=0$ is the equilibrium of systems (1). Based on the Lemma 2, we get the conditions of the existence of the equilibrium for system (1) as follows.

*Theorem 1.* If Assumptions 1-4 are satisfied, and $\Pi=D-(|A|+|B|+|\alpha|+|\beta|)L$ is an M- matrix, then for each *J*, system (1) has a unique equilibrium point.

*Proof.* In order to prove that systems (1) have a unique equilibrium point $x^*$, it is only need to prove that $H(x)$ is a homeomorphism on $R^n$. In the following, we shall prove that map $H(x)$ is a homeomorphism in two steps.

*Step 1.* We prove that $H(x)$ is an injective on $\mathrm{R}^n$. For purposes of contradiction, suppose that there exist $x,y\in R^n$ with $x\ne y$, such that $H(x)=H(y)$, i.e,

$$\begin{aligned}c_i(x_i)-c_i(y_i)&=\sum_{j=1}^{n}(a_{ij}+b_{ij})[f_j(x_j)-f_j(y_j)]+\bigwedge_{j=1}^{n}\alpha_{ij}f_j(x_j)-\bigwedge_{j=1}^{n}\alpha_{ij}f_j(y_j)\\&\quad+\bigvee_{j=1}^{n}\beta_{ij}f_j(x_j)-\bigvee_{j=1}^{n}\beta_{ij}f_j(y_j),\qquad i=1,2,\cdots n\,.\end{aligned}$$

We have

$$|c_i(x_i)-c_i(y_i)|\le|\sum_{j=1}^{n}(a_{ij}+b_{ij})[f_j(x_j)-f_j(y_j)]|+|\bigwedge_{j=1}^{n}\alpha_{ij}f_j(x_j)-\bigwedge_{j=1}^{n}\alpha_{ij}f_j(y_j)|$$
$$+|\bigvee_{j=1}^{n}\beta_{ij}f_j(x_j)-\bigvee_{j=1}^{n}\beta_{ij}f_j(y_j)|,\ i=1,2,\cdots n.$$

From Lemma 3, and Assumption 1-3, for all $i=1,2,\cdots n$, we get

$$d_i|x_i-y_i|\le\sum_{j=1}^{n}(|a_{ij}|+|b_{ij}|)L_j|x_j-y_j|+\sum_{j=1}^{n}|\alpha_{ij}|L_j|x_j-y_j|+\sum_{j=1}^{n}|\beta_{ij}|L_j|x_j-y_j|.$$

Rewriting the above inequalities as matrix form, we have

$$[D-(|A|+|B|+|\alpha|+|\beta|)L]|x-y|\le 0\ . \tag{3}$$

Because of $\Pi$ being an M-matrix, from Lemma 1, we know that all elements of $(D-(|A|+|\alpha|+|\beta|)L)^{-1}$ are nonnegative. Therefore $|x-y|\le 0$, i.e., $x=y$. From the supposition $x\ne y$, thus this is a contradiction. So map $H(x)$ is injective.

*Step 2.* We prove that $\|H(x)\|\to\infty$ as $\|x\|\to\infty$.

Let $\overline{H}(x)=H(x)-H(0)$. From (2), we get

$$\overline{H}_i(x_i)=-[c_i(x_i)-c_i(0)]+\sum_{j=1}^{n}(a_{ij}+b_{ij})[f_j(x_j)-f_j(0)]+\bigwedge_{j=1}^{n}\alpha_{ij}f_j(x_j)-\bigwedge_{j=1}^{n}\alpha_{ij}f_j(0)$$
$$+\bigvee_{j=1}^{n}\beta_{ij}f_j(x_j)-\bigvee_{j=1}^{n}\beta_{ij}f_j(0)\ \ (i=1,2,\cdots n). \tag{4}$$

Since $D-(|A|+|B|+|\alpha|+|\beta|)L$ is an M-matrix, from Lemma 1, there exists a diagonal matrix $T=\text{diag}\{T_1,T_2,\cdots,T_n\}>0$, such that

$$[T(-D+(|A|+|B|+|\alpha|+|\beta|)L)]^s\le-\varepsilon E_n<0\ , \tag{5}$$

where $\varepsilon>0$ and $E_n$ is the identity matrix. From equation (4) and Lemma 3, we get

$$[Tx]^{\mathrm{T}}\overline{H}(x)=\sum_{i=1}^{n}x_iT_i\{-[c_i(x_i)-c_i(0)]+\sum_{j=1}^{n}(a_{ij}+b_{ij})[f_j(x_j)-f_j(0)]$$
$$+\bigwedge_{j=1}^{n}\alpha_{ij}f_j(x_j)-\bigwedge_{j=1}^{n}\alpha_{ij}f_j(0)+\bigvee_{j=1}^{n}\beta_{ij}f_j(x_j)-\bigvee_{j=1}^{n}\beta_{ij}f_j(0)\}$$
$$\le\sum_{i=1}^{n}T_i\{-d_ix_i^2+|x_i|\sum_{j=1}^{n}(|a_{ij}|+|b_{ij}|)|f_j(x_j)-f_j(0)|$$
$$+|x_i|\sum_{j=1}^{n}|\alpha_{ij}||f_j(x_j)-f_j(0)|+|x_i|\sum_{j=1}^{n}|\beta_{ij}||f_j(x_j)-f_j(0)|\}$$
$$\le\sum_{i=1}^{n}T_i\{-d_ix_i^2+|x_i|\sum_{j=1}^{n}(|a_{ij}|+|b_{ij}|)L_j|x_j|+|x_i|\sum_{j=1}^{n}|\alpha_{ij}|L_j|x_j|$$
$$+|x_i|\sum_{j=1}^{n}|\beta_{ij}|L_j|x_j|\}$$
$$\le|x|^{\mathrm{T}}[T(-D+(|A|+|B|+|\alpha|+|\beta|)L)]^s|x|\le-\varepsilon\|x\|^2\ . \tag{6}$$

Using Schwarz inequality, and from (6), we get $\varepsilon\|x\|^2 \le \|T\|\|x\|\|\overline{H}(x)\|$, so $\|\overline{H}(x)\| \ge \varepsilon\|x\|/\|T\|$. Therefore, $\|\overline{H}(x)\| \to +\infty$, i.e., $\|H(x)\| \to +\infty$ as $\|x\| \to +\infty$.

Based on Lemma 2, from steps 1 and 2, we know $\overline{H}(x)$ is a homeomorphism and for every *J*, map $H(x)$ is a homeomorphism on $R^n$. So system (1) has a unique equilibrium point. The proof is completed.

## 4 Global Exponential Stability of the Equilibrium Point

*Theorem 2.* If Assumptions 1-4 are satisfied and $\Pi = D-(|A|+|B|+|\alpha|+|\beta|)L$ is an M-matrix, then for each $J$, system (1) has a unique equilibrium point, which is globally exponentially stable.

*Proof.* Since $\Pi$ is an M-matrix, from Theorem 1, system (1) has a unique equilibrium $x^*$. Let $y(t)=x(t)-x^*$, we have

$$\begin{aligned}\dot{y}_i(t) = \theta_i(y(t)+x^*)[&-c_i(y_i(t)+x_i^*)+c_i(x_i^*) + \sum_{j=1}^{n} a_{ij}(f_j(y_j(t)+x_j^*)-f_j(x_j^*))\\ &+\sum_{j=1}^{n} b_{ij}(f_j(y_j(t-\tau_{ij}(t))+x_j^*)-f_j(x_j^*))\\ &+\bigwedge_{j=1}^{n}\alpha_{ij}\int_{-\infty}^{t}k_{ij}(t-s)f_j(y_j(s)+x_j^*)\mathrm{d}s - \bigwedge_{j=1}^{n}\alpha_{ij}f_j(x_j^*)\\ &+\bigvee_{j=1}^{n}\beta_{ij}\int_{-\infty}^{t}k_{ij}(t-s)f_j(y_j(s)+x_j^*)\mathrm{d}s - \bigvee_{j=1}^{n}\beta_{ij}f_j(x_j^*)] \quad (i=1,2,\cdots,n). \end{aligned} \tag{7}$$

The initial conditions of equation (7) are $\Psi(s)=\phi(s)-x^*$, $s\in(-\infty,0]$. Systems (7) have a unique equilibrium at $y=0$.

Let

$$V_i(t)=\mathrm{e}^{\lambda t}|y_i(t)|, \tag{8}$$

where $\lambda$ is a constant to be given. Calculating the upper right derivative of $V_i(t)$ along the solutions of (7), we have

$$\begin{aligned} D^+(V_i(t)) &= \mathrm{e}^{\lambda t}\operatorname{sgn}(y_i(t))[\dot{y}_i(t)+\lambda y_i(t)]\\ &\le \mathrm{e}^{\lambda t}\{\theta_i(y(t)+x^*)\,[-\operatorname{sgn}(y_i)(c_i(y_i(t)+x_i^*)-c_i(x_i^*))\\ &+\sum_{j=1}^{n}|a_{ij}\|f_j(y_j(t)+x_j^*)-f_j(x_j^*)| + \sum_{j=1}^{n}|b_{ij}\|f_j(y_j(t-\tau_{ij}(t))+x_j^*)-f_j(x_j^*)|\\ &+|\bigwedge_{j=1}^{n}\alpha_{ij}\int_{-\infty}^{t}k_{ij}(t-s)f_j(y_j(s)+x_j^*)\mathrm{d}s - \bigwedge_{j=1}^{n}\alpha_{ij}f_j(x_j^*)|\\ &+|\bigvee_{j=1}^{n}\beta_{ij}\int_{-\infty}^{t}k_{ij}(t-s)f_j(y_j(s)+x_j^*)\mathrm{d}s - \bigvee_{j=1}^{n}\beta_{ij}f_j(x_j^*)|] + \lambda|y_i(t)|\}\\ &\le \mathrm{e}^{\lambda t}\{\theta_i(y(t)+x^*)\,[-d_i|y_i(t)| + \sum_{j=1}^{n}|a_{ij}|L_j|y_j(t)| + \sum_{j=1}^{n}|b_{ij}|L_j|y_j(t-\tau_{ij}(t))| \end{aligned}$$

$$+\sum_{j=1}^{n}|\alpha_{ij}||\int_{-\infty}^{t}k_{ij}(t-s)f_j(y_j(s)+x_j^*)ds - f_j(x_j^*)|$$

$$+\sum_{j=1}^{n}|\beta_{ij}||\int_{-\infty}^{t}k_{ij}(t-s)f_j(y_j(s)+x_j^*)ds - f_j(x_j^*)|] + \lambda|y_i(t)|\}$$

$$\le \theta_i(y(t)+x^*)\,[-d_iV_i(t) + \sum_{j=1}^{n}|a_{ij}|L_jV_j(t) + \sum_{j=1}^{n}|b_{ij}|e^{\lambda\tau}L_jV_j(t-\tau_{ij})$$

$$+\sum_{j=1}^{n}(|\alpha_{ij}|+|\beta_{ij}|)\int_{-\infty}^{t}k_{ij}(t-s)e^{\lambda(t-s)}L_jV_j(s)ds] + \lambda V_i(t) \quad (i=1,2,\cdots,n).$$

From Assumption 3, we know that $0<\sigma_i \le \theta_i(y(t)+x^*)$, so $\theta_i(y(t)+x^*)/\sigma_i \ge 1$. Thus, from Assumption 1 and Lemma 3, we get

$$D^+(V_i(t)) \le \theta_i\{(-d_i+\lambda/\sigma)V_i(t) + \sum_{j=1}^{n}L_j[|a_{ij}|V_j(t) + e^{\lambda\tau}|b_{ij}|V_j(t-\tau_{ij})$$

$$+(|\alpha_{ij}|+|\beta_{ij}|)\int_{-\infty}^{t}k_{ij}(t-s)e^{\lambda(t-s)}V_j(s)ds]\} \ . \qquad (9)$$

Due to $\Pi$ is an M-matrix, from the Lemma 1, there exist positive constant numbers $\xi_i, i=1,2,\cdots n$, satisfying

$$-\xi_i d_i + \sum_{j=1}^{n}\xi_j(|a_{ij}|+|b_{ij}|+|\alpha_{ij}|+|\beta_{ij}|)L_j < 0 \quad (i=1,2,\cdots n).$$

It is obvious that there exists a constant $\lambda>0$ such that

$$-\xi_i(d_i-\lambda/\sigma) + \sum_{j=1}^{n}\xi_j[|a_{ij}|+e^{\lambda\tau}|b_{ij}|+(|\alpha_{ij}|+|\beta_{ij}|)p_{ij}(\lambda)]L_j < 0\,(i=1,2,\cdots n)\,. \quad (10)$$

Define the curve $\gamma=\{z(l): z_i=\xi_i l, l>0, \quad i=1,2,\cdots,n\}$ and the set $\Omega(z)=\{u: 0\le u\le z, z\in\gamma\}$ . Let $\xi_M = \max_{i=1,\ldots,N}\xi_i$ , $\xi_m = \min_{i=1,\ldots,N}\xi_i$ , taking $l_0=(1+\delta)e^{\lambda\tau}\|\Psi\|/\xi_m$, where $\delta>0$ be a constant.

Defining set

$$O=\{V: V=e^{\lambda s}\|\Psi_1(s)\|,\cdots,\|\Psi_n(s)\|)^{\mathrm{T}}, -\infty<s\le 0\}.$$

So, $O\subset\Omega(z_0(l_0))$, namely

$$V_i(s) \le e^{\lambda s}\|\Psi_i(s)\| < \xi_i l_0, -\infty<s\le 0, \quad (i=1,2,\cdots n). \qquad (11)$$

In the following, we shall prove

$$V_i(t) < \xi_i l_0,\ t>0,\ (i=1,2,\cdots n). \qquad (12)$$

If (12) is not true, then from (11), there exist $t_1>0$ and some index $i$ such that

$$V_i(t_1)=\xi_i l_0,\ D^+(V_i(t_1))\ge 0,\ V_j(t)\le \xi_j l_0,\ t\in(-\infty,t_1],\ j=1,2,\cdots n\ . \qquad (13)$$

However, from (9), and (10), we get

$$D^+(V_i(t_1)) \le \theta_i\{-\xi_i(d_i-\lambda/\sigma) + \sum_{j=1}^{n}\xi_j[|a_{ij}|+e^{\lambda\tau}|b_{ij}|+p_{ij}(\lambda)(|\alpha_{ij}|+|\beta_{ij}|)]L_j\}l_0 < 0 .$$

This is a contradiction. So $V_i(t) < \xi_i l_0$, for $t > 0$ ($i = 1,2,\cdots n$). Furthermore, from (8), and (12), we obtain

$$|y_i(t)| \le \xi_i l_0 e^{-\lambda t} \le (1+\sigma)e^{\lambda\tau}\xi_M/\xi_m \|\Psi\| e^{-\lambda t} \le M\|\Psi\| e^{-\lambda t},\ t \ge 0\ (i = 1,2,\cdots n),$$

where $M = (1+\sigma)e^{\lambda\tau}\xi_M/\xi_m$. Thus

$$|x_i(t) - x_i^*| \le M\|x_i(t) - x_i^*\| e^{-\lambda t},$$

and the equilibrium point of (1) is globally exponentially stable. The proof is completed.

## 5 Conclusions

In this paper, we extend the Cohen–Grossberg neural networks from classical to fuzzy sets, and propose the fuzzy Cohen–Grossberg neural networks (FCGNN). We analyze the existence, uniqueness, and global exponential stability of the equilibrium point of FCGNN with variable delays and distributed delays. Applying the idea of vector Liapunov function method, by constructing proper nonlinear integro-differential inequalities involving both variable delays and distributed delays, we obtain sufficient conditions for global exponential stability independent of the delays. The conditions are explicit and easy to test for designing neural networks.

**Acknowledgments.** This work is supported by National Program for New Century Excellent Talents in University (No.NCET-04-0889), Natural Science Foundation of China (No. 50525518), and Youth Science Foundation of Sichuan (No. 05ZQ026-015).

## References

1. Cohen, M.A., Grossberg, S.: Absolute Stability and Global Pattern Formation and Parallel Memory Storage by Competitive Neural Networks. IEEE Trans. Syst., Man, Cybern., 13(1983) 815-826.
2. Arik, S.: An Improved Global Stability Result for Delayed Cellular Neural Networks. IEEE Trans. Circ. Syst. 49 (2002) 1211-1214
3. Forti, M., Tesi, A.: New Conditions for Global Stability of Neural Networks with Application to Linear and Quadratic Programming Problems. IEEE Trans. Circ. Syst.-I 42 (1995) 354-366
4. Zhang, J.: Globally Exponential Stability of Neural Networks with Variable Delays. IEEE Trans. Circ. Syst.-I 50(2003) 288-291
5. Yucel, E., Arik, S., New Exponential Stability Results for Delayed Neural Networks with Time Varying Delays, Physica D, 191(2004) 314–322.

6. Xu, D., Zhao, H., Zhu, H.: Global Dynamics of Hopfield Neural Networks Involving Variable Delays. Computers and Mathematics with Applications 42(2001) 39-45
7. Zhang, J., Suda, Y., Iwasa, T.: Absolutely Exponential Stability of A Class of Neural Networks with Unbounded Delay, Neural Networks, 17(2004) 391-397.
8. Wang, L.: Stability of Cohen-Grossberg Neural Networks with Distributed Delays, Applied Mathematics and Computation, 160(2005), 93-110.
9. Chen, T., Rong, L.: Delay-independent Stability Analysis of Cohen-Grossberg Neural Networks, Physics Letters A, 317(2003), 436-449.
10. Wang, C.C., Cheng, C.J., Liao, T.L.: Globally Exponential Stability of Generalized Cohen-Grossberg Neural Networks with Delays, Physics Letters A, 319(2003) 157-166.
11. Chen, T., Rong, L.: Robust Global Exponential Stability of Cohen- Grossberg Neural Networks with Time-Delays. IEEE Transactions on Neural Networks 15(2004) 203-206.
12. Xiong, W., Cao, J.: Absolutely Exponential Stability of Cohen-Grossberg Neural Networks with Unbounded Delays. Neurocomputing 68(2005) 1-12
13. Song, Q., Cao, J.: Stability Analysis of Cohen–Grossberg Neural Network with both Time-Varying and Continuously Distributed Delays, Journal of Computational and Applied Mathematics 197 (2006) 188-203
14. Zhang, J., Suda, Y., Komine, H.: Global Exponential Stability of Cohen-Grossberg Neural Networks with Variable Delays. Physics Letter A 338(2005) 44-50
15. Yang, T., Yang, L.B.: Exponential Stability of Fuzzy Cellular Neural Networks with Constant and Time-Varying Delays. IEEE Trans. Circ. Syst.-I 43 (1996) 880-883
16. Yang, T., Yang, L.B.: Fuzzy Cellular Neural Networks: A New Paradigm for Image Processing. Int. J. Circ. Theor. Appl. 25 (1997) 469-481
17. Liu Y., Tang, W.: Exponential Stability of Fuzzy Cellular Neural Networks with Constant and Time-Varying Delays. Physics Letters A 323 (2004) 224-233
18. Zhang, J., Ren, D., Zhang, W.: Global Exponential Stability of Fuzzy Cellular Neural Networks with Variable Delays. Lecture Notes in Computer Science 3971 (2006) 236-242
19. Yuan, K., Cao, J., Deng, J.: Exponentially Stability and Periodic Solutions of Fuzzy Cellular Neural Networks with Time-Varying Delays. Neurocomputing 69(2006) 1619-1627.

# Global Exponential Synchronization of a Class of Chaotic Neural Networks with Time-Varying Delays

Jing Lin and Jiye Zhang

National Traction Power Laboratory, Southwest Jiaotong University,
Chengdu 610031, China
jyzhang@home.swjtu.edu.cn

**Abstract.** This paper aims to present a synchronization scheme for a class of chaotic neural networks with time-varying delays, which covers the Hopfield neural networks and cellular neural networks. Using the drive-response concept, a control law of two identical chaotic neural networks is derived to achieve the exponential synchronization. Furthermore, based on the idea of vector Lyapunov function, and M-matrix theory, the sufficient conditions for global exponential synchronization of a class of chaotic neural networks are obtained. The synchronization condition is easy to verify and removed some restriction on the chaotic neural networks. Finally, some chaotic neural networks with time-varying delays are given as examples for illustration.

**Keywords:** Exponential synchronization, Lyapunov function, chaos.

## 1 Introduction

Over past two decades, much research effort has been devoted to the study of control, synchronization and application of chaotic system [1-4]. Since the drive-response concept of couple chaotic systems introduced by Pecora and Carroll in their pioneering work [5], the synchronization of couple chaotic systems has been received considerable attention in the last decade due to its potential applications including secure communication systems and signal-processing systems [6-9]. There are several different approaches including some conventional linear control techniques and advanced nonlinear control schemes to achieve synchronization of the chaotic systems have been proposed in the literature [10-16]. More precisely, state variables of a given chaotic drive system are used as input to drive a response system that is the same as the drive system. Using the restrictive condition, the response system is to synchronize to that of the drive system. In [17], synchronization control of stochastic neural networks with time-varying delays was studied by linear matrix inequality approach.

Our objective in this paper is to study the global exponential synchronization problem of a class of chaotic neural networks with time- varying delays. This class of chaotic neural networks includes several well-known neural networks, such as Hopfield neural networks and cellular neural networks which have been studied extensively over past two decades [14-17]. Based on the vector Lyapunov function, M-matrix theory [18] and drive-response synchronization concept, a control law with an

D.-S. Huang, L. Heutte, and M. Loog (Eds.): ICIC 2007, LNAI 4682, pp. 75–82, 2007.

appropriate gain matrix is derived to achieve synchronization of the drive-response-based chaotic neural networks with time-varying delays.

We first give some notations which will be used throughout this paper. $x = (x_1, ..., x_n)^T \in R^n$ denotes a column vector (the symbol $(^T)$ denotes transpose). $|x|$ denotes the absolute-value vector given by $|x| = (|x_1|, ..., |x_n|)^T$. For a matrix $A = (a_{ij})_{n\times n}$, $|A|$ denotes absolute-value matrix defined by $|A| = (|a_{ij}|)_{n\times n}$.

## 2 Systems Description and Synchronization Problem

A class of chaotic neural networks in this paper is described by the delayed differential equations:

$$\dot{x}_i(t) = -g_i(x_i(t)) + \sum_{j=1}^{n} a_{ij} f_j(x_j(t)) + \sum_{j=1}^{n} b_{ij} f_j(x_j(t - \tau_{ij}(t)) + J_i \ , (i = 1, ..., n\ ), \quad (1)$$

where $n \geq 2$ denotes the number of neurons, $x_i$ is the state of neuron $i$, $i = 1, ..., n$, $g_i(x_i(t))$ is an appropriately behaved function, $f_i$ is the activation function of the neurons. The feedback matrix $A = (a_{ij})_{n\times n}$ indicates the strength of the neuron interconnections within the network. $B = (b_{ij})_{n\times n}$ indicates the strength of the neuron interconnections within the network with time-varying delay parameter $\tau_{ij}(t)$, $(i, j = 1, ..., n)$ ($\tau = \max_{1\leq i,j\leq n, t\in R}\{\tau_{ij}(t)\}$), $J = (J_1, ..., J_n)^T$ is the constant input vector. The initial conditions of (1) are of the form $x_i(s) = \phi_i(s)$, $s \in [-\tau, 0]$, where $\phi_i$ is bounded and continuous on $[-\tau, 0]$.

We consider the functions of the neurons satisfying the following assumptions.

**Assumption 1.** For each function $g_i$: $R \to R$, $i = 1, ..., n$, there exists constant $G_i > 0$ such that

$$\frac{g_i(u_i) - g_i(v_i)}{u_i - v_i} \geq G_i > 0 \quad \text{for } u_i \neq v_i .$$

**Assumption 2.** Each function $f_i$ : $R \to R$, $i = 1, ..., n$, is globally Lipschitz with Lipschitz constant $L_i$, i.e.,

$$|f_i(u_i) - f_i(v_i)| \leq L_i |u_i - v_i| \text{ for all } u_i, v_i .$$

Let $G = \text{diag}\{G_1, ..., G_n\}$, $L = \text{diag}\{L_1, ..., L_n\}$.

Thc class of neural networks can describe several well-known neural networks such as Hopfield neural network [17] and cellular neural network [14,15].

If the system's matrix $A$ and $B$ as well as the delay parameter $\tau_{ij}(t)$ are suitably chosen, the system (1) will display a chaotic behavior [14,15]. In this paper, we are concerned with the synchronization problem of this class of chaotic neural networks.

Based on the drive-response concept, synchronization behavior for two chaotic neural networks is studied. The drive and response system are described by the following equations, respectively:

$$\dot{x}_i(t) = -g_i(x_i(t)) + \sum_{j=1}^{n} a_{ij} f_j(x_j(t)) + \sum_{j=1}^{n} b_{ij} f_j(x_j(t-\tau_{ij}(t))) + J_i\,,\ (i = 1,\dots,n)\,, \quad (2)$$

and

$$\dot{z}_i(t) = -g_i(z_i(t)) + \sum_{j=1}^{n} a_{ij} f_j(z_j(t)) + \sum_{j=1}^{n} b_{ij} f_j(z_j(t-\tau_{ij}(t))) + J_i - u_i\,,\ (i = 1,\dots,n)\,, \quad (3)$$

with initial condition $z_i(s) = \varphi_i(s)$ , $s \in [-\tau, 0]$ , where it is usually assumed that $\varphi_i \in C([-\tau,0], R)$, $i$=1,…,$n$, in which $u_i$ denotes the external control input.

*Definition 1*. The system (2) and the uncontrolled system (3) (i.e. $u \equiv 0$ ) are said to be globally exponentially synchronized if there exist constant $M$>0 and $\lambda$ >0 such that

$$\|x(t) - z(t)\| \le M \|\phi(s) - \varphi(s)\| e^{-\lambda t}$$

for all $t \ge 0$ , where $\|\phi(s) - \varphi(s)\| = \max_{1 \le i \le n} \sup_{s \in [-\tau,0]} |\phi_i(s) - \varphi_i(s)|$ , $\lambda$ is the exponential synchronization rate.

*Definition 2.*[18] A real matrix $A = (a_{ij})_{n \times n}$ is said to be an M-matrix if $a_{ij} \le 0$ , $i$, $j$=1,2,…,$n$, $i \ne j$ , and all successive principal minors of $A$ are positive.

## 3 Main Results

If the drive and response system with same system's parameter but the differences in initial conditions, it is studied that how to deal the control input $u_i$ with the state-feedback for the purpose of global exponential synchronization.

### 3.1 Controller Design

Defining the synchronization error signal $\beta_i(t) = x_i(t) - z_i(t)$ , $i = 1,\dots,n$ , where $x_i(t)$ and $z_i(t)$ are the state variable of the drive and response neural networks, respectively. $\beta \to 0$ means that the drive and response system are synchronized. The error dynamics between (2) and (3) can be written as

$$\dot{\beta}_i(t) = -[g_i(\beta_i(t) + z_i) - g_i(z_i)] + \sum_{j=1}^{n} a_{ij} [f_j(\beta_j(t) + z_j) - f(z_j)]$$
$$+ \sum_{j=1}^{n} b_{ij} [f_j(\beta_j(t - \tau_{ij}) + z_j) - f_j(z_j)] + u_i\ , \quad (4)$$

or the following compact form:

$$\dot{\beta}(t) = -Q(\beta(t)) + AP(\beta(t)) + BP(\beta(t-\tau(t))) + u(t) \ , \tag{5}$$

where $\beta(t) = [\beta_1(t),...,\beta_n(t)]^{\mathrm{T}}$, $u(t) = [u_1(t),...,u_n(t)]^T$ denotes the input vector;

$$P(\beta) = [p_1(\beta_1),...,p_n(\beta_n)]^T = [f_1(\beta_1 + z_1) - f(z_1),...,f_n(\beta_n + z_n) - f_n(z_n)]^T \ ;$$
$$Q(\beta) = [q_1(\beta_1),...,q_n(\beta_n)]^T = [g_1(\beta_1 + z_1) - g_1(z_1),...,g_n(\beta_n + z_n) - g_n(z_n)]^T \ .$$

Using the state variable of the two systems to drive the response system, the control input vector with state feedback is designed as follows:

$$\begin{bmatrix} u_1(t) \\ \vdots \\ u_n(t) \end{bmatrix} = \begin{bmatrix} \omega_{11} & \cdots & \omega_{1n} \\ \vdots & \cdots & \vdots \\ \omega_{n1} & \cdots & \omega_{nn} \end{bmatrix} \begin{bmatrix} x_1(t) - z_1(t) \\ \vdots \\ x_n(t) - z_n(t) \end{bmatrix} = \Omega \begin{bmatrix} \beta_1(t) \\ \vdots \\ \beta_n(t) \end{bmatrix} , \tag{6}$$

where $\Omega$ is the controller gain matrix. The following compact form is as follows:

$$\dot{\beta}(t) = -Q(\beta(t)) + AP(\beta(t)) + BP(\beta(t-\tau(t))) + \Omega\beta(t) \ . \tag{7}$$

### 3.2 Global Exponential Synchronization Condition

In the following, we give a condition ensuring the global exponential synchronization.

*Main theorem.* For the drive-response chaotic neural networks (2) and (3) which satisfy assumptions (A1)-(A2), if $G - (|A| + |B|)L - \Omega^*$ is an M-matrix, where $\Omega^* = (\omega_{ij}^*)_{n \times n}$, $\omega_{ij}^* = |\omega_{ij}|$ ( $i \neq j$ ), $\omega_{ii}^* = \omega_{ii}$ ( $i, j = 1,2,...,n$ ), then for each $J \in R^n$, the global exponential synchronization of system (2) and system (3) is obtained.

*Proof.* Since $G - (|A| + |B|)L - \Omega^*$ is an M-matrix [18], there exists $\xi_i > 0$ $i = 1,...,n$ satisfying

$$-\xi_i G_i + \sum_{j=1}^{n} \xi_j [(|a_{ij}| + |b_{ij}|)L_j + \omega_{ij}^*] < 0 \ \ (i = 1,...,n).$$

Defining function as follows:

$$F_i(\mu) = -\xi_i(G_i - \mu) + \sum_{j=1}^{n} \xi_j [(|a_{ij}| + \mathrm{e}^{\mu\tau} |b_{ij}|)L_j + \omega_{ij}^*], \ \ i = 1,...,n \ .$$

We know that $F_i(0) < 0$. So, there exists a constant $\lambda > 0$ such that

$$-\xi_i(G_i - \lambda) + \sum_{j=1}^{n} \xi_j [(|a_{ij}| + \mathrm{e}^{\lambda\tau} |b_{ij}|)L_j + \omega_{ij}^*] < 0 \ , \ i = 1,...,n \ . \tag{8}$$

Here, $\tau$ is a fixed number according to assumption of chaotic neural networks (1). Let $V_i(t) = \mathrm{e}^{\lambda t} |\beta_i(t)|$, $i = 1,...,n$. It can casily be verified that $V_i$ is a non-negative function over $[-\tau, +\infty)$ and that it is radially unbounded, i.e. $V \to +\infty$ as $\|\beta\| \to +\infty$. Calculating the upper right derivation $D^+ V_i$ of $V_i$ along the solution of (7), we get

$$D^+V_i = e^{\lambda t}\operatorname{sgn}\beta_i(t)\{-q_i(\beta_i(t))+\sum_{j=1}^{n}[a_{ij}p_j(\beta_j(t))+\omega_{ij}\beta_j(t)$$
$$+b_{ij}p_j(\beta_j(t-\tau_{ij}(t)))]\}+\lambda e^{\lambda t}|\beta_i(t)|$$
$$\le e^{\lambda t}\{-|q_i(\beta_i(t))|+\sum_{j=1}^{n}[|a_{ij}||p_j(\beta_j(t))|+\omega_{ij}^*|\beta_j(t)|$$
$$+|b_{ij}||p_j(\beta_j(t-\tau_{ij}(t)))|]\}+\lambda e^{\lambda t}|\beta_i(t)|$$
$$\le e^{\lambda t}\{(\lambda-G_i)|\beta_i(t)|+\sum_{j=1}^{n}[|a_{ij}||p_j(\beta_j(t))|$$
$$+\sum_{j=1}^{n}|b_{ij}||p_j(\beta_j(t-\tau_{ij}(t)))|]+\sum_{j=1}^{n}\omega_{ij}^*|\beta_j(t)|\}$$
$$\le(\lambda-G_i)V_i(t)+\sum_{j=1}^{n}[L_j(|a_{ij}|V_j(t)+e^{\lambda\tau_{ij}(t)}|b_{ij}|e^{\lambda(t-\tau_{ij}(t))}|\beta_j(t-\tau_{ij}(t))|)+\omega_{ij}^*V_j(t)]$$
$$\le(\lambda-G_i)V_i(t)+\sum_{j=1}^{n}[L_j(|a_{ij}|V_j(t)+e^{\lambda\tau}|b_{ij}|\sup_{t-\tau<s<t}V_j(s))+\omega_{ij}^*V_j(t)]\,. \quad (9)$$

Defining the curve: $\gamma=\{z(l): z_i=\xi_i l,\ l>0,\ i=1,\dots,n$ and the set: $\kappa(z)=\{u: 0\le u\le z, z\in\gamma\}$. Let $\xi_{\min}=\min_{1\le i\le n}\{\xi_i\}$, $\xi_{\max}=\max_{1\le i\le n}\{\xi_i\}$. Taking $l_0=\delta\|\beta(t)\|/\xi_{\min}$, where $\delta>1$ is a constant, then $\{V: V=e^{\lambda\tau}|\beta(s)|, -\tau\le s\le 0\}\subset\kappa(z_0(l_0))$, namely $V_i(s)<\xi_i l_0, -\tau\le s\le 0$, $i=1,\dots,n$.

We claim that $V_i(t)<\xi_i l_0$ for $t\in[0,+\infty]$, $i=1,\dots,n$. If it is not true , then there exist some $i$ and $t_1$ （$t_1>0$） , such that $V_i(t_1)=\xi_i l_0$, $D^+V_i(t_1)\ge 0$ and $V_j(t)\le\xi_j l_0$ for $-\tau\le t\le t_1$, $j=1,\dots,n$. However, from (9) and (8), we get

$$D^+V_i\le\{\xi_i(\lambda-G_i)+\textstyle\sum_{j=1}^{n}\xi_j[(|a_{ij}|+e^{\lambda\tau}|b_{ij}|)L_j+\omega_{ij}^*]\}l_0<0\,.$$

This is a contradiction. So, $V_i(t)<\xi_i l_0$ for $t\in[0,+\infty]$, furthermore

$$|\beta_i(t)|<\xi_i l_0 e^{-\lambda t}\le\delta\|\beta(t)\|\xi_{\max}/\xi_{\min}e^{-\lambda t}=M\|\beta(t)\|e^{-\lambda t}\text{ for }t\ge 0\,,$$

where $M=\delta\xi_{\max}/\xi_{\min}$. From Definition 1, the $\beta$ converges to zero exponentially, which in turn implies that system (2) and system (3) also converges global exponential synchronization. The proof is completed.

*Remark.* The sufficient condition for global exponential synchronization of systems (2) and (3) is independent of the delay parameter but relay on the system's parameter and the controller gain.

## 4 Illustrative Example

The sufficient condition for global exponential synchronization is demonstrated by following delayed neural network.

**Example.** Consider a chaotic Hopfield neural network (HNN) with variable delay [16,17]:

$$\begin{bmatrix}\dot{x}_1\\ \dot{x}_2\end{bmatrix}=-\begin{bmatrix}x_1(t)\\ x_2(t)\end{bmatrix}+\begin{bmatrix}2 & -0.1\\ -5 & 3\end{bmatrix}\begin{bmatrix}f_1(x_1(t))\\ f_2(x_2(t))\end{bmatrix}+\begin{bmatrix}-1.5 & -0.1\\ -0.2 & -2.5\end{bmatrix}\begin{bmatrix}f_1(x_1(t-\tau_1(t)))\\ f_2(x_2(t-\tau_2(t)))\end{bmatrix}, \qquad (10)$$

where $g_i(x_i)=x_i$ ,and $f_i=\tanh(x_i)$ , $i$=1,2. $\tau_1(t)=\tau_2(t)=1+0.1\sin t$ . The feedback matrix and the delayed feedback matrix are specified as

$$A=(a_{ij})_{2\times2}=\begin{bmatrix}2 & -0.1\\ -5 & 3\end{bmatrix},\quad B=(b_{ij})_{2\times2}=\begin{bmatrix}-1.5 & -0.1\\ -0.2 & -2.5\end{bmatrix},$$

respectively. The system satisfies assumptions (A1)-(A2) with $L_1=L_2=1$ and $G_1=G_2=1$. The system (10) possesses a chaotic behavior.

Now the response chaotic cellular neural network is designed as follows:

$$\begin{bmatrix}\dot{z}_1\\ \dot{z}_2\end{bmatrix}=-\begin{bmatrix}z_1(t)\\ z_2(t)\end{bmatrix}+\begin{bmatrix}2 & -0.1\\ -5 & 3\end{bmatrix}\begin{bmatrix}f_1(z_1(t))\\ f_2(z_2(t))\end{bmatrix}+\begin{bmatrix}-1.5 & -0.1\\ -0.2 & -2.5\end{bmatrix}\begin{bmatrix}f_1(x_1(t-\tau_1(t)))\\ f_2(x_2(t-\tau_2(t)))\end{bmatrix}-u(t)\ . \qquad (11)$$

The controller gain matrix is chosen as $\Omega=(\omega_{ij})_{2\times2}=\begin{bmatrix}-12 & 4\\ 4 & -20\end{bmatrix}$. It can be easily verified that $G-(|A|+|b|)L-\Omega^*$ is M matrix. Fig.1 depicts the synchronization error with the initial condition $[x_1(s)\quad x_2(s)]^T=[0.45\quad 0.65]^T$ and $[z_1(s)\quad z_2(s)]^T$ = $[0.5\quad 0.6]^T$ for all $-\tau\le s\le 0$, respectively.

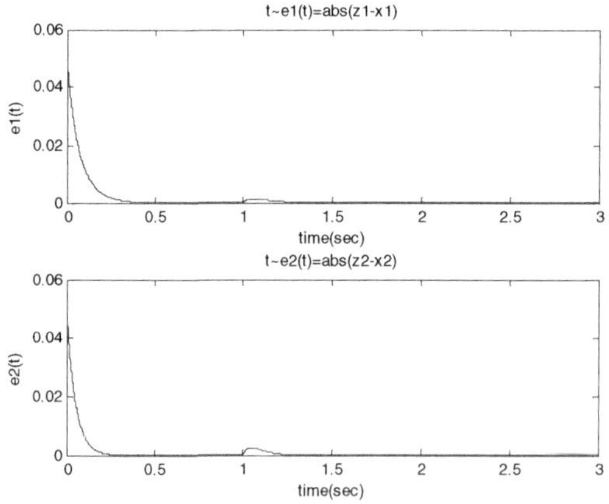


**Fig. 1.** The synchronization error

## 5 Conclusions

Applying the idea of vector Liapunov function and M-matrix theory, this paper presented a sufficient condition to guarantee the global exponential synchronization for a class of chaotic neural networks including Hopfield neural networks and cellular neural networks with time-varying delays.

**Acknowledgments.** This work is supported by National Program for New Century Excellent Talents in University (No.NCET-04-0889), Youth Science Foundation of Sichuan (No. 05ZQ026-015).

## References

1. Wu, C.W., Chua, L.O.: On Adaptive Synchronization and Control of Nonlinear Dynamical Systems. Int. JBC, 6 (1996)455- 461
2. Gilli, M.: Strange Attractors in Delayed Cellular Neural Networks. IEEE Trans Circ Syst, 40(11) (1993)849–853
3. Bondarenko, V.E.: Control and 'Anticontrol' of Chaos in an Analog Neural Network with Time Delay. Chaos Solitons Fract, 13 (2002)139–154
4. Chen, G., Dong, X.: On Feedback Control of Chaotic Continuous-Time Systems. IEEE Trans Circ Syst, 40 (1993)591-601
5. Pecora, L.M., Carroll, T.L.: Synchronization in Chaotic Systems. Phys Rev Lett, 64 (1990)821-824
6. Zhang, Y.F., Chen, G.R., Zhu, C.Y.: A System Inversion Approach to Chaos-Based Secure Speech Communication. Int. J.B.C, 15 (2005)2569-2572.
7. Lian, K.Y., Chiang, T.S., Chiu, C.S., Liu, P.: Synthesis of Fuzzy Model-Based Designs to Synchronization and Secure Communications for Chaotic Systems. IEEE Trans Circ Syst , 31 (2001)66-68
8. Oppenheim, A.V., Womell, C.W., Sabelle, S.H.: Signal Processing in the Context of Chaotic Signals. In Proc. IEEE. ICASSP (1992)117-120
9. Short, K.M.: Steps Toward Unmasking Secure Communications. Int. JBC, 4 (1994) 959- 977
10. Liao, T.L., Tsai, S.H.: Adaptive Synchronization of Chaotic Systems and Its Application to Secure Communications. Chaos, Solitons & Fractals, 11 (2000)1387-1396
11. Itoh, M., Murakami, H.: New Communication Systems via Chaotic Synchronizations and Modulation. IEICE Trans. Fundamentals, E78-A (1995)285–290
12. Lu, H.T.: Chaotic Attractors in Delayed Neural Networks. Phys. Lett. A, 298 (2002)109–116
13. Kocarev, L., Halle, K.S., Eckert, K., Chua, L.O., Parlitz, U.: Experimental Demonstration of Secure Communications via Chaotic Synchronization. Int. J. Bifurc. Chaos, 2 (1992)709–713
14. Chen, G., Zhou, J., Liu, Z.: Global Synchronization of Coupled Delayed Neural Networks with Application to Chaotic CNN Models. Int J Bifurcat Chaos, 14 (2004)2229–2240
15. Jankowski, S., Londei, A., Lozowski, A., Mazur, C.: Synchronization and Control in a Cellular Neural Network of Chaotic Units by Local Pinnings. Int. J. Circuit Theory Applicat., 24 (1996)275-281

16. Hopfield, J.J.: Neural Networks and Physical Systems with Emergent Collective Computational Abilities. Proc. Nat. Acad. Sci, 79 (1982)2554-2558
17. Yu, W., Cao, J.: Synchronization Control of Stochastic Delayed Neural Networks. Physica A, 373(2006) 252-260.
18. Zhang, J., Suda, Y., Komine, H.: Global Exponential Stability of Cohen–Grossberg Neural Networks with Variable Delays. Physica Letters A, 338(2005)44-50

# Grinding Wheel Topography Modeling with Application of an Elastic Neural Network

Błażej Bałasz, Tomasz Szatkiewicz, and Tomasz Królikowski

The Technical University of Koszalin, Department of Fine Mechanics,
75-256 Koszalin, ul. Racławicka 15-17, Poland
blazej.balasz@tu.koszalin.pl

**Abstract.** The article presents an application of a two-dimensional elastic neuron network for the generation of the surfaces of abrasive grains with macro-geometric parameters set. In the neuron model developed, the output parameters are the number of the grain vertices, the apex angle and the vertex radius. As a result of the work of the system, a random model of a grain with set parameters is obtained. The neuron model developed is used as a generator of the surface of the model of abrasive grains in the system of modeling and simulation of grinding processes.

## 1 Introduction

The efficiency and quality of abrasive machining processes has a decisive influence on the costs and quality of elements produced as well as whole products. The machining potential of abrasive tools is used insufficiently. One of more important reasons for an insufficient use of the machining potential is a slow development of new abrasive tools – development work focuses more on the improvement of the known technologies and not so much on the creation of new abrasive tools. Also, due to high costs of research into tools from ultra-hard materials concerning new tools, such research has not made a sufficient progress.

The use of the machining potential of tools depends of the optimization of the loading of abrasive grains, while typical empirical research allows solely for the designation of the global features of the process and not local ones, and temporary working conditions of active abrasive grains. The development of new modeling methods and the simulation of generation processes will facilitate a substantial progress in the creation of the basis of the system under development [1, 2, 4] and additionally it will enable to set assumptions for the creation of new abrasive tools with parameters to facilitate obtaining the expected results of machining, an increase of the productivity of the process and a much better use of the machining potential of grinding wheels.

## 2 Modeling of the Abrasive Grain Surface

Abrasive grains applied in machining can be divided into monocrystal, polycrystal and aggregate ones. The grain's geometrical parameters play a vital part on the

D.-S. Huang, L. Heutte, and M. Loog (Eds.): ICIC 2007, LNAI 4682, pp. 83–90, 2007.

machining process as it is its shape that micro-machining processes are dependent on. Precise defining of the grain's shape is very difficult due to a great variety of geometrical forms of grains created in the generating process. By a mutual agreement, abrasive grains are divided into groups defined as isometric, plate-like, pillar-like, sword-tail-like and needle-like. There are numerous methods to asses the grain's shape in such a manner, so that it should be possible, apart from geometrical features, to additionally assess in an indirect manner other features of grains, such as bulk density, abrasive ability or mechanical strength.

In order to make a complete assessment of an abrasive grain one should also determine the number and parameters of abrasive tool points located on the grain's surface. This assessment is conducted through the measurement of the corner radius of tool point $\rho$, as well as of apex angle $2\varepsilon$ which determines the grain's sharpness. The nose radius of the grain has a substantial influence on the machining process. Its size is closely related with the apex angle of the tool point, but the values of the radius for the same apex angles differ depending of the type of the abrasive material. Its change occurs also a result of the wear of the grains following the contact in the machining zone. It increases when there occurs a wear of the grain's vertex, and it decreases when the fragments of the abrasive grain break up. While making an assessment of the usable features of grains one should also consider the structure of their surfaces (the surface morphology). Due to the fact that the penetration of a single abrasive grain in the material machined does not constitute more than 5% of its largest size [5], an important part is played by the features of the surface morphology of the grain, such as micro- and macro-cracks, notches in the surface, the number of vertices and their location (cf. fig. 1). All these factors play an influence on the nature of the grain's work during the machining process, as well as its wear and ability to self-sharpen.

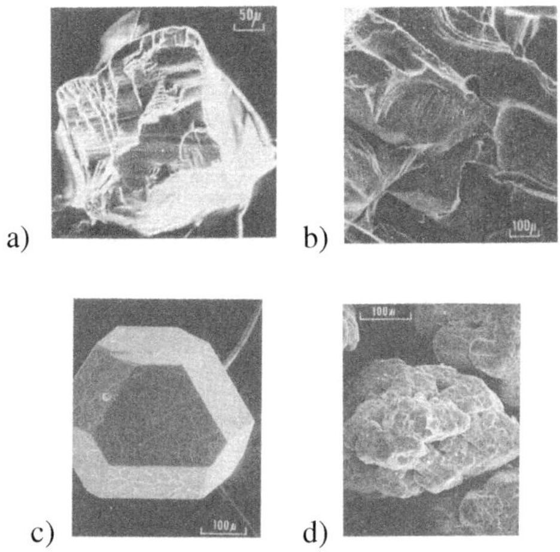


**Fig. 1.** Pictures of abrasive grains taken with the use of a scanning microscope: a) monocrystalic Al2O3 b) green SiC, c) diamond, d) diamond covered with copper [2]

Analyses of the stereometry of real grains on the basis of research results quoted in literature [5] formed the basis for the development of models of abrasive grains. In the simulation method developed it was assumed that what is important for the machining process is the grains' contours protruding over the surface of the grinding wheel as well as their shape and size above the level of the binding material, as it is only those fragments of the grain that have an influence on the grain's contact with the material and its wear. For this reason, the models develop describe only the stereometry of the part of the grain located above the geometrical surface of the grinding wheel. It was assumed in the modeling of the grains that the shape of the grain is described on a convex solid, with the local concavities of the grain's surface being taken into account and modeled in the form of micro-roughness on the surface. In the model developed, the grain's surface is described by a function whose components determine the grain's shape $f_{shape}(x,y)$ and its micro-topography (irregularities of the shape) $f_{mtp}(x,y)$. The components of the function are combined in an additive or a multiplicative manner (1).

$$z_k(x,y) = f_{shape}(x,y) + f_{mtp}(x,y) \tag{1}$$

A numerical notation of the shape of the grain obtained is done with the use of a matrix of real numbers $z_k$ (2), whose size [m, n] is determined on the basis of assumptions concerning the size of the grain modeled. The size of the matrix increases together with the growth of the grain's sizes.

$$Z_k(x,y) = \begin{bmatrix} z_{11} & z_{12} & \dots & z_{1n} \\ z_{21} & z_{22} & \dots & z_{2n} \\ \dots & \dots & \dots & \dots \\ z_{m1} & z_{m2} & \dots & z_{mn} \end{bmatrix} \tag{2}$$

where: $z_k(x_i, y_j) = f_{shape}(x_i, y_j) + f_{mtp}(x_i, y_j)$

A numerical notation of the shape of the grain's topography facilitates a modification of its shape during the simulation process caused by the grain's contacts with the material machined, and also as a result of the dressing process of the grinding wheel.

The further part of the article presents the application of an elastic neuron network for the modeling of the surfaces of abrasive grains.

## 3 Modeling of the Surface of Abrasive Grains with the Use of an Elastic Neuron Network

In the neuron model developed, the output parameters are the number of the grain vertices, the apex angle and the vertex radius. As a result of the work of the system, a random model of a grain with set parameters is obtained. In the network developed, the weights of individual neurons represent the coordinates of points on the surface of the grain generated. The work of this neuron network consists in the change of the values of neuron weights, as a result of which the coordinates of points describing the surface of the modeled grain are obtained.

The proposed elastic neuron network consists of *N* neurons,

$$A = \{n_1, n_2, ..., n_N\} \quad (3)$$

where each one of them has a vector of weights assigned

$$w_n \in R^N \quad (4)$$

to determine its location in the space of possible states $R^N$. Between the neurons in the network, there exists a system of elastic connections

$$C \subset A \times A \quad (5)$$

These connections are symmetric.

$$c(i, j) = c(j, i) \quad (6)$$

For each neuron n, a set of neurons is assigned with which it is directly connected, also called adjacent neurons.

$$N_N = \{i \in A | (c, i) \in C\} \quad (7)$$

Each connection $c(i, j)$ is assigned function $f_e(d)$, called the function of elasticity. This function depends on the distances of weight vectors of connected neurons $n_i$ and $n_j$:

$$d(n_i, n_j) = |w_i - w_j| \quad (8)$$

in accordance with the agreed space metric $R^N$. Function $f_e(d)$ is most often of a linear nature and is the same for all the connections $c(i, j) \in C$) (if the network is to be homogeneous). The value of function $f_e(d)$ constitutes the quantity of the attractive force occurring between two adjacent neurons.

The network described, after its initiation, has the form of a rectangular grid, and so each neuron initially has 4 neighbors, with the exception of utmost neurons, which posses 2 or 3 neighbors each. At the same time, in this specific application for the simulation of the abrasive grain's surfaces, the weights of utmost neurons are blocked. It means they do not change during the adaptation process.

The system of M nodes constituting characteristic points on the grain's surface constitutes the output data for the network:

$$L = \{l_1, l_2, ... l_N\} \in R^N \quad (9)$$

where each one of them has a vector of weights assigned to them:

$$w_m \in R^N \quad (10)$$

to determine its location in space $R^N$. These nodes, in the case in question, constitute a system of characteristic points of the surface of the simulated abrasive grain.

During the network's adaptation process (cf. fig. 2), the weight vectors of individual nodes affect simultaneously all the neurons located in the neighborhood determined by a certain radius. With the progress of the network's adaptation both the neighborhood radius and the impact factor are subject to a reduction to lead to the network's stabilization.

The purpose of the network's adaptation in the case in question is to obtain such a final form of the network, i.e. such vectors of the weights of neurons $w_n$ and such vectors of connections $N_N$ that it should map the abrasive grain's surface (cf. fig. 2d).Two types of forces act on individual neurons: an attractive force from adjacent neurons and a force from the nodes, i.e. from the input data fed to the input of the network. For this reason, the following rule of the changes of the weights of neurons can be derived:

$$\forall c(n,j) \neq 0 \left[ \Delta w_n = \beta(\sum_m \Lambda^m(n) + f_e(\kappa \sum (w_n - w_j))) \right] \tag{11}$$

where:

$c(n,j)$ – connection between neurons n and j,

$\beta$ – coefficient of network's learning,

$f_e()$ – the elasticity function accepted,

$\kappa$ – elasticity coefficient, variable in the duration of the network's adaptation process, proportional to the network's temperature $\beta$,

$\Lambda^m(n)$ – coefficient of an impact of node $l_m$ on neuron $n_n$ expressed with the following formula:

$$\Lambda^m(n) = \frac{\exp(-|w_m - w_n|^2 / 2\sigma^2}{\sum_m \exp(-|w_j - w_n|^2 / 2\sigma^2)} \tag{12}$$

where: $\sigma$ – effective range of the impact of nodes on the neurons.

In equation (11), the first expression is the force attracting every neuron $n_n$ in the direction of the node (a characteristic point on the grain's surface) $l_m$ with the coefficient of impact $\Lambda^m(n)$. The second expression is the total elasticity force, which attracts every neuron in the direction of the adjacent neurons. The whole expression depends of the parameter of learning coefficient $\beta$.

As it can bee seen from Fig. 2, the network in its first phase, right after its initiation (cf. fig. 2a) gradually maps the space of input signals. The elastic impact simulated results in the fact that the network during an expansion behaves like an elastic membrane and evolves like an equipotential surface in a certain vector field. The effect of the network's work is a random surface of the abrasive grain with set parameters, which is then transformed into matrix $Z_k$ (2), used in the system of the simulation of the machining process.

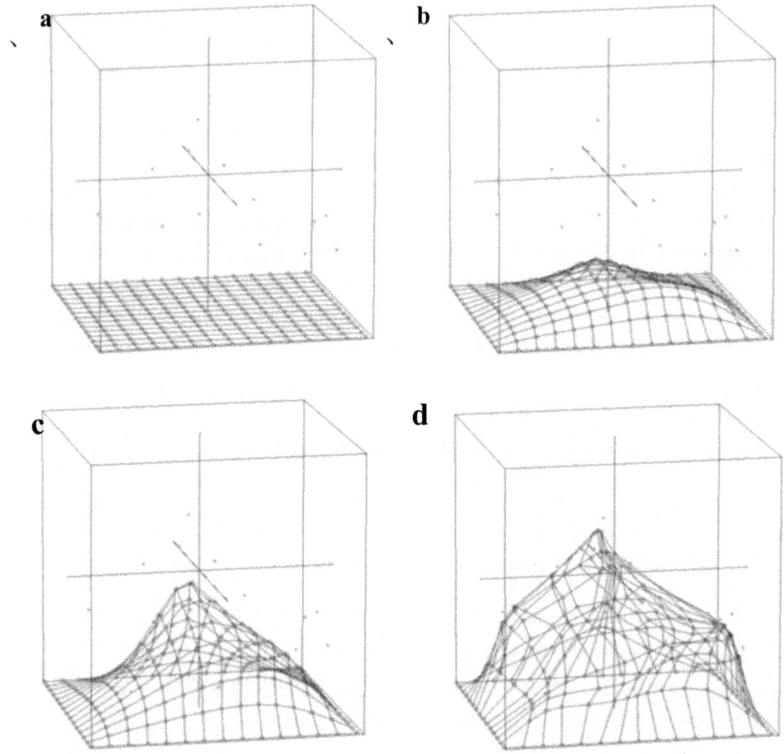


**Fig. 2.** Individual stages of the adaptation process of an elastic neuron network: a) initiation of network in a grid form, b-c) adaptation of network, d) network's final form depicting the surface of the simulated abrasive grain

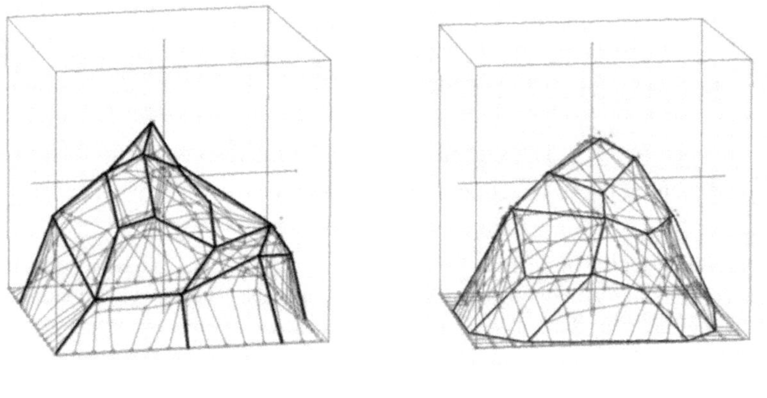

**Fig. 3.** Sample final surface of simulated abrasive grains with crystalic edges marked

## 4 Modeling of the Grinding Tool Surface

Structure of grinding tool is composed of grains located randomly on its surface. Both a grain size and its locations have a great influence on quality of machined surface. In the developed model of grinding tool surface, one of the most significant factor of optimization of grinding process is achieved with optimal location of grinding grains on the surface. In the process of modeling a grinding tool surface, every single grain is randomly located on the surface with specified grain concentration (cf. fig 4a).

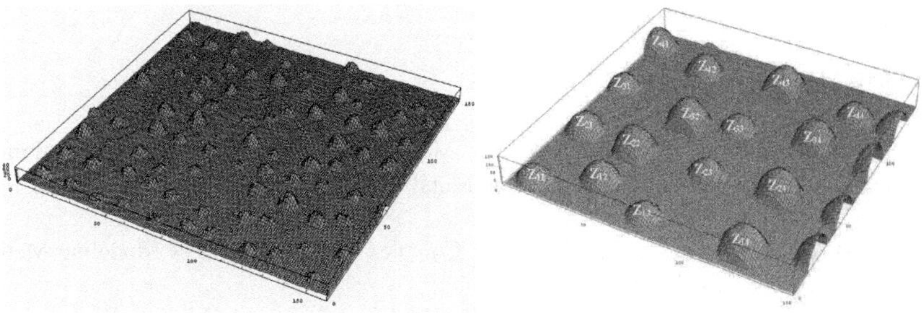

**Fig. 4.** Grinding wheel topography a) model, b) indexes of single grains

With every generated grain there is associated vector of grain parameters, describing temporal states of the grain during the whole process (e.g. number of contacts with workpiece material, volume of removed material, normal and tangential forces etc.). After grain generation, the working surface of the grinding wheel is generated by the aggregation of single grains into one surface, where each grain has a unique index (cf. fig. 4b). Thanks to that, the characteristic of behavior of contact during the process could be thoroughly discovered. On generated surface the model of the bond is placed on. As a completion to this task, models of grain displacement and removal and the dressing process are also elaborated.

## 5 Conclusion

The models developed reveal features which enable a generation of the surface of grains with properties statistically compliant with specified types of grains from different abrasive materials. The models of abrasive grains designed underwent an empirical verification. Due to the fact that the basic features which have an influence on the nature of the work of the grain (type of contact) are the parameters of the abrasive tool point, a comparative analysis was conducted in the range of checking the compatibility of the apex angle of the tool point $2\varepsilon$, the radius of the nose radius $\rho$ of the model grains and the proportion of the height $h_w$ of the grain to the width of its base $b$, as regards the real grains. The verification process consisted in determining the geometrical parameters of the models of grains generated. The values of the apex angle $2\varepsilon$ and of the rounding angle of the vertex $\rho$ were determined for various penetration depths of the tool point. The verification of the shape of the grains served to determine boundary values of the coefficients of the shape for individual types of grains, owing to which geometrical correctness of the modeled grains is ensured in the duration of the simulation process.

## Acknowledgements

This work was supported by grant: KBN Nr 4 T07D 033 28 form Polish Ministry of Science and Higher Education.

## References

1. Bałasz, B., Królikowski, T., Kacalak, W.: Method of Complex Simulation of Grinding Process. Third International Conference on Metal Cutting and High Speed Machining Metz, France (2001) 169-172
2. Bałasz, B. Królikowski, T..:Utility of New Complex Grinding Process Modeling Method. PAN Koszalin, (2002)93-109
3. Brinksmeier, E., et al.: Advances in Modeling and Simulation of Grinding Processes. Annals of the CIRP, vol. 55/2.(2006), 667-696
4. Królikowski,T., Bałasz,B.,Kacalak,W.:The Influence of Micro- And Macrotopography of the Active Grinding Surface on the Energy Consumption in the Grinding Process. 15th European Simulation Multiconference, Prague, Czech Republic, (2001) 339-341
5. Shaw, M.: Principles of Abrasive Processing, Oxford University Press, Oxford (1996)
6. Stępień, P., Bałasz, B.: Simulation of the Formation Process of Regular Grooves on Surface Ground, Industrial Simulation Conference, Palermo, Italy (2006) 269 – 276

# Hybrid Control of Hopf Bifurcation for an Internet Congestion Model*

Zunshui Cheng[1,3], Jianlong Qiu[2,3], Guangbin Wang[1], and Bin Yu[1]

[1] School of Mathematics and Physics, Qingdao University of Science and Technology
Qingdao 266061, China
[2] Department of Mathematics, Linyi Normal University
Linyi, Shandong 276005, China
[3] Department of Mathematics, Southeast University
Nanjing 210096, China
chengzunshui@gmail.com

**Abstract.** In this paper, the problem of Hopf bifurcation control for an Internet congestion model with time delays is considered by using a new hybrid control strategy, in which state feedback and parameter perturbation are used. It is well known that for the system without control, as the positive gain parameter of the system passes a critical point, Hopf bifurcation occurs. To control the Hopf bifurcation, a hybrid control strategy is proposed and the onset of an inherent bifurcation is delayed (advanced) when such bifurcation is undesired (desired). Furthermore, the dynamic behaviors of the controlled system can also be changed by choosing appropriate control parameters. Numerical simulation results confirm that the new control strategy is efficient in controlling Hopf bifurcation.

## 1 Introduction

Bifurcation control refers to the task of designing a controller to suppress or reduce some existing bifurcation dynamics of a given nonlinear system, thereby achieving some desirable dynamical behaviors [5]. Aim of bifurcation control is to delay the onset of an inherent bifurcation, change the parameter value of an existing bifurcation point, stabilize a bifurcated solution or branch, etc [5]-[6]. In recent years, researchers from various disciplines were attracted to bifurcation control and various methods of bifurcation control can be found (see, for example, [6]-[11]). In [11], a new hybrid control strategy was proposed, in which state feedback and parameter perturbation were used to control the bifurcations.

In this paper, a hybrid control strategy is used to control bifurcations for an Internet model with a single link and single source. The model can be described as:

$$\frac{\mathrm{d}x(t)}{\mathrm{d}t} = k[w - x(t-D)p(x(t-D))], \tag{1}$$

* This work was jointly supported by the Science and Technology Plans of the Department of Education, Shandong Province under Grant J06P04, the Youth Framework Teacher Subsidize Item of Henan Province under Grant 20050181, and the Natural Science Foundation of Henan Province, China under Grant 0611055100.

D.-S. Huang, L. Heutte, and M. Loog (Eds.): ICIC 2007, LNAI 4682, pp. 91–97, 2007.

where $k$ is a positive gain parameter and $x(t)$ is the rate at which a source sends packets at time $t$. In the Internet, the communication delay is comprised of propagation delay and queuing delay. As the router hardware and network capacity continue to improve rapidly, the queuing delay becomes smaller compared to the propagation delay. $D$ is the sum of the forward and returning delays, that is, the time during which the packet makes a round trip from a sender to a receiver, and back to the sender. As a result, the sum of the forward and returning delays is fixed for resources on a given route. $w$ is a target (set-point), and $p(\cdot)$ is the congestion indication function. When a resource within the network becomes overloaded, one or more packets are lost, and the loss of a packet is taken as an indication of congestion. The congestion indication function is assumed to be increasing, non-negative, and not identically zero [3,6]. We will show, with a Hopf bifurcation controller, that one can increase the critical value of positive gain parameter. Furthermore, the stability and direction of bifurcating periodic solutions can also be changed by choosing appropriate parameters.

The remainder of this paper is organized as follows. The existence of Hopf bifurcation parameter is determined in Section 2. In Section 3, based on the normal form method and the center manifold theorem introduced by Hassard et al. [13], the direction, orbitally stability and the period of the bifurcating periodic solutions are analyzed. To verify the theoretic analysis, numerical simulations are given in Section 4. Finally, Section 5 concludes with some discussions.

## 2 Existence of Hopf Bifurcation

In this section, we focus on designing a controller in order to control the Hopf bifurcation arising from the Internet congestion model. The following conclusions for the uncontrolled system (1) are needed at first [4]:

**Lemma 1.** *When the positive gain parameter $k$ passes through the critical value $k^* = \frac{\pi}{2D(p(x^*)+x^*p'(x^*))}$, there is a Hopf bifurcation of system (1) at its equilibrium $x^*$.*

**Lemma 2.** *The Hopf bifurcation for the Internet congestion model (1) is determined by the parameters $\mu_2$, $\beta_2$ and $\tau_2$, where $\mu_2$ determines the direction of the Hopf bifurcation: the Hopf bifurcation is supercritical (subcritical ) when $\mu_2 > 0$ ($\mu_2 < 0$), and the bifurcating periodic solutions exist (do not exist) if $\mu > \mu^*$ ($\mu < \mu^*$); $\beta_2$ determines the stability of the bifurcating periodic solutions: the solutions are orbitally stable (unstable) if $\beta_2 < 0$ ($\beta_2 > 0$); and $\tau_2$ determines the period of the bifurcating periodic solutions: the period increases (decreases) if $\tau_2 > 0$ ($\tau_2 < 0$).*
where

$$\mathrm{Re}\lambda'(0) = \frac{-b_1^* D[p(x^* + x^* p'(x^*))]}{1 + (b_1^* D)^2},$$

$$\mathrm{Im}\lambda'(0) = \frac{p(x^* + x^* p'(x^*))}{1 + (b_1^* D)^2}.$$

$$
\begin{aligned}
C_1(0) &= \frac{\mathrm{i}}{2\omega_0}\left(g_{20}g_{11} - 2|g_{11}|^2 - \frac{1}{3}|g_{02}|^2\right) + \frac{g_{21}}{2}, \\
\mu_2 &= -\frac{\mathrm{Re}\{C_1(0)\}}{\mathrm{Re}\lambda'(0)}, \\
\beta_2 &= 2\mathrm{Re}\{C_1(0)\}, \\
\tau_2 &= -\frac{\mathrm{Im}\{C_1(0)\} + \mu_2 \mathrm{Im}\lambda'(0)}{\omega_0}\,.
\end{aligned} \tag{2}
$$

in which

$$
\begin{aligned}
\omega_0 &= \frac{\pi}{2D}, \\
g_{20} &= g_{02} = -g_{11} = \frac{-2b_2^*}{1 + b_1^* D e^{-i\omega_0 D}}\,, \\
g_{21} &= \frac{2i}{1 + b_1^* D e^{-i\omega_0 D}}[2\frac{2b_2^* - g_{11} - \bar{g}_{11}}{b_1^*} + \frac{(-g_{20} - \bar{g}_{02} + 2b_2^*)b_2^*}{b_1^* - 2i\omega_0} - 3b_3^*]\,, \\
b_1^* &= -k^*[p(x^*) + x^* p'(x^*)], \\
b_2^* &= -\frac{k^*}{2}[2p'(x^*) + x^* p''(x^*)], \\
b_3^* &= -\frac{k^*}{6}[3p''(x^*) + x^* p'''(x^*)].
\end{aligned} \tag{3}
$$

We now turn to study how to control the Hopf bifurcation to achieve desirable behaviors through control parameters. The controller system is designed as follows:

$$
\frac{dx(t)}{dt} = (1-\alpha)k[w - x(t-D)p(x(t-D))] + \alpha(x(t) - x^*), \tag{4}
$$

where $x^*$ is the equilibrium point of system (1) and $\alpha$ are parameters, which can be used to control the Hopf bifurcation.

Expanding the right-hand side of system (4) into first, second and third terms around $x^*$, we have

$$
\frac{dv(t)}{dt} = r_1 v(t-D) + r_2 v^2(t-D) + r_3 v^3(t-D)\,, \tag{5}
$$

where

$$
\begin{aligned}
v(t) &= x(t) - x^*, \\
r_1 &= \alpha - k(1-\alpha)[p(x^*) + x^* p'(x^*)], \\
r_2 &= -\frac{k}{2}(1-\alpha)[2p'(x^*) + x^* p''(x^*)], \\
r_3 &= -\frac{k}{6}(1-\alpha)[3p''(x^*) + x^* p'''(x^*)].
\end{aligned} \tag{6}
$$

The linear equation of system (5) is

$$
\frac{dv(t)}{dt} = r_1 v(t-D)\,, \tag{7}
$$

whose characteristic equation is

$$\lambda - r_1 e^{-\lambda D} = 0\,. \tag{8}$$

We first examine when the characteristic equation (8) has pairs of pure imaginary roots. If $\lambda = \pm i\omega$ with $\omega > 0$, then we have

$$r_1 \cos(\omega D) = 0\,, \tag{9}$$

$$\omega + r_1 \sin(\omega D) = 0\,. \tag{10}$$

It has been shown by Li et al. [4] that the characteristic equation does not have roots with positive real parts unless $\omega_0 = \frac{\pi}{2D}$. Thus, we obtain

$$\frac{\pi}{2D} + r_1 = 0, \tag{11}$$

or

$$\frac{\pi}{2D} + \alpha - k(1-\alpha)[p(x^*) + x^* p'(x^*)] = 0, \tag{12}$$

which lead to

$$k^* = \frac{\pi}{2D(1-\alpha)[p(x^*) + x^* p'(x^*)]} + \frac{\alpha}{2D(1-\alpha)[p(x^*) + x^* p'(x^*)]}\,. \tag{13}$$

In order to create a Hopf bifurcation from the bifurcation point, the following *transversality condition* is needed

$$\frac{\mathrm{d}(\mathrm{Re}(\lambda))}{\mathrm{d}k}|_{k=k^*} \neq 0\,. \tag{14}$$

Letting $\lambda = \mathrm{Re}(\lambda) + \mathrm{Im}(\lambda)\mathrm{i}$, and then substituting $\lambda$ into the characteristic equation (8), we have

$$\mathrm{Re}(\lambda) - e^{-\mathrm{Re}(\lambda)D} r_1 \cos(\mathrm{Im}(\lambda)D) = 0\,,$$
$$\mathrm{Im}(\lambda) + e^{-\mathrm{Re}(\lambda)D} r_1 \sin(\mathrm{Im}(\lambda)D) = 0\,.$$

Thus we get

$$\begin{aligned}\frac{\mathrm{dRe}(\lambda)(k^*, \omega_0)}{\mathrm{d}k} &= -\frac{2D\xi^2 r_1}{(1-\alpha)[p(x^*) + x^* p'(x^*)][1 + r_1^2 D^2]} \\ &= \frac{\pi \xi^2}{(1-\alpha)[p(x^*) + x^* p'(x^*)][1 + r_1^2 D^2]} > 0\,.\end{aligned} \tag{15}$$

Therefore, the final condition for the occurrence of a Hopf bifurcation in nonlinear model (4) is indeed satisfied. Then we have the following theorem.

**Theorem 3.** *For the controlled system (4), there exists a Hopf bifurcation emerging from its equilibrium $x^*$, when the positive parameter, $k$, passes through the critical value,*

$$k^* = \frac{\pi}{2D(1-\alpha)[p(x^*) + x^* p'(x^*)]} + \frac{\alpha}{2D(1-\alpha)[p(x^*) + x^* p'(x^*)]}\,,$$

*where the equilibrium point $x^*$ is kept unchanged.*

*Remark 1.* Theorem 3 can be applied to system (4) for the purpose of control and anti-control of bifurcations. One can delay or advance the onset of a Hopf bifurcation without changing the original equilibrium points by choosing an appropriate value of $\alpha$ (see Section 4).

## 3 Direction and Stability of Hopf Bifurcation

From this section we know that one can also change the stability and direction of bifurcating periodic solutions by choosing appropriate values of $\alpha$.

The bifurcating periodic solutions $v(t, \mu(\epsilon))$ of (4) (where $\epsilon > 0$ is a small parameter) have amplitude $\mu(\epsilon)$, period $\tau(\epsilon)$ and nonzero Floquet exponent $\beta(\epsilon)$, where $\mu$, $\tau$ and $\beta$ have the following (convergent) expansions:

$$\mu(\epsilon) = \mu_2\epsilon^2 + \mu_4\varepsilon^4 + \cdots$$
$$\tau(\epsilon) = \tau_2\epsilon^2 + \tau_4\varepsilon^4 + \cdots$$
$$\beta(\epsilon) = \beta_2\epsilon^2 + \beta_4\varepsilon^4 + \cdots .$$

By Li et al. [4] as well as in the textbook [13], we have the following theorem for the controlled Internet congestion model.

**Theorem 4.** *The Hopf bifurcation exhibited by the controlled Internet congestion model (4) is determined by the parameters $\mu_2$, $\beta_2$ and $\tau_2$, where $\mu_2$ determines the direction of the Hopf bifurcation, if $\mu_2 > 0 (< 0)$, then the Hopf bifurcation is supercritical (subcritical) and the bifurcating periodic solutions exist for $k > k^*$ ($< k^*$); $\beta_2$ determines the stability of the bifurcating periodic solutions: the bifurcating periodic solutions are orbitally stable (unstable) if $\beta_2 < 0$ ($> 0$), and $\tau_2$ determines the period of the bifurcating periodic solutions: the period increase (decreases) if $\tau_2 > 0$ ($\tau_2 < 0$).*

The parameters $\mu_2$, $\beta_2$ and $\tau_2$ can be found using the following formulas:

$$\begin{aligned} C_1(0) &= \frac{\mathrm{i}}{2\omega_0}\left(g_{20}g_{11} - 2|g_{11}|^2 - \frac{1}{3}|g_{02}|^2\right) + \frac{g_{21}}{2} \\ \mu_2 &= -\frac{\mathrm{Re}\{C_1(0)\}}{\mathrm{Re}\lambda'(0)} \\ \beta_2 &= 2\mathrm{Re}\{C_1(0)\} \\ \tau_2 &= -\frac{\mathrm{Im}\{C_1(0)\} + \mu_2\mathrm{Im}\lambda'(0)}{\omega_0} . \end{aligned} \tag{16}$$

in which

$$\omega_0 = \frac{\pi}{2D},$$

$$g_{20} = g_{02} = -g_{11} = \frac{-2r_2^*}{1 + r_1^* De^{-i\omega_0 D}} ,$$

$$g_{21} = \frac{2i}{1 + r_1^* De^{-i\omega_0 D}}[2\frac{2r_2^* - g_{11} - \bar{g}_{11}}{r_1^*} + \frac{(-g_{20} - \bar{g}_{02} + 2r_2^*)r_2^*}{r_1^* - 2i\omega_0} - 3r_3^*] ,$$

$$
\begin{aligned}
r_1^* &= \alpha - k^*(1-\alpha)[p(x^*) + x^* p'(x^*)], \\
r_2^* &= -\frac{k^*}{2}(1-\alpha)[2p'(x^*) + x^* p''(x^*)], \\
r_3^* &= -\frac{k^*}{6}(1-\alpha)[3p''(x^*) + x^* p'''(x^*)]. \qquad (17)
\end{aligned}
$$

*Remark 2.* From Theorem 4, we can change the the parameters $\mu_2$, $\beta_2$ and $\tau_2$ by choosing appropriate control parameters $\alpha$, thereby change the stability and direction of bifurcating periodic solutions.

## 4 Numerical Examples

In this section, we present numerical results to verify the analytical predictions obtained in the previous section, using the hybrid control strategy to control the Hopf bifurcation of Internet congestion model (1). These numerical simulation results constitute excellent validations of our theoretical analysis. For a consistent comparison, we choose the same function, $p(x) = x/(20-3x)$ and $\delta = 1$ used in Li et al. [4]. The dynamical behavior of this uncontrolled model

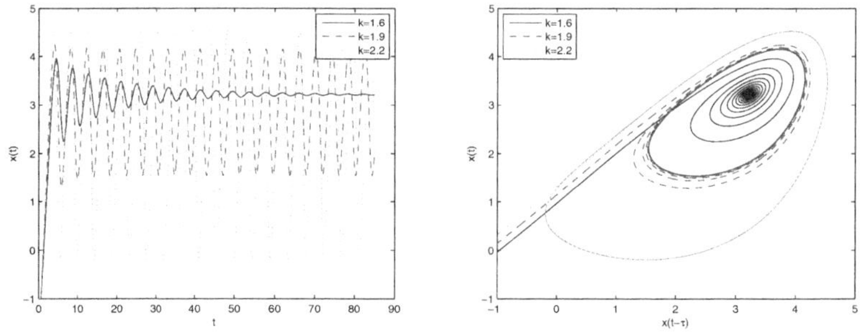


**Fig. 1.** Waveform plot and phase portrait of model (1) for $k = 1.6,\ 1.9,\ 2.2$, respectively

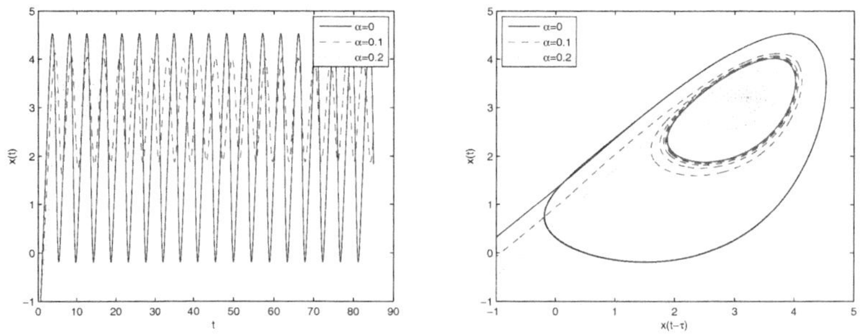


**Fig. 2.** Waveform plot and phase portrait of model (4) for $k = 2.2$ and $\alpha = 0,\ 0.1,\ 0.2$, respectively

is illustrated in Fig. 1. It is shown that when $k < k^* \approx 1.7231$, trajectories converge to the equilibrium point, while as $k$ is increased to pass $k^*$, $x^*$ loses its stability and a Hopf bifurcation occurs (see Fig. 1).

Now we choose appropriate values of $\alpha$ to control the networks. For $k = 2.2$, by choosing $\alpha = 0,\ 0.1,\ 0.2$, respectively, the periodic solution disappeared and $x^*$ become stable. That is, the onset of the Hopf bifurcation is delayed (see Fig. 2).

## 5 Conclusions

In this paper, the problem of Hopf bifurcation control for a small-world network model with time delays has been studied. To control the Hopf bifurcation, a time-delayed feedback controller has been proposed. This controller can delay the onset of an inherent bifurcation when such bifurcation is undesired. Furthermore, this controller can effectively control the amplitude of the bifurcated limit cycle. Numerical results have been presented to verify the analytical predictions.

## References


1. Kelly, F. P., Maulloo, A., Tan, D. K. H.: Rate Control in Communication Networks: Shadow Prices, Proportional Fairness, and Stability. J. Oper. Res. Soc. **49** (1998) 237–252
2. Kelly, F. P.: Models for a Self-managed Internet. Philos. Trans. Roy. Soc. A **358** (2000) 2335–2348
3. Johari, R., Tan, D. K. H.: End-to-end Congestion Control for the Internet: Delays and Stability. IEEE/ACM Trans. Networking **9** (2001) 818–832
4. Li, C., Chen, G.: Hopf Bifurcation in an Internet Congestion Control Model. Chaos, Solitons & Fractals **19** (2004) 853–862
5. Chen, G., Moiola, J. L., Wang, H. O.: Bifurcation Control: Theories, Methods, and Applications. Int. J. Bifur. Chaos **10** (2000) 511–548
6. Chen, Z., Yu, P.: Hopf Bifurcation Control for an Internet Congestion Model. Int. J. Bifur. Chaos **15** (2005) 2643–2651
7. Berns, D. W., Moiola, J. L., Chen, G.: Feedback Control of Limit Cycle Amplitudes from a Frequency Domain Approach. Automatica **34** (1998) 1567–1573
8. Ott, E., Grebogi, C., Yorke, J. A.: Controlling Chaos. Phys. Rev. Lett. **64** (1990) 1196–1199
9. Bleich, M. E., Socolar, J. E. S.: Stability of Periodic Orbits Controlled by Time-delay Feedback. Phys. Lett. A **210** (1996) 87–94.
10. Berns, D. W., Moiola, J. L., Chen, G.: Feedback Control of Limit Cycle Amplitudes from a Frequency Domain Approach. Automatica **34** (1998) 1567–1573.
11. Liu, Z., Chung, K. W.: Hybrid Control of Bifurcation in Continuous Nonlinear Dynamical Systems. Int. J. Bifur. Chaos **15** (2005) 3895–3903.
12. Wang, X. F.: Complex networks: Topology, Dynamics and Synchronization. Int. J. Bifur. Chaos **12** (2002) 885–916.
13. Hassard, B. D., Kazarinoff, N. D., Wan, Y. H.: *Theory and Applications of Hopf Bifurcation.* Cambridge University Press, Cambridge, 1981

# MATLAB Simulation of Gradient-Based Neural Network for Online Matrix Inversion

Yunong Zhang, Ke Chen, Weimu Ma, and Xiao-Dong Li

Department of Electronics and Communication Engineering
Sun Yat-Sen University, Guangzhou 510275, China
ynzhang@ieee.org

**Abstract.** This paper investigates the simulation of a gradient-based recurrent neural network for online solution of the matrix-inverse problem. Several important techniques are employed as follows to simulate such a neural system. 1) Kronecker product of matrices is introduced to transform a matrix-differential-equation (MDE) to a vector-differential-equation (VDE); i.e., finally, a standard ordinary-differential-equation (ODE) is obtained. 2) MATLAB routine "ode45" is introduced to solve the transformed initial-value ODE problem. 3) In addition to various implementation errors, different kinds of activation functions are simulated to show the characteristics of such a neural network. Simulation results substantiate the theoretical analysis and efficacy of the gradient-based neural network for online constant matrix inversion.

**Keywords:** Online matrix inversion, Gradient-based neural network, Kronecker product, MATLAB simulation.

## 1 Introduction

The problem of matrix inversion is considered to be one of the basic problems widely encountered in science and engineering. It is usually an essential part of many solutions; e.g., as preliminary steps for optimization [1], signal-processing [2], electromagnetic systems [3], and robot inverse kinematics [4]. Since the mid-1980's, efforts have been directed towards computational aspects of fast matrix inversion and many algorithms have thus been proposed [5]-[8]. It is known that the minimal arithmetic operations are usually proportional to the cube of the matrix dimension for numerical methods [9], and consequently such algorithms performed on digital computers are not efficient enough for large-scale online applications. In view of this, some $O(n^2)$-operation algorithms were proposed to remedy this computational problem, e.g., in [10][11]. However, they may be still not fast enough; e.g., in [10], it takes on average around one hour to invert a 60000-dimensional matrix. As a result, parallel computational schemes have been investigated for matrix inversion.

The dynamic system approach is one of the important parallel-processing methods for solving matrix-inversion problems [2][12]-[18]. Recently, due to the in-depth research in neural networks, numerous dynamic and analog solvers

D.-S. Huang, L. Heutte, and M. Loog (Eds.): ICIC 2007, LNAI 4682, pp. 98–109, 2007.

based on recurrent neural networks (RNNs) have been developed and investigated [2][13]-[18]. The neural dynamic approach is thus regarded as a powerful alternative for online computation because of its parallel distributed nature and convenience of hardware implementation [4][12][15][19][20].

To solve for a matrix inverse, the neural system design is based on the equation, $AX - I = 0$, with $A \in \mathbb{R}^{n \times n}$. We can define a scalar-valued energy function such as $\mathcal{E}(t) = \|AX(t) - I\|^2/2$. Then, we use the negative of the gradient $\partial\mathcal{E}/\partial X = A^T(AX(t) - I)$ as the descent direction. As a result, the classic linear model is shown as follows:

$$\dot{X}(t) = -\gamma \frac{\partial \mathcal{E}}{\partial X} = -\gamma A^T(AX(t) - I), \quad X(0) = X_0 \tag{1}$$

where design parameter $\gamma > 0$, being an inductance parameter or the reciprocal of a capacitive parameter, is set as large as the hardware permits, or selected appropriately for experiments.

As proposed in [21], the following general neural model is an extension to the above design approach with a nonlinear activation-function array $\mathcal{F}$:

$$\dot{X}(t) = -\gamma A^T \mathcal{F}(AX(t) - I) \tag{2}$$

where $X(t)$, starting from an initial condition $X(0) = X_0 \in \mathbb{R}^{n \times n}$, is the activation state matrix corresponding to the theoretical inverse $A^{-1}$ of matrix $A$. Like in (1), the design parameter $\gamma > 0$ is used to scale the convergence rate of the neural network (2), while $\mathcal{F}(\cdot) : \mathbb{R}^{n \times n} \to \mathbb{R}^{n \times n}$ denotes a matrix activation-function mapping of neural networks.

## 2 Main Theoretical Results

In view of equation (2), different choices of $\mathcal{F}$ may lead to different performance. In general, any strictly-monotonically-increasing odd activation-function $f(\cdot)$, being an element of matrix mapping $\mathcal{F}$, may be used for the construction of the neural network. In order to demonstrate the main ideas, four types of activation functions are investigated in our simulation:

- linear activation function $f(u) = u$,
- bipolar sigmoid function $f(u) = (1 - \exp(-\xi u))/(1 + \exp(-\xi u))$ with $\xi \geqslant 2$,
- power activation function $f(u) = u^p$ with odd integer $p \geqslant 3$, and
- the following power-sigmoid activation function

$$f(u) = \begin{cases} u^p, & \text{if } |u| \geqslant 1 \\ \frac{1+\exp(-\xi)}{1-\exp(-\xi)} \cdot \frac{1-\exp(-\xi u)}{1+\exp(-\xi u)}, & \text{otherwise} \end{cases} \tag{3}$$

  with suitable design parameters $\xi \geqslant 1$ and $p \geqslant 3$.

Other types of activation functions can be generated by these four basic types.

Following the analysis results of [18][21], the convergence results of using different activation functions are qualitatively presented as follows.

**Proposition 1.** *[15]-[18][21] For a nonsingular matrix $A \in \mathbb{R}^{n\times n}$, any strictly monotonically-increasing odd activation-function array $\mathcal{F}(\cdot)$ can be used for constructing the gradient-based neural network (2).*

1. *If the linear activation function is used, then the global exponential convergence is achieved for neural network (2) with convergence rate proportional to the product of $\gamma$ and the minimum eigenvalue of $A^TA$.*
2. *If the bipolar sigmoid activation function is used, then the superior convergence can be achieved for error range $[-\delta,\delta]$, $\exists\delta \in (0,1)$, as compared to the linear-activation-function case. This is because the error signal $e_{ij} = [AX - I]_{ij}$ in (2) is amplified by the bipolar sigmoid function for error range $[-\delta,\delta]$.*
3. *If the power activation function is used, then the superior convergence can be achieved for error ranges $(-\infty,-1]$ and $[1,+\infty)$, as compared to the linear-activation-function case. This is because the error signal $e_{ij} = [AX - I]_{ij}$ in (2) is amplified by the power activation function for error ranges $(-\infty,-1]$ and $[1,+\infty)$.*
4. *If the power-sigmoid activation function is used, then superior convergence can be achieved for the whole error range $(-\infty,+\infty)$, as compared to the linear-activation-function case. This is in view of Properties 2) and 3).*

In the analog implementation or simulation of the gradient-based neural networks (1) and (2), we usually assume that it is under ideal conditions. However, there are always some realization errors involved. For example, for the linear activation function, its imprecise implementation may look more like a sigmoid or piecewise-linear function because of the finite gain and frequency dependency of operational amplifiers and multipliers. For these realization errors possibly appearing in the gradient-based neural network (2), we have the following theoretical results.

**Proposition 2.** *[15]-[18][21] Consider the perturbed gradient-based neural model*

$$\dot{X} = -\gamma(A + \Delta_A)^T\mathcal{F}\left((A + \Delta_A)X(t) - I\right),$$

*where the additive term $\Delta_A$ exists such that $\|\Delta_A\| \leqslant \varepsilon_1$, $\exists\varepsilon_1 \geqslant 0$, then the steady-state residual error $\lim_{t\to\infty}\|X(t) - A^{-1}\|$ is uniformly upper bounded by some positive scalar, provided that the resultant matrix $A + \Delta_A$ is still nonsingular.*

For the model-implementation error due to the imprecise implementation of system dynamics, the following dynamics is considered, as compared to the original dynamic equation (2).

$$\dot{X} = -\gamma A^T\mathcal{F}\left(AX(t) - I\right) + \Delta_B, \tag{4}$$

where the additive term $\Delta_B$ exists such that $\|\Delta_B\| \leqslant \varepsilon_2$, $\exists\varepsilon_2 \geqslant 0$.

**Proposition 3.** *[15]-[18][21] Consider the imprecise implementation (4), the steady state residual error $\lim_{t\to\infty}\|X(t) - A^{-1}\|$ is uniformly upper bounded by some positive scalar, provided that the design parameter $\gamma$ is large enough (the so-called design-parameter requirement). Moreover, the steady state residual error $\lim_{t\to\infty}\|X(t) - A^{-1}\|$ can be made to zero as $\gamma$ tends to positive infinity .*

As additional results to the above lemmas, we have the following general observations.

1. For large entry error (e.g., $|e_{ij}| > 1$ with $e_{ij} := [AX - I]_{ij}$), the power activation function could amplify the error signal ($|e_{ij}^p| > \cdots > |e_{ij}^3| > |e_{ij}| > 1$), thus able to automatically remove the design-parameter requirement.
2. For small entry error (e.g., $|e_{ij}| < 1$), the use of sigmoid activation functions has better convergence and robustness than the use of linear activation functions, because of the larger slope of the sigmoid function near the origin.

Thus, using the power-sigmoid activation function in (3) is theoretically a better choice than other activation functions for superior convergence and robustness.

## 3 Simulation Study

While Section 2 presents the main theoretical results of the gradient-based neural network, this section will investigate the MATLAB simulation techniques in order to show the characteristics of such a neural network.

### 3.1 Coding of Activation Function

To simulate the gradient-based neural network (2), the activation functions are to be defined firstly in MATLAB. Inside the body of a user-defined function, the MATLAB routine "nargin" returns the number of input arguments which are used to call the function. By using "nargin", different kinds of activation functions can be generated at least with their default input argument(s).

The linear activation-function mapping $\mathcal{F}(X) = X \in \mathbb{R}^{n\times n}$ can be generated simply by using the following MATLAB code.

```
function output=Linear(X)
output=X;
```

The sigmoid activation-function mapping $\mathcal{F}(\cdot)$ with $\xi = 4$ as its default input value can be generated by using the following MATLAB code.

```
function output=Sigmoid(X,xi)
if nargin==1, xi=4; end
output=(1-exp(-xi*X))./(1+exp(-xi*X));
```

The power activation-function mapping $\mathcal{F}(\cdot)$ with $p = 3$ as its default input value can be generated by using the following MATLAB code.

```
function output=Power(X,p)
if nargin==1, p=3; end
output=X.^p;
```

The power-sigmoid activation function defined in (3) with $\xi = 4$ and $p = 3$ being its default values can be generated below.

```
function output=Powersigmoid(X,xi,p)
if nargin==1, xi=4; p=3;
elseif nargin==2, p=3;
end
output=(1+exp(-xi))/(1-exp(-xi))*(1-exp(-xi*X))./(1+exp(-xi*X));
i=find(abs(X)>=1);
output(i)=X(i).^p;
```

### 3.2 Kronecker Product and Vectorization

The dynamic equations of gradient-based neural networks (2) and (4) are all described in matrix form which could not be simulated directly. To simulate such neural systems, the Kronecker product of matrices and vectorization technique are introduced in order to transform the matrix-form differential equations to vector-form differential equations.

- In general case, given matrices $A = [a_{ij}] \in \mathbb{R}^{m\times n}$ and $B = [b_{ij}] \in \mathbb{R}^{p\times q}$, the Kronecker product of $A$ and $B$ is denoted by $A \otimes B$ and is defined to be the following block matrix

$$A \otimes B := \begin{pmatrix} a_{11}B & \dots & a_{1n}B \\ \vdots & \vdots & \vdots \\ a_{m1}B & \dots & a_{mn}B \end{pmatrix} \in \mathbb{R}^{mp\times nq}.$$

  It is also known as the direct product or tensor product. Note that in general $A \otimes B \neq B \otimes A$. Specifically, for our case, $I \otimes A = \mathrm{diag}(A, \dots, A)$.
- In general case, given $X = [x_{ij}] \in \mathbb{R}^{m\times n}$, we can vectorize $X$ as a vector, i.e., $\mathrm{vec}(X) \in \mathbb{R}^{mn\times 1}$, which is defined as

$$\mathrm{vec}(X) := [x_{11}, \dots, x_{m1}, x_{12}, \dots, x_{m2}, \dots, x_{1n}, \dots, x_{mn}]^T.$$

  As stated in [22], in general case, let $X$ be unknown, given $A \in \mathbb{R}^{m\times n}$ and $B \in \mathbb{R}^{p\times q}$, the matrix equation $AX = B$ is equivalent to the vector equation $(I \otimes A)\,\mathrm{vec}(X) = \mathrm{vec}(B)$.

Based on the above Kronecker product and vectorization technique, for simulation proposes, the matrix differential equation (2) can be transformed to a vector differential equation. We thus obtain the following theorem.

**Theorem 1.** *The matrix-form differential equation (2) can be reformulated as the following vector-form differential equation:*

$$\mathrm{vec}(\dot{X}) = -\gamma(I \otimes A^T)\mathcal{F}\big((I \otimes A)\,\mathrm{vec}(X) - \mathrm{vec}(I)\big), \tag{5}$$

*where activation-function mapping $\mathcal{F}(\cdot)$ in (5) is defined the same as in (2) except that its dimensions are changed hereafter as $\mathcal{F}(\cdot) : R^{n^2\times 1} \to R^{n^2\times 1}$.*

*Proof.* For readers' convenience, we repeat the matrix-form differential equation (2) here as $\dot{X} = -\gamma A^T \mathcal{F}(AX(t) - I)$.

By vectorizing equation (2) based on the Kronecker product and the above vec(·) operator, the left hand side of (2) is vec($\dot{X}$), and the right hand side of equation (2) is

$$\begin{aligned} &\operatorname{vec}\left(-\gamma A^T \mathcal{F}(AX(t) - I)\right) \\ &= -\gamma \operatorname{vec}\left(A^T \mathcal{F}(AX(t) - I)\right) \\ &= -\gamma(I \otimes A^T)\operatorname{vec}(\mathcal{F}(AX(t) - I)). \end{aligned} \tag{6}$$

Note that, as shown in Subsection 3.1, the definition and coding of the activation function mapping $\mathcal{F}(\cdot)$ are very flexible and could be a vectorized mapping from $\mathbb{R}^{n^2\times 1}$ to $\mathbb{R}^{n^2\times 1}$. We thus have

$$\begin{aligned} &\operatorname{vec}(\mathcal{F}(AX(t) - I)) \\ &= \mathcal{F}(\operatorname{vec}(AX(t) - I)) \\ &= \mathcal{F}(\operatorname{vec}(AX) + \operatorname{vec}(-I)) \\ &= \mathcal{F}\big((I \otimes A)\operatorname{vec}(X) - \operatorname{vec}(I)\big). \end{aligned} \tag{7}$$

Combining equations (6) and (7) yields the vectorization of the right hand side of matrix-form differential equation (2):

$$\operatorname{vec}\left(-\gamma A^T \mathcal{F}(AX(t) - I)\right) = -\gamma(I \otimes A^T)\mathcal{F}\big((I \otimes A)\operatorname{vec}(X) - \operatorname{vec}(I)\big).$$

Clearly, the vectorization of both sides of matrix-form differential equation (2) should be equal, which generates the vector-form differential equation (5). The proof is thus complete.

*Remark 1.* The Kronecker product can be generated easily by using MATLAB routine "kron"; e.g., $A\otimes B$ can be generated by MATLAB command `kron(A,B)`. To generate vec($X$), we can use the MATLAB routine "reshape". That is, if the matrix $X$ has $n$ rows and $m$ columns, then the MATLAB command of vectorizing $X$ is `reshape(X,m*n,1)` which generates a column vector, $\operatorname{vec}(X) = [x_{11}, \ldots, x_{m1}, x_{12}, \ldots, x_{m2}, \ldots, x_{1n}, \ldots, x_{mn}]^T$.

Based on MATLAB routines "kron" and "vec", the following code is used to define a function returns the evaluation of the right-hand side of matrix-form gradient-based neural network (2). In other words, it also returns the evaluation of the right-hand side of vector-form gradient-based neural network (5). Note that $I \otimes A^T = (I \otimes A)^T$.

```
function output=GnnRightHandSide(t,x,gamma)
if nargin==2, gamma=1; end
A=MatrixA; n=size(A,1); IA=kron(eye(n),A);
% The following generates the vectorization of identity matrix I
vecI=reshape(eye(n),n^2,1);
% The following calculates the right hand side of equations (2) and (5)
output=-gamma*IA'*Powersigmoid(IA*x-vecI);
```

Note that we can change "Powersigmoid" in the above MATLAB code to "Sigmoid" (or "Linear") for using different activation functions.

## 4 Illustrative Example

For illustration, let us consider the following constant matrix:

$$A = \begin{bmatrix} 1 & 0 & 1 \\ 1 & 1 & 0 \\ 1 & 1 & 1 \end{bmatrix}, \quad A^T = \begin{bmatrix} 1 & 1 & 1 \\ 0 & 1 & 1 \\ 1 & 0 & 1 \end{bmatrix}, \quad A^{-1} = \begin{bmatrix} 1 & 1 & -1 \\ -1 & 0 & 1 \\ 0 & -1 & 1 \end{bmatrix}.$$

For example, matrix $A$ can be given in the following MATLAB code.

```
function A=MatrixA(t)
A=[1 0 1;1 1 0;1 1 1];
```

The gradient-based neural network (2) is thus in the following specific form

$$\begin{bmatrix} \dot{x}_{11} & \dot{x}_{12} & \dot{x}_{13} \\ \dot{x}_{21} & \dot{x}_{22} & \dot{x}_{23} \\ \dot{x}_{31} & \dot{x}_{32} & \dot{x}_{33} \end{bmatrix} = -\gamma \begin{bmatrix} 1 & 1 & 1 \\ 0 & 1 & 1 \\ 1 & 0 & 1 \end{bmatrix} \mathcal{F} \left( \begin{bmatrix} 1 & 0 & 1 \\ 1 & 1 & 0 \\ 1 & 1 & 1 \end{bmatrix} \begin{bmatrix} x_{11} & x_{12} & x_{13} \\ x_{21} & x_{22} & x_{23} \\ x_{31} & x_{32} & x_{33} \end{bmatrix} - \begin{bmatrix} 1 & 0 & 0 \\ 0 & 1 & 0 \\ 0 & 0 & 1 \end{bmatrix} \right).$$

### 4.1 Simulation of Convergence

To simulate gradient-based neural network (2) starting from eight random initial states, we firstly define a function "GnnConvergence" as follows.

```
function GnnConvergence(gamma)
tspan=[0 10]; n=size(MatrixA,1);
for i=1:8
  x0=4*(rand(n^2,1)-0.5*ones(n^2,1));
  [t,x]=ode45(@GnnRightHandSide,tspan,x0,[],gamma);
  for j=1:n^2
    k=mod(n*(j-1)+1,n^2)+floor((j-1)/n);
    subplot(n,n,k); plot(t,x(:,j)); hold on
  end
end
```

To show the convergence of the gradient-based neural model (2) using power-sigmoid activation function with $\xi = 4$ and $p = 3$ and using the design parameter $\gamma := 1$, the MATLAB command is `GnnConvergence(1)`, which generates Fig. 1(a). Similarly, the MATLAB command `GnnConvergence(10)` can generate Fig. 1(b).

To monitor the network convergence, we can also use and show the norm of the computational error, $\|X(t) - A^{-1}\|$. The MATLAB codes are given below, i.e., the user-defined functions "NormError" and "GnnNormError". By calling "GnnNormError" three times with different $\gamma$ values, we can generate Fig. 2. It shows that starting from any initial state randomly selected in $[-2, 2]$, the state matrices of the presented neural network (2) all converge to the theoretical

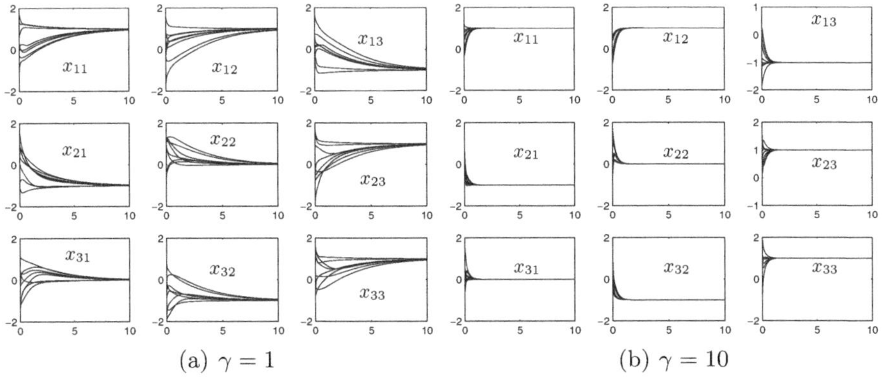


**Fig. 1.** Online matrix inversion by gradient-based neural network (2)

inverse $A^{-1}$, where the computational errors $\|X(t) - A^{-1}(t)\|$ all converge to zero. Such a convergence can be expedited by increasing $\gamma$. For example, if $\gamma$ is increased to $10^3$, the convergence time is within 30 milliseconds; and, if $\gamma$ is increased to $10^6$, the convergence time is within 30 microseconds.

```
function NormError(x0,gamma)
tspan=[0 10]; options=odeset();
[t,x]=ode45(@GnnRightHandSide,tspan,x0,options,gamma);
Ainv=inv(MatrixA);
B=reshape(Ainv,size(Ainv,1)^2,1);
total=length(t); x=x';
for i=1:total, nerr(i)=norm(x(:,i)-B); end
plot(t,nerr); hold on
```

```
function GnnNormError(gamma)
if nargin<1, gamma=1; end
total=8; n=size(MatrixA,1);
for i=1:total
    x0=4*(rand(n^2,1)-0.5*ones(n^2,1));
    NormError(x0,gamma);
end
text(2.4,2.2,['gamma=' int2str(gamma)]);
```

### 4.2 Simulation of Robustness

Similar to the transformation of the matrix-form differential equation (2) to a vector-form differential equation (5), the perturbed gradient-based neural network (4) can be vectorized as follows:

$$\mathrm{vec}(\dot{X}) = -\gamma(I \otimes A^T)\mathcal{F}\big((I \otimes A)\,\mathrm{vec}(X) - \mathrm{vec}(I)\big) + \mathrm{vec}(\Delta_B). \tag{8}$$

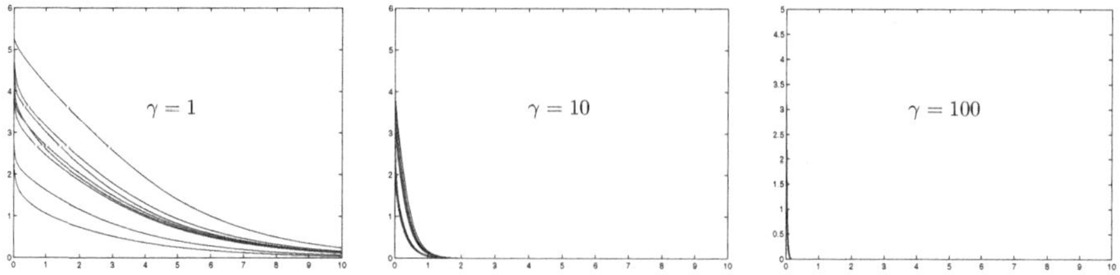


**Fig. 2.** Convergence of $\|X(t) - A^{-1}\|_F$ using power-sigmoid activation function

To show the robustness characteristics of gradient-based neural networks, the following model-implementation error is added in a sinusoidal form (with $\varepsilon_2 = 0.5$):

$$\Delta_B = \varepsilon_2 \begin{bmatrix} \cos(3t) & -\sin(3t) & 0 \\ 0 & \sin(3t) & \cos(3t) \\ 0 & 0 & \sin(2t) \end{bmatrix}.$$

The following MATLAB code is used to define the function "GnnRightHandSideImprecise" for ODE solvers, which returns the evaluation of the right-hand side of the perturbed gradient-base neural network (4), in other words, the right-hand side of the vector-form differential equation (8).

```
function output=GnnRightHandSideImprecise(t,x,gamma)
if nargin==2, gamma=1; end
e2=0.5;
deltaB=e2*[cos(3*t) -sin(3*t) 0; 0 sin(3*t) cos(3*t);0 0 sin(2*t)];
vecB=reshape(deltaB,9,1);
vecI=reshape(eye(3),9,1);
IA=kron(eye(3),MatrixA);
output=-gamma*IA'*Powersigmoid(IA*x-vecI)+vecB;
```

To use the sigmoid (or linear) activation function, we only need to change "Powersigmoid" to "Sigmoid" (or "Linear") in the above MATLAB code. Based on the above function "GnnRightHandSideImprecise" and the function below (i.e.,"GnnRobust"), MATLAB commands `GnnRobust(1)` and `GnnRobust(100)` can generate Fig. 3.

```
function GnnRobust(gamma)
tspan=[0 10]; options=odeset(); n=size(MatrixA,1);
for i=1:8
    x0=4*(rand(n^2,1)-0.5*ones(n^2,1));
    [t,x]=ode45(@GnnRightHandSideImprecise,tspan,x0,options,gamma);
    for j=1:n^2
      k=mod(n*(j-1)+1,n^2)+floor((j-1)/n);
      subplot(n,n,k); plot(t,x(:,j)); hold on
    end
end
```

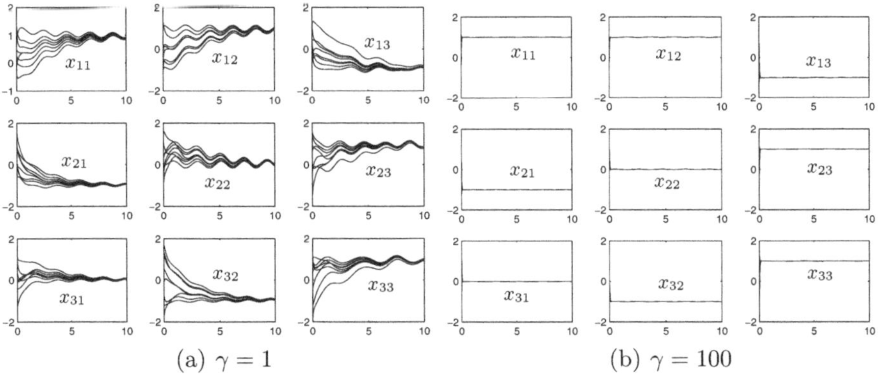


**Fig. 3.** Online matrix inversion by GNN (4) with large implementation errors

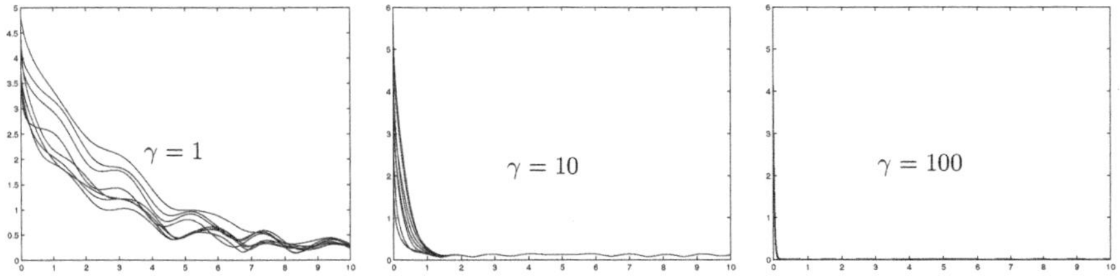


**Fig. 4.** Convergence of computational error $\|X(t) - A^{-1}\|$ by perturbed GNN (4)

Similarly, we can show the computational error $\|X(t) - A^{-1}\|$ of gradient-based neural network (4) with large model-implementation errors. To do so, in the previously defined MATLAB function "NormError", we only need change "GnnRightHandSide" to "GnnRightHandSideImprecise". See Fig. 4. Even with imprecise implementation, the perturbed neural network still works well, and its computational error $\|X(t) - A^{-1}\|$ is still bounded and very small. Moreover, as the design parameter $\gamma$ increases from 1 to 100, the convergence is expedited and the steady-state computational error is decreased. It is worth mentioning again that using power-sigmoid or sigmoid activation functions has smaller steady-state residual error than using linear or power activation functions. It is observed from other simulation data that when using power-sigmoid activation functions, the maximum steady-state residual error is only $2 \times 10^{-2}$ and $2 \times 10^{-3}$ respectively for $\gamma = 100$ and $\gamma = 1000$. Clearly, compared to the case of using linear or pure power activation functions, superior performance can be achieved by using power-sigmoid or sigmoid activation functions under the same design specification. These simulation results have substantiated the theoretical results presented in previous sections and in [21].

## 5 Conclusions

The gradient-based neural networks (1) and (2) have provided an effective online-computing approach for matrix inversion. By considering different types of activation functions and implementation errors, such recurrent neural networks have been simulated in this paper. Several important simulation techniques have been introduced, i.e., coding of activation-function mappings, Kronecker product of matrices, and MATLAB routine "ode45". Simulation results have also demonstrated the effectiveness and efficiency of gradient-based neural networks for online matrix inversion. In addition, the characteristics of such a negative-gradient design method of recurrent neural networks could be summarized as follows.

- From the viewpoint of system stability, any monotonically-increasing activation function $f(\cdot)$ with $f(0) = 0$ could be used for the construction of recurrent neural networks. But, for the solution effectiveness and design simplicity, the strictly-monotonically-increasing odd activation-function $f(\cdot)$ is preferred for the construction of recurrent neural networks.
- The gradient-based neural networks are intrinsically designed for solving time-invariant matrix-inverse problems, but they could also be used to solve time-varying matrix-inverse problems in an approximate way. Note that, in this case, design parameter $\gamma$ is required to be large enough.
- Compared to other methods, the gradient-based neural networks have an easier structure for simulation and hardware implementation. As parallel-processing systems, such neural networks could solve the matrix-inverse problem more efficiently than those serial-processing methods.

**Acknowledgements.** This work is funded by National Science Foundation of China under Grant 60643004 and by the Science and Technology Office of Sun Yat-Sen University. Before joining Sun Yat-Sen University in 2006, the corresponding author, Yunong Zhang, had been with National University of Ireland, University of Strathclyde, National University of Singapore, Chinese University of Hong Kong, since 1999. He has continued the line of this research, supported by various research fellowships/assistantship. His web-page is now available at http://www.ee.sysu.edu.cn/teacher/detail.asp?sn=129.

## References

1. Zhang, Y.: Towards Piecewise-Linear Primal Neural Networks for Optimization and Redundant Robotics. Proceedings of IEEE International Conference on Networking, Sensing and Control (2006) 374-379
2. Steriti, R.J., Fiddy, M.A.: Regularized Image Reconstruction Using SVD and a Neural Network Method for Matrix Inversion. IEEE Transactions on Signal Processing, Vol. 41 (1993) 3074-3077
3. Sarkar, T., Siarkiewicz, K., Stratton, R.: Survey of Numerical Methods for Solution of Large Systems of Linear Equations for Electromagnetic Field Problems. IEEE Transactions on Antennas and Propagation, Vol. 29 (1981) 847-856

4. Sturges Jr, R.H.: Analog Matrix Inversion (Robot Kinematics). IEEE Journal of Robotics and Automation, Vol. 4 (1988) 157-162
5. Yeung, K.S., Kumbi, F.: Symbolic Matrix Inversion with Application to Electronic Circuits. IEEE Transactions on Circuits and Systems, Vol. 35 (1988) 235-238
6. El-Amawy, A.: A Systolic Architecture for Fast Dense Matrix Inversion. IEEE Transactions on Computers, Vol. 38 (1989) 449-455
7. Neagoe, V.E.: Inversion of the Van Der Monde Matrix. IEEE Signal Processing Letters, Vol. 3 (1996) 119-120
8. Wang, Y.Q., Gooi, H.B.: New Ordering Methods for Space Matrix Inversion via Diagonaliztion. IEEE Transactions on Power Systems, Vol. 12 (1997) 1298-1305
9. Koc, C.K., Chen, G.: Inversion of All Principal Submatrices of a Matrix. IEEE Transactions on Aerospace and Electronic Systems, Vol. 30 (1994) 280-281
10. Zhang, Y., Leithead, W.E., Leith, D.J.: Time-Series Gaussian Process Regression Based on Toeplitz Computation of $O(N^2)$ Operations and $O(N)$-Level Storage. Proceedings of the 44th IEEE Conference on Decision and Control (2005) 3711-3716
11. Leithead, W.E., Zhang, Y.: $O(N^2)$-Operation Approximation of Covariance Matrix Inverse in Gaussian Process Regression Based on Quasi-Newton BFGS Methods. Communications in Statistics - Simulation and Computation, Vol. 36 (2007) 367-380
12. Manherz, R.K., Jordan, B.W., Hakimi, S.L.: Analog Methods for Computation of the Generalized Inverse. IEEE Transactions on Automatic Control, Vol. 13 (1968) 582-585
13. Jang, J., Lee, S., Shin, S.: An Optimization Network for Matrix Inversion. Neural Information Processing Systems, American Institute of Physics, NY (1988) 397-401
14. Wang, J.: A Recurrent Neural Network for Real-Time Matrix Inversion. Applied Mathematics and Computation, Vol. 55 (1993) 89-100
15. Zhang, Y.: Revisit the Analog Computer and Gradient-Based Neural System for Matrix Inversion. Proceedings of IEEE International Symposium on Intelligent Control (2005) 1411-1416
16. Zhang, Y., Jiang, D., Wang, J.: A Recurrent Neural Network for Solving Sylvester Equation with Time-Varying Coefficients. IEEE Transactions on Neural Networks, Vol. 13 (2002) 1053-1063
17. Zhang, Y., Ge, S.S.: A General Recurrent Neural Network Model for Time-Varying Matrix Inversion. Proceedings of the 42nd IEEE Conference on Decision and Control (2003) 6169-6174
18. Zhang, Y., Ge, S.S.: Design and Analysis of a General Recurrent Neural Network Model for Time-Varying Matrix Inversion. IEEE Transactions on Neural Networks, Vol. 16 (2005) 1477-1490
19. Carneiro, N.C.F., Caloba, L.P.: A New Algorithm for Analog Matrix Inversion. Proceedings of the 38th Midwest Symposium on Circuits and Systems, Vol. 1 (1995) 401-404
20. Mead, C.: Analog VLSI and Neural Systems. Addison-Wesley, Reading, MA (1989)
21. Zhang, Y., Li, Z., Fan, Z., Wang, G.: Matrix-Inverse Primal Neural Network with Application to Robotics. Dynamics of Continuous, Discrete and Impulsive Systems, Series B, Vol. 14 (2007) 400-407
22. Horn, R.A., Johnson, C.R.: Topics in Matrix Analysis, Cambridge University Press, Cambridge (1991)

# Mean Square Exponential Stability of Uncertain Stochastic Hopfield Neural Networks with Interval Time-Varying Delays

Jiqing Qiu[1], Hongjiu Yang[1], Yuanqing Xia[2], and Jinhui Zhang[2]

[1] College of Sciences, Hebei University of Science and Technology, Shijiazhuang, 050018, China
qiujiqing@263.net, yanghongjiu@126.com
[2] Department of Automatic Control, Beijing Institute of Technology, Beijing 100081, China
xia_yuanqing@163.net, jinhuizhang82@gmail.com

**Abstract.** The problem of mean square exponential stability of uncertain stochastic Hopfield neural networks with interval time-varying delays is investigated in this paper. The delay factor is assumed to be time-varying and belongs to a given interval, which means that the derivative of the delay function can exceed one. The uncertainties considered in this paper are norm-bounded and possibly time-varying. By Lyapunov-Krasovskii functional approach and stochastic analysis approach, a new delay-dependent stability criteria for the exponential stability of stochastic Hopfield neural networks is derived in terms of linear matrix inequalities(LMIs). A simulation example is given to demonstrate the effectiveness of the developed techniques.

## 1 Introduction

The Hopfield neural networks were first introduced by Hopfield [1]. In recent years, it has been investigated extensively and successful applications in pattern recognition, image processing, optimization problems, and so on. Since time delays may lead to instability and oscillation of the Hopfield neural network model, the issue on the stability analysis of Hopfield neural networks with time delays has received more and more attention. It is well known that the delay-dependent criteria type is less conservative than the delay-independent one as for example [7, 8, 11].

As far as we know, there are systems which are stable with some nonzero delays, but are unstable without delay. Therefore, it is important to analyze the stability of systems with nonzero delays, and the nonzero delay can be placed into a given interval such as [6]. There are also many stochastic perturbations which affect the stability of neural networks. Therefore, it is necessary to consider stochastic effects on the stability property of the delayed Hopfield neural networks (for example in [2, 5, 9, 12]). The exponential stability of neural networks has been considered in [3, 4, 10, 13, 14]. But to the best of our knowledge,

D.-S. Huang, L. Heutte, and M. Loog (Eds.): ICIC 2007, LNAI 4682, pp. 110–119, 2007.

very few papers consider the stochastic exponential stability problem of Hopfield neural networks with interval time-varying delays.

The problem of mean square exponential stability of uncertain stochastic Hopfield neural networks with interval time-varying delays is considered in this paper. The uncertainties are norm-bounded and the derivative of the delay function can exceed one. Based on Lyapunov-Krasovskii functional approach and stochastic analysis approach, a new stability criteria in terms of linear matrix inequalities is proposed which can be solved numerically using the Matlab LMI control toolbox. A numerical example is given to illustrate the feasibility and effectiveness of the proposed technique.

*Notations*: The notation in this paper is quite standard. The superscript "T" denotes matrix transposition; $\mathbf{R}^n$ and $\mathbf{R}^{n\times n}$ denote an $n$-dimensional Euclidean space and the set of all $n \times n$ real matrices, respectively; the notation $X > Y$ ( $X \geq Y$ ) means that the matrix $X-Y$ is positive definite ($X-Y$ is semi-positive definite, respectively); $\lambda_{min}(\cdot)$ and $\lambda_{max}(\cdot)$ denote the minimum and maximum eigenvalue of a real symmetric matrix, respectively; $I$ is the identity matrix of appropriate dimension; $diag\{\cdots\}$ denote the block diagonal matrix; $\|\cdot\|$ is the Euclidean vector norm, and the symmetric terms in a symmetric matrix are denoted by $\star$.

## 2 Problem Formulation

In this section, the following uncertain Hopfield neural networks with time-varying and distributed delays is investigated:

$$\begin{aligned} dx(t) = {} & [-A(t)x(t) + W(t)f(x(t-\tau(t)))]dt \\ & + [H_0(t)x(t) + H_1(t)x(t-\tau(t))]d\omega(t) \end{aligned} \tag{1}$$

where $x = [x_1,\ x_2,\ \cdots,\ x_n]^\top \in \mathbf{R}^n$ is the neural state vector, and $f(x) = [f_1(x_1),\ f_2(x_2),\ \cdots,\ f_n(x_n)]^\top \in \mathbf{R}^n$ denotes the neural activation function, $\omega(t) = [\omega_1(t),\ \omega_2(t),\ \cdots,\ \omega_m(t)] \in \mathbf{R}^m$ is $m$-dimensional Brownian motion defined on a complete probability space $(\Omega,\ \mathbf{F},\ P)$. The matrices $A(t) = A+\Delta A(t)$, $W(t) = W + \Delta W(t)$, $H_0(t) = H_0 + \Delta H_0(t)$ and $H_1(t) = H_1 + \Delta H_1(t)$, where $A = diag\{a_1, a_2, ..., a_n\}$ is a positive diagonal matrix, $W \in \mathbf{R}^{n\times n}$ is the connection weight matrix, $H_0 \in \mathbf{R}^{n\times n}$ and $H_1 \in \mathbf{R}^{n\times n}$ are known real constant matrices, $\Delta A(t)$, $\Delta W(t)$, $\Delta H_0(t)$ and $\Delta H_1(t)$ are parametric uncertainties, which are assumed to be of the following form

$$[\Delta A(t),\ \Delta W(t),\ \Delta H_0(t),\ \Delta H_1(t)] = DF(t)[E_1,\ E_2,\ E_3,\ E_4] \tag{2}$$

where $D$, $E_1$, $E_2$, $E_3$ and $E_4$ are known real constant matrices with appropriate dimensions, and $F(t)$ is the time-varying uncertain matrix satisfying

$$F^\top(t)F(t) \leq I. \tag{3}$$

And $\tau(t)$ is the time-varying delay satisfies $0 < \tau_m \leq \tau(t) \leq \tau_M$, where $\tau_m$ and $\tau_M$ are positive constants. In this paper, it is denoted that $\tau_0 = \frac{1}{2}(\tau_M + \tau_m)$ and $\delta = \frac{1}{2}(\tau_M - \tau_m) = \tau_M - \tau_0 = \tau_0 - \tau_m$.

*Remark 1.* Obviously, when $\delta = 0$ i.e., $\tau_m = \tau_M$, then $\tau(t)$ denotes a constant delay, which is investigated in [10]; the case when $\tau_m = 0$, i.e., $\tau_0 = \delta = \frac{\tau_M}{2}$, it implies that $0 < \tau(t) < \tau_M$, which is investigated in [7].

**Definition 1.** *The equilibrium point of the delayed neural networks (1) is said to be globally robustly exponentially stable for all admissible uncertainties satisfying (3)-(4) in the mean square if there exist positive constants $\alpha > 0$ and $\mu > 0$, such that the following condition holds:*

$$\mathbf{E}\{\|x(t)\|\} \leq \mu e^{-\alpha t} \sup_{-k\leq s\leq 0} \mathbf{E}\{\|x(s)\|\}, \forall t > 0. \tag{4}$$

Before ending this section, the following lemma is cited to prove our main results in the next section.

**Lemma 1.** [6] *For any positive define matrix $M \in \boldsymbol{R}^{n\times n}$, two vectors $a$ and $b$ with appropriate dimension, the following inequality holds:*

$$2a^\top Mb \leq a^\top Ma + b^\top Mb.$$

**Lemma 2.** [15] *For any constant matrix $M \in \boldsymbol{R}^{n\times n}$, $M = M^\top > 0$, scalar $\gamma > 0$, vector function $\omega : [0, \gamma] \longrightarrow \boldsymbol{R}^n$ such that the integrations are well defined, the following inequality holds:*

$$\left(\int_0^\gamma \omega(s)ds\right)^\top M \left(\int_0^\gamma \omega(s)ds\right) \leq \gamma \int_0^\gamma \omega^\top(s)M\omega(s)ds.$$

**Lemma 3.** [16]. *For some given matrices $Y$, $G$ and $E$ of appropriate and with $Y$ symmetric, then for all $F(t)$ satisfying $F^\top(t)F(t) \leq I$ and*

$$Y + GF(t)E + E^\top F^T(t)G^\top \leq 0,$$

*if and only if there exists scalar $\alpha > 0$ such that*

$$Y + \alpha GG^\top + \alpha^{-1}E^\top E \leq 0.$$

## 3 Main Results

This section will perform global robust stability analysis for uncertain Hopfield neural networks (1). Based on Lyapunov-Krasovskii stability theorem, the following result is carried out.

**Theorem 1.** *The uncertain neural networks (1) is robustly asymptotically stable, if there exist symmetric positive definite matrices $P$, $Q$, $R_1$, $R_2$, $M$ and scalars $\alpha_i > 0$, $i = 1, 2$ such that the following LMI holds:*

$$\Sigma = \begin{bmatrix} \Gamma_{11} & \alpha_2 E_3^T E_4 & PW + \alpha_1 E_1^T E_2 & 0 & H_0^T P & PD & 0 \\ \star & -Q + \alpha_2 E_4^T E_4 & 0 & 0 & H_1^T P & 0 & 0 \\ \star & \star & -M + \alpha_1 E_2^T E_2 & M & 0 & 0 & 0 \\ \star & \star & \star & \Gamma_{44} & 0 & 0 & 0 \\ \star & \star & \star & \star & -P & 0 & PD \\ \star & \star & \star & \star & \star & -\alpha_1 I & 0 \\ \star & \star & \star & \star & \star & \star & -\alpha_2 I \end{bmatrix} < 0, \tag{5}$$

*where*

$$\Gamma_{11} = -2PA + Q + \tau_0 R_1 + 2\delta R_2 + \alpha_1 E_1^T E_1 + \alpha_2 E_3^T E_3,$$
$$\Gamma_{44} = -M - \frac{1}{\tau_0} R_1 - \frac{1}{\delta} R_2$$

*Proof.* First of all, we define the following positive define Lyapunov-Krasovskii functional,

$$V(x(t),\ t) = x^\top(t)Px(t) + \int_{t-\tau(t)}^{t} x^\top(s)Qx(s)ds + \int_{t-\tau_0}^{t}\int_{s}^{t} x^\top(v)R_1x(v)dvds$$
$$+2\delta \int_{t-\tau_0+\delta}^{t} x^\top(s)R_2x(s)ds + \int_{t-\tau_0-\delta}^{t-\tau_0+\delta}\int_{s}^{t-\tau_0+\delta} x^\top(v)R_2x(v)dvds.$$

By Itô's differential formula, the stochastic derivative of $V(x(t),\ t)$ along the trajectory (1) can be obtained as follows:

$$\begin{aligned}
&dV(x(t),\ t)\\
&\leq \{2x^\top(t)P[-A(t)x(t) + W(t)f(x(t-\tau(t)))] - x^\top(t-\tau(t))Qx(t-\tau(t))\\
&+x^\top(t)Qx(t) + \tau_0 x^\top(t)R_1x(t) - \int_{t-\tau_0}^{t} x^\top(s)R_1x(s)ds + 2\delta x^\top(t)R_2x(t)\\
&- \int_{t-\tau_0-\delta}^{t-\tau_0+\delta} x^\top(s)R_2x(s)ds + [H_0(t)x(t) + H_1(t)x(t-\tau(t))]^\top P[H_0(t)x(t)\\
&+H_1(t)x(t-\tau(t))]\}dt + \{2x^\top(t)P[H_0(t)x(t) + H_1(t)x(t-\tau(t))]\}d\omega(t).
\end{aligned}$$

From Lemma 2, we have

$$-\int_{t-\tau_0}^{t} x^\top(s)R_1x(s)ds \leq -\frac{1}{\tau_0}\left[\int_{t-\tau_0}^{t-\tau(t)} x(s)ds\right]^\top R_1 \left[\int_{t-\tau_0}^{t-\tau(t)} x(s)ds\right],$$

and

$$-\int_{t-\tau_0-\delta}^{t-\tau_0+\delta} x^\top(s)R_2x(s)ds \leq -\frac{1}{\delta}\left[\int_{t-\tau_0}^{t-\tau(t)} x(s)ds\right]^\top R_2 \left[\int_{t-\tau_0}^{t-\tau(t)} x(s)ds\right].$$

From Lemma 1, it can be obtained that

$$\begin{aligned}
&2f^\top(x(t-\tau(t)))M\int_{t-\tau_0}^{t-\tau(t)} x(s)ds\\
&\leq f^\top(x(t-\tau(t)))Mf(x(t-\tau(t))) + \left[\int_{t-\tau_0}^{t-\tau(t)} x(s)ds\right]^\top M \left[\int_{t-\tau_0}^{t-\tau(t)} x(s)ds\right].
\end{aligned}$$

Substituting above inequalities into $dV(x(t),\ t)$, we have

$$dV(x(t),\ t) \leq \{\xi_1^T(t)\Sigma_0\xi_1(t)\}dt + \{2x^\top(t)P[H_0(t)x(t) + H_1(t)x(t-\tau(t))]\}d\omega(t)$$

where

$$\Sigma_0 = \begin{bmatrix} (1,1) & H_0^\top(t)PH_1(t) & PW(t) & 0 \\ \star & -Q + H_1^\top(t)PH_1(t) & 0 & 0 \\ \star & \star & -M & M \\ \star & \star & \star & -M - \frac{1}{\tau_0}R_1 - \frac{1}{\delta}R_2 \end{bmatrix},$$

$$\xi_1^T(t) = \left[ x^\top(t),\ x^\top(t-\tau(t)),\ f^\top(x(t-\tau(t))),\ \left[\int_{t-\tau_0}^{t-\tau(t)} x(s)ds\right]^\top \right]$$

with

$$(1,1) = -2PA(t) + Q + \tau_0 R_1 + 2\delta R_2 + H_0^\top(t)PH_0(t).$$

Utilizing Shur complement $\Sigma_0 < 0$ can be changed to

$$\Sigma_1 = \begin{bmatrix} (1,1) & 0 & PW(t) & 0 & H_0^\top(t)P \\ \star & -Q & 0 & 0 & H_1^\top(t)P \\ \star & \star & -M & M & 0 \\ \star & \star & \star & -M - \frac{1}{\tau_0}R_1 - \frac{1}{\delta}R_2 & 0 \\ \star & \star & \star & \star & -P \end{bmatrix} < 0$$

where

$$(1,1) = -2PA(t) + Q + \tau_0 R_1 + 2\delta R_2.$$

Using Lemma 3, taking into account (2) and (3) the matrix $\Sigma_1 < 0$ can be changed to

$$\begin{aligned} \Sigma_2 &= \Upsilon + \eta_1^\top F(t)\eta_2 + \eta_2^\top F^\top(t)\eta_1 + \eta_3^\top F(t)\eta_4 + \eta_4^\top F^\top(t)\eta_3 \\ &\le \Upsilon + \alpha_1^{-1}\eta_1^\top\eta_1 + \alpha_1\eta_2^\top\eta_2 + \alpha_3^{-1}\eta_3^\top\eta_3 + \alpha_2\eta_4^\top\eta_4 < 0, \end{aligned}$$

where

$$\begin{aligned} \eta_1 &= \left[ D^T P\, 0\, 0\, 0\, 0 \right], & \eta_2 &= \left[ -E_1\, 0\, E_2\, 0\, 0 \right], \\ \eta_3 &= \left[ 0\, 0\, 0\, 0\, D^T P \right], & \eta_4 &= \left[ E_3\, E_4\, 0\, 0\, 0 \right], \end{aligned}$$

and

$$\Upsilon = \begin{bmatrix} -2PA + Q + \tau_0 R_1 + 2\delta R_2 & 0 & PW & 0 & H_0^\top P \\ \star & -Q & 0 & 0 & H_1^\top P \\ \star & \star & -M & M & 0 \\ \star & \star & \star & -M - \frac{1}{\tau_0}R_1 - \frac{1}{\delta}R_2 & 0 \\ \star & \star & \star & \star & -P \end{bmatrix}.$$

Utilizing Shur complement again, the matrix $\Sigma_2 < 0$ can be changed to $\Sigma < 0$.

It is obvious that for $\Sigma < 0$, there must exist a scalar $\gamma > 0$ such that $\Sigma + diag\{\gamma I\ 0\ 0\ 0\ 0\} < 0$, which indicates that

$$dV(x(t),t) = -\gamma \|x(t)\|^2 dt + \{2x^\top(t)P[H_0(t)x(t) + H_1(t)x(t-\tau(t))]\}d\omega(t). \quad (6)$$

Taking the mathematical expectation of both sides of (4), we have

$$\frac{d\mathbf{E}V(x(t),t)}{dt} = -\gamma \mathbf{E}\|x(t)\|^2 dt, \tag{7}$$

which indicates from the Lyapunov stability the dynamics of hopfield neural neural networks (1) is globally robustly asymptotically stable in the mean square. In the following, we will show the global exponential stability for the delayed hopfield neural neural networks (1).

Considering $V(x(t))$, it is easy to get that

$$\begin{aligned} V(x(t),t) \le \lambda_{max}(P)\|x(t)\|^2 &+ \lambda_{max}(Q)\int_{t-\tau_M}^{t} \|x(\alpha)\|^2 d\alpha \\ &+\lambda_{max}(R_1)\int_{t-\tau_0}^{t}\int_{s}^{t} \|x(\alpha)\|^2 d\alpha ds + 2\delta\cdot\lambda_{max}(R_2)\int_{t-\tau_M}^{t} \|x(\alpha)\|^2 d\alpha \\ &+\lambda_{max}(R_2)\int_{t-\tau_0-\delta}^{t-\tau_0+\delta}\int_{s}^{t-\tau_0+s} \|x(\alpha)\|^2 d\alpha ds \end{aligned}$$

Note that

$$\int_{t-\tau_0}^{t}\int_{s}^{t} \|x(\alpha)\|^2 d\alpha ds \le \int_{t-\tau_0}^{t}\int_{-\tau_0}^{0} \|x(\alpha)\|^2 du d\alpha \le \tau_0\int_{t-\tau_M}^{t} \|x(\alpha)\|^2 d\alpha,$$

and

$$\int_{t-\tau_0-\delta}^{t-\tau_0+\delta}\int_{s}^{t-\tau_0+s} \|x(\alpha)\|^2 d\alpha ds \le (\tau_0+\delta)\int_{t-\tau_M}^{t} \|x(\alpha)\|^2 d\alpha,$$

Then, it follows

$$V(x(t),t) \le a\left(\|x(t)\| + \int_{t-\tau_M}^{t} \|x(\alpha)\|^2 d\alpha\right). \tag{8}$$

where $a = \max\{\lambda_{max}(P),\ \lambda_{max}(Q) + \tau_0\lambda_{max}(R_1) + (\tau_0+3\delta)\lambda_{max}(R_2)\}$.

Let $Y(x(t),t) = e^{\theta t}V(x(t),t)$, where $\theta$ is to be determined. Then, we have

$$\begin{aligned} dY(x(t),t) \le &\left\{e^{\theta t}\left[(\theta a-\gamma)\|x(t)\|^2 + \theta a\int_{t-\tau_M}^{t} \|x(\alpha)\|^2 d\alpha\right]\right\} dt \\ &+\{2x^\top(t)P[H_0(t)x(t) + H_1(t)x(t-\tau(t))]\}d\omega(t) \end{aligned} \tag{9}$$

Integrating both sides of (9) from 0 to $T > 0$ and then taking the mathematical expectation results in

$$\begin{aligned} &\mathbf{E}\{e^{\theta T}V(x(T),T) - V(x(0),0)\} \\ &\le \mathbf{E}\left\{(\theta a-\gamma)\int_0^T e^{\theta t}\|x(t)\|^2 dt + \theta a\int_0^T\int_{t-\tau_M}^{t} e^{\theta t}\|x(\alpha)\|^2 d\alpha dt\right\} \end{aligned}$$

Observe that

$$\int_0^T \int_{t-\tau_M}^t e^{\theta t}\|x(\alpha)\|^2 d\alpha dt < \tau_M e^{\theta \tau_M} \int_{-\tau_M}^T e^{\theta \alpha}\|x(\alpha)\|^2 d\alpha \tag{10}$$

Now, choose $\theta > 0$ satisfying $\theta a - \gamma + \theta a \tau_M e^{\theta \tau_M} = 0$. This together with (7) implies

$$\mathbf{E}\{e^{\theta T} V(x(T),T)\} \leq \mathbf{E}\left\{\theta a \tau_M e^{\theta \tau_M} \int_0^{-\tau_M} e^{\theta t}\|x(t)\|^2 dt + V(x(0),0)\right\}. \tag{11}$$

By (11) and (8), it is obtained that

$$\mathbf{E}\{V(x(T),T)\} \leq 2e^{-\theta T}(a + a\tau_M + \theta a \tau_M^2 e^{\theta \tau_M}) \sup_{-\tau_M \leq \theta \leq 0} \{\mathbf{E}\{\|x(t)\|\}\} \tag{12}$$

which implies that

$$\mathbf{E}\{\|x(t)\|\} \leq \mu e^{-\delta T} \sup_{-\tau_M \leq \theta \leq 0} \{\mathbf{E}\{\|x(t)\|\}\} \tag{13}$$

where

$$\mu = \sqrt{\frac{2(a + a\tau_M + \theta a \tau_M^2 e^{\theta \tau_M})}{\lambda_{min}(P)}}, \qquad \delta = \frac{\theta}{2}.$$

Therefore, by Definition 1, It is easy to see an equilibrium point of the delayed Hopfield neural network in (1) is globally exponentially stable.

**Theorem 2.** *The uncertain Hopfield neural networks (1) with $F(t) = 0$ is robustly asymptotically stable, if there exist symmetric positive definite matrices $P$, $Q$, $R_1$, $R_2$ and $M$ such that the following LMI holds:*

$$\begin{bmatrix} \Gamma_{11} & 0 & PW & 0 & H_0^T P \\ \star & -Q & 0 & 0 & H_1^T P \\ \star & \star & -M & M & 0 \\ \star & \star & \star & \Gamma_{44} & 0 \\ \star & \star & \star & \star & -P \end{bmatrix} < 0, \tag{14}$$

*where*

$$\Gamma_{11} = -2PA + Q + \tau_0 R_1 + 2\delta R_2, \ \Gamma_{44} = -M - \frac{1}{\tau_0}R_1 - \frac{1}{\delta}R_2.$$

## 4 Numerical Examples

*Example 1.* Consider the following norm-bounded uncertain hopfield neural networks with time-varying delays:

$$\begin{aligned} dx(t) = {} & [-A(t)x(t) + W(t)f(x(t - \tau(t)))]dt \\ & + [H_0(t)x(t) + H_1(t)x(t - \tau(t))]d\omega(t) \end{aligned} \tag{15}$$

where

$$A = \begin{bmatrix} 1.2 & 0 \\ 0 & 1.15 \end{bmatrix}, \ W = \begin{bmatrix} 0.4 & -1 \\ -1.4 & 0.4 \end{bmatrix}, \ H_0 = \begin{bmatrix} -0.2 & 0 \\ 0 & 0.1 \end{bmatrix}, \ H_1 = \begin{bmatrix} 0.1 & 0 \\ 0 & -0.3 \end{bmatrix},$$

$$D = \begin{bmatrix} 0.1 & 0 \\ 0 & -0.5 \end{bmatrix}, \ E_1 = \begin{bmatrix} 0.6 & 0 \\ 0 & 0.6 \end{bmatrix}, \ E_2 = E_3 = E_4 = \begin{bmatrix} 0.2 & 0 \\ 0 & 0.2 \end{bmatrix},$$

and the delay function $\tau(t) = 0.06 + 1.01 sin^2(t)$, it is easy to see that $\dot{\tau}(t) = 1.01 sin(2t)$ which can be larger than one. Using Theorem 1 and LMI control toolbox in Matlab, we can find that the neural networks (14) is asymptotically stable and the solution of the LMI (5) is given as follows:

$$P = \begin{bmatrix} 0.4597 & -0.1851 \\ -0.1851 & 0.3430 \end{bmatrix}, \ Q = \begin{bmatrix} 0.0504 & 0.0299 \\ 0.0299 & 0.0774 \end{bmatrix}, \ R_1 = \begin{bmatrix} 0.7225 & -0.4123 \\ -0.4123 & 0.4324 \end{bmatrix},$$

$$R_2 = \begin{bmatrix} 0.0366 & 0.02 \\ 0.02 & 0.0132 \end{bmatrix}, \ M = \begin{bmatrix} 52.0692 & -36.3296 \\ -36.3296 & 36.3820 \end{bmatrix}, \ \alpha_1 = 0.3235, \ \alpha_2 = 0.207.$$

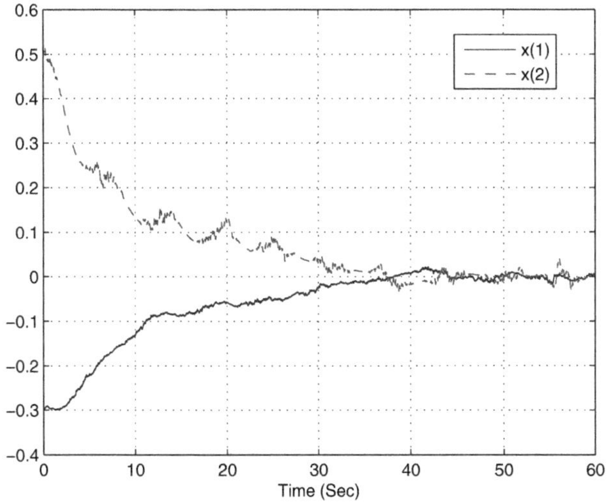


**Fig. 1.** The dynamical behavior of the Hopfield neural networks (14)

# 5 Conclusions

In this paper, the robust mean square exponential stability of uncertain stochastic Hopfield neural networks with interval time-varying delays is investigate. The delay factor was assumed to be time-varying and belongs to a given interval, which means that the derivative of the delay function can exceed one. The uncertainties considered in this paper are norm-bounded and possibly time-varying.

Based on the Lyapunov-Krasovskii functional approach and stochastic analysis approach, a new delay-dependent stability criteria for the exponential stability of the uncertain stochastic Hopfield neural networks with interval time-varying delays is derived in terms of linear matrix inequalities(LMIs). The efficiency of our method was demonstrated by the numerical example.

**Acknowledgements.** The work of Yuanqing Xia was supported by the National Natural Science Foundation of China under Grant 60504020 and Excellent young scholars Research Fund of Beijing Institute of Technology 2006y0103.

## References

1. Hopfield, J. J.: Neural Networks And Physical Systems with Emergent Collect Computational Abilities. Proc. Nat. Acad. Sci. USA, 79(2) (1982) 2554-1558
2. Wang, Z., Shu, H., Fang, J. and Liu, X.: Robust Stability For Stochastic Hopfield Neural Networks With Time Delays. Nonlinear Analysis: Real World Applications, 7(5) (2006) 1119-1128
3. Liao, X., Wong, K. and Li, C.: Global Exponential Stability For A Class Of Generalized Neural Networks With Distributed Delays. Nonlinear Analysis: Real World Applications, 5(3) (2004) 527-547
4. Song, Q. and Wang, Z.: An Analysis On Existence And Global Exponential Stability Of Periodic Solutions For BAM Neural Networks With Time-Varying Delays. Nonlinear Analysis: Real World Applications, in press, (2006)
5. V, Singh.: On Global Robust Stability Of Interval Hopfield Neural Networks With Delay. Chaos, Solitons and Fractals, 33(4) (2007) 1183-1188
6. Qiu, J., Yang, H., Zhang, J. and Gao, Z.: New Robust Stability Criteria For Uncertain Neural Networks With Interval Time-Varying Delays. Chaos, Solitons and Fractals, in press, (2007)
7. Qiu, J. and Zhang J.: New Robust Stability Criterion For Uncertain Fuzzy Systems With Fast Time-Varying Delays, Lecture Notes in Computer Sciences. 223(4) (2006) 41-44
8. Qiu, J., Zhang, J. and Shi, P.: Robust Stability Of Uncertain Linear Systems With Time-Varying Delay And Nonlinear Perturbations. Proceedings of The Institution of Mechanical Engineers Part I: Journal of Systems and Control Engineering, 220(5) (2006) 411-416
9. Lou, X. and Cui, B.: Delay-Dependent Stochastic Stability Of Delayed Hopfield Neural Networks With Markovian Jump Parameters. Journal of Mathematical Analysis and Applications, 328(1) (2007) 316-326
10. Wang Z., Liu Y., Yu L. and Liu X.: Exponential Stability Of Delayed Recurrent Neural Networks With Markovian Jumping Parameters. Physics Letters A, 356(4-5) (2006) 340-352
11. Zhang, J., Shi P. and Qiu, J.: Robust Stability Criteria For Uncertain Neutral System With Time Delay and Nonlinear Uncertainties. Chaos, Solitons and Fractals, in press, (2006)
12. Wang Z., Liu Y., Fraser K. and Liu X.: Stochastic Stability Of Uncertain Hopfield Neural Networks With Discrete And Distributed Delays. Physics Letters A, 354(4-5) (2006) 288-297

13. Ou, O.: Global Robust Exponential Stability Of Delayed Neural Networks: An LMI Approach. Chaos, Solitons and Fractals, 32(5) (2007) 1742-1748
14. Mohamad, S.: Exponential Stability In Hopfield-Type Neural Networks With Impulses. Chaos, Solitons and Fractals, 32(2) (2007) 456-467
15. Jiang, X. and Han, Q.-L.: On $H_\infty$ Control For Linear Systems With Interval Time-Varying Delay. Automatica, 41(12) (2005) 2099-2106
16. Barmish, B.R.: Necessary And Sufficient Conditions For Quadratic Stability Of An Uncertain System. Journal of Optimal Theory Apply, 46(12) (2004) 2147-2152

# New Stochastic Stability Criteria for Uncertain Neural Networks with Discrete and Distributed Delays

Jiqing Qiu, Zhifeng Gao, and Jinhui Zhang

College of Sciences, Hebei University of Science and Technology, Shijiazhuang, 050018, China
qiujiqing@263.net, gaozhifeng80@126.com

**Abstract.** This paper is concerned with robust asymptotic stability for uncertain stochastic neural networks with discrete and distributed delays. The parameter uncertainties are assumed to be time-varying and norm-bounded. We removed the traditional monotonicity and smoothness assumptions on the activation function, by utilizing a Lyapunov-Krasovskii functional and conducting stochastic analysis, a new stability criteria is provided, which guarantees uncertain stochastic neural networks is robust asymptotical stable and depends on the size of the distributed delays, The criteria can be effectively solved by some standard numerical packages. A numerical example is presented to illustrate the effectiveness of the proposed stability criteria.

**Keywords:** Robust asymptotic stability, Stochastic neural networks, Norm-bounded uncertainties.

## 1 Introduction

In the past two decades, neural networks have received considerable research attentions, and found successful applications in all kinds of areas such as pattern recognition, associate memory and combinatorial optimization. The dynamical behaviors of various neural networks, such as the stability, the attractivity, the oscillation, have been hot research topics that have drawn much attention from mathematicians, physicists and computer scientists, a large amount of results have been available in the recent literatures.

Axonal signal transmission delays often occur in various neural networks, and may cause undesirable dynamic network behaviors such as oscillation and instability. Therefore, there has been a growing research interest on the stability analysis problems for delayed neural networks, and a large amount of literature has been available. sufficient conditions, either delay-dependent or delay-independent, have been proposed to guarantee the asymptotic or exponential stability for neural networks, see [1-6] for some recent results.

Generally speaking, there are two kinds of disturbances to be considered when one models the neural networks. They are parameter uncertainties and

D.-S. Huang, L. Heutte, and M. Loog (Eds.): ICIC 2007, LNAI 4682, pp. 120–129, 2007.

stochastic perturbations, which are unavoidable in practice. For the parameter uncertainties, there have been a great deal of robust stability criteria proposed by some researchers, see [7-9] for some recent results. For the stability analysis of stochastic neural networks, some results related to this problem have been published, see [10-14]. As far as we known, in most published papers, the stochastic analysis problems and the robust stability analysis problems have been treated separately. Up to now, the robust stability analysis problem for stochastic neural networks with parameter uncertainties has not been fully studied. Therefore, it is important and challenging to get some useful stability criteria for uncertain stochastic neural networks.

In this paper, we consider the problem of robust asymptotic stability for uncertain stochastic neural networks with discrete and distributed delays. We have removed the traditional monotonicity and smoothness assumptions on the activation function. By utilizing Lyapunov-Krasovskii functional and conducting stochastic analysis, a new stability criteria is presented in terms of linear matrix inequalities to guarantee uncertain stochastic neural networks to be robust asymptotical stable. A numerical example is presented to illustrate the feasibility of the proposed stability criteria.

*Notation:* The symmetric terms in a symmetric matrix are denoted by $*$.

## 2 Problem Formulation

In this section, we consider the following uncertain stochastic neural networks with discrete and distributed delays:

$$\begin{aligned} dx(t) = [&-(A+\Delta A(t))x(t) + (W_0+\Delta W_0(t))F(x(t)) \\ &+ (W_1+\Delta W_1(t))G(x(t-\tau)) \\ &+ (W_2+\Delta W_2(t))\int_{t-h}^{t} H(x(\alpha))d\alpha]dt + \sigma(x(t),x(t-\tau),t)d\omega(t) \end{aligned} \quad (1)$$

where $x(t) = [x_1(t),\ x_2(t),\ \cdots,\ x_n(t)]^T \in \mathbf{R}^n$ is the neural state vector, $A = diag\{a_1, a_2, \cdots, a_n\}$ is a diagonal matrix, where $a_i > 0$, $i = 1, \cdots, n$. The matrices $W_0 \in \mathbf{R}^{n\times n}$, $W_1 \in \mathbf{R}^{n\times n}$ and $W_2 \in \mathbf{R}^{n\times n}$ are the connection weight matrix, the discretely delayed connection weight matrix, and the distributively delayed connection weight matrix, respectively, $\Delta A(t)$, $\Delta W_0(t)$, $\Delta W_1(t)$, $\Delta W_2(t)$ are the time-varying parameters uncertainties. $\tau > 0$ is the discrete delay, and $h > 0$ is the distributed delay. $F(x(t) = [f_1(x_1(t)),\ \cdots,\ f_n(x_n(t))]^T \in \mathbf{R}^n$, $G(x(t-\tau)) = [g_1(x_1(t-\tau)),\ \cdots,\ g_n(x_n(t-\tau))]^T \in \mathbf{R}^n$, $H(x) = [h_1(x_1(\alpha)),\ \cdots,\ h_n(x_n(\alpha))]^T \in \mathbf{R}^n$ are the neuron activation function. $\omega(t) = [\omega_1(t), \omega_2(t), \cdots, \omega_m(t)]^T \in \mathbf{R}^m$ is a $m$-dimensional Brownian motion defined on a complete probability space $(\Omega, \mathcal{F}, P)$. Assume that $\sigma : \mathbf{R}^+ \times \mathbf{R}^n \times \mathbf{R}^n$, is local Lipschitz continuous satisfies the linear growth condition. For convenience, we denote that $A(t) = A + \Delta A(t)$, $W_0(t) = W_0 + \Delta W_0(t)$, $W_1(t) = W_1 + \Delta W_1(t)$, $W_2(t) = W_2 + \Delta W_2(t)$.

*Remark 1.* The motivation we consider system (1) containing uncertainties $\Delta A(t)$, $\Delta W_0(t)$, $\Delta W_1(t)$ and $\Delta W_2(t)$ stems from the fact that, in practice, it is almost impossible to get an exact mathematical model of a dynamic system owing to the complexity of the systems, environmental noises, etc. Indeed, it is reason able and practical that the model of the controlled system contain some type of uncertainties.

In order to obtain our main result, the assumptions are always made.

**Assumption 1.** *For $i \in \{1, 2, \cdots, n\}$, the activation function $F(x)$, $G(x)$, $H(x)$ in (1) satisfy the following condition:*

$$l_i^- \leq \frac{f_i(s_1) - f_i(s_2)}{s_1 - s_2} \leq l_i^+, \quad m_i^- \leq \frac{g_i(s_1) - g_i(s_2)}{s_1 - s_2} \leq m_i^+,$$
$$n_i^- \leq \frac{h_i(s_1) - h_i(s_2)}{s_1 - s_2} \leq n_i^+. \tag{2}$$

*where $l_i^-$, $l_i^+$, $m_i^-$, $m_i^+$, $n_i^-$, $n_i^+$, are some constant.*

**Assumption 2.** *The admissible parameter uncertainties are assumed to be the following form:*

$$[\Delta A(t) \;\; \Delta W_0(t) \;\; \Delta W_1(t) \;\; \Delta W_2(t)] = DF(t)[E_1 \;\; E_2 \;\; E_3 \;\; E_4] \tag{3}$$

*where $D$, $E_i$ ($i$=1, $\cdots$, 4), are known real constant matrices with appropriate dimensions, and $F(t)$ is the time-varying uncertain matrix which satisfies that*

$$F^T(t)F(t) \leq I. \tag{4}$$

Let $x(t, \xi)$denote the state trajectory of the neural network (1) from the initial data $x(\theta) = \xi(\theta)$ on $-\tau \leq \theta \leq 0$ in $\xi \in L^2_{\mathcal{F}_0}([-\tau, 0]; R^n)$, It can be easily seen that the system (1) admits a trivial solution $x(t; 0) \equiv 0$ corresponding to the initial data $\xi = 0$, see [2,10].

Before ending this section, we recall the following definition and lemmas which will be used in the next section.

**Definition 1.** *For the neural network (1) and every $\xi \in L^2_{\mathcal{F}_0}([-\tau, 0]; R^n)$, the trivial solution (equilibrium point) is robust asymptotical stable in the mean square if, for all admissible uncertainties satisfying (3), the following holds:*

$$\lim_{t\to\infty} \mathbb{E}|x(t;\xi)|^2 = 0.$$

**Lemma 1.** *[9] For given matrices $D$, $E$ and $F$ with $F^T F \leq I$ and scalar $\varepsilon > 0$, the following inequality holds:*

$$DFE + E^T F^T D^T \leq \varepsilon DD^T + \varepsilon^{-1} E^T E$$

**Lemma 2.** *[15] For any constant matrix $M \in \mathbf{R}^{n\times n}$, $M = M^T > 0$, positive scalar $\sigma > 0$, vector function $\omega : [0, \sigma] \longrightarrow \mathbf{R}^n$ such that the integrations are well defined, the following inequality holds:*

$$\left[\int_0^\sigma \omega(s)ds\right]^T M \left[\int_0^\sigma \omega(s)ds\right] \leq \sigma \int_0^\sigma \omega^T(s)M\omega(s)ds$$

For presentation convenience, in the following, we denote that

$$L_1 = diag(l_1^+ l_1^-, \cdots, l_n^+ l_n^-), \qquad L_2 = diag(\frac{l_1^+ + l_1^-}{2}, \cdots, \frac{l_n^+ + l_n^-}{2}), \tag{5}$$

$$M_1 = diag(m_1^+ m_1^-, \cdots, m_n^+ m_n^-), \quad M_2 = diag(\frac{m_1^+ + m_1^-}{2}, \cdots, \frac{m_n^+ + m_n^-}{2}), \tag{6}$$

$$N_1 = diag(n_1^+ n_1^-, \cdots, n_n^+ n_n^-), \qquad N_2 = diag(\frac{n_1^+ + n_1^-}{2}, \cdots, \frac{n_n^+ + n_n^-}{2}). \tag{7}$$

## 3 Main Results

In this section, we will perform the robust asymptotic stability analysis for uncertain stochastic neural networks (1). Based on the Lyapunov-Krasovskii stability theorem and stochastic analysis approach, we have the following main theorem which can be expressed as the feasibility of a linear matrix inequality.

**Theorem 1.** *Assume that there exist matrix $P_1 > 0$, $C_i \geq 0 (i = 1, \cdots 4)$ such that $trace[\sigma^T P_1 \sigma] \leq x^T(t) C_1 x(t) + x^T(t-\tau) C_2 x(t-\tau) + F^T(x(t)) C_3 F(x(t)) + G^T(x(t-\tau)) C_4 G(x(t-\tau))$, system (1) is robust asymptotical stable, if there exist symmetric positive definite matrices $P_2$, $P_3$, $P_4$, diag real matrices $K_1 = diag\{\mu_1, \cdots, \mu_n\}$, $K_2 = diag\{\lambda_1, \cdots, \lambda_n\}$, $K_3 = diag\{\beta_1, \cdots, \beta_n\}$, and positive scalars $\varepsilon_1 > 0$, such that the following LMI holds:*

$$\Xi = \begin{bmatrix} \Xi_{11} & 0 & \Xi_{13} & K_2 M_2 & \Xi_{15} & K_3 N_2 & \Xi_{17} & P_1 D \\ * & -P_2 + C_2 & 0 & 0 & 0 & 0 & 0 & 0 \\ * & * & \Xi_{33} & 0 & \varepsilon_1 E_2^T E_3 & 0 & \varepsilon_1 h E_2^T E_4 & 0 \\ * & * & * & P_3 - K_2 & 0 & 0 & 0 & 0 \\ * & * & * & * & \Xi_{55} & 0 & \varepsilon_1 h E_3^T E_4 & 0 \\ * & * & * & * & * & hP_4 - K_3 & 0 & 0 \\ * & * & * & * & * & * & \Xi_{77} & 0 \\ * & * & * & * & * & * & * & -\varepsilon_1 I \end{bmatrix} < 0 \tag{8}$$

*where* $\Xi_{11} = -P_1 A - A^T P_1 + P_2 + C_1 - K_1 L_1 - K_2 M_1 - K_3 N_1 + \varepsilon_1 E_1^T E_1$, $\Xi_{13} = P_1 W_0 + K_1 L_2 - \varepsilon_1 E_1^T E_2$, $\Xi_{15} = P_1 W_1 - \varepsilon_1 E_1^T E_3$, $\Xi_{17} = h P_1 W_2 - \varepsilon_1 h E_1^T E_4$, $\Xi_{33} = -K_1 + C_1 + \varepsilon_1 E_2^T E_2$, $\Xi_{55} = -P_3 + C_4 + \varepsilon_1 E_3^T E_3$, $\Xi_{77} = -hP_4 + \varepsilon_1 h^2 E_4^T E_4$.

*Proof.* Using famous Schur complement, $\Xi < 0$ (8) implies that

$$\tilde{\Xi} = \begin{bmatrix} \Xi_1 & 0 & P_1 W_0 + K_1 L_2 & K_2 M_2 & P_1 W_1 & K_3 N_2 & h P_1 W_2 \\ * & -P_2 + C_2 & 0 & 0 & 0 & 0 & 0 \\ * & * & -K_1 + C_3 & 0 & 0 & 0 & 0 \\ * & * & * & -K_2 + P_3 & 0 & 0 & 0 \\ * & * & * & * & -P_3 + C_4 & 0 & 0 \\ * & * & * & * & * & -K_3 + hP_4 & 0 \\ * & * & * & * & * & * & -hP_4 \end{bmatrix}$$

$$+ \varepsilon_1^{-1} \eta_1 \eta_1^T + \varepsilon_1 \eta_2^T \eta_2 < 0. \tag{9}$$

where

$$\Xi_1 = -P_1A - A^TP_1 + P_2 - K_1L_1 - K_2M_1 - K_3N_1$$
$$\eta_1 = \begin{bmatrix} D^TP_1 \; 0\;0\;0\;0\;0\;0 \end{bmatrix}^T, \quad \eta_2 = \begin{bmatrix} -E_1 \; 0 \; E_2 \; 0 \; E_3 \; 0 \; hE_4 \end{bmatrix}.$$

Then, noting that (3) and (4) and using Lemma 1, we have

$$\begin{bmatrix} -P_1\Delta A(t) - \Delta A^T(t)P_1 & 0 & P_1\Delta W_0(t) & 0 & P_1\Delta W_1(t) & 0 & hP_1\Delta W_2(t) \\ * & 0 & 0 & 0 & 0 & 0 & 0 \\ * & * & 0 & 0 & 0 & 0 & 0 \\ * & * & * & 0 & 0 & 0 & 0 \\ * & * & * & * & 0 & 0 & 0 \\ * & * & * & * & * & 0 & 0 \\ * & * & * & * & * & * & 0 \end{bmatrix}$$

$$= \eta_2^TF^T(t)\eta_1^T + \eta_1F(t)\eta_2 \le \varepsilon_1^{-1}\eta_1\eta_1^T + \varepsilon_1\eta_2^T\eta_2. \tag{10}$$

From (9) and (10), we have the following inequality,

$$\begin{bmatrix} [1.1] & 0 & P_1W_0(t) + K_1L_2 & K_2M_2 & P_1W_1(t) & K_3N_2 & hP_1W_2(t) \\ * & -P_2 + C_2 & 0 & 0 & 0 & 0 & 0 \\ * & * & -K_1 + C_3 & 0 & 0 & 0 & 0 \\ * & * & * & -K_2 + P_3 & 0 & 0 & 0 \\ * & * & * & * & -P_3 + C_4 & 0 & 0 \\ * & * & * & * & * & hP_4 - K_3 & 0 \\ * & * & * & * & * & * & -hP_4 \end{bmatrix} < 0 \tag{11}$$

where $[1.1] = -P_1A(t) - A^T(t)P_1 + P_2 + C_1 - K_1L_1 - K_2M_1 - K_3N_1$.

Constructing positive definite Lyapunov-Krasovskii functional $V(x(t), t) \in \mathcal{C}^{2,1}(\mathbf{R}^+ \times \mathbf{R}^n; \mathbf{R}^+)$ as follows:

$$V(x(t),t) = x^T(t)P_1x(t) + \int_{t-\tau}^{t} x^T(\alpha)P_2x(\alpha)d\alpha + \int_{t-\tau}^{t} G^T(x(\alpha))P_3G(x(\alpha))d\alpha$$
$$+ \int_{-h}^{0}\int_{t+s}^{t} H^T(x(\alpha))P_4H(x(\alpha))d\alpha ds \tag{12}$$

where $P_2 > 0$, $P_3 > 0$, and $P_4 > 0$ are the solutions of LMI (8).

By Itô's differential formula, utilizing Lemma 2, the stochastic derivative of $V(x(t), t)$ along the trajectory of system (1) is

$$dV(x(t),t) \le \{2x^T(t)P_1[-A(t)x(t) + W_0(t)F(x(t)) + W_1(t)G(x(t-\tau))$$
$$+W_2(t)\int_{t-h}^{t} H(x(\alpha))d\alpha] + x^T(t)P_2x(t) - x^T(t-\tau)P_2x(t-\tau)$$
$$+G^T(x(t))P_3G(x(t)) - G^T(x(t-\tau))P_3G(x(t-\tau)) + H^T(x(t))[hP_4]H(x(t))$$
$$-\left(\frac{1}{h}\int_{t-h}^{t} H(x(\alpha)d\alpha\right)^T [hP_4]\left(\frac{1}{h}\int_{t-h}^{t} H(x(\alpha))d\alpha\right) + x^T(t-\tau)C_2x(t-\tau)$$

$$
\begin{aligned}
&+x^T(t)C_1x(t) + F^T(x(t))C_3F(x(t)) + G^T(x(t-\tau))C_4G^T(x(t-\tau))\}dt \\
&+\{2x^T(t)P_1\sigma(x(t), x(t-\tau), t)\}d\omega(t). \\
&= \xi^T(t)\Theta\xi(t)dt + \{2x^T(t)P_1\sigma(x(t), x(t-\tau), t)\}d\omega(t)
\end{aligned}
$$

where

$$
\xi^T(t) = [x^T(t) \;\; x^T(t-\tau) \;\; F^T(x(t)) \;\; G^T(x(t)) \;\; G^T(x(t-\tau)) \;\; H^T(x(t)) \;\; \left(\frac{1}{h}\int_{t-h}^{t} H(x(\alpha))d\alpha\right)^T]
$$

$$
\Theta = \begin{bmatrix}
\Theta_{11} & 0 & P_1W_0(t) & 0 & P_1W_1(t) & 0 & hP_1W_2(t) \\
* & -P_2+C_2 & 0 & 0 & 0 & 0 & 0 \\
* & * & C_3 & 0 & 0 & 0 & 0 \\
* & * & * & P_3 & 0 & 0 & 0 \\
* & * & * & * & -P_3+C_4 & 0 & 0 \\
* & * & * & * & * & hP_4 & 0 \\
* & * & * & * & * & * & -hP_4
\end{bmatrix}
$$

$$
\Theta_{11} = -P_1A(t) - A^T(t)P_1 + P_2 + C_1.
$$

From (2), we have

$$
\begin{aligned}
&\left(\frac{f_i(x_i(t))}{x_i(t)} - l_i^+\right)\left(\frac{f_i(x_i(t))}{x_i(t)} - l_i^-\right) \le 0, \quad i = 1, \cdots n \\
&\left(\frac{g_i(x_i(t))}{x_i(t)} - m_i^+\right)\left(\frac{g_i(x_i(t))}{x_i(t)} - m_i^-\right) \le 0, \quad i = 1, \cdots n \\
&\left(\frac{h_i(x_i(t))}{x_i(t)} - n_i^+\right)\left(\frac{h_i(x_i(t))}{x_i(t)} - n_i^-\right) \le 0, \quad i = 1, \cdots n
\end{aligned}
$$

From the above three inequalities, we can get the following inequalities:

$$
\begin{aligned}
&(f_i(x_i(t)) - l_i^+x_i(t))(f_i(x_i(t)) - l_i^-x_i(t)) \le 0, \quad i = 1, \cdots n \\
&(g_i(x_i(t)) - m_i^+x_i(t))(g_i(x_i(t)) - m_i^-x_i(t)) \le 0, \quad i = 1, \cdots n \\
&(h_i(x_i(t)) - n_i^+x_i(t))(h_i(x_i(t)) - n_i^-x_i(t)) \le 0, \quad i = 1, \cdots n
\end{aligned}
$$

which are equivalent to the following:

$$
\begin{bmatrix} x(t) \\ F(x(t)) \end{bmatrix}^T \begin{bmatrix} l_i^+l_i^-e_ie_i^T & -\frac{l_i^++l_i^-}{2}e_ie_i^T \\ -\frac{l_i^++l_i^-}{2}e_ie_i^T & e_ie_i^T \end{bmatrix} \begin{bmatrix} x(t) \\ F(x(t)) \end{bmatrix} \le 0, \quad i = 1, \cdots n
$$

$$
\begin{bmatrix} x(t) \\ G(x(t)) \end{bmatrix}^T \begin{bmatrix} m_i^+m_i^-e_ie_i^T & -\frac{m_i^++m_i^-}{2}e_ie_i^T \\ -\frac{m_i^++m_i^-}{2}e_ie_i^T & e_ie_i^T \end{bmatrix} \begin{bmatrix} x(t) \\ G(x(t)) \end{bmatrix} \le 0, \quad i = 1, \cdots n
$$

$$
\begin{bmatrix} x(t) \\ H(x(t)) \end{bmatrix}^T \begin{bmatrix} n_i^+n_i^-e_ie_i^T & -\frac{n_i^++n_i^-}{2}e_ie_i^T \\ -\frac{n_i^++n_i^-}{2}e_ie_i^T & e_ie_i^T \end{bmatrix} \begin{bmatrix} x(t) \\ H(x(t)) \end{bmatrix} \le 0, \quad i = 1, \cdots n
$$

where $e_i$ denotes the unit column vector having '1' element on its $i$th row and zeros elsewhere. Consequently, we have the following:

$$\begin{aligned}
&\xi^T(t)\Theta\xi(t) - \sum_{i=1}^{n}\mu_i \begin{bmatrix} x(t) \\ F(x(t)) \end{bmatrix}^T \begin{bmatrix} l_i^+ l_i^- e_i e_i^T & -\frac{l_i^+ + l_i^-}{2} e_i e_i^T \\ -\frac{l_i^+ + l_i^-}{2} e_i e_i^T & e_i e_i^T \end{bmatrix} \begin{bmatrix} x(t) \\ F(x(t)) \end{bmatrix} \\
&- \sum_{i=1}^{n}\lambda_i \begin{bmatrix} x(t) \\ G(x(t)) \end{bmatrix}^T \begin{bmatrix} m_i^+ m_i^- e_i e_i^T & -\frac{m_i^+ + m_i^-}{2} e_i e_i^T \\ -\frac{m_i^+ + m_i^-}{2} e_i e_i^T & e_i e_i^T \end{bmatrix} \begin{bmatrix} x(t) \\ G(x(t)) \end{bmatrix} \\
&- \sum_{i=1}^{n}\beta_i \begin{bmatrix} x(t) \\ H(x(t)) \end{bmatrix}^T \begin{bmatrix} n_i^+ n_i^- e_i e_i^T & -\frac{n_i^+ + n_i^-}{2} e_i e_i^T \\ -\frac{n_i^+ + n_i^-}{2} e_i e_i^T & e_i e_i^T \end{bmatrix} \begin{bmatrix} x(t) \\ H(x(t)) \end{bmatrix} \\
&= \xi_1^T(t)\Theta\xi_1(t) + \begin{bmatrix} x(t) \\ F(x(t)) \end{bmatrix}^T \begin{bmatrix} -K_1L_1 & K_1L_2 \\ K_1L_2 & -K_1 \end{bmatrix} \begin{bmatrix} x(t) \\ F(x(t)) \end{bmatrix} + \begin{bmatrix} x(t) \\ G(x(t)) \end{bmatrix}^T \\
&\begin{bmatrix} -K_2M_1 & K_2M_2 \\ \Gamma M_2 & -K_2 \end{bmatrix} \begin{bmatrix} x(t) \\ G(x(t)) \end{bmatrix} + \begin{bmatrix} x(t) \\ H(x(t)) \end{bmatrix}^T \begin{bmatrix} -K_3N_1 & K_3N_2 \\ \Delta N_2 & -K_3 \end{bmatrix} \begin{bmatrix} x(t) \\ H(x(t)) \end{bmatrix} \\
&= \xi^T(t)\Psi\xi(t)
\end{aligned}$$

where

$$\Psi = \begin{bmatrix}
\Psi_{11} & 0 & P_1W_0(t) + K_1L_2 & K_2M_2 & P_1W_1(t) & K_3N_2 & hP_1W_2(t) \\
* & -P_2 + C_2 & 0 & 0 & 0 & 0 & 0 \\
* & * & -K_1 + C_3 & 0 & 0 & 0 & 0 \\
* & * & * & -K_2 + P_3 & 0 & 0 & 0 \\
* & * & * & * & -P_3 + C_4 & 0 & 0 \\
* & * & * & * & * & -K_3 + hP_4 & 0 \\
* & * & * & * & * & * & -hP_4
\end{bmatrix}$$

with $\Psi_{11} = -P_1A(t) - A^T(t)P_1 + P_2 + C_1 - K_1L_1 - K_2M_1 - K_3N_1$

From (11), it is obvious that for $\Psi < 0$, there exists a scalar $\gamma > 0$ such that

$$\Psi + diag\{\gamma I, 0, 0, 0, 0, 0, 0\} < 0$$

which indicates that

$$dV(x(t), t) \leq -\gamma\|x(t)\|^2 dt + \{2x^T(t)P_1\sigma(x(t), x(t-\tau), t)\}d\omega(t) \tag{13}$$

Taking the mathematical expectation of both sides of (13), we have

$$\frac{d\mathbb{E}V(x(t), t)}{dt} \leq -\gamma\mathbb{E}\|x(t)\|^2 \tag{14}$$

which indicates from the Lyapunov stability theory that the dynamics for uncertain stochastic neural networks (1) is robust asymptotic stable. This completes the proof.

*Remark 2.* It should be noted that the condition (8) is given as linear matrix inequality, therefore, by using the Matlab LMI Toolbox, it is straightforward to check the feasibility of (8) without tuning any parameters.

Based on the proof of Theorem 1, if there are no parameter uncertainties in $A(t)$, $W_0(t)$, $W_1(t)$ and $W_2(t)$, the neural networks (1) is simplified to the following form:

$$dx(t) = [-Ax(t) + W_0F(x(t)) + W_1G(x(t-\tau)) + W_2\int_{t-h}^{t} H(x(\alpha))d\alpha]dt \\ + \sigma(x(t), x(t-\tau), t)d\omega(t) \tag{15}$$

then we have the following corollary.

**Corollary 1.** *Assume there exists matrix $P_1 > 0$, $C_i \geq 0 (i = 1, \cdots 4)$, such that $trace[\sigma^T P_1 \sigma] \leq x^T(t)C_1x(t) + x^T(t-\tau)C_2x(t-\tau) + F^T(x(t))C_3F(x(t)) + G^T(x(t-\tau))C_4G(x(t-\tau))$, system (15) is global asymptotical stable, if there exist symmetric positive definite matricces $P_2$, $P_3$, $P_4$, diag real matrices $K_1 = diag\{\mu_1, \cdots, \mu_n\}$, $K_2 = diag\{\lambda_1, \cdots, \lambda_n\}$, $K_3 = diag\{\beta_1, \cdots, \beta_n\}$, and positive scalar $\varepsilon_1 > 0$, such that the following LMI holds:*

$$\begin{bmatrix} [1.1] & 0 & P_1W_0 + K_1L_2 & K_2M_2 & P_1W_1 & K_3N_2 & hP_1W_2 \\ * & -P_2 + C_2 & 0 & 0 & 0 & 0 & 0 \\ * & * & -K_1 + C_3 & 0 & 0 & 0 & 0 \\ * & * & * & -K_2 + P_3 & 0 & 0 & 0 \\ * & * & * & * & -P_3 + C_4 & 0 & 0 \\ * & * & * & * & * & hP_4 - K_3 & 0 \\ * & * & * & * & * & * & -hP_4 \end{bmatrix} < 0$$

*where* $[1.1] = -P_1A - A^TP_1 - K_1L_1 - K_2M_1 - K_3N_1 + P_2 + C_1$.

## 4 Numerical Examples

In this section, we provide a numerical example to demonstrate the effectiveness of the proposed stability criteria.

*Example 1.* In this example, we consider uncertain stochastic neural networks with discrete and disturbuted delays (1), and the parameter matrices as follow:

$$A = \begin{bmatrix} 1.6 & 0 \\ 0 & 1.7 \end{bmatrix}, \quad W_0 = \begin{bmatrix} 1.2 & 1.4 \\ 1.5 & 1.4 \end{bmatrix}, \quad W_1 = \begin{bmatrix} 1.8 & 2.4 \\ 1.6 & 2.1 \end{bmatrix}, \quad W_2 = \begin{bmatrix} 1.6 & 2.3 \\ 1.8 & 2.4 \end{bmatrix},$$

$$D = \begin{bmatrix} 0.3 & 0 \\ 0 & 0.3 \end{bmatrix}, \quad L_2 = M_2 = N_2 = \begin{bmatrix} 0.3 & 0 \\ 0 & 0.3 \end{bmatrix}, \quad L_1 = M_1 = N_1 = 0,$$

$$E_1 = \begin{bmatrix} -0.1 & 0 \\ 0 & 0.1 \end{bmatrix}, \quad E_2 = \begin{bmatrix} 0.2 & 0 \\ 0 & -0.2 \end{bmatrix}, \quad E_3 = \begin{bmatrix} -0.3 & 0 \\ 0 & 0.3 \end{bmatrix}, \quad E_4 = \begin{bmatrix} 0.4 & 0 \\ 0 & -0.4 \end{bmatrix},$$

$$C_1 = \begin{bmatrix} 0.5 & 0 \\ 0 & 0.5 \end{bmatrix}, \quad C_2 = \begin{bmatrix} 3.0 & 0 \\ 0 & 3.0 \end{bmatrix}, \quad C_3 = \begin{bmatrix} 2.0 & 0 \\ 0 & 2.0 \end{bmatrix}, \quad C_4 = \begin{bmatrix} 0.3 & 0 \\ 0 & 0.3 \end{bmatrix}.$$

We choose $P_1 = 0.1I$, using Matlab LMI Control Toolbox, by Theorem 1 in this paper, we find that system (1) is robust asymptotical stable, obtain the maximum distributed time delay $h = 0.4216$, the solutions of LMI (8) as follow:

$$P_2 = \begin{bmatrix} 0.9807 & -0.0659 \\ -0.0659 & 0.9695 \end{bmatrix}, \; P_3 = \begin{bmatrix} 0.7036 & 0.0138 \\ 0.0138 & 0.7540 \end{bmatrix}, \; P_4 = \begin{bmatrix} 1.6474 & -0.0045 \\ -0.0045 & 1.6722 \end{bmatrix},$$
$$K_1 = \begin{bmatrix} 1.1129 & 0 \\ 0 & 1.0375 \end{bmatrix}, \quad K_2 = \begin{bmatrix} 0.2490 & 0 \\ 0 & 0.2197 \end{bmatrix}, \; K_3 = \begin{bmatrix} 0.2844 & 0 \\ 0 & 0.2469 \end{bmatrix},$$

$\varepsilon_1 = 0.9129.$

The dynamical behavior of stochastic neural network systems in this example is shown in Fig. 1. The simulation result implies the stochastic neural networks in this example is indeed robust asymptotical stable.

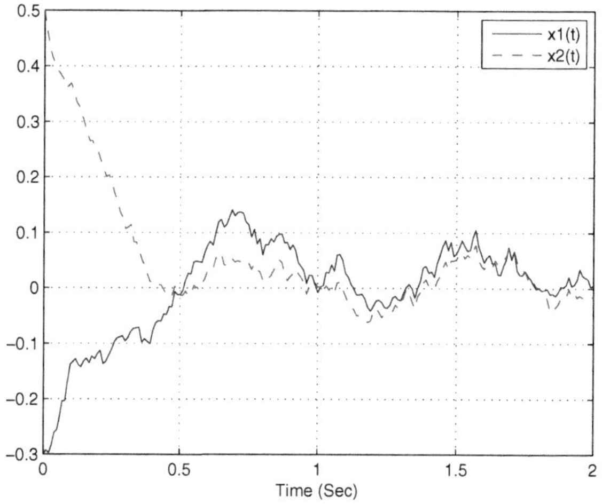


**Fig. 1.** The trajectories for the state of stochastic neural network systems

## References

1. Zhao, H.: Global Stability of Bidirectional Associative Memory Neural Networks with Distributed Delays. Physics Letters A, 297 (2002) 182-190
2. Wang, Z.D., Liu, Y.R., Liu, X.H.: On Global Asymptotic Stability of Neural Networks with Discrete and Distributed Delays. Physics Letters A, 345 (2005) 299-308
3. Liu, Y.R., Wang, Z.D., Liu, X.H.: Global Exponential Stability of Generalized Recurrent Neural Networks with Discrete and Distributed Delays. Chaos, Solitons & Fractals, 28 (2006) 793-803
4. Singh, V.: Global Robust Stability of Delayed Neural Networks: An LMI Approach. IEEE Transactions on Circuits and Systems, 52 (2005) 33-36

5. Wan, A., Qiao, H., Peng, J., Wang, M.: Delay-Independent Criteria for Exponential Stability of Generalized Cohen-Grossberg Neural Networks with Discrete Delays. Physics Letters A, 353 (2006) 151-157
6. Zhang, J.: Globally Exponential Stability of Neural Networks with Variable Delays. IEEE Transactions on Circuits and Systems, 50 (2003) 288-291
7. He, Y., Wang, Q.G., Zhang, W.: Global Robust Stability for Delayed Neural Networks with Polytopic Type Uncertainties. Chaos, Solitons & Fractals, 26(2005) 1349-1354
8. Wang, L., Gao, Y.: Global Exponential Robust Stability of Reaction-Diffusion Interval Neural Networks with Time-varying Delays. Physics Letters A, 350(2006) 342-348
9. Qiu, J.Q., Zhang, J.H., Shi, P.: Robust Stability of Uncertain Linear Systems with Time-Varying Delay and Nonlinear Perturbations. Proceedings of the Institution of Mechanical Engineers, Part I, Journal of Systems and Control Engineering, 220 (2006) 411-416
10. Wang, Z.D., Lauria, S.,Fang, J.A., Liu, X.P.: Exponential Stability of Uncertain Stochastic Neural Networks with Mixed Time-Delays. Chaos, Solitons & Fractals, 32 (2007) 62-72
11. Liu, Y.R., Wang, Z.D., Liu, X.H.: On Global Exponential Stability of Generalized Stochastic Neural Networks with Mixed Time-Delays. Neurocomputing, 70 (2006) 314-326
12. J. Hu, S. Zhong and L. Liang, Exponential Stability Analysis of Stochastic Delayed Cellular Neural Network. Chaos, Solitons & Fractals, 27 (2006) 1006-1010
13. Huang, H., Ho, D.W.C., Lam, J.: Stochastic Stability Analysis of Fuzzy Hopfield Neural Networks with Time-Varying Delays. IEEE Transactions on Circuits and Systems, 52 (2005) 251-255
14. Wan, L., Sun, J.: Mean Square Exponential Stability of Stochastic Delayed Hopfield Neural Networks. Physics Letters A, 343 (2005) 306-318
15. Gu, K.: An Integral Inequality In the Stability Problem of Time-Delay systems. Proceedings of 39th IEEE Conference on Decision and Control. Sydney Australia (2000) 2805-2810

# Novel Forecasting Method Based on Grey Theory and Neural Network

Cheng Wang and Xiaoyong Liao

College of Mathematics and Information Science,
Huanggang Normal University, Huanggang 438000, Hubei, China
wangc80@163.com

**Abstract.** In this paper, a new forecasting model named GGNNM(1,1) model is presented.First of all, a generalized GM(1,1) model based on the traditional GM (1,1) model is established, then the generalized GM(1,1) model and the theory of neural network are combined to establish the GGNNM(1,1) model.Furthermore, the algorithm for solving this new model is given. Finally, a forecasting example is given to demonstrate the feasibility and rationality of this new model.

**Keywords:** forecast, generalized GM(1,1) model, neural network, Generalized Grey Neural Network Model (GGNNM(1,1)).

## 1 Introduction

Nowadays,many scholars research on the grey models and neural network models, and they conclude that these two kinds of models can be combined to present more advanced and more applied forecasting models, such as the CGNN model in [1], the GNNM (1,1) model and GNNM (2,1) model in [2], the PGNN model, SGNN model and IGNN model in [3] and so on. According to the application effects of these models, we can get two conclusions. The first one is that the computing of grey neural network (GNN) model is simpler than neural network model's, and the forecasting precision of GNN model is higher than that of neural network model in the condition of little data. The second one is that the GNN model has the advantages of high forecasting precision and error controllability compared with grey forecasting models.

However, the applicable range of GNN model is limited in practical application. In essence, this disadvantage is due to the limit of applicable range of the traditional AGO GM(1,1) model. Based on the traditional GM (1,1) model and all kinds of improved GM (1,1) models[4-9], this paper presents a new generalized GM(1,1) model firstly, then combines the generalized GM(1,1) model and the theory of neural network to establish the GGNNM(1,1) model, and gives the algorithm for solving this new model. Finally, in order to demonstrate the feasibility and superiority of this new model, a forecasting example is given, and the ideal forecasting results are obtained.

D.-S. Huang, L. Heutte, and M. Loog (Eds.): ICIC 2007, LNAI 4682, pp. 130–136, 2007.

# 2 Establishment of the GGNNM(1,1) Model

## 2.1 Generalized GM(1,1) Model

Firstly, a new generalized GM(1,1) model is established based on the traditional GM(1,1) model.

The usual AGO matrix is an upper triangular matrix, of which elements are all 1 in the principal diagonal and above the principal diagonal, and its elements of every row are not decreasing from left to right. Abstracting these characters, we give the definition of GAGO as follows:

**Definition 1.** Let $A$ be a n-th order upper triangular matrix, and

$$A = \begin{pmatrix} \alpha_1 & \alpha_1 & \alpha_1 & \cdots & \alpha_1 \\ 0 & \alpha_2 & \alpha_2 & \cdots & \alpha_2 \\ 0 & 0 & \alpha_3 & \cdots & \alpha_3 \\ \cdots & \cdots & \cdots & \cdots & \cdots \\ 0 & 0 & 0 & \cdots & \alpha_n \end{pmatrix} \tag{1}$$

where $\alpha_i > 0, \ i = 1, 2, \cdots, n$. Then $A$ is called a GAGO matrix ( or called GAGO).

**Definition 2.** Let $x^{(0)} = (x^{(0)}(1), x^{(0)}(2), \cdots, x^{(0)}(n))$ be raw series, $A$ be a GAGO matrix. By $x^{(1)} = x^{(0)}A$, we have a new series

$$\begin{aligned} x^{(1)} &= (x^{(1)}(1), x^{(1)}(2), \cdots, x^{(1)}(n)) \\ &= (a_1 x^{(0)}(1), \ \sum_{i=1}^{2} a_i x^{(0)}(i), \cdots, \ \sum_{i=1}^{n-1} a_i x^{(0)}(i), \sum_{i=1}^{n} a_i x^{(0)}(i)) \end{aligned} \tag{2}$$

then $x^{(1)}$ is called a GAGO series of $x^{(0)}$.

Secondly, we present a new generalized GM(1,1) model. Based on expression (1) and (2), the generalized GM(1,1) model can be expressed as

$$\tau_k x^{(0)}(k) + a z^{(1)}(k) = b \tag{3}$$

where

$$\tau_k = \begin{cases} \alpha_2 & k = 2 \\ \alpha_3 & k = 3 \\ \cdots & \\ \alpha_{n-1} & k = n-1 \\ \alpha_n & k = n \end{cases}$$

$$z^{(1)}(k) = 0.5x^{(1)}(k) + 0.5x^{(1)}(k-1), \quad k = 2, 3, \cdots, n$$

According to existing research results of GM(1,1) model, the generalized GM(1,1) model (3) at least includes the following known models, i.e. the traditional GM(1,1) model [4], the PGAGO GM(1,1) forecasting model[8], the MGAGO GM(1,1) forecasting model[9], the generalized models of PGAGO GM(1,1) model and MGAGO GM(1,1)model.

By the above analysis, we can conclude that the generalized GM(1,1) model (3) is a more universal model than the existing GM(1,1) models.

Referring to the methods of parameter identification in [8] and [9], we can get the following conclusions.

**Theorem 1.** Let $C, D, E, F$ be the intermediate parameters of the generalized GM(1,1) model (3), where

$$\begin{aligned} C &= \sum_{k=2}^{n} z^{(1)}(k), \\ D &= \sum_{k=2}^{n} \alpha_k x^{(0)}(k), \\ E &= \sum_{k=2}^{n} \alpha_k z^{(1)}(k) x^{(0)}(k), \\ F &= \sum_{k=2}^{n} (z^{(1)}(k))^2 \end{aligned}$$

then the expression of parameters $A$ and $B$ in model (3) can be expressed as

$$\begin{aligned} a &= \frac{CD-(n-1)E}{(n-1)F-C^2} \\ b &= \frac{DF-CE}{(n-1)F-C^2} \end{aligned} \tag{4}$$

**Proof.** Suppose that

$$\begin{aligned} Y &= [\alpha_2 x^{(0)}(2),\ \alpha_3 x^{(0)}(3),\ \cdots, \alpha_n x^{(0)}(n)]^T, \\ T &= [a, b]^T, \\ B &= \begin{bmatrix} -z^{(1)}(2) & -z^{(1)}(3) & \cdots & -z^{(1)}(n) \\ 1 & 1 & 1 & 1 \end{bmatrix}^T \end{aligned}$$

Substituting $k = 2, 3, \cdots, n$ into model (3), we have

$$\begin{cases} \alpha_2 x^{(0)}(2) + a z^{(1)}(2) = b \\ \alpha_3 x^{(0)}(3) + a z^{(1)}(3) = b \\ \cdots \\ \alpha_n x^{(0)}(n) + a z^{(1)}(n) = b \end{cases} \tag{5}$$

System of equation (5) can be denoted as $Y = BT$.

Replacing $\alpha_k x^{(0)}(k)$ with $-az^{(1)}(k) + b,\ k = 2, 3, \cdots, n$, so the error of series can be expressed as $\varepsilon = Y - BT$. Suppose that

$$\begin{aligned} e &= \varepsilon^T \varepsilon = (Y - BT)^T (Y - BT) \\ &= \sum_{k=2}^{n} (\alpha_k x^{(0)}(k) + a z^{(1)}(k) - b)^2 \end{aligned}$$

When the value of $e$ is taken the minimum, the parameters $a, b$ should satisfy the following condition

$$\begin{cases} \frac{\partial e}{\partial a} = 2 \sum_{k=2}^{n} (\alpha_k x^{(0)}(k) + a z^{(1)}(k) - b) \cdot z^{(1)}(k) = 0 \\ \frac{\partial e}{\partial a} = -2 \sum_{k=2}^{n} (\alpha_k x^{(0)}(k) + a z^{(1)}(k) - b) = 0 \end{cases} \tag{6}$$

Solving the system of equation (6) based on the expressions of $C, D, E, F$, we can get the result of (4).

**Theorem 2.** The white response of the generalized GM(1,1) model (3) can be expressed as

$$\hat{x}^{(1)}(k+1) = (\alpha_1 x^{(0)}(1) - \frac{b}{a})e^{-ak} + \frac{b}{a}, \quad k = 0, 1, 2, \cdots \tag{7}$$

and the forecasting formulas can be expressed as

$$\hat{x}^{(0)}(k) = \begin{cases} \frac{\hat{x}^{(1)}(k) - \hat{x}^{(1)}(k-1)}{\alpha_k}, & k = 2, 3, \cdots n \\ \frac{\hat{x}^{(1)}(k) - \hat{x}^{(1)}(k-1)}{\alpha_n}, & k = n+1, n+2, \cdots \end{cases} \tag{8}$$

**Proof.** The proof method is similar to Ref. [10].

### 2.2 Generalized Grey Neural Network Model (GGNNM(1,1))

Based on the advantages of neural network in intelligent computation, we integrate the neural network into the generalized GM(1,1) model and establish a new forecasting model, i.e. GGNNM(1,1) model. The main modeling steps are given as follows.

(1) Mapping the white response expression (7) into a BP neural network

First of all, the expression (7) is transformed as

$$\begin{aligned} \hat{x}^{(1)}(k+1) &= (\alpha_1 x^{(0)}(1) - \tfrac{b}{a})e^{-ak} + \tfrac{b}{a} \\ &= [(\alpha_1 x^{(0)}(1) - \tfrac{b}{a}) \cdot \tfrac{e^{-ak}}{1+e^{-ak}} + \tfrac{b}{a} \cdot \tfrac{1}{1+e^{-ak}}] \cdot (1 + e^{-ak}) \\ &= [(\alpha_1 x^{(0)}(1) - \tfrac{b}{a}) \cdot (1 - \tfrac{1}{1+e^{-ak}}) \\ &\quad + \tfrac{b}{a} \cdot \tfrac{1}{1+e^{-ak}}] \cdot (1 + e^{-ak}) \\ &= [(\alpha_1 x^{(0)}(1) - \tfrac{b}{a}) - \alpha_1 x^{(0)}(1) \cdot \tfrac{1}{1+e^{-ak}} \\ &\quad + 2 \cdot \tfrac{b}{a} \cdot \tfrac{1}{1+e^{-ak}}] \cdot (1 + e^{-ak}) \end{aligned} \tag{9}$$

Then the expression (9) is mapped into a BP neural network.

(2) Determining the node weight value and threshold value of BP neural network

The value assignment of node weight value are as follows:

$$\begin{aligned} W_{11} &= a, \\ W_{21} &= -\alpha_1 x^{(0)}(1), \\ W_{22} &= \tfrac{2b}{a}, \\ W_{31} &= W_{32} = 1 + e^{-ak} \end{aligned}$$

and the threshold value is taken as

$$\theta_{y_1} = (1 + e^{-ak})(\frac{b}{a} - \alpha_1 x^{(0)}(1))$$

(3) Determining the activation function of every nerve cell in BP neural network

By expression (9), the activation function of nerve cell for layer LB is taken as

$$f(x) = \frac{1}{1+e^{-x}},$$

and the activation function of nerve cell for layer LA , LC , LD are all taken as

$$f(x) = x$$

(4) Computing the output value of every node

By step 2 and step 3, we have

$$\begin{aligned}
a_1 &= k \cdot W_{11} = ak, \\
b_1 &= f(a_1) = f(ak) = \tfrac{1}{1+e^{-ak}}, \\
c_1 &= W_{21}b_1 = -\alpha_1 x^{(0)}(1) \cdot \tfrac{1}{1+e^{-ak}}, \\
c_2 &= W_{22}b_1 = \tfrac{2b}{a} \cdot \tfrac{1}{1+e^{-ak}}, \\
d_1 &= W_{31}c_1 + W_{32}c_2 - \theta_{y_1} = (1+e^{-ak}) \cdot (-\tfrac{\alpha_1 x^{(0)}(1)}{1+e^{-ak}}) \\
&\quad + (1+e^{-ak}) \cdot \tfrac{2b}{a} \cdot \tfrac{1}{1+e^{-ak}} - (1+e^{-ak})(\tfrac{b}{a} - \alpha_1 x^{(0)}(1)) \\
&= [(\alpha_1 x^{(0)}(1) - \tfrac{b}{a}) - \alpha_1 x^{(0)}(1) \cdot \tfrac{1}{1+e^{-ak}} \\
&\quad + 2 \cdot \tfrac{b}{a} \cdot \tfrac{1}{1+e^{-ak}}] \cdot (1+e^{-ak}) = \hat{x}^{(1)}(k+1), \\
d_1 &= y_1 = \hat{x}^{(1)}(k+1)
\end{aligned}$$

(5) Training the network

Using the algorithm of Back Propagation[3] to train the network. When this network is convergent, the coefficients of relevant equation are extracted in the trained BP neural network, so a whitenization differential equation is obtained. Then we can solve this equation and forecast the future.

## 3 An Application Example

Now we know the raw data series as follows:

$$x^{(0)} = (0.727, 0.761, 0.646, 0.735)$$

(1) Establishing the generalized GM(1,1) model

Obviously, $x^{(0)}(3) = 0.646$ is a jump point in $x^{(0)}$, so we can establish a PGAGO GM(1,1) model.

Let the PGAGO matrix be

$$A = \begin{pmatrix} \alpha & \alpha & \alpha & \alpha \\ 0 & \alpha & \alpha & \alpha \\ 0 & 0 & \beta & \beta \\ 0 & 0 & 0 & \alpha \end{pmatrix}$$

Referring to the modeling method in [10], we can get the values of $\alpha$ and $\beta$ as follows:

$$\alpha = 0.78830, \quad \beta = 0.9127$$

**Table 1.** The forecasting results of three forecasting models

| k | $x^{(0)}$ | GM(1,1) | GNNM(1,1) | GGNNM(1,1) |
|---|---|---|---|---|
| | | Forcasting value | Forcasting value | Forcasting value |
| 1 | 0.727 | 0.727 | 0.727 | 0.727 |
| 2 | 0.761 | 0.728 | 0.761 | 0.760 |
| 3 | 0.646 | 0.714 | 0.741 | 0.645 |
| 4 | 0.735 | 0.700 | 0.722 | 0.735 |
| Mean Error | | 6.533 | 5.51 | 0.0024 |

Substituting the values of $\alpha$ and $\beta$ into formula (4), we have

$$a = 0.01738, \quad b = 0.6151$$

Substituting the values of $a$ and $b$ into formula (7), we can get the white response of GM(1,1) as follows

$$\hat{x}^{(1)}(k+1) = -34.8171e^{-0.01738k} + 35.3901, \quad k = 0, 1, 2, \cdots \qquad (10)$$

(2) Establishing the GGNNM(1,1) model

Based on the white response expression (10), we use the above modeling steps of GGNNM(1,1) model to establish a GGNNM(1,1) model, the last forecasting results are listed in table 1.

In addition, we also use the traditional 1-AGO GM(1,1) model and GNNM(1,1) model in [5] for forecasting, the forecasting values are listed in table 1.

According to the results in table 1, we can conclude that the forecasting effect of GGNNM(1,1) model is optimal, and its precision is up to 99.9976%. Therefore, the GGNNM(1,1) model is feasible and advanced in forecasting.

## 4 Conclusions

This paper presents a new GGNNM(1,1) model. According to the forecasting results of the application example, we can conclude that the GGNNM(1,1) model has some advantages,they are represented as the following three aspects. (1)GGNNM(1,1) model is a new model which combines the generalized GM(1,1) model and the method of neural network. The white response of the generalized GM(1,1) model is mapped into a BP neural network, in the process of training the network, the weight values of node are amended gradually and the values of grey parameters $a$ and $b$ are kept improving, so the forecasting effect of the generalized GM(1,1) model is improved gradually in this process. Therefore, GGNNM(1,1) model can further improve the forecasting precision based on the generalized GM(1,1) model. (2)The activation function of nerve cell for layer LB is taken as the Sigmoid function, which is a S-type function and exists a high gain area, so it can ensure the network to reach the stable state, which means the network can reach the convergent state by training. (3)On the one hand, we use the GAGO series to establish the GGNNM(1,1) model, so the randomness of the raw data is weakened and the change rule of data is found easily. On the other hand, we make full use of the BP neural network's advantages of parallel computation, distributed information storage, strong fault-tolerance capability and self-adaptive learning to establish the GGNNM(1,1) model. In a word, the GGNNM(1,1) model synthesizes the advantages of the generalized GM(1,1) model and the method of neural network, it has better forecasting effect,and has great theoretic value and applied value in practice.

## Acknowledgments

This work is supported by the National Natural Science Foundation of China Grant (No.70671050) and the Key Project of Hubei Provincial Department of Education (No. D200627005).

## References

1. Ma, X., Hou, Z., Jiang, C.: Electricity Forward Price Forecasting Based on Combined Grey Neural Network Model. Journal of Shanghai Jiaotong University **9** (2003) 14–23
2. Shang, G., Zhong L., Yan,J.: Establishment and Application of Two Grey Neural Network Model. Journal of Wuhan University of Technology **12** (2002) 78–81
3. Chen, S., wang, W.: Grey Neural Network Forcasting for Traffic Flow. Journal of Southeast University (Natural Science Edition) **4** (2004) 541–544
4. Deng, J.: The Foundation of Grey Theory. Wuhan, Huazhong University of Science and Technology Press (2002)
5. Hung, C., Lu, M.: Two Stage GM(1,1) Model: Grey Step Model. The Journal of Grey System **1** (1997) 9–24
6. Geng, J., Sun, C.: Grey Modeling via Jump Trend Series. The Journal of Grey System **4** (1998) 351–354
7. Chen, C.: A New Method for Grey Modeling Jump Series, The Journal of Grey System **2** (2002) 123–132
8. Rao, C., Xiao, X., Peng, J.: A GM(1,1) Control Model with Pure Generalized AGO Based on Matrix Analysis, Proceedings of the 6th World Congress on intelligent control and automation **1** (2006) 574–577
9. Rao, C., Xiao, X., Peng, J.: A New GM(1,1) Model for Prediction Modeling of Step Series. Dynamics of Continuous Discrete and Impulsive Systems-Series B-Applications and Algorithms **1** (2006) 522–526

# One-Dimensional Analysis of Exponential Convergence Condition for Dual Neural Network

Yunong Zhang[1] and Haifeng Peng[2]

[1] Department of Electronics and Communication Engineering
School of Information Science and Technology
[2] School of Life Science
Sun Yat-Sen University, Guangzhou 510275, China
ynzhang@ieee.org

**Abstract.** In view of its fundamental role arising in numerous fields of science and engineering, the problem of online solving quadratic programs (QP) has been investigated extensively for the past decades. One of the state-of-the-art recurrent neural network (RNN) solvers is dual neural network (DNN). The dual neural network is of simple piecewise-linear dynamics and has global convergence to optimal solutions. Its exponential-convergence property relies on a so-called exponential convergence condition. Such a condition often exists in practice but seems difficult to be proved. In this paper, we investigate the proof complexity of such a condition by analyzing its one-dimensional case. The analysis shows that in general the exponential convergence condition often exists for dual neural networks, and always exists at least for the one-dimensional case. In addition, the analysis is very complex.

**Keywords:** Quadratic programming, Redundant systems, Dual neural network, Online solution, Exponential convergence, Proof complexity.

## 1 Introduction

In view of its fundamental role arising in numerous fields of science and engineering, the problem of solving quadratic programs has been investigated extensively for the past decades. For example, about the recent research based on recurrent neural networks (specifically, the Hopfield-type neural networks), we can refer to [1]-[4] and the references therein. The neural network (NN) approach is now thought to be a powerful tool for online computation, in view of its parallel distributed computing nature and convenience of hardware implementation.

Motivated by the engineering application of quadratic-programs in robotics [2][3][5], the following general problem-formulation has been preferred in our research frequently:

$$\text{minimize} \quad x^T W x/2 + q^T x, \tag{1}$$

$$\text{subject to} \quad Jx = d, \tag{2}$$

$$Ax \leqslant b, \tag{3}$$

$$\xi^- \leqslant x \leqslant \xi^+, \tag{4}$$

D.-S. Huang, L. Heutte, and M. Loog (Eds.): ICIC 2007, LNAI 4682, pp. 137–147, 2007.

**Fig. 1.** Human limbs are also redundant systems similar to robot manipulators

where $W \in R^{n \times n}$ is assumed to be positive-definite in this paper. In performance index (1), $q \in R^n$. In equality constraint (2), $J \in R^{m \times n}$ and $d \in R^m$. In inequality constraint (3), $A \in R^{l \times n}$ and $b \in R^l$. In bound constraint (4), $\xi^- \in R^n$ and $\xi^+ \in R^n$.

In the context of robotic research, the QP problem formulation (1)-(4) can be used to solve the inverse-kinematic problem of redundant robotic manipulators [2]-[8]. Redundant manipulators are robots having more degrees-of-freedom (DOF) than required to perform a given end-effector primary task (usually no more than 6DOF). The inverse-kinematic problem is that, given the desired Cartesian trajectory $r(t) \in R^m$ at the manipulator end-effector, how could we generate the corresponding joint trajectory $\theta(t) \in R^n$ in real time $t$? Note $m \leqslant n$. Such an inverse-kinematic problem can be effectively converted into the QP problem formulation (1)-(4), where the physical meaning and utility of each equation/term are shown clearly in the literature [5][6].

In addition to the above-mentioned inverse-kinematic problem of redundant robot manipulators, it is worth mentioning here that our human limbs are also such redundant systems [9]. See Fig. 1. As simply extended from the robotic research, the general QP formulation (1)-(4) and its online dynamic-system solution (e.g., a dual neural network to be introduced in the ensuing sections) might be generalized to the diversity analysis of human body/limb movements [9]-[11]. This is in view of the facts that our human body/limbs are also redundant systems and there might be natural mechanisms for the involved inverse-kinematic online solution.

## 2 Dual Neural Network

The dual neural network is an online QP solver in the form of a hardware-implementable dynamic system. For other types of recurrent neural networks (and/or other authors' related works) which can solve QP or linear-programming (LP) problems in real time $t$, please refer to [6] and the references therein.

To solve online the QP problem (1)-(4) via a dual neural network, the following design procedure is presented.

Firstly, we treat the equality and inequality constraints in (2) and (3) as a special case of bound constraints [1]:

$$\zeta^- := \begin{pmatrix} d \\ -\varpi 1_{\text{v}} \\ \xi^- \end{pmatrix}, \ \zeta^+ := \begin{pmatrix} d \\ b \\ \xi^+ \end{pmatrix}, \ H := \begin{pmatrix} J \\ A \\ I \end{pmatrix},$$

where constant $\varpi \gg 0$ is sufficiently large to represent $+\infty$ for simulation and hardware-implementation purposes, and $1_{\text{v}}$ denotes an appropriately-dimensioned vector made of ones; e.g., here $1_{\text{v}} := [1, \cdots, 1]^T \in R^l$. In this sense, the QP problem in (1)-(4) is converted to the following bounded QP problem:

$$\text{minimize} \quad x^T W x/2 + q^T x, \tag{5}$$
$$\text{subject to} \quad \zeta^- \leqslant Hx \leqslant \zeta^+. \tag{6}$$

Secondly, we could then reformulate the Karush-Kuhn-Tacker (KKT) optimality/complementarity conditions of (5)-(6) to a system of piecewise-linear equations [1][2]. That is,

$$\begin{cases} \mathcal{P}_\Omega(HMH^T u - HMq - u) - (HMH^T u - HMq) = 0, \\ x = MH^T u - Mq, \end{cases} \tag{7}$$

where $M$ denotes the inverse, $W^{-1}$. The auxiliary vector, $u \in R^{m+l+n}$, represents the dual decision variable vector corresponding to augmented constraint (6). Note that the set $\Omega$ of piecewise-linear projection operator $\mathcal{P}_\Omega(\cdot) : R^{m+l+n} \to \Omega \subseteq R^{m+l+n}$ uses boundaries $[\zeta^-, \zeta^+]$ here [1]-[4].

Thirdly, based on solving (7), we could thus have the following dynamic equations of the dual neural network to solve QP (1)-(4) in real time [1]:

$$\begin{aligned} \dot{u} &= \kappa\{\mathcal{P}_\Omega\left(HMH^T u - HMq - u\right) - \left(HMH^T u - HMq\right)\}, \\ x &= MH^T u - Mq, \end{aligned} \tag{8}$$

where $\kappa > 0$ is the design parameter used to adjust the network-convergence rate. Furthermore, assuming the existence of optimal solution $x^*$ to QP (1)-(4), we have the following lemmas.

**Lemma 1.** *Starting from any initial state $u(0)$, the dual neural network (8) is convergent to an equilibrium point $u^*$, of which the network output $x^* = MH^T u^* - Mq$ is the optimal solution to QP (1)-(4) [1][2].*

**Lemma 2.** *Starting from any initial state $u(0)$, the dual neural network (8) can exponentially converge to an equilibrium point $u^*$, provided that there exists a constant $\rho > 0$ such that $\|\mathcal{P}_\Omega(HMH^T u - HMq - u) - (HMH^T u - HMq)\|^2 \geqslant$*

*$\rho\|u - u^*\|^2$, where the exponential convergence rate is proportional to $\kappa\rho$. In addition, if such an exponential convergence condition (ECC) holds true, then the network output $x(t) = MH^T u(t) - Mq$ will also globally exponentially converge to the optimal solution $x^* = MH^T u^* - Mq$ of QP (1)-(4) [1][2].*

Before ending this section, we may ask our fellow researchers the following question: could the above dynamic system or its variants be the natural mechanisms for handling the inverse-kinematic problem inside our human body/limbs?

## 3 Exponential Convergence Analysis

The exponential convergence condition (ECC) in Lemma 2 is an unsolved problem. The mentioned exponential convergence implies an arbitrarily fast convergence of the dual neural network; otherwise, it could only be of asymptotical convergence. For a better understanding on the significance of this research, we show Fig. 2 so as to give a very clear comparison between asymptotical convergence and exponential convergence. Asymptotical convergence here implies that network-output $x(t)$ approaches the theoretical solution $x^*$ as time $t$ goes to $+\infty$, which may not be accepted in practice: who could wait for an infinitely-long time-period to get the answer?

So, in this paper, we have to focus on the exponential convergence and exponential convergence condition of dual neural network (8). Now, in this section, by analyzing the one-dimensional case, we will investigate the proof complexity of the above exponential convergence condition, which has been an unsolved prob-

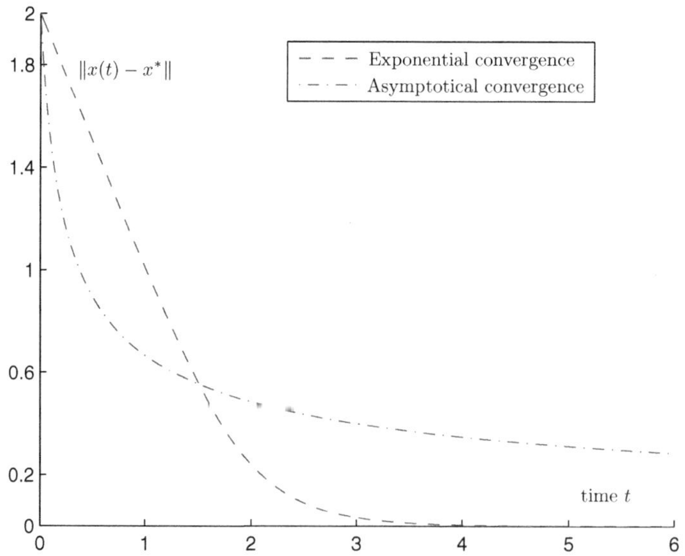


**Fig. 2.** Comparison between asymptotical convergence and exponential convergence

lem since 2001. We will also show that such a one-dimensional case of exponential convergence condition (ECC) always holds true.

For the exponential-convergence condition, in the one-dimensional case, we define $HMH^T := \gamma$ and $\mathcal{P}_\Omega(\cdot) := g(\cdot)$ for simplicity, and assume $q = 0$. Therefore the condition becomes that there exists a constant $\rho > 0$ such that

$$|g(\gamma u - u) - \gamma u|^2 \geqslant \rho |u - u^*|^2, \tag{9}$$

where the equilibrium $u^*$ satisfies

$$g(\gamma u^* - u^*) - \gamma u^* = 0. \tag{10}$$

We have five cases by discussing $\gamma = 0$, $\gamma > 1$, $\gamma = 1$, $0 < \gamma < 1$ and $\gamma < 0$.

**CASE 1 of $\gamma = 0$:**

Equation (10) becomes $g(-u^*) = 0$, and we have four sub-cases (for simplicity, we may use the word "subcases" or "subcase" without using the hyphen):

$$\text{subcase 1.1} \begin{cases} \xi^- < u^* < \xi^+ \\ u^* = 0 \end{cases}, \quad \text{subcase 1.2} \begin{cases} \xi^- < \xi^+ = 0 \\ u^* \in \{R^-, 0\} \end{cases},$$

$$\text{subcase 1.3} \begin{cases} \xi^- = 0 < \xi^+ \\ u^* \in \{R^+, 0\} \end{cases}, \quad \text{subcase 1.4} \begin{cases} \xi^- = \xi^+ = 0 \\ u^* \in R \end{cases},$$

where, in this paper, $u^* \in R^-$ means $u^* < 0$, while $u^* \in R^+$ means $u^* > 0$.

For the subcase 1.1, we have $|g(\gamma u - u) - \gamma u|^2 = |g(-u)|^2 := \varrho |u|^2$, where

$$\varrho = \begin{cases} 1 & \text{if } \xi^- \leqslant u \leqslant \xi^+, \\ |\xi^-/u|^2 & \text{if } u > \xi^+ \geqslant -\xi^-, \\ 1 & \text{if } -\xi^- \geqslant u > \xi^+, \\ |\xi^+/u|^2 & \text{if } u < \xi^- \leqslant -\xi^+, \\ 1 & \text{if } -\xi^+ \leqslant u < \xi^-. \end{cases}$$

Clearly, it follows from the proved convergence property of the dual neural network (i.e., Lemma 1) that there exists a constant $\rho > 0$ such that $\varrho \geqslant \rho$ and thus $|g(\gamma u - u) - \gamma u|^2 = \varrho |u|^2 \geqslant \rho |u|^2$ [which is exactly the one-dimensional ECC (9)], as the initial state $u(0)$ is not equal to $\pm\infty$.

For the subcase 1.2, we have $|g(\gamma u - u) - \gamma u|^2 = |g(-u)|^2 := \varrho |u|^2$, where

$$\varrho = \begin{cases} 1 & \text{if } \xi^- < -u \leqslant \xi^+, \\ |\xi^-/u|^2 & \text{if } -u \leqslant \xi^-. \end{cases}$$

Note that when $-u > \xi^+$, any $u$ has been the equilibrium $u^*$ according to (10) and the definition of subcase 1.2. It follows from the convergence property (i.e., Lemma 1) that there exists a constant $\rho > 0$ such that $\varrho \geqslant \rho$ and ECC (9) holds true, as the initial state $u(0)$ is not equal to $+\infty$.

For the subcase 1.3, we have

$$\varrho = \begin{cases} 1 & \text{if } \xi^- \leqslant -u < \xi^+, \\ |\xi^+/u|^2 & \text{if } -u \geqslant \xi^+. \end{cases}$$

Note that when $-u < \xi^-$, any $u$ has been the equilibrium $u^*$ according to (10) and the definition of subcase 1.3. It follows from the convergence property (i.e., Lemma 1) that there exists a constant $\rho > 0$ such that $\varrho \geqslant \rho$, as $u(0) \neq -\infty$.

For the subcase 1.4, note that $u^* \in R$ (i.e., any $u$ has been the equilibrium), and $\rho$ can be viewed as 1. This means that the one-dimensional ECC (9) holds true for this subcase either.

Before ending the discussion of CASE 1, we would like to interpret the physical meaning of such an analysis in the context of solving QP (1)-(4) and QP (5)-(6). In the CASE 1 of $\gamma = 0$, the definitions of $\gamma = HMH^T = 0$ and $M = W^{-1} > 0$ implies $H = 0$, which further implies that QP (1)-(4) and QP (5)-(6) reduce to an unconstrained quadratic minimization problem of $x^T Wx/2$. Clearly, $x^* = 0$ is the optimal solution. Accordingly, dual neural network (8) reduces to $\dot{u} = \kappa \mathcal{P}_\Omega(-u)$ and $x = 0$, which gives the optimal solution $x = x^* = 0$ in no time! In this case, global exponential convergence (actually a much superior convergence to it) certainly holds, which has substantiated the above analysis of one-dimension ECC (9) when $\gamma = 0$.

**CASE 2 of $\gamma > 1$:**

Equation (10) becomes $g((\gamma-1)u^*) - \gamma u^* = 0$, and we have six sub-cases:

$$\text{subcase 2.1} \quad \begin{cases} \xi^- < (\gamma-1)u^* < \xi^+ \\ (\gamma-1)u^* = \gamma u^* \end{cases} \quad \Longrightarrow \quad u^* = 0,$$

$$\text{subcase 2.2} \quad \begin{cases} (\gamma-1)u^* > \xi^+ \\ \xi^+ = \gamma u^* \end{cases} \quad \Longrightarrow \quad \xi^+ < 0, \ u^* = \xi^+/\gamma < 0,$$

$$\text{subcase 2.3} \quad \begin{cases} (\gamma-1)u^* < \xi^- \\ \xi^- = \gamma u^* \end{cases} \quad \Longrightarrow \quad \xi^- > 0, \ u^* = \xi^-/\gamma > 0,$$

$$\text{subcase 2.4} \quad \begin{cases} \xi^- < (\gamma-1)u^* = \xi^+ \\ \xi^+ = \gamma u^* \end{cases} \quad \Longrightarrow \quad \xi^+ = 0, \ u^* = 0,$$

$$\text{subcase 2.5} \quad \begin{cases} \xi^- = (\gamma-1)u^* < \xi^+ \\ \xi^- = \gamma u^* \end{cases} \quad \Longrightarrow \quad \xi^- = 0, \ u^* = 0,$$

$$\text{subcase 2.6} \quad \begin{cases} \xi^- = (\gamma-1)u^* = \xi^+ \\ \xi^- = \gamma u^* = \xi^+ \end{cases} \quad \Longrightarrow \quad \xi^- = \xi^+ = 0, \ u^* = 0.$$

For the subcase 2.1, we have

$$|g((\gamma-1)u) - \gamma u|^2 = \begin{cases} |u|^2 & \text{if } \xi^- \leqslant (\gamma-1)u \leqslant \xi^+ \\ |\xi^+ - \gamma u|^2 & \text{if } (\gamma-1)u > \xi^+ \\ |\xi^- - \gamma u|^2 & \text{if } (\gamma-1)u < \xi^- \end{cases} := \varrho |u|^2,$$

where we have $|\xi^+ - \gamma u|^2 > |\xi^+/(r-1)|^2 > 0$ when $(\gamma-1)u > \xi^+$, and $|\xi^- - \gamma u|^2 > |-\xi^-/(r-1)|^2 > 0$ when $(\gamma-1)u < \xi^-$, resulting in

$$\varrho \geqslant \begin{cases} 1 & \text{if } \xi^- \leqslant (\gamma-1)u \leqslant \xi^+, \\ |\xi^+/[(r-1)u]|^2 & \text{if } (\gamma-1)u > \xi^+, \\ |\xi^-/[(r-1)u]|^2 & \text{if } (\gamma-1)u < \xi^-. \end{cases}$$

Thus, it follows from the proved convergence property of the dual neural network (i.e., Lemma 1) that there exists a constant $\rho > 0$ such that $\varrho \geqslant \rho$, as $u(0) \neq \pm\infty$.

For the subcase 2.2, we have $|u - u^*|^2 = |u - \xi^+/\gamma|^2 = (1/\gamma)^2|\gamma u - \xi^+|^2$ and

$$|g((\gamma-1)u) - \gamma u|^2 = \begin{cases} |u|^2 & \text{if } \xi^- \leqslant (\gamma-1)u \leqslant \xi^+ < 0 \\ |\xi^+ - \gamma u|^2 & \text{if } (\gamma-1)u > \xi^+ \\ |\xi^- - \gamma u|^2 & \text{if } (\gamma-1)u < \xi^- < 0 \end{cases} := \varrho|u - u^*|^2,$$

where because of $-u \geqslant -\xi^+/(\gamma-1) > 0$ when $\xi^- \leqslant (\gamma-1)u \leqslant \xi^+$, and $\xi^- - \gamma u > -\xi^-/(\gamma-1) > 0$ when $(\gamma-1)u < \xi^-$, we have

$$\varrho \geqslant \begin{cases} |\xi^+|^2/|[(\gamma-1)(u-\xi^+/\gamma)]|^2 > 0 & \text{if } \xi^- \leqslant (\gamma-1)u \leqslant \xi^+ < 0, \\ \gamma^2 & \text{if } (\gamma-1)u > \xi^+, \\ |\xi^-/[(r-1)u]|^2 > 0 & \text{if } (\gamma-1)u < \xi^-. \end{cases}$$

That is, there exists a constant $\rho > 0$ such that $\varrho \geqslant \rho$, as $u(0) \neq -\infty$.

For the subcase 2.3, since it is symmetric to the case 2.2, we have

$$\varrho \geqslant \begin{cases} |\xi^-|^2/|[(\gamma-1)(u-\xi^-/\gamma)]|^2 > 0 & \text{if } 0 < \xi^- \leqslant (\gamma-1)u \leqslant \xi^+, \\ |\xi^+/[(r-1)u]|^2 > 0 & \text{if } (\gamma-1)u > \xi^+, \\ \gamma^2, & \text{if } (\gamma-1)u < \xi^-. \end{cases}$$

Thus there exists a constant $\rho > 0$ such that $\varrho \geqslant \rho$, as $u(0) \neq +\infty$.

For the subcase 2.4, in view of $\xi^+ = 0$ and $u^* = 0$, we have

$$|g((\gamma-1)u) - \gamma u|^2 = \begin{cases} |u|^2 & \text{if } \xi^- \leqslant (\gamma-1)u \leqslant \xi^+ \\ |-\gamma u|^2 & \text{if } (\gamma-1)u > \xi^+ \\ |\xi^- - \gamma u|^2 & \text{if } (\gamma-1)u < \xi^- \end{cases} := \varrho|u|^2,$$

where we have $\xi^- - \gamma u > -\xi^-/(r-1) > 0$ when $(\gamma-1)u < \xi^-$, and thus

$$\varrho \geqslant \begin{cases} 1 & \text{if } \xi^- \leqslant (\gamma-1)u \leqslant \xi^+, \\ \gamma^2 & \text{if } (\gamma-1)u > \xi^+, \\ |\xi^-/[(r-1)u]|^2 & \text{if } (\gamma-1)u < \xi^-. \end{cases}$$

It follows that there exists a constant $\rho > 0$ such that $\varrho \geqslant \rho$, as $u(0) \neq -\infty$.

For the subcase 2.5 (symmetric to the subcase 2.4), we have

$$\varrho \geqslant \begin{cases} 1 & \text{if } \xi^- \leqslant (\gamma-1)u \leqslant \xi^+, \\ |\xi^+/[(r-1)u]|^2 & \text{if } (\gamma-1)u > \xi^+, \\ \gamma^2 & \text{if } (\gamma-1)u < \xi^-. \end{cases}$$

It follows that there exists a constant $\rho > 0$ such that $\varrho \geqslant \rho$, as $u(0) \neq +\infty$.

For the subcase 2.6, it is clear that

$$\varrho = \begin{cases} \gamma^2 & \text{if } (\gamma-1)u > \xi^+, \\ \gamma^2 & \text{if } (\gamma-1)u < \xi^-. \end{cases}$$

Note that when $u = \xi^+$ or $u = \xi^-$, the $u = 0$ has been the equilibrium $u^*$ and $\varrho := 1$. It follows that there exists a constant $\rho > 0$ such that $\varrho \geqslant \rho$, as $u(0) \neq \pm\infty$.

**CASE 3 of $\gamma = 1$:**

Equation (10) becomes $g(0) - u^* = 0$, which includes six sub-cases:

subcase 3.1 $\begin{cases} \xi^- < 0 < \xi^+ \\ u^* = 0 \end{cases}$, subcase 3.2 $\begin{cases} 0 < \xi^- < \xi^+ \\ u^* = \xi^- \end{cases}$,

subcase 3.3 $\begin{cases} \xi^- < \xi^+ < 0 \\ u^* = \xi^+ \end{cases}$, subcase 3.4 $\begin{cases} \xi^- < \xi^+ = 0 \\ u^* = 0 \end{cases}$,

subcase 3.5 $\begin{cases} 0 = \xi^- < \xi^+ \\ u^* = 0 \end{cases}$, subcase 3.6 $\begin{cases} \xi^- = \xi^+ = 0 \\ u^* = 0 \end{cases}$.

Clearly, it is seen in the subcases 3.1, 3.4, 3.5 and 3.6 that $g(0) = 0$ and $u^* = 0$, thus $\varrho = |0 - \gamma u|^2/|u - 0|^2 = \gamma^2 = 1$ for those subcases. But for the subcases 3.2 and 3.3, a little detailed discussion has to be given as follows. For the subcase 3.2, since $g(0) = \xi^-$ and $u^* = \xi^-$, we have

$$\varrho = |g(0) - u|^2/|u - u^*|^2 = |\xi^- - u|^2/|u - \xi^-|^2 = 1.$$

So is the subcase 3.3. Thus, in the case of $\gamma = 1$, there also exists a constant $\rho > 0$ such that $\varrho \geqslant \rho$ (actually here $\rho \equiv 1$).

**CASE 4 of $0 < \gamma < 1$:**

Equation (10) is $g((\gamma-1)u^*) - \gamma u^* = 0$, which includes the same six sub-cases as CASE 2. Moreover, we have derived the same conclusion as in CASE 2 that there exists a constant $\rho > 0$ such that $\varrho \geqslant \rho$, as the initial state $u(0)$ is not equal to $\pm\infty$. The difference between CASE 4 and CASE 2 is that the derivation of CASE 4-involved inequalities makes use of $\gamma - 1 < 0$, while the CASE 2 derivation makes use of $\gamma - 1 > 0$.

**CASE 5 of $\gamma < 0$:**

Equation (10) is $g((\gamma-1)u^*) - \gamma u^* = 0$, and similar to CASE 2, we have the following six sub-cases (with sign differences in sub-cases 5.2 and 5.3):

$$\text{subcase 5.1} \quad \begin{cases} \xi^- < (\gamma-1)u^* < \xi^+ \\ (\gamma-1)u^* = \gamma u^* \end{cases} \quad \Longrightarrow u^* = 0,$$

$$\text{subcase 5.2} \quad \begin{cases} (\gamma-1)u^* > \xi^+ \\ \xi^+ = \gamma u^* \end{cases} \quad \Longrightarrow \xi^+ > 0,\ u^* = \xi^+/\gamma < 0,$$

$$\text{subcase 5.3} \quad \begin{cases} (\gamma-1)u^* < \xi^- \\ \xi^- = \gamma u^* \end{cases} \quad \Longrightarrow \xi^- < 0,\ u^* = \xi^-/\gamma > 0,$$

$$\text{subcase 5.4} \quad \begin{cases} \xi^- < (\gamma-1)u^* = \xi^+ \\ \xi^+ = \gamma u^* \end{cases} \quad \Longrightarrow \xi^+ = 0,\ u^* = 0,$$

$$\text{subcase 5.5} \quad \begin{cases} \xi^- = (\gamma-1)u^* < \xi^+ \\ \xi^- = \gamma u^* \end{cases} \quad \Longrightarrow \xi^- = 0,\ u^* = 0,$$

$$\text{subcase 5.6} \quad \begin{cases} \xi^- = (\gamma-1)u^* = \xi^+ \\ \xi^- = \gamma u^* = \xi^+ \end{cases} \quad \Longrightarrow \xi^- = \xi^+ = 0,\ u^* = 0.$$

For the subcase 5.1, we have

$$\varrho \geqslant \begin{cases} 1 & \text{if } \xi^- \leqslant (\gamma-1)u \leqslant \xi^+, \\ |\xi^+/[(1-r)u]|^2 > 0 & \text{if } (\gamma-1)u > \xi^+, \\ |-\xi^-/[(1-r)u]|^2 > 0 & \text{if } (\gamma-1)u < \xi^-. \end{cases}$$

For the subcase 5.2, we have

$$\varrho \geqslant \begin{cases} |\xi^+|^2/|[(1-\gamma)(u-\xi^+/\gamma)]|^2 > 0 & \text{if } \xi^- \leqslant (\gamma-1)u \leqslant \xi^+ < 0, \\ \gamma^2 & \text{if } (\gamma-1)u > \xi^+, \\ |\xi^-/[(1-r)u]|^2 > 0 & \text{if } (\gamma-1)u < \xi^-. \end{cases}$$

For the subcase 5.3, we have

$$\varrho \geqslant \begin{cases} |\xi^-|^2/|[(1-\gamma)(u-\xi^-/\gamma)]|^2 > 0 & \text{if } 0 < \xi^- \leqslant (\gamma-1)u \leqslant \xi^+, \\ |\xi^+/[(1-r)u]|^2 > 0 & \text{if } (\gamma-1)u > \xi^+, \\ \gamma^2 & \text{if } (\gamma-1)u < \xi^-. \end{cases}$$

For the subcase 5.4, we have

$$\varrho \geqslant \begin{cases} 1 & \text{if } \xi^- \leqslant (\gamma-1)u \leqslant \xi^+, \\ \gamma^2 & \text{if } (\gamma-1)u > \xi^+, \\ |\xi^-/[(1-r)u]|^2 & \text{if } (\gamma-1)u < \xi^-. \end{cases}$$

For the subcase 5.5, we have

$$\varrho \geqslant \begin{cases} 1 & \text{if } \xi^- \leqslant (\gamma-1)u \leqslant \xi^+, \\ |\xi^+/[(1-r)u]|^2 & \text{if } (\gamma-1)u > \xi^+, \\ \gamma^2 & \text{if } (\gamma-1)u < \xi^-. \end{cases}$$

For the subcase 5.6, we have

$$\varrho = \begin{cases} \gamma^2 & \text{if } (\gamma-1)u > \xi^+, \\ \gamma^2 & \text{if } (\gamma-1)u < \xi^-. \end{cases}$$

That is, there always exists a constant $\rho > 0$ such that $\varrho \geqslant \rho$, as $u(0) \neq \pm\infty$.

In summary, we have discussed all the possible cases (including sub-cases) of the one-dimensional form of the original exponential-convergence condition (note that $q = 0$ is assumed in the analysis for simplicity and clarity). By applying the proved convergence property of the dual neural network (i.e., Lemma 1: $u(t) \to u^*$ as $t \to +\infty$), it has been shown that there always exist a lower bound $\rho > 0$ such that one-dimensional ECC (9) holds true, provided that the initial state $u(0)$ is not selected at $\pm\infty$ (this always hold in mathematics and in practice). In addition, the basic tools in the proof are equilibrium equation (10) and the piecewise-linearity of projection operator $g(\cdot)$.

Before ending this section, we would like to point out that the two-dimensional or higher-dimensional analysis of such an exponential convergence condition (9) will be much more complex than the above one-dimensional analysis of ECC (9), and that $q \neq 0$ will further complicate the analysis. However, as shown in Fig. 2, global exponential convergence/stability is one of the most desirable properties of recurrent neural networks or engineering systems. From the viewpoint of real applications, we thus have to work on it. Moreover, as one of the reviewers said, from mathematical viewpoint, this topic is interesting as well. If it could be of the mathematician readers' interest to further explore the general existence of such a condition, this might be another contribution of this paper.

## 4 Conclusions

To solve QP (1)-(4) in real time and in an error-free parallel-computing manner, dual neural network (8) has been proposed. Being globally exponentially stable, dual neural networks can converge to their optimal solution most rapidly. The global exponential stability/convergence relies on a so-call exponential convergence condition (ECC). In our research of nearly six years, we have numerically observed that this exponential-convergence condition always/often exists in practice, but it is hard to be proved. To be mathematically rigorous, it has been formulated in this research as a condition instead of an assumption. This paper has investigated the proof complexity of such an exponential-convergence condition by analyzing its one-dimensional case (with $q = 0$). The analysis results are that the one-dimensional case of ECC (9) always holds true, and that the proof is quite complex with many sub-cases. Future research directions may lie in the proof of general existence of such a condition and its equivalence/conversion/link to other types of conditions found by other researchers.

**Acknowledgements.** This work is funded by National Science Foundation of China under Grant 60643004 and by the Science and Technology Office of Sun

Yat-Sen University. Before joining Sun Yat-Sen University in 2006, the corresponding author, Yunong Zhang, had been with National University of Ireland, University of Strathclyde, National University of Singapore, Chinese University of Hong Kong, since 1999. He has continued the line of this research, supported by various research fellowships/assistantship. His web-page is now available at http://www.ee.sysu.edu.cn/teacher/detail.asp?sn=129.

## References


1. Zhang, Y., Wang, J.: A Dual Neural Network for Convex Quadratic Programming Subject to Linear Equality and Inequality Constraints. Physics Letters A, Vol. 298 (2002) 271-278
2. Zhang, Y., Wang, J., Xu, Y.: A Dual Neural Network for Bi-Criteria Kinematic Control of Redundant Manipulators. IEEE Transactions on Robotics and Automation, Vol. 18 (2002) 923-931
3. Zhang, Y., Ge, S.S., Lee, T.H.: A Unified Quadratic Programming Based Dynamical System Approach to Joint Torque Optimization of Physically Constrained Redundant Manipulators. IEEE Transactions on Systems, Man, and Cybernetics, Vol. 34 (2004) 2126-2133
4. Zhang, Y.: On the LVI-Based Primal-Dual Neural Network for Solving Online Linear and Quadratic Programming Problems. Proceedings of American Control Conference (2005) 1351-1356
5. Zhang, Y.: Minimum-Energy Redundancy Resolution Unified by Quadratic Programming. The 15th International Symposium on Measurement and Control in Robotics, Belgium (2005)
6. Zhang, Y.: Towards Piecewise-Linear Primal Neural Networks for Optimization and Redundant Robotics. Proceedings of IEEE International Conference on Networking, Sensing and Control (2006) 374-379
7. Zhang, Y.: Inverse-Free Computation for Infinity-Norm Torque Minimization of Robot Manipulators. Mechatronics, Vol. 16 (2006) 177-184
8. Zhang, Y.: A Set of Nonlinear Equations and Inequalities Arising in Robotics and its Online Solution via a Primal Neural Network. Neurocomputing, Vol. 70 (2006) 513-524
9. Latash, M.L.: Control of Human Movement. Human Kinetics Publisher, Chicago (1993)
10. Zhang, X., Chaffin, D.B.: An Inter-Segment Allocation Strategy for Postural Control in Human Reach Motions Revealed by Differential Inverse Kinematics and Optimization. Proceedings of IEEE International Conference on Systems, Man, and Cybernetics (1997) 469-474
11. Iqbal, K., Pai, Y.C.: Predicted Region of Stability for Balance Recovery: Motion at the Knee Joint can Improve Termination of Forward Movement. Journal of Biomechanics, Vol. 33 (2000) 1619-1627
12. Zhang, Y., Wang, J.: Global Exponential Stability of Recurrent Neural Networks for Synthesizing Linear Feedback Control Systems via Pole Assignment. IEEE Transactions on Neural Networks, Vol. 13 (2002) 633-644

# Stability of Stochastic Neutral Cellular Neural Networks

Ling Chen[1,2] and Hongyong Zhao[2]

[1] Basic Department, Jinling Institute of Technology, Nanjing 210001, China
chensxa@126.com
[2] Department of Mathematics, Nanjing University of Aeronautics and Astronautics, Nanjing 210016, China
zhaohy@nuaa.edu.cn

**Abstract.** In this paper, we study a class of stochastic neutral cellular neural networks. By constructing a suitable Lyapunov functional and employing the nonnegative semi-martingale convergence theorem we give some sufficient conditions ensuring the almost sure exponential stability of the networks. The results obtained are helpful to design stability of networks when stochastic noise is taken into consideration. Finally, two examples are provided to show the correctness of our analysis.

## 1 Introduction

Recently, the analysis of the dynamics of delayed cellular neural networks has attracted much attention due to their applicability in pattern recognition, image processing, speed detection of moving objects, optimization problems and so on [1,2]. Many important results have been reported in the prior literature, see [3-12] and the references therein. However, due to the complicated dynamic properties of the neural cells, in many cases the existing delayed neural networks models cannot characterize the properties of a neural reaction process precisely. To describe further the dynamics for such complex neural reactions, a new type of model called neutral neural networks is set up and analyzed. It is reasonable to study neutral neural networks. For example, in the biochemistry experiments, neural information may transfer across chemical reactivity, which results in a neutral-type process [13]. A different example is proposed in [14,15], where the neutral phenomena exist in large-scale integrated circuits. There exist some results on the stability of neutral neural networks, we refer to Refs. [16, 17]. However, most neutral neural networks models proposed and discussed in existing literature are deterministic. A real system is usually affected by external perturbations which in many cases are of great uncertainty and hence may be treated as random, as pointed out by Haykin [18] that in real nervous systems, synaptic transmission is a noise process brought on by random fluctuations from the release of neurotransmitters, and other probabilistic causes. Under the effect of the noise, the trajectory of system becomes a stochastic process. There are various kinds convergence concepts to describe limiting behaviors of stochastic

D.-S. Huang, L. Heutte, and M. Loog (Eds.): ICIC 2007, LNAI 4682, pp. 148–156, 2007.

processes, see for example [19]. The almost sure exponential stability is the most useful because it is closer to the real situation during computation than other forms of convergence (see [20, 21] for the detailed discussions). Therefore, it is significant to study almost sure exponential stability for stochastic neutral cellular neural networks. To the best of the authors' knowledge, the almost sure exponential stability analysis problem for stochastic neutral cellular neural networks has not been fully investigated, and remains important and challenging.

Motivated by the above discussion, our objective in this paper is to study stochastic neutral cellular neural networks, and give some sufficient conditions ensuring the almost sure exponential stability by constructing a suitable Lyapunov functional and applying the nonnegative semi-martingale convergence theorem. It is easy to apply these conditions to the real networks.

## 2 Preliminary

$R^n$ and $C[X,Y]$ denote the n–dimensional Euclidean space and a continuous mapping set from the topological space $X$ to the topological space $Y$, respectively. Especially, $C \triangleq C[[-\tau,0],R^n]$, where $\tau > 0$.

Consider the following stochastic neutral cellular neural networks model:

$$\begin{cases} d(x_i(t)-G_i(x_i(t-\tau_i))) = [-c_i x_i(t) + \sum\limits_{j=1}^{n} a_{ij} f_j(x_j(t)) + \sum\limits_{j=1}^{n} b_{ij} g_j(x_j(t-\tau_j)) \\ \qquad\qquad + J_i]dt + \sum\limits_{j=1}^{n} \sigma_{ij}(x_j(t), x_j(t-\tau_j))d\omega_j(t), \quad t \geq 0 \\ x_i(t) = \phi_i(t), \quad t \in [-\tau, 0] \end{cases}, \tag{1}$$

where $i = 1, \cdots, n$. $n$ denotes the number of neurons in the neural networks; $x_i(t)$ denotes the state of the $i$th neuron at time $t$; $c_i > 0$ is the neuron firing rate; $\tau_i$ represents transmission delays with $0 \leq \tau_i \leq \tau$; $a_{ij}$ and $b_{ij}$ denote the delayed connection weight and neutral delayed connection weight, respectively; $G_i(\cdot)$, $f_j(\cdot)$ and $g_j(\cdot)$ are the activation functions; $J_i$ is the external input; $\sigma(\cdot,\cdot) = (\sigma_{ij}(\cdot,\cdot))_{n\times n}$ is the diffusion coefficient matrix and $\omega(\cdot) = (\omega_1(\cdot), \cdots, \omega_n(\cdot))^T$ is an $n-$dimensional Brownian motion; Assume, throughout this paper, that $\sigma_{ij}(\cdot,\cdot)$ is locally Lipschitz continuous and satisfies the linear growth condition as well. So it is known that Eq.(1) has a unique global solution on $t \geq 0$, which is denoted by $x(t)$, where $x(t) = (x_1(t), \cdots, x_n(t))^T$. $\phi_i(t)$ is the initial function, and is assumed to be continuous and bounded on $[-\tau, 0]$.

Throughout the paper, we always assume that:

$(H_1)$. There are positive constants $k_i \in (0,1)$, $\lambda_j$ and $\mu_j (i,j = 1, \cdots, n)$, such that $k_i = \sup\limits_{u\neq v} |\frac{G_i(u)-G_i(v)}{u-v}|$, $\lambda_j = \sup\limits_{u\neq v} |\frac{f_j(u)-f_j(v)}{u-v}|$ and $\mu_j = \sup\limits_{u\neq v} |\frac{g_j(u)-g_j(v)}{u-v}|$, for all $u, v \in R$ .

$(H_2)$. There are a set of positive constants $d_1, \cdots, d_n$, such that

$$2d_i c_i k_i + \sum_{j=1}^{n} |a_{ji}| d_j \lambda_i + \sum_{j=1}^{n} |b_{ji}| d_j \mu_i < d_i c_i, \quad i = 1, \cdots, n \ . \tag{2}$$

For any $x(t) = (x_1(t), \cdots, x_n(t))^T \in R^n$, we define the norm $||x(t)|| = \sum_{i=1}^{n}\{d_i|x_i(t)|\}$. For any $\phi(t) = (\phi_1(t), \cdots, \phi_n(t))^T \in C$, we define $\| \phi \|_\tau = \sum_{i=1}^{n}\{d_i \| \phi_i \|_\tau\}$, where $\| \phi_i \|_\tau = \sup_{-\tau \le t \le 0} | \phi_i(t) |$ .

**Lemma 1.** *(Semi-martingale Convergence Theorem [22]) Let $A(t)$ and $U(t)$ be two continuous adapted increasing processes on $t \ge 0$ with $A(0) = U(0) = 0$ a.s.. Let $M(t)$ be a real-valued continuous local martingale with $M(0) = 0$ a.s.. Let $\zeta$ be a nonnegative $F_0$-measurable random variable with $E\zeta < \infty$. Define*

$$X(t) = \zeta + A(t) - U(t) + M(t), \quad \text{for } t \ge 0 \ .$$

*If $X(t)$ is nonnegative, then*

$$\{\lim_{t\to\infty} A(t) < \infty\} \subset \{\lim_{t\to\infty} X(t) < \infty\} \cap \{\lim_{t\to\infty} U(t) < \infty\} \quad a.s. \ ,$$

*where $B \subset D$ a.s. means $P(B \cap D^c) = 0$. In particular, if $\lim_{t\to\infty} A(t) < \infty$ a.s., then for almost all $\omega \in \Omega$*

$$\lim_{t\to\infty} X(t) < \infty \quad \text{and} \quad \lim_{t\to\infty} U(t) < \infty \ ,$$

*that is both $X(t)$ and $U(t)$ converge to finite random variables.*

**Lemma 2.** [23] *Assume that $G : R^n \to R^n$ is a Borel measurable function such that for some $l \in (0, 1)$*

$$\|G(y)\| \le l\|y\|, \quad \text{for all } y \in R^n \ .$$

*Let $\varphi(t)$, $-\tau \le t \le \infty$, be a Borel measurable $R^n$-valued function. Let $\alpha > 0$ and $K > 0$. Assume*

$$e^{\alpha t}\|\varphi(t) - G(\varphi(t - \tau))\|^2 \le K, \quad \text{for all } t \ge 0 \ .$$

*Then*

$$\limsup_{t\to\infty} \frac{1}{t} \log \|\varphi(t)\| \le -\frac{\alpha \bigwedge \beta}{2} \ ,$$

*where*

$$\beta = -\frac{2}{\tau} \log l > 0 \ .$$

## 3 Main Results

For the deterministic system

$$\begin{cases} d(x_i(t) - G_i(x_i(t - \tau_i))) = [-c_i x_i(t) + \sum_{j=1}^{n} a_{ij} f_j(x_j(t)) + \sum_{j=1}^{n} b_{ij} g_j(x_j(t - \tau_j)) \\ \qquad\qquad\qquad\qquad +J_i]dt, \quad t \ge 0 \\ x_i(t) = \phi_i(t), \quad t \in [-\tau, 0] \ , \end{cases} \tag{3}$$

we have the following result.

**Theorem 1.** *If* $(H1)$ *and* $(H_2)$ *hold, then system (3) has a unique equilibrium point* $x^* = (x_1^*, \cdots, x_n^*)^T$.

*Proof.* The proof is similar to that of [3]. So we omit it here.

In the paper, we assume that
$(H_3)$. $\quad \sigma_{ij}(x_j^*, x_j^*) = 0, \quad i,j = 1, \cdots, n$ .
Thus, system (1) admits an equilibrium point $x^* = (x_1^*, \cdots, x_n^*)^T$.

For the sake of simplicity in the stability proof of system (1), we make the following transformation for system (1):

$$y_i(t) = x_i(t) - x_i^*, \; \varphi_i(t) = \phi_i(t) - x_i^* \; ,$$

where $y(t) = (y_1(t), \cdots, y_n(t))^T, G(y(t-\tau)) = (G_1(y_1(t-\tau_1)), \cdots, G_n(y_n(t-\tau_n)))^T$. Under the transformation, it is easy to see that system (1) becomes:

$$\begin{cases} d(y_i(t) - G_i(y_i(t-\tau_i))) = [-c_i y_i(t) + \sum\limits_{j=1}^n a_{ij} f_j(y_j(t)) + \sum\limits_{j=1}^n b_{ij} g_j(y_j(t-\tau_j))]dt \\ \qquad\qquad + \sum\limits_{j=1}^n \sigma_{ij}(y_j(t), y_j(t-\tau_j)) d\omega_j(t), \quad t \geq 0 \\ y_i(t) = \varphi_i(t), \quad t \in [-\tau, 0] \; , \end{cases} \tag{4}$$

where $f_j(y_j(t)) = f_j(x_j(t)) - f_j(x_j^*)$, $g_j(y_j(t-\tau_j)) = g_j(x_j(t-\tau_j)) - g_j(x_j^*)$, $G_i(y_i(t-\tau_i)) = G_i(x_i(t-\tau_i)) - G_i(x_i^*)$ and $\sigma_{ij}(y_j(t), y_j(t-\tau_j)) = \sigma_{ij}(x_j(t), x_j(t-\tau_j)) - \sigma_{ij}(x_j^*, x_j^*)$ .

Clearly, the equilibrium point $x^*$ of (1) is almost surely exponentially stable if and only if the equilibrium point $O$ of system (4) is almost surely exponentially stable. Thus in the following, we only consider the almost surely exponential stability of the equilibrium point $O$ for system (4).

**Theorem 2.** *Suppose that* $(H_1)-(H_3)$ *hold. Then system (4) has an equilibrium point* $O$ *which is almost surely exponentially stable.*

*Proof.* It follows from $(H_2)$ that there exists a sufficiently small constant $0 < \lambda < \min\limits_i c_i$ $(i = 1, \cdots, n)$ such that

$$\lambda d_i - d_i c_i + \sum_{j=1}^n |a_{ji}| d_j \lambda_i + e^{\lambda\tau}(\lambda d_i k_i + 2 d_i c_i k_i + \sum_{j=1}^n |b_{ji}| d_j \mu_i) \leq 0 \; . \tag{5}$$

Define the following Lyapunov functional:

$$V(y(t) - G(y(t-\tau)), t) = e^{\lambda t} \sum_{i=1}^n d_i |y_i(t) - G_i(y_i(t-\tau_i))| \; ,$$

and applying *Itô*'s formula to $V(y(t) - G(y(t-\tau)), t)$, we have

$$V(y(t) - G(y(t-\tau)), t) = \xi + \int_0^t \lambda e^{\lambda s} \sum_{i=1}^n d_i |y_i(s) - G_i(y_i(s-\tau_i))| ds$$

$$
\begin{aligned}
&+ \int_0^t e^{\lambda s} \sum_{i=1}^n d_i \mathrm{sgn}[y_i(s) - G_i(y_i(s-\tau_i))][-c_i y_i(s) \\
&+ \sum_{j=1}^n a_{ij} f_j(y_j(s)) + \sum_{j=1}^n b_{ij} g_j(y_j(s-\tau_j))]ds \\
&+ M(\omega) \\
&\leq \xi + \int_0^t \lambda e^{\lambda s} \sum_{i=1}^n d_i(|y_i(s)| + |G_i(y_i(s-\tau_i))|)ds \\
&+ \int_0^t e^{\lambda s} \sum_{i=1}^n d_i \mathrm{sgn}[y_i(s) - G_i(y_i(s-\tau_i))][-c_i y_i(s) \\
&+ c_i G_i(y_i(s-\tau_i)) - c_i G_i(y_i(s-\tau_i)) \\
&+ \sum_{j=1}^n a_{ij} f_j(y_j(s)) + \sum_{j=1}^n b_{ij} g_j(y_j(s-\tau_j))]ds \\
&+ M(\omega) \\
&\leq \xi + \int_0^t \lambda e^{\lambda s} \sum_{i=1}^n d_i(|y_i(s)| + k_i|y_i(s-\tau_i)|)ds \\
&+ \int_0^t e^{\lambda s} \sum_{i=1}^n d_i[-c_i|y_i(s) - G_i(y_i(s-\tau_i))| + c_i k_i |y_i(s-\tau_i)| \\
&+ \sum_{j=1}^n |a_{ij}|\lambda_j|y_j(s)| + \sum_{j=1}^n |b_{ij}|\mu_j|y_j(s-\tau_j)|]ds + M(\omega) \\
&\leq \xi + \int_0^t \lambda e^{\lambda s} \sum_{i=1}^n d_i(|y_i(s)| + k_i|y_i(s-\tau_i)|)ds \\
&+ \int_0^t e^{\lambda s} \sum_{i=1}^n d_i[-c_i|y_i(s)| + c_i|G_i(y_i(s-\tau_i))| + c_i k_i |y_i(s-\tau_i)| \\
&+ \sum_{j=1}^n |a_{ij}|\lambda_j|y_j(s)| + \sum_{j=1}^n |b_{ij}|\mu_j|y_j(s-\tau_j)|]ds + M(\omega) \\
&\leq \xi + \int_0^t \lambda e^{\lambda s} \sum_{i=1}^n d_i(|y_i(s)| + k_i|y_i(s-\tau_i)|)ds \\
&+ \int_0^t e^{\lambda s} \sum_{i=1}^n d_i[-c_i|y_i(s)| + 2c_i k_i |y_i(s-\tau_i)| \\
&+ \sum_{j=1}^n |a_{ij}|\lambda_j|y_j(s)| + \sum_{j=1}^n |b_{ij}|\mu_j|y_j(s-\tau_j)|]ds + M(\omega) \ , \qquad (6)
\end{aligned}
$$

where

$$\xi = \sum_{i=1}^{n} d_i |y_i(0) - G_i(y_i(-\tau_i))|$$

and

$$M(\omega) = \int_0^t e^{\lambda s} \sum_{i=1}^{n} d_i sgn[y_i(s) - G_i(y_i(s-\tau_i))] \sum_{j=1}^{n} \sigma_{ij}(y_j(s), y_j(s-\tau_j)) d\omega_j(s) \ .$$

Note that

$$\int_{t-\tau_i}^{t} e^{\lambda s}|y_i(s)|ds = \int_{-\tau_i}^{t} e^{\lambda s}|y_i(s)|ds - \int_0^t e^{\lambda(s-\tau_i)}|y_i(s-\tau_i)|ds \ .$$

So

$$\int_0^t e^{\lambda s}|y_i(s-\tau_i)|ds = e^{\lambda\tau_i}\int_{-\tau_i}^{t} e^{\lambda s}|y_i(s)|ds - e^{\lambda\tau_i}\int_{t-\tau_i}^{t} e^{\lambda s}|y_i(s)|ds \ ,$$

that is

$$\int_0^t e^{\lambda s}|y_i(s-\tau_i)|ds \le e^{\lambda\tau_i}\int_{-\tau_i}^{t} e^{\lambda s}|y_i(s)|ds \le e^{\lambda\tau}\int_{-\tau}^{t} e^{\lambda s}|y_i(s)|ds \ .$$

Following from (6) we have

$$\begin{aligned} V(y(t) - G(y(t-\tau)), t) \le \xi &+ \int_0^t e^{\lambda s} \sum_{i=1}^{n} (\lambda d_i - d_i c_i + \sum_{j=1}^{n} |a_{ji}| d_j \lambda_i)|y_i(s)|ds \\ &+ \int_0^t e^{\lambda(s+\tau)} \sum_{i=1}^{n} (\lambda d_i k_i + 2 d_i c_i k_i + \sum_{j=1}^{n} |b_{ji}| d_j \mu_i)|y_i(s)|ds \\ &+ \eta + M(\omega) \ , \end{aligned} \qquad (7)$$

where

$$\eta = \int_{-\tau}^{0} e^{\lambda(s+\tau)} \sum_{i=1}^{n} (\lambda d_i k_i + 2 d_i c_i k_i + \sum_{j=1}^{n} |b_{ji}| d_j \mu_i)|y_i(s)|ds \ .$$

It is obvious that $M(\omega)$ is a nonnegative semi-martingale. Applying Lemma 1, one derives

$$e^{\lambda t}\sum_{i=1}^{n} d_i |y_i(t) - G_i(y_i(t-\tau_i))| < +\infty, \quad t \ge 0 \ . \qquad (8)$$

This, together with Lemma 2, implies

$$\limsup_{t\to\infty} \frac{1}{t} \log \|y(t)\| \le -\frac{\lambda \bigwedge \beta}{2} \ , \qquad (9)$$

where $\beta = -\frac{2}{\tau}\log(\max_i k_i) > 0$. This proof is complete.

**Corollary 1.** *Assume* $(H_1)$ *and* $(H_3)$ *holds. Moreover, if the following inequality holds*

$$2c_ik_i + \sum_{j=1}^{n} |a_{ji}|\lambda_i + \sum_{j=1}^{n} |b_{ji}|\mu_i < c_i, \quad i = 1, \cdots, n \ . \tag{10}$$

*Then system (4) has an equilibrium point O which is almost surely exponentially stable.*

## 4 Examples

*Example 1.* Let $n = 1$. Consider the following stochastic neutral cellular neural networks:

$$\begin{aligned} d(x(t) - G(x(t-\tau))) = [&-cx(t) + af(x(t)) + bg(x(t-\tau)) \\ &+J]dt + \sigma(x(t), x(t-\tau))d\omega(t), \quad t \geq 0 \ . \end{aligned} \tag{11}$$

Choose $G(x) = \frac{1}{8}(x+\cos x-1)$, $f(x) = \sin x$, $g(x) = \frac{1}{4}x+1$, $J = -1$, $\sigma(x(t), x(t-\tau)) = x(t)$. Clearly, $(H_1)$ and $(H_3)$ hold, and $k = \frac{1}{4}$, $\lambda = 1$, $\mu = \frac{1}{4}$. Let $c = 2$, $a = \frac{1}{2}$, $b = 1$. By simple calculation, we easily see that $(H_2)$ holds. Thus, system (11) has an equilibrium point $O$ which is almost surely exponentially stable.

*Example 2.* Let $n = 2$. Consider the following stochastic neutral cellular neural networks:

$$\begin{aligned} d(x_i(t) - G_i(x_i(t-\tau_i))) = [-c_ix_i(t) + \sum_{j=1}^{n} a_{ij}f_j(x_j(t)) + \sum_{j=1}^{n} b_{ij}g_j(x_j(t-\tau_j)) \\ + J_i]dt + \sum_{j=1}^{n} \sigma_{ij}(x_j(t), x_j(t-\tau_j))d\omega_j(t), \ t \geq 0 \ , \end{aligned} \tag{12}$$

where $i = 1, 2$. Choose $G_1(x_1) = \frac{1}{12}(|x_1 + 1| - |x_1 - 1|)$, $G_2(x_2) = \frac{1}{8}(|x_2 + 1| - |x_2 - 1|)$, $f_j(x_j) = x_j$, $g_j(x_j) = 5 + \cos x_j$, $J_1 = -4$, $J_2 = -3$, $\sigma_{ij}(x_j(t), x_j(t-\tau_j)) = x_j(t)(i, j = 1, 2)$. Clearly, $(H_1)$ and $(H_3)$ hold, and $k_1 = \frac{1}{6}$, $k_2 = \frac{1}{4}$, $\mu_j = \lambda_j = 1(j = 1, 2)$. Let $c_1 = 2$, $c_2 = 3$, $a_{11} = \frac{1}{2}$, $a_{12} = a_{21} = \frac{1}{4}$, $a_{22} = \frac{1}{3}$, $b_{11} = \frac{1}{2}$, $b_{12} = \frac{1}{6}$, $b_{21} = b_{22} = \frac{1}{4}$. Take $d_1 = 5$, $d_2 = 3$. By simple calculation, we easily see that $(H_2)$ holds. Thus, system (12) has an equilibrium point $O$ which is almost surely exponentially stable.

## 5 Conclusions

In this paper, stochastic neutral cellular neural networks model has further been investigated. Some sufficient conditions ensuring the almost surely exponential stability are obtained by constructing a suitable Lyapunov functional and employing the nonnegative semi-martingale convergence theorem. These conditions obtained have important leading significance in the designs and applications of neural networks.

**Acknowledgement.** This research was supported by the Grant of "Qing-Lan Engineering" Project of Jiangsu Province, and the Science Foundation of Nanjing University of Aeronautics and Astronautics.

## References

1. Chua, L., Yang, L.: Cellular Neural Networks: Theory and Applications. IEEE Transactions on Circuits and Systems I **35** (1988) 1257–1290
2. Chua, L., Roska, T.: Cellular Neural Networks and Visual Computing. Cambridge, UK: Cambridge University Press, (2002)
3. Zhao, H., Cao, J.: New Conditions for Global Exponential Stability of Cellular Neural Networks with Delays. Neural Networks **18** (2005) 1332–1340
4. Zhao, H.: Globally Exponential Stability and Periodicity of Cellular Neural Networks with Variable Delays. Phys. Lett. A **336** (2005) 331–341
5. Chen, A., Cao, J., Huang, L.: Global Robust Stability of Interval Cellular Neural Networks with Time-varying Delays. Chaos, Solitons and Fractals **23** (2005) 787–799
6. Hu, J., Zhong, S., Liang, L.: Exponential Stability Analysis of Stochastic Delayed Cellular Neural Network. Chaos, Solitons and Fractals **27** (2006) 1006–1010
7. Cao, J., Ho, D.: A General Framework for Global Asymptotic Stability Analysis of Delayed Neural Networks Based on LMI Approach. Chaos, Solitons and Fractals **24** (2005) 1317–1329
8. Li, C., Liao, X., Zhang, R., Prasad, A.: Global Robust Exponential Stability Analysis for Interval Neural Networks with Time-varying Delays. Chaos, Solitons and Fractals **25** (2005) 751–757
9. Xu, D., Yang, Z.: Impulsive Delay Differential Inequality and Stability of Neural Networks. J. Math. Anal. Appl. **305** (2005) 107–120
10. Zhang, J.: Global Stability Analysis in Delayed Cellular Neural Networks. Computers and Mathematics with Applications **45** (2003) 1707–1727
11. Zhang, J., Suda, Y., Iwasa, T.: Absolutely Exponential Stability of A Class of Neural Networks with Unbounded Delay. Neural Networks **17** (2004) 391–397
12. Zhao, H.: A Comment on "Globally Exponential Stability of Neural Networks with Variable Delays". IEEE Transactions on Circuits and Systems II **53** (2006) 77–78
13. Curt, W.: Reactive Molecules: The Neutral Reactive Intermediates in Organic Chemistry. Wiley Press, New York (1984)
14. Salamon, D.: Control and Observation of Neutral Systems. Pitman Advanced Pub. Program, Boston (1984)
15. Shen, Y., Liao, X.: Razumikhin-type Theorems on Exponential Stability of Neutral Stochastic Functional Differential Equations. Chinese Science Bulletin **44** (1999) 2225–2228
16. He, H., Liao, X.: Stability Analysis of Neutral Neural Networks with Time Delay. Lecture Notes in Computer Science **3971** (2006) 147–152
17. Xu, S., Lam, J., Ho, D., et al.: Delay-dependent Exponential Stability for A Class of Neural Networks with Time Delays. Journal of Computational and Applied Mathematics **183** (2005) 16–28
18. Haykin, S.: Neural Networks. Prentice-Hall, NJ (1994)
19. Hasminskii, R.: Stochastic Stability of Differential Equations. D. Louvish, Thans., Swierczkowski, ED (1980)

20. Yang, H., Dillon, T.: Exponential Stability and Oscillation of Hopfield Graded Response Neural Network. IEEE Trans. On Neural Networks **5** (1994) 719–729
21. Liao, X., Mao, X.: Exponential Stability and Instability of Stochastic Neural Networks. Stochast. Anal. Appl. **14** (1996) 165–185
22. Mao, X.: Stochastic Differential Equation and Application. Horwood Publishing, Chichester (1997)
23. Liao, X., Mao, X.: Almost Sure Exponential Stability of Neutral Stochastic Differential Difference Equations. Journal of Mathematical Analysis and Applications **212** (1997) 554–570

# Synchronization of Neural Networks by Decentralized Linear-Feedback Control

Jinhuan Chen[1,2], Zhongsheng Wang[1], Yanjun Liang[1], Wudai Liao[1], and Xiaoxin Liao[3]

[1] College of Electronics and Information, Zhongyuan University of Technology Zhengzhou, P.R. China, 450007
zswang3@263.net
[2] Department of mathematics ,Zhengzhou University,Zhengzhou, P.R. China,450002
[3] Department of Control Science and Engineering, Huazhong University of Science and Technology, Hubei, Wuhan, P.R. China, 430074

**Abstract.** The problem of synchronization for a class of neural networks with time-delays has been discussed in this paper.By using of the Lyapunov stability theorem, a novel delay-independent and decentralized linear-feedback control law is designed to achieve the exponential synchronization. The controllers can be more easily designed than that obtained. The illustrative examples show the effectiveness of the presented synchronization scheme.

## 1 Introduction

In recent years, neural networks has attracted the attention of the scientists, due to their promising potential for the tasks of classification, associate memory and parallel computation, communication such as secure communication through the chaotic system, etc., those neural networks have been applied to describe complex nonlinear dynamical systems, and have become a field of active research over the past two decades[1-10 ]. It is known that the finite speed of amplifiers and the communication time of neurons may induce time delays in the interaction between the neurons when the neural networks were implemented by very large-scale integrated(VLSI) electronic circuits. Many researchers have devoted to the stability analysis of this kind of neural networks with time-delays. The chaotic phenomena in Hopfield neural networks and cellular neural networks with two or more neurons and differential delays have also been found and investigated[11,12]. Neural networks are nonlinear and high-dimensional systems consisting many neurons. To such systems, the centralized control method is hard to implement. In this paper, the decentralized control method is discussed for the synchronization problem of a class of chaotic systems such as Hopfield neural networks and cellular neural networks with time-delays. By using of the Lyapunov stability theorem, a novel delay-independent and decentralized linear control law is designed to achieve the exponential synchronization. The controllers can be more easily designed than that obtained in [12]. The illustrative examples show the effectiveness of the presented synchronization scheme.

D.-S. Huang, L. Heutte, and M. Loog (Eds.): ICIC 2007, LNAI 4682, pp. 157–163, 2007.

## 2 Synchronization Problem and Lemma

We consider the neural networks with time delay described by the differential delayed equation of the form

$$\begin{aligned}\dot{x}_i(t) = -d_i(c_i(x_i(t)) - \textstyle\sum_{j=1}^{n} a_{ij} f_j(x_j(t)) \\ - \textstyle\sum_{j=1}^{n} b_{ij} f_j(x_j(t-\tau_j)) + J_i), i = 1, ..., n\end{aligned} \tag{1}$$

Where $n \geq 2$ denotes the number of neurons in the networks, $x_i$ is the state variable associated with the $ith$ neurons, $d_i > 0$ represents an amplification gain,and $c_i(x_i)$is an appropriately behaved function remaining the solution of model(1) bounded. Feedback matrix $A = (a_{ij})_{n\times n}$ and the delayed feedback matrix $B = (b_{ij})_{n\times n}$indicate the interconnection strength among neurons without and withe time delay $\tau_j$, respectively. The activation function $f_i$ describes the manner in which the neurons respond to each other. Moreover,$f_i$ satisfies $0 < f_i^{'} \leq \overline{M_i}, i = 1, 2, ..., n$, in (1), it is assumed that $0 \leq \tau_j^* = max(\tau_j)$ for $1 \leq j \leq n$. $J_i$ is a external constant input. The initial conditions of (1) are given by $x_i(t) = \psi_i(t) \in C([-\tau_j^*, 0], R)$, where $C([-\tau_j^*, 0], R)$ denotes the set of all continuous functions from $[-\tau_j^*, 0]$ to $R$. The system (1) is called the master system.

A couple of chaotic neural networks are described by the following equation which is called the slave system

$$\begin{aligned}\dot{z}_i(t) = -d_i(c_i(z_i(t)) - \textstyle\sum_{j=1}^{n} a_{ij} f_j(z_j(t)) \\ - \textstyle\sum_{j=1}^{n} b_{ij} f_j(z_j(t-\tau_j)) + J_i) + u_i, i = 1, ..., n\end{aligned} \tag{2}$$

with the initial conditions of (2) are given by $z_i(t) = \phi_i(t) \in C([-\tau_j^*, 0], R)$, where $u_i$ is the appropriate control input that will be designed to obtain a certain control objective. Although the system's parameters are same, the initial condition on the (1) is different from that of the system (2). In fact,even the infinitesimal difference in the initial condition in (1) and (2) will lead to different chaotic phenomenon in those system. Let us define the synchronization error vector $e(t)$ as $e(t) = [e_1(t), e_2(t), ..., e_n(t)]^T$, where $e_i(t) = x_i(t) - z_i(t)$. We make the following assumption for the functions $c_i(x_i)$ and the activation function $f_i$.

**Assumption 1.** Function $c_i(x_i)$ and $(c_i(x_i))^{-1}$,where $i \in [1, 2, ..., n]$, are globally Lipschitz continuous. Moreover, $c_i^{'}(x_i) = \frac{dc_i(x_i)}{x_i} \geq \gamma_i > 0$

**Assumption 2.** Each function $f_i : R \to R, i \in [1, 2, ..., n]$, is bounded,and satisfies the Lipschitz condition with a Lipscgitz constant $L_i$, that is ,$|f_i(u) - f_i(v)| \leq L_i|u - v|$ for all $u, v \in R$.

**Definition 1**[12]**.** The system (1) and the uncontrolled system (2)( i.e. $u \equiv 0$) are said to be exponentially synchronized if there exist constants $\eta \geq 1$ and $\theta > 0$

such that $|x_i(t) - z_i(t)| \leq \eta max_{-\tau^* \leq s \leq 0}|x_i(s) - z_i(s)|exp(-\theta t)$ for all $t \geq 0$. Moreover, the constant $\theta$ is defined as the exponential synchronization rate.

Before we give the Lemma ,we consider the following differentiable inequality

$$D^+x_i \leq \sum_{j=1}^{n} c_{ij}x_j + \sum_{j=1}^{n} d_{ij}x_j(t-\tau) \tag{3}$$

where $i, j \in [1, 2, ..., n], x_i, d_{ij} \in C(R, R_+), c_{ij} \in C(R, R_+)(i \neq j), c_{ii} \in C(R, R)$, $R_+ = [0, +\infty)$

**Lemma** [7]**:** If there exists a $\eta < 0$ such that for any $i \in [1, 2, ..., n]$,

$$(c_{ii} - \eta) + \sum_{j=1, j\neq i}^{n} c_{ij} + \sum_{j=1}^{n} d_{ij}exp(-\eta\tau) < 0$$

then, any solution $x_i(t)$ of inequality (3) satisfies

$$x_i(t) \leq \overline{x}(t_0)exp(\eta(t - t_0))$$

The aim of this paper is to design the decentralized linear-feedback control $u_i$ associated only with the state error $e_i$ for the purpose of achieving the exponential synchronization between system (1) and (2) with the same system's parameters but the differences in initial conditions.

## 3 Decentralized Linear-Feedback Controller Design and Main Result

The error dynamical between system(1) and(2) can be expressed by the following equation

$$\begin{aligned} \dot{e_i}(t) = &-d_i(c_i(e_i(t) + z_i(t)) - c_i(z_i(t)) \\ &- \sum_{j=1}^{n} a_{ij}[f_j(e_j(t) + z_j(t)) - f_j(z_j(t))] \\ &- \sum_{j=1}^{n} b_{ij}[f_j(e_j(t - \tau_j) + z_j(t - \tau_j)) \\ &- f_j(z_j(t - \tau_j))]]) - u_i, i = 1, ..., n \end{aligned} \tag{4}$$

or expressed by the following compact form:

$$\begin{aligned} \dot{e_i}(t) = &-d_i(\beta_i(e_i(t)) - \sum_{j=1}^{n} a_{ij}\phi_j(e_j(t)) \\ &- \sum_{j=1}^{n} b_{ij}\phi_j(e_j(t - \tau_j))) - u_i, i = 1, ..., n \end{aligned} \tag{5}$$

where

$$\beta_i(e_i) = c_i(e_i + z_i) - c_i(z_i), \phi_j(e_j(t)) = f_j(e_j(t) + z_j(t)) - f_j(z_j(t)) \in R$$

**Main Theorem.** For system (1)and (2)which satisfy Assumption1 and 2, if the control input $u_i$ designed as

$$u_i(t) = K_i e_i(t)$$

where $K_i$ satisfies

$$-d_i r_i - K_i + \theta + \sum_{j=1}^{n} d_i |a_{ij}| L_j + \sum_{j=1}^{n} d_i |b_{ij}| L_j exp(\theta \tau_i) < 0$$

synchronization of system (1) and (2) can be obtained with synchronization rate $\theta$.

**Remark:** The constructer of the controllers is more simpler than that obtained in[12], and the synchronization rate $\theta$ can be selected.

**Proof.** In order to confirm the origin of (5) is globally exponentially stable, we construct the Lyapunov function V as

$$V = (|e_1(t)|, |e_2(t)|, ..., |e_n(t)|) = (V_1(t), V_2(t), ..., V_n(t))$$

Using the definition of $\phi_j(e_j(t))$ and the Assumption 2 yields

$$|\phi_j(e_j(t))| \leq L_j |e_j(t)|$$

$$|\phi_j(e_j(t - \tau_j))| \leq L_j |e_j(t - \tau_j)|$$

Taking the time derivative of $V$ along the trajectory of (5) :

$$\begin{aligned} D^+ V_i(t) &= \dot{e}_i(t) sign(e_i(t)) \\ &= [-d_i(\beta_i(e_i(t)) - \textstyle\sum_{j=1}^{n} a_{ij} \phi_j(e_j(t)) \\ &\quad - \textstyle\sum_{j=1}^{n} b_{ij} \phi_j(e_j(t - \tau_j))) - u_i] sign(e_i(t)) \end{aligned} \tag{6}$$

Since

$$\begin{aligned} -d_i(\beta_i(e_i(t)) sign(e_i(t)) &\leq -d_i r_i e_i(t) sign(e_i(t)) = -d_i r_i |e_i(t)| \\ &= -d_i r_i V_i(t) \end{aligned} \tag{7}$$

$$\begin{aligned} d_i \textstyle\sum_{j=1}^{n} a_{ij} \phi_j(e_j(t)) sign(e_i(t)) &\leq \textstyle\sum_{j=1}^{n} d_i |a_{ij}| L_j |(e_j(t))| \\ &= \textstyle\sum_{j=1}^{n} d_i |a_{ij}| V_j(t) \end{aligned} \tag{8}$$

$$\begin{aligned} d_i \textstyle\sum_{j=1}^{n} b_{ij} \phi_j(e_j(t - \tau_j)) sign(e_i(t)) &\leq \textstyle\sum_{j=1}^{n} d_i |b_{ij}| L_j |e_j(t - \tau_j)| \\ &= \textstyle\sum_{j=1}^{n} d_i |b_{ij}| L_j V_j(t - \tau_j) \end{aligned} \tag{9}$$

$$-u_i sign(e_i(t) = -K_i e_i(t) sign(e_i(t)) = -K_i V_i(t) \tag{10}$$

Then from (7)-(10) ,we can obtain

$$\begin{aligned} D^+ V_i(t) \leq &-(d_i r_i + K_i) V_i(t) + \textstyle\sum_{j=1}^{n} d_i |a_{ij}| V_j(t) \\ &+ \textstyle\sum_{j=1}^{n} d_i |b_{ij}| L_j V_j(t - \tau_j) \end{aligned} \tag{11}$$

by the Lemma and the condition of the main Theorem,the proof can be completed.

# 4 Illustrative Example

Consider the delayed Hopfied neural networks with two neurons as below[12]

$$\begin{bmatrix}\dot{x}_1\\ \dot{x}_2\end{bmatrix} = -\begin{bmatrix}1 & 0\\ 0 & 1\end{bmatrix}(\begin{bmatrix}x_1\\ x_2\end{bmatrix} - \begin{bmatrix}2 & -0.1\\ -5 & 3\end{bmatrix}\begin{bmatrix}f_1(x_1(t))\\ f_2(x_2(t))\end{bmatrix} - \begin{bmatrix}-1.5 & -0.1\\ -0.2 & -2.5\end{bmatrix}\begin{bmatrix}f_1(x_1(t-\tau_1))\\ f_2(x_2(t-\tau_2))\end{bmatrix}), \tag{12}$$

where $d_i = [1,1]^T, c_i(x_i) = x_i, \tau_i = 1$ and $f_i(x_i) = tanh(x_i), i = 1,2$. The feedback matrix and the delayed feedback matrix are specified as

$$A = (a_{ij})_{2\times 2} = \begin{bmatrix}2 & -0.1\\ -5 & 3\end{bmatrix}, B = (b_{ij})_{2\times 2} = \begin{bmatrix}-1.5 & -0.1\\ -0.2 & -2.5\end{bmatrix},$$

respectively. The system satisfies Assumption 1,2 with $L_1 = L_2 = 1$ and $r_1 = r_2 = 1$. The response chaotic Hopfield neural networks with delays is designed by

$$\begin{bmatrix}\dot{z}_1\\ \dot{z}_2\end{bmatrix} = -\begin{bmatrix}1 & 0\\ 0 & 1\end{bmatrix}(\begin{bmatrix}z_1\\ z_2\end{bmatrix} - \begin{bmatrix}2 & -0.1\\ -5 & 3\end{bmatrix}\begin{bmatrix}f_1(z_1(t))\\ f_2(z_2(t))\end{bmatrix} - \begin{bmatrix}-1.5 & -0.1\\ -0.2 & -2.5\end{bmatrix}\begin{bmatrix}f_1(z_1(t-\tau_1))\\ f_2(z_2(t-\tau_2))\end{bmatrix}) + \begin{bmatrix}u_1(t)\\ u_2(t)\end{bmatrix}. \tag{13}$$

Taking $\theta = 1$,, it follows from the Main Theorem that if the control input $u_i(t)$ chosen as

$$u_1(t) = 7e_1(t), u_2(t) = 16e_2(t)$$

then the system (12) and (13) can be synchronized with the exponential convergence rate $\theta = 1$, Fig.1 depicts the synchronization error of the state variables between the drive system(12) and the response system (13) with the initial condition $[x_1(s), x_2(s)] = [0.3, 0.4]^T$, and $[z_1(s), z_2(s)] = [0.1, 0.3]^T$, respectively.

Taking $\theta = 3$,, it follows from the Main Theorem that if the control input $u_i(t)$ chosen as

$$u_1(t) = 37e_1(t), u_2(t) = 65e_2(t)$$

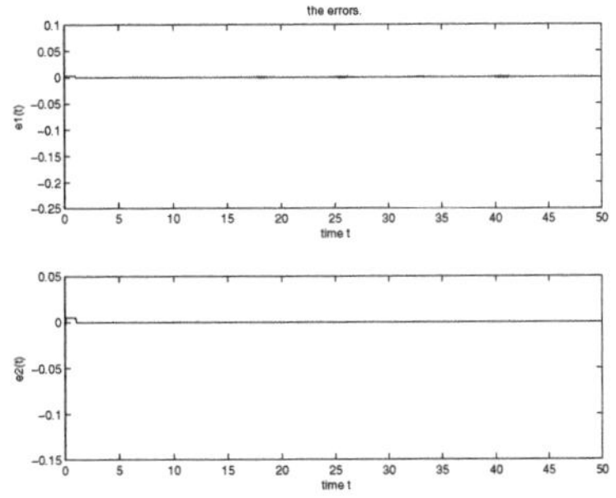


**Fig. 1.** The graphs of state $e_1, e_2$when $K_1 = 7, K_2 = 16$

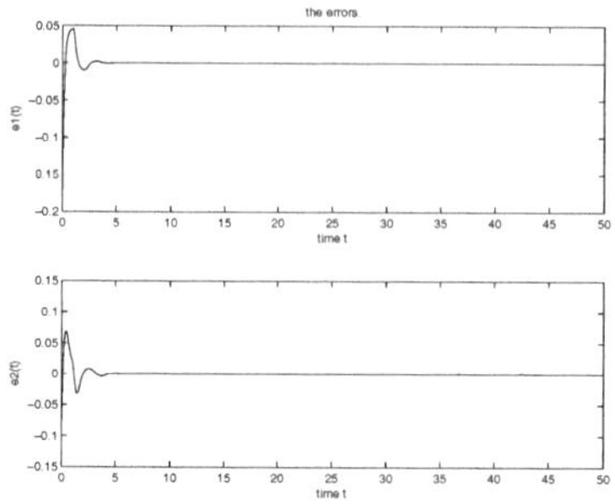


**Fig. 2.** Waveform graphs of $e_1, e_2$ when $K_1 = 37, k_2 = 65$

then the system (12) and (13) can be synchronized with the exponential convergence rate $\theta = 3$, Fig.2 depicts the synchronization error of the state variables between the drive system(12) and the response system (13) with the initial condition $[x_1(s), x_2(s)] = [0.4, 0.7]^T$, and $[z_1(s), z_2(s)] = [0.15, 0.55]^T$, respectively.

## 5 Conclusion

The synchronization problem for a class of Hopfield neural networks has been discussed in this paper, a novel decentralized linear-feedback control has been designed.The controllers associated only with the current state error can be constructed easily. The illustrative examples show the effectiveness of the presented synchronization scheme.

## Acknowledgements

This work is supported by National Natural Science Foundation of China (No.60274007), Foundation of Ph.D candidate of Zhengzhou University (No.20040907) and Foundation of Young Bone Teacher of Henan Province(No. 2004240).

## References

1. Liang X. B., Wu L. D.: Globally Exponential Stability of Hopfield Neural Networks and Its Applications, Sci. China (series A), (1995) 523-532
2. Forti M. and Tesi A.: New Conditions for Global Stability of Neural Networks with Application to Linear and Quadratic Programming Problems , IEEE Trans. on Circ. and Sys. I: Fundanmetntal Theory and Application, (1995) 354-366
3. Liao X. X., Xiao D. M.: Globally Exponential Stability of Hopfield Neural Networks with Time-Varying Delays, ACTA Electronica Sinica, (2000) 1-4
4. Marco M. D., Forti M. and Tesi A.: Existence and Characterization of Limit Cycles in Nearly Symmetric Neural Networks, IEEE Trans. on Circ. and Sys. I: Fundanmetntal Theory and Application, (2002) 979-992

5. Forti M.: Some Extensions of a New Method to Analyze Complete Stability of Neural Networks, IEEE Trans. on Neural Networks, **13** (2002) 1230-1238
6. Zeng Z. G., Wang J. and Liao X. X.: Global Exponential Stability of a General Class of Recurrent Neural Networks with Time-Varying Delays, IEEE Trans. on Circ. and Sys. I: Fundanmetntal Theory and Application, **50** (2003) 1353-1358
7. Zeng Z.G., Wang J. and Liao X. X.: Global Asymptotic Stability and Global Exponetial Stability of Networks with Unbounded Time-Varying Delays, IEEE Trans. Circuits Syst. II:express briefs , **52** (2005) 168-173
8. Cao J. D., Huang D. S.,Qu Y. Z.:Global Robust Stability of Delayed Recurrent Neural Networks, Chaos,Solitons and Fractals, **23** (2005)221-229
9. Fantacci R., Forti M., Marini M.,and etc.,:A Neural Network for Constrained Optimaization with Application to CDMA Communication Systems, IEEE Trans. on Circ. and Sys. II:Analog and Digital Signal Processing , **50** (2003) 484-487
10. Zhou S. B., Liao X. F. Yu J. B.,and etc.: Chaos and Its Synchronization in Two-Neuron Systems with Discret Delays, Chaos,Solitons and Fractals, **21** (2004) 133-142
11. Cao, J. D. : Global Stability Conditionso for Delayed CNNs. Hopfield Neural Networks and Its Applications. IEEE Trans. Circuits Syst. I, **48** (2001) 1330-1333
12. Cheng C. J., Liao T. L., Yan J. J and etc.,: Synchronization of Neural Networks by Decentralized Feedback Control. Physics Letters A , **338** (2005) 28-35

# Synchronous Pipeline Circuit Design for an Adaptive Neuro-fuzzy Network

Che-Wei Lin, Jeen-Shing Wang, Chun-Chang Yu, and Ting-Yu Chen

Department of Electrical Engineering,
National Cheng Kung University
Tainan 701, Taiwan, R.O.C.
jeenshin@mail.ncku.edu.tw

**Abstract.** This paper presents an efficient synchronous pipeline hardware implementation procedure for a neuro-fuzzy (NF) circuit. We decompose the NF circuit into a feedforward circuit and a backpropagation circuit. The concept of pre-calculation to share computation results between the feedforward circuit and backpropagation circuit is introduced to achieve a high throughput rate and low resource usage. A novel pipeline architecture has been adopted to fulfill the concept of pre-calculation. With the unique pipeline architecture, we have successfully enhanced the throughput rate and resource sharing between modules. Particularly, the multiplier usage has been reduced from 7 to 3 and the divider usage from 3 to 1. Finally, we have implemented the NF circuit on FGPA. Our experimental results show a superior performance than that of an asynchronous pipeline design approach and the NF system implemented on MATLAB®.

**Keywords:** Synchronous pipeline design, neuro-fuzzy circuit, and FPGA.

## 1 Introduction

Intelligent systems have combined with different knowledge, techniques, and methods in the area of science and have been regarded as effective tools to solve complex and real-world problems over a long period of time. These systems usually have self-adaptive capacity and clear decision procedures for solving problems in specific areas such as general human professional knowledge in various environments. Within these systems, neuro-fuzzy (NF) systems are one of the representatives. A NF system consists of a neural network (NN) and a fuzzy logic system under the same structure [1]. The fuzzy logic system uses the fuzzy inference rule (IF-THEN rules) to transform the linguistic terms into mathematical functions that can be computed. The neural network provides the ability to learn and adapt, and also to ensure that the NF system will still keep working well in changing circumstances [2], [3].

Although an adaptive neuro-fuzzy system has already been developed over a very long period of time, there are still some difficulties in practical applications. One of the main reasons for this is that its algorithm for updating parameters is so complicated that it spends a lot of time in computation. If the NF networks can be implemented into

D.-S. Huang, L. Heutte, and M. Loog (Eds.): ICIC 2007, LNAI 4682, pp. 164–173, 2007.

hardware, their value will be improved greatly because of the high-speed computation ability of hardware. In the past few years, many hardware implementations have been realized through analog or digital methods. In the field of neural network chips and fuzzy controller chips, many researchers [5-11] have shown that either digital or analog technology can be utilized to meet different demands and specifications. In [5], Wang et al. introduced a hardware structure of a single perceptron that serves as the basic nerve cell and its implementation method with FPGA. Porrmann et al. [6] proposed the implementation of three different artificial neural networks on reconfigurable hardware accelerators. Vitabile et al. [7] proposed an efficient multilayer perceptron (MLP) digital implementation on FPGA. Togai and Watanabe first proposed a digital hardware FLC in [8]. Jou et al. [9] designed an adaptive fuzzy logic controller by VLSI. In [10], an online adaptive temperature control with adaptive recurrent fuzzy controller chips was implemented by FPGA. Juang and Chen proposed a hardware implementation of the Takagu-Sugeno-Kan (TSK)-type recurrent fuzzy network (TRFN-H) for water bath temperature control in [11].

In this paper, we focus on the pipelined hardware design of a neuro-fuzzy circuit with on-chip learning capability. The research topics of this paper include the computation analysis of a neuro-fuzzy network, dataflow analysis, and pipeline structuring design. The main design idea lies not only in using fewer resources but also in giving high operation efficiency. By simplifying the network computation and avoiding the computation of multiplication and division, we can make each multiplier and divider reach parallel processing or operate at the same time.

## 2 Computational Procedures of Neuro-Fuzzy Networks

The network computation can be separated into feedforward and backpropagation procedures. In the feedforward procedure, the input variables are fed to the network and go through fuzzification, fuzzy rule inference engine, and defuzzification operations to obtain the corresponding output variables. The obtained outputs are then compared with the desired outputs to generate an error signal for tuning the network adjustable parameters in the backpropagation procedure. The operations involved in these two procedures are introduced in the following two subsections.

### 2.1 Feedforward Procedure

The operations of the nodes in each layer are as follows:

- Layer 1: The node in this layer only transmits input values to layer 2.
- Layer 2: Each node in this layer represents fuzzification. The output of the node generates the firing strength corresponding to the input values transmitted from Layer 1. Considering the simplicity of hardware implementation, we adopt isosceles triangular functions as the membership functions. The membership grade of the triangular membership function is expressed by (1), where $x_i$ denotes the $i^{th}$ input, and $a_{ij}$ and $b_{ij}$ are the center and width of the $j^{th}$ triangular membership function for the $i^{th}$ input, respectively. $M$ represents the total number of fuzzy rules.

$$\mu_{ij}^{(2)}(x_i)=1-2\frac{|x_i-a_{ij}|}{b_{ij}},\ i=1,2...,n \text{ and } j=1,2...,M. \tag{1}$$

- Layer 3: The node in this layer executes the function of the fuzzy inference. The node integrates the firing strengths of the corresponding fuzzification functions, and its mathematical expression is as (2).

$$\mu_k^{(3)}(\mathbf{x})=\prod_{i=1}^{n}\mu_{ij}^{(2)},\ j\in\left\{\mu_{ij}^{(2)} \text{ with connection to } k^{th} \text{ node}\right\}. \tag{2}$$

- Layer 4: The output node plays a weighted-average defuzzification as in (3).

$$y=\frac{\sum_{l=1}^{M}\mu_k^{(3)}w_k^{(3)}}{\sum_{l=1}^{M}\mu_k^{(3)}}. \tag{3}$$

## 2.2 Backpropagation Procedure

A backpropagation learning algorithm is utilized to update the centers and width of the fuzzification layer and the weights of the output layer. First, the error function is defined as (4), where $y$ is the current output and $y^d$ is the desired output.

$$E=\frac{1}{2}(y-y^d)^2. \tag{4}$$

(5), (6) and (7) express the corresponding error signal of adjustable parameters.

$$\frac{\partial E}{\partial a_{ij}}=\frac{\partial E}{\partial \mu_l^{(3)}}\frac{\partial \mu_l^{(3)}}{\partial \mu_{ij}^{(2)}}\frac{\partial \mu_{ij}^{(2)}}{\partial a_{ij}}=\left((y-y^d)\frac{1}{ACC}\right)\times\left(\sum_l w_l\mu_l^{(3)}-y\sum_l \mu_l^{(3)}\right)\times\frac{1}{b_{ij}}\times\frac{2}{\mu_{ij}^{(2)}}\times sign(x_i-a_{ij}). \tag{5}$$

$$\frac{\partial E}{\partial b_{ij}}=\frac{\partial E}{\partial \mu_l^{(3)}}\frac{\partial \mu_l^{(3)}}{\partial \mu_{ij}^{(2)}}\frac{\partial \mu_{ij}^{(2)}}{\partial b_{ij}}=\left((y-y^d)\frac{1}{ACC}\right)\times\left(\sum_l w_l\mu_l^{(3)}-y\sum_l \mu_l^{(3)}\right)\times\frac{1}{b_{ij}}\times(\frac{1}{\mu_{ij}^{(2)}}-1). \tag{6}$$

$$\frac{\partial E}{\partial w_k}=\frac{\partial E}{\partial y}\frac{\partial y}{\partial w_k}=\left((y-y^d)\frac{1}{ACC}\right)\times\mu_k^{(3)}. \tag{7}$$

The update rules of adjustable parameter are described as (8), (9) and (10). $\eta$ is the learning rate.

$$a_{ij}(t+1)=a_{ij}(t)-\eta\frac{\partial E}{\partial a_{ij}}. \tag{8}$$

$$b_{ij}(t+1)=b_{ij}(t)-\eta\frac{\partial E}{\partial b_{ij}}. \tag{9}$$

$$w_k(t+1)=w_k(t)-\eta\frac{\partial E}{\partial w_k}. \tag{10}$$

# 3 Hardware Design and Implementation of NF Networks

We will introduce the procedures of NF network design in this section. The design procedures include dataflow analysis, pipeline structure design, resource allocation, control circuit design.

## 3.1 Dataflow of Feedforward Circuit

There are two parts in the feedforward circuit design: 1) To generate modules with respect to the operation of each layer, 2) To rearrange each module after the analysis of pre-calculated terms in an update rule. First, we partition our design into three primary modules: fuzzification (FC), inference engine (IE), and defuzzification (DF), with respect to the function of each layer in the NF network. The second step is to modify the DFG of three modules in the feedforward circuit. The modifications include: 1) adding an operation of necessary pre-calculation terms in the update rule, and 2) combining some operation procedures due to the particular architecture of the NF network. These modifications may accelerate the execution speed of the backpropagation circuit, prevent redundant memory saving, and retrieve between operations of the IE module and the DF module.

The final DFG after modification of each module in the feedforward circuit is stage 1, stage 2, and a combination of stage 3, and pre-backward as shown in Fig.1. We name the three modified modules as fuzzification, Mf2 and Mf3. Two sub-blocks with dotted-line circles are the pre-calculation for minor terms in the backpropagation circuit. In our design, we implement a two-input-one-output NF network. All inferred

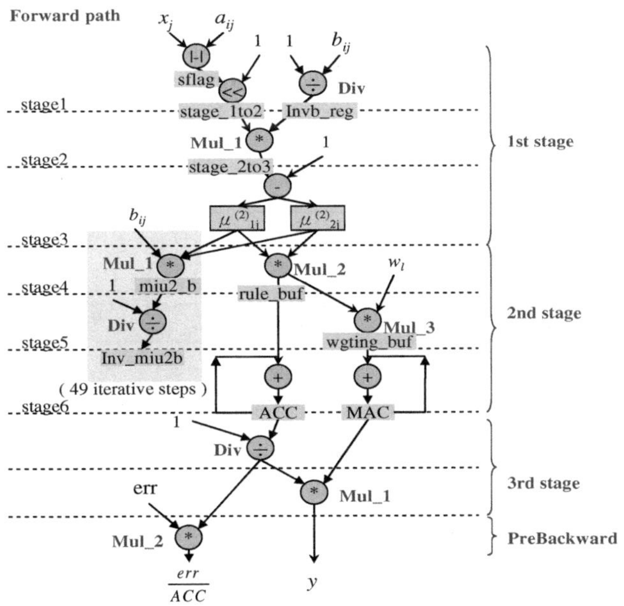


**Fig. 1.** Three primary modules in the feedforward circuit

results in the inference layer are directly sent to a single node in the defuzzification layer. Thus, we combine some operation procedures originally in the DF module with the IE module in order to prevent unnecessary memory saving and retrieval.

## 3.2 Dataflow of Backpropagation Circuit

There are three steps in the backpropagation circuit design. The first step is to analyze and label the terms that can be pre-calculated in the feedforward circuit. The second step is to generate a data flow graph (DFG) of the backpropagation circuit and perform integer linear programming (ILP) to achieve optimal scheduling.

An analysis of pre-calculated terms on the update rule reduces the control steps and resource usage in the backpropagation circuit. For example, the term, $\left(\sum_l w_l\mu_l^{(3)} - y\sum_l \mu_l^{(3)}\right)\times\frac{1}{b_{ij}}\times\frac{1}{\mu_{ij}^{(2)}}$ in update rule of $a_{ij}$ and $b_{ij}$ is complicated to implement in the backpropagation circuit, while it is easy to implement if some minor terms in the feedforward circuit are pre-calculated. We analyze the update rule and partition it into several minor terms such as labels *wgting_buf*, *rule_buf*, *Inv_miu2b* in Fig. 2. Labels *wgting_buf* and *rule_buf* can be easily achieved in the feedforward circuit, and label *Inv_miu2b* can be achieved by extending the operation in the inference engine module of the feedforward circuit. An effective analysis of pre-calculated terms reduces the control steps in our design from 10 to 5 steps. The sharing terms such as label *mod_err* occurs in each update rule. The pre-calculation of *mod_err* can also be arranged into the feedforward circuit. All pre-calculated terms in the feedforward circuit store in memory indexed by *i*, *j* and *l*. These terms will be retrieved from memory during the operation of the backpropagation circuit.

The second step is to generate the DFG of the backpropagation circuit and perform an integer linear programming (ILP) to achieve optimal scheduling. The final DFG of the backpropagation circuit is shown in Fig. 3.

$$\frac{\partial E}{\partial a_{ij}} = \frac{\partial E}{\partial \mu_l^{(3)}}\frac{\partial \mu_l^{(3)}}{\partial \mu_{ij}^{(2)}}\frac{\partial \mu_{ij}^{(2)}}{\partial a_{ij}} = \left\{\left((y-y^d)\frac{1}{ACC}\frac{1}{\mu_{ij}^{(2)}}\right)\times\left(\sum_l w_l\mu_l^{(3)} - y\sum_l \mu_l^{(3)}\right)\right\}\frac{\partial \mu_{ij}^{(2)}}{\partial a_{ij}}$$

$$= \left\{\underbrace{\boxed{\left((y-y^d)\frac{1}{ACC}\right)}}_{\text{mod\_err}}\times\left(\underbrace{\boxed{\sum_l w_l\mu_l^{(3)}}}_{\text{wgting\_buf}} - y\underbrace{\boxed{\sum_l \mu_l^{(3)}}}_{\text{rule\_buf}}\right)\times\frac{1}{b_{ij}}\right\}\times\underbrace{\boxed{\frac{2}{\mu_{ij}^{(2)}}}}_{\text{Inv\_miu2b}}\times\underbrace{\boxed{sign(x_i-a_{ij})}}_{\text{sflag}}$$

$$\frac{\partial E}{\partial b_{ij}} = \underbrace{\boxed{\left\{\left((y-y^d)\frac{1}{ACC}\right)\times\left(\sum_l w_l\mu_l^{(3)} - y\sum_l \mu_l^{(3)}\right)\times\frac{1}{b_{ij}}\right\}\times\frac{1}{\mu_{ij}^{(2)}}}}_{\text{buffer\_a}}\times(1-\mu_{ij}^{(2)})$$

$$\frac{\partial E}{\partial w_k} = \underbrace{\boxed{\left((y-y^d)\frac{1}{ACC}\right)}}_{\text{mod\_err}}\times\mu_k^{(3)}$$

**Fig. 2.** Learning rules for sharing computation results

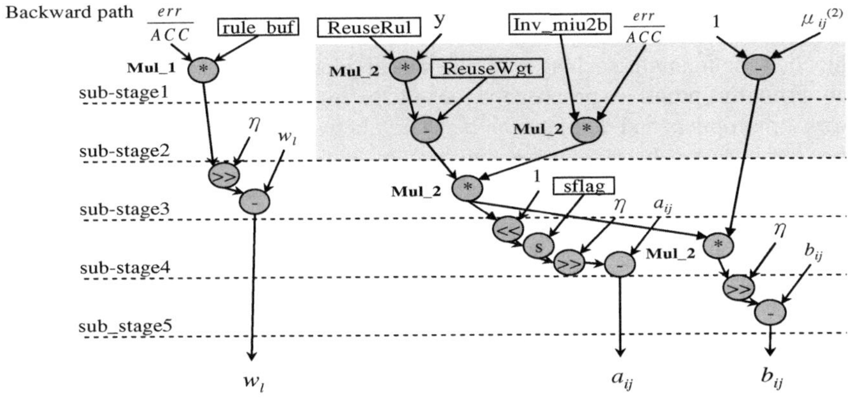


**Fig. 3.** The DFG of the backpropagation circuit

### 3.3 Pipeline Architecture of Feedforward Circuit and Backpropagation Circuit

This section illustrates the different pipeline strategies of the feedforward circuit and the backpropagation circuit. The fine-grain pipeline is adopted in three modules of the feedforward circuit to increase the throughput rate. There are various numbers of nodes in each layer of the NF network. The data dependency between different layers in the NF network yields the three modules that are impossible to execute concurrently. Based on this property, three modules are designed as three asynchronous islands and are communicated by a handshaking signal. We integrate the synchronous and asynchronous design methodology in the feedforward circuit. We named this architecture a globally-asynchronous-locally-synchronous (GALS) architecture. The architecture of GALS in the feedforward circuit is shown in Fig. 4.

The backpropagation circuit is realized by an ordinary pipeline structure but with a different pipeline stage and latency for various adjustable parameters. The reason is that the data of updating $w_l$ executes 49 times in the backpropagation circuit while $a_{ij}$ and $b_{ij}$ executes only 14 times. The overall control cost of the backpropagation circuit is determined by updating $w_l$,; thus, we increase the pipeline latency for updating $a_{ij}$ and $b_{ij}$ to reduce resource usage. The datapath of updating $w_l$ is designed into two pipeline stages with no clock latency, and $a_{ij}$ and $b_{ij}$ are designed into three pipeline stages with two clock latency.

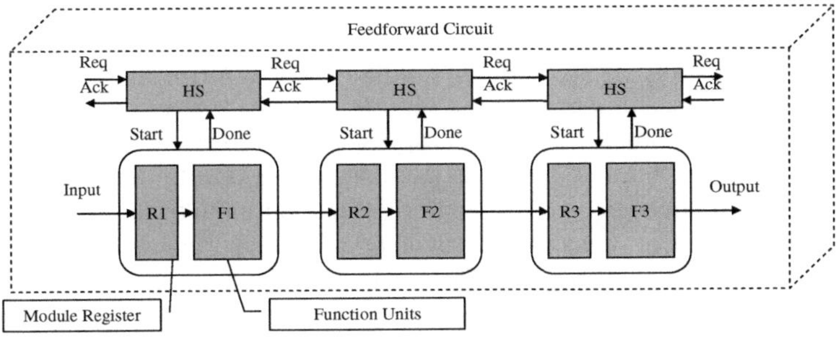


**Fig. 4.** Globally asynchronous locally synchronous architecture

### 3.4 Resource Allocation

Arithmetic function units such as multipliers or dividers are area expensive in a digital circuit. From this point, we propose the idea of sharing multipliers and dividers. In NF networks, multipliers and dividers can be shared between the feedforward circuit and the backpropagation because these two circuits never execute concurrently. Three modules in the feedforward circuit can also share multipliers and dividers because the three modules also never execute concurrently in the feedforward circuit under the GALS architecture. From the above concept, we can determine the minimum hardware usage by finding the maximum usage of each module. Table 1 lists the resource consumption (multipliers and dividers) of each module in our design. Finally, we adopt 3 multipliers and 1 divider and share them in both the feedforward and backpropagation circuits. In Table 1, we have successfully reduced the usage of the multipliers from 7 to 3 and the dividers from 3 to 1.

**Table 1.** Resource usage of multipliers and dividers in each module

| Module name | Multiplier | Divider |
|---|---|---|
| Fuzzification | 1 | 1 |
| Mf2 | 3 | 1 |
| Mf3 | 1 | 1 |
| Backpropagation circuit | 2 | 0 |

### 3.5 Control Circuit Design

The control circuit realized by finite-state machines (FSMs) not only coordinates the executions on the datapath, but also generates a great quantity of control signals such as propagation indexes and special flags. In other words, FSMs produce the signals that fetch data from memory, load and read the contents to/from registers, direct signals through multiplexers, and control the operations of function units. Based on the previous datapath, we have designed the control unit consisting of six components: a main FSM, a function FSM, two forward sub-FSMs, and two backpropagation sub-FSMs. The structure of the control unit is shown in Fig. 5. The main FSM takes the responsibility of enabling other machines through generating control signals, such as en_back, fw_run, run_err, firemem, run_fwb. The interface between the FSMs is also illustrated in Fig. 5. The computation of the NF network circuit is iterative and many indexes require special arrangements to take the sequence of operations procedure by FSMs into account. The control signals produced by the feedforward and backpropagation sub-FSMs dominate the operations of the feedforward and backpropagation modules, respectively. The forward sub-FSMs and backpropagation sub-FSMs also generate control signals to the main FSM for indicating the transition progress. In addition, function units of the circuit should be coordinated by control units. The calculation FSM generates signals to select multipliers and a division operator. That is, the selection signals enable multipliers and a division operation at specific control states. These signals also control the multiplexers and de-multiplexers of function units.

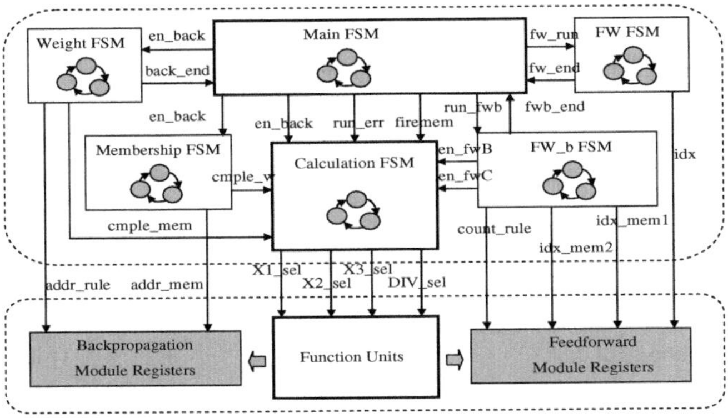


**Fig. 5.** Architecture of the control circuit

# 4 Hardware Verification

The proposed circuit has been coded in Verilog and synthesized by Synopsys™ Design Compiler. A throughput rate comparison among the proposed pipeline NF network, asynchronous pipeline, GALS structure, and MATLAB® is discussed in this section.

In our previous research, we proposed the asynchronous pipeline design and the GLAS structure design of the same structure as that of our NF network. We downloaded the asynchronous circuit, GALS structure circuit and proposed circuit to the same FGPA device (Clock: 50MHz) to compare the throughput rate. The detailed execution performance of the feedforward circuit and the backpropagation circuit are listed in Table 2. A throughput rate comparison in Table 2 shows that the proposed circuit outperforms the asynchronous circuit and the GLAS structure circuit about 10.12, 1.312 times respectively. In general, to establish an NF network with the MATLAB® simulation platform is the most typical method. Table 2 also lists the throughput rate of MATLAB®. The throughput rate of the proposed design is 2203.9 times faster than the same implementation established in MATLAB®. Table 3 is the area report of each sub-module of the proposed design. The high cost of the multiplier is shown in Table 3; the area of the multiplier is almost as large as that of the backpropagation circuit.

**Table 2.** Throughput rate comparison

| Circuits | Proposed design (KHz) | Asynchronous pipeline design (KHz) | GALS structure (KHz) | MATLAB® (KHz) |
|---|---|---|---|---|
| Feedforward circuit | 308.64 | 37.74 | 308.64 | 0.438 |
| Backpropagation circuit | 510.21 | 38.28 | 322.58 | 0.1 |
| Overall | 192.31 | 19.00 | 146.63 | 0.087 |

**Table 3.** Area report of the proposed design

| Sub-module name | Area ($\mu m^2$) |
|---|---|
| Divider | 79530.710938 |
| Multiplier_1 | 77917.515625 |
| Multiplier_2 | 77917.515625 |
| Multiplier_3 | 77917.515625 |
| Multiplexer (To select the input of multipliers/dividers) | 7018.685547 |
| De-Multiplexer (To select the output of multipliers/dividers) | 5518.506348 |
| Reuse Register (Storage of the pre-calculated terms) | 54972.214844 |
| Backpropagation circuit | 82674.773438 |
| Three primary modules in the feedforward circuit | 820007.250000 |
| Control circuit | 10351.767578 |
| Total | 1295469.875000 |

## 5 Conclusion

This paper presents an efficient synchronous pipeline hardware implementation procedure for a neuro-fuzzy (NF) circuit. The proposed idea of pre-calculation terms greatly reduces the control steps and resource usage of the backpropagation circuit. Resource sharing between various modules reduces the multiplier usage from 7 to 3 and the divider usage from 3 to 1. Even we share multipliers and dividers; the throughput rate is still maintained at a high speed (192.31 KHz). We attribute these merits to our proposed synchronous pipeline architecture. The effectiveness and superiority of the proposed design approach has been validated through the comparison with an asynchronous pipeline design approach and the NF system implemented on MATLAB®.

## References


1. Wang, J.-S., Lee, C.-S.G.: Self-Adaptive Neuro-Fuzzy Inference Systems for Classifications Applications. IEEE Trans. on Fuzzy Systems, 10 (6) (2002) 790-802
2. Wang, J.-S., Lee, C.-S.G.: Self-Adaptive Recurrent Neuro-Fuzzy Control of an Autonomous Underwater Vehicle. IEEE Trans. on Robotics and Automation, 19 (2) (2003) 283-295
3. Rubaai, A., Kotaru, R., Kankam, M. D.: A Continually Online-Trained Neural Network Controller for Brushless DC MotorDrives. IEEE Trans. on Industry Applications, 36 (2) (2000) 475-483
4. Micheli, G.-D.: Synthesis and Optimization of Digital Circuits. McGraw-Hill, Newyork (1994)
5. Wang, Q., Yi, B., Xie, Y., Liu, B.: The Hardware Structure Design of Perceptron with FPGA Implementation. Proc. of the IEEE Int. Conf. on Systems, Man and Cybernetics, 1 (2003) 762-767
6. Porrmann, M., Witkowski, U., Kalte, H., Ruckert, U.: Implementation of Artificial Neural Networks on a Reconfigurable Hardware Accelerator. Proc. of 10th Euromicro Workshop on Parallel, Distributed and Network-based Processing, (2002) 243-250

7. Vitabile, S., Conti, V., Gennaro, F., Sorbello, F.: Efficient MLP Digital Implementation on FPGA. Proc. of 8th Euromicro Conf. on Digital System Design, (2005) 218-222
8. Togai, M., Watanabe, H.: Expert System on a Chip: An Engine for Real-Time Approximate Reasoning. IEEE Expert, 1 (3) (1986) 55-62
9. Jou, J.-M., Chen, P.-Y., Yang, S.-F.: An Adaptive Fuzzy Logic Controller: Its VLSI Architecture and Applications. IEEE Trans. on VLSI Systems, 8 (1) (2000) 52-60
10. Juang, C.-F., Hus, C.-H.: Temperature Control by Chip-Implemented Adaptive Recurrent Fuzzy Controller Designed by Evolutionary Algorithm. IEEE Trans. on Circuits and Systems,52 (11) (2005) 2376-2384
11. Juang, C.-F., Chen, J.-S.: Water Bath Temperature Control by a Recurrent Fuzzy Controller and its FPGA Implementation. IEEE Trans. on Industrial Electronics, 53 (3) (2006) 941-949
12. Hwang, C.-T., Lee, J.-H., Hsu, Y.-C.: A Formal Approach to the Scheduling Problem in High Level Synthesis. IEEE Trans. on Computer-Aided Design,10 (4) (1991) 464-475
13. Paulin, P.G., Knight, J.P.: Algorithm for High-Level Synthesis. IEEE Trans. on Design & Test of Computers, 6 (6) (1989) 18-31
14. Gajski, D., Wu, A., Dutt, N., Lin, S.: High-level Synthesis: Introduction to Chip and System Design. Kluwer Academic, Boston (1992)
15. Mitra, S., Hayashi, Y.: Neuro-Fuzzy Rule Generation: Survey in Soft Computing Framework. IEEE Trans. on Neural Networks,11 (3) (2000) 748-768

# The Projection Neural Network for Solving Convex Nonlinear Programming

Yongqing Yang and Xianyun Xu

School of Science, Southern Yangtze University, Wuxi 214122, China
yyq640613@gmail.com, xxy630417@163.com

**Abstract.** In this paper, a projection neural network for solving convex optimization is investigated. Using Lyapunov stability theory and LaSalle invariance principle, the proposed network is showed to be globally stable and converge to exact optimal solution. Two examples show the effectiveness of the proposed neural network model.

## 1 Introduction

The convex programming problems arise often in scientific research and engineering application. The traditional numerical methods for solving convex programming problems involve a complex iterative process and have longer computational time. This may limit their usage in large-scale or real-line optimization such as in regression analysis, image and signal progressing, parameter estimation, filter design, robot control, etc. It is well-known that the neural network can solve optimization problems in real time. Recently, the studies of constructing neural networks for optimization have been a new focus point. Some neural networks for solving convex optimization were proposed based on gradient method and dual theorem and projection method [1]-[15]. Kennedy and Chua [2] proposed a neural network for nonlinear programming. The network contains a finite penalty parameter, so it converges an approximation optimal solution only. Chen *et al.* [3] proposed a neural network for solving convex nonlinear programming problems based on primal-dual method. Its distinguishing features are that the primal and dual problems can be solved simultaneously. But the number of state variables becomes more, which enlarges the scale of network. Based on projection method and Karush-Kuhn-Tacker (KKT) optimality conditions of convex programming, Friesz *et al*[15], Xia and Wang [4] proposed a projection neural network. However, for some convex programming problems, the stability of Friesz neural networks can not be guaranteed (see example 2).

Motivated by the above discussions, in this paper, we present a new projection neural network for solving convex programming. This new projection neural network improved the Friesz projection network. The global stability and convergence are proved using Lyapunov stability theory and LaSalle invariance principle.

The organization of the paper is as follows: In Section 2, we will construct a neural network model based on projection theorem and KKT conditions. In Section 3, the global stability and convergence will be proved. In Section 4, two

D.-S. Huang, L. Heutte, and M. Loog (Eds.): ICIC 2007, LNAI 4682, pp. 174–181, 2007.

illustrative examples and simulation results will be given to show the effectiveness of the proposed network. Conclusions are given in Section 5.

## 2 Preliminaries

In the paper, we consider the following convex programming problems

$$\min_{x\in\Omega} f(x)$$
$$\text{s.t.} \quad g(x) \le 0, \tag{1}$$

where $f(x)$, $g(x) = (g_1(x), ..., g_s(x))$ are twice continuous differentiable convex functions.

It is well known that if a point $x^* \in R^n$ is the optimal solution of (1), then there exists $\lambda^* = (\lambda_1^*, ..., \lambda_s^*)^T \in R_+^s$, such that $(x^*, \lambda^*)$ satisfies the following variational inequalities

$$\begin{cases} (x-x^*)^T(\nabla f(x) + \nabla g(x)y) = 0, \ \forall x \in \Omega \\ (\lambda - \lambda^*)^T(-g(x^*)) \ge 0, \ \forall \lambda \ge 0 \end{cases} \tag{2}$$

where $\nabla f(x) = (\partial f(x)/\partial x)^T$, $\nabla g(x) = (\nabla g_1(x), ..., \nabla g_s(x))$. $x^*$ is called a KKT point of (1) and $\lambda^*$ is called the Lagrangian multiplier vector corresponding to $x^*$. Moreover, if $f$ and $g_i$, $i = 1, ..., s$ are all convex, then $x^*$ is an optimal solution of (1) if and only if $x^*$ is a KKT point of (1).

Based on Friesz projection neural network, we have

$$\begin{cases} \frac{dx}{dt} = -(x - P_\Omega[x - \nabla f(x) - \nabla g(x)^T\lambda]), \\ \frac{d\lambda}{dt} = -(\lambda - [\lambda + g(x)]^+). \end{cases} \tag{3}$$

Unfortunately, for some convex programming problems, the neural networks (3) is unstable (see example 2).

In this paper, we construct a new projection neural network model for solving (1). For simplicity, we denote $u(t) = (x^T, \lambda^T)^T \in R^{n+s}$, $D = \Omega \times R_+^s$, $\bar{\lambda} = [\lambda + g(x)]^+$, $\bar{x} = P_\Omega[x - \nabla f(x) - \nabla g(x)^T\lambda]$, and $D^*$ is the optimal point set of (1).

$$\begin{cases} \frac{dx}{dt} = -(x - P_\Omega[x - \nabla f(x) - \nabla g(x)^T\bar{\lambda}]), \\ \frac{d\lambda}{dt} = -(\lambda - [\lambda + g(x)]^+)/2. \end{cases} \tag{4}$$

## 3 Stability and Converge Analysis

In the section, we will study the stability and convergence of neural network (4). Before proving the theorem, we first introduce an Lemma.

**Lemma 1** [16]: Assume that the set $\Omega \subset R^n$ is a closed convex set. Then

$$(v - P_\Omega(v))^T(P_\Omega(v) - u) \ge 0, \quad v \in R^n, \quad u \in \Omega, \tag{5}$$

and

$$\| P_\Omega(u) - P_\Omega(v) \| \le \| u - v \|, \quad u, v \in R^n. \tag{6}$$

**Theorem 1:** For any initial point $u(t_0) = (x(t_0)^T, \lambda(t_0)^T)^T \in R^{n+s}$, there exists a unique continuous solution $u(t) = (x(t)^T, \lambda(t)^T)^T$ for system (4). Moreover, $x(t) \in \Omega$ and $\lambda(t) \geq 0$, provided that $x(t_0) \in \Omega$ and $\lambda(t_0) \geq 0$.

**Proof:** The projection mappings $P_\Omega(\cdot)$ and $(\cdot)^+$ are nonexpansive. Since $\nabla f(x)$ and $\nabla g_i(x)$, $i = 1, 2, ..., s$ are continuously differentiable on an open convex set $D \subseteq R^{n+s}$ including $\Omega \times R^s_+$, therefore $x - P_\Omega[x - \nabla f(x) - \nabla g(x)^T \bar{\lambda}]$ and $\lambda - [\lambda + g(x)]^+$ are locally Lipschitz continuous. According to the local existence of ordinary differential equations, the initial value problem of the system (4) has a unique solution.

Let initial point $x_0 = x(t_0) \in \Omega$ and $\lambda_0 = \lambda(t_0) \geq 0$. Since

$$\begin{cases} \frac{dx}{dt} + x = P_\Omega[x - \nabla f(x) - \nabla g(x)^T \bar{\lambda}], \\ \frac{d\lambda}{dt} + \lambda = [\lambda + g(x)]^+/2. \end{cases} \tag{7}$$

$$\begin{cases} \int_{t_0}^t \left(\frac{dx}{dt} + x\right) e^t dt = \int_{t_0}^t P_\Omega[x - \nabla f(x) - \nabla g(x)^T \bar{\lambda}] e^t dt, \\ \int_{t_0}^t \left(\frac{d\lambda}{dt} + \lambda\right) e^t dt = \int_{t_0}^t [\lambda + g(x)]^+ e^t/2dt. \end{cases} \tag{8}$$

thus

$$\begin{cases} x(t) = e^{-(t-t_0)} x_0 + e^{-t} \int_{t_0}^t e^t P_\Omega[x - \nabla f(x) - \nabla g(x)^T \bar{\lambda}] dt, \\ \lambda(t) = e^{-(t-t_0)} \lambda_0 + e^{-t} \int_{t_0}^t e^t [\lambda + g(x)]^+/2dt. \end{cases} \tag{9}$$

By the integration mean value theorem, we have

$$\begin{cases} x(t) = e^{-(t-t_0)} x_0 + \left(1 - e^{-(t-t_0)}\right) P_\Omega[\hat{x} - \nabla f(\hat{x}) - \nabla g(\hat{x})^T \hat{\lambda}], \\ \lambda(t) = e^{-(t-t_0)} \lambda_0 + \left(1 - e^{-(t-t_0)}\right) [\hat{\lambda} + g(\hat{x})]^+/2. \end{cases} \tag{10}$$

It follows $x(t) \in \Omega$ and $\lambda(t) \geq 0$ from $x(t_0) \in \Omega$ and $\lambda(t_0) \geq 0$. This completes the proof.

**Theorem 2:** Assume that $f(x)$, $g_i(x)$, $i = 1, 2, ..., s, x \in R^n$ are convex differentiable on an open convex set $D \subseteq R^{n+s}$ including $\Omega \times R^s_+$, then neural network (4) is globally stable in the sense of Lyapunov and, for any initial point $(x(t_0)^T, \lambda(t_0)^T)^T \in R^{n+s}$, the solution trajectory of (4) will converge to a point in $D^*$. In particular, neural network (4) is asymptotically stable when $D^*$ has only a point.

**Proof:** By Theorem 1, $\forall (x_0^T, \lambda_0^T)^T \in \Omega \times R^s_+$, there exists a unique continuous solution $(x(t)^T, \lambda(t)^T)^T \subseteq \Omega \times R^s_+$ for system (4).

Define a Lyapunov function as follow

$$\begin{aligned} V(x, \lambda) = f(x) - f(x^*) + \frac{1}{2}[(\bar{\lambda})^2 - (\lambda^*)^2] - (x - x^*)^T (\nabla f(x^*) + \nabla g(x^*)^T \lambda^*) \\ -(\lambda - \lambda^*)^T \lambda^* + \frac{1}{2} \parallel x - x^* \parallel^2 + \frac{1}{2} \parallel \lambda - \lambda^* \parallel^2 . \end{aligned} \tag{11}$$

Noting that $\| \bar{\lambda} \|^2 = \sum_{i=1}^{s} [(\lambda_i + g_i(x))^+]^2$ and

$$[(\lambda_i + g_i(x))^+]^2 = \begin{cases} [(\lambda_i + g_i(x))]^2, & \text{if } \lambda_i + g_i(x) \geq 0, \\ 0, & \text{otherwise,} \end{cases} \tag{12}$$

we have

$$\nabla \| \bar{\lambda} \|^2 = \nabla \Big( \sum_{i=1}^{s} [(\lambda_i + g_i(x))^+]^2 \Big) = \begin{pmatrix} 2\nabla g(x)^T \bar{\lambda} \\ 2\bar{\lambda} \end{pmatrix}. \tag{13}$$

Calculating the derivative of $V(t)$ along the trajectory of system (4) and by $-\lambda + \bar{\lambda} = g(x) - (\lambda + g(x))^-$, one has

$$\begin{aligned}
\frac{dV(x,\lambda)}{dt} &= [\nabla f(x) + \nabla g(x)^T \bar{\lambda} - \nabla f(x^*) - \nabla g(x^*)^T \lambda^* + x - x^*]^T (-x + \bar{x}) \\
&\quad + (\bar{\lambda} + \lambda - 2\lambda^*)^T (-\lambda + \bar{\lambda})/2 \\
&= - \| x - \bar{x} \|^2 \\
&\quad + [\nabla f(x) + \nabla g(x)^T \bar{\lambda} - \nabla f(x^*) - \nabla g(x^*)^T \lambda^* + \bar{x} - x^*]^T (-x + \bar{x}) \\
&\quad - \frac{1}{2} \| \lambda - \bar{\lambda} \|^2 + (\bar{\lambda} - \lambda^*)^T (-\lambda + \bar{\lambda}) \\
&= - \| x - \bar{x} \|^2 - [\nabla f(x) - \nabla f(x^*)]^T (x - x^*) \\
&\quad - (\nabla f(x^*) + \nabla g(x^*)^T \lambda^*)^T (\bar{x} - x^*) \\
&\quad - [x - \nabla f(x) - \nabla g(x)^T \bar{\lambda} - \bar{x}]^T (\bar{x} - x^*) \\
&\quad - (\nabla g(x)^T \bar{\lambda} - \nabla g(x^*)^T \lambda^*)^T (x - x^*) - \frac{1}{2} \| \lambda - \bar{\lambda} \|^2 \\
&\quad + (\bar{\lambda} - \lambda^*)^T [g(x) - (\lambda + g(x))^-] \\
&= - \| x - \bar{x} \|^2 - \frac{1}{2} \| \lambda - \bar{\lambda} \|^2 - [\nabla f(x) - \nabla f(x^*)]^T (x - x^*) \\
&\quad - [\nabla f(x^*) + \nabla g(x^*)^T \lambda^*]^T (\bar{x} - x^*) \\
&\quad - [x - \nabla f(x) - \nabla g(x)^T \bar{\lambda} - \bar{x}]^T (\bar{x} - x^*) \\
&\quad - \bar{\lambda}^T [-\nabla g(x)(x^* - x) - g(x) + g(x^*)] + \bar{\lambda}^T g(x^*) - \bar{\lambda}(\lambda + g(x))^- \\
&\quad + (\lambda^*)^T [\nabla g(x^*)(x - x^*) - g(x) + g(x^*)] \\
&\quad - (\lambda^*)^T g(x^*) + (\lambda^*)^T (\lambda + g(x))^- \\
&= - \| x - \bar{x} \|^2 - \frac{1}{2} \| \lambda - \bar{\lambda} \|^2 - [\nabla f(x) - \nabla f(x^*)]^T (x - x^*) \\
&\quad - [\nabla f(x^*) + \nabla g(x^*)^T \lambda^*]^T (\bar{x} - x^*) \\
&\quad - [x - \nabla f(x) - \nabla g(x)^T \bar{\lambda} - \bar{x}]^T (\bar{x} - x^*) \\
&\quad - \bar{\lambda}^T [g(x^*) - g(x) - \nabla g(x)(x^* - x)] + \bar{\lambda}^T g(x^*) \\
&\quad - (\lambda^*)^T [g(x) - g(x^*) - \nabla g(x^*)(x - x^*)] + (\lambda^*)^T (\lambda + g(x))^- \\
&\leq - \| x - \bar{x} \|^2 - \frac{1}{2} \| \lambda - \bar{\lambda} \|^2 - [\nabla f(x) - \nabla f(x^*)]^T (x - x^*) \\
&\quad - [\nabla f(x^*) + \nabla g(x^*)^T \lambda^*]^T (\bar{x} - x^*)
\end{aligned}$$

$$
\begin{aligned}
&-[x - \nabla f(x) - \nabla g(x)^T \bar{\lambda} - \bar{x}]^T (\bar{x} - x^*) \\
&-\bar{\lambda}^T [g(x^*) - g(x) - \nabla g(x)(x^* - x)] \\
&-(\lambda^*)^T [g(x) - g(x^*) - \nabla g(x^*)(x - x^*)]. \qquad (14)
\end{aligned}
$$

In the inequality of Lemma 1, let $v = x - \nabla f(x) - \nabla g(x)^T \bar{\lambda}$ and $u = x^*$, we obtain

$$(x - \nabla f(x) - \nabla g(x)^T \bar{\lambda} - \bar{x})^T (\bar{x} - x^*) \geq 0 \qquad (15)$$

From the differentiable convexities of $f(x)$ and $g(x)$, $\forall x \in \Omega$, we have

$$\begin{cases} [\nabla f(x) - \nabla f(x^*)]^T (x - x^*) \geq 0 \\ g(x^*) - g(x) - \nabla g(x)(x^* - x) \geq 0 \\ g(x) - g(x^*) - \nabla g(x^*)(x - x^*) \geq 0 \end{cases} \qquad (16)$$

Substituting (2), (15) and (16) into (14), one has

$$\frac{dV(x,\lambda)}{dt} \leq - \parallel x - \bar{x} \parallel^2 - \frac{1}{2} \parallel \lambda - \bar{\lambda} \parallel^2 \leq 0 \qquad (17)$$

This means neural network (4) is globally stable in the sense of Lyapunov. Next, since

$$V(x,\lambda) \geq \frac{1}{2}(\parallel x - x^* \parallel^2 + \parallel \lambda - \lambda^* \parallel^2).$$

That is, $V(x,\lambda)$ is positive definite and radially unbounded. Thus, there exists a convergent subsequence

$$\{(x(t_k)^T, \lambda(t_k)^T)^T | t_0 < t_1 < ... < t_k < t_{k+1} < ...\}, \text{ and } t_k \to \infty \text{ as } k \to \infty$$

such that $\lim_{k\to\infty} (x(t_k)^T, \lambda(t_k)^T)^T = (\hat{x}^T, \hat{\lambda}^T)^T$, where $(\hat{x}^T, \hat{\lambda}^T)^T$ satisfies

$$\frac{dV(x,\lambda)}{dt} = 0,$$

which indicates that $(\hat{x}^T, \hat{\lambda}^T)^T$ is an $\omega-$limit point of $\{(x(t)^T, \lambda(t)^T)^T | t \geq t_0\}$. From the LaSalle Invariant Set Theorem, one has that $\{(x(t)^T, \lambda(t)^T)^T \to M\}$ as $t \to \infty$, where $M$ is the largest invariant set in $K = \{(x(t)^T, \lambda(t)^T)^T | \frac{dV(x,\lambda)}{dt} = 0\}$. From (4) and (17), it follows that $\frac{dx}{dt} = 0$ and $\frac{d\lambda}{dt} = 0 \Leftrightarrow \frac{dV(x,\lambda)}{dt} = 0$. Thus, $(\hat{x}^T, \hat{\lambda}^T)^T \in D^*$ by $M \subseteq K \subseteq D^*$.

Substituting $x^* = \hat{x}$ and $\lambda^* = \hat{\lambda}$ in (11), we define another Lyapunov function

$$
\begin{aligned}
\hat{V}(x,\lambda) &= F(x) - F(\hat{x}) + \frac{1}{2}[(\bar{\lambda})^2 - (\hat{\lambda})^2] - (x - \hat{x})^T (\nabla F(\hat{x}) + \nabla g(\hat{x})\hat{\lambda}) - (\lambda - \hat{\lambda})^T \hat{\lambda} \\
&+ \frac{1}{2} \parallel x - \hat{x} \parallel^2 + \frac{1}{2} \parallel \lambda - \hat{\lambda} \parallel^2 \qquad (18)
\end{aligned}
$$

Then $\hat{V}(x,\lambda)$ is continuously differentiable and $\hat{V}(\hat{x},\hat{\lambda}) = 0$.

Noting that $\lim_{k\to\infty}(x(t_k)^T, \lambda(t_k)^T)^T = (\hat{x}^T, \hat{\lambda}^T)^T$, we have

$$\lim_{k\to\infty} \hat{V}(x(t_k)^T, \lambda(t_k)^T)^T = \hat{V}(\hat{x}, \hat{\lambda}) = 0.$$

So, $\forall\varepsilon > 0$, there exists $q > 0$ such that for all $t > t_q$, we have $\hat{V}(x, \lambda) < \varepsilon$.

Similar to the above analysis, we can prove that function $\frac{d\hat{V}(x,\lambda)}{dt} \leq 0$. It follows that for $t \geq t_q$

$$\| x(t) - \hat{x} \|^2 /2 + \| \lambda(t) - \hat{\lambda} \|^2 /2 \leq \hat{V}(x, \lambda) \leq \varepsilon.$$

This is, $\lim_{t\to\infty} x(t) = \hat{x}$, and $\lim_{t\to\infty} \lambda(t) = \hat{\lambda}$. So, the solution trajectory of the neural network (4) is globally convergent to an equilibrium point $(\hat{x}^T, \hat{\lambda}^T)^T$, i.e. $(\hat{x}^T, \hat{\lambda}^T)^T$ is also an optimal solution of (1).

In particular, if $D^* = \{((x^*)^T, (\lambda^*)^T)^T\}$, then for each $x_0 \in \Omega$ and $\lambda_0 \geq 0$, the solution $(x^T, \lambda^T)^T$ with initial point $(x_0^T, \lambda_0^T)^T$ will approach to $((x^*)^T, (\lambda^*)^T)^T$ by the analysis above. That is, the neural network (4) is globally asymptotically stable. This completes the proof.

## 4 Simulation Examples

In this section, two simulation examples are given to demonstrate the feasibility and efficiency of the proposed neural networks for solving the convex nonlinear programming problems. The simulation is conducted on Matlab, and the ordinary differential equation is solved by Runge-Kutta method.

**Example 1:** Consider the following nonlinear programming

$$\min \frac{1}{2}[(x_1 - x_2)^4 + (x_2 + x_3)^2 + (x_1 + x_3)^2],$$

$$\text{subject to } \begin{cases} x_1^2 + x_2^4 - x_3 \leq 0, \\ (2 - x_1)^2 + (2 - x_2)^2 - x_3 \leq 0, \\ 2e^{-x_1+x_2} - x_3 \leq 0, \\ x_1^2 + x_2^2 - 2x_1 + x_2 \leq 4, \\ |x_1| \leq 2, \quad |x_2| \leq 2, \quad x_3 \geq 0. \end{cases} \tag{19}$$

This problem has an optimal solutions $x^* = (1.0983, 0.9037, 1.9565)^T$. Using neural network (4) to solve the problem (19), all simulation results show the trajectory of neural network (4) converge to the optimal solution. The corresponding transient behavior is shown in Fig.1.

**Example 2:** Consider the following nonlinear programming

$$\min \frac{1}{4}(x_1 + x_2)^4 - 16x_2,$$

$$\text{subject to } \begin{cases} -x_1 + x_2 \leq 0, \\ x \geq 0. \end{cases} \tag{20}$$

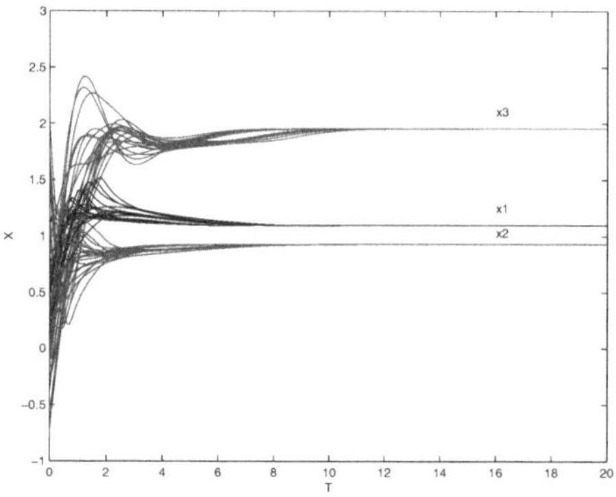


**Fig. 1.** Trajectories of network (4)

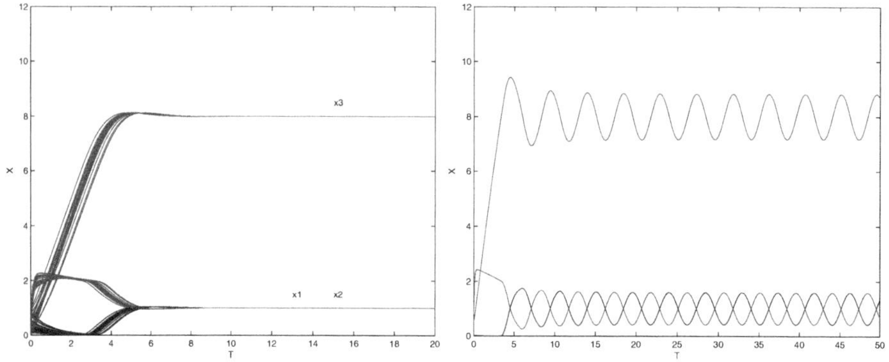


**Fig. 2.** (a)Trajectories of network (4), (b) Trajectories of network (3)

The nonlinear programming problem has an optimal solution $x^* = (1,1)^T$. Using neural network (4) to solve the problem, all simulation results show the trajectory of neural network (4) converge to the optimal solution of problem (20). To make a comparison, we solve the problem (20) by using neural network (3). Simulation result shows the trajectory of neural network (20) is not stable. The corresponding transient behavior is shown in Fig.2(a) and (b).

# 5 Conclusions

In the paper, we have investigated a convex nonlinear programming problem with nonlinear inequality constraints. Based on projection theorem, a new projection neural network model was constructed. This network improved the Friesz projection network and was proved to be globally stable in the sense of Lynapunov and the solution trajectory can converge to an optimal solution of original optimization problem. Two illustrative examples were given to show the effectiveness of the proposed neural network. Thus, we can conclude that the proposed projection neural network is feasible.

## References

1. Tank, D. W., Hopfield, J. J.: Simple 'Neural' Optimization Network: An A/D Converter, Signal Decision Circuit and A Linear Programming Circuit, IEEE Trans. Circuits Syst., **33**, (1986), 533-541
2. Kennedy, M. P., Chua, L. O.: Neural Networks for Nonlinear Programming, IEEE Trans. Circuits Syst., **35**, (1988), 554-562
3. Chen,K. Z., Leung, Y., Leung, K. S., Gao, X. B.: A Neural Network for Solving Nonlinear Programming Problem, Neural Comput. and Applica., **11**, (2002), 103-111
4. Xia, Y., Wang, J.: A Recurrent Neural Networks for Nonlinear Convex Optimization Subject to Nonlinear Inequality Constraints, IEEE Trans. Circuits Syst.-I, **51**, (2004) 1385-1394
5. Gao, X. B.: A Novel Neural Network for Nonlinear Convex Programming, IEEE Trans. Neural Networks, **15**, (2004), 613-621
6. Zhang, Y., Wang, J.: A Dual Neural Network for Convex Quadratic Programming Subject to Linear Equality and Inequality Constraints, Phys. Lett. A, **298**, (2002), 271-278
7. Tao, Q., Cao, J., Xue, M., Qiao, H.: A High Performance Neural Network for Solving Nonlinear Programming Problems with Hybrid Constraints, Phys. Lett. A, **288**, (2001), 88-94
8. Liu, Q., Cao, J., Xia, Y.: A Delayed Neural Network for Solving Linear Projection Equations and Its Analysis, IEEE Trans. Neural Networks, **16**, (2005), 834-843
9. Yang, Y., Cao, J.: Solving Quadratic Programming Problems by Delayed Projection Neural Network, IEEE Trans. Neural Networks, **17**, (2006), 1630-1634
10. Yang, Y., Cao, J.: A Delayed Neural Network Method for Solving Convex Optimization Problems, Intern. J. Neural Sys. **16**, (2006), 295-303
11. Yang, Y., Xu, Y., Zhu, D.: The Neural Network for Solving Convex Nonlinear Programming Problem, Lecture Note Compu. Sci., **4113**, (2006), 494-499
12. Xia, Y., Feng, G., Wang, J.: A Recurrent Neural Networks with Exponential Convergence for Solving Convex Quadratic Program and Related Linear Piecewise Equations, Neural Networks, **17**, (2004), 1003-1015
13. Liu, Q., Wang, J., Cao, J.: A Delayed Lagrangian Network for Solving Quadratic Programming Problems with Equality Constraints, Lecture Note Comput. Sci., **3971**, (2006), 369-378
14. Hu, X., Wang, J.: Solving Pseudomonotone Variational Inequalities and Pseudoconvex Optimization Problems Using the Projection Neural Network, IEEE Trans. Neural Networks, **17**, (2006), 1487-1499
15. Friesz, T. L., Bernstein, D. H., Mehta, N. J., Tobin, R. L., Ganjlizadeh, S.: Day-today Dynamic Network Disequilibria and Idealized Traveler Information Systems, Opera. Rese., **42**, (1994), 1120-1136
16. Kinderlehrer, D., Stampcchia, G.: An Introduction to Variational Inequalities and Their Applications, New York: Academic, 1980

# Usage of Hybrid Neural Network Model MLP-ART for Navigation of Mobile Robot

Andrey Gavrilov and Sungyoung Lee

Department of Computer Engineering, Kyung Hee University, 1, Soechen-ri, Giheung-eop, Yongin-shi, Gyeonggi-do, 449-701, Korea
Andr_gavrilov@yahoo.com, sylee@oslab.khu.ac.kr

**Abstract.** We suggest to apply the hybrid neural network based on multi layer perceptron (MLP) and adaptive resonance theory (ART-2) for solving of navigation task of mobile robots. This approach provides semi supervised learning in unknown environment with incremental learning inherent to ART and capability of adaptation to transformation of images inherent to MLP. Proposed approach is evaluated in experiments with program model of robot.

**Keywords:** neural networks, mobile robot, hybrid intelligent system, adaptive resonance theory.

## 1 Introduction

Usage of neural networks for navigation of mobile robots is a very popular area at last time. This tendency was born in works of N.M.Amosov [1] and R.Brooks [2]. Short review of this topic may be found in [3]. This interest of using neural networks for this task is explained by that a key challenge in robotics is to provide the robots to function autonomously in unstructured, dynamic, partially observable, and uncertain environments. The problem of navigation may be divided on following tasks: map building, localization, path planning, and obstacle avoidance.

Many attempts to employ different neural networks models for solving of navigation tasks are known.

Usage of multi layer perceptrons (MLP) with error back propagation learning algorithm has some disadvantages most of them are complexity or even impossibility to relearn, slow training and orientation on supervised learning. In [4] was made the attempt to overcome some of these shortcomings by development of multi layer hybrid neural network with preprocessing with principle component analysis (PCA). This solution allows some reduce the time of learning. But rest disadvantages of MLP are remained.

In [5] A.Billard and G.Hayes suggested architecture DRAMA based on recurrent neural network with delays. This system is interesting as probably first attempt to develop universal neural network based control system for behavior in uncertain dynamic environment. However it was oriented on enough simple binary sensors detecting any events.

D.-S. Huang, L. Heutte, and M. Loog (Eds.): ICIC 2007, LNAI 4682, pp. 182–191, 2007.

We suppose that most perspective approach is usage of unsupervised learning based on adaptive resonance theory [6]. In [7] usage of this approach for building of map for navigation was proposed. The attempt of employ of model ART-2 for solving of navigation task of robot oriented on interaction by natural language was carried out [8]. But this model is dealing with primary features of images and so is sensitive to its transformations. This disadvantage leads to impossibility to use it in dynamic unknown environment for solving of such task as avoidance of obstacles using real time information from sensors. To overcome this drawback in [9] was employed multi-channel model and evaluated for solving of minefield navigation task. But in this model for every category is needed to use separate ART model. This feature limits availability of such approach, essentially in case of using visual-like sensor information.

We suggest employing of hybrid model MLP-ART2, proposed by authors and evaluated in processing of visual information [10, 11]. In this model multi-layer perceptron with error back propagation algorithm as preprocessor is used for reducing of sensitivity of ART to transformations of images from sensors. In this paper we propose usage of the model MLP-ART2 for solving of one high level task of navigation namely recognition of situation in environment with respect to position of obstacles and target and decision making about changing of direction of movement. This task is solved in combination with avoidance of obstacles solved by simple deterministic algorithms.

## 2 Hybrid Neural Network MLP-ART2

In our model of neural network (figure 1) the first several layers of neurons are organized as MLP. Its outputs are the inputs of model ART-2. MLP provides conversion of primary feature space to secondary feature space with lower dimension. Neural network ART-2 classifies images and uses secondary features to do it. Training of MLP by EBP (with limited small number of iterations) provides any movement of an output vector of MLP to centre of recognized cluster of ART-2 in feature space.

In this case the weight vector (center) of recognized cluster is desired output vector of MLP. It could be said that the recognized class is a context in which system try to

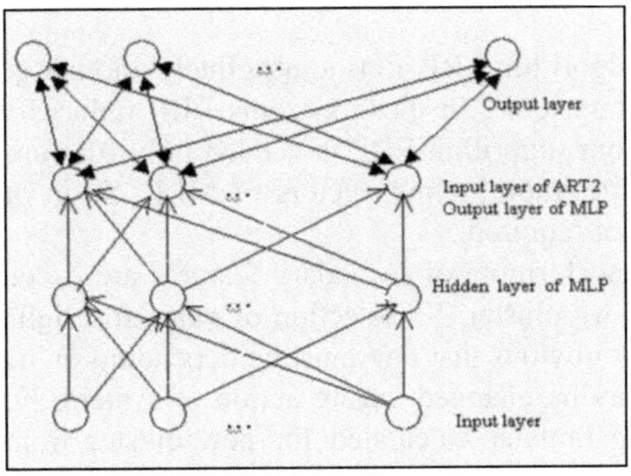


**Fig. 1.** Structure of hybrid neural network

recognize other images like previous, and in some limits the system "is ready to recognize" its by this manner. In other words neural network "try to keep recognized pattern inside corresponding cluster which is recognizing now".

Action of the suggested model is described by the following unsupervised learning algorithm:

1. In MLP let the weights of connections equal to *1/n,* where $n$ is quantity of neurons in the previous layer (number of features for first hidden layer). The quantity of output neurons $N_{out}$ of ART-2 is considered equal zero.

2. The next example from training set is presented to inputs of MLP. Outputs of MLP are calculating.

3. If $N_{out}$=0, then the output neuron is formed with the weights of connections equal to values of inputs of model ART-2 (the outputs of MLP).

4. If $N_{out}$> 0, in ART-2 the algorithm of calculation of distances between its input vector and centers of existing clusters (the weight vectors of output neurons) is executing using Euclidian distance:

$$d_j = \sqrt{\sum_i (y_i - w_{ij})^2} \,,$$

where: $y_i$ – $i^{th}$ feature of input vector of ART-2, $w_{ij}$ – $i^{th}$ feature of weight vector of $j^{th}$ output neuron (the center of cluster). After that the algorithm selects the output neuron-winner with minimal distance. If the distance for the neuron-winner is more than defined a vigilance threshold or radius of cluster $R$, the new cluster is created as in step 3.

5. If the distance for the neuron-winner is less than $R$, then in model ART-2 weights of connections for the neuron-winner are updating by:

$$w_{im} = w_{im} + (y_i - w_{im})/(1 + N_m)\,,$$

where: $N_m$ – a number of recognized input vectors of $m^{th}$ cluster before. Also for MLP an updating of weights by standard error back propagation algorithm (EBP) is executing. In this case a new weight vector of output neuron-winner in model ART-2 is employed as desirable output vector for EBP, and the quantity of iterations may be small enough (e.g., there may be only one iteration).

6. The algorithm repeats from step 2 while there are learning examples in training set.

Note that in this algorithm EBP aims at absolutely another goal different from that in usual MLP-based systems. In those systems EBP reduces error-function to very small value. But in our algorithm EBP is needed only for some decreasing distance between actual and desirable output vectors of MLP. So in our case the long time learning of MLP is not required.

Algorithm EBP and forming of secondary features are executed only when image "is captured" by known cluster. So selection of value for vigilance threshold is very important. Intuitively obvious that one must be depending on transformation speed of input images and may be changed during action of system. For our architecture we used value of this parameter calculated for new cluster from distance of neuron-winner by formula

$$r = K \min d_i ,$$

where $K$ is a coefficient between 1 and 2. In our experiments it was selected 1.2.

## 3 Simulation and Experiments

To evaluate the proposed model for selection of direction of movement with respect to position of robot, obstacles and target, experiments are conducted based on program simulation of mobile robot in 2D space for solving of navigation task, i.e. moving to target avoiding the obstacles. These experiments were provided by special program MRS developed in Delphi for simulation of mobile robots in two-dimensional simplified environment.

In our simulation following base primitives are assumed to be applied for interaction of robot with environment:

1) *dist(i)* – value of distance getting from i-th range sensor ( one of 12 sensors);
2) *target_dist* – distance from target;
3) *target_dir* – direction to target (in degrees);
4) *robot_dir* – direction of robot's movement (in degrees);
5) *move* – command to robot "move forward in one step";
6) *turn(a)* – command to robot "turn on angle *a* (in degrees)";
7) *stop* – command to robot to halt;
8) *intersection* – situation when the target is not looked by robot directly because obstacles;
9) *target_orientation* – command to robot "turn to target direction";
10) *input* – input vector for neural network consisting of values 1 for 12 sensors, 2, 3 and 4. Length of this vector is equal 15;
11) *work_NN(input)* – start of neural network *with associative memory*, returns value of needed turn of robot in degree. The value 0 is corresponding to retain of current direction of movement, TARGET is corresponding to turn to target;
12) *ask* – prompt value of angle for rotation of robot in degree. One of possible value is SAME. It means that user agrees with value proposed by robot;
13) *current_state* – last recognized cluster or selected number of direction of movement;
14) *direction(i)* – direction corresponding to *i*th recognized cluster.

Set of distance sensors, performance of robots and obstacles are shown in fig. 4.

**Algorithm of simulation of robot behavior**

```
While (target_dist > 20) and not stop
   move;
   get values from sensors;
   delta = 0;
   min_distance = min(dist(0),dist(11));
   if min_distance <25 then
      if dist(0)<dist(11) then
         delta = 30
      else
```

```
            delta = -30;
        end if
    end if
  if min_distance < 5 then
        stop
    end if
    if abs(delta)=30 then
        turn(delta)
      else
        if intersection then
            Preparing vector input for NN;
            delta = work_NN(input);
            if delta = TARGET then
                target_orientation
              else
                if delta <> 0 then
                    turn(delta)
                end if
            end if
          else
            target_orientation
        end if
    end if
end while
```

**End of algorithm of simulation of robot behavior**

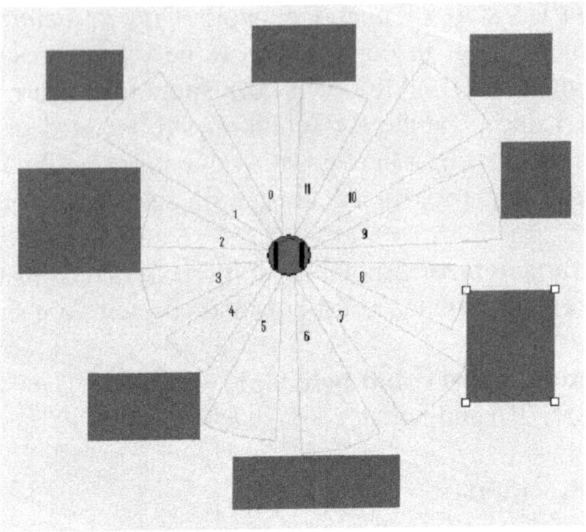

**Fig. 2.** Distance sensors of robot

We can see that in this algorithm two kinds of making decision are implemented – simple rules and neural network. Neural network is not used when robot may see directly target without obstacles and when robot is too closely in front of obstacle.

Otherwise, we utilized the neural network with associative memory (table) for storing of direction corresponding to cluster. In this case creating of new cluster causes the storing in this table of association between cluster (any situation) and corresponding appropriate action in this situation (selected direction of movement).

**Algorithm *work_NN***

Input: input vector consisting of normalized values distance *dist(i)* from 1 for 12 sensors, *target_dist, target_dir, robot_dir.* Length of this vector is equal 15. Vigilance threshold *r*.

Output: value of angle for rotation of robot (direction).

```
Calculation of outputs of MLP and outputs of ART-2 (distances between input
vector of ART-2 and centers of existing clusters);
If minimal value of outputs of ART-2 > r then
    Delta = Ask;
    r = 1.2 * minimal value of outputs of ART-2;
    If delta <> SAME then
        Creation of new cluster (with number i) with center equal input
            vector of ART-2 (output vector of MLP);
        Direction(i) = Delta;
    End if
  Else
    Delta = direction from i-th row of associative memory, where i is
        number of recognized cluster;
    Update weights of ART-2;
End if
If (minimal value of outputs of ART-2 ≤ r)
      or (delta=SAME) then
   Update weights of MLP.
   If current_state = i then
        Delta = 0;
      Else
        Delta = Direction(i); Current_state = i;
   End if
End if
```

**End of algorithm *work_NN***

The experiments were conducted with two kinds of neural network – ART-2 and MLP-ART2. Respectively in first case in algorithm *work_NN* the calculation of outputs and the updating of weights for MLP are absent. The lot of experiments were conducted with different values of vigilance parameter *r* and number of iterations of EBP in MLP. Parameters of MLP are as follows: number of hidden neurons is 10, number of output neurons is 5 and the activation function is exponential sigmoid with parameter 1.

Below some screenshots of experiments are presented. The following notations are employed: 1) the trajectory of robot moving from left start point to right point which is position of target, 2) obstacle as green rectangle and 3) yellow positions of robot means that it could not select direction from associative memory itself (could not recognize known cluster) and requested prompting (supervised learning).

In fig. 3 the comparison between behavior of robot with standard model ART-2 (left) and model MLP-ART2 (right) is shown for case with one obstacle.

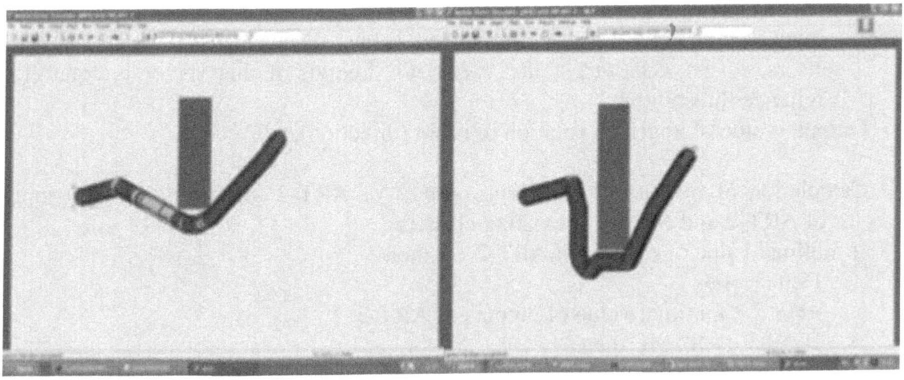

**Fig. 3.** The behavior of robot using ART-2 (left) and MLP-ART2 (right)

The conducting experiments show that in case of using model ART-2 without previous processing of signals from sensors the robot often asks user "what to do". In contrast to it the model MLP-ART2 reduces number of such situations essentially after some learning and filling of associative memory by associations between created cluster and appropriate action. For configuration of environment with one obstacle and determined position of target shown in figures just 5-7 clusters are creating during learning and it is enough for practically autonomous behavior of robot independent on start position. And in this case just one iteration of EBP algorithm is enough.

Figures 4, 5, and 6 show series of screenshots obtained during sequence of experiments for case with multiple obstacles. Every experiment of series is movement of robot to target from any arbitrary point after learning during previous experiments of this series.

**Fig. 4.** The behavior of robot using the model MLP-ART2 at the first experiment of series

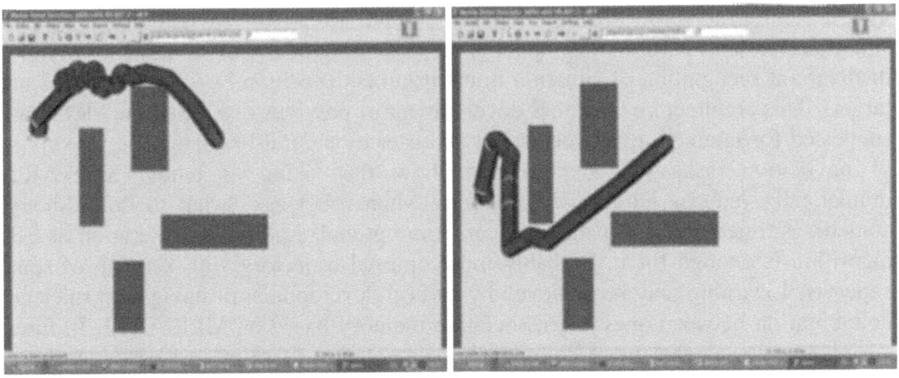

**Fig. 5.** The behavior of robot using the model MLP-ART2 at the 3rd and the 4th experiments of series

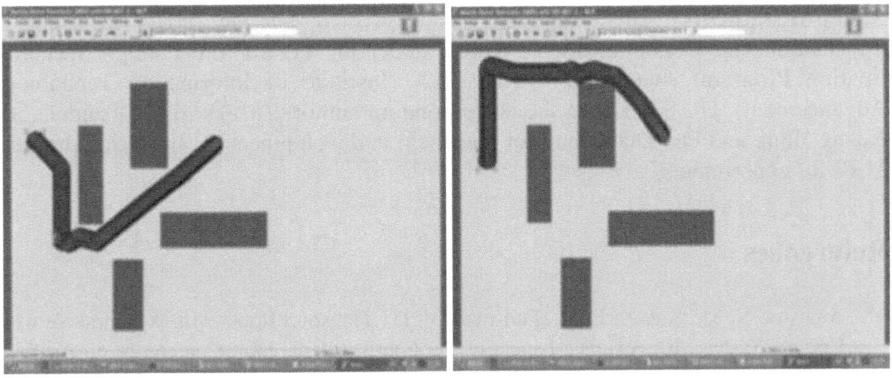

**Fig. 6.** The behavior of robot using the model MLP-ART2 at the 7th and the 10th experiments

Results of experiments with multiple obstacles show the decreasing of number of confusions when the robot demands assistance of operator although sometimes the robot requires help even after training before that (Figure 5, left and right). This fact may be result of not enough careful training before. Experiments show that sometimes the trajectory of movement is far from optimal essentially when environment includes many obstacles (Figure 6, right). Sometimes we can see "deadlock" when the robot can not to return from circle motion.

To overcome these disadvantages it is possible to improve logical part of control or introduce more sophisticated relations between logical rules and neural network MLP-ART2, for example, similar to proposed in [12] for hybrid expert systems. It is goal of our further researches.

## 4 Conclusions

In this paper we suggest and experimentally evaluate the novel approach to development of control system for navigation task of mobile robot. It is based on hybrid neural network MLP-ART2 and simple rules for navigation in specific situations. Role of MLP in this model is preprocessing of sensor signals for providing of invariant recognition of situation in environment (position of robot, obstacles and target). This architecture is further development of previous one based on ART-2 and suggested for interaction between robot and user by natural-like language for solving of navigation tasks [8]. Experiments show that using of model MLP-ART2 dramatically reduces number of situations when robot ask "what to do" although sometimes trajectory of movement is far from optimal. And just one iteration of EBP algorithm is enough for it. Probably more optimal trajectory with keeping of semi-supervised learning may be achieved by careful development of navigation rules and collaboration between ones and associative memory based on MLP-ART2. In future we plan to investigate more complex implementation of rules as knowledge based system cooperated with model MLP-ART2 through black board similar to mechanism proposed in [12]. Furthermore we plan to continue investigation the influence of parameters of our hybrid model on navigation efficiency.

**Acknowledgments.** This research was supported by the MIC (Ministry of Information and Communication), Korea, under the ITFSIP (IT Foreign Specialist Inviting Program) supervised by the IITA (Institute of Information Technology Advancement). Dr. S.Y.Lee is the corresponding author. Thanks to PhD students Le Xuang Hung and Pho Duc Giang for their help in development of simulation program MRS for experiments.

## References

1. Amosov, N. M., Kussul, E. M., Fomenko, V. D.: Transport Robot with A Neural Network Control System. In: Advance papers of the Fourth Intern. Joint Conference on Artificial intelligence 9 (1975) 1-10
2. Brooks, R.: A Robust System Layered Control System for a Mobile Robot. In: IEEE Trans. on robotics and automation RA-2 (1986) 14-23
3. Zou, A.M., Hou, Z.G., Fu, S.Y., Tan,M.: Neural Network for Mobile Robot Navigation: A Survey. In: Proceedings of International Symposium on Neural Networks ISNN-2006, LNCS 3972, Springer-Verlag, Berlin Heidelberg New York (2006) 1218-1226
4. Janglova, D.: Neural Networks in Mobile Robot Motion. In: International Journal of Advanced Robotic Systems 1(1) (2004) 15-22
5. Billard, A., Hayes, G.: DRAMA, A Connectionist Architecture for Control and Learning in Autonomous Robots. In: Adaptive Behavior 7(1) (1999) 35-63
6. Carpenter, G., A., Grossberg, S. Pattern Recognition by Self-Organizing Neural Networks, Cambridge, MA, MIT Press (1991)
7. Rui, A.: Prune-able Fuzzy ART Neural Architecture for Robot Map Learning and Navigation in Dynamic environment. In: IEEE Trans. on Neural Networks 17(5) (2006) 1235-1249
8. Gavrilov, A.V., Gubarev, V.V., Jo K.-H., Lee H.-H. In: Hybrid Neural-based Control System for Mobile Robot. In: Proceedings of 8th Korea-Russia International Symposium on Science and Technology KORUS-2004, Vol. 1, TPU, Tomsk, (2004) 31-35

9. Tan, A.H.: FALCON: A Fusion Architecture for Learning, Cognition and Navigation. In: Proceedings of IEEE International Joint Conference on Neural Networks IJCNN04, Vol. 4, 3297-3302
10. Gavrilov, A.V.: Hybrid Neural Network Based on Models Multi-Layer Perceptron and Adaptive Resonance Theory. In: Proceedings of 9$^{th}$ International Russian-Korean Symposium KORUS-2005, Novosibirsk (2005) 604-606
11. Gavrilov, A.V., Lee,Y.K., Lee, S.Y.: Hybrid Neural Network Model based on Multi-Layer Perceptron and Adaptive Resonance Theory. In: Proceedings of International Symposium on Neural Networks ISNN06, China, Chengdu, LNCS 3972, Shpringer-Verlag, Berlin Heidelberg New York (2006) 707-713
12. Gavrilov, A.V., Chistyakov, N.A.: An Architecture of the Toolkit for Development of Hybrid Expert Systems. In: International Conference IASTED ACIT-2005, Novosibirsk, (2005)

# Using a Wiener-Type Recurrent Neural Network with the Minimum Description Length Principle for Dynamic System Identification

Jeen-Shing Wang[1], Hung-Yi Lin[1], Yu-Liang Hsu[1], and Ya-Ting Yang[2]

[1] Department of Electrical Engineering,
[2] Institute of Education
National Cheng Kung University
Tainan 701, Taiwan, R.O.C.
jeenshin@mail.ncku.edu.tw

**Abstract.** This paper presents a novel Wiener-type recurrent neural network with the minimum description length (MDL) principle for unknown dynamic nonlinear system identification. The proposed Wiener-type recurrent network resembles the conventional Wiener model that consists of a dynamic linear subsystem cascaded with a static nonlinear subsystem. The novelties of our approach include: 1) the realization of a conventional Wiener model into a simple connectionist recurrent network whose output can be expressed by a nonlinear transformation of a linear state-space equation; 2) the state-space equation mapped from the network topology can be used to analyze the characteristics of the network using the well-developed theory of linear systems; and 3) the overall network structure can be determined by the MDL principle effectively using only the input-output measurements. Computer simulations and comparisons with some existing recurrent networks have successfully confirmed the effectiveness and superiority of the proposed Wiener-type network with the MDL principle.

**Keywords:** Wiener model, recurrent neural network, minimum description length.

## 1 Introduction

Good system identification performance relies on a suitable selection of a model representation. Diverse model representations have been proposed for different nonlinear system identification problems [11]. The Wiener model consisting of a dynamic linear block cascaded with a static nonlinear block is one of the notable block-oriented (BO) representations. The main advantage of Wiener models is that the well-known nonlinear/linear system theories can be used to deal with the nonlinear and linear blocks separately. A large number of research studies have indicated the superior capability and effectiveness of Wiener models in nonlinear dynamic system identification and control [6], [12]. Various selections of linear and nonlinear blocks can be found in [6]. Recently, neural networks together with some

D.-S. Huang, L. Heutte, and M. Loog (Eds.): ICIC 2007, LNAI 4682, pp. 192–201, 2007.

linear models have been utilized to construct Wiener models. To name a few, Fang and Chow [2] proposed a Wiener model that consists of an orthogonal wavelet-based neural network (OWNN) as the nonlinear block and a linear autoregressive moving average (ARMA) model as the linear block. The identification algorithm combines the OWNN with traditional least squares method. Janczak [5] compared four gradient-learning algorithms for neural-network-based Wiener models. The neural-network-based Wiener models, composed of different dynamic linear subsystems and static nonlinear subsystems, are used to identify a single-input-single-output (SISO) system. The backpropagation (BP), sensitivity method (SM) and backpropagation through time (BPTT) were applied to train the models and compared for the identification performance. This approach was further extended to MIMO series-parallel and MIMO parallel Wiener models in [6].

From the literature review, we found that trial-and-error processes were frequently used to find suitable network structures when model orders were unknown [12], [20]. To exempt from trial-and error processes for model order selection, we proposed to employ the minimum description length (MDL) principle to determine the model order. The MDL principle, first proposed by Rissanen [13], was derived from the concept of the data compression. Nowadays, the MDL principle has been widely used in many scientific domains such as statistical modeling [1], signal detection problems [17], noise reduction [14], and shape description problems [7]. The MDL principle has been used to deal with the structure as well as parameter selection of neural networks. To name a few, Leonardis [9] used the MDL principle not only to construct the radius basis functions (RBF) network but also to select the parameters of the network. Lappalainen [8] used the MDL principle as the cost function for selecting a feedforward neural network. Small and Tse [16] applied the MDL principle as a criterion to construct a feedforward neural network which can use a minimum number of neurons to mimic a nonlinear system well.

In this paper, we integrate the network construction for dynamic nonlinear system identification problems into an integral task. First, we developed a Wiener type recurrent neural network. The advantages of this network include: 1) the realization of a traditional Wiener model into a simple recurrent neural network whose output can be expressed by a nonlinear transformation of a linear state-space equation; 2) the state-space equation mapped from the network topology can be used to analyze the characteristics of the network using the well-developed theory of linear systems; and 3) the overall network structure can be determined by the MDL principle effectively using only the input-output data. Based on the above advantages, we have developed a system identification algorithm to obtain optimal identification performance. The network construction is implemented by the MDL principle as a stopping criterion that helps us to get a reasonable network size and suitable initial parameters as well. Subsequently, a recursive recurrent learning algorithm derived based on the ordered derivatives [18] with a moment term is applied to obtain better performance. Finally, computer simulations on benchmark examples of nonlinear dynamic applications have successfully validated the effectiveness of the proposed network and algorithm in constructing a quality network with satisfactory performance.

The rest of this paper is organized as follows. In Section 2, the Wiener-type recurrent network is introduced. The MDL-based learning algorithm for establishing a recurrent network is presented in Section 3. Section 4 provides computer simulations of dynamic applications to validate the effectiveness of the proposed network and algorithm. Finally, Section 5 is devoted to conclusions.

## 2 Structure of Wiener-Type Recurrent Model

Consider the conventional Wiener model shown in Fig. 1. The Wiener model is composed of a linear dynamic element cascaded with a nonlinear static element. One of the advantages of the Wiener model is that the complexity of system dynamics is contained in the linear element, whereas the complexity of nonlinearity is only in the static element. We integrate the two cascaded elements into a simple recurrent neural network. The network structure consists of one input layer, one hidden layer (dynamic layer), and one output layer. The input layer and the dynamic layer form the linear dynamic subsystem and the output layer acts as the nonlinear static subsystem. The input layer conveys the input value to the neurons of the hidden layer. The dynamic layer integrates the current input information from the input layer and the state history stored in the memories of the neurons in the dynamic layer to infer the current states of the network. The neurons of the output layer perform a nonlinear transformation for the state variables with different link weights. Fig. 2(a) shows the proposed recurrent structure, which can be expressed by a block diagram illustrated in Fig. 2(b). The actual output $y(k)$ and state variables $\mathbf{x}(k)$ are obtained by calculating the activities of all nodes on each layer and corresponding functions are summarized as

$$\mathbf{x}(k)=\sum_{j=1}^{J}((\sum_{i=1}^{J}a_{ji}x_i(k-1))+(\sum_{h=1}^{p}b_{jh}u_h(k-1))) \ , \tag{1}$$

$$s=\mathbf{C}\mathbf{x}(k)=\sum_{j=1}^{J}c_j x_j(k) \ , \tag{2}$$

$$y(k)=n(s)=\frac{\exp(s)-\exp(-s)}{\exp(s)+\exp(-s)} \ . \tag{3}$$

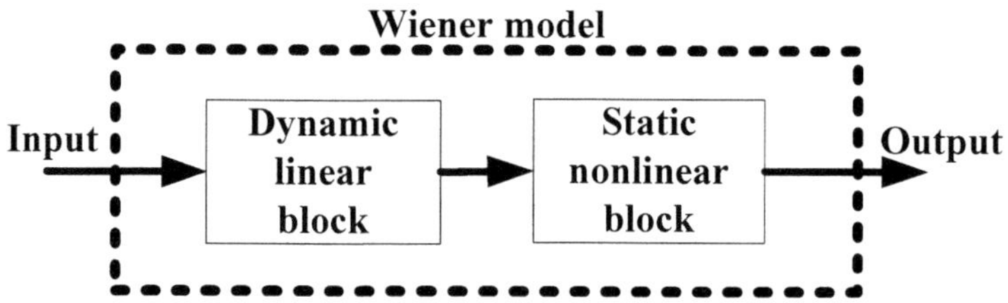


**Fig. 1.** The block diagram of the Wiener model

Based on (1)-(3), the following state space equations can be used to express the proposed connectionist network.

$$\begin{aligned}\mathbf{x}(k+1)&=\mathbf{A}\mathbf{x}(k)+\mathbf{B}\mathbf{u}(k),\\ \mathbf{y}(k)&=\mathbf{N}(\mathbf{C}\mathbf{x}(k))\ ,\end{aligned} \tag{4}$$

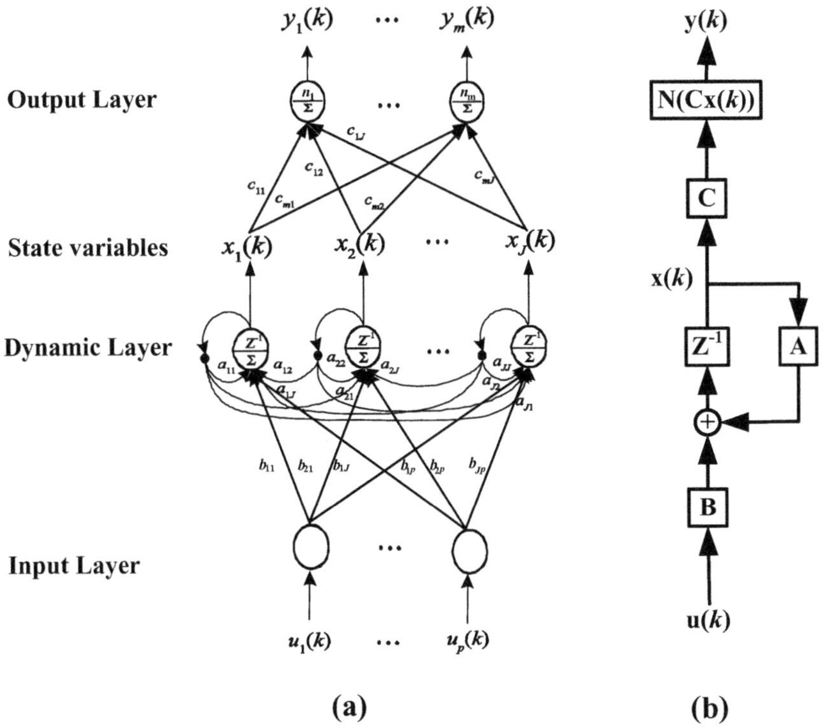


**Fig. 2.** (a) The topology of the proposed Wiener-type recurrent neural network. (b) The block diagram of the proposed network.

where the elements of the state matrix $\mathbf{A} \in \mathbb{R}^{J \times J}$ are the weights of the self-feedback connections and stand for the degrees of the inter-correlations among the state variables. The elements of the input link matrix $\mathbf{B} \in \mathbb{R}^{J \times p}$ are the weights between the input layer and the dynamic layer. The elements of the output link matrix $\mathbf{C} \in \mathbb{R}^{m \times J}$ are the weights of the states. $\mathbf{u} = [u_1, \ldots, u_p]^T$ is the input vector, $\mathbf{x} = [x_1, \ldots, x_J]^T$ is the state vector, $\mathbf{y} = [y_1, \ldots, y_m]^T$ is the output vector, $\mathbf{N} = [n_1, \ldots, n_m]^T$ is the nonlinear activation function vector, and $J$, $p$ and $m$ are the total number of state variables, the numbers of the inputs and outputs, respectively. Based on the proposed Wiener-type recurrent network, we have developed a system identification algorithm. The algorithm consists of two procedures: 1) using the MDL principle as a stop criterion to determine a recurrent neural network size and its initial parameters, and 2) the online parameter tuning algorithm based on the concept of ordered derivatives.

## 3 MDL-Based Learning Algorithm

In this paper, we have presented the proposed Wiener-type recurrent neural network with a systematic identification algorithm to perform the identification task from the input-output measurements of the nonlinear system. The algorithm is composed of the procedures of network construction and recursive parameter learning. We now introduce the two procedures in detail as follows.

### 3.1 Network Construction Algorithm

Typically, determining the structure of neural networks usually takes a trial-and-error approach. In this paper, we employ the MDL principle as a stop criterion to determine the size of the proposed Wiener-type recurrent neural network because of its good performance in nonlinear systems. In the MDL criterion, the basic principle is to estimate the value of *M*(*J*) and *E*(*J*), where *M*(*J*) is called the cost function of describing the model and *E*(*J*) is called the error function of the model prediction errors. Let *J* be the size of the network, so the description length (DL) of the particular model can be represented as follows:

$$D(J) = M(J) + E(J) \; . \tag{5}$$

Obviously, when the network size *J* increases, the *E*(*J*) increases, and the *M*(*J*) decreases. The MDL principle states that the optimal model is the one for which *D*(*J*) is minimal. In [16], *M*(*J*) is expressed as

$$M(J) = \sum_{j=1}^{J} \ln \frac{\gamma}{\delta_j} \; , \tag{6}$$

where $\gamma$ is a constant and represents the number of bits required in the exponent of the floating point representation. In (6), $\delta_j$ is interpreted as the optimal precision of the network parameters and $(\delta_1, \ldots, \delta_J)$ are defined as the solution of

$$\left( Q \begin{bmatrix} \delta_1 \\ \delta_2 \\ \delta_3 \\ \cdot \\ \cdot \\ \cdot \\ \delta_J \end{bmatrix} \right)_j = \frac{1}{\delta_j} \; , \tag{7}$$

where *Q* is the second derivative of *E*(*J*) with respect to the model parameters. Rissanen [13] has shown the *E*(*J*) to be the negative logarithm of the likelihood of the errors $\mathbf{e} = \{e_i\}_{i=1}^{V}$ under the assumed distribution of those errors

$$E(J) = -\ln \text{Prob}(\mathbf{e} \mid w_J) \; , \tag{8}$$

where $w_J$ is the parameters of the model and *V* is the number of data. For regression problems, the probability function can be represented as [3]

$$\text{Prob}(\mathbf{e} \mid w_J) = \left( \frac{1}{\sqrt{2\pi\sigma}} \right)^V \exp\left(-\frac{\mathbf{e}^T \mathbf{e}}{2\sigma^2}\right) \; , \tag{9}$$

where

$$\sigma^2 = \sum_{i=1}^{V} e_i^2 \Big/ V \ . \tag{10}$$

The algorithm of finding the neural network structure using a minimum number of neurons is represented as follows:

*Step 1.* Generate a set of candidate neurons randomly, and calculate the value of $err_g = \sum_{i=1}^{V} e_i h_{ig}$ for each candidate neuron, where $g$ = 1,..., $R$. $R$ is the number of candidate neurons, $V$ is the number of data, $e_i$ is the error of the current network, and $h_{ig}$ is the output of the candidate neuron.

*Step 2.* Find which candidate neuron occurs at the maximum value of $H_g$. $H_g = \left|\sum_{\psi=1}^{J} \mathbf{e}^T \mathbf{h}_\psi\right| + \left|err_g\right|$, where $\mathbf{e}$ is the error of the current network, $\mathbf{h}_\Psi$ is the hidden neuron output of the current network, and $J$ is the number of hidden neurons of the current network. Add the neuron to the current network.

*Step 3.* Calculate the value of $L_\psi = \left|\mathbf{e}^T \mathbf{h}_\psi\right|$, where $\Psi$ = 1,..., $J$, and find which hidden neuron causes the minimum value. Delete the neuron from the current network. In addition, if the neuron is added in this iteration, keep it in the network.

*Step 4.* Find the value of DL. If the MDL criterion is reached (generally, we define the value of DL is minimum if it is smaller than the following ten DL values), then stop. Otherwise, go to Step 1.

Upon completion of the network construction and parameter initialization with the MDL principle, we can establish our recurrent network preliminarily. To closely emulate the dynamic behavior of the unknown system, we have derived the update rule based on the ordered derivatives to fine-tune the parameters of the network further.

### 3.2 Recursive Recurrent Learning Algorithm

In this section, we derive a recursive learning algorithm based on the ordered derivatives with momentum terms to improve the network performance. The momentum terms with a proper learning rate usually accelerate the parameter learning convergence for each parameter update rules. To ease our discussion, the optimization target is characterized to minimize the following error function with respect to the adjustable parameters (**w**) of a MISO network.

$$E_{MDL}(\mathbf{w},k) = \tfrac{1}{2}(y_d(k) - y(k))^2 = \tfrac{1}{2} e_{MDL}(k)^2 \ , \tag{11}$$

where $y_d(k)$ and $y(k)$ are the desired output and the actual output, respectively, and **w** is the adjustable parameters. The update rule is presented as follows:

$$\Delta w(k) = -\xi(\frac{\partial^+ E_{MDL}}{\partial w}) + \alpha\Delta w(k-1) \ , \tag{12}$$

$$w(k) = w(k-1) + \Delta w(k)\,, \tag{13}$$

where $\xi$ is the learning rate and $\partial^+ E_{MDL}/\partial w$ is the ordered derivative which considers the direct and indirect effects on changing the parameter involved in the current state and previous states. $\alpha\Delta w$ is the momentum term, where $\alpha\in[0, 1]$ is the learning rate. The adjustable parameters **w** of the proposed Wiener-type recurrent network include the weights between the state variables and the output layer, $\mathbf{C}\in\mathbb{R}^{m\times J}$, the elements of the state matrix, $\mathbf{A}\in\mathbb{R}^{J\times J}$, and the weights between the input layer and the dynamic layer, $\mathbf{B}\in\mathbb{R}^{J\times p}$. To ease our discussion, we only derive the update rule of the weights between state variables and output layer $c_j$ is:

$$c_j(k) = c_j(k-1) + (-\xi_{ac}\frac{\partial^+ E_{MDL}(k)}{\partial c_j} + \alpha\Delta c_j(k-1))\,, \tag{14}$$

where $\xi_{ac}$ is the learning rate for adjusting $c_j$ and $a_{ji}$. According to (2) and (3), $\partial^+ E_{MDL}/\partial c_j = \partial E_{MDL}/\partial c_j = -e_{MDL}(k)\left(4/(\exp(s)+\exp(-s))^2\right)x_j(k)$ is an ordinary derivative because there is no indirect effect on changing $c_j$. To update the rest of the parameters ($a_{ji}$ and $b_{jh}$), we have to propagate the current error signal to not only the current state but also the previous states. Similarly, the update rules for the rest adjustable parameters can be derived based on the same procedure of the above derivations. For more detailed derivations, please refer to [19].

## 4 Simulation Results

In the following examples, we will demonstrate the capability of our Wiener-type recurrent network and the proposed identification algorithm in identifying MIMO and SISO systems using minimal network sizes and less training time.

*Example 1*: The following MIMO plant shown in (15) was adopted from [10] and the training procedure in [4] was used.

$$\begin{aligned} y_{p1}(k+1) &= 0.5\left[\frac{y_{p1}(k)}{(1+y_{p2}^2(k))} + u_1(k)\right], \\ y_{p2}(k+1) &= 0.5\left[\frac{y_{p1}(k)y_{p2}(k)}{(1+y_{p2}^2(k))} + u_2(k)\right]. \end{aligned} \tag{15}$$

A total of 11000 time steps, including 6000 time steps of two *i.i.d* uniform sequences within the limits [-2, 2] and sinusoid signals given by $\sin(\pi k/45)$ for the remaining training time, were generated to train the proposed network. The first 500 time steps were used to determine the network size and initialize the network parameters. The remaining time steps were employed to optimize the parameters by the recursive recurrent learning algorithm. In the learning phase, we selected the learning rates $\xi_{ac} = 0.0006$ and $\xi_b = 0.006$ for adjustable parameters $a_{ji}$, $c_j$ and $b_{jh}$. Subsequently, we used the same testing input signal as (16) in [4] to verify the identification performance of the proposed recurrent network after the network was trained.

$$u(k)=\begin{cases}\sin(\pi k/25), & 0\le k<250\\ 1.0, & 250\le k<500\\ -1.0, & 500\le k<750\\ 0.3\sin(\pi k/25)+0.1\sin(\pi k/32) & \\ +0.6\sin(\pi k/10). & 750\le k<1000\end{cases} \quad (16)$$

In the beginning of the system identification algorithm, the network construction and parameter initialization algorithm with the MDL principle were used to determine the network size and initial parameters of the proposed network. We obtained the network size as equal to 2. To validate our better performance in dynamic system identification problems using our recurrent network, we compared the performance of the network construction by the MDL principle without the proposed recursive learning algorithm, denoted as Wiener I, and after the network structure was determined, we used the proposed recursive learning algorithm, denoted as Wiener II. Further, we also compared the same simulation with the two existing recurrent networks. From Table 1, we can see that the performance of Wiener I and Wiener II with fewer parameters is better than that of the two existing networks.

**Table 1.** Identification Performance Comparisons of the Proposed Recurrent Network with Existing Recurrent Networks for Example 1

| Network Type | No. of Parameters | Training Time (time steps) | MSE |
|---|---|---|---|
| Wiener I | 12 | 500 | $y_1$=5.8×$10^{-3}$<br>$y_2$=9.4×$10^{-3}$ |
| Wiener II | 12 | 11,000 | $y_1$=1.6×$10^{-3}$<br>$y_2$=9.3×$10^{-3}$ |
| RSONFIN [4] | 77 | 11,000 | $y_1$=1.24×$10^{-2}$<br>$y_2$=1.97×$10^{-2}$ |
| MNN [15] | 131 | 77,000 | $y_1$=1.86×$10^{-2}$<br>$y_2$=3.27×$10^{-2}$ |

*Example 2*: The following SISO plant is adopted from [10] and the training procedure is similar to the previous example.

$$y(k+1)=f[y(k),y(k-1),y(k-2),u(k),u(k-1)], \quad (17)$$

where

$$f[x_1,x_2,x_3,x_4,x_5]=\frac{x_1x_2x_3x_5(x_3-1)+x_4}{1+x_3^2+x_2^2}. \quad (18)$$

A total of 9000 time steps, including 5000 time steps of an *i.i.d* uniform sequences within the limits [-2, 2] and a sinusoid signals given by 1.05 × sin($\pi k$/45) for the remaining training time, were generated to train the proposed network. We set the learning rates $\xi_{ac} = 0.0001$ and $\xi_b = 0.001$. In the beginning, the first 500 time steps were employed to decide the network size and initial values of the whole network parameters by the MDL principle. The network size was selected as equal to 3. The remaining time steps were used to optimize the parameters by the recursive recurrent learning algorithm. The MSE of the Wiener I and Wiener II for the same testing input signal given in (16) are $6.04\times10^{-2}$ and $2.79\times10^{-2}$, respectively. We also compared the simulation results with those of two existing recurrent networks. From Table 2, we can see that the performance of Wiener I and Wiener II with fewer parameters is better than those of the two existing networks.

**Table 2.** Identification Performance Comparisons of the Proposed Recurrent Network with Existing Recurrent Networks for Example 2

| Network Type | No. of Parameters | Training Time (time steps) | MSE |
|---|---|---|---|
| Wiener I | 15 | 500 | $6.04\times10^{-2}$ |
| Wiener II | 15 | 9,000 | $2.79\times10^{-2}$ |
| RSONFIN [4] | 38 | 9,000 | $4.41\times10^{-2}$ |
| MNN [15] | 81 | 620,000 | $7.52\times10^{-2}$ |

## 5 Conclusion

A novel Wiener-type recurrent neural network with the minimum description length principle has been proposed for nonlinear unknown dynamic system identification problems. The advantages of our approach include: 1) the realization of a conventional Wiener model into a simple connectionist recurrent network whose output can be expressed by a nonlinear transformation of a linear state-space equation; 2) the overall network structure can be determined by the MDL principle effectively using only the input-output patterns; 3) the trained network topology can be translated into a state-space equation that can be directly used to analyze the characteristics of the network using the well-developed theory of linear systems; and 4) the proposed network is capable of accurately identifying nonlinear dynamic systems using fewer parameters with higher accuracy. Finally, several computer simulations on nonlinear unknown dynamic examples have been successfully validated the effectiveness and superiority of our proposed approach.

## References


1. Barron, A., Rissanen, J., Yu, B.: The Minimum Description Length Principle in Coding and Modeling. IEEE Trans. Information Theory. 44 (6) (1998) 2743–2760
2. Fang, Y., Chow, T.W.S.: Orthogonal Wavelet Neural Networks Applying to Identification of Wiener Model. IEEE Trans. Circuits and Systems-I. 47 (4) (2000) 591–593
3. Grunwald, P.D., Myung, I.J., Pitt, M.A.: Advances in Minimum Description Length. The MIT Press (2005)
4. Juang, C.-F., Lin, C.-T.: A Recurrent Self-Organizing Neural Fuzzy Inference Network. IEEE Trans. Neural Networks. 10 (4) (1999) 828–845

5. Janczak, A.: Comparison of Four Gradient-Learning Algorithms for Neural Network Wiener Models. International Journal of Systems Science. 34 (1) (2003) 21–35
6. Janczak, A.: Identification of Nonlinear Systems using Neural Networks and Polynomial Models. Springer-Verlag, New York (2005)
7. Li, M.: Minimum Description Length Based 2D Shape Description. IEEE 4$^{th}$ International Conf. Computer Vision. (1993) 512–517
8. Lappalainen, H.: Using an MDL-based Cost Function with Neural Networks. IEEE Conf. Neural Networks. 3 (1998) 2384–2389
9. Leonardis, A., Bischof, H.: An Efficient MDL-based Construction of RBF Networks. Neural Networks. 11 (5) (1998) 963–973
10. Narendra, K.S., Parthasarathy, K.: Identification and Control of Dynamical Systems using Neural Network. IEEE Trans. Neural Networks. 1 (1) (1990) 4–27
11. Nelles, O.: Nonlinear System Identification. Springer-Verlag, New York (2001)
12. Nagammai, S., Sivakumaran, N., Radhakrishnan, T.K.: Control System Design for a Neutralization Process using Block Oriented Models. Instrumentation Science and Technology. 34 (2006) 653–667
13. Rissanen, J.: Stochastic Complexity in Statistical Inquiry. World Scientific (1989)
14. Rissanen, J.: MDL Denoising. IEEE Trans. Information Theory. 46 (7) (2000) 2537–2543
15. Sastry, P.S., Santharam, G., Unnikrishnan, K.P.: Memory Neuron Networks for Identification and Control of Dynamical Systems. IEEE Trans. Neural Networks. 5 (2) (1994) 306–319
16. Small, M., Tse, C.-K.: Minimum Description Length Neural Networks for Time Series Prediction. Physical Review E - Statistical Physics, Plasmas, Fluids, and Related Interdisciplinary Topics. 66 (62) (2002) 066701/1–066701/12
17. Valaee, S., Champagne, B., Kabal, P.: Sinusoidal Signal Detection using the Minimum Description Length and the Predictive Stochastic Complexity. International Conf. Digital Signal Processing. 2 (1997) 1023–1026
18. Werbos, P.: Beyond Regression: New Tools for Prediction and Analysis in the Behavior Sciences. Ph.D. dissertation, Harvard Univ., Cambridge, MA (1974)
19. Wang, J.-S., Chen, Y.-P.: A Fully Automated Recurrent Neural Network for Unknown Dynamic System Identification and Control. IEEE Trans. Circuits and Systems-I. 53 (6) (2006) 1363–1372
20. Xu, M., Chen, G., Tian, Y.-T.: Identifying Chaotic Systems using Wiener and Hammerstein Cascade Models. Mathematical and Computer Modeling. 33 (2001) 483–493

# A Parallel Independent Component Implement Based on Learning Updating with Forms of Matrix Transformations

Jing-Hui Wang[1], Guang-Qian Kong[2], and Cai-Hong Liu[3]

[1] TianJin University of Technology, TianJin,300191, China
[2] GuiZhou University,GuiYan,550025, China
[3] Northwest Minorities University, Lanzhou 730030, China

**Abstract.** PVM (Parallel virtual machine) library is a tool which used processes large amounts of data sets. This paper wants to achieve a high performance solution that exploits PVM library and parallel computers to solve ICA (Independent Component Analysis) problem. The paper presents parallel power ICA implementations to decomposition data sets. Power iteration (PI) is an algorithm for independent component analysis, which has some desired features. It has higher performance and data capacity than current sequential implementations. This paper, we show the power iteration algorithm which learning updating is in the form of matrix transformation . From power iteration algorithm, we develop parallel power iteration algorithm and implement parallel component decomposition solution. At last, experimental results, analysis and future plans are presented.

**Keywords:** Independent Component Analysis, Parallel Virtual Machine, Parallel Program.

## 1 Introduction

Independent component analysis (ICA) has been extensively investigated in its theory, implementation and applications. Up to now, there have already been many algorithms for ICA ([1, 2]). The ICA algorithms can be roughly categorized as gradient-based, Newton's, i.e, the second order gradient-based and diagonalization-based, etc. ICA algorithm requires a huge number of calculations[3].

Our work wants to solve these limitations by increasing the computational speed. One way of increasing the computational speed is by using multiple processors operating together on a single problem. There have been several software packages for workstation cluster parallel programming. We want to use Parallel Virtual Machine (PVM) implement ICA algorithm.

Recently, a power iteration algorithm ([4,5,6]) was introduced. Those papers deal with power iteration (PI) algorithm and its performance. This paper we develop power iteration algorithm to parallel power iteration and analysis parallel algorithm performance. The paper describes the parallel power iteration algorithm solves the ICA problem.

D.-S. Huang, L. Heutte, and M. Loog (Eds.): ICIC 2007, LNAI 4682, pp. 202–211, 2007.

The iteration algorithm has at least two desired features[6]: 1) the algorithm does not include any predetermined parameter such as the learning step size as in the gradient-based algorithm which unknown in ICA applications; 2) in its iteration, the updating of ICA matrix is fully-multiplicative, i.e. ,

$$W^{(n+1)} = T^{(n)} W^{(n)} \tag{1}$$

where $W^{(n)}$ is the estimation of separation matrix at $n$-th iteration, and $T^{(n)}$ is a transformation matrix that may, or may not, be near to or equal to the unit matrix $I$. The updating will be terminated when $T^{(n)} = I$ .$T^{(n)}$ will be determined by a ICA criterion or a cost function. Compared with (1), in a gradient-based algorithm, the updating is in the form as

$$W^{(n+1)} = W^{(n)} - u\Delta W^{(n)} \tag{2}$$

where $u$ is the predetermined learning step size, and $\Delta W^{(n)}$ is the updating amount at $n$-th iteration. $\Delta W^{(n)}$ will be determined by a ICA criterion or a cost function. Because of feature (1), the power iteration algorithm does not belong to any categories of conventional algorithms.

The updating in the form of (1) is more natural than that in (2).In fact, $T^{(n)}$ is a transform upon to a matrix (here it is $W^{(n)}$ and $T^{(n)} \in GL(N)$ , the $N$ - dimensional general linear group. If $T^{(n)}$ is near to $I$ , it is easy to show that (1) can be transferred into (2) approximately. That is,

$$T^{(n)} = \exp(I - u\Delta) \approx I - u\Delta \tag{3}$$

if $\Delta$ is sufficient small operator(i.e., if its principal eigenvalue is sufficient small). However, if $T^{(n)}$ is not near to $I$ , this approximation relation does not hold anymore. A remarkable benefit of updating in the form (1) is that it allows a finite updating that is not near to $I$ as well.

In paper ([6]) show how it is possible to perform ICA in the updating form (1). The criterion of ICA is based on a diagonalization of a non-linearized covariance matrix that is defined by ICA outputs and their non-linear mappings. The activation function, which features the probability distribution of sources, may be chosen as such a nonlinear map. The covariance matrix was termed as activation function mapped covariance matrix. This criterion can also be derived from the minimization of the Kullback-Leibler divergence, as performed in the most conventional ICA algorithms.

## 2 Problem Formulation

The problem is from blind source separation (BSS), assume the existence of $M$ zero-mean statistically independent sources $s(t) = [s_1(t),..., s_M(t)]^T$ , where t is the time instant. The original sources $s_i(t)$ are unknown, and we observe $N$ possibly noisy but different linear, instantaneous mixture $x(t) = [x_1(t),..., x_N(t)]^T$ of the sources.

The constant mixing coefficients are also unknown. That is, the mixing model can be written in a matrix form as

$$x(t) = As(t) + n(t) \tag{4}$$

where $A$ is a constant, full-rank $N \times M$ mixing matrix whose elements are the unknown coefficients of the mixtures. Here $n(t)$ is additive noise vector with a same rank to $x(t)$.

In ICA, the task is to find $M$ output waveforms that are as independent as possible. Denoting the output waveforms by

$$y(t) = [y_1(t),..., y_M(t)]^T \quad , \quad y(t) = W(t)x(t) \tag{5}$$

where $W(t)$ is $N \times M$ matrix that is estimated making the independencies between the output waveforms.

In several conventional ICA algorithms, the observation vector $x(t)$ is pre-processed by whitening them through a linear Transformation $V$ so that the covariance matrix $E\{x(t)x(t)^T\}$ becomes the $N$ -rank unit matrix $I_N$.

Therefore, if $x(t)$ is whitened signal vector, the ICA network can be modeled as

$$y(t) = B(t)x(t) \tag{6}$$

where $B(t)$ is an orthogonal $N \times M$ matrix.

## 3 The ICA Criterion Based on Activation Function Mapped Covariance Matrix

Assume $\varphi(.)$ is the non-linear activation function. Let us define the activation function mapped covariance matrix as

$$G(B) \equiv E\{\varphi(y)y^T\} = E\{\varphi(Bx)x^T B^T\} \tag{7}$$

where the symbol $E\{\varphi(x(t))\}$ denotes statistical expectation of the component and sample-wise function $\varphi(.)$ over the distribution of random vector $x$.

The non-linear function $\varphi(.)$ is required to satisfy the following conditions[4]:

(1) The non-linear function is odd;
(2) The non-linear function is, at least approximately, activation functions such that $\varphi(y_i) \sim y_i$, for $y_i -> 0$ and $i = 1,2..M$ ; As is well known[1], the natural gradient updating rule for ICA matrix can be written as

$$\Delta B = u(I_M - E\{\varphi(y)y^T\})B \tag{8}$$

where $u$ denotes the learning step size. This rule means that the minimization can be arrived when $E\{x(t)x(t)^T\} = I_M$ . Since the scale indeterminacy of ICA, this can be relaxed as $E\{x(t)x(t)^T\} = \Lambda$, where $\Lambda$ is a M-rank diagonal matrix. Therefore the off-diagonal norm of the AFMC matrix can be taken as the cost function

$$J(G,B) = \sum_{1\le i\ne j\le M} |G(B)_{ij}|^2 \tag{9}$$

It is worthy of notice that the principle is different that of joint diagonalization of multi-lagged covariance matrices in blind source separation (BSS) [9, 10]. The main justification for using the non-linear activation function $\varphi(.)$ is that it introduces higher-order statistics into the cost function. If the activation function is chosen as the non-linear function[7,11,12,13], the output independence in all orders of statistics can be reached.

## 4 Diagonalization Principle and Method of Activation Function Mapped Covariance Matrix for ICA

In reference [4,5,6,7], an algorithm for the diagonalization was performed by a learning rule based on the gradient searching for the minimization of the cost function (9). In paper[6], they propose a novel algorithm for more effective realization of the diagonalization. The proposed algorithm is not initialed from a consideration of the minimization of the cost function (9). Instead, they directly consider the diagonalization of the AFMC matrix (7).Assume that at $n$-th iteration the separation matrix is $B^{(n)}$ and the activation function mapped covariance matrix is

$$G(B^{(n)}) = E\{\varphi(B^{(n)}x)x^T B^{(n)T}\}.$$

The posed ICA problem is, for an arbitrary given initial matrix $B_0$, to find a series of matrix Transformations $T^{(n)}$; for $n = 1,2,\ldots M$ such that

$$B^{(n+1)} = T^{(n)}B^{(n)} \tag{10}$$

and the activation function mapped covariance matrix

$$G(B^{(n+1)}) = E\{\varphi(B^{(n+1)}x)x^T B^{(n+1)T}\} \tag{11}$$

satisfies the following conditions: (1) $G(B^{(n+1)})$ becomes more or equal diagonal if $G(B^{(n)})$ is not diagonal, i.e., $J(G,B_{n+1}) \le J(G,B_n)$ (2) $G(B^{(n+1)})$ keeps diagonal if $G(B^{(n)})$ has been diagonal. The iteration will be terminated and ICA is reached by $y = B^{(n+1)}x$. The activation function mapped covariance matrix (7) has been diagonalized,it can express this as

$$G(\hat{B}) = \Lambda \tag{12}$$

where $\Lambda = diag(\lambda_1, \lambda_2, \ldots, \lambda_M)$ is an $N \times M$ real diagonal matrix. Here $\hat{B}$ is the final estimation of $B$ that makes the independencies between the components of $\hat{s} = \hat{B}x$, where $\hat{s}$ is the estimation of s. That is,

$$G(\hat{B}) = E\{\varphi(\hat{B}x)x^T\hat{B}^T\} = E\{\varphi(\hat{s})\hat{s}^T\} \quad (13)$$

It is worthy of mention that $\Lambda$ in (12) can be found by the eigenvalue problem as

$$G(\hat{B})q_k = \lambda_k q_k \quad (14)$$

for $k = 1,2,\ldots P$, where $P$ is the number of non-zero eigenvalues. Here $q_k$ is k-th eigenvector corresponding to the eigenvalue $\lambda_k$. Therefore, the problem to find $\lambda_k$ can be attributed as an eigenvalue problem. Although to solve the eigenvalue problem (14) is very classical problem and there have already been a lot of methods for it, instead, the purpose is to find $\hat{B}$ but not the eigenvector. Of course that discussion[4,5] is approximately valid. Here an exact analysis.

$$E\{\hat{s}_i\hat{s}_j\} = E\{\varphi(\hat{s}_i)\hat{s}_j\} \quad (15)$$

where the subscripts $i, j$ denoted independent (white) stochastic processes. At first rewritten equation (15) in matrix form

$$E\{\hat{s}\hat{s}^T\} = E\{\varphi(\hat{s})\hat{s}^T\} \quad (16)$$

$$\hat{B}E\{xx^T\}\hat{B}^T = E\{\varphi(\hat{B}x)x^T\hat{B}^T\} \quad (17)$$

From equations (14) and (17),

$$\hat{B}E\{xx^T\}\hat{B}^T q_k = \lambda_k q_k; \forall k \in (1,2,\ldots, P\} \quad (18)$$

Notice that equation (18) is exact, rather than approximate in [6, 5]. Since the matrix $\hat{B}E\{xx^T\}\hat{B}^T$ is real Hermitian,there are $N$ orthogonal eigenvectors, i.e $q_i^T q_k = \delta_{ik}, \forall i,k \in \{1,2,\ldots, N\}$ and $P=N$. Here $\delta_{ik}$ is the Kronecker delta function. As we have assumed that the observations are whitened, $E\{xx^T\} = I_N$. Substituting this and the orthogonality of $q_k$ into (18), obtain

$$\hat{B}\hat{B}^T q_k = \lambda_k q_k \quad (19)$$

For solving equation (19), define $\hat{B} = [\hat{b}_1, \hat{b}_2, \ldots, \hat{b}_M]^T$, where $\hat{b}_k$ is a $N$ -rank column vector. Some papers[4,5,6] find a solution for equation (19) as

$$\hat{b}_k = \lambda_k^{1/2} q_k \quad (20)$$

Equation (20) means that once find the eigenvalue and eigenvectors of the matrix $G(\hat{B})$, from those papers,we can construct the ICA matrix $\hat{B}$.

Suppose that we have arranged the eigenvalues and eigenvectors in an order such that $|\lambda_1| > |\lambda_2| > \ldots > |\lambda_p|, 1 \le P \le M$ . Then we can obtain the ICA vector from (20). Although the algorithm can work for any $P \le M$ , for simplicity, the algorithm here for the case of $P = M$.

$$\begin{aligned}
\tilde{Q}^{(n+1)} &= G(B^{(n)})Q^{(n)} \\
Q^{(n+1)} &= \tilde{Q}^{(n+1)}(\tilde{Q}^{(n+1)T}\tilde{Q}^{(n+1)})^{-1/2} \\
\lambda_k^{(n+1)} &= q_k^{(n+1)T}G(B^{(n)})q_k^{(n+1)}, \forall k = 1,\ldots, M \\
B^{(n+1)} &= (\Lambda^{(n+1)})^{1/2}Q^{(n+1)}
\end{aligned} \tag{21}$$

where $n = 0,1,\ldots$ , denotes the iteration index; $Q \equiv [q_1, q_2, \ldots, q_M]^T$ ; and $\hat{Q} \equiv [\hat{q}_1, \hat{q}_2, \ldots, \hat{q}_M]^T$ . Here, $Q^{(0)}$ and $B^{(0)}$ are initial guesses of $Q$ and $B$; each of them can be an arbitrary $N * M$ matrix. Equation (21) denotes the orth-normalization. Since $Q$ is orthogonal, it means $B$ is also orthogonal. Thus both the rows and the columns become normalized.

We show that the updating (21) can be cast into the form of (10). Indeed, since

$$B^{(n+1)} = (\Lambda^{(n+1)})^{1/2}G(B^{(n)})(\Lambda^{(n)})^{-1/2}B^{(n)} \tag{22}$$

we obtain

$$T^{(n)} = (\Lambda^{(n+1)})^{1/2}G(B^{(n)})(\Lambda^{(n)})^{-1/2} \tag{23}$$

The proposal algorithm is based on (21)-(23).

The PowerICA Algorithm[4]

Initialization: $Q^{(0)} \in R^{M \times N}$ ,

where $Q^{(0)}Q^{(0)T} = I_N ; B^{(0)T} = Q^{(0)}$ .

n = 0, where n is the iteration number.

Do until convergence

$n \leftarrow n + 1$

$y^{(n)}(t) \leftarrow B^{(n)}x(t)$

$\tilde{Q}^{(n+1)} \leftarrow G(B^{(n)})Q^{(n)}$

$Q^{(n+1)} \leftarrow \tilde{Q}^{(n+1)}(\hat{Q}^{(n+1)T}\hat{Q}^{(n+1)})^{-1/2}$

Do k=1 through M

$\lambda_k^{(n+1)} \leftarrow q_k^{(n+1)T}G(B^{(n)})q_k^{(n+1)}$

```
    End
    B^(n+1) ← diag((λ_k^(n+1))^(1/2))Q^(n+1)
  End
```

Here $diag\ (x)$ denotes the diagonal matrix formed from a vector $x \in R^{M \times 1}$.

## 5 Parallel ICA Algorithm Based on Learning Updating with Forms of Matrix Transformations and the Diagonalization Principle

In PVM[14,15,16], the programmer decomposes the problem into separate programs. Programs communicate by message passing using PVM library routines such as *pvm_send()* and *pvm_recv()*,which are embedded into the programs prior to compilation. All PVM send routines are nonblocking (or asynchronous in PVM terminology) while PVM receive routines can be either blocking (synchronous) or nonblocking. The key operations of sending and receiving data are done through message buffers. Once a send buffer is loaded with data, a PVM send routine is used to initiate sending the contents of the buffer through the network to a destination receive buffer, from which it can be picked up with a PVM receive routine. PVM uses a *message tap(msgtag)*, attached to a message to differentiate between types of messages being sent. These messages can include data that other processors may require for their computations.

We need one master processor and $L$ slave processors. The parallel ICA code could be of the form:

**Master (Processor 0)**

Initialization: $Q^{(0)} \in R^{M \times N}$ ,

where $Q^{(0)}Q^{(0)T} = I_N ; B^{(0)T} = Q^{(0)}$ .

n = 0, where n is the iteration number.

Do until convergence

$n \leftarrow n+1$

$y^{(n)}(t) \leftarrow B^{(n)}x(t)$

$\tilde{Q}^{(n+1)} \leftarrow G(B^{(n)})Q^{(n)}$

$Q^{(n+1)} \leftarrow \tilde{Q}^{(n+1)}(\hat{Q}^{(n+1)T}\hat{Q}^{(n+1)})^{-1/2}$

```
    Send(&x,Pi);          /* send x to processor i*/
    Send(&Q,Pi);          /* send Q to processor i*/
  Send(&B,Pi);          /* send B to processor i*/
  Wait(...)
```

```
  Recv(Pi,..);          /*receive λ_k^(n+1) from processor i*/
```

$B^{(n+1)} \leftarrow diag\ ((\lambda_k^{(n+1)})^{1/2})Q^{(n+1)}$

```
end
```

**Slave(Processor 1,...Processor L)**

```
 Initialization: recv(&x,Pi); /* receive x to processor
0*/
  Do unitl M/L λ_k^(n+1) compute
    recv(&Q,Pi);          /*receive Q to processor 0*/
    recv(&B,Pi);          /*receive B to processor 0*/
```

$G(B) \equiv E\{\varphi(y)y^T\} = E\{\varphi(Bx)x^T B^T\}$

$\lambda_k^{(n+1)} \leftarrow q_k^{(n+1)T} G(B^{(n)}) q_k^{(n+1)}$

```
    Send(&λ_k^(n+1),P0);   /* send λ_k^(n+1) to processor 0*/
  end.
```

# 6 Analysis and Conclusion

## 6.1 Communication and Computation Time Analysis

We verified the validity and efficiency implementations of parallel power iteration algorithms with a series of tests. The validity tests ensured that the algorithms gave the correct answer. We have compared the results of the EEGLAB of Infomax with the parallel power iteration results. The tested platforms processors are Inter Pentium processors which frequency is 1.73GHz.

The number of channels used in these experiments varied from 64 to 256.

In those programs, the master sends *Q*, *B* to the slaves. The master waits for any slave to respond. Each slave will receive *Q*,*B* and send $\lambda_k^{(n+1)}$, giving a communication time of $t_{comm} = t_{startup} + t_{x\_time} + t_{QB\_time} + \left\lceil \frac{M}{L} \right\rceil t_{\lambda\_time}$ ,Once the slaves receive, they spend $t_{comp}$ which has a time complexity of $O(n^4)$ .When $n$ increased, $t_{comp} >> t_{comm}$ . A measure of relative performance between multiprocessor system and a single processor system is the speedup factor.

The performance of the parallel powerICA algorithm is more constrained. One reason is that the amount of parallel work available in PVM parallel regions is not large in relation to the sequential computation. One possible reason is that the algorithm suffers caching effects due to the composition of blocks from random selection of samples. As the number of samples increase, we should see better speedups for larger numbers of channels.

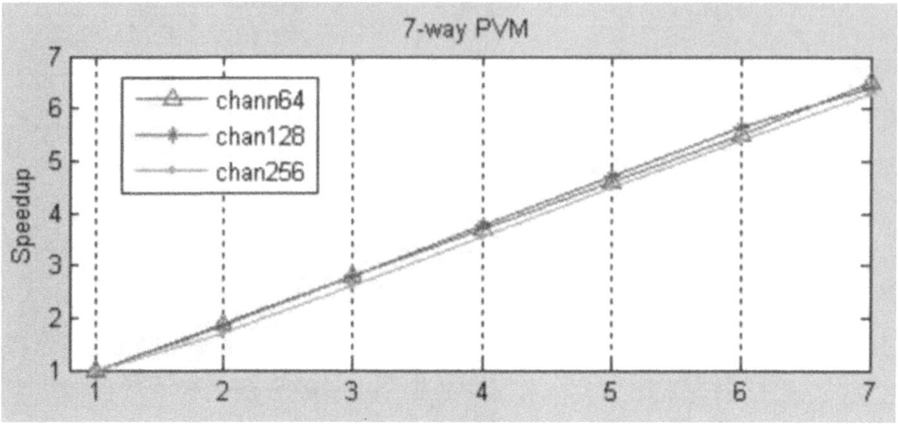


**Fig. 1.** 7-way PVM

### 6.2 Conclusions

PowerICA algorithm is very processor intensive, especially with large data sets. We described the PowerICA techniques and parallel implementations PowerICA algorithms. The method increased process speed compared to the sequential implementations. The ability handles data set sizes larger than the sequential implementations.

We need to further investigate ways to increase the portion of the algorithm that can operate in parallel. This includes minor changes such as adjusting the block size used during the training of a weight vector, and major changes such as allowing each worker thread to work on its own block in parallel, merging the learned weights after each step. The mathematical legitimacy of these optimizations must be analyzed.

Application-Specific Integrated Circuits (ASIC) has more advantages than computer networks. We have complete single TMS320C6713 DSP board to process high speed ICA problem. In future, we will implement the more complex parallel PowerICA algorithm on multiprocessor.

## References

1. Cichocki, A., Amari ,S.: Adaptive Blind Signal and Image Processing. John Wiley, LTD(2003)
2. Hyvarinen, A., Karhun, J., Oja, E.: Independent Component Analysis. John Wiley & Sons, LTD(2001)
3. Tucker, D.: Spatial Sampling Of Head Electrical Fields: The Geodesic Sensor Net. Electroencephalography and Clinical Neurophysiology(1993)145–163
4. Ding , S.: Independent Component Analysis without Predetermined Learning Parameters. In Proc 2006 IEEE International Conference on Computer and Information Technology(CIT 2006), Seoul, Korea(2006)

5. Ding, S.: A Power Iteration Algorithm for ICA Based on Diagonalizations of Nonlinearized Covariance Matrix. Proc 2006 International Conference on Innovative Computing, Information and Control, Beijing (2006)
6. Ding, S. : Independent Component Analysis Based on Learning Updating with Forms of Matrix Transformations and the Diagonalization Principle, Proceedings of the Japan-China Joint Workshop on Frontier of Computer Science and Technology (FCST'06)(2006)
7. Cichocki, A., Umbehauen, R., Rummert, E.: Robust learning for blind separation of signals. Electronics Letters(1994)1386–1387
8. Golub, G. H., Loan, C. F. V.: Matrix Computations. The Johns Hopkins University Press, Third Edition(1996).
9. Cardoso ,J. A., Souloumiac, J. :Angles for Simultaneous Diagonalization. SIAM Journal of Matrix Analysis and Applications(1996)161–164
10. Molgedey, L., Schuster,H. G.: Separation of a mixture of independent signals using time delayed correlations. Physical Review Letter(1994)3634–3637
11. Bell, A. ,Sejnowski, T.: An Information Maximization Approach to Blind Separation and Blind Deconvolution. Neural Computation, (1995)1129–1159
12. Jutten, C., Herault,J.: Blind Separation of Sources, Part I: An Adaptive Algorithm Based on Neuromimetic Architecture.Signal Processing(1991)1–10
13. Fiori, S.: Fully Multiplicative Orthogonal Group ICA Neural Algorithm. Electronics Letters ( 2003)1737 – 1738
14. Wilkinson, Barry, Michael ,A.: Parallel Programming, Techniques and Applications, Pearson Educatio,Reading, Massachusetts(2002)
15. Aho, A.V., Hopcroft, J.E., Ullman, J.D.: The Design and Analysis of Computer Algorithms, Addison-Wesley Publishing Company.
16. Sunderam,V.: PVM: A Framework for Parallel Distributed Computing, Concurrency Practice&Experience Vol. 2.315-339

# Application Study on Monitoring a Large Power Plant Operation

Pingkang Li[1], Xun Wang[2], and Xiuxia Du[1]

[1] Beijing Jiaotong University, 100044, Beijing, China
[2] Intelligent Systems and Control Group, Queen's University of Belfast, BT9 5AH, U.K.

**Abstract.** Upon close examination of a set of industrial data from a large scale power plant, time varying behavior are discovered. If a fixed model is applied to monitor this process, false alarms will be inevitable. This paper suggests the use of adaptive models to cope with such situation. A recently proposed technique, fast algorithm for Moving Window Principal Component Analysis (MWPCA) was employed because of its following strength: (i) the ability in adapting process changes, (ii) the conceptual simplicity, and (iii) its computational efficiency. Its advances in fault detection is demonstrated in the paper by comparing with the conventional PCA. In addition, this paper proposed to plot the scaled variables in conjunction with MWPCA for fault diagnosis, which is proved to be effective in this application.

**Keywords:** Model Adaptation, Process Monitoring, Moving Window, Principal Component Analysis, Power Plant.

## 1 Introduction

To model and monitor modern industrial processes, where a huge number of variables are frequently recorded, Multivariate Statistical Process Control (MSPC) techniques have been widely recognized and applied.[1] They can establish models using a reduced number of "artificial variables", due to the relationships among the original process variables. By plotting and observing the monitoring statistics generated from the models, fault detection and diagnosis became much more efficient than using the plots of individual process variables as in the conventional way. Among the MSPC approaches, Principal Component Analysis (PCA) has probably received the widest attention for its simplicity. The idea of PCA dates back to 1901, when Pearson described it mathematically as a method for obtaining the "best-fitting" straight line or hyper-plane to data in two or higher dimensional space.[2] Jackson summarized pioneering work on the use of PCA for statistical process control.[3]

As Gallagher *et al.* pointed out that most industrial processes are time-varying and that the monitoring of such processes requires the adaptation on models to accommodate this behavior.[4] However, the updated model must still be able to detect abnormal behavior with respect to statistical confidence limits, which themselves may also have to vary with time.[5, 6] There are two techniques that allow such an adaptation of the PCA model, i.e. Moving Window PCA (MWPCA) and

D.-S. Huang, L. Heutte, and M. Loog (Eds.): ICIC 2007, LNAI 4682, pp. 212–221, 2007.

Recursive PCA (RPCA). The pros and cons of these two methods have been reviewed, which led to the proposal of a fast MWPCA approach.[6]

The principle behind the moving window is well known. As the window slides along the data, a new process model is generated based on the data selected within the current window. This allows older samples to be discarded in favor of newer ones that are more representative of the current process operation. It has to be noted that a sufficient number of data points should be included in the window in order to present adequate process variation for modeling and monitoring purposes. However, this causes the computational speed of MWPCA drop significantly, particularly in cases where the process has a large number of variables. When an on-line process monitoring is required, MWPCA may become inapplicable due to this drawback. The fast MWPCA overcomes the difficulty with large window size.[6] This method relies on the combined use of RPCA and MWPCA to enhance the application of adaptive condition monitoring. Applied to the power plant data considered in this paper, it can detect the fault spontaneously. Upon detecting the fault, it is important to trace the root cause of the fault in order to take immediate actions. This is a particularly difficult task in modern industry that presents large number of variables. Various techniques could be applied, such as contribution charts. This paper proposed the use of the scaled values of all variables. Since the scaling factors are updated along with moving window, process changes can manifest themselves to affect the values of the scaled variables. By comparing the significance of all variables at the same time instance after the occurrence of the fault, emphasis can be laid on those variables with large values.

The next section gives a brief review of the conventional PCA and the fast MWPCA algorithm. Section 3 introduces the power plant and analyses the data set, where a PCA model is used to demonstrate the time-varying behavior. Successful application of the fast MWPCA to detect and diagnose the fault is shown and explained in Section 4. The conclusions appear in Section 5.

## 2 Review of PCA and Fast MWPCA

### 2.1 Generating a PCA Model

A PCA model can be constructed from the correlation matrix of the original process data matrix, $\mathbf{X}_k^0 \in \mathbb{R}^{k\times m}$, which includes *m* process variables collected from time instant *1* until *k*, the mean and standard deviation are given by $\mathbf{b}_k$ and $\boldsymbol{\Sigma}_k = diag\{\sigma_k(1) \quad \cdots \quad \sigma_k(m)\}$. The matrix $\mathbf{X}_k^0$ is then scaled using these two factors to produce $\mathbf{X}_k$, such that each variable now has zero mean and unit variance. The correlation matrix, $\mathbf{R}_k$, of the scaled data set is given by

$$\mathbf{R}_k = \frac{1}{k-1}\mathbf{X}_k^T\mathbf{X}_k \tag{1}$$

By carrying out the eigenvalue-eigenvector decomposition, $\mathbf{R}_k$ is decomposed into a product of two matrices, denoted as score matrix $\mathbf{T}_k$ and loading matrix $\mathbf{P}_k$, as highlighted in Equation 2.

$$\mathbf{R}_k = \mathbf{T}_k \cdot \mathbf{P}_k^T \tag{2}$$

The loading matrix provides the PCA model for further process monitoring tasks.

## 2.2 Fast MWPCA Models

RPCA updates the correlation matrix by adding a new sample to its current value. Conventional MWPCA operates by first discarding the oldest sample from the correlation matrix and then adding a new sample to the matrix. The details of this two-step procedure are shown in Figure 1 for a window size *L*. The fast MWPCA algorithm is based on this, but incorporating the adaptation technique in RPCA. The three matrices in Figure 1 represent the data in the previous window (Matrix I), the result of removing the oldest sample $\mathbf{x}_k^0$ (Matrix II), and the current window of selected data (Matrix III) produced by adding the new sample $\mathbf{x}_{k+L}^0$ to Matrix II.

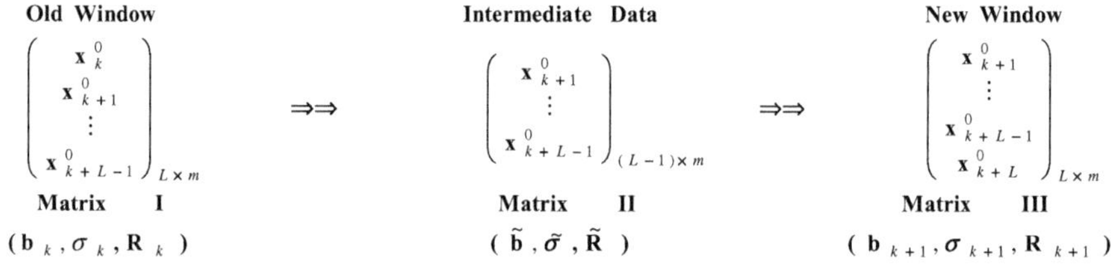


**Fig. 1.** Two-step adaptation to construct new data window [6]

The procedure of updating the correlation matrix is provided in Table 1 for convenience.

## 2.3 Monitoring Procedure Using Fast MWPCA

The monitoring scheme used in this paper is based on one-step-ahead prediction, which calculates the monitoring statistics of time *k* based on the previous PCA model obtained at time *(k-1)*. The use of *N*-step-ahead prediction has been proposed for cases, where the window size is small or the faults are gradual.[6]

The one-step-ahead prediction is now described in more detail. SPE statistic is employed to describe the fitness of the model, which for the $k^{th}$ sample is defined as:

$$SPE_k = \mathbf{x}_k^T \left(\mathbf{I} - \mathbf{P}_{k-1}\mathbf{P}_{k-1}^T\right)\mathbf{x}_k \tag{3}$$

Note that $\mathbf{P}_{k-1}$ is the loading matrix of the $(k-1)^{th}$ model, while $\mathbf{x}_k$ is the $k^{th}$ process sample scaled using the mean and variance for that model.

**Table 1.** Procedure to update correlation matrix for the fast MWPCA approach [6]

| STEP | EQUATION | DESCRIPTION |
|---|---|---|
| 1 | $\tilde{\mathbf{b}} = \frac{1}{L-1}\left(L\mathbf{b}_k - \mathbf{x}_k^0\right)$ | Mean of Matrix II |
| 2 | $\Delta\tilde{\mathbf{b}} = \mathbf{b}_k - \tilde{\mathbf{b}}$ | Difference between means |
| 3 | $\mathbf{x}_k = \mathbf{\Sigma}_k^{-1}\left(\mathbf{x}_k^0 - \mathbf{b}_k\right)$ | Scale the discarded sample |
| 4 | $\mathbf{R}^* = \mathbf{R}_k - \mathbf{\Sigma}_k^{-1}\Delta\tilde{\mathbf{b}}\Delta\tilde{\mathbf{b}}^T\mathbf{\Sigma}_k^{-1} - \frac{1}{L-1}\mathbf{x}_k\mathbf{x}_k^T$ | Bridge over Matrix I and III |
| 5 | $\mathbf{b}_{k+1} = \frac{1}{L}[(L-1)\tilde{\mathbf{b}} + \mathbf{x}_{k+L}^0]$ | Mean of Matrix III |
| 6 | $\Delta\mathbf{b}_{k+1} = \mathbf{b}_{k+1} - \tilde{\mathbf{b}}$ | Difference between means |
| 7 | $\sigma_{k+1}^2(i) = \sigma_k^2(i) + \Delta b_{k+1}^2(i) - \Delta\tilde{b}^2(i)$ $\frac{\left[x_{k+L}^0(i) - b_{k+1}(i)\right]^2 - \left[x_k^0\right.}{L-1}$ | Standard deviation of Matrix III |
| 8 | $\mathbf{\Sigma}_{k+1} = diag\{\sigma_{k+1}(1) \quad \cdots \quad \sigma_{k+1}(m)\}$ | |
| 9 | $\mathbf{x}_{k+L} = \mathbf{\Sigma}_{k+1}^{-1}\left(\mathbf{x}_{k+L}^0 - \mathbf{b}_{k+1}\right)$ | Scale the new sample |
| 10 | $\mathbf{R}_{k+1} = \mathbf{\Sigma}_{k+1}^{-1}\mathbf{\Sigma}_k\mathbf{R}^*\mathbf{\Sigma}_k\mathbf{\Sigma}_{k+1}^{-1} + \mathbf{\Sigma}_{k+1}^{-1}\Delta\mathbf{b}_{k+1}\Delta\mathbf{b}_{k+1}^T$ $\frac{1}{L-1}\mathbf{x}_{k+L}\mathbf{x}_{k+L}^T$ | Correlation matrix of Matrix III |

In this paper, the confidence limits are also calculated using a moving window technique. The window size employed is the same as that for updating the PCA model. The monitoring charts presented in the paper employ 95% and 99% confidence limits.

As shown in the 9th step in Table 1, the newly included sample $\mathbf{x}_{k+L}^0$ is scaled using the new scaling factors, $\mathbf{b}_{k+1}$ and $\mathbf{\Sigma}_{k+1}$, to calculate $\mathbf{x}_{k+L}$. If an abnormal event happens at time ($k$+$L$), $\mathbf{x}_{k+L}^0$ should present noticeable variation from the historical data. However, it is not fair to compare variables of different nature using their un-scaled values. With the MWPCA approach, although the newly updated scaling factors are "corrupted" by including the faulty sample $\mathbf{x}_{k+L}^0$, the scaled values in $\mathbf{x}_{k+L}$ are still able to show distinction from the former samples, given that the window size is sufficiently large. This forms the foundation for the proposed fault diagnosis technique in this paper.

# 3 Power Plant and the Application of PCA

## 3.1 Description of the Process

The power plant is a boiler–turbine–generator unit of 600 MW capacity, shown in a simple schematic diagram in Figure 2. External fans are provided to give sufficient air for combustion. The Forced Draft (FD) fan takes air from atmosphere and injects it through an airpreheater to the air nozzles on the boiler furnace to give hot air for better combustion. The Induced Draft (ID) fan sucks out or draws out the combustible gases from the furnace to assist FD fan and to maintain always slightly negative pressure in the furnace to avoid backfiring through any opening. The pretreated coal is conveyed by hot air injectors through coal pipes into the furnace to give a swirling action for proper mixing of coal powder and the hot air from FD fans. This steam ejected from the boiler above the furnace passes the superheated pipes to reach sufficient combustion temperature. The turbine-generator unit takes the prepared steam to its high pressure turbine and low pressure turbine to generator power. A reheater is fitted between the turbines to guarantee a satisfactory combustion temperature.

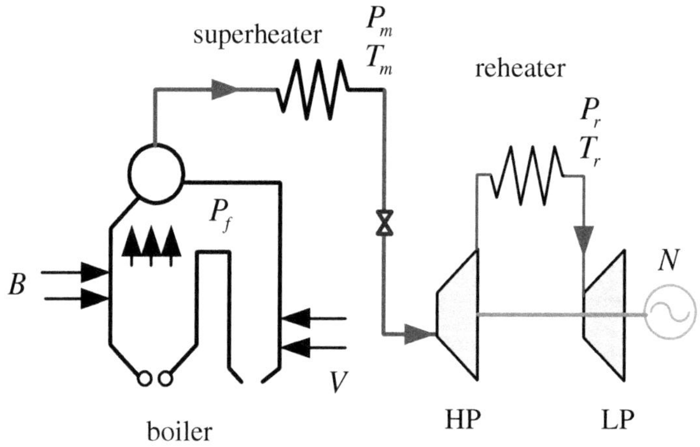


**Fig. 2.** A schematic diagram of the power plant unit

Since this paper concerns a fault in the boiler unit, the monitoring and alarm system for this unit are of particular interest. Checking through the major problems experienced with this unit (not limited to the plant under investigation), a few times furnace explosions have occurred due to wrong operation. In one case the boiler suffered a very bad shock that even stay girders got bent, in addition to good number of tube ruptures. It has also happened that large amount of fuel got sucked into the turbine boiler cycle during normal operation of the unit, indicated by all drains showing foaming. Therefore, the safety aspects and the normal procedures have to be looked into at all stages of operation. Manual intervention is unavoidable; however, much the system is made automatic. In view of this necessary protection, monitoring with alarms for out of limit parameters, and auto and manual control equipment are provided on the operators' console, both on mechanical and electrical equipment.

### 3.2 Available Data and the Process Fault

There are a total of 9 variables recorded from the power plant, as listed in Table 2. They were recorded for a period of about one and half hours, resulting in a total of 2500 variables at the sampling interval of 2 seconds. All available data are plotted in Figure 3.

**Table 2.** Variables measured from the power plant

| Variable no. | Symbol | Description | unit |
|---|---|---|---|
| 1 | *N* | Generator load | MW |
| 2 | *Pm* | Main steam pressure | Mpa |
| 3 | *V* | Total air flow | $km^3/h$ |
| 4 | *Pf* | Furnace pressure | Mpa |
| 5 | *B* | Total fuel flow | t/h |
| 6 | *Dp* | Differential pressure (furnace/big air box) | Pa |
| 7 | *Pr* | Reheater steam pressure | Mpa |
| 8 | *Tm* | Main steam temperature | ℃ |
| 9 | *Tr* | Reheater steam temperature | ℃ |

The working condition before the occurrence of the fault is: power load 550MW, main steam pressure 16.5 Mpa, total air flow 1500$km^3$/h, main steam and reheat temperatures are 536℃. The fault was triggered by a trip out of a FD fan, noticeable from the 1286$^{th}$ sample in the figure. The trip out caused the total air flow into the furnace (variable 3) to decrease spontaneously. Other variables related with air, differential pressure between inside furnace air and air entering furnace (variable 6) and reheater steam pressure (variable 7), also presented immediate sharp drops in value. As these 3 variables are directly affected by the FD fan air flow, they represent the most significant changes after the fault. Because of the existence of the controller for keeping the coal combustion in the furnace, the ID fan (for furnace pressure control) and Runback (RB) took action to cope with the fault. Due to the intervention of the controllers, the furnace pressure (variable 4) and total fuel flow (variable 5)

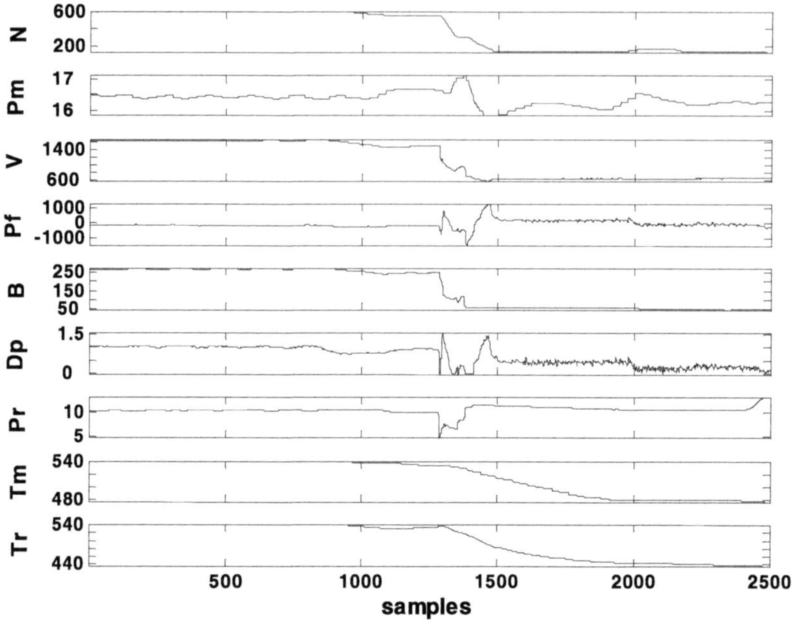


**Fig. 3.** Original variables from plant

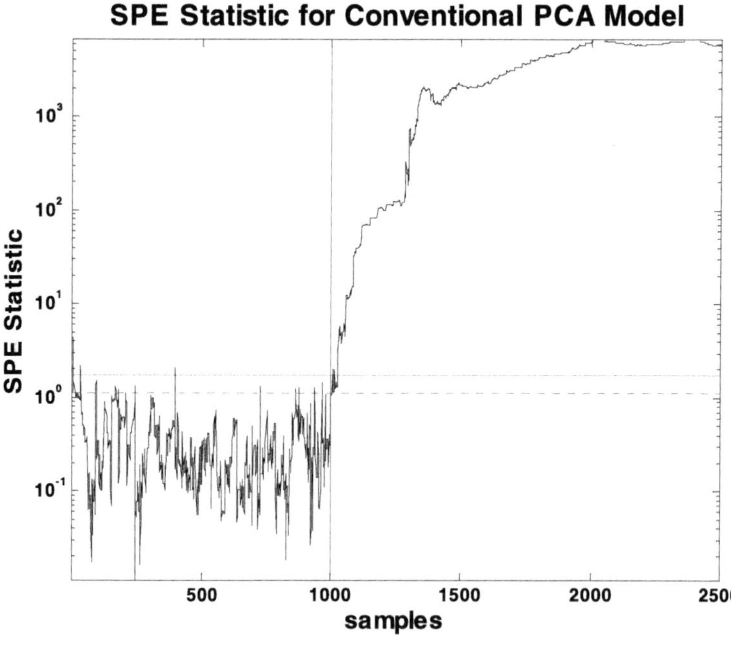


**Fig. 4.** SPE statistic by applying conventional PCA model

have a more gradual drop, and to an even less degree was the drop with the steam temperatures (variables 8 and 9). On notifying the fault, manual adjustments were also carried out. As the ultimate result, the power generated (variable 1) suffered a decrease to below 200 MW.

It can be noticed that the main steam pressure (variable 2) did not decrease its value immediately after the fault. This shows that the furnace and the boiler managed to operate normally by all the actions taken, although the operating condition was not as desired for its highest possible performance.

### 3.3 Application of PCA

By selecting the first 1000 samples as training data, a PCA model was built using 5 Principal Components (PCs). The training data was scaled to zero mean and unit variance. These scaling factors are saved to be used by any other data that this model is tested on. The PCA model is then applied to the rest of the data. The SPE values of all samples are plotted in Figure 4. The 99% and 95% control limits were calculated using the values of SPE statistic from the training data. It is obvious that as soon as the PCA model is used to analyze any other data from the training data, alarms were raised, even without an occurrence of a real fault. This could be a sign of the time varying behavior of the process. It is apparent that a fixed PCA model can not be used to monitor the data.

## 4 Application of MWPCA

The fast MWPCA is now applied for detecting and diagnose the fault. By setting window length as 200, the fault can be detected precisely on time as shown in Figure 5. Before the fault, the statistic did not present excessive false alarms. There is a major violation at the 1286$^{th}$ sample.

Besides its advantage in fault detection over conventional PCA, the fast MWPCA offers higher computational efficiency comparing than conventional MWPCA. By calculating the floating point operations, the fast version of MWPCA is almost 6 times faster.

It should be noted that after a fault is detected, it does not make sense to continue the MWPCA approach any further, as the model has already been corrupted by the fault. The detected fault should be diagnosed and fixed. Only when the process starts to operate normally, can the MWPCA be continued. Figure 5 shows the results of running the moving window through the entire data set for demonstration purpose only.

The scaled values of all variables for the 1286$^{th}$ sample are plotted in Figure 6. This fault diagnosis suggests that, the 3$^{rd}$, 6$^{th}$ and 7$^{th}$ variables are the most dominating ones to the fault, which coincides with the previous description of the fault.

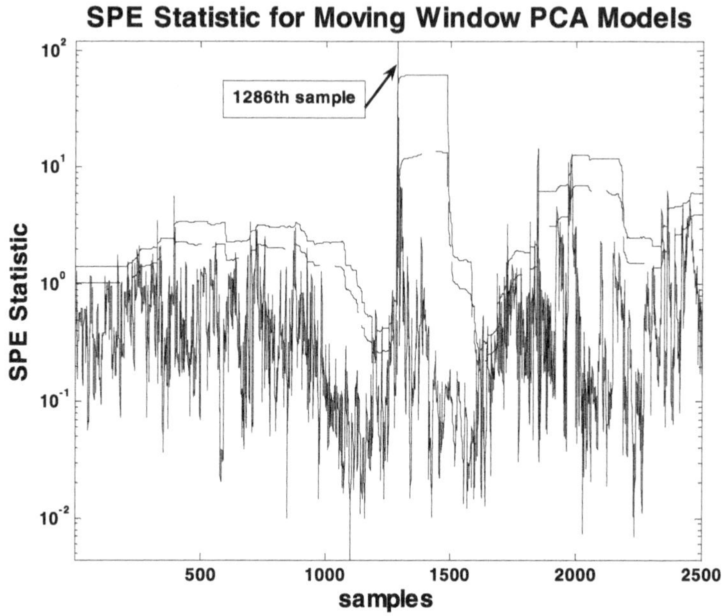


**Fig. 5.** SPE statistic by applying fast MWPCA

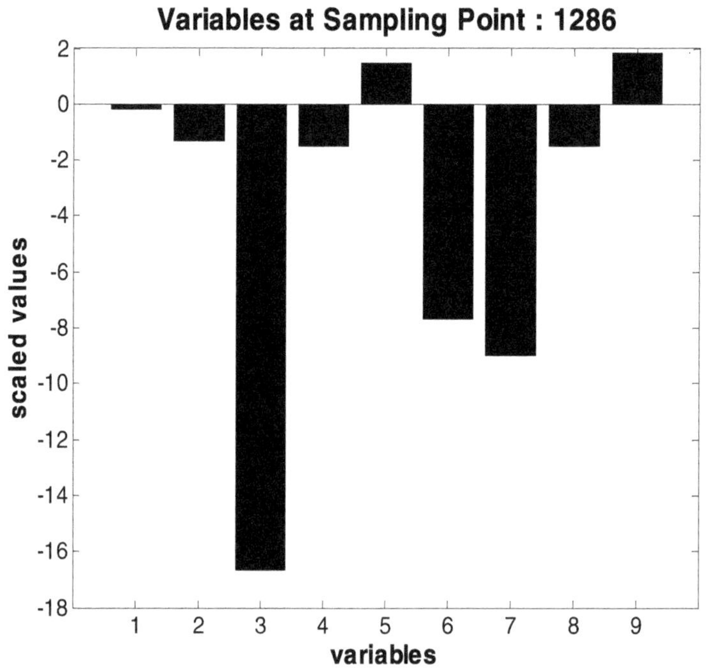


**Fig. 6.** Scaled values of all variables for the 1286$^{th}$ sample

# 5 Conclusions

This paper focused on the detection and diagnosis of an abnormal event recorded from an industrial power plant. By applying the conventional PCA approach, a great number of false alarms occurred when the model was tested on unknown data. This phenomenon suggests the use of adaptive models to monitor this process. By applying the fast MWPCA, the false alarms were eliminated, while the fault could still be detected. Due to its computational efficiency, this approach can be applied on-line to monitor future operation. Its potential in fault diagnosis was further explored, where correct information was extracted from plotting out the scaled values of all variables.

The success of applying MWPCA to the power plant demonstrated its strength in monitoring such processes. Future work can be continued on applying MWPCA on some minor faults, which may require the monitoring scheme based on *N*-step-ahead prediction.

**Acknowledgement.** Dr Xun Wang would like to acknowledge financial support from the U.K. Engineering and Physical Science Research Council (Grant No. EP/C005457).

# References

1. Wise, B.M., Gallagher, N.B.: The Process Chemometrics Approach to Process Monitoring and Fault Detection. Journal of Process Control. 1996, 6(6), 329-348
2. Pearson, K.: On Lines and Planes of Closest Fit to Systems of Points in Space, Phil. Mag., 1901, 2(11), 559-572
3. Jackson, J.E.: Principal Components and Factor Analysis: part 1 – principal analysis, J. Qual. Technol., 1980, 12, 201-213
4. Allagher, N..B., Wise, B..M., Butler, S.W., White, D.D., Barna, G.G.: Development and Benchmarking of Multivariate Statistical Process Control Tools for a Semiconductor Etch Process: Improving Robustness Through Model Updating. Proc. ADCHEM 97, Banff, Canada, 1997, 78-83
5. Wang, X., Kruger, U., Lennox, B.: Recursive Partial Least Squares Algorithms for Monitoring Complex Industrial Processes, Control Engineering Practice, 2003, 11(6), 613-632
6. Wang, X., Kruger, U., Irwin, G.W.: Process Monitoring Approach Using Fast Moving Window PCA, Industrial & Engineering Chemistry Research, 2005, 44(15), 5691-5702

# Default-Mode Network Activity Identified by Group Independent Component Analysis*

Conghui Liu[1,2,**], Jie Zhuang[4], Danling Peng[2], Guoliang Yu[1], and Yanhui Yang[3]

[1] Institute of Psychology, Renmin University of China, Beijing, 100872, China
[2] State Key Laboratory of Cognitive Neuroscience and Learning, Beijing Normal University, Beijing, 100875, China
[3] Department of Radiology, Xuanwu Hospital, Capital University of Medical Sciences, Beijing, 100053, China
[4] Department of Experimental Psychology, University of Cambridge, Cambridge, CB23EB, UK
liuconghui2005@hotmail.com

**Abstract.** Default-mode network activity refers to some regional increase in blood oxygenation level-dependent (BOLD) signal during baseline than cognitive tasks. Recent functional imaging studies have found co-activation in a distributed network of cortical regions, including ventral anterior cingulate cortex (vACC) and posterior cingulate cortex (PPC) that characterize the default mode of human brain. In this study, general linear model and group independent component analysis (ICA) were utilized to analyze the fMRI data obtained from two language tasks. Both methods yielded similar, but not identical results and detected a resting deactivation network at some midline regions including anterior and posterior cingulate cortex and precuneus. Particularly, the group ICA method segregated functional elements into two separate maps and identified ventral cingulate component and fronto-parietal component. These results suggest that these two components might be linked to different mental function during "resting" baseline.

**Keywords:** fMRI, default mode network, independent component analysis.

## 1 Introduction

Typical functional magnetic resonance imaging (fMRI) technology is often applied to study the changes in blood oxygenation level-dependent (BOLD) signal driven by stimuli presented in experimental tasks. Recently, increased attention has been directed at investigating default mode network or task-induced deactivation (TID) [1, 2]. TID refers to greater BOLD signal during "passive" or "resting" baseline condition than during any experimental task [3]. It has been suggested that the fluctuations in BOLD signal during "passive" baseline reflect the neuronal baseline activity of the brain [4].

* Contract grant sponsor: National Science Foundation of China (30570614, 30670705).
** Corresponding author.

D.-S. Huang, L. Heutte, and M. Loog (Eds.): ICIC 2007, LNAI 4682, pp. 222–233, 2007.

Numerous studies on default mode activity have been conducted, reporting a consistent mode across different task, stimuli [5, 6]. Common areas of default mode mainly include ventral anterior cingulate cortex and medial frontal cortex (often extending into the rectus and orbital frontal cortex), posterior cingulate cortex (often extending into the precuneus, angular gyrus, superior occipital cortex and supramarginal gyrus) [1, 5, 7]. The "passive" baseline is a complex state that might include attention or anxiety or memory, however, the precise mental processes supported by the default mode network remain to be elucidated [8].

Several theories have been proposed to explain the nature of TID mode network. The "vascular steal" hypothesis [9] showed that the decrease might be the result of a redistribution of cerebral blood flow to regions that are active from adjacent areas. However, little evidence supported this theory [7]. A more popular point of view is that the decrease is caused by interruption of ongoing internal information processing during the passive or resting state [5, 6, 10]. Some researchers also suggested that the TID mode network play a role in the attention to internal and external stimuli [11]. In addition, Simpson et al. [12] found that some parts of TID mode network might reflect the relationship between attention and anxiety.

In terms of the relationship between TID mode network and task difficulty, some researches [10] suggested that TID mode network is completely task-independent since no difference was oberved across tasks with different difficulties. Others [7] argued that the amplitude of neural deactivation varied according to different task difficulties within the same region of interest. D'Esposito et al. [28] even found that deactivation was extended to adjacent brain areas when the task became more memory demanding.

There is no consent on what kinds of mental processes were involved in TID mode network till now, although most studies agreed with that the baseline state of brain is dynamic and coherent activity. It is very difficult to resolve this question by using the traditional subtraction approach.

In the current study, we used two complementary methods, one applying the general linear model and the other method derived from adapted independent component analysis (ICA) to derive the default mode network from data of multiple subjects. Most of studies employed ICA approach to analyze the data from one subject in a single estimation [13, 14]. It is a method that attempts to separate out linearly mixed spatially or temporally independent components, not only from stimulation, which subjects receive during fMRI experiments, but also signals from other sources, such as "slowly varying" sources and head movements [13]. In our application, we assume independence of the hemodynamic source location from the fMRI data resulting in maps for each of these regions, as well as the time course representing the fMRI hemodynamics. Currently, this approach becomes general tool for detecting default mode [15].

Recently, some researchers extended ICA to allow for the analysis of multiple subjects [16, 17, 18]. This analysis can simultaneously decompose group fMRI data into different component maps. It has been demonstrated that the group ICA approach can analyze the activation data from all subjects in a single ICA estimation [16]. In this study, we applied this method to our data from two fMRI experiments: one with Chinese verb generation task, the other with English verb reading task, with the aim of identifying default or TID coherencies that are consistent across subjects, stimuli and tasks. In addition, we compared the results of fMRI imaging data processed with ICA with the results obtained with conventional hypothesis-driven analysis.

# 2 Method

## 2.1 Subjects

Twenty four right-handed healthy, under graduate students were recruited from a university campus Beijing, China., participated in the study (8 males and 16 females; 18~23 years old with and average age of 21 years, standard error 1.8 years). The subjects are native speakers of Mandarin Chinese with English as their second language. All subjects have normal or corrected-to-normal visual acuity.

## 2.2 Materials and Procedures

Forty 3~9 letters English words and 40 double-characters Chinese words were common nouns and used in everyday life. The experiment includes two runs, with 40 Chinese words in one run and 40 English words in the other. The sequence of them was counterbalanced such that half subjects performed the English task first and the other half performed the Chinese task first. Each run lasted 4min 48s and consisted of 4 blocks. Of the 4 blocks, 2 were experimental blocks and 2 were control blocks. Experimental and control blocks were always presented one after the other. The order and length of the experimental and control blocks was displayed in Fig.1. Each experimental block consisted of 40 trials. There is a 2s instruction before each experimental and control block. The stimuli was programmed with DMDX (http://www.u.arizona.edu/Bkforster/dmdx/dmdx.htm) on a notebook computer and presented by a projector onto a translucent screen. Subjects viewed the stimuli through a mirror attached to the head coil.

In this study, we used a verb generation task and noun reading task based on that described by Petersen and his colleagues [19]. During the experimental condition, each Chinese or English noun was presented for 150 ms, followed by a "+" blank screen for 3850 ms (see Fig.1). Subjects are required to speak the Chinese verb associated with the presented Chinese nouns on the computer screen as quickly and correctly as possible. For example, in Chinese task, the subject might speak the verb "吃" (eat) if the noun, "苹果" (apple), is presented. In other run, subjects are asked to read the English noun presented on the screen. The subjects were asked to silently fixate on "+" passively without any response in control block. The subjects were instructed to speak the verb silently, in order to minimize the motion artifact of speech. During the control condition, the stimuli is "+". The subjects were asked to view the "+"passively.

## 2.3 fMRI Apparatus

This study was performed on a 1.5T (Siemens Magnetom Sonata Maestro Class, Germany) whole-body MRI scanner. Functional scans were obtained using a

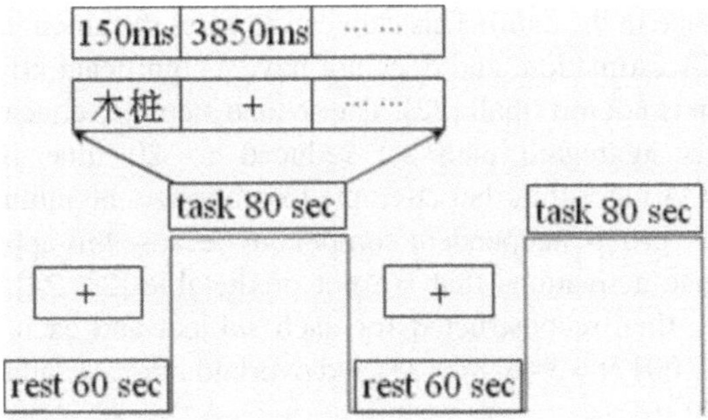


**Fig. 1.** An example and arrangement of materials in Chinese verb generation task. Task =Chinese verb generation or English noun reading; rest = baseline condition.

T2-weighted gradient echo EPI sequence (20 contiguous axial slices, slice thickness=6mm, inter-slice gap=1.8mm, in-plane resolution=3.6mm×3.6mm, TR/TE/θ=2000ms/50ms/90°; FOV=230×230mm2, matrix=64×64). 288 data sets were collected using a task/rest block paradigm, with a total time of 576 sec. The high-resolution anatomical images were acquired using an axial multi-slice T1-weighted FLASH sequence (96 sagittal slices, slice thickness=1.7mm, inter-slice gap=0.85mm; TR=1970ms, TE=3.93ms, flip angle=15°; FOV=250×235mm2, matrix=179×235).

### 2.4 fMRI Data Analysis

Image processing and statistical analysis were carried out using SPM2 software (www.fil.ion.ucl.ac.uk/spm) [20]. The first four volumes of each run were discarded to allow for signal stabilization, the remaining functional images were realignment to the first volume. No subject had more than 2.0mm of head movement in any plane. Co-registered images were normalized into the standard space [21], and then smoothed to decrease spatial noise (8mm FWHM Gaussian filter).

The general linear model was used to estimate the condition effect of individual subject. Firstly, individual results were acquired by defining four effects of interests (baseline minus Chinese verb generation, baseline minus English noun reading,) for each subject with the relevant parameter estimates. The threshold for significant activation was $P<0.001$ (uncorrected). The group averaging effects were computed with a random-effects model. The cluster with more than 10 voxels activated above a thresholds of $P<0.001$ (uncorrected) were considered as significant. We calculated contrasts comparing control conditions to experimental (baseline-Chinese verb generation, baseline-English noun reading).

### 2.5 Independent Component Analysis

The group ICA was carried out using GIFT (Group ICA of fMRI Toolbox) software (http://icatb.sourceforge.net/). The smoothed data from each subject were reduced from 144 to 40 time points using principal component analysis (PCA) (representing greater than 99% of variance in the data). This step can reduces the amount of memory require to perform the ICA estimation, and does not have a significant affect on the results if the number chosen is not too small [22]. The second step is to concatenate data from all subjects, and this aggregate data set reduced to 20 time points using PCA. Informax-based ICA algorithm that attempts to minimize the mutual information was used to estimate the group independent component because this approach appears to be suited to investigate activations that are not predictable [23, 24]. Time courses and spatial maps were then reconstructed for each subject and each group images was thresholded at $P$<0.001 ($t = 3.48$, $df = 23$) and overlaid onto a standard SPM anatomical template brain [16].

# 3 Result

## 3.1 Behavioral and Physiological Data

In the behavioral experiment, we collected the response times (RTs) and error rate. The average RT and error rate of responses were 420 ms and 2.1% for English noun reading

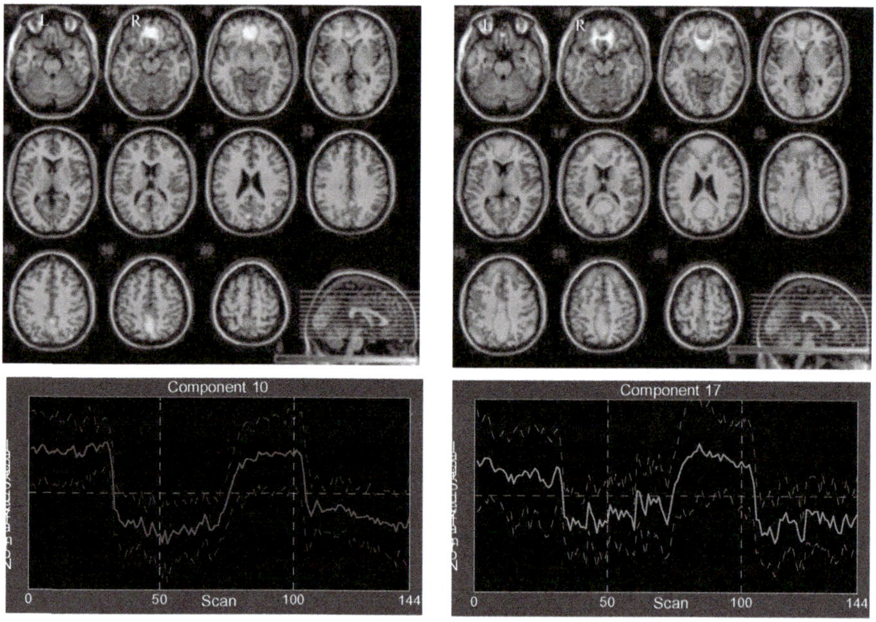


**Fig. 2.** Group averaged axial t-maps ($P$ < 0.001, uncorrected) for baseline minus English noun word reading (A), Group averaged component 12 (red) and 18(green) t-maps ($P$ < 0.001, uncorrected) in English noun word reading task (B). Group averaged time courses for component 12 (C) and component 18 (D) are presented. Standard deviation across the group is indicated for each time course with dotted lines. The images were superimposed on a standard SPM anatomical template brain in neurological convention with Z coordinate for each slice shown in Talairach Space. The color bar shows t value, ranging from 0 to 10.

task, 817ms and 5.5% for the Chinese verb generation task, respectively. The reaction times were significantly faster ($P < 0.001$, two-tailed paired t test) during Chinese verb generation task compared with the English noun reading task. The difference of error rate between the two tasks did not reach significant level ($P = 0.16$).

## 3.2 fMRI Data

The SPM group analysis (Fig 2A, Fig 3A and Table 1, 2, 3, 4) revealed that a similar network were deactivated significantly in English noun word reading task and Chinese verb generation task, including anterior cingulate gyrus (BA 32) and medial frontal cortex (BA 10/11/25) as well as precuneus (BA 7/31). Notably, there is stronger deactivation in Chinese verb generation task than English noun word reading task.

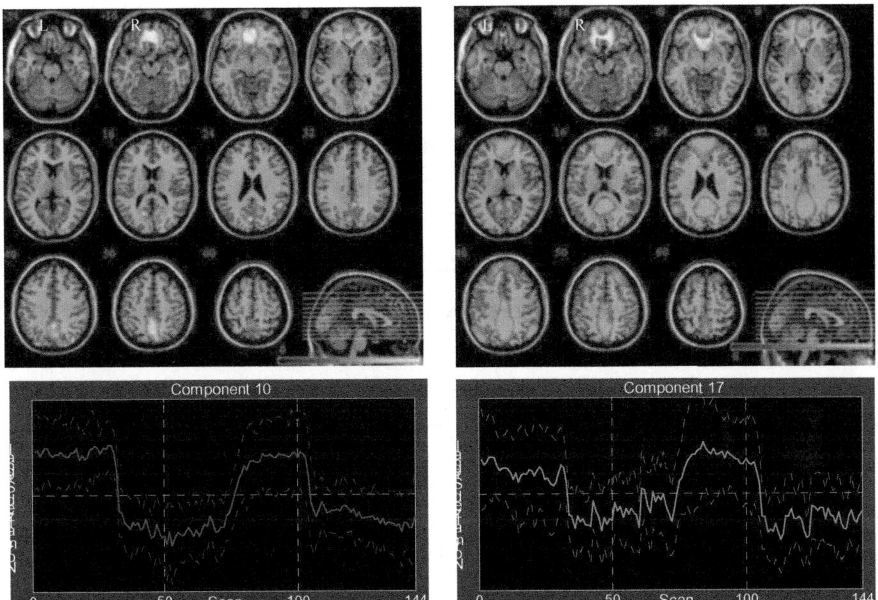


**Fig. 3.** Group averaged axial t-maps ($P < 0.001$, uncorrected) for baseline minus Chinese verb generation task (A), Group averaged component 10 (red) and 17(green) t-maps ($P < 0.001$, uncorrected) in Chinese verb generation task (B). Group averaged time courses for component 10 (C) and component 17 (D) are presented. Standard deviation across the group is indicated for each time course with dotted lines. The images were superimposed on a standard SPM anatomical template brain in neurological convention with Z coordinate for each slice shown in Talairach Space. The color bar shows t value, ranging from 0 to 10.

The ICA group results are depicted in Fig 2B, Fig 3B and Table 1, 2, 3, 4. Twenty components were estimated for each subject after reducing the data. The resulting 20 time courses were sorted according to their correlation with the design matrix in SPM.

Of the 20 components, only two largest components were selected. In English task, the correlation coefficient of component 12 and 18 is 0.78 and 0.46 respectively. In Chinese task, the correlation coefficient of component 10 and 17 is 0.79 and 0.60 respectively. The group-averaged time courses for the fixation-task paradigm (with the standard deviation across the 24 subjects) for component 12 and 18, component 10 and 17, are presented in Fig 2C, D respectively.

The activation pattern of component 12 and component 18 in English task resemble the activation maps of component 10 and 17 in Chinese task respectively. Fox example,

**Table 1.** Brain activation for baseline vs. English noun word reading

| Brain Region | BA | SPM | BA | ICA(C12) |
|---|---|---|---|---|
| | | Max T(x,y,z) | | Max T(x,y,z) |
| L/R Anterior Cingulate | 32(R) | 8.30(8,35,-5) | 32(R),25(R) | 18.35(6,35,-5) |
| L/R Medial Frontal Gyrus | 10 | 7.91(8,44,-9) | 10,11,25(R) | 18.14(4,30,-12) |
| L/R Inferior Frontal Gyrus | | | 11(L) | 8.41(-20,34,-22) |
| L/R Superior Frontal Gyrus | 8(R) | 5.16(24,29,43) | 10(R) | 7.56(8,62,-3) |
| L/R Middle Frontal Gyrus | 8(R) | 5.90(30,39,46) | | |
| L/R Precuneus | 7(L),31(R) | 8.58(-2,-66,44) | | |
| L/R Superior Parietal Lobule | 7(R) | 11.63(34,-75,44) | | |
| L/R Posterior Cingulate | 30(R) | 8.01(14,-54,10) | | |

$P < 0.001$, uncorrected, voxel = 10

L, left hemisphere; R, right hemisphere; BA, Brodmann's area

**Table 2.** Brain activation for baseline vs. English noun word reading

| Brain Region | BA | SPM | BA | ICA(C18) |
|---|---|---|---|---|
| | | Max T(x,y,z) | | Max T(x,y,z) |
| L/R Anterior Cingulate | 32(R) | 8.30(8,35,-5) | 32(R) | 6.63(8,43,-2) |
| L/R Medial Frontal Gyrus | 10 | 7.91(8,44,-9) | 9,10 | 10.78(-2,52,-6) |
| L/R Middle Frontal Gyrus | 8(R) | 5.90(30,39,46) | 8(R) | 5.95(28,33,43) |
| L/R Superior Frontal Gyrus | 8(R) | 5.16(24,29,43) | 8(R),9(L) | 6.66(-16,48,34) |
| L/R Precuneus | 7(L),31(R) | 8.58(-2,-66,44) | 7,31 | 17.12(4,-53,32) |
| L/R Superior Parietal Lobule | 7(R) | 11.63(34,-75,44) | | |
| L/R Posterior Cingulate | 30(R) | 8.01(14,-54,10) | | |
| L/R Inferior Parietal Lobule | | | 40 | 8.39(51,-58,36) |
| L/R Angular Gyrus | | | 39(L) | 7.84(-44,-68,29) |
| L/R Supramarginal Gyrus | | | 40(R) | 6.91(55,-51,28) |

$P < 0.001$, uncorrected, voxel = 10

L, left hemisphere; R, right hemisphere; BA, Brodmann's area

**Table 3.** Brain activation for baseline vs. Chinese verb generation

| Brain Region | BA | SPM | BA | ICA(C10) |
|---|---|---|---|---|
| | | Max T(x,y,z) | | Max T(x,y,z) |
| L/R Anterior Cingulate | 32(L) | 5.78(-8,23,-8) | 32(L),23(L) | 14.78(-4,35,-5) |
| L/R Medial Frontal Gyrus | 11,25(R) | 13.30(-4,36,-12) | 10(L),11(R),25(L) | 15.46(2,32,15) |
| L/R Inferior Frontal Gyrus | 47 | 7.36(-16,23,-15) | 47(L) | 6.06(-22,12,-24) |
| L/R Middle Frontal Gyrus | 8(R) | 5.25(28,35,42) | | |
| L/R Precuneus | 7,31(L) | 8.62(4,-46, 43) | | |
| L/R Supramarginal Gyrus | 40(R) | 7.81(50,-49,25) | | |
| L/R Posterior Cingulate | 29(R) | 6.95(14,-50,6) | | |
| L/R Parahippocampal Gyrus | 37(L) | 6.60(-26,-45,-11) | | |

$P < 0.001$, uncorrected, voxel = 10
L, left hemisphere; R, right hemisphere; BA, Brodmann's area

**Table 4.** Brain activation for baseline vs. Chinese verb generation

| Brain Region | BA | SPM | BA | ICA(C17) |
|---|---|---|---|---|
| | | Max T(x,y,z) | | Max T(x,y,z) |
| L/R Anterior Cingulate | 32(L) | 5.78(-8,23,-8) | 32 | 12.53(-8,41,5) |
| L/R Medial Frontal Gyrus | 11,25(R) | 13.30(-4,36,-12) | 10(L),11(L) | 15.25(-2,56,1) |
| L/R Middle Frontal Gyrus | 8(R) | 5.25(28,35,42) | 8(R) | 10.20(26,29,43) |
| L/R Inferior Frontal Gyrus | 47 | 7.36(-16,23,-15) | | |
| L/R Superior Frontal Gyrus | | | 10(L),8(L) | 11.78(-18,57,19) |
| L/R Precuneus | 7,31(L) | 8.62(4,-46, 43) | 31,7(L),19(L) | 23.54(0,-59,32) |
| L/R Posterior Cingulate | 29(R) | 6.95(14,-50,6) | 30 | 24.02(6,-50,17) |
| L/R Supramarginal Gyrus | 40(R) | 7.81(50,-49,25) | | |
| L/R Cingulate Gyrus | | | 31,24(L) | 12.89(6,-37,31) |
| L/R Angular Gyrus | | | 39 | 13.00(46,-64,33) |
| L/R Parahippocampal Gyrus | 37(L) | 6.60(-26,-45,-11) | | |
| L/R Middle Temporal Gyrus | | | 21,39(L) | 11.58(-51,-61,25) |

$P < 0.001$, uncorrected, voxel = 10, L, left hemisphere; R, right hemisphere; BA, Brodmann's area

the component 12 and 10 overlap heavily in the ventral anterior cingulate and medial frontal regions. The component 18 and 17 also have similar spatial pattern of brain activation in midline regions and precuneus. The Talariach coordinates of the maxima of each region with in the maps are presented in Table 1, 2, 3, 4. However, the Chinese verb generation produced greater and stronger deactivation than did the English word reading task. Additionally, many of regional locations identified in ICA analysis corresponded well with the SPM2 analysis.

## 4 Discussion

In this study, we used two methods to investigate the default or TID mode of two different tasks. The results produced by conventional hypothesis-driven method (general linear model analysis) showed that English noun word reading and Chinese verb generation deactivated a similar neural network which includes anterior cingulate cortex, posterior cingulate cortex and precuneus (Fig 2A, Fig 3 A and Table 1, 2, 3, 4). These areas are very consistent with results of previous studies [5, 8, 18]. For example, Shulman et al. [5] performed a large PET meta-analysis study involving 97 subjects in several different processing tasks, in which two midline regions, the posterior cingulate cortex and ventral anterior cingulate cortex, consistently demonstrate TID in several cognitive tasks. We found that default network exhibit high spatial consistency across subjects, stimuli and tasks.

We found stonger task-related deactivation in Chinese verb generation task than in English noun word reading task, which is consistent with behavioral data that RTs decreased as tasks became more difficult. The results are against the task difficulty-independent opinion of TID [10], but in favor of task difficulty-dependent point of view [7]. It might indicate that more difficult levels of the task required greater processing and cognitive resources. Alternatively, it is possible that the default mode neural network is active during the "passive" baseline, less disrupted during English word reading with low cognitive demand, but more disrupted during verb generation task with high cognitive demand [8].

Our result from group ICA analysis showed that ICA and conventional methods agree in most cases, which is consistent with the findings reported by Quigley and his colleagues [25]. The default or TID maps of brain in English and Chinese tasks were decomposed by group ICA into several different components. We calculated the correlation with the standard reference function as a means to rank independent components. Only two components were selected on the basis of their temporal dynamic mode and spatial activation pattern. Fig.2B and Fig 3B show ventral anterior cingulate cortex in one component (component 12 and 10). The other component involves mainly anterior cingulate cortex, posterior cingulate cortex and precuneus (component 17 and 18, see Fig.2B, Fig.3B). The activation pattern of the only two components is very similar in English and Chinese tasks. The Images processed with group ICA resembled images processed with conventional means. However, ICA analysis found the two different components. This suggested that the default mode network might recruit multiple neural systems to support different mental activities. The medial prefrontal cortex and anterior cingulate cortex (component 10, 12, see Fig.2B, Fig 3B) is linked primarily to paralimbic regions associated with affective processes. Many fMRI studies revealed that the ventral cingulate cortex might interact with other cortical structures as a part of the circuits involved in the regulation of emotional activity [26, 27]. Simpson et al. [12] found that blood flow reduction in medial prefrontal cortex might reflect a dynamic balance between attention and anxiety. The anterior cingulate cortex and posterior cingulate cortex (component 17, 18) is consistent with recent studies that the midline default network played an important role in attending to environmental stimulus [11]. Although we found similar TID mode network in English reading and Chinese verb generation tasks, stronger activation is showed in component 17 in Chinese task, compared to component 18 in

English task. It might be resulted from different attentional resources acquired by different tasks. Some investigators demonstrated that the default mode network was constant without changes during a simple task which required little attentional resources [8]. In a previous similar ICA analysis of default mode network, Esposito et al [28] found that the greater extension of the anterior and lesser extension of the posterior cingulate region were detected when the task was switched from low to high working memory loads. We did not find the less activation in posterior cingulate cortex in Chinese task, compared to English task. Such differences may be biased by the specific task engagement. Esposito et al. [28] used the n-back memory task while we selected language task to investigate the default mode network. Both language tasks required subjects to retrieve more language rather than memory related information. The verb generation task need more semantic search than noun reading task [19], however, the difficulty difference between them might not be large enough as the n-back memory task to show the same result pattern as Esposito et al. [28]. Another similar explanation [1] is that posterior cingulate cortex in default network is intensively involved in memory processing as Alzheimer patients showed reduced connectivity in posterior cingulate cortex.

## 5 Conclusion

In conclusion, there were three main findings in this study. First, the traditional GLM and ICA detect similar default network across different tasks, stimuli, and subjects, which mainly include midline regions. Second, default mode network might recruit two main neural systems, each of which might have different deactivation pattern and mental function. The anterior default mode activity system (ventral anterior cingulate cortex) might be linked to the emotional regulation, whereas the posterior default mode activity mode system (midline regions and precuneus) might be involved to attention system and memory processing. Third, the amplitude of neural deactivation is manipulated by task difficulty. More difficult task (e.g., Chinese verb generation task) produced stronger deactivation than relatively easier task (e.g., English noun reading task).

## References

1. Greicius, M.D., Srivastava, G., Reiss, A.L., Menon, V.: Default-mode Network Sctivity Fistinguishes Alzheimer's Fisease from Healthy Sging: Evidence from functional MRI. Proceedings of the National Academy of Sciences, U.S.A., 101, (2004)4637-4642
2. Damoiseaux, J.S., Rombouts, S.A.R.B., Barkhof, F., Scheltens, P., Stam, C.J., Smith, S.M., Beckmann, C.F.: Consistent Testing-state Networks Scross Healthy Dubjects. Proceedings of the National Academy of Sciences, U.S.A., 103: (2006)13848-13853
3. Ogawa, S., Lee, T.M., Kay, A.R., Tank, D.W.: Brain Magnetic Resonance Imaging with Contrast Dependent on Blood Oxygenation. Proceedings of the National Academy of Sciences, U.S.A., 87, (1990)9868-9872
4. Laufs, H., Krakow, K., Sterzer, P., Eger, E., Beyerle, A., Salek-Haddadi, A., Kleinschmidt. Electroencephalographic Signatures of Attentional and Cognitive Default Modes in Spontaneous Brain Activity Fluctuations at rest. Proceedings of the National Academy of Sciences, U.S.A., 100, (2003)11053-11058

5. Shulman, G.L., Fiez, J.A., Corbetta, M., Buckner, R.L., Miezin, F.M., Raichle, M.E., Pertersen, S.E. :Common Blood Flow Changes Across Visual Tasks: II. Decreases in Cerebral Cortex. Journal of Cognitive Neuroscience, 9, (1997)648-663
6. Blinder, J.R., Frost, J.A., Hammeke, T.A., Bellgowan, P.S.F., Rao, S.M., Cow, R.W. :Conceptual Processing During the Conscious Resting State: A Functional MRI Study. Journal of Cognitive Neuroscience, 11(1), (1999) 80-93
7. McKiernan, K.A., Kaufman, J.N., Kucer-Thompson, J., Binder, J.R.: A Parametric Manipulation of Factors Affecting Task-induced Deactivation in Function Neuron Imaging. Journal of Cognitive Neuroscience. 15(3), (2003)394-408
8. Greicius, M.D., Krasnow, B., Reiss, A.L., Menton, V. : Functional Connectivity in the Resting Brain: A Network Analysis of the Default Mode Hypothesis. Proceedings National Academy Sciences, U.S.A., 100,(2003)253-256
9. Shmuel, A., Yacoub, T., Pfeuffer, J., Van De Moortele, P.-F., Adriany, G., Ugurbil, K., Hu, X.: Negative BOLD Response and Its Coupling to the Positive Response in the Human Brain. International Conference on Functional Mapping of the Human Brain, NeuroImage, 13(6), (2001)1005
10. Gusnard, D.A., Raichle, M.E. :Searching for Baseline: Functional Imaging and the Resting Human Brain. Nature Reviews Neuroscience, 2, (2001)685-694
11. Raichle, M.E., Macleod, A.M., Snyder, A.Z., Powers, W.J., Gusnard, D.A., Shulman, G.L.: A Default Mode of Brain Function. Proceedings National Academy Sciences USA. 98, (2001)676-682
12. Simpson, J.R., Drevets, W.C., Snyder, A.Z., Gusnard, D.A., Raichle, M.E.: Emotion-induced Changes in Human Medial Prefrontal Cortex: II. During Anticipatory Anxiety. Proceedings National Academy Sciences, U.S.A., 98, (2001)688-693
13. McKeown, M.J., Makeig, S., Brown, G.G., Jung, T.P., Kindermann, S.S., Bell, A.J., Sejnowski, T.J.: Analysis of fMRI Data by Blind Separation into Independent Spatial Components. Human Brain Mapping, 6, (1998)160-188
14. Mckeown M.J., Hansen, L.K., Sejnowski, T.J. :Independent Component Analysis of Functional MRI: What is Signal And What is Noise? Current Opinion in Neurobiology, 13, (2003)1-10
15. Ma, L., Wang, B., Chen, X., Xiong, J. Detecting Functional Connectivity in the Resting Brain: A Comparison Between ICA and CCA. Magnetic Resonance Imaging, 25(1), (2007)47-56
16. Calhoun, V.D., Pekar, J.J., McGinty, V.B., Adali, T., Watson, T.D., Pearlson, G.D.: Different Activation Dynamics in Multiple Neural Systems During Simulated Driving. Human Brain Mapping, 16, (2002)158-167
17. Svensen, M., Kruggel, F., Benali, H.: ICA of fMRI Group Study Data. NeuroImage, 16, (2002)551-563
18. Calhoun, V.D., Adali, T., Stevens, M.C., Kiehl, K.A., Pekar, J.J.:Semi-blind ICA of fMRI: A Method for Utilizing Hypothesis-derived Time Courses in A Spatial ICA Analyais. NeuroImage, 25, (2005)527-538
19. Petersen, S.E., Fox, P.T., Posner, M.I., Mintun, M., Raichle, M.E.: Positron Emission Tomographic Studies of the Cortical Anatomy of Single-word Processing. Nature, 331, (1988)585-589
20. Friston, K.J., Holmes, A.P., Poline, J.B., Grasby, P.J., Williams, S.C., Frackowiak, R.S., Tumer, R.: Analysis of fMRI Time-series Revisited. NeuroImage, 2, (1995)45-53
21. Talairach, J., and Tournoux, P.: A Co-planar Stereotaxic Atlas of a Human Brain. Stuttgart: Thieme, (1988)

22. Calhoun, V.D., Adali, T., Pearlson, G.D., Pekar, J.J. : A Method for Making Group Inferences from fMRI Data Using Independent Component Analysis. Human Brain Mapping, 14, (1995)140-151
23. Bell, A.J., Sejnowski, T.J.: An Information–maximization Approach to Blind Separation and Blind Deconvolution. Neural Comput 7, (1995)1004–1034
24. Esposito, F., Formisano, E., Seifritz, E., Goebel, R., Morrone, R., Tedeschi, G., Salle, F.D. Spatial Independent Component Analysis of Functional MRI Time-series: to what Extent Do Results Depend on the Algorithm Used? Human Brain Mapping. 16, (2002)146-157
25. Quigley, M.A., Haughton, V.M., Carew, J., Cordes, D., Moritz, C.H., Meyerand, M.E.: Comparison of Independent Component Analysis and Conventional Hypothesis-driven Analysis for Clinical Functional MR Image Processing. American Journal of Neuroradiology, 23, (2002)49-58
26. Greicius, M.D., Flores, B.H., Menon, V., Glover, G.H., Solvason, H.B., Kenna, H., Reiss, A.L., Schatzberg, A.F.: Resting-state Functional Connectivity in Major Depression: Abnormally Increased Contributions from Subgenual Cingulate Cortex and Thalamus. Biology Psychiatry, In Press, (2007)
27. Bush, G., Luu, P., Posner, M.I.: Cognitive and Emotional Influences in Anterior Cingulate Cortex. Trends in Cognitive Sciences, 4(6), (2000)215-222

# Mutual Information Based Approach for Nonnegative Independent Component Analysis

Hua-Jian Wang[1], Chun-Hou Zheng[2], and Li-Hua Zhang[1]

[1] College of Electric Information and Automotion, Qufu Normal University, 276826 Rizhao, Shandong, China
[2] School of Information and Communication Technology, Qufu Normal University
zhch@qrnu.edu.cn

**Abstract.** This paper proposes a novel algorithm for nonnegative independent component analysis, which is based on minimizing the mutual information of the separated signals, and is truly insensitive to the particular underlying distribution of the source data. The unmixing system culminates to a novel neural network model. Compared with other algorithms for nonnegative ICA, the method proposed in this paper can work efficiently even in the case that the source signals are not well grounded, and that pre-whiting process is not needed. Finally, the experiments were performed on both simulating signals and mixtures of image data, the results indicate that the algorithm is efficient and effective.

## 1 Introduction

Independent component analysis (ICA) has become an important research area in recent years [1,2,3,22]. Considering $n$ statistically independent random variables $s_i$ (the sources), which form a random vector $\mathbf{s} = (s_1, \cdots, s_n)^T$, and assuming that we have an observation vector $\mathbf{x}$ generated according to

$$\mathbf{x} = \mathbf{As} \tag{1}$$

The task of ICA is then to discover the source $\mathbf{s}$ and mixing matrix $\mathbf{A}$ given just the observation, using the assumption of independence between different variables of $\mathbf{s}$. Hence, the common methods for ICA is to construct an unmixing matrix $\mathbf{B} = \mathbf{RA}^{-1}$ giving

$$\mathbf{y} = \mathbf{Bx} = \mathbf{BAs} = \mathbf{Rs} \tag{2}$$

In traditional methods for ICA, the observation $\mathbf{x}$ are often assumed to be zero-mean, or transformed to be so, and are commonly prewhitened by a matrix $\mathbf{V}$:

$$\mathbf{z} = \mathbf{Vx} \tag{3}$$

so that $E\{\mathbf{zz}^{\mathrm{T}}\} = \mathbf{I}$ holds before an optimization algorithm. However, in many real-world problems, we know that the sources $s_i$ must be nonnegative. We call a source

D.-S. Huang, L. Heutte, and M. Loog (Eds.): ICIC 2007, LNAI 4682, pp. 234–244, 2007.

$s_i$ *nonnegative* if $\Pr(s_i < 0) = 0$, where $\Pr(\bullet)$ is probability function, i.e., the sources must be either zero or positive [5]. Moreover, the combination of these constraints on the sources $s_i$ is referred to as *nonnegative independent component analysis*. Also, Plumbley *et. al* gave out some useful algorithms for nonnegative ICA [12, 13, 14]. Compared with traditional ICA, these algorithms for nonnegative ICA did not remove the mean of the data in the whitening transform of eqn.(3), since in doing so would lose information about the nonnegativity of the sources. Another assumption of these algorithms is that the sources $s_i$ are well grounded except for independence and non-negativity. We call a source $s_i$ *well grounded* if $\Pr(s_i < \delta) > 0$ for any $\delta > 0$, i.e., $s_i$ has nonzero probabilistic density function (pdf) all the way down to zero.

However, in practical applications, many real-world nonnegative sources are not well grounded, e.g., images. In this paper, we will propose a new algorithm for non-negative ICA even in the case that the sources are not well grounded, which is an approach based on minimizing mutual information.

## 2 Mutual Information Criterion for ICA

### 2.1 Independence Criterion

Since statistical independence of the sources is the main assumption, any separation structure of ICA is tuned so that the components of its output $\mathbf{y}$ become statistically independent. In practice, there are several sensible measures of mutual dependence, but one of the best ones is Shannon's mutual information. It can be defined as

$$I(\mathbf{y}) = \sum_i H(y_i) - H(\mathbf{y}) \tag{4}$$

where $H(v) = -\int p(v)\log p(v)dv$ denotes the entropy of variable $v$, and $p(\bullet)$ is probabilistic density function (pdf). The mutual information $I(\mathbf{y})$ measures the amount of information that is shared by the components of $\mathbf{y}$. In fact, it is always non-negative and vanishes iff the components of $\mathbf{y}$ are mutually independent, i.e.

$$p(\mathbf{y}) = \prod_i p(y_i) \tag{5}$$

### 2.2 Minimizing Mutual Information

The minimizing mutual information method has been proposed in literatures [6,7] for performing ICA based on a principle of maximizing information preservation. It uses a network structure depicted in Fig.1, where B is a separating matrix in ICA, $y_i$ are extracted independent components, and blocks $\psi_i$ are auxiliary, being used only during the optimization phase.

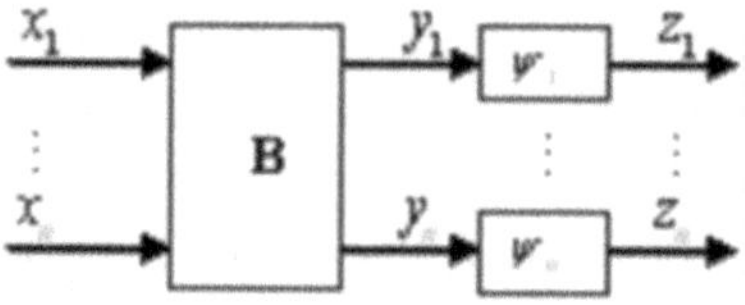


**Fig. 1.** Structure of the ICA system

Assume that each function $\psi_i$ is the cumulative probability function (CPF) of the corresponding component $y_i$, i.e.

$$z_i = \psi_i(y_i) = \int p(y_i)dy_i \tag{6}$$

then

$$p(z_i) = \frac{p(y_i)}{|\partial z_i / \partial y_i|} = \frac{p(y_i)}{p(y_i)} = 1 \tag{7}$$

That is to say, $z_i$ are uniformly distributed in [0, 1], so that $H(z_i) = 0$. Therefore, we have

$$I(\mathbf{y}) = I(\mathbf{z}) = \sum_i H(z_i) - H(\mathbf{z}) = -H(\mathbf{z}) \tag{8}$$

where $H(\mathbf{z})$ is the joint entropy of random vector $\mathbf{z}$. Therefore, maximizing the output entropy is equivalent to minimizing the mutual information of the extracted components $y_i$. In literatures [6,9], the nonlinear functions $\psi_i$ are chosen *a priori*. However, the method will fail if there is a strong mismatch between the $\psi_i$ and the true CPFs of $y_i$.

It has been proved in literature [8] that given the constraints placed on the functions $\psi_i$ so $z_i$ is bounded to [0, 1]. Besides, given that $\psi_i$ is also constrained to be an increasing function (see Section 4), then maximizing the output entropy $H(\mathbf{z})$ will lead the functions $\psi_i$ to become the estimates of the CPFs of the corresponding $y_i$ components.

## 3 Algorithm for Nonnegative ICA

### 3.1 Algorithm Architecture

**Lemma 1.** Let $\mathbf{s} = (s_1, \cdots, s_n)^T$ be an $n$-dimensional random vector of real-valued independent sources which have nongaussian distributions, $\mathbf{A}$ and $\mathbf{B}$ be nonsingular $n \times n$ real mixing matrices, and $\mathbf{x} = \mathbf{As}$ be linear mixing model of $\mathbf{s}$, $\mathbf{y} = \mathbf{Bx} = \mathbf{BAs} = \mathbf{Rs}$ be linear unmixing model of $\mathbf{x}$. Then the mutual information

$I(\mathbf{y})$ is minimized if and only if $\mathbf{R} = \mathbf{\Lambda P}$, where $\mathbf{\Lambda}$ is a diagonal matrix and $\mathbf{P}$ is a permutation matrix.

*Proof:* In the case where $\mathbf{Q} = \mathbf{\Lambda P}$, $\mathbf{y}$ is simply a permutation of the independent source vector $\mathbf{s}$ with just sign and scale ambiguity, then the mutual information $I(\mathbf{y})$ is zero. Proving the converse is relatively complicated, but interested readers can refer to literature [1].

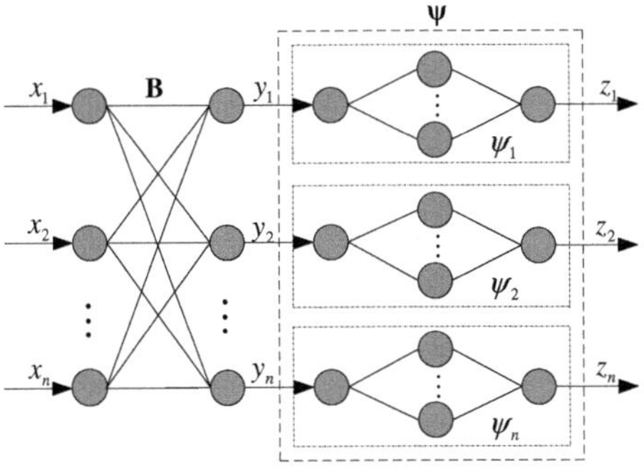


**Fig. 2.** Structure of nonnegative ICA unmixing system proposed in this paper

We shall assume, without loss of generality, $\mathbf{R}$ to be a diagonal matrix, i.e., $r_{ij} = \delta_{ij}$, then we have $y_i = r_{ii}s_i$. So $y_i$ is a duplicate of $s_i$ with just sign and scale ambiguity. Moreover, by considering the sources $s_i$ to be nonnegative in this paper, we will see that $y_i$ is either nonnegative or non-positive corresponding respectively to a positive $r_{ii}$ or a negative one. Consequently, we can eliminate the sign ambiguity by taking absolute value of $y_i$, i.e., $|y_i|$, as the recovered signals.

According to the theory given above, the unmixing system of nonnegative ICA can be constructed as shown in Fig.2. where B is the unmixing matrix in ICA, $y_i$ are the extracted independent components, and $\psi_i$ are some nonlinear mappings. The basic problem that we have to solve is to optimize the networks by maximizing the output entropy $H(\mathbf{z})$, which will be equivalent to minimizing the mutual information of the extracted components $y_i$, so $y_i$ will be the duplicates of $s_i$ according to lemma 1. In next section, we shall discuss the algorithm proposed in this paper in detail.

### 3.2 Learning Algorithm

With respect to the separation structure of this paper, the joint probabilistic density function of the output vector $\mathbf{z}$ can be calculated as:

$$p(\mathbf{z}) = \frac{p(\mathbf{x})}{|\det(\mathbf{B})| \prod_{i=1}^{n} |\psi'_i(\boldsymbol{\varphi}_i, y_i)|} \tag{9}$$

where $\psi'_i(\boldsymbol{\varphi}_i, y_i)$ is the derivative of $\psi_i(\boldsymbol{\varphi}_i, y_i)$ with respect to $y_i$, and $\boldsymbol{\varphi}_i$ is the parameters contained in nonlinear function $\psi_i$. From Eqn. (9), we can immediately achieve the following expression of the joint entropy:

$$H(\mathbf{z}) = H(\mathbf{x}) + \log\left|\det(\mathbf{B})\right| + \sum_{i=1}^{n} E\left(\log\left|\psi'_i(\boldsymbol{\varphi}_i, y_i)\right|\right) \tag{10}$$

The minimization of $I(\mathbf{y})$, which is equal to maximizing $H(\mathbf{z})$ here, requires the computation of its gradient with respect to the separation structure parameters $\mathbf{B}$ and $\boldsymbol{\varphi}$.

Since $H(\mathbf{x})$ does not contain any parameters of the separating system, it becomes null when taking gradient with respect to the parameters. We thus have the following gradient expressions:

$$\frac{\partial H(\mathbf{z})}{\partial \mathbf{B}} = \frac{\partial \log\left|\det(\mathbf{B})\right|}{\partial \mathbf{B}} + \frac{\partial\left(\sum_{i=1}^{n} E\left(\log\left|\psi'_i(\boldsymbol{\varphi}_i, y_i)\right|\right)\right)}{\partial \mathbf{B}} \tag{11}$$

$$\frac{\partial H(\mathbf{z})}{\partial \boldsymbol{\varphi}_k} = E\left(\frac{\partial \log\left|\psi'_k(\boldsymbol{\varphi}_k, y_k)\right|}{\partial \boldsymbol{\varphi}_k}\right) \tag{12}$$

Of course, their computation depends on the structure of the parametric nonlinear mapping function $\boldsymbol{\psi}$.

In this paper, we use multilayer perceptrons (MLP) [10] with a single hidden layer to model the nonlinear parametric functions $\psi_k(\boldsymbol{\varphi}_k, y_k)$, thus they can be written as:

$$\psi_k(\boldsymbol{\varphi}_k, y_k) = \psi_k(\boldsymbol{\alpha}_k, \boldsymbol{\beta}_k, \boldsymbol{\mu}_k, y_k) = \sum_{j=1}^{M_k} \alpha_j^k \tau(\beta_j^k y_k - \mu_j^k) \tag{13}$$

where $\boldsymbol{\alpha}$ and $\boldsymbol{\beta}$ are the weight matrixes of input layer and output layer, respectively, $\boldsymbol{\mu}$ is the hidden unit's bias term, and $\tau(\bullet)$ is the active function of hidden layer.

From Eqs. (11)- (13), we can easily calculate the gradients of $H(\mathbf{z})$ with respect to each parameter, and then optimize the network accordingly.

## 4 Experimental Results and Discussions

In this section, four experiments are carried out to verify the efficacy and effectiveness of the proposed method. The first three experiments have similar settings but differ in terms of the source signals being used. In the last experiment, image data are used to complete the investigation for the algorithm.

### 4.1 Supergaussian and Subgaussian Data

In this experiment, three nonnegative source signals are generated synthetically as the original sources. The three source signals can be expressed as:

$$\mathbf{s}=\begin{bmatrix} s_1 \\ s_2 \\ s_3 \end{bmatrix}=\begin{bmatrix} ((\sin(t/3))+1)+\lambda \\ ((\mathrm{rem}(t,23)-11)/9)^5+2.8)+\lambda \\ ((\mathrm{rem}(t,27)-13)/9+1.5)+\lambda \end{bmatrix}$$

where λ is a nonnegative constant used to control the well-grounded degree of the source signals, the function rem$(u,v)$ represents the remainder of $u$ divided by $v$. The seconded one is a supergaussian signal and the other two are subergaussian. Figure 3 shows the three source signals in the case of λ=0 and λ=0.3, respectively. Clearly, they are all nonnegative and not well-grounded when λ=0.3, but approximate well-grounded when λ=0. In this experiment, three source signals (λ=0.3) are mixed using a $3\times3$ mixing matrix

$$\mathbf{A}=\begin{bmatrix} 0.7412 & -0.4513 & 0.1234 \\ 0.1864 & 0.8015 & -0.3241 \\ 0.5123 & 0.2314 & 0.8234 \end{bmatrix}$$

which was chosen randomly, thus the sources are nonnegative, while the mixing matrix is of mixed sign.

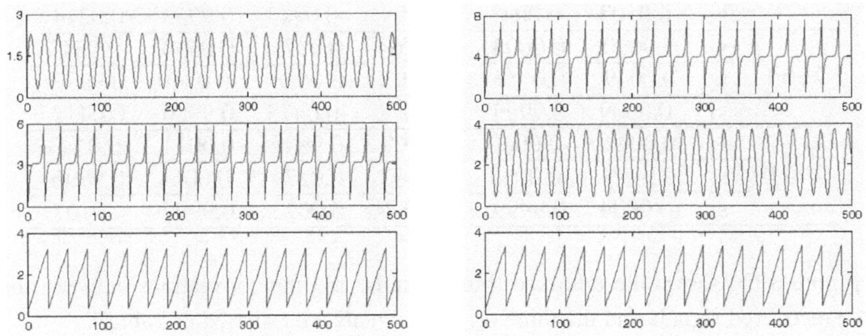

(a) (b)

**Fig. 3.** (a) Original source signals( λ=0.3), (b) Recovered signals (λ=0.3)

Some explanations should be given here. One is that to improve the efficiency of the algorithm, the momentum and adaptive step sizes with error control have been used. And, in this experiment, the parameter *Mi* is set as 4 (the number of the hidden layer neurons of $\psi_i$ blocks). Finally, to implement the constraints on the $\psi_i$ functions, which is the increasing functions with values in a finite interval, the sigmoids of the hidden units of $\psi_i$ blocks were chosen as increasing functions, and the vector of weights leading from the hidden units to the output units was normalized at the end of each epoch, also all weights in each $\psi_i$ block were initialized to positive values, which results in an increasing $\psi_i$ function.

**Table 1.** Correlations between recovered signals and the original signals

| | | Method in this paper | | | Method in literature [11] | | |
|---|---|---|---|---|---|---|---|
| | | $y_1$ | $y_2$ | $y_3$ | $y_1$ | $y_2$ | $y_3$ |
| λ=0.5 | $s_1$ | 0.9999 | 0.0039 | -0.0016 | 0.9742 | 0.2069 | -0.0902 |
| | $s_2$ | 0.0030 | 0.9997 | 0.0100 | 0.0898 | 0.0761 | 0.9931 |
| | $s_3$ | -0.0092 | -0.0168 | 1.0000 | -0.2086 | 0.9766 | -0.0516 |
| λ=0.4 | $s_1$ | 1.0000 | -0.0003 | 0.0014 | 0.9804 | 0.1746 | -0.0907 |
| | $s_2$ | 0.0055 | 0.9998 | 0.0175 | 0.0967 | 0.0516 | 0.9940 |
| | $s_3$ | -0.0016 | -0.0115 | 0.9999 | -0.1744 | 0.9842 | -0.0298 |
| λ=0.3 | $s_1$ | 0.0046 | 1.0000 | -0.0052 | 0.9862 | 0.1390 | -0.0903 |
| | $s_2$ | 0.9999 | 0.0072 | 0.0096 | 0.1015 | 0.0243 | 0.9945 |
| | $s_3$ | -0.0073 | -0.0012 | 0.9999 | -0.1366 | 0.9906 | -0.0059 |

**Table 1.** Correlations between recovered signals and the original signals (continued)

| | | Method in this paper | | | Method in literature [11] | | |
|---|---|---|---|---|---|---|---|
| | | $y_1$ | $y_2$ | $y_3$ | $y_1$ | $y_2$ | $y_3$ |
| λ=0.2 | $s_1$ | 0.0047 | 0.9999 | -0.0086 | 0.9917 | 0.0993 | -0.0813 |
| | $s_2$ | 0.9999 | 0.0001 | 0.0003 | 0.0962 | -0.0079 | 0.9953 |
| | $s_3$ | -0.0103 | -0.0041 | 0.9999 | -0.0943 | 0.9953 | 0.0214 |
| λ=0.1 | $s_1$ | 1.0000 | 0.0003 | -0.0064 | 0.9971 | 0.0533 | -0.0543 |
| | $s_2$ | 0.0049 | 0.0153 | 0.9997 | 0.0698 | -0.0340 | 0.9970 |
| | $s_3$ | -0.0009 | 0.9999 | -0.0072 | -0.0475 | 0.9980 | 0.0417 |
| λ=0.0 | $s_1$ | 0.0044 | -0.0060 | 0.9999 | 0.9999 | 0.0014 | -0.0143 |
| | $s_2$ | 0.9999 | -0.0057 | 0.0067 | 0.0279 | -0.0064 | 0.9996 |
| | $s_3$ | -0.0004 | 0.9999 | -0.0027 | 0.0023 | 0.9999 | 0.0107 |

Figure 3 (b) shows the unmixed three signals, and the correlations between these three recovered signals and the three original signals are reported in Table 1.

From Figure 3 (b) we can see that the unmixed signals are all nonnegative and they are very similar to the original signals shown in Figure 3 (a).

For comparison, we also use another non-negative ICA algorithm proposed in literature [11] to conduct the same experiment. The correlations between the recovered signals and the original sources are also reported in Table1. From Table 1, it can be seen that the separated signals using the method proposed in this paper is more similar to the original signals than the other one.

To compare the two methods systemically, we also have done some other experiments when λ was set to other values, the results are also shown in Table 1. From Table 1, we can find that the results of our method are very steady, yet the results of the algorithm in literature [11] become bad clearly when the source signals are away from well-grounded. Moreover, we can also find that the two methods are very similar when the source signals are well grounded (λ=0), whereas the method proposed in this paper is more efficient when the sources are not well grounded. And, the more away from well grounded the sources are, the bigger the difference of the two methods is. These are mainly because the nonnegative algorithm in literature [11] as well as algorithms in literature [4,5], are all based on the assumption that the original signals are well grounded.

### 4.2 Subgaussian Data

In the seconded experiment, three subgaussian signals are expressed as

$$\mathbf{s}=\begin{bmatrix} s_1 \\ s_2 \\ s_3 \end{bmatrix}=\begin{bmatrix} (\sin(\pi t/10)+1)+\lambda \\ (\sin(600\pi t/10000+6\cos(120\pi t/10000))+1)+\lambda \\ \text{Uniformly distributed signal}+\lambda \end{bmatrix}$$

The three signals are mixed by the mixing matix

$$\mathbf{A}=\begin{bmatrix} 0.5412 & -0.4513 & 0.1234 \\ 0.2864 & 0.7015 & -0.1241 \\ 0.3123 & 0.4314 & 0.8234 \end{bmatrix}$$

**Table 2.** Correlations between recovered signals and the original signals

| | | Method in this paper | | | Method in literature [11] | | |
|---|---|---|---|---|---|---|---|
| | | $y_1$ | $y_2$ | $y_3$ | $y_1$ | $y_2$ | $y_3$ |
| λ=0.25 | $s_1$ | 0.0008 | 1.0000 | -0.0052 | 0.9712 | 0.1797 | -0.1565 |
| | $s_2$ | -0.0678 | -0.0008 | 1.0000 | 0.1907 | -0.1923 | 0.9626 |
| | $s_3$ | 0.9987 | 0.0038 | -0.0247 | -0.1410 | 0.9686 | 0.2046 |
| λ=0.20 | $s_1$ | 0.0045 | 1.0000 | -0.0036 | 0.9826 | 0.1378 | -0.1249 |
| | $s_2$ | -0.0649 | -0.0027 | 1.0000 | 0.1466 | -0.1612 | 0.9760 |
| | $s_3$ | 0.9988 | 0.0012 | -0.0233 | -0.1117 | 0.9804 | 0.1620 |
| λ=0.15 | $s_1$ | -0.0060 | -0.0005 | 1.0000 | 0.9906 | 0.0994 | -0.0937 |
| | $s_2$ | 0.9999 | -0.0679 | -0.0009 | 0.1063 | -0.1297 | 0.9858 |
| | $s_3$ | -0.0274 | 0.9986 | 0.0040 | -0.0825 | 0.9890 | 0.1225 |

The correlations between the recovered signals and the original sources corresponding to different situations are reported in Table2. From Table 2, we can reach similar conclusions as from the first experiment.

### 4.3 Supergaussian Data

To test the method proposed in this paper systemically, we also do the unmixing experiment where three source signals expressed as

$$\mathbf{s}=\begin{bmatrix} s_1 \\ s_2 \\ s_3 \end{bmatrix}=\begin{bmatrix} \text{Laplacian distributed signal} \\ (\text{rem}(t,23)-11)/9)^5 \\ \text{Impulsive noise} \end{bmatrix}$$

are all supergaussian. Where the impulsive noise is generated by $(2(r_1(t)<0.5)-1)\circ(\log(r_2(t)))$, '$\circ$' denotes the Hadamard product, $r_1(t)$ and $r_2(t)$ are the uniform distributed signals. The three signals are mixed by the mixing matrix

$$\mathbf{A} = \begin{bmatrix} -0.6412 & 0.3511 & 0.3234 \\ 0.1864 & 0.6013 & -0.2244 \\ 0.4123 & -0.1314 & 0.7233 \end{bmatrix}$$

Table 3 presents the correlations between the recovered signals and the original sources.

**Table 2.** Correlations between recovered signals and the original signals (continued)

| | | Method in this paper | | | Method in literature [11] | | |
|---|---|---|---|---|---|---|---|
| | | $y_1$ | $y_2$ | $y_3$ | $y_1$ | $y_2$ | $y_3$ |
| λ=0.10 | $s_1$ | 1.0000 | -0.0069 | 0.0021 | 0.9958 | -0.0671 | 0.0627 |
| | $s_2$ | -0.0056 | 0.9999 | -0.0677 | -0.0590 | 0.0555 | 0.9967 |
| | $s_3$ | 0.0007 | -0.0256 | 0.9987 | 0.0763 | 0.9948 | -0.0672 |
| λ=0.05 | $s_1$ | -0.0047 | 1.0000 | -0.0013 | 0.9991 | -0.0328 | 0.0285 |
| | $s_2$ | 1.0000 | -0.0002 | -0.0690 | -0.0291 | -0.0162 | 0.9994 |
| | $s_3$ | -0.0239 | 0.0044 | 0.9985 | 0.0378 | 0.9993 | 0.0010 |
| λ=0.00 | $s_1$ | 1.0000 | -0.0032 | -0.0023 | 1.0000 | -0.0024 | 0.0000 |
| | $s_2$ | 0.0003 | 1.0000 | -0.0676 | -0.0001 | -0.0129 | 0.9999 |
| | $s_3$ | 0.0048 | -0.0210 | 0.9982 | 0.0075 | 1.0000 | -0.0034 |

**Table 3.** Correlations between recovered signals and the original signals

| | Method in this paper | | | Method in literature [11] | | |
|---|---|---|---|---|---|---|
| | $y_1$ | $y_2$ | $y_3$ | $y_1$ | $y_2$ | $y_3$ |
| $s_1$ | 0.0738 | 0.9998 | 0.0341 | 0.9574 | -0.0763 | 0.2786 |
| $s_2$ | 0.9999 | 0.0653 | 0.0311 | 0.2318 | 0.9445 | -0.2329 |
| $s_3$ | 0.0122 | 0.0453 | 0.9998 | -0.2015 | 0.2943 | 0.9342 |

We can see from Table 3 that the difference between the two methods is evident. This is because the first signal and third one are not well grounded due to its supergaussian distributing. Of course, supergaussion distributing is not necessarily related to not well grounded, e.g. the second source signal. Yet it is do that in the first source signal and the third one.

### 4.4 Image Data

In order to test the efficacy of the proposed scheme in practical term, we also applied the algorithm to unmix image data in this experiment. For this task, three image patches of size 60×60 were used. The first image is a women face and the second one is a picture of natural scene, the third image is an artificial one containing only noisy signal. Each of the images was treated as one source, with its pixel values representing the 60×60=3600 samples.

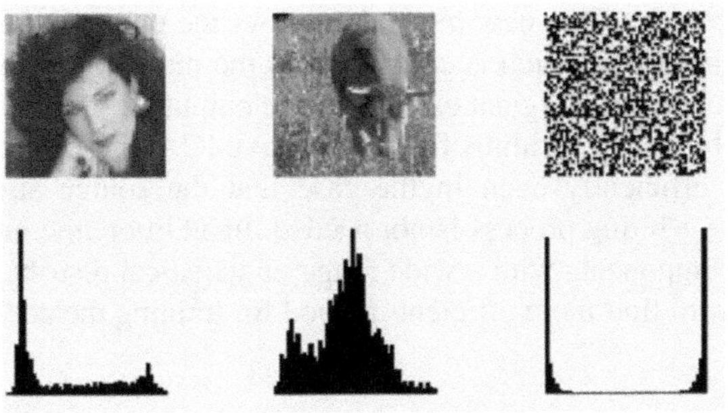

**Fig. 4.** The original source images and their histograms

Figure 4 shows the original images and their histograms. Note that the histograms indicate that the first source image is not well-grounded, moreover, the distributing of the first two images are nonsymmetric.

In this experiment, no special pre-processing was performed on the mixed image data, other than divided by a constant, so they were appropriate for use with our networks (input values are roughly between -2 and 2).The correlations between these three recovered images and the three original images are reported in Table 4.The source-to-output matrix $\mathbf{R}=\mathbf{BA}$ was

$$\mathbf{R}=\begin{bmatrix} -0.0000 & -0.0000 & -1.6390 \\ -1.0299 & 0.0113 & -0.0004 \\ 0.0753 & 1.0043 & 0.0141 \end{bmatrix}$$

**Table 4.** Correlations between recovered images and the original images

| | Method in this paper | | | Method in literature [11] | | |
|---|---|---|---|---|---|---|
| | $y_1$ | $y_2$ | $y_3$ | $y_1$ | $y_2$ | $y_3$ |
| $s_1$ | -0.0143 | 1.0000 | -0.0862 | -0.0359 | 0.9874 | -0.1540 |
| $s_2$ | -0.0107 | -0.1030 | 0.9932 | 0.0803 | -0.0401 | 0.9960 |
| $s_3$ | 1.0000 | -0.0135 | 0.0237 | 0.9959 | 0.0075 | -0.0907 |

Clearly, the algorithm is able to separate the images reasonably well.In addition, we also conducted the same experiment by using the algorithm proposed in literature [11] too. The correlations between the recovered images and the original sources are also reported in Table 4. Clearly, our method is more efficient. Furthermore, we can also find that the distinction of the two experimental results is small. The reason of this phenomenon is that the source images are approximately well grounded, especially the last two ones.

## 5 Conclusions

This paper considered the task of nonnegative independent component analysis and proposed a new unmixing scheme to separate mixed signals using a neural network with a special structure. This new method employs the output entropy of the network as the objective function, which is equivalent to the mutual information criterion but needs not to calculate the marginal entropy of the output.

Compared with other algorithms for nonnegative ICA, the method proposed in this paper can work efficiently even in the case that the source signals are not well grounded, and pre-whiting process is not needed. In addition, the method can separate the mixtures of components with a wide range of statistical distributions.In future, we will focus on how to find more efficient method for training the network.

## References

1. Hyvärinen, A., Karhunen, J., Oja, E.: Independent Component Analysis. J. Wiley, (2001)
2. Hyvärinen, A.: Fast and Robust Fixed-Point Algorithms for Independent Component Analysis. IEEE Trans. Neural Neworks, 10(3) (1999)626–634
3. Lee, D.D., Seung, H.S.: Learning the Parts of Objects by Nonnegative Matrix Factorization. Nature, 401 (1999) 788–791
4. Plumbley, M.D., Oja, E.: A 'Nonnegative PCA' Algorithm for Independent Component Analysis. IEEE Trans.Neural Networks, 15 (2004)66–76
5. Plumbley, M.D.: Algorithms for Nonnegative Independent Component Analysis. IEEE Trans. Neural Networks, 14 (2003)534–543
6. Bell, A., Sejnowski, T.: An Information-Maximization Approach to Blind Separation and Blind Deconvolution. Neural Computation, 7 (1995) 1129–1159
7. Lee, T.W., Girolami, M., Sejnowski, T.: Independent Component Analysis Using an Extended Infomax Algorithm for Mixed Sub-Gaussian and Super-Gaussian Sources. Neural Computation, 11 (1999) 417–441
8. Almeuda, L.B.: Linear and Nonlinear ICA Based On Mutual Information –the MISEP Method. Signal Processing, 84 (2004) 231-245
9. Zheng, C.H., Huang, D.S., Sun, Z.L., Shang, L.: Post-nonlinear Blind Source Separation Using Neural Networks with Sandwiched Structure. Lecture Notes in Computer Science, 3497 (2005) 478 - 483
10. Huang, D.S.: Systematic Theory of Neural Networks for Pattern Recognition. Publishing House of Electronic Industry of China, Beijing, (1996)
11. Plumbley, M.D.: Optimization using Fourier Expansion over a Geodesic for Non-Negative ICA. In Proceedings of the International Conference on Independent Component Analysis and Blind Signal Separation (ICA 2004), (2004) 49-56

# Modeling of Microhardness Profile in Nitriding Processes Using Artificial Neural Network

Dariusz Lipiński and Jerzy Ratajski

Koszalin University of Technology, Faculty of Mechanical Engineering,
Racławicka 15-17, 75-620 Koszalin, Poland
{dariusz.lipinski, jerzy.ratajski}@tu.koszalin.pl

**Abstract.** A artificial neural network was assigned to modeling of hardness profiles in the nitrided layer. In the model developed, a feed-forward neural network was applied. The designed network possesses good capacities to generalize knowledge included in experiential data. Matching the model with the training data made it possible to determine, with a good approximation, hardness profiles, which make up a set of verifying data.

**Keywords:** nitriding, microhardness, neural network, modeling.

## 1 Introduction

Growing significance of gaseous nitriding process is being noticed contemporarily, in spite of intense development of new technologies of forming of superficial layers. It happens so since the process is very efficient both in the mass and long-run production. Moreover, nowadays nitriding process is very often used in so called duplex processes, i.e. the sequential application of two established surface technologies to produce a surface composite with combined properties which are unobtainable through any individual surface technology [1]. A typical duplex process involves combined nitriding process (gaseous or plasma) and PVD ceramic coating treatment of steels. The basic condition however of its universal application possibilities is the obtaining of nitrided layer with demanded hardness and thickness.

Nitriding process produces a relatively thick (300–500μm) and hard (900-1200 HV) diffusion zone, and at the same time a thin iron (carbo)nitride compound layer is formed at the surface. In the latest solutions of the controlled nitriding process, the software of the control system is based upon an assumed algorithm of changes of nitrogen potential $K_N = pNH_3/pH_2^{3/2}$ in the function of time and process temperature. However, the complex relation between mentioned nitriding parameters and the composition and phase constitution of the compound layer [2] limits the tailoring of properties of the nitrided case and the controlling of growth kinetics in the diffusion zone. One of the way for finding right relation between the process parameters and the structure of the layer is the development of mathematical

D.-S. Huang, L. Heutte, and M. Loog (Eds.): ICIC 2007, LNAI 4682, pp. 245–252, 2007.

models to simulate surface engineering processes and to predict the service behavior of the resultant surface engineered systems [3–4].

More and more often, artificial intelligence methods are applied for the modelling of surface treatment processes. Intelligent data base and expert system are developed [5-6]. These are complementary tools, which are intended to enable a comprehensive simulation of the process and as a result of it, to allow for the development of software for control systems focused on obtaining optimal results of the process. The results obtained indicate that models can be effectively used in the optimisation of properties on the macro and micro levels.

In particular, in order to fully realize the maximum uptake of the benefits available from a nitrided layer, it is essential to select optimum grade of steel and nitriding parameters which ensure desired profile of micro-hardness in the diffusion zone. In view of this, a neural network models [7-8] which simulates the nitriding process has been successfully developed.

## 2 Experimental Data

The results of experimental research conducted during the last decades at Koszalin University of Technology and Radom Institute for Sustainable Technologies have been used in micro-hardness process modeling (Table 1).

**Table 1.** Characteristics of experimental data used for modeling

| | steel | number of samples | parameters range | | | | | |
|---|---|---|---|---|---|---|---|---|
| | | | first stage | | | second stage | | |
| | | | T [°C] | t [min] | $K_H$ | T [°C] | t [min] | $K_H$ |
| 1 | 18H2N2 | 10 | - | - | - | 530-570 | 120-780 | 3.2-6 |
| 2 | 18HGT | 85 | 480-580 | 120-270 | 1.15-30 | 480-580 | 25-3090 | 0.4-30 |
| 3 | 20MnCr5 | 23 | - | - | - | 530-590 | 120-1920 | 2.75-4.5 |
| 4 | 25H3M | 10 | 530 | 240-270 | 5.25-11.5 | 530 | 120-3090 | 0.6-11.5 |
| 5 | 25H5M | 2 | 530 | 270 | 5.25-7 | 530 | 780-3090 | 0.8-5.25 |
| 6 | 30CrMoV9 | 32 | - | - | - | 530-590 | 120-1920 | 2.75-4.5 |
| 7 | 33H3MF | 35 | 530-570 | 270 | 1.5-30 | 530-570 | 120-3090 | 0.8-30 |
| 8 | 35CrAl5 | 3 | - | - | - | 530-570 | 480 | 2.75-4.4 |
| 9 | 35HGSA | 2 | 540 | 360 | 20 | 540-550 | 3240 | 0.7-0.9 |
| 10 | 36CrAl6 | 22 | - | - | - | 530-570 | 120-1920 | 3-4.5 |
| 11 | 36H3M | 6 | - | - | - | 550 | 720 | 1.5-30 |
| 12 | 38CrMoV21,14 | 32 | - | - | - | 530-590 | 120-1920 | 2.75-4.5 |
| 13 | 38HMJ | 44 | 480-580 | 120-480 | 1.1-30 | 480-580 | 25-3240 | 0.4-30 |
| 14 | 40H | 43 | 500-570 | 60-270 | 1.5-30 | 500-570 | 120-3090 | 0.8-30 |
| 15 | 40H2MF | 2 | 530 | 270 | 5.25-7 | 530 | 780-3090 | 0.8-5.25 |
| 16 | 40HM | 4 | - | - | - | 560-570 | 240-540 | 0.85-3.1 |
| 17 | 4140 | 15 | 530-570 | 270-450 | 0.85-7 | 530-570 | 120-3090 | 0.8-6.1 |
| 18 | 4340 | 30 | 480-570 | 120-270 | 3.2-20 | 480-570 | 25-3090 | 0.8-20 |
| 19 | 45 | 50 | 500-590 | 270 | 3-7 | 500-590 | 120-3090 | 0.8-4.75 |
| 20 | IMPACTO | 8 | - | - | - | 530-570 | 120-720 | 3.2-6.1 |
| 21 | NC10 | 3 | - | - | - | 530-570 | 480 | 2.75-4.4 |
| 22 | SPS | 8 | - | - | - | 530-570 | 120-720 | 3.2-6.1 |
| 23 | SW3S2 | 5 | 530 | 240 | 9.25 | 530 | 120-1920 | 0.6 |
| 24 | WCL | 16 | 530-570 | 240-270 | 1.5-30 | 530-570 | 120-3090 | 0.6-30 |

### 2.1 Data Description

Experimental data specified the influence of main process parameters (temperature T, time t, and nitrogen potential $K_H$) and values describing material composition (atomic concentration of 11 nitride forming elements) on microhardness profile.

Microhardness profile in diffusion zone has been estimated for every sample subjected to nitrating. In each case, the value of microhardness in several characteristic points has been tested (Fig. 1). As a result of the experiment a dataset D has been obtained:

$$D=[F\ M\ x\ \Delta HV(x)]^T . \tag{1}$$

where F is the process parameters vector, M is the vector of material composition (atomic concentration of 11 nitride forming elements), x is the depth [μm], ΔHV(x) is the value of microhardness increase dependent on core hardness in a depth x.

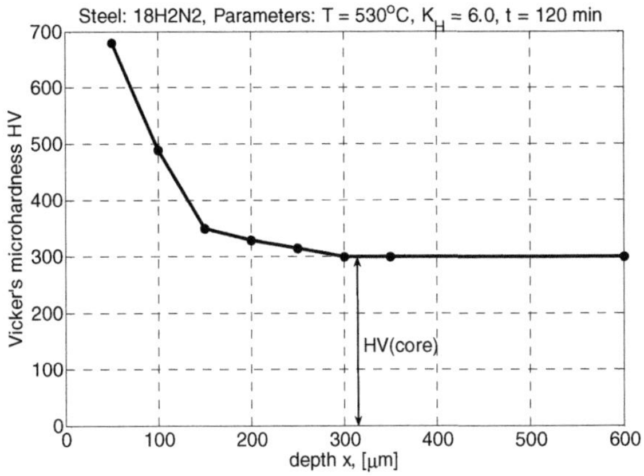


**Fig. 1.** Exemplary microhardness profile of 18H2N2 steel, nitriding parameters: T = 530°C, t=120min, $K_H$ =6

### 2.2 Data normalization

The values in dataset D had different dimensions and ranges. Variable values $d_i$ $(d_i \in D)$ used for modeling have been normalized according to a relation:

$$d_{in} = (d_i - d_{imean}) / d_{istd} , \tag{2}$$

where $d_{in}$ is the normalized value of i-th parameter in dataset, $d_i$ is the measured value for i-th parameters, $d_{imean\backslash}$ and $d_{istd}$ is the mean and the standard deviation values in the dataset for i-th parameter, respectively.

### 2.3 Reflection of Nitriding Process Multistaging in Experimental Data

Nitriding process can be conducted as a single or double stage process. A single stage process is described by $F_I$ = $[T_2\ Np_2\ t_2]$ parameters while a double stage one by

$F_{II}$ = [$T_1$ $Np_1$ $t_1$ $T_2$ $Np_2$ $t_2$]. In order to include process multistaging in one neural model values which have not been represented in the $F_I$ set have been substituted with zeros (which corresponds to average values of the first stage normalized parameters):

$$F_I = [0\ 0\ 0\ T_2\ t_2\ Np_2\ F_s]^T;\ F_{II} = [T_1\ t_1\ Np_1\ T_2\ t_2\ Np_2\ F_s]^T. \tag{3}$$

where: $F_s$ stands for number of stages of the process, $F_s = 0$ for single stage process, $F_s = 1$ for double stage process.

As a result of the above transformations, sets of input P and output T values of a model have been obtained:

$$P=[F_n\ M_n\ x_n]^T;\ T = \Delta HV_n(x_n)\ , \tag{4}$$

where: $F_n$ - normalized process properties including multistaging, $M_n$ - normalized material composition, $x_n$ - normalized depth, $\Delta HV_n(x_n)$ - normalized microhardness increment in a normalized depth x.

A subset (being 1/3 of the input dataset) of test data used for verifying prediction accuracy in the profile of microhardness has been randomly isolated from the dataset.

# 3 Neural Network Model

The task of modeling the increment of microhardness in a depth x, and for known process parameters and material properties is based on finding an unknown function g(.) (Fig. 2).

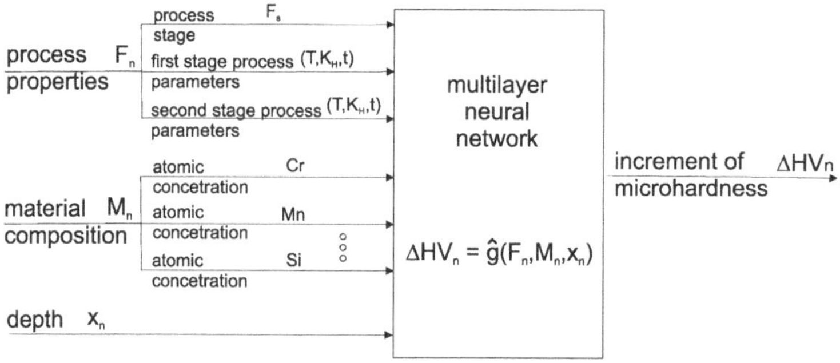


**Fig. 2.** Schematic model of artificial neural network for modeling of microhardness profile after nitriding

## 3.1 Training Algorithm

The Levenberg-Marquardt algorithm [9] was used to adjust the network weights and biases w in order to minimize the performance function:

$$F_e = \frac{1}{N}\sum_{i=1}^{N}(y_i - \hat{y}_i)^2\ , \tag{5}$$

where: $y_i$ – network output expected value, $\hat{y}_i$ – network response, N – number of cases in the input set.

In order to improve network generalization, the Bayesian Regularization [10] was used. The performance function was modified by adding an additional term $F_w$:

$$F = \alpha \cdot F_e + (1-\alpha) \cdot F_w \ , \tag{6}$$

where: $F_w = \frac{1}{N}\sum_{i=1}^{N} w_i^2$ , $\alpha$ - objective function parameters.

Regularization of network connections has been applied in order to provide a higher capacity of generalization of data in the output set. Application of combination of algorithms enables obtaining lower network connection values, which, consequently, ensures lower network susceptibility to excessive matching to the training dataset.

### 3.2 Model Predictive Capacity Estimation

Optimization of a neural network structure was based on selection of: (i) the number of network hidden layers ($n_h$=1..2) and (ii) the number of neurons in each hidden layers ($k^{\{nh\}}$=1..20). In order to estimate model predictive capacity, for each of the analyzed architectures the following relations shall be determined:

$$MAD = \frac{1}{n}\sum_{i=1}^{n}|y_i - \hat{y}_i|, \quad MAPE = \left(\frac{1}{n}\sum_{i=1}^{n}\left|\frac{y_i - \hat{y}_i}{y_i}\right|\right)\cdot 100\% \ ,$$

$$MRSE = \frac{1}{n}\sqrt{\sum_{i=1}^{n}(|y_i - \hat{y}_i|)^2} \ , \quad R^2_{prediction} = 1 - \frac{\sum_{i=1}^{n}(y_i - \hat{y}_i)^2}{\sum_{i=1}^{n} y_i^2} \ . \tag{7}$$

where: n – number of samples in testing dataset, $y_i$ – expected output value for i-th sample, $\hat{y}_i$ - modeled output value for i-th sample.

420 different neural network architectures have been tested. For every analyzed neural network before mentioned parameters providing predictive capacity estimation have been determined. The results of predictive capacity estimation in selected models have been presented in table 2.

**Table 2.** Results of predictive capacity estimation in selected models

| Network architecture (19-$k^{\{1\}}$-$k^{\{2\}}$-1) | MAD | MAPE | MRSE | $R^2_{pred}$ [%] |
|---|---|---|---|---|
| 19-18-19-1 | 0.161 | 79.363 | 0.0078 | 90.85 |
| **19-18-3-1** | **0.154** | 83.872 | **0.0074** | **91.87** |
| 19-17-1 | 0.157 | **78.595** | 0.0076 | 91.39 |
| 19-16-7-1 | 0.167 | 89.977 | 0.0075 | 91.49 |
| 19-13-4-1 | 0.161 | 85.189 | 0.0075 | 91.67 |
| 19-18-12-1 | 0.158 | 85.446 | 0.0076 | 91.59 |

The best value of MAD, MAPE and $R^2_{prediction}$ have been obtained for neural network with two hidden layers. Basing upon the above dependences a 19-18-3-1 structure of neural network proves to be optimal.

The accuracy of data representation in training and testing sets of a selected model has been shown in figure 3.

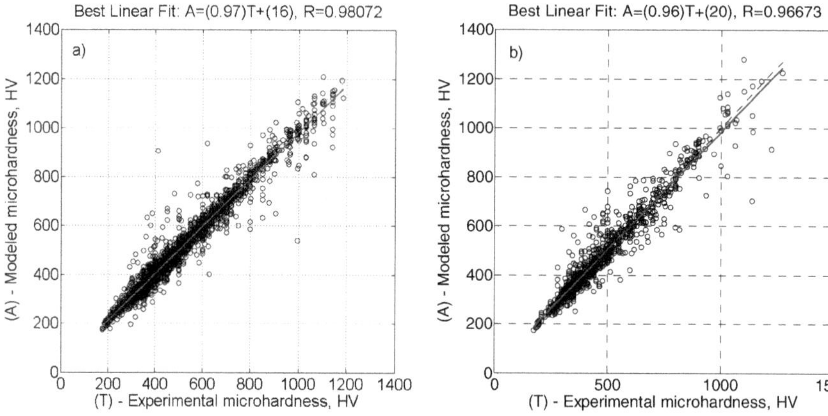


**Fig. 3.** Accuracy of data representation with a: (a) training and (b) testing set in a 19-18-3-1 structure of neural network

# 4 Modeling Results

The established neural model enables rating of influence of the material properties and nitration process parameters including staging, on micro-hardness profile. Exemplary results of microhardness profile modeling in the diffusion zone have been shown on figure 4.

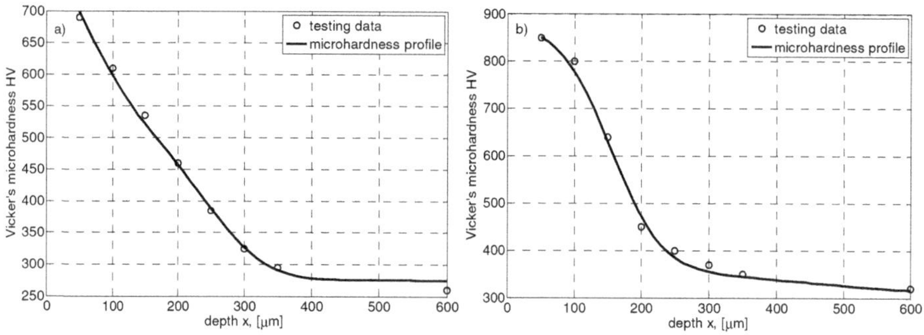


**Fig. 4.** Exemplary results of microhardness profiles modeling with the use of neural network model: a) steel 18HGT, process parameters: (first stage) T=580°C, $K_H$ = 10, t=120min, (second stage) T=580°C, $K_H$ = 0.4, t=240min; b) steel 36H3M, process parameters: T=550°C , $K_H$ = 17, t = 720 min

For the future application of the elaborated neural network, the presented model has been tested for the ability of generalization. On figure 5 a profiles of microhardness for parameters not being included in the set of experimental data has been shown.

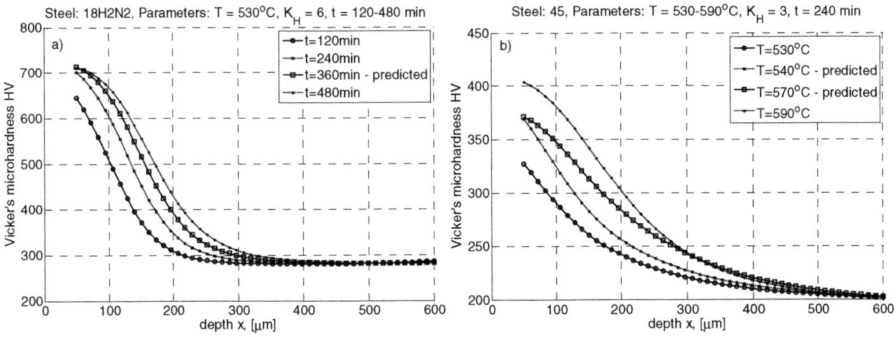


**Fig. 5.** Exemplary results of microhardness profiles prediction with the use of neural network model: a) steel 18H2N2, process parameters: T=530°C, $K_H$ = 6, different periods of time, b) steel 45, process parameters: $K_H$ = 3, t = 240 min, different temperatures

Created model has a good ability of knowledge generalization based on training dataset. It proves that presented model can be applied for prediction of microhardness profile as well as selection of nitrining process parameters in order to obtain a expected microhardness profile.

## 5 Summary

Elaborated neural network model constitute tool to the simulation of nitriding process The model correctly describes the profiles of microhardness in the nitrided layer - predicted results showed relatively low scatter with experimental results.

In particularly, the model can be used for:

- Prediction of the microhardness profile for any steel and nitriding conditions;
- Comparison and analysis of microhardness profiles at different conditions, prediction of the process parameters for given grade of steel and desired profile of microhardness,
- Selecting optimum grade of steel and nitriding parameters which ensure optimal profile of micro-hardness in diffusion zone for duplex process.

The model is open for constant upgrade and improvement and also can be applied in a control system and in visualization of the process course.

**Acknowledgments.** Scientific work carried out within the project "Development of nanotechnologies in surface engineering" in The Multi-Year Programme "Development of innovativeness systems of manufacturing and maintenance 2004-2008".

## References

1. Bell, T., Dong, H., Sun, Y.: Realising the Potential of Duplex Surface Engineering. Tribology International, Vol. 31 (1998) 127–137
2. Ratajski, J., Tacikowski, J., Somers, M.A.J.: Development of Compound Layer of Iron (carbo) Nitrides during Nitriding of Steel. Surface Engineering, Vol. 19 (2003) 285
3. Ratajski, J.: Model of Growth Kinetics of Nitrided Layer in the Binary Fe-N System. Zeitschrift fur Metallkunde, 95 (2004) 9, 23
4. Bell, T., Sun, Y., Mao, K., Buchhagen, P.: Mathematical Modelling of the Plasma Nitriding Process and the Resultant Load Bearing Capacity. Advanced Materials and Processes, No. 4(1996) 40Y–40BB
5. Dobrzański, L.A., Madejski, J., Malina, W., Sitek, W.: The Prototype of an Expert System for the Selection of High-speed Steels for Cutting Tools. Journal of Materials Processing Technology, 56 (1996) 873-881
6. Kumar, S., Singh, R.: A Short Note on an Intelligent System for Selection of Materials for Progressive Die Components. Journal of Materials Processing Technology, 182 (2007) 456-461
7. Zhecheva, A., Malinov, S., Sha, W.: Simulation of Microhardness Profiles of Titanium Alloys after Surface Nitriding Using Artificial Neural Network. Surface and Coatings Technology, 200 (2005) 2332-2342
8. Genel, K.: Use of Artificial Neural Network for Prediction of Iron Nitrided Case Depth in Fe-Cr alloys. Materials and Design, 24 (2003) 203-207
9. Hagan, M.T., Menhaj, M.: Training Feedforward Networks with the Marguardt Algorithm. IEEE Transaction of Neural Networks, 5 (1994) 989-993
10. Forsee, F.D., Hagan, M.T.: Gauss-Newton Approximation to Bayesian Learning. Proceedings of the International Joint Conference on Neural Networks (1997) 1930-1935

# A Similarity-Based Approach to Ranking Multicriteria Alternatives

Hepu Deng

School of Business Information technology, RMIT University,
GPO Box 2476V, Melbourne, 3000, Victoria, Australia
hepu.deng@rmit.edu.au

**Abstract.** This paper presents a similarity-based approach to ranking multicriteria alternatives for solving discrete multicriteria problems. The approach effectively makes use of the ideal solution concept in such a way that the most preferred alternative should have the highest degree of similarity to the positive ideal solution and the lowest degree of similarity to the negative-ideal solution. The overall performance index of each alternative across all criteria is determined based on the concept of the degree of similarity between each alternative and the ideal solution using alternative gradient and magnitude. An example is presented to demonstrate the applicability of the proposed approach. A comparative analysis between the proposed approach and the technique for order preference by similarity to ideal solution is conducted for demonstrating the merits of the proposed approach for solving discrete multicriteria analysis problems.

**Keywords:** Multicriteria analysis; discrete optimization; Decision making.

## 1 Introduction

Many decision problems in real world settings require simultaneous consideration of several aspects rather than of a single criterion [1, 3, 5, 15]. Decision making that deals with several aspects of a finite set of available alternatives in a given situation is often referred to as multicriteria analysis. Multicriteria analysis is distinguished from single-criterion decision making and from multi-objective decision making, in which alternatives are not explicitly enumerated but implicitly defined by constraints on decision variables [3, 16, 19, 21].

Tremendous efforts have been spent and numerous approaches have been developed for effectively addressing general multicriteria analysis decision problems, leading to many successful applications of these approaches in the literature [10, 11, 14, 16, 18]. One of the mostly commonly used approaches in this regard is the technique for order preference by similarity to ideal solution (TOPSIS) [6, 11, 12 19]. The TOPSIS approach is developed based on the perception that a preferred alternative should be as close to the positive ideal solution as possible and as far from the negative ideal solution as possible which is simple and understandable [3, 6, 11]. As a result, numerous applications of such an approach have been reported in the

D.-S. Huang, L. Heutte, and M. Loog (Eds.): ICIC 2007, LNAI 4682, pp. 253–262, 2007.

literature for addressing various practical multicriteria analysis problems in the real world setting.

The process of actually calculating the performance index for each alternative across all criteria using the TOPSIS approach, however, may need further consideration [2, 3]. Under some circumstances counter intuition outcomes may occur while comparing two alternatives (vectors) just simply based on the distance between them and the ideal solution. Mathematically, the relative similarity (closeness) between each alternative and the ideal solution is better represented by the magnitude of the alternatives and the degree of conflict between them [4, 14].

To avoid the concern that the TOPSIS approach has, this paper presents a similarity based approach for solving the general multicriteria analysis problem. The approach effectively makes use of the ideal solution concept in such a way that the most preferred alternative should have the highest degree of similarity to the positive ideal solution and the lowest degree of similarity to the negative-ideal solution. The overall performance index of each alternative across all criteria is determined based on the concept of the degree of similarity between each alternative and the ideal solution using alternative gradient and magnitude. An example is presented to demonstrate the applicability of the proposed approach. A comparative analysis between the proposed approach and the TOPSIS approach is conducted for demonstrating the merits of the proposed approach for solving discrete multicriteria analysis problems.

In what follows, we first formulate the general multicriteria analysis problem to pave the way for the development of the multicriteria analysis approach, followed by the introduction of the concept of the degree of conflict and the degree of similarity between alternatives. A multicriteria analysis approach is then presented by combining the concept of the degree of similarity and the ideal solution together with the illustration of a case study.

## 2 Formulating the Multicriteria Analysis Problem

Multicriteria analysis is used to assist the decision maker (DM) in prioritizing or selecting one or more alternatives from a finite set of available ones with respect to multiple, usually conflicting criteria. The general multicriteria analysis problem usually consists of a number of alternatives $A_i$ $(i = 1, 2, ..., n)$ to be evaluated against a set of criteria $C_j$ $(j = 1, 2, ..., m)$. To determine the overall ranking of all alternatives across all criteria, the DM is usually required (a) to assess the performance of each alternative $A_i$ with respect to each criterion, denoted as $x_{ij}$, and the relative importance of the each criterion, represented as $w_j$, with respect to the overall objective of the problem and (b) to aggregate the performance ratings of the alternatives and the criteria weights for calculating the overall performance index for each alternative across all criteria. As a result, the decision matrix $X = (x_{ij})$ and the weighting vector $W = (w_1, w_2, ..., w_m)$ for the multicriteria analysis problem can be determined respectively as follows:

$$X = \begin{bmatrix} x_{11} & x_{12} & \dots & x_{1m} \\ x_{21} & x_{22} & \dots & x_{2m} \\ \dots & \dots & \dots & \dots \\ x_{n1} & x_{n2} & \dots & x_{nm} \end{bmatrix} \tag{1}$$

$$W = (w_1, w_2, \dots, w_m) \tag{2}$$

To facilitate the development of the multicriteria analysis approach, all decision criteria are assumed to be benefit criteria in the current discussion. This simply means the larger the value that an alternative has on a criterion, the more preferable the alternative [11, 20]. If a criterion is not a benefit one, necessary transformation processes, such as a reversal of the original criterion value, can be carried out in the decision matrix for consistency.

Given the decision matrix and the weight vector described as above, the overall objective of solving the multicriteria analysis problem is to prioritize all decision alternatives with respect to their overall performance across all criteria. To pave the way for the development of the multicriteria analysis approach, the concept of the degree of conflict and the degree of similarity between alternatives are discussed next.

## 3 Degree of Conflict

Real-world decision making problems are very often large, multi-dimensional, conflicting and non-commensurable. There is no exception for multicriteria analysis problems [6, 9, 13, 14, 17]. Conflict is a fundamental nature of multicriteria analysis problems which constitutes the core of each decision situation. A multicriteria analysis problem in which the performances of the alternatives in all evaluation criteria are in complete concordance, does not present any interest, as the choice is evident [2, 3].

There are various ways to represent the conflict between two alternatives in multicriteria analysis problems [2, 3, 9, 20]. Among them, the concept of alternative gradient to represent the conflict of decision alternatives in multicriteria analysis problems is the mostly common one [4]. Using this method, a conflict index between two alternatives is calculated to show the degree of conflict between the alternatives.

Assuming that $A_i$ and $A_j$ are the two alternatives concerned in a given multicriteria analysis problem, these two alternatives can be considered as two vectors in the m-dimensional real space. The angle between $A_i$ and $A_j$ in the $m$-dimensional real space is a good measure of conflict between them. As shown in Figure 1, $A_i$ and $A_j$ is in no-conflict if $\theta_{ij} = 0$, the conflict is possible if $\theta_{ij} \neq 0$, *i.e.* $\theta_{ij} \in (0, \pi/2)$. This is so because when $\theta_{ij} = 0$ the gradients of both the alternatives $A_i$ and $A_j$ are simultaneously in the same increasing direction and there is no conflict between them. The situation of conflict occurs when $\theta_{ij} \neq 0$, i.e. when the gradients of $A_i$ and $A_j$ are not coincident.

The degree of conflict between alternatives $A_i$ and $A_j$ is determined by

$$\cos\theta_{ij} = \frac{\sum_{k=1}^{m} x_{ik} x_{jk}}{\left[ (\sum_{k=1}^{m} x_{ik}^2)(\sum_{k=1}^{m} x_{jk}^2) \right]^{1/2}} \quad (3)$$

where $\theta_{ij}$ is the angle between the gradients of the two alternatives, and $(x_{i1}, x_{i2}, ..., x_{in})$ and $(x_{j1}, x_{j2}, ..., x_{jn})$ are the gradients of two alternatives $A_i$ and $A_j$ respectively.

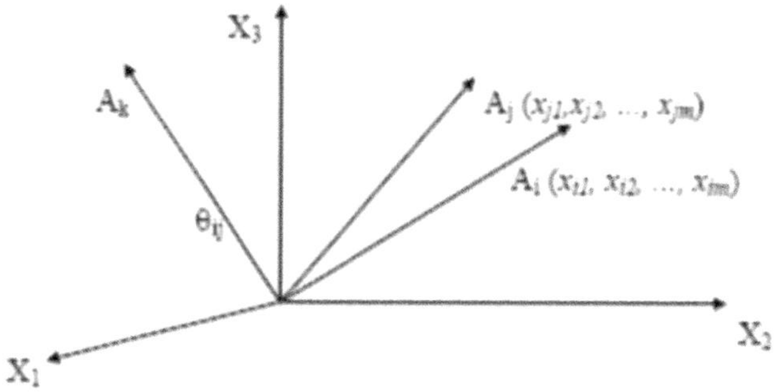


**Fig. 1.** Degree of conflict between alternatives by gradients

The conflict index equals to one characterized by $\theta_{ij} = 0$ as the corresponding gradient vectors lie in the same direction of improvement. Similarly, the conflict index is zero characterized by $\theta_{ij} = \pi/2$ which indicates that their gradient vectors have the perpendicular relationship between each other.

Based on the degree of the conflict between the alternatives, the degree of similarity between the two alternatives can be calculated. The degree of similarity between alternative $A_i$ and $A_i$, denoted as $S_{ij}$, measures the relative similarity (closeness) alternative $A_j$ to $A_i$, given as:

$$S_{ij} = \frac{(\sum_{k=1}^{m} x_{ik}^2)^{1/2} \cos\theta_{ij}}{(\sum_{k=1}^{m} x_{jk}^2)^{1/2}} \quad (4)$$

where $\theta_{ij}$ is the angle between alternative $A_i$ and alternative $A_j$. represented the degree of conflict as discussed above. The larger the $S_{ij}$ is, the higher the degree of similarity between alternative $A_i$ and to $A_j$.

## 4 The Multicriteria Analysis Approach

Given the problem structure defined as above and the concept introduced, this section proposes a multicriteria analysis approach to ranking multicriteria alternatives by combining the alternative gradient and magnitude. The concept of the ideal solution is used in such a way that the most preferred alternative should have the highest degree of similarity to the positive ideal solution and the lowest degree of similarity to the negative-ideal solution.

The ranking approach starts by normalizing the decision matrix as in (1) to ensure all the criteria involved are benefit ones based on (5), described as

$$x_{ij}^{'} = \frac{x_{ij}}{(\sum_{k=1}^{n} x_{ik}^{2})^{1/2}} \tag{5}$$

As a result, a normalized decision matrix can be determined as

$$X' == \begin{bmatrix} x'_{11} & x'_{12} & \dots & x'_{1m} \\ x'_{21} & x'_{22} & \dots & x'_{2m} \\ \dots & \dots & \dots & \dots \\ x'_{n1} & x'_{n2} & \dots & x'_{nm} \end{bmatrix} \tag{6}$$

The weighted performance matrix which reflects the performance of each alternative with respect to each criterion is determined by multiplying the normalized decision matrix in (6) by the weight vector described in (2), given as

$$y = \begin{bmatrix} w_1 x'_{11} & w_2 x'_{12} & \dots & w_m x'_{1m} \\ w_1 x'_{21} & w_2 x'_{22} & \dots & w_m x'_{2m} \\ \dots & \dots & \dots & \dots \\ w_1 x'_{n1} & w_2 x'_{n2} & \dots & w_m x'_{nm} \end{bmatrix} = \begin{bmatrix} y_{11} & y_{12} & \dots & y_{1m} \\ y_{21} & y_{22} & \dots & y_{2m} \\ \dots & \dots & \dots & \dots \\ y_{n1} & y_{n2} & \dots & y_{nm} \end{bmatrix} \tag{7}$$

The positive (or negative) ideal solution consists of the best (or worst) criteria values attainable from all the alternatives if each criterion takes monotonically increasing or decreasing values [6, 11]. This concept has been widely used in various multicriteria analysis models for solving practical decision problems [7, 8, 17]. This is due to (a) its simplicity and comprehensibility in concept, (b) its computational efficiency, and (c) its ability to measure the relative performance of the decision alternatives in a simple mathematical form. Based on this concept, the positive ideal solution and the negative ideal solution can be determined from the performance matrix in (7), given as

$$\begin{aligned} A^{+} &= (y_1^{+}, y_2^{+}, \dots, y_m^{+}) \\ A^{-} &= (y_1^{-}, y_2^{-}, \dots, y_3^{-}) \end{aligned} \tag{8}$$

Where

$$\begin{aligned} y_j^{+} &= \max_{i=1,2,\dots,n} y_{ij}^{'} \\ y_j^{-} &= \min_{i=1,2,\dots,n} y_{ij}^{'} . \end{aligned} \tag{9}$$

The degree of conflict between each alternative $A_i$ and the positive ideal solution (the negative ideal solution) can be determined based on (3), given as

$$\cos\theta_{i+} = \frac{\sum_{j=1}^{m} y'_{ij}\, y^{+}_{j}}{\left(\sum_{j=1}^{m} y'^{2}_{ij} \sum_{j=1}^{m} y^{+2}_{j}\right)^{1/2}}$$
$$\cos\theta_{i-} = \frac{\sum_{j=1}^{m} y'_{ij}\, y^{-}_{j}}{\left(\sum_{j=1}^{m} y'^{2}_{ij} \sum_{j=1}^{m} y^{-2}_{j}\right)^{1/2}}. \tag{10}$$

As a consequence, the degree of similarity between each alternative $A_i$ and the positive ideal solution and the negative ideal solution can be determined by

$$S_i^{+} = \frac{\left(\sum_{k=1}^{m} y'^{2}_{ik}\right)^{1/2} \cos\theta_{i+}}{\left(\sum_{j=1}^{m} y^{+2}_{j}\right)^{1/2}}$$
$$S_i^{-} = \frac{\left(\sum_{k=1}^{m} y'^{2}_{ik}\right)^{1/2} \cos\vartheta_{i-}}{\left(\sum_{j=1}^{m} y^{-2}_{j}\right)^{1/2}}, \tag{11}$$

An overall performance index for each alternative across all criteria can then be calculated based on the concept of the degree of similarity of alternative $A_i$ relative to the ideal solution as

$$P_i = \frac{S_i^{+}}{S_i^{+} + S_i^{-}}, \; i = 1, 2, \ldots, n. \tag{12}$$

The larger the index value, the more preferred the alternative.

Summarizing the discussion as above, the proposed multicriteria analysis approach can be presented in a algorithmically form as follows:

Step 1. Determine the decision matrix as in (1).
Step 2. Determine the weighting vector as in (2).
Step 3. Normalize the decision matrix as in (6) obtained by Step 1 by (5)
Step 4. Calculate the performance matrix as expressed in (7)
Step 5. Determine the positive ideal solution and the negative ideal solution by (8) and (9).
Step 6. Calculate the conflict index between the alternatives and the positive ideal solution and the negative-ideal solution using (4).
Step 7. Calculate the degree of similarity of the alternatives between each alternative and the positive ideal solution and the negative-ideal solution by (11).
Step 8. Calculate the overall performance index for each alternative across all criteria by (12).
Step 9. Rank the alternatives in the descending order of the index value.

## 5 An Example

A country has decided to purchase a fleet of jet fighters from the U.S. The Pentagon officials offered the characteristic information of the four models ($A_1$, $A_2$, $A_3$, $A_4$) which may be sold to that country. The air force analyst team of that country agreed that six characteristics (criteria) should be considered They are (a) maximum speed $C_1$, (b) ferry range $C_2$, (c) maximum payload $C_3$, (d) purchasing cost $C_4$, (e) reliability $C_5$, and (f) maneuverability $C_6$. The team has assessed the performance of the four alternatives with respect to each of the six criteria attributes. Table 1 presents the performance assessments. This case example is adopted from Hwang and Yoon [16].

**Table 1.** A fighter aircraft selection problem

| | $C_1$ | $C_2$ | $C_3$ | $C_4$ | $C_5$ | $C_6$ |
|---|---|---|---|---|---|---|
| $A_1$ | 2.0 | 1500 | 20000 | 5.5 | average | very high |
| $A_2$ | 2.5 | 2700 | 18000 | 6.5 | low | average |
| $A_3$ | 1.8 | 2000 | 21000 | 4.5 | high | high |
| $A_4$ | 2.2 | 1800 | 20000 | 5.0 | average | average |

As a result, the decision matrix for the fighter aircraft selection problem can be determined as

$$X = \begin{array}{c} \\ A_1 \\ A_2 \\ A_3 \\ A_4 \end{array} \begin{array}{cccccc} C_1 & C_2 & C_3 & C_4 & C_5 & C_6 \\ \hline 2.0 & 1500 & 20000 & 5.5 & \text{average} & \text{very high} \\ 2.5 & 2700 & 18000 & 6.5 & \text{low} & \text{average} \\ 1.8 & 2000 & 21000 & 4.5 & \text{high} & \text{high} \\ 2.2 & 1800 & 20000 & 5.0 & \text{average} & \text{average} \end{array}$$

By quantifying the non-numerical assessments on the criteria of $C_5$ and $C_6$ based on a ten-point scale (Saaty, 1980), the decision matrix is adjusted as

$$X = \begin{array}{c} \\ A_1 \\ A_2 \\ A_3 \\ A_4 \end{array} \begin{array}{cccccc} C_1 & C_2 & C_3 & C_4 & C_5 & C_6 \\ \hline 2.0 & 1500 & 20000 & 5.5 & 5 & 9 \\ 2.5 & 2700 & 18000 & 6.5 & 3 & 5 \\ 1.8 & 2000 & 21000 & 4.5 & 7 & 7 \\ 2.2 & 1800 & 20000 & 5.0 & 5 & 5 \end{array}$$

All criteria except $C_4$ are benefit ones. Therefore, transformation of criterion $C_4$ into a benefit one is necessary by using the reversal of the original criterion value. The corresponding decision matrix is given as

$$X = \begin{array}{c} \\ A1 \\ A2 \\ A3 \\ A4 \end{array} \begin{array}{cccccc} C_1 & C_2 & C_3 & C_4 & C_5 & C_6 \\ \hline 2.0 & 1500 & 20000 & .1818 & 5 & 9 \\ 2.5 & 2700 & 18000 & .1538 & 3 & 5 \\ 1.8 & 2000 & 21000 & .2222 & 7 & 7 \\ 2.2 & 1800 & 20000 & .2000 & 5 & 5 \end{array}$$

By normalized the decision matrix by (5), the decision matrix is given as

$$X = \begin{array}{c} \\ A_1 \\ A_2 \\ A_3 \\ A_4 \end{array} \begin{array}{cccccc} C_1 & C_2 & C_3 & C_4 & C_5 & C_6 \\ \hline .4671 & .3662 & .5056 & .4757 & .4811 & .6708 \\ .5839 & .6591 & .4550 & .4024 & .2887 & .3727 \\ .4204 & .4882 & .5308 & .5814 & .6736 & .5217 \\ .5139 & .4392 & .5056 & .5233 & .4811 & .3727 \end{array}$$

The weight vector of the attribute is given by the DM as $w = (w_1, w_2, w_3, w_4, w_5, w_6) = (.2, .1, .1, .1, .2, .3)$. By multiplying the normalized decision matrix by the weight vector as expressed in (7), the weighted performance matrix is obtained as

$$Y = \begin{matrix} A_1 \\ A_2 \\ A_3 \\ A_4 \end{matrix} \overset{\begin{matrix} C_1 & C_2 & C_3 & C_4 & C_5 & C_6 \end{matrix}}{\begin{bmatrix} .0934 & .0366 & .0506 & .0476 & .0962 & .2012 \\ .1168 & .0659 & .0455 & .0402 & .0577 & .1118 \\ .0841 & .0488 & .0531 & .0581 & .1347 & .1565 \\ .1028 & .0439 & .0506 & .0523 & .0962 & .1118 \end{bmatrix}}$$

The positive ideal solution and the negative-ideal solution are determined by (8) and (9) as

$$A^+ = ( .1168, .0659, .0531, .0581, .1347, .2012),$$
$$A^- = ( .0841, .0366, .0455, .0402, .0577, .1118).$$

Therefore, the degree of conflict between each alternative and the positive ideal solution and the negative ideal solution is calculated by (10) as

$$\cos\theta_{1+} = .992, \qquad \cos\theta_{2+} = .936,$$
$$\cos\theta_{3+} = .975, \qquad \cos\theta_{4+} = .963$$
$$\cos\theta_{1-} = .976, \qquad \cos\theta_{2-} = .981,$$
$$\cos\theta_{3-} = .924 \qquad \cos\theta_{4-} = .981$$

The degree of similarity between each alternative and the positive ideal solution and the negative ideal solution is determined by (11) as

$$S_1^+ = .862, \qquad S_2^+ = .619,$$
$$S_3^+ = .804, \qquad S_4^+ = .655,$$
$$S_1^- = 1.485, \qquad S_2^- = 1.137,$$
$$S_3^- = 1.335, \qquad S_4^- = 1.678.$$

An overall performance index for each alternative across all criteria can be determined by (12). Table 2 shows the results. For the sake of comparison, the ranking outcomes by using the TOPSIS approach are also included.

**Table 2.** Alternatives rankings between TOPSIS approach and the proposed MA approach

| | The TOPSIS Approach | | The Multicriteria Analysis Approach | |
|---|---|---|---|---|
| | Index | Ranking | Index | Ranking |
| $A_1$ | .643 | 1 | .367 | 2 |
| $A_2$ | .268 | 4 | .353 | 3 |
| $A_3$ | .613 | 2 | .376 | 1 |
| $A_4$ | .312 | 3 | .281 | 4 |

Table 2 shows slightly different ranking outcomes between the TOPSIS approach and the proposed multicriteria analysis approach. The proposed multicriteria analysis approach is believed to have provided a better ranking outcome. However, it is very difficult to evaluate whether the approach is more appropriate practically although there are sound theoretically ground to support the new approach.

## 6 Conclusions

This paper presents a new approach using the concept of alternative gradient and magnitude for effectively solving the general multicriteria analysis problem. The proposed approach is capable of addressing the concern of the TOPSIS approach that the comparison of the alternatives cannot be determined solely by the distance between the alternatives. The concept of the degree of similarity between the alternatives and the ideal solution is combined to derive an overall performance index of each alternative for the general multicriteria analysis problem which has shown some potential. The underlying concept of this approach is simple and easy to understand. The computation process is very easy to handle. As a consequence, the proposed multicriteria analysis approach is of practical use in solving real multicriteria analysis decision problems.

## References

1. Bryson, N.: Group Decision-making and the Analytic Hierarchy Process: Exploring the Consensus-Relevant Information Content. Computers and Operations Research. 23 (1) (1996) 27-35
2. Carlsson, C., Fuller, R.: Multiple Criteria Decision Making: The Case for Interdependence. Computers and Operations Research. 22 (3) (1995) 251-260
3. Chen, S. J., Hwang, C.L.: Fuzzy Multiple Attribute Decision Making: Methods and Applications. Springer-Verlag, New York (1992)
4. Cohon, J.L.: Multi-objective Programming and Planning. Academic Press, New York (1978)
5. Deng, H., Yeh, C.H.: Simulation-based Evaluation of Defuzzification-based Approaches to Fuzzy Multiattribute Decision Making. IEEE Transactions on Systems, Man, and Cybernetics. 36 (5) (2005) 968-977
6. Deng, H., Yeh, C.H., Willis, R.J.: Inter-company Comparison using Modified TOPSIS with Objective Weights. Computers and Operations Research. 27 (2000) 963-973
7. Deng, H.: Multicriteria Analysis with Fuzzy Pairwise Comparison. International Journal of Approximate Reasoning. 21 (3) (1999) 215-231
8. Deng, H., Yeh, C.H.: Ranking Multi-criteria Alternatives under Uncertainty. Proceedings of the International Conference on Computational Intelligence and Multimedia Applications. World Scientific, Singapore (1998) 504-509
9. Diakoulaki, D., Mavrotas, G., Papayannakis, L.: Determining Objective Weights in Multiple Criteria Problems: the CRITIC Method. Computers and Operations Research. 22 (7) (1995) 763-770
10. Hwang, C.L., Lai, Y.J., Liu, T.Y.: A New Approach for Multiple Objective Decision Making. Computers and Operations Research. 20 (9) (1993) 889-899
11. Hwang, C.L., Yoon, K.S.: Multiple Attribute Decision Making: Theory and Applications. Springer-Verlag, New York (1981)

12. Mohanty, B.K., Vijayaraghavan, T.-A.S.: A Multi-objective Programming Problem and its Equivalent Goal Programming Problem with Appropriate Priorities and Aspiration Levels: A Fuzzy Approach. Computers and Operations Research. 22 (8) (1995) 771-778
13. Olson, D.L.: Decision Aids for Selection Problems. Springer-Verlag, New York (1996)
14. Roy, B., Vincke, P.: Multicriteria Analysis: Survey and Promising Directions. European Journal of Operational Research. 8 (1981) 207-218
15. Saaty, T.L.: The Analytic Hhierarchy Process. McGraw-Hill, New York (1980)
16. Saaty, T.L.: How to Make A Decision: the Analytic Hierarchy Process. Interfaces 24 (1994) 19-43
17. Shipley, M.F., de Korvin, A., Obid, R.: A Decision Making Model for Multi-Attribute Problems Incorporating Uncertainty and Bias Measures. Computers and Operations Research. 18 (1991) 335-342
18. Stewart, T.J.: A Critical Survey on the Status of Multiple Criteria Decision Making: Theory and Practice. Omega 20 (1992) 569-586
19. Yeh, C.H., Deng, H., Pan, H.: Multi-criteria Analysis for Dredger Dispatching under Uncertainty. Journal of the Operational Research Society. 50 (1999) 35-43
20. Zeleny, M.: Multiple Criteria Decision Making: Eight Concepts of Optimality. Human Systems Management. 17 (2) (1998) 97-107
21. Zionts, S.: A Multiple Criteria Method for Choosing among Discrete Alternatives. European Journal of Operational Research. 7 (1981) 143-147

# Algorithms for the Well-Drilling Layout Problem*

Aili Han[1,2], Daming Zhu[2], Shouqiang Wang[2], and Meixia Qu[1]

[1] Dept. of Comput. Sci. and Tech., Shandong University, Weihai 264209, China
[2] Sch. of Comput. Sci. and Tech., Shandong University, Jinan 250061, China
hanal@sdu.edu.cn

**Abstract.** Given some discrete points in a plane, ones move a grid to maximize the number of the points that can be used. This is the well-drilling layout problem. If only consider the translation motion, we present an algorithm with time complexity of $O(n^2r)$ to compute the translation location instead of the previous algorithms with time complexity of $O(n^2r^2)$, where $n$ is the number of the discrete points and $r$ is the radius of error-round. In consideration of the rotation and translation motion, we present an algorithm with time complexity of $O(n^3d)$ to compute the rotation angle and the translation location instead of the previous algorithms with time complexity of $O(n^3r^2d)$, where $d$ is the maximum distance between any two discrete points.

## 1 Introduction

When prospecting for oil, the tentative wells are firstly drilled to ascertain the distribution of oil field and the content of oil. The tentative drilling is a random one that randomly selects some points where the wells are drilled to obtain the original data. And then the formal drilling is carried out, which is a grid one. According to the data obtained from the tentative drilling, a grid layout is marked out and the well is drilled at each node of the grid. If a tentative well locates in a circle with center being a node and radius being $r$, it is used as a formal well. Obviously, the tentative wells should be sufficiently used to reduce the cost. This problem is called the well-drilling layout problem, which has also important signification in other fields.

For the well-drilling layout problem, let a set of points $\{P_i \mid P_i=(a_i, b_i), i=1,2,\ldots,n\}$ represent the tentative wells, a set of points $\{N_i \mid N_i=(X_i, Y_i), i\in Z\}$ represent the nodes of grid, and $h$ represent the side-length of a unit of grid. If a tentative well locates in a circle with center $N_i$ and radius $r(r<h)$, it is called near the node $N_i$, or it can be used as a formal one. Thus, the well-drilling layout problem can be formalized as follows:

*Instance*: A set of points $\{P_i \mid P_i=(a_i, b_i), i=1,2,\ldots,n\}$, a positive integer $h$ and a positive real number $r$ $(r<h)$.

---

* Supported by the Science and Technology Development Foundation from Shandong University at Weihai; the National Natural Science Foundation of China under Grant No.60573024; the National Grand Fundamental Research 973 Program of China under Grant No.2005CCA04500.

D.-S. Huang, L. Heutte, and M. Loog (Eds.): ICIC 2007, LNAI 4682, pp. 263–271, 2007.

*Objective*: Maximize the number of the points near the nodes. That is, $\max\{P_i=(a_i, b_i) \mid \exists N_j(X_j, Y_j)$ to satisfy $(a_i-X_j)^2+(b_i-Y_j)^2 \le r^2\}$.

It is easy to known that all nodes of grid are fixed if any two nodes are fixed. Suppose that the original location of grid is $\{(ih, jh) \mid i,j \in Z\}$, where $(ih, jh)$ represents the coordinates of any node. The well-drilling layout problem can also be described as follows: Given a set of points and the original location of grid, how to move the grid to maximize the number of the points near the nodes.

The movement of grid corresponds to the counter-movement of all the given points. That is, moving the grid clockwise is equivalent to moving all the given points counter-clockwise. If only consider the translation motion, we present an algorithm with time complexity of $O(n^2r)$, where $n$ is the number of the given points and $r$ is the radius of error-round. If consider the rotation and translation motion, we give an algorithm with time complexity of $O(n^3d)$, where $d$ denotes the maximum distance between any two given points.

In order to be convenient for description, the grid is considered as a $L\times W$ one, denoted by $Gr(L,W)$, and all nodes of it are denoted by $N(Gr)$. Since all of the given points locate in a plane where the formal drillings are carried out, we can assume that $L$ and $W$ are so great that $Gr(L,W)$ can cover all of the given points when it moves. Here, the values of $L$ and $G$ don't affect the number of the points that can be used as formal wells, so $L$ and $G$ are considered as unlimited and $Gr(L,W)$ is briefly written as $Gr$ in the following.

## 2 Algorithm for Translation Motion

### 2.1 Principle of the Translation Algorithm

*Definition* 1. For an instance of the well-drilling layout problem, a roundlet with center $P_i$ and radius $r$ is called the *error-round* of $P_i$. A point $P_j$ locating in an error-round is also called the error-round *cover* the point $P_j$.

*Definition* 2. For an instance of the well-drilling layout problem, let $a_i'=a_i \bmod h$ and $b_i'=b_i \bmod h$. The obtained point $P_i'(a_i', b_i')$ is called the *image point* of $P_i(a_i, b_i)$, and the original point $P_i(a_i,b_i)$ is called the *source point* of $P_i'(a_i', b_i')$. The procedure from the source points to the image points is called *coordinate mapping*.

*Theorem* 1. For an instance of the well-drilling layout problem, the sufficient and necessary condition of $P_i$ and $P_j$ being used as formal wells is that the image points $P_i'$ and $P_j'$ can be covered by an error-round.

*Proof.* Let $N(Gr)=\{(ph, qh) \mid p,q \in Z\}$. (i) Necessity: Suppose that $P_i(a_i, b_i)$ and $P_j(a_j, b_j)$ can be used as formal wells and they locate in the error-rounds of $(p_1h, q_1h)$ and $(p_2h, q_2h)$, respectively. That is, $(a_i-p_1h)^2+(b_i-q_1h)^2 \le r^2$, $(a_j-p_2h)^2+(b_j-q_2h)^2 \le r^2$. Let $a_i'=a_i-p_1h$, $b_i'=b_i-q_1h$, $a_j'=a_j-p_2h$, $b_j'=b_j-q_2h$, then $a_i'^2+b_i'^2 \le r^2$, $a_j'^2+b_j'^2 \le r^2$. This means that the image points $(a_i', b_i')$ and $(a_j', b_j')$ locate in the error-round of (0, 0). (ii) Sufficiency: Suppose that the image points $P_i'$ and $P_j'$ can be covered by an error-round. That is, exist $(x_t,y_t)$ satisfying $(a_i'-x_t)^2+(b_i'-y_t)^2\leq r^2$ and $(a_j'-x_t)^2+(b_j'-y_t)^2\leq r^2$. Translate the grid to let $x'=x+x_t$ and $y'=y+y_t$. Here, $\{(ph+x_t, qh+y_t) \mid p,q\in Z\}$ are the new nodes of grid. Let $a_i'=a_i-p_3h$, $b_i'=b_i-q_3h$, $a_j'=a_j-p_4h$, $b_j'=b_j-q_4h$. According to the assumption, we can conclude that $((a_i-p_3h)-x_t)^2+((b_i-q_3h)-y_t)^2\leq r^2$ and $((a_j-p_4h)-x_t)^2+ ((b_j-q_4h)-y_t)^2\leq r^2$, or $(a_i-(p_3h+x_t))^2+(b_i-(q_3h+y_t))^2\leq r^2$ and $(a_j-(p_4h+x_t))^2+(b_j-(q_4h+y_t))^2\leq r^2$. Thus, $P_i(a_i, b_i)$ and $P_j(a_j, b_j)$ can be used as formal wells.

*Inference* 1. For an instance of the well-drilling layout problem, the sufficient and necessary condition of $P_{i1},P_{i2},\ldots,P_{ik}$ being used as formal wells is that the image points of $P_{i1},P_{i2},\ldots,P_{ik}$ can be covered by an error-round.

According to Inference 1, the well-drilling layout can be computed as follows: All of the coordinates are firstly mapped to let the image points locate in the unit $[0, h]*[0, h]$. And then, move an error-round in the unit $[0, h]*[0, h]$ to cover the most image points. If an image point locates in the border region of $[0, h]*[0, h]$, the error-round is likely to span the border, as shown in Fig. 1. Thus, the image points close to other vertices or edges of the unit $[0, h]*[0, h]$ haven't been covered by the error-round. In order to avoid this case, the border problem should be firstly dealt with.

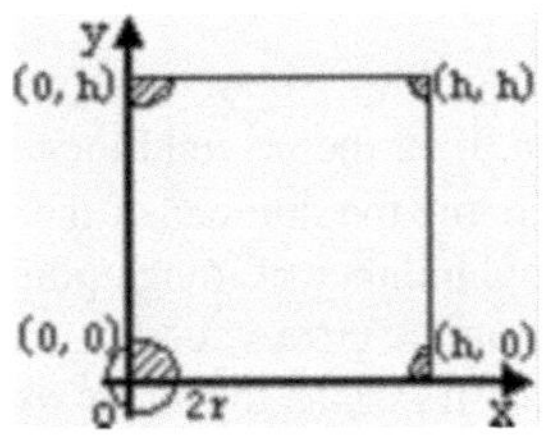


**Fig. 1.** A case of error-round spanning the borders

## 2.2 Method of Dealing with the Border Problem

For an instance of the well-drilling layout problem, the method of dealing with the border problem is given as follows.

(1) The image points in region $[h-r, h] *[0, h]$ are copied to $[-r, 0]*[0, h]$.
(2) The image points in region $[0, r] *[0, h]$ are copied to $[h, h+r]*[0, h]$.
(3) The image points in region $[-r, h+r] *[h-r, h]$ are copied to $[-r, h+r]*[-r, 0]$.
(4) The image points in region $[-r, h+r] *[0, r]$ are copied to $[-r, h+r]*[h, h+r]$.

Thus, the region of the image points lying in is enlarged from $[0, h]*[0, h]$ to $[-r, h+r]*[-r, h+r]$, as shown in Fig. 2.

After dealing with the border problem, each image point close to the vertices of $[0, h]*[0, h]$ has four images in the enlarged unit, shown as * in Fig. 2; each image point close to the edges of $[0, h]*[0, h]$ has two images, shown as # in Fig. 2. Thus, an error-round of covering the most image points can be obtained through moving the error-round in the enlarged unit.

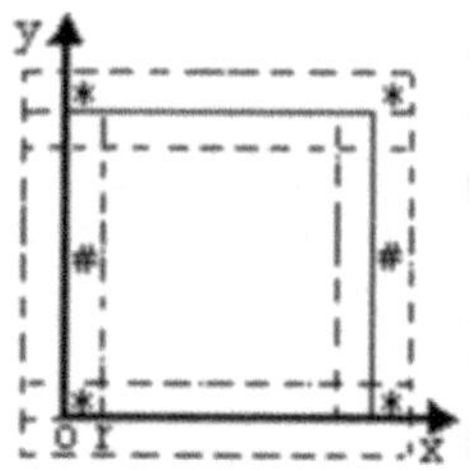


**Fig. 2.** An enlarged unit of grid

### 2.3 The Translation Algorithm

Given an instance of the well-drilling layout prob-lem. If only consider the translation motion, the proposed algorithm is given as follows.

*Algorithm* 1. The translation algorithm

(1) For each discrete point, the coordinates are mapped to let the image point locate in the unit $[0, h]*[0, h]$.

(2) Deal with the border problem to let the image points lie in the enlarged unit $[-r, h+r]*[-r, h+r]$.

(3) For $i$=1 to $n$ do

(3.1) For the points $P_i'(a_i', b_i')$, let the original location of an error-round is $(x_e, y_e)$, where $x_e=a_i'$, $y_e=b_i'-r$. Let $pp$ denote the number of the image points lying in the error-round, $pp_{\max}$ denote the maximal number of image points lying in the error-round, and $S_{PE}$ denote the set of the points on the error-round.

(3.2) For each point $P \in S_{PE}$, translate the error-round to let $P$ locate at $P_i'$ and compute: $pp$=the number of image points lying in the error-round. Let $pp_{\max}=\max\{pp, pp_{\max}\}$. The center of the error-round of covering $pp_{\max}$ image points is marked as $P_{\max}$.

(4) Let $P_{\max}=(x_{\max}, y_{\max})$. The grid is translated as follows: $(ih, jh) \xrightarrow{translation} (ih+x_{\max}, jh+y_{\max})$. The source points corresponding to $pp_{\max}$ image points in the error-round are the solution of the well-drilling layout problem.

## 3 Algorithm for Rotation Motion

For an instance of the well-drilling layout problem, the algorithm in consideration of the rotation and translation motion is given as follows.

### 3.1 Principle of the Rotation Algorithm

*Definition* 3. For an instance of the well-drilling layout problem, the circle with center $N_i$ and radius $2r$ is called the *analysis circle* of $N_i$. The circle with center $P_i$ and radius $d_{ij}$, where $d_{ij}$ is the distance between $P_i$ and $P_j$, is called the *rotation circle* of $P_j$.

*Theorem* 2. For an instance of the well-drilling layout problem, let $d_{ij}$ denote the distance between the points $P_i$ and $P_j$. The sufficient and necessary condition of $P_i$ and

$P_j$ being used as formal wells in consideration of the rotation and translation motion is that exist two nodes $N_i(X_i,Y_i)$ and $N_j(X_j,Y_j)$ satisfying the equation set

$$\begin{cases}(x-X_i)^2+(y-Y_i)^2=d_{ij}^{\ 2}\\(x-X_j)^2+(y-Y_j)^2=(2r)^2\end{cases}$$

*Proof.* Fix $P_i$ to the node $N_i$. It is easy to known that the rotation circle of $P_j$ intersecting the analysis circle of $N_j$ is equivalent to that the equation set has solution. (i) Sufficiency: Suppose that the rotation circle of $P_j$ intersects the analysis circle of $N_j$ at $M_1$ and $M_2$, as shown in Fig. 3. Fix $P_j$ to a point on the arc $M_1M_2$ of the rotation circle of $P_j$. Here, $P_i$ and $P_j$ can be used as formal wells through translating the grid. (ii) Necessity: suppose that $P_i$ and $P_j$ can be used as formal wells, and $P_i$ and $P_j$ locate in the error-rounds of $N_i$ and $N_j$, respectively. When $P_i$ is translated to $N_i$, $P_j$ will locate in the analysis circle of $N_j$, that is, the equation set has solution.

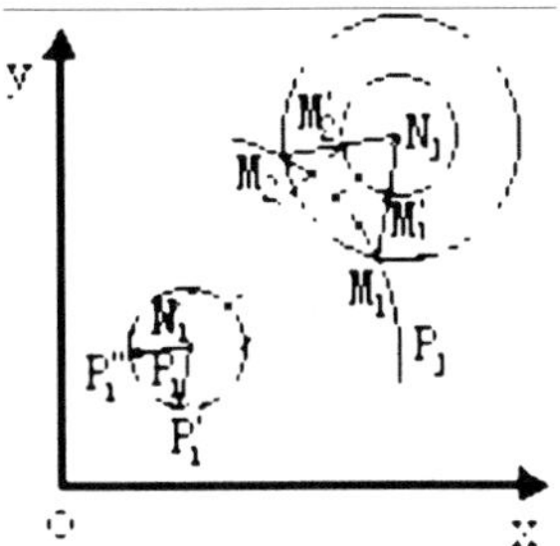

**Fig. 3.** The rotation circle of Pj intersects the analysis circle of Nj at M1 and M2

*Definition* 4. Let $P_i$ locate at a node of grid and the rotation circle of $P_j$ intersects the analysis circle of $N_j$ at $M_1$ and $M_2$. If the rotation angle is $\theta_{j1}$ when $P_j$ is rotated to $M_1$ from the original location and the rotation angle is $\theta_{j2}$ when $P_j$ is rotated to $M_2$ from the original location, the angle interval $[\theta_{j1},\theta_{j2}]$ is called the *rotation interval* of $P_j$ near the node $N_j$.

According to definition 4, $P_j$ locates in the analysis circle of $N_j$ if the rotation angle $\phi\in[\theta_{j1},\theta_{j2}]$. Here, according to theorem 2, $P_i$ and $P_j$ can be used as formal wells.

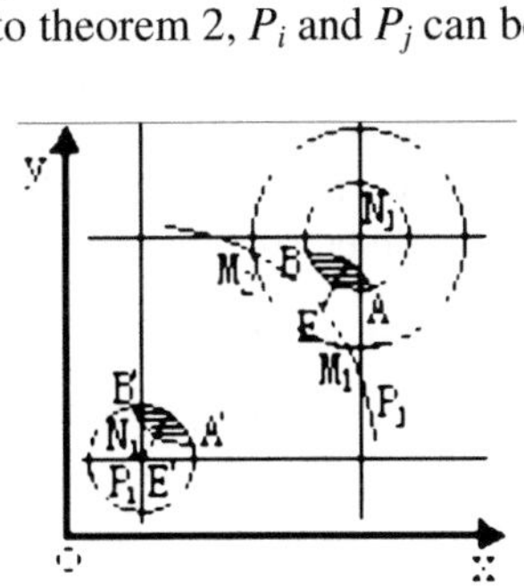

**Fig. 4.** The translation region of Pj

*Definition* 5. Let $P_i$ locate at a node of grid and the rotation circle of $P_j$ intersects the analysis circle of $N_j$ at $M_1$ and $M_2$. For a point $E$ on the arc $M_1M_2$ of the rotation circle of $P_j$, if a circle with center $E$ and radius $r$ intersects the error-round of $N_j$ at $A$ and $B$, the region between $A$ and $B$ is called the *translation region* of $P_j$, as shown in Fig. 4. The translation region of $P_j$ is translated to let $E$ locate at $P_i$. The corresponding region in the error-round of $N_i$ is called the *mapping region* of $P_j$.

*Theorem* 3. For an instance of the well-drilling layout problem, let $P_i$ locate at a node of grid. The sufficient and necessary condition of $P_i$, $P_j$ and $P_k$ being used as formal wells is that the rotation intervals of $P_j$ and $P_k$ intersect and there exists an angle $\phi$ in the common interval $[\phi_1,\phi_2]$ to make the mapping regions of $P_j$ and $P_k$. overlap each other.

*Proof*. Necessity is obvious. Sufficiency: Suppose that $P_j$ locate at $E$ and $P_k$ locate at $F$ when the rotation angle is $\phi\in[\phi_1,\phi_2]$, and the common part of the mapping regions of $P_j$ and $P_k$ is marked as $H$, as shown in Fig. 5. When $P_i$ is translated from $N_i$ to a point $G$ in the region $H$, the point $P_j$ will be translated to the error-rounds of $N_j$ and the point $P_k$ will be translated to the error-rounds of $N_k$. Thus, $P_i$, $P_j$ and $P_k$ can be used as formal wells.

*Inference* 2. For an instance of the well-drilling layout problem, let $P_{i1}$ locate at a node of grid. The sufficient and necessary condition of $P_{i1}$, $P_{i2}$, $P_{i3}$,...,$P_{ik}$ being used as formal wells is that the rotation intervals of $P_{i2}$, $P_{i3}$,...,$P_{ik}$ intersect and there exists an angle $\phi$ in the common interval $[\phi_1,\phi_2]$ to make the mapping regions of $P_{i2}$, $P_{i3}$,...,$P_{ik}$ overlap each other.

According to Inference 2, the well-drilling layout problem in consideration of the rotation and translation motion is changed into the problem of seeking the most points whose rotation intervals overlap and the translation regions also overlap.

## 3.2 Computing the Rotation Angle

The rotation angle can be obtained in the interval [0°, 90°] since the grid is composed of squares. Fix $P_i$ to a node of grid. $P_j$ is rotated to obtain its rotation interval. For each angle in the rotation interval, compute the number of the points being used as formal wells. And then, the optimal layout can be obtained from the maximal number of the points being used as formal wells.

The rotation circle of $P_j$ may intersect several analysis circles, so the corresponding nodes need to be analyzed in turn when computing the rotation interval of $P_j$. Fix $P_i(a_i, b_i)$ to $N_i$. If the rotation circle of $P_j$ and the analysis circle of $N_j(X_j,Y_j)$ intersect, the rotation interval of $P_j$ near $N_j$ can be computed through the following algorithm 2.

*Algorithm* 2. Computing the rotation interval of $P_j$

(1) The coordinates $(x_1, y_1)$ and $(x_2, y_2)$ are obtained by solving the equations set

$$\begin{cases}(x-a_i)^2+(y-b_i)^2=d_{ij}^{\ 2}\\(x-X_j)^2+(y-Y_j)^2=(2r)^2\end{cases}$$

(2) According to the formula $\mathrm{tg}\theta_0=(b_j-b_i)/(a_j-a_i)$, $\mathrm{tg}\theta_1=(y_1-b_i)/(x_1-a_i)$ and $\mathrm{tg}\theta_2=(y_2-b_i)/(x_2-a_i)$, the angles $\theta_0,\theta_1,\theta_2\in[-90°, 90°]$ are computed, respectively.

(3) Let $\phi_1=\theta_1-\theta_0$, $\phi_2=\theta_2-\theta_0$. If $\phi_1,\phi_2\notin[0°,90°]$, let $\phi_1=\phi_1 \bmod 90$, $\phi_2=\phi_2 \bmod 90$. Here, the rotation interval of $P_j$ is $[\phi_1, \phi_2]$.

Now, determine the analysis circles that may intersect with the rotation circle of $P_j$. Fix a node $N_i$ of grid to $P_i(a_i,b_i)$ and another node $N_j$ to $(a_i+ph, b_i+qh)$. If the rotation circle of $P_j$ intersects with the analysis circle of $N_j(a_i+ph, b_i+qh)$, the distance between the two centers $N_i$ and $N_j$ satisfy:

$$d_{ij}-2r\le\sqrt{(ph)^2+(qh)^2}\le d_{ij}+2r$$

Let $c=\sqrt{(d_{ij}-2r)^2-(qh)^2}$ and $d=\sqrt{(d_{ij}+2r)^2-(qh)^2}$. According to the above formula, the values of $q$ and $p$ are as follows:

$$-(d_{ij}+2r)/h \le q \le (d_{ij}+2r)/h$$

$$c/h \le p \le d/h, \text{ or } -d/h \le p \le -c/h$$

Owing to $d-c<h$, or $d/h-c/h<1$, for a certain value of $q$, while $c/h\le p\le d/h$, the rotation circle of $P_j$ may intersect with one analysis circle. Since $p$ is an integer and $c/h\ge 0$, let $p_1=\lfloor c/h\rfloor$, $p_2=\lfloor c/h\rfloor+1$. In the same way, while $-d/h\le p\le -c/h$, let $p_1=-\lfloor c/h\rfloor$, $p_2=-\lfloor c/h\rfloor-1$. Thus, the rotation circle of $P_j$ may intersect with one of the analysis circles of $(p_1h, qh)$ and $(p_2h, qh)$. It cannot intersect with the other analysis circle.

The rotation intervals of $P_j$ are in the interval [0°, 90°] since the grid is composed of squares. Consequently, the rotation intervals of $P_j$ can be computed through the following algorithm 3.

*Algorithm* 3. Computing the rotation intervals

FOR $q$:=0 TO $[(d_{ij}+2r)/h]+1$ DO

{$c:=\sqrt{(d_{ij}-2r)^2-(qh)^2}$ ;

IF $b_i\le b_j$

THEN IF $a_i\le a_j$

THEN {$q_1:=q$; $p_1:=\lfloor c/h\rfloor$; $p_2:=\lfloor c/h\rfloor+1$}

ELSE {$q_1:=q$; $p_1:=-\lfloor c/h\rfloor$; $p_2:=-\lfloor c/h\rfloor-1$}

ELSE IF $a_i\le a_j$

THEN {$q_1:=-q$; $p_1:=\lfloor c/h\rfloor$; $p_2:=\lfloor c/h\rfloor+1$}

ELSE {$q_1:=-q$; $p_1:=-\lfloor c/h\rfloor$; $p_2:=-\lfloor c/h\rfloor-1$};

IF the rotation circle of $P_j$ intersects with the analysis circle of $(a_i+p_1h, b_i+q_1h)$ or $(a_i+p_2h, b_i+q_1h)$

THEN the rotation interval of $P_j$ near the node is computed through algorithm 2.

}

### 3.3 Computing the Translation Region

After the rotation intervals are computed through algorithm 2, the translation region of $P_j$ need to be computed. Since the translation region of $P_j$ is not easy to be represented and overlapped, we need to look for a substitute for it. Obviously, the translation region of $P_j$ is one-to-one mapping to the mapping region of $P_j$, and the mapping region of $P_j$ can be expressed through the corresponding arc at the analysis circle of $N_i$, as shown in Fig. 4. Thus, the maximal number of the points being used can be obtained through overlapping the arcs.

Let $P_i$ locate at the node $N_i$ and $P_j$ locate at any point $E(x_0,y_0)$ in the analysis circle of $N_j(X_j,Y_j)$ when the rotation angle is $\varphi$. The arc corresponding to the mapping region of $P_j$ can be computed through the following algorithm 4.

*Algorithm* 4. Computing the arc corresponding to the mapping region of $P_j$

(1) The coordinates $(x_1,y_1)$ and $(x_2,y_2)$ of the intersections $A$ and $B$ are computed through the following equations set

$$\begin{cases}(x-x_0)^2+(y-y_0)^2=r^2\\(x-X_j)^2+(y-Y_j)^2=r^2\end{cases}$$

(2) The three points $E(x_0,y_0)$, $A(x_1,y_1)$ and $B(x_2,y_2)$ are translated to let $E$ locate at $P_i$. Here, the images $A'(x_1',y_1')$ and $B'(x_2',y_2')$ are the ends of the arc corresponding to the mapping region of $P_j$.

### 3.4 The Rotation Algorithm

In consideration of the rotation and translation motion, the algorithm of solving the well-drilling layout problem is given as follows.

*Algorithm* 5. The rotation algorithm

(1) for $i$=1 to $n$ do

(2) for $j$=1 to $i$-1 do

(2.1) Fix $P_i$ to a node of grid. Computing the rotation intervals through algorithm 3.

(2.2) For each angle in the rotation intervals, compute the locations of other points and the corresponding arcs in the error-round of $N_i$ through algorithm 4. And then, the number of points being used is recorded.

(3) The rotation intervals and the translation regions of the most points being used can be obtained from steps (1) and (2), which corresponds to the solution of the well-drilling layout problem.

## 4 Comparison with Other Methods

For an instance of the well-drilling layout problem, if only consider the translation motion, the previous translation algorithms [1,2,3] are with time complexity of $O(n^2r^2)$, where $n$ is the number of the given points and $r$ is the radius of error-round. The time complexity of the proposed translation algorithm is analyzed as follows. Steps 1, 2 and 4 are with time complexity of $O(n)$. In step 3, the set of points on an error-round is $\lfloor 2\pi r \rfloor$; judging the number of the image points in the error-round needs to do at most $n$ computations, and the times of judgment is at most $n$. Thus, step 3 is with time complexity of $O(n^2r)$. Therefore, the time complexity of the translation algorithm is $O(n^2r)$.

In consideration of the rotation and translation motion, the previous rotation algorithms [1,2,3] are with time complexity of $O(n^3r^2d)$ , where $d$ is the maximum distance between any two given points. The time complexity of the proposed rotation algorithm is analyzed as follows. Step (2.1) is with time complexity of $O(\lfloor(d_{ij}+2r)/h\rfloor)$. In steps (2.2), the rotation angle of $P_j$ in the worst case is [0°, 90°], which is with time complexity of $O(\lfloor\pi \times d_{ij}/2\rfloor \times n)$. Owing to $\lfloor\pi \times d_{ij}/2\rfloor \times n$ is above to $\lfloor(d_{ij}+2r)/h\rfloor$ in the general case, the rotation algorithm is with time complexity of $O(n^3d)$.

## 5 Experimental Results

Let 100 be the side-length of a unit of grid and 5 be the radius of error-round, or $h$=100, $r$=5. The coordinates of the given points are as follows:

| $i$ | 1 | 2 | 3 | 4 | 5 | 6 | 7 | 8 | 9 | 10 | 11 | 12 |
|---|---|---|---|---|---|---|---|---|---|---|---|---|
| $a_i$ | 50 | 141 | 300 | 337 | 340 | 472 | 472 | 543 | 757 | 838 | 898 | 950 |
| $b_i$ | 200 | 350 | 150 | 351 | 5<br>50 | 200 | 624 | 410 | 201 | 450 | 341 | 80 |

According to the translation algorithm, the most points that can be used as formal wells are 2, 4, 5, 10. And according to the rotation algorithm, the most points that can be used as formal wells are 1, 6, 7, 8, 9, 11.

## 6 Conclusion

We discuss the methods of solving the well-drilling layout problem in this paper. For any instance of the well-drilling layout problem, if only consider the translation motion, we gave a new algorithm with time complexity of $O(n^2r)$ to maximize the number of the given points that can be used as formal wells. In consideration of the rotation and translation motion, we present an algorithm with time complexity of $O(n^3d)$ to maximize the number of the given points that can be used as formal wells. The proposed algorithms are with less time complexity than the previous ones.

## References

1. Chen, G., Cheng, G. L., Wu, T. B.. Location Arrangement Model of Drilling Well. Mathematics in Practice and Theory, 30(1) (2000) 46-54
2. Xu, S. Y., Chen, S., Jin, H.. Well-Drilling Lay-out. Mathematics in Practice and Theory, 30 (1) (2000) 55-59
3. Hu, H., Y., Chen, J., Lu, X.,. The Mathematical Model of Borehole Layout. Mathematics in Practice and Theory, 30 (1) (2000) 60-66
4. Han, A. L.: Complexity Analysis for the HEWN Algorithm. Journal of Software, 2002, 13 (12) (2002) 2337~2342
5. Han, A. L.: A Study on the Solution of 9-room Diagram by State Space Method. Journal of Shandong University (Engineering Science), 34 (4) (2004) 51~54
6. Han, A. L, Zhu, D. M.: A Network Layout Algorithm Based on the Principle of Regular Hexagons Covering a Plane. Journal of Information and Computational Science, 3 (4) (2006) 753-759
7. Dorit S. H.: Approximation Algorithms for NP-hard Problems. PWS Publishing Company, (1997)

# Application of Dynamic Programming to Solving K Postmen Chinese Postmen Problem

Rong Fei, Duwu Cui, Yikun Zhang, and Chaoxue Wang

School of Computer Science and Engineering, Xi'an university of technology, Xi'an, China
annyfei@hotmail.com

**Abstract.** In this paper, Dynamic Programming is used to solve K postmen Chinese postmen problem for the first time. And a novel model for decision-making of KPCPP and the computation models for solving the whole problem are proposed. The arcs of G are changed into the points of G' by CAPA, and the model is converted into another one, which applies to Multistep Decision Process, by MDPMCA. On the base of these two programs, Dynamic Programming algorithm KMPDPA can finally solve the NPC problem-KPCPP. An illustrative example is given to clarify concepts and methods. The accuracy of these algorithms and the relative theories are verified by mathematical language.

**Keywords:** Dynamic Programming, KPCPP, CAPA, MDPMCA, KMDPA.

## 1 Introduction

K Postmen Chinese Postmen Problems is presented on the basis of Chinese Postmen Problems [5] [6] [7]. In reference [4], this problem is defined as KPCPP.

It has been proved by reference [4] that KPCPP is NPC. In general, KPCPP can be described with graph theory [1] as follows:

G=⟨ V,A;W⟩ is undirected graph, and let weight denote the length of every line. All the postmen (the number of postmen ≥2), start from one vertex of G, and run k lines at the same time, When they go back, every arc should be passed through at least once. These k lines are called the delivery routes, the length of which is the sum of all the arcs they passed through, and the routes group using the least delivery time is called optimal delivery routes. W(G) is the total weight of these optimal routes.

Dynamic programming [2] has two essences: the thought of ruling separately and the solution of redundancy. Now, we present a Dynamic programming algorithm system in order to solve KPCPP (k Postmen Chinese Postmen Problems), in which k equals the number of the edges of start vertex. CAPA is presented for making the model of KPCPP apply to decision-making, then, we put forward MDPMCA to make this model meet the demand of the Multistep Decision Process [10]. Pro tan to, we can use KMDPA to solve KPCPP. It's for the first time that KPCPP is solved by the thought of dynamic programming.

This paper is divided into four sections. In Section 2, functional definitions of KPCPP、 elements of a dynamic programming model and the problem are introduced.

D.-S. Huang, L. Heutte, and M. Loog (Eds.): ICIC 2007, LNAI 4682, pp. 272–281, 2007.

In Section 3, three new algorithms are proposed in 3 parts separately, every algorithm and related theorems are discussed, at the same time, the examples for proving the accuracy of algorithms is given. In Section 4, conclusion.

# 2 KPCPP

## 2.1 Background

**Definition 1.** $v_i$ is the node, $1 \le i \le n$, give a set of nodes V as V= $\{v_1, v_2..., v_n\}$ ; if there is an arc $a_{ij}$ between $v_i$ and $v_j$, then give a set of arcs A as A= $\{a_{ij} | 1 \le i \le n, 1 \le j \le n, i \ne j\}$ ; the length of $a_{ij}$ is $w_{ij}$, and give a set of length W as W=$\{ w_{ij} | 1 \le i < j \le n, 1 \le j \le n, i \ne j\}$; use $s_w = \sum_{\substack{i=1, j=1 \\ i \ne j}}^{n} w_{ij}$ to represent the set of length ;

**Definition 2.** $x_{km}$ is the state variable of the kth step, $X_k$ is the set of admissible state of the kth step, that is to say, $X_k = \{ x_{km} | 1 \le m \le n\}$;

**Definition 3.** Let $x_{n+1}$ be a terminal variable, $X_{n+1}$ be the set of terminal variable;

**Definition 4.** Let (k, $x_k$, $u_k(x_k)$, dk, $f_k(x_k)$) be a model of dynamic programming function, which has five parts: the number of steps represented by k, which is divided according to processes; the states represented by $x_k$, which is decided by the position of every step; the direction starting from every state represented by $u_k(x_k)$; objective function represented by $d_k(x_k, u_k(x_k))$, which is the distance between two adjacent states; the optimal value function represented by $f_k(x_k)$, which is the shortest distance between $x_k$ and terminal. The basic equations are defined as follows:

$x_{k+1} = u_k(x_k)$ ;

$d_{kn}(x_k, u_k, x_{k+1}, x_{n+1}) = \sum_{j=k}^{n} d_j(x_j, u_j)$ ;

$f_k(x_k) = \min[d_k(x_k, u_k(x_k)) + f_{k+1}(x_{k+1})], k=n,\dots,1$ ;

$f_{n+1}(x_{n+1}) = 0$. [2][3]

The optimal policy $p_{kn}^* = \{u_k^*, \dots, u_n^*\}$ [2] [3] is defined as which has the optimal value of the objective function $d_{kn}$, $p_{kn}^*$ is the optimal policy in whole course. Starting from the first state $x_1^*$ $(=x_1^*)$, the optimal trajectory $\{x_1^*, x_2^*, \dots, x_n^*\}$ is derived from $p_{kn}^*$ and the equation of state transition. Now, the question is defined as follows:

**Definition 5.** In KPCPP, the relatively optimal policy $p_{kn}^* = \{u_k^*, \ldots, u_n^*\}$is defined as the one that has the relatively optimal value of the objective function $d_{kn}$, $p_{1n}^*$ is the relatively optimal policy in whole course. Starting from the first state $x_1^*$ (=$x_1^*$), the relatively optimal trajectory{ $x_1^*$, $x_2^*$, …, $x_n^*$} is derived from $p_{1n}^*$ and the equation of state transition.

**Definition 6.** In KPCPP, we define M as threshold of $d_{kn}$, and the range of M is restricted by W(G), W(G)≥M. Now, the relatively optimal trajectory group meets the demand that $|d_{kn} - M|$ is minimum.

### 2.2 The Description of Problem

Based on the above definitions, we define the issue as follows:

G= ⟨V, A ; W⟩ is an undirected graph, $a_{ij}$ is an arc, whose length is $w_{ij}$, and $w_{ij} \geq 0$. If we start from $v_0$, $v_0 \in V$, and k equals the number of the edges of start vertex, we should walk along k paths at the same time, traveling every arc at least one time, then go back to the start point after completing our own task. The k lines using the least delivery time is called the optimal delivery routes.

## 3 Algorithms and Related Theorems

The standard dynamic programming [3] has obviously divided steps and equation of state transition. However, it's not distinct for most of the problems to divide the steps [8] [9]. To solve the question, we approximately transform it to a standard dynamic programming model by: CAPA and MDPMCA, and then, the new Dynamic Programming algorithm — KMDPA can be used to solve KPCPP.

### 3.1 Change Arc into Point Algorithm(CAPA)

**A CAPA**

Step 0. Convert $a_{ij}$ to the function $a_k(v_i,v_j) = w_{ij}$, $1 \leq k \leq m$,m is the number of arcs of G;

Step 1. k=1, k++, conversion of G→G' is completed until all the arcs which have the common nodes are connected, otherwise, go to step2;

Step 2. Seek the function $a_s(v_i,v_j) = w_{il}$ which has the common node, then connect them to create an arc. Let this arc be $v_{ks}$, the length of which is $e_{ks}$ =0;

Now, we give a new definition to G', G' =⟨ A,V;E⟩, the set of node A={ $a_k(v_i,v_j)|1 \leq i \leq n$ , $1 \leq j \leq n$, $i \neq j$, $1 \leq k \leq m$} , the set of arcs V={$v_{ij}|1 \leq i \leq m$, $1 \leq j \leq m$, $i \neq j$}, and the set of length E={ $e_{ls}$ $|1 \leq l \leq m, 1 \leq s \leq m$, $l \neq s$} ;

We give a sample example for CAPA, Using CAPA, we convert Fig 1 into Fig 2:

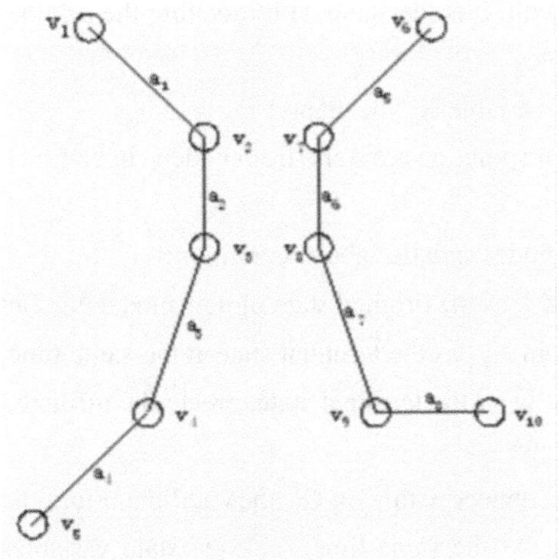

Fig. 1. A sample graph G

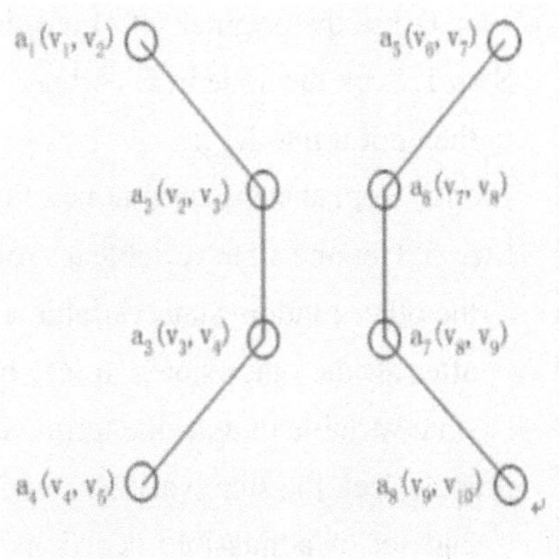


Fig. 2. The conversion graph G'

### B The algorithmic property

**Property 1.** if the algorithm is over, it is impossible in G' to have an arc $v_{ij}$ which is connected by two nodes: $a_i$ and $a_j$, those corresponding arcs in the graph G are not connected, $1 \leq i < m$, $1 \leq j < m$, $i \neq j$.

**Proof.** It is known that $a_i$ and $a_j$ are not connected in G. Suppose an arc $v_{ij}$, $1 \leq i < m$, $1 \leq j < m$, $i \neq j$, satisfy the nodes $a_i$ and $a_j$ in G' to be connected by an arc, when the algorithm is over, according to Step3 in CEAP, there always is a common node between $a_i$ and $a_j$, that is to say, the two arcs in G which are the origins of functions $a_i$ and $a_j$ is connected. It is conflict to the given condition. Therefore, when the algorithm is over, for any created $a_i$ in G', $1 \leq i < m$., if it hasn't the common node between the arcs $a_i$ and $a_j$ in graph G, it is impossible for G' to exist an arc $v_{ij}$, $1 \leq j < m$, $i \neq j$, which connects nodes $a_i$ and $a_j$.

## 3.2 Multistep Decision Process Model Convert Algorithm (MDPMCA)

### A MDPMCA

CAPA converts the arc of G into the point of G', which makes the problem of traveling arcs be one of searching points. This algorithm is universal for G', but not suitable for the multistep decision process, so we give MDPMCA to make this model meet its demand. The involved accords with, it is the same with traveling the original state variable or the affiliated state variables.

Step 0. Let the original set of terminal variable $X_{n+1}$ be empty;

Step 1. Seek the node in G', whose corresponding arc starts from node $v_i$ in graph G, then put it into $X_{n+1}$;

Step 2. Repeat step1 until none of the nodes satisfies above condition;

Step 3. Get one state variable $a_k$ from $X_{n+1}$ as the original state of new model $N_k$. Get the other random state variable $a_j$ from $X_{n+1}$ as the terminal state at the same time; after all the other states in $X_{n+1}$ have been the terminal state, create the affiliated state variable of $a_k$ as the terminal state;

Step 4. Seek the state variables which connect with $a_k$ in G', then add them into the 2nd set of admissible decisions set. At the same time, seek the state variables which connect with the terminal state, then add them into the admissible decisions set before the final decisions set.

Step 5. do step4, until the number of admissible decision sets is 2(m-2);

Step 6. According to the condition of G', connect all the steps. If the affiliated state variable has the same attribute, that is, they have the own parent, connect them with arc. The modeling of $N_k$ is over;

Step 7. repeat step3-6, after every state variable of $X_{n+1}$ has ever been the original state, then, all the models have been created. The algorithm is over.

Thus, all the models $N_k$ have been created by MDPMCA.

## B The algorithmic property

Now we describe KPCPP as follows:

$N_k$ =<A,V; E> is the decision-making model. Let z equal 2m(m-1)+2 ( m is the number of nodes of graph G') , in the model, the parameters are described as: $A=\{a_p(v_i,v_j)|1\le i\le n,\ 1\le j\le n,\ i\ne j,\ 1\le p\le z\}$, $V=\{v_{ij}|1\le i\le z,\ 1\le j\le z,\ i\ne j\}$, $E=\{e_{ls}|1\le l\le z,\ 1\le s\le z,\ l\ne s\}$; Find out the relatively optimal trajectory group which has k trajectory; k equals the number of the state variables of which $v_0$ is the vertex. It is when start from the state of $X_{n+1}$ and then come back to the state of $X_{n+1}$ that the time for traveling all the original state variables is shortest

When algorithm2 is over, according to Step2, $N_k$ has 4 independent admissible decision sets; according to Step4, there exists 2(m-2) admissible decisions sets in $N_k$, and there isn't any access in every set. Now, we know there are 2m independent sets of admissible decision. So, we can divide the model into k steps by the principle of multistep decision where k is 2m-1. **Therefore, we deduce the following property:**

**Property 2.** Any model $N_k$ can be divided into 2m-1 steps.

### 3.3 K Postmen Decision Process Algorithm (KMDPA)

#### A Theorization

**Lemma 1.** $s_w/k$ is always the topmost threshold for any W (G).

**Proof.** Proof by contradiction. Suppose there exists $m > s_w/k$, which satisfies the condition of the topmost limited testing, according to the definition 6, $W(G) \geq km$ is true. When W (G) = $s_w$, that is, all the arcs aren't repetitious and the length of every k paths is equal, there exists a plus L= $s_w/k$ which makes W (G) be kill. According to definition 6, it's right, that is, L can be the threshold. It is known that L>m, so, the suppose is wrong, this completes the proof...

Here, let M be $s_w/k$.

**Theorem 1.** For graph G, the if-and-only-if statement of the relatively optimal trajectory group is that the limit of the max function $d_{kn}$ of this group is $s_w/k$, that is, $|\max(d_{kn}) - s_w/k|$ is minimum.

**Proof.** The if part: If there is a relatively optimal trajectory group for the original graph G, $|\max(d_{kn}) - s_w/k|$ must be minimum, otherwise, we can find a max $d_{kn}$ corresponding to other groups. $d_{kn}$ Satisfies $\lim d_{kn} = s_w/k$, which contradicts definition 6.

The only-if part: If $/\max(d_{kn}) - s_w/k/$is minimum, that is, when $d_{kn}$ which corresponds to the group is max, $\lim d_{kn} = s_w/k$, now, we get $W(G) \leq k\, d_{kn}$, that is, $\lim W(G) = s_w$.If there is another relatively optimal trajectory group for the original graph G, its $/\max(d_{kn}) - s_w/k/$ isn't the minimum, relatively, there exist two conditions in W(G)' of this group:

1) $d_{kn}' > d_{kn}$, for $d_{kn}'$ there exists $d_{kn}$ which uses less time, but the group of $d_{kn}'$ doesn't consume the least time. It contradicts definition 6;

2) $d_{kn}' \bullet d_{kn}$, for $d_{kn}'$, $\lim d_{kn}' \neq s_w/k$, then:

if $d_{kn}' > s_w/k$, it is known that $d_{kn}' < d_{kn}$, so $d_{kn}'$ is closer to $s_w/k$ than $d_{kn}$, which contradicts the known.;

if $d_{kn}' < s_w/k$, it is inexistent according to the lemma 1.

This completes the proof.

We can get a deduce from the theorem 1:

**Deduce:** (the statement of optimizing of KPCPP) For every node of $N_k$, if $a_p(v_i,v_j)$ is the node, suppose $Group_k((a_1,\ldots),\ldots,(a_k,\ldots))$ is the trajectory group which starts from $v_0$ and comes back to $v_0$ in original graph G, the if-and -only-if statement of consuming least time for trajectory group is that, for all the trajectories of the group, $/\max(d_{kn})-s_w/k/$ is less than other trajectory groups.

According to the theorem 1 and its deduce, we present a algorithm KMDPA for KPCPP, which bases on decision-making thought of dynamic programming. Then we will give some properties of this algorithm and prove the correctness of the KMDPA.

**B KMDPA**

Step 0. Let the original set $X_l$ be empty. Pick out state variables from $X_{n+1}$ except $a_k$ and other states which are connected with those ones, then put them into $X_l$;

Step 1. i=1, search the decision-making model $N_k$ which starts from $a_k$;

Step 2. start from $a_k$ which is the start point of the 1st step;

1) if: some state variable of the i+1th set of admissible state has been traveled less than 2 times, and it's not the affiliated state variables of $a_k$, it satisfies $f_k(x_k)=\min[d_k(x_k,u_k(x_k))+f_{k+1}(x_{(k+1)})]$. We choose this state variable to be the next direction from $u_k(x_k)$;

2) if: All states of the i+1th set of admissible state have been traveled 2 times, choose the affiliated state variables to be the next direction from $u_k(a_k)$;

Step 3. if: the common point of $a_k$ and former state variable $a_k$ is the terminal point of $a_k$, then: $e_{kl}= a_k$, note $a_k$ or $a_l$ has been traveled respectively as 1,

else: $e_{kl}=0$, note the time which $a_l$ has been traveled as 1, $d_i= e_{kl} + a_k$;

Step 4. i++; when $i\le 2(m+1)$, do step 2-3, otherwise, go to 5;

Step 5. get the decision which has the $\min d_{kn}$, and find out its trajectory;

Step 6. compare $d_{kn}$ with $s_w/k$, if $d_{kn} = s_w/k$, over,

Else if: the former $d_{kn}=0$, go to 6;else: Compare $d_{kn}$ with the former one:

1) they all $>s_w/k$ or they all $<s_w/k$:

if: There exist some variables which aren't noted in $X_l$,

if: $|\text{this value}-s_w/k| \geq |\text{the former} d_{kn}-s_w/k|$, put the state variable which has been taken out back to the model, and connect the arcs of it., then go to 7;

else, keep this $d_{kn}$ and its trajectory, and then go to 7;

else: after all the state variables of $X_l$ have been noted, keep the $d_{kn}$ which is closer to $s_w$ /k and its trajectory, then go to 8;

2) one $>s_w$ /k, the other $<s_w$ /k:

if: after all the state variables of $X_l$ have been noted, keep the $d_{kn}$ which is closer to $s_w$ /k and its trajectory, then go to 8;

else: there exist the variables which haven't been noted, then go to 7;

Step 7. for $N_k$, take a state variable that has been noted out of $X_l$ and its arcs, and note this state has been taken out. Reset up $N_k$', then do 1-6 for it;

Step 8. judge whether this model group has been traveled entirely,

If: not, keep the trajectory of $d_{kn}$. For the state variables which haven't been traveled, note them as having never been taken out, and revert other state variables as having never been noted, reset the times of traveling, do 1-7;

else: go to 9;

Step 9. judge whether all the groups have been traveled entirely,

If: not, adjust the order of $N_k$, and repeat 1-8; else: go to 10;

Step 10. compare the results of every group and the max $d_{kn}$ of every result, and then select min $d_{kn}$ of them, its trajectory is the relatively optimal trajectory.

### C The proof of the algorithmic correctness

**Theorem 2.** When the algorithm is over, for the relatively optimal trajectory group, ensure the integrity of its points, and, at the same time, satisfy theorem1.

**Proof.** The algorithm is over, if there exist some points which have never been visited, they have surely been taken out and not been traveled during the course of resetting models. According to Step8, We know that every state variable must not be taken out during the course of resetting model if it's not traveled. According to step2-3, they must be traveled during the next course of traveling, so, when the algorithm is over, there isn't any point which hasn't been visited.

When algorithm is over, for relatively optimal trajectory group, if $/\max d_{kn} - s_w/k/$ isn't minimum, according to Step6, we know that the condition of judgment is true, the algorithm can't be over, which contradicts the situation that algorithm is over, so, it satisfies the statement of optimizing of KPCPP. This completes the proof.

### D The proving of the algorithmic validity:

The original graph is given by Fig 3:

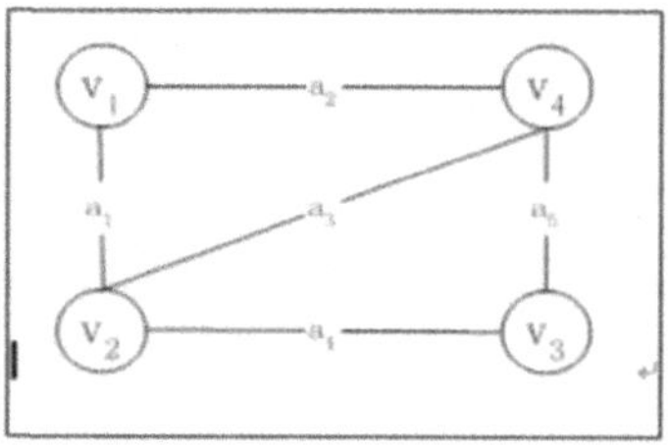


**Fig. 3.** A four points graph

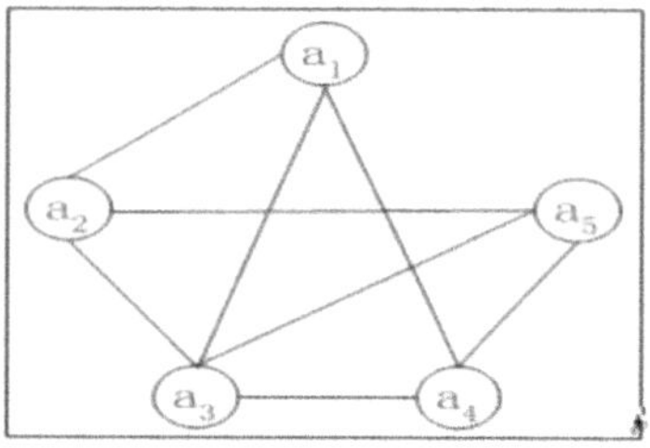


**Fig. 4.** Conversion graph

Using above algorithm system, firstly, the graph is changed into Fig 4. suppose the post office can stay any point, we get different results as Table 1 shows:

**Table 1.** Results from different starting point

| vertex | k | relatively optimal trajectory group |
|---|---|---|
| $v_0$ | 2 | $v_0$-$v_1$-$v_2$-$v_3$-$v_0$, $v_0$-$v_1$-$v_3$-$v_0$ |
| $v_1$ | 3 | $v_1$-$v_0$-$v_3$-$v_1$, $v_1$-$v_3$-$v_1$, $v_1$-$v_2$-$v_3$-$v_1$ |
| $v_2$ | 2 | $v_2$-$v_1$-$v_3$-$v_2$, $v_2$-$v_1$-$v_0$-$v_3$-$v_2$ |
| $v_3$ | 3 | $v_3$-$v_0$-$v_1$-$v_3$, $v_3$ -$v_1$-$v_3$, $v_3$-$v_2$-$v_1$-$v_3$ |

## 4 Conclusions

This paper shows how to solve KPCPP with dynamic programming algorithm. For the first time, it presents an algorithm KMDPA based on dynamic programming, and solves the problem of relatively optimal trajectory of $N_k$ model group. This algorithm system can be used in computer network communication, traffic and transport, etc. It ensures the integrality of paths during the course of model conversion, but it also has the problem that the speed can't be enhanced when the dimension is too much. In the future research, we should pay more attention to this drawback.

## References

1. Bondy, J.A., Murty, U.S.R.: Graph Theory with Applications. The Macmillan Press LTD, London, England (1976)
2. Bellman, R.E, Dreyfus, S.E.: Applied Dynamic Programming. Princeton University Press, Princeton, New Jersey (1962)
3. Bellman, R.E.: Dynamic Programming. Princeton University Press, Princeton, New Jersey (1957)
4. Wang S.: Many Postmen Chinese Postmen Problems, Journal of University of Science and Technology of China( in Chinese) , Beijing, China, (1995) 4, 454~460

5. Edmonds, J., Johnson E.L, Matching.: Enley Tours and the Chinese Postman, Math Programming, (1973) 5, 88-124
6. Even, S.: Graph Algorithms, Computer Science Press, Beijing, China (1979)
7. K.M.Koh, H.H.The.: On Directed PostMan Problem. Nanyang University Journal, Vol Ⅷ &IX (1974/75), 14-25
8. Papadimition, C.H., Steiglitz, K.: Combinatorial Optimization: Algorithms and Complexity, Printice-Hall, New Jersey: U.S.A (1982)
9. Goldberg, D.E.: Genetic Algorithms in Optimization, and Machine Learning. Addison-Wesley. New York, U.S.A (1989)
10. Wang S.T.: Fuzzy Heuristic Search Algorithm FDA* For Fuzzy Multi-stage Decision Problems, Journal of Computer Research and Development (in Chinese), Beijing, China. (1998) 35(7): 652~656

# Choices of Interacting Positions on Multiple Team Assembly

Chartchai Leenawong and Nisakorn Wattanasiripong

Department of Mathematics and Computer Science
King Mongkut's Institute of Technology Ladkrabang
Bangkok, 10520 Thailand
klchartc@kmitl.ac.th, nisakorn_narak@hotmail.com

**Abstract.** This paper proposes a new method for choosing interacting positions that affect team performance on multiple team assembly in an organization. Various approaches for replacing team members are also reviewed and adjusted so that the resulting team obtained will be as effective and efficient as possible. This multiple team assembly is a combinatorial optimization problem that focuses on examining complexity in an organization. The objective of the problem is to achieve maximum performance of the team while at the same time trying to reduce the expected number of replacements and the expected number of trials needed to arrive at that performance level. Computer simulation is used to implement and demonstrate the proposed ideas.

**Keywords:** Combinatorial Optimization, Complexity, Computer Simulation, Organizational Behavior.

## 1 Introduction

The study of complex systems is bringing new vitality to many areas of science. The word "complex systems" is therefore often used as a broad term encompassing a research approach to problems in many diverse disciplines [1][2][9][11] including neuroscience, meteorology, chemistry, physics, computer science, psychology, artificial life, evolutionary computation, economics, and so on.

In general, a *single* complex system is a system consisting of a finite numbers of parts, each of which can be filled by one of the interchangeable components available for that part. The objective of this problem is to achieve the best system. However, the interaction among the components in the system is one difficulty in measuring the "best". To understand the problem more clearly, *the NK model* proposed by Kauffman [2] in chromosome evolution, is adapted [4][10]. A *multiple* complex system is then defined as a system having more than one subsystem, each of which is a complex system itself. The interaction factors in this extended system become the interaction among the components both from within the same subsystem and also from other subsystems. A generalized mathematical model for studying the multiple-complex-system problem is called *the NKC model* [3][5]. Note that both the single and multiple complex-system problems have been proved to be *NP*-complete problems [5][10].

D.-S. Huang, L. Heutte, and M. Loog (Eds.): ICIC 2007, LNAI 4682, pp. 282–291, 2007.

Application of the multiple complex systems can be found in many different areas as well. One example of interest here is a study of multiple team assembly in an organization with an objective of accomplishing highest performance possible.

In this paper, to make the model more realistic, a new method for choosing the interacting positions is proposed. Computer simulation is used to show the effects of the way the interacting positions are chosen on multiple team assembly. More precisely, the new and existing methods are applied to the *NKC* model using also various replacement heuristics algorithms previously proposed [6][8] for replacing a current team member with the other candidate for that position. Simulation results on the performance of the team, the expected number of replacements and the expected number of trials to a local maximum team are to be presented.

From a managerial viewpoint, the *NKC* model and all of the methodology for replacing the team members and for selecting the interacting positions can be interpreted in the following way. Not surprisingly, every team manager would like the team to be most effective. By interchanging the team members, the *NKC* model attempts to search for a team with the performance as good as it can be under several limitations. At the same time, the team manager would also favor keeping the costs and efforts of obtaining such a team as low as possible. Those costs and efforts are reflected in the expected number of replacements and the expected number of trials because the former involves the process of firing a team member and hiring a replacement one whereas the latter involves the process of interviewing the candidates for each job position. Hence it is useful to examine different methods contributing to the efficiency of those processes.

A review of the *NKC* model for studying multiple teams together with different replacement algorithms is given in Section 2. A proposed method for choosing the interacting positions and the modification of some replacement algorithms are presented in Section 3. Computer simulation results and their discussions are followed in Section 4. Finally, conclusions of this work are provided in Section 5.

## 2 The *NKC* Model and Replacement Heuristics Algorithms

It is assumed, throughout this paper, that a multiple team consists of two subteams. The *NKC* model tries to find a team that has the best overall performance. Let a pair of two binary *N*-vectors (**x**,**y**) represent a multiple team where **x** is one feasible subteam and **y** is the other feasible subteam. In general, for position $i$ of both subteams **x** and **y**, there are $2^{K+C+1}$ possible combinations of choices for the team members at the $K$+$C$+1 positions that affect the contribution of the team member in position $i$.

The value of the contribution to performance of team **x** is defined as $f_i(x_i^K, y_i^C)$ and of team **y** as $f_i(y_i^K, x_i^C)$. Each value is chosen from a list of $2^{K+C+1}$ uniform 0–1 random numbers that corresponds to the combination of team members in position $i$, the $K$/2 positions on either side of position $i$ in the same subteam, and $C$/2 positions on either side of position $i$ in the other subteam. The performance of subteam **x** affected by subteam **y**, $f(\mathbf{x}^K, \mathbf{y}^C)$, is then an average of these contributions. Similarly, the performance of team **y** affected by team **x**, $f(\mathbf{y}^K, \mathbf{x}^C)$, is also an average of the corresponding contributions. The overall performance, $f(\mathbf{x},\mathbf{y})$, of the multiple team $(\mathbf{x},\mathbf{y})$ is then an average of the average performances of the two subteams as follows:

$$f(\mathbf{x},\mathbf{y}) = \frac{f(\mathbf{x}^K,\mathbf{y}^C) + f(\mathbf{y}^K,\mathbf{x}^C)}{2}. \quad (1)$$

It was shown, by reduction, that the *NKC* problem is also an *NP*-complete problem [5]. Computer experiments using $C^{++}$ programming were conducted to study the effects of the interaction among the team members. The results show that the *complexity catastrophe* still exists in the *NKC* model.

The replacement algorithms previously used in determining which order the team members should be replaced in the *NKC* model are divided into two groups. One is with no effects from the interaction among the team members [6] and the other is with the effects [8]. All of the replacement algorithms are briefly explained here.

*Optimal Performance Policy (OPP).* In each search for a better subteam, the algorithm tries to obtain the best subteam chosen from among the current subteam and all of its corresponding neighboring subteams.

*Random Improvement Policy (RIP).* The algorithm randomly obtains a new subteam with better performance than the current subteam. Similar to OPP, this approach considers every single position of the subteam and then randomly identifies the position to be replaced which results in higher performance.

*First Come First Serve (FCFS).* This approach considers each position in a given order of the subteam of interest, the first position that causes improved performance when replaced with the other candidate will be chosen and the algorithm then moves to the other subteam.

*Sorted First Come First Serve (S/FCFS).* This approach is similar to FCFS except that the positions are first reordered in increasing order according to their individual contributions, after which S/FCFS is applied.

*Sorted First Come First Serve based on K (SK/FCFS).* This approach as well as the next one takes into accounts the interaction effects. In this approach, it still involves reordering the positions of the current subteam but not just by their individual contributions. The approach reshuffles the positions in the order of the total contributions, each of which is a summation of the individual contribution of the considered position and all contributions of those positions on the same subteam that are affected by that considered position. After that, FCFS can again be applied.

*Sorted First Come First Serve based on K and C (SKC/FCFS).* This approach is similar to SK/FCFS except that it now adds into each previously-defined total contribution the contributions of those positions from the other subteam that affect the concerned position.

For each value of *N*, *K*, and *C*, 500 randomly generated problems are generated using $C^{++}$ programming. The results show that all of these different replacement algorithms do not have any significant impact on the expected performance of a local maximum team. As for the expected number of replacements to a local maximum, the best replacement algorithm is OPP and the worst is FCFS. Last but not least, in terms of the expected number of trials to a local maximum team, the best replacement algorithm is SKC/FCFS and the worst is RIP.

# 3 A Proposed Method for Choosing the Interacting Positions

In the *NKC* model, the current method of choosing the interacting positions is based on the neighboring positions of the concerned position. More specifically, the numbers of interacting positions both from within the same subteam ($K$) and from the other subteam ($C$) are split in half to the left and the right sides (rounded up to the left in case of odd numbers) of the considered position. For future references, this method will be called the Left-Right method (LR).

However, in a more realistic scenario, the interacting positions may be chosen from any other positions. The proposed method will employ this idea by randomly choosing the interacting positions, and hence be called the RANDOM method. In this method, the interacting positions within the same subteam can be any positions *other than* the concerned position whereas the interacting positions from the other subteam can be any arbitrary positions.

In RANDOM, some modifications on certain replacement algorithms, namely, SK/FCFS and SKC/FCFS will be needed because now the numbers of positions affected by each considered position may not be equal as they are in LR. Comparing the total contributions of all positions calculated by the summation of all associated positions would not be fair. The details of the modified SK/FCFS and SKC/FCFS algorithms are presented now.

## 3.1 SK/FCFS – An Average Approach (or SK/average)

This replacement approach is modified from SK/FCFS. It still involves reordering the positions of the current subteam but is in increasing order of $h_i(\mathbf{x})$, an average over contributions of the team member in position $i$ and all contributions of those team members in the positions in the same subteam that are affected by that team member. Afterwards, the first-come first-serve rule can be applied to the team.

In particular, at first, the team members of the current subteam $\mathbf{x}$ are sorted in increasing order of their average contributions defined above. Note that all of the positions in subteam $\mathbf{y}$ will be reordered accordingly as well. Then, sequentially consider replacing the team members based on this order in an attempt to find a first subteam $\mathbf{x}'$ with $f(\mathbf{x}',\mathbf{y}) > f(\mathbf{x},\mathbf{y})$. Repeat the process for subteam $\mathbf{y}$ and continue in this manner until a local optimal team is reached.

For each position $i = 1, 2, \ldots, N$ in a given subteam $\mathbf{x}$, let

$$h_i(\mathbf{x}) = [f_i(x_i^K, y_i^C) + \Sigma f_j(x_j^K, y_j^C)]/[\text{Number of affected positions} + 1] \quad (2)$$

where $j$ = the positions in subteam $\mathbf{x}$, affected by position $i$.

## 3.2 SKC/FCFS – An Average Approach (or SKC/average)

Similar to SK/average, this approach is modified from the original SKC/FCFS. The only change is similar to that change in SK/average, that is, the total contributions used in the reordering process are now the averages over the contributions of the affected positions both within the same subteam and in the other subteam. Note again that the positions in the other team will be reordered accordingly.

## 4 Computer Simulation Results and Discussions

Computer simulation has long been used to project the behavior of organizations too complex for analytical calculation [12]. Although increasingly important, modeling organization performance is more difficult than modeling individual performance because of the complexities and dynamics inherent in organization performance.

In this section, computer simulation results of the multiple-team assembly problem when the new method of choosing the interacting positions, namely, RANDOM, is used are presented. It is applied on each and every replacement algorithm previously stated. To observe the effectiveness and the efficiency of the problem, three characteristics resulted from each replacement approach will be investigated. They are the expected performance of a local maximum team, the expected number of replacements needed to reach a local maximum, and the expected number of trials needed for a local maximum.

These computer simulations are conducted using $C^{++}$ programming. For a fixed team size $N$ = 40, some fixed amount of internal interaction $K$, and the amount of external interaction $C$ varying from 0 to $N$–1, 500 independent problems are generated randomly. Moreover, for a fixed team size $N$ = 40, some fixed amount of external interaction $C$, and the amount of internal interaction $K$ varying from 0 to $N$–1, another set of 500 independent problems are generated randomly. The computer simulation results are now presented with regard to each problem characteristic mentioned above.

### 4.1 The Expected Performance of a Local Maximum Team

When RANDOM is used in the multiple team setting for various replacement algorithms including the two modified ones, the expected performances of local maximum teams are shown in Fig. 1 as a function of $C$ when $K$ is fixed at the value of 0. The results imply that, for a large team, the complexity catastrophe still exists in all of the replacement algorithms used here even though the SKC/FCFS, SK/average, and SKC/average curves show a slightly slower decrease in the expected performance as $C$ increases.

In particular, when $K$ = 0 and $C$ = 0, the expected performance is approximately 0.66. As $C$ increases toward $N$–1, the performance decreases, theoretically, to 0.5 in Fig. 1 for a larger team [5]. The patterns of the curves for other values of $N$, $K$, and $C$ are comparable to Fig. 1, although they are not shown in this paper. In addition, the conclusions drawn in this section are similar to those in [7] when LR was the method for choosing the interacting positions instead. Other selection methods may be needed if the objective is to reduce the complexity catastrophe.

### 4.2 The Expected Number of Replacements to a Local Maximum Team

In terms of the expected number of replacements to a local maximum, this value is shown in Fig. 2 to Fig. 4 as a function $C$ or $K$ for different replacement algorithms when RANDOM is used. Note that a qualified replacement is when a current subteam is replaced with one of its neighbors, keeping the other subteam unchanged. The process repeatedly alternates between the two subteams until a local optimum is reached. The lower this value gets, the more efficient the algorithm is.

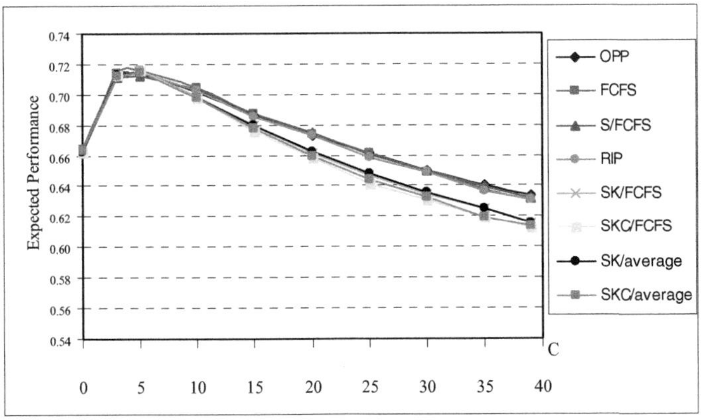


**Fig. 1.** The expected performance of a local maximum as a function of $C$ for different replacement algorithms when $N = 40$, $K = 0$, and RANDOM is used

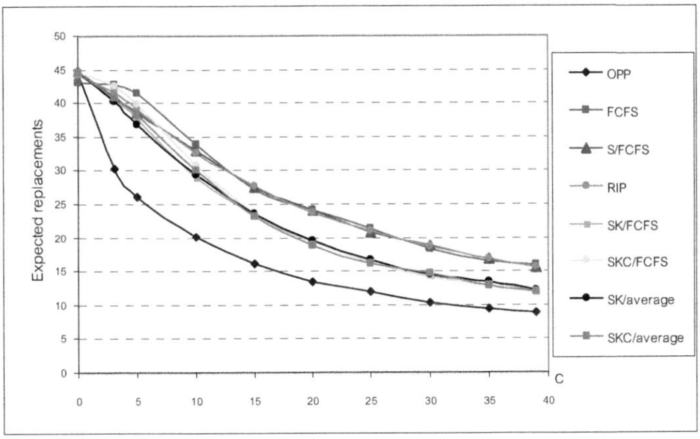


**Fig. 2.** The expected number of replacements to a local maximum as a function of $C$ for different replacement algorithms when $N = 40$, $K = 0$, and RANDOM is used

Fig. 2 indicates that when there is no interaction between the team members in the same subteam, OPP is the most preferred replacement algorithm. FCFS, S/FCFS, and RIP are among the least efficient algorithms. Though not most efficient, all the four interaction-based replacement algorithms are relatively good. Comparatively, when there is no external interaction, though the results are not shown in this paper, the curves are still declining. The only difference is that the SK/FCFS and SKC/FCFS curves are shifted up to the least efficient group.

Fig. 3 and Fig. 4 show the expected numbers of replacements to a local maximum as a function of $K$ or $C$, either of which is fixed at a positive value. The two figures can lead to similar conclusions of the previous cases. In summary, according to the expected number of replacements, OPP is the most efficient algorithm because, at each iteration, it moves from a current subteam to one of its neighbors that gives the

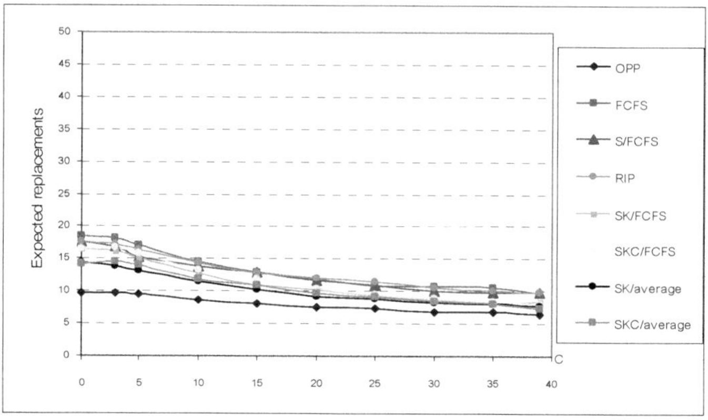


**Fig. 3.** The expected number of replacements to a local maximum as a function of $C$ for different replacement algorithms when $N = 40$, $K = 20$, and RANDOM is used

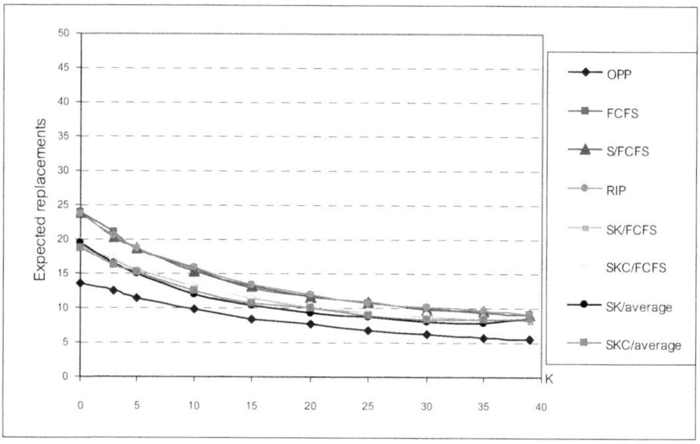


**Fig. 4.** The expected number of replacements to a local maximum as a function of $K$ for different replacement algorithms when $N = 40$, $C = 20$, and RANDOM is used

highest performance. On the contrary, all other replacement algorithms may not select the best next subteam from among all the current subteam's neighbors. As opposed to LR already examined in [7], RANDOM has only little effects on this efficiency indicator especially after some modification on the interaction-based replacement algorithms.

### 4.3 The Expected Number of Trials to a Local Maximum Team

As explained earlier in Section1, this value is another efficiency indicator. It is counted when the algorithm considers replacing the team member in a position with the other candidate available for that position. When looked at the overall team, each and every *different* trial will be counted as one. Note that the values reported in this section are the expected values of the *total* numbers of trials needed for achieving a local maximum.

Fig. 5 reports the expected total number of trials as a function of $C$, when $N = 40$, $K = 0$, and RANDOM is used in the process of choosing the interacting positions, for different replacement algorithms. It reveals that at the beginning when external interaction is low, the numbers of trials of OPP and RIP are relatively high compared to those of other algorithms. Nonetheless, as external interaction increases, these values seem to fall down faster. Similarly, for the case when there is no external interaction, results comparable to Fig. 5 can be obtained for the expected number of trials as a function of $K$.

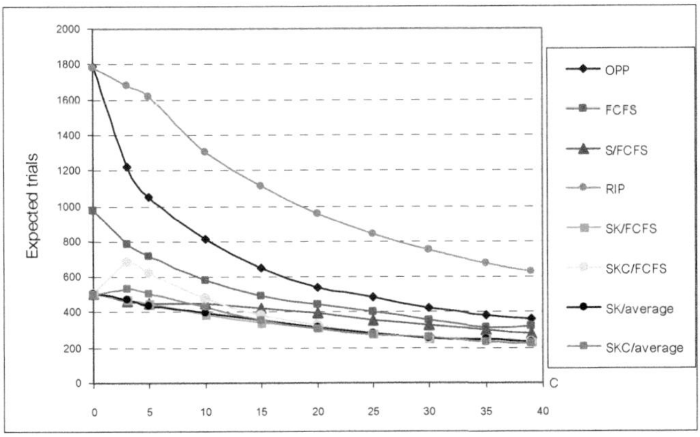


**Fig. 5.** The expected number of trials to a local maximum as a function of $C$ for different replacement algorithms when $N = 40$, $K = 0$, and RANDOM is used

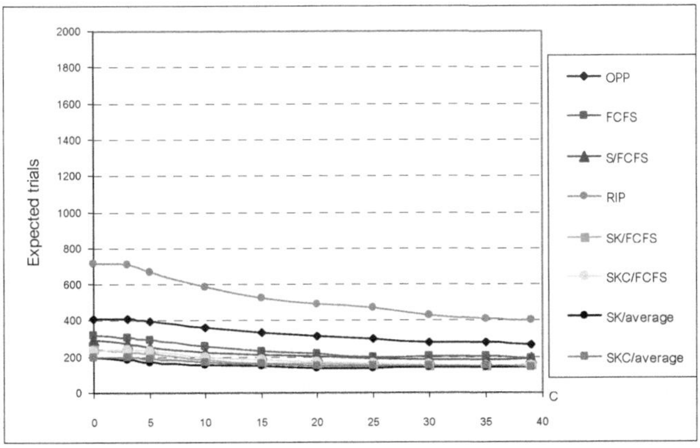


**Fig. 6.** The expected number of trials to a local maximum as a function of $C$ for different replacement algorithms when $N$=40, $K$=20, and RANDOM is used

In Fig. 6 and Fig. 7, when both internal and external interactions are present, the patterns of the curves for all the replacement algorithms are similar to Fig. 5 but it is

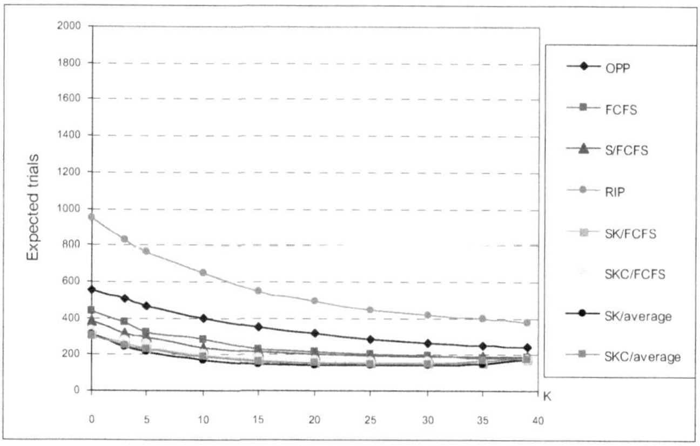


**Fig. 7.** The expected number of trials to a local maximum as a function of *K* for different replacement algorithms when *N*=40, *C*=20, and RANDOM is used

clearer now that the two modified interaction-based algorithms, namely, SK/average and SKC/average outperform all other algorithms including their associated original ones.

In summary, except OPP and RIP, the values of this expected number of trials for all algorithms indicate insignificant differences especially when the amount of interaction is high. The OPP and RIP curves being somewhat away from the others conform to the fact that OPP and RIP have to spend so much time checking every position before finally making the decision for a replacement.

## 5 Conclusions

This paper has proposed a new method for choosing the interacting positions that affect the individual contribution to team performance in the multiple-team assembly problem with an effort of making the model more realistic. The RANDOM method used in the paper arbitrarily select the internal and the external interacting positions based on the values of *K* and *C*, respectively. This proposed method is applied to the *NKC* model using various algorithms in the process of replacing team members to achieve a more effective team. In addition, due to the random nature of this proposed method, the numbers of positions affected by a position of interest are unequal.

More appropriately, therefore, proposed modifications on some replacement algorithms, especially the ones with interaction effects, namely, SK/FCFS and SKC/FCFS have been presented. Computer simulation is used in implementing the proposed ideas. The simulation results on the effectiveness and efficiency aspects of the local maximum team obtained have been reported. This is a proof that the *NKC* model for studying multiple team assembly is robust. However, the two methods of choosing the interacting positions used in this paper has only one-way effect, i.e., they affect the performance contribution of the considered position but not the other way around. Hence, it is natural to assume such two-way effect for future research work.

## References


1. Derrida, B.: Random-Energy Model an Exactly Solvable Model of Disordered Systems. Physical Review B. 24 (1981) 2613–2620
2. Kauffman, S.A.: The Origins of Order. Oxford University Press, Oxford (1993)
3. Kauffman, S.A., Johnsen, S.: Convolution to the Edge of Chaos: Coupled Fitness Landscapes, Poised States, and Coevolutionary Avalanches. Journal of Theoretical Biology. 149 (1991) 476–505
4. Leenawong, C.: On Modeling a Complex System with Interacting Components. KMITL Science Journal. 3 (2003) 107–115
5. Leenawong, C., Maneechai, S.: Combinatorial Optimization Model for Studying Multiple Complex Systems. Proceedings of the International Conference on Computing, Communications and Control Technologies, Austin, TX. (2004) 88–96
6. Leenawong, C., Wattanasiripong, N.: Replacement Algorithms for the Multiple Complex-System Model, KMITL Science Journal. 5 (2005) 329–338
7. Leenawong C., Wattanasiripong, N.: Simulations of Interaction-based Replacement Algorithms for the Multiple Complex System Model. Proceedings of the 2006 International Conference on Business, Honolulu, HI. (2006) 1732–1740
8. Leenawong, C., Wattanasiripong, N., Netisopakul, P.: Interaction-Based Algorithms for Replacing Components in the Multiple Complex-System Model. Proceedings of the International Technical Conference on Circuits/Systems, Computers and Communications, Jeju, Korea. (2005) 1143–1144
9. Levinthal, D.A.: Adaptation on Rugged Landscapes. Management Science. 43 (1997) 934–950
10. Solow, D., Burnetas, A.N., Tsai, M., Greenspan, N.: On the Expected Performance of Systems with Complex Interactions among Components. Complex Systems. 12 (2000) 423-456
11. Westhoff, F.H., Yarbrough, B.V., Yarbrough, R.M.: Complexity, Organization, and Start Kauffman's the Origins of Order. Economic Behavior and Organization 29 (1996) 1–25
12. William, B., Kenneth, R.: Organization Simulation. Wiley-Interscience, Hoboken New Jersey (2005)

# Genetic Local Search for Optimum Multiuser Detection Problem in DS-CDMA Systems

Shaowei Wang and Xiaoyong Ji

Department of Electronics Science and Engineering, Nanjing University, Nanjing, Jiangsu, 210093, P.R. China
{wangsw,jxy}@nju.edu.cn

**Abstract.** Optimum multiuser detection (OMD) in direct-sequence code-division multiple access (DS-CDMA) systems is an NP-complete problem. In this paper, we present a genetic local search algorithm, which consists of an evolution strategy framework and a local improvement procedure. The evolution strategy searches the space of feasible, locally optimal solutions only. A fast iterated local search algorithm, which employs the proprietary characteristics of the OMD problem, produces local optima with great efficiency. Computer simulations show the bit error rate (BER) performance of the GLS outperforms other multiuser detectors in all cases discussed. The computation time is polynomial complexity in the number of users.

## 1 Introduction

In direct-sequence code-division multiple access (DS-CDMA) communication systems, transmitters multiply each user's signal by a distinct code waveform. Detectors receive a signal composed of the sum of all active users' signals, which overlap in time and frequency. A particular user's signal is detected by correlating the entire received signal with that user's code waveform without regard for the other users, which inevitably yields multiple access interference (MAI) at the output of matched filter. MAI is the main factor limiting performance in DS-CDMA systems. While optimum multiuser detection (OMD) [1] scheme is the most promising technique for mitigating MAI, its computational complexity increases exponentially with the number of active users [1], which leads to its implementation impractical.

The OMD is based on the maximum-likelihood sequence-estimation rule and searches exhaustively for all possible combinations of the users' entire transmitted bit sequence that maximizes the log-likelihood function [1] related to the outputs of matched filters. For an asynchronous DS-CDMA system which has $K$ active users and the packet size of each user is $M$, the possible bit sequence combinations are $2^{MK}$. The computational complexity of the OMD can be reduced to $2^K$ by exploiting the Viterbi algorithm [1], but it still increases exponentially with the number of active users. From a combinatorial optimization viewpoint, the OMD problem is NP-complete [2]. Due to the exponential computational complexity of the OMD, some researchers have concentrated their effort on designing heuristics

D.-S. Huang, L. Heutte, and M. Loog (Eds.): ICIC 2007, LNAI 4682, pp. 292–299, 2007.

yielding sum-optimal solutions which can satisfy the practical demand. Earlier works on applying heuristics to the OMD problem can be found in [3-7].

In this paper, we propose a genetic local search (GLS) algorithm [8] for the OMD problem. The GLS consists of the application of genetic operators to a population of local optima produced by a special local search procedure. The process is iterated until a maximal number of generations is reached. Simulation results show that the GLS based multiuser detector can converge to the (near) optimum solution rapidly. The bit error rate (BER) performance of the GLS is superior to other heuristic multiuser detectors in all cases considered. The remainder of this paper is organized as following. Section 2 introduces the DS-CDMA system model and constructs the OMD problem. The GLS based multiuser detector is described in section 3. In section 4, all simulation results are given with comparison between the proposed GLS and other detectors, followed by a short conclusion in section 5.

## 2 System Model and Problem Formulation

Assume a binary phase shift keying (BPSK) transmission through an additive-white-Gaussian-noise (AWGN) channel shared by $K$ active users with packet size $M$ in an asynchronous DS-CDMA system. The real-valued base-band signal received can be expressed as [9]

$$r(t) = \sum_{k=1}^{K} A_k \sum_{m=1}^{M} b_k(m) s_k(t - mT_b - \tau_k) + n(t) \tag{1}$$

where $A_k$ is the signal amplitude of the $k$th user, $b_k(m)$ is the $m$th transmitted bit of the $k$th user, $s_k(t)$ is the normalized signature waveform of the $k$th user, $T_b$ is the bit duration, $\tau_k \in [0, T_b]$ is the transmission delay of the $k$th user, $n(t)$ is the white Gaussian noise with power spectral density $N_0/2$. Without loss of generality, the transmission delay $\tau$ is assumed to satisfy $0 = \tau_1 < \tau_2 < \ldots < \tau_K$ and $(\tau_k - \tau_{k-1}) = T_c$, where $T_c$ is chip duration time. The sufficient statistics for demodulation of the transmitted bits $\mathbf{b}$ are given by the $MK$ length vector generated by matched filter banks [10]

$$\mathbf{y} = \mathbf{RAb} + \mathbf{n} \tag{2}$$

where $\mathbf{y} = [y_1(1), y_2(1), \ldots, y_K(1), \ldots, y_1(M), y_2(M), \ldots, y_K(M)]^T$, $\mathbf{b} = [b_1(1), b_2(1), \ldots, b_K(1), \ldots, b_1(M), b_2(M), \ldots, b_K(M)]^T$. $\mathbf{A}$ is the $MK \times MK$ diagonal matrix whose $k + iK$ diagonal element is the $k$th user's signal amplitude $A_k$ and $i = 1, 2, \ldots, M$. $\mathbf{R} \in \mathbb{R}^{MK \times MK}$ is the signature correlation matrix and can be written as

$$\mathbf{R} = \begin{bmatrix} \mathbf{R[0]} & \mathbf{R}^T\mathbf{[1]} & \mathbf{0} & \cdots & \mathbf{0} & \mathbf{0} \\ \mathbf{R[1]} & \mathbf{R[0]} & \mathbf{R}^T\mathbf{[1]} & \cdots & \mathbf{0} & \mathbf{0} \\ \mathbf{0} & \mathbf{R[1]} & \mathbf{R[0]} & \cdots & \mathbf{0} & \mathbf{0} \\ \vdots & \vdots & \vdots & \vdots & \vdots & \vdots \\ \mathbf{0} & \mathbf{0} & \mathbf{0} & \cdots & \mathbf{R[1]} & \mathbf{R[0]} \end{bmatrix} \tag{3}$$

where $\mathbf{R[0]}$ and $\mathbf{R[1]}$ are $K \times K$ matrices defined by

$$R_{jk}[0] = \begin{cases} 1, & \text{if } j = k; \\ \rho_{jk,} & \text{if } j < k; \\ \rho_{kj,} & \text{if } j > k; \end{cases} \tag{4}$$

$$R_{jk}[1] = \begin{cases} 0, & \text{if } j \geq k; \\ \rho_{jk,} & \text{if } j < k. \end{cases} \tag{5}$$

$\rho_{jk}$ denotes the partial crosscorrelation coefficient between the $j$th user and the $k$th user. $\mathbf{n}$ is a real-valued zero-mean Gaussian random vector with a covariance $\frac{N_0}{2}\mathbf{H}$.

The optimum multiuser detection problem is to generate an estimation sequence $\hat{\mathbf{b}} = [\hat{b}_1(1), \hat{b}_2(1), ..., \hat{b}_K(1), ..., \hat{b}_1(M), \hat{b}_2(M), ..., \hat{b}_K(M)]^T$ to maximize the objective function

$$f(\mathbf{b}) = 2\mathbf{y}^T\mathbf{Ab} - \mathbf{b}^T\mathbf{Hb} \tag{6}$$

$\mathbf{H} = \mathbf{ARA}$. It means to search $2^{MK}$ possible bit sequences exhaustively and is an NP-complete problem [2].

Obviously a synchronous DS-CDMA system can be seen as a special case of an asynchronous one (for the case $M = 1$). On the other hand, an asynchronous DS-CDMA system can be interpreted as an equivalent synchronous system too [9]. In the following we only consider the synchronous case to simplify analysis without loss of generality.

## 3 Genetic Local Search for the OMD Problem

Evolutionary algorithms (EAs), such as genetic algorithm, evolutionary programming and evolution strategy, are known to be robust optimization techniques that have already been successfully applied to many combinatorial optimization problems. Previous works in multiuser detection domain are found in [3-5]. In [3] and [4], genetic algorithms are used to detect transmission sequences. In [5], an evolutionary programming based multiuser detector is proposed and shows better performance than genetic algorithm. On the other hand, some local search methods have been used to solve the OMD problem too. In [6], a gradient guided search algorithm, which is essentially a 1-opt local search, is proposed. An efficient k-opt local search algorithm is given in [7]. They show lower computational complexity than that of the EAs.

**Procedure GLS**

**Initialization:** $\mathbf{b} = \mathbf{b}_{\text{initial}} = sign(\mathbf{y}) \in \{-1,+1\}^K$ ;

$t := 0$ ;

**Repeat:**

$f(\mathbf{b}) = 2\mathbf{b}^T\mathbf{A}\mathbf{y} - \mathbf{b}^T\mathbf{H}\mathbf{b}$ ;

**for** $i = 1, 2, ..., \lambda$

$\mathbf{b}_i := sign(\mathbf{b} + \mathbf{N}(0, \sigma^2))$ ;

$f(\mathbf{b}_i) = 2\mathbf{b}_i^T\mathbf{A}\mathbf{y} - \mathbf{b}_i^T\mathbf{H}\mathbf{b}_i$ ;

**endfor**

$f(\mathbf{b}_i) = \max\{f(\mathbf{b}_1), f(\mathbf{b}_2), \cdots, f(\mathbf{b}_\lambda)\}$ ;

Perform FILS on $\mathbf{b}_i$ and produce local optima $\mathbf{b}_{\text{opt}}$ ;

**if** $f(\mathbf{b}_{\text{opt}}) \geq f(\mathbf{b})$

$\mathbf{b} := \mathbf{b}_{\text{opt}}$ ;

**else**

$\mathbf{b} := \mathbf{b}$ ;

**endif**

$t := t + 1$ ;

**Until Pre-assigned number of iteratons;**

**Return:** $\mathbf{b}$ .

**Fig. 1.** Procedure of the GLS algorithm, $\mathbf{N}(0, \sigma^2)$ represents a Gaussian random variable with mean 0 and variance $\sigma^2$ , $\lambda$ is the offspring population size

Generally, EAs based mutluser detectors can approach the OMD bound [5] when the number of users is relatively small, but the computational complexity of these detectors is much higher than other suboptimum algorithms, such as the multistage detector (MSD) [11]. Local search multiuser detectors require computation of $O(n^2)$ , which is much lower than EAs based ones, but its BER performance decreases dramatically when the active users increase. For example, the BER performance of the k-opt detector [7] decreases about 3dB when the number of users increases from 10 to 20.

It is reasonable to combine EAs with local search to achieve a compromise between the computational complexity and the BER performance. Here we propose a genetic local search (GLS) [8] multuser detector. The framework of the GLS proposed is $(1+\lambda)$ evolution strategy (ES) [12]. After each generation, the $(1+\lambda)$ ES selects the best offspring as the current solution and a fast iterated local search (FILS) is performed on it to obtain local optima. The main procedure of the GLS is shown in Fig.1.

In principle, any local search algorithm can be applied in GLS, but the performance of the GLS algorithm with respect to solution quality and computation speed strongly depends on this choice. It has been widely accepted that the specific information of a problem can speedup random search algorithm dramatically. The OMD problem differs from general combinatorial optimization problems in the following characteristics. First, simple multiuser detectors, such as conventional matched filters, can provide solutions close to the optimum in most of the cases in DS-CDMA systems. Second, the epistasis [13] of the objective function given in equation (6) is weaker. In other words, each bit of a possible solution contributes its own part to the fitness of the solution almost independently [14].

Based on the first characteristic, we can take the output of conventional detector as the initial solution to speedup the search process. The second characteristic indicates that greedy strategy can be efficient to exploit the fitness landscape [15] of the OMD problem. Here we propose a fast iterated local search (FILS) to produce local optima, which employs greedy strategy and flips a bit as long as the associated gain in improvement occurs. The basic procedure of iterated local search can be found in [16]. The details of the FILS are following.

Denote the current solution vector at the $t$th generation of iterated local search by $\mathbf{b}_0^t = [b_1^t, b_2^t, ..., b_n^t]^T$, where $n = K$. $\mathbf{b}_j^t$ is the solution with only the $j$th bit different from $\mathbf{b}_0^t$, where $\mathbf{b}_j^t = [b_1^t, ..., -b_j^t, ..., b_n^t]^T$, the associated gain $g_j^t$ from $\mathbf{b}_{j-1}^t$ to $\mathbf{b}_j^t$

$$g_j^t = f(\mathbf{b}_j^t) - f(\mathbf{b}_{j-1}^t),\ 1 \le j \le n \tag{7}$$

By flipping $b_1^t$ of the $\mathbf{b}_0^t$, the greedy local search begins and the current solution $\mathbf{b}_1^t$ is created, $\mathbf{b}_1^t = [-b_1^t, b_2^t, ..., b_n^t]^T$. While FILS running, the current solution vector $\mathbf{b}_j^t$ is updated as

$$\mathbf{b}_j^t = \begin{cases} \mathbf{b}_j^t, & g_j^{t-1} > 0 \\ \mathbf{b}_{j-1}^t, & \text{otherwise} \end{cases},\ 1 \le j \le n. \tag{8}$$

When the $n$th associated gain is calculated, the local optimal solution $\mathbf{b}_n^t$ is produced and taken as the current solution of the next generation, $\mathbf{b}_0^{t+1} = \mathbf{b}_n^t$. The local search terminates until there is no associated gain after $n$ flips in a generation. The procedure of FILS is illustrated in Fig.2.

Unlike other local search algorithms, such as gradient guided [6] and k-opt [7] local search, searching for a flip with the highest associated gain in improvement in each iteration, the proposed FILS flips a bit as long as the positive associated gain of this bit exists. The advantage of this method is that there is no need to search all neighborhood of the current solution exhaustively to determine flipping a bit or not, as the gradient guided search and k-opt algorithms do. The FILS also differs from the general ILS. It takes the local optimal found in the previous iteration as the start search point of the current generation, not performing a perturbation on the current optimal.

**Procedure FILS**

**Initialization:** $\mathbf{b} := \mathbf{b}_i$ ;

**Repeat:**

$\mathbf{b}_0^t := \mathbf{b} = \left[b_1^t,\ldots,b_i^t,\ldots,b_n^t\right]^T$ ;

**for** $i = 1, 2, \ldots, n$

  Let $\mathbf{b}_i^t = [b_1^t,\ldots,-b_i^t,\ldots,b_n^t]^T$ ;

  Calculate gain $g_i^t := f(\mathbf{b}_i^t) - f(\mathbf{b}_{i-1}^t)$ ;

  **if** $g_i^t > 0$

    $\mathbf{b}_i^t := \mathbf{b}_i^t$ ;

  **else**

    $\mathbf{b}_i^t = \mathbf{b}_{i-1}^t$ ;

  **endif**

**endfor**

$\mathbf{b} := \mathbf{b}_n^t$ ;

**Until no associated gain after** $n$ **flips;**

**Return:** $\mathbf{b}$ .

**Fig. 2.** Procedure of the FILS algorithm

# 4 Simulation Results

Consider a synchronous DS-CDMA system with perfect power control and random binary sequences with length $L = 127$ are employed as spreading sequences. The outputs of the conventional detector are used to initialize the start solutions of the GLS.

The BER performance of the conventional detector (CD), EP [5], MSD [11], k-opt [7], and the proposed GLS is illustrated in Fig.3(a) and (b) by the curves of BER versus SNR. The number of users is 30 and 40 respectively. Because of the limitation of computational time, the population size of EP is set to 60 (30 users) and 100 (40 users) and runs for 30 generations. Since there is no improvement after 10 stages for the MSD, the MSD runs for 10 generations. The k-opt local search is carried out as [7] described. From Fig.3 we can see that the proposed GLS obviously outperforms the CD, EP, MSD and k-opt in two cases discussed. The performance of CD is very poor because the MAI is heavy, especially in the case $K = 40$ . The EP detector

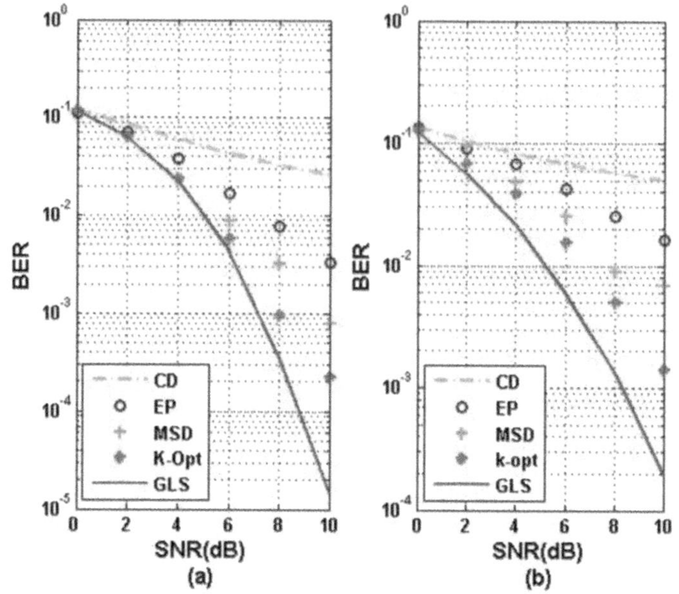


**Fig. 3.** BER as a function of SNR for: (a) $K = 30$ and (b) $K = 40$

performs poor because of the small population size and number of iterations. The k-opt is inferior to GLS because it can't step out local optima effectively.

The computational complexity of the GLS is estimated by curve fitting techniques. A personal computer with 2.66-GHz CPU and 512MB of RAM is used to perform all procedures under MATLAB programming environment. The average CPU time is approximated as

$$C_{\text{OMD}} = 2.32\times10^{-4}2^{K} \quad (9)$$

$$C_{\text{GLS}} = 4.71\times10^{-3}K^{3} \quad (10)$$

Additionally, the associated gain ($g_j^t = f(\mathbf{b}_j^t) - f(\mathbf{b}_{j-1}^t)$) of flipping the $j$th bit of $\mathbf{b}_j^t$ for the FILS can be calculated by an efficient method proposed in [17]. Instead of recalculating the fitness function, the associated gain can be updated by only calculating the difference of the gains. So the computational complexity of the GLS can be reduced more in this way.

## 5 Conclusions

We propose a genetic local search algorithm for the optimum multiuser detection problem in DS-CDMA systems. The GLS adopt an efficient iterated local search method to improve the quality of the solution produced by $(1+\lambda)$ evolution strategy, which can explore the search space effectively. Simulation results show the GLS has better performance than other heuristics based multiuser detectors, such as evolutionary programming and k-opt local search. The average computation time is polynomial in the number of users.

## References


1. Verdu, S.: Minimum Probability of Error for Asynchronous Gaussian Multiple-access Channels. IEEE Transactions on Information Theory, 32 (1986) 85-96
2. Verdu, S.: Computational Complexity of Optimal Multiuser Detection. Algorithmica, 4 (1989) 303–312
3. Wang, S., Zhu, Q., Kang, L.: $(1+\lambda)$ Evolution Strategy Method for Asynchronous DS-CDMA Multiuser Detection. IEEE Communications Letters, 10(6) (2006)423-425
4. Adebi, S., Tafazolli, R.: Genetically Modified Multiuser Detection for Code Division Multiple Access Systems. IEEE Journal of Selected Areas in Communications, 20 (2002) 463-473
5. Lim, H., Rao, M., Alan, W. Chuah, H.: Multiuser Detection for DS-CDMA Systems Using Evolutionary Programming. IEEE Communications Letters, 7 (2003) 101-103
6. Hu, J., Blum, R.S.: A Gradient Guided Search Algorithm for Multiuser Detection. IEEE Communications Letters, 4 (2000) 340-342
7. Lim, H., Venkatesh, B.: An Efficient Local Search Heuristics for Asynchronous Multiuser Detection. IEEE Communications Letters, 7 (2003) 299-301
8. Merz, P., Freisleben, B.: Genetic Local Searchfor the TSP: New results. IEEE International Conference on Evolutionary Computation, IEEE Press, (1997) 159–164
9. Proakis, J.G.: Digital Communications. 4th edn., McGraw-Hill, USA. (2001)
10. Verdu, S.: Multiuser Detection. Cambridge University Press, Cambridge, U.K. (1998)
11. Varanasi, M. K., Aazhang, B.: Multi-stage Detection in Asynchronous Code-Division Multiple Access Communications. IEEE Transactions on Communications, 38 (1990) 509-519
12. Beyer, H. G., Schwefel, H. P.: Evolution Strategies: A Comprehensive Introduction. Natural Computing, 1 (2002) 3-52
13. Bart, N., Leila, K.: A Comparison of Predictive Measure of Problem Difficulty in Evolutionary Algorithms. IEEE Transactions on Evolutionary Computations, 4 (2000) 1-15
14. Wang, S., Zhu, Q., and Kang, L.: Landscape Properties and Hybrid Evolutionary Algorithm for Optimum Multiuser Detection Problem. Lecture Notes in Computer Science, 3991 (2006)340-347
15. Weinberger, E.D.: Correlated and Uncorrelated Fitness Landscapes and How to Tell the Difference. Biological Cybernetics, 63 (1990) 325-336
16. Holger, H., Thomas, S.: Stochastic Local Search: Foundations and Applications. Morgan Kaufmann Publishers/Elsevier, San Fransisco, USA, 2004
17. Merz, P., Freisleben, B.: Greedy and Local Search Heuristics for Unconstrained Binary Quadratic Programming. Journal of Heuristics, 8 (2002) 197-213

# Motion Retrieval with Temporal-Spatial Features Based on Ensemble Learning

Jian Xiang

School of Information and Electronic Engineering, ZheJiang University of Science and Technology, 310023, Hangzhou, China
Freexiang@Gmail.com

**Abstract.** Along with the development of Motion Capture technique, more and more 3D motion database become available. In this paper, a novel method is presented for motion retrieval based on Ensemble HMM learning. First 3D temporal-spatial features and their keyspaces of each human joint are extracted for training data of Ensemble HMM learning. Then each action class is learned with one HMM. Since ensemble learning can effectively enhance supervised learners, ensembles of weak HMM learners are built. Experimental results show that our approaches are effective for motion data retrieval.

**Keywords:** Motion Capture, Temporal-Spatial, Ensemble Learning, HMM.

## 1 Introduction

Now more and more motion capture systems are used to acquire realistic human motion data. Therefore an efficient motion data recognition and retrieval technique is needed to support motion data processing, such as motion morph, edition and synthesis, etc. At present, most of motion data are stored in Mocap database with different length of motion clips, which is convenient for manipulating in animation authoring systems and retrieval based on keyword or content.

To resolve above-mentioned challenges, the temporal-spatial feature is defined in this paper first, which describes 3D space relationship of each joint. Comparing with the aforementioned motion features [1] [2] that is made up of 2D mathematic features such as joints positions, angles, speed and angular velocity, etc., temporal-spatial features are 3D features based on 3D time and space of each joint. Because conventional motion features are 2D, a complete motion must be described by 2D motion features of all joints. But for 3D temporal-spatial features, each joint's features can represent a part of the whole motion independently. Conventional motion features are extracted from original motion data, which has high time and space complexity with high dimension, so these methods need some dimension reduction algorithms. And 3D temporal-spatial features can avoid contacting with original motion data and eliminate "curse of dimensionality".

When temporal-spatial features are extracted, for each feature, the dynamics of one action class is learned with on continuous Hidden Markov Model (HMM) with outputs modeled by a mixture of Gaussian. HMM is a kind of temporal training

D.-S. Huang, L. Heutte, and M. Loog (Eds.): ICIC 2007, LNAI 4682, pp. 300–308, 2007.

models used successfully in speech recognition [3], and it has been applied to the video content analysis in constrained conditions [4]. Lv[5] use HMM to recognize and segment motion data.During the past years, diverse ensemble learning algorithms have been developed, such as Bagging[6], AdaBoost[7]. In [8], an integration called "boosted HMM" is proposed for lip reading. In this paper, Adaboost is used for Ensemble HMM learning.

## 2 3D Temporal-Spatial Features

In this paper, a simplified human skeleton model is defined, which contains 16 joints that are constructed in the form of tree. Joint *root* is root of the tree and those paths from root to all endmost joints in human skeletal model from sub-trees of root.

World coordinate of each joint can be represented as follow:

$$M = \{F(1), F(2),..., F(t),..., F(n)\} \quad F(t) = \{p(t), q_1(t),..., q_m(t)\} \tag{1}$$

where $F(t)$ is the *t-th* frame in motion clip M, $p(t)$ is the rotation of the root joint and $q_i(t)$ is the rotation of joint $i$ at frame t. m is the number of joints used in human skeleton.

All of the motions used by us are performed by a real actor and recorded by an optical motion capture system at frame rate 120. Each motion is presented by the same skeleton with 51 DOFs(corresponding to 16 joints of human body.

According to Equation (1), we can calculate world coordinate of each joint and get 48 dimensional data.

Given a motion M consisting of n sampling frames, each motion can be represented as follow:

$$M_s = (F_1, F_2,..., F_n) ; F_i = (p_{i1}, p_{i2},..., p_{ij},..., p_{i16}) ; p_{ij} = (x, y, z) \tag{2}$$

where n is the number of frames of motion data, $p_{ij}$ is world coordinate of joint j at $i^{th}$ frame.

Now space transformations of each joint are calculated. Firstly, we define a space transformation set of upper body $S_{up}$, and a space transformation set of lower body $S_{down}$ as following:

$S_{ui} \in S_{up}$, i=1,2...m; $S_{dj} \in S_{down}$, j=1,2...m; where m is the number of spaces in space transformation set, $S_{up}$ and $S_{down}$ have the same number of spaces. If we take Root as benchmark, then space transformations of joints above Root belong to $S_{up}$, and others belong to $S_{down}$, if a joint on upper body enters into space $S_{ui}$, its space transformation is $S_{ui}$.

$$front(N_i,N_j)=\begin{cases}1, N_i \text{ in front of } N_j\\0, N_i \text{ behind of } N_j\end{cases} \qquad left(N_i,N_j)=\begin{cases}1, N_i \ leftto\ N_j\\0, N_i \ rightto\ N_j\end{cases}$$

$$high(N_i,N_j)=\begin{cases}1, N_i \text{ above } N_j\\0, N_i \text{ below } N_j\end{cases} \qquad far(N_i,N_j)=\begin{cases}1, N_i \text{ distancefrom } N_j > \lambda\\0, N_i \text{ distancefrom } N_j < \lambda\end{cases}$$

Four space partition rules are defined as above. where rules of front, left and high depend on space relationship of up/down and left/right between joint $N_i$ and $N_j$, rule of far depends on range of motion. As usual, in rules of front and left, $N_j$ is Root, but in rules of high and far, $N_j$ on upper and lower body are different. $N_i$, $N_j$ are both at the same sampling frame.

Now we define motion space transformations:

$$(B = S_1, S_2, ..., S_{16})' , S_i = (S_{i1}, S_{i2}, ..., S_{in}) \tag{3}$$

where $S_i$ is space transformation vector of joint i, n is the number of frames, $s_{ip}$ is space transformation of joint i at p $^{th}$ frame. Suppose $S_a$ is space transformation vector of joint a on lower body, $S_a$ =( $s_{a1}$ , $s_{a2}$ … $s_{aj}$ … $s_{an}$ ):

**Table 1.** Space rule table to calculate Saj, N aj is joint a at j-th frame, Nrj is joint root at j-th frame, Nkj is joint knee at j-th frame

| $S_{aj}$ | $front(N_{aj},N_{rj})$ | $left(N_{aj},N_{rj})$ | $high(N_{aj},N_{kj})$ | $far(N_{aj},N_{kj})$ |
|---|---|---|---|---|
| $S_{aj}=S_{d1}$ | 1 | 1 | 1 | 1 |
| $S_{aj}=S_{d2}$ | 0 | 1 | 1 | 1 |
| … | … | … | … | … |
| $S_{aj}=S_{dm}$ | 0 | 0 | 0 | 0 |

In Table 1, some rules can be concluded:

If $s_{aj} = s_{d1}$ $\Leftrightarrow$ rule:

$$front(N_{aj}, N_{rj}) \wedge left(N_{aj}, N_{rj}) \wedge ; high(N_{aj}, N_{kj}) \wedge far(N_{aj}, N_{kj})$$

The rules cited above are calculated by 48 dimension data from Equation (2). Because these rules are all calculated at same frame, time and space complexity are not high. Moreover, space transformations of each joint are independent.

For example, we extract local space transformations of motion run's (see Fig.1) left foot and right foot as following: $S_{leftfoot}$ =($S_{dk}$ ,$S_{dj}$ ,$S_{dk}$ ,$S_{dj}$ ,…); $S_{rightfoot}$ =($S_{di}$ ,$S_{dl}$ ,$S_{di}$ ,$S_{dl}$ ,…).

Up to now, motion's space transformations are extracted, which is a kind of the reflection of motion spatial characteristic. But first of all, a complete motion is a group of time series data. Without time property, temporal-spatial features cannot represent motion clearly.

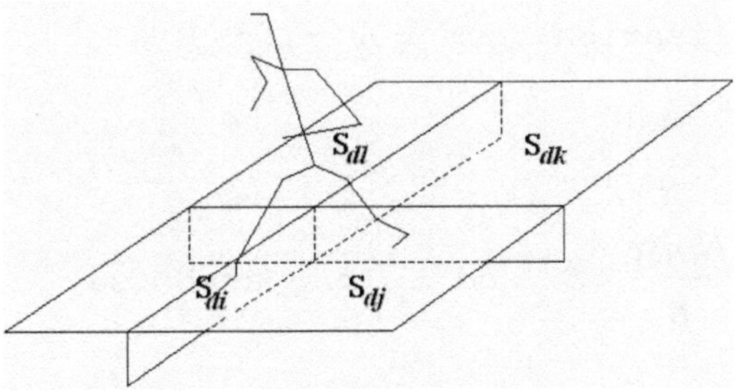


**Fig. 1.** Space transformations of run's feet

So the time property of motion is calculated as a part of temporal-spatial features.

The first time property is space transformation speed. Because of independence of each joint's space transformations, space transformation speed is independent either. The algorithm can be summarized as follow:

Procedure SpaceSpeed()

Input: local space transformation vector of $k^{th}$ joint $s_k = (s_{k1}, s_{k2}, \dots, s_{kn})$, n is the number of frames.

Output: $SP_k = (SP_{k1}, \dots, SP_{ki}, \dots)$, $SP_{ki}$ is space transformation $S_{ki}$'s speed of $k^{th}$ joint.

(1) Initialization: $num_j = 0, i=1, j=0, L=S_{ki}$

(2) if $s_{ki} \neq s_{k(i+1)}$, {$spacespeed_{kl} = num_j$, $l=S_{k(i+1)}$, j=j+1}

else $num_j = num_j + 1$;

(3) i=i+1, if meet the end of frames goto (4) else goto (2)

(4) return $SP_k$

This spacespeed is actually the speed of a joint moving from a space to another. The weighted sum of every joints' spacespeeds consists of the whole motion's spacespeed.

During similarity measure, because of irregularity and contingency of human motion, there are odd space transformations that cannot be matched. Therefore spacenoise is defined to measure some odd space transformations.

```
Procedure SpaceNoise()
Input: local space transformation vector of k^th joint
         s_k = (s_k1, s_k2, ..., s_kn), n is the number of frames
Output: SpaceNoise_k
(1) Initialization: num_j=0,i=1,j=0,l=1
(2) if s_ki ≠ s_k(i+1)
        Noise= num_j, j=j+1,
      if Noise/n < ε  add S_ki to SpaceNoise_k
      else
         num_j=num_j+1;
(3) i=i+1,if meet the end of frames goto (4) else goto
    (2)
(4) return SpaceNoise_k
```

As space transformations, spacespeeds and spacenoises of 16 joints are gotten, complete temporal-spatial features are formed through the merger of them.

# 3 Ensemble HMM Learning

## 3.1 Weak HMM Classifier

We choose a hidden Markov mode (HMM) to capture the dynamic information in the feature vectors as experience shows HMM to be more powerful than models such as Bayesian network or DTW. The basic theory of HMM was presented in the late 1960s and early 1970s. Widespread understanding and application of the theory of HMMs to speech processing has occurred within the past several years.

Until now, a N×D matrix of HMMs is formed by each feature of each type of motions and $\text{HMM}_{i,j}$ is the HMM model of *i*-th motion type and *j*-th feature and its corresponding parameters is $\lambda_{i,j}$. All HMMs in column *j* correspond the classifier for feature *j*.

Given one observation sequence O, we compute $P(O \mid \lambda)$ for each HMM using the Forward Backward algorithm. Motion type classification based on feature *j* can be solved by finding action class *i* that has the maximum value of $P(O \mid \lambda)$. As shown in :

$$\text{Action}(O) = \arg\max_{i:i=1,\dots,N}(P(O \mid \lambda) \tag{4}$$

The training and classification algorithm of HMM classifiers are listed as follows:

```
Procedure: HMM classification algorithm
Input: M motion samples((x_1,y_1),...,(x_M,y_M)), x_k is a clip
       with motion type y_k, y_k ∈ {1,…,N}, k=1,…,M, N is
       the number of types of motions ,
       An observation clip O = O_1 O_2 ... O_T
Output: Motiontype(O)
(1) Classifier these samples into N classes , each
     class contains the same type of motion.
(2) for i=1 to M
       Train HMM for each feature of each motion type
       (using Baum-Welch algorithm)
(3) for j=1 to N
       Computer P(O | λ_{j,i})
(4) return  Motiontype(O)= argmax_{i:i=1,...,N}(P(O | λ) .
```

Since our HMM mode has 3 states with a 3-component mixture Gaussian and Baum-Welch algorithm usually converges in less than 10 iterations, the complexity of all HMM models training is O(DM), M is the total length of training samples of all motion types. And the complexity of classification is O(NT). So the complexity of the whole procedure is O(DM+NT).

### 3.2 Ensemble HMM Learning

Ensemble methods are learning algorithms that construct a set of classifiers and then classify new data points by taking a (weighted) vote of their predictions. The most popular ensemble learning method is the boosting that is implementing on weighted training sets. In weighted training sets, each example has its weight $w_j \geq 0$. When the weight is higher, the example is more important during learning. The following shows the algorithm.

```
Procedure AdaBoost
Input: examples,set of N labeled examples
       ((x_1,y_1),...,(x_N,y_N)); L,a learning algorithm,M,the
       number of hypotheses in the ensemble
Output: a weighted-majority hypothesis
Local variables: w, a vector of N example
weights,initially 1/N
h,a vector of M hypotheses
z,a vector of M hypothesis weights
for m=1 to M do
       h[m]=L(examples,w)
       error=0
       for j=1 to N do
```

```
        if h[m](x_j)=y_j then w[j]=w[j]*error/(1-
            error);
    for j=1 to N do
        if h[m](x_j)≠y_j then error=error+w[j]
    w=Normalize(w)
    z[m]=log(1-error)/error
```

For all motion clips, initializing $w_j$=1. When the first hypothesis is given, the weights of motion clips in wrong classes increase and the weights in correct class reduce. Then a new weighted training set is created and new hypothesis is given based on this new training set repeatedly.

For the ensemble HMMs, a methodology for assessing prediction quality is used to estimate our method. First we collect a large set of example and divide it into two disjoints sets: the training set and the test set, then use the proposed method with training set as examples to generate a hypothesis H and measure the percentage of example in the test set that are correctly classified by H. At last, steps above are repeated for different sizes of training sets and different randomly selected training sets of each size.

## 4 Experimental Results and Analysis

We implement out algorithm in matlab. It is more than 1000 motion clips with 15 common types in the database for test. Most of the typical human motions are performed by actors, such as walking, running, kicking, punching, jumping, washing floor, etc.

Comparing ensemble HMMs with the individual HMM learners, Fig.2 show that the performance of the ensemble HMMs is higher. Results show that after combine 5 features of HMM learners by the adaboost algorithm, the final learner

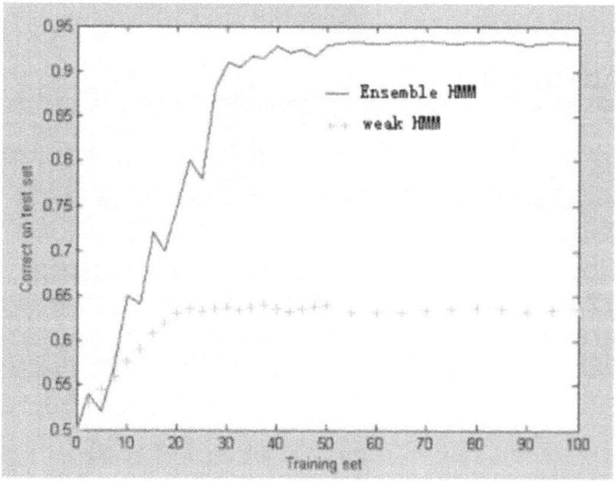


**Fig. 2.** Comparison of the performance of Ensemble HMM with weak HMM classifier

archieves a recognition rate of 93.2% on the motion type run, showing the effectiveness of the algorithm.

During motion recognition and retrieval, query examples always belong to common motion type. Given a query example, we compute P(O| $\lambda$ ) for each motion type and found the type which has argmax(P(O| $\lambda$ ). So out motion retrieval avoids a great deal of motion similarity measuring and matching and become more efficient. Table 2 shows the comparison of retrieval efficiency between these two methods. And Table 3 shows that HMM learning based on temporal-spatial features can save a great deal of time for HMM training.

**Table 2.** Recall and Precision

| Motion clips | Recall | | Precision | |
|---|---|---|---|---|
| | Conventional method | Our method | Conventional method | Our method |
| Walk | 0.79 | 0.96 | 0.91 | 0.97 |
| Run | 0.71 | 0.97 | 0.82 | 0.96 |
| Jump | 0.61 | 0.93 | 0.71 | 0.92 |
| Punch | 0.49 | 0.89 | 0.59 | 0.91 |

Results show that the ensemble HMM learning based on temporal-spatial features is efficient and accurate for motion retrieval in large human motion capture database.

**Table 3.** Training time for HMM

| Training data | Training time | | | |
|---|---|---|---|---|
| | Walk | Run | Jump | Punch |
| Originalmotion features | 59.2144s | 65.8490s | 77.1392s | 66.1121s |
| Temporal-spatial features | 4.3135s | 6.9182s | 8.2942s | 6.1631s |

## 5 Conclusion

In this paper, an Ensemble HMM learning method is proposed. Before learning, temporal-spatial features are extracted with out dimensionality reduction which describe 3D space relationship of each joint Then HMM models of some common motion types for each low-dimensional space feature are learned and ensemble learning method: adaboost is applied to combine weak HMM learner for each feature to form a strong learners for motion recognition. At last, the whole motion database is automatically built and efficiently and accurately indexed. The motion retrieval system is also sped up significantly.

## References

[1] Liu, F., Zhuang, Y.T., Wu, F., Pan, Y.H.: 3D Motion Retrieval with Motion Index Tree[J]. Computer Vision and Image Understanding, 92(2-3) (2003) 265-284
[2] Chui, Y., Chao, S.P., Wu, M.Y., Yang, S.N., Lin, H.C.: Content-based Retrieval for Human Motion Data. Journal of Visual Communication and Image Representation, 16(3) (2006) 446-466
[3] Rabiner, L.: A Tutorial on Hidden Markov Models and Selected Applications in Speech Recognition. Proceedings of the IEEE, 77(2) (1989) 257-286
[4] Starner, T.: Visual Recognition of American Sign Language Using Hidden Markov Modles. Master's Thesis, MIT Media Laboratory (1995) 189-194
[5] Lv, F.J., Nevatia, R.: Recognition and Segmentation of 3-D Human Action using HMM and Multi-Class AdaBoost. Proceedings of 9th European Conference on Computer Vision (ECCV), (2006) 359-372
[6] Breiman, L.: Bagging Predictors. Machine Learning, 24 (1996) 123-140
[7] Freund, Y., Schapire, R.E.: A Decision-Theoretic Generalization of On-Line Learning and An Application to Boosting. Proceedings of the 2nd European Conference on Computational Learning Theory, (1995) 23-37
[8] Yin, P., Essa, I., Rehg, J.M.: Asymmetrically Boosted HMM for Speech Reading. Proceedings of CVPR, (2004) 755-761

# The Study of Pavement Performance Index Forecasting Via Improving Grey Model

Ziping Chiang[1], Dar-Ying Jan[1], and Hsueh-Sheng Chang[2]

[1] Assistant Professor, Department of Logistics Management,
Leader University, 709 Taiwan, China
[2] Assistant Professor, Department of Local Development and Management,
Leader University, 709 Taiwan, China
{ziping,dyj,chs}@mail.leader.edu.tw

**Abstract.** This paper proposed a time series forecasting approach based on improving grey model (IGM). This method is based on fitting difference equation and yields better predictive result than the traditional one, and is demonstrated by forecasting pavement performance index as international roughness index. The results show that this approach can minimizes the error based on the traditional grey model, adaptive α, and grey rolling model with 19.4%, 17.7%, and 9.5%.

**Keywords:** Grey Model, Time Series Forecasting, Pavement Performance.

## 1 Introduction

Predicting algorithm is very important in pavement management system (PMS). Butt (1987), Easa (1989), and Lee (1993) stated the advantages of pavement-forecasting model obtain at least 2 capabilities [1-3]. (1) Predicting future pavement performance. (2) Reasoning pavement deterioration model. Thus PMS may establish the optimal strategy to distribute funds reasonably. Especially in Taiwan, previous attempts for developing pavement index forecasting models met with a lot of difficulties. There are two reasons; first is the performance indexes were affected by many dependent parameters which cause the hardness to build multi-variables regression model (MVRM). The second one is very difficult to collect the performance data completely.

Butt et al. (1987) used Markov process to build the transition matrix for modeling pavement future performance in United States [1]. Lee (1993) and Paterson (1989) tried to connect the correlation with the index and factors by MVRM [3,4]. In Taiwan, Niu (1995) also used MVRM to build a model for explaining the cause-result relationship of local pavement deterioration [5]. Huang (1997) used Markov process model to build pavement condition [6]. Hung (2000) based on the fuzzy regression to rebuild pavement performance prediction model [7]. Meantime, time series forecasting method was studied for the simulate pavement system. Shahin et al. (1987) developed the simple time regression method (STRM) to model pavement deterioration [8]. Lu et al. (1992) forecasted the pavement roughness with adaptive

D.-S. Huang, L. Heutte, and M. Loog (Eds.): ICIC 2007, LNAI 4682, pp. 309–314, 2007.

filter model (AFM) [9]. But STRM, AFM, and autoregressive integrated moving average (ARIMA) need considerable data to formulate the model and are not suitable to Taiwan. This paper proposed an improving GM (IGM) for forecasting pavement future condition.

# 2 Description of Traditional Grey Forecasting Model

## 2.1 Nomenclature

| | |
|---|---|
| $x^{(0)}(k)$ | represent the original series, k=1,2,3,…,n. |
| $x^{(1)}(k)$ | the first-order accumulated generating operation. |
| $z^{(1)}(k)$ | the mean generating operation. |
| $\alpha$ | the parameter for mean generating operation. |
| a and b | the undetermined parameters of the grey difference equation. |
| $\Delta$(k) | the time gap of original series, k=2,3,…,n. |

## 2.2 Grey Forecasting Model

Deng (1982) developed the grey model (GM) for time series forecasting based on grey differential equation [10]. Huang et al. (1996) integrated the fuzzy and GM; the results are very satisfy [11]. Liang et al. (2001) used GM to evaluating the carbonation damage to concrete bridges [12]. The consequence showed that GM is a good forecasting model for limited data. GM modeling process will be showing as follow:

Given original positive discrete data $\{x^{(0)}(k); x^{(0)}(k) > 0, 1 \le k \le n\}$, and apply the accumulated generating operation (AGO) $x^{(1)}(k) \equiv \sum_{P=1}^{k} x^{(0)}(P)$ to transfer $x^{(0)}(k)$ to a new space $\{x^{(1)}(k); x^{(1)}(k) > 0, 1 \le k \le n\}$. Easily to see that $x^{(1)}$(k) is a positive monotonic increasing form. Obviously, the governing equation will be $\frac{dx^{(1)}(k)}{dk} + a \cdot x^{(1)}(k) = b$, where $a$ and $b$ are the undetermined parameters for the system.

Based on the difference operation yields

$$x^{(0)}(k) + a \cdot z^{(1)}(k) = b. \tag{1}$$

where $z^{(1)}(k) = \alpha \cdot x^{(1)}(k) + (1-\alpha) \cdot x^{(1)}(k-1)$, $\alpha$ actually is the weighting factor for two adjacent data within [0,1]. The solution of Eq. (1) will be

$$\hat{x}^{(1)}(k+1) = (x^{(0)}(1) - \frac{b}{a})e^{-ak} + \frac{b}{a}, \qquad k = 2,3,\ldots,n. \tag{2}$$

# 3 Improving Grey Rolling Model

Deng (1993) suggested $\alpha$ =0.5 (equal weighting case), which it is quite suitable for the monotonic and smoothing data [13]. Otherwise is doubtful. Wen et al. (1999) and

Wen et al. (2000) studied the $\alpha$ of GM [14,15]. The results of Wen's study concluded that the criterion for $\alpha$ is applicable to minimize the predicted error easily, we developed grey rolling model (GRM) where the parameters $\alpha$ (*k*) are rebuild as

$$z^{(1)}(k) = \alpha(k) \cdot x^{(1)}(k) + (1-\alpha(k))x^{(1)}(k-1), \qquad 2 \le k \le n. \tag{3}$$

GRM can also provide more adjustable value for the weights and its performance is better than traditional GM and adaptive $\alpha$. This paper extended GRM into a general form of $z^{(1)}(k)$ as

$$z^{(1)}(k) = \alpha_{(j)}^{(k)} \cdot x^{(1)}(k) + \alpha_{(j-1)}^{(k)} \cdot x^{(1)}(k-1) + ... + \alpha_{(1)}^{(k)} \cdot x^{(1)}(1), \tag{4}$$

where $k = 2,3,...,n$, and $\alpha_{(j)}^{(k)} + \alpha_{(j-1)}^{(k)} + ... + \alpha_{(1)}^{(k)} = 1$.

If the range of the rolling interval for $n \ge 4$, then the outcome of system can be forecasted to $\{\hat{x}^{(0)}(k), k \ge n+1\}$. The processes are described as the following:

Step 1: Original data series are $\{x^{(0)}(k), 1 \le k \le n\}$, and the time-gap series are $\{\Delta k, 2 \le k \le n\}$.

Step 2: Generate $\{x^{(1)}(k), 1 \le k \le n\}$ by AGO.

Step 3: Build $z^{(1)}(k)$ series as Eq. (4).

Step 4: Estimate the parameters $a$ and $b$ by the Least Square Method:

$$\begin{bmatrix} a \\ b \end{bmatrix} = (B^T \cdot B)^{-1} \cdot B^T \cdot y \cdot \tag{5}$$

$$\text{where } B = \begin{bmatrix} -z^{(1)}(2) \cdot \Delta(2) & \Delta(2) \\ -z^{(1)}(3) \cdot \Delta(3) & \Delta(3) \\ \vdots & \\ -z^{(1)}(k) \cdot \Delta(\mathrm{k}) & \Delta(\mathrm{k}) \end{bmatrix}. \tag{6}$$

$$y = \begin{bmatrix} x^{(0)}(2) \\ x^{(0)}(3) \\ \vdots \\ x^{(0)}(k) \end{bmatrix}. \tag{7}$$

Step 5: The predicted value $\hat{x}^{(0)}(k), k \ge n+1$ is obtained by

$$\hat{x}^{(0)}(k) = \hat{x}^{(1)}(k) - \hat{x}^{(1)}(k-1), \text{ where } \hat{x}^{(1)}(k) = (\hat{x}^{(0)}(1) - \frac{b}{a})e^{-a(k-1)} + \frac{b}{a} \cdot \tag{8}$$

## 4 Model Implementation

In order to verify the pavement forecasting based on IGM, two cases has been studied via four time series analysis models (traditional GM ($\alpha$=0.5), adaptive α, GRM, and

IGM) to predict pavement performance index as international roughness index (IRI). Roughness is one of important pavement performance indexes and Paterson (1989) used it to determine the performance of pavement [4]. Data in the table 1 are surveyed by material laboratory, department of civil engineering, N.C.U.

**Table 1.** Test Data in Case 1

| Pavement section | Time-gap (month) | | | |
|---|---|---|---|---|
| | 0 | 2.8 | 2.5 | 3.5 |
| Section 1 | 2.47 | 2.88 | 2.86 | 2.77 |
| Section 2 | 2.42 | 2.73 | 2.3 | 4.45 |
| Section 3 | 3.05 | 2.9 | 2.74 | 2.76 |
| Section 4 | 2.54 | 2.84 | 2.66 | 3.36 |
| Section 5 | 2.91 | 3.22 | 3.1 | 3.37 |

In table 2, the data are collected by the Central District Project Office, National Freeway Bureau, Taiwan. The root mean square (RMS) technique and total-error-comparison (TEC) are used for the error analysis in this research and defined as Eq. (9) and Eq. (10).

**Table 2.** Test Data in Case 2

| Pavement section | Time-gap (month) | | | |
|---|---|---|---|---|
| | 0 | 3 | 11 | 6 |
| Section 6 | 1.85 | 1.54 | 1.45 | 3.18 |
| Section 7 | 2.42 | 2.58 | 2.44 | 2.48 |
| Section 8 | 2.68 | 2.58 | 2.81 | 3.32 |
| Section 9 | 1.22 | 1.37 | 1.92 | 2.32 |
| Section 10 | 1.81 | 1.90 | 1.58 | 3.08 |

$$RMS = \sqrt{\frac{\sum_{k=1}^{n} (\hat{x}^{(0)}(k) - x^{(0)}(k))^2}{n-1}} . \tag{9}$$

$$TEC_i = \frac{(\sum_{j=1}^{n} RMS_{ij} - \sum_{j=1}^{n} RMS_{ij})}{\sum_{j=1}^{n} RMS_{ij}} \times (-100\%) . \tag{10}$$

where $i$ is compared method, $j$ is section number. The results are presented in the table 3 and table 4.

**Table 4.** RMS of the Four Methodologies

| | Traditional grey model (α=0.5) | Adaptiveα | GRM | Improving GM |
|---|---|---|---|---|
| Section 1 | 1.7487 | 1.6992 | 1.6709 | 1.6197 |
| Section 2 | 2.4207 | 2.3511 | 2.3386 | 2.0686 |
| Section 3 | 1.7198 | 1.6675 | 1.6421 | 1.5842 |
| Section 4 | 1.9409 | 1.9282 | 1.9174 | 1.8417 |
| Section 5 | 2.0359 | 1.9957 | 1.9717 | 1.9024 |
| Section 6 | 2.2760 | 2.2167 | 1.9331 | 1.6666 |
| Section 7 | 2.1659 | 2.1627 | 1.6486 | 1.4598 |
| Section 8 | 2.6813 | 2.6491 | 2.1915 | 1.8951 |
| Section 9 | 1.8040 | 1.7738 | 1.5780 | 1.3592 |
| Section 10 | 2.2859 | 2.2059 | 1.8813 | 1.6011 |
| Σ | 21.0791 | 20.6499 | 18.7732 | 16.9984 |

**Table 5.** Total error comparison of the four methodologies

| | Traditional grey model (α=0.5) | Adaptiveα | GRM | IGM |
|---|---|---|---|---|
| RMS Σ | 21.0791 | 20.6499 | 18.7732 | 16.9984 |
| TEC 1 | - | 2.0% | 10.9% | 19.4% |
| TEC 2 | -2.1% | - | 9.1% | 17.7% |
| TEC 3 | -12.3% | -10.0% | - | 9.5% |
| TEC 4 | -24.0% | -21.5% | -10.4% | - |

# 5 Discussion and Conclusions

This approach can model pavement deterioration merely need four survey data. The RMS error obtained by four individual calculation based on the traditional GM, adaptive $\alpha$, GRM, and IGM presented in table 3. The data related to the comparison of IGM with the other three methods is in table 4. The results showed IGM is much improved with 19.4%, 17.7%, and 9.5%. This approach can consecutively adjust the model according to the new input data, and also avoids the rectification of pavement conditions after maintaining in MVRM. Based on IGM, one can forecast pavement performance index, establish an optimal strategy to distribute funds reasonably, and the best serviceability condition is provided for the entire network level system in Taiwan.

# References


1. Butt, A.A., Shahin, M.Y. , Feighan, K.J., Carpenter, S.H.: Pavement Performance Prediction Model Using the Markov Process, Transportation Research Record 1123 (1987) 12-19
2. Easa, S., Kikuchi, S.: Pavement Performance Prediction Models: Review and Evaluation. Delaware Transportation Center (1989)

3. Lee, Y. H.: Development of Pavement Prediction Models, Ph.D. Thesis, University of Illinois, Urbana (1993)
4. Paterson W.D.O.: A Transferable Causal Model for Predicting Roughness Progression in Flexible Pavements, Transportation Research Record 1215 (1989) 70-84
5. Niu, W.Y.: The Study of Processing Build for Flexible Pavement Performance Forecasting Model, Master thesis, National Taiwan University (1995)
6. Huang, C.C.: Development of Freeway Pavement Performance Prediction Model Using Markov Chain, Master thesis, Tamkang University (1997)
7. Hung, C.T.: The Study on Establishing the Present Serviceability Index and Predictive Model of Flexible Pavement, Master thesis, National Central University (2000)
8. Shahin, M.Y., Nunez, M.M., Broten, M.R., Carpenter, S.H., Sameh, A.: New Techniques for Modeling Pavement Deterioration, Transportation Research Record 1123 (1987) 40-46
9. Lu, J., Bertrand, C., Hudson, W.R., McCullough, B.F.: Adaptive Filter Forecasting System for Pavement Roughness, Transportation Research Record 1344 (1992) 124-129
10. Deng, J.L.: Control Problems of Grey System, Systems Control Letter 5 (1) (1982) 288-294
11. Huang, Y.P., Huang, C.C.: The Integration and Application of Fuzzy and Grey Modeling Methods, Fuzzy Sets and Systems 78 (1) (1996) 107-119
12. Liang, M.T., Zhao, G.F., Chang C. W., Liang, C.H.: Evaluating the Carbonation Damage to Concrete Bridges Using a Grey Forecasting Model Combined with a Statistical Method, Journal of the Chinese Institute of Engineers 24 (1) (2001) 85-94
13. Deng, J.L.: Grey Differential Equation, The Journal of Grey System 5(1) (1993) 1-14.
14. Wen, K.L., Chang, T.C., Chang, H.T., You, M.L.: The Adaptive $\alpha$ in GM(1,1) Model, The Proceeding of IEEE SMC International Conference (1999) 304-308
15. Wen, J.C., Huang, K.H., Wen, K.L.: The Study of $\alpha$ in GM(1,1) Model, Journal of the Chinese Institute of Engineers 23 (5) (2000) 583-589

# An Adaptive Recursive Least Square Algorithm for Feed Forward Neural Network and Its Application

Xi-hong Qing[1], Jun-yi Xu[1], Fen-hong Guo[2], Ai-mu Feng[3], Wei Nin[4], and Hua-xue Tao[1]

[1] College of Geo-Information Science and Engineering, Shandong University of Science and Technology, 271019,Qingdao, Shandong,China
[2] College of Applied mathematics, Guangdong University of Technology, 510090,Guangzhou, Guangdong, China
[3] Daqing Oilfield NO.2 Oil Production Company, 163414, Heilongjiang daqing, China
[4] Shandong Agricultural University,271018,Taian Shandong, China
qingyin@sdust.edu.cn, xjyxxl@yahoo.com.cn, guofenhong@163.com, fengaimu@petrochina.com.cn

**Abstract.** In high dimension data fitting, it is difficult task to insert new training samples and remove old-fashioned samples for feed forward neural network (FFNN). This paper, therefore, studies dynamical learning algorithms with adaptive recursive regression (AR) and presents an advanced adaptive recursive (AAR) least square algorithm. This algorithm can efficiently handle new samples inserting and old samples removing. This AAR algorithm is applied to train FFNN and makes FFNN be capable of simultaneously implementing three processes of new samples dynamical learning, old-fashioned samples removing and neural network (NN) synchronization computing. It efficiently solves the problem of dynamically training of FFNN. This FFNN algorithm is carried out to compute residual oil distribution.

**Keywords:** feed forward neural network, adaptive recursive regression, least square algorithms, dynamical learning, residual oil, Voronoi graph.

## 1 Introduction

Dynamical learning of feed forward neural network (DLFFNN) is always related to surface reconstruction and fitting. There are a lot of methods to reconstruct surface from unorganized points using neural network (NN), such as geometry modeling algorithms [1][2][3]. In recent years, a lot of researchers have considered NN algorithms to fit scattered data [4]. People are paying much attention to the prediction of scattered data's space properties using NN [5].

Feed forward neural network (FFNN) is often thought of as recursive weighted least squared algorithms or extended Kalman filters [6][7][8][9][10]. And it is feasible to train NN using adaptive recursive regression (AR) [8][9][10]. Moving window is

D.-S. Huang, L. Heutte, and M. Loog (Eds.): ICIC 2007, LNAI 4682, pp. 315–323, 2007.

usually a circle or hyper-sphere to fit or localize scattered data in isotropic space. However, the samples data in moving window is variable [11]. For example, it needs to insert new training samples and remove old-fashioned training samples. Therefore it is useful to solve the problem of dynamical learning for NN during high dimension data fitting by studying moving windows and variable steps. Consequently, this paper presents an advanced adaptive recursive regression (AAR) algorithm with any new training samples inserting and old-fashioned samples removing. Using the AAR algorithm, a dynamical learning algorithm is designed to train the weight vector of FFNN. The results of residual oil distribution show the efficiency of our approach.

## 2 Dynamical Learning Process of FFNN

In this paper, we only study the three-layered FFNN. Let $\mathbf{x}=(x_1,x_2,\ldots,x_m)$ be the input data vector, $u_i$ be the state of neuron, $\theta$ be the threshold. The hard-limit transfer function $f(u)$ of the neuron is defined as

$$f(u)=\begin{cases}1, u>\theta \\ 0, u\le\theta\end{cases}. \tag{1}$$

The relation of neuron's output $z_i$, the input data $\mathbf{x}$ and the transfer function $f(u)$ is

$$u_i=\sum_{j=1}^{m}x_j w_{i,j}+b_i=(\mathbf{x},1)\begin{pmatrix}\mathbf{w}_i\\ b_i\end{pmatrix},\ z_i=f(u_i)=\begin{cases}1, u_i>\theta \\ 0, u_i\le\theta\end{cases}. \tag{2}$$

Here $b_i$ is a bias, $\mathbf{w}_i=(w_{i,1},\ldots,w_{i,m})^T$ is weight vector. If there are multi input samples matrix $\mathbf{x}_k=(x_{k,1},x_{k,2},\ldots,x_{k,m})$, the target is $y_k$, $1\le k\le n$ ,then the output of $i^{\text{th}}$ neuron is

$$u_{k,i}=\sum_{j=1}^{m}x_{k,j}w_{i,j}+b_i=(\mathbf{x}_k,1)\begin{pmatrix}\mathbf{w}_i\\ b_i\end{pmatrix},\ \mathbf{u}_i=\mathbf{A}\tilde{\mathbf{w}}_i. \tag{3}$$

Where

$$\mathbf{u}_i=(u_{1,i},\cdots,u_{n,i})^T,\mathbf{A}=\mathbf{A}_{n\times(m+1)}=\begin{pmatrix}\mathbf{x}_1^T & \cdots & \mathbf{x}_n^T\\ 1 & \cdots & 1\end{pmatrix}^T,\tilde{\mathbf{w}}_i=\begin{pmatrix}\mathbf{w}_i\\ b_i\end{pmatrix}. \tag{4}$$

Let the input layer have $m$ neurons, the hidden layers have $p$ neurons, then for all neuron in the hidden layer, we have

$$\begin{aligned}&\mathbf{U}=\mathbf{AW},Vec(\mathbf{U})=(\mathbf{I}\otimes\mathbf{A})Vec(\mathbf{W})\\ &\mathbf{Z}_p=f(\mathbf{U}),y_k=f(\frac{1}{p}\sum_{i=1}^{p}z_{k,i}),1\le k\le n\end{aligned}, \tag{5}$$

Where $\mathbf{U}=\mathbf{U}_{n\times p}=(\mathbf{u}_1,\mathbf{u}_2,\ldots,\mathbf{u}_p)$, $\mathbf{W}=\mathbf{W}_{(m+1)\times p}=(\tilde{\mathbf{w}}_1\tilde{\mathbf{w}}_2,\cdots,\tilde{\mathbf{w}}_p)$, $(z_{i,j})=(\mathbf{z}_1,\ldots,\mathbf{z}_p)=\mathbf{Z}_p=f(\mathbf{U})=(f(\mathbf{u}_1),\ldots,f(\mathbf{u}_p))$, $f(\mathbf{u}_i)=(f(\mathrm{u}_{1,i}),f(\mathrm{u}_{2,i}),\ldots,f(\mathrm{u}_{n,i}))^T$, $\mathbf{I}=\mathbf{I}_{p\times p}$ is a identity matrix, $Vec(*)$ is a operator of matrix vectorization and $\otimes$ is the Kronecker product.

# 3 Algorithm

## 3.1 AAR Algorithm to Synchronize Variable Data Sets

The existing AR algorithms have no power of simultaneously implementing variable new training samples inserting and any old-fashioned samples removing. This is our motivation to study the AAR algorithm with new samples inserting and old samples removing.

Let $\mathbf{R}^m$, $\mathbf{R}^{n\times m}$ represent $m$ and $n\times m$ dimensional linear space respectively. Linear regression model is written in the following expression:

$$\mathbf{y}_n = \mathbf{A}_{n\times m}\mathbf{b}_n + \boldsymbol{\varepsilon}_n, \mathbf{y}_n, \boldsymbol{\varepsilon}_n \in \mathbf{R}^n, \mathbf{b}_n \in \mathbf{R}^m, \mathbf{A}_{n\times m} \in \mathbf{R}^{n\times m}. \tag{6}$$

Its least square regression is

$$\hat{\mathbf{b}}_n = (\mathbf{A}_{n\times m}^T \mathbf{A}_{n\times m})^{-1} \mathbf{A}_{n\times m}^T \mathbf{y}_n . \tag{7}$$

Where $\hat{\mathbf{b}}_n$ is the $\mathbf{b}_n$'s parameter solution. It is the weight vector in FFNN. $\boldsymbol{\varepsilon}_n$ is Gaussian noise. $n$ is number of samples. Let { $\mathbf{M}_{n,t}$, $\mathbf{Y}_{n,t}$ } and { $\mathbf{D}_{n,r}$, $\mathbf{Y}'_{n,r}$ } be $t$ new training and $r$ old-fashioned samples set. In addition, the following notations are used:

$$\begin{gathered}
\mathbf{x}_{n_i} = (x_{n_i,1}, x_{n_i,2}, \cdots, x_{n_i,m})_{1\times m}, \ \mathbf{M}_{n,t} = (\mathbf{x}_{n_1}^T, \mathbf{x}_{n_2}^T, \cdots, \mathbf{x}_{n_t}^T)^T, \\
\mathbf{y}_{n,t} = (y_{n_1}, y_{n_2}, \cdots, y_{n_t})^T, \ \mathbf{A}_{n\times m} = (\mathbf{x}_1^T, \cdots, \mathbf{x}_n^T)^T, \\
\mathbf{x}_j = (x_{j,1}, \cdots, x_{j,m}), 1 \le j \le n, \mathbf{P}_n = (\mathbf{A}_{n\times m}^T \mathbf{A}_{n\times m})^{-1}, \\
\hat{\mathbf{b}}_n = \mathbf{P}_n \mathbf{A}_{n\times m}^T \mathbf{y}_n, \ \hat{\mathbf{y}}_{n,t} = \mathbf{M}_{n,t}\hat{\mathbf{b}}_n, \ \Delta\mathbf{y}_{n,t} = \mathbf{y}_{n,t} - \hat{\mathbf{y}}_{n,t}, \\
\mathbf{d}'_{n_i} = (x'_{n_i,1}, x'_{n_i,2}, \cdots, x'_{n_i,m})_{1\times m}, \ \mathbf{D}_{n,r} = (\mathbf{d}_{n_1}^T, \mathbf{d}_{n_2}^T, \cdots, \mathbf{d}_{n_r}^T)^T, \\
\mathbf{y}'_{n,r} = (y'_{n_1}, y'_{n_2}, \cdots, y'_{n_r})^T, \ \mathbf{D}_{n,r}\hat{\mathbf{b}}_n = \hat{\mathbf{y}}'_{n,r}, \ \Delta\mathbf{y}'_{n,r} = \mathbf{y}'_{n,r} - \hat{\mathbf{y}}'_{n,r}
\end{gathered} \tag{8}$$

Let $\mathbf{O}$=zero matrix, $\mathbf{I}$=identity matrix, $\boldsymbol{\Gamma}_{n,j} = diag(\rho_{1,i}, \rho_{2,i}, \cdots, \rho_{n,i}) > \mathbf{O}$ be the weighted diagonal matrix, which weight to samples. Let $n+t-r$ be the samples number after inserting new $t$ samples and removing old $r$ samples, $n+t$ denote the samples number after inserting new $t$ samples, $n-r$ represent the samples number after removing old $r$ samples. It is easy to prove the following theorem 1 and inference 1.

**Theorem 1:** Dynamical memory recursive regression with any new training samples inserting and any old samples removing is given by.

$$\begin{aligned}
\mathbf{P}_{n+t} &= \mathbf{P}_n - \mathbf{P}_n \mathbf{M}_{n,t}^T (\mathbf{I} + \mathbf{M}_{n,t}\mathbf{P}_n\mathbf{M}_{n,t}^T)^{-1} \mathbf{M}_{n,t}\mathbf{P}_n \\
\mathbf{P}_{n+t-r} &= \mathbf{P}_{n+t} + \mathbf{P}_{n+t}\mathbf{D}_{n,r}^T (\mathbf{I} - \mathbf{D}_{n,r}\mathbf{P}_{n+t}\mathbf{D}_{n,r}^T)^{-1} \mathbf{D}_{n,r}\mathbf{P}_{n+t} \ , \\
\hat{\mathbf{b}}_{n+t-r} &= \hat{\mathbf{b}}_n + \mathbf{P}_{n+t-r}(\mathbf{M}_{n,t}^T \Delta\mathbf{y}_{n,t} - \mathbf{D}_{n,r}^T \Delta\mathbf{y}'_{n,r})
\end{aligned} \tag{9}$$

**Inference 1:** Dynamical memory weighted recursive regression with any new training samples inserting and any new old samples removing is

$$\begin{aligned}\mathbf{P}_{n+t} &= \mathbf{P}_n - \mathbf{P}_n\mathbf{M}_{n,t}^T(\mathbf{\Gamma}_{n,t}^{-2} + \mathbf{M}_{n,t}\mathbf{P}_n\mathbf{M}_{n,t}^T)^{-1}\mathbf{M}_{n,t}\mathbf{P}_n \\ \mathbf{P}_{n+t-r} &= \mathbf{P}_{n+t} + \mathbf{P}_{n+t}\mathbf{D}_{n,r}^T(\mathbf{\Gamma}_{n,r}^{-2} - \mathbf{D}_{n,r}\mathbf{P}_{n+t}\mathbf{D}_{n,r}^T)^{-1}\mathbf{D}_{n,r}\mathbf{P}_{n+t} , \\ \hat{\mathbf{b}}_{n+t-r} &= \hat{\mathbf{b}}_n + \mathbf{P}_{n+t-r}(\mathbf{M}_{n,t}^T\mathbf{\Gamma}_{n,t}^2\Delta\mathbf{y}_{n,t} - \mathbf{D}_{n,r}^T\mathbf{\Gamma}_{n,r}^2\Delta\mathbf{y}'_{n,r})\end{aligned} \tag{10}$$

We call theorem 1 and inference 1 AAR algorithm.

**Proof of theorem1:** let $t$ and $r$ be the number of new training samples and of removing old samples set: $\{ (\mathbf{x}_{n_1}, y_{n_1}),(\mathbf{x}_{n_2}, y_{n_2}),\cdots,(\mathbf{x}_{n_t}, y_{n_t}) \},\{ (\mathbf{d}_{n_1}, y'_{n_1}),(\mathbf{d}_{n_2}, y'_{n_2}),\cdots,$ $(\mathbf{d}_{n_r}, y'_{n_r}) \}$. Let $\mathbf{A}_{n_j}^T = (\mathbf{x}_{1+n_j}^T,\cdots,\mathbf{x}_{n_{j+1}}^T)$ and insert $\mathbf{M}_{n,t}$ into $\mathbf{A}_n = (\mathbf{A}_{n_1}^T,\mathbf{A}_{n_2}^T,\cdots,\mathbf{A}_{n_t}^T,\mathbf{A}_{n_{t+1}}^T)^T$ :

$$\begin{aligned}\mathbf{A}_{n+t} &= (\mathbf{x}_1^T,\cdots,\mathbf{x}_{n_1}^T,\mathbf{x}_{n_1}^T,\mathbf{x}_{1+n_1}^T,\cdots,\mathbf{x}_{n_t}^T,\mathbf{x}_{n_t}^T,\mathbf{x}_{1+n_t}^T,\cdots,\mathbf{x}_n^T)^T \\ &= (\mathbf{A}_{n_1}^T,\mathbf{x}_{n_1}^T,\mathbf{A}_{n_2}^T,\mathbf{x}_{n_2}^T,\cdots,\mathbf{A}_{n_t}^T,\mathbf{x}_{n_t}^T,\mathbf{A}_{n_{t+1}}^T)^T\end{aligned} , \tag{11}$$

Then

$$\mathbf{A}_{n+t}^T\mathbf{A}_{n+t} = \sum_{j=1}^{r+1}\mathbf{A}_{n_j}\mathbf{A}_{n_j}^T + \sum_{j=1}^{r}\mathbf{x}_{n_j}\mathbf{x}_{n_j}^T = \mathbf{A}_n^T\mathbf{A}_n + \mathbf{M}_{n,t}^T\mathbf{M}_{n,t} . \tag{12}$$

Let $\mathbf{D}_{n+t,r} = (\mathbf{O}_1^T,\mathbf{d}_{n_1}^T,\cdots,\mathbf{O}_r^T,\mathbf{d}_{n_r}^T,\mathbf{O}_{r+1}^T)^T$ denote the positions of vector $\mathbf{d}_{n_j}$ in $\mathbf{A}_{n+t}$, and $\mathbf{A}_{n+t-r}$ be the result after inserting $t$ new samples $\mathbf{M}_{n,t}$ and removing $r$ old samples $\mathbf{D}_{n,r}$ from $\mathbf{A}_n$. Then

$$\begin{aligned}&\mathbf{A}_{n+t-r} = \mathbf{A}_{n+t} - \mathbf{D}_{n+t,r} \\ &= (\mathbf{A}_{n_1}^T,\mathbf{x}_{n_1}^T,\cdots,\mathbf{A}_{n_t}^T,\mathbf{x}_{n_t}^T,\mathbf{A}_{n_{t+1}}^T)^T - (\mathbf{O}_1^T,\mathbf{d}_{n_1}^T,\cdots,\mathbf{O}_r^T,\mathbf{d}_{n_r}^T,\mathbf{O}_{r+1}^T)^T , \\ &\mathbf{A}_{n+t-r}^T = (\mathbf{A}_{n_1},\mathbf{x}_{n_1},\cdots,\mathbf{A}_{n_t},\mathbf{x}_{n_t},\mathbf{A}_{n_{t+1}}) - (\mathbf{O}_1,\mathbf{D}_{n_1},\cdots,\mathbf{O}_r,\mathbf{D}_{n_r},\mathbf{O}_{r+1})\end{aligned} \tag{13}$$

Therefore

$$\begin{aligned}&\mathbf{A}_{n+t-r}^T\mathbf{A}_{n+t-r} = \sum_{j=1}^{r+1}\mathbf{A}_{n_j}\mathbf{A}_{n_j}^T + \sum_{j=1}^{r}\mathbf{x}_{n_j}\mathbf{x}_{n_j}^T - 2\sum_{j=1}^{r}\mathbf{d}_{n_j}\mathbf{d}_{n_j}^T + \sum_{j=1}^{r}\mathbf{d}_{n_j}\mathbf{d}_{n_j}^T \\ &= \sum_{j=1}^{r+1}\mathbf{A}_{n_j}\mathbf{A}_{n_j}^T + \sum_{j=1}^{r}\mathbf{x}_{n_j}\mathbf{x}_{n_j}^T - \sum_{j=1}^{r}\mathbf{d}_{n_j}\mathbf{d}_{n_j}^T , \\ &= \mathbf{A}_n^T\mathbf{A}_n + \mathbf{M}_{n,t}^T\mathbf{M}_{n,t} - \mathbf{D}_{n,r}^T\mathbf{D}_{n,r} = \mathbf{A}_{n+t}^T\mathbf{A}_{n+t} - \mathbf{D}_{n,r}^T\mathbf{D}_{n,r}\end{aligned} \tag{14}$$

By $\hat{\mathbf{y}}'_{n,r} = \mathbf{D}_{n,r}\hat{\mathbf{b}}_n$, $\Delta\mathbf{y}'_{n,r} = \mathbf{y}'_{n,r} - \hat{\mathbf{y}}'_{n,r}$ we have

$$\begin{aligned}
&\mathbf{P}_{n+t} = (\mathbf{A}_{n+t}^T \mathbf{A}_{n+t})^{-1} = (\mathbf{A}_n^T \mathbf{A}_n + \mathbf{M}_{n,t}^T \mathbf{M}_{n,t})^{-1} \\
&= \mathbf{P}_n - \mathbf{P}_n \mathbf{M}_{n,t}^T (\mathbf{I} + \mathbf{M}_{n,t} \mathbf{P}_n \mathbf{M}_{n,t}^T)^{-1} \mathbf{M}_{n,t} \mathbf{P}_n \\
&\mathbf{P}_{n+t-r} = (\mathbf{A}_{n+t-r}^T \mathbf{A}_{n+t-r})^{-1} = (\mathbf{A}_{n+t}^T \mathbf{A}_{n+t} - \mathbf{D}_{n,r}^T \mathbf{D}_{n,r})^{-1} \\
&= \mathbf{P}_{n+t} + \mathbf{P}_{n+t} \mathbf{D}_{n,r}^T (\mathbf{I} - \mathbf{D}_{n,r} \mathbf{P}_{n+t} \mathbf{D}_{n,r}^T)^{-1} \mathbf{D}_{n,r} \mathbf{P}_{n+t} \\
&\mathbf{A}_{n+t-r}^T \mathbf{y}_{n+t-r} = \mathbf{A}_{n+t-r}^T \mathbf{A}_{n+t-r} \hat{\mathbf{b}}_{n+t-r} \\
&= (\mathbf{A}_n^T \mathbf{A}_n + \mathbf{M}_{n,t}^T \mathbf{M}_{n,t} - \mathbf{D}_{n,r}^T \mathbf{D}_{n,r}) \hat{\mathbf{b}}_{n+t-r} \qquad , \\
&= \mathbf{A}_n^T \mathbf{A}_n \hat{\mathbf{b}}_{n+t-r} + \mathbf{M}_{n,t}^T \mathbf{M}_{n,t} \hat{\mathbf{b}}_{n+t-r} - \mathbf{D}_{n,r}^T \mathbf{D}_{n,r} \hat{\mathbf{b}}_{n+t-r}
\end{aligned} \tag{15}$$

According to $\mathbf{A}_{n+t-r}^T \mathbf{y}_{n+t-r} = \mathbf{A}_n^T \mathbf{y}_n + \mathbf{M}_{n,t}^T \mathbf{y}_{n,t} - \mathbf{D}_{n,r}^T \mathbf{y}'_{n,r}$, we have

$$\begin{aligned}
&\mathbf{A}_{n+t-r}^T \mathbf{A}_{n+t-r} (\hat{\mathbf{b}}_{n+t-r} - \hat{\mathbf{b}}_n) \\
&= \mathbf{A}_{n+t-r}^T \mathbf{y}_{n+t-r} - (\mathbf{A}_n^T \mathbf{A}_n + \mathbf{M}_{n,t}^T \mathbf{M}_{n,t} - \mathbf{D}_{n,r}^T \mathbf{D}_{n,r}) \hat{\mathbf{b}}_n \\
&= \mathbf{M}_{n,t}^T (\mathbf{y}_{n,t} - \hat{\mathbf{y}}_{n,t}) - \mathbf{D}_{n,r}^T (\mathbf{y}'_{n,r} - \hat{\mathbf{y}}'_{n,r}) \\
&= \mathbf{M}_{n,t}^T \Delta \mathbf{y}_{n,t} - \mathbf{D}_{n,r}^T \Delta \mathbf{y}'_{n,r}, \\
&\hat{\mathbf{b}}_{n+t-r} = \hat{\mathbf{b}}_n + (\mathbf{A}_{n+t-r}^T \mathbf{A}_{n+t-r})^{-1} (\mathbf{M}_{n,t}^T \Delta \mathbf{y}_{n,t} - \mathbf{D}_{n,r}^T \Delta \mathbf{y}'_{n,r}) \\
&= \hat{\mathbf{b}}_n + \mathbf{P}_{n+t-r} (\mathbf{M}_{n,t}^T \Delta \mathbf{y}_{n,t} - \mathbf{D}_{n,r}^T \Delta \mathbf{y}'_{n,r})
\end{aligned} \tag{16}$$

■

**Proof of inference1:** Let us suppose new samples and old samples are weighted input vectors. We can rewrite input matrix **A** as **ΓA**, where weighted matrix **Γ** is diagonal matrix. Let $\mathbf{\Gamma}_{n,t}$ and $\mathbf{\Gamma}_{n,r}$ be the new samples and old samples diagonal weighted matrix . According to theorem 1, we obtain

$$\begin{aligned}
&\mathbf{P}_{n+t} = \mathbf{P}_n - \mathbf{P}_n \mathbf{M}_{n,t}^T \mathbf{\Gamma}_{n,t} (\mathbf{I} + \mathbf{\Gamma}_{n,t} \mathbf{M}_{n,t} \mathbf{P}_n \mathbf{M}_{n,t}^T \mathbf{\Gamma}_{n,t})^{-1} \mathbf{\Gamma}_{n,t} \mathbf{M}_{n,t} \mathbf{P}_n \\
&= \mathbf{P}_n - \mathbf{P}_n \mathbf{M}_{n,t}^T ((\mathbf{I} + \mathbf{\Gamma}_{n,t} \mathbf{M}_{n,t} \mathbf{P}_n \mathbf{M}_{n,t}^T \mathbf{\Gamma}_{n,t}) \mathbf{\Gamma}_{n,t}^{-1})^{-1} \mathbf{\Gamma}_{n,t} \mathbf{M}_{n,t} \mathbf{P}_n \\
&= \mathbf{P}_n - \mathbf{P}_n \mathbf{M}_{n,t}^T (\mathbf{\Gamma}_{n,t}^{-2} + \mathbf{M}_{n,t} \mathbf{P}_n \mathbf{M}_{n,t}^T))^{-1} \mathbf{M}_{n,t} \mathbf{P}_n .
\end{aligned} \tag{17}$$

And

$$\begin{aligned}
&\mathbf{P}_{n+t-r} = \mathbf{P}_{n+t} + \mathbf{P}_{n+t} \mathbf{D}_{n,r}^T \mathbf{\Gamma}_{n,r} (\mathbf{I} - \mathbf{\Gamma}_{n,r} \mathbf{D}_{n,r} \mathbf{P}_{n+t} \mathbf{D}_{n,r}^T \mathbf{\Gamma}_{n,r})^{-1} \mathbf{\Gamma}_{n,r} \mathbf{D}_{n,r} \mathbf{P}_{n+t} \\
&= \mathbf{P}_{n+t} + \mathbf{P}_{n+t} \mathbf{D}_{n,r}^T ((\mathbf{I} - \mathbf{\Gamma}_{n,r} \mathbf{D}_{n,r} \mathbf{P}_{n+t} \mathbf{D}_{n,r}^T \mathbf{\Gamma}_{n,r}) \mathbf{\Gamma}_{n,r}^{-1})^{-1} \mathbf{\Gamma}_{n,r} \mathbf{D}_{n,r} \mathbf{P}_{n+t} \\
&= \mathbf{P}_{n+t} + \mathbf{P}_{n+t} \mathbf{D}_{n,r}^T (\mathbf{\Gamma}_{n,r}^{-2} - \mathbf{D}_{n,r} \mathbf{P}_{n+t} \mathbf{D}_{n,r}^T))^{-1} \mathbf{D}_{n,r} \mathbf{P}_{n+t}, \\
&\hat{\mathbf{B}}_{n+t-r} = \hat{\mathbf{B}}_n + (\mathbf{A}_{n+t-r}^T \mathbf{A}_{n+t-r})^{-1} (\mathbf{M}_{n,t}^T \mathbf{\Gamma}_{n,t} \mathbf{\Gamma}_{n,t} \Delta \mathbf{Y}_{n,t} - \mathbf{D}_{n,r}^T \mathbf{\Gamma}_{n,r} \mathbf{\Gamma}_{n,r} \Delta \mathbf{Y}'_{n,r}) \\
&= \hat{\mathbf{B}}_n + \mathbf{P}_{n+t-r} (\mathbf{M}_{n,t}^T \mathbf{\Gamma}_{n,t}^2 \Delta \mathbf{Y}_{n,t} - \mathbf{D}_{n,r}^T \mathbf{\Gamma}_{n,r}^2 \Delta \mathbf{Y}'_{n,r})
\end{aligned} \tag{18}$$

Where $\mathbf{\Gamma}^{\beta}_{p,q} = diag(\rho^{\beta}_{p,q}, \cdots, \rho^{\beta}_{1,q}), \beta = \pm 1, \pm 2$. ■

### 3.2 An AAR Algorithm for FFNN to Synchronize Variable Data Sets

(1) Initial step:

1. Let the training samples be

$$\mathbf{x}_j = (x_{j,1}, x_{j,2}, \cdots, x_{j,m-1}, 1)_{1\times m}, y_j, 1 \le j \le n,$$
$$\mathbf{A}_{n\times m} = (\mathbf{x}_1^T, \mathbf{x}_2^T, \cdots, \mathbf{x}_n^T)^T, \mathbf{y}_n = (y_1, y_2, \cdots, y_n)^T, \tilde{\mathbf{w}}_i \overset{d}{=} \tilde{\mathbf{b}}_i, \tag{19}$$

2. The initial weight vector of the $i^{\text{th}}$ neuron is

$$\mathbf{y}_n = \mathbf{u}_i = \mathbf{A}_{n\times m}\tilde{\mathbf{b}}_i, \hat{\mathbf{b}}_i = (\mathbf{A}^T_{n\times m}\mathbf{A}_{n\times m})^{-1}\mathbf{A}^T_{n\times m}\mathbf{y}_n, \tag{20}$$

(2) Learning and computing step:

1. Input the initial samples $\mathbf{C} = (c_1, c_2 \ldots c_k)_{1\times k}$.
2. Input the new training samples $(\mathbf{M}_{n,t}, \mathbf{Y}_{n,t})$ which will be inserted into the network

$$\mathbf{M}_{n,t} = (\mathbf{x}^T_{n_1}, \mathbf{x}^T_{n_2}, \cdots, \mathbf{x}^T_{n_t})^T, \mathbf{y}_{n,t} = (y_{n_1}, y_{n_2}, \cdots, y_{n_t})^T. \tag{21}$$

3. Input the old samples ($\mathbf{D}_{n,r}$, $\mathbf{Y}'_{n,r}$), which will be removed from the network.

$$\mathbf{D}_{n,r} = (\mathbf{D}^T_{n_1}, \mathbf{D}^T_{n_2}, \cdots, \mathbf{D}^T_{n_r})^T, \mathbf{y}'_{n,r} = (y'_{n_1}, y'_{n_2}, \cdots, y'_{n_r})^T. \tag{22}$$

4. Update the weight vector. The new weight value $\hat{\mathbf{b}}_{n+t-r}$ is given by the formula 1 or 2, where the output of $i^{\text{th}}$ neuron is

$$\hat{y}_i = f(u_i) = \begin{cases} 1, \text{if } \theta > 0 \\ 0, \text{else} \end{cases}. \tag{23}$$

We ask the output of the $i^{\text{th}}$ neuron is $\hat{\mathbf{y}}_i = \mathbf{x} \cdot \mathbf{w}_i^T = f(\mathbf{x} \cdot \mathbf{w}_i^T)$. The gradient is

$$\left(\frac{\partial f(\mathbf{x} \cdot \mathbf{w}_i^T)}{\partial \mathbf{w}_i}\right)_{\mathbf{w}(j)=\tilde{\mathbf{w}}(j)} = \frac{\partial(\mathbf{x} \cdot \mathbf{w}_i^T)}{\partial \mathbf{w}_i} = \mathbf{x}_i^T. \tag{24}$$

By the formula 1 or 2 and (25) Network will update weight value $\hat{\mathbf{b}}_{n+t-r}$.
(3) Simulation:

Input $\mathbf{x}_0 = (x_1, x_2, \cdots, x_{m-1}, 1)_{1\times m}$, and the output of NN is

$$u_i = \sum_{j=1}^{m} x_j w_{i,j} + b_i = (\mathbf{x}_0, 1)\begin{pmatrix} \mathbf{w}_i \\ b_i \end{pmatrix}, \mathbf{Z}_p = f(\mathbf{U}), \hat{y}_0 = f(\frac{1}{p}\sum_{i=1}^{p} z_i) \tag{25}$$

## 4 Experimental Results

Our approach is carried out to compute residual oil distribution. We use a block data in a sample oil field. The well number is 1813, the grid size of the coordinate $(x_1,x_2)$ is 10379×3367,and the threshold $\theta$=0.5. Let the local data fitting function be the following polynomial:

$$y = y(x_1, x_2) = \sum_{i=0,j=0}^{1,1} a_{i,j} x_1^i x_2^j , \mathbf{x} = (1, x_1, x_2, x_1 x_2) \tag{26}$$

The input layer has 4 nodes. The forgetting factor is

$$\rho_i = \frac{\| \mathbf{p}_i - \mathbf{p}_{i0} \|}{(1 + | y_i - y_{i0} |)(1 + \| \mathbf{p}_i - \mathbf{p}_0 \|)^{2.5}} , \quad \Gamma = diag(\rho_1, \cdots, \rho_n), i = 1,2,\cdots,n \tag{27}$$

Where $\mathbf{p}_0 = (x_1^{(0)}, x_2^{(0)})$ is the planar coordinate of the evaluated value, $\mathbf{p}_i = (x_1^{(i)}, x_2^{(i)})$ is the $i^{\text{th}}$ well coordinate, $\mathbf{p}_{i,0} = (x_1^{(i,0)}, x_2^{(i,0)})$ is the nearest well coordinate of the point $\mathbf{p}_i$, $y_i$ is the height of the point $\mathbf{p}_i$, $y_{i,0}$ is the height of the point $\mathbf{p}_{i,0}$, ||*|| is the Euclidean distance norm, |*| is a absolute operator. Fig.1 shows the process of the new training samples inserting and old samples removing. Fig.2 shows the residual oil distribution by our approach.

A comparison experiment of our approach with simple Kriging (Surfer software) is also carried out. The results are shown in Fig.3. From Fig.3, we conclude that the two methods are similar, value and shape-preserved. However, the connectivity is

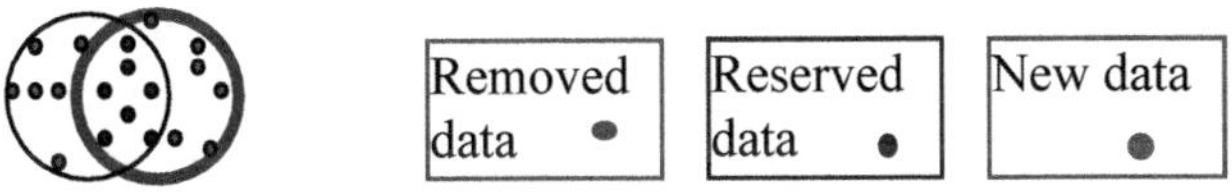


**Fig. 1.** The process of the fitting window moving and data updating

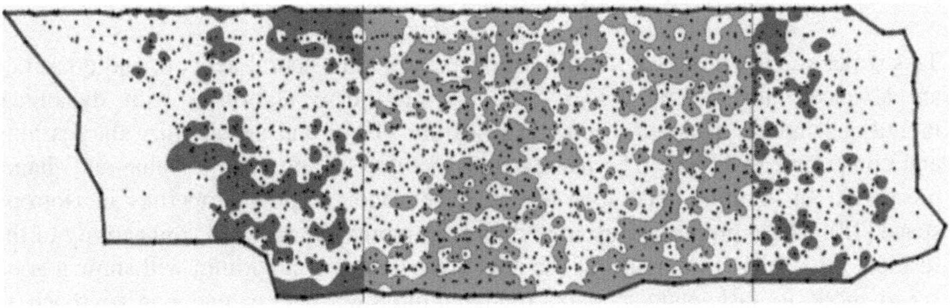

**Fig. 2.** Residual oil distributions by our approach

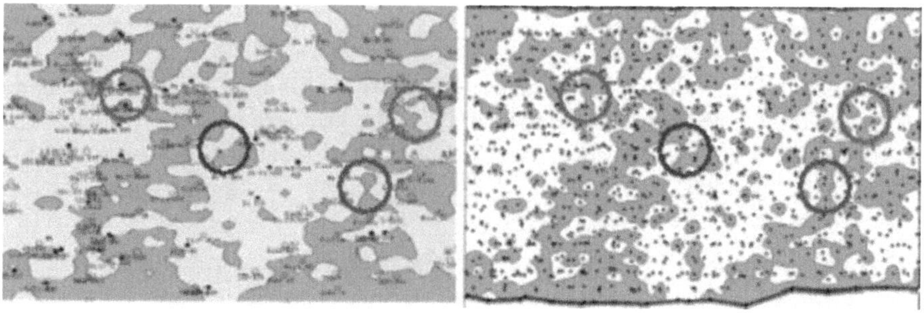

a. Residual oil distribution by Kriging method.

b. Residual oil distribution by our approach.

**Fig. 3.** Comparison of our approach with Kriging

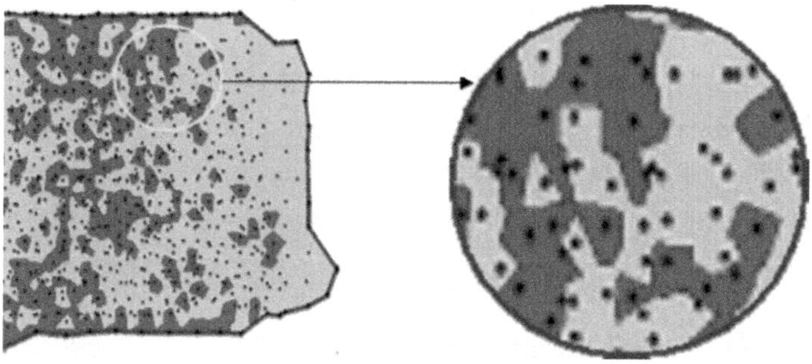

**Fig. 4.** Voronoi graph computed by our dynamical learning neural network

different. Our approach is of better connectivity than the Kriging. This connectivity is important to decide how and where to develop oilfield. Our approach is in accord with the engineer's estimation for the prediction of residual oil distribution. Fig.4 shows the Voronoi graph result [12]. This is the optimal result if there is no anisotropy.

## 5 Conclusion and Future Work

This paper studied the AAR algorithm for feed forward neural network and presented an advanced adaptive recursive (AAR) least square algorithm with dynamical inputting window. This approach can fast train FFNN and synchronize the learning and computing in FFNN. The results showed that our approach is value and shape-preserved. In addition, our approach is of Voronoi graph's properties in isotropy space [12]. These properties are important to compute the regional connectivity of the residual oil. With the increase of the new input data, our algorithm will show a good speed merit to fast evaluate data. In the future, we will extend our approach to implement the fast moving fitting of GPS terrain surface.

## References


1. Hoffmann, M. , Kovács, E.: Developable Surface Modeling by Neural Network. Mathematical and Computer Modelling, 38(2003) 849-853
2. Hoffmann, M., Kohonen.: Neural Network for Surface Reconstruction. Publ. Math. 54 Suppl (1999) 857-864
3. Yu, Y.: Surface Reconstruction from Unorganized Points using Self-organizing Neural Networks. In IEEE Visualization 99,Conference Proceedings (1999) 61–64
4. Várady, L., Hoffmann, M. , Kovács, E.: Improved Free-form Modelling of Scattered Data by Dynamic Neural Networks. Journal for Geom. and Graph, 3 (1999) 177-183
5. Wu, A, Hsieh, W.W., Tang, B.: Neural Network Forecasts of the Tropical Pacific Sea Surface Temperatures. Neural Networks: the Official Journal of the International Neural Network Society, 19 (2) (2006) 145-54
6. Zadeh, L. A.: From Circuit Theory to System Theory. Proc. IRE, 50(5) (1962) 856-865
7. Eykhoff, P.: System Identification – Parameter and State Estimation. John Wiley & Sons, INC.(1974)
8. Palmieri, F., *et al.*: Sound Localization with a Neural Network Trained with the Multiple Extended Kalmann Algorithm. Proc IJCNN, (1991)125-131
9. Azimi-Sadjadi, M. R. , Liou, R. J.: Fast Learning Process of Multi-Layer Neural Networks Using RLS Technique. IEEE Trans. on Signal Processing, SP-40(2) (1992)446-450
10. Shah, S., Palmieri, F., Datum, M.: Optimal Filtering Algorithms for Fast Learning in Feedforward Neural Networks. Journal for Neural Networks, 5 (5) (1992)779-787
11. Li, A.G., Qin, Z.: Moving Windows Quadratic Autoregressive Model for Predicting Nonlinear Time Series. Chinese Journal of Computers, 27 (7) (2004) 1004-1008
12. Amenta, N., Bern, M., Kamvysselis, M.: A New Voronoi–based Surface Reconstruction Algorithm. In SIGGRAPH 98, Conference Proceedings (1998) 415–422

# BOLD Dynamic Model of Functional MRI

Ling Zeng, Yuqi Wang, and Huafu Chen*

School of Applied Mathematics, School of Life Science & Technology, University of Electronic Science and Technology of China, Chengdu 610054, China
chenhf@uestc.edu.cn

**Abstract.** Blood oxygenation level dependent (BOLD)contrast based functional magnetic resonance imaging (fMRI)can be used to detect brain neural activities. In this paper, a new procedure is presented which allows the estimation of the hemodynamic approach from BOLD responses. The procedure is based on Friston proposed dynamic model and Agnes Aubert proposed a correlation model between activation and motabolism, in this case, adopted to characterize hemodynamic responses in functional magnetic resonance imaging (fMRI). This work represents a fundamental improvement over existing approaches to system identification using nonlinear hemodynamic models. The model can simulate the change of oxygen motabolism, de-oxyhemoglobine, cerebral blood flow and volume to brain activation.

## 1 Introduction

Blood oxygenation level dependent (BOLD)contrast based functional magnetic resonance imaging (fMRI)can be used to detect brain neural activities The physiological mechanisms underlying the relationship between synaptic activation and vascular/metabolic controlling systems have been widely reported. Hence, some authors have attempted to model the BOLD signal at the macroscopic level by differential equations systems, relating the hemodynamical variations to relative changes in a set of variables with physiological sense. The Balloon approach, based on the mechanically compelling model of an expandable venous compartment [1] and the standard Windkessel theory [2], has become an established idea. Friston et al have extended the Balloon approach, named in this paper simply the hemodynamic approach, to include interrelationships between physiological (i.e., neuronal synaptic activity and a flow-inducing signal) and hemodynamic processes[3]. In the hemodynamic approach, a set of four nonlinear and nonautonomous ordinary differential equations governs the dynamics of the intrinsic variables: the flow-inducing signal, the Cerebral Blood Flow (CBF), the Cerebral Blood Volume (CBV) and the total de-oxyhemoglobine (dHb). This dynamic system is, in effect, nonautonomous due to the time-varying dependence of the synaptic activity, which will be referred to henceforth as the input sequence. Though this theoretical model could have a tremendous impact on fMRI analysis, there is little work done in fitting and validating it from actual data. The most important attempt to date has been presented by Friston [3] using a Volterra

* Corresponding author.

D.-S. Huang, L. Heutte, and M. Loog (Eds.): ICIC 2007, LNAI 4682, pp. 324–329, 2007.

series expansion to capture nonlinear effects on the output of the model produced by predefined input sequences. In that work, the Volterra kernels were explicitly computed for the hemodynamic approach, after a set of assumptions that forced the original deterministic and continuous differential equations system to have a bilinear form. An EM implementation of a Gauss–Newton search method, in the context of maximum a posteriori mode estimation, was used to determine the hemodynamic parameters. Even though this methodology theoretically allows the computation of Volterra kernels of any order, in practice, a finite truncation of the series must be carried out, limiting the representation of higher order nonlinear dynamics. The estimation of the states and parameters of the hemodynamic approach from blood oxygenation level dependent (BOLD) responses are reported [6]. On other hand, model of the coupling between brain electrical activity and metabolism [7,8], and model of the hemodynamic response and oxygen delivery to brain [9] are reported. Further, the BOLD signal model is applied to study brain functional activation [10-11], and the fMRI data analysis method is improved to better locate brain functional activation[12-13].

In this paper, an extended BOLD dynamic model are firstly presented based on Friston dynamic and Aubert model to simulation BOLD dynamic process, which included CBF,CBV, bHb, oxygenation metabolism. Finally, the model dynamic response is analyzed by using gamma input function.

## 2 An Extended BOLD Dynamic Model

In this section we describe a hemodynamic model that mediates between synaptic activity and measured BOLD responses. This model essentially combines the Balloon model and a simple linear dynamical model of changes in regional cerebral blood flow (rCBF) caused by neuronal activity.

### 2.1 The Balloon Component

This component links rCBF and the BOLD signal as described in Buxton *et al.*([1]. All variables are expressed in normalized form, relative to resting values. The BOLD signal $y(t) = \lambda(v, q, E_0)$ is taken to be a static nonlinear function of normalized venous volume (v), normalized total deoxyhemoglobin voxel content (q) and resting net oxygen extraction fraction by the capillary bed ( $E_0$ )

$$y(t) = V_0(k_1(1-q) + k_2(1-q/v) + k_3(1-v)) \tag{1}$$
$$k_1 = 7E_0$$
$$k_2 = 2$$
$$k_3 = 2E_0 - 0.2$$

where $V_0$ is resting blood volume fraction. This signal comprises a volume-weighted sum of extra- and intra-vascular signals that are functions of volume and deoxyhemoglobin content. The latter are the state variables whose dynamics need specifying. The rate of change of volume is simply

$$\dot{V} = f_{in} - f_{out} \tag{2}$$

Equation (2) says that volume changes reflect the difference between inflow $f_{in}$ and outflow $f_{out}$ from the venous compartment with a time constant.

Note that outflow is a function of volume. This function models the balloon-like capacity of the venous compartment to expel blood at a greater rate when distended. We model it with a single parameter a based on the windkessel model

$$f_{out} = V^{1/\alpha} \tag{3}$$

At steady state empirical results from PET suggest a 0.38.

The change in deoxyhemoglobin $\dot{q}$ reflects the delivery of deoxyhemoglobin into the venous compartment of deoxyhemoglobin into the venous compartment

$$\dot{q} = f_{in} \frac{E(f_{in}, E_0)}{E_0} - f_{out}(v) q / V \tag{4}$$

where $E(f_{in}, E_0)$ is the fraction of oxygen extracted from the inflowing blood. This is assumed to depend on oxygen delivery and is consequently flow-dependent. A reasonable approximation for a wide range of transport conditions is [1].

$$E(f_{in}, E_0) = 1 - (1 - E_0)^{1/f_{in}} \tag{5}$$

## 2.2 rCBF Component

Friston suggests that the observed nonlinearities enter into the translation of rCBF into a BOLD response (as opposed to a nonlinear relationship between synaptic activity and rCBF) in the auditory cortices [3]. Under the constraint that the dynamical system linking synaptic activity and rCBF is linear we have chosen the most parsimonious model

$$\dot{f}_{in} = s \tag{6}$$

where s is some flow inducing signal defined, operationally, in units corresponding to the rate of change of normalized flow.

The signal is assumed to subsume many neurogenic and diffusive signal subcomponents and is generated by neuronal activity u(t)

$$\dot{s} = \varepsilon u(t) - s / \tau_s - (f_{in} - 1) / \tau_f \tag{7}$$

$\varepsilon, \tau_s, \tau_f$ are the three unknown parameters that determine the dynamics of this component of the hemodynamic model. They represent the efficacy with which neuronal activity cause an increase in signal [4].

### 2.3 Oxygen Extraction

We assume that the average concentration of oxygen present inside the capillary is

$$O_{2c} = (O_{2\bar{c}} + O_{2a})/2 \tag{8}$$

where $O_{2a}$ is the arterial oxygen concentration, and $O_{2\bar{c}}$ the oxygen concentration at the end of the capillaries. The results obtained using this simple expression are close to those obtained with more complex ones, derived by integrating oxygen extraction along capillary segment, provided that the oxygen extraction fraction [7,8]

$$E = 1 - O_{2\bar{c}} / O_{2a} \tag{9}$$

is less than 0.8

Then mass balance leads to the equation Capillary oxygen

$$\frac{dO_{2c}}{dt} = V_{O2C} - V_{O2m}\frac{V_i}{V_{cap}} \tag{10}$$

where the rate of oxygen in capillary $V_{O2C}$ is

$$V_{O2C} = \frac{2F_0 f_{in}(t)}{V_{cap}}(O_{2a} - O_{2C}) \tag{11}$$

$V_{O2m}$ is the rate of net oxygen transport across the blood–brain barrier per unit intracellular volume. $Vcap$ is capillary volumes。

Combining equation(10) and (11), the new similar equation can be acquired

$$\frac{dO_{2c}}{dt} = (f_{in}(t) - f_{out}(v,\alpha))(O_{2a} - O_{2c})/Vcap \tag{12}$$

## 3 Result

Prior BOLD model did not discuses model input function $u(t)$ ([3-5]). But it is important to stable of the dynamic mode. The gamma functions are discussed as model input as follows.

Stimulus input function $u(t)$ is supposed as gamma function

$$u(t) = \frac{c}{\tau_h m!}\left(\frac{t - t_d}{\tau_h}\right)^m e^{-(t-t_d)/\tau_h} \tag{13}$$

where $t_d$ shows the time delay, $\tau_h$ signifies the blurring effect, $m$ is a response scale which will affect the shape of h(t), and $c$ is an amplitude factor of the response which does not affect the shape of the function[9-11]). We can get dynamic model result using Equation (12) in Figure1. CBF, CBV, de-oxyhemoglobine (dHb) , the oxygen extraction fraction and BOLD are simulated to fit the physiological characteristic.

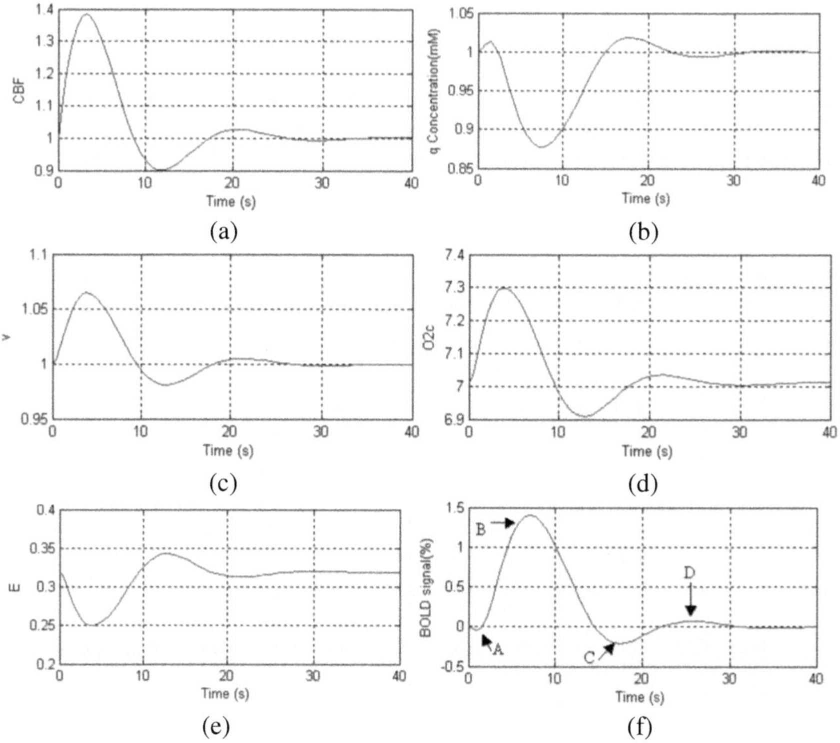


**Fig. 1.** Dynamic model simulation result based gamma function, where $t_d = 1.5$ , $\tau_h = 1.5$ , $c = 54$ , $t_s = 0.8$ $m = 20$ , $E_0 = 0.319$ .(a) is CBF simulation result. (b) is the de-oxyhemoglobine simulation result. (c) is CBV simulation result. (d) is the simulation result of the rate of oxygen in capillary. (e). is the simulation result of the oxygen extraction fraction. (f) is the simulation result of BOLD signal.

## 4 Conclusion

In this paper, the extended dynamic model is proposed based on Friston's BOLD dynamic model and Agnes Aubert's brain electricity blood dynamic model, Which can link between metabolize and CBF, CBF, CBV. Our model suggests that the BOLD response seem have rebound response after the end of the post-stimulus undershoot (D part in Fig1.(f)). It is clear that more study needs to be undertaken to further delineate the precise physical and biological mechanisms leading to these pattern.

## Acknowledgment

Supported in NSFC# 30570507, 30525030, New Century Excellent Talents in University(NECT-05-0809). Key research project of science and technology of MOE(107097).

## References

1. Buxton, R.B., Wong ,E.C., Frank, L.R. : Dynamics of Blood Flow and Oxygenation Changes During Brain Activation: the Balloon Model. Magnetic Resonance in Medicine 39(1998) 855-864
2. Mandeville, J.B., Marota, J.J.A, Ayata, C., Zaharchuk, G., Moskowitz, M.A., Rosen, B.R., Weisskoff,R.M.: Evidence of Cerebrovascular Postarteriole Windkessel with Delayed Compliance. J. Cereb. Blood Flow Metab 19 (1999) 679– 689
3. Friston ,K.J.: Bayesian Estimation of Dynamical Systems: An Application to fMRI. Neuroimage( 2002) 513-530
4. Friston, K. J., Josephs, O., Rees, G., Turner, R.: Nonlinear Event-related Responses in fMRI. Magn. Reson Med. 39(1998) 41–52
5. Friston, K.J., Mechelli, A., Turner, R., Price, C.J.: Nonlinear Responses in fMRI the Balloon Model, Volterra Kernels, and Other Hemo-Dynamics. NeuroImage 12(2000) 466– 477
6. Riera, J.J., Watanabe, J., Kazuki I. ,Naoki M. ,Aubert E., Ozaki ,T., Kawashima, R.A.: State-Space Model of the Hemodynamic Approach: Nonlinear Filtering of BOLD Signals. NeuroImage 21 (2004) 547–567
7. Aubert, A., Costalat, R.: A Model of the Coupling between Brain Electrical Activity, Metabolism and Hemodynamic: Application to the Interpretation of Functional Neuroimaging. NeuroImage. 17(2002) 1162- 1181
8. Aubert , Costalat, R., Valabrègue ,R.: Modeling of The Coupling Between Brain Electrical Activity and Metabolism. Acta Biotheoretica 49(2001)301-326
9. Zheng, Y., Martindale, J., Johnston, D., Jones, M., Berwick, J., Mayhew, J.: A Model of Hemodynamic Response and Oxygen Delivery to Brain. NeuroImage 16(2002) 617–637
10. Chen, H, Yao,D, Liu, Z:Analysis of the fMRI BOLD Response of Spatial Visual by Analysis of the fMRI BOLD Response . Brain Topography17( 2004) 39-46
11. Chen, H, Yao, D , Liu, Z: A Comparison of Gamma and Gaussian Dynamic Convolution Models of the fMRI BOLD Response. Magnetic Resonance Imaging 23 (2005) 83-88
12. Chen, H, Yuan., H, Yao., D, Chen, L. , Chen, W.: An Integrated Neighborhood Correlation and Hierarchical Clustering Approach of Functional MRI. IEEE Trans ,Biomedical Engineering ,53(2006) 452-258
13. Chen, H., Yao, D., Chen, W., Chen, L.: Delay Correlation Subspace Decomposition Algorithm and Its Application in fMRI IEEE Trans, Medical Imaging(2005)1647-1650

# Partial Eigenanalysis for Power System Stability Study by Connection Network

Pei-Hwa Huang and Chao-Chun Li

Department of Electrical Engineering, National Taiwan Ocean University
Peining Road, Keelung 20224, Taiwan
B0104@mail.ntou.edu.tw

**Abstract.** Power system small signal stability concerns the ability of the power system to maintain stable subject to small disturbances. The method of frequency-domain analysis, namely the analysis of system eigenstructure, is commonly employed for the study of small signal stability. However, we often face high-order system matrix due to the large number of generating units so that it will be undesirable to calculate and analyze the whole system eigenstructure. The main purpose of this paper is to present an algorithm to find out the eigenvalue of the worst-damped electromechanical mode or the eigenvalues of all unstable electromechanical modes, i.e. to figure out those eigenvalues of critical oscillatory modes. The proposed algorithm takes advantage of the specific feature of the parallel structure of connection networks for calculating the eigenvalues. Numerical results from performing eigenvalue analysis on a sample power system are demonstrated to verify the proposed method.

**Keywords:** Connection Network, Artificial Neural Network, Power System Stability, Eigenvalue Calculation, Partial Eigenstructure.

## 1 Introduction

Power system small signal stability concerns the ability of the power system to maintain stable subject to small disturbances. [1-12]. There are generally two kinds of approaches for analyzing the power system small signal stability, namely the time domain analysis and the frequency domain analysis. The time domain simulation method is first to apply small disturbances in the system, and then to find the solutions of the state equations as well as to observe the variations of the state variables of the power system to determine the stability of the system. The major disadvantage of the time domain approach is that the procedure is time consuming and several tests might be required. Besides, the system response is the composite response of several oscillating modes; it is hard to determine the damping of the individual oscillating mode. On the other hand, in the frequency domain approach, the problem of small signal stability of the power system is focused on finding the system eigenstructure, namely the eigenvalues and the corresponding eigenvectors. Because the small signal stability concerns the system to remain stable operation under small

D.-S. Huang, L. Heutte, and M. Loog (Eds.): ICIC 2007, LNAI 4682, pp. 330–339, 2007.

disturbances, the original nonlinear system can be linearized at the operating point to obtain the state equations of the linearized system. Therefore, we can use the linear system theory to find the system eigenstructure and based on which we are able to determine whether the power system is stable or not [13,14]. However, we often face high-order system matrix due to the large number of generating units in the system so that it will be time consuming to calculate and to analyze the whole system eigenstructure.

The main purpose of this paper is to present an algorithm to calculate the eigenvalue of the worst-damped electromechanical mode or those eigenvalues of all unstable oscillatory modes, i.e. to figure out those eigenvalues associated with critical oscillatory modes instead of figuring out all the system eigenvalues. The proposed calculation method takes advantages of the specific feature of the parallel structure of the connection network (artificial neural network) [15-19], combined with the operations of matrix shifting and inversion [20-24], to figure out the subset of eigenvalues associated with the most unstable oscillatory mode, which is the mode with lowest damping, and/or to all unstable oscillatory modes of the power system. Numerical results from performing eigenvalue analysis on a sample power system are demonstrated to verify the proposed method.

## 2 Small Signal Stability Analysis

Power system small signal stability is often referred to as power system dynamic stability and it focuses on the ability of the system to maintain stable subject to small disturbances [1-3]. Instead of employing the time domain approach of applying various small disturbances on the system to observe the dynamic behaviors of the system, the frequency domain approach, i.e. performing eigen-analysis by calculating the eigenvalue/eigenvector of the system matrix of the linearized system under study, has been widely adopted in the industry for power system small signal stability analysis. Eigenanalysis is primarily based on modal expansion theory (modal analysis) [13,14].

Consider the linear unforced system described in (1):

$$\dot{x}(t) = Ax(t), \quad x(0) = x_0 \tag{1}$$

where $x(t)$, $x_0$, and $A$ denote the $n\times1$ state vector, the $n\times1$ initial states, and the $n\times n$ system matrix, respectively. The solution of (1) is

$$x(t) = e^{At}x_0 \,. \tag{2}$$

We are to use the concept of eigenvalue/eigenvector to further analyze the system described in (1). The eigenvalues of a $n\times n$ matrix $A$ are the $n$ scalars, denoted by $\lambda_i$, $i = 1,2,\cdots,n$ and each associated with a corresponding $n\times1$ vector $v_i$, satisfying (3)

$$Av_i = \lambda_i v_i, \quad i = 1,2,\cdots,n \,. \tag{3}$$

Note that $\lambda_i$ is the $i$th eigenvalue and $v_i$ is the eigenvector corresponding to $\lambda_i$.

Assume that all eigenvalues are distinct and thus $\{v_1, v_2, \cdots, v_3\}$ is a set of linearly independent vectors. Define the modal matrix $M$ :

$$M = \begin{bmatrix} v_1 & v_2 & \cdots v_n \end{bmatrix}. \tag{4}$$

The inverse matrix $M^{-1}$ exists because $\det(M) \neq 0$. Consider a new state vector $z$ defined by the transformation

$$x = Mz, \quad z = M^{-1}x. \tag{5}$$

The system in (1) can be rewritten as

$$\dot{z}(t) = M^{-1} A M \, z(t), \quad z_0 = M^{-1} x_0 \tag{6}$$

and $M^{-1}AM$ is a diagonal matrix with eigenvalues as its diagonal elements. Define

$$\Lambda = M^{-1} A M = \text{diag}\begin{bmatrix} \lambda_1 & \lambda_2 & \cdots \lambda_n \end{bmatrix}. \tag{7}$$

Therefore,

$$e^{\Lambda t} = \text{diag}\left[ e^{\lambda_1 t} \;\; e^{\lambda_2 t} \cdots e^{\lambda_n t} \right]. \tag{8}$$

The solution of (6) is thus

$$z(t) = e^{\Lambda t} z_0 . \tag{9}$$

Then we can obtain the original state vector $x(t)$ as

$$x(t) = M e^{\Lambda t} M^{-1} x_0 . \tag{10}$$

Consider the modal matrix $M$ in (4). Denotes the $i$th row of $M^{-1}$ as $\ell_i$, that is

$$M^{-1} = \begin{bmatrix} \ell_1 \\ \ell_2 \\ \vdots \\ \ell_n \end{bmatrix}. \tag{11}$$

The row vector $\ell_i$ is of dimension $1 \times n$ and is named the left eigenvector of matrix $A$, and the earlier-mentioned $n \times 1$ column vector $v_i$ is often referred to as the right eigenvector. Hence $x(t)$ can be further expressed as

$$x(t) = M e^{\Lambda t} M^{-1} x_0 = \begin{bmatrix} v_1 \cdots v_n \end{bmatrix} \cdot \text{diag}\left[ e^{\lambda_1 t} \cdots e^{\lambda_n t} \right] \cdot \begin{bmatrix} \ell_1 \\ \vdots \\ \ell_n \end{bmatrix} \cdot x_0 \tag{12}$$

Define a $1 \times n$ vector $\alpha$

$$\alpha = [\alpha_1 \ \alpha_2 \cdots \alpha_n]^T = M^{-1} x_0 \tag{13}$$

in which the scalar element $\alpha_i = \ell_i x_0$. Finally the state vector $x(t)$ is obtained as

$$x(t) = \sum_{i=1}^{n} \alpha_i e^{\lambda_i t} v_i . \tag{14}$$

Equation (14) is referred to as the Modal Expansion Theory [13,14]. From (14), the unforced system response $x(t)$ depends upon $\lambda_i$, $v_i$ and $\alpha_i$. Each term of $\exp(\lambda_i t) v_i$ is referred to as a *mode* and $x(t)$ is a composite response formed from the linear combination of every mode $\exp(\lambda_i t) v_i$ with the initial state related scalar term $\alpha_i$ as the coefficients. A real eigenvalue corresponds to a non-oscillatory mode. A positive real eigenvalue represents an aperiodic unstable mode, and a negative real eigenvalue represents a decaying mode. On the other hand, complex eigenvalues occur in conjugate pairs and each pair corresponds to an oscillatory mode. A pair of complex eigenvalues $\lambda = \sigma \pm j\omega$ include a real part $\sigma$ and an imaginary part $\omega$. The imaginary part $\omega = 2\pi f$ gives the frequency $f$ of the corresponding oscillatory mode. The real part $\sigma$ reveals the damping of the associated oscillatory mode: a positive value means a negative damping while a negative value represents a positive damping. A real part with zero value implies there is no damping with the mode. A linear system is stable if every eigenvalue of its system matrix has a negative real part.

Since power system small signal stability concerns the ability of the system to maintain stable subject to small disturbances, eigenanalysis is suitable for the studies of small signal stability. The original system under study is first linearized at the operating point to derive the linear state equations from which the system matrix is obtained. Then the eigenvalues of the system matrix are calculated for checking the small signal stability. If all the real parts of eigenvalues are negative, the system will be classified as stable in the sense of small signal stability. It is noted that the system to be studied usually has high order system matrix due to the large number of generators so that it will be time consuming to calculate and to analyze all the eigenvalues. It is most desirable in system planning and operation to find out only those eigenvalues corresponding to the mode with lowest damping, and/or to all unstable oscillatory modes, instead of all the system eigenvalues for fast determination of system stability.

## 3 Connection Network

The connection network or the artificial neural network is a data processing system which simulates the functions and operations of the human brain. A typical neural network consists of a set of processing units, the neurons, and the neurons communicate with each other through weighted links. The neurons process their input values in parallel and independently of each other. The output of one neuron becomes the input of other neurons and the connection between any pair of neurons sets up the structure of the neural network [15-19]. The connection network is used for the calculation of eigenvalues and eigenvectors in this paper.

A simple neuron is shown in Fig. 1 where $x_i$ stands for the value of the $i$th input of the neuron, $w_i$ is the weight associated with the link between the $i$th input and the neuron, $y$ is the output of the neuron, and $f(\cdot)$ is the activation function.

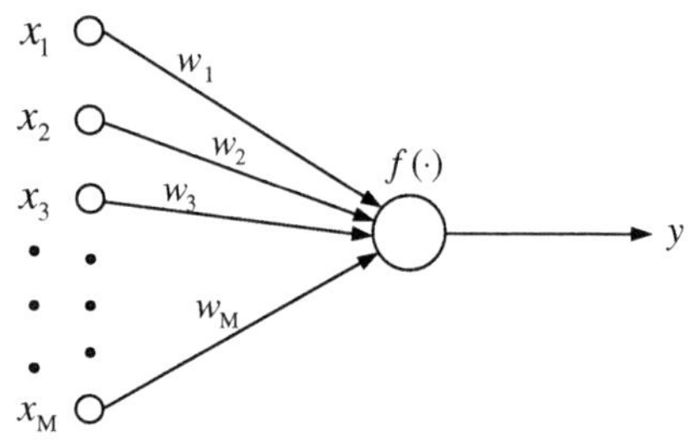


**Fig. 1.** Structure of a simple neuron

In Fig. 1, the network input of the neuron is the weighted sum of all input values

$$u = \sum_{i=1}^{m} w_i x_i \tag{15}$$

and the output of the neuron is

$$y = f(u). \tag{16}$$

The neurons process their input values in parallel and independently of each other and thus the structure of the connection network is adopted to perform parallel processing. In this paper the connection network is employed for the calculation of eigenvalues. Define the connection vector $w$ and the input vector $x$ for the network in Fig. 1 as

$$w = [w_1 \ \ w_2 \ \cdots \ w_m], \tag{17}$$

$$x = [x_1 \ \ x_2 \ \cdots \ x_m]. \tag{18}$$

and (15) can be represented as

$$u = w^T x. \tag{19}$$

In a connection network, the weights between any pair of neurons can be modified by using a learning rule. A Hebbian learning rule in (20) can be used for determining the values of the weights. Note that $\gamma$ is a constant between 0 and 1.

$$w(t+1) = w(t) + \gamma u(t) x(t) \tag{20}$$

In this paper the structure of the connection network is utilized for finding eigenvalues and eigenvectors. Consider the network structure in Fig. 2 in which $v_i$ stands for the output of the $i$th neuron and $w_{ij}$ represents the weight of the link between the $i$th and the $j$th neurons [17].

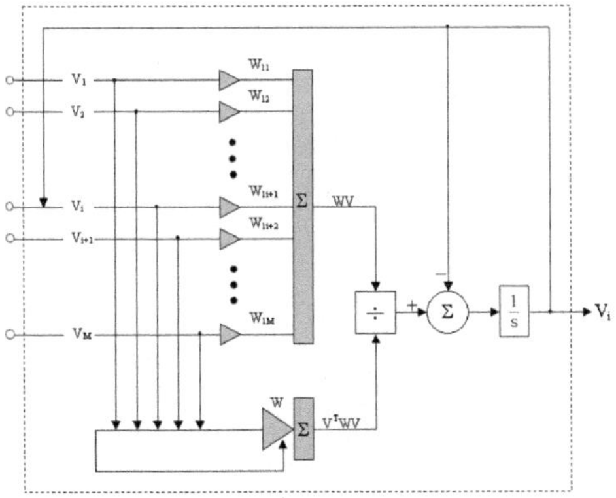

**Fig. 2.** Connection network structure for finding eigenvector

Denote the eigenvalues of the weight matrix $W=[w_{ij}]$ as $\lambda_1, \lambda_2, \cdots, \lambda_M$, with the decreasing order of magnitude as $|\lambda_1|>|\lambda_2|>\cdots>|\lambda_M|$, and the corresponding eigenvectors $V_1, V_2, \cdots, V_M$, respectively. The input-output dynamic relationship of each neuron is

$$\frac{d}{dt}V_i(t) = V(t) + k \cdot \left[ \frac{\sum_{j=1}^{M} W_{ij} V_j(t)}{\sum_{i\leq j}^{M} W_{ij} V_i(t) V_j(t)} - V_i(t) \right] \tag{21}$$

where $k$ is a constant. Rearranging (21) in the vector form will yield

$$\frac{d}{dt}V(t) = V(t) + k \cdot \left[ \frac{W\,V(t)}{V^T(t)\,W\,V(t)} - V(t) \right] \tag{22}$$

After finding the solution of $V(t)$ in (22) and substituting $V(t)$ by $V_1(t)$, the eigenvalue with the largest magnitude, $\lambda_1$, can be obtained as

$$\lambda_1 = \frac{V_1^T W V_1(t)}{V_1^T V_1(t)}. \tag{23}$$

The symmetric matrix $W$ in (24) represents the relationship between eigenvalues and eigenvectors

$$W = \sum_{i=1}^{M} \lambda_i V_i V_i^T . \tag{24}$$

Define a transform matrix $T$ as

$$T^{(1)} = I - V_1 V_1^T . \tag{25}$$

Multiplying (25) with (24) will get

$$W^{(1)} = T^{(1)} W = (I - V_1 V_1^T) \sum_{i=1}^{M} \lambda_i V_i V_i^T \tag{26}$$

and since $V_i^T V_j = 0, i \neq j$ we have

$$W^{(1)} = \sum_{i=1}^{M} \lambda_i V_i V_i^T \tag{27}$$

Note that $\lambda_1 = 0$ in (27) and this will make the eigenvector associated with the eigenvalue with the second largest magnitude become the eigenvector corresponding to the eigenvalue with the largest magnitude for $W^{(1)}$. If $W^{(1)}$ is used as the weight matrix for the connection network shown in Fig. 2 and the output of the network will be the eigenvector corresponding to the original eigenvalue with the second largest magnitude. Then likewise, we can further define a transform matrix

$$T^{(i-1)} = (I - V_{i-1} V_{i-1}^T) . \tag{28}$$

The new weight matrix can be found as

$$W^{(i-1)} = T^{(i-1)} W \tag{29}$$

where $\lambda_{i-1} = 0$. Substituting $\lambda_{i-1} = 0$ back into the network in Fig. 2 will result in the output as the eigenvector corresponding to the $i$th eigenvalue, i.e. the eigenvalue with the $i$th largest magnitude. In this way, we can find the eigenvalues with the largest down to the smallest magnitudes and their corresponding eigenvectors. Such process forms the foundation for the calculation of critical eigenvalues in power system small signal stability analysis.

## 4 Calculation of Critical Eigenvalues

When the above-mentioned connection network based eigenvalue/eigenvector calculation process is used for power system small signal stability analysis, the operations of matrix shifting and inversion are included to devise a systematic procedure suitable for the calculation of power system critical eigenvalues.

The following steps comprise the procedure for calculating critical eigenvalues of the power system.

(1) Make a matrix shifting operation for the original system matrix by $A' = A - \beta I$ where matrix $A'$ is the shifted matrix obtained from performing matrix shifting on the original system matrix $A$ and $\beta$ is a complex number for shifting operation. Normally $\beta$ is chosen to be a location in the right half plane of the complex plane, e.g. $30 + j5$.

(2) Find the inverse of the shifted matrix $A'$. That is, we are to find the matrix $(A')^{-1} = (A - \beta I)^{-1}$. Denote the eigenvalues of $A'$ as $\lambda_1', \lambda_2', \cdots, \lambda_M'$ for which $|\lambda_1'| > |\lambda_2'| > \cdots > |\lambda_M'|$.

(3) Use the connection network eigenvalue/eigenvector calculation process to compute $\lambda_1'$ which is the eigenvalue with the largest magnitude among those eigenvalues of $A'$.

(4) Calculate $\lambda_1 = \beta + 1/\lambda_1'$ where $\lambda_1$ is the most unstable eigenvalue of the system matrix $A$.

(5) If $\mathrm{Re}(\lambda_1) < 0$, then the system under study is stable. If $\mathrm{Re}(\lambda_1) \geq 0$, the system under study is unstable; then we go to step (3) and repeat the eigenvalue/eigenvector calculation process for the next iteration to obtain the eigenvalue with magnitude next in order until a stable eigenvalue is figured out.

The proposed algorithm as described in the above five steps will be employed for calculating critical eigenvalues in power system small signal stability analysis.

## 5 Analysis of Sample Example

A sample power system described in [25] is adopted as the study system for testing the proposed approach. The single-line diagram of the study system is shown in Fig. 3. The

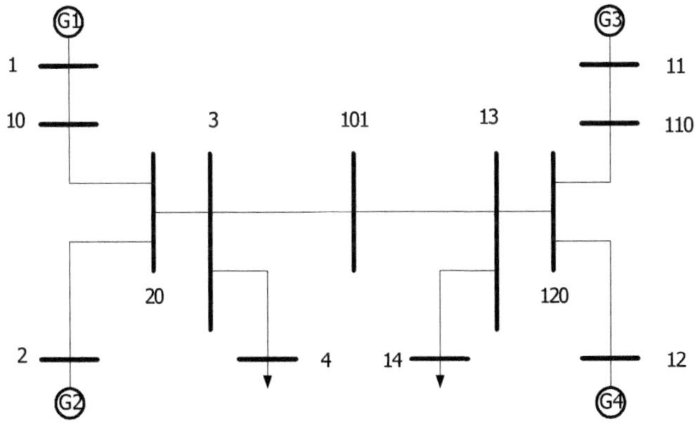


**Fig. 3.** Single line diagram of the study system

study system is a system with thirteen buses and four generators. After the linearization process, a system state matrix of the order of 57×57 is obtained. Then the most unstable eigenvalues are found to be 0.066632±j3.2429 and another unstable eigenvalue is computed as 0.000015102. Because the eigenvalue of the mode with the lowest damping falls on the right half plane of the complex plane, the system is an unstable system. All eigenvalues of the state matrix are shown in Table 1. It is worth noting that the error of this calculation is less than $1\times10^{-10}$, as compared to the solution from the Matlab software. The computer time for the calculation is 0.078 second.

**Table 1.** All eigenvalues of the study system

| | | | |
|---|---|---|---|
| 0.000015102 | -0.19468 | -0.19861 | -0.19862 |
| -0.58289 | -0.37474+j0.45428 | -0.37474-j0.45428 | -0.38627+j0.44995 |
| -0.38627-j0.44995 | -0.68001 | -0.24917+j0.64503 | -0.24917-j0.64503 |
| -1.5913 | -0.50516+j1.7217 | -0.50516-j1.7217 | -1.8933 |
| -2.0008 | -2.0011 | -1.2625+j1.9041 | -1.2625 –j1.9041 |
| -2.7687+j0.0054216 | -2.7687-j0.0054216 | 0.066632 +j3.2429 | 0.066632-j3.2429 |
| -3.3805 | -3.4764 | -4.4515 | -4.4702 |
| -0.49102+j6.8639 | -0.49102-j6.8639 | -0.49142+j6.9059 | -0.49142-j6.9059 |
| -10.07 | -10.07 | -10.1 | -10.11 |
| -14.24 | -14.248 | -14.479 | -14.628 |
| -20 | -20 | -20 | -20 |
| -27.611+j5.0277 | -27.611-j 5.0277 | -29.513 | -33.566 |
| -34.566 | -35.848 | -36.052 | -37.12 |
| -37.167 | -99.998 | -99.998 | -99.999 |
| -99.999 | | | |

## 6 Conclusion

The main purpose of this paper is to discuss an algorithm for the analysis of power system small signal stability in order to compute the eigenvalues of the worst-damped oscillatory mode or the eigenvalues of all unstable electromechanical modes, i.e. to figure out those eigenvalues of critical oscillatory modes. The proposed method takes advantages of the specific feature of the parallel structure of the connection network (the neural network), along with the operations of matrix shifting and inversion, for finding the partial eigenstructure corresponding to the most unstable oscillatory mode, i.e. the mode with lowest damping, and/or all unstable oscillatory modes of the system. Numerical results from performing eigenanalysis on a sample power system are demonstrated and it is found the proposed approach is suitable for the analysis of power system small signal stability.

## References


1. Anderson, P.M., Fouad, A.A.: Power System Control and Stability. IEEE Press (1994)
2. Kundur, P.: Power System Stability and Control. McGraw-Hill (1994)
3. Rogers, G.: Power System Oscillations. Kluwer Academic Publishers (2000)
4. Campagnolom, J.M., Martins, L., Lima, T.G.: Fast Small-Signal Stability Assessment Using Parallel Processing. IEEE Trans. on Power Systems, 9 (1994) 949-956
5. Angelidis, G., Semlyen, A.: Efficient Calculation of Critical Eigenvalue Clusters in The Small Signal Stability Analysis of Large Power System. IEEE Trans. on Power Systems,10 (1995) 427-432
6. Campagnolo, J.M., Martins, N.D., Falcao, M.: An Efficient and Robust Eigenvalue Method for Small-Signal Stability Assessment in Parallel Computers. IEEE Trans. on Power Systems,10 (1995) 506-511
7. Lima, T.G., Bezerra, H., Martins, L.: New Mcthods for Fast Small-Signal Stability Assessment of Large Scale Power System. IEEE Trans. on Power Systems, 10 (1995) 1979-1985
8. Angelidis, G. Semlyen, A.: Improved Methodologies for the Calculation of Critical Eigenvalues in Small Signal Stability Analysis. IEEE Transactions on Power Systems, 11 (1996) 1209-1217
9. Makarov, Y.V., Dong, Z.Y., Hill, D.J.: A General Method for Small Signal Stability Analysis. IEEE Trans. on Power Systems, 13 (1998) 979-985
10. Wang, K.W., Chung, C Y., Tse, C.T., Tsang, K.M.: Multimachine Eigenvalues Sensitivities of Power System Parameters. IEEE Trans.on Power Systems, 15 (2000) 741-747
11. Gomes, S., Martins, N., Portela, C.: Computing Small-Signal Stability Boundaries for Large-Scale Power Systems. IEEE Trans. on Power Systems, 18 (2003) 747-752
12. Zhang, X., Shen, C.: A Distributed-computing-based Eigenvalue Algorithm for Stability Analysis of Large-scale Power Systems. Proceedings of 2006 International Conference on Power System Technology (2006) 1-5
13. Kailath, T.: Linear Systems. Prentice-Hall (1980)
14. Ogata, K.: System Dynamics. 4th edn. Prentice Hall (2003)
15. Oja, E.: A Simplified Neuron Model as a Principle Components Analyzer. Journal of Mathematical Biology,15 (1982) 267-273
16. Lau, C.: Neural Networks-Theoretical Foundations and Analysis. IEEE Press (1992)
17. Li, T.Y.: Eigen-decompositioned Neural Networks for Beaming Estimation. M.Sc. Thesis, National Taiwan Ocean University (1994)
18. Nauck, D., Klawonn, F., Kruse, R: Neuro-Fuzzy System. John Wiley & Sons (1997)
19. Haykin, S.: Neural Network. Prentice-Hall (1999)
20. Golub, G.H., van Loan, C.F.: Matrix Computations. 2nd edn. The Johns Hopkins University Press (1989)
21. Goldberg, J.L.: Matrix Theory with Applications. McGraw-Hill (1992)
22. Datta, B.N.: Numerical Linear Algebra and Applications. Brooks/Cole (1995)
23. Anton, H., Rorres, C.: Elementary Linear Algebra Application. John Wiley & Sons, Inc. (2000)
24. Leon, S.J.: Linear Algebra with Applications. 6th edn. Prentice Hall (2002)
25. Yu, Y.N., Siggers, C.: Stabilization and Optimal Control Signal for Power Systems. IEEE Trans. on Power Apparatus and Systems, 90 (1971) 1469-1481

# A Knowledge Navigation Method for the Domain of Customers' Services of Mobile Communication Corporations in China

Jiangning Wu and Xiaohuan Wang

Institute of Systems Engineering, Dalian University of Technology
Dalian, Liaoning, 116024, P.R. China
jnwu@dlut.edu.cn, xhwang@student.dlut.edu.cn

**Abstract.** Rapidly increasing amount of mobile phone users and types of services leads to a great accumulation of complaining information. How to use this information to enhance the quality of customers' services is a big issue at present. To handle this kind of problem, the paper presents an approach to construct a domain knowledge map for navigating the explicit and tacit knowledge in two ways: building the Topic Map-based explicit knowledge navigation model, which includes domain TM construction, a semantic topic expansion algorithm and VSM-based similarity calculation; building Social Network Analysis-based tacit knowledge navigation model, which includes a multi-relational expert navigation algorithm and the criterions to evaluate the performance of expert networks. In doing so, both the customer managers and operators in call centers can find the appropriate knowledge and experts quickly and exactly. The experimental results show that the above method is very powerful for knowledge navigation.

**Keywords:** Topic Map, Social Network Analysis, Knowledge Navigation, Explicit Knowledge, Tacit Knowledge.

## 1 Introduction

With the rapid development of China's economy and communication technologies, the number of mobile phone users in China is greatly increasing year by year. Meanwhile, the Mobile Communication Corporations (MCCs) in China are providing more types of services now than before. Consequently, more and more complaining information come forth. So there is a great need for effective tools that can quickly find useful information and then extract interesting knowledge. Topic Map (TM) as an effective knowledge organization and navigation tool is adopted in the study for navigating the explicit knowledge. With respect to the tacit knowledge, a tool namely social network analysis (SNA) is introduced. In the domain of Customers' Services in MCCs of China, the explicit knowledge refers to the customers' complaining pieces in the form of document, and the tacit knowledge there refers to the person (expert) who owns more practical experiences in problem solving.

D.-S. Huang, L. Heutte, and M. Loog (Eds.): ICIC 2007, LNAI 4682, pp. 340–349, 2007.

In order to navigate both explicit knowledge and tacit knowledge simultaneously, the paper presents an approach to build a knowledge map in the given domain that consists of two parts: TM-based explicit knowledge navigation and SNA-based tacit knowledge navigation. Such knowledge map brings more benefits for customer managers and operators in call centers. The experimental results show that both TM and SNA respectively are powerful tools for explicit and tacit knowledge navigation.

# 2 TM-Based Explicit Knowledge Navigation

## 2.1 TM Construction

According to the TM structure [1], there are three main phases involved in the TM construction process: *topic selection*, *occurrence appending*, and *association analysis*. In the study, the data is collected from a MCC of a certain city as well as the official website of MCC of a certain province in China. The test data are 500 pieces of customers' complaining documents in the form of Excel.

In *topic selection* phase, topics are selected in the following ways, shown in Fig. 1.

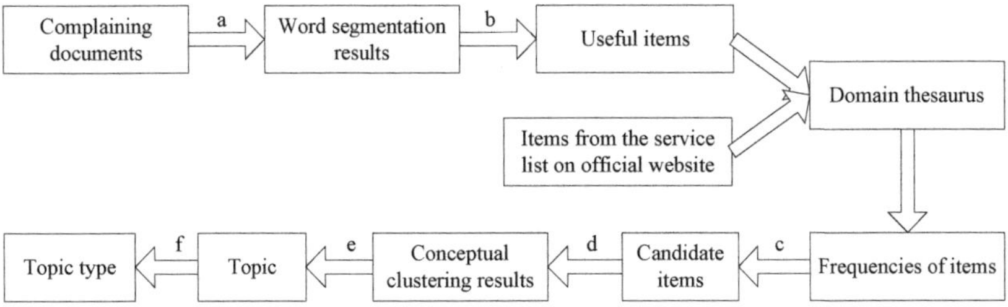


**Fig. 1.** Process of topic and topic type selection

Where, (a) 500 pieces of complaining documents are segmented based on the algorithm in Ref. [2]; (b) and (c) 5228 segmented items are obtained, of which 420 items are selected according to the following principle: the item with 2-6 characters and appearing above 5 times has a good description about the certain domain, and then 210 candidate items are selected in terms of their frequencies; (d) 110 items are selected after conceptual clustering process based on the algorithm in Ref. [3]; (e) 89 items are chosen among the above items and named as topics, which are quite relative to the given domain and can describe the domain well; (f) All topics are classified into 5 topic types, which are *Service*, *Customer*, *Network*, *System*, and *Dealer*.

In *occurrence appending* phase, occurrences are appended in the following steps:

Step 1: Map the multi-dimension space namely knowledge level of domain TM into the one-dimension space. Each topic type is considered as a topic concept tree, which can then be transformed into one dimension vector. See an example in Fig. 2.

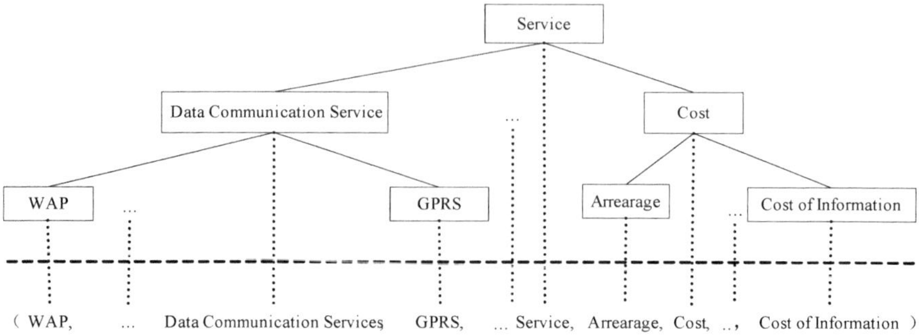


**Fig. 2.** A part of domain TM and the mapping results

Considering different topics at different levels present different importance and the topics at the lower level of the topic concept tree are more important to the users, therefore, different topics should be given different weights, defined in Equation (1):

$$\beta = h/H \ . \tag{1}$$

Where $\beta$ denotes the weight of the current topic, $h$ denotes the height of the current topic, and $H$ denotes the hierarchical height of the branch where the current topic exists in.

For all topics belonging to the same level, in this case, the leaf topics are much more concrete than the non-leaf topics. So the distances between leaf topics and non-leaf topics should be considered. Therefore, the weight definition is modified as:

$$w = \beta / \mathrm{K}^r = h / H \cdot \mathrm{K}^r \ . \tag{2}$$

Where $w$ denotes the modified weight of the current topic, $r$ denotes the distance between the current topic and the leaf topic of the same branch to which the current topic is belonging, K is a constant, normally, K=2. Here, the root topic is at level 0.

In this paper, topic type is defined as $T_i$, $T_{ij}$ is hyponymy topic of $T_i$, $T_{ijk}$ is hyponymy topic of $T_{ij}$, and so forth, down to the leaf topics. Correspondingly, $w_i$, $w_{ij}$, $w_{ijk}$ are weights of $T_i$, $T_{ij}$, $T_{ijk}$, and $i, j, k$ are natural numbers. Finally, the topic map can be represented as one-dimension vector, $\boldsymbol{T} = \{w_1, w_{11}, \ldots, w_{1i}, \ldots, w_{1jk}, \ldots, w_i, \ldots, w_{ij}, \ldots, w_{ijk}, \ldots\}$.

Step 2: Construct an M×N Topic-Document Matrix, where the number of complaining documents is M, and the number of topics is N. Element $t_{m,ijk}$ in the matrix denotes whether topic $T_{ijk}$ has appeared in document $D_m$; if present, then mark 1; otherwise, mark 0. If both hypernymy topic and hyponymy topic appear in the same document, then append the document to the hyponymy topic. Actually, one document is permitted to connect with more than one topics belonging to different topic types. Here, $t_{m,ijk}$ is defined as follows:

$$\begin{aligned} &\forall m \in \mathrm{M}, \text{ if } T_i \in D_m, \text{ then } t_{m,i} = 1; \text{ or } T_i \in D_m, \text{ and } T_{ij} \in D_m, \text{ then } t_{m,i} = 0, t_{m,ij} = 1 \ . \\ &\forall m \in \mathrm{M}, \text{ if } T_{ij} \in D_m, \text{ then } t_{m,ij} = 1; \text{ or } T_{ij} \in D_m, \text{ and } T_{ijk} \in D_m, \text{ then } t_{m,ij} = 0, t_{m,ijk} = 1 \ . \end{aligned} \tag{3}$$

Step 3: Appending occurrences back to the TM. Occurrences in TM are customers' complaining documents. The matrix obtained in Step 2 shows the relations between topics and documents. Therefore, the complaining documents can be appended back to each topic according to the Topic-Document Matrix.

In *association analysis* phase, relations between topics and topic types are analyzed. These relations are associations of the domain TM. From 500 pieces of complaining documents, 3 kinds of associations, viz. *Contain*, *Influence-on* and *Complain*, with 6 kinds of association roles, viz. *Hypernymy/Hyponymy*, *Customer/Complaining object* and *Cause/Result*, are extracted manually.

Up to now, the whole TM in the domain of Customers' Services for MCCs has been constructed.

## 2.2 TM Maintenance

Managers and operators can use the developed TM to serve users and make some improvements. As time moves on, the quality of current services would become satisfying; meanwhile, new kinds of services would cause new problems. Therefore, the TM has to be modified timely, and two ways are presented in this section.

Adding a topic: A newly happened problem, which is not belonging to the current complaining type, cannot be solved very well at the beginning. Later, if the number of complaining documents towards this kind of problem is large enough, the new problem should be considered as a new complaining type. The adding condition is shown in the following expressions:

$$n_{t+k} \le n_{t+k+1}, \quad k = 0, 1, 2, 3 \ldots .$$

$$\frac{n_{t+k} + n_{t+k+1}}{2} \le \frac{n_{t+k+1} n_{t+k+2}}{2}, \quad k = 0, 1, 2, 3 \ldots . \tag{4}$$

$$\frac{\frac{n_{t+k} + n_{t+k+1}}{2} + \frac{n_{t+k+1} + n_{t+k+2}}{2}}{2} \le \frac{\frac{n_{t+k+1} + n_{t+k+2}}{2} + \frac{n_{t+k+2} + n_{t+k+3}}{2}}{2}, \quad k = 0, 1, 2, 3 \ldots .$$

Until the last piece of document.

Removing a topic: If an old problem, which is belonging to a complaining type, is fully solved, the number of complaining documents would become smaller and smaller. Under this circumstance, the topic can be removed off the domain TM. The removing condition is shown in the expressions below:

$$n_{t+k} \ge n_{t+k+1}, \quad k = 0, 1, 2, 3 \ldots .$$

$$\frac{n_{t+k} + n_{t+k+1}}{2} \ge \frac{n_{t+k+1} n_{t+k+2}}{2}, \quad k = 0, 1, 2, 3 \ldots . \tag{5}$$

$$\frac{\frac{n_{t+k} + n_{t+k+1}}{2} + \frac{n_{t+k+1} + n_{t+k+2}}{2}}{2} \ge \frac{\frac{n_{t+k+1} + n_{t+k+2}}{2} + \frac{n_{t+k+2} + n_{t+k+3}}{2}}{2}, \quad k = 0,1,2,3 \ldots .$$

Until the last piece of document.

Where, $n_t$ denotes the number of complaining documents at the time of t, $k$ denotes the number of months after time t, and t is a constant.

### 2.3 TM Usages

There are two main usages of TM in the knowledge navigation system: knowledge browse and information retrieval.

Knowledge browse: People are able to find certain knowledge by browsing the knowledge level of domain TM, and address the information resources by browsing the information level.

Information retrieval: A semantic topic expansion algorithm is proposed for this usage, which is described in detail in Algorithm 1, in which queries are obtained from the given lists by choosing target topics, associations and occurrence types.

**Algorithm 1.** Semantic based topic expansion algorithm

Input: Target topics, target associations, target occurrence types
Output: Extended sub-TM, viz. relevant topics, associations, and occurrence types
Step 1: Choose target $topic_i$ from the topic list (multiple selections are possible); list all topics which are associated with $topic_i$ by "contain", and "contain" associations themselves. Then $graph_1$, whose apex is $topic_i$, is obtained;
Step 2: Choose target $association_j$ from the association list (multiple selections are possible); list all topics which are associated with topics in $graph_1$ by $association_j$, and $association_j$ themselves. Then $graph_2$ is obtained;
Step 3: List all topics which are associated with topics in $graph_2$ by "contain", and "contain" association themselves. Then $graph_3$ is obtained;
Step 4: List all topics which are associated with topics in $graph_3$ by $association_j$, and $association_j$ themselves. Then $graph_4$ is obtained;
Step 5: Repeat Steps 3 and 4 until no more $association_j$ appears; then $graph_s$ is obtained, $s \geq 2$.
Step 6: Choose target $occurrence\ type_k$ from the occurrence type list (multiple selections are possible); list all occurrences of $graph_s$ belonging to this type.
Then the expanded sub-TM is obtained.

This algorithm can not only reveal semantic relations between topics, but also can realize some reasoning processes. Take Fig. 3 into account, the given query is "What influence on the signals of a mobile phone?" in which (a) shows that the general retrieval process can only find out "*Network* influences on *Signal*"; but (b) shows that Algorithm 1 helps to find out there is an influence-on relationship between *System* and *Network*; and (c) shows that *Equipment* belongs to *System*. Then the conclusion can be made, i.e., "*Equipment* is the real reason influencing on the *Signals*".

Although the users are able to obtain some results related to the given query by the Algorithm 1, to get more satisfying results, similarities between the complaining documents and the given query should be calculated. In this paper, the similarity is calculated by the cosine measure based on VSM [4]. Here, the topic weights are defined

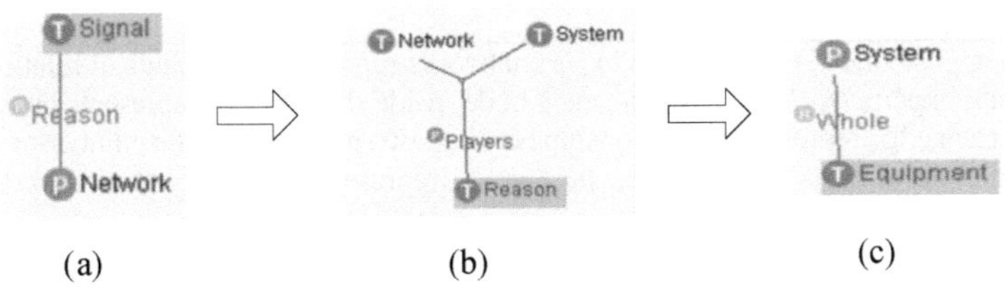


**Fig. 3.** Process of topic expansion (Created by *TM4J*, available at: http://compsci.wssu.edu/iis/nsdl/download.html)

as in Section 2.1. Therefore each complaining document can be represented as $\boldsymbol{D}_s=\{w_{s1}, w_{s11}, \ldots w_{s1i}, \ldots, w_{s1jk}, \ldots, w_{si}, \ldots, w_{sij}, \ldots, w_{sijk}, \ldots\}$; and the query can be represented as $\boldsymbol{Q}=\{w_{q1}, w_{q11}, \ldots w_{q1i}, \ldots, w_{q1jk}, \ldots, w_{qi}, \ldots, w_{qij}, \ldots, w_{qijk}, \ldots\}$ in the same way. The similarity between $\boldsymbol{D}_s$ and $\boldsymbol{Q}$ is given by Equation (6):

$$sim(\boldsymbol{D}_s, \mathrm{Q}) = \cos\theta = \frac{w_{s1}w_{q1} + \ldots + w_{si}w_{qi} + \ldots + w_{sij}w_{qij} + \ldots + w_{sijk}w_{qijk}}{\sqrt{w_{s1}^2 + \ldots + w_{sijk}^2}\sqrt{w_{q1}^2 + \ldots + w_{qijk}^2}}. \tag{6}$$

Then a threshold is set to limit the relevant result outputs.

The developed retrieval system is very friendly for users by providing fixed query lists, such as lists of topics, associations, and occurrence types, which avoids troubles in inputting the queries. The system performance is evaluated by precision and recall.

For both TM-based and keyword-based information retrieval systems, the average precisions are 84.64% and 72.92% respectively, while the average recalls are 69.68% and 61.10% respectively. Apparently, TM-based information retrieval system has a better performance than keyword-based one.

## 3 Social Network Based Tacit Knowledge Navigation

Tacit knowledge is difficult in coding and spreading abroad, because it is always in forms of experiences, techniques, etc; and it is mainly stored in humans' brains [5]. So navigating tacit knowledge is transformed into navigating experts. Since there are many experts existing inside MCCs, how to find out the appropriate experts to solve problems quickly and reasonably becomes a hot topic. Social Network Analysis (SNA) is a powerful tool to deal with people's relationships [6], and it is also helpful to enhance the effect and efficiency of tacit knowledge navigation [7].

### 3.1 Multi-relational Expert Navigation Method

Since there are many kinds of relationships existed among experts, we should consider this multi-relational fact when searching experts. First of all, a multi-relational expert navigation algorithm is proposed to realize experts' navigation.

**Algorithm 2.** Multi-relational expert navigation algorithm

---

Suppose that (1) There are R expert networks representing R kinds of relationship among experts $N_1$, $N_2$… $N_r$… $N_R$; each node inside the network represents an expert, and each edge represents a relationship between two experts; (2) The number of nodes in each network are $n_1$, $n_2$… $n_r$… $n_R$; (3) $a_{r,ij}$ represents the edge between $node_i$ and $node_j$ in network $N_r$; (4) $\lambda_{r,ij}$ represents the weight of edge $a_{r,ij}$.

Suppose that (1) The new network is $N$; (2) The number of nodes is $n$; (3) $a_{ij}$ represents the edge between $node_i$ and $node_j$ in network $N$; (4) $\lambda_{ij}$ represents the weight of edge $a_{ij}$. Here, $1 \le r \le R$; $1 \le i, j \le n_r$.

Then map networks $N_1, N_2, \ldots, N_r, \ldots, N_R$ into network $N$, with no changes towards the nodes and edges, but the weight of each edge is changed as follows:

$$\forall a_{ij} \in N, \lambda_{ij} = \min\{\lambda_{r,ij}\} \ . \tag{7}$$

In doing so, a new expert network with different edge weights, named as multi-relational expert network, is built.

Suppose that the number of expert navigating routes is M, each route between two experts is represented as $S_1$, $S_2$…$S_m$…$S_M$; $s_m$ is the length of route $S_m$.

If $a_{ij} \in S_m$, then $a_{ij}=1$; otherwise, $a_{ij}=0$.Then,

$$s_m(i,j) = \sum_{i}^{j} a_{ij}\lambda_{ij} \ , \ 1 \le m \le M \ . \tag{8}$$

Rank all $s_m$. To the end, $S_m$ with the smallest $s_m$ is the best expert navigating route. Moreover, *E.D.Dijkstra* method can be used to provide navigating routes as well.

---

In the study, the relationships are extracted out from questionnaires, and there are two questions involved in the survey. One is "Have you ever cooperated with expert Ei?" Another is "Would you like to work with expert Ei? or Do you think expert Ei is a reliable person?" Ten experts are participating in this survey, and according to their answers, two kinds of expert networks are obtained, as shown in Fig. 4.

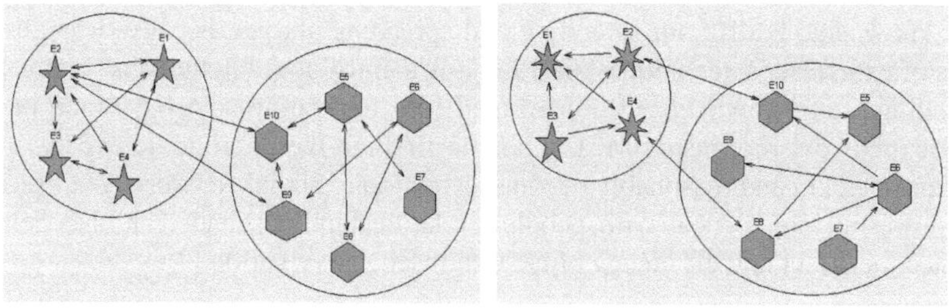


**Fig. 4.** Two kinds of graphs of experts' relationship networks

Suppose that all the relationships in the same network are equal to each other, and two networks have different weights λ1, λ2, hereλ1=2λ2. Then both graphs are mapped into a new expert network based on Algorithm 2. Then the result is shown in Fig. 5.

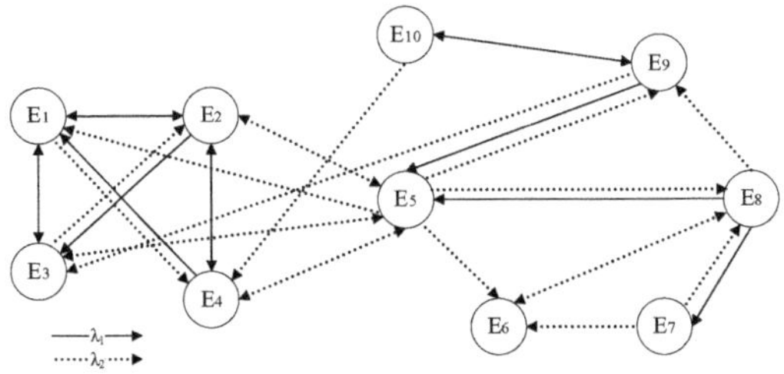


**Fig. 5.** Graph of 2-relational expert network

Based on *E.D.Dijkstra* method [8], the shortest navigating route between $E_1$ and $E_6$ is $E_1 \xrightarrow{\lambda_2} E_4 \xrightarrow{\lambda_2} E_5 \xrightarrow{\lambda_2} E_6$, and the length is $3\lambda_2$. That is to say, $E_1$, $E_4$, $E_5$, $E_6$ are navigated based on the second kind of relationship.

### 3.2 Criterions to Evaluate the Performance of Expert Network

Many SNA tools provide criterions to evaluate the performance of expert network, such as InFlow [9], UCINET, Netdraw [10], KeyPlayer, SociometryPro [11], etc. In this study, SociometryPro tool is adopted for this purpose.

SociometryPro provides two kinds of criterions, group index, which includes Density, Cohesion, Stability, and Intensity; and individual index, which includes Weight, Emotional effusiveness, Satisfaction, and Status. Take the right graph in Fig. 4 as an example, the evaluation results are shown in Fig. 6.

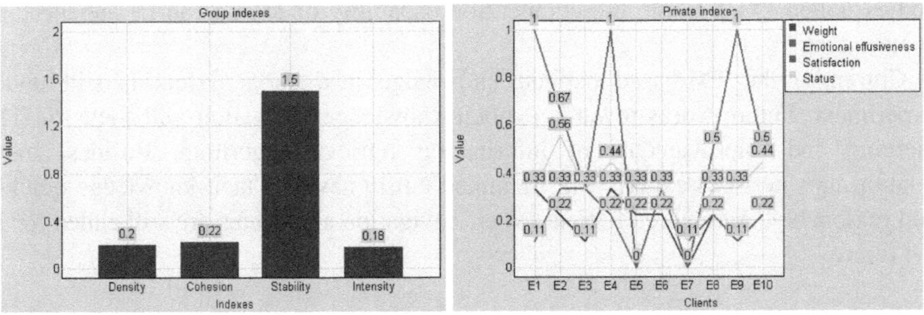


**Fig. 6.** Results of evaluation (Created by *SociometryPro2.3*, available at: http://www.allworldsoft.com/download/16-578-sociometrypro_download.htm)

From the above results, some conclusions can be summarized as follows:

(1) The expert network is weak to some extent in terms of *Stability*. It points out minimal part of the group must be removed to divide the group into unrelated parts. Here, the value of *Stability* is 1.5, which means someone' leaving might result in the group's disjunction. In that case, communications between people should be enhanced, and more opportunities for people to get to know each other should be created.

(2) The values of *weight* for $E_1$ and $E_4$ are the same 0.33, which are the highest among all the experts. It implies that $E_1$ and $E_4$ play very important roles in the corporation. In fact, they do improve the communication between experts and accordingly enhance navigating tacit knowledge inside the corporation.

(3) The values of *satisfaction* for $E_1$, $E_4$, and $E_9$ are all 1.0, which are the highest among all the experts. It implies that $E_1$, $E_4$, and $E_9$ are very satisfied with their partners and vice versa. Actually, they have potential possibilities to improve the communications and realize the tacit knowledge navigation as best as they can.

(4) The values of *satisfaction* for $E_5$ and $E_7$ are both 0.0, which are the lowest among all the experts. It implies that $E_5$ and $E_7$ are not satisfied with their partners and vice versa. As a matter of fact, they are very likely to cumber the communications of tacit knowledge. Approaches should be proposed to improve their attitudes towards work and relationships with others.

Now, the status of the corporations can be easily viewed. So the corresponding decisions should be made to improve the tacit knowledge navigation.

## 4 Conclusions and Future Works

The paper presents a domain knowledge map with which both explicit knowledge and tacit knowledge involved in the customers' services of MCCs can be navigated efficiently. The knowledge map is composed of two models: TM-based explicit knowledge navigation model and SNA-based tacit knowledge navigation model. By means of these two models, knowledge inside the corporations can be well managed and exploited. As a result, the competition capability of MCCs can be enhanced to some extent.

Currently, the TM-based explicit knowledge navigation system is still under experiment. Future works towards explicit knowledge navigation will focus on TM merging and improvements of information retrieval algorithm. Besides, more relationships between experts will be mined out to navigate tacit knowledge quickly and reasonably. And more efficient expert navigation algorithms are still called for in the future.

**Acknowledgements.** This study is sponsored by the National Natural Science Foundation of China (NSFC), Grant Nos. 70431001 and 70620140115.

## References

1. Steven, P.: The TAO of Topic Map: Finding the Way on the Age of Infoglut. [Online] Available at: http://www.ontopia.net/ topicmaps/meterials/ tao.html
2. Jiang, S.H.: Segmentation Algorithm for Chinese Text Based on Length Descending and String Frequency Statistics. Vol. 25, No. 1 (2006) 74-79 (in Chinese)

3. Wu, J.N., Tian, H.Y., Yang, G.F.: A Multilayer Topic-Map-Based Model Used for Document Resources Organization. In De-Shuang Huang, Kang Li, George William Irwin (Eds.): Lecture Notes in Control and Information Sciences, Vol. 344. Springer-Verlag, Berlin Heidelberg (2006) 753-758
4. Salton, G., Wong, A., Yang, C.S.: A Vector Space Model for Automatic Indexing. Communications of the ACM, Vol.18, No.11 (1975) 613-620
5. Polanyi, M.: Personal Knowledge. Routledge, London (1958)
6. Liu, J.: Introduction to Social Network Analysis. Social Science Literature Publishing House, Beijing (2004) (in Chinese)
7. Social network analysis-KM toolkit: inventory of tools and techniques-knowledge management. [Online] Available at: http://www.nelh.nhs.uk/knowledge_management/km2/social_network.asp
8. Hu, Y.Q.: Introduction to operations research. Harbin institute of technology press, Harbin (1998) (in Chinese)
9. InFlow: [Online] Available at: http://www.orgnet.com/inflow3.html
10. NewDraw: [Online] Available at: http://www.analytictech.com/downloadnd.htm
11. Social Network Analysis: Introduction and Resources: [Online] Available at: http://lrs.ed.uiuc.edu/tse-portal/analysis/social-network-analysis/#portals

# A Method for Building Concept Lattice Based on Matrix Operation

Kai Li[1], Yajun Du[1], Dan Xiang[1], Honghua Chen[1], and Zhenwen Liao[2]

[1] School of Mathematics & Computer Science, Xihua University, Chengdu, Sichuan, 610039, China
[2] Chengdu Center, China Geological Survey, Chengdu 610081, China
likai0148@hotmail.com, duyajun@mail.xhu.edu.cn, jintianxd@163.com, setfuzzy@yahoo.com.cn

**Abstract.** As a power tool for analyzing data, concept lattice has been extensively applied in several areas such as knowledge discovery, software engineering and case-based reasoning. However, building concept lattice is time-consuming and complicated; it becomes the bottleneck of application. Therefore, a simple and efficient method for building concept lattice is proposed in this paper. We transform binary formal context into matrix at first, and then discuss how to build concept lattice based on basic concepts and added concepts, which the two concepts can be got from matrix operation. We also present a fast algorithm BCLMO (Building Concept Lattice based on Matrix Operation) for building concept lattice, and analyze the time complexity of BCLMO. The method we proposed could remarkably reduce the time complexity and improve the efficiency of building concept lattice.

**Keywords:** BCLMO; Concept Lattice; Matrix Operation; Formal Concept Analysis.

## 1 Introduction

Formal Concept Analysis (FCA) is a mathematical method for analyzing binary relations, it's a power tool which used to analyze data and extract knowledge from formal context by concept lattice. 1982, concept lattice was first introduced by Wille [1], it established on the basis of FCA in theory. In FCA, each element in the concept lattice is a formal concept, and the corresponding graph (Hasse diagram) is considered as the generalization/specialization relationship between concepts. At present, FCA has been extensively applied in several areas such as knowledge discovery [2], software engineering [3] and case-based reasoning [4].

There are many algorithms for building concept lattice. Bordat [5] and CBO [6] use trees for storing concepts, which allows efficient search for a concept when the diagram constructed. Nourine [7] algorithm constructs a tree of concepts and searches for every newly generated concept. Qiao algorithm [8] derived all the concepts of the context, when database updates, it is suitable for added some new

D.-S. Huang, L. Heutte, and M. Loog (Eds.): ICIC 2007, LNAI 4682, pp. 350–359, 2007.

objects into the concept lattice. Missaoui and Godin [9] proposed an algorithm based on a hash function, which makes it possible to distribute concepts among 'buckets' and reduce search. In [10,11], the method of constructing increment concept lattice based on multi-valued formal context is presented. Literature [10], uses rough set theory to reduce the attributes in formal context, thereby reducing the time of building concept lattice. LCA [11] uses a support degree $\varepsilon$ to measure the quality of concept lattice and reduce the number of formal concept. Iceberg concept lattices are proposed in [12,13], which can be constructed by TITANIC.

In this paper, we propose a new method for building concept lattice, which based on the matrix operation. In the following section, we recall some basic definitions related to the concept lattice. Section 3 introduces how to extract formal concepts based on the matrix operation. The algorithm BCLMO is described in section 4 for building concept lattice. We conclude our work in section 5 with a look in the future.

## 2 Basic Notions

In this section, we will recall some necessary basic notions used in our paper. The detail description about concept lattice can be found in [1,14,15,16].

**Definition 1.** *A formal context is a triple* $\boldsymbol{K}$*:= (G,M,I) where G and M are sets and* $I \subseteq G \times M$ *is a binary relation. The elements of G are called objects and the elements of M are called attributes. The inclusion (g, m)* $\in I$ *is read "object g has attribute m". For* $A \subseteq G$*, we define*

$$A^{\uparrow} := \{m \in M | \forall g \in A : (g, m) \in I\};$$

*and for* $B \subseteq M$*, we define dually*

$$B^{\downarrow} := \{g \in G | \forall m \in B : (g, m) \in I\}.$$

*In this paper, we assume that all sets are finite, especially G and M.*

**Definition 2.** *A formal concept is a pair (A,B) with* $A \subseteq G$*,* $B \subseteq M$*,* $A^{\uparrow} = B$ *and* $B^{\downarrow} = A$*. (This is equivalent to* $A \subseteq G$ *and* $B \subseteq M$ *being maximal with* $A \times B \subseteq I$*.) A is called extent and B is called intent of the concept.*

**Definition 3.** *The set* $\underline{\boldsymbol{B}}(\boldsymbol{K})$ *of all concepts of a formal context* $\boldsymbol{K}$ *together with the partial order* $(A_1,B_1) \leq (A_2,B_2) \iff A_1 \subseteq A_2$ *(which is equivalent to* $B_2 \subseteq B_1$*) is called concept lattice of* $\boldsymbol{K}$*.*

Example. Table 1 describes a binary formal context. $G = \{1, 2, 3\}$, $M = \{a, b, c, d, e, f\}$, $I$ depicts objects in G have attributes in M. Fig. 1 depicts the concept lattice that corresponds to the context in Table 1.

**Table 1.** A binary formal context

| | a | b | c | d | e | f |
|---|---|---|---|---|---|---|
| 1 | × | × | | × | × | × |
| 2 | × | | × | × | × | × |
| 3 | | × | × | × | × | × |

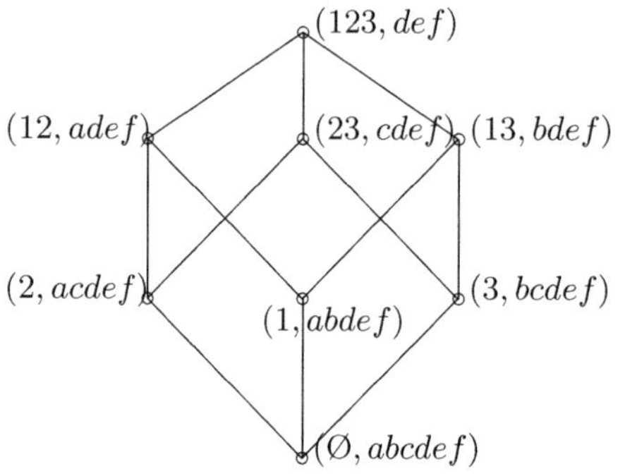


**Fig. 1.** The concept lattice that corresponds to the context in Table 1

## 3 Extracting Concepts from Formal Context Based on Matrix Operation

As we known, extraction of formal concept is the core of constructing concept lattice, the main contribution of our present work is proposing a distinct method for extracting formal concept. In this section, we divide the formal concepts into basic concepts and added concepts, which the two concepts can be acquired from the matrix operation. The following definitions and theorems are defined for explaining how to acquire the two concepts.

**Definition 4.** *In a binary formal context, give m objects,* $G = \{g : 1 \ldots m\}$, *and n attributes,* $M = \{m : 1 \ldots n\}$. *We produce a* $m \times n$ *matrix from the formal context:* $a_{ij} = 1$ *(*$a_{ij}$ *denotes the element in ith row and jth column of a matrix) iff a cell contains* ×, *and the other elements are set 0.*

For example, the matrix that corresponds to Table 1 is shown in Fig. 2.

$$T = \begin{pmatrix} 1\,1\,0\,1\,1\,1 \\ 1\,0\,1\,1\,1\,1 \\ 0\,1\,1\,1\,1\,1 \end{pmatrix}$$

**Fig. 2.** The matrix that corresponds to Table 1

$T'$ is the transpose of $T$:

$$T' = \begin{pmatrix} 1\,1\,0 \\ 1\,0\,1 \\ 0\,1\,1 \\ 1\,1\,1 \\ 1\,1\,1 \\ 1\,1\,1 \end{pmatrix}$$

**Fig. 3.** The transpose of $T$

**Definition 5.** *For a formal context* $\boldsymbol{K}$:= (G,M,I), *let* $g_i$ *denotes the ith object in G,* $m_j$ *denotes the jth attribute in M,* $a_{ij}$ *denotes the element in the ith row and the jth column of matrix* $\boldsymbol{A}$ *which corresponds to the formal context. Iff* $a_{ij}$=1, *then* $(g_i, m_j) \in I$, *i.e.* $g_i I m_j$.

**Definition 6.** *A* $m \times n$ *matrix* $\boldsymbol{A}$ *corresponds to a formal context,* $\boldsymbol{A}'$ *denotes the transpose of* $\boldsymbol{A}$, *let* $C = \boldsymbol{A} \otimes \boldsymbol{A}'$, *then* $c_{ij} = \{m_k \in M | a_{ik} = 1, a'_{kj} = 1, k = 1 \ldots n\}$ *(i=1...m, j=1...m,* $c_{ij}$ *denotes the elements in matrix C,* $a'_{kj}$ *denotes the elements in matrix* $\boldsymbol{A}'$*).*

For example, according to definition 6, then:

$$M \otimes M' = \begin{pmatrix} abdef & adef & bdef \\ adef & acdef & cdef \\ bdef & cdef & bcdef \end{pmatrix}$$

**Fig. 4.** The result of $M \otimes M'$

**Corollary 1.** *A* $m \times n$ *matrix* $\boldsymbol{A}$ *corresponds to a formal context,* $\boldsymbol{A}'$ *denotes the transpose of* $\boldsymbol{A}$, *let* $C = \boldsymbol{A} \otimes \boldsymbol{A}'$, *then* $c_{ij}$ *denotes the common attributes of the ith object and the jth object.*

*Proof.* According to definition 6, $c_{ij} = \{m_k \in M | a_{ik} = 1, a'_{kj} = 1, k = 1 \ldots n\}$, $a'_{kj}$ denotes the elements in matrix $\boldsymbol{A}'$, so $a'_{kj} = a_{jk}$, $c_{ij} = \{m_k \in M | a_{ik} = 1, a_{jk} = 1, k = 1 \ldots n\}$. Therefore, $c_{ij}$ denotes the common attributes of the ith object and the jth object.

**Theorem 1.** *A* $m \times n$ *matrix* $\boldsymbol{A}$ *corresponds to a formal context,* $\boldsymbol{A}'$ *denotes the transpose of* $\boldsymbol{A}$, *let* $C = \boldsymbol{A} \otimes \boldsymbol{A}'$, $X = \{x \in G | xIc_{ij}\}$, *then* $(X, c_{ij})$ *is called basic concept.*

*Proof.* $X \subseteq G$, $X^{\uparrow} = \{m \in M | \forall x \in X : (x, m) \in I\} = c_{ij}$; $c_{ij} \subseteq M$, $c_{ij}^{\downarrow} := \{x \in G | \forall m \in c_{ij} : (x, m) \in I\} = X$. then $(X, c_{ij})$ is a concept.

According to theorem 1, the basic concepts extracted from Fig. 4 are $(1, abdef)$, $(12, adef)$, $(13, bdef)$, $(2, acdef)$, $(23, cdef)$, $(3, bcdef)$.

**Theorem 2.** *If* $(X_1, B_1)$ *and* $(X_2, B_2)$ *are concepts, then* $((X_1 \cup X_2)^{\uparrow\downarrow}, B_1 \cap B_2)$ *is a concept, which called added concept if it is not a basic concept.*

*Proof.* Because $(X_1, B_1)$ and $(X_2, B_2)$ are concepts, so $X_1 = B_1^{\downarrow}$, $X_1^{\uparrow} = B_1$, $X_2 = B_2^{\downarrow}$, $X_2^{\uparrow} = B_2 \Longrightarrow (X_1 \cup X_2)^{\uparrow\downarrow\uparrow} = (B_1^{\downarrow} \cup B_2^{\downarrow})^{\uparrow} = B_1 \cap B_2$; $(B_1 \cap B_2)^{\downarrow} = (X_1^{\uparrow} \cap X_2^{\uparrow})^{\downarrow} = (X_1 \cup X_2)^{\uparrow\downarrow}$. Therefore, $((X_1 \cup X_2)^{\uparrow\downarrow}, B_1 \cap B_2)$ is a concept.

According to theorem 2, we can extract the added concepts is $(123, def)$.

**Theorem 3.** $\forall(X, Y)$ *is a concept, then* $(X, Y)$ *is a basic concept or added concept.*

*Proof.* $(X, Y)$ is a concept, let $X = \{X_{i_1} \cup X_{i_2} \cdots \cup X_{i_n}\}^{\uparrow\downarrow}$, $X_{i_j}$ is an object. According to definition 6, theorem 1 and theorem 2, Y is the common attributes set of $X_{i_j}$, i.e., $Y = \{c_{i_1 i_2} \cap c_{i_1 i_3} \cdots \cap c_{i_1 i_n}\}$, furthermore, $c_{i_1 i_j}$ can be extracted from the matrix. So $(X, Y)$ is a basic concept or added concept.

## 4 Algorithm for Building Concept Lattice

In this section, the algorithm BCLMO (Building Concept Lattice based on Matrix Operation) is proposed for building concept lattice. The following steps introduce how to build concept lattice by using BCLMO:

1. Transforming binary formal context into 0-1matrix. (definition 4)
2. Getting a new matrix by matrix operation. (definition 6)
3. Getting basic concepts. (theorem 1)
4. Getting add concepts: (theorem 2)
   - **4.1.** Utilizing theorem 2 to examine the basic concepts. If there produce new concepts, they are added concepts.
   - **4.2.** Utilizing theorem 2 to examine the added concepts (if there are more than one added concepts), if there produce new concepts, they are added concepts.
   - **4.3.** Continuously executing step 4.2 until there does not produce new added concepts.
5. Concept lattice denotes the relations among the formal concepts which consist of basic concepts and added concepts; it is constructed by depth-first method.
6. Using a graph structure to store the nodes and edges in concept lattice.

To better implement BCLMO, we divide BCLMO into a main-algorithm and three sub-algorithms. According to definition 6, it is easy to get the needed matrix by matrix operation. Therefore, BCLMO will focus on how to extract concepts from the new matrix and build concept lattice.

**Main-algorithm**
*//Matrix C can be got by definition 6.*
*//conceptset is the set including all concept.*
01 BEGIN

02 conceptset ← Ø ;
03 conceptset ← BasicConcept(C);
04 AddConcept(C);
05 Enter(queue,conceptset);
06 WHILE queue≠ Ø DO
07 BEGIN
08 $(X, X^{\uparrow})$ ← queue.concept;
09 conceptset ← conceptset - $\{(X, X^{\uparrow})\}$;
10 SubNodes ← FindSubNodes $(X, X^{\uparrow})$ ;
11 IF SubNodes ≠ Ø THEN
12 FOR $(Y, Y^{\uparrow}) \in$ SubNodes DO
13 $(X, X^{\uparrow})$.Edge ← $(Y, Y^{\uparrow})$ ;
14 IF SubNodes = Ø THEN
15 $(X, X^{\uparrow})$.Edge ← $(\emptyset, D)$ ;
16 END WHILE
17 END BEGIN

First, conceptset is NULL. In step 03, main-algorithm calls sub-algorithm BasicConcept(C) for getting basic concepts. In step 04, main-algorithm calls sub-algorithm AddConcept(C) for getting added concepts. In step 05, all concepts are stored in a FIFO queue. Step 06-16 use a while-loop to construct concept lattice.

**Sub-algorithm BasicConcept(C)**

19 BEGIN
20 conceptset ← Ø ;
21 FOR i ← 1 to | $O$ | DO
22 FOR j ← 1 to | $O$ | DO
23 IF $(c_{ij}^{\downarrow}, c_{ij}) \notin$ conceptset then
24 conceptset ← conceptset ∪ $(c_{ij}^{\downarrow}, c_{ij})$;
25 RETURN conceptset;
26 END.

In the above sub-algorithm, $|O|$ records the number of objects. Step 21-24 use a double for-loop to get basic concepts. Step 23 avoids extracting the repeated concept.

**Sub-algorithm AddConcept(C)**

28 BEGIN
29 conceptset1 ← conceptset;
30 conceptset2 ← Ø;
31 DO
32 BEGIN
33 FOR $(X_1, Y_1)$, $(X_2, Y_2)$ in conceptset1 DO
34 BEGIN
35 Y ← $(Y_1 \cap Y_2)$;
36 IF $(Y^{\downarrow}, Y) \notin$ conceptset then

```
37          BEGIN
38             conceptset ← conceptset ∪ (Y↓, Y);
39             conceptset2 ← conceptset2 ∪ (Y↓, Y);
40          END IF;
41       END FOR;
42       conceptset1 ← conceptset2;
43       conceptset2 ← Ø;
44    END DO
45    UNTIL conceptset1=Ø;
46 END.
```

We use the above sub-algorithm to get all added concepts. Step 33-41 extract added concepts from conceptset1. In step 45, this sub-algorithm will halt when conceptset1 is NULL.

**Sub-algorithm FindSubNodes**$(X, X^{\uparrow})$

```
48 BEGIN
49    SubNodes ← Ø ;
50    FOR ∀(Y, Y↑) ∈ conceptset DO
51    BEGIN
52       IF Y ⊂ X && X↑ ⊂ Y↑ THEN
53       BEGIN
54          Flag=False;
55          FOR ∀(Z, Z↑) ∈ SubNodes DO
56          BEGIN
57             IF Z ⊂ Y THEN
58             BEGIN
59                SubNodes ← SubNodes - (Z, Z↑);
60                SubNodes ← SubNodes ∪(Y, Y↑);
61                Flag=True;
62             END
63             ELSE
64                Flag=True;
65          END FOR
66          IF NOT Flag THEN
67             SubNodes ← SubNodes ∪(Y, Y↑);
68       END IF
69    END FOR
70    RETURN SubNodes;
71 END.
```

We use this sub-algorithm to search the son concepts of $(X, X^{\uparrow})$. In step 52, the concept $(Y, Y^{\uparrow})$ which satisfies both $Y \subset X$ and $X^{\uparrow} \subset Y^{\uparrow}$ is found. Note that $(Y, Y^{\uparrow})$ may be not a son concept of $(X, X^{\uparrow})$. In step 55-step 65, if $\forall(Z, Z^{\uparrow}) \in$ SubNodes and $Z \subset Y$, then $(Z, Z^{\uparrow})$ is not a son concept of $(X, X^{\uparrow})$.

## 5 Algorithm Analysis

In the following equations, $\mid O \mid$, $\mid D \mid$, $\mid L \mid$, $\mid L_1 \mid$ and $\mid L_2 \mid$ denotes the number of objects, attributes, all concepts, basic concepts, added concepts, respectively.

1. For the matrix operation, the time complexity is $\mid O \mid^2 \times \mid D \mid$;
2. The time complexity of generating basic concepts: Because all basic concepts can be got from matrix directly, so the time complexity of extracting a basic concept is $\mid O \mid^2$. So the time complexity of extracting all basic concepts is $\mid O \mid^2 \times \mid L_1 \mid$;
3. The time complexity of generating added concepts: According to theorem 3'proof, we can regard an added concept as the intersection of two concepts, so the time complexity of extracting an added concept is $r \times C^2_{|O|}$, r is the number of iteration. So the time complexity of extracting all added concepts is $r \times C^2_{|O|} \times \mid L_2 \mid \leq r \times \mid O \mid^2 \times \mid L_2 \mid$.

To sum up, the time complexity of our algorithm is $O(\mid O \mid^2 \times (\mid D \mid + \mid L \mid))$. Table 2 shows BCLMO in comparison with other algorithms.

**Table 2.** Time complexity comparison of building concept lattice

| | Algorithm | Time complexity |
|---|---|---|
| 1 | Bordat [5] | $O(\mid D \mid^2 \times \mid O \mid \times \mid L \mid)$ |
| 2 | CBO [6] | $O(\mid D \mid \times \mid O \mid^2 \times \mid L \mid)$ |
| 3 | Nourine [7] | $O((\mid O \mid + \mid D \mid) \times \mid O \mid \times \mid L \mid)$ |
| 4 | S. Y. Qiao [8] | $O((\mid O \mid + \mid D \mid) \times \mid D \mid \times \mid L \mid)$ |
| 5 | Chein [17] | $O(\mid D \mid \times \mid O \mid^3 \times \mid L \mid)$ |
| 6 | Norris [18] | $O(\mid O \mid^2 \times \mid D \mid \times \mid L \mid)$ |
| 7 | BCLMO | $O(\mid O \mid^2 \times (\mid D \mid + \mid L \mid))$ |

## 6 Conclusions

FCA has shown have many advantages in the field of knowledge discovery, concept lattice is a convenient tool and has been applied in data analysis and knowledge discovery. However, the complexity of building concept lattice becomes the bottleneck of application.

In this paper, we proposed a simple and efficient method for building concept lattice. As we known, extraction of formal concept is the core of constructing concept lattice, the main contribution of our present work is proposing a distinct method for extracting formal concept. We divide the formal concepts into basic concepts and added concepts, and define a series of definitions and theorems to explain how to acquire the two concepts.

Based on the matrix operation, the algorithm BCLMO is proposed for building concept lattice. By algorithm analysis, we compare our algorithm with some classical algorithms, and the time complexity of our algorithm has remarkably decreased.

For future work, we will apply BCLMO to some classical datasets, and do experiments by comparing with some classical algorithms. We will also research how to apply BCLMO to the multi-value formal context.

## Acknowledgments

This work is supported by the Education Department Foundation of Sichuan Province (Grant No.2006A086), the Application Foundation of Sichuan Province (Grant No.2006J13-056), the Cultivating Foundation of Science and Technology of Xihua University (Grant No.R0622611), the cultivating foundation of the science and technology leader of sichuan province.

## References

1. Wille, R.: Restructuring Lattice Theory: an Approach Based on Hierarchies of Concepts, in: I. Rival (Ed.), Ordered Sets, Reidel, Dordrecht, Boston, (1982) 445-470
2. Stumme, G., Wille, R., Wille, U.: Conceptual Knowledge Discovery in Databases Using Formal Concept Analysis Methods, in: Proceedings of the 2nd European Symposium on Principles of Data Mining and Knowledge Discovery, (1998) 450-458.
3. Tilley, T., Cole, R., Becker, P., Eklund, P.: A Survey of Formal Concept Analysis Support for Software Engineering Activities, in: Proceedings of the First International Conference on Formal Concept Analysis. (2003)
4. Di az-Agudo, B., Gonza lez-Calero, P.A.: Classification-Based Retrieval using Formal Concept Analysis, in: Proceedings of the 4th International Conference on Case-Based Reasoning, (2001) 173-188
5. Bordat, J.P.: Calcul Partique Du Treillis de Galois dune correspondence, Math. Sci. Hum. (1986) 96:31-47
6. Kuznetsov, S.O.: A Fast Algorithm for Computing All Intersections of Objects in a Finite Semi-lattice, Automatic Documentation and Mathematical Linguistics. (1993) 27(5):11-21
7. Nourine, L., Raynaud, O.: A Fast Algorithm for Building Lattices. Information Processing Letters. (1999) 71: 199-204
8. Qiao, S.Y., Wen, S.P., Chen, C.Y., Li, Z.G.: A Fast Algorithm for Building Concept Lattice. Proceedings of the Second International Conference on Machine Learning and Cybernetics, Xi'an. (2003) 163-167
9. R Godin, R Missaoui, H Alaoui. Incremental concept formation algorithms based on Galois (concept) lattices. Computational Intelligence, (1995) 11(2): 246-267
10. Wang, Z.H., Hu, K.Y., Hu, X.G., Liu, Z.T., Zhang, D.C.: General and Incremental Algorithms of Rule Extraction Based on Concept Lattice. Computer Journal. (1999) 22(1): 66-70
11. Hu, K.Y., Lu, Y.C., Shi, C.Y.: An Integrated Mining Approach for Classification and Association Rule Based on Concept Lattice. Journal of software. (2000) 11(11): 1479-1484
12. Stumme, G., Taouil, R., Bastide, Y., Lakhal, L.: Conceptual Clustering with Iceberg Concept Lattices. In: Proceedings of GIFachgruppentreffen Maschinelles Lernen01, Universität Dortmund, vol. 763, October 2001.

13. Stumme, G., Taouil, R., Bastide, Y., Pasqier, N., Lakhal, L.: Computing Iceberg Concept Lattices with Titanic. J. on Knowledge and Data Engineering (KDE). (2002) 42(2) : 189-222
14. Ganter, B., Wille, R.: Formal Concept Analysis: Mathematical Foundations, Springer, Berlin, (1999)
15. Wrobel, S., Morik, K., Joachims, T.: Maschinelles lernen und data mining, in: G. Grz, C.-R. Rollinger, J.Schneeberger (Eds.), Handbuch der Knstlichen Intelligenz, vol. 3, Auflage, Oldenbourg, Munchen, Wien, (2000) 517-597
16. Sergei, O.K.: Complexity of learning in concept lattices from positive and negative examples, in: Discrete Applied Mathematics 142, (2004) 111-125
17. Chein, M.: Algorithm De Recherche Des Sous-Matrices Premiresdune Matrice, Bull. Math. Soc. Sci. Math. R.S. Roumanie. (1969) 13:21-25
18. Norris, E. M.: An Algorithm for Computing the Maximal Rectangles in a Binary Relation,Revue Roumaine de Mathematiques Pures et Appliques. (1978) 23(2):243-250

# A New Method of Causal Association Rule Mining Based on Language Field

Kaijian Liang[1,2], Quan Liang[2], and Bingru Yang[2]

[1] Department of Computer, Hunan Institute of Engineering, Xiangtan 411101
[2] School of Information and Engineering, University of Science and Technology Beijing, Beijing, 100083
liangquanlq@126.com

**Abstract.** Aiming at the research that using more new knowledge to develope knowledge system with dynamic accordance, and under the background of using Fuzzy language field and Fuzzy language values structure as description framework , the generalized cell Automation that can synthetically process fuzzy indeterminacy and random indeterminacy and generalized inductive logic causal model is brought forward. On this basis, the paper provides a kind of the new methods that can discover causal association rules. According to the causal information of Standard Sample Space and Commonly Sample Space,through constructing its state (abnormality) relation matrix, causal association rules can be gained by using inductive reasoning mechanism.The estimate of this algorithm complexity is given,and its validity is proved through case.

**Keywords:** knowledge discovery, language field, language value structure, generalized cell automation, causal association rule.

## 1 Introduction

In the research of intricate system control and complicated affair reasoning, the problem of mechanism and computational model of reasoning has become a very important issue in the academic world. Thus the research of indeterminacy inductive automatic reasoning mechanism is more important.

In the development of current logic science, an important trend that the research of logic thought and method merged into logic language has taken place. Thus the intelligence reasoning procedure is regarded as a procedure of the intelligence language's reasoning, quantifying, composing and transforming in the language information field.

The language field offers us a framework for the quantitative description of model and mechanism of reasoning flow and the generalized inductive logic causal model offers us a logic underground of inductive reasoning mechanism. Only on this basis it is possible to establish a computational model and automatic reasoning mechanism of indeterminacy causal inductive reasoning. The research of computational model of

D.-S. Huang, L. Heutte, and M. Loog (Eds.): ICIC 2007, LNAI 4682, pp. 360–366, 2007.

reasoning has theoretical importance and wide prospect of application in expert system, automatic inference, knowledge engineering, intelligent control and neutral network.

## 2 Language Field and Language Value Structure

### 2.1 Basic Concept

The language field and language value structure with sketchy outline established here will supply a framework for the description of computational model of reasoning. An initial discussion about the framework is done as follows.

**Definition 1.** We call U = <X, N, $\psi$, D> as a normal structure of state description, if :

① $X = \bigcup_{i=1}^{n} x_i$, X is called state space, $x_i$ is called a state class, each state class(a set of state description the same thing) is regard as a state language variable;
② $N = \{N_i | i \in I_+\}$, $N_i$ is called a language value set;
③ $\psi : X \to N$, for each $x_i$ there is $\psi(x_i) = N_j (i = 1, 2, \cdots, n;\ j = 1, 2, \cdots, m)$;
④ $D \subseteq R$ is called the possible universe of discourse and it is usually a real interval in real world corresponding with state language variables.

**Definition 2.** Given a sequence of n real interval, if every two adjacent intervals $L_i$ and $L_j$ do not contain each other and $L_i \cap L_j \neq \Phi$, then we call the sequence as a overlapping interval sequence.

Obviously, regarding to state language variable $x_i$, all the realistic quantity interval corresponding with language value (in real domain) compose a overlapping interval sequence.

**Definition 3.** As for set E which contains n real intervals composing an overlapping interval sequence, we can get the binary relation " $\leq$ " which is to any two intervals $[X_1, Y_1] \in E$ and $[X_2, Y_2] \in E$ : $[X_1, Y_1] \leq [X_2, Y_2] \Leftrightarrow (X_1 \leq X_2) \wedge (Y_1 \leq Y_2)$

Obviously, the binary relation " $\leq$ " defined on E is a complete ordering relation.

**Definition 4.** The middle point of basic variable sub-interval corresponding to language value and interval value of its $\varepsilon$-neighborhood are called standard value (usually $\varepsilon$ is reasonable error). The sample of standard value is called standard sample, otherwise it is called non-standard sample. The standard sample space and the nonstandard sample space which is separately composed by standard sample and the nonstandard sample are called general sample space.

**Definition 5.** In state description standard structure U ,we call C=<E, I, N, $\leq_N$> as a language field, if :

① E is the set of overlapping closed interval on the R (in basic variable domain);
② N is a finite set of language value, and not empty;

③ $\leq_N$ is complete ordering relation in N;
④ I: N→E, mapping from language value to its standard value, is a standard value mapping, and satisfies order-preserving.

**Definition 6.** In state description standard structure U, for the language field C=<E, I, N, $\leq_N$>, F=<C, W, K> is a language value structure of C, if:

① C satisfies definition 5;
② K is natural number;
③ W: N$\rightarrow R^K$, it satisfies the following conditions:
$\forall n_1, n_2 \in N\ (n_1 \leq_N n_2 \rightarrow W(n_1) \leq_{dic} W(n_2))$,
$\forall n_1, n_2 \in N\ (n_1 \neq n_2 \rightarrow W(n_1) \neq W(n_2))$.

Where, $\leq_{dic}$ is lexicographic order in $R^K$.

In Fuzzy state description standard structure U, when R is defined to [0, 1], Definition 5 and Definition 6 defines Fuzzy language field and language value structure respectively.

### 2.2 Basic Frameworks

**Definition 7.** Given two language fields $C_1$ and $C_2$, if there are 1-1 mappings f: $E_1 \rightarrow E_2$, g: $N_1 \rightarrow N_2$, it satisfies the following conditions:

① $f$ is monotone;
② $\forall n_1 \in N_1$, $f(I(n_1)) = I_2(g(n_1))$;

Where $C_1 = <E_1, I_1, N_1, \leq_{N_1}>$, $C_2 = <E_2, I_2, N_2, \leq_{N2}>$.
Then $C_1$ is called an extension of $C_2$.

**Theorem 1.** If language field $C_1$ is the extension of $C_2$, then g: $N_1 \rightarrow N_2$ must be monotonic mapping , that is to say , if $n_1 \leq_{N_1} n_1'$ then $g(n_1) \leq_{N2} g(n_1')$, where $n_1$, $n_1' \in N_1$, (proof is omited)

**Definition 8.** If $C_1 = <E_1, I_1, N_1, \leq_{N_1}>$, $C_2 = <E_2, I_2, N_2, \leq_{N2}>$, $|N_1| = |N_2|$, then $C_1$ and $C_2$ are the same type language fields.

## 3 Construction of Causal Association Rule

### 3.1 Indeterminacy Causal Association Rule Under Standard Sample Space

(1) In generalized inductive logic causal model , given the causes A , B , C , … that lead to the effect S , when the state(abnormality) relation between cause and effect of standard sample space at moment t was described by generalized causal cell automaton,

the language value description and the corresponding discrete vector expression of all kinds of states(abnormalities) of cause and effect can first be gotten . For example, the causes corresponding to 5 language values "the change is very small ", " the change is small ", " the change is not great and not small ", " the change is great ", " the change is very great " can be expressed as

$$A_t^{(i)}=(a_i,b_i,c_i,d_i,e_i)_t \quad (i=1,2,3,4,5).$$

It is called A's state(abnormality) standard vector at moment $t$. In the same way the effect S's state(abnormality) standard vector $S_{t'}^{(j)}=(p_j,q_j,\ldots,r_j)_{t'}$ $(j=1,2,3,4,5)$ at moment $t'$ can be gotten.

## 3.2 Indeterminacy Causal Association Rules Under Commonly Sample Space and Single Language Field

(1) In Commonly Sample Space, For cause A , the input vector of cause state (abnormalities)(i.e. $\alpha_t$ (non-standard vectors)) can be gained using interpolation formula according to standard vectors of adjacent cause state (abnormalities). That is:

$$\alpha_t = A_t \cdot \left(1-\frac{|t_i - t_{i_0}|}{l_i}\right) + A_{adjacent} \cdot \frac{|t_i - t_{i_0}|}{l_i}$$

Where, $t_i$ is input data of $i$ interval, $t_{i_0}$ is middle point data of $i$ interval, $l_i$ is length of $i$ interval, $A_t$ is standard vector of cause state (abnormality) of $i$ interval, $A_{adjacent}$ is standard vector of cause state (abnormality) of the adjacent of left or right which is determined according to the point that t fall on.

(2) **Definition 9.** In generalized inductive logic causal model and the same language value structure, the measurement of cause state (abnormalities) input vector $a_t$ and standard vector $a_t^{(i)}$ can be confirmed by the following formula:

$$d_H(a_t, A_t^{(i)}) = \sum_{i=1}^{k} |\overline{\mu a_t^{(i)}} - \overline{\mu A_t^{(i)}}|$$

$\overline{\mu a_t^{(i)}}$ and $\overline{\mu A_t^{(i)}}$ are their corresponding mark respectively. (Definition of the state (abnormalities) corresponding measurement is analogous). According to this definition, For cause A, the measure of $\alpha_t$ and any state standard vector of A is calculated by following formula, then the cause state (abnormality) type (language value) which $\alpha_t$ belongs to is determined according to the minimum of the measure.

(3) In the construction of generalized inductive logic causal model and in non-standard sample space of the possible causal world, by means of determining the type of cause state (abnormality)(such as $A_t^{(w)}$ type) which is the input vector of cause

state (abnormality) $\alpha_t$ belongs to and determining the type of local major premise,we can find its sole matching knowledge matrix ( $M_\sigma^*$ ) through self-organizing in the state (abnormality) knowledge of standard sample space . Under the background ( major premise ) of $M_\sigma^*$, the effect state (as a conclusion) which results from cause A at a certain state (abnormality) can be gained according to the automaton reasoning rule as follows:

$$\begin{array}{ll} M_\sigma^* & \text{( major premise)} \\ \quad \alpha_t & \text{( minor premise)} \\ \hline S^* \triangleq \alpha_t \circ M_\sigma^* & \text{(conclusion)} \end{array}$$

That is to say, the conclusion $S^*$ can be gained through secondary composition.

(4) Type accumulation: Measure of $S^*$ and standard vector that has known effect state (abnormality) is calculated and the effect state (abnormality) type (language value) which $S^*$ belongs to is determined according to the minimum of the measure. then the causal association rule is gained $A_t^* \xrightarrow{\Theta} S$

### 3.3 Indeterminacy Causal Association Rules Under Commonly Sample Space and Comprehensive Language Field

At the aspect of algorithm complexity,this algorithm flow chart does not increase the top value of complexity additionally,thus not multiple complexity or increase it exponentially.The complexity of this algorithm is the linear sum of originals only.While value of $N1$,$N2$ and $N3$ is very large,or while they run to $\infty$ in Sample Space,this algorithm only has a **O**(n) complexity. Summation of this algorithm complexity lies on others such as Knowledge base,Compound principle and so on,which exist already.So,this algorithm itself is available.

### 3.4 Case Verifying

While verifying this algorithm,we use partial data from result database of a certain American state society investigation in 1991.The database content includes many items of investigated object such as occupation,marriage,education years,annual earning,etc.Record number of the database is 1500.Education as premise,and annual earning as effect,we study to find some reasonable and available rules.

In causal language field,language variable is education years which can be divided into 5 language value such as very short education years ($A1$),short education years($A2$),moderate education years($A3$),long education years($A4$),very long education years($A5$).Max value is 20 (Unit:year) and the minimal is 0.The standard sample point and radius that each language value correspond to is confirmed by experts or users,and here,let them be $A1(1,2)$,$A2$ (8.2,1), $A3$ (11.8,1), $A4$ (15,1), $A5$ (15,1) separately,and others can be gained by fuzzy switch.Let $A2 = (1\ 0.8\ 0.6\ 0.4\ 0.2)$,$A4 = (0.2\ 0.4\ 0.6\ 0.8\ 1)$,

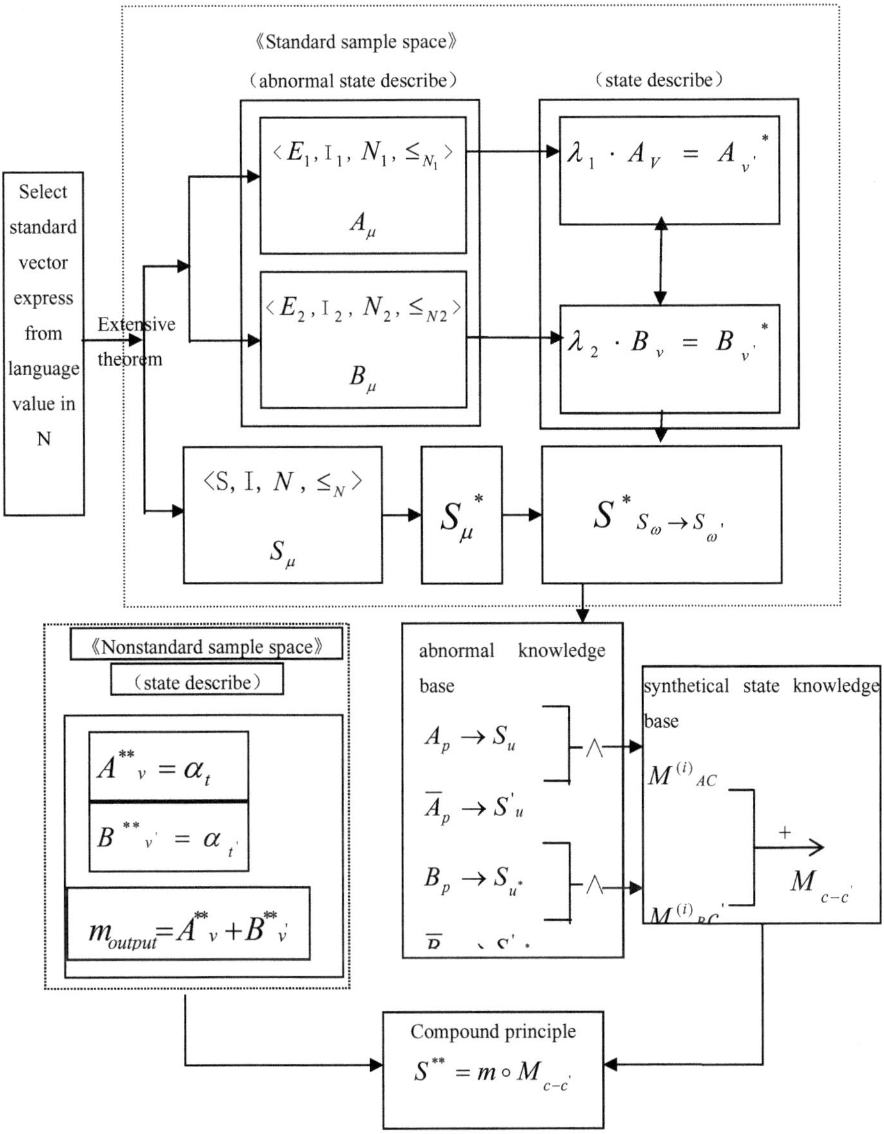


**Fig. 1.** Algorithm flow chart

$A1 = (A2)^2 = (1\ 0.64\ 0.36\ 0.16\ 0.04)$, $A5 = (A4)^2 = (0\ 0.04\ 0.16\ 0.36\ 0.64)$, $A3 = (1-A2) \wedge (1-A4) = (0\ 0.2\ 0.4\ 0.2\ 0)$,all these values can be gained according to data distributing or experience.Be the same reason,to process like that in language field,we can get relevant standard sample points,radius and standard vectors,which come from 5 language value of annual earning separately,viz. very little annual earning (S1),little annual earning

(S2),moderate annual earning (S3),much annual earning (S4) and very much annual earning (S5).

After all these precedures,2 causal association rules can be gained and represented as $R1$:[$A4$]$\overset{\ominus}{\longrightarrow}$[$S4$] and $R1$:[$A4$] $\overset{\ominus}{\longrightarrow}$ [$S4$].The first rule $R1$ represents:long education year is one cause of much annual earning,but not the direct cause.Obviously,this result matches people's experience well.So,the algorithm validity is proved through this case.

## 4 Conclusion

Using language field as description framework and under the background of generalized inductive logic causal model, we have discussed the rule and algorithm of indeterminacy causal inductive automation reasoning mechanism based on fuzzy state description and given feasible and judgment solution to solve the problem of causal disturbance correspondence in the Causal state (abnormality).That is to say, according to model and corresponding algorithm, we can gained the corresponding effect information and on the base gained more new knowledge automatically to development knowledge system with dynamic accordance. The research results discussed in this paper are very important to constructing comprehensive knowledge discovery systems.

## References


1. Heckerman, D.: Bayesian Networks for Data Mining. Data Mining & Knowledge Discovery, 1(1997) 79-119
2. Jagielska, I., Matthews, W.: An Investigation into the Application of Neural Networks, Fuzzy Logic, Genetic Algorithms, and Rough Sets to Automated Knowledge Acquisition for Classification Problems. Neurocomputing, 24(1999) 37-54
3. Wang, Y.T., Wu, B.R.: Inductive Logic and Artificial Intelligence. Beijing: the Publishing House of the Textile University of China, (1995)
4. Shi, C. Y.: Development of Qualitative Reasoning, CJCAI, (1992)
5. Yoon, J., Kerschberg, L.A.: Framework for Knowledge Discovery and Evolution in Databases. IEEE Transactions on Knowledge and Data Engineering, 5(6)(1993) 973-979
6. Agrawal, R., Srikant, R.: Mining Generalized Association Rules. In Proc of the 21st VL DB. Zurich, Switzerland, (1995) 407-419

# A Particle Swarm Optimization Method for Spatial Clustering with Obstacles Constraints

Xueping Zhang[1,2,3], Jiayao Wang[2], Zhongshan Fan[4], and Xiaoqing Li[1]

[1] School of Information Science and Engineering, Henan University of Technology, Zhengzhou 450052, China
[2] School of Surveying and Mapping, PLA Information Engineering University, Zhengzhou 450052, China
[3] Geomatics and Applications Laboratory, Liaoning Technical University, Fuxin 123000, China
[4] Henan Academy of Traffic Science and Technology, Zhengzhou 450052, China
zhang_xpcn@yahoo.com.cn

**Abstract.** Spatial clustering is an important research topic in Spatial Data Mining (SDM). In this paper, we propose a particle swarm optimization (PSO) method for Spatial Clustering with Obstacles Constraints (SCOC). In the process of doing so, we first use PSO algorithm via MAKLINK graphic to get the optimal obstructed path, and then we developed PSO K-Medoids SCOC (PKSCOC) algorithm to cluster spatial data with obstacles constraints. The experimental results demonstrate the effectiveness and efficiency of the proposed method, which can not only give attention to higher local constringency speed and stronger global optimum search, but also get down to the obstacles constraints and practicalities of spatial clustering.

**Keywords:** Spatial Clustering, Obstacles Constraints, Particle Swarm Optimization, K-Medoids Algorithm.

## 1 Introduction

Spatial clustering is not only an important effective method but also a prelude of other task for Spatial Data Mining (SDM). Many methods have been proposed in the literature, but few of them have taken into account constraints that may be present in the data or constraints on the clustering. These constraints have significant influence on the clustering results. Spatial clustering with constraints has two kinds of forms [1]. One kind is Spatial Clustering with Obstacles Constraints (SCOC), such as bridge, river, and highway etc. whose impact on the result should be considered in the clustering process. As an example, Fig.1 shows clustering spatial data with physical obstacle constraints. Ignoring the constraints leads to incorrect interpretation of the correlation among data points. The other kind is Spatial Clustering with Handling Operational Constraints [2], it consider some operation limiting conditions in the clustering process. SCOC is mainly discussed in this paper.

To the best of our knowledge, only three clustering algorithms for SCOC have been proposed very recently, that is COD-CLARANS [3], AUTOCLUST+ [4], and

D.-S. Huang, L. Heutte, and M. Loog (Eds.): ICIC 2007, LNAI 4682, pp. 367–376, 2007.

DBCluC [5]-[8], and many questions exist in them. COD-CLARANS computes obstructed distance using visibility graph costly and is unfit for large spatial data. In addition, it only gives attention to local constringency. AUTOCLUST+ builds a Delaunay structure for solving SCOC costly and is also unfit for large spatial data. DBCluC cannot run in large high dimensional data sets etc. We developed Genetic K-Medoids SCOC (GKSCOC) based on Genetic algorithms (GAs) and Improved K-Medoids SCOC (IKSCOC) in [9], however, GKSCOC has a comparatively slower speed in clustering.

Particle Swarm Optimization (PSO) is a population-based optimization method first proposed by Kennedy and Eberhart in 1995 [10, 11]. Compared to GAs, the advantages of PSO are that it is easier to implement and there are fewer parameters to be adjusted, and it can be efficiently used on large data sets.

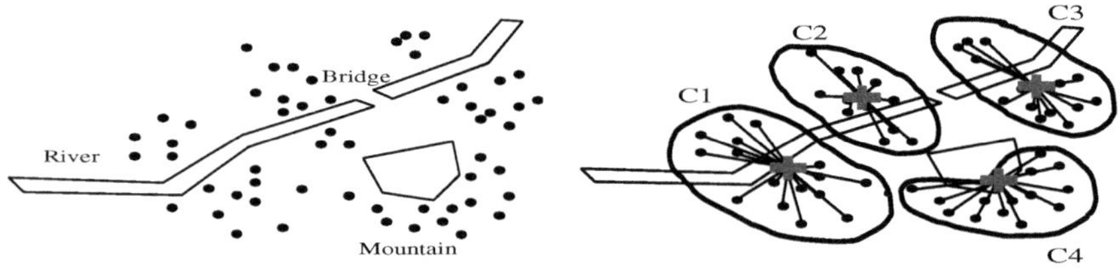


(a) Data objects and obstacles constraints (b) Clusters ignoring obstacle constraints

**Fig. 1.** Clustering data objects with obstacles constraints

In this paper, we propose a PSO method for SCOC. In the process of doing so, we first use PSO algorithm via MAKLINK graphic to get the optimal obstructed path, and then we developed PSO K-Medoids SCOC (PKSCOC) algorithm to cluster spatial data with obstacles constraints. The experimental results demonstrate the effectiveness and efficiency of the proposed method, which can not only give attention to higher local constringency speed and stronger global optimum search, but also get down to the obstacles constraints and practicalities of spatial clustering.

The remainder of the paper is organized as follows. Section 2 introduces PSO. Using PSO to get the obstructed distance is discussed in Section 3. Section 4 presents PKSCOC. The performances of PSO method for SCOC on real datasets are showed in Section 5, and Section 6 concludes the paper.

## 2 Particle Swarm Optimization

Particle Swarm Optimization (PSO) is a population-based optimization method first proposed by Kennedy and Eberhart [10, 11]. In order to find an optimal or near-optimal solution to the problem, PSO updates the current generation of particles (each particle is a candidate solution to the problem) using the information about the best solution obtained by each particle and the entire population. The mathematic description of PSO is as the following. Suppose the dimension of the searching space is D, the number of the particles is n. Vector $X_i = (x_{i1}, x_{i2}, \ldots, x_{iD})$ represents the position of the $i^{th}$ particle and $pBest_i = (p_{i1}, p_{i2}, \ldots, p_{iD})$ is its best position searched by now, and the whole particle swarm's best position is represented as $gBest = (g_1, g_2, \ldots, g_D)$ .Vector $V_i = (v_{i1}, v_{i2}, \ldots, v_{iD})$ is the position change rate of the $i^{th}$ particle. Each particle updates its position according to the following formulas:

$$v_{id}(t+1) = w * v_{id}(t) + c_1 * rand() * [p_{id}(t) - x_{id}(t)] + c_2 * rand() * [g_d(t) - x_{id}(t)] \quad (1)$$

$$x_{id}(t+1) = x_{id}(t) + v_{id}(t+1) \ , \ 1 \le i \le n, \ 1 \le d \le D \quad (2)$$

where $w$ is the inertia weight, $c_1$ and $c_2$ are positive constant parameters, and $Rand()$ is a random function with the range [0, 1]. Equation (1) is used to calculate the particle's new velocity, then the particle flies toward a new position according to Equation (2).The various range of the $d^{th}$ position is $[XMINX_d, XMAXX_d]$ and the various range $[-VMAXX_d, VMAXX_d]$. If the value calculated by equations (1) and (2) exceeds the range, set it as the boundary value. The performance of each particle is measured according to a predefined fitness function, which is usually proportional to the cost function associated with the problem. This process is repeated until user-defined stopping criteria are satisfied.

PSO is effective in nonlinear optimization problems and it is easy to implement. In addition, only few input parameters need to be adjusted in PSO. Because the update process in PSO is based on simple equations, PSO can be efficiently used on large data sets. A disadvantage of the global PSO is that it tends to be trapped in a local optimum under some initialization conditions [12].

## 3 Using PSO to Get the Obstructed Distance

### 3.1 Obstructed Distance

To derive a more efficient algorithm for SCOC, obstructed distance is first introduced.

**Definition 1.** *(Obstructed Distance)* Given point $p$ and point $q$ , the obstructed distance $d_o(p,q)$ is defined as the length of the shortest Euclidean path between two points $p$ and $q$ without cutting through any obstacles.

### 3.2 Obstacles Modeling

Path planning with obstacles constraints is the key to computing obstructed distance. Here, we adopt a simple model of obstacles called MAKLINK graphic [13] for path planning with obstacles constraints, which can reduce the complicacy of the model and get the optimized path. An example is shown in Fig.2. Further explanations and detail on how to construction MAKLINK graphic can be found in [13].

### 3.3 Using PSO to Get the Optimal Obstructed Path

In this paper, path planning with obstacles constraints is divided into two stages. Firstly, we can use Dijkstra algorithm to found out the shortest path from the start point to the goal point in the MAKLINK graph. The simulation result is in Fig.2 and the black solid line represents the shortest path we got. And then, we adopt PSO algorithm to optimize the shortest path and get the best global path, which is inspired by [14].

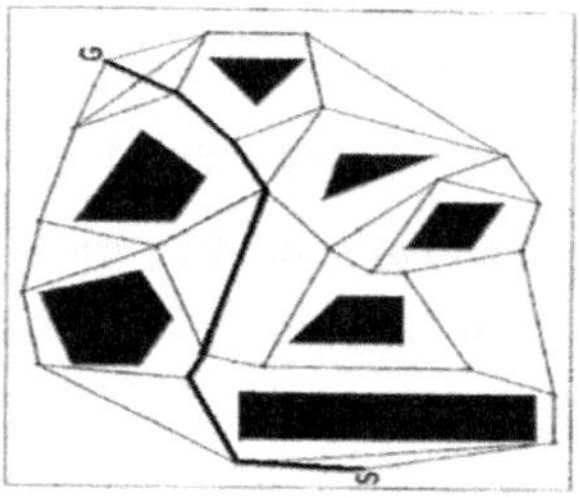


**Fig. 2.** MAKLINK and shortest path

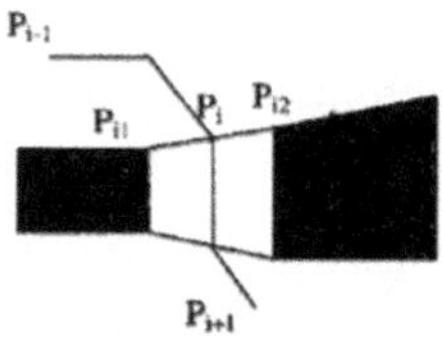


**Fig. 3.** Path coding

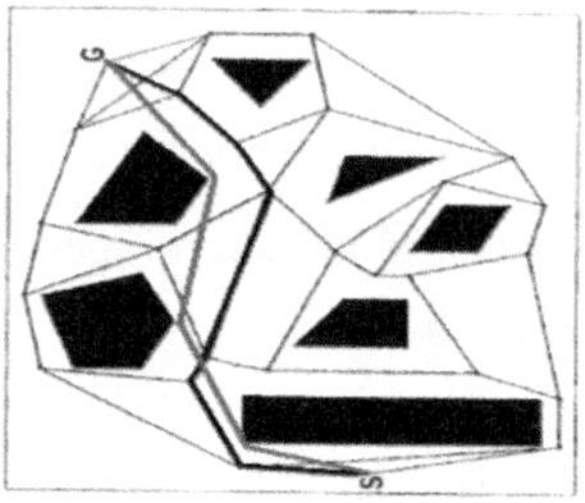


**Fig. 4.** Optimal obstructed path

Suppose the shortest path of the MAKLINK graph that we get by Dijkstra algorithm is $P_0, P_1, P_2, \ldots, P_D, P_{D+1}$, where $P_0 = start$ is the start point and $P_{D+1} = goal$ is the goal point. $P_i (i = 1, 2, \ldots, D)$ is the midpoint of the free link. The optimization task is to adjust the position of $P_i$ to shorten the length of path and get the optimized (or acceptable) path in the planning space. The adjust process of $P_i$ is shown in Figure 3. The position of $P_i$ can be decided by the following parametric equation:

$$P_i = P_{i1} + (P_{i2} - P_{i1}) \times t_i, t_i \in [0,1], i = 1, 2, \ldots D \tag{4}$$

Each particle $X_i$ is constructed as: $X_i = (t_1 t_2 \ldots t_D)$ .Accordingly, the $i^{th}$ particle's fitness value is defined as:

$$f(X_i) = \sum_{k=1}^{D+1} P_{k-1} P_k, i = 1, 2, \ldots, n \tag{5}$$

where $P_{k-1} P_k$ is the direct Euclidean distance between the two points and $P_k$ can be calculated according to equation (5). Thus the smaller the fitness value, the better the solution.

Here, the PSO is adopted as follows.

1. Initialize particles at random, and set $pBest_i = X_i$ ;
2. Calculate each particle's fitness value according to equation (5) and label the particle with the minimum fitness value as $gBest$ ;
3. For $t_1 = 1$ to $t_{1\max}$ do {
4. For each particle $X_i$ do {
5. Update $v_{id}$ and $x_{id}$ according to equations (1) and (2);
6. Calculate the fitness according to equation (5) ;}
7. Update $gBest$ and $pBest_i$ ;

8. If $\|v\| \le \varepsilon$, terminate ;}
9. Output the obstructed distance.

where $t_{1}\max$ is the maximum number of iterations, $\varepsilon$ is the minimum velocity. The simulation result is in Fig.4 and the red solid line represents the optimal obstructed path obtained by PSO.

# 4 PKSCOC Based on PSO and K-Medoids

This section first introduces IKSCOC in section 4.1, and then presents the PKSCOC algorithm in section 4.2.

## 4.1 IKSCOC Based on K-Medoids

There are three typical Partitioning-base algorithms: K-Means, K-Medoids and CLARANS. K-Medoids algorithm is adopted for SCOC to avoid cluster center falling on the obstacle.

The clustering quality is estimated by an object function. Square-error function is adopted here, and it can be defined as:

$$E = \sum_{j=1}^{N_c} \sum_{p \in C_j} (d(p, m_j))^2 \tag{6}$$

where $N_c$ is the number of cluster $c_j$, $m_j$ is the cluster centre of cluster $c_j$, $d(p,q)$ is the direct Euclidean distance between the two points $p$ and $q$.

To handle obstacle constraints, accordingly, criterion function for estimating the quality of spatial clustering with obstacles constraints can be revised as:

$$E_o = \sum_{j=1}^{N_c} \sum_{p \in C_j} (d_o(p, m_j))^2 \tag{7}$$

where $d_o(p,q)$ is the obstructed distance between point $p$ and point $q$.

The method of IKSCOC is adopted as follows [9].

1. Select $N_c$ objects to be cluster centers at random;
2. Distribute remain objects to the nearest cluster center;
3. Calculate $E_o$ according to equation (7);
4. Do {let current $E = E_o$;
5. Select a not centering point to replace the cluster center $m_j$ randomly;
6. Distribute objects to the nearest center;
7. Calculate $E$ according to equation (6);
8. If $E >$ current $E$, go to 5;
9. Calculate $E_o$;
10. If $E_o <$ current $E$, form new cluster centers;
11. } While ( $E_o$ changed).

While IKSCOC still inherits two shortcomings because it is based on standard partitioning algorithm. One shortcoming is that selecting initial value randomly may cause different results of the spatial clustering and even have no solution. The other is that it only gives attention to local constringency and is sensitive to an outlier.

### 4.2 PKSCOC Based on PSO and K-Medoids

Particle Swarm Optimization (PSO) has been applied to data clustering [15-18]. In the context of clustering, a single particle represents the $N_c$ cluster centroid. That is, each particle $X_i$ is constructed as follows:

$$X_i = (m_{i1}, \ldots, m_{ij}, \ldots, m_{iN_c}) \tag{8}$$

where $m_{ij}$ refers to the $j^{th}$ cluster centroid of the $i^{th}$ particle in cluster $C_{ij}$. Here, the objective function is defined as follows:

$$f(X_i) = \frac{1}{J_i} \tag{9}$$

$$J_i = \sum_{j=1}^{N_c} \sum_{p \in C_{ij}} d_o(p, m_j) \tag{10}$$

Spatial Clustering with Obstacles Constraints based on PSO and K-Medoids (PKSCOC), which is inspired by the K-means PSO hybrid [16], is adopted as follows.

1. Execute the IKSCOC algorithm to initialize one particle to contain $N_c$ selected cluster centroids;
2. Initialize the other particles of the swarm to contain $N_c$ selected cluster centroids at random;
3. For $t = 1$ to $t_{\max}$ do {
4. For each particle $X_i$ do {
5. For each object $p$ do {
6. Calculate $d_o(p, m_{ij})$;
7. Assign object $p$ to cluster $C_{ij}$ such that $d_o(p, m_{ij}) = \min \forall c = 1, \ldots, N_c \{d_o(p, m_{ic})\}$;
8. Calculate the fitness according to equation (9) ;}}
9. Update *gBest* and $pBest_i$;
10. Update the cluster centroids according to equation (1) and equation (2);
11 If $\|v\| \le \varepsilon$, terminate;
12. Optimize new individuals using the IKSCOC algorithm ;}

where $t_{\max}$ is the maximum number of iteration, $\varepsilon$ is the minimum velocity. STEP 1 is to overcome the disadvantage of the global PSO which tends to be trapped in a local optimum under some initialization conditions. STEP 12 is to improve the local constringency speed of the global PSO.

## 5 Results and Discussion

We have made experiments separately by K-Medoids, IKSCOC, GKSCOC, and PKSCOC. $n = 50, w = 0.72, c_1 = c_2 = 2, V_{max} = 0.4, t_{max} = 100, \varepsilon = 0.001$.

Fig.5 shows the results on synthetic Dataset1. Fig.5 (a) shows the original data with simple obstacles. Fig.5 (b) shows the results of 4 clusters found by K-Medoids without considering obstacles constraints. Fig.5(c) shows 4 clusters found by IKSCOC. Fig.5 (d) shows 4 clusters found by GKSCOC. Fig.5 (e) shows 4 clusters found by PKSCOC. Obviously, the results of the clustering illustrated in Fig.5(c), Fig.5 (d) and Fig.5 (e) all have better practicalities than that in Fig.5 (b). And the one in Fig.5 (e) is superior to the one in Fig.5(c) but is less inferior to the one in Fig.5 (d).

Fig.6 shows the results on synthetic Dataset2. Fig.6 (a) shows the original data with various obstacles. Fig.6 (b) shows 4 clusters found by K-Medoids. Fig.6 (c) shows 4 clusters found by PKSCOC. Obviously, the result of the clustering illustrated in Fig.6(c) has better practicalities than the one in Fig.8 (b).

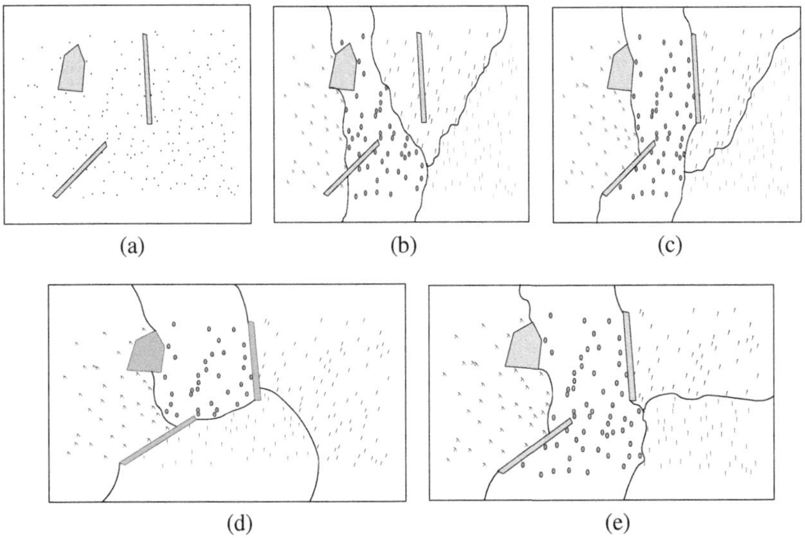


**Fig. 5.** Clustering dataset Dataset1

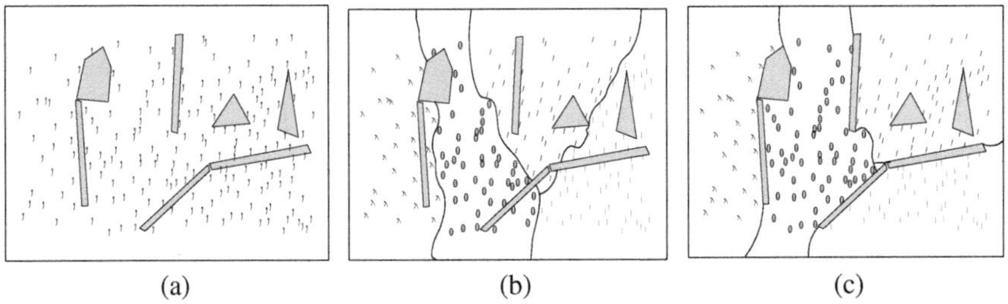


**Fig. 6.** Clustering dataset Dataset2

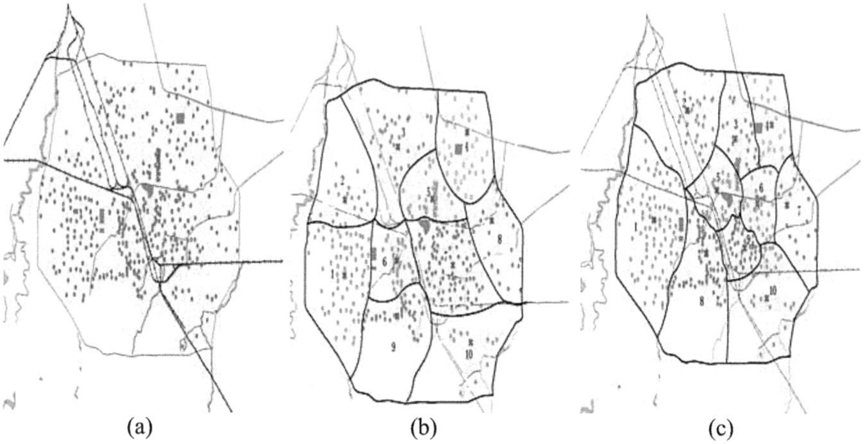


**Fig. 7.** Clustering dataset Dataset3

Fig.7 shows the results on real Dataset3 of residential spatial data points with river and railway obstacles in facility location on city parks. Fig. 7(a) shows the original data with river and railway obstacles. Fig. 7(b) and Fig. 7(c) show 10 clusters found by K-Medoids and PKSCOC respectively. Obviously, the result of the clustering illustrated in Fig. 7(c) has better practicalities than the one in Fig. 7(b). So, it can be drawn that PKSCOC is effective and has better practicalities

Fig.8 is the constringency speed in one experiment on Dataset1. It is showed that PKSCOC constringes in about 12 generations while GKSCOC constringes in nearly 25 generations. So, we can draw the conclusion that PKSCOC is effective and has higher constringency speed than GKSCOC.

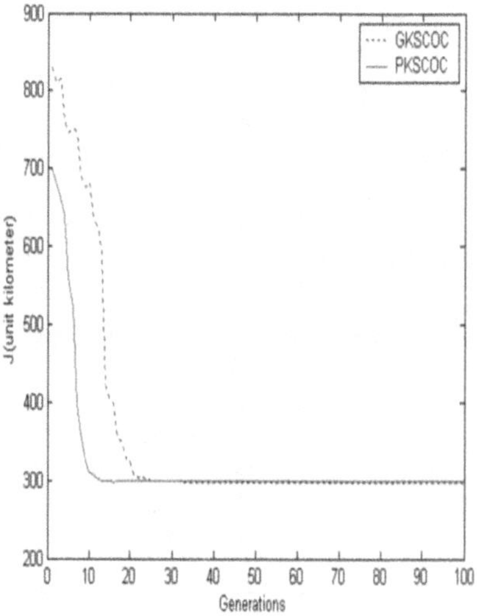


**Fig. 8.** PKSCOC vs. GKSCOC

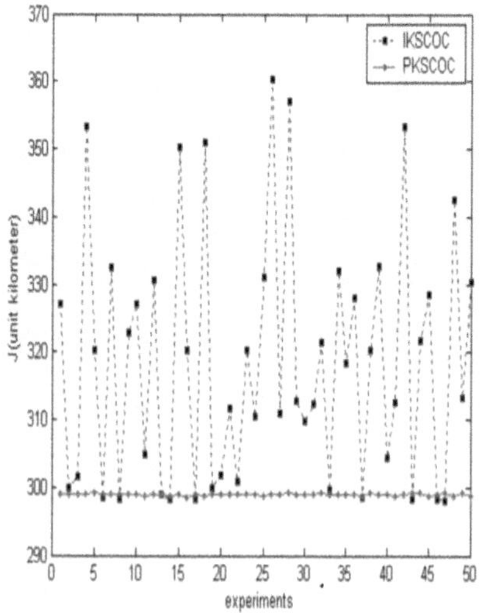


**Fig. 9.** PKSCOC vs. IKSCOC

Fig.9 is the value of $J$ showed in every experiment on Dataset1. It is showed that IKSCOC is sensitive to initial value and it constringes in different extremely local optimum points by starting at different initial value while PKSCOC constringes nearly in the same optimum points each time. Therefore, we can draw the conclusion that PKSCOC has stronger global constringent ability comparing with IKSCOC.

## 6 Conclusions

Spatial clustering has been an active research area in the data mining community. Classic clustering algorithms have ignored the fact that many constraints exit in the real world and could affect the effectiveness of clustering result. This paper proposes a PSO method for SCOC. In the process of doing so, we first use PSO algorithm via MAKLINK graphic to get the optimal obstructed path, and then we developed PKSCOC algorithm to cluster spatial data with obstacles constraints. The experimental results demonstrate the effectiveness and efficiency of the proposed method, which can not only give attention to higher local constringency speed and stronger global optimum search, but also get down to the obstacles constraints and practicalities of spatial clustering. But the drawback of this method is that using the PSO algorithm based MAKLINK graphic to obtain the best obstructed path is unfit for irregular shape obstacles.

**Acknowledgments.** This work is partially supported by the Natural Sciences Fund Council of China (Number: 40471115), the Natural Sciences Fund of Henan( Number:0511011000, Number: 0624220081), and the Open Research Fund Program of the Geomatics and Applications Laboratory, Liaoning Technical University (Number: 2004010).

## References

1. Tung, A.K.H., Han, J., Lakshmanan, L.V.S., Ng, R.T.: Constraint-Based Clustering in Large Databases. In Proceedings of the International Conference on Database Theory (ICDT'01). London U.K. (2001) 405-419
2. Tung, A.K.H., Ng, R.T., Lakshmanan, L.V.S., Han, J.: Geospatial Clustering with User-Specified Constraints. In Proceedings of the International Workshop on Multimedia Data Mining (MDM/KDD 2000). Boston USA (2000) 1-7
3. Tung, A.K.H., Hou, J., Han, J.: Spatial Clustering in the Presence of Obstacles. In Proceedings of International Conference on Data Engineering (ICDE'01). Heidelberg Germany (2001) 359-367
4. Estivill-Castro, V., Lee, I.J.: AUTOCLUST+: Automatic Clustering of Point-Data Sets in the Presence of Obstacles. In Proceedings of the International Workshop on Temporal, Spatial and Spatial-Temporal Data Mining. Lyon France (2000) 133-146
5. Zaïane, O.R., Lee, C.H.: Clustering Spatial Data When Facing Physical Constraints. In Proceedings of the IEEE International Conference on Data Mining (ICDM'02). Maebashi City Japan (2002) 737-740
6. Wang, X., Hamilton, H.J.: DBRS: A Density-Based Spatial Clustering Method with Random Sampling. In Proceedings of the 7th PAKDD. Seoul Korea (2003) 563- 575

7. Wang, X., Rostoker, C., Hamilton, H.J.: DBRS+: Density-Based Spatial Clustering in the Presence of Obstacles and Facilitators. Ftp.cs.uregina.ca/Research/Techreports/2004-09.pdf. (2004)
8. Wang, X., Hamilton, H.J.: Gen and Data Generators for Obstacle Facilitator Constrained Clustering. Ftp.cs.uregina.ca/Research/Techreports/2004-08.pdf. (2004)
9. Zhang, X.P., Wang, J.Y., Wu, F., Fan, Z.S, Li, X.Q.: A Novel Spatial Clustering with Obstacles Constraints Based on Genetic Algorithms and K-Medoids. In Proceedings of the Sixth International Conference on Intelligent Systems Design and Applications (ISDA 2006), Jinan Shandong China (2006) 605-610
10. Eberhart, R., Kennedy, J.: A New Optimizer Using Particle Swarm Theory. In Proceedings of the Sixth International Symposium on Micro Machine and Human Science, Nagoya Japan (1995) 39-43
11. Kennedy, J., Eberhart, R.: Particle Swarm Optimization. In Proceedings of IEEE International Conference on Neural Networks, Vol. 4. Perth Australia (1995) 1942-1948
12. Van den Bergh, F.: An Analysis of Particle Swarm Optimizers. Ph.D. thesis, University of Pretoria. (2001)
13. Habib, M.K., Asama, H.: Efficient Method to Generate Collision Free Paths for Autonomous Mobile Robot Based on New Free Space Structuring Approach. In Proceedings of International Workshop on Intelligent Robots and Systems, Japan, November, (1991) 563-567
14. Qin, Y.Q., Sun, D.B., Li, N., Cen, Y.G.: Path Planning for Mobile Robot Using the Particle Swarm Optimization with Mutation Operator. In Proceedings of the Third International Conference on Machine Learning and Cybernetics. Shanghai China (2004) 2473-2478
15. Xiao, X., Dow, E.R., Eberhart, R., Miled, Z.B., Oppelt, R.J.: Gene Clustering Using Self-Organizing Maps and Particle Swarm Optimization. In Proceedings of the International Conference on Parallel and Distributed Processing Symposium (IPDPS). (2003)
16. Vander, M. D.W., Engelbrecht, A.P.: Data Clustering Using Particle Swarm Optimization. In Proceedings of IEEE Congress on Evolutionary Computation 2003. (2003) 215-220
17. Omran, M.G.H.: Particle Swarm Optimization Methods for Pattern Recognition and Image Processing. Ph.D. thesis, University of Pretoria. (2005)
18. Cui, X.H., Potok, T.E., Palathingal, P.: Document Clustering Using Particle Swarm Optimization. In Proceedings of IEEE on Swarm Intelligence Symposium (SIS 2005). (2005) 185-191

# A PSO-Based Classification Rule Mining Algorithm

Ziqiang Wang, Xia Sun, and Dexian Zhang

School of Information Science and Engineering, Henan University of Technology,
Zheng Zhou 450052, China
wzqagent@126.com

**Abstract.** Classification rule mining is one of the important problems in the emerging field of data mining which is aimed at finding a small set of rules from the training data set with predetermined targets. To efficiently mine the classification rule from databases, a novel classification rule mining algorithm based on particle swarm optimization (PSO) was proposed. The experimental results show that the proposed algorithm achieved higher predictive accuracy and much smaller rule list than other classification algorithm.

## 1 Introduction

The current information age is characterized by a great expansion in the volume of data that are being generated and stored. Intuitively, this large amount of stored data contains valuable hidden knowledge, which could be used to improve the decision-making process of an organization. With the rapid growth in the amount of information stored in databases, the development of efficient and effective tools for revealing valuable knowledge hidden in these databases becomes more critical for enterprise decision making. One of the possible approaches to this problem is by means of data mining or knowledge discovery from databases (KDD)[1]. Through data mining, interesting knowledge can be extracted and the discovered knowledge can be applied in the corresponding field to increase the working efficiency and to improve the quality of decision making.

Classification rule mining is one of the important problems in the emerging field of data mining which is aimed at finding a small set of rules from the training data set with predetermined targets[2]. There are different classification algorithms used to extract relevant relationship in the data as decision trees that operate a successive partitioning of cases until all subsets belong to a single class. However, this operating way is impracticable except for trivial data sets. There are many other approaches for data classification, such as statistical and roughest approaches and neural networks. These classification techniques require significant expertise to work effectively but do not provide intelligible rules though they are algorithmically strong.

The classification problem becomes very hard when the number of possible different combinations of parameters is so high that algorithms based on exhaustive

D.-S. Huang, L. Heutte, and M. Loog (Eds.): ICIC 2007, LNAI 4682, pp. 377–384, 2007.

searches of the parameter space become computationally infeasible rapidly. The self-adaptability of evolutionary algorithms based on population is extremely appealing when tackling the tasks of data mining. Thus it is natural to direct attention to heuristic approaches to find a "good-enough" solution to combat the classification problem. In recent years, evolutionary algorithms(such as genetic algorithm,immune algorithm and ant colony algorithm) have emerged as promising techniques to discover useful and interesting knowledge from databases[3]. Especially, there are numerous attempts to apply genetic algorithms(GAs) in data mining to accomplish classification tasks. In addition, the particle swarm optimization (PSO) algorithm[4], which has emerged recently as a new meta-heuristic derived from nature, has attracted many researchers' interests. The algorithm has been successfully applied to several minimization optimization problems and neural network training. Nevertheless, the use of the algorithm for mining classification rule in the context of data mining is still a research area where few people have tried to explore.

Recently, Eberhart and Kennedy suggested a particle swarm optimization (PSO) based on the analogy of swarm of bird[4]. The algorithm, which is based on a metaphor of social interaction, searches a space by adjusting the trajectories of individual vectors, called "particles " as they are conceptualized as moving points in multidimensional space. The individual particles are drawn stochastically toward the position of their own previous best performance and the best previous performance of their neighbors. The main advantages of the PSO algorithm are summarized as: simple concept, easy implementation, robustness to control parameters, and computational efficiency when compared with mathematical algorithm and other heuristic optimization techniques. The original PSO has been applied to a learning problem of neural networks and function optimization problems, and efficiency of the method has been confirmed. In this paper, the objective is to investigate the capability of the PSO algorithm to discover classification rule with higher predictive accuracy and a much smaller rule list.

The rest of the paper is organized as follows. In the next section, we give a brief problem description about mining classification rule. In section 3, we present the basic idea and key techniques of the PSO algorithm. In section 4,the PSO-based classification rule mining algorithm is proposed. Section 5 reports experimental results when comparing with Ant-Miner[5] and GA-based classification algorithm across six data sets. Finally, the paper ends with conclusions and future research directions.

## 2 Classification Rule Problem Description

In general, the problem on mining classification rules can be stated as follows. There is a large database $D$, in which each tuple consists of a set of n attributes (features), $\{A_1, A_2, \ldots, A_n\}$. For example, attributes could be name, gender, age, salary range, zip code, etc. Our purpose is to assign each case(object, record, or instance) to one class out of a set of predefined classes based on the values of some attributes(called predictor attributes) for the case.

In the classification task, the discovered knowledge is usually represented in the form of IF -THEN prediction rules, which have the advantage of being of a high-level and symbolic knowledge representation contributing to the comprehensibility of the discovered knowledge. In this paper, knowledge is presented as multiple IF-THEN rules in a classification rules list. Such rules state that the presence of one or more items (antecedents) implies or predicts the presence of other items(consequents). A typical rule has the following form: IF term1 AND term2 AND ... THEN class, where each term of the rule antecedent is a triple <attribute, operator, value>,such as <age=70>. The rule consequent(THEN part) specifies the class predicted for cases whose predictor attributes satisfy all the terms specified in the rule antecedent. This kind of classification rule representation has the advantage of being intuitively comprehensible for the user.

Classification rule mining is one of the important data mining technique. Many classification algorithms have been proposed, such as statistical based, distance based, neural network base and decision tree based,have been constructed and applied to discover knowledge from data in different applications, yet many suffer from poor performance in prediction accuracy in many practical domains. While it seems unlikely to have an algorithm to perform best in all the domains, it may well be possible to produce classifiers that perform better on a wide variety of real-world domains. To achieve this objective,a novel classification rule mining algorithm based on particle swarm optimization (PSO) is proposed. The experimental results show that the proposed algorithm achieved higher predictive accuracy and much smaller rule list than other classification algorithm.

## 3 The Particle Swarm Optimization Algorithm

PSO is a relatively new population-based evolutionary computation technique[4]. In contrast to genetic algorithms (GAs)which exploit the competitive characteristics of biological evolution. PSO exploits cooperative and social aspects, such as fish schooling, birds flocking, and insects swarming. Resembling the social behavior of a swarm of bees to search the location with the most flowers in a field, the optimization procedure of PSO is based on a population of particles that fly in the solution space with velocity dynamically adjusted according to its own flying experience and the flying experience of the best among the swarm. In the past several years, PSO has been successfully applied in many different application areas due to its robustness and simplicity. In comparison with other stochastic optimization techniques like genetic algorithms (GAs), PSO has fewer complicated operations and fewer defining parameters, and can be coded in just a few lines. Because of these advantages, the PSO has received increasing attention in data mining community in recent years. PSO is applied to classification rule mining in this work.

The PSO definition is described as follows. Let $s$ denote the swarm size. Each individual particle $i(1 \leq i \leq s)$ has the following properties: a current position $x_i$ in search space, a current velocity $v_i$, and a personal best position $p_i$ in the search space, and the global best position $p_{gb}$ among all the $p_i$. During each iteration, each particle in the swarm is updated using the following equation.

$$v_i(t+1) = k[w_i v_i(t) + c_1 r_1(p_i - x_i(t)) + c_2 r_2(p_{gb} - x_i(t))] \ , \tag{1}$$

$$x_i(t+1) = x_i(t) + v_i(t+1) \ , \tag{2}$$

where $c_1$ and $c_2$ denote the acceleration coefficients, and $r_1$ and $r_2$ are random numbers uniformly distributed within [0,1].

The value of each dimension of every velocity vector $v_i$ can be clamped to the range $[-v_{max}, v_{max}]$ to reduce the likelihood of particles leaving the search space. The value of $v_{max}$ chosen to be $k \times x_{max}$(where $0.1 \leq k \leq 1$). Note that this does not restrict the values of $x_i$ to the range $[-v_{max}, v_{max}]$. Rather than that, it merely limits the maximum distance that a particle will move.

Acceleration coefficients $c_1$ and $c_2$ control how far a particle will move in a single iteration. Typically, these are both set to a value of 2.0, although assigning different values to $c_1$ and $c_2$ sometimes leads to improved performance. The inertia weight $w$ in Equation (1) is also used to control the convergence behavior of the PSO. Typical implementations of the PSO adapt the value of $w$ linearly decreasing it from 1.0 to near 0 over the execution. In general, the inertia weight $w$ is set according to the following equation[6]:

$$w_i = w_{max} - \frac{w_{max} - w_{min}}{iter_{max}} \cdot iter, \tag{3}$$

where $iter_{max}$ is the maximum number of iterations, and $iter$ is the current number of iterations.

In order to guarantee the convergence of the PSO algorithm, the constriction factor $k$ is defined as follows:

$$k = \frac{2}{|2 - \varphi - \sqrt{\varphi^2 - 4\varphi}|} \ , \tag{4}$$

where $\varphi = c_1 + c_2$ and $\varphi > 4$.

The PSO algorithm performs the update operations in terms of Equation (1) and (2) repeatedly until a specified number of iterations have been exceeded, or velocity updates are close to zero. The quality of particles is measured using a fitness function which reflects the optimality of a particular solution. Some of the attractive features of the PSO include ease of implementation and the fact that only primitive mathematical operators and very few algorithm parameters need to be tuned. It can be used to solve a wide array of different optimization problems, some example applications include neural network training and function minimization. However, the use of the PSO algorithm for mining classification rule in the context of data mining is still a research area where few people have tried to explore. In this paper,a PSO-based classification rule mining algorithm is proposed in later section.

## 4 The PSO-Based Classification Rule Mining Algorithm

The steps of the PSO-based classification rule mining algorithm are described as follows.

Step1: Initialization and Structure of Individuals. In the initialization process, a set of individuals(i.e.,particle) is created at random. The structure of an individual for classification problem is composed of a set of attribute values. Therefore, individual $i's$ position at iteration 0 can be represented as the vector $X_i^0 = (x_{i1}^0, \ldots, x_{in}^0)$ where $n$ is the number of attribute numbers in attribute table. The velocity of individual $i$(i.e.,$V_i^0 = (v_{i1}^0, \ldots, v_{in}^0)$)corresponds to the attribute update quantity covering all attribute values,the velocity of each individual is also created at random. The elements of position and velocity have the same dimension.

Step2: Evaluation Function Definition. As in all evolutionary computation techniques there must be some function or method to evaluate the goodness of a position. The fitness function must take the position in the solution space and return a single number representing the value of that position. The evaluation function of PSO algorithm provides the interface between the physical problem and the optimization algorithm. The evaluation function used in this study is defined as follows:

$$F = 1 - \frac{N(R)}{M} + \frac{TP}{TP+FN} \cdot \frac{TN}{TN+FP}, \tag{5}$$

where $(1 - \frac{N(R)}{M})$ denotes the comprehensibility metric of a classification rule, $N(R)$ is the number of conditions in the rule $R$, $M$ denotes the allowable maximal condition number of the rule $R$. In general the smaller the rule, the more comprehensible it is.

In addition,$(\frac{TP}{TP+FN} \cdot \frac{TN}{TN+FP})$ denotes the quality of rule $R$,where TP(true positives) denotes the number of cases covered by the rule that have the class predicted by the rule,FP(false positives) denotes the number of cases covered by the rule that have a class different from the class predicted by the rule,FN(false negatives) denotes the number of cases that are not covered by the rule but that have the class predicted by the rule,TN(true negatives) denotes the number of cases that are not covered by the rule and that do not have the class predicted by the rule.

Therefore, $F's$ value is within the range [0,1] and the larger the value of $F$ , the higher the comprehensibility and quality of the rule.

Step3: Personal and Global Best Position Computation. Each particle $i$ memorizes its own $F's$ value and chooses the maximum one, which has been better so far as personal best position $p_i^t$. The particle with the best $F's$ value among $p_i^t$ is denoted as global position $p_{gb}^t$,where $t$ is the iteration number. Note that in the first iteration,each particle $i$ is set directly to $p_i^0$, and the particle with the best $F's$ value among $p_i^0$ is set to $p_{gb}^0$.

Step4: Modify the velocity of each particle according to Equation(1). If $v_i^{(t+1)} > V_i^{max}$, then $v_i^{(t+1)} = V_i^{max}$. If $v_i^{(t+1)} < V_i^{min}$, then $v_i^{(t+1)} = V_i^{min}$.

Step5: Modify the position of each particle according to Equation(2).

Step6: Rule pruning. The main goal of rule pruning is to remove irrelevant terms that might have been unduly included in the rule. Morever,rule pruning can increase the predictive power of the rule, helping to improve the simplicity of the rule. The process of rule pruning is as follows: a)Compute a rule quality value using Equation(5); b) Check the attribute pairs in the reverse order in which they were selected to see if a pair can be removed without causing the rule quality to decrease. If yes, remove it. This process is repeated until no pair can be removed.

Step7: If the best evaluation value $p_{gb}$ is not obviously improved or the iteration number $t$ reaches the given maximum,then go to Step8. Otherwise, go to Step2.

Step8: The particle that generates the best evaluation value $F$ is the output classification rule.

## 5 Experimental Results

To thoroughly investigate the performance of the proposed PSO algorithm, we have conducted experiment with it on a number of datasets taken from the UCI repository[7]. In Table 1, the selected data sets are summarized in terms of the number of instances, and the number of the classes of the data set. These data sets have been widely used in other comparative studies. All the results of the comparison are obtained on a Pentium 4 PC(CPU 2.2GHZ,RAM 256MB).

**Table 1.** Dataset Used in the Experiment

| Data Set | Instances | Classes |
|---|---|---|
| Ljubljana Breast Cancer | 282 | 2 |
| Wisconsin Breast Cancer | 683 | 2 |
| Tic-Tac-Toe | 958 | 2 |
| Dermatology | 366 | 6 |
| Hepatitis | 155 | 2 |
| Cleveland Heart Disease | 303 | 5 |

In all our experiments,the PSO algorithm uses the following parameter values. Inertia weight factor $w$ is set by Equation (3),where $w_{max} = 0.9$ and $w_{min} = 0.4$. Acceleration constant $c_1 = c_2 = 2$. The population size in the experiments was fixed to 20 particles in order to keep the computational requirements low. Each run has been repeated 50 times and average results are presented.

We have evaluated the performance of PSO by comparing it with Ant-Miner [5], OCEC(a well-known genetic classifier algorithm)[8]. The first experiment was carried out to compare predictive accuracy of discovered rule lists by well-known ten-fold cross-validation procedure[9]. Each data set is divided into ten partitions, each method is run ten times, using a different partition as test set and the other nine partitions as the training set each time. The predictive accuracies of the ten runs are averaged as the predictive accuracy of the discovered rule list. Table 2 shows the results comparing the predictive accuracies of PSO, Ant-Miner and OCEC, where the symbol " ± " denotes the standard deviation of the corresponding predictive accuracy. It can be seen that predictive accuracies of PSO is higher than those of Ant-Miner and OCEC.

**Table 2.** Predictive Accuracy Comparison

| Data Set | PSO(%) | Ant-Miner(%) | OCEC(%) |
|---|---|---|---|
| Ljubljana Breast Cancer | 78.56±0.24 | 75.28±2.24 | 76.89±0.18 |
| Wisconsin Breast Cancer | 98.36±0.28 | 96.04±0.93 | 95.42±0.02 |
| Tic-Tac-Toe | 98.89±0.13 | 73.04±2.53 | 92.51±0.15 |
| Dermatology | 98.24±0.26 | 94.29±1.20 | 93.24±0.12 |
| Hepatitis | 95.75±0.31 | 90.00±3.11 | 91.64±0.23 |
| Cleveland Heart Disease | 79.46±0.34 | 57.48±1.78 | 76.75±0.16 |

In addition, We compared the simplicity of the discovered rule list by the number of discovered rules. The results comparing the simplicity of the rule lists discovered by PSO,Ant-Miner and OCEC are shown in Table 3. As shown in those tables, taking into number of rules discovered, PSO mined rule lists much simpler(smaller) than the rule lists mined by Ant-Miner and OCEC.

**Table 3.** Number of Rules Discovered Comparison

| Data Set | PSO | Ant-Miner | OCEC |
|---|---|---|---|
| Ljubljana Breast Cancer | 6.05±0.21 | 7.10±0.31 | 16.65±0.21 |
| Wisconsin Breast Cancer | 4.23±0.13 | 6.20±0.25 | 15.50±0.13 |
| Tic-Tac-Toe | 6.45±0.37 | 8.50±0.62 | 12.23±0.25 |
| Dermatology | 6.39±0.24 | 7.30±0.47 | 13.73±0.18 |
| Hepatitis | 3.01±0.26 | 3.40±0.16 | 10.73±0.35 |
| Cleveland Heart Disease | 7.15±0.23 | 9.50±0.71 | 15.37±0.42 |

At last, we also compared the running time of PSO with Ant-Miner and OCEC. The experimental results are reported Table 4,as expected,we can see that PSO's running time is fewer than Ant-Miner's and OCEC's in all data sets. The main reason is that PSO algorithm is conceptually very simple and requires only primitive mathematical operators codes. In addition,PSO can be implemented in a few lines of computer codes,those reduced PSO's running time.

In summary, PSO algorithm needs to tune very few algorithm parameters, taking into account both the predictive accuracy and rule list simplicity criteria, the proposed PSO-based classification rule mining algorithm has shown promising results.

**Table 4.** Running Time Comparison

| Data Set | PSO | Ant-Miner | OCEC |
|---|---|---|---|
| Ljubljana Breast Cancer | 31.25 | 55.28 | 46.37 |
| Wisconsin Breast Cancer | 42.35 | 58.74 | 45.25 |
| Tic-Tac-Toe | 38.65 | 61.18 | 52.38 |
| Dermatology | 27.37 | 49.56 | 37.23 |
| Hepatitis | 38.86 | 56.57 | 42.89 |
| Cleveland Heart Disease | 31.83 | 48.73 | 35.26 |

## 6 Conclusions

Classification rule mining is one of the most important tasks in data mining community because the data being generated and stored in databases are already enormous and continues to grow very fast.In this paper, a PSO-based algorithm for classification rule mining is presented. Compared with the Ant-Miner and OCEC in public domain data sets,the experimental results show that the proposed algorithm achieved higher predictive accuracy and much smaller rule list than Ant-Miner and OCEC.

## References

1. Fayyad, U.M., Piatetsky-Shapiro, G., Smyth, P.: From Data Mining to Knowledge Discovery: an Overview.In Advances in Knowledge Discovery & Data Mining, MIT Press(1996)1–34
2. Quinlan, J.R.: Induction of Decision Trees. Machine Learning,1(1986)81–106
3. Freitas, A.A.: Data Mining and Knowledge Discovery with Evolutionary Algorithms. Springer-Verlag,Berlin(2002)
4. Eberhart, R.C., Kennedy,J.: A New Optimizer using Particle Swarm Theory.In:Proc. 6th Symp.Micro Machine and Human Science,Nagoya,Japan(1995)39–43
5. Parpinelli, R.S., Lopes, H.S., Freitas, A.A.: Data Mining with an Ant Colony Optimization Algorithm.IEEE Transactions on Evolutionary Computing 6(2002)321–332
6. Kennedy, J.: The particle Swarm:Social Adaptation of Knowledge.In: Proc. IEEE Int. Conf. Evol. Comput., Indianapolis,IN(1997)303-308
7. Hettich, S., Bay, S.D.: The UCI KDD Archive. URL:http://kdd.ics.uci.edu (1999)
8. Liu, J., Zhong, W.-C., Liu, F., Jiao, L.-C.: Classification Based on Organizational Coevolutionary Algorithm. Chinese Journal of Computers 26(2003)446–453
9. Weiss, S.M., Kullkowski, C.A.: Computer Systems that Learn.Morgan Kaufmann Press,San Mateo,CA(1991)

# A Similarity Measure for Collaborative Filtering with Implicit Feedback

Tong Queue Lee[1], Young Park[2], and Yong-Tae Park[3]

[1] Dept. of Mobile Internet Dongyang Technical College
62-160 Gocheok-dong, Guro-gu, Seoul 152-714, Korea
queue@dongyang.ac.kr
[2] Dept. of Computer Science & Information Systems
Bradley University, W. Bradley Ave., Peoria, IL 61625, USA
young@bradley.edu
[3] Dept. of Industrial Engineering, Seoul National University
San 56-1, Sillim-dong, Gwanak-gu, Seoul 151-742, Korea
parkyt@cybernet.snu.ac.kr

**Abstract.** Collaborative Filtering(CF) is a widely accepted method of creating recommender systems. CF is based on the similarities among users or items. Measures of similarity including the Pearson Correlation Coefficient and the Cosine Similarity work quite well for explicit ratings, but do not capture real similarity from the ratings derived from implicit feedback. This paper identifies some problems that existing similarity measures have with implicit ratings by analyzing the characteristics of implicit feedback, and proposes a new similarity measure called Inner Product that is more appropriate for implicit ratings. We conducted experiments on user-based collaborative filtering using the proposed similarity measure for two e-commerce environments. Empirical results show that our similarity measure better captures similarities for implicit ratings and leads to more accurate recommendations. Our inner product-based similarity measure could be useful for CF-based recommender systems using implicit ratings in which negative ratings are difficult to be incorporated.

**Keywords:** E-commerce, recommender system, collaborative filtering, implicit feedback, similarity measure, recommendation accuracy.

## 1 Introduction

Today users face the problem of choosing the right products or services within a flood of information. A variety of recommender systems help users select relevant products or services. Among these recommender systems, collaborative filtering-based recommender systems are effectively used in many practical areas [1,2]. A hybrid method is also used by using item content information in addition to user feedback data [3].

Collaborative filtering determines the user's preference from the user's rating data. In general, rating data is generated by explicit feedback from users. Obtaining explicit feedback is not always easy and sometimes unfeasible. Users tend to be reluctant to

D.-S. Huang, L. Heutte, and M. Loog (Eds.): ICIC 2007, LNAI 4682, pp. 385–397, 2007.

partake in the intrusiveness of giving explicit feedback. In some cases, users give arbitrary ratings, leading to incorrect recommendations. There has been research on constructing rating data by using implicit feedback such as Web log instead of explicit feedback [4,5,6,7].

Once user rating data is established, collaborative filtering computes similarity among users or items using some similarity measure. A number of similarity measures have been used. The Pearson Correlation Coefficient and Cosine Similarity are two popular measures of similarity. These measures do not distinguish between explicit and implicit rating data. These measures work quite well with explicit ratings, but do not capture the real similarity of implicit ratings because the rating data derived from implicit feedback is different from explicit rating data.

In this paper we look at the characteristics of implicit feedback and propose a new similarity measure. We investigate the effectiveness of the proposed measure by conducting some experiments on real data in e-commerce environments. Our similarity measure could be used for collaborative filtering-based recommender systems using only implicit ratings, in which negative ratings are difficult to be incorporated.

The rest of this paper is organized as follows: Section 2 describes the characteristics of implicit ratings compared with explicit ratings. Some problems of existing similarity measures with implicit ratings are discussed in Section 3. In Section 4, a new similarity measure for implicit ratings is proposed. Experiments and empirical results are described in Section 5. Section 6 concludes with future work.

## 2 Deriving Ratings from Implicit Feedback

User preference is the basis of collaborative filtering. There are two ways of finding user preferences – explicit feedback and implicit feedback.

Ratings and reviews are popular forms of explicit feedback. Ratings are easily quantifiable and thus are used as the basis of collaborative filtering in practice called rating-based CF. For example, consider explicit ratings for movies using a scale of 1 (negative preference) to 5 (positive preference) as shown in Table 1.

**Table 1.** Explicit Movie Ratings (scales 1-5)

| | Movie 1 | Movie 2 | Movie 3 | Movie 4 |
|---|---|---|---|---|
| User A | 5 | | 3 | 1 |
| User B | | 1 | 5 | |
| User C | 1 | | | 4 |

User A's preference to Movie 1 and User B's preference to Movie 3 are high, meaning they like those movies. User A's preference to Movie 4, User B's preference to Movie 2 and User C's preference to Movie 1 are very low, meaning they dislike those movies.

With explicit feedback, users can clearly express positive or negative preferences. However, it is not always easy to obtain explicit feedback. It is practically impossible in some situations, such as mobile e-commerce environments. In this case, recommender systems should rely on implicit feedback.

Implicit feedback includes purchase patterns, page visits, page viewing times, and Web surfing paths. This data is usually obtained by analyzing the Web log. This approach needs preprocessing in order to build implicit ratings by extracting meaningful data from the whole Web log. The amount of meaningful data in the Web log is usually small. Collaborative filtering based on this data is called log-based CF [8,9].

With implicit feedback, users cannot clearly express negative preferences. Implicit ratings constructed from implicit feedbacks do not include negative preferences. For example, consider the implicit ratings for items as shown in Table 2. They are constructed by using the number of item's Web page visits.

**Table 2.** Implicit Ratings from the Number of Item's Web Page Visits

| | Item 1 | Item 2 | Item 3 | Item 4 |
|---|---|---|---|---|
| User A | 15 | | 7 | 3 |
| User B | | 2 | 13 | |
| User C | 4 | | | 12 |

From Table 2, we infer that User A has high preference to Item 1 and User B has high preference to Item 3. We can also view that User A's preference to Item 4 and User B's preference to Item 2 are relatively low. However, it is rather difficult to conclude that they do not like those items. Implicit values are derived from implicit feedback. Lower values do not necessarily correspond to lower preferences.

As another example, consider the implicit ratings for items by using the purchase of items (Table 3).

**Table 3.** Implicit Ratings from the Purchase of Items

| | Item 1 | Item 2 | Item 3 | Item 4 |
|---|---|---|---|---|
| User A | 1 | | 1 | 1 |
| User B | | 1 | 1 | |
| User C | 1 | | | 1 |

In Table 3, 1 indicates that the user purchased the item. In this case, we can infer that the user likes the purchased item. However, we cannot conclude that the user dislikes all the items that were not purchased.

## 3 Similarity Problems with Implicit Ratings

A similarity measure is used in collaborative filtering in order to determine the similarity between two users or items using users' item ratings. The Pearson Correlation Coefficient and the Cosine Similarity are two popular measures of similarity. These two measures work quite well with explicit user ratings. However, there are some problems when these measures are applied to implicit ratings.

### 3.1 Pearson Correlation Coefficient

The Pearson Correlation Coefficient is one of the most widely used similarity measures from the early days of collaborative filtering to the present [1].

The Pearson Correlation Coefficient is defined as follows:

$$P\_sim(a,b) = \frac{\sum_j (P_{aj} - \overline{P_a})(P_{bj} - \overline{P_b})}{\sqrt{\sum_j (P_{aj} - \overline{P_a})^2} \sqrt{\sum_j (P_{bj} - \overline{P_b})^2}} \quad (1)$$

Here, a and b are users, $P_{aj}$ is the current preference of user a on item j, $P_{bj}$ is the current preference of user b on item j, $\overline{P_a}$ is the average current preference of user a, and $\overline{P_b}$ is the average current preference of user b.

The Pearson Correlation Coefficient considers the differences in users' average preferences by subtracting the average preference from the current preference. By dividing by the standard deviations it also considers the differences in user rating values.

For instance, consider the explicit ratings for movies using a scale of 1 (negative preference) to 5 (positive preference). An example matrix is shown in Table 4.

**Table 4.** Explicit Ratings (scale 1-5)

| | Movie 1 | Movie 2 | Movie 3 |
|---|---|---|---|
| User A | 1 | 2 | 3 |
| User B | 5 | 4 | 3 |

From Table 4, we see that the rating trends of User A and User B are opposite. When we compute the Pearson Correlation Coefficient between User A and User B, it is negative and thus shows that these two users are dissimilar as shown in Fig. 1. The Pearson Correlation Coefficient appears to be a good similarity measure for explicit ratings given by users.

However, the Pearson Correlation Coefficient does not capture the real similarity between users from implicit ratings. For example, consider the number of web page visits. Table 5 shows an example implicit rating matrix.

Note that Table 5 looks similar to Table 4, but it contains the number of visits rather than actual rating values. Thus, like Table 4, the Pearson Correlation Coefficient between User A and User B is negative, which implies that these two users are dissimilar. However, because the values in the implicit rating matrix do not indicate any negative preferences, it is difficult to conclude that two users are

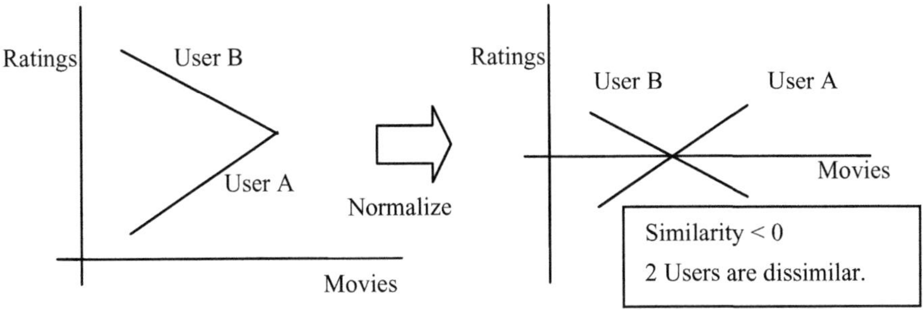


**Fig. 1.** Similarity using Pearson Correlation Coefficient

**Table 5.** Implicit Ratings based on Page Visit Counts

| | Page 1 | Page 2 | Page 3 |
|---|---|---|---|
| User A | 2 | 4 | 6 |
| User B | 10 | 8 | 6 |

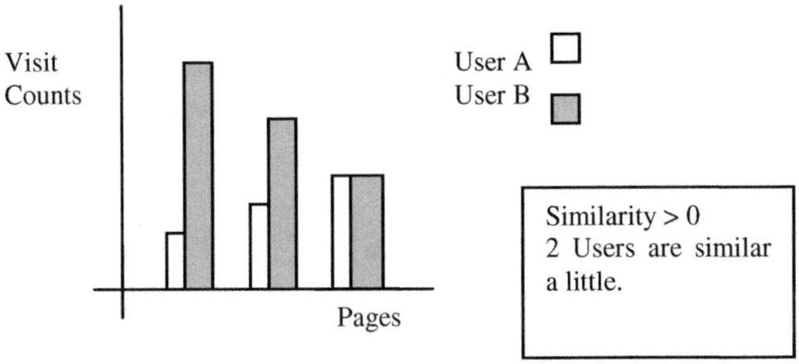


**Fig. 2.** Similarity with Implicit Feedback

dissimilar. Smaller numbers of visits do not necessarily correlate to negative preferences. In fact, User A and User B may have very similar preference trends as shown in Fig. 2.

### 3.2 Cosine Similarity

The Cosine Similarity is also one of the similarity measures that are widely used in collaborative filtering. The Cosine Similarity is defined as follows:

$$C\_sim(a,b) = \frac{\sum_j (P_{aj})(P_{bj})}{\sqrt{\sum_j (P_{aj})^2}\sqrt{\sum_j (P_{bj})^2}}. \qquad (2)$$

Here, a and b are users, $P_{aj}$ is the current preference of user a on the item j, $P_{bj}$ is the current preference of user b on the item j.

The Cosine Similarity between user u1 and user u2 can be viewed as the angle between u1's preference vector and u2's preference vector. The smaller the angle is, the greater the degree of similarity between the users is. For example, consider the explicit ratings for articles using a scale of 1 (negative preference) to 5 (positive preference). Consider an example matrix as shown in Table 6.

**Table 6.** Explicit Ratings (scales 1-5)

| | Article 1 | Article 2 |
|---|---|---|
| User A | 2 | 3 |
| User B | 1 | 2 |
| User C | 2 | 4 |

The Cosine Similarity between User A and User B is the same as the Cosine Similarity between User A and User C. Considering User C's rating values are proportionately larger than User B's, we infer that User B and User C are equally similar to User A. The Cosine Similarity normalizes rating values of a user in order to incorporate the user's trends on the rating values. Thus, as shown in Fig. 3, the Cosine Similarity seems reasonable for explicit ratings.

Like the Pearson Correlation Coefficient, however, the Cosine Similarity is problematic in capturing the real similarity between users from implicit ratings. For example, consider the page viewing time. Table 7 shows an example implicit rating matrix.

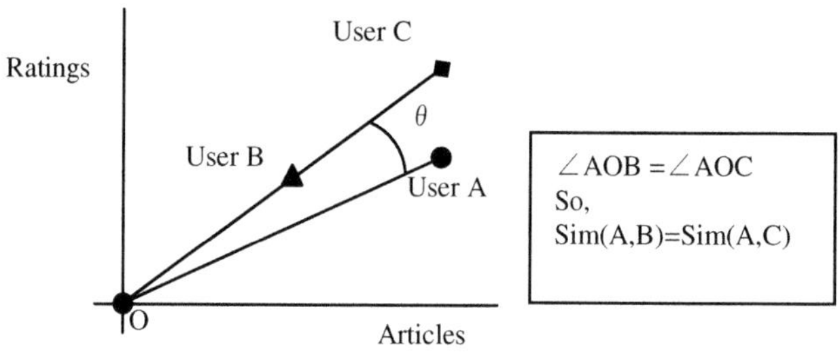


**Fig. 3.** Similarity using Cosine Similarity

**Table 7.** Implicit Ratings based on View Time (seconds)

| | Article 1 | Article 2 |
|---|---|---|
| User A | 20 | 30 |
| User B | 10 | 20 |
| User C | 20 | 40 |

Note that Table 7 looks similar to Table 6, but it contains the viewing duration in seconds rather than actual rating values. The Cosine Similarity between User A and User B is the same as the Cosine Similarity between User A and User C.

Still, it is difficult to conclude that User B and User C have the same extent of similarity with respect to User A because the values in the implicit rating matrix are not preference values. It is more natural that the values in the implicit rating matrix themselves without normalization should be viewed as preferences. User C spent more time viewing the articles than User B. Thus, as shown in Fig. 4, it could be that User C is more similar to User A than User B is to User A.

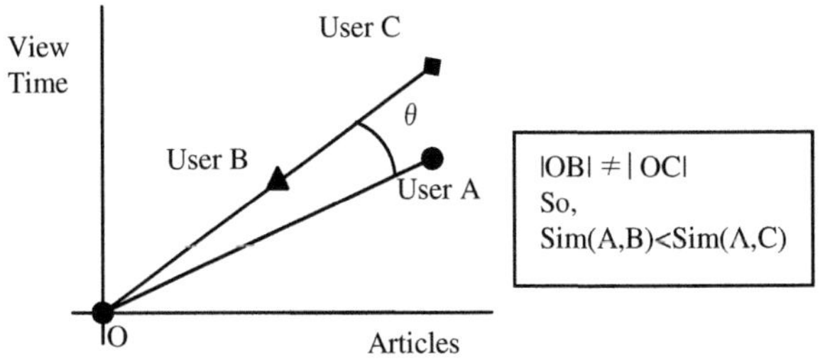


**Fig. 4.** Similarity with Implicit Feedback

# 4 A New Similarity Measure for Implicit Ratings

We propose a new similarity measure for implicit ratings. Our similarity measure solves the problems of negative preferences and normalization in implicit ratings that are constructed from various implicit feedbacks. The new similarity measure is called Inner Product and is defined as follows:

$$IP\_sim(a,b) = \vec{P_a} \bullet \vec{P_b} = \sum_j (P_{aj})(P_{bj}) \,. \tag{3}$$

Here, a and b are users, $\vec{P_a}$ is the preference vector of user a, $\vec{P_b}$ is the preference vector of user b, $P_{aj}$ is the current preference of user a on item j, and $P_{bj}$ is the current preference of user b on item j.

Compared with the Pearson Correlation Coefficient and the Cosine Similarity, the Inner Product measure better captures real similarity among users from implicit ratings. For example, consider the example implicit ratings based on page visit counts (Table 5). The implicit rating values indicate only positive preferences. The similarity value between User A and User B using the Pearson Correlation Coefficient is -1, which implies that the two users are dissimilar. However, the similarity value is 88 when using the Inner Product measure. This indicates that the two users have very similar preferences. The Inner Product measure reflects users' real preferences (shown by implicit ratings) better than the Pearson Correlation Coefficient.

We cannot compute similarity using the Pearson Correlation Coefficient if the standard deviation of User A or User B is 0 because the denominator becomes 0. However, we can compute the Inner Product-based similarity value regardless of the user's standard deviation.

Consider, for example, the example implicit rating matrix in Table 7 based on page view time. The similarity value using the Cosine Similarity between User A and User B is $8/\sqrt{65}$. The similarity value using Cosine Similarity between User A and User C is also $8/\sqrt{65}$. Because User C spent more time viewing the articles than User B, however, User C seems to be more similar to User A than User B is to User A. When the Inner Product measure is used, the similarity value between User A and User B is 800. But, the similarity value between User A and User C is 1600, which is twice the similarity value between User A and User B. The Inner Product measure reflects similarity between users more accurately than the Cosine Similarity in the context of implicit ratings.

The proposed Inner Product measure has the following improvements over the major existing similarity measures:

- The Inner Product measure solves the negative preference problem with the Pearson Coefficient.
- The Inner Product measure also solves the normalization problem with the Cosine Similarity.
- The Inner Product measure solves the problem with the Pearson Coefficient when the standard deviation is 0.

## 5 Experiments and Results

In order to investigate the effectiveness of our similarity measure, we conducted experiments on user-based collaborative filtering-based recommender systems using two data sets - data set 1 and data set 2.

The data set 1 is the set of purchase transactions of character images in a mobile environment provided by SKT in 2004. SKT is one of leading mobile service companies in Korea. The number of users that purchased at least once is 1,922. The number of character images is 9,131. The total number of transactions is 65,101. The data set 2 is the web-log of an on-line cosmetics store "H" in 2005. "H" is an Internet shopping mall in Korea. The number of users is 208. The number of items is 1,682. The total number of transactions is 16,959.

We used 80% of the transaction data as training data. The remaining 20% of the transaction data was used to test the accuracy of user-based collaborative filtering-based recommender systems. We used 10 neighbors to find the nearest neighbors. We recommended 10 items. Simulation was done by using VBA (Visual Basic for Applications) on the Excel worksheet containing the data.

### 5.1 Experiment I-A: Using Implicit Ratings from Purchase Information of Data Set 1

In Experiment I-A, we constructed implicit ratings from purchase information. When someone purchased an item, we assigned the rating value 1 to the user-item pair.

An example user-item rating matrix is shown in Table 8. Here, User A purchased Item 1, 3 and 4, User B purchased Item 2 and 3, and User C purchased Item 1 and 4.

**Table 8.** Implicit Rating Matrix Example Based on Purchase Information Only

| | Item 1 | Item 2 | Item 3 | Item 4 |
|---|---|---|---|---|
| User A | 1 | | 1 | 1 |
| User B | | 1 | 1 | |
| User C | 1 | | | 1 |

In order to evaluate accuracy, we compared the number of items actually purchased from the items recommended by the user-based collaborative filtering-based recommender systems using the Pearson Correlation Coefficient, Cosine Similarity, and Inner Product. The empirical results of Experiment I-A are summarized in Table 9.

**Table 9.** Empirical Results with Purchase Information Only of Data Set 1

| | Pearson Correlation Coefficient | Cosine Similarity | New IP Similarity |
|---|---|---|---|
| # of items purchased from recommended items | 123 | 127 | 118 |
| # of items per user | 0.11 | 0.11 | 0.11 |

The Pearson Correlation Coefficient showed 123 actual purchases and Cosine Similarity showed 127 purchases. Our Inner Product measure resulted in 118 actual purchases from the recommended list. As shown in Fig. 5, the three similarity measures all showed similar accuracy.

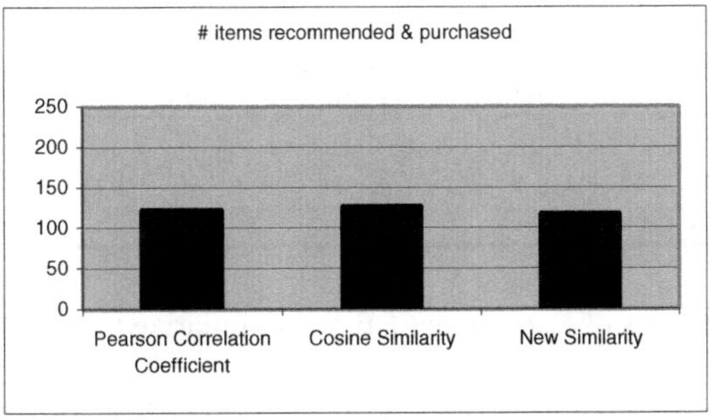


**Fig. 5.** Comparison of Similarity Measures with Purchase Information Only of Data Set 1

Cosine Similarity showed slightly better accuracy than the other two. Our Inner Product measure showed slightly worse accuracy. This is because implicit ratings from solely purchase information are binary.

## 5.2 Experiment I-B: Using Implicit Ratings from Purchase and Time Information of Data Set 1

In Experiment I-B, we constructed the implicit ratings from both purchase and time information. We used two kinds of time information: item launch time and user purchase time.

Item launch time was used to improve the scalability and accuracy of collaborative filtering-based recommender systems [10]. User purchase time was also used in order to improve recommendation accuracy [11].

Original rating values are weighted by assigning more weight to recent launch time and recent purchase time. We divided the launch time and purchase time into three groups, respectively. We then gave more weight to recent groups. The weight scheme used is given in Table 10.

**Table 10.** Weight Scheme for Time information

| | Old purchase group | Middle purchase group | Recent purchase group |
|---|---|---|---|
| Old launch group | 0.7 | 1.7 | 2.7 |
| Middle launch group | 1 | 2 | 3 |
| Recent launch group | 1.3 | 2.3 | 3.3 |

For example, consider the example user-item rating matrix shown in Table 8. Assume the following time information: Item 1 belongs to the Old launch group, Item 2 and 3 belong to the Middle launch group and Item 4 belongs to the Recent launch group. Suppose that the User A-Item 1 purchase, the User B-Item 2 purchase and the User A-Item 4 purchase belong to the Old purchase group, the User A-Item 1 purchase and the User B-Item 3 purchase belong to the Middle purchase group, and the User C-Item 1 purchase, the User A-Item 3 purchase and the User C-Item 4

**Table 11.** Implicit Rating Matrix Example Based on Purchase and Time Information

| | Item 1 | Item 2 | Item 3 | Item 4 |
|---|---|---|---|---|
| User A | 0.7 | | 3 | 1.3 |
| User B | | 1 | 2 | |
| User C | 2.7 | | | 3.3 |

The empirical results of Experiment I-B are summarized in Table 12.

**Table 12.** Empirical Results with Purchase and Temporal Information of Data Set 1

| | Pearson Correlation Coefficient | Cosine Similarity | New IP Similarity |
|---|---|---|---|
| # of items purchased from recommended items | 180 | 170 | 229 |
| # of items per user | 0.16 | 0.15 | 0.21 |

purchase belong to the Recent purchase group. The corresponding user-item rating matrix for this case is shown in Table 11.

The Pearson Correlation Coefficient resulted in 180 actual purchases and Cosine Similarity showed 170 actual purchases. Our Inner Product measure showed 229 actual purchases from the recommended list. Fig. 6 depicts the accuracy of three similarity measures.

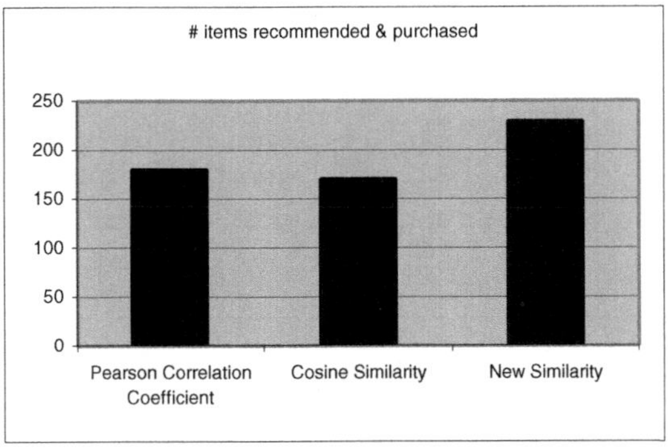


**Fig. 6.** Comparison of Similarity Measures with Purchase and Temporal Information of Data Set 1

Our Inner Product measure showed 27% increase in accuracy over the Pearson Correlation Coefficient and 35% increase in accuracy over Cosine Similarity.

### 5.3 Experiment II: Using Implicit Ratings from Web Log of Data Set 2

In Experiment II, we constructed implicit ratings from web-log information as follows. When someone clicked an item, we assigned the rating value 1 to the user-item pair. When someone put an item in the shopping cart, we assigned the rating value 2 to the user-item pair. When someone actually purchased an item, we assigned the rating value 3 to the user-item pair. We compared accuracy of recommendation using MAE (Mean Absolute Error).

The empirical results of Experiment II are summarized in Table 13, and the accuracy comparison of three similarity measures is shown in Fig. 7.

**Table 13.** Empirical Results with Web Log of Data Set 2

| | Pearson Correlation Coefficient | Cosine Similarity | New IP Similarity |
|---|---|---|---|
| Mean Absolute Error | 0.483 | 0.472 | 0.418 |

Our Inner Product measure showed 13% increase in accuracy over the Pearson Correlation Coefficient and 11% increase in accuracy over Cosine Similarity.

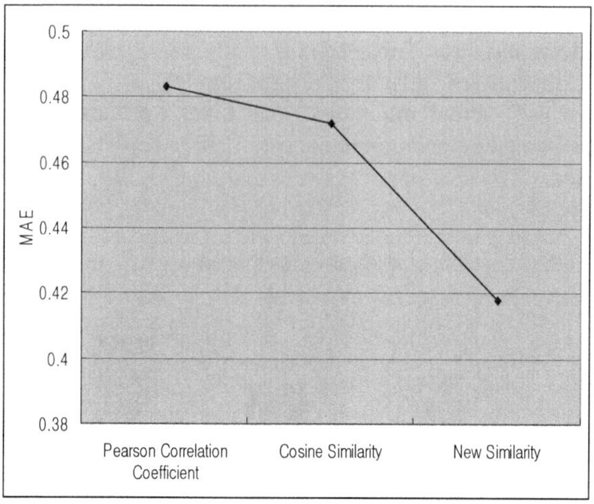


**Fig. 7.** Comparison of Similarity Measures with Web Log of Data Set 2

## 6 Conclusion and Future Work

We have presented a new similarity measure suitable for implicit ratings. It is based on inner product and resolves some problems associated with the existing similarity measures (including the Pearson Correlation Coefficient and the Cosine Similarity) with regard to implicit ratings in collaborative filtering. Empirical data from two e-commerce environments (including a mobile environment) showed that user-based collaborating filtering using the proposed similarity measure resulted in more accurate recommendations. Our inner product similarity measure could be useful for collaborative filtering-based recommender systems using implicit ratings, in which negative ratings are not readily incorporated.

Future work will focus on conducting more experiments with a variety of implicit rating data. Further research is also needed to incorporate the factors of rating scales and rating average shifts into the inner product measure.

**Acknowledgments.** We would like to thank Dr. Y. H. Cho for permitting us to share the Data Set. This work is supported in part by Dongyang Technical College Academy Research Expenses and the Caterpillar Research Fellowship.

## References

1. Resnick, P., Iacovou, N., Suchak, M., Bergstrom, P., Riedl, J. GroupLens: An Open Architecture for Collaborative Filtering of Netnews. Proceedings of CSCW '94 (1994) 175-186
2. Linden, G., Smith, B., York, J.: Amazon.com Recommendations. Item-to-Item Collaborative Filtering. IEEE Internet Computing (2003)

3. Melville, P., Mooney, R. J., Nagarajan, R.: Content-Boosted Collaborative Filtering for Improved Recommendations. Proceedings of Eighth National Conference an Artificial Intelligence (2002) 187-192
4. Caglayan, A., Snorrason, M., Jacoby,J., Mazzu, J.,Jones, R., Kumar, K.: Learn Sesame - A Learning Agent Engine. Applied Artificial Intelligence. Vol. 11 (1997) 393-412
5. Middleton, S.E., Shadbolt, N.R., de Roure, D.C.: Ontological User Profiling in Recommender Systems. ACM Trans. Information Systems. Vol. 22. no. 1 (2004) 54-88
6. Oard, D.W., Kim, J.: Implicit Feedback for Recommender Systems. Proceedings of Recommender Systems 1998 Workshop (1998)
7. Schein, A.I., Popescul, A., Ungar, L.H., Pennock, D.M.: Methods and Metrics for Cold-Start Recommendations. Proceedings of Ann. Int'l ACM SIGIR Conf. (2002)
8. Mobasher, B., Dai, H., Luo, T., Sun, Y., Zhu, J.: Automatic Personalization Based on Web Usage Mining. Communications of the ACM. Vol. 43(8) (2000) 142-151
9. Anderson, C. R., Domingos, P., Weld, D. S.: Personalizing Web Sites for Mobile Users. Proceedings of the 10th Conference on the World Wide Web (2001)
10. Tang, T. Y., Winoto, P., Chan, K. C. C.: Scaling Down Candidate Sets Based on the Temporal Feature of Items for Improved Hybrid Recommendations. Intelligent Techniques in Web Personalization. LNAI 3169 (2003) 169-185
11. Ding Y., Li, X. Orlowska M.: Recency-Based Collaborative Filtering, Australian Computer Science Communications. Vol. 28 No 2. Australasian Database Conference. ACM Digital Library (2006) 99-107

# An Adaptive $k$-Nearest Neighbors Clustering Algorithm for Complex Distribution Dataset

Yan Zhang[1], Yan Jia[1], Xiaobin Huang[2], Bin Zhou[1], and Jian Gu[1]

[1] School of Computer, National University of Defense Technology, 410073 Changsha, China
jane325@tom.com
[2] Department of Information Engineering, Air Force Radar Academy, 430019 Wuhan, China
hxbtougao@gmail.com

**Abstract.** To resolve the shortage of traditional clustering algorithm when dealing data set with complex distribution, a novel adaptive $k$-Nearest Neighbors clustering(AKNNC) algorithm is presented in this paper. This algorithm is made up of three parts: (a)normalize data set; (b)construct initial patterns; (c)merge initial patterns. Simulation results show that compared with classical FCA, our AKNNC algorithm not only has better clustering performance for data set with Complex distribution, but also can be applied to the data set without knowing cluster number in advance.

## 1 Introduction

The target of clustering is to classify similar objects into a single class. Clustering is a very important preprocessing technology for Pattern Recognition, Image Processing, Medical Diagnosis, etc. Clustering usually can be classified into two classes [1]: Hierarchical Clustering and Dynamic Clustering. Dynamic Clustering has been attracting many researchers and a lot of clustering algorithms have been proposed in this area. These dynamic clustering methods can be mainly classified into four kinds: 1)center clustering methods, such as FCA [1]; 2)center clustering methods based on neural network, such as LVQ algorithm [1]; 3)clustering methods based on characteristics of data distribution [2,3]; 4)graphic clustering methods [4]. In real applications, the pattern number of processed data set is unknown and its distribution is very complicated, which make many clustering algorithms failure. So, it is extremely important to research a new clustering algorithm suitable for this situation. Much work has been done to resolve this problem [5,6]. With the help of pioneer work, a novel adaptive $k$-Nearest Neighbors clustering(AKNNC) algorithm is presented in this paper. The AKNNC algorithm processes the data set through three phases which are 1)normalize data set; 2)construct initial patterns using the adaptive $k$-Nearest Neighbors searching. The number of initial patterns is usually larger than that of final patterns; 3)with the help of connected graph theory, these initial patterns are merged into final patterns. Simulation results show that compared with classical FCA, this

D.-S. Huang, L. Heutte, and M. Loog (Eds.): ICIC 2007, LNAI 4682, pp. 398–407, 2007.

AKNNC algorithm not only has better clustering performance for data set with complex distribution, but also can be applied to the data set without knowing cluster number in advance.

The contents of this paper is arranged as follows: the AKNNC algorithm is introduced in Section 2; in Section 3, we give the time complexity analysis for AKNNC, in Section 4, simulation results with two complex distribution data sets are presented; the discussion in Section 5 summaries our work.

## 2 The AKNNC Algorithm

Usually, FCA gets clustering centre by minimizing the following objective function

$$J = \sum_{j=1}^{C} \sum_{i=1}^{n} [\mu_j(\boldsymbol{s}_i)]^b \left\| \boldsymbol{s}_i - \boldsymbol{c}_j \right\|^2 \tag{1}$$

Please refer to [1] for the meaning of parameters in formula (1). FCA has good performance when the data set has spherical-shape distribution, because it minimizes the distances between samples and class centre which the samples belong to. Obviously, FCA explores data set distribution in global scope, which will be ineffective in the case of complex data distribution. However, if we detect the local data distribution first, and use local information to determine the final patterns, then we can get better clustering performance. According to this idea, the AKNNC algorithm is proposed in this paper. This algorithm consists of three parts which will be introduced in detail as follows.

### 2.1 Normalize Data Set

Assume having $N$ $M$-dimensional samples, denoted by $\hat{\boldsymbol{S}} = \{\hat{\boldsymbol{s}}_1, \hat{\boldsymbol{s}}_2, \cdots, \hat{\boldsymbol{s}}_h, \cdots, \hat{\boldsymbol{s}}_N\}$, where $\hat{\boldsymbol{s}}_h$ is the $h$th sample$(1 \le h \le N)$, $\hat{s}_{hi}$ is the $i$th component of the sample $\hat{\boldsymbol{s}}_h (1 \le i \le M)$. Using formula (2) to get the normalized data set $\boldsymbol{S} = \{\boldsymbol{s}_1, \boldsymbol{s}_2, \cdots, \boldsymbol{s}_N\}$.

$$s_{hi} = \frac{\hat{s}_{hi}}{\max_{1 \le i \le M}[\max_{1 \le h \le N}(\hat{s}_{hi}) - \min_{1 \le h \le N}(\hat{s}_{hi})]} \tag{2}$$

where $s_{hi}$ is the $i$th component of the normalized sample $\boldsymbol{s}_h$. This preprocessing makes all samples locate in unit space, which will not destroy the distribution characteristic.

### 2.2 Construct Initial Patterns

The constitution of initial patterns is in fact exploring the local structure of data set. Studies show that the $k$-Nearest Neighbors strategy is an efficient method for analyzing the local structure of data set [1]. But $k$ is fixed in the classical $k$-Nearest Neighbors algorithm, which will not be fit for complex data set. For example, in the case where the patterns' densities are uneven, if $k$ is larger than the amount of the "small" pattern, it will make the initial patterns error; if $k$ is too small, then the number of initial patterns is too large which increases computational burden for the latter stage. So how to choose the value of $k$ adaptively according to the pattern's density is key to resolving this problem. Prior work [1] shows that the trace of within-classes covariance is a good indicator for pattern density. The pattern is denser when the trace of its within-classes covariance is smaller. Furthermore, the trace of within-classes covariance has rotation-invariant property.

Before introducing the adaptive $k$-Nearest Neighbors clustering algorithm, we define:

a) $E_p : (\boldsymbol{l}_p, k_p\text{-NN of } \boldsymbol{l}_p, \boldsymbol{c}_p, T_p)$ — the $p$th initial class
b) $\boldsymbol{l}_p$ — the kernel sample of initial pattern $E_p$
c) $k_p$-NN of $\boldsymbol{l}_p$ — $k_p$ nearest neighbor nodes of $\boldsymbol{l}_p$
d) $\boldsymbol{c}_p$ — the centre of $k_p$+1 nodes, and its computing formula is $\boldsymbol{c}_p = \frac{\sum_{\boldsymbol{s}_k \in E_p} \boldsymbol{s}_k}{k_p+1}$
e) $T_p$ — the trace of within-classes covariance matrix of $k_p$+1 samples

The initial classes formation is implemented through 5 steps shown as follows:

**Step 1:** Set $p = 1$ and use formula (3) to get the distance matrix $\boldsymbol{D} = [d_{ij}]_{N \times N}$ of data set $\boldsymbol{S}$.

$$d_{ij} = \|\boldsymbol{s}_i - \boldsymbol{s}_j\|^2 = \sqrt{\sum_{k=1}^{M} (s_{ik} - s_{jk})^2} \tag{3}$$

Where $d_{ij}$ denotes the distance between $\boldsymbol{s}_i$ and $\boldsymbol{s}_j$.
**Step 2:** If $p$=1, $\boldsymbol{l}_1$ is obtained with the following restrictions:

a) $\boldsymbol{l}_1 \in E_1$.
b) $\boldsymbol{l}_1$ is the farthest sample from the global centre $\boldsymbol{c}_0$.
c) $\boldsymbol{c}_0 = \frac{\sum_{\boldsymbol{s}_k \in \boldsymbol{S}} \boldsymbol{s}_k}{N}$.

If $p \geq 2$, $\boldsymbol{l}_p$ is obtained with the following restrictions:

a) $\boldsymbol{l}_p \in E_p$ and $\boldsymbol{l}_p \notin E_\alpha, 1 \leq \alpha < p$.
b) $\boldsymbol{l}_p$ is the farthest sample from the local centre $\boldsymbol{c}_{p-1}$.

**Step 3:** Obviously $\boldsymbol{l}_p \in \boldsymbol{S}$, so we might assume it as $\boldsymbol{s}_l$. Extract the $l$th row elements from the distance matrix $\boldsymbol{D}$ to form the distance vector $\{d_{l1}, d_{l2}, \cdots, d_{lN}\}$, then delete the elements that denotes the distance between $\boldsymbol{s}_l$ and $\boldsymbol{s}_j$ where $\boldsymbol{s}_j \in E_\alpha$. So we get the abridged distance vector $\Psi = \{d_{lj} | 1 \leq j \leq N, \boldsymbol{s}_j \notin E_\alpha\}$, sort $\Psi$ increasingly as:

$$\Psi' = \{d_{lj_1}, d_{lj_2}, \cdots, d_{lj_\beta} | d_{lj_m} \leq d_{lj_n}, 1 \leq m \leq n \leq \beta\} \tag{4}$$

where $\beta$ denotes there are $\beta$ samples which have not been put into the initial patterns.
**Step 4:** Use the following algorithm to get the $k_p$ Nearest Neighbors of $\boldsymbol{l}_p$ and the trace of within-classes covariance matrix $T_p$. For accelerating computation, a recursive method to calculate trace of within-classes covariance [7] can be used.

```
Ω = l_p;
set the threshold of the trace of within-classes covariance matrix;
for i=1 to β begin
    Ω = {Ω, s_{j_i} | s_{lj_i} ∈ Ψ'};
    t_i = trace(conv(Ω));
    if t_i ≤ threshold
        T_i = t_i;
    else begin
        Ω = Ω\s_{j_i};
        break;
    end.
end.
```

**Step 5:** If there are still samples that have not been put into the initial patterns, then go to step 2, otherwise the constitution of initial patterns is finished.

## 2.3 Merge Initial Patterns

Firstly we can get distance matrix between initial patterns, and then transform distance matrix into a binary matrix by using a distance threshold. If we consider the initial patterns as "dots", then these initial patterns make up of a graph, and the binary matrix will be the connected matrix of this graph. Because the connected graph can classify the dots, so the merging of initial patterns can be done with connected graph theory. The detail process is depicted as follows.

**Step 1:** Use formula (5) to obtain the distance matrix $\boldsymbol{W} = [w_{ij}]_{Q\times Q}$,

$$w_{ij} = \min_{\boldsymbol{s}_m \in E_i, \boldsymbol{s}_n \in E_j} \|\boldsymbol{s}_m - \boldsymbol{s}_n\|_2 \tag{5}$$

where $Q$ is the number of initial patterns, $w_{ij}$ is the distance between the initial $i$th pattern and $j$th pattern.

**Step 2:** Construct the binary matrix $\boldsymbol{A} = [a_{ij}]_{Q\times Q}$ by using the following formula.

$$a_{ij} = \begin{cases} 1, & \text{if } w_{ij} \le \delta \\ 0, & \text{if } w_{ij} > \delta \end{cases} \tag{6}$$

where $\delta$ is the distance threshold.

**Step 3:** Compute $\boldsymbol{A}^{Q-1}$ in boolean algebra [1], then use the following two lemmas [1] to get the number of the connected subgraphs and the amount of dots in each connected subgraph.

a) The order of $\boldsymbol{A}^{Q-1}$ is the number of the connected graphs;
b) Get the linearly dependent row vectors from $\boldsymbol{A}^{Q-1}$, then those dots, whose sequence numbers are the row numbers of the vectors, belong to the same connected subgraph.

**Step 4:** Use the mapping relationship between initial pattern and the "dots", we can get the final clustering result.

# 3 Time Complexity Analysis

Suppose there are $N$ samples in the data set and $Q$ initial patterns after initial pattern construction. In initial pattern construction phase, the first step for calculating distance matrix $\boldsymbol{D}$ has $O(N^2)$ time complexity. Step two to Step five build up a dual loop. Obviously, the run times of inner loop and outer loop are both less than $N$, so the time complexity of these steps is also $O(N^2)$, then we can get the time complexity of initial pattern construction phase is $O(N^2)$. In combination step, the first step is also calculating distance matrix, so the time complexity is $O(Q^2)$. The second step is actually a loop operation, and its time complexity is also $O(Q^2)$. Carefully analyzing the third step, we can find it is a triple loop operation, and run times for each loop are less than $Q$, so this step's time complexity is $O(Q^3)$. In real applications, $Q$ is always less than $\sqrt{N}$ , so the time complexity of combination phase is $O(N^{3/2})$. With the time complexity analysis for the two phase, we can conclude that the time complexity of AKNNC is $O(N^2)$.

# 4 Simulation and Discussion

## 4.1 Simulation

We use two data sets shown in Fig.1 and Fig.2 in this experiment. The first data set has 60 samples which present linear distribution, and obviously they can be divided into seven classes; the second data set has 100 samples which present semicircle distribution, and they can be divided into two classes. We compare the classic FCA with our AKNNC method in 20 experiments. Fig.3 and Fig.4 give the FCA clustering results for the first data set, Fig.5 and Fig.6 give the AKNNC clustering results for the second data set. Notice that randomly setting the clustering centre in initial phase for FCA, the 20 experiments have different results with this method, so we select one of the best result. However, with our AKNNC method, we get the same results in all experiments. The parameters setting for these two methods are shown in Tab.1, Tab.2 shows the initial patterns for these two data set, Tab.3 compares the clustering performance for the two methods.

## 4.2 Discussion

From the FCA clustering results shown in Fig.3 and Fig.4, classes can be overlaid with a serial of circles which overlap each other very little, in that FCA is only fit for the spherical shape distribution data set. So the FCA clustering results is badly disaccord with the actual classes. Furthermore, because FCA randomly selects the clustering centre in its initial phase, each experiment may have different clustering result for complex distribution data set. However, our AKNNC method firstly uses trace of within-classes covariance matrix to construct the initial patterns which can successfully detect the local data structure, then merges

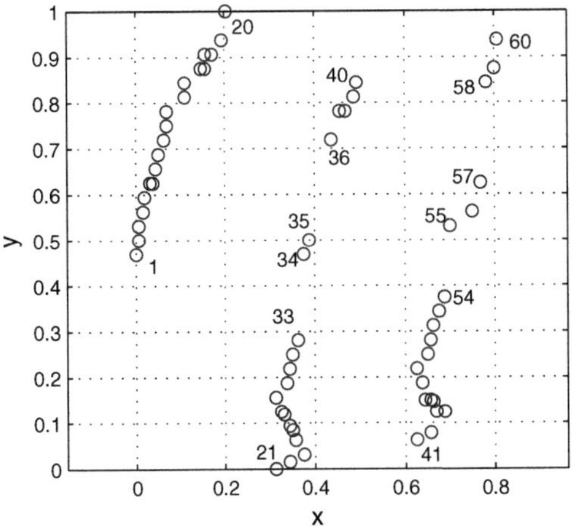


**Fig. 1.** Distribution of first data set. There are 60 samples in this data set, each sample has two components, namely x and y. Obviously, there should be seven classes in this data set, 1-20 is the first class, 21-33 the second class, 34-35 the third class, 36-40 the forth class, 41-54 the fifth class, 55-57 the sixth class and 58-60 the seventh class.

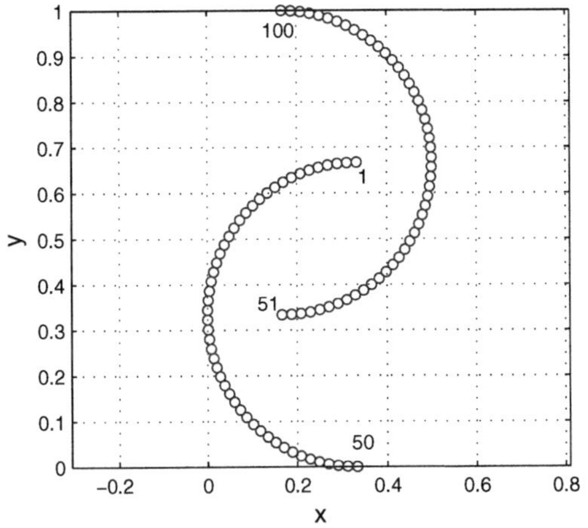


**Fig. 2.** Distribution of second data set. There are 100 samples in this data set, each sample has two components, namely x and y. Obviously, there should be two classes in this data set, 1-50 is the first class and 51-100 the second class.

these local initial patterns into the final classes. With this two-phase operation, AKNNC ensures that each experiment has the same clustering result.

We should notice that AKNNC method needs two parameters, namely *threshold* and $\delta$, but FCA needs only one parameter(number of cluster). Maybe you

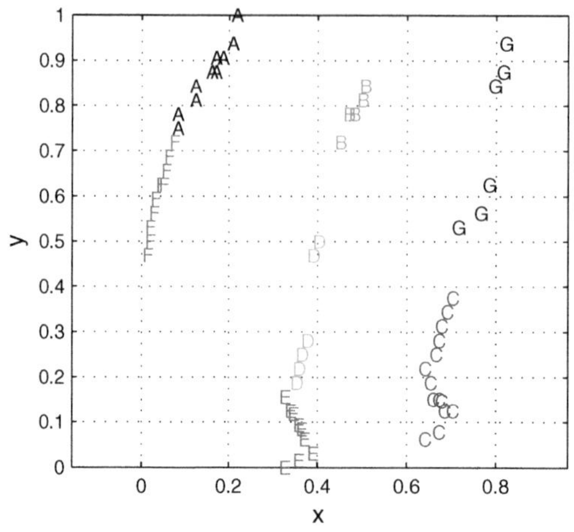


**Fig. 3.** Clustering result for first data set with FCA(one experiment). The same letters belong to the same class. Obviously, FCA badly destroys the initial data structure, because it is only fit for the spherical shape distribution data sets.

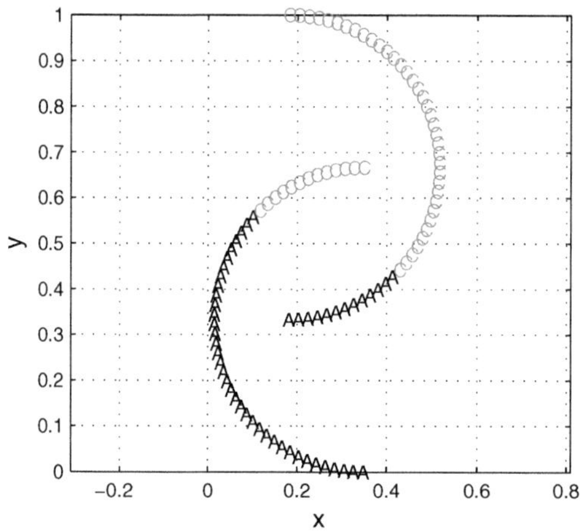


**Fig. 4.** Clustering result for second data set with FCA(one experiment). The same letters belong to the same class. Again FCA badly destroys the initial data structure.

will think that the AKNNC method needs more priori information than FCA, but from the experiments, we observe that *threshold* and $\delta$ parameters little influence the clustering result, whereas clustering result of FCA strongly rely on the selection of number of cluster.

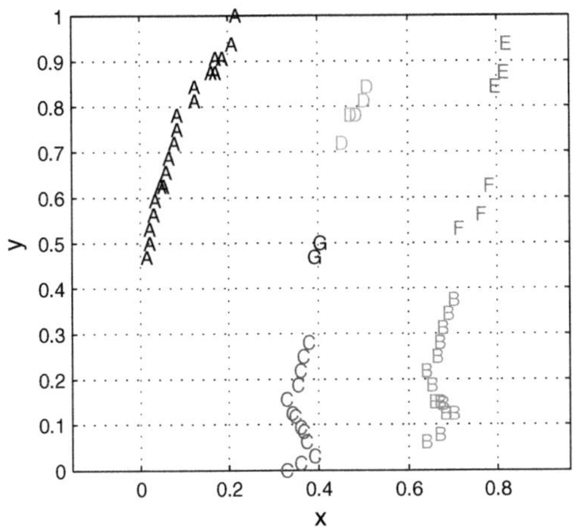


**Fig. 5.** Clustering result for first data set with AKNNC. The same letters belong to the same class. AKNNC successfully detects the local linear structure for this data set, so it divides the data set into correct seven classes.

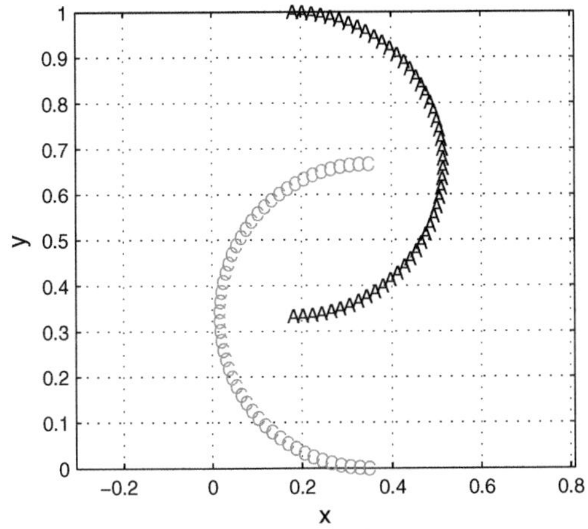


**Fig. 6.** Clustering result for second data set with AKNNC. The same letters belong to the same class. Again AKNNC successfully detects the local semicircle structure for this data set, so it divides the data set into correct two classes.

**Table 1.** Parameters Setting of FCA and AKNNC Algorithms

| Algorithm | Parameters setting | |
|---|---|---|
| | first data set | second data set |
| FCA(cluster number $C$) | $C = 7$ | $C = 2$ |
| AKNNC(*threshold* and $\delta$) | $threshold = 0.01, \delta = 0.1$ | $threshold = 0.01, \delta = 0.1$ |

**Table 2.** Initial patterns of the AKNNC algorithm

| Serial number of samples in each initial pattern (initial pattern: samples serial number) | |
|---|---|
| first data set | second data setc |
| 1: 1, 2,10 | 1: 1, 2,15 |
| 2: 11,12,20 | 2: 16,17,35 |
| 3: 21,22,29 | 3: 36,37,50 |
| 4: 29,30,33 | 4: 51,52,65 |
| 5: 34,35 | 5: 66,67,85 |
| 6: 36,37,40 | 6: 86,87,100 |
| 7: 41,42,49 | |
| 8: 50,51,54 | |
| 9: 55,56, 57 | |
| 10: 68,59,60 | |

**Table 3.** Clustering performace comparsion of FCA and AKNNC algorithms

| Algorithm | Error rate | |
|---|---|---|
| | first data set | second data set |
| FCA | 28.33% | 26% |
| AKNNC | 0 | 0 |

# 5 Conclusion

A novel AKNNC algorithm is presented in this paper for complex data set without knowing patterns number. And we analyze time complexity for it in detail. Use clustering validity index to evaluate the clustering result for AKNNC is our future work. Also our method is a useful building block, which can be applied to many fields. We have already used our AKNNC algorithm for construction a new resource discovery method in grid environment, related work can be viewed at http://blog.xiaobing.org/.

**Acknowledgements.** This work is supported by 973 project (No. 2005CB321800) of China, and 863 project (No. 2006AA01Z198) of China.

# References

1. Fukunaga, K.: Introduction to Statistical Pattern Recognition. Academic Press, San Diego, CA (1990)
2. Baraldi, F., Parmiggiani, F.: Fuzzy-shell Clustering and Applications to Circle Detection in Digital Images. Int. J. General Syst, **16**(1995) 343-355
3. Frigui, H., Krishnapuram, R.: A Comparison of Fuzzy Shell-clustering Method for the De-tection of Ellipses. IEEE Transactions on Fuzzy System, **4**(1996) 193-199
4. Hubert, L.J.: Some Applications of Graph Theory to Clustering. Psychonmetrika, **4**(1974) 435-475

5. Liu, Y.T., Shiueng, B.Y.: A Genetic Algorithm for Data with Non-spherical-shape Clusters. Pattern Recognition, **33**(2000) 1251-1259
6. Patrick, K.S.: Fuzzy Min-Max Neural Networks-Part1: Classification. IEEE Transactions On Neural Networks, **3**(1992) 776-786
7. Huang, X.B., Wan, J.W., Wang, Z.: A Recursive Algorithm for Computing the Trace of the Sample Covariance Matrix. Pattern Recognition and Artificial Intelligence, **17**(2004) 497-501

# Defining a Set of Features Using Histogram Analysis for Content Based Image Retrieval

Jongan Park[1], Nishat Ahmad[1], Gwangwon Kang[1], Jun H. Jo[3], Pankoo Kim[1], and Seungjin Park[2]

[1] Dept of Information & Communications Engineering
Chosun University, Kwangju, South Korea
japark@chosun.ac.kr
[2] Dept of Biomedical Engineering,
Chonnam National University Hospital, Kwangju, South Korea
[3] School of Information and Communication Technology Griffith University, Australia
j.jo@griffith.edu.au

**Abstract.** A new set of features are proposed for Content Based Image Retrieval (CBIR) in this paper. The selection of the features is based on histogram analysis. Standard histograms, because of their efficiency and insensitivity to small changes, are widely used for content based image retrieval. But the main disadvantage of histograms is that many images of different appearances can have similar histograms because histograms provide coarse characterization of an image. Hence we further refine the histogram using the histogram refinement method. We split the pixels in a given bucket into several classes just like histogram refinement method. The classes are all related to colors and are based on color coherence vectors. After the calculation of clusters using histogram refinement method, inherent features of each of the cluster is calculated. These inherent features include size, mean, variance, major axis length, minor axis length and angle between x-axis and major axis of ellipse for various clusters.

## 1 Introduction

Research in content based image retrieval is an active discipline and its expanding in length & breadth. The deeper problems in computer vision, databases and information retrieval are being emphasized with the maturation of content based image retrieval technology. The web has huge collection of digital media which contains all sorts of digital content including still images, video, audio, graphics, animation etc. We concentrate on the visual content especially on still images.

One of the most effective ways of accessing visual data is Content-based image retrieval (CBIR). The visual content such as color, shape and image structure is considered for the retrieval of images instead of an annotated text method. However, one major problem with CBIR is the issue of predicting the relevancy of retrieved images. This retrieval is based on various image features. Our objective is the selection of such features which can provide accurate and precise query results.

D.-S. Huang, L. Heutte, and M. Loog (Eds.): ICIC 2007, LNAI 4682, pp. 408–417, 2007.

## 2 Related Work

The best review of CBIR till 2000 is provided by Arnold et. al. [3]. They reviewed 200 references in content based image retrieval. They discussed the working conditions of content-based retrieval: patterns of use, types of pictures, the role of semantics, and the sensory gap.

Histogram refinement method was first proposed by Pass and Zabih [2]. They partition histogram bins by the spatial coherence of pixels. They further refine it by using additional feature, i.e., the center of the image. The center of the image is defined as the 75% centermost pixels. An unsupervised learning network to incorporate a self-learning capability into image retrieval systems was proposed by Paisarn [3]. The adoption of a self-organizing tree map (SOTM) is introduced, to minimize the user participation in an effort to automate interactive retrieval.

Zhang [4] discussed a generic Fourier descriptor (GFD) to overcome the drawbacks of existing shape representation techniques. Special emphasis was made on content-based indexing and retrieval by Djeraba [5]. They try to add the generalization capability for indexing and retrieval. JongAn, Bilal et al. [6] provided shape description based on histogram based chain codes. One of the problems is the search in large collections of heterogeneous images. Vasileios [7] presented an image retrieval methodology for this problem.

## 3 Pre-processing

After the image acquisition, the image needs to be pre-processed before feature extraction process. We consider the grayscale images for feature extraction. Therefore, first the image is converted to grayscale image using threshold. The RGB image is changed to grayscale image, also known as the intensity image, which is a single 2-D matrix containing values from 0 to 255.

For grayscale image, we do not consider all the 256 levels. Hence after the conversion from RGB to grayscale image, we perform quantization to reduce the number of levels in the image. We reduce the 256 levels to 16 levels in the quantized image. For reducing the number of levels from 256 to 16, we use uniform quantization. Figure 1 shows the block diagram of the algorithm. The steps in the pre-processing stage can be observed from the first three blocks in figure 1.

## 4 Selection of Features

### 4.1 Coherency and Incoherency

First we find out the coherent pixels and incoherent pixels. We use color refinement method for calculation of coherency and incoherency among pixels. Color refinement is based on histogram refinement [2] method. The histogram refinement method provides that the pixels within a given bucket be split into classes based upon some local property and these split histograms are then compared on bucket by bucket basis and the pixels within a bucket are compared.

Color histogram buckets are partitioned based on spatial coherence just like computed by Pass and Zabih [2]. A pixel is coherent if it is a part of some sizable similar colored region, otherwise it is incoherent. So the pixels are classified as coherent or incoherent within each color bucket. If a pixel is part of a large group of pixels of the same color which form at least five percent of the image then that pixel is a coherent pixel and that group is called the coherent group or cluster. Otherwise it is incoherent pixel and the group is incoherent group or cluster.

Then two more properties are calculated for each bin. First the numbers of clusters are found for each case, i.e., coherent and incoherent case in each of the bin. Secondly, the average of each cluster is computed. So for each bin, there are six values: one each for percentage of coherent pixels and incoherent pixels, number of coherent clusters and incoherent clusters, average of coherent cluster and incoherent cluster. This is shown in block diagram in figure 1.

For each discretized color j, let the number of coherent pixels as $\alpha_j$, the number of coherent connected components as $C_{\alpha j}$ and the average of coherent connected component as $\mu_{\alpha j}$. Similarly, let the number of incoherent pixels as $\beta_j$, the number of incoherent connected components as $C_{\beta j}$ and the average of incoherent connected component as $\mu_{\beta j}$. For each discretized color j, the total number of pixels are $\alpha_j+\beta_j$ and the color histogram summarizes the image as $<\alpha_1+\beta_1,\ldots,\alpha_n+\beta_n>$.

### 4.2 Features from Coherent Clusters

Coherent clusters are considered only for the additional features. At this stage, incoherent clusters are ignored. The reason for selecting coherent clusters only is based on the assumption that objects of significant size are considered only, i.e., cluster size is equal to or greater than 5% of the image.

Four features are selected among the coherent clusters. Three of them are based on the size of the clusters while one is statistical in nature. They are; (i) Size of largest cluster in each bin, (ii) Size of median cluster in each bin, (iii) Size of smallest cluster in each bin, and (iv) Variance of clusters in each bin.

Let us denote the largest cluster in each bin as $L_{\alpha j}$, the median cluster in each bin as $M_{\alpha j}$, the smallest cluster in each bin as $S_{\alpha j}$ and variance of clusters in each bin as $V_{\alpha j}$. These features are shown in figure 1.

### 4.3 Additional Features Based on Size of Cluster

Again these additional features are based on the coherent clusters only. The following features are selected for retrieval for each of the largest cluster, median cluster and smallest cluster in each of the bin; (i) Major axis length, (ii) Minor axis length, and (iii) Angle between x-axis and major axis of ellipse.

Let us denote the major axis length of the largest cluster in each bin as $MAL_{\alpha Lj}$, the minor axis length of the largest cluster in each bin as $MIL_{\alpha Lj}$ and angle as $Ang_{\alpha Lj}$. Similarly, let us denote the major axis length of the median cluster in each bin as $MAL_{\alpha Mj}$, the minor axis length of the median cluster in each bin as $MIL_{\alpha Mj}$, the angle of median cluster as $Ang_{\alpha Mj}$, the major axis length of the smallest cluster in each bin as $MAL_{\alpha Sj}$, the minor axis length of the smallest cluster in each bin as $MIL_{\alpha Sj}$ and the angle of smallest cluster as $Ang_{\alpha Sj}$. This is shown in figure 1.

# 5 The Retrieval Method

Image retrieval is done in 3 stages hence we can call it incremental retrieval approach.

## 5.1 Stage 1

The features obtained in section 4.1 are used for retrieval at first level. We use the $L_1$ distance to compare two images I and I′.

$$\Delta_1 = |(\alpha_j-\alpha'_j)| + |(\beta_j-\beta'_j)|,\ \Delta_2 = |(C_{\alpha j}-C'_{\alpha j})| + |(C_{\beta j}-C'_{\beta j})|,\ \Delta_3 = |(\mu_{\alpha j}-\mu'_{\alpha j})| + |(\mu_{\beta j}-\mu'_{\beta j})|$$

## 5.2 Stage 2

This level of retrieval is used for further refining the result obtained in section 4.1. The additional features obtained in section 4.2 are used at this level of retrieval. Again we use the $L_1$ distance to compare two images I and I′.

$$\Delta_4 = |(L_{\alpha j} - L'_{\alpha j})|,\ \Delta_5 = |(M_{\alpha j} - M'_{\alpha j})|,\ \Delta_6 = |(S_{\alpha j} - S'_{\alpha j})|,\ \Delta_7 = |(V_{\alpha j} - V'_{\alpha j})|$$

## 5.3 Stage 3

This level of retrieval is used for final retrieval of images from the result obtained in section 4.2. The additional features obtained in section 4.3 are used at this level of retrieval. Again we use the $L_1$ distance to compare two images I and I′.

$$\Delta_8 = |(MAL_{\alpha Lj} - MAL'_{\alpha Lj})|,\ \Delta_9 = |(MIL_{\alpha Lj} - MIL'_{\alpha Lj})|,\ \Delta_{10} = |(Ang_{\alpha Lj} - Ang'_{\alpha Lj})|$$

$$\Delta_{11} = |(MAL_{\alpha Mj} - MAL'_{\alpha Mj})|,\ \Delta_{12} = |(MIL_{\alpha Mj} - MIL'_{\alpha Mj})|,\ \Delta_{13} = |(Ang_{\alpha Mj} - Ang'_{\alpha Mj})|$$

$$\Delta_{14} = |(MAL_{\alpha Sj} - MAL'_{\alpha Sj})|,\ \Delta_{15} = |(MIL_{\alpha Sj} - MIL'_{\alpha Sj})|,\ \Delta_{16} = |(Ang_{\alpha Sj} - Ang'_{\alpha Sj})|$$

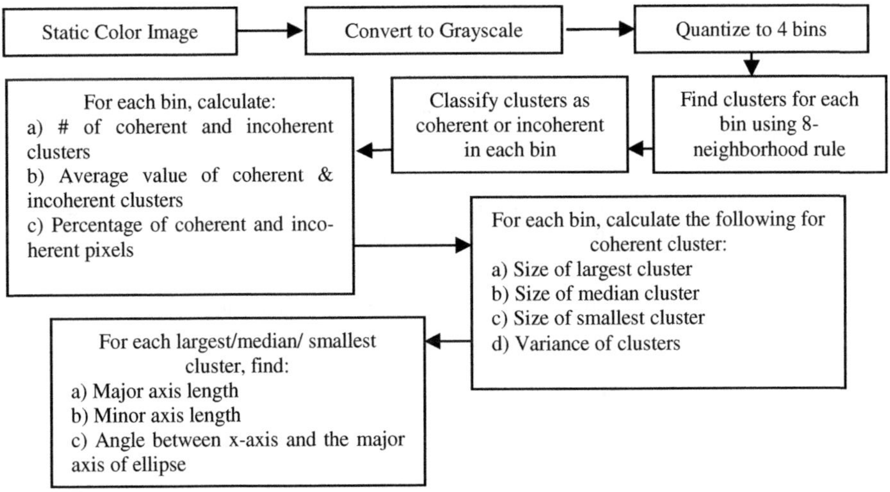


**Fig. 1.** Block diagram of the feature extraction algorithm

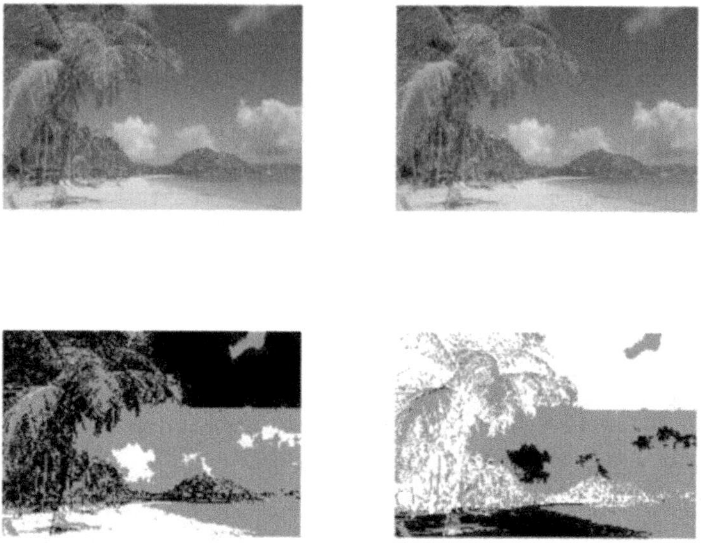

**Fig. 2.** One of the image from the database, converted to grayscale & quantized

## 6 Results and Discussion

We used the database provided by James S. Wang et. al [8, 9] to test the proposed method. First the images were preprocessed and converted to grayscale images. Then the images were quantized and the features described in section 4.1 were calculated based on coherent and incoherent clusters. Then, the features described in section 4.2 were calculated for coherent clusters only. Finally the features defined in section 4.3 were calculated based on the size of clusters. These features were calculated and stored for each of the images. Figure 2 shows one of the image from the database, its corresponding grayscale image and then the corresponding quantized images.

Consider table 1. Table 1 provides the parameter values related with the incoherent clusters. The parameters include percentage of incoherent pixels ($\beta_j$), number of incoherent clusters ($C_{\beta j}$) and average of incoherent cluster ($\mu_{\beta j}$) for each jth bucket or bin. As an example, we show the results for 4 bins of one of the images from the database in table 1. Figure 3 shows the corresponding incoherent clusters.

**Table 1.** Example of parameter values for Incoherent pixels

| | $\beta_j$ | $C_{\beta j}$ | $\mu_{\beta j}$ |
|---|---|---|---|
| **Bin 1** | 0.78% | 38 | 1.1053 |
| **Bin 2** | 7.02% | 64 | 5.8438 |
| **Bin 3** | 31.02% | 86 | 19.209 |
| **Bin 4** | 61.18% | 105 | 31.048 |

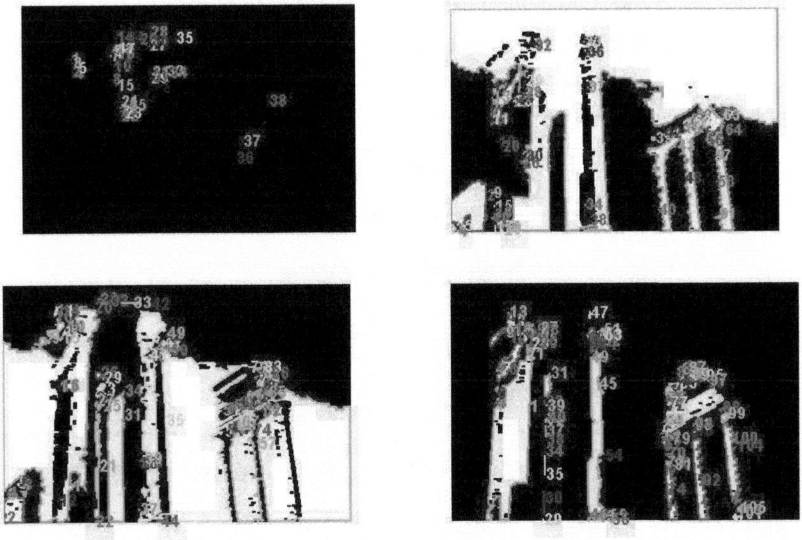

**Fig. 3.** Incoherent clusters in 4 different bins

**Table 2.** Example of parameter values for coherent pixels

| | $\alpha_j$ | $C_{\alpha j}$ | $\mu_{\alpha j}$ |
|---|---|---|---|
| **Bin 1** | 0 | 0 | 0 |
| **Bin 2** | 50.61% | 2 | 26689 |
| **Bin 3** | 41.68% | 4 | 10990 |
| **Bin 4** | 7.71% | 3 | 2712 |

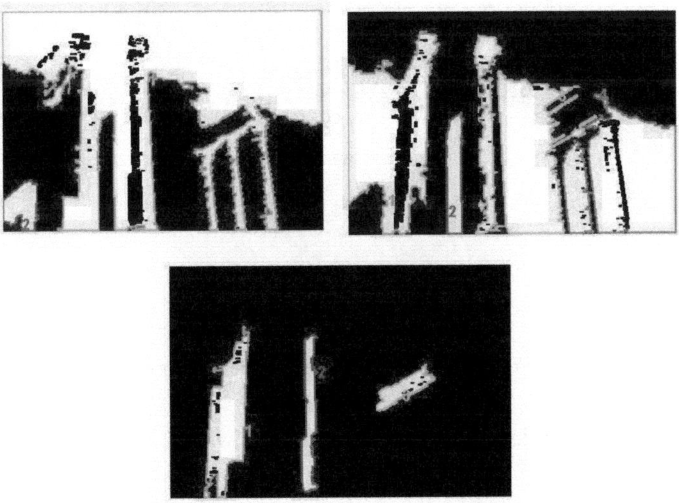

**Fig. 4.** Coherent clusters in 3 different bins

**Table 3.** Additional parameter values for coherent pixels

| | $L_{\alpha j}$ | $M_{\alpha j}$ | $S_{\alpha j}$ | $V_{\alpha j}$ |
|---|---|---|---|---|
| **Bin 1** | 0 | 0 | 0 | 0 |
| **Bin 2** | 51606 | 0 | 1772 | 1.24E+09 |
| **Bin 3** | 14553 | 12637 | 2340 | 34021226 |
| **Bin 4** | 4996 | 2025 | 1115 | 4119517 |

(a) Query Image

(b) Stage 1

(c) Stage 2

(d) Stage 3

**Fig. 5.** Image Retrieval from the database

Consider table 2. Table 2 provides the parameter values related with the coherent clusters. The parameters include percentage of coherent pixels ($\alpha_j$), number of coherent clusters ($C_{\alpha j}$) and average of coherent cluster ($\mu_{\alpha j}$) for each jth bucket or bin. As

**Table 4.** Features based on largest coherent cluster

| | $MAL_{\alpha Lj}$ | $MIL_{\alpha Lj}$ | $Ang_{\alpha Lj}$ |
|---|---|---|---|
| **Bin 1** | 0 | 0 | 0 |
| **Bin 2** | 413 | 277 | 2.88 |
| **Bin 3** | 231 | 107 | -81.62 |
| **Bin 4** | 170 | 42 | 80.58 |

(a) Query Image

(b) Stage 1

(c) Stage 2

(d) Stage 3

**Fig. 6.** Another example of image retrieval from the database an example, we show the results for 4 bins of one of the images in the database in table 2. Figure 4 shows the corresponding coherent clusters.

Consider table 3. Table 3 provides the additional parameter values related with the coherent clusters. The parameters include size of largest cluster in each bin ($L_{\alpha j}$), size of median cluster in each bin ($M_{\alpha j}$), size of smallest cluster in each bin ($S_{\alpha j}$) and Variance of coherent clusters in each bin ($V_{\alpha j}$). As an example, we show the results for 4 bins of one of the images in table 3.

Consider table 4. Table 4 provides the additional parameter values based on the various sizes of the coherent clusters. Although there are nine parameters as defined in section 4.3 but as an example, table 4 shows only 3 of the 9 features for the largest cluster. Hence, the parameters include the major axis length of the largest cluster ($MAL_{\alpha Lj}$), the minor axis length of the largest cluster ($MIL_{\alpha Lj}$) and angle ($Ang_{\alpha Lj}$). As an example, we show the results for 4 bins of one of the images in table 4.

The results were compared with the $L_1$ distance as described in section 5. Consider figure 5 and figure 6. Both the figures show the query images and the first 3 results obtained by using the above described algorithm. On inspection of all the images of the database, we found that this was the closest result. Similar query results were obtained for various query images.

## 7 Conclusions

This paper is based on the concept of coherency and incoherency and all the features are defined on top of this core concept. We have shown that the features obtained using color refinement algorithm is quite useful for relevant image retrieval queries. The feature selection is based on the number, color and shape of objects present in the image. The grayscale values, mean, variance, various sizes of the objects, axis length and angle of ellipses are considered as appropriate features for retrieval.

For retrieval of images based on queries, we proposed a three tier incremental approach. At first stage, the initial set of features described in section 4.1 is used for image retrieval. At next stage, the additional features described in section 4.2 are considered for retrieval. At the final stage, features described in section 4.3 are considered for retrieval. Hence, this approach is computationally efficient and provides refined result. The results are refined incrementally based on user's choice.

**Acknowledgements.** This study was supported by Ministry of Culture & tourism and Culture & Content Agency in Republic of Korea.

## References

1. Arnol, W.M., Worring, M., Santini, S., Gupta, A., Jain, R.: Content Based Image Retrieval at the End of the Early Years. IEEE Transactions on Pattern Analysis and Machine Intelligence, 22 (2000) 1349-1380
2. Greg, P., Ramin, Z.: Histogram Refinement for Content–based Image Retrieval. In IEEE Workshop on Applications of Computer Vision, (1996) 96-102

3. Paisarn, M., Ling, G.: Automatic Machine Interactions for Content Based Image Retrieval using a Self Organizing Tree Map Architecture. IEEE Transactions on Neural Networks, 13 (2002) 821-834
4. Zhang, D.S., Lu, G.J.: Shape Based Image Retrieval using Generic Fourier Descriptor. Signal Processing: Image Communication, 17 (2002) 825-842
5. Chabane, D.: Association and Content Based Retrival. IEEE Transactions on Knowledge and Data Engineering, 15 (2003) 118-135
6. Park, J.A., Chang, M.H., Choi, T.S., Muhammad, B.A.: Histogram based Chain Codes for Shape Description. IEICE Trans. On Communications, E86-B (2003) 3662-3665
7. Vasileios, M., Kompatsiaris, I., Strintzis, M.G.: Region-based Image Retrieval Using an Object Ontology and Relevance Feedback. EURASIP Journal on Applied Signal Processing, 6 (2004) 886–901
8. Wang, J.Z., Li, J., Wiederhold, G.: SIMPLIcity: Semantics-sensitive Integrated Matching for Picture Libraries. IEEE Trans. on Pattern Analysis and Machine Intelligence, 23 (2001) 947-963
9. Li, J., Wang, J.Z.: Automatic Linguistic Indexing of Pictures by a Statistical Modeling Approach. IEEE Transactions on Pattern Analysis and Machine Intelligence, 25 (2003) 1075-1088

# Determine the Kernel Parameter of KFDA Using a Minimum Search Algorithm

Yong Xu[1], Chuancai Liu[2], and Chongyang Zhang[2]

[1] Department of Computer Science & Technology, Shenzhen Graduate School, Harbin Institute of Technology, Shenzhen, China
[2] Department of Computer Science & Technology, Nanjing University of Science & Technology, Nanjing, China
laterfall@hitsz.edu.cn, chuancailiu@yahoo.com.cn, zcy603@163.com

**Abstract.** In this paper, we develop a novel approach to perform kernel parameter selection for Kernel Fisher discriminant analysis (KFDA) based on the viewpoint that optimal kernel parameter is associated with the maximum linear separability of samples in the feature space. This makes our approach for selecting kernel parameter of KFDA completely comply with the essence of KFDA. Indeed, this paper is the first paper to determine the kernel parameter of KFDA using a search algorithm. Our approach proposed in this paper firstly constructs an objective function whose minimum is exactly equivalent to the maximum of linear separability. Then the approach exploits a minimum search algorithm to determine the optimal kernel parameter of KFDA. The convergence properties of the search algorithm allow our approach to work well. The algorithm is also simple and not computationally complex. Experimental results illustrate the effectiveness of our approach.

**Keywords:** Kernel Fisher discriminant analysis (KFDA); parameter selection; Linear separability.

## 1 Introduction

Kernel Fihser discriminant analysis (KFDA) [1-7] is a well-known and widely used kernel method. This method roots in Fisher discriminant analysis (FDA) [8-11]. FDA aims at achieving the optimal discriminant direction that is associated with the best linear separability. We can say that two procedures are implicitly contained in the implementation of KFDA. The first procedure maps the original sample space i.e. input space into a new space i.e. feature space and the second procedure carries out FDA in the feature space. Note that the feature space induced by KFDA is usually equivalent to a space obtained through a nonlinear transform. As a result, KFDA might produce linear-separable features for such data that are from the input space and have bad linear separability. On the other hand, FDA is not capable of doing so.

A kernel function is associated with KFDA and the parameter in the function is called kernel parameter. When we carry out KFDA, we should specify the value of the

D.-S. Huang, L. Heutte, and M. Loog (Eds.): ICIC 2007, LNAI 4682, pp. 418–426, 2007.

kernel parameter. Because different parameter values usually produce different feature extraction performances, to select a suitable value for the kernel parameter is significant. An expectation maximization algorithm developed by T. P. Centeno et al. determined the kernel parameter and the regularization coefficient through the maximization of the margin- likelihood of data [12]. Note that the optimization procedure in [12] is not guaranteed to find the global minimum. S. Ali and K. A. Smith have proposed an automatic parameter learning approach using Bayes inference [13]. The cross-validation criterion was also used to select free parameters in KFDA [14]. Though a nonlinear programming algorithm [15] can be applied to determine kernel and weighting parameters of a support vector machine, the effect depends on the choice of initial values of parameters. The DOE (design of experiments) technique was also used to select parameters for SVM machines [16]. These parameter selection approaches can be classified into two classes. The first class of approach usually determins the parameter value by maximizing the likelihood and the second class of approach is based on a criterion with respective to the relation between samples.

We can consider that the kernel parameter that results in the largest Fisher criterion is the optimal parameter. The rationale is as follows: first, the larger the Fisher criterion is, the greater linear separability different classes in the feature space have. Second, greater linear separability may allow higher classification performance to be produced. With this paper, we develop a novel kernel parameter selection approach for KFDA. This approach takes the maximization of Fisher criterion value as the target of parameter selection and uses a search algorithm. As far as the knowledge of the authors, no any other researcher has proposed the same parameter selection idea as ours. The theoretic property of the search algorithm can guarantee that the parameter selection approach has good performance. The moderate computation complexity allows parameter selection to be implemented efficiently. Moreover, the developed parameter approach does obtain good experimental results and gains performance improvement for KFDA. The other parts of this paper are organized as follows: KFDA are introduced briefly in Section 2. The idea and the algorithm of parameter selection are presented in Section 3. Experimental results are shown in Section 4. In Section 5 we offer our conclusion.

## 2 KFDA

KFDA [1], [2], can be derived formally from FDA as follows. Let $\{x_i\}$ denote the samples in the input space and let $\phi$ be a nonlinear function that transforms the input space into the feature space. Consequently, Fisher criterion in the feature space is

$$J(w)=\frac{w^T S_b^\phi w}{w^T S_w^\phi w} \tag{1}$$

where $w$ is a discriminant vector, $S_b^\phi$ and $S_w^\phi$ are respectively between-class and within-class scatter matrixes in the feature space. Suppose that there are two classes,

$c_1$ and $c_2$, and the numbers of samples in $c_1$ and $c_2$ are $N_1$ and $N_2$, respectively. Then the total number of the samples is $N = N_1 + N_2$. $x_j^1, j = 1,2,\dots,N_1$ denotes the $j-th$ sample in $c_1$. $x_j^2, j = 1,2,\dots,N_2$ means the $j-th$ sample in $c_2$. If the prior probabilities of the two classes are equal, then we have,

$$S_b^\phi = (m_1^\phi - m_2^\phi)(m_1^\phi - m_2^\phi)^T \tag{2}$$

$$S_w^\phi = \sum_{i=1,2} \sum_{j=1,2,\dots,N_i} (\phi(x_j^i) - m_i^\phi)(\phi(x_j^i) - m_i^\phi)^T \tag{3}$$

where $m_i^\phi = \frac{1}{N_i} \sum_{j=1,N_i} \phi(x_j^i), i = 1,2$. According to the theory of reproducing kernels, $w$ can be an expressed in terms of all the training samples, i.e.

$$w = \sum_{i=1}^{N} \alpha_i \phi(x_i) \tag{4}$$

where each $\alpha_i$ is a scalar. We introduce a kernel function $k(x_i, x_j)$ to denote the dot production $\phi(x_i) \cdot \phi(x_j)$ and define $M_1, M_2$ and $Q$ as follows:

$$(M_i)_j = \frac{1}{N_i} \sum_{s=1}^{N_i} k(x_j, x_s^i), j = 1,2,\dots,N, i = 1,2 \tag{5}$$

$$Q = \sum_{i=1}^{2} K_i (I - I_{N_l}) K_i^T \tag{6}$$

where $I$ is the identity, $I_{N_l}$ is an $N_i \times N_i$ matrix whose each element is $1/N_i$ , $K_i$ is a $N \times N_i$ matrix, $(K_n)_{i,j} = k(x_i, x_j^n), i = 1,2,\dots,N, j = 1,2,\dots,N_n, n = 1,2$.

Then we introduce notation $M$ to mean the following formula:

$$M = (M_1 - M_2)(M_1 - M_2)^T \tag{7}$$

Note that Fisher criterion in the feature space can be expressed in terms of

$$J(\alpha)=\frac{\alpha^T M\alpha}{\alpha^T Q\alpha} \tag{8}$$

where $\alpha=[\alpha_1 \quad . \quad . \quad .\alpha_N]^T$.

As a result, the problem for obtaining the optimal discriminant vector $w$ in the feature space can be converted into the problem for solving optimal $\alpha$, which is associated with the maximum $J(\alpha)$. On the other hand, the optimal $\alpha$ will be obtained by solving the following eigenequation

$$M\alpha=\lambda Q\alpha\,. \tag{9}$$

After $\alpha$ is obtained, we can use it to extract features for samples. For detail please see [7]. Because the method presented above is defined on the basis of kernel function and Fisher discriminant analysis, it is called kernel Fisher discriminant analysis (KFDA). Note that the use of the kernel function allows KFDA to have a much lower computational complexity than an ordinary nonlinear Fisher discriminant analysis that implements explicitly FDA in the feature space obtained using a real mapping procedure. In addition, KFDA is able to obtain linear separable features for non-linear separable data whereas FDA cannot do so.

## 3 Select the Parameter Using a Search Algorithm

### 3.1 General Description of the Parameter Selection Scheme

As indicated in the context above, large Fisher criterion value means that the feature space has greater linear separability and higher classification accuracy can be expected. On the other hand, different parameter values of the kernel function will produce different Fisher criterion values. Consequently, the maximization of Fisher criterion (8) can be regarded as the objective of parameter selection. Note that the maximum of (8) coincides with the minimum of the following formual

$$J_2(\alpha)=\frac{\alpha^T Q\alpha}{\alpha^T M\alpha} \tag{20}$$

Thus, if a kernel parameter corresponds to a $\alpha$ that results in the minimum of (10), then the kernel parameter is the optimal parameter. In practice, if the $M, Q$ associated with different kernel parameters are known, then the kernel parameter that is able to result in the minimum of (10) can be taken as the optimal parameter.

The Nelder-Mead simplex algorithm [17] is an enormously popular search algorithm for unconstrained minimization and it usually performs well in practice. The convergence properties of this search algorithm has been studied[18]. In fact, J. G. Lagarias has proved clearly that the algorithm converges to a minimizer for dimension 1. Moreover, the search algorithm is simple and not computationally complex.

### 3.2 Procedure of Parameter Selection

The following procedure can carry out the parameter selection scheme described in subsection 3.1:

Step 1. Set an initial value for the kernel parameter
Step 2. Calculate $M, Q$ using (5), (6) and (7)
Step 3. Solving the smallest eigenvalue of $Q\alpha = \lambda M\alpha$.

Note that step 2 and step 3 will be repeatedly performed and will not be terminated until the convergence occurs. What the search algorithm does is to lead the computation to the convergence and to obtain the optimal kernel parameter that results in the minimum of (10).

### 3.3 Introduction to the Nelder-Mead Simplex Algorithm

The Nelder-Mead algorithm [18] focuses on minimizing a real-valued function f(x) for $x \in R^n$. Four scalar parameters exist in this method: coefficients of reflection ($\rho$), expansion ($\chi$), contraction ($\gamma$), and shrinkage ($\sigma$). These parameters satisfy

$$\rho > 0, \chi > 1, \chi > \rho, 0 < \gamma, \sigma < 1. \tag{31}$$

At the beginning of the $k$ th iteration, a nondegenerate simplex $\Delta_k$ is given, along with its $n + 1$ vertices, each of which is a point in $R^n$. Assume that iteration $k$ begins by ordering and labeling these vertices as $x_1^{(k)}, x_2^{(k)}, \ldots, x_{n+1}^{(k)}$, such that $f_1^{(k)} \le f_2^{(k)} \le \ldots \le f_{n+1}^{(k)}$, where $f_i^{(k)} = f(x_i^{(k)})$. Note that the $k$th iteration generates $n + 1$ vertices that define a different simplex for the next iteration, so that $\Delta_{k+1} \ne \Delta_k$. The result of each iteration must be either of the follows: (i) a single new vertex, i.e. the accepted point, which replaces $x_{n+1}$ in the set of vertices for the next iteration. (ii) if a shrink is performed, a set of $n$ new points that, together with $x_1$, form the simplex at the next iteration. The Nelder-Mead algorithm can be implemented by the following iteration procedure [18]:

Step 1 (order). Order the $n + 1$ vertices to satisfy $f(x_1) \le f(x_2) \le \ldots \le f(x_{n+1})$ using the tie-breaking rules given below.

Step 2 (reflection). Calculate the reflection point $x_r$ using $x_r = \bar{x} + \rho(\bar{x} - x_{n+1}) = (1+\rho)\bar{x} - \rho x_{n+1}$, where $\bar{x}$ denotes the mean of all vertices except for $x_{n+1}$. If $f_1 \le f_r < f_n$, the reflected point $x_r$ should be accepted and then the iteration is terminated.

Step 3 (expansion). If $f_r < f_1$, we compute the expansion point $x_e$ using $x_e = \bar{x} + \gamma(x_r - \bar{x}) = (1+\rho\chi)\bar{x} - \rho\chi x_{n+1}$. Then $f_e = f(x_e)$. If $f_e < f_r$, $x_e$ will be accepted and the iteration will be terminated; otherwise, $x_r$ will be accepted and the iteration will be terminated.

Step 4 (contraction). If $f_r \geq f_n$, we conduct a contraction between $\bar{x}$ and the better of $x_{n+1}$ and $x_r$ as follows.

(i) If $f_n \leq f_r < f_{n+1}$, let $x_c = \bar{x} + \gamma(x_r - \bar{x}) = (1+\rho\gamma)\bar{x} - \rho\gamma x_{n+1}$ and $f_c = f(x_c)$. If $f_c \leq f_r$, we accept $x_c$ and terminate the iteration; otherwise, go to step 5.

(ii) If $f_r \geq f_{n+1}$, let $x_c = \bar{x} - \gamma(\bar{x} - x_{n+1}) = (1-\gamma)\bar{x} + \gamma x_{n+1}$ and $f_c = f(x_c)$. If $f_c < f_{n+1}$, we accept $x_c$ and terminate the iteration; otherwise, go to step 5.

Step 5 (shrinkage). Evaluate $f$ at the n points $v_i = x_1 + \sigma(x_i - x_1), i = 2,3,\ldots,n+1$. Then the vertices of the simplex at the next iteration will be $x_1, v_2, \ldots, v_{n+1}$. The following rules are the so-called tie-breaking rules, which assign to the new vertex the highest possible index consistent with the relation $f(x_1^{(k+1)}) < f(x_2^{(k+1)}) \leq \ldots \leq f(x_{n+1}^{(k+1)})$.

(i) Nonshrink ordering rule. When a nonshrink step occurs, we discard the worst vertex $x_{n+1}^{(k)}$. Then the accepted point created during iteration $k$, denoted by $v^{(k)}$ becomes a new vertex and takes position $j+1$ in the vertices of $\Delta_{k+1}$, where $j = \max_{0 \leq l \leq n}\{l \mid f(v^{(k)}) < f(x_{l+1}^{(k)})\}$. All other vertices retain their relative ordering from iteration $k$.

(ii) Shrink ordering rule. If a shrink step occurs, the only vertex carried over from $\Delta_k$ to $\Delta_{k+1}$ is $x_1^{(k)}$. Only one tie-breaking rule is specified, for the case in which $x_1^{(k)}$ and one or more of the new points are tied as the best point: if $\min\{f(v_2^{(k)}), f(v_3^{(k)}), \ldots, f(x_{n+1}^{(k+1)})\} = f(x_1^{(k)})$, then $x_1^{(k+1)} = x_1^{(k)}$. A notation change index $k^*$ of iteration $k$ is defined as the smallest index of a vertex that differs between iterations $k$ and $k+1$. When Nelder-Mead algorithm terminates in step 2, $1 < k^* \leq n$; for termination in step 3, $k^* = 1$; for termination in step 4, $1 \leq k^* \leq n+1$; and for termination in step 5, $k^*$ is 1 or 2.

## 4 Experiments

We conducted experiments on several benchmark datasets to compare naive KFDA and KFDA with the parameter selection scheme. The kernel function employed in KFDA is the Gaussian kernel function $k(x_i, x_j) = \exp(\| x_i - x_j \|^2 / \eta)$. The minimum distance classifier was used for classification. For naive KFDA, the kernel parameter $\eta$ are respectively set to be the norm of the covariance matrix of the training samples and its three times. For KDA with the parameter selection scheme, $\eta$ are also respectively initially set to be the two values. Since each dataset has 100 training subsets and testing subsets, we conducted training and testing for every couple of training subset and testing subset. That is, if training was performed on the first training subset, testing would be carried out for the first test subset, and so on. As a result, for one dataset, we obtained 100 classification error rates respectively associated with 100 subsets. Then the mean and the standard deviation of the error rates were calculated. Table 1 indicates characteristics of these datasets. Table 2 and Table 3 respectively show classification results of naive KFDA and the KFDA model obtained using our parameter selection approach. Note that using the parameter selection scheme, KFDA obtained lower classification error rates.

**Table 1.** Characteristics of the datasets

| | Banana | Diabetis | Heart | thyroid |
|---|---|---|---|---|
| Dimension of the sample vector | 2 | 8 | 13 | 5 |
| Number of classes | 2 | 2 | 2 | 2 |
| Sample number of each training subset | 400 | 468 | 170 | 140 |

**Table 2.** Mean and standard deviation of classification error rates of naive KFDA on the subsets of one dataset. The first and second percentages denote the mean and standard deviation of classification error rates, respectively (the second percentage is written in the bracket). $\eta = \text{var}$ means that $\eta$ is set the norm of the covariance matrix of the training samples.

| | Banana | Diabetis | Heart | thyroid |
|---|---|---|---|---|
| $\eta = \text{var}$ | 12.99%(0.7%) | 30.45%(2.2%) | 23.16%(4.0%) | 5.28 (3.0%) |
| $\eta = 3 \cdot \text{var}$ | 12.96%(0.8%) | 27.15%(2.3%) | 23.14%(3.5%) | 5.39%(2.4%) |

**Table 3.** Mean and standard deviation of classification error rates of our approach on the subsets of one dataset. The first and second percentages denote the mean and standard deviation of classification error rates, respectively (the second percentage is written in the bracket). $\eta = \mathrm{var}$ means that the initial value of $\eta$ is set the norm of the covariance matrix of the training samples.

| | banana | Diabetis | Heart | thyroid |
|---|---|---|---|---|
| $\eta = \sigma$ | 11.35%(0.6%) | 26.40%(2.1%) | 20.86%(3.6%) | 5.08 %(2.2%) |
| $\eta = 3\sigma$ | 12.33%(0.7%) | 25.92%(1.9%) | 18.94%(3.2%) | 5.10%(2.6%) |

## 5 Conclusion

Our kernel parameter selection approach, which relates the optimal kernel parameter selection issue of KFDA with the Fisher-criterion maximization issue, is perfectly subject to the nature of FDA. This makes our approach be distinctive from all other parameter selection approaches. Additionally, one can understand easily the underlying reasonableness and rationality of our approach.

The underlying principle of our parameter selection approach is as follows: the optimal parameter should produce the best linear separability that is associated with the largest Fisher criterion value. Based on the defined objective function, whose minimum coincides with the maximum of Fisher-criterion, the approach developed in this paper can determine effectively the optimal kernel parameter by using a minimum search algorithm. In fact, the evidenced convergence property of the minimum search algorithm provides theoretical reasonability and practical feasibility with the parameter selection approach. Moreover, the fact that the search algorithm is simple and not computationally complex allows our approach to be carried out efficiently. Experimental results show that our approach allows the performance of KFDA to be greatly improved.

**Acknowledgements.** This work was supported by Natural Science Foundation of China （No. 60602038） and Natural Science Foundation of Guangdong Province, China （No. 06300862） .

## References

1. Mika, S., Rätsch, G., Weston, J., et al.: Fisher Discriminant Analysis with Kernels. In: Y H Hu, J Larsen, E Wilson, S Douglas eds. Neural Networks for Signal Processing IX, IEEE, (1999) 41-48
2. Muller, K.-R., Mika, S., Ratsch, G., Tsuda, K., Scholkopf, B.: An Introduction to Kernel-based Learning Algorithms. IEEE Trans. On Neural Network, 12(1) (2001) 181-201
3. Billings, S.A., Lee, K.L.: Nonlinear Fisher Discriminant Analysis Using a Minimum Square Error Cost Function and the Orthogonal Least Squares Algorithm. Neural Networks, 15(1) (2002) 263-270
4. Yang, J., Jin, Z.H., Yang, J.Y., Zhang, D., Frangi, A.F.: Essence of Kernel Fisher Discriminant: KPCA plus LDA. Pattern Recognition 37(10) (2004) 2097-2100

5. Xu, Y., Yang, J.-Y., Lu, J., Yu, D.J.: An Efficient Renovation on Kernel Fisher Discriminant Analysis and Face Recognition Experiments. Pattern Recognition, 37 (2004) 2091-2094
6. Xu, Y., Yang, J.-Y., Yang, J.:A Reformative Kernel Fisher Discriminant Analysis. Pattern Recognition, 37 (2004) 1299-1302
7. Xu,Y., Zhang, D., Jin, Z., Li, M., Yang J.-Y.: A Fast Kernel-based Nonlinear Discriminant Analysis for Multi-class Problems. Pattern Recognition 39(6) (2006) 1026-1033
8. Duda, R., Hart, P.: Pattern Classification and Scene Analysis. New York: Wiley (1973)
9. P. Belhumeur, J. Hespanha, D. Kriegman.: Eigenface vs. Fisherface: Recognition Using Class Specific Linear Projection, IEEE Trans. Pattern Anal. And Mach. Intelligence, vol. 19, no. 10 (1997) 711-720
10. Xu, Y., Yang, J.Y., Jin, Z.:Theory Analysis on FSLDA and ULDA. Pattern Recognition, 36(12) (2003) 3031-3033
11. Xu, Y., Yang, J.-Y., Jin, Z.: A Novel Method for Fisher Discriminant Analysis. Pattern Recognition, 37(2) (2004) 381-384
12. Tonatiuh Peña Centeno, Neil D,Lawrence.: Optimising Kernel Parameters and Regularisation Coefficients for Non-linear Discriminant Analysis, Journal of Machine Learning Research 7 (2006) 455–491
13. Shawkat, Ali., Kate, A. Smith.: Automatic Parameter Selection for Polynomial Kernel, Proceedings of the IEEE International Conference on Information Reuse and Integration, USA (2003) 243-249
14. Volker, Roth.: Outlier Detection with One-class Kernel Fisher Discriminants. In Lawrence K. Saul, Yair Weiss, and Léon Bottou, editors, Advances in Neural Information Processing Systems 17, Cambridge, MA, MIT Press (2005) 1169–1176
15. Schittkowski, K.: Optimal Parameter Selection in Support Vector Machines, Journal of Industrial and Management Optimization, Vol. 1, No. 4, (2005) 465-476
16. Carl, Staelin.: Parameter Selection for Support Vector Machines, Technical report, HP Laboratories Israel (2003)
17. McKinnon, K.I.M.: Convergence of the Nelder-Mead to a No Stationary Point[J].SIAM Journal Optimization, 9 (1998) 148-158
18. Lagarias, J. G., Reeds, J. A., Wright, M.H., et al.: Convergence Properties of the Nelder-Mead Simplex Method in Low Dimensions, SIAM Journal of Optimization, 9(1) (1998) 112-147

# Hidden Markov Models with Multiple Observers

Hua Chen, Zhi Geng*, and Jinzhu Jia

School of Mathematical Sciences, Peking University, Beijing 100871, China
zgeng@math.pku.edu.cn

**Abstract.** Hidden Markov models (HMMs) usually assume that the state transition matrices and the output models are time-invariant. Without this assumption, the parameters in a HMM may not be identifiable. In this paper, we propose a HMM with multiple observers such that its parameters are local identifiable without the time-invariant assumption. We show a sufficient condition for local identifiability of parameters in HMMS.

**Keywords:** Multiple observers, Hidden Markov models, Identifiability.

## 1 Introduction

Hidden Markov models (HMMs) are widely applied to pattern recognition, computational molecular biology, computer vision and so on [1]. HMMs usually assume that the state transition matrices and the output models are not dependent on time. Without this time-invariant assumption, the models are more complicate and the parameters in a HMM may not be identifiable. This assumption, however, may not be true in many applications.

Some literatures have discussed parameter identifiability under the time-varying assumption in HMMs. For continuous variables, Gaussian HMMs with time-varying transition probabilities depending on exogenous variables through a logistic function were discussed in [3]. Spezia proposed Markov chain Monte Carlo algorithms for model selection and parameter estimation. For discrete variables, Van de Pol *et al.* proposed multiple-group analysis which can be only used with time-constant covariates [4]. Vermunt *et al.* proposed a flexible logit regression approach under discrete-time discrete-state HMMs with time-constant and time-varying covariates [5].

In this paper, suppose all variables are discrete and there are no covariates. We propose a HMM with multiple observers such that its parameters are identifiable even without the time invariant assumption. These models are reasonable in some applications. For example, every subject is scored or observed independently by multiple experts or observers at the same time and the observed states are subject to measurement error. Then the observed transitions between two points in time will include both true change and spurious change caused by measurement error. Then we can apply our method to these cases. Moreover, such

* Corresponding author.

D.-S. Huang, L. Heutte, and M. Loog (Eds.): ICIC 2007, LNAI 4682, pp. 427–435, 2007.

a HMM with multiple observers and without time-invariant assumption can be used to analyze association or relationship among hidden variables, which may represent different unobservable variables even with different domains.

Section 2 describes notation and HMMs with multiple observers. In Section 3, we discuss identifiability of parameters in HMMs. Section 4 shows simulation to illustrate and evaluate our approach. Finally we will summary our results in section 5.

## 2 Notation and Definitions

Let $X_1, \ldots, X_T$ denote $T$ hidden variables, where $T$ may or may not represent the number of time points. Suppose that $K$ observers simultaneously observe each individual. Let $Y_{1t}, \ldots, Y_{Kt}$ denote $K$ manifest variables with respect to the hidden variable $X_t$, which are observed by $K$ observers respectively. Assume that $Y_{1t}, \ldots, Y_{Kt}$ are mutually and conditionally independent given $X_t$ and that $X_1, \ldots, X_T$ satisfy the Markov property: $X_{t+1}$ is conditionally independent of $X_{t-1}$ given $X_t$, see Fig. 1. We assume that all variables are discrete with multiple categories. Let $J_t$ be the number of $X_t$'s categories and $I_{kt}$ be the number of $Y_{kt}$'s categories.

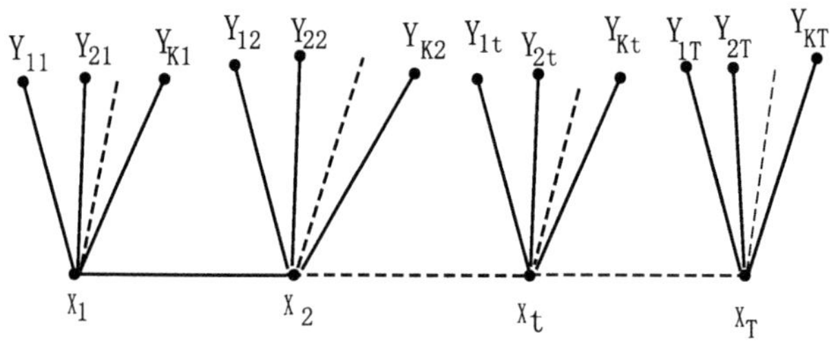


**Fig. 1.** A HMM with $K$ observers

Under the hidden Markov model with multiple observers, the joint probability can be written as

$$\pi_{y_{11}\cdots y_{K1}\cdots y_{1T}\cdots y_{KT}x_1\cdots x_T}^{Y_{11}\cdots Y_{K1}\cdots Y_{1T}\cdots Y_{KT}X_1\cdots X_T} = \pi_{x_1}^{X_1} \prod_{k=1}^{K} \pi_{y_{k1}|x_1}^{Y_{k1}|X_1} \prod_{t=2}^{T} \left( \pi_{x_t|x_{t-1}}^{X_t|X_{t-1}} \prod_{k=1}^{K} \pi_{y_{kt}|x_t}^{Y_{kt}|X_t} \right) \quad (1)$$

where $\pi_u^U$ denotes the probability of $U = u$ and $\pi_{u|v}^{U|V}$ denotes the conditional probability of $U = u$ given $V = v$. Then the marginal probability of manifest variables

$$\pi_{y_{11}\cdots y_{K1}\cdots y_{1T}\cdots y_{KT}} = \sum_{x_1} \cdots \sum_{x_T} \pi_{y_{11}\cdots y_{K1}\cdots y_{1T}\cdots y_{KT}x_1\cdots x_T}^{Y_{11}\cdots Y_{K1}\cdots Y_{1T}\cdots Y_{KT}X_1\cdots X_T}. \quad (2)$$

The vector of parameters is denoted as

$$\pi = \{\pi_{x_1}^{X_1}, \pi_{x_2|x_1}^{X_2|X_1}, \ldots, \pi_{x_T|x_{T-1}}^{X_T|X_{T-1}}, \pi_{y_{11}|x_1}^{Y_{11}|X_1}, \ldots, \pi_{y_{K1}|x_1}^{Y_{K1}|X_1}, \ldots, \pi_{y_{1T}|x_T}^{Y_{1T}|X_T}, \ldots, \pi_{y_{KT}|x_T}^{Y_{KT}|X_T}\},$$

and let $\hat{\pi}$ denote its maximum likelihood estimate (MLE).

If $\pi$ is uniquely determined by the joint probability $\pi_{y_{11}\ldots y_{K1}\ldots y_{1T}\ldots y_{KT}}$ of manifest variables, then we say that the parameters of the HMM is *identifiable*, or simply the HMM is identifiable. If $\pi$ is uniquely determined by $\pi_{y_{11}\ldots y_{K1}\ldots y_{1T}\ldots y_{KT}}$ within some neighborhood of $\pi$, we say that the parameters of the HMM are *locally identifiable*, or simply the HMM is locally identifiable.

## 3 Identification of Parameters in Hidden Markov Models with Multiple Observers

In this section, we discuss conditions for local identification of parameters in the HMM. We discuss identifiability for the cases with two hidden variables $X_1$ and $X_2$ at first. Then we discuss the cases with multiple hidden variables. Below we give an obvious necessary condition.

From (1) and (2), we get

$$\pi_{y_{11}\ldots y_{K1}\ldots y_{1T}\ldots y_{KT}} = \sum_{x_1}\cdots\sum_{x_T} \pi_{x_1}^{X_1} \prod_{t=2}^{T} \pi_{x_t|x_{t-1}}^{X_t|X_{t-1}} \prod_{t=1}^{T}\left(\prod_{k=1}^{K} \pi_{y_{kt}|x_t}^{Y_{kt}|X_t}\right) \tag{3}$$

Formula (3) describes a set of functions that map free parameters in $\pi$ into the probability $\pi_{y_{11}\ldots y_{K1}\ldots y_{1T}\ldots y_{KT}}$ of manifest variables. The number of free parameters in $\pi$ is

$$J_1 - 1 + \sum_{i=2}^{T} J_{i-1}(J_i - 1) + \sum_{i=1}^{T}\sum_{k}(I_{ki} - 1)J_i$$

since

$$\sum_{x_1} \pi_{x_1}^{X_1} = \sum_{x_2} \pi_{x_2|x_1}^{X_2|X_1} = \ldots = \sum_{x_T} \pi_{x_T|x_{T-1}}^{X_T|X_{T-1}} = \sum_{y_{kt}} \pi_{y_{kt}|x_t}^{Y_{kt}|X_t} = 1. \tag{4}$$

The set of these free parameters is called the basic set. The number of observed frequencies is $\prod_{kt} I_{kt}$. A necessary condition of identifiability is that the number of observed frequencies is larger than the number of free parameters in $\pi$. In the cases that all variables are binary, if there is only one observer, then the parameters are not identifiable. For example, in the case of $T = 2$, the number of free parameters is 7 but the number of observed frequencies is only 4. It can be shown that there must be at least three observers for the case with only one hidden variable to satisfy the necessary condition and that there must be at least two observers for more hidden variables.

### 3.1 Local Identifiability for HMMs with Two Hidden Variables

Goodman [2] showed a sufficient condition for local identifiability of parameters in latent class models which has only one hidden variable. In this subsection, we extend Goodman's approach to show a sufficient condition for local identifiability of parameters of models with two hidden variables, and we discuss the case with more hidden variables in the next subsection. For the case with two hidden variables $X_1$ and $X_2$, the joint probability of hidden and manifest variables is

$$\pi^{Y_{11}\ldots Y_{K1}Y_{12}\ldots Y_{K2}X_1X_2}_{y_{11}\cdots y_{K1}y_{12}\cdots y_{K2}x_1x_2} = \pi^{X_1}_{x_1}\pi^{X_2|X_1}_{x_2|x_1}\left(\prod_{k=1}^{K}\pi^{Y_{k1}|X_1}_{y_{k1}|x_1}\right)\left(\prod_{k=1}^{K}\pi^{Y_{k2}|X_2}_{y_{k2}|x_2}\right), \quad (5)$$

and the marginal probability of manifest variables is

$$\pi^{Y_{11}\ldots Y_{K1}Y_{12}\ldots Y_{K2}}_{y_{11}\cdots y_{K1}y_{12}\cdots y_{K2}} = \sum_{x_1,x_2}\pi^{X_1}_{x_1}\pi^{X_2|X_1}_{x_2|x_1}\left(\prod_{k=1}^{K}\pi^{Y_{k1}|X_1}_{y_{k1}|x_1}\right)\left(\prod_{k=1}^{K}\pi^{Y_{k2}|X_2}_{y_{k2}|x_2}\right). \quad (6)$$

**Lemma 1.** *The sufficient condition for local identifiability is that the rank of the derivative matrix of the function $\pi_{y_{11}\cdots y_{K1}\cdots y_{1T}\cdots y_{KT}}$ in (6) with respect to the parameters in the basic set, equals the number of columns of the derivative matrix.*

*Example 1.* The model with two hidden variables $X_1$ and $X_2$ and three observers is shown in Fig. 2, where all variables are binary.

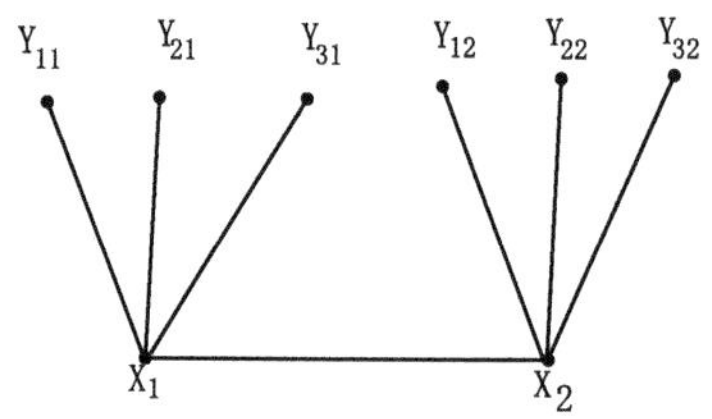


**Fig. 2.** A hidden Markov model with two hidden variables and three observers

The marginal probability of manifest variables is

$$\pi_{y_{11}y_{21}y_{31}y_{12}y_{22}y_{32}} = \sum_{x_1}\sum_{x_2}\pi^{Y_{11}Y_{21}Y_{31}Y_{12}Y_{22}Y_{32}X_1X_2}_{y_{11}y_{21}y_{31}y_{12}y_{22}y_{32}x_1x_2}. \quad (7)$$

The vector of parameters is denoted as

$$\pi = \{\pi^{X_1}_{x_1}, \pi^{X_2|X_1}_{x_2|x_1}, \pi^{Y_{11}|X_1}_{y_{11}|x_1}, \pi^{Y_{21}|X_1}_{y_{21}|x_1}, \pi^{Y_{31}|X_1}_{y_{31}|x_1}, \pi^{Y_{12}|X_2}_{y_{12}|x_2}, \pi^{Y_{22}|X_2}_{y_{22}|x_2}, \pi^{Y_{32}|X_2}_{y_{32}|x_2}\}.$$

The derivative matrix has 63 rows and 15 columns, and it can be calculated as follows:

$$\frac{\partial \pi_{y_{11}y_{21}y_{31}y_{12}y_{22}y_{32}}}{\partial \pi_1^{X_1}}$$

$$= \sum_{x_2} \left( \pi_{x_2|1}^{X_2|X_1} \prod_{k=1}^{3} \pi_{y_{k1}|1}^{Y_{k1}|X_1} - \pi_{x_2|0}^{X_2|X_1} \prod_{k=1}^{3} \pi_{y_{k1}|0}^{Y_{k1}|X_1} \right) \prod_{k=1}^{3} \pi_{y_{k2}|x_2}^{Y_{k2}|X_2},$$

$$\frac{\partial \pi_{y_{11}y_{21}y_{31}y_{12}y_{22}y_{32}}}{\partial \pi_{1|x_1}^{X_2|X_1}}$$

$$= \pi_{x_1}^{X_1} \prod_{k=1}^{3} \pi_{y_{k1}|x_1}^{Y_{k1}|X_1} \left( \prod_{k=1}^{3} \pi_{y_{k2}|1}^{Y_{k2}|X_2} - \prod_{k=1}^{3} \pi_{y_{k2}|0}^{Y_{k2}|X_2} \right),$$

$$\frac{\partial \pi_{y_{11}y_{21}y_{31}y_{12}y_{22}y_{32}}}{\partial \pi_{1|x_1}^{Y_{11}|X_1}}$$

$$= \begin{cases} \sum_{x_2=0}^{1} \pi_{x_1}^{X_1} \pi_{x_2|x_1}^{X_2|X_1} \prod_{k=2}^{3} \pi_{y_{k1}|x_1}^{Y_{k1}|X_1} \prod_{k=1}^{3} \pi_{y_{k2}|x_2}^{Y_{k2}|X_2}, & y_{11} = 1, \\ -\sum_{x_2=0}^{1} \pi_{x_1}^{X_1} \pi_{x_2|x_1}^{X_2|X_1} \prod_{k=2}^{3} \pi_{y_{k1}|x_1}^{Y_{k1}|X_1} \prod_{k=1}^{3} \pi_{y_{k2}|x_2}^{Y_{k2}|X_2}, & y_{11} = 0, \end{cases}$$

$$\frac{\partial \pi_{y_{11}y_{21}y_{31}y_{12}y_{22}y_{32}}}{\partial \pi_{1|x_1}^{Y_{21}|X_1}}$$

$$= \begin{cases} \sum_{x_2=0}^{1} \pi_{x_1}^{X_1} \pi_{x_2|x_1}^{X_2|X_1} \pi_{y_{11}|x_1}^{Y_{11}|X_1} \pi_{y_{31}|x_1}^{Y_{31}|X_1} \prod_{k=1}^{3} \pi_{y_{k2}|x_2}^{Y_{k2}|X_2}, & y_{21} = 1, \\ -\sum_{x_2=0}^{1} \pi_{x_1}^{X_1} \pi_{x_2|x_1}^{X_2|X_1} \pi_{y_{11}|x_1}^{Y_{11}|X_1} \pi_{y_{31}|x_1}^{Y_{31}|X_1} \prod_{k=1}^{3} \pi_{y_{k2}|x_2}^{Y_{k2}|X_2}, & y_{21} = 0, \end{cases}$$

$$\frac{\partial \pi_{y_{11}y_{21}y_{31}y_{12}y_{22}y_{32}}}{\partial \pi_{1|x_1}^{Y_{31}|X_1}}$$

$$= \begin{cases} \sum_{x_2=0}^{1} \pi_{x_1}^{X_1} \pi_{x_2|x_1}^{X_2|X_1} \pi_{y_{11}|x_1}^{Y_{11}|X_1} \pi_{y_{21}|x_1}^{Y_{21}|X_1} \prod_{k=1}^{3} \pi_{y_{k2}|x_2}^{Y_{k2}|X_2}, & y_{31} = 1, \\ -\sum_{x_2=0}^{1} \pi_{x_1}^{X_1} \pi_{x_2|x_1}^{X_2|X_1} \pi_{y_{11}|x_1}^{Y_{11}|X_1} \pi_{y_{21}|x_1}^{Y_{21}|X_1} \prod_{k=1}^{3} \pi_{y_{k2}|x_2}^{Y_{k2}|X_2}, & y_{31} = 0, \end{cases}$$

$$\frac{\partial \pi_{y_{11}y_{21}y_{31}y_{12}y_{22}y_{32}}}{\partial \pi_{1|x_2}^{Y_{12}|X_2}}$$

$$= \begin{cases} \sum_{x_1=0}^{1} \pi_{x_1}^{X_1} \pi_{x_2|x_1}^{X_2|X_1} \prod_{k=1}^{3} \pi_{y_{k1}|x_1}^{Y_{k1}|X_1} \pi_{y_{22}|x_2}^{Y_{22}|X_2} \pi_{y_{32}|x_2}^{Y_{32}|X_2}, & y_{12} = 1, \\ -\sum_{x_1=0}^{1} \pi_{x_1}^{X_1} \pi_{x_2|x_1}^{X_2|X_1} \prod_{k=1}^{3} \pi_{y_{k1}|x_1}^{Y_{k1}|X_1} \pi_{y_{22}|x_2}^{Y_{22}|X_2} \pi_{y_{32}|x_2}^{Y_{32}|X_2}, & y_{12} = 0, \end{cases}$$

$$\frac{\partial \pi_{y_{11}y_{21}y_{31}y_{12}y_{22}y_{32}}}{\partial \pi_{1|x_2}^{Y_{22}|X_2}}$$

$$= \begin{cases} \sum_{x_1=0}^{1} \pi_{x_1}^{X_1} \pi_{x_2|x_1}^{X_2|X_1} \prod_{k=1}^{3} \pi_{y_{k1}|x_1}^{Y_{k1}|X_1} \pi_{y_{12}|x_2}^{Y_{12}|X_2} \pi_{y_{32}|x_2}^{Y_{32}|X_2}, & y_{22} = 1, \\ -\sum_{x_1=0}^{1} \pi_{x_1}^{X_1} \pi_{x_2|x_1}^{X_2|X_1} \prod_{k=1}^{3} \pi_{y_{k1}|x_1}^{Y_{k1}|X_1} \pi_{y_{12}|x_2}^{Y_{12}|X_2} \pi_{y_{32}|x_2}^{Y_{32}|X_2}, & y_{22} = 0, \end{cases}$$

$$\frac{\partial \pi_{y_{11}y_{21}y_{31}y_{12}y_{22}y_{32}}}{\partial \pi_{1|x_2}^{Y_{32}|X_2}}$$

$$= \begin{cases} \sum_{x_1=0}^{1} \pi_{x_1}^{X_1} \pi_{x_2|x_1}^{X_2|X_1} \prod_{k=1}^{3} \pi_{y_{k1}|x_1}^{Y_{k1}|X_1} \pi_{y_{12}|x_2}^{Y_{12}|X_2} \pi_{y_{22}|x_2}^{Y_{22}|X_2}, & y_{32} = 1, \\ -\sum_{x_1=0}^{1} \pi_{x_1}^{X_1} \pi_{x_2|x_1}^{X_2|X_1} \prod_{k=1}^{3} \pi_{y_{k1}|x_1}^{Y_{k1}|X_1} \pi_{y_{12}|x_2}^{Y_{12}|X_2} \pi_{y_{22}|x_2}^{Y_{22}|X_2}, & y_{32} = 0. \end{cases}$$

Note that this lemma is also a sufficient condition for local identifiability of parameters of models with multiple hidden variables. But the lemma is not convenient for use in practice because we have to deduce a huge derivative matrix when the model is complex. Even there are only two latent binary variables and three observed binary variables corresponding to every latent variable, we must compute a 65-by-15 matrix. In the next subsection we consider HMMs with multiple hidden variables.

### 3.2 Local Identifiability for HMMs with Multiple Hidden Variables

In this subsection, we use the result obtained in the previous subsection and the Markov property of HMMs to show sufficient condition for local identifiability of parameters in HMMs with multiple hidden variables.

**Theorem 1.** *A HMM with multiple hidden variables is locally identifiable if each of its sub-models composed of $X_t$ and $X_{t+1}$ is locally identifiable.*

*Proof.* From (1), we have the marginal probability for a sub-models composed of $X_t$ and $X_{t+1}$ as follows

$$
\begin{aligned}
&\pi^{Y_{1t}\ldots Y_{Kt}Y_{1t+1}\ldots Y_{Kt+1}X_tX_{t+1}}_{y_{1t}\cdots y_{Kt}y_{1t+1}\cdots y_{Kt+1}x_tx_{t+1}} \\
&= \sum_{x_i,y_{1i},\ldots,y_{Ki},i\neq t,t+1} \pi^{X_1}_{x_1} \prod_{k=1}^{K} \pi^{Y_{k1}|X_1}_{y_{k1}|x_1} \prod_{m=2}^{T} \left( \pi^{X_m|X_{m-1}}_{x_m|x_{m-1}} \prod_{k=1}^{K} \pi^{Y_{km}|X_m}_{y_{km}|x_m} \right) \\
&= \pi^{X_t}_{x_t} \pi^{X_{t+1}|X_t}_{x_{t+1}|x_t} \left( \prod_{k=1}^{K} \pi^{Y_{kt}|X_t}_{y_{kt}|x_t} \right) \left( \prod_{k=1}^{K} \pi^{Y_{kt+1}|X_{t+1}}_{y_{kt+1}|x_{t+1}} \right). \qquad (8)
\end{aligned}
$$

Especially for $t = 1$,

$$
\pi^{Y_{11}\ldots Y_{K1}Y_{12}\ldots Y_{K2}X_1X_2}_{y_{11}\cdots y_{K1}y_{12}\cdots y_{K2}x_1x_2} = \pi^{X_1}_{x_1} \pi^{X_2|X_1}_{x_2|x_1} \left( \prod_{k=1}^{K} \pi^{Y_{k1}|X_1}_{y_{k1}|x_1} \right) \left( \prod_{k=1}^{K} \pi^{Y_{k2}|X_2}_{y_{k2}|x_2} \right). \qquad (9)
$$

Then if all of sub-models are locally identifiable by Lemma 1, we obtain all of parameters are locally identifiable from (8) and (9). □

*Example 2.* For a HMM with three hidden variables $X_1$, $X_2$ and $X_3$ and three observers where all variables are binary, the marginal probability of manifest variables is

$$
\pi_{y_{11}y_{21}y_{31}\ldots y_{33}} = \sum_{x_1,x_2,x_3} \pi^{Y_{11}Y_{21}Y_{31}\ldots Y_{33}X_1X_2X_3}_{y_{11}y_{21}y_{31}\ldots y_{33}x_1x_2x_3}, \qquad (10)
$$

where

$$
\begin{aligned}
&\pi^{Y_{11}Y_{21}Y_{31}\ldots Y_{33}X_1X_2X_3}_{y_{11}y_{21}y_{31}\cdots y_{33}x_1x_2x_3} \\
&= \pi^{X_1}_{x_1} \prod_{k=1}^{3} \pi^{Y_{k1}|X_1}_{y_{k1}|x_1} \pi^{X_2|X_1}_{x_2|x_1} \prod_{k=1}^{3} \pi^{Y_{k2}|X_2}_{y_{k2}|x_2} \pi^{X_3|X_2}_{x_3|x_2} \prod_{k=1}^{3} \pi^{Y_{k3}|X_3}_{y_{k3}|x_3}. \qquad (11)
\end{aligned}
$$

By Theorem 1 we only need that the following sub-models are locally identify:

$$\pi^{Y_{11}Y_{21}Y_{31}\ldots Y_{32}X_1X_2}_{y_{11}y_{21}y_{31}\ldots y_{32}x_1x_2} = \pi^{X_1}_{x_1}\pi^{X_2|X_1}_{x_2|x_1}\left(\prod_{k=1}^{3}\pi^{Y_{k1}|X_1}_{y_{k1}|x_1}\right)\left(\prod_{k=1}^{3}\pi^{Y_{k2}|X_2}_{y_{k2}|x_2}\right), \quad (12)$$

and

$$\pi^{Y_{12}Y_{22}Y_{32}\ldots Y_{33}X_2X_3}_{y_{12}y_{22}y_{32}\ldots y_{33}x_2x_3} = \pi^{X_2}_{x_2}\pi^{X_3|X_2}_{x_3|x_2}\left(\prod_{k=1}^{3}\pi^{Y_{k2}|X_2}_{y_{k2}|x_2}\right)\left(\prod_{k=1}^{3}\pi^{Y_{k3}|X_3}_{y_{k3}|x_3}\right). \quad (13)$$

## 4 Simulation

In this section, we use a hidden Markov model with three hidden variables $X_1$, $X_2$ and $X_3$ whose true parameters are given in Table 1. First, we consider identifiability of the HMM. According to the result in Section 3.1, we can show that the rank of the derivative matrix for the HMM with two hidden variables $X_1$ and $X_2$ is 15 which is equal to the number of parameters in the basic set, and thus the HMM with $X_1$ and $X_2$ is locally identifiable. Similarly, we can show that the HMM with two hidden variables $X_2$ and $X_3$ is also locally identifiable. Thus by Theorem 1, we obtain that the HMM with three hidden variables $X_1$, $X_2$ and $X_3$ is locally identifiable.

Next we evaluate the maximum likelihood estimates (MLEs) obtained by using the expectation-maximization (EM) algorithm. We generate a sample from the multinomial distribution with a sample size 800 and parameters $\{\pi_{y_{11}y_{21}y_{31}\ldots y_{33}}\}$ obtained by formulas (10) and (11) and the true values in Table 1, and then we use

**Table 1.** True parameters, initial values, and means and standard errors of MLEs

| Parameter | True | Init. | Mean | Std. Err. | Parameter | True | Init. | Mean | Std. Err. |
|---|---|---|---|---|---|---|---|---|---|
| $\pi^{X_1}_{1}$ | 0.55 | 0.5 | 0.5514586 | 0.0592352 | | | | | |
| $\pi^{X_2\|X_1}_{1\|0}$ | 0.18 | 0.5 | 0.1782013 | 0.0721840 | $\pi^{X_2\|X_1}_{1\|1}$ | 0.70 | 0.5 | 0.7029848 | 0.0761647 |
| $\pi^{X_3\|X_2}_{1\|0}$ | 0.65 | 0.5 | 0.6242690 | 0.0924664 | $\pi^{X_3\|X_2}_{1\|1}$ | 0.40 | 0.5 | 0.3746524 | 0.0895316 |
| $\pi^{Y_{11}\|X_1}_{1\|0}$ | 0.15 | 0.1 | 0.1478441 | 0.0564458 | $\pi^{Y_{11}\|X_1}_{1\|1}$ | 0.80 | 0.9 | 0.8009310 | 0.0485992 |
| $\pi^{Y_{21}\|X_2}_{1\|0}$ | 0.15 | 0.1 | 0.1456320 | 0.0470327 | $\pi^{Y_{21}\|X_2}_{1\|1}$ | 0.80 | 0.9 | 0.8033483 | 0.0587846 |
| $\pi^{Y_{31}\|X_3}_{1\|0}$ | 0.15 | 0.1 | 0.1606069 | 0.0700405 | $\pi^{Y_{31}\|X_3}_{1\|1}$ | 0.80 | 0.9 | 0.8203083 | 0.0746082 |
| $\pi^{Y_{12}\|X_1}_{1\|0}$ | 0.25 | 0.1 | 0.2510445 | 0.0432321 | $\pi^{Y_{12}\|X_1}_{1\|1}$ | 0.70 | 0.9 | 0.6983308 | 0.0376889 |
| $\pi^{Y_{22}\|X_2}_{1\|0}$ | 0.25 | 0.1 | 0.2486549 | 0.0361424 | $\pi^{Y_{22}\|X_2}_{1\|1}$ | 0.70 | 0.9 | 0.6993636 | 0.0426023 |
| $\pi^{Y_{32}\|X_3}_{1\|0}$ | 0.25 | 0.1 | 0.2534798 | 0.0561735 | $\pi^{Y_{32}\|X_3}_{1\|1}$ | 0.70 | 0.9 | 0.7153454 | 0.0516805 |
| $\pi^{Y_{13}\|X_1}_{1\|0}$ | 0.35 | 0.1 | 0.3498366 | 0.0351648 | $\pi^{Y_{13}\|X_1}_{1\|1}$ | 0.60 | 0.9 | 0.5969391 | 0.0318683 |
| $\pi^{Y_{23}\|X_2}_{1\|0}$ | 0.35 | 0.1 | 0.3459355 | 0.0299180 | $\pi^{Y_{23}\|X_2}_{1\|1}$ | 0.60 | 0.9 | 0.6035706 | 0.0372744 |
| $\pi^{Y_{33}\|X_3}_{1\|0}$ | 0.35 | 0.1 | 0.3570095 | 0.0372560 | $\pi^{Y_{33}\|X_3}_{1\|1}$ | 0.60 | 0.9 | 0.6060694 | 0.0336440 |

'Init.' denotes the initial values used in the EM algorithm.

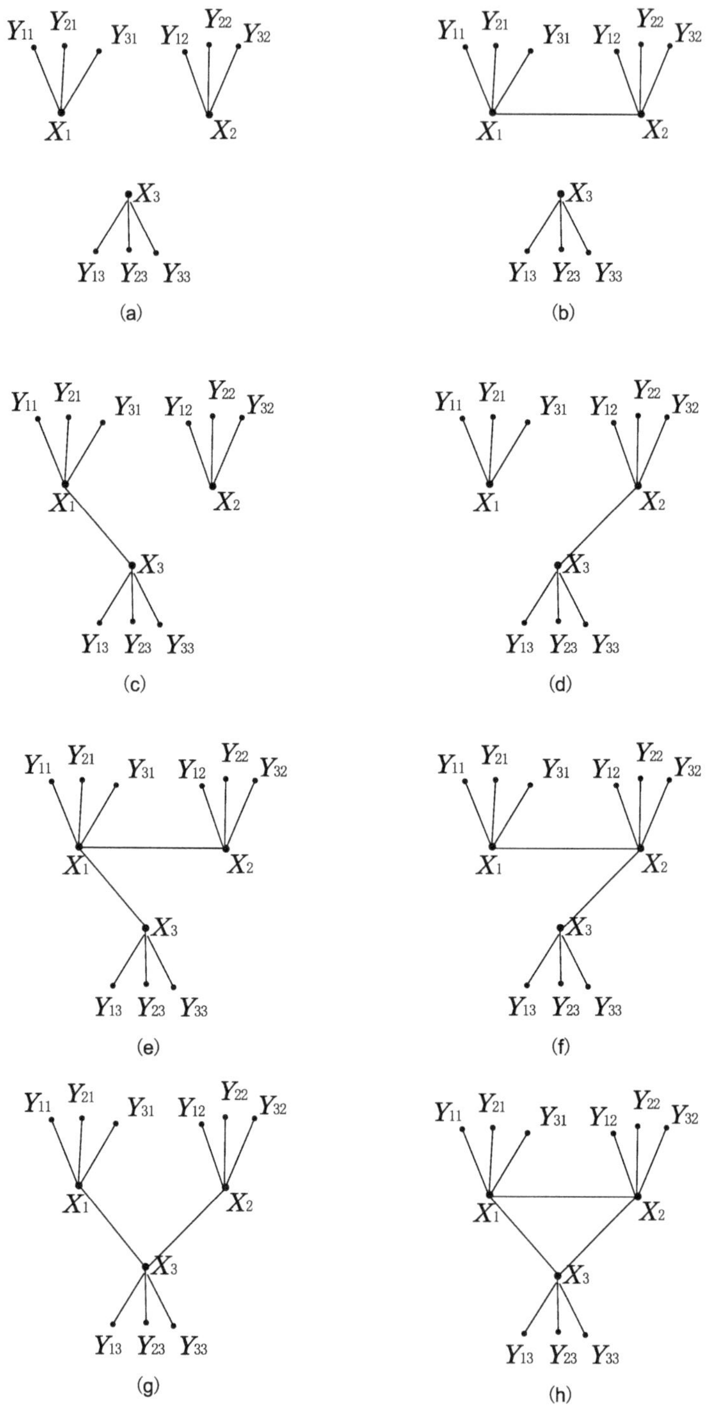


**Fig. 3.** 8 possible graphical models over $X_1$, $X_2$ and $X_3$

the EM algorithm to find MLEs. Repeat this process 200 times, and find means and variances of estimates as shown in Table 1. It can be seen that the estimates are quite close to the true vales.

Finally we illustrate the model selection. Given three hidden model $X_1$, $X_2$ and $X_3$, there are 3 possible edges between them, and thus there are 8 possible graphical models over $X_1$, $X_2$ and $X_3$, see Fig. 3. We generate a sample with size 800 from the true model $X_1 - X_2 - X_3$, and we select a model which has the least value of BIC. Repeat this process 100 times. We correctly selected the true model 99 times, and the other selected $X_1 - X_3 - X_2$ incorrectly.

## 5 Summary

We focused on the identifiability of parameters in discrete-time discrete-status HMMs with multiple observers. We first discussed local identifiability of the cases of two latent variables in lemma 1. Then we gave the identifiable results of the cases of multiple hidden variables which satisfy Markov property in theorem 1. For identifiable models, we proposed to find the maximum likelihood estimates by the EM algorithm. At last we tried to apply our method to analysis relationship of hidden variables which may not satisfy Markov property.

## Acknowledgements

This research was supported by NSFC, NBRP 2003CB715900 and NBRP 2005CB523301.

## References

1. Ghahramani, Z.: An Introduction to Hidden Markov Models and Bayesian Networks. Hidden Markov Models: Applications in Computer Vision, (2001) 9-42
2. Goodman, L.A.: Exploratory Latent Structure Analysis Using Both Identifiable and Unidentifiable Models: Biometrika, **61** (1974) 215-231
3. Spezia, L.: Bayesian Analysis of Non-homogeneous Hidden Markov Models: Journal of Statistical Computation and Simulation, **76** (2006) 713-725
4. Van de Pol, F., Langeheine, R.: Mixed Markov Latent Class Models. In C.C. Clogg(Ed.), Sociological Methodology Oxford: Blackwell. (1990)
5. Vermunt, J. K., Langeheine, R., Bockenholt, U.: Discrete-time Discrete-state Latent Markov Models with Time-constant and Time-varying Covariates: Journal of Educational and Behavioral Statistics, **24** (1999) 179-207

# K-Distributions: A New Algorithm for Clustering Categorical Data

Zhihua Cai[1], Dianhong Wang[2], and Liangxiao Jiang[3]

[1] Faculty of Computer Science, China University of Geosciences
Wuhan, Hubei, P.R.China, 430074
zhcai@cug.edu.cn
[2] Faculty of Electronic Engineering, China University of Geosciences
Wuhan, Hubei, P.R.China, 430074
wangdh@cug.edu.cn
[3] Faculty of Computer Science, China University of Geosciences
Wuhan, Hubei, P.R.China, 430074
ljiang@cug.edu.cn

**Abstract.** Clustering is one of the most important tasks in data mining. The K-means algorithm is the most popular one for achieving this task because of its efficiency. However, it works only on numeric values although data sets in data mining often contain categorical values. Responding to this fact, the K-modes algorithm is presented to extend the K-means algorithm to categorical domains. Unfortunately, it suffers from computing the dissimilarity between each pair of objects and the mode of each cluster. Aiming at addressing these problems confronting K-modes, we present a new algorithm called K-distributions in this paper. We experimentally tested K-distributions using the well known 36 UCI data sets selected by Weka, and compared it to K-modes. The experimental results show that K-distributions significantly outperforms K-modes in term of clustering accuracy and log likelihood.

**Keywords:** K-means, K-modes, K-distributions, clustering, categorical data sets, log likelihood.

## 1 Introduction

Clustering [1] is one of the most important tasks in data mining. The goal of clustering is to partition a set of objects into clusters of similar objects. Thus, a cluster is a collection of objects that are similar to one another within the same cluster and are dissimilar to the objects in other clusters. Different from classification, clustering doesn't rely on predefined classes and class-labelled training data. For this reason, it is a kind of typical unsupervised learning based on observation. Clustering analysis has been widely used in in many real-world data mining applications. For example, in business, clustering analysis may help marketers discover distinct groups in their customer bases and characterize customer groups based on purchasing patterns.

D.-S. Huang, L. Heutte, and M. Loog (Eds.): ICIC 2007, LNAI 4682, pp. 436–443, 2007.

The K-means algorithm [2] is the most popular one for clustering because of its efficiency. However, it works only on numeric values because it need to minimize a cost function by calculating the means of clusters. This limits its use in data mining because data sets in data mining often contain categorical values. The whole algorithm can be described as follow.

**Algorithm.** K-means(**D**, K)
**Input** : a data set **D** containing $n$ objects, the number of clusters K
**Output** : a set of K clusters
**Method** : the K-means algorithm is implemented as follows

1. Partition all objects into K nonempty and mutually exclusive subsets randomly, and each subset is treated as a cluster.
2. Compute each cluster's mean and assign each object to the cluster whose mean is the nearest to it according to the standard Euclidean distance.
3. Repeat 2 until no more new assignment.

Responding to this fact, the K-modes algorithm [3] is presented to extend the K-means algorithm to categorical domains whilst preserving the efficiency of the k-means algorithm. In the K-modes algorithm, three major modifications have been made to the K-means algorithm: Using different dissimilarity measures, replacing k means with k modes, and using a frequency based method to update modes. The whole K-modes algorithm[1] can be described as follow.

**Algorithm.** K-modes(**D**, K)
**Input** : a data set **D** containing $n$ objects, the number of clusters K
**Output** : a set of K clusters
**Method** : the K-modes algorithm is implemented as follows

1. Partition all objects into K nonempty and mutually exclusive subsets randomly, and each subset is treated as a cluster.
2. Compute each cluster's mode and assign each object to the cluster whose mode is the nearest to it according to the simple dissimilarity measure (the number of different attribute values).
3. Repeat 2 until no more new assignment.

Although K-modes is successful in extending K-means to categorical domains, it suffers from computing the dissimilarity between each pair of objects and the mode of each cluster. Aiming at addressing these problems confronting K-modes, we present a new algorithm called K-distributions in this paper. The experimental results in Section 3 show that K-distributions significantly outperforms K-modes in term of accuracy and log likelihood.

The rest of the paper is organized as follows. In Section 2, we present a new algorithm simply called K-distributions. In Section 3, we describe the experimental setup and results in detail. In Section 4, we draw conclusions and outline our main directions for future research.

---

[1] This algorithm is a little different from Huang's [3].

## 2 K-distributions: A New Algorithm for Clustering Categorical Data

Categorical data as referred to in this paper is the data describing objects which have only categorical attributes, which is identical with the data defined in K-modes [3]. Assume that $D(X_1, X_2, \ldots, X_n)$ is a categorical data set consisting of $n$ categorical objects and $A_1, A_2, \ldots, A_m$ are $m$ categorical attributes of each categorical object $X$, then the categorical object $X$ is represented by a vector $< a_1, a_2, \ldots, a_m >$, where $a_i$ is the value of the attribute $A_i$.

Just as shown before, K-modes suffers from computing the dissimilarity between each pair of objects and the mode of each cluster. This fact raises the question of whether a clustering algorithm without computing the dissimilarity between each pair of objects and the mode of each cluster can perform even better. Responding to this question, we present a new clustering algorithm simply called K-distributions in this paper. Our motivation is to develop a new algorithm to efficiently and effectively cluster categorical data. Our new algorithm can be described as follow.

**Algorithm.** K-distributions(**D**, K)
**Input** : a data set **D** containing $n$ objects, the number of clusters K
**Output** : a set of K clusters
**Method** : the K-distributions algorithm is implemented as follows

1. Partition all objects into K nonempty and mutually exclusive subsets randomly, and each subset is treated as a cluster.
2. For each object $< a_1, a_2, \ldots, a_m >$, compute each cluster's joint probability $P(a_1, a_2, \ldots, a_m)$ and assign this object to the cluster which has maximal joint probability.
3. Repeat 2 until no more new assignment.

Seen from the K-distributions algorithm, we only need to compute each cluster's joint probability $P(a_1, a_2, \ldots, a_m)$ for each object $< a_1, a_2, \ldots, a_m >$. Out of question, estimating the optimal joint probability of $P(a_1, a_2, \ldots, a_m)$ from a set of categorical data is NP-hard problem.

To simplify the computation, we assume that all attributes are fully independent within each cluster. Then the resulting joint probability can be simplified as $\prod_{i=1}^{m} P(a_i)$. Just as we all know, the value of each item $P(a_i)$ can be easily estimated from a data set by calculating the related frequency. We estimates the base probabilities $P(a_i)$ using a special m-estimate as follows:

$$P(a_i) = \frac{F(a_i) + \frac{1}{|A_i|}}{N + 1.0} \tag{1}$$

where $F(a_i)$ is the frequency that $A_i = a_i$ appears in this cluster, $|A_i|$ is the number of values of attribute $A_i$, $N$ is the number of objects in this cluster.

Like the K-means algorithm and the K-modes algorithm, the K-distributions algorithm also produces locally optimal solutions that are dependent on the initial partition.

## 3 Experimental Methodology and Results

We ran our experiments on 36 UCI data sets [4] selected by Weka [5], which represent a wide range of domains and data characteristics listed in Table 1. In our experiments, we adopted the following five preprocessing steps.

**Table 1.** Description of data sets used in the experiments. All these data sets are the whole 36 UCI data sets selected by Weka. We downloaded these data sets in format of *arff* from main web site of weka.

| No. | Dataset | Instances | Attributes | Classes | Missing | Numeric |
|---|---|---|---|---|---|---|
| 1 | anneal | 898 | 39 | 6 | Y | Y |
| 2 | anneal.ORIG | 898 | 39 | 6 | Y | Y |
| 3 | audiology | 226 | 70 | 24 | Y | N |
| 4 | autos | 205 | 26 | 7 | Y | Y |
| 5 | balance-scale | 625 | 5 | 3 | N | Y |
| 6 | breast-cancer | 286 | 10 | 2 | Y | N |
| 7 | breast-w | 699 | 10 | 2 | Y | N |
| 8 | colic | 368 | 23 | 2 | Y | Y |
| 9 | colic.ORIG | 368 | 28 | 2 | Y | Y |
| 10 | credit-a | 690 | 16 | 2 | Y | Y |
| 11 | credit-g | 1000 | 21 | 2 | N | Y |
| 12 | diabetes | 768 | 9 | 2 | N | Y |
| 13 | Glass | 214 | 10 | 7 | N | Y |
| 14 | heart-c | 303 | 14 | 5 | Y | Y |
| 15 | heart-h | 294 | 14 | 5 | Y | Y |
| 16 | heart-statlog | 270 | 14 | 2 | N | Y |
| 17 | hepatitis | 155 | 20 | 2 | Y | Y |
| 18 | hypothyroid | 3772 | 30 | 4 | Y | Y |
| 19 | ionosphere | 351 | 35 | 2 | N | Y |
| 20 | iris | 150 | 5 | 3 | N | Y |
| 21 | kr-vs-kp | 3196 | 37 | 2 | N | N |
| 22 | labor | 57 | 17 | 2 | Y | Y |
| 23 | letter | 20000 | 17 | 26 | N | Y |
| 24 | lymph | 148 | 19 | 4 | N | Y |
| 25 | mushroom | 8124 | 23 | 2 | Y | N |
| 26 | primary-tumor | 339 | 18 | 21 | Y | N |
| 27 | segment | 2310 | 20 | 7 | N | Y |
| 28 | sick | 3772 | 30 | 2 | Y | Y |
| 29 | sonar | 208 | 61 | 2 | N | Y |
| 30 | soybean | 683 | 36 | 19 | Y | N |
| 31 | splice | 3190 | 62 | 3 | N | N |
| 32 | vehicle | 846 | 19 | 4 | N | Y |
| 33 | vote | 435 | 17 | 2 | Y | N |
| 34 | vowel | 990 | 14 | 11 | N | Y |
| 35 | waveform-5000 | 5000 | 41 | 3 | N | Y |
| 36 | zoo | 101 | 18 | 7 | N | Y |

**Table 2.** Experimental results for comparing K-modes and K-distributions in term of clustering accuracy. The symbols v and * denotes statistically significant improvement and degradation respectively over K-modes using two-tailed t-test with a 95% confidence level. The average value and the $w/t/l$ value are summarized at the bottom of the table.

| Datasets | K-modes | K-distributions | Result of T-Test |
|---|---|---|---|
| anneal | 36.86 | 36.41 | * |
| anneal.ORIG | 37.53 | 39.76 | v |
| autos | 48.78 | 36.59 | * |
| balance-scale | 41.92 | 37.28 | * |
| breast-cancer | 73.43 | 71.68 | * |
| breast-w | 96.85 | 97.42 | v |
| colic | 64.67 | 66.03 | v |
| colic.ORIG | 57.61 | 54.08 | * |
| credit-a | 54.93 | 83.91 | v |
| credit-g | 61.5 | 62.8 | v |
| diabetes | 55.6 | 62.89 | v |
| glass | 35.98 | 41.59 | v |
| heart-c | 80.86 | 81.85 | v |
| heart-h | 67.01 | 74.15 | v |
| heart-statlog | 76.3 | 82.59 | v |
| hepatitis | 80.65 | 74.84 | * |
| hypothyroid | 45.55 | 51.67 | v |
| ionosphere | 60.68 | 74.36 | v |
| iris | 49.33 | 72.67 | v |
| kr-vs-kp | 50.97 | 51.16 | v |
| labor | 73.68 | 57.89 | * |
| lymph | 38.51 | 52.7 | v |
| mushroom | 58.32 | 83.7 | v |
| segment | 53.2 | 53.55 | v |
| sick | 56.84 | 75.77 | v |
| sonar | 66.83 | 52.88 | * |
| soybean | 56.52 | 60.76 | v |
| splice | 41.97 | 70.47 | v |
| vehicle | 39.36 | 35.22 | * |
| vote | 87.36 | 87.82 | v |
| vowel | 19.9 | 24.04 | v |
| waveform-5000 | 58.42 | 52.6 | * |
| zoo | 72.28 | 73.27 | v |
| **Mean** | 57.58 | 61.65 | 23/0/10 |

1. Hiding class attribute values: Clustering is a typical unsupervised learning. So, we need to hide class attribute values but use the number of classes as the number of clusters during learning and recur them during evaluation.
2. Ignoring three multi-classes data sets: For saving the time of running experiments, we ignore three data set whose number of clusters are above 20. They are "audiology", "letter", "primary-tumor" in turn.

**Table 3.** Experimental results for comparing K-modes and K-distributions in term of log likelihood. The symbols v and * denotes statistically significant improvement and degradation respectively over K-modes using two-tailed t-test with a 95% confidence level. The average value and the $w/t/l$ value are summarized at the bottom of the table.

| Datasets | K-modes | K-distributions | Result of T-Test |
|---|---|---|---|
| anneal | -14.17 | -13.52 | v |
| anneal.ORIG | -9.58 | -9.48 | v |
| autos | -28.9 | -29.59 | * |
| balance-scale | -6.62 | -6.6 | v |
| breast-cancer | -9.24 | -9.11 | v |
| breast-w | -11.4 | -11.28 | v |
| colic | -23.1 | -22.71 | v |
| colic.ORIG | -26.9 | -26.68 | v |
| credit-a | -13.07 | -13.08 | * |
| credit-g | -21.58 | -21.36 | v |
| diabetes | -12.64 | -12.27 | v |
| glass | -11.2 | -10.68 | v |
| heart-c | -15.06 | -14.82 | v |
| heart-h | -12.43 | -12.11 | v |
| heart-statlog | -15.43 | -15.15 | v |
| hepatitis | -15.49 | -15.27 | v |
| hypothyroid | -10.17 | -9.37 | v |
| ionosphere | -59.56 | -56.44 | v |
| iris | -7.57 | -6.89 | v |
| kr-vs-kp | -14.16 | -13.52 | v |
| labor | -15.58 | -15.81 | * |
| lymph | -13.96 | -13.7 | v |
| mushroom | -19.61 | -18.43 | v |
| segment | -17.98 | -15.09 | v |
| sick | -10.39 | -9.97 | v |
| sonar | -111.75 | -110.47 | v |
| soybean | -15.28 | -14.26 | v |
| splice | -81.54 | -80.8 | v |
| vehicle | -26.57 | -25.62 | v |
| vote | -7.72 | -7.69 | v |
| vowel | -21.75 | -21.15 | v |
| waveform-5000 | -71.35 | -68.89 | v |
| zoo | -6.16 | -6.24 | * |
| **Mean** | -22.97 | -22.37 | 29/0/4 |

3. Replacing missing attribute values: We used the unsupervised filter named *ReplaceMissingValues* in Weka to replace all missing attribute values in each data set, because we don't handle missing attribute values.
4. Discretizing numeric attribute values: We used the unsupervised filter named *Discretize* in Weka to discretize all numeric attribute values in each data set, because we don't handle numeric attribute values.

5. Removing useless attributes: Apparently, if the number of values of an attribute is almost equal to the number of instances in a data set, it is a useless attribute. Thus, we used the unsupervised filter named *Remove* in Weka to remove this type of attributes. In these 36 data sets, there are only three such attributes: the attribute "Hospital Number" in the data set "colic.ORIG", the attribute "instance name" in the data set "splice" and the attribute "animal" in the data set "zoo".

We conducted our experiments to compare K-modes and K-distributions in terms of clustering accuracy and log likelihood [6,7,8]. We implemented these two algorithms within the Weka system [5]. In all experiments, each algorithm's clustering accuracy and log likelihood on each data set was obtained via 10 repeated runs. Finally, we conducted two-tailed $t$-test with a 95% confidence level [9] to compare K-modes and K-distributions.

Table 2 and Table 3 respectively shows each algorithm's clustering accuracy and log likelihood on each data set, and the symbols v and * in the table denotes statistically significant improvement and degradation respectively over K-modes. The average value and the $w/t/l$ value (wins in $w$ data sets, ties in $t$ data sets, and loses in $l$ data sets) are summarized at the bottom of the tables. The experimental results show that K-distributions significantly outperforms K-modes. Now, we summarize the highlights as follows:

1. In term of clustering accuracy, K-distributions significantly outperforms K-modes. Compared to K-modes, in the 33 data sets we test, K-distributions wins in 23 data sets, and only loses in 10 data sets. In addition, the average accuracy of K-distributions is 61.65, much higher than K-modes' 57.58.
2. In term of log likelihood, K-distributions also significantly outperforms K-modes. Compared to K-modes, in the 33 data sets we test, K-distributions wins in 29 data sets, and only loses in 4 data sets. In addition, the average accuracy of K-distributions is -22.37, much higher than K-modes' -22.97.

## 4 Conclusions and Future Work

K-modes is a popular algorithm for clustering categorical data sets in data mining. However, it suffers from computing the dissimilarity between each pair of objects and the mode of each cluster. In this paper, we present another new clustering algorithm simply called K-distributions. Our motivation is to develop a new algorithm to efficiently and effectively cluster categorical data without the troubles confronting K-modes. The experimental results show that K-distributions significantly outperforms K-modes in term of clustering accuracy and log likelihood.

In K-distributions, how to estimate joint probability $P(a_1, a_2, \ldots, a_m)$ is crucial. Currently, we assume that all attributes are fully independent within each cluster. So, the resulting joint probability can be simplified as $\prod_{i=1}^{m} P(a_i)$. We believe that relaxing this unrealistic assumption could further improve the performance of the current K-distributions algorithm and make its advantage stronger. This is one of our main directions for future research.

## References

1. Jain, A. K., Murty, M. N., Flynn, P. J.: Data Clustering: A Review. ACM Computing Surveys (CSUR). 31 (1999) 264-323
2. MacQueen, J. B.: Some Methods for Classification and Analysis of Multivariate Observations. In: Proceedings of the 5th Berkeley Symposium on Mathematical Statistics and Probability. Univ.of California, Berkeley, USA: Statistics and Probability (1967) 281-297
3. Huang, Z.: A fast Clustering Algorithm to Cluster Very Large Categorical Data Sets in Data Mining. In: Proceedings of the SIGMOD Workshop on Research Issues on Data Mining and Knowledge Discovery. Tucson, Arizona, USA (1997) 146-151
4. Merz, C., Murphy, P., Aha, D.: UCI Repository of Machine Learning Databases. In Dept of ICS, University of California, Irvine (1997) http://www.ics.uci.edu/ mlearn/MLRepository.html
5. Witten, I. H., Frank, E.: Data Mining: Practical Machine Mearning Tools and Techniques. 2nd Edition, Morgan Kaufmann, San Francisco (2005) http://prdownloads.sourceforge.net/weka/datasets-UCI.jar
6. Friedman, N., Geiger, D., Goldszmidt, M.: Bayesian Network Classifiers. Machine Learning. 29 (1997) 131-163
7. Grossman, D., Domingos, P.: Learning Bayesian Network Classifiers by Maximizing Conditional Likelihood. In: Proceedings of the Twenty-First International Conference on Machine Learning. Banff, Canada. ACM Press (2004) 361-368
8. Guo, Y., Greiner, R.: Discriminative Model Selection for Belief Net Structures. In: Proceedings of the Twentieth National Conference on Artificial Intelligence. AAAI Press (2005) 770-776
9. Nadeau, C., Bengio, Y.: Inference for the Generalization Error. In: Advances in Neural Information Processing Systems. MIT Press, 12 (1999) 307-313

# Key Point Based Data Analysis Technique

Su Yang* and Yong Zhang

Department of Computer Science and Engineering,
Fudan University, Shanghai 200433, P.R. China
suyang@fudan.edu.cn

**Abstract.** In this paper, a new framework for data analysis based on the "key points" in data distribution is proposed. Here, the key points contain three types of data points: bridge points, border points, and skeleton points, where our main contribution is the bridge points. For each type of key points, we have developed the corresponding detection algorithm and tested its effectiveness with several synthetic data sets. Meanwhile, we further developed a new hierarchical clustering algorithm SPHC (Skeleton Point based Hierarchical Clustering) to demonstrate the possible applications of the key points acquired. Based on some real-world data sets, we experimentally show that SPHC performs better compared with several classical clustering algorithms including Complete-Link Hierarchical Clustering, Single-Link Hierarchical Clustering, KMeans, Ncut, and DBSCAN.

## 1 Introduction

The rapid development of information technologies over the past few decades has led to continual collection and fast accumulation of data in repositories [6]. However, data is not equivalent to information (or knowledge) [2]. Data analysis plays an important role in data mining applications [2].

The aim of data analysis lies in knowledge discovery, which is a non-trivial process [2]. For this purpose, many techniques such as classification, clustering, association rule mining, and outlier analysis have been developed in data mining field [6].

If the underlying technique is ignored, data analysis approaches can be divided into three categories: classical analysis, Bayesian analysis, and exploratory analysis [1]. The difference is the sequence and focus of intermediate steps (Fig. 1).

Different from the three data analysis approaches discussed above, in this paper, we propose a new framework for data analysis based on the "key points" in the data distribution. We refer to it as KPDA (*Key Point based Data Analysis*). For KPDA, we do not require model imposition. The conclusions (or knowledge) can be revealed by the "key points" directly or further analysis performed over the acquired "key points". Note that KPDA is based on the observation that "key points" are more useful than model in revealing knowledge sometimes. Take border points for example. This set of

* Corresponding author.

D.-S. Huang, L. Heutte, and M. Loog (Eds.): ICIC 2007, LNAI 4682, pp. 444–455, 2007.

points may denote a subset of population that should have developed certain diseases. Special attention is certainly warranted for this set of people since they may reveal some interesting characteristics of the disease [6].

**Classical analysis:**
Problem→Data→Model→Analysis→Conclusion
**Bayesian analysis:**
Problem→Data→Model→Prior Distribution→Analysis→Conclusion
**Exploratory analysis:**
Problem→Data→Analysis→Model→Conclusion

**Fig. 1.** Different processes of three popular data analysis approaches: classical analysis, Bayesian analysis, and exploratory analysis

In this paper, we mainly concern with three types of "key points": bridge points, border points, and skeleton points. Accordingly, we propose three algorithms BPF (*Bridge Point Filter*), BPD (*Border Point Detection*), and SPE (*Skeleton Point Extraction*) to detect the corresponding "key points", respectively. In addition, we further develop a novel hierarchical clustering algorithm SPHC (*Skeleton Point based Hierarchical Clustering*) to test the effectiveness of the key points acquired. The main contribution of this paper is the introduction of bridge point as well as the corresponding detection algorithm BPF. To the best of our knowledge, BPF is the first work.

The remainder of the paper is organized as follows: Section 2 presents different "key points" and corresponding detection algorithms. Section 3 describes the application of the "key points" to clustering analysis (SPHC). Section 4 presents the experiment results. Finally, Section 5 concludes the whole paper.

## 2 Key Points

### 2.1 Preliminary

Throughout this paper, we use $p$, $q$, and $r$ to denote data points in a data set. We use the notation $d(p,q)$ to denote the distance (Euclidean distance if no further explanation) between point $p$ and $q$.

Since bridge point and border point are both detected based on the neighborhood of a data point, we must select an appropriate neighborhood diagram first. There exist many kinds of neighborhood diagrams, among which kNN diagram, ε-diagram, and Delaunay diagram [8] are used frequently in related works.

Compared with kNN diagram and ε-diagram, the key advantage of Delaunay diagram lies in that it is parameter-free. In contrast, its drawback is also apparent: Although the algorithm is efficient for 2 or 3-dimensional data sets, it becomes inefficient rapidly for large-scale data sets when the dimensionality $n$ is higher than 4 due to the high time complexity $O(m^{n/2})$, where $m$ is the number of data points. On the other hand, the time complexity of constructing kNN diagram or ε-diagram is not sensitive to the dimensionality but the specification of $k$ or ε parameter may be difficult sometimes.

In this paper, we adopt Delaunay diagram for very low dimensional situations (e.g., $n \le 3$) and kNN or ε-diagram for other circumstances. For the sake of simplicity, we just use kNN diagram to describe the algorithms although different diagrams can be adopted according to the dimensionality.

### 2.2 Bridge Point Filter

For supervised learning like classification, the data points at the boundary of two or more classes do affect the final decision result, since these data points are always error-prone. Many techniques have been developed to process even remove these data points so as to achieve better results [3]. For unsupervised learning like clustering, these data points also affect the final clustering result. In this paper, we just refer to these points of interest as bridge points; the formal definition is as follows:

**Definition 1** (Bridge Point):
A bridge point $p$ is a data point that is at the boundary between two or more (potential, for unsupervised learning) classes.

To the best of our knowledge, there exists no formal definition of bridge point yet. Note that the above definition is also an abstract description, which needs further concretions in different algorithms.

In following, we present the corresponding algorithm for detecting bridge points, and we refer to it as BPF (*Bridge Point Filter*). Here, note that BPF is based on the following observation: If we build a local neighborhood diagram over all the data points, the shortest paths connecting every pair of data points should pass through the bridge points more times than other data points.

**Algorithm 1.** Bridge Point Filter
**Input:**
The data set $S = \{x_1, x_2, \cdots, x_m\}$, where $x_i$, $1 \le i \le m$, denotes an $n$-dimensional column vector
The kNN parameter $K$
The tuning parameter $\lambda$
**Output:**
The acquired bridge point set $BPS$
**Steps:**
Step 1 Build kNN neighborhood diagram $KD$ over the data set $S$
Step 2 Set $BPS = \varnothing$, $CPN[i]=0$, $1 \le i \le m$, where $CPN[i]$ denotes the number of such paths that pass through point $x_i$
Step 3 Apply Floyd algorithm to find all the shortest paths connecting every data point pair and save the result as $P = \{P_{ij}\}$, where $P_{ij}$ denotes the shortest path between point $x_i$ and $x_j$, which can be regarded as a point sequence $x_i x_{k1} x_{k2} \cdots x_j$

Step 4 For every path $P_{ij}$ in $P$ , do:
For every intermediate point $x_k$ in $P_{ij}$ , do:
$CPN[k] = CPN[k] + 1$
Step 5 Compute $\overline{CPN} = (\sum_{i=1}^{m} CPN[i]) / m$
Step 6 For every data point $x_i$ in *S*, do
If $CPN[i] > \lambda * \overline{CPN}$ , then add $x_i$ into *BPS* otherwise continue
Step 7 Return *BPS*

Our previous definition is not applicable to such data set that only contains a single class. If we apply BPF to the data set containing only the points from just one class, intuitively, the data points deep in the cluster are more likely to be labeled as bridge points. The experimental result shown in Fig. 2 (a) confirms our expectation well. Meanwhile, Fig. 2 (b-d) shows the detection results for applying BPF to two data sets containing two or three classes respectively.

(a) (b) (c) (d)

**Fig. 2.** The detection results of BPF on four synthetic data sets

Here, three issues should be noticed regarding the above algorithm. First, we detect bridge points on the basis of intuitive observation. Although good results are obtained, we still believe that a more thorough study of the algorithm from mathematic viewpoint is necessary. We leave it for future study. Second, the time complexity of BPF algorithm will be O($m^3$) due to the computation of all the shortest paths using Floyd algorithm. However, there exist many techniques to reduce the complexity to O($m^2\log m$) [4]. Besides, since we only build edges between neighboring points, the computational cost can thus be reduced further. Third, the neighborhood diagram construction requires the specification of the kNN parameter, which is sometimes difficult, especially for some data sets without any prior knowledge.

## 2.3 Border Point Detection

Usually, border points are data points that are at the margin of densely distributed data such as a cluster. They are useful in many fields like data mining, image processing, pattern recognition, etc.

As an active research direction, border point detection has been drawing much attention from different researchers. In image processing field, there exist various algorithms for border point detection [5]. In addition, there are also many techniques [6-8] developed to detect general border points. For example, in [6], Chenyi Xia et al develop a method called BORDER that utilizes the special property of the reverse k nearest neighbor (RkNN) and employs the state-of-the-art database technique – the Gorder kNN join to find boundary points in a data set. In [8], the authors utilize the Delaunay diagram to detect boundary points of clusters.

In our opinion, [7] captures the typical characteristics of border points, "Border points are not surrounded by other points in all directions while the interior points are".Different from [7], in this paper, we interpret this observation from a novel viewpoint. For a interior data point, surrounded by its neighboring points in nearly all directions usually means homogeneousness. On the other hand, the distribution of the neighboring points of a border point is usually biased. In other word, we can detect border points through homogeneousness measurement. Here, the key problem lies in the measurement of homogeneousness for the neighborhood of a data point. Intuitively, more homogeneous distribution means higher symmetry degree and vice versa. Thus, homogeneousness measurement can be achieved with the help of symmetry degree measurement. To measure the symmetry (or asymmetry) degree of a given data set, a simple method is to compare the original data set and its symmetric image just like in [9].

Based on the above discussion, we present the detailed algorithm, BPD(*Border Point Detection*), as follows:

**Algorithm 2.** Border Point Detection

**Input:**

The data set $S = \{x_1, x_2, \cdots, x_m\}$, where $x_i$, 1≤*i*≤m, is an *n*-dimensional column vector

The kNN parameter *K*

The tuning parameter $\lambda$

**Output:**

The acquired border point set *BPS*

**Steps:**

Step 1 Build kNN neighborhood diagram *KD* over the data set *S*

Step 2 Set *BPS* = ∅ ,*AD*[i]=0, 1≤*i*≤m , where *AD*[i] denotes the asymmetry degree of the neighborhood of point $x_i$

Step 3 For every data point $x_i$ in *S*, do:

Step 3.1 Determine the kNN neighborhood $N_k(x_i)$ of $x_i$

Step 3.2 For every point *p* in $N_k(x_i)$, do:

Compute $$d(p, x_i, N_k(x_i)) = \min_{q \in N_k(x_i)} d(p^*, q)$$

where $p^*$ is the image point of *p* with respect to point $x_i$ and set $AD[i] = AD[i] + d(p, x_i, N_k(x_i))$

Step 4 Compute $$\overline{AD} = (\sum_{i=1}^{m} AD[i]) / m$$

Step 5 For every data point $x_i$ in *S*, do

If $AD[i] > \lambda * \overline{AD}$, then add $x_i$ into *BPS* otherwise continue

Step 6 Return *BPS*

The detection results are illustrated in Fig. 3.

(a) (b) (c) (d)

**Fig. 3.** Detection results of BPD on four synthetic data sets

The time complexity of BPD is $O(mk^2)$ due to the computation of the asymmetry degree for every data point in *S*. In most circumstances, *k* value is far smaller than *m*. Hence, the computational cost of BPD is linear with the number of the data points, which is tractable even for some large-scale data sets.

### 2.4 Skeleton Point Extraction

Skeleton points, also called representative points, are often used to represent the underlying structure of the original data set. They can find applications in data compression, data clustering, pattern classification, and statistical parameter estimation. In the literature of pattern recognition and statistical analysis, there exist many approaches regarding skeleton point extraction [10-14].

For the integrity of key point based data analysis, skeleton points are also indispensable parts. As we know, if a data set is hyper spherical in shape, then the center of the data set can represent the whole data set well. On the other hand, any elongated or nonconvex data set can be considered as the union of a few distinct hyper spherical clusters. Based on this consideration, in this paper, we intend to pack the whole data set with different spheres. Then, the centers of all the spheres constitute the skeleton point set. In order to determine the number and radiuses of such spheres, it is essential to find out the border points at first.

Similar to [7], we also use border points to detect the shape of a cluster and hence determine the number of spheres required. However, we adopt BPD as the underlying algorithm to detect border points.

The detailed algorithm, SPE(*Skeleton Point Extraction*), is presented as follows:

**Algorithm 3.** Skeleton Point Extraction

**Input:**

The data set $S = \{x_1, x_2, \cdots, x_m\}$, where $x_i$, $1 \le i \le m$, is an *n*-dimensional column vector

The threshold *tn* and *td*

**Output:**

The acquired skeleton point set *SPS*

**Steps:**

Step 1 Initialize current sample set *curS* = S, *SPS* = $\varnothing$

Step 2 Apply Parsen window method to estimate the probability density for every data point in *curS*

Step 3 Apply BPD to detect the border point set $B$ of *curS*

Step 4 Find the point with the highest probability density estimated, say $p$, and add p into *SPS*

Step 5 Compute

$$maxb = \max_{q\in B}\|q-p\| \; minb = \min_{q\in B}\|q-p\|, \; fb = maxb - minb$$

Step 6 If $fb \le td$ , go to Step 8, else go to Step 7

Step 7 Remove the point $q$ in *curS* satisfying the following condition: $q \in S_0, S_0 = \{q \mid \|q-p\| \le minb\}$. If $|curS| - |S_0| < tn$, go to Step 8. Else, go to Step 3

Step 8 Return *SPS*

For the above algorithm, we set the data points with the local highest probability density values estimated as the centers of the spheres required (Step 2, Step 4). Meanwhile, we determine the number of spheres required and the corresponding radii of these spheres through the border points detected (Step 5, Step 7).

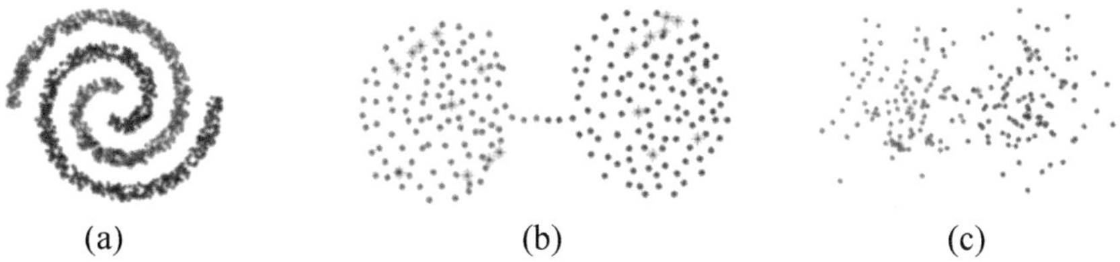


**Fig. 4.** Detection results of SPE on the three synthetic data sets

Fig. 4 illustrates the detection results, which demonstrates the effectiveness of SPE. In addition, there are two issues regarding SPE that should be noticed. The time complexity of SPE is approximately O($m$), which is tractable even for some large-scale data sets.

## 3 Application to Clustering Analysis

As mentioned earlier, the key points (bridge point, border point, and skeleton point) can reveal knowledge about the underlying data set directly or be used as intermediate steps for further analysis. The key points can find applications in various fields like data classification, clustering, outlier detection, etc. In this section, we develop a new hierarchical clustering algorithm, SPHC (*Skeleton Point based Hierarchical Clustering*), to illustrate the potential application of the key points acquired.

The basic idea of SPHC is very simple. We intend to perform traditional hierarchical clustering algorithm like Complete-Link hierarchical clustering algorithm over the skeleton points extracted from the data set instead of the original data set so as to obtain clearer cluster boundaries and reduce the computational cost. The remaind data points are assigned to the skeleton point by nearest neighbor rule.

**Algorithm 4.** Skeleton Point based Hierarchical Clustering
**Input:**
The data set $S = \{x_1, x_2, \cdots, x_m\}$, where $x_i$, $1 \le i \le m$, is an $n$-dimensional column vector
The required class number $K$
**Output:**
The labels for every point in $S$
**Steps:**
Step 1 Apply BPF algorithm to remove the bridge points and obtain the modified data set *ms*
Step 2 Apply SPE algorithm to obtain the skeleton point set *SK* from data set *ms*
Step 3 Perform Complete-Link hierarchical clustering over *SK*, and form $K$ clusters
Step 4 For every data point $p$ in $S$, do:
Find $sk_0$ satisfying $\|p - sk_0\| = \min_{q \in SK} \|q - p\|$ and then set the label of $sk_0$ as the label of $p$, i.e., label($p$) = label($sk_0$)
Step 5 Return the labels for every point in $S$

There are two issues should be noticed about the above algorithm. First, we must specify several parameters (like the kNN parameter, the tuning parameter, etc) for SPHC due to its underlying BPF, SPE, and BPD algorithms. However, we design SPHC just as an example to demonstrate the application of the key points acquired. More work should be done to automate the determination of the required parameters if we want to make it a practical algorithm. Second, the time complexity of SPHC will be $O(m^3)$ due to the detection of bridge points. For some large-scale data sets, SPHC will become intractable. However, if we do not use BPF as the preprocessing stage to filter the bridge points, the time complexity will be reduced to $O(m)$.

# 4 Experiment

## 4.1 Data Sets and Evaluation Criterion

In order to present the results of key points detection visually, we mainly tested BPF, BPD, and SPE over several 2-dimensional synthetic data sets.

- Data set Ⅰ. A single class contains 167 data points.
- Data set Ⅱ. Two densely distributed clusters, which are connected by a narrow bridge, where each cluster contains 115 data points.
- Data set Ⅲ. Two clusters (Gaussian distribution) which have partial points overlapped, where each cluster contains 100 data points.

- Data set IV. Three densely distributed clusters (685 data points) with some outliers (74 data points).
- Data set V. A two-spiral structure which contains 1500 data points.

As for SPHC algorithm, we also tested its effectiveness on several real-world data sets in addition to the above synthetic data sets. All the real-world data sets were obtained from the UCI repository [15]. Table. 1 summarizes the properties of these data sets: The number of instances, the number of dimensions (attributes), and the number of classes.

**Table 1.** The properties of the real-world data sets

| Data sets | #Instances | #Attributes | #Classes |
|---|---|---|---|
| Iris | 150 | 4 | 3 |
| Balance-scale | 625 | 4 | 3 |
| Wdbc | 569 | 30 | 2 |
| Wpbc | 194 | 33 | 2 |
| Glass | 214 | 9 | 6 |
| House | 506 | 13 | 5 |
| Iono | 351 | 34 | 2 |
| Pima | 768 | 8 | 2 |

In addition, Rand Index [16] was adopted to evaluate the performance of different clustering algorithms. Let $n_s$, $n_d$ be the number of point pair that are assigned to the same/different cluster(s) in both partitions respectively. The Rand Index is defined as the ratio of ($n_s$+ $n_d$) to the total number of point pairs, $m(m\text{-}1)/2$, where $m$ denotes the number of data points in the given data set. The Rand Index lies between 0 and 1, and when the two partitions are consistent completely, the Rand Index will be 1.

### 4.2 Evaluation of BPF, BPD and SPE

The corresponding detection results of BPF, BPD, and SPE over the synthetic data sets are presented in Fig. 2, Fig. 3, and Fig. 4, respectively.

For different data sets, we utilize different neighborhood diagrams. The details can be found in Table.2. For ε-diagram, we set ε value as the average of the minimum and maximum pair-wise distance of the given data set. As mentioned earlier, SPE algorithm also utilizes BPD algorithm to detect the border points. Here, we apply Delaunay diagram and set the tuning parameter $\lambda$ as 1.5 uniformly. For Parsen window method, we set the required parameter h1 as the average of the minimum and maximum pair-wise distance of the given data set.

### 4.3 Evaluation of SPHC

For all the data sets (synthetic and real-world), we uniformly set the required parameters as follows: For underlying BPF algorithm, we set the kNN parameter $K$ =

**Table 2.** The parameter settings for the three detection algorithms

| Data sets | BPF | | BPD | | SPE | |
|---|---|---|---|---|---|---|
| | diagram | $\lambda$ | diagram | $\lambda$ | *tn* | *td* |
| Data set Ⅰ | Delaunay | 1.5 | ε-diagram | 1.4 | | |
| Data set Ⅱ | kNN(k=5) | 2.5 | Delaunay | 1.0 | 5 | 2 |
| Data set Ⅲ | Delaunay | 2.0 | ε-diagram | 1.4 | 5 | 0.1 |
| Data set □ | kNN(k=189) | 2.0 | | | | |
| Data set □ | | | kNN(k=10) | 1.0 | 5 | 0.1 |

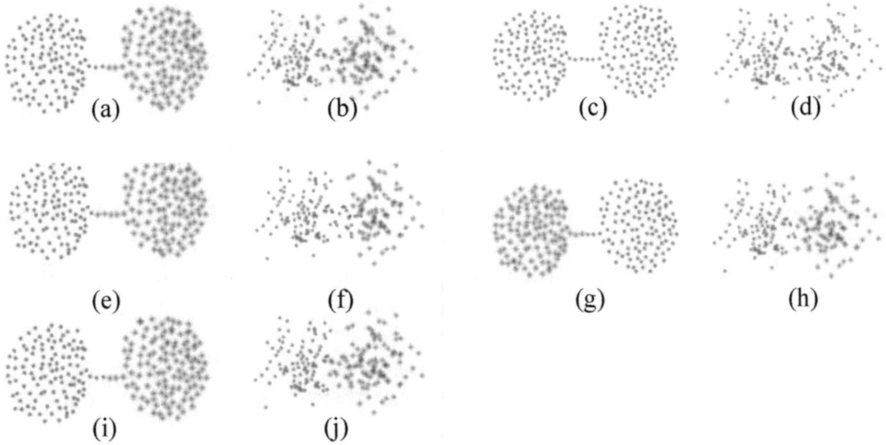


**Fig. 5.** The clustering results of different algorithms over the two synthetic data sets, where (a-b): original data set distribution, (c-d): SLHC, (e-f): CLHC, (g-h): KMeans, (i-j): SPHC

10, the tuning parameter $\lambda$= 2.0; for underlying SPE algorithm, we set the two thresholds *tn* = 10, *td* = 1.5*the minimum pair-wise distance of the given data set. Besides, during the extracting process, we also need to detect the border points of current sample set continuously. Here, for BPD algorithm, we apply kNN diagram. We set the kNN parameter as 0.05*the number of samples in current sample set and set the tuning parameter $\lambda$ = 1.0. For Parsen window method, the required parameter h1 is set to 2.0 for all the data sets.

The results indicate that SPHC achieves better clustering results for most data sets compared with other traditional clustering techniques. Although it seems that the clustering results of SLHC, KMeans, and SPHC in Fig.5 do not differ much, SPHC achieves much better results compared with traditional Complete-Link Hierarchical Clustering algorithm. Meanwhile, for the real-world data sets, SPHC also performs better than SLHC and KMeans in most cases. In this sense, the effectiveness of the key points extracted is confirmed.

**Table 3.** The clustering results of SPHC compared with Complete-Link Hierarchical Clustering (CLHC), Single-Link Hierarchical Clustering (SLHC), KMeans, Ncut, and DBSCAN algorithms over 8 real-world data sets

| Data sets | CLHC | SLHC | KMeans | Ncut | DBSCAN | SPHC |
|---|---|---|---|---|---|---|
| Iris | 0.8368 | 0.7766 | 0.8597 | 0.8115 | 0.7763 | **0.8859** |
| Balance-scale | **0.6039** | 0.4329 | 0.5977 | 0.5837 | 0.4299 | 0.5911 |
| Wdbc | 0.5521 | 0.5326 | 0.7004 | 0.7479 | 0.5317 | **0.7605** |
| Wpbc | 0.5335 | **0.6418** | 0.5335 | 0.5705 | 0.6363 | 0.5745 |
| Glass | 0.5822 | 0.2970 | 0.6064 | 0.5867 | 0.5871 | **0.6350** |
| House | 0.5906 | 0.5108 | 0.5364 | 0.5376 | 0.5500 | **0.5929** |
| Iono | 0.5684 | 0.5401 | 0.5089 | **0.6232** | 0.5385 | 0.5706 |
| Pima | 0.5443 | 0.5458 | 0.4507 | **0.6219** | 0.5419 | 0.5443 |

## 5 Conclusion

In this paper, we introduce a new data analysis framework KPDA based on the key points in the data set, where the key points are referred to as bridge points, border points, and skeleton points. For each type of key points, we propose the corresponding detection algorithms, respectively. The detection results with several synthetic data sets demonstrate their effectiveness.

In order to illustrate the possible application of the key points acquired, we further develop a new hierarchical clustering algorithm SPHC based on the key points. The comparison results with some other traditional algorithms indicate that SPHC usually performs better than the others.

There are some limitations that should be noticed. First, the time complexity of BPF is $O(m^3)$, where $m$ is the number of the data points in the given data set. This is not tractable for some large-scale data sets. Second, we must specify the required parameters for every algorithm proposed in this paper. This may be difficult for common users. These will be the future possible research directions.

For synthetic data sets, we compared the proposed SPHC algorithm with Complete-Link Hierarchical Clustering (CLHC), Single-Link Hierarchical Clustering (SLHC), and KMeans algorithms. For real-world data sets, we also compared SPHC with two other algorithms Ncut [17] and DBSCAN [18]. We set the parameter *MinPts* of DBSCAN to 10 and $\varepsilon$ as default.

Fig. 5 shows the clustering results of different algorithms on the two synthetic data sets, and Table. 3 summarize the results over the real-world data sets.

## Acknowledgement

This work is supported in part by Natural Science Foundation of China under grant 60305002 and China/Ireland Science and Technology Research Collaboration Fund under grant CI-2004-09.

## References


1. NIST/SEMATECH e-Handbook of Statistical Methods, http://www.itl.nist.gov/div89/handbook/ (2006)
2. Han, J.W., Kamber, M.: Data Mining: Concepts and Techniques. Beijing: China Machine Press, (2003)
3. Wilson, D. R., Martinez, T. R.: Instance Pruning Techniques. Proceedings of the 14th International Conference on Machine Learning, San Francisco, CA, USA, Morgan Kaufmann Publishers Inc,(1997) 403-411
4. Moat, A., Takaoka, T.: An All Paris Shortest Path Algorithm with Expected Time $O(n^2 logn)$. SIAM Journal on Computing, Vol. 16, No. 6, (1987)1023-1031
5. Gonzalez, R. C., Woods, R. E.: Digital Image Processing, Second Edition. Publishing House of Electronics Industry, Beijing (2003)
6. Xia, C. Y., Hsu, W., Lee, M. L., BORDER, B. C.: Efficient Computation of Boundary Points. IEEE Transactions on Knowledge and Data Engineering, Vol. 18, No. 3, (2006)289-303
7. Chaudhuri, D., Chaudhuri. B. B.: A Novel Nonhierarchical Data Clustering Technique. IEEE Transactions on Systems, Man, and Cybernetics—part B: Cybernetics, Vol. 27, No. 5, (1997)871-877
8. Estivill-Castro, V., Lee, I. AutoClust: Automatic Clustering via Boundary Extraction for Massive Point-Data Sets. In Proceedings of the 5th International Conference on Geocomputation (2000)
9. Colliot, O., Tuzikov, A. V., Cesar, R. M., Bloch, I.: Approximate Reflectional Symmetries of Fuzzy Objects with An Application in Model-based Object Recognition. Fuzzy Sets and Systems 147: (2004)141-163
10. Chaudhuri, D., Murthy, C. A., Chaudhuri, B. B.: Finding a Subset of Representative Points in a Data Set. IEEE Transactions on Systems, Man, and Cybernetics, Vol. 24, No. 9, (1994)1416-1424
11. Mitra, P., Murthy, C. A., Pal, S. K.: Density-Based Multiscale Data Condensation. IEEE Transactions on Pattern Analysis and Machine Intelligence, Vol. 24, No. 6, (2002)734-747
12. Ansari, N., Huang, K. W.: Non-Parametric Dominant Point Detection. SPIE Vol. 1606 Visual Communications and Image Processing: Image Processing, (1991)31-42
13. Yao, Y. H., Chen, L. H., Chen, Y. Q.: Using Cluster Skeleton as Prototype for Data Labeling. IEEE Transactions on Systems, Man, and Cybernetics—part B: Cybernetics, Vol. 30, No. 6, (2000)895-904
14. Choi, W. P., Lam, K. M., Siu, W. C.: Extraction of the Euclidean Skeleton Based on a Connectivity Criterion. Pattern Recognition, 36: (2003)721-729
15. Blake, L., Merz, J.: UCI Repository of Machine Learning Databases. http://www.ics.uci.edu/~mlearn/MLRepository.html(1998)
16. Xu, R., Wunsch, D.: Survey of Clustering Algorithms. IEEE Transaction on Neural Networks, Vol. 16, No. 3, (2005)645-678
17. Shi, J.B., Malik, J.: Normalized Cuts and Image Segmentation. IEEE Transactions on Pattern Analysis and Machine Intelligence. Vol. 22, No. 8, (2002)888-905
18. Ester, M., Kriegel, H.P., Sander, J., Xu, X.W.: A Density–Based Algorithm for Discovering Clusters in Large Spatial Databases with Noise. International Conference on Knowledge Discovery and Data Mining, (1996)226-231

# Mining Customer Change Model Based on Swarm Intelligence

Peng Jin[1,2] and Yunlong Zhu[1]

[1] Shenyang Institute of Automation of the Chinese Academy of Sciences, Shenyang, 110016, China
[2] Graduate School of the Chinese Academy of Sciences Beijing, 100039, China
{jinpeng,ylzhu}@sia.cn

**Abstract.** Understanding and adapting to changes of customer behavior is an important aspect of surviving in a continuously changing market environment for a modern company. The concept of customer change model mining is introduced and its process is analyzed in this paper. A customer change model mining method based on swarm intelligence is presented, and the strategies of pheromone updating and items searching are given. Finally, an examination on two customer datasets of a telecom company illuminates that this method can achieve customer change model efficiently.

**Keywords:** Data Mining, Customer Change Mode, Swarm Intelligence, Rule Change Mining.

## 1 Introduction

With the development of new business models such as e-business, the market environments become more and more complex, and the demands of customers are changing all the time. Understanding and adapting to changes of customer behaviors is an important aspect of surviving in a continuously changing environment. For a modern company, knowing what is changing and how it has been changed is of crucial importance because it allows businesses to provide the right products and services to suit the changing market needs [1]. For examples, most decision makers in many companies need to know the answers to following questions: Which customer group's sales are gradually increasing? Which customer group's favorite products or services have been changed? What has been changed about customer behavior and how it has been happened? The answers can be found out through customer change model mining.

Swarm intelligence is a general designation of algorithms or distributed problem-solving devices inspired by the collective behavior of social insect colonies and other animal societies. Individuals with simple structure compose the swarm, and they interact directly or indirectly by some simple rules. The complex collective behaviors of the swarm can emerge out of simple rules [5]. A single customer record is similar

D.-S. Huang, L. Heutte, and M. Loog (Eds.): ICIC 2007, LNAI 4682, pp. 456–464, 2007.

with an individual of a swarm. It has simple structure, and can't provide significant customer model. But the customer model acquire from many approximate customer records can reflect the common characteristics of this customer cluster. On the other hand, Data mining can discover implicit and valuable knowledge and rules with automatic or semiautomatic methods. The method based on swarm intelligence and data mining is adopted to analyze customer change model.

IF-THEN rules are widely used in expression of customer models. The results of association rules analysis, classification and predication, and clustering analysis can be described with rules. So it is significative to analyze and mining the change of rules. The existing researches have focus on the change of rule structures, but can not find the change of samples covered by the rule. It is not enough for customer analysis because it is need to know that where the customers come from and where they go in the changed rules. The difficulties in rule change mining are: 1) the rule structures are not all the same and can not be compared directly; 2) how to estimate what kinds of change and how many changes have occurred, and the reason of changes. In this paper, each customer data is considered as an agent, and the customer change model mining method based on swarm intelligence is adopted to search and match rules in two rule sets. The change of rules and the characteristics of corresponding customer clusters can be found. This method discovers the change of rules not from the aspect of rule structures but the change of customers, so it can support the decision making more effectively.

The rest of this paper is organized as follows. Section 2 introduces the definition of customer change mode and existing researches, and illuminates the parameters and symbols used in this paper. Section 3 presents the customer change model mining method based on swarm intelligence. Section 4 reports an experiment to illuminate the performance of this method. Finally, Section 5 concludes the paper and points expectation for future research.

## 2 Research on Customer Change Model

Customer model, namely customer consumption model or customer behavior model, describes the characteristics of corresponding customer cluster. With the continuously changing of market environment, the demands and behaviors of customers also change at times. So the concept of customer change model is introduced. It is defined as the kind and degree of change in customer model and the reason of change. The task of customer change model mining is to support decision making.

Mining customer change model can use the methods of rules change analysis. Existing researches on comparing or analyzing different datasets or rule sets are clustered as seven categories as follows [2].

1. Rule maintenance. The purpose of these studies is improving accuracy in changing environment, but these techniques do not provide any changes for the user, they just maintain existing knowledge.

2. Emerging patterns discovering. Emerging patterns can capture emerging trends in time stamped databases, or useful contrasts between data classes, but they do not consider the structural changes in the rules.

3. Unexpected rules mining. This technique can not be used for detecting changes, as its analysis only compares each newly generated rule with each existing rule to find degrees of difference, and it does not find which aspects have changed, what kinds of changes have taken place and how much change has occurred.

4. Mining from time series data. These studies focus on the detection of regularity rather than irregularity from data.

5. Mining class comparisons. These techniques can only detect change about the same structured rule.

6. Change mining of decision trees. This technique can not detect complete sets of change or provide any information for the degree of change.

7. Rules change mining. These techniques focus on the change of rule structure, but can not find the change of samples covered by the rule.

Solving the problems existing in these researches, the customer change model mining method based on swarm intelligence is adopted in this paper. This method considers the aspect of customer switching, and discovers that where the customers come from and where they go in the changed rules, how many changes have been occurred and the reason of customer changes. The results of customer change model mining can help company to make appropriate market strategies.

The parameters and symbols used in this paper are illuminated as follows.

$R^t$ customer model set for time $t$;

$R^{t+k}$ customer model set for time $t+k$;

$r_i^t$ a customer model in $R^t$, $r_i^t \in R^t$;

$r_j^{t+k}$ a customer model in $R^{t+k}$, $r_j^{t+k} \in R^{t+k}$;

$|M_i^t|$ the number of attributes in the conditional parts for $r_i^t$;

$|M_j^{t+k}|$ the number of attributes in the conditional parts for $r_j^{t+k}$;

$|N_i^t|$ the number of attributes in the consequent parts for $r_i^t$;

$|N_j^{t+k}|$ the number of attributes in the consequent parts for $r_j^{t+k}$;

$A_{ij}$ the set of attributes included in conditional part for both $r_i^t$ and $r_j^{t+k}$;

$|A_{ij}|$ the number of attributes in $A_{ij}$;

$B_{ij}$ the set of attributes included in consequent part for both $r_i^t$ and $r_j^{t+k}$;

$|B_{ij}|$ the number of attributes in $B_{ij}$;

$X_{ijp}$ a binary variable, where $X_{ijp}=1$, if the $p$th attribute in $A_{ij}$ has the same value for $r_i^t$ and $r_j^{t+k}$, otherwise $X_{ijp}=0$, $p=1,2,\ldots,|A_{ij}|$;

$Y_{ijq}$ a binary variable, where $Y_{ijq}=1$, if the $q$th attribute in $B_{ij}$ has the same value for $r_i^t$ and $r_j^{t+k}$, otherwise $Y_{ijq}=0$, $q=1,2,\ldots,|B_{ij}|$;

$RulePair_{ij}$ the rule pair composed with $rule_i$ and $rule_j$;

*RulePairsSet* the set of candidate items composed with rule pairs;

$ListofRulePair_{ij}$ the list of customers covered by $rule_i$ and $rule_j$ in $RulePair_{ij}$;

$c$ the number of customers;

$a$ the number of rule pairs;
$\rho$ the coefficient of pheromone decay.

# 3 Customer Change Model Mining Based on Swarm Intelligence

## 3.1 The Process of Customer Change Model Mining

The goal of customer change model mining is to predicate or evaluate market strategies through discovering the change of customer and its reason. On one hand, when a new market strategy has been made, customer change model under this strategy need be predicated. On the other hand, customer change model can be obtained by mining from the datasets collected before and after a strategy execution to evaluate the effect of this strategy. The main problem of customer change model mining is analyzing two or more customer datasets from different periods to find out the change of customers. The process of customer change model mining is shown in figure 1.

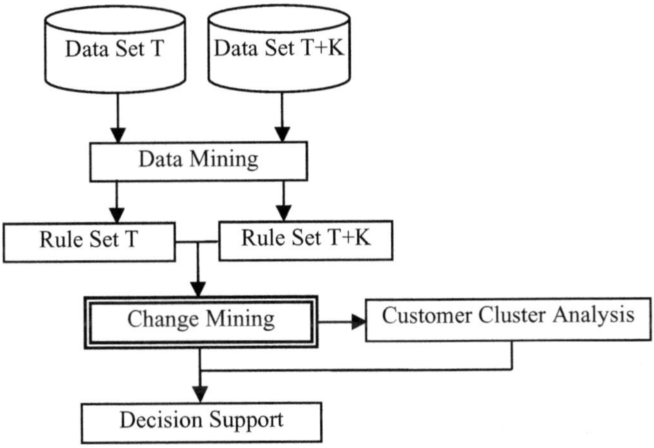


**Fig. 1.** The process of customer change model mining

At first, data mining methods, such as classification and clustering analysis, are applied to analyze two or more customer datasets from different periods. The rule sets obtained from data mining are expressed as customer models. Then the customer change model mining method is implemented to discover what kinds of customer change model has occurred, where the customers come from and where they go in the changed rules, how many changes have been occurred and the reason of customer changes. Finally, the results of customer change model mining are used to help company to make appropriate market strategies. The key step is the rule change mining, so it will be discussed as follows in detail.

## 3.2 High-Level Description of the Algorithm

Algorithm 1. The Customer Change Model Mining Algorithm Based on Swarm Intelligence

$RulePairsSet = \{(r_i^t, r_j^{t+k}) | r_i^t \in R^t, r_j^{t+k} \in R^{t+k}\}$

```
for (n = 1; n <= c; n++) {
  Initialize RulePairsSet;
  for (m = 1; m <= a; m++) {
    Choose RulePair_ij according to chosen probability;
    if (Customer_n • RulePair_ij) {
      Add Customer_n to ListofRulePair_ij;
      Update pheromone of RulePair_ij;
      break;
    }
    else if (Customer_n ∈ r_i^t)
      Preserve the items including r_i^t in RulePairsSet;
      else if (Customer_n ∈ r_j^{t+k})
        Preserve the items including r_j^{t+k} in RulePairsSet;
      else
        Remove all items including r_i^t or r_j^{t+k} in RulePairsSet;
  }
  if (Customer_n ∉ ∀ RulePair_ij)
    Assign Customer_n to appropriate change model;
}
```

Find customer change model according to the number of customers in $ListofRulePair_{ij}$ and threshold value.

## 3.3 Particular Discussion of the Algorithm

### 3.3.1 Preprocessing and Initializing

For the items which include two rules with the same structure from two rule sets, there must be some customers matching them. So these items can not provide any useful knowledge. In the phase of initialization, the rules with the same structure from different rule sets are distinguished according to the following formula, and the customer data covered by these rules are removed.

$$\begin{cases} |A_{ij}| = |M_i^t| = |M_j^{t+k}| \\ |B_{ij}| = |N_i^t| = |N_j^{t+k}| \\ \sum_p X_{ijp} \times \sum_q Y_{ijq} = 1 \end{cases} \tag{1}$$

The following operations will be executed in this phase. The rest rules in these two rule sets compose the set of rule pair $(r_i^t, r_j^{t+k})$ named an item. The numbers of customers covered by each rule are counted for repeatedly using in the algorithm. The

*ListofRulePair*$_{ij}$ for each item is built and it is empty initially. The pheromone of each item is initialized as $\tau(0)=1/a$.

### 3.3.2 The Strategy of Pheromone Updating and Item Searching

The strategy of pheromone updating and item searching used in this paper is based on ant colony optimization method [5]. This method is inspired by behaviors of ant colonies finding the shortest path between their nest and a food source. According to the characteristics of rule change mining, the strategy of pheromone updating is presented as follow.

The pheromone of the item which has been used by customer$_n$ should be increased for simulating that ants leave pheromone on the trail passed by. The formula of pheromone updating is:

$$\tau_{ij}(t+1)=\tau_{ij}(t)+\eta_{ij}\cdot\tau_{ij}(t) \tag{2}$$

The pheromone of the item which has not been used by customer$_n$ should be decreased for simulating the decay of pheromone. The formula of pheromone updating is:

$$\tau_{ij}(t+1)=\tau_{ij}(t)-\rho\cdot\tau_{ij}(t) \tag{3}$$

The heuristic function based on the support rate of rule is adopted for effective convergence of the algorithm. The formula of the heuristic function is:

$$\eta_{ij}=(s_i+s_j)/2 \tag{4}$$

where $s_i$ and $s_j$ express the support rate of $r_i^t$ and $r_j^{t+k}$, i.e. the proportion of the number of customers covered by the rule and the whole number of customers.

Based on above computations, the chosen probability of item *RulePair*$_{ij}$ can be achieved with following formula:

$$p_{ij}(t)=\frac{\tau_{ij}(t)\eta_{ij}}{\sum_{\text{RulePairsSet}}\tau_{ij}(t)\eta_{ij}} \tag{5}$$

### 3.3.3 Process of Item Searching

For enhancing the running efficiency of the algorithm, some judgment conditions have been used in the process of item searching to avoid computing all of items. If customer$_n$ match the item, then customer$_n$ is added to the list of this item and the pheromone is updated. Otherwise, if customer$_n$ matches $r_i^t$, remove items which do not include $r_i^t$ from *RulePairsSet*. Otherwise, if customer$_n$ matches $r_j^{t+k}$, remove items which do not include $r_j^{t+k}$ from *RulePairsSet*. If customer$_n$ matches neither $r_i^t$ nor $r_j^{t+k}$, remove items which include $r_i^t$ or $r_j^{t+k}$ from *RulePairsSet*. These judgment conditions can decrease the number of items in candidate set and computational quantity.

### 3.3.4 Finding the Customer Change Model

The customer change model can be found according the results at the end of algorithm. If a change of rule satisfies the threshold value designated in advance, for example, if the support rate is higher than a certain value, it can be considered as a customer change model. The threshold value is designated by the experts of application domains.

## 4 Experimental Results

In this section, we experiment on two customer datasets of a telecom company. The interval of these two datasets is three months. The attributes of customers used in our experiment are shown in table 1.

**Table 1.** Data attributes used in experiment

| Variable name | Description |
|---|---|
| Regular_dur | Minutes of call in regular time |
| Discount_dur | Minutes of call in discount time |
| Local_dur | Minutes of local call |
| Domestic_dur | Minutes of domestic call |
| Svc_sms | Times of short message service |
| Svc_type | Number of service types |
| Svc_time | Number of service times |
| Card_type | Type of cell card |
| Disc_type | Type of discount |
| Age | Customer age |
| Gender | Customer gender |
| Arrearage_time | Times of arrearage |
| ARPU | Average Revenue Per User |
| Churn | Customer is churning or not |

The experiment is implemented with SIMiner, a self-development data mining software system based on swarm intelligence. According to the process of customer change model mining presented in this paper, we analyzed the two datasets using the classification mining method in [4] and obtained two rule sets. Then the customer change model mining method based on swarm intelligence was adopted to mining rule changes in these two rule sets. The output of rule change set is illuminated in figure 2.

It can be found from this figure that different kinds of customer change model have been occurred. The first two are unexpected model, i.e. the consequent parts of two rules are the same but the conditional parts are different. The third is a perished model, i.e. the rule exists in $R^t$ but not $R^{t+k}$. The fourth is added model, i.e. the rule exists in $R^{t+k}$ but not $R^t$.

The results of analyzing each kind of customer change model can support decision-making effectively. For example, from change model 1, we can find that a part of customers using card A have more call in discount time instead of regular time. This illuminates that the discount strategy of card A has worked. Furthermore, analyzing the characteristics of customer cluster covered by each change model can help company to understand the reason of change, and support decision making more effectively.

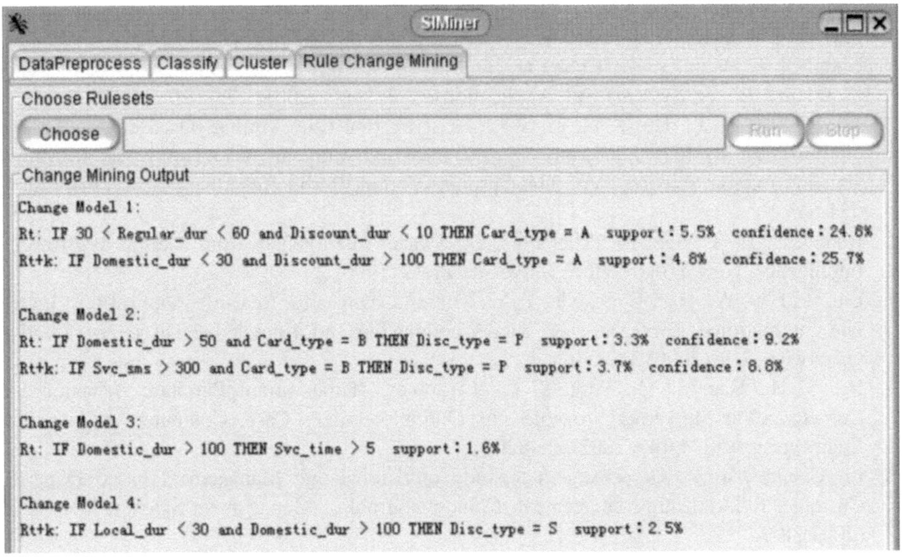


**Fig. 2.** The results of customer change model mining

# 5 Conclusions

With the development of new business models and continuous change of customer demand and behavior model, the dynamic analysis of customer data and customer relationship management have to face new challenges. The concept of customer change model has been introduced in this paper. The process of customer change model mining is analyzed, and a customer change model mining method based on swarm intelligence is presented to discover the change of rules not from the aspect of rule structure but the change of customers. The results of experiment illuminates that this method can support the decision making effectively.

In the future research, the measure method of four kinds of customer change model, namely emerging model, added model, perished model, and unexpected model, should be studied. The computational method of change degree in each change model should be improved based on the distinguishing of four kinds of change model.

# Acknowledgements

This work is supported by the National Natural Science Foundation of China (Grant No. 70431003).

## References

1. Liu, B., Hsu, W., Han, H. S., Xia, Y.: Mining Changes for Real-Life Applications. In: Second International Conference on Data Warehousing and Knowledge Discovery (2000) 337-346
2. Song, H. S., Kim, J. K., Kim, S. H.: Mining the Change of Customer Behavior in an Internet Shopping Mall. Expert Systems with Applications. Vol. 21(3) (2001)157-168
3. Chen, M. C., Chiu, A. L., Chang H. H.: Mining Changes in Customer Behavior in Retail Marketing. Expert Systems with Applications. Vol. 28(4) (2005) 773-781
4. Jin, P., Zhu, Y. L., Hu, K.Y., Li, S. F.: Classification Rule Mining Based on Ant Colony Optimization. In: ICIC 2006: Intelligent Control and Automation. Lecture Notes in Control and Information Sciences, Vol 344. Springer-Verlag, Berlin Heidelberg New York (2006) 654-663
5. Bonabeau, E., Dorigo, M., Theraulaz, G.: Swarm Intelligence: from Natural to Artificial Intelligence. New York: Oxford University Press (1999)
6. Liu, B., Hsu, W., Han, H. S., Xia, Y. Y.: Mining Changes for Real-life Applications. In: the 2nd International Conference on data Warehousing and Knowledge Discovery, London Greenwich, UK (2000) 337-346
7. Ha, S. H., Bae, S. M., Park, S. C.: Customer's Time-variant Purchase Behavior and Corresponding Marketing Strategies: an Online Retailer's Case. Computers & Industrial Engineering. Vol. 43(4) (2002)801-820
8. Li, C. Q., Xu, Y. G., Zhang, Y.: Study on Knowledge Management based Dynamic Customer Relationship Management. Chinese Journal of Management Science. Vol. 12(2) (2004)88-94

# New Classification Method Based on Support-Significant Association Rules Algorithm

Guoxin Li[1] and Wen Shi[2,*]

[1] School of Management, Harbin Institute of Technology, 150001, China
liguoxin@hit.edu.cn
[2] Department of Computer Science, Northeast Agricultural University 150030, China
Tel.: +86-451-55191146
shiw@neau.edu.cn

**Abstract.** One of the most well-studied problems in data mining is mining for association rules. There was also research that introduced association rule mining methods to conduct classification tasks. These classification methods, based on association rule mining, could be applied for customer segmentation. Currently, most of the association rule mining methods are based on a support-confidence structure, where rules satisfied both minimum support and minimum confidence were returned as strong association rules back to the analyzer. But, this types of association rule mining methods lack of rigorous statistic guarantee, sometimes even caused misleading. A new classification model for customer segmentation, based on association rule mining algorithm, was proposed in this paper. This new model was based on the support-significant association rule mining method, where the measurement of confidence for association rule was substituted by the significant of association rule that was a better evaluation standard for association rules. Data experiment for customer segmentation from UCI indicated the effective of this new model.

**Keywords:** data mining, classification method, association rule mining, customer segmentation.

## 1 Introduction

The term "data mining" has been applied to a broad range of activities that attempt to discover interesting information from existing data, where usually the original information was gathered for a purpose entirely different from the way it is used for data mining. Typically the applications involve large-scale data banks such as data warehouses or datacubes. One of the most well-studied problems in data mining is the search for association rules from database.

Association rule mining was first proposed by Agrawal, etc. in 1993. An association rule, which is measured via support and confidence is primarily intended to identify rules of the type, "A customer purchasing item X is likely to also purchase item Y". There was also research (Bing Liu, 1998) that introduced association rule

* Corresponding author.

D.-S. Huang, L. Heutte, and M. Loog (Eds.): ICIC 2007, LNAI 4682, pp. 465–474, 2007.

mining methods to conduct classification mining tasks. These association classification rule mining methods could be used for establishing customer segmentation models. Currently, most of the association rule mining methods are based on a support and confidence structure, where rules satisfied both minimum support and minimum confidence were returned as strong association rules to the analyzer. But, these types of association rule mining methods lack of rigorous statistic guarantee, sometimes even caused misleading. So, new methods for association rule mining with strict statistic support are expected. The classical association rule mining methods and its shortcomings under support and confidence structure were discussed in section 2. New association rule mining method for classification was proposed in section 3and Section 4. Section 5 was experiment and result. Section 6 was conclusion and summary.

## 2 Classification Method Under Support-Confidence Structure

### 2.1 Support-Confidence Structure of Association Rule Mining

Let $I=\{i_1,i_2,\cdots,i_m\}$ be a set of items. Let *D*, the task-relevant data, be a set of database transactions where each transaction *T* is a set of items such that $T\subseteq I$. Each transaction is associated with an identifier, called TID. Let A be a set of items. A transaction T is said to contain A if and only if $A\subseteq T$. An association rule is an implication of the form $A\Rightarrow B$, where $A\subset I, B\subset I$ ,and $A\cap B=\phi$. The rule $A\Rightarrow B$ holds in the transaction set *D* with *support s,* where ***s*** is the percentage of transactions in D that contain $A\cup B$. This is taken to be the probability, $P(A\cup B)$. The rule $A\Rightarrow B$ has *confidence c* in the transaction set *D* if ***c*** is the percentage of transactions in D containing A that also contain B. This is taken to be the conditional probability, $P(B\mid A)$. That is,

$$Support(A\Rightarrow B)=P(A\cap B),\ Confidence(A\Rightarrow B)=P(B\mid A)$$

Rules that satisfy both a minimum support threshold and a minimum confidence threshold are called *strong Rules.* A set of items is referred to as an itemset. The occurrence frequency of an itemset is the number of transactions that contain the itemset. This is also known, simply, as the frequency, support count, or count of the itemset. An itemset satisfies minimum support if the occurrence frequency of the itemset is greater than or equal to the product of minimum support and the total number of transactions in D. The mumber of transactions required for the itemset to satisfy minimum support is therefore referred to as the minimum support count. If an itemset satisfies minimum support, then it is called *frequent itemset.* Usually an association rule mining process contains the following two steps:

(1) Find all frequent itemsets: By definition, each of these itemsets will occur at least as frequently as a pre-determined minimum support count.

(2) Generate strong association rules from the frequent itemsets: By definition, these rules must satisfy minimum support and minimum confidence.

Up to now, most of the researches are focus on the first step, i.e. the algorithms of finding frequent itemsets.

## 2.2 Classification Method Under the Association Rule Mining Algorithm

Let D be the dataset. Let I be the set of all items in D, and Y be the set of class labels. We say that a data case d_ $\in$ D contains X$\subseteq$I, a subset of items, if X$\subseteq$d. A classification association rule (CAR) is an implication of the form X$\rightarrow$y, where X$\subseteq$I, and y$\in$ Y. A rule X$\rightarrow$y holds in D with confidence c if c% of cases in D that contain X are labeled with class y. The rule X$\rightarrow$ y has support s in D if s% of the cases in D contain X and are labeled with class y.

The algorithm is given in by Bing Liu (1998) in paper [1].

## 2.3 The Shortage of Support and Confidence Structure Mining Methods

(1) Not all of the strong rules are interesting

Example A: Suppose we are interested in analyzing transactions at an online sports store OSS with respect to the purchase of ping-pong ball and badminton. Let PB refer to the account of transactions containing ping-pong ball, and BM refer to the account of those containing badminton. Of the 10000 transactions analyzed, the data showed that 6000 of the customer transactions included ping-pong ball, while 7500 included badminton, and 4000 included both ping-pong ball and badminton. Suppose that a data mining program for discovering association rules were run on the data, using a minimum support of, say, 30% and a minimum confidence of 60%. The following association rule would be discovered:

$$buys(X,"ping-pong\quad ball") \Rightarrow buys(X,"bad\min ton")$$
$$[support = 40\%,\ confidence = 66\%]$$

**Table 1.** Data of the ping-pong ball (PB) and badminton (BM) Selling in a store

| | The number of customers bought *PB* | The number of customers didn't buy PB |
|---|---|---|
| The number of customers bought *BM* | 4000 | 3500 |
| The number of customers did not buy BM | 2000 | 500 |

However, consider now the fact that the apriori probability that a customer buys badminton is 75%. In other words, a customer who is known to buy ping-pong ball is less likely to buy badminton than a customer about whom we have no information. Of course, it may still be interesting to know that such a large number of people who buy ping-pong ball also buy badminton, but stating that rule by itself is at best incomplete information and at worst misleading. The truth here is that there is a negative correlation between buying ping-pong ball and buying badminton. Without fully under standing this phenomenon, one could make improper business decisions based on the rules derived.

(2) Omitting Useful Rules
Example B: There is a hospital attached to a certain chemical plant. The relationship between a kind of occupational disease X and a certain occupation A in that plant could be analyzed by the association rule method from the database of health checkup. The data is in table 2 (given Sup_min=2%, Conf_min=20%).

**Table 2.** Health Checkup Data in a Chemical Plant Attached Hospital

| | With Disease X | Without Disease X | Total |
|---|---|---|---|
| The number of workers in occupation A | 52 | 636 | 688 |
| The number of workers in other occupation | 11 | 1800 | 1811 |
| Total | 63 | 2436 | 2499 |

Rule of Candidate: *Job ( w| "With occupation A") ⇒ Disease ( d\ "Cause disease X")*

According to the data in table-2 the support (Sup) and confidence (Conf) of this rule of candidate is Sup=2.08% (>Min_Sup); Conf=7.56% (<Min_Conf). It doesn't satisfy the minimum confidence, so this will not be a strong rule to return to the decision maker. But we noticed, that the average ratio of disease X suffered workers among the workers with occupation A (7.56%) is three times that of other occupation workers (2.52%). The figure difference between disease ratios is *EXTREME SIGNIFICANT* in terms of statistics standard (p<0.01). So, actually this rule is a very important rule to research the relationship between occupation and occupational disease. But it can't be found out by the support-confident structure of association rule mining methods.

(3) lack of rigorous statistic guarantee
The classical mining methods for association rules under support and confidence structure are experiential methods. Though the support-confidence structure methods were widely used, those methods were lack of rigorous statistic guarantee. Also there isn't a universal standard for confidence or support among difference databases.

# 3 Mining Association Rules Under Support-Significant Structure

To solve the above problems, a new method of association rule mining was proposed. In this new method, .the classic statistic method of t-testing was introduced.

## 3.1 T-Testing for the Comparison of Proportions in Statistics

Suppose there is a sample S of $n$ size (i.e. S has $n$ objects). Among the $n$ objects, $n_K$ objects have the property K. That is to say, the proportion of objects with property K is $p = n_K / n$ in S. When we want to know whether or not the difference between $p$ and a certain ratio $\pi$ is significant, t-testing could be used.

$$t = \frac{p-\pi}{\sigma_p}, \quad \text{where } \sigma_P = \sqrt{\frac{\pi(1-\pi)}{n}}$$

The difference between $p$ and $\pi$ is $\begin{cases} not \quad significant, |t| < t_{0.05}(n) \\ significant, t_{0.05}(n) < |t| < t_{0.01}(n) \\ extremely \quad significant, |t| > t_{0.01}(n) \end{cases}$.

This t-testing method could be introduced into the evaluation of association rules in data mining and to build a "support and T-value" mining structure.

### 3.2 Data Mining Based on T-Testing

To avoid the shortage of "support & confidence" data mining structure, we bring forward a new data mining structure, "support & T-value" mining structure.

Let $I = \{i_1, i_2, \cdots, i_m\}$ be a set of items. Let *D*, the task-relevant data, be a set of database transactions where each transaction *T* is a set of items such that $T \subseteq I$. Each transaction is associated with an identifier, called TID. Let A be a set of items. A transaction T is said to contain A if and only if $A \subseteq T$. An association rule is an implication of the form $A \Rightarrow B$, where $A \subset I, B \subset I$ ,and $A \cap B = \phi$. The rule $A \Rightarrow B$ holds in the transaction set *D* with *support s,* where *s* is the percentage of transactions in D that contain $A \cup B$. This is taken to be the probability, $P(A \cup B)$.

$$Support(A \Rightarrow B) = P(A \cup B)$$

If an itemset satisfies minimum support, then it is called *frequent itemset.* $t_{A \Rightarrow B} = \frac{P(B \mid A) - P(B)}{\sigma_p}$ named the *significant* of B from A or $Sig_{A \Rightarrow B}$, where $\sigma_P = \sqrt{\frac{P(B)(1-P(B))}{n_A}}$

If $Sig_{A \Rightarrow B} > t_\alpha(n_A)$, then the rule $A \Rightarrow B$ is called significant rule. (Because we want only to know whether or not p is greater than $\pi$, so instead of absolute value there is the original value of $t_{A \Rightarrow B}$). Where, $t_\varepsilon(n_A)$, *named minimum significant,* is the threshold $t$ value under $\alpha$ significant level with $n_A$ freedom dimensions in $T$ distribution. Usually $n_A$ have a very large value so $t_\varepsilon(n_A) \approx u_\alpha$, $u_\alpha$ is the $u$ threshold of $\alpha$ significant level in normal distribution.

This new association rule mining method includes two steps.

(1) Find all frequent itemsets.
The proportion of objects with a certain itemset greater than a minimum support would be a frequent itmeset. This step could be conducted by the algorithms like apriori, etc.

(2) Generating significant association rules from the frequent itemsets.
Rules would be derived from frequent itemsets. Choose a significant level $\alpha$ (It is comparable between different databases). The rules with $Sig_{A \Rightarrow B}$ greater than the threshold $t_{\alpha}(df)$ (*minimum significant*) would be returned to the analyzer as significant rules under $\alpha$ significant level.

## 4 Association Classification Method Based on Support-Significant Structure of Association Rule Mining

To avoid the shortcoming of current association classification methods, we proposed a new association classification method. The basic idea of this new classification method is to replace the traditional support-confidence structure of association rule mining model by the support-significant structure of association rule mining model in the classification process. This new method consisted of a classification rule mining algorithm and a classifier composing algorithm. The working principle of an association classifier is as in figure 1.

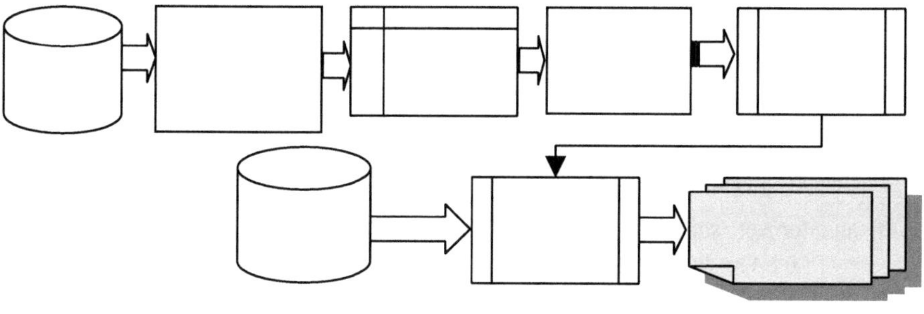

**Fig. 1.** Working principle of an Association Classifier

In the algorithm of support-significant structure of association rule mining method, only significant rule would be add to the rule set.

Modifying to the CBA algorithm of Liu (1998) etc., we proposed a new association rule classification algorithm *Classification Based on Significant Associations rule (CBSA).*

Definition 1. Association rule like condset$\Rightarrow$y is classification association rule, when condset is conditional item set, y$\in$ Y is class lable, <condset,y> is called rule item set.

Definition 2. For a classification association rule of condset$\Rightarrow$y, the count of record with conditional item set condset in the database is called conditional support count (consupCount) . The conditional item set is called frequent conditional item set, if the conditional support count equal to or greater than the minimum support count.. The number of record with rule item set <condset,y> in the database is called rule support count (rulesupCount). If the rule support count of a rule item set equal to or greater than the minimum support count, then the rule item set is called frequent rule item set.

When the record in database with a class label is greater than the minimum support count, the class indicated by this class label is called frequent class.

Definition 3. Rules were called significant classification association rules (SCARs) when the classification rules represent by a frequent rule item set satisfied minimum significant condition.

In the following classification association rule mining method, suppose D is a database contained relational data form. There are $\boldsymbol{N}$ records with $\boldsymbol{l}$ discrete attributes in the database. The $N$ records were divided into $q$ categories with corresponding class label.

We proposed the association rule classification algorithm *Classification Based on Significant Associations rule (*CBSA*)* to generate significant classification association rules as following:

Algorithm: SCARs generation algorithom ——CBSA
Input: Relational data base D, minimum support Min_sup, minimum significant Min_Sig
Output: significant classification association rules SCAR

```
procedure CBSA(D, Min_sup, Min_sig)
1) Public Integer n=|D|
2) F_0=find_frequent_class(D)
3) For each class-i∈F_0 P_c-i=Sup(class-i)
4) F_1={frequent 1-ruleitems}
5) SCAR_1=SignRule_Gen (F_1)
6) prSCAR_1=pruneRules(SCAR_1)
7) for (k=2;F_k-1•∅;k++) do{
8)   C_k=Candidate_Gen(F_k-1•Min_sup)
9)   for each datacase d∈D do{
10)    C_d=ruleSubset(C_k,d);
11)   for each candidate c∈C_d do{
12)     c.condsupCount ++;
13)     if d.class=c.class then c.rulesupCount ++;
14)   }  // end of for 11)
15)  }  // end of for 9)
16)  F_k={c∈C_k |c.rulsupCount≧min_sup};
17)  SCAR_k= SignRule_Gen(F_k);
18)  pr CAR_k=pruneRules(F_k);
19) } // end of for 7)
20) SCARs=∪_k SCAR_k;
21) Ruturn SCARs
procedure SignRule_Gen (F)
1) SignRules=null;
2) for each ruleitemsets i∈F{
3)          sig_i=((count_ruleitemset_i/count_condset_i)-p_y)/
SQRT(p_y*( 1-p_y)/n);
4)          if sig_i>min_sig then add ruleitemset-i to
SignRules;
5) } // end of for
6) return SignRules
```

```
procedure Candidate_Gen(F_{k-1}•Min_sup)
1) for each itemset i_1∈ F_{k-1}{
2)        for each itemset i_2∈ F_{k-1}{
3)     if (i_1[1]= i_2[1])∧ (i_1[2]= i_2[2])∧…∧ (i_1[k-2]=
i_2[k-2])
∧ (i_1[k-1]< i_2[k-1]) then{
4)     c=i_1[1]i_1[2]…i_1[k-2]i_1[k-1]i_2[k-1];
5)     if has_infrequent_subset(c, F_{k-1}) then
6)        delete c ;
7)     else add c to C_k;
8)     }     // end of if
9)   } // end of for 2)
10)}  // end of for 1)
11) return C_k;
```

## 5 Experiment and Result

We applied the above methodology onto the *Census Income* data obtained from the Machine Learning Repository in the University of California at Irvine (UCI, http://www.ics.uci.edu/~mlearn/MLRepository.html).

The people were divided into two class according there income. Class 1 is lower income (<=50K). Class 2 is high income (>50K). The attributes include *range of age, occupation, education experience, marriage status, job position, family, race, gender, nationality*.

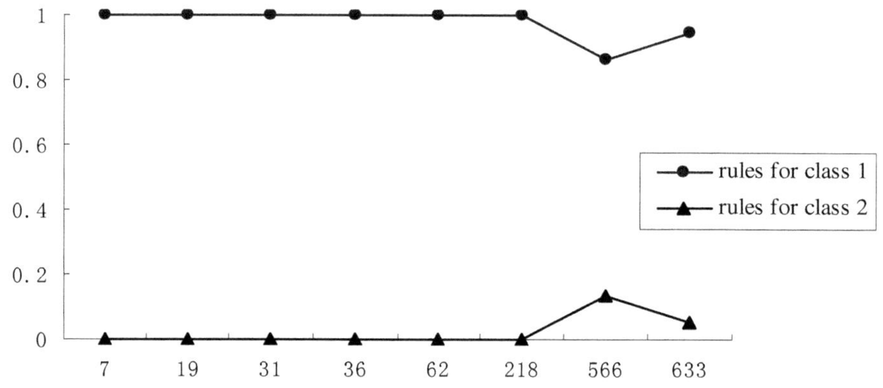


**Fig. 2.** Classification rules generated by traditional algorithm

Rules for classification these people by their attributes could be generated by applying training algorithms on the training data. There would be two types of rules to be generated. Some of rules were for identification the lower income people, and the other rules were for identification higher income people. We applied both the common association classification algorithm and our new algorithm on the training data with several different min_support, min_confident or min_significant levels. The numbers of both types of rules generated by common algorithm were in figure 2. Figure 3 showed the numbers of two types of rules generated by the new algorithm proposed by this study. The figures in X axis present the numbers of total rules generated in one experiment. The figures in Y axis present the percentage of rules for identifying certain class of people. We can find that the new algorithm is better than the traditional algorithm, for the traditional algorithm have very poor ability in identifying the high income people (class 2), i.e. asymmetrical between classes, but the new algorithm have good ability for identifying both lower income and higher income people.

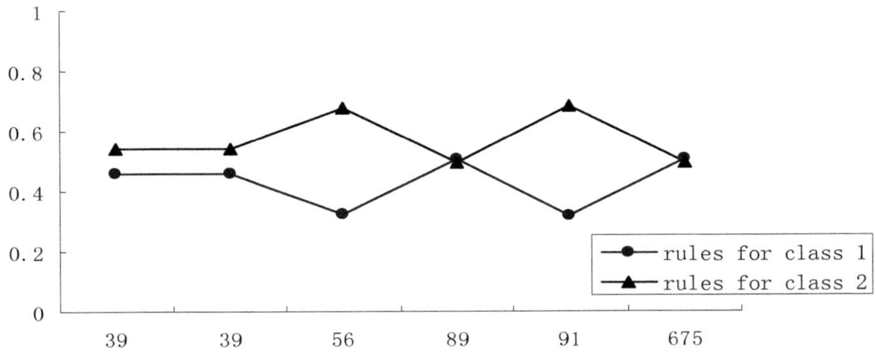


**Fig. 3.** Classification rules generated by the new algorithm

## 6 Conclusion and Summary

Traditional classification association rule mining methods under support and confidence structure lack of strict statistic support. It might cause misleading in the decisions making process, for not all of the strong rules are interesting. A classical statistic method, t-test was introduced into the classification association rule mining process to build a support-significant structure of classification mining method. This new mining structure consisted two steps: 1) find all frequent itemset;2) Generate significant association classification rules from the frequent itemsets with t-test. With rigorous statistic support, the rules mined from this t-test based mining structure would be more meaningful and useful. Data experiment indicated that the proposed new algorithm have better ability in generating classification rules.

**Acknowledgments.** The work was partially supported by the National Science Foundation of China (Grant No. 70501009) and Heilongjiang Natural and Science Fund Project (G0304). This work was performed at the National Center of Technology, Policy and Management (TPM) (Grant No. htcsr06t04), Harbin, China.

## References


1. Liu, B., Hsu, W., Ma, Y.M.: Integrating Classification and Association Rule Mining. Proceedings of the Fourth International Conference on Knowledge Discovery and Data Mining, New York (1998) 1-7
2. Ha n, E.H., Karypis, G., Kumar, V.: Scalable Parallel Data Mining for Association Rules. Knowledge and Data Engineering, IEEE Transactions (2000) 337-352
3. Han, J.W., Karmbr. M.: Data Mining: Concepts and Techniques. Morgan Kaufmann Publishers (2001) 225~330
4. Agrawal, R., Imielinski, T., Swami, A.: Mining Association Rules Between Sets of Items in Large Databases. In Proc. 1993 ACM-SIGMOD Int. Conf. Management of Data (SIGMOD'93). (1993) 207~216
5. Brin, S., Motwani, R., Silverstein, C.: Beyond Market Baskets: Generalizing Association Rules to Correlations. ACM SIGMOD Record, Proceedings of the ACM SIGMOD international conference on Management of data. June (1997) 265 - 276
6. Ye, Q., Li, Y.J., Zhang, J.: Improved Method in Association Rule Mining. Proceeding of The 8th Asia Pacific Management Conference (2002) 1-8
7. Adomavicius, G., Tuzbilin, A.: Using Data Mining Methods to Build Customer Profiles. Computer (2001) 74-82
8. Yin, X., Han J.: CPAR: Classification Based on Predictive Association Rules. Proceedings of the SDM(2003) 80-86
9. Tsay, Y.J., Chiang, J.Y.: CBAR: An Efficient Method for Mining Association Rules. Knowledge Based Systems, (2005) 432–444
10. Mielikäinen, T.: Frequency-based Views to Pattern Collections. Discrete Applied Mathematics, Vol. 154. 7(2006) 1113-1139

# Scaling Up the Accuracy of Bayesian Network Classifiers by M-Estimate

Liangxiao Jiang[1], Dianhong Wang[2], and Zhihua Cai[3]

[1] Faculty of Computer Science, China University of Geosciences
Wuhan, Hubei, P.R. China, 430074
ljiang@cug.edu.cn

[2] Faculty of Electronic Engineering, China University of Geosciences
Wuhan, Hubei, P.R. China, 430074
wangdh@cug.edu.cn

[3] Faculty of Computer Science, China University of Geosciences
Wuhan, Hubei, P.R. China, 430074
zhcai@cug.edu.cn

**Abstract.** In learning Bayesian network classifiers, estimating probabilities from a given set of training examples is crucial. In many cases, we can estimate probabilities by the fraction of times the events is observed to occur over the total number of opportunities. However, when the training examples are not enough, this probability estimation method inevitably suffers from the zero-frequency problem. To avoid this practical problem, Laplace estimate is usually used to estimate probabilities. Just as we all know, m-estimate is another probability estimation method. Thus, a natural question is whether a Bayesian network classifier with m-estimate can perform even better. Responding to this question, we single out a special m-estimate method and empirically investigate its effect on various Bayesian network classifiers, such as Naive Bayes (NB), Tree Augmented Naive Bayes (TAN), Averaged One-Dependence Estimators (AODE), and Hidden Naive Bayes (HNB). Our experiments show that the classifiers with our m-estimate perform better than the ones with Laplace estimate.

**Keywords:** Bayesian network classifiers, m-estimate, Laplace estimate, probability estimation, classification.

## 1 Introduction

A Bayesian network consists of a structural model and a set of conditional probabilities. The structural model is a directed graph in which nodes represent attributes and arcs represent attribute dependencies. Attribute dependencies are quantified by conditional probabilities for each node given its parents. Bayesian networks are often used for classification problems, in which a learner attempts to construct a classifier from a given set of training examples with class labels. Assume that $A_1, A_2, \cdots, A_n$ are $n$ attributes (corresponding to attribute nodes in a Bayesian network). An example $E$ is represented by a vector $(a_1, a_2, , \cdots, a_n)$,

D.-S. Huang, L. Heutte, and M. Loog (Eds.): ICIC 2007, LNAI 4682, pp. 475–484, 2007.

where $a_i$ is the value of $A_i$. Let $C$ represent the class variable (corresponding to the class node in a Bayesian network). We use $c$ to represent the value that $C$ takes and $c(E)$ to denote the class of $E$. The Bayesian network classifier represented by a Bayesian network is defined in Equation 1.

$$c(E) = \arg\max_{c \in C} P(c) \prod_{i=1}^{n} P(a_i|\Pi_{a_i}), \tag{1}$$

where $\Pi_{a_i}$ is the set of parents of $A_i$.

In learning a Bayesian network classifier, we need to estimate probabilities of $P(c)$ and conditional probabilities $P(a_i|\Pi_{a_i})$ from a given set of training examples. In many cases, we can estimate probabilities by the fraction of times the events is observed to occur over the total number of opportunities. However, when the training examples are not enough, this probability estimation method inevitably suffers from the zero-frequency problem. In order to avoid this practical problem, Laplace estimate is usually used to estimate probabilities. Just as we all know, m-estimate is another probability estimation method. Thus, a natural question is whether a Bayesian network classifier with m-estimate can perform even better. Responding to this question, we single out a special m-estimate method in this paper.

The rest of the paper is organized as follows. In section 2, we introduce we introduce four Bayesian network classifiers studied in this paper. In section 3, we single out a special m-estimate method after that we simply introduce Laplace estimate and m-estimate. In section 4, we describe the experimental setup and results in detail. In section 5, we make a conclusion and outline our main directions for future research.

## 2 Bayesian Network Classifiers

Theoretically, learning an optimal Bayesian network is intractable [1]. Moreover, it has been observed that learning an unrestricted Bayesian network classifier seems to not necessarily lead to a classifier with good performance. For example, Friedman et al. [3] observed that unrestricted Bayesian network classifiers do not outperform naive Bayes, the simplest Bayesian network classifier, on a large sample of benchmark data sets. One major reason is that the resulting network tends to have a complex structure, and thus has high variance because of the inaccurate probability estimation caused by the limited amount of training examples. So in practice, learning restricted Bayesian network classifiers is a more realistic solution.

Naive Bayes (simply NB) [2] is based on an assumption that all attributes are independent given the class. In NB, each attribute node has the class node as its parent, but does not have any parent from attribute nodes. Figure 1 shows an example of naive Bayes. The corresponding naive Bayes classifier is defined as follows.

$$c(E) = \arg\max_{c \in C} P(c) \prod_{i=1}^{n} P(a_i|c) \tag{2}$$

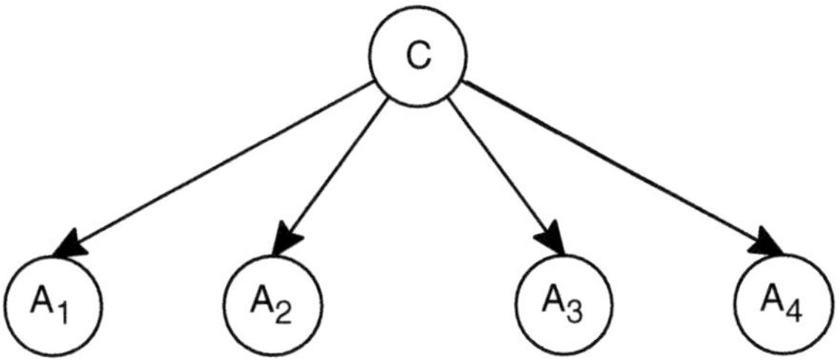


**Fig. 1.** An example of naive Bayes

Tree augmented naive Bayes (simply TAN) [3] extends naive Bayes by allowing each attribute has at most one attribute parent. Figure 2 shows an example of TAN. The corresponding TAN classifier is defined as follows.

$$c(E) = \arg\max_{c \in C} P(c) \prod_{i=1}^{n} P(a_i|p_{ai}, c) \tag{3}$$

where $p_{ai}$ is the attribute parent of $A_i$.

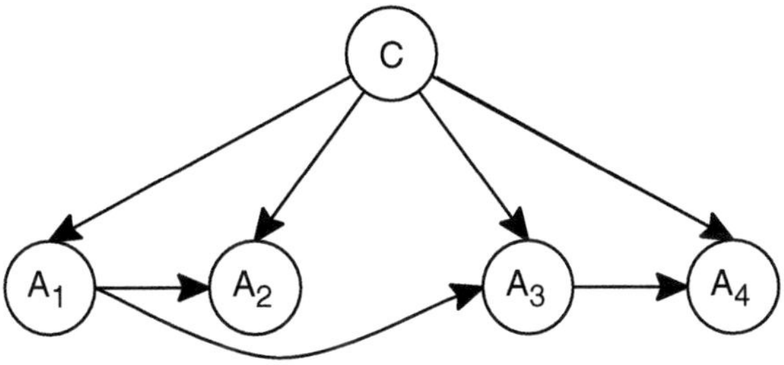


**Fig. 2.** An example of TAN

Averaged One-Dependence Estimators (simply AODE) [4] is an ensemble of one-dependence classifiers and produces the prediction by aggregating the predictions of all qualified one-dependence classifiers. More precisely, in AODE, a special TAN is built for each attribute, in which the attribute is set to be the parent of all other attributes. AODE classifies an instance using Equation 4.

$$c(E) = \arg\max_{c \in C}\left(\frac{\sum_{i=1 \land F(a_i) \geq m}^{n} P(c)P(a_i|c)\prod_{j=1, j \neq i}^{n} P(a_j|a_i, c)}{numParent}\right) \tag{4}$$

where $F(a_i)$ is the number of training examples having attribute-value $a_i$, $m$ is a constant, $numParent$ is the number of the root attributes, which satisfy the condition that the training data contain more than $m$ examples with the value $a_i$ for the parent attribute $A_i$. Figure 3 shows an example of the aggregate of AODE. Hidden naive Bayes (HNB) [5] is another extension of naive Bayes, in

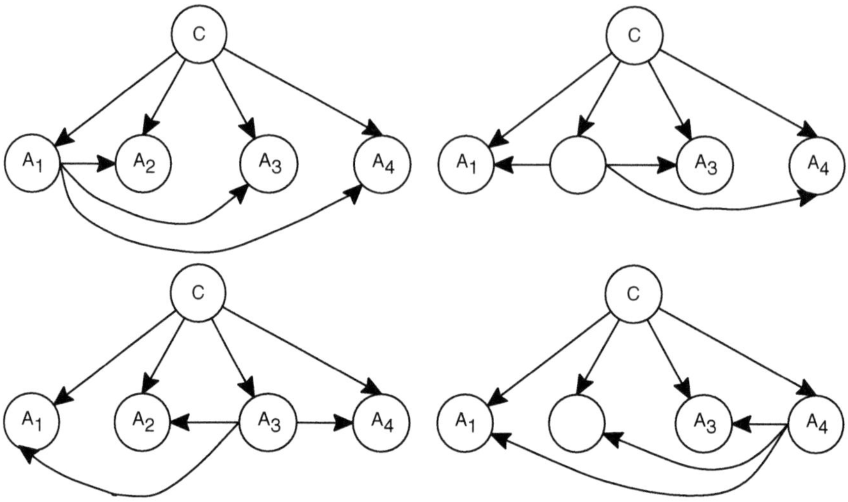


**Fig. 3.** An example of the aggregate of AODE

which a hidden parent $A_{hp_i}$ is created for each attribute $A_i$ to integrate the influences from all other attributes. Figure 4 shows the structure of HNB. HNB classifies an instance using Equation 5.

$$c(E) = \arg\max_{c \in C} P(c) \prod_{i=1}^{n} P(a_i | a_{hp_i}, c) \tag{5}$$

where

$$P(a_i | a_{hp_i}, c) = \sum_{j=1, j \neq i}^{n} w_{ij} * P(a_i | a_j, c) \tag{6}$$

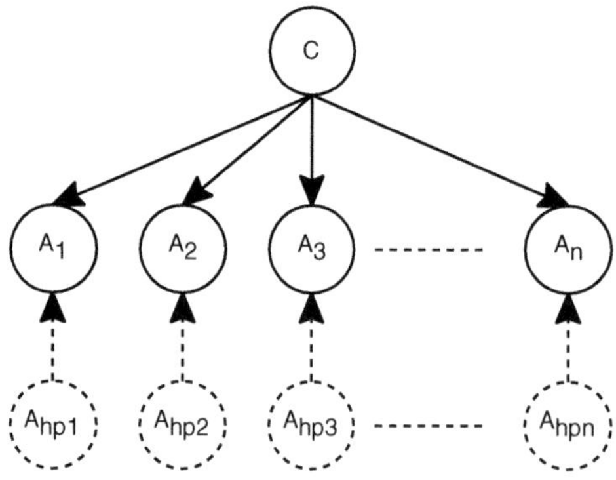


**Fig. 4.** The structure of HNB

and

$$w_{ij} = \frac{I_P(A_i; A_j|C)}{\sum_{j=1, j \neq i}^{n} I_P(A_i; A_j|C)} \tag{7}$$

In Equation 7, $I_P(A_i; A_j|C)$ is the conditional mutual information between $A_i$ and $A_j$ given $C$. It can be defined as:

$$I_P(A_i; A_j|C) = \sum_{a_i, a_j, c} P(a_i, a_j, c) log \frac{P(a_i, a_j|c)}{P(a_i|c)P(a_j|c)} \tag{8}$$

## 3 Laplace Estimate and M-Estimate

If we adopt Laplace estimate to estimate probabilities of $P(c)$ and conditional probabilities $P(a_i|\Pi_{a_i})$, then

$$P(c) = \frac{F(c) + 1.0}{N + |C|} \tag{9}$$

$$P(a_i|\Pi_{a_i}) = \frac{F(a_i, \Pi_{a_i}) + 1.0}{F(\Pi_{a_i}) + |A_i|} \tag{10}$$

where $F(\bullet)$ is the frequency with which a combination of terms appears in the training examples, $N$ is the number of training examples, $|C|$ is the number of classes, $|A_i|$ is the number of values of attribute $A_i$.

M-estimate [6] is another method to estimate probability, which can be defined as follows.

$$P(c) = \frac{F(c) + mp}{N + m} \tag{11}$$

$$P(a_i|\Pi_{a_i}) = \frac{F(a_i, \Pi_{a_i}) + mp}{F(\Pi_{a_i}) + m} \tag{12}$$

where $m$ and $p$ are two parameters. $p$ is the prior estimate of the probability we wish to determine. $m$ is a constant called the *equivalent sample size*, which determines how heavily to weight $p$ relative to the observed data. In fact, m-estimate can be comprehend as augmenting the actual observations by an additional $m$ virtual samples distributed according to $p$.

Since $m$ can be an arbitrary natural number, such as 1,2,3,$\cdots$, we set it to 1 in our implementation. In estimating probabilities of $P(c)$, we set $p$ to an uniform distribution. Namely, $p = \frac{1}{|C|}$. In estimating conditional probabilities of $P(a_i|\Pi_{a_i})$, we set $p$ to $P(a_i)$. Namely, $p = P(a_i)$, where $P(a_i)$ can be estimated by m-estimate again. So $P(a_i)$ can be defined as follows.

$$P(a_i) = \frac{F(a_i) + mp}{N + m} \tag{13}$$

where $m = 1$ and $p = \frac{1}{|A_i|}$.

Now, let's rewrite two equations used to estimate probabilities of $P(c)$ and conditional probabilities $P(a_i|\Pi_{a_i})$ as follows.

$$P(c) = \frac{F(c) + 1.0\frac{1}{|C|}}{N + 1.0} \tag{14}$$

$$P(a_i|\Pi_{a_i}) = \frac{F(a_i, \Pi_{a_i}) + 1.0\frac{F(a_i)+1.0\frac{1}{|A_i|}}{N+1.0}}{F(\Pi_{a_i}) + 1.0} \tag{15}$$

## 4 Experimental Methodology and Results

We conducted experiments under the framework of Weka [7] to study the effect of m-estimate on the performance of Bayesian network classifiers. We ran our experiments on 36 UCI data sets [8] selected by Weka [7], which represent a wide range of domains and data characteristics listed in Table 1. In our experiments, we adopted the following three preprocessing steps.

1. Replacing missing attribute values: We don't handle missing attribute values. Thus, we used the unsupervised filter named *ReplaceMissingValues* in Weka to replace all missing attribute values in each data set.
2. Discretizing numeric attribute values: We don't handle numeric attribute values. Thus, we used the unsupervised filter named *Discretize* in Weka to discretize all numeric attribute values in each data set.
3. Removing useless attributes: Apparently, if the number of values of an attribute is almost equal to the number of instances in a data set, it is a useless attribute. Thus, we used the unsupervised filter named *Remove* in Weka to remove this type of attributes. In these 36 data sets, there are only three such attributes: the attribute "Hospital Number" in the data set "colic.ORIG", the attribute "instance name" in the data set "splice" and the attribute "animal" in the data set "zoo".

We empirically investigated four Bayesian network classifiers: NB [2], TAN [3], AODE [4], and HNB [5], in terms of classification accuracy. We implemented TAN and HNB within the Weka framework and used the implementation of NB and AODE in Weka. In all experiments, the classification accuracy of classifiers on a data set was obtained via 10 runs of 10-fold cross validation. Runs with the various algorithms were carried out on the same training sets and evaluated on the same test sets. Finally, we conducted a two-tailed $t$-test with a 95% confidence level [9] to compare the classifiers with m-estimate and the ones with Laplace estimate.

Table 2 and Table 3 show the classification accuracy and standard deviation of each classifier on each data set. The symbols v and * in the tables respectively denotes statistically significant improvement and degradation with a 95% confidence level.

Our experiments show that the classifiers with our m-estimate perform overall better than the classifiers with Laplace estimate. We summarize the highlights briefly as follows:

**Table 1. Description of data sets used in the experiments.** All these data sets are the whole 36 UCI data sets selected by Weka. We downloaded these data sets in format of arff from main web of Weka.

| No. | Dataset | Instances | Attributes | Classes | Missing | Numeric |
|---|---|---|---|---|---|---|
| 1 | anneal | 898 | 39 | 6 | Y | Y |
| 2 | anneal.ORIG | 898 | 39 | 6 | Y | Y |
| 3 | audiology | 226 | 70 | 24 | Y | N |
| 4 | autos | 205 | 26 | 7 | Y | Y |
| 5 | balance-scale | 625 | 5 | 3 | N | Y |
| 6 | breast-cancer | 286 | 10 | 2 | Y | N |
| 7 | breast-w | 699 | 10 | 2 | Y | N |
| 8 | colic | 368 | 23 | 2 | Y | Y |
| 9 | colic.ORIG | 368 | 28 | 2 | Y | Y |
| 10 | credit-a | 690 | 16 | 2 | Y | Y |
| 11 | credit-g | 1000 | 21 | 2 | N | Y |
| 12 | diabetes | 768 | 9 | 2 | N | Y |
| 13 | Glass | 214 | 10 | 7 | N | Y |
| 14 | heart-c | 303 | 14 | 5 | Y | Y |
| 15 | heart-h | 294 | 14 | 5 | Y | Y |
| 16 | heart-statlog | 270 | 14 | 2 | N | Y |
| 17 | hepatitis | 155 | 20 | 2 | Y | Y |
| 18 | hypothyroid | 3772 | 30 | 4 | Y | Y |
| 19 | ionosphere | 351 | 35 | 2 | N | Y |
| 20 | iris | 150 | 5 | 3 | N | Y |
| 21 | kr-vs-kp | 3196 | 37 | 2 | N | N |
| 22 | labor | 57 | 17 | 2 | Y | Y |
| 23 | letter | 20000 | 17 | 26 | N | Y |
| 24 | lymph | 148 | 19 | 4 | N | Y |
| 25 | mushroom | 8124 | 23 | 2 | Y | N |
| 26 | primary-tumor | 339 | 18 | 21 | Y | N |
| 27 | segment | 2310 | 20 | 7 | N | Y |
| 28 | sick | 3772 | 30 | 2 | Y | Y |
| 29 | sonar | 208 | 61 | 2 | N | Y |
| 30 | soybean | 683 | 36 | 19 | Y | N |
| 31 | splice | 3190 | 62 | 3 | N | N |
| 32 | vehicle | 846 | 19 | 4 | N | Y |
| 33 | vote | 435 | 17 | 2 | Y | N |
| 34 | vowel | 990 | 14 | 11 | N | Y |
| 35 | waveform-5000 | 5000 | 41 | 3 | N | Y |
| 36 | zoo | 101 | 18 | 7 | N | Y |

1. NB-M significantly outperforms NB-L. Compared to NB-L, in the 36 data sets we test, NB-M wins in 8 data sets, loses in 0 data sets, and ties in all the others.
2. TAN-M is competitive with TAN-L. Compared to TAN-L, in the 36 data sets we test, TAN-M wins in 5 data sets, loses in 5 data sets, and ties in all the others.

**Table 2. The detailed experimental results on classification accuracy and standard deviation.** NB-L: Naive Bayes with Laplace estimate; NB-M: Naive Bayes with m-estimate; TAN-L: Tree Augmented Naive Bayes with Laplace estimate; TAN-M: Tree Augmented Naive Bayes with m-estimate. v, * : statistically significant improvement or degradation with a 95% confidence level.

| Datasets | NB-L | NB-M | TAN-L | TAN-M |
|---|---|---|---|---|
| anneal | 94.32±2.23 | 96.94±1.60 v | 96.75±1.73 | 98.37±1.28 v |
| anneal.ORIG | 88.16±3.06 | 88.12±3.22 | 90.48±2.16 | 91.65±2.77 |
| audiology | 71.4±6.37 | 77.16±9.13 | 65.3±6.81 | 72.48±9.18 v |
| autos | 63.97±11.35 | 66.9±11.19 | 72.59±9.64 | 79.02±8.86 v |
| balance-scale | 91.44±1.3 | 91.44±1.29 | 85.97±2.95 | 86.5±2.91 |
| breast-cancer | 72.94±7.71 | 72.17±7.96 | 69.53±6.55 | 67.54±7.8 |
| breast-w | 97.3±1.75 | 97.38±1.73 | 95.52±2.38 | 95.95±2.23 |
| colic | 78.86±6.05 | 78.75±6.09 | 80.03±5.99 | 80.06±6.24 |
| colic.ORIG | 74.21±7.09 | 73.42±6.54 | 67.76±6.07 | 64.96±7.1 |
| credit-a | 84.74±3.83 | 84.23±3.85 | 84.19±4.15 | 82.26±3.95 |
| credit-g | 75.93±3.87 | 75.68±3.95 | 74.84±3.86 | 73.46±4 |
| diabetes | 75.68±4.85 | 75.01±5.07 | 76.04±4.85 | 74.98±4.92 |
| glass | 57.69±10.07 | 57.86±9.35 | 58.64±9.06 | 58.26±9.14 |
| heart-c | 83.44±6.27 | 82.29±6.69 | 79.83±8.55 | 75.99±8.35 * |
| heart-h | 83.64±5.85 | 83.02±6.23 | 81.2±5.97 | 77.08±6.19 * |
| heart-statlog | 83.78±5.41 | 82.11±6.1 | 79.59±5.87 | 75.70±7.34 * |
| hepatitis | 84.06±9.91 | 85.87±9.08 | 83±9.11 | 82.67±10.0 |
| hypothyroid | 92.79±0.73 | 92.77±0.74 | 93.35±0.59 | 92.83±0.73 * |
| ionosphere | 90.86±4.33 | 90.74±4.34 | 91.4±4.5 | 92.77±4.13 |
| iris | 94.33±6.79 | 94.13±6.65 | 94.07±5.68 | 92.6±7.15 |
| kr-vs-kp | 87.79±1.91 | 87.81±1.91 | 92.86±1.47 | 92.85±1.46 |
| labor | 96.7±7.27 | 95.3±9.13 | 89±12.39 | 83.43±14.1 |
| letter | 70.09±0.93 | 70.75±0.95 v | 82.67±0.8 | 83.85±0.72 v |
| lymph | 85.97±8.88 | 84±9.05 | 84.51±9.39 | 79.89±9.92 |
| mushroom | 95.52±0.78 | 98.89±0.36 v | 99.99±0.05 | 100±0.02 |
| primary-tumor | 47.2±6.02 | 47.15±6.06 | 44.8±6.74 | 44.37±6.37 |
| segment | 89.03±1.66 | 90.48±1.55 v | 93.91±1.57 | 94.53±1.46 |
| sick | 96.78±0.91 | 97.17±0.78 v | 97.69±0.69 | 97.61±0.72 |
| sonar | 76.35±9.94 | 75.34±10.2 | 75.39±9.47 | 72.17±9.68 |
| soybean | 92.2±3.23 | 93.54±2.92 v | 94.93±2.44 | 94.6±2.59 |
| splice | 95.42±1.14 | 95.52±1.13 | 94.87±1.23 | 95.14±1.21 |
| vehicle | 61.03±3.48 | 61.11±3.65 | 73.34±3.8 | 74.09±3.95 |
| vote | 90.21±3.95 | 90.28±3.93 | 94.43±3.34 | 94.64±3.29 |
| vowel | 66.09±4.78 | 67.92±4.56 v | 91.87±2.8 | 94.52±2.68 v |
| waveform-5000 | 79.97±1.46 | 79.89±1.52 | 80.41±1.82 | 77.80±1.75 * |
| zoo | 94.37±6.79 | 97.83±4.35 v | 96.63±5.84 | 96.83±6.47 |

3. AODE-M significantly outperforms AODE-L. Compared to AODE-L, in the 36 data sets we test, AODE-M wins in 8 data sets, loses in 1 data sets, and ties in all the others.

**Table 3. The detailed experimental results on classification accuracy and standard deviation.** AODE-L: Averaged One-Dependence Estimators with Laplace estimate; AODE-M: Averaged One-Dependence Estimators with m-estimate; HNB-L: Hidden Naive Bayes with Laplace estimate; HNB-M: Hidden Naive Bayes with m-estimate. v, * : statistically significant improvement or degradation with a 95% confidence level.

| Datasets | AODE-L | AODE-M | HNB-L | HNB-M |
|---|---|---|---|---|
| anneal | 96.74±1.72 | 97.88±1.44 v | 97.74±1.28 | 98.39±1.33 |
| anneal.ORIG | 88.79±3.17 | 88.8±3.13 | 89.87±2.2 | 91.82±2.74 v |
| audiology | 71.66±6.42 | 77.91±9.13 v | 69.04±5.83 | 80.99±8.68 v |
| autos | 73.38±10.24 | 77.91±9.63 | 75.49±9.89 | 79.02±9.3 |
| balance-scale | 89.78±1.88 | 89.39±1.96 | 89.14±2.05 | 89.59±2.48 |
| breast-cancer | 72.53±7.15 | 71.8±6.7 | 73.09±6.11 | 70.3±6.69 |
| breast-w | 97.11±1.99 | 96.64±2.21 | 95.67±2.33 | 96.74±1.96 |
| colic | 80.9±6.19 | 80.95±6.3 | 81.44±6.12 | 81.15±6.34 |
| colic.ORIG | 75.3±6.6 | 76.2±7.2 | 75.66±5.19 | 76.88±6.69 |
| credit-a | 85.91±3.78 | 85.06±3.9 | 85.8±4.1 | 84.58±4.6 |
| credit-g | 76.42±3.86 | 75.85±4.05 | 76.29±3.45 | 76.82±3.74 |
| diabetes | 76.37±4.35 | 76.11±4.7 | 76±4.6 | 75.62±4.73 |
| glass | 61.13±9.79 | 58.08±9.54 | 59.02±8.67 | 59.1±8.82 |
| heart-c | 82.48±6.96 | 80.96±7.08 | 82.31±6.82 | 81.2±7.59 |
| heart-h | 84.06±5.85 | 82.97±5.72 | 83.21±5.88 | 79.95±5.82 |
| heart-statlog | 83.67±5.37 | 81.15±6.21 | 82.7±5.89 | 81.11±6.24 |
| hepatitis | 84.82±9.75 | 86.2±8.29 | 83.92±9.43 | 82.19±10.2 |
| hypothyroid | 93.53±0.62 | 93.28±0.63 | 93.49±0.47 | 93.29±0.55 |
| ionosphere | 92.08±4.24 | 92.77±3.94 | 92±4.32 | 92.82±3.86 |
| iris | 94.47±6.22 | 94.47±6.29 | 93.93±5.92 | 93.33±7.03 |
| kr-vs-kp | 91.01±1.67 | 91.29±1.56 | 92.36±1.3 | 92.35±1.3 |
| labor | 95.3±8.49 | 92.87±10.9 | 92.73±11.16 | 90.9±12.04 |
| letter | 85.54±0.68 | 88.33±0.56 v | 84.68±0.74 | 86.11±0.70 v |
| lymph | 86.25±9.43 | 83.99±8.04 | 83.9±9.31 | 81.69±8.02 |
| mushroom | 99.95±0.07 | 99.96±0.06 | 99.94±0.1 | 99.96±0.06 |
| primary-tumor | 47.67±6.3 | 47.68±6.03 | 47.66±6.21 | 47.55±5.86 |
| segment | 92.94±1.4 | 95.16±1.30 v | 93.72±1.5 | 94.77±1.42 v |
| sick | 97.51±0.73 | 97.91±0.64 v | 97.77±0.68 | 97.67±0.76 |
| sonar | 79.04±9.42 | 79.34±10.0 | 81.75±8.4 | 79.6±8.95 |
| soybean | 93.28±2.84 | 94.58±2.33 v | 93.88±2.47 | 94.76±2.41 |
| splice | 96.12±1 | 96.32±0.97 | 95.84±1.1 | 96.13±0.99 v |
| vehicle | 71.62±3.6 | 72.79±3.81 | 72.15±3.41 | 73.37±3.94 |
| vote | 94.52±3.19 | 94.53±3.17 | 94.43±3.18 | 94.36±3.2 |
| vowel | 89.52±3.12 | 93.39±2.42 v | 91.34±2.92 | 92.63±2.66 |
| waveform-5000 | 84.24±1.59 | 83.49±1.65 * | 83.79±1.54 | 83.39±1.61 |
| zoo | 94.66±6.38 | 98.03±3.97 v | 97.73±4.64 | 98.62±3.44 |

4. HNB-M significantly outperforms HNB-L. Compared to HNB-L, in the 36 data sets we test, HNB-M wins in 5 data sets, loses in 0 data sets, and ties in all the others.

## 5 Conclusions and Future Work

In learning Bayesian network classifiers, how to estimate probabilities from a given set of training examples is crucial problem. Responding to this question, we single out a special m-estimate method and empirically investigate its effect on various Bayesian network classifiers, such as Naive Bayes (NB) [2], Tree Augmented Naive Bayes (TAN) [3], Averaged One-Dependence Estimators (AODE) [4], and Hidden Naive Bayes (HNB) [5]. Our experiments show that the classifiers with our m-estimate perform better than the ones with Laplace estimate.

In principle, our m-estimate could be used to improve the probability estimation of other classification models, such as decision trees [10]. It is our main direction for future research.

## References

1. Chickering, D. M.: Learning Bayesian Networks is NP-Complete. In: Fisher, D. and Lenz, H., editors: Learning from Data: Artificial Intelligence and Statistics. Springer-Verlag, New York (1996) 121-130
2. Langley, P., Iba, W., Thomas, K.: An Analysis of Bayesian Classifiers. In: Proceedings of the Tenth National Conference of Artificial Intelligence. AAAI Press (1992) 223-228.
3. Friedman, N., Geiger, D., Goldszmidt, M.: Bayesian Network Classifiers. Machine Learning. 29 (1997) 131-163
4. Webb, G. I., Boughton, J., Wang, Z.: Not so Naive Bayes: Aggregating One-Dependence Estimators. Machine Learning. 58 (2005) 5-24
5. Zhang, H., Jiang, L., Su, J.: Hidden Naive Bayes. In: Proceedings of the 20th National Conference on Artificial Intelligence. AAAI Press (2005) 919-924
6. Mitchell, T. M.: Machine learning. McGraw-Hill (1997)
7. Witten, I. H., Frank, E.: Data Mining: Practical Machine Mearning Tools and Techniques. 2nd Edition, Morgan Kaufmann, San Francisco (2005) http://prdownloads.sourceforge.net/weka/datasets-UCI.jar
8. Merz, C., Murphy, P., Aha, D.: UCI Repository of Machine Learning Databases. In Dept of ICS, University of California, Irvine (1997) http://www.ics.uci.edu/ mlearn/MLRepository.html
9. Nadeau, C., Bengio, Y.: Inference for the Generalization Error. In: Advances in Neural Information Processing Systems. MIT Press, 12 (1999) 307-313
10. Quinlan, J. R.: C4.5: Programs for Machine Learning. Morgan Kaufmann, San Mateo, CA (1993)

# Similarity Computation of Fuzzy Membership Function Pairs with Similarity Measure

Dong-hyuck Park, Sang H. Lee, Eui-Ho Song, and Daekeon Ahn

School of Mecatronics, Changwon National University
9 Sarim-dong, Changwon, Gyeongnam, 641-773, Korea
{gurehddl, leehyuk, ehsong, niceahn}@changwon.ac.kr

**Abstract.** The similarity computations for fuzzy membership function pairs are carried out. Similarity measure is proposed for the general fuzzy sets. Obtained similarity measure has the inverse meaning of fuzzy entropy, and the proposed similarity measure is also constructed through distance measure. Finally similarity computation results are computed for the various membership function pairs.

**Keywords:** Similarity measure, distance, fuzzy number.

## 1 Introduction

Computation of similarity between two or more informations is very interesting for the fields of decision making, pattern classification, or *etc*.. Until now the research of designing similarity measure has been made by numerous researchers[1-6]. Most studies are focussed on designing similarity measure based on membership function. Hence the studies are mainly carried out for the triangular or trapezoidal fuzzy sets. With the previous results it is vague to obtain degree of similarity between general fuzzy sets, and furthermore crisp set and crisp set or crisp set and fuzzy set. In this paper with our previous similarity measure results we try to compute the similarity measure of two fuzzy membership functions, and analyze the result of degree of similarity between fuzzy set and crisp set. First we introduce the similarity measure which is previously derived from fuzzy number, and derive similarity measure via well known-Hamming distance. We explain the similarity measure with the certainty and uncertainty point of view. The larger area of coinciding certainty or uncertainty, the better similarities are. Two similarity measures that are derived from fuzzy number and distance measure are compared with computation of fuzzy membership function pairs. Two similarity measures have their own strong points, fuzzy number methods is simple and easy to compute similarity if membership function is trapezoidal or triangular. Whereas similarity with distance method needs more time and consideration, however that can be applied to the general membership function. At this point, it is interesting to study and compare two similarity measure for the fuzzy set and crisp set.

In the next section, preliminary results about fuzzy number, center of gravity, and the similarity measure are introduced. In Section 3, similarity measures with distance measure and fuzzy number are derived and proved. Also two similarity measures are compared and discussed in Section 4. In the example, we obtain similarity measure

D.-S. Huang, L. Heutte, and M. Loog (Eds.): ICIC 2007, LNAI 4682, pp. 485–492, 2007.

values that have proper meaning. Conclusions are followed in Section 5. Notations of Liu's are used in this paper [7].

## 2 Similarity Measure Preliminaries

In this section, we introduce some preliminary results for the degree of similarity. Fuzzy number, center of gravity, and axiomatic definitions of similarity measure are included.

### 2.1 Fuzzy Number, and Center of Gravity

A generalized fuzzy number $\tilde{A}$ is defined as $\tilde{A}=(a,b,c,d,\omega)$, where $0<\omega\leq 1$ and $a,b,c$ and $d$ are real numbers [1,2]. Trapezoidal membership function $\mu_{\tilde{A}}$ of fuzzy number $\tilde{A}$ satisfies the following conditions[4]:

1) $\mu_{\tilde{A}}$ is a continuous mapping from real number to the closed interval $[0,1]$
2) $\mu_{\tilde{A}}(x)=0$, where $-\infty<x\leq a$
3) $\mu_{\tilde{A}}(x)$ is strictly increasing on $[a,b]$
4) $\mu_{\tilde{A}}(x)=\omega$ , where $b\leq x\leq c$
5) $\mu_{\tilde{A}}(x)$ is strictly decreasing on $[c,d]$
6) $\mu_{\tilde{A}}(x)=0$ , where $d\leq x<\infty$.

If $b=c$ is satisfied, then it is natural to satisfy triangular type. Four fuzzy number operations are also found in literature [4].

Traditional center of gravity(COG) is defined by

$$x^{*}_{\tilde{A}}=\frac{\int x\mu_{\tilde{A}}(x)dx}{\int \mu_{\tilde{A}}(x)dx}$$

where $\mu_{\tilde{A}}$ is the membership function of the fuzzy number $\tilde{A}$, $\mu_{\tilde{A}}(x)$ indicates the membership value of the element $x$ in $\tilde{A}$, and generally, $\mu_{\tilde{A}}(x)\in[0,1]$ . Chen and Chen presented a new method to calculate COG point of a generalized fuzzy number [4]. They derived the new COG calculation method based on the concept of the medium curve. These COG points play an important role in the calculation of similarity measure with fuzzy number. We will introduce more in Section 3.

### 2.2 Similarity Measure

Liu suggested axiomatic definition of similarity measure as follows [7]. By this definition, we study the meaning of similarity measure.

Definition 2.1 [7] A real function : $s:F^{2}\rightarrow R^{+}$ is called a similarity measure, if $s$ has the following properties:

(S1) $s(A,B)=s(B,A), \forall A,B\in F(X)$
(S2) $s(D,D^c)=0 \ \forall D\in P(X)$
(S3) $s(C,C)=\max_{A,B\in F} s(A,B), \ \forall C\in F(X)$
(S4) $\forall A,B,C\in F(X)$ ,if $A\subset B\subset C$ ,then $s(A,B)=s(A,C)$ and $s(B,C)=s(A,C)$.

Where , $R^+=[0,\infty)$, $X$ is the universal set, $F(X)$ is the class of all fuzzy sets of $X$, $P(X)$ is the class of all crisp sets of $X$ , and $D^C$ is the complement of $D$. Fuzzy normal similarity measure on $F$ is also obtained by the division of $\max_{C,D\in F} s(C,D)$.

## 3 Similarity Measure by Fuzzy Number and Distance Measure

In this section we introduce the degree of similarity which are contained in the previous literatures [1-4]. Which are all based on the fuzzy number. And the similarity measure construction with the distance measure is contained in subsection 3.2, and proved.

### 3.1 Similarity Measure Via Fuzzy Number

In the literatures [1-4], degrees of similarities are derived through membership function fuzzy number and center of gravity. We introduce the conventional fuzzy measure that is based on the fuzzy number. Chen introduced the degree of similarity for trapezoidal or triangular fuzzy membership function of $\tilde{A}$ and $\tilde{B}$ as [1]

$$s(\tilde{A},\tilde{B})=1-\frac{\sum_{i=1}^{n}|a_i-b_i|}{4} \tag{1}$$

where $s(\tilde{A},\tilde{B})\in[0,1]$. If $\tilde{A}$ and $\tilde{B}$ are trapezoidal or triangular fuzzy numbers, then the $n$ can be 4 or 3, respectively. For trapezoidal membership function fuzzy number satisfy $\tilde{A}=(a_1,a_2,a_3,a_4,1)$ and $\tilde{B}=(b_1,b_2,b_3,b_4,1)$.

Hsieh *et. al.* also proposed similarity measure for the trapezoidal and triangular fuzzy membership function as follows [2]:

$$s(\tilde{A},\tilde{B})=\frac{1}{1+d(\tilde{A},\tilde{B})} \tag{2}$$

where $d(\tilde{A},\tilde{B})=|P(\tilde{A})-P(\tilde{B})|$, and if $\tilde{A}$ and $\tilde{B}$ are triangular fuzzy number, then the graded mean integration of $\tilde{A}$ and $\tilde{B}$ are defined as follows:

$$P(\tilde{A})=\frac{a_1+4a_2+a_3}{6} \text{ and } P(\tilde{B})=\frac{b_1+4b_2+b_3}{6},$$

if $\tilde{A}$ and $\tilde{B}$ are trapezoidal fuzzy number, then the graded mean integration of $\tilde{A}$ and $\tilde{B}$ are also defined as follows:

$$P(\tilde{A})=\frac{a_1+2a_2+2a_3+a_4}{6} \text{ and } P(\tilde{B})=\frac{b_1+2b_2+2b_3+b_4}{6}.$$

Lee derived the trapezoidal similarity measure using fuzzy number operation and norm definition. That is

$$S(\tilde{A},\tilde{B}) = 1 - \frac{\|\tilde{A}-\tilde{B}\|_{l_p}}{\|U\|} \times 4^{-1/p} \tag{3}$$

where $\|\tilde{A}-\tilde{B}\|_{l_p} = \left(\sum_i (|a_i - b_i|)\right)^{1/p}$, $\|U\| = \max(U) - \min(U)$, and $p$ is the natural number greater or equal 1, finally $U$ is the universe of discourse.

Chen and Chen propose similarity measure to overcome the drawbacks of existing similarity:

$$S(\tilde{A},\tilde{B}) = [1 - \frac{\sum_i |a_i - b_i|}{4}] \times (1 - |x^*_{\tilde{A}} - x^*_{\tilde{B}}|)^{B(S_{\tilde{A}},S_{\tilde{B}})} \times \frac{\min(y^*_{\tilde{A}}, y^*_{\tilde{B}})}{\max(y^*_{\tilde{A}}, y^*_{\tilde{B}})} \tag{4}$$

where $(x^*_{\tilde{A}}, y^*_{\tilde{A}})$ and $(x^*_{\tilde{B}}, y^*_{\tilde{B}})$ are the COG of fuzzy number $\tilde{A}$ and $\tilde{B}$, $S_{\tilde{A}}$ and $S_{\tilde{B}}$ are expressed by $S_{\tilde{A}} = a_4 - a_1$ and $S_{\tilde{B}} = b_4 - b_1$ if they are trapezoidal. $B(S_{\tilde{A}}, S_{\tilde{B}})$ is denoted by 1 if $S_{\tilde{A}} + S_{\tilde{B}} > 0$, and 0 if $S_{\tilde{A}} + S_{\tilde{B}} = 0$. In (4), $B(S_{\tilde{A}}, S_{\tilde{B}})$ is used to determine whether we consider the COG distance or not.

### 3.2 Similarity Measure with Distance Function

To design the similarity measure via distance, first we introduce distance measure [7].

**Definition 3.1.** A real function : *d* is called a distance measure on *F(X)*, if *d* satisfies the following properties:

(D1) $d(A,B) = d(B,A)$, $\forall A,B \in F(X)$

(D2) $d(A,A) = 0$, $\forall A \in F(X)$

(D3) $d(D,D) = \max_{A,B \in P} d(A,B)$, $\forall D \in P(X)$

(D4) $\forall A,B,C \in F(X)$, if $A \subset B \subset C$, then $d(A,B) \le d(A,C)$ and $d(B,C) \le d(A,C)$.

Hamming distance is commonly used as distance measure between fuzzy sets $A$ and $B$,

$$d(A,B) = \frac{1}{n}\sum_{i=1}^{n} |\mu_A(x_i) - \mu_B(x_i)|$$

where $X = \{x_1, x_2, \cdots x_n\}$, $|\kappa|$ is the absolute value of $\kappa$. $\mu_{\tilde{A}}$ is the membership function of $A \in F(X)$. With Definition 3.1, we propose the following theorem as the similarity measure.

**Theorem 3.1.** For any set $A, B \in F(X)$ or *P(X)*, if $d$ satisfies Hamming distance measure, then

$$s(A,B)=2-d((A\cap B),[1])-d((A\cup B),[0]) \tag{5}$$

is the similarity measure between set $A$ and set $B$.

Proof. We prove that the eq. (5) satisfies the Definition 3.1. (S1) means the commutativity of set and , hence it is clear from (5) itself. For (S2),

$$s(D,D^C)=2-d((D\cap D^C),[1])-d((D\cup D^C),[0])$$

then $d((D\cap D^C),[1])$ and $d((D\cup D^C),[0])$ become 1.

For arbitrary sets $A$, $B$ inequality of (S3) is proved by

$$\begin{aligned}s(A,B)&=2-d((A\cap B),[1])-d((A\cup B),[0])\\&\leq 2-d((C\cap C),[1])-d((C\cup C),[0])\\&=s(C,C).\end{aligned}$$

Inequality is satisfied from $d((A\cap B),[1])\geq d((C\cap C),[1])$ and $d((A\cup B),[0])-d((C\cup C),[0])$.

Finally, (S4) is $\forall A,B,C\in F(X)$, $A\subset B\subset C$ ,

$$\begin{aligned}s(A,B)&=2-d((A\cap B),[1])-d((A\cup B),[0])\\&=2-d(A,[1])-d(B,[0])\\&\geq 2-d(A,[1])-d(C),[0])\\&=s(A,C)\end{aligned}$$

also

$$\begin{aligned}s(B,C)&=2-d((B\cap C),[1])-d((B\cup C),[0])\\&=2-d(B,[1])-d(C,[0])\\&\geq 2-d(A,[1])-d(C),[0])\\&=s(A,C)\end{aligned}$$

is satisfied. Inequality is also satisfied from the facts of $d(B,[0])\leq d(C,[0])$ and $d(B,[1])\leq d(A,[1])$. Therefore proposed similarity measure (5) satisfies modified similarity measure. In the following Section 4 we compute the degree of similarity between membership functions. The results are compared with two similarity measures.

## 4 Computation of Similarity Measures

In [4], Chen and Chen computed degree of similarity for the following 12 membership function sets. 12 pairs contain fuzzy-fuzzy sets, crisp-crisp sets, and fuzzy-crisp set.

They proposed 7 descriptions compare to the existing method. One of descriptions is represented as follows

1) From Set 1, we can see that $\tilde{A}$ and $\tilde{B}$ are different generalized fuzzy number. However, from Table 1, we can see that if we apply Hsieh and Chen's method, it has the same degree of similarity[4].

The other 6 description also pointed out the same degree of similarity of other method[4]. Main characteristics of the Chen and Chen's are 10 sets are all different

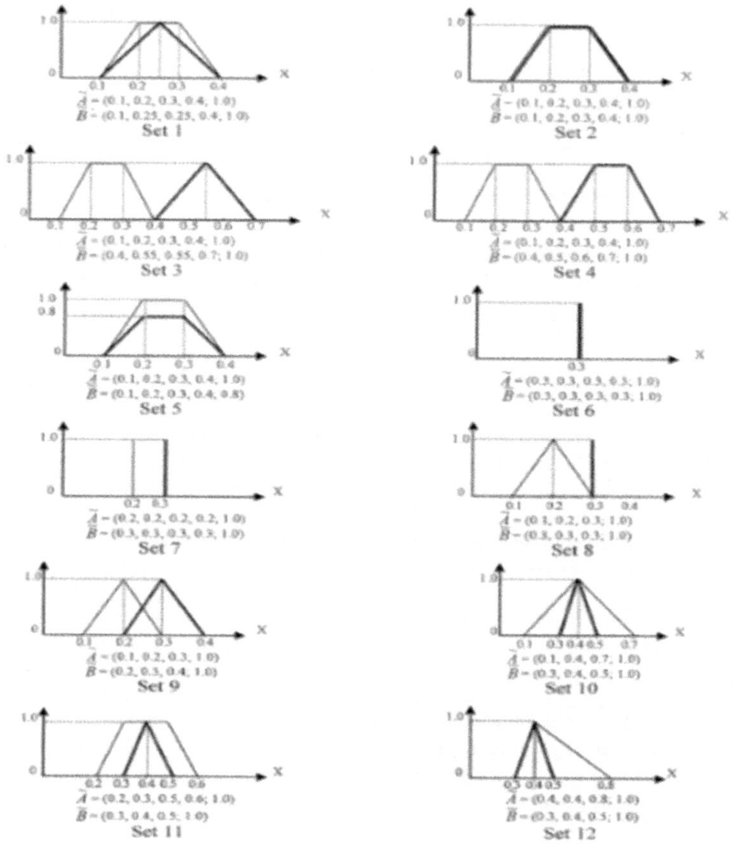

**Fig. 1.** Twelve sets of fuzzy numbers[4]

**Table 1.** Comparison with the result of Chen and Chen

| | Set1 | Set2 | Set3 | Set4 | Set5 | Set6 | Set7 | Set8 | Set9 | Set10 | Set11 | Set12 |
|---|---|---|---|---|---|---|---|---|---|---|---|---|
| Lee[3] | 0.9167 | 1 | 0.5 | 0.5 | 1 | * | 0 | 0.5 | 0.6667 | 0.8333 | 0.75 | 0.8 |
| Hsieh and Chen[2] | 1 | 1 | 0.7692 | 0.7692 | 1 | 1 | 0.909 | 0.909 | 0.909 | 1 | 1 | 0.9375 |
| Chen [1] | 0.975 | 1 | 0.7 | 0.7 | 1 | 1 | 0.9 | 0.9 | 0.9 | 0.9 | 0.9 | 0.9 |
| Chen and Chen[4] | 0.8357 | 1 | 0.42 | 0.49 | 0.8 | 1 | 0.9 | 0.54 | 0.81 | 0.9 | 0.72 | 0.78 |
| The proposed Method | 0.839 | 1 | 0.426 | 0.344 | 0.871 | 1 | 0 | 0.476 | 0.516 | 0.672 | 0.512 | 0.618 |

except Set 2 and 6. We compute 12 sets with our similarity measure (5). In our computation, same results with those of Chen and Chen are obtained, *i.e* different similarity degrees between 10 sets except Set 2 and Set 6. Similarity computation results are illustrated in Table 1.

From now we will compute one of sets in Fig. 1, Set 8. With (4) Chen and Chen compute the degree of similarity as follows

$$S(\tilde{A},\tilde{B})=[1-\frac{0.2+0.1+0}{3}]\times(1-0.1)^{1}\times\frac{\min(1/3,0.5)}{\max(1/3,0.5)}=0.54\cdot$$

Where as we needed computation conditions as

Universe of discourse : 0.1~0.8
Data points : 70
Sample distance : 0.01

In Set 8, for fuzzy set $A$, domain can be from 0.1 to 0.3 among universe of discourse, whereas crisp set $B$ has value only on 0.3. With similarity measure (5), similarity computation is 0.476.

Finally one more interesting comparison is the result of Set 7 similarity comparison. Chen and Chen compute as follows

$$S(\tilde{A},\tilde{B})=[1-\frac{0.4}{4}]\times(1-|0.1|)^{B(S_{\tilde{A}},S_{\tilde{B}})}\times\frac{\min(0.5,0.5)}{\max(0.5,0.5)}$$
$$=[1-0.1]=0.9\ \cdot$$

This computation keep the rule of (4), hence the result is obtained. However there can be another way of approach to the similarity between crisp sets. With our similarity measure we compute Set 7 pair similarity as follows.

$$\begin{aligned}s(A,B)&=2-d((A\cap B),[1])-d((A\cup B),[0])\\&=2-d([0],[1])-d([1],[0])\\&=2-1-1=0.\end{aligned}$$

Where $(A\cap B)$ means the $\min(A(x_i),B(x_i))$, hence it satisfies $[0]$. Also $(A\cup B)$ represents the maximum value of between $A(x_i)$ and $B(x_i)$. By inspection of Set 7, two variables 0.2 and 0.3 have the corresponding membership value 1. Therefore

$$\begin{aligned}d((A\cup B),[0]_X)&=\frac{1}{2}\sum_{i=1}^{2}\{|\mu_{A\cup B}(0.2)-0|+|\mu_{A\cup B}(0.3)-0|\}\\&=\frac{1}{2}\sum_{i=1}^{2}\{|1-0|+|1-0|\}=1\end{aligned}$$

is satisfied.

Similarity satisfy zero with (4), if it satisfy

$$[1-\frac{\sum_{i=1}^{4}|a_i-b_i|}{4}]=0 \text{ or } (1-|x^*_{\tilde{A}}-x^*_{\tilde{B}}|)=0\cdot$$

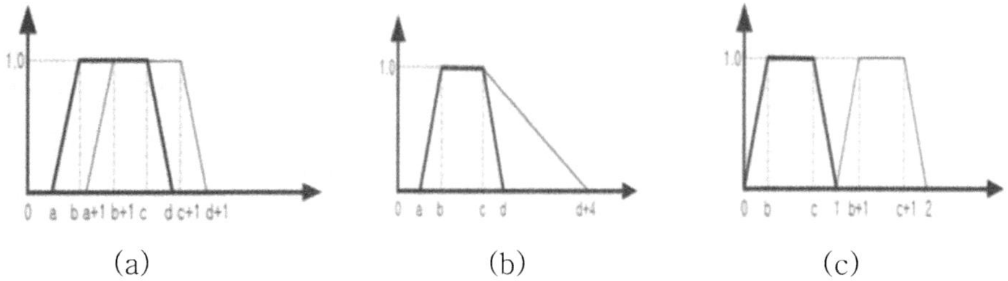


**Fig. 2.** Similarity zero membership function pairs

For this satisfaction, summation of all difference satisfies 4 for trapezoidal case, or difference of $x$-COG also satisfies 1. Fig. 2 can be the similarity zero cases.

3 cases are not proper for the normalized universe of discourse. If we do not consider normalized cases, Fig. 2 membership functions may not have the zero degree of similarity.

## 5 Conclusions

We have introduced the fuzzy number and the similarity measure that is derived from fuzzy number. These results are easy to compute, however the result is strictly limited for the trapezoidal or triangular membership functions. Whereas with similarity measure we also compute the similarity measure, and results are generally applied for the arbitrary shape of membership functions. The usefulness of proposed similarity measure is proved. By the comparison with previous example, we can see that proposed similarity measure can be applied to the general types of fuzzy membership functions.

## References

1. Chen, S.M.: New Methods for Subjective Mental Workload Assessment and Fuzzy Risk Analysis, Cybern. Syst. : Int. J., vol 27, no. 5, (1996) 449-472
2. Hsieh, C.H., Chen, S.H.: Similarity of Generalized Fuzzy Numbers with Graded Mean Integration Representation, in Proc. 8th Int. Fuzzy Systems Association World Congr., vol 2, (1999) 551-555
3. Lee, H.S.: An Optimal Aggregation Method for Fuzzy Opinions of Group Decision, Proc. 1999 IEEE Int. Conf. Systems, Man, Cybernetics, vol. 3, (1999) 314-319
4. Chen S.J., Chen, S.M.: Fuzzy Risk Analysis Based on Similarity Measures of Generalized Fuzzy Mumbers, IEEE Trans. on Fuzzy Systems, vol. 11, no. 1, (2003) 45-56
5. Lee, S.H., Cheon, S.P., Jinho, K.: Measure of certainty with fuzzy entropy function, LNAI, Vol. 4114, (2006) 134-139
6. Lee, S.H., Kim, J.M., Choi, Y.K.: Similarity Measure Construction Using Fuzzy Entropy and Distance Measure, LNAI Vol.4114, (2006) 952-958
7. Liu, X.: Entropy, Distance Measure and Similarity Measure of Fuzzy Sets and Their rRelations, Fuzzy Sets and Systems, 52, (1992) 305-318
8. Fan, J.L., Xie, W.X.: Distance Measure and Induced Fuzzy Entropy, Fuzzy Set and Systems, 104, (1999) 305-314
9. Fan, J.L., Ma, Y.L., Xie, W.X.: On Some Properties of Distance Measures, Fuzzy Set and Systems, 117, (2001) 355-361

# Spatial Selectivity Estimation Using Cumulative Density Wavelet Histogram

Byung Kyu Cho

Department of Computer Science, Chungju National University, Korea
bkcho@chungju.ac.kr

**Abstract.** The purpose of selectivity estimation is to minimize the error of estimated value and query result using the summary data maintained on small memory space. Many works have been performed to estimate accurately selectivity. However, the existing works require a large amount of memory to retain accurate selectivity. In order to solve this problem, we propose a new technique cumulative density wavelet histogram, called CDW Histogram which is able to compress summary data and get an accurate selectivity in small memory space. The proposed method is based on the sub-histograms created by CD histogram and the wavelet transformation technique. The experimental results showed that the proposed method is superior to the existing selectivity estimation technique.

**Keywords:** Spatial Selectivity Estimation, CD Histogram, Wavelet, Histogram compression.

## 1 Introduction

There are several components in a spatial database management system that requires reasonably accurate estimates of the result size for spatial queries [6,7,9,11]. For example, cost-based query optimizers use it to evaluate the costs of different query execution plans and choose the preferred one. Also, query profilers use them to provide quick feedback to users as a mean to detect some forms of semantic misconceptions before queries are actually executed [4].

Several techniques have been proposed in the literature to estimate query result sizes, including histograms, sampling and parametric techniques [1,2,4]. Of these, histograms approximate the frequency distribution of an attribute by grouping attribute values into buckets and approximating true attribute values and their frequencies in data based on summary statistics maintained in each bucket [8,10,12,13]. The main advantages of histograms over other techniques are that they incur almost no run-time overhead; they do not require data to fit a probability distribution or a polynomial one for real-world databases.

This paper focuses on estimating the selectivity of range queries on rectangular objects. Rectangular objects incur multiple-count problem when they span across the several buckets. To solve this problem, the CD and Euler histograms are proposed in the literature [7,11]. Those techniques can give very accurate results for range queries

D.-S. Huang, L. Heutte, and M. Loog (Eds.): ICIC 2007, LNAI 4682, pp. 493–504, 2007.

on rectangular objects. The CD histogram can give a good result on both point and rectangular objects while the Euler histogram can be just applied to rectangular objects. Especially, although the CD histogram gives very accurate results on spatial datasets, they have the problem which they require a large amount of memory to maintain sub-histogram for four-corner points of objects. If such method is used in given small memory capacity, good selectivity cannot be obtained. Also recent advancements in computing and mobile technology make it possible to provide information services on the user's position and geography using small size database, thus increasing the importance, in practical as well as in theoretical aspects, of selectivity estimation method for small database.

Motivated by the above reasoning, we propose a novel technique cumulative density wavelet histogram, called CDW histogram that requires a small memory space over CD histogram. The proposed technique take advantage of strong points of cumulative density histogram and Haar wavelet transform technique - high accuracy provided by the former and economization of memory space supported by the latter. Consequently, our technique is able to support exact estimation, high compression effect.

The rest of this paper is organized as follows. In the next section we summarize related work. The proposed technique is presented in section 3. In section 4 we describe the strengths and weakness of the proposed method through experiments. Finally, we draw conclusions and give a future work in Section 5.

## 2 Related Works

Selectivity estimation is a well-studied problem for traditional data types such as integer. Histograms are most widely used forms for doing selectivity estimation in relational database systems. Many different histograms have been proposed in the literature and some have been deployed in commercial RDBMSs. In case selectivity estimation in terms of spatial data, some techniques for range queries have been proposed in the literature [6,7,9,11]. Most of spatiotemporal histogram focuses on point object [10,12,13,14,16], and some techniques just focus on rectangular object[6,7,11].

In [6], Acharya et. al. proposed the MinSkew algorithm. The MinSkew algorithm starts with a density histogram of the dataset, which effectively transforms region objects to point data. The density histogram is further split into more buckets until the given bucket count is reached or the sum of the variance in each bucket cannot be reduced by additional splitting. In result, the MinSkew algorithm constructs a spatial histogram to minimize the spatial-skew of spatial objects.

The CD histogram technique is proposed in [7]. Typically when building a histogram for region objects, an object may be counted multiple times if it spans across several buckets. The algorithm of CD histogram addresses this problem by keeping four sub-histogram to store the number of corresponding corner points that fall in the buckets, so even if a rectangle spans several buckets, it is counted exactly once in a each sub-histogram. The Euler Histogram technique is proposed in [11]. The mathematical foundation of the Euler Histogram is based on Euler's Formula in graph theory, hence the name Euler Histogram. As in the CD Histogram, Euler Histogram also addresses the multiple-count problem. Though these techniques are efficient methods to approximate range query selectivity estimation in spatial databases. These techniques require a large amount of memory for better accuracy.

To compress the summary information in databases, in [3,5,8,9,15] Matias et al. introduce a new type of histograms, called wavelet-based histograms, based upon multidimensional wavelet decomposition. Wavelet decomposition is performed on the underlying data distribution, and most significant wavelet coefficients are chosen to compose the histogram. In other words, the data points are compressed into a set of numbers via a sophisticated multi-resolution transformation. Those coefficients constitute the final histogram. This approach can be extended very naturally to efficiently compress the joint distribution of multiple attribute.

## 3 Cumulative Density Wavelet Histogram

In order to reduce the restriction on memory space of cumulative density histogram, we apply wavelet transformation method to the histogram. The proposed technique, CDW histogram, is combination method taking advantage of strong point of CD histogram and wavelet transformation. Table 1 and table 2 show symbols that are used to describe the CDW histogram.

**Table 1.** Symbols for wavelet transformation

| Parameters | Description |
|---|---|
| $A_i$ , $W_i$ | Input data array and wavelet coefficient array |
| $B_{i,}$ $D_i$ | Bucket and data value for grid cell i |
| $O_i$ | Recovery value of cell i |
| $r_i$ | Resolution level of wavelet coefficient i |
| $Wav.coeff_i$ | Non-normalized wavelet coefficient |
| $Norm.coeff_i$ | Normalized wavelet coefficient |
| $Retained\ coeff_i$ | The number of Retained wavelet coefficient |

**Table 2.** Symbols for CDW histogram

| Parameters | Description |
|---|---|
| *Q* | Query window with (qxl,qyl,qxh,qyh) coordinate value |
| *BQ* | Bucket intersected with query Q |
| *xBucket* | x axis size of bucket |
| *yBucket* | y axis size of bucket |
| *Llp* | Lower-left corner point of object |
| *Lrp* | Lower-right corner point of object |
| *Ulp* | Upper-left corner point of object |
| *Urp* | Upper-right corner point of object |
| *Hll(i,j)* | Llp number cumulated from cell(0,0) to (i,j) |
| *Hlr( i,j)* | Lrp number cumulated from cell(0,0) to (i,j) |
| *Hul( i,j)* | Ulp number cumulated from cell(0,0) to (i,j) |
| *Hur( i,j)* | Urp number cumulated from cell(0,0) to (i,j) |

### 3.1 Construction of CDW Histogram

The construction procedure for CDW histogram consists of the following three stages.

**Construction of cumulative density histogram stage:** Divide the entire space $|DX| * |DY|$ into a same size of gird cells, and determine the size of bucket $B_i$ for each grid cell. Determine the position for each corner point (*Llp*, *Lrp*, *Ulp* and *Urp*) of objects, and then construct four sub-histograms through by accumulating each corner point for objects. Figure 1 shows the *Hll* histogram accumulating *Llp* of objects. CD histogram has following structure.

***CD Histogram = < bucket range, Hll, Hlr, Hul, Hur >***

- *bucket range* = < PL∈{xl, yl}, PU∈ {xh, yh} >
- *{xl, yl},{xh, yh}* : the pair of lower left and upper right cell of each bucket
- *bucket range* : the range of each bucket
- *Hll, Hlr, Hul, Hur* : cumulative density for each corner point of object

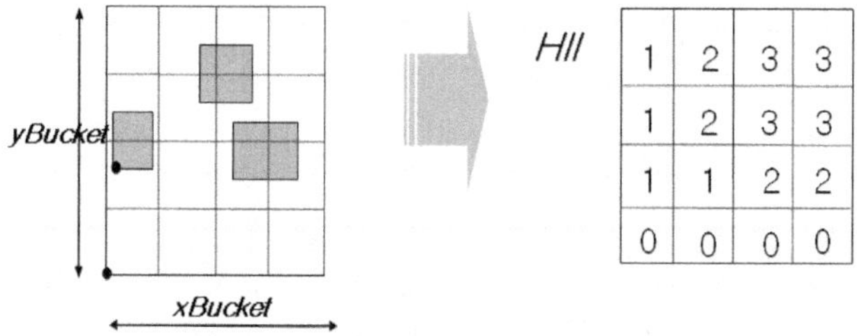


**Fig. 1.** Sub-histogram for lower-left-corner point

**Wavelet transformation stage:** Transform two dimensional buckets for four corner points(*Llp, Lrp, Ulp, Urp*) into one dimensional buckets using space-ordering method, and then generate wavelet synopsis $W_i$ after applying one dimensional Haar wavelet to the domain of each bucket $B_i$, i.e, $B_i$ transforms into $W_i$.

**Wavelet coefficient reduction stage:** Reduce the number of coefficients to be kept in each wavelet synopsis $W_i$ until the limited storage space is completely filled. Each bucket has following structure.

***B = < Wavelet synopsis ∈{coefficient, coefficient index} >***

Where, Wavelet Synopsis is a set of preserved wavelet coefficient and index.

#### 3.1.1 Construction Cumulative Density Histogram

The cumulative density histogram is summary information which is made by using MBR of rectangle object. It is constructed by following procedure. First, partition the whole space into the same size of gird cells, and then assign each gird cell to bucket. Each bucket keeps sub-histogram information. It is represented as follows:

***CDH(i,j)={Spatial MBR, Hll(i,j), Hlr(i,j), Hul(i,j), Hur(i,j)}***

Where, *Spatial MBR* represents spatial range of each bucket for x and y axis, the four of information for rectangle object mean as follows:

- *Hll(i,j)* keeps the counts of lower-left corner point of the objects. It can be calculated by using the following equation, where *BS(i,x)* is the number of rectangles whose lower-left corner point lie in the range *(0,x)* to *(i,x)*.

$$Hll(i,j) = \sum_{x=0}^{x=j} BS(i,x) \tag{1}$$

- *Hlr(i,j)* keeps the counts of lower-right corner point of the objects. It can be calculated by using the following equation, where *BE(i,x)* is the number of rectangles whose lower-right corner point lie in the range *(0,x)* to *(i,x)*.

$$Hlr(i,j) = \sum_{x=0}^{x=j} BE(i,x) \tag{2}$$

- *Hul(i,j)* keeps the counts of upper-left corner point of the objects. It can be calculated by using the following equation, where *US(i,x)* is the number of rectangles whose upper-left corner point lie in the range *(0,j)* to *(i,x)*.

$$Hul(i,j) = \sum_{x=0}^{x=j} US(i,x) \tag{3}$$

- *Hur(i,j)* keeps the counts of upper-right corner point of the objects. It can be calculated by using the following equation, where *UE(i,x)* is the number of rectangles whose upper-right corner point lie in the range *(0,x)* to *(i,j)*.

$$Hur(i,j) = \sum_{x=0}^{x=j} UE(i,x) \tag{4}$$

### 3.1.2 Haar Wavelet Transformation

After composing cumulative density histogram, compress the generated sub-histogram using wavelet transformation technique. First procedure is to transform two dimensional grid cell arrays for bucket into one dimensional array. This process is accomplished using a space-ordering method. When the values of the adjacent domain are similar, wavelet transformation technique generate a lot of coefficients close to 0, increasing the compression effects further. In this paper, we use Z-mirror ordering method considering the compression effect of wavelet transformation. Second step is to transform the one dimensional array into a wavelet synopsis by Haar wavelet, and then remove coefficients whose values are zero. Figure 2 shows the procedure of wavelet transformation.

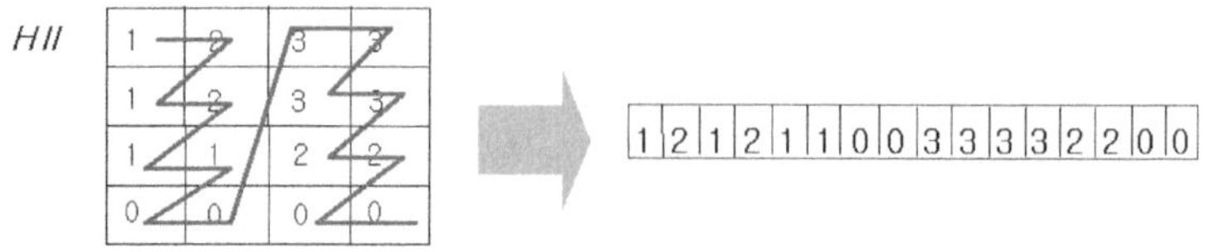


(a) Transformation two dimensional array into one dimensional array

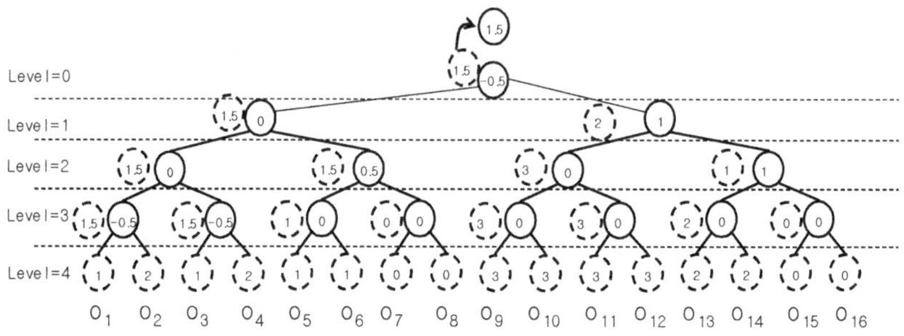


(b) Error tree for wavelet Transformation

**Fig. 2.** Wavelet Transformation for Cumulative Density Wavelet Histogram

Figure 2(a) is an example transforming two dimensional gird cell array of *Hll* histogram for lower left corner point into one dimensional array using Z-odering method. Figure 2(b) shows the error tree made by wavelet transformation. For example, the average of source data O1 and O2 is (1+2)/2 = 1.5, and detail coefficient is (1-2)/2 = -0.5. The average of O3 and O4 is (1+2)/2 = 1.5, and detail coefficient is (1-2)/2 = -0.5. The error tree is construed by performing repeatedly that the average and detail coefficient of upper level (i.e., level 3) is computed by using the average value of previous level (i.e., level 4). In figure 2(b), since the number of wavelet coefficient and original data is same, require the process of wavelet coefficient reduction to get the compression effect. Wavelet technique can get the compression effect by changing coefficients near to zero into zero, because coefficients with zero value do not have influence on data recovery.

### 3.1.3 Wavelet Coefficient Reduction

Compression effects are obtained by assigning zero to all non-retained coefficients. The goal of coefficient threshold is to determine the best subset of coefficient to retain, so that some overall error measure in the approximation is minimized. Conventional coefficient threshold is a deterministic threshold that typically retains the largest wavelet coefficients of all absolute normalized values. This deterministic process minimizes the overall root-mean-squared error(ie., L2–norm average error) in reconstructing all the data values. Namely, deterministic threshold retains the wavelet coefficient with largest absolute value after normalization. The table 3 shows the wavelet synopsis for the data array in Figure 2. In the table 3, the normalized coefficient is obtained by using the deterministic threshold. If given memory size = 8, we retain the coefficients {1.5, -0.5, 0.5, 1, 0.5, 1, -0.5, -0.5} by using deterministic threshold. Figure 3 shows the wavelet error tree for wavelet coefficient in table 3.

**Table 3.** Wavelet synopsis

| | | | | | | | | |
|---|---|---|---|---|---|---|---|---|
| Index i | 0 | 1 | 2 | 3 | 4 | 5 | 6 | 7 |
| | 8 | 9 | 10 | 11 | 12 | 13 | 14 | 15 |
| Wav.coeff | 1.5 | -0.5 | 0.5 | 1 | 0 | 0.5 | 0 | 1 |
| | -0.5 | -0.5 | 0 | 0 | 0 | 0 | 0 | 0 |
| Level coeff | 0 | 0 | 1 | 1 | 2 | 2 | 2 | 2 |
| | 3 | 3 | 3 | 3 | 3 | 3 | 3 | 3 |
| Norm coeff | 1.5 | 0.5 | 0.5 | 1 | 0 | 0.5 | 0 | 1 |
| | $\frac{0.5}{2\sqrt{2}}$ | $\frac{0.5}{2\sqrt{2}}$ | 0 | 0 | 0 | 0 | 0 | 0 |

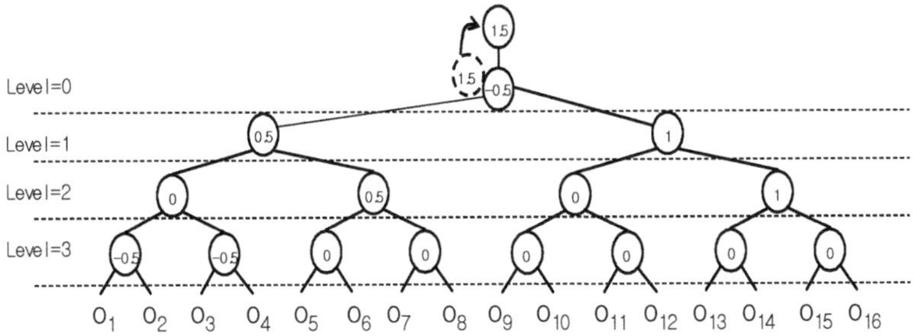


**Fig. 3.** Wavelet error tree of memory size = 8

## 3.2 Selectivity Estimation

If query Q(qxl, qxh, qyl, qyh) is given, first, a bucket index for query Q find in the one dimensional array of each sub-histogram transformed by space-ordering method, and then the original value of bucket index is recovered by wavelet recovering process. The selectivity is obtained by using the recovered original data. Thus in case of the proposed method, it takes *longN+1* time more compared with the existing cumulative density histogram to recovery wavelet coefficient. However, the proposed method has the high memory space efficiency than the existing method.

Figure 4 shows an example of query Q and sub-histogram. The bucket index for each sub-histogram *Hll[qxh,qyh]*, *Hlr[qxl-1,qyh]*, *Hul[qxh,Wyl-1]* and *Hur[qxl-1,qyl-1]* find in the one dimensional array transformed by space-ordering method.

Figure 5(a) shows the one dimensional array for *Hll* histogram, the index $O_{10}$ is the index of *Hll[qxh,qyh]* for query Q. Figure 5(b) shows the recovery process of original data for the index $O_{10}$. If this data is in the left node from starting the root the coefficient gets (+). Otherwise, if this data is right node, the coefficient gets (-). That is, to get the original data of $O_{10}$, we can recover it by calculating all the existing nodes within the path ($O_{10}$).

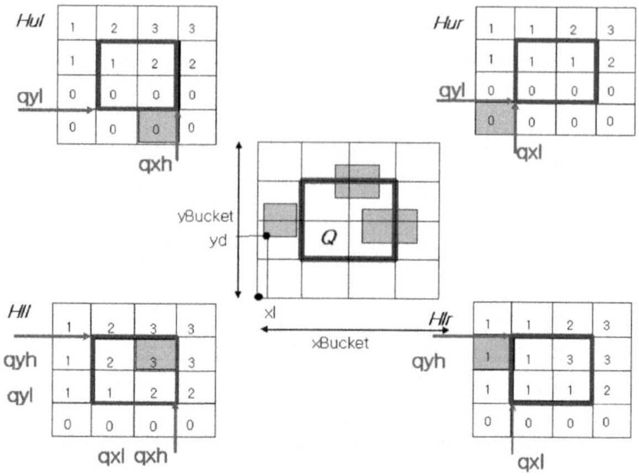


**Fig. 4.** Example of query and sub-histogram

| | | | | | | | | | | | | | | | | |
|---|---|---|---|---|---|---|---|---|---|---|---|---|---|---|---|---|
| Count value | 1 | 2 | 1 | 2 | 1 | 1 | 0 | 0 | 3 | 3 | 3 | 3 | 2 | 2 | 0 | 0 |
| OrIndex Oi | $O_0$ | $O_1$ | $O_2$ | $O_3$ | $O_4$ | $O_5$ | $O_6$ | $O_7$ | $O_8$ | $O_9$ | $O_{10}$ | $O_{11}$ | $O_{12}$ | $O_{13}$ | $O_{14}$ | $O_{15}$ |

(a) Data array of *Hll* by Z-Mirror Order

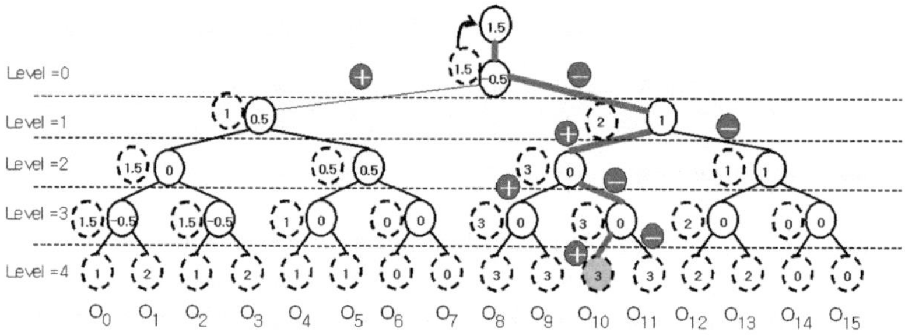


(b) Error tree of original data

**Fig. 5.** Recovery of error tree for estimating selectivity

In case of $O_{10}$, it is recovered as Path($O_{10}$) = 1.5 - (-0.5) + 1 - 0 + 0 = 3 . For Each sub-histogram, we can obtain *Hll[qxh,qyh]= 3, Hlr[qxl-1,qyh] = 1, Hul[qxh,Wyl-1] = 0, and Hur[qxl-1,qyl-1] = 0* by recovering the bucket count value as above. Finally, selectivity is obtained as following.

$$\textit{Selectivity} = Hll\,[O_{10}] - Hlr\,[O_2] - Hul\,[O_{14}] + Hur\,[O_6] = 2$$

## 4 Experiment and Performance Evaluation

In this section, we evaluate the accuracy with which the designed method estimates using actual data, alternating various factors. Our experiments were conducted on

Intel Pentium Ⅳ 2GHz PC with following three rectangle datasets: 1) dataset of commercial building located in Seoul Korea(D1) which contains 11,000, 2) dataset of California taken from TIGER/LINE(D2) which contains the MBRs of 2,249,727 roads, 3) polygon dataset(Level 1) taken from Sequoia 2000 Benchmark(D3) which is one of the polygon datasets from the SEQUIOA 2000 Benchmark, and consists of 22,288 number of urban and built-up land features. We have considered different query window sizes (5%, 10%, 15%, and 20% of spatial extent).

In order to evaluating the average relative error according to memory space, we changed storage space to 25~50% of total space. If the number of bucket is 100, the required memory space of CD histogram is 800. In case of CDW histogram, the storage size of CDW1, CDW2, and CDW3 is 400, 266 and 200. Namely, we compared the average relative error of CD histogram with CDW histogram assigned 50%, 33%, 25% space size of CD histogram. We took the average value of 10 queries with equal size, and compared with the estimated result. Average relative error(Er), defined as follows, and was used to estimate the accuracy of the estimation.

$$Er = \sum_{q \in Q}^{n} \frac{|Nq - Nq'|}{Nq} \cdot 100 \qquad (5)$$

where *Nq* is actual size of the result, *Nq'* is estimated size of the result.

## 4.1 Experimental Results

Figure 6 shows the experimental result for CDW1~CDW3 and CD histogram. It is the average relative error for 5%~ 20% queries. As shown this figure, generally, the accuracy of selectivity increases as the size of the query increases. This is because in

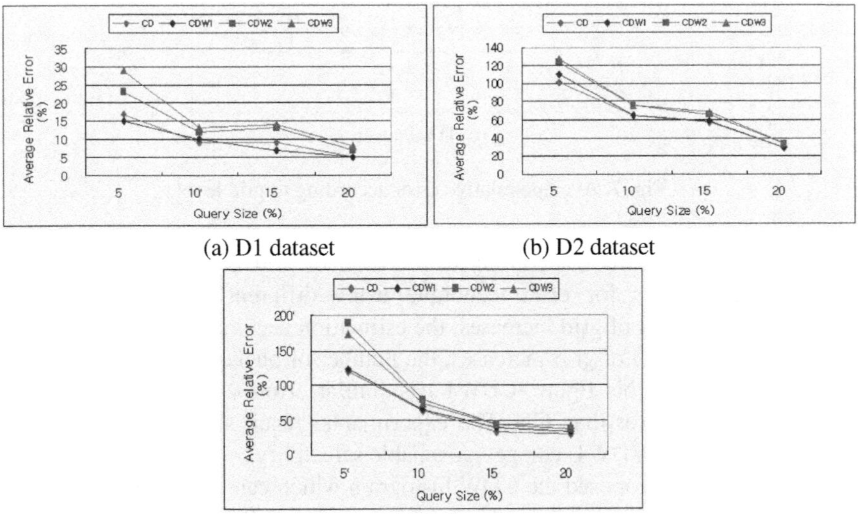


(a) D1 dataset (b) D2 dataset

(c) D3 dataset

**Fig. 6.** Average relative error according to query size

case of small query, the intersecting number of buckets is small, thus the error rate preferably increases; conversely the case of large query may get the high accuracy against the small query.

The experimental result showed that CDW1 which has 50% storage space of CD has similar error with CD, but CDW2 and CDW3 has higher error than CD. If small storage space is used, memory space which keep wavelet coefficient is saved, and the wavelet recovering time is also decreased because the number of coefficient to be used in recovery reduce. However, the coefficient ignoring by wavelet compression make the error when performing wavelet recovery. Therefore, wavelet compression should be performed so that the accuracy increases storage space decrease. In this experiment, CDW1 which has 50% of CD storage size showed that the proposed technique can maintain more information even with small storage space.

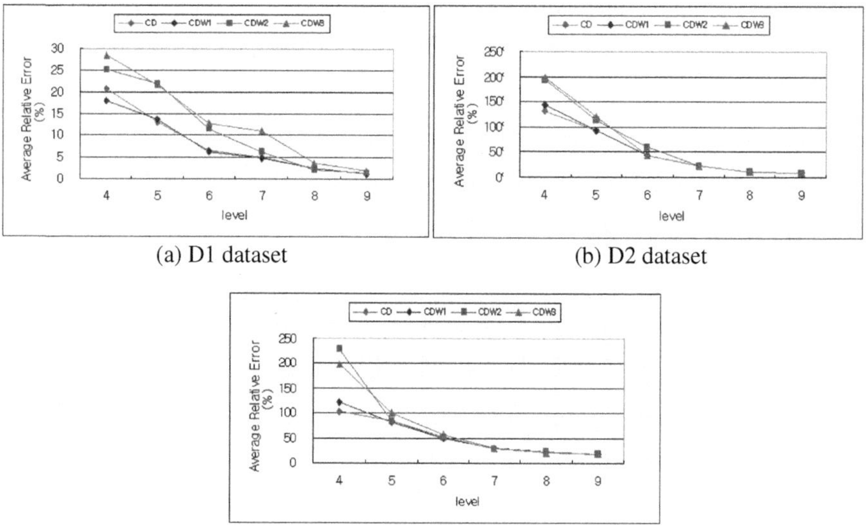


(a) D1 dataset

(b) D2 dataset

(c) D3 dataset

**Fig. 7.** Average relative error according to grid level

The estimation accuracy according to the level of grid is shown in figure 7. We have obtained results for each technique using different levels (h=4,5,6,7,8,9). Generally, as the level of grid increases, the estimation accuracy improves. This is the reason that as the level of grid increases, the number of bucket included in query also increases. As shown this figure, CDW1 has similar error with CD, and CDW2 and CDW3 has higher error than CD. The experimental result shows that the proposed technique, especially CDW1, can get reasonable selectivity.

In this paper, we proposed the CDW histogram which can maintain synopsis in the small storage space and can obtain high accuracy. Especially, CDW1 which use 50% storage space of CD is proved the high accuracy through the various experimental evaluations. We showed that CDW2 or CDW3 also has reasonable selectivity in case of very restrictive storage space.

# 5 Conclusion and Future Works

Selectivity estimation is used in query optimization and decision of optimal access path cardinally. Until now, several techniques of spatial selectivity estimation have been proposed. These techniques are focused on obtaining high accuracy and fast response time. However, they require very large memory space to maintain high accuracy of selectivity if spatial domain is also large. Therefore, we proposed a new method called CDW histogram that could get reasonable selectivity with small memory size. CDW histogram combined cumulative density histogram technique with Haar wavelet transformation so that we obtained maximum compression effects consequently. Based on our experimental analysis, we showed that the proposed technique which called CDW histogram can obtain maximum compression effects and reasonable selectivity simultaneously.

In the future, we need to analyze our histogram to improve much experimental evaluation. We also will extend our histogram to do work easily about dynamic insertion.

# References

1. Ioannidis, Y. E., Poosala, V.: Histogram-Based Solutions to Diverse Database Estimation Problems, IEEE Data Engineering Bulletin, Vol.18, No.3 (1995) 10-18
2. Poosala, V., Haas, P. J., Ioannidis, Y. E.: Improved Histograms for Selectivity Estimation of Range Predicates, ACM SIGMOD (1996) 294-305
3. Stollnitz, E., DeRose, T., Salesin, D.: Wavelet for Computer Graphics Theory and Applications, Morgan Kaufmann(1996)
4. Ioannidis, Y. E.: Query Optimization, ACM Computing Surveys, Vol.28, No.1(1996) 121-123
5. Vitter, J. S., and Wang, M.: Approximate Computation of Multidimensional Aggregates of Sparse Data using Wavelets, ACM SIGMOD(1999) 193-204
6. Acharya, S., Poosala , V., Ramaswamy, S.: Selectivity Estimation in Spatial Databases, ACM SIGMOD(1999) 13-24
7. Jin, J., An, N., and Sivasubramaniam, A.: Analyzing Range Queries on Spatial Data, ICDE(2000) 525-534
8. Matias, Y., Vitter, J. S., Wang, M.: Dynamic Maintenance of Wavelet-Based Histograms, The VLDB Journal(2000) 101-110
9. Wang, M., Vitter, J. S., Lim, L., Padmanabhan, S.: Wavelet-based Cost Estimation for Spatial Queries, SSTD(2001) 175-196,
10. Choi, Y. J., Chung, C. W.: Selectivity Estimation for Spatio-Temporal Queries to Moving Objects, ACM SIGMOD(2002) 440-451
11. Sun,C., Agrawal, D., Abbadi, A. El.:Selectivity for spatial joins with geometric selections, EDBT(2002) 609-626

12. Hadjieleftheriou, M., Kollios, G., Tsotras, V.: Performance Evaluation of Spatio-Temporal Selectivity Estimation Techniques, SSDB(2003) 202-211
13. Tao, Y., Sun, J., Papadias, D.:Selectivity Estimation for Predictive Spatio-Temporal Queries, ICDE(2003) 417-428
14. Zhang, Q., Lin, X.,: Clustering Moving Objects for Spatio-Temporal Selectivity Estimation, ADC(2004) 123-130
15. Chi, J. H., Kim, S. H., Ryu, K. H.: Spatial Selectivity Estimation using Compressed Histogram Information, APWeb(2005) 489-494
16. Elmongui, H. G., Mokbel, M. F., Aref, W. G.: Spatio-temporal Histogram, SSTD(2005) 19-36

# Image Segmentation Based on Chaos Immune Clone Selection Algorithm

Junna Cheng, Guangrong Ji, and Chen Feng

Electronic Department, Information College, Ocean University of China, 238 Hao, Songling Road, Laoshan Area, Qingdao, 266100, China
chengjunna@yahoo.com.cn

**Abstract.** Image segmentation is a fundamental step in image processing. Otsu's threshold method is a widely used method for image segmentation. In this paper, a novel image segmentation method based on chaos immune clone selection algorithm (CICSA) and Otus's threshold method is presented. By introducing the chaos optimization algorithm into the parallel and distributed search mechanism of immune clone selection algorithm, CICSA takes advantage of global and local search ability. The experimental results demonstrate that the performance of CICSA on application of image segmentation has the characteristic of stability and efficiency.

**Keywords:** Otsu's threshold method, Immune clone selection algorithm, Chaos optimization algorithm.

## 1 Introduction

Image segmentation is the process of separating objects of interest from background. It is an essential preliminary step in image processing. Over the past decades a great deal of image segmentation technique has emerged, including Edge Detection, clustering, thresholding, region growing, region splitting and merging. One of the most commonly used methods for segmenting images is thresholding, such as Otsu's threshold method, Chow-Kaneko's adaptive thresholding, Capur's maximum entropy method and so on [1][2]. Otsu's threshold method is an automatic unsupervised segmentation method. Due to its relatively simple calculation, and in most cases a satisfactory segmentation result can be achieved, it becomes a widely used method for image segmentation.

During recent years, artificial immune systems have become the research focus. It consists of three typical intelligent computational algorithms termed negative selection, clone selection and immune network theory [10]. They have been successfully applied to optimization, pattern recognition, machine learning and other engineering problems.

Immune clone selection algorithm takes the parallel and distributed search mechanism, thus it has nice global search capability and efficiency. But its local search ability is weak. The chaos optimization algorithm (COA) is a new kind of

D.-S. Huang, L. Heutte, and M. Loog (Eds.): ICIC 2007, LNAI 4682, pp. 505–512, 2007.

searching method. When the solution space is not very large, COA has nice global and local search capability. But it is not efficient while the solution space is large. In this paper, taking advantages of the two algorithms, a novel chaos immune clone selection algorithm is presented and is applied to search the optimal thresholds of image based on Otsu's threshold method. Experimental results demonstrate that the hybrid algorithm can obtain a good segmentation of the image and has the characteristic of stability and efficiency.

This paper is organized as follows: in Section 2, the basic idea of chaos immune clone selection algorithm (CICSA ) is described. In Section 3, main steps of CICSA to Otsu's threshold segmentation are presented. In Section 4, experimental results are shown and the performance of CICSA on image segmentation is verified.

## 2 Chaos Immune Clone Selection Algorithm

Immune clone selection is the theory used to explain how an immune response is mounted by a B-cell of Vertebrate immune system. When some B-cell receptors recognize a special kind of invading antigen such as viruses and bacteria with a certain affinity, these B cells are selected to proliferate. The proliferation rate of each immune cell is proportional to its affinity with the selective antigen. The B-cell clones also suffer mutation. The mutation rate of each B cell is inversely proportional to the affinity. During the process of selection, proliferation and mutation, B cells with the highest affinity for the antigen are generated. The highest affinity B cells release soluble forms of B-cell receptors, which are termed as antibodies to bind to antigens leading to the elimination of the antigen.

Inspired from the process of selection, proliferation and mutation of the immune system, a clone selection algorithm (CLONALG) is proposed [3]. The basic steps of CLONALG can be described as follows: 1 Initialize a Population of antibodies randomly. 2 calculate the affinity of each antibody in the Population with the specific antigen. 3 Select $n1$ of the highest affinity antibodies and generate copies of these antibodies proportionally to their affinity with the antigen. Mutate all these copies with a rate in inverse proportion to their affinity. Replace some low affinity antibodies by random antibodies. 4 Select a few antibodies to be kept as memory colony. 5 Repeat Steps 2 to 5 until a stop criterion is met.

The chaos optimization algorithm is a new kind of searching method[4].The procedure of chaos search includes two steps[5]. First, search the whole limited space by serial chaos iteration and find the current optimum point; then take the current optimum point as the center, more subtle search is performed by imposing a tiny chaos disturbance to find the final optimal point. Due to the ergodic and dynamic properties of chaos variables, chaos search is more capable of hill-climbing and escaping from local optima than random search [6].

In this paper, the chaos search mechanism is integrated with CLONALG and a novel chaos immune clone selection algorithm (CICSA) is developed. The initial Population of CLONALG is generated randomly; and in order to keep the diversity of every Population a mount of fresh antibodies are also produced by random. CICSA takes place the randomcity by the ergodic chaos sequence to improve the global explore ability of CLONALG. And after certain generations of evolution, when optimal solution will not progress, a current optimal point is get and tiny chaos disturbance is imposed on it to search in its neighborhood. In addition, to make full use of the information of the memory colony, tiny chaos disturbance is also performed on the individuals of the memory. Thus the local exploit ability of CICSA is improved too. CICSA integrates the virtue of parallel and distributed search mechanism and the excellent local search capability.

## 3 Image Segmentation Based on CICSA and Otsu's Method

Otsu's threshold method for image segmentation is a histogram-based method. Assuming the grey level of image is ranged within [0, 1,⋯,$k-1$], for single threshold segmentation, suppose that a threshold $t$ is chosen and the whole image is divided into two classes: $C_0$ is the set of pixels with levels [0,1, ⋯, $t$] and $C_1$ is the set of pixels with levels [$t$+1,$t$+2, ⋯, $k-1$]

We can get the probability distribution of all grey levels by :

$$p_i = \frac{n_i}{N} \quad (p_i \ge 0, \sum_{i=0}^{k-1} p_i = 1) \tag{1}$$

where $n_i$ is the number of pixels that have grey level $i$, $N$ is the total number of pixels in the image.

Define $w_0$ and $w_1$ as the probability of $C_0$ and $C_1$ respectively:

$$w_0 = P(C_0) = \sum_{i=0}^{t} p_i \quad w_1 = P(C_1) = \sum_{i=t+1}^{k-1} p_i \tag{2}$$

Define $u_0$ and $u_1$ as the mean grey level of $C_0$ and $C_1$ respectively, $u_T$ is the mean grey level of the whole image:

$$u_0 = \sum_{i=0}^{t} i\frac{p_i}{w_0} \quad u_1 = \sum_{i=t+1}^{k-1} i\frac{p_i}{w_1} \quad u_T = \sum_{i=0}^{k-1} ip_i \tag{3}$$

The optimal threshold value $t^*$ is the one that maximizes *between-class variance* $\sigma_B{}^2$:

$$\begin{cases} t^* = Arg \quad Max\, \sigma_B{}^2 \\ \sigma_B{}^2 = w_0(u_0 - u_T)^2 + w_1(u_1 - u_T)^2 \end{cases} \tag{4}$$

Otsu's method can be extended to multiple thresholds segmentation. Assume $M$ is the number of thresholds, the *between-class variance* $\sigma_B{}^2$ is defined as:

$$\sigma_B{}^2 = \sum_{j=0}^{M} w_j (u_j - u_T)^2 \tag{5}$$

Image segmentation based on Otsu's threshold method can be modeled as the following optimization problem:

$$\begin{cases} \max \quad f(x_1, x_2, \cdots, x_r) \\ s.t. \quad x_i \in [a, b\ ], \quad i = 1, 2, \cdots r \end{cases} \tag{6}$$

where $r$ is the number of optimization variables, $x_i$ is the optimization variable corresponding to threshold value of image, $[a,b\ ]$ is the range of grey level of an image, $f$ is the objective function corresponding to Eq.5.

Image to be segmented is regarded as the antigen. Optimization variables ($x_1, x_2, \cdots, x_r$) is expressed by an antibody and encoded as a binary code. Take objective function as the evaluation function of the affinity of the antibody.

### 3.1 Main Steps of CICSA to Otsu's Threshold Segmentation

Step 1: initialize Population and Memory colony. Generate $N$ antibodies of Population and $M$ antibodies of Memory colony by chaos.

Step 2: calculate the affinity of each antibody in the Population and sort them by their affinities in descending order. If evolutionary stop criterion is met, the current optimum antibody is achieved and go to step 5; else go to step 3.

Step 3: update the Memory colony based on compositive affinity of the antibodies in the Memory colony.

Step 4: evolve the current Population. First, P highest affinity antibodies of current Population are selected and cloned proportionally to their affinity with the antigen: The higher the affinity, the more the number of copies, and vice-versa. Then the copies of the P antibodies are mutated with a rate in inverse proportion to their affinity: the higher the affinity, the smaller the mutation rate, and vice-versa. After clone and mutation, the P highest affinity antibodies are selected and kept to next generation of Population. Second, take the Elitist strategy: the best antibody in the current generation enters the next generation directly. Third, produce $H$ antibodies by chaos iteration and add them to the next generation. Go to step 2.

Step 5: impose tiny chaos disturbance on the current optimum antibody and the individuals of the Memory colony to get the optimal thresholds. When stop criterion for chaos iteration is met, the algorithm is terminated.

### 3.2 Generate Antibody by Chaos

The chaos system is produced by the following famous Logistic mapping:

$$z^{k+1} = \mu z^k (1 - z^k), \quad z^k \in [0,1], \qquad k = 1, 2, \cdots \tag{7}$$

where $z$ is the chaos variable, $k$ is the iteration times, $z^k$ is the value of chaos variable $z$ at the $k$th iteration times. $\mu$ is the chaotic attractor and when $\mu = 4$ the system is entirely in chaos situation. Given a initial value $z^0$, chaos variable $z$ can go through every state during chaos space [0,1]according to their own regularity without repetition and produce chaotic sequence[9]. Chaos sequence has the characteristics of ergodicity, randomicity and extreme sensitivity to the initial value.

In order to generate an antibody by chaos, $r$ chaos variables each corresponds to an optimization variable should be conducted by Eq.8.

$$z_i^{k+1} = \mu z_i^k (1 - z_i^k), \quad z_i^k \in [0,1], \qquad i = 1,2,\cdots,r \qquad k = 1,2,\cdots \tag{8}$$

where $r$ is the total number of chaos variables, $z_i$ is the $i$th chaos variable. Ergodic space of the chaos variable is [0, 1], while the space of optimization variables is [a, b]. Thus the $r$ chaos variables should be mapped to the $r$ optimization variables $x_i$ by:

$$x_i = a + (b - a) z_i \quad i = 1,2,\cdots,r \tag{9}$$

Given $r$ different initial value of the $r$ chaos variables: $z_1^0, z_2^0, \cdots, z_r^0$, after each iteration by Eq.8 and mapping by Eq.9, an antibody is generated.

### 3.3 Update the Memory Colony Based on Compositive Affinity

The individuals in the Memory colony are used to be imposed tiny chaos disturbance to achieve the final optimal thresholds. In order to keep the diversity of Memory colony, an updating method based on compositive affinity is adopted. The antibodies in the Memory colony should have high affinities with the antigen, while great similarities between individuals should be avoided. Similarity between every two antibodies $S_{ij}$ is defined as:

$$S_{ij} = \frac{1}{1 + H(Ag_i, Ag_j)} \qquad i = 1,2,\cdots,M \quad j = 1,2,\cdots,M \tag{10}$$

where $M$ is the total number of individuals in Memory colony, $H(Ag_i, Ag_j)$ is the entropy [8] between $Ag_i$ and $Ag_j$.

Assume $d_i$ as the density of an antibody $Ag_i$, which is defined by Eq.11.

$$d_i = \frac{N_i}{M} \quad d_i \in [0,1] \tag{11}$$

where $N_i$ is the number of antibodies which similarity with $Ag_i$ is above a threshold[8].

The compositive affinity $CAff_i$ of antibody $Ag_i$ is defined by Eq.12.

$$CAff_i = \frac{Aff_i}{1+\lambda d_i} \quad \lambda > 0 \tag{12}$$

where $Aff_i$ is the affinity of $Ag_i$,, $d_i$ is the density of $Ag_i$ , $\lambda$ is adjustive parameter.

During the evolution of Population, the highest affinity antibody of every generation is selected and added to Memory colony. When a new antibody is put to the Memory colony, compositive affinity of each antibody is calculated by Eq.12 and $M$ individuals with the highest compositive affinity $CAff_i$ are selected to constitute the new generation of Memory colony.

### 3.4 Chaos Disturbance Mode

After certain generations of evolution of CICSA, when the optimal solution is in a state of stagnant, it is considered that the current optimal thresholds value is obtained. The left wok is done by tiny chaos disturbance to get improved local search ability.

The chaos disturbance mode [7] used in this paper is defined by Eq.13.

$$Y^k = (1-\beta)Z^* + \beta Z^k \quad \beta \in (0,\ 0.5) \tag{13}$$

where $Z^*$ is the chaos variable vector corresponds to current optimum point which is mapped from the current optimal thresholds value. $Z^k$ is the chaos variable vector iterated by Eq.8, $\beta Z^k$ is the tiny chaos disturbance imposed on the current optimum point $Z^*$ , $Y^k$ is the chaos variable vector corresponds to a point near $Z^*$ after chaos disturbance. $\beta$ is an adjustive parameter.

## 4 Experimental Results

To verify the performance of image segmentation based on CICSA, it is used to segment the standard test image of Lenna. Lenna's original image and its histogram of grey level are shown in Fig.1 (a), (b). The experimental result of single threshold segmentation and two thresholds segmentation based on CICSA and Otsu's method are shown in Fig.1 (c), (d).

To compare the performance of CICSA with CLONALG, Each algorithm run 30 times to reduce the stochastic influences. The experimental results are given in Table1, and the average evolutionary curves for two thresholds segmentation of Lenna's image are shown in Fig.2. From Table 1, we can see that the performance of CICSA is very stable and has the 100% convergence probability. From Fig.2, we can see that: CICSA achieves the maximum objective function value after 600 evaluations[9] and CLONALG reaches the maximum objective function value after 900 evaluations. The convergence speed of CICSA is quicker than that of CLONALG.

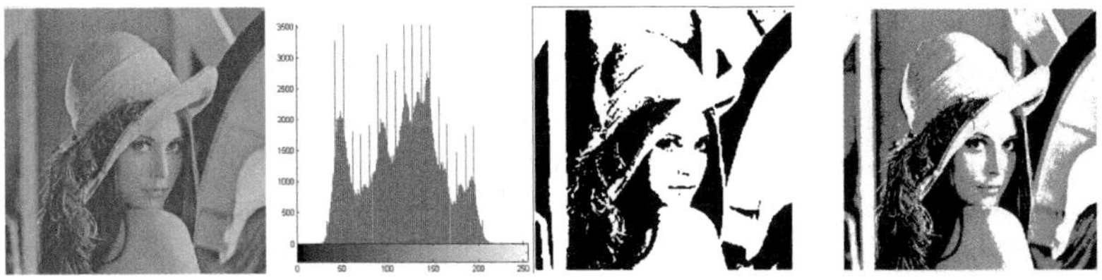

(a) (b) (c) (d)

**Fig. 1.** (a) Original image of Lenna (b) The histogram of grey level of Lenna (c) Otus's single threshold segmentation image by CICSA (d) Otus's two thresholds segmentation image by CICSA

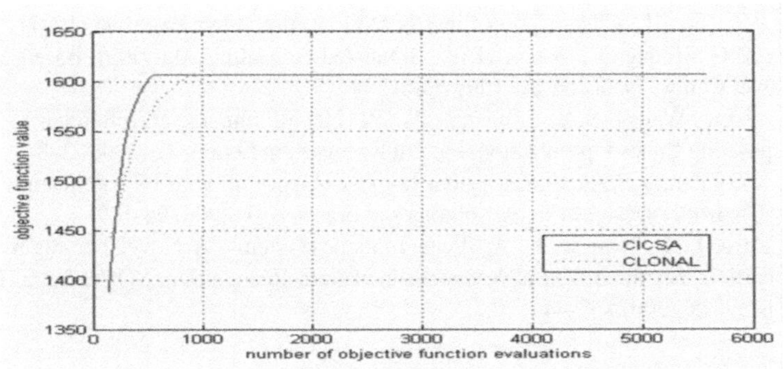


**Fig. 2.** Evolutionary cave for two thresholds segmentation of Lenna by CICSA and CLONALG

**Table 1.** Performance of CICSA and CLONALG for segmentation of Lenna

| | Single threshold | | Two threshold | |
|---|---|---|---|---|
| | CICSA | CLONALG | CICSA | CLONALG |
| best threshold | 110 | 110 | 87, 140 | 87,140 |
| worst threshold | 110 | 109 | 87, 139 | 88,141 |
| average threshold | 110 | 109.9 | 87, 139.9 | 87.2,140.3 |
| Average number of objective function evaluations | 460 | 530 | 650 | 910 |
| Convergence probability | 100% | 100% | 100% | 100% |

# 5 Conclusion

Taking advantages of the ergodic and stochastic properties of chaotic variables and the parallel and distributed search mechanism of immune clone selection, CICSA achieves powerful global and local search ability. Its application on image segmentation has the characteristic of stability and efficiency.

## References

1. Sahoo, P.K., Soltani, S., Wong, A.: A Survey of Thresholding Techniques. Computer Vision, Graphics and Image 41 (1988) 233-260
2. Spirkovska, L.: A Summary of Image Segmentation Techniques. NASA Technical Memorandum 104022 (1993)
3. De Castro, L.N., Von Zuben, F.J.: The Clonal Selection Algorithm with Engineering Applications. *GECCO'00 – Workshop Proceedings,* (2000) 36-37
4. Li, B., Jiang W.S.: Chaos Optimization Method and Its Application. Control Theory and Applications, 14 (1997) 613-615
5. Yao, J.F., Mei, C., Peng, X.Q.: The Application Research of The Chaos Genetic Algorithm (CGA) and Its Evaluation of Optimization Efficiency. Acta Automat Sinica 28 (2002) 935–942
6. Zhou, C., Chen, T.: Chaotic Annealing for Optimisation. Phys Rev E 55 (1997) 2580–2587
7. Wang, Z.C., Zhang, T., Wang, H.W.: Simulated Annealing Algorithm Based on Chaotic Variable. Control Decision 14 (1999) 382–384
8. Guo, Z.L., Wang, S.A., Zhuang, J.: A Novel immune Evolutionary Algorithm Incorporating Chaos Optimization. Pattern Recognition Letters 27 (2006) 2–8
9. Zuo, X.Q., Fan, Y.S.: A Chaos Search Immune Algorithm with Its Application to Neuro-fuzzy Controller Design. Chaos, Solitons and Fractals 30 (2006) 94-109
10. De Castro, L.N., Timmis, J.: Artificial Immune Systems: A Novel Paradigm to Pattern Recognition. Artificial Neural Networks in Pattern Recognition, SOCO-2002, University of Paisley UK (2002) 67-84

# Research a Novel Integrated and Dynamic Multi-object Trade-Off Mechanism in Software Project

Weijin Jiang and Yuhui Xu

Department of computer, Hunan business college, Changsha 410205, P.R. China
jiangwj_nudt@163.com

**Abstract.** Aiming at practical requirements of present software project management and control, the paper presented to construct integrated multi-object trade-off model based on software project process management, so as to actualize integrated and dynamic trade-oil of the multi-object system of project. Based on analyzing basic principle of dynamic controlling and integrated multi-object trade-off system process, the paper integrated method of cybernetics and network technology, through monitoring on some critical reference points according to the control objects, emphatically discussed the integrated and dynamic multi- object trade-off model and corresponding rules and mechanism in order to realize integration of process management and trade-off of multi-object system.

**Keywords:** Software item management; Software project; Process management; Dynamic trade-off; Multi – object.

## 1 Introduction

The project of developing a large and complicated software is a multi-object system. Horizontally, there is multi-project participate in different objects of subject respectively; and vertically, "top three controls" including project progress, cost and quality are important control objects of each subject. All of these form the integrated and dynamic multi-object system frame. Especially, "top three controls" objects, which are interactional and interrestrict[1,2], vertically make up an organic indivisible dialectical entail. To effectually actualize the management and control of software project, it is necessary to consider horizontally the harmonious communication between related subjects, and to trade-off synthetically and optimize multi-object control system vertically.

On one hand, the horizontal harmonious communication between every subject is mainly concerting the problems of organizing mechanism and managing method reform. Through introducing the dynamic organizing and managing method[3] of software projects and advocating a kind of thinking mode which is result-oriented and emphasizes the process interface integration management, we can change the conventional thinking mode which is process-oriented and neglects wholly and harmonious control of projects. Technically supported by the integrated management system and the information network platform, every subject organizes and manages

D.-S. Huang, L. Heutte, and M. Loog (Eds.): ICIC 2007, LNAI 4682, pp. 513–524, 2007.

interface and every processing interface ingrates intellectually. All these form a information sharing and uniform harmonious management and control mechanism, which provides all software project participants with a efficient communicating and cooperating environment, and helps to realize horizontal harmonious communication between software projects[4-6]. Furthermore, on the base of perfecting related prompting measure, by enhancing the contract management, it is possible that every participant corresponds with project owner on project wholly benefit while pursuits max interest of himself.

On the other hand, however, vertical multi-object integrated trade-off mainly concerns the problem of management technology and implement method. Every subject, especially, the project owner in critical position, must be integrally analyzed according to project condition, project organization, function requirement and technology complexity[7-9]. Only when top three control objects are dynamically overall trade-off, the highest constructing speed, the least investment and the best outcome would be possible and the software project construction would be completed quickly, well and economically[10].

Now, the study of software project control mainly concerns two kind of method. One is network technology method including many problems, such as the decomposition of network plan, synthesis and control, comanagement of process and cost, which copes with object detailing to working procedures. Another is cybernetics, which emphasizes macro-aspect study, whose object is to realize phase objects of the software project. For instance, literature[5] based on PERT technology uses the method of system analogue to make the random variable, which accords to job time of prescribed distribution, for each working procedures, and analogizes statistical index of optimized schedule, cost and quality. Literature[11] studies the balance relationship of schedule, cost and quality through three linear layout models. Literature[12] applies the linear models in literature[13] to appraise this method's practicability for a factory information system constructing project. Literature dedicates to control investment effectively and presents the nonlinear motility model, so to realize the intellectual management of project's multi-object. Literature[14] and [15], based on network plan technology, respectively uses multi-property effect function and takes cohesion function as the target function to build up the software project management resources balanced and optimized model. Both design corresponding inherited arithmetic to solve the models, hoping to get the satisfying plan. These control methods, based on network technology, still consider the balanced and optimized problem between software project process, cost and quality control objects from the angle of plan, and aim for planning to determine the reasonable project time limit and the lowest investment on the condition of quality requests[16-18]. As for the problem of how to implement dynamic trade-off according to plan control object is hardly studied during project implement. Furthermore, according to our practical investigation, during the implements of some domestic large software project, the inspection and trade-off of control object still depend on the human experience judgment and subjective decision without the decision-making support provided by corresponding DSS. This postmortem control method is difficult to avoid the condition of exceeding project time and budget, and must be improved through strengthening underway control and even aforehand control. Literature[19] and [20] point out the validity of the monitor system for controlling the project cost and enhancing project management performance. Currently, the control for project implement process and corresponding monitor systems also basically base on the critical path method (CPM) and the process control technology. This paper integrates cybernetics and the network technology, and emphasizes the integrated trade-off of each phase object control during the project management from the angle of the software project dynamic trade-off, introduces the theory of constraints (TOC) based on the critical path method and imposes real-time monitor on control objects through monitoring on some critical reference points, emphatically discusses the rules and mechanism of integrated dynamic trade-off between "top three controls"

## 2 The Basic Principle of Dynamic Controlling Process in Software Project

As a dynamic, uncertain and inconclusive real-time system, the software project management, in a brief, has basic contents as "plan + control". Process's uncertain makes "plan" became the necessary foundation and precondition of the project management. At the same time, it's the existence of uncertain that make the project management must base on "control". By controlling, the project process, cost and quality are limited in the plan object. This is also the essence of the project management. In the meantime, saw from the continuity of the software project implement, the "input" of next process must be the "output" of previous process. But the traditional software project management is guided by the independent process management controlling. The controlling functions of each phase processes are separated, the controlling objects are disjointed; the overall trade-off and control are neglected. It will eventually influence the overall controlling object of project, and can not adapt the realistic request of the software project management. So it inevitably requests to control the system dynamically based on the overall integrated trade-off and management of software project. Furthermore, new interferential factors are continuously produced during the project progress and results in new deviation. So the project controlling is a kind of dynamic circle which is "...identify the deviation — adjust the controlling — implement the development — trace and check — compare and analyze...". The basic principle is shown in Figure 1.

## 3 The Process of the Integrated Multi-object Trade-Off System

As shown in Figure 1(a), the process management in software project is the base of object controlling. Every process has input and output, and the "input" of next process must be the "output" of the previous process. The inputs and outputs between the processes constitute the interfaces between the processes. Furthermore, no only the adjacent processes have information relationship, but also prophase processes and anaphase processes have information relationship. Therefore, to control the process in software project means to control the software interface and to control the information stream flowing through the software interfaces. As the process of the integrated multi-object trade-off system based on process management, it inevitably presents in a great

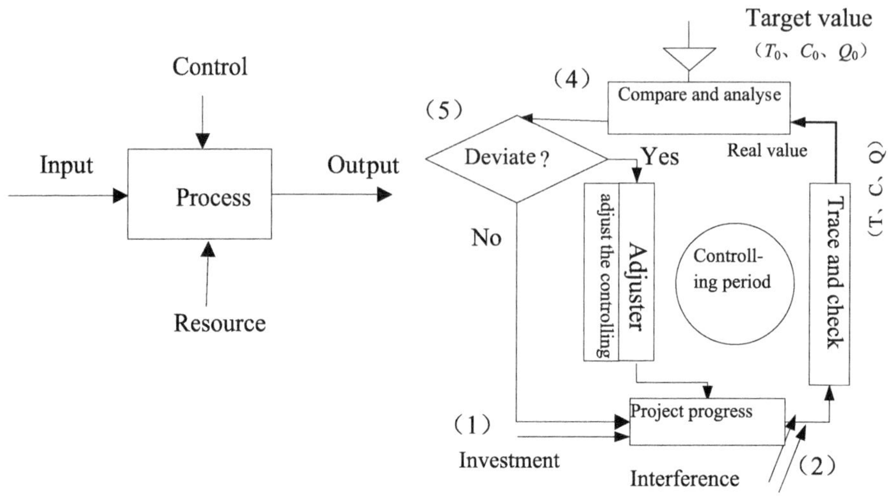


(a)The principle model of procress management and controlling

(b) The principle of dynamic controlling based on the process management

**Fig. 1.** The basic principle of dynamic controlling process in software project

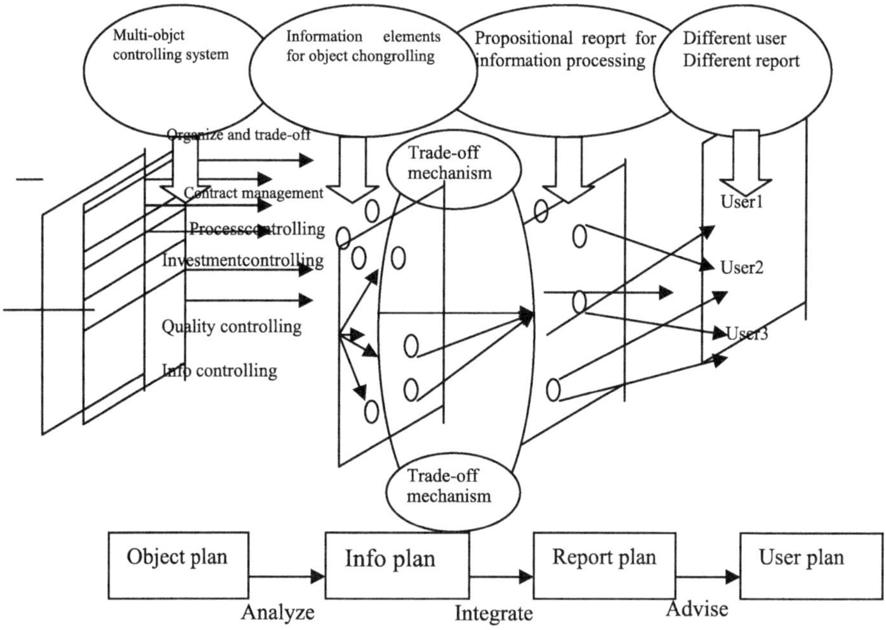


**Fig. 2.** Process of the integrated multi-object trade-off system based on process management

extent as the system process including the collecting, the processing and the analysis of the project information. According to the states and mapping relationship of information in the overall process of the software project management, this trade-off system process can be devided into four plans, namely the object plan, the info plan, the report plan and the user plan. It is shown in Figure 2.

(1) The object plan is, based on the overall process in software project, the integrated overall multilayer reticulate object controlling system after analyzing the project object.
(2) The info plan is the necessary information elements for the object controlling, such as the uniform information classifying and coding, the uniform rules for using the central database and the computer network, the time and the content that each subjects report the information, the related standard information of the object controlling schedule and so on.
(3) The report plan is the decision-making information for object controlling. The process of system information processing from the info plan to the report plan is using IT technology, by comparing and analyzing with the standard information of the object controlling schedule, to find the deviation, and according to the related trade-off mechanism to make propositional report aiming at the multi-object controlling decision-making to provide the support for the user decision-making.
(4) The user plan.

## 4 Integrated and Dynamic Multi-object Trade-Off Mechanism

The pluralism of the project object controlling in Figure 2 relate to the all kinds of aspects in the overall project process. And in the "top three controlling" objects of every subject, the project quality controlling object is the base and the process controlling is in the position of the relative core. Therefore, taking the quality controlling as the precondition, the progress controlling as the head, to leade the investment cost controlling is a main line to realize the multi-object trade-off in software project[21]. At the same time, the realization of the quality controlling object can be embodied by the realization of the schedule controlling object and the investment controlling object. Because if the quality is not satisfied, it is need to rework or repair. Thus it will doubtless delay the project development progress and increase investment. So, in the condition of insuring the software project quality request, the relative coordination and balance between "the top three controlling" objects can be realized by making the reasonable schedule and confining the reasonable investment. If the schedule is delayed and the cost is increase because of quality, it will need to seek the integrated balance between the schedule and the cost in the condition of the project quality by reworking and repair[22]. This kind of the dynamic monitor and trade-off mechanism of "the top three controlling objects" in the overall project progress is shown in Figure 3.

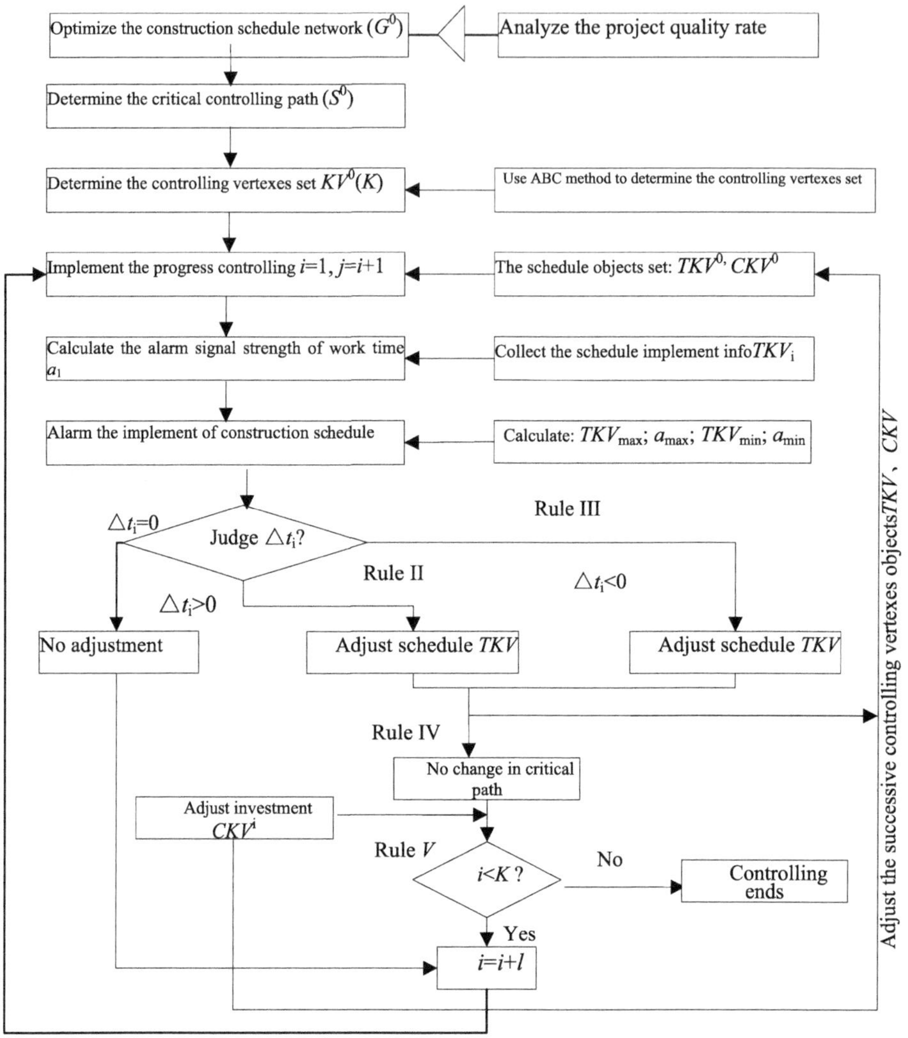


**Fig. 3.** The integrated and dynamic multi-object trade-off mechanism model in software project management

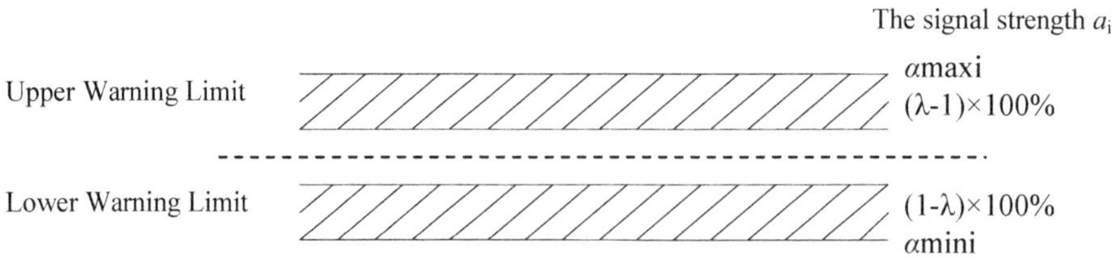


**Fig. 4.** The progress monitoring alarm zone of controlling vertexes

(1) Because the software project quality object must be insured and is certain, namely the software project qualification rate must reach 100%. If there are quality problems in the software development, namely the qualification rate can not reach 100%, the plan quality object should be realized by prolonging the work time for project and increasing the investment. The dynamic balance between "the top three controlling objects" should be reach considering the optimization of the two aspects, the schedule and the investment controlling.

(2) The optimization of network schedule $G^0=(V^0, E^0)$ is the punctuality in the condition that the total work time is not more than the contract time limit $T_0$, the total cost is least and satisfies the schedule of resource configuration. Among them, $V^0$ is the network schedule vertexes set, and $E^0$ is the network schedule activities set.

(3) The controlling points are set on the certain critical path $S_M^0=\{MV^0, ME^0(L, P)|L, P\in MV^0\}$ of the optimized network schedule $G^0$. Among them, $MV^0$ is the vertexes set of the critical controlling path, $ME^0(L, P)$ is the activities set on the critical path. Using the ABC method to determine the K controlling vertexes set on the critical controlling path $S_M^0$ is $KV^0=\{ KV_0^i \mid KV_0^i \in MV^0, i=1, 2, \ldots, K, K\geq 2 \}$.

(4) The controlling object is divided into the controlling vertex work time object $TKV^0$, the controlling vertex cost object $CKV_0$ and the controlling vertex quality rate object ($Q_0$=100%). Among them, $TKV^0=\{TKV_i^0 | KV_i^0 \in KV^0, i=1, 2,...,K, K\geq 2\}$, in this formula, $TKV_i^0$ is the scheduled work time of the controlling vertex $KV_i^0$. $TKV_i^0$ is calculated from the time parameter of the optimized network schedule $G^0$. There is $TKV_i^0 = TL_{-i}^0$ , $TL_{-i}^0$ is the latest implement time of the controlling vertex $KV_i^0$ on the network schedule $G^0$, and satisfy $TL_{-i}^0 \leq T_0$.

At the same time, $CKV^0=\{CKV_i^0 \mid KV_i^0 \in KV^0, i=1, 2, \ldots, K, K\geq 2\}$ , in this formulation, the development cost controlling object of the controlling vertex $KV_i^0$ is $CKV_i^0 = \sum_{\substack{L,P\in MV^0 \\ TKV_P^0 \leq TKV_i^0}} \mathrm{N}_u^0(m, n)\, t^0(m, n)\, r^0(m, n)$ , in this formulation, $\mathrm{M}_u^0(L, P)$ and $\mathrm{N}_u^0(m, n)$ are respectively the object work time of the vertex P on the critical controlling path $S_M^0$ and the object work time of the vertex $n$ on the non-critical controlling path; $r^0(m, n)$ is the discount coefficient of effective work time, and there is

$$r^0(m, n) = \begin{cases} 1 & \text{when } TNV_n^0 \leq TKV_i^0 \\ \dfrac{\mathrm{TNV}_n^0 - \mathrm{TKV}_i^0}{\mathrm{TNV}_n^0 - \mathrm{TNV}_m^0} & \text{when } TNV_m^0 \leq TKV_i^0 \leq TNV_n^0 \end{cases} \tag{1}$$

(5) The monitoring signal strength of the development progress is $a_i= \frac{TNV_i - TKV_i^0}{\lambda - TNV_i^0} \times 100\%$, in this formula, $a_i$ is the monitoring signal strength of the controlling vertex $KV_i^0$; $TKV_i$ is the actual work time of the controlling vertex $KV_i^0$; λ is the permissible floating coefficient of the progress controlling, commonly $1.00 \leq \lambda \leq 1.05$. So the progress monitoring alarm zone to the controlling vertex $KV_i^0$ is shown as the Figure 4.

In the figure, $a_{maxi}$ and $a_{mini}$ are respectively the maximum and minimum of the work time monitoring signal strength of the controlling vertex $KV_i^0$:

$$a_{maxi}=\frac{TKV_{maxi} - TKV_i^0}{TKV_i^0} \times 100\%, \quad a_{mini}=\frac{TKV_{mini} - TKV_i^0}{TKV_i^0} \times 100\% \tag{2}$$

In the formula (2), $TKV_{max}$ and $TKV_{min}$ are the latest implement time and the earliest implement time of the controlling vertex $KV_i^0$ that are respectively computed from the the longest duration $t_{max\,(i,\,j)}$ and the shortest duration $t_{min\,(i,\,j)}$ of each activities in the development schedule network $G^0$, and $i, j \in V^0$.

(6) After dynamically tracing the development schedule and computing the monitoring signal strength $a_i$ of corresponding controlling vertex, the adjustive value $\Delta t_i$ of implement time of controlling vertex $KV_i^0$ can be computed by following formula:

$$\Delta t_i = TKV_i - TKV_i^0 = \begin{cases} > 0 when (\lambda - 1) \times 100\% < ai \leq a_{\max i} \\ \geq 0 when (1 - \lambda) \times 100\% \leq ai \leq (\lambda - 1) \times 100\% \\ < 0 when a_{\min i} \leq ai \leq (1 - \lambda) \times 100\% \end{cases} \tag{3}$$

Combining with the alarm zone figure of controlling vertexes work time, according to the process management fact in software project, in the above precondition of quality controlling, lead by the progress controlling, and driving the trade-off main line of the cost controlling, the following rules are established:

(1) Rule I: the fixed work time rule

When $\Delta t_i=0$, it is not necessary to adjust the plan to insure that the object work time of the successive controlling vertex $KV_j^0$ is fixed.

(2) Rule II: the work time delay rule

When $\Delta t_i>0$, noting the real work time of controlling vertex $KV_i^0$ lags behide the object work time, it is necessary to adjust the object work time of the successive controlling vertex $KV_j^0$. In another word, by compressing the critical activities' durations between the controlling vertexes $KV_i^0$ and $KV_j^0$, the anaphase procedures schedule and the total project work time object are insured, and the following conditions are satisfied:

$$\begin{cases} TKV_j^1 = TKV_j^0 + [\Delta t_i - \sum_{KV_i^0 \le L,P \le KV_J^0} \Delta(L, P) \ ] \\ \Delta(L, P) \ \le \Delta_{max}(L, P) \end{cases} \quad (4)$$

In formula (4), $TKV_j^0$ and $TKV_j^1$ are respectively the pre-adjust and post-adjust object work time of the controlling vertex $KV_j^0$; $\Delta(L, P)$ are respectively the compression and the max compression of duration of the critical activities (*L-P*) from the controlling vertex $KV_i^0$ to $KV_j^0$. Among them, $\Delta(L, P) = t^0(L, P) - t_{min}(L, P)$.

(3) Rule III: the work time advance rule

When $\Delta t_i < 0$, because the object work time of the controlling vertex $KV_j^0$ can be realized ahead of time, the implement schedule network generally is not adjusted. But the advancement of work time can increase the cost and resource consumption. For insuring the balance between the controlling objects of the controlling vertexes, we can adjust the necessary development network schedule. In the condition that the object work time of controlling vertex $KV_j^0$ keeps constant, the duration of the critical activities between the controlling vertex $KV_j^0$ and $KV_j^0$ are prolonged properly. The prolongation $|\Delta t_i|$ must satisfy:

$$\begin{cases} |\Delta ti| = \sum_{KV_i^0 \le L,P \le KV_J^0} \overline{\Delta}(L, P) \ ] \\ \overline{\Delta}(L, P \ ) \le \overline{\Delta}_{max}(L, P) \end{cases} \quad (5)$$

In the fomula (5), $\overline{\Delta}(L,P)$ and $\overline{\Delta}_{max}(L,P)$ are respectively the prolongation and the maximum prolongation of the duration of the critical activities (*L-P*) between the controlling vertex $KV_i^0$ and $KV_j^0$. Among them, $\overline{\Delta}_{max}(L,P) = t_{max}(L, P) - t^0(L, P)$.

(4) Rule IV: the fixed critical controlling path rule

If the durations of critical activities are compressed to adjust the successive controlling vertexes' object work time according to Rule II, it is possible that the original cirtical controlling path $S_M^0$ becames the non-critical path. If we want to continue tracing and controlling the development progress, it is necessary to reselect the critical controlling path and reset the new controlling vertexes set, so increasing the complexity of network schedule. And in course of the progress controlling practice in software prject, controlling the critical path of network schedule sometimes corresponds to controlling the critical portion of software project, and the controlling line and the critical controlling vertexes are basically changeless. Therefor, for keeping the original critical controlling path $S_M^0$ and the original controlling set $KV^0$ no change, it is necessary to compress the duration of other corresponding activities on the critical path paralleling to the critical activities between the controlling vertex $KV_i^0$ and $KV_j^0$. And it satisfies:

$$\begin{cases} \sum\limits_{TKV_i^0 \le TK_u, TK_\upsilon \le TKV_j^0} \Delta(u, \upsilon) = \sum\limits_{KV_i^0 \le L, P \le TKV_j^0} \Delta(u, \upsilon) \\ \Delta(L, P) \le \Delta_{max}(L, P) \\ \Delta(u, \upsilon) \le \Delta_{max}(u, \upsilon) \end{cases} \tag{6}$$

In formula (6), $\Delta(L, P)$ and $\Delta(u, \upsilon)$ are respectively the duration compression of the activities $(L, P)$ on the critical controlling path $S_M^0$ and the duration compression of the activities $(u, \upsilon)$ on the other critical path paralleling to the former activities; $TK_u$ and $TK_\upsilon$ are respectively the latest implement time of vertex $u$ and $\upsilon$. Among them, $\Delta_{max}(u, \upsilon) = t^0(u, \upsilon) - t_{min}(u, \upsilon)$.

(5) Rule V: the controlling vertex cost trade-off rule[23]

After adjusting the implement schedule, the cost and the resource consumption should be adjusted properly to insure the economy and the balance of implement schedule.

From the controlling vertex $KV_i^0$ to vertex $KV_j^0$, after adjusting the duration of the critical activities, the development cost increase to:

$$\Delta CKV_j^0 = CKV_j^1 - CKV_j^0 = \sum_{KV_i^0 \le L, P \le TKV_j^0} M_u^0(L,P) \times \Delta(L,P) - \sum_{TKV_i^0 \le TKV_n^0 \le TKV_j^0} N_u^0(m,n) \overline{\Delta}_N^0(m,n) \tag{7}$$

and satisfies: $\overline{\Delta}_N^0(m,n) \le \min\{TF(m, n), FF(m, n)\}$. So the development cost object of the successive controlling vertex $KV_j^0$ is corrected as: $CKV_j^1 = CKV_j^0 + \Delta CKV_j^0$.

In formula (7), $\sum M_u^0(L, P) \times \Delta(L, P)$ denotes the cost increment caused by compressing the duration of the critical activities $(L\text{-}P)$ between the controlling vertex $KV_i^0$ and $KV_j^0$. $\sum N_u^0(m, n) \overline{\Delta}_N^0(m, n) r^0(m, n)$ is the cost decrement result from prolonging the duration of the non-critical activities between the controlling vertex $KV_i^0$ and $KV_j^0$. $TF(m, n)$ and $FF(m, n)$ are respectively the total time difference and free time difference of the activity $(m, n)$.

## 5 Conclusions

(1) The multi-object trade-off system based on software project, especially the relationship between "the three top controlling object" as the quality, process and investment, explains the necessary of the integrated multi-object trade-off on the each phase of the overall software project management. By analyzing the basic principle of the dynamic trade-off in the software project process, the integrated multi-object trade-off system process based on the process management from the angle of information processing and controlling mechanism.

(2) The dynamic process management controlling based on the alarm monitoring is a kind of active controlling method, helps to do beforehand and afterhand controlling well and realize the anticipate object of the project management.

(3) By introducing TOC based on the critical path method, selecting the critical controlling vertexes on the critical path, real-time monitoring the controlling object, the management cost can be reduced and the management efficiency can be improved.

(4) Using the quality alarm point to determine whether it is necessary to modify the controlling object plan, when it is sure to modify, the modification is in the feasible and reasonable range, and the realization of the anticipate object is insured as possible.

(5) Fusing the cybernetics and the network technology not only helps to the unparalleled advantage of the network technology on the aspects that depicts the difference between the schedule executive state and the original object, but also can exerts the positive function of the cybernetics on the aspects of macroscopically trade-off, and helps to realize the unification of the progress controlling, the object controlling and the optimized controlling.

(6) Studying the ideal can provide guidance for the management and controlling practice in software project and development of the corresponding integrated decision support system.

(7) The limitation of this study is that it only considerate the condition that the software quality is 100% and ignores the condition that the quality level is allowed to fluctuate in a certain range.

**Acknowledgments.** This paper is supported by the Natural Science Foundation of Hunan Province of China No. 06JJ2033, and the science and technology of Department of Education of Hunan Province of China No. 06C268.

## References

1. Yang F.q., Hong, M., Jian, L., Zhi, Z.: Some Discussion on the Development of Software Technology, Acta electronica sinica. (2002) 1901-1906
2. Roger Atkinson. Project management: cost, time and quality two best guesses and a phenomenon, it's time to accept other success criteria [J]. International Journal of Project Management, (1999) 337-342
3. Duan, G.J., Chin, K.S., Tang, X.: QA Panoramic Review and Vision on "Integration" for Quality Managent. The Asia Journal on Quality, (2002) 93-112
4. Kaganov, M.: A Quality Manual for the Transition and Beyond. Quality Progress, (2003) 27-31

5. Metri, B.A, Srividya, A.: IT-driven Quality Benchmarking for Competitive Advantage. IETE Technical Review (Institution of Electronics and Telecommunication Engineers, India), (2001) 17-21
6. Babu, A.J.G., Suresh, N.: Project Management with Time, Cost, and Quality onside Rations[J]. European Journal of Operational Research, (1996) 320-327
7. Do, B.K., Yin, M.M.: Time, Cost and Quality Tradeoff in Software Project Management: a Case Study[J]. International Journal of software Project Management, (1999) 15-114
8. Chang, J., Scott, C.: Agent-based Workflow: TRP Support Environment (TSE) [J]. Computer Networks and ISDN Systems, (1996) 1501-1511
9. Zeng, L., Ngu, A., Benatallah, B.: Agent-based Approach for Supporting Cross-enterprise Workflows [A]. In: proceedings of the 12th Australasian Database Conference, Queensland, Australasian Database Conference, Queensland, Australia, (2001) 123-130
10. Mei, H., Huang, G., Xing,Y., Peng, F.: An Introduction to Feature Interaction Problem, Acta electronica sinica. (2002) 1923-1927
11. I-Jibouri, S.H.A.: Monitoring Systems and Their Effectiveness for Project Cost Control in Construction[J].International Journal of Project Management, (2003) 145-154
12. Crawford, P., Bryce, P.: Project Monitoring and Evaluation: a Method for Enhancing the Efficiency and Effectveness of Aid Project Implementation[J]. International Journal of Project Management, (2003) 363-373
13. Cheung, S.O., Suen, H.C.H., Cheung, K.K.W.: PPMS:a Web-based Construction Project Performance Monitoring System[J].Automation in Construcetion, (2004) 361-376
14. Abeid, J., Allouche, E., Arditi, D., Hayman, M.: PHOTONET II:a Computer-based Monitoring System Applied to Project Management[J].Aautomation in Construction, (2003) 12(5):603-616
15. Shih, H.M., Tseng, M.M.: Workflow Technology-based Monitoring and Control for Business Process and Project Management[J]. International Journal of Project Management, (1999) 373-378
16. Sadiq, S., Sadiq, W., Orlowska M.: Pockets of Flexibility in Workflow Specifications [A]. In: proceeding of the 20th International Conference on Conceptual Modeling, Yokohama, (2001)
17. Heinl, P., Horn, S., Jablonskis, et al.: A Comprehensive Approach to Flexibility in Workflow Management Systems [A]. In: Proceedings of the International Joint Conference on Work Activities Coordination and Collaboration, San Francisco (1999) 79-89
18. Jiang, W.J.: Research on Diagnosis Model Distributed Intelligence and Key Technique Based on MAS [J]. Control Throe and Applications, (2004) 82-88
19. Jiang, W.J.: Research on Key Technologies of Virtual Enterprise and Dynamic Modeling Based on MA & BP. Information and Control, (2002) 329-335
20. Mei, H., Huang, G., Xing, Y., Peng, F.: An Introduction to Feature Interaction Problem, Acta electronica sinica. (2002) 1923-1927
21. Mei, H., Gan, H.: Twards Self-Healing Systems via Dependable Architecture and Reflective Middleare, invited paper, to appear in IEEE International Workshop on Object Oriented Real-time and Dependable Systems (WORDS), Arizona, USA ( 2005)
22. Hang, G., wang Q.X., mei, H., yang, F.Q.: Research on Architecture-Based Reflective Middleware, Journal of Software. (2003) 1819-1826
23. Jiang,w.j.: Research on Diagnosis Model Distributed Intelligence and Key Technique Based on MAS [J]. Control Throe and Applications, (2004) 82-88

# A Swarm-Based Learning Method Inspired by Social Insects*

Xiaoxian He[1,2], Yunlong Zhu[1], Kunyuan Hu[1], and Ben Niu[1,2]

[1] Shenyang Institute of Automation, Chinese Academy of Sciences, Shenyang
[2] Graduate school of the Chinese Academy of Sciences, Beijing
{hexiaoxian, ylzhu}@sia.cn

**Abstract.** Inspired by cooperative transport behaviors of ants, on the basis of Q-learning, a new learning method, Neighbor-Information-Reference (NIR) learning method, is present in the paper. This is a swarm-based learning method, in which principles of swarm intelligence are strictly complied with. In NIR learning, the *i*-interval neighbor's information, namely its discounted reward, is referenced when an individual selects the next state, so that it can make the best decision in a computable local neighborhood. In application, different policies of NIR learning are recommended by controlling the parameters according to time-relativity of concrete tasks. NIR learning can remarkably improve individual efficiency, and make swarm more "intelligent".

**Keywords:** Neighbor-Information-Reference (NIR) learning, *i*-interval neighbor, discounted reward, Q-learning, swarm intelligence.

## 1 Introduction

In recent years, more and more researchers are interested in an exciting way of achieving a form of artificial intelligence, namely swarm intelligence. It is inspired by collective behaviors of social insects, and can solve problems by groups of simple individuals. Eric Bonabeau [1] and J. Kennedy [2] have given comprehensive descriptions for swarm intelligence. Researchers have good reasons to find swarm intelligence appealing: when the world is becoming so complex that no single human being can understand it, when tools and software systems become so intractable that they can no longer be controlled by a few persons, swarm intelligence offers an alternative way of designing "intelligent" systems, in which autonomy, emergence, and distributed functioning replace control, preprogramming and centralization. Up to now, however, researches mainly focused on optimization in this field [3] [4]. Dwelling on the subfield, though advantageous, doesn't consequentially make us closer to the goals of designing swarm intelligence systems.

Machine learning is the essence of machine intelligence. When we have systems that learn, we will have true artificial intelligence. There exist many machine learning strategies and methods [5] [6] [7]. More recently, agent learning becomes a booming

* This work is supported by the National Natural Science Foundation, China (No. 70431003) and the National Basic Research Program, China (2002CB312204).

D.-S. Huang, L. Heutte, and M. Loog (Eds.): ICIC 2007, LNAI 4682, pp. 525–533, 2007.

field [8]. In the state of the art, however, only a few results about swarm learning are reported. For example, Keiki Takadama et al. developed a novel organizational learning model for group of swarm adaptive robots [9], James F. et al. proposed an approach to realize swarm learning and multi-agent cooperation [10]. These methods are more adaptive for multi-agent than for swarm. Agent learning, including multi-agent learning, is not necessarily feasible in swarm systems. This is because: (1) a single individual doesn't have a route map to the goal in swarm systems; (2) a single individual can not make sure that its behavior is helpful without other individuals' information in dynamic environments; (3) a single individual can not get any global information.

Fortunately, social insects present metaphors for solving these problems. Inspired by cooperative transport behaviors of social insects, on the basis of Q-learning, a swarm-based learning method is proposed in this paper. The rest of this paper is organized as follows. Section 2 describes the cooperative transport behaviors in ants. Section 3 introduces Q-learning which is mostly applied in agent technology. The new method, called Neighbor-Information-Reference (NIR) learning method, is present in section 4. Section 5 analyzes different policies for concrete tasks, and discusses the effect of NIR learning on swarm intelligence. Finally, section 6 outlines some conclusions.

## 2 Cooperative Transport in Ants

Ants of many species are capable of collectively retrieving large prey that are impossible for a single ant to retrieve, which has been reported in several species of ants: weaver ants *Oecophylla smaragdina* [11] and *Oecophylla longinada* [12], army ants *Eciton burchelli* [13], African driver ants *Dorylus* [14] and some other species (Fig. 1). Although these ant species are distributed in different areas of the world and have different living habits, they surprisingly exhibit the same behavioral patterns in solitary and group transports [15] at the beginning stage. This includes: (1) when an ant finds a prey, it tries to carry it; (2) if the ant does not succeed in moving the prey, it tries to drag it in various directions; (3) if the prey does not move, the ant grasps the prey differently, then tries and drags it in various directions; (4) if the prey still does not move, the ant starts recruiting nest mates. Firstly, it releases a secretion in the air in order to attract nearby ants (short range recruitment). If the number of recruited ants is not enough to move the prey, the ant goes back to the nest leaving a pheromone trail on the ground. Such trail will lead other ants to the prey (long range recruitment). The recruitment phase stops as soon as the group is able to move the prey.

The large prey, however, is not always transported smoothly. Especially at the beginning of cooperative retrieving, ants often are stagnant to fulfill the tasks for a long time according to observing results. They may tumble, rotate, and even move for wrong directions. Some ants may be efficient draggers, while some may be not. There may be some ants that do useless or adverse works. Some naughty ants can even crawl on the moving preys. These phenomenon are easily observed in natural world, and also proved by experimental results [16]. Fig. 2 indicates that at the prophase of cooperative transport, ants only can move big prey slowly. By adjusting, the velocity becomes obviously higher and keeps relatively stable at last.

**Fig. 1.** Cooperative transport (large prey retrieving) in ant colony

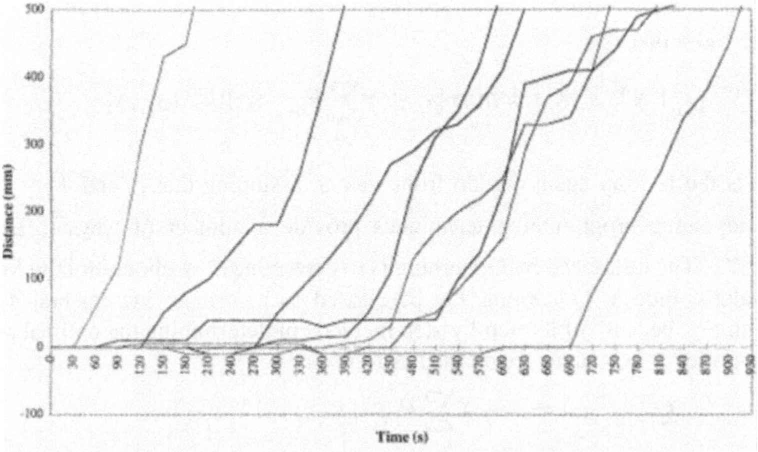


**Fig. 2.** Distance over which a larva of *Tenebrio molitor* has been transported by *Formica polyctena* ants as a function of time, eight experiments are shown [16]

From what described above, it is reasonable to believe that in the beginning phase of cooperative transport, ants must have learned to realign themselves and cooperate with others. We have already known that in swarm systems like ant colony, a single ant can not evaluate rewards of its action sequence with its own limited information. The environment and the goal for a specific ant are changing in dynamic. Ants only can interact with its neighbors and get updated local information. Therefore, they can only learn from their own experiences and neighbor's information, adjust themselves in time according to what they learn, so that they can move the big prey smoothly. This is a heuristic for us to design swarm-based learning methods.

## 3 Q-Learning

Q-learning [17] is an approach of reinforcement learning which is widely applied. Because it assigns rewards to a state-action pair, the agent is therefore not required to predict the future state and does not require a model. The task facing the agent is that of determining an optimal policy $\pi^*$, which selects actions that maximize the long-term measure of reinforcement given the current state. Normally the measure used is the total discounted expected reward. By discounted reward, we mean that future rewards are worth less than rewards received now, by a factor of $\gamma^s (0 < \gamma < 1)$. Under a policy $\pi$, the value of state *s* is

$$V^{\pi} \equiv r_a + \gamma \sum_{s_{t+1}} P_{s_t s_{t+1}} [\pi(s_t)] V^{\pi}(s_{t+1}) \tag{1}$$

The agent expects to receive reward *r* immediately for performing the action $\pi$ recommends, and then moves to a state that is "worth" $V^{\pi}(s_{t+1})$ to it with probability $P_{s_t s_{t+1}}[\pi(s_t)]$. The theory assures us that there is at least one stationary policy $\pi^*$ such that

$$V^*(s_t) \equiv V^{\pi^*}(s_t) = \max_a \{r_a + \gamma \sum_{s_{t+1}} P_{s_t s_{t+1}} [a] V^{\pi^*}(s_{t+1})\} \tag{2}$$

This is the best an agent can do from state *s*. assuming that $r_a$ and $P_{s_t s_{t+1}}[a]$ are known, dynamic programming techniques provide a number of ways to calculate $V^*$ and $\pi^*$. The task faced by Q-learning is to determine $\pi^*$ without initially knowing these values. Indeed, Q-learning can be classed as a form of incremental dynamic programming, because of its step-by-step methods of determining the optimal policy.

For a policy $\pi$, Q values are defined as follows:

$$Q^{\pi}(s_t, a) = r_a + \gamma \sum_{s_{t+1}} P_{s_t s_{t+1}} [\pi(s_t)] V^{\pi}(s_{t+1}) \tag{3}$$

In other words, the *Q* value is the expected discounted reward for executing action *a* at state *x* and following policy $\pi$ thereafter. The object in Q-learning is to estimate the *Q* values for an optimal policy.

## 4 The Framework of Neighbor-Information-Reference (NIR) Learning Method in Swarm Systems

### 4.1 Individual

A state is the description of an individual that captures all the information relevant to its decision-making process at a particular time. In swarm systems, individuals share the same state set $S = \{s_1, s_2, \ldots, s_N\}$ available to them because of their homogeneous nature. In other words, each state $s_i (i = 1,2,\ldots,N)$ can be experienced by any individual at an appropriate time step. All individuals comply with some simple rules in swarm systems. As a result, there are only a few states for individuals, namely $N$ is a relatively small number, which make it possible that individual can transit to any state from its current state in a short time period. For simplification, we assume that individuals are capable of transiting to any state in one step by taking actions, so that we can pay most of our attention to state transition instead of actions individuals take.

### 4.2 Interaction with Neighbors

In swarm systems, each individual is self-autonomous according to some rules [18]. They can obtain local information, and interact with their geographical neighbors. They can also change the local environments or mark in the local environments to interact with the remote individuals indirectly, namely stigmergy. Complex collective and self-organizing behaviors emerge from the interaction of individuals. In the past years, the information interaction of the individuals was ignored to some extend. In fact it is very important [19] for solving problems in swarm intelligence.

In this work, $Ind_i$ is employed to denote the $i^{th}$ individual. All individuals who can interact directly with $Ind_i$ are called 1-interval neighbors of $Ind_i$. Individuals who can only interact directly with 1-interval neighbors are denoted by 2-interval neighbors of $Ind_i$, and so on. Individual can get its 1-interval neighbors' performance information directly. It also can get its $j$-interval $(1 < j)$ neighbors' information indirectly by interacting with $l$-interval $(1 \le l < j)$ neighbors if there is enough time. If one of neighbors gets very good performance, or this neighbor can get good performance with a high probability in the near future, it is likely for the individual to select this neighbor's state as its next state. The neighbor relationship is showed in Fig. 3.

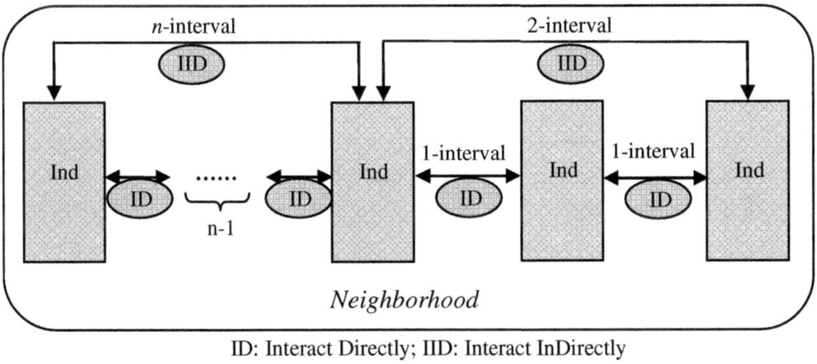


**Fig. 3.** Interaction relationships of individuals in neighborhood

### 4.3 Environments and Rewards

The goal of a swarm is implied in the environment information. Individuals only can achieve their goal by cooperative behaviors. A single individual contributes only a little to the whole performance. When $Ind_i$ is at state $s_j$ at time $t$, it gets reward $r_{s_j}^t$. What must be mentioned is that the value of $r_{s_j}^t$ only depends on $(t, j)$. In other words, if more than two individuals are in the same state $s_j$ at time $t$, they get the same rewards. Because individuals are distributed in different areas, and each individual makes decisions according to local information it gets, it is impossible to assure that each one can do right work at any time. Taking cooperative transport for example, many ants have trouble in understanding what they should do at the beginning. Even when the task is being fulfilled smoothly, there still are idle and naughty individuals. So, not everyone can get positive rewards.

### 4.4 Neighbor-Information-Reference Learning

Consider an individual moving around some discrete, finite world in computational environments, choosing one from a finite collection of states at every time step. At time $t$, the individual is equipped to register the state $s^t (\in S)$ of the world, and can receive a reward $r_{s^t}$ immediately. It transits to the state $s^{t+1}$ of its neighbors according to a policy $\pi$. The task facing the individual is that of determining an optimal policy $\pi^*$, that transits states that maximize the long-term measure of reinforcement. In this work the measure used is the total discounted expected reward. Under policy $\pi$, the discounted value of $s^t$ is

$$V^{\pi}(s^t) = r_{s^t} + \gamma \sum_{s_{1-\mathrm{int}-Nei}} P_{s^t, s_{1-\mathrm{int}-Nei}} V^{\pi}(s_{1-\mathrm{int}-Nei}) \tag{4}$$

where 1-int-Nei denotes 1-interval neighbor of this individual, $P_{s^t, s_{1-\mathrm{int}-Nei}}$ is the probability of the individual transits to $s_{1-\mathrm{int}-Nei}$ at time $t$+1 from $s^t$ according to policy $\pi$. Since each individual has its 1-interval neighbors, we set

$$V^{\pi}(s_{R-\mathrm{int}-Nei}) = r_{R-\mathrm{int}-Nei} \tag{5}$$

where $R$ is a control parameter. Namely, when the information of $R$-interval neighbor is referenced, use its immediate reward instead of discounted value. $R$ has its physical meaning that the information can be directly transferred for $2 \times R$ times between individuals in an allowed time period. There is at least one policy $\pi^*$ such that

$$V^{\pi^*}(s^t) \equiv V^*(s^t) = \max\{r_{s^t} + \gamma \sum_{s_{1-\mathrm{int}-Nei}} P_{s^t, s_{1-\mathrm{int}-Nei}} V^{\pi^*}(s_{1-\mathrm{int}-Nei})\} \tag{6}$$

when $R$ is given. This is the best an individual can do from state $s^t$. Because $r$ and $P$ are known, $V^*$ is computable according (4) and (5). It can be written as a formation of Q-learning as follows:

$$Q^{\pi}(s^t) = r_{s^t} + \gamma \sum_{s_{1-\mathrm{int}-Nei}} P_{s^t, s_{1-\mathrm{int}-Nei}} V^{\pi}(s_{1-\mathrm{int}-Nei}) \tag{7}$$

In NIR learning, the aim is to estimate individual's neighbors' $Q$ values. If one of its 1-interval neighbors has the maximum $Q$, then transiting to this neighbors' state in the next time is optimal for the individual at this time step. The NIR algorithm is described as in Fig. (4), where $\alpha$ and $\gamma$ are learning rate and discounted factor, respectively.

1. Initiate arbitrarily all Q(s) values;
2. Repeat (for each episode):
   1) Put all individuals on the working area;
   2) Repeat (for each individual):
      ① Choose random (initial) states of all individuals;
      ② $i$=1;
      ③ Repeat(for each step in the episode) :
         i. Receive immediate reward $r$ of current state, observe the 1-interval neighbors' state $s_{1-\mathrm{int}-Nei}$;
         ii. $Q(s) \leftarrow Q(s) + \alpha(r + \gamma \max_{s_{1-\mathrm{int}-Nei}} Q(s_{1-\mathrm{int}-Nei}) - Q(s))$;
         iii. $s \leftarrow s_{1-\mathrm{int}-Nei}$;
         iv. $Ind \leftarrow Ind(s_{1-\mathrm{int}-Nei})$;
         v. $i \leftarrow i+1$
         Until $i = R+1$;
      Until each individual is computed;
   Until the desired number of episodes have been investigated.

**Fig. 4.** The NIR learning algorithm

# 5 Analysis and Discussion

NIR learning is a model-free learning method. Individuals need not build maps of their environments. They need not predict the future states either. In NIR learning, each individual has its neighbor's information as referencing information, so that to improve their own performance. By discounted rewards, individuals are assured to make optimal decision in $R$-interval neighborhood area. In fact, $R$ denotes the radius of effective area in one-step decision-making for an individual in NIR learning. The value of $R$ is assigned according to time-relativity of the task swarm fulfills. If the task is urgent, the value of $R$ must be small because there is not much time to interact. When $R$ =1, individual only gets 1-interval neighbors' information, and can make its decision in the shortest time. On the contrary, when the value $R$ is large enough, individual can get global optimal information. Assuming that there are $n$ individuals, and each individual can interact directly with $m$ neighbors, if

$$R \geq \log_m n \tag{8}$$

then an individual can get information of every individual at one time step.

In swarm systems, everyone is equally important to the swarm. When a few individual's states are abnormal, the swarm's efficiency is almost not influenced. Furthermore, swarm is capable of self-repairing by easily accepting new individuals. In this view, a single individual is not important. On the other hand, however, it is important to the swarm. Collective functions of the swarm are carried out based on every individual's work. John Holland [20] believes that there is an "emergence" procedure from individual simplicity to collective complexity. Although mechanisms underlying emergence are unclear, some researchers believe that if the adaptive individuals are properly designed in simulation, and make them cooperate according to a few rules, new swarm functions will emerge from evolution of self-organized individuals. Experimental results, however, were not as exciting as researchers anticipated. Except for unforeseen reasons, the lack of appropriate learning methods for swarm systems has markedly reduced individual efficiency and swarm performance. So, the application of NIR learning will undoubtedly make swarm more "intelligent".

## 6 Conclusions

Most of traditional learning methods do not fit for swarm intelligence. The idea for multi-agent, though looks similar with swarm, do not work when treating on swarm learning in applications. In this paper, a new learning method, NIR learning method, is present on the basis of Q-learning. This is a swarm-based learning method in which the principles of swarm intelligence are strictly complied with. According to time-relativity of concrete tasks, different policies are analyzed under the method framework. The application of NIR learning will not only improve individual efficiency, but also make swarm more "intelligent". In future works we will pay more attention to simulation and computation of this method.

## References

1. Bonabeau, E., Dorigo, M., Theraulaz, G.: Swarm Intelligence: from Natural to Artificial System. Oxford University Press, New York (1999)
2. Kennedy, J., Eberhart, R.C., Shi, Y.: Swarm Intelligence. Morgan Kaufmann Publishers, San Francisco (2001)
3. Dorigo, M., Gianni, D.C.: Ant Algorithms for Discrete Optimization. Artificial Life, 5(3) (1999) 137–172
4. Kennedy, J., Eberhart, R.C.: Particle Swarm Optimization. Proceedings of IEEE International Conference on Neural Networks, Piscataway, NJ (1995) 1942–1948
5. Vapnik, V.: The Nature of Statistical Learning Theory. 2nd Edition, Springer, New York (2000)
6. Poggio, T., Sung, K.K.: Example-Based Learning for View-Based Human Face Detection. Proceedings of the ARPA Image Understanding Workshop (II) (1994) 843–850
7. Mitra, P., Murthy, C.A., Pal, S.K.: A Probabilistic Active Support Vector Learning Algorithm. IEEE Trans. on PAMI 26(3) (2004) 413–418
8. Tillotsona, P.R.J., Wu, Q.H., Hughes, P.M.: Multi-agent Learning for Routing Control within an Internet Environment. Engineering Applications of Artificial Intelligence, 17(2) (2004) 179–185

9. Takadama, K., Hajiri, K., Nomura, T., okada, M., Nakasuka, S., Shimohara, K.: Learning Model for Adaptive Behaviors as an Organized Group of Swarm Robots. Artificial Life Robotics, 2 (1998) 123–128
10. James, F.P., Henry, C.: Reinforcement Learning in Swarms that Learn. Proceedings of the 2005 IEEE/WIC/ACM International Conference on Intelligent Agent Technology (IAT'05), Compiegne, France (2005) 400–406
11. Hölldobler, B.: Territorial Behavior in the Green Tree Ant (Oecophylla smaragdina). Biotropica, 15 (1983) 241–250
12. Wojtusiak, J., Godzinska, E. J., Dejean, A.: Capture and Retrieval of Very Large Prey by Workers of the African Weaver Ant Oecophylla loginada. Tropical Zool. 8 (1995) 309–318
13. Franks, N. R., Gomez, N., Goss, S., Deneubourg, J.-L.: The Blind Leading the Blind in Army Ant Raid Patterns: Testing a Model of Self-Organization (Hymenoptera: Formicidae). Insect behav. 4 (1991) 583–607
14. Moffett, M.W.: Cooperative Food Transport by an Asiatic ant. National Geog. Res. 4 (1988) 386–394
15. Martino, G.D.S., Cardillo, F.A., Starita, A.: A New Swarm Intelligence Coordination Model Inspired by Collective Prey Retrieval and Its Application to Image Alignment. Lecture Notes in Computer Science, Vol. 4193 (2006) 691–700
16. Kube, C.R., Bonabeau, E.: Cooperative Transport by Ants and Robots. Robotics and Autonomous Systems 30 (2000) 85–101
17. Watkins, C., Dayan, P.: Technical Note: Q-Learning. Machine Earning, 8 (1992) 279–292
18. He, X., Zhu, Y., Wang, M.: Knowledge Emergence and Complex Adaptability in Swarm Intelligence. The Proceedings of the China Association for Science and Technology, 3 (2007) 310–316
19. He, X., Zhu, Y., Hu, K., Niu, B.: Information Entropy and Interaction Optimization Model Based on Swarm Intelligence. Lecture Notes in Computer Science, Vol. 4222 (2006) 136–145
20. John, H.: Emergence: from Chaos to Order. Oxford University Press (1998)
21. Williams, R.J.: Simple statistical gradient-following algorithms for connectionist reinforcement learning. Machine Learning 8 (1992) 229–256
22. Berny, A.: Statistical Machine Learning and Combinatorial Optimization. Theoretical Aspects of Evolutionary Computing, Springer (2001) 287-306

# A Genetic Algorithm for Shortest Path Motion Problem in Three Dimensions*

Marzio Pennisi[1], Francesco Pappalardo[1,2], Alfredo Motta[3], and Alessandro Cincotti[4]

[1] Department of Mathematics and Computer Science, University of Catania
[2] Faculty of Pharmacy, University of Catania
mpennisi@dmi.unict.it,francesco@dmi.unict.it
[3] Politecnico di Milano
Milano, Italy
motta.lrd@gmail.com
[4] School of Information Science
Japan Advanced Institute of Science and Technology, Japan
cincotti@jaist.ac.jp

**Abstract.** We present an evolutionary approach to search for near-optimal solutions for the shortest path motion problem in three dimensions (between a starting and an ending point) in the presence of obstacles. The proposed genetic algorithm makes use of newly defined concepts of crossover and mutation and effective, problem optimized, methods for candidate solution generation. We test the performances of the algorithm on several test cases.

## 1 Introduction

The application of genetic algorithms (GA) [10] to problems where the search space is particularly wide and complex can produce good results [11],[4]. One of the problems we can adapt to GA search is the shortest path motion problem in three dimensions (between a starting and an ending point) in the presence of obstacles. It has been proved that if the obstacles are constituted by polyhedrons, the problem is NP hard [2],[9]. In particular, the problem is exponential in the number of vertices.Various approximating algorithms have been proposed in literature. The interested reader can find additional informations in [1],[8],[3].

The complexity of this kind of problem in its general form is not strictly and directly connected to the complexity of the obstacles. It has been proved [15] that the Euclidean shortest path motion problem in three dimensions remains NP hard if the obstacles are disjoint parallelepipeds parallel to the axes. This is the case that we are going to analyze in this paper.

This problem can be found in many practical applications and has investigated in many different environments. "Path planning research" is a common problem

* F.P. and M.P. acknowledge partial support from IMMUNOGRID project, under EC contract FP6-2004-IST-4, No. 028069.

D.-S. Huang, L. Heutte, and M. Loog (Eds.): ICIC 2007, LNAI 4682, pp. 534–542, 2007.

in robot automation: in order to move a robot arm, all the obstacles have to be avoided and obtaining shortest path is preferable. This problem has been explored for example in [6],[7].

Another topic that founds application can be represented by routing in communication networks. In this case shortest path can represent best route and obstacles can be link failure or congestion. A Neural Network for this particular case has been presented in [5], while some AntNet algorithms have been shown in [13],[14].

Automated driving systems in presence of obstacles can also represent another practical application of the problem.

The paper is organized as follows. In section 2, we describe in depth our algorithm. Section 3 provides computational results, while Section 4 is devoted to conclusions and final remarks.

## 2 The Genetic Algorithm

We start our description with the formal definition of the problem we intend to study.

- Instance: two points $s$ and $t$ and $N_p$ parallelepipeds pairwise-disjoint in an Euclidean three-dimensional space;
- solution: the shortest path between $s$ and $t$ that avoids the interior of the given obstacles;
- measure: the total length of the path in Euclidean metric.

With term *world* we intend a subset of the Euclidean space $\mathbf{R}^3$, while we consider an *obstacle* as parallelepiped parallel to the Cartesian axes. A *gene* is defined as a tern of real values $< x, y, z >$. It describes the position of a point in the Euclidean space. Genes are subject to particular constraints:

- Every gene is positioned on one of the 12 edges of the $N_p$ parallelepipeds, limiting considerably the search space.
- Two consecutive genes in a chromosome cannot be connected by a segment intersecting an obstacle.

A *chromosome* is a sequence of genes representing the vertices of the broken line. $s$ and $t$ are identified as terns of real values $< x, y, z >$ representing the starting and the ending points. They are respectively connected to the first and the last gene of every chromosome by a segment that does not intersect any parallelepiped. Therefore every chromosome represents a candidate solution to the problem.

The number of chromosomes and the maximum dimension of a chromosome are fixed and we call these values respectively $N_c$ and $D_c$.

The algorithm resembles classical GAs and can be briefly described by the pseudo-code showed on Procedure 1.

During the generation of chromosomes, we proceed in the following way: let $f$ be the segment connecting $s$ to $t$. If there is not a direct connection from $s$

**Procedure 1.** GeneticAlgorithm for SPMP3D

Generate the initial population
Compute the fitness of each individual
**while** no. of desired iterations is **not** reached **do**
  Select best-ranked individuals from population
  Execute **crossover** and **mutation** operators to obtain new offsprings
  Compute the fitness of the offsprings
  Replace worst-ranked individuals with offsprings
**end while**
**return** best-ranked individual

to $t$, it will intersect at least an obstacle. If this is the case, we proceed with a corrective approach eliminating all the "errors", i.e. all intersections between $s$ and $t$.

To gain individuals diversity, for other chromosomes, we choose parallel segments $[s', t']$ to $f$. After the initial corrections, we substitute $s', t'$ with $s, t$. After that we make corrections only on the final parts. For choosing $s', t'$ we use the following method to find acceptable points: let $< x_1, y_1, z_1 >$ and $< x_2, y_2, z_2 >$ be respectively the coordinates of $s$ and $t$. Let $< X_m, Y_m, Z_m >$ be the tern indicating the max dimensions of the world, we define a "validity range" $r$ as follows:

$$r = min(x_1, y_1, z_1, x_2, y_2, z_2, X_m - x_1, Y_m - y_1, Z_m - z_1, X_m - x_2, Y_m - y_2, Z_m - z_2).$$

Three random values $v_x, v_y, v_z \in (0, r)$ are chosen. $s'$ and $t'$ will be defined respectively as $< x_1 + v_x, y_1 + v_y, z_1 + v_z >$ and $< x_2 + v_x, y_2 + v_y, z_2 + v_z >$. Figure 1 shows the method.

For correcting the "errors" we act in two different ways, according with the position of the ingoing and the outgoing points on the obstacle. If the ingoing and outgoing points are on adjacent faces, the gene will be positioned on a random point of the edge shared by the two faces. If they are on parallel faces, we need two genes. First we choose a face that can minimize the path between the two points. Then we choose randomly:

1. one point on the edge shared by this face and the face containing the ingoing point,
2. one point on the edge shared by this face and the face containing the outgoing point.
   The process is shown on Figure 2.

Fitness function $F$ is defined as follows. Let $c$ be a chromosome; let $g_i$ be the $i$-th gene of the chromosome $c$; let $g_0 = s$ and $g_{D_c} = t$. We have:

$$F(c) = \sum_{i=1}^{D_c} p(g_i, g_{i-1}),$$

where $p(g_i, g_{i-1})$ represents the Euclidean metric distance between $g_i$ and $g_{i-1}$.

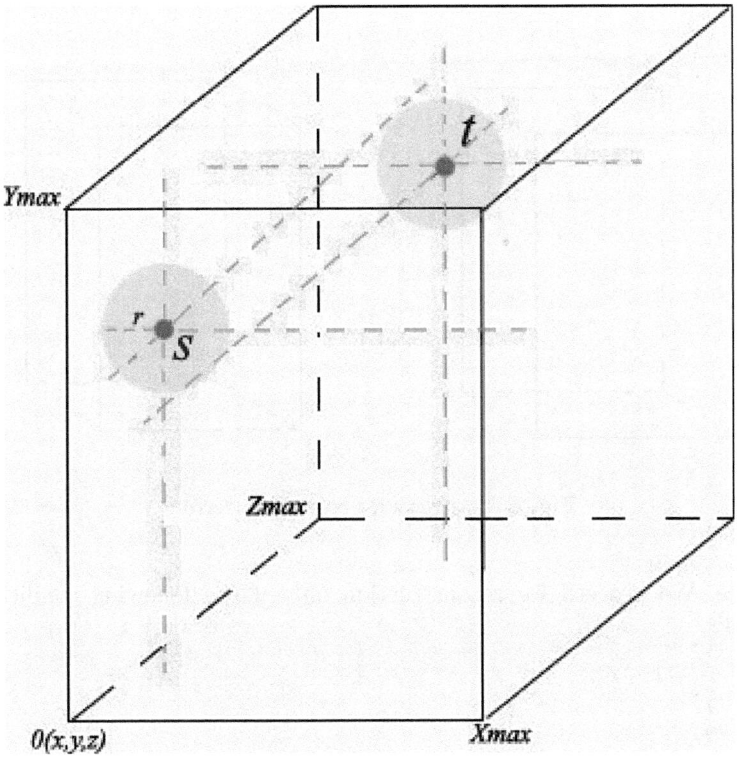


**Fig. 1.** Example of validity range in an Euclidean space $XYZ$

A "roulette wheel" selection method is used to select the chromosomes which will take part to the crossover process: the chance of a chromosome of getting selected is proportional to its fitness.

Elitism on the best chromosome is implemented: the chromosome with the best fitness will be preserved and it will be a member of the next population.

We proceed with a modified single-point crossover: every gene in a chromosome is placed on an edge of parallelepiped.

Given $g_p$ a randomly chosen crossover point, the first part from beginning of chromosome to $g_p$ is copied from the first parent, the rest is copied from the second parent. If $g_p = g_{p+1}$ or the segment $(g_p, g_{p+1})$ intersects no obstacles the chromosome is accepted and no more work has to be done.

We else allow, to permit chromosome acceptance, the reconstruction of a subpart of the chromosome in a such way that no obstacles are intersected. To avoid total chromosome reconstruction we introduced a fixed threshold $C_x$ indicating the maximum number of genes that can be replaced (Figure 3).

Starting from $g_p$ and proceeding towards the ends we compute the sub-part of ends $(g_{i_1}, g_{i_k})$ whose exclusion avoids the repeating of parallelepipeds in the

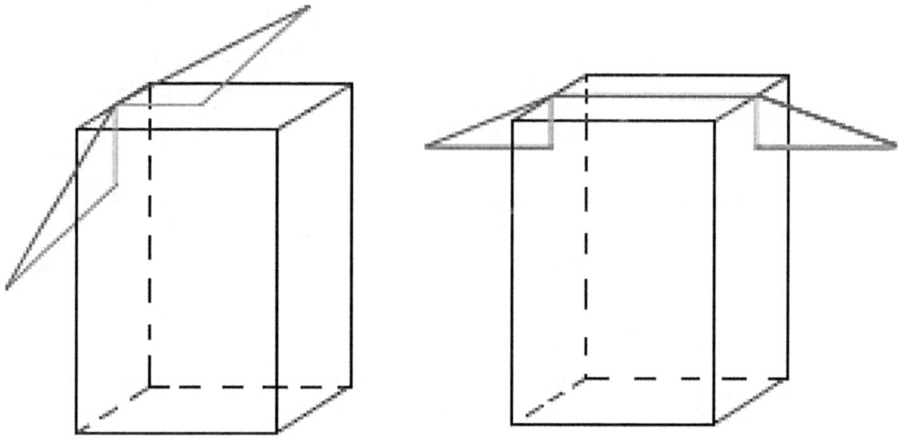

**Fig. 2.** Two ways for correcting "errors"

sequence. We proceed for reconstruction only if the following conditions are satisfied:

1. $i_k - i_1 + 1 \leq C_x$
2. $i_1 > 1$
3. $i_k < D_c$.

For rebuilding the remaining sub-part of the chromosome under the compatible threshold, we have to recalculate the missing genes of the new offspring from $g_{i_1}$ to $g_{i_k}$.

We proceed in the following way: let $f$ be the straight line connecting $s$ to $t$. Consider $D_c$ equidistant points in $f$ so that the number of these points is equal to the number of genes in a chromosome. We assimilate the $i$-th gene of a chromosome to the $i$-th point of $f$ supposing that in most cases a good chromosome contains genes whose position is not too far from the indicated points. We therefore proceed during the initialization of the algorithm building a $[D_c \times N_p]$ matrix so that the $(i, j)$ cell contains the $j$-th obstacle closer to the $i$-th point. Let $r$ be the index of gene we need to recalculate, we choose an integer value $y$ between 0 and $N_p - 1$ using the following law:

$$y = \lfloor(((k_1^x - 1)/k_2) \cdot N_p)\rfloor$$

where $k_1 = 10, k_2 = 9$ are two constants and $x \in [0, 1[\subset \mathbf{R}$ is a randomly chosen value. We finally choose to position the $r$-th gene on a random edge of the obstacle contained in the $(r, y)$ cell.

From experimental results we observed that the presented law tends toward closest obstacles without excluding the distant ones.

Mutation process can happen in different ways. Due to particular chromosome structure and constraints, a canonical mutation process was unusable. It

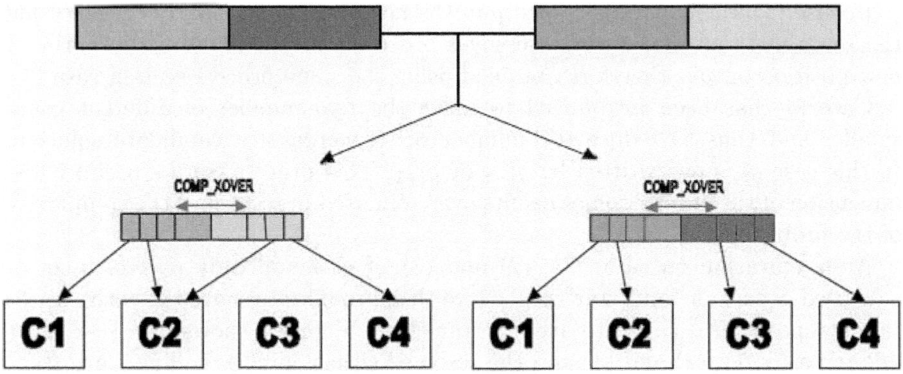


**Fig. 3.** An example of crossover: inside the first offspring, $c_2$ represents a parallelepiped out of the threshold $C_x$. The first offspring will be rejected. The second offspring will pass the test and will be accepted.

was instead necessary to take into account particular chromosome meaning. We therefore decided to allow specific mutation in four different ways, each way with a specific probability $p_i$, $\sum_{i=1}^{4} p_i = 1$.

When a gene $g_i$ of a chromosome is selected for mutation, a random real number $p \in (0, 1)$ is generated. Let $g_i'$ be the mutated gene, $g_i'$ is obtained from $g_i$ using one of the following mutation processes:

1. shift $g_i$ on the same edge ($0 \leq p < p_1$);
2. move $g_i$ on different edge of the same parallelepiped ($p_1 \leq p < p_1 + p_2$);
3. move $g_i$ on a parallelepiped in the neighborhood ($p_1 + p_2 \leq p < p_1 + p_2 + p_3$);
4. collapse $g_i$ on the previous or subsequent gene ($p_1 + p_2 + p_3 \leq p < 1$).

We chose $p_1 \geq p_2 \geq p_3 \geq p_4$ to favor mutations which alter less the chromosome. It's clear that only one mutation process is chosen at a time (*mutually exclusive events*). Two mutations cannot occur to the same gene in the same time step.

If the mutated chromosome does not respect the constraints, $p$ is regenerated and the entire process will be repeated for no more than $N_t$ times (where $N_t$ is a positive integer value). If the number of tries exceeds the threshold $N_t$, the mutation process will fail and the chromosome will not be modified.

For the case (1), the best results have been obtained limiting the length of the range where a new position has to be chosen: $g_i'$ is obtained choosing a random position on the same edge where $g_i$ is placed in a such way that the distance between $g_i'$ and the segment $[s, t]$ will be not greater to than the distance between $g_i$ and the same segment.

If we are in case (2), $g_i'$ is obtained first choosing a random edge $e$ of the parallelepiped where $g_i$ is located, and then choosing a random position on $e$.

In case (3) a new parallelepiped $p$ in the neighborhood is firstly chosen using the same method seen during crossover process for rebuilding of sub-parts. A new position on $p$ is therefore obtained using the same process seen in case (2).

Case (4) has been introduced to make the real number of different genes smaller and thus to reduce the number of segments of a candidate solution. In this case $g_i$ is overwritten by $g_{i-1}$ or $g_{i+1}$. This process can be useful if the dimension of the chromosomes results overestimated in respect of the complexity of the problem.

After mutation on cases (1), (2) and (3), if $g$ was already part of a set of collapsed genes, an "anti-star" procedure that provides to move the entire set to the new position is called to avoid a star effect. Figure 4 shows us a star effect due to cases (1), (2) and (3) and the resolved situation after calling "anti-star" procedure.

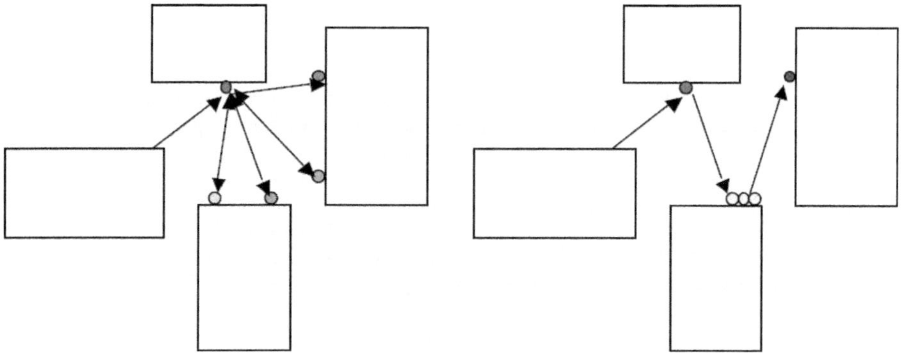

**Fig. 4.** An example without (left side) and with (right side) "anti-star" procedure

We also use some auxiliary and optimization procedures for obtaining best results. The first one is called after the crossover process. This procedure selects a chromosome with a given probability $P_v$ from those have not taken part to the crossover process and overwrites it with a new generated one. In this way diversity of the population is maintained and local minimums should be avoided.

The second procedure looks into the chromosomes for two non consecutive genes $[g_{i1}, g_{ik}]$ placed on the same obstacle $p$ and will collapse, if necessary, the entire sequence in a such way that all the constraints are respected. It is used to avoid that a candidate solution passes from a point $g_{i1}$ on an obstacle $p$ and, after a loop, it returns to $p$ (Figure 5).

## 3 Computational Results

To our best knowledge there are no test suites available for the problem. For that reason we tested the algorithm in two different ways.

On the first 6 cases we used worlds with a well-known solution, created "ad-hoc" for testing purposes. On the other ones we used random bigger worlds

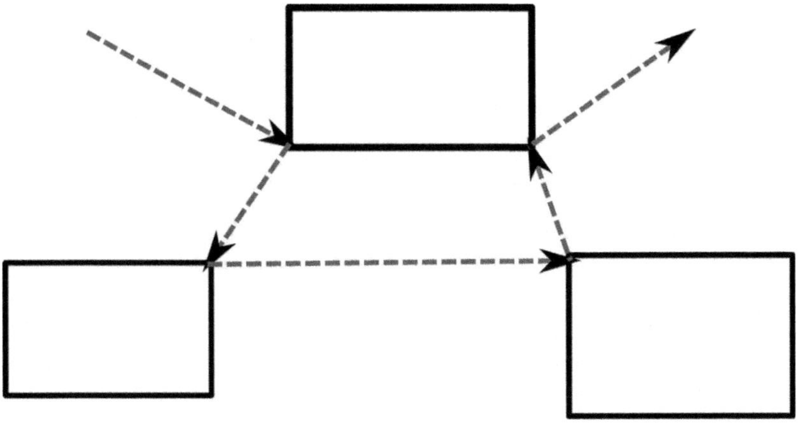

**Fig. 5.** An example of loop

without knowing the best solution. The algorithm has been repeated 20 times for every case.

We have set $N_c$ and $D_c$ to a congruous value for every case. We use "2-decimal" precision for the results except for the standard deviation that uses a "5-decimal" precision.

**Table 1.** Path lengths for different test cases

| Obstacles | Best | Best result % | Mean | Standard deviation |
|---|---|---|---|---|
| 2 | 137.33 | 100% | 137.33 | 0 |
| 3 | 194.47 | 100% | 194.47 | 0 |
| 4 | 109.44 | 100% | 109.44 | 0 |
| 5 | 168.65 | 100% | 168.65 | 0 |
| 14 | 190.85 | 100% | 190.85 | 0 |
| 28 | 368.47 | 100% | 368.47 | 0 |
| **Obstacles** | **Best Found** | **Best Found %** | **Mean** | **Standard deviation** |
| 20 | 342.68 | 60% | 342.70 | 0.03996 |
| 20 | 288.08 | 70% | 288.08 | 0.00113 |
| 20 | 147.05 | 100% | 147.05 | 0 |
| 40 | 240.05 | 5% | 240.07 | 0.00570 |
| 40 | 295.56 | 45% | 295.80 | 0.49896 |
| 40 | 221.80 | 85% | 221.80 | 0.00185 |
| 60 | 363.41 | 10% | 363.99 | 0.26146 |
| 60 | 533.03 | 10% | 534.79 | 2.38956 |
| 60 | 371.37 | 5% | 373.78 | 1.34352 |
| 80 | 349.98 | 70% | 350.87 | 2.73466 |
| 80 | 399.70 | 5% | 403.74 | 2.94401 |
| 80 | 549.32 | 5% | 551.23 | 0.45042 |

From Table 1 we can see that the given algorithm is able to find always the best solution for less-populated words where the optimal solution is known. Further analysis and comparison of the remaining cases with approximated algorithms will be examined in future work.

## 4 Conclusion and Future Work

We have presented an evolutionary algorithm to find effective near-optimal solution for the shortest path motion problem in three dimensions.

One of the major novelties of our algorithm, is the usage of particularly adapted optimization procedures, like new defined crossover and mutation.

Future work will see our GA compared to approximate algorithms and to be adapted in worlds where the position of obstacles changes with the passing of the time.

## References

1. Papadimitriou, C.H.: An Algorithm for Shortest-Path Motion in Three Dimensions.Inform Process. Lett20.(1985)259-263
2. Canny, J.,Reif,J.H.: Lower Bound for Shortest Paths and Related Problems. In Proceedings of 28th Annual Symposium on Foundations of Computer Science (1987)49-60
3. Clarkson, K.L.: Approximation algorithms for shortest path motion planning. Proceedings of 19th Annual ACM Symposium on Theory of Computing (1987)56-65
4. Goldberg,D.E.: Genetic Algorithms in Search. Optimization and Machine Learning, Addison-Wesley(1989)1-88
5. Zhang, L., Thomopoulos, S.C.A: Neural Network Implementation of the Shortest Path Algorithm for Traffic Routing in Communication Networks. International Joint Conference on Neural Networks.Vol2. 591(1989)
6. Fujimura,K., Samet,H.: Planning A Time-Minimal Motion among Moving Obstacles. Algorithmica, Vol.10.(1993)41-63
7. Fujimura, K.: Motion Planning Amid Transient Obstacles. International Journal of Robotics Research, Vol.13.No.5.(1994)395-407
8. Choi,J., Sellen,J., Chee,K.Y.: Approximate Euclidean Shortest Path in 3-space. Annual Symposium on Computational Geometry Archive. Proceedings of the Tenth Annual Symposium on Computational Geometry. (1994)41-48
9. Reif,J.H., Storer,J.A.: A Single-Exponential Upper Bound for Finding Shortest Paths in Three Dimensions. J. ACM (1994)1013-1019
10. Whitley, D.: A Genetic Algorithm Tutorial. Statistics and Computing (1994)65-85
11. Chambers, L.: Practical Handbook of Genetic Algorithms,Applications Vol.1. CRC Press (1995)143-172
12. Mitchell, M.: An Introduction to Genetic Algorithms. The Mit Press (1996)
13. Baran, B., Sosa,R.: A new approach for AntNet routing. Proceedings. Ninth International Conference on Computer Communications and Networks (2000)303-308
14. Baran,B.: Improved AntNet routing. ACM SIGCOMM Computer Communication Review, Vol.31.Issue 2 Supplement(2001)42-48
15. Mitchell, J.S.B., Sharir, M.: New Results on Shortest Paths in Three Dimensions. Annual Symposium on Computational Geometry Archive Proceedings of the Twentieth Annual Symposium on Computational Geometry (2004)124-133

# A Hybrid Electromagnetism-Like Algorithm for Single Machine Scheduling Problem

Shih-Hsin Chen[1], Pei-Chann Chang[2], Chien-Lung Chan[2], and V. Mani[2]

[1] Department of Industrial Engineering and Management, Yuan Ze University
[2] Department of Information Management, Yuan Ze University,
135 Yuan Tung Road, Ne-Li, Tao-Yuan, Taiwan, R.O.C., 32026
[3] Department of Aerospace Engineering, Indian Institute of Science
Bangalore, 560-012, India
iepchang@saturn.yzu.edu.tw

**Abstract.** Electromagnetism-like algorithm (EM) is a population-based meta-heuristic which has been proposed to solve continuous problems effectively. In this paper, we present a new meta-heuristic that uses the EM methodology to solve the single machine scheduling problem. Single machine scheduling is a combinatorial optimization problem. Schedule representation for our problem is based on random keys. Because there is little research in solving the combinatorial optimization problem (COP) by EM, the paper attempts to employ the random-key concept enabling EM to solve COP in single machine scheduling problem. We present a hybrid algorithm that combines the EM methodology and genetic operators to obtain the best/optimal schedule for this single machine scheduling problem, which attempts to achieve convergence and diversity effect when they iteratively solve the problem. The objective in our problem is minimization of the sum of earliness and tardiness. This hybrid algorithm was tested on a set of standard test problems available in the literature. The computational results show that this hybrid algorithm performs better than the standard genetic algorithm.

## 1 Introduction

Single-machine scheduling problems are one of the well-known combinatorial optimization problems and the earliness/tardiness problem is shown in literature that this problem is NP-hard (Lenstra et al., 1977). The results not only provide the insights into the single machine problem but also for more complicated environment (Pinedo, 2002). In this paper, we consider the single machine scheduling problem with the objective of minimizing the sum of earliness and tardiness penalties. Earlier studies on single machine scheduling with the objective of minimizing the sum of earliness and tardiness penalties are studied by several researchers (Belouadah et al., 1992; Hariri and Potts, 1983; Kim et al., 1994; Akturk and Ozdemir, 2000, 2001; Valente and Alves, 2003).

EM type algorithm has been used for optimization problems, which starts with a randomly selected points from the feasible region for a given optimization problem.

D.-S. Huang, L. Heutte, and M. Loog (Eds.): ICIC 2007, LNAI 4682, pp. 543–552, 2007.

EM employs an attraction-repulsion mechanism to move points (particles) towards the optimal solution. Each point (particle) is treated as a solution and has a charge. A better solution contains a stronger charge. The charge of each point relates to the objective function value we like to optimize. EM method has been tested on available test problems in Birbil and Fang (2003). In this study, it is shown that EM is able to converge to the optimal solution in less number of function evaluations without any first or second order derivative information. A theoretical study of this EM analysis and a modification for convergence to the optimal solution is presented in Birbil et al. (2004).

Hence, in this study we use the random-key approach to represent a schedule and incorporate the EM methodology to solve the single machine scheduling problem.

## 2 Literature Review

There are some researcher extended EM algorithm or applied EM to solve different problems. Debels et al. (2006) integrated a scatter search with EM for the solution of resource constraint project scheduling problems. It is the first paper that includes an EM type methodology for the solution of a combinatorial optimization problem. Birbil and Feyzioglu (2003) used EM type algorithms solving fuzzy relation equations, and Wu et al. (2005) obtained fuzzy if-then rules.

Though EM algorithm is designed for solving continuous optimization problems with bounded variables, the algorithm can be extended to solve combinatorial optimization problem (COP). When we extend the EM algorithm to COPs, the first important step is the representation of a solution. Bean (1994) introduced a random-key (RK) approach for real-coded GA for solving sequencing problem. Subsequently, numerous researchers show that this concept is robust and can be applied for the solution of different kinds of COPs (Norman and Bean, 1999; Snyder and Daskin, 2006]. The random key approach is used to solve single machine scheduling problems and permutation flowshop problems using particle swarm optimization (PSO) algorithm by (Tasgetiren et al., 2007).

Hence, in our study we use the random-key approach to represent a schedule and incorporate the EM methodology to solve the single machine scheduling problem. In our algorithm, the EM procedures are modified to obtain better solution quality effectively. For example, the local search operator perturbs the best solution and to replace the worst one when the objective value is better than the worst solution. In addition, Debels et al. (2006) proposed a new method in calculating the particle charge and exertion force. Both of them are adopted in the research.

According to our experimental results, EM algorithm provides good solution diversity because there are few solutions are overlapped or redundant. Consequently, a hybrid framework is proposed that EM algorithm is combined with GA which is able to converge quickly by its selection and crossover operator.

The rest of the paper is organized as follows: section 3 presents the original EM-like algorithm in solving the continuous problem; the methodology is described in section 4. The experimental result is carried out in section 5, which compared EM with Genetic Algorithms (GAs). Section 6 draws the discussion and conclusions.

## 3 Electromagnetism-Like Algorithm

EM simulates the attraction-repulsion mechanism of electromagnetism theory which is based on Coulomb's law. Each particle represents a solution and the charge of each particle relates to its solution quality. The better solution quality of the particle, the higher charge the particle has. Moreover, the electrostatic force between two point charges is directly proportional to the magnitudes of each charge and inversely proportional to the square of the distance between the charges[1]. The fixed charge of particle $i$ is shown as follows:

$$q^i = \exp\left( -n \frac{f(x^i) - f(x^{best})}{\sum_{k=1}^{m} (f(x^k) - f(x^{best}))} \right), \forall i. \tag{1}$$

where $q^i$ is the charge of particle $i$, $f(x^i)$, $f(x^{best})$, and $f(x^k)$ denote the objective value of particle $i$, the best solution, and particle $k$. Finally, $m$ is the population size.

The solution quality or charge of each particle determines the magnitude of an attraction and repulsion effect in the population. A better solution encourages other particles to converge to attractive valleys while a bad solution discourages particles to move toward this region. These particles move along with the total force and so diversified solutions are generated. The following formulation is the force of particle $i$.

$$F^i = \sum_{j \neq i}^{m} \left\{ \begin{array}{lll} (x^j - x^i) \dfrac{q^i q^j}{\left\| x^j - x^i \right\|^2} & if & f(x^j) < f(x^i) \\ (x^i - x^j) \dfrac{q^i q^j}{\left\| x^j - x^i \right\|^2} & else & f(x^j) \geq f(x^i) \end{array} \right\}, \forall i. \tag{2}$$

The fundamental procedures of EM include initialize, local search, calculating total force, and moving particles. The generic pseudo-code for the EM is as follows:

```
Algorithm 1. EM()
  initialize()
  while (hasn't met stop criterion) do
     localSearch()
     calculate total force F()
     move particle by F()
     evaluate particles()
  End While
```

[1] http://en.wikipedia.org/wiki/Coulomb's_law

# 4 Methodology

This paper proposes a hybrid framework that combines EM-like algorithm and genetic operator for solving scheduling problems. The fundamental method is the random-key technique that enables EM to solve this kind of problems. Because the time-complexity is high and to obtain better solution quality for EM-like meta-heuristic with RK approach, some procedures like local search, particle charge, and electrostatic force are modified. The purpose of this hybrid framework is to take the advantage of EM, which yields a high diversity population, and GA operator let the algorithm converge faster. Since the random-key technique is a fundamental method in this paper, it is introduced in the beginning and the later sections describe the detailed approaches of the hybrid framework and modified EM procedures.

## 4.1 A Random-Key Method

In order to enable EM to solve scheduling problems, the random-key technique is introduced. The concept of RK technique is simple and can be applied easily. When we obtain a *k*-dimension solution, we sort the value corresponding to each dimension. Any sorting algorithm can be used in the method and the paper uses quick sort because its time-complexity is O(*nlogn*). After having a sequence, we can use it to compute the objective function value of this sequence.

Figure 2 demonstrates a 10-dimension solution. The value of dimension 1 is 0.5, value 9.6 is at dimension 2, dimension 3 represents 3.0, and etc. Then, we apply the random-key method to sort these values in ascending order. Thus sequence at position 1 is 8 that mean we schedule job 8 in the beginning and job 2 is scheduled at the last position. By the random-key method, continuous EM algorithm is able to solve all kinds of sequencing problem.

Activities

| | 1 | 2 | 3 | 4 | 5 | 6 | 7 | 8 | 9 | 10 |
|---|---|---|---|---|---|---|---|---|---|---|
| Before | 0.5 | 9.6 | 3.0 | 2.9 | 2.2 | 8.0 | 4.2 | 0.1 | 7.1 | 5.6 |

(a) Value of activities

| | | | | | | | | | | |
|---|---|---|---|---|---|---|---|---|---|---|
| After | 8 | 1 | 5 | 4 | 3 | 7 | 10 | 9 | 6 | 2 |

(b) Schedule list

**Fig. 1.** An example of attract-repulse effect on particle number 3

## 4.2 A Hybrid Framework Combines the Modified EM and Genetic Operators

The hybrid framework includes modified EM procedures and genetic operators, which adopts selection and mating. The selection operator is binary tournament and uniform crossover operator is applied in the framework. Generic EM provides an excellent diversity while GA is able to converge to a better solution quickly. Thus the hybrid method takes the advantage of both sides.

The hybrid system starts with determining which particle is moved by EM or mated by GA crossover operator. In a paper by Debels et al. (2006), they suggested that a new solution can be obtained from crossing between a better solution selected by a binary tournament method. And EM is used to move the inferior solution to a new position. This hybrid approach may encourage solutions converging toward better region quickly and to prevent from trapping into local optima by maintaining the population diversity. Algorithm 1 is the pseudo code of the main procedures of the hybrid framework.

Algorithm 2. A Hybrid Framework

```
initialize()
while (hasn't met stop criterion) do
   localSearch()
   avg ← calcAvgObjectiveValues()
   for i = 1 to m do
      if i ≠ best and f(x^i) < avg then
         j ← a selected particle to mate particle i by binary tournament()
         uniformCrossover(x^i, x^j)
      else if f(x^i) > avg then
         CalcF and Move(x^i)
      end if
   end for
   find sequence by random-key method()
   evaluate particles()
end while
```

According to algorithm 1 (line 1), we initiate the particles in the population. Then, the local search procedure is implemented before the EM procedures and genetic operators. To determine which solution is good or inferior, an average objective value *avg* is calculated. Then, if the solution is better than *avg*, this solution is mated by the other better solution obtained by binary tournament (line 7-8). Otherwise, this solution is moved by modified EM algorithm (line 10). After these particles are mated or moved along with their own total force, the next step is to generate corresponding sequences by random-key technique. As soon as the sequence is obtained, we can obtain objective value of the solution. Finally, because the initialization, local search, particle charge, calculated total force, and move are modified, we discuss them in the following sections.

### 4.3 Particle Charges, Electrostatic Force and Move

The study uses the total force algorithm proposed by Debels et al. (2006), which determines the force exerted on particle *i* by point *j* that does not use the fixed charge of $q_i$ and $q_j$. Instead, $q_{ij}$ depends on the relative deviation of $f(x^i)$ and $f(x^j)$. Thus this particle charge is calculated as follows:

$$q_{ij} = \frac{f(x^i) - f(x^j)}{f(x^{worst}) - f(x^{best})} \tag{3}$$

If the objective value $f(x^i)$ is larger than $f(x^j)$, particle *j* will attract particle *i*. On the other hand, when $f(x^i) < f(x^j)$, a repulsion effect is occurred. There is no action when $f(x^i) = f(x^j)$ because $q_{ij}$ is equal to zero. After the $q_{ij}$ is obtained, the force on particle *i* by particle *j* is

$$F^{ij} = (x^j - x^i) \cdot q_{ij} \tag{4}$$

Thus the particle $x^i$ moves to $x^i + F^{ij}$ in the direction of particle $x^j$. This method is similar to the path relinking method [13] which gradually moves from one point to another (Debels et al., 2006).

## 5 Experimental Results

This study proposed a hybrid framework that combines modified EM meta-heuristic and genetic operator in solving the single machine problem in minimizing the earliness and tardiness penalty. In order to evaluate the performance of this hybrid framework, it is compared with GA which is a well known meta-heuristic. Across these experiments, we adopt the scheduling instances of Sourd and Sidhoum (2005) whose job size are 20, 30, 40, and 50[2]. Each experiment is replicated 30 times and the stopping criterion is to fix the number of examined solutions that is set to 100,000.

Before we validate these methods and to compare the performance between the proposed algorithm and GA, a Design of Experiment (DOE) is carried out to examine the parameter settings of the hybrid framework. The DOE result of it is shown in section 5.1. Then, we compare the performance of the hybrid framework with GA under the job-dependent due date. It is presented in section 5.2.

[2] The name of each instance for 20, 30, 40, and 50 jobs are sks222a, sks322a, sks422a, and sks522a, respectively.

### 5.1 Design of Experiment for EM in Single Machine Scheduling Problems

There are two parameters that should be tuned in EM algorithm. In continuous EM, Birbil and Fang (2003) suggested a population size that is four times the dimensions. However, since there is no result for this problem, this experiment fills up the gap which identifies the appropriate population size. Secondly, the local search method is modified and the number of local search is unknown. Thus the number of local search is considered in the DOE experiment.

Except for the parameter setting of EM algorithm, the study includes the comparison of the performance of hybrid model and the modified EM algorithm that works alone. The parameter setting is shown in table 1 and DOE is applied to select the parameters. The final parameter setting of this hybrid framework is shown in table 2.

**Table 1.** The parameter settings of the EM algorithm

| Factor | Treatments |
|---|---|
| Population Size (popSize) | 50 and 100 |
| Number of Local Search (LS) | 10 and 25 |
| Methods | 1. Modified EM algorithm<br>2. Hybrid Model (Modified EM algorithm and genetic operators) |
| Job Instance (Size) | 20, 30, 40, 50 |
| Number of examined solutions | 100,000 |

**Table 2.** The parameter settings of the hybrid algorithm

| Factor | Treatments |
|---|---|
| Population Size (popSize) | 50 |
| Number of Local Search (LS) | 25 |
| Methods | Hybrid Model (Modified EM algorithm and genetic operators) |

### 5.2 The Comparison Between Hybrid Framework and GAs

We consider the scheduling problem under the job-dependent due date without learning consideration first. The proposed hybrid framework is compared with Genetic Algorithm. The parameter of GA includes crossover rate, mutation rate, and population size, which are set to 0.8, 0.3, and 100, respectively. Above GA parameter settings and experimental result of GA are adopted from our previous research Mani et al. The comparison results are presented in table 3 and the hybrid framework outperforms GA in average across all instances. On other hand, the hybrid model spends more computational effort than GA.

**Table 3.** The comparison between hybrid algorithm and GA

| | GA | | | | Hybrid Framework | | | |
|---|---|---|---|---|---|---|---|---|
| Job | *Min* | *Mean* | *Max* | *Secs* | *Min* | *Mean* | *Max* | *Secs* |
| 20 | 5286 | 5401.7 | 5643 | 1.0573 | 5287 | 5331.8 | 5464 | 1.9542 |
| 30 | 11623 | 12066 | 12916 | 1.6838 | 11584 | 11794 | 12223 | 2.8208 |
| 40 | 25656 | 26211 | 27462 | 2.4548 | 25706 | 25933 | 26294 | 3.3386 |
| 50 | 29485 | 30623 | 32340 | 3.5406 | 29490 | 29902 | 30447 | 4.1182 |

## 6 Discussion and Conclusions

Owing to the random-key method, continuous EM is able to solve sequencing problem now. To improve the performance of EM algorithm, a hybrid framework is proposed which combines EM algorithm and genetic operators. The purpose of this hybrid framework is to take the advantage of EM algorithm and genetic operator, which provides better solution diversity in population and good convergence ability, respectively. A DOE shows the performance of hybrid method is better than to use EM algorithm alone.

According to the comparison between hybrid framework and GA in single machine scheduling problem, the proposed method may be better than GA. However, since RK technique sorts out each solution to generate a sequence, it needs O(*nlogn*) time-complexity to do it while GA is able to provide a sequence representation directly. As a result, the computational effort of hybrid framework is higher than GA. For future research, a better local search such as Variable Neighborhood Search (VNS) can be applied into EM which may improve solution quality. Furthermore, since EM can be extended to multi-objective algorithm, it is an entirely new research area.

## References

1. Abdul-Razaq, T., Potts, C.N.: Dynamic Programming State-Space Relaxation for Single Machine Scheduling, Journal of the Operational Research Society, 39 (1988) 141-152
2. Akturk, M.S., Ozdemir, D.: An Exact Approach to Minimize Total Weighted Tardiness with Release Date, IIE Tranactions, 32 (2000) 1091-1101
3. Akturk, M.S., Ozdemir, D.: A New Dominance Rule to Ninimize Total Weighted Tardiness with Unequal Release Dates, European Journal of Operational Research, 135 (2001) 394-412
4. Azizoglu, M., Kondakci, S., Omer, K.: Bicriteria Scheduling Problem Involving Total Tardiness and Total Earliness Penalties, International Journal of Production Economics, 23 (1991) 17-24.
5. Bauman, J., Józefowska, J.: Minimizing the Earliness–Tardiness Costs on a Single Machine, Computers & Operations Research, 33(11) (2006)3219-3230
6. Bean, J.C.: Genetic Algorithms and Random Keys for Sequencing and Optimization, ORSA Journal on Computing, 6(2) (1994) 154-160

7. Belouadah, H., Posner, M.E., Potts, C.N.: Scheduling with Release Dates on a Single Machine to Minimize Total Weighted Completion Time, Discrete Applied Mathematics, 36 (1992) 213-231
8. Birbil, S.I., Fang, S.C.: An Electromagnetism-like Mechanism for Global Optimization, Journal of Global Optimization, 25 (2003) 263–282.
9. Birbil, S.I., Fang, S.C., Sheu, R.L.: On the Convergence of a Population-Based Global Optimization Algorithm, Journal of Global Optimization, 30 (2004) 301-318
10. Birbil, S. I., Feyzioglu, O.: A Global Optimization Method for Solving Fuzzy Relation Equations, Lecture Notes in Artificial Intelligence, 2715 (2003) 718-724
11. Chang, P.C.: A Branch and Bound Approach for Single Machine Scheduling with Earliness and Tardiness Penalties, Computers and Mathematics with Applications, 37 (1999) 133-144
12. Debels, D., Reyck, B.D., Leus, R., Vanhoucke, M.: A Hybrid Scatter Search/Electromagnetism Meta-Heuristic for Project Scheduling, European Journal of Operational Research , 169 (2006) 638–653
13. Glover, F., Laguna, M., Marti, R.: Fundamentals of Scatter Search and Path Relinking, Control and Cybernetics, 39 (2000) 653–684
14. Hariri, A.M.A., Potts, C.N.: Scheduling with Release Dates on a Single Machine to Minimize Total Weighted Completion Ttime, Discrete Applied Mathematics, 36 (1983) 99-109
15. Kim, Y.D., Yano, C.A.: Minimizing Mean Tardiness and Earliness in Single-Machine Scheduling Problems with Unequal Due Dates, Naval Research logistics, 41 (1994) 913-933
16. Lenstra, J.K., RinnooyKan, A.H.G., Brucker, P.: Complexity of Machine Scheduling Problems, Annals of Discrete Mathematics, 1 (1977) 343-362
17. Li, G.: Single Machine Earliness and Tardiness Scheduling, European Journal of Operational Research, 96 (1997) 546-558
18. Liaw, C.F.: A Branch and Bound Algorithm for the Single Machine Earliness and Tardiness Scheduling Problem, Computers and Operations Research, 26 (1999) 679-693
19. Mani, V., Chang P.C., Chen, S.H.: Single Machine Scheduling: Genetic Algorithm with Dominance Properties, Submitted to International Journal of Production Economics (2006)
20. Norman, B.A., Bean, J.C.: A Genetic Algorithm Methodology for Complex Scheduling Problems, Naval Research Logistics, 46 (2) (1999) 199-211
21. Ow, P.S., Morton, E.T.: The Single Machine Early/Tardy Problem, Management Science, 35 (1989) 171-191
22. Pinedo, M., Scheduling: Theory, Algorithms, and Systems, Prentice Hall, Upper Saddle River, NJ (2002)
23. Snyder, L.V., Daskin, M.S.: A Random-Key Genetic Algorithm for the Generalized Traveling Salesman Problem, European Journal of Operational Research, 174(1) (2006)38-53
24. Sourd, F., Sidhoum, S.K.: An Efficient Algorithm for the Earliness/Tardiness Scheduling Problem, Working paper - LIP6 (2005)
25. Su, L.H., Chang, P.C.: A Heuristic to Minimize a Quadratic Function of Job Lateness on a Single Machine, International Journal of Production Economics, 55 (1998) 169-175
26. Su, L.H., Chang, P.C.: Scheduling n Jobs on One Machine to Minimize the Maximum Lateness with a Minimum Number of Tardy Jobs, Computers and Industrial engineering, 40 (2001) 349-360

27. Tasgetiren, M.F., Sevkli, M., Liang, Y.C., Gencyilmaz, G.: Forthcoming, Particle Swarm Optimization Algorithm for Makespan and Total Flowtime Minimization in Permutation Flowshop Sequencing Problem, Accepted to the EJOR Special Issue on Evolutionary and Meta-Heuristic Scheduling by European Journal of Operational Research
28. Valente, J.M.S., Alves, R.A.F.S.: Heuristics for the Early/Tardy Scheduling Problem With Release Dates, Working paper, 129, Faculdade de Economia do porto, Portugal (2003)
29. Wu, P., Yang, K.J., Hung, Y.Y.: The Study of Electromagnetism-Like Mechanism Based Fuzzy Neural Network for Learning Fuzzy If-Then Rules, Lecture Notes in Computer Science, 3684 (2005) 382-388
30. Wu, S.D., Dtorer, R.H., Chang, P.C.: One Machine Heuristic with Efficiency and Stability as Criteria, Computers and Operations Research, 20 (1993) 1-14

# A Self-adaptive Evolutionary Algorithm for Multi-objective Optimization

Ruifen Cao[1], Guoli Li[2], and Yican Wu[1]

[1] Institute of Plasma Physics, Chinese Academy of Sciences,
Hefei of Anhui. Prov., 230031, China
[2] School of Electrical Engineering and Automation, Hefei
University of Technology, Hefei 230009, China
{rfcao, lgli ,ycwu}@.ipp.ac.cn.

**Abstract.** Evolutionary algorithm has gained a worldwide popularity among multi-objective optimization. The paper proposes a self-adaptive evolutionary algorithm (called SEA) for multi-objective optimization. In the SEA, the probability of crossover and mutation, $\boldsymbol{P_c}$ and $\boldsymbol{P_m}$, are varied depending on the fitness values of the solutions. Fitness assignment of SEA realizes the twin goals of maintaining diversity in the population and guiding the population to the true Pareto Front; fitness value of individual not only depends on improved density estimation but also depends on non-dominated rank. The density estimation can keep diversity in all instances including when scalars of all objectives are much different from each other. SEA is compared against the Non-dominated Sorting Genetic Algorithm (NSGA-II) on a set of test problems introduced by the MOEA community. Simulated results show that SEA is as effective as NSGA-II in most of test functions, but when scalar of objectives are much different from each other, SEA has better distribution of non-dominated solutions.

**Keywords:** Multi-objective optimization, evolutionary algorithm, SEA; non-dominated.

## 1 Introduction

Some real world problems usually consist of many objectives which conflict with each other. As there are several possibly contradicting objectives to be optimized simultaneously, there is no longer a single optimal solution but rather a whole set of possible solutions of equivalent quality. To obtain the optimal solution, there will be a set of optimal trade-offs between the objectives.

In recent years, evolutionary algorithm is popular with multi-objective optimization, because it is characterized by a population of solution candidates and could obtain a set of approximate solutions in a simulated run. During the past decade, various multi-objective evolutionary algorithms (MOEAs) have been proposed and applied in multi-objective optimization problem (MOP). A representative collection of these algorithms includes the vector evaluated genetic algorithm (VEGA) by Schaffer[2], the niched pareto genetic algorithm (NPGA) by

D.-S. Huang, L. Heutte, and M. Loog (Eds.): ICIC 2007, LNAI 4682, pp. 553–564, 2007.

Horn et.al.[3], the non dominated sorting genetic algorithm (NSGA) by Srinivas and Deb [4], the non-dominated sorting genetic algorithm II (NSGA-II) by Deb et al.[1], the strength pareto evolutionary algorithm (SPEA) by Zitzler and Thiele[5], the strength Pareto evolutionary algorithm II (SPEA-II) by Zitzler et.al.[6], the pareto archived evolution strategy (PAES) by Knowles and Corne[7] and the memetic PAES (M-PAES) by Knowles and Corne[8] et.al. Although these MOEAs differ from each other in both exploitation and exploration, they share the common purpose of searching for a near-optimal, well-extended and uniformly diversified Pareto-optimal front for a given MOP.

In this work, a novel MOEA called self-adaptive evolutionary algorithm (SEA) is formulated and developed in section 3. Some conceptions and definitions about multi-objective optimization are introduced in section 2. SEA was tested against NSGA-II on a set of suitably chose test problems in section 4. Lastly, concluding remarks are given in section 5.

## 2 Multi-objective Optimization

A general multi-objective optimization problem is expressed by

$$\begin{aligned} \min f(x) &= (f_1(x),\ f_2(x),\ldots,\ f_m(x)) \\ s.t.\ \ & X \in S \\ & x = (x_1, x_2, \ldots, x_n) \in X \end{aligned} \tag{1}$$

where $(f_1(x), f_2(x), \ldots, f_m(x))$ are the $m$ objective functions, $(x_1, x_2, \ldots, x_n)$ are the $n$ optimization parameters and $S \in R^n$ is the solution or parameter feasible space.

**Definition 1(Dominate).** Let $x1 \in S$, $x2 \in S$, $x1$ dominates $x2$ ($x1 \succ x2$), if satisfies $(f_j(x1) \le f_j(x2)$ for all $j=1, 2, \ldots, m$ and $(f_j(x1) < f_j(x2)$ for at least one objective function $f_j$.

**Definition 2(Pareto solution).** $x^*$ is said to be a Pareto optimal solution of MOP, if there is no other feasible solution $x$ dominates $x^*$. All the Pareto solutions form Pareto-optimal Front.

The objective of MOP is searching for a near-optimal, well-extended and uniformly diversified Pareto-optimal Front.

## 3 SEA Algorithm

The difference between single objective optimization and multi-objective optimization is that it is difficulty to evaluate the solutions, which is exactly a difficulty for multi-objective evolutionary algorithm. In order to alleviate the above difficulty, SEA develops a formula to calculate fitness value, including dummy fitness based on fast non-dominated rank [1] and density fitness based on improved density estimation. The dummy fitness can guide the searching process to true Pareto Front and the density fitness can preserve diversity in all instances. Based on above fitness assignment, SEA introduces self-adaptive crossover and mutation to evolutionary process according to fitness values of solutions. In the following, we present a number of different modules that form SEA.

### 3.1 Fast Non-dominated Sorting

First, for each solution $\boldsymbol{i}$ we calculate two entities: 1) $\boldsymbol{n_i}$ , the number of solutions that dominate the solution $\boldsymbol{i}$ , and 2)$\boldsymbol{S_i}$, a set of solutions that are dominated by $\boldsymbol{i}$. Then, we identify all those solutions whose $\boldsymbol{n_i=0}$ and put them in a list $\boldsymbol{H_1}$. We call $\boldsymbol{H_1}$ the current front. For each solution $\boldsymbol{i}$ in the current front, we visit each member $\boldsymbol{j}$ in its set $\boldsymbol{S_i}$ and reduce $\boldsymbol{n_j}$ count of the member $\boldsymbol{j}$ by one. By doing so, for any member $\boldsymbol{j}$, if the count $\boldsymbol{n_j=0}$, it will be put in another list $\boldsymbol{H_2}$. We repeat the process of $\boldsymbol{i}$ until all members of the current front are checked. Now $\boldsymbol{H_2}$ is the current front. For the current list $\boldsymbol{H_i}$ ($\boldsymbol{i}$=2…), we continue the process like $\boldsymbol{H_1}$, until all the solutions are identified, and the subscript $\boldsymbol{i}$ of $\boldsymbol{H_i}$ is the non-dominated rank number of every individual in $\boldsymbol{H_i}$.

### 3.2 Density Estimation

In order to keep diversity of population, we get an estimate of the density surrounding a given point in the population. Differing from NSGA-II, SEA takes the average relative distance of the two points on either side of this point (relative to the distance of two border points) along each of the objectives (Fig. 1.(b)). The quantity of $i_{\text{distance}}$ serves as the relative average side-length of the largest cuboid enclosing the point $i$ without including any other point in the population (we call this the crowding distance). The following algorithm is used to calculate the crowding.

*Crowding-distance-assignment:*

```
l=|L|                          //number of solutions in L
for each i, set L[i]distance=0     //initialize distance
  for each objective m
   L=sort(L,m)                 //sort using each objective
                               //according ascending

   if L[0]==L[l-1]
     For i=0 to i=l-1
       L[i]distance=L[i]distance+0
   else
       L[0]distance=L[l-1]distance=1 //boundary points
                                     //are always selected
       For i=1 to i=l-2
         L[i]distance=L[i]distance+ (L[i+1].m-L[i-1].m)
                 /(L[l-1].m-L[0].m)
L[i]distance=L[i]distance/m
```

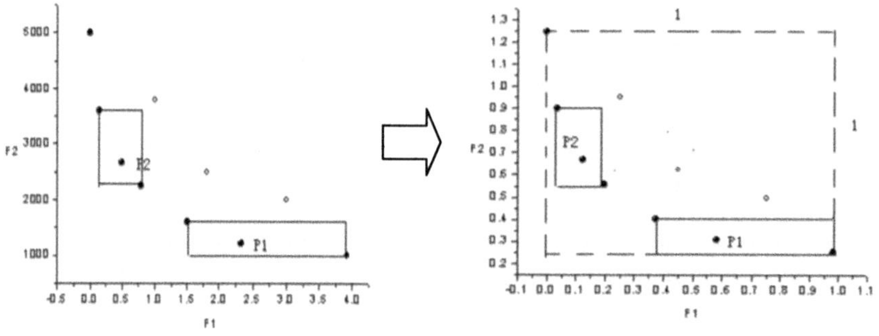


(a) (b)

**Fig. 1.** The comparison of crowding distance

SEA uses the relative value instead of absolute value used by literature [1] as density estimated value, so that it can keep diversity in all instances, even though scalars of all objectives are very different from each other. The figure 1 shows $P_1$ and $P_2$ have the same non-dominated rank number, but the density estimation value of $P_1$ is actually more than $P_2$. Since scalar of ***F2*** is much bigger than ***F1***, if the absolute value is used (Fig.1.(a)), ***F2*** will predominate and density value of $P_2$ will be bigger than $P_1$. If $P_1$ and $P_2$ tourney, $P_2$ will be selected and most of the non-dominated solutions will lean to ***F2*** at the end of evolution. To avoid the instance, SEA converts density estimation from (a) to (b) in Fig.1.and selects $P_1$ into next generation. So it can obtain a better (uniform) distribution Pareto Front than NSGA-II. The tested example 6 proves.

### 3.3 Fitness Assignment

Fitness assignment scheme of SEA obeys two guidelines that are the design objectives of every evolutionary algorithm: guiding the direction of evolution to the true Pareto Front and keeping the diversity of population. Firstly, SEA assigns a dummy fitness (called dumfit) $1-i_{rank} / rank_{max}$ for each individual $i$ in population according to its non-dominated rank number $i_{rank}$ , $rank_{max}$ is the max of all individual rank numbers. The smaller the rank number is, the bigger the $i_{dumfit}$ will be, and individuals in the same non-dominated rank will have the same $i_{dumfit}$. Secondly, SEA gives a density fitness (called denfit) $(1/ rank_{max}) \times i_{distance}$ for each individual according to its density estimation value $i_{distance}$. Lastly, the fitness of each individual is computed as:

$$i_{fitness} = i_{dumfit} + i_{denfit}$$

or

$$i_{fitness} = 1 - \frac{i_{rank}}{rank_{max}} + \frac{1}{i_{rank}} \times i_{distance} \tag{2}$$

Because the distance of individual isn't bigger than 1, individuals with larger rank numbers impossibly have the same fitness as those with small rank numbers, even if

$i_{\text{distance}}$of them are very large. Individuals with same non-dominated rank number will have different fitness because of their different density. In competition, individuals with smaller non-dominated rank numbers will win, and for those with same rank numbers, individuals with larger $i_{\text{distance}}$ will win. By doing that, fitness assignment considered the diversity and non-dominated rank of solutions at the same time, SEA not only ensures the evolutionary process towards true Pareto Front, but also obtains equable distribution of solutions.

### 3.4 Self-adaptive Crossover and Mutation

SEA introduces self-adaptive crossover and mutation into evolutionary process, which can self-adaptively adjust crossover probability $P_c$ and mutation probability $P_m$, according to fitness values of solutions (equation 3-4). $P_c$ and $P_m$ are important elements for maintaining the diversity of population and sustaining the convergence capacity of evolutionary algorithm. To obtain a good set of ($P_c$, $P_m$) for the given problem, if current general multi-objective evolutionary algorithms are used, it is necessary for user to adjust ($P_c$, $P_m$) again and again, which is very troublesome. SEA provides self-adaptively a best set of $P_c$ and $P_m$ to certain solution. Solutions with high fitness are protected, while solutions with subavage fitness are totally disrupted; when fitness values of all individuals in populations become similar or approach a local optimum, the $P_c$ and $P_m$ of solutions will be big; when fitness values of all individuals are dispersive, the $P_c$ and $P_m$ of solutions will be small.

$$P_c = \begin{cases} P_{c1} - \dfrac{(P_{c1} - P_{c2})(fit' - fit_{avg})}{fit_{\max} - fit_{avg}}, & fit' \geq fit_{avg} \\ P_{c1}, & fit' < fit_{avg} \end{cases} \tag{3}$$

$$P_m = \begin{cases} P_{m1} - \dfrac{(P_{m1} - P_{m2})(fit_{\max} - fit)}{fit_{\max} - fit_{avg}}, & fit \geq fit_{avg} \\ P_{m1}, & fit < fit_{avg} \end{cases} \tag{4}$$

Where $P_{c1}$=0.9, $P_{c2}$=0.6, $P_{m1}$=0.1, $P_m$=0.01, $fit_{avg}$ and $fit_{max}$ are average fitness and max fitness of population respectively, $fit'$ is the larger fitness among two individuals that would cross, $fit$ is the fitness of individual that would mutate.

### 3.5 The Main Loop

Initially, a random parent population $P_0$ of size $N$ is generated. The population is sorted based on the non-domination rank (according to 3.1). Density of each solution is computed (according to 3.2). Each solution is assigned fitness value (according to 3.3). Thus, maximization of fitness is adopted. Binary tournament selection, crossover, and mutation operators are used to create a child population $Q_0$ of size $N$. Then population $P_0$ and $Q_0$ are combined to form population $R_0$. $R_0$ is sorted according to 3.1, density estimated (3.2), fitness assigned (3.3) and sorted according to fitness, and then $N$ individuals with the maximal fitness are selected from $R_0$ into population $P_1$. $P_1$will repeat the above process of $P_0$ and create population $Q_1$, $P_1$ and

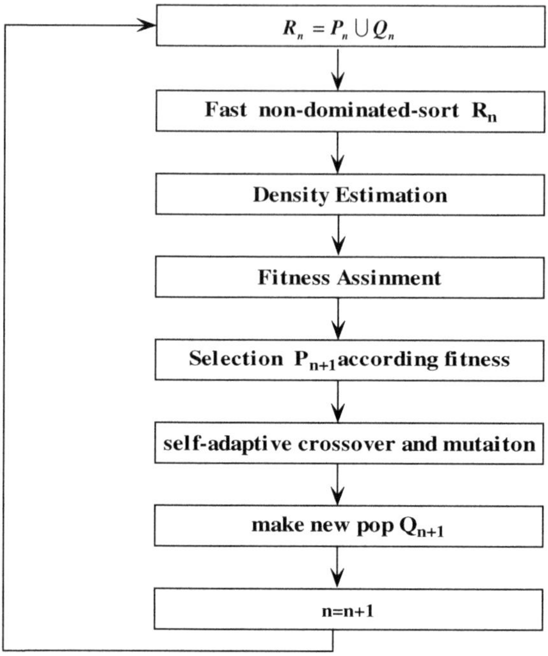


**Fig. 2.** The main Loop

$Q_1$ will be combined to form another population $R_1$ as well, repeat the same loop until the iterative times equal to given value. The iteration could be shown as fig.2:

SEA implements two elitist strategies: i) creating a mating pool by combining the parent and child populations for selecting and ii) Selecting N individuals with maximal fitness into next generation. So the best individual will be kept down and won't be lost.

# 4 Numerical Testing and Analysis

SEA was tested and compared with NSGA-II, which is one of the most successful MOEAs in the literature. In other experimental comparative studies [9], SPEA2 had been shown to be as effective as NSGA-II. Here NSGA-II was chosen as it is more efficient and simple to implement.

For all test problems and per algorithm, the best outcomes of ten runs were adopted. We used a population size of 100, the max generation of 250. The variables were treated as real numbers; the simulated binary crossover (SBX) and the real-parameter mutation operator have been used. Besides, NSGA-II used a crossover probability of 0.8 and a mutation probability of ***1/n*** (***n*** is the number of variables) [1].

In order to test the performance of SEA, the test was divided into two steps. First, we used the same functions in which NSGA-II has been better than the other MOEAs. Then, we compared the distribution of NSGA-II with SEA, using a problem in which scalars of each objective are much different. It showed that the NSGA-II fell short of uniform diversity for the problem in which scalar of all objectives were different, while SEA could obtain uniform distribution solutions. The step and result are shown as follows:

**1) Test Performance in Which NSGA-II Is Better Than Other MOEAs**

The test functions used in this part are exactly those used by Deb et al. 2001[1] when NSGA-II was firstly proposed. In fact, the popularity of the algorithm has started after it outperformed other MOEAs on these test problems. Due to space limit, the reader could refer to [1] for a complete and detailed listing of these tests suite functions.

In order to compare the performance of SEA with NSGA-II, we made a quantitative analysis for the results. A Common Pareto Front (CPF) was filtered from both algorithms: CPF = ND (SEA $\cup$ NSGA – II). Then two main performance values, including the percentage of a Pareto Front in the common archive (PF) and the relative covering index (CS) [11], were computed. Also, the number of solutions of each MOEA that are in the CPF was also obtained: MOEAPareto (MP) = MOEA $\cap$ CPF. Let MOEA $\in$ {SEA, NSGA – II}. The indexes above are computed as:

$$PF = \frac{|\mathbf{MOEAPareto}|}{|\mathbf{CPF}|} \tag{5}$$

$$\mathbf{CS(SEAPareto,NSGA\text{-}IIPareto)} = \frac{|x \in \mathbf{NSGA\text{-}IIPareto};\ \exists x' \in \mathbf{SEAPareto}: x' \succ x|}{|\mathbf{NSGA\text{-}IIPareto}|} \tag{6}$$

$$\mathbf{CS(NSGAPareto,SEAPareto)} = \frac{|x \in \mathbf{SEAPareto};\ \exists x' \in \mathbf{NSGA\text{-}IIPareto}: x' \succ x|}{|\mathbf{SEAPareto}|} \tag{7}$$

Table 1 shows the number of Pareto solutions, CPF, MP, PF and CS of NSGA-II and SEA. CS of NSGA-II is CS (NSGA-IIPareto, SEAPareto); CS of SEA is CS (SEAPareto, NSGA-IIPareto). It is clear that an algorithm with bigger MP, PF and CS is better, in terms of its ability to approach the true Pareto Front. From the table as we can see SEA seems to be a little better in MP, PF, and CS than NSGA-II on the MOP2, EC4 and EC6.

**Table 1.** Comparison of relative covering index

| Problem | MOP2 | | MOP3 | | MOP4 | | EC4 | | EC6 | |
|---|---|---|---|---|---|---|---|---|---|---|
| MOEA | NSGA-II | SEA | NSGA-II | SEA | NSGA-II | SEA | NSGA-II | SEA | NSGA-II | SEA |
| Pareto | 100 | 100 | 100 | 100 | 100 | 100 | 100 | 100 | 100 | 100 |
| CPF | 163 | 163 | 192 | 192 | 18 | 18 | 168 | 168 | 175 | 175 |
| MP | 78 | 85 | 96 | 96 | 9 | 9 | 71 | 97 | 85 | 90 |
| PF | 47.9% | 52% | 50% | 50% | 50% | 50% | 42% | 59% | 48.6% | 51.4% |
| CS | 0.15 | 0.22 | 0.04 | 0.04 | 1 | 1 | 0.03 | 0.29 | 0.1 | 0.15 |

In order to have a better understanding of how these algorithms are able to spread solutions over the non-dominated front, we present the entire non-dominated front found by NSGA-II and SEA in three of the above test problems (MOP2, EC4, and EC6), results of the other two problems (MOP3 and MOP4) obtained by NSGA-II and SEA are similar in distribution and indexes above (Table 1).

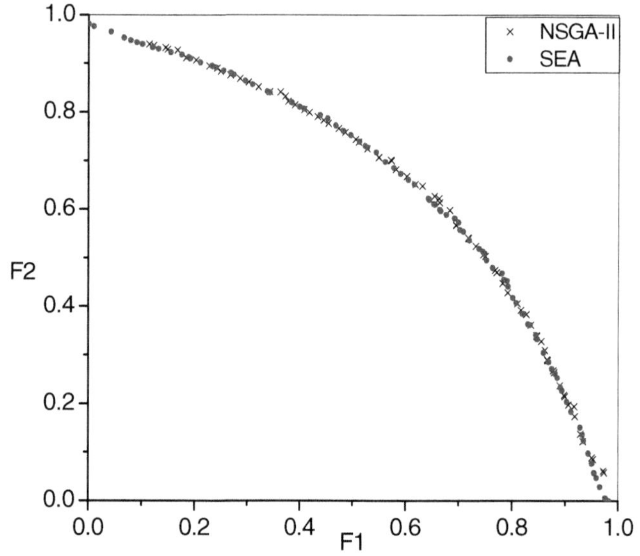


**Fig. 3.** The non-dominated solutions obtained by SEA and NSGA-II on MOP2

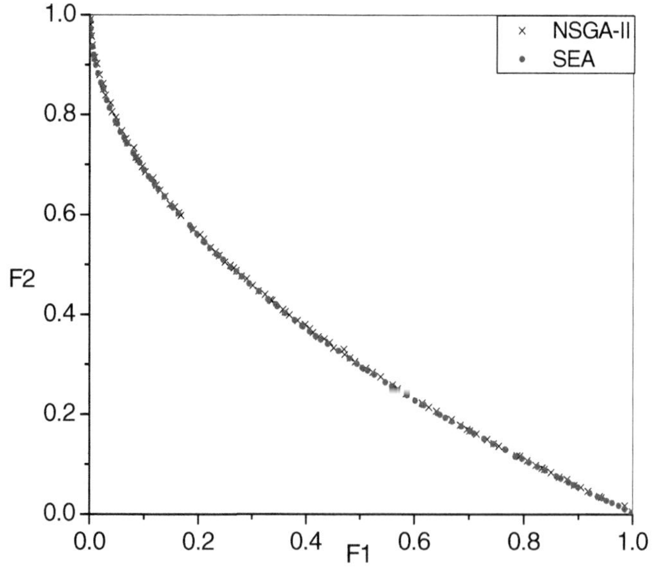


**Fig. 4.** The non-dominated solutions obtained by SEA and NSGA-II on EC4

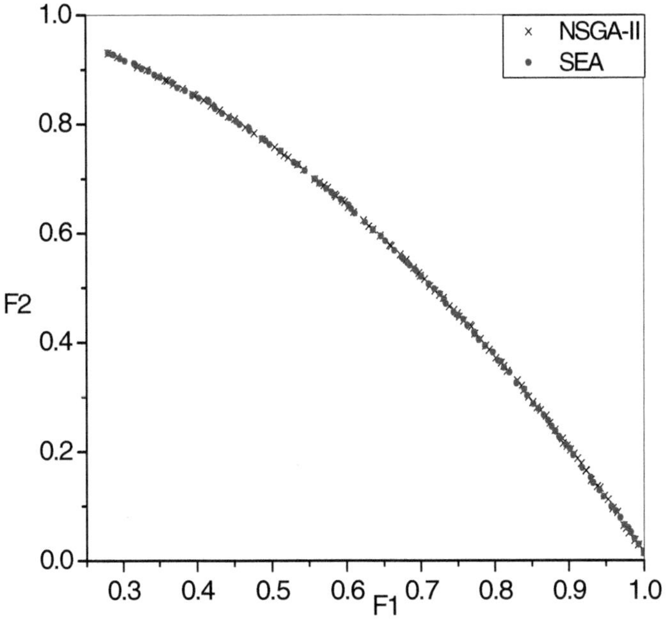


**Fig. 5.** The non-dominated solutions obtained by SEA and NSGA-II on EC6

From the figure 3, we can see the range of the result obtained by SEA is a little larger than NSGA-II; from the figure 4-5, we can see the Pareto Front of SEA has spread over the Pareto Front surface of NSGA-II. It means that the SEA has the similar distribution in range and diversity as NSGA-II. But from Table 1. EEA seems better than NSGA-II in CS,MP and PF.

**2) Test the Distribution of SEA and NSGA-II When Scalars of Objectives Are Very Different**

In the first part we have tested the usual problem. In this part we will test the result distribution when objectives have much different scalars. The test function (Test Problem 6) is described as:

$$\begin{aligned}&\text{Minimize } \boldsymbol{F} = (f_1(x), f_2(x))\\&\text{Where } \ f_1(x) = x^2\\&f_2(x) = 1000 + (x-2)^2 \times 1000\\&-10^5 \le x_1, x_2 \le 10^5 .\end{aligned} \tag{8}$$

As the true Pareto Front (PFture) of the problem can be obtained easily, the result of SEA and NSGA-II are compared with PFtrue respectively in Figure6-7. Since the diversity among optimized solutions is an important matter in multi-objective optimization, we devised a measure based on Crowding Distance (3.2). The $D_{max}$ and

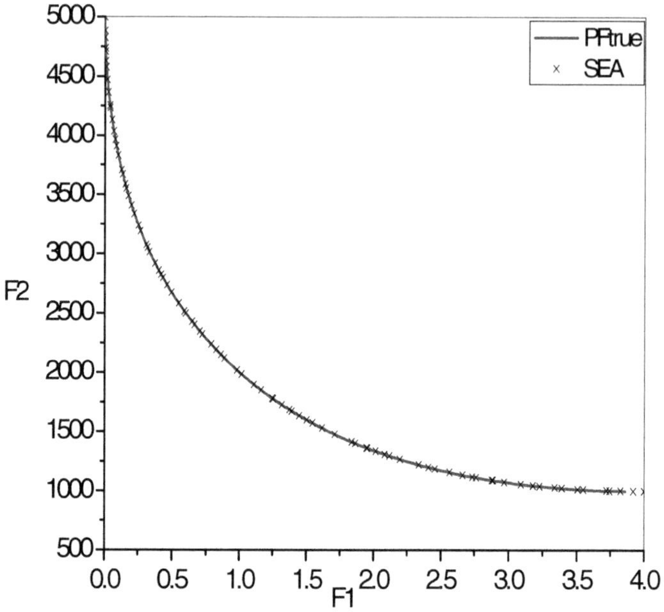


**Fig. 6.** The true Pareto front and non-dominated solutions obtained by SEA

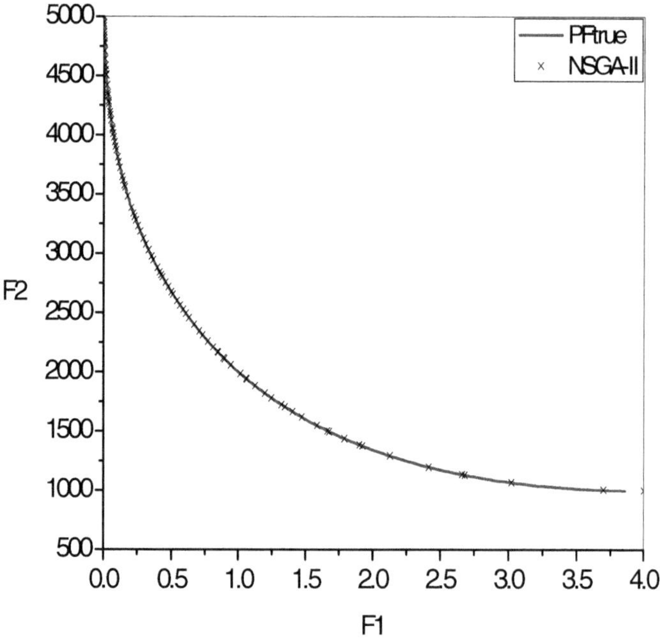


**Fig. 7.** The true Pareto front and non-dominated solutions obtained by SEA

$D_{min}$ are the max and min crowding distance among the solutions of the best non-dominated front in the final population, if the $D_{max}$ equals the $D_{min}$, the distribution of result is the best, namely, uniform distribution.

**Table 2.** Comparison of crowding distance and other indexes

| MOEA | NSGA-II | SEA |
|---|---|---|
| Pareto | 100 | 100 |
| MP | 98 | 99 |
| PF | 49.7% | 50.3% |
| CS | 0.01 | 0.02 |
| $D_{max}$ | 0.184821 | 0.049413 |
| $D_{min}$ | 0 | 0 |

Figure 6-7 show true Pareto Front and the non-dominated solutions obtained by SEA and NSGA-II for the test problem 6. Both of the results can approach the true Pareto, but SEA is able to distribute its population along the true front better than NSGA-II. From table 2, SEA also seems to be able to find a distribution of solutions close to a uniform distribution along the non-dominated front, but the result of NSGA-II leans to F2 axis.

## 5 Conclusions

In this paper a self-adaptive multi-objective evolutionary algorithm (SEA) is proposed. Introduction of self-adaptive crossover and mutation operator makes it simple for application; new fitness assignment and improved density estimation make it very effective in convergence and diversity keeping. In addition, the fitness assignment enables multi-objective optimization to use some effective operators of single objective optimization to improve the performance of algorithm. SEA was compared against NSGA-II by using the same test functions in which NSGA-II has excelled. The test results show that the SEA has near-optimal and better distribution along the true Pareto front than NSGA-II when scalars of the objectives are very different. SEA could have many applications in multi-objective optimization problems, such as inverse planning for Intensity Modulation Radiation Therapy, optimization design and so on.

## References

1. Deb K., Pratap, A., Agarwal, S., and Meyarivan, T.: A Fast and Elitist Multiobjective Genetic Algorithm: NSGA-II. IEEE Transactions on Evolutionary Computation, vol 6(2), (2001) 182-197
2. Schaffer, J.D.: Multiple Objective Optimization with Vector Evaluated Genetic Algorithms, in: J.J. Grefenstette et al. (Eds.), Genetic algorithms and their applications, In: Proceedings of the 1st International Conference on Genetic Algorithms, Lawrence Erlbaum, Mahwah, NJ, (1985) 93–100

3. Horn, J., Nafpliotis, N., Goldberg, D.E.: A Niched Pareto Genetic Algorithm for Multiobjective Optimization, in: J.J. Grefenstette et al.(Eds.), IEEE World Congress on Computational Intelligence, In: Proceedings of the 1st IEEE Conference on Evolutionary Computation, IEEE Press, Piscataway, NJ, (1994) 82–87
4. Srinivas, N., Deb, K.: Multiobjective Optimization Using Nondominated Sorting in Genetic Algorithms, Evolutionary Computation 2 (3) (1999) 221–248.
5. Zitzler, E., Thiele, L.: Multiobjective Optimization Using Evolutionary Algorithms: A Comparative Case Study, in: A.E. Eiben, T. Back,M. Schoenauer, H.P. Schwefel, (Eds.), Fifth International Conference on Parallel Problem Solving from Nature (PPSN-V), Berlin,Germany, (1998) 292–301
6. Zitzler, E., Laumanns, M., Thiele, L.: Improving the Strength Pareto Evolutionary Algorithm for Multiobjective Optimization.In: Proceedings of Evolutionary methods for design, optimization and control with applications to industrial problems, EUROGEN2001, Athens, Greece, (2001)
7. Knowles, J.D., Corne, D.W.: The Pareto Archived Evolution Strategy: A New Baseline Algorithm for Multiobjective Optimization, in: In:Proceedings of the 1999 Congress on Evolutionary Computation, IEEE Press, Piscataway, NJ, (1999) 98–105
8. Knowles, J.D., Corne, D.W., M-PAES: A Memetic Algorithm for Multiobjective Optimization, in: In: Proceedings of the 2000 Congresson Evolutionary Computation, IEEE Press, Piscataway, NJ, (2000) 325–332
9. Zitzler, E., Laumanns, M., Thiele, L.: SPEA2: Improving the Strength Pareto Evolutionary Algorithm. Technical report, Swiss Federal Institute of Technology (2001)
10. Burke, E.K., Landa Silva, J.D.: The Influence of the Fitness Evaluation Method on the Performance of Multiobjctive Optimisers, European Journal of Operational Research, Volume 169 issue 3, (2006) 875-897
11. Cui, Y.: Multiobjective Evolutionary Algorithms and their Application. Beijing: National defense industry press, (2006) 161-162

# An Adaptive Immune Genetic Algorithm for Edge Detection

Ying Li, Bendu Bai, and Yanning Zhang

School of Computer Science, Northwest Polytechnical University,
Xi'an, 710072, China
lybyp@nwpu.edu.cn

**Abstract.** An adaptive immune genetic algorithm (AIGA) based on cost minimization technique method for edge detection is proposed. The proposed AIGA recommends the use of adaptive probabilities of crossover, mutation and immune operation, and a geometric annealing schedule in immune operator to realize the twin goals of maintaining diversity in the population and sustaining the fast convergence rate in solving the complex problems such as edge detection. Furthermore, AIGA can effectively exploit some prior knowledge and information of the local edge structure in the edge image to make vaccines, which results in much better local search ability of AIGA than that of the canonical genetic algorithm. Experimental results on gray-scale images show the proposed algorithm perform well in terms of quality of the final edge image, rate of convergence and robustness to noise.

## 1 Introduction

Edge detection is an important task in computer processing. Most classical edge detection operators such as the gradient operator, the Laplacian operator or the Laplacian-of-Gaussian operator are based on the derivatives of the pixel intensity values. In spite of simplicity of these operators, they are only suitable for detecting limited types of edges and are highly susceptible to noise often resulting in fragmented edges. Recently, a class of detection techniques [1~3] based on cost function optimization has been present. These approaches all cast the edge detection problem as one of minimizing the cost of an edge image firstly, and then exploit different technique to optimize the cost function. The edges detected by all these approaches are expected to be well localized, continuous and thin.

This paper presents an adaptive immune genetic algorithm (AIGA) based on cost minimization technique for edge detection. The immune genetic algorithm is a novel evolutionary algorithm which combining the immune mechanism and evolutionary mechanism. IGA is further improved in this paper, and used in the context of edge detection.

## 2 Cost Function Evaluation

The cost function of an edge image is defined in terms of the enhanced image. Therefore, the first step in the detection process is dissimilarity enhancement where the

D.-S. Huang, L. Heutte, and M. Loog (Eds.): ICIC 2007, LNAI 4682, pp. 565–571, 2007.

pixels in the image that are likely candidates for edge pixels are selectively enhanced. The enhanced image $D=\{d(i,j);\ 1\le i\le M, 1\le j\le N\}$ is a collection of pixels where each pixel value is proportional to the degree of region dissimilarity that exists at that pixel site. The pixel values in $D$ lie in the range [0, 1]. The enhanced image $D$ is obtained using the same procedure as that of [1~3].

The edge cost function at each pixel site $(i,j)$ is a weighted sum of the following terms

$$F(i,j)=\sum_i w_i C_i\ ,\ i\in\{d,t,c,f,e\}\ , \tag{1}$$

where the $C_i$'s are the cost factors which are similar to the ones used by Bhandarkar *et al*[3], and $w_i$'s empirically predetermined weights assigned to the respective terms.

The edge cost function for an entire image of size $M\times N$ pixels is given by

$$F=\sum_{i=1}^{M}\sum_{j=1}^{N}F(i,j)\ . \tag{2}$$

## 3 Cost Function Minimization Based on AIGA

Genetic algorithms (GAs) are optimization techniques which are based on natural selection, crossover and mutation operations. Compared to the traditional optimization methods, GAs are robust, global and can be generally applied without recourse to domain-specific heuristics. But GAs are easy trapped into the local optimum or premature when they are used to solve the problems with the high order, long length building blocks. This drawback is particularly prominent in the context of image edge detection where the solution space is very large. On the other hand, there are many basic and obvious characteristics or knowledge in a pending problem. However the crossover and mutation operator in GA lack the capability of meeting an actual situation, so that some torpidity appears when solving problems, which is conductive to the universality of the algorithm but neglects the assistant function of the characteristics or knowledge. The loss due to the negligence is sometimes considerable in dealing with some complex problems. With a view toward alleviating these shortcomings in GA, the immune GA (IGA) presented in [4] leads the immune concepts and methods into the canonical GA. On condition of preserving GA's advantage, IGA utilizes some characteristics and knowledge in the pending problems for restraining the degenerative phenomena during evolution, so as to improve the algorithmic efficiency. IGA is further improved in this paper, and used in the context of edge detection. The presented algorithm named AIGA recommends the use of adaptive probabilities of crossover, mutation and immune operation. Furthermore, it effectively exploits some prior knowledge of pending problem and the information of evolved individual's past history to make vaccines. The AIGA-based edge detection algorithm can be implemented as the following procedure.

1. Generate an initial population and evaluate the fitness for each individual.
2. Abstract vaccines according to the prior to knowledge.
3. If the current population contains the optimal individual, then the course halts; or else, continues.
4. Select n individuals as parent generation from the present population.
5. Perform the crossover and mutation operation on the parents to obtain the offspring generation.
6. Perform the immune operation on the offspring generation to generate the next population, and go to step 3.

### 3.1 Encoding Scheme and Fitness Evaluation

Each chromosome of the population is represented by two-dimensional binary array of 1s or 0s which corresponds to an edge image. The fitness of the *i*-th individual in the current generation is computed as

$$fitness[i] = \left(F[worst] - F[i]\right)^n , \tag{3}$$

where $F[worst]$ is the cost associated with the worst individual and $F[i]$ the cost associated with the $i$ th individual in the current generation. Both $F[worst]$ and $F[i]$ are computed using (1). During the earlier phases of evolution, we set $n = 2$. After the solutions converge to a certain extent, we make n successively n successively larger—up to $n = 5$.

### 3.2 Selection Mechanism

A pair of individuals is selected from the current population for mating using the rank based selection mechanism [5]. Let M sorted individuals be numbered as 0, 1,…, *M*-1, with the zero-*th* being the fittest. Then the (*M-j*)- *th* individual is selected with probability

$$P(M - j) = \frac{j}{\sum_{k=1}^{M} k} \tag{4}$$

### 3.3 Crossover and Mutation Operator

Crossover is applied to the newly selected (parents) individuals to generate two offsprings. Since our representation is two dimensional, two-point crossover is employed. And the mutation operator is performed by flipping the bit value at a randomly chosen position in the bit string. In our AIGA implementation, a high probability is assigned to the crossover operator in the initial stages of the AIGA run and the crossover probability is decreased by a small amount with every generation. The initial values of the crossover and mutation probabilities and the corresponding decrement and increment values respectively were chosen empirically after several experiments. The rationale here is to enable the AIGA in the later stages of evolution to focus on local search via mutation while forgoing exploration of large regions of the search space via crossover.

### 3.4 Immune Operator

An immune operator is composed of the following two operations:

1) *The Vaccination*: A vaccination means modifying the genes on some bits in accordance with priori knowledge so as to gain higher fitness with greater probability. A vaccine is abstracted form the prior knowledge of the pending problem, whose information amount and validity play an important role in the performance of the algorithm. In the context of edge detection, the vaccines are selected and performed based on the examination of the local neighborhood in a 3×3 window centered at a randomly chosen pixel location. In particular, the valid two-neighbor local edge structures, the most frequently encountered valid local edge structures in an edge images, are mainly used as the vaccines. The vaccination probabilities are determined by the following guidelines. Vaccines that result in straight local edge structures are assigned a higher probability; vaccines that result in local edge structures that turn by 45° are assigned a higher probability than that those that turn by more than 45°; vaccines resulting valid local edge structures are more favored than those resulting invalid local edge structures. Fig. 1 shows some vaccines used for edge detection. The vaccination operation is characterized by two parameters: $p_1$ which denotes the fraction of individuals in the current binary solutions $P(t)$ to be subject to vaccination and $p_2$ which denotes the number of pixels in the chosen individual to be subject to vaccination. Both $p_1$ and $p_2$ are incremented by a small amount after each generation. The initial values of $p_1$ and $p_2$ and the corresponding increment values were chosen empirically after several experiments.

2) *The Immune Selection*: This operation is accomplished by the following two steps. The first one is the immune test. If the fitness of the vaccinated individual is smaller than that of the parent, the parent will participate in the next competition by replacing the vaccinated individual; the second one is the annealing selection, i.e. selecting an individual $x_i$ in the present offspring $E_k = (x_1, \ldots x_{n_0})$ to join in the new parents with the probability as follows:

$$P(x_i) = \frac{e^{f(x_i)/T_k}}{\sum_{i=1}^{n_0} e^{f(x_i)/T_k}}, \quad (5)$$

where $f(x_i)$ is the fitness of the individual $x_i$ and the set $\{T_K\}$ is called an annealing temperature schedule.

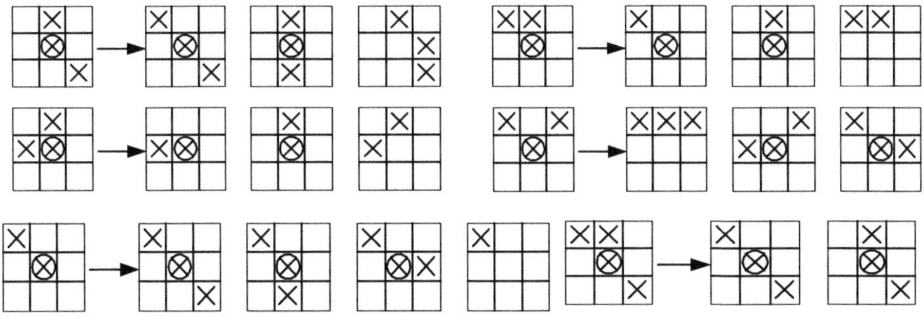

**Fig. 1.** Some vaccines used for edge detection

## 4 Experimental Results

In this section, we present some experimental results of edge detection based on the cost minimization approach using the proposed AIGA. In the experiments, the weights used in the cost function were set to $w_c = 0.5$, $w_d = 2$, $w_e = 1$, $w_f = 3$, and $w_t = 6.51$. Figure 2(a) is the original telephone image, and the edge image detected by AIGA is shown in Fig. 2(b). Fig. 3 shows the progress of the cost function found by AIGA and the conventional GA with the elitism strategy over 200 generations. AIGA is shown to have a much faster convergence rate than GA due to its better local search ability.

(a) (b)

**Fig. 2.** Original image and detected edges (a) Original telephone image (b) edges detected using AIGA

In order to test the robustness of noise of the AIGA for edge detection, the ring image was corrupt with additive Gaussian noise with zero mean and stardard variances 55 shown as Fig. 4(a). The detected edges for the noisy image using the Canny operator, and the AIGA approach are shown in the same figure. The experimental result shows that the AIGA has good robustness to noise.

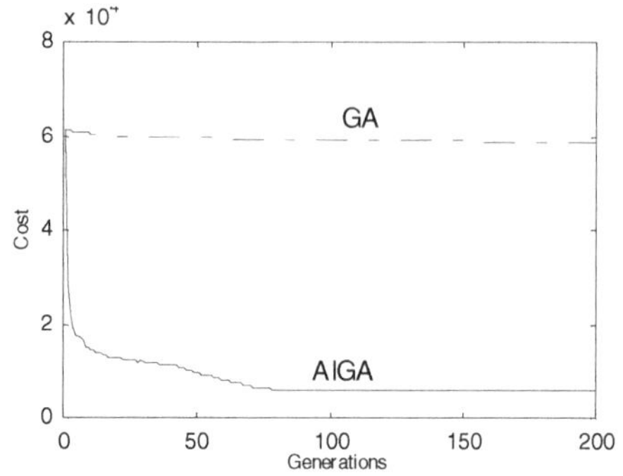


**Fig. 3.** Comparison of cost function between GA and AIGA

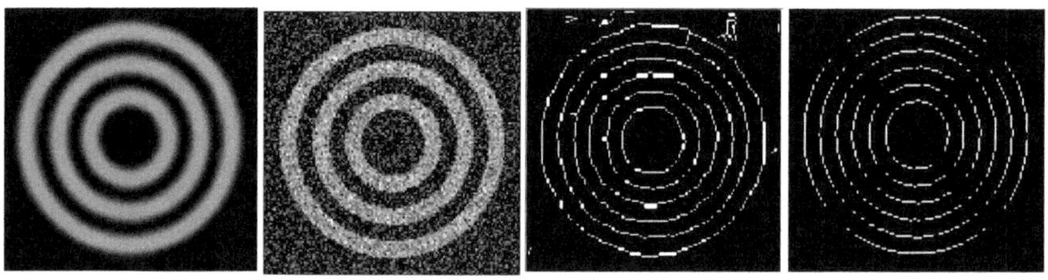

(a) (b) (c) (d)

**Fig. 4.** Noisy image and detected edges. (a) original ring image. (b) noisy image. (c) edges detected using Canny operator. (d) edges detected using AIGA.

## 5 Conclusion

Based on cost minimization technique, this paper proposed an adaptive immune genetic algorithm (AIGA) for edge detection. The edge detection problem was cast as one of minimizing the cost of an edge image, and the desired edge image was deemed to be one that corresponds to the global minimum of the cost function. The proposed AIGA used the adaptive probabilities of crossover, mutation and immune operation, and a geometric annealing schedule in the immune operator. Furthermore, AIGA can effectively exploit some prior knowledge and information of the local edge structure in the edge image to make vaccines, which are shown to improve the local search ability.

Future research will investigate various refinements of the basic AIGA operators including crossover operator, mutation operator, and immune operator in the context of edge detection. How to obtain more effective encoding scheme of chromosome will also be investigated.

**Acknowledgment.** This work is supported by the National Natural Science Foundation of China (60472072), the Natural Science Foundation of Shaanxi Province(No. 2006F05), the Aeronautical Science Foundation (No.05I53076), and Specialized Research Found for the Doctoral Program of Higher Education (20040699034).

## References


1. Tan, H.L., Gelfand, S.B., Delp, E.J.: A Comparative Cost Function Approach to Edge Detection. IEEE Trans. System, Man and Cybernetic. 16 (1989) 1337-1349
2. Tan, H.L., Gelfand, S.B., Delp, E.J.: A Cost Minimization Approach to Edge Detection Using Simulated Annealing. IEEE Trans. Pattern Anal. Machine Intel. 14 (1991) 3-18
3. Bhandarkar, S.M., Zhang, Y., Potter, W.D.: An Edge Detection Technique using Genetic Algorithm-based Optimization. Pattern Recog. 27 (1994) 1159-1180
4. Jiao, L.C., Wang, L.: A Novel Genetic Algorithm based on Immunity. IEEE Trans. System Man Cybernetic. 30 (2000) 552-561
5. Yao, X., Liu, Y.: A New Evolutionary System for Evolving Artificial Neural Networks. IEEE Trans. on Neural Networks. 8 (1997) 694-713

# An Improved Nested Partitions Algorithm Based on Simulated Annealing in Complex Decision Problem Optimization*

Yan Luo[1] and Changrui Yu[2]

[1] Institute of System Engineering, Shanghai Jiao Tong University, 200052 Shanghai, China
[2] School of Information Management and Engineering, Shanghai University of Finance and Economics, Shanghai 200433, China
{yanluo, yucr}@sjtu.edu.cn

**Abstract.** This paper introduces the main ideas of the nested partitions (NP) method, analyses its efficiency theoretically and proposes the way to improve the optimization efficiency of the algorithm. Then the paper introduces the simulated annealing (SA) algorithm and incorporates the ideas of SA into two of the arithmetic operators of NP algorithm to form the combined NP/SA algorithm. Moreover, the paper presents the explicit optimization procedure of the combined algorithm NP/SA and explains the feasibility and superiority of it. The NP/SA algorithm adopts the global optimization ability of NP algorithm and the local search ability of SA algorithm so that it improves the optimization efficiency and the convergence rate. This paper also illustrates the NP/SA algorithm through an optimization example.

## 1 Introduction

The solution of many complex decision problems involves combinatorial optimization, i.e., obtaining the optimal solution among a finite set of alternatives. Such optimization problems are notoriously difficult to solve. One of the primary reasons is that in most applications the number of alternatives is extremely large and only a fraction of them can be considered within a reasonable amount of time. As a result, heuristic algorithms, such as evolutionary algorithms, tabu search, and neural networks, are often applied in combinatorial optimization. All of these algorithms are sequential in the sense that they move iteratively between single solutions or sets of solutions. However, in some applications to the complex decision it may be desirable to maintain a more global perspective, that is, to consider the entire solution space in each iteration. In this paper we propose a new optimization algorithm to address this difficult class of problems. The new method combines the nested partitions (NP) method and the simulated annealing (SA) method. It converges to a global optimum for combinatorial optimization problems in finite time, and effectively reduces the number of times backtracking occurs in the nested partitioning. Numerical results demonstrate the effectiveness of our proposed method.

* This research work is supported by the Natural Science Fund of China (# **70501022**).

D.-S. Huang, L. Heutte, and M. Loog (Eds.): ICIC 2007, LNAI 4682, pp. 572–583, 2007.

The remainder of the paper is organized as follows. In Section 2 we review the general procedure of the NP method and analyse its optimization efficiency in detail. In Section 3 we present a combined NP/SA algorithm, i.e. an improved NP algorithm enhanced with simulated annealing. In Section 4 we give a numerical example to illustrate the hybrid method, and Section 5 contains some concluding remarks and future research directions.

## 2 The Nested Partitions Method

The NP method, an optimization algorithm proposed by L. Shi and S. Ólafsson [1], may be described as an adaptive sampling method that uses partitioning to concentrate the sampling effort in those subsets of the feasible region that are considered the most promising. It combines global search through global sampling of the feasible region, and local search that is used to guide where the search should be concentrated. This method has been found to be promising for difficult combinatorial optimization problems such as: the traveling salesman problem [2], buffer allocation problem [3], product design problem [4] [5], and production scheduling problems [6].

Suppose the finite feasible region of a complex decision problem is $\Theta$. Our objective is to optimize the objective performance function $f$: $\Theta \rightarrow \mathbf{R}$, that is, to solve:

$$\max_{\theta \in \Theta} f(\theta),$$

where $|\Theta| < \infty$. Also, to simplify the analysis, we assume that there exists a unique solution $\theta_{opt} \in \Theta$ to the above problem, which satisfies $f(\theta_{opt}) > f(\theta)$ for all $\theta \in \Theta \setminus \{\theta_{opt}\}$.

**Definition 1.** A region partitioned using a fixed scheme is called a valid region. In a discrete system a partitioned region with a singleton is called a singleton region. The collection of all valid regions is denoted by $\Sigma$. Singleton regions are of special interest in the process of optimization, and $\Sigma_0 \subset \Sigma$ denotes the collection of all such valid regions.

The optimization process of the NP method is a sequence of set partitions using a fixed partitioning scheme, with each partition nested within the last. The partitioning is continued until eventually all the points in the feasible region correspond to a singleton region.

**Definition 2.** The singleton regions in $\Sigma_0$ are called regions of maximum depth. More generally, we define the depth, $dep: \Sigma \rightarrow N_0$, of any valid region iteratively with $\Theta$ having depth zero, subregions of $\Theta$ having depth one, and so forth. Since they cannot be partitioned further, we call the singleton regions in $\Sigma_0$ regions of maximum depth.

**Definition 3.** If a valid region $\sigma \in \Sigma$ is formed by partitioning a valid region $\eta \in \Sigma$, then $\sigma$ is called a subregion of region $\eta$, and region $\eta$ is called a superregion of region $\sigma$. We define the superregion function $s:\Sigma \to \Sigma$ as follows. Let $\sigma \in \Sigma \backslash \Theta$. Define $s(\sigma) = \eta \in \Sigma$, if and only if $\sigma \subset \eta$ and if $\sigma \subseteq \xi \subseteq \eta$ then $\xi = \eta$ or $\xi = \sigma$. For completeness we define $s(\Theta) = \Theta$.

A set performance function $I:\Sigma \to \mathbf{R}$ is defined and used to select the most promising region and is therefore called the promising index of the region.

In the *k*-th iteration of the NP method there is always a region $\sigma(k) \subseteq \Theta$ that is considered the most promising, and as nothing is assumed to be known about location of good solutions before the search is started, $\sigma(0) = \Theta$. The most promising region is then partitioned into $M_{\sigma(k)}$ subregions, and what remains of the feasible region $\sigma(k)$ is aggregated into one region called the surrounding region. Therefore, in the *k*-th iteration $M_{\sigma(k)} + 1$ disjoint subsets that cover the feasible region are considered. Each of these regions is sampled using some random sampling scheme, and the samples used to estimate the promising index for each region. This index is a set performance function that determines which region becomes the most promising region in the next iteration. If one of the subregions is found to be best, this region becomes the most promising region. If the surrounding region is found to be best, the method backtracks to a larger region. The new most promising region is partitioned and sampled in a similar fashion.

## 2.1 The NP Algorithm

The NP method comprises four basic arithmetic operators during the four steps respectively: partitioning the solution space, obtaining the sampling points, selecting a promising index function, and backtracking.

**Step 1: Partitioning.** After the *k*-th iteration ($k > 0$), the most promising region $\sigma(k)$ is further partitioned into $M_{\sigma(k)}$ subregions $\sigma_1(k),\ldots,\sigma_{M_{\sigma(k)}}(k)$. What remains of the feasible region $\sigma(k)$, i.e., $\Theta \backslash \sigma(k)$, is aggregated into the surrounding region $\sigma_{M_{\sigma(k)}+1}(k)$. Then, $M_{\sigma(k)} + 1$ partitioned regions are obtained. When the first partition starts, the whole feasible region $\Theta$ is considered the most promising region, i.e., $\sigma(0) = \Theta$.

Since the feasible region $\Theta$ is finite, the partitioned regions we obtain will eventually be singleton regions, i.e., $M_{\sigma(k)} = 1$. Then two regions are obtained: $\sigma(k)$ and $\Theta \backslash \sigma(k)$.

**Step 2: Random sampling.** The next step of the algorithm is to randomly select $N_j$ samples $\theta_1^{(j)}, \theta_2^{(j)},\ldots,\theta_{N_j}^{(j)}$, $j = 1,2,\ldots,M_{\sigma(k)} + 1$, from each of the subregions $\sigma_j(k)$ obtained by the partitioning operator. Because of the openness of the NP method, various random sampling methods can be adopted with a requirement that the possibility of each point in each region being selected is more than zero [7].

**Step 3: Calculation of promising index.** Given a promising index function $I:\Sigma \to \mathbf{R}$, sample each region $\sigma_j(k)$, where $j = 1,2,\ldots,M_{\sigma(k)} + 1$, according to the fixed sampling strategy and estimate the promising index value of each region. For example, assume that the promising index value is the maximal objective function value of each region,

$$I(\sigma_j(k)) = \max_{\theta \in \sigma_j(k)} f(\theta), \ j = 1,2,\ldots,M_{\sigma(k)} + 1.$$

Estimate the promising index value of each region $\sigma_j(k)$,

$$\hat{I}(\sigma_j(k)) = \max_{i=1,2,\ldots,N_j} f(\theta_i^{(j)}), \ j = 1,2,\ldots,M_{\sigma(k)} + 1.$$

Notice that $\hat{I}(\sigma_j(k))$ is a random variable.

As long as the promising index corresponds to the performance function in singleton region, it can adopt any form. That is to say, when $\sigma_j(k)$ is the region of maximum depth, i.e., $\sigma_j(k) = \{\theta\}$, $I(\sigma_j(k))$ must equal to $f(\theta)$, i.e., $I(\sigma_j(k)) = f(\theta)$. Except for this restriction the NP method does not have restrictions on the selection of promising index function, which indicates the openness of the NP methods. Then, the promising index values of the $M_{\sigma(k)} + 1$ regions are compared, and the most promising region is determined:

$$\hat{j}_k = \arg\max \hat{I}(\sigma_j(k)), \ j = 1,2,\ldots,M_{\sigma(k)} + 1$$

If $\hat{j}_k \leq M_{\sigma(k)}$, i.e., one of the subregions of the current most promising region is found to have the maximum promising index, then this subregion is the most promising region in the next iteration. If $\hat{j}_k = M_{\sigma(k)} + 1$, then the most promising region in the next iteration is determined by the backtracking operator.

**Step 4: Backtracking.** If the entire region except $\sigma(k)$ is found to be the most promising region, the algorithm backtracks to a larger region that contains the current most promising region $\sigma(k)$. The backtracking rules can be determined by the requirements. An obvious backtracking method is to make the superregion of the current most promising region the backtracking objective. The selection of the present most promising region is denoted as

$$\sigma(k+1) = \begin{cases} \sigma_{\hat{j}_k}(k) & if \quad \hat{j}_k \leq M_{\sigma(k)} \\ s(\sigma(k)) & otherwise \end{cases}$$

Certainly, the entire finite feasible region $\Theta$ can be considered the backtracking objective, i.e., $\sigma(k+1) = \Theta$. Starting from the new most promising region $\sigma(k+1)$, the algorithm continues with the above-mentioned steps of partitioning, sampling, promising indices, and backtracking. Then, a sequence of partitioned regions is obtained. Finally, the algorithm comes to an end when the points in all feasible regions correspond to the singleton regions. The point in the singleton that has been considered the most promising regions for the most times can be considered the global optimal solution.

## 2.2 The Analysis on Optimization Efficiency of the NP Method

### 2.2.1 The Significance of the Number of Times Backtracking Is Implemented to the Optimization Efficiency of the NP Method

During the optimization process using the NP method, if the current most feasible region is proved to be unsatisfactory by sampling and calculation of promising index, backtracking is then necessary. This implicates that the last time partitioning, sampling, and promising indices are invalid. The algorithm should backtrack to the last iteration and continue with sampling and promising indices. Therefore, backtracking implies the decrease of calculation efficiency.

In the $k$-th iteration of the NP method if the surrounding regions of $\sigma(k)$ is considered the most promising, it then backtracks to the superregion $s(\sigma(k))$ of the current most promising region and makes $s(\sigma(k))$ the most promising region for the next partitioning. In the condition that the partitioning and sampling schemes are fixed, each backtracking results in two more times of partitioning and $2N(M_{\sigma(k)}+1)$ more points in the feasible regions are sampled, where $M_{\sigma(k)}$ is the number of feasible regions for partitioning with a fixed partitioning scheme, and $N$ is the number of sampled points in each feasible region. Calculating the promising index at these points requires $2N(M_{\sigma(k)}+1)$ performance functions of the promising index. The backtracking rate of the NP method is tightly related to the optimization efficiency indexes such as the convergence rate. If the backtracking is reduced once, $2N(M_{\sigma(k)}+1)$ performance functions of the promising index are reduced, which consequently shortens the optimization route, reduces optimization time, and speeds up the convergence. Thus, the number of times backtracking occurs is an important criterion for measuring the efficiency of this simulated optimization method.

### 2.2.2 The Analysis on Optimization Probability of the NP Method

L. Shi and S. Ólafsson improved that the NP method converges to a global optimal solution with probability one [1]. Let $\eta_l \in \Sigma$ be a feasible region obtained by nested partitions, $\theta_l^*$ be the optimum we get after introducing some other local optimization algorithms (such as SA, tabu search, etc.) into sampling of the NP method, and $\theta_l^{'}$ be the optimum we get using the other simple random sampling methods. Although we cannot assure that $\theta_l^*$ is the global optimum of the feasible region, the probability of $\theta_l^*$ being the global optimum is greater than the probability of $\theta_l^{'}$ being the global optimum in that these local optimization algorithms are capable of avoiding getting trapped in the local optima, i.e.,

$$P\{\theta_l^* \textit{ is the global optimum of } \eta_l\} > P\{\theta_l^{'} \textit{ is the global optimum of } \eta_l\}.$$

Suppose the global optimal solution to the original problem $\theta^* \in \eta_l \in \Sigma$, i.e., $\eta_l$ is the feasible region that contains the global optimal solution. Then, in the process of nested partitioning, $\eta_l$ is unavoidable in the way to the global optimal solution. Compare the promising index of $\eta_l$ is compared with those of the other regions $\eta_i\left(i=1,\cdots,M_{\sigma(k)}+1, i\neq l\right)$. If $\eta_l$ is selected to be the most promising region, the backtracking is reduced for at least once. Therefore, we can infer that, if the probability of $\eta_l$ being selected to be the most promising region is increased, the efficiency of the algorithm will be improved. The probability of $\eta_l$ being selected to be the most promising region is:

$$P\left\{f(\theta_l^*) > f(\theta_1^*),\ldots, f(\theta_l^*) > f(\theta_{M_{\sigma(k)}+1}^*)\right\} = \prod_{\substack{i=1\\ i\neq l}}^{M_{\sigma(k)}+1} P\left\{f(\theta_l^*) > f(\theta_i^*)\right\},$$

where $P\left\{f(\theta_l^*) > f(\theta_i^*)\right\} = \omega\rho + \psi(1-\rho) = \rho + \psi(1-\rho)$, $\omega$ is the probability of $f(\theta_l^*) > f(\theta_i^*)$ under the condition that $\theta_l^*$ is the global optimal solution, $\rho$ is the probability of $\theta_l^*$ being the global optimal solution, and $\psi$ is $\theta_l^*$ is the probability of $f(\theta_l^*) > f(\theta_i^*)$ under the condition that $\theta_l^*$ is the local optimal solution. As the $l$-th feasible region contains the global optimum, $\omega = 1$. The above probability function is shown as Fig. 2.

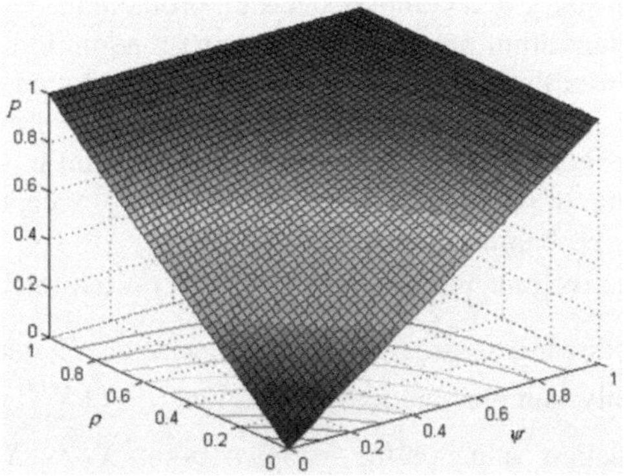


**Fig. 2.** The figure of the probability function

Therefore, the above probability equals the weighted average of 1 and $\psi$. And because $\psi \in (0,1)$, we have

$$\frac{\partial P}{\partial \rho} = 1 - \psi > 0 .$$

If the probability $\rho$ of $\theta_l^*$ being the global optimal solution is increased greatly, the above probability will correspondingly be increased. If the random sampling operator of the NP algorithm is changed and the probability of obtaining the global optimal solution in each region is increased, the convergence will be sped up and the efficiency of the algorithm will be improved greatly. The probability that the point we obtain using the local search of the SA method is the global optimal solution is much greater than the probability that the points we get using other simple randomized sampling methods are the global optima. Hence, the ideas of SA can be introduced into the NP method in order to increase the probability that $\eta_l$ is selected properly, decrease the number of times that backtracking in the NP method is implemented, speed up the convergence, and eventually improve the optimization efficiency. In the next section we present a new algorithm combining NP and SA.

# 3 The Combined NP/SA Algorithm

## 3.1 The Simulated Annealing Method

The simulated annealing algorithm (SA) is essentially a heuristic algorithm. The technique has been widely applied to a variety of problems including many complex decision problems. The term simulated annealing derives from the roughly analogous physical process of heating and then slowly cooling a substance to obtain a strong crystalline structure [8].

Often the solution space of a complex decision problem has many local minima. A simple local search algorithm proceeds by choosing random initial solution and generating a neighbor from that solution. The neighboring solution is accepted if it is a cost decreasing transition. Such a simple algorithm has the drawback of often converging to a local minimum. The SA method, though by itself it is a local search algorithm, avoids getting trapped in a local minimum by accepting cost increasing neighbors with some probability.

To solve the objective function Z: $\max_{s \in \Theta} f(s)$, over a feasible region $\Theta$, SA is implemented in the following steps. Firstly, at temperature $T$, starting from an initial point $X^{(0)}$, randomly sample the feasible region. If $f(X^{(k)}) \geq f(X^{(0)})$, where $f(X^{(k)})$ is the function value of the sampled point $X^{(k)}$, $X^{(k)}$ is accepted and taken as the initial point $X^{(0)}$ to continue the optimization; otherwise, if $f(X^{(k)}) < f(X^{(0)})$, $X^{(k)}$ is accepted with a probability of $\exp((f(X^{(k)}) - f(X^{(0)}))/T)$. Then, beginning from the initial annealing temperature $T_0$, the annealing temperature is lowered at a fixed temperature interval of $\Delta T$. At each annealing temperature $N$ points are randomly sampled. The above process is implemented repeatedly until the temperature reaches the final annealing one $T_f$ [9][10] and the algorithm converges to the global optimum.

## 3.2 The Combined NP/SA Algorithm

For a given feasible region the SA method focuses on searching for feasible points. It is capable of obtaining the global optima with a great probability and has a very strong local search ability. Applying the ideas of SA to the random sampling of the NP algorithm will greatly improve the ability of global optimization of the NP algorithm and the ability of local optimization of the SA method; hence the efficiency of the NP algorithm is improved greatly. Merging the SA method into the NP algorithm, we get the combined NP/SA algorithm.

Note that NP/SA is not simply merging the whole SA into the random sampling of the NP algorithm, but combining the basic optimization idea of SA with the complete optimization process of the NP algorithm properly in order to improve the optimization efficiency of the NP algorithm.

### 3.2.1 The Implementation Procedure of NP/SA

Similar to the preparatory work of SA implementation, firstly we need to set the initial annealing temperature $T_0$, the final annealing temperature $T_f$, and the number $N$ of random samples at each annealing temperature. NP/SA is an improvement of the NP algorithm. It has the same operations in partitioning, calculation of promising indices and backtracking. The random sampling of NP/SA is improved. Actually, NP/SA does not implement a complete annealing process in every sampled region to obtain an optimal solution over the region. Instead, NP/SA carry out the optimization according to the same annealing temperature over the feasible regions at the same depth. According to the maximum depth $dep(\sigma)$ ($\sigma \in \Sigma_0$) of singleton region in the feasible region, the annealing speed $\Delta T = (T_0 - T_f)/dep(\sigma)$ is set.

Respectively optimize the uncrossed $M_{\sigma(k)} + 1$ feasible regions obtained through the $k$-th partitioning at the annealing temperature $T_k = T_0 - dep(\sigma(k)) \cdot \Delta T$ according to the SA method. That is to say, starting from a certain initial point $X^{(0)}$, randomly sample the feasible regions. If $f(X^{(k)}) \geq f(X^{(0)})$, where $f(X^{(k)})$ is the function value of the sampled point $X^{(k)}$, $X^{(k)}$ is accepted and taken as the initial point $X^{(0)}$ to continue the optimization; otherwise, if $f(X^{(k)}) < f(X^{(0)})$, $X^{(k)}$ is accepted with a probability of $\exp((f(X^{(k)}) - f(X^{(0)}))/T)$ and taken as the initial point $X^{(0)}$ to continue the optimization. When $N$ points are sampled, the function value $f(X^{(0)})$ at the optimal point is used as the promising index function of each feasible region to fix the next most feasible region. The pseudo-code of the optimization process is following.

```
  (k)= ;
d( (k))=0;
Repeat
Partition the current promising region (k) into Mσ(k)
subregions.
T(k)=T(0)-dep( (k))* T
```

```
For i=1 to M_σ(k) +1 do
For j=1 to N do
Generate_state_x(j);
 =f(x(j))-f(x(k));
if >0 then k=j
   else if random(0,1)<exp(-  /T(k)) then
      k=j;
      Promising(i)=f(x(k));
End
if promising(i)>promising(m) then m=i;
if m<=M_σ(k) then  (k+1)=subregion(m);
        dep(  (k))= dep(  (k))+1;
   else backtrack(  (k-1));
        dep(  (k))= dep(  (k))-1;
until it reaches the maximum depth and stabilizes.
```

We may notice that the same annealing temperature is applied to the sampling operation of $M_{\sigma(k)}+1$ feasible regions at the same depth. When the depth of the feasible region is low, the annealing temperature is high and the probability of the worse solutions being accepted in sampling is also high. As the partitioning is moved on and the depth of the feasible region is increased, the annealing temperature used is comparatively low. At this temperature the probability of the worse solutions being accepted in sampling is hence low. NP/SA does not implement the complete annealing process of SA over every feasible region to be sampled.

### 3.2.2 Feasibility Analysis on NP/SA

The openness of the NP algorithm allows for the introduction of other algorithm and thoughts. The NP algorithm implicitly contains a requirement: the modifications to the operators of the NP algorithm are allowed so long as two conditions are satisfied. They are: (*a*) the probability of each point in the feasible region being sampled is larger than 0, and (*b*) the promising index corresponds with the performance function of the singleton region.

Although NP/SA is different from the pure NP algorithm in fixing the optima in the partitioned regions, its essential sampling method is still random sampling. This ensures that the probability of each point in the feasible region being sampled is larger than 0. Therefore, NP/SA completely satisfies condition (*a*) of the NP algorithm. When the partitioning process of the NP/SA algorithm moves on to singleton, there is only one feasible point in the feasible region and only one point is obtained through sampling. The promising index at this point is the function value of this point; hence it corresponds with the performance function over the singleton. Thus, NP/SA satisfies the condition (*b*) of the NP algorithm.

In all, the introduction of SA into the NP algorithm satisfies the openness of the latter one, which ensures that NP/SA converges to the global optimal solution with a probability of 1.

### 3.2.3 Superiority Analysis on NP/SA

As the NP algorithm evolves, the sequence of most promising regions $\{\sigma(k)\}_{k=1}^{\infty}$ forms a Markov chain with state space $\Sigma$. The singleton regions with the global optima are denoted as the absorbing states. In literature [1] and [2], L. Shi and S. Ólafsson proved that, the expected number of nested partitioning when the NP algorithm converges to the optimal solution is given by the following equation:

$$E[Y]=1+\sum_{\eta\in\Sigma_1}\frac{1}{P_\eta[T_{\sigma_{opt}}<T_\eta]}+\sum_{\eta\in\Sigma_2}\frac{P_\Theta[T_\eta<\min\{T_\Theta,T_{\sigma_{opt}}\}]}{P_\eta[T_\Theta<T_\eta]\cdot P_\Theta[T_{\sigma_{opt}}<\min\{T_\Theta,T_\eta\}]},$$

where $T_\eta$ is the hitting time of state $\eta\in\Sigma$, i.e. the first time that the Markov chain visits the state, $P_\eta[\cdot]$ denotes the probability of an event given that the chain starts in state $\eta\in\Sigma$, $\sigma_{opt}$ is the region corresponding to the unique global optimum, and $\Sigma_1=\{\eta\in\Sigma\setminus\{\sigma_{opt}\}|\sigma_{opt}\subseteq\eta\}$, $\Sigma_2=\{\eta\in\Sigma|\sigma_{opt}\not\subseteq\eta\}$ and $\Sigma=\{\sigma_{opt}\}\cup\Sigma_1\cup\Sigma_2$ are disjoint state spaces.

NP/SA introduces SA into the NP algorithm, which increases the probability of obtaining the global optima in the sampled regions and further increases the probability of the state of the Markov chain changes in the correct direction. Consequently, probability $P_\eta[T_{\sigma_{opt}}<T_\eta]$ at time $\eta\in\Sigma_1$, probability $P_\eta[T_\Theta<T_\eta]$ at time $\eta\in\Sigma_2$, and probability $P_\Theta[T_{\sigma_{opt}}<\min\{T_\Theta,T_\eta\}]$ are increased while probability $P_\Theta[T_\eta<\min\{T_\Theta,T_{\sigma_{opt}}\}]$ at time $\eta\in\Sigma_2$ is decreased. The combined effect of these factors reduces the expected number of nested partitioning when the NP algorithm converges to the global optima, and thus speeds up the convergence of the algorithm.

## 4 A Numerical Example

In this section we consider a numerical example to illustrate the combined NP/SA method. In order to prove the optimization efficiency of the NP/SA method, we respectively implement the NP algorithm using the traditional random sampling as well as the NP/SA method for the minimization problem of Schaffer's $f_6$ function. Then we represent numerical results that compare the computation efficiency of the NP/SA method to a pure NP implementation.

The Schaffer's $f_6$ function is designed to have its global optimum at 0, surrounded by circular "valleys" designed to trap methods based on local search, see Fig. 3 [11].

The function is given by

$$f(x_1,x_2)=0.5-\frac{\left(\sin\sqrt{x_1^2+x_2^2}\right)^2-0.5}{\left[1.0+0.001(x_1^2+x_2^2)\right]^2},$$

where $x_1,x_2\in[-4,4]$. It is commonly used to test global optimization algorithms. In order to maintain the original purpose of this function, we minimize it.

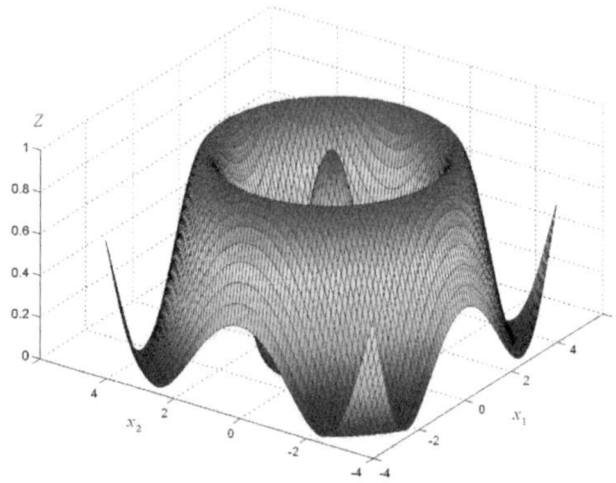


**Fig. 3.** The Schaffer's $f_6$ function

To calculate its optimal solution, firstly we implement the pure NP method with the traditional random sampling operator. The following scheme is used. In each iteration the most promising region is partitioned into nine subregions. 30 points in each subregion is randomly sampled. The algorithm terminates at the tenth iteration. Secondly, we implement the NP/SA method and use the same nested partitioning scheme as the pure NP method. Moreover, the initial annealing temperature is 10 and the final annealing temperature is 0.0001. The comparison between the results of the two methods is shown in Table 1.

**Table 1.** Comparison of NP and NP/SA Performance

| Algorithm | Result | Number of Times Backtracking Occurs |
|---|---|---|
| NP | 0 | 6 |
| NP/SA | 0 | 0 |

As a result, after the adoption of the NP/SA method, the saved number of times we need calculate the performance function values is

$$\Delta C = 2N(M_{\sigma(k)} + 1)H = 2 \times 30 \times (9+1) \times 6 = 3600 \text{,}$$

where $H$ is the reduced number of times backtracking occurs after NP/SA is adopted.

These results give a strong indication that the NP/SA method obtained by introducing SA into the NP algorithm is very useful in combining the global optimum search capability of the NP algorithm and the local search capability of the SA algorithm, reducing the number of times backtracking occurs in the nested partitioning, and making great improvements in calculation efficiency.

## 5 Conclusions

We have presented a new optimization algorithm that combines the NP algorithm and the SA algorithm. The resulting algorithm NP/SA retains the benefits of both algorithms, i.e., the global perspective and convergence of the NP algorithm and the powerful local search capabilities of the SA.

Since the random sampling operator of the NP algorithm is changed and the probability of obtaining the global optimal solution in each region is increased, the convergence is sped up, the number of times backtracking occurs in the nested partitioning is reduced, and hence the optimization efficiency is improved.

However, further theoretical and empirical development is needed for the algorithm. The NP/SA algorithm can be enhanced in several aspects. For example, we can use more elaborate partitioning, sampling and backtracking schemes if we have more knowledge of the specific decision problem. If we know that some solutions with certain properties are better than other solutions, we can add more weights on the regions containing these points. Future work will also focus on more numerical experiments and implementing the algorithm for complex decision problems in many fields to improve the current solving methods.

## References

1. Shi, L., Ólafsson, S.: Nested Partitions Method for Global Optimization. Operations Research. 4008 (2000) 390-407
2. Shi, L., Ólafsson, S., Sun, N.: New Parallel Randomized Algorithms for the Traveling Salesman Problem. Computers & Operations Research. 26 (1999) 371-394
3. Shi, L., Men, S.: Optimal Buffer Allocation in Production Lines. IIE Transactions. 35 (2003) 1-10
4. Shi, L., Ólafsson, S., Chen, Q.: A New Hybrid Optimization Algorithm. Computers & Industrial Engineering. 36 (1999) 409-426
5. Shi, L., Ólafsson, S., Chen, Q.: An Optimization Framework for Product Design. Management Science. 47 (2001) 1681-1692
6. Ólafsson, S., Shi, L.: A Method for Scheduling in Parallel Manufacturing Systems with Flexible Resources. IIE Transactions. 32 (1998) 135-146
7. Ólafsson, S., Gopinath, N.: Optimal Selection Probability in the Two-stage Nested Partition Method for Simulation-based Optimization. Proceedings of the 2000 Winter Simulation Conference (2000) 736-742
8. Kirkpatrick, S., Gelatt, Jr., C., Vecchi, M.: Optimization by Simulated Annealing. Science. 220 (1983) 671-680
9. Barretto, R.P., Chwif, L., Eldabi, T., et al.: Simulation Optimization with the Linear Move and Exchange Move Optimization Algorithm. Proceedings of the 1999 Winter Simulation Conference (1999) 806-811
10. Ahmed, M.A., Alkhamis, T.M.: Simulation-based Optimization Using Simulated Annealing with Ranking and Selection. Computers & Operations Research. 29 (2002) 387-402
11. Battiti, R., Brunato, M., Pasupuleti, S.: Do Not Be Afraid of Local Minima: Affine Shaker and Practicle Swarm. Technical Report # DIT-05-049, Department of Computer Science and Telecommunications, University of Trento, Italy (2005)

# DE and NLP Based QPLS Algorithm

Xiaodong Yu, Dexian Huang, Xiong Wang, and Bo Liu

Department of Automation, Tsinghua University, Beijing 100084, P.R. China
yuxd06@mails.tsinghua.edu.cn

**Abstract.** As a novel evolutionary computing technique, Differential Evolution (DE) has been considered to be an effective optimization method for complex optimization problems, and achieved many successful applications in engineering. In this paper, a new algorithm of Quadratic Partial Least Squares (QPLS) based on Nonlinear Programming (NLP) is presented. And DE is used to solve the NLP so as to calculate the optimal input weights and the parameters of inner relationship. The simulation results based on the soft measurement of diesel oil solidifying point on a real crude distillation unit demonstrate that the superiority of the proposed algorithm to linear PLS and QPLS which is based on Sequential Quadratic Programming (SQP) in terms of fitting accuracy and computational costs.

**Keywords:** DE, NLP, QPLS, application.

## 1 Introduction

As a robust multivariate linear regression technique for the analysis and modeling of noisy and highly correlated data, Partial Least Squares (PLS) has been successfully applied in the modeling, prediction and statistical control of the behavior of a wide variety of linear processes. However, when dealing with non-linear complex problems especially in chemical engineering, linear PLS cannot always capture the underlying model structure. To account for this non-linear factor, several attempts have been made to produce a Nonlinear Partial Least Squares (NPLS) algorithm which retains the orthogonality properties of the linear methodology while the nonlinear features could also be incorporated [1], including QPLS [1], Spline PLS (SPLS) [2], Neural Networks PLS (NNPLS) [3, 4], Fuzzy PLS (FPLS) [5]. For instance, Wold et al.[1] proposed a nonlinear (polynomial) PLS regression algorithm which retains the framework of the linear PLS algorithm, and modifies the linear inner relation between the predictor and the response latent variables to a nonlinear relationship. In particular, they proposed a quadratic polynomial relation for the inner mapping. Wold also proposed updating the weights of the input outer relationship by means of a Newton-Raphson linearization of the inner relation, i.e. a first-order Taylor series expansion of the quadratic inner relationship, and solving it with respect to the weights increments. However, the present algorithm (QPLS) [1] is fairly complicated and converges slowly when the data lack structure. Baffi et al. [6] proposed an error-based QPLS algorithm. By Tu and Tian

D.-S. Huang, L. Heutte, and M. Loog (Eds.): ICIC 2007, LNAI 4682, pp. 584–592, 2007.

et al. [7] modified the error-based QPLS algorithm where the procedure of updating the weights could be formulated as a nonlinear programming (NLP) problem.

Recently, a new evolutionary technique, Differential Evolution (DE) [8, 9], has been proposed as an alternative to genetic algorithm (GA) and Particle Swarm Optimization (PSO) [10] for unconstrained continuous optimization problems. Although the original objective in the development of DE was for solving the Chebychev polynomial problem, it has been found to be an efficient and effective solution technique for complex functional optimization problems. In a DE system, a population of solutions is initialized randomly, which is evolved to find optimal solutions through the mutation, crossover, and selection operation procedures. DE uses simple differential operator to create new candidate solutions and one-to-one competition scheme to greedily select new candidate, which work with real numbers in natural manner and avoid complicated generic searching operators in GA. It has memory, so knowledge of good solutions is retained in current population, whereas in standard GA, previous knowledge of the problem is destroyed once the population changes and in PSO, a secondary archive is needed. It also has constructive cooperation between individuals, individuals in the population share information between them. Due to the simple concept, easy implementation and quick convergence, nowadays DE has attracted much attention and wide applications in different fields mainly for various continuous optimization problems [8, 9, 10, 11]. However, to the best of our knowledge, there is no research on DE for NPLS problems.

In this paper, a new algorithm of Quadratic Partial Least Squares (QPLS) based on Nonlinear Programming (NLP) is presented. And DE algorithm is used to solve the NLP problem so as to calculate the optimal input weights and the parameters of inner relationship. The following paper is organized as follows. In section 2, QPLS algorithm is briefly introduced. Subsequently NLP-based QPLS algorithm is introduced in section 3, where highlighting the modification of the updating procedure of the weights as a NLP problem. Section 4 provides an overview of DE algorithm. Then the proposed algorithm is applied to measurement of diesel oil solidifying point on a real crude distillation unit in Section 5.

## 2 QPLS Modeling Method

Basically, the PLS method is a multivariable linear regression algorithm that can handle correlated inputs and limited data [1, 5]. The algorithm reduces the dimension of the predictor variables (input matrix, X) and response variables (output matrix, Y) by projecting them to the directions (input weight w and output weight c) that maximize the covariance between input and output variables. The decomposition of X and Y by score vectors is formulated as follows:

$$X = \sum_{h=1}^{m} t_h p_h^T + E \tag{1}$$

$$Y = \sum_{h=1}^{m} u_h q_h^T + F \tag{2}$$

where p and q are loading vectors, and E and F are residuals. This relation is known as the PLS outer relation. The relation between score vectors t and u is known as the inner relation which is formulated as follows:

$$u = f(t) + r \tag{3}$$

where r is a vector of residuals.

The original PLS algorithm was developed as linear regression method that uses a linear inner relation on the latent space. In the present work the linear PLS model is extended to the case when the inner model relating the score vectors t and u is nonlinear. By now various nonlinear PLS algorithms have been proposed to cope with the problems of the regression coefficients of the inner relationship. The QPLS algorithm was proposed by Wold et al. [1] for the model where the inner relation is a quadratic polynomial in 1989:

$$u = c_0 + c_1 t + c_2 t^2 + e \tag{4}$$

# 3 NLP-Based QPLS Algorithm

## 3.1 Updating the Weights Based on NLP

The procedure of updating the weights could be considered as a NLP problem with constraints [7]. . We can use NLP to work out the optimal input weight and the coefficient of the quadratic polynomial for only one time rather than iteration. Therefore the computing speed can be greatly improved. The objective function is formulated as follows:

$$\min\{(u-\hat{u})^T (u-\hat{u})\} \qquad s.t.\|w\| = 1 \tag{5}$$

where

$$\hat{u} = [1 \quad t \quad t^2]^T \qquad c, t = X \cdot w \tag{6}$$

The above NLP problem should be solved by calculating optimal w and c minimizing the objective function while satisfying the constraints[12].

## 3.2 NLP-Based QPLS Algorithm

The basic procedure of NLP-based QPLS algorithm is summarized as follows:

Step1: Mean centre and scale X and Y.
Step2: Set the output scores u equal to a column of Y.
Step3: Regress columns of X on u.

$$w' = u'X / u'u \tag{7}$$

Step4: Normalize w to unit length.

$$w' = w' / \|w'\| \tag{8}$$

Step5: Calculate the input scores.

$$t = Xw / w'w \tag{9}$$

Step6: Fit the nonlinear inner relation.

$$c \leftarrow fit[u = f(t) + r] \tag{10}$$

Step7: Calculate the nonlinear prediction of u.

$$\hat{u} = f(t, c) \tag{11}$$

Step8: Find the optimal input weight w and coefficient c according to. In this paper, we use DE, which will be introduced in section 4, to find the optima.
Step9: Calculate the input scores again according to Eq. (9).
Step10: Calculate the X loadings.

$$p' = t'X / t't \tag{12}$$

Step11: Normalize p to unit length.

$$p' = p' / \|p'\| \tag{13}$$

Step12: Calculate new nonlinear prediction of u according to Eq. (11).
Step13: Regress the columns of Y on $\hat{u}$.

$$q' = \hat{u}'Y / \hat{u}'\hat{u} \tag{14}$$

Step14: Normalize q to unit length.

$$q' = q' / \|q'\| \tag{15}$$

Step15: Calculate the input residual matrix.

$$E_h = E_{h-1} - t_h p_h'; X = E_0 \tag{16}$$

Step16: Calculate the output residual matrix.

$$F_h = F_{h-1} - \hat{u}_h' q_h'; Y = F_0 \tag{17}$$

Step17: If additional PLS dimensions are necessary, X and Y are replaced by E and F, and steps 2-17 are repeated.

In comparison with the procedure proposed by Wold which is based on Newton-Raphson linearization of the inner relation, the basic procedure introduced above is less complex [7]. Therefore we can obtain the optimal input weights and the parameters of inner relationship directly by means of NLP, which requires less computing complexity.

## 4 Brief Introduction to DE

As a branch of evolutionary algorithm for optimization problems over continuous domains, DE has gained much attention and wide applications in many fields due to its attractive advantages. Starting from the random initialization of a population of individuals in the search space, it can find the global optima by dynamically altering the differentiation's direction and step length. At each generation, the mutation and crossover operators are applied to individuals to generate a new population. Then, selection takes place and the population is updated [13, 14, 15]. The basic scheme of DE can be described as follows:

Step1: Initialize such parameters as NP, which donates the size of the population, $F \in [0,2]$, which is constant called scaling factor which controls amplification of the differential variation, NM, which donates the maximal number of the mutation, and $CR \in [0,1]$, which is also constant called crossover parameter that controls the diversity of the population.

Step2: Randomly generate the initial population $W^0 : \{w_i^0 (i = 1,2,\cdots, NP)\}$.

Step3: Evaluate $PE(w_i^G)$ which are the objective values of all individuals, and determine the best individual $w_b^G$ which has the best objective value.

Step4: Perform mutation operation for each individual, $w_i^G (i = 1,2,\cdots, NP)$, according to Eq. (18) in order to obtain each individual's corresponding mutation vector $\hat{w}_i^{G+1}$,

$$\hat{w}_i^{G+1} = w_i^G + F(w_b^G + w_j^G - w_k^G - w_i^G) \tag{18}$$

where $1 \leq j,k,l \leq NP$ are randomly chosen and mutually different and also different from the current index $i$.

Step5: Perform crossover operation between each individual and its corresponding mutation vector according to Eq. (19) in order to obtain each individual's trial vector.

$$w_{ij}^G \begin{cases} w_{ij}^G (RandomNumber > CR) \\ \hat{w}_{ij}^{G+1} (otherwise) \end{cases} \tag{19}$$

Step6: Evaluate the objective values of the trial vectors.

Step7: Perform selection operation by means of one greedy selection criterion between each individual and its corresponding trial vector according to Eq. (20) so as to generate the new individual for the next generation.

$$w_i^{G+1} = \begin{cases} w_i^{G+1} \left[PE(w_i^{G+1} \leq PE(w_i^G))\right] \\ w_i^G (otherwise) \end{cases} \tag{20}$$

Step8: Determine the best individual of the current new population with the best objective value. If the objective value of the current best individual is better than that of $w_b^G$, then update $w_b^G$ and its objective value.

Step9: If a stopping criterion is met, then output $w_b^G$ and its objective value, otherwise go back to Step 3.

## 5 Simulation Results

In this section, a soft measurement of diesel oil solidifying point on a real crude distillation unit is considered as a test example. Our proposed algorithm are compared with the traditional linear PLS and the QPLS algorithm based on Sequential Quadratic Programming (SQP) [16, 17], which is used to regress the optimal coefficients of the inner relationship.

The factors affecting the diesel oil solidifying point, one of real industrial qualitative index about diesel ,include the flow rate and temperature of the feed, the top pressure and temperature, the characteristic of the crude oil, and so on. According to the real industrial technic, we choose 12 variables as the inputs, such as the top pressure, the top temperature, the temperature and flow rate of the 3rd draw, the temperature and flow rate of the feed, etc., while choosing the diesel oil solidifying point as the single output. To build the PLS models, data corresponding to roughly about eight months of plant operation (and featuring a full range of acceptable disturbances) was collected, filtered and down-sampled to give 600 data points, which was split into two sets: one set of 400 points for model building (training and cross validation [18]) and a set of 200 points for model testing. Table 1 shows the model performance of different modeling algorithms by the means of the number of principal component and Sum of Squared Errors (SSE) of the predicted output.

**Table 1.** The model performance of different modeling algorithms

| | Number of principal component | SSE of the predicted output(200) |
|---|---|---|
| Traditional linear PLS | 6 | 12.7784 |
| SQP and NLP based QPLS | 4 | 11.4712 |
| DE and NLP based QPLS | 4 | 7.214 |

From Table 1, it can be seen that the QPLS algorithm is capable of modeling nonlinear systems greatly better than traditional linear PLS algorithm. And it is approved clearly again that when dealing with complex systems such as chemical distillation columns which contain strong nonlinear characteristics, NPLS algorithms show better performances in comparison with traditional linear PLS algorithms.

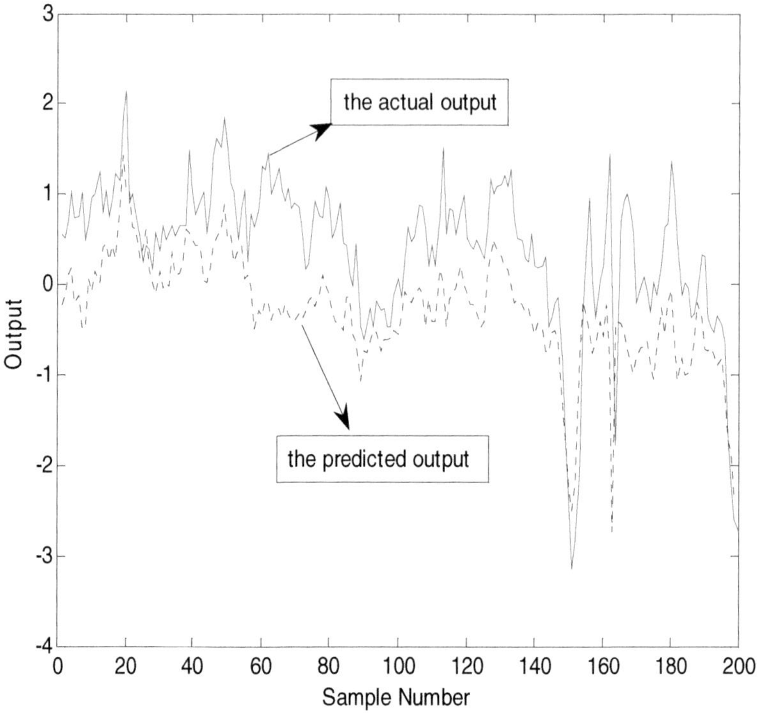


**Fig. 1.** Results for the validation data set using the traditional linear PLS algorithm

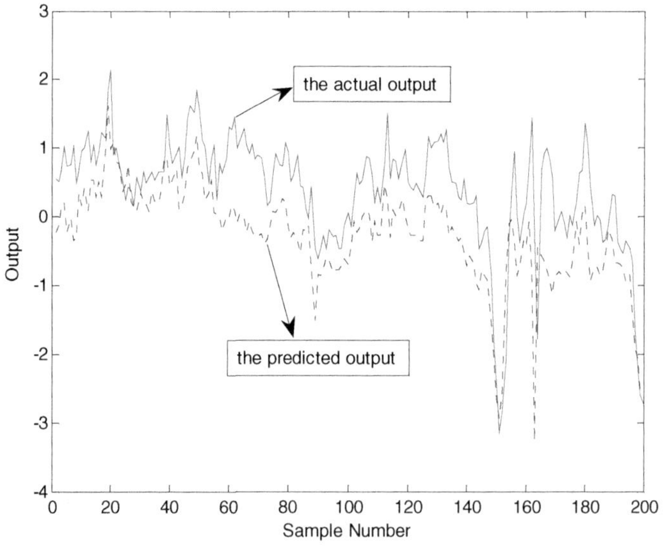


**Fig. 2.** Results for the validation data set using the SQP and NLP based QPLS algorithm

Meanwhile, the computational results of the proposed algorithm also confirm a significant improvement over the SQP and NLP based QPLS algorithm, demonstrating that the proposed algorithm can improve the fitting accuracy of the model and decrease the computation burden greatly, which is significantly important in chemical industry. Beside that, the model is less sensitive to the initial values while using DE.

The actual and the predicted output for the validation data set using traditional linear PLS algorithm, SQP based QPLS algorithm and DE based QPLS algorithm are shown in Fig. 1, Fig. 2 and Fig. 3, respectively.

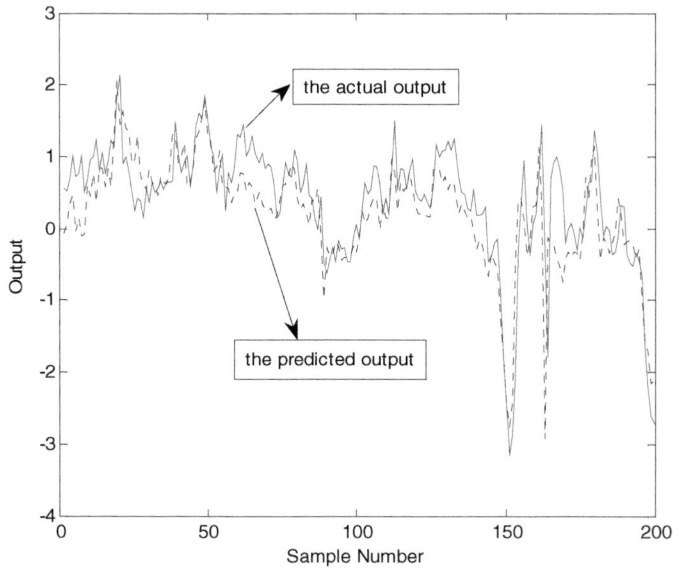


**Fig. 3.** Results for the validation data set using the DE and NLP based QPLS algorithm

# 6 Conclusions

To the best of our knowledge, this is the first paper to apply DE for NPLS problems. The proposed model uses a QPLS framework while considering the procedure of updating the weights as a NLP problem. And we use DE to calculate the optimal input weights and the parameters of inner relationship. Compared with the traditional linear PLS and the SQP and NLP based QPLS, the simulation results demonstrated that the proposed algorithm can improve the fitting accuracy of the model and decrease the computation burden and the sensitivity to the initial values. Meanwhile the proposed algorithm is also robust, simple and easy to implement.

**Acknowledgement.** The authors wish to thank three anonymous referees for a number of constructive comments on the earlier manuscript of this paper. This research is partially supported by National Science Foundation of China (Grant No. 60574072) as well as the National high tech. project of China(863/CIMS 2006AA04Z168).

## References


1. Wold, S., Wold, N.K., Skagerberg, B.: Nonlinear PLS Modeling. In Chemometrics Int. Lab. System. 11( 7) (1989) 53-65
2. Wold, S.: Nonlinear Partial Least Square Modeling (□) Spline Inner Function. In Chemometrics Int. Lab. System. 14 (1) (1992) 71-84
3. Qin, S.J., McAvoy, T.J.: Nonlinear PLS Modeling using Neural Networks. In Comput. Chem. Eng. 16(4) (1992) 379-391
4. Baffi, G., Martin, E.B., Morris, A.J.: Non-linear Projection to Latent Structures Revisited (the Neural Network PLS Algorithm). In Comput. Chem. Eng. 23 (1999) 1293-1307
5. Yoon, H.B., Chang, K.Y., Lee, I.: Nonlinear PLS Modeling with Fuzzy Inference System. In Chemometrics Int. Lab. System.. 64(2) (2003) 137-155
6. Baffi, G., Martin, E.B., Morris, A.J.: Non-linear Projection to Latent Structures Revisited: the Quadratic PLS Algorithm. In Comput. Chem. Eng. 23 (1999) 395-411
7. Ling, Tu., Tian, X.: Quadratic PLS Algorithm Based on Nonlinear Programming. In Control Engineering of China. 11 (supplement) (2004) 117-119
8. Storn, R., Price, K.: Differential Evolution – A Simple Evolution Strategy for Fast Optimization. In Dr. Dobb's Journal. 22 (4) (1997) 18-24
9. Lampinen, J.: A Bibliography of Differential Evolution Algorithm. http://www.lut.fi/~jlampine/debiblio.htm, 2002
10. Liu, B., Wang, L., Jin, Y.H.: Advances in Particle Swarm Optimization Algorithm. In Control and Instruments in Chemical Industry. 32(3) (2005) 1-6
11. Liu, B., Wang, L., Jin, Y.H.: Advances in Differential Evolution. In Control and Decision. (in press)
12. Wang, G., Li, X.: Nonlinear Programming Algorithm and Its Convergence Rate Analysis. In Chinese Quarterly Journal of Mathematics. 13(1) (1998) 8-13
13. Fang, Q., Cheng, D., Yu, H.: Eugenic Strategy and its Application to Chemical Engineering. In Journal of Chemical Industry and Engineering (China). 55(4)(2004) 598-602
14. Storn, R.: On the Usage of Differential Evolution for Function Optimization. In Proceedings of Biennial Conference of the North American. (1996) 519-523
15. Cheng, S., Hwang, C.: Optimal Approximation of Linear Systems by a Differential Evolution Algorithm. In IEEE Transactions on Systems, Man and Cybernetics, Part A. 31(6) (2001) 698-707
16. Shi, R., Pan, L.: Modified Method of Nonlinear PLS and its Application-Based on Chebyshev Polynomial. In Control Engineering of China. 10(6) (2003) 506-508
17. Fu, L., Wang, H.: A Comparative Research of Polynomial Regression Modeling Method. In Application of Statistics and Management. 23(1) (2004) 48-52
18. Zhang, J., Yang, X.H.: Multivariate Statistical Process Control. The Chemical Industry Press. (2000)

# Fuzzy Genetic Algorithm Based on Principal Operation and Inequity Degree

Fachao Li[1,2] and Chenxia Jin[2]

[1] School of Economy and Management, Hebei University of Science and Technology, Shijiazhuang, Hebei, 050018, China
[2] School of Science, Hebei University of Science and Technology, Shijiazhuang, Hebei, 050018, China
lifachao@tsinghua.org.cn, jinchenxia2005@126.com

**Abstract.** In this paper, starting from the structure of fuzzy information, by distinguishing principal indexes and assistant indexes, give comparison of fuzzy information on synthesizing effect and operation of fuzzy optimization on principal indexes transformation, further, propose axiom system of fuzzy inequity degree from essence of constraint, and give an instructive metric method; Then, combining genetic algorithm, give fuzzy optimization methods based on principal operation and inequity degree (denoted by BPO&ID-FGA, for short); Finally, consider its convergence using Markov chain theory and analyze its performance through an example. All these indicate, BPO&ID-FGA can not only effectively merge decision consciousness into the optimization process, but possess better global convergence, so it can be applied to many fuzzy optimization problems.

**Keywords:** Fuzzy optimization, fuzzy inequity degree, principal index, fuzzy genetic algorithm, BPO&ID-FGA, Markov chain.

## 1 Introduction

The theory of fuzzy numbers is very popular in describing uncertain phenomena in actual problems. Its trace can be found in many domains such as fuzzy control, fuzzy optimization, fuzzy data analysis and fuzzy time serial etc. For fuzzy optimization, good results both in theory and in application mainly focus on fuzzy linear optimization [1-5], which were mostly obtained by transforming a fuzzy linear optimization problem to a classical one according to the structure properties of fuzzy numbers. With the development of computer science and evolutionary computation theory, evolutionary computation methods have entered into the field of vision of scholars those are interested in fuzzy optimization problems. For instance, genetic algorithms were used to processing the optimization problems with fuzzy coefficients but real variables in [6] and [7], and evolutionary computation were used to the linear optimization problems with fuzzy coefficients and fuzzy variables in [8], the essence of which are transforming a fuzzy linear optimization problem to an ordinary one. Up to

D.-S. Huang, L. Heutte, and M. Loog (Eds.): ICIC 2007, LNAI 4682, pp. 593–604, 2007.

now, there is still no effective and common method for general fuzzy optimization problems, in which the bottleneck is presented by the following aspects: ① The ordering of fuzzy information; ② The judgment of fuzzy constraints; ③ The operable description of fuzzy information; ④ The operation of optimization process.

In ranking fuzzy information, many systemic research findings have been already achieved [9-15], but the other three aspects can not be still solved effectively. In this contribution, for the general optimization problems with fuzzy coefficients, fuzzy variables and fuzzy constraint, we have the following findings: 1) By distinguishing principal indexes and assistant indexes, give comparison method of fuzzy information on synthesizing effect and description method of fuzzy information on principal indexes; 2) Starting from the structure characteristic of fuzzy information and essence of constraint, propose axiom system of fuzzy inequity degree, and give an instructive metric method; 3) Establish a kind of broad and operable fuzzy optimization model, and combining the transform strategy by penalizing for problems with constraints, a new kind of fuzzy genetic algorithm based on principal operation and inequity degree is proposed (denoted by BPO&ID-FGA, for short); 4) Give the concrete implementation step and the crossover ,mutation strategy; 5) Consider its global convergence under the elitist reserved strategy using Markov chain theory; 6) Further analyze the performance of BPO&ID-FGA through an example.

## 2 Preliminaries

Fuzzy numbers, with the feature of both fuzzy sets and numbers, are the most common tool for describing fuzzy information in real problems. In the following, the definition of fuzzy number is introduced.

**Definition 1** [16]. Let $A$ be a fuzzy set on the real number field R, $A_\lambda = \{x \mid A(x) \geq \lambda\}$ be the $\lambda-$cuts of $A$. If $A_1 = \{x \mid A(x) = 1\} \neq \phi$, $A_\lambda$ are closed intervals for each $\lambda \in (0,1]$, $\text{supp}A = \{x \mid A(x) > 0\}$ is bounded, then $A$ is called a fuzzy number. The class of all fuzzy numbers is called fuzzy number space, which is denoted by $E^1$. Particularly, if there exists real number $a, b, c$ such that $A(x) = (x-a)/(b-a)$ for each $x \in [a, b)$, $A(b) = 1$, $A(x) = (x-c)/(b-c)$ for each $x \in (b, c]$, and $A(x) = 0$ for each $x \in (-\infty, a) \cup (c, +\infty)$, then we say $A$ is a triangular fuzzy number, and written as $A = (a, b, c)$ for short.

The operations of fuzzy numbers, established based on Zadeh's extension principle, should be the foundation for optimization problems. For the arithmetic operations of fuzzy numbers, we have the following Theorem.

**Theorem 1** [16]. Let $A, B \in E^1$, $k \in$ R, $f(x, y)$ be a continuous binary function, $A_\lambda$, $B_\lambda$ be the $\lambda-$cuts of $A$ and $B$, respectively. Then $f(A, B) \in E^1$, and $(f(A, B))_\lambda = f(A_\lambda, B_\lambda)$ for each $\lambda \in (0,1]$.

Fuzzy numbers have many good analytical properties, we can see ref. [16] for the concrete content.

# 3 Compound Quantification Description of Fuzzy Information

## 3.1 Basic Idea of Compound Quantification

Ranking fuzzy numbers, as the main components of fuzzy number theory, is the key for fuzzy optimization problems. Up to now, usually, by appropriate transformation, each fuzzy number can be mapped onto a real number, through which the comparison and ranking of fuzzy number can be realized.

**Definition 2** [17]. For uncertain information *A*, real number *a* (called the principal value of *A*) denotes the centralized quantification value under a certain consciousness, sequence $a_1, a_2, \cdots, a_s$ denote the assistant quantity indexes describing the connection between *a* and *A* from different sides. The whole constituted by *a* and $a_1, a_2, \cdots, a_s$ is said to be a compound quantification value, written as $(a; a_1, a_2, \cdots, a_s)$ for short.

In fuzzy optimization problem, the assistant index play the role of supplement and constraint for principle index, we may obtain specific quantitative values by acting the assistant index in $(a; a_1, a_{2,} \cdots, a_s)$ of compound quantification of fuzzy information into its principle index through an effect synthesizing function, through which the size comparison of fuzzy values can be realized from global view.

## 3.2 Compound Quantification Based on Level Effect Function

**Definition 3.** Say $L(\lambda): [0,1] \to [a,b] \subset [0,\infty)$ a level effect function, if $L(\lambda)$ is piecewise continuous and monotone non-decreasing. For $A \in E^1$, Let

$$I(A) = \frac{1}{L^*} \int_0^1 L(\lambda) M_\theta(A_\lambda) \mathrm{d}\lambda , \tag{1}$$

$$CD(A) = \int_0^1 L(\lambda) m(A_\lambda) \mathrm{d}\lambda . \tag{2}$$

Then $I(A)$ is called the centralized quantification value of *A*, $CD(A)$ is called the concentration degree of *A*. Particularly, if $L^* = 0$, $I(A)$ is defined as the midpoint of $A_1$, $CD(A)$ the length of $A_1$. Where, $L^* = \int_0^1 L(\lambda)\mathrm{d}\lambda$ ; $M_\theta([a,b]) = a + \theta(b-a)$ , $\theta \in [0,1]$ , and *m* is the Lebesgue measure.

Obviously, in the sense of level effect function $L(\lambda)$ and risk parameter $\theta$, $I(A)$ is the centralized quantification value, also the principle index describing the position of *A*, while $CD(A)$ is assistant index further describing the reliability of $I(A)$ , so $(I(A); CD(A))$ is the compound quantification value of *A*. And in the implementation process of BPO&ID-FGA, we select $S(I(A), CD(A)) = I(A)/(1+\beta CD(A))^\alpha$ as the synthesizing effect function, here, $\alpha, \beta \in (0,+\infty)$ all represents some kind of decision consciousness.

## 4 Compound Quantification Description of Fuzzy Constraint

Generally, the constraints of fuzzy optimization problems have some uncertainty, how to judge the satisfaction is the main factor, in which the most common used is the method based on the order relation of fuzzy information. Owing to the essential differences between fuzzy number and real number, there exists weakness for the current methods. For this, references [18,19] defined the degree $D(A \le x)$ of fuzzy number $A$ not exceeding real number $x$ by the location relationship of all level cuts and $x$, then give the definition $D(A \le B)$ (that is, the degree of fuzzy number $A$ not exceeding fuzzy number $B$) by $D(A - B \le 0)$, further, combine a given threshold $\beta \in (0, 1]$, tell whether $A \le B$ is right through whether $D(A \le B) \ge \beta$ is right. For any fuzzy number, because the addition operation and subtraction operation are not inverse, the degree of $A \le B$ defined by the degree of $A - B \le 0$ is not reasonable, which directly embody that, if $A_\lambda$ $(0 < \lambda < 1)$ is not single-point set, then $D(A \le A) = 0.5$.

From the above analysis, the current methods of testing fuzzy constraints all exist a certain weakness. Because the fuzzy numbers do not have the ordering like real numbers, by adopting some quantification strategy under a certain consciousness, the comparison of fuzzy information can be realized, which is the basic method of processing fuzzy constraints. To establish general rules, the axiom system of fuzzy inequity degree is introduced as follows:

**Definition 4.** Let $D(A, B)$ be function on $E^1 \times E^1$ (denoted by $D(A \le B)$ for short), $D$ is called the fuzzy inequity degree on $E^1$, if $D$ satisfies the following conditions:

1) *Normality*: $0 \le D(A \le B) \le 1$ for any $A, B \in E^1$;
2) *Reflexivity*: $D(A \le A) = 1$ for any $A \in E^1$;
3) *Monotonicity*: $D(A^{(1)} + A^{(2)} \le B^{(1)} + B^{(2)}) = 1$ for any $A^{(1)}, A^{(2)}, B^{(1)}, B^{(2)} \in E^1$ with $D(A^{(1)} \le B^{(1)}) = D(A^{(2)} \le B^{(2)}) = 1$;
4) *Semi-linearity*: $D(kA \le kB) = D(A \le B)$ for any $A, B \in E^1$ and $k \in (0, \infty)$;
5) *Translation invariance*: $D(a + A, a + B) = D(A, B)$ For any $A, B \in E^1$ and $k \in \mathrm{R}$.

In Definition 4, 0 and 1 separately denote absolute dissatisfaction state and satisfaction state. Obviously, each requirement reflects the basic characteristic of no excess relationship from different aspects. For given $\alpha \in [0, 1]$, Let

$$D(A \le B) = H(M_\theta(B_\alpha) - M_\theta(A_\alpha)). \tag{3}$$

Where, $M_\theta([a, b]) = a + \theta(b - a)$, $\theta \in [0, 1]$; and $H(x) = 1$ for each $x \in [0, +\infty)$, and $H(x) = 0$ for each $x \in (-\infty, 0)$. According to Definition 4, it is easy to verify that formula (3) is fuzzy inequity degree on $E^1$.

From (3), this kind of fuzzy inequity degree contains the no excess relationship $\le$, but it doesn't make full use of the location relationship of $A$ and $B$ under all levels. To establish more perfect model describing fuzzy inequity degree, we introduce the following formula (4).

$$D(A \le B) = \frac{1}{L^*} \int_0^1 L(\lambda) H(M_\theta(B_\lambda) - M_\theta(A_\lambda)) \mathrm{d}\lambda. \tag{4}$$

Here, $L(\lambda)$ is level effect function, $L^* = \int_0^1 L(\lambda)d\lambda$ ; and we say if $L^* = 0$ , then $D(A \le B) = H(M_\theta(B_1) - M_\theta(A_1))$ . Through the above analysis, we can obtain the following conclusion.

**Theorem 2.** $D(A \le B)$ defined by formula (4) is fuzzy inequity degree on $E^1$ .

This theorem can be proved by the properties of fuzzy number and integral and Definition 4.

In the optimization and decision process of many real problems, importance degree to the studied problems varies with different levels, so the influence of the degree of $A_\lambda \le B_\lambda$ under different levels on the global degree of $A \le B$ is not same. In (4), level effect function $L(\lambda)$ is a kind of decision parameter describing effect value under levels, therefore, (4) is essentially instructive measure method reflecting fuzzy information $A$ no exceeding $B$.

## 5 The Solution Model of Fuzzy Optimization Problem Based on Inequity Degree

In this paper we will consider the following optimization problems in which both objective function and constraints are with fuzzy uncertainty, the general form of the mathematical model can be expressed as:

$$\begin{cases} \max f(x), \\ \text{s. t. } g_i(x) \lesssim b_i, i = 1, 2, \cdots, m. \end{cases} \tag{5}$$

Where, $x = (x_1, x_2, \cdots, x_n)$ , $f$ and $g_1, g_2, \cdots, g_m$ are all $n$-dimensional fuzzy value function, $\lesssim$ denotes the inequality relationship in the fuzzy sense, $x_i \in E^1$ the optimized variable or decision variable, $b_i \in E^1$ the given fuzzy number.

Because the fuzzy numbers do not have the comparability like real numbers, model (5) is just a formal model, and can't be easily solved. According to the above compound quantification strategy and fuzzy inequity degree, it can be converted into the following model (6) by synthesizing effect function.

$$\begin{cases} \max E(f(x)), \\ \text{s. t. } D(g_i(x)) \le b_i) \ge \beta_i, i = 1, 2, \cdots, m. \end{cases} \tag{6}$$

Where, $E(f(x))$ denotes the synthesizing effect value of $f(x)$ , $D(g_i(x)) \le b_i)$ denotes the degree of $g_i(x) \le b_i$ , $\beta_i \in (0, 1]$ denotes minimum requirement for satisfying $g_i(x) \le b_i$ . If (1) and (2) are taken as the compound quantification description of fuzzy information, $S(a, b)$ as the synthesizing effect operator, (4) as measure method of inequity degree, then we have

$$E(f(x)) = S(I((f(x)), CD((f(x))) , \tag{7}$$

$$D(g_i(x) \le b_i) = \frac{1}{L^*} \int_0^1 L(\lambda) H(M_\theta((g_i(x)_\lambda) - M_\theta((b_i)_\lambda)) d\lambda . \tag{8}$$

Obviously, model (6) have the feature of optimization operation, but it is not conventional optimization problem, and can't be solved by existing methods, its bottleneck lies that it is hard to describe the changing way of fuzzy information in detail. Considering that triangular fuzzy numbers are often used to describe the fuzzy information in practical problems, we previously arrange that optimized variables and coefficients are all triangular fuzzy number in this article.

Owing to the intrinsic difference with the real number in operations, the corresponding optimization problem is not still solved by analytical methods even if triangular fuzzy numbers are strong in description. For this, we can establish concrete solution methods by combining genetic algorithm and compound quantification strategy of fuzzy information (denoted by BPO&ID-FGA, for short).

## 6 Fuzzy Genetic Algorithm Based on Principal Operation and Inequity Degree

Genetic algorithms [20] possess the features of easy operation and strong flexibility, which help itself become one of the most common used method in many fields. In this section ,we will focus on the structure of BPO&ID-FGA. And the basic operation strategy of BPIO-FGA includes the following three aspects:

1) For decision variable $A=(a, b, c)$, we see $b$ as the principle index describing the size position of $A$, $a$ and $c$ the assistant indexes. In the optimization process, we first consider the change of $b$, and then by combining the lengths of $[a, b]$ and $[b, c]$ and the change result of $b$, determine the change results of $a$ and $c$ by random supplement strategy. Given the change result $A'=(a', b', c')$ of $A=(a, b, c)$ largely depends on the principle index $b$ in this kind of operational strategy, this strategy is one of the main background we name our algorithm as what we do.

2) For the problems of the evaluation of the objective function, we take the effect synthesizing value of the compound quantification description of fuzzy information constituted by (1) and (2) as the main criteria of operation. From what we discussed in previous section 3, we are involved in the concept of principle index and assistant index as well, which becomes another main background we name our algorithm as what we do.

3) For the satisfaction of the fuzzy constraints, we take fuzzy inequity degree (4) as the main criteria, which becomes the third background we name our algorithm as what we do.

Owing to the nonnegativity of the object function value in real problems, in the following we assume that: 1) $E(f(x)) \geq 0$, if not, we can convert it into $M+E(f(x))$ by selecting appropriate large $M$ ; 2) the optimization problem is the maximum one, and the minimum optimization problem $\min f(x)$ can be converted into the maximum optimization problem by $\max[M-E(f(x))]$, where, $M$ is appropriate large positive number.

### 6.1 Coding

Coding is the most basic component of genetic algorithm. In BPO&ID-FGA, for fuzzy number ($a$, $c$, $b$), we have adopted three equal lengths 0, 1 code to separately represent the principle index $c$ and the left, right assistant indexes $a$ and $b$.

### 6.2 Crossover and Mutation

The crossover and mutation operations are the specific strategies to find the optimal or satisfied solution. In BPO&ID-FGA, we only act the crossover and mutation operations on the middle section of fuzzy variables. And the two ends of coding string can be obtained by random complement or definite complement strategy. The details are given below.

**Crossover Operation.** For two given fuzzy numbers $A^{(1)}=(a_1,b_1,c_1)$ and $A^{(2)}=(a_2,b_2,c_2)$, cross the two strings representing $b_1$ and $b_2$ separately, and take one of the obtained strings $b$ as the crossover result of $b_1$ and $b_2$, then the left and right assistant indexes $a$ and $c$ can be determined by the following methods (here, both $r_1$ and $r_2$ are random numbers in specified scope) :

① $a=b-r_1 b$, $c=b+r_2 b$; ② $a=b-r_1$, $c=b+r_2$;
③ $a=b-r_1(b_1-a_1)-r_2(b_2-a_2)$, $c=b+r_1(c_1-b_1)-r_2(c_2-b_2)$.

**Mutation Operation.** For any given fuzzy number $A=(a,b,c)$, mutate the string representing $b$, and obtain the mutation results $b'$, then the left and right assistant indexes $a'$ and $c'$ can be determined by the following methods (here, both $r_1$ and $r_2$ are random numbers in specified scope):

① $a'=b'-r_1 b'$, $c'=b'+r_2 b'$; ② $a'=b'-r_1$, $c'=b'+r_2$;
③ $a'=b'-r_1(b-a)$, $c'=b'+r_1(c-b)$.

In this paper, we choose ①s as the method of crossover and mutation.

### 6.3 Replication

In designing genetic algorithm, penalty strategy is commonly used to eliminate constraints in optimization process. Its purpose is to convert infeasible solution into feasible solution by adding penalty item in the objective function, by which, the chance of infeasible solution selected for evolution is lowed according to some rules. In BPO&ID-FGA, we use the following fitness function with some penalty strategy.

$$F(x)=E(f(x))\cdot p(x). \tag{9}$$

And, take (9) as the basis of proportional selection. Where, $E(f(x))$ is synthesizing effect value of object function $f(x)$, $p(x)$ is penalty factor, the basic form as follows: if all the constraints are satisfied, then $p(x)=1$; if the constraints are not completely satisfied, then $0 \le p(x) \le 1$. In general, exponential function can be used as penalty function as follows:

$$p(x) = \exp\{-K \cdot \sum_{i=1}^{m} \alpha_i \cdot r_i(x)\}. \tag{10}$$

Here, $K \in (0, \infty]$, $\alpha_i \in (0, \infty]$, $r_i(x) \in [0, \infty)$, $0 \cdot \infty = 0$.

Obviously, $K = \infty$ implies decision result must satisfy all the constraints, $\alpha_i = \infty$ implies decision result must satisfy the $i^{\text{th}}$ constraints, and $0 < \alpha_i, K < \infty$ implies the decision result can break $i^{\text{th}}$ constraint. In the following example, let $\alpha_i = 1$, $K = 0.01$, $r_i(x)$ be the difference of synthesizing effect value between two sides of $i^{\text{th}}$ constraints.

# 7 Convergence of BPO&ID-FGA

We can know from the discussion above that, the process of crossover, mutation and selection in BPO&ID-FGA is only relevant to current state of populations, but has nothing to do with the former one. Thus the BPO&ID-FGA is still a Markov chain, and its convergence could be analyzed by the Markov chain theory.

**Lemma 1.** Genetic sequence $\{X(t)\}_{t=1}^{\infty}$ of BPO&ID-FGA is a Markov chain which is homogenous and mutually attainable.

**Lemma 2.** Genetic sequence $\{X(t)\}_{t=1}^{\infty}$ of BPO&ID-FGA is an ergodic Markov chain. The above results can be directly proved according to the structure of BPO&ID-FGA and the definition of Markov chain.

**Theorem 3.** BPO&ID-FGA using the elitist preservation strategy in replication process is global convergent.

**Proof.** Because it is used the elitist protection strategy, there are some changes happened on the nature of Markov chain. When the GA evolves to a new generation (for example generation *j*), the most superior individual of previous generation (generation *j*-1) will replace the worst individual of this generation (namely generation *j*). At the same time, we suppose that generation *i* be one of the previous generations of generation *j*, and there produced a more superior new individual in the evolution process from generation *i* to generation *j*. It is obvious that $P_{ij}^{(n)} > 0$ by now, which is to say it is reachable from *i* to *j*; but it is not reachable from *j* to *I*, that is, $P_{ji}^{(n)} = 0$, which is because the individual of generation *j* is forced to be replaced by the most superior individual of the previous generation. In the evolution process, for *i* and *j* are arbitrary, we may obtain that: The BPO&ID-FGA using the most superior individual protection strategy is a non-return evolution process, and it will finally converge to the global optimal solution.

# 8 Application Example

Consider the following fuzzy nonlinear programming

$$\max\ f(x_1,x_2) = -(0.1, 0.3, 0.8)x_1^2 - (0.2, 0.4, 0.7)x_2^2 + (16.1, 17, 17.3)x_1 + (17.7, 18, 18.6)x_2,$$
$$\text{s.t.}\quad (1.4, 2, 2.6)x_1 + (2.7, 3, 3.3)x_2 \mathrel{\tilde{\leqq}} (47, 50, 51),$$
$$(3.8, 4, 4.4)x_1 + (1.6, 2, 2.2)x_2 \mathrel{\tilde{\leqq}} (40, 44, 47),$$
$$(2.6, 3, 3.2)x_1 + (1.6, 2, 2.2)x_2 \cong (32, 36, 40),$$
$$x_1, x_2 \mathrel{\tilde{\geqq}} 0.$$

For this optimization problem, when both coefficients and variables are real numbers, the optimal solutions are $x_1 = 4.8333, x_2 = 10.75$, $\max f(x_1,x_2) = 222.4329$.

Let the size of population be 80, (1) be the centralized quantification value, (2) be the concentration degree of $A$, $S(I(A), CD(A)) = I(A)/[1+0.001 \cdot CD(A)]^{0.5}$ be the synthesizing effect function, and $L(\lambda) = \lambda$ be the level effect function. By using BPO&ID-FGA with 20 bits of binary coding, we can get the optimal value shown on Fig. 1 after 100 times of iterations (taking the times of iteration as $x$-coordinate, and the synthesizing effect value of fuzzy minimum value as y-coordinate). The optimal solutions are $x_1 = (4.6595, 4.9902, 5.3576)$, $x_2 = (10.5398, 11.0000, 11.4577)$, and the synthesizing effect value of fuzzy maximum value is 222.1152.

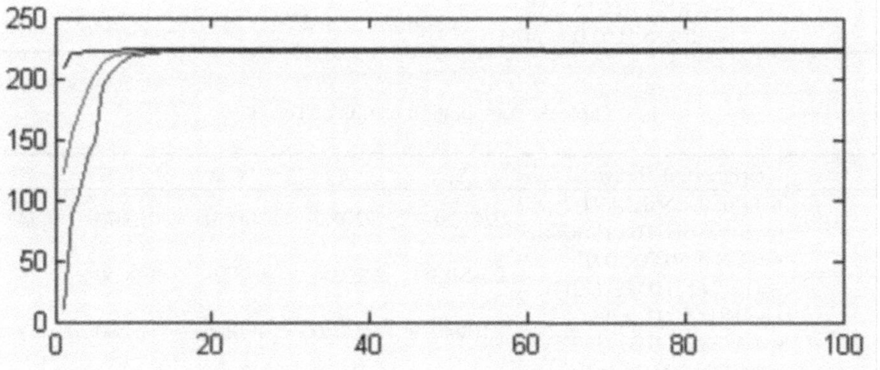


**Fig. 1.** 100 iteration results for Example 1

In order to further analyze the performance of BPO&ID-FGA, for different synthesizing effect functions and level effect functions, we separately make tests from the following three aspects:

*Test 1* For $L(\lambda) = \lambda$ and $S(I(A), CD(A)) = I(A)/(1+\beta \cdot CD(A))^{\alpha}$, and $(\alpha, \beta)$ takes (0.5, 0.1), (0.5, 1), (2,0.1) and (2, 1) separately, the computation results are stated in Table 1.

*Test 2* For $S(I(A), CD(A)) = I(A)/(1+0.01 \cdot CD(A))^{0.5}$, and $L(\lambda)$ be $\lambda, \lambda^2, \lambda^{0.5}$, separately, the computation results are stated in Table 2.

*Test 3* For $S(I(A), CD(A)) = I(A)/(1+0.001 \cdot CD(A))^{0.5}$ and $L(\lambda) = \lambda$, the results of 10 experiments separately are stated in Table 3.

**Table 1.** Computation results of Test 1

| | $(\alpha, \beta)$ | Optimization solutions | $Y_1$ | $Y_2$ | C.D. | C.T. | C |
|---|---|---|---|---|---|---|---|
| 1 | (0.5,0.1) | $x_1$=(4.7628,5.0000,5.3213)<br>$x_2$=(10.7847,10.9785,11.2183) | 224.5967 | 137.9064 | 9.9930 | 21.5160 | 21 |
| 2 | (0.5,1) | $x_1$=(4.9370,5.0000,5.1036)<br>$x_2$=(10.8373,11.0000,11.0212) | 224.2314 | 49.4265 | 9.2659 | 18.8130 | 22 |
| 3 | (2,0.1) | $x_1$=(4.5064,4.9756,5.0580)<br>$x_2$=(8.4635,8.7527,9.0867) | 201.4342 | 42.0757 | 7.8990 | 18.6250 | 21 |
| 4 | (2,1) | $x_1$=(1.8102,2.2385,2.7164)<br>$x_2$=(3.2860,3.3118,3.6266) | 92.2965 | 2.4456 | 3.2300 | 20.7970 | 19 |

**Table 2.** Computation results of Test 2

| | $L(\lambda)$ | Optimization solutions | $Y_1$ | $Y_2$ | C.D. | C.T. | C |
|---|---|---|---|---|---|---|---|
| 1 | $\lambda$ | $x_1$=(4.9150,5.0000,5.3484)<br>$x_2$=(10.5883, 11.0000, 11.1770) | 224.8013 | 213.1663 | 9.9593 | 20.5630 | 14 |
| 2 | $\lambda^2$ | $x_1$=(4.5137,4.9853,5.0258)<br>$x_2$=(10.6904, 11.0000, 11.4684) | 224.6494 | 217.2436 | 6.7096 | 21.8130 | 21 |
| 3 | $\lambda^{0.5}$ | $x_1$=(4.8342, 5.0000, 5.2254)<br>$x_2$=(10.9886, 11.0000, 11.2472) | 224.4106 | 210.2351 | 12.1189 | 22.9060 | 20 |

**Table 3.** Computation results of Test 3

| | Optimization solutions | $Y_1$ | $Y_2$ | C.D. | C.T. | C |
|---|---|---|---|---|---|---|
| 1 | $x_1$=(4.5100,4.9951,5.1611)<br>$x_2$=(10.895,511.0000,11.2828) | 224.1051 | 222.0662 | 10.0664 | 21.5470 | 14 |
| 2 | $x_1$=(4.8933,5.0000,5.0607)<br>$x_2$=(10.9749,11.0000,11.2311) | 224.5483 | 222.2854 | 9.3827 | 19.8750 | 13 |
| 3 | $x_1$=(4.8844,5.0000,5.3914)<br>$x_2$=(10.6385,11.0000,11.1725) | 224.8849 | 222.4760 | 10.0166 | 17.3590 | 17 |
| 4 | $x_1$=(4.8038,5.0000,5.1821)<br>$x_2$=(10.7043,11.0000,11.4831) | 224.8755 | 221.9763 | 10.0919 | 24.6250 | 16 |
| 5 | $x_1$=(4.9157,5.0000,5.2617)<br>$x_2$=(10.7470,11.0000,11.1360) | 224.6761 | 222.4789 | 9.6900 | 24.0320 | 18 |
| 6 | $x_1$=(4.5845,5.0000,5.4958)<br>$x_2$=(10.6656,11.0000,11.0513) | 224.4286 | 222.1919 | 10.3633 | 24.3900 | 16 |
| 7 | $x_1$=(4.8672,4.9902,5.3289)<br>$x_2$=(10.8135,11.0000,11.1829) | 224.7948 | 222.2969 | 9.8394 | 25.3750 | 19 |
| 8 | $x_1$=(4.7202,4.9951,5.4219)<br>$x_2$=(10.7028,11.0000,11.0075) | 224.4004 | 222.1757 | 10.0315 | 23.2500 | 19 |
| 9 | $x_1$=(4.8632,4.9951,5.2407)<br>$x_2$=(10.6387,11.0000,11.0100) | 224.1772 | 222.1572 | 9.6592 | 24.9530 | 18 |
| 10 | $x_1$=(4.8818,5.0000,5.2377)<br>$x_2$=(10.7529,11.0000,11.1897) | 224.6644 | 222.3695 | 9.7466 | 25.9070 | 16 |
| A.V. | $x_1$=(4.7924,4.9976,5.2782)<br>$x_2$=(10.7534,11.0000,11.1747) | 224.5555 | 222.2474 | 9.8888 | 23.1313 | 16.6 |

In the Table 1~3, $Y_1$ denotes the centralized quantification value of the maximum value, $Y_2$ the synthesizing effect value of the maximum value, C.D. the concentration degree, C. the convergence generation, C.T. the computation times, and A.V. the average value

All the calculations above are based on Matlab 6.5 and 2.00 GHz Pentium 4 processor and worked out under WindowsXP Professional Edition platform.

From the results above we can see that: ① Computational results are related to the level effect function and synthesizing effect function, and the difference is obvious (for case 1 and case 4 in Test 1), which shows BPO&ID-FGA can effectively merge decision consciousness into decision process; ② Despite of variation of parameters, the convergence time is about 20 seconds, and the convergence generation is about 20, also, the rate of getting the optimal result is almost more than 80% , which shows the algorithm have higher computational efficiency and good convergence performance; ③ Though the computational complexity is a bit larger than that of conventional algorithms, the difference is not great under high-performance parallel computing environment, so BPO&ID-FGA has good practicability; ④ BPO&ID-FGA, with the feature of good interpretability and strong operability, have good structure.

Synthesizing the computation results above and the theoretical analysis of section 7, we can see that BPO&ID-FGA is of stronger robust and good convergence, and suitable for the optimization problems under uncertain environment.

## 9 Conclusion

In this paper, on the basis of distinguishing principal indexes and assistant indexes and the restriction and supplementation relation between them, give comparison method of fuzzy information on synthesizing effect and description method of fuzzy information on principal indexes; using the structure characteristic of fuzzy information and essence of constraint, propose axiom system of fuzzy inequity degree, and give an instructive metric method; a new kind of fuzzy genetic algorithm based on the principal operation and inequity degree for the general optimization problems with fuzzy coefficients, fuzzy variables and fuzzy constraint is proposed(denoted by BPO&ID-FGA, for short); consider its convergence using Markov chain theory and analyze its performance through simulation, which indicate that this kind of algorithm not only merge decision consciousness effectively into optimization process, but posses many interesting advantages such as strong robust, faster convergence, less iterations and less chance trapping into premature states, so it can be applied to many fuzzy fields such as artificial intelligence, manufacture management and optimization control etc.

**Acknowledgements.** This work is supported by the National Natural Science Foundation of China (70671034) and the Natural Science Foundation of Hebei Province (F2006000346) and the Ph. D. Foundation of Hebei Province (05547004D-2, B2004509).

## References


1. Tang, J.F., Wang, D.W.: Fuzzy Optimization Theory and Methodology Survey. Control Theory and Application 17 (2000) 159–164
2. Cadenas, J.M., Verdegay, J.L.: Using Ranking Functions in Multiobjective Fuzzy Linear Programming. Fuzzy Sets and Systems 111 (2000) 47–531
3. Maleki, H.R., Tala, M., Mashinchi, M.: Linear Programming with Fuzzy Variables. Fuzzy Sets and Systems 109 (2000) 21–33
4. Tanaka, H.: Fuzzy Data Analysis by Possibillistic Linear Models. Fuzzy Sets and Systems 24 (1987) 363–375
5. Kuwano, H.: On the Fuzzy Multi-objective Linear Programming Problem: Goal Programming Approach. Fuzzy Sets and Systems 82 (1996) 57–64
6. Leu, S.S., Chen, A.T., Yang, C.H.: A GA-Based Fuzzy Optimal Model For Construction Time-Cost Trade-Off. International Journal of Project Management 19 (2001) 47–58
7. Tang, J.F., Wang, D.W., Fung, R.Y.K.: Modeling and Method Based on GA For Nonlinear Programming Problems With Fuzzy Objective and Resources. International Journal of System Science 29 (1998) 907–913
8. Buckley, J.J., Feuring, T.: Evolutionary Algorithm Solution to Fuzzy Problems: Fuzzy Linear Programming. Fuzzy Sets and Systems 109 (2000) 35–53
9. Zhang, K.L., Hirota, K.: On Fuzzy Number-Lattice $(\tilde{R}, \leq)$. Fuzzy Sets and Systems 92 (1997) 113–122
10. Liu, M., Li, F.C., Wu, C.: The Order Structure of Fuzzy Numbers Based on The Level Characteristic and Its Application in Optimization Problems. Science in China (Series F) 45 (2002) 433–441
11. Kim, K., Park, K.S.: Ranking Fuzzy Numbers with Index of Optimism. Fuzzy Sets Systems 35 (1990) 143–150
12. Wang, H.L.-K., Lee, J.-H.: A Method for Ranking Fuzzy Numbers and Its Application to Decision- Making. IEEE Transactions on Fuzzy Systems 7 (1999) 677–685
13. Tseng, T.Y., Klein, C.M.: New Algorithm for the Ranking Procedure in Fuzzy Decision Making. IEEE Trans. Syst. Man and Cybernetics 19 (1989) 1289–1296
14. Yager, R.R.: Procedure for Ordering Fuzzy Subsets of the Unit Interval. Information Science 24 (1981) 141–161
15. Cheng, C.H.: A New Approach for Ranking Fuzzy Numbers by Distance Method. Fuzzy Sets and Systems 95 (1998) 307–317
16. Diamond, P., Kloeden, P.: Metric Space of Fuzzy Set: Theory and Applications. Singapore: Word Scientific (1994)
17. Li, F.C., Yue, P.X., Su, L.Q.: Research on the Convergence of Fuzzy Genetic Algorithms Based on Rough Classification. Proceedings of the Second International Conference on Natural Computation and the Third International Conference on Fuzzy Systems and Knowledge Discovery (2006) 792–795
18. Ishbuchi, H., Tanaka, H.: Formulation and Analysis of Linear Programming Problem with Interval Coefficients. Journal of Japan Industrial Management Association 40 (1989) 320–329
19. Li, F.C., Liu, M., Wu, C.: Fuzzy Optimization Problems Based on Inequality Degree. IEEE International Conference on Machine Learning and Cybernetics, Vol. 3. Beijing (2002) 1566–1570
20. Holland, J.H.: Genetic Algorithms and the Optimal Allocations of Trials. SIAMJ of Computing 2 (1973) 8–105

# Immunity-Based Adaptive Genetic Algorithm for Multi-robot Cooperative Exploration

Xin Ma[1,2], Qin Zhang[1], Weidong Chen[2], and Yibin Li[1]

[1] School of Control Science and Engineering, Shandong University, 73 Jingshi Road, Jinan, 250061, China
[2] School of Electronic, Information and Electrical Engineering, Shanghai Jiao Tong University, 800 Dongchuan Road, Shanghai, 200240, China
maxin@sdu.edu.cn

**Abstract.** The key to multi-robot exploration is how to select appropriate targets for robots to avoid collision and overlap. However, the distribution of targets for multiple robots is an NP hard problem. This paper presents a multi-robot cooperative exploration strategy based on the immune genetic algorithm. With its random global searching and parallel processing, genetic algorithm is applied for multi-robots multiple targets combinatorial distribution. With its antibody diversity maintaining mechanism, the immune algorithm is used to get over the premature convergence of genetic algorithm. The selection probability is computed based on the similarity vector distance to guarantee the antibody's diversity. The crossover and mutation probability are adjusted based on the fitness of antibody to decrease the possibility of local optimal. The extensive simulations demonstrate that the immunity-based adaptive genetic algorithm can effectively distribute the targets to multiple robots in various environments. The multiple robots can explore the unknown environment quickly.

**Keywords:** Exploration, Genetic algorithm, Immunity, Multi-robot.

## 1 Introduction

With the development of the robotics, mobile robots have been applied from known structural environment to unknown dynamic unstructured environment. In order to accomplish some intelligent tasks in unknown dynamic environment effectively, the robots need to explore the unknown environment. It is a fundamental problem in mobile robotics.

Obviously, there are many advantages for exploration with multiple robots compared to a single robot. Multiple robots can explore environment faster and more tolerant [1]. However the premise for realizing the advantages is a good exploration strategy which is much difficult for coordinating multiple robots to maximize the utility of the whole system and acquire the information of the environment effectively.

The exploration strategy had been only simple and passive walk-following or random wandering before Yamauchi presented frontier-based exploration method [2].

D.-S. Huang, L. Heutte, and M. Loog (Eds.): ICIC 2007, LNAI 4682, pp. 605–616, 2007.

Frontier was defined as a boundary between the open area and the unknown area in grid maps. With searching for new frontiers, the robot could explore the unknown environment actively and effectively. Frontier-based exploration method was extended to multiple robots in [3]. The multiple robots shared information with each other and explored independently, which made the system not efficient due to the absence of the coordination. More than one robot may explore a same frontier which caused collisions. The key for the effective coordination of multi-robot exploration is how to assign the frontiers to the multiple robots. It was shown that the optimal allocation is an NP-hard problem, even in known environments [5].

Many researchers have recently investigated market-based approaches, in particular, auctions, to solve the coordination of the multiple robots. With auction algorithm, robots were regarded as bidders, while frontiers were regarded as goods. A central executive integrated the local maps to create a consistent global map, received the 'bids' of each local robot and made global decisions to assign the frontiers to each robot based on their bids and try to maximize overall utility. A single-item auction method was applied to assign the frontiers to robots [4], [5]. However, single-item auctions can result in highly suboptimal allocations if there are strong synergies between the items for the bidders. Combinatorial auction was used for multi-robot coordinated exploration to remedy the disadvantages of single-item auctions by allowing bidders to bid on bundles of items [6]. In theory, the method could produce the optimal solution and improve the exploration efficiency largely and collision could be avoided. Since the number of bundles increase exponentially with the number of frontiers, bid valuation, communication, and auction would become intractable. The method is infeasible for large number of frontiers. Moreover, the bidding strategies are still open problem. Generally, bids are computed based on utilities and costs. The cost of reaching the current frontier cell is proportional to the distance between the current position of robot and the frontier. The utility computation of frontier cells is more difficult. The actual new information that can be gathered by moving to the frontier cell is impossible to predict. Burgard et al. presented a technique that estimates the expected utility of a frontier call based on the distance and visibility to cells that are assigned to other robots [7]. The utility of a target location depends on the probability that this location is visible from target locations assigned to other robots. A decision theoretic approach is presented to explicitly coordinate multiple robots [7] by maximizing the overall utility and minimizing the potential for overlap in information gain amongst the multiple robots. The method simultaneously considers the utility of unexplored areas and the cost for reaching these areas. Coordination among the multiple robots is achieved in a very elegant way by balancing the utilities and the cost and further reducing the utilities according to the number of robots that ready to move toward this area. An iterative approach was used to determine appropriate target points for all robots. The complexity of the algorithm is $O(n^2T)$ where n is the number of robots and T is the number of frontier cells. The computation burden for distributing target cells to multiple robots will become very large if there are many frontiers in complex environment. The robots have to spend so much time to wait for receiving commands about their target cells. The multi-robot coordinated exploration can not be finished effectively. Market-based approach was improved with computing costs in the condition of connection without adding extra communication [8]. The exploration efficiency could be improved in the open or office environment with the method. However the improvement is limited for the complex environments.

Aiming at the problem, we applied genetic algorithm for distributing the frontier cells to multiple robots based on its characteristics of random global search and parallel processing [12]. On the basis of Burgard's work, the minus between the utility of some target for some robot and the cost for the robot reaching to the target is defined as the fitness function. Some possible assignments are randomly selected as initial population. A near-optimal assignment can be acquired after many generations through selection, crossover and mutation operations. The genetic algorithm-based multi-robot exploration strategy can reduce the computation time for distributing the targets to multiple robots. However, the selection, crossover and mutation operations are randomly carried out in the sense of probability. The traditional genetic algorithm has its disadvantage. Premature convergence can result in suboptimal solution. Moreover, the diversity of population decreases very quickly. The immune genetic algorithm combines the immunity principle with genetic algorithm to improve the performance of the algorithm. In this paper, the antibody's diversity maintaining mechanism of artificial immunity algorithm is applied into the genetic algorithm to get over the premature convergence. The antibody's diversity is guaranteed with the selection probability computed on the basis of the distance of similarity vector. On the basis of the immunity-based genetic algorithm, the crossover and mutation probability can be adjusted adaptively based on the fitness of antibody to decrease the possibility of local optimal. The extensive simulation experiments demonstrate that the immunity-based adaptive genetic algorithm can improve the exploration efficiency of multi-robot system.

The article is organized as follows: Section 2 gives a brief description of the immunity-based genetic algorithm. Section 3 gives a detail presentation of the immunity-based adaptive genetic algorithm for distributing multiple targets to multiple robots. Section 4 presents extensive simulation experiments and result analysis. Section 5 provides conclusions and future work.

## 2 The Immune Genetic Algorithm

### 2.1 Genetic Algorithm

Genetic algorithm is a random global search and optimization method which is developed from imitating the biologic genetic mechanism in nature. The parametric encoding character strings are operated by reproduction, crossover and mutation genetic operations. Each character string is corresponding to a possible solution. The genetic operations are carried out for many possible solutions. There are several advantages: the parallel searching is carried out in objective function space with the colony manner, the information can be exchanged between the possible solutions and some new possible solutions can be produced with crossover and mutation operations, and the individual is evaluated only by the fitness function. The direction of searching is guided with the variance rule of probability, which guarantees the robustness of the searching. However, the traditional genetic algorithm has some disadvantages. The single encoding can not represent the constraints of some optimization problems, the solution is apt to premature convergence, and the searching process may be sluggish in the end due to the individual diversity decreases quickly.

## 2.2 The Immunity Algorithm

The immunity algorithm is developed from the natural biologic immunity principle. The problem is corresponding to an antigen, and a solution to the problem is corresponding to an antibody. Many antibodies can be produced to resist various antigens in biological immunity systems. Thus many solutions can be guaranteed for solving problems. Moreover, the immunity algorithm has ability to maintain the immunity balance. The number of solutions can be adjusted adaptively with adjusting adaptively the number of antibodies suppressing and stimulating the antibodies.

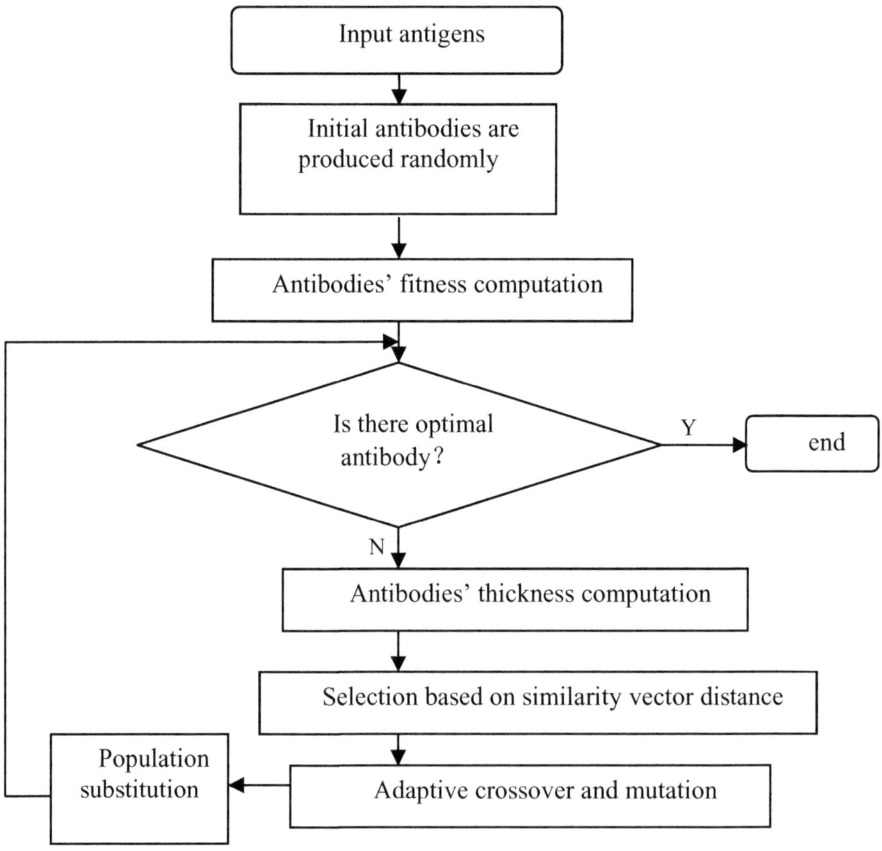


**Fig. 1.** The flow of the immune genetic algorithm

### 2.3 The Immune Genetic Algorithm

The immune genetic algorithm combines the natural biological immune system's self-adaptability and the ability to eliminate the antigen that invades into the body with the genetic algorithm. It introduces the characteristics of the immune system, that is, learning, memory, diversity and identification, to the genetic algorithm. In the context of some practical problems, the objective function and the constraints are treated as antigen inputs. Some initial antibody population is produced. Through reproduction, crossover and mutation operations and the computation of antibodies' similarity, some antibody corresponding to the antigen, that is, the solution to the problem can be found while maintaining the antibody diversity. For multi-robot exploration application, the antigen corresponds to the problem, that is, how to assign the multiple targets to multiple robots. The antibody corresponds to the possible targets-robots assignments. The antibodies similarity describes the similarity of loca between the two antibodies, that is, the similarity between some two target-robot assignments. The detail of the immune genetic algorithm is described as Fig.1.

## 3 Immunity-Based Adaptive Genetic Algorithm (IAGA) for Multi-target Multi-robot Assignment

In this section, the immunity-based adaptive genetic algorithm for multi-target multi-robot assignment is presented in detail.

### 3.1 Chromosome Encoding and the Initial Population

The chromosome is encoded with the decimal codes. Each chromosome corresponds to a target-robot assignment. The value of each locus is the robot's number that is assigned to the corresponding the target. The length of the chromosome is equal to the number of the targets. The initial population is produced randomly with forty assignments.

### 3.2 The Fitness Function

The genetic algorithm carries out its evolution by searching individual's fitness of the population. In the context of multi-robot exploration, the fitness function is defined as the objective function for optimization. The input of antigen is:

$$fitness = utility - \gamma \cdot \cos t \,. \tag{1}$$

where utility represents the possible new information if the robot reaches to the target, cost represents the cost for the robot to reach for the target. $\gamma$ weighs the relative importance of utility to cost. The experiments showed that the exploring time almost was similar for $\gamma \in [0.01, 50]$. Moreover, if $\gamma$ is too large or near to zero, the coordination between robots will be weakened and the exploring time will increase [7]. In our experiments, $\gamma = 0.1$.

### 3.3 The Three Operations

**Selection Probability Based on Similarity Vector Distance**

In general genetic algorithm, the selection probability usually is proportional to the fitness of the individual in the population. The number of individual with similar fitness increases quickly, which results in the local optimal. In order to get over the problem, we define the selection probability based on similarity vector distance by taking the similarity between the antibodies' encoding into account.

The antibodies' similarity is defined as the Euclidean distance of their encoding. The Euclidean distance between the antibody $a_1, a_2, \cdots, a_n$ and the antibody $b_1, b_2, \cdots, b_n$ is:

$$d = \sqrt{\sum_{1 \le i \le n} (a_i - b_i)^2} \ . \tag{2}$$

The larger $d$, the less similar the two antibodies. The concentration of the antibody $i$ is defined as:

$$C_i = \frac{\textit{the number of antibodies which similarity with } i \textit{ is less than } \lambda}{N} \ . \tag{3}$$

where, $N$ is the number of the antibody population. $\lambda$ is the defined threshold.

The selection probability based on the similarity vector distance is [9]:

$$P_s(x_i) = \alpha \frac{\rho(x_i)}{\sum_{i=1}^{N} \rho(x_i)} + (1-\alpha)\frac{1}{N} e^{-\frac{C_i}{\beta}} \ . \tag{4}$$

where, $\alpha$ and $\beta$ are constant adjusting factor. $0 \le \alpha \le 1$ , $0 \le \beta \le 1$ . $x_i$ is antibody. $f(x_i)$ is fitness function. $\rho(x_i) = \sum_{j=1}^{N} \left| f(x_i) - f(x_j) \right|$ is the vector distance of the antibody.

It can be seen that not only is the fitness of antibody related with selection probability, but also the antibody's similarity. To some extent, the similarity vector distance-based selection probability can maintain the antibody's diversity and get over the problem of local optimal solution.

**The Crossover and Mutation Operations**

The crossover operation can prevent the premature convergence to make the searching in solution space more robust. The mutation operation changes some loca of individuals of the population to improve the local searching ability of the genetic algorithm.

The crossover probability $P_c$ and the mutation probability $P_m$ are the keys that influence the behaviors and performance of genetic algorithm. They directly influence the convergence of the genetic algorithm. Too small or too large $P_c, P_m$ will go against the convergence. In our paper, we apply the adaptive genetic algorithm on the basis of the immunity-based genetic algorithm. $P_c, P_m$ changes with the fitness of antibody [10]. $P_c, P_m$ increases if all individuals of the population have similar fitness or in local optimal, decreases if all individuals of the population have dispersing fitness. Moreover, the individual which fitness is larger than the average fitness of the population has lower $P_c, P_m$ to protect them into the next generation. The individual which fitness is less than the average fitness corresponds to larger $P_c, P_m$. Thus they will be eliminated in the next generation.

$$P_c = \begin{cases} P_{c1} - \dfrac{(P_{c1} - P_{c2})(f' - f_{avg})}{f_{\max} - f_{avg}} & f' \geq f_{avg} \\ P_{c1} & f' < f_{avg} \end{cases} . \tag{5}$$

$$P_m = \begin{cases} P_{m1} - \dfrac{(P_{m1} - P_{m2})(f - f_{avg})}{f_{\max} - f_{avg}} & f \geq f_{avg} \\ P_{m1} & f < f_{avg} \end{cases} . \tag{6}$$

where, $f_{\max}, f_{avg}$ are the maximum and average fitness of the population respectively. $f'$ is the larger fitness of the two antibodies for crossover operation. $f$ is the fitness of the individual for mutation operation. $P_{c1}, P_{m1}$ are the largest crossover and mutation probability respectively defined in advanced. $P_{c2}, P_{m2}$ are the lowest crossover and mutation probability respectively defined for the individual which has the largest fitness value. $P_{c1} = 0.9$ , $P_{c2} = 0.06$ , $P_{m1} = 0.1$ , $P_{m2} = 0.001$ . Thus the crossover and mutation probability $P_c$ and $P_m$ can be adjusted adaptively to decrease the possibility of getting into the local optimal.

### 3.4 The Immunity-Based Adaptive Genetic Algorithm for Multi-robot Exploration

The above immunity-based adaptive genetic algorithm is applied to assigning multiple targets to multiple robots for exploring unknown environment effectively. The basis idea of exploration is "frontier cells", that is, the targets for robots getting new information in the near future [3]. When the robots find the frontiers, the frontier cells will be assigned among the robots for cooperatively exploration. The detail description about the immunity-based adaptive genetic algorithm multi-target multi-robot assignment strategy is as follows.

1. Input the objective functions, which will be discussed in the next section, as antigens, and initialize the population, the number of evolutionary generation, crossover and mutation probability.

2. Produce the initial antibodies.

Identify the antigens. Extract the minimum value of the optimized variables from the immune memory database. The initial parent antibodies are produced by adding some random variables on the minimum value. Then the maximum and average fitness $f_{\max}$, $f_{avg}$ are computed respectively. The optimal individual of the parent generation is marked.

3. The fitness of each antibody is evaluated. If there existing the individual fits the requirement in the current population, then end. Otherwise, go to the next step.

4. Select operation. Some individuals are selected on the basis of the similarity vector distance to get into the next generation according to Equation (4).

5. Crossover and mutation operation. The crossover and mutation probability $P_c, P_m$ are adjusted adaptively on the basis of the fitness of each antibody according to Equation (5) and (6).

6. Update the population and return to step 3.

## 4 The Simulation Experiments and the Result Analysis

Extensive simulation experiments are done with MATLAB. The environment is represented with occupancy grid map. Each grid has a value, which represents the posterior occupied probability. In simulated environment, each robot scans its surrounding with simulated sonar model. After scanning, robots will find several frontier cells, which are targets be assigned among the multiple robots. The detail flow for the multi-robot exploration strategy is described as follows:

1. A set of target, frontier cells, are obtained after scanning.

2. Compute the cost $V_t^i$ for each robot $i$ reaching to each target $t$.

3. Compute the utility $U_t$ of each target $t$, taking the influence of the assigned targets into account.

4. Define the objective function $U_t - \beta V_t^i$ as the fitness function. Randomly select some possible assignments as initial population.

5. According to the immunity-based adaptive genetic algorithm described in the above section, some optimal assignment can be acquired after several generations.

6. Each robot will go to the assigned targets.

7. At the new positions, all robots scan the environment. Then further exploration begins.

Three kinds of virtual environments are shown in Fig. 2.

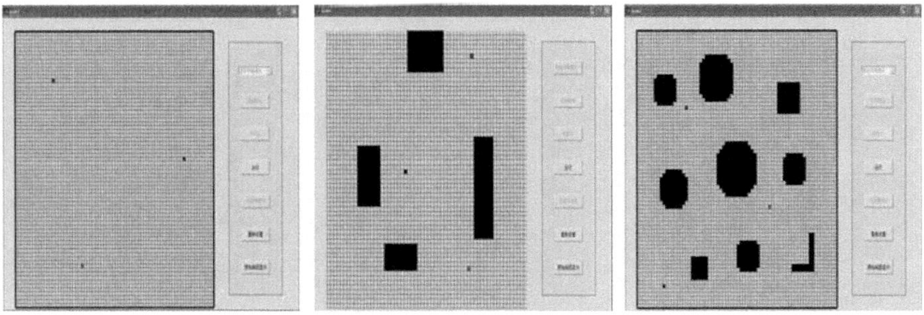

(a) Open environment (b) Office environment (c) Complex environment

**Fig. 2.** Three kinds of virtual environments

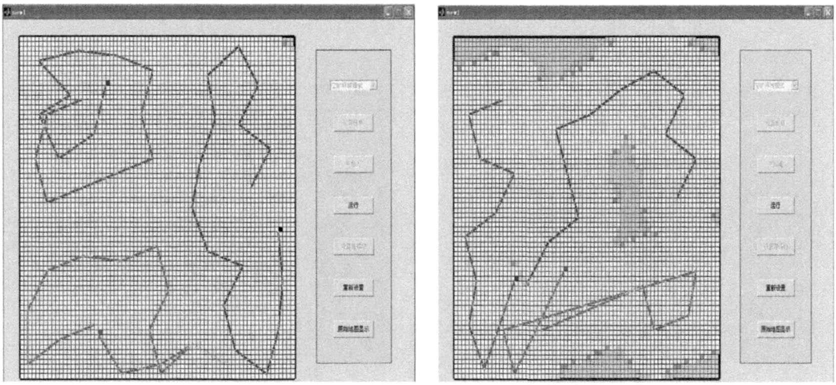

(a) IAGA (Immunity-based Adaptive Genetic Algorithm) (b) Burgard's approach

**Fig. 3.** Multi-robot coordinated exploration in open environment

In order to make the system simple and easy to be realized and compared with the method in [7], we define there are three robots in the system noted with yellow, blue and red color respectively. We assume that the robots mount with sonar. The purpose of the simulator is to study the multi-robot exploration strategy.

At the beginning, we define the robots' initial locations and the environment. Assume that the locations of all robots and their information about environment can be known by each other during their exploration. The three robots apperceive their surrounding environment with sonar. The simulated sonar data can be acquired with the sonar model [11]. The simulated sonar data is fused by the Dempster-Shafter evidential method. The local map is obtained. Then the frontiers are acquired. The information about the frontiers includes their sizes and locations. The three robots cooperatively explore the environment with the immunity-based adaptive genetic algorithm, which is described in the above section, and the approach proposed by Burgard in [7] respectively. Extensive simulation experiments have been done with different initial positions of robots. The results are shown in Fig.3, 4 and 5.

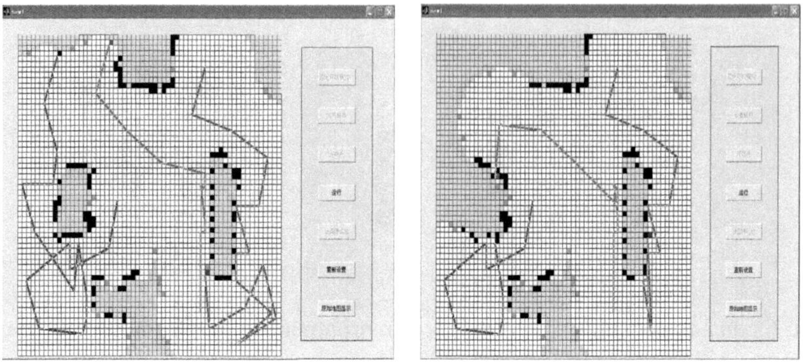

(a) IAGA (Immunity-based Adaptive Genetic Algorithm) (b) Burgard's approach

**Fig. 4.** Multi-robot coordinated exploration in office environment

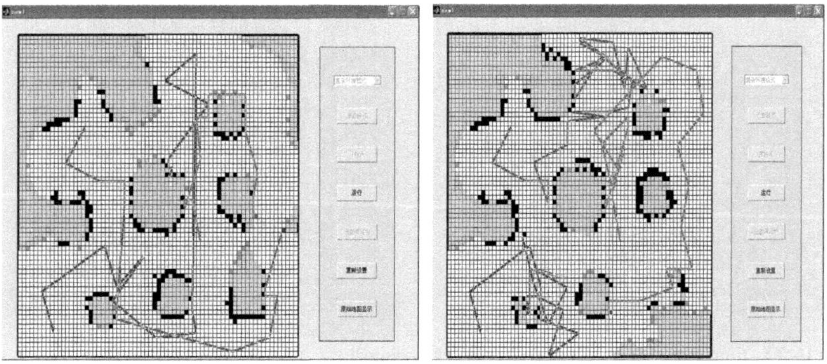

(a) IAGA (Immunity-based Adaptive Genetic Algorithm) (b) Burgard's approach

**Fig. 5.** Multi-robot coordinated exploration in complex environment

The immunity-based genetic algorithm is applied for distributing frontier cells to multiple robots. We focus our attention on the improvement on the time spent on distributing frontier cells. The results are shown in Table 1.

**Table 1.** Comparison of time spent for distributing targets

| Environment | IAGA | Burgard's method |
|---|---|---|
| Open | 3.2s | 12.1s |
| Office | 4.1s | 53.4s |
| Complex | 5.9s | 28.3s |

From the results shown in Fig. 3-5, we can see that the immunity-based adaptive genetic algorithm could distribute frontier cells to multiple robots effectively. The path length for exploring the whole environment reduced obviously. And the useless repeated exploring in the corner area is avoided. Combining the random global search and parallelism characteristics of the genetic algorithm with the antibody diversity mechanism of the immune system, the immunity-based adaptive genetic algorithm is more effective than Burgards' approach in [7]. Table 1 shows that the time spent for distributing frontier cells during multi-robot cooperative exploration is largely reduced.

## 5 Conclusion

In this paper, we present an immunity-based adaptive genetic algorithm for assigning multiple targets among multiple robots for effective multi-robot cooperative exploration. Combining the random global search and parallelism characteristics of the genetic algorithm with the antibody diversity mechanism of the immune system, the immunity-based adaptive genetic algorithm is more effective than Burgards' approach in [7]. The selection probability based on the similarity vector distance and the crossover and mutation probability adjusted adaptively improved the antibody diversity to guarantee the global optimal assignment solution furthermore. From the simulation results, it can be found that the algorithm is feasible and the computation time required for distributing frontier cells to multiple robots is reduced. The multi-robot coordinated exploration can be finished very effectively, especially for the situation that there are many robots exploring unknown complex environment.

**Acknowledgments.** This work was supported in part by the CHINA Ministry of Education under Grant 20060400649 Postdoctoral Research Award, Shandong Provincial Department of Science and Technology under Grant 2006GG3204018 and Shandong Provincial Information Development Plan under Grant 2006R00048.

## References

1. Zlot, R. , Stentz, A., Dias, M. B., Thayer, S.: Multi-robot Exploration Controlled by a Market Economy. In Proceedings of the 2002 IEEE International Conference on Robotics & Automation, Washington DC, (2002) 3016-3023
2. Yamauchi, B.: A Frontier-based Approach for Autonomous Exploration. In Proceedings of the 1997 IEEE International Symposium on Computational Intelligence in Robotics and Automation, Monterey, CA, (1997) 146-151
3. Yamauchi, B.: Frontier-Based Exploration Using Multiple Robots. In Proceedings of Second International Conference on Autonomous Agents, Minneapolis MN, (1998) 47-53
4. Lagoudakis, M. G., Berhault, M., Koenig S., Keskinocak, P., Kleywegt, A. J.: Simple Auctions with Performance Guarantees for Multi-robot Task Allocation. In Proceedings of 2004 IEEE/RSJ International Conference on Intelligent Robots and Systems, (2004) 698-705
5. Simmons, R., Apfelbaum, D., Burgard, W., Fox, D., Thrun, S., Younes, H.: Coordination for Multi-Robot Exploration and Mapping. In Proceedings of the National Conference on Artificial Intelligence, AAAI, (2000) 852-858
6. Berhault, M. , Huang, H., Keskinocak, P., Koenig, S., Elmaghraby, W., Griffin P., Kleywegt, A.: Robot Exploration with Combinatorial Auctions. Conference on Intelligent Robots and Systems, (2003)1957-1962

7. Burgard, W., Moors, M., Schneider, F.: Coordinated Multi-robot Exploration. IEEE Transactions on Robotics, 21(3) (2005) 376-378
8. Zhang, F., Chen, W. D., Xi, Y.: Improving Collaboration through Fusion of Bid Information for Market-based Multi-robot Exploration. In Proceedings of the IEEE International Conference on Robotics and Automation, Barcelona, Spain, (2005) 1157-1162
9. Zheng, R., Mao, Z.Y., Luo, X. X.: Artificial Immune Algorithm Based on Euclidean Distance and King-crossover. Control and Decision, 20(2) (2005)161-164
10. Srinivas, M., Patnaik, L. M.: Adaptive Probabilities of Crossover and Mutation in Genetic Algorithms. IEEE Trans on Systems, Man and Cybernetics, 24( 4) (1994) 656-667
11. Ma, X., Liu, W., Li, Y. B., Song, Rui.: LVQ Neural Network Based Target Differentiation Method for Mobile Robot. In Proceedings of IEEE 12th International Conference on Advanced Robotics, Seattle, USA, (2005) 680-685
12. Ma, X., Zhang, Q., Li, Y. B.: Genetic Algorithm-based Multi-robot Cooperative Exploration. In Proceedings of IEEE International Conference on Control and Automation, Guangzhou, CHINA, (2007) 1018-1023

# Improved Genetic Algorithms to Fuzzy Bimatrix Game

RuiJiang Wang[1], Jia Jiang[1], and XiaoXia Zhu[2]

[1] College of Economics and Management, Hebei University of Science and Technology, Shijiazhuang, 050018, China
[2] College of Science, Hebei University of Science and Technology, Shijiazhuang, 050018, China
ruijiang_wang@126.com

**Abstract.** According to the features of fuzzy information, we put forward the concept of level effect function $L(\lambda)$, established a very practical and workable measurement method $I_L -$ which can quantify the location of fuzzy number intensively and globally, and set up the level of uncertainty for measurement $I_L -$ under the level effect function $L(\lambda)$. Thus we can improve the fuzzy bimatrix game. For this problem, after establishing the model involving fuzzy variable and fuzzy coefficient for each player, we introduced the theory of modern biological gene into equilibrium solution calculation of game, then designed the genetic algorithm model for solving Nash equilibrium solution of fuzzy bimatrix game and proved the validity of the algorithm by the examples of bimatrix game. It will lay a theoretical foundation for uncertain game under some consciousness and have strong maneuverability.

**Keywords:** bimatrix game, fuzzy, level effect function, IL-metric, LU-level of uncertainty, genetic algorithm, Nash equilibrium solution.

## 1 Introduction

In recent years, game theory has been attached more and more importance in the economic field. With building the game model, people have studied prisoners' dilemma, oligarch competition and evolvement of biological species, etc. Nash proved the existence of game equilibrium solution, but he didn't develop general algorithm for solving Nash equilibrium. At present, there are many algorithms for solving Nash equilibrium, such as geometric algorithm, Lemke-Howson algorithm, and emulation algorithm, etc. [1-4] but each method has its limitations. For geometric algorithm, it is intuitionist and concise, while it is unworkable when game matrix above three orders. For Lemke-Howson algorithm, it can convert solving equilibrium problem into linear programming problem involving multiple steps, but it is very hard to get the result. For emulation algorithm, it applies computer to develop emulation by simulating the biological evolvement, which stands for a new way of calculation. [5-6] while using the above three algorithms, we often have some difficulties in solving

D.-S. Huang, L. Heutte, and M. Loog (Eds.): ICIC 2007, LNAI 4682, pp. 617–628, 2007.

game problems. The following are the main reasons: First, there are many kinds of game problems. The different features and forms of equilibrium solution for each game problem cause some difficulties in solving. Second, the complexity of discussing the existence of solution for a game problem limits the application of many solving methods. Third, there may be several equilibrium solutions for one game problem. We can select the equilibrium solution of the anticipating result only with ensuring the complete accessing in solving as well as comparing each solution, which is a very high demand for the algorithms.

Since J.P.Aubin firstly studied fuzzy game in 1974, the research of fuzzy game has developed very quickly. For fuzzy two-person game problem, two-person zero-sum fuzzy game was studied in bibliography [7]. The basic thoughts are the following: With treating game value as a clear variable, solution was carried out by linear programming method involving fuzzy coefficient. When considering that the game cannot ensure determinate level game value, we blurred the constraint condition and then reflect the comparing relation among fuzzy numbers based on number features (e.g. median point and mean) of fuzzy numbers. Then according to different fuzzy number features, we set up different assistant model of solving fuzzy matrix game problem.

Based on the above mentioned, this paper discussed the following aspects: a) According to the features of fuzzy information, the author put forward the concept of level effect function describing fuzzy information processing, set up the pool quantification method for fuzzy information with the significance of extensive guidance, suggested the uncertainty measurement model for the value of pool quantification, and discussed related operation property. b) The author established the solving model involving fuzzy variable and fuzzy coefficient of fuzzy matrix game with extensive maneuverability, and designed the concurrent selection genetic algorithm solving Nash equilibrium solution of fuzzy double-matrix game on the basis of measurement of level effect function and the gene theory. c) The author proved the workability of this method by referring to the data in the examples of bibliography [7].

## 2 Preliminaries

In the following, let $R$ be the real number field, $F(R)$ the family of all fuzzy sets over $R$ .For any $A \in F(R)$, the membership function of $A$ is written as $A(x)$, the $\lambda-$cuts of $A$ as $A_\lambda = \{x \mid A(x) \geq \lambda\}$, and the support set of A as $\mathrm{supp}A = \{x \mid A(x) > 0\}$. In what follows, we introduce the definition of fuzzy number and its basic operation properties.

**Definition 1** [4,8]**.** $A \in$ F(R) is called a fuzzy number if it satisfies the following conditions: 1) For any given $\lambda \in (0,1]$, $A_\lambda$ are closed intervals; 2) $A_1 = \{x \mid A(x) = 1\} \neq \phi$; 3) supp$A$ is bounded. The class of all fuzzy numbers is called fuzzy number space, which is denoted by $E^1$. In particular, if there exists $a, b, c \in$ R such that $A(x) = (x-a)/(b-a)$ for each $x \in [a, b)$, and $A(b)=1$, and $A(x) = (x-c)/(b-c)$ for each $x \in (b, c]$, and $A(x) = 0$ for each $x \in (-\infty, a) \bigcup (c, +\infty)$, then we say that A is a triangular fuzzy number, and written as $A = (a, b, c)$ for short.

For convenience sake, in what follows we denote the closure of supp$A$ by $A_0$. Clearly, $A\in E^1$ implies that $A_0$ is a closed interval.

For $A=(a,b,c)$, it is easy to obtain that $A_\lambda=[a+(b-a)\lambda,\ c-(c-b)\lambda]$ for each $\lambda\in(0,1]$ by directly verification.

Obviously, if we regard real number $a$ as a fuzzy set whose membership function is $a(x)=1$ for $x=a$ and $a(x)=0$ for each $x\neq a$, then fuzzy number can be thought as an extension of real number, so fuzzy numbers possess the properties of both numbers and sets, which is the widest description of fuzzy information in many practical domains.

In many applied fields, the algebraic operation of fuzzy numbers is the most basic operation and also the most common tool for dealing with optimization problems. The widely received operation method used now is established based on Zadeh's extension principle.

**Theorem 1** [6]**.** Let $A,B\in E^1$, $k\in\mathrm{R}$, $f(x,y)$ be a continuous binary function, and $A_\lambda=[\underline{a}(\lambda),\overline{a}(\lambda)]$, $B_\lambda=[\underline{b}(\lambda),\overline{b}(\lambda)]$ be the $\lambda-$cuts of $A$ and $B$, respectively. Then for $f(A,B)\in E^1$ and each $\lambda\in(0,1]$, $(f(A,B))_\lambda=f(A_\lambda,B_\lambda)$. Particularly, the following conclusions always hold:

1) $A+B=B+A$, $A\cdot B=B\cdot A$, $k(A\pm B)=kA\pm kB$;
2) $(A+B)_\lambda=A_\lambda+B_\lambda=[\underline{a}(\lambda)+\underline{b}(\lambda),\overline{a}(\lambda)+\overline{b}(\lambda)]$
   $(A-B)_\lambda=A_\lambda-B_\lambda=[\underline{a}(\lambda)-\overline{b}(\lambda),\overline{a}(\lambda)-\underline{b}(\lambda)]$;
3) $(A\times B)_\lambda=A_\lambda\times B_\lambda=[\underline{a}(\lambda)\times\underline{b}(\lambda),\ \overline{a}(\lambda)\times\overline{b}(\lambda)],\ \underline{a}(\lambda)\geq 0,\underline{b}(\lambda)\geq 0$;
4) $(A\div B)_\lambda=A_\lambda\div B_\lambda=[\underline{a}(\lambda)\div\overline{b}(\lambda),\ \overline{a}(\lambda)\div\underline{b}(\lambda)],\underline{a}(\lambda)\geq 0,\underline{b}(\lambda)>0$;
5) For $A=(a_1,b_1,c_1)$, $B=(a_2,b_2,c_2)$,
   $A+B=(a_1+a_2,b_1+b_2,c_1+c_2)$; $A-B=(a_1-c_2,b_1-b_2,c_1-a_2)$;
6) For $A=(a_1,b_1,c_1)$, if $k\geq 0$, then $kA=(ka_1,kb_1,kc_1)$, if $k<0$, then $kA=(kc_1,kb_1,ka_1)$.

Fuzzy numbers have many good analytical properties and perfect theory systems. We can see ref. [6] for the detailed contents.

## 3 IL-Metric for Fuzzy Number

### 3.1 Conception and Properties of $I_L$-Metric

The decomposition theorem of fuzzy sets provides us a basic method to realize and deal with fuzzy information, but in many real problems, we often depend on the global features of fuzzy information to make the decision. It is easy to see that the individual with different membership characteristics will play different role during the process of decision-making. To establish a general theoretical model for this problem, we introduce the concept of level effect function.

**Definition 2.** Say $L(\lambda):[0,1]\to[a,b]\subset[0,\infty)$ a level effect function, if $L(\lambda)$ is piecewise continuous and monotone non-decreasing. For $A\in E^1$, let $A_\lambda=[\underline{a}(\lambda),\overline{a}(\lambda)]$ be the $\lambda-$cuts of A, and $L^*=\int_0^1 L(\lambda)\mathrm{d}\lambda$. Then

$$I_L(A) = \frac{1}{2L^*}\int_0^1 L(\lambda)(\underline{a}(\lambda)+\overline{a}(\lambda))\mathrm{d}\lambda\,, \tag{1}$$

is called the $I_L$-metric of $A$, particularly, if $L^* = 0$, we define $I_L(A) = [\underline{a}(1)+\overline{a}(1)]/2$.

In Definition 2, if we interpret the level effect function as the description for the confidence degree of information with different levels, $A_\lambda$ as the intrinsic information of A and $L(\lambda)$ as a kind of decision parameter, then $I_L(A)$ is just a method centralized quantifying A. Obviously, by the IL-metric values of fuzzy numbers, we can establish an order relation on $E^1$, which is denoted as $(E^1, I_L)$.

**Definition 3.** Let $A, B \in E^1$. If $I_L(A) < I_L(B)$, then we say A is less than B with respect to the IL-metric, and written as $A < B$; If $I_L(A) = I_L(B)$, then we say A is equal to B with respect to the IL-metric, and written as $A = B$; If $I_L(A) \le I_L(B)$, then we say A is not more than B with respect to the IL-metric, and written as $A \le B$.

**Remark 1.** Order structure $(E^1, I_L)$ provides a kind of model for describing the sequence feature of fuzzy information, and has favorable interpretability and operability, moreover, it is very typical, and almost all existing ranking methods for fuzzy numbers can be seen as its special cases. For example, $(E^1, I_L)$ keeps the order relation $\le_1$ defined by level cuts of fuzzy numbers (here $A \le_1 B \Leftrightarrow A_\lambda \le B_\lambda$ for each $\lambda \in [0,1]$, and $[a,b] \le [c,d] \Leftrightarrow a \le c, b \le d$), that is $I_L(A) \le I_L(B)$) if $A \le_1 B$; When $L(\lambda) \equiv 1$, $(E^1, I_L)$ just coincides with the order relation proposed in ref. [5].

**Theorem 2.** Let $A, B \in E^1$, $k \in \mathrm{R}$. Then: 1) $I_L(A \pm B) = I_L(A) \pm I_L(B)$; 2) $I_L(kA) = kI_L(A)$.

**Proof.** Let $A_\lambda = [\underline{a}(\lambda), \overline{a}(\lambda)]$, $B_\lambda = [\underline{b}(\lambda), \overline{b}(\lambda)]$ be the $\lambda$-cuts of $A$, $B$ respectively. Using the properties [6] of fuzzy numbers, we have $(A+B)_\lambda = [\underline{a}(\lambda)+\underline{b}(\lambda), \overline{a}(\lambda)+\overline{b}(\lambda)]$ and $(A-B)_\lambda = [\underline{a}(\lambda)-\overline{b}(\lambda), \overline{a}(\lambda)-\underline{b}(\lambda)]$ for each $\lambda \in [0,1]$, $(kA)_\lambda = [k\underline{a}(\lambda), k\overline{a}(\lambda)]$ for each $\lambda \in [0,1]$ and all $k \ge 0$, and $(kA)_\lambda = [k\overline{a}(\lambda), k\underline{a}(\lambda)]$ for each $\lambda \in [0,1]$ and all $k < 0$. So, the following can be got from the above and the properties of Lebesgue integral:

$$\begin{aligned}
I_L(A+B) &= \frac{1}{2L^*}\int_0^1 L(\lambda)[\underline{a}(\lambda)+\underline{b}(\lambda)+\overline{a}(\lambda)+\overline{b}(\lambda)]\mathrm{d}\lambda \\
&= \frac{1}{2L^*}\int_0^1 L(\lambda)[\underline{a}(\lambda)+\overline{a}(\lambda)]\mathrm{d}\lambda + \frac{1}{2L^*}\int_0^1 L(\lambda)[\underline{b}(\lambda)+\overline{b}(\lambda)]\mathrm{d}\lambda = I_L(A)+I_L(B); \\
I_L(A-B) &= \frac{1}{2L^*}\int_0^1 L(\lambda)[\underline{a}(\lambda)-\overline{b}(\lambda)+\overline{a}(\lambda)-\underline{b}(\lambda)]\mathrm{d}\lambda \\
&= \frac{1}{2L^*}\int_0^1 L(\lambda)[\underline{a}(\lambda)+\overline{a}(\lambda)]\mathrm{d}\lambda - \frac{1}{2L^*}\int_0^1 L(\lambda)[\underline{b}(\lambda)+\overline{b}(\lambda)]\mathrm{d}\lambda = I_L(A)-I_L(B); \\
I_L(kA) &= \frac{1}{2L^*}\int_0^1 L(\lambda)[k\,\underline{a}(\lambda)+k\overline{a}(\lambda)]\mathrm{d}\lambda = \frac{k}{2L^*}\int_0^1 L(\lambda)[\underline{a}(\lambda)+\overline{a}(\lambda)]\mathrm{d}\lambda = kI_L(A).
\end{aligned}$$

### 3.2 LU-Level of Uncertainty on IL-Metric

For order structure $(E^1, I_L)$, when $I_L(A) = I_L(B)$, it is not adequate for us to do further comparison between fuzzy numbers A and B only with IL-metric. Due to the decision processing in practical problems, we not only consider the decision solution itself, but also the reliability degree. In order to abstract the quantity feature of fuzzy information more objectively, we introduce the concept of uncertain level on IL-metric.

**Definition 4.** Let $u:[0,\infty)\rightarrow[0,1]$, $u$ is called uncertainty basis function, if it satisfies the following conditions: 1) $u(0)=0$, $\lim_{x\rightarrow\infty}u(x)=1$; 2) $u(x)$ is monotone non-decreasing.

**Definition 5.** Let $A\in E^1$, $\theta\in(0,\infty)$, $A_\lambda=[\underline{a}(\lambda),\overline{a}(\lambda)]$ be the $\lambda-$cuts of $A$, $L(\lambda)$ be a level effect function, and $u$ be a uncertainty basis function. Denote

$$\delta=\int_0^1 L(\lambda)(\overline{a}(\lambda)-\underline{a}(\lambda))d\lambda\ , \tag{2}$$

then $LU(A)=u(\delta(A))$ is called the $LU$-uncertainty degree on $I_L$-metric based on $L(\lambda)$, for short, we call it $LU$- level of uncertainty of $A$.Let $A^{(i)}\in E^1$, $[\overline{a}_i(\lambda),\underline{a}_i(\lambda)]$ be the $\lambda-$cuts of $A^{(i)}$,

$$\delta_i=\delta(A^{(i)})=\int_0^1 L(\lambda)(\overline{a}_i(\lambda)-\underline{a}_i(\lambda))d\lambda\, , i=1,2,\cdots,n\ ,\quad , \tag{3}$$

then by using the properties of integral and fuzzy numbers, we can get $LU(A^{(1)}+A^{(2)}+\cdots+A^{(n)})=u(\delta_1+\delta_2+\cdots+\delta_n)$.

By the implication of integral, we know that $LU(A)$ is just the synthetic measurement for the uncertain feature of $A$ under the level decision consciousness $L(\lambda)$, a description of uncertainty of $A$. The smaller $LU(A)$ is, the smaller the uncertain level of $I_L(A)$; and the bigger $LU(A)$ is, the bigger the uncertain level of $I_L(A)$. During the process of dealing with fuzzy information, $I_L$-metric and $LU$-level of uncertainty can constraint and complement each other. Generally speaking, in considering maximal (or minimum) fuzzy optimization problems, decision-makers always hope that the compound quantification of objective function is as great (or small) as possible, the corresponding $LU$-level of uncertainty is as small as possible simultaneously, which is the basis for solution transition of fuzzy programming.

## 4 Bimatrix Games with Fuzzy Payoffs

In this section, we define a fuzzy expected payoff in a bimatrix game with fuzzy payoffs.

**Definition 6** [6,9,10]**.** Let $I=\{1,2,\cdots,m\}$ denote a set of pure strategies of Player I and $J=\{1,2,\cdots,n\}$ denote that of Player II. Mixed strategies of Players I and II are represented by probability distributions to pure strategies of them, i.e.,

$x=(x_1,x_2,\cdots,x_m)^T\in X=\{x\in\Re_+^m \mid \sum_{i=1}^m x_i=1\}$ is a mixed strategy of Player I, and

$y=(y_1,y_2,\cdots,y_n)^T\in Y=\{y\in\Re_+^n \mid \sum_{j=1}^n y_j=1\}$ is a mixed strategy of Player II,

where $\Re_+^m=\{a\in\Re^m \mid a_i\geq 0, i=1,2\cdots,m\}$ and where $x^T$ is the transposition of $x$.

Payoffs of Players I and II are $U_1(i,j)=\tilde{a}_{ij}$ and $U_2(i,j)=\tilde{b}_{ij}$, respectively when Player I chooses a pure strategy $i\in I$ and Player II chooses a pure strategy $j\in J$. Then a non-zero-sum two-person game in normal form is represented as a pair of $m\times n$ payoffs matrices

$$\tilde{A}=\begin{bmatrix}\tilde{a}_{11} & \cdots & \tilde{a}_{1n}\\ \vdots & \ddots & \vdots\\ \tilde{a}_{m1} & \cdots & \tilde{a}_{mn}\end{bmatrix},\ \tilde{B}=\begin{bmatrix}\tilde{b}_{11} & \cdots & \tilde{b}_{1n}\\ \vdots & \ddots & \vdots\\ \tilde{b}_{m1} & \cdots & \tilde{b}_{mn}\end{bmatrix}$$

The game is defined by $(\tilde{A},\tilde{B})$ and is also referred to as a fuzzy bimatrix game.

When Player I choose a mixed strategy $x\in X$ and Player II chooses a mixed strategy $y\in Y$, expected payoffs of Players I and II are

$$E_I=\sum_{i=1}^{m}\sum_{j=1}^{n}\tilde{a}_{ij}x_iy_j=X\tilde{A}Y^T,\quad E_{II}=\sum_{i=1}^{m}\sum_{j=1}^{n}\tilde{b}_{ij}x_iy_j=X\tilde{B}Y^T$$

respectively.

**Definition 7** [10, 11, 12]**.** For a fuzzy bimatrix game $(\tilde{A},\tilde{B})$, a Nash equilibrium solution is a pair of strategies *m*-dimensional column vector $x^*$ and *n*-dimensional column vector $y^*$ if, for any other mixed strategies *x* and *y*,

$$x^{*T}\tilde{A}y^*\ge x^T\tilde{A}y,\ x^{*T}\tilde{B}y^*\ge x^T\tilde{B}y$$

Where $(X^*\tilde{A}Y^{*T},X^*\tilde{B}Y^{*T})$ is defined Nash equilibrium value of the Fuzzy Bimatrix game.

**Lemma 1** [5,13]**.** A pair of strategies $(x^*,y^*)$ is an equilibrium solution to the aforementioned bimatix game with fuzzy goals, if and only if $(x^*,y^*)$ satisfies the following conditions: for players' payoffs of the fuzzy bimatrix game:

$$f_1=\sum_{j=1}^{n}\max\{\tilde{A}_iY^T-\tilde{A}_jY^T\mid 1\le i\le m\},\ f_2=\sum_{i=1}^{m}\max\{X\tilde{B}_i-X\tilde{B}_j\mid 1\le j\le n\}$$

respectively, such that function $f=f_1+f_2$ gets minimum.

Because the fuzzy numbers do not have the comparability like real numbers, so above model is just a formal model, and can't be directly used for solving operation, for that, we can convert fuzzy information into centralized numerical value; furthermore, solvable transformation can be realized.

Based on the above analysis, according to IL-metric and LU-level of uncertainty in section 3, under some decision consciousness, we can convert the model of fuzzy bimatrix game to the following nonlinear programming: A pair of strategies $(x^*,y^*)$ is a Nash equilibrium solution to the aforementioned bimatrix game with fuzzy goals, if and only if $(x^*,y^*)$ satisfies the following conditions:

$$f_1=\sum_{j=1}^{n}\max\{I_L(\tilde{A}_i)Y^T-I_L(\tilde{A}_j)Y^T\mid 1\le i\le m, LU(\tilde{A}_i)\le\varepsilon, LU(\tilde{A}_i)\le\varepsilon\},$$

$$f_2=\sum_{i=1}^{m}\max\{XI_L(\tilde{B}_i)-XI_L(\tilde{B}_j)\mid 1\le j\le n, LU(\tilde{B}_i)\le\eta, LU(\tilde{B}_j)\le\eta\}$$

respectively, such that function $f=f_1+f_2$ gets minimum.

# 5 Genetic Algorithms to Fuzzy Bimatrix Games

In order to introduce genetic algorithm into the calculation of Nash equilibrium solution, we first suppose such corresponding relation: comparing each mixture condition to an organism in the nature, and comparing mix strategy of each player to different chromosome of organism. Just like the character of organism is related to the genes of chromosome group, equilibrium solution will be the best mixture condition in the process of algorithm, thus the Nash equilibrium solution of games is obtained. Genetic algorithm is an effective method of solving the problem of combination optimization and intelligence optimization at present. In the process of genetic algorithm solving, with the start of an initial group, we seek the optimization solution or satisfaction solution of the problem from generation to generation until meets convergence or pre-established degree of iteration. The basic genetic operation includes selection, crossover and variation.

The key contents of genetic algorithm consists of parameter coding, initial group setting, function of adaptation level designing, genetic operator fixing, and controlling parameter selecting. The following are the specific implement strategy of $GA_{FBMG.}$

## 5.1 Coding

Combining the feature of game, in this paper, we use multidimensional chromosome multi-parameter mapping coding, that is, to the mix strategy $x_i$ of each player, each parameter $x_{ij}$ $(1 \le j \le m_i)$ is binary coded to obtain a substring, all which are integrated into a complete chromosome of $x_i$ . Then, the mix strategy coding of different players constitute $n$ chromosomes. Therefore, the whole mix situation corresponds to a binary $n$-dimensional chromosome. Suppose one has five pure strategies, and each is coded for 00000000-11111111, then the coding for mix strategy is forty bits. But mix strategy should satisfy $x_{ij} \ge 0$ , $\sum x_{ij} = 1$ , so the above coding exists a certain redundancy, it is necessary to employ normalization strategy to coding after real value transformation.

## 5.2 Fitness Function

We use fitness assignment based ranking, that is, sort objective function value in decreasing sequence; the individual with the smallest fitness value is placed the first position; the most optimal is placed position Nind(the size of population). Each fitness value is calculated according to the position $g$ , namely,

$$Fit(g) = \frac{Nind \times X^{g-1}}{\sum_{i=1}^{Nind} X_i}$$

## 5.3 Crossover Operator and Mutation Operator

The old population can generate new population with the crossover and mutation operations in the process of living beings evolution. In order to avoid generating unfeasible solutions, in the paper, we use the crossover operation MCUOX(Multicomponent uniform order-based crossover) in [14] and discrete mutation operation with probability $p_m$ . If $p_m$ is omission, then we suppose $p_m = 0.7 / Lind$ (Here, Lind is the length of chromosomes), the value of $p_m = 0.7 / Lind$ promises that the mutation probability of each individual can approach to 0.5. However, with the simulation of living beings evolution, make the chromosome with highest fitness or mixed situation be reserved, and then it can approach to the Nash Equilibrium. Considering the nature of hybrid strategy, we apply normalized treatment to the chromosome coding after the crossover and mutation operations.

### 5.4 Selection Operator

Selection operation is applied for operating individuals in the population according to the principle, that is, the individuals with higher fitness value possess bigger probability of surviving in next generation, and the individuals with lower fitness value possess smaller probability of surviving in next generation. Roulette wheel selection is used for genetic algorithm, it is just a proportional strategy based on fitness value, with the property that better individuals possess bigger survival probability in proportional way, and all the individuals in the population have the opportunity to be selected.

### 5.5 Forced Reserved Strategy

Forced reserved strategy is a reservation way, which assure that the optimal solution can be got as soon as possible. Its operation method is that taking the optimal individual and suboptimal individual as the result of generations in the evolutionary process. The operation procedures are as follows:

a) For two parent individuals $X_1$ and $X_2$, generating $X_1'$ and $X_2'$ through crossover;
b) For child individuals $X_1'$ and $X_2'$, generating $X_1''$ and $X_2''$ through mutation;
c) Compare the fitness value of parent individuals $X_1, X_2$ with child individuals $X_1'', X_2''$, reserve the two individuals with the biggest and the second biggest fitness value. For example, if $F(X_1) = 0.6$, $F(X_2) = 0.8$, $f(X_1'') = 0.5$, $f(X_2'') = 0.9$, then we take $X_2$ and $X_2''$ as the evolution results of $X_1$ and $X_2$.

## 6 Performance Analysis of GAFBMG

To analyze the performance of $GA_{FBMG}$ theoretically, we first give the definitions of Markov chain and the convergence of genetic algorithm.

**Definition 6** [9]. Let $X(n) = \{X_1(n), X_2(n), \cdots, X_N(n)\}$ be the $n^{th}$ population of genetic algorithm, $Z_n$ denote the optimal value in the population $X(n)$, that is $Z_n = \max\{f(X_i(n)) \mid i = 1, 2, \cdots, N\}$. If $\lim_{n\to\infty} P\{Z_n = f^*\} = 1$, then we say the genetic sequence $\{X(n)\}_{n=1}^{\infty}$ converges. Here, $f^* = \max\{f(X) \mid X \in S\}$ denotes the global optimal value of individuals.

**Definition 7** [10]. Let the random sequence $\{X(n)\}_{n=1}^{\infty}$, which can only take countable values $I = \{i_0, i_1, \cdots\}$, satisfy the conditions: for arbitrary natural number $n$ and $\{i_0, i_1, \cdots, i_n\} \subset I$, when $P\{X(0) = i_0, X(1) = i_1, \cdots, X(n) = i_n\} > 0$, we have

$$P\{X(n+1) = i_{n+1} \mid X(0) = i_0, X(1) = i_1, \cdots, X(n) = i_n\} = P\{X(n+1) = i_{n+1} \mid X(n) = i_n\},$$

then we say $\{X(n)\}_{n=1}^{\infty}$ is a Markov chain with discrete time and discrete states, and say Markov chain for short.

**Definition 8** [10]. For Markov chain $\{X(n)\}_{n=1}^{\infty}$, if the transition probability starting from state $i$ to state $j$

$$p_{ij}(t)=p\{X(t+1)=j|X(t)=i\}=p_{ij}\ (i,j\in I)$$

is irrelevant to initiation time $t$, then $\{X(n)\}_{n=1}^{\infty}$ is called homogeneous Markov chain.

**Theorem 4.** The genetic sequence $\{X(n)\}_{n=1}^{\infty}$ of $GA_{FBMG}$ is a homogeneous Markov chain.

**Proof.** Through symbolic coding, the size of the population is $s=n!$ (here, $n$ be natural number). We may know from the constructive process of $GA_{FAP}$ that the $N^{th}$ population $X(N)$ in the evolutionary process is merely relevant to the $N$-1$^{th}$ population $X(N-1)$ and genetic operators, irrelevant to $X(N-2)$, $X(N-3),\cdots,X(0)$. So $P\{X(N)=i_N\mid X(0)=i_0,\ X(1)=i_1,\ \cdots,\ X(N-1)=i_{N-1}\}=P\{X(N)=i_N\mid X(N-1)=i_{N-1}\}$, this implies that $\{X(n)\}_{n=1}^{\infty}$ is a Markov chain.

Let $p_{ij}^{n}(m)=P\{X_{m+n}=j\mid X_m=i\}$ denote the transition probability of state $i$ to $j$ after $n$ steps from $m^{th}$ time, from the above operation, the transition probability of each generation is only relevant to the crossover probability, the mutation probability as well as the population of this generation, and it does not alter with time (e.g. evolution generation), that is, $p_{ij}^{n}(m)$ is irrelevant to $m$, so $\{X(n)\}_{n=1}^{\infty}$ is a homogeneous Markov chain.

**Theorem 4.** $GA_{FBMG}$ can converge to the global optimal solution.

**Proof.** Because forced reserved strategy is used in $GA_{FBMG}$, there are some changes happened on the nature of Markov chain. When the genetic algorithm evolves to a new generation (for example generation $N$), compare all the parent population (generation $N$-1) which take part in evolution with the child population generated; the most superior individual of previous generation will replace the worst individual (generation $N$) of this generation. At the same time, we suppose that generation $M$ be one of the previous generations of generation $N$, and there produced a more superior new individual in the evolution process from generation $M$ to generation $N$, it is very obvious that $P_{MN}>0$, which is to say it is reachable from $M$ to $N$, but it is not reachable from $N$ to $M$, that is, $P_{NM}=0$, which is because the individual of generation $N$ is forced to be replaced by the most superior individual of the previous generations. In the evolution process, for $M$ and $N$ are arbitrary, so the simple fuzzy genetic algorithm using the forced reserved strategy is a non-return evolution process, and it will converge to the global optimal solution.

## 7 Number Simulation

In this section, we will take an example to analyze the performance of solving algorithm for the fuzzy bimatrix game. For the sake of simplicity, we suppose all the elements in efficiency matrixes to be triangular fuzzy numbers.

**Example** [4]**.** We take a fuzzy bimatrix game into account as following:

$$\tilde{A}=\begin{bmatrix}(4.0,4.5,4.8) & (6.4,7.0,7.6) & (8.7,9.3,9.7)\\ (4.9,5.4,6.0) & (6.3,6.9,7.2) & (8.1,8.4,8.9)\\ (5.5,6.1,6.7) & (7.0,7.5,7.9) & (7.9,8.6,9.1)\end{bmatrix}\quad \tilde{B}=\begin{bmatrix}(5.8,6.4,7.1) & (4.9,5.5,6.1) & (3.0,3.6,4.1)\\ (5.9,6.5,7.0) & (6.3,6.75,7.1) & (6.0,6.6,7.2)\\ (6.1,6.7,7.4) & (6.8,7.1,7.9) & (7.1,7.7,8.2)\end{bmatrix}$$

For the sake of the specificity of level effect function, we may consider as following:

Firstly, let level effect function $L(\lambda)=\lambda$, for triangular fuzzy numbers $A=(a,b,c)$, from (1)、(2) and $A_\lambda=[a+(b-a)\lambda, c-(c-b)\lambda]$, according to the properties of integral, we can obtain that $I_L(A)=(a+4b+c)/6$, $\delta(A)=(c-a)/6$, so the matrixes of $\delta(\tilde{A})$, $\delta(\tilde{B})$ and $I_L(\tilde{A})$, $I_L(\tilde{B})$ are follows:

$$I_L(\tilde{A})=\begin{bmatrix}4.47 & 7.00 & 9.27\\ 5.42 & 6.45 & 8.43\\ 6.10 & 7.48 & 8.57\end{bmatrix},\ \delta(\tilde{A})=\begin{bmatrix}0.1333 & 0.1333 & 0.1667\\ 0.1833 & 0.1500 & 0.1333\\ 0.1333 & 0.1500 & 0.2000\end{bmatrix}$$

$$I_L(\tilde{B})=\begin{bmatrix}6.42 & 5.50 & 3.58\\ 6.48 & 6.73 & 6.57\\ 6.68 & 7.12 & 7.68\end{bmatrix},\ \delta(\tilde{B})=\begin{bmatrix}0.2167 & 0.2000 & 0.1833\\ 0.1833 & 0.1333 & 0.2000\\ 0.2167 & 0.1333 & 0.1500\end{bmatrix}$$

According to the structure of $GA_{FBMG}$, if setting the genetic parameters as follows: the size of population is 80, the number of evolution generation is 100, crossover probability $p_c=0.6$, mutation probability $p_m=0.1$, then for $\varepsilon=\eta=0.6$, we can discuss this problem from two aspects:

**Case 1.** when $u(x)=x/(5+x)$, the optimal solution can be:

$$(x^*,y^*)=(0.4323,0.2338,0.3339;0.4739,0.4489,0.0772),\ f^*=3.6111$$

Fig.1 can show the variation of the optimal value

**Case 2.** when $u(x)=x/(10+x)$, the optimal solution can be:

$$(x^*,y^*)=(0.2446,0.03581,0.3972;0.1593,0.7890,0.0517),\ f^*=3.6614$$

Fig.2 can show the variation of the optimal value

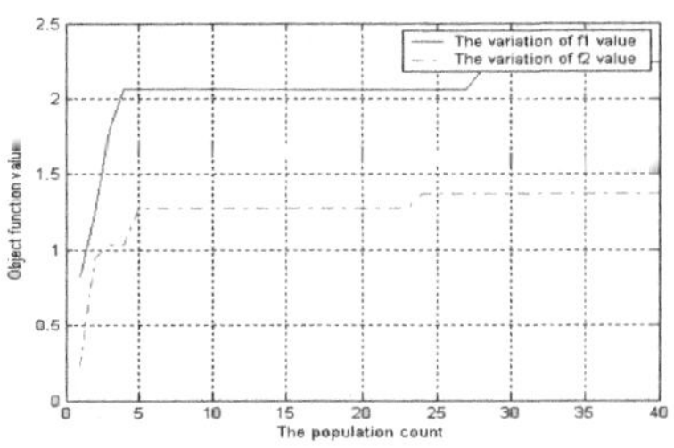


**Fig. 1.**

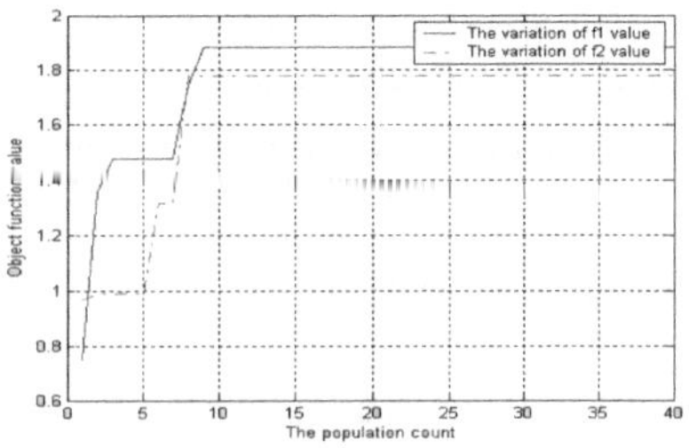


**Fig. 2.**

Secondly, Let level effect function $L(\lambda)=\lambda^2$, for $A=(a,b,c)$, from (1)、(2) and $A_\lambda=[a+(b-a)\lambda, c-(c-b)\lambda]$, according to the properties of integral, we can obtain that $I_L(A)=(6b+a+c)/72$, $\delta(A)=(c-a)/12$, $\varepsilon=0.4, \eta=0.7$ .We can get solutions as following:

**Case 3.** when $u(x)=x/(5+x)$, the optimal solution can be:

$$(x^*, y^*)=(0.0243,0.2929,0.6828;0.4400,0.34720,0.2128)\quad f^*=0.3890$$

The variation of the optimal value can be shown by Fig.3

**Case 4.** when $u(x)=x/(10+x)$, the optimal solution can be:

$$(x^*, y^*)=(0.4403,0.4989,0.0.0607;0.4043,0.5143,0.0814)\quad f^*=0.3767$$

The variation of the optimal value can be shown by Fig.4

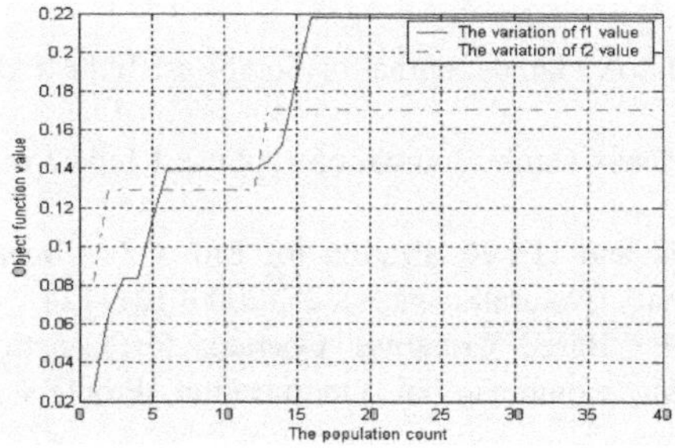


Fig. 3.

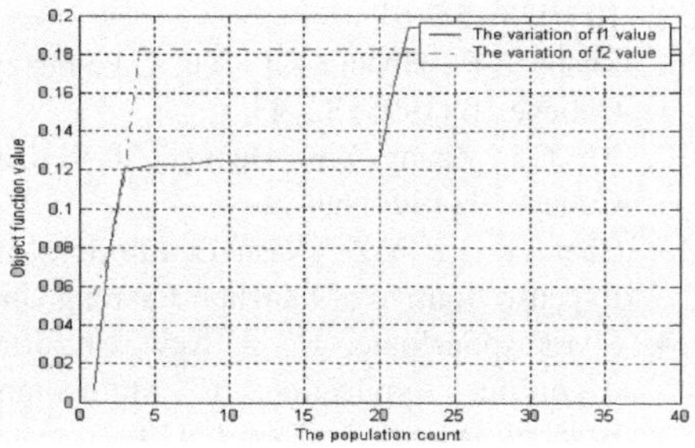


Fig. 4.

## 8 Conclusion

Considering the fuzzy features of fuzzy bimatrix game, we treat the game value as fuzzy variable, and establish the model involving fuzzy variable and fuzzy coefficient with the corresponding fuzzy bimatrix game problem. Due to the shortage of reflecting the comparing relation among fuzzy numbers with fuzzy number features, we use the comparing relation of fuzzy number based on the level effect function, and convert the original fuzzy bimatrix game problem into common bimatrix game problem. Therefore, we successfully solved the fuzzy bimatrix game problem.

**Acknowledgement.** The National Natural Science Foundation of China (70671034) and the Natural Science Fund of Hebei Province (F2006000346) and the Science Fund of Hebei University of Science and Technology (XL2006035) and the Ph. D. Fund of Hebei Province (05547004D-2) support this work.

## References

1. Wang, J.H.: Game Theory. Beijing: Tsinghua University Press (1986)
2. NairK, G.G., Tanjith, G.: Solution of 3×3 Games Using Graphic a Method . European Journal of Operational Research, 112 (1999) 472–478

3. Liu, D., Huang, Z.G.: Game Theory and Application. Changsha: National University of Defence Technology Press (1994)
4. Shi, X.: A Algorithm of Solving Nash Equilibrium Solution. Systems Engineering, Vol. 16. (1998)
5. Chen, S.J., Sun, Y.G., Wu, Z.X.: A Genetic Algorithm of Nash Equilibrium Solution. Systems Engineering, 19 (2001) 67–70
6. Nishizaki, I., Sakawa, M.: Equilibrium Solutions in Multiobjective Bimatrix Games with Fuzzy Payoffs and Fuzzy Goals. Fuzzy Sets and Systems, 111 (2000) 99–116
7. Campos, L.: Fuzzy Linear Programming Models to Solve Fuzzy Matrix Games. Fuzzy Sets and Systems, 32 (1989) 275–289
8. Li, F.C., Wu, C.X., Qiu, J.Q.: Platform Fuzzy Number and Separtability of Fuzzy Number Space. Fuzzy Sets and Systems, 117 (2001) 347–353
9. Diamond, P., Kloeden, P.: Metric Space of Fuzzy Set: Theory and Applications. Word Scientific, Singapore (1994)
10. Zhang, Z.F., Huang, Z.L., Yu, C.J.: Fuzzy Matrix Game. Fuzzy System and Mathmatics, 10 (1996) 55–61
11. Zhang Z.F., Huang, Z.L, Yu, C.J.: Fuzzy Matrix Game. Journal of Southwest Industrial College,10 (1995) 32–43
12. Yu ,C.J., Zhang, Z.F., Huang, Z.L.: Fuzzy Matrix Game. Journal of Southwest Industrial College, 9 (1994) 69–74
13. Chen, J., Li, Y.Z.: Nash Equilibrium Model and GA Realization for Bid of No Bear Expense. Journal of Lanzhou Jiaotong University (NaturalSciences), 25 (2006) 121–124
14. SivrikayaSerfoglu, F.: A New UniformOrder Based Crossover Operator for Genetic Algorithm Applications to Multicomponent Combinatorial Optimization Problems. Istanbul: BobaziciUniversity (1997)

# K$\oplus$1 Composite Genetic Algorithm and Its Properties

Fachao Li[1,2] and Limin Liu[2]

[1] College of Economics and Management, Hebei University of Science and Technology, Shijiazhuang Hebei 050018, China
[2] College of Science, Hebei University of science and technology, Shijiazhuang Hebei 050018, China
lifachao@tsinghua.org.cn, liulm2001@163.com

**Abstract.** In view of the slowness and the locality of convergence for Simple Genetic Algorithm (SGA for short) in solving complex optimization problems, K$\oplus$1 Composite Genetic Algorithm (K$\oplus$1-CGA for short), as an improved genetic algorithm, is proposed by reducing the optimization-search range gradually, the structure and the implementation steps of K$\oplus$1-CGA are also given; then consider its global convergence under the elitist preserving strategy using Markov chain theory, and analyze its performance from different aspects through simulation. All these indicate that the new algorithm possesses interesting advantages such as better convergence, less chance trapping into premature states. So it can be widely used in many optimization problems with large-scale and high- accuracy.

**Keywords:** Genetic Algorithm, Convergence, Markov Chain, Optimization, K$\oplus$1 Composite Genetic Algorithm (K$\oplus$1-CGA).

## 1 Introduction

Genetic Algorithm[1] (GA for short), proposed by Holland in 1975, is a kind of optimization search algorithm based on the theory of evolution and the genetic mutation theory of Mendel. Recently, it has been a hot spot[2-4] in many fields such as data mining, optimization control, artificial Intelligence etc., and the applications have been achieved in many corresponding fields. Genetic Algorithm, with the evolutionary theory and coding strategy, possess the feature of no consideration the complex mathematics characteristic of real problems and no restriction on objective function, which can be described as follows: ① Generate randomly an initial population from feasible solution space; ② Evaluate the population through some norms (say it a fitness function); ③ Generate the new population by selection, crossover, mutation operations on the basis of ②; ④ Repeat the process above until some pre-conditions are satisfied .

Despite the advantage of being easy and direct in operation of GA, there still exist some shortcomings of premature phenomenon and lower convergence precision, especially for optimization problems with large-scale and high-accuracy. In recent years, many authors have proposed a variety of improved GA, but most of them focus on the value of selection, crossover, mutation probability and the selection of fitness

D.-S. Huang, L. Heutte, and M. Loog (Eds.): ICIC 2007, LNAI 4682, pp. 629–640, 2007.

function[5,6] and have strong points, and can't essentially make up for the deficiency of algorithm above. Combining the mechanism of GA, in this contribution, a kind of improved genetic algorithm, K⊕1 composite genetic algorithm (K⊕1-CGA for short), is proposed, and consider its convergence using Markov chain theory and analyze its performance through simulation. All the results indicate that the improved genetic algorithm possesses the interesting advantages such as better convergence under the elitist preserving strategy and less chance trapping into premature states, and could be widely used in many fields such as optimization problems with large-scale and high-accuracy, complex system numerical optimization etc..

## 2 Structure of K⊕1-CGA

### 2.1 Basic Idea of K⊕1-CGA

No matter what optimization problems whether in complex optimization system or other related fields in actual life, it possesses high application value for the research of precision. Theoretically speaking, for a given optimization problem, the accurate optimal solution must be found if it exists. However, for practical optimization problems, the satisfaction solution is usually focused because of the existing of theoretical error of the model, the information error in the data and the cognition bias. Generally, the varying region of variable has close relation with the precision of solution, so it is difficult to find the optimal solution with large-scale and high-accuracy of the optimization problems. Accordingly, for optimization problems with large-scale and high-accuracy, it is obviously to help us to find the optimal solution or satisfactory solution by reducing the optimization-search range gradually without losing the optimal solutions.

K⊕1-CGA is just followed the way mentioned above, which is divided into two phases: optimal pre-judgement phase and optimal searching phase. The optimal pre-judgement phase is made up by K times (which is independence each other) genetic searching for its objective of determining the basic features of the optimal solutions or satisfactory solutions under some strategies, which based on the relative satisfaction solutions obtained from each time. Further, by combing some methods such as statistics laws and reseau theory etc. to reduce the optimization-search range. The optimal searching phase is to search the higher precision of satisfactory solution based on the reduced range on optimal pre-judgement phase. Obviously, if K=1, then K⊕1-CGA be the simple genetic algorithm, this indicates that K⊕1-CGA is extendability and perfection of SGA. In what following, we first give the implementation steps of K⊕1-CGA.

### 2.2 The Implementation Steps of K⊕1-CGA

Based on the analysis above, we may design the implementation steps of K⊕1-CGA as following:

**Step 1.** Choose the encoded mode of individuals;
**Step 2. (Optimal pre-judgement)** Repeat the operations K times independently as following: Generate randomly an initial population including $N$ individuals, then apply the genetic operations to them according to the pre-setting generations and write down each individual and its fitness of each time;
**Step 3. (Reducing search range)** According to some strategy and the results from step2, determine the relative satisfactory spaces, and reduce the search range by combining the encoded mode of individual;
**Step 4. (Optimal searching)** Implement genetic search on the basis of the range from step3;
**Step 5. (Termination test)** If the stopping condition is satisfied, stop it; otherwise return to step2 based on the search range from Step 4.

# 3 The Strategies of Reducing Search Range

The key link for K⊕1-CGA is to reduce search range, when fulfilling the concrete methods, we should combine the properties of optimization problems and the encoded form of individuals. Generally speaking, it can be generalized the following two methods:

**Method 1:** For symbolic coding and binary coding, reducing search range by determining the important genes or unimportant genes.
**Method 2:** For real coding, reducing search range by shortening the bound;

Generally, to the $K \cdot N$ individuals obtained from optimal pre-judgement phase, whose search range can be reduced using the following flow diagram.

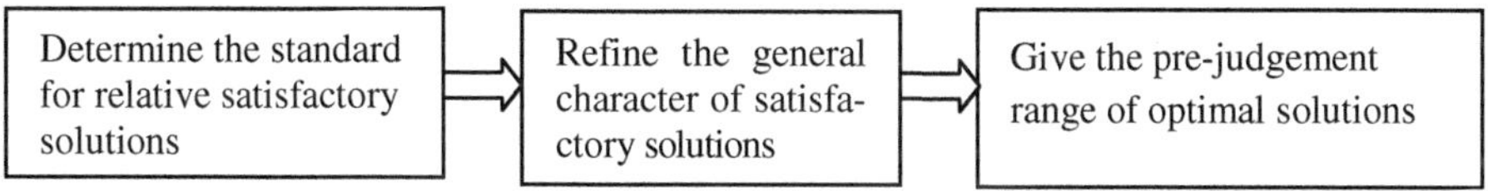


In what following, we will give some concrete methods for reducing search range.

## 3.1 The Methods of Reducing Search Range Based on Statistics Law

We know from the statistics theory that the statistics rules are reliable only if the data is much enough. In this case, when $K$ is a bit bigger and $K \cdot N$ individuals possess the general characters on pre-judgement phase, we can reduce the search range with the strategy as follows.

1) Determining the relative satisfaction solutions ***C***, the commonly used methods are as follows:

① Determine ***C*** by ratio $\alpha$ $(0 < \alpha \le 1)$, that is, taking $\text{int}(K \cdot N)$ individuals with bigger fitness as the relative satisfaction solutions ***C***.

② Take the biggest fitness $W$ of $K \cdot N$ individuals as the standard, and determine ***C*** by relative optimal satisfactory level $\beta$ $(0 < \beta < 1)$, that is, determine ***C*** by selecting the individuals whose fitness $w$ satisfy the condition $(W - w)/W \le \beta$.

2) Giving the pre-judgement range, the commonly used methods are as follows:

① Determine the important genes using stable rate of genes, that is to take the genes whose stable rate exceed $\beta$ $(0 < \beta \le 1)$ as the important genes, which is suitable for the situation with not real coding.

② Determine the pre-judgement range using the method of symmetric points, that is to determine it by using symmetric points $\beta$ $(0 < \beta \le 1)$ of distribution of solutions based on the probability distribution of satisfaction solutions.

### 3.2 The Min-Max Method for Reducing Search Range

From the statistics theory, the reduced range with high reliability can not be obtained if there is not obvious general characters for $K \cdot N$ individuals from pre-judgement phase. Under the circumstance of not losing the optimal solution information as possible, in order to achieve the goal of reducing search range, we can use the following Min-Max strategy.

**Step1.** Determine the relative satisfaction solutions $\boldsymbol{C}$ according to some rules, for, abandoning the bad individuals by proportion or by relative satisfaction.
**Step2.** Based on relative satisfaction solutions, take separately smallest fitness and biggest fitness of individuals in $\boldsymbol{C}$ as the infimum and supremum of reduced range.

### 3.3 Several Remarks

*Remark 1.* The objective on optimal pre-judgement phase is to reducing hunting region gradually without losing optimal solutions, so we use reserving optimal individuals in the process of genetic operations in order to obtain more optimal solutions information.

*Remark 2.* The value of $K$ has direct relation with K$\oplus$1-CGA, if $K$ is too big, the result will be bad, such as time and efficiency, if $K$ is small, the result will be distortion. It can be determined by combining encoding mode of solutions, population space in pre-judgement phase and the strategy to reduce hunting region. Generally speaking, for the method of reducing hunting region based on statistics law, it is better to take 4 to 10, and for the method of Min-Max, it is better to take 3 to 6.

*Remark 3.* Since the main objectives of the two phases are different, we should select appropriate parameters on each phase. Generally, the mutation probability on the pre-judgement phase should be a bit larger than the mutation probability on the searching phase, and the genetic generation on the pre-judgement phase should be smaller than the genetic generation on the searching phase.

## 4 Convergence of K$\oplus$1-CGA

Since the population $X(t+1)$ of generation $t+1$ is only relate to population $X(t)$ of generation $t$ in process of genetic iteration, and the transition probability of each generation is irrelative with the origin time, then genetic sequence $\{X(t)\}_{t=1}^{\infty}$ can be regarded as a homogeneous Markov chain. In what follows, we use Markov chain to analyze the performance for K$\oplus$1-CGA.

### 4.1 Convergence and Other Related Concepts

The convergence of genetic algorithms usually refers to that the iterative population (or distribution) generated by GA converges to a steady state (or distribution), or that the maximum or average value of its fitness function drives to the optimal solution of the optimization problem as the iteration progresses.

**Definition 1 [7].** Let $X(n)=\{X_1(n), X_2(n),\cdots, X_N(n)\}$ be the $n^{\text{th}}$ population of GA, $Z_n = \max\{ f(X_i(n)) \mid i=1,2,\cdots,N\}$ denote the optimal value in population $X(n)$, $f^* = \max\{f(X) \mid X \in S\}$ be the global optimal value. If $\lim_{n\to\infty} P\{Z_n = f^\bullet\}=1$, then we say the genetic sequence $\{X(n)\}_{n=1}^{\infty}$ is convergent.

**Definition 2 [8].** Let $\{X(t)\}_{t=1}^{\infty}$ be a Markov chain, $P_{ij}^{(t)}$ be the transition probability from state *i* to state *j*. ① For any states *i* and *j*, if there exists a natural number *n* such that $P_{ij}^{(n)} > 0$, then we say $\{X(t)\}_{t=1}^{\infty}$ is irreducible; ② For any states *i* and *j*, if $D=\{n: n\ge 1, P_{ii}^{(n)} > 0\}$ is not empty and the greatest common divisor of it is 1, then we say $\{X(t)\}_{t=1}^{\infty}$ is nonperiodic. ③ For any state $j$, if $\sum_{n=1}^{\infty} P_{jj}^{(n)} = 1$, then we say state *j* is common return. ④ If $\sum_{n=1}^{\infty} P_{jj}^{(n)} < 1$, then we say state *j* is seldom return.

**Definition 3 [8].** For common return state *i* of Markov chain $\{X(t)\}_{t=1}^{\infty}$, if $u_i = \sum_{n=1}^{\infty} tP_{ii}^{(t)} < \infty$, then we say the state *i* is positive common return; if for any state *j*, *j* is positive common return and nonperiodic, then we say Markov chain $\{X(t)\}_{t=1}^{\infty}$ is ergodic.

### 4.2 Two Propositions on K⊕1-CGA

**Proposition 1.** The genetic sequence $\{X(n)\}_{n=1}^{\infty}$ of K⊕1-CGA is a homogeneous Markon chain.

**Proof.** By the operating process of K⊕1-CGA, we know the $n^{\text{th}}$ population $X(n)$ is merely depend on the $(n\text{-}1)^{\text{th}}$ population $X(n-1)$, and it is irrelevant to $X(n-2)$, $X(n-3)\cdots, X(0)$, so

$$P\{X(n)=i_n \mid X(0)=i_0, X(1)=i_1,\cdots, X(n-1)=i_{n-1}\}$$
$$= P\{X(n)=i_n \mid X(n-1)=i_{n-1}\}.$$

By Definition 2, we can know that $\{X(n)\}_{n=1}^{\infty}$ is a Markov chain.

Let $P_{ij}^{(n)}(m) = P\{X_{m+n} = j \mid X_m = i\}$ denote the transition probability of state *i* to *j* after *n* steps from $n^{\text{th}}$ population. Because the transition probability of each generation in K⊕1-CGA is only relevant to the crossover probability, the mutation probability as well as the population of this generation, and it does not alter with time (e.g. evolution generation), that is, $P_{ij}^{(n)}(m)$ is irrelevant to *m*, so $\{X(n)\}_{n=1}^{\infty}$ is a homogeneous Markov chain.

**Proposition 2.** The genetic sequence $\{X(t)\}_{t=1}^{\infty}$ of K⊕1-CGA is an ergodic Markov chain.

**Proof.** Because the genetic sequence $\{X(t)\}_{t=1}^{\infty}$ of K ⊕ 1-CGA is not only a homogenous, but also a mutually attainable Markov chain, so $\{X(t)\}_{t=1}^{\infty}$ is an irreducible, positive recurrent and non-periodic Markov chain. Using the theory of stochastic process (See [7]) we can know that the genetic sequence $\{X(t)\}_{t=1}^{\infty}$ is an ergodic Markov chain, and its stationary probability distribution exists, that is, as $n \to \infty$, there exists a probability distribution $\lim_{n\to\infty} P_{ij}^{(n)} = p_j \ (j = 1, 2, \cdots)$ which is irrelevant to the original states and satisfies $P_j > 0$ and $\sum_{j=1}^{\infty} P_j = 1$.

### 4.3 Two Main Theorems

**Theorem 1.** The genetic sequence $\{X(n)\}_{n=1}^{\infty}$ of K⊕1-CGA is not convergent to the global optimal solution.

**Proof.** Since K⊕1-CGA is ergodic, that is to say, all the probability $P_j = \lim_{n\to\infty} P_{ij}^{(n)}$ starting from any original state *i* with any state *j* as its limiting state are bigger than 0, and $\sum_{j=1}^{\infty} P_j = 1$. Accordingly, the probability with the optimal state $f^*$ as its limiting state is smaller than 1, that is $\lim_{n\to\infty} P\{Z_t = f^*\} < 1$, which implies that K⊕1-CGA is not convergent to the global optimal solution in probability.

**Theorem 2.** The genetic sequence $\{X(n)\}_{n=1}^{\infty}$ of K⊕1-CGA that includes the strategy of reserving the optimal individual is convergent to the global optimal solution.

**Proof.** Suppose that when the population evolves to a new generation (for example generation *j*), the most superior individual of previous generation (generation $j-1$) will replace the worst individual (for instance the individual at position *k*) of this generation (namely generation *j*). At the same time, we suppose that generation *i* be one of the previous generation of generation *j*, and there produced a more superior new individual is produced in the evolution process from generation *i* to generation *j* (namely the most superior individual of generation *j* is more outstanding than the most superior individual of generation *i* ). It is very obvious that $P_{ij}^{(n)} > 0$ by now, which is to say, it is reachable from *i* to *j*; simultaneously, we also obtain that $P_{ji}^{n} = 0$, which is because the individual at position *k* of generation *j* is forced to be replaced by the most superior individual of the previous generation which is definite and unmodifiable, and can not be the same with the individual at position *k* of generation *j* (for there does not exist so outstanding individual in generation *i*, namely it is inaccessible from *j* to *i*. In above analysis, since *i* and *j* are arbitrary, we may obtain that K⊕1-CGA using the most superior individual protection strategy is a non-return evolution process, so the genetic sequence $\{X(n)\}_{n=1}^{\infty}$ of K⊕1-CGA will finally converge to the global optimal solution.

*Remark 4.* From the structure of K⊕1-CGA, we can obtain the genetic sequences are all Markov chain in corresponding state space whether using real coding or other coding. The main differences between them are the state space with real coding is infinite while the state space with other coding is finite. In this way, the convergence analysis above is still true if we make appropriate change on state space.

## 5 Application Examples

This section, in order to analyze the performance of K⊕1-CGA further, we use two difficult functions which are usually used to test the performance for algorithms to analyze and discuss. And all experiments are based on MATLAB 6.5 and 2.0 GHz Pentium 4 processor and worked out under windows 2000 Professional Edition Platform.

**Example 1.** Consider the minimum value of Shaffer function (See [9, 10]):

$$f(x_1,x_2)=0.5-\frac{\sin^2\sqrt{x_1^2+x_2^2}-0.5}{[1+0.001(x_1^2+x_2^2)]^2},\quad -100\le x_1,\,x_2\le 100.$$

This function has only one global maximal point $(0,0)$, and the maximal value is $f(0,0)=1$. In what following, we make the experiments by using K⊕1-CGA in this paper and SGA based on real coding, respectively. Here, the parameters setting of optimal pre-judgement phase and optimal searching phase and SGA are as follows:

SGA: The size of population 80, the maximal times of iteration 100, the crossover probability $p_c=0.6$, the mutation probability $p_m=0.001$.

K⊕1-CGA: ① Optimal pre-judgement phase: The size of population 80, the maximal times of iteration 40, the times for pre-judgement $K=5$, the crossover probability $p_c=0.6$, the mutation probability $p_m=0.002$; ② Optimal searching phase: The size of population 80, the maximal times of iteration 100, the crossover probability $p_c=0.6$, the mutation probability $p_m=0.001$; ③ Using the symmetrical point $\beta$ ( $\beta=0.1$ ) based on probability distribution of satisfaction solutions on pre-judgement phase to reduce the search range.

Fig.1 and Fig.2 denote the evolution curve of 100 iterations for SGA and 5⊕1-CGA; Fig.3 and Fig.4 are the distribution of optimal solutions for 5⊕1-CGA on pre-judgement phase.

We may see from Fig.1 that SGA is not well convergent to the global optimal solution by any means, however, using 5⊕1-CGA in this paper, we may see from Fig.2 the population converges to the global satisfaction solution only after 10 generations. The results indicate that the convergence precision of 5⊕1-CGA is much better than SGA. Also, we can obtain from Fig.3 and Fig.4 the satisfaction solutions on the pre-judgement phase are around the optimal solution with high probability, it indicates the method of reducing search range in section 3 of this paper is feasible.

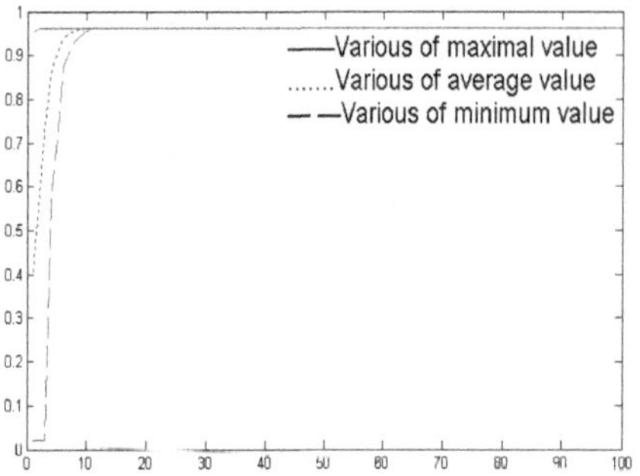


**Fig. 1.** The result of iterations for SGA

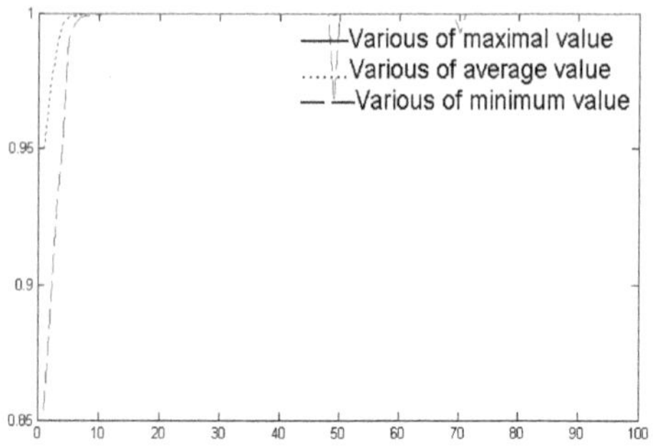


**Fig. 2.** The result of iterations for 5⊕1-CGA

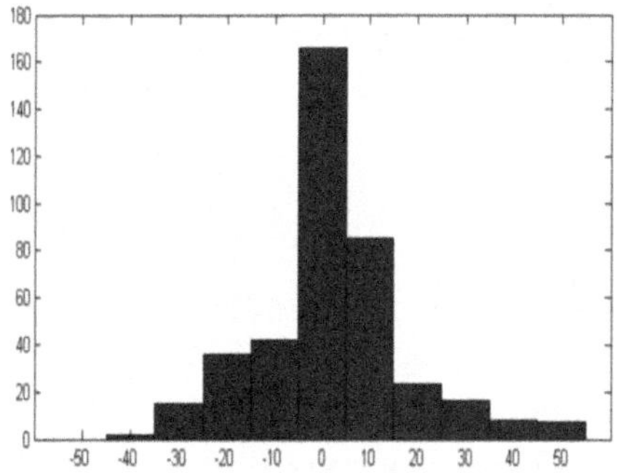

**Fig. 3.** Probability distribution of $x_1$

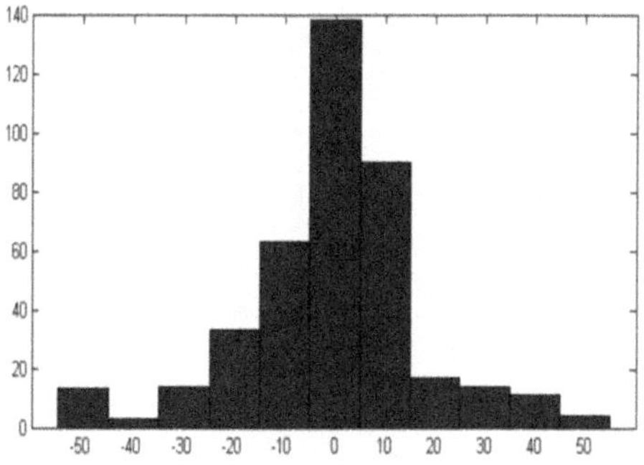

**Fig. 4.** Probability distribution of $x_2$

Further, in order to analyze the convergence performance of 5⊕1-CGA, we made 10 times simulation testing by using real coding and binary coding between 5⊕1-CGA and SGA based on parameters given above, respectively, and the results of the testing shown on Table1. Here, the strategies for reducing search range are as following:

For real coding: Using the symmetrical point $\beta$ ( $\beta = 0.1$ ) based on probability distribution of satisfaction solutions on pre-judgement phase to reduce the search range.

For binary coding: Reduce the search range by determining the important genes based on the individuals on pre-judgement phase.

In Table 1, C.V. denotes convergence value, C.G. the convergence generation, C.T. convergence time and A.V. average value.

We can obtain the following from table 1: 1) The 5⊕1-CGA possesses the global convergence performance whether using real coding or binary coding; 2) The convergence generation and convergence time of 5⊕1-CGA with real coding are better than 5⊕1-CGA with binary coding. The results indicate that it is better to use K⊕1-CGA with real coding for optimization problems with large-scale and high-accuracy.

**Table 1.** The comparison of convergence results between real coding and binary coding

| | Real coding | | | | | | Binary coding | | | | | |
|---|---|---|---|---|---|---|---|---|---|---|---|---|
| | 5⊕1-CGA | | | SGA | | | 5⊕1-CGA | | | SGA | | |
| | C.V. | C.G | C.T. | C.V. | C.G | C.T. | C.V. | C.G. | C.T. | C.V. | C.G. | C.T. |
| 1 | 1.0000 | 9 | 0.7780 | 0.8484 | 10 | 0.5780 | 0.9949 | 13 | 2.9060 | 0.8380 | 12 | 1.0940 |
| 2 | 0.9966 | 8 | 0.7000 | 0.9508 | 12 | 0.5620 | 0.9959 | 15 | 2.4530 | 0.8235 | 11 | 1.2030 |
| 3 | 0.9993 | 10 | 0.8030 | 0.9137 | 11 | 0.5320 | 0.9969 | 12 | 2.7190 | 0.8364 | 14 | 1.1880 |
| 4 | 0.9990 | 9 | 0.7180 | 0.8563 | 10 | 0.5940 | 0.9962 | 13 | 1.9370 | 0.9875 | 12 | 1.0780 |
| 5 | 0.9983 | 11 | 0.7350 | 0.8443 | 11 | 0.5780 | 0.9949 | 15 | 1.9840 | 0.8381 | 12 | 1.0340 |
| 6 | 0.9910 | 8 | 0.6720 | 0.8150 | 11 | 0.5710 | 0.9968 | 12 | 2.3750 | 0.8377 | 13 | 1.2810 |
| 7 | 0.9989 | 12 | 0.6400 | 0.9597 | 12 | 0.5160 | 0.9969 | 13 | 2.0780 | 0.8332 | 11 | 1.2350 |
| 8 | 0.9963 | 11 | 0.6250 | 0.8672 | 9 | 0.5310 | 0.9900 | 13 | 1.9460 | 0.8381 | 11 | 1.0780 |
| 9 | 0.9971 | 10 | 0.6720 | 0.8730 | 12 | 0.5630 | 0.9967 | 10 | 1.8910 | 0.9075 | 14 | 1.2190 |
| 10 | 1.0000 | 10 | 0.6560 | 0.8217 | 11 | 0.5620 | 0.9959 | 14 | 2.4060 | 0.9544 | 13 | 1.0930 |
| A.V | 0.9976 | 9.8 | 0.6999 | 0.8750 | 10.9 | 0.5587 | 0.9955 | 13.00 | 2.2695 | 0.8694 | 12.3 | 1.1503 |

**Example 2.** Consider the minimum value of Six-Hump Camel Back Function (See [7, 8]):

$$f(x_1,x_2)=(4-2.1x_1^2+\frac{1}{3}x_1^4)x_1^2+x_1x_2+(-4+4x_2^2)x_2^2, \qquad -100\le x_1,x_2\le 100 .$$

For this function, there are six local minimum points, but only (-0.0898, 0.7126) and (0.0898, -0.7126) are global minimum points, and the minimum value is -1.0326. In what following, we make the experiment by using K⊕1-CGA in this paper and SGA based on real coding, , respectively. Here, the parameters setting of K⊕1-CGA and SGA are as follows:

SGA: The size of population 80, the maximal times of iteration 100, the crossover probability $p_c = 0.6$, the mutation probability $p_m = 0.001$.

K⊕1-CGA: ① Optimal pre-judgement phase: The size of population 80, the maximal times of iteration 40, the times for optimal pre-judgement $K = 5$, the crossover probability $p_c = 0.6$, the mutation probability $p_m = 0.002$; ② Optimal searching phase: The size of population 80, the maximal times of iteration 100, the crossover probability $p_c = 0.6$, the mutation probability $p_m = 0.001$; ③ Using the symmetrical point $\beta$ ( $\beta = 0.1$ ) based on probability distribution of satisfaction solutions on pre-judgement phase to reduce the search range.

Fig.5 and Fig.6 denote the evolution curve of 100 iterations for SGA and 5⊕1-CGA; Fig.7 and Fig.8 denote the distribution of optimal solutions for 5⊕1-CGA on pre-judgement phase.

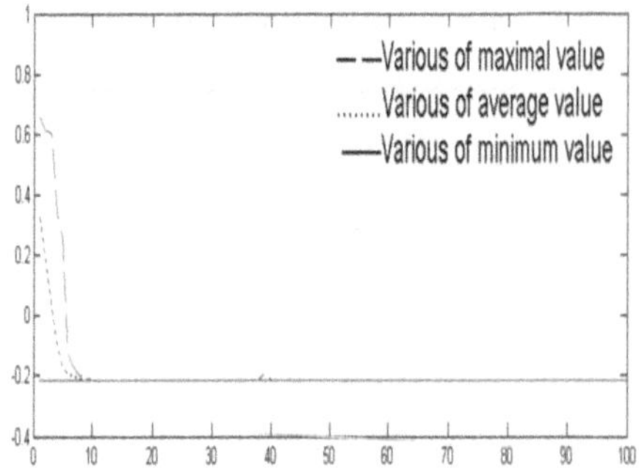


**Fig. 5.** The result of iterations for SGA

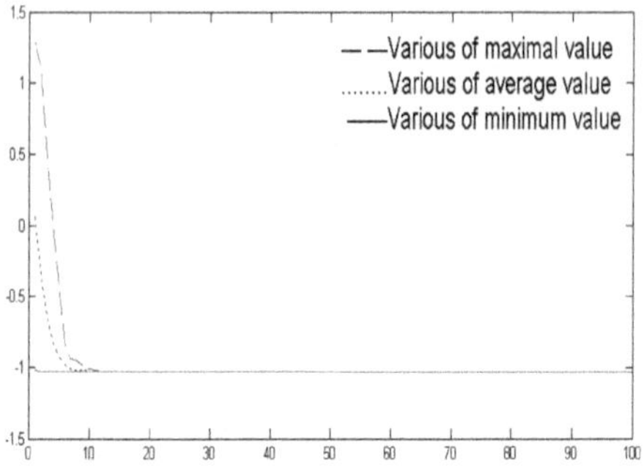


**Fig. 6.** The result of iterations for 5⊕1-CGA

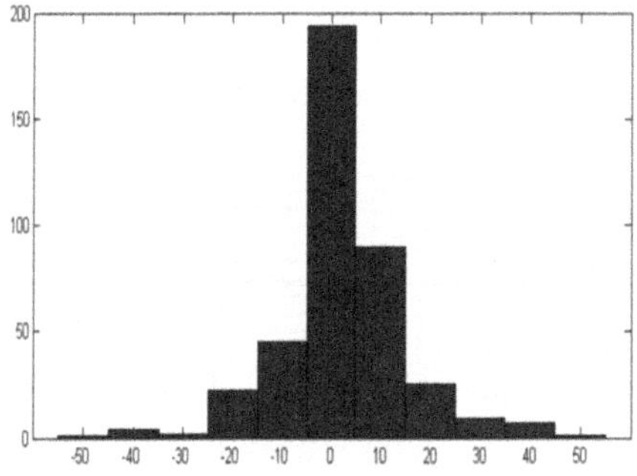

**Fig. 7.** Probability distribution of $x_1$

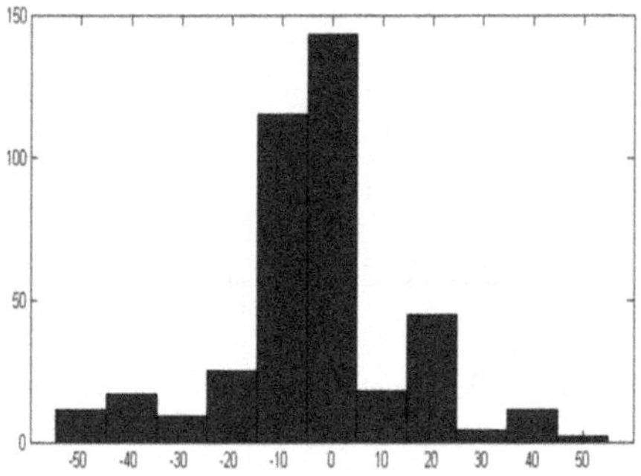

**Fig. 8.** Probability distribution of $x_2$

We can obtain from Fig.5 and Fig.6 that the convergence value of SGA is -0.2014 with the deviation 0.8312, while it is -0.1322 of 5⊕1-CGA with the deviation 0.0004. It obviously that 5⊕1-CGA is better than SGA in convergence precision. Fig.7 and Fig.8 demonstrate the satisfaction solutions on the pre-judgement phase are around the optimal solution (0.0898, -0.7126) with high probability, it indicates the method of reducing search range in section 3 of this paper is feasible.

In order to analyze the performance of K⊕1-CGA in the whole, we make 10 times experiments based on the setup parameters above under parameters K with different value 0, 2, 4 and 6. The results are shown on Table 2.

In Table 2, C.V. denotes convergence value, C.G. the convergence generation, C.T. convergence time and A.V. average value.

We can obtain from table 2: 1) Despite of variation of parameter K, K⊕1-CGA possess better convergence stability, such as convergence time and convergence generation; 2) The convergence precision of K⊕1-CGA will be improved gradually with the augment of parameter K; 3) The computational results will not be changed when parameter K is big enough.

Synthesizing the analysis and discussion above, K⊕1-CGA can not only avoid the premature phenomenon, but also possesses the global convergence performance.

**Table 2.** The computational result under parameters K with different value

| | SGA | | | 2⊕1-CGA | | | 4⊕1-CGA | | | 6⊕1-CGA | | |
|---|---|---|---|---|---|---|---|---|---|---|---|---|
| | C.V. | C.G. | C.T. | C.V. | C.G. | C.T. | C.V. | C.G. | C.T. | C.V. | C.G. | C.T. |
| 1 | -0.2111 | 8 | 0.5620 | -1.0047 | 10 | 0.6280 | -1.0277 | 10 | 0.7810 | -1.0324 | 12 | 0.8440 |
| 2 | -0.4988 | 9 | 0.6400 | -1.0152 | 11 | 0.5310 | -1.0314 | 12 | 0.7190 | -1.0321 | 11 | 0.8720 |
| 3 | -0.3450 | 8 | 0.6090 | -1.0208 | 11 | 0.7340 | -1.0300 | 10 | 0.7190 | -1.0326 | 12 | 0.7560 |
| 4 | -0.2711 | 7 | 0.6100 | -0.9753 | 8 | 0.5530 | -1.0303 | 11 | 0.7340 | -1.0323 | 12 | 0.7340 |
| 5 | -0.0018 | 6 | 0.5940 | -1.0232 | 10 | 0.5780 | -1.0302 | 13 | 0.7500 | -1.0317 | 14 | 0.7810 |
| 6 | -0.1432 | 10 | 0.5570 | -1.0216 | 13 | 0.6090 | -1.0316 | 11 | 0.8590 | -1.0320 | 12 | 0.7810 |
| 7 | -0.1360 | 8 | 0.5250 | -1.0295 | 11 | 0.6100 | -1.0278 | 13 | 0.7970 | -1.0322 | 12 | 0.6880 |
| 8 | -0.1021 | 11 | 0.5400 | -1.0280 | 11 | 0.6250 | -1.0291 | 10 | 0.7340 | -1.0316 | 10 | 0.7660 |
| 9 | -0.1142 | 10 | 0.5410 | -1.0250 | 13 | 0.5940 | -1.0305 | 12 | 0.6560 | -1.0317 | 11 | 0.7560 |
| 10 | -0.2333 | 9 | 0.5720 | -1.0280 | 12 | 0.6250 | -1.0298 | 12 | 0.7340 | -1.0326 | 12 | 0.8120 |
| A.V. | -0.2057 | 8.600 | 0.5750 | -1.0171 | 11.000 | 0.6087 | -1.0298 | 11.400 | 0.7483 | -1.0321 | 11.800 | 0.7790 |

# 6 Conclusion

In view of the slowness and the locality of convergence for Simple Genetic Algorithm (SGA for short), combine the analysis of solving mechanism of genetic algorithm, K⊕1Composite Genetic Algorithm (K⊕1-CGA for short), based on the optimal pre-judgement and optimal searching, is proposed. The implementation steps of K⊕1-CGA are also given, and the convergence performance is analyzed by the methods of Markov chain and simulation technology. All results indicate that the new type of algorithm enrich and perfect the evolutional computational theory and methods in mechanization. It can not only avoid the premature phenomenon in process of evolutionary, but also possesses the stability global convergence and better accountability and strong operability. It will be appropriate to optimization problems with large-scale and high-accuracy and possesses vast application prospect in complex system optimization, manufacture management etc.

**Acknowledgments.** This work is supported by the National Natural Science Foundation of China (70671034) and the Natural Science Foundation of Hebei Province (F2006000346) and the Ph. D. Foundation of Hebei Province (05547004D-2, B2004509).

# References

1. Holland, J.H.: Adaptive of Natural and Artificial Systems. Michigan: The University of Michigan Press (1975)
2. Srinivas, M, Patnaik, M.: Genetic algorithm: A survey. IEEE Computer 27 (1994) 17–26

3. Foge, D.B.: An Introduction to Simulated Evolutionary Optimization. IEEE Trans.on SMC 24 (1999) 3–14
4. Atmar, W.: Noteson the Simulation of Evolution. IEEE Trans. on SMC 24 (1994) 130–147
5. Gong, D.W., Sun, X.Y., Guo, X.J.: A New Kind of Survival of the Fittest Genetic Algorithm of. Control and Decision 11 (2002) 908–912
6. Han, W.L.: Improvement of Genetic Algorithm. Journal of China University of Mining & Technology 3 (2001) 102–105
7. Fang, Z.B., Miu, B.Q.: Random Process. University of Science and Technology of China Press (1993)
8. Zhang, W.X., Liang, Y.: Mathematical Foundation of Genetic Algorithms. Xi'an: Xi'an Jiao Tong University Press (2003)
9. Wang, X.P., Cao, L.M.: Theory of Genetic Algorithm, Application and Software implemented. Xi'an: Xi'an Jiao Tong University Press (2002)
10. Chen, G.L.: Genetic Algorithm and its application. Beijing: Posts and Telecom Press (1996)

# Parameter Tuning for Buck Converters Using Genetic Algorithms

Young-Kiu Choi and Byung-Wook Jung

School of Electrical Engineering, Pusan National University
Changjeon-dong, Geumjeong-gu, Busan 609-735, Korea
{ykichoi,wooroogy}@pusan.ac.kr

**Abstract.** The buck converter is one of DC/DC converters that are often used as power supplies. This paper presents parameter tuning methods to obtain circuit element values for the buck converter to minimize the output voltage variation under load changing environments. The conventional method using the concept of the phase margin is extended to have optimal phase margin that gives slightly improved performance in the output voltage response. For this, the phase margin becomes the tuning parameter that is optimized with the genetic algorithm. Next, the circuit element values are directly considered as the tuning parameters and optimized using the genetic algorithm to have very improved performance in the output voltage control of the buck converter.

**Keywords:** buck converter, output voltage control, genetic algorithm.

## 1 Introduction

DC/DC converters are equipments that transform some DC voltages into required DC voltages. DC/DC converters are usually classified into buck, boost, buck-boost and Cúk converters. DC/DC converters with rectifier stage on the AC side are used as power supplies that should maintain constant DC output voltages[1-3]. Even though the loads of DC/DC converters often change abruptly, DC/DC converters should keep constant output voltages with some forms of feedback control. A design method proposed by Venable[4,5] using the concept of phase margins has been widely used. It has voltage feedback controllers with error amplifiers composed of OP-Amps, resistors and capacitors. Other design methods using the root locus[6], PI control[7] and robust control[8] were also proposed for the output voltage control of DC/DC converters. These design approaches essentially have some design parameters such as phase margins and gains. The performance of feedback controllers for output voltages is closely related to those design parameters; however, these parameters usually rely on designers' experience. So, we have optimization problems for DC/DC converters with respect to those parameters and the problems may be efficiently solved by genetic algorithms[9].

In this paper, the conventional design method based on the phase margin[5] is optimized with the genetic algorithm for the buck converter that is one of DC/DC converters; the phase margin is the tuning parameter that is optimized with the genetic

D.-S. Huang, L. Heutte, and M. Loog (Eds.): ICIC 2007, LNAI 4682, pp. 641–647, 2007.

algorithm to have some improved output voltage responses. Next, resistances and capacitances of the voltage feedback controllers are directly regarded as the tuning parameters and they are optimized using the genetic algorithm to have very improved response of output voltage in the buck converter.

## 2 System Configuration of the Buck Converter

Since the output voltages of DC/DC converters are influenced by the change of loads, voltage feedback controls are required to maintain constant output voltages. Fig.1 shows a circuit diagram of the buck converter with the voltage control loop.

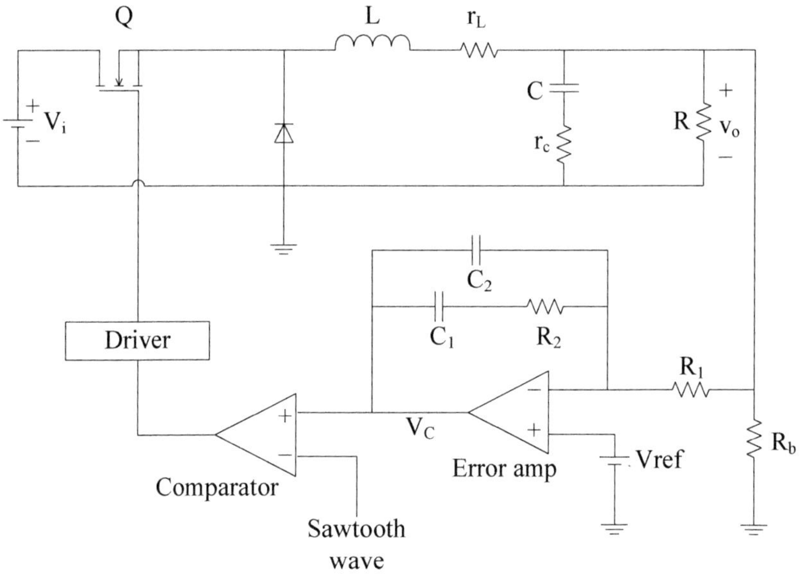


**Fig. 1.** Buck converter with voltage control loop

Let $G_P(s)$ be a transfer function relating the output voltage $v_O(s)$ to the control voltage $v_C(s)$. Then we have

$$G_P(s) = \frac{V_i / V_P}{LC} \left[ \frac{1 + s r_C C}{s^2 \left(1 + \frac{r_C}{R}\right) + s \left( \frac{1}{RC} + \frac{r_C}{L} + \frac{(r_C + R) r_L}{RL} \right) + \frac{(r_L + R)}{RLC}} \right]. \tag{1}$$

Where $V_i$ is the input source voltage, $V_P$ is the peak voltage of PWM circuits, and $R$ is the load resistance. $L$ is the inductance of the inductor coil, $r_L$ is the resistance of the inductor coil, $C$ is the capacitance of the capacitor, and $r_C$ is the series equivalent resistance of the capacitor.

We should have proper values of circuit elements of $R_1$, $R_2$, $C_1$ and $C_2$ of the error amplifier in Fig. 1 to minimize the variation of the converter output voltage caused by the change of the load resistance $R$. The conventional procedure to select the proper values of the circuit elements is as follows[5].

i) Plot the Bode diagram of $G_P(s)$

ii) Select a desired bandwidth $\omega_{CO}$ $(=\omega_S/10 \sim \omega_S/5)$, where $\omega_S$ is the switching frequency. Find $R_1$ and $R_2$ such that $G_P(j\omega_{CO}) = R_1/R_2$.

iii) Choose a proper phase margin (*PM*) that should be usually greater than or equal to 45°. Calculate the following equations:

$$\varphi_{CO} = PM - \angle G_P(j\omega_{CO}) - 180°. \quad (2)$$

$$K^2 - 2\tan(\varphi_{CO} + 90°)K - 1 = 0. \quad (3)$$

iv) Find the zero frequency $\omega_Z$ and pole frequency $\omega_P$:

$$\omega_Z = \omega_{CO}/K, \quad \omega_P = K\,\omega_{CO}. \quad (4)$$

v) Finally, $C_1$ and $C_2$ are obtained as follows:

$$C_1 = 1/(R_2\omega_Z), \quad C_2 = 1/(R_2\omega_P). \quad (5)$$

## 3 Parameter Tuning Method Using Genetic Algorithms

In the conventional procedure previously stated, the phase margin should be chosen to minimize the variation of the output voltage of the converter caused by the load change; however, the optimum value of the phase margin is not known. In this paper, the phase margin is considered as the tuning parameter and the genetic algorithm is applied to optimize the phase margin to find the values of $R_1$, $R_2$, $C_1$ and $C_2$ of the error amplifier minimizing the output voltage variation. To improve further the circuit performance beyond the conventional procedure based on the phase margin, we have $R_1$, $R_2$, $C_1$ and $C_2$ themselves as the tuning parameters, i.e., the chromosomes of the genetic algorithm. The chromosomes are encoded to be binary forms of 28 bits. The cost function $J$ and the fitness $F$ for the genetic algorithm are defined as below:

$$J = \int_0^{T_f} |e(t)|\,dt. \quad (6)$$

Where $e(t)$ is the output error voltage that is the difference between the reference voltage $V_{ref}$ and the output voltage $v_o(t)$. $T_f$ is the final time for evaluation of the cost function.

$$F = \frac{1}{1+\alpha J}. \quad (7)$$

Where $\alpha$ is a weighting factor for the fitness value.

Fig. 2 shows the total flow chart for parameter tuning with the genetic algorithm. PM denotes the phase margin.

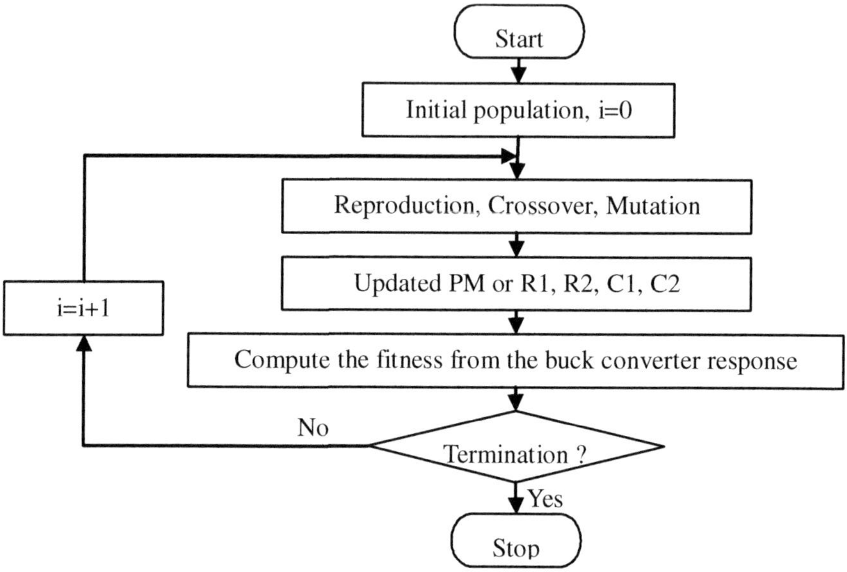


**Fig. 2.** Flow chart of the parameter tuning algorithm

## 4 Simulation Results and Discussion

Let the buck converter in Fig. 1 have the following values: $V_i = 20V$, $V_{ref} = 8V$, $L = 100\mu H$, $r_L = 0.5\Omega$, $C = 80\mu F$, $r_C = 0.6\Omega$, $V_P = 3V$, $\omega_{CO} = 2\pi \times 10^4 [rad/s]$.

The load resistance $R$ is set to be $5\,\Omega$ in the time interval 0 ~ 0.6*ms*, is changed to be $2.5\,\Omega$ in the time interval 0.6*ms* ~ 1*ms*, and is set to be $5\,\Omega$ again in the time interval 0.6*ms* ~ 1*ms*. $T_f$ in eq.(6) is 1.5*ms* and $\alpha$ in eq.(7) is $2\times10^5$.

Given the phase margin 46° that is arbitrarily chosen, the conventional procedure previously stated for the buck converter generates the following element values: $R_1 = 20k\Omega$, $R_2 = 33.04k\Omega$, $C_1 = 1.4254nF$, and $C_2 = 162.75pF$. The cost function $J$ is $7.7923\times10^{-5}$ and the output voltage of the buck converter is shown in Fig. 3.

Next, the genetic algorithm is applied to optimize the phase margin. The phase margin is regarded as binary chromosomes, the population size is 100, the crossover rate is 0.75, the mutation rate is 0.008, and the number of generations is 10. The load resistance $R$ is changed in the same way as before. As a result, the cost function $J$ is $7.725\times10^{-5}$ and the phase margin is 51.55°. Fig. 4 shows the output voltage response of the buck converter that is slightly improved compared to that in the case of the phase margin 46°.

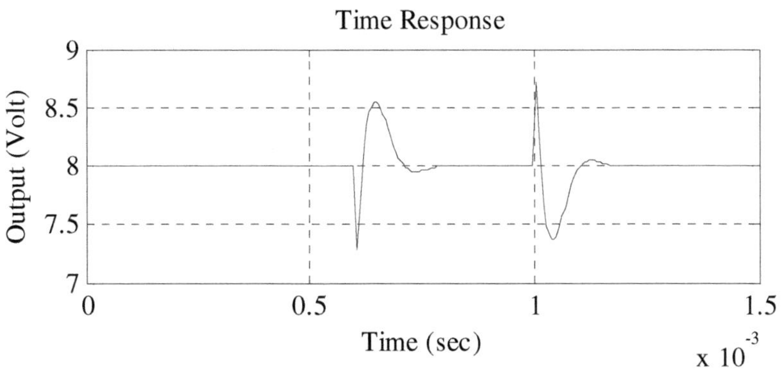


**Fig. 3.** Output voltage of the buck converter with the phase margin 46°

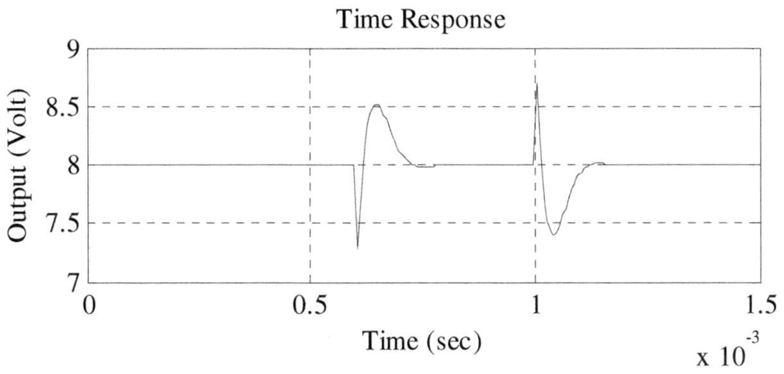


**Fig. 4.** Output voltage of the buck converter with the phase margin 51.55°

The circuit element values are also a little bit changed: $R_1 = 20k\Omega$, $R_2 = 33.04k\Omega$, $C_1 = 1.6914nF$, $C_2 = 137.15pF$.

To improve the output voltage response further, $R_1$, $R_2$, $C_1$ and $C_2$ are directly regarded as the tuning parameters and encoded in the form of binary chromosomes, and then the genetic algorithm is applied to tune the parameters. The population size is 100, the crossover rate is 0.75, the mutation rate is 0.008, and the number of generations is 20. The load resistance $R$ is changed in the same way as before. The cost function $J$ is so much decreased to be $1.953\times10^{-5}$ and the circuit parameters are $R_1 = 10k\Omega$, $R_2 = 39k\Omega$, $C_1 = 0.2nF$ and $C_2 = 10pF$. Fig. 5 shows the output voltage response of the buck converter that seems very improved when compared to that of the phase margin 51.55° in the sense of the magnitude and duration of the transient response: the magnitude decreased 34.1% and the duration also decreased 57.3%.

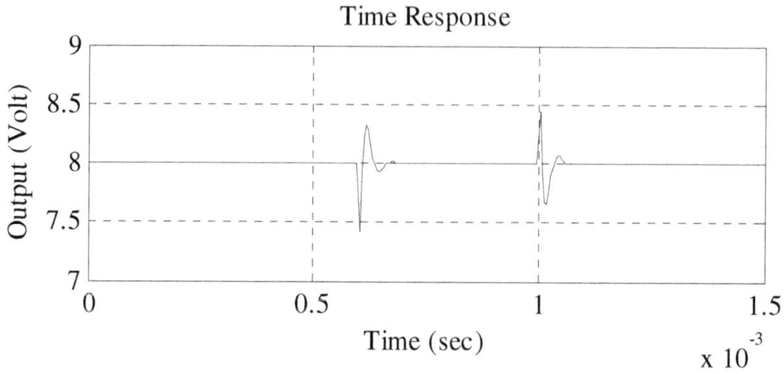


**Fig. 5.** Output voltage of the buck converter in the final case

# 5 Conclusions

The buck converter is one of DC/DC converters that are often used as power supplies with precise voltage regulation. This paper presents a parameter tuning method using the genetic algorithm to obtain circuit element values to minimize the output voltage variation under various load conditions. First, an optimal phase margin for the conventional procedure has been obtained using the genetic algorithm; however, it ensures only a little bit improvement over the phase margin 46° that was arbitrarily chosen. Second, two resistances and two capacitances of the error amplifier are considered as the tuning parameters, and the genetic algorithm is applied. The optimal parameters give us very improved control performances for the output voltage of the buck converter.

# Acknowledgement

This work was supported for two years by Pusan National University Research Grant.

# References

1. Mohan, N., Undeland, T.M., Robbins, W.P.: Power Electronics. 3rd edn. John Wiley & Sons, Inc. (2003)
2. Chen, Y.M., Liu, Y.C., Lin, S.H.: Double-Input PWM DC/DC Converter for High-/Low-Voltage Sources. IEEE Trans. on Industrial Electronics, vol. 53, no. 5 (2006) 1538-1545
3. Wei, S., Lehman, B.: Current-Fed Dual-Bridge DC-DC Converter. IEEE Trans. on Power Electronics, vol. 22, no. 2 (2007) 461-469
4. Venable, D.: The K Factor: A New Mathematical Tool for Stability Analysis and Synthesis. Proceedings Powercon, Vol. 10 (1983)
5. Hart, D.W.: Introduction to Power Electronics, Prentice-Hall (1996)

6. Guo, L., Hung, J.Y., Nelms, R.M.: Digital Controller Design for Buck and Boost Converters Using Root Locus. Proceedings IEEE IECON (2003) 1864-1869
7. Guo, H., Shiroishi., Y., Ichinokura, O.: Digital PI Controller for High Frequency Switching DC/DC Converters Based on FPGA. Proceedings IEEE INTELEC (2003) 536-541
8. Higuchi, K., Nakano, K., Kajikawa, T., Takegami, E., Tomioka, S., Watanabe, K.: Robust Control of DC-DC Converter by High-Order Approximate 2-Degree-of-Freedom Digital Controller. Proceedings IEEE IES (2004) 1839-1844
9. Michalewicz, Z.: Genetic Algorithms + Data Structures = Evolution Programs. 3rd edn. Springer-Verlag, Berlin Heidelberg New York (1996)

# Research a New Dynamic Clustering Algorithm Based on Genetic Immunity Mechanism

Yuhui Xu and Weijin Jiang

Department of Computer, Hunan Business College, Changsha 410205, P.R.China
jiangwj_nudt@163.com

**Abstract.** A novel dynamic evolutionary clustering algorithm is proposed in this paper to overcome the shortcomings of fuzzy modeling method based on general clustering algorithms that fuzzy rule number should be determined beforehand. This algorithm searches for the optimal cluster number by using the improved genetic techniques to optimize string lengths of chromosomes; at the same time, the convergence of clustering center parameters is expedited with the help of Fuzzy C-Means algorithm. Moreover, by introducing memory function and vaccine inoculation mechanism of immune system, at the same time, dynamic evolutionary clustering algorithm can converge to the optimal solution rapidly and stably. The proper fuzzy rule number and exact premise parameters are obtained simultaneously when using this efficient dynamic evolutionary clustering algorithm to identify fuzzy models. The effectiveness of the proposed fuzzy modeling method based on dynamic evolutionary clustering algorithm is demonstrated by simulation examples, and the accurate non-linear fuzzy models can be obtained when the method is applied to the thermal processes.

**Keywords:** Dynamic clustering algorithm, Immune mechanism, Genetic algorithm, Fuzzy model.

## 1 Introduction

Alone with the improvement of capacity and parameter of modern electric power production (power-plant) system and the complication of equipment system, it leads to a higher demand to the automatic control of electric power production process [1] in order to make sure that the electric power equipment can run economically and stably. Generally, many systems in electric power production process has a set of characteristics including high rank inertia, pure delay, non-linearity and time varying. The control quality can be affected and users are even unable to operate normally when it comes a big change on the processing operation based on the control system of conventional linear model. Therefore, Establishing accurate global non-linear model of the thermal process was the foundation to enhance the performance of control system [2-3]. In recent years, the fuzzy modeling has become of a research hotspot [4] of non-linear modeling. Compared to other non-linear modeling methods, the merit of fuzzy modeling is that it is constituted by the if-then rule which it allows

D.-S. Huang, L. Heutte, and M. Loog (Eds.): ICIC 2007, LNAI 4682, pp. 648–659, 2007.

the model structure and the physical meaning of parameter easily to understand; Moreover, the fuzzy model not only can use the survey data but also can fully use the experience and knowledge which are described by language. T-S fuzzy model is one kind of fuzzy modeling which is able to only use quite few rule numbers to describe a given unknown system and its conclusion can be described by linear model. It can be very convenient to adopt the conventional control theory to design controller. A kind of fuzzy molding which has a similar structure to the T-S fuzzy model's is proposed in this paper.

It needs two steps to operate the fuzzy modeling--structure identification and parameter identification, and the structure identification takes much more trouble. Generally, the methods of identifying structure are average partition algorithm, hill climbing algorithm and clustering algorithm. We only can identify the system by the input-output data [5-6] unless we have sufficient information about it. The clustering number and the clustering center represent the model rule number and some of the model parameters when we use the clustering algorithm to deal with modeling, therefore, partitioning the global system means to figure out the proper clustering number and the precise clustering center. Like many clustering algorithms mentioned by other reference such as the C-Means value algorithm, FCM (Fuzzy C-Means) which develops based on C-average value algorithm, PCM (Possibilistic C-Means) algorithm and G-K (Gustafson-Kessel) algorithm, they are all belong to the static clustering which the clustering number should be determined beforehand [7-9]. However the proper rule number (clustering number) are generally unable to be determined beforehand in practical, accordingly, we use the clustering algorithm to figure out the clustering center through continuously changing the determined clustering munber, and find out the optimal clustering number according to a certain clustering validity criterion [10]. Obviously the quantity of calculation which using the method of iteration-trial to collect sample number is remarkable when the sample number is quite large. And the clustering algorithm has its own problem, such as it is sensitive to the starting value and easy to sink into local minimum and so on, the clustering center won't be the optimum, thus it leads to affect the accuracy of modeling.

Some researchers use the genetic algorithm [11] and the immune evolution algorithm [12] to overcome the shortcomings of general clustering algorithm is easy to sink into the local minimum and sensitive to the initialization. But these improved clustering algorithms are still static which means that it is unable to identify the clustering number directly. Therefore a novel variable lengths of chromosomes genetic algorithm is proposed in this paper to deal with the dynamic clustering, the optimal clustering number can be determined dynamically as well as the clustering center can be determined accurately. In this novel algorithm, different string lengths of chromosomes represent different number of clustering center. In order to adapt to this encoding method, we improved the conventional crossover operation in this paper, at the same time, in order to make sure the system can be optimized more rapidly and stably, we used the local search capacity based on FCM algorithm, and also introduced memory cells and vaccine inoculation mechanism of immune system. A kind of fuzzy model identifying method based on this highly effective dynamic clustering algorithm is proposed in this paper; this method can simultaneously identify the premise structures and parameters of the non-linear system fuzzy model. As the simulation example indicates, this kind of identification has the merit of simple calculation, only few fuzzy rule number needed and higher accuracy.

## 2 New Dynamic Evolutionary Clustering Algorithm

Generally, clustering contains 3 sections: selecting clustering validity criterion function, determining clustering center and selecting clustering algorithm. The clustering is static if the clustering number is determined beforehand, on the contary, the clustering which its clustering number can be determined in the course of clustering is dynamic clustering. Given the X={x1,x2,...,xn} ⊂ Rp reprents the to-be-classified sample, V={v1,v2,...,vn} ⊂ Rp represents the clustering number, let c (1<c<n) be the classified number, U=[uik] (1≤i≤c, 1≤k≤n, ) be the partition matrix, uik represents the membership of the xk to the ith class. It is concluded that the clustering process is to optimizing the clustering criterion function. It ultimately aims to obtain the clustering number V and the partition matrix U of the sample X.

In the general genetic algorithm, the lengths of different chromosomes of the evolutionary population are all the same, the central position can change but the clustering number is unable to change, therefore it can not be practical in dealing with dynamic clustering problem. Therefore a newly variable lengths of chromosomes genetic immunity algorithm is proposed in this papper, in this algorithm, different chromosomes may have different lengths and they represent clustering center with different number, moreover the number and the position of clustering center can continuously change until they are optimal. In order to enhance local search capacity, we use the mechanism of searching for optimum of FCM algorithm to update each chromosome continuously while the global search capacity can be further improved with help of the introduction of the immune function.

The process of dynamic clustering based on variable lengths of chromosomes genetic immunity algorithm and FCM algorithm displayed as fig. 1 below, here is the explanation of the main approach.

(1) Encoding the variable-length chromosomes and initializing population
There are usually two encoding approaches based on genetic algorithm to deal with clustering problem. One is based on partition matrix U [7], which the search space is variable along with the data set sample number, the search space increase rapidly when the data set sample number increase, it is quite difficult to search for the optimum. Therefore we here introduce another kind of real number encoding approach based on clustering center which the search space is unvaried while the data set sample number changes.

Chromosome Si represents p-dimensional space with ci( 1<ci<n) clustering center(s), the encoding format : Si=( g1,g2, … ,gp,gp+1, …, gl) , which let the length of chromosome be l=ci×p, the first p values in Si represent the first one clustering centers, the other different-length chromosomes represent the different number of clustering center by analogy.

Starting to generate a random population: randomly selecting N data in sample data set, the number of the i( 1≤i≤N) th data is ci which represents ci clustering centers. Each data string code is a ci×p bit individual, randomly selected N data to form a initial population with N individuals.

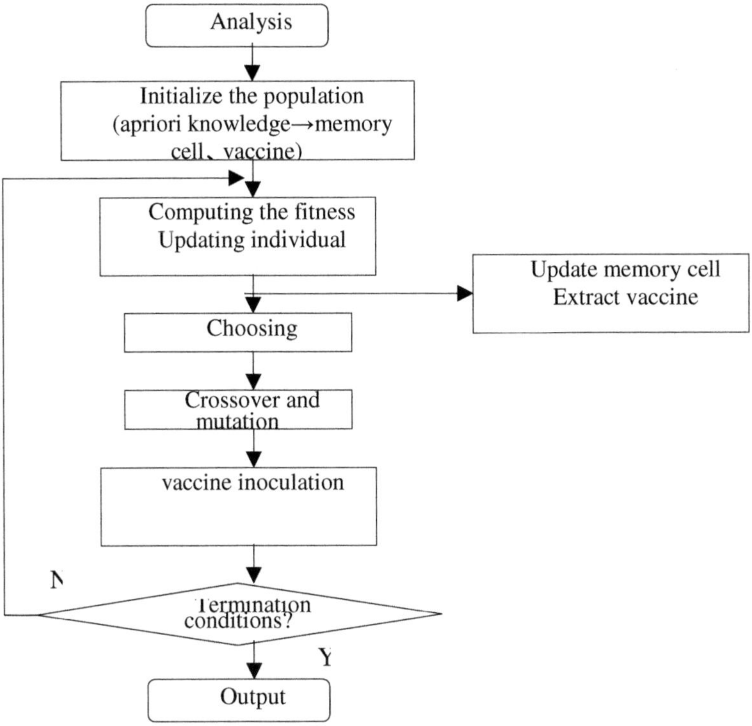


**Fig. 1.** Flow chart of dynamic clustering based on immune genetic algorithm

(2) Fitness function and individual updating based on FCM algorithm
Fitness function is a criterion to evaluate the fitness of individual to the population environment, the probabity of that a certain individual succeeding to its offspring can be obtained by individual fitness value,. Fitness function can be obtained by some kind of mapping transformation from the objective function, but regarding to clustering, the objective function is the clustering validity criterion function, it must be able to manifest the performance of classification. It made a comparison of some clustering validity evaluation index such as DB( Davies-Bouldin) 、XB( Xie-Beni) and a newly proposed index in the reference [9], the result indicated that the effectiveness of cluttering is more expressive by using the newly index, it is defined as below:

$$I(c) = (\frac{1}{c} \times \frac{E_1}{J_c} \times D_c)^r \tag{1}$$

$$J_c = \sum_{i=1}^{c}\sum_{k=1}^{n} u_{ik} \| x_k - v_i \| ;\ Dc = \max_{i,j=1}^{c} \| v_i - v_j \| ;\ \|\cdot\|$$

represents the Euclidean norm; c represents the clustering number, exponential r≥1 represents real number; E1 represents a constant to a given data set, it's main role is standardization and avoiding to obtain a minimum. Jc represents the sum of the distance from the points to the center in each class, Dc represents the maximum distance from each class center, the optimal classification should be that the point inside the class can converge together and the distance from each class should be as far as possible, therefore the bigger value I could be means the better performance of clustering.

This newly index represents the objective function. Given one individual with length = c×p, c clustering center(s) {v1,v2,...,vc} can be obtained by decoding this individual, the formula of the fitness function of this individual as below:

$$f = \{\frac{1}{c} \times \frac{E_1 \times \max_{i,j=1}^{c} \| v_i - v_j \|}{\sum_{i=1}^{c}\sum_{k=1}^{n} u_{ik} \| x_k - v_i \|}\}^2 \tag{2}$$

By using the following formula to partition matrix $U=[u_{ik}]_{c}\times_{p}$:

$$\begin{cases} u_{ik} = \dfrac{1}{\sum_{j=1}^{c}\left(\dfrac{\| v_i - x_k \|}{| v_j - x_k \|}\right)^{2/(m-1)}}, & I_k = \varphi \\ u_{ik} = 0, \forall i \in \bar{I}_k, \sum_{i \in I_k} u_{ik} = 1, & I_k = \varphi \end{cases} \tag{3}$$

Ik={i|1 ≤ i ≤ c, ||vi-xk||=0}, $\bar{I}_k$ ={1, 2, ..., c}- Ik; $m \in [1, \infty]$ represents the index of fuzzy degree, usually m=2. After computing on the fitness of one individual, then we use the local search capacity based on FCM to update each clustering center by the following formula:

$$v_i = \frac{\sum_{k=1}^{n} (u_{ik})^m x_k}{\sum_{k=1}^{n} (u_{ik})^m}, 1 \le i \le c \tag{4}$$

We form a new individual which replace the current individual by using the updated c clustering center(s)

(3) The newly crossover operation

The selection of general genetic algorithms is still practical in this application, such as the proportion of fitness method and roulette wheel selection. The selected individual(s) carry out the crossover operation with probability Pc. wether crossover is good or not, to a large extent, decide the performance of genetic algorithm. As the variable lengths of chromosomes encoding method is proposed the general crossover methods of genetic algorithm are no longer applicable in this paper. Therefore two newly crossover methods are proposed in this paper, the selected individuals' crossover with equal probability by one of these two crossover methods.

One is the gene string matching crossover based on nearest neighbor method. Given S1=( a1a2…ac1) and S2=( b1b2bc2) are the two individuals, c1≤c2, Each one of S1 represents a gene string of clustering center $a_i = (a_i^{(1)}, a_i^{(2)}, \cdots, a_i^{(p)})$, as in the gene string matching process that selecting the one which is the nearest to the ai, the matched gene string no longer participated in the following matching. Rearranging the c1 gene strings in S2 according to the sequence of gene string matching, and then randomly select one point in 0~ci×p to obtain two general new individual S3 and S4. by single point crossover on S1 and S2 . Rearranging gene string aims to let the different individuals have similar clustering centers in the same position, thus can avoided the population degradation caused by two elitism parent-individuals generate mutated offspring individuals after crossover.

The offspring individual copy the clustering center of its parents by using the first crossover method. To obtain the offspring individuals of different clustering centers as well as maintaining the variety of chromosomes, the second crossover method based on cross-cut and cloning is introduced as follows. The gene string/set is seen to be inseparable in this crossover method, the crossover point can only be selected between different gene string/set. Take the uncrossovered S1 and S2 as an example to demonstrate crossover process as follows:

S1 (ci×p) ={a1 a2…at1-1 at1||at1+1 … ac1}
S2 (ci×p) ={b1 b2…bt2-1 bt2||bt2+1 … bc2}

Crossover them to obtain two new individuals:

S5 ((c2+t1-t2) ×p= {a1 a2…at1-1 at1 || bt2+1 … bc2}
S6 ((c1+t2-t1) ×p= {b1 b2 … bt2-1 bt1 || at2+1 … ac2}

tl and t2 are the crossover points of chromosomes S1 and S2, the two new individuals S5 and S6 after crossover represent the number of clustering centers (c2+tl-t2) and (cl+t2-t1) .

The mutation operation will be taken by gene bit on the new after-crossover individuals, each one of the floating number of the gene bit will be mutated randomly with the mutation probability Pm, the mutated gene bit is replaced by one other uniform distributed random number.

(4) The immune operation

The introduction of the immune memory and vaccine inoculation in the algorithm aims to improve the global search capacity.

1)Immune memory
As for the apriority knowledge, the initial value can be seen as the initial memory cells, with randomly generated individuals, constitute an initial population. In the evolutionary process of continuously updating memory cells: rank the chromosomes by fitness descending order, add the Nm high fitness individuals into the memory cells warehouse, as the limited number of memory cells, the one which is the closest to the original individual is replaced by the newly antibody

2)Vaccine inoculation
Vaccine refer to some basic characteristics information extracted from the evolutionary environment or the apriority knowledge based on the issues, it is an estimation to the optimal chromosome gene, Vaccine inoculation is a measure of revising chromosome gene according to the vaccine, it aims to improve individual fitness. the information and accuracy of the vaccine play an very important role of the algorithm performance. There are usually two methods of extracting the immune vaccine [10], one is to make a specific analysis of the issues to collect characteristics information to produce immune vaccine; another is to extract the equal information from the gene of the optimal individual to produce immune vaccine in the evolutionary process. It is usually limited for people to acquire sufficient apriority knowledge and it is much difficult to extract equal characteristics information to obtain the proper immune vaccine, therefore people tend to use the second method of adaptive extracting vaccine.

Given the population of the kth generation Pk={Sk1,Sk2,...,SkN} has obtained the optimal individual which is the highest fitness individual Sk_opt, decomposing Sk_opt to obtain the gene string{v1,v2,...,vc} as to be the immune vaccines, population Pk overstated genetically to be pk_new. Inoculate vaccine to pk_new with probability Pv, given the individual S=( d1d2...dc3) is selected, extract a vaccine v from the vaccine warehouse, then find out a gene string which is the nearest one to vaccine v in S , and this gene string is replaced by the v to obtain a new individual Snew. Using the nearest neighbor switchover rule in order to avoid that one individual has similar center

The termination conditions of the algorithm adopt a combined method of limiting the times of iteration and terminating computation if the optimal solution which is obtained by several times iterating (eg. 5 times iteration) is still unable to improve, when it meet the termination conditions, the process of searching for the optimum is terminated and then output the optimal solution.

A clustering example

One clustering example presented in this paper verify the effectiveness of this newly clustering algorithm.

Fig. 2(a) is a two-dimensional original data set contains four classes and a total number of 100 data points. The parameters of clustering algorithm are selected as follows: N=50, Pc=0.9, Pm=0.05, Pv=0.1, Nm=5, the evolutionary algebra is 40. the clustering results as fig. 2(b) below, the Original data set is divided into four classes, the little circles on the fig.2(b) represent the clustering centers' position. So the algorithm can be implemented accurately and effectively.

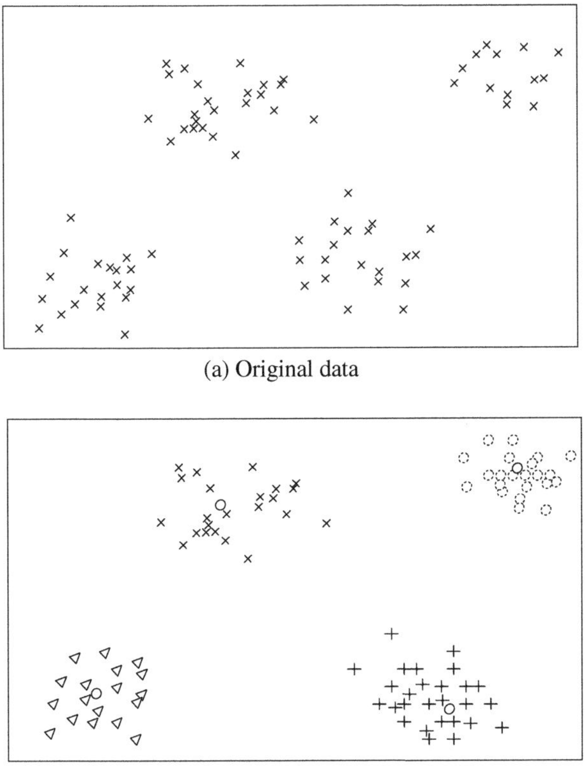

(a) Original data

(b) Clustering result

**Fig. 2.** Test for clustering algorithm

## 3 Identification of Fuzzy Model Based on Dynamic Clustering

The MIMO system can be decomposed into several MISO subsystems to identify, therefore it maintains its universality, so we only discuss the identification of MISO system in the paper.

The first step of fuzzy modeling is to identify the structure of model, including partition the input-output space and determine the fuzzy rule number. The basic idea of using clustering algorithm to identify structure is classifying the samples which have similar characteristics based on some kind of comparability measurement or distance information, each class represents one fuzzy set, the parameters of memberships function are determined by the characteristic function of class. By using the dynamic clustering method, attention is paid to an new approach of fuzzy structure identification based on dynamic clustering in this paper, the optimal fuzzy rule number and the precise premise parameter can be obtained simultaneously, and it overcomes the shortcomings of the rule number should be determined beforehand, and the identification process can be then well simplified.

Regarding a MISO system, the sample set consists of the input-output data of the system, given the sample as follows: $\{(\varphi_i, y_i), i=1,2,\ldots,n\}$, $\varphi_i$ represents the vector of affecting system output, namely the general input vector. Generally, selecting the current and past input-output data as its component of vector. yi represents the system output. Define the $z_i = (\varphi_i, y_i)$, the sample set can be then represented as $Z = \{z_1, z_2, \cdots, z_n\} \subset R^{r+1}$.

Given c classes are obtained by the convergence of the sample set using the dynamic clustering method mentioned above, the centers are $V = \{v_1, v_2, \cdots, v_c\} \subset R^{r+1}$, remove the output components of the clutering centers to obtain the clustering centers of the input space: $V^x = \{v_1^x, v_2^x, \cdots, v_c^x\} \subset R^r$, the clustering center $v_j^x$ represents the center of fuzzy set Bj, and the c sub-rule models which represent each clustering center can be obtained by the clustering result as follows, Rj: if $\varphi \in B_j(v_j^x, \sigma_j)$ then

$$y^j = p_0^j + p_1^j\varphi(1) + p_2^j\varphi(2) + \cdots p_r^j\varphi(r) \quad (j=1,2,\ldots,c) \tag{5}$$

Rj represents the jth fuzzy rule; $\varphi$ represents the input vector of fuzzy model; Bj represents the jth sample set of input vector; yj represents the jth rule output; $p_i^j$ (i=0,1...,r) represents the conclusion parameter. The global model of system can be figured out by the fuzzy weighted operation between these c submodels.

$$y = f(\varphi) = \sum_{j=1}^{c} w^j y^j \tag{6}$$

As for $w^j = \mu^j / \sum_{j=1}^{c} \mu^j$, $\mu^j$ represents the membership of the output to fuzzy set Bj . The gauss function serve as the membership function of fuzzy set Bj in this paper

$$\mu_{B_j} = \exp[-(\frac{\| \varphi - v_j^x \|}{\sigma_j})^2] \tag{7}$$

$\sigma_j$ represents the width of membership function which can be evaluated by the clustering result. Using the nearest neighbor domain method to determine $\sigma_j$, which means the width of fuzzy set Bj is determined by the average distance from the center to its nearest neighbor k centers.

$$\sigma_j = \frac{1}{\beta}(\frac{1}{k}\sum_{l=1}^{k} \| v_j - v_l \|) \tag{8}$$

vl(l=1,2,…,k) represents the k centers which are the nearest one to No. j center vj , k =1 or k=2 when the rule number is few; $\beta$ represents a constant, and $\beta$=4.

When it demands the higher accuracy of system modeling, we can use the Gradient Descent method to implement a further adjustment of the center and width of fuzzy set. But in practical, the following simulation test indicate that the approach of building system model introduced in this paper has already acquired much higher accuracy. After the premise structure and parameter are determined by the dynamic clustering method, then we can use the conventional Least- Square method to obtain the optimal conclusion parameter. The strange matrix can be solved by the SVD decomposition approach.

As to measure the accuracy of modeling, the performance criterion is then defined as follows:

$$\varepsilon_{\text{MSE}} = \frac{1}{n}\sum_{k=1}^{n}[y(k) - y_m(k)]^2 \tag{9}$$

n represents the total number of data points; y(k) represents the practical output of the kth point; ym(k) represents the output of model. The less of the model error the higher accuracy of the model.

## 4 Case Testing

Box-Jenkins gas furnace data [1] is a typical example of identifying system which has been adopted by many fuzzy identification researchers. The data consists of 296 pairs of intput-output observations, it is a dynamic SISO system, intput u(t) represents the methane gas feedrate, output y(t) represents the C02 concentration. We will demonstrate three pairs of intput variables in order to easily compared with: ① y(t-1), u(t-4); ② y(t-1), y(t-2), u(t-3), u(t-4); ③ y(t-1), y(t-2), y(t-3), u(t-u), u(t-2), u(t-3). The parameters of clustering algorithm are selected as: N=50, Pc=0.9, Pm=0.05, Nm=5, let the evolutionary algebra be 50.

**Table 1.** Comparison of gas furnace modeling results based on different fuzzy identifying methods

| Reference | Input variables | Rule number | $\varepsilon_{\text{MSE}}$ |
|---|---|---|---|
| Reference [13] | *y*(*t*-1),*u*(*t*-4) | 25 | 0.328 |
| Reference [14] | *y*(*t*-1),*u*(*t*--4) | 3 | 0.2678 |
| Reference [15] | *y*(*t*-1),*y*(*t*-2), *u*(*t*-3),*u*(*t*-4) | 5 | 0.248 |
| Reference [16] | *y*(*t*-1),*y*(*t*-2), *y*(*t*-3), *u*(*t*-1),*u*(*t*-2),*u*(*t*-3) | 2 | 0.068 |
| | *y*(*t*-1), *u*(*t*-4) | 3 | 01515 |
| This paper | *y*(*t*-1),*y*(*t*-2), *u*(*t*-3),*u*(*t*-4) | 2 | 0.0636 |
| | *y*(*t*-1), *y*(*t*-2), *y*(*t*-3), *u*(*t*-1),*u*(*t*-2),*u*(*t*-3) | 2 | 0.0607 |

The identifying result are displayed as tab. 1. Tab. 1 makes a comparison of the accuracy of the method presented in this paper and other identifying methods in the equal performance criterion.

## 5 Conclusions

A variable length of chromosome genetic algorithm is proposed in this paper to deal with the dynamic clustering problem in fuzzy modeling process. The convergence of the algorithm is expedited with the help of the local search capacity of FCM algorithm and the introduction of memory cells and vaccine inoculation mechanism of immune system. Identifying the premise parameter and structure of nonlinear system fuzzy rule model based on this high effective clustering algorithm. Using the Least-Square method to obtain the solution on model conclusion parameter. The expeditiousness, accuracy and effectiveness of the proposed fuzzy modeling method based on DECA is demonstrated by simulation examples. Using this fuzzy model identification method in the application of building nonlinear model in the electric power process enables that the rule number no longer need to be determined beforehand, fewer quantity of calculation and high accuracy of the identified model, thus establishing a model foundation for the global optimizing control in the electric power production process.

**Acknowledgments.** This paper is supported by the science and technology of Department of Education of Hunan Province of China No. 06C268, and the Natural Science Foundation of Hunan Province of China No. 06JJ2033.

## References

1. Bertotti, G.: Identification of the Damping Coefficient in Landau-Lifshitz Equation. Physical B (2001) 102–105
2. Mau, S T.: A Subspace Modal Superposition Method for Non-classical Damping Systems. Earthquake Eng Struct Dyn (1998) 931–942
3. Zhang, T.J., Lu, J.H., Yu, K.J.: A New Approach for Predictive Control Based on Fuzzy Decision-making and Its Application to Thermal Process. Proceedings of the CSEE (2004) 179–184
4. Liu, Z.Y., Lu, J.H., Chen, L.J.: A Novel RBF Neural Network and Its Application in Thermal Processes Modeling. Proceedings of the CSEE (2002) 8-122
5. Feng, W.X., Chen, X.: Amethod for Estimating of Damping Matrix of Structural Dynamic Systems. J of Guangdong University of Technology (2001) 6-11
6. Maulik U., Bandyopdhyay, S.: Performance Evaluation of Some Clustering Algorithms and Validity Indices. IEEE Trans on Pattern Analysis and Machine Intelligence (2002) 1650-1654
7. Hou, Y.W., Shen, J., Li, Y.G.: A Simulation Study on Load Modeling of A Thermal Power Unit Based on Wavelet Neural Networks. Proceedings of the CSEE (2003) 220-224
8. Jiang, W.J.: Research on the Learning Algorithm of BP Neural Networks Embedded in Evolution Strategies. WCICA'2005 (2005) 222-227

9. Xu, Q., Wen, X.R.: High Precision Direct Integration Scheme for Structural Dynamic Load Identification. Chinese J of Computational Mechanics (2002) 53-57
10. Wu, J.Y., Wang, X.C.: A Parallel Genetic Design Method With Coarse Grain. Chinese J of Computational Mechanics (2002) 148-153
11. Li, S.J., Liu, Y.X.: Identification of Structural Vibration Parameter Based on Genetic Algorithm. J of Chinese University of Mining Science and Technology (2001) 256-260
12. Gomez-Skarmeta A.F., Delgado, M., Vila, M.A.: About the Use of Fuzzy Clustering Techniques for Fuzzy Model Identification. Fuzzy Sets and Systems (1999) 179-188
13. Furukawa, T.: An Automated System for Simulation and Parameter Identification of Inelastic Constitute Models. Computer Methods Appl. Mech. Eng. (2002) 2235-2260
14. Deng, H., Sun, Z.Q., Sun, F.C.: The Fuzzy Cluster Identification Algorithm.Control Theory and Applications (2001) 171-175
15. Zhao, L., Tsujimura, Y., Gen M.: Genetic Algorithm for Fuzzy Clustering. Proceedings of IEEE International Conference on Evolutionary Computation (1996)
16. Liu, J., Zhong, W.C., Liu, F. et al. A Novel Clustering Based on the Immune Evolutionary Algorithm. ACTA Electronic SWICA (2001) 1860-1072

# Applying Hybrid Neural Fuzzy System to Embedded System Hardware/Software Partitioning

Yue Huang and YongSoo Kim*

Department of Computer Science in Kyungwon University,
Sujeong-Gu, Seongnam-Si, Gyeonggi-Do, 461-701, Korea
huangyue@ku.kyungwon.ac.kr, kimys@kyungwon.ac.kr

**Abstract.** Hardware/Software (HW/SW) partitioning is becoming an increasingly crucial step in embedded system co-design. To cope with roughly assumed parameters in system specification and imprecise benchmarks for judging the solution's quality, researchers have been trying to find methods for a semi-optimal partitioning scheme in HW/SW partitioning for years. This paper proposes an application of a hybrid neural fuzzy system incorporating Boltzmann machine to the HW/SW partitioning problem. The hybrid neural fuzzy system's architecture and performance estimation against simulated annealing algorithm are evaluated. The simulation result shows the proposed system outperforms other algorithm in cost and performance.

**Keywords:** hardware/software partitioning, hybrid neural fuzzy system, Boltzmann machine, embedded system.

## 1 Introduction

An embedded system is some combination of computer hardware and software, specifically designed to perform a particular function. It is widely used in industry machines, medical equipments, automobiles, airplanes, robots, and other various fields. In the early times, the way of embedded system design was mainly classified into two categories: one is developing some software for given hardware; the other is implementing specific hardware architecture for existent software. These methods couldn't assure the attention to the requirements of both hardware and software. The partitioning scheme could be evaluated only when hardware and software were all prepared, thus it was difficult to find the hardware and software compatibility problem at the early time of the development cycle. Nowadays embedded system becomes more functional and complex. To meet the more complicated design requirements, smaller size and shorter product life cycle with performance, cost, and reliability goals, a new way of embedded system design was needed. HW/SW co-design [1] method is introduced to fulfill this requirement.

Different from the traditional methods, HW/SW co-design focuses more on the cooperation of both hardware and software designers. It considers and balances the

* Corresponding author.

D.-S. Huang, L. Heutte, and M. Loog (Eds.): ICIC 2007, LNAI 4682, pp. 660–669, 2007.

hardware and software components from the time when system specification is proposed. The designs of hardware and software are carried out in parallel. They influence each other though the whole design procedure. A HW/SW partitioning technique is used to adjust and evaluate the partitioning schemes and finally choose the best solution from all the candidates. Therefore HW/SW partitioning is one of the foremost issues in HW/SW co-design [2].

HW/SW partitioning is a process of determining what part of the system is suitable for a hardware implementation (i.e. as ASIC) and what part is better implemented as software (i.e. as code for a microprocessor) [3]. The "suitable" and "better" mean the implementation needs less area consumption and less executing time compared to the other way in the coordinate condition. As VLSI and IT technology advance, the hardware and software resources can be chosen in increasingly various ways and the boundary between them isn't as obvious as before. On one hand it is not surprising to see that some complicated arithmetic is implemented in hardware, on the other hand it is also quite common to see that software in RISC implements a function that has been done in hardware. Thus we urgently need a methodology that guides us to precise partitioning for the agile development and best performance.

## 2 Related Work

HW/SW partitioning is not a new problem. Many researches have been addressing on it since 1970s [4]. Till now, there have been many achievements in this field. In [3], E. Barros introduced a HW/SW partitioning approach supported by the use of UNITY, a specification language based on static transitions. Genetic algorithm is a search technique widely used in computing an approximate solution for optimization problem, thus it is also suitable for solving HW/SW partitioning problem [5] [6]. In [7], the proposed HW/SW partitioning algorithm constructed an efficient Branch-and-Bound approach to partition the hot path selected by path profiling techniques, and communication overhead was taken into account. Paper [8] investigated HW/SW partitioning problem solved by simulated annealing algorithm, a generic probabilistic mete-algorithm for the global optimization problem. Paper [9] presented two heuristics for automatic HW/SW partitioning of system level specifications, simulated annealing algorithm and tabu algorithm.

Most of these published works divided partitioning flow into two steps: mapping system models to hardware and software sets and estimating the system cost according to the partitioning scheme. Through the iterative process, optimal solution could be obtained. This methodology has been proved to be practical and effective, therefore our proposed method for HW/SW partitioning also follows this way.

In the rest of this paper, we propose a new idea for HW/SW partitioning using a Boltzmann machine incorporated hybrid neural fuzzy system. Section 3 explains the partitioning model formalization and section 4 illustrates the hybrid system's architecture. Section 5 analyzes its performance by simulation.

## 3 Specification of HW/SW Partitioning Problem

HW/SW partitioning is NP-Hard problem. According to [5], [6], [9] and [10], it can be expressed as a set of communicating nodes, represented by a Directed Acyclic Graph (DAG) in form of G = (N, E). N is the node set, $N=\{n_0,n_1,......n_{k-1}\}$. It is composed of two parts, $N=N_s+N_h$, where $N_s$ is the subset of the nodes that is implemented by software and $N_h$ is the subset of the nodes that is implemented by hardware, $|N_s|+|N_h|=k$, $N_s \cap N_h = \varnothing$. E is the edge set, $E=\{e_{ij}\}$, $0 \le i, j \le k-1$, $E \subseteq N \times N$.

For each node $n_i \in N$, it represents an atomic unit of functionality to be executed in the system, which we call as "module". In the rest of this paper, we will call the elements in formula (1) as main properties of module *i*.

$$n_i = (ah_i, th_i, as_i, ts_i, \tau_i, F_i), \tag{1}$$

where $ah_i$ indicates the area consumption for module $i$ when it is implemented by hardware, which may represent the number of FPGA(Field-Programmable Gate Array) or ASIC (Application Specific Integrated Circuits) modules;

$th_i$ is the time consumption when module $i$ is implemented by hardware;
$as_i$ is the area consumption when module $i$ is implemented by software, which may represent the amount of memory for program logic;
$ts_i$ is the time consumption when module $i$ is implemented by software;
$\tau_i$ is the number of times the module $i$ has executed;
$F_i$ is a binary mapping function, where $F_i$=1 means node $n_i$ should be mapped into hardware set and $F_i$=0 means node $n_i$ should be mapped into software set. Initially, its value is 0 for every $n_i$.

For each edge $e_{ij} \in E$, it indicates some kind of communication between two modules. It is meaningful when two modules are implemented in different ways. If they are implemented in the same way, both by hardware or both by software, we assume there is no time consumption between them.

$$e_{ij} = t_{ij} \times (1-F_i) \times F_j + t_{ij} \times F_i \times (1-F_j), \tag{2}$$

where $e_{ij}$ indicates the input communication of node $n_j$ from node $n_i$, or output communication of node $n_i$ to node $n_j$, which may present the time spent on transferring data between output and input module though bus;
$t_{ij}$ is the time consumption on data transfer when module $i$ is implemented by hardware and module $j$ is implemented by software, or module $i$ is implemented by software and module $j$ is implemented by hardware.

We suppose some preparatory works have been done before our research begins, such as doing HW/SW specification and verification, estimating $ah_i$ and $th_i$ using ASAP (As-Soon-As-Possible) scheduling algorithm, estimating $as_i$ and $ts_i$ though DFG liner technology and selection of different micro-process, etc. Our algorithm has to be applied after these important parameters have been calculated. We can use them directly to infer the value $F_i$ for each module. As we mentioned, HW/SW partitioning is a NP-Hard problem. If there are $k$ nodes in the DAG, there will be $2^k$ candidate HW/SW partitioning schemes. What we intend to do is finding out a scheme from them giving the least time consumption ($TC$) and the least area consumption ($AC$) under the time cost constraint $TC_{con}$ and the area cost constraint $AC_{con}$. Based on the study in [5] and [10], we modified the formulae which are used to denote the partitioning problem as in formula (3) to (7):

$$TC = ST - \sum_{i=0}^{k-1} \tau_i \times (ts_i - th_i) \times F_i + \frac{1}{2} \times \sum_{i=0}^{k-1} \sum_{j=0, j \neq i}^{k-1} \tau_i \times [t_{ij} \times (1 - F_i) \times F_j + t_{ij} \times F_i \times (1 - F_j)] \quad (3)$$

$$ST = \sum_{i=0}^{k-1} \tau_i \times ts_i \quad (4)$$

$$AC = \sum_{i=0}^{k-1} ah_i \times F_i + \sum_{i=0}^{k-1} as_i \times (1 - F_i) \times \alpha \quad (5)$$

$$TC \leq TC_{con} \quad (6)$$

$$AC \leq AC_{con} \quad (7)$$

Hitherto, area consumption of a software module has not been taken into the calculation of the system's area consumption. But we regard $as_i$ a valuable factor in calculating $AC$ and included it in formula (5). Size becomes one of the most important parameters when we evaluate an embedded system. To decrease the system's size, designer always tries to use as less registers as possible. If a module implemented by software uses a lot of registers which are unnecessary for other modules, we'd better reconsider its implementation mode carefully. Based on this consideration, we redefine formula (5) and take $as_i$ with a coefficient $\alpha$ in it. $\alpha$ is correlative with the importance of the number of registers and the size of the memory. Heuristically, $\alpha \in [0.1, 0.5]$. Finally, the cost function can be defined as:

$$cost = \min(AC + TC) \quad (8)$$

## 4 Architecture of the Hybrid Neural Fuzzy System

The interest in hybrid neural fuzzy system has grown rapidly since it came out. Researchers have applied it to data classification, image retrieval, controlling, decision making and many other fields because of its great ability. It takes the merits of both of fuzzy logic and neural network, and avoids their drawbacks [11]. As time goes on, its potential applicability will be found in more domains. In this paper, we apply it in HW/SW partitioning for embedded system.

Our proposed hybrid neural fuzzy system is composed of two parts: a classification network and an operation network. The classification network is used to generate an initial scheme for the HW/SW partitioning problem by using fuzzy logic. Its outputs are transferred into the operation network to generate the optimal partitioning scheme under the constraints. In an overview, the whole system is working as a fuzzy system, while in terms of its topological structure, it is a network composed of nodes and weights. Therefore it combines the advantages of both fuzzy system and neural network with a great mapping and self-learning ability. The whole system's architecture is shown in Figure 1.

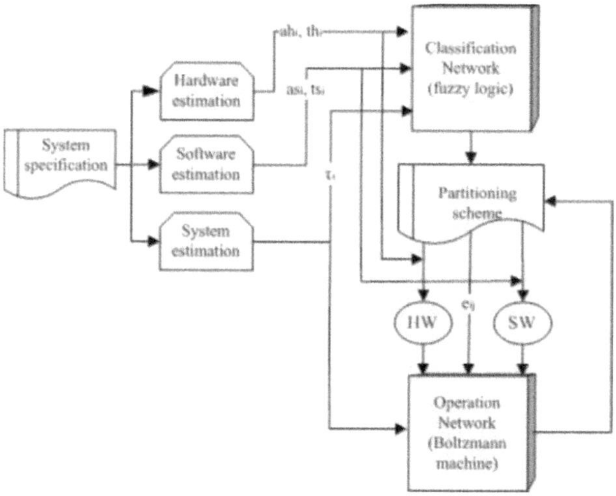


**Fig. 1.** Architecture of the hybrid neural fuzzy system

### 4.1 The Structure of Classification Network

We propose five layers in the classification network. Neurons in the input layer are assumed to be the nodes in the DAG. The input data are the vectors composed of elements in formula (1) which indicate the main properties of each node in the DAG. Three layers between input and output layers are hidden in the black box.

The first hidden layer is called fuzzy layer. A membership function is used to classify the vector element into several classes, which are the fuzzy sets. Supposing each vector element data can be classify into $s$ classes, there are $c \times s$ neurons in this layer in total.

The second hidden layer is called rule layer. The outputs of the fuzzy layer are transferred into this layer to match the rules from domain experts. If there are $r$ rules, there will be $r$ neurons in this layer.

Because every rule neuron in the rule layer gives out a prediction and the predictions may be conflicting, we define the third hidden layer as a confirm layer to calculate the confidence of the final prediction, which is represent as in formula (9) and (10),

$$\text{confidence}=\begin{cases}\frac{a}{r} & \text{if } a>\frac{r}{2}\\ 1-\frac{a}{r} & \text{if } 0\le a\le\frac{r}{2}\end{cases} \tag{9}$$

$$a=\sum_{i=0}^{r-1} 1\times O_{\text{rulelayer-i}} \tag{10}$$

The last layer is output layer. Each input neuron has a corresponding output neuron to show the final predication for the input vector as in formula (11), where result 0 indicates input neuron should be implemented by software and result 1 indicates input neuron should be implemented by hardware.

$$O_i=\begin{cases}1 & \text{if } a>\frac{r}{2}\\ 0 & \text{if } 0\le a\le\frac{r}{2}\end{cases} \tag{11}$$

The structure of the network is shown in Figure 2. To make the network's architecture clear and easy to see, we only show the neurons and connections just for one input neuron $n_i$.

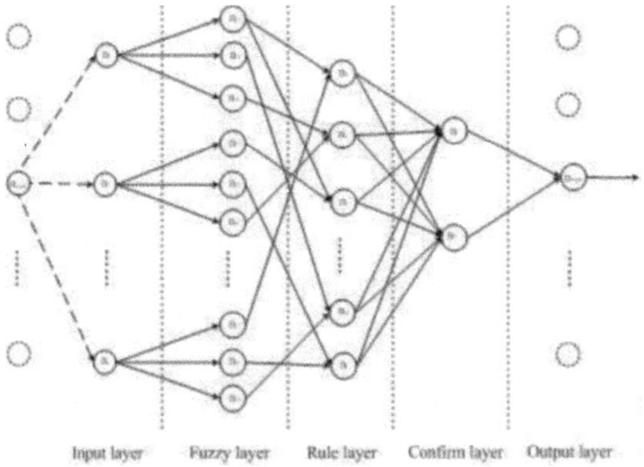


**Fig. 2.** Structure of the classification network

After running this classification network, we can get an initial partitioning scheme: every node is marked by an alternative number 0 or 1, indicating the implementation mode.

### 4.2 The Structure of Operation Network

The operation network is a neural network incorporated with Boltzmann machine. Boltzmann machine is put forward to solve the local minimum problem of both BP neural network (which means the multi-layer network trained via BP training algorithm) and Hopfield neural network [12] [13]. Similar to BP neural network,

Boltzmann machine also has layers, called visual layer and hidden layer. And visual layer can be divided into input part and output part. The difference between Boltzmann machine and other multiple layer networks is that the layers in Boltzmann machine have no obvious circumscription and the connection between two neurons is bidirectional. Like Hopfield neural network, all the neurons in the network are connected and the change of network state obeys the probability distribution. This combination helps Boltzmann machine to avoid falling into the local minimum of error function or energy function and finally get the optimal solution of the problem. Figure 3 shows the structure of two simple Boltzmann machine networks.

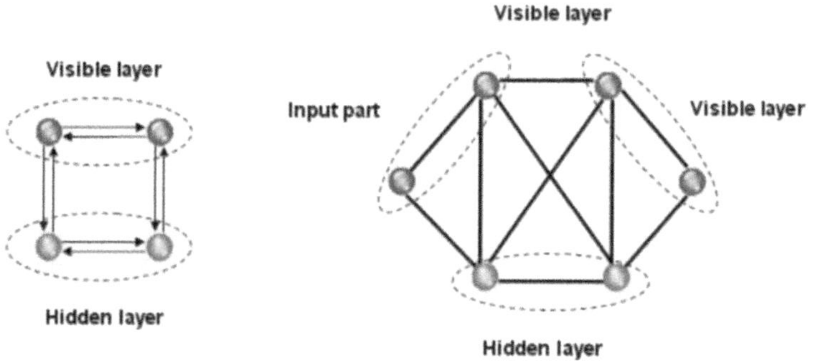


**Fig. 3.** Structure of simplele Boltzmann machine

The input data for operation network are the character properties which are the same as classification network's. The output is a sequence composed of binary number 0 or 1. Initial partitioning scheme gotten from the classification network is the initial state of the operation network. If $state_i$=1, $ah_i$, $th_i$, $t_{ij}$, where $n_i \in N_h \cap n_j \in N_s$, are enabled and taken into the calculation of *cost*; if $state_i$ =0, $as_i$, $ts_i$, $t_{ij}$, where $n_i \in N_s \cap n_j \in N_h$, are enabled and considered. For every iteration, the network tries to get a balanced state for all the neurons under the controlling parameter *T*. As iteration goes on, both parameter *T* and energy *cost* decreases. When *T* becomes small enough and the network reaches a balanced state, the network's output is the optimized solution for the problem.

## 5 Performance Evaluation

The theory behind our system indicates that if the Boltzmann machine receives more crisp meaningful inputs, it will improve the overall output and quality of its prediction. To verify the availability of our proposed system, we adopt the similar simulation method introduced in [14] and [15]. For there is no standard test data in this domain, we generate a few DAGs randomly and arrange the attributes to each node and edge. We assume these are the practical data gotten from previous work, such as system performance analysis.

### 5.1 Embedded System Architecture

During our study on designing an embedded system, we always keep a viewpoint in mind that embedded system is still a kind of computer system but the processor's architecture and its software are a little different from the typical conventional computer. Therefore for simulation purpose, we assume the system model has nine basic modules like a real embedded system typically has, as shown in Table 1. Some of the modules are HW-bound, some are SW-bound, and some are mixed. Nodes in the DAG are distributed into these modules. To make the simulation more reflective of the real system, we allocate the properties for each node based on its module.

**Table 1.** System modules

| HW-bound | HW-SW | SW-bound |
|---|---|---|
| CPU | Driver | File System |
| RAM | Loader | Scheduling |
| Flash Memory | Memory Management | Inter-Processor Call |

### 5.2 Simulation Data, Fuzzy Rules and the Neural Network

For simulation, we generate five DAGs randomly by using GVF (Graph Visualization Framework), with 20, 50, 100, 200, 400 nodes respectively. Each edge and node in the graph has different character properties. To enlarge the simulation data base, we add the Gaussian noise at 10 different levels to the 5 groups of sample data. As a result, there are 7700 sample data for testing in total.

Since currently there are no standard fuzzy rules from the domain experts available to use, we choose 200 sample nodes from the simulation data base, whose properties vectors are with obvious hardware implementation characteristic or software implementation characteristic, as the learning data. An open-source data mining software WEKA is used to get the fuzzy rules for our hybrid neural fuzzy system. WEKA can work on the learning data and generate fuzzy rules automatically.

Properties vectors are the input data fed into our hybrid system and the partitioning scheme, which is a 0-1 sequence, is the output data. The main logic in classification network and operation network is implemented using MATLAB. In classification network, we choose sigmoid membership function to change original data into fuzzy sets. We take the coefficient $\alpha$ in formula (5) as $\alpha$ =0.2, and the controlling parameter $T$ in Boltzmann machine as $T_0$=200 (the initial value of $T$), $T_{final}$=0.001 (the final value of $T$). According to the simulation results with different coefficients, we found when $\alpha$ is more than 0.5, sometimes AC couldn't be under the constraint $AC_{con}$. And also, when $\alpha$ is less than 0.1, $as_i$ becomes too small to be meaningfull. That's the reason why we recommend the value of $\alpha$ should be between 0.1 and 0.5 and we choose $\alpha$=0.3 in our simulation.

### 5.3 Simulation Results

Among such a huge number of methods for solving HW/SW partitioning problem, simulated annealing algorithm (SAA) is most similar to Boltzmann machine network.

So we take the comparison of our proposed hybrid neural fuzzy system (HNFS) with SAA. The following Table 2 shows the results. The value of *AC*, *TC* and *cost* are listed. According to the data in table, for any set of nodes, the *cost* of HNFS is less than that of SAA, but the difference is not obvious, sometimes even the same.

**Table 2.** Performance comparison of HNFS with SAA

| Number of nodes | $AC_{con}$ | $TC_{con}$ | HNFS | | | SAA | | |
|---|---|---|---|---|---|---|---|---|
| | | | AC | TC | cost | AC | TC | cost |
| 20 | 210 | 650 | 202 | 622 | 824 | 186 | 647 | 833 |
| 50 | 560 | 2000 | 555 | 1939 | 2494 | 550 | 1947 | 2497 |
| 100 | 1200 | 4100 | 1197 | 4026 | 5223 | 1200 | 4059 | 5259 |
| 200 | 2500 | 8500 | 2490 | 8462 | 10952 | 2493 | 8482 | 10975 |
| 400 | 5100 | 17800 | 5088 | 17741 | 22829 | 5080 | 17774 | 22854 |

Compared to SAA, HNFS's advantage mainly represents in running time. Figure 4 shows the running time of these two methods. The curve illustrates that the time HNFS spends in getting the final result is obviously less than that of SAA, even when the size of the sample data is not large. Because the initial state of HNFS is trained by fuzzy logic, it is more meaningful than SAA's random initial state.

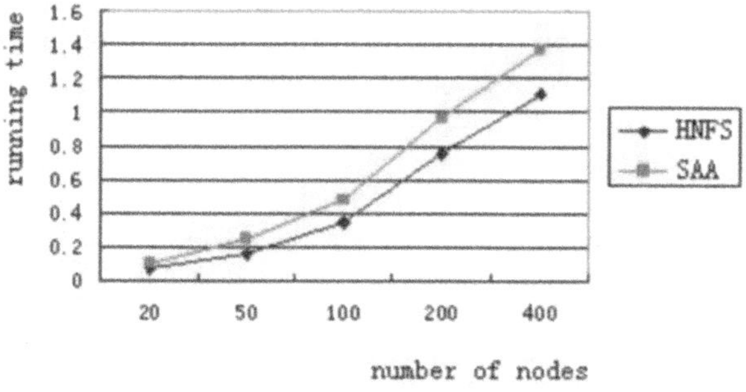


**Fig. 4.** Running time curve

## 6 Conclusion

This paper introduces a method of applying hybrid neural fuzzy system into the HW/SW partitioning problem. Firstly, we use fuzzy logic mechanism to match the character properties to the expert rules and generate an initial partitioning scheme. Then the initial scheme is fed into a Boltzmann machine network together with the character properties. Neuron in the network changes its state as iteration goes on, and correspondingly the partitioning scheme is changed. When the network reaches a balanced state and the control parameter becomes small enough, the output of the operation network is regarded as the final scheme for HW/SW partitioning problem. The simulation result has demonstrated our method for HW/SW partitioning is viable and has a better performance than some of the current methods in this research domain.

## References


1. Wolf, W.: Hardware-software Co-design of Embedded System, In: Proc. of the IEEE, Volume 82, Issue 7 (1994) 967-989
2. Staunstrup, J., Wolf, W.: Hardware/Software Co-Design: Principles and Practice, In: Kluwer Academic Publishers (1997)
3. Barros, E., Rosenstiel W.: A Method for Hardware Software Partitioning, In: Proc. of CompEuro '92 Computer Systems and Software Engineering (1992) 580-585
4. Estrin, G.: A Methodology for Design of Digital Systems Supported by SARA at Age One, In: National Computer Conference (1978)
5. Saha, D., Mitra, R.R., Basu, A.: Hardware Software Partitioning Using Genetic Algorithm, In: Proc. of the Tenth International Conference on VLSI Design (1997) 155-160
6. Arato, P., Juhasz, S., Mann, Z.A., Orban, A. Papp, D.: Hardware-software Partitioning in Embedded System Design, In: Proc. of Intelligent Signal Processing 2003 IEEE International Symposium (2003) 197-202
7. Wu, J., Srikanthan, T.: A Branch-and-Bound Algorithm for Hardware/Software Partitioning, In: Proc. of IEEE Symposium on Signal Processing and Information Technology (ISSPIT) (2004) 526-529
8. Henkel, J., Ernst, R.: An approach to automated hardware/software partitioning using a flexible granularity that is driven by high-level estimation techniques, In: IEEE Trans. VLSI Syst. 9 (2) (2001) 273-289
9. Eles, P., Peng, Z., Kuchcinski, K., Doboli, A.: System Level Hardware/Software Partitioning Based on Simulated Annealing and Tabu Search, In: Journal on Design Automation for Embedded System, 2(1) (1997) 5-32
10. Ma, T.Y., Li, Z.Q., Yang, J.: A Novel Neural Network Search for Energy-Efficient Hardware-Software Partitioning, In: Machine Learning and Cybernetics, 2006 International Conference (2006) 3053-3058
11. Nauck, D., Klawonn, F., Kruse, R.: Choosing Appropriate Neuro-Fuzzy Models, In: Proc. of EUFIT'94, Aachen (1994) 552-557
12. Lin, C.T., Lee, C.S.G.: A multi-valued Boltzman machine, In: Systems Man and Cybernetics, IEEE Transaction on Volume 25, Issue 4 (1995) 660-669
13. Ma, H.: Pattern Recognition Using Boltzmann Machine, In: Proc. of Southeastcon '95 Visualize the Future (1995) 23-29
14. Xiong, Z.H., Chen, J.H., Li, S.K.: Hardware/Software partitioning for platform-based design method, In: Proc. of Asia and South Pacific-Design Automation Conference, Volume 2 (2005) 691-696
15. Wang, G., Gong, W.R., Kastner, R.: A New Approach for Task Level Computational Resource Bi-partitionging, In: Proc. of the IASTED Int'l Conf. on Parallel and Distributed Computing and Systems (PDCS), ACTA Press (2003) 434-444

# Design of Manufacturing Cells for Uncertain Production Requirements with Presence of Routing Flexibility

Ozgur Eski and Irem Ozkarahan

Dokuz Eylul University, Faculty of Engineering, Industrial Engineering Dept., 35100 Bornova-Izmir, Turkey
{ozgur.eski,irem.ozkarahan}@deu.edu.tr

**Abstract.** Cellular manufacturing has been seen as an effective strategy to the changing worldwide competition. Most of the existing cell design methods ignore the existence of stochastic production requirements and routing flexibility. In this study, a simulation based Fuzzy Goal Programming model is proposed for solving cell formation problems considering stochastic production requirements and routing flexibility. The model covers the objectives of minimizing inter-cell movements, maximizing system utilization, minimizing mean tardiness and minimizing the percentage of tardy jobs. The simple additive method and max-min method are used to handle fuzzy goals. A tabu search based solution methodology is used for solution of the proposed models and the results are presented.

**Keywords:** Cellular manufacturing, fuzzy goal programming, tabu search, simulation, routing flexibility.

## 1 Introduction

Cellular manufacturing (CM) today is a well known strategy for improving the productivity of batch production system. In CM systems, the parts that have similar operations are grouped and manufactured in a dedicated production area called manufacturing cell.

Cell formation problem is a well defined problem in literature. In general, there are two main objectives in the cell formation procedure: (1)Part family and machine cells formation, (2)The allocation of families to the machine cells.

Mathematical programming methods have been widely used for solving cell formation problem. They generally focus on multiple objectives such as minimization of material handling costs, minimization of setup costs, minimization of workload imbalances, maximization of utilization etc[11]. Simulation studies performed in cell formation literature have indicated the importance of other performance measures such as machine utilization, workload balance etc. in determining manufacturing cells. Moreover the performance measures such mean tardiness, percentage of tardy jobs etc. are important especially for manufacturing systems which operates with “Just in Time” manufacturing philosophy. However, mathematical presentation of some objectives such as system utilization, average time spent in system, average

D.-S. Huang, L. Heutte, and M. Loog (Eds.): ICIC 2007, LNAI 4682, pp. 670–681, 2007.

throughput, mean tardiness, number of tardy jobs etc. is difficult. Analytic representation of such objectives leads computationally complex models which are not practical for real applications.

Deterministic models are generally used in most of the studies. However real manufacturing systems have stochastic nature. Simulation is an useful tool for analyzing such systems. Since simulation is not an optimization tool simulation studies performed in cell formation literature are generally focused on analyzing the performance of CM systems.

In most of the studies performed in cell formation literature, parts are assumed to have only one process plan(route). There is no loading problem under such an assumption. However in real life, parts can have different processing plans. Part loading problem arises when the parts have alternative routes. The loading problem can be summarized as searching for the good routing among different alternative routes.

Real manufacturing systems tend to have uncertainty or vagueness in system parameters. Fuzzy set theory gives the opportunity to deal with uncertainty and vagueness. Fuzzy clustering techniques has been applied to the cell formation problems. However Fuzzy clustering is different from Fuzzy mathematical programming(FMP) problem. In fuzzy clustering problem the fuzzy membership of a machine with respect to a cell is defined and hierarchical clustering is performed for designing cells. In FMP, linguistic vagueness in information related with many other design parameters may be modeled[12]. Flexibility is an important feature especially for production systems in which demand rate and part types are not stable. In such systems, a cell developed according to precise goal values may become completely infeasible when production requirements are changed. Allowing for vague aspirations of decision maker brings flexibility to the cell formation process.

The aim of this study is to develop a cell design methodology considering the existence of stochastic production requirements and routing flexibility. In this study a hybrid analytic-simulation Fuzzy Goal Programming (FGP) model is proposed in order to support the cell formation process. In proposed hybrid analytic-simulation model, the stochastic nature of the production system is represented by a simulation model. The part processing times, intercellular part movement times and the part arrivals are all stochastic. The objectives of maximizing system utilization, minimizing mean tardiness and minimizing the percentage of tardy jobs which are difficult to represent analytically are obtained by simulation model. The other objective (min. of inter-cell movements) is obtained by an analytical equation. The fuzzy goals are handled by using both additive method and max-min method. A tabu search based solution methodology is used for solution of the proposed models.

The paper is structured as follows: In Section 2, detailed description of proposed FGP model is given. Section 3 describes the solution methodology. In section 4, the proposed methodology is applied to a case problem. Simple additive method and max-min method are used to handle fuzzy goals and the results are presented.

## 2 A FGP Model for Cell Formation

Goal programming (GP) is one of the most powerful, multi-objective decision making approaches in practical decision making. This method requires the decision maker

(DM) to set goals for each objective that he/she wishes to attain. In a standard GP formulation, goals and constraints are defined precisely. However, one of the major drawbacks for a DM in using GP is to determine precisely the goal value of each objective function. Applying fuzzy set theory (FST) into GP has the advantage of allowing for the vague aspirations of a DM. In this study a simulation based fuzzy goal programming model is developed and used for cell formation. The following notation is used in the development of the mathematical model:

**Indices**

| | |
|---|---|
| i =1,2,...,I | Jobs |
| o =1,2...O | Operations |
| c=1,2,...C | Cells |
| m=1,2,...M | Machines |

**Parameters**

$$P_{iom} = \begin{cases} 1 & \text{if oth operation of job } i \text{ can be performed on machine } m \\ 0 & \text{Otherwise} \end{cases}$$

K: A big number
$M_{min}$: Min number of machines in order to form a cell
$M_{max}$: Max number of machines that can be included in a cell.
$goal_{int}$: Aspiration level for goal_1
$goal_{util}$: Aspiration level for goal_2
$goal_{tardiness}$: Aspiration level for goal_3
$goal_{tardyjobs}$: Aspiration level for goal_4

**Decision variables**

$$Q(c) = \begin{cases} 1 & \text{if cell is formed} \\ 0 & \text{Otherwise} \end{cases} \quad (1)$$

$$Z_{ioc} = \begin{cases} 1 & \text{if oth op. of job } i \text{ is performed in cell } c \\ 0 & \text{Otherwise} \end{cases} \quad (2)$$

$$D1_{ioc} = \begin{cases} 1 & \text{if oth oper. of job } i \text{ is performed in another cell} \\ 0 & \text{Otherwise} \end{cases} \quad (3)$$

$$X_{iocm} = \begin{cases} 1 & \text{if oth op. of job } i \text{ is assigned to machine } m \text{ in cell } c \\ 0 & \text{Otherwise} \end{cases} \quad (4)$$

$$Y_{cm} = \begin{cases} 1 & \text{if mahine } m \text{ is assigned to cell } c \\ 0 & \text{Otherwise} \end{cases} \quad (5)$$

**Goals:**

$$Goal\_1: \sum_i \sum_o \sum_c D_{1ioc} \prec goal_{int} \tag{6}$$

$$Goal\_2: system \quad utilization \succ goal_{util} \tag{7}$$

$$Goal\_3: mean\ tardiness \prec goal_{tardiness} \tag{8}$$

$$Goal\_4: percentage\ of\ tardy\ jobs \prec goal_{tardyjobs} \tag{9}$$

In formulation (6-9), the symbols "$\prec$" and "$\succ$" denote the fuzzified versions of "$\leq$" and "$\geq$" and can be read as "approximately less (greater) than or equal to". The objectives of the mathematical model are minimizing inter-cell movements(6), maximizing system utilization(7), minimizing mean tardiness(8) and minimizing the percentage of tardy jobs (9). The inter-cell movements should be substantially smaller than $goal_{int}$, system utilization should be substantially greater than $goal_{util}$, the mean tardiness should be substantially smaller than $goal_{tardiness}$ and the percentage of tardy jobs should be substantially smaller than $goal_{tardyjobs}$.

In cellular manufacturing systems, it is desired to complete all operations of a part in the same cell. However, in real applications, parts can visit different cells when it requires processing on a machine that is not available in the allocated cell of the part. The inter-cell movements results in extra transportation costs. This also requires more coordinating effort between cells. Thus, inter-cell movements are undesirable and the objective of minimizing inter-cell movements is essential for designing manufacturing cells.

Utilization is another important performance measure in determining cell formation. Since the set-up times are decreased, the effective capacity of the machines are increased thus leading to lower utilization. Demand fluctuations can also lead to lower utilizations. The general level of utilization of cells is of the order of 60-70%. Hence the maximization of system utilization is an important objective for cellular manufacturing systems.

Minimization of mean tardiness objective is important when customers tolerate smaller tardiness but become rapidly and progressively more upset for larger ones. The objective of minimizing the percentage of tardy jobs is important when customers simply refuse to accept tardy jobs, so that the order is lost. These performance measures are important especially for manufacturing systems work with just in time manufacturing philosophy.

In proposed model, the first objective is determined by an analytical equation whereas other three objectives which are difficult to obtain analytically are determined by a simulation model.

**Constraints:**

$$\sum_c \sum_m X_{iocm} P_{iom} = 1 \tag{10}$$

$$\sum_c Y_{cm} = 1 \qquad \forall m \tag{11}$$

$$X_{iocm} \leq K.Y_{cm} \qquad \forall i,o,c,m \tag{12}$$

$$X_{iocm} \leq K.Z_{ioc} \qquad \forall i,o,c,m \tag{13}$$

$$\sum_{c} Z_{ioc} = 1 \qquad \forall i,o \tag{14}$$

$$Z_{ioc} - Z_{ioc-1} = D_{1ioc} - D_{ioc} \qquad \forall i,o,c \tag{15}$$

$$\sum_{m} Y_{cm} \leq M_{\max} Q_c \qquad \forall c \tag{16}$$

$$\sum_{m} Y_{cm} \geq M_{\min} Q_C \qquad \forall c \tag{17}$$

$$X_{iocm}, Y_{cm}, Z_{ioc}, Q_c, D_{1ioc}, D_{ioc} = [0,1] \tag{18}$$

Equation (10) ensures that the operation of a job is assigned to a machine in a cell. Equation (11) ensures that each machine is assigned to only one manufacturing cell. Equation (12) indicates that if an operation in a cell c is assigned to a machine m, this machine is assigned to cell c. Equation(13) indicates that if an operation is assigned to a machine m in a cell c, this operation is assigned to cell c. Equation (14) ensures that an operation is assigned to only one cell. Equation (15) controls whether the consecutive two operations of a job is performed in the same cell. Equation (16-17) constraints the number of machines assigned to each cell if it is formed.

The membership functions of objectives are given in Equations (19-22)

$$\mu_1 = \begin{cases} 1 & \text{if } f_1 \leq L_1 \\ \dfrac{U_1 - f_1}{U_1 - L_1} & \text{if } L_1 \leq f_1 \leq U_1 \\ 0 & \text{if } f_1 \geq U_1 \end{cases} \tag{19}$$

$$\mu_2 = \begin{cases} 1 & \text{if } f_2 \geq U_2 \\ \dfrac{f_2 - L_2}{U_2 - L_2} \text{ if} & L_2 \leq f_2 \leq U_2 \\ 0 & \text{if } f_2 \leq L_2 \end{cases} \tag{20}$$

$$\mu_3 = \begin{cases} 1 & \text{if } f_3 \leq L_3 \\ \dfrac{U_3 - f_3}{U_3 - L_3} \text{ if} & L_3 \leq f_3 \leq U_3 \\ 0 & \text{if } f_3 \geq U_3 \end{cases} \tag{21}$$

$$\mu_4 = \begin{cases} 1 & if\ f_4 \le L_4 \\ \dfrac{U_4 - f_4}{U_4 - L_4}\ if & L_4 \le f_4 \le U_4 \\ 0 & if\ f_4 \ge U_4 \end{cases} \tag{22}$$

Where, $f_i$ is the value of *ith* objective function; $U_i$ and $L_i$ are min.-max. limits of objectives. The shapes of the membership functions are given in Figure 1.

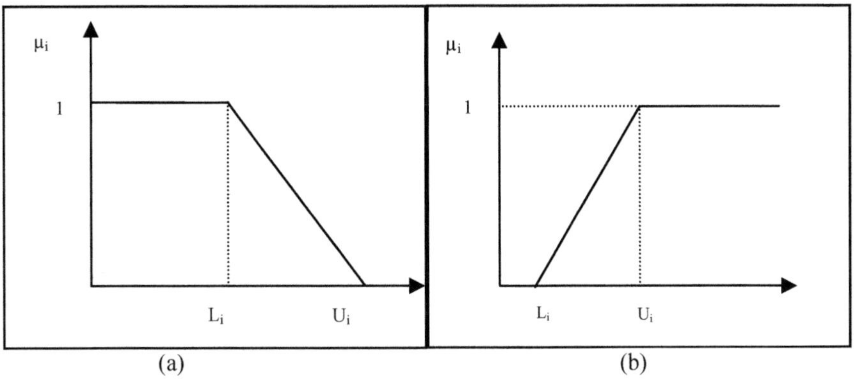


**Fig. 1.** (a) The shape of membership function for objectives 1,3 and 4(minimization) (b) The shape of membership function for objective 2 (maximization)

Using the additive method[13], standard goal programming formulation can be equivalently transformed as:

$$MaxZ = \sum_{i=1}^{4} \mu_i \tag{23}$$

$$\mu_1, \mu_2, \mu_3, \mu_4 \ge 0 \tag{24}$$

And other constraints (10-18).

In additive method, the sum of the membership values of the goals ($\Sigma\mu$) is maximized. The use of additive method can obtain the maximum sum of goals' achievement degrees[ 4]

Using the max-min operator[3] $\lambda$, which is the overall satisfactory level of compromise, the standard goal programming formulation can be equivalently transformed as:

$$MaxZ = \lambda \tag{25}$$

$$\mu_1, \mu_2, \mu_3, \mu_4 \ge \lambda \tag{26}$$

And other constraints (10-18).

In the next section, a tabu search based solution methodology is presented for the solution of proposed models.

## 3 Solution of Fuzzy Goal Programming Models Using Tabu Search

Tabu search (TS) [7],[8] is a global optimization heuristic and can handle any type of objective function and any type of constraints. The solution process of TS involves working with more than one solution (neighborhood solutions) at a time. Baykasoglu[1],[2],[9] noted that this feature of TS gives a great opportunity to deal with multiple objectives or goals and proposed a TS based solution methodology for multi-objective optimization problems. The steps of Tabu search algorithm used for solving FGP models are as follows:

*Initial solution:* Algorithm starts with a randomly generated feasible solution vector. Starting with a known good solution vector can decrease the computation time.

*Generation of neighborhoods:* Different move strategies have been presented in TS literature. The move strategies depend on the type of variables. Since all variables are 0-1 variables in our model, the move strategy given below is used for generation of neighborhoods.

$$x_i^* = \begin{cases} 1 & if \quad x_i = 0 \\ 0 & if \quad x_i = 1 \end{cases} \tag{27}$$

Where $x_i$ = Value of the $i$ th variable prior to the neighborhood move, $x_i$* = Value of the $i$th variable after the neighborhood move.

*Selection:* For simple additive method, the membership values of goals are calculated and summed. The the neighbor with the highest sum ($\Sigma\mu$) is selected as the current best solution. When the max-min method is used, the $\lambda$ values are calculated for each neighborhood solution. The solution with the highest $\lambda$ value is selected as a new seed.

*Updating Tabu list and current best solution list:* The current best solution list is updated when a better solution is obtained. A predefined number of previous moves are recorded in tabu list. Tabu list is updated in each iteration. When it is full, the first item of the list is removed and replaced with a new one.

*Termination:* If a predefined number of iteration is reached or if there is no improvement in the current best move list in the last $t$ iterations the algorithm terminates.

The above algorithm is applied to the test problems selected from literature[1],[4],[5],[6]. The results showed that tabu search algorithm can solve Fuzzy Goal programming problems efficiently.

Since the proposed mathematical model for cell formation is a hybrid analytic-simulation optimization model, it is needed to integrate the mathematical model

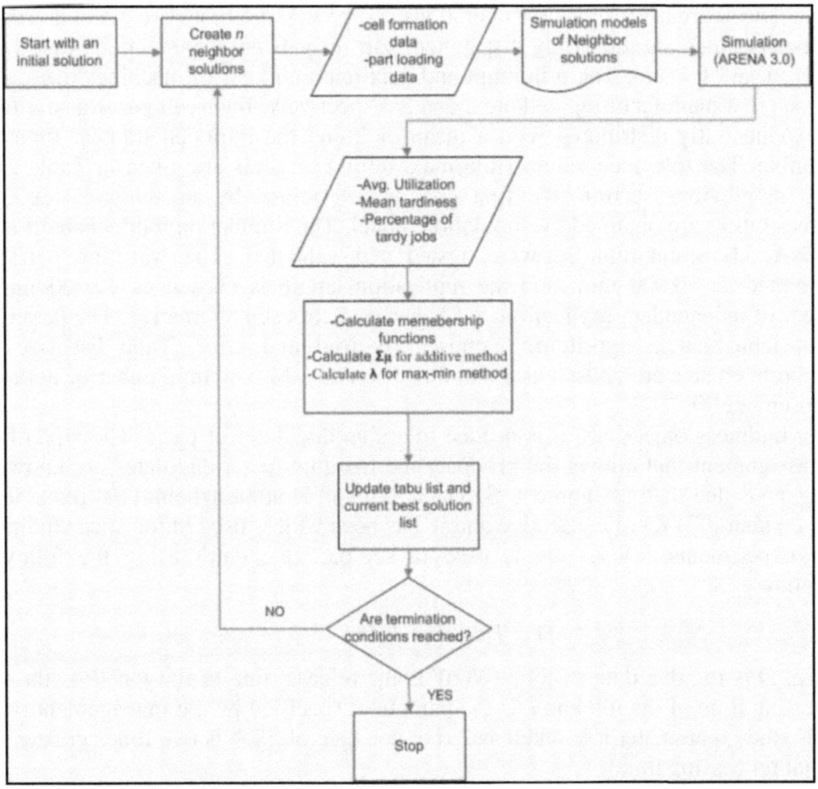


**Fig. 2.** Tabu search algorithm for solving hybrid analytic-simulation FGP model

and simulation model. Simulation models of manufacturing system are built in SIMAN-ARENA 3.0 simulation software. A computer program is coded in Turbo C and used for implementing Fuzzy Goal Programming Model which is integrated with simulation model. The flowchart of C program is illustrated in Figure 2. As seen in Figure 2, simulation model uses the loading data and cell formation data. When tabu search algorithm creates neighborhood solutions, simulation models are created automatically and ARENA Simulation Software runs and provides the system utilization level, mean tardiness and percentage of tardy jobs for each neighbor solution. Then the membership functions and $\Sigma\mu$ values ($\lambda$ values for max-min method) each solution are calculated. The solution with the highest $\Sigma\mu$ value (highest $\lambda$ value for max-min method) is chosen as a new seed. The procedure is terminated when termination conditions are reached.

## 4 Experimental Work

The proposed hybrid analytic-simulation FGP model is applied to a case problem. The manufacturing system under consideration consists of 6 machines and performs 6 different jobs. Each jobs consists of 3 sub-operations and can have alternative process routes. The alternative routes for operations and processing times are given in Table1. Processing times are uniformly distributed, part arrivals are exponentially distributed with a mean of 7 min. The minimum and maximum number of machines that can be assigned to a manufacturing cell are 2 and 3 respectively. Inter-cell part transfer times are exponentially distributed with a mean of 2 and the intra-cell transfer times are negligible. The tolerance values (min-max. limits) of goals are given in Table 2. As stated in previous section, the first goal is determined by an analytical equation whereas others are obtained by simulation model. The Simulation model is built using ARENA 3.0 Simulation Software, tested and validated. The warm-up period is determined as 10.000 min. and the replication length is chosen as 40.000min.The number of independent replications is chosen as 5 for each alternative. The parameter set of tabu search algorithm is chosen by trial and error. Tabu list size and neighborhood size are chosen as 8 and 5 respectively. Maximum number of iterations is chosen as 500.

For tardiness objectives, it is needed to assign due dates of parts. The type of due date assignment that allows the producer the freedom to set due dates are known as endogenous due date assignment. Sabuncuoğlu and Hommertzheim [10] found Total work content (TWK) rule effective and it has been widely used in job shop studies. In these experiments, TWK rule is used to set part due dates using the following definition:

$$D = TNOW + k.P \tag{28}$$

Where, $D$ is the due date of job, $TNOW$ is the release time of the job, $P$ is the total processing time of the job and $k$ is the parameter specified by the management ($k \geq 1$). In this study, parameter k is taken as 2 (i.e due date of a job is two times greater than its total processing time).

**Table 1.** Alternative routes(process plans) and Processing times of operations

| Job | Operation | Alternative process plan | Processing time (min.) |
|---|---|---|---|
| | A1 | 1, 5, 6 | Unif(6,7) |
| JOB1 | A2 | 3, 4 | Unif(5,8) |
| | A3 | 1, 5 | Unif(4,7) |
| | B1 | 5, 6 | Unif(5,6) |
| JOB2 | B2 | 1, 2, 3 | Unif(5,6) |
| | B3 | 5 | Unif(6,7) |
| | C1 | 4, 5 | Unif(5,8) |
| JOB3 | C2 | 1, 4 | Unif(3,4) |
| | C3 | 1, 2, 5 | Unif(5,7) |
| | D1 | 2 | Unif(7,8) |
| JOB4 | D2 | 2, 3 | Unif(5,6) |
| | D3 | 3 | Unif(6,7) |
| | E1 | 1, 2 | Unif(5,7) |
| JOB5 | E2 | 3 | Unif(7,9) |
| | E3 | 1, 4 | Unif(6,8) |
| | F1 | 4, 6 | Unif(7,8) |
| JOB 6 | F2 | 6 | Unif(4,5) |
| | F3 | 3 | Unif(4,6) |

**Table 2.** The tolerance values of goals

| Goal | Min-Max limits | |
|---|---|---|
| Inter-cell movements | 2 | 5 |
| System utilization | 0.30 | 0.75 |
| Mean tardiness | 0 | 7 |
| Percentage of tardy jobs | 10 | 30 |

First, the proposed methodology is applied to the above case using simple additive method. The best solution is found at 133rd iteration. The best $\Sigma\mu$ value is found as 3.28. The solution is summarized in Table 3. According to the solution vector, 2 cells are formed. The first cell is composed of machines 1-4-5 and the second cell is composed of machines 2-3 and 6. There are 2 inter-cell movement and the satisfaction level of the first goal $\mu_1$ = 1. The system utilization level is found as 0.5820 and $\mu_2$=0.6267. The mean tardiness is found as 1.9541 min. and $\mu_3$=0.7208. The percentage of tardy jobs is found as 11,35% and $\mu_4$=0.9325.

**Table 3.** Machine Cells formation and part assignments according to the solution(Simple additive method)

| CELL 1 | | CELL 2 | |
|---|---|---|---|
| Machines | Operations | Machines | Operations |
| 1 | A3, B2, C3 | 2 | D1, D2, E1 |
| 4 | A2, C2, E3 | 3 | D3, E2, F3 |
| 5 | A1, B3, C1 | 6 | B1, F1, F2 |

Then the methodology is applied to the same case using max-min method. The best solution is found in 93rd iteration. The solution is summarized in Table 4. According to the solution using max-min method two cells formed. The first machine cell consists of machines 1-4-6 and the second cell composed of machines 2-3-5.The best $\lambda$ value is found as 0.6517($\mu_1$=1; $\mu_2$=0.6517; $\mu_3$= 0.6611; $\mu_4$= 0.8325).

**Table 4.** Machine Cells formation and part assignments according to the solution (max-min method)

| CELL 1 | | CELL 2 | |
|---|---|---|---|
| Machines | Operations | Machines | Operations |
| 1 | A3, C2, C3 | 2 | D1, D2, E1,E3 |
| 4 | A2, C1, F1 | 3 | B2, D3,E2, F3 |
| 6 | A1, F2 | 5 | B1, B3 |

Based on the solution obtained by simple additive method, the achievement degrees of the second goal (max. of system utilization) is small(0.6267) because it is difficult to achieve. However the achievement levels of other goals are between 0.7208 and 1. According to the solution obtained by max-min method, the achievement level of goal_2 (0.6517) is higher than simple additive method. However the achievement levels of goal_3 and goal_4 are lessen. In simple additive method, the achievement levels of some goals will not decrease because of a particular goal that is difficult to achieve. This advantage makes the simple additive method appealing. As a whole, the sum of achievement levels of goals in the solution obtained by simple additive method is greater than max-min method.

It is obvious that a decision maker can find different cell configurations by using different tolerance limit sets. The changes in the part arrival rates or part processing times also lead different cell configurations. Since the proposed model is based on a parametric simulation model, the system can be easily adapted for different production requirements. For example when the part arrivals are exponentially distributed with a mean of 5 (high demand) instead of 7 the manufacturing cells would form as in table 5. In this case, $\Sigma\mu$ value is found as 2.9094 ($\mu_1$= 1; $\mu_2$=0.8742; $\mu_3$= 0.4408; $\mu_4$= 0.5944). It is obvious that, in high demand case, the achievement level of system utilization is increased whereas achievement levels of mean tardiness and percentage of tardy jobs are decreased.

**Table 5.** Machine Cells formation and part assignments according to the solution.(High Demand Case)

| CELL 1 | | CELL 2 | |
|---|---|---|---|
| Machines | Operations | Machines | Operations |
| 1 | A1, C3 | 2 | D1, D2, E1 |
| 4 | A2, C2,E3 | 3 | B2, D3, E2,F3 |
| 5 | A3, B3, C1 | 6 | B1,F1, F2 |

# 5 Conclusion

Cell formation decisions are made based on several factors such as machining times, utilization, workload, alternative routings, capacities, operation sequences. Most of the existing procedures for cell formation ignores the existence of stochastic production requirements and the existence of alternative process plans.

In this paper, a hybrid analytic-simulation Fuzzy Goal programming Model is proposed for cell formation problem considering stochastic production requirements and alternative routes. The model covers the objectives of minimizing inter-cell movements, maximizing system utilization, minimizing mean tardiness and minimizing the percentage of tardy jobs. The first objective is calculated by an analytical equation whereas others are obtained by simulation model. A tabu search based solution methodology is used for both simple additive model and max-min model and the results are presented. A computer program is also developed using Turbo C in order to implement proposed methodology.

# References

1. Baykasoglu, A.: Solution of Goal Programming Models Using a Basic Taboo Search Algorithm. Journal of the Operational Research Society, 50. (1999) 960-973
2. Baykasoglu, A., Gokcen, T.: A Tabu Search Approach to Fuzzy Goal Programs and an Application to Aggregate Production Planning, Engineering Optimization, 31 (2006) 155-177

3. Bellman, R. E., Zadeh, L. A.: Decision-making in a Fuzzy Environment. Management Science, 17 (1970) 141-64
4. Chen, L. H., Tsai, F. C.: Fuzzy Goal Programming with Different Importance Priorities. European Jour. Of. Operational Res., 133 (2001) 548-556
5. Fang, H. C., Teng, C. J., Li, S. Y.: A Fuzzy Goal Programming Approach to Multi-objective Optimization Problem with Priorities. European Jour. Of. Operational Res., 176.(2007) 1319-1333
6. Gen, M., Ida, K., Tsujimura, Y., Kim, C. E.: Large Scale 0-1 Fuzzy Goal Programming and Its Application to Reliability Optimization Problem. Computers Ind. Engineering, 24 (1993) 539-549
7. Glover, F.:Tabu Search-Part 1.ORSA Journal on Computing, 1 (1989) 190-206
8. Glover, F.:Tabu Search-Part 2.ORSA Journal on Computing, 2 (1990) 4-32
9. Saad, S.M., Baykasoglu, A., Gindy, G.: An Integrated Framework for Reconfiguration of Cellular Manufacturing Systems Using Virtual Cells. Production Planning&Control, 13 (2002) 381-393
10. Sabuncuoglu, I., Hommertzheim, D. L.: Dynamic Dispatching Algorithm for Scheduling Machines and Automated Guided Vehicles in a Flexible Manufacturing System. International Journal of Production Research, 30 (1989) 1059-1079
11. Selim, H., Askin, R. G., Vakharia, A.J.: Cell Formation in Group Technology: Review, Evaluation and Directions for Future Research. Computers and Ind. Eng., .34 (1998) 3-20
12. Shanker, R., Vrat, P.: Some Design Issues in Cellular Manufacturing Using the Fuzzy Programming Approach. Int. Journal of Prod. Research, 37 (1997) 2545-2563
13. Tiwari, R. N., Dharmar, S., Rao, J. R.: A Fuzzy Goal Programming-an Additive Model. Fuzzy Sets and Systems, 24 (1987) 27-34

# Developing a Negotiation Mechanism for Agent-Based Scheduling Via Fuzzy Constraints

K. Robert Lai[1], Menq-Wen Lin[2], and Bo-Ruei Kao[1]

[1] Department of Computer Science & Engineering Yuan Ze University
[2] Department of Information Management, Ching Yun University,
Chung-Li, Taiwan 32026, R.O.C.
krlai@cs.yzu.edu.tw, linmw@cyu.edu.tw, s909402@mail.yzu.edu.tw

**Abstract.** The paper presents a negotiation mechanism for agent-based scheduling via fuzzy constraints. Scheduling is considered as a global consistency enforcing via iterative constraint adjustment and relaxation by agents. Fuzzy constraints, in this way, are used not only to represent the temporal relations that jobs being scheduled must satisfy, but also to specify the possibilities prescribing to what extent the solutions are suitable for scheduling to rank the solutions. The negotiation mechanism based on fuzzy constraint provides a systematic method to gradually relax the temporal constraints to generate a proposal, and then utilizes possibility functions to select an alternative schedule that is subject to the others' acceptability. Thus, each agent, who is in charge of different aspects of the problem, not only distributively solves its problems to maximize its local objectives, but also works together with other agents to attain a globally beneficial schedule. Experimental results suggest the proposed approach provides superior performance in all criteria to the contract net protocol and auction-based approaches.

**Keywords:** Scheduling, Multi-Agent Systems, Fuzzy Constraints, Negotiation.

## 1 Introduction

Scheduling is understood as the problem of suitable assignment of resources to tasks/jobs within a specified time window and coping with a set of constraints. Most of researches emphasize the optimization of scheduling based on a centralized, monolithic model in the past. However, real-world scheduling problems are often inherently distributed, highly combinatorial aspects, and usefulness for practical applications, the focus of scheduling research has been shifted to emphasize scheduling flexibility [16,12] and toward solving problems in distributed environments [11,15].

Agent-based approaches, which are essentially distributed, efficient, and adaptable to dynamic environment, have been widely applied to the practical scheduling problem. Several negotiation models of agent-based approach have been proposed to solving scheduling[13]. Among them, the contract net protocol (CNP),

D.-S. Huang, L. Heutte, and M. Loog (Eds.): ICIC 2007, LNAI 4682, pp. 682–692, 2007.

a commonly used negotiation model, involves a process of task announcement, bidding, and awarding to establish a deal among agents [14]. Relying on this protocol, several bidding-based or auction-based approaches have demonstrated a flexible manner for resources selection and allocation [6,9,12]. Though these negotiation models emphasize the flexibility and responsiveness over optimality of solutions, the weaknesses of these approaches are inferior quality of the schedules generated and the unpredictability of the system performance.

To address these problems, agents have to act from their individual perspectives to negotiate with others and trade off local performances to improve global performance. Meanwhile, the decision scheme of agents depends primarily on individual knowledge, shared information and the negotiation mechanisms. As a result, combination of rich knowledge bases, individual reasoning mechanism and negotiation mechanisms in agent-based approach for scheduling are very important.

The paper presents a negotiation mechanism for agent-based scheduling via fuzzy constraints. Each agent, who is in charge of different aspects of the scheduling, can be represented as a fuzzy constraint satisfaction problem. Scheduling via agent negotiation is considered as a global consistency enforcing via iterative constraint adjustment and relaxation. Fuzzy constraints, in this way, are used not only to represent the temporal relations that jobs being scheduled must satisfy, but also to specify the possibilities prescribing to what extent the solutions are suitable for scheduling to rank the solutions [1,2]. The negotiation mechanism based on fuzzy constraint provides a systematic method to gradually relax the temporal constraints to generate a proposal, and then utilizes possibility functions to select an alternative schedule that is subject to the others' acceptability [4,5,8]. Thus, each agent, who is in charge of different aspects of the problem, not only distributively solves its problems to maximize its local objectives, but also works together with other agents to attain a globally beneficial schedule. Experimental results suggest the proposed approach provides superior performance in all criteria to the contract net protocol and auction-based approaches.

The remainder of this paper is organized as follows. Section 2 introduces the theoretical basis of modeling distributed scheduling as agent negotiation. Section 3 presents the negotiation process for obtaining the scheduling solutions. Section 4 demonstrates the effectiveness of the proposed approach followed by some conclusions in Section 5.

## 2 Constraint-Directed Negotiation Mechanism in Agent-Based Scheduling

Planning a schedule among a set of entities can be modeled as agent negotiation in that finding a satisfactory scheduling solution in a distributed environment is the same as reaching an acceptable agreement in agent negotiation. Furthermore, fuzzy constraints have also been used to represent the requirements that jobs being scheduled must satisfy [2,3]. Thus, a distributed scheduling problem can be viewed as a distributed fuzzy constraint satisfaction problem (DFCSP) and its graphical representation, a distributed fuzzy constraint network (DFCN) adapted from [5].

A distributed fuzzy constraint network is consisted of a set of fuzzy constraint networks $\{\mathfrak{N}^1, ..., \mathfrak{N}^L\}$. Each fuzzy constraint networks (FCN) $\mathfrak{N}^k$ involves a set of objects, $\mathbf{X}^k=\{X^k_{1^k}, \ldots, X^k_{n^k}\}$, a set of constraints, $\mathbf{C}^k$, over these objects, and a universe of discourse $\mathcal{U}^k$ for an FCN $\mathfrak{N}^k$. An FCN $\mathfrak{N}^k$ is connected to other FCNs by a set of external fuzzy constraints $\mathbf{C}^{k_e}$ which referring to at least one object in $\mathbf{X}^k$ and another in other FCNs. That is, each individual fuzzy constraint network $\mathfrak{N}^k = (\mathcal{U}^k, \mathbf{X}^k, \mathbf{C}^k)$ in a DFCN can represent the job, resource, or some other forms of agents in a distributed scheduling problem. Then, the task of distributed scheduling is to attain a schedule that can satisfy all the fuzzy constraints in DFCN simultaneously. The job and resource agents can be further described as follows.

A job agent $k^J_i$, which involves a set of activities $a_{i1}$,...,$a_{im_i}$ required by job $J_i$ and concerns with temporal, precedence, and required resource constraints, can be represented as a fuzzy constraint network $\mathfrak{N}^{k^J_i} = (\mathcal{U}^{k^J_i}, \mathbf{X}^{k^J_i}, \mathbf{C}^{k^J_i})$. In FCN $\mathfrak{N}^{k^J_i}$, $\mathbf{X}^{k^J_i}$ includes a start time $s_{ij}$, process time $p_{ij}$, and required resource $r_{ij}$ associated to activity $a_{ij}$, where $a_{ij} \in J_i$; $\mathbf{C}^{k^J_i}$ includes the temporal constraint $\mathbf{C}_{tmp}$, precedence constraint $\mathbf{C}_{pre}$, resource constraint $\mathbf{C}_{req}$, in which

- $\mathbf{C}_{tmp}$ represents the temporal constraint which the job has to be started after the release date and finished before the deadline. For job $J_i$, $\mathbf{C}_{tmp(i)}$ implies start time $s_i$ has to be later than release date $\widetilde{R_i}$, and end time $e_i$ has to be earlier than due-date $\widetilde{D_i}$. Release date and due-date are often subject to preference and are modeled by fuzzy number.

$$\mathbf{C}_{tmp(i)} : s_i \in [\widetilde{R_i}, +\infty) \ \wedge e_i \in (0, \widetilde{D_i}],$$
$$s_i = \min_{j=1,\ldots,m_i} s_{ij},\ e_i = (\max_{j=1,\ldots,m_i} s_{ij} + p_{ij}) \tag{1}$$

- $\mathbf{C}_{pre}$ represents the precedence constraint which defines the preceding restriction between two activities. $\mathbf{C}_{pre(ij->iq)}$ implies that activity $a_{ij}$ has to be performed before activity $a_{iq}$.

$$\mathbf{C}_{pre(ij->iq)}:\ s_{ij} + p_{ij} \leq s_{iq} \tag{2}$$

- $\mathbf{C}_{req}$ represents the required resource constraint which defines the set of possible resources are required by an activity. $\mathbf{C}_{req(ij,H)}$ implies that activity $a_{ij}$ can be performed by a set of alternative resources $H_{ij}$. For each candidate resource $R_h \in H_{ij}$ is held by resource agent $k^R_h$.

$$\mathbf{C}_{req(ij,H)}:\ r_{ij} = R_h,\ R_h \in H_{ij} \tag{3}$$

On the other side, a resource agent $k^R_h$, which holds resource $R_h$ and concerns with processing time, capacity, and problem-specific constraints, can be represented as a fuzzy constraint network $\mathfrak{N}^{k^R_h} = (\mathcal{U}^{k^R_h}, \mathbf{X}^{k^R_h}, \mathbf{C}^{k^R_h})$. In FCN $\mathfrak{N}^{k^R_h}$, $\mathbf{X}^{k^R_h}$ includes a start time $s_{hj}$, and processing time $p_{hj}$ associated to activity $a_{jh}$ which requires resource $R_h$; $\mathbf{C}^{k^R_h}$ includes the processing time constraint $\mathbf{C}_{pro}$, and capacity constraints $\mathbf{C}_{cap}$ in which

- $\mathbf{C}_{pro}$ represents processing time constraint which defines the possible duration of processing time for an activity. Durations are determined by tuning the machine or allocating the amount of resources. $\mathbf{C}_{pro(hj)}$ implies that the processing time $p_{hj}$ of activity $a_{hj}$ are bounded by possible duration $\widetilde{P_{hj}}$ which is represented as a fuzzy number.
- $\mathbf{C}_{cap}$ represents the capacity constraint which limits the available capacity of resource over time. $\mathbf{C}_{cap(hj,hq)}$ implies that the processing times of activities $a_{hj}$ and $a_{hq}$ which are performed on the resource $R_h$, cannot overlap on times.

$$\mathbf{C}_{cap(hj,hq)}\colon s_{hj} \geq s_{hq} + p_{hq} \ \wedge \ s_{hq} \geq s_{hj} + p_{hj} \tag{4}$$

When planning a schedule, both the job agents and resource agents govern activity by maintaining the consistency of inter-constraints $C_{jr(ijh)}$, which requires that activity $a_{ijh}$, performed by a resource agent $k_h^R$, has to start at $s_{ij}$ assigned by a job agent $k_i^J$, where

$$\mathbf{C}_{jr(ijh)}\colon s_{hj} = s_{ij} \tag{5}$$

However, maintaining the consistency of activities by job agents may incur constraint violations for resource agents and vice-versa. Thus, negotiation mechanism is employed to resolve the conflicts among the agents. Each agent determines its actions considering the trade-offs of local and other agents' preferences and revises them by getting feed-back from other agents. However, how does the agent negotiate with other agents to decide its local scheduling solution to reach an agreement that benefits all agents with a high satisfaction degree of fuzzy constraints, and move toward the deal more quickly?

To that end, the negotiation strategies are adopted by agents to determine the negotiation process in scheduling. These strategies determine how agents evaluate and generate local schedules to reach an agreement that is most in their self-interest or perform global goals. Agents exchange local schedules throughout the negotiation according to their own negotiation strategies. Whenever an local schedule is not acceptable by other agents, they make counter-offers by making concessions or by finding new alternatives to move toward an agreement. Hence, a concession strategy is presented, and a trade-off strategy is proposed to find alternatives.

A concession is a revision of a previous position that has been held and justified publicly. In a scheduling process, agents employ the concession strategy to compromise their private schedules which are movable. Agents attempt to entice one another into agreement by manipulating the ranges associated with a given constraint in a scheduling problem. Hence, the set of feasible concession scheduling proposals for agent $k$ at a threshold $\alpha_i^k$ is defined as follows.

**Definition 1.** ***(Set of feasible concession scheduling proposals):*** *Given the latest scheduling offer* $\mathbf{u}$ *and a threshold* $\alpha_i^k$ *of agent* $k$*, the set of feasible concession scheduling proposals at the threshold* $\alpha_i^k$ *for the next offer of agent* $k$*, denoted by* ${}_{\alpha_i^k}\mathfrak{C}_{\mathbf{u}}^k$*, can be defined as*

$${}_{\alpha_i^k}\mathfrak{C}_{\mathbf{u}}^k = \left\{\mathbf{v} \mid \left(\mu_{\mathbf{C}^k}(\mathbf{v}) \geq \alpha_i^k\right) \wedge \left(\Psi^k(\mathbf{v}) = \Psi^k(\mathbf{u}) - r\right)\right\}, \tag{6}$$

*where* $r$ *is the concession value.*

Concessions are always expected in negotiation, but negotiators nevertheless try to move away from their preferences as little as possible. The agent's concession value $r$ for its next offer may be determined from the agent's mental state and the opponent's responsive state.

A trade-off strategy is an approach by which an agent generates an alternative without reducing requirements. In a scheduling process, agents employ the trade-off strategy to reschedule the private schedules without reducing satisfactions. Agents attempt to entice one another into agreement by reconciling their constraints. An agent may respond to a borderline unacceptable cost through extending the range of a due-date constraint which did not exist previously. Hence, the set of feasible trade-off scheduling proposals is defined as follows.

**Definition 2.** ***(Set of feasible trade-off scheduling proposals)****: Given the latest scheduling offer* $\mathbf{u}$ *and a threshold* $\alpha_i^k$ *of agent* $k$*, the set of feasible trade-off scheduling proposals at the threshold* $\alpha_i^k$ *for the alternatives of agent* $k$*, denoted by* ${}_{\alpha_i^k}\mathfrak{T}_{\mathbf{u}}^k$*, is defined as*

$$ {}_{\alpha_i^k}\mathfrak{T}_{\mathbf{u}}^k = \left\{ \mathbf{v} \mid \left(\mu_{\mathbf{C}^k}(\mathbf{v}) \geq \alpha_i^k\right) \wedge \left(\Psi^k(\mathbf{v}) = \Psi^k(\mathbf{u})\right) \right\}. \quad (7) $$

A normalized Euclidean distance can be applied in establishing a trade-off strategy to measure the similarity between alternatives, and thus generate the best possible scheduling offer. This function tends to distinguish options whose satisfaction values are relatively close. Hence, a similarity function is defined as follows.

**Definition 3.** ***(Similarity function)****: Assuming that* $\mathbf{U}' = (\mathbf{u}_1', \ldots, \mathbf{u}_n')$ *is the set of offers proposed by* $n$ *other agents, and* $\mathbf{V} = (\mathbf{v}_1, \ldots, \mathbf{v}_n)$ *is a feasible trade-off scheduling proposal of agent* $k$ *for* $n$ *other agents, the similarity function between* $\mathbf{V}$ *and* $\mathbf{U}'$ *on the negotiated issues for agent* $k$*, denoted by* $\Theta^k(\mathbf{V}, \mathbf{U}')$*, is defined as*

$$ \Theta^k(\mathbf{V}, \mathbf{U}') = 1 - \frac{1}{n} \sum_{j=1}^{n} \left( \frac{1}{m} \left( \sum_{i=1}^{m} \left( \mu_{C_i^k}(\mathbf{v}_j) - \mu_{C_i^k}(\mathbf{u}_j') + p_{C_i^k}(\mathbf{u}_j') \right)^2 \right)^{\frac{1}{2}} \right), \quad (8) $$

*where* $m$ *is the number of fuzzy constraints of agent* $k$ *on issues,* $\mu_{C_i^k}(\mathbf{v}_j)$ *and* $\mu_{C_i^k}(\mathbf{u}_j')$ *denote the satisfaction degree of the* $i^{th}$ *(weighted) fuzzy constraint associated with the* $\mathbf{v}_j$ *and the* $\mathbf{u}_j'$ *for agent* $k$ *to agent* $j$*, and* $p_{C_i^k}(\mathbf{u}_j')$ *denotes the penalty from the* $i^{th}$ *dissatisfied (weighted) fuzzy constraint associated with the offer* $\mathbf{u}_j'$ *made by agent* $k$*.*

For each feasible trade-off scheduling proposal $\mathbf{v}$ of an agent, a fuzzy similarity between any $\mathbf{v}$ and the scheduling offer $\mathbf{u}'$ proposed by the opponent can be defined as a fuzzy set in which the membership grade of any particular $\mathbf{v}$ represents the similarity between $\mathbf{v}$ and $\mathbf{u}'$. Hence, the expected trade-off scheduling proposal $\mathbf{U}^*$ that benefits all parties can be defined as follows.

**Definition 4.** ***(Expected trade-off scheduling proposal)***: *Assuming that agent $k$ proposes a scheduling offer $\mathbf{U}$ to its opponents, and that the opponents subsequently proposes a set of scheduling counter-offer $\mathbf{U}'$ to agent $k$, the expected trade-off scheduling proposal $\mathbf{U}^*$ for the next scheduling offer by agent $k$ is defined as*

$$\mathbf{U}^* = \arg_{\mathbf{V}} \left( \max_{\mathbf{V} \in {}_{\alpha_i^k}\mathfrak{B}_{\mathbf{u}}^k} \Theta^k \left( \mathbf{V}, \mathbf{U}' \right) \right), \tag{9}$$

*where $\alpha_i^k$ is the highest possible threshold such that ${}_{\alpha_i^k}\mathfrak{B}_{\mathbf{u}}^k \neq \{\}$ and $\Theta^k \left( \mathbf{V}, \mathbf{U}' \right) > \Theta^k \left( \mathbf{U}, \mathbf{U}' \right)$.*

The constraint $\Theta^k \left( \mathbf{V}, \mathbf{U}' \right) > \Theta^k \left( \mathbf{U}, \mathbf{U}' \right)$ is used to ensure that the next scheduling solution is better than the previous solution. Thus, based on the fuzzy similarity, an agent can use a trade-off strategy to generate a scheduling proposal that may benefit all parties without lowering the agent's requirements. Thus, by trade-off negotiation in a scheduling problem, agents can reallocate their initially assigned resources whenever timing of the jobs is undesirable.

Different combinations of strategies can be applied to cooperative or competitive situations. Hence, the trade-off strategy and/or concession strategy can be further meshed and ordered into a meta strategy $\mathfrak{M}$ over the whole scenario of negotiation.

## 3 Negotiation Process

A solution of agent-based scheduling can be obtained by maintaining the satisfiability of both inter-agent and intra-agent constraints. (A solution of distributed scheduling can be obtained via fuzzy constraint-based agent negotiation by maintaining the satisfiability of both inter-agent and intra-agent constraints.) Agents take turns to propose local schedules to explore potential global schedules, thereby moving the negotiation toward a consensus. The process of each agent's behavior for scheduling is shown in Fig. 1.

Given the local schedule (the time interval of activities) $\mathbf{U}' = \{\mathbf{u}_{k^1}, ..., \mathbf{u}_{k^J}\}$ from agents $K' = \{k^1, ..., k^J\}$, each agent $k$ finds the solution concurrently and independently for obtaining the feasible solution. As shown with reference to the meta strategy of the negotiation, Deal and Failure are flags that indicate whether agents have made a deal. Accept($\mathbf{U}'$,$k$,$K'$) means that agent $k$ accepts the local schedules $\mathbf{U}'$ sent by the opponent agents $K'$. Fail ($k$,$K'$) means that agent $k$ cannot propose any solution. Tell($\mathbf{U}^*$,$k$,$K'$) means that agent $k$ proposes the local schedules to agents $K'$. In the negotiation process, an agent will interpret messages sent by the opponent. If an agent receives its opponents' Accept message, it sets Deal=True and the negotiation succeeds (in lines 3 and 4). If an agent receives anyone of its opponent's Fail message, it sets Failure=True and the negotiation fails (in lines 5 and 6). If an agent receives its opponents' local schedules $\mathbf{U}'$, the agent determines whether to accept the local schedules (in lines 8 to 10 or lines 32 to 34) or generates new preferred local schedules (in lines 12 to 36).

```
i = 1; $\Psi_{th}^{k} = 1.0$; $\alpha_i^k = 1.0$; Deal ← False; Failure ← False;
repeat
  if Receive "Accept(U',K',k)" then
    Deal ← True; end if;
  if Receive "Fail( k',k)" then
    Deal ← False;
    exit; end if;
  if Receive "Tell(U',K',k)" then
    if $\Psi^k$(U') ≥ $\Psi^k$(U) and $\mu_{C^k}$(U') ≥ $\alpha_i^k$ then
      Deal ← True;
      Accept(u', k, K');
    else
      $\mathfrak{M} \leftarrow PS_n^k$;
      while(True)
        if ${}_{\alpha_i^k}\mathfrak{B}_u^k \neq \{\}$ then
          U* ← $Sel\_offer({}_{\alpha_i^k}\mathfrak{B}_u^k, \mathfrak{M})$;
          if U* ≠ {} then
            exit; end if;
        else
          if $Chk\_con(\mathfrak{M})$ = True then
            $\Psi_{th}^k = \Psi_{th}^k - r$;
            ${}_{\alpha_i^k}\mathfrak{B}_u^k = LocalSch(\Psi_{th}^k, \alpha_i^k$, U', ∅); end if;
          if $Chk\_tra(\mathfrak{M})$ = True then
            ${}_{\alpha_i^k}\mathfrak{B}_u^k = LocalSch(\Psi_{th}^k, \alpha_i^k$, U', ∅); end if;
          if ${}_{\alpha_i^k}\mathfrak{B}_u^k = \{\}$ then
            $\alpha_i^k \leftarrow \alpha_{i+1}^k$;
            if $\alpha_i^k < \delta^k$ then
              Failure ← True;
              Fail(k,k');
              exit; end if; end if;
      end while;
      if $\Psi^k$(U') ≥ $\Psi^k$(U*) and $\mu_{C^k}$(U') ≥ $\alpha_i^k$ then
        Deal ← True;
        Accept(u', k, K');
      else
        Tell(U*,K,K');end if;end if;
until Deal = True or Failure = True;
```

**Fig. 1.** Agent behavior for scheduling

Following min-conflict principle/max similarity principle by Definition 4, a local schedule $\mathbf{U}^* = \{\mathbf{u}^*_{k^1}, ..., \mathbf{u}^*_{k^J}\}$ would be selected from the feasible schedules $\mathfrak{B}^k_{\mathbf{u}}$ and proposed to the corresponding agents $K'$ (in line 16). To ensure that the next local schedule solution $\mathbf{U}^*$ is better than the previous solution $\mathbf{U}$ for gradually converge, the constraint $\Theta^k(\mathbf{U}', \mathbf{U}^*) > \Theta^k(\mathbf{U}', \mathbf{U})$ has to be satisfied. If no solution found (in lines 19 to 30), agent $k$ will relax the constraint to the next acceptable threshold $\Psi^k_{i+1}$ to create a new feasible solution space $\mathfrak{B}^k_{\mathbf{u}}$ ($\mathfrak{B}^k_{\mathbf{u}} = \mathfrak{C}^k_{\mathbf{u}}$, in lines 20 to 22) by concession strategy (Definition 1); or will create a new alternative solution space $\mathfrak{B}^k_{\mathbf{u}}$ ($\mathfrak{B}^k_{\mathbf{u}} = \mathfrak{T}^k_{\mathbf{u}}$, in lines 23 and 24) by trade-off strategy (Definition 2). Solution space $\mathfrak{B}^k_{\mathbf{u}}$ is obtained from *LocalSch*.

According to the meta strategy $\mathfrak{M}$, which is determined by agent $k$'s mental state (in line 13), the functions, *Chk_con* and *Chk_tra* are used to verify whether the meta strategy will support a concession or choose a new alternative (in lines 20 to 24). In the concession strategy (Definition 1), $r$ is the agent's concession value, which is determined by the negotiator's mental state and the opponent's responsive state. In the trade-off strategy (Definition 2), *LocalSch* returns a feasible solution space $\mathfrak{B}^k_{\mathbf{u}}$ without reducing the agent's demands (in line 24).

If agent $k$ faces no feasible proposal that matches the expected satisfaction value at the threshold $\alpha^k_i$, with the capability of self-relaxation, the agent lowers its threshold of acceptability to the next threshold $\alpha^k_{i+1}$ until it generates an expected offer $\mathbf{U}^*$ or the threshold is less than $\delta^k$ (in line 19 or 23) in which case the negotiation fails and terminates.

If the expected local schedules $\mathbf{U}^*$ is generated, then the agent compares the opponents' local schedule $\mathbf{U}'$ with the selected solution $\mathbf{U}^*$ to determine whether to accept the opponents' local schedule $\mathbf{U}'$ (in line 32 to 34) or to propose the selected solution $\mathbf{U}^*$ through a Tell message to its opponents $K'$ (in line 36). Finally, the negotiation process terminates when either Deal=True or Failure=True (in line 37).

## 4 Experiments

In what follows, we conduct several experiments to examine the performances of the fuzzy constraint-based agent negotiation model (FCAN) for scheduling problem. The experiments demonstrate the proposed approach, based on fuzzy constraint theory and negotiation mechanism, provides the predictability of the system performance, and well quality of the schedules generated. Results obtained using the proposed approach were compared with those obtained from the CNP with priority dispatching strategies, and the auction-based approach using market-model [7,10]. The priority dispatching strategies used for comparison are Shortest Processing Time (SPT), Longest Processing Time (LPT), Earliest Due Date (EDD), Slack, and Critical Ratio (CR). To evaluate the qualities of schedule, the average of flow time, tardiness and resource utilization are chosen as performance measures.

The approaches are evaluated on a benchmark of job shop scheduling problems where parameters, such as number of jobs/activities, range of due dates and

**Table 1.** Comparison of FCAN approach with CNP and auction-based approaches over flow time

| Number of jobs/ activities | CNP with priority dispatching strategies | | | | | Auction | FCAN | Imp. over Best CNP (%) | Imp. over Auction (%) |
|---|---|---|---|---|---|---|---|---|---|
| | SPT | LPT | EDD | SLACK | CR | | | | |
| 10/5 | 100.81 | 125.75 | 97.36 | 111.40 | 129.65 | 84.51 | 80.29 | 17.53 | 5.00 |
| 20/5 | 181.63 | 227.21 | 181.04 | 204.85 | 233.53 | 128.66 | 116.37 | 35.72 | 9.56 |
| 30/5 | 257.22 | 309.12 | 255.43 | 288.56 | 326.76 | 199.30 | 173.84 | 31.94 | 12.77 |
| 40/5 | 335.74 | 409.42 | 332.98 | 380.19 | 417.63 | 230.92 | 190.87 | 42.68 | 17.34 |
| 50/5 | 409.02 | 511.92 | 405.47 | 470.59 | 521.65 | 280.29 | 214.84 | 47.01 | 23.35 |

**Table 2.** Comparison of FCAN approach with CNP and auction-based approaches over tardiness

| Number of jobs/ activities | CNP with priority dispatching strategies | | | | | Auction | FCAN | Imp. over Best CNP (%) | Imp. over Auction (%) |
|---|---|---|---|---|---|---|---|---|---|
| | SPT | LPT | EDD | SLACK | CR | | | | |
| 10/5 | 235 | 405 | 252 | 353 | 410 | 208 | 205 | 12.77 | 1.44 |
| 20/5 | 1911 | 2761 | 1999 | 2424 | 2873 | 1577 | 1357 | 28.99 | 13.95 |
| 30/5 | 5085 | 6589 | 5162 | 6097 | 7111 | 4830 | 4050 | 20.35 | 16.15 |
| 40/5 | 9887 | 12787 | 9926 | 11744 | 13113 | 7580 | 6062 | 38.69 | 20.03 |
| 50/5 | 16005 | 21109 | 15994 | 19173 | 21594 | 10531 | 8417 | 47.37 | 20.07 |

activity durations are varied to generate a broad range of problem instances. Each job has a linear process routing specifying a sequence. For each activity of the job, it is equipped with predefined resource and processing time is deterministic. The processing time of the activity is uniformly generated from 1 to 10 time units. The results of each set of parameters are averaged over 100 different randomly generated data sets. For simplicity, all agents employ a fixed concession strategy with 0.1 urgency value.

The experiments illustrate the comparisons among FCAN, CNP and auction-based approaches across of a range of problem sizes over the criteria of flow time, tardiness, and resource utilization are shown in Table 1, Table 2 and Table 3, respectively. The performances of CNP with priority dispatching strategies are presented in columns 2 through 6, and the results of the auction-based approach and FCAN approach are in columns 7 and 8, respectively. The improvements in using the proposed approach, as compared with the results obtained from the best of different priority dispatching strategies and the auction-based approach, are presented in columns 9 and 10, respectively.

Comparing FCAN approach and CNP with priority dispatching strategies over different problem sizes, the results obtained from the best of priority dispatching strategies have been improved from 17.53% to 47.01% on the criterion of average flow time, 12.77% to 47.37% on the tardiness, and 11.57% to 53.16% on the resource utilization. Further, when FCAN approach compared with the

**Table 3.** Comparison of FCAN approach with CNP and auction-based approaches over resource utilization

| Number of jobs/ activities | CNP with priority dispatching strategies | | | | | Auction | FCAN | Imp. over Best CNP (%) | Imp. over Auction (%) |
|---|---|---|---|---|---|---|---|---|---|
| | SPT | LPT | EDD | SLACK | CR | | | | |
| 10/5 | 41.21 | 41.30 | 34.33 | 33.28 | 41.57 | 45.26 | 46.38 | 11.57 | 2.47 |
| 20/5 | 41.50 | 41.92 | 33.64 | 33.30 | 36.17 | 43.86 | 47.39 | 11.52 | 8.07 |
| 30/5 | 42.46 | 43.34 | 33.71 | 33.59 | 35.38 | 47.43 | 51.30 | 18.36 | 8.15 |
| 40/5 | 42.38 | 43.09 | 33.80 | 33.40 | 35.42 | 52.20 | 58.48 | 35.72 | 12.03 |
| 50/5 | 42.77 | 42.36 | 33.86 | 33.28 | 34.85 | 59.18 | 65.51 | 53.16 | 10.69 |

auction-based approach over different problem sizes, the global performances of schedule have been improved from 5.0% to 23.35% on the average flow time, 1.44% to 20.07% on the tardiness, and 2.47% to 12.03% on the resource utilization.

From the observation of the experimental results, CNP with priority dispatching strategies has inferior performance of scheduling for all problem sizes. Through iterative bidding, the auction-based approach is more aware about resource contention and performs better than the formers. However, these approaches with local decision cannot guarantee the overall system performance.

In the proposed approach, each agent keeps more local schedules in bounded solution space which satisfaction level is not less than the threshold. Negotiation with a prior knowledge also provides a guideline for agents to further restrict solution space and move towards a consistent agreement that benefits all agents. The experimental results show the proposed approach provides superior performance in all criteria to the contract net protocol and auction-based approaches. Meanwhile, as problem sizes is growing, the performance improved by the FCAN approach increases more significantly.

## 5 Conclusions

A negotiation mechanism applied fuzzy constraints is proposed to govern agent-based scheduling. The proposed approach enables an agent to maximize its own objectives with a different perspective, and attains the common agreement that benefits all agents at the high satisfaction degree. The concession and trade-off strategy is presented to ensure agents move towards a consistent agreement more quickly since their searches focus only on feasible and bounded solution space. The gradual relaxation and evaluation method with iterative negotiation process enables participants in distributed scheduling to progressively move toward a globally satisfactory schedule.

Our experimental results illustrate the proposal approach incorporates activities' demands and guides the local schedule procedure by the society of interacting agents, facilitating rapid convergence to a feasible and global beneficial solution. While the proposed model yielded some promising results, considerable work remains to be done, such as designing a learning model, applying to other forms of scheduling problems, and studying coherence of negotiation strategies in various scheduling problems.

## References


1. Dubois, D., Fargier, H., and Prade, H.: Fuzzy Constraints in Job-shop Scheduling. *Journal of Intelligent Manufacturing* 6 (1995) 215-234.
2. Dubois, D., Fragier, H., and Philippe Fortemps.: Fuzzy Scheduling: Modeling flexible Constraints vs. Coping with Incomplete Knowledge. *European Journal of Operational Research* 147 (2003) 231-252
3. Lai, K. R.: *Fuzzy Constraint Processing.* Ph.D. thesis, NCSU, Raleigh, N. C. (1992)
4. Lai, K. R., and Lin, M. W.: Agent Negotiation as Fuzzy Constraint Processing. *FUZZ-IEEE'02. Proceedings of the 2002 IEEE International Conference on Fuzzy Systems* 2(12-17) (2002) 1021-1026
5. Lai, K. R., and Lin, M. W.: Modeling Agent Negotiation via Fuzzy Constraints in e-Business. *Computational Intelligence* 20(4) (2004) 624-642
6. Lee, Y., and Kumara, S. R., and Chatterjee, K.: Multiagent Based Dynamic Resource Scheduling for Distributed Multiple Projects Using a Market Mechanism. *Journal of Intelligent Manufacturing* 14 (5) (2003) 471-484
7. Lin, G. Y. J., and Solberg, J. J.: Integrated Shop Floor Control Using Autonomous Agents. *IIE Trans.: Des. Manuf.* 24(3) (1992) 57-71
8. Luo, X.D., Nicholas R. Jennings., Nigel Shadbolt, Ho-fung Leung, and Jimmy Homan Lee.: A Fuzzy Constraint Based Model for Bilateral Multi-issue Negotiations in Semi-competitive Environments. *Artificial Intelligence* 148 (2003) 53-102
9. McDonnell, P., Smith, G. Joshi, S., and Kumara, S. R. T.: A Cascading Auction protocol as a Framework for Integrating Process Planning and Heterarchical Shop Floor Control. *Int. J. Flexible Manuf. Syst.* 11(1) (1999) 37-62
10. Macchiaroli, R., and Riemma, S.: A Negotiation Scheme for Autonomous. Agents in Job Shop Scheduling. *International Journal of Computer Integrated. Manufacturing* 15(3) (2002) 222-232
11. Miyashita, K.: CAMPS: A Constraint-Based Architecture for Multi-Agent Planning and Scheduling. *Journal of Intelligent Manufacturing* 9(2) (1998) 147-154.
12. Siwamogsatham, T., Saygin, C.: Auction-based Distributed Scheduling and Control Scheme for Flexible Manufacturing Systems. *International Journal of Production Research* 42(3), (2004) 547-572
13. Shen, W., Wang, L., and Hao, Q.: Agent-based Distributed Manufacturing Process Planning and Scheduling: A State-of-the-art Survey. *IEEE Trans. on Systems, Man, and Cybernetics, Part C: Applications and Reviews* 36(4) (2006) 563-577
14. Smith, R. G.: The Contract Net Protocol: High-level Communication and Control in a Distributed Problem Solver. *IEEE Trans. Comput.* C-29(12) (1980) 1104-1113
15. Wang, L., Shen, and W.: DPP: An Agent-based Approach for Distributed Process Planning. *J. Intell. Manuf.* 14(5) (2003) 429-440
16. Zukin, M., and Young, R. E.: Applying Fuzzy Logic and Constraint Networks to a Problem of Manufacturing Flexibility. *International Journal of Production Research* 39(14) (2001) 3253-3273

# Lyapunov Stability of Fuzzy Discrete Event Systems*

Fuchun Liu[1,2] and Daowen Qiu[1,**]

[1] Department of Computer Science, Zhongshan University, Guangzhou 510275, China
[2] Faculty of Applied Mathematics, Guangdong University of Technology, Guangzhou 510090, China

liufch@gdut.edu.cn, issqdw@mail.sysu.edu.cn
http://www.sysu.edu.cn

**Abstract.** *Fuzzy discrete event systems* (FDESs) as a generalization of (crisp) discrete event systems (DESs) may better deal with the problems of fuzziness, impreciseness, and subjectivity. Qiu, Cao and Ying, Liu and Qiu interestingly developed the theory of FDESs. As a continuation of Qiu's work, this paper is to deal with the Lyapunov stability of FDESs, some main results of crisp DESs are generalized. We formalize the notions of the reachability of fuzzy states defined on a metric space. A linear algorithm of computing the $r$-reachable fuzzy state set is presented. Then we introduce the definitions of stability and asymptotical stability in the sense of Lyapunov to guarantee the convergence of the behaviors of fuzzy automaton to the desired fuzzy states when system engages in some illegal behaviors which can be tolerated. In particular, we present a necessary and sufficient condition for stability and another for asymptotical stability of FDESs.

**Keywords:** Discrete event systems, Lyapunov stability, asymptotical stability, fuzzy finite automata, metric space.

## 1 Introduction

*Discrete event systems* (DESs) are dynamical systems whose evolution in time is governed by the abrupt occurrence of physical events at possibly irregular time intervals. Up to now, the theory of DESs has been significantly applied to many practical systems such as automated manufacturing systems, interaction telecommunication networks and communication networks [1, 2]. In most of engineering applications, DESs are modeled by finite state automata with crisp states and crisp events. However, such crisp DESs are not sufficient for some

* This work was supported in part by the National Natural Science Foundation under Grant 90303024 and Grant 60573006, and the Research Foundation for the Doctorial Program of Higher School of Ministry of Education under Grant 20050558015 of China.

** Corresponding author.

D.-S. Huang, L. Heutte, and M. Loog (Eds.): ICIC 2007, LNAI 4682, pp. 693–701, 2007.

practical fields such as biomedical applications in which the patient's health status and state transition are always somewhat uncertain and vague. For example, it is vague to describe a patient's health condition to be "good", and, it is imprecise to say at what point exactly the patient has changed from "poor" to "good". In order to more effectively cope with the real-world problems of vagueness, Lin and Ying [3, 4] initiated the study of fuzzy discrete event systems (FDESs) by combining fuzzy set theory with crisp DESs. Notably, FDESs have been successfully applied to biomedical control for HIV/AIDS treatment planning [5, 6] and robotic control for intelligent information processing [7, 8].

As Lin and Ying [4] pointed out, a comprehensive theory of FDESs still needs to be set up, including many important concepts, methods and theorems, such as controllability, observability, and optimal control. These issues have been partially investigated in [9-12]. Qiu [9], Liu and Qiu [10] investigated the supervisory control and decentralized supervisory control of FDESs, respectively; and independently, Cao and Ying [11, 12] significantly developed the theory of FDESs from different aspect. In particular, Qiu [9] first devised an algorithm of checking the existence of supervisors for classical DESs.

On the other hand, there has been much interest recently in studying the stability properties of DESs, and several definitions for stability and methods for stability analysis have been proposed [13-17]. Among these works, the Lyapunov approach is considered as a general characterization of the stability properties of DESs, which has been applied to deal with the load balancing problem in computer networks [15] and stability analysis in petri nets [16, 17].

As a continuation of Qiu's work [9], this paper is to deal with the Lyapunov stability of FDESs. We formalize the notions of reachability, stability and asymptotical stability in the framework of FDESs. We first define a metric space on fuzzy systems, and the reachability of fuzzy states on this metric space is formalized. We further present an algorithm that is linear in the number of states of system to compute the $r$-reachable fuzzy state set. Then we introduce the notions of stability and asymptotical stability in the sense of Lyapunov to guarantee the convergence of the behaviors of fuzzy system to the desired fuzzy states. In particular, we present a necessary and sufficient condition for stability and another for asymptotical stability of FDESs.

## 2 Preliminaries of Fuzzy Discrete Event Systems

In the setting of FDESs, a fuzzy state is represented as a vector $[a_1, a_2, \cdots, a_n]$ that stands for the possibility distributions over crisp states, i.e., each $a_i \in [0, 1]$ represents the possibility that the system is in the $i$th crisp state. Similarly, a fuzzy event is denoted by a matrix $\tilde{\sigma} = [a_{ij}]_{n\times n}$, where each $a_{ij} \in [0, 1]$ means the possibility of system transferring from the $i$th crisp state to the $j$th crisp state when event $\sigma$ occurs, and $n$ is the number of all possible crisp states.

*Definition 1:* A *fuzzy finite automaton* is formally defined as a fuzzy system

$$\tilde{G} = (\tilde{Q}, \tilde{\Sigma}, \tilde{\delta}, \tilde{Q}_0),$$

where $\tilde{Q}$ is the set of fuzzy state vectors; $\tilde{Q}_0 \subseteq \tilde{Q}$ is the set of initial fuzzy states; $\tilde{\Sigma}$ is the set of fuzzy event matrices; $\tilde{\delta} : \tilde{Q} \times \tilde{\Sigma} \to \tilde{Q}$ is a transition function which is defined by $\tilde{\delta}(\tilde{q}, \tilde{\sigma}) = \tilde{q} \odot \tilde{\sigma}$ for $\tilde{q} \in \tilde{Q}$ and $\tilde{\sigma} \in \tilde{\Sigma}$, where $\odot$ denotes the *max-product* [3, 4] or *max-min* [9] operation: for matrices $A = [a_{ij}]_{n\times m}$ and $B = [b_{ij}]_{m\times k}$, then $A \odot B = [c_{ij}]_{n\times k}$, where $c_{ij} = \max_{l=1}^{m} a_{il} \times b_{lj}$ after max-product operation, or $c_{ij} = \max_{l=1}^{m} \min\{a_{il}, b_{lj}\}$ after max-min operation.

*Remark 1:* Transition function $\tilde{\delta}$ can be extended to $\tilde{Q} \times \tilde{\Sigma}^*$ in the usual manner:

$$\tilde{\delta}(\tilde{q}, \lambda) = \tilde{q}, \qquad \tilde{\delta}(\tilde{q}, \tilde{s}\tilde{\sigma}) = \tilde{\delta}(\tilde{\delta}(\tilde{q}, \tilde{s}), \tilde{\sigma}),$$

where $\tilde{\Sigma}^*$ is the Kleene closure of $\tilde{\Sigma}$, $\lambda$ denotes the empty string, $\tilde{q} \in \tilde{Q}$, $\tilde{\sigma} \in \tilde{\Sigma}$ and $\tilde{s} \in \tilde{\Sigma}^*$. Moreover, without loss of generality, $\tilde{\delta}$ can be regarded as a partial transition function in practice.

We define a set valued function $d : \tilde{Q} \to 2^{\tilde{\Sigma}}$ to represent the set of all possible fuzzy events defined at each fuzzy state. That is, for $\tilde{q} \in \tilde{Q}$,

$$d(\tilde{q}) = \{\tilde{\sigma} \in \tilde{\Sigma} : \exists \tilde{q}' \in \tilde{Q}(\tilde{q} \odot \tilde{\sigma} = \tilde{q}' \wedge \tilde{\delta}(\tilde{q}, \tilde{\sigma})!)\}, \tag{1}$$

where the notation "!" is used to denote "is defined".

A finite string of fuzzy states $\tilde{p} = \tilde{q_1}\tilde{q_2} \cdots \tilde{q_j}$ is called to be *a state trajectory* from fuzzy state $\tilde{q_1}$ if $\tilde{q}_{i+1} = \tilde{\delta}(\tilde{q_i}, d(\tilde{q_i}))$ for all $i = 1, 2, \cdots, j-1$, where

$$\tilde{\delta}(\tilde{q}, d(\tilde{q})) = \bigcup_{\tilde{\sigma} \in d(\tilde{q})} \{\tilde{q} \odot \tilde{\sigma} : \tilde{\delta}(\tilde{q}, \tilde{\sigma})!\}. \tag{2}$$

Similarly, string $\tilde{s} = \tilde{\sigma_1}\tilde{\sigma_2} \cdots \tilde{\sigma_j}$ is called to be *a valid event trajectory* from $\tilde{q}$, if $\sigma_1 \in d(\tilde{q})$ and $\tilde{\sigma}_{i+1} = d(\tilde{q} \odot \tilde{\sigma_1} \odot \cdots \tilde{\sigma_i})$ for $i = 1, 2, \cdots, j-1$.

Denote $L(\tilde{G}, \tilde{q})$ and $L_a(\tilde{G})$ as the set of all possible valid event trajectories from $\tilde{q}$ and the set of all allowed event trajectories in $\tilde{G}$, respectively. $L_a(\tilde{G}, \tilde{q})$ stands for the set of all allowed event trajectories from $\tilde{q}$. Denote $\tilde{\Sigma}^0 = \{\lambda\}$, and $\tilde{\Sigma}^k$ is the set of all fuzzy event string with the length of $k$, i.e.,

$$\tilde{\Sigma}^k = \{\tilde{\sigma_1}\tilde{\sigma_2} \cdots \tilde{\sigma_k} : \tilde{\sigma_i} \in \tilde{\Sigma}, i = 1, 2, \cdots, k\}. \tag{3}$$

## 3 Reachability of Fuzzy States

In order to investigate the stability of FDESs, we first consider the problem of the reachability from fuzzy states in FDESs defined on a metric space.

*Definition 2:* Let $\tilde{G} = (\tilde{Q}, \tilde{\Sigma}, \tilde{\delta}, \tilde{Q_0})$ be an FDES modeled by a fuzzy finite automaton. *A metric* $\rho : \tilde{Q} \times \tilde{Q} \to [0, +\infty)$ is defined as: for $\tilde{q_1}, \tilde{q_2} \in \tilde{Q}$,

$$\rho(\tilde{q_1}, \tilde{q_2}) = \frac{1}{n} \sum_{i=1}^{n} |a_i - b_i|, \tag{4}$$

where $\tilde{q_1} = [a_1, \cdots, a_n]$ and $\tilde{q_2} = [b_1, \cdots, b_n]$. Obviously, $(\tilde{Q}, \rho)$ is a metric space.

Let $\tilde{Q}_z \subseteq \tilde{Q}$, the distance from a fuzzy state $\tilde{q}$ to set $\tilde{Q}_z$ is

$$\rho(\tilde{q}, \tilde{Q}_z) = inf\{\rho(\tilde{q}, \tilde{q}') : \tilde{q}' \in \tilde{Q}_z\}. \tag{5}$$

*Definition 3:* For $r \geq 0$, the *r-neighborhood* of the set $\tilde{Q}_z$ is

$$S(\tilde{Q}_z; r) = \{\tilde{q} \in \tilde{Q} : \rho(\tilde{q}, \tilde{Q}_z) \leq r\}. \tag{6}$$

A fuzzy state $\tilde{q}'$ is said to be *reachable* from a fuzzy state $\tilde{q}$ if there is $\tilde{s} \in \tilde{\Sigma}^*$ such that $\tilde{\delta}(\tilde{q}, \tilde{s})!$ and $\tilde{q} \odot \tilde{s} = \tilde{q}'$. In order to more effectively describe the vagueness of the fuzzy systems, we introduce the notion of $r$-reachability. A fuzzy state $\tilde{q}'$ is said to be $r$*-reachable* from a fuzzy state $\tilde{q}$ if there is $\tilde{s} \in \tilde{\Sigma}^*$ such that $\tilde{\delta}(\tilde{q}, \tilde{s})!$ and $\tilde{q} \odot \tilde{s} \in S(\{\tilde{q}'\}; r)$, which is denoted by $\tilde{q} \rightarrow_r^* \tilde{q}'$. We use

$$R_r(\tilde{G}, \tilde{q}) = \{\tilde{q}' \in \tilde{Q} : \tilde{q} \rightarrow_r^* \tilde{q}'\} \tag{7}$$

to represent all fuzzy states that are $r$-reachable from fuzzy state $\tilde{q}$.

Now we present an approach to compute the accessibility of $\tilde{G}$ which is a DFA whose state space is made up of all $r$-reachable states.

*Definition 4:* The accessibility of $\tilde{G}$ is defined as

$$\tilde{G}_r = (\tilde{Q}_r, \tilde{\Sigma}, \tilde{\delta}_r, \tilde{Q}_0), \tag{8}$$

where $\tilde{\Sigma}$ and $\tilde{Q}_0$ are the same as those of $\tilde{G}$; $\tilde{Q}_r$ is the set of all $r$-reachable fuzzy states from $\tilde{Q}_0$; and the transition function $\tilde{\delta}_r : \tilde{Q}_r \times \tilde{\Sigma} \rightarrow \tilde{Q}_r$ is defined as follows: for $\tilde{q} \in \tilde{Q}_r$ and $\tilde{\sigma} \in \tilde{\Sigma}$,

$$\tilde{\delta}_r(\tilde{q}, \tilde{\sigma}) = \tilde{q}' \quad \text{iff} \quad \tilde{q} \odot \tilde{\sigma} = \tilde{q}' \text{ and } \tilde{\delta}(\tilde{q}, \tilde{\sigma})! \text{ in } \tilde{G}. \tag{9}$$

Denote

$$R_r(\tilde{G}, \tilde{Q}_0) = \bigcup_{\tilde{q} \in \tilde{Q}_0} R_r(\tilde{G}, \tilde{q}). \tag{10}$$

Notice that $\tilde{Q}_r$ is a fixed point of function $g : 2^{\tilde{Q}} \rightarrow 2^{\tilde{Q}}$, where for $\tilde{Q}_z \subseteq \tilde{Q}$,

$$g(\tilde{Q}_z) = R_r(\tilde{G}, \tilde{Q}_z \odot \tilde{\Sigma}) = \bigcup_{\tilde{q} \in \tilde{Q}_z} R_r(\tilde{G}, \tilde{q} \odot d(\tilde{q})). \tag{11}$$

Thus, we can give an algorithm to compute $\tilde{Q}_r$ of all $r$-reachable fuzzy states.

*Algorithm:*

- Let $R_0 = P_0 = \tilde{Q}_0$.
- Iterate:
$$R_{k+1} = R_k \bigcup g(P_k), \qquad P_{k+1} = R_{k+1} \bigcap \overline{R}_k, \tag{12}$$
  where overline denotes the complement operator in set theory.
- Terminate when $R_{k+1} = R_k$.

*Theorem 1:* Above algorithm can compute the set $\tilde{Q}_r$ of all $r$-reachable fuzzy states from $\tilde{Q}_0$, and it has complexity $O(n)$, where $n = |\tilde{Q}|$.

*Proof:* Clearly, the algorithm terminates in a finite number of steps, say $t$. From Eqs.(11, 12), we know that each fuzzy state of $R_t$ is $r$-reachable from one of initial fuzzy state, so $R_t \subseteq \tilde{Q}_r$.

On the other hand, for any $\tilde{q} \in \tilde{Q}_r$, there is $\tilde{q}_0 \in \tilde{Q}_0$ and $\tilde{s} \in \tilde{\Sigma}^*$ such that $\tilde{\delta}(\tilde{q}_0, \tilde{s})!$ and $\tilde{q}_0 \odot \tilde{s} \in S(\{\tilde{q}\}; r)$. We prove $\tilde{q} \in R_t$ by induction on the length of $\tilde{s}$. If $\tilde{s} = \lambda$, then the result holds clearly. Assume that $\tilde{s} = \tilde{s}_1\tilde{\sigma}$ where $\tilde{q}_0 \odot \tilde{s}_1 \in R_k$ ($k \leq t$)). By the step of iteration of the algorithm, we have $\tilde{q}_0 \odot \tilde{s} \in R_k$ or $\tilde{q}_0 \odot \tilde{s} \in g(P_k)$. Thus, $\tilde{q} \in R_{k+1}$ and $\tilde{q} \in R_t$. That is, $\tilde{Q}_r \subseteq R_t$.

Since each fuzzy state is visited at most only once, the complexity of the algorithm is $O(n)$. □

*Remark 2:* For max-min fuzzy automata, Qiu [9] provided a good approach to calculate all fuzzy states reachable from the initial state by means of computing tree. Our algorithm presented above is suitable for calculating $r$-reachability for both max-min and max-product fuzzy automata. Furthermore, the case of $r = 0$ coincides with Qiu's approach.

## 4 Lyapunov Stability of Fuzzy Discrete Event Systems

In this section, we generalize the main results of stability from crisp DESs [15] to FDESs, and establish the theory of Lyapunov stability in the framework FDESs. As we know that stability can be thought of as error recovery that allows the system to engage in some illegal behaviors, but it must go to one of the desired states after a finite number of transitions. The invariant set $\tilde{Q}_m$ is denoted those desired fuzzy states generated by the legal event trajectories.

*Definition 5:* The set $\tilde{Q}_m \subseteq \tilde{Q}$ is called to be *invariant* with respect to (w.r.t.) $\tilde{G}$ if for any $\tilde{q} \in \tilde{Q}_m$ and any $\tilde{s} \in \tilde{\Sigma}^k$, we have

$$\tilde{\delta}(\tilde{q}, \tilde{s})! \quad \Rightarrow \quad \tilde{q} \odot \tilde{s} \in \tilde{Q}_m. \tag{13}$$

*Proposition 2:* $\tilde{Q}_m \subseteq \tilde{Q}$ is invariant w.r.t. $\tilde{G}$ if and only if $R_0(\tilde{G}, \tilde{Q}_m) = \tilde{Q}_m$.

*Proof:* It is clearly proved by Definition 5 and Eq. (10). □

*Definition 6:* For a given $\tilde{q} \in \tilde{Q}$ and $k$, *a motion function* $f_{\tilde{q}}^k : \tilde{\Sigma}^k \to \tilde{Q}$ is defined as a partial function: for any $\tilde{s} \in \tilde{\Sigma}^k$, if $\tilde{\delta}(\tilde{q}, \tilde{s})!$, then $f_{\tilde{q}}^k(\tilde{s}) = \tilde{q} \odot \tilde{s}$, which is called a motion; otherwise, $f_{\tilde{q}}^k(\tilde{s})$ is not defined.

*Proposition 3:* For any $\tilde{q} \in \tilde{Q}$, we have $f_{\tilde{q}}^0(\lambda) = \tilde{q}$. Furthermore, for any $\tilde{s_1} \in \tilde{\Sigma}^{k_1}$ and $\tilde{s_2} \in \tilde{\Sigma}^{k_2}$, if $\tilde{\delta}(\tilde{q}, \tilde{s_1}\tilde{s_2})!$, then

$$f^{k_2}_{f_{\tilde{q}}^{k_1}(\tilde{s_1})}(\tilde{s_2}) = f_{\tilde{q}}^{k_1+k_2}(\tilde{s_1}\tilde{s_2}). \tag{14}$$

*Proof:* It can be obtained easily from Definition 6. □

*Definition 7:* Let $L_a(\tilde{G})$ be the set of allowed event trajectories. An invariant set $\tilde{Q}_m$ is said to be *stable in the sense of Lyapunov* w.r.t. $L_a(\tilde{G})$, if for any $\epsilon > 0$, there is $\eta > 0$ such that when $\rho(\tilde{q}, \tilde{Q}_m) < \eta$, we have

$$\rho(f_{\tilde{q}}^k(\tilde{s}), \tilde{Q}_m) < \epsilon \tag{15}$$

for all $k$ and $\tilde{s} \in \tilde{\Sigma}^k \cap L_a(\tilde{G})$, where $\tilde{\delta}(\tilde{q}, \tilde{s})!$.

Intuitively, for a fuzzy system equipped with a stable state set $\tilde{Q}_m$, in order to make system transfer to a state that is sufficiently near to $\tilde{Q}_m$ after $k$ transitions, we only need ensure that the original state is suitably close to $\tilde{Q}_m$.

*Definition 8:* An invariant set $\tilde{Q}_m$ is said to be *asymptotically stable in the sense of Lyapunov* w.r.t. $L_a(\tilde{G})$, if $\tilde{Q}_m$ is stable in the sense of Lyapunov, and it is possible that there is $\eta > 0$ such that when $\rho(\tilde{q}, \tilde{Q}_m) < \eta$, we have

$$\lim_{k \to \infty} \rho(f_{\tilde{q}}^k(\tilde{s}), \tilde{Q}_m) = 0 \tag{16}$$

for all $\tilde{s} \in \tilde{\Sigma}^k \cap L_a(\tilde{G})$, where $\tilde{\delta}(\tilde{q}, \tilde{s})!$.

Intuitively, a desired fuzzy state set $\tilde{Q}_m$ being asymptotically stable means that, if the original state is close enough to $\tilde{Q}_m$, then the system will asymptotically transfer to a desired fuzzy state finally along the legal behaviors.

*Theorem 4:* Let $\tilde{G} = (\tilde{Q}, \tilde{\Sigma}, \tilde{\delta}, \tilde{Q}_0)$ be an FDES. The invariant set $\tilde{Q}_m \subseteq \tilde{Q}$ is stable in the sense of Lyapunov w.r.t. $L_a(\tilde{G})$, if and only if, in a sufficiently small neighborhood $S(\tilde{Q}_m; r)$, there is a function $V : S(\tilde{Q}_m; r) \to (0, +\infty)$ with the following conditions:

(1) For any constant $c_1 > 0$ and any $\tilde{q} \in S(\tilde{Q}_m; r)$, there is a constant $c_2 > 0$ such that when $\rho(\tilde{q}, \tilde{Q}_m) > c_1$, we have $V(\tilde{q}) > c_2$.

(2) For any constant $c_4 > 0$ and any $\tilde{q} \in S(\tilde{Q}_m; r)$, there is a constant $c_3 > 0$ such that when $\rho(\tilde{q}, \tilde{Q}_m) < c_3$, we have $V(\tilde{q}) < c_4$.

(3) $V(f_{\tilde{q}}^k(\tilde{s}))$ is a nonincreasing function of $k$, where $\tilde{q} \in S(\tilde{Q}_m; r)$, $\tilde{s} \in \tilde{\Sigma}^k \cap L_a(\tilde{G})$, $\tilde{\delta}(\tilde{q}, \tilde{s})!$ and $f_{\tilde{q}}^k(\tilde{s}) \in S(\tilde{Q}_m; r)$.

*Proof: Necessity:* Assume that $\tilde{Q}_m$ is stable in the sense of Lyapunov w.r.t. $L_a(\tilde{G})$. Then from Definition 7, for any $\epsilon > 0$, there is $\eta > 0$ such that

$$\rho(\tilde{q}, \tilde{Q}_m) < \eta \qquad \Rightarrow \qquad \rho(f_{\tilde{q}}^k(\tilde{s}), \tilde{Q}_m) < \epsilon \tag{17}$$

for all $k$ and $\tilde{s} \in \tilde{\Sigma}^k \cap L_a(\tilde{G})$, where $\tilde{\delta}(\tilde{q}, \tilde{s})!$.

Define function $V : S(\tilde{Q}_m; r) \to (0, +\infty)$ as follows: for $\tilde{q} \in S(\tilde{Q}_m; r)$,

$$V(\tilde{q}) = sup\{\rho(f_{\tilde{q}}^k(\tilde{s}), \tilde{Q}_m) : \text{for all } k \text{ and } \tilde{s} \in \tilde{\Sigma}^k \cap L_a(\tilde{G}),\ \tilde{\delta}(\tilde{q}, \tilde{s})!\}. \tag{18}$$

It is not difficult to verify $V$ satisfying conditions (1, 2) of the theorem. In the following, we prove that $V(f_{\tilde{q}}^k(\tilde{s}))$ is a nonincreasing function of $k$.

From the definition of $V$ and Proposition 3, we have

$$\begin{aligned} V(f_{\tilde{q}}^{k}(\tilde{s})) &= sup\{\rho(f_{f_{\tilde{q}}^{k}(\tilde{s})}^{k'}(\tilde{s}'), \tilde{Q}_m) : \tilde{s}' \in \tilde{\Sigma}^{k'} \cap L_a(\tilde{G}),\ \ \tilde{\delta}(\tilde{q}, \tilde{s}\tilde{s}')!\} \\ &= sup\{\rho(f_{\tilde{q}}^{k+k'}(\tilde{s}\tilde{s}'), \tilde{Q}_m) : \tilde{s}' \in \tilde{\Sigma}^{k'} \cap L_a(\tilde{G}),\ \ \tilde{\delta}(\tilde{q}, \tilde{s}\tilde{s}')!\}. \end{aligned} \tag{19}$$

Similarly,

$$V(f_{\tilde{q}}^{k+1}(\tilde{s}\tilde{\sigma})) = sup\{\rho(f_{\tilde{q}}^{k+1+k'}(\tilde{s}\tilde{\sigma}\tilde{s}'), \tilde{Q}_m) : \tilde{s}' \in \tilde{\Sigma}^{k'} \cap L_a(\tilde{G}),\ \ \tilde{\delta}(\tilde{q}, \tilde{s}\tilde{\sigma}\tilde{s}')!\}. \tag{20}$$

Therefore, $V(f_{\tilde{q}}^{k}(\tilde{s})) \geq V(f_{\tilde{q}}^{k+1}(\tilde{s}\tilde{\sigma}))$.

*Sufficiency:* Suppose that function $V : S(\tilde{Q}_m; r) \to (0, +\infty)$ satisfies conditions (1, 2, 3). We prove $\tilde{Q}_m$ being stable by contradiction. Assume that there are $\epsilon > 0$, $\tilde{q} \in S(\tilde{Q}_m; r)$, and $\tilde{s} \in \tilde{\Sigma}^k \cap L_a(\tilde{G})$ (without loss of generality, let $\epsilon < r$) such that $\tilde{\delta}(\tilde{q}, \tilde{s})!$. Then for any $\eta > 0$, when $\rho(\tilde{q}, \tilde{Q}_m) < \eta$, we have

$$\rho(f_{\tilde{q}}^{k}(\tilde{s}), \tilde{Q}_m) \geq \epsilon. \tag{21}$$

Denote that $\mu = inf\{V(\tilde{q}) : \tilde{q} \in \triangle\}$, where

$$\triangle = \{\tilde{q} \in S(\tilde{Q}_m; r) : \rho(\tilde{q}, \tilde{Q}_m) \geq \epsilon\}. \tag{22}$$

From condition (1), we have $\mu > 0$. Furthermore, by condition (2), there is $\eta > 0$ such that $V(\tilde{q}) < \mu$ when $\tilde{q} \in S(\tilde{Q}_m; r)$ and $\rho(\tilde{q}, \tilde{Q}_m) < \eta$. By condition (3), we know that $f_{\tilde{q}}^{k}(\tilde{s}) \in S(\tilde{Q}_m; r)$ and $V(f_{\tilde{q}}^{k}(\tilde{s})) \leq V(\tilde{q})$. Therefore,

$$V(f_{\tilde{q}}^{k}(\tilde{s})) < \mu. \tag{23}$$

However, from $f_{\tilde{q}}^{k}(\tilde{s}) \in S(\tilde{Q}_m; r)$ and Ineq. (21), we have $f_{\tilde{q}}^{k}(\tilde{s}) \in \triangle$. That is,

$$V(f_{\tilde{q}}^{k}(\tilde{s})) \geq inf\{V(\tilde{q}) : \tilde{q} \in \triangle\} = \mu, \tag{24}$$

which is in conflict with Ineq. (23). □

*Theorem 5:* Let $\tilde{G} = (\tilde{Q}, \tilde{\Sigma}, \tilde{\delta}, \tilde{Q}_0)$ be an FDES. The invariant set $\tilde{Q}_m \subseteq \tilde{Q}$ is asymptotically stable in the sense of Lyapunov w.r.t. $L_a(\tilde{G})$, if and only if, in a sufficiently small neighborhood $S(\tilde{Q}_m; r)$, there is a function $V : S(\tilde{Q}_m; r) \to (0, +\infty)$ with conditions (1, 2, 3) of Theorem 4, and, furthermore,

$$\lim_{k\to\infty} V(f_{\tilde{q}}^{k}(\tilde{s})) = 0 \tag{25}$$

for $\tilde{s} \in \tilde{\Sigma}^k \cap L_a(\tilde{G})$, where $\tilde{\delta}(\tilde{q}, \tilde{s})!$ and $f_{\tilde{q}}^{k}(\tilde{s}) \in S(\tilde{Q}_m; r)$.

*Proof: Necessity:* Assume that the invariant set $\tilde{Q}_m$ is asymptotically stable. Then, $\tilde{Q}_m$ is stable, and, the function $V : S(\tilde{Q}_m; r) \to (0, +\infty)$ constructed in Theorem 4 satisfies the conditions (1,2,3) of Theorem 4. Furthermore, from Definition 8, there is $\eta > 0$ such that

$$\rho(\tilde{q}, \tilde{Q}_m) < \eta \quad \Rightarrow \quad \lim_{k\to\infty} \rho(f_{\tilde{q}}^{k}(\tilde{s}), \tilde{Q}_m) = 0 \tag{26}$$

for all $\tilde{s} \in \tilde{\Sigma}^k \cap L_a(\tilde{G})$, where $\tilde{\delta}(\tilde{q}, \tilde{s})!$. That is, for any $\epsilon > 0$, there is $N \in \mathbf{N}$ such that when $k \geq N$, we have $\rho(f_{\tilde{q}}^k(\tilde{s}), \tilde{Q}_m) < \epsilon$. Therefore, when $k \geq N$,

$$V(f_{\tilde{q}}^k(\tilde{s})) == sup\{\rho(f_{\tilde{q}}^{k+k'}(\tilde{s}\tilde{s}'), \tilde{Q}_m) : \tilde{s}' \in \tilde{\Sigma}^{k'} \cap L_a(\tilde{G}), \;\; \tilde{\delta}(\tilde{q}, \tilde{s}\tilde{s}')!\} < \epsilon. \quad (27)$$

*Sufficiency:* Suppose that the conditions of this theorem are satisfied. From Theorem 4, $\tilde{Q}_m$ is stable. That is, for any $\epsilon > 0$, there is $\eta > 0$ such that for all $k$ and $\tilde{s} \in \tilde{\Sigma}^k \cap L_a(\tilde{G})$ where $\tilde{\delta}(\tilde{q}, \tilde{s})!$, when $\rho(\tilde{q}, \tilde{Q}_m) < \eta$, we have

$$\rho(f_{\tilde{q}}^k(\tilde{s}), \tilde{Q}_m) < \epsilon. \quad (28)$$

In the following, we show that the above $\eta$ can be chosen to make Eq. (16) hold. Otherwise, there exist infinitely many $k$ such that

$$\rho(f_{\tilde{q}}^k(\tilde{s}), \tilde{Q}_m) > c_1 \quad (29)$$

for some $c_1 > 0$. Then, from the condition (1) of this theorem, we have that there is $c_2 > 0$ such that

$$V(f_{\tilde{q}}^k(\tilde{s})) > c_2 \quad (30)$$

for infinitely many $k$, which is in conflict with Eq. (25). Therefore, the above $\eta$ can be chosen to make Eq. (16) hold, i.e., $\tilde{Q}_m$ is asymptotically stable. □

## 5 Concluding Remarks

As a continuation of Qiu's work [9], this paper is concerned with the stability of FDESs. We formalized the notions of reachability, stability and asymptotical stability in the sense of Lyapunov, which guarantee the convergence of the behaviors of system to the desired states when system engages in some illegal behaviors. In particular, a necessary and sufficient condition for stability and another for asymptotical stability of FDESs are presented. As we know, Lyapunov stability has important applications in the load balancing problem of computer networks [15] and petri nets [16, 17]. Therefore, a further issue worthy of consideration is to use Lyapunov stability of FDESs presented in this paper to deal with the load balancing problem in fuzzy petri nets.

## References

1. Cassandras, C. G., Lafortune, S.: Introduction to Discrete Event Systems. Boston, MA, Kluwer (1999)
2. Lin, F., Wonham, W. M.: On Observability of Discrete Event Systems. Inform. Sci., Vol. 44 (1988) 173-198
3. Lin, F., Ying, H.: Fuzzy Discrete Event Systems and Their Observability. Pro. Joint Int. Conf. 9th Int. Fuzzy Systems Assoc. World Congr. 20th North Amer. Fuzzy Inform. Process. Soci., Canada (2001) 25-28
4. Lin, F., Ying, H.: Modeling and Control of Fuzzy Discrete Event Systems. IEEE Trans. Syst., Man, Cybern. B, Vol. 32, No. 4 (2002) 408-415

5. Lin, F., Ying, H., Luan, X., MacArthur, R.D., Cohn, J.A., Barth-Jones, D.C., Crane, L.R.: Fuzzy Discrete Event Systems and Its Applications to Clinical Treatment Planning. Proceedings of 43rd IEEE Conf. Decision and Control, Budapest, Hungary (2004) 197-202
6. Lin, F., Ying, H., Luan, X., MacArthur, R.D., Cohn, J.A., Barth-Jones, D.C., Crane, L.R.: Theory for A Control Architecture of Fuzzy Discrete Event System for Decision Making. 44th Conference on Decision and Control and European Control Conference ECC (2005)
7. Huq, R., Mann, G. K. I., Gosine, R. G.: Distributed Fuzzy Discrete Event System for Robotic Sensory Information Processing. Expert Systems, Vol. 23, No. 5 (2006) 273-289
8. Huq, R., Mann, G. K. I., Gosine, R. G.: Behavior-Modulation Technique in Mobile Robotics Using Fuzzy Discrete Event System. IEEE Trans. Robotics, Vol. 22, No. 5 (2006) 903-916
9. Qiu, D. W.: Supervisory Control of Fuzzy Discrete Event Systems: A Formal Approach. IEEE Trans. Syst., Man, Cybern. B, Vol. 35, No. 1 (2005) 72-88
10. Liu, F. C., Qiu, D. W.: Decentralized Supervisory Control of Fuzzy Discrete Event Systems. European Control Conference, Kos, Greece (2007)
11. Cao, Y., Ying, M.: Supervisory Control of Fuzzy Discrete Event Systems. IEEE Trans. Syst., Man, Cybern. B, Vol. 35, No. 2 (2005) 366-371
12. Cao, Y., Ying, M.: Observability and Decentralized Control of Fuzzy Discrete-Event Systems. IEEE Trans. Fuzzy Syst., Vol. 14, No. 2 (2006) 202-216
13. Ozveren, C. M., Willsky, A. S., Antsaklis,P. J.: Stability and Stabilizability of Discrete Event Dynamic Systems. Journal of the Association for Computing Machinery, Vol. 38, No. 3 (1991) 730-752
14. Zubov, V.I.: Methods of A.M. Lyapunov and Their Applications. The Netherlands: Noordhoff (1964)
15. Passino, K. M., Burgess, K. L.: Stability Analysis of Discrete Event Systems. Wiley, New York (1998)
16. Passino,K. M., Michel, A. N., Antsaklis, P. J.: Lyapunov Stability of A Class of Discrete Event Systems. IEEE Trans. Automat. Contr., Vol. 39, No. 2 (1994) 269-279
17. Tzafestas, S. G., Rigatos, G. G.: Stability Analysis of An Adaptive Fuzzy Control System Using Petri Nets and Learning Automata. Mathematics and Computers in Simulation, Vol. 51 (2000) 315-339

GPSR Compliance

*The European Union's (EU) General Product Safety Regulation (GPSR) is a set of rules that requires consumer products to be safe and our obligations to ensure this.*

*If you have any concerns about our products, you can contact us on ProductSafety@springernature.com*

In case Publisher is established outside the EU, the EU authorized representative is:

Springer Nature Customer Service Center GmbH
Europaplatz 3
69115 Heidelberg, Germany

Batch number: 10755036

Printed by Printforce, the Netherlands

Lecture Notes in Artificial Intelligence 4682

Edited by J. G. Carbonell and J. Siekmann

Subseries of Lecture Notes in Computer Science

De-Shuang Huang Laurent Heutte
Marco Loog (Eds.)

# Advanced Intelligent Computing Theories and Applications

## With Aspects of Artificial Intelligence

Third International Conference on
Intelligent Computing, ICIC 2007
Qingdao, China, August 21-24, 2007
Proceedings

Springer

Series Editors

Jaime G. Carbonell, Carnegie Mellon University, Pittsburgh, PA, USA
Jörg Siekmann, University of Saarland, Saarbrücken, Germany

Volume Editors

De-Shuang Huang
Chinese Academy of Sciences
Institute of Intelligent Machines, China
E-mail: dshuang@iim.ac.cn

Laurent Heutte
Université de Rouen
Laboratoire LITIS
76800 Saint Etienne du Rouvray, France
E-mail: Laurent.Heutte@univ-rouen.fr

Marco Loog
University of Copenhagen
Datalogical Institute
2100 Copenhagen Ø, Denmark
E-mail: loog@diku.dk

Library of Congress Control Number: 2007932602

CR Subject Classification (1998): I.2.3, I.2, F.4.1, F.1, I.5, F.2, G.2, I.4

LNCS Sublibrary: SL 7 – Artificial Intelligence

ISSN 0302-9743
ISBN-10 3-540-74201-8 Springer Berlin Heidelberg New York
ISBN-13 978-3-540-74201-2 Springer Berlin Heidelberg New York



Springer is a part of Springer Science+Business Media

springer.com


Printed in Germany

Typesetting: Camera-ready by author, data conversion by Scientific Publishing Services, Chennai, India
Printed on acid-free paper SPIN: 12107902 06/3180 5 4 3 2 1 0

# Preface

The International Conference on Intelligent Computing (ICIC) was formed to provide an annual forum dedicated to the emerging and challenging topics in artificial intelligence, machine learning, bioinformatics, and computational biology, etc. It aims to bring together researchers and practitioners from both academia and industry to share ideas, problems and solutions related to the multifaceted aspects of intelligent computing.

ICIC 2007, held in Qingdao, China, August 21-24, 2007, constituted the Third International Conference on Intelligent Computing. It built upon the success of ICIC 2006 and ICIC 2005 held in Kunming and Hefei, China, 2006 and 2005, respectively.

This year, the conference concentrated mainly on the theories and methodologies as well as the emerging applications of intelligent computing. Its aim was to unify the picture of contemporary intelligent computing techniques as an integral concept that highlights the trends in advanced computational intelligence and bridges theoretical research with applications. Therefore, the theme for this conference was "**Advanced Intelligent Computing Technology and Applications**". Papers focusing on this theme were solicited, addressing theories, methodologies, and applications in science and technology.

ICIC 2007 received 2875 submissions from 39 countries and regions. All papers went through a rigorous peer review procedure and each paper received at least three review reports. Based on the review reports, the Program Committee finally selected 496 high-quality papers for presentation at ICIC 2007, of which 430 papers have been included in three volumes of proceedings published by Springer: one volume of *Lecture Notes in Computer Science* (LNCS), one volume of *Lecture Notes in Artificial Intelligence* (LNAI), and one volume of *Communications in Computer and Information Science* (CCIS). The other 66 papers will be included in four international journals.

This volume of *Lecture Notes in Artificial Intelligence* (LNAI) includes 139 papers.

The organizers of ICIC 2007, including the Ocean University of China and the Institute of Intelligent Machines of the Chinese Academy of Science, made an enormous effort to ensure the success of ICIC 2007. We hereby would like to thank the members of the ICIC 2007 Advisory Committee for their guidance and advice, the members of the Program Committee and the referees for their collective effort in reviewing and soliciting the papers. We would like to thank Alfred Hofmann, executive editor from Springer, for his frank and helpful advice and guidance throughout and for his support in publishing the proceedings. In particular, we would like to thank all the authors for contributing their papers. Without the high-quality submissions from the authors, the success of the conference would not have been possible. Finally, we are especially grateful to the IEEE Computational Intelligence Society, the International Neural Network Society and the National Science Foundation of China for their sponsorship.

June 2007

De-Shuang Huang
Laurent Heutte
Marco Loog

# ICIC 2007 Organization

**General Co-chairs**

De-Shuang Huang, China
Luonan Chen, Japan

**International Advisory Committee**

Moonis Ali, USA
Shun-Ichi Amari, Japan
Zheng Bao, China
John L. Casti, USA
Guoliang Chen, China
Diane J. Cook, USA
Ruwei Dai, China
John O Gray, UK
Aike Guo, China
Fuchu He, China
Xingui He, China
Tom Heskes, Netherlands
Mustafa Khammash, USA
Okyay Knynak, Turkey
Yanda Li, China
Marios M. Polycarpou, USA
Songde Ma, China
Erke Mao, China
Michael R. Lyu, Hong Kong
Yunyu Shi, China
Harold Szu, USA
Stephen Thompson, UK
Mathukumalli Vidyasagar, India
Shoujue Wang, China
Paul Werbos, USA
George W. Irwin, UK
DeLiang Wang, USA
Youshou Wu, China
Xin Yao, UK
Nanning Zheng, China
Yixin Zhong, China
Mengchu Zhou, USA
Qingshi Zhu, China
Xiang-Sun Zhang, China

**Steering Committee Co-chairs**

Sheng Chen, UK
Xiao-Ping Zhang, Canada
Kang Li, UK

**Program Committee Chair**

Laurent Heutte, France

**Organizing Committee Co-chairs**

Guo Chen, China
Ming Lv, China
Guangrong Ji, China
Ji-Xiang Du, China

**Publication Chair**

Marco Loog, Denmark

**Special Session Chair**

Wanquan Liu, Australia

**International Liaison Chair**

Prashan Premaratne, Australia

**Tutorial Chair**

Robert Hsieh, Germany

## Publicity Co-chairs

Liyanage C. De Silva , New Zealand
Vitoantonio Bevilacqua, Italy
Kang-Hyun Jo, Korea
Jun Zhang, China

## Exhibition Chair

Bing Wang, China

## International Program Committee

Andrea Francesco Abate, Italy
Waleed H. Abdulla, New Zealand
Shafayat Abrar, Pakistan
Parag Gopal Kulkarni, UK
Vasily Aristarkhov, Russian Federation
Masahiro Takatsuka, Australia
Costin Badica, Romania
Soumya Banerjee, India
Laxmidhar Behera, India
Vitoantonio Bevilacqua, Italy
Salim Bouzerdoum, Australia
David B. Bracewell, Japan
Toon Calders, Belgium
Vincent C S Lee, Australia
Gianluca Cena, Italy
Pei-Chann Chang, Taiwan
Wen-Sheng Chen, China
Hong-Qiang Wang, Hong Kong
Rong-Chang Chen, Taiwan
Geoffrey Macintyre, Australia
Weidong Chen, China
Chi-Cheng Cheng, China
Ziping Chiang, Taiwan
Min-Sen Chiu, Singapore
Tommy Chow, Hong Kong
Mo-Yuen Chow, USA
Rasoul Mohammadi Milasi, Canada
Alexandru Paul Condurache, Germany
Sonya Coleman, UK
Pedro Melo-Pinto, Portugal
Roman Neruda, Czech Republic
Gabriella Dellino, Italy
Grigorios Dimitriadis, UK
Mariagrazia Dotoli, Italy
Minh Nhut Nguyen, Singapore
Hazem Elbakry, Japan
Karim Faez, Iran
Jianbo Fan, China
Minrui Fei, China
Mario Koeppen, Japan
Uwe Kruger, UK
Fausto Acernese, Italy
Qing-Wei Gao, China
Takashi Kuremoto, Japan
Richard Lathrop, USA
Agostino Lecci, Italy
Marco Loog, Denmark
Choong Ho Lee, Korea
Jinde Cao, China
Kang Li, UK
Peihua Li, China
Jin Li, UK
Xiaoli Li, UK
Chunmei Liu, USA
Paolo Lino, Italy
Ju Liu, China
Van-Tsai Liu, Taiwan
Wanquan Liu, Australia
Brian C. Lovell, Australia
Hongtao Lu, China
Mathias Lux, Austria
Sheng Chen, UK
Jinwen Ma, China
Yongjun Ma, China
Guido Maione, Italy
Vishnu Makkapati, India
Filippo Menolascina, Italy
Damien Coyle, UK
Cheolhong Moon, Korea
Angelo Ciaramella, Italy
Tark Veli Mumcu, Turkey
Michele Nappi, Italy
Kevin Curran, UK
Giuseppe Nicosia, Italy
Kenji Doya, Japan
Ahmet Onat, Turkey
Ali Ozen, Turkey
Sulin Pang, China
Antonino Staiano, Italy
David G. Stork, USA
Fuchun Sun, China
Zhan-Li Sun, Hong Kong
Maolin Tang, Australia
John Thompson, UK
Amir Atiya, Egypt
Anna Tramontano, Italy
Jose-Luis Verdegay, Spain
Sergio Vitulano, Italy
Anhua Wan, China
Chengxiang Wang, UK
Bing Wang, China
Kongqiao Wang, China
Zhi Wang, China
Hong Wang, China
Hong Wei, UK
Xiyuan Chen, China
Chao-Xue Wang, China
Yong Wang, Japan
Xue Wang, China
Mike Watts, New Zealand
Ling-Yun Wu, China

## Reviewers

Dong Cao, Alessandra Carbotti, Jesus Ariel Carrasco-Ochoa, Deborah Carvalho, Roberto Catanuto, Xiujuan Chai, Kap Luk Chan, Chien-Lung Chan, Ram Chandragupta, Hong Chang, Hsueh-Sheng Chang, Clément Chatelain, Dongsheng Che, Chun Chen, Chung-Cheng Chen, Hsin-Yuan Chen, Tzung-Shi Chen, Xiaohan Chen, Y.M. Chen, Ying Chen, Ben Chen, Yu-Te Chen, Wei-Neng Chen, Chuyao Chen, Jian-Bo Chen, Fang Chen, Peng Chen, Shih-Hsin Chen, Shiaw-Wu Chen, Baisheng Chen, Zhimin Chen, Chun-Hsiung Chen, Mei-Ching Chen, Xiang Chen, Tung-Shou Chen, Xinyu Chen, Yuehui Chen, Xiang Cheng, Mu-Huo Cheng, Long Cheng, Jian Cheng, Qiming Cheng, Ziping Chiang, Han-Min Chien, Min-Sen Chiu, Chi Yuk Chiu, Chungho Cho, Sang-Bock Cho, Soo-Mi Choi, Yoo-Joo Choi, Wen-Shou Chou, T Chow, Xuezheng Chu, Min Gyo Chung, Michele Ciavotta, Ivan Cibrario Bertolotti, Davide Ciucci, Sonya Coleman, Simona Colucci, Patrick Connally, David Corne, Damien Coyle, Cuco Cristi, Carlos Cruz Corona, Lili Cui, Fabrizio Dabbene, Weidi Dai, Thouraya Daouas, Cristina Darolti, Marleen De Bruijne, Leandro De Castro, Chaminda De Silva, Lara De Vinco, Carmine Del Mondo, Gabriella Dellino, Patrick Dempster, Da Deng, Yue Deng, Haibo Deng, Scott Dexter, Nele Dexters, Bi Dexue, Wan Dingsheng, Banu Diri, Angelo Doglioni, Yajie Dong, Liuhuan Dong, Jun Du, Wei-Chang Du, Chen Duo, Peter Eisert, Mehdi El Gueddari, Elia El-Darzi, Mehmet Engin, Zeki Erdem, Nuh Erdogan, Kadir Erkan, Osman Kaan Erol, Ali Esmaili, Alexandre Evsukoff, Marco Falagario, Shu-Kai Fan, Chin-Yuan Fan, Chun-I Fan, Lixin Fan, Jianbo Fan, Bin Fang, Yikai Fang, Rashid Faruqui, Markus Fauster, Guiyu Feng, Zhiyong Feng, Rui Feng, Chen Feng, Yong Feng, Chieh-Chuan Feng, Francisco Fernandez Periche, James Ferryman, Mauricio Figueiredo, Vítor Filipe, Celine Fiot, Alessandra Flammini, Girolamo Fornarelli, Katrin Franke, Kechang Fu, Tiaoping Fu, Hong Fu, Chaojin Fu, Xinwen Fu, Jie Fu, John Fulcher, Wai-keung Fung, Zhang G. Z., Sebastian Galvao, Junying Gan, Zhaohui Gan, Maria Ganzha, Xiao-Zhi Gao, Xin Gao, Liang Gao, Xuejin Gao, Xinwen Gao, Ma Socorro Garcia, Ignacio Garcia-del-Amo, Lalit Garg, Shuzi Sam Ge, Fei Ge, Xin Geng, David Geronimo, Reza Ghorbani, Paulo Gil, Gustavo Giménez-Lugo, Tomasz Gingold, Lara Giordano, Cornelius Glackin, Brendan Glackin, Juan Ramón González González, Jose-Joel Gonzalez-Barbosa, Padhraig Gormley, Alfredo Grieco, Giorgio Grisetti, Hanyu Gu, Xiucui Guan, Jie Gui, Aaron Gulliver, Feng-Biao Guo, Ge Guo, Tian-Tai Guo, Song Guo, Lingzhong Guo, Yue-Fei Guo, P Guo, Shwu-Ping Guo, Shengbo Guo, Shou Guofa, David Gustavsson, Jong-Eun Ha, Risheng Han, Aili Han, Fengling Han, Hisashi Handa, Koji Harada, James Harkin, Saadah Hassan, Aboul Ella Hassanien, Jean-Bernard Hayet, Hanlin He, Qingyan He, Wangli He, Haibo He, Guoguang He, Pilian He, Yanxiang He, Pawel Herman, Francisco Herrera, Jan Hidders, Grant Hill, John Ho, Xuemin Hong, Tzung-Pei Hong, Kunjin Hong, Shi-Jinn Horng, Lin Hou, Eduardo Hruschka, Shang-Lin Hseih, Chen-Chiung Hsieh, Sun-Yuan Hsieh, Jih-Chang Hsieh, Chun-Fei Hsu, Honglin Hu, Junhao Hu, Qinglei Hu, Xiaomin Hu, Xiaolin Hu, Chen Huahua, Xia Huang, Jian Huang, Xiaojing Huang, Gan Huang, Weitong Huang, Jing Huang, Weimin Huang, Yufei Huang, Zhao Hui, Sajjad Hussain, Thong-Shing Hwang, Giorgio Iacobellis, Francesco Iorio, Mohammad Reza Jamali, Horn-Yong Jan, Dar-Yin Jan, Jong-Hann Jean, Euna Jeong, Mun-Ho Jeong, Youngseon Jeong, Zhen Ji, Qing-Shan Jia, Wei Jia, Fan Jian, Jigui Jian, Peilin Jiang, Dongxiang Jiang, Minghui Jiang, Ping Jiang, Xiubao Jiang, Xiaowei Jiang, Hou Jiangrong, Jing Jie, Zhang Jihong, Fernando Jimenez, Guangxu Jin, Kang-Hyun Jo,

Guillaume Jourjon, Jih-Gau Juang, Carme Julià, Zhou Jun, Dong-Joong Kang, Hee-Jun Kang, Hyun Deok Kang, Hung-Yu Kao, Indrani Kar, Cihan Karakuzu, Bekir Karlik, Wolfgang Kastner, John Keeney, Hrvoje Keko, Dermot Kerr, Gita Khalili Moghaddam, Muhammad Khurram Khan, Kavi Umar Khedo, Christian Kier, GwangHyun Kim, Dae-Nyeon Kim, Dongwon Kim, Taeho Kim, Tai-hoon Kim, Paris Kitsos, Kunikazu Kobayashi, Sarath Kodagoda, Mario Koeppen, Nagahisa Kogawa, Paul Kogeda, Xiangzhen Kong, Hyung Yun Kong, Insoo Koo, Marcin Korze, Ibrahim Kucukdemiral, Petra Kudova, Matjaz Kukar, Parag Kulkarni, Saravana Kumar, Wen-Chung Kuo, Takashi Kuremoto, Janset Kuvulmaz, Jin Kwak, Lam-For Kwok, Taekyoung Kwon, Marcelo Ladeira, K. Robert Lai, Darong Lai, Chi Sung Laih, Senthil Kumar Lakshmanan, Dipak Lal Shrestha, Yuk Hei Lam, M. Teresa Lamata, Oliver Lampl, Peng Lan, Vuokko Lantz, Ana Lilia Laureano-Cruces, Yulia Ledeneva, Vincent C S Lee, Narn-Yih Lee, Malrye Lee, Chien-Cheng Lee, Dong Hoon Lee, Won S Lee, Young Jae Lee, Kyu-Won Lee, San-Nan Lee, Gang Leng, Agustin Leon Barranco, Chi Sing Leung, Cuifeng Li, Fuhai Li, Chengqing Li, Guo-Zheng Li, Hongbin Li, Bin Li, Liberol Li, Bo Li, Chuandong Li, Erguo Li, Fangmin Li, Juntao Li, Jinshan Li, Lei Li, Ming Li, Xin Li, Xiaoou Li, Xue li, Yuan Li, Lisa Li, Yuancheng Li, Kang Li, Jun Li, Jung-Shian Li, Shijian Li, Zhihua Li, Zhijun Li, Zhenping Li, Shutao Li, Xin Li, Anglica Li, Wanqing Li, Jian Li, Shaoming Li, Xiaohua Li, Xiao-Dong Li, Xiaoli Li, Yuhua Li, Yun-Chia Liang, Wei Liang, Wuxing Liang, Jinling Liang, Wen-Yuan Liao, Wudai Liao, Zaiyi Liao, Shizhong Liao, Vicente Liern, Wen-Yang Lin, Zhong Lin, Chih-Min Lin, Chun-Liang Lin, Xi Lin, Yu Chen Lin, Jun-Lin Lin, Ke Lin, Kui Lin, Ming-Yen Lin, Hsin-Chih Lin, Yu Ling, Erika Lino, Erika Lino, Paolo Lino, Erika Lino, Shiang Chun Liou, Ten-Yuang Liu, Bin Liu, Jianfeng Liu, Jianwei Liu, Juan Liu, Xiangyang Liu, Yadong Liu, Yubao Liu, Honghai Liu, Kun-Hong Liu, Kang-Yuan Liu, Shaohui Liu, Qingshan Liu, Chen-Hao Liu, Zhiping Liu, Yinyin Liu, Yaqiu Liu, Van-Tsai Liu, Emmanuel Lochin, Marco Loog, Andrew Loppingen, Xiwen Lou, Yingli Lu, Yao Lu, Wen-Hsiang Lu, Wei Lu, Hong Lu, Huijuan Lu, Junguo Lu, Shangmin Luan, Jiliang Luo, Xuyao Luo, Tuan Trung Luong, Mathias Lux, Jun Lv, Chengguo Lv, Bo Ma, Jia Ma, Guang-Ying Ma, Dazhong Ma, Mi-Chia Ma, Junjie Ma, Xin Ma, Diego Magro, Liam Maguire, Aneeq Mahmood, Waleed Mahmoud, Bruno Maione, Agostino Marcello Mangini, Weihua Mao, Kezhi Mao, Antonio Maratea, Bogdan Florin Marin, Mario Marinelli, Urszula Markowska-Kaczmar, Isaac Martin, Francesco Martinelli, Jose Fco. Martínez-Trinidad, Antonio David Masegosa Arredondo, Louis Massey, Emilio Mastriani, Marco Mastrovito, Kerstin Maximini, Radoslaw Mazur, Daniele Mazzocchi, Malachy McElholm, Gerard McKee, Colin McMillen, Jian Mei, Belen Melian, Carlo Meloni, Pedro Melo-Pinto, Corrado Mencar, Luis Mesquita, Jianxun Mi, Pauli Miettinen, Claudia Milaré, Rasoul Milasi, Orazio Mirabella, Nazeeruddin Mohammad, Eduard Montseny, Inhyuk Moon, Hyeonjoon Moon, Raul Morais, J. Marcos Moreno, José Andrés Moreno, Philip Morrow, Santo Motta, Mikhal Mozerov, Francesco Napolitano, David Naso, Wang Nengqiang, Mario Neugebauer, Yew Seng Ng, Wee Keong Ng, Tam Nguyen, Quang Nguyen, Thang Nguyen, Rui Nian, James Niblock, Iaobing Nie, Eindert Niemeijer, Julio Cesar Nievola, Haijing Niu, Qun Niu, Changyong Niu, Asanao Obayashi, Kei Ohnishi, Takeshi Okamoto, Jose Angel Olivas, Stanley Oliveira, Kok-Leong Ong, Chen-Sen Ouyang, Pavel Paclik, Tinglong Pan, Sanjib Kumar Panda, Tsang-Long Pao, Emerson Paraiso, Daniel Paraschiv, Giuseppe

Patanè, Kaustubh Patil, Mykola Pechenizkiy, Carlos Pedroso, Zheng Pei, Shun Pei, Chang Pei-Chann, David Pelta, Jian-Xun Peng, Sheng-Lung Peng, Marzio Pennisi, Cathryn Peoples, Eranga Perera, Alessandro Perfetto, Patrick Peursum, Minh-Tri Pham, Phuong-Trinh Pham-Ngoc, Lifton Phua, Son Lam Phung, Alfredo Pironti, Giacomo Piscitellei, Elvira Popescu, Girijesh Prasad, Prashan Premaratne, Alfredo Pulvirenti, Lin Qi, HangHang Qi, Yu Qiao, Xiaoyan Qiao, Lixu Qin, Kai Qin, Jian-long Qiu, Ying-Qiang Qiu, Zhonghua Quan, Thanh-Tho Quan, Chedy Raïssi, Jochen Radmer, Milo Radovanovi, Bogdan Raducanu, Humera Rafique, Thierry Rakotoarivelo, Nini Rao, Ramesh Rayudu, Arif Li Rehman, Dehua Ren, Wei Ren, Xinmin Ren, Fengli Ren, Orion Reyes, Napoleon Reyes, Carlos Alberto Reyes-Garcia, Alessandro Rizzo, Giuseppe Romanazzi, Marta Rosatelli, Heung-Gyoon Ryu, Hichem Sahbi, Ying Sai, Paulo Salgado, Luigi Salvatore, Nadia Salvatore, Saeid Sanei, Jose Santos, Angel Sappa, Heather Sayers, Klaus Schöffmann, Bryan Scotney, Carla Seatzu, Hermes Senger, Murat Sensoy, Carlos M.J.A. Serodio, Lin Shang, Li Shang, XiaoJian Shao, Andrew Shaw, Sheng Yuan Shen, Yanxia Shen, Yehu Shen, Linlin Shen, Yi Shen, Jinn-Jong Sheu, Mingguang Shi, Chaojian Shi, Dongfeng Shi, June-Horng Shiesh, Yen Shi-Jim, Zhang Shuhong, Li Shundong, Nanshupo Shupo, Oliver Sinnen, Sukree Sinthupinyo, Silvia Siri, Ernest Sithole, Nicolas Sklavos, Stanislav Slusny, Pilar Sobrevilla, Ignacio Solis, Anthony Solon, Andy Song, Liu Song, Qiankun Song, Zheng Song, Yinglei Song, Nuanwan Soonthornphisaj, Aureli Soria-Frisc, Jon Sporring, Kim Steenstrup Pedersen, Domenico Striccoli, Juhng Perng Su, Shanmugalingam Suganthan, P. N. Suganthan, Youngsoo Suh, Yonghui Sun, Xinghua Sun, Ning Sun, Fuchun Sun, Lily Sun, Jianyong Sun, Jiande Sun, Worasait Suwannik, Roberto T. Alves, Tele Tan, Taizhe Tan, Xuan Tan, Xiaojun Tan, Hong Zhou Tan, Feiselia Tan, Hong Tang, Chunming Tang, David Taniar, Michele Taragna, David M.J. Tax, Ziya Telatar, Zhi Teng, John Thompson, Bin Tian, Ching-Jung Ting, Fok Hing Chi Tivive, Alexander Topchy, Juan Carlos Torres, Ximo Torres, Joaquin Torres-Sospedra, Hoang Hon Trinh, Chia-Sheng Tsai, Chieh-Yuan Tsai, Huan-Liang Tsai, Wang-Dauh Tseng, Yuan-Jye Tseng, Yifeng Tu, Biagio Turchiano, Cigdem Turhan, Anna Ukovich, Muhammad Muneeb Ullah, Nurettin Umurkan, Mustafa Unel, Daniela Ushizima, Adriano Valenzano, Pablo A. Valle, Bram Van Ginneken, Christian Veenhuis, Roel Vercammen, Enriqueta Vercher, Silvano Vergura, Brijesh Verma, Raul Vicente Garcia, Boris X. Vintimilla Burgos, Gareth Vio, Stefano Vitturi, Aristeidis Vlamenkoff, John Wade, Manolis Wallace, Li Wan, Shijun Wang, Xiaodong Wang, Xue Wang, Zhi Wang, Bing Wang, Chih-Hung Wang, Chao Wang, Da Wang, Jianying Wang, Le Wang, Min Wang, Rui-Sheng Wang, Sheng Wang, Jiahai Wang, Guanjun Wang, Linshan Wang, Yanyan Wang, Xuan Wang, Xiao-Feng Wang, Yong Wang, Zidong Wang, Zhongsheng Wang, Zhengyou Wang, Yen-Wen Wang, Shiuh-Jeng Wang, Shouqi Wang, Ling Wang, Xiang Wang, Lina Wang, Qing-Guo Wang, Yebin Wang, Dingcheng Wang, Dianhui Wang, Meng Wang, Yi Wang, Bao-Yun Wang, Xiaomin Wang, Huazhong Wang, Jeen-Shing Wang, Haili Wang, Haijing Wang, Jian Wang, Yoshikazu Washizawa, Yuji Watanabe, Wiwat Watanawood, Michael Watts, Richard Weber, Lisheng Wei, Zhi Wei, Yutao Wei, Hong Wei, Li Weigang, Dawid Weiss, Hou Weiyan, Guo-Zhu Wen, Brendon Woodford, Derek Woods, Lifang Wu, Zikai Wu, Ke Wu, Xinan Wu, Hsien-Chu Wu, QingXiang Wu, Shiqian Wu, Lihchyau Wuu, Jun-Feng Xia, Li Xia, Xiao Lei Xia, Zhiyu Xiang, Kui Xiang, LiGuo Xiang, Tao Xiang, Jing Xiao, Min Xiao, Liu

Xiaodong, Zhao Xiaoguang, Xiangpeng Xie, Zhijun Xie, Shaohua Xie, Jiang Xie, Hong Xie, Rui Xing, Li Xinyu, Wei Xiong, Huan Xu, Jiangfeng Xu, Jianhua Xu, Yongjun Xu, Jun Xu, Hongji Xu, Bingji Xu, Yu Xue, Yun Xue, Mehmet Yakut, Xing Yan, Jiajun Yan, Hua Yan, Yan Yang, Hsin-Chang Yang, Tao Yang, Chengfu Yang, Banghua Yang, Ruoyu Yang, Zhen Yang, Zhichun Yang, Wu-Chuan Yang, Ming Yang, Cheng-Zen yang, Shouyi Yang, Ming-Jong Yao, Kim-Hui Yap, Hao Ye, Chia-Hsuan Yeh, James Yeh, Jun-Heng Yeh, Shwu-Huey Yen, Sang-Soo Yeo, Yang Yi, Tulay Yildirim, PeiPei Yin, Junsong Yin, Lin Ying, Ling Ying-Biao, Yang Yongqing, Kaori Yoshida, Tomohiro Yoshikawa, Qi Yu, Wen Yu, Wen-Shyong Yu, Kun Yuan, Kang Yuanyuan, Chen Yuepeng, Li Yun, Kun Zan, Chuanzhi Zang, Ramon Zatarain-Cabada, Faiz ul Haque Zeya, Zhihui Zhan, Changshui Zhang, Yongping Zhang, Jie Zhang, Jun Zhang, Yunchu Zhang, Zanchao Zhang, Yifeng Zhang, Shihua Zhang, Ningbo Zhang, Junhua Zhang, Jun Zhang, Shanwen Zhang, Hengdao Zhang, Wen-sheng Zhang, Haoshui Zhang, Ping Zhang, Huaizhong Zhang, Dong Zhang, Hua Zhang, Byoung-Tak Zhang, Guohui Zhang, Li-Bao Zhang, Junping Zhang, Junpeng Zhang, Jiye Zhang, Junying Zhang, JingRu Zhang, Jian Zhang, Duanjin Zhang, Xin Zhang, Huaguang Zhang, Guo Zhanjie, Jizhen Zhao, Zhong-Qiu Zhao, Li Zhao, Ming Zhao, Yinggang Zhao, Ruijie Zhao, Guangzhou Zhao, Liu Zhaolei, Fang Zheng, Ying Zheng, Chunhou Zheng, Cong Zheng, Guibin Zheng, Qinghua Zheng, Wen-Liang Zhong, Jinghui Zhong, Jiayin Zhou, Jie Zhou, Xiaocong Zhou, Fengfeng Zhou, Chi Zhou, Sue Zhou, Mian Zhou, Zongtan Zhou, Lijian Zhou, Zhongjie Zhu, Xinjian Zhuo, Xiaolan Zhuo, Yanyang Zi, Ernesto Zimmermann, Claudio Zunino, Haibo Deng, Wei Liu.

# Table of Contents

## Neural Networks

## Independent Component Analysis and Blind Source Separation

## Combinatorial and Numerical Optimization



## Neural Computing and Optimization

## Fuzzy Systems and Soft Computing

## Particle Swarm Optimization and Niche Technology



## Swarm Intelligence and Optimization

## Kernel Methods and Support Vector Machines



## Fine Feature Extraction Methods

## Intelligent Fault Diagnosis

## Traffic Optimization



## Intelligent Mobile and Wireless Sensor Networks



## Intelligent Prediction and Time Series Analysis

## Intelligent Computing in Neuroinformatics

# Managing Target Cash Balance in Construction Firms Using Novel Fuzzy Regression Approach

Chung-Fah Huang[1], Morris H.L. Wang[2], and Cheng-Wu Chen[3,⋆]

[1] Department of Civil Engineering, National Kaohsiung University of Applied Sciences, Kaohsiung 807, Taiwan, R.O.C.
[2] Department of Civil Engineering, Vanung University, Chung-li, Taiwan 320, R.O.C.
[3] Department of Logistics Management, Shu-Te University, Yen Chau, Kaohsiung, Taiwan 82445, R.O.C
cwchen@mail.stu.edu.tw

**Abstract.** This study presented the cash portion of working capital management (WCM) by using the concept of target cash balance and developed a practical model for construction firms in Taiwan for rationalizing the amount of cash and current assets which should be possessed in any point of time. The model developed by Miller and Orr is introduced here for understanding the issues involved. Because the S-curve has unique merits to represent the relationship between project duration and complete progress in practical usage of construction management, based on the technique of Takagi-Sugeno (T-S) fuzzy model, the fuzzy S-curve regression is hereby constructed in this paper.

**Keywords:** T-S fuzzy model, working capital management.

## 1 Introduction

General contractors play a prominent role in the construction industry, causing the supply chain to respond to a variety of construction needs submitted by the demand chain. Along the demand chain, public owners, private property developers, banking institutions, and all shareholders of those business entities are directly or indirectly involved in the demand chain. As generally known, a supply market is overwhelmed by a large number of construction companies with relatively comparable backgrounds and capabilities. In this situation, nearly every playerinvolved in the demand chain must evaluate the background of the general contractor(s) before entering into the contract stage.

One of the principle criteria for evaluating a general contractor is the liquidity of his firm. A healthy liquidity greatly improves the firm's solvency and is generally a sign of energetic operating capability. Basically, a firm's liquidity can be fully reflected by his working capital management which covers both short term assets and debts. As carrying too much cash would bear the firm unnecessary financial costs and too little risks of bankruptcy, this study attempts to

⋆ Corresponding author.

D.-S. Huang, L. Heutte, and M. Loog (Eds.): ICIC 2007, LNAI 4682, pp. 702–711, 2007.

are two logical steps involved, including identification of working capital and determination of target cash balance, discussed below.

In a construction firm, the needs for working capital may be motivated by transaction concerns, precautionary concerns and speculative concerns. Each of the motive categories is briefly described below.

1. Transaction motive
This is the most common cause for a firm to hold current assets, mostly cash. To a construction firm, the main category of transactions are (1) for outflow of cash, subcontractors, material vendors, equipment leases and direct-hire workers and (2) for inflow of cash, mainly construction clients or their representatives. Salary for internal employees however may not be regarded as this motive.

2. Precautionary motive
As cash outflows and inflows may vary as planned, a firm has to prepare for itself a handful sum of current assets for unexpected shortage of debt payments. By definition, this category is for precaution of short-term insolvency.

3. Speculative motive
A firm may encounter opportunities of price negotiation in procurement of service or material. A lucrative price discount may be offered, if the firm is able to transact with the opposite by cash or the equivalent. For this reason or the like, the firm may be willing to accept the cost of borrowing in hope of high speculative return.

Once the need for working capital is identified, a firm is underway of figuring out its most appropriate level of cash balance or the target cash balance. Any diversion to the target level bears the firm a penalty. When a firm holds superfluous current assets, its penalty is the excess interest payments. On the contrary, if the firm is in need of cash for debt payments, its penalty is the cost of trading notes with cash. Further, if the firm has drained out all current assets, the additional penalty is the opportunity cost for arranging the borrowing in a short time. Obviously, it exists a balance between the two extremes, as depicted in Fig. 1.

This study seeks to understand this balance in a construction firm by incorporating the popular Miller-Orr Model illustrated in Fig. 2 [13]; [14]. This model argues that the irregular pattern of cash needs along various times can be best handled by the idea of dual control limits. In other words, a firm can use its operating characteristics and credit conditions as a basis for constructing a lower cash balance limit. Similarly, the firm can construct an upper cash balance limit and the target cash balance by using its transaction costs, variance of cash flow, and opportunity cost of holding cash. After identifying the upper/lower limits, it is convenient for the firm to discern timing for investing cash in marketable securities or trading notes for cash. In short, the Miller-Orr model reduces the difficulty of working capital management into finding the target cash balance and the associated limits. The model states that

$$C^* = L + (\frac{3}{4} \times F \times \frac{\sigma}{R})\frac{1}{3} \tag{1}$$

where

$$U = 3 \times C \times -2 \times L \qquad (2)$$

L: Lower cash balance limit F: Transaction costs of trading valuable notes for cash or arranging short-term loans m2: Variance of cash flow R: Opportunity costs, equivalent to interest rate of loans or security notes.

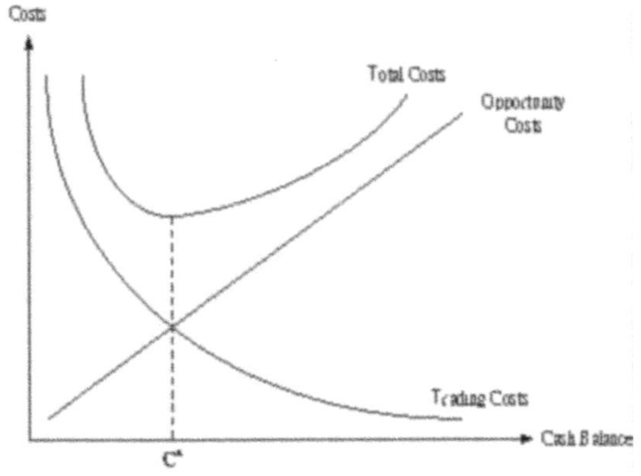


**Fig. 1.** Balance between superfluous or shortage of working capital

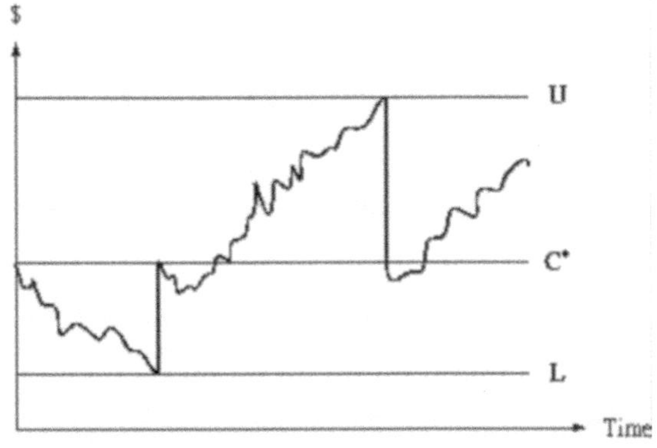


**Fig. 2.** Dual control of Miller-Orr model

## 4 Applying Miller-Orr Model in Construction

The cash flow of a firm is dependent upon its operating cycle, which begins at procurement of service or material and ends at sales and inflow of revenues. More relevant to the firm, however, is often the cash cycle, which strictly relates to all cash outflows for procurement and the inflows of sales. Although logically connected, in practice, the two cycles may be quite different from each other. For a

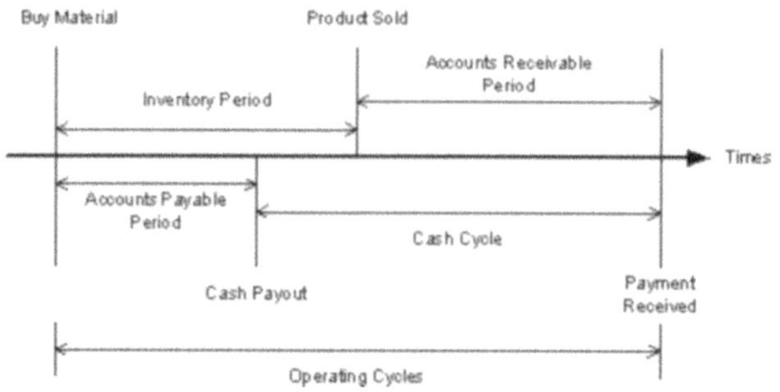


**Fig. 3.** Relationship between operating cycles and cash cycles

construction firm, the danger of insolvency often occurs when there is a long delay or considerable gap between a cash outflow and the expected inflow. Markedly, it is impossible to know a firm's cash cycle directly based on its published financial statements. The amount of details involved is enormous. Rather, it is more useful to first peek into a firm's operating cycle and then, by subtracting the accounts payable period, to measure the cash cycle, as depicted in Fig. 3.

## 5 Classic S-Curve Theory

An S-shaped curve is often used to reflect the phenomena in biology and social economy. It means that the trend of growth gets slow first and finally saturation rapidly. In other words, the typical S-shaped curve is generally a build-up period first, then a relatively steady load period, with a final tail-off period. The characteristic that the build-up and tail-off periods vary from slow to steep depends

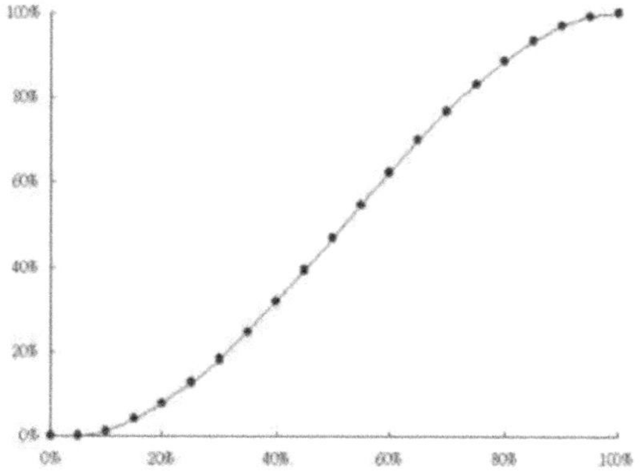


**Fig. 4.** Typical S-curve figure

on the type of project, for example the typical shape of construction activity within a project is a quick build-up period, a steady load period and a slow tail-off period. The relationship between budgets and time limit for a project can be represented via S-curve fitting. A typical S-curve figure is shown in Fig. 4. The x-axis and y-axis denote project duration and complete progress, respectively.

[3] proposed an S-curve equation which can be used in a variety of applications related to project control. The S-curve model is of the following form:

$$P = \frac{3^T}{2} sin[\frac{\pi(1-T)}{2}] sin(\pi T) log(\frac{T+(1.5-T_p)}{T_p+T}) - 2T^3 + 3T^2 \quad (3)$$

where P denotes percentage completion of a project or an activity; T denotes time at any point of the duration of a project or an activity; TP is shape factor. Fig. 5 is plotted with various values of TP between T = 0 and T = 100% duration and the envelope of curves for TP = 0 and TP = 100% in Eq. (3).

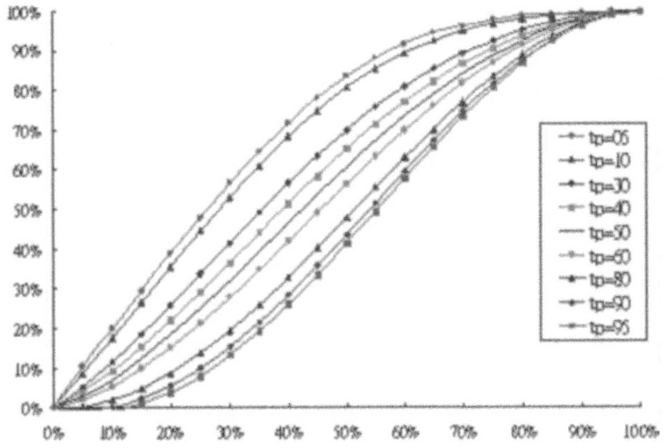


**Fig. 5.** Miskawi S-curve model

Here we suppose we can exactly get all observed data taking part in the problems, but, actually, we may not know exact values rather some approximation [9]. For this reason, the traditional fitting method may not be quite suitable and [8], [9] hence proposed an S-shaped curve regression model for fitting data that exist fuzziness or uncertainty. However, the S-curve fitting model by data of large-scale engineering must be different with that of small-scale engineering. In order to let an S-curve model be generally used in capital management for construction firms, Takagi-Sugeno (T-S) fuzzy model is utilized to develop a practical S-curve model. That is to say, the fuzzy regression curve, obtained for project control of large-scale or small-scale engineering, is smoothly connected by the T-S fuzzy model in the following.

## 6 Fuzzy S-Curve Via T-S Fuzzy Model

The T-S fuzzy model was developed primarily from the pioneering work of [15], to represent the nonlinear relation of multiple input and output data, according to the format of fuzzy reasoning. Namely, the resulting overall fuzzy regression model, nonlinear in general, is achieved by fuzzy blending of each individual input-output realization [16].

Before constructing fuzzy regression model, we are used to choosing the following polynomial equation when k order curve fitting is adopted:

$$y = a_k x^k \tag{4}$$

by choosing the order k we can represent nonlinear relations. Parameters are determined so that the distance (or error) between an observed data point and its corresponding point on the polynomial will be minimal.

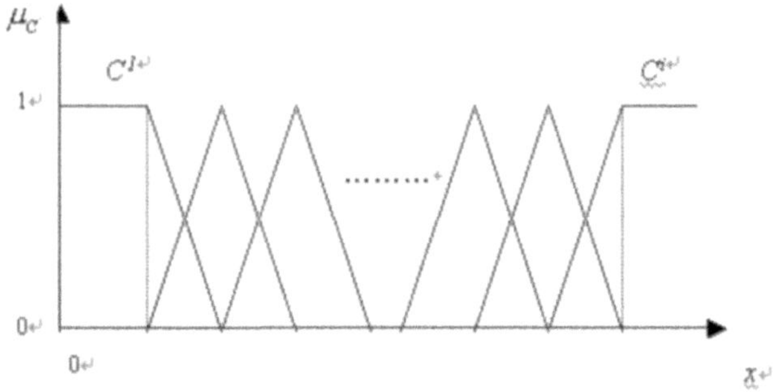

**Fig. 6.** Fuzzy sets to represent low and high cost

In this paper, we distribute the data clusters of cost into some overlapping regions to represent the outlays of engineering constructions such as shown by the membership functions of fuzzy sets C1, C2 K Ci in Fig.6. Therefore, the ith rule of fuzzy inference is described by a set of fuzzy IF-THEN rules in the following form[15],[16], [18], [19]:

$$\begin{aligned} &Rule1 : \text{IF x is } C^1 \text{ THEN } y_1 = a_{1k} x_1^k \\ &Rule2 : \text{IF x is } C^2 \text{ THEN } y_2 = a_{2k} x_2^k \\ &\ldots\ldots\ldots \\ &Rulei : \text{IF x is } C^i \text{ THEN } y_i = a_{ik} x_i^k \end{aligned} \tag{5}$$

where in this case x, input, represents the cost and $y_i$ (i = 1, 2), output, stand for progress of work.$i = 1, 2 \ldots r$ ; in which r is the number of IF-THEN rules and x is the premise variable. Using the center of gravity defuzzification, product inference, and single fuzzifier, the final output is inferred as follows:

$$y = \frac{\sum_{i=1}^{r} w_i y_i}{\sum_{i=1}^{r} w_i} = \sum_{i=1}^{r} h_i y_i \tag{6}$$

It is assumed that$w_2 \geq 0$,i=1,2,... r ; $\sum_{i=1}^{r} w_i > 0$. Therefore, $h_i \geq 0$ and $\sum_{i=1}^{r} h_i = 1$ .

*Remark 1.* $w_i$ is the degree of membership belonging to either the low (i = 1) or high (i = 2) fuzzy sets. When x is smaller than CL, the regression model of rule 1 is solely applied. Contrarily, when x is greater than CH, only regression model of rule 2 is applied. When x is in between, both equations are employed with the continuously varying degree of weight . For instance, as the value of x falls in higher in the interval of [CL, CH], more weight is given to the regression model of rule 1, and less weight to the regression model of rule 2.

*Remark 2.* (Wang and Chiu 1999)[17]: the resultant fuzzy number is the same type as the original fuzzy numbers after the operation of addition, subtraction or multiplication. Namely, If A and B are the fuzzy numbers with the same type of membership function, then A+B, A-B and KA, K$\in$ R , are also the same type as A and B.

## 7 Conclusions

We propose here a fuzzy S-curve regression method for a better understanding of the issues involved. The aim is to develop a practical model for construction firms in Taiwan to rationalize the amount of cash and current assets possessed in certain time of duration. A simplified case is also introduced for demonstrating the concept and steps of applying the conceptual model.

## References

1. Halpin, D. W., Woodhead, R. W.: Construction Management. Wiley, New York (1998)
2. Kim, Y. H., Srinivasan, V.: Advances in Working Capital Management. JAI Press, Greenwich (1988)
3. Miskawi, Z.: An S-curve equation for project control. Construction Management and Economics, Vol.7, (1989)115-124
4. Romie, T. J.: A restatement of the s-curve hypothesis. Review of Development Economics, Vol.3, (1999) 207-214
5. Rudolf, E.: The S-curve Relation between Per-capita Income and Insurance Penetration. Geneva Papers on Risk and Insurance V Issues and Practice, Vol.25, (2000) 396-406
6. Peters, G.:Fuzzy Linear Regression with Fuzzy Intervals. Fuzzy Sets Syst., Vol.63, (1994) 45-55
7. Tanaka, H., Uejima, S., Asai, K.: Linear Regression Analysis with Fuzzy model. IEEE Trans. Syst., Man, Cybern., Vol.12, (1982) 903-907
8. Xu, R.: A Linear Regression Model in Fuzzy Environment. Adv. Modeling Simulation, Vol.27, (1991) 31-40
9. Xu, R.: SVCurve Regression Model in Fuzzy Environment. Fuzzy Sets Syst., Vol.90, (1997) 317-326

10. Yang, M. S.: Fuzzy least-squares Linear Regression Analysis for Fuzzy Input-output Data. Fuzzy Sets Syst., Vol.126(2002) 389-399
11. Hwee, N. G., Tiong, R. L. K. , Model on Cash Flow Forecasting and Risk Analysis for Contracting Firms. Int. J. Project Management, Vol.20, (2002) 351-363
12. Navon, R.:Company-level Cash-flow Management. J. Construction Engineering and Management, ASCE, Vol.122,(1996) 22-29
13. Juang, J. L.: The research on Working Capital Investment. Journal Nan-Tai College Bullet, Vol.20, (1994) 93-97
14. Ross, S. A., Westerfield, R. W., Jordan, B. D.: Fundamentals of Corporate Finance. Richard D. Irwin Inc., New York (1995)
15. Takagi, T., Sugeno, M.: Fuzzy Identification of Systems and its Applications to Modeling and Control. IEEE Trans. Syst., Man, Cybern., Vol.15, (1985) 116-132
16. Wang, H. O., Tanaka, K., Griffin, M. F.: An Approach to Fuzzy Control of Nonlinear Systems: Stability and Design Issues. IEEE Trans. Fuzzy Syst., Vol.4, (1996) 14-23
17. Wang, W. J., Chiu, C. H.: Entropy Variation on the Fuzzy Numbers with Arithmetic Operations. Fuzzy Sets Syst., Vol.103 (1999) 443-456
18. Hsieh,T.Y., Wang, M. H. L., Chen, C.W. , Chen, C.Y., Yu, S.E., Yang, H.C., Chen, T,H.: A New Viewpoint of S-Curve Regression Model and its Application to Construction Management. International Journal on Artificial Intelligence Tools, Vol.15, No. 2, (2006) 131-142
19. Chen, C.W.: Stability Conditions of Fuzzy Systems and Its Application to Structural and Mechanical Systems. Advances in Engineering Software, Vol. 37. No. 9, (2006) 624-629

# Medical Diagnosis System of Breast Cancer Using FCM Based Parallel Neural Networks

Sang-Hyun Hwang, Dongwon Kim, Tae-Koo Kang, and Gwi-Tae Park

Department of Electrical Engineering, Korea University, 1, 5-ka, Anam-dong, Seongbuk-ku, Seoul 136-701, Korea
{tomcroze,upground,tkkang,gtpark}@korea.ac.kr

**Abstract.** In this paper, a new methodology for medical diagnosis based on fuzzy clustering and parallel neural networks is proposed. Intelligent systems have various fields. Breast cancer is one of field to be targeted, which is the most common tumor-related disease among women. Diagnosis of breast cancer is not task for medical expert owing to many attributes of the disease. So we proposed a new method, FCM based parallel neural networks to handle difficult. FCM based parallel neural networks composed of two parts. One is classifying breast cancer data using Fuzzy c-means clustering method (FCM). The other is designing the multiple neural networks using classified data by FCM. The proposed methodology is experimented, evaluated, and compared the performance with other existed models. As a result we can show the effectiveness and precision of the proposed method are better than other previous models.

**Keywords:** Fuzzy c-means clustering, parallel neural networks, lookup table.

## 1 Introduction

An important problem in the medical science is attaining the correct diagnosis in the preprocessing of medical cure. In the modern medical science, the various tests have been performed to a patient for the ultimate diagnosis. However, making a correct and accurate diagnosis is not easy to medical expert. In addition, the more various tests performed a patient, the more medical expert complicated diagnosis of a disease. Specifically, medical experts have difficulties in diagnosing some disease such as breast cancer which has many attributes. Also, man's eye could not exactly classify a tumor of breast cancer whether it is a malignant or benign. Therefore, many medical experts and scientists are concerned about computerizing tools to diagnose a disease in the medical science.

Computerizing tools intended to aid the medical expert in making sense out of the welter of data. A well-designed computerized diagnosis system of breast cancer could be used to directly attain the ultimate diagnosis with artificial intelligent algorithms which perform roles as classifier.

D.-S. Huang, L. Heutte, and M. Loog (Eds.): ICIC 2007, LNAI 4682, pp. 712–719, 2007.

There had been previous research works to an automatic ultimate diagnostic system with breast cancer database. Neural networks, adaptive boosting, genetic algorithm, fuzzy inference system, adaptive Neuro-Fuzzy hybrid model have been applied to this problem [4, 5, 6, 11, 12]. The performances of methodologies were evaluated with calculating the degree of correctness in predicted results against diagnosed results represented as Positive Predicted Value (PPV) [5]. Most of performances are about 95% to previous methodologies.

In this paper, we present the Fuzzy c-means based Parallel Neural Networks (FbPNN) for solving the breast cancer problem. This methodology is a combination of Neural Networks and Fuzzy c-means clustering, which improves each of performance and decreases as for problem of number of sampling data.

The remainder of this paper is organized as follows: Section Ⅱ describes the WBCD to be targeted. Section Ⅲ proposes system using FbPNN for medical diagnosis. Section Ⅳ shows the experiments performed using the FbPNN classifier. Finally, conclusions are described in section Ⅴ.

## 2 Breast Cancer Data

The breast cancer is the most common tumor-related disease among women in Korea and throughout the world. It is considered to be the major cause to death to women. That's why we should be interested in the disease.

We have used well-known WBCD which is the result from efforts provided by the University of Wisconsin Hospital based on microscopic examination of breast masses with fine needle aspirate tests. The WBCD problem involves classifying a presented case whether it is benign or malignant. The WBCD database consists of nine measures represented an integer value between 1 and 10. In our experiments, the WBCD database separates out training set and testing set. We normalized the WBCD database between 0 and 1. They are :

1. Clump Thickness : 1-10
2. Uniformity of Cell Size : 1-10
3. Uniformity of Cell Shape : 1-10
4. Marginal Adhesion : 1-10
5. Single Epithelial Cell Size : 1-10
6. Bare Nuclei : 1-10
7. Bland Chromatin : 1-10
8. Normal Nucleoli : 1-10
9. Mitoses : 1-10

The database consists of 683 except 16 data which has involved a missing value.

**Table 1.** WBCD database

| | Case | $X_1$ | $X_2$ | $X_3$ | $\cdots$ | $X_9$ | Diagnostics |
|---|---|---|---|---|---|---|---|
| Training data | 1 | 5 | 1 | 1 | $\cdots$ | 1 | Benign |
| | 2 | 3 | 2 | 2 | $\cdots$ | 1 | Malignant |
| | $\vdots$ | $\vdots$ | $\vdots$ | $\vdots$ | | $\vdots$ | $\vdots$ |
| | 400 | 4 | 8 | 8 | $\cdots$ | 1 | Malignant |
| Test data | 401 | 6 | 6 | 6 | $\cdots$ | 2 | Benign |
| | 402 | 4 | 8 | 2 | $\cdots$ | 1 | Benign |
| | $\vdots$ | $\vdots$ | $\vdots$ | $\vdots$ | | $\vdots$ | $\vdots$ |
| | 683 | 4 | 8 | 8 | $\cdots$ | 1 | Malignant |

As can be seen from Table 1, any information for cancer diagnostics is not provided whether tumor of breast cancer is malignant or benign. And there are no relationship between measured values and diagnostics. Therefore, a correct diagnosis is so difficulty with original data even for medical experts. The proposed methodology can be assisted to medical experts.

# 3 Diagnosis System Using FCM Based Parallel Neural Networks

## 3.1 Overall Medical Diagnosis System

The overall system largely divides into two parts. One is the fuzzy c-means clustering for classifying the WBCD database, which is used as making a several subnetworks. A number of clusters treat the next section. The other is the decision analyzer for

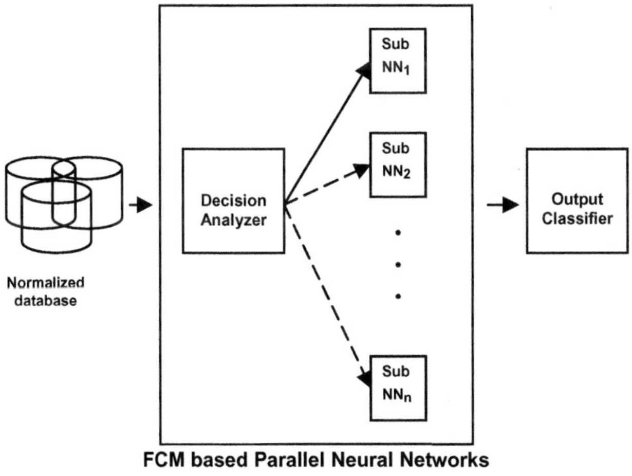


**Fig. 1.** Overall system architecture

selecting an optimal network model. The overall system is constructed as illustrated in Fig.1. Fuzzy c-means clustering is used in training for a consisting the parallel neural networks, and the decision analyzer is used in testing for selecting an optimal network model, respectively.

### 3.2 Breast Cancer Data Clustering Using FCM

In this paper, we used Fuzzy c-means clustering method for classifying the WBCD database each of subset which has similar attributes.

Fuzzy c-means (FCM) [13] is a method of clustering which allows one piece of data to belong to two or more clusters. This method (developed by Dunn in 1973 and improved by Bezdek in 1981) is frequently used in pattern recognition. It is based on minimization of the following objective function:

$$J_m = \sum_{i=1}^{N}\sum_{j=1}^{C} u_{ij}^m \left\| x_i - c_j \right\|^2 \quad , \quad 1 \le m < \infty \tag{1}$$

Where $m$ is any real number greater than 1, $u_{ij}$ is the degree of membership of $x_i$ in the cluster $j$, $x_i$ is the $i$ th of d-dimensional measured data, $c_j$ is the d-dimension center of the cluster, and $\|*\|$ is any norm expressing the similarity between any measured data and the center.

Fuzzy partitioning is carried out through an iterative optimization of the objective function shown above, with the update of membership $u_{ij}$ and the cluster centers $c_j$ by:

$$u_{ij} = \frac{1}{\sum_{k=1}^{c}\left(\frac{\left\| x_i - c_j \right\|}{\left\| x_i - c_k \right\|}\right)^{\frac{2}{m-1}}} \quad , \quad c_j = \frac{\sum_{i=1}^{N} u_{ij}^m x_i}{\sum_{i=1}^{N} u_{ij}^m} \tag{2}$$

This iteration will stop when $\max_{ij}\left\{\left| u_{ij}^{(k+1)} - u_{ij}^{(k)} \right|\right\} < \varepsilon$ ,where is a termination criterion between 0 and 1, whereas $k$ are the iteration steps. This procedure converges to a local minimum or a saddle point of $J_m$.

In this paper, we use the FCM to classify the WBCD database for performance of neural networks. We set the number of initial cluster as 7. This value was determined empirically. In order to determine the number of initial cluster, we experiment Fuzzy c-means clustering method procedures several times.

### 3.3 Parallel Neural Networks

If high dimensions and number of sampling data are used in training of neural networks, local minimum problem which not converges the optimal values are occurred by feature of learning method, minimum gradient.

In this paper, the parallel neural networks is composed of a 3 layer back-propagation type of neural networks. Each layer of the neural networks has a 3~5 neurons for hidden layer and 9 neurons for input layer. Neurons in the input layers are connected with 9 features of breast cancer. The Parallel Neural Networks is constructed by empirically determined the number of initial cluster. Whereas a small number of initial cluster decreases the performance of diagnosis, increasing the number of initial cluster improves the performance of the diagnosis. If the Parallel Neural Networks construct the number of initial cluster, the diagnosis system will be fall into a local minimum. Therefore number of initial cluster must be set to 7. Proposed the parallel neural networks is illustrated in Fig. 2.

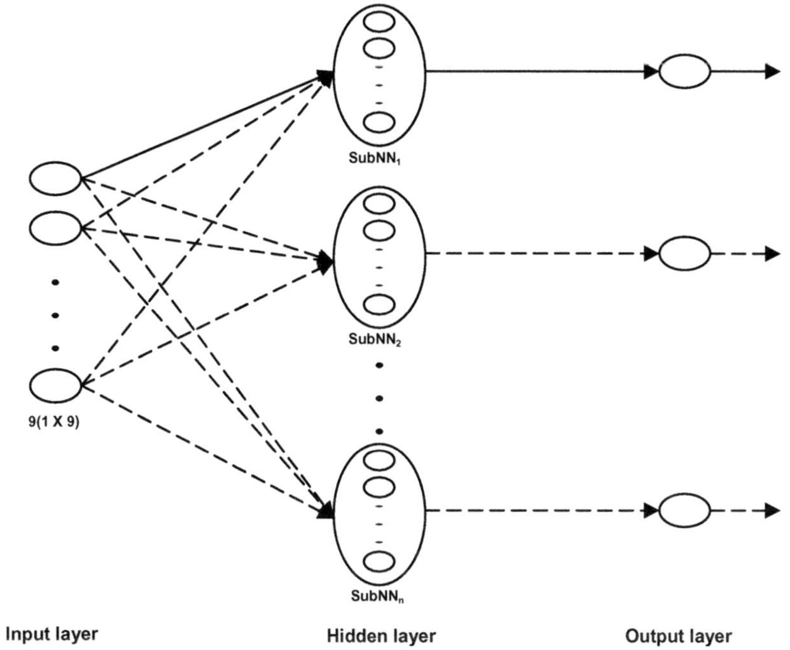


**Fig. 2.** Structure of parallel neural networks

The WBCD database is classified in 7 subnetworks by FCM so that data which have similar attributes will be used for the neural networks training data.

### 3.4 Optimal Neural Networks Model Decision by FCM Value

The proposed methodology has a one problem to find optimal neural networks model. The input data will be entered in neural networks which is in the closet convergence of learning due to training after grouping using Fuzzy c-means clustering. When test data enter system as input data, it ought to find optimal neural networks model. If it does not find the optimal neural networks model, on the contrary the proposed structure of parallel neural networks may decreases performances. In this paper, we present the decision method, which decision analyzer to search an optimal neural networks model. By implementing the FCM, we can obtain FCM center values in training procedure. We call these values CV (Criteria Values), in this work. Decision analyzer uses those MSE (Mean Square Error) between CV and input neuron to classify each of testing data. Therefore each of testing data enters optimal subnetworks.

## 4 Experiments and Discussion Results

Our experiments have been done as 3 steps. First step is that the WBCD database is classified with Fuzzy c-means clustering. Second step is making a lookup table to create the decision analyzer which selects an optimal neural networks. Third step is parallel neural networks simulate test data. To explain result, Each of step has a table and comment.

In our experiments, we experimented with 683 individuals in the WBCD database. Experiments were simulated with 400 training data and 283 testing data. Getting the number of initial cluster, we performed several experiments. Therefore, the number of initial cluster is set to 7. (Table 2) shows FCM center values. All experiments were conducted with Pentium4™ 3.0Ghz, 1Gb memory system using MATLAB™R2006a.

**Table 2.** Values of FCM center

| | $X_1$ | $X_2$ | $X_3$ | $X_4$ | $X_5$ | $X_6$ | $X_7$ | $X_8$ | $X_9$ |
|---|---|---|---|---|---|---|---|---|---|
| 1st Cluster | 0.7309 | 0.6972 | 0.6913 | 0.5836 | 0.5944 | 0.7839 | 0.5888 | 0.6774 | 0.3110 |
| 2nd Cluster | 0.7253 | 0.7498 | 0.7325 | 0.6334 | 0.6220 | 0.8024 | 0.6158 | 0.7102 | 0.3336 |
| 3rd Cluster | 0.7318 | 0.6802 | 0.6763 | 0.5673 | 0.5834 | 0.7817 | 0.5790 | 0.6541 | 0.3029 |
| 4th Cluster | 0.7117 | 0.5133 | 0.5419 | 0.4329 | 0.4624 | 0.7818 | 0.4909 | 0.4655 | 0.2184 |
| 5th Cluster | 0.3982 | 0.1345 | 0.1439 | 0.1294 | 0.2142 | 0.1331 | 0.2532 | 0.1301 | 0.1082 |
| 6th Cluster | 0.1511 | 0.1109 | 0.1188 | 0.1125 | 0.1978 | 0.1154 | 0.2274 | 0.1114 | 0.1045 |
| 7th Cluster | 0.7118 | 0.5138 | 0.5423 | 0.4334 | 0.4628 | 0.7888 | 0.4912 | 0.4635 | 0.2118 |

Those values in Table 2 are used for classifying the WBCD database. Also those values are used for input data to select the optimal subnetworks among network models in testing procedure. Table 3 shows similar index which be calculated by MSE (Mean Square Error) of CV and test data. Calculated similarity indexes are used as lookup table for selecting the optimal subnetworks.

**Table 3.** Lookup table using similar index

| | mse1 | mse2 | mse3 | mse4 | mse5 | mse6 | mse7 | Similar index |
|---|---|---|---|---|---|---|---|---|
| Test data 1 | 0.014 | 0.145 | 0.161 | 0.002 | 0.267 | 0.150 | 0.262 | 4th NNs |
| Test data 2 | 0.164 | 0.036 | 0.042 | 0.142 | 0.087 | 0.037 | 0.085 | 2nd NNs |
| Test data 3 | 0.003 | 0.145 | 0.161 | 0.002 | 0.267 | 0.150 | 0.262 | 4th NNs |
| ⋮ | ⋮ | ⋮ | ⋮ | ⋮ | ⋮ | ⋮ | ⋮ | ⋮ |
| Test data 283 | 0.002 | 0.196 | 0.212 | 0.013 | 0.324 | 0.200 | 0.300 | 1st NNs |

To evaluate the correctness of the proposed system, PPV (Positive Predicted Value) [5] was computed in each case. Here, following table shows that the more the number of initial cluster increases, the more performance improves.

**Table 4.** FbPNN performances by the number of initial cluster

| | Positive Predicted Value |
|---|---|
| No cluster | 95.5842 |
| 2 cluster | 96.0342 |
| 3 cluster | 98.1284 |
| 4 cluster | 98.2462 |
| 5 cluster | 99.0459 |
| 6 cluster | 99.2933 |
| 7 cluster | 99.5289 |

Table 5 shows the comparison between result of FbPNN and results of other methods.

**Table 5.** Experimental results of previous works

| | Positive Predicted Value | Reference |
|---|---|---|
| ANFIS | 97.95 | [5] |
| Fuzzy-Genetic | 97.07 | [6] |
| ILFN | 97.23 | [7] |
| Fuzzy | 96.71 | [8] |
| ILFN & Fuzzy | 98.13 | [9] |
| SANFIS | 96.07~96.3 | [10] |
| NNs | 97.95 | [4] |

In our experiments FbPNN shows dramatically better performance than other methods on breast cancer diagnosis problem. If the proposed system has a significantly higher advantage in storage, it would be a better method to be implemented in real situations. Therefore, the proposed method may be appropriate method for the problem of medical diagnosis including breast cancer diagnosis.

## 5 Conclusions

In this paper, method of automatic breast cancer diagnosis system with FCM based Parallel Neural Networks is proposed for correct a diagnosis. The Wisconsin breast cancer diagnosis (WBCD) database is divided into several groups to have a similarity based on fuzzy c-means clustering for improving the performance of diagnosis. Also method to enter optimal subnetworks, we proposed decision analyzer. By using this method, correct diagnosis rate of over 99% is obtained, which is better than some other results.

Our experiments indicate a way to have higher performance of diagnosis with these powerful classification algorithms. The proposed method using FbPNN would be improving performances not only medical diagnosis but also classification problems which have high complexity and nonlinear system with huge data.

## References

1. Bazanov, P., Kim, T., Kee, S., Lee, S.: Hybrid And Parallel Face Classifier Based on Artificial Neural Networks and Principal Component Analysis. Proceedings of IEEE International Conference on Image Processing. (2002) 22-25
2. Yuan, X., Lu, J., Yahagi, T.: A Personal Identification System Based on Fuzzy Clustering and Parallel Neural Network. Proceedings of International Symposium on Communications and Information Technologies 2004 (2004) 383-388
3. Husain, H., Khalid, M., Yusof, R.: Automatic Clustering of Generalized Regression Neural Network By Similarity Index based Fuzzy C-Means Clustering. Proceedings of TENCON 2004.2004 IEEE Region 10 Conference 2 (2004) 302-305
4. Arulampalam, G., Bouzerdoum, A.: Application of shunting inhibitory artificial neural networks to medical diagnosis. The Seventh Australian and New Zealand 2001, Intelligent Information Systems Conference. (2001) 89-94
5. Song, H., Lee, S., Kim, D., Park, G.: New Methodology of Computer Aided Diagnostic System on Breast Cancer. Proceedings of International Symposium on Neural Networks 2005. (2005) 780-789
6. Pena-Reyes, C., Sipper, M.: Designing Breast Cancer Diagnostic System via a Hybrid Fuzzy-Genetic Methodology. Proceedings of IEEE International Fuzzy Systems Conference 1 (1999) 135-139
7. Meesad, P., Yen, G,G. : Combined Numerical and Linguistic Knowledge Representation and Its Application to Medical Diagnosis. Proceedings of IEEE Transactions on Systems, Man and Cybernatics 2 (2003) 206-222
8. Wang, J., Lee, G.: Self-Adaptive Neuro-Fuzzy Inference Systems for Classification Applications. Proceedings of IEEE Transactions and Fuzzy Systems 10 (2002) 790-802
9. Setiono, R.: Generating Concise and Accurate Classification Rules for Breast Cancer Diagnosis. Artificial Intelligence in Medicine (2000)
10. Lovel, B. C., Bradley, A.: The multiscale Classifier. Proceedings of IEEE Transactions on Pattern Analysis and Machine Intelligence 18 (1996)
11. Jang, J.: ANFIS : Adaptive-Network Based Fuzzy Inference System. Proceedings of IEEE Transactions on Systems, Man and Cybernatics 3 (1993) 665-685
12. Freund, Y., Schapire, R.: Experiments with a New Boosting Algorithm. Machine Learning : Proceedings of the Thirteenth International Conference (1996) 148-156
13. Dunn., J.: A Fuzzy Relative of the ISODATA Process and Its Use in Detecting Compact Well-Separated Clusters, Journal of Cybernetics 3 (1973) 32-57

# Optimal Sizing of Energy Storage System in Solar Energy Electric Vehicle Using Genetic Algorithm and Neural Network

Shiqiong Zhou[1], Longyun Kang[2,1], MiaoMiao Cheng[1], and Binggang Cao[1]

[1] School of Mechanical Engineering, Xi'an Jiaotong University, Xi'an 710049, China
[2] School of Automotive Engineering,South China University of Technology, Guangzhou 510640, China

**Abstract.** Owing to sun's rays distributing randomly and discontinuously and load fluctuation, energy storage system is very important in Solar Energy Electric Vehicle (SEEV). The combinatorial optimization by genetic algorithm and neural network was used to optimize the energy storage system (including storage batteries and flywheel).In the optimization design, the operation strategy of the system was fixed and used to instruct the simulation about the system's operation. And the optimal objective was selected as minimizing the total capital cost of the energy storage system, subject to the main constraint of the Loss of Power Supply Probability (LPSP). Studies have proved that the combinatorial optimization by genetic algorithm and neural network converges well, lessen calculation time and it is feasible.

**Keywords:** battery flywheel, genetic algorithm, neural network.

## 1 Introduction

Petroleum lack and public environment protection consciousness improving, that compel scientists and carmakers design new type vehicle to the best of our abilities to reduce exhaust gas to let[1]. By research and experiment for a few years, a new type of no-pollution vehicle based on the solar offering the power, namely Solar Energy Electric Vehicle (SEEV), was developed by scientists recently. As a novel type of green vehicle, SEEV takes on many virtues, such as zero-letting, low-noise, and energy source collecting expediently and so on.

SEEV is composed of photovoltaic (PV) arrays, maximum PV power tracking (MPPT), storage system (including storage batteries and flywheel), motor driver controller and direct current motors. And system structure of SEEV is showed in Fig.1.

In SEEV system, solar energy takes on randomicity and discontinuousness, which results in the system power supplied by PV arrays changing randomly and acutely. But the vehicle motor requires steady power supplied by generate electricity, so the storage system is the indispensable components, and their capacities are selected suitably that affect directly the economy benefit and the reliability of the SEEV

D.-S. Huang, L. Heutte, and M. Loog (Eds.): ICIC 2007, LNAI 4682, pp. 720–729, 2007.

system. In order to use the renewable energy as much as possible, decrease the dependence on the common grid, ensure the reliability and stability of the system and to the best of our abilities to low the cost, the storage settings capacities must be designed with optimization.

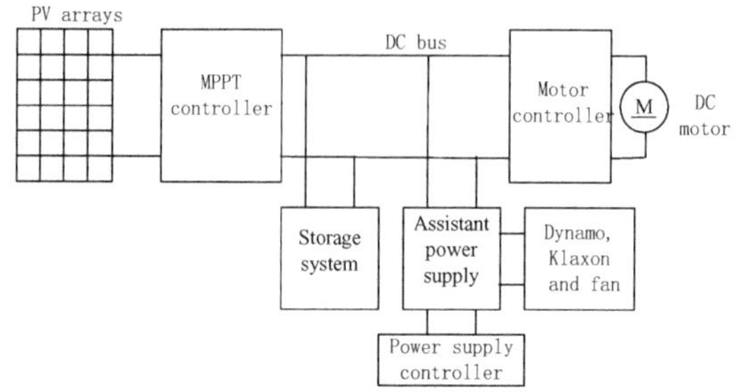


**Fig. 1.** The system structure of SEEV

On the basis of literature 2, in this paper, the combination optimization by genetic algorithm and neural network was used to optimize the total capital cost of the energy storage system in SEEV, namely to minimize the capacities of storage battery and the mass of flywheel.

## 2 Mathematical Model Building

### 2.1 PV Cell

Photovoltaic engineering model is as follow [2]:

$$I = I_{SC}\{1 - C_1[\exp(\frac{V - \Delta V}{C_2 V_{OC}}) - 1]\} + \Delta I \tag{1}$$

Here, $I_{SC}$ is the short-circuit current, $V_{OC}$ is the open-circuit voltage, $C_1$、$C_2$、$\triangle I$、$\triangle V$ can be expressed as follows:

$$C_1 = (1 - \frac{I_m}{I_{sc}})\exp(-\frac{V_m}{C_2 V_{oc}}), \qquad C_2 = (\frac{V_m}{V_{oc}} - 1)[\ln(1 - \frac{I_m}{I_{sc}})]^{-1}$$

$$\Delta V = -\beta \times \Delta T - R_s \times \Delta I, \qquad \Delta I = \alpha \frac{S}{S_{ref}} \bullet \Delta T + (\frac{S}{S_{ref}} - 1) I_{sc}$$

Where, $I_m$ is maximum power current, $V_m$ is maximum power voltage $V_m$, $I_{SC}$ is short circuit , current, $V_{OC}$ is open circuit voltage , $S$ is the solar radiation intensity, $R_s$ is series-wound resistance of PV cell α is current temperature coefficient under reference solar radiation intensity, (A/℃),$\beta$ is temperature coefficient of voltage under the reference solar irradiance, V/℃, $\alpha = 0.0012 I_{sc}$ ( A/℃ ), $\beta = 0.005 V_{oc}$ ( V/℃ ) ,$T = T_{air} + 0.03 H_T$ , and $\Delta T = T - T_{ref}$ .

Maximum-power operation is used in this paper. Assume the photovoltaic cells are always working in maximum-power point. That is $V=V_m$. Under normal conditions, given the solar irradiance is $H_{ST}$ , cell temperature is $T_{ST}$ , maximum-power current is $I_{mo}$ , maximum-power voltage is $V_{mo}$ . Actual solar irradiance is $H_T$ , Actual cell temperature is $T$ , then,

$$V_m = V_{mo} \bullet [1 + 0.0539 \lg [\frac{H_T}{H_{ST}}]] + \beta_0 \bullet (T - T_{ST}) \tag{2}$$

According to the equation (1), we can get equation as follows:

$$I_m = I_{SC} \{1 - C_1 [\exp(\frac{V_m - \Delta V}{C_2 V_{OC}}) - 1]\} + \Delta I \tag{3}$$

Then, under any condition, output of the cell is expressed as follows:

$$P_m = V_m I_m \tag{4}$$

## 2.2 Battery Model

KiBaM dynamic discharge and charge model is used in this paper. This model is a real-time simulation model, and it can reflect the real-time relation of battery capacity with discharge and charge current[3]. Assume that in a step size Δt , rated voltage of the system is *U*, the power output of photovoltaic generator is *b*, load power is *h*, then, $P_e = b - h$, when $P_e$>0, battery charges, Charge current: $I_c = P_e / U$

Maximum charge current $I_{c\max}$ :

$$I_{c\max} = \left(-kcq_{\max} + kq_{10}e^{-k\Delta t} + q_0 kc\left(1 - e^{-k\Delta t}\right)\right) / \left(1 - e^{-k\Delta t} + c\left(k\Delta t - 1 + e^{-k\Delta t}\right)\right) \tag{5}$$

$P_d = h - b$ , when $P_d$ >0, battery discharges, Discharge current: $I_d = P_d / U$

Maximum discharge current $I_{d\max}$ :

$$I_{d\max} = \left(kq_{10}e^{-k\Delta t} + q_0 kc\left(1 - e^{-k\Delta t}\right)\right) / \left(1 - e^{-k\Delta t} + c\left(k\Delta t - 1 + e^{-k\Delta t}\right)\right) \tag{6}$$

Where, *c* is the ratio of available charge handling capacity to total capacity, $q_{10}$ is the available charge handling capacity at the beginning ofΔt, Ah, $q_0$ is the charge handling capacity at the beginning of Δt, Ah, k is the ratio coefficient, $h_{rs}^{-1}$, $q_{\max}$ is the maximum capacity, Ah.

## 2.3 Flywheel Model

The available energy stored in the flywheel is calculated as follows:

$$\Delta E = J(\omega_{max}^2 - \omega_{min}^2)/2 \tag{7}$$

Where, $J$ is the moment of inertia, $\omega_{max}$ is the maximum angular rate, $\omega_{min}$ is the minimum angular rate.

We can see that, the capacity of flywheel is relevant with its angular rate and its moment of inertia. Here, flywheel's maximum angular rate is limited by its material and structure, and the ratio of the maximum angular rate and minimum angular rate is 1.6:1 in SEEV [4]. Flywheel's moment of inertia is decided by its mass and geometry, while its geometry is usually limited by the space. So flywheel's mass is chosen as the optimum object in this paper.

Flywheel's specific energy, that is energy stored per unit mass, is decided by the following:

$$e = k\left(\frac{\sigma}{\rho}\right) \tag{8}$$

Here, $e$ is specific energy; $k$ is appearance coefficient, $\rho$ is material's density, $\sigma$ is material's strength.

## 3 Storage System Optimization

### 3.1 Objective Function

The object is to minimize the total capital cost of the energy storage system with performance indices are satisfied[2]:

$$\min C_b P_b + C_f P_f \tag{9}$$

$C_b, C_f$ --- battery and flywheel's unit price

$P_b, P_f$ --- battery and flywheel's rated capacity.

### 3.2 Constraint Function

Assume that, the power output of photovoltaic is b, motor power is h, and then, constraint functions are:

$$\begin{cases} E(b) = E(h) \\ \Pr\{h - b - UI_c - P_F \le 0\} \ge \alpha, (\alpha = 0.5 \sim 1), I_c \ge 0, 0 \le P_F \le \dfrac{me}{60} \\ \int_0^t bdt - \int_0^t hdt \ge P_b + P_f \end{cases} \tag{10}$$

Expressions (10) is explained as follows:

The first equation reflects the system's reasonableness. The second equation reflects the system's reliability. That is, when the system is unavailable (no irradiance), the energy storage section can provide energy to ride through these periods reliably. Ic is actual discharge and charge current; PF is flywheel's actual discharge and charge power. Where, assume the available energy stored in the flywheel can be discharged in a minute.

The third equation reflects the system's practicability. That is when load is low, the battery and flywheel can charge into full capacity. Here, $P_b = 10UI_b$ , $P_f = 64me/39$ . $I_b$ and $m$ are the optimum objects, they are separately charge current of battery and flywheel's mass.

# 4 The Combinatorial Optimization by Genetic Algorithm and Neural Network

## 4.1 Combinatorial Opitmization

Genetic algorithm is simulating the course of biology inheriting and evolving. There exist three major processes, namely selection, crossover and mutation. Genetic algorithm based on stochastic simulation is very effective for the solvable general chance constrained programming, and the optimization of energy source system in SEEV is a typical stochastic programming[5] (showed in Fig.2).

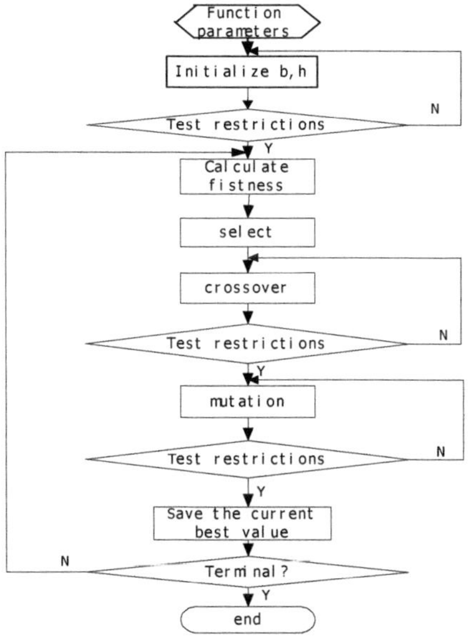


**Fig. 2.** GA flow chart

GA flow chart is explained as follows:

1) The function parameters, such as population size, crossover rate, mutation rate and genetic generation and so on, are defined.
2) Real-number encoding is used to describe at first hand the question and improve on the operation rate of crossover and mutation.
3) Test the restrictions and divest the invalid random number.
4) Fitness is calculated and then selection, crossover and mutation are used until meet the scheduled maximum genetic generation or reach to the precision required.

Individual performance influences directly the efficiency of colony evolving. In fact, to test the feasibility of individuals is to evaluate the performance of individuals. In the course of individuals testing, more time is cost on the optimization along with the group number increasing. And it restricts the optimization efficiency. Obviously, to test the feasibility of the individuals is a classifying problem. Here, the Artificial Neural Network (ANN) is constructed to fulfill the classifying problem. In the GA, transfer the network trained to test the feasibility of individuals, and the colony propagation is limited to the feasibility field of individuals and it fasten the rate of search the best value. This is the combinatorial optimization by genetic algorithm and neural network.

## 4.2 Training and Applying ANN

Firstly, chromosomes' feasibility is tested, which is a classifying problem. And in this paper, the BP arithmetic that is in common use is used to work out this problem. MATLAB is utilized to realize the arithmetic. We can take full advantages of the particular predominance of MATLAB in matrix calculation because ANN is involved with a plenty of matrix calculation. At the same time, the ANN toolbox is supplied by MATLAB6.5, which brings many conveniences to ANN calculation. And it can be divided into four steps: fixing on network structure, preparing stylebook, training network and checking up network[5].

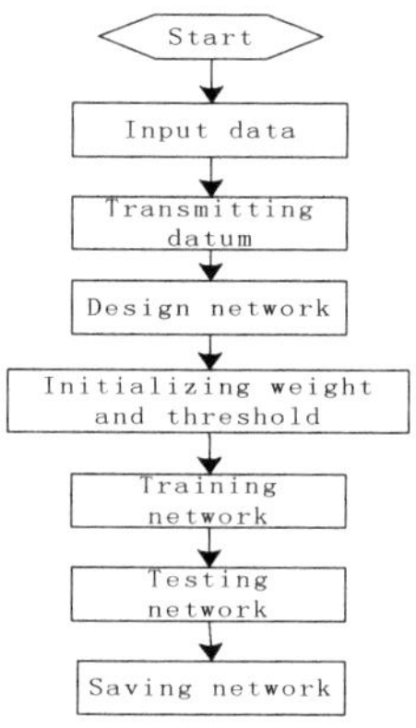


**Fig. 3.** Training net flow chart

Training and applying network are separate and the flow charts are showed in Fig.3 and Fig.4. In the GA transferring, the program section of application network is transferred viz. the network trained is used and the network is not needed to be trained again. And the operating time of program is reduced.

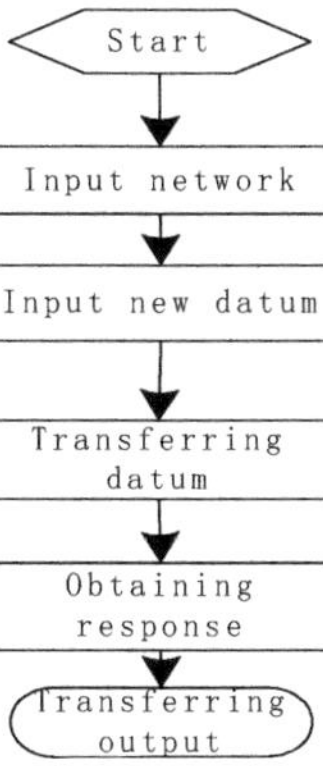


**Fig. 4.** Transferring net flow chart

# 5 Example and Analysis

## 5.1 Datum Resource

In this paper, the experiment datum obtained from the existing SEEV run on Silk Road the Silk Road , including solar radiant intensity, the voltage and current of load, derive from a cooperation project by Xi'an Jiaotong University and Osaka Sangyo University in Oct, 2005. These datum reflect the route and the weather status on the course of the SEEV running. The output power of PV arrays and the consumed power by load (motor) are showed separately in Fig.5 and Fig.6 .

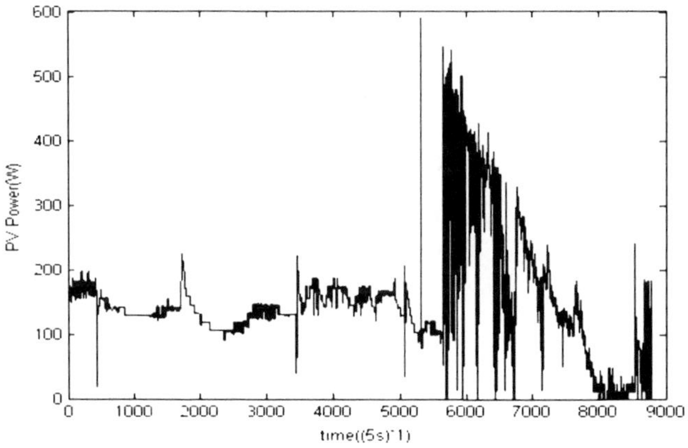


**Fig. 5.** The output power of PV arrays (datum are obtained on 19, 20,22,23,24,27,Oct,2005)

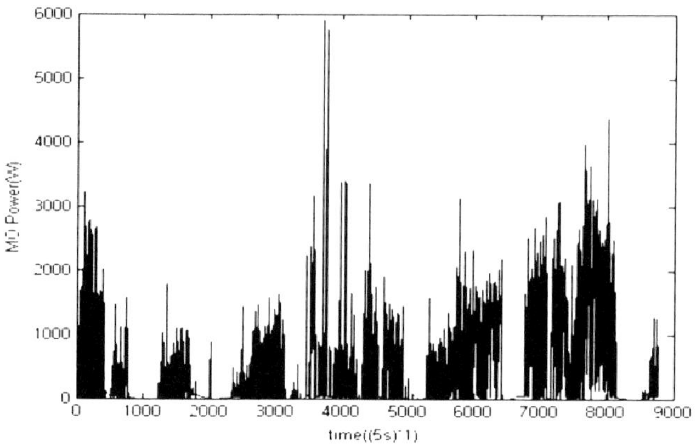


**Fig. 6.** The consumed power by load (motor) (datum are obtained as the former)

### 5.2 Result and Analysis

When the radiant intensity is less, the output power calculated is negative because PV arrays must get over interior all kinds losses. In fact, the bounce-back diodes are selected and then the value is zero.

In this paper, the combinatorial optimization by genetic algorithm and neural network was used to optimize the energy storage system (including storage batteries and flywheel). Assume that battery's unit price is 0.8yuan/Wh; flywheel is made of steel 45#, of which the unit price is 4.1yuan/kg, and available specific energy is 5Wh/kg [2] . Assume that the selection rate is 0.8 and the mutation rate is 0.85, Population size is 30 and genetic generate is 20. Studies have proved that GA converge stably and can offer gist to design. In GA, the neural network trained is used to replace the processing section of restrictions, namely the feasibility test section of chromosome individual that saves the operating time of program. The result of neural network trained is showed in Fig.7. We can see that only 11 steps are required to reach to desire error and complete the training neural network.

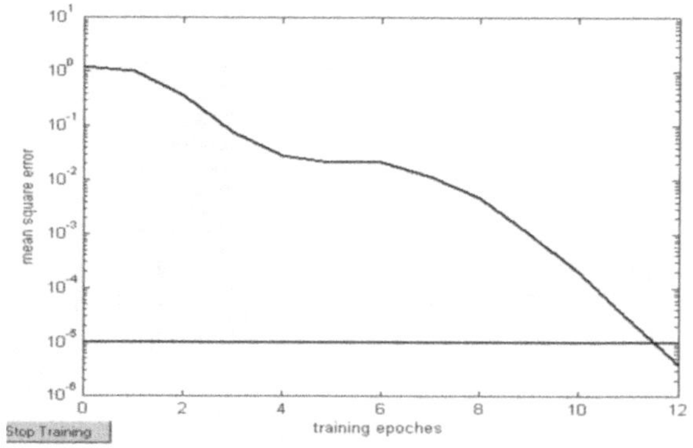


**Fig. 7.** The result of neural network trained

The optimization results arc showed in Fig.8 , Fig.9 and Fig.10.

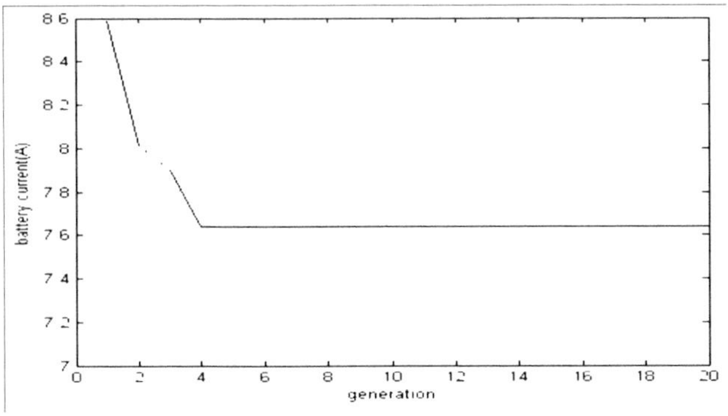


**Fig. 8.** Battery charge current

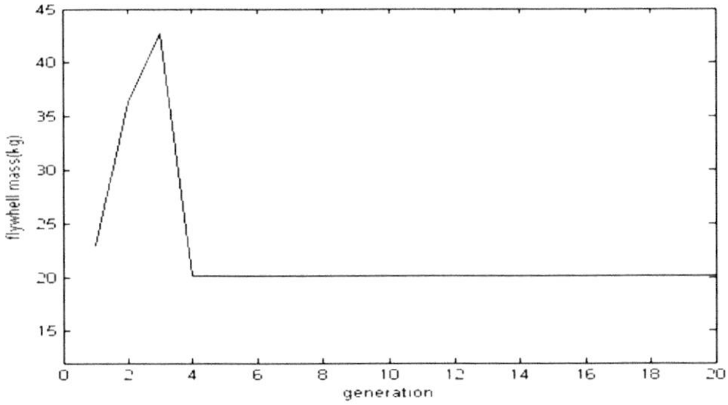


**Fig. 9.** Flywheel mass

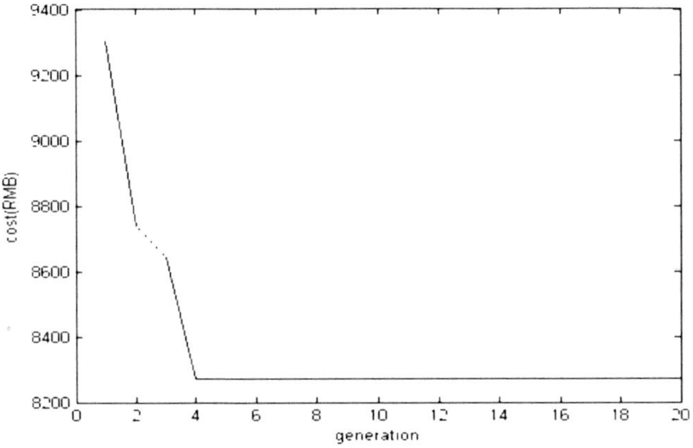


**Fig. 10.** The simulation result of cost

From the optimization result by genetic algorithm and neural network, we can find that, with generation increase, the combination of battery's current and flywheel's mass reach gradually to optimization. However, the sum of their investment cost is decreasing, as Fig.10. In this example, the battery's minimum charge current is 7.639A or 8A by ceiling of as integer. The flywheel's minimum mass is 20.122kg or 21kg by ceiling of as integer. And the sum cost of the storage system in SEEV is RMB:8272.5.

## 6 Conclusion

The combinatorial optimization by genetic algorithm and neural network was used to optimize the energy storage system in SEEV. The optimal result, satisfied with the load requirement, can be obtained and the algorithm can converge stably, if the population size and genetic generation are sufficient. Besides, for battery and flywheel can be complementary as storage section, when designing SEEV system, the object may be utilizing solar irradiance energy as more as possible, not worrying about the power's waste too much. It will be positive to the utilization of renewable energy.

## References

1. Xiong, Q., Tang, D. H.: Research Progress on Supercapacitor in Hybrid Electric Vechicle. ACTA Scientiarum Naturalium Universitis Sunyatseni Vol.42 (2003)
2. Cheng, M.M., Kang, L.Y., Xu Daming.: Optimal Capacity of Energy-Storing Section in PV/wind Hybrid System. International Symposium on Mechanical &Aerospace Engineering 2005. August 22~25, 2005 Xi'an , China
3. Zuo,W.: Simulation of Wind Energy and Solar Energy for distributed Generation System [D].Xi'an : Xi'an Jiaotong University , (2004)
4. Mao, M.Q., Yu, S.J., Su, J.H., Shen, Y.L.: Research on Variable Structure Simulation Modeling for Wind-Solar Hybrid Power Systems [J] Journal of System Simulation, Vol.5. (2003) 361-364
5. Chen, Z.C. , Lou, J.N. , Zhu, B.X.: Genetic Algorithm and Neural Network Structure Optimization stategy [J] . Nanking Chymistry Industry University Transaction , 1999

# Research on Error Compensation for Oil Drilling Angle Based on ANFIS

Fan Li, Liyan Wang, and Jianhui Zhao

School of instrument science & opto-electronics engineering,
Beihang university, Beijing 100083, China
lifan@buaa.edu.cn, yanzididi007@163.com,
zhaojianhui@buaa.edu.cn

**Abstract.** Gyro survey technique has applied and played an important role in many areas, such as offshore oil drilling, directional drilling and so on. Considering the influence of the compensation for the large surveying azimuth error, in this paper, the principle of the gyro survey system is described, and the ANFIS architecture is employed to model the survey azimuth error, predict a chaotic current, all yielding remarkable results according to the gyro survey principle and the data sampled from the two-axis turntable. From the simulation and the testing result, we can see that ANFIS is an effective and feasible way to model and compensate the azimuth error, and the precision of the ANFIS method is higher than the methods of the bilinear interpolation and the radius basis function (RBF), so it is available and advisable in engineering.

**Keywords:** ANFIS, Error compensation, Gyro survey, Bilinear interpolation, RBF.

## 1 Introduction

Gyro survey technique plays an important role in directional survey field, and has applied in many areas, especially in offshore oil drilling, etc.

In this paper, the error compensation technique of inertia gyro survey system based on ANFIS is studied.

As for the directional survey technique in oil and other industry, the survey based on inertia technique is more accurate and steady. Using the dynamically tuned gyroscopes (DTG) to sensitize the rotational angular velocity of the earth and using the accelerometers to sensitize the gravity, the strapdown inertia navigation system (SINS) can obtain parameters such as inclination, azimuth and tool angles. Therefore, the precision of the survey system is largely depending on the precision of the inertial measure components (a two-axis DTG and two force-feedback accelerometers). We know that the error formed by the system, the influence of some physical factors and other outside interference all influence the accuracy. So error compensation is the key technique to improve the precision.

ANFIS is often referred to as the neural network-based fuzzy modeling because the parameters of fuzzy membership functions are identified embedding the fuzzy inference system into a framework of adaptive networks. For training, ANFIS

D.-S. Huang, L. Heutte, and M. Loog (Eds.): ICIC 2007, LNAI 4682, pp. 730–737, 2007.

employs a hybrid learning procedure combining the gradient decent in the backward pass usually called backpropagation and the least-square method in the forward pass. These contents can be seen in [1]. This hybrid learning allows speed up of the learning process substantially by decreasing the dimension of the search space. Besides, ANFIS is successfully used in the prediction of time array and is becoming one of the most attractive research fields recently for the machine learning.

Firstly, after analyzing the principle of the gyro surveying, the inclination and azimuth angles are sampled from the two-axis turntable under the condition of the given ideal inclination and azimuth angles. Then the azimuth error model is set up based on the ANFIS and it leads to higher accuracy and a better result than the models based on both of the bilinear interpolation and the RBF.

## 2 The Inertial Survey Theory

The well bore survey system in this paper includes a DGT and two accelerometers, the rotation axis of DGT coincides with the axis of this survey system as shown in Fig.1, those two accelerometers are arranged in two mutually orthogonal directions, the plane which two output axis locates is vertical with the axis of this survey system, the direction of those two accelerometers' output axis ( $X_a$, $Y_a$ ) are identical with the direction of gyro's output axis ( $X_g$ , $Y_g$ ).

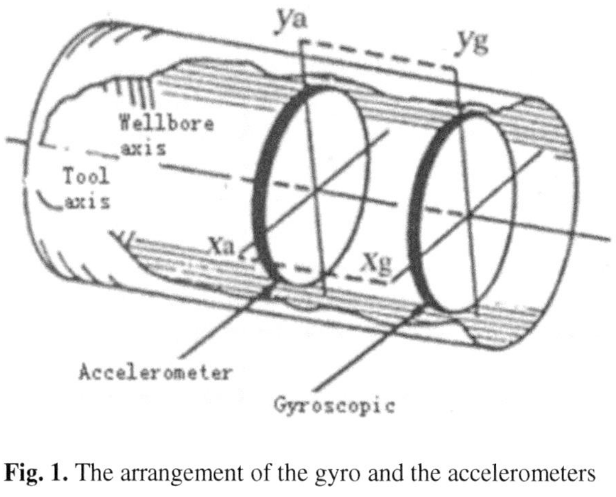


**Fig. 1.** The arrangement of the gyro and the accelerometers

From calculation, we can get the attitude value from the relationship among geocentric coordinate system, terrestrial coordinates system, geographic coordinate system, body coordinate system. The angle functions of the surveying system are as show in (1) (2) (3). where $\omega_e$ is the rotational angular velocity of the earth, $\varphi$ is local attitude, $g$ is gravitational acceleration, $A$ , $I$ , $T$ are azimuth angle, inclination angle, tool angle we want to know, $\alpha_x$ , $\alpha_y$ , $\alpha_z$ are projection of $g$ on body coordinate system's every axis, $\omega_x$ , $\omega_y$ , $\omega_z$ are projection of $\omega_e$ on body coordinate system's every axis.

$$\text{Tool angle: } T = -\arctan\frac{\alpha_y}{\alpha_x} \tag{1}$$

$$\text{Inclination: } I = \arcsin\frac{\sqrt{\alpha_x^2 + \alpha_y^2}}{g} \tag{2}$$

$$\text{Azimuth: } A = \arctan\frac{(\alpha_x \omega_y - \alpha_y \omega_x)\cos I}{\alpha_x \omega_x + \alpha_y \omega_y - g\omega_e \sin\varphi \sin^2 I} \tag{3}$$

Limited by the length of the paper, the detailed compute process is elliptical and can refer [2] to see more.

## 3 Algorithm of the ANFIS Network [3]

Adaptive neuron-fuzzy inference systems (ANFIS) represents a neural network approach to the design of fuzzy inference systems. Since their introduction, ANFIS networks have been widely considered in the technical literature and successfully applied to classification tasks, rule-based process controls, pattern recognition problems, and so on. An ANFIS network makes use of a supervised learning algorithm to determine a nonlinear model of the input–output function [4], which is represented by a training set of numerical data. Since under proper conditions it can be used as a universal approximator, an ANFIS network is particularly suited for solving function approximation problems in several engineering fields.

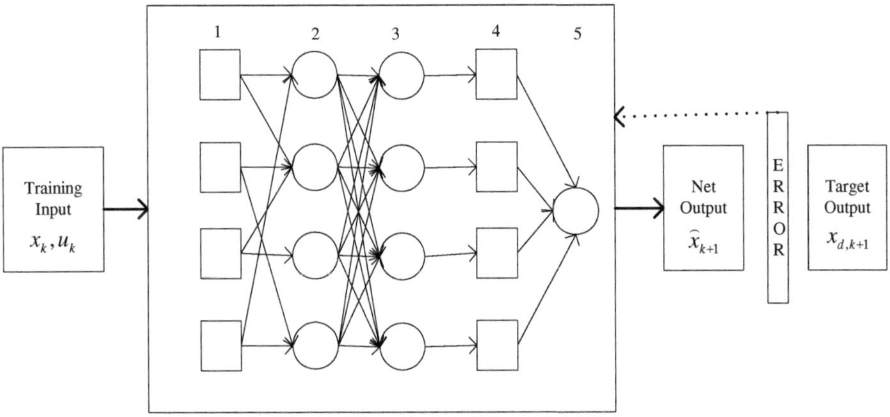


**Fig. 2.** A schematic diagram of the ANFIS model

A dynamical system in discrete time can be modeled by the equation

$$x_{K+1} = f(x_k, u_k) \tag{4}$$

where $x \in R^m$ and $u \in R^n$ are the state system output and control input respectively. For training, the error is defined as

$$e_k = x_{d,k} - \hat{x}_k \tag{5}$$

where $\hat{x}_k$ and $x_{d,k} \in R^m$ are the net model output and the training target output respectively. Adaptive network-based fuzzy inference system (ANFIS) developed by Jang is a first order Sugeno-type fuzzy inference system represented by the structure and parameters of adaptive networks. ANFIS based identification model has been demonstrated to be superior to the back-propagation neural networks and other methods. An ANFIS model for Takagi-Sugeno type fuzzy inference system, where two membership functions are assigned to each input variable and four if-then rules are employed, is illustrated in Fig. 2. Layer 1 in the model consists of a set of variables of input membership functions known as premise parameters. For example, the generalized bell membership function is defined as

$$\mu(x) = \frac{1}{1+[(\frac{x-c}{a})^2]^b} \tag{6}$$

Where $a$, $b$ and $c$ are adaptable premise parameters. In layer 2, the nodes with T-norm operators known as node functions produce the firing strength of each rule simply by multiplying the incoming signals. The firing strength from layer 2 is normalized by layer 3. In layer 4, The adaptable variables in layer 4 called consequent parameters are multiplied by the output of layer 3. The single node in layer 5 sums up all the incoming values and produces an adaptive network output. Hybrid learning procedure combining the gradient method and the least squares estimate is achieved by a forward pass and a backward pass of the adaptive network. In forward pass, while holding the premise parameters, the consequent parameters in layer 4 are identified by the least squares method. In backward pass, on the contrary, the consequent parameters are held fixed and the error rates calculated after output node are back-propagated then the premise parameters in input nodes are updated by the gradient method. The details of other forms of ANFIS architecture and learning procedure can be found in [4].

# 4 The Prediction Based on ANFIS

## 4.1 Data Acquisition

After study the operation principle of this well bore survey system, we can establish the reference attitude via two-axis turntable, set desired inclination angle and Azimuth angle at two freedom of turntable, sample the response output of inertial measurement unit, calculate the actual inclination angle and Azimuth angle which includes error signal. Every group of data is sampled after the gyro and the accelerometer is stable, and samples each data five times to get the mean value to avoid random error. Thus the main source of error is the instrument error and calculation tolerance.

To ensure the reliability of modeling, the range of testing points must cover the whole scope to show the character of this survey system. Because the error of azimuth angle is the largest, so we compensate this angle in this paper, the data to be modeling via experiment is shown in table 1 and table 2. There are desired azimuth angle, desired inclination angle, actual azimuth angle after calculation and the error. The selected testing points of inclination angle range from 0° to 70° are 1°, 3°, 5°,10°, 20°, 30°, 40°, 50°, 60°, 70° separately. The selected testing points of azimuth angle range from 0° to360°, the equal intervals is set to 20°. The data in table 2 which is also got through experiment is sampled at the points where the azimuth error is large in table 1, and is prepared to verify the effect of modeling .

**Table 1.** The Model Points

| Ideal I / Ideal A / Test A | 1° | 3° | 5° | 10° | 20° | … | 50° | 60° | 70° |
|---|---|---|---|---|---|---|---|---|---|
| 0° | 352.5 | 354 | 354.2 | 354.5 | 353.6 | … | 351.5 | 351.6 | 348 |
| 20° | 12.8 | 16.2 | 14.8 | 16 | 15.6 | … | 16.6 | 23.4 | 21.6 |
| 40° | 33.3 | 36.2 | 36.8 | 36.4 | 36.2 | … | 42.2 | 46.4 | 45.7 |
| … | … | … | | | | … | … | … | … |
| 320° | 310.8 | 312.9 | 313.3 | 312.7 | 310.7 | … | 300.3 | 292.1 | 282.1 |
| 340° | 334.4 | 331.8 | 333.7 | 333.6 | 331.5 | … | 326.5 | 320.3 | 311.4 |
| 360° | 352.4 | 353.8 | 353.7 | 354.3 | 352.9 | … | 351.8 | 351.3 | 347.8 |

**Table 2.** The Test Points

| | 1 | 2 | 3 | 4 | … | 42 | 43 | 44 | 45 | 46 |
|---|---|---|---|---|---|---|---|---|---|---|
| Test I(°) | 69.01 | 59.48 | 59.56 | 49.77 | … | 9.87 | 4.97 | 4.85 | 2.56 | 2.59 |
| Test A(°) | 209.91 | 208.87 | 130.5 | -7.07 | … | 212.9 | -4.1 | 174.89 | 86.45 | 175.5 |

### 4.2 The Modeling Results

Adopting the ANFIS toolbox under the environment of Matlab and adjusting the parameter, we can get the azimuth error model based on ANFIS.

Substituting the data in table1 into the learning system of ANFIS for training, and it is different from the learning mechanism of neural network. ANFIS gets the parameters automatically under the environment of Matlab. Through rectifying the changing parameters continually, we got a set of parameters to acquire the optimal result as shown in fig.3.

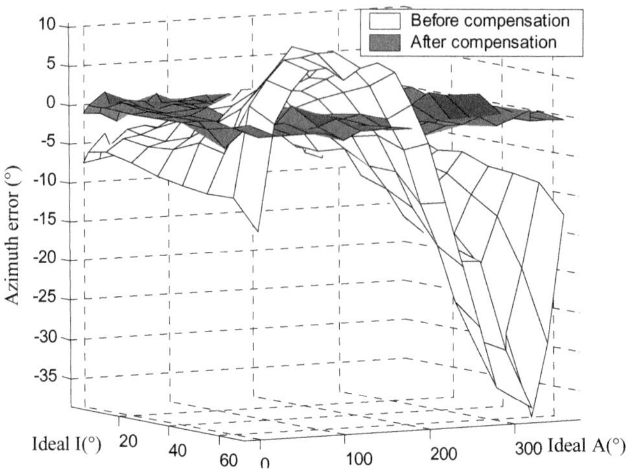


**Fig. 3.** The compensated effect of the model based on ANFIS

## 4.3 The Simulation Results

To see the effect of ANFIS model, we use modeling error to see the effect of modeling, use testing error to verify the prediction ability of this model. The modeling error and testing error of ANFIS method, bilinear method and RBF neural network is calculated and compared, refer to fig. 4 ~ fig. 6 to see the verify effect picture of all methods. The performance parameter of each method is shown in table 3.

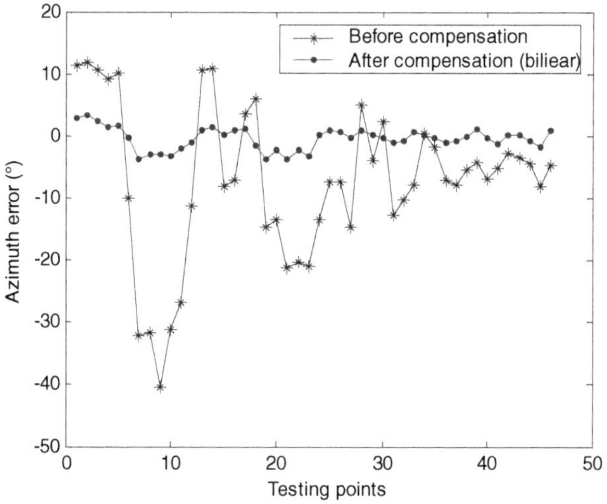


**Fig. 4.** The error before and after the compensation based on bilinear interpolation

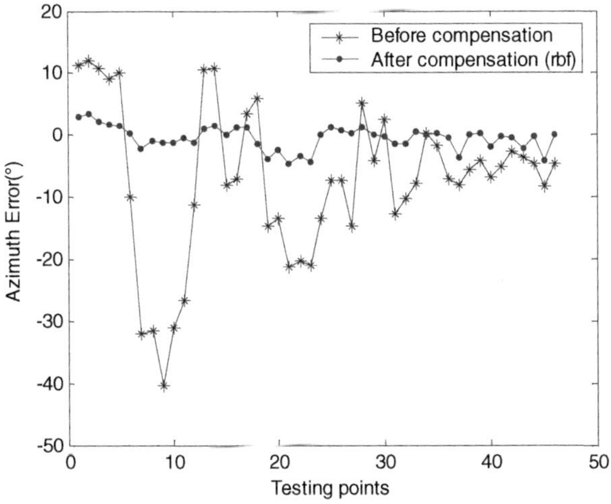


**Fig. 5.** The error before and after the compensation based on RBF

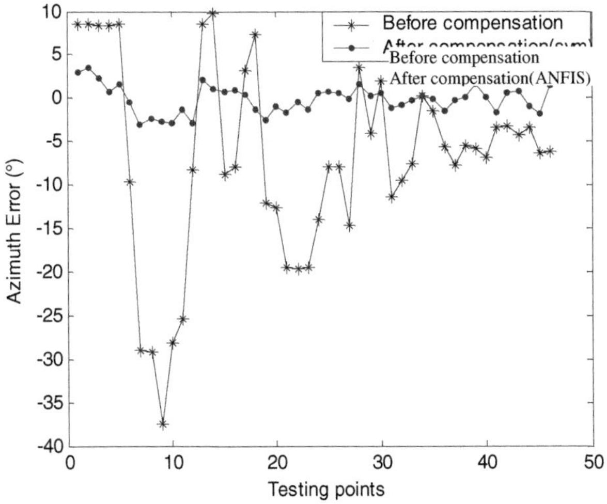


**Fig. 6.** The error before and after the compensation based on ANFIS

**Table 3.** The Performance of the Three Models

| Model | Modeling | | | Test | | |
|---|---|---|---|---|---|---|
| | Mean E(°) | Max E(°) | RMSE(°) | Mean E(°) | Max E(°) | RMSE(°) |
| Bilinear | 0.12 | 2.1 | 0.812 | 0.4130 | 3.8 | 1.7686 |
| RBF | 0.0047 | 1 | 0.0053 | 0.4476 | 4.5688 | 1.7461 |
| ANFIS | 0.0217 | 1.92 | 0.4218 | 0.1875 | 3.2053 | 1.4120 |

From the comparison of the compensation results in table 3, we can come to a conclusion that the performance parameter of ANFIS method is better than that of the other two methods. All the parameter performances are better than that of the bilinear method which is usually used in engineering. Although the modeling errors are not smaller but even larger than that of RBF method, the testing error is enhanced distinctly. And that's just of great significance in both the field of theory and engineering.

## 5 Conclusion

Different methods for the compensation of the azimuth error have been implemented and compared on the basis of the maximum and mean errors according the data obtained from a gyro survey system. Through the verifying result, we can see that the azimuth error compensation based on ANFIS is feasible and effective. Compared with the models based on RBF and bilinear interpolation, ANFIS fits with high accuracy. Besides, the results illustrated in the paper encourage to a further development of the error compensation method.

## Acknowledgments

This work was supported by the National Natural Science Foundation of China under grant 50674005, CNPC Innovation Fund and Electronic Test Technology Key Laboratory Foundation under grant 51487040105HK0101 to Jianhui Zhao.

## References

1. Massimo, P., Antonio, G.: An Input-output Clustering Approach to the Synthesis of ANFIS Network. IEEE transactions on fuzzy system 13(1) (2005) 69–79
2. Zhang H. J.: Error Analysis & Simulation Research of the Gyroscopic-Survey Instrument in the Continuous Mode. Beijing university of aeronautics and astronautics, Beijing (2000)
3. Hho, K., Agarwal, R.K.: Fuzzy Logic Model-based Predictive Control of Aircraft Dynamics Using ANFIS. $39^{th}$ AIAA Aerospace Sciences Meeting & Exhibit 8-11, RENO, NV (2001)
4. Jang, J. R.: ANFIS: Adaptive-Network Based Fuzzy Inference System. IEEE Transactions on System, Man, And Cybernetics 23 (1993) 1134-1141

# Rough Set Theory of Shape Perception

Andrzej W. Przybyszewski

Department of Psychology, McGill University Montreal, Canada
Dept of Neurology, University of Massachusetts Medical Center, Worcester MA USA
przy@ego.psych.mcgill.ca

**Abstract.** Humans can easily recognize complex objects even if values of their attributes are imprecise and often inconsistent. It is not clear how the brain processes uncertain visual information. We have tested electrophysiological activity of the visual cortex (area V4), which is responsible for shape classifications. We formulate a theory in which different visual stimuli are described through their attributes and placed into a decision table, together with the neural responses to them, which are treated as decision attributes. We assume that the brain interprets sensory input as bottom-up information which is related to hypotheses, while top-down information is related to predictions. We have divided neuronal responses into three categories: (a) Category 0 - cell response is below 20 spikes/s, which indicates that the hypothesis is rejected, (b) Category 1 - cell activity is higher than 20 spikes/s, which implies that the hypothesis is accepted, 3. Category 2 - cell response is above 40 spikes/s, which means that the hypothesis and prediction are valid. By comparing responses of different cells we have found equivalent concept classes. However, many different cells show inconsistency between their decision rules, which may suggest that parallel different decision logics may be implemented in the brain.

**Keywords:** visual brain, imprecise computation, bottom-up, top-down processes, neuronal activity.

## 1 Introduction

Imprecise reasoning is a characteristic of natural languages and is related to human decision-making effectiveness [1]. However, natural language used by humans is related to awareness and description is connected to an object of attention. Therefore it is a serial process on top of many other sensory and motor processes. These other processes are preattentive. These so-called early processes extract and integrate into many parallel channels basic features of the environment. In this work, we concentrate on early preattentive processes in the visual system. Our work is related to the constitution of decision rules extracting basic features from the visual stream.

Our eyes constantly perceive changes in light colors and intensities. From these sensations our brain extracts features related to different objects. So-called "basic features" were identified in psychophysical experiments as elementary features which can be extracted in parallel. Evidence of parallel extraction comes from the fact that their extraction time is independent of the number of objects. Other features need

D.-S. Huang, L. Heutte, and M. Loog (Eds.): ICIC 2007, LNAI 4682, pp. 738–749, 2007.

serial search, so that the time needed to extract them is proportional to the number of objects. We would like to find relationships between decision rules detected in the neurological data from V4 and the basic features found from psychophysics.

The brain, in contrast to the computer, constantly integrates many asynchronous parallel streams of information [2], which help in its adaptation to the environment. Most of our knowledge about function of the brain is based on electrophysiological recordings from single neurons. In this paper we will describe properties of cells from the visual area V4. This intermediate area of the ventral stream mediates shape perception, but different laboratories propose different often-contradictory hypotheses about properties of V4 cells. We propose the use of rough set theory (Pawlak, [3]) to classify concepts as related to different stimuli attributes. We will show several examples of our method.

## 2 Method

Most of our analysis will be related to data from Pollen et al [4]. As mentioned above we have divided all cell responses in V4 into three ranges. Activity below 20 spikes/s is defined as a category 0 cell response. Activity above 20 spikes/s is defined as category 1, and activity above 40 spikes/s as category 2.

The reason for choosing the minimum significant cell activity of 20 spikes/s is as follows. During normal activity our eyes are constantly moving. The fixation periods are between 100 and 300ms, similar to those of monkeys. Assuming that a single neuron, in order to give reliable information about an object, must fire a minimum of 2-3 spikes during the eye fixation period, we obtain a minimum frequency of 20 spikes/s. We assume that these discharges are determined by the bottom-up information (hypothesis testing) and that they are related to the sensory information about object's form. The brain is constantly making predictions which are verified by comparing them with sensory information. These tests are performed in a positive feedback loop (Przybyszewski et al. [5], Przybyszewski and Kon, [6]). If prediction is in agreement with the hypothesis, we assume that activity of the cell increases approximately twofold similarly to the strength of the feedback from V1 to LGN [5]. This increased activity is related to category 2 (neuronal discharges of 40 spikes/s).

We will represent data from Pollen et al. [4] in the following table. In the first column are neural measurements. Neurons are identified using numbers related to a collection of figures in the previous paper [4]. Different measurements of the same cell are denoted by additional letters (a, b,...). For example, 11a denotes the first measurement of a neuron numbered 1 Fig. 1 of [4], 11b the second measurement, etc. Stimulus properties (see Fig 1) are as characterized as follows:

1. orientation in degrees appears in the column labeled *o*, and orientation bandwidth is labeled by *ob*.
2. spatial frequency is denoted as *sf*, and spatial frequency bandwidth is *sfb*
3. x-axis position is denoted by *xp* and the range of x-positions is *xpr*
4. y-axis position is denoted by *yp* and the range of y-positions is *ypr*
5. x-axis stimulus size is denoted by *xs*

6. y-axis stimulus size is denoted by *ys*
7. stimulus shape is denoted by *s*, values of s are following: for grating *s=1*, for vertical bar *s= 2*, for horizontal bar *s= 3*, for disc *s= 4*, for annulus *s=5*

Cell responses (*r*) are divided into 3 ranges:

*category 0* : activity below 20 sp/s labeled by $r_0$;
*category 1*: activity above 20sp/s labeled by $r_1$;
*category 2*: activity above 40sp/s labeled by $r_2$.

Thus the full set of stimulus attributes is expressed as $B = \{o, ob, sf, sfb, xp, xpr, yp, ypr, xs, ys, s\}$.

After Pawlak [3], we define an information system as $S = (U, A)$, where $U$ is a set of objects and $A$ is set of attributes. If $a \in A$ *and* $u \in U$, the value $a(u)$ is a unique element of $V$ (a value set). The *indiscernibility* relation of any subset $B$ of $A$, or $IND(B)$, *is* defined [3] as the equivalence relation whose elements are the sets $\{u: b(u) = v\}$ as $v$ varies in $V$, and $[u]_B$ is the equivalence class of $u$. The concept $X \subseteq U$ is *B-definable* if for each $u \in U$ either $[u]_B \subseteq X$ or $[u]_B \subseteq U \backslash X$. $\underline{B}X = \{u \in U: [u]_B \subseteq X\}$ is a lower approximation of $X$. The concept $X \subseteq U$ is *B-indefinable* if exists such $u \in U$ such that $[u]_B \cap X \neq \phi \}$. $\overline{B}X = \{u \in U: [u]_B \cap X \neq \phi\}$ is an upper approximation of $X$. The set $BN_B(X) = \overline{B}X - \underline{B}X$ will be referred to as the *B-boundary region* of. If the boundary region of X is the empty set than X is *exact* (*crisp*) with respect to *B*; otherwise if $BN_B(X) \neq \phi$ X is not *exact* (*rough*) with respect to *B*.

In our work universe $U$ is defined as all visual patterns that are characterized by their attributes $A$. The purpose of our research is to find how these objects are classified in the brain. Therefore we are looking to determine visual patterns (shapes) with indiscernible attributes $B \subseteq A$ on the basis of a single neuron recording from the visual area in the brain.

## 3 Results

We have analyzed the experimental data from several neurons recorded in the monkey's V4 [4]. One example of the V4 cell responses to the vertical (horizontal) bars in different horizontal - x (vertical –y) positions is shown in the upper (lower) right parts of Fig. 1. Cell responses show two maxima for the bar position along the x-axis and two maxima for the bar position along the y-axis. It was found that most of V4 cells show local extreme that was the reason to divide receptive field into several smaller subfields [4]. In the next figure (Fig. 2) the receptive field of the other V4 cell was divided into four subfields, which were independently stimulated. Horizontal lines in plots of both figures divide cell responses into the three categories $r_0$, $r_1$, $r_2$, which were related to the response strength (see Methods). Stimuli attributes and cell responses classified into categories are shown in the table 1 for cell in Fig. 1 and in table 2 for cell in Fig. 2. Our figures are modified in comparison to [4] because they also show a schematic of the optimal stimulus. These schematics were made on the basis of the decision tables (Table 1, Table 2). Fig. 1 (left side) shows the cell's responses to the stimulus, which was a long narrow bar with vertical (Fig.1 C) or horizontal (Fig.1 D) orientation. The schematic representation on the top right side of Fig. 1 shows positions of the bars in the cell receptive field when cell responses were above 20 sp/s (category 1). Therefore these bar positions represent equivalence class of stimuli related to the concept 1. The schematic in the lower right side of Fig. 1 is characterized by cell responses above 40sp/s (category 2) and this configuration represents concept 2 stimuli.

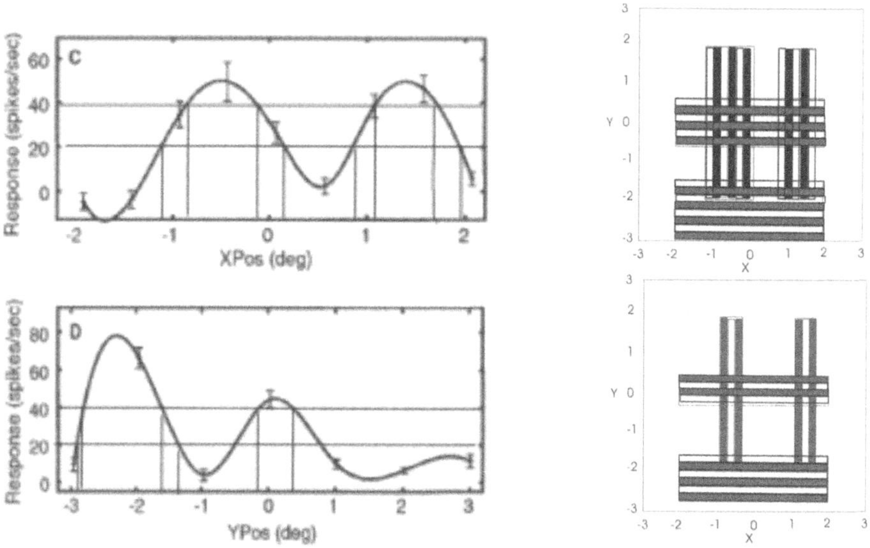


**Fig. 1.** Curves represent approximated responses of a cell from area V4 to vertical (C), and horizontal (D) bars. Bars change their position along x-axis (Xpos) or along y-axis (Ypos). Responses of the cell are measured in spikes/sec. Mean cell responses ± SE are marked in the figures. Cell responses are divided into three ranges (concepts) by two horizontal lines. On the right is a schematic representation of cell response on the basis of Table 1. Vertical and horizontal bars in certain x- and y-positions gave strong (concept 1 – upper schematic) or very strong (concept 2 – lower schematic) responses.

**Table 1.** Decision table for the cell shown in Fig. 1. Attributes ob, sf, sfb were constant and are not presented in the table.

| Cell | o | xp | xpr | yp | ypr | xs | ys | s | r |
|---|---|---|---|---|---|---|---|---|---|
| 12a | 90 | -0.6 | 1.2 | 0 | 0 | 0.4 | 4 | 2 | 1 |
| 12a1 | 90 | -0.6 | 0.6 | 0 | 0 | 0.4 | 4 | 2 | 2 |
| 12a2 | 90 | 1.3 | 1 | 0 | 0 | 0.4 | 4 | 2 | 1 |
| 12a3 | 90 | 1.3 | 0.5 | 0 | 0 | 0.4 | 4 | 2 | 2 |
| 12b | 0 | 0 | 0 | -2.2 | 1.6 | 4 | 0.4 | 3 | 1 |
| 12b1 | 0 | 0 | 0 | -2.2 | 1.2 | 4 | 0.4 | 3 | 2 |
| 12b2 | 0 | 0 | 0 | 0.15 | 1.3 | 4 | 0.4 | 3 | 1 |
| 12b3 | 0 | 0 | 0 | 0.15 | 0.7 | 4 | 0.4 | 3 | 2 |

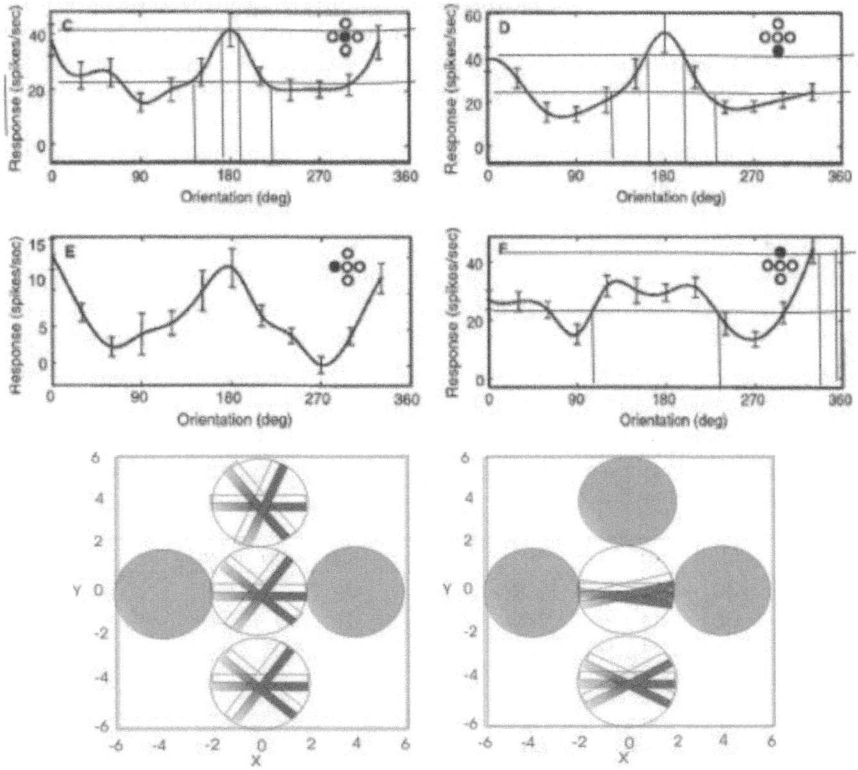


**Fig. 2.** Modified plots on the basis of [4] (upper plots), and their representation on the basis of table 2 (lower plots). C-F Curves represent V4 cell responses to different orientations of grating patches. This cell has a 6 degree dimension receptive field. Stimuli have a 2 degree dimension and are two degrees away from each other. Their relative dimensions and positions are shown in each plot. Lower plots: Gray circles indicate cell response below 20 spikes/s in the left schematic, and responses below 40 spikes/s in the right schematic. Plots on the left are related to stimulus concept 1, and plots on the right to stimulus concept 2.

We assign the narrow ($xpr_n$), medium ($xpr_m$), and wide ($xpr_w$) x position ranges as follows: $xpr_n$ *if* ($xpr$: $0 < xpr \leq 0.6$), medium $xpr_m$ *if* ($xpr$: $0.6 < xpr \leq 1.2$), wide $xpr_w$ *if* ($xpr$: $xpr > 1.2$). We assign the narrow ($ypr_n$), medium ($ypr_m$), and wide ($ypr_w$) y position range: $ypr_n$ *if* ($ypr$: $0 < ypr \leq 1.2$), medium $ypr_m$ *if* ($ypr$: $1.2 < xpr \leq 1.6$), wide $ypr_w$ *if* ($ypr$: $ypr > 1.6$). Notice that there is an asymmetry in the cell responses for the bar position along the horizontal and the vertical axis (Fig. 1).

Our results from the two-bar study can be presented as the following rules:

**Decision rules:**

**DR1:** $o_{90} \wedge xpr_n \wedge (xp_{-0.6} \vee xp_{1.3}) \wedge xs_{0.4} \wedge ys_4 \rightarrow r_2$

**DR2:** $o_0 \wedge ypr_n \wedge (yp_{-2.2} \vee yp_{0.15}) \wedge xs_4 \wedge ys_{0.4} \rightarrow r_2$

**DR3:** $o_{90} \wedge xpr_m \wedge (xp_{-0.6} \vee xp_{1.3}) \wedge xs_{0.4} \wedge ys_4 \rightarrow r_1$

**DR4:** $o_0 \wedge ypr_m \wedge (yp_{-2.2} \vee yp_{0.15}) \wedge xs_4 \wedge ys_{0.4} \rightarrow r_1$

**DR5:** $(o_{90} \wedge xpr_w) \vee (o_0 \wedge ypr_w) \rightarrow r_0$

These decision rules can be interpreted as follows: the narrow vertical or narrow horizontal bar evokes strong response in certain positions, medium size bars evoke medium responses in certain positions, and wide horizontal or vertical bars evoke no responses. We say that such a cell is tuned to narrow vertical and narrow horizontal bars.

The decision table (Table 2) describes properties of stimuli placed in four positions when the stimulus orientation varied (Fig. 2 c, d, e, f: cells 3c* to 3e) and when the stimulus spatial frequency varied (from Fig. 5 in [4], cells 5a to 5c*) as a function of response strength. This table is converted into two schematics (lower part of Fig. 2), which show areas of cell responses related to category 1 (left part) and to category 2 (right part). Gray areas are related to the subfields where responses were below threshold for the concept 1 (left schematic) or concept 2 (right schematic) stimuli. White and black bars show schematically the range of possible bar orientations which give response concept 1 or 2 in each subfield.

**Table 2.** Decision table for one cell shown in Fig. 3. Attributes xpr, ypr, s are constant and are not presented in the table.

| Cell | o | ob | sf | sfb | xp | yp | r |
|---|---|---|---|---|---|---|---|
| 3c | 172 | 105 | 2 | 0 | 0 | 0 | 1 |
| 3c1 | 10 | 140 | 2 | 0 | 0 | 0 | 1 |
| 3c2 | 180 | 20 | 2 | 0 | 0 | 0 | 2 |
| 3d | 172 | 105 | 2 | 0 | 0 | -2 | 1 |
| 3d1 | 5 | 100 | 2 | 0 | 0 | -2 | 1 |
| 3d2 | 180 | 50 | 2 | 0 | 0 | -2 | 2 |
| 3e | 180 | 0 | 2 | 0 | -2 | 0 | 0 |
| 3f | 170 | 100 | 2 | 0 | 0 | 2 | 1 |
| 3f1 | 10 | 140 | 2 | 0 | 0 | 2 | 1 |
| 3f2 | 333 | 16 | 2 | 0 | 0 | 2 | 2 |
| 5a | 180 | 0 | 2.3 | 2.6 | 0 | -2 | 1 |
| 5b | 180 | 0 | 2.5 | 3 | 0 | 2 | 1 |
| 5c | 180 | 0 | 2.45 | 2.9 | 0 | 0 | 1 |
| 5c1 | 180 | 0 | 2.3 | 1.8 | 0 | 0 | 2 |

We assign the narrow ($ob_n$), medium ($ob_m$), and wide ($ob_w$) orientation bandwidth as follows: $ob_n$ *if* ($ob$: $0<ob\leq 50$), medium $ob_m$ *if* ($ob$: $50<ob<100$), wide $ob_w$ *if* ($ob$: $ob\geq 100$). We assign the narrow ($sfb_n$), medium ($sfb_m$), and wide ($sfb_w$) spatial frequency bandwidth: $sfb_n$ *if* ($sfb$: $0<sfb\leq 2$), medium $sfb_m$ *if* ($sfb$: $2<sfb\leq 2.5$), wide $sfb_w$ *if* ($sfb$: $sfb>2.5$ ).

Our results from the separate subfields stimulation study can be presented as the following rules:

**Decision rules:** **DR6:** $ob_n \wedge (o_{180} \vee o_{333}) \wedge xp_0 \wedge (yp_{-2} \vee yp_0 \vee yp_2) \rightarrow r_2$,
**DR7:** $ob_w \wedge xp_0 \wedge (yp_{-2} \vee yp_0 \vee yp_2) \rightarrow r_1$,
**DR8:** $sfb_n \wedge xp_0 \wedge yp_0 \rightarrow r_2$,
**DR9:** $sfb_w \wedge xp_0 \wedge (yp_{-2} \vee yp_0 \vee yp_2) \rightarrow r_1$.

These decision rules can be interpreted as follows: narrowly oriented discs in the horizontal middle of the receptive field but at different vertical positions evoke strong responses. Similarly, a narrowly tuned disc in spatial frequency in the middle of the receptive field evokes strong cell responses. Stimuli with wide bandwidths of orientations or spatial frequencies in similar positions evoke medium cell responses. We say that such a cell is tuned to vertical discs with narrow orientations and narrow spatial frequencies.

Notice that Figs 2 and 4 show possible configurations of the optimal stimulus. However, they do not take into account interactions between several stimuli, when more than one subfield is stimulated. Therefore in addition we should take into account interactions between effects of different stimuli:

**Subfield Interaction Rules:**

**SIR1:** facilitation when stimulus consists of multiple bars with small distances (0.5-1deg) between them, and inhibition when distance between bars is 1.5 -2 deg.

**SIR2:** inhibition when stimulus consists of multiple similar discs with distance between them ranging from 0 deg (touching) to 3 deg.

**SIR3:** Center-surround interaction, which is described below in detail.

We will concentrate on the center-surround interaction **SIR3**. We will make a decision table for nine different cells tested with discs or annuli (Pollen et al. [4] Fig. 10). If the center was stimulated with a stimulus different from that in the surround then the surround inhibitory mechanism was weak (Fig. 9B in [4]). In order to compare different cells, we have normalized their optimal orientation and denoted it as 1, and removed them from the table.

We assign the spatial frequency: low ($sf_l$), medium ($sf_m$), and high ($sf_h$) as follows: $sf_l$ *if* $(sf: 0<sf\leq 1)$, medium $sf_m$ *if* $(sf: 1 <sf\leq 4)$, wide $sf_h$ *if* $(sf: sf>4)$. On the basis of this definition we calculate for each row in Table 3 the spatial frequency range by taking into account the spatial frequency bandwidth (*sfb*) e. g. cell 107: *sf*: 0.375 – 0.657 c/deg which means $sf_l$, 107b: *sf*: 0.25 – 3.95 c/deg which means that this cell gives response *r*2 to the stimulus with frequencies $sf_l$ and $sf_m$ , etc. Therefore we have to split case 107a to 107al and 107am, 108a to 108al and 108am, and 108b to 108bl, 108bm, 108bh.

Stimuli used in these experiments can be placed in the following six categories:

$Y_o = |sf_l\, xo_7\, xi_0\, s_4| = \{101, 105\};$
$Y_1 = |sf_l\, xo_7\, xi_2\, s_5| = \{101a, 105a\};$
$Y_2 = |sf_l\, xo_8\, xi_0\, s_4| = \{102, 104\};$
$Y_3 = |sf_l\, xo_8\, xi_3\, s_5| = \{102a, 104a\};$
$Y_4 = |sf_l\, xo_6\, xi_0\, s_4| = \{103, 106, 107, 108, 20a, 20b\};$
$Y_5 = |sf_l\, xo_6\, xi_2\, s_5| = \{103a, 106a, 107al, 108bl\};$
$Y_6 = |sf_l\, xo_4\, xi_0\, s_4| = \{108al\};$
$Y_7 = |sf_m\, xo_6\, xi_2\, s_5| = \{107am, 108bm\};$
$Y_8 = |sf_m\, xo_4\, xi_0\, s_4| = \{107b, 108am\};$
$Y_9 = |sf_h\, xo_6\, xi_2\, s_5| = \{108bh\}.$

These are equivalence classes for stimulus attributes, which means that in each class they are indiscernible *IND*(*B*). We have normalized orientation bandwidth to 0 in {20a, 20b} and spatial frequency bandwidth to 0 in cases {107, 107a, 108a, 108b}.

**Table 3.** Decision table for eight cells comparing the center-surround interaction. All stimuli were concentric discs or annuli with xo –outer diameter, xi – inner diameter. All stimuli were localized around the middle of the receptive field, so that xp = yp = xpr = ypr = 0 were constant and we did not put them in the table.

| Cell | sf | sfb | xo | xi | s | r |
|---|---|---|---|---|---|---|
| 101 | 0.5 | 0 | 7 | 0 | 4 | 0 |
| 101a | 0.5 | 0 | 7 | 2 | 5 | 1 |
| 102 | 0.5 | 0 | 8 | 0 | 4 | 0 |
| 102a | 0.5 | 0 | 8 | 3 | 5 | 0 |
| 103 | 0.5 | 0 | 6 | 0 | 4 | 0 |
| 103a | 0.5 | 0 | 6 | 2 | 5 | 1 |
| 104 | 0.5 | 0 | 8 | 0 | 4 | 0 |
| 104a | 0.5 | 0 | 8 | 3 | 5 | 2 |
| 105 | 0.5 | 0 | 7 | 0 | 4 | 0 |
| 105a | 0.5 | 0 | 7 | 2 | 5 | 1 |
| 106 | 0.5 | 0 | 6 | 0 | 4 | 1 |
| 106a | 0.5 | 0 | 6 | 2 | 5 | 2 |
| 107 | 0.5 | 0.25 | 6 | 0 | 4 | 2 |
| 107a | 2.1 | 3.8 | 6 | 2 | 5 | 2 |
| 107b | 2 | 0 | 4 | 0 | 4 | 1 |
| 108 | 0.5 | 0 | 6 | 0 | 4 | 1 |
| 108a | 0.9 | 0.9 | 4 | 0 | 4 | 2 |
| 108b | 5 | 9 | 6 | 2 | 5 | 2 |
| 20a | 0.5 | 0 | 6 | 0 | 4 | 1 |
| 20b | 0.5 | 0 | 6 | 0 | 4 | 2 |

There are three ranges of responses, denoted as $r_o$, $r_1$, $r_2$:

$| r_o | =$ {*101, 102, 102a, 103, 104, 105*};
$| r_1 | =$ {*101a, 103a, 105a, 107b, 108, 20a*};
$| r_2 | =$ {*104a, 106a, 107, 107al, 107am, 108al, 108am, 108bl, 108bm, 108bh, 20b*};

which are denoted as $X_o$, $X_1$, $X_2$.

We want to find out whether equivalence classes of the relation *IND*{*r*} form the union of some equivalence relation *IND*(*B*), or whether $B \Rightarrow \{r\}$.

We will calculate the lower and upper approximation [3] of the basic concepts in terms of stimulus basic categories:

$\underline{B} X_o = Y_o \cup Y_2 =$ {*101, 105, 102, 104*},

$\overline{B} X_o = Y_o \cup Y_2 \cup Y_3 \cup Y_4 =$ {*101, 105, 102, 104, 102a, 104a, 103, 106, 107, 108, 20a, 20b*},

$\underline{B} X_1 = Y_1 = \{101a, 105a\}$,

$\overline{B} X_1 = Y_1 \cup Y_5 \cup Y_6 \cup Y_4 = \{101a, 105a, 103a, 107al, 108b, 106a, 20b, 107b, 108a, 103, 107, 106, 108, 20a\}$,

$\underline{B} X_2 = Y_7 \cup Y_9 = \{107am, 108bm, 108bh\}$,

$\overline{B} X_2 = Y_7 \cup Y_9 \cup Y_8 \cup Y_3 \cup Y_4 \cup Y_5 \cup Y_6 = \{107am, 108bm, 108bh, 107b, 108am, 102a, 104a, 103a, 107a, 108bl, 106a, 20b, 103, 107, 106, 108, 20a, 108al\}$

Concept 0, 1, and 2 are roughly *B-definable*, which means that only with some approximation we found that stimuli that do not evoke response, or evoke weak or strong response in the area V4 cells. Certainly such stimulus as $Y_0$ or $Y_2$ does not evoke a response in all our examples, in cells 101, 105, 102, 104. Also stimulus $Y_1$ evokes a weak response in all our examples: 101a, 105a. We are interested in stimuli, which evoke a strong response because they are specific for the area V4 cells. We have found two such stimuli: $Y_7$ and $Y_9$. While other stimuli such as $Y_3$, $Y_4$ evoke no response or weak or strong responses in our data.

We can find quality [3] of our experiments by comparing properly classified stimuli $POS_B(r)=\{101, 101a, 105, 105a, 102, 104, 107, 109\}$ to all stimuli and responses:

$$\gamma\{r\} = |\{101, 101a, 105, 105a, 102, 104, 107, 109\}|/|\{101, 101a, ..., 20a, 20b\}| = 0.3.$$

We can also ask what percentage of cells we have fully classified. We obtain consistent responses from 2 of 9 cells, which means that $\gamma\{cells\} = 0.22$. This is related to the fact that for some cells we have tested more than two stimuli. What is also important from an electrophysiological point of view that there are negative cases. There are many negative instances for the stimuli of the concept 0, which means that in most instances cells in this brain area respond to our stimuli; even if our concepts are still only roughly defined.

Our results from the center-surround interaction study can be presented as the following rules:

**Decision rules:** **DR10:** $sf_l \wedge x_{o7} \wedge x_{i2} \wedge s_5 \rightarrow r_1$,
**DR11:** $sf_l \wedge x_{o7} \wedge x_{i0} \wedge s_4 \rightarrow r_0$,
**DR12:** $sf_l \wedge x_{o8} \wedge x_{i0} \wedge s_4 \rightarrow r_0$,
**DR13:** $(sf_m \vee sf_l) \wedge xo_6 \wedge xi_2 \wedge s_5 \rightarrow r_2$.

They can be interpreted as the statement that for the stimuli modulated with a low spatial frequency grating, a large annulus (s5) evokes weak response, but a large disc (s4) evokes no response. However, little bit smaller annulus but containing medium or high spatial frequency grating, evokes strong responses. It is unexpected that certain stimuli evoke inconsistent responses in different cells, for example:

103: $sf_l \wedge x_{o6} \wedge x_{i0} \wedge s_4 \rightarrow r_0$,
106: $sf_l \wedge x_{o6} \wedge x_{i0} \wedge s_4 \rightarrow r_1$,
107: $sf_l \wedge x_{o6} \wedge x_{i0} \wedge s_4 \rightarrow r_2$,
103a: $sf_l \wedge x_{o6} \wedge x_{i2} \wedge s_5 \rightarrow r_1$,
106a: $sf_l \wedge x_{o6} \wedge x_{i2} \wedge s_5 \rightarrow r_2$.

The same disc with not very large dimension containing a low spatial frequency grating can evoke no response (103), a small response (106), or a strong response (107).

## 4 Discussion

The purpose of our study has been to determine how different categories of stimuli and particular concepts, as related to the responses of a single cell. We test our theory on a set of data from David et al. [7], shown in Fig.3.

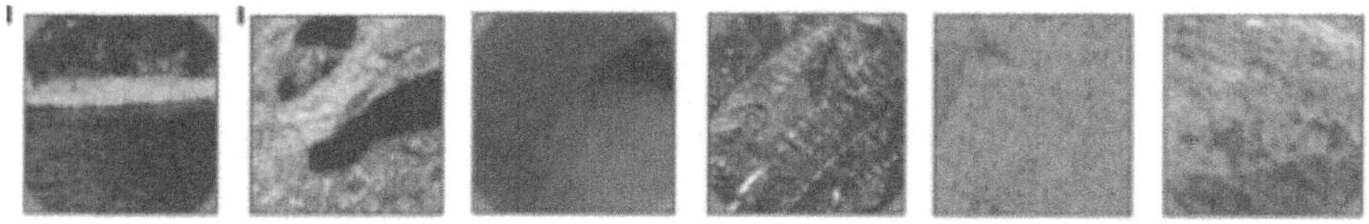

**Fig. 3.** In their paper David et al. [7] stimulated V4 neurons (medium size of their receptive fields was 10.2 deg) with natural images. Several examples of their images are shown above. We have divided responses of cells into three concept categories. The two images on the left represent cells, which give strong responses related to stimulus concept 2. The two images in the middle evoke responses above 20 spikes/s; they are related to stimulus concept 1. Two images on the right gave very weak responses; they are related to stimulus concept 0.

We assume that the stimulus configuration in the first image on the left is similar to that proposed in Fig. 1; the dominating object in the stimulus is a horizontal narrow bar, so that we can apply the decision rule *DR2*. The second image from the left can be divided into central and surround parts. The stimulus in the central disc is similar to that from Fig. 2: narrow in orientation and in spatial frequency bandwidth (*DR6, DR8*). Stimuli in the upper and right parts of the surround have a common orientation and larger orientation bandwidth in comparison with the center (Fig. 3). These differences make weak interactions between discs as in *SIR2* or between center-surround as in *SIR3*. This means that these images will be related to stimulus concept 2. Two middle images show significant differences between their central and surround parts. Assuming that the center and surround are tuned to a feature of the object in the images, we believe that these images would also give significant responses. However, in the left image in the middle part of Fig. 5, stimuli in the center and in the surround consist of many orientations ($ob_w$) and many spatial frequencies ($sfb_w$); therefore rules *DR7*, and *DR9* can be applied. The right middle image shows an interesting stimulus but with a narrow range of orientations but a wider range of spatial frequencies. There are small but significant differences between center and surround parts of the image. This image can be seen as a group of medium x position range bars (bars of medium width), which means using the *DR3* decision rule. Even if this image shows differences between its central and surround parts, they have also many similar features like orientation or spatial frequencies. Therefore even if the center and surround alone would give strong cell responses, their interactions will be inhibitory (rule *SIR3*). In consequence, both middle images are related to stimulus concept 1. In the two images on the right there is no significant difference between the stimulus in the center and the surround. Therefore the response will be similar to that obtained when a single disc covers the whole receptive field: *DR11, DR12*. In most cells such stimuli class will be equivalent to a stimulus concept 0.

In the following paragraph we discuss the meaning our analysis in a context of the psychophysical experiments related the human visual system.

The conventional view based mostly on the psychophysical experiments is that the perception proceeds along at least two stages:

1. low-level parallel visual processing, largely unconscious, rapid, global high efficiency categorization of items and events,
2. high-level serial visual processing, attentional stage, identification, integration, and consolidation of items with a conscious report.

However, recent experiments of Thorpe et al. [8] found that primates (human and non-human) are capable of rapid and accurate categorization of the briefly flashed natural images. Human observers are very good at deciding whether a novel image contains an animal. The underlying visual processing reflecting the decision that there was a target present are under 150ms [8]. These finding seems to be in contradiction to the classical view that only the "basic features" can be processed in parallel [9]. Certainly natural scenes contain more complex stimuli than "simple" geometric shapes. Treisman [9] showed that instances of disjunctive set of at least four basic features can be detected through parallel processing. Other labs gave evidence for parallel detection of more complex features, as for example shape from shading [10], or features of intermediate complexity that can be learned by experience [11].

It seems that the conventional two-stage perception processing model needs correction, because to the "basic feature" we have to add a set of unknown intermediate features. We propose that at least some intermediate features are related to the receptive field properties of the area V4. By using multi-valued categorization of the V4 neuron responses we have differentiated between the bottom-up information (hypothesis testing) that are related to the sensory input, and predictions, some of which can be learned and are related to the positive feedback from the higher areas. If prediction is in agreement with the hypothesis object classification will change from the category 1 to the category 2. We suggest that such decisions can be made very effectively during pre-attentive, parallel processing in the multiple visual areas. In addition, we found that the decision rules of different neurons could be inconsistent. This inconsistency could help to process different aspects of the complex object properties. The principle is similar to that observed in the primary visual cortex orientation tuning cells. Neurons in V1 with overlapping receptive fields show different preferred orientations. It is assumed that this help to extract the local orientations in different parts of the object. However, it is still not clear which cell will dominate if several cells with overlapping receptive fields are tuned to different attributes of the stimulus.

*In summary*, we have showed that using rough set theory we can divide stimulus attributes in relationships to neuronal responses into different concepts. Even if most of our concepts were very rough, they determine rules on whose basis we can predict neural responses to new, natural images.

**Acknowledgements.** I thank M. Kon for useful discussion and comments.

## References

1. Zadeh, L.A.: Toward a Perception-based Theory of Probabilistic Reasoning with Imprecise Probabilities. Journal of Statistical Planning and Inference 105 (2002) 233 -264
2. Przybyszewski, A.W., Linsay, P.S., Gaudiano, P., Wilson, C.: Basic Difference Between Brain and Computer: Integration of Asynchronous Processes Implemented as Hardware Model of the Retina. IEEE Trans Neural Networks 18 (2007) 70-85
3. Pawlak, Z.: Rough Sets - Theoretical Aspects of Reasoning about Data. Kluwer Academic Publishers,, Boston, London, Dordrecht. (1991)
4. Pollen, D.A., Przybyszewski, A.W., Rubin, M.A., Foote, W.: Spatial Receptive Field Organization of Macaque V4 Neurons. Cereb Cortex 12 (2002) 601-16
5. Przybyszewski, A.W., Gaska, J.P., Foote, W., Pollen, D.A.: Striate Cortex Increases Contrast Gain of Macaque LGN Neurons. Vis Neurosci. 17 (2000) 485-94
6. Przybyszewski, A.W., Kon, M.A.: Synchronization-based Model of the Visual System Supports Recognition.. Program No. 718.11. 2003 Abstract Viewer/Itinerary Planner. Washington, DC: Society for Neuroscience, (2003) Online.
7. David, S.V., Hayden, B.Y., Gallant, J.L.: Spectral Receptive Field Properties Explain Shape Selectivity in Area V4. J Neurophysiol. 96 (2006) 3492-505
8. Thorpe, S., Faze, D., Merlot, C.: Speed of Processing in the Human Visual System Nature 381 (1996) 520-522
9. Treisman, A.: Features and Objects: The Fourteenth Bartlett Memorial Lecture. Q J Exp Psychol A. 40 (1988) 201-37
10. Ramachandran, V.S.: Perception of Shape From Shading. Nature 331 (1988) 163-6.
11. Ullman, S, Vidal-Naquet, M., Sali, E.: Visual Features of Intermediate Complexity and Their Use in Classification Nature Neuroscience 5 (2002) 682-687

# Stability Analysis for Floating Structures Using T-S Fuzzy Control

Chen-Yuan Chen[1], Cheng-Wu Chen[2,*], Ken Yeh[3], and Chun-Pin Tseng[4]

[1] Department of Management Information System, Yung-Ta institute of Technology and Commerce, Pingtung, Taiwan
[2] Department of Logistics Management, Shu-Te University, Yen Chau, Kaohsiung, Taiwan, R.O.C
cwchen@mail.stu.edu.tw
[3] Department of Civil Engineering, De-Lin Institute of Technology, Tucheng, Taipei, Taiwan, R.O.C
[4] Department of Civil Engineering, National Central University, Jhongli City, Taoyuan County 32001, Taiwan

**Abstract.** This study constructs a mathematical model of an ocean environment in which wave-induced flow fields cause structural surge motion. The solutions corresponding to the mathematical model are derived analytically. In this study, a fuzzy control technique is developed to mitigate structural vibration. The Takagi-Sugeno (T-S) fuzzy model is employed to approximate the oceanic structure and a parallel distributed compensation (PDC) scheme is utilized in the controller design procedure to reduce structural response.

**Keywords:** fuzzy model, floating structure.

## 1 Introduction

In recent decades, cylindrical piles beneath coastal and marine structures (e.g., breakwaters, wharfs, quays, lighthouses, artificial islands and platforms) have been used extensively for academic research and petroleum extraction. Emerged and submerged porous cylindrical structures are used by fisheries for offshore and coastal aquaculture projects in relatively shallow and intermediate water depths. The primary reason for using cylindrical piles is to reduce the interaction between sea waves and marine structures. Additionally, many innovative floating offshore structures have been proposed for cost savings for deep-water offshore oil and gas exploration. To minimize wave-induced motion, the natural frequency of these proposed offshore structures is designed far from peak frequency of the force power spectra. A spar platform is an offshore floating structure utilized in deep water for drilling, production, processing, storage, and offloading of ocean deposits. A spar platform consists of a vertical cylinder that floats vertically in water. The structure floats extremely deep in water. Consequently, the force of wave action at the surface is minimized by the counter balance effect of the structure's weight. The semi-submerged tension leg platform

* Corresponding author.

D.-S. Huang, L. Heutte, and M. Loog (Eds.): ICIC 2007, LNAI 4682, pp. 750–758, 2007.

(TLP) system can be comprised of different materials with a top of working floating body. In the design of cylindrical piles, the effectiveness of various platform parameters is considered when attempting to reduce vibration in an offshore floating platform system. These platform parameters, which affect platform resonant frequency and mitigate amplitude, include the platform mass, floating barrel diameter, platform draft, and platform dimensions. Application of a tuned liquid column damper significantly reduces the peak value of the platform. The effectiveness of a tuned liquid column damper is significant when the ratio of its width to length is high [1].

Simplified models for the surge motion of an impermeable marine structure with linearly elastic, pre-tensioned legs have been developed [2],[3],[4] for investigating wave energy dissipation. The response of a floating body subjected to wave force has recently been studied numerically. This study presents a novel approach for analyzing the dynamics of marine platforms. Since Zadeh [5] first proposed a linguistic approach to simulate human thought processes and judgment, many studies have explored this field further ([6],[7],[8] and the references therein). However, all these studies lack mathematical theories and systematic designs. In 1985, Takagi and Sugeno [9] proposed a fuzzy inference system called the Takagi-Sugeno (T-S) fuzzy model. As the concepts underlying this fuzzy model are natural and simple, many applications have been proposed (for example, [10],[11],[12] and the references therein). The T-S-type fuzzy models combine linguistic rule descriptions with traditional functional descriptions of system operation. This local description of system operation, which is linear description, is relatively easy to identify. Any complex nonlinear system surfaces can be represented by a set of flat linear segments and these segments can be described using one rule in the T-S fuzzy model. Due to the design of linear controllers in a well-developed procedure, the T-S model is suitable for describing nonlinear controllers, especially for describing working points (operating sectors). Based on the use of T-S fuzzy models, the parallel distributed compensation (PDC). The fuzzy controllers created are based on the use of T-S fuzzy models. The PDC approach using a model-based design procedure was first proposed by Sugeno and Kang [13]. The PDC is a technique for designing a fuzzy controller based on a given T-S fuzzy model, meaning that each control rule is constructed from the corresponding rule in a T-S fuzzy model. Therefore, real systems, such as mechanical systems, chaotic systems, and resonant systems, can been represented by T-S fuzzy models, and PDC controllers are designed to achieve the stability and stabilization. Each fuzzy implication is expressed by a linear system model such that linear feedback control can be utilized, as in the case of feedback stabilization. Because a linear feedback control is designed for each local linear model, the resulting overall nonlinear controller is a fuzzy blending of each individual linear controller. This so-called T-S fuzzy model and PDC control technique were analyzed in [14],[15]. Analytical results indicate that the T-S fuzzy model and PDC technique are suited to stability analysis of control systems. Moreover, stability analysis and control design problems can be reduced to linear matrix inequality (LMI) problems ([16],[17] and the references therein). Controller design and stability problems can be transformed into LMI problems. To represent the nonlinearity exactly, this study used numerous linear segments to derive a fuzzy model without simplifying the original nonlinear model. Using the local approximation idea, Tanaka and Wang [18] proved that the T-S fuzzy model and PDC technique are approximate nonlinear terms by judiciously choosing linear terms. This procedure results in a reduction in the number of linear model rules. The number of linear model rules is directly correlated with the complexity of stability analysis and solving LMI conditions because the rule number is generally the combination of the fuzzy model and fuzzy control rules. Due to the local linear structure of the representation of nonlinear systems, the stability of plant and controlled systems, which is described in the T-S form, is easily proven.

## 2 Mathematical Formulation

### 2.1 Initial Boundary Value Problem of Fluid–Structure Interaction

A stationary cubical element is shown in Fig. 1a. The mass inside a fixed surface, bounding the closed volume will increase if mass flows into the volume and decrease if it flows out. The inflow and outflow process is shown in Fig.1b. For incompressible fluids the fluid density is a constant throughout the flow field. Thus [19],

$$\nabla \cdot \vec{V} = 0 \tag{1}$$

the fluid considered is inviscid, and the flow is assumed to move from rest such that it is irrotational. Therefore, fluid velocity $\mathrm{V}$ can be described by the gradient of velocity potential $\Phi(x,z,t)$ in the fluid domain, i.e., $\mathrm{V} = \nabla\, \Phi(x,z,t)$. The governing equation for this problem satisfies the Laplace equation for velocity potential, i.e.

$$\nabla^2 \Phi(x,z,t) = 0 \tag{2}$$

The derivations of fluid domain equations are based on the following assumptions:

1. The fluid considered is inviscid.
2. The flow is incompressible and irrotational, and surface tension effects can be neglected.
3. A scalar velocity potential describes the flow, satisfying the Laplace equation within the fluid domain.
4. No breaking waves occur on the sea surface.

Consider a wave-induced flow field system in which a Cartesian coordinate system $oxz$ is employed. As shown in the sketch of a 2D numerical wave flume, plane $z$=0 coincides with the undisturbed still water level, and the $z$-axis is directed vertically upward. The vertical elevation of any point on the free surface can be defined by the function $z=\eta(x,y,t)$, in which surface tension is negligible. As depicted in Fig. 2, $-\infty < x < -b$, the total velocity potential $\Phi_I$ in Region I consists of incident waves $\Phi_i$, scattered waves $\Phi_{IS}$, and motion radiation waves $\Phi_{IW}$. In Region II, $-b < x < b$, and in Region III, $b < x < \infty$; the total velocity potentials $\Phi_{II}$ and $\Phi_{III}$ consist of both scattered, $\Phi_{IIS}$ and $\Phi_{IIIS}$, and radiated waves, $\Phi_{IIW}$, $\Phi_{IIIW}$. Subscript s denotes the scattering problem, and subscript w denotes the wave-maker (i.e., primitive radiation) problem induced by platform surge motion. Displacement of the surge motion with an unknown amplitude $S$ is given by $X = Se^{-i\sigma t}$, and platform deformation on the $x$-axis is defined as $S$.

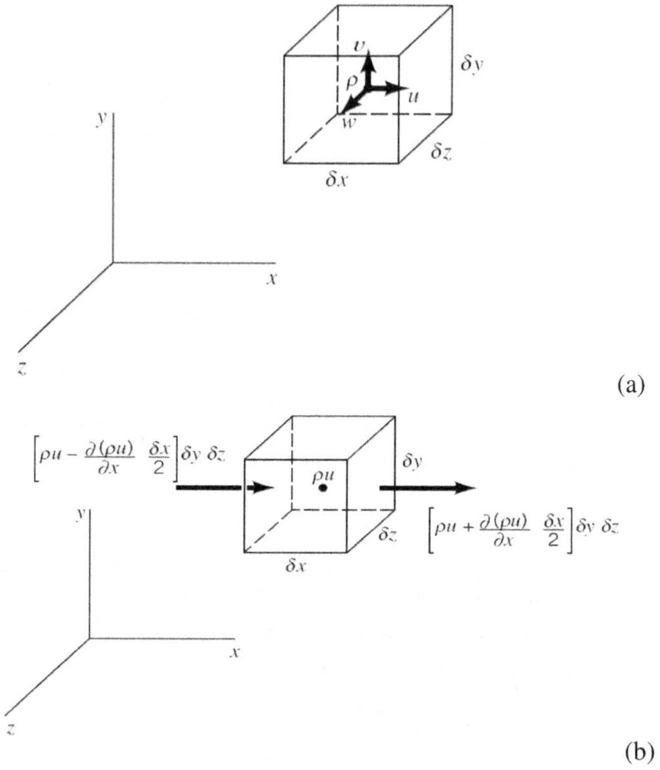


**Fig. 1.** A differential element for the development of conservation of mass equation

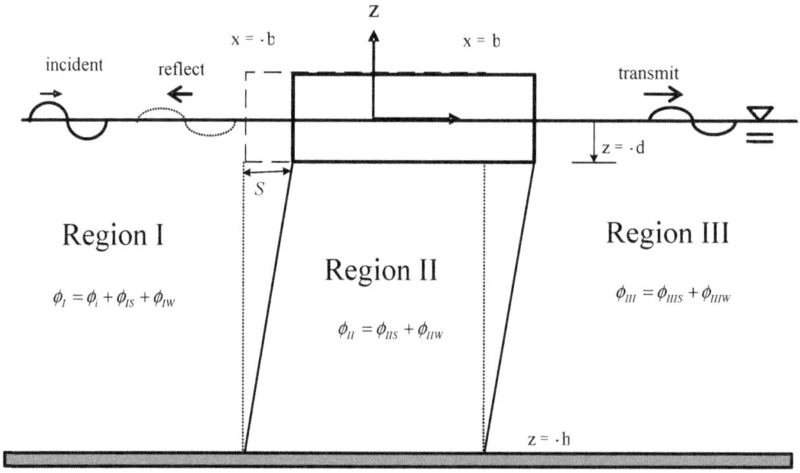


**Fig. 2.** Definition sketch of a deformable tension leg platform subjected to wave force

No flow across an interface is assumed for any fluid interface, indicating that fluid particles can only move in a direction tangential to a fluid interface. The required kinematic boundary conditions (see Appendix A) are as follows:

$$\frac{\partial \eta}{\partial t} = \frac{\partial \Phi}{\partial z} - \frac{a}{\lambda}\frac{\partial \Phi}{\partial x}\frac{\partial \eta}{\partial x} \quad \text{on the surface} \tag{3}$$

$$\frac{\partial \Phi}{\partial n} = U_n \quad \text{on the rigid boundaries} \tag{4}$$

where $a << \lambda$ for small-amplitude waves can neglect the non-linear convective term, and $n$ is the outward normal to the boundary. Furthermore, applying the linearized condition at $z$=0 instead of $z= \eta$ results in the kinematic boundary condition $\partial \eta / \partial t = w$, suggesting that the vertical velocity component of the fluid at the interface must equal the interface velocity. When the rigid boundaries are stationary on the seabed, the normal velocity component $U_n$ becomes zero.

The dynamic boundary condition (see Appendix B) on the free surface is utilized to calculate the dynamic pressure and horizontal fluid velocity. The dynamic conditions on the free surface are derived based on the conservation of linear momentum. Briefly, the discontinuity in the normal stress is proportional to the mean curvature of the free surface caused by surface tension.

$$\frac{\partial \Phi}{\partial t} = C - g\eta - \frac{1}{2}\left[\left(\frac{\partial \Phi}{\partial x}\right)^2 + \left(\frac{\partial \Phi}{\partial z}\right)^2\right] - \frac{P}{\rho} \tag{5}$$

where $C$ is the Bernoulli constant. When atmospheric pressure is taken as zero, term $P$ will also equal zero. In free-surface problems, nonlinearity in the potential flow problem is only derived from free-surface boundary conditions when inviscid and incompressible fluid and irrotational flow assumptions are made. For small amplitude waves, the high order terms in the free surface boundary conditions given by Eqs. (2-3) and (2-5) are ignored, and the resulting conditions are applied at the undisturbed water level $z$=0 with $C$=0; the following expressions are obtained.

$$\eta = -\frac{1}{g}\frac{\partial \Phi}{\partial t} \tag{6}$$

The Sommerfeld radiation condition is utilized as an outflow boundary condition with no interference inside the computational domain.

$$\lim_{x\to\pm\infty}\left[\frac{\partial \phi_{IS/IIIS}}{\partial x} \pm \frac{1}{c}\frac{\partial \phi_{IS/IIIS}}{\partial t}\right] = 0 \tag{7}$$

### 2.2 Kinematic Boundary Condition

Various kinematic boundary conditions are required at an interface. No flow across an interface is assumed at any fluid interface. In other words, fluid particles can only move tangentially to a fluid interface. To express this condition mathematically, we must consider the interface between the two fluids in more detail. The interface location, $z_j$ is defined by a mathematical expression in the form

$$F(x, z, t) = z_j - \eta(x, t) = 0 \tag{8}$$

Since the interface itself is also a streamline, hence $F = 0$. Writing this Lagrangian reference in the Eulerian frame gives

$$\left.\frac{\partial F}{\partial t} + u \cdot \nabla F\right|_{F(x,z,t)=0} = 0, \text{ or } \frac{\partial F}{\partial t} + u \cdot \nabla F = 0, \text{ at } z = \eta \tag{9}$$

To simplify this equation we may recast it in a non-dimensional form using several transformation variables,

$$x = \lambda x^* \;, \; z = az^* \;, \; u = U_0 u^* \;, \; w = U_0 w^* \;, t = \frac{a}{U_0} t^*$$

where $\lambda$ is a wavelength, $a$ the wave amplitude, $U_0$ a characteristic fluid velocity and the variables denoted by asterisk (*) are the new non-dimensional variables.

Denoting $F$ by, equation (2-6) may be converted into non-dimensional form,

$$\frac{\partial \eta}{\partial t} + \frac{a}{\lambda} u \frac{\partial \eta}{\partial x} = w \tag{10}$$

where asterisks for non-dimensional variables are dropped, to simplify the expressions. For small-amplitude waves, $a << \lambda$, we can neglect the non-linear convective term. Further, applying the linearized condition at $z = 0$ instead of $z = \eta$, leads to the kinematic boundary condition $\partial \eta / \partial t = w$. This suggests the vertical velocity component of the fluid at the interface must equal to the velocity of the interface. Since we have vertical velocities in both layers, this condition can be applied twice giving

$$\frac{\partial \phi}{\partial z} = \frac{\partial \eta}{\partial t} \tag{11}$$

## 2.2 Dynamic Boundary Condition

We must supply the initial conditions to satisfy the principle of conserving momentum. This provides a dynamic boundary condition at the fluid interface: the normal stress of the fluid is continuous across the interface. For an inviscid fluid, this implies the pressure is continuous on the interface. Taking the inviscid momentum equation in the vertical direction, we have

$$\frac{Dw}{Dt} = -\frac{1}{\rho}\frac{\partial p}{\partial z} - g \tag{12}$$

where $D/Dt$ is the material derivative operator, $\rho$ is fluid density, $p$ is fluid pressure, $g$ is the acceleration due to gravity. Substituting the velocity potential and linearizing the convective terms of the material derivative renders

$$\frac{D}{Dt}\left(\frac{\partial \phi}{\partial z}\right) = -\frac{1}{\rho}\frac{\partial p}{\partial z} - g \tag{13}$$

Upon integrating equation (2-13) with respect to $z$ and after dividing it by $\gamma$, we have

$$\frac{1}{g}\frac{\partial \phi}{\partial t} + \frac{p}{\gamma} + z = C(t) \tag{14}$$

where $C(t)$ is a function of the time. This equation is also known as the unsteady Bernoulli equation, in which $C(t)$ is a constant along a streamline.

## 3 Analytical Solutions

### 3.1 Vibration Radiation Problem

The momentum equation can be obtained from the floating structure motion, which is derived from Newton's second law. Assume the momentum equation of a TLP system is controlled by actuators that can be characterized by the following differential equation:

$$M\ddot{\overline{X}}(t) = \overline{B}U(t) - M\overline{r}\phi(t) \tag{15}$$

where $\overline{X}(t) = [\overline{x}_1(t), \overline{x}_2(t) \cdots \overline{x}_n(t)] \in R^n$ is an n-vector; $\ddot{\overline{X}}(t)$, $\dot{\overline{X}}(t)$, $\overline{X}(t)$ are acceleration, velocity, and displacement vectors, respectively; $\overline{B}$ is a $(n \times m)$ matrix denoting the locations of m control forces; $U$(t) corresponds to the actuator forces (e.g., generated via the active tendon system or an active mass damper).

The overall closed-loop controlled system is as follows:

$$\dot{X}(t) = \sum_{i=1}^{r}\sum_{l=1}^{r} h_i(t)h_l(t)[(A_i - B_i K_l)X(t)] + E_i\phi(t) \tag{16}$$

**Theorem 1.** The equilibrium point of fuzzy control system is stable in the large if there exist a common positive definite matrix $P$ and feedback gains $K$ such that the following two inequalities are satisfied:

$$(A_i - B_i K_i)^T P + P(A_i - B_i K_i) + \frac{1}{\eta^2} P E_i E_i^T P < 0$$

and

$$\left(\frac{(A_i - B_i K_l) + (A_l - B_l K_i)}{2}\right)^T P + P\left(\frac{(A_i - B_i K_l) + (A_l - B_l K_i)}{2}\right) + \frac{1}{\eta^2} P E_i E_i^T P < 0$$

with $P = P^T > 0$, for $i < l \le r$ and $i = 1, 2, \cdots, r$

# 4 Conclusion

Since tendon length and water collapse pressure increase proportionally to water depth, tendon pipes with high resistance are critical, especially in deep-water environments. Instead of the tether drag effects used in previous studies, this study presented a novel concept of control force to stabilize a TLP system. This proposed system can improve the limitation of steel performance for maximum water depth attainable with a TLP system. Dependence of wave motion and structural surge motion was demonstrated, from which the analytical solutions demonstrate that the response of a floating structure can be calculated using certain parameters, including structural properties and wave characteristics. Subjecting a floating structure to wave force results in vibration that can be mitigated rapidly using fuzzy controllers.

**Acknowledgments.** This work is partially supported by the National Science Council of Republic of China under Grant No. NSC 95-2221-E-366-001.

# References

1. Weng, S.H.: The Dynamic Analysis and Vibration Suppression of TLP with Tuned Liquid Column Damper. Master Thesis, National Sun Yat-sen University, Taiwan, (2000)
2. Lee, C.P., Lee, J.F.: Interaction between Waves and Tension Leg Platform. ASCE Eng. Mechanics Conf. on Mechanics Computing in 1990's and Beyond, 1991.
3. Yamamoto, T., Yoshida, A., Ijima, T.: Dynamics of Elastically Moored Floating Objects. Dynamic Analysis of Offshore Structures. (1982) 106-113
4. Mei, C.C.: Numerical Methods in Water Wave Diffraction and Radiation, Ann. Rev. Fluid Mech. 10, 393, 1978.
5. Zadeh, L. A.: Fuzzy Sets. Inform. and Contr, Vol.8 (1965) 338-353
6. Chang, S.S.L., Zadeh, L.A.: On Fuzzy Mapping and Control. IEEE Trans. Syst. Man Cybern, Vol.2 (1972) 30-34
7. Kickert, W.J.M., Mamdani, E.H.: Analysis of a Fuzzy Logic Controller. Fuzzy Sets Syst, Vol.1 (1978) 29-44
8. Braae, M., Rutherford, D.A.: Theoretical and Linguistic Aspects of the Fuzzy Logic Controller. Automatica, Vol.15 (1979) 553-577
9. Takagi, T., Sugeno, M.: Fuzzy Identification of Systems And its Applications to Modeling and Control. IEEE Trans. Syst. Man Cybern, Vol.15 (1985) 116-132
10. Hsieh, T.Y., Wang, M.H.L., Chen, C.W. et al.: A new viewpoint of s-curve regression model and its application to construction management. Int. J. Artif. Intell. Tools, Vol.15 (2006) 131-142
11. Cococcioni, M., Guasqui, P., Lazzerini, B.: Identification of Takagi-Sugeno Fuzzy Systems Based on Multi-Objective Genetic Algorithms. Lect. Note Artif. Int., Vol.3849 (2006) 172-177
12. Zhang, Z.Y., Zhou, H.L., Liu, S.D. et al.: An Application of Takagi-Sugeno Fuzzy System to the Classification of Cancer Patients Based on Elemental Contents in Serum Samples. Chemometr. Intell. Lab. Syst., Vol.82 (2006) 294-299
13. Sugeno, M., Kang, G..T.: Fuzzy Modeling and Control of Multilayer Incinerator. Fuzzy Sets Syst., Vol.18 (1986) 329-346

14. Tanaka, K. Sugeno, M.: Stability Analysis and Design of Fuzzy Control Systems. Fuzzy Sets Syst., Vol.45 (1992) 135-156
15. Wang, H.O., Tanaka, K., Griffin, M.F.: Parallel Distributed Compensation of Nonlinear Systems by Tanaka-Sugeno Fuzzy Model. Proc. FUZZ IEEE/IFES'95, (1995) 531-538
16. Chen, C.W., Chiang, W.L., Hsiao, F.H.: Stability Analysis of T-S fuzzy Models for Nonlinear Multiple Time-Delay Interconnected Systems. Math. Comput. Simul., Vol.66 (2004) 523-537
17. Chen, C.W., Chiang, W.L., Tsai, C.H.: Fuzzy Lyapunov Method for Stability Conditions of Nonlinear Systems. Int. J. Artif. Intell. Tools, Vol.15 (2006) 163-171
18. Tanaka, K., Wang, H.O.: Fuzzy Control Systems Design and Analysis. John Wiley & Sons. Inc, New York (2001)
19. Munson, B.R., Young, D.F., Okiishi, T.H.: Fundamentals of Fluid Mechanics. 4th Edition. John Wiley & Sons Inc (2002) 299-308

## Notations

$a$: wave amplitude
$b$: platform width
$C$: Bernoulli constant
$d$: platform draft
$E$: Young's modulus
$g$: gravitational acceleration
$k$: wave number ($=2\pi/\lambda$)
$M$: platform mass
$S$: platform amplitude
$T$: wave period
$\Phi$: velocity potential
$\rho$: fluid density
$\lambda$: wavelength
$\sigma$: wave frequency
$\eta$: vertical elevation

# Uncertainty Measures of Roughness of Knowledge and Rough Sets in Ordered Information Systems

Wei-Hua Xu[1], Hong-zhi Yang[2], and Wen-Xiu Zhang[3]

[1] Institute for Information and System Sciences,
Xi'an Jiaotong University, Xi'an, Shaan'xi 710049, P.R. China
datongxuweihua@126.com
[2] He'nan Pingyuan University, Xinxiang 453003, P.R. China
Institute for Information and System Sciences,
Xi'an Jiaotong University, Xi'an, Shaan'xi 710049, P.R. China
yzxz@163.com
[3] Faculty of Science, Institute for Information and System Sciences,
Xi'an Jiaotong University, Xi'an, Shaan'xi 710049, P.R. China
wxzhang@mail.xjtu.edu.cn

**Abstract.** Rough set theory has been considered as a useful tool to deal with inexact, uncertain, or vague knowledge. However, in real-world, most of information systems are based on dominance relations, called ordered information systems, in stead of the classical equivalence for various factors. So, it is necessary to find a new measure to knowledge and rough set in ordered information systems. In this paper, we address uncertainty measures of roughness of knowledge and rough sets by introducing rough entropy in ordered information systems. We prove that the rough entropy of knowledge and rough set decreases monotonously as the granularity of information becomes finer, and obtain some conclusions, which is every helpful in future research works of ordered information systems.

**Keywords:** Rough set, Information systems, Dominance relation, Rough entropy, Rough degree.

## 1 Introduction

The rough set theory, proposed by Pawlak in the early 1980s[1], is an extension of the classical set theory for modeling uncertainty or imprecision information. The research has recently roused great interest in the theoretical and application fronts, such as machine learning, pattern recognition, data analysis, and so on [2-6].

In Pawlak's original rough set theory, partition or equivalence (indiscernibility relation) is a important and primitive concept. However, partition or equivalence relation, as the indiscernibility relation in Pawlak's original rough set theory, is still restrictive for many applications. To address this issue, several interesting

D.-S. Huang, L. Heutte, and M. Loog (Eds.): ICIC 2007, LNAI 4682, pp. 759–769, 2007.

and meaningful extensions to equivalence relation have been proposed in the past, such as tolerance relations [17], similarity relations [16], others [18-20]. Particularly, Greco, Matarazzo, and Slowinski[7-11] proposed an extension rough sets theory, called the dominance-based rough sets approach(DRSA) to take into account the ordering properties of attributes. This innovation is mainly based on substitution of the indiscernibility relation by a dominance relation. In DRSA condition attributes and classes are preference ordered. And many studies have been made in DRSA[12-15].

On the other hand, the concept of entropy, originally defined by Shannon in 1948 for communication theory, gives a measure of uncertainty about the structure of a system. It has been useful concept for characterizing information content in a great diversity of models and applications. Attempts have been made to use Shannon's entropy to measure uncertainty in rough set theory [21-24]. Moreover, information entropy is introduced into incomplete information systems, and a kind of new rough entropy is defined to describe the incomplete information systems and roughness of rough set. While, most of information systems are based on dominance relations, i.e., ordered information systems. Hence, consideration of the uncertain measure about entropy in ordered information systems is needed. This paper discussed the problem mainly.

In this paper, we address uncertainty measures of roughness of knowledge and rough sets by introducing rough entropy in ordered information systems. We prove that the rough entropy of knowledge and rough set decreases monotonously as the granularity of information becomes finer, and obtain some conclusions, which is every helpful in future research works of ordered information systems.

## 2 Rough Sets and Ordered Information Systems

The following recalls necessary concepts and preliminaries required in the sequel of our work. Detailed description of the theory can be found in [4,15].

The notion of information system (sometimes called data tables, attribute-value systems, knowledge representation systems etc.) provides a convenient tool for the representation of objects in terms of their attribute values.

An information system is an ordered quadruple $\mathcal{I} = (U, A, F)$, where $U = \{x_1, x_2, \cdots, x_n\}$ is a non-empty finite set of objects called the universe, and $A = \{a_1, a_2, \cdots, a_p\}$ is a non-empty finite set of attributes, such that there exists a map $f_l : U \to V_{a_l}$ for any $a_l \in A$, where $V_{a_l}$ is called the domain of the attribute $a_l$, and denoted $F = \{f_l | a_l \in A\}$.

In an information systems, if the domain of a attribute is ordered according to a decreasing or increasing preference, then the attribute is a criterion.

**Definition 2.1.** An information system is called an ordered information system(OIS) if all condition attributes are criterions.

Assumed that the domain of a criterion $a \in A$ is complete pre-ordered by an outranking relation $\succeq_a$, then $x \succeq_a y$ means that $x$ is at least as good as $y$ with respect to criterion $a$. And we can say that $x$ dominates $y$. In the following, without any loss of generality, we consider criterions having a numerical domain, that is, $V_a \subseteq \mathcal{R}$($\mathcal{R}$ denotes the set of real numbers).

We define $x \succeq y$ by $f(x,a) \geq f(y,a)$ according to increasing preference, where $a \in A$ and $x, y \in U$. For a subset of attributes $B \subseteq A$, $x \succeq_B y$ means that $x \succeq_a y$ for any $a \in B$, and that is to say $x$ dominates $y$ with respect to all attributes in $B$. Furthermore, we denote $x \succeq_B y$ by $xR_B^{\geq} y$. In general, we denote a ordered information systems by $\mathcal{I}^{\succeq} = (U, A, F)$. Thus the following definition can be obtained.

**Definition 2.2.** Let $\mathcal{I}^{\succeq} = (U, A, F)$ be an ordered information, for $B \subseteq A$, denote

$$R_B^{\succeq} = \{(x,y) \in U \times U | f_l(x) \geq f_l(y), \forall a_l \in B\};$$

$R_B^{\succeq}$ are called dominance relations of ordered information system $\mathcal{I}^{\succeq}$.

Let denote

$$\begin{aligned} [x_i]_B^{\succeq} &= \{x_j \in U | (x_j, x_i) \in R_B^{\succeq}\} \\ &= \{x_j \in U | f_l(x_j) \geq f_l(x_i), \forall a_l \in B\}; \\ U/R_B^{\succeq} &= \{[x_i]_B^{\succeq} | x_i \in U\}, \end{aligned}$$

where $i \in \{1, 2, \cdots, |U|\}$, then $[x_i]_B^{\succeq}$ will be called a dominance class or the granularity of information, and $U/R_B^{\succeq}$ be called a classification of $U$ about attribute set $B$.

The following properties of a dominance relation are trivial by the above definition.

**Proposition 2.1.** Let $R_A^{\succeq}$ be a dominance relation.

(1) $R_A^{\succeq}$ is reflexive,transitive, but not symmetric, so it is not a equivalence relation.

(2) If $B \subseteq A$, then $R_A^{\succeq} \subseteq R_B^{\succeq}$.

(3) If $B \subseteq A$, then $[x_i]_A^{\succeq} \subseteq [x_i]_B^{\succeq}$

(4) If $x_j \in [x_i]_A^{\succeq}$, then $[x_j]_A^{\succeq} \subseteq [x_i]_A^{\succeq}$ and $[x_i]_A^{\succeq} = \cup\{[x_j]_A^{\succeq} | x_j \in [x_i]_A^{\succeq}\}$.

(5) $[x_j]_A^{\succeq} = [x_i]_A^{\succeq}$ iff $f(x_i, a) = f(x_j, a)$ $(\forall a \in A)$.

(6) $|[x_i]_B^{\succeq}| \geq 1$ for any $x_i \in U$.

(7) $U/R_B^{\succeq}$ constitute a covering of $U$, i.e., for every $x \in U$ we have that $[x]_B^{\succeq} \neq \phi$ and $\bigcup_{x \in U} [x]_B^{\succeq} = U$.

where $|\cdot|$ denote cardinality of the set.

For any subset $X$ of $U$, and $A$ of $\mathcal{I}^{\succeq}$ define

$$\underline{R_A^{\succeq}}(X) = \{x \in U | [x]_A^{\succeq} \subseteq X\}; \quad \overline{R_A^{\succeq}}(X) = \{x \in U | [x]_A^{\succeq} \cap X \neq \phi\},$$

$\underline{R_A^{\succeq}}(X)$ and $\overline{R_A^{\succeq}}(x)$ are said to be the lower and upper approximation of $X$ with respect to a dominance relation $R_A^{\succeq}$. And the approximations have also some properties which are similar to those of Pawlak approximation spaces.

**Proposition 2.2.** Let $\mathcal{I}^{\succeq} = (U, A, F)$ be an ordered information systems and $X, Y \subseteq U$, then its lower and upper approximations satisfy the following properties.

(1) $\underline{R_A^{\succeq}}(X) \subseteq X \subseteq \overline{R_A^{\succeq}}(X)$.

(2) $\overline{R_A^{\succeq}}(X \cup Y) = \overline{R_A^{\succeq}}(X) \cup \overline{R_A^{\succeq}}(Y)$;
$\underline{R_A^{\succeq}}(X \cap Y) = \underline{R_A^{\succeq}}(X) \cap \underline{R_A^{\succeq}}(Y)$.

(3) $\underline{R_A^{\succeq}}(X) \cup \underline{R_A^{\succeq}}(Y) \subseteq \underline{R_A^{\succeq}}(X \cup Y)$;
$\overline{R_A^{\succeq}}(X \cap Y) \subseteq \overline{R_A^{\succeq}}(X) \cap \overline{R_A^{\succeq}}(Y)$.

(4) $\underline{R_X^{\succeq}}(\sim X) = \sim \overline{R_A^{\succeq}}(X)$; $\overline{R_A^{\succeq}}(\sim X) = \sim \underline{R_A^{\succeq}}(X)$.

(5) $\underline{R_A^{\succeq}}(U) = U$; $\overline{R_A^{\succeq}}(\phi) = \phi$.

(6) $\underline{R_A^{\succeq}}(X) \subseteq \underline{R_A^{\succeq}}(\underline{R_A^{\succeq}}(X))$; $\overline{R_A^{\succeq}}(\overline{R_A^{\succeq}}(X)) \subseteq \overline{R_A^{\succeq}}(X)$.

(7) If $X \subseteq Y$, then $\underline{R_A^{\succeq}}(X) \subseteq \underline{R_A^{\succeq}}(Y)$ and $\overline{R_A^{\succeq}}(X) \subseteq \overline{R_A^{\succeq}}(Y)$.

where $\sim X$ is the complement of $X$.

**Definition 2.3.** For a ordered information system $\mathcal{I}^{\succeq} = (U, A, F)$ and $B, C \subseteq A$.

(1) If $[x]_B^{\succeq} = [x]_C^{\succeq}$ for any $x \in U$, then we call that classification $U/R_B^{\succeq}$ is equal to $R/R_C^{\succeq}$, denoted by $U/R_B^{\succeq} = U/R_C^{\succeq}$.

(2) If $[x]_B^{\succeq} \subseteq [x]_C^{\succeq}$ for any $x \in U$, then we call that classification $U/R_B^{\succeq}$ is finer than $R/R_C^{\succeq}$, denoted by $U/R_B^{\succeq} \subseteq U/R_C^{\succeq}$.

(3) If $[x]_B^{\succeq} \subseteq [x]_C^{\succeq}$ for any $x \in U$ and $[x]_B^{\succeq} \neq [x]_C^{\succeq}$ for some $x \in U$, then we call that classification $U/R_B^{\succeq}$ is properly finer then $R/R_C^{\succeq}$, denoted by $U/R_B^{\succeq} \subset U/R_C^{\succeq}$.

For a ordered information system $\mathcal{I}^{\succeq} = (U, A, F)$ and $B \subseteq A$, it is obtained that $U/R_A^{\succeq} \subseteq U/R_B^{\succeq}$ directly by Proposition 2.1(3) and above definition. And an ordered information systems $\mathcal{I}^{\succeq} = (U, A, F)$ be regarded as knowledge representation system $U/R_A^{\succeq}$ or knowledge $A$, as is same to classical rough set based on equivalence relation.

**Example 2.1.** Given an ordered information system in Table 1.

Table 1 An ordered information system

| $U \times A$ | $a_1$ | $a_2$ | $a_3$ |
|---|---|---|---|
| $x_1$ | 1 | 2 | 1 |
| $x_2$ | 3 | 2 | 2 |
| $x_3$ | 1 | 1 | 2 |
| $x_4$ | 2 | 1 | 3 |
| $x_5$ | 3 | 3 | 2 |
| $x_6$ | 3 | 2 | 3 |

If denote $B = \{a_1, a_2\}$, from the table we have

$$[x_1]_A^{\succeq} = \{x_1, x_2, x_5, x_6\};$$

$$[x_2]_A^{\succeq} = \{x_2, x_5, x_6\};$$
$$[x_3]_A^{\succeq} = \{x_2, x_3, x_4, x_5, x_6\};$$
$$[x_4]_A^{\succeq} = \{x_4, x_6\};$$
$$[x_5]_A^{\succeq} = \{x_5\};$$
$$[x_6]_A^{\succeq} = \{x_6\};$$

and

$$[x_1]_B^{\succeq} = \{x_1, x_2, x_5, x_6\};$$
$$[x_2]_B^{\succeq} = \{x_2, x_5, x_6\};$$
$$[x_3]_B^{\succeq} = \{x_1, x_2, x_3, x_4, x_5, x_6\};$$
$$[x_4]_B^{\succeq} = \{x_2, x_4, x_5, x_6\};$$
$$[x_5]_B^{\succeq} = \{x_5\};$$
$$[x_6]_B^{\succeq} = \{x_5, x_6\}.$$

Thus, it is obviously that $U/R_A^{\succeq} \subseteq U/R_B^{\succeq}$, i.e., classification $U/R_A^{\succeq}$ is finer than classification $U/R_B^{\succeq}$.

For simple description, in the following information systems are based on dominance relations generally, i.e., ordered information systems.

## 3 Rough Entropy of Knowledge in Ordered Information Systems

In classical rough set theory, knowledge be regarded as partition of set of objects to an information system. However, it is known that equality relations is replaced by dominance relations in an ordered information system. Thus, knowledge be regarded as classification of set of objects to an ordered information system by section 2.

In this section, we will introduce rough entropy of knowledge and establish relationships between roughness of knowledge and rough entropy in ordered information systems.

Firstly, let give the concept of rough entropy of knowledge in ordered information systems.

**Definition 3.1.** Let $\mathcal{I}^{\succeq} = (U, A, F)$ be an ordered information systems and $B \subseteq A$. The rough entropy of knowledge $B$ is defined as follows:

$$E(B) = \sum_{i=1}^{|U|} \frac{|[x_i]_B^{\succeq}|}{|U|} \cdot \log_2 |[x_i]_B^{\succeq}| \,,$$

where $|\cdot|$ is the cardinality of sets.

**Example 3.1.** In Example 2.1, the rough entropy of knowledge $A = \{a_1, a_2, a_3\}$ can be calculated by above definition, which is

$$\begin{aligned} E(A) &= \frac{4}{6} \cdot \log_2 4 + \frac{3}{6} \cdot \log_2 3 + \frac{5}{6} \cdot \log_2 5 + \\ &\quad \frac{2}{6} \cdot \log_2 2 + \frac{1}{6} \cdot \log_2 1 + \frac{1}{6} \cdot \log_2 1 \\ &= \frac{2}{3} \cdot 2 + \frac{1}{2} \cdot \log_2 3 + \frac{5}{6} \cdot \log_2 5 + \frac{1}{3} \\ &= 4.39409 \end{aligned}$$

**Proposition 3.1.** Let $\mathcal{I}^{\succeq} = (U, A, F)$ be an ordered information systems and $B \subseteq A$. The following hold.

(1) $E(B)$ can obtain its maximum $|U| \cdot \log_2 |U|$, iff $U/R_B^{\succeq} = U$.

(2) $E(B)$ can obtain its minimum 0, iff $U/R_B^{\succeq} = \{\{x_1\}, \{x_2\}, \cdots, \{x_{|U|}\}\}$.

*Proof.* It is straightforward by Definition 3.1.

From Proposition 3.1, it can be concluded that information quantity provided by knowledge $B$ is zero when its rough entropy reaches maximum, and its cannot distinguish any two objects in $U$, when the classification of ordered information systems is no meaning. When the rough entropy of knowledge $B$ obtains its minimum, the information quantity is the most and every objects can be discriminated by $B$ in the ordered information systems.

**Theorem 3.1.** Let $\mathcal{I}^{\succeq} = (U, A, F)$ be an ordered information systems and $B_1, B_2 \subseteq A$. If $U/R_{B_1}^{\succeq} \subset U/R_{B_2}^{\succeq}$, then $E(B_1) < E(B_2)$.

*Proof.* Because of $U/R_{B_1}^{\succeq} \subset U/R_{B_2}^{\succeq}$, we have that $[x_i]_{B_1}^{\succeq} \subseteq [x_i]_{B_2}^{\succeq}$ for every $x_i \in U$. Thus there exists some $x_j \in U$ such that $|[x_j]_{B_1}^{\succeq}| < |[x_j]_{B_2}^{\succeq}|$. Hence, by the Proposition 2.1 and Definition 3.1 we can obtain

$$\sum_{i=1}^{|U|} |[x_i]_{B_1}^{\succeq}| \cdot \log_2 |[x_i]_{B_1}^{\succeq}| < \sum_{i=1}^{|U|} |[x_i]_{B_2}^{\succeq}| \cdot \log_2 |[x_i]_{B_2}^{\succeq}|,$$

i.e.,

$$E(B_1) < E(B_2).$$

From Theorem 3.1, we can find that rough entropy of knowledge decreased monotonously as the granularity of information became smaller through finer classifications of objects set $U$.

**Corollary 3.1.** Let $\mathcal{I}^{\succeq} = (U, A, F)$ be an ordered information systems and $B_1, B_2 \subseteq A$. If $B_2 \subseteq B_1$, then $E(B_1) \leq E(B_2)$.

**Theorem 3.2.** Let $\mathcal{I}^{\succeq} = (U, A, F)$ be an ordered information systems and $B_1, B_2 \subseteq A$. If $U/R_{B_1}^{\succeq} = U/R_{B_2}^{\succeq}$, then $E(B_1) = E(B_2)$.

*Proof.* Since $U/R_{B_1}^{\succeq} = U/R_{B_2}^{\succeq}$, we have that $[x_i]_{B_1}^{\succeq} = [x_i]_{B_2}^{\succeq}$ for every $x_i \in U$. Thus, it is obtain $E(B_1) = E(B_2)$ directly.

The theorem states that two equivalence knowledge representation systems have same rough entropy.

**Theorem 3.3.** Let $\mathcal{I}^{\succeq} = (U, A, F)$ be an ordered information systems and $B_1, B_2 \subseteq A$. If $U/R^{\succeq}_{B_1} \subseteq U/R^{\succeq}_{B_2}$ and $E(B_1) = E(B_2)$, then $U/R^{\succeq}_{B_1} = U/R^{\succeq}_{B_2}$.

*Proof.* Since $E(B_1) = E(B_2)$, it follows that

$$\sum_{i=1}^{|U|} |[x_i]^{\succeq}_{B_1}| \cdot \log_2 |[x_i]^{\succeq}_{B_1}| = \sum_{i=1}^{|U|} |[x_i]^{\succeq}_{B_2}| \cdot \log_2 |[x_i]^{\succeq}_{B_2}|. \qquad (*)$$

From $U/R^{\succeq}_{B_1} \subseteq U/R^{\succeq}_{B_2}$, we have that $[x_i]^{\succeq}_{B_1} \subseteq [x_i]^{\succeq}_{B_2}$ for every $x_i \in U$. This show that $1 \leq |[x_i]^{\succeq}_{B_1}| \leq |[x_i]^{\succeq}_{B_2}|$. Thus, it is true that

$$|[x_i]^{\succeq}_{B_1}| \cdot \log_2 |[x_i]^{\succeq}_{B_1}| \leq |[x_i]^{\succeq}_{B_2}| \cdot \log_2 |[x_i]^{\succeq}_{B_2}|.$$

By the formula $(*)$, it follows that

$$|[x_i]^{\succeq}_{B_1}| \cdot \log_2 |[x_i]^{\succeq}_{B_1}| = |[x_i]^{\succeq}_{B_2}| \cdot \log_2 |[x_i]^{\succeq}_{B_2}|.$$

So, we easily obtain $|[x_i]^{\succeq}_{B_1}| = |[x_i]^{\succeq}_{B_2}|$, for every $x_i \in U$.

On the other hand, $[x_i]^{\succeq}_{B_1} \subseteq [x_i]^{\succeq}_{B_2}$, we get $[x_i]^{\succeq}_{B_1} = [x_i]^{\succeq}_{B_2}$ for every $x_i \in U$.

Hence, $U/R^{\succeq}_{B_1} = U/R^{\succeq}_{B_2}$.

Theorem 3.3 states that if two knowledge representation systems exists inclusion relation and their rough entropy are equal, then two knowledge representation systems is equivalent.

**Corollary 3.2.** Let $\mathcal{I}^{\succeq} = (U, A, F)$ be an ordered information systems and $B_1, B_2 \subseteq A$. If $B_2 \subseteq B_1$ and $E(B_1) = E(B_2)$, then $U/R^{\succeq}_{B_1} = U/R^{\succeq}_{B_2}$.

**Example 3.2.** We had got that $U/R^{\succeq}_{A} \subseteq U/R^{\succeq}_{B}$, if denote $B = \{a_1, a_2\}$ in the ordered information system of Example 2.1. Moreover, $E(B)$ cab be calculated easily, which is

$$\begin{aligned} E(B) &= \frac{4}{6} \cdot \log_2 4 + \frac{3}{6} \cdot \log_2 3 + \frac{6}{6} \cdot \log_2 6 + \\ &\quad \frac{4}{6} \cdot \log_2 4 + \frac{1}{6} \cdot \log_2 1 + \frac{2}{6} \cdot \log_2 2 \\ &= \frac{2}{3} \cdot 4 + \frac{1}{2} \cdot \log_2 3 + \log_2 6 + \frac{1}{3} \\ &= 6.37744 \end{aligned}$$

On the other hand, by Example 3.1, we obtained $E(A) = 4.39409$.

Thus, it is obvious that $E(A) \leq E(B)$.

However, if denote $B' = \{a_1\}$ and $B'' = \{a_2\}$ in the system of Example 2.1, we have that

$$[x_1]^{\succeq}_{B'} = [x_3]^{\succeq}_{B'} = \{x_1, x_2, x_3, x_4, x_5, x_6\};$$

$$[x_2]^{\succeq}_{B'} = [x_5]^{\succeq}_{B'} = [x_6]^{\succeq}_{B'} = \{x_2, x_5, x_6\};$$
$$[x_4]^{\succeq}_{B'} = \{x_2, x_4, x_5, x_6\},$$

and

$$[x_1]^{\succeq}_{B''} = [x_2]^{\succeq}_{B''} = [x_6]^{\succeq}_{B''} = \{x_1, x_2, x_5, x_6\};$$
$$[x_3]^{\succeq}_{B''} = [x_4]^{\succeq}_{B''} = \{x_1, x_2, x_3, x_4, x_5, x_6\};$$
$$[x_5]^{\succeq}_{B''} = \{x_5\}.$$

Furthermore, we can obtain that $E(B') = 8.88071$ and $E(B'') = 9.16993$, which show $E(B') < E(B'')$. While, $U/R^{\succeq}_{B'} \subseteq U/R^{\succeq}_{B''}$ doesn't hold. So, it can be concluded that the converse proposition of Theorem 3.1 does not hold.

## 4 Rough Entropy of Rough Sets in Ordered Information Systems

In rough set theory, the roughness of a rough set can be measured by its rough degree. So we give the rough degree of a rough set in ordered information systems.

**Definition 4.1.** Let $\mathcal{I}^{\succeq} = (U, A, F)$ be an ordered information systems and $B \subseteq A$. The rough degree of a rough set $X \subseteq U$ about knowledge $B$ is defined as follows:

$$\rho_B(X) = 1 - \frac{|\underline{R^{\succeq}_B}(X)|}{|\overline{R^{\succeq}_B}(X)|},$$

where $|\cdot|$ is the cardinality of sets.

From the above definition and Proposition 2.2, it is obvious to $0 \leq \rho_B(X) \leq 1$, and the following property can be obtained easily.

**Theorem 4.1.** Let $\mathcal{I}^{\succeq} = (U, A, F)$ be an ordered information systems and $B_1, B_2 \subseteq A$. If $U/R^{\succeq}_{B_1} \subseteq U/R^{\succeq}_{B_2}$, then $\rho_{B_1}(X) \leq \rho_{B_2}(X)$, for any rough set $X \subseteq U$.

**Example 4.1.** In Example 2.1, we have known $U/R^{\succeq}_A \subseteq U/R^{\succeq}_B$, i.e., classification $U/R^{\succeq}_A$ is finer than classification $U/R^{\succeq}_B$ in the system of Table 1.

For $X = \{x_4, x_5, x_6\}$, we have

$$\underline{R^{\succeq}_A}(X) = \{x_4, x_5, x_6\}, \quad \overline{R^{\succeq}_A}(X) = U;$$
$$\underline{R^{\succeq}_B}(X) = \{x_5, x_6\}, \quad \overline{R^{\succeq}_B}(X) = U.$$

Thus, by calculating, the rough degrees of $X$ about knowledge $B$ and $A$ can be obtained respectively, which are

$$\rho_A(X) = \frac{1}{2}; \quad \rho_B(X) = \frac{2}{3};$$

Obviously, $\rho_A(X) \leq \rho_B(X)$.

From Theorem 4.1 and Example 4.1, we can get that coarser is the classification of ordered information systems, smaller is not the rough degree of a rough set of the system.

However, it can be find that the uncertainty measure, i.e., rough degree, of a rough set is not exact in ordered information systems by the following example.

**Example 4.2.** Let $X' = \{x_3, x_5, x_6\}$ in Example 4.1, we get

$$\underline{R_A^{\succeq}}(X') = \underline{R_B^{\succeq}}(X') = \{x_5, x_6\};$$
$$\overline{R_A^{\succeq}}(X') = \overline{R_B^{\succeq}}(X') = U.$$

So have

$$\rho_A(X) = \rho_B(X) = \frac{1}{3}\,.$$

In other words, the uncertainty of knowledge $B$ is larger than that of $A$ in Example 4.2, but $X'$ has the same rough degree. Therefore, it is necessary to find a new and more accurate uncertainty measure for rough sets in ordered information systems.

**Definition 4.2.** Let $\mathcal{I}^{\succeq} = (U, A, F)$ be an ordered information systems and $B \subseteq A$. The rough entropy of a rough set $X \subseteq U$ about knowledge $B$ is defined as follows:

$$E_B(X) = \rho_B(X)E(B).$$

From Definition 4.2, the rough entropy of rough sets is related not only to its own rough degree, but also to the uncertainty of knowledge in the ordered information systems.

**Example 4.3.** The rough entropy of $X'$ in Example 4.2 is calculated about knowledge $B$ and $A$ respectively, which are

$$E_B(X') = \rho(X')E(B) = \frac{1}{3} \times 6.37744 = 2.12579;$$
$$E_A(X') = \rho(X')E(A) = \frac{1}{3} \times 4.39409 = 1.46468.$$

Thus, we have

$$E_A(X') < E_B(X').$$

By this example, it is obvious that the rough entropy of rough sets is more accurate than the rough degree to measure the roughness of rough sets in ordered information systems.

Furthermore, the following property can be obtained about the entropy of rough sets.

**Theorem 4.2.** Let $\mathcal{I}^{\succeq} = (U, A, F)$ be an ordered information systems and $B_1, B_2 \subseteq A$. If $U/R_{B_1}^{\succeq} \subset U/R_{B_2}^{\succeq}$, then $E_{B_1}(X) < E_{B_2}(X)$, for any $X \subseteq U$.

*Proof.* It is straightforward by Theorem 3.1 and Theorem 4.1.

**Corollary 4.1.** Let $\mathcal{I}^{\succeq} - (U, A, F)$ be an ordered information systems and $B_1, B_2 \subseteq A$. If $B_2 \subseteq B_1$, then $E_{B_1}(X) \leq E_{B_2}(X)$ for any $X \subseteq U$.

It can be deduced from the above propositions that the rough entropy of a rough set monotonously decreases as the classification becomes finer in ordered information systems.

## 5 Conclusions

Rough set theory is a new mathematical tool to deal with vagueness and uncertainty. Development of a rough computational method is one of the most important research tasks. While, in practise, ordered information system confines the applications of classical rough set theory. In this article, a measure to knowledge and its important properties are established by proposed rough entropy in ordered information systems. We prove that the rough entropy of knowledge and rough set decreases monotonously as the granularity of information becomes finer, and obtain some conclusions, which is every helpful in future research works of ordered information systems.

## References


1. Pawlak, Z.: Rough sets, Internantional Journal of Computer and Information Science, 11(5)(1982) 341-356
2. Kryszkiewicz, M.: Comprative Studies of Alternative Type of Knowledge Reduction in Inconsistent Systems. International Journal of Intelligent Systems, 16(2001) 105-120
3. Leuang,Y., Wu, W.Z., Zhang, W.X.: Knowledge Acquisition in Incomplete Information Systems: A Rough Set Approach, European Journal of Operational Research, 168(1)(2006) 164-180
4. Zhang, W.X., Wu, W.Z., Liang, J.Y., Li, D.Y.: Theory and Method of Rough Sets, Science Press, Beijing (2001)
5. Wu, W.Z., Leuang, Y., Mi, J.S.: On Characterizations of (I,T)-Fuzzy Rough Approximation Operators. Fuzzy Sets and Systems, 154(1)(2005) 76-102
6. Wu,W.Z., Zhang, M., Li, H.Z., Mi, J.S.: Knowledge Reduction in Random Information Systems Via Dempster-Shafer Theory of EvidenceInformation Sciences, 174(3-4)(2005) 143-164
7. Greco, S., Matarazzo, B., Slowingski, R.: Rough Approximation of A Preference Relatioin by Dominance Relatioin. ICS Research Report 16 / 96, Warsaw University of Technology 1996 and in Eru H Oper Res, 117(1999) 63-83
8. Greco, S., Matarazzo, B., Slowingski, R.: A New Rough Set Approach to Multicriteria and Multiattribute Classificatioin. In: Polkowsik L, Skowron A, Editors. Rough Sets and Current Trends in Computing (RSCTC'98), Lecture Notes in Artificial Intelligence, Vol 1424. Springer-Verlag ,Berlin(1998) 60-67
9. Greco, S., Matarazzo, B., Slowingski, R.: A New Rough Sets Approach to Evaluation of Bankruptcy Risk. In: Zopounidis X, Editor. Operational Tools in the Management of Financial Risks. Dordrecht: Kluwer, (1999) 121-136
10. Greco, S., Matarazzo, B., Slowingski, R.: Rough Sets Theory for Multicriteria Decision Analysis. Europe Journal Operational Research, 129(2001) 11-47

11. Greco, S., Matarazzo, B., Slowingski, R.: Rough Sets Methodology for Sorting Problems in Presence of Multiple Attributes and Criteria, Europe Journal Operational Research, 138(2002) 247-259
12. Dembczynski, K., Pindur, R., Susmaga, R.: Generation of Exhaustive Set of Rules within Dominance-Based Rough Set Approach. Electronic Notes of Theory Computer Sciences, 82(4)(2003)
13. Dembczynski, K., Pindur, R., Susmaga, R.: Dominance-based Rough Set Classifier without Induction of Decision Rules. Electronic Notes of Theory Computer Sciences, 82(4)(2003)
14. Sai, Y., Yao, Y.Y., Zhong, N.: Data Analysis and Mining in Ordered Information Tables. Proceeding 2001 IEEE International Conference on Data Mining, IEEE Computer Society Press (2001) 497-504
15. Shao, M.W., Zhang, W.X.: Dominance Relation and Relus in an Incomplete Ordered Information System. International Journal of Intelligent Sytems, 20(2005) 13-27
16. Cattaneo, G.: Abatract Approximation spaces for Rough Theories, in: Polkowski, Skoeron (Eds) Rough Sets in Knowledge Discovery 1: Methodolgy and Applications, Physica-Verlag, Heidelberg, (1998) 59-98
17. Skowron, A., Stepaniuk, J.: Tolerance Approxximation Space, Fundamental Information, 27 (1996) 245-253
18. Yao, Y.Y.: Relational Interpretations of Neighborhood Operators and Rough Set Approximation Operators. Information, Sciences, 101 (1998) 239-259
19. Yao, Y.Y.: Constructive and Algebraic Methods of Theory of Rough Sets, Information. Sciences, 109 (1998) 21-47
20. Yao, Y.Y.: Probabilistic Approaches to Rough Sets, Expert Systems, 2003,20(5)287-297
21. Beaubouef, T., Petry, F.E., Arova, G.: Information-Theoretic Measures of Uncertainty for Rough Sets and Rough Relational databases, Information Sciences,109(1998) 185-195
22. Wierman, M.J.: Measureing Uncertainty in Rough Set Theory.International Journal of General System, 28(1999) 283-297
23. Duntsch, I.,Gediga, G.: Uncertainty Measures of Rough Set Prediction, Artificial Intelligent, 106(1998) 109-137
24. Duntsch, I.,Gediga, G.: Roughian: Rough Information Analysis, International Journal of Intelligent System, 16(2001) 121-147

# Particle Swarm Optimization with Dynamic Step Length

Zhihua Cui[1,2], Xingjuan Cai[1], Jianchao Zeng[1], and Guoji Sun[2]

[1] Division of System Simulation and Computer Application
[2] Taiyuan University of Science and Technology, 030024, P.R. China
[3] State Key Laboratory for Manufacturing Systems Engineering
Xi'an Jiaotong University, Xi'an, 710049,P.R. China
cuizhihua@gmail.com, cai-xing-juan@sohu.com,
zengjianchao@263.net, gjsun@sei.xjtu.edu.cn

**Abstract.** Particle swarm optimization (PSO) is a robust swarm intelligent technique inspired from birds flocking and fish schooling. Though many effective improvements have been proposed, however, the premature convergence is still its main problem. Because each particle's movement is a continuous process and can be modelled with differential equation groups, a new variant, particle swarm optimization with dynamic step length (PSO-DSL), with additional control coefficient- step length, is introduced. Then the absolute stability theory is introduced to analyze the stability character of the standard PSO, the theoretical result indicates the PSO with constant step length can not always be stable, this may be one of the reason for premature convergence. Simulation results show the PSO-DSL is effective.

**Keywords:** Particle swarm optimization, dynamic step length, absolute stability theory.

## 1 Introduction

Due to the increased large industrial requirements, many non-differential, high dimensional optimization problems need to be answered. Natural inspired computation is a wide class random research algorithms simulated with the real-world's complex systems, such as evolutionary computation-the biological process of evolution[1], artificial neural networks-the function of neurons in the brain[2], and swarm techniques-the behavior of social insects[3][4][5],etc. All of these algorithms are population-based, and show a potential capabilities to solve the complex,non-differential, high dimensional problems.

Particle swarm optimization (PSO)[6][7] is a new swarm technique simulated with animal social behaviors such as birds flocking, fish schooling and insects herding. The method of this algorithm is simple and effective. Suppose all particles live in a $D-$dimensional problem space. They share the information and adapt their search patterns by collaborative manner to seek the food. Once a particle finds a food source, others will fly to that location from every directions.

D.-S. Huang, L. Heutte, and M. Loog (Eds.): ICIC 2007, LNAI 4682, pp. 770–780, 2007.

If the symbol $\overrightarrow{x}_j(t)$ represents the position of $j^{th}$ particle at time $t$, then in the next time it's position $\overrightarrow{x}_j(t+1)$ will change as follows:

$$\overrightarrow{x}_j(t+1) = \overrightarrow{x}_j(t) + \overrightarrow{v}_j(t) \tag{1}$$

where $\overrightarrow{v}_j(t)$ denotes moving speed of particle $j$ at time $t$. $w$ is inertia weight between 0 and 1, accelerator coefficients $c_1$ and $c_2$ are known as constants, $r_1$ and $r_2$ are two random numbers generated with uniform distribution within $(0, 1)$.

It is found that the velocity $\overrightarrow{v}_j(t)$ is consisted with three different parts[6]: the previous inertia, personal experience and others' experiences. Therefore, the velocity vector of $j^{th}$ particle is updated as follows:

$$\overrightarrow{v}_j(t+1) = w\overrightarrow{v}_j(t) + c_1r_1(\overrightarrow{p}_j(t) - \overrightarrow{x}_j(t)) + c_2r_2(\overrightarrow{p}_g(t) - \overrightarrow{x}_j(t)) \tag{2}$$

where $\overrightarrow{p}_j(t)$ represents the particle's thinking itself, $\overrightarrow{p}_g(t)$ is the best location found by the entire swarm. To make the particle is roaming within the problem space, a predefined constant $v_{max}$ is used to control the size of velocity.

Although the PSO algorithm is easy implementation and the fast convergent speed, it still easily traps in a local optimum. Many researchers works on this subject to improve the performance, thereby several interesting variants are proposed such as fitness estimation particle swarm optimization[8], Kalman particle swarm optimization[9], adaptive particle swarm optimization with velocity feedback[10], comprehensive learning particle swarm optimization[11], guaranteed convergence particle swarm optimization[12], particle swarm optimization with time-varying acceleration coefficient[13], etc. All of these variants uses the discrete model of PSO methodology. However, as we known, the searching food process of bird is a continuous process in nature, that means the corresponding simulation model - particle swarm optimization shall be a differential equations. Follow this idea, we investigate the differential model of PSO, and find the step length inside the update equation (2) is an important parameter to affect the algorithm's performance escaping from a local optimum.

The rest of paper is organized as follows. In section 2, the differential model and concept of step length are introduced, as well as the analysis of absolute stability are provided in section 3 . Then, the details of the particle swarm optimization with dynamic step length (PSO-DSL) is designed. Finally, Several benchmark functions are used to testify the new algorithm's efficiency.

## 2 Differential Model and Step Length

In this paper, only the unconstrained optimization problem is considered:

$$\min \quad f(\overrightarrow{x}) \quad x \in [L, U]^D \subseteq R^D \tag{3}$$

Substituting (1) into (2), and using the $k^{th}$ dimensional value to replace the vector formula resulting

$$x_{jk}(t+1) = wv_{jk}(t) + \varphi_1 p_{jk}(t) + \varphi_2 p_{gk}(t) + (1-\varphi)x_{jk}(t) \tag{4}$$

where$\varphi_1$ and $\varphi_2$ denote $c_1 r_1$ and $c_2 r_2$, respectively, $\varphi$ is the sum of $\varphi_1$ and $\varphi_2$.

Suppose

$$\frac{dv_{jk}}{dt} = v_{jk}(t+1) - v_{jk}(t) \tag{5}$$

and

$$\frac{dx_{jk}}{dt} = x_{jk}(t+1) - x_{jk}(t) \tag{6}$$

both with step length one, then the formula (2) and (4) are changed as follows:

$$\frac{dv_{jk}}{dt} = (w-1)v_{jk} - \varphi x_{jk} + (\varphi_1 p_{jk} + \varphi_2 p_{gk}) \tag{7}$$

$$\frac{dx_{jk}}{dt} = wv_{jk} - \varphi x_{jk} + (\varphi_1 p_{jk} + \varphi_2 p_{gk}) \tag{8}$$

In the following analysis, $\varphi$,$\varphi_1$ and $\varphi_2$ are regarded as constants. We also regard $p_{jk}$ and $p_{gk}$ as constants, although $p_{jk}$ and $p_{gk}$ are, in practice, dynamic in a search process.

Formula (7) and (8) are the corresponding differential model. If the step length is taken to one and Euler integral method is used, the update equation (1) and (2) are obtained. This means the update equations of the standard particle swarm optimization implies the step length is a constant with one. Because many experimental and theoretical tests have been down and results show the standard PSO is not effective and efficient in some cases, the constant step length setting may be one of the reasons resulting bad computational efficiency. However, few papers are concerned about it. Therefore, in this paper, we investigate the affection of step length.

Suppose symbol $h$ denotes the step length, making differences of formula (7) and (8) using Euler method with the same step length $h$ may result in

$$v_{jk}(t+1) = [1 + h(w-1)]v_{jk}(t) - h\varphi x_{jk} + h(\varphi_1 p_{jk} + \varphi_2 p_{gk}) \tag{9}$$

$$x_{jk}(t+1) = hwv_{jk} + (1-h\varphi)x_{jk} + h(\varphi_1 p_{jk} + \varphi_2 p_{gk}) \tag{10}$$

This model is our main contribution that incorporates the step length explicitly. The following part will explain the relationship between step length and other three parameters.

## 3 Selection Principle of Step Length

Since step length $h$ is a key factor to reflect the differences of the corresponding differential model, the absolute stability condition should be necessary considered.

Firstly, we consider the absolute stability conditions of velocity differential equation (7). Suppose

$$\frac{dv_{jk}}{dt} = f_1(t, v_{jk}, x_{jk}) \tag{11}$$

where

$$f_1(t, v_{jk}, x_{jk}) = (w-1)v_{jk} - \varphi x_{jk} + (\varphi_1 p_{jk} + \varphi_2 p_{gk}) \tag{12}$$

According to the numerical treatments of differential modal, the absolute stability with Euler method is defined as follows.

$$|1 + \lambda h| < 1 \tag{13}$$

where

$$\lambda = \frac{\partial f_1}{\partial v_{jk}} = w - 1 \tag{14}$$

Substituting (14) into (13), resulting

$$|1 + (w-1) \times h| < 1 \tag{15}$$

So the absolute stability interval of step length $h$ of $k^{th}$ variable $v_{jk}$ of velocity vector of particle $j$ is defined with:

$$0 < h < \frac{2}{1-w} \tag{16}$$

According to the definition of absolute stability, parameter $\lambda = \frac{\partial f_1}{\partial v_{jk}} < 0$, it is true if and only if $w < 1$. In one word, the absolute stability conditions of differential modal of velocity vector are:

$$w < 1 \; and \;\; 0 < h < \frac{2}{1-w} \tag{17}$$

The same method is implemented with the absolute stability conditions of position differential equation. Formula (8) can be represented with

$$\frac{dx_{jk}}{dt} = f_2(t, v_{jk}, x_{jk}) \tag{18}$$

where

$$f_2(t, v_{jk}, x_{jk}) = wv_{jk} - \varphi x_{jk} + (\varphi_1 p_{jk} + \varphi_2 p_{gk}) \tag{19}$$

and coefficient $\lambda$ is satisfied with

$$\lambda = \frac{\partial f_2}{\partial x_{jk}} = -\varphi < 0 \tag{20}$$

Thus the absolute interval of position differential equation is

$$0 < h < \frac{2}{\varphi} \tag{21}$$

From the above mentioned, the absolute stability conditions of differential modal with step length $h$ is

$$0 < h < min\{\frac{2}{1-w}, \frac{2}{\varphi}\} \tag{22}$$

It means

$$\begin{cases} 0 < h < \frac{2}{1-w}, & \text{if } (\frac{2}{1-w} < \frac{2}{\varphi}) , \\ 0 < h < \frac{2}{\varphi} \quad , & \text{otherwise.} \end{cases} \tag{23}$$

condition $\frac{2}{1-w} < \frac{2}{\varphi}$ is true if and only if the following formula is true.

$$\varphi + w < 1 \tag{24}$$

Since then, we hvae

$$\begin{cases} 0 < h < \frac{2}{1-w}, & \text{if } w + \varphi < 1 , \\ 0 < h < \frac{2}{\varphi} \quad , & \text{otherwise.} \end{cases} \tag{25}$$

Because the coefficient $\varphi$ is a random variable, while the condition $\frac{2}{1-w} < \frac{2}{\varphi}$ is not always true. Thus, the step length $h$ is kept as a random variable and selected as follows.

Case 1: $\varphi + w < 1$

The absolute stability interval is $(0, \frac{2}{1-w})$, the step length $h$ is uniformly generated within this interval.

Case 2: $\varphi + w \geq 1$

The step length $h$ is generated with uniform distribution within $(0, \frac{2}{\varphi})$.

Now, let us consider a special issue that the step length is taken as one. Suppose $h = 1$, applied formula (22), we have

$$1 < \frac{2}{1-w} < \frac{2}{\varphi} .or. 1 < \frac{2}{\varphi} < \frac{2}{1-w} \tag{26}$$

It is true if and only if the following conditions is satisfied:

$$|w| < 1 \ .and. \ \varphi < 2 \tag{27}$$

Many researchers uses PSO with coefficients setting $c_1 = c_2 = 2.0$, though the expect value of random variable $\varphi$ is computed with

$$E(\varphi) = E(c_1 r_1 + c_2 r_2) = \frac{c_1 + c_2}{2} = 2.0 \tag{28}$$

So the condition $\varphi < 2$ is true only with probability $P(\varphi < 2) = 0.5$. It means the accelerator selection principles of standard PSO can guarantee absolute stability with 50 percent probability only, though in other cases it can not guaranteed absolute stability. This may provide an roughly explanation for the premature convergence. Therefore, we propose a variant of PSO methodology with dynamic step length (PSO-DSL) so that it is always stable. From the stability point of view, PSO-DSL employs a power global exploration ability.

The detail steps of particle swarm optimization with dynamic step length (PSO-DSL) are listed as follows.

Step1. Initializing each coordinate $x_{jk}(0)$ and $v_{jk}(0)$ sampling within $[x_{min}, x_{max}]$ and $[-v_{max}, v_{max}]$, respectively.

Step2. Computing the fitness of each particle.

Step3. For each dimension $k$ of particle $j$, the personal historical best position $p_{jk}(t)$ is updated as follows.

$$p_{jk}(t) = \begin{cases} x_{jk}(t), & \text{if } f(x_j(t)) < f(p_j(t-1)) \ , \\ p_{jk}(t-1) & , \quad \text{otherwise.} \end{cases} \tag{29}$$

Step 4. For each dimension $k$ of particle $j$, the global best position $p_{gk}(t)$ is updated as follows.

$$p_{gk}(t) = \begin{cases} p_{jk}(t), & \text{if } f(p_j(t)) < f(p_g(t-1)) \ , \\ p_{gk}(t-1) & , \quad \text{otherwise.} \end{cases} \tag{30}$$

Step5. Computing the step length $h$ value of each particle according to formula (25).

Step6. Updating the velocity and position vectors with equations (7) and (8).

Step7. If the criteria is satisfied, output the best solution; otherwise, goto step 2.

# 4 Simulation Results

## 4.1 Test Functions

In order to certify the efficiency of the PSO-DSL, we select six famous benchmark functions to testify the performance, and compare PSO-DSL with standard PSO (SPSO), Modified PSO with time-varying accelerator coefficients (MPSO-TVAC)[13], and comprehensive learning PSO (CLPSO)[11].

Sphere Modal:

$$f_1(x) = \sum_{j=1}^{n} x_j^2$$

where $|x_j| \leq 100.0$, and

$$f_1(x^*) = f_1(0, 0, ..., 0) = 0.0$$

Schwefel Problem 2.22:

$$f_2(x) = \sum_{j=1}^{n} |x_j| + \prod_{k=1}^{n} |x_k|$$

where $|x_j| \leq 10.0$, and

$$f_2(x^*) = f_2(0, 0, ..., 0) = 0.0$$

Schwefel Problem 2.26:

$$f_3(x) = -\sum_{j=1}^{n} (x_j \sin(\sqrt{|x_j|}))$$

where $|x_j| \leq 500.0$, and

$$f_3(x^*) = f_3(420.9687, 420.9687, ..., 420.9687) \approx -12569.5$$

Ackley Function:

$$f_4(x) = -20exp(-0.2\sqrt{\frac{1}{n}\sum_{j=1}^{n} x_j^2}) - exp(\frac{1}{n}\sum_{k=1}^{n} \cos 2\pi x_k) + 20 + e$$

where $|x_j| \leq 32.0$, and

$$f_4(x^*) = f_4(0, 0, ..., 0) = 0.0$$

Penalized Function1:

$$f_5(x) = \frac{\pi}{30}\{10\sin^2(\pi y_1) + \sum_{i=1}^{n-1} (y_i - 1)^2[1 + \sin^2(\pi y_{i+1})] + (y_n - 1)^2\} + \sum_{i=1}^{n} u(x_i, 10, 100, 4)$$

where $|x_j| \leq 50.0$, and

$$f_5(x^*) = f_5(1, 1, ..., 1) = 0.0$$

Penalized Function2:

$$f_6(x) = 0.1\{\sin^2(3\pi x_1) + \sum_{i=1}^{n-1} (x_i - 1)^2[1 + \sin^2(3\pi x_{i+1})] + (x_n - 1)^2[1 + \sin^2(2\pi x_n)]\} + \sum_{i=1}^{n} u(x_i, 5, 100, 4)$$

where $|x_j| \leq 50.0$, and

$$u(x_i, a, k, m) = \begin{cases} k(x_i - a)^m, & \text{if } x_i > a \\ 0, & \text{if } -a \leq x_i \leq a \\ k(-x_i - a)^m, & \text{if } x_i < -a \end{cases}$$

$$y_i = 1 + \frac{1}{4}(x_i + 1)$$

$$f_6(x^*) = f_6(1, 1, ..., 1) = 0.0$$

Sphere Model and Schwefel Problem 2.22 are unimodel functions, whereas Schwefel Problem 2.26, Ackley function and two Penalized Functions are multi-model functions with many local minima.

### 4.2 Parameter Setting

The coefficients of SPSO,MPSO-TVAC and PSO-DSL are set as follows:

The inertia weight $w$ is decreased linearly from 0.9 to 0.4. Two accelerator coefficients $c_1$ and $c_2$ are both set to 2.0 with SPSO and PSO-DSL, as well as in MPSO-TVAC, $c_1$ decreased from 2.5 to 0.5,while $c_2$ increased from 0.5 to 2.5. For CLPSO, the parameter $c$ is chosen as 1.49445, as well as the selection probability $p_c$ is increased from 0.05 up to 0.5[11].

Total individuals are 100, and $v_{max}$ is set to the upper bound of domain. The dimensions are set to 30. Each experiment the simulation run 30 times while each time the largest evolutionary generation is 1500.

### 4.3 Performance Analysis

Table 1 is the comparison results of the benchmark functions under the same evolution generations respectively. The average mean value and average standard deviation of each algorithm are computed with 30 runs and listed as follows.

For unimodal functions such as Sphere and Schwefel problem 2.22, CLPSO is superior to other three variants no matter the mean value and the standard deviation. The performance of PSO-DSL is nearly equal to that of MPSO-TVAC. From Fig.1 and 2, we can find the search pattern of CLPSO, MPSO-TVAC and PSO-DSL are not the same. For MPSO-TVAC and PSO-DSL, in the first stage they exhibit a powerful search ability, however, for CLPSO, it's search speed is much slow than other two variants in the first stage. This provides a potential search ability in the latter period.

For other four multi-modal functions with many local optima, PSO-DSL shows its superior power to detect the global optimum in complex problems except for Ackley. For Ackley, the performance of PSO-DSL and CLPSO is equivalent. One interesting phenomenon is the performance of PSO-DSL for two penalized functions, they are very unstable for many previous variants, however, for PSO-DSL, it is very stable contrarily.

**Table 1.** Comparison Results

| Function | Algorithm | Average Mean Value | Average Standard Deviation |
|---|---|---|---|
| $f_1$ | SPSO | 5.000018276888388e-010 | 6.602992640198069e-010 |
| $f_1$ | MPSO_TVAC | 4.781370708169390e-028 | 2.131351038064648e-027 |
| $f_1$ | CLPSO | 2.389585420687056e-040 | 1.306584448941780e-040 |
| $f_1$ | PSO-DSL | 2.251666904639589e-030 | 3.443127420553516e-030 |
| $f_2$ | SPSO | 1.004081284537217e-007 | 5.554568029985854e-008 |
| $f_2$ | MPSO_TVAC | 8.312001736590997e-009 | 3.089620613600377e-008 |
| $f_2$ | CLPSO | 6.881755625527024e-022 | 2.435108597937871e-022 |
| $f_2$ | PSO-DSL | 6.103484918221024e-015 | 1.305899099430203e-014 |
| $f_3$ | SPSO | -6.611115600258316e+003 | 9.449995477386343e+002 |
| $f_3$ | MPSO_TVAC | -6.622503709486737e+003 | 6.152232639474220e+002 |
| $f_3$ | CLPSO | -3.953923148429431e+003 | 2.690564797326014e+002 |
| $f_3$ | PSO-DSL | -1.083633599477462e+004 | 3.078960158313349e+002 |
| $f_4$ | SPSO | 6.958953254354583e-006 | 6.106325746998641e-006 |
| $f_4$ | MPSO_TVAC | 6.519229600598920e-014 | 8.535472145016816e-014 |
| $f_4$ | CLPSO | 8.881784197001252e-016 | 0 |
| $f_4$ | PSO-DSL | 1.101341240428155e-014 | 4.504940039743684e-015 |
| $f_5$ | SPSO | 3.110078447844952e-002 | 4.874122225306644e-002 |
| $f_5$ | MPSO_TVAC | 5.183451012524805e-003 | 2.318109764409115e-002 |
| $f_5$ | CLPSO | 2.488923030568993e-001 | 2.627405698009492e-002 |
| $f_5$ | PSO-DSL | 1.929423896797247e-021 | 8.628645943849921e-021 |
| $f_6$ | SPSO | 1.641873384803907e-007 | 4.493491018898719e-007 |
| $f_6$ | MPSO_TVAC | 7.931743741404794e-023 | 2.507187260938738e-022 |
| $f_6$ | CLPSO | 1.768231420217310e+000 | 1.675090075786802e-001 |
| $f_6$ | PSO-DSL | 1.349783804395671e-032 | 2.808011502358267e-048 |

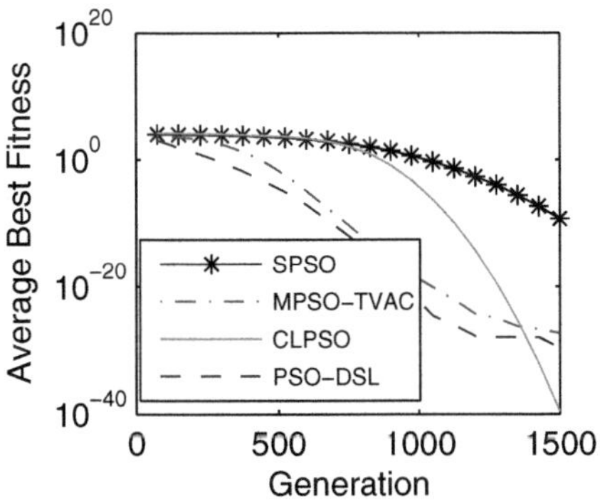


**Fig. 1.** Dynamic Comparison of $f_1$

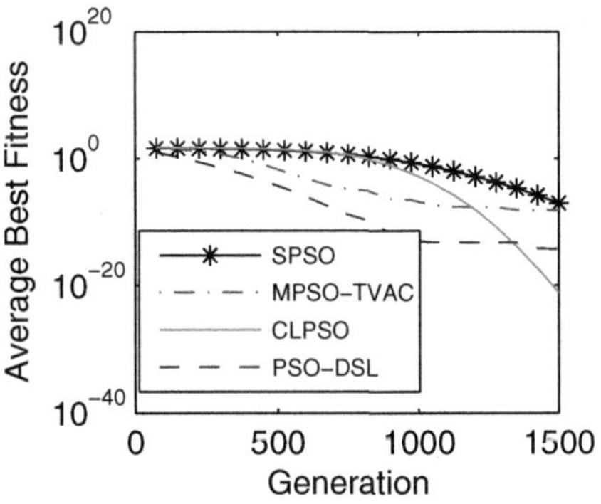


**Fig. 2.** Dynamic Comparison of $f_2$

In all, from the test suit, PSO-DSL is very satisfied with the multi-modal functions with many local optima. However, for unimodal functions, its performance is less than CLPSO.

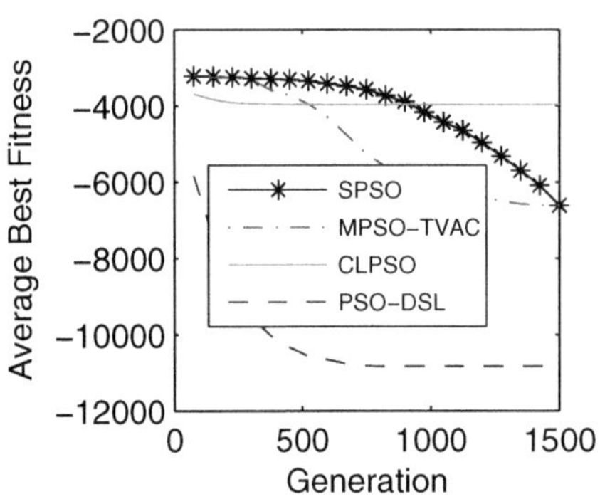


**Fig. 3.** Dynamic Comparison of $f_3$

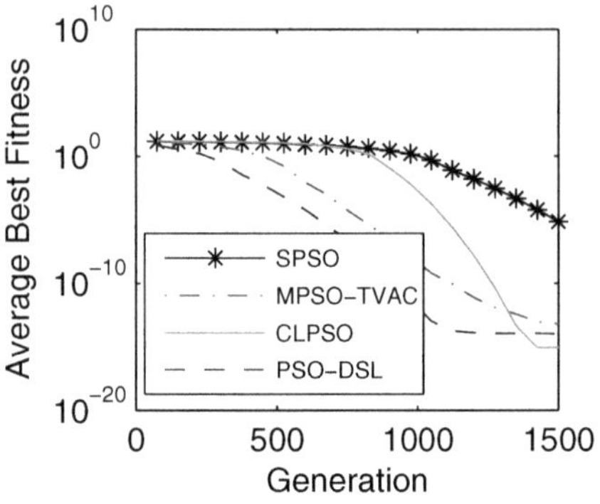


**Fig. 4.** Dynamic Comparison of $f_4$

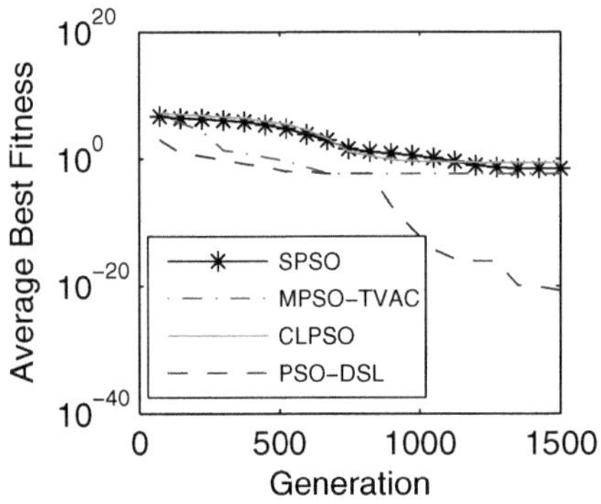


**Fig. 5.** Dynamic Comparison of $f_5$

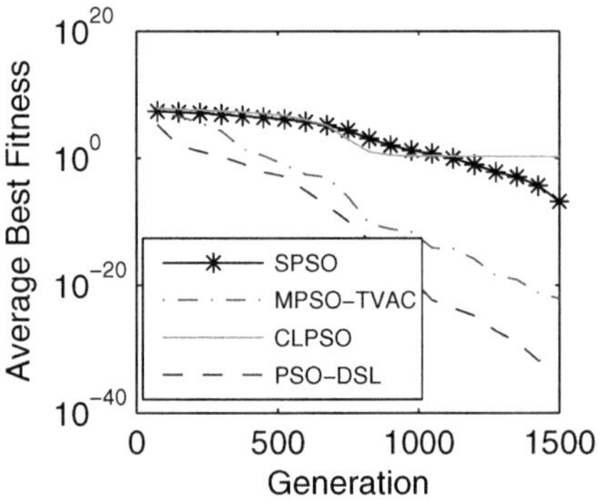


**Fig. 6.** Dynamic Comparison of $f_6$

## 5 Conclusion

This paper introduces the concept of step length, and the corresponding selection strategy with the absolute stability theory. Simulation results show the proposed particle swarm optimization with dynamic step length is effective with multi-modal benchmarks. The further research is to apply this new version of PSO into discrete area.

## Acknowledgment

This paper was supported by National Natural Science Foundation of China under Grant No.60674104, and Shanxi Educational Department Science and Technology Funds of China under Grant No.20051310.

## References

1. Bäck, T.: Evolutionary Algorithms in Theory and Practice, Oxford University Press, New York(1996)
2. Anderson, J.: A Simple Neural Network Generating an Interactive Memory, *Mathematical Biosciences*, 14 (1972) 197-220
3. Bonabeau, E., Dorigo, M., Theraulaz, G.: Swarm Intelligence: From Natural to Artificial Systems, Santa Fe Institute Publications(1999)
4. Abraham, A., Grosan, C., Ramos, V.: Swarm Intelligence and Data Mining, Studies in Computational Intelligence, Springer(2006)
5. Andries,G.,Engelbrecht P.: Fundamentals of Computational Swarm Intelligence, Wiley Publishing(2006)
6. Kennedy, J.,Eberhart, R.C.: Particle Swarm Optimization, Proceedings of IEEE International Conference on Neural Networks. (1995) 1942-1948
7. Eberhart, R.C.,Kennedy, J.: A New Optimizer Using Particle Swarm Theory, Proceedings of $6^{th}$ International Symposium on Micro Machine and Human Science, (1995) 39-43
8. Cui, Z.H., Zeng, J.C., Sun, G.J.: A Fast Particle Swarm Optimization, *International Journal of Innovative Computing, Information and Control*, 2 (2006) 1365-1380
9. Monson, C.K., Seppi, K.D.: The Kalman Swarm: A New Approach to Particle Motion in Swarm Optimization, Proceedings of the Genetic and Evolutionary Computation Conference, (2004) 140-150
10. Iwasaki, N.,Yasuda, K.: Adaptive Particle Swarm Optimization Using Velocity Feedback, *International Journal of Innovative Computing, Information and Control*, 1 (2005) 369-380
11. Liang, J.J., Qin, A.K., Suganthan, P.N., Baskar, S.: Comprehensive Learning Particle Swarm Optimizer for Global Optimization of Multimodal Functions, *IEEE Transactions on Evolutionary Computation*, 10 (2006) 281-295
12. Cui, Z.H., Zeng, J.C.: A Guaranteed Global Convergence Particle Swarm Optimizer, Lecture Notes in Artificial Intelligence, vol.3066, Sweden, (2004) 762-767
13. Ratnaweera A., Halgamuge S.K., Watson H.C.: Self-Organizing Hierarchical Particle Swarm Opitmizer with Time-Varying Acceleration Coefficients, *IEEE Transactions on Evolutionary Computation*, 8(2004) 240-255

# Stability Analysis of Particle Swarm Optimization

Jinxing Liu[1,2], Huanbin Liu[1], and Wenhao Shen[1]

[1] State Key Laboratory of Pulp and Paper Engineering,
South China University of Technology,
510640, Guangzhou, Guangdong, China
[2] Qufu Normal University
sdcavell@126.com

**Abstract.** This paper explores how the particle swarm optimization algorithm works inside and how the values of $\beta$ influence the behavior of the particle. According to Lyapunov Stability theorem, the stability of the PSO algorithm is analyzed. It is found that when $\beta < 4$, the PSO algorithm is stable; when $\beta > 4$, the PSO algorithm is unstable; when $\beta = 4$, the PSO algorithm is sensitive to the initial value and the system is chaotic. The experiment validated the above conclusions.

**Keywords:** Stability, Particle Swarm Optimization, Lyapunov Stability Theorem.

## 1 Introduction

Particle swarm adaptation has been shown to successfully optimize a wide range of continuous functions. Particle swarm optimization (PSO) is a stochastic population based optimization approach, first published by Kennedy and Eberhart in 1995 [1], [2]. Since its first publication, a large body of research has been done to study the performance of PSO, and to improve its performance [3], [4]. From these empirical studies, many efforts have been invested to obtain a better understanding of the convergence properties of PSO. From these empirical studies it can be concluded that the PSO is sensitive to control parameter choices.

To gain a better, general understanding of the behavior of a particle, in-depth theoretical analyses of particle trajectories are necessary. Till now, a few theoretical studies of particle trajectories have been done, which concentrate on PSO systems [5], [6], [7].

The particle swarm is an algorithm for finding optimal regions of complex search spaces through the interaction of individuals in a population of particles. Even though the algorithm has been shown to perform well, researchers have not adequately explained how it works. To ensure the algorithm executing efficiently, we will study how to control the trajectories of particles. Furthermore, according to Lyapunov

D.-S. Huang, L. Heutte, and M. Loog (Eds.): ICIC 2007, LNAI 4682, pp. 781–790, 2007.

stability theorem, this paper provides a formal proof that the algorithm is stable while the attraction coefficient is less than four.

## 2 Stability Algorithm for PSO

### 2.1 Lyapunov Stability Theorems

Lyapunov stability theorems [8], [9], [10], [11] are the basic theories for stability analysis of a system. The following is a stability theorem of linear time-invariant discrete-time systems deduced from the Lyapunov stability theorems.

**Theorem.** Consider the discrete-time system

$$X(k+1) = GX(k)\,, \tag{1}$$

where $X$ is a state vector (an n-vector) and $G$ is an $n \times n$ constant nonsingular matrix. A necessary and sufficient condition for the equilibrium state $X = 0$ to be asymptotically stable in the large is that, given any positive-definite Hermitian (or any positive definite real symmetric) matrix $Q$, there exists a positive-definite Hermitian (or a positive definite real symmetric) matrix $P$ such that

$$G^T PG - P = -Q\,, \tag{2}$$

where $G^T$ is the transposition of $G$. The scalar function $X^T PX$ is a Lyapunov function for this system [8].

For a test of positive definiteness of an $n \times n$ matrix, **Sylvester's criterion** can be applied, which states that a necessary and sufficient condition for the matrix to be positive definite is that the determinants of all the successive principal minors of the matrix be positive. Consider, for example, the following $n \times n$ Hermitian matrix $P$ (if the elements of $P$ are all real, then the Hermitian matrix becomes a real symmetric matrix):

$$P = \begin{bmatrix} p_{11} & p_{12} & \cdots & p_{1n} \\ \overline{p}_{12} & p_{22} & \cdots & p_{2n} \\ \vdots & \vdots & & \vdots \\ \overline{p}_{1n} & \overline{p}_{2n} & \cdots & p_{nn} \end{bmatrix}, \tag{3}$$

where $\overline{p}_{ij}$ denotes the complex conjugate of $p_{ij}$. The matrix $P$ is positive definite if all the successive principal minors are positive, that is,

$$p_{11} > 0, \begin{vmatrix} p_{11} & p_{12} \\ \overline{p}_{12} & p_{22} \end{vmatrix} > 0, \cdots, \begin{vmatrix} p_{11} & p_{12} & \cdots & p_{1n} \\ \overline{p}_{12} & p_{22} & \cdots & p_{2n} \\ \vdots & \vdots & & \vdots \\ \overline{p}_{1n} & \overline{p}_{2n} & \cdots & p_{nn} \end{vmatrix} > 0\,. \tag{4}$$

## 2.2 Particle Swarm Optimization (PSO) Algorithm

### 2.2.1 Standard Algorithm [1], [2]

The basic PSO algorithm can be described in vector notation as follows:

$$\begin{cases} \vec{v}_{k+1} = \vec{v}_k + \vec{\beta}_1(\vec{p}_l - \vec{x}_k) + \vec{\beta}_2(\vec{p}_g - \vec{x}_k) \\ \vec{x}_{k+1} = \vec{v}_{k+1} + \vec{x}_k \end{cases}. \tag{5}$$

At iteration $k$, the velocity $\vec{v}_k$ is updated based on its current value and on a term which attracts the particle towards previously found best positions: its own previous best position ( $\vec{p}_l$.)and global best position in the whole swarm ( $\vec{p}_g$ ). The strength of attraction is given by the coefficients $\vec{\beta}_1$ and $\vec{\beta}_2$, which are the vectors of random numbers. They are usually selected as uniform random numbers in the range $[0, \beta_{max}]$ The particle position $\vec{x}_k$ is updated using its current value and the newly computed velocity $\vec{v}_{k+1}$.

### 2.2.2 One-Dimensional Algorithm

It appears from Eq. (5) that each dimension is updated independently from the others. The only link between the dimensions of the problem space is the global best position ( $\vec{p}_g$ ) found so far. Thus, without loss of generality, the algorithm description can be reduced for analysis purpose to the one-dimensional case:

$$\begin{cases} v_{k+1} = v_k + \beta_1(p_l - x_k) + \beta_2(p_g - x_k) \\ x_{k+1} = v_{k+1} + x_k \end{cases}. \tag{6}$$

Let

$$\beta = \beta_1 + \beta_2, \quad p = \frac{\beta_1 p_l + \beta_2 p_g}{\beta_1 + \beta_2}. \tag{7}$$

Then, Eq. (6) can be simplified as:

$$\begin{cases} v_{k+1} = v_k - \beta x_k + \beta p \\ x_{k+1} = v_k + (1-\beta)x_k + \beta p \end{cases}. \tag{8}$$

The newly introduced attraction coefficient $\beta$ is the combination of the local and global attraction coefficients $\beta_1$ and $\beta_2$, so $\beta$ is selected in the range $[0, 2\beta_{max}]$. The attraction point $p$ is the weighted average of its own previous best position ( $p_l$.)and global best position in the whole swarm ( $p_g$ ).

## 3 Dynamic Analysis

To be convenient for the dynamic analysis, let

$$y_k = x_k - p \,, \tag{9}$$

Eq. (8) can be simplified as:

$$\begin{cases} v_{k+1} = v_k - \beta y_k \\ y_{k+1} = v_k + (1-\beta) y_k \end{cases}. \tag{10}$$

Let

$$X_k = \begin{bmatrix} v_k \\ y_k \end{bmatrix}, \tag{11}$$

then Eq. (10) can be written in matrix-vector notation as follow:

$$X_{k+1} = GX_k \,, \tag{12}$$

where

$$G = \begin{bmatrix} 1 & -\beta \\ 1 & (1-\beta) \end{bmatrix}. \tag{13}$$

According to the theorem in section 2.1,

$$G^T PG - P = -I \,, \tag{14}$$

where

$$P = \begin{bmatrix} p_{11} & p_{12} \\ p_{12} & p_{22} \end{bmatrix}, \quad I = \begin{bmatrix} 1 & 0 \\ 0 & 1 \end{bmatrix}, \tag{15}$$

$G^T$ is the transposition of G.

Substituting Eq. (13) and Eq. (15) into Eq. (14), the Eq. (16) is obtained:

$$\begin{bmatrix} 2p_{11} + p_{22} & -\beta p_{11} - 2\beta p_{12} + (1-\beta) p_{22} \\ -\beta p_{11} - 2\beta p_{12} + (1-\beta) p_{22} & \beta^2 p_{11} + 2(\beta^2 - \beta) p_{12} + (\beta^2 - 2\beta) p_{22} \end{bmatrix} = \begin{bmatrix} -1 & 0 \\ 0 & -1 \end{bmatrix}. \tag{16}$$

Then, the following system of equations is obtained:

$$\begin{cases} 2p_{11} + p_{22} = -1 \\ -\beta p_{11} - 2\beta p_{12} + (1-\beta) p_{22} = 0 \\ \beta^2 p_{11} + 2(\beta^2 - \beta) p_{12} + (\beta^2 - 2\beta) p_{22} = -1 \end{cases}. \tag{17}$$

From Eq. (17), the Eq. (18) can be obtained:

$$\begin{cases} p_{12} = (\beta p_{11} + \beta - 1)/2 \\ p_{22} = \beta p_{11} - \beta \end{cases}. \tag{18}$$

According to Sylvester's criterion, the necessary and sufficient condition is that matrix P must be positive definite:

$$\begin{cases} p_{11} > 0 \\ p_{11}p_{22} - p_{12}p_{21} > 0 \end{cases}. \tag{19}$$

Let

$$\begin{aligned} f &= p_{11}p_{22} - p_{12}p_{21} \\ &= (\beta - \frac{\beta^2}{4})p_{11}^2 - \frac{\beta^2+1}{2}p_{11} - \frac{1}{4}(\beta-1)^2, \end{aligned} \tag{20}$$

and

$$a = \beta - \frac{\beta^2}{4}, \quad b = -\frac{\beta^2+1}{2}, \quad c = -\frac{1}{4}(\beta-1)^2. \tag{21}$$

It is obvious that b and c are not greater than zero. Here,

$$\Delta = b^2 - 4ac = \beta[\beta^2 + (\beta-1)^2]. \tag{22}$$

For $\beta > 0$, by all appearances, $\Delta > 0$, so the equation f = 0 has two different real roots which can be expressed as:

$$\lambda_1 = \frac{-b+\sqrt{\Delta}}{2a}, \quad \lambda_2 = \frac{-b-\sqrt{\Delta}}{2a}. \tag{23}$$

To satisfy condition of the Eq. (19), condition f > 0 must be satisfied.

- Case $a > 0$, that is, $0 < \beta < 4$, so $\lambda_1 > 0 > \lambda_2$.

While $p_{11} > \lambda_1 > 0$, then f > 0, thus matrix P is positive definite. Therefore, while $0 < \beta < 4$, the algorithm of PSO is large-scale asymptotically stable.

- Case $a < 0$, that is, $\beta > 4$ here,

$$b^2 - \Delta = \frac{1}{4}\beta(\beta-4)(\beta-1)^2 > 0, \tag{24}$$

And b < 0, so

$$\lambda_1 > 0, \lambda_2 = \frac{-b-\sqrt{\Delta}}{2a} < 0. \tag{25}$$

To satisfy $f>0$, $p_{11}$ must be bound in the range $(\lambda_1,\lambda_2)$. According to Sylvester's criterion, matrix P is negative definite. Therefore, while $\beta>4$, the algorithm of PSO is unstable.

- Case a = 0, that is, $\beta=4$.

In this situation,

$$G=\begin{bmatrix}1 & -4\\ 1 & -3\end{bmatrix}. \tag{26}$$

In this particular case, the eigenvalues are both equal to -1 and there is just one family of eigenvectors, generated by

$$V=\begin{bmatrix}2\\ 1\end{bmatrix}. \tag{27}$$

So, $GV=-V$.

Fig. 1 shows the line which is fixed by both point (0, 0) and point (2, 1).

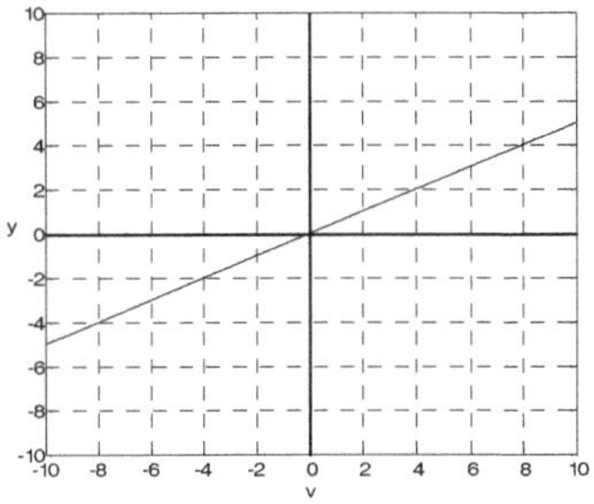


**Fig. 1.** The line y = x / 2.

Thus, if $P_0$ is an eigenvector and proportional to V (that is to say, if $P_0$ lies in the line $y=x/2$, see Fig. 1), for

$$P_{t+1}=\pm\begin{bmatrix}2y_0\\ y_0\end{bmatrix}=-P_t, \tag{28}$$

there are just two symmetrical points $(2y_0,y_0)$ and $(-2y_0,-y_0)$. In this case, the system is neither divergent nor convergent. So it is an undamped oscillating system which is stable in the sense of Lyapunov.

In the case where $P_0$ is not an eigenvector, it can be directly computed how $\|P_t\|$ decreases and/or increases.

Define $\Delta_t=\|P_{t+1}\|^2-\|P_t\|^2$, where $\|P_t\|$ is the Euclidean norm.

Then select the initial point $P_0(2y_0+\varepsilon,y_0)$ above the line $y=x/2$, where $\varepsilon>0$. By recurrence, the following is derived:

$$\begin{aligned} \Delta_0 &= -10y_0\varepsilon + \varepsilon^2 \\ \Delta_1 &= -10y_0\varepsilon + 11\varepsilon^2 \\ \Delta_2 &= -10y_0\varepsilon + 21\varepsilon^2 \\ &\vdots \end{aligned} \tag{29}$$

$$\Delta_t = -10y_0\varepsilon + (10t+1)\varepsilon^2 . \tag{30}$$

As long as $\Delta_t < 0$, that is to say, $(10t+1)\varepsilon^2 < 10y_0\varepsilon$, then,

$$t < Integer(\frac{y_0}{\varepsilon} - \frac{1}{10}) + 1 , \tag{31}$$

where $\varepsilon > 0$ and $y_0 > \varepsilon/10$, $\|P_t\|$ decreases.

When

$$t > Integer(\frac{y_0}{\varepsilon} - \frac{1}{10}) + 1 , \tag{32}$$

$\|P_t\|$ increases.

From Eq. (31), if $\varepsilon > 0$ and $y_0 \le \varepsilon/10$, $\|P_t\|$ directly increases.

Similarly, if $\varepsilon < 0$ and $y_0 > \varepsilon/10$, Eq. (31) is satisfied, $\|P_t\|$ decreases. After that, $\|P_t\|$ increases. If $\varepsilon > 0$ and $y_0 \le \varepsilon/10$, $\|P_t\|$ directly increases.

Thus, it can be concluded that $\|P_t\|$ decreases/increases when t satisfied Eq. (31). In particular, even if $\|P_t\|$ begins to decrease, it will tend to increase in the end. So the system is unstable.

# 4 Experiment

## 4.1 Experimental Result

Fig. 2 shows a two-dimensional representation of the restrictions $(v, y)$ of a particle with different values of $\beta$ and different initial values $(v_0, y_0)$. The value of $\beta$ and initial values $(v_0, y_0)$ are shown on the top of each figure. Fifty iterations are executed in Fig. 2(a) ~ Fig. 2(j). Thirty iterations are executed in Fig. 2(k) and Fig. 2(m), two-hundred iterations in Fig. 2(l) and Fig. 2(n).

## 4.2 The Result Analysis

From Fig. 2, it can be found that:

(I) the particle converges toward a nontrivial attractor [12] when $\beta < 4$ (See Fig. 2(a), (b), (c) and (d)); quick divergence when $\beta > 4$ (See Fig. 2(g) and (h)); yet difficult convergence when $\beta \approx 4$ (See Fig. 2(e), (f), (i), (j), (k), (l), (m) and (n)).

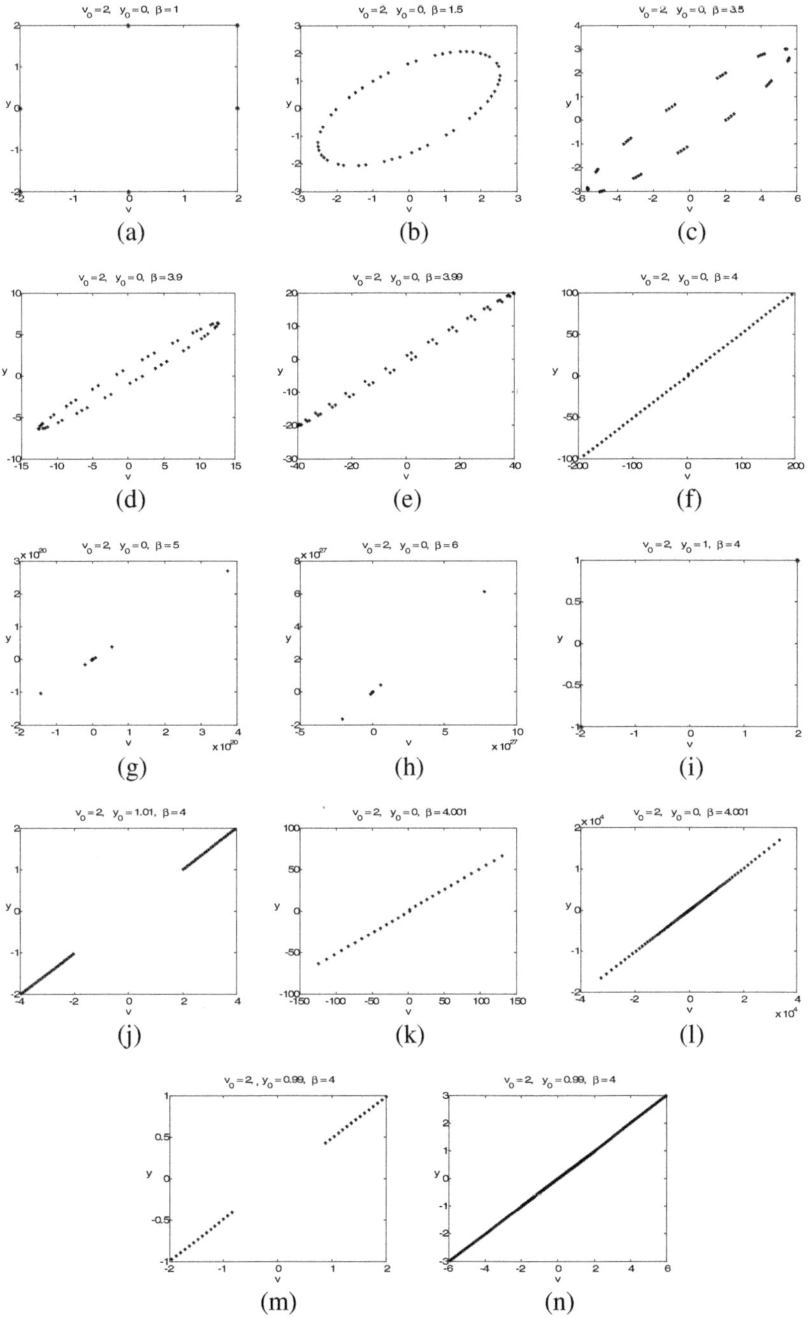


**Fig. 2.** Trajectories of a particle in two-dimensional space with different values of $\beta$ and different initial values $(v_0, y_0)$. The initial values $(v_0, y_0)$ and $\beta$ are shown on the top of each figure.

(II) Case $\beta < 4$ : From Fig. 2(a), (b), (c), (d) to (e), we can see that the trajectories of the particle are approximately elliptical. In addition, with the increase of the values of $\beta$ (where $\beta < 4$ ), the slope of macro axis gradually changes from negative infinite to 1/2, the radius of the macro axis becomes longer and the radius of the minor axis becomes shorter. That is to say, the trajectories are gradually close to the line $y = x/2$ and stretched out on the line.

(III) Case $\beta > 4$ : From Fig. 2(k), (l), (g) to (h), it can be seen that the trajectories of the particles are close to some lines. With the times increasing, the trajectories of the particle rapidly deviated from the origin point. In addition, the larger the value of $\beta$ , the rapider the deviation.

(IV) Case $\beta = 4$ : From Fig. 2(f), (i), (j), (m) and (n), we can see that the trajectories are near the line $y = x/2$ , but, sensitive to the initial values $(v_0, y_0)$ . From Fig. 2(i), with the initial value $(2,1)$ , the trajectory is two points $(2,1)$ and $(-2,-1)$ . From Fig. 2(j), for the initial value $(2,1.01)$ , the trajectory is approximate-linearly away from the origin point. For the initial value $(2,0.99)$ , Fig. 2(m) shows that the trajectory is close to the origin point. Otherwise, for the same initial value, Fig. 2(n) shows that the trajectory is away from the origin point. From Eq. (31), we can calculate the value of t equal to 50. While the number of iteration is less than 50, the trajectory is close to the origin point. Otherwise, from Eq. (32), when the number of iteration times greater than 50, the trajectory is away from the origin point. Both Fig. 2(m) and (n) validate these points. From Fig. 2(f), we can see that when the initial value is not proportional to $(2,1)$ , the trajectories are divergent approximately along the line $y = x/2$ .

In addition, when the value of $\beta$ is less than four, the system is convergent. When $\beta$ is greater than four, the system is divergent. When $\beta$ is equal to four, the chaos appears.

## 5 Conclusions

According to Lyapunov Stability theorem, it is found that when $\beta < 4$ , the PSO algorithm is stable; when $\beta > 4$ , the PSO algorithm is unstable; when $\beta = 4$ , the PSO algorithm is sensitive to the initial value.

How the different values of $\beta$ (where $\beta < 4$ ) influence the behavior of a particle is a challengeable topic for a future paper. The research on the chaos ( $\beta = 4$ ) is an interesting topic in future.

## References

1. Kennedy, J., Eberhart, R.C. : Particle Swarm Optimization, in: Proceedings of the IEEE International Joint Conference on Neural Networks, IEEE Press, (1995) 1942–1948
2. Eberhart, R.C., Kennedy, J.: A New Optimizer Using Particle Swarm Theory, in: Proceedings of the Sixth International Symposium on Micromachine and Human Science, Nagoya, Japan, (1995) 39–43

3. Liu, B., Wang, L.,et al.: Improved Particle Swarm Optimization combined with Chaos, Chaos, Solitons and Fractals 25 (2005) 1261–1271
4. Yi, D., Ge, X.: An Improved PSO-based ANN with Simulated Annealing Technique, Neurocomputing 63 (2005) 527–533
5. Clerc, M., Kennedy, J.: The Particle Swarm-Explosion, Stability, and Convergence in a Multidimensional Complex Space, IEEE Transactions on Evolutionary Computation 6 (1) (2002) 58–73
6. Trelea, I.C.: The Particle Swarm Optimization Algorithm: Convergence Analysis and Parameter Selection, Information Processing Letters 85 (6) (2003) 317–325
7. Yasuda, K., Ide, A., Iwasaki, N.: Adaptive Particle Swarm Optimization, in: Proceedings of the IEEE International Conference on Systems, Man, and Cybernetics, (2003) 1554–1559
8. Ogata,K.: Discrete-time Control Systems, Prentice-Hall International, Inc. (1987) 545 – 561
9. Chen, C. T.: Linear System Theory and Design, Holt, Rinehart and Winstom, (1984) 412 – 425
10. Strejc, U.: State Space Theory of Discrete Linear Control, John, Wiley and Sons, (1981) 197 – 204
11. Lehnigk, S. H.: Stability Theorems for Linear Motions with an Introduction to Lyapunov's Direct method, Pentic-hall Inc. N.J. (1966) 25 – 71
12. Skowronski, J.M.: Nonlinear Lyapunov dynamics, World Scientific Publishing, (1990) 254 – 267

# A Novel Discrete Particle Swarm Optimization Based on Estimation of Distribution

Jiahai Wang

Department of Computer Science, Sun Yat-sen University,
No.135, Xingang West Road, Guangzhou 510275, P.R. China
wjiahai@hotmail.com

**Abstract.** The philosophy behind the original PSO is to learn from individual's own experience and best individual experience in the whole swarm. Estimation of distribution algorithms sample new solutions from a probability model which characterizes the distribution of promising solutions in the search space at each generation. In this paper, a novel discrete particle swarm optimization algorithm based on estimation of distribution is proposed for combinatorial optimization problems. The proposed algorithm combines the global statistical information collected from local best solution information of all particles and the global best solution information found so far in the whole swarm. To demonstrate its performance, experiments are carried out on the knapsack problem, which is a well-known combinatorial optimization problem. The results show that the proposed algorithm has superior performance to other discrete particle swarm algorithms as well as having less parameters.

**Keywords:** Discrete particle swarm optimization, estimation of distribution, knapsack problem, combinatorial optimization problem.

## 1 Introduction

The PSO is inspired by observing the bird flocking or fish school [1]. A large number of birds/fishes flock synchronously, change direction suddenly, and scatter and regroup together. Each individual, called a particle, benefits from the experience of its own and that of the other members of the swarm during the search for food. Comparing with genetic algorithm, the advantages of PSO lie on its simple concept, easy implementation and quick convergence. The PSO has been applied successfully to continuous nonlinear function [1], neural network [2], nonlinear constrained optimization problems [3], etc. Most of the applications have been concentrated on solving continuous optimization problems [4].

To solve discrete (combinatorial) optimization problems, Kennedy and Eberhart [5] also developed a discrete version of PSO (DPSO), which however has seldom been utilized. DPSO essentially differs from the original (or continuous) PSO in two characteristics. First, the particle is composed of the binary variable. Second, the velocity must be transformed into the change of probability, which is the chance of the binary variable taking the value one. Furthermore, the

D.-S. Huang, L. Heutte, and M. Loog (Eds.): ICIC 2007, LNAI 4682, pp. 791–802, 2007.

relationships between the DPSO parameters differ from normal continuous PSO algorithms [6] [7]. Although Kennedy and Eberhart [5] have tested the robustness of the discrete binary version through function optimization benchmark, few applications for combinatorial optimization have ever been developed based on their work.

Though it has been proved the DPSO can also be used in discrete optimization as a common optimization method, it is not as effective as in continuous optimization. When dealing with integer variables, PSO sometimes are easily trapped into local minima [5]. Therefore, Yang et al. [8] proposed a quantum particle swarm optimization (QPSO) for discrete optimization in 2004. Their simulation results showed that the performance of the QPSO was better than DPSO and genetic algorithm. Recently, Yin [9] proposed a genetic particle swarm optimization (GPSO) with genetic reproduction mechanisms, namely crossover and mutation to facilitate the applicability of PSO to combinatorial optimization problem, and the results showed that the GPSO outperformed the DPSO for combinatorial optimization problems.

In the last decade, more and more researchers tried to overcome the drawbacks of usual recombination operators of evolutionary computation algorithms. Therefore, estimation of distribution algorithms (EDAs) [10] have been developed. These algorithms, which have a theoretical foundation in probability theory, are also based on populations that evolve as the search progresses. EDAs use probabilistic modeling of promising solutions to estimate a distribution over the search space, which is then used to produce the next generation by sampling the search space according to the estimated distribution. After every iteration, the distribution is re-estimated.

The philosophy behind the original PSO is to learn from individual's own experience and best individual experience in the whole swarm. Estimation of distribution algorithms sample new solutions from a probability model which characterizes the distribution of promising solutions in the search space at each generation. In this paper, a discrete particle swarm optimization algorithm based on estimation of distribution is proposed for combinatorial optimization problems. The proposed algorithm combines the global statistical information collected from local best solution information of all particles and the global best solution information found so far in the whole swarm. To demonstrate its performance, experiments are carried out on the knapsack problem, which is a well-known combinatorial optimization problem. The results show that the proposed algorithm has superior performance to other discrete particle swarm algorithms as well as having less parameter.

## 2 Particle Swarm Optimization

PSO is initialized with a group of random particles (solutions) and then searches for optima by updating each generation. In every iteration, each particle is updated by following two best values. The first one is the local best solution (fitness) a particle has obtained so far. This value is called personal best solutions.

Another best value is that the whole swarm has obtained so far. This value is called global best solution. The philosophy behind the original PSO is to learn from individual's own experience (personal best solution) and best individual experience (global best solution) in the whole swarm, which can be described by Fig.1.

Denote by $N$ particle number in the swarm. Let $X_i(t) = (x_{i1}(t), \cdots, x_{id}(t), \cdots, x_{iD}(t))$, be particle $i$ with $D$ bits at iteration $t$, where being treated as a potential solution. Denote the velocity as $V_i(t) = (v_{i1}(t), \cdots, v_{id}(t), \cdots, v_{iD}(t))$, $v_{id}(t) \in R$. Let $PBest_i(t) = (pbest_{i1}(t), \cdots, pbest_{id}(t), \cdots, pbest_{iD}(t))$ be the best solution that particle $i$ has obtained until iteration $t$, and $GBest(t) = (gbest_1(t), \cdots, gbest_d(t), \cdots, gbest_D(t))$ be the best solution obtained from $PBest_i(t)$ in the whole swarm at iteration $t$. Each particle adjusts its velocity according to previous velocity of the particle, the cognition part and the social part. The algorithm is described as follows [1]:

$$v_{id}(t+1) = v_{id}(t) + c_1 \cdot r_1 \cdot (pbest_{id}(t) - x_{id}(t)) + c_2 \cdot r_2 \cdot (gbest_d(t) - x_{id}(t)), \quad (1)$$

$$x_{id}(t+1) = x_{id}(t) + v_{id}(t+1), \quad (2)$$

where

$c_1$ is the cognition learning factor and $c_2$ is the social learning factor;

$r_1$ and $r_2$ are the random numbers uniformly distributed in [0,1].

Most of the applications have been concentrated on solving continuous optimization problems. To solve discrete (combinatorial) optimization problems, Kennedy and Eberhart [5] also developed a discrete version of PSO (DPSO), which however has seldom been utilized. DPSO essentially differs from the original (or continuous) PSO in two characteristics. First, the particle is composed of the binary variable. Second, the velocity must be transformed into the change of probability, which is the chance of the binary variable taking the value one. The velocity value is constrained to the interval [0, 1] using the following sigmoid function:

$$s(v_{id}) = \frac{1}{1 + \exp(-v_{id})}, \quad (3)$$

where $s(v_{id})$ denotes the probability of bit $x_{id}$ taking 1. Then the particle changes its bit value by

$$x_{id} = \begin{cases} 1 & \text{if } \mathrm{rand}() \le s(v_{id}) \\ 0 & \text{otherwise} \end{cases}, \quad (4)$$

where rand() is a random number selected from a uniform distribution in [0,1]. To avoid $s(v_{id})$ approaching 1 or 0, a constant $V_{\max}$ as a maximum velocity is used to limit the range of $v_{id}$, that is, $v_{id} \in [-V_{\max}, V_{\max}]$. The basic flowchart of PSOs (including continuous PSO and discrete PSO) is shown by Fig.1.

## 3 Estimation of Distribution Algorithms (EADs)

Evolutionary Algorithms that use information obtained during the optimization process to build probabilistic models of the distribution of good regions in the

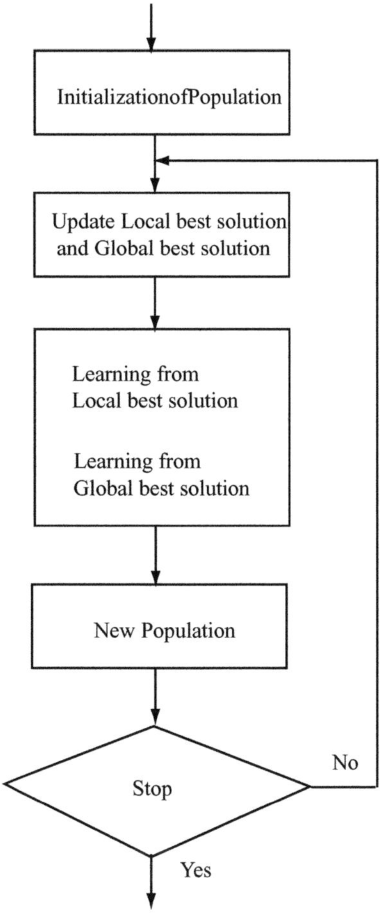


**Fig. 1.** The basic flowchart of PSO

search space and that use these models to generate new solutions are called estimation of distribution algorithms (EDAs) [11]. These algorithms, which have a theoretical foundation in probability theory, are also based on populations that evolve as the search progresses. EDAs use probabilistic modeling of promising solutions to estimate a distribution over the search space, which is then used to produce the next generation by sampling the search space according to the estimated distribution. After every iteration, the distribution is re-estimated.

An algorithmic framework of most EDAs can be described as:

InitializePopulation( ) /*Initialization*/
**While** Stopping criteria are not satisfied do /*Main Loop*/
$P_{sel}$=Select($P$)/*Selection*/

$P(x) = P(x|P_{sel})$=EstimateProbabilityDistribution( ) /*Estimation*/
$P$=SampleProbabilityDistribution( ) /*Sample*/
**EndWhile**

An EDA starts with a solution population $P$ and a solution distribution model $P(x)$. The main loop consists of three principal stages. The first stage is to select the best individuals (according to some fitness criteria) from the population. These individuals are used in a second stage in which the solution distribution model $P(x)$ is updated or recreated. The third stage consists of sampling the updated solution distribution model to generate new solutions offspring. EDAs are based on probabilistic modelling of promising solutions to guide the exploration of the search space instead of using crossover and mutation like in the well-known genetic algorithms (GAs). The basic flowcharts of EDAs and GAs is illustrated by Fig.2, which is also shows the difference between GAs and EDAs.

There has been a growing interest for EDAs in the last years. More comprehensive presentation of the EDA field can be found in Refs. [12] [13].

## 4 Novel Discrete PSO Based on EDA

Several different probability models have been introduced in EDAs for modeling the distribution of promising solutions. The univariate marginal distribution (UMD) model is the simplest one and has been used in univariate marginal distribution algorithm [14], population-based incremental learning (PBIL) [15], compact GA [16]. In this section, we describe the proposed discrete algorithm which uses global statistic information gathered from the local best solutions of all particles during the optimization process to guide the search.

In the proposed algorithm, as defined in the previous section, denote by $N$ the number of particles in the swarm. Let $X_i(t) = (x_{i1}(t), \cdots, x_{id}(t), \cdots, x_{iD}(t))$, $x_{id}(t) \in \{0, 1\}$, be particle $i$ with $D$ bits at iteration $t$, where $X_i(t)$ being treated as a potential solution. Firstly, all the local best solutions are selected; then, the UMD model is adopted to estimate the distribution of good regions over the search space based on the selected local best solutions. The UMD uses a probability vector $P = (p_1, \cdots, p_d, \cdots, p_D)$ to characterize the distribution of promising solutions in the search space, where $p_d$ is the probability that the value of the $d$-th position of a promising solution is 1. New offspring solutions are thus generated by sampling the updated solution distribution model. The probability vector $P = (p_1, \cdots, p_d, \cdots, p_D)$ guides a particle to search in binary 0-1 solution space in the following way:

**If** rand() $< \beta$
if rand() $< p_d$, set $x_{id}(t+1) = 1$, otherwise set $x_{id}(t+1) = 0$;
**Otherwise** $x_{id}(t+1) = gbest_d(t)$.

In the sample process above, a bit is sampled from the probability vector $P$ randomly or directly copied from the global best, which is controlled or balanced

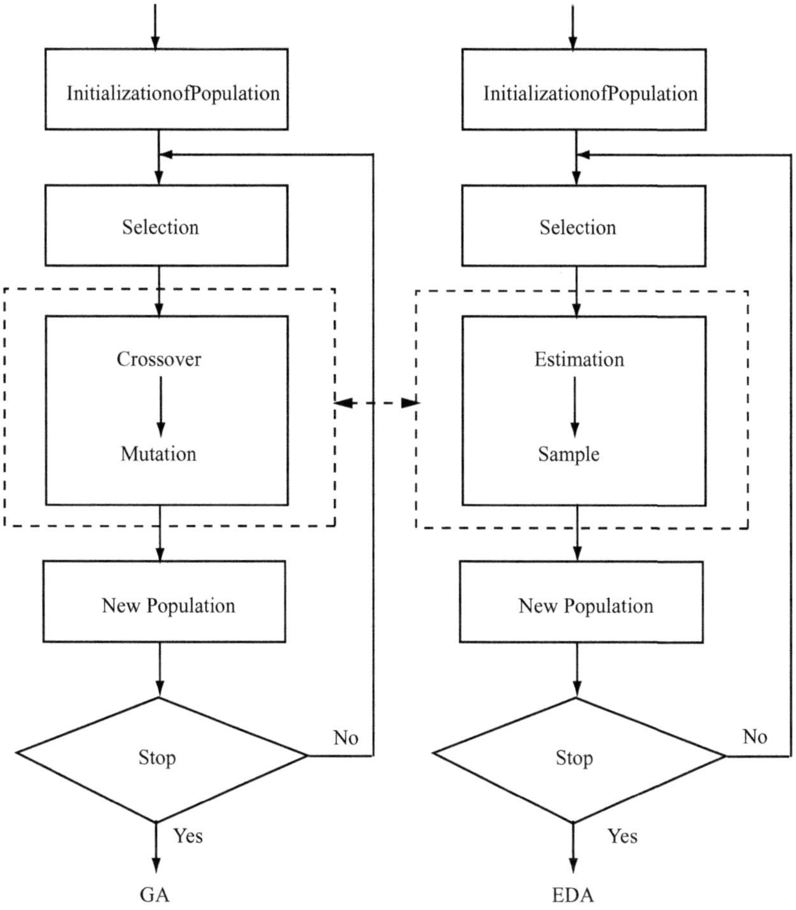


**Fig. 2.** The basic flowcharts of GA and EDA

by a parameter $\beta$. The larger $\beta$ is, the more elements of $X_i(t)$ are sampled from the vector $P$.

The probability vector $P$ is initialized by the following rule:

$$p_d = \frac{\sum_{i=1}^{N} pbest_{id}}{N}, \tag{5}$$

$p_d$ is the percentage of the binary strings with the value of the $d$-th element being 1. $P$ can also be regarded as the center of the personal best solutions of all the particles.

The probability vector in the proposed algorithm can be learned and updated at each iteration for modeling the distribution of promising solutions. Since some elements of the offspring are sampled from the probability vector $P$, it can be expected that should fall in or close to a promising area. The sampling mechanism can also provide diversity for the search afterwards. At each iteration $t$ in the proposed algorithm, the personal best solutions of all the particles are selected and used for updating the probability vector $P$. Therefore, the probability vector $P$ can be updated in the same way as in the PBIL algorithm [15]:

$$p_d = (1 - \lambda)p_d + \lambda \frac{\sum_{i=1}^{N} pbest_{id}}{N}, \tag{6}$$

where $\lambda \in (0, 1]$ is the learning rate. As in PBIL [15], the probability vector $P$ is used to generate the next set of sample points; the learning rate also affects which portions of the problem space will be explored. The setting of the learning rate has a direct impact on the trade-off between exploration of the problem space and exploitation of the exploration already conducted. For example, if the learning rate is 0, there is no exploitation of the information gained through search. As the learning rate is increased, the amount of exploitation increases, and the ability to search large portions of the problem space diminishes. In order to balance the exploration and exploitation ability in the proposed algorithm, the probability vector is updated as:

$$p_d = (1 - \lambda_d)p_d + \lambda_d \frac{\sum_{i=1}^{N} pbest_{id}}{N}, \tag{7}$$

where $\lambda_d$ is a random number selected from a uniform distribution in (0,1]. In this equation, different dimensions of the probability vector adopt different random learning rates, which is a random way to balance the exploration and exploitation ability.

In order to keep the diversities in particle swarm, a mutation operator is also incorporated into the proposed algorithm. After each bit is decided in accordance with estimated marginal distribution, the mutation operator independently flips the bit of an individual with a mutation probability.

The basic flowchart of the proposed algorithm is illustrated by Fig.3. From Fig.1–3, we can see that the proposed algorithm is different from the pure EDAs or PSO. Pure EDAs extract global statistical information from the previous search and then represent it as a probability model, which characterizes the distribution of promising solutions in the search space. New solutions are generated by sampling from this model. However, the location information of the locally optimal solutions found so far, for example, the global best solution in the population, has not been directly used in the pure EDAs. In the proposed algorithm, the location information of the locally optimal solutions found so far, global best solution, has been directly used, therefore a particle can learn not only from the global statistical information collected by the historical personal best solutions of all the particles, but also from the global best solution found so far in the whole swarm. Further, in contrast to that a particle can only learn from oneself best experience in the original PSO, a particle can learn from the global statistical information collected by the personal best experiences of all the particles in the proposed algorithm. That is, all particles can potentially contribute to a particle's search via the probability vector $P$, which can be seen as a

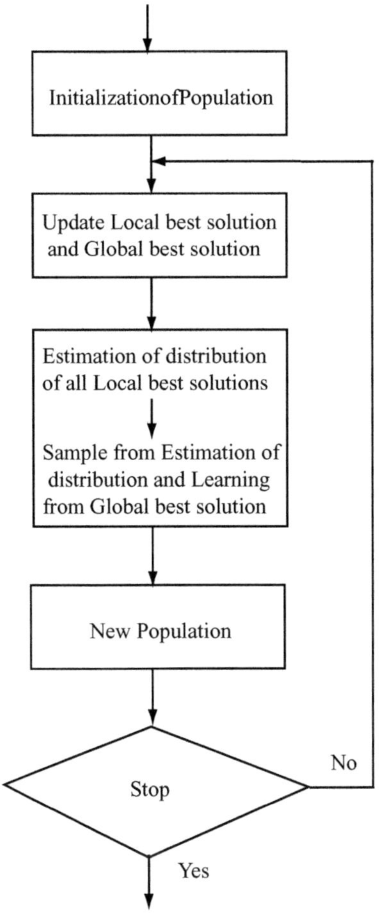


**Fig. 3.** The basic flowchart of the proposed algorithm

kind of comprehensive learning ability. Moreover, a bit-wise mutation operation is incorporated into the proposed algorithm. The evolution mechanism of the proposed algorithm keeps to the philosophy behind the original PSO.

## 5 Simulation Results

To demonstrate the performance of the proposed algorithm, experiments are carried out on the knapsack problem, which is a well-known combinatorial optimization problem. The classical knapsack problem is defined as follows: We are given a set of $n$ items, each item $i$ having an integer profit $p_i$ and an integer weight $w_i$. The problem is to choose a subset of the items such that their total profit is maximized, while the total weight does not exceed a given capacity

$C$. We may formulate the problem to maximize the total profit $f(X)$ as the following [17]:

$$f(X) = \sum_{i=1}^{n} p_i x_i,$$

subject to

$$\sum_{i=1}^{n} w_i x_i \leq C,$$

where the binary decision variables $x_i$ are used to indicate whether item $i$ is included in the knapsack or not. Without loss of generality it may be assumed that all profits and weights are positive, that all weights are smaller than the capacity $C$, and that the total weight of the items exceeds $C$.

In all experiments, strongly correlated sets of data were considered:

$$w_i = \text{uniformly random}[1, R],$$

$$p_i = w_i + R/10,$$

and the following average knapsack capacity was used:

$$C = \frac{1}{2} \sum_{i=1}^{n} w_i.$$

The traditional test instances with small data range are too easy to draw any meaningful conclusions; therefore we test the proposed algorithm on a class of difficult instances with large coefficients [17]. That is, the weights are uniformly distributed in a large data rang $R = 10^6$. This makes the dynamic programming algorithms run slower, but also the upper bounds get weakened since the gap to the optimal solution is scaled and cannot be closed by simply rounding down the upper bound to the nearest smaller integer [17]. Five knapsack problems with 100, 500, 1000, 5000, and 10000 items were considered.

In the proposed algorithm, the parameter $\beta = 0.95$, and mutation probability is set 0.06. For comparison, QPSO, DPSO and GPSO for this problem were also implemented. All the algorithms were implemented in C on a DELL–PC (Pentium4 2.80 GHz). In the QPSO [8], the parameters $\alpha = 0.1$, $\beta = 0.9$, $c_1 = c_2 = 0.1$, and $c_3 = 0.8$ are used. In the DPSO, the parameters, two acceleration coefficients $c_1 = c_2 = 1.2$, and velocity limit $V_{max} = 4$ were used in the simulations. In the GPSO, the standard parameters are adopted from Ref. [9]: the value of $w_1$ is dynamically tuned from 0.9 to 0.4 according to the number of generations and $w_2 = 0.2w_1 + 0.8$. The bit mutation probability $p_m$ is set to 0.001. In all algorithms, the population size and maximum iteration number are set to 40 and 1500, respectively. In all algorithms, the greedy repair is adopted to handle the constraint of knapsack problems [18].

Table 1 shows simulation results. The best total profit ("Best") and average total profit ("Av.") produced by QPSO, DPSO, GPSO and the proposed algorithm respectively within 20 simulation runs are shown. Simulation results

**Table 1.** Simulation results of 5 test problems

| Algorithm | | $n = 100$ | $n = 500$ | $n = 1000$ | $n = 5000$ | $n = 10000$ |
|---|---|---|---|---|---|---|
| QPSO | Best | 34885926 | 155155453 | 323632482 | 1601064252 | 3208126993 |
| | Av. | 34855727.7 | 154988339.05 | 323187543.7 | 1598475266.6 | 3200372726.95 |
| GPSO | Best | 34885971 | **155431535** | 325342751 | **1607304363** | 3214945549 |
| | Av. | 34885853.7 | 155431513.1 | 325342740.2 | 1607145495 | 3214588236.15 |
| DPSO | Best | 34885545 | 155431349 | 325142736 | 1607159843 | 3214745648 |
| | Av. | 34885545 | 155431349 | 325142736 | 1606874334.95 | 3214032772.4 |
| Proposed algorithm | Best | **34885973** | **155431535** | **325342752** | **1607304363** | **3215245262** |
| | Av. | **34885899.2** | **155431528.7** | **325342745.9** | **1607145515** | **3215161845.7** |

**Table 2.** Computation time of the algorithm of QPSO, DPSO, GPSO and the proposed algorithm (seconds)

| Algorithm | $n = 100$ | $n = 500$ | $n = 1000$ | $n = 5000$ | $n = 10000$ |
|---|---|---|---|---|---|
| QPSO | 6.32 | 13.81 | 28.57 | 144.9 | 293.17 |
| GPSO | 2.58 | 12.51 | 24.99 | 125.86 | 255.68 |
| DPSO | 3.48 | 28.5 | 56.48 | 287.47 | 565.73 |
| Proposed algorithm | 3.69 | 16.68 | 35.24 | 175.02 | 338.30 |

show that the proposed algorithm can obtain better solutions than the other particle swarm optimization algorithm. Furthermore, all the average solutions of the proposed algorithm are better than the best solutions of QPSO and DPSO algorithms. The better average performance of the proposed algorithm shows that the proposed algorithm is of a certain robustness for the initial solutions. Bold figures indicate the best results among the four algorithms.

In the QPSO, a particle based on on the quantum bit is defined. The value of a quantum bit is in essential a probabilistic value which represents the probability of this bit being 0. In order to guarantee that the computation results of a quantum bit is in $[0.0, 1.0]$, the parameters in the QPSO updating rule, $c_1 + c_2 + c_3 = 1$, must be satisfied. In addition, the probabilistic value of a quantum bit is computed by the simple adding and sum, which is lack good theory foundation. In the DPSO, the non-monotonic shape of the changing probability function Eq.(3) of a bit causes a problem: Eq.(3) has a concave shape so that for some bigger velocity values the changing probability will decrease. This seems to be an unusual probability function, because a higher changing probability is expected as the velocity increases. The updating rule of the GPSO is analogue to the genetic algorithm with crossover and mutation operators, therefore there are many parameters that are not easy to be tuned. The proposed algorithm is based on the sound theory foundation with less control parameters, which greatly contributes to its good performance.

Table 2 shows the comparison of computation time which is the average of 20 simulations. The proposed algorithm requires a little more CPU computational time than QPSO and GPSO because the proposed algorithm spends time in computing and updating the estimation of distribution at each iteration. DPSO is the slowest algorithm because it spends a lot time in computing the nonlinear sigmoid function at each iteration. Therefore, we can conclude that the proposed algorithm can search better solution within reasonable time.

## 6 Conclusions

To our best knowledge, this is the first report of combining the philosophy of particle swarm optimization and estimation of distribution algorithm to form a hybrid discrete particle swarm optimization algorithm for combinatorial optimization problems. The proposed algorithm combines the global statistical information collected from local best solution information of all particles and the global best solution information found so far in the whole swarm. To demonstrate its performance, experiments are carried out on the knapsack problem, which is a well-known combinatorial optimization problem. The results show that the proposed algorithm has superior performance to other discrete particle swarm algorithms as well as having less parameter. The future work is to use the high order probability models to estimation the distribution of promising solutions and their applications to different kinds of hard combinatorial optimization problems.

**Acknowledgments.** The Project was supported by the Scientific Research Foundation for Outstanding Young Teachers, Sun Yat-sen University.

## References

1. Kennedy, J., Eberhart, R.C.: Particle Swarm Optimization. Proceedings of IEEE International Conference on Neural Networks. NJ: Piscataway, (1995) 1942–1948
2. Van den Bergh, F., Engelbrecht, A.P.: Cooperative Learning in Neural Network Using Particle Swarm Optimizers. South African Computer Journal, **26** (2000) 84–90
3. El-Galland, AI., El-Hawary, ME., Sallam, AA.: Swarming of Intelligent Particles for Solving the Nonlinear Constrained Optimization Problem. Engineering Intelligent Systems for Electrical Engineering and Communications, **9** (2001) 155–163
4. Parsopoulos, K.E. and Vrahatis, M.N.: Recent approaches to global optimization problems through Particle Swarm Optimization. Natural Computing, **1**(2–3) (2002) 235–306
5. Kennedy, J., Eberhart R.C.: A Discrete Binary Version of the Particle Swarm Algorithm. Proceedings of the World Multiconference on Systemics, Cybernetics and Informatics. NJ: Piscatawary, (1997) 4104–4109

6. Franken, N., Engelbrecht, A.P.: Investigating Binary PSO Parameter Influence on the Knights Cover Problem. IEEE Congress on Evolutionary Computation, **1** (2005) 282–289
7. Huang, Y.-X., Zhou, C.-G., Zou, S.-X., Wang, Y.: A Hybrid Algorithm on Class Cover Problems. Journal of Software (in Chinese), **16**(4) (2005) 513–522
8. Yang, S.Y., Wang, M., Jiao, L.C.: A Quantum Particle Swarm Optimization. Proceeding of the 2004 IEEE Congress on Evolutionary Computation, **1** (2004) 320–324
9. Yin, P.Y.: Genetic Particle Swarm Optimization for Polygonal Approximation of Digital Curves. Pattern Recognition and Image Analysis, **16**(2) (2006) 223–233
10. Mühlenbein, H., Paa$\beta$, G.: From Recombination of Genes to the Estimation of Distributions. in Proceedings of the 4th Conference on Parallel Problem Solving from Nature-PPSN IV, H.-M. Voigt, W. Ebeling, I. Rechenberg, and H.-P. Schwefel, Eds. Lecture Notes in Computer Science. Springer, Berlin, **1411** (1996) 178–187
11. Pelikan, M., Goldberg, D.E., Lobo, F.: A Survey Of Optimization by Building and Using Probabilistic Models. Computational Optimization and Applications, **21**(1) (2002) 5–20
12. Kern, S., Muller, S.D., Hansen, N., Buche, D., Ocenasek, J., Koumoutsakos, P.: Learning Probability Distributions in Continuous Evolutionary Algorithms-A Comparative Review. Natural Computing, **3**(1) (2004) 77–112
13. Larrañaga, P., Lozano, J.: Estimation of Distribution Algorithms: A New Tool for Evolutionary Computation. Genetic Algorithms and Evolutionary Computation. Springer, **2** (2001)
14. Müehlenbein, H.: The Equation for Response to Selection and Its Use for Prediction. Evol. Comput., **5**(3) (1997) 303–346
15. Baluja, S.: Population-Based Incremental Learning: A Method for Integrating Genetic Search Based Function Optimization and Competitive Learning. School of Comput. Sci., Carnegie Mellon Univ., Pittsburgh, PA, Tech. Rep. CMU-CS-94-163, (1994)
16. Harik, G. R., Lobo, F. G. and Goldberg, D. E.: The Compact Genetic Algorithm. IEEE Trans. Evol. Comput., **3**(4) (1999) 287–297
17. Pisinger, D.: Where Are the Hard Knapsack Problem? Computer & Operations Research, **32** (2006) 271–2284
18. Michalewicz, Z.: Genetic Algorithm+ Data Structure=Evolution Programs. Beijng: Science Press, (2000) 59–65

# An Improved Particle Swarm Optimization for Traveling Salesman Problem

Xinmei Liu[1], Jinrong Su[2], and Yan Han[1]

[1] School of information and communication engineering, North University of China, Tai Yuan 030051, China
[2] Department of information engineering, business college of Shan Xi University, Tai Yuan 030031, China
sjr_su@tom.com

**Abstract.** In allusion to particle swarm optimization being prone to get into local minimum, an improved particle swarm optimization algorithm is proposed. The algorithm draws on the thinking of the greedy algorithm to initialize the particle swarm. Two swarms are used to optimize synchronously. Crossover and mutation operators in genetic algorithm are introduced into the new algorithm to realize the sharing of information among swarms. We test the algorithm with Traveling Salesman Problem with 14 nodes and 30 nodes. The result shows that the algorithm can break away from local minimum earlier and it has high convergence speed and convergence ratio.

**Keywords:** Particle swarm optimization, Traveling salesman problem, Greedy algorithm, Crossover, Mutation.

## 1 Introduction

Particle swarm optimization (PSO) is an evolutionary computation technique that was introduced by Eberhart and Kennedy in 1995[1, 2]. It developed out of work simulating the behaviors of bird flocking involving the scenario of a group of birds randomly looking for food in an area. PSO shares many features with Genetic Algorithms (GA). Similar to GA, PSO is a population based optimization tool that search for optima by updating generations, but PSO has lesser parameters than GA and PSO has no operators such as crossover and mutation in GA. PSO has proved to be efficient at solving global optimization and engineering problems [3,4].

After the nearly ten years development, PSO is applied widely in many fields such as function optimization, artificial neural network training, fuzzy systems control, printed circuit board assembly, combinatorial optimization and decision making dispatching, etc [6-13]. The performance of PSO is known to do well in the early iterations of the search process, but has problems in reaching a near optimal solution, which leads to the shortcomings of low convergence speed and not easy to converge to global optima.

TSP is described as: give n cities and the distances between two arbitrary cities, and seek an optimal tour of n cities, visiting each city exactly once with no sub tours. TSP is

D.-S. Huang, L. Heutte, and M. Loog (Eds.): ICIC 2007, LNAI 4682, pp. 803–812, 2007.

a famous combinatorial optimization problem, and is also NP-Hard problem. It is always used to test and verify the validity of intelligent heuristic algorithms [15].

In reference [15] Lan Huang constructed an algorithm of a kind of special particle swarm optimization via presenting the concepts of swap operator and swap sequence and then applied it to a 14-node traveling salesman problem successfully. There is great disparity in the ability and speed to solve the problem between basic PSO and other classical algorithms solving TSP, but it's a new attempt to solve TSP with PSO [16]. Basic PSO is improved in this paper. We initialize the particle swarm drawing on the thinking of the greedy algorithm. Two swarms are used to search synchronously. PSO algorithm is carried out independently inside the two swarms. Particles in the two swarms mutate according to a certain mutation probability. The individual optima of particles in two swarms cross according to a certain probability, which improves the population quality and increases information sharing. Simulation results show that the improved PSO is of high convergence speed and convergence rate.

## 2 Standard Particle Swarm Optimization

PSO algorithm conducts search using a population of particles. The population of PSO is called a *swarm* and each individual in the population of PSO is called a *particle*. Each particle represents a candidate solution to the problem at hand. In an *N*-dimensional searching space the *i*th particle at iteration □has two attributes: a current position $X_i^k = (x_1^k, \cdots, x_n^k, \cdots, x_N^k)$ and a current velocity $V_i^k = (v_1^k, \cdots, v_n^k, \cdots, v_N^k)$. where $x_n^k \in [l_n, u_n]$, $1 \le n \le N$, $l_n$ and $u_n$ is lower and upper bound for the *n*th dimension, $V_i^k$ is bounded by a maximum velocity $V_{\max}^k$ and a minimum velocity $V_{\min}^k$. The position and velocity of swarm update by the following equations[1]:

$$V_i^{k+1} = \omega V_i^k + c_1 r_1 (P_i^k - X_i^k) + c_2 r_2 (P_g^k - X_i^k) \tag{1}$$

$$X_i^{k+1} = X_i^k + V_i^{k+1} \tag{2}$$

Where $P_i^k$ is the best previous position of the *i*th particle (also known as *pbest*) in the $k$ th iteration. $P_g^k$ is the best position among all the particles in the swarm in the $k$ th iteration (also known as *gbest* ). The variables $c_1$ and $c_2$ are acceleration constants, which adjust the relative significance of $P_i^k$ and $P_g^k$. $r_1$ and $r_2$ are elements from two uniform random sequences in the range $(0,1)$. $\omega$ is an inertia weight that is a key factor to affect the convergence of PSO [1],[5],[14].

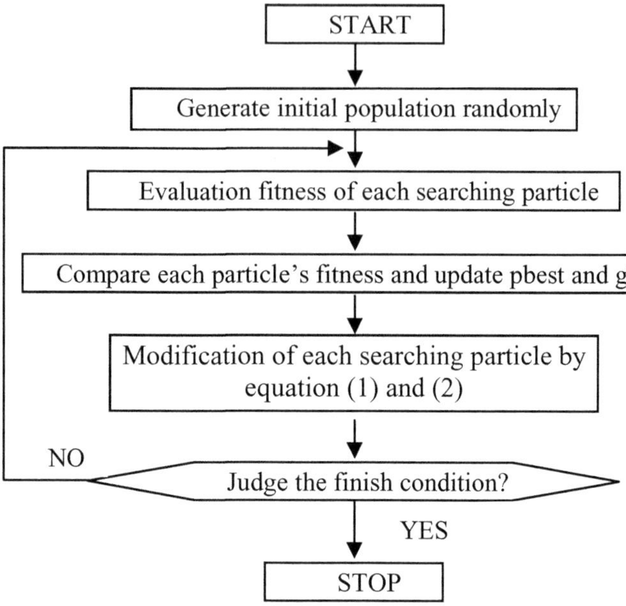


**Fig. 1.** Flowchart of standard particle swarm optimization

The process of the standard PSO algorithm can be expressed as follows (see Fig.1):

Step1: Generate a population of particles with random positions and velocities in *N*-dimensional searching space.

Step2: Evaluate the fitness of each particle by using the objective functions of the target problem.

Step3: Compare each particle's fitness with its previous best fitness (*pbest*) for every iteration. If the current value is better than the *pbest*, then replace *pbest* with the current value and the *pbest* location equal to the current location.

Step4: Compare *pbest* of each particle and update the swarm global best position with the greatest fitness (*gbest*).

Step5: Change the position and velocity of each particle according to equation (1) and (2) respectively.

Step6: Judge the finish condition. If satisfy the condition, then stop and output the result. If not then repeat steps 2 to 5 until satisfy the finish condition.

## 3 Improved Particle Swarm Optimization

Reference [14] redefined the concepts of velocity and position of standard PSO by introducing the concepts of swap operator and swap sequence. The updating equations of velocity and position are as follows:

$$V'_{id} = V_{id} \oplus \alpha(P_{id} - X_{id}) \oplus \beta(P_{gd} - X_{id}) \tag{3}$$

$$X'_{id} = X_{id} + V'_{id} \tag{4}$$

In equation (3), $\alpha$ and $\beta$ are elements from two uniform random sequences in the range $(0,1)$. $(P_{id} - X_{id})$ is swap sequence of the *i*th particle and *pbest*, which is reserved by the probability $\alpha$. $(P_{gd} - X_{id})$ is swap sequence of the *i*th particle and *gbest*, which is reserved by the probability $\beta$. Particles update their positions by equation (4).

Basic PSO algorithm is prone to fall into local optimum when it solves TSP, which affects the algorithm on convergence speed and convergence rate. In view of these shortcomings we now modify the algorithm appropriately.

Firstly, drawing on the idea of greedy algorithm, we take local optimum every step when initialize swarm. In this way the global best position of the initial swarm is fairly close to solution of problem, therefore, we can save searching time and improve convergence speed.

Secondly, particles mutate by a certain probability. The particles of initial swarm produced greedily are of high quality, but it's diversity, which affect the algorithm on it's global exploration performance greatly, is poorer than that produced randomly. Therefore, in the process of evolution, mutation is introduced to increase the diversity of particles so as to enhance the global exploration ability of PSO. Whether to mutate depends on the mutation threshold value *m*. *m* is a fix number that is larger than the least error *e*. If difference of two evolutions' fitness is smaller than *m* also is bigger than the least error *e*, this indicates that particles does not have obvious evolution also does not satisfy the conclusion condition, so now carry on mutation to particles to increase the diversity of swarms.

Thirdly, Two swarms are used to search synchronously. PSO algorithm is carried out independently inside the two swarms. Meanwhile the individual optimums of particles in the two swarms cross by a certain probability. Two swarms optimizing may reduce the probability of algorithm to fall into local best position. Crossing between two swarms' individual optimum can strengthen information sharing between two swarms as well as particles and transmit the optimum value information duly, thus enhance the particles' speed to get to a better solution. Choosing the best individual positions of swarms to cross is equal to the mating of minority excellent individuals in the biosphere, which is of advantage to produce fine descendant and can guarantee multiplication of the previous generation's choiceness character. This is consistent with the evolution mechanism that superior wins and the inferior washes out in biosphere, therefore, particles produced by crossing is advantageous to swarm's evolution.

Moreover, we make speed of particles update by equation (5), in which there is inertia weight $\omega$. $\omega$ will be decreased linearly during evolution.

$$V'_{id} = \omega V_{id} \oplus \alpha(P_{id} - X_{id}) \oplus \beta(P_{gd} - X_{id}) \tag{5}$$

The step of improved PSO algorithm is as follows:

Step 1: Produce two initial swarms A and B greedily, produce two basic swap sequences randomly as initial velocities of the two swarms, and set inertia weight $\omega$, mutation threshold *m* as well as least error *e*;

Step 2: Evaluate fitness of each searching particle;

Step 3: Carry on mutation if swarms satisfy mutation condition;

Step 4: Cross the best individual positions of swarm A and swarm B to produce new best individual positions;

Step 5: Update particles' position and velocity by equation (4) and equation (5) respectively inside the two swarms, produce two new global best positions and then go to step2;

Step 6: Stop and demonstrate result as long as one swarm satisfies stop condition (difference of fitness between two iterations be less than the least error *e*). If no swarm satisfies stop condition repeat step 2 to step 6.

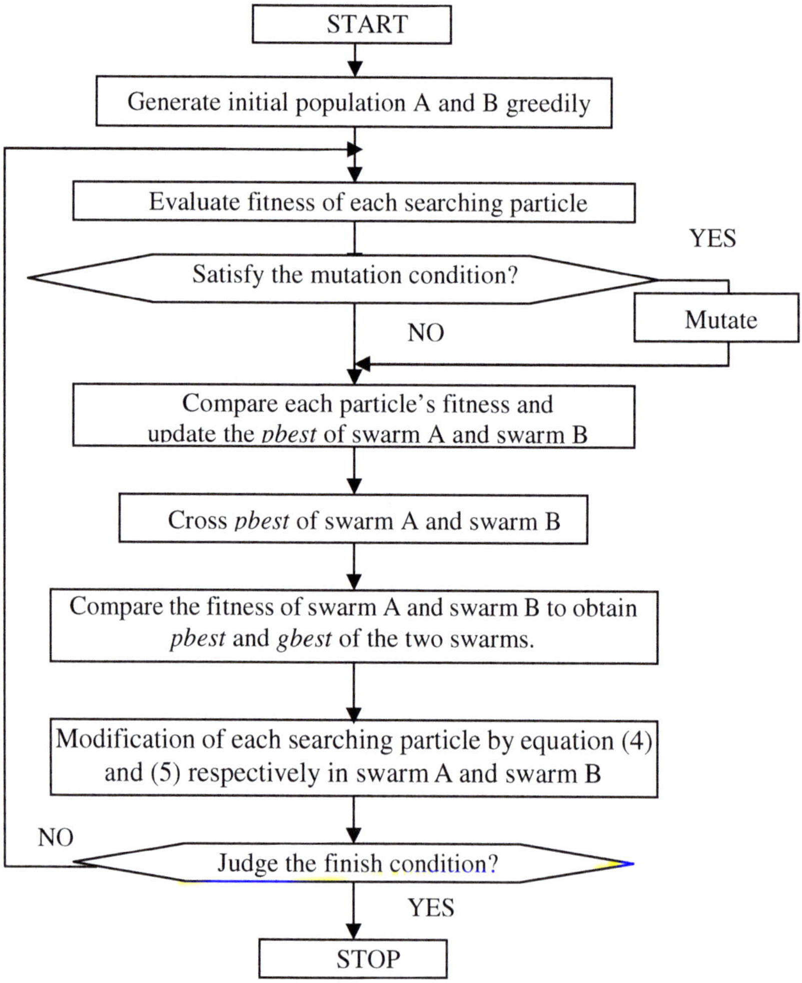


**Fig. 2.** Flowchart of improved particle swarm optimization

# 4 Simulation Experiment

We used 14-node and Benchmark 30-node TSP standard data to test the validity of improved PSO algorithm (the question origin: http://elib.zib.de/pub/Packages/mp-testdata/tsp/tsplib/tsp/). Table 1 describes the 14-node traveling salesman problem. The best solution of 14-node TSP known at present is 30.8785 and the tour path is 1-10-9-11-8-13-7-12-6-5-4-3-14-2-1. The best solution of 30-node TSP known at present is 423.741 and the path is 1-2-3-9-18-19-20-21-10-11-7-8-14-15-24-25-26-27-28-29-16-17-22-23-30-12-13-4-5-6.

## 4.1 Parameter Setting in Experiments

The experiment environment is Pentium IV, 2.93GHz CPU, 512M RAM, Windows XP system and program tool is Matlab6.5. In order to be convenient for comparison, we use basic PSO and genetic algorithm (GA) together with the improved PSO presented in this paper. We repeated experiments for 30 runs continuously for every algorithm. The population size of all algorithms used in our experiments was set at 50. In GA we used ordered cross operator and fixed cross probability at 0.9 and mutation probability at 0.1. In PSO algorithms, the swap operator of swap sequence was 7 at most. $\alpha$ and $\beta$ are elements from two uniform random sequences in the range $(0,1)$ .The inertia weight $\omega$ linearly decreased from 1.4 to 0.5. A fixed number of maximum generations 2000 is applied to algorithms. The mutation threshold $m=1e\text{-}3$, the least error $e=1e\text{-}10$ were adopted. The cross probability was set at 0.9. Experimental results are listed in Table 2 and Table 3. Figure 3 shows the best tour path of 14-node TSP and Fig.4 shows the path of 30-node TSP we obtained. They are consistent with the best solutions known at present.

## 4.2 Analyzing of Experiment Result

In table 2, the convergence ratio is the ratio of times that the algorithms restrained to the best result 30.8785 to testing times 30. The data in table 2 shows that compare with basic PSO the improved PSO have distinct enhancement in convergence speed and convergence ratio. Compared with GA the improved PSO also advances a little in convergence speed. Figure 3 is the best tour path we obtained in experiment, the length of which is 30.8785.

In 30 runs, for 14-node TSP the improved PSO presented in this article has fallen into the local optima 31.2088 in an early time but jumped out at about 50 generations, and then got the best result 30.8785. It is obvious that the improved algorithm can overcome the shortcoming of being easy to fall into local optima to a certain extent. In addition, the initial swarm produced greedily played an important part in quickening algorithm convergence steps. The algorithm once has restrained in the 2nd generation to the best result 30.8785. Table 3 shows that the performance of improved PSO didn't deteriorate a lot as the increasing of problem's scale. Table 4 is the comparison of initial swarm global best position average produced greedily and randomly. We can see that *gbest* reduces gradually with the increase of population size. The number of iterations was reduced to a great extent because of greedily initializing and the convergence speed was heightened consequently.

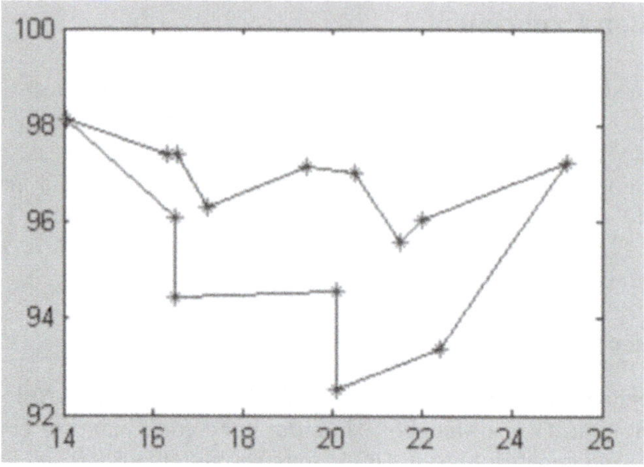


**Fig. 3.** Best path of the 14-node TSP we obtained

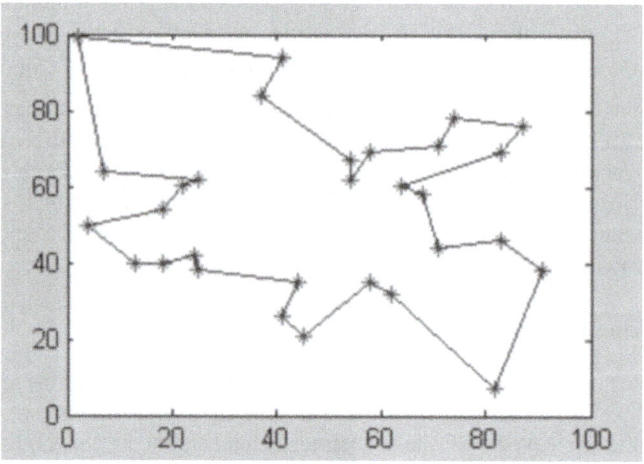


**Fig. 4.** Best path of the 30-node TSP we obtained

**Table 1.** Data of 14-node TSP

| Node | 1 | 2 | 3 | 4 | 5 | 6 | 7 | 8 | 9 | 10 | 11 | 12 | 13 | 14 |
|---|---|---|---|---|---|---|---|---|---|---|---|---|---|---|
| X | 16.47 | 16.47 | 20.09 | 22.39 | 25.23 | 22.00 | 20.47 | 17.20 | 16.30 | 14.05 | 16.53 | 21.52 | 19.41 | 20.09 |
| Y | 96.10 | 94.44 | 92.54 | 93.37 | 97.24 | 96.05 | 97.02 | 96.29 | 97.38 | 98.12 | 97.38 | 95.59 | 97.13 | 94.55 |

**Table 2.** Experimental results of 14-node TSP

| Algorithms | Best solution | Worst solution | Convergence ratio | Average generations |
|---|---|---|---|---|
| Basic PSO | 30.8785 | 31.8194 | 20% | 231.6 |
| GA | 30.8785 | —— | 100% | 82.4 |
| Improved PSO | 30.8785 | —— | 100% | 43.5 |

**Table 3.** Experimental results of 30-node TSP

| Algorithms | Best solution | Worst solution | Convergence ratio | Average generations |
|---|---|---|---|---|
| Basic PSO | 432.7617 | 482.6495 | 0 .00 | —— |
| GA | 423.7406 | 431.3098 | 36.67% | 1781.00 |
| Improved PSO | 423.7406 | 424.6918 | 76.67% | 1186.60 |

**Table 4.** Comparison of initial swarm produced by two methods

| Nodes of TSP | Population size | randomly | greedily |
|---|---|---|---|
| 14-node TSP | 20 | 49.4320 | 32.4179 |
| | 30 | 49.1652 | 32.0122 |
| | 50 | 46.7954 | 31.8825 |
| 30-node TSP | 30 | 1097.8000 | 480.9184 |
| | 40 | 1082.7000 | 472.3400 |
| | 50 | 1063.6000 | 468.8726 |

**Table 5.** Comparison of initial swarm variability

| Algorithms | node of TSP | Population size | Average fitness | variability |
|---|---|---|---|---|
| A | 14 | 20 | 0.0167 | 0.8250 |
| B | 14 | 20 | 0.0273 | 0.4500 |
| C | 14 | 20 | 0.0267 | 0.6125 |
| A | 14 | 50 | 0.0162 | 0.6560 |
| B | 14 | 50 | 0.0277 | 0.2040 |
| C | 14 | 50 | 0.0268 | 0.4200 |
| A | 30 | 30 | 7.576e-4 | 1.0000 |
| B | 30 | 30 | 1.889e-3 | 0.1583 |
| C | 30 | 30 | 1.836e-3 | 0.2083 |
| A | 30 | 50 | 7.658e-4 | 1.0000 |
| B | 30 | 50 | 1.879e-3 | 0.4920 |
| C | 30 | 50 | 1.824e-3 | 0.6360 |

We define variability of swarm as the ratio of number of different fitness value to the population size of swarm. Table 5 is the comparison of initial swarm variability of three different methods. “A” means the swarm is born randomly. In condition “B” the swarm is born greedily. Swarm in “C” is greedily initialized and then mutate at a mutation probability. Referring to Table 5, it is observed that the variability of “A” is the best and the average fitness of “B” is best no matter what the node of TSP and the population size is. But if consider fitness and variability together the swarm in “C” is best. The method “C” is the very way to initialize swarm in the PSO improved in this paper. The quality of swarm in “C” is higher than “A” and “B”, which improves the performance of PSO a lot.

## 5 Conclusions

PSO algorithm is a novel intelligent optimization algorithm developed in ten years. Being simple and easy to realize, it is successfully applied in many fields. It is a new attempt to solve TSP with PSO. But the performance of basic PSO is not very satisfying when solve TSP. In this article we have made appropriate improvements to basic PSO by initializing the swarm greedily and introducing cross and mutation operation. As a result the convergence speed and convergence ratio were enhanced. The subsequent simulation experiments have also proven the validity of improved PSO. The strategy to initialize swarm put forward in this article is effective. It also will be effective in other intelligent algorithms based on swarm. The PSO algorithm is in the initial period of research at present, but in view of its application effect, this algorithm has tremendous potential and it will be applied in more widespread domains.

## References


1. Kennedy, J., Eberhart, R.C.: Particle Swarm Optimization. In: Proceedings IEEE International Conference on Neural Networks, Volume: 4, 27 November-1 December (1995) 1942-1948
2. Kennedy, J., Eberhart, R.C.: Swarm Intelligence. Morgan Kauffman publishers, San Francisco, CA.ISBN 1-55860-595-9
3. Parsopoulos, K.E., Plagianakos, V.P., Magoulas, G.D., Vrahatis, M.N.: Stretching Technique for Obtaining Global Minimizers Through Particle Swarm Optimization. In: Proc. Particle Swarm Optimization Workshop, (2001) 22-29
4. Parsopoulos, K.E., Vrahatis, M.N.: Modification of the Particle Swarm Optimizer for Locating All the Global Minima. Artificial Neural Networks and Genetic Algorithms, (2001) 324-327
5. Shi, Y., Eberhart, R.C.: A Modified Particle Swarm Optimizer. In: Proceedings of the IEEE International Conference on Evolutionary Computation. PiscatawayNJ: IEEE Press, (1998) 69-72
6. Xie, X.F., Zhang, W.J., Yang, Z.L.: Overview of Particle Swarm Optimization. Control and Decision. Vol.18 No.2. 129-134
7. Salman, A., Ahmad, M., Al-Madani, S.: Particle Swatm Optimization for Task Assignment Problem. In: Microprocessors and Microsystems 26 (2002) 363-371
8. Yo, S.H., Kawata, K., Fukuyama, Y.: A Particle Swarm Optimization for Reactive Power and Voltage Control Considering Voltage Security Assessment. Trans of the Institute of Electrical Engineers of Japan, (1999) 1462-1469
9. Jiang, C.W., Bompard, E.: A Hybrid Method of Chaotic Particle Swarm Optimization and Linear Interior for Reactive Power Optimization. Mathematics and Computers in Simulation 68 (2005) 57-65
10. Da, Y., Ge, X.R.: An Improved PSO-Based ANN with Simulated Annealing Technique. Neurocomputing. 63 (2005) 527-533
11. Ghoshal, S.P.: Optimization of PID Gains by Particle Swarm Optimizations in Fuzzy Based Automatic Generation Control. Electric Power Systems Research. 72 (2004): 203-212
12. Zhang, H., Li, X.D., Li, H., Huang, F.L.: Particle Swarm Optimization-Based Schemes for Resource-Constrained Project Scheduling. Automation in Construction 14 (2005): 393-404
13. Chen, Y.M., Lin, C.T.: A Particle Swarm Optimization Approach to Optimize Component Placement in Printed Circuit Board Assembly. Springer-Verlag London Limited 2006, DOUI 10.1007/s00170-006-0777-y

14. Zeng, J.C.: Particle Swarm Optimization. Scientific publishing company. 2004: 50-53
15. Wang, K.P., Huang, L., Zhou, C.G.: Particle Swarm Optimization for Traveling Salesman Problems. In: Proceedings of the 2nd International Conference on Machine Learning and Cybernetics. Xi'an, Nov 2003,1583-158
16. Gao, H.C., Feng, B.Q., Zhu, L.: Reviews of the Meta-heuristic Algorithm for TSP. Control and Decision[J]. Vol.21,No.3 : 241-247

# An Improved Swarm Intelligence Algorithm for Solving TSP Problem

Yong-Qin Tao[1,2], Du-Wu Cui[1], Xiang-Lin Miao[2], and Hao Chen[1]

[1] School of Computer Science and Engineering ,Xi'an University of Technology
Xi'an 710048 China
[2] School of Electronic and Information Engineering ,Xi'an Jiaotong University
Xi'an 710049 China
Tyq_123@126.com, cuidw@xaut.edu.cn, miaoxianglin@126.com

**Abstract.** Traveling Salesman Problem (TSP) is a typical NP—Complete problem. This paper, through finding the solution of TSP, combining the use of high—efficiency gene regulatory algorithm , particle swarm optimization and ant colony optimization, proposes a kind of improved swarm intelligence algorithm GRPSAC. The GRPSAC overcomes the disadvantages of several algorithms through the use of the crossover, the mutation and the gene regulation. The experimental results indicate that GRPSAC not only has a high-efficiency, but also induces better optimal results

**Keywords:** swarm intelligence algorithm, gene regulation, tsp.

## 1 Introduction

Since the bionics was established in the 20 centuries 50's middle, people had got enlightenment from the living evolutionary mechanism, and had proposed the many new algorithms of solving the complicated problem, such as: genetic algorithm, evolutionary rule, evolutionary strategy, etc. The swarm intelligence algorithm which is used as new developing technique in evolved calculation has become the focus of concern among more and more researchers. The colony is described in the swarm intelligence as a group of agent which can communicate directly and indirectly each other. The agent of group can cooperate and solve distributed problem. Currently, there are two kinds of Algorithms in the swarm intelligence theories and research: They are Ant Colony Optimization (ACO) and Particle Swarm Optimization (PSO). The swarm intelligence is an algorithm in probability search. It is different from the other optimization. This paper deeply studies the principle of intelligence Algorithms, analyzing idea and result of intelligence Algorithm from the different angle, and combines with gene engineering, then put forward an improved algorithm of different mechanism which is called as Gene Regulation Particle Swarm Ant Colony Optimization (GRPSAC). Its outstanding characteristics are that the construction is more simple, the speed of running is faster, the number of compute is smaller, global searching capability is stronger when GRPSAC is used to solve TSP problem.

D.-S. Huang, L. Heutte, and M. Loog (Eds.): ICIC 2007, LNAI 4682, pp. 813–822, 2007.

## 2 Traveling Salesman Problem

The TSP can be described as follows:

In the graph G = (V, E), V is the set of nodes, or cities, E is the set of edges, E={ (a, b)| a, b ∈ V}. The Euclidean distance between a and b is $D_{ab}$, supposing $D_{ab}= D_{ba}$. The object of TSP is to find a minimal length closed tour during which he visits each city once and only once and the closed tour is also called Hamiltonian Cycle. TSP has been proven to be a NP-complete problem[2]. This problem is the problem of searching shortest route which go through n cites. Its mathematics describing is a array which search the subset of natural number X ={1,2,...,n}, (the element of X shows serial number of n cities) C={$c_1$,$c_2$,..., $c_n$ }, and makes $T_d$ get minimum.

$$T_d = \sum_{i=1}^{n-1} d(c_i, c_{i+1}) + d(c_1, c_n) \quad (1)$$

d($c_i$,$c_i$ ) expresses the distance from the $C_i$ city to $C_{i+1}$ city in the Formula (1). It has been proved NP (nondeterministic polynomial) entirety problem, namely: nondeterministic algorithm can get the solution of problem in polynomial time. Total of possible route is n!/2n to n cities, With the growth of n the route will increase at the speed of the exponent blast. For example: When n is 20 it will need 350 years to computer at a hundred million per second according to the limit search.

ACO has been proven to be effective for TSP in Swarm Intelligence. It is good at resolving the problem of discrete optimized. However, PSO is adopted at resolving problem of successive optimization. The paper proposes GRPSAC algorithm to solve TSP problem through the use of the crossover, the mutation and the gene regulation operation. The method has been proved more precise and more approximate excellent

## 3 The Analysis of the Improved Swarm Intelligence Algorithm (GRPSAC)

### 3.1 Ant Colony Optimization (ACO)

ACO is a novel approach of evolutionary computation. It was first introduced by Italy scholars Dorigo and Maniezzo in 1992. The ACO has been proposed by the research elicitation on real ant colony's foraging behavior like ant, particle etc. The behavior of the single insect is easy. But constituted colony represents the very complicated behavior. Through the large quantity of the research the bionics experts discover that the ant individual delivers information by the substance called the ectohormone that the ant leaves a kind of volatility secretion (called pheromone)on his passing path. The pheromone will be vaporized and disappear gradually along with the time proceeding. The ant can apperceive this kind of material existence and its strengths in the process of looking for food and guide its own moving direction with this method to incline toward the direction of the high substance strength, namely: the probability of choosing path is the proportion at strength of this path then. The path with more strength information is selected by more many ants. If pheromone left on that path is larger, more ants will be attracted. Consequently the positive feedback will be formed. The ant can discover the best path through the feedback finally and result in most ants to walk this path[1].

### 3.2 Particle Swarm Optimization(PSO)

Particle Swarm Optimization (PSO) is an algorithm based stochastic optimization technique developed by Dr. Eberhart and Dr. Kennedy in 1995, inspired by social behavior of bird flocking or fish schooling. In the PSO system, the swarm is made up of a certain number of particles. In each iteration, particles fly around in a multidimensional search space to find the global optima. The velocity and the position of each particle are adjusted by the following formulas(2) and (3)respectively.[2]

$$V_i^{t+1} = \omega V_i^t + C_1 * Random()(P_i^t - X_i^t) + C_2 * Random()(P_g^t - X_i^t) \quad (2)$$

$$X_i^{t+1} = X_i^t + V_i^{t+1} \quad (3)$$

Here $\omega$ is called the inertia weight. $C_1$and $C_2$ are acceleration constants, which are often called the cognitive confidence coefficients. Random () are random values between 0 and 1.Variable i denotes the ith particle in the swarm. Variable t represents the iteration number, $V_i$ is the velocity vector of the ith particle, $X_i$ is the position vector of the ith particle. During flight, each particle adjusts its position $P_{i,}$ ,which is the local best position that the ith particle had reached according to its own experience, and according to the experience of a neighboring particle, itself and its neighbor encounter the best position $P_g$ ,which is the global best position that all particles had reached .

## 4 Application of the GRPSAC Algorithm in TSP

ACO makes use of pheromone to deliver the information, but PSO makes use of three information such as oneself information, individual extremum information, global extremum information to guide particle next iterative position. ACO makes use of the positive feedback principle and combines with some heuristic algorithm organically. ACO easily emerges precocious phenomenon and plunges to local superior solution. The mixed thought is to let ant have character of particle. First of all, ACO random products some group of better solution to build distributing of pheromone. Then, some groups of these solutions are searched by ACO according to total renewing pheromone. Finally the operations of the crossover and the mutation algorithm are carried through by PSO. Through leading into gene regulation operator in gene engineering the control method of regulation switch valve can decide the operation of regulation operator to increase colony variety, and can lead flock to evolution quickly. The ant again processes the auto-adjustment according to location superior solution and global superior solution until getting most effective solution.

### 4.1 The Flow Chart of GRPSAC Algorithm

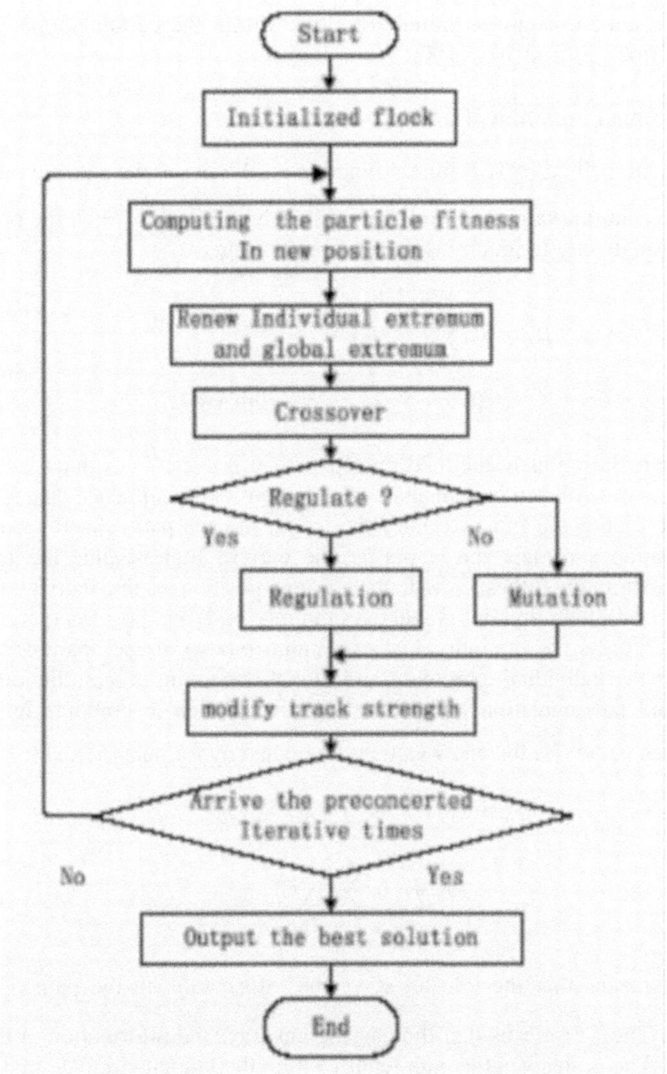


**Fig. 1.** The flow chart of GRPSAC Algorithm

### 4.2 Coding Scheme

The paper adopts real number coding and uses order to express cities. For example: [4,7,6,5,9,1,2,8,10,3] denotes a path which starts out from 4 city via 7-6-5-9-1-2-8-10-3 cities and finally return to city 4.

### 4.3 Initialized Flock

A flock is a particle. Each particle is an individual inside the flock. The individual with certain condition grows initializes flock within the certain scope. The size of initializing flock is from 50 to 100.

### 4.4 The Fitness Function of TSP

The computed method of TSP fitness function is: When t=0 the ants are putted in each city. Suppose initial value of information in each path is $\tau_{ij}(0)$ = C, the probability of each ant from ith city to jth city is decided by formula (4)

$$p_{ij}^{k} = \begin{cases} \dfrac{\tau_{ij}^{\alpha}\eta_{ij}^{\beta}(t)}{\sum\limits_{s\in allowd_k}\tau_{is}^{\alpha}\eta_{is}^{\beta}(t)} & j \in allowed_k \\ 0 & \text{otherwise} \end{cases} \tag{4}$$

$\tau_{ij}(t)$ Means how much ants had passed path (i,j) ever; $\eta_{ij}$ explains the probability that nearer city are picked out.αandβare used to control the influence degree that choose the ant and to compute fitness value ( the length of each path) ltsp0 according to the current position ( namely :basic path) .The current fitness value are token as the individual extremum (ptbest). And, the current position is the individual extremum position (pcbest), then find out global extremum (gtbest) and position (gcbest) of global extremum according to the individual extremum (ptbest) of each particle. The current solutions of the individual extremum and global extremum process the crossover, the mutation and the regulation operation respectively, then to compute information of renewed each path $\tau_{ij}(t)$. Its renew value is computed by formula (5) and (6):

$$\tau_{ij}(t+n) = \rho \cdot \tau_{ij}(t) + \Delta\tau_{ij} \tag{5}$$

$$\Delta\tau_{ij} = \sum_{k=1}^{m} \Delta\tau_{ij}^{k} \tag{6}$$

The $\Delta\tau_{ij}^{k}$ means that the kth ant stays the information in the path ij among this circulation. The $\Delta\tau_{ij}$ means that the passing ant stays the information in the i,j in this circulation. The $L_k$ denotes the path length which the kth ant circulate in a circulation. The Q expresses the constant.

$$\Delta\tau_{ij}^{k} = \begin{cases} \dfrac{Q}{L_k} & \text{When ant k pass i j} \\ 0 & \text{When ant k do not pass} \end{cases} \tag{7}$$

The last requests the best short path $\min L_k$ in the circulation. This process repeats continuously until the circulation number attains the tallest number $NC_{\max}$ that set up in advance.

### 4.5 The Crossover and the Mutation Operator

The crossover and the mutation operator are put forward by literature [3]. The crossover operator is to choose a crossover district randomly in the second string as the crossover district:6543, old2 crossed district add the old1 position of city 6 and the last deletes the city that had appeared in the old2 among the old1.

For example:

Two father and son string is: old1=1 2 3 4 5 6 7 8 9,
old2=9 8 7|6 5 4 3| 2 1,
After crossover it is : new1=1 2 6 5 4 3 7 8 9

The mutation operator is to choose j1 and j2 visitorial city randomly between the first and the nth. cities. Supposing the j1< j2, the j1 visitorial city is arranged before j2 visitorial in the path $C_0$ . The rest is constant, the path is $C_1$.here

For example:

$C_0$=2 3 4 1 5 7 9 8 6, j1=2 , j2=7, follow
$C_1$=2 4 1 5 7 3 9 8 6

### 4.6 The Design of the Gene Regulation Operator

Operator theory of the operation was first proposed by France scholars Monord and Jacob in 1961.The gene decides evolution. The decision function of gene is regulated by the other gene. There are the structure gene, the enlightened gene and the operation gene in the operator of operation .The enlightened gene locates in front of operation gene, and the two link each other closely, while the structure gene is regulated by two switch genes—operation gene and enlightened gene. Only when the two switches turn on, can structure gene be activated. This paper leads the regulation function of the gene into swarm intelligence and constructs a regulated operator to guide the flock evolution, so as to improve the property of swarm intelligence. Its parameter (variety or density) is as follows:

$$\delta = \frac{\mathrm{Ni}}{\mathrm{N}} \times 100\% \tag{8}$$

The N is a community scale. The $N_i$ is the ith independence ( dissimilarity ) individual number. The $\varepsilon$ is the threshold of the regulation switch. The $P_m$ is the mutation probability. The operation of regulation operator is as follows:

(1)When $\delta < \varepsilon$ in addition to individual that attend the normal mutation, in the individual of the flock surplus the new individual that is generated in adding $N_0$ random individual is

$$N_0 = | \mathrm{N} \times \mathrm{Pm} - \mathrm{int}[\mathrm{N} \times (\varepsilon - \delta)] | ; \tag{9}$$

(2) When $\delta \geq \varepsilon$ The $P_m$ keeps constantly [6].

### 4.7 The Step of GRPSAC Algorithm

The step of GRPSAC is as follows:

(1) nc= 0 ( the nc is iterative step or manhunt times), initialization, producing a path ( such as 100) and choosing more excellent path (such as 30), making these paths leave pheromone, placing the m ant on the n top.

(2) computing fitness value ( the length of each path ) ltsp0 according to the current position, establishing the current fitness value as the individual extremum (ptbest). The current position is the individual extremum position (pcbest). Finding out global extremum (gtbest) and position ( gcbest ) of global extremum according to the individual extremum (ptbes) of each particle.

(3) placing the start of each ant into the gather of the current solution. Moving each ant k ( k=1,2, … , m) to the next acme j according to probability $p_{ij}^{k}$. Placing acme j into gather in the front solution.

(4) proceeding operate to each ant is as follows, the jth ant crosses with gcbest to get the $C_1^1$( j) through the path $C_0$( j) , the $C_1^1$ ( j) crosses with pcbest to get the $C_1^n$ ( j), $C_1^n$ ( j)mutates into $C_1$( j) at the certain probability, then computing the length of the path (ltsp1) according to the current position. If the new target function become better the new value is accepted, otherwise Refused. The path $C_1$( j) of the jth particle is still $C_0$( j) , then finding out the individual extremum (ptbest) and the extremum position (pcbest) of the each ant newly, finding out global extremum (gtbest) and global extremum positon (gcbest).

(5) comparing and process the crossover , the mutation and the regulation operation..

(6) computing the path length $L_k$( 1,2,…,m) of each ant, recording the current best solution.

(7) if the path length $L_k$ is less than settled path, to modify track strength with the renewal formula(7).

(8) nc =nc+1.

(9) if nc< preconcerted iterative times, having no the degeneration behavior (namely : finding out the same solution all), then turning to the step 2.

(10) outputting the current best solution.

## 5 Analysis of the Test and Result

For the sake of the usefulness of the testing algorithm we take Olive30 and Att48 as experiment example respectively and compare GRPSAC algorithm with Simulated Annealing (SA) , Genetic Algorithm (GA), Ant Colony Optimization(ACO) , Particle Swarm Optimization-Ant Colony Optimization (PSAC). Thereinto, the results of SA and GA are from the literature [3]' While The results of PSAC and GRPSAC algorithm are edited respectively with MATLAB7.1. Their parameters are $\alpha$=1.5, m=30, $\beta$=2,$\rho$=0.9. Each algorithm tests 20 times respectively. The origination temperature of SA is T=100000. The temperature of end is $T_0$=1. The velocity of

**Table 1.** The experimental result of these algorithms

| Algorithm | Oliver30 | | | Att48 | | |
|---|---|---|---|---|---|---|
| | Average | Best | Worst | Average | Best | worst |
| SA | 438.522 | 424.692 | 479.831 | 35176 | 34958 | 40536 |
| GA | 483.457 | 467.684 | 502.574 | 38732 | 38541 | 42458 |
| ACO | 550.035 | 491.958 | 599.933 | 36532 | 35876 | 42234 |
| PSAC | 436.458 | 423.949 | 457.316 | 35032 | 34672 | 40348 |
| RGPSAC | 423.576 | 413.564 | 451.287 | 34692 | 34253 | 36137 |

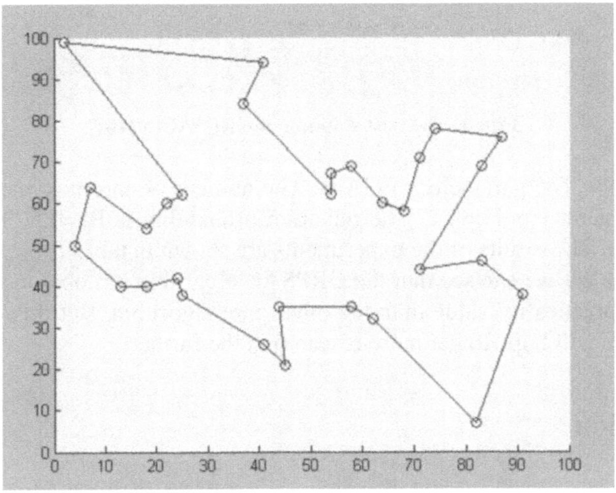


**Fig. 2.** The best solution of GRPSAC(Oliver30)

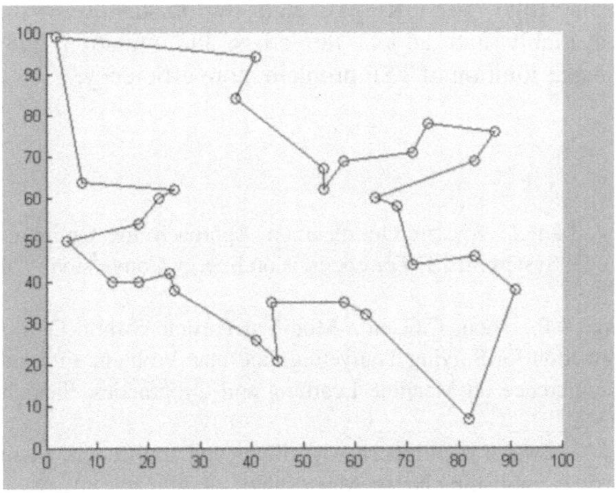


**Fig. 3.** The best solution of PSAC(Oliver30)

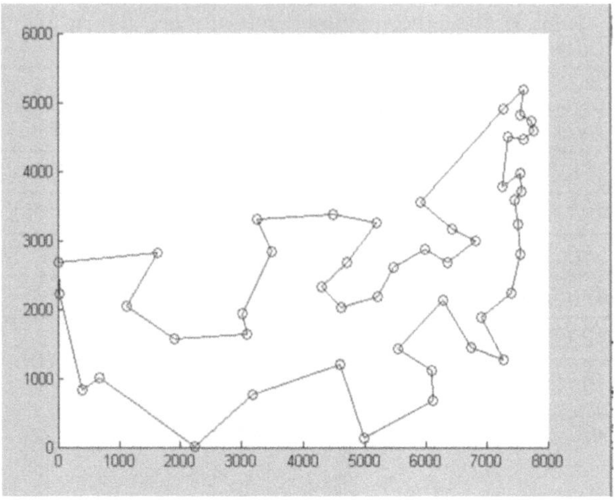


**Fig. 4.** The best solution of GRPSAC(att48)

anneal is a=0.99. The parameter of GA is: The number of chromosome is N=30. The crossover probability is Pc=0.2. The mutation probability is $P_m$=0.5. The number of iterative is 100. The results of the experiments are shown in table 1.

From the table 1 we can see that the GRPSAC algorithm has obviously better result and the more practicality value than the other pure algorithm. But this method is still not mature, We will hope to get more research in the further.

## 6 Conclusion

The ACO appears easily precocious phenomenon and local superior solution. While PSO is more simple and its capability of looking for the excellent solution is stronger. GRPSAC algorithm combines ACO with PSO organically and adds gene regulation operator at the same time. The GRPSAC algorithm makes use of the advantage of two algorithm availably and adopts the cross, the mutation and the regulation operator, which make solution of TSP problem more efficiency.

## References

1. Gaing, Z.L.: A Particle Swarm Optimization Approach for Optimum Design of PID Controller in AVR System. IEEE Transactions on Energy Conversion, Vol. 19 No.2 ( 2004) 384-391
2. Pang,W., Wang, K.P., Zhou, C.G.,etc.: Modified Particle Swarm Optimization Based on Space Transformation for Solving Tranveling Salesman Problem. Proceedings of the Third International Conference on Machine Learning and Cybernetics, Shanghai, 26-29 August (2004) 2342-2346
3. Gao,S., Yang,J.Y.: Swarm Intelligence Algorithms and application. Beijing Chinese water and water electricity publishing house, May ( 2006)

4. Maeda,Y., Kuratani,T.: Simultaneous Perturbation Particle Swarm Optimization. 2006 IEEE Congress on Evolutionary Computation Sheraton Vancouver Wall Centre Hotel, Vancouver, BC, Canada 16-21 July( 2006) 672-675
5. Gao,H.C., Feng,B.Q., Zhu,L.::Reviews of the Meta-heuristic Algorithms for TSP. Control and Decision , Vol.21 No.3. (2006) 241-246
6. Wang,Y.,DongYe,G.S.,Wang,L.G.:Improve of Genetic Algorithm Based Gene Regulation .Journal of Jian University (Sci.&Tech) , Vol. 20 No.2 (2006) 144-147

# MAS Equipped with Ant Colony Applied into Dynamic Job Shop Scheduling

Kai Kang, Ren feng Zhang, and Yan qing Yang

School of Management, Hebei University of Technology, Tianjin, 300401, China
kkang2000@sina.com,
gerry_hai@sohu.com,
happyyyq@163.com

**Abstract.** This paper presents a methodology adopting the new structure of MAS(multi-agent system) equipped with ACO(ant colony optimization) algorithm for a better schedule in dynamic job shop. In consideration of the dynamic events in the job shop arriving indefinitely schedules are generated based on tasks with ant colony algorithm. Meanwhile, the global objective is taken into account for the best solution in the actual manufacturing environment. The methodology is tested on a simulated job shop to determine the impact with the new structure.

**Keywords:** Dynamic Scheduling, Multi-Agent System, Ant Colony Optimization.

## 1 Introduction

Scheduling is always the key part in manufacturing and becoming more essential in recent years. Although the classical scheduling algorithms have been applied and studied widely over the years, the result is not satisfied in practice. The reason is that the classical scheduling algorithm is mostly to aim at the problems in the static environment, but the actual environment that will influence the effect of the scheduling is filled with dynamic events such as the arrival of the new orders, the cancellation of the original orders and the malfunction of machines. Consequently, the dynamic job shop scheduling is more needed.

Previous research on dynamic job shop scheduling tried to construct a new schedule so recently arrived jobs can be integrated into the schedule soon after they arrive. Wooseung Jang[1] presents a heuristic based on a myopically optimal solution to construct dynamic scheduling for stochastic jobs. The problem with this strategy is that constantly changing the production schedule can induce instability and the performance is not satisfied. A rescheduling methodology is proposed that schedules are generated at each rescheduling point using a genetic local search algorithm[2]. A periodic policy with a frozen interval is adopted to increase stability on the shop floor using a genetic algorithm to find a schedule so that both production idle time and

D.-S. Huang, L. Heutte, and M. Loog (Eds.): ICIC 2007, LNAI 4682, pp. 823–835, 2007.

penalties on tardiness and earliness of both original orders and new orders are minimized at each rescheduling point[3]. But how to find the best rescheduling point and scale the interval should be given more research. If the interval is too long, the emergent tasks may be delayed. If the interval is too short, the instability of the system is induced consequently. Products are assembled with lots of parts, which means that one product can not be completed until the last part was processed. So we define these parts as one task having the same due date. In this research, a methodology based on tasks to improve DJSS(dynamic job shop scheduling) is presented adopting the new structure of MAS(multi-agent system) and ACO(ant colony optimization) algorithm.

## 2 Basic Concepts and Notation

The job shop scheduling problem deals with the processing of a finite set of jobs on a finite set of machines. The process plan defines the selection and sequencing of operations for each job. Generally, the objective of the problem is mainly to find a sequence that minimizes the makespan, that is, the maximum completion time of the jobs. In this section we introduce the performance criteria of the dynamic job shop scheduling and the notation that we use. The performance criteria of the dynamic job shop scheduling is introduced in Section 2.1. In Section 2.2 we deal with the notation and present the objective function of the scheduling problems.

### 2.1 Performance Criteria of the Dynamic Job Shop Scheduling

The actual job shop scheduling problems are complex, dynamic and stochastic. Meanwhile, there are lots of restrictions such as the capabilities of the machines and the processing orders of the parts. And the dynamic job shop scheduling is multi-objective, which means that it considers more than one optimization objective such as the minimum of the makespan and the maximum of the profits simultaneously (or the minimum of the costs). More often the different objectives conflict, so we need to coordinate them according to the specific situation.

The performance criteria of the dynamic job shop scheduling is classified as time criteria, economical criteria and systemic criteria. The time criteria involves the minimum of the makespan, the due date of the tasks, mean flow time of the parts, the completion time of the tasks and so on. More research efforts have been spent on the minimum of the makespan and mean flow time of the parts. Tasks are determinate in previous static scheduling, so it is easier to find the best solution than that in dynamic scheduling. The economical criteria involves costs, penalties for tardiness, the inventory costs for the early completion and etc. The mode of JIT (just in time) is used widely in manufacturing firms recent years, which means that we should try our best to make the completion time of the processing come close to the due date in order to cut down the inventory costs and reduce the risk. The systemic criteria involves the utilization rate of the machines, productivity and so on. The proper allocation of the tasks on the machines can increase the efficiency of the processing and maintain the good performance of the machines. One job shop scheduling can not satisfy all the criteria at one time. We should choose the proper criteria according to our needs. Here we mainly consider the minimum of the makespan and the minimum of the prices for the tardiness as our optimization objectives.

## 2.2 The Notation and the Objective Function

The job shop scheduling problem can be characterized as n jobs to be processed on m machines. Generally, it is using a finite set of resources where resources are mainly consisted of machines and basic tasks are called jobs. Each job is a request for scheduling a set of operations according to a process plan which considers precedence restrictions. We have

a set of machines: $M_1,M_2,\ldots, M_m$
a set of parts: $P_1,P_2,\ldots, P_n$
a set of operations: $O_1,O_2,\ldots, O_n$

Every part is consisted of operations and every operation has to be processed on a given machine for a given time. For each operation there is a part to which it belongs, a machine on which it has to be processed and a processing time of the operation. $C_{ij}$ is the completion time for part i ($1 \le i \le$ n) processing on machine j ($1 \le j \le$ m). This set of parts consists of a task that is classified by the due date and one task is sequenced as a unit. We define $Fi$ as the penalties for the tardiness of the taski. Here we have some assumptions about the problem. The processing order of each part has to be maintained and each machine can only process one part at the same time; no part can be preempted; once an operation starts it must be completed; there is no precedence restriction on the operations of different parts.

Our aim is to find the starting times of all operations so that the completion time of the very last operation is minimal even if the dynamic events arrive. Meanwhile, this sequence leads to the least penalties when the tasks can not be finished in the due date. So we choose objective function as formula 1:

$$\begin{cases} Min \max\limits_{1\le j\le m}\left\{\max\limits_{1\le i\le n} C_{ij}\right\} \\ MinF_i = r_i * (C_i - D_i) \end{cases} \quad (1)$$

$C_i$ : the completion time of the $task_i$;
$D_i$ : the due date of the $task_i$;
$r_i$ : the penalty rate of the $task_i$

# 3 Multi-Agent Systems (MAS)

The advantages of Multi-Agent Systems (MAS) have been widely realized in manufacturing because of its flexibility and re-configurability in recent years. A decentralized multi-agent system is based on the idea that several distributed agents who are dependent decision-makers according to the information acquired can cooperate and interact together in order to obtain globally optimal performances[4]. In various kinds of applications such as distributed resource allocation, contract-net based negotiation mechanism is mostly adopted and has played an important role to achieve outstanding performance. But some subsequent problems that influence the effect of the negotiation to some extent come out. In section 3.1 some demerits of CNP (contract-net protocol) are displayed. And we introduce a new structure of MAS not adopting contract-net based negotiation mechanism in section 3.2. Solutions for several different dynamic events are presented with the new MAS in section 3.3.

## 3.1 Demerits of CNP

In the simple MAS architecture, two types of agents—part agents (PA) and resource agents (RA), are used to represent parts and resources, respectively. In consideration of the scheduling requirements and availability of manufacturing resources, the processing plan will be established through negotiation between the PAs and RAs. Most of the negotiation protocols are based on the renowned contract-net protocol (CNP)[5]. A contract-net based negotiation protocol is carried out according to the fictitious cost which reflects the objective of optimization. There have been lots of research efforts trying to extend the original CNP. One modification is to support biddings between multiple managers and multiple contractors[6]. A hybrid contract-net protocol (HCNP) is proposed to support a multi-task many-to-many negotiation[7]. While there are some unsurpassable obstacles in this negotiation protocol.

A great deal of communications come out between PAs and RAs because of the negotiation, which leads to lots of resources of agents occupied. So the capabilities of agents are strongly restricted and the results of scheduling are influenced consequently. Meanwhile, precedence restrictions of the tasks are not considered in CNP, which may need more coordination and cooperation between agents. Every task can not be understood totally by the agents because of the local sight and every agent can be a manager or a contractor, which induced the instability of the system.

## 3.2 The Proposed Structure of MAS

In consideration of the merits of MAS and demerits of CNP, we present a new structure of MAS not adopting CNP to assign the tasks. TAs(task agents, we define parts as one task having the same due date) exist as the tasks arrive, one TA corresponding to one task. RAs still represent the machines. It is difficult for the local agents TAs and RAs to reach a globally optimum solution. So we need a global controller—MA(management agent) to coordinate the local agents TAs and RAs. The MA is empowered to access full information and status of all the agents in the system ensuring that the global objective is being observed.

MA is the global control center depending on the computational ability and the system requirement trying to control the time, qualities, quantities and resources of the processing. When the new tasks arrive, MA makes a decision whether to take them according to the computational profits and the actual status of processing. And MA takes responsibilities to construct a new schedule for new tasks because of its global perspective. It is possible to consider one or several different global objectives such as minimizing jobs' makespan, minimizing jobs' tardiness, or balancing machines' loading etc. Then MA gives birth to TAs based on the tasks arrived and TAs are queued according to the priorities of their own. MA distinguishes the different priorities of TAs on account of the due dates of TAs and significances of customers to us. Meanwhile, MA transmits the information of tasks to TAs. All the activities are supervised by MA and the timing and frequency of intervention are also determined by the MA.

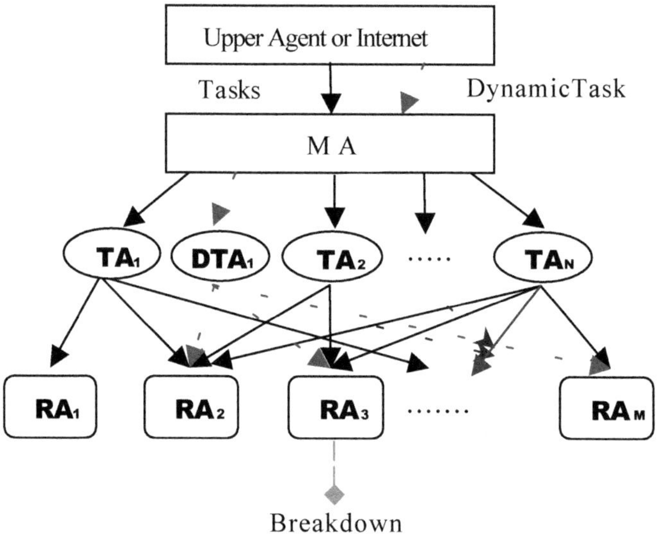


**Fig. 1.** The new structure of MAS

Every TA builds best schedule for the common task with ant colony algorithm and transmits the information to RAs. Then RAs mark "occupied" on the machines when assigned and return the information to TAs, which can be considered for TAs in the future schedules. RAs always choose tasks of high priorities when TAs transmit the processing messages to them. The new structure of MAS is showed in Figure 1.

### 3.3 Solutions for Dynamic Events

The actual job shop scheduling problem is more difficult because it is filled with dynamic events such as the arrival of the new orders, the cancellation of the original orders and the malfunction of machines. In this section, we will present the solutions considering some typical dynamic events in the proposed MAS.

#### 3.3.1 Arrival of the New Orders

With the arrival of the new orders, MA gives birth to DTAs and put them into the queue according to their priorities. We have assumptions that DTAs must be finished once taken by MA because MA has considered the global objective according to its computational ability. Here we present the solution for the typical situation as an example. When the dynamic task($DTA_1$) arrives, we queued it between $TA_1$ and $TA_2$ according to the priority of $DTA_1$. Meanwhile, the due date of $TA_2$ is behind the due date of $DTA_1$. In Figure 1 the red line represents $DTA_1$. MA constructs a hybrid sequence for $DTA_1$ and the remained operations of $TA_1$ with ant colony algorithm on premise that $DTA_1$ must be finished before the due date of $DTA_1$. Then MA transmits the best solution chosen from the alternative sequences to $TA_1$ and $DTA_1$.

#### 3.3.2 Cancellation of the Original Orders

Some of the original orders may be cancelled because of the market or the design in the actual manufacturing. Accordingly, MA will cancel the TAs and RAs assigned by these TAs will be reassigned for the new assignment. Meanwhile, TAs that have not started being processed will build new schedules on more RAs for a better performance.

#### 3.3.3 The Malfunction of Machines

MA will write off the RAs that have malfunctions from the database and tries to find some replaceable RAs for TAs that have assigned processing tasks on these abnormal RAs. Then TAs will construct a new schedule on these RAs according to their priorities as is shown in Figure 1.

## 4 ACO Applied into MAS

Ant colony optimization has become an increasingly popular method mimicing behaviour of processes that exist in nature. Ants in nature can always find the shortest path from their nest to the food source. The information is communicated through any chemical or set of chemicals produced by ants, called pheromone by a process called stigmergy, a particular form of indirect communication used by social insects to coordinate their activities. All the ants secrete this pheromone while walking and pheromone is volatile and evaporates quickly. A strong pheromone concentration on a path will stimulate the ants to move in that direction. While ants using a shorter path back to the nest, it must be faster than ants taking a longer path because the quantity of pheromone laid down on the shorter path grows faster than on the longer ones. Meanwhile, there will be some stray ants that may take the longer paths, which may explore other new routes to the food and back to the nest. The choice of path seems almost probabilistic in nature. The artificial ants can be furnished with some oddities that real ants do not have, for instance a local heuristic function to guide their search through a set of feasible solutions and an adaptive tabulist so that they can remember visited nodes.

The original ant algorithm was introduced by Marco Dorigo in his doctoral thesis[8] and was called an ant system (AS). AS was applied into job shop scheduling and proved to be a noteworthy method in a paper by Colorni et al[9]. Ying et al.[10] applied the ant colony system to permutation flow-shop sequencing and effectively solved the n/m/P/Cmax problem. Gajpal and Rajendran[11] used a new ACO algorithm to minimize the completion-variance of jobs, showing that work with ACO algorithms is an ongoing process to modify and improve the original AS and apply it to a variety of scheduling problems. J. Heinonen, F. Pettersson[12] applied the ant colony optimization with a postprocessing algorithm in the job shop scheduling and the performance approves that the new method is a noteworthy competitor to existing scheduling approaches. In our new structure TAs construct schedules for the original tasks and MA constructs schedules for dynamic tasks with ACO algorithm.

### 4.1 ACO

Some similarities can be found in JSSP(job shop scheduling problem) and TSP. A connected graph G = (N,A,E) is constructed with weights on the edges between the nodes referring to the processing. The nodes denote the operations, and the edge weight is the local heuristic function we choose according to the global objective. This graph is constructed based on parts for the feasibility of processing. And we have two dummy nodes as the starting node and the finishing node. The goal is to find a tour in G that connects all operations(from the starting node to the finishing node) so that the overall time is minimal. All ants are initially put in the starting node, and move to a node in their feasible list. Each edge eij has the pheromone value $\tau_{ij}$ associated to it. When located at a node i an ant k uses the pheromone trails $\tau_{ij}$ to compute the probability of choosing node j as the next node in the formula 2:

$$\begin{cases} p_{ij}^k = \dfrac{\tau_{ij}^{\alpha}\eta_{ij}^{\beta}}{\sum \tau_{ij}^{\alpha}\eta_{ij}^{\beta}} & ,\text{if } i \in N_i^k \\ p_{ij}^k = 0 & ,\text{if } i \notin N_i^k \end{cases} \tag{2}$$

$N_i^k$ is the allowed neighborhood of ant k when in node i, that is, the list of operations that ant k has not yet visited. If $q \le q_0$, choose the next node j according to

$\max_{j \in allowed_k} \{\tau_{ij}^{\alpha}\eta_{ij}^{\beta}\}$, otherwise choose the next node j according to the method as above. q is a random number in [0,1], and $q_0$ is a parameter that determines the relative influence of the new trail and the heuristic information.

It is not entirely straightforward what visibility is as the distance of TSP and what effect it has on computations with regard to ACO on schedules. So an additional problem when working with this method is that of visibility. Some various approaches to ACO-visibility in schedules are undertaken and studied. The parameter $\eta_{ij}$ is the measure of visibility and in problems with an appearance like TSP the meaning is clear and all values of $\eta_{ij}$ can be computed according to the distances between nodes. Here we use SPT(shortest processing time) as visibility that ranks the operations according to length of their processing time, shorter processing time means a higher probability of being chosen. So the parameter is computed as formula 3:

$$\eta_{ij} = 1/T_{ij} \tag{3}$$

where $T_{ij}$ is the processing time of $O_{ij}$ . The parameters $\alpha$ and $\beta$ are two parameters which determine the relative influence of the pheromone trail and the heuristic information. When the ant moves from node i to j, local pheromone gets updated when the ant passed as formula 4:

$$\begin{cases} \tau_{ij} \leftarrow (1-\rho)\bullet\tau_{ij} + \rho \bullet \Delta\tau_{ij} \\ \tau_0 = (nT_r)^{-1} \end{cases} \tag{4}$$

where $\tau_0$ is the initial pheromone,$T_i$ is the stochastic completion time, $n$ is the number of the nodes. $\Delta\tau_{ij}$ is the pheromone deposited by an ant. $0 < \rho \le 1$ is the evaporation rate, which enables the algorithm to ''forget'' previous bad decisions and avoids unlimited accumulation on the edges[13]. Once all ants have completed their tour the best global trail get updated to guide the search for ants as formula 5 :

$$\begin{cases} \tau_{ij} \leftarrow (1-\rho)\bullet\tau_{ij} + \rho\bullet\Delta\tau_{ij} \\ \Delta\tau_{ij} = (T_{gb})^{-1} & \text{,if (i, j)} \in \text{best globalroute} \\ \Delta\tau_{ij} = 0 & \text{,if (i, j)} \notin \text{best globalroute} \end{cases} \tag{5}$$

where $T_{gb}$ is the shortest time in the iteration. When the system expresses stagnation behaviour, pheromone values are reinitialized to encourage ant exploration. The ACO algorithm also keeps tracks of visited nodes, meaning that

**Table 1.** ACO parameter settings

| Parameter | Value | Meaning |
|---|---|---|
| N | 20 | Number of ants |
| $\alpha$ | 1.0 | Influence of the pheromone trail |
| $\beta$ | 2.0 | Influence of heuristic information |
| $\rho$ | 0.4 | Evaporation rate |
| $q_0$ | 0.9 | Influence of new trail and heuristic information |

**Table 2.** The processing information of $TA_1$

| Part | Operation 1 | | Operation 2 | | Operation 3 | | Operation 4 | |
|---|---|---|---|---|---|---|---|---|
| | Process | Time | Process | Time | Process | Time | Process | Time |
| Part 1 | 3 | 12 | 1 | 4 | 2 | 6 | 4 | 7 |
| Part 2 | 2 | 8 | 3 | 6 | 1 | 10 | 4 | 10 |
| Part 3 | 3 | 6 | 4 | 4 | 2 | 8 | 1 | 12 |

**Table 3.** The information of RAs

| Machine | Process |
|---|---|
| 1 | 1 |
| 2 | 2 |
| 3 | 3 |
| 4 | 4 |

the ants have a memory which helps them select the next node from a list of possible choices.

### 4.2 Simulation

Experiments were conducted on Matlab 7.0 and the program was executed on the test problem using a PC with Pentium4 processor running at 1.70GHZ and 256MB of RAM. All the ACO parameter settings can be seen in Table 1 and we use SPT(shortest processing time) as visibility. MA gives birth to TAs based on the tasks arrived and TAs are queued according to the priorities of their own. The processing information of TA1 can be seen in Table 2. And the due date of $TA_1$ is 70. The information of RAs can be seen in Table 3. After 30 rounds of calculations in 4.106 seconds best found result is 43 and the original schedule with ACO for $TA_1$ can be seen in Table 4.

The actual environment is filled with dynamic events that will exert an influence on the constructed schedule. And high prices will be paid if not properly solved. Here

**Table 4.** The original schedule of $TA_1$

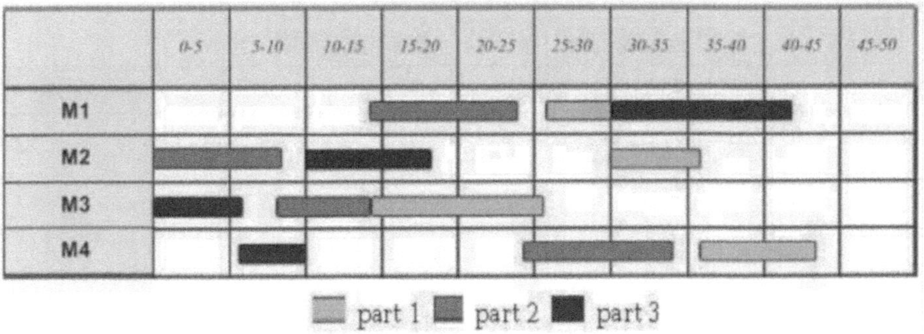


**Table 5.** The processing information of $DTA_1$

| Part | Operation 1 | | Operation 2 | | Operation 3 | | Operation 4 | |
|---|---|---|---|---|---|---|---|---|
| | Process | Time | Process | Time | Process | Time | Process | Time |
| Part 4 | 4 | 5 | 2 | 18 | 1 | 24 | 3 | 3 |
| Part 5 | 4 | 12 | 2 | 8 | 3 | 8 | - | - |

we mainly consider the arrival of emergent tasks coming in different time. The emergent task comes indefinitely, which means that it may come at any point of the processing. We deal with two typical situations that $DTA_1$ arrives at 0 and 20 for the general solution. The processing information of $DTA_1$ can be seen in Table 5 and the due date of $DTA_1$ is 70.

When $DTA_1$ that has higher priority than $TA_1$ arrives at 0(the processing of $TA_1$ has not started yet), the schedule for $DTA_1$ and $TA_1$ is displayed in Table 6 and the best result is 69(60 rounds of calculations in 22.993 seconds). When $DTA_1$ arrives at 20(the processing of $TA_1$ has started), the schedule for $DTA_1$ and $TA_1$ is displayed in Table 7 and the best result is 70. Dynamic events happened at other time can be solved similarly according to the specific situation.

**Table 6.** The schedule for $DTA_1$ arriving at 0

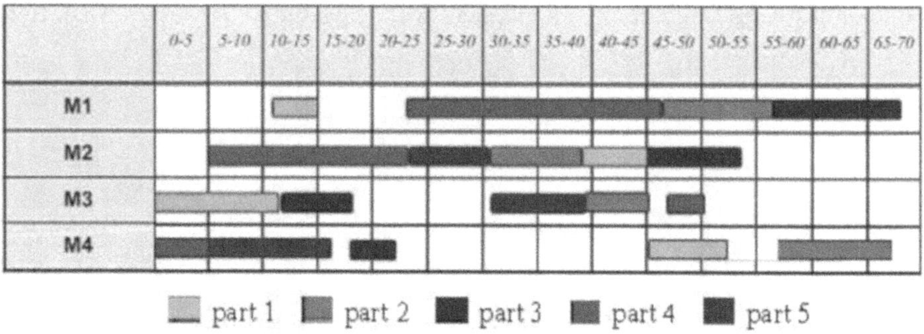


**Table 7.** The schedule for $DTA_1$ arriving at 20

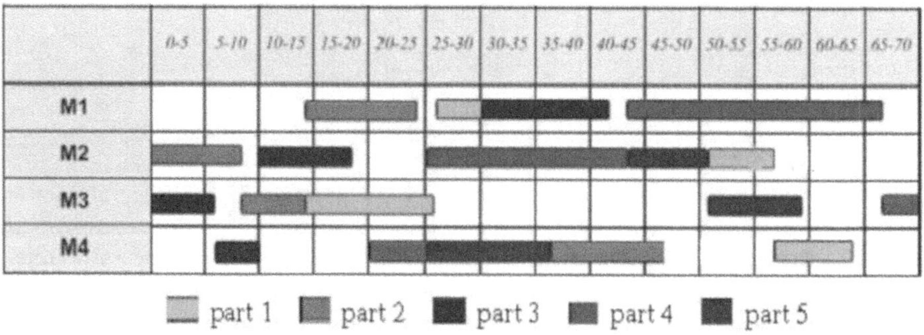


From the table we can see that all the tasks are scheduled properly and the results are satisfied. Excellent performance has been made with this structure based on the arrival of new tasks. Meanwhile, this method can deal with the dynamic events effectively and efficiently.

## 5 Conclusions

This paper presents a methodology based on tasks to improve DJSS adopting the new structure of MAS(multi-agent system) equipped with ACO(ant colony optimization) algorithm. The proposed structure is to aim at supporting a better schedule for the dynamic job shop. The numerical results confirm that the proposed methodology can improve the schedule. Besides, the global objective can be considered first when constructing a new schedule, so the structure is feasible and attractive in the actual shop floor.

## References

1. Jang, W.: Dynamic Scheduling of Stochastic Jobs on a Single Machine. European Journal of Operational Research, 138 (2002) 518–530
2. Rangsaritratsamee, R., Ferrell, W., Kurz, M. B.: Dynamic Rescheduling that Simultaneously Considers Efficiency and Stability. Computers & Industrial Engineering, Vol. 46 (2004) 1–15
3. Chen, K.J., Ji, P.: A Genetic Algorithm for Dynamic Advanced Planning and Scheduling (Daps) with A Frozen Interval. Expert Systems with Applications (2006)
4. Durfee, E.H.: Distributed Problem Solving and Planning. In G. Weiss(Ed.), Multiagent systems: A Modern Approach to Distributed Artificial Intelligence, Cambridge, MA: MIT Press (1999) 121–164
5. Smith, R.G..: The Contract Net Protocol: High-Level Communication and Control in a Distributed Problem Solver. IEEE Transactions on Computers, Vol. 29(12) (1980) 1104–1113
6. Calosso, T., Cantamessa, M., Vu, D., Villa, A.: Production Planning and Order Acceptance in Business to Business Electronic Commerce. International Journal of Production Economics, Vol. 85(2) (2003) 233–249
7. Wong, T.N., Leung, C.W., Mak, K.L., Fung, R.Y.K.: Dynamic Shopfloor Scheduling in Multi-Agent Manufacturing Systems. Expert Systems with Applications, Vol. 31 (2006) 486–494
8. Dorigo, M.: Optimization, Learning and Natural Algorithms. Ph.D. Thesis, Dipartemento di Elettronica, Politecnico di Milano (1992)
9. Gutjahr, W.J., Rauner, M.S.: An ACO Algorithm for A Dynamic Regional Nurse-Scheduling Problem in Austria, Comp. Oper. Res. 41 (3) (2007) 642–666
10. Ying, K.C., Liao, C.J.: An Ant Colony System for Permutation Flow-Shop Sequencing. Comput. Oper. Res. 31 (2004) 791–801
11. Gajpal, Y., Rajendran, C.: An Ant-Colony Optimization Algorithm for Minimizing the Completion-Time Variance of Jobs in Flowshops. Int. J. Prod. Econom. 101 (2006) 259–272

12. Heinonen, J., Pettersson, F.: Hybrid Ant Colony Optimization and Visibility Studies Applied to A Job-Shop Scheduling Problem. Applied Mathematics and Computation (2006)
13. Dorigo, M., Bonabeau, E., Theraulaz, G.: Ant Algorithms and Stigmergy. Future Generation Computer Systems, Vol. 16 (2000) 851–871

# Optimizing the Selection of Partners in Collaborative Operation Networks

Kai Kang, Jing Zhang, and Baoshan Xu

School of Management, Hebei University of Technology, Tianjin, 300401, China
kkang2000@sina.com, zhangjing1219@gmail.com, xubaoshan202@163.com

**Abstract.** Through the situation today it is necessary that small and medium sized enterprises collaborate in so-called collaborative operation network. In the center of interest is the development of a virtual enterprise model which is based on small collaborative cells in so-called operation centers. Thus, the concentration on the core competences is supported and the market power is increased by the help of the collaborative operation networks. The automated selection of the partners is the major problems in virtual enterprises. In the paper, a method for choosing the most capable operation centers for every order is designed. The selected operation centers fulfill the tasks of a value chain particularly well. Within the approach, the problem will be solved by Ant Colony Optimization in combination with the Analytical Hierarchy Process.

**Keywords:** Ant colony optimization, Collaborative operation network, Operation center, Partner selection, Virtual enterprise.

## 1 Introduction

The continual development of modern communication technologies and the quickly increasing globalization force enterprises to re-think their economic behavior. The classical image of the enterprise does not completely meet the modern economic reality anymore. Thereby, the concentration on core competences implies the increase in enterprise-spanning cooperation having the objective of releasing cost reduction potentials and of being present on global market places [1]. The pressure to face those challenges considerably increases especially for small and medium-sized enterprises (SMEs) in order to secure their own survival. In the paper, a virtual enterprise model is developed in order to improve the competitiveness of SMEs, so called *collaborative operation network (CON)*. This is based on very small collaborative cells, so called *operation centers (OCs)*.

The detailed discussion of the OCs and the operating mechanisms of CONs are far beyond the scope of this paper. Due to the space limitations, the paper will restrict our description in the selection of partners in collaborative operation networks. There are two problems of the selection of partners. On the one hand, the OC has to be evaluated. The difficulty consists in the fact that several criteria, which are different from one

D.-S. Huang, L. Heutte, and M. Loog (Eds.): ICIC 2007, LNAI 4682, pp. 836–850, 2007.

another in terms of content, need to be included in the evaluation. On the other hand, there is a great variety of alternatives for manufacturing one product, so a complete enumeration is not possible [2].

The contribution introduces a possibility of generating a solution for the shown problem by *Ant Colony Optimization (ACO)* (for the selection of partners) in combination with *Analytic Hierarchy Process (AHP)* (for the computation of the objective function value) [3], [4], [5].

## 2 Problem Description

The idea of forming a CON out of a pool of autonomous OCs follows this pattern: Every OC in the pool is able to attract a customer's production task. Once the production planning exists, the OC has to search for suitable partners for all work steps which it is not able to fulfill [1]. Figure 1 depicts the attraction of OCs out of a pool to a customer's production task.

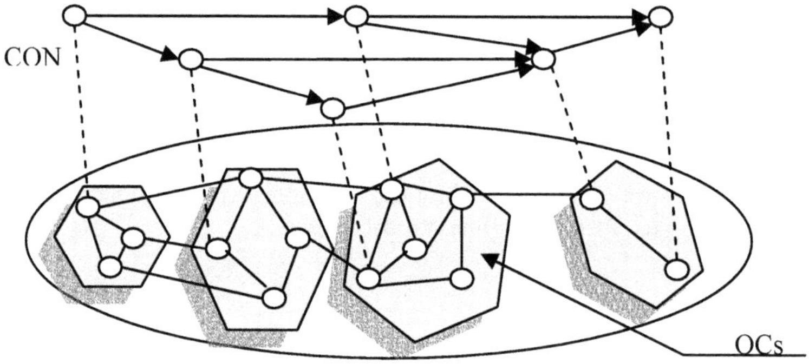


**Fig. 1.** Working Model: collaborative operation network

The customer's production tasks are described exactly by the inquiry vectors (IV). Those define the necessary work plans in order to complete an intermediate product. According to the IVs, corresponding OCs are searched for all elements, which are potentially able to manufacture the intermediate product. That means, the offer vectors (OV) of the OCs have to equal to the IVs to a certain degree. The principal structure of an OC-offer vector is illustrated in figure 2.

The *specialist competence (SpC)* within the OV is reflected in two measures similarity and saving of time. The parameter similarity conceals the percentage of the conformity of IV and OV. That value depends on the order. A high value (max.100%) is aimed at, but not stringently necessary. The term potential of saving of time refers to a possible saving of time achieved by an adaptation of the production system in terms of intensity.

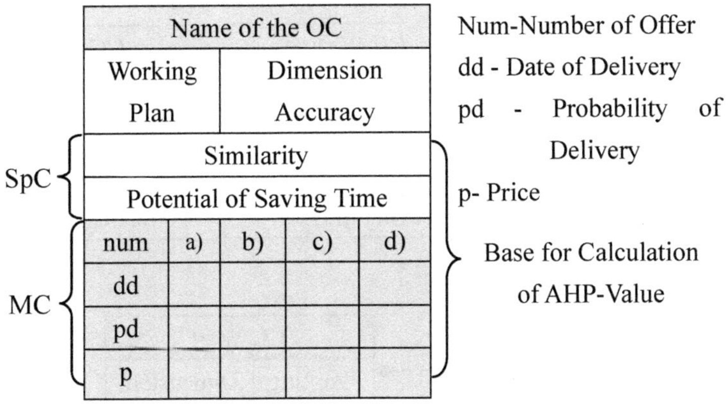


**Fig. 2.** Offer vectors of an OC

Within the *method competence (MC)*, the OV is provided with a number of value-triples consisting of the date of delivery, the probability of delivery and the resulting cost. The starting point of that consideration is the date of delivery desired by the final customer. This is followed by the calculation of a time slot via a backward scheduling.

Within the scope of the search for an optimal manufacturing variant, Soft-facts are considered as marks of the *social competence (SoC)* included in the network. Thereby, it is dealt with qualitative parameters for the description of the social feature of OCs such as confidence. The parameters connectivity and eccentricity, calculated by the help of the polyhedral analysis, are included in the social evaluation of every work variant in the OC-offer network [6].

## 3 Modeling

For the optimization by the help of an algorithm, all work variants within the OC-offer network are illustrated as a directed graph when the pairs of nodes *(i, j)* attached to an edge are arranged. Thereby, *i* is the first node-the initial node, and *j* is the second node-final node of the pairs *(i, j)*. Therefore, it is necessary to insert an initial node, a so-called *source*, for all nodes, which are in the beginning of the value chain. Starting from that, all OC-offers are integrated in the graph according to the sequence in the single process variants. After the last work step of all variants, the alternatives meet in a final node of the graph, the final product. That point is called *drain*. The objective is the maximization of the cumulated AHP-values of the OCs.

Figure 3 illustrates a part of the modeling as a direct graph for a simple value chain. For every step of production between the intermediate products, several manufacturing variants exist, out of whom the best has to be selected. It has to be considered that not all potential OC-offers for the processing step $i+1$ can be attained by every OC-offer in

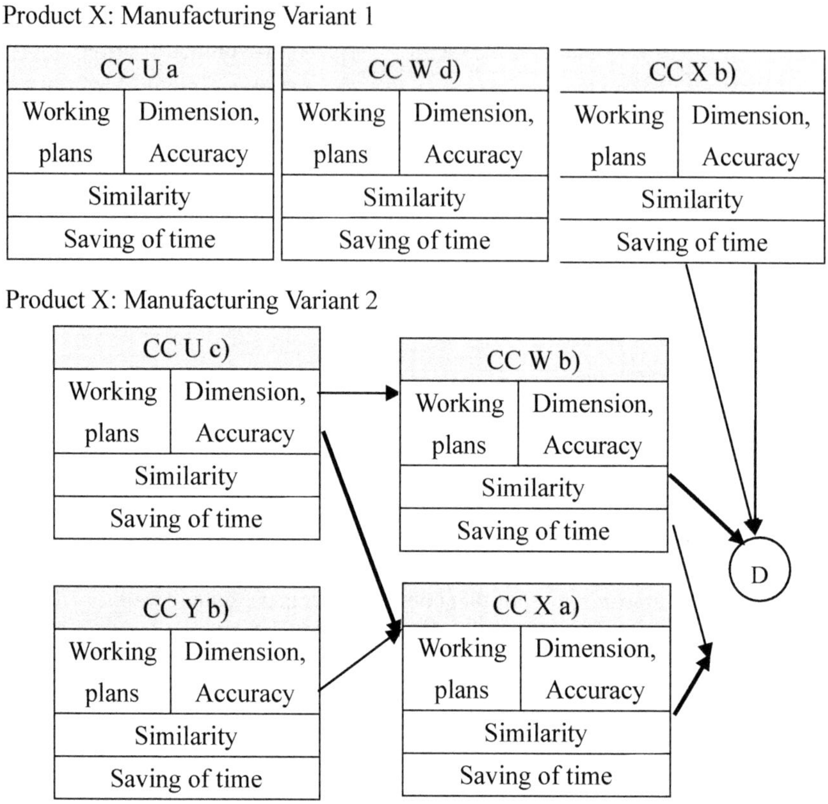


**Fig. 3.** Illustration of the problem (not parallel, no converging production)

the manufacturing step *i*. The reason for that is to be found in the overlapping of dates of delivery and the latest beginning date of the following OC. An example is the missing link between offer OC *Y b)* and offer OC *W b)* in Figure 3. The emphasized route represents a concrete, realizable manufacturing alternative.

The problem illustrated in Figure 3 is simplified and does not comprise a converging production (assembly of parts). If real products are to be produced in the network, one can not act on the assumption. Therefore, it is absolutely necessary to consider branches. Before the assembly, the components are manufactured independently and are not included in the final product before the time of assembly. Figure 4 expands the graph in Figure 3 by branches. Furthermore, it needs to be recognized that there are two strongly emphasized route from the source via the offers of the first process step OC *Wa)* and OC *Vd)* up to the offer of the second process step OC *Wc)*. That means that both ways are necessary in the manufacturing variant [7][8]. Generally, the involved OC-offers OC of a manufacturing variant $k$ are stored in $\psi$ and have the objective function value $L_k$. The attractiveness of an OC-offer is determined by the according AHP-value and it is constant during the whole search and it is independent from the predecessor or the successor.

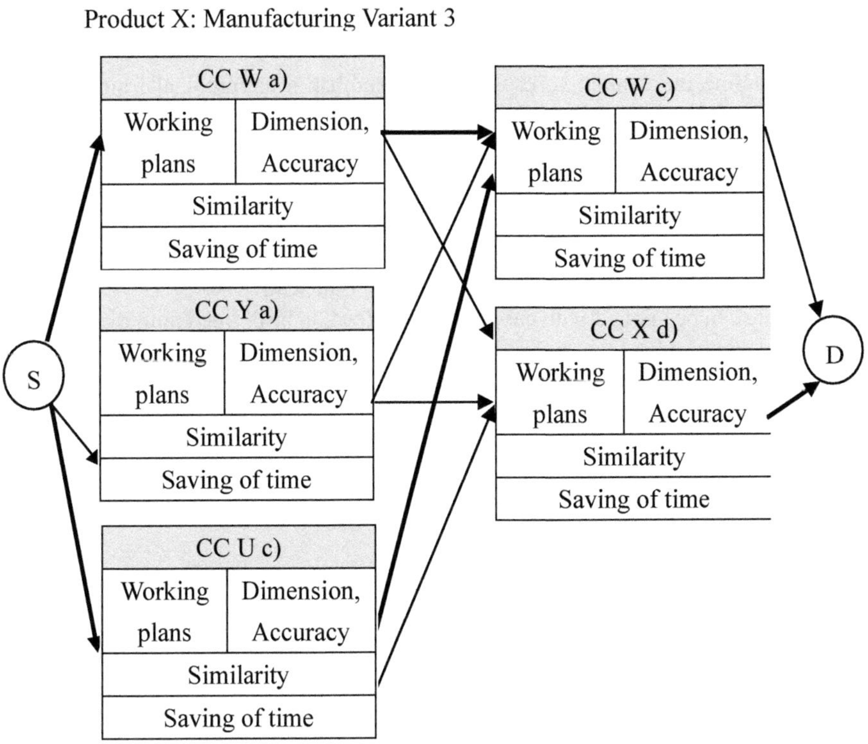


**Fig. 4.** Illustration of the problem (parallel, converging production)

In theory, the ACO demonstrated its suitability in a great variety of different problems. One for example found good results for problems [11], [12], [13], etc. All the practical problems have in common, that they dispose of a great number of nodes, that these are dynamic and that they evade themselves from other solution methods by difficulty manageable restrictions of an efficient modeling.

For the use of the ACO for the selection of OCs in networks, it is a pre-condition that it can also be applied for converging production (real value chains-see Figure 4). The necessary adaptations are described in the following paragraph.

## 4 The Computation of the Objective Function Value

The AHP-methodology is a method for decision formulation and analysis. It is a multi-criteria decision procedure which arranges a final number of alternatives by the help of a linear preference index which has been successfully applied to a diverse array of problems. In the paper, the objective function value is computed by the help of AHP. The values within the specialist competence, method competence and social competence form the basis for the evaluation of the OC-offers. In this section, we use an example to show the application of the algorithm for the calculation of the OCs'attractiveness. The process comprises the following steps [4] [5].

Step1: Define the evaluative criteria, and establish a hierarchical framework (see figure 5).

Step2: Establish each factor of the pair-wise comparison matrix.

In the step, the elements of a particular level are compared pair-wise, with respect to a specific element in the immediate upper level. Saaty (1980) suggests the use of a 9-point scale to transform the verbal judgments into numerical quantities representing the values of $aij$. Table 1 lists the definition of 9-point scale.

This scale can be applied with ease to criteria that can be defined numerically as well as to those can not be defined numerically. Relative importance scale is presented. The decision maker is supposed to specify their judgments of the relative importance of each contribution of criteria towards achieving the overall goal. Table 2 presents the main criteria as the sample.

Step3: Calculate the eigenvalue and eigenvector.

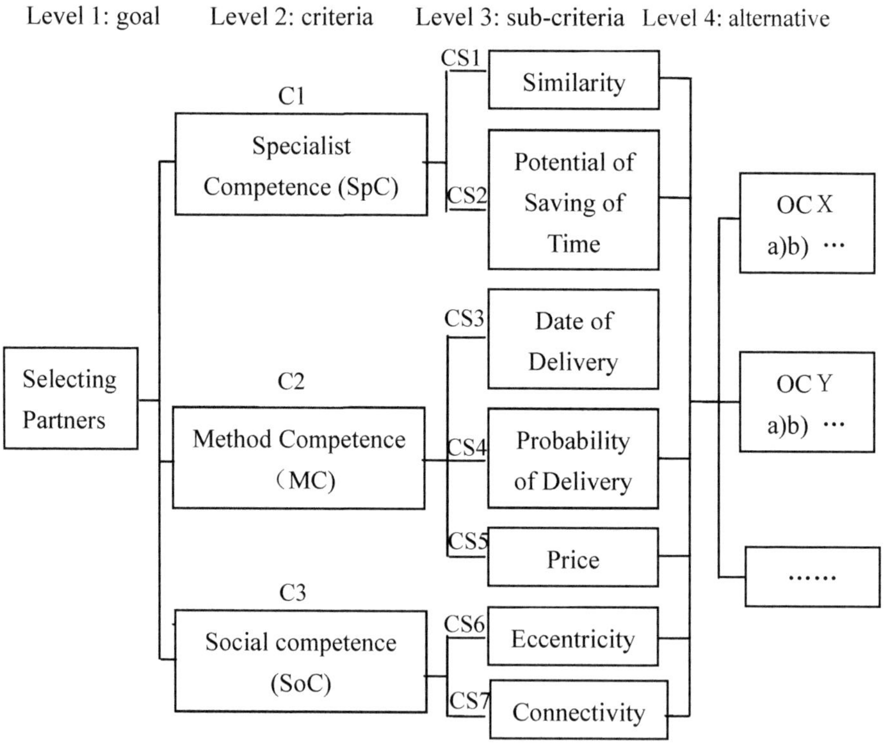


**Fig. 5.** Hierarchical structure to select the partners

Saaty's method computes *W* as the principal right eigenvector of the matrix *A*:

$$AW = \lambda_{max} W. \tag{1}$$

$$(A - \lambda_{max} I) = 0. \tag{2}$$

Here, using the comparison matrix (such as in Table 2) the eigenvectors were calculated by equation (1) and (2). Table 3 summarizes the results of the eigenvectors for criteria, sub-criteria and three OCs. Besides, the results for each level relative weight of the elements are showed in table 3.

Step4: Perform the consistency test. The eigenvector method yields a natural measure of consistency. Saaty defined the consistency index (CI) and consistency ratio (CR):

$$CI = \frac{(\lambda_{max} - n)}{(n-1)}. \tag{3}$$

$$CR = \frac{CI}{RI}. \tag{4}$$

**Table 1.** The pair-wise comparison scale (Saaty, 1980)

| Intensity of importance | Definition |
|---|---|
| 1 | Equal importance both element |
| 3 | Weak importance one element over another |
| 5 | Essential or strong importance one element |
| 7 | Demonstrated importance one element over |
| 9 | Absolute importance one element over |
| 2,4,6,8 | Intermediate values between two adjacent |

**Table 2.** Aggregate pair-wise comparison matrix for criteria of level 2

| Goal | C1 | C2 | C3 |
|---|---|---|---|
| C1 | 1 | 1.582 | 1.622 |
| C2 | 0.632 | 1 | 1.026 |
| C3 | 0.616 | 0.975 | 1 |
| $\lambda_{max} = 3.149056; CI = 0.074528; RI = 0.90; CR = 0.082808 \leq 0.1$ | | | |

A value of the consistency ratio *CR≤0.1* is considered acceptable. Large values of *CR* require the decision maker to revise his judgments.

Step5: Calculate the level hierarchy weight and the attractiveness to every OC-offer.

The composite priorities of the alternatives are showed in Table 4. According to table 4, attractiveness is calculated.

**Table 3.** Weight of the criteria, sub-criteria and three OCs

| Criteria | Weight | Subcriteria | Weight | Synthesis | OC X | OC X | OC X |
|---|---|---|---|---|---|---|---|
| C1 | 0.369 | CS1 | 0.598 | 0.221 | 0.242 | 0.390 | 0.282 |
| | | CS2 | 0.402 | 0.148 | 0.314 | 0.440 | 0.332 |
| | | Synthesis value | | | 0.278 | 0.415 | 0.307 |
| C2 | 0.321 | CS3 | 0.250 | 0.056 | 0.327 | 0.430 | 0.243 |
| | | CS4 | 0.348 | 0.078 | 0.117 | 0.464 | 0.359 |
| | | CS5 | 0.402 | 0.091 | 0.361 | 0.347 | 0.293 |
| | | Synthesis value | | | 0.287 | 0.408 | 0.303 |
| C3 | 0.310 | CS6 | 0.453 | 0.095 | 0.345 | 0.418 | 0.236 |
| | | CS7 | 0.547 | 0.116 | 0.352 | 0.355 | 0.293 |
| | | Synthesis value | | | 0.349 | 0.390 | 0.261 |

**Table 4.** Selection of the partner in terms of precision

| Criteria | Weight | OC X a) | OC X b) | OC X c) |
|---|---|---|---|---|
| C1 | 0.369 | 0.278 | 0.415 | 0.307 |
| C2 | 0.321 | 0.287 | 0.408 | 0.303 |
| C3 | 0.310 | 0.349 | 0.390 | 0.261 |
| Result | Attractiveness | 0.305 | 0.404 | 0.291 |

## 5 The Selection of OCs by the Help of ACO

In general, the problem described in that contribution can also be interpreted as a Traveling Salesman Problem (TSP)[11]. That means, a way has to be found from the source to the drain, which provides a high objective function value. An important difference consists in the face that not all nodes (OC-offers) need to be visited in the graph.

One of the most important assumptions is the disregard of the distance between the OCs. The length of the edges between the nodes, which could be interpreted as transport costs, is already included in the costs within the OC-offers. For that reason, $\eta_{ij}$ does not represent the length of the way between two OCs, but it indicates the attractiveness (to select) of the OC-offer node at the end of the edge. Thereby, the AHP-value calculated from saving the of time, similarity, date, probability of delivery, costs, eccentricity and connectivity is applied. The heuristics value remains constant. Because of the objective of maximizing the AHP-value, the heuristics value is calculated from the reciprocal, as in TSP, but it is immediately set equal to the AHP-value.

The pheromone value $\tau ij$ on the edge $(i \rightarrow j)$ remains the leading variable of the search. It is responsible for the improvement of the solutions in the course of time. The reason is the dependence of the objective function values $L_k$ on the solutions $\psi_k$. In case, a solution is qualitatively good, that means, $L_k$ is high, all edges $(i \rightarrow j \in \psi_k)$ receive extra pheromone ($\triangle\tau$). This increases the attractiveness of the edges for following ants and iterations. If, in contrast, the quality of the found complete solution is weak ($L_k$ is lower), the single edges do not receive few pheromone. Which calculating the transition rule, the pheromone values $\tau_{ij}$ on the edge $(ij)$ as well as the AHP values of the possible OC-offers $j \in N_i^k$ (nodes) are included.

Further adaptations are necessary because of the branches in the graph (converging production, assembly processes). Within the original algorithm for the TSP, each ant finds exactly one way through the graph. However, in case necessary branches (see Figure 4) are existing, different parallel way result, which are all part of the solution. The solution of the ant in contrast might only include one branch. A further, easier realizable possibility consists in the addition of new ants at branches. Normally, the search of the ants happens from the source to the drain. Thereby, it can not be recognized that a branches takes place before a node where two ways are united. As oppose, if one takes the reverse way from the drain (final product) to be source, one recognizes the branches before deciding for a branch (route). In that case as many ants as branches are set onto the node where the branches start. Then, those ants respectively move one branch and search for a solution. The ant up to the branch as well as the new (added) ants on behalf of their contents belongs to a solution and form a family. Thus, now all the OC-offers of a family, from which the objective function value $L_k$ rises, are stored in $k$. For reasons of performance, the transformation of the graph is neglected in the following.

Figure 6 illustrates the developed program procedure as pseudo code. After constructing the *problem structure*, that means storing the concrete graph from the information of the central database, the search of the ants is started. It lasts until the fixed break condition is achieved. Parameter $m$ regulates the size (amount of ant families) of a colony. This needs to be chosen dependent on the size of the problem.

Begin

Initialization ( $problem\_structure$ );

$i = drain$ ;

While not ( $exit\_condition$ ) do

For $k := 0$ to m steps 1 do

While $(N_i^k \neq \phi) \cap (i \neq source)$ do

Procedure ( handle_branching )

Random ( $z$ );

if $z \leq q$ then $p_{ij}^k(t) = [\tau_{ij}(t)]^\alpha [\eta_{ij}]^\beta$ ;

$$p_{ij}(t) = \frac{[\tau_{ij}(t)]^\alpha [\eta_{ij}]^\beta}{\sum_{i \in N_i^k} [\tau_{ij}(t)]^\alpha [\eta_{ij}]^\beta}$$

Else ;

Decide $(j)$ ;

$\Psi_k = \Psi_k \cup OC(\max(AHP(j)))$;

/* local Pheromone update*/

$\tau_{ij}(t+1) = (1-\rho') * \tau_{ij} + \rho' * \Delta\tau_{ij}; i := j$

End

End

/*global Pheromone update*/

$\tau_{ij}(t+1) \leftarrow (1-\rho) * \tau_{ij}(t) + \Delta\tau_{ij}(t) \forall \tau_{ij}$;

$\max - \min - rule\, \tau_{ij;}$

$Decide(\psi_k)$;

End

$\Psi_k \in M : \forall \Psi_k mit L_k \succ \kappa * L_k^* \quad 0 \leq \tau \leq 1$

Decide $(\Psi_k^{\max} : \max(aggregation(MK_k, SK_K)))$

End

**Fig. 6.** Procedure of the algorithm

The procedure in case of a branch, that means the inclusion of additional ants and the establishment of the ants-family, happens in line 7 by the procedure *handle branching*.

The calculation of the transition rule $p_{ij}^k(t)$ for all alternative nodes j is carried through by the two formulas in line9 and 10. After an ant *k* has decided for a node *j* (OC-offer), that offer is transferred into the solution $\psi_k$ of the current ant *k*. After an ant has reached the source, the corresponding temporal objective function value $L_k$, aggregated from costs, times and probabilities of delivery, can be calculated by the help of the sequence in $\psi_k$.

The formula carrying through the local pheromone update in line 14 is applied. The same is valid for the global pheromone update in line 18. The pheromone values are limited by upper and under barriers corresponding line 19. In case, the pheromone value of an edge $\tau_{ij}$ is higher than the upper barrier $\tau_{max}$ or lower than the under barrier $\tau_{min}$ , the started value is correspondingly adapted. Thereby, it is dealt with the MAX-MIN Ant system.

After attaining the exit-condition, the search is stopped. By the help of the objective function value $L_k$, the quality of the solutions can be evaluated. This happens dependent on all the solutions found $\psi_k$ and their level of the objective function value $L_k$ by ranking. Subsequently, the x-best solutions have to be chosen from all the solutions. Several possibilities exist for the determination of the boundary value. On the one hand, a fixed number can be determined or a minimal objective function value can be applied. That work used the second approach. Only solutions $\psi_k$□M are further considered whose $L_k$ achieve at least $\kappa$*100% of the maximum objective function value $L_k^*$. The following formula makes clear that coherence:

$$\Psi_k \in M : \forall \Psi_k withL_k \succ \kappa * \cdot L_k^* \; 0 \leq \kappa \leq 1$$

For the remaining solutions, the eccentricity values and the connectivity value are subsequently calculated by the help of the polyhedral analysis. It is the aim to evaluate the good solutions concerning the social competence of the OCs involved and to give statements about the quality of the team-work.

## 6 Computational Experience

To illustrate the application of the algorithm for partner selection, twenty sets of data are generated randomly, in which the attractiveness should be calculated actually by AHP proposed before. The precedence relationship between the work steps can be described by a directed graph (a network) with each subtask as a node and it is shown in Figure 7. The figure is an acyclic digraph.

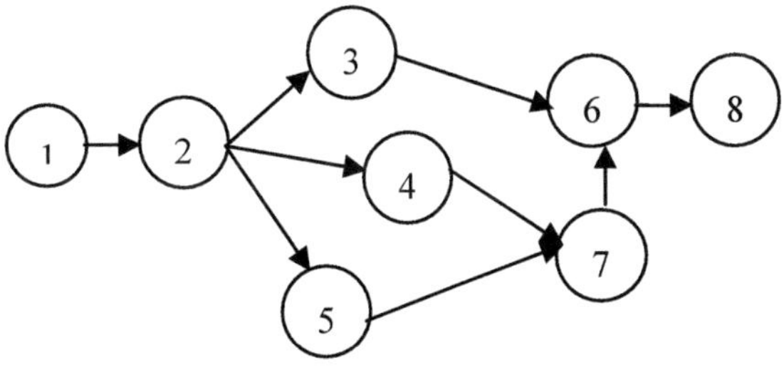


**Fig. 7.** Relation of the work steps

The number of ant $m$ regulates the size of a colony, this need to be chosen dependent on the size of the problem. In the illustrative example, $m$ is equal to 4. The parameters studied are $\alpha$, $\beta$ and $\rho$. The algorithm is tested on small test problems with default value set $\alpha=\beta=1$ and $\rho=0.75$. The parameters decide the trade-off between the importance of the trail intensity and visibility. With $\alpha \in \{0,0.25,0.33,0.5,1,2,3,4\}$ and $\beta \in \{0.5,1,2,4,8\}$, the values of objective functions are observed to select the best combination of partners. The best value was found to be $\alpha=2$ and $\beta=1$ (see Figure 8).

Another important parameter in the algorithm is $\rho$. A too high value of parameter $\rho$ in the original algorithm results in a situation called stagnation. Stagnation denotes the undesirable situation in which all ants construct the same solution over and over again, making further exploration of newer paths almost impossible. While a very low value of $\rho$ results in little information conveying from previous solutions and the algorithm becomes a randomized greedy search procedure. A study of $\rho$ behavior is done with set of values {0.1, 0.3, 0.5, 0.7, 0.9}. A value of 0.7 for $\rho$ renders minimum computation time as shown in Figure 8.

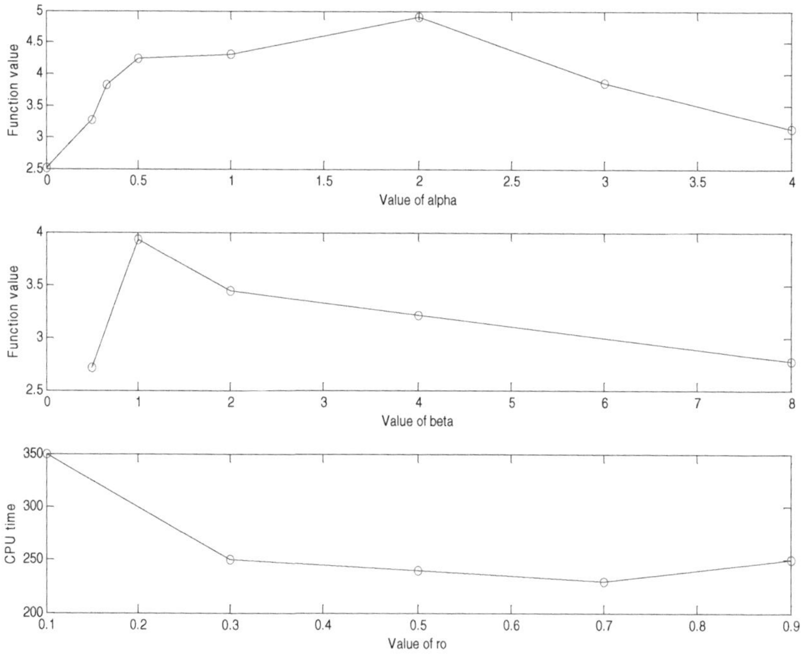


**Fig. 8.** Behavior of parameter α, β and ρ

The proposed algorithm was coded in C++ and was compiled through Microsoft Visual Studio 6.0. The program was executed on the test problem using a Lenovo compatible PC with Pentium4 processor running at 2.93GHZ and 512MB of RAM. The values of the objective function and computational time taken for 20 problems can be seen in Figure 9.

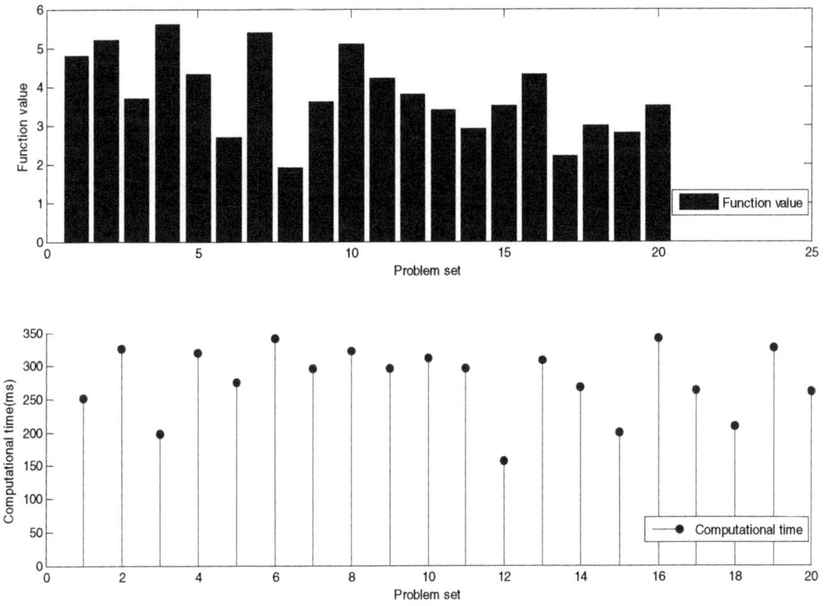


**Fig. 9.** Objective function values and computational time

Compared to the solution obtained by the enumeration algorithm and the solution obtained by the Genetic algorithm (GA) and Rule-based Genetic algorithm (R-GA)[21], it can be seen that the solution is optimal and requires least time ( see table 5) .

**Table 5.** Comparison of the different algorithms

| Algorithms | Best result | Run time | CPU time (ms) | Rate of best result |
|---|---|---|---|---|
| AHP+ACO | 5.56 | 20 | 284 | 95% |
| GA | 2.93 | 20 | 580 | 50% |
| R-GA | 3.81 | 20 | 396 | 65% |
| Enumeration | 5.86 | 20 | 2962 | 100% |

## 7 Conclusion

This paper introduced an approach for the selection of final manufacturing partners (OCs). That approach is to be found within the network controlling and can be understood as a decision supporting tool. Thereby, the approach includes economic parameters as well as social factors by applying the AHP-method.

After the description of the optimization problem, the method was selected. Besides classical procedures, numerous iterative improvement procedures, which formerly were used for similarly complicating, but in terms of content different situations, could have been chosen. After carrying through an extensive analysis of the problem and the resulting implications for the optimization model, one decided for the ACO. The following modeling, implementation and various tests proved the desired efficiency of the procedure for the selection of the partners in networks.

## References


1. Neubert, R., Langer, O., Görlitz, O., Benn, W.: Virtual Enterprises - Challenges from a Database Perspective. In: M.E.Orlowska, M.Yoshikawa (Eds.): Proc. of the Workshop on Information Technology for Virtual Enterprises ITVE, 23(2001)
2. Tich, T., Zschorn, L.: Management of Production Networks-A New Approach to Work with Probabilities of Delivery. In: Dresden.: Germany Proceedings of the 12th International Conference on Flexible Automation & Intelligent Manufacturing, 12 (2002) 762-771
3. Teich, T., Fischer, M., A New Ant Colony Algorithm for the Job Shop Scheduling Problem. In: Francisco, S.: California Proceedings of the Genetic and Evolutionary Computation Conference, (2001) 803-812
4. Saaty, TL.: The Analytic Hierarchy Process. New York, NJ:McGraw-Hill (1980)
5. Atkin, R., Casti, J.: Polyhedral Dynamics and Geometry of Systems. Laxenburg, Austria. International Institute for Applied Systems Analysis (IIASA), (1977) 77-106
6. Ho, C.T.: Strategic Evaluation of Emerging Technologies in the Semiconductor Foundry Industry. Portland State University, (2004) 251–278
7. Saaty, T.L.: How to Mark a Decision: the Analytic Hierarchy Process. European Journal of Operational Research, 48 (1990) 9-26
8. Yurdakul, M.: AHP as a Strategic Decision Making Tool to Justify Machine Tool Selection. Journal of Materials Processing technology, 146 (2004) 365-376
9. Dorigo, M., DiCaro, G.: The Ant Colony Optimization Meta-heuristic. In: New ideas in optimization. New York (1999)
10. Bonabeau, E., Dorigo, M.: Swarm Intelligence-From Natural to Artificial Systems. New York, Oxford University press (1999)
11. Dorigo, M., Gambardella, LM.: Ant Colonies for the Traveling Salesman Problem. BioSystems, (1997) 73–81
12. Maniezzo, V., Colorni, A.: The Ant System Applied to the Quadratic Assignment Problem. IEEE Trans Knowledge Data Eng, (1999) 769–778
13. Stuetzle, T., Dorigo, M.: ACO Algorithms for the Quadratic Assignment Problem. In: Corne, D., Dorigo, M., Glover, F.: New Ideas Optimzation. New York: McGraw-Hill(1999)
14. Bullnheimer, B., Hartl, RF., Strauss, C.: Applying the Ant System to the Vehicle Routing Problem. In: Voss, S., Martello, S., Osman, IH., Roucairol, C.: Meta-heuristics: Advances and Trends in Local Search Paradigms for Optimization. Dordrecht: Kluwer, (1999) 285–296
15. Gambardella, LM., Taillard, E., Agazzi, G.: MACS-VRPTW a Multiple Ant Colony System for Vehicle Routing Problems with Time Windows. In: Corne, D., Dorigo, M., Glover, F.: New Ideas in Optimization. New York: McGraw-Hill (1999) 63–76

16. Caro, G., Dorigo, M.: Ant Colonies for Adaptive Routing in Packetswitched Communication Networks. Presented at fifth International Conference on Parallel Problem Solving from Nature (PPSN V), Amsterdam, The Netherlands (1998)
17. Costa, D., Hertz, A.: Ants can Color Graphs. J Oper Res Soc, (2003) 295–305
18. Schoofs, L., Naudts, B.: Ant Colonies are Good at Solving Constraint Satisfaction Problems. Presented at Proceedings of 2000 Congress on Evolutionary Computation, San Diego, USA (2000)
19. Wagner, IA., Bruckstein, AM.: Hamiltonian(t)—an Ant Inspired Heuristic for Recognizing Hamiltonian Graphs. Presented at Proceedings of 1999 Congress on Evolutionary Computation, Washington (2003)
20. Besteb, M., Stutzle, T., Dorigo, M.: Ant Colony Optimization for the Total Weighted Tardiness Problem. Presented at sixth International Conference on Parallel Problem Solving from Nature (PPSN VI), Berlin (2000)
21. Zhao, Fuqing., Hong, Yi., Yu, Dongmei.: A Multi-objective Optimization Model of the Partner Selection Problem in a Virtual Enterprise and Its Solution with Genetic Algorithms. In: Advantage Manufacture Technology, 28 (2006) 1246-1253

# Quantum-Behaved Particle Swarm Optimization with Generalized Local Search Operator for Global Optimization

Jiahai Wang and Yalan Zhou

Department of Computer Science, Sun Yat-sen University,
No.135, Xingang West Road, Guangzhou 510275, P.R. China
wjiahai@hotmail.com

**Abstract.** In this paper, we propose a local quantum-behaved particle swarm optimization (LQPSO) as a generalized local search operator. The LQPSO is incorporated into a main quantum-behaved particle swarm optimization (QPSO), which leads to a hybrid QPSO scheme QPSO-LQPSO, with enhanced searching qualities. The main QPSO performs global exploration search while the LQSPO exploits a neighborhood of the current solution provided by the main QPSO to search better solutions. The proposed QPSO-LQPSO scheme is tested on a test set. Simulation results demonstrate the efficiency of the proposed QPSO-LQPSO scheme. For the same number of fitness evaluations, QPSO-LQPSO exhibited a significantly better performance than other particle swarm optimization algorithms.

**Keywords:** Quantum-behaved particle swarm optimization, generalized local search operator, global optimization.

## 1 Introduction

The particle swarm optimization (PSO) is inspired by observing the bird flocking or fish school [1]. A large number of birds/fishes flock synchronously, change direction suddenly, and scatter and regroup together. Each individual, called a particle, benefits from the experience of its own and that of the other members of the swarm during the search for food. Comparing with genetic algorithm, the advantages of PSO lie on its simple concept, easy implementation and quick convergence. The PSO has been applied successfully to continuous nonlinear function [1], neural network [2], nonlinear constrained optimization problems [3], etc. Lots of the applications have been concentrated on solving continuous optimization problems [4].

However, the evolution equation of standard PSO (SPSO) cannot guarantee the algorithm to find out the global optimum with probability 1, that is, SPSO is not a global optimization algorithm, as F. van den Bergh has demonstrated [5]. Sun et al. [6] [7] proposed a global convergence-guaranteed search technique, quantum-behaved particle swarm optimization algorithm (QPSO), whose performance is superior to the standard PSO (SPSO). The proposed QPSO algorithm,

D.-S. Huang, L. Heutte, and M. Loog (Eds.): ICIC 2007, LNAI 4682, pp. 851–860, 2007.

kept to the philosophy of PSO, is based on Delta potential well and depicted only with the position vector without velocity vector, which is a simpler algorithm. And the results show that QPSO performs better than SPSO on several benchmark test functions and is a promising algorithm due to its global convergence guaranteed characteristic. Therefore, QPSO is applied to some problems in practice, for example, clustering problem [8], and multi-period financial planning problem [9].

In order to further improve the QPSO search ability, two improved QPSO algorithms, QPSO with simulated annealing scheme (QPSO-SA) and QPSO with mutation operator (QPSO-mutation) also be proposed by Sun [10] [11]. In the QPSO-SA, simulated annealing (SA), as a selection operator, is introduced to QPSO, which effectively employs the ability to jump from the local minima in SA and the capacity of global search in QPSO algorithm. This combination of the two different optimization mechanism algorithms QPSO and SA enriched the search behavior greatly during the search process and increase its search capacity and efficiency in global and local area. In the QPSO-mutation, the mutation mechanism is introduced into QPSO to increase the diversity of the swarm and then effectively escape from local minima to increase its global search ability.

In general, the search performed by a metaheuristic approach should both intensively explore areas of the search space with high quality solutions, and to move to unexplored areas of the search space when necessary. These two requirements, exploitation and exploration, are conflicting, therefore a proper balance of exploitation and exploration ability is a crucial issue in heuristics. The QPSO, as a global optimization metaheuristic approach, has powerful global search ability. At the same time, it also needs a local search mechanism in order to provide an effective local search.

In this paper, we propose a local quantum-behaved particle swarm optimization (LQPSO) as a generalized local search operator. The LQPSO is incorporated into a main quantum-behaved particle swarm optimization (QPSO), which leads to a hybrid QPSO scheme QPSO-LQPSO, with enhanced searching qualities. The main QPSO performs global exploration search while the LQSPO exploits a neighborhood of the current solution provided by the main QPSO to search better solutions. The proposed QPSO-LQPSO scheme is conducted on a test set. Simulation results demonstrate the efficiency of the proposed QPSO-LQPSO scheme. For the same number of fitness evaluations, QPSO-LQPSO exhibited a significantly better performance in terms of solution accuracy and robustness.

## 2 PSO and Quantum-Behaved PSO

### 2.1 Stardand Particle Swarm Optimization (SPSO)

PSO is initialized with a group of random particles (solutions) and then searches for optima by updating each generation. In every iteration, each particle is updated by following two best values. The first one is the local best solution (fitness) a particle has obtained so far. This value is called personal best solutions. Another best value is that the whole swarm has obtained so far. This value is called global best solution. The philosophy behind the original PSO is to learn from individual's own experience (personal best solution) and best individual experience (global best solution) in the whole swarm.

Denote by $N$ particle number in the swarm. Let $X_i(t) = (x_{i1}(t), \cdots, x_{id}(t), \cdots, x_{iD}(t))$, be particle $i$ with $D$ bits at iteration $t$, where being treated as a potential solution. Denote the velocity as $V_i(t) = (v_{i1}(t), \cdots, v_{id}(t), \cdots, v_{iD}(t))$, $v_{id}(t) \in R$. Let $PBest_i(t) = (pbest_{i1}(t), \cdots, pbest_{id}(t), \cdots, pbest_{iD}(t))$ be the best solution that particle $i$ has obtained until iteration $t$, and $GBest(t) = (gbest_1(t), \cdots, gbest_d(t), \cdots, gbest_D(t))$ be the best solution obtained from $PBest_i(t)$ in the whole swarm at iteration $t$. Each particle adjusts its velocity according to previous velocity of the particle, the cognition part and the social part. The algorithm is described as follows [1]:

$$v_{id}(t+1) = v_{id}(t) + c_1 \cdot r_1 \cdot (pbest_{id}(t) - x_{id}(t)) + c_2 \cdot r_2 \cdot (gbest_d(t) - x_{id}(t)), \quad (1)$$

$$x_{id}(t+1) = x_{id}(t) + v_{id}(t+1), \quad (2)$$

where

$c_1$ is the cognition learning factor and $c_2$ is the social learning factor;

$r_1$ and $r_2$ are the random numbers uniformly distributed in [0,1].

In [12], Clerc and Kennedy analyze the trajectory and prove that each particle in the PSO system converges to its local point $g$, whose coordinates are $g_d = (\varphi_{1d} pbest_{id} + \varphi_{2d} gbest_d)/(\varphi_{1d} + \varphi_{2d})$ so that the best previous position of all particles will converge to an exclusive global position with $t \longrightarrow \infty$, where $\varphi_{1d}$, and $\varphi_{2d}$ are random numbers distributed uniformly on [0,1].

## 2.2 Quantum-Behaved PSO (QPSO)

SPSO is not a global convergence-guaranteed optimization algorithm, as F. van den Bergh has demonstrated [5]. Sun et al. [6] [7] proposed a global convergence-guaranteed search technique, quantum-behaved particle swarm optimization algorithm (QPSO), whose performance is superior to the standard PSO (SPSO). The proposed QPSO algorithm, kept to the philosophy of PSO, is based on Delta potential well and depicted only with the position vector without velocity vector, which is a simpler algorithm. In the quantum model of a PSO, the state of a particle is depicted by wavefunction $\psi(x,t)$, instead of position and velocity. The dynamic behavior of the particle is widely deferent from that of the particle in traditional PSO systems in that the exact values of position and velocity cannot be determined simultaneously. We can only learn the probability of the particle's appearing in position $x$ from probability density function $|\psi(x,t)|^2$, the form of which depends on the potential field the particle lies in. The particles move according to the following iterative equation [6] [7]:

$$x_{id}(t+1) = \begin{cases} p_{id} - \beta \cdot |mbest_{id} - x_{id}(t)| \cdot \ln(1/u) & \text{if } \mathrm{rand}() \geq 0.5 \\ p_{id} + \beta \cdot |mbest_{id} - x_{id}(t)| \cdot \ln(1/u) & \text{otherwise} \end{cases}, \quad (3)$$

where

$$p_{id} = \varphi \cdot pbest_{id} + (1 - \varphi) \cdot gbest_d,$$

and

$$mbest_{id} = \frac{1}{N} \sum_{i=1}^{N} pbest_{id},$$

$mbest$ (mean best position) is defined as the mean value of all particles' the best position, $\varphi$ and $u$ are random number distributed uniformly on [0,1] respectively. The parameter, $\beta$, called contraction-expansion coefficient, is the only parameter in QPSO algorithm. From the results of stochastic simulations, QPSO has relatively better performance by varying the value of $\beta$ from 1.0 at the beginning of the search to 0.5 at the end of the search to balance the exploration and exploitation [13][14]. The value of $\beta$ is dynamically tuned from 1.0 to 0.5 according to the number of generations such that more exploration search is pursued during the early generations and the exploitation search is emphasized afterward.

The basic flowcharts of SPSO and QPSO are described by Fig.1.

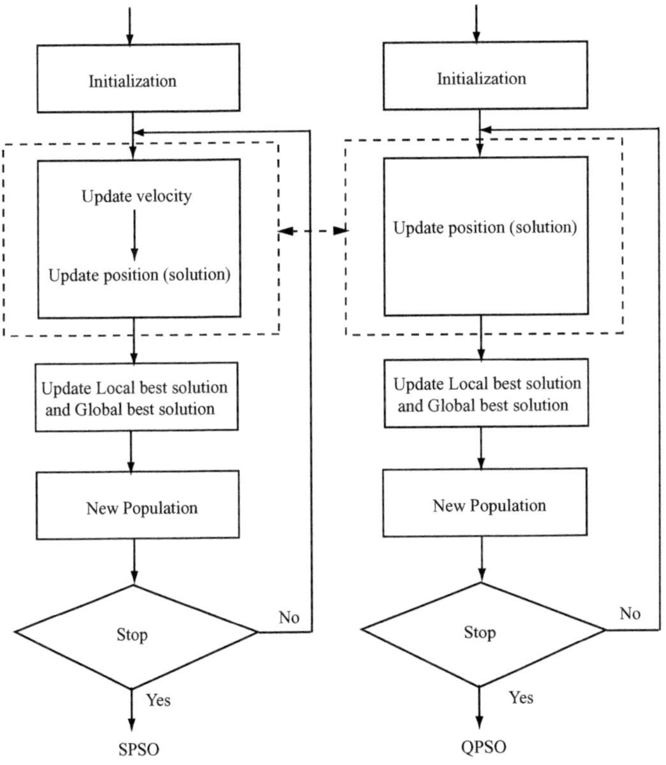


**Fig. 1.** The basic flowcharts of SPSO and QPSO

# 3 QPSO with Local QPSO (QPSO-LQPSO)

## 3.1 Generalized Local Search Operator-Local QPSO (LQPSO)

The QPSO, as a global optimization metaheuristic approach, has powerful global search ability. At the same time, it also needs a local search mechanism in order to provide an effective local search. In this section, we propose a generalized local search operator for QPSO. The generalized local search operator uses a reduced version of a QPSO, a local quantum-behaved particle swarm optimization (LQPSO), as a means for searching the neighborhood of the current global swarm best solution provided by the main QPSO.

We consider the neighborhood of global best solution $Gbest$ found by the QPSO at a particular generation, which is formulated as follows:

$$N(Gbest, radius) = gbest_d \cdot \text{rand}(1 - radius, 1 + radius) \tag{4}$$

where $radius \in [0, 1]$ represents the search radius, rand() produces random number in the given interval. Hence, $radius$ can be regarded as a measure that defines the size of the local search area.

The objective of the LQPSO is to explore $N(Gbest, radius)$. The LQPSO, as a reduced but complete version of QPSO, is carried out with smaller population size and less generation of evolution than main QPSO. Further, since the LQPSO focus on local search, the value of $\beta$ in LQSPO is set to 0.5 to emphasize the exploitation search.

## 3.2 QPSO-LQPSO

In this section, we attempt to incorporate the LQPSO to the main QPSO to improve its performance. The procedure of QPSO-LQPSO is described as follows:

1. Initialize.
   - **1.1** Generate $N$ particles at random, and determine local and global best solutions.
2. Repeat until a given maximal number of iterations ($MaxIter$) is achieved.
   - **2.1** Update particle position using Eq.(3).
   - **2.2** Evaluate each particle.
   - **2.3** Determine the local best solution.
   - **2.4** Determine the global best solution.
   - **2.5** Compute the neighborhood of global best solution Using Eq.(4).
   - **2.6** Perform the LQPSO on the neighborhood of global best solution.

The hybrid QPSO-LQPSO scheme, which is illustrated by Fig.2, provides an integrated means for solving a wide variety of optimization problems. Its effectiveness results from the synergetic contribution of QPSO and the LQPSO. The main QPSO performs the global search to explore the entire search space. On the other hand, the LQPSO operator performs the local search on the neighborhood of the global best solution obtained by main QPSO.

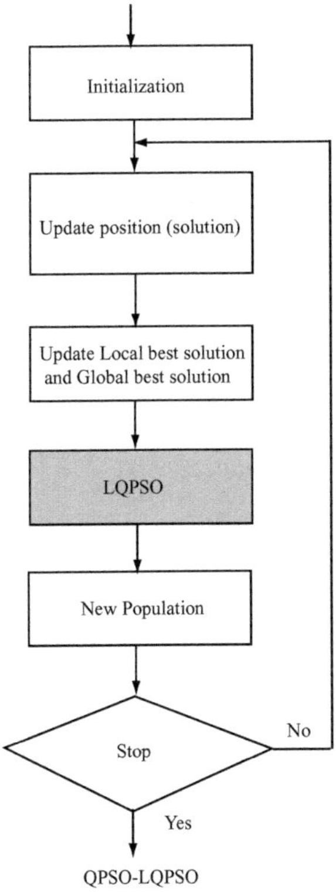


**Fig. 2.** The basic flowcharts of QPSO-LQPSO

Support the population size of main QPSO is $N$, and the population size and generation run limit of LQPSO is $sN$ and $sGen$, respectively. The fitness evaluations performed by the main QPSO and the LQPSO are $N$ and $sN \times sGen$, respectively, per generation of the main QPSO. The total number of fitness evaluations performed per generation in hybrid QPSO-LQPSO is $N + sN \times sGen$. Given a total number of fitness evolutions performed per generation in the hybrid algorithm, there is a fitness allocation scheme problem. This problem relates to the sharing of the search activity between the main QPSO and the LQPSO in the hybrid QPSO-LQPSO scheme. In other words, it consists of dividing the available computation effort between the global search (main QPSO) and the local search (LQPSO). The computational burden is measured on the basis of the fitness evaluations performed at each generation of the main QPSO. Given a total number of fitness evolutions, we can not allocate too much fitness evolution to LQPSO, or else only a small portion of the available computations remain for the evolution of main QPSO, which degrades the performance of the overall algorithm. Experience indicates that the portion $(sN \times sGen)/(N + sN \times sGen)$ should be kept to a reasonably low level, that is, the percentage amount $(sN \times sGen)/(N + sN \times sGen)$ should vary in the range 0.3-0.5. Obviously, if the portion is too small or even equal to 0 then the hybrid algorithm QPSO-LQPSO degenerates to the algorithm with little local search or the pure QPSO without any local search ability.

## 4 Simulation Results

In order to investigate the effectiveness of the QPSO-LQPSO method and its capability to attain near optimum solutions, three multimodal benchmark functions [10] [11] are tested. All functions are tested on 10, 20 and 30 dimensions. The properties and the formulas of these functions are presented below.
1) Rosenbrock function

$$f_1 = \sum_{i=1}^{n} (100(x_{i+1} - x_i^2)^2 + (x_i - 1)^2)$$

2) Rastrigrin function

$$f_2 = \sum_{i=1}^{n} (x_i^2 - 10\cos(2\pi x_i) + 10)$$

3) Griewank function

$$f_3 = \frac{1}{4000}\sum_{i=1}^{n} (x_i - 100)^2 + \prod_{i=1}^{n} \cos(\frac{x_i - 100}{\sqrt{i}}) + 1$$

Rosenbrock function can be treated as a multimodal problem. Even for $n = 2$, the Rosenbrock function surface exhibits narrow ridges, which make it difficult to approach the optimum. This situation is even more complicated when larger values of are considered. We have chosen $n$ =10, 20, 30. Rastrigin function is a complex multimodal problem with a large number of local optima. When attempting to solve Rastrigin function, algorithms may easily fall into a local optimum. Hence, an algorithm capable of maintaining a larger diversity is likely to yield better results. Griewank function has a component $\prod_{i=1}^{n} \cos(\frac{x_i - 100}{\sqrt{i}})$ causing linkages among variables, thereby making it difficult to reach the global optimum. An interesting phenomenon of Griewank function is that it is more difficult for lower dimensions than higher dimensions [15]. Table 1 shows the initialization range and the corresponding limits to the search space and the global minima of all the test function. Biased initializations are used for the functions whose global minimum is at the centre of the search range.

As in [10] [11], for each function, three different dimension sizes, 10, 20 and 30 are tested. The corresponding maximum generations are 1000, 1500 and 2000 respectively in the SPSO, QPSO, QPSO-SA and QPSO-mutation (global best mutation) algorithms. And the population size is also set to 20, 40 and 80. In the QPSO-LQPSO, the fitness evaluations are performed by the main QPSO (global search) is equal to the evaluations are performed by the LQPSO operator (local search) per generation of the main QPSO, that is, $N = sN \times sGen$. Therefore,

**Table 1.** Benchmark configuration for simulations

| Function | Global minimum | Search Range | Initialization Range |
|---|---|---|---|
| $f_1$ | 0 | [-100,100] | [15,30] |
| $f_2$ | 0 | [-10,10] | [2.56,5.12] |
| $f_3$ | 0 | [-600,600] | [300,600] |

**Table 2.** Simulation results for Rosenbrock function

| *N* | *D* | *Gen* | SPSO Mean (St.Dev) | QPSO Mean (St.Dev) | QPSO-SA Mean (St.Dev) | QPSO-mutation Mean (St.Dev) | QPSO-LQPSO Mean (St.Dev) |
|---|---|---|---|---|---|---|---|
| 20 | 10 | 1000 | 94.1276 | 59.4764 | 25.5521 | 21.2081 | **4.6132** |
| | | | (194.3648) | (153.0842) | (58.8220) | (60.0583) | **(1.2655)** |
| | 20 | 1500 | 204.336 | 110.664 | 98.9765 | 61.9268 | **14.0711** |
| | | | (293.4544) | (149.5483) | (122.2852) | (92.9440) | **(0.2531)** |
| | 30 | 2000 | 313.734 | 147.609 | 112.0748 | 86.1195 | **23.9154** |
| | | | (547.2635) | (210.3262) | (54.0904) | (127.6446) | **(0.391129)** |
| 40 | 10 | 1000 | 70.0239 | 10.4238 | 10.7750 | 8.1828 | **3.5882** |
| | | | (174.1108) | (14.4799) | (12.5061) | (8.3604) | **(1.1132)** |
| | 20 | 1500 | 179.291 | 46.5957 | 38.1721 | 40.0749 | **13.4845** |
| | | | (377.4305) | (39.5360) | (33.4951) | (68.4074) | **(0.65829)** |
| | 30 | 2000 | 289.593 | 59.0291 | 47.9188 | 65.2891 | **23.0864** |
| | | | (478.6273) | (63.4940) | (39.2296) | (79.4420) | **(0.631055)** |
| 80 | 10 | 1000 | 37.3747 | 8.63638 | 6.7566 | 7.3686 | **3.31003** |
| | | | (57.4734) | (16.6746) | (6.7435) | (8.4972) | **(1.5225)** |
| | 20 | 1500 | 83.6931 | 35.8947 | 59.2269 | 30.1607 | **13.3877** |
| | | | (137.2637) | (36.4702) | (99.7291) | (33.2090) | **(0.7065)** |
| | 30 | 2000 | 202.672 | 51.5479 | 41.6666 | 38.3036 | **22.3896** |
| | | | (289.9728) | (40.8490) | (29.9889) | (27.4658) | **(0.67088)** |

when $N$ in the main QPSO is 20, 40 and 80, respectively, the corresponding $sN \times sGen$ in LQPSO is set to $4 \times 4$, $5 \times 8$ and $8 \times 10$, respectively. The corresponding maximum generations are 500, 750 and 1000, respectively; therefore the total number of fitness evolutions in the QPSO-LQPSO is the same to the other QPSO algorithms. The search radius $radius = 0.2$ is used in the QPSO-LQPSO.

A total of 50 runs for each experimental setting are conducted for all of the algorithms. The mean values and standard deviations for 50 runs of each test function are recorded in Table 2 and Table 3. The numerical results in Table 2 and Table 3 show that the solutions found by the QPSO-LQPSO are statistically significantly better than other PSO algorithms for Rosenbrock and Rastrigin function. For the Griewank function, all PSO algorithms produced comparable results (all near global minimum of the function) and there is no statistically significant difference among those algorithms, therefore these results do not be given because of page limit. More result details of these PSO algorithms can be found in Ref. [10] [11]. The better results on Rosenbrock and Rastrigin function show that QPSO-LQPSO both can identify and follow narrow ridges of arbitrary

Table 3. Simulation results for Rastrigrin function

| $N$ | $D$ | $Gen$ | SPSO Mean (St.Dev) | QPSO Mean (St.Dev) | QPSO-SA Mean (St.Dev) | QPSO-mutation Mean (St.Dev) | QPSO-LQPSO Mean (St.Dev) |
|---|---|---|---|---|---|---|---|
| 20 | 10 | 1000 | 5.5382 (3.0477) | 5.2543 (2.8952) | 4.9388 (2.6520) | 4.2976 (2.5325) | **0** (**0**) |
| | 20 | 1500 | 23.1544 (10.4739) | 16.2673 (5.9771) | 13.6808 (4.6682) | 14.1678 (4.9272) | **0** (**0**) |
| | 30 | 2000 | 47.4168 (17.1595) | 31.4576 (7.6882) | 29.5396 (7.6264) | 25.6415 (6.6575) | **0** (**0**) |
| 40 | 10 | 1000 | 3.5778 (2.1384) | 3.5685 (2.0678) | 2.7779 (1.3363) | 3.2046 (3.0587) | **0** (**0**) |
| | 20 | 1500 | 16.4337 (5.4811) | 11.1351 (3.6046) | 10.8366 (4.5036) | 9.5793 (2.8107) | **0** (**0**) |
| | 30 | 2000 | 37.2796 (14.2838) | 22.9594 (7.2455) | 21.1007 (5.0758) | 20.5479 (5.0191) | **0** (**0**) |
| 80 | 10 | 1000 | 2.5646 (1.5728) | 2.1245 (1.1772) | 2.1476 (1.3866) | 1.7166 (1.3067) | **0** (**0**) |
| | 20 | 1500 | 13.3829 (8.5137) | 10.2759 (6.6244) | 8.5381 (6.4073) | 7.2041 (2.4822) | **0** (**0**) |
| | 30 | 2000 | 28.6293 (10.3431) | 16.7768 (4.4858) | 15.1721 (3.9442) | 15.0393 (4.1800) | **0** (**0**) |

direction within the search space, and can escape from local minima. Therefore, we can conclude that the proposed algorithm has superior global and local search ability for global optimization.

## 5 Conclusions

This paper proposes a LQPSO as a generalized local search operator. It explores the neighborhood of the best swarm solution provided by the main QPSO with the aim of finding a better solution. By combining LQPSO with a QPSO, this paper proposes a hybrid QPSO-LQPSO, where the main QPSO performs the global search while the local search is accomplished by the LQPSO. Simulation results show that the QPSO-LQPSO is superior to other PSO algorithms for global optimization.

**Acknowledgments.** The Project was supported by the Scientific Research Foundation for Outstanding Young Teachers, Sun Yat-sen University.

## References

1. Kennedy, J., Eberhart, R.C.: Particle Swarm Optimization. Proceedings of IEEE International Conference on Neural Networks, NJ: Piscataway, (1995) 1942–1948
2. Van den Bergh, F., Engelbrecht, A.P.: Cooperative Learning In Neural Network Using Particle Swarm Optimizers. South African Computer Journal, **26** (2000) 84–90

3. El-Galland, AI., El-Hawary, ME., Sallam, AA.: Swarming Of Intelligent Particles For Solving The Nonlinear Constrained Optimization Problem. Engineering Intelligent Systems for Electrical Engineering and Communications, **9** (2001) 155–163
4. Parsopoulos, K.E. and Vrahatis, M.N.: Recent Approaches To Global Optimization Problems Through Particle Swarm Optimization. Natural Computing, **1**(2-3) (2002) 235–306
5. Van den Bergh, F.: An Analysis of Particle Swarm Optimizers. PhD Thesis, University of Pretoria, Nov. (2001)
6. Sun, J. et al: Particle Swarm Optimization with Particles Having Quantum Behavior. Proc. Congress on Evolutionary Computation, (2004) 325–331
7. Sun, J. et al: A Global Search Strategy of Quantum-Behaved Particle Swarm Optimization. Proc. IEEE Conference on Cybernetics and Intelligent Systems, (2004) 111–116.
8. Sun, J., Xu, W., Ye, B.: Quantum-Behaved Particle Swarm Optimization Clustering Algorithm. Proceedings Advanced Data Mining and Applications, Lecture Notes in Artificial Intelligence, **4093** (2006) 340–347
9. Sun, J., Xu, W., Fang, W.: Solving Multi-Period Financial Planning Problem via Quantum-Behaved Particle Swarm Algorithm. Proceedings Computational Intelligence, Lecture Notes In Artificial Intelligence, **4114** (2006) 1158–1169
10. Liu, J., Sun, J., Xu, W.: Improving Quantum-Behaved Particle Swarm Optimization by Simulated Annealing. Proceedings Computational Intelligence and Bioinformatics, Lecture Notes in Computer Science, **4115** (2006) 130–136
11. Liu, J., Sun, J., Xu, W.: Quantum-Behaved Particle Swarm Optimization with Adaptive Mutation Operator. Advances in Natural Computation, Lecture Notes in Computer Science, **4221** (2006) 959–967
12. Clerc, M. and Kennedy, J.: The Particle Swarm: Explosion, Stability and Convergence in a Multi-Dimensional Complex Space. IEEE Transaction on Evolutionary Computation, **6** (2002) 58–73
13. Xu, W., Sun, J.: Adaptive Parameter Selection of Quantum-Behaved Particle Swarm Optimization on Global Level. Proceedings Advances in Intelligent Computing, Lecture Notes in Computer Science, **3644** (2005) 420–428
14. Sun, J., Xu, W., Liu, J.: Parameter Selection of Quantum-Behaved Particle Swarm Optimization. Proceedings Advances in Natural Computation, Lecture Notes in Computer Science, **3612** (2005) 543–552

# Kernel Difference-Weighted k-Nearest Neighbors Classification

Wangmeng Zuo[1], Kuanquan Wang[1], Hongzhi Zhang[1], and David Zhang[2]

[1] School of Computer Science and Technology, Harbin Institute of Technology
150001 Harbin, China
cswmzuo@gmail.com
[2] Department of Computing, the Hong Kong Polytechnic University
Kowloon, Hong Kong

**Abstract.** Nearest Neighbor (NN) rule is one of the simplest and most important methods in pattern recognition. In this paper, we propose a kernel difference-weighted *k*-nearest neighbor method (KDF-WKNN) for pattern classification. The proposed method defines the weighted KNN rule as a constrained optimization problem, and then we propose an efficient solution to compute the weights of different nearest neighbors. Unlike distance-weighted KNN which assigns different weights to the nearest neighbors according to the distance to the unclassified sample, KDF-WKNN weights the nearest neighbors by using both the norm and correlation of the differences between the unclassified sample and its nearest neighbors. Our experimental results indicate that KDF-WKNN is better than the original KNN and distance-weighted KNN, and is comparable to some state-of-the-art methods in terms of classification accuracy.

**Keywords:** *k*-nearest neighbor, pattern classification, kernel methods.

## 1 Introduction

Nearest Neighbor rule is one of the oldest, simplest and most important methods for pattern classification and case-based-reasoning. Nowadays NNs have been widely exploited in varieties of artificial intelligence tasks, such as pattern recognition, data mining, posterior probabilities estimation, similarity-based query, computer vision, and bioinformatics. Various aspects of NN development are investigated from algorithmic innovations, computational concerns, theoretical analysis and visualization.

Assume that we are given a set of training data, $\mathbf{X} = \{\mathbf{x}_1, \cdots, \mathbf{x}_m\} \subseteq \mathbb{R}^N$ with corresponding labels $\mathbf{y} = \{y_1, \cdots, y_m\} \subseteq \{\omega_1, \omega_2, \cdots, \omega_C\}$. Given a sample $\mathbf{x}$, the *k*-Nearest Neighbor (KNN) rule first identifies the *k* nearest neighbors of the *m* training samples, identifies the number of samples $k_i$ that belong to class $\omega_i$, and then assign $\mathbf{x}$ to the class $\omega_i$ with the maximum $k_i$ of samples. To enforce the computational and classification performance of KNN, it has recently been proposed a variety of prototype editing, distance measure, and weighted NN techniques. Prototype editing techniques are used to efficiently reduce a large training set to a small, representative prototype set

D.-S. Huang, L. Heutte, and M. Loog (Eds.): ICIC 2007, LNAI 4682, pp. 861–870, 2007.

while maintaining the classification performance by instance filtering, instance abstraction, or their combination. Distance measure technique intends to work out a dissimilarity measure or metric with a better class discriminative capability. Weighted NN aims to weigh the discriminative capability of different nearest neighbors (sample-weighted KNN) or different features (weighted metric).

Here we focus on the development of weighted Nearest Neighbor to achieve better classification performance. We roughly grouped all the current weighted NN strategies into two major categories, weighted metric and sample-weighted KNN.

## 1.1 Weighted Metric for NN Classification

Generally, weighted metric can be defined as a distance between a unclassified sample $\mathbf{x}$ and a training sample $\mathbf{x}_i$ as $d(\mathbf{x},\mathbf{x}_i)=(\mathbf{x}-\mathbf{x}_i)^T\Xi_i(\mathbf{x}-\mathbf{x}_i)^T$. If all $\Xi_i$=$\Xi$, the metric is a global metric, otherwise it is a local metric. Short and Fukunaga proposed to use a local adaptive distance metric for a two-class problem [12]. This method obtains the first $k$ nearest neighbor $\{\mathbf{x}_1^{NN},\cdots,\mathbf{x}_k^{NN}\}$ using the Euclidean distance. Let $\{y_1^{NN},\cdots,y_k^{NN}\}$, $y\in\{\omega_1,\omega_2\}$ be the class labels of the nearest neighbors. The local metric first defines $\mathbf{M}_1=\sum_{y_i^{NN}=\omega_1}(\mathbf{x}_i^{NN}-\mathbf{x})/k_1$ and $\mathbf{M}_0=\sum_i(\mathbf{x}_i^{NN}-\mathbf{x})/k$, where $k_1$ is the number of $\mathbf{x}_{j_i}$ in class $\omega_1$, and then calculate the distance $d(\mathbf{x},\mathbf{x}_i^{NN})=\left|(\mathbf{M}_1-\mathbf{M}_0)^T(\mathbf{x}-\mathbf{x}_i^{NN})\right|$.

Subsequently, Fukugana and Flick presented an optimal global metric for NN and developed a nonparametric $\Xi$ estimator [5]. Later Hastie and Tibshirani proposed a discriminant adaptive nearest neighbor metric which estimates $\Xi$ as a product of weighted local within and between sum-of-squares matrices [6]. Recently, Domeniconi and Peng further developed DANN into an adaptive metric NN algorithm [3].

Other strategies are investigated to learn the weighted metric for nearest neighbor classification. Ricci and Avesani proposed to learn a local distance measure from a viewpoint of data compression [11]. One main advantage of this method is that it can simultaneously learn the weighted metric and reduce the number of prototypes. Most recently, Paredes and Vidal present a method to learn the local weighted metrics by minimizing the leave-one-out (LOO) classification error [10].

## 1.2 Sample-Weighted KNN Classification

The KNN classifier assigns to a test sample the class label which has the maximum number of nearest neighbors. This rule, however, neglects that the nearest neighbor close to the unclassified sample should contribute more to classification. To weigh the samples nearby more heavily in making the decision, Dudani proposed a distance-weighted KNN method (DS-WKNN) [4]. Let $\{\mathbf{x}_1^{NN},\cdots,\mathbf{x}_k^{NN}\}$ denote the $k$ nearest neighbors of $\mathbf{x}$ arranged in an increasing order according to the distance between $\mathbf{x}_i^{NN}$ and $\mathbf{x}$, $d(\mathbf{x},\mathbf{x}_i^{NN})$. DS-WKNN assign to each nearest neighbor $\mathbf{x}_i^{NN}$ a weight $w_i$ as a function of $d(\mathbf{x},\mathbf{x}_i^{NN})$, $w_i=[d(\mathbf{x},\mathbf{x}_k^{NN})-d(\mathbf{x},\mathbf{x}_i^{NN})]/[d(\mathbf{x},\mathbf{x}_k^{NN})-d(\mathbf{x},\mathbf{x}_1^{NN})]$. In the same paper, Dudani also discussed several alterative weighting functions. Later Keller *et al.* extended the fuzzy set concepts to KNN and proposed a fuzzy KNN method which can be considered as an alterative weighted KNN rule [7].

Theoretical analysis of DS-WKNN was pioneered by Bailey and Jain, whose study indicated that, given an infinite set of training samples, the asymptotic error rate of unweighted KNN is always lower than that of any weighted KNN [1]. This conclusion, however, would not apply when the number of training samples is finite [8]. Experimental results also showed that, it is very possible that weighted KNN achieves lower error rate than unweighted KNN in the finite-training-sample case [14].

In this paper, we propose another sample-weighted KNN method for classification. Unlike DS-WKNN, the approach we propose is a difference-weighted KNN method. Difference-weighted KNN first obtains the $k$ nearest neighbors $\{\mathbf{x}_1^{NN},\cdots,\mathbf{x}_k^{NN}\}$ of an unclassified sample **x**, and then calculates the difference between nearest neighbors and **x**. Based on the difference, we solve a constrained least-squares optimization problem to compute the weight of each nearest neighbor, and then we can predict the classification result of the sample **x** using the difference-weighted KNN rule.

The remainder of the paper is organized as follows: Section 2 first presents a definition of the objective function in assigning weights for KNN classification, and then solves a constrained least-squares optimization problem for difference-weighted KNN. Section 3 further presented the kernel DF-KNN method. In Section 4, experiments are used to evaluate KDF-KNN. Finally, Section 5 concludes this paper.

## 2 Difference-Weighted KNN Rule

In this section, we propose a difference-weighted KNN rule. First we present the formulation of the difference-weighted KNN rule as a constrained least-squares optimization problem, and then propose our solution to this problem for determining the weight of each nearest neighbor.

### 2.1 Problem Formulation

Let $\{(\mathbf{x}_1, y_1), \cdots, (\mathbf{x}_m, y_m)\}$ be a training set, where $\mathbf{x}_i$ is the $i$th training sample, and $y_i$ is the corresponding class label. Given a unclassified sample **x**, a distance metric (e.g., Euclidean) is used to obtain the first $k$ nearest neighbors $\{\mathbf{x}_1^{NN},\cdots,\mathbf{x}_k^{NN}\}$ and their corresponding class labels $\{y_1^{NN},\cdots,y_k^{NN}\}$, $y_i^{NN}\in\{\omega_1,\cdots,\omega_C\}$. A KNN is defined as a sample-weighted KNN classifier if it consists of the next two steps:

(1) Assign each nearest neighbor $\mathbf{x}_i^{NN}$ a weight $w_i$ using a weight algorithm;

(2) Assign class label $\omega_i$ to sample **x** using the following rule

$$\omega_{j_{\max}} = \arg\min_{\omega_j}\Big(\sum_{y_i^{NN}=\omega_j} w_i\Big) . \qquad (1)$$

So far, there are some methods in assigning the weights. For example, Dudani (1976) proposed a distance-weighted KNN method which assigns to the $i$th nearest neighbor a weight $w_i$ as a function of their distance $d(\mathbf{x}_i, \mathbf{x})$. This approach, however, is intuitively derived and lacks of theoretical or geometric evidence. In this section, we define the difference-weight assignment as a constrained optimization problem of sample reconstruction from its neighborhood.

**Problem.** The weights of the nearest neighbors $\mathbf{w}=[w_1, \cdots, w_k]^T$ is defined as a vector corresponding to the constrained optimal reconstruction of $\mathbf{x}$ using $\mathbf{X}=[\mathbf{x}_1^{NN},\cdots,\mathbf{x}_k^{NN}]$,

$$\mathbf{w} = \arg\min_{\mathbf{w}} \left\|\mathbf{x}-\mathbf{w}^T\mathbf{X}\right\|_2 \text{ s. t. } \sum_i w_i = 1 . \tag{2}$$

## 2.2 Solution of the Constrained Least-Squares Optimization Problem

In the constrained least-squares optimization problem defined in Section 2.1, the objective is a quadratic function while the constraint is a linear, thus is formulated as a quadratic programming (QP) problem. We then use the Lagrangian multiplier method to seek a simple and efficient solution to this QP problem.

Let $\mathbf{D}=[\mathbf{x}-\mathbf{x}_1^{NN},\cdots,\mathbf{x}-\mathbf{x}_k^{NN}]$. The optimization problem of Eq. (2) can be rewritten as

$$\mathbf{w} = \arg\min_{\mathbf{w}} \mathbf{w}^T\mathbf{D}\mathbf{D}^T\mathbf{w} \text{ s. t. } \sum_i w_i = 1 . \tag{3}$$

From Eq. (3), the optimal weights $\mathbf{w}$ is only dependent on the difference and its $k$ nearest neighbors, $\{\mathbf{x}-\mathbf{x}_1^{NN},\cdots,\mathbf{x}-\mathbf{x}_k^{NN}\}$. Thus we name our method rule as difference-weghted KNN (DF-WKNN), to distinguish from Dudani's distance -weighted KNN (DS-WKNN).

The Lagrangian function for the constrained optimization problem is

$$L(\mathbf{w},\lambda) = \frac{1}{2}\mathbf{w}^T\mathbf{D}\mathbf{D}^T\mathbf{w} - \lambda(\mathbf{w}^T\mathbf{1}_k - 1) , \tag{4}$$

where $\mathbf{1}_k$ is a $k\times1$ vector with each element equal 1. Let $\mathbf{G} = \mathbf{D}\mathbf{D}^T$, and $\nabla_{\mathbf{w}} L(\mathbf{w}, \lambda) = 0$, $\nabla_{\lambda} L(\mathbf{w}, \lambda) = 0$, we obtain the next linear equations

$$\mathbf{G}\mathbf{w} - \lambda\mathbf{1}_k = 0 , \tag{5}$$

$$\mathbf{w}^T\mathbf{1}_k - 1 = 0 . \tag{6}$$

We further rewrite these linear equations as

$$\begin{bmatrix} \mathbf{G} & -\mathbf{1}_k \\ -\mathbf{1}_k^T & 0 \end{bmatrix}\begin{bmatrix} \mathbf{w} \\ \lambda \end{bmatrix} = \begin{bmatrix} 0 \\ -1 \end{bmatrix} . \tag{7}$$

If the matrix $\mathbf{G}$ is invertible, the inverse of the matrix $\mathbf{A}=\begin{bmatrix} \mathbf{G} & -\mathbf{1}_k \\ -\mathbf{1}_k^T & 0 \end{bmatrix}$ is existed. So we can compute the Lagrangian multiplier $\lambda$ and the weights $\mathbf{w}$ by solving the linear system of equations.

In fact, we can use a more efficient and numerically stable approach to compute the weights $\mathbf{w}$. This approach first solves the system of linear equations $\mathbf{G}\mathbf{w}_0=\mathbf{1}_k$, and then rescale the weights using $\mathbf{w}=\mathbf{w}_0/(\mathbf{w}_0^T\mathbf{1}_k)$. It is simple to see that this approach would yield the same weight vector $\mathbf{w}$ as the Langrangian multiplier method.

In some cases, the inverse of the matrix **G** is not unique if **G** is singular (e.g., the number of nearest neighbors, $k>N$, the dimension of the sample). To avoid this case, we adopt a regularization method by adding a small multiple of the identity matrix,

$$\mathbf{G} = \mathbf{G} + \eta \mathrm{tr}(\mathbf{G})/k , \tag{8}$$

where tr(**G**) denotes the trace of the matrix **G**, and $\eta=10^{-0}\sim10^{-3}$ is the regularization parameter.

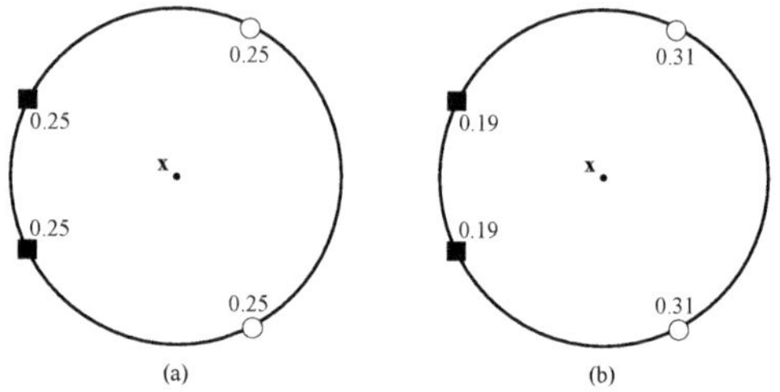


**Fig. 1.** Assignment of weights to the nearest neighbors using (a) the distance-weighted KNN rule and (b) the difference-weighted KNN rule

Figure 1 illustrates an example of DF-WKNN and DS-WKNN in assigning weights. The DF-WKNN utilizes both the norm and correlation of the differences $\mathbf{D}=[\mathbf{x}-\mathbf{x}_1^{NN},\cdots,\mathbf{x}-\mathbf{x}_k^{NN}]$ to determine the weights **w**, while DS-WKNN only uses the distance between **x** and its nearest neighbors. Thus in some cases, DF-WKNN may achieve better classification performance than DS-WKNN.

We briefly summary the main steps of DF-WKNN. Given a unclassified sample **x**, DF-WKNN first obtain the first $k$ NNs $\{\mathbf{x}_1^{NN},\cdots,\mathbf{x}_k^{NN}\}$ and their corresponding class labels $\{y_1^{NN},\cdots,y_k^{NN}\}$, and then calculate the difference of **x** and its $k$ nearest neighbors, $\mathbf{D}=[\mathbf{x}-\mathbf{x}_1^{NN},\cdots,\mathbf{x}-\mathbf{x}_k^{NN}]$. Finally the weights **w** of $k$ nearest neighbors are determined by solve the system of linear equations $[\mathbf{D}\mathbf{D}^T+\eta\mathrm{tr}(\mathbf{D}\mathbf{D}^T)/k]\mathbf{w}=\mathbf{1}_k$.

## 3 Kernel DF-KNN Rule

Using the kernel trick, we extend DF-WKNN to its nonlinear version, kernel DF-WKNN. DF-WKNN uses a linear method, QP, to assign weights to nearest neighbors, which can not utilize the nonlinear structure information. Extension to kernel DF-WKNN will provide a strategy to circumvent this restriction. This extension includes two steps: extension to kernel distance and to the kernel Gram matrix.

The Euclidean distance can be extended to its corresponding kernel distance measure. Given two samples **x** and **x**′, we define a kernel function $k(\mathbf{x}, \mathbf{x}')=(\Phi(\mathbf{x})\cdot\Phi(\mathbf{x}'))$. Using the kernel function, data **x** are implicitly mapped into a higher dimensional or infinite dimensional *feature space* $\mathbb{F}: \mathbf{x}\rightarrow\Phi(\mathbf{x})$ and the inner product in *feature space* can be easily computed using the kernel function $k(\mathbf{x}, \mathbf{x}')=(\Phi(\mathbf{x})\cdot\Phi(\mathbf{x}'))$. Two popular kernel functions are radial basis function (RBF) kernel $k(\mathbf{x}, \mathbf{x}')=\exp(-\|\mathbf{x}-\mathbf{x}'\|^2/2)$ and polynomial kernel $k(\mathbf{x}, \mathbf{x}')=(1+\mathbf{x}\cdot\mathbf{x}')^d$. The kernel distance in the feature space is then defined as

$$d(\mathbf{x},\mathbf{x}') = \left\|\Phi(\mathbf{x}) - \Phi(\mathbf{x}')\right\|^2 = \mathrm{k}(\mathbf{x},\mathbf{x}) - 2\mathrm{k}(\mathbf{x},\mathbf{x}') + \mathrm{k}(\mathbf{x}',\mathbf{x}') . \tag{9}$$

The matrix **G** can also be extended to its kernel version by constructing the kernel Gram matrix $\mathbf{G}^{\mathrm{k}}$. In the data space, the element $g_{ij}$ of the matrix **G** is defined as

$$g_{ij} = ((\mathbf{x} - \mathbf{x}_i^{NN}) \cdot (\mathbf{x} - \mathbf{x}_j^{NN})) , \tag{10}$$

where $\mathbf{x}_i^{NN}$ is the *i*th nearest neighbor of the unclassified sample **x**. Analogously, we define the element $g_{ij}^{\mathrm{k}}$ of the kernel gram matrix $\mathbf{G}^{\mathrm{k}}$ as

$$g_{ij}^{\mathrm{k}} = ((\Phi(\mathbf{x}) - \Phi(\mathbf{x}_i^{NN})) \cdot (\Phi(\mathbf{x}) - \Phi(\mathbf{x}_j^{NN}))) . \tag{11}$$

Using the kernel trick, $g_{ij}^{\mathrm{k}}$ can be calculated explicitly

$$g_{ij}^{\mathrm{k}} = \mathrm{k}(\mathbf{x},\mathbf{x}) - \mathrm{k}(\mathbf{x},\mathbf{x}_i^{NN}) - \mathrm{k}(\mathbf{x},\mathbf{x}_j^{NN}) + \mathrm{k}(\mathbf{x}_i^{NN},\mathbf{x}_j^{NN}) . \tag{12}$$

We can further derive a more compact expression of the kernel matrix $\mathbf{G}^{\mathrm{k}}$,

$$\mathbf{G}^{\mathrm{k}} = \mathbf{K} + \mathbf{1}_{kk}\mathrm{k}(\mathbf{x},\mathbf{x}) - \mathbf{1}_k\mathbf{k}_c^T - \mathbf{k}_c\mathbf{1}_k^T , \tag{13}$$

**Table 1.** Summary of data sets and their characteristics

| Data Set | Instances | Classes | Features | |
|---|---|---|---|---|
| | | | Category | Numeric |
| balance | 625 | 3 | 0 | 4 |
| bupa liver | 345 | 2 | 0 | 6 |
| ecoli | 336 | 8 | 0 | 7 |
| glass | 214 | 6 | 0 | 9 |
| haberman | 306 | 2 | 0 | 3 |
| ionosphere | 351 | 2 | 0 | 34 |
| image | 2310 | 7 | 0 | 19 |
| iris | 150 | 3 | 0 | 4 |
| letter | 20000 | 16 | 0 | 26 |
| optdigit | 5620 | 10 | 0 | 64 |
| page block | 5473 | 5 | 0 | 10 |
| pendigit | 10992 | 10 | 0 | 16 |
| spam | 4601 | 2 | 0 | 57 |
| wine | 178 | 3 | 0 | 13 |
| vehicle | 846 | 4 | 0 | 18 |
| abalone | 4177 | 3 | 1 | 7 |
| cmc | 1473 | 3 | 7 | 2 |
| dermatology | 366 | 6 | 1 | 33 |
| heart | 270 | 2 | 6 | 7 |
| monk1 | 556 | 2 | 6 | 0 |
| monk2 | 601 | 2 | 6 | 0 |
| monk3 | 554 | 2 | 6 | 0 |
| nursery | 12960 | 5 | 8 | 0 |
| shuttle land | 279 | 2 | 6 | 0 |
| statlog DNA | 3186 | 3 | 180 | 0 |
| tae | 151 | 6 | 1 | 4 |
| tic-tac-toe | 958 | 2 | 9 | 0 |
| thyroid | 7200 | 3 | 15 | 6 |
| vote | 435 | 2 | 16 | 0 |
| zoo | 101 | 2 | 15 | 1 |

where $\mathbf{K}$ is a $k\times k$ Gram matrix with the element $k_{ij}=\mathrm{k}(\mathbf{x}_i^{NN},\mathbf{x}_j^{NN})$, $\mathbf{1}_{kk}$ is a $k\times k$ matrix with each element equals 1, $\mathbf{1}_k$ is a $k\times 1$ vector with each element equals 1, $\mathbf{k}_c$ is a $k\times 1$ vector with the $i$th element equals $\mathrm{k}(\mathbf{x},\mathbf{x}_i^{NN})$. After obtaining the kernel matrix $\mathbf{G}^{\mathrm{k}}$, we can assign the weights to the nearest neighbors by solve the linear system of equations $[\mathbf{G}^{\mathrm{k}}+\eta\mathrm{tr}(\mathbf{G}^{\mathrm{k}})/k]\mathbf{w}=\mathbf{1}_k$.

# 4 Experimental Results and Discussion

In this section, we evaluate the classification performance of KDF-WKNN using the UCI Machine Learning Repository (http://www.ics.uci.edu/mlearn/MLRepository.html). First, we investigate the performance of DF-WKNN and KDF-WKNN. Second, we compare the classification performance of DF-WKNN with several state-of-the-art methods.

**Table 2.** Hyper parameters and ACR (%) using DF-WKNN and KDF-WKNN

| Data Set | DF- WKNN | | KDF- WKNN | |
|---|---|---|---|---|
| | $k$ | Accuracy | $[k, \sigma]$ | Accuracy |
| balance | 31 | **91.20±0.19** | [31, 8] | 91.18±0.18 |
| bupa liver | 151 | **73.91±0.73** | [151, 2] | 73.51±1.01 |
| ecoli | 101 | 87.11±0.62 | [101, 0.5] | **87.47±0.53** |
| glass | 25 | 70.51±1.46 | [25, 3] | **71.21±2.23** |
| haberman | 81 | 75.56±0.65 | [81, 8] | 75.56±0.68 |
| ionosphere | 51 | **92.71±0.70** | [51, 10] | 92.56±0.57 |
| image | 25 | 97.32±0.13 | [25, 1] | **97.53±0.10** |
| iris | 47 | 97.93±0.58 | [47, 4] | 97.93±0.49 |
| letter | 5 | 96.63±0.11 | [5, 1] | **96.68±0.11** |
| optdigit | 17 | 99.20±0.04 | [17, 3] | **99.24±0.04** |
| page block | 65 | 96.90±0.07 | [65, 10] | 96.90±0.07 |
| pendigit | 81 | 99.68±0.01 | [81, 1] | **99.71±0.02** |
| spam | 41 | 92.20±0.17 | [41, 0.25] | **93.16±0.14** |
| wine | 91 | 98.93±0.17 | [91, 2] | **99.38±0.39** |
| vehicle | 35 | 82.21±0.80 | [35, 12] | **82.23±0.87** |
| abalone | 201 | 65.94±0.17 | [201, ] | 65.94±0.08 |
| cmc | 51 | **47.98±0.85** | [51, 4] | 47.93±0.75 |
| dermatology | 31 | 95.64±0.57 | [31, 1] | **97.21±0.32** |
| heart | 201 | 83.70±0.69 | [201, 16] | **83.74±1.00** |
| monk1 | 25 | 99.78±0.18 | [25, 2] | **99.91±0.12** |
| monk2 | 9 | 84.09±1.70 | [9, 4] | **84.36±1.68** |
| monk3 | 51 | 98.88±0.07 | [51, 6] | 98.88±0.07 |
| nursery | 15 | **94.38±0.12** | [15, 6] | 94.36±0.12 |
| shuttle land | 31 | 96.26±0.66 | [31, 4] | **96.37±0.72** |
| statlog DNA | 71 | 90.52±0.28 | [71, 6] | **93.00±0.16** |
| tae | 1 | 64.83±2.72 | [1, 1] | 64.83±2.72 |
| tic-tac-toe | 3 | 100.0±0.00 | [3, 1] | 100.0±0.00 |
| thyroid | 41 | 95.38±0.06 | [41, 6] | 95.38±0.06 |
| vote | 101 | **96.57±0.20** | [101, 8] | 96.44±0.24 |
| zoo | 9 | 96.93±0.31 | [9, 3] | 96.93±0.31 |
| Average | | 88.76 | | 88.98 |

### 4.1 The Experimental Settings

DF-KWNN and KDF-KWNN are tested on 30 benchmark data sets from the UCI Repository. Table 1 summarizes the information on the number of numeric and categorical features, number of classes $C$, and number of total instances $m$ for each data set. These data sets include 12 2-class problems, 6 3-class problems and 12 multi-class problems, and cover a wide range of applications such as medical diagnosis, and image analysis. We describe the experimental settings as follows:

(1) Distance measure. Features of some data may be categorical variables. In these cases, each categorical variable is converted into a vector of 0 or 1 variables. If a categorical variable $x$ takes $l$ values $\{c_1, c_2, \ldots, c_l\}$, it is replaced by a ($l$-1)-dimensional vector $[x^{(1)}, x^{(2)}, \ldots, x^{(l-1)}]$ such as $x^{(i)} = 1$ if $x = c_i$ and $x^{(i)} = 0$ otherwise, for $i = 1, \ldots, l\text{-}1$. If $x = c_l$, all the variables of the vector are zeros.

(2) Each of the data sets we randomly split into 10 folds, and use a 10-fold cross validation (cv) method to determine the classifier parameters and classification rate. To reduce bias in evaluating the performance, we calculate the average and standard deviation of the classification rates of the 10 runs of 10-fold cv.

(3) Normalization. For all the data sets, each of all the input features is normalized to values within [0, 1].

(4) Performance evaluation. To compare the differences in the performances of multiple classifiers, it is usual to select a number of data sets to test the individual performance scores (e.g., classification rate). It is then possible, based on these individual performance scores, such as the average classification rate (ACR) over all data sets, to evaluate the overall performance of a classifier.

### 4.2 Comparisons with the KNN Classifiers

Before applying DF-WKNN to a classification task, we should always determine one hyper parameter, the number of the nearest neighbors $k$. Further, KDF-WKNN will introduce other kernel parameters (Gaussian kernel $\sigma$). In our experiments, the optimal values of these hyper parameters are determined using 10-fold cv. Table 2 lists the optimal values of hyper parameters, classification rates, and standard deviation of DF-WKNN and KDF-WKNN. KDF-WKNN's ACR is 88.98%, is slightly higher than that of DF-WKNN, 88.76%. We further count the number of data sets for which KDF-WKNN performs better than DF-WKNN, 15 (*win* record), the number of data sets for which KDF-WKNN and DF-WKNN have the same classification rate, 9 (*draw* record), and the number of data sets for which DF-WKNN performs better than KDF-WKNN, 6 (*loss* record). For most data sets, KDF-WKNN is able to achieve classification rates competitive to or better than DF-WKNN.

Using ACR, we compare the classification performance of KDF-WKNN with that of other KNN classifiers, such as KNN and DS-WKNN. Table 3 shows the average classification rate of KNN, DS-WKNN, and KDF-WKNN on each of the test data sets, and the overall average classification rates of each method over all data sets. The average classification rate of KDF-WKNN over all data sets is 88.98%, which is higher than those of KNN at 86.56%, and DS-WKNN at 86.66%.

### 4.3 Comparisons of Multiple Classifiers

In this section, we evaluate KDF-WKNN by comparing it with multiple state-of-the-art classifiers, such as SVM and the reduced multivariate polynomial model (RM):

(1) SVM is a recently developed nonlinear classification approach and has achieved great success in many application tasks [9]. In this section, we use the OSU-SVM toolbox (http://svm.sourceforge.net/docs/3.00/api/) with RBF kernel.

(2) The RM model, which transforms the original data into a reduced polynomial feature space, has performed well in classification tasks that involve few features and many training data [13].

**Table 3.** Comparisons of average classification rates (%) obtained using different methods on the 30 data sets

| Data Set | KDF-WKNN | KNN | DS-WKNN | SVM | RM |
|---|---|---|---|---|---|
| balance | 91.18±0.18 | 88.86±0.82 | 89.89±0.22 | **99.89±0.16** | 91.74±0.15 |
| bupa liver | 73.51±1.01 | 63.48±1.42 | 64.64±1.09 | 63.80±0.79 | **72.58±0.77** |
| ecoli | 87.47±0.53 | 87.32±0.49 | 87.29±0.50 | 87.45±0.55 | **87.61±0.64** |
| glass | 71.21±2.23 | 69.39±1.60 | 66.07±0.85 | **71.31±1.00** | 62.66±1.77 |
| haberman | **75.56±0.68** | 72.88±0.72 | 74.74±0.65 | 72.85±0.49 | 75.35±0.63 |
| ionosphere | 92.56±0.57 | 86.89±0.53 | 86.72±0.52 | **95.04±0.62** | 88.54±0.82 |
| image | **97.53±0.10** | 97.10±0.13 | 97.16±0.14 | 96.44±0.18 | 94.11±0.20 |
| iris | **97.93±0.49** | 95.93±0.46 | 95.60±0.46 | 97.00±0.47 | 96.83±0.44 |
| letter | 96.68±0.11 | 96.18±0.46 | 96.32±0.06 | **97.34±0.15** | 74.14±0.05 |
| optdigit | **99.24±0.04** | 98.82±0.05 | 98.89±0.06 | 98.82±0.06 | 95.37±0.08 |
| page block | 96.90±0.07 | 96.04±0.10 | 96.12±0.11 | 96.34±0.06 | 95.49±0.06 |
| pendigit | **99.71±0.02** | 99.39±0.02 | 99.44±0.02 | 99.42±0.02 | 95.68±0.05 |
| spam | **93.16±0.14** | 90.92±0.18 | 90.92±0.18 | 91.03±0.05 | 92.85±0.16 |
| wine | **99.38±0.39** | 96.24±0.75 | 97.58±0.59 | 88.39±0.94 | 98.88±0.30 |
| vehicle | 82.23±0.87 | 71.38±0.49 | 71.54±0.49 | 81.58±0.96 | **83.10±0.46** |
| abalone | 65.94±0.08 | 64.09±0.32 | 64.43±0.40 | **66.46±0.04** | **66.46±0.12** |
| cmc | 47.93±0.75 | 45.42±0.68 | 46.24±0.99 | 48.90±0.79 | **54.25±0.47** |
| dermatology | **97.21±0.32** | 96.90±0.24 | 96.42±0.28 | 97.01±0.37 | 97.14±0.39 |
| heart | 83.74±1.00 | 81.22±0.35 | 79.67±0.78 | 69.81±1.22 | **83.86±0.91** |
| monk1 | 99.91±0.12 | 96.55±0.60 | 98.15±0.42 | **100.0±0.00** | 98.71±0.79 |
| monk2 | 84.36±1.68 | 81.05±2.20 | 81.06±2.20 | **100.0±0.00** | 75.91±1.57 |
| monk3 | **98.88±0.07** | 96.01±0.71 | 94.19±0.97 | 97.87±0.07 | 91.57±0.79 |
| nursery | 94.36±0.12 | 93.22±0.11 | 93.62±0.13 | **100.0±0.00** | 91.02±0.05 |
| shuttle land | 96.37±0.72 | 94.57±0.88 | 94.75±0.76 | **98.89±0.48** | 95.98±0.34 |
| statlog DNA | 93.00±0.16 | 88.00±0.20 | 88.93±0.13 | **96.23±0.10** | 95.09±0.10 |
| tae | **64.83±2.72** | **64.83±2.72** | 64.83±2.72 | 62.24±1.87 | 56.82±2.31 |
| tic-tac-toe | **100.0±0.00** | **100.0±0.00** | **100.0±0.00** | 99.76±0.13 | 98.33±0.05 |
| thyroid | 95.38±0.06 | 93.88±0.07 | 93.99±0.08 | **95.43±0.05** | 94.32±0.05 |
| vote | **96.44±0.24** | 93.54±0.38 | 93.82±0.46 | 95.35±0.26 | 95.43±0.07 |
| zoo | **96.93±0.31** | 96.73±0.66 | 96.73±0.66 | 96.55±0.82 | 96.25±1.48 |
| ACR | 88.98 | 86.56 | 86.66 | 88.70 | 86.54 |

Table 3 lists the classification rates and standard deviations of KDF-WKNN and the other two classifiers. The overall average classification rate of KDF-WKNN is 88.98%, which is higher than the classification rates of SVM (88.70) and RM (86.54).

## 5 Conclusion

In this paper we proposed a kernel difference-weighted KNN method for pattern classification. Given an unclassified sample **x**, KDF-WKNN use the difference between **x** and its neighborhood to weigh the influence of each neighbor, and then use weighted KNN rule to classify **x**. Compared with distance-weighted KNN, KDF-WKNN has a distinct geometric explanation as an optimal constrained reconstruction problem. Experimental results show that, in terms of classification performance, KDF-WKNN is better than KNN and distance weighted KNN, and is comparable to or better than several state-of-the-art methods, such as SVM and RM. In the future, systemic experiments [2] will be carried out to evaluate KDF-KNN.

## Acknowledgments

The work is partially supported by the NSFC foundation under the contracts No. 60332010 and No. 60571025, the 863 project under the contracts No. 2006AA01Z308.

## References

1. Bailey, T., Jain, A.K.: A Note on Distance-Weighted K-Nearest Neighbor Rules. IEEE Trans. Systems, Man, and Cybernetics, 8(1978) 311-313
2. Demšar, J.: Statistical Comparisons of Classifiers over Multiple Data Sets. Journal of Machine Learning Research, 7(2006) 1-30
3. Domeniconi, C., Peng, J., Gunopulos, D.: Locally Adaptive Metric Nearest Neighbor Classification. IEEE Trans. PAMI, 24(2002) 1281-1285
4. Dudani, S.A.: The Distance-Weighted K-Nearest-Neighbor Rule. IEEE Trans. Systems, Man, and Cybernetics, 6(1976) 325-327
5. Fukunaga, K., Flick, T.E.: An Optimal Global Nearest Neighbor Metric. IEEE Trans. PAMI, 6(1984) 314-318
6. Hastie, T., Tibshirani, R.: Discriminant Adaptive Nearest Neighbor Classification. IEEE Trans. PAMI, 18(1996) 607-616
7. Keller, J.M., Gray, M.R., Givens, Jr., J.A.: A Fuzzy K-Nearest Neighbor Algorithm. IEEE Trans. Systems, Man, and Cybernetics, 15(1985) 580-585
8. Macleod, J.E.S., Luk, A., Titterington, D.M.: A Re-examination of the Distance-Weighted *K*-Nearest Neighbor Classification Rule. IEEE Trans. SMC, 17(1987) 689-696
9. Müller, K.R., Mika, S., Rätsch, G., Tsuda, K., Schölkopf, B.: An Introduction to Kernel-based Learning Algorithms. IEEE Trans. Neural Networks, 12(2001) 181-202
10. Paredes, R., Vidal, E.: Learning Weighted Metrics to Minimizing Nearest-Neighbor Classification Error. IEEE Trans. PAMI, 28(2006) 1100-1110
11. Ricci, F., Avesani, P.: Data Compression and Local Metrics for Nearest Neighbor Classification. IEEE Trans. PAMI, 21(1999) 380-384
12. Short, R.D., Fukunaga, K.: The Optimal Distance Measure for Nearest Neighbor Classification. IEEE Trans. Information Theory, 27(1981) 622-627
13. Toh, K.A., Tran, Q.L., Srinivasan, D.: Benchmarking a Reduced Multivariate Polynormial Pattern Classifier. IEEE Trans. PAMI, 26(2004) 740-755
14. Wang, H.: Nearest Neighbors by Neighborhood Counting. IEEE Trans. PAMI, 28(2006) 942-953

# Novel Design of Decision-Tree-Based Support Vector Machines Multi-class Classifier

Liaoying Zhao[1], Xiaorun Li[2], and Guangzhou Zhao[2]

[1] Institute of Computer Application Technology, HangZhou Dianzi University, Hangzhou 310018, China
[2] College of Electrical Engineering, Zhejiang University, Hangzhou 310027, China
Zhaoly@hdu.edu.cn

**Abstract.** Designing the hierarchical structure is a key issue for the decision-tree-based (DTB) support vector machines multi-class classification. Inter-class separability is an important basis for designing the hierarchical structure. A new method based on vector projection is proposed to measure inter-class separability. Furthermore, two different DTB support vector multi-class classifiers are designed based on the inter-class separability: one is in the structure of DTB-balanced branches and another is in the structure of DTB-one against all. Experiment results on three large-scale data sets indicate that the proposed method speeds up the decision-tree-based support vector machines multi-class classifiers and yields higher precision.

**Keywords:** Pattern classification, Support vector machines, Vector projection, Inter-class separability.

## 1 Introduction

Support vector machines (SVMs), motivated by statistical learning theory, is a new machines learning technique proposed recently by Vapnik and co-workers [1]. The main feature of SVMs is that they use the structural risk minimization rather than the empirical risk minimization. The SVMs has been successful as a high performance classifier in several domains including pattern recognition [2, 3], fault diagnosis [4], and bioinformatics [5]. It has strong theoretical foundations and good generalization capability.

The SVMs approach was originally developed for two-class or binary classification. Practical classification applications are multi-class problems commonly. Forming a multi-class classifier by combining several binary classifiers is the way commonly used, methods such as one-against-all (OAA) [6] , one-against-one (OAO) [7] and DAG (decision directed acyclic graph) support vector machines [8] are all based on binary classifications. Decision-tree-based SVMs (DTBSVMs) [9-12] which combine SVMs and decision tree is also a good way for solving multi-class problems. However, additional work is required to effectively design the hierarchical structure of the DTBSVMs.

D.-S. Huang, L. Heutte, and M. Loog (Eds.): ICIC 2007, LNAI 4682, pp. 871–880, 2007.

The classification performances of DTBSVMs multi-class classifier with different hierarchical structure differ a lot. The inner-class separability is an important basis for designing the hierarchical structure. In this paper, a new method based on vector projection is proposed to measure inter-class separability, and two ways are presented to design the hierarchical structure of the multi-class classifier based on the inter-class separability.

This paper is organized as follow. In section 2, the structure of decision-tree-based SVMs is briefly described; in section 3, the seperability measure is defined based on vector projection. Two algorithms for design DTBSVMs are given in section 4, and the simulation experiments and results are given in section 5.

## 2 The Structure of Decision-Tree-Based SVMs Classifier

The DTBSVMs classifier decomposes the *C*-class classification problem into *C*-1 sub-problems, each separating a pair of micro-classes. Two structures of the DTBSVMs classifier for a 4-class classification problem are shown in Fig.1. Fig.1(a) is partial binary tree structure, also called DTB-one against all (DTB-OAA), represents a simplification of the OAA strategy obtained through its implementation in a hierarchical context; Fig.1(b) is the DTB-balanced branches (DTB-BB) structure. The DTBSVMs classifier discussed in paper [9]、[10] and [11] are all based on the DTB-OAA strategy, while in [12], a DTB-BB strategy is described. In this paper, we investigate a new design method of the two different DTB hierarchies.

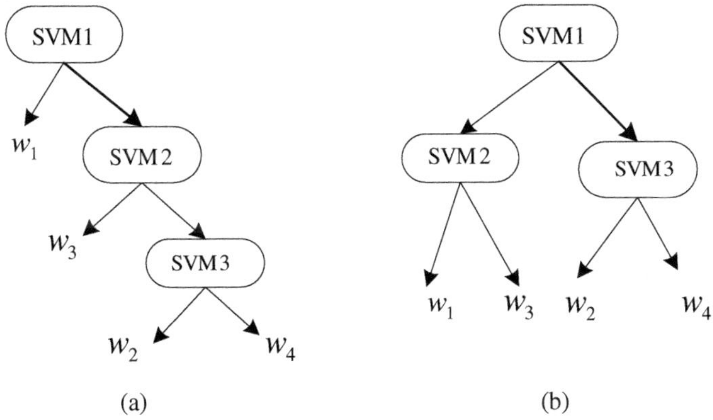


**Fig. 1.** Structures of DBTSVMs classifier

The distance between the separating hyperplane and the closed data points of training set is called margin. The following lemma [13] gives the relation between the margin and the generalization error of the classifier.

**Lemma 1.** Suppose we are able to classify an $m$ sample of labeled examples using a perceptron decision tree and suppose that the tree obtained contains $k$ decision nodes with margin $\gamma_i$ at node $i$, $i = 1,2,\cdots,k$, then we can bound the generalization error with probability greater than 1- $\delta$ to be less than

$$\frac{130R^2}{m}[D'\log(4em)\log(4m)+\log\frac{(4m)^{k+1}\binom{2k}{k}}{(k+1)\delta}] \tag{1}$$

where , $D' = \sum_{i=1}^{k}\frac{1}{\gamma_i^2}$ , $\delta > 0$ and $R$ is radius of a sphere containing the support of the unknown (but fixed) distribution $P$.

According to lemma1, for a given set of train samples, the less the number of nodes, the smaller of generalization error of the classifier, and the larger the margin, the higher generalization ability of the classifier. Thus, in order to get better generalization ability, the margin in the DTB is an important basis for designing the hierarchical structure.

Different classes have different domains in the sample space. If the domains of two classes are not intersected, the margin is larger and the two classes are more separable. While the margin is smaller if the domains of two classes are intersected, and the larger ratio of the intersected samples to the total number of the two classes leads to more difficulties in separating.

Now the problem is how to judge two classes intersect or not and how to estimate the separability between two classes.

## 3 The Inter-class Separabilty Measure

This section will mainly discuss that how to measure the inter-class separability between two classes. In order to be comprehensible, we first discuss the seperability measure in linear space and then generalize it to nonlinear feature space.

### 3.1 The Seperabiliy Measure in Linear Space

First we give some definitions.

**Definition 1.** (sample center $\boldsymbol{m}_i$)Consider the set of samples $\boldsymbol{X}_i = \{\boldsymbol{x}_1, \boldsymbol{x}_2, \cdots, \boldsymbol{x}_n\}$, the sample center of class-$i$ is defined by

$$\boldsymbol{m}_i = \frac{1}{n}\sum_{j=1}^{n}\boldsymbol{x}_j \tag{2}$$

**Definition 2.** ( feature direction ) Define the direction of vector $\overrightarrow{\boldsymbol{m}_1\boldsymbol{m}_2}$ as the feature direction of pattern-1 , and the direction of vector $\overrightarrow{\boldsymbol{m}_2\boldsymbol{m}_1}$ as the feature direction of pattern-2。

**Definition 3.** ( feature distance ) Let $\boldsymbol{x}_i \in \boldsymbol{X}_1 = \{\boldsymbol{x}_1, \boldsymbol{x}_2, \cdots, \boldsymbol{x}_n\}$, $\boldsymbol{x}_i^o$ be the projection of data $\boldsymbol{x}_i$ to the feature direction of pattern-1, $\boldsymbol{m}_1$ be the sample center of $\boldsymbol{X}_1$, the feature distance of $\boldsymbol{x}_i$ can be defined as

$$\left\|\overrightarrow{\boldsymbol{m}_1\boldsymbol{x}_i^o}\right\|_2 = \left\|\boldsymbol{m}_1 - \boldsymbol{x}_i^o\right\|_2 \tag{3}$$

It is easy to proof the following theorem by reduction to absurdity.

**Theorem 1.** Suppose $d = \|\boldsymbol{m}_1 - \boldsymbol{m}_2\|$ is the sample centers distance of data set $\boldsymbol{X}_1 = \{\boldsymbol{x}_1, \boldsymbol{x}_2, \cdots, \boldsymbol{x}_{l_1}\}$ and $\boldsymbol{X}_2 = \{\boldsymbol{y}_1, \boldsymbol{y}_2, \cdots, \boldsymbol{y}_{l_2}\}$, calculate the feature distance of data $\boldsymbol{x}_i$ as $\left\|\overrightarrow{\boldsymbol{m}_1\boldsymbol{x}_i^o}\right\|_2$ and $\boldsymbol{y}_j$ as $\left\|\overrightarrow{\boldsymbol{m}_2\boldsymbol{y}_j^o}\right\|_2$ respectively, let

$$r_1 = \max_{\boldsymbol{x}_i \in \boldsymbol{X}_1}\left(\left\|\overrightarrow{\boldsymbol{m}_1\boldsymbol{x}_i^o}\right\|_2\right) \tag{4}$$

$$r_2 = \max_{\boldsymbol{y}_j \in \boldsymbol{X}_2}\left(\left\|\overrightarrow{\boldsymbol{m}_2\boldsymbol{y}_j^o}\right\|_2\right) \tag{5}$$

then the data domains of data set $\boldsymbol{X}_1$ and $\boldsymbol{X}_2$ are not intersected if $r_1 + r_2 < d$, while if the data domains of $\boldsymbol{X}_1$ and $\boldsymbol{X}_2$ are intersected, it is surely that $r_1 + r_2 \geq d$.

According to theorem 1, the inter-class seperability measure can be defined on the principle that the smaller measure value, the larger margin.

**Definition 4.** If $r_1 + r_2 < d$, then the inter-class seperability is defined as

$$se_{12} = se_{21} = -d \tag{6}$$

If $r_1 + r_2 \geq d$, assume the number of data in $\boldsymbol{X}_1$ that satisfied $d - r_2 \leq \left\|\overrightarrow{\boldsymbol{m}_1\boldsymbol{x}_i^o}\right\|_2 \leq r_1$ is $tr_1$, the number of data in $\boldsymbol{X}_2$ that satisfied $d - r_1 \leq \left\|\overrightarrow{\boldsymbol{m}_2\boldsymbol{y}_j^o}\right\|_2 \leq r_2$ is $tr_2$, the inter-class seperability is defined as

$$se_{12} = se_{21} = (tr_1 + tr_2)/(l_1 + l_2) \tag{7}$$

### 3.2 The Sepearability Measure in Nonlinear Space

The following lemma [14] gives the formula of Euclidean distance between two vectors in the feature space.

**Lemma 2.** If two vectors $\boldsymbol{x}=(x_1,x_2,\cdots,x_n)$ and $\boldsymbol{y}=(y_1,y_2,\cdots,y_n)$ are projected into a high-dimension feature space by a nonlinear map $\boldsymbol{\Phi}(\bullet)$ , the Euclidean distance between vector $\boldsymbol{x}$ and $\boldsymbol{y}$ in the corresponding feature space is given by

$$d^H(\boldsymbol{x},\boldsymbol{y})=\sqrt{k(\boldsymbol{x},\boldsymbol{x})-2k(\boldsymbol{x},\boldsymbol{y})+k(\boldsymbol{y},\boldsymbol{y})} \tag{8}$$

where , the function $k(\boldsymbol{x},\boldsymbol{y})=\boldsymbol{\Phi}(\boldsymbol{x})\boldsymbol{\Phi}(\boldsymbol{y})$ is a kernel function.

According lemma2 , the center distance between class-*i* and class-*j* is

$$d^H=\left\|\boldsymbol{\Phi}(\boldsymbol{m}_i)-\boldsymbol{\Phi}(\boldsymbol{m}_j)\right\|_2=\sqrt{k(\boldsymbol{m}_i,\boldsymbol{m}_i)-2k(\boldsymbol{m}_i,\boldsymbol{m}_j)+k(\boldsymbol{m}_j,\boldsymbol{m}_j)} \tag{9}$$

**Lemma 3.** Consider three vectors $\boldsymbol{x}=(x_1,x_2,\cdots,x_n)$ , $\boldsymbol{y}=(y_1,y_2,\cdots,y_n)$ and $\boldsymbol{z}=(z_1,z_2,\cdots,z_n)$ , suppose $\boldsymbol{\Phi}(\bullet)$ is a feature map function , let $\overrightarrow{\boldsymbol{\Phi}(\boldsymbol{x})\boldsymbol{\Phi}(z^o)}$ be the projection of vector $\overrightarrow{\boldsymbol{\Phi}(\boldsymbol{x})\boldsymbol{\Phi}(z)}$ onto vector $\overrightarrow{\boldsymbol{\Phi}(\boldsymbol{x})\boldsymbol{\Phi}(\boldsymbol{y})}$, then the feature distance is given by

$$\left\|\overrightarrow{\boldsymbol{\Phi}(\boldsymbol{x})\boldsymbol{\Phi}(z^o)}\right\|_2=\frac{k(z,\boldsymbol{y})-k(z,\boldsymbol{x})-k(\boldsymbol{x},\boldsymbol{y})+k(\boldsymbol{x},\boldsymbol{x})}{\sqrt{k(\boldsymbol{x},\boldsymbol{x})-2k(\boldsymbol{x},\boldsymbol{y})+k(\boldsymbol{y},\boldsymbol{y})}} \tag{10}$$

The inter-class seperability measure in nonlinear space can be defined as the definition in linear space.

**Definition 5.** Suppose $d^H=\left\|\boldsymbol{\Phi}(\boldsymbol{m}_1)-\boldsymbol{\Phi}(\boldsymbol{m}_2)\right\|$ is the sample centers distance of data set $\boldsymbol{X}_1=\{\boldsymbol{x}_1,\boldsymbol{x}_2,\cdots,\boldsymbol{x}_{l_1}\}$ and $\boldsymbol{X}_2=\{\boldsymbol{y}_1,\boldsymbol{y}_2,\cdots,\boldsymbol{y}_{l_2}\}$ in the feature space, calculate the feature distance of data $\boldsymbol{x}_i$ as $\left\|\overrightarrow{\boldsymbol{\Phi}(\boldsymbol{m}_1)\boldsymbol{\Phi}(\boldsymbol{x}_i^o)}\right\|_2$ and $\boldsymbol{y}_j$ as $\left\|\overrightarrow{\boldsymbol{\Phi}(\boldsymbol{m}_2)\boldsymbol{\Phi}(\boldsymbol{y}_j^o)}\right\|_2$ respectively, let

$$r_1=\max_{\boldsymbol{x}_i\in\boldsymbol{X}_1}\left(\left\|\overrightarrow{\boldsymbol{\Phi}(\boldsymbol{m}_1)\boldsymbol{\Phi}(\boldsymbol{x}_i^o)}\right\|_2\right) \tag{11}$$

$$r_2=\max_{\boldsymbol{y}_j\in\boldsymbol{X}_2}\left(\left\|\overrightarrow{\boldsymbol{\Phi}(\boldsymbol{m}_2)\boldsymbol{\Phi}(\boldsymbol{y}_j^o)}\right\|_2\right) \tag{12}$$

If $r_1 + r_2 < d^H$, the inter-class seperability is defined as

$$se_{12} = se_{21} = -d^H \tag{13}$$

if $r_1 + r_2 \geq d^H$, assume the number of data in $X_1$ that satisfied $d^H - r_2 \leq \left\| \overrightarrow{\Phi(m_1)\Phi(x_i^{\,o})} \right\|_2 \leq r_1$ is $tr_1$, the number of data in $X_2$ that satisfied $d^H - r_1 \leq \left\| \overrightarrow{\Phi(m_2)\Phi(y_j^{\,o})} \right\|_2 \leq r_2$ is $tr_2$, the inter-class seperability is defined as

$$se_{12} = se_{21} = (tr_1 + tr_2)/(l_1 + l_2) \tag{14}$$

## 4 Construct DTBSVMs Classifier

In classification of DTBSVMs classifier, starting from the top of the decision tree, we calculate the value of the decision function for input data $x$ and according to the value we determine which node to go to. We iterate this procedure until we reach a leaf node and classify the input data into the class associated with the node. According to this classification procedure of DTBSVMs classifier, not all the decision functions need to be calculated, and the more the data are misclassified at the upper node of the decision tree, the worse the classification performance becomes. Therefore, the classes that are easily separated need to be separated at the upper node of the decision tree.

Suppose $S_j, j = 1,2,\cdots,c$ are sets of $l$ pairs training data included in $c$ classes, and $y_i = j$ if $x_i \in S_j$. The new design procedures of DTB-OAA and DTB-BB are described respectively.

### 4.1 DTB-OAA

For DTB-OAA classifier, one class is separated from the remaining classes at the hyperplane corresponding to each SVMs of the decision tree. For the sake of convenience for realization, taking an array $L$ to keep the markers of the classes according their seperability in descend. The algorithm of DTB-OAA is proposed as follows.

Step1. Calculate the separability measure in feature space $se_{ij}$, $se_{ij} = se_{ji}$, $i, j = 1,2,\cdots,c$, $i \neq j$, construct a symmetric matrix of separability measures

$$SE = \begin{bmatrix} 0 & se_{12} & \cdots & se_{1,c-1} & se_{1,c} \\ se_{12} & 0 & \cdots & se_{2,c-1} & se_{2,c} \\ \vdots & \vdots & \vdots & \vdots & \vdots \\ se_{c-1,1} & se_{c-1,2} & \cdots & 0 & se_{c-1,c} \\ se_{c,1} & se_{c,2} & \cdots & se_{c,c-1} & 0 \end{bmatrix}$$

Step2. Define array $D\_no$ =[1,2,…,*c*], let *i*=1, and $SE(k,:)$ indicate the row *k* of $SE$, for $j=1$ to $c-2$, repeat the following procedure to get the most easily separated class from the remaining classes:

1 ) Calculate $k_0 = \arg\min_{k=1,\cdots,c+1-j} sum(SE(k,:))$ , $L(i) = D\_no(k_0)$ . If $k_0$ exists for plural k, regard the one got first as minimization;

2 ) Set $SE(k_0,:)$=null, $SE(:,k_0)$=null, and $D\_no$ ( $k_0$ )=null, *i*=*i*+1.

Step3. $L(c-1) = D\_no(1)$ , $L(c) = D\_no(2)$ .

Step4. Define structure array $node$ to keep the information of each node (including support vector, weight $\alpha$ and , threshold *b* et al). For $j$=1 to $c$-1, repeat the following procedure to construct the classifier: regard class- $L(j)$ as the plus samples of SVMs-*j*, and union the rest class $L(j+1),\cdots,L(c)$ as the negative samples of SVMs-j. Training SVMs-*j* to get the structure information of $node(j)$.

### 4.2 DTB-BB

In the DTB-BB strategy, the tree is defined in such a way that each node (SVMs) discriminate between two groups of classes with maximum margin. The algorithm that implements the DTB-BB strategy is described as follows:

Step1、2、3 is the same as DTB-OAA to get array $L$.

Step4. Define a binary tree structure $\theta = \{node(i)\}$ . The structure variable $node(i)$ keeps the information of each node (including support vector, weight $\alpha$ and , threshold *b* etc). Let $node(i).I$ keep the markers of the classes included in $node(i)$ and variable *endnodes* be the number of leaf nodes. Set $i=1$ , $node(1).I = L$ , $t=1, j=1$, $endnodes=0$.

Step5. If $length(node(i).I)$=1, then go to Step9.

Step6. Let $num = length(node(i).I)$ , divide classes in $node(i)$ into two groups in such a way that $node(i).pl = j+1$ , $node(i).pr = j+2$ ,

$$node(j+1).I = node(i).I(1,\cdots,[num/2]) ,$$

$$node(j+2).I = node(i).I([num/2]+1,\cdots,num)$$

Step7. Regard the classes in $node(t).pl$ as the plus samples and the classes in $node(t).pr$ as the negative samples of classifier- $t$ , train the SVMs to get the information of $node(t)$ .

Step8. Set $i = i+1$, $j = j+1$ and $t = t+1$, go to Step5.

Step9. Set $endnodes = endnodes + 1$, if $endnodes = c$ then Stop, otherwise, set $i = i+1$, go to Step5.

## 5 Experimental Results

The experiments reported in this section have been conducted to evaluate the performance of the two DTBSVMs multi-class classifier proposed in this paper, in comparison with the OAO algorithm. The experiments focus on the following three issues: classification accuracy, execution efficiency and the number of support vectors.

The kernel function used in the experiments is the radial basis function kernel $k(x,y) = \mathbf{exp}(-\|x-y\|^2/\gamma)$. Table 1 lists main characteristics for the three large dataset used in our experiments. The data sets are from the UCI repository (http://www.ics.uci.edu/~mlearn/MLRepository.html). In these experiments, the SVMs software used is SVM_V0.51 [15] with the radial basis kernel. Cross validation has been conducted on the training set to determine the optimal parameter values to be used in the testing phase. Table 2 is the optimal parameters for each data set, where $C$ is the castigatory coefficient of SVMs, ones(1,$n$) denotes an all 1s vector of size $1 \times n$ .

**Table 1.** Benchmark data sets used in the experiments

| Date set | # trainging samples | # testing samples | #class numbers | # attribute numbers |
|---|---|---|---|---|
| Letter | 15 000 | 5 000 | 26 | 16 |
| Satimage | 4 435 | 2 000 | 6 | 36 |
| Shuutle | 43 500 | 14 500 | 7 | 9 |

Table 3 compares the results delivered by alternative classification algorithms with the three large benchmark data sets, where Tc/s is the testing time in second, Tx/s is the training time in second, #SVs denotes the number of all support vectors (with intersection), u_SVs denotes the number of different support vectors, and CRR denotes the correct recognition rate. As Table 3 shows that the two DTBSVMs classifiers and the OAO classifier basically deliver the same level of accuracy. The OAO needs more support vector in training, but the numbers of different support vectors are approximately equal. For **letter**, the test time of OAO is much higher than that of DTB-OAA and that of DTB-BB. For **satimage**, the test time of OAO is more than twice of that of DTB-OAA and almost triple of that of DTB-BB. For **shuttle**, the test time of OAO is approximate to that of DTB-OAA and almost twice of that of DTB-BB.

Table 3 also shows that DTB-BB is more efficient than DTB-OAA both in accuracy and speed. This is consistent with the theoretic analyse in paper [12].

**Table 2.** The optimal parameters for each data set

| Date set | $C$ | $\gamma$ | | |
|---|---|---|---|---|
| | | OAO | DTB-OAA | DTB-BB |
| Letter | 8 | 64 | 64×ones(1, 25) | 64×ones(1, 25) |
| Satimage | 1.5 | 3048 | 3048×ones(1,5) | 3048×ones(1,5) |
| Shuutle | $2^{12}$ | 4096 | [4096, 1024, 1024, 1024, 1024, 1024] | [4096, 1024, 1024, 1024, 1024, 1024] |

**Table 3.** Comparison of the results

| Date set | OAO | | | DTB-OAA | | | DTB-BB | | |
|---|---|---|---|---|---|---|---|---|---|
| | Tx/s<br>Tc/s | #SVs<br>u_SVs | CRR<br>% | Tx/s<br>Tc/s | #SVs<br>u_SVs | CRR<br>% | Tx/s<br>Tc/s | #SVs<br>u_SVs | CRR<br>% |
| Letter | 397<br>348 | 33204<br>7750 | 97.4 | 3916<br>58 | 7389<br>5087 | 96.4 | 2068<br>18 | 8489<br>5475 | 96.5 |
| Satimage | 60<br>35 | 3404<br>1510 | 91.8 | 43<br>17 | 2191<br>1428 | 91.2 | 53<br>13 | 2208<br>1529 | 92 |
| Shuutle | 7182<br>26 | 1239<br>382 | 99.9 | 15452<br>28 | 1219<br>499 | 99.8 | 6807<br>14 | 703<br>417 | 99.9 |

## 6 Conclusion

In this paper, we proposed new formulation of SVMs for a multi-class problem. A novel inter-class separability measure is given based on vector projection, and two algorithms are presented to design the DTBSVMs multi-class classifier based on the inter-class separability. Classification experiments for three large-scale data sets prove that the two DTBSVMs classifiers basically deliver the same level of accuracy as the OAO classifier, and the executing time is shortened.

Based on the study presented in this paper, there are several issues that deserve further studies. The first issue is the experiment on other benchmark data sets or some real data sets such as remote sensing images with the proposed algorithms to verify their effectiveness. The second issue is a more reasonable design for the structure of DTB-BB classifier. The third issue is the choice of parameters of kernel function.

**Acknowledgments.** This work is supported by Natural Science Basic Research Plan in Zhejiang Province of China Grant Y106085 to L.Y.Zhao.

## References

1. Vapnik ,V.: The Nature of Statistical Learning Theory. New York: Springer (1995)
2. Ma, C., Randolph, M.A., Drish, J.: A Support Vector Machines-Based Rejection Technique for Speech Recognition. Proceeding of IEEE Int. Conference on Acoustics, Speech, and Signal Processing (2001) 381-384
3. Brunelli, R.: Identity Verification Through Finger Matching: A Comparison of Support Vector Machines and Gaussian Basis Functions Classifiers. Pattern Recognition Letters 27 (2006) 1905-1915
4. Ma, X.X., Huang, X.Y., Chai, Y.: 2PTMC Classification Algorithm Based on Support Vector Machines and Its Application to Fault Diagnosis. Control and Decision 18 (2003) 272-276
5. Jin, B., Tang, Y.C., Zhang, Y.Q.: Support Vector Machines with Genetic Fuzzy Feature Transformation for Biomedical Data Classification. Information Sciences 177 (2007) 476-489
6. Bottou, L., Cortes, C., Denker, J.: Comparison of Classifier Methods: A Case Study in Handwriting Digit Recognition. Proceedings of the 12th IAPR International Conference on Pattern Recognition, Jerusalem: IEEE (1994) 77-82
7. Kebel, U.: Pairwise Classification and Support Vector Machines. Advances in Kernel Methods-Support Vector Learning, MIT, Cambridge (1999) 255-258
8. Platt, J., Cristianini, N., Shawe-Taylor, J.: Large Margin DAG's for Multiclass Classification. Advances in Neural Information Processing Systems 12, MA, Cambridge (2000) 547-553
9. Hsu, C. W., Lin, C. J.: A Comparison of Methods for Multi-Class Support Vector Machines. IEEE Transaction on Neural Network 13 (2002) 415-425
10. Wang, X.D., Shi, Z.W., Wu, C.M. Wang, W.: An Improved Algorithm for Decision-tree-based SVM. Proceedings of the 6th World Congress on Intelligent Control and Automation, Dalian, China (2006) 4234-4237
11. Sahbi, H., Geman, D., Perona, P.: A Hierarchy of Support Vector Machines for Pattern Detection. Journal of Machine Learning Research 7 (2006) 2087-2123
12. Zhao, H., Rong, L.L., Li, X.: New Method of Design Hierarchical Support Vector Machine Multi-class Classifier. Application Research of Computers 23 (2006) 34-37
13. Bennet, K.P., Cristianini, N., Shaue T.J.: Enlarging the Margins of Perceptron Decision Trees. Machine Learning 3 (2004) 295-313
14. Li, Q., Jiao, L.C., Zhou, W.D.: Pre-Extracting Support Vector for Support Vector Machine Based on Vector Projection, Chinese Journal of Computers 28 (2005) 145-152
15. Platt, J.C.: Fast Training of Support Vector Machines Using Sequential Minimal Optimization. http://research.microsoft.com/~jplatt

# Tuning Kernel Parameters with Different Gabor Features for Face Recognition


Linlin Shen[1], Zhen Ji[1], and Li Bai[2]

[1] Faculty of Information and Engineering, ShenZhen University, 518060, China
{llshen,jizhen}@szu.edu.cn
[2] School of Computer Science and Information Technology, University of Nottingham, Nottingham NG8 1BB, UK
bai@cs.nott.ac.uk



**Abstract.** Kernel methods like support vector machine, kernel principal component analysis and kernel fisher discriminant analysis have recently been successfully applied to solve pattern recognition problems such as face recognition. However, most of the papers present the results without giving kernel parameters, or giving parameters without any explains. In this paper, we present an experiments based approach to optimize the performance of a Gabor feature and kernel method based face recognition system. During the process of parameter tuning, the robustness of the system against variations of kernel function, kernel parameters and Gabor features are extensively tested. The results suggest that the kernel method based approach, with tuned parameters, achieves significantly better results than other algorithms available in literature.



**Keywords:** Kernel methods, Gabor features.


## 1 Introduction

Face recognition has been widely used in commercial and law-enforcement applications such as surveillance, security, telecommunication and human-computer interaction. Many face recognition algorithms have been reported in the literature such as the Eigenface method based on Principle Component Analysis (PCA) [1], Fisherface method based on Linear Discriminant Analysis (LDA) [2], Hidden Markov Models [3], and neural network approaches [4].

Whilst PCA projection aims at a subspace that maximizes the overall data variance, LDA projection on the other hand aims at a subspace that maximizes between-class variance and minimizes within-class variance. It is observed that variations between the face images of the same person (within-class scatter) due to illumination and pose are almost always larger than that due to facial identity (between-class scatter) [5]. As a result, LDA based Fisherface methods have been proven to perform better than PCA based Eigenface approaches [2], when sufficient training samples are available. However, both PCA and LDA are linear methods. Since facial variations are mostly nonlinear, PCA and LDA projections could only


D.-S. Huang, L. Heutte, and M. Loog (Eds.): ICIC 2007, LNAI 4682, pp. 881–890, 2007.

provide suboptimal solutions for face recognition tasks [6]. Recently, kernel methods have been successfully applied to solve pattern recognition problems [7-10] because of their capacity in handling nonlinear data. Support Vector Machines (SVM) are typical kernel methods and have been successfully applied to face detection [11], face recognition [12] and gender classification [13]. By mapping sample data to a higher dimensional feature space, effectively a nonlinear problem defined in the original image space is turned into a linear problem in the feature space. PCA or LDA can subsequently be performed in the feature space and thus Kernel Eigenface (KPCA) [8] and Kernel Fisher discriminant analysis (KFDA) [14]. In literature, a number of variations of KFDA have also been proposed [15-17]. Experiments show that KPCA and KFDA are able to extract nonlinear features and thus provide better recognition rates in applications such as character [18] and face recognition [10, 14].

While a large number of the kernel methods use raw pixel values as features for face recognition [8] [14], some works do apply more complicate and robust features, e.g. Gabor features [10]. The combination of Gabor features with kernel methods has been shown to achieve significantly better results than the systems using raw pixel values and linear subspace methods [19]. While the robustness of Gabor features has been proved by a number of research works, the feature extraction process is, however, quite complex and computationally costive. To tackle this problem, we have proposed to use boosting algorithm to simplify the feature extraction process [20]. In the paper, a variation of KFDA, Generalized Discriminant Analysis (GDA) [7] was applied to the selected Gabor features for face recognition. While efficiency has been substantially improved, the system still achieves similar accuracy with approaches using conventional feature extraction process.

Though quite a number of nonlinear kernel methods have been proposed and successfully applied to pattern recognition problems, few researches have been done on how to choose kernel functions and tune related parameters. Most of the papers present the results without giving parameters, or give parameters without any explains. In this paper, we will discuss related parameters when different kernel functions, e.g. Radial Basis Function (RBF) and polynomial function etc. are used. Following the discussion, we will present the effects of different parameters and different Gabor features on performance of the GDA based face recognition system, and present an experiment based kernel parameter tuning approach. By tuning the kernel parameters and subspace dimension, the GDA based system has shown significantly better accuracy than other methods such as PCA, LDA and KPCA. We have also shown that GDA become much more robust against the variations of kernel functions and kernel parameters when the boosting selected Gabor features are used.

## 2 Gabor Feature Representation

### 2.1 Gabor Wavelets

In the space domain, the 2D Gabor wavelet is a Gaussian kernel modulated by a sinusoidal plane wave [21]:

$$g(x,y) = w(x,y)s(x,y) = e^{-(\alpha^2 x'^2 + \beta^2 y'^2)} e^{j2\pi f x'}$$
$$x' = x\cos\theta + y\sin\theta \qquad (1)$$
$$y' = -x\sin\theta + y\cos\theta$$

where $f$ is the central frequency of the sinusoidal plane wave, $\theta$ is the anti-clockwise rotation of the Gaussian and the plane wave, $\alpha$ is the sharpness of the Gaussian along the major axis parallel to the wave, and $\beta$ is the sharpness of the Gaussian minor axis perpendicular to the wave. To keep the ratio between frequency and sharpness constant, $\gamma = \frac{f}{\alpha}$ and $\eta = \frac{f}{\beta}$ are defined and the Gabor wavelets can now be rewritten as:

$$\varphi(x,y) = \frac{f^2}{\pi\gamma\eta} g(x,y) = \frac{f^2}{\pi\gamma\eta} e^{-(\alpha^2 x'^2 + \beta^2 y'^2)} e^{j2\pi f x'} \qquad (2)$$

## 2.2 Downsampled Gabor Features

The Gabor wavelet representation of a face image is the convolution of the image with the family of Gabor wavelets as defined by (1). The convolution of image $I(\vec{x})$ and a Gabor wavelet $\varphi_{u,v}(\vec{x})$ can be defined as follows:

$$G_{u,v}(\vec{x}) = (I * \varphi_{u,v})(\vec{x}) \qquad (3)$$

where $G_{u,v}(\vec{x})$ denote the convolution result corresponding to the Gabor wavelet at orientation $u$ and scale $v$. As a result, the image $I(\vec{x})$ can be represented by a set of Gabor wavelet coefficients $\{G_{u,v}(\vec{x}), v = 0,...,4; u = 0,..,7\}$.

When the convolution results $G_{u,v}(\vec{x})$ over each pixel of the image could be concatenated to from an augmented feature vector, the size of the vector could be very large. Take an image with size 24×24 for example, the convolution result will give 24×24×5×8=23,040 features. To make the following kernel methods applicable to such a huge dimensional feature, each $G_{u,v}(\vec{x})$ is firstly downsampled by a factor $r$, normalized to zero mean and unit variance, and then transformed to a vector $x^r_{u,v}$ by concatenating its rows [19]. Therefore, a downsampled Gabor feature (DGF) vector $x^r$ can be derived to represent the image $I$ by concatenating those vectors $x^r_{u,v}$:

$$x^r = ((x^r_{0,0})^t \ (x^r_{0,1})^t \ \cdots \ (x^r_{4,7})^t)^t \qquad (4)$$

## 2.3 The Optimized Gabor Features

While important information could be lost during the downsampling process, the feature dimension, after downsampling, could still be large. As a result, a better approach is required to reduce the feature dimension. we have recently developed a boosting based algorithm to identify the most significant Gabor features for face recognition [20]. In this work, the task of a multi-class face recognition problem was transformed to a two-class problem: selecting Gabor features that are effective for intra- and extra-person space discrimination. Such selected Gabor features should be robust for face recognition, as intra- and extra-person space discrimination is one of the major difficulties in face recognition.

By using the boosting algorithm, the most significant Gabor features are selected one by one, in sequence. Upon completion of *T* boosting iterations, *T* most significant Gabor features for face recognition will be identified. Fig. 1 shows the 12 Optimized Gabor Features (OGF) and the first 200 positions identified by the boosting algorithm for feature extraction. The results suggest that the locations around eyes, eyebrows and nose seem to be more important for face recognition.

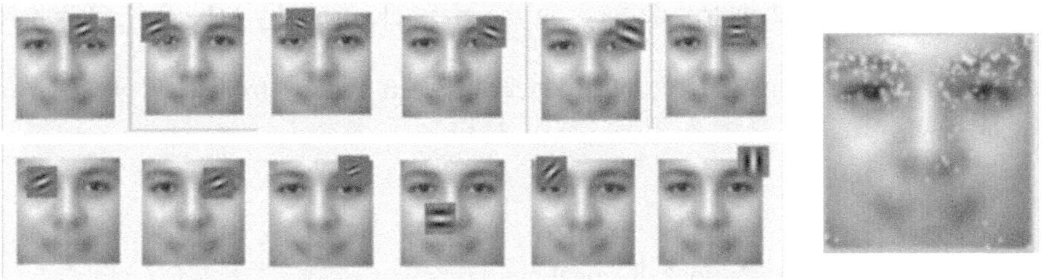

**Fig. 1.** The first 12 Gabor features and the 200 positions for feature extraction

## 3 Generalized Discriminant Analysis

Similar to LDA, the purpose of GDA [14] is to maximize the quotient between the inter-classes inertia and the intra-classes inertia. Considering a *C*-class problem and letting $N_c$ be the number of samples in class *c*, a set of training patterns from the *C* classes can be defined as $\{x_{ck}, c=1,2,...C; k=1,2,...,N_c\}, N=\sum_{c=1}^{C} N_c$ . Given a nonlinear mapping $\phi: R^N \to F$ , the set of training samples in the mapped feature space can be represented as $\{\phi(x_{ck}), c=1,2,...C; k=1,2,...,N_c\}$. The $S_b$ and $S_w$ of the training set can be computed as:

$$S_w = \frac{1}{C}\sum_{c=1}^{C}\frac{1}{N_c}\sum_{k=1}^{N_c}\phi(x_{ck})\phi(x_{ck})^T \tag{5}$$

$$S_b = \frac{1}{C}\sum_{c=1}^{C}(\mu_c - \mu)(\mu_c - \mu)^T \tag{6}$$

GDA finds the eigenvalues $\lambda \geq 0$ and eigenvectors $v \in F \setminus \{0\}$ satisfying

$$\lambda S_w v = S_b v , \tag{7}$$

where all solutions $\mathbf{v}$ lie in the span of $\phi(x_{11})$, ..., $\phi(x_{ck})$, ... and there exist coefficients $\alpha_{ck}$ such that

$$v = \sum_{c=1}^{C} \sum_{k=1}^{N_c} \alpha_{ck} \phi(x_{ck}) \tag{8}$$

Using kernel techniques, the dot product of a sample $i$ from class $p$ and the other sample $j$ from class $q$ in the feature space, denoted as $(k_{ij})_{pq}$, can be calculated by a kernel function as below:

$$(k_{ij})_{pq} = \phi(x_{pi}) \cdot \phi(x_{qj}) = k(x_{pi}, x_{qj}) \tag{9}$$

Let $\mathbf{K}$ be a $M \times M$ matrix defined on the class elements by ($(\mathbf{K}_{pq})_{\substack{p=1,...C \\ q=1,...C}}$), where $\mathbf{K}_{pq}$ is a matrix composed of dot products between vectors from class $p$ and $q$ in feature space:

$$\mathbf{K}_{pq} = (k_{ij})_{\substack{i=1,...,N_p \\ j=1,...,N_q}} \tag{10}$$

We also define a $M \times M$ block diagonal matrix:

$$\mathbf{U} = (\mathbf{U}_c)_{c=1,...,C} \tag{11}$$

where $\mathbf{U}_c$ is $N_c \times N_c$ a matrix with terms all equal to $\frac{1}{N_c}$.

By substituting (5), (6) and (8) into (7) and taking inner-product with vector $\phi(x_{ij})$ on both sides, the solution of (8) can be achieved by solving:

$$\lambda \mathbf{KK}\boldsymbol{\alpha} = \mathbf{KUK}\boldsymbol{\alpha} \tag{12}$$

where $\boldsymbol{\alpha}$ denotes a column vector with entries $\alpha_{ck}, c = 1,...C, k = 1,..., N_c$. The solution of $\boldsymbol{\alpha}$ in equation (13) is equivalent to find the eigenvectors of the matrix $(\mathbf{KK})^{-1}\mathbf{KUK}$. However, similar to the small sample size, the matrix $\mathbf{K}$ might not be reversible. GDA find the eigenvector $\boldsymbol{\alpha}$ by first diagonalising matrix $\mathbf{K}$ (see [14] for more details). Once the first $L$ significant eigenvectors are found, a projection matrix can be constructed as:

$$\mathbf{W} = [\boldsymbol{\alpha}_1 \ \boldsymbol{\alpha}_2 \ ... \boldsymbol{\alpha}_L] \tag{13}$$

The projection of $x$ in the $L$-dimensional GDA space is given by:

$$\mathbf{y} = \mathbf{k}_{\mathbf{x}} \mathbf{W} \tag{14}$$

where

$$\mathbf{k}_{\mathbf{x}} = [k(x, x_{11}) ... k(x, x_{ck}) ... k(x, x_{CN_C})] \tag{15}$$

As suggested in [19], normalized correlation distance measure and the nearest neighbor classifier is used thereafter for the GDA based face recognition system.

## 4 Kernel Functions and Parameters Tuning

While GDA differs with other KFDA methods in solving the eigen decomposition problem in discriminant analysis, different GDA implementations might also vary in the kernel functions to be applied. Among them, polynomial function $k(x,y)=(x\cdot y)^d$ and RBF function $k(x,y)=e^{-\|x-y\|^2/r}$ are the most widely used. As seen from the equations, degree $d$ and RBF parameter $r$ need to be decided for polynomial function and RBF function, respectively. To apply GDA for face recognition, the dimension $L$ of learned GDA subspace has to be decided as well. Given certain Gabor features, i.e. DGF and OGF, a GDA based face recognition system need to tune subspace dimension $L$ and kernel parameter, i.e. degree $d$ or RBF parameter $r$ for the best performance. In this paper, we find the optimal kernel parameter and subspace dimension using the following process:

1. Give an initial guess on the kernel parameter, e.g. degree $d_{ini}$ or RBF parameter $r_{ini}$;
2. increase the value of subspace dimension with a small step, test the performance of the system and find the optimal dimension: $L_{opt}$;
3. set the subspace dimension as $L_{opt}$, vary the value of kernel parameter with a reasonable step, test the performance of the system and find the optimal degree $d_{opt}$ or RBF parameter $r_{opt}$.

In the following section, we will perform the process to find the optimal space dimension and kernel parameters for both the DGF and OGF, and test their effects on performance of the GDA based face recognition system.

## 5 Experimental Results

### 5.1 The Database

The FERET database is used to evaluate the performance of the proposed method for face recognition. The database consists of 14051 eight-bit grayscale images of human heads with views ranging from frontal to left and right profiles. 600 frontal face images corresponding to 200 subjects are extracted from the database for the experiments - each subject has three images of size 256×384 with 256 gray levels. The images were captured at different photo sessions so that they display different illumination and facial expressions. The following procedures were applied to normalize the face images prior to the experiments:

- The centers of the eyes of each image are manually marked,
- Each image is rotated and scaled to align the centers of the eyes,

- Each face image is cropped to the size of 64×64 to extract facial region,
- Each cropped face image is normalized to zero mean and unit variance.

Of the 600 face images, two images of each subject, totally 400 face images, will be randomly selected for training. The remaining 200 images, one image per subject, will be used for testing.

### 5.2 The Results

Following the process described in section 4, we will first test the effects of RBF parameter $r$ and subspace dimension $L$ on recognition accuracy of the GDA based system, when different Gabor features are used. While 200 OGF are selected using the boosting algorithm, the dimension of the DGF is set as 10,240 in our experiments, with downsample rate set as 16. As a result, the maximum subspace dimensions of GDA (with RBF kernel) for DGF and OGF are 70 and 199, respectively. Once the value of $r$ is increased by a pre-set step, the GDA subspace will be retrained using the training set and tested using the 200 test images. Fig. 2a gives the performance of GDA with RBF kernel (we initially set $r = 2\times10^3$) when different Gabor features are used. It can be observed that OGF based GDA achieves the best result with $L_{opt}$ =40, while DGF based GDA achieves the highest accuracy with $L_{opt}$ =180. Fig. 2b shows recognition rate as a function of the value of RBF kernel parameter ( $r$ ) when subspace dimension is fixed as $L_{opt}$ , the optimal RBF parameter $r_{opt}$ is found to be $8\times10^4$ and $12\times10^3$ for DGF and OGF, respectively. The recognition rate of GDA with optimal kernel parameters and subspace dimensions are 98% for OGF, and 97% for DGF. Even when significantly fewer features are used, OGF based GDA still achieves a higher recognition rate than DGF based GDA. The inferiority of DGF could be caused by loss of useful information during the downsampling process. One can also observe from the figure that, when OGF is used, the performance of GDA with RBF kernel is much more stable against the variation of kernel parameter $r$ .

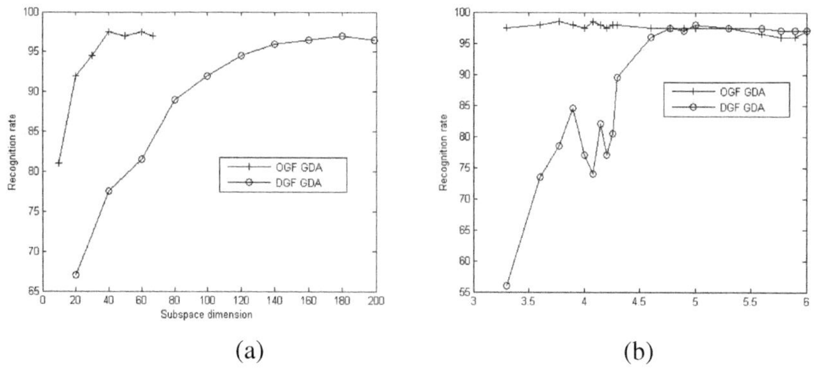


**Fig. 2.** Performance of GDA with RBF kernel using different Gabor features. (a) recognition rate as function of subspace dimension; (b) recognition rate as function of the logarithm of $r$ .

While Fig. 3a shows the performance of GDA with different polynomial kernels for DGF, Fig. 3b gives the result of OGF based GDA with different polynomial kernels. Both figures suggest that the polynomial kernel with degree 2 ( $d_{opt}$ =2) achieves the best results. While 91% accuracy is achieved for DGF based GDA with $L_{opt}$ =140, 97% is achieved for OGF based GDA $L_{opt}$ =60. Note that we test polynomial kernels with degree 2, 3 and 4 only in this paper, as polynomial kernels with higher degrees are not widely used. However, the parameter tuning process could be easily applied to test the performance of polynomial kernel based GDA with higher degrees.

The robustness of OGF against variation of kernel functions can also be proved by comparing the results obtained using polynomial kernels with that of RBF kernels. While the accuracy of DGF based GDA with polynomial kernel ( $d_{opt}$ =2, $L_{opt}$ =140) is 6% lower than that of DGF based GDA with RBF kernel ( $r_{opt} = 8 \times 10^4$ , $L_{opt}$ =180), the difference has been reduced to only 1% when OGF based GDA is concerned.

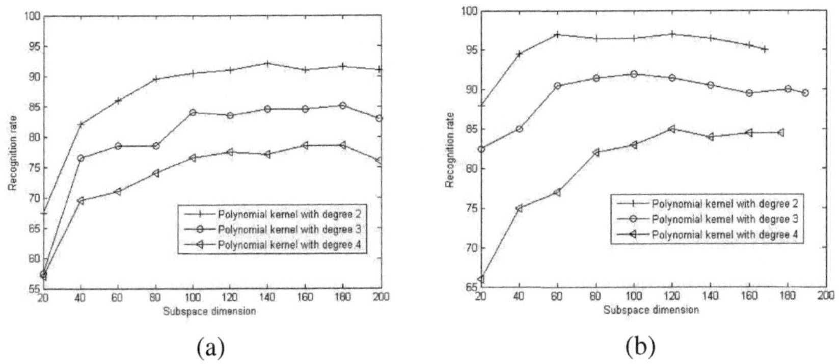


**Fig. 3.** Performance of GDA with polynomial kernel using (a) DGF; (b) OGF

**Table 1.** Comparative results with other approaches

| *Method* | *Recognition Accuracy* |
|---|---|
| DGF PCA | 80.0% |
| DGF LDA | 92.0% |
| DGF KPCA | 80.0% |
| DGF GDA | 97.0% |
| OGF PCA | 93.5% |
| OGF LDA | 77.0% |
| OGF KPCA | 93.5% |
| **OGF GDA** | **98.0%** |

We have also applied other subspace methods such as PCA, LDA and KPCA to both DGF and OGF, for evaluation. As summarized in Table 1, the results suggest that OGF GDA achieves significantly better accuracy than other approaches and when OGF is used, PCA, KPCA and GDA achieve better accuracy. However, the performance of LDA drops from 92% to as low as 77%, which suggests that when the input features are discriminate enough, LDA may not necessarily generate a more discriminative space. As a kernel version of LDA, GDA is obviously more robust. All of the results were obtained by optimizing the parameters for the best performance, as described in the previous section.

## 5 Conclusions

We have presented in this paper an experiment based approach for tuning kernel parameters. The approach has been successfully applied to optimize a Gabor feature and GDA based face recognition system. Different kernel functions, e.g. RBF function and polynomial function have been tested and effects of variant kernel parameters are demonstrated. Two different Gabor features, i.e. DGF and OGF are tested and the results show that OGF based GDA are much more robust against the variations of kernel functions and parameters. By eliminating redundant information and keeping important features, OGF based GDA shows advantages on both efficiency and accuracy over DGF based GDA. With the tuned parameters, OGF based GDA has also been shown to perform significantly better than PCA, LDA and KPCA when the FERET database is used for testing.

**Acknowledgments.** Research funded by SZU R/D Fund 200746.

## References

1. Turk, M., Pentland, A.: Eigenfaces for Recognition. Journal of Cognitive Neuroscience 3 (1991) 71-86
2. Belhumeur, P.N., Hespanha, J.P., Kriegman, D.J.: Eigenfaces vs. Fisherfaces: Recognition Using Class Specific Linear Projection. IEEE Transactions on Pattern Analysis and Machine Intelligence 19 (1997) 711-720
3. Samaria, F., Young, S.: Hmm-Based Architecture for Face Identification. Image and Vision Computing 12 (1994) 537-543
4. Er, M.J., Wu, S.Q., Lu, J.W., Toh, H.L.: Face Recognition With Radial Basis Function (RBF) Neural Networks. IEEE Transactions on Neural Networks 13 (2002) 697-710
5. Adini, Y., Moses, Y., Ullman, S.: Face Recognition: The Problem of Compensating for Changes in Illumination Direction. IEEE Transactions on Pattern Analysis and Machine Intelligence 19 (1997) 721-732
6. Gupta, H., Agrawal, A.K.: An Experimental Evaluation of Linear and Kernel-Based Methods for Face Recognition. Proceedings of the Sixth IEEE Workshop on Applications of Computer Vision, (WACV 2002) (2002) 13-18
7. Baudat, G., Anouar, F.E.: Generalized Discriminant Analysis Using a Kernel Approach. Neural Computation 12 (2000) 2385-2404
8. Kim, K.I., Jung, K., Kim, H.J.: Face Recognition Using Kernel Principal Component Analysis. IEEE Signal Processing Letters 9 (2002) 40-42
9. Liu, Q.S., Huang, R., Lu, H.Q., Ma, S.D.: Kernel-Based Nonlinear Discriminant Analysis for Face Recognition. Journal of Computer Science and Technology 18 (2003) 788-795

10. Shen, L., Bai, L.: Face Recognition Based on Gabor Reatures Using Kernel Methods. Proc. of the 6th IEEE Conference on Face and Gesture Recognition, Korea (2004) 170-175
11. Osuna, E., Freund, R., Girosit, F.: Training Support Vector Machines: An Application to Face Detection. Proceedings of IEEE Computer Society Conference on Computer Vision and Pattern Recognition (1997) 130-136
12. Guo, G.D., Li, S.Z., Chan, K.L.: Support Vector Machines for Face Recognition. Image and Vision Computing 19 (2001) 631-638
13. Moghaddam, B., Yang, M.: Gender Classification with Support Vector Machines. Proceedings. Fourth IEEE International Conference on Automatic Face and Gesture Recognition (2000) 306-311
14. Yang, M.: Kernel Eigenfaces vs. Kernel Fisherfaces: Face Recognition Using Kernel Methods. Proc. of the Fifth IEEE International Conference on Automatic Face and Gesture Recognition, Washington, D.C. (2002) 205-211
15. Liu, Q.S., Lu, H.Q., Ma, S.D.: Improving Kernel Fisher Discriminant Analysis for Face Recognition. IEEE Transactions on Circuits and Systems for Video Technology 14 (2004) 42-49
16. Xu, Y., Yang, J.Y., Lu, J.F., Yu, D.J.: An Efficient Renovation on Kernel Fisher Discriminant Analysis and Face Recognition Experiments. Pattern Recognition 37 (2004) 2091-2094
17. Yang, J., Frangi, A.F., Yang, J.Y.: A New Kernel Fisher Discriminant Algorithm With Application to Face Recognition. Neurocomputing 56 (2004) 415-421
18. Scholkopf, B., Smola, A., Muller, K.R.: Nonlinear Component Analysis as A Kernel Eigenvalue Problem. Neural Computation 10 (1998) 1299-1319
19. Shen, L., Bai, L., Fairhurst, M.: Gabor Wavelets and General Discriminant Analysis for Face Identification and Verification. Image and Vision Computing 25 (2007) 553-563
20. Shen, L., Bai, L.: MutualBoost Learning for Selecting Gabor Features for Face Recognition. Pattern Recognition Letters 27 (2006) 1758-1767
21. Shen, L., Bai, L.: A Review on Gabor Wavelets for Face Recognition. Pattern Analysis and Applications 9 (2006) 273-292

# Two Multi-class Lagrangian Support Vector Machine Algorithms

Hua Duan[1,2], Quanchang Liu[2], Guoping He[2], and Qingtian Zeng[2]

[1] Department of Mathematics, Shanghai Jiaotong University, Shanghai 200240, P.R. China
[2] College of Information Science and Engineering, Shandong University of Science and Technology, Qingdao 266510, P.R. China

**Abstract.** Support vector machines (SVMs) were designed for two-class classification problems, and multi-class classification problems have been solved by combining independently produced two-class decision functions. In this paper, we propose two multi-class Lagrangian Support Vector Machine(LSVM) algorithms using the quick and simple properties of LSVM. The experimental results in the linear and nonlinear cases indicate that the CPU running time of these two algorithms is shorter than that of the standard support vector machines, and their training correctness and testing correctness are almost identical.

## 1 Introduction

Support vector machines(SVMs) proposed by [1][2] were designed for two-class classification problems. However, the number of applications that require multi-class classification problems are immense. A few examples for such applications are text and speech categorization, natural language processing tasks such as part-of-speech tagging, gesture and object recognition in machine vision[10]. An effective extension from two-class to multi-class classification problems has different types that can be divided into two kinds. One is by constructing and combining several two-class classifiers while the other's by directly considering all data in one optimization formulation [1][8][9][11]. Methods for solving multi-class classification problems using two-class SVMs include one-vs-one[1], one-vs-all[1], error-correcting codes[7][10][13], directed acyclic graph[12], and pairwise coupling[6]. For these methods above, the resulting set of two-class decision functions must be combined in some way after the two-class classification problems have been solved[4]. There is variables proportional to the number of classes in the optimization formulation to solve multi-class SVM problems in one step. Hence multi-class SVM problems are computationally more expensive than two-class SVM problems with the same number of data. An interesting comparison of multi-class methods is presented in [5].

Lagrangian support vector machine (LSVM) proposed by Mangasarian and Musicant is a quick and simple classification method [3] which is trained by solving an iteration scheme of a simple linear convergence. In this paper we

D.-S. Huang, L. Heutte, and M. Loog (Eds.): ICIC 2007, LNAI 4682, pp. 891–899, 2007.

discuss an extension of LSVM to the multi-class case. We only focus on the most two popular methods that are one-vs-all and one-vs-one.

This paper is organized as follows: Section 2 presents LSVM, Section 3 gives one-vs-all multi-class LSVM, section 4 gives one-vs-one multi-class LSVM, section 5 gives experiments, and section 6 concludes the whole paper and gives the discussions.

## 2 Lagrangian Support Vector Machines

We first give a description of the two-class LSVM. Let $T = \{(x_i, y_i)|x_i \in R^n, i = 1, \cdots, m\}$ be the training set of a classification problem, where $x_i$ is the sample point of an $n$-dimensional space, represented by $A^T_{m\times n} = (x_1, \ldots, x_m)$, and $y_i \in \{\pm 1\}$ be the labels of the positive and negative class as to $x_i$ where $i = 1, \cdots, m$, represented by a diagonal matrix $D_{m\times m} = diag(y_1, \ldots, y_m)$. The LSVM with a linear **kernel** is given by the following quadratic program:

$$\begin{array}{l} \min \frac{1}{2}(\|w\|^2 + b^2) + \frac{C}{2}\xi^T\xi \\ s.t.\ \ y_i((w \cdot x_i) + b) + \xi_i \geq 1 \end{array} \tag{1}$$

where $C > 0$ is the penalty parameter. And, its Lagrangian function is:

$$L = \frac{1}{2}(\|w\|^2 + b^2) + \frac{C}{2}\xi^T\xi - \sum_{i=1}^{m} \alpha_i(y_i((w \cdot x_i) + b) + \xi_i - 1)$$

where $\alpha_i \geq 0$ is the Lagrangian multiplier. After derivation, $w = A^T D\alpha$, $b = e^T D\alpha$, and $\xi = \frac{\alpha}{C}$, where $e$ is a vector of ones of the appropriate dimension. The linear classifier is:

$$f(x) = sgn(g(x)) = sgn(\alpha^T DAx + b)$$

The dual problem is:

$$\min_{0\leq\alpha\in R^m} \frac{1}{2}\alpha^T Q\alpha - e^T\alpha \tag{2}$$

where $Q = \frac{I}{C} + HH^T$, $H = D[A \ \ -e]$.

The optimization KKT condition of its dual problem is $0 \leq \alpha \perp Q\alpha - e \geq 0$. By using the identity between any two real numbers (or two vectors) $a$ and $b$:

$$0 \leq a \perp b \geq 0 \Longleftrightarrow a = (a - \lambda b)_+, \quad \lambda > 0$$

where $(x)_+$ denotes the vector in $R^n$ in which all of its negative components are set to zero.

The iteration formula given by LSVM algorithm is

$$\alpha^{i+1} = Q^{-1}(e + ((Q\alpha^i - e) - \lambda\alpha^i)_+), i = 0, 1, \ldots, \lambda > 0. \tag{3}$$

While $0 < \lambda < \frac{2}{C}$, the algorithm is the global linear convergence from any starting point [3]. The inversion of $m$ matrix $Q$ changes to the inversion of

$n+1(n \ll m)$ matrix by using SMW identity. This leads to process large data sets feasibly, and the computation time is reduced.

The SMW identity is:

$$\left(\frac{I}{C}+HH^T\right)^{-1}=C\left(I-H(\frac{I}{C}+H^TH)^{-1}H^T\right)$$

where $C>0$ and $H$ is an $m \times n$ matrix. SMW identity was also used in [17], [18], and [19] to reduce computation time of algorithm.

To obtain LSVM nonlinear classifier, we use nonlinear **kernel**. A typical kernel is the Gaussian Radial Basis Kernel $K(x,y)=\exp(-\|x-y\|^2/2\sigma^2)$, where exp is the base of natural logarithms. The only price paid for this nonlinear kernel is that problems with large datasets cannot be handled using the SMW identity. Nevertheless LSVM may be a useful tool for classification with nonlinear kernels because of its extreme simplicity. The nonlinear classifier is:

$$f(x)=sgn(g(x))=sgn(\alpha^T DK(A,x)+b)$$

where $\alpha$ is the solution of the dual problem with $Q$ re-defined for a nonlinear kernel as follows:

$$G=[A \ \ -e], \ \ Q=\frac{I}{C}+DK(G,G^T)D$$

The iterative scheme and convergence of linear case remain valid, with $Q$ re-defined as above. Nonlinear classifier cannot handle very large problem because SMW identity can not be applied for the inversion of $Q$.

## 3 One-vs-All Multi-class Lagrangian Support Vector Machines

For multi-class classification problems, we consider a given training set $T=\{(x_1,y_1),\cdots,(x_m,y_m)\}$, where $x_i \in R^n$, $y_i \in \{1,\cdots,k\}, i=1,\cdots,m$, and $k$ is the number of classes. The multi-class classification problem is to construct a decision function $f(x)$, which classifies a new sample point $x$.

The earliest used implementation for multi-class classification SVM maybe the one-vs-all method[14][5]. It constructs $k$ two-class SVM models. First, several notations are given for convenience. $A_l \in R^{m_l \times n}$ denotes $m_l$ sample points in class $l$, $l \in \{1,\cdots,k\}$, and $A^T=\left[A_1^T \cdots A_k^T\right]$. To extend two-class classification to $k$-class, we need separate class $l$ from the rest $k-1$ classes as follow:

$$A_{+1}=A_l, \quad A_{-1}^T=\left[A_1^T \cdots A_{l-1}^T \ A_{l+1}^T \cdots A_k^T\right] \quad l \in \{1,\cdots,k\} \tag{4}$$

here, the $m \times m$ label diagonal matrix $D$ is :

$$\begin{array}{l} D_{ii}=\ \ 1 \ \ for \ \ x_i^T \in A_l \\ D_{ii}=-1 \ \ for \ \ x_i^T \notin A_l \end{array} \quad l \in \{1,\cdots,k\} \tag{5}$$

With $A$ and $D$ defined as above, $k$ classification problems are solved by iteration formula (3). Then $k$ linear decision functions:

$$f^l(x) = sgn(g^l(x)) = sgn({\alpha^l}^T DAx + b^l) \qquad l = 1, \cdots, k \tag{6}$$

A new input point $x \in R^n$ is assigned to class $r$, where $r$ is the superscript of the maximum of $g^1(x), \ldots, g^k(x)$, that is:

$$g^r(x) = \max_{l=1,\cdots,k} g^l(x) \tag{7}$$

Based on the above analysis one-vs-all linear multi-class LSVM Algorithm be presented.

**Algorithm 1: (One-vs-All linear multi-class LSVM)**

Step 1: Let training set $T = \{(x_1, y_1), \cdots, (x_m, y_m)\}$, where $x_i \in R^n$, $y_i \in \{1, \cdots, k\}, i = 1, \cdots, m$, $k$ is the number of classes.

Step 2: For $l = 1, \cdots, k$, the class $l$ is regarded as a positive class and the rest $k-1$ classes are negative class. The decision functions presented in (6) are solved using LSVM iteration formula (3).

Step 3: To judge a new input point $x \in R^n$ belongs to class $r$ or not according to (7).

We extend the linear results to the nonlinear LSVM. The matrix $Q$ is different from that of linear case. In the computation, the $m \times m$ kernel matrix $K(G, G^T)$ is replaced by the rectangular kernel $K(G, \overline{G}^T)$, where $\overline{G} \in R^{\overline{m} \times (n+1)}$ is a subset chosen randomly from $G$(Typically $\overline{m}$ is 1% to 10% of $m$)[16]. This leads to reduce computation time. As in the linear case, we extend two-class classification to $k$-class classification. Obtaining $k$ nonlinear decision functions:

$$f^l(x) = sgn(g^l(x)) = sgn({\alpha^l}^T DK(A, x) + b^l), \quad l = 1, \cdots, k \tag{8}$$

A new input point $x \in R^n$ is assigned to class $r$, where $r$ is the superscript of the maximum of $g^1(x), \ldots, g^k(x)$, presented in equation(7).

The one-vs-all nonlinear multi-class LSVM Algorithm is presented as follows.

**Algorithm 2: (One-vs-All nonlinear multi-class LSVM)**

Step 1: Let training set $T = \{(x_1, y_1), \cdots, (x_m, y_m)\}$, where $x_i \in R^n$, $y_i \in \{1, \cdots, k\}, i = 1, \cdots, m$, $k$ is the number of classes.

Step 2: For $l = 1, \cdots, k$, the class $l$ is regarded as a positive class and the rest $k-1$ classes are negative class. The decision functions presented in (8) are solved using LSVM iteration formula (3).

Step 3: To judge a new input point $x \in R^n$ belongs to class $r$ or not according to (7).

## 4 One-vs-One Lagrangian Support Vector Machines

One-vs-one method was proposed in [15], and the first use of this method in SVM was in [6][20]. The method constructs $k(k-1)/2$ decision functions where each one is trained on data from two classes. For the training data from the $i$th and $j$th classes, i.e. $(i, j) \in \{(i, j)|i \leq j, i, j = 1, \ldots, k\}$, which form a training set $T_{i-j} = \{(x_l, y_l)|y_l = i \ or \ j, l = 1, \ldots, m\}$. In this case, $A$ and $Q$ defined in section 2 are necessary to be redefined.

$$A^{ij} = \begin{bmatrix} A^i \\ A^j \end{bmatrix} \begin{matrix} D^{ij}_{ll} = 1 \ for \ (x_l, y_l) \in T_{i-j} \ and \ y_l = i \\ D^{ij}_{ll} = -1 \ for \ (x_l, y_l) \in T_{i-j} \ and \ y_l = j \end{matrix} \quad i, j = 1, \ldots, k \tag{9}$$

For the linear case:$H^{ij} = D^{ij}[A^{ij} \ -e], Q^{ij} = \frac{I}{C} + H^{ij}{H^{ij}}^T$. The $i-j$ linear decision function is obtained using the iteration formula (3) :

$$f^{ij}(x) = sgn(g^{ij}(x)) = sgn({\alpha^{ij}}^T D^{ij} A^{ij} x + b^{ij}) \quad i, j = 1, \ldots, k \tag{10}$$

For nonlinear case:$G^{ij} = [A^{ij} \ -e]$, $Q^{ij} = \frac{I}{C} + D^{ij} K(G^{ij}, {G^{ij}}^T) D^{ij}$. The $i-j$ nonlinear decision function is obtained using the iteration formula (3):

$$f^{ij}(x) = sgn(g^{ij}(x)) = sgn({\alpha^{ij}}^T D^{ij} K(A^{ij}, x) + b^{ij}) \quad i, j = 1, \ldots, k \tag{11}$$

After constructing all the $k(k-1)/2$ decision functions, we need to judge which class a new point $x$ belongs to. We use the following voting strategy[20]: if $f^{ij}(x)$ says $x \in R^n$ is in the class $i$ , then the vote for the class $i$ is added by one. Otherwise, the class $j$ is increased by one. And then $x$ is assigned to the class with the largest vote.

Based on the above analysis one-vs-one linear and nonlinear multi-class LSVM Algorithm be presented.

**Algorithm 3: (One-vs-One linear and nonlinear multi-class LSVM)**

Step 1: Let training set $T = \{(x_1, y_1), \cdots, (x_m, y_m)\}$, where $x_i \in R^n$, $y_i \in \{1, \cdots, k\}, i = 1, \cdots, m$, $k$ is the number of classes.

Step 2: For $\forall i, j \in \{1, \cdots, k\}$, the training set is $T_{i-j} = \{(x_l, y_l)|y_l = i \ or \ j, l = 1, \ldots, m\}$. The class $i$ is regarded as a positive class and the class $j$ is negative class. The decision functions presented in (10)(for nonlinear case is (11)) are solved using LSVM iteration formula (3).

Step 3: If $f^{ij}(x)$ says a new input point $x \in R^n$ is in the class $i$ , then the vote for the class $i$ is added by one. Otherwise, the class $j$ is increased by one.

Step 4: A new input point $x \in R^n$ assigned to class with the largest vote.

## 5 Experiment

In order to evaluate the performances of the algorithms presented in this paper, the experiments are given based on five groups data sets. The experiments are implemented by Mathlab 7.0, and they run on PC environment. The main configurations of the PC are: (1) CPU: Pentium IV 2.0G, (2) Memory: 256M, and (3) OS: Windows XP.

In the following discussions, in order to save space, we denote

- **OALSVM:** One-vs-all classifier using Lagrangian support vector machines for every two-class classification problems.
- **OOLSVM:** One-vs-one classifier using Lagrangian support vector machines for every two-class classification problems.
- **OASVM:** One-vs-all classifier using a standard support vector machines quadratic programming for every two-class classification problems.
- **OOSVM:** One-vs-one classifier using a standard support vector machines quadratic programming for every two-class classification problems.

The parameters $C$ and $\sigma$ in each of those methods are chosen by using a tuning set extracted from the training set. First, we compare the performances of OALSVM, OOLSVM, OASVM and OOSVM in the linear case, and their experimental results are shown in Table 1. According to the results shown in Table 1, we can see that the CPU running time of OALSVM and OOLSVM is much shorter than OASVM and OOSVM, respectively, although their training correctness and testing correctness are almost identical. It indicates that OALSVM and OOLSVM can reduce the running time of CPU efficiently, so to reduce the CPU running time is one of the most advantages of the two algorithms proposed in this paper.

In the non-linear case, the kernel function is Gaussian Radial Basis Kernel $K(x,y) = \exp(-\|x-y\|^2/2\sigma^2)$. The experimental results of multi-class in the

**Table 1.** The experimental results of multi-class LSVM and SVM in the linear case

| Dataset | Methods | $C$ | Training correctness | Testing correctness | CPU Sec. |
|---|---|---|---|---|---|
| Iris | OALSVM | 10 | 95.00% | 86.12% | 0.2598 |
| train size:100*4 | OOLSVM | 10 | 97.00% | 72.00% | 0.1617 |
| test size:50*4 | OASVM | 10 | 96.00% | 86.00% | 3.1562 |
| classes: 3 | OOSVM | 10 | 92.00% | 70.00% | 1.5670 |
| Wine | OALSVM | 10 | 100% | 86.21% | 0.1790 |
| train size:120*13 | OOLSVM | 10 | 100% | 86.21% | 1.2499 |
| test size:58*13 | OASVM | 100 | 100% | 88.48% | 4.9749 |
| classes: 3 | OOSVM | 100 | 86.67% | 82.14% | 2.1345 |
| Glass | OALSVM | 100 | 87.93% | 73.00% | 0.0129 |
| train size:114*9 | OOLSVM | 100 | 91.44% | 72.00% | 0.2391 |
| test size:100*9 | OASVM | 1000 | 84.54% | 72.12% | 9.9256 |
| classes: 7 | OOSVM | 1000 | 81.41% | 73.23% | 3.3008 |
| Vehicle | OALSVM | 100 | 82.96% | 76.75% | 0.2691 |
| train size:446*18 | OOLSVM | 100 | 85.87% | 76.03% | 0.2262 |
| test size:400*18 | OASVM | 100 | 81.17% | 80.75% | 17.9729 |
| classes: 4 | OOSVM | 100 | 80.25% | 72.03% | 3.9876 |
| Segment | OALSVM | 0.1 | 92.48% | 91.20% | 1.9240 |
| train size:1500*19 | OOLSVM | 0.1 | 96.33% | 96.17% | 0.6088 |
| test size:810*19 | OASVM | 100 | 81.24% | 78.91% | 23.3311 |
| classes: 7 | OOSVM | 100 | 77.32% | 73.46% | 19.9567 |

**Table 2.** The experimental results of multi-class LSVM and SVM in the nonlinear case

| Dataset | Methods | $(C,\sigma)$ | Training correctness | Testing correctness | CPU Sec. |
|---|---|---|---|---|---|
| Iris | OALSVM | (10,0.5) | 100% | 84.00% | 2.3962 |
| train size:100*4 | OOLSVM | (10,0.5) | 100% | 86.00% | 0.1153 |
| test size:50*4 | OASVM | (100,0.5) | 98.00% | 83.00% | 3.5363 |
| classes: 3 | OOSVM | (10,0.5) | 96.00% | 83.00% | 1.5995 |
| Wine | OALSVM | (10,0.5) | 100% | 85.86% | 2.9486 |
| train size:120*13 | OOLSVM | (100,0.5) | 100% | 88.28% | 0.3079 |
| test size:58*13 | OASVM | (100,0.1) | 100% | 89.66% | 5.2229 |
| classes: 3 | OOSVM | (100,0.1) | 96.50% | 84.76% | 2.3150 |
| Glass | OALSVM | (100,0.1) | 100% | 88.00% | 4.4421 |
| train size:114*9 | OOLSVM | (10,0.5) | 100% | 93.41% | 0.1474 |
| test size:100*9 | OASVM | (10,0.1) | 93.86% | 89.24% | 10.1690 |
| classes: 7 | OOSVM | (100,0.1) | 79.59% | 76.23% | 3.4126 |
| Vehicle | OALSVM | (100,0.5) | 100% | 75.75% | 18.4495 |
| train size:446*18 | OOLSVM | (10,0.5) | 100% | 75.75% | 0.4956 |
| test size:400*18 | OASVM | (100,0.5) | 100% | 75.25% | 17.1813 |
| classes: 4 | OOSVM | (50,0.5) | 84.68% | 73.75% | 4.2732 |
| Segment | OALSVM | (10,0.5) | 92.48% | 92.48% | 1.7855 |
| train size:1500*19 | OOLSVM | (10,0.5) | 100% | 87.65% | 10.5883 |
| test size:810*19 | OASVM | (10,0.5) | 85.75% | 73.24% | 20.1352 |
| classes: 7 | OOSVM | (100,0.5) | 80.13% | 70.03% | 16.4451 |

nonlinear case are shown in Table 2. According to the results shown in Table 2, the similar conclusions as in the linear case can also be obtained.

## 6 Conclusion

In this paper, we propose two simple and efficient classification algorithms for one-vs-all and one-vs-one multi-class LSVMs, respectively. It is required to solve $k$ iteration schemes in OALSVM, and $k(k-1)/2$ iteration schemes are required in OOLSVM, where $k$ is the number of classes. In contrast, OASVM and OOSVM require to solve the more costly quadratic program. Through the experiments, it indicates that the CPU running time of OALSVM and OOLSVM is much shorter than OASVM and OOSVM in the linear and nonlinear cases, respectively, and their training correctness and testing correctness are almost identical. It shows that OALSVM and OOLSVM proposed in this paper can reduce the running time of CPU efficiently.

We only pay our attention on the general multi-class classification of Lagrangian support vector machines. The future research work will be the incremental multi-class classification for large data sets.

**Acknowledgements.** This work is supported partially by national science foundation of China (10571109 and 60603090).

## References


1. Vapnik, V.: The Nature of Statistical Learning Theory, Springer-Verlag, New-York, (1995)
2. Vapnik, V.: Statistical Learning Theory. New York: Wiley, (1998)
3. Mangasarian, O.L., Musicant, D.R.: Lagrangian Support Vector Machines. Journal of Machine Learning Research, (2001) 167-177
4. Duan, K., Keerthi, S. S.: Which Is the Best Multiclass SVM Method? An Empirical Study. Proc. Multiple Classifier Systems, (2005) 278-285
5. Hsu, C.-W., Lin. C.-J.: A Comparison of Methods for Multi-class Support Vector Machines. IEEE Trans. on Neural Networks, (2002) 415-425
6. Kre, U. H.-G. et. al.: Pairwise Classification and Support Vector Machines. In B. Schŏlkopf, C. J. C. Burges, A. J. Smola (Eds.), Advances in kernel methods: Support Vector Learning. Cambridge, MA: MIT Press. (1999) 255-268
7. Dietterich, T. G., Bakiri, G.: Solving Multiclass Learning Problems via Error-correcting Output Codes. Journal of Artificial Intelligence Research, (1995) 263-286
8. Weston, J., Watkins, C.: Multi-class Support Vector Machines. In M.Verleysen, editor, Proceedings of ESANN 99, Brussels, D. Facto Press, (1999)
9. Bredensteiner, E.J., Bennett, K.P.: Multicategory Classification by Support Vector Machines. Computational Optimization and Applications, (1999) 53-79
10. Suykens, J.A.K., Vandewalle, J.: Multiclass LS-SVMs: Moderated Outputs and Coding-decoding Schemes. In Proceedings of IJCNN, Washington D.C., (1999)
11. Suykens, J.A.K., Vandewalle, J.: Multiclass Least Squares Support Vector Machines. In: Proc. International Joint Conference on Neural Networks (IJCNN 99), Washington DC, (1999)
12. Platt, J.C., Cristianini, N., Shawe-Taylor, J.: Large Margin DAGs for Multiclass Classification. In Advances in Neural Information Pressing Systems, MIT Press. (2000) 547-553
13. Kindermann, J., Leopold, E., Paass, G.: Multi-class Classification with Error Correcting Codes. In E.Leopold and M.Kirsten, editors, Treffen der GI-Fachgruppe 1.1.3, Maschinelles Lernen, GMD Report 114, (2000)
14. Bottou, L., Cortes, C., Denker, J., Drucker, H., et. al.: Comparison of Classifier Methods: a Case Study in Handwriting Digit Recognition. In International Conference on Pattern Recognition. IEEE Computer Society Press, (1994) 77-87
15. Knerr, S., Personnaz, L., Dreyfus, G.: Single-layer Learning Revisited: a Stepwise Procedure for Building and Training a Neural Network. In J. Fogelman, editor, Neurocomputing: Algorithms, Architectures and Applications. Springer-Verlag, (1990)
16. Lee, Y.-J., Mangasarian, O. L.: RSVM: Reduced Support Vector Machines. Technical Report 00-07, Data Mining Institute, Computer Sciences Department, University of Wisconsin, Madison, Wisconsin, July(2000)
17. Ferris, M.C., Munson, T.S.: Interior Point Methods for Massive Support Vector Machines. Technical Report 00-05, Computer Sciences Department, University of Wisconsin, Madison, May(2000)

18. Fung, G.,Mangasarian, O.L.: Proximal Support Vector Machine Classifiers. In F.Provost and R.Srikant, editors, Proceedings KDD-2001: Knowledge Discovery and Data Mining, New York, (2001) 77-86
19. Fung, G., Mangasarian, O. L.: Finite Newton Method for Lagrangian Support Vector Machine ClassiFication. Technical Report 02-01, Data Mining Institute, Computer Sciences Department, University of Wisconsin, Madison, Wisconsin, (2002)
20. Friedman, J. H.: Another Approach to Polychotomous Classification. Technical report, Department of Statistics, Stanford University, (1996)

# Research on On-Line Modeling of Fed-Batch Fermentation Process Based on ν-SVR

Yongjun Ma

College of Computer Science and Information Engineering,
Tianjin University of Science and Technology, Tianjin, China
yjma@ustc.edu

**Abstract.** The fermentation process is very complex and non-linear, many parameters are not easy to measure directly on line, soft sensor modeling is a good solution. This paper introduces $\nu$-support vector regression ($\nu$-SVR) for soft sensor modeling of fed-batch fermentation process. $\nu$-SVR is a novel type of learning machine. It can control the accuracy of fitness and prediction error by adjusting the parameter $\nu$. An on-line training algorithm is discussed in detail to reduce the training complexity of $\nu$-SVR. The experimental results show that $\nu$-SVR has low error rate and better generalization with appropriate $\nu$.

## 1 Introduction

The fermentation process is complex and non-linear, some key parameters are difficult to measure on line, such as biomass concentration, substrate concentration, production concentration. It is impractical to analysis the fermentation process by using analytic model. Artificial neural network (ANN) is used for modeling fermentation process, and it has shown better performance than analytic model method. However, it is a hard work to collect enough experimental data in fermentation process, even in off-line condition. Furthermore, ANN has its own defects, for example, the net parameters are not easy to tune, the structure is difficult to determine [1-2].

$\nu$-SVR is a novel type of learning machine, which based on statistical learning theory (SLT). It introduces a new parameter $\nu$ to control the fitness and predication accuracy. $\nu$-SVR has shown to provide a better generalization performance than traditional techniques, including neural networks [3].

In this paper $\nu$-SVR based modeling algorithm is proposed for fed-batch fermentation process, and an on-line training algorithm is discussed in detail to reduce the training complexity of $\nu$-SVR.

This paper is organized as follows. In section 2 we discuss the construction of $\nu$-SVR. Section 3 shows how to use $\nu$-SVR to construct soft sensor modeling of fermentation process. The on-line training algorithm based on $\nu$-SVR is proposed in this section. The obtained experimental results are illustrated in section 4. Finally, Section 5 summarizes the conclusions that can be drawn from the presented research.

D.-S. Huang, L. Heutte, and M. Loog (Eds.): ICIC 2007, LNAI 4682, pp. 900–908, 2007.

## 2 ν-SV Regression

ν-SVR seeks to estimate functions

$$f(x) = (w \cdot x) + b \tag{1}$$

$$\text{where } w,\ x \in R^N, b \in R \tag{2}$$

based on independent identically distributed data

$$(x_1, y_1), \ldots\ldots, (x_\ell, y_\ell) \in \chi \times R \tag{3}$$

Here, $\chi$ is the space in which the input patterns live.

To estimate functions (1) from empirical data (3), we can obtain a small risk by solve the following constrained optimization problem:

$$\tau(w, \xi^{(*)}, \varepsilon) = \frac{1}{2} \| w \|^2 + C \cdot (\upsilon\varepsilon + \frac{1}{\ell} \sum_{i=1}^{\ell} (\xi_i + \xi_i^*)) \tag{4}$$

$$((w \cdot X_i) + b) - y_i \le \varepsilon + \xi_i \tag{5}$$

$$y_i - ((w \cdot x_i) + b) \le \varepsilon + \xi_i^*,\ \xi_i^* \ge 0, \xi_i \ge 0 \tag{6}$$

where C is a constant determining the trade-off.

At each point xi, an error of ε is allowed. Everything above ε is captured in slack variables $\xi_i^{(*)}$, which are penalized in the objective function via a regularization constant C, chosen a priori. The size of ε is traded off against model complexity and slack variables via a constant $\nu > 0$.

Constructing Lagrangian

$$L_\nu(\boldsymbol{w}, \boldsymbol{\xi}, b, \rho, \boldsymbol{\alpha}, \boldsymbol{\beta}, \delta) = \frac{1}{2} \|\boldsymbol{w}\|^2 - \nu\rho + \frac{1}{n} \sum_{i=1}^{n} \xi_i \tag{7}$$

$$- \sum_{i=1}^{n} \alpha_i \{ y_i [(\boldsymbol{w} \cdot \boldsymbol{x}_i) + b] - \rho + \xi_i \} + \sum_{i=1}^{n} \beta_i \xi_i - \delta\rho$$

$$\text{where } \alpha_i, \beta_i, \delta \ge 0 \tag{8}$$

At the saddle point, *L* has a minimum, thus we can write

$$\boldsymbol{w} = \sum_{i=1}^{n} \alpha_i y_i \boldsymbol{x}_i \tag{9}$$

$$\alpha_i + \beta_i = \frac{1}{n} \tag{10}$$

$$\sum_{i=1}^{n} \alpha_i y_i = 0 \tag{11}$$

$$\sum_{i=1}^{n} \alpha_i - \delta = v \tag{12}$$

Considering Karush-Kuhn-Tucker (KKT) conditions and dual problem, the v-SVR regression estimate then takes the form

$$Q_v(\boldsymbol{\alpha}) = -\frac{1}{2}\sum_{i,j=1}^{n} \alpha_i \alpha_j y_i y_j k(\boldsymbol{x}_i, \boldsymbol{x}_j) \tag{13}$$

$$\text{Subject to } \quad 0 \le \alpha_i \le \frac{1}{n}, \sum_{i=1}^{n} \alpha_i y_i = 0, \sum_{i=1}^{n} \alpha_i \ge v \tag{14}$$

The decision function becomes

$$f(x) = \sum_{i=1}^{\ell} (\alpha_i * - \alpha_i) k(x_i, x) + b \tag{15}$$

where $v \geq 0$, $C > 0$, $\alpha_i^{(*)}$ is multiplier, $k(x, y)$ is kernel function. b (and ε) can be computed by taking into account that (6) and (7) become equalities with $\xi_i^{(*)} = 0$, for points with $0 < \alpha_i^{(*)} < C / l$, respectively, due to the KKT conditions.

From [3] we also know that v is an upper bound on the fraction of errors, so we can control the error by deciding v. We can use it to control the prediction accuracy during the fermentation process. This is the reason why we select v-SVR instead of SVR.

# 3 Soft Sensor Modeling of Fermentation Process Based on *ν*-SVR

## 3.1 The Construction of Model Based on *ν*-SVR

The fermentation process is complex and non-linear, many parameters are not easy to measure, such as biomass concentration, substrate concentration, production concentration. It is impractical to analysis the fermentation process using analytic model [5-6]. We introduce *ν*-SVR as the soft sensor model. We took the following function as the model description (see (7)).

Radical basis function (RBF) is chosen as kernel function:

$$K(\mathbf{x}, \mathbf{x}_i) = \exp\left(-\frac{|\mathbf{x} - \mathbf{x}_i|^2}{\sigma^2}\right) \tag{16}$$

It is critical to select the type of kernel function and the parameters such as $\nu$ and $C$ during the modeling process. Cross validation is used to determine the optimal parameters.

## 3.2 *ν*-SVRM : The Online Training Algorithm Based on *ν*-SVR

Cross validation is used to determine the parameters and the type of kernel function, but it can not be used to fine-tune the model online on line. So a new on-line fine-tune algorithm of model is proposed which is named $\nu$-SVRM ($\nu$-SVR for Modelling).

Firstly select $n$ input samples and build up a training set for the training of $\nu$-SVRM. The optimal parameters are selected as the model parameters after validation. Secondly add new sample and renew the set according to some rules. Finally fine-tune the parameters of models.

The detailed steps are as following:

Step1. Normalize the working set $w = \{(x_1, y_1), (x_2, y_2), \cdots (x_n, y_n)\}$

Step2. Train $\nu$-SVRM model $f(x)$ using cross validation method.

Step3. Use $f(x)$ to predict a new sample $(x_{n+1}, y_{n+1})$

Step4. If $\left|\dfrac{f(x_{n+1}) - y_{n+1}}{y_{n+1}}\right| > v$, add $(x_{n+1}, y_{n+1})$ to the working set.

Step5. Remove a non-SV sample to form a new working set

Step6. If there are still new samples, go to step2, else go to the end.

Partial Matlab code which use LibSVM as training algorithms

```
P=[P1;P2;P3;P5]; % Training set
T=[T1;T2;T3;T5]; % Testing set
p_test=P4;
T_test=T4
s = sprintf('-s %d -n %.4g -p %.7g -t 2 -c %d -g
%d',s,n,p,c,g);
model=svmtrain(T,P,s);% use vSVM
[predict_label,accuracy,decision_values]=svmpredict(T_t
est,p_test,model);
e=(decision_values-T_test).*(decision_values-T_test);
E=sum(e);% Compute error
```

# 4 Experiments

## 4.1 Experimental Conditions

In the experiments we take polylysine batch fermentation and feed the fermentor with 2.5L materials each time. There are many parameters which can influence the polylysine fermentation process. We choose some key factors as the input set of model, which are temperature, PH value, dissolved oxygen (DO), stirring speed, fermentation time and the biomass concentration last period. We take the biomass concentration as the model output [4]. The total batch is 5.

The experimental equipment is an intelligent fermentation process control system designed by ourselves. The software platform is PIV2.66GHz/1G Memory/ WindowsXP/ Matlab7.0/ VC++6.0.

The fermentation equipment is as following:

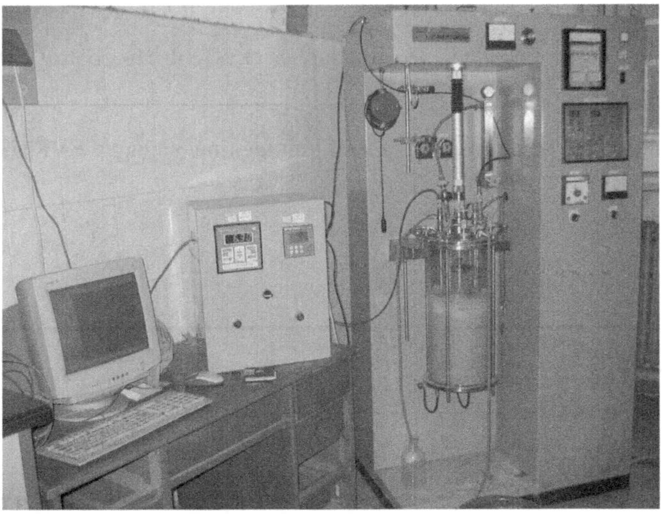

**Fig. 1.** Experimental equipment

## 4.2 Experimental Results

The following table are the experimental data.

**Table 1.** Partial experimental data

| Predict data | Input data | | |
|---|---|---|---|
| | 1th Column: time | 2th column: PH | 3th colum: DO |
| 0.0231 | 0.0000 | 0.2611 | 0.2956 |
| 0.0363 | 0.0143 | 0.2541 | 0.2493 |
| 0.0662 | 0.0286 | 0.2413 | 0.2134 |
| 0.1044 | 0.0429 | 0.2258 | 0.1971 |
| 0.1424 | 0.0571 | 0.2107 | 0.1945 |
| 0.1717 | 0.0714 | 0.1989 | 0.1929 |
| 0.1838 | 0.0857 | 0.1936 | 0.1919 |
| 0.1842 | 0.1000 | 0.1927 | 0.1909 |
| 0.1848 | 0.1286 | 0.1919 | 0.1898 |
| 0.1864 | 0.1429 | 0.1943 | 0.1912 |
| 0.1936 | 0.1714 | 0.2016 | 0.1950 |
| 0.2000 | 0.1857 | 0.2022 | 0.1945 |
| 0.2087 | 0.2000 | 0.2026 | 0.1933 |

In the experiments we select RBF kernel function. $\sigma$ is the width coefficient, little value can get the good fitness, but too little value will lead to poor generation. Penalty parameter $C$ can punish the error, the increasing of $C$ will decrease the fitness error and the prediction error. But when $C$ becomes too big, it will get into trouble of over fitness. $v$ is an upper bound on the fraction of errors, it can control the prediction accuracy by adjusting $v$.

Table 2 shows the comparison results of training time among $v$-SVRM, $v$-SVR and ε -SVR.

**Table 2.** Comparison of training time (c=250, σ=15)

| Model | $v$-SVR ($v$ = 0.10) | $v$-SVRM ($v$ = 0.10) | $v$-SVR ($v$ = 0.30) | $v$-SVRM ($v$ = 0.30) | ε -SVR |
|---|---|---|---|---|---|
| Training time (s) | 3.84 | 2.69 | 2.14 | 1.98 | 3.73 |

From the table above we can conclude that $v$-SVRM need shorter training time with the same value of parameters $v$.

The following table 2 is the experimental results of all 5 batches. It indicates that $v$-SVRM has fine-tuning ability. With the increase of experimental data, $v$-SVRM shows better prediction accuracy.

**Table 3.** On-line predictive error of biomass concentration (C=250 , σ =15)

| Batch data | RMSE: predictive error of each batch | | | | |
|---|---|---|---|---|---|
| | 1th batch | 2th batch | 3th batch | 4th batch | 5th batch |
| 2 batches | 0.00753 | 0.00816 | — | — | — |
| 3 batches | 0.00623 | 0.00511 | 0.00693 | — | — |
| 4 batches | 0.00508 | 0.00531 | 0.00542 | 0.00499 | — |
| 5 batches | 0.00512 | 0.00494 | 0.00501 | 0.00489 | 0.00457 |

The 5th batch experimental results are as following figures ($v$ = 0.10).

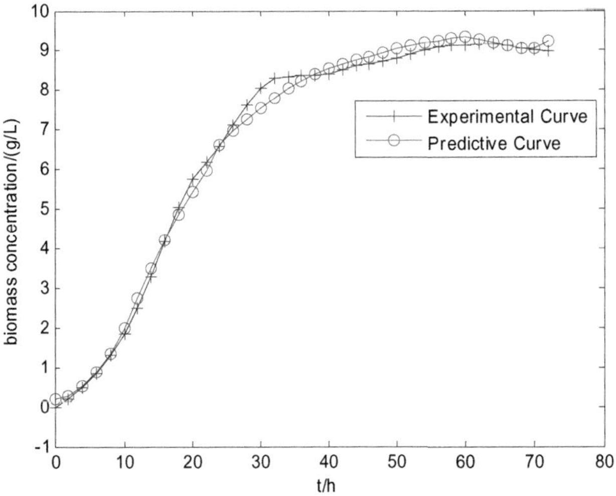


(a) *v*-SVRM ( RMSE=0.00457)

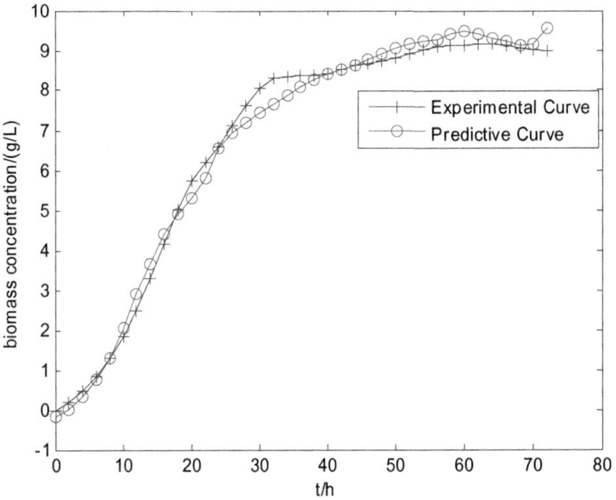


(b) *v*-SVR ( RMSE=0.00716)

**Fig. 2.** Comparison among v-SVRM, v-SVR, SVR and BP

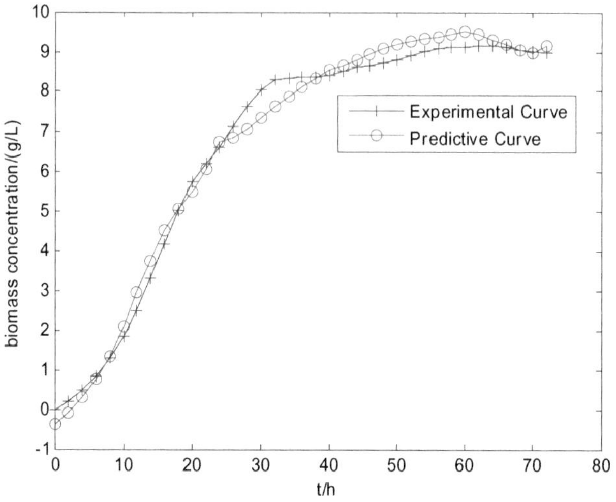


(c) ε -SVR ( RMSE=0.00608)

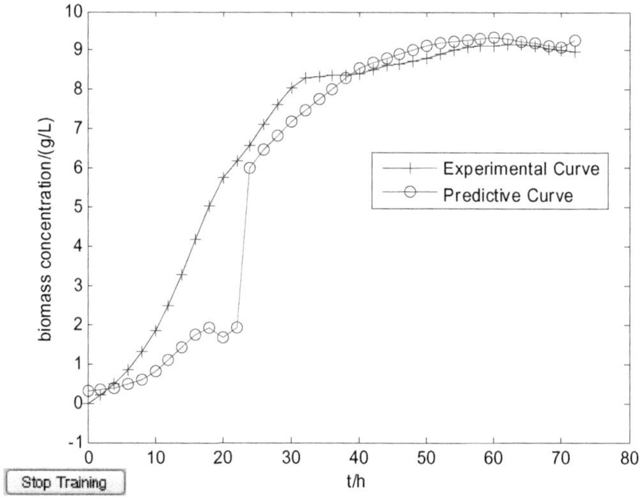


(d) BP net ( RMSE=0.0289)

**Fig. 2.** (*continued*)

Figure 1 (a), (b) and (c) show that *v*-SVRM, *v*-SVR, SVR has similar predictive accuracy.

The predicting results of BP net are not satisfying (RMSE=0.0289), the main reason is artificial neural network is based on traditional statistics, which need a large amount of training samples. Actually it is difficult to get enough samples in the fermentation process. SVR can get better performance in such a case.

## 5 Conclusions

In experiments $\nu$-SVR shows good performance for soft sensor modeling of fed-batch fermentation process, the on-line training algorithm $\nu$-SVRM is discussed which can reduce the training complexity of $\nu$-SVR. The experimental results show that v-SVR has low error rate and better generalization by adjusting the parameter $\nu$.

**Acknowledgement.** This research is sponsored by a grant of Tianjin Science&Technology Development Foundation of High School under contract 20061011, and partly sponsored by a grant of Tianjin Key Technologies R&D Program under contract 04310951R.

## References

1. Ma,Y.J., Kong,B.: A Study of Object Detection based on Fuzzy Support Vector Machine and Template Matching. *IEEE Proceedings of the 5th World Congress on Intelligent Control and Automation*, vol.5, Hangzhou, P.R. China, ( 2004), .4137-4140
2. Ma, Y.J., Fang, K., Fang, T.J. :A Study of Classification based on Support Vector Machine and Distance Classification for Texture Image (Chinese). *Journal of Image and Graphics,* Vol. 7(A), no.11, (2002),1151-1155
3. Scholkopf,B.,.Smolad, A.J.: New Support Vector Algorithms. neurocolt2 nc2-tr-1998-031. Technical report, GMD First and Australian National University, (1998)
4. Liu,Y.M., Meng,Z.P., Yu,H.W., et al: The Realization of Fermentation Process Status Pre-estimate Model Based on BP NN (In: Chinese). *Journal of Tianjin university of light industry,* vol.18, no. 3. (2003)35~38
5. Xiong,Z.H., Zhang,J.C., Shao,H.H.: GP-based Soft Sensor Modeling. *Journal of system simulation.* Vol. 17, no. 4, (2005) 793~800
6. Wang, J.L., Yu, T.: Research Progress in Soft Sensor Techniques for On-Line Biomass Estimation. Modern chemistry industry. Vol.25, no.6, ( 2005) 22~25

# Kernel Generalized Foley-Sammon Transform with Cluster-Weighted

Zhenzhou Chen

Computer School, South China Normal University,
Guangzhou 510631, China
chenzhenzhou@scnu.edu.cn

**Abstract.** KGFST (Kernel Generalized Foley-Sammon Transform) has been proved very successfully in the area of pattern recognition. By the kernel trick, one can calculate KGFST in input space instead of feature space to avoid high dimensional problems. But one has to face two problems. In many applications, when $n$ (the number of samples) is very large, it not realistic to store and calculate serval $n \times n$ metrics. Another problem is the complexity for the eigenvalue problem of $n \times n$ metrics is $O(n^3)$. So a new nonlinear feature extraction method CW-KGFST (KGFST with Cluster-weighted) based on KGFST and Clustering is proposed in this paper. Through Cluster-weighted, the number of samples can be reduced, the calculate speed can be higher and the accuracy can be preserved simultaneously. Lastly, our method is applied to digits and images recognition problems, and the experimental results show that the performance of present method is superior to the original method.

**Keywords:** Foley-Sammon Transform, Kernel, Cluster-weighted.

## 1 Introduction

Fisher discriminant based Foley-Sammon Transform (FST)[1] has great influence in the area of pattern recognition. Guo et al.[2] proposed a generalized Foley-Sammon transform (GFST) based on FST. GFST is a linear feature extraction method, but the linear discriminant is not always optimal. By kernel trick, a feature extraction method KGFST (Kernel Generalized Foley-Sammon Transform) is proposed[3].

By the kernel trick[4,5], one can calculate KGFST in input space instead of feature space to avoid high dimensional problems. But one has to face two problems. In many applications, when $n$ (the number of samples) is very large, it not realistic to store and calculate serval $n \times n$ metrics efficiently. Another problem is the complexity for the eigenvalue problem of $n \times n$ metrics is $O(n^3)$ although there exist many efficient off-the-shelve eigensolvers or Cholesky packages which could be used to optimize. So a new nonlinear feature extraction method CW-KGFST (KGFST with Cluster-weighted)based on KGFST and Clustering[6] is proposed in this paper.

The remainder of the paper is organized as follows: Section 2 gives a brief review of KGFST. Section 3 shows how to combine KGFST method and clustering

D.-S. Huang, L. Heutte, and M. Loog (Eds.): ICIC 2007, LNAI 4682, pp. 909–918, 2007.

and proves that the CW-KGFST also can get good performance as KGFST does. Section 4 provides some experiments of CW-KGFST and KGFST. Finally, section 5 gives a brief summary of the present method.

## 2 A Review of Kernel Generalized Foley-Sammon Transform

Let $Z = \{(x_1, y_1), ..., (x_n, y_n)\} \subseteq R^m \times \{\omega_1, \ldots, \omega_C\}$. The number of samples in each class $\omega_i$ is $n_i$. The Fisher's linear discriminant [1] in feature space $H$ is given as:

$$J(\mathbf{a}) = \frac{\mathbf{a}^T M \mathbf{a}}{\mathbf{a}^T N \mathbf{a}}, \tag{1}$$

where $M$ and $N$ are $n \times n$ matrices.

Let $\mathbf{a}_1$ be the vector which maximizes $J(\mathbf{a})$ and $\mathbf{a}_1^T K \mathbf{a}_1 = 1$, then $\mathbf{a}_1$ is the first vector of KGFST optimal set of discriminant vectors, the $i$th vector $(\mathbf{a}_i)$ of KGFST optimal discriminant set can be calculated by optimizing the following problem[3]:

$$\begin{cases} \max[J(\mathbf{a}_i)], \quad see \ (1) \\ s.t. \quad \mathbf{a}_i^T K \mathbf{a}_j = 0, \quad j = 1, \cdots, i-1 \ . \\ \mathbf{a}_i^T K \mathbf{a}_i = 1 \end{cases} \tag{2}$$

First let's rewrite the dicriminant criterion of KGFST:

$$J(\mathbf{a}_i) = \frac{\sum_{j=1}^{i-1} \mathbf{a}_j^T M \mathbf{a}_j + \mathbf{a}_i^T M \mathbf{a}_i / \mathbf{a}_i^T K \mathbf{a}_i}{\sum_{j=1}^{i-1} \mathbf{a}_j^T N \mathbf{a}_j + \mathbf{a}_i^T N \mathbf{a}_i / \mathbf{a}_i^T K \mathbf{a}_i} = \frac{\mathbf{a}_i^T \tilde{M}_i \mathbf{a}_i}{\mathbf{a}_i^T \tilde{N}_i \mathbf{a}_i},$$

where

$$\tilde{M}_i = (\sum_{j=1}^{i-1} \mathbf{a}_j^T M \mathbf{a}_j) K + M, (\tilde{M}_1 = M)$$

$$\tilde{N}_i = (\sum_{j=1}^{i-1} \mathbf{a}_j^T N \mathbf{a}_j) K + N, (\tilde{N}_1 = N).$$

The Lagrangian for the discriminant vector $\mathbf{a}_i$ is:

$$L(\mathbf{a}_i, \lambda) = \mathbf{a}_i^T \tilde{M}_i \mathbf{a}_i - \lambda(\mathbf{a}_i^T \tilde{N}_i \mathbf{a}_i - 1) - \sum_{j=1}^{i-1} \mu_j \mathbf{a}_i^T K \mathbf{a}_j.$$

Just like above, on the saddle point, the following condition must be satisfied:

$$\frac{\partial L(\mathbf{a}_i, \lambda)}{\partial \mathbf{a}_i} = 2\tilde{M}_i \mathbf{a}_i - 2\lambda \tilde{N}_i \mathbf{a}_i - \sum_{j=1}^{i-1} \mu_j K \mathbf{a}_j = 0. \tag{3}$$

If both sides of (3) multiply $\mathbf{a}_k^T K \tilde{N}_i^{-1} (k < i)$, one can get:

$$2\mathbf{a}_k^T K \tilde{N}_i^{-1} \tilde{M}_i \mathbf{a}_i - \sum_{j=1}^{i-1} \mu_j \mathbf{a}_k^T K \tilde{N}_i^{-1} K \mathbf{a}_j = 0, \ k = 1, \cdots i-1. \tag{4}$$

Let

$$\begin{aligned} \boldsymbol{u} &= [\mu_1, \cdots, \mu_{i-1}]^T, \\ D &= [\mathbf{a}_1, \cdots, \mathbf{a}_{i-1}]^T, \end{aligned}$$

then (4) can be rewritten as

$$2DK\tilde{N}_i^{-1}\tilde{M}_i \mathbf{a}_i = DK\tilde{N}_i^{-1} K D^T \boldsymbol{u} \ ,$$

i.e.

$$\boldsymbol{u} = 2(DK\tilde{N}_i^{-1}KD^T)^{-1} DK\tilde{N}_i^{-1}\tilde{M}_i \mathbf{a}_i. \tag{5}$$

We know that in (3): $\sum_{j=1}^{i-1} \mu_j K \mathbf{a}_j = KD^T \boldsymbol{u}$. Substituting $\boldsymbol{u}$ of (3) with (5), then the following formula is obtained:

$$P\tilde{M}_i \mathbf{a}_i = \lambda \tilde{N}_i \mathbf{a}_i, \tag{6}$$

where

$$P = I - KD^T (DK\tilde{N}_i^{-1} KD^T)^{-1} DK\tilde{N}_i^{-1}.$$

So $\mathbf{a}_i$ is the eigenvector corresponding to the largest eigenvalue of the generalized eigenvalue problem (6).

After $\mathbf{a}_i$ has been obtained, one should normalize $\mathbf{a}_i$ with $\mathbf{a}_i^T K \mathbf{a}_i = 1$.

## 3 KGFST with Cluster-Weighted

### 3.1 GFST with Cluster-Weighted

Let $Z = \{(x_1, y_1), ..., (x_n, y_n)\} \subseteq R^m \times \{\omega_1, \ldots, \omega_C\}$. The number of samples in each class $\omega_i$ is $n_i$. Suppose the mean vector, the covariance matrix and a priori probability of each class $\omega_i$ are $m_i$, $S_i$, $P_i$, respectively. The global mean vector is $m_0$. Then the between-class scatter matrix $S_B$ and the within-class scatter matrix $S_W$ are determined by the following formulae:

$$S_B = \sum_{i=1}^{C} P_i(m_i - m_0)(m_i - m_0)^T,$$

$$S_W = \sum_{i=1}^{C} P_i S_i.$$

Let $Z_c$ be the clustering result of $Z$. $Z_c = \{(x_1^c, y_1), \cdots, (x_l^c, y_l)\} \subseteq X_c \times Y$, $X_c \in R^m$, $Y = \{\omega_1, \cdots, \omega_C\}$, the number of $Z_c$ is $l$, the number of $\omega_i$ is $l_i(l_i/n_i = l/n)$, sample $x_i^c$ represent $q_i$ original samples. Suppose the mean vector, the covariance matrix and a priori probability of each class $\omega_i$ are $m_{ci}$, $S_{ci}$, $P_{ci}(P_{ci} = P_i)$, respectively. The global mean vector is $m_{c0}$. Then the between-class scatter matrix $S_{cB}$ and the within-class scatter matrix $S_{cW}$ on $Z_c$ are determined by the following formulae:

$$S_{cW} = \sum_{i=1}^{C} P_{ci} S_{ci},$$

$$(S_{ci} = \frac{1}{n_i} \sum_{j=1}^{l_i} q_{ij}(x_{ij}^c - m_{ci})(x_{ij}^c - m_{ci})^T),$$

$$S_{cB} = \sum_{i=1}^{C} P_{ci}(m_{ci} - m_{c0})(m_{ci} - m_{c0})^T,$$

where $x_{ij}^c$ is the $jth$ clustering sample of $\omega_i$, $q_{ij}$ (weight) is the number of the original samples represented by the $jth$ clustering sample of $\omega_i$.

It is easy to prove that

$$\begin{aligned} m_{ci} &= \frac{1}{n_i} \sum_{j=1}^{l_i} q_{ij} x_{ij}^c \\ &= \frac{1}{n_i}(x_{ij} + \cdots + x_{in_i}) \\ &= \frac{1}{n_i} \sum_{j=1}^{n_i} x_{ij} = m_i. \end{aligned}$$

For the same reason, one can get $m_0 = m_{c0}$. So one can draw the following conclusion:

$$S_B = S_{cB}.$$

For the within-class scatter matrices $S_W$ and $S_{cW}$, one should only compare $S_i$ with $S_{ci}$(for $P_i = P_{ci}$).

For

$$S_i = E[(X - m_i)(X - m_i)^T]$$
$$= \frac{1}{n_i}\sum_{j=1}^{n_i}(x_{ij} - m_i)(x_{ij} - m_i)^T$$
$$= \frac{1}{n_i}(\sum_{j=1}^{n_i} x_{ij}x_{ij}^T + \sum_{j=1}^{n_i} m_i m_i^T - 2\sum_{j=1}^{n_i} x_{ij}m_i^T).$$

and

$$S_{ci} = \frac{1}{n_i}\sum_{j=1}^{l_i} q_{ij}(x_{ij}^c - m_{ci})(x_{ij}^c - m_{ci})^T$$
$$= \frac{1}{n_i}(\sum_{j=1}^{l_i} q_{ij}x_{ij}^c(x_{ij}^c)^T + \sum_{j=1}^{l_i} q_{ij}m_{ci}m_{ci}^T - 2\sum_{j=1}^{l_i} q_{ij}x_{ij}^c m_{ci}^T).$$

then

$$S_i - S_{ci} = \frac{1}{n_i}(\sum_{j=1}^{n_i} x_{ij}x_{ij}^T - \sum_{j=1}^{l_i} q_{ij}x_{ij}^c(x_{ij}^c)^T).$$

So we know that $S_i \approx S_{ci}$ and $S_W \approx S_{cW}$.

### 3.2 KGFST with Cluster-Weighted

As showed in GFST with Clustering, if we use clustering in feature space for KGFST, the only thing we should do is how to calculate the matrices $M_c$, $N_c$ and $K_c$ corresponding to $M$, $N$ and $K$.

Let $Z_c^{\Phi} = \{(\Phi_c(t_1), y_1), \cdots, (\Phi_c(t_l), y_l)\}$ be the clustering result of $Z$ in feature space. The scale of $Z_c^{\Phi}$ is $l$, the scale of $\omega_i$ is $l_i(l_i/n_i = l/n)$, sample $\Phi_c(t_i)$ represent $q_i$ samples in feature space. Then the between-class scatter matrix $S_{cB}^{\Phi}$ and the within-class scatter matrix $S_{cW}^{\Phi}$ on $Z_c^{\Phi}$ are determined by the following formulae:

$$S_{cW}^{\Phi} = \sum_{i=1}^{C} P_{ci}S_{ci}^{\Phi},$$
$$(S_{ci}^{\Phi} = \frac{1}{n_i}\sum_{j=1}^{l_i} q_{ij}(\Phi_c(t_{ij}) - m_{ci}^{\Phi})(\Phi_c(t_{ij}) - m_{ci}^{\Phi})^T),$$
$$S_{cB}^{\Phi} = \sum_{i=1}^{C} P_{ci}(m_{ci}^{\Phi} - m_{c0}^{\Phi})(m_{ci}^{\Phi} - m_{c0}^{\Phi})^T.$$

We can easy know that:

$$m_{ci}^{\Phi} = \frac{1}{n_i}\sum_{j=1}^{l_i} q_{ij}\Phi_c(t_{ij})$$

$$= \frac{1}{n_i}\sum_{j=1}^{n_i}\Phi(x_{ij})$$
$$= m_i^{\Phi},$$
$$m_{c0}^{\Phi} = \sum_{i=1}^{C} P_{ci} m_{ci}^{\Phi}$$
$$= \frac{1}{n}\sum_{i=1}^{n}\Phi(x_i)$$
$$= m_0^{\Phi},$$

Let $w_c = \sum_{i=1}^{l} a_i \Phi_c(t_i)$, then

$$w_c^T m_{ci}^{\Phi} = \mathbf{a}^T M_i^c,$$
$$(M_i^c)_j = \frac{1}{q_j n_i}\sum_{p=1}^{q_j}\sum_{k=1}^{n_i} k(x_{jp}^c, x_{ik}), j = 1, \cdots, l.$$
$$w_c^T m_{c0}^{\Phi} = \mathbf{a}^T M_0^c,$$
$$(M_0^c)_j = \frac{1}{q_j n}\sum_{p=1}^{q_j}\sum_{k=1}^{n} k(x_{jp}^c, x_k), j = 1, \cdots, l.$$

where $x_{jp}^c$ is the $pth$ clustering sample of the $jth$ class, $x_{ik}$ is the $ith$ sample of the $kth$ class and $x_k$ is the $kth$ sample of the whole samples.

Then we can get the following formulae:

$$w_c^T S_{cB}^{\Phi} w_c = \mathbf{a}^T M_c \mathbf{a},$$

where

$$M_c = \sum_{i=1}^{C} P_i (M_i^c - M_0^c)(M_i^c - M_0^c)^T.$$

According to the results above and the definition of $S_{cW}^{\Phi}$, we can get:

$$w_c^T S_{cW}^{\Phi} w_c = \mathbf{a}^T N_c \mathbf{a},$$

where

$$N_c = \sum_{i=1}^{C} P_i (N_i^c - N_0^c)(N_i^c - N_0^c)^T,$$
$$(N_i^c)_j = \frac{1}{q_j q_{im}}\sum_{p=1}^{q_j}\sum_{k=1}^{q_{im}} k(x_{jp}^c, x_{imk}^c),$$
$$(N_0^c)_j = \frac{1}{q_j n_i}\sum_{p=1}^{q_j}\sum_{k=1}^{n_i} k(x_{jp}^c, x_{ik}).$$

$q_{ij}$ (weight) is the number of samples represented by the $jth$ clustering sample of $\omega_i$ and $x^c_{imk}$ is the $kth$ original sample of the $mth$ clustering of $\omega_i$.

The Kernel matrix $K_c$ also can be calculate easily:

$$\begin{aligned}(K_c)_{ij} &= \Phi_c(t_i) \cdot \Phi_c(t_j) \\ &= \frac{\Phi(x^c_{i1}) + \cdots + \Phi(x^c_{iq_i})}{q_i} \cdot \frac{\Phi(x^c_{j1}) + \cdots + \Phi(x^c_{jq_j})}{q_j} \\ &= \frac{1}{q_i q_j} \sum_{p=1}^{q_i} \sum_{k=1}^{q_j} (x^c_{ip}, x^c_{jk}).\end{aligned}$$

Once we get the $l \times l$ matrices $K_c$, $M_c$ and $N_c$, we can easily solve the problem of CW-KGFST (KGFST with Cluster-weighted) according to KGFST[3].

## 4 Computational Comparison and Applications

In this section, we compare the performance of KFGST against CW-KGFST. We implemented all these methods in Matlab R2006 and ran them on a 1.70G MHz PM machine.

### 4.1 The Datasets and Algorithms

The following datasets are used in our experiments:

*Dataset A:* The "Optdigits" database from the UCI repository. Optdigits is a Optical-based recognition problem of handwritten digits ($0 \sim 9$). The digits written by 30 writers are used for training and the digits written by other 13 writers are used for testing. Each pattern contains one class attribute and 64 input features and each feature value is between 0 and 1. We produce a series subsets of Pendigits A$i(i = 3, ..., 10)$ which is a classification problem of $i$ classes.

*Dataset B:* The "Pendigits" database from the UCI repository. Pendigits is a Pen-based recognition problem of handwritten digits ($0 \sim 9$). The digits written by 30 writers are used for training and the digits written by other 14 writers are used for testing. Each pattern contains one class attribute and 16 input features. We also produce a series subsets of Pendigits B$i(i = 3, ..., 10)$ which is a classification problem of $i$ classes.

To compare the methods above, we use linear support vector machines (SVM) [7] and K-nearest neighbors (KNN)[8] algorithm as classifiers.

### 4.2 Results and Analysis

Table 1 and 2 discrible the relationship of the project vectors got by KGFST and CW-KGFST on dataset A and dataset B. $w_1, w_2, \cdots$ are the project vectors got by KGFST and $w_{c1}, w_{c2}, \cdots$ are the project vectors got by CW-KGFST.

**Table 1.** Relationship of project vectors got by methods above on dataset A(RBF:0.3)

| A3 | $w_{c1}$ | $w_{c2}$ | A4 | $w_{c1}$ | $w_{c2}$ | $w_{c3}$ | A5 | $w_{c1}$ | $w_{c2}$ | $w_{c3}$ | $w_{c4}$ |
|---|---|---|---|---|---|---|---|---|---|---|---|
| $w_1$ | 0.889 | 0.005 | $w_1$ | 0.899 | 0.097 | 0.083 | $w_1$ | 0.902 | 0.067 | 0.076 | 0.099 |
| $w_2$ | 0.033 | 0.893 | $w_2$ | 0.056 | 0.709 | 0.498 | $w_2$ | 0.069 | 0.848 | 0.004 | 0.241 |
| | | | $w_3$ | 0.133 | 0.504 | 0.733 | $w_3$ | 0.071 | 0.039 | 0.882 | 0.043 |
| | | | | | | | $w_4$ | 0.135 | 0.217 | 0.04 | 0.864 |

**Table 2.** Relationship of project vectors got by methods above on dataset B(RBF:2)

| B3 | $w_{c1}$ | $w_{c2}$ | B4 | $w_{c1}$ | $w_{c2}$ | $w_{c3}$ | B5 | $w_{c1}$ | $w_{c2}$ | $w_{c3}$ | $w_{c4}$ |
|---|---|---|---|---|---|---|---|---|---|---|---|
| $w_1$ | 0.930 | 0.062 | $w_1$ | 0.933 | 0.102 | 0.105 | $w_1$ | 0.939 | 0.021 | 0.076 | 0.039 |
| $w_2$ | 0.048 | 0.877 | $w_2$ | 0.113 | 0.842 | 0.418 | $w_2$ | 0.033 | 0.898 | 0.206 | 0.189 |
| | | | $w_3$ | 0.042 | 0.402 | 0.782 | $w_3$ | 0.065 | 0.223 | 0.575 | 0.619 |
| | | | | | | | $w_4$ | 0.026 | 0.023 | 0.681 | 0.596 |

According to table 1 and 2, we can see that the product of the main corresponding project vectors constituting KGFST and CW-KGFST approximate 1. That is to say, the main project directions are coincident.

Table 3 and 4 describe the running speed of KGFST and CW-KGFST and the classification accuracy of KNN and SVM on dataset A and dataset B.

**Table 3.** The running speed of KGFST and CW-KGFST and the classification accuracy on dataset A(RBF:0.3)

| dataset | KGFST | | | CW-KGFST | | |
|---|---|---|---|---|---|---|
| | Times | Accuracy | | Times | Accuracy | |
| | | KNN | SVM | | KNN | SVM |
| A3 | 22.86s | 99.4382% | 99.4382% | 5.312s | 99.4382% | 99.8250% |
| A4 | 71.53s | 99.0237% | 99.1632% | 8.703s | 98.7448% | 98.6053% |
| A5 | 184.38s | 98.9989% | 98.8877% | 12.58s | 98.2202% | 98.3315% |
| A6 | 775.58s | 98.3225% | 98.4157% | 17.30s | 97.2041% | 97.3905% |
| A7 | 41m | 98.4051% | 98.4051% | 23.28s | 97.4482% | 99.0994% |
| A8 | 1.5h | 98.1882% | 98.3275% | 30.06s | 97.5610% | 96.8641% |
| A9 | 7.7h | 97.2136% | 97.0279% | 37.95s | 96.0372% | 95.7276% |
| A10 | —— | —— | —— | 48.03s | 94.8247% | 91.7641% |

According to table 3 and 4, for the same dataset, the classification accuracy by KGFST is approximate to that by CW-KGFST. But the running times on the same dataset by KGFST and CW-KGFST are very different. For example, on dataset A9, the running time of KGFST is 7.7 hours while the running time of CW-KGFST is 37.95s. That is to say that the running speed of CW-KGFST is higher to that of KGFST while preserving the classification ability of project vectors for a dataset.

**Table 4.** The running speed of KGFST and CW-KGFST and the classification accuracy on dataset B(RBF:2)

| dataset | KGFST | | | CW-KGFST | | |
|---|---|---|---|---|---|---|
| | Times | Accuracy | | Times | Accuracy | |
| | | KNN | SVM | | KNN | SVM |
| B3 | 213.78s | 99.8069% | 99.8069% | 11.78s | 99.8069% | 99.7104% |
| B4 | 30m | 99.4898% | 99.4898% | 28.16s | 99.4898% | 99.3440% |
| B5 | 5.85h | 98.0415% | 97.8687% | 32.68s | 97.6959% | 97.8687% |
| B6 | —— | —— | —— | 48.87s | 97.2844% | 97.0939% |
| B7 | —— | —— | —— | 233.3s | 97.0767% | 96.7925% |
| B8 | —— | —— | —— | —— | —— | —— |
| B9 | —— | —— | —— | —— | —— | —— |
| B10 | —— | —— | —— | —— | —— | —— |

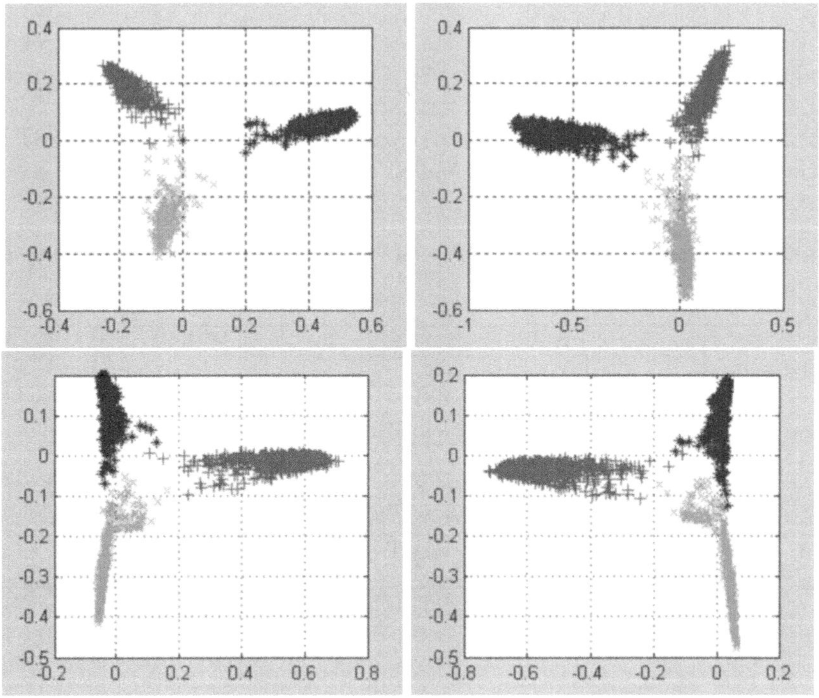


**Fig. 1.** (First)space distribution of A3 on the features extracted by KGFST; (Second) space distribution of A3 on the features extracted by CW-KGFST; (Third) space distribution of B3 on the features extracted by KGFST; (Fourth) space distribution of B3 on the features extracted by CW-KGFST;

Figure 1 describes the space distributions of A3 and B3 on the features extracted by KGFST and CW-KGFST. Form Figure 1, we can see that the space distributions of A3 and B3 on the features extracted by KGFST is approximate to that extracted by CW-KGFST.

From the results above, we can see that the product of the corresponding project vectors constituting KGFST and CW-KGFST approximate 1, the space distributions of A3 and B3 on the features extracted by KGFST is approximate to that extracted by CW-KGFST and the running speed of CW-KGFST is higher to that of KGFST while preserving the classification ability of project vectors.

## 5 Conclusion

In this paper, a new nonlinear feature extraction method CW-KGFST (KGFST with Cluster-weighted) based on KGFST and Clustering is proposed. By the Cluster-weighted, the number of samples can be reduced, the calculate speed can be higher and the accuracy can be preserved simultaneously. Lastly, our method is applied to digits and images recognition problems, and the experimental results show that the performance of present method is superior to the original method.

## References


1. Foley, D.H., Sammon, J.W.: An Optimal Set of Discriminant Vectors. IEEE Trans on Computers, 24(1975)281–289
2. Guo, Y.F., Li, S.J., et al.: A generalized Foley-Sammon Transform Based on Generalized Fisher Discriminant Criterion and its Application to Face Recognition. Pattern Recognition Letters, 24(2003)147–158
3. Chen, Z.Z., Li, L.: Generalized Foley-Sammon Transform with Kernels. Advances in Neural NetworksCISNN 2005: Second International Symposium on Neural Networks, Part II(2005)817–823
4. Mika, S., Schölkopf, B., et al.: Kernel PCA and De-noising in Feature Spaces. In M. S. Kearns, S. A. Solla, and D. A. Cohn, editors, *Advances in Neural Information Processing Systems 11*, MIT Press(1999)536–542
5. Bach, F.R., Jordan, M.I.: Kernel Independent Component Analysis. (Kernel Machines Section) 3(2002)1–48
6. Bradley P., Fyyad U., ReinaC.: Scaling Clustering Algorithms to Large Databases. In Proc. 1998 Int. Conf. Knowledge Discovery and Data Mining (KDD'98)(1998) 9–15
7. Burges, C.J.C.: A Tutorial on Support Vector Machines for Pattern Recognition. Data Mining and Knowledge Discovery, 2(1998)955–974
8. Arya, S., Mount, D.M., Netanyahu, N.S., Silverman, R. and Wu, A.: An Optimal Algorithm for Approximate Nearest Neighbor Searching. In Proc. 5th ACM-SIAM Sympos. Discrete Algorithms(1994)573–582

# Supervised Information Feature Compression Algorithm Based on Divergence Criterion

Shiei Ding[1,2] Wei Ning[3], Fengxiang Jin[4], Shixiong Xia[1], and Zhongzhi Shi[2]

College of Computer Science and Information Engineering,
Tianjin University of Science and Technology, Tianjin, China
yjma@ustc.edu dingsf@cumt.edu.cn

**Abstract.** The fermentation process is very complex and non-linear, many parameters are not easy to measure directly on line, soft sensor modeling is a good solution. This paper introduces $\nu$-support vector regression ($\nu$-SVR) for soft sensor modeling of fed-batch fermentation process. $\nu$-SVR is a novel type of learning machine. It can control the accuracy of fitness and prediction error by adjusting the parameter $\nu$. An on-line training algorithm is discussed in detail to reduce the training complexity of $\nu$-SVR. The experimental results show that $\nu$-SVR has low error rate and better generalization with appropriate $\nu$.

## 1 Introduction

# Supervised Information Feature Compression Algorithm Based on Divergence Criterion

The fermentation process is complex and non-linear, some key parameters are difficult to measure on line, such as biomass concentration, substrate concentration, production concentration. It is impractical to analysis the fermentation process by using analytic model. Artificial neural network (ANN) is used for modeling fermentation process, and it has shown better performance than analytic model method. However, it is a hard work to collect enough experimental data in fermentation process, even in off-line condition. Furthermore, ANN has its own defects, for example, the net parameters are not easy to tune, the structure is difficult to determine [1-2].

$\nu$-SVR is a novel type of learning machine, which based on statistical learning theory (SLT). It introduces a new parameter $\nu$ to control the fitness and predication accuracy. $\nu$-SVR has shown to provide a better generalization performance than traditional techniques, including neural networks [3].

In this paper $\nu$-SVR based modeling algorithm is proposed for fed-batch fermentation process, and an on-line training algorithm is discussed in detail to reduce the training complexity of $\nu$-SVR.

This paper is organized as follows. In section 2 we discuss the construction of $\nu$-SVR. Section 3 shows how to use $\nu$-SVR to construct soft sensor modeling of fermentation process. The on-line training algorithm based on $\nu$-SVR is proposed in this section. The obtained experimental results are illustrated in section 4. Finally, Section 5 summarizes the conclusions that can be drawn from the presented research.

D.-S. Huang, L. Heutte, and M. Loog (Eds.): ICIC 2007, LNAI 4682, pp. 919 – 927, 2007.

## 2 ν-SV Regression

ν-SVR seeks to estimate functions

$$f(x) = (w \cdot x) - b \tag{1}$$

$$\text{where } w, \quad x \in R^N, b \in R \tag{2}$$

based on independent identically distributed data

$$(x_1, y_1), \ldots\ldots, (x_\ell, y_\ell) \in \chi \times R \tag{3}$$

Here, $\chi$ is the space in which the input patterns live.

To estimate functions (1) from empirical data (3), we can obtain a small risk by solve the following constrained optimization problem:

$$\tau(w, \xi^{(*)}, \varepsilon) = \frac{1}{2} \| w \|^2 + C \cdot (\upsilon\varepsilon + \frac{1}{\ell} \sum_{i=1}^{\ell} (\xi_i + \xi_i^{*})) \tag{4}$$

$$((w \cdot X_i) - b) - y_i \le \varepsilon - \xi_i \tag{5}$$

$$y_i - ((w \cdot x_i) + b) \le \varepsilon + \xi_i^{*}, \ \xi_i^{*} \ge 0, \xi_i \ge 0 \tag{6}$$

where C is a constant determining the trade-off.

At each point xi, an error of ε is allowed. Everything above ε is captured in slack variables $\xi_i^{(*)}$, which are penalized in the objective function via a regularization constant C, chosen a priori. The size of ε is traded off against model complexity and slack variables via a constant $\nu > 0$.

Constructing Lagrangian

$$L_\nu(\boldsymbol{w}, \xi, b, \rho, \boldsymbol{\alpha}, \boldsymbol{\beta}, \delta) = \frac{1}{2}\|\boldsymbol{w}\|^2 - \nu\rho + \frac{1}{n}\sum_{i=1}^{n}\xi_i \tag{7}$$

$$-\sum_{i=1}^{n}\alpha_i\{y_i[(\boldsymbol{w} \cdot \boldsymbol{x}_i) + b] - \rho + \xi_i\} + \sum_{i=1}^{n}\beta_i\xi_i - \delta\rho$$

$$\text{where } \alpha_i, \beta_i, \delta \ge 0 \tag{8}$$

At the saddle point, *L* has a minimum, thus we can write

$$\boldsymbol{w} = \sum_{i=1}^{n} \alpha_i y_i \boldsymbol{x}_i \tag{9}$$

$$\alpha_i + \beta_i = \frac{1}{n} \tag{10}$$

$$\sum_{i=1}^{n} \alpha_i y_i = 0 \tag{11}$$

$$\sum_{i=1}^{n} \alpha_i - \delta = v \tag{12}$$

Considering Karush-Kuhn-Tucker (KKT) conditions and dual problem, the v-SVR regression estimate then takes the form

$$Q_v(\boldsymbol{\alpha}) = -\frac{1}{2}\sum_{i,j=1}^{n} \alpha_i \alpha_j y_i y_j k(\boldsymbol{x}_i, \boldsymbol{x}_j) \tag{13}$$

$$\text{Subject to } \; 0 \le \alpha_i \le \frac{1}{n}, \sum_{i=1}^{n} \alpha_i y_i = 0, \sum_{i=1}^{n} \alpha_i \ge v \tag{14}$$

The decision function becomes

$$f(x) = \sum_{i=1}^{\ell} (\alpha_i * - \alpha_i) k(x_i, x) + b \tag{15}$$

where v≥0, C > 0, $\alpha_i^{(*)}$ is multiplier, $k(x, y)$ is kernel function. b (and ε) can be computed by taking into account that (6) and (7) become equalities with $\xi_i^{(*)} = 0$, for points with $0 < \alpha_i^{(*)} < C / l$, respectively, due to the KKT conditions.

From [3] we also know that v is an upper bound on the fraction of errors, so we can control the error by deciding v. We can use it to control the prediction accuracy during the fermentation process. This is the reason why we select v-SVR instead of SVR.

# 3 Soft Sensor Modeling of Fermentation Process Based on ν-SVR

## 3.1 The Construction of Model Based on ν-SVR

The fermentation process is complex and non-linear, many parameters are not easy to measure, such as biomass concentration, substrate concentration, production concentration. It is impractical to analysis the fermentation process using analytic model [5-6]. We introduce ν-SVR as the soft sensor model. We took the following function as the model description (see (7)).

Radical basis function (RBF) is chosen as kernel function:

$$K(\mathbf{x}, \mathbf{x}_i) = \exp\left(-\frac{|\mathbf{x} - \mathbf{x}_i|^2}{\sigma^2}\right) \tag{16}$$

It is critical to select the type of kernel function and the parameters such as $v$ and $C$ during the modeling process. Cross validation is used to determine the optimal parameters.

## 3.2 ν-SVRM : The Online Training Algorithm Based on ν-SVR

Cross validation is used to determine the parameters and the type of kernel function, but it can not be used to fine-tune the model online on line. So a new on-line fine-tune algorithm of model is proposed which is named $\nu$-SVRM ($\nu$-SVR for Modelling).

Firstly select $n$ input samples and build up a training set for the training of $\nu$-SVRM. The optimal parameters are selected as the model parameters after validation. Secondly add new sample and renew the set according to some rules. Finally fine-tune the parameters of models.

The detailed steps are as following:

Step1. Normalize the working set $w = \{(x_1, y_1), (x_2, y_2), \cdots (x_n, y_n)\}$

Step2. Train $\nu$-SVRM model $f(x)$ using cross validation method.

Step3. Use $f(x)$ to predict a new sample $(x_{n+1}, y_{n+1})$

Step4. If $\left|\frac{f(x_{n+1}) - y_{n+1}}{y_{n+1}}\right| > \nu$, add $(x_{n+1}, y_{n+1})$ to the working set.

Step5. Remove a non-SV sample to form a new working set

Step6. If there are still new samples, go to step2, else go to the end.

Partial Matlab code which use LibSVM as training algorithms

```
P=[P1;P2;P3;P5]; % Training set

T=[T1;T2;T3;T5]; % Testing set

p_test=P4;

T_test=T4

s = sprintf('-s %d -n %.4g -p %.7g -t 2 -c %d -g
%d',s,n,p,c,g);

model=svmtrain(T,P,s);% use vSVM

[predict_label,accuracy,decision_values]=svmpredict(T_t
est,p_test,model);

e=(decision_values-T_test).*(decision_values-T_test);

E=sum(e);% Compute error
```

# 4 Experiments

## 4.1 Experimental Conditions

In the experiments we take polylysine batch fermentation and feed the fermentor with 2.5L materials each time. There are many parameters which can influence the polylysine fermentation process. We choose some key factors as the input set of model, which are temperature, PH value, dissolved oxygen (DO), stirring speed, fermentation time and the biomass concentration last period. We take the biomass concentration as the model output [4]. The total batch is 5.

The experimental equipment is an intelligent fermentation process control system designed by ourselves. The software platform is PIV2.66GHz/1G Memory/ WindowsXP/ Matlab7.0/ VC++6.0.

The fermentation equipment is as following:

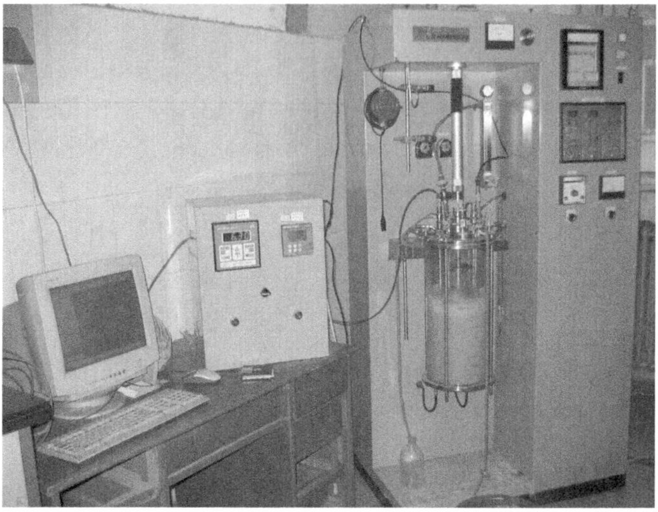

**Fig. 1.** Experimental equipment

## 4.2 Experimental Results

The following table are the experimental data.

**Table 1.** Partial experimental data

| Predict data | Input data | | |
|---|---|---|---|
| | 1th Column: time | 2th column: PH | 3th colum: DO |
| 0.0231 | 0.0000 | 0.2611 | 0.2956 |
| 0.0363 | 0.0143 | 0.2541 | 0.2493 |
| 0.0662 | 0.0286 | 0.2413 | 0.2134 |
| 0.1044 | 0.0429 | 0.2258 | 0.1971 |
| 0.1424 | 0.0571 | 0.2107 | 0.1945 |
| 0.1717 | 0.0714 | 0.1989 | 0.1929 |
| 0.1838 | 0.0857 | 0.1936 | 0.1919 |
| 0.1842 | 0.1000 | 0.1927 | 0.1909 |
| 0.1848 | 0.1286 | 0.1919 | 0.1898 |
| 0.1864 | 0.1429 | 0.1943 | 0.1912 |
| 0.1936 | 0.1714 | 0.2016 | 0.1950 |
| 0.2000 | 0.1857 | 0.2022 | 0.1945 |

| | | | |
|---|---|---|---|
| 0.2087 | 0.2000 | 0.2026 | 0.1933 |

In the experiments we select RBF kernel function. $\sigma$ is the width coefficient, little value can get the good fitness, but too little value will lead to poor generation. Penalty parameter $C$ can punish the error, the increasing of $C$ will decrease the fitness error and the prediction error. But when $C$ becomes too big, it will get into trouble of over fitness. $v$ is an upper bound on the fraction of errors, it can control the prediction accuracy by adjusting $v$.

Table 2 shows the comparison results of training time among $v$-SVRM, $v$-SVR and ε -SVR.

**Table 2.** Comparison of training time (c=250, σ=15)

| Model | $v$-SVR ($v = 0.10$) | $v$-SVRM ($v = 0.10$) | $v$-SVR ($v = 0.30$) | $v$-SVRM ($v = 0.30$) | ε -SVR |
|---|---|---|---|---|---|
| Training time (s) | 3.84 | 2.69 | 2.14 | 1.98 | 3.73 |

From the table above we can conclude that $v$-SVRM need shorter training time with the same value of parameters $v$.

The following table 2 is the experimental results of all 5 batches. It indicates that $v$-SVRM has fine-tuning ability. With the increase of experimental data, $v$-SVRM shows better prediction accuracy.

**Table 3.** On-line predictive error of biomass concentration (C=250 , σ =15)

| Batch data | RMSE: predictive error of each batch | | | | |
|---|---|---|---|---|---|
| | $1^{th}$ batch | $2^{th}$ batch | $3^{th}$ batch | $4^{th}$ batch | $5^{th}$ batch |
| 2 batches | 0.00753 | 0.00816 | ___ | ___ | ___ |
| 3 batches | 0.00623 | 0.00511 | 0.00693 | ___ | ___ |
| 4 batches | 0.00508 | 0.00531 | 0.00542 | 0.00499 | ___ |
| 5 batches | 0.00512 | 0.00494 | 0.00501 | 0.00489 | 0.00457 |

The $5^{th}$ batch experimental results are as following figures ($v = 0.10$).

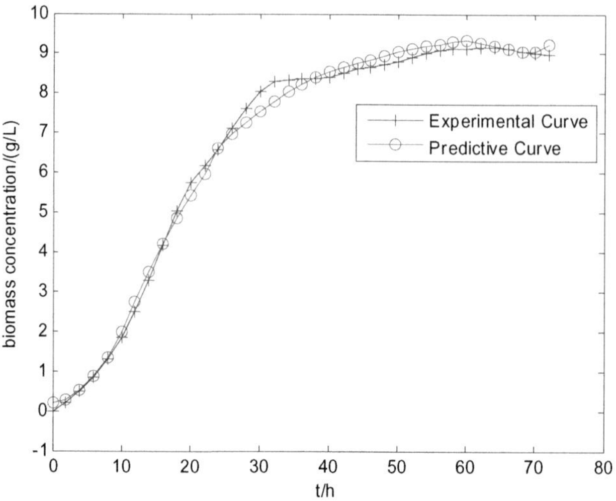


(a) $v$-SVRM ( RMSE=0.00457)

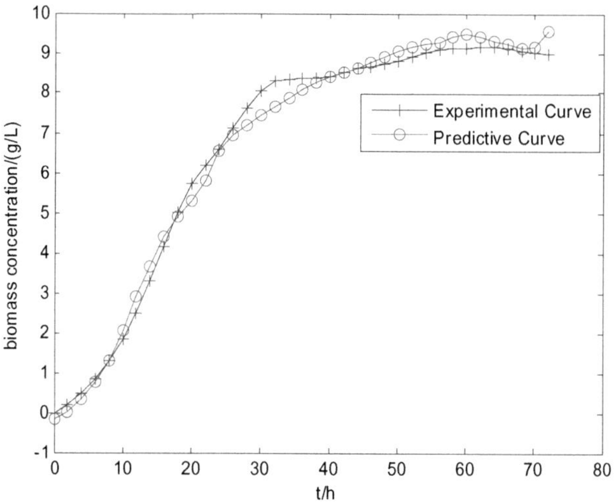


(b) $v$-SVR ( RMSE=0.00716)

**Fig. 2.** Comparison among v-SVRM, v-SVR, SVR and BP

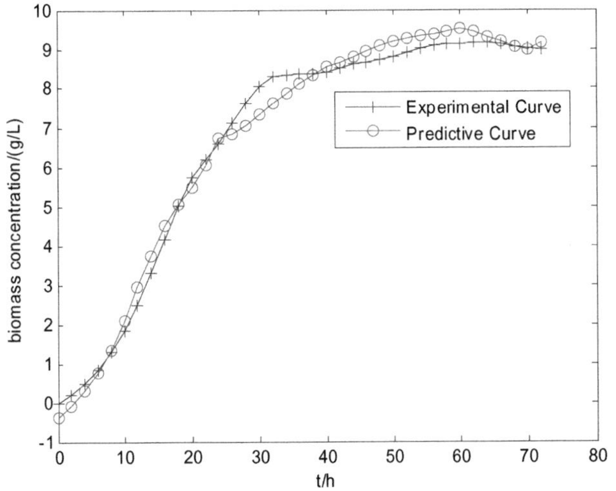


(c) ε -SVR ( RMSE=0.00608)

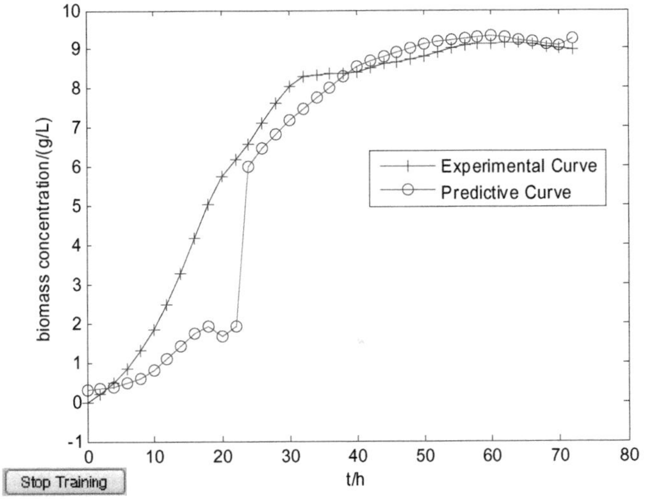


(d) BP net ( RMSE=0.0289)

**Fig. 2.** *(continued)*

Figure 1 (a), (b) and (c) show that *v*-SVRM, *v*-SVR, SVR has similar predictive accuracy.

The predicting results of BP net are not satisfying (RMSE=0.0289), the main reason is artificial neural network is based on traditional statistics, which need a large amount of training samples. Actually it is difficult to get enough samples in the fermentation process. SVR can get better performance in such a case.

## 5 Conclusions

In experiments $\nu$-SVR shows good performance for soft sensor modeling of fed-batch fermentation process, the on-line training algorithm $\nu$-SVRM is discussed which can reduce the training complexity of $\nu$-SVR. The experimental results show that v-SVR has low error rate and better generalization by adjusting the parameter $\nu$.

**Acknowledgement.** This research is sponsored by a grant of Tianjin Science&Technology Development Fundation of High School under contract 20061011, and partly sponsored by a grant of Tianjin Key Technologies R&D Program under contract 04310951R.

## References

1. Ma,Y.G., Kong,B.: A Study of Object Detection based on Fuzzy Support Vector Machine and Template Matching. *IEEE Proceedings of the 5th World Congress on Intelligent Control and Automation*, vol.5, Hangzhou, P.R. China, ( 2004), .4137-4140
2. Ma, Y.G., Fang, K., Fang, T.J. :A Study of Classification based on Support Vector Machine and Distance Classification for Texture Image (Chinese). *Journal of Image and Graphics,* Vol. 7(A), no.11, (2002),1151-1155
3. Scholkopf,B.,.Smolad, A.J.: New Support Vector Algorithms. neurocolt2 nc2-tr-1998-031. Technical report, GMD First and Australian National University, (1998)
4. Liu,Y.M., Meng,Z.P., Yu,H.W., et al: The Realization of Fermentation Process Status Pre-estimate Model Based on BP NN (In: Chinese). *Journal of Tianjin university of light industry,* vol.18, no. 3. (2003)35~38
5. Xiong,Z.H., Zhang,J.C., Shao,H.H.: GP-based Soft Sensor Modeling. *Journal of system simulation.* Vol. 17, no. 4, (2005) 793~800
6. Wang, J.L., Yu, T.: Research Progress in Soft Sensor Techniques for On-Line Biomass Estimation. Modern chemistry industry. Vol.25, no.6, ( 2005) 22~25

# The New Graphical Features of Star Plot for K Nearest Neighbor Classifier

Jinjia Wang[1,2], Wenxue Hong[1], and Xin Li[1]

[1] Department of Biomedicine Engineer, Yanshan University, Qinhuangdao 066004
[2] Information Colleges, Yanshan University, Qinhuangdao 066004

**Abstract.** The graphical representation or graphical analysis for multi-dimensional data in multivariate analysis is a very useful method. But it rarely is used to the pattern recognition field. The paper we use the stat plot to represent one observation or sample with multi variances and extract the new graphical features of star plot: sub-area features and sub-barycentre features. The new features are used for the K nearest neighbor classifier (KNN) with leave one out cross validation. Experiments with several standard benchmark data sets show the effectiveness of the new graphical features.

**Keywords:** star plot, graphical features, features extraction, K nearest neighbor classifier.

## 1 Introduction

The feature selection and extraction is the key question for the pattern recognition [1, 2]. Because in many practical application the most important features is often difficult to find out, or is difficult to measure owing to the limited conditions. The question is pay attention to more and more. One often utilize the physical and structural features to recognize the object, as these features are easily found out by the vision, hearing, touch and other feeling organ. But it is some complex for these features to construct the pattern recognition system using the computer. In general, it is very complex to simulate the human feeling organ using the hardware. But the capacity of extracting the mathematic features using computer, such as statistical mean, correlation, eigenvalue and eigenvector of sample covariance, is more superior to human. The keystone in pattern recognition is that how the mathematic features are selected and extracted by the learning samples.

Glyphs provide a means of displaying items of multivariate data by representing individual units of sample as icon-graphical objects [3]. Such glyphs may help to uncover specific clusters of both simple relations and interactions between dimensions. One commonly used glyph form is the 'star plot', in which the profile lines are placed on spokes so that the profile plot looks a bit like a star. Each dimension is represented by a line segment radiating from a central point. The ends of the line segment are joined. The length of the line segment indicates the value of the corresponding dimension. A second interesting form of glyph is 'Chernoff faces', which display data using cartoon faces by relating different dimension to facial features. We here use the star plot.

D.-S. Huang, L. Heutte, and M. Loog (Eds.): ICIC 2007, LNAI 4682, pp. 926–933, 2007.

From the star plot of a multivariate observation, we see the irregular polygonal shape by encircled the variance value on spokes. Based on the shape, we propose the sub-area features and sub-barycentre features for each observation, the number of which is both the same as the dimension of the observation. These new graphical features extend the basic feature concept. Moreover, these new graphical features establish a relation of the physical feature and mathematic feature. That is to say, these new graphical features not only are regard as the features found by human feeling organ, but also as the features mathematically calculated by the compute. This is our contribution.

The new graphical features are evaluated by the K nearest neighbor classifier, which is compared with original sample data. The reason of selecting the K nearest neighbor classifier is that it is a simple, yet useful approach to pattern recognition [4, 5]. The error rate of the KNN has been proven to be asymptotically at most twice that of the Bayesian error rate. The most important factor impacting the performance of KNN is the distance metric. We use the Euclidean distance. The evaluation of the consequent classifier is done through leave one out cross validation procedure repeated ten times. Experiments with several standard benchmark data sets show the effectiveness of the new graphical features.

## 2 Approach

### 2.1 Star Plot

The star plot is a simple means of multivariate visualization, which represents the value of an attribute through the length of lines radiating from the icon's center. Figure 1 displays star plots of the IRIS data. Each symbol displays all four variables. It is created by the Matlab function glyphplot(X), which creates a star plot from the multivariate data in the n-by-p matrix X. Rows of X correspond to observations, columns to variables. A star plot represents each observation as a "star" whose i-th spoke is proportional in length to the i-th coordinate of that observation. glyphplot standardizes X by shifting and scaling each column separately onto the interval [0,1] before making the plot, and centers the glyphs on a rectangular grid that is as close to square as possible. glyphplot treats NaNs in X as missing values, and does not plot the corresponding rows of X.

This method provides an overall impression of change of variable values across subjects. However, when there are too many variables and observations, a star plot will no longer be appropriate. This visual approach shows all data and thereby, it is considered a noisy technique. A star plot is not effective in examining multivariate relationships in a still mode, due to the difficulty for us to picture so many changes across subjects, especially when there are many observations. However, if individual stars are put together as a movie, the animated star can present a clear picture of how the values of multiple variables vary across subjects or over time relative to each other. From the vector data, we should not be limited in the only data graphical representation, but should full utilize data graphical analysis. That is, we should look for a method to mining the vector features of the star plot. So we propose the graphical features of data star plot: sub-area features and sub-barycentre features.

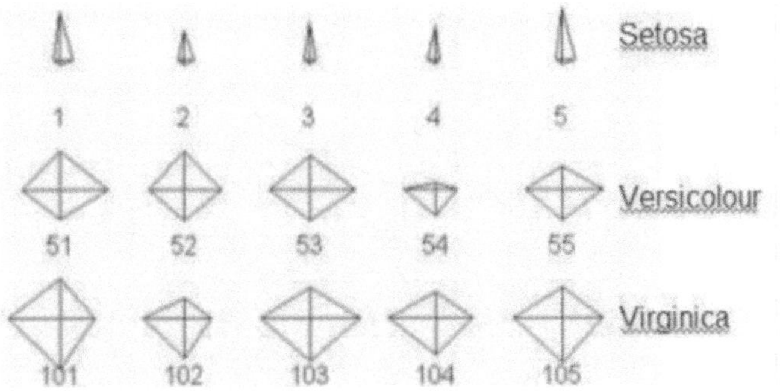


**Fig. 1.** Star plots of some IRIS data with four variables and there class

## 2.2 Graphical Features

To construct a star plot, we first rescale each variable to range from c to 1, where c is the desired length of the smallest ray relative to the largest. c may be zero. If $x_{ij}$ is the j-th observation of the i-th variable, then the scaled variable is

$$x_{ij}^{*} = c + (1 - c)\frac{x_{ij} - \min_{j} x_{ij}}{\max_{j} x_{ij} - \min_{j} x_{ij}} \tag{1}$$

To display n variables, we choose n rays whose directions are equally spaced around the circle, so that the i-th ray is at an angle $W_i = 2\pi(i - 1)/n$ from the horizontal line, for i = 1,...,n. Then for the j-th rescaled observation $(x^*_{1j},\ldots, x^*_{nj})$, we draw a star whose i-th ray is proportional to $x^*_{ij}$ in the direction $W_i$. In other words, if we want the maximum radius to be R, then the required star is obtained by computing and connecting the n points $P_{ij}$, for $i = 1,\ldots,n$. We need to repeat $i = 1$ at the end to close the star. Figure 2 displays a star plot of the j-th observation.

$$P_{ij} = (x_{ij}^{*} R \cos \omega_i , x_{iji}^{*} R \sin \omega_i) \tag{2}$$

When there are many variables involved in a star plot, there is a serious question as to whether a viewer can get a visual impression of the behavior of a particular variable, or of the joint behavior of two variables. One of the main purposes of such a scheme is to obtain a star with a distinctive shape for each observation, so that the viewer can look for pairs or groups of stars with similar shapes, or individual observations that are very different from the rest.

The sub-area graphical features are designed as the following. For one observation with n dimension variance, its star plot include n triangle, which is a visional shape feature. Each triangle has an area value $S_{i,,}$ and a whole star plot has n dimensional area value. So the sub-area graphical features with n dimension variance can calculated as the following equation

$$S_i = \frac{1}{2} r_i \bullet r_{i+1} \bullet \sin \omega_i , i = 1,\cdots,n \tag{3}$$

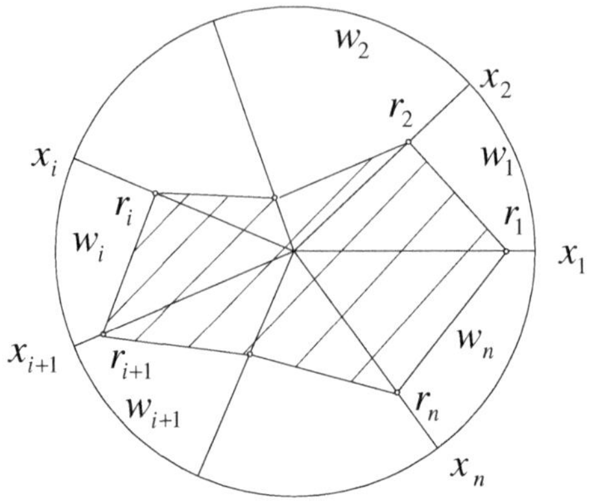


**Fig. 2.** A star plot of the j-th observation used to calculate sub-area graphical features, where n is the variance number of a observation, $r_i$ is rescaled observation to [0 1], Wi= $2\pi(i-1)/n$ is an angle

So based on star plots, the original data are changed to the sub-area graphical features with the same size.

The sub-barycentre graphical features are considered as the following. For one observation with n dimension variance, its star plot include n triangle, which is a visional shape feature. Each triangle has an barycentre $G_i$=( $abs_i$ , $angle_i$ ), and a whole star plot has n barycentre with n amplitude value $abs_i$ and n angle value $angle_i$ . So the sub-barycentre graphical features with n amplitude value and n angle value can calculated as the following equation

$$\begin{cases} abs_i = \sqrt{(\frac{r_i}{3}\sin\omega_i)^2 + ((r_i\cos\omega_i - \frac{r_{i+1}}{2})/3 + \frac{r_{i+1}}{2})^2} \\ angle_i = ar\sin(\frac{\frac{r_i}{3}\sin w_i}{abs}) \end{cases} ,i = 1,\cdots,n \quad (4)$$

So based on star plots, the original data are changed to the sub-barycentre graphical features with the double size, which is shown as Fig.3. For simplification or dimension reduction, we only consider the n amplitude value $abs_i$ as the sub-barycentre graphical features for a star plot. Finally the original data are changed to the sub-barycentre graphical features with the same size.

### 2.3 K Nearest Neighbor Classifier

The KNN method is a simple yet effective method for classification in the areas of pattern recognition, machine learning, data mining, and information retrieval. It has been successfully used in a variety of real-world applications. KNN can be very competitive with the state-of-the-art classification methods. A successful application of KNN depends on a suitable distance function and a choice of K. IF K=1, KNN

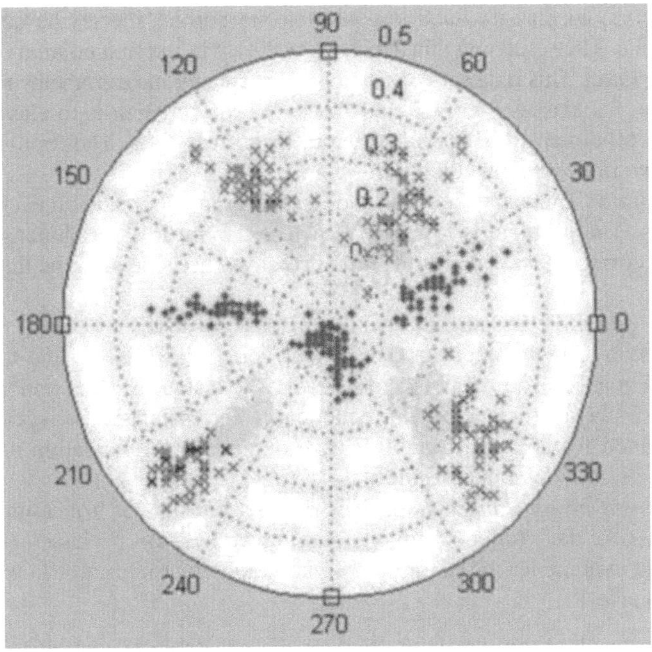


**Fig. 3.** the sub-barycentre graphical features of star plots for IRIS data sets with 4 dimensions, 150 observations and 3 class( iris setosa, iris versicolor and iris virginica are corresponding to the color of red, yellow and blue

classifier becomes the Nearest Neighbor classifier (1NN). The distance function puts data points in order according to their distance to the query and k determine show many data points are selected and used as neighbors. Classification is usually done by voting among the neighbors. There exist many distance functions in the literature. No distance function is known to perform consistently well, even under some conditions; no value of k is known to be consistently good, even under some circumstances. In other words, the performance of distance functions is unpredictable. This makes the use of KNN highly experience-dependent. The Euclidean distance function is probably the most commonly used in any distance-based algorithm.

## 3 Experiments and Results

### 3.1 Experments

Several standard benchmark corpora from the UCI Repository of Machine Learning Databases and Domain Theories (UCI) have been used[1]. A short description of these corpora is given below:

*1) Iris data: This* data set consists of 4 measurements made on each of 150 iris plants of 3 species. The two species are iris setosa, iris versicolor and iris virginica.

[1] http://www.ics.uci.edu/mlearn/MLRepository.html

The problem is to classify each test point to its correct species based on the four measurements. The results on this data set are shown in the first column of Table 1.

*2) Sonar data:* This data set consists of 60 frequency measurements made on each of 208 data of 2 classes ("mines" and "rocks"). The problem is to classify each test point in the 60-dimensional feature space to its correct class. The results on this data set are shown in the second column of Table 1.

*3) Liver data:* This data set consists of 6 measurements made on each of 345 data of 2 classes. The problem is to classify each test point in the 6-dimensional feature space to its correct class. The results on this data set are shown in the third column of Table 1.

*4) Vote data:* This data set includes votes for each of the U.S. House of Representatives Congressmen on the 16 key votes identified by the Congressional Quarterly Almanac. The data set consists of 232 instances after removing missing values, and 2 classes (democrat and republican). The instances are represented by 16 Boolean valued features. The average leave-one-out cross validation error rates are shown in the fourth column of Table 1

*5) Wisconsin breast cancer data:* This data set consists of 9 measurements made on each of 683 data (after removing missing values) of 2 classes (malignant or benign). The average leave-one-out cross validation error rates are shown in the fifth column of Table 1.

Besides, our algorithm has been tried on the vegetable oil data [6]. This data set collects 95 samples from seven different classes: pumpkin oils; sunflower oils; peanut oils; oliver oils; soybean oils; rapeseed oils and corn oils. A 7-dimensional fatty acid feature of each sample is measured which is Palmitic, Stearic, Oleic, Linoleic, Linolenic, Eicosanoic and Eicosenoic. The average leave-one-out cross validation error rates are shown in the sixth column of Table 1.

For the 1NN, KNN classifier we use PRTOOLS toolbox [7], and the most best K value selected by leave-one-out cross validation method. The 1NN, KNN classifier was also explicitly compared with SVM with radial basis kernels. We used SVMlight toolbox [8], and set the kernel scale value of in equal to the optimal one determined via cross validation. Also the value of C for the soft-margin classifier is optimized via cross validation.

**Table 1.** Average classification error rates for real data(%)

| | Iris | Sonar | Liver | Vote | breast cancer | oil |
|---|---|---|---|---|---|---|
| 1NN | 4.7 | 12.5 | 34.4 | 4.4 | 4.3 | 5.3 |
| KNN | 4.0 | 12.5 | 26.1 | 3.0 | 2.6 | 4.2 |
| SVM | 2.6 | 14.4 | 32.5 | 7.8 | 3.7 | 3.2 |
| 1NN a | 4.0 | 12.0 | 25.3 | 3.0 | 3.2 | 0 |
| KNN a | 4.0 | 11.3 | 24.6 | 3.0 | 2.9 | 0 |
| 1NN b | 3.3 | 11.7 | 21.2 | 2.8 | 4.3 | 0 |
| KNN b | 3.3 | 10.9 | 20.6 | 2.6 | 2.5 | 0 |

[a] with sub-area graphical features

[b] with sub-barycentre graphical features

### 3.2 Results

From Table I, the performance of KNN is superior to that of 1NN, as K is selected by leave-one-out cross validation method with minimum error rate. The performance of SVM with radial basis kernels is superior to that of 1NN, which is not surprised. For the optimized KNN and the optimized SVM with radial basis kernels, each has his strong point. The different of the two methods depend on the data set.

From Table I, the performance of KNN with sub-area graphical features is not superior to that of the performance of KNN with sub-barycentre graphical features. The performance of 1NN with sub-area graphical features is not superior to that of the performance of 1NN with sub- barycentre graphical features. These indicate the better class separability of sub-barycentre graphical features.

From Table I, the performance of KNN with graphical features is superior to that of the performance of KNN without graphical features. Even sometimes the performance of 1NN with graphical features is superior to that of the performance of KNN or SVM without graphical features.

Note this result only depend on the six data sets.

## 4 Conclusion

Based on the concept of the graphical representation, this paper proposes the concept of the graphical features and gives two graphical features based on star plot: the sub-area features and sub-barycentre features.

The effectiveness of the two graphical features was tested using six data sets. The results shows that the proposed graphical features can achieve high classification accuracy, even compared the best SVM classifier.

To fully investigate the potential of the graphical features, more comprehensive experiments can be performed. One possible future direction is the improved sub-barycentre features which make the class reparability. Another possible future direction is that star plots succeed in displaying high-dimensional data without any dimension reduction. But they also suffer from a problem: The order of attributes has an impact on the resulting overall shape and therefore on how the data is perceived.

**Acknowledgments.** This work was supported by National Natural Science Foundation of China (No60504035, No60474065, and No60605006). The work was also partly supported by the Science Foundation of Yanshan University for the Excellent Ph.D Students.

## References

1. Jain. A. K., Duin R., Mao Jianchang.: "Statistical Pattern Recognition: A Review," IEEE Transactions on Pattern Analysis and Machine Intelligence, 22(1),(2000) 4-37
2. Duda R.O., Hart P.E., Stork D.G.: Pattern Classification and Scene Analysis. 2nd ed, New York: John Wiley & Sons. (2000)
3. Anscombe, F.J.,: Graphs in Statistical Analysis, the American Statistician, 27, 17–21

4. Cover, T.M., Hart, P.E.: Nearest Neighbor Pattern Classification . IEEE Trans. Information Theory, vol. 13, no. 1, (1967) 21-27
5. Paredes, R., Vidal, E.: A Class-Dependent Weighted Dissimilarity Measure for Nearest Neighbor Classification Problems," Patter Recognitin Letters, 21 (2000)1027-1036
6. Darinka, B. V., Zdenka, C. K., Marjana N.: Multivariate Data Analysis in Classification of Vegetable Oils Characterized by the Content of Fatty Acids. Chemometrics and Intelligent Laboratory Systems, 75 (2005) 31– 43
7. Duin, R.P.W., Juszczak, P., Paclik, P., Pekalska, E., Deridder, D. , Tax, D.M.J.: PRTools4, A Matlab Toolbox for Pattern Recognition, Delft University of Technology(2004)
8. Joachims, T.: Making Large-Scale SVM Learning Practical. Advances in Kernel Methods - Support Vector Learning, B. Schölkopf and C. Burges and A. Smola (ed.), MIT-Press, (1999) Available: http://svmlight.joachims.org/

# A Mixed Algorithm of PCA and LDA for Fault Diagnosis of Induction Motor

Wook Je Park, Sang H. Lee, Won Kyung Joo, and Jung Il Song

School of Mecatronics, Changwon National University
9 Sarim-dong, Changwon, Gyeongnam, 641-773, Korea
{leehyuk, parkwj, nom2479, jisong}@changwon.ac.kr

**Abstract.** In this paper, we propose a feature extraction method and fusion algorithm which is constructed by PCA and LDA to detect a fault state of the induction motor that is applied over the whole field of a industry. After yielding a feature vector from current signal which is measured by an experiment using PCA and LDA, we use the reference data to produce matching values. In a diagnostic step, two matching values which are respectively obtained by PCA and LDA are combined by probability model, and a faulted signal is finally diagnosed. As the proposed diagnosis algorithm brings only merits of PCA and LDA into relief, it shows excellent performance under the noisy environment. The simulation is executed under various noisy conditions in order to demonstrate the suitability of the proposed algorithm and it showed more excellent performance than the case just using conventional PCA or LDA

**Keywords:** PCA, LDA, induction motor, fault diagnosis.

## 1 Introduction

For the reduction of maintenance cost and preventing the unscheduled downtimes of the induction motor, fault detection techniques for the induction motors have been studied by the numerous researchers [1-7]. Faults of an induction machine are classified by bearing fault, coupling and rotor bar faults, air gap, rotor, end ring and stator faults, *etc.*. Various measurements, vibration signal, stator currents, lights, sound, heat, *etc.* are required to monitor the status of the motor or to detect the faults. It is well known that the current signal is useful to detect the faults because of cost reduction. To detect the faults of induction motor, we can derive analytically or heuristically. In both cases, features of the faulty or healthy motor are needed.

In this paper, we focus about the characteristic extraction of the healthy and faulted induction motor. Obtaining characteristic values from the stator current can be used in the frequency domain and time domain approach. In frequency domain, Fourier and Wavelet transformation of the signal have good points for the obtaining characteristics. However, these methods have not complete result alone, hence other methods, PCA(principal component analysis) and LDA(linear discriminant analysis) applied to obtain characteristics. In the Section 2, we combine PCA and LDA. That mixed algorithm have the robust characteristics over the noise condition. Proposed algorithm has the advantages of each methods, and reveals good performance

D.-S. Huang, L. Heutte, and M. Loog (Eds.): ICIC 2007, LNAI 4682, pp. 934–942, 2007.

compared to the individual results. In Section 3, properness of the algorithm is checked under the various noise condition. Finally, conclusions follow in Section 4.

## 2 LDA and PCA

The By linear transformation, PCA presents projecting the high-dimensional data onto a lower dimensional space[8-10]. This approach seeks a projection that best separates the data in a least-square sense. However components that are obtained by the PCA have not discrimination characteristic between data in different classes. Next, we find an orientation for which the projected samples are well separated. This is exactly the goal of LDA. PCA and LDA methods is applied to the determination of healthy and faulty induction motor. Procedure is illustrated in Fig. 1. Ratings and specifications of experimental motor are illustrated in Table 1.

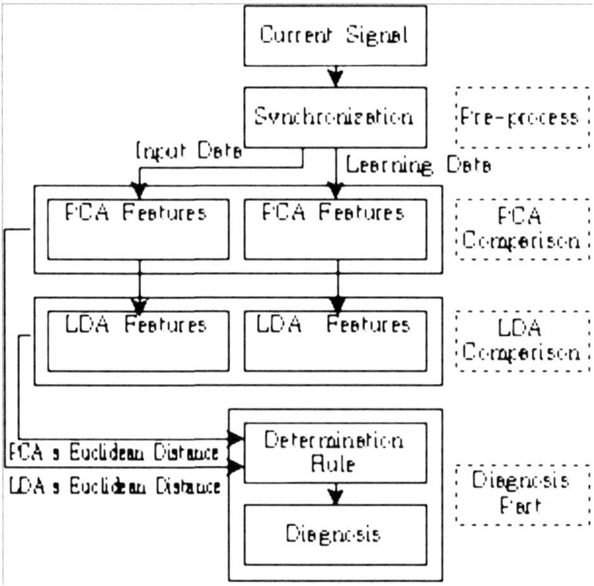


**Fig. 1.** Fault diagnosis system for induction motor

**Table 1.** Ratings and specifications of experimental motor

| Motor rating | | Motor spec. | |
|---|---|---|---|
| Rated voltage | 220V | No. of slots | 34 |
| Rated speed | 3450 rpm | No. of poles | 4 |
| Rated power | 0.5 HP | No. of rotor bars | 24 |

Considering faulty conditions are 5 cases, which are bearing fault, bowed rotor bar, broken rotor bar, static and dynamic eccentricity. In addition, healthy condition is included. In this paper, total 6 cases patterns are classified by PCA and LDA method.

### 2.1 Principal Component Analysis(PCA)

We consider $n$ dimensional samples $x_1, \cdots, x_n$, by a single vector $x_0$. Suppose that we want to find a vector $x_0$ such that the sum of the squared distances between $x_0$ and the various $x_k$ is as small as possible. Then, $x_0$ becomes sample mean. Data $x_k$ denotes

$$x_k = m + a_k e. \tag{1}$$

Where, $m$ is sample mean, $e$ be a unit vector in the direction of the line. Optimal set of coefficients $a_k$ by minimizing the squared-error criterion function

$$\begin{aligned} J_1(a_1, \cdots, a_n, e) &= \sum_{k=1}^{n} \|(m + a_k e) - x_k\|^2 \\ &= \sum_{k=1}^{n} a_k^2 \|e\|^2 - 2\sum_{k=1}^{n} a_k e^t (x_k - m) + \sum_{k=1}^{n} \|x_k - m\|^2 \end{aligned} \tag{2}$$

where, $\|\cdot\|$ is the 2-norm, $\|e\| = 1$.

Also $\dfrac{\partial J_1}{\partial a_k} = 0$ gives

$$a_k = e^t (x_k - m) \tag{3}$$

where, $a_k$ is the basis or feature vector of $x$, and, principal component. In order to find $e$, first we define Scatter matrix

$$S = \sum_{k=1}^{n} (x_k - m)(x_k - m)^t. \tag{4}$$

Substituting (3) into (2), we derive following equation.

$$\begin{aligned} J_1(e) &= \sum_{k=1}^{n} a_k^2 - 2\sum_{k=1}^{n} a_k + \sum_{k=1}^{n} \|x_k - m\|^2 \\ &= -\sum_{k=1}^{n} \left[e^t (x_k - m)\right]^2 + \sum_{k=1}^{n} \|x_k - m\|^2 \\ &= -\sum_{k=1}^{n} e^t (x_k - m)(x_k - m)^t e + \sum_{k=1}^{n} \|x_k - m\|^2 \\ &= -e^t S e + \sum_{k=1}^{n} \|x_k - m\|^2 \end{aligned}$$

Clearly, the vector $e$ that minimizes $J_1$ also maximizes $e^t Se$. Lagrange multipliers to maximize $e^t Se$ subject to the constraint that $\|e\|=1$. Let $\lambda$ be the undetermined multiplier,

$$L = e^t Se - \lambda(e^t e - 1)$$

we differentiate with respect to $e$. Setting $\frac{\partial L}{\partial e}$ = 0, we see that $e$ must be an eigenvector of the Scatter matrix:

$$Se = \lambda e \tag{5}$$

*$\lambda$ is the eigenvalue of $S$ and $e$ is the eigenvector corresponding to the $\lambda$. Because $e^t Se = \lambda e^t e$, it follows that to maximize $e^t Se$, we want to select the eigenvector corresponding to the largest eigenvalue of the Scatter matrix $S$. Now we consider principal value $a_k$ into the characteristic value to classify pattern between healthy and faulty condition. From (3), principal value $a_k$ of known vector $x$ is calculated, and unknown vector principal value $a_k^*$ is also obtained.*

## 2.2 Linear Discriminant Analysis(LDA)

LDA seeks directions that are efficient for discrimination. For this discrimination analysis, we first define Between-Class Scatter matrix (BCS) $S_B$ and Within-Class Scatter(WCS) $S_W$ by

$$S_B = \sum_{i=1}^{c} n_i (m_i - m)(m_i - m)^T \tag{6}$$

$$S_W = \sum_{i=1}^{c} \sum_{x \in C_i} (x - m_i)(x - m_i)^t \tag{7}$$

where, $c$ is the number of class, $m_i$ is the average value of the samples in class $c_i$. Average of the total samples is $m$, $n_i$ is the number of signal in class $c_i$.

In terms of $S_W$ and $S_B$, the criterion can be written as

$$J(W) = \frac{\left|W^T S_B W\right|}{\left|W^T S_W W\right|} \tag{8}$$

where, $W = [w_1, w_2, \cdots, w_{c-1}]$. Rectangular matrix $W$ : maximizes (7). The columns of $W$ are the generalized eigenvectors that correspond to the largest eigenvalues in

$$S_B w_i = \lambda_i S_W w_i, i = 1, 2, \cdots, c-1. \tag{9}$$

Conventional eigenvalue problem requires an unnecessary computation of the inverse of $S_w$. Instead, with the eigenvalues as the roots of the characteristic polynomial $\lfloor S_B - \lambda_i S_W \rfloor = 0$, eigenvectors are directly solved.

$$(S_B - \lambda_i S_W) w_i = 0, i = 1, 2, \cdots, c-1. \quad (10)$$

For training data $x_i s$, feature vector $T_i$ can be obtained as follows

$$T_i = W^T a_i = W^T e^t (x_i - m). \quad (11)$$

PCA feature vector $a_i$ is projected to the LDA space by the matrix $W$. Generally, training data $c$ is less than the data points of signal, WCS matrix $S_W$ becomes singular. This means that projection matrix $W$ has to be chosen properly. Next, we compute the distance of training PCA feature vector $a_i$ and test PCA feature vector $a_i'$ as $D_{PCA}$. Furthermore LDA feature distance is also computed as follows

$$D_{PCA} = \sqrt{(a_i - a_i')^T (a_i - a_i')} \quad (12)$$

$$D_{LDA} = \sqrt{(T_i - T_i')^T (T_i - T_i')} \quad (13)$$

where $T_i$ and $T_i'$ are the training LDA feature and test feature vector, respectively. When the Euclidean distance satisfied $min(D_{LDA}) \langle T_{th}$, where $T_{th}$ is the predetermined threshold value, then fault detection process carry out. We choose value of $T_{th}$ that $D_{LDA}$ is larger than $D_{PCA}$ as the noise rises with the iterative experiments. Whereas in the case of $min(D_{LDA}) \rangle T_{th}$, new distance $D_{SUM}$ is calculated by $D_{PCA}$ and $D_{LDA}$ each cases.

$$D_{SUM} = D_{PCA} + D_{LDA} \quad (14)$$

In order to get more reliable data, we apply Bootstrap method to the $D_{SUM}$, then we get Gaussian distribution over each fault cases. With this result we regard minimum distance of Health (1), Fault (2), and Fault (N) as the fault condition.

Signal has 128 data points. Training vectors are 54(9×6cases), and mean of $x_i$, $m$ represent [1×128] size. Sampling frequency is 3kHz and sampling time is 0.13(=1/(60×128))[ms]. Fig. 3 shows the result without noise case, and Fig. 4 is the result of SNR=5 case. As shown in Fig.'s, it is hard to discriminate when there is noise. Discrimination results are compared with those of LDA later. Under noise free

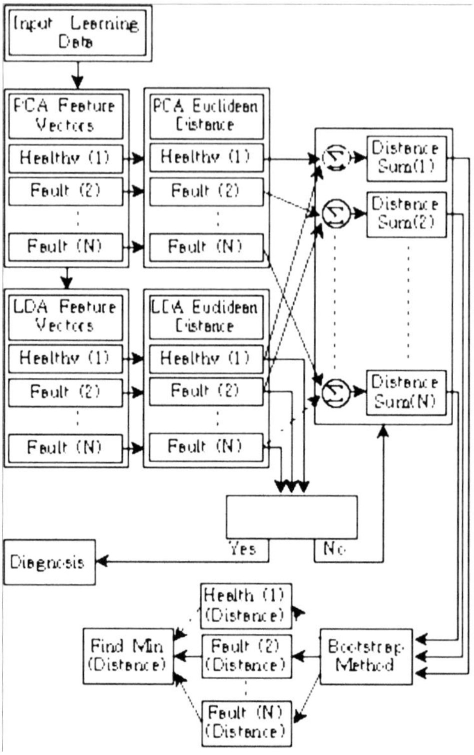


**Fig. 2.** Fusion algorithm for a fault diagnosis

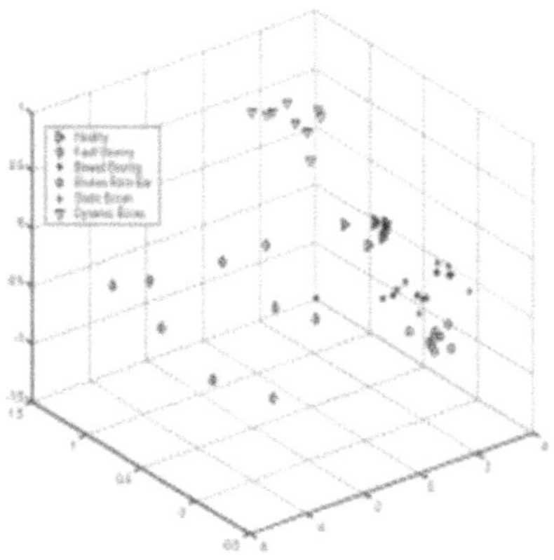

**Fig. 3.** Feature vectors(by PCA)

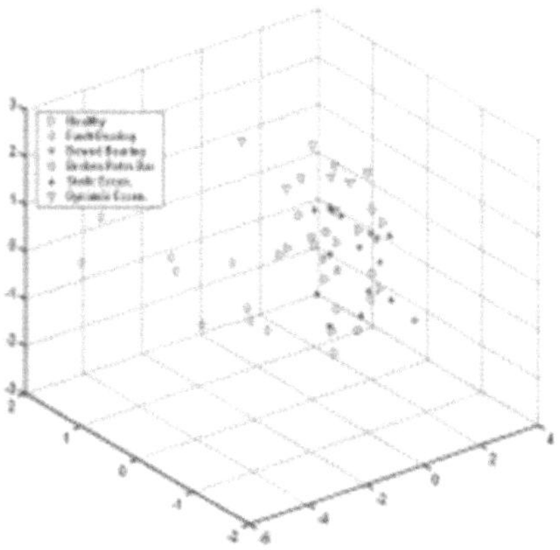

**Fig. 4.** Feature vectors(by PCA, SNR=5)

condition, Fig. 5 show superior to Fig. 3. When the SNR is 5, PCA and LDA results are illustrated in Fig. 4 and Fig. 6. Both cases can not discriminate faults. Hence we use the above mixed algorithm.

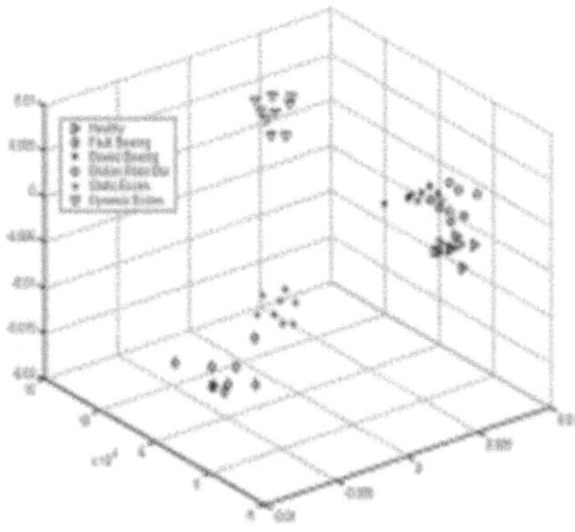

Fig. 5. Feature vectors(by LDA)

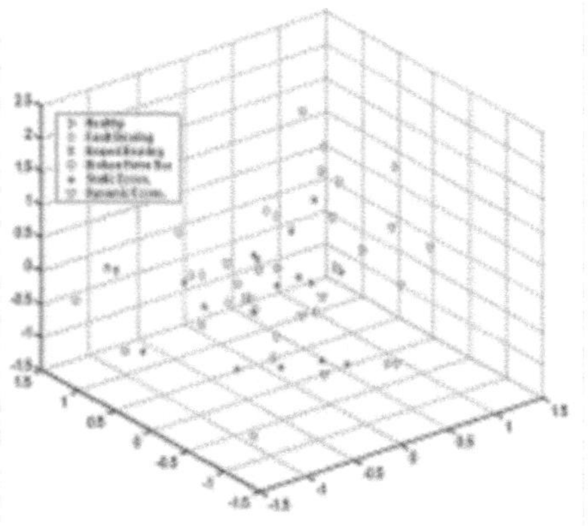

Fig. 6. Feature vectors (by LDA, SNR=5)

## 3 Experimental Results

For the extraction of current characteristics, we consider 3-phases induction motor with 220V, 5hp, 4 poles. Experimental system is illustrated in Fig. 7. System contains 5kw Permanent Magnet Synchronous Motor, induction motor, PWM inverter and PWM converter, furthermore digital board containing TMS320VC33 DSP chip. Data acquisition device of NI co. is equipped to obtain many data.

Fig. 7. Experimental system

We tried noise free case and SNR(signal noise ratio) from 5 to 35. 9 signals per one fault, total 54 cases are tested. For the case of noise free, LDA show perfectly. Noise free result is illustrated in Table 2. As in Table 2, bowed rotor and static eccentricity case has 4 detection errors. Hence LDA has the advantage over the noise free case because of maximizing discrimination of each cases. Recognition results under the noise condition are carried out in Table 3. Over SNR = 40, there are no changes.

**Table 2.** Recognition result

| Driving Condition | LDA | | PCA | |
|---|---|---|---|---|
| | Recognition | Error | Recognition | Error |
| Healthy Condition | 9 | 0 | 9 | 0 |
| Faulted Bearing | 9 | 0 | 9 | 0 |
| Bowed Rotor | 9 | 0 | 7 | 2 |
| Broken Rotor Bar | 9 | 0 | 9 | 0 |
| Static Eccentricity | 9 | 0 | 7 | 2 |
| Dynamic Eccentricity | 9 | 0 | 9 | 0 |

**Table 3.** Recognition result according to noise variation

| SNR | Recognition Ratio | | |
|---|---|---|---|
| | PCA | LDA | Proposed |
| 35 | 92.6 | 100 | 100 |
| 30 | 91.3 | 98.7 | 98.7 |
| 25 | 92.22 | 95.17 | 95.17 |
| 20 | 88.89 | 85.56 | 90.74 |
| 15 | 82.78 | 67.96 | 84.82 |
| 10 | 72.78 | 51.23 | 77.59 |
| 5 | 60.56 | 38.52 | 62.96 |

Results of Table 3 are indicate that LDA performance is better than PCA when SNR is from 100 to 25. However error rate deteriorate rapidly as noise larger(SNR< 25). At SNR is 5, recognition rate of LDA is 22% less than that of PCA result. As a result, proposed algorithm reveals over 4.8% and 26% compared to PCA and LDA only, respectively.

## 4 Conclusions

Mixed algorithm based on the PCA and LDA methods is proposed for the detection of faulty induction motor. LDA has the good result when noise free case. When there is noise, mixed PCA/LDA algorithm is proposed to raise recognition rate. With total 108 data of the 6 cases, we applied 54 data to PCA and LDA respectively. Remaining 54 data are tested to verify that the proposed approach have better result than the individual with or without noise.

## References

1. B, 1. P.: Vas, Parameter Estimation, Condition Monitoring, and Diagnosis of Electrical Machines, Clarendron Press, Oxford, (1993)
2. Nejjari, H., Benbouzid, M.E.H.: Monitoring and Diagnosis of Induction Motors Electrical Faults Using a Current Park's Vector Pattern Learning Approach, IEEE Transactions on Industry Applications, Vol. 36, Issue 3, (2000) 730-735

3. Bellini, A., Filippetti, F., Franceschini, G., Tassoni, C., Kliman, G.B.: Quantitative Evaluation of Induction Motor Broken Bars by Means of Electrical Signature Analysis, IEEE Transactions on Industry Applications, Vol. 37, Issue 5, (2000) 1248-1255
4. Kyusung, K., Parlos, A.G., Mohan Bharadwaj, R.: Sensorless Fault Diagnosis of Induction Motors, IEEE Transactions on Industrial Electronics, Vol. 50 Issue 5, (2003) 1038-1051
5. Zidani, F., El Hachemi Benbouzid, M., Diallo, D., Nait-Said, M.S.: Induction Motor Stator Faults Diagnosis by a Current Concordia Pattern-based Fuzzy Decision System, IEEE Transactions on Energy Conversion, Vol. 18, Issue 4, (2003) 469-475
6. Haji, M., Toliyat, H.A.: Pattern Recognition-a Technique for Induction Machines Rotor Broken Bar Detection, IEEE Trans. on, Energy Conversion, Vol. 16, Issue 4, (2001) 312 - 317
7. Trzynadlowski, A.M., Ritchie, E.: Comparative Investigation of Dagnostic Media for Induction Motors: a Case of Rotor Cage Faults, IEEE Trans. on, Industrial Electronics, Vol. 47, Issue 5, (2000) 1092 -1099
8. Turk, M., Pentland, A., Face Recognition Using Eigenfaces, Proc. IEEE Conf. on Computer Vision and Pattern Recognition, (1991) 586-591
9. Belhumeur, P.N., Hespanha, J.P., Kriegmaqn, D.J.: Eigenfaces vs. Fisherfaces : Recognition Using Class Specific Linear Projection, IEEE Trans. on Pattern Analysis and Machine Intell., 19(7), (1997) 711-720
10. Richard, O.D., Peter, E.H., David, G. S.: Pattern Classification, JOHN WILEY&SONS, INC. Second Edition, 2002

# A Test Theory of the Model-Based Diagnosis

XueNong Zhang[1,2], YunFei Jiang[1], and AiXiang Chen[1]

[1] Institute of Software Research, Zhongshan University
[2] Network Center, GuangDong Pharmaceutical University
zxnxlq@163.com

**Abstract.** For finding the actual diagnosis of the faulty system, this paper discusses the relationship between the candidate of diagnosis and the set of the actual faulty components. Then we define the notion of adoptability of the diagnostic system and prove that the consistency-based diagnosis is adoptable. On the basis of the above works, a test theory of the diagnosis based on consistency is proposed, which indicates that how tests provide information about the current space of diagnoses.

**Keywords:** model-based diagnosis, adoptability, test theory.

## 1 Introduction

Due to its generality and its dramatic importance in many application domains, automated diagnosis has long been an active research area of Artificial Intelligence.

In 1987, a logical theory of diagnosis was proposed by Reiter[1], and it is usually called the theory of consistency-based diagnosis. Its main idea is to establish a model of the normal structure and behavior of the diagnosed objects. Diagnosis is then modeled as finding a discrepancy between the normal behavior predicted from the model and the actually observed abnormal behavior. The discrepancy in this approach is formalized as logical inconsistency.

The classical model describes the system's structure and behavior usually by the first-order language. Luca Chittaro et al. [2] proposed a hierarchical model which can represent multiple behavioral modes of one component in its various states. P.Baroni et al.[3] proposed a dynamic system model based on the finite-state automata. Console et al.[4] described the diagnostic problem based on the process algebra.

The computational complexity of diagnosis is one of the well-known problems that need to be tackled in order to deploy the real-world applications of model-based diagnosis. Several relevant contributions can be found in the references [5-9].

However, for a given diagnostic problem, there are lots of candidates of diagnosis. Therefore, we must test them for finding the actual diagnosis. In general, for a given faulty system, we may adopt different diagnostic methods and standards to find different diagnoses. In our view, if a diagnostic method is adoptable, then the actual diagnosis, which is the set of the actual faulty components of the system, should be included in the set of candidates of diagnosis, based on the related principle of the diagnostic system. Otherwise, testing of the diagnoses is worthless.

D.-S. Huang, L. Heutte, and M. Loog (Eds.): ICIC 2007, LNAI 4682, pp. 943–951, 2007.

Hence, our works allow attention to be focused on the relationship between the candidate of the diagnosis and the set of the actual faulty components of the considered system before and after the execution of a test.

This paper is structured as follows: The classical method of model-based diagnosis is introduced in Section 2; Section 3 discusses the relationship between the diagnosis and the set of the actual faulty components. Then, we define the adoptability of diagnostic system and prove that the consistency-based diagnosis is adoptable; On the basis of the above works, a test theory of the diagnosis based on consistency is proposed in Section 4. Related research works are discussed in Section 5 and the conclusions are drawn in Section 6.

## 2 Model-Based Diagnosis

In this section, we simply introduce the classical method of model-based diagnosis which is proposed by Reiter[1], including the definition of model-based diagnosis and the process of generating the consistency-based diagnoses. Reiter's definition of diagnosis is based on logical consistency.

**Definition 1.** consistency-based (minimal) diagnosis
The consistency-based (minimal) diagnosis of the diagnostic problem (*SD*, *COMPS*, *OBS*) is a (minimal) set $D \subseteq COMPS$ which satisfies that $SD \cup OBS \cup \{ \neg ab(c) \mid c \in COMPS - D\}$ is consistent, where:

*SD*, the system description, is a finite set of first-order sentences;
*COMPS*, the system components, is a finite set of constants;
*OBS* is a finite set of first-order formulas which describe the system observations;
*ab* is a unary predicate, interpreted to mean "abnormal". When component *c* is abnormal, *ab*(*c*) is true.

**Definition 2.** causality-based diagnosis
The causality-based (minimal) diagnosis of the diagnostic problem (*SD*, *COMPS*, *OBS*) is a (minimal) set $D \subseteq COMPS$ which satisfies $SD \cup \{ \neg ab(c) \mid c \in COMPS - D\}$ is consistent and $SD \cup \{\neg ab(c) \mid c \in COMPS\text{-}D\} \vdash OBS$.

Based on the above definition, the approach for finding diagnoses with the structure and observations of the system can be proposed as following: first, generating each subset *D* of the *COMPS*; second, testing the consistency of $SD \cup OBS \cup \{\neg ab(c) \mid c \in COMPS - D\}$. Obviously, this naive method is too complex to deploy the real-world applications. Hence, for finding the minimal diagnoses, all minimal conflict sets must be computed firstly, then, the minimal diagnoses can be obtained by computing the hitting set of the conflict sets.

**Definition 3.** (minimal) conflict set
The (minimal) conflict set of the diagnostic problem (*SD*, *COMPS*, *OBS*) is a (minimal) set $\{C_1, C_2, \dots, C_k\} \subseteq COMPS$ satisfied that $SD \cup OBS \cup \{\neg ab(C_1), \dots, \neg ab(C_k)\}$ is inconsistent.

**Definition 4.** (minimal) hitting set
The (minimal) hitting set is a (minimal) set $H \subseteq \cup_{S \in C} S$ satisfied $H \cap S \neq \emptyset$ for any $S \in C$.

**Theorem 1.** Suppose *D* is a subset of *COMPS*, *D* is a minimal diagnosis of (*SD*, *COMPS*, *OBS*) if and only if *D* is a minimal hitting set of all of the minimal conflict sets of (*SD*, *COMPS*, *OBS*).

## 3 Diagnosis and the Actual Faulty Components

Before discussing the relationship between the diagnosis and the actual faulty components, we first give a set of symbols. We note diagnostic problem as *M* = (*SD*, *COMPS*, *OBS*). Every diagnostic system can resolve the diagnostic problem by some method. We note the diagnostic system which adopt consistency-based diagnostic method as *CD*, *CD*(*M*) express the set of the consistency-based diagnoses of *M*, and $CD_{min}(M)$ express the set of the minimal consistency-based diagnoses of *M*; *AD* is the causality diagnostic system, *AD*(*M*) is the set of the causality diagnoses of *M*, and $AD_{min}(M)$ is the set of the minimal causality diagnoses of *M*.

**Definition 5.** compare of the diagnostic systems
Given diagnostic systems *R*1 and *R*2, for any diagnostic problem *M*, if $R1(M) \supseteq R2(M)$, then we said that *R*1 is not stronger than *R*2. It is noted as $R1 < R2$.

**Theorem 2.** $CD < AD$
**Proof**
For any diagnostic problem *M*, if a diagnosis $D \in AD(M)$, then $SD \cup \{\neg ab(c) \mid c \in COMPS - D\}$ is consistent and $SD \cup \{\neg ab(c) \mid c \in COMPS\text{-}D\} \vdash OBS$. Therefore, $SD \cup OBS \cup \{\neg ab(c) \mid c \in COMPS\text{-}D\}$ is consistent. Based on definition 1, $D \in CD(M)$. Hence, $CD < AD$ is valid.

Theorem 2 details the relationship of the consistency-based diagnosis and the causality diagnosis. For the same diagnostic problem, every causality diagnosis is also a consistency-based diagnosis.

The aim of diagnosis is finding the actual faulty components of the system (the actual diagnosis). Therefore, we hope that the set of diagnoses includes the actual diagnosis. Thus, it is possible that we can eliminate all diagnoses but the actual diagnosis by testing them. The notion of the adoptability of diagnostic system is proposed as following.

**Definition 6.** adoptability of the diagnostic system
The diagnostic system *R* is adoptability if and only if $RealDiag \in R(M)$ for any diagnostic problem *M*, where *RealDiag* is the actual diagnosis.

**Theorem 3.** *CD* is adoptable.
**Proof**
For any given *M* = (*SD*, *COMPS*, *OBS*), let *RealDiag* is the actual diagnosis, all components of the system have the actual values of the inputs and the output, which is

the reasonable interpretation of the current observation. Therefore, $SD \cup OBS \cup \{\neg ab(c) \mid c \in COMPS\text{-}RealDiag\}$ is consistent. It is valid that $RealDiag \in CD(M)$ based on the definition 1. Hence, *CD* is adoptable.

In general, because there are lots of candidates of diagnosis, we only find all the minimal diagnoses. For example, adopting Reiter's method[1], there are 26 diagnoses, but only 4 minimal diagnoses in the example 1.

For the set of minimal diagnoses, we hope that there exists at least one minimal diagnosis satisfied that all the components in it are abnormal. The following theorem details that the set of the minimal diagnoses obtained by the adoptability diagnostic system satisfies the above requirement.

**Theorem 4.** If diagnostic system *R* is adoptable, then for any diagnostic problem *M*, at least exist a minimal diagnosis $D \in R(M)$, which satisfies $D \subseteq RealDiag$.

**Proof**

Because diagnostic system *R* is adoptable, for any diagnostic problem *M*, $RealDiag \in R(M)$. If *RealDiag* is a minimal diagnosis, let $D = RealDiag$, the theorem is valid. If *RealDiag* is not a minimal diagnosis, we can obtain a minimal diagnosis *D* by delete some components from *RealDiag*. Therefore, the theorem is also hold.

**Example 1**

The poly-box system, depicted in Fig.1, contains five components: *M*1, *M*2 and *M*3 are multipliers; *A*1 and *A*2 are adders. System's inputs are: $A = 3$, $C = 3$, $E = 3$, $B = 2$, $D = 2$, $F = 2$; output are: $G = 10$, $H = 12$. Suppose *M*1 and *A*1 are actual faulty components, and the actual value of *X* is 5, the actual values of *Y* and *Z* are 6. Therefore, *RealDiag* is $\{M1, A1\}$.

Because the output *G* of the system is abnormal, there arises a diagnostic problem and four consistency-based minimal diagnoses will be found: $\{M1\}$, $\{A1\}$, $\{M2, A2\}$ and $\{M2, M3\}$.

As said in theorem 4, there exists at least one minimal diagnosis which is the subset of the actual diagnosis: $\{M1\} \subseteq RealDiag$ and $\{A1\} \subseteq RealDiag$.

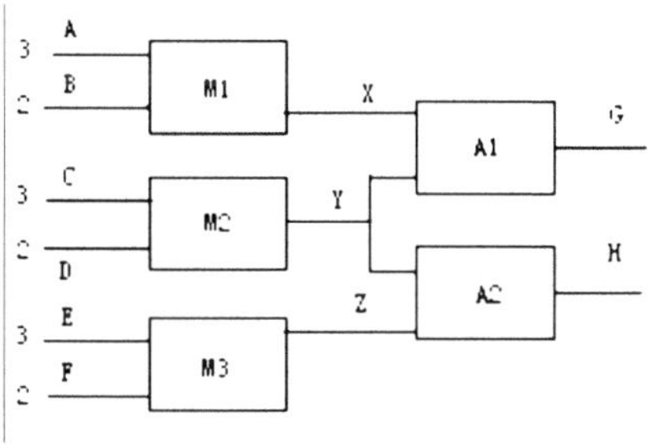


**Fig. 1.** Poly-box system

Not all diagnostic systems are adoptable. If the knowledge about the considered system is not complete, the causality diagnostic system is not adoptable. Because the model of system has not enough knowledge to explain the current behavior of the considered system, it is possible that the actual diagnosis has been lost when we obtained the causality diagnoses. It is noticeable that the consistency-based diagnosis is equal to the causality diagnosis when the knowledge about the considered system is complete[10].

Generally, experience-based diagnostic system is also not adoptable. There are some abnormal behaviors of the system which overstep the range of the current experience we have. Therefore, it is impossible to find the actual diagnosis by the current experience.

On the basis of the above works, in the following contents, we will discuss the test of the set of minimal diagnoses, which is a small part of the whole diagnosis space. Again, for finding the actual diagnosis, we only consider the consistency-based diagnostic system because it is adoptable.

## 4 Consistency-Based Test

Informally, the notion of a test provides for certain initial conditions which may be established by the tester, together with the specification of an observation whose outcome determines what the test conclusions are to be. McIlraith et al.[11] provide for a formal definition of a test by distinguishing a subset of literals of the propositional language, called the *achievable literals*. These will specify the initial condition for a test. In addition, a distinguished subset of the propositional symbols of the language called the *observables* is required. These will specify the observations to be made as part of a test.

**Definition 7.** test
A test is a pair $(A, o)$ where $A$ is a conjunction of achievable literals and $o$ is an observable.

A test specifies some initial condition $A$ which the tester establishes, and an observation $o$ whose truth value the tester is to determine.

**Definition 8.** outcome of a test
The outcome of a test $(A, o)$ is one of $o$, $\neg o$.

As a result of performing the test in the real world, the truth value of o is observed. If $o$ is observed to be true, the outcome of the test is $o$, otherwise $\neg o$.

In this paper, we only discuss the simple test, where $A$ = *true* and the test does not change the state of the considered system. Therefore, a test is a pair (*true*, $o$). In fact, the main results of the reference [11] are also for the test (*true*, $o$).

**Definition 9.** confirmation and refutation
For a given $M = (SD, COMPS, OBS)$, outcome $\alpha$ of the test (*true*, $o$) confirms (refutes) $D \in CD(M)$ iff $SD \wedge COMPS \wedge OBS \wedge HD \wedge \alpha$ is satisfiable (unsatisfiable), where $HD = \neg ab(c_1) \wedge \ldots \wedge \neg ab(c_i) \wedge ab(c_{i+1}) \wedge \ldots \wedge ab(c_j)$ , $\{c_1, \ldots, c_i\} = COMPS\text{-}D$ and $\{c_{i+1}, \ldots, c_j\} = D$.

**Definition 10.** prime implicate
A prime implicate of a propositional formula $\Sigma$ is a clause $C$ such that $\Sigma \vdash C$, and for no proper subclause $C'$ of $C$ does $\Sigma \vdash C'$.

**Definition 11.** discriminating test
A test (*true*, *o*) is a discriminating test for the $CD(M)$ iff there exists $D_i$, $D_j \in CD(M)$ such that the outcome $\alpha$ of test (*true*, *o*) refutes either $D_i$ or $D_j$.

In other words, a discriminating test must refute at least one diagnosis in the diagnosis space.

**Theorem 5.** For a given $M = (SD, COMPS, OBS)$, suppose $SD \cup OBS$ has at least two prime implicates of the form $\neg HD_i' \vee o$ and $\neg HD_j' \vee \neg o$ where $HD_i'$ and $HD_j'$ are subconjuncts of $HD_i$ and $HD_j$ respectively, for some $HD_i$, $HD_j \subset CD(M)$. Then (*true*, *o*) is a discriminating test for the $CD(M)$.

**Proof**
We prove the result in the case that $o$ is the outcome of (*true*, *o*). A symmetrical proof applies when the outcome is $\neg o$.

Since $\neg HD_i' \vee o$ is a prime implicate of $SD \cup OBS$, we have $SD \cup OBS \cup HD_i' \vdash o$. in addition, $HD_i'$ is subconjuncts of $HD_i$ , we have $SD \cup OBS \cup HD_i \vdash o$. Therefore, $SD \cup OBS \cup HD_i \cup o$ is satisfiable. Similarly, $SD \cup OBS \cup HD_j \cup \neg o$ is satisfiable.

Hence $o$ confirms $HD_i$ and refutes $HD_j$.

In what follows, we discuss the execution of a single test and its impact on the set of minimal diagnoses.

**Theorem 6.** For a given problem $M = (SD, COMPS, OBS)$, let the set of the diagnoses in the $CD_{min}(M)$ which refuted by outcome $\alpha$ of the test (*true*, *o*) is noted as *REF*, let $M'=(SD, COMPS, OBS \cup \alpha)$, we have

(1) $CD_{min}(M) - REF \subseteq CD_{min}(M')$;

(2) $REF \cap CD_{min}(M') = \varnothing$;

(3) for any new diagnosis $D' \in CD_{min}(M') - (CD_{min}(M) - REF)$, there exists a diagnosis $D \in REF$ satisfied $D \subset D'$.

**Proof**
Based on the definition 9 and definition 1, we have $CD_{min}(M) - REF \subseteq CD_{min}(M')$ and $REF \cap CD_{min}(M') = \varnothing$.

For any diagnosis $D' \in CD_{min}(M') - (CD_{min}(M) - REF)$, $SD \cup OBS \cup \alpha \cup \{\neg ab(c) \mid c \in COMPS\text{-}D'\}$ is consistent. Obviously, $SD \cup OBS \cup \{\neg ab(c) \mid c \in COMPS\text{-}D'\}$ is also consistent. Let $D$ is a minimal subset of $D'$ which satisfies that $SD \cup OBS \cup \{\neg ab(c) \mid c \in COMPS\text{-}D\}$ is consistent, we have $D \in REF$. Because test (*true*, *o*) refutes the diagnosis $D$, we have $D \neq D'$. Therefore $D \subset D'$ .

Theorem 6 indicates that the test may eliminate some diagnoses and/or is helpful of obtaining more information about diagnosis (see example 2).

**Example 2**
Consider the poly-box system in example1, system's inputs are: $A = 3$, $C = 3$, $E = 3$, $B = 2$, $D = 2$, $F = 2$; output are: $G = 10$, $H = 12$. Suppose $M1$ and $A1$ are actual faulty components, and the actual value of $X$ is 5, the actual values of $Y$ and $Z$ are 6. *RealDiag* is $\{M1, A1\}$ and $CD_{min}(M)=\{\{M1\}, \{A1\}, \{M2, A2\},\{M2, M3\}\}$.

If we choice variable $Y$ to test, and the result of the test is $Y=6$, then the test refutes the minimal diagnoses $\{M2, A2\}$ and $\{M2, M3\}$, $CD_{min}(M')=\{\{M1\}, \{A1\}\}$.

Again, if we test the value of $X$ and the result is $X=5$, then the test refutes all of the minimal diagnoses in $CD_{min}(M')$, we can compute that $CD_{min}(M'')=\{\{M1, A1\}\}$, Obviously, $\{M1\}\subset\{M1, A1\}$ and $\{A1\}\subset\{M1, A1\}$.

As said in example 2, the first test eliminated two diagnoses, the rest diagnoses are close to the actual diagnosis; after the execution of the second test, we found the actual diagnosis using the results of the test.

**Theorem 7.** Given a diagnostic problem $M = (SD, COMPS, OBS)$, suppose that the outcome of the test (*true*, *o*) is $\alpha$. Let $D'\in CD_{min}(M')$ and $D' \subseteq RealDiag$, where $M'=(SD, COMPS, OBS\cup \alpha)$, there exists a diagnosis $D\in CD_{min}(M)$ satisfied $D \subseteq D'$.

**Proof**
On the basis of the theorem 6, if $D' \in CD_{min}(M)- REF$, let $D = D'$, the theorem is valid, else $D' \in CD_{min}(M')- (CD_{min}(M)- REF)$, there exists a diagnosis $D\in CD_{min}(M)$ satisfied $D\subset D'$. therefore, the theorem is also valid.

**Theorem 8.** If there is a minimal diagnosis $D\in CD_{min}(M)$ which satisfies $D= RealDiag$, then not exists any test refutes the diagnosis $D= RealDiag$.

**Proof**
For any given $M = (SD, COMPS, OBS)$, if there is a minimal diagnosis $D\in CD_{min}(M)$ which satisfies $D= RealDiag$, then $SD \cup OBS \cup \{\neg ab(c) \mid c\in COMPS\text{-}RealDiag\}$ is consistent. All components of the system have the actual values of the inputs and the output, which is the reasonable interpretation of the current observation. The results of any test are actual values of the system. Therefore, there not exists any test refutes the diagnosis $D= RealDiag$.

The above results construct a theoretical foundation of test for finding the actual diagnosis. Theorem 7 indicates that the obtained information about the actual diagnosis is always increased after the execution of the test, until the actual diagnosis is included in the set of minimal diagnoses. Theorem 8 tells that once the actual diagnosis is included in the set of minimal diagnoses, it would be confirmed by any test which may refute other diagnoses.

## 5 Related Works

Reference [12] provided a probabilistic analysis to decide what measurement to take next for discriminating the diagnoses. The DART system[13] was capable of proposing inputs and observations to be made in order to confirm or refute a possible diagnosis. The systematic study of the design and role of tests in the area of diagnosis has been proposed in reference [11]. Aimin Hou[14] studied the test of diagnosis and the generation of the tests with conflicts. However, the above works did not discuss the relationship between the diagnoses and the set of the actual faulty components. Therefore, if the considered diagnostic system is not adoptable, although we eliminated all candidates of diagnosis but one, it may be not the actual diagnosis. Again, the scale of the whole diagnoses space is very huge. Based on Reiter's diagnostic theories, reference [15] investigated into the replacement tests of component by combining replacing with the theories of model-based diagnosis. In fact, some treatments play a dual role as treatment and test.

Not for all diagnoses, we only discuss the test's impact on the set of minimal diagnoses, which is a small part of the whole diagnoses space. In addition, our works are focused on the relationship between the diagnoses and the set of the actual faulty components of the considered system before and after the execution of a test. Therefore, we only consider the adoptable diagnostic system such as consistency-based diagnostic system. After a sequence of test and diagnosis, the actual diagnosis must be found.

## 6 Conclusions and the Future Works

Firstly, this paper defines the notion of adoptability of the diagnostic system on the basis of the relationship between diagnosis and the set of the actual faulty components and proves that the consistency-based diagnosis is adoptable.

On the basis of the above works, a test theory of the diagnosis based on consistency is proposed, which indicates that how tests provide information about the current space of diagnoses. First, the obtained information about the actual diagnosis is always increased after the execution of the test, until the actual diagnosis is included in the set of minimal diagnoses. Again, once the actual diagnosis is included in the set of minimal diagnoses, it would be confirmed by any tests which may refute other diagnoses.

In this paper, we only discuss the simple test (*true*, *o*). Differential diagnosis for arbitrary test (*A*, *o*) is more difficult to characterize because the realization of initial conditions *A* could have side effects in the world which would change the truth value of previous observations. Differential diagnosis for arbitrary test (*A*, *o*) is our future works.

## References

1. Reiter.: A Theory of Diagnosis from First Principles. Artificial Intelligence, (1987) 57-96
2. Chittaro, L.: Hierarchical Model-Based Diagnosis Based on Structural Abstraction. Artificial Intelligence (2004) 147-182
3. Baroni, P.: Diagnosis of Large Active Systems Artificial Intelligence (1999) 135-183
4. Console, L., Picardi, G., Ribaudo, M.: Process Algebras for Systems Diagnosis. Artificial Intelligence (2002) 19–51
5. Pencole, Y.: A Formal Framework for the Decentralized Diagnosis of Large Scale Discrete Event Systems and its Application to Telecommunication Networks. Artificial Intelligence, (2005) 121-170

6. Fattah,Y., Dechte, R.: Diagnosing Tree-decomposable Circuits. In: Proceedings of International Joint Conference on Artificial Intelligence, Montreal, Canada (1995) 572-578
7. Portinale, L., Magro, D., Torasso, P.: Multi-modal Diagnosis Combining Case-based and Model-based Reasoning: a Formal and Experimental Analysis. Artificial Intelligence, (2004) 109–153
8. Console, L.: Temporal Decision Trees: Model-based Diagnosis of Dynamic Systems On-board Journal of Artificial Intelligence Research, (2003) 469-512
9. Milde, H.: Integrating Model-based Diagnosis Techniques into Current Work Processes-Three Case Studies from the INDIA Project AI Communications, (2000) 99-123
10. Console, L., Torasso, P.: A Spectrum of Logical Definitions of Model-based Diagnosis. Computational Intelligence, (1991) 133-141
11. McIlraith., Reiter.: On Test for Hypothetical Reasoning. Readings in Model-based Diagnosis, Morgan Kaufmann Publishers. (1992) 89-96
12. Kleer., Williams.: Diagnosing Multiple Faults. Artificial Intelligence, (1987) 97-130
13. Genesereth, M.R.: The Use of Design Descriptions in Automated Diagnosis. Artificial Intelligence, (1984) 411-436
14. Hou, A. M.: A Theory of Measurement in Diagnosis from First Principles. Artificial Intelligence, (1994) 281-328
15. Jiang, Y.F., Li, Z.S.: On Component Replacing and Replacement Tests for Model-Based Biagnosis. Chinese Journal of Computers, (2001) 666-672

# Bearing Diagnosis Using Time-Domain Features and Decision Tree

Hong-Hee Lee, Ngoc-Tu Nguyen, and Jeong-Min Kwon

School of Electrical Engineering, University of Ulsan, Ulsan, Korea
hhlee@mail.ulsan.ac.kr, nntu@hcmut.edu.vn,
ruee006@mail.ulsan.ac.kr

**Abstract.** Bearing fault detection with the aid of the vibration signals is presented. In this paper, time-domain features are extracted to indicate bearing fault, which collected from tri-axial vibration signal. Decision tree is chosen as an effective diagnostic tool to obtain bearing status. The paper also introduces principal component analysis (PCA) algorithm to reduce training data dimension and remove irrelevant data. Both original data and PCA-based data are used to train C4.5 decision tree models. Then, the result of PCA-based decision tree is compared with normal decision tree to get the best performance of classification process.

**Keywords:** bearing diagnosis, decision tree, vibration, principal component analysis.

## 1 Introduction

Bearing defects are the most popular type of machinery fault. Nowadays, most of diagnostic methods are based on measurement of vibration, acoustic noise, stator currents, or temperature. The vibration measurement is commonly method used in industry, because it is relatively cheaper and more reliable than the others. Vibration measurement methods can be based on time-domain, frequency-domain vibration signals, or both of them. Frequency domain bearing diagnosis method often monitors the fundamental frequencies generated by the defective bearing: rotating frequency, fundamental train frequency, ball pass frequency of the outer race, ball pass frequency of the inner race, ball spin frequency and their harmonics. Meanwhile time-domain bearing diagnosis method is using simple processing to analyze the time waveform characteristics.

Recently, the time signal analysis method for fault diagnosis has been introduced in many researches, such as proximal support vector machine [1], artificial neural network [2] for bearing diagnosis. Applications of decision tree [3], support vector machine [4] for motor diagnosis, etc. have been shown as the effective way on machine fault diagnosis field. However, there are still few projects of decision tree in bearing condition monitoring in particular. Therefore, this paper presents a bearing fault detection method by developing a decision tree, which is based on C4.5 algorithm. Compare to other methods such as neural network, fuzzy system, etc., decision tree has construction that users can understand easily, and have very fast learning rate.

D.-S. Huang, L. Heutte, and M. Loog (Eds.): ICIC 2007, LNAI 4682, pp. 952–960, 2007.

J. S. Sohre in his knowledge chart for rotating machine diagnosis [9] showed that the direction of vibration caused by bearing damage is 30% at axial, 40% at horizontal, and 30% at vertical direction. For this reason, in order to gather as much as possible of bearing information that needed for diagnosing, a tri-axial accelerometer is installed to collect vibration signals in three directions: axial, horizontal, and vertical in this work. Then, PCA is chosen as the feature extraction algorithm to reduce data dimension and remove the useless and irrelevant information. Finally, decision tree is studied with this processed data set to illustrate the advantage of PCA method.

## 2 Decision Tree Algorithm

The decision tree is a diagnostic tool that builds the knowledge-based system by the inductive inference from case histories. A decision tree contains:

- Leaf nodes (or answer nodes) which contain class name.
- Decision nodes ( or non-leaf nodes) that specifies some test to be carried out on a single attribute value, with one branch and sub-tree for each possible outcome of the test.

Structure of decision tree highly depends on how a test is selected as the root of the tree. The criterion for selecting the root of the tree is Quilan's information theory (information gain) [5]. This criterion means the information that is conveyed by a message depends on its probability and can be measured in bits as minus the logarithm to base 2 of that probability. The construction of decision tree bases on a training set T, which is a set of cases. Each case specifies values for a collection of attributes and for a class. Let the classes be denoted as {C1, C2, ..., Ck}. Suppose we have a possible test with n outcomes that partitions the training set T into subsets T1, T2, ... , Tn. Let S is any set of cases, freq(Ci , S) is the number of cases in S that belong to class Ci , |S| is the number of cases in set S. If we select one case at random from set S and announce that it belongs to class Cj. This message has probability

$$freq\left(C_j, S\right) / |S| \,. \tag{1}$$

and so the information it conveys is

$$-\log_2 \frac{freq(C_j, S)}{|S|} \text{ bits} \,. \tag{2}$$

The expected information needed to identify the class of case in S is

$$\text{info(S)} = -\sum_{j=1}^{k} \frac{freq(C_j, S)}{|S|} \times \log_2 \left( \frac{freq(C_j, S)}{|S|} \right) \text{ bits} \,. \tag{3}$$

When (3) is applied to the set of training cases, *info(T)* measures the average amount of information needed to identify the class of a case in T.

A similar measurement after T has been partitioned in accordance with n outcomes of a test X:

$$\mathrm{info}_X(T) = \sum_{i=1}^{n} \frac{|T_i|}{|T|} \times \mathrm{info}(T_i) \text{ bits} . \tag{4}$$

The quantity

$$\mathrm{Gain}(X) = \mathrm{info}(T) - \mathrm{info}_X(T) . \tag{5}$$

measures the information that is gained by partitioning T in accordance with the test X. The gain criterion selects a test to maximize this information gain.

## 3 Time Signal Features

In order to prepare data inputs for decision tree classifier, features are calculated from three-dimension vibration signal and extracted by PCA algorithm.

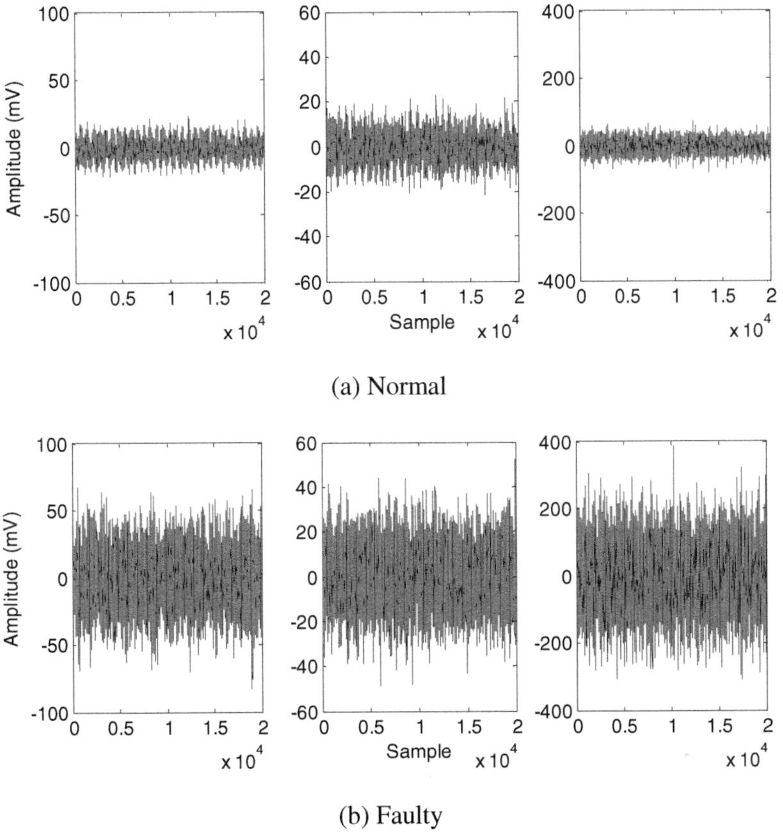


**Fig. 1.** a. Normal bearing (horizontal, axial, and vertical); b. Defective bearing time signals

### 3.1 Bearing Features

Time domain features are extracted to diagnose the bearing status such as root mean square (rms), variance, skewness, kurtosis, crest factor and maximum value.

- Skewness is a measure of symmetry, or the lack of symmetry of signal.
- Kurtosis is a measure of whether the data are peaked or flat relative to a normal distribution.
- Crest factor is a measure of how much impacting is occurring in time waveform. Impacting in time waveform may indicate rolling element wear or cavitations.
- Variance is a measure of the dispersion of a waveform about its mean, or is called the second moment of signal.
- Maximum amplitude and rms of signal indicate the severity of bearing defect.

Tri-axial accelerometer is installed to measure vibration signals in three directions, 6 features are extracted from each signal. The feature set which has total 18 features is used to train decision tree.

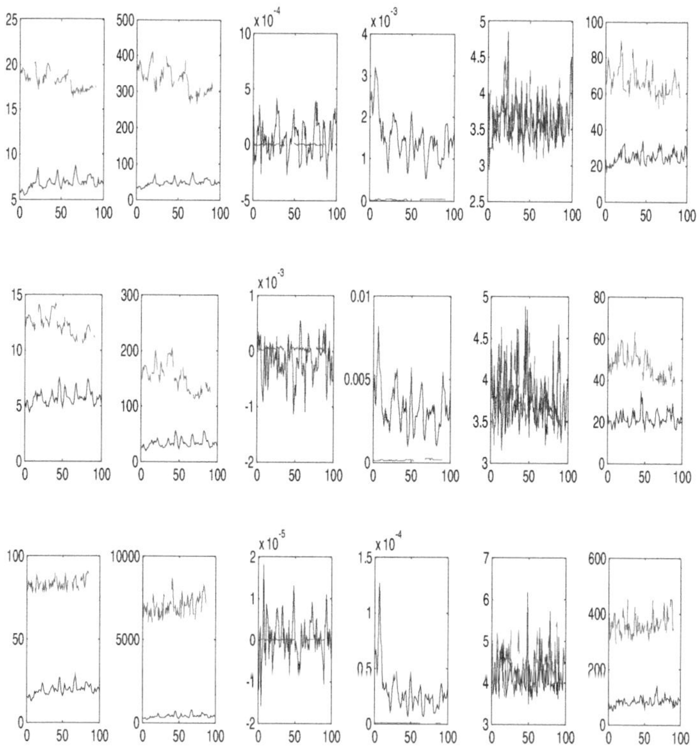

**Fig. 2.** Extracted features, (-, blue) normal and (--, red) defective bearing

Dataset is formed as follows: *rms(a), variance(a), skewness(a), kurtosis(a), crest factor(a), maximum(a), rms(h), variance(h), skewness(h), kurtosis(h), crest factor(h), maximum(h), rms(v), variance(v), skewness(v), kurtosis(v), crest factor(v), maximum(v).* Where the terms (a), (h), and (v) represent to axial, horizontal, and vertical direction, respectively.

Fig. 1 shows vibration signals in 3 directions of both normal and fault bearings. From these data, features are extracted as plots in Fig. 2, which shows 18 features in normal and defective cases. The dataset uses to train decision tree has about 1300 samples measured from 5 bearings with different conditions. A dataset consist about 190 samples, which collected from other bearings, is used to test the decision tree.

### 3.2 Feature Reduction

Principal component analysis (PCA) is a technique for simplifying the data by extracting the most relevant information from original dataset and forming a new lower dimension data for analysis. A N-dimensional (zero mean) dataset $x_i$ (i=1, 2, ..., m, N<m) is projected on the eigenvectors of its covariance matrix

$$v = U^T x_i \ . \tag{6}$$

where U is an orthogonal matrix which contains the eigenvectors of the data covariance matrix C

$$C = \frac{1}{m}\sum_{i=1}^{m} x_i x_i^T \ . \tag{7}$$

The eigenvalues of C are computed and sorted in decreased order to form matrix U

$$\lambda_1 \geq \lambda_2 \geq \ldots \geq \lambda_k \ldots \geq \lambda_N \ . \tag{8}$$

By keeping only the most significant eigenvectors which corresponding to k largest eigenvalues, we can reduce the data dimension, while preserving most of information in the data. In order to choose k, the following criterion is used

$$\left(\sum_{i=1}^{k} \lambda_i\right) / \left(\sum_{i=1}^{N} \lambda_i\right) > \textbf{\textit{threshold}} \ . \tag{9}$$

The dataset used to train bearing diagnostic decision tree has 18 dimensions. After processed by PCA algorithm, we can choose the first 9 eigenvalues (k=9), then the percentage of information after the projection is about 100% (threshold ≈ 1.0). If we choose the first 4 eigenvalues (k=4), then threshold value is about 99.9%.

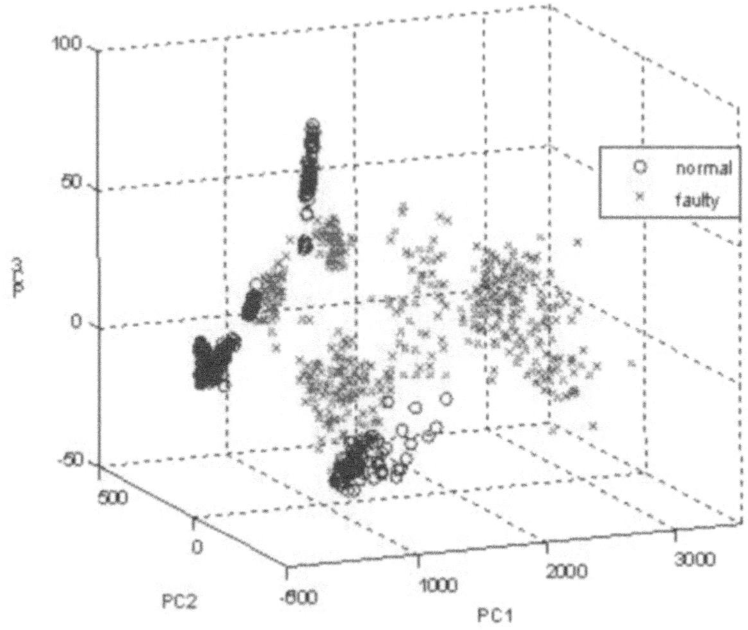


**Fig. 3.** The projection of the training data on the first three axes

# 4 Results

In this paper, the C4.5 algorithm [5] is used to train decision tree. Training set consists 1372 samples is used to train classifier, and test set which consists 191 samples is used to test validity of the classifier. Decision trees are proposed corresponding to normal and PCA-based ones as shown in Fig. 4-6.

The term F in Fig. 4-6 is the abbreviation of feature, e.g. F1 is feature number 1. The trees can rewrite in a collection of rules, one for each leaf in the tree. There are 13 rules for trees in Fig. 4 and Fig. 6, 12 rules for tree in Fig. 5. Each rule is an if-then statement which traces from the root to a leaf. A case (yes or no) is classified by a leaf when all the conditions of a rule are satisfied along the path from the root to that leaf.

Table 1 summarizes the performance of result decision trees tested with training data and test data. As shown in this table, the classification performance of trees with PCA feature extraction is 100% accuracy. It is better than the classification of tree without feature extraction (normal decision tree).

**Table 1.** Compare the performance of normal decision tree and PCA based decision tree

| Type | Size | Evaluation on training data | Evaluation on test data |
|---|---|---|---|
| Normal decision tree | 25 | 99.6% | 95.8% |
| PCA-based decision tree ( 4 new features) | 25 | 99.9% | 100% |
| PCA-based decision tree ( 9 new features) | 23 | 99.9% | 100% |

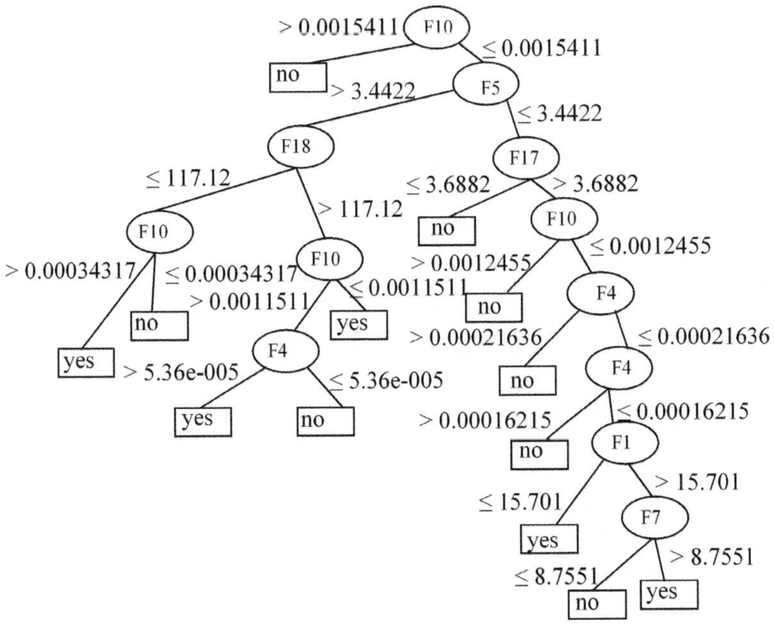


**Fig. 4.** Decision tree without feature extraction

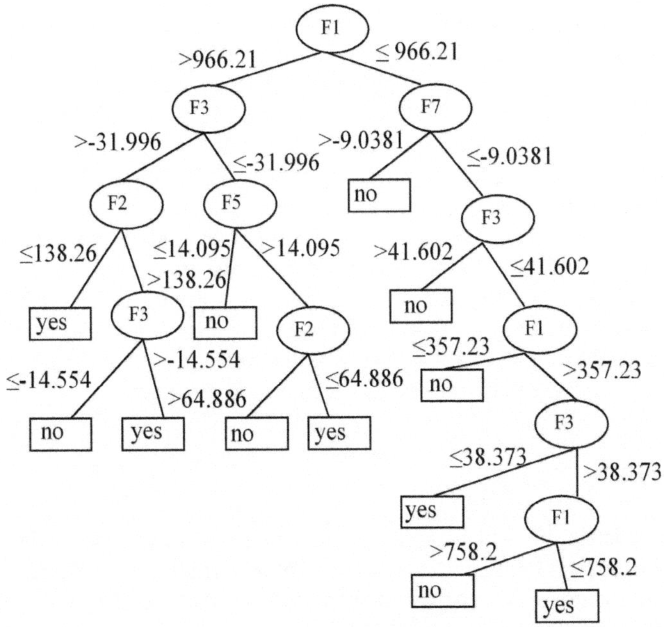


**Fig. 5.** PCA-based decision tree with 9 new features

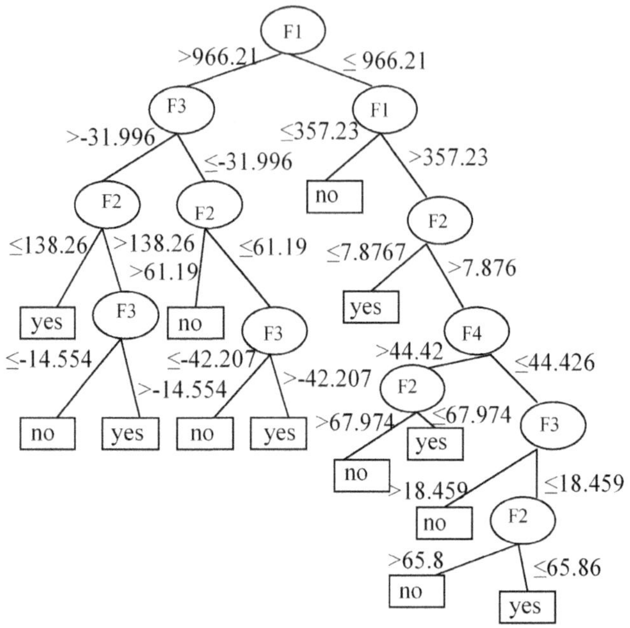


**Fig. 6.** PCA-based decision tree with 4 new features

The accuracy of PCA-based decision trees when evaluated with training and test set increase compare with the normal tree. Without PCA processing, the decision tree uses 7 of 18 features (F1, F4, F5, F7, F10, F17, F18), while the PCA-based decision trees uses 5 in 9 new features (F1, F2, F3, F5, F7) in Fig. 5 and all 4 new features (F1, F2, F3, and F4) in Fig. 6 but they only have maximum 6 nodes depth compare to 8 of normal decision tree. Smaller depth level makes PCA-based decision tree more compact and faster than normal one. Beside that, selecting all the variables of the system features will bring in much irrelevant information that can spoil the performance of classification system and increase the error ratio. In Table 1, when evaluating the classification system with test set, the normal tree with all features has lower performance accuracy compare to PCA-based ones. This also is illustrated by looking at the performance of two PCA-based trees, where the one with only 4 features has almost same performance as the tree with 9 features.

## 5 Conclusion

Decision tree has established for bearing fault diagnosis with and without reducing features in this paper. Only two states (fault and normal) of bearing condition are considered but it is possible to apply this method with multi bearing fault types.

The drawback of this method is discrete output, so that decision tree cannot give the severity level of bearing fault. This requires the system has to have continuous output. Another problem is the sensitivity to noise of decision tree, if a small amount of noise is added to attribute values then the tree can give the wrong results. But beside all that weak points, decision tree has simple construction that can be understood easily. In this paper, the decision trees have very high accuracy when evaluated with the test set.

**Acknowledgments.** The authors would like to thank University of Ulsan, the Ministry of Commerce, Industry and Energy (MOCIE) and Ulsan Metropolitan City that partly supported this research through the Network-based Automation Research Center (NARC).

## References

1. Sugumaran, V., Muralidharan, V., Ramachandran, K.I.: Feature Selection Using Decision Tree and Classification Through Proximal Support Vector Machine for Fault Diagnostics of Roller Bearing, Mechanical Systems and Signal Processing. 21 (2007) 930-942
2. Samanta, B., Al-Balushi, K.R.: Artificial Neural Network Based Fault Diagnostics of Rolling Element Bearings Using Time-domain Features, Mechanical Systems and Signal Processing. 17 (2003) 317-328
3. Sun, W., Chen, J., Li, J.: Decision Tree and PCA-based Fault Diagnosis of Rotating Machinery, Mechanical Systems and Signal Processing (2006)
4. Widodo, A., Yang, B.S., Han, T.: Combination of Independent Component Analysis and Support Vector Machines for Intelligent Faults Diagnosis of Induction Motors, Expert Systems with Applications. 32 (2007) 299-312
5. Quinlan, J.R.: C4.5 : Programs for Machine Learning, Morgan Kaufmann Publisher, Inc. (1993)
6. Yang, B.S., Park, C.H., Kim, H.J.: An Efficient Method of Vibration Diagnostics for Rotating Machinery using a Decision Tree, International Journal of Rotating Machinery. 1 (2000) 19-27
7. Lim, D.S., Yang, B.S., Kim, D.J.: An Expert System for Vibration Diagnosis of Rotating Machinery using Decision Tree, International Journal of COMADEM. (2000) 31- 36
8. Samanta, B., Al-Balushi, K. R., Al-Araimi, S.A.: Artificial Neural Networks and Genetic Algorithm for Bearing Fault Detection, Soft Coput. (2006) 264-271
9. Rao, J.S.: Vibratory Condition Monitoring of Machines. Alpha Science International Ltd. (2000) 364-373
10. Casimir, R., Boutleux, E., Clerc, G., Yahoui, A.: The Use of Features Selection and Nearest Neighbors Rule for Faults Diagnostic in Induction Motors. Engineering Applications of Artificial Intelligence. (2006) 169-177
11. Yang, J., Zhang, Y., Zhu, Y.: Intelligent fault diagnosis of rolling element bearing based on SVMs and fractal dimension. Mechanical Systems and Signal Processing (2006), doi: 10.1016/j.ymssp.2006.10.005
12. Purushotham, V., Narayanan, S., Suryanarayana, Prasad, A.N.: Multi-fault Diagnosis of Rolling Bearing Elements Using Wavelet Analysis and Hidden Markov Model based Fault Recognition, NDT&E International. 38 (2005) 654-664
13. Widodo, A., Yang, B.S.: Application of Nonlinear Feature Extraction and Support Vector Machines for Fault Diagnosis of Induction Motors. Expert System with Application (2006)
14. Rojas, A., Asoke, Nandi, K.: Practical Scheme for Fast Detection and Classification of Rolling-element Bearing Faults Using Support Vector Machines, Mechanical Systems and Signal Processing. 20 (2006) 1523-1536

# CMAC Neural Network Application on Lead-Acid Batteries Residual Capacity Estimation

Chin-Pao Hung and Kuei-Hsiang Chao

National Chin-Yi University of Technology, Department of Electrical Engineering, 35, 215 Lane, Sec. 1, Chung Shan Road, Taiping, Taichung, Taiwan
cbhong@ncut.edu.tw, chaokh@ncut.edu.tw

**Abstract.** This paper proposed a novel residual capacity estimation method for lead-acid battery. Generally, the battery residual capacity is related to open-circuit-voltage (OCV) and inner resistance. Due to complexities, delay, coupling and nonlinearity on the relation described above, the residual capacity estimation is difficult to be obtained by the voltage or inner resistance measurement only. In this paper, by observing and recording the constant current discharging process, we generated the residual capacity pattern for different capacity levels and built a CMAC (Cerebellar Model Articulation Controller) neural network to estimate the lead-acid battery residual capacity. The characteristic of self-learning and generalization, like the cerebellum of human being, a CMAC NN estimation scheme enables powerful, straightforward, and efficient battery residual capacity estimation. With application of this scheme to the experimental data test, the estimation results demonstrate the new scheme with high accuracy and high noise rejection ability.

**Keywords:** CMAC, neural network, batteries, capacity estimation.

## 1 Introduction

Lead-acid batteries have a broad application on portable device today because of its sealed structure and low cost. But portable device usually needs to detect battery residual capacity to evaluate the operating time. Hard to measure capacity becomes its major drawback. Many researchers proposed different schemes to achieve the capacity measurement, such as open-circuit voltage (OCV), electrolyte specific gravity, loaded voltage, and inner resistance measure method. However, the recover time delay, extra gravimeter sensor, nonlinear discharging curve or inner resistance, made the capacity estimation lost its accuracy or the measure cost is expensive.

Also, many researchers proposed intelligent scheme to estimate the battery capacity, such as multi-layer neural network architecture by EBP (error back propagation) learning method [1-2]. Some successful results demonstrated the feasibility using neural network to estimation the residual capacity. However, the local minimum problem and slower learning speed are its major drawbacks. To solve these drawbacks described above, we proposed a CMAC neural network

D.-S. Huang, L. Heutte, and M. Loog (Eds.): ICIC 2007, LNAI 4682, pp. 961–970, 2007.

(CMAC NN) estimation system to evaluate the battery residual capacity. Albus first proposed CMAC NN and applied it to control system because of its on-line learning ability [3].

In our recently researches, we applied it to the fault diagnosis of transformer and others and many successful cases are obtained, such as air-conditioning system [4], generator sets [5], power transformer system [6] and water circulation system [7]. CMAC, like the models of human memory, perform a reflexive processing [9]. By pattern collection and converging learning, the diagnosis performance outperforms than other intelligent schemes such as fuzzy methodology [10], multi-layer neural network [11] and wavelet neural network [12]. Main advantages conclude fast learning speed, noise rejection ability and high accuracy. In this paper, we apply it to the estimation field first. Firstly, we measure and record the discharge curves include the variation of OCV and inner resistance for lead-acid battery. Then we transformed the recorded curve as the pattern of residual capacity and built a CMAC NN to learn the feature of discharge curve. Finally, the trained CMAC NN can estimate the battery residual capacity.

## 2 The Configuration of CMAC NN Battery Residual Capacity Estimation System

To estimate the battery residual capacity, we first observed and recorded the constant current (1A) discharging process for lead-acid battery. The curves of OCV and inner resistance are relative to capacity and recorded as Figure 1. It is evident the nonlinearity appears in the discharging process especially in the low residual capacity.

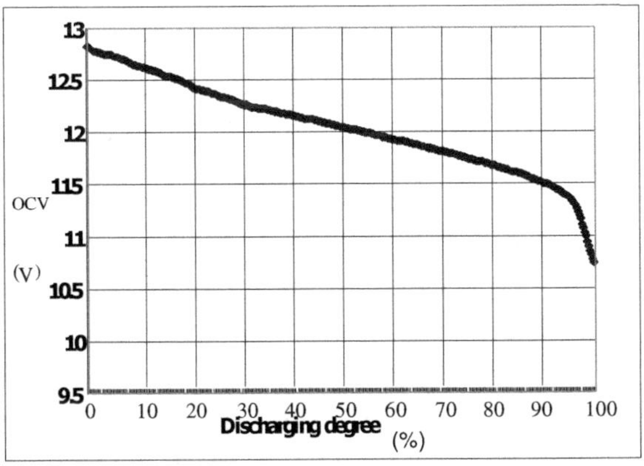


**Fig. 1(a).** OCV and battery discharging degree plot

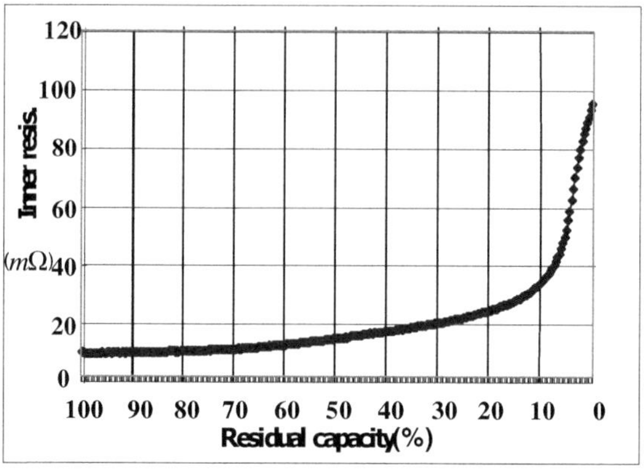


**Fig. 1(b).** Inner resistance and battery residual capacity plot

From these observations, we transformed the residual capacity estimation as pattern recognition problem. For example, we define the residual capacity as 10 levels and rearrange the recorded data as Table 1. Then Table 1 will be the patterns of different residual capacity level. Noted $V_{oc} / R_{in}$ represents the short current.

**Table 1.** Classification for battery residual capacity

| | |
|---|---|
| K1: 90% residual capacity. | K6: 40% residual capacity. |
| K2: 80% residual capacity. | K7: 30% residual capacity. |
| K3: 70% residual capacity. | K8: 20% residual capacity. |
| K4: 60% residual capacity. | K9: 10% residual capacity. |
| K5: 50% residual capacity. | K0: 0% residual capacity. |

## 2.1 The Development of CMAC NN Fault Diagnosis System

Depending on the above definition, the CMAC based battery estimation system is shown as Figure 2. As described above, inner resistance $R_{in}$, OCV $V_{oc}$ and short current $V_{oc} / R_{in}$ are used as the input states. The output contains 10 parallel memory layers and every memory layer has one output node. Every memory layer remembers one residual capacity feature. E.g. layer 1 stores the features of class *K1* of Table 2, layer 2 stores the features of class *K2* of Table 2, etc. Inputting $R_{in}, V_{oc}$ and $V_{oc} / R_{in}$ to CMAC, through a series of mapping, the input data will generate one group excited

**Table 2.** Inner resistance(Ω) ,OCV (volt) and short current (A)range

| | | | |
|---|---|---|---|
| **K1** | **$R_{in}$ : 11.19~11.82.** | **K6** | **$R_{in}$ : 15.96~20.18** |
| | $V_{oc}$ :12.71~12.82 | | $V_{oc}$ :11.87~12.18 |
| | $V_{oc} / R_{in}$ :1.081~1.142 | | $V_{oc} / R_{in}$ :0.588~0.763 |
| K2 | $R_{in}$ : 11.83~12.16. | K7 | $R_{in}$ : 20.19~25.24. |
| | $V_{oc}$ :12.59~12.69 | | $V_{oc}$ :11.70~11.86 |
| | $V_{oc} / R_{in}$ :1.036~1.073 | | $V_{oc} / R_{in}$ :0.466~0.587 |
| K3 | $R_{in}$ : 12.18~12.64. | K8 | $R_{in}$ : 25.27~36.65 |
| | $V_{oc}$ :12.49~12.58 | | $V_{oc}$ :11.59~11.74 |
| | $V_{oc} / R_{in}$ :0.988~1.033 | | $V_{oc} / R_{in}$ :0.316~0.465 |
| K4 | $R_{in}$ : 12.65~13.69. | K9 | $R_{in}$ : 33.66~46.50. |
| | $V_{oc}$ :12.33~12.46 | | $V_{oc}$ :11.48~11.60 |
| | $V_{oc} / R_{in}$ :0.901~0.985 | | $V_{oc} / R_{in}$ :0.227~0.316 |
| K5 | $R_{in}$ : 13.70~15.95. | K0 | $R_{in}$ : 46.70~96.70 |
| | $V_{oc}$ :12.18~12.32 | | $V_{oc}$ :11.05~11.47 |
| | $V_{oc} / R_{in}$ :0.775~0.941 | | $V_{oc} / R_{in}$ :0.119~0.246 |

memory addresses. To sum the excited memory addresses of each layer, output node will obtain a value to express the possibility of class *Kn*.

## 2.2 The Training Mode of CMAC NN Residual Capacity Estimation System

The proposed scheme recorded the discharging data as the training pattern. Assuming each level capacity is piecewise continuous, the training data can be regenerated to replace limited pattern. It will benefit the CMAC NN to learn the feature for each residual capacity level. For example, class *K1* with ( $R_{in}$ , $V_{oc}$ , $V_{oc} / R_{in}$ ) = ([11.19~11.82], [12.71~12.82], [1.081~1.142]). Using the program 1(designed by MATLAB), the training data can be generated. In program 1 the step value *STEP_X* determines the resolution of training data. High resolution will cause long training time. The training data then send to the CMAC network, through

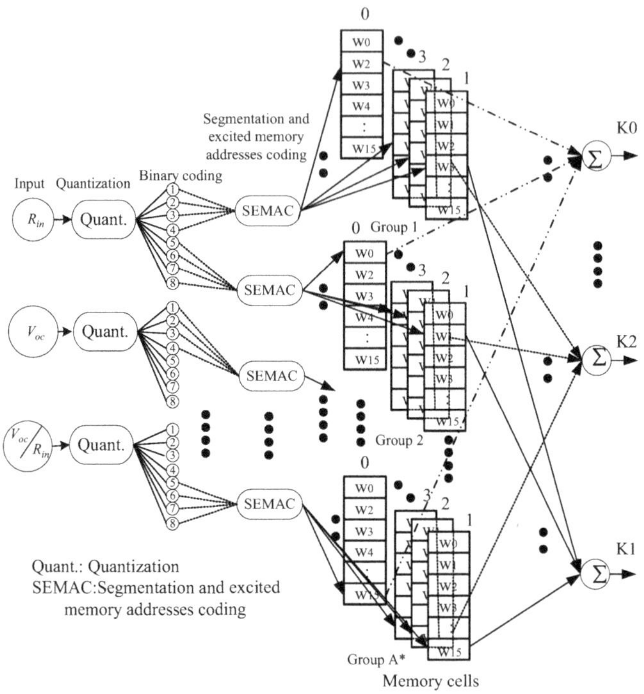


**Fig. 2.** CMAC based battery residual capacity estimation system

the quantization, excited addresses coding, and sums the fired memory cells weights to obtain an output value. Compare the desired output 1, then the error used to tune the excited memory weights. The details will describe as follows:

```
for Rin=11.19:STEP_1:11.82
  for Voc=12.71:STEP_2:12.82
    for Voc_Rin=1.081:STEP_3:1.142
    %quantization, excited addresses coding, summation;
    end
  end
end
```

**Program 1.** Training data regeneration

### 2.2.1 Quantization Mapping

The input values send to the CMAC network, it is first though the quantization mapping $Q$ to produce a quantization level output. The quantization output can be described as follows [9].

$$q_i = Q(x_i, x_{i\min}, x_{i\max}, q_{i\max}), \quad i = 1, \cdots n \tag{1}$$

where $n$ is the input numbers. The resolution of this quantization depends on the expected maximum and minimum inputs, $x_{i\max}$ and $x_{i\min}$, and on the number of quantization levels, $q_{i\max}$. High resolution will have good generalization ability but more memory size is needed. Assuming the maximum quantization level $q_{\max}$ is chosen as 16, then the quantization mapping diagram can be shown as Fig. 3. Figure 3 shows that the input state between $x_{i\max}$ and $x_{i\min}$ are quantized as 0 to 15 orderly. In this paper, because the states distributed in a wide range, we choose $q_{\max}$ as 256 to increase the differentiability.

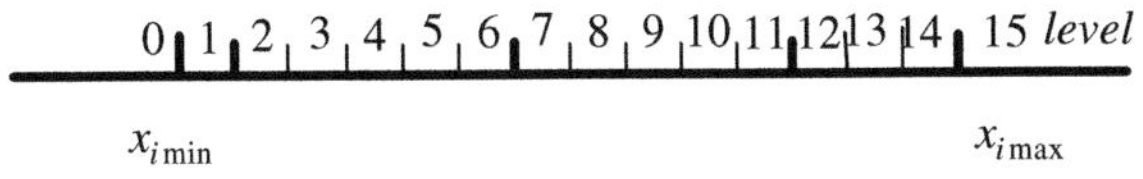


**Fig. 3.** Quantization mapping diagram

### 2.2.2 Excited Addresses Coding

Assuming the quantization level of each input signal is calculated as ( $R_{in}$ , $V_{oc}$ , $V_{oc} / R_{in}$ )=(7,0,14). Then the quantization level can be rewritten in binary form as 00000111, 00000000 and 00001110. Concatenating the binary bits will obtain the following binary string.

000001110000000000001110

By taking four bits to code an excited memory address, the excited memory addresses will be 0,7,0,0,0,14. I.e. the features of specific level capacity will be distributed store on six memory addresses and group number $A^*$ is 6 in Figure 2. To add the weights of excited memory addresses will produce the CMAC output. The output of CMAC can be expressed as

$$y = \sum_{i=1}^{A*} w_i^{ai} \;, i=1,\ldots,A^* \tag{2}$$

where $w_i^{ai}$ denotes the $ai$-th addresses of group $i$.

### 2.2.3 Learning Rule

Assuming the memory layer $i$ ($i$=1,…,9,0) output one denotes the residual capacity $Ki$ is confirmed, then one can be thought as the teacher and the supervised learning algorithm can be described as [5,9]

$$w_{i\,(new)}^{ai} \leftarrow w_{i\,(old)}^{ai} + \beta \frac{Y_d - Y}{A*}, \quad i = 1,2,\ldots,A^* \;\ldots\;\ldots \tag{3}$$

where $w^{a_i}_{i(new)}$ are the new weight values after the weights tuning, $w^{a_i}_{i(old)}$ are the old weight values before weight tuning, and *ai* denotes the excited memory addresses, β the learning gain($0<\beta<1$), $Y_d=1$ the desired output, $Y$ the real output.

#### 2.2.4 Learning Convergence and Performance Evaluation

From [8], the convergence of a supervised learning algorithm can be guaranteed. Assuming the *i*-th (*i*=1,…,9,0) layer outputs one denotes the system has capacity level *Ki*, and the number of training patterns is $N_p$. $y_i$ is the CMAC output for pattern *i*. Let the performance index be

$$E=\sum_{i=1}^{N_p}(y_i-1)^2 \tag{4}$$

When $E<\varepsilon$, stop the training process. ($\varepsilon$ is a small positive constant).

### 2.3 Estimation Mode

When the training mode is finished, the CMAC battery capacity estimation system can be utilized to evaluate the battery residual capacity level. Inputting the measuring data ($R_{in}, V_{oc}, V_{oc}/R_{in}$) to the estimation system, the operations of CMAC NN are same as the training mode. But in estimation mode, the same excited memory addresses weights of every memory layer are summed up and each layer has one output value. If the input signal is the same as the training patterns of *Ki*, it will excite the same memory addresses of layer $i$ and layer $i's$ output near one denotes the residual capacity level is *Ki*. But other layer's output, generally, far away from one expresses a low possibility of capacity *Kj* ($j\neq i$).

### 2.4 Training and Estimation Algorithm

Based on the configuration of Figure 2, the training and estimation algorithm are summarized as follows.

#### 2.4.1 Training Mode

step 1. Build the configuration of CMAC NN estimation system. It includes 3 input signals, 10 parallel memory layers and 10 output nodes.

step 2. Input the training patterns, through quantization, excited memory addresses coding, and summation of excited memory addresses weights to produce the node output.

step 3. Calculate the difference of actual output and the desired output ($y_d$=1) and using equation (3) to update the weight values.

step 4. Training performance evaluation. If $E<\varepsilon$, the training is finished. Save the memory weights. Otherwise, go to step 2.

### 2.4.2 Estimation Mode

step 5. Load the up to date memory weights from the saved file.

step 6. Inputting the measured data ( $R_{in}, V_{oc}, V_{oc} / R_{in}$ ).

step 7 Quantization, excited memory address coding, and summation of the excited memory weights using equation (2).

step 8. Outputting the estimation results.

## 3 Case Studies and Discussions

To demonstrate the effectiveness of the proposed CMAC NN scheme, twenty experimental data are tested after the training mode. The training patterns are generated from Program 1. Inputting the test data of Table 3 into the estimation system, the nodes output in Table 6, shown all the estimation results are correct except 19$^{th}$ row. However, we found the 19$^{th}$ row test data also located at the class *K9* range of Table 2. Rigorous speaking, the estimated result is right. Observing the nodes output of 9 and 0, the residual capacity can be identifies as *K9* or *K0*. These results are caused by the nonlinearity of battery discharging process as described above. If necessary, this problem can be solved by increasing the residual capacity level, increasing $q_{i\max}$ or calibrating the training pattern interval more precisely.

To test the noise rejection ability, we also added 10% to 50% noise to the original data, the diagnosis results still confirms what residual capacity level is. It guarantees the proposed estimation scheme with high feasibility, high accuracy and high noise rejection ability. The related CMAC parameters are listed in Table 5 and some weights distribution plots are shown in Figure 4.

**Table 5.** The CMAC NN parameters of case study

| Parameter | Value |
|---|---|
| Learning time | 10 |
| Class level | 10 |
| Step_Rin | $(RinMax - RinMin)/10$ |
| Step_Voc | $(VocMax - VocMin)/10$ |
| Step_Voc_Rin | $(Voc\_RinMax - Voc\_RinMin)/10$ |
| $\beta$ | 0.95 |
| $q_{max}$ | 256 |
| A* | 6 |
| $\varepsilon$ | 0.1 |

**Table 6.** Detail outputs of CMAC NN method for test data

| No. | $R_{in}$ | $V_{OC}$ | $R_{in}/V_{OC}$ | Each node output(estimation output) | | | | | | | | | | Real class |
|---|---|---|---|---|---|---|---|---|---|---|---|---|---|---|
| | | | | 1 | 2 | 3 | 4 | 5 | 6 | 7 | 8 | 9 | 0 | |
| 1 | 11.55 | 12.78 | 1.106 | 0.92 | 0.51 | 0.40 | 0.43 | 0.51 | 0.31 | 0.13 | 0.23 | 0.20 | 0.18 | K1 |
| 2 | 11.78 | 12.74 | 1.081 | 1.00 | 0.79 | 0.61 | 0.48 | 0.51 | 0.43 | 0.04 | 0.11 | 0.20 | 0.17 | K1 |
| 3 | 12.13 | 12.65 | 1.042 | 0.78 | 1.00 | 0.70 | 0.40 | 0.45 | 0.37 | 0.08 | 0.10 | 0.12 | 0.08 | K2 |
| 4 | 12.09 | 12.67 | 1.048 | 0.82 | 1.0 | 0.7 | 0.48 | 0.57 | 0.24 | 0.21 | 0.06 | 0.13 | 0.33 | K2 |
| 5 | 12.54 | 12.53 | 0.999 | 0.47 | 0.49 | 0.94 | 0.78 | 0.51 | 0.39 | 0.09 | 0.11 | 0.12 | 0.20 | K3 |
| 6 | 12.62 | 12.51 | 0.991 | 0.43 | 0.49 | 0.99 | 0.78 | 0.45 | 0.36 | 0.13 | 0.19 | 0.12 | 0.09 | K3 |
| 7 | 12.85 | 12.42 | 0.967 | 0.43 | 0.32 | 0.88 | 0.99 | 0.58 | 0.41 | 0.13 | 0.17 | 0.06 | 0.36 | K4 |
| 8 | 13.42 | 12.35 | 0.920 | 0.47 | 0.33 | 0.40 | 0.98 | 0.95 | 0.39 | 0.13 | 0.06 | 0.06 | 0.36 | K4 |
| 9 | 14.26 | 12.28 | 0.861 | 0.39 | 0.33 | 0.40 | 0.64 | 0.94 | 0.42 | 0.07 | 0.23 | 0.13 | 0.20 | K5 |
| 10 | 15.19 | 12.22 | 0.804 | 0.47 | 0.35 | 0.40 | 0.49 | 0.92 | 0.60 | 0 | 0.12 | 0.05 | 0 | K5 |
| 11 | 16.38 | 12.08 | 0.737 | 0.16 | 0.03 | 0.05 | 0 | 0 | 0.91 | 0.38 | 0.06 | 0 | 0.21 | K6 |
| 12 | 18.68 | 11.96 | 0.640 | 0.07 | 0.01 | 0.06 | 0.13 | 0.06 | 1.09 | 0.45 | 0 | 0.18 | 0.17 | K6 |
| 13 | 21.81 | 11.82 | 0.541 | 0.20 | 0.09 | 0.16 | 0 | 0.05 | 0.07 | 0.91 | 0.56 | 0.06 | 0.30 | K7 |
| 14 | 22.59 | 11.81 | 0.523 | 0.17 | 0.18 | 0 | 0.09 | 0 | 0.12 | 1.02 | 0.80 | 0.15 | 0.09 | K7 |
| 15 | 28.50 | 11.69 | 0.410 | 0 | 0.05 | 0.17 | 0.09 | 0 | 0.15 | 0.41 | 0.94 | 0.12 | 0.09 | K8 |
| 16 | 34.07 | 11.61 | 0.341 | 0.03 | 0.03 | 0.15 | 0.04 | 0.12 | 0.13 | 0.28 | 0.97 | 0.38 | 0.20 | K8 |
| 17 | 39.70 | 11.58 | 0.292 | 0.07 | 0.09 | 0.12 | 0.14 | 0.05 | 0.10 | 0 | 0.35 | 0.85 | 0.52 | K9 |
| 18 | 42.70 | 11.54 | 0.270 | 0.03 | 0.03 | 0.02 | 0.03 | 0.11 | 0.05 | 0.22 | 0.45 | 1.00 | 0.42 | K9 |
| 19 | 48.5 | 11.46 | 0.236 | 0.03 | 0 | 0.05 | 0.03 | 0.09 | 0.37 | 0 | 0.11 | 0.91 | 0.81 | K0 |
| 20 | 68.6 | 11.29 | 0.165 | 0.03 | 0.05 | 0.05 | 0.03 | 0.20 | 0.23 | 0.14 | 0.05 | 0.12 | 0.71 | K0 |

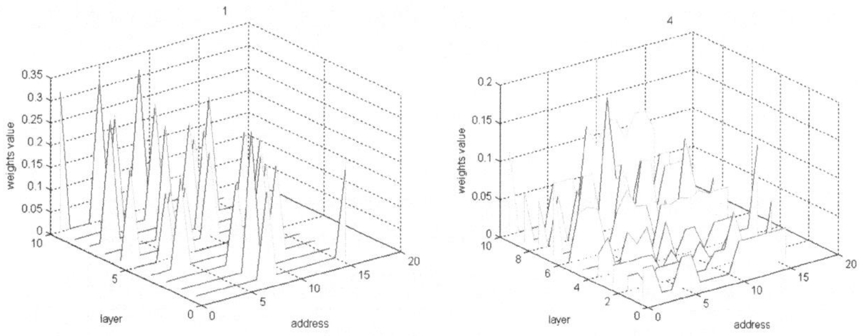


**Fig. 4.** Memory weights distribution plots of group 1,4

# 4 Conclusions

This paper presents a novel CMAC NN residual capacity estimation method for lead-acid battery. Using the characteristic of generalization, local reflexive action and self-learning ability, the proposed scheme achieves at least the following merits: 1) High estimation accuracy is obtained. 2) High noise rejection ability. 3) Suit to non-training data and associate the most similar residual capacity class. 4) Don't require any expert experience to train the CMAC neural network. The tested data demonstrate the success of proposed scheme. However, how to design an optimal memory size, quantization level, and associated cells numbers to obtain more efficient application are our future work and under studying.

## Acknowledgments

The authors gratefully acknowledge the support of National Science Council, Taiwan, R.O.C., under the grant NO. NSC 95-2221-E-167-024.

## References

1. Liu, Q.: Estimating SOC of MH/Ni Batteries Based on Artificial Neural Network. Journal of WuHan University of Technology 3 (2006)
2. Hu, R., Han, Z., Wang, K.: Estimation of Resting Batteries' remaining Capacity Based on BP Neural Networks. Battery bimonthly 1 (2006)
3. Albus, J. S: A New Approach to Manipulator Control: the Cerebeller Model Articulation Controller (CMAC)[1] . Trans. ASME J. Dynam., Syst., Meas., and Contr. 97 (1975) 220-227
4. Hung, C. P., Wang, M. H.: Fault Diagnosis of Air-conditioning System Using CMAC Neural Network Approach. Advances in Soft Computing - Engineering, Design and Manufacturing. Springer (2003)
5. Hung, C. P., Wang, M., Cheng, C., Lin, W.: Fault Diagnosis of Steam Turbine-generator Using CMAC Neural Network Approach. International Joint Conference on Neural Network 4 (2003) 2988 - 2993
6. Hung, C. P., Wang, M.: Diagnosis of Incipient Faults in Power Transfers Using CMAC Neural Network Approach. Electric Power Systems Research 71 (2004) 235-244
7. Hung, C. P., Lin, Y., Liu, W.: PIC Microcontroller Based Fault Diagnosis Apparatus Design for Water Circulation System Using CMAC Neural Network Approach. WSEAS TRANS. On Information Science & Application 4(2) (2007) 393-399
8. Wong, Y. F., Sideris, A.: Learning Convergence in the Cerebellar Model Articulation Controller. IEEE Trans. on Neural Network 3(1) (1992) 115-121
9. Handeiman, D. A., Lane, S. H., Gelfand, J. J.: Integrating Neural Networks and Knowledge-based Systems for Intelligent Robotic Control. IEEE Control System Magazine (1990) 77-86
10. Su, Q., Mi, C., Lai, L. L., Austin, P.: A Fuzzy Dissolved Gas Analysis Method for the Diagnosis of Multiple Incipient Faults in a Transformer. IEEE Trans. on Power Systems 15(2) (2000) 593-598
11. Zhang, Y., Ding, X., Liu, Y., Griffin, P. J.: An Artificial Neural Network Approach to Transformer Fault Diagnosis. IEEE Trans. on PWRD. 11(4) (1996) 1836-1841
12. LI, H., Sun, C. X., Hu, X.S., Yue, G., Tang, N.F., Wang, K.: Study of Method on Adaptive Wavelets for Vibration Fault Diagnosis of Steam Turbine-generator Set. Journal of Electrical Engineering (China) 15(3) (2000)

# Diagnosing a System with Value-Based Reasoning

XueNong Zhang[1,2], YunFei Jiang[1], and AiXiang Chen[1]

[1] Institute of Software Research, Zhongshan University
[2] Network Center, GuangDong Pharmaceutical University
zxnxlq@163.com

**Abstract.** This paper presents a value propagation model and redefines the notion of diagnosis. On the basis of the value propagation model, an algorithm for finding a minimal diagnosis is proposed. This algorithm need not compute the minimal conflicts and provides a reasonable interpretation for the minimal diagnosis. In addition, we present a method for repairing the faulty system by integrating diagnoses and test.

**Keywords:** Model-based diagnosis, Value propagation, Minimal diagnosis.

## 1 Introduction

Due to its generality and its dramatic importance in many application domains, model-based diagnosis has been receiving considerable attention in the field of AI research. It addresses the systems whose nominal behaviors can be specified as a mapping from their input variables to their output variables. The classical method is built on the well-known consistency-based theory and the classical model[1] describes the system's structure and behavior usually by the first-order language. Luca Chittaro[2] et al. proposed a hierarchical model which can represent multiple behavioral modes of one component in its various states. Baroni[3] et al. proposed a dynamic system model based on the finite-state automata. Console[4] et al. described the diagnostic problem based on the process algebra. In order to deploy the real-world applications of Model-based diagnosis, several relevant contributions were proposed in the literatures [5,6,7,8].

In our view, diagnosis of a physical system requires the interpretation of what happened to it, based on the related observations and models. The interpretation of the system refers to the output values and the state of each component. Therefore, our proposed method lends itself to the notion of explanatory diagnosis, according to which diagnosis is the explanation of the behavior of the considered system, rather than the mere identification of a set of faulty components. Hence, our approach allows attention to be focused on the value propagation of the components and the system.

This paper is structured as follows: The value propagation diagnostic method is informally described in Section 2; Section 3 defines the concepts of value propagation model and asserts a theorem of value propagation diagnostic method; Section 4 details the diagnosis algorithms and provides the complexity analysis; Section 5 discusses the relational works; Conclusions are drawn in Section 6.

D.-S. Huang, L. Heutte, and M. Loog (Eds.): ICIC 2007, LNAI 4682, pp. 971–981, 2007.

## 2 Informal Presentation of the Approach

In this section, we are devoted to the informal presentation of the value propagation diagnostic method, which is then formalized in Section 3. Consider two assertions:

(1). Giving a system and its observation, for each component, there are actual input values and output value corresponding to the observation and the actual state of the system.

(2). If values of input and output of a component are not consistent with its function, this component is abnormal.

Therefore, if we establish a reasonable hypothesis of the input values and output value of all components corresponding to the observation, a candidate of diagnosis can be obtained.

To establish an above hypothesis reasonably, the value propagation reasoning is a feasible approach. Usually, for a given component, there are two types of value propagation. One is the positive value propagation, which determines the output value of a component based on its input values. The other one is the inverse value propagation, which determines an input value of a component based on its output value and its other partial input values.

To find a diagnosis, we first suppose that some of the components are normal. Then, based on the observation and the normal component's behavior, we can determine all the values of the input and output of the system components by value propagation reasoning. If input values and output value of a component are not consistent with its function, we call that this component is abnormal. The following examples demonstrate the process of diagnosing a system with value propagation reasoning.

**Example 1**

The poly-box system, depicted in Fig.1, contains five components: *M1*, *M2* and *M3* are multipliers, whose reliability is 0.97, 0.99 and 0.97, respectively. *A1* and *A2* are adders, whose reliability is 0.98 and 0.975, respectively. System's input are: $A = 3$, $C = 3$, $E = 3$, $B = 2$, $D = 2$, $F = 2$; output are: $G = 10$, $H = 12$.

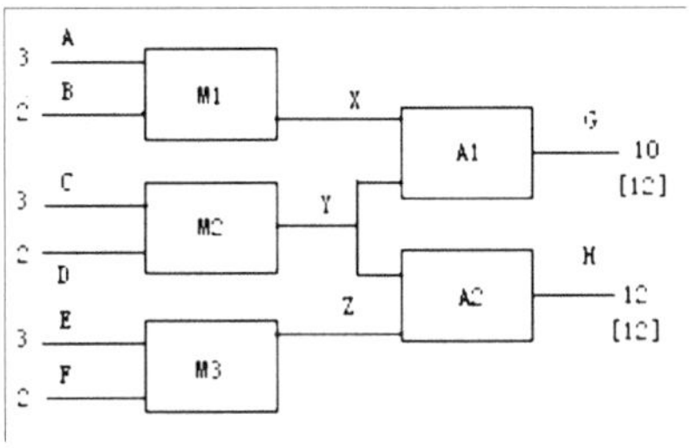


**Fig. 1.** Poly-box system

There are four candidates of minimal diagnoses: {*M1*}, {*A1*}, {*M2*, *A2*}and {*M2*, *M3*}. These minimal diagnoses can be found by value propagation reasoning. First, supposing *M2* is normal, whose reliability is higher than other components, its output *Y* can thus be determined: $Y = C \times D = 6$. Then, supposing *A1* is normal, we can determine that $X = G - Y = 4$. Finally, supposing *A2* is normal, we can determine that $Z = H - Y = 6$. At this time, all the input and output values of the given system have been determined. Checking the rest components of the system, we can find that the input values and output value of *M3* are consistent with its function, but *M1*'s are not. Hence, we find a minimal diagnosis: {*M1*}. Obviously, the minimal diagnosis explains the behavior of the given system reasonably: because it is abnormal, *M1* outputs a false value $X = 4$, which makes the system output the false value $G = 10$.

Once a minimal diagnosis has been obtained, we can test it. If the test result indicates that *M1* is abnormal, we repair it or replace it with a new one. Otherwise, *M1* is normal and we have: $X = A \times B = 6$. Then, supposing *M2* is normal, whose reliability is higher than other components, its output value can thus be determined: $Y = 6$. Then, supposing *A2* is normal, we can determine that $Z = 6$. Thus, all the values of the given system have been determined. Checking the rest components, we find a minimal diagnosis: {*A1*}.

If *A1* is also normal, similarly, we can find other minimal diagnoses: {*M2, M3*}, {*M2, A2*}. □

As mentioned in example 1, each reasoning process based on value propagation can find a minimal diagnosis without computing the conflict sets. By diagnosis generation, testing and repairing, we can find the actual diagnosis and repair the system gradually. However, two problems arise: the first one is that whether the found diagnosis must be a minimal diagnosis; the second one is that whether we can always find at least one diagnosis by our method.

## 3 Formal Presentation of the Value Propagation Approach

In this section, the value propagation model is formally given. Before defining the model of value propagation, we first formally define the system model, which describes the system's structure by the directed graph, models system's behavior by the value constraints.

### 3.1 System Model

**Definition 1.** (model of system)
The model of system Σ is a directed graph $M_\Sigma = G(V, E)$:

(1). $V = I(\Sigma) \cup O(\Sigma) \cup COMPS(\Sigma)$,
(2). $E = \{e_{ij}\ i,j \in COMPS(\Sigma) \mid$ output of $i$ is an input of $j$ $\}$,

where $I(\Sigma)$ is the set of input vertexes, $O(\Sigma)$ is the set of output vertexes, and $COMPS(\Sigma)$ is the set of the system components. Function $f_c$: $D(x_1) \times \ldots \times D(x_{n-1}) \rightarrow D(x_n)$ describes the normal behavior of the component $c$, where $x_1,\ldots,x_{n-1}$ are the input variables of $c$, $x_n$ is the output variable of $c$, $D(x_i)$ is the range of $x_i$. If component $c$ is in a normal state, its input and output values satisfy the constraint $R_{normal}(c) = \{(a_1,\ldots,a_n) \mid f_c(a_1,\ldots,a_{n-1}) = a_n \text{ and } a_i \in D(x_i)\}$. If component $c$ is abnormal, its input and output values satisfy the constraint $R_{abnormal}(c) = D(x_1) \times \ldots \times D(x_n) - R_{normal}(c)$. Each edge $e_{ij} \in E$ has been marked by a variable.

**Example 2**
Considering component *M1* of the poly-box system depicted in Fig.1, its input variables are *A* and *B*, output variable is *X*, function is $A \times B = X$. When *M1* is normal, its corresponding constraint is:

$$R_{normal}(M1)=\{(a_1, a_2, a_3) \mid a_1 \times a_2 = a_3 \text{ and } a_1 \in D(A),\ a_2 \in D(B),\ a_3 \in D(X)\}. \qquad \Box$$

Because the static system has no feedback, *G* (*V*, *E*) is a directed acyclic graph. Fig 2 is a graph of the poly-box system depicted in Fig 1. In this paper, a component can have several inputs but only one output, but one output may acts as an input of many components respectively.

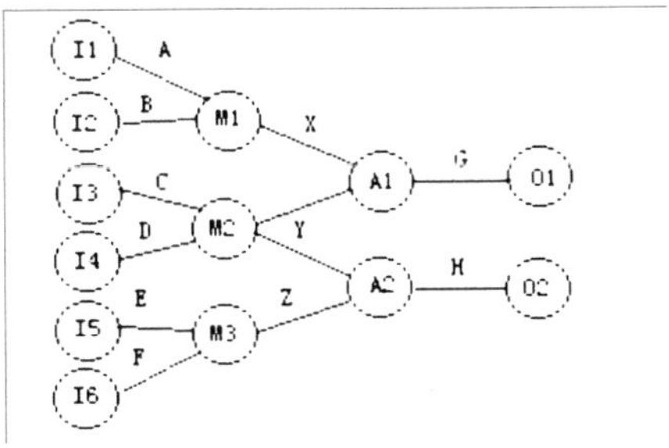


**Fig. 2.** A graph of the poly-box system

**Definition 2.** (system assignment)
System assignment is a pair ($I_V$, $I_E$), where $I_V$ assigning a state $s \in \{normal, abnormal\}$ to every component $c \in COMPS(\Sigma)$, and $I_E$ assigning a value $a \in D(x)$ to every variable *x*. If $I_{VP} \subseteq I_V$ and $I_{EP} \subseteq I_E$, then ($I_{VP}$, $I_{EP}$) is a partial system assignment.

**Definition 3.** (consistent system assignment)
System assignment ($I_V$, $I_E$) is consistent if and only if any component $c \in COMPS(\Sigma)$ satisfies $(I_E(x_1), \ldots, I_E(x_n)) \in R_m(c)$, where $x_1,\ldots,x_n$ are variables of component *c*, and $m = I_V(c)$. Partial system assignment ($I_{VP}$, $I_{EP}$) is consistent if and only if any component $c \in COMPS(\Sigma)$ satisfies $(I_{EP}(x_1), \ldots, I_{EP}(x_n)) \in R_m(c)$, where $x_1,\ldots,x_n$ are variables of component *c*, and $m = I_{VP}(c)$.

**Definition 4.** (diagnostic problem)
A diagnostic problem of the system $\Sigma$ is a triple ($M_\Sigma$, *IN*, *OUT*), where $M_\Sigma$ is the system model, *IN* is the set of the assignment of all input variables of the system, and *OUT* is the set of assignment of all output variables of the system.

**Definition 5.** (system diagnosis)
System diagnosis is a consistent system assignment ($I_V$, $I_E$), which satisfies $I_E(x) = IN(x)$ and $I_E(y) = OUT(y)$ for any input variable *x* and output variable *y* of the given system.

System diagnosis is a reasonable explanation of the current behavior of the given system. If $I_V$ indicates that all components are normal, then $I_E$ describes the normal behavior of the given system. If $I_V$ indicates that some of the components are abnormal, then $I_E$ describes the abnormal behavior of the given system: because some abnormal components export false values which propagate through other components, the system thereby exports the current value which is different from that should be.

In fact, the definition of diagnosis in this paper is equivalent to Reiter's definition[1]. Because system diagnosis is a reasonable explanation of the current behavior of the given system, it surely satisfies the logical consistency, and vice versa.

**Definition 6.** (minimal diagnosis)
System diagnosis $(I_V, I_E)$ is minimal if and only if there does not exist any system diagnosis $(I_V', I_E')$ which satisfies $\{c \in COMPS(\Sigma) \mid I_V'(c) = abnormal\} \subset \{c \in COMPS(\Sigma) \mid I_V(c) = abnormal\}$.

We can simply define a minimal diagnosis $(I_V, I_E)$ as a set of faulty components $D=\{c\in COMPS(\Sigma) \mid I_V(c)=abnormal\}$, which resembles Reiter's definition.

### 3.2 Value Propagation Model

**Definition 7.** (component value propagation)
For any normal component $c$, its value propagation, $c[(x_1/a_1,\ldots,x_k/a_k)\rightarrow(y_1/b_1,\ldots,y_j/b_j)]$,is a process to determine the values of some variables $y_1,\ldots,y_j$ based on other known variables $x_1,\ldots,x_k$ and constraint $R_{normal}(c)$: if $x_1 = a_1,\ldots,x_k = a_k$ and $c$ is normal, then $y_1 = b_1,\ldots,y_j = b_j$.

Usually, the positive value propagation is to determine the output value of a component based on its input values, and the inverse value propagation is to determine an input value of a component based on its output value and its other partial input values. For example, determining $X = 6$ by $A = 3$ and $B = 2$ is a positive value propagation of component $M1$; determining $Z = 6$ by $H = 12$ and $Y = 6$ is an inverse value propagation of component $A2$.

**Definition 8.** (system value propagation)
System value propagation is a sequence $S=(I_{VP1}, I_{EP1}), \ldots, (I_{VPn}, I_{EPn})$, which satisfies the following conditions:

(1) For any adjacent elements $(I_{VPi}, I_{EPi})$ and $(I_{VPi+1}, I_{EPi+1})$, there exists $c[(x_1/a_1,\ldots,x_k/a_k) \rightarrow (y_1/b_1,\ldots,y_j/b_j)]$, which satisfies $\{x_1/a_1,\ldots,x_k/a_k\}\subseteq I_{EPi}$, $I_{VPi} \cup\{c/normal\}= I_{VPi+1}$ and $I_{EPi} \cup\{ y_1/b_1,\ldots,y_j/b_j \}= I_{EPi+1}$.

(2) $I_{VP1}=\varnothing$, $I_{EP1} =IN\cup OUT$, $EPn=E$.

Let $I_{VPn} = I_{VP}$, then $(I_{VP}, I_E)$ is the result of value propagation, where $VP$ is the set of the components which participate in the process of value propagation.

Actually, system value propagation is a process to assign values to all the variables and assign states to the components which participate in the process of value propagation. It is noticeable that any $(I_{VPi}, I_{EPi})$ in sequence $S$ is consistent. (see theorem 1)

**Theorem 1**
For a given system value propagation $S=(I_{VP1}, I_{EP1}), \ldots, (I_{VPn}, I_{EPn})$, any partial system assignment $(I_{VPi}, I_{EPi})$ in sequence $S$ is consistent..

**Proof**
For definition 3 and definition 8, $<I_{VP1},I_{EP1}>=<\varnothing, IN\cup OUT>$ is consistent. Supposing that $<I_{VPi},I_{EPi}>$ is consistent, we only need to prove that $<I_{VPi+1},I_{EPi+1}>$ is also consistent.

Because of $I_{VPi} \cup \{c/normal\} = I_{VPi+1}$, $I_{EPi} \cup \{ y_1/b_1,\ldots,y_j/b_j \} = I_{EPi+1}$, and $<I_{VPi}, I_{EPi}>$is constent, we only need to prove that $<I_{EPi+1}(x_1),\ldots, I_{EPi+1}(x_n)> \in R_{normal}(c)$. For definition 8, $c[(x_1/a_1,\ldots,x_k/a_k) \rightarrow (y_1/b_1,\ldots,y_j/b_j)]$, thus, $<I_{EPi+1}(x_1),\ldots,I_{EPi+1}(x_n)> \in R_{normal}(c)$. Therefore, $<I_{VPi+1}, I_{EPi+1}>$ is also consistent. □

In other words, system value propagation establishes a reasonable hypothesis, which has been mentioned in section 2, of the input values and output value of all components corresponding to the observation. The following example describes a possible process of system value propagation.

**Example 3**
Considering the poly-box system depicted in Fig.1, sequence *u1u2u3u4* is a value propagation of system and *u4* is a value propagation result, where:

*u1*=( ∅, {*A*/3, *C*/3, *E*/3, *B*/2, *D*/2, *F*/2, *G*/10, *H*/12})
*u2*=({*M2*/*normal*}, { *A*/3, *C*/3, *E*/3, *B*/2, *D*/2, *F*/2, *G*/10, *H*/12, *Y*/6})
*u3*=({ *M2*/*normal* , *A1*/*normal*},{ *A*/3, *C*/3, *E*/3, *B*/2, *D*/2, *F*/2, *G*/10, *II*/12, *Y*/6, *X*/4})
*u4*=({ *M2*/*normal*, *A1*/*normal*, *A2*/*normal*}, { *A*/3, *C*/3, *E*/3, *B*/2, *D*/2, *F*/2, *G*/10, *H*/12, *Y*/6, *X*/4, *Z*/6})
*u1u2* corresponds to *M2*[(*C*/3,*D*/2) → (*Y*/6)]
*u2u3* corresponds to *A1*[(*G*/10,*Y*/6) → (*X*/4)]
*u3u4* corresponds to *A2*[(*H*/12,*Y*/6) → (*Z*/6)]. □

**Theorem 2**
Given the inputs and outputs of the system, $N$ is the set of the normal components and $I$ is the actual assignment[1] of the variables of the system, if there is a result of value propagation ($I_{VP}$, $I_E$) which satisfies $VP \subseteq N$, then $I_E = I$.

**Proof**
Suppose that $I_{EP}$ ($I_{EN}$) is the set of the assignments of the variables which have been determined by the positive (inverse) value propagation, so $IN \cup OUT \cup I_{EP} \cup I_{EN} = I_E$ and $I_{EP} \cap I_{EN} = \varnothing$. Corresponding to $I_{EP}$ and $I_{EN}$, we divide $I$ into $I_P$ and $I_N$, then we have $IN \cup OUT \cup I_P \cup I_N = I$ and $I_P \cap I_N = \varnothing$.

Because $VP \subseteq N$, it holds that $I_P = I_{EP}$.

If $I_{EN} = \varnothing$, then $I_E = I$, else because $G$ ($V$, $E$) is a directed acyclic graph, there exists a component $c$ which has performed inverse value propagation and whose output variable is assigned by $OUT$. Let $I_{EN}$ ($x$) be an assignment of the input variable which has been determined by the value propagation of $c$. Based on the definition 7, we have $I_N(x) = I_{EN}(x)$. If the output of component $c'$ is the input of $c$ and $I_{EN}(x')$ is an assignment of the input variable which has been determined by the inverse value propagation of $c'$. We have $I_N(x') = I_{EN}(x')$. By analogy, we have $I_N = I_{EN}$. Therefore, $I_E = I$. □

As said in theorem 2, the intuitional meaning of the result of value propagation is: given the input and output of the system and ensure some components are normal, if there is a value propagation process with those normal components, then the assignments of the variables with the result of value propagation describe the current behavior of the considered system (see example 4).

---

[1] The actual assignment of the variables of the system is assigning an actual value to each variable of the system. It is a description of the actual behavior of the considered system.

**Example 4**
The poly-box system, depicted in Fig.1, suppose *M*1 is an actual faulty component, and the actual values of the system are: *A*=3, *C*=3, *E*=3, *B*=2, *D*=2, *F*=2, *G*=10, *H*=12, *Y*=6, *X*=4, *Z*=6. In other words, the actual assignment *I* ={ *A*/3, *C*/3, *E*/3, *B*/2, *D*/2, *F*/2, *G*/10, *H*/12, *Y*/6, *X*/4, *Z*/6}. In addition, suppose that we know that *M2*, *A1* and *A2* are normal components.

As said in example 3, sequence *u1u2u3u4* is a value propagation of system and *u4* is a value propagation result ($I_{VP}$, $I_E$), where *VP* = {*M2*, *A1*, *A2*}. Obviously, as said in theorem 2, $I_E = I$. □

### 3.3 Value Propagation and Minimal Diagnosis

In this section, we will discuss the relationship between the value propagation and the minimal diagnosis.

**Theorem 3**
Given a diagnostic problem ($M_\Sigma$, *IN*, *OUT*), there exists a system value propagation which gets a result of value propagation ($I_{VP}$, $I_E$).

**Proof**
For any normal component, given its input values, its output value can be definitely determined. Therefore, given a diagnostic problem ($M_\Sigma$ *IN*, *OUT*), if there is a variable which has not been assigned a value, it can be assigned by the positive value propagation, although it may also be assigned by the inverse value propagation.

Hence, theorem 3 is confirmed. □

**Theorem 4**
Given a diagnostic problem ($M_\Sigma$, *IN*, *OUT*), if there is a result of value propagation ($I_{VP}$, $I_E$), then there exists a minimal diagnosis ($I_V$, $I_E$ ) corresponding to ($I_{VP}$, $I_E$).

**Proof**
For $<I_{VP}, I_E>$, we structure assignment *I* as follows: For every $c \in V/VP$, if $<I_E(x_1),\ldots,I_E(x_n)> \in R_{normal}(c)$ then *I* (*c*) =*normal*, else *I* (*c*) =*abnormal*. Let $I_V = I \cup I_{VP}$, for definition 5 and theorem 1, $<I_V, I_E>$ is a diagnosis.

Suppose $<I_V, I_E>$ is not a minimal diagnosis, then there exists a diagnosis which satisfies $D' \subset D$, where:

$$D = \{c \in COMPS\ (\Sigma) \mid I_V\ (c) = abnormal\};$$
$$D' = \{c' \in COMPS\ (\Sigma) \mid I_V'\ (c') = abnormal\}.$$

Consider a system $\Sigma'$, which has a set $D'$ of the abnormal components. Given the same inputs as $\Sigma$'s, $\Sigma'$ will export the same result as $\Sigma$'s. Obviously, $I_E'$ is consistent with the functions of the components of $D/D'$. Because $VP \subset COMPS(\Sigma')/ D'$, the same value propagation process will produce the result of value propagation $<I_{VP}, I_E>$. $I_E$ is inconsistent with the functions of the components of $D / D'$. Because $VP \subset COMPS(\Sigma')/D'$, based on theorem 2, it holds that $I_E = I_E'$, which contradicts to the previous result. Therefore, *D* is a minimal diagnosis. □

Above mentioned theorems illuminate the relationship between the value propagation and the minimal diagnosis. Theorem 3 indicates that the result of system value propagation can be definitely obtained by a value propagation process. Theorem 4 tells that the result of value propagation corresponds to a minimal diagnosis. Therefore, here is the following deduction.

**Corollary 1**
Given a diagnostic problem ($M_\Sigma$, *IN*, *OUT*), there exists a system value propagation which can result in a minimal diagnosis ($I_V$, $I_E$ ).

**Proof**
It can be derived straightforwardly from theorem 3 and theorem 4. □

Thus, to resolve the problems mentioned in section 2, we construct a theoretical foundation of the diagnostic method based on the value propagation model. The following example describes a possible process for finding a minimal diagnosis.

**Example 5**
Considering the poly-box system depicted in Fig.1, sequence *u1u2u3u4* (see example 3) is a system value propagation which gets the result of value propagation ($I_{VP}$, $I_E$):

$I_{VP}$ ={ *M2/normal*, *A1/normal*, *A2/normal*}
$I_E$ ={ *A*/3, *C*/3, *E*/3, *B*/2, *D*/2, *F*/2, *G*/10, *H*/12, *Y*/6, *X*/4, *Z*/6}.
It corresponds to a minimal diagnosis ($I_V$, $I_E$ ), simply noted as *D* = {*M1*}:
$I_V$={ *M2/normal*, *A1/normal*, *A2/normal*, <u>*M1/abnormal*</u>, *M3/normal*}
$I_E$ ={ *A*/3, *C*/3, *E*/3, *B*/2, *D*/2, *F*/2, *G*/10, *H*/12, *Y*/6, *X*/4, *Z*/6}. □

# 4 Diagnostic Approach

As mentioned in section 3, a minimal diagnosis can be found by the value propagation reasoning. Therefore, given a faulty system, we can repair it by integrating diagnoses and test:

(1) find a minimal diagnosis;
(2) test the diagnosis, if there are faulty components, repair them or replace them with new components;
(3) generate a new diagnostic problem, repeat the above process until the system's output is normal.

Step (1) is the most important one in the above process. In subsection 4.1, we will propose an algorithm for finding a minimal diagnosis.

## 4.1 Algorithm for Finding a Minimal Diagnosis

**Algorithm 1**
**FIND-ONE-DIAG** ($M_\Sigma$, *IN*, *OUT*)
**Begin**
$I_V$ =∅; $I_E$ = *IN*∪*OUT*;
*K* ← the set of all components which can perform value propagation;
**while** ($K \neq \emptyset$)
**do**{ choose a component *c* from *K*;
perform $c[(x_1/a_1,\ldots,x_k/a_k)\rightarrow(y_1/b_1,\ldots,y_j/b_j)]$ and mark *c*;
$I_V = I_V \cup \{c/normal\}$;
$I_E \leftarrow I_E \cup \{ y_1/b_1,\ldots,y_j/b_j \}$;
$K \leftarrow (K/\{c\}) \cup \{$ new components which can perform value propagation$\}$; }
**for** (each unmarked component *c'*)

**do**{**if** (input and output values are inconsistent with the component function)
  **then** $I_V$=$I_V$∪{ *c'/abnormal* }**else** $I_V$=$I_V$∪{ *c'/normal* };}
**return** ($I_V$, $I_E$);
**End**

Actually, algorithm 1 performs a practical task of system value propagation and returns a minimal diagnosis. We have the following theorem.

**Theorem 5**
Given a diagnostic problem ($M_\Sigma$, *IN*, *OUT*), algorithm 1 will terminate and return a minimal diagnosis consequentially.

**Proof**
As said in the proof of theorem 3, it holds that if there exists variables which have not been assigned a value, $K \neq \varnothing$. Again, the number of variables and components is finite. Therefore, algorithm 1 will terminate with a system assignment ($I_V$, $I_E$). For theorem 4, the system assignment ($I_V$, $I_E$) is a minimal diagnosis. □

In fact, we can adopt certain strategy for choosing the value propagating component from *K*. For example, in general, we always choose the most reliable component in set *K* to perform value propagation. As a result, a more possible diagnosis can usually be achieved.

Specially, if the algorithm returns a null set, then the system's behavior is normal.

### 4.2 Complexity Analysis

Without loss of generality, we suppose that any component has *k* variables at most, and each variable has *m* possible values. Therefore, the complexity of determining weather can a component perform value propagation is $O(m^k)$. For a system which has *n* components, we can demonstrate that the complexity of algorithm 1 is $O(n^2m^k)$:

(1) Finding a set *K* of all components which can perform value propagation requires at most *n* computations;
(2) Updating *K* requires at most *n* computations;
(3) The algorithm requires at most *n* value propagations.

It is noticeable that, for many systems, the number of the variables of the component is a constant, and the number of the possible values of the variable is also a constant. Therefore, in this case, the time cost of once component value propagation is a constant, and the complexity of algorithm 1 is $O(n^2)$.

As mentioned in above discusses, our algorithm can find a minimal diagnosis rapidly, even for a large system. Table 1 is the experimental results[2] of the algorithm.

---

[2] For the poly-box system, determining weather can a component perform value propagation is very easy. Any component which has only one unknown variable can perform value propagation. But for other system, the algorithm may cost more time to find the component which can perform value propagation. It is depend on the function of the component.

**Table 1.** The experimental results on the poly-box system

| Number of the components | Time cost (millisecond) |
|---|---|
| 5 | 1 |
| 15 | 10 |
| 120 | 75 |
| 500 | 300 |
| 1000 | 750 |

## 5 Related Works

Reiter[1] proposed a method of finding diagnosis by computing the conflict sets, which complexity is exponential time cost. Davis[9], de Kleer and Williams[10] proposed constrain propagation method to compute the conflict sets. Unlike Reiter's method, our approach can find the minimal diagnosis without computing the conflict sets.

Fattah and Dechter[5], Stumptner and Wotawa[11] respectively proposed a polynomial-time algorithm for diagnosing the tree-structured systems. However, because many actual systems are not tree-structured, their approach only has a limited application field. Gerhard Friedrich[12] also presented a polynomial-time approach for finding a minimal diagnosis. Based on the information of system structure, literature [13] proposed a method of diagnosing a physical system without computing the conflict sets. For some specially structured system, it can find all the minimal diagnoses within polynomial time.

Console [7] and Milde [8] proposed a method of diagnosing an actual system by generating a decision tree based on the system model. But for large-scale system, the scale of the decision tree is very huge.

## 6 Conclusions and Future Works

Our approach allows attention to be focused on the value propagation of the components and the system. We present a system model based on the value propagation and redefine the notion of diagnosis, which is equivalent to Reiter's classical definition. On the basis of the above job, an algorithm for finding a minimal diagnosis is proposed. This algorithm need not compute the minimal conflicts. The resulting diagnosis is just the explanation of the behavior of the considered system, rather than the mere identification of a set of faulty components. The experimental results indicate that the algorithm can find a minimal diagnosis rapidly, even for a large system which involves lots of components. Finally, this paper presents a framework for repairing the system by integrating diagnoses and test.

For a given diagnostic problem, there exist many possible value propagations which can obtain many different value propagation results respectively. Therefore, we can get many different diagnoses based on those value propagation results. But for a large-scale system, its number of the possible value propagations is very big. We will discuss this problem in the future works.

## References


1. Reiter, R.: A Theory of Diagnosis from First Principles. Artificial Intelligence, 32 (1987) 57-96
2. Chittaro, L.: Hierarchical Model-Based Diagnosis Based on Structural Abstraction. Artificial Intelligence, 155 (2004) 147-182
3. Baroni, P .: Diagnosis of Large Active Systems. Artificial Intelligence,110 (1999) 135-183
4. Console, L., Picardi, C., Ribaudo, M.: Process Algebras for Systems Diagnosis. Artificial Intelligence, 142 (2002) 19–51
5. Fattah,Y. E., Dechter, R..: Diagnosing Tree-Decomposable Circuits. In: Proceedings of International Joint Conference on Artificial Intelligence, Montreal, Canada, (1995)572-578
6. Portinale, L., Magro, D., Torasso, P.: Multi-Modal Diagnosis Combining Case-Based and Model-Based Reasoning: A Formal and Experimental Analysis. Artificial Intelligence, 158 (2004) 109–153
7. Console, L.: Temporal Decision Trees: Model-Based Diagnosis of Dynamic Systems On-Board. Journal of Artificial Intelligence Research, 19 (2003) 469-512
8. Milde, H.: Integrating Model-Based Diagnosis Techniques into Current Work Processes-Three Case Studies from the INDIA Project. AI Communications, 13 (2000) 99-123
9. Davis, R.: Diagnostic Reasoning Based on Structure and Behavior. Artificial Intelligence, 24 (1984) 347-410
10. de Kleer, J., Williams, B. C.: Diagnosing Multiple Faults. Artificial Intelligence, 32 (1987) 97-130
11. Stumptner, M., Wotawa, F.: Diagnosing Tree-Structured Systems. Artificial Intelligence, 127 (2001) 1-29
12. Friedrich, G., Gottlob, G., Nejdl, W.: Physical Impossibility Instead of Fault Models. AAAI (1990) 331-336.
13. Luan, S. M., Dai, G. Z.: An Approach to Diagnosing a System with Structure Information. Chinese Journal of Computers, 28 (2005) 801-808

# Modeling Dependability of Dynamic Computing Systems

Salvatore Distefano and Antonio Puliafito

University of Messina, Engineering Faculty,
Contrada di Dio, S. Agata, 98166 Messina, Italy
{salvatdi,apulia}@ingegneria.unime.it

**Abstract.** A trend actually characterizing any technological and application field is the use of control and computing devices. This moves the attention of the insiders on the management of more complex processes and systems, modifying requirements, tasks and skills. In this scenario the dependability evaluation becomes strategic. The increasing complexity of systems requires a serious revision of the techniques and the methodologies exploited for evaluating the system dependability. Notations as reliability block diagrams (RBD), fault trees (FT) or reliability graphs (RG) and their extensions, such as dynamic FT (DFT), could not adequately model dynamic reliability/availability aspects and behaviors as the units' interactions. To overcome those problems we developed a new formalism derived from RBD: the dynamic RBD (DRBD). DRBD allow to represent dynamic reliability/availability behaviors by specifying the dependency concept. The dependency is the building block for modeling dynamic aspects in DRBD. In this paper we detail the DRBD modeling technique thorough an example of distributed computing system taken from literature. By this we also compare our approach with the DFT one, providing proofs of the DRBD potentialities in dynamic reliability/availability modeling.

**Keywords:** Dependability Modeling, Dynamic Systems, Dynamic Reliability Block Diagrams, Dynamic Fault Trees.

## 1 Introduction

There are several approaches to represent system reliability/availability. In particular the *combinatorial models* are specific high-level reliability/availability modeling formalisms. *Reliability block diagrams* (RBD) [1], *fault trees* (FT) [2] and *reliability graphs* (RG) [3] belong to this class. Although RBD, RG and FT provide a view of the system close to the modeler, more readable and understandable than any other formalism, they are defined on the *stochastic independence* assumption of the units. They do not provide any elements or capabilities to model reliability interactions among components or subsystems, or to represent system configuration changes, aspects characterized as *dynamic*.

From the reliability/availability point of view, it could be possible that, generally, a subsystem has some influence to other subsystems, characterizing the *system dynamics*.

D.-S. Huang, L. Heutte, and M. Loog (Eds.): ICIC 2007, LNAI 4682, pp. 982–991, 2007.

Examples of such behaviours are: *load-sharing, standby redundancy, interferences, dependencies, probabilistic and common mode failures,* .... Also the configuration of a system could vary: a failed component/subsystem could be repaired (maintenance, imperfect fault coverage, reliability growth), the system could be *phased-mission*, and so on. In particular these remarks concern computing systems: load sharing phenomena could affect the network availability; dependability resources-optimization requirements could be translated into maintenance and/or redundancy schemes and policies; interference or inter-dependence among components could arise (wireless devices, sensors, ...); common cause failures could group electric devices, in case of power jumps or sudden changes of temperature (melting point of transistors, ...).

These argumentations awakened the scientific community to the need of new formalisms as the *dynamic fault trees* (DFT) [4,5]. DFT extend static FT to enable modeling of time dependent failures, introducing new dynamic gates and elements. DFT add a temporal notion to FT: system failures can depend on the order of component failures. DFT are a good attempt in dynamic reliability modeling, but they cannot adequately represent some of the dynamic behaviours previously listed. To overcome the problems or lacks in dynamic reliability/availability modeling, in [6,7,8] we have defined a new reliability/availability modeling notation named *dynamic reliability block diagrams* (DRBD), by extending the RBD formalism.

In this paper we face the problem of modeling by DRBD notation in practice. We base the approach's description on an example of a multiprocessor distributed computing system. During the modeling explanation, the guidelines of a methodology for modeling system reliability by DRBD are provided. Moreover a comparison with the DFT approach is investigated showing how it is possible to map from the existing model types to the DRBD model type. By this way, the capabilities, the flexibility, and the other features of the DRBD formalism in modeling system reliability/availability could be adequately pointed out.

The reminder of the paper is organized as follows: first of all some background concepts of the notations utilized in the paper are provided in section 2. Then, the modeling section 3 describes the DRBD modeling approach, highlighting the comparison between the DFT and the DRBD notations by exploiting the multiprocessor distributed computing system taken as example. Lastly, section 4 provides some final considerations on the DRBD methodology.

## 2 Background

This section briefly introduces the notations object of interest of the paper, used and compared in the following. In subsection 2.1 an overview of fault trees and dynamic fault trees is provided. Then, in subsection 2.2, an outline of the reliability block diagrams notation is summarized in order to better introduce the dynamic reliability block diagrams, discussed in subsection 2.3. More details on the DRBD formalism can be found in [6,7,8].

### 2.1 FT and DFT

Fault trees provide a compact, graphical method to analyze system reliability. Static trees [2] use Boolean gates to represent how component failures combine to produce system failure. Dynamic trees ([4,5,9]) add a temporal notion, since system failures can depend on the order of component failures. Although static fault trees are more commonly used, they are limited in modeling systems that have no sequential relationships among their component failures. Component failure sequences are best captured using dynamic fault trees that extend static trees to enable modeling of time dependent failures. They can model dynamic replacement of failed components from pools of spares (CSP, WSP and HSP gates); failures that occur only if others occur in certain orders (PAND gates); dependencies that propagate the failure of one component to others (FDEP gates); and specification of constraints on failure orders that simplify analysis computations (SEQ gates).

### 2.2 RBD

The main virtue of a RBD is that it is easy to read. In a RBD, the logic diagram is arranged to indicate which combinations of component failures result in the failure of the system, or which combinations of properly working components keep the system operational. A block in RBD represents the working physical component, and the failure of this component is indicated by the removal of the corresponding block. If enough blocks are removed in an RBD to interrupt the connection between the input and output points, the system fails. Generally three main types of connections, series, parallel and non series-parallel, can be established between two or more components [1]. The blocks in either a series, parallel and non series-parallel structure can be merged into a new block. Using such combinations, any system can be modeled as a block and its reliability can be easily computed by repeatedly using the structure equations [1].

### 2.3 DRBD

DRBD enhance the RBD formalism in the systems' dynamics representation. The original RBD syntax is extended with some specific elements modeling dynamic aspects. Two are the key points of DRBD: the *component's dynamics* characterization and the specification of the *dependency* concept as the building block of dynamic reliability modeling.

In a DRBD model each component is characterized by a variable *state* identifying its operational condition at a given time. The evolution of a component's state (*component's dynamics*) is characterized by the *events* occurring to it, as depicted in Fig. 1. The states a generic DRBD component can assume are: *active* if the component works without any problem, *failed* if component is not operational, following up its failure, and *standby* if it is operable but not committable. Active components participate actively to perform the work, the task of the system, while standby components do not contribute to this, they do not interact with the other components as consequence of a *dependency* application. But, at the same time, a component in standby is not failed, it just performs its internal activities. An event

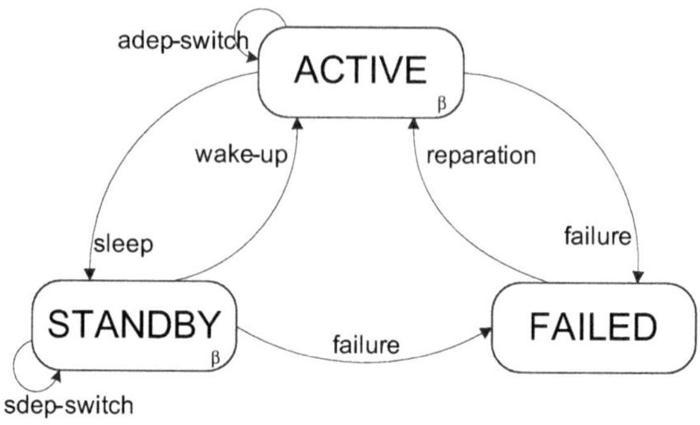


**Fig. 1.** DRBD component's states-events Finite State Automata

represents the transition from a components' state to another one: the *failure* event models a states changing from active or standby to the failed state, the *wake-up* switches from standby to active states, the *sleep* from active to standby states, the *reparation* from failed to active state, the *adep-switch* represents the transition between two active states and *sdep-switch* between two standby states. These two latter events are related to the concurrency property of dependencies. For details see [6,7].

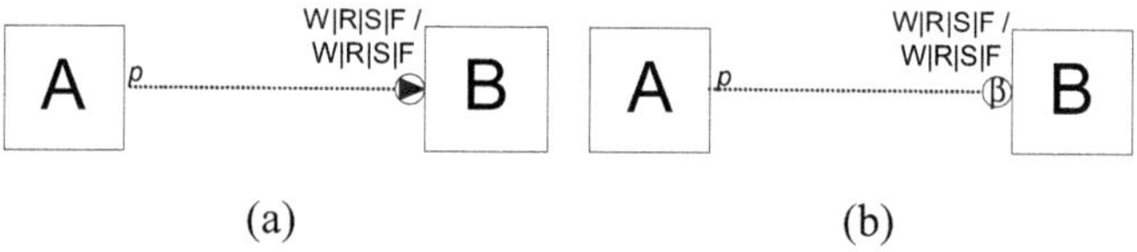


**Fig. 2.** DRBD order (a) and strong (b) dependencies representation

The main enhancement introduced by DBRD is the capability to model dependencies among subsystems or components concerning their reliability behaviours. A dependency establishes a reliability relationship between two components or subsystems, a *driver* and a *target*. Informally a dependency works as follow: when a specified event, named *action* or *trigger*, occurs to the driver, the dependency condition is applied to the target. This condition is associated to a specific target event, named reaction. When the satisfied dependency condition becomes unsatisfied, the target component comes back to the fully active state. A dependency could model two kinds of different relationships between a couple of driver-target components/subsystems: the *order* (Fig. 2 (a)) establishes the sequence order between trigger and reaction events, the *strong* (Fig. 2 (b)) forces the target component to react when a trigger event occurs. A dependency is also characterized by the action (trigger) and the reaction events. Four types of trigger and reaction events can be identified: wake-up (W), reparation (R), sleep (S) and failure (F). Combining action and reaction, 16 types of dependencies are identified, 32 considering the two relationships. In the examples shown in Fig. 2, A is the driver component and B the target. The dependency action or trigger event is indicated by a letter (W, R, S or F as above), and the reaction by another letter (from the same set), separated from the action by a slash. The total string is placed near the circle. In case of order dependencies an arrow is placed inside the circle (Fig. 2 (a)), while a number characterizes strong dependencies (the $\beta$ of Fig. 2 (b)): it indicates the *dependency rate*. This latter characterizes strong dependencies with wake-up (W) and/or sleep (S) reaction, weighting, in terms of reliability, the dependence of the target component from the driver. More details and the formal specification of the DRBD notation could be found in [8].

The concept of dependence is exploited in DRBD as the basis to represent dynamic reliability behaviors. For example, redundancy can be easily represented by combining dependencies among units. Other possible applications of dependencies in dynamic system reliability modeling could be load sharing, common cause failure, reparation, and so on. The modeling of such aspects will be illustrated in the following by means of the multiprocessor computing system example described in subsection 3.1. Further details and more formal specifications of these interesting capabilities of DRBD can be found in [6,7,8].

## 3 The Modeling

In this section we describe the process of modeling the multiprocessor computing system described in subsection 3.1 by means of DFT (in subsection 3.2) and DRBD (subsection 3.3), in order to compare the two modeling approaches.

### 3.1 Motivating Example: A Multiprocessor Distributed Computing System

In order to better describe our methodology, a step by step approach focused on the reliability modeling of a multiprocessor system taken as case study is adopted. It refers to a multiprocessor computing system drawn from literature ([10,11]), represented in Fig. 3. It is composed by two computing modules: $CM_1$ and $CM_2$. Each of them contains one processor ($P_1$ and $P_2$ respectively), one memory ($M_1$ and $M_2$) and two hard disks: a primary ($D_{11}$ and $D_{21}$) and a backup disk ($D_{12}$ and $D_{22}$). Initially, the primary disk is accessed by the computing modules, while the backup disk contains the copy of the information inside the primary disk, and is accessed only periodically

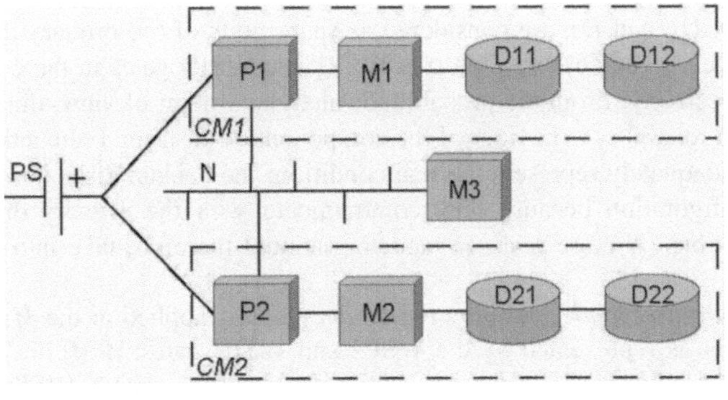


**Fig. 3.** Schematic representation of the Multiprocessor Distributed Computing System

for update operations. If the primary disk fails, it is replaced by the backup disk. When the primary disk is operational, the reliability of the backup disk increases as consequence of the lower workload condition.

The computing modules are connected by means of the bus *N* and are energized by the power supply *PS*.

$M_3$ is a spare memory replacing $M_1$ or $M_2$ in the case of failure, kept in a partly-loaded standby: when $M_1$ or $M_2$ fail, $M_3$ substitutes the failed unit, if it is not already failed.

In order to work properly the multiprocessor computing system of Fig. 3 requires that at least one computing module ($CM_1$ or $CM_2$), the power supply *PS* and the bus *N* are operating correctly. Moreover a computing module ($CM_1$ and $CM_2$) is operational if the processor ($P_1$ and $P_2$ respectively), one between the local memory ($M_1$ and $M_2$) and the shared memory $M_3$ and one disk ($D_{11}$ or $D_{21}$ for $CM_1$ and $D_{12}$ or $D_{22}$ for $CM_2$) are not failed.

The reported example does not represent/implement the optimal system's management configuration available in terms of reliability. To improve the system reliability, a better solution could be to consider the two computing modules $CM_1$ and $CM_2$ as redundant subsystems and therefore managed such as. Another lack of the model is that no load sharing effects are considered between the computing modules: the fact that both $CM_1$ and $CM_2$ are operating could increase the reliability of each computing module splitting the workload with the other one. If only a computing module works, it must process all the incoming requests, before elaborated in collaboration with the other *CM*, for this reason it could be more stressed and therefore its probability to fail could increase. These behaviours cannot be modeled by a DFT because they involve dependencies composition and/or modeling of uncovered aspects as the load sharing. In the following sections we explain how to represent the redundancy aspects by DRBD, considering the two computing modules as redundant subsystems.

### 3.2 The DFT Model

In [10,11], the DFT modeling the multiprocessor computing system is composed by a FDEP and four WSP, as shown in Fig. 4. The FDEP gate models the functional dependency among the power supply *PS* and the two processors $P_1$ and $P_2$. The backup disks $D_{12}$ and $D_{22}$ are considered as spare units of the primary disks $D_{11}$ and $D_{21}$ respectively, thus $D_{11}$ and $D_{21}$ drive WSP1 and WSP4 gates in the control of $D_{12}$ and $D_{22}$ respectively. From the probabilistic/analytical point of view, this choice well describes the reliability behaviour of the components, but, from a semantic viewpoint, it does not adequately represent the real condition: the backup disks do not assume a standby configuration because they communicate with the primary disks, making active operations. A more realistic modeling should therefore take into account this fact.

The *warm/partly loaded standby redundancy policy* applied to the $M_1$, $M_2$ and $M_3$ memory units, is represented by the WSP2 and WSP3 gates: if $M_1$ or $M_2$ fail, $M_3$ is activated. The other DFT gates are static: the internal events $DISK_1$ and $DISK_2$

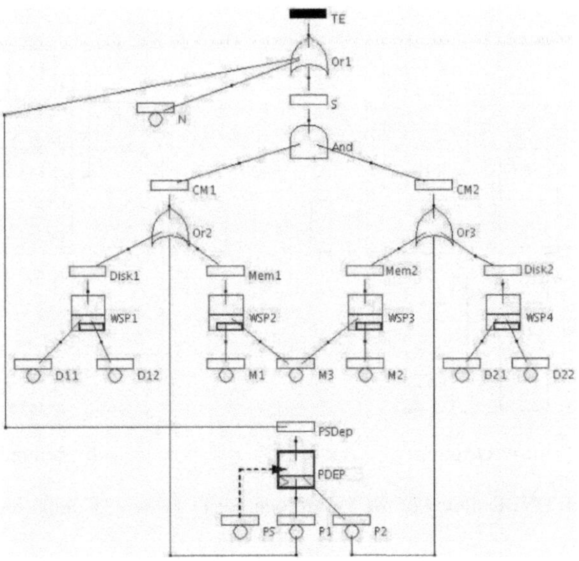


**Fig. 4.** DFT models of the Multiprocessor Distributed Computing System

represent the failures of the storage blocks, similarly $MEM_1$ and $MEM_2$ represent the computing modules' memory blocks failures. The failure of the processor ($P_1$ and $P_2$) or of the memory block ($MEM_1$ and $MEM_2$) or of the disk block ($DISK_1$ and $DISK_2$) drives to the failure of the corresponding computing module ($CM1$ and $CM2$ internal events). Finally, if both the computing modules fail, or the power supply *PS* goes down, or the bus *N* fails, the overall system fault occurs, represented in the DFT as the top event *TE*.

### 3.3 The DRBD Model

The DRBD model reported in Fig. 5(a) represents the multiprocessor computing system object of this study. Since the power supply *PS* energizes the $P_1$ and $P_2$ processors, the failure of *PS* does not imply the failure of $P_1$ and $P_2$. Thus we represent this behaviour in the DRBD model by a simple series between each processor and *PS*.

The disks management policy is represented by a DRBD wake-up/wake-up strong dependency: when the primary disks $D_{11}$ and/or $D_{21}$ are operational, the backup disks $D_{12}$ and/or $D_{22}$ respectively are partially active, maintaining the backup. When a primary disk fails, the corresponding backup disk that substitutes it becomes fully active and fully energized since the dependency condition becomes unsatisfied. The level of activity of the dependent components is numerically translated into the DRBD by the dependency rate $\beta$, that, in this case, assumes a value in the range between 0 and 1 representing decreasing failure rate conditions. In this case $\beta$ is related to the DFT dormancy factor $\alpha$ by the relationship $\beta = 1 - \alpha$.

Wake-up/standby strong dependencies are instead exploited to model the redundancy policy managing the memories. These represent the disabling conditions

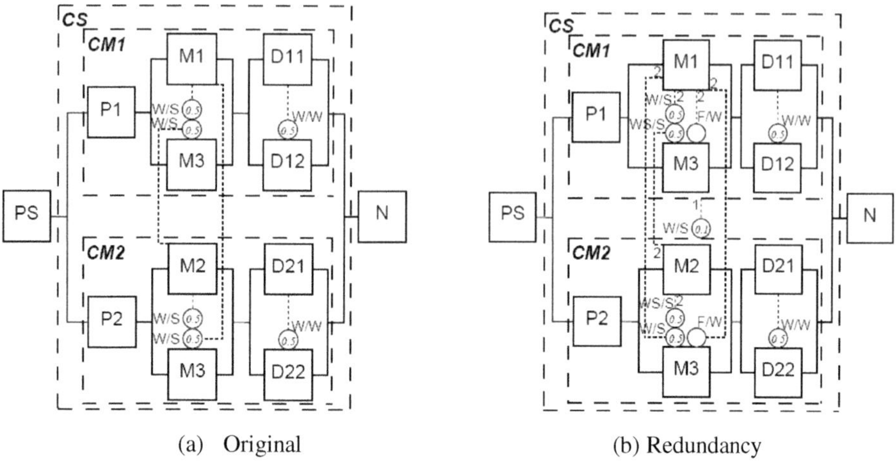


**Fig. 5.** DRBD model of the Multiprocessor Distributed Computing System

applied by $M_1$ and $M_2$, when they are operational, on $M_3$, keeping this latter in standby. The overall dependency must be applied to $M_3$ if and only if both $M_1$ and $M_2$ are at the same time operational; when one of these fails, $M_3$ must switch to the fully active state. To realize this condition the two wake-up/standby strong dependencies, from $M_1$ to $M_3$ and from $M_2$ to $M_3$ are *series composed*: when both are simultaneously satisfied the component $M_3$ is kept in standby, otherwise $M_3$ is active. This composed dependency is duplicated in the DRBD model for the sake of clarity, identifying and separating the two computing modules $CM_1$ and $CM_2$.

To demonstrate the effectiveness of the DRBD methodology, a standby redundancy policy for managing the two computing modules is introduced in the DRBD model of Fig. 5(b). This is represented by the wake-up/standby strong dependency between the two modules $CM_1$ and $CM_2$: when $CM_1$ is operational $CM_2$ is kept in standby, activated if the former fails. But this dependency conflicts with the dependencies modeling the memories' redundancy policy. This aspect needs further deepening. When $M_1$ fails, $M_3$ must be activated even though the $CM_2$ block is disabled. This behaviour is represented by the failure/wake-up strong dependency between $M_1$ and $M_3$. Moreover, if component $M_2$ is in standby, the memory dependency is not anymore applied. It is necessary to include this case in the memory dependency condition, for this reason, in the model of Fig. 5(b), there is a wake-up-standby/standby strong dependency between $M_2$ and $M_3$ that keeps $M_3$ in standby if also $M_2$ is in standby. But such modification does not resolve, alone, the modeling problem. By this way, as introduced above, dependencies conflicts arise: which dependency must be applied on $M_3$ when $M_1$ is active and $CM_2$ is in standby according to the computing modules redundancy policy? In this condition, $M_2$ is in standby and both the memory dependency and the computing modules dependency are applied to $M_3$. This problem can be solved by establishing a priority order among the conflicting dependencies. If $CM_2$ is in standby $M_3$ must be listening if $M_1$ or $M_2$ fail. In case of $M_1$ failure, $M_3$ must be activated. If $M_2$ fails from the active state, $M_3$ is activated, if $M_2$

**Table 1.** States of $M_3$ in relation to $M_1$ and $M_2$

| $M_1$ | $M_2$ | $M_3$ |
|---|---|---|
| A | S | S1 |
| A | F | S2 |
| F | * | A |

fails from the standby state due to the computing modules redundancy policy, $M_3$ replaces $M_2$ and it is placed in the $CM_1$ standby until $M_1$ or $CM_1$ are not failed. Table 1 summarizes the behaviour of $M_3$.

In Table 1, A stands for active state, F for failed state, S1 represents the $M_3$ standby state due to the memory dependency, S for $M_2$ and S2 for $M_3$ represent the standby states reached following up the computing modules dependency. To implement this behaviour it is necessary to fix the priority of the memory dependencies greater than the computing modules dependency one. In the DRBD model of Fig. 5(b), priority 1 is assigned to this latter and 2 to the memory dependencies that are considered priority. It is important to remark that no conflicts arise between the computing modules dependency and the $CM_2$ disks dependency because they are not overlapped: when the former is applied, $D_{21}$ and $D_{22}$ are in standby state. When $CM_2$ is activated, the disks dependency is applied and $D_{22}$ is partially activated according to the dependency rate specified by the dependency triggered by $D_{21}$. These behaviors, and, more in general, the composition among dependencies, are difficult to model by DFT and sometimes they cannot be represented at all, as in this case. In DFT there are no elements to distinguish among dependencies (priority) and therefore to model dependency concurrency.

## 4 Conclusions

In this paper, the DRBD methodology is explained by exploiting an example regarding a multiprocessor computing system taken from literature. The strength expressiveness of DRBD is further highlighted by extending the multiprocessor computing system with a standby redundancy management policy, usually not representable by DFT and other specific reliability/availability formalisms due to the lack of dependencies composition features.

By such examples, the effectiveness of the DRBD notation is demonstrated, also in comparison with other formalism as the DFT one. So, it is possible to affirm that, considering their flexibility and the modeling power, DRBD are a valid alternative in dynamic reliability/availability evaluation scenario.

## References

1. Rausand, M., Hoyland, A.: System Reliability Theory: Models, Statistical Methods, and Applications. 3rd edn. Wiley-IEEE (2003)
2. Vesely, W.E., Goldberg, F.F., Roberts, N.H., Haasl, D.F.: Fault Tree Handbook. U. S. Nuclear Regulatory Commission, NUREG-0492, Washington DC (1981)

3. Sahner, R., Trivedi, K., Puliafito, A.: Performance and Reliability Analysis of Computer Systems: An Example-based Approach Using the SHARPE Software Package. Kluwer Academic Publisher (1996)
4. Boyd, M.A.: Dynamic Fault Tree Models: Techniques for Analysis of Advanced Fault Tolerant Computer Systems. PhD thesis, Duke University, Department of Computer Science (1991)
5. Dugan, J.B., Doyle, S.A.: New results in Fault-tree Analysis. In: Reliability and Maintainability Symposium. (1996) 568-573 Tutorial Notes.
6. Distefano, S., Puliafito, A.: Dynamic Reliability Block Diagrams vs Dynamic Fault Trees. In: Proceedings of the 53rd Annual Reliability and Maintainability Symposium (RAMS07), IEEE (2007)
7. Distefano, S., Puliafito, A.: System Modeling with Dynamic Reliability Block Diagrams. In: Proceedings of the Safety and Reliability Conference (ESREL06), ESRA (2006)
8. Distefano, S.: System Dependability and Performances: Techniques, Methodologies and Tools. PhD thesis, University of Messina (2005)
9. Dugan, J.B., Bavuso, S., Boyd, M.: Dynamic Fault Tree Models for Fault Tolerant Computer Systems. IEEE Transactions on Reliability 41(3) (1992) 363-377
10. Malhotra, M., Trivedi, K.S.: Dependability Modeling using Petri-nets. IEEE Transaction on Reliability 44(3) (1995) 428-440
11. Montani, S., Portinale, L., Bobbio, A., Raiteri, D.C.: Automatically Translating Dynamic Fault Trees into Dynamic Bayesian Networks by means of a Software Tool. In: Proceedings of the The First International Conference on Availability, Reliability and Security, ARES 2006, IEEE Computer Society (2006) 804-809

# Particle Swarm Trained Neural Network for Fault Diagnosis of Transformers by Acoustic Emission

Cheng-Chien Kuo

Department of Electrical Engineering, Saint John's University
499, Sec. 4, Tam King Road, Tamsui, Taipei, Taiwan
cckuo@mail.sju.edu.tw

**Abstract.** A top-down experimental procedure for defect type recognition of epoxy-resin transformers by Partial Discharge (PD) is proposed. Most of the PD detection methods could be performed only at the shutdown period of equipments. By using Acoustic Emission (AE), the real-time and online detection could be reachable. Therefore, this paper conducted high voltage test of pre-faulty transformers and measured those PD signals for recognition needed. Afterward, the selected features that proposed in this paper can be extracted from these collected PD signals. According to these features, effective identification of their faulty types can be done using the proposed particle swarm optimization combined with neural network. Finally, with a view to apply in field, this research adds different noise levels to distort the original data. These distorted data are entered for subsequent testing. Research shows encouraging results that with 30% noise per discharge count, an 80% successful recognition rate can be achieved.

**Keywords:** Acoustic Emission, Partial Discharge, Particle Swarm Optimization, Neural Network, Transformer.

## 1 Introduction

There is a growing demand on energy resources and power consumption with rocketing industrial and trade development. In addition to power supply, the stability of electric power shall also be improved accordingly. Degraded insulating property of electric equipments will lead to serious accident and great loss for the utilities and customers. As the performance of power supply is closely related to electric equipments, it is possible to improve considerably the stability of power supply if real-time status of electric equipments is made available. Transformer is one of electric equipments most closely associated with power consumption. The lifespan of transformer depends upon the insulating strength of insulation materials. So, real-time awareness of the insulation state is helpful to prevent failure. At present, PD detection has extremely high diagnosis accuracy for insulating medium. Thus, PD detection is an important index to prevent equipment trouble [1-2].

In the event of equipment failure and local discharge, energy release will occur in tune with supersonic wave. If the supersonic wave can be analyzed and processed, it

D.-S. Huang, L. Heutte, and M. Loog (Eds.): ICIC 2007, LNAI 4682, pp. 992–1003, 2007.

should be possible to identify the nature of failure, thus making breakthrough contribution to power detection. Any breakdown of running transformer will lead to dramatic loss and negative influence. In an attempt to prevent and identify occasional accident, degrading diagnosis and failure identification of transformer are urgently required based on a safe, real-time and accurate solution. PD detection is basically considered as a pulse phenomenon of energy release, and subsequent sound release is named Acoustic Emission (AE) from the perspective of supersonic wave. So, an ideal way to judge the performance of transformer is based on computer-aided AE method for PD detection. The following are three highlights of AE methods in PD detection for transformer:[3-6]

(1) On-line real-time supervision : AE can be detected on-line without switching-off, or any special arrangements.
(2) High safety : AE is detected in a contact-free manner, thereby helping to improve greatly the safety of maintenance personnel.
(3) Perfect trouble-shooting : In the case of new or unknown troubles of electric equipments, only AE data shall be required for future identification and prevention.

The major purpose of this paper is to set up a real-time online detection unit based on AE detection method, and also provide the end users with an approach to obtain the status of objects as well as relevant information and reference for failure prevention and diagnosis of any electric equipment. Based upon the transformers with discharge failure and a trouble-free transformer, this paper analyzed the data on the MATLAB platform and artificial neural network, and then identified the presence of or otherwise the nature of failure. To put it bluntly, "AE" represents supersonic wave of PD within transformer.

## 2 AE Technology

AE refers to natural phenomenon in many situations, such as earthquake, rock burst, thunder, crack of tree branches, fracture of glass or pottery, etc. In ancient times, the quality of pottery ware was identified with AE principle; at present, the maximum load of automobile components is tested with the same principle. The ASTM standard E610-82 gave a detailed description of AE as "AE is a transient elastic waves generated by the rapid release of energy form localized sources within a material." Therefore, this paper intends to apply AE principle to analyze PD of transformers that is difficult to diagnosis by other method. The whole procedure for the proposed method is detail stated below.

### 2.1 Acquisition of AE Data

PD data, if obtained from computer simulation testing, may differ from those derived from field test research. In this paper we use field experiment, and employs epoxy-resin transformers with designated defects to get the defect type pattern of PD. Five types of defects designed which include model A- improper solder on high voltage side, model B and C- improper soft copper wire on grounding and high voltage panel, model D and E- improper bubble within insulating material on low and high voltage side. To reduce interference from noise and to make a very high level of dependability, the experiment is conducted mainly in a control room and a magnetically shielded room. The conceptual diagram of equipment configuration is shown in Fig. 1, and the proposed process is in Fig. 2.

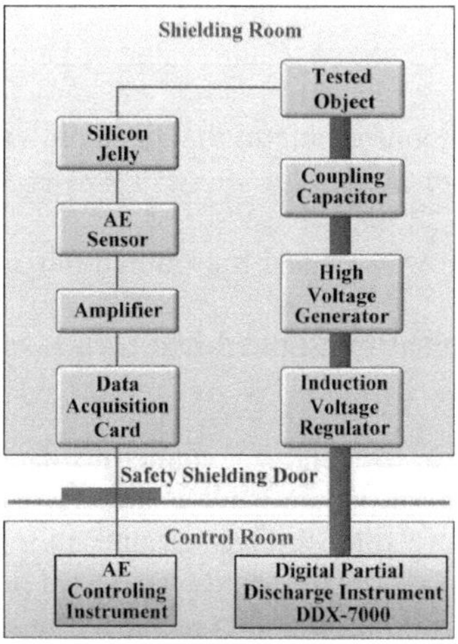


**Fig. 1.** The hardware structure of PD measurement for AE

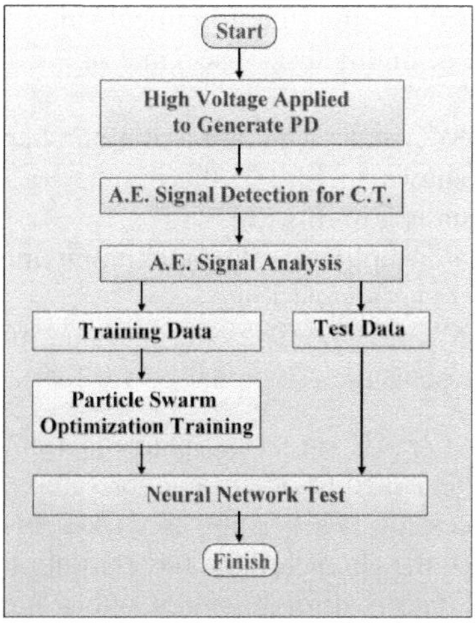


**Fig. 2.** Flowchart of the proposed approach for defect types recognition by PD signal

When the test sample (epoxy-resin transformer) generates PD and AE phenomenon, the voltage value obtained from sensor will be transmitted to amplifier and then AE data capture card for A/D conversion. Finally, the digital data pattern is stored in database.

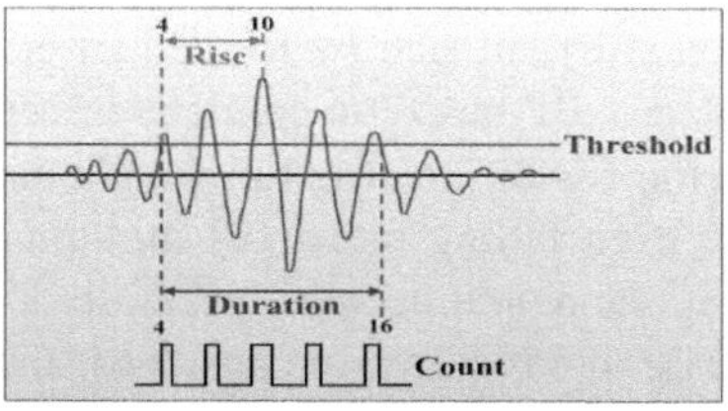


**Fig. 3.** Explanatory drawing of feature value

## 2.2 Feature Selection

The feature values are selected according to (a) feature value of the same nature with little difference, (b) feature value of different natures with identifiable difference. Therefore, five feature values are taken as the identification data in this paper: Rise Time, Count, Duration and RMS. These values are represented in a statistical way as shown in Fig. 3:

(1) Rise Time : The time from generation of AE signal to the peak value of AE signal. It is assumed that the time to threshold value is 4, and the time to peak value is 10, as shown in Fig. 3. In this test, Rise Time of AE wave is 10-4=6 uS, so feature value of Rise is 6.

(2) Duration : The time from beginning to the end of a complete AE signal, such that the magnetite is bigger than threshold value and then fallen below it. It is assumed that the time to critical value is 4, and the time to peak hour is 16, as shown in Fig. 3. In this test, Rise Time of AE wave is 16-4=12 uS, so feature value of Duration is 12.

(3) Count : The extraction number of AE wave conforming to the threshold value. As shown in Fig. 3: the number of exceeding threshold value of AE wave is 5, so Count is 5 in this test.

(4) Amplitude : The maximum amplitude of AE wave signal transmitted into computer capture card.

(5) Energy : The energy value of AE wave signal transmitted into the sensor. It represents the AE signal power for each hit.

After AE signal is obtained, it is also impossible to identify if numerous data are not analyzed to obtain feature value. So, acquisition of feature value is the key to data processing. This paper uses fixed mode threshold to filter the background noise and judge the characteristic representation. A passive sensor is triggered by external pulse, and data is directly stored into a memory. The filtering methods include: threshold filtering, frequency filtering, sound characteristics filtering and contact surface filtering methods. Finally, fiver feature values Rise Time, Count, Duration Amplitude and Energy are selected.

## 3 PSO Neural Network

The weight and bias training of neural network involves actually complex continuous optimization of parameters. The feed-forward network being widely developed is based on Back Propagation (BP) initiated by Rumelhart et al. Notwithstanding it has simple and flexible advantages, BP based on gradient descent is extremely sensitive to initial weight and bias vector. So, different initial weight and bias vectors may lead to totally different results or even falling into local minimum. To prevent the outputs turning into local optimum, the neural network generally allows for setting randomly the initial value. In such case, it's required to consider if the weight and bias are optimum one after training. Trial and error method has been applied by some researchers to find out better initial weight settings [7]. In addition, selection of dynamic optimal learning rate, etc, depends upon test and experience. The network cannot be converged if the value is improperly set, or even in the case of convergence, the slow convergence speed may lead to longer training and local optimum without optimal weight and bias distribution.

Recently, the use of the global optimization technique called PSO [8], to solve real world problems have aroused researchers' interest due to its flexibility and efficiency. Limitations regarding the form of the objective function employed and the continuity of variables used for the classical greedy search technique can be completely eliminated. Considering the complexity of error curve of BP neural network and its sensitivity to initial weight, as well as the global searching advantage of PSO, this paper applied PSO as the training method to get the weight and bias of neural network. PSO algorithm is mainly used to train the neural network. When the weight and bias are set, the feed forward neural network is used for recognition. By combining these advantages of two algorithms, higher accuracy and quicker convergence is possible. This algorithm combining PSO and neural network together is called PSO-NN hybrid algorithm.

## 3.1 Brief Review of PSO

Particle swarm optimization (PSO), first introduced by Kennedy and Eberhart, is one of the modern heuristic algorithms. It was developed through simulation of a simplified social system, and has been found to be robust in solving continuous nonlinear optimization problems [8-9]. The PSO technique can generate a high-quality solution within shorter calculation time and stable convergence characteristic than other stochastic methods [10-12]. Much research is still in progress for proving the potential of the PSO in solving complex power system operation problems. Researchers including Yoshida et al. have presented a PSO for reactive power and voltage control considering voltage security assessment. The feasibility of their method is compared with the reactive tabu system and enumeration method on practical power system, and has shown promising results [13]. Naka et al. have presented the use of a hybrid PSO method for solving efficiently the practical distribution state estimation problem [14].

Searching procedures by PSO based on the above concept can be described as follows: a flock of individuals optimizes a certain objective function. Each individual knows its best value *Pbest* so far and its position. Moreover, each individual knows the best value in the group *Gbest* among *Pbest*, namely the best value so far of the group. The modified velocity of each individual can be calculated using the current velocity and the distance from *Pbest* and *Gbest* as shown below:

$$v_i^{k+1} = \omega v_i^k + c_1 Rand() \times (Pbest_i^k - s_i^k) + c_2 Rand() \times (Gbest^k - s_i^k) \tag{1}$$

$$s_i^{k+1} = s_i^k + v_i^{k+1} \tag{2}$$

where:

$v_i^k$ : current velocity of individual i at iteration k,
$v_i^{k+1}$ : modified velocity of individual i at iteration k+1,
$Rand()$ : random number between 0 and 1,
$s_i^k$ : current position of individual i at iteration k,
$Pbest_i^k$ : *Pbest* of individual i until iteration k,
$Gbest^k$ : *Gbest* of the group until iteration k,
$\omega$ : weight function for velocity of individual ,
$c_i$ : weight coefficients for each term.

The constants $c_1$ and $c_2$ represent the weighting of the stochastic acceleration terms that pull each individual toward the *Pbest* and *Gbest* positions. Low values allow individual to roam far from the target regions before being tugged back. On the other hand, high values result in abrupt movement toward target regions. Hence, the acceleration constants $c_1$ and $c_2$ were often set to be 2.0 according to simulation experiences. Suitable selection of inertia weight $\omega$ in (3) provides a balance between global and local explorations, thus requiring less iteration on average to find a sufficiently optimal solution. As originally developed $\omega$ often decreases linearly from about 0.9 to 0.4 during a run. In general, the inertia weight $\omega$ is set according to the following equation:

$$\omega = \omega_{\max} - \frac{\omega_{\max} - \omega_{\min}}{Iter_{\max}} \times Iter \tag{3}$$

where:

$\omega_{\max}$ : initial weight,
$\omega_{\min}$ : final weight,
$Iter_{\max}$ : maximum number of iterations
$Iter$ : current number of iterations.

## 3.2 Representation of Individual String

Implementation of a problem in the PSO framework starts from the parameter encoding, i.e., the representation of the problem. In this study, weight and bias of neural network are chosen for each particle. The individual string structure is represented in Fig. 4. The parameter $W_{i,h}^{IH}$ describes the weight value between the connection of the *i*th input and the *h*th hidden layers (represent by upper notes *IH*). The parameter $W_{h,o}^{HO}$ describes the weight value between the connection of the *h*th hidden and the *o*th output layers (represent by upper notes *HO*). Also, The parameter $B_i^I$ , $B_h^H$ and $B_o^O$ describe the bias values of the *i*th input, *h*th hidden and *o*th output neurons respectively.

| | | | | | | | | | | | | | | | |
|---|---|---|---|---|---|---|---|---|---|---|---|---|---|---|---|
| $W_{1,1}^{IH}$ | .... | $W_{i,h}^{IH}$ | $W_{1,1}^{HO}$ | .... | $W_{h,o}^{HO}$ | $B_1^I$ | .... | $B_i^I$ | $B_1^H$ | .... | $B_h^H$ | $B_1^O$ | .... | $B_o^O$ |

**Fig. 4.** The individual string structure

## 3.3 Evaluation Function

Implementation of an optimization problem in PSO is realized within the evolutionary process of an evaluation function. The function adopted is the MSE of training period that given below.

### 3.4 The Processing Steps of PSO-NN Hybrid Algorithm

The following describes the detail steps of PSO-NN process :

1. Treating each weight and bias of multi-layer neural networks as the particle of PSO.
2. Initialize N particles using random generator.
3. Evaluate each particle by MSE.
4. Modify *Gbest* and *Pbest* by simply comparison their fitness values.
5. Calculate velocities for each particle by Equ. (1).
6. Renew each particle to the new position by Equ. (2) that including random search phenomenon
7. If the evolution process reach to a satisfy condition (or maximum evolution number is reached) then go on to step 9, else modify the inertia weight $\omega$ by Equ. (3) and go back to step 5.
8. Using the best particle as the weight and bias for multi-layer neural network.
9. Performing test data to evaluate the performance of this PSO-NN.

## 4 Implementation

In the past, heavy-duty electric equipments were often detected during the scheduled shutdown, which was a time-consuming process against performance of power supply. In such case, a real-time online detector will extend significantly the service life of electric devices, but also improve dramatically the performance of power supply to meet the growing demand. Hence, this paper intends to detect and identify PD for fault diagnosis of epoxy-resin transformers based on AE sensing theory.

Aiming the testing object, the testing specification is epoxy-resin transformers EWF-20DB. The Rated maximum voltage is 23kV, the impulse voltage withstand is 125kV. This testing involved five epoxy-resin transformer models which are model A- improper solder on high voltage side, model B and C- improper soft copper wire on grounding and high voltage panel, model D and E- improper bubble within insulating material on low and high voltage side. Therefore the particular defect must be made before case-resin in manufacture procedure.

### 4.1 Test of PD

Fig. 5 illustrates the experiment procedure. High voltage controller, via switch control panel, orders high voltage generator to generate a high voltage, which delivers 34.5kV, 1.5 times as high as the maximum voltage of the transformer, within 50 seconds, and holds for 1 minute and subsequently drops to 23kV in 20 seconds. At the point, the detector starts to gauge PD for a session of 2 minutes. 20 minutes after the detecting finishes, the transformers recuperate and the next stage of the experiment unfolds. It is for the purpose of exciting PD that 34.5kV is held for one minute. Voltage drops to 23kV in order for PD to continue and for the PD detector to gauge and record. Each experiment renders a set of data. There are five kinds of discharge models; each model is experimented on 120 times. Firstly, test the qualified products, and

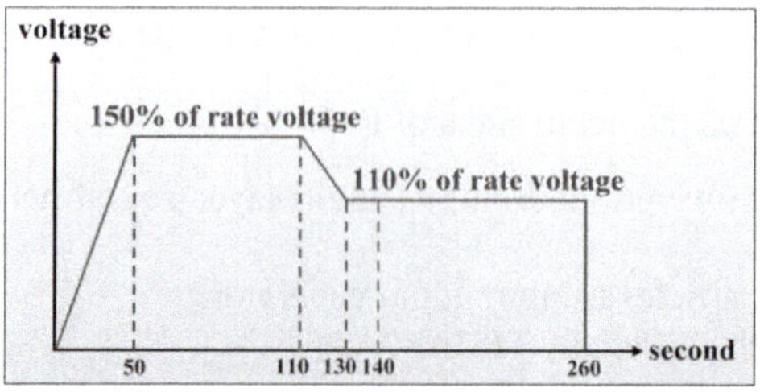


**Fig. 5.** The voltage applying procedure for PD testing

set the critical value without signal and other five with AE signal. Then, acquire AE data according to five faults. This experiment produces 120 sets of data, 80 and 40 of which is for training and for testing.

### 4.2 Apply to PSO Neural Network Algorithm

(1) Setup of Artificial Neural Network

In this paper, multi-layer neural network is used for identification on MATLAB platform with high-speed mathematical operating capability for numerous data [15-18]. The neural network architecture has three layers (input layer, hidden layer and output layer), while the learning rule is based on the proposed PSO-BP algorithm and the neuron of input and output layer are decided by the users as shown in Fig. 6. The number of hidden layer will be discussed later in (3) and the results shows that five neurons will be the optimal choose. [19]

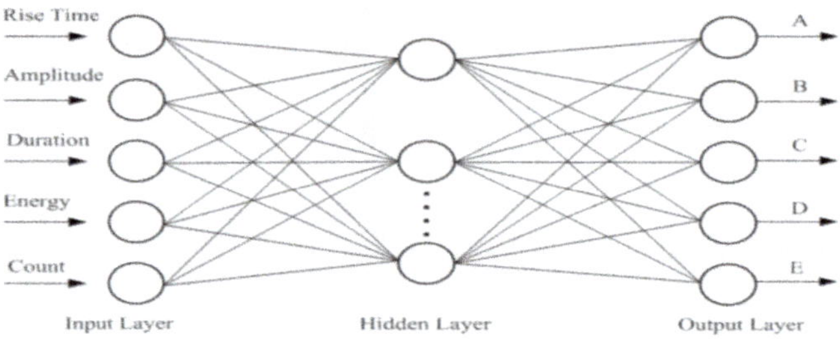


**Fig. 6.** Topological structure of neural network

(2) Applying PSO to get the weight and bias.

PSO is applied to get the optimal or near optimal weight and bias of neural network. 1000 times iteration (or moves in PSO) is performed and the convergence approach is shown in Fig. 7. It is obviously that the PSO is converged after 1000 times iteration and the final MSE is about 8.5.

(3) Number of Neurons on Hidden Layer

We are interested in knowing how the number of neurons on the hidden layer influences the MSE value during the training process. Fig. 8 shows the simulation results after 1000 times recurrent training period. It represents that the NN

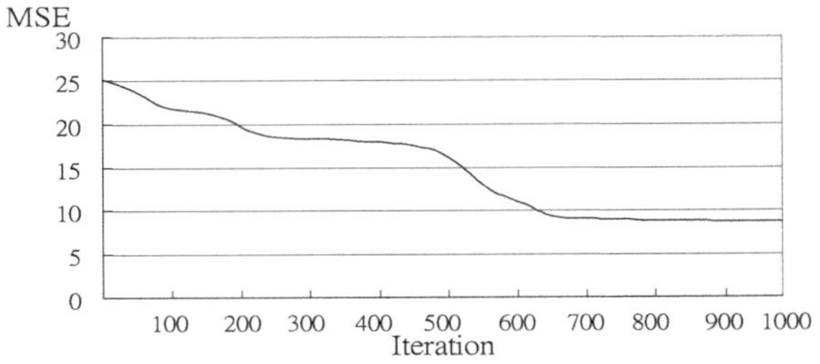


**Fig. 7.** The converging status for each iteration of PSO

prediction accuracy increases with the hidden layer neurons number in the beginning but reverse at the end. However, the optimum value is the one that allows reaching quickly the desired MSE during the training process. This computation time of training process depends partially on the size of NN. Then, by observing Fig. 8, we can notice that five neurons are enough in our study to obtain good identification.

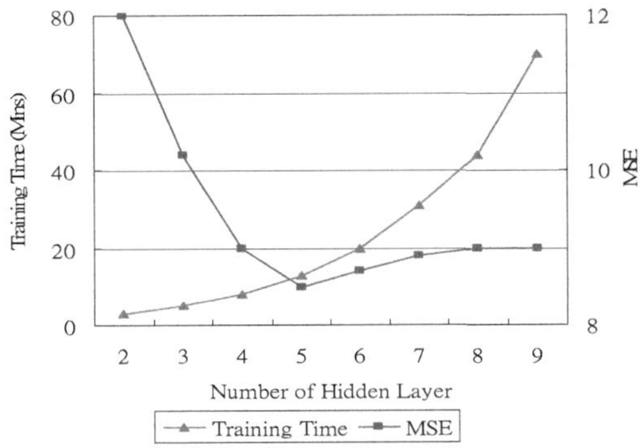


**Fig. 8.** MSE and Training time versus the number of hidden layer's neuron

(4) Setup of Training Group and Test Group

120 data samples are selected randomly from each defect types, each of which is divided into two groups. The former 80 data samples are used as training data of PSO-NN, and latter 40 data samples as test data. Therefore, there are totally 400 training data and 200 test data. Five defect types of training objects (desire outputs of neural network) are set below:

1 0 0 0 0= improper solder on high voltage side
0 1 0 0 0= improper soft copper wire on grounding panel
0 0 1 0 0= improper soft copper wire on high voltage panel
0 0 0 1 0= improper bubble within insulating material on low voltage side
0 0 0 0 1= improper bubble within insulating material on high voltage side

## 5 Results and Discussion

The research results from PSO-NN neural network are divided into two different parts and described below:

### 5.1 Noise-Free Recognition Results

600 sets of data obtained in a shielded room without noise distortion undergo PSO-BP recognition, 400 of which are for training purposes, 200 for testing. Two different methods that include BP and PSO are applied to the same PD recognition problem. The recognition rates results and comparison figure are shown in Table 1 and Fig. 9. There are obviously that PSO-NN has the best percentage among these three methods. Although the training time consuming by PSO-NN may be a little bit greater than BP, it still not affect the feasibility about this method. Because, training time is only needed once, feed forward neural network that is very fast is applied in the real field. From this recognition results, noise-free recognition rate is as high as 82% to 90% by the proposed approach. This encourage rates should be high enough for practical use.

**Table 1.** The recognition rates for different methods and defect types

| Types / Training | A | B | C | D | E |
|---|---|---|---|---|---|
| BP | 86 | 80 | 88 | 80 | 76 |
| PSO | 90 | 82 | 88 | 90 | 86 |

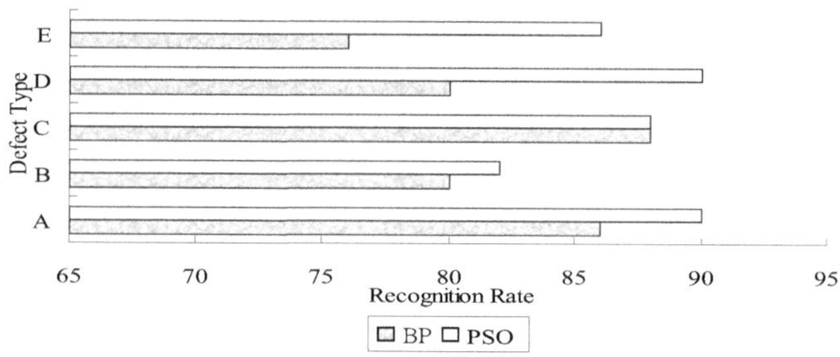


**Fig. 9.** The comparison figure of recognition rate with different methods

### 5.2 Noise-Added Recognition Results

In this section, white noise with 10% to 30% respect to the amplitude of original AE signals are added on test data set to simulate the practical environment on field testing. These distorted signals are then applying to the neural network to evaluate the recognition rate under noise condition. Results are displayed in Fig. 10, including comparison for different level of noise distortion. It is clearly that even as high as 30% noise disturb the original AE signal, this proposed method can still have 80% recognition rate to correctly distinguish failure type.

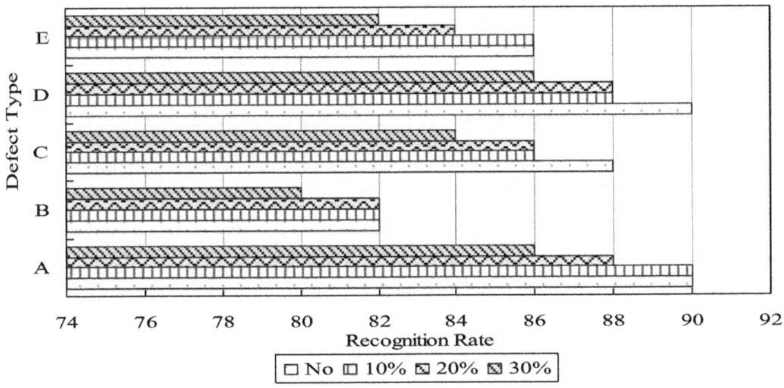


**Fig. 10.** Recognition rate with different noise levels for five experimental models

# 6 Conclusions

Most of the PD detection methods can be applied on offline period of equipments only. By using Acoustic Emission, the real-time and online detection could be easily performed. Therefore, a top-down experimental procedure of epoxy-resin transformers for defect type recognition by PD is developed in this paper. High voltage test for pre-faulty transformers is conducted and PD signals are measured for defect recognition and type classification. Five selected features are selected in this paper to increase the successful probability of recognition rate. A PSO-NN approach is also provided for better recognition rate. The whole proposed procedure can be easily extended to other electrical equipments for on-line defect detection and type classification. To show the effectiveness of the proposed method, several classification and identification experiments are used to evaluate. In a magnetically shielded room experiment, noise-free recognition rate is as high as 82% to 90%. It should be high enough for practical use. Although, noise may interfere with or overlap the PD pattern, and likewise influence the recognition rate. In our proposed approach, even on 30% noise-added situation, at least 80% recognition rate can be guaranteed. These encouraging results show that this paper can provide a feasible and effective way to early detect the possible failure of transformers and can also determine the types of failure to help utility for maintenance needed.

**Acknowledgments.** Support for this research by the National Science Council of the Republic of China under Grant No. NSC95-2221-E-129-022 is gratefully acknowledged.

# References

1. Kweon, D.J., Chin, S.B., Kwak, H.R., Kim, J.C., Song, K.B.: The Analysis of Ultrasonic Signals by Partial Discharge and Noise from the Transformer, IEEE Transactions on Power Delivery, Vol.20, No.3, (2005) 1976-1983
2. Yue, B., Chen, X., Cheng, Y., Song, J., Xie, H.: Diagnosis of Stator Winding Insulation of Large Generator Based on Partial Discharge Measurement, IEEE Transactions on Energy Conversion, Vol.21, No.2, (2006) 387-395

3. Grossmann E., Feser K.: Sensitive Online PD-Measurements of Onsite Oil/Paper-Insulated Devices by Means of Optimized Acoustic Emission Techniques, IEEE Transactions on Power Delivery, Vol.20, No.1, (2005) 158-162
4. Chen, L.J., Tsao, T.P., Lin, Y.H.: New Diagnosis Approach to Epoxy Resin Transformer Partial Discharge Using Acoustic Technology, IEEE Transactions on Power Delivery, Vol.20, No.4, (2005) 2501–2508
5. Boczar, T., Zmarzly, D.: Application of Wavelet Analysis to Acoustic Emission Pulses Generated by Partial Discharges, IEEE Transactions on Dielectrics and Electrical Insulation, Vol.11, No.3, (2004) 433-449
6. Lundgaard, L.E.: Particles in GIS Characterization from Acoustic Signatures, IEEE Transactions on Dielectrics and Electrical Insulation, Vol.8, No.6, (2001) 1064-1074
7. Amir, A., Chuanyi, J.: How Initial Conditions Affect Generalization Performance in Large Networks, IEEE Transaction on Neural Networks, Vol.8, NO.2, (1997) 448-451
8. Kennedy, J., Eberhart, R.: Particle Swarm Optimization, Proceedings of the IEEE International Conference on Neural Networks, (1995) 1942–1948
9. Clerc, M., Kennedy, J.: The Particle Swarm-Explosion, Stability, and Convergence in a Multidimensional Complex Space. IEEE Transactions on Evolution. Computing. Vol. 6, No. 1, (2002) 58–73
10. Eberhart, R.C., Shi, Y.: Comparison between Genetic Algorithms and Particle Swarm Optimization. IEEE International Conference Evolution Computing, (1998)611–616
11. Gaing, Z.L.: Particle Swarm Optimization to Solving the Economic Dispatch Considering the Generator Constraints. IEEE Transactions on Power Systems, Vol. 18, (2003)1187–1195
12. Shi, Y., Eberhart, R.C.: Empirical Study of Particle Swarm Optimization. Proceedings of the Conference Evolution Computing, NJ, (1999)1945–1950
13. Yoshida, H., Kawata, K., Fukuyama, Y., Takayama, S., Nakanishi, Y.: A Particle Swarm Optimization for Reactive Power and Voltage Control Considering Voltage Security Assessment. IEEE Transactions on Power Systems, Vol. 15, (2000)1232–1239
14. Naka, S., Genji, T., Yura, T., Fukuyama, Y.: Practical Distribution State Estimation Using Hybrid Particle Swarm Optimization. Proceedings of the IEEE Power Engineering Society Winter Meeting, Vol. 2, (2001)815–820
15. Salama, M.M.A., Bartnikas, R.: Determination of Neural Network Topology for Partial Discharge Pulse Pattern Recognition, IEEE Transactions on Neural Networks, Vol.13, No.2, (2002) 446–456
16. Candela, R., Mirelli, G., Schifani, R.: PD Recognition by Means of Statistical and Fractal Parameters and a Neural Network, IEEE Transactions on Dielectrics and Electrical Insulation, Vol.7, No.1, (2000) 87–94
17. Haykin, S.: Neural Networks, a Comprehensive Foundation, 2nd ed. Upper Saddle River, NJ: Prentice-Hall, 1999
18. Mazroua, A.A., Salama, M.M.A., Bartnikas, R.: PD Pattern Recognition with Neural Networks Using Multi-layer Perceptron Technique, IEEE Transactions on Dielectrics and Electrical Insulation, Vol.28, No.3, (1993) 1082–1089
19. Kélouwani, S., Agbossou, K.: Nonlinear Model Identification of Wind Turbine With a Neural Network, IEEE Transactions on Energy Conversion, Vol.19, NO3, (2004) 607-612

# Prediction of Chatter in Machining Process Based on Hybrid SOM-DHMM Architecture

Jing Kang, Chang-jian Feng, Qiang Shao, and Hong-ying Hu

Department of mechanical engineering Dalian Nationalities University
Dalian (116600), China
kj@dlnu.edu.cn

**Abstract.** To distinguish chatter gestation, a new method of chatter prediction based on hybrid SOM/DHMM is proposed for dynamic patterns of chatter gestation in cutting process. At first FFT features are extracted from the vibration signal of cutting process, then FFT vectors are presorted and coded into code book of integer numbers by SOM(Self-Organizing Feature Map), and these code books are introduced to DHMM (Discrete Hidden Markov Models), for machine learning and classification. Finally the results of chatter gestation recognition and chatter prediction experiments are presented and show that the method proposed is effective.

**Keywords:** discrete hidden Markov model (DHMM), chatter, self-organizing feature map(SOM), prediction, vector coding.

## 1 Introduction

If chatter is being gestated in the cutting process, information of chatter will be shown in the vibration signal of cutting process. We predict the latent chatter in the process of chatter gestation so that reasonable method can be taken to eliminate chatter in the early period of chatter gestation. Thus, recognition of chatter gestation is always paid attention to by experts, and there have been researches on features extracted from the vibration signal to predict latent chatter, chatter prediction and recognition based on dynamic signal of cutting force, and chatter monitor using AE sensor, etc[1]. These methods are based on recognition of narrow-band feature from the spectrum. The availability of these methods highly relies on the rightness of feature selection and the rationality of judging threshold [1,2]. Reference [3] and [4] present the fault diagnosis method based on DHMM, and its experiment yields good results. Based on the hybrid SOM-DHMM architecture, we presort and code the spectrum vector by SOM (vector quantification) and recognize chatter gestation by DHMM theory. Thus, a new recognition method of chatter gestation is presented in this paper.

## 2 Method of SOM Vectors Coding

### 2.1 SOM Topology Architecture

Self Organizing Feature Map (SOM) is a kind of neural networks, whose feature of input data is acquired by self-organized, unsupervised and competitive neighborhood

D.-S. Huang, L. Heutte, and M. Loog (Eds.): ICIC 2007, LNAI 4682, pp. 1004–1013, 2007.

learning. As shown in fig.1, SOM consists of two layers of neurons, input layer and competitive layer. There are n neurons on input layer, so the input can be an n-dimension vector $\boldsymbol{x}$ . The competitive layer is a two-dimension array of neurons. However, for each input vector, calculated by competitive algorithm, only one neuron on the competitive layer is excitatory and yields one output yi. We adopt SOM to code the FFT spectrum vector, and take the serial number i as coding feature. This is because SOM preserves the topology in vector projection. That is to say, similar vectors in n-dimension space generate similar serial numbers of excitatory neuron after mapping.

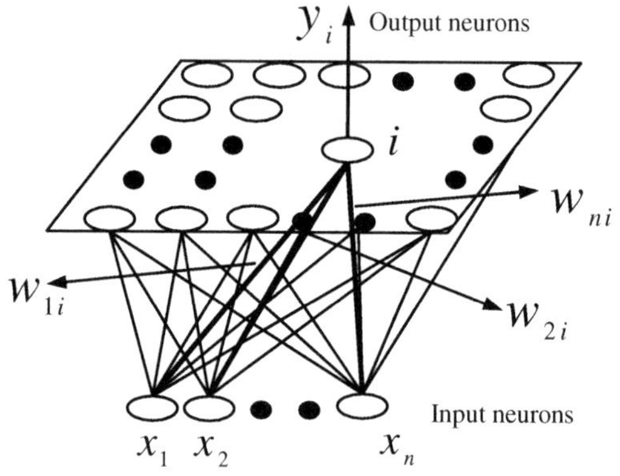


**Fig. 1.** Architecture of SOM network

## 2.2 Feature of SOM Training Arithmetic and SOM Coding

The domain of the prevailing neuron will become smaller by self learning and iterative calculating until the error is within the desired range. This is the standout feature of SOM. The algorithm is summarized as below:

Calculating a small stochastic number $\boldsymbol{w}_{ij}$, which is the weight of mapping the input neuron $\boldsymbol{i}$ to the feature node $\boldsymbol{j}$ in initialization .

Offering a new input mode *x(t).*

Calculating the Euclidean distance $\boldsymbol{d}_j$, this is distance from input sample to each output neuron $\boldsymbol{j}$.

$$d_j = \sum_{i=1}^{n} [x(t) - w_{ij}(t)]^2 \tag{1}$$

Then, a neuron which has minimum distance is selected from all output neurons as the prevailing neuron $\boldsymbol{j}^*$.

Correcting output neuron $\boldsymbol{j}^*$ and weight matrix of neuron in the domain where $\boldsymbol{j}^*$ is, according to equation (2).

$$w_{ij}(t+1) = w_{ij}(t) + \eta(t)[x_i(t) - w_{ij}(t)] \tag{2}$$

where, $N$ is nodes number of output neurons, $n$ is dimension of input modes, and $\eta$(t) $\in$(0,1) is regulating exponential of weight $w$; which is a plus term and gradually converges at 0 as time passes.

Offering a new sample and repeating the learning process mentioned above until the error is within the desired range.

Through competitive neighborhood learning, the positions of output neurons of SOM network are close if input samples are similar. Therefore, SOM can be used widely in mode recognition and vector clustering analysis. In this paper, SOM is used in code of vector clustering. After certain input FFT spectrum vectors pass through SOM, only one output neuron on the competitive layer is excitatory, and then we take this sequence number of neuron node as feature of clustering code.

## 3 Basic Theory of HMM

A HMM is a kind of statistic modeling tool for time sequence, whose application in speech recognition, faults diagnosis and computational biology processing is successful. HMM accomplishes statistic of non-stationary signal and parameterized modeling, and it can be used easily in probability reasoning. Therefore, HMM is a useful tool for dynamic pattern classification. In the cutting process when chatter gestates, the vibration features change with the stages. The changes of vibration signal are represented statistically by HMM so that the whole vibration pattern of chatter gestating process will be clear. Then we classify the different states of chatter in method of HMM pattern classification, which is a good method of chatter monitor without doubt. There are two kind of HMM due to different values observed: continuous HMM (CHMM) where the observed values are continuous and discrete HMM (DHMM) where the value sequence is within a limited range and represented by symbols or codes.

### 3.1 Structure of DHMM

HMM comprises two parts: Markov chain and stochastic process. Markov chain whose output is a sequence of state, can be described by $\pi$ and A, while stochastic process whose output is a sequence of observed values, is described by B. The structure of HMM is shown in fig.2, where $T$ is time length of observed sequence.

A DHMM is described by parameters defined below:

*1)* $N$: state number of Markov chain.$\theta_1$, $\theta_2$, ..., $\theta_N$ are the possible states of Markov chain and $q_t$ is the state at time t, so we get $q_t \in(\theta_1, \theta_2, ..., \theta_N)$.

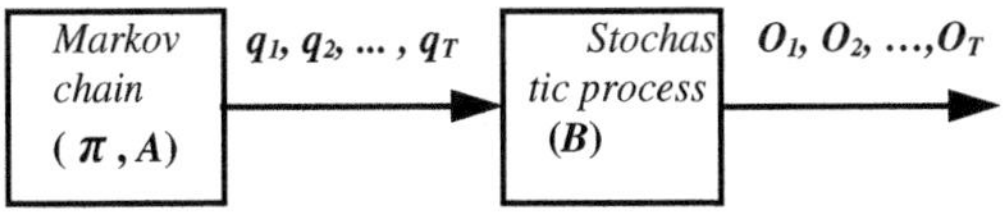


**Fig. 1.** Structure of HMM

*2) M*: possible number of observed value in each state. If $v_1, v_2, \cdots, v_M$ are the observed values and $o_t$ is the value at time t, then $o_t \in (v_1, v_2, \cdots, v_M)$.

*3)* $\boldsymbol{\pi}$ : initial probability distribution vector, where $\pi_i = P(q_t = \theta_i),\ 1 \le i \le N$

*4) A:* state shift probability matrix,

$$\boldsymbol{A} = (a_{ij})_{N\times N} \tag{3}$$

where: $a_{ij} = P(q_{t+1} = \theta_j / q_t = \theta_i)\ \ 1 \le i, j \le N$

$$\boldsymbol{B} = (b_{jk})_{N\times M} \tag{4}$$

*5) B:* probability matrix of observed values,
where: $b_{jk} = P(o_t = v_k / q_t = \theta_j), 1 \le j \le N, 1 \le k \le M$

Thus, a DHMM can be described as:

$$\lambda = (N, M, \pi, \boldsymbol{A}, \boldsymbol{B}) \tag{5}$$

According to the model defined, a three-state DHMM can be illustrated as fig.3. In fig.3, $s_1$, $s_2$ and $s_3$ are unobservable states which are described by probabilities of observable symbols, A, B, C, D, E and F in this example. The arrows illustrate the possible change between every two states, while the symbols at the arrows are change probabilities.

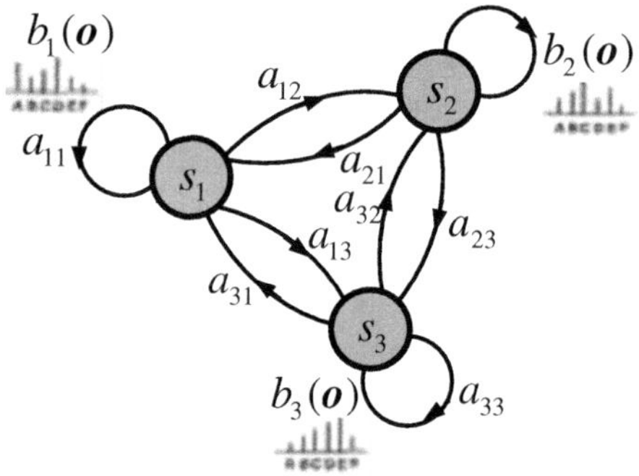


**Fig. 3.** A three-state DHMM

### 3.2 Basic Algorithm of DHMM

Let us denote a sequence of observations $\{\boldsymbol{O}_t\}$ and a sequence of hidden states $\{S_t\}$, where t=1,…,T. A HMM assumes two sets of conditional independence relations: (1) the observation $\boldsymbol{O}_t$ is independent of all other observation and states given $S_t$; (2) the state St depends on only $S_{t-1}$, i.e., states satisfy the first order Markov property. It

$$P(s_1, \cdots, s_T, o_1, \cdots, o_T) = P(s_1)P(o_1 \mid s_1)\prod_{t=2}^{T} P(s_t \mid s_{t-1})P(o_t \mid s_t) \tag{6}$$

follows from this conditional independence relation that the joint probability distribution of states and observations can be factorized as:

A HMM assumes that hidden states variables are discrete-valued, i.e., St ∈ {1,…,K} The states vector $S_{t-1}$ is K-dimensional vector with only one elements being unity and the rest of elements being zeros. In the other words, which element of the state vector is unity, depends on which state value is active. Then P(St|St-1)can be represented by $K \times K$ state transition matrix that is noted by A. $P(S_1)$is a K-dimensional vector for initial state probability that is denoted by .

Learning HMM consists of two steps: (1) inference step where the posterior distribution over hidden states is calculated; (2) learning step where parameters (such as initial state probability, state transition probability, and emission probability) are identified. The well-known forward-backward recursion allows us to infer the posterior over hidden states efficiently. Forward variable is defined as

$$\alpha_t(i) = P(o_1, o_2, \cdots, o_t, q_t = \theta_i \mid \lambda),\ 1 \le t \le T \tag{7}$$

thus,

$$P(\boldsymbol{O} \mid \lambda) = \sum_{i=1}^{N} \alpha_T(i) \tag{8}$$

Backward variable is defined as

$$\beta_t(i) = P(o_{t+1}, o_{t+2}, \cdots, o_T \mid q_t = \theta_i, \lambda),\quad 1 \le t \le T-1 \tag{9}$$

Backward algorithm is

$$P(\boldsymbol{O} / \lambda) = \sum_{i=1}^{N} \beta_t(i) \tag{10}$$

The next problem is to work out an optimal state sequence. The optimal state sequence $\boldsymbol{Q}^*$ here refers to the state sequence when the probability $P(\boldsymbol{Q}|\boldsymbol{O},\lambda)$ reaches the maximum. The problem can be solved by **Viterbi** algorithm

$$\boldsymbol{P}^* = \max_{1 \le i \le N}[\delta_T(i)] \tag{11}$$

$$q_T^* = \arg\max_{1 \le i \le N}[\delta_T(i)] \tag{12}$$

Optimal state sequence is

$$q_t^* = \varphi_{t+1}(q_{t+1}^*), t = T-1, T-2, \cdots, 1 \tag{13}$$

To make the model more robust, we need multiple samples to train the model so that the re-estimation formulas have to be adapted. Suppose the sequence of K observed samples is

$$\boldsymbol{O} = [\boldsymbol{O}^{(1)}, \boldsymbol{O}^{(2)}, \cdots, \boldsymbol{O}^{(K)}] \tag{14}$$

where $O^{(k)} = [o_1^{(k)}, o_2^{(k)}, \cdots o_t^{(k)}, o_{T_k}^{(k)}]$ is the K st sample whose sample length is $T_k$, we know that every sample is an ordered set of a vector series. In modeling of vibration signal we use samples of equal length, $T_k=T$. Suppose that each sample is independent of others, the goal of model training is to adapt the parameter λ to yield a maximum of the value below

$$P(O \mid \lambda) = \prod_{k=1}^{K} P(O^{(k)} \mid \lambda) = \prod_{k=1}^{K} P_k \tag{15}$$

The re-estimation formulas are based on the frequency under which different observed event occurs. Therefore, re-estimation formulas of multiple observed samples can be obtained by adding every frequency under which an event occurs independently. The adapted Baum-Welch re-estimation formulas with k sample training can be obtained, which are reasoned out in reference [4]. More details on HMM can be found in [3, 5, 7].

## 4 Feature Vector Extraction and Quantified Coding

Chatter prediction using HMM is based on recognition of statistic mode, so feature information should be recognized [3, 4]. When chatter gestates, we get feature information by calculating FFT of vibration acceleration [3, 4, and 8]. The peak points of the spectrum compose a spectrum vector which manifests current cutting state and quantifying the feature vector yields training parameters, or observed sequence $O$.

The observed variable of DHMM need to be discrete, so the spectrum should be discretized in frequency domain and the vector quantified. So we divided the amplitude of signal into N-1 parts and make ascending sort. The index value of each area of signal x is defined as:

$$index(x) = \begin{cases} 1, x \le partition(1) \\ i+1, partition(i) < x \le partition(i+1) \\ N, partition(N-1) < x \end{cases} \tag{16}$$

SOM to code the FFT spectrum vector, and take the serial number i as coding feature. This is because SOM preserves the topology in vector projection. That is to say, similar vectors in n-dimension space generate similar serial numbers of excitatory neuron after mapping

## 5 State Recognition Method for Chatter

For SOM-DHMM state recognition of chatter, besides a DHMM for normal state, models for all possible cutting state, e.g. chatter gestation and explosion, should be trained. Thus, a model library is formed. To recognize cutting states, pretreatment of vibration signal (FFT, feature extraction and vector coding) is needed. Then we calculate the probability of each output of models in the library, and choose the maximum, so cutting process is recognized.

According to the theory of DHMM, the process from stable cutting, chatter gestating to chatter explosion can be simulated by an enantiomorphous three-state DHMM, illustrated in Fig.4, because the spectrum vectors of different states are obviously different. In the figure, circles with numbers stand for three states the relation between states and variables beside which $a_{ij}$, i, j=1,2,3 is the state change probability; $\boldsymbol{o}_t$ is the observed code at time t, or the serial number of excitatory neuron which is the output of FFT spectrum vector after mapping.

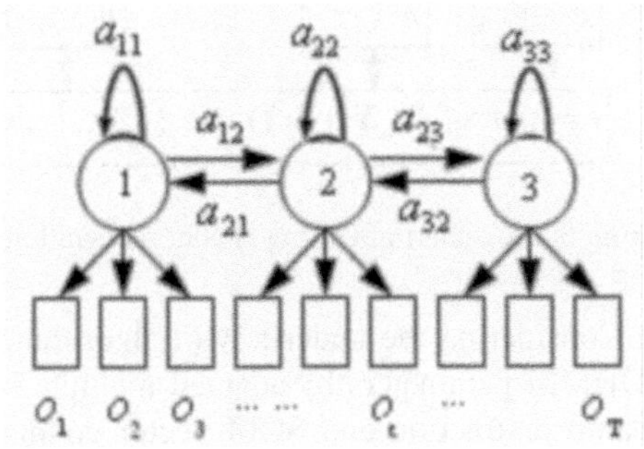


**Fig. 4.** DHMM Structure of cutting chatter

Furthermore, the enantiomorphous DHMM reflects well the exchangeable relation in time domain, so it is suitable for chatter state recognition. After modeling DHMM for every cutting process, we can recognize the cutting process by the steps in fig.5

Method of feature extraction of vibration data in fig.5 can be selected in light of requirement. c is the total number of cutting process types, and it can be increased when necessary. FFT-DHMM recognition of chatter can be described as below:

(1)Acquisition of spectrum sequence To be used in training, several cutting processes are needed. In the k st process, T frames of vibration signal are acquired. Spectrum feature is extracted from every frame, and the spectrum vector sequence of the whole process can be written as

$$X^{(k)} = [x_1^{(k)}, x_2^{(k)}, \cdots x_t^{(k)}, x_{T_k}^{(k)}] \tag{17}$$

where $x_t^{(k)}$ is the t st spectrum frame of the k st process.

(2) Quantification of observed sequence Every spectrum is normalized and coded, so a discrete set of codes is formed. Suppose that the quantified sequence of spectrum is ot , where1≤t≤T. Thus, the coded vector of the t st spectrum frame of the k st process is $\boldsymbol{o}_t^{(k)}$. The set of observed sequence after quantifying the feature sample of the k st process is

$$\boldsymbol{O}^{(k)} = [\boldsymbol{o}_1^{(k)}, \boldsymbol{o}_2^{(k)}, \cdots \boldsymbol{o}_t^{(k)}, \boldsymbol{o}_{T_k}^{(k)}] \tag{18}$$

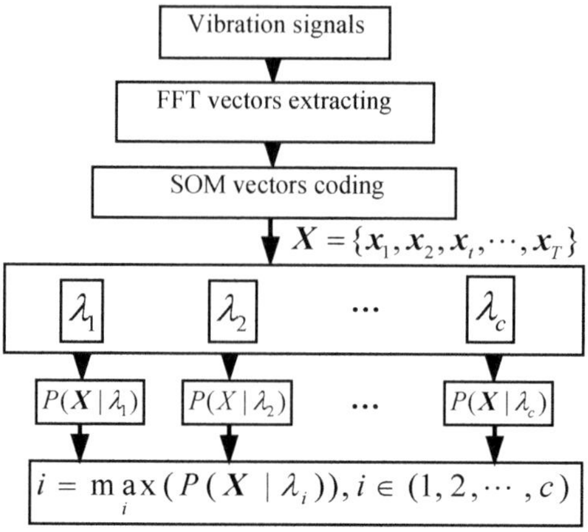


**Fig. 5.** Flowchart of cutting chatter recognition

(3)DHMM training Considering the underflow of algorithm, iterative algorithm is used to calculate the DHMM parameters by adapted multiple sequence re-estimation formulas [1]. After feature extraction and SOM vector coding, DHMM for normal cutting, chatter gestation and chatter explosion can be modeled.

(4)Recognition of chatter gestation After modeling DHMM of different cutting state, Viterbi algorithm can be used to calculate and recognize chatter gestation. The current observed sequence is submitted into DHMM of different state. Then we calculate the reasoning probability $\boldsymbol{P}$ $(\boldsymbol{O} \mid \lambda)$ with which current sequence turns up in the state by Viterbi algorithm. Finally, we know the current state whose probability $\boldsymbol{P}$ $(\boldsymbol{O} \mid \lambda)$ is the largest. In this way, chatter gestation can be recognized.

## 6 Experiment of Chatter Prediction

To confirm the availability of pattern recognition in chatter prediction and monitor, we did cutting test on CNC lathe (CK6150) during which DHMM for normal cutting, chatter gestation and chatter explosion were modeled. The conditions are: n=15~20r /s, f=0.4mm/s, ap=0.12mm, d=35~40mm. In the test, we sampled cutting force signal Fz(t) and acceleration signal of cutter vibration a(t) of each state and extracted feature signal after FFT. Data acquisition was accomplished by variable rate sampling of complete period. The sample rate is 128 times of the rate of the rotary workpiece. To compare the spectrum of different states, we normalized the feature signal, coded the vectors and obtained observed sequence O. The results of normalization and coding are shown in tab.1.

In the training of DHMM, we use 10 test samples of every cutting state. The trained DHMM for normal cutting, chatter gestation and chatter explosion are written as λ1, λ2, , λ3. When current observed sequence Ot is submitted into each state, we get its probability P(Ot| λi) where i=1,2,3. If the state is chatter gestation, we will have

$$P(O_t|\lambda_2)=\max_{t=1,2,\dots,T}[P(O_t|\lambda_i)]. \tag{19}$$

In 20 tests of normal cutting recognition, there has been one false report; in 20 tests of chatter gestation recognition there has been one false report and it failed to report once. The results of recognition are shown in tab.2.

**Table 1.** Normalization Amplitude And Coding In Different Stage

| Feature sample | | $f_1$ | $f_2$ | $f_3$ | $f_4$ | $f_5$ |
|---|---|---|---|---|---|---|
| Normal cutting | Amp. | 0.03 | 0.07 | 0.11 | 0.21 | 0.28 |
| | code | 0 | 2 | 6 | 12 | 14 |
| Chatter gestation | Amp. | 0.08 | 0.2 | 0.45 | 0.20 | 0.14 |
| | code | 2 | 9 | 21 | 12 | 6 |
| Chatter explosion | Amp. | 0.12 | 1 | 0.15 | 0.03 | 0.04 |
| | code | 6 | 38 | 8 | 0 | 1 |

**Table 2.** Result of chatter by DHMM

| Cutting stage | Normal cutting | Chattergestation | Chatter explosion |
|---|---|---|---|
| Normal cutting recognition | 19 | 1 | 0 |
| Chatter gestation recognition | 1 | 18 | 1 |
| Chatter explosion recognition | 0 | 0 | 20 |

## 7 Conclusion

A. The vibration signal of cutting process from normal cutting to chatter explosion is non-stationary. This non-stationary process can be described by statistic model DHMM and its states can be sorted in method of probability. Tab.2 shows it effective to recognize different cutting states in method of vector quantified DHMM.

B. Training of DHMM is simple. Because SOM preserves the topology in vector projection, we transform the observed FFT spectrum vector into SOM quantified code.

C. In tests of chatter recognition, failing to report and false report occurs. It is because FFT is fit for transform of stationary signal and information of spectrum vector loses when spectrum vector is quantified. Therefore, feature extraction method of non-stationary signal needed to be adapted.

D. The results show that chatter explosion will happen around 600~1200ms after its report in recognition of chatter gestation based on DHMM theory. According to reference [2], the time between report and chatter explosion is 1000~1500ms. Thus, a better algorithm can lengthen the time

## References

1. FEI, R., WANG, M.: Recent Development in Machining Chatter Control and Prediction. J. CHINA MECHANICAL ENGNEERING. Vol. 12, No. 9, 1075-1079
2. YU, J., ZHOU, X.: Prediction & Control of Cutting Chatter, J. CHINA MECHANICAL ENGNEERING. Vol. 10, No. 9, 1028-1032
3. FENG, C.: HMM Dynamical Pattern Recognition Theories, Methods and Applications in Faults Diagnosis of Rotating Machine, Doctoral Degree Dissertation of Zhejiang University, Zhejiang University, Nov. 2002
4. Ding, Q., Feng, C., Li, Z., Wu, Z.: Study on DHMM Fault Diagnostic Methods for Rotating Machinery during Run-up, J. Journal of Vibration Engineering, Vol. 16 no. 1,Jan. 2003
5. Rabiner, L.R., "A tutorial on Hidden Markov Models and Selected Application in Speech Recognition". Proc. IEEE, 77(2) (1989) 257-286
6. HE, Q., MAO, S., ZHANG, Y.: Re-estimation of Continuous Hidden Markov Model with Multiple Observation without Overflow, J. ACTA ELECTRONICA SINICA, Vol28 (10) 98-101
7. Varsta, M., Heikkonen, J., Lampinen, J., and Millán, J. del R.: Temporal Kohonen map and recurrent Self-Organizing Map: Analytical and Experimental Comparision. Neural Processing Letters, 13(3) (2001) 237-251
8. Feng, C., Ding, Q., Wu, Z.: Hybrid SOM & HMM Method in Fault Diagnosis Applications of Rotating Machinery Overall Speed Ascend & Descend Process. China Mechanical Engineering, vol.13, no.20
9. Feng, C., Wu, B., Kang, J., Shao, Q.: Faults Diagnosis of Rotor Machine in Whole Running-up Process Based on Hybrid SVM/HMM Architecture. Proceedings of IEEE ICCA07 Guangzhou China, May, 2007

# Research of the Fault Diagnosis Method for the Thruster of AUV Based on Information Fusion

Yu-Jia Wang, Ming-Jun Zhang, and Juan Wu

College of Mechanical and Electrical Engineering Harbin Engineering University
Harbin, Heilongjiang 150001, China
97771@163.com

**Abstract.** Aiming at the problem of thruster fault diagnosis of AUV, the motion condition model of AUV based on the improved dynamic recursive Elman neural network, and the performance model of thruster based on the Radial Basis Function network were established. And the fault fusion diagnosis method was proposed according to the overall and local fault detection. Through comparing the output value of motion condition model with the measured value of actual speed and angle, it obtained the overall fault information. Also, it obtained the direct fault information through analyzing the residual which was produced by comparing the output of the performance model with the measured value of the actual voltage and current of the each thruster. According to the decision level information fusion of two kinds of information, it realized the fault diagnosis of thrusters and analyzed the fault degree and reliability. The results of the fault-simulation experiment show that the proposed fault fusion diagnosis method for the thruster of AUV is feasible and effective.

**Keywords:** Underwater robot, Thruster fault diagnosis, Neural network, Information fusion.

## 1 Introduction

With the technology development of underwater robot, they are extensively used in scientific, commercial, military mission for various purposes. But because of the complexity of the working environment and the multiplicity of the task, underwater robot are liable to faults or failure during underwater mission, the reliability of its components needs more strict requirement and challenge. It is necessary to embed fault diagnosis system into AUVs to find fault in time and increase the reliability of the vehicle and decrease serious loss. The thruster is the key component in realizing the motion of robots, and it is significant to detect the faults in time to ensure the AUV's work safety, as in [1] and [2], it is also the one of the most common and most important source of faults.

In recent years, the problem of fault diagnosis of underwater robot has been made a great deal of researches, but many of them only are the basic researches, in the project application are very few. And many method of fault diagnosis are all based on the analytical model, we all known, the accurate mathematic model of underwater robot is

D.-S. Huang, L. Heutte, and M. Loog (Eds.): ICIC 2007, LNAI 4682, pp. 1014–1023, 2007.

difficult to obtain due to various reasons, even if we obtained the model, it also omitted the many terms according to need. It can't completely describe characteristic of underwater robot, so the method of artificial intelligence has been received attention of researchers. In this paper, based on the control system of underwater robot and the characteristic of the thruster, the motion condition model and thruster performance model of underwater robot were designed by improved dynamic recursion Elman network and RBF neural network respectively, and the fault fusion diagnosis method with local and overall information is proposed. We verify the method by computer fault simulation, and the simulation result shown the method is feasible. It proposed a new thought for fault diagnosis of underwater robot, and it also can be application to fault diagnosis of the other strong non-line system.

## 2 Test-Platform of Underwater Robot

It is necessary to offer a certain objective for FDD (fault detection and diagnosis) according to the general FDD theory. Based on the underwater robot test-platform, which is developed by the researchers of Harbin Engineering University in 2005, it establishes the overall fault diagnosis module and the local fault diagnosis module. The structure of the test-platform is shown in Fig.1, and the detail structure parameters are described in reference [3].

**Fig. 1.** Test-platform of underwater robot

The propulsion system is comprised of two main thrusters in horizontal direction and two vertical thrusters in vertical direction, which realize the motion in the two planes respectively. For detecting the condition of the thruster, it loads DC motor electricity detection sensors to measure the voltage and current of thruster motor armature.

The motion parameters of the test-platform, such as yaw, speed and depth, are obtained from reading the digital compass, velocity measurement system and depth gauge. These motion parameters reflect the change of running condition when the thruster is failure. It uses the scroll windows mode, which has the function of catch gradual fault condition, to realize the signal processing.

# 3 Fault Diagnosis System Structure for Underwater Robot

According to the control system of the test-platform which uses PD controller as its kernel, the fault diagnosis system structure is established and shown in Fig.2.

The Underwater robot motion condition model block and the Overall fault detection block constitute the overall fault diagnosis module and the thruster' performance model group block and the Local fault detection block constitute the local fault diagnosis module.

Here $E_1$ represents the residual of the motion condition model output and actual running information, such as speed , yaw and depth, $E_2$ represents the residual of the performance model output and actual thruster electric motor armature information, such as voltage and current as in [4].

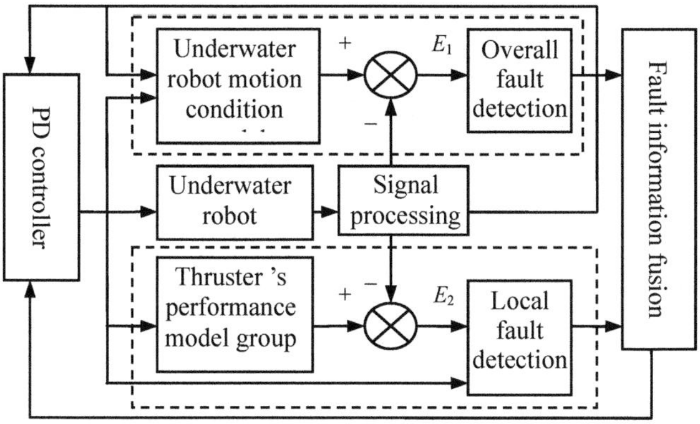


**Fig. 2.** The structure of underwater robot fault diagnosis system

## 3.1 Underwater Robot Motion Condition Models

The thruster is the primary actuator for the motion of underwater robot. By using differential control mode to operate the two horizontal or vertical thrusters, it can go ahead, fall back, turn around, rise and descend, described in [5].

In the case of sensors working well, the motion condition reflects the health status of thrusters, for example if one horizontal propeller is wrapped, the thruster motor revolution will decrease, and further result is the speed fall and course deflexion. It obtains the theoretic motion condition by establishing underwater robot forward model, and gains the fault character residual by comparing the theoretic output and actual condition output, which realizes the fault detection. It is difficult to establish a precise dynamic model for the underwater robot because of the complexity of motion [6], so an improved dynamic recursion Elman neural network structure is proposed to establish robot motion condition model in horizontal and vertical respectively. The network structure is shown in Fig.3.

The inputs of the network (horizontal or vertical) include last time thruster control voltage $\boldsymbol{U}(k\text{-}1)$ and measured speed or yaw angle-speed information $\boldsymbol{V}(k\text{-}1)$, the output of the network is the current time theoretic motion condition $\boldsymbol{Y}(k)$. Comparing with the structure only input the control value, the improved network images the time character of robot motion more accurate.

Based on the input-output pairs of actual control system in normal condition, it trains the network with BP algorithm, adjusts the weights between layers, and satisfies the structure and parameters of the dynamic forward model. Then it carries on the overall fault diagnosis based the residual $\boldsymbol{E}_1$, shown in fig.2, which obtained through comparing the model output and the actual condition measured value.

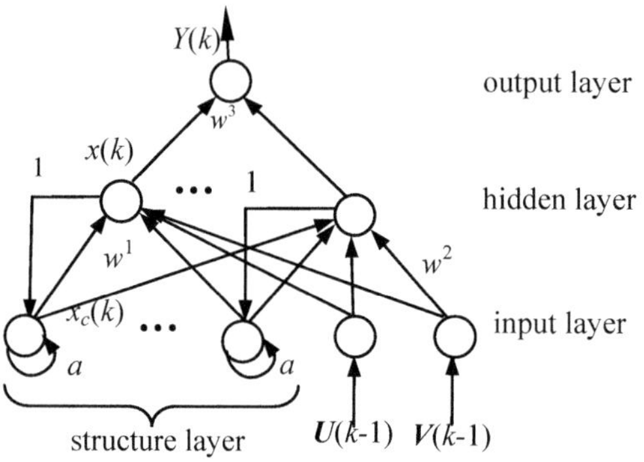


**Fig. 3.** Improved Elman network structure

### 3.2 Thruster Performance Model

The propulsion system includes drivers, thrust motor and screw propeller. The control voltage dominates the revolution of the thruster DC motor, and the joint propeller circumrotates to bring thrust [7].

According to relative theory and integrate with a great deal of experiment data of fault simulation, it summarizes the voltage and current character of typical fault, such as drivers failure, motor coil turnoff, propeller wrap, block, disrepair and racing, which is shown in Table 1.

**Table 1.** Electricity Character in Fault Occurring

| Voltage | Current | Revolution | Fault type |
|---|---|---|---|
| Zero | zero | zero | Drivers failure |
| Normal | zero | zero | motor coil turnoff |
| a little low | a little high | a little low | propeller wrap |
| Normal | a little low | a little high | propeller disrepair |
| Low | high | low | propeller block |
| very low | zero | high | motor racing |

Based on the character described above, it establishes thruster performance model by RBF neural network, whose hidden layer neurons adopt local activation function and have faster learning speed and higher approach precision compared with BP neural network, as in [8]. The structure of performance model comprises voltage model and current model with SISO mode, the input is thruster control voltage, but the outputs are theoretic voltage and current respectively. It takes the measured value of normal condition and typical fault as the training sample of the network to tune the center and width of the hidden layer neurons. The structure of the local fault detection module, which is established based on the performance model, is shown in Fig.4.

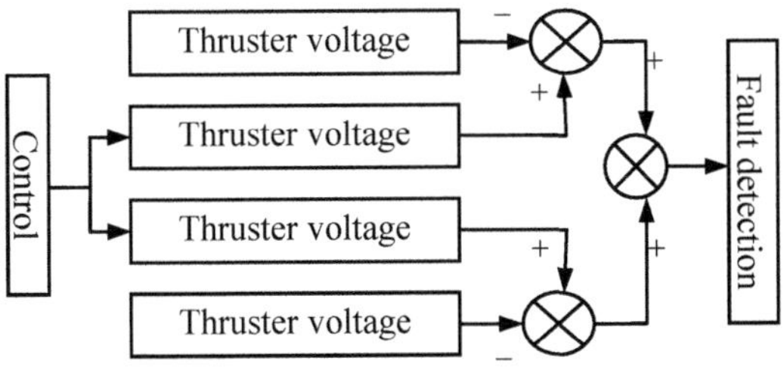


**Fig. 4.** Single thruster fault detection module

### 3.3 Fault Fusion Diagnosis

According to the thought of information fusion, we carried on distributional diagnosis to each sub-model of thrust fault detection in order to fully use the information provided by detection value, and fused the correlative diagnosis result to obtain the partial diagnosis result. This process was called the partial fusion, then further fused each partial diagnosis result and whole diagnosis result got the final result of system fault diagnosis, it be named overall fusion [8]. Take the underwater robot horizontal plane movement as the example to describe the fusion diagnosis process, as shown in Fig. 5.

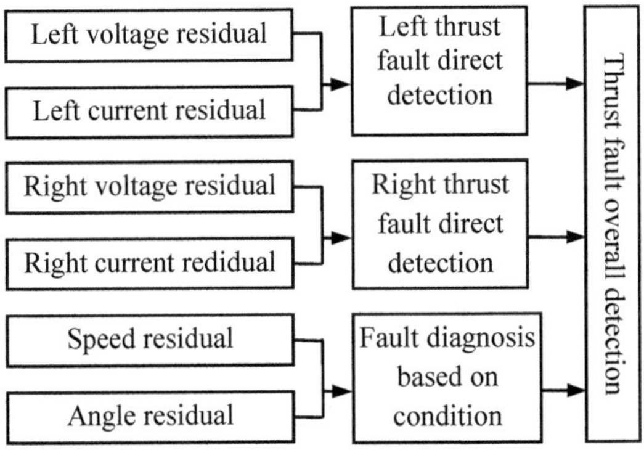


**Fig. 5.** The frame of information fusion fault diagnosis process

The fault partial detection was executed according to the residual whether excess the set threshold. To direct fault detection, the real-time fault classification was executed according to list characteristic of Table I. And the overall fault detection based on condition has some schedule effect, the speed residual can indirect effect the degree of fault, And the angle residual further indicate the part of fault; the fault overall detection according to the information of partial detection to fusion, increased the reliability of diagnosis and decreased the misdiagnosis condition because the thruster instantaneous disturbance is aroused by disturbance in the direct fault detection.

### 3.4 Fault Degree and Reliability Analysis

In order to increase higher reliability and stronger observability of fault diagnosis result, according to D-S proof theory and considered fault diagnosis system structure of underwater robot, the reliability of fault diagnosis can be defined as:

In the certain interval, the probability of residual excess the threshold range was called the reliability of fault, expressed as $\delta$, namely

$$\delta = q / h. \tag{1}$$

Where $q$ is the step of the residual excess the threshold range in the certain interval, and $h$ is the whole step amount in the certain interval.

According to above definition, the fault character information reliability of underwater robot motion condition whole fault detection sub-model can be defined with $\delta_1$, and the fault character information reliability of thruster direct fault diagnosis sub-model can be defined with $\delta_2$. Considering the relativity of each fault character, we took the average reliability as the reliability of fault diagnosis result after information fusion, namely

$$\delta = (\delta_1+\delta_2) / 2. \tag{2}$$

Because the residual size has reflected the degree of fault to a certain extent, therefore the mean square sum of residual in the scheduled step divide the mean square sum of theory output of model, and then evolution, namely the proportion of the fault relative to the high-point case, which be taken as the standard of judgment fault, namely

$$\sigma = \sqrt{\sum_{i=1}^{h} e_i^2 \Big/ \sum_{i=1}^{h} M_i^2}\,. \tag{3}$$

Where e is the difference between model output and actual output, M is the model output.

According to fault diagnosis system structure of underwater robot and demand of experiment to the thruster performance, the high-point case of motion condition fault residual is 1/2 of model theory output, so obtained the following judgment way of fault degree:

Motion condition fault degree index:

$$\sigma_1 = \sqrt{\sum_{t=1}^{h} e_{1t}^2 \Big/ \sum_{t=1}^{h} (0.5 * m_t)^2}\,. \tag{4}$$

Thruster fault degree index:

$$\sigma_2 = \sqrt{\sum_{t=1}^{h} e_{2i_t}^2 \Big/ \sum_{t=1}^{h} r_{i_t}^2} \,. \tag{5}$$

Where, $t$ is step, $h$ is known detection step number, $m$, $r$ are speed, angle of model output and voltage, current information respectively. The fault degree judgment index of final fault diagnosis result can be defined based on the thought of information fusion.

$$\sigma = (\sigma_1+\sigma_2) / 2, \sigma \in [0,1] \,. \tag{6}$$

Thus, we can fuse the fault information of underwater robot fault detection model output at different time and space, and realized fault recognition, orientation and judgment of fault degree.

## 4 Fault Simulation Experiment

In order to prove the presented method of thruster fault fusion diagnosis, according to the actual equipment of the certain type underwater robot and the pool experiment environment, the motion experiment of underwater robot in the horizontal plane at simulating the right thruster enwind fault is designed. The underwater robot used the double close loop p control of speed and angle, the control period is 0.2s, and the type of experiment motion is the speed track direct navigation, the aim speed is shown as Fig. 6. The model output and actual measure are shown as Fig. 7- Fig. 9.

In Fig.9, the current condition residual of thruster excess the threshold (1.0) range at step 38 and following 20 step, combined with voltage character at figure 7, the thruster fault direct diagnosis sub-model has detected fault. But form figure 6 to look, underwater robot motion condition overall detection sub-model has not found fault.

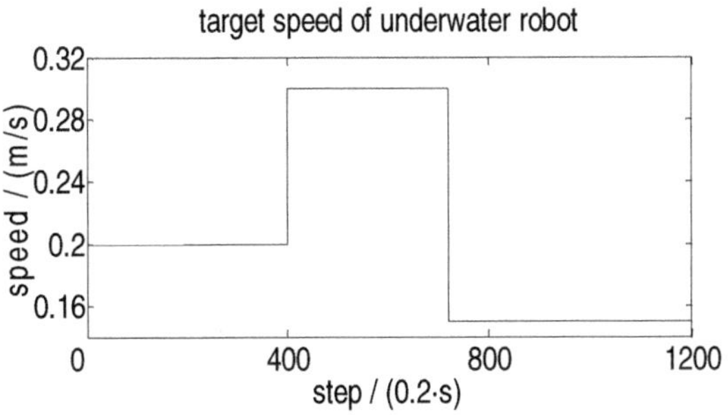


**Fig. 6.** The aim speed of underwater robot motion

Obviously when robot movement at low speed, thruster provides thrust under outside thing enwind disturbance still be able to satisfy the movement request.

In figure 6, at step 559 and following 20 steps, motion speed residual excess the threshold (0.02) rang. At the same step, the thrust fault diagnosis sub-system also obtained the fault information, through fusion can ensure appear the fault.

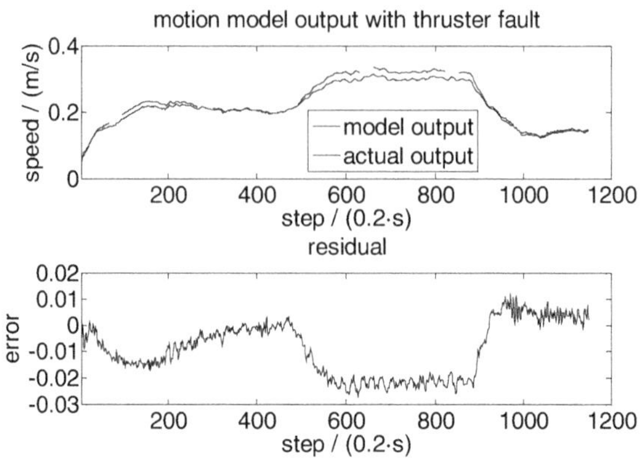


**Fig. 7.** The robot motion condition and model output in thruster enwind fault

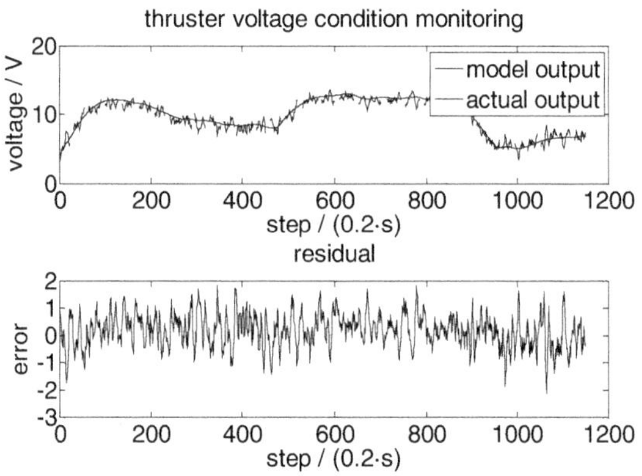


**Fig. 8.** The right thruster voltage model output

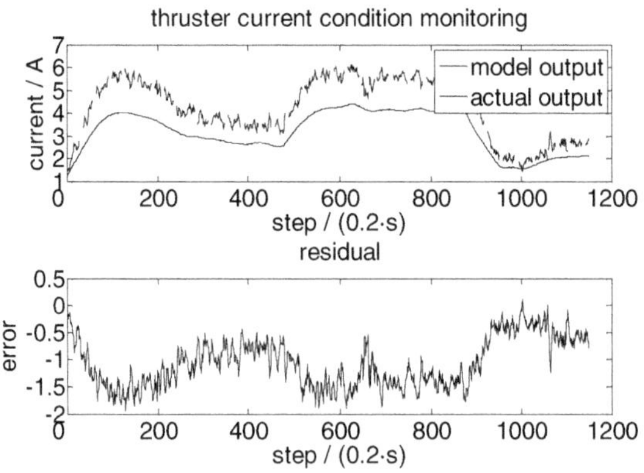


**Fig. 9.** The right thruster current model output

**Table 2.** Residual and Model Output

| step | Speed residual $e_I$ | Voltage residual $e_{Vi}$ | Current residual $e_{Ai}$ |
|---|---|---|---|
| 559 | -0.023411 | 0.70479 | -1.4242 |
| 560 | -0.022969 | 0.90943 | -1.427 |
| 561 | -0.020885 | 0.48819 | -1.505 |
| 562 | -0.020662 | 0.50094 | -1.607 |
| 563 | -0.021014 | 0.18184 | -1.6498 |
| 564 | -0.020702 | 0.20457 | -1.8122 |
| 565 | -0.02015 | 0.60977 | -1.5474 |
| 566 | -0.021086 | 1.1786 | -1.5342 |
| 567 | -0.022045 | 1.4182 | -1.3042 |
| 568 | -0.020822 | 1.286 | -1.3058 |
| 569 | -0.019087 | 1.6438 | -1.1406 |
| 570 | -0.019385 | 1.3694 | -1.3182 |
| 571 | -0.020208 | 0.68522 | -1.4946 |
| 572 | -0.020006 | 0.42521 | -1.6782 |
| 573 | -0.020073 | 0.9151 | -1.5534 |
| 574 | -0.021176 | 0.73659 | -1.4314 |
| 575 | -0.022251 | 0.62829 | -1.4398 |
| 576 | -0.022073 | 0.8842 | -1.3146 |
| 577 | -0.021623 | 0.86703 | -1.2874 |
| 578 | -0.022252 | 0.35903 | -1.5802 |

Under the effect of angle controller, because the thrust of fault thruster was compensated, it residual relative less, and the residual of left thruster performance model output and actual measure quantity of electric output is inconspicuous, therefore we not discuss at here. The residual information at step 559 and following 20 steps is shown as Table 2.

According to the produced method of fault degree and reliability estimation, the motion condition of actual sensor can be calculated.

Motion condition fault degree: $\sigma_1 = 0.0681$.

The thruster fault degree: $\sigma_2 = 0.4231$.

Fault degree of final fault diagnosis result:

$\sigma = (\sigma_1 + \sigma_2) / 2 = 0.2456$.

The reliability of diagnosis result:

$\delta = (\delta_1 + \delta_2) / 2 = (0.9+1) / 2 = 0.95$.

We can see the adjustment of controller make the fault degree relative less which obtained from motion condition fault, but the direct fault information of thruster is obviously, after fusion the two residuals can exact diagnosis the fault.

## 5 Conclusion

We have presented the information fusion method Underwater robot motion condition model based on improved Elman neural network of input information was built can exact response nonlinear dynamic characteristic, the thruster performance model based on RBF neural network was built can approach it voltage, current etc. quantity of electric characteristic. The overall fault detection model which take the underwater robot motion condition model and the thruster performance model as the core and direct fault detection sub-model through comparing the model output with actual output obtained the residual which require for fusion diagnosis, carried on partial and overall fault fusion, can detect the thruster fault availably. The analytic method of fault degree and reliability quantified the thruster fault case, finished the thruster fault diagnosis. Through analysis the result of pool experiment, the presented fault fusion diagnosis method is proved and it is feasible and reliable. Simultaneity, it has the practical meaning for executer fault diagnosis of other strong nonlinear time-delay system.

## Acknowledgments

The author wishes to thank Zhixian Jin for his zealous help. Thanks are also due to Mingjun Zhang and Juan Wu for their valuable suggestions.

## References

1. Ni, L.L.: Fault-Tolerant Control of Unmanned Underwater Vehicles, Dissertation of Ph. D for Virginia Polytechnic Institute and State University, June, (2001) 37-62
2. Omerdic, E., Roberts, G.: Thruster Fault Diagnosis and Accommodation for Open-frame Underwater Vehicles. Control Engineering Practice, 12(12) (2004) 1575–1598
3. Wang, Y.J., Zhang, M.J.: Research on the Test-platform and Condition Monitoring Method for AUV. IEEE International Conference on Mechatronics and Automation, 3 (2006) 2404-2409
4. Kimmich, F., Schwarte, A., Isermann, R.: Fault Detection for Modern Diesel Engines using Signal- and Process Model-based Methods. Control Engineering Practice, 13 (2) (2005)189–203
5. Jiang, X.S.: Underwater robot, Liaoning: Technology publisher house of Liaoning, (2000) 292-303
6. Jason, E., Meyer, N.: Dynamics Modeling and Performance Evaluation of an Autonomous Underwater Vehicle. Ocean Engineering, 31(14) (2004) 1835-1858
7. Tang, Y.Q.: Motor Technology. Harbin: Harbin University of Science and Technology, (2002) 128-156
8. Ye, S.W. , Shi, Z.Z.: The Theory of Neural Nnetwork. Beijing: Mechanical Industry Publishing House, (2004) 305-356
9. Wang, Z.S.: Intelligent Fault Diagnosis and Fault Tolerant Control. Xian: the Northwest Industry University Publishing House, (2005) 212-233

# Synthesized Fault Diagnosis Method Based on Fuzzy Logic and D-S Evidence Theory*

Guang Yang[1,2] and Xiaoping Wu[1]

[1] Department of Information Security, Naval University of Engineering, 430033 Wuhan, China
sunyoung0783@sina.com
[2] Department of Watercraft Engineering, Zhenjiang Watercraft College, 212003 Zhenjiang, China
sunyoung0783@163.com

**Abstract.** Based on fuzzy logic theory, the information fusion technique at decision-making level is introduced for fault diagnosis system. A comprehensive fault diagnosis method based on fuzzy logic and D-S evidence theory is presented by making use of multi-sensor information fusion technique. In this method, the basic reliability distribution of evidence theory is obtained by using fuzzy membership function, which can significantly improve the accuracy of the fault diagnosis through taking full advantages of redundant and complementary fault information from all sensors. Finally the method is applied for fault diagnosis of ship diesel engine. Diagnostic results indicate that the technique is effective, greatly improving the efficiency of fault diagnosis.

**Keywords:** fault diagnosis, fuzzy logic, information fusion, D-S evidence theory.

## 1 Introduction

In the past few years, sensor technique developed fast, and many complex multi-sensor information systems emerged. In these systems, the diversity of information representation, information capability and information processing speed etc. have greatly exceeded the information synthesis ability of human brains, so information fusion technique emerged as required. As information fusion system has potential characters such as robustness, wide spatial-temporal coverage, high measurement dimension, high resolution in object space, strong fault tolerance and system reconfiguration capability and so on, information fusion arouses much attention in many fields since its inception [1-3].

With many types of state monitoring and fault diagnosis information, each type may contain information reflecting running state. So it is incomprehensive, uncertain and even wrong to diagnose running state of the whole equipment only using few signals and single sensor. Reliable results can only be obtained by studying in detail

* This work is partially supported by CNSF Grant #70471031.

D.-S. Huang, L. Heutte, and M. Loog (Eds.): ICIC 2007, LNAI 4682, pp. 1024–1031, 2007.

all kinds of information and consequently comprehensively utilizing redundant and complementary fault information from all sensors [4]. In this paper, to avoid information loss in the fault diagnosis system, the information fusion technique at decision-making level is introduced for fault diagnosis system and a comprehensive fault diagnosis method based on fuzzy logic and D-S evidence theory is presented. The method is applied to fault diagnosis of ship diesel engine, and diagnostic results indicate that it can significantly increase the degree of confidence in fault diagnosis and improve the efficiency of fault diagnosis.

## 2 Comprehensive Diagnosis Model

### 2.1 Structure of Comprehensive Diagnosis System

The comprehensive diagnosis system consists of decision layer and characteristic layer, with the former one being main while the latter being complementary. Fig. 1 is the structure framework of the multi-sensor information fusion system. Firstly dynamic signals of diagnostic objects are measured by all sensors, and then the faults are identified by appropriate eigenvalues of all sensors' signals using the feature extraction method. In the measured faults of the system, some extracted eigenvalues from different sensors sometimes are approximate, so wrong decisions can be made. What's more, when a certain sensor goes wrong, wrong decisions can also be made possibly. If these sensors identify faults independently, paradoxical results may be produced when a wrong feature is obtained. Aimed at these problems, fuzzy logic theory is introduced. As for every sensor, fuzzy membership functions are used to express the probability of which type the measured faults belong to. There still exists some problems here. For example, the membership functions of a certain fault belonging to every type are different and even big deviation and paradoxical condition may appear. In order to solve these problems, the fuzzy logic and D-S evidence theory is combined for information fusion. In the process of fault diagnosis, membership functions of the measured faults belonging to every type are used as the correlation coefficients of the basic possibility assignment of every sensor in D-S evidence theory. Fusing the basic possibility assignment of every sensor by using fusion rule of D-S evidence theory, the basic possibility assignment after fusion is obtained. Then according to fusion results and decision rules, faults are identified and decided.

### 2.2 Diagnosis Fusion Algorithm

Basic Possibility Assignment (BPA). Basic possibility assignment (or measurement function assignment) denotes the reasoning for believability of object assumption and a judgment, which is affected by all kinds of factors. Different ideas can form different formulae of basic possibility assignment [5]. Every sensor becomes an evidence body, and the target faults constitute the frame of discernment $\Theta = \{F_1, F_2, \cdots, F_n\}$. Define correlation coefficient between evidence body $E_i$ and

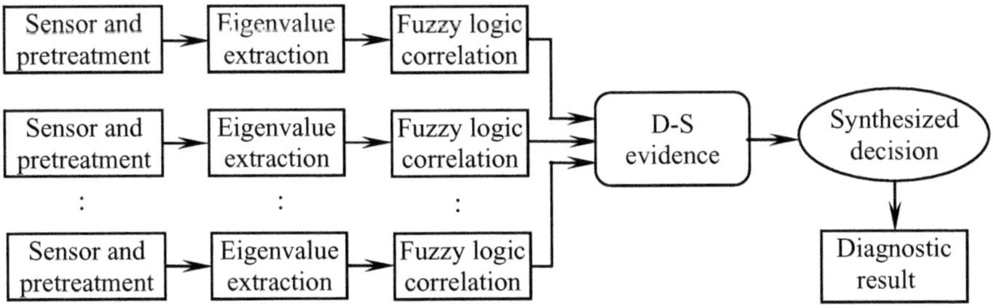


**Fig. 1.** Structure framework of comprehensive diagnosis system

proposition $F_j$ in the frame of discernment as $C_i(F_j)$, and then evidence body $E_i$ endows target object $F_j$ with the basic possibility assignment $m_i(F_j)$, as is shown in (1) and (2).

$$m_i(F_j) = \frac{C_i(F_j)}{\sum_j C_i(F_j) + R_i} \tag{1}$$

$$m_i(\Theta) = \frac{R_i}{\sum_j C_i(F_j) + R_i} \tag{2}$$

$$R_i = \omega_i (1-\alpha_i)(1-\beta_i) \tag{3}$$

$$\alpha_i = C_i(F_m) - \max_{j \neq m}\{C_i(F_j)\} \tag{4}$$

$$C_i(F_m) = \max_j\{C_i(F_j)\} \tag{5}$$

$$\mu_i = \frac{1}{M-1} \sum_{j=0, j\neq m}^{M} C_i(F_i) \tag{6}$$

$$\beta_i = \sqrt{\frac{1}{M-1} \sum_{j=1}^{M} (C_i(F_j) - \mu_i)^2} \tag{7}$$

Where, $R_i$ denotes the whole uncertainty of diagnostic process. $\alpha_i$ reflects the outstanding degree of the proposition which has the biggest correlation coefficient in the frame of discernment. The value of $\alpha_i$ reflects diagnosis reliability from the aspect of the outstanding degree of the proposition correlativity during the course of information fusion; $\mu_i$ is mean of all propositions' correlation coefficients except the biggest one. $\beta_i$ is the standard deviation of all propositions' correlation coefficients except the biggest one in the frame of discernment, which reflects the concentrative degree of other propositions' correlation coefficients. The value of $\beta_i$ reflects diagnosis reliability from the aspect of the concentrative degree of the proposition's correlation coefficient during the course of information fusion. $\omega_i$ is the weight of $E_i$. In the process of multi-sensor information fusion fault diagnosis, because of the difference in sensitivity, reliability and precision, the factor $\omega_i$ is given to improve correctness of decision results.

In view of subjectivity and inconsistency in selecting correlation coefficient $C_i(F_j)$, fuzzy membership function $\mu_{ij}$ is presented to substitute correlation coefficient $C_i(F_j)$. Obviously in physical meaning, both parameters denote the probability (i.e. the correlation between them) of measured object belonging to a certain target object according to a sensor's measuring value, which is in accordance with brain's thinking habit and has scientism and feasibility. In practical application, it has much difficulty in describing probability density function of the sensor's output values with precise mathematics expression for they are influenced by mechanism, temperature and pressure [6]. From many experimental data and observations, a trapezoid function $\mu_{ij}$, shown in Fig.2, is selected as the fuzzy membership function for each fault eigenvalue. It is defined as:

$$\mu_{ij}=\begin{cases}0 & x_i\le x_{0ij}-t_{ij}\\ \dfrac{x_i-x_{0ij}+t_{ij}}{t_{ij}-e_{ij}} & x_{0ij}-t_{ij}\le x_i\le x_{0ij}-e_{ij}\\ 1 & x_{0ij}-e_{ij}\le x_i\le x_{0ij}+e_{ij}\\ -\dfrac{x_i-x_{0ij}-t_{ij}}{t_{ij}-e_{ij}} & x_{0ij}+e_{ij}\le x_i\le x_{0ij}+t_{ij}\\ 0 & x\ge x_{0ij}+t_{ij}\end{cases} \tag{8}$$

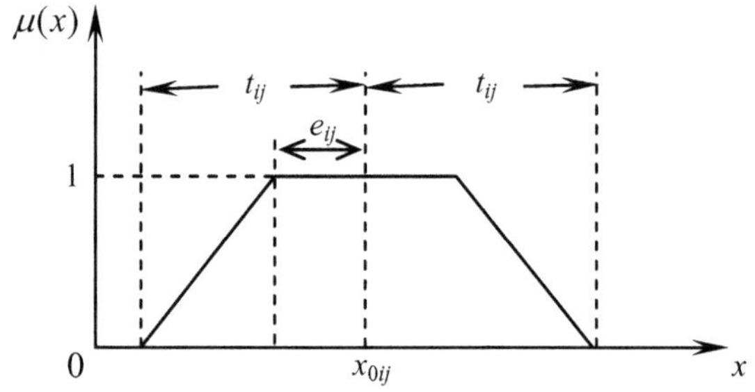


**Fig. 2.** Fuzzy membership function of eigenvalue

Where, $x_i$ denotes practical measuring eigenvalue of the sensor $i$. $\mu_{ij}$, $x_{0ij}$, $e_{ij}$ and $t_{ij}$ denote membership function, standard eigenvalue, eigenvalue error and the biggest deviation of eigenvalue of diagnosed fault's sort belonging to $j$th sort respectively.

**D-S Evidence Theory.** D-S evidence theory begins Dempster-Shafer's multi-value map work in the 1960s, where belief function is kept relationship with upper limit and lower limit of probability and then a general frame of uncertain reasoning model is presented [1,7].

(1). Let $\Theta$ denote a universe set of all possible values of X and all elements in $\Theta$ are disjoint, then $\Theta$ is defined as the frame of discernment of X.

(2). Define BPA m as set function $m: 2^{\Theta} \to [0,1]$ in frame of discernment $\Theta$, then: $m(\phi) = 0$; $\sum_{A \subset \Theta} m(A) = 1$.

(3). Define function $Bel: 2^{\Theta} \to [0,1]$ which defined by $Bel(A) = \sum_{B \subset A} m(B)$ $(\forall A \subset \Theta)$ as belief function in $\Theta$; define $pl: 2^{\Theta} \to [0,1]$ as $pl(A) = 1 - Bel(\overline{A}) = \sum_{B \cap A \neq \phi} m(B)$ and $pl$ is named plausibility function ; $[Bel(A), pl(A)]$ denotes uncertain interval of proposition A.

(4). Dempster-Shafer's rule of combination: let $Bel_1, Bel_2, \cdots, Bel_n$ be belief functions in the same frame of discernment $\Theta$ and $m_1, m_2, \cdots, m_n$ be corresponding BPA respectively, then BPA after combination can be denoted as:

$$m(A) = \begin{cases} 0 & A = \phi \\ \dfrac{\sum_{\cap A_n = A} \prod_{n=1}^{N} m_n(A_n)}{\sum_{\cap A_n \neq \phi} \prod_{n=1}^{N} m_n(A_n)} & A \neq \phi \end{cases} \tag{9}$$

Because different sensors offer different information sources and different information sources can be regarded as different evidences in D-S evidence theory, D-S combination rule provides the method fusing different evidences, i.e., fusing different sensors. Therefore, it is feasible and necessary for D-S evidence theory to be applied for the comprehensive fault diagnosis of multi-sensor information fusion.

**Decision Rule for Comprehensive Diagnosis.** After belief interval $[Bel,pl]$ and evidence's uncertainty $m_i(\Theta)$ of all propositions in frame of discernment $\Theta$ are gained, diagnostic result $F_a$ can be ascertained according to the following rules:

*Rule 1:* $Bel(F_a) = \max_j \{Bel(F_j)\}$;

*Rule 2:* $Bel(F_a) - Bel(F_j) > \varepsilon$, $Bel(F_a) - m(\Theta) > \varepsilon$, $\varepsilon \in R$ and $\varepsilon > 0$;

*Rule 3:* $m(\Theta) < \gamma$, $\gamma \in R$ and $\gamma > 0$.

Rule 1 indicates that the diagnostic result must be the proposition with the biggest possibility; rule 2 indicates that the possibility of diagnostic result must be $\varepsilon$ bigger than all other propositions' possibilities and evidence's uncertainty; rule 3 indicates that evidence's uncertainty must be less than $\gamma$. Where, $\varepsilon$ and $\gamma$ can be chosen according to practical application condition. With the premises of achieving the above three rules, fault $F_a$ can just be ascertained. If it can't be ascertained, new frame of discernment should be constructed or more evidence bodies should be chosen to carry out fusion computation.

## 3 Diagnostic Example

The warship of type X is the main force of surface warship formation. Being the key part of the warship, power equipment affects campaign efficiency directly. The comprehensive fault diagnosis ability is also an important embodiment of battlefield rush-repair and comprehensive support ability. Thus, real-time fault monitoring and diagnosis should be processed for its power system. In this paper, the relation between running faults and symptoms of the warship's main power system (the diesel engine of type Y) is analyzed and discussed, through which we can research on the presented diagnosis method in the application of ship power equipment fault diagnosis.

The common running faults of this warship diesel engine are regarded as the frame of discernment $\Theta$ of fusion-decision diagnosis, denoted as $\Theta=\{F_0,F_1,F_2,F_3,F_4,F_5,F_6,F_7,F_8,F_9\}$={normal, the injector's nozzle enlarges, the injector's nozzle is jammed, the air inlet is jammed, the valve seat of injector leaks, fuel is sprayed too late, fuel is sprayed too early, the vent-pipe is jammed, the injection pressure is too high, the quality of fuel is bad}. Sensors of the synthesized diagnosis system are made up of the vibration sensor of cylinder head, the pressure sensor of cylinder and the temperature sensor of exhaust. Each sensor constitutes an evidence body which is denoted as $E_1$, $E_2$, $E_3$ respectively. According to experts' experience, the weight of evidence $E_1$, $E_2$, $E_3$ can be selected as $(\omega_1,\omega_2,\omega_3)=(0.400,0.300,0.300)$. Feature extraction is processed with the measured signals of sensors, and the normalized correlation coefficient $\mu_{ij}=C_i(F_j)$ between each evidence and decision objects can be attained, as is shown in Table 1. Then the basic probability assignments of every evidence can be obtained, as is shown in Table 2. The values of $\varepsilon$ and $\gamma$ (decision parameters) are selected as 0.25 and 0.10 respectively. Compute every proposition's belief interval $[Bel(A), pl(A)]$ by single evidence and by combination of evidences, and diagnostic results can be gotten using decision rules, as is shown in Table 3 and Table 4.

**Table 1.** Correlation coefficient between each evidence and decision objects

| Evidence | $C_i(F_j)$ | | | | | | | | | |
|---|---|---|---|---|---|---|---|---|---|---|
| | $F_0$ | $F_1$ | $F_2$ | $F_3$ | $F_4$ | $F_5$ | $F_6$ | $F_7$ | $F_8$ | $F_9$ |
| $E_1$ | 0.040 | 0.032 | 0.306 | 0.085 | 0.153 | 0.093 | 0.084 | 0.095 | 0.093 | 0.028 |
| $E_2$ | 0.060 | 0.066 | 0.285 | 0.063 | 0.074 | 0.172 | 0.078 | 0.081 | 0.076 | 0.045 |
| $E_3$ | 0.050 | 0.070 | 0.360 | 0.060 | 0.059 | 0.082 | 0.184 | 0.030 | 0.090 | 0.015 |

**Table 2.** Basic probability assignments of every evidence

| Evidence | $m_i(\Theta)$ | $m_i(F_j)$ $F_0$ | $F_1$ | $F_2$ | $F_3$ | $F_4$ | $F_5$ | $F_6$ | $F_7$ | $F_8$ | $F_9$ |
|---|---|---|---|---|---|---|---|---|---|---|---|
| $E_1$ | 0.2367 | 0.0305 | 0.0244 | 0.2336 | 0.0649 | 0.1168 | 0.0710 | 0.0641 | 0.0725 | 0.0710 | 0.0214 |
| $E_2$ | 0.1972 | 0.0482 | 0.0530 | 0.2288 | 0.0506 | 0.0594 | 0.1381 | 0.0626 | 0.0650 | 0.0610 | 0.0361 |
| $E_3$ | 0.1811 | 0.0409 | 0.0573 | 0.2948 | 0.0491 | 0.0483 | 0.0671 | 0.1507 | 0.0246 | 0.0737 | 0.0123 |

**Table 3.** Belief intervals and diagnostic results by single evidence

| Evidence | $[Bel(F_j),pl(F_j)]$ $F_0$ | $F_1$ | $F_2$ | $F_3$ | $F_4$ | $F_5$ | $F_6$ | $F_7$ | $F_8$ | $F_9$ | $m_i(\Theta)$ | Diagnostic results |
|---|---|---|---|---|---|---|---|---|---|---|---|---|
| $E_1$ | 0.0305 | 0.0244 | 0.2336 | 0.0649 | 0.1168 | 0.0710 | 0.0641 | 0.0725 | 0.0710 | 0.0214 | 0.2367 | Uncertain |
| | 0.2672 | 0.2611 | 0.4703 | 0.3016 | 0.3535 | 0.3077 | 0.3008 | 0.3092 | 0.3077 | 0.2581 | | |
| $E_2$ | 0.0482 | 0.0530 | 0.2288 | 0.0506 | 0.0594 | 0.1381 | 0.0626 | 0.0650 | 0.0610 | 0.0361 | 0.1972 | Uncertain |
| | 0.2454 | 0.2502 | 0.4260 | 0.2478 | 0.2566 | 0.3353 | 0.2598 | 0.2622 | 0.2582 | 0.2333 | | |
| $E_3$ | 0.0409 | 0.0573 | 0.2948 | 0.0491 | 0.0483 | 0.0671 | 0.1507 | 0.0246 | 0.0737 | 0.0123 | 0.1811 | Uncertain |
| | 0.2220 | 0.2384 | 0.4759 | 0.2302 | 0.2294 | 0.2482 | 0.3318 | 0.2057 | 0.2548 | 0.1934 | | |

**Table 4.** Belief intervals and diagnostic results by combination of evidences

| Evidence | $[Bel(F_j),pl(F_j)]$ $F_0$ | $F_1$ | $F_2$ | $F_3$ | $F_4$ | $F_5$ | $F_6$ | $F_7$ | $F_8$ | $F_9$ | $m_i(\Theta)$ | Diagnostic results |
|---|---|---|---|---|---|---|---|---|---|---|---|---|
| $E_1$&$E_2$ | 0.0400 | 0.0395 | 0.3257 | 0.0595 | 0.0933 | 0.1197 | 0.0667 | 0.0729 | 0.0695 | 0.0287 | 0.0989 | Uncertain |
| | 0.1389 | 0.1384 | 0.4246 | 0.1584 | 0.1922 | 0.2186 | 0.1656 | 0.1718 | 0.1684 | 0.1276 | | |
| $E_1$&$E_3$ | 0.0349 | 0.0411 | **0.3839** | 0.0563 | 0.0811 | 0.0711 | 0.1208 | 0.0440 | 0.0754 | 0.0150 | 0.0909 | $F_2$ |
| | 0.1258 | 0.1320 | **0.4748** | 0.1472 | 0.1720 | 0.1620 | 0.2117 | 0.1349 | 0.1663 | 0.1059 | | |
| $E_2$&$E_3$ | 0.0421 | 0.0537 | **0.3747** | 0.0479 | 0.0519 | 0.1066 | 0.1133 | 0.0409 | 0.0675 | 0.0211 | 0.0801 | $F_2$ |
| | 0.1222 | 0.1338 | **0.4548** | 0.1280 | 0.1320 | 0.1867 | 0.1934 | 0.1210 | 0.1476 | 0.1012 | | |
| $E_1$&$E_2$&$E_3$ | 0.0337 | 0.0394 | **0.4807** | 0.0484 | 0.0683 | 0.0949 | 0.0967 | 0.0455 | 0.0653 | 0.0177 | 0.0468 | $F_2$ |
| | 0.0805 | 0.0862 | **0.5275** | 0.0952 | 0.1151 | 0.1417 | 0.1435 | 0.0923 | 0.1121 | 0.0645 | | |

From table 3 and table 4, it is obvious that belief measurement is lower if only a single evidence is used to diagnose and identify fault state, i.e., faults can't be identified accurately. When multi-evidence fusing information is used, the identification rate of faults is significantly improved. For example, when the fault $F_2$ (the injector's nozzle is jammed) happens, if single evidence is used, the uncertainty $m_i(\Theta)$ of the 3 evidences is too big and $\gamma$ greater than the selected value of 0.10, so the running state can't be judged from using each single evidence because 3 decision rules are not satisfied simultaneously. While multi-evidences is fused to decide, the belief interval and evidence's uncertainty decreases quickly, and the value of belief function $Bel(F_2)$ of diagnostic result increases obviously, which indicates that diagnostic precision and reliability are improved with evidence fusion. Therefore, the identification and decision ability in diagnosis system is improved.

## 4 Conclusions

A comprehensive diagnosis method using multi-sensor fault characteristic information fusion is presented based on fuzzy logic and D-S evidence theory.

Construction method of basic possibility assignment (BPA) based on fuzzy membership function and diagnostic decision rules are presented. The fault diagnosis example indicates that the calculated BPA can preferably reflects each proposition's credibility of frame of discernment and that uncertainty $m_i(\Theta)$ is much less.

After fusing multi-fault characteristic information, the credibility of diagnostic results increases obviously and uncertainty decreases evidently, which shows the validity of the fault diagnosis method based on fuzzy logic and D-S evidence theory.

**Acknowledgments.** Many thanks for my professor Xiaoping Wu and associate professor Yexin Song for their detailed and helpful comments. And I am also grateful to my colleague Yinchun Chen for his good advices to this paper.

## References

1. Waltz, E., Lilnas, A.: Multisensor Data Fusion. Artech House, Boston (1990)
2. Hall, D. L., Lilnas, J.: An Introduction to Multisensor Data Fusion. In: Proc IEEE, Vol. 85, no 1 (1997) 6–23
3. Varshney, P. K.: Multisensor Data Fusion. J Eclec Commu Eng, Vol. 9, no.6 (1997) 245–253
4. Luo, R. C., Gonzalez, R. C.: Data Fusion and Sensor Integration. State-of-the-art 1990s, Data Fusion in Robotics and Machine Intelligence, Edited by Abidi, M. A., Academic Press, Inc.(1992) 7–136
5. Liu, L J., Yang, J. Y.: Model-based object classification using fused data. SPIE Vol. 1611. (1991) 65–73
6. Han, J., Tao, Y. G.: Data Fusion Algorithm of Multi-sensor Based on D-S Evidential Theory and Fuzzy Mathematic. Chinese Journal of Scientific Instrument, Vol. 21, no.6 (2000) 644–647
7. He, Y., Wang, G. H., Lu, D. J.: Multi-sensor Information Fusion With Application. Publishing House of Electronics Industry, Beijing (2000)

# Test Scheduling for Core-Based SOCs Using Genetic Algorithm Based Heuristic Approach

Chandan Giri[1], Soumojit Sarkar[2], and Santanu Chattopadhyay[1]

[1] Department of E & ECE, IIT Kharagpur, India
{chandan,santanu}@ece.iitkgp.ernet.in
[2] Department of CSE, IIT Kanpur, India
soumojitsarkar@gmail.com

**Abstract.** This paper presents a Genetic algorithm (GA) based solution to co-optimize test scheduling and wrapper design for core based SOCs. Core testing solutions are generated as a set of wrapper configurations, represented as rectangles with width equal to the number of TAM (Test Access Mechanism) channels and height equal to the corresponding testing time. A locally optimal best-fit heuristic based bin packing algorithm has been used to determine placement of rectangles minimizing the overall test times, whereas, GA has been utilized to generate the sequence of rectangles to be considered for placement. Experimental result on ITC'02 benchmark SOCs shows that the proposed method provides better solutions compared to the recent works reported in the literature.

**Keywords:** SOC testing, test scheduling, wrapper design, test access mechanism.

## 1 Introduction

In order to meet the demand of low time-to-market, core-based SOC design is becoming a widely used paradigm for integrated circuit design. However, integrating reusable cores into an SOC involves complicated design and test issues. As one of the objectives of SOC testing is to reduce the test application time, scheduling test solutions for embedded IP cores can be facilitated to reduce it. Hence, test scheduling problem is one of the major issues in test integration. A general problem for SOC test integration consists of the design of test access mechanism (TAM) architecture, that transports test data between SOC pin and core wrapper. Wrapper provides an interface between TAM and the core and can be operated in several modes [1]. So test scheduling is a process that determines the start and ending time of testing each core in the SOC such that the overall test application time is minimized.

Several recent works considered various aspects of the test scheduling problem. Earlier works propose methods to solve wrapper design and test scheduling as separate problems. Recently [2] and [3] proposed an integer linear programming based solution for co-optimization of wrapper design and test scheduling for SOCs. Huang et al [4] formulated the problem of SOC pin allocation to cores and test

D.-S. Huang, L. Heutte, and M. Loog (Eds.): ICIC 2007, LNAI 4682, pp. 1032–1041, 2007.

scheduling using 2-D bin-packing or rectangle packing approach. Same authors also proposed 3-D bin-packing approach [5] considering power constraints. Several other works [6-11] also consider SOC power dissipation constraints during scheduling. A heuristic approach using the sequence pair representation for test scheduling was considered in [12]. Zou et al [13] proposed test scheduling algorithm based on simulated annealing (SA) using the sequence pair representation. A B*-tree based approach has been proposed in [14] to get the test schedule. Recently in [15] SOC test scheduling with reconfigurable core wrappers is used. Ant colony optimization (ACO) based approach [16] considers the rectangle packing for test scheduling solution. A two-stage genetic algorithm (GA) based algorithm was proposed in [17] where each solution is represented by a sequence pair.

In this paper, we present a genetic algorithm (GA) based approach for the SOC test scheduling and wrapper design co-optimization problem to achieve minimization of SOC test application time. We represent wrapper design as a rectangle with width being the number of TAM channels and height being the test time. To get the complete test scheduling we used the two-dimensional bin-packing approach based on best-fit heuristics. It uses the sequence of rectangles obtained from GA and placed one rectangle after another each time optimizing the placement based on the local scenario reached by the placement of all the previous rectangles. For a particular rectangle, the placement algorithm evaluates the possible placement time instants on the basis of two cost parameters, TAM width utilization (U) and increase in total test time (T). Experimental results show that the proposed method obtains better test time results for the SOCs with larger number of cores than the recently proposed works.

The remainder of this paper is organized as follows. Section 2 describes the overall SOC test scheduling problem. Section 3 discusses the wrapper design optimization methods used. Section 4 discusses the test scheduling algorithm with the proposed bin-packing approach. Proposed GA formulation for selection of one rectangle from the set of rectangles of each core generated by wrapper design is presented in Section 5. We present our experimental result based on the ITC'02 benchmark [18] SOCs in Section 6. Section 7 concludes the paper.

## 2 Problem for Mulation

Let the SOC design consist of N cores, and each core $C_i$ ($1 \leq i \leq N$) has $n_i$ input terminals, $m_i$ output terminals, $b_i$ bi-directional I/Os, $sin_i$ scan inputs and $sot_i$ scan outputs. Let the total width of TAMs be K and each core must be tested with $P_i$ patterns. So the overall problem that we have to solve is as follows.

*Given a set of N cores, their specific test parameters and the number of I/O pins for an SOC, design the test schedule with wrapper designs for all wrapper-based cores such that overall testing time is minimized.*

So there exist basically two steps to solve the problem. First we generate possible optimized wrapper configurations for each core under specified TAM width. In the next step solve the test scheduling problem using the sets of optimized wrapper solutions.

## 3 Core Wrapper Design

A test wrapper is the interface between the TAM and the core. Since larger cores typically have hundreds of core terminals and the total number of TAM channels available depends on the limited number of SOC pins, wrappers facilitate test width adaptation when the TAM width is not equal to the number of core terminals. In this paper we consider only internal testing. There are two kinds of wrapper designs, balanced and unbalanced wrapper design [1]. For cores having no internal scan chains (that is, containing input and output pins only), unbalanced wrapper design is preferred since it can obtain a lower test time than the balanced one [14]. To calculate test time, T for a wrapper for different TAM widths we use the well-known formula [1] given below.

$$T= \{ 1 + max\ (S_i,S_o)\} \times P + min((S_i,S_o) \tag{1}$$

Where P is the number of test patterns and $S_i(S_o)$ denotes the length of longest wrapper scan chain used during scan-in(out) for a core. We consider two different approaches to the design of wrapper for cores with internal scan chains and for cores that do not have any internal scan chains. For cores with internal scan chain we use the *Design_wrapper* algorithm proposed in [3] and for cores without internal scan chain we use the scheme proposed in[13].

Using the wrapper design method for each core we can generate a set of wrapper configurations with the TAM wire usage of 1 to K, where K is maximum number of TAM channels allocated to test the SOC. Hence each wrapper configuration can be considered as a rectangle with width equal to the test time and height corresponding to the TAM channels used. So each configuration is represented by a two tuple $(W_{ij}, T(W_{ij}))$ – where, core i has been assigned a TAM width $W_{ij}$ resulting in a test time $T(W_{ij})$. From all the wrapper configurations for a core i a smaller set of wrapper configuration can be considered in the test scheduling. It is based on pareto-optimal design [3] principle – for a range of TAM widths, test time remain unchanged.

## 4 Test Scheduling Problem

Suppose a SOC with N cores is to be tested using K TAM channels. Each core $C_i$ ( $1 \leq i \leq N$) is represented by a set of $R_i$ wrapper configurations. Each wrapper configuration is represented by a pair $(W_{ij}, T(W_{ij}))$, where $W_{ij}$ stands for the wrapper width of the j-th wrapper configuration for core $C_i$ and $T(W_{ij})$ is the test time of the core $C_i$ with the wrapper width $W_{ij}$. So the objective is the assignment of the core wrapper pins to the pins of SOC and obtains the test starting time and finishing time for each core such that overall test time is minimized.

This problem can be transformed into the well-known rectangle-packing problem, in which the SOC is represented by bin with width K and a set of $R_i$ SOC wrappers for each core represented by a set of $R_i$ rectangles with width $W_{ij}$ and test time $T(W_{ij})$. We want to choose one rectangle from each set of rectangles for a core $C_i$ and pack all the rectangles in the bin, so that height of the bin is minimized.

However, the test schedule of a SOC is feasible if no two cores are assigned to the same SOC pin at the same time instant and for each core all its wrapper pins are assigned to SOC pins for the entire time needed to test that core. Therefore in the process of bin-packing no rectangle should be overlapped. It should be noted that a core does not need to be assigned consecutive SOC pins. So the wrapper configuration for that particular core can be split into several smaller rectangles as long as those rectangles have the same test start time. But, splitting rectangles complicates the problem. To avoid that splitting we use the following constraint.

***Constraint:*** *At any instant of time within entire test time of the SOC the sum of the widths of all the rectangles that are tested is equal to or smaller than the total TAM width provided to test the given SOC.*

To satisfy this constraint it needs to satisfy at the test start time $t_{si}$ of each core i. If this is fulfilled at every core test start time $t_{si}$, it will hold for all the time instances from 0 to $T_{end}$, where $T_{end}$ is the time needed to complete testing of the given SOC. The following section discusses the proposed rectangle packing algorithm:

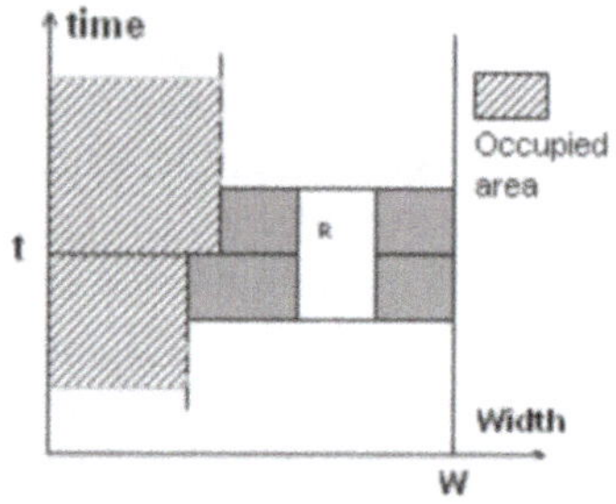


**Fig. 1.** An optimum placement of the incoming rectangle

### 4.1 Placement Algorithm

We used a best-fit heuristic based placement or packing algorithm satisfying the above mentioned constraint for placing the rectangles in a bin of width W (maximum allowable TAM width) whose height is infinite (∞). The algorithm tries to place one rectangle after another each time optimizing the placement based on a cost function. The cost is rectangle specific and is calculated considering the existing profile of the already placed rectangles. The profile provides the information regarding TAM channels used in various time instants. Hence cost function depends on two parameters, TAM width utilization (U) and increase in test time (T). For a particular rectangle, the algorithm evaluates the possible placement points (time instants) on the basis of cost parameters and the point with lowest cost value is selected for placing the rectangle. For finding that optimal position we restrict our search only to the points where the profile function changes value. As shown in the Fig. 1 for the next rectangle R to be placed if we move the rectangle upwards then we are surely going to reduce the area (solid shaded region) formed by the free TAM channels. And this will continue till we reach at the time instant marked as t. Similarly for downward movement also the same time instant will provide the reduced area. So for each possible placement point (time instant) we consider two types of orientations – up and down, which denotes whether the rectangle is placed upwards (or downwards). These two possible placements are shown in Fig. 2.

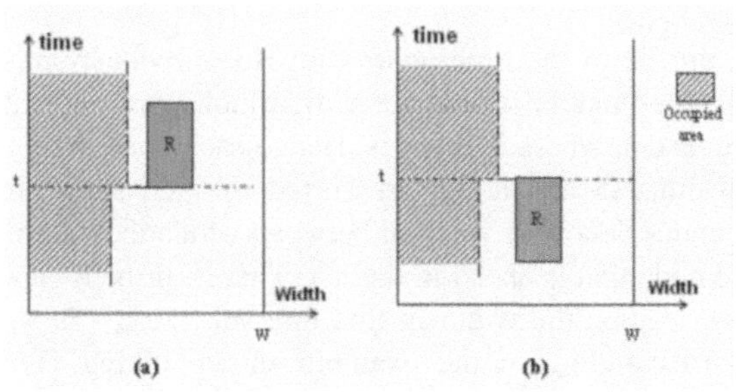


**Fig. 2.** Upward and downward oriented placements

We use weighted sum of these two parameters to determine the final cost of a particular placement. This is

$$Cost=wt \times u(U) + (1\text{-}wt) \times t(T), \quad (0 \leq wt \leq 1) \tag{2}$$

Where *wt* is the weight in favour of the utilization parameter and $u(U)=U/U_{max}$ and $t(T)=T/T_{max}$. The utilization parameter is a measure of utilization of TAM lines during the entire height of the rectangle being used. It is maximized by minimizing the number of unused TAM lines. The variable U represents the sum of unutilized TAM lines at each time instant during the testing time of the core, that is, the height of the rectangle. Suppose that the rectangle is placed between the time instants $t_1$ and $t_1+h$. In this interval, there will be subintervals during which the number of unutilized lines does not change. The changes occur only at the boundaries of subintervals. For instance, may be the TAM lines used in the duration $t_1$ and $t_1+\Delta$ be $w_1$. If W be the total TAM width used, the total number of unutilized TAM lines for this subinterval is $(W - w_1)\times \Delta$. For each such subinterval, the number of unutilized TAM lines is calculated and summed up to get the total unutilized TAM lines U. Once all such U values have been calculated, for different possible placements $U_{max}$ is taken to be the maximum of U values for all these placements. Thus, $u(U)=U/U_{max}$ gives a normalized value of unutilized TAM lines. The second parameter increase in test time (T) is an obvious attempt to attach some penalty with those possible placements that result in putting the upper edge of the rectangle in consideration beyond the test time required by all the rectangles placed before it. This inclusion of such a parameter can be justified on the basis of the simple fact that it is the testing time, which we are trying to minimize. The value of $wt=0.8$ has been found to work well with the ITC'02 benchmark SOCs.

For each of the possible placement points we check whether the constraint is satisfied or not and the cost is evaluated. The point with lowest cost is selected for placement of the rectangle. This process is continued for all the cores and the last time instant provides the final test time for the selected sequence of rectangle obtained from GA.

**Data Structure used:**

1. The sequence of representative rectangles for each core generated by GA is in the form of a queue $L_{CORE}$.

2. The current profile of the time-width plot (for previously placed rectangles) is maintained as a double-linked list time_instants, which stores the value of the width at the particular time instant where it changes. Hence each node in this list is a tuple like <time_instant(t),width(w)> and nodes are sorted by increasing time_instant values. Two consecutive nodes denote an interval between which the width is fixed by the w-value stored in the node that is ahead in the list. That is, if the $i^{th}$ and $(i+1)^{th}$ tuples are $<t_i,w_i>$ and $<t_{i+1},w_{i+1}>$ then the width in that interval $[t_i , t_{i+1})$ is $w_i$ and beyond $t_{i+1}$, width is $w_{i+1}$. Corresponding to the example shown in Fig. 3 the list would be $\{<0,w_2>, <t_1,w_4>, <t_2,w_1>, <t_3,w_3>, <t_4,0>\}$. Note that $[t_i , t_{i+1})$ is in close interval at left side.

3. Apart from all the above we also use a list to store the locations and orientation of the placed rectangles.

The placement procedure is represented in Algorithm1.

**Algorithm 1:**

---

**Input:** Ordered list of cores $L_{CORE}$ in descending order according to their average area obtained from GA.

**Output:** Final Schedule after placing all the cores in the list $L_{CORE}$ satisfying constraint.

---

Step 1. Pick first core $C_0$ from $L_{CORE}$ and place at $0^{th}$ time instant with 'up' orientation.
  Time_instants = $\{<0,C_{0w}>, <C_{0h},0>\}$ // $C_{0w}$ and $C_{0h}$ are the width and height of the core $C_0$ respectively.

Step 2. While ($L_{CORE}$ is NOT empty)

**Begin**

- Pick core $C_i$ from $L_{CORE}$ in the order of the list.
- Find possible time instants $t_k$'s satisfying TAM width constraints for both the orientations and calculate cost for each of these possible time instants according to equation (2)
- Place the core $C_i$ at time instant $t_k$ giving minimum cost (in case of ties, the placement with the lowest $t_k$ - value and at such level a 'down' orientation is preferred).
- Update time_instants list.

**End**

---

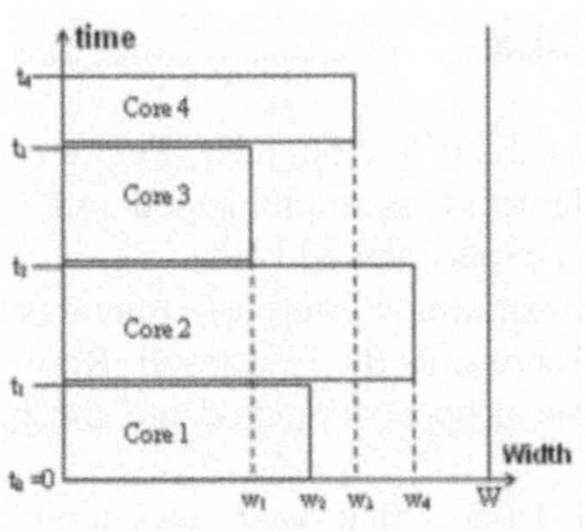


**Fig. 3.** A possible profile of the time-width plot

# 5 GA Formulation

In this paper GA formulation is used to get sequence of rectangles generated by the wrapper design method. For each core in the SOC we have to select one rectangle from the set of rectangles (wrapper configurations) generated for that core so that total testing time will be less. The order in which we select the rectangle for each core depends on the average area of the rectangles obtained from wrapper design algorithm used. For this purpose, cores are sorted in terms of decreasing average rectangle area. Throughout the running of the algorithm, this order is maintained. Next we define the individual representation of chromosomes, mutation and crossover operators, and fitness function and selection procedure.

## 5.1 GA Formulation to Get a Sequence of Rectangles

For this problem, chromosome structure is encoded with floating point values $<f_1,f_2,\ldots,f_N>$ of length $N$ depending on the number of cores present in an SOC, where $f_i$'s are floating point value within range of 0 to 1. Each $f_i$ is used to select one rectangle from the set of rectangles obtained for the i-th core using wrapper design process. The value $f_i$ is multiplied by the number of rectangles for core i to identify the rectangle selected for it. For each chromosome, fitness value is calculated using the best-fit heuristic based placement algorithm discussed in Section 4. This fitness value is calculated after getting sequence of representative rectangles from each core as discussed.

To move towards the promising region of search space different evolutionary operators are used. To create populations for new generation 20% best fit chromosomes are directly copied and remaining 80% chromosomes are created using crossover and mutation operators. In our GA approach we used parameterized mutation and crossover operation [19]. The algorithm is run for 150-200 generations with 3000-5000 population sizes depending on the number of cores present in the SOC. Chromosome with lowest fitness value provides the solution for the concerned problem.

## 6 Experimental Results

All the tests were conducted on a 2.66 GHz Pentium IV machine with 512 MB memory. The proposed algorithm is implemented in C++ and we used the ITC'02 SOC benchmarks [18] to be experimented with.

Table 1 shows the comparison of our test time results with recently proposed methods. Number in bold represents the best result. Row "Our" indicates our method. The results reported for the proposed method are the best among five runs of the procedure.

It can be noted from the Table 1 that our results significantly improve upon earlier approaches. For SOCs with larger number of cores our proposed method provides better results for most of the cases of different TAM widths.

**Table 1.** Comparison of test scheduling times (in clock cycles) for different ITC'02 SOC benchmarks

| SOC | Algo | Number of TAM wires | | | | | | |
|---|---|---|---|---|---|---|---|---|
| | | 16 | 24 | 32 | 40 | 48 | 56 | 64 |
| D695 | Our | 41604 | 27767 | 20957 | 16913 | 14273 | 12084 | 10723 |
| | [16] | 41737 | 28080 | 21098 | 17075 | 14310 | 12110 | 10783 |
| 10 | [17] | 40691 | 28060 | 20977 | 16894 | 14129 | 11453 | 10573 |
| cores | [14] | **39489** | **26203** | **19773** | **16149** | **13649** | **11285** | **9885** |
| P22810 | Our | 428852 | **286352** | **216570** | **175946** | **147898** | 127071 | 112498 |
| | [16] | **424889** | 289190 | 218035 | 177214 | **147898** | 130479 | 115791 |
| 28 | [17] | 438619 | 288565 | 216747 | 177633 | 148832 | **123857** | **103321** |
| cores | [14] | 438619 | 287999 | 216747 | 178223 | 149592 | 129624 | 115406 |
| P34392 | Our | 939855 | **626122** | **544579** | **544579** | **544579** | **544579** | **544579** |
| 19 | [16] | **931588** | 631035 | **544579** | **544579** | **544579** | **544579** | **544579** |
| cores | [14] | 935649 | 635237 | **544579** | **544579** | **544579** | **544579** | **544579** |
| P93791 | Our | 1750830 | **1170620** | **877073** | **704272** | **587117** | **506176** | **442455** |
| 32 | [16] | **1747504** | 1175988 | 891103 | 716112 | 598286 | 517692 | 452951 |
| cores | [14] | 1782067 | 1190565 | 890092 | 707664 | 609580 | 517017 | 452245 |

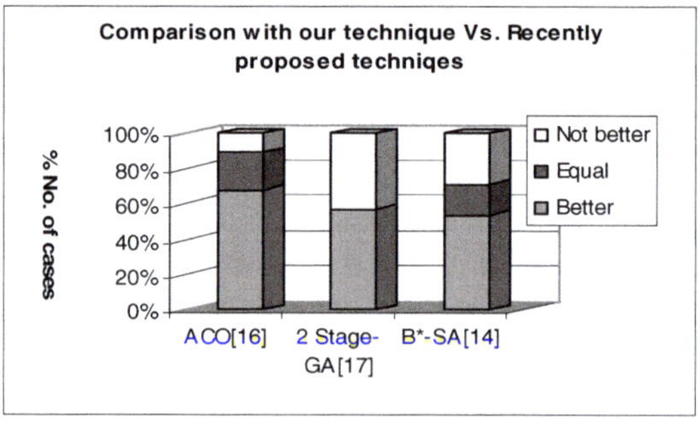


**Fig. 4.** Comparison with our technique Vs. recently proposed technique for different number of cases

Fig. 4 shows for how many cases (taking each TAM width value as one case for each core) our results are better, equal or not better than the recently proposed schemes. It is seen from the Fig. 4 that our proposed approach is better than ACO[16] for 67% of the cases, more than 50% of the cases compared to 2-Stage GA[17], more than 53% of the cases compared to B*-SA[14].

## 7 Conclusion

In this paper we proposed SOC test scheduling algorithm based on 2-dimensional rectangle bin packing approach considering best-fit heuristic. A genetic algorithm scheme is also proposed for selection of representative chromosome for each core present in an SOC. The test scheduling results are obtained for ITC'02 benchmark circuits and it is seen that our proposed method provides better results than the recently proposed techniques.

**Acknowledgments.** This work is supported in part by DST, India, under grant SR/S3/EECE/19/2003-SERC-ENGG, dated 05-05-2004.

## References

1. Marinissen, E. J., Goel, S. K., Lousberg, M.: Wrapper Design for Embedded Core Test, in Proc. ITC (2000) 911-920
2. Chakrabarty, K.: Test Scheduling for Core-Based Systems Using Mixed-Integer Linear Programming, in IEEE TCAD,(2000) 1163-1174
3. Iyengar, V., Chakrabarty, K., Marinissen, E. J.: Test Wrapper and Test Access Mechanism Co-Optimization for System-On-Chip, In Journal of Electronic Testing: Theory and Applications 18 (2002) 18
4. Huang Y. et al, Resource Allocation and Test Scheduling for Concurrent Test of Core-Based SOC Design, In proc. ATS (2001) 265-270
5. Huang, Y., Reddy, S. M., Cheng, W.-T. ,Reuter, P.: Optimal Core Wrapper Width Selection and SOC Test Scheduling Based On 3D-bin Packing Algorithm, In Proc. Intl. Test Conference(ITC) Baltimore (2002) 74-82
6. Ravikumar C. P., Verma A., Chandra G.: A Polynomial Time Algorithm for Power Constrained Testing of Core-Based Systems, In proc. ATS (1999) 107-112
7. Xia Y., Chrzanowska-Jeske M., Wang B., Jeske M.: Using a Distributed Bin-Packing Approach for Core-based SOC Test Scheduling with Power Constraints, In Proc. ICCAD (2003 ) 100-105
8. Su, C.-P.,Wu, C.-W.: A Graph-Based Approach to Power-Constrained SOC Test Scheduling, In Journal of Electronic Testing: Theory and Applications (2004) 45-60
9. Pouget, J. , Larsson, E., Peng, Z.: Multiple-Constraint Driven System-On-Chip Test Time Optimization, Journal of Electronic Testing:Theory and Applications, 21 (2005) 599-611
10. 10. Zhao, D., Upadhyay, S.: A Generic Resource Distribution and Test Scheduling Scheme for Embedded Core-Based SOCs, IEEE Trans.. Instrumentation and Measurement, 53 (2004) 318-329
11. Iyengar, V., Chakrabarty, K.: Precedence-Based, Preemptive, and Power Constrained Test Scheduling for System-On-Chip, In Proc. VLSI Test Symposium, (2001) 368-374

12. Koranne, S., Iyengar, V.: On the Use of k-Tuples for SOC Test Schedule Representation, In Proc. ITC, (2002) 539-548
13. Zou, D., Reddy, S. M., Pomeranz, I. ,Huang, Y.: SOC Test Schdeuling Using Simulated Annealing, In Proc. VLSI Test Symposium, (2003) 325-330
14. Wuu, J. –Yi. , Chen, T.-C. ,Chang, Y.-W.: SOC Test Scheduling Using B*-Tree Based Floor Planning Technique, In Proc. ASP- DAC, (2005) 1188-1191
15. Larsson E., Fujiwara H.: System-On-Chip Test Scheduling with Reconfigurable Core Wrappers, In IEEE Trans. VLSI Systems, 14 (2006) 305-309
16. Ahn, J.H., Kang, S.: SoC Test Scheduling Algorithm Using ACO-Based Rectangle Packing, In Proc. ICIC, (2006) 655-660
17. Yu Y, Peng X Y, Peng Y.: A Test Scheduling Algorithm Based on Two-Stage GA, In Proc. International Symposium on Instrumentation Science and Technology, (2006) 658-662
18. Marinissen, E. J., Iyengar, V.,Chakrabarty, K.: A Set of Benchmarks for Modular Testing of SOCs, In Proc. Int. Test Conf.,(2002) 519–528
19. Melanie Mitchell: An Introduction to Genetic Algorithm, Prentice Hall India

# The Design of Finite State Machine for Asynchronous Replication Protocol

Yanlong Wang[1], Zhanhuai Li[1], Wei Lin[1], Minglei Hei[1], and Jianhua Hao[2]

[1] School of Computer Science, Northwestern Polytechnical University, No.127 West Youyi Road, Xi'an, Shaanxi, China 710072
[2] Department of Finance Shandong University of Technology, Zibo, China 255049
{wangyl,linwei}@mail.nwpu.edu.cn, lizhh@nwpu.edu.cn, mythay@126.com

**Abstract.** Data replication is a key way to design a disaster tolerance system and to achieve reliability and availability. It is difficult for a replication protocol to deal with the diverse and complex environment. This means that data is less well replicated than it ought to be. To reduce data loss and to optimize replication protocols, we (1) present a finite state machine, (2) run it to manage an asynchronous replication protocol and (3) report a simple evaluation of the asynchronous replication protocol based on our state machine. It's proved that our state machine is applicable to guarantee the asynchronous replication protocol running in the proper state to the largest extent in the event of various possible events. It also can helpful to build up replication-based disaster tolerance systems to ensure the business continuity.

## 1 Introduction

With the widespread use of computers, data is becoming more and more important in human life. But all kinds of accidents and disasters occur frequently. For example, hardware breaks, software has defects, viruses propagate, buildings catch fire, power fails and people make mistakes [1]. Data corruption and data loss by those disasters have become more dominant, accounting for up to 80% [2] of data loss. Many data disaster tolerance technologies have been employed to increase the availability of data and to reduce the damage caused by data loss, corruption and disaster [2] [4].

Replication [5] is a key technology for disaster tolerance and is different from the static and periodical data backup. It replicates business data on a primary system to some remote backup systems in primary-backup architecture dynamically and on-line. It can not only retain the replicas at remote sites, but also make one of the backup systems take over the primary system in the event of a disaster. Replication protocol is the focus of replication research. Traditional replication protocols can be broadly classified into the two categories of synchronous and asynchronous [6]. Both of the two replication protocols have been used as a part of disaster tolerance solutions, such as IBM's Peer-to-Peer Remote Copy (PPRC) [7], EMC's Symmetrix Remote Data Facility (SRDF) [8], Hitachi's Remote Copy [9] and VERITAS's Volume Replicator (VVR) [10]. In addition, some optimized replication protocols are presented, such as Seneca [6], SnapMirror [11] and semi-synchronous [12].

D.-S. Huang, L. Heutte, and M. Loog (Eds.): ICIC 2007, LNAI 4682, pp. 1042–1053, 2007.

From these replication protocols and related commercial products, we observe the two phenomena as follows:

- Synchronous replication protocol replicates every data block written to the primary system to the backup system before the local write operation returns, as well as asynchronous replication protocol allows a divergence between the primary and backup systems. Therefore, asynchronous replication protocol is used more widely than synchronous replication protocol.
- Both synchronous and asynchronous replication protocols primarily run under normal conditions. It is rarely discussed about how to deal with various complicated internal or external events. That means that the replication process may lack the ability in response to abnormal events, which maybe results in data inconsistency.

To handle various diverse and complex events [13], we present a finite state machine (FSM) and use it to run an asynchronous replication protocol. FSM can not only deal with all of the normal, failure and recovery events to achieve data consistency, but also guarantee different event-handling strategies to achieve data consistency.

The remainder of the paper is outlined as follows. In Section 2, we put our work into perspective by considering related work. In Section 3, we provide a typical asynchronous replication protocol and then present a finite state machine for it. In Section 4, we describe a taxonomy of the events that all of current asynchronous replication systems have to face, and discuss the actions of the asynchronous replication protocol triggered by all kinds of normal and abnormal events. In Section 5, we implement the FSM to schedule the asynchronous replication protocol and report a simple evaluation of the FSM based on our logical volume replicator (LVR). Finally, in Section 6 we conclude the paper by highlighting the advantages of this work and discussing future research directions.

## 2 Related Works

While computer systems and networks are becoming more and more complicated, they are becoming less reliable. This motivates the study of all kinds of state machines in different areas and at various times because they have practical importance and theoretical interest [14]. In recent years, state machine replication comes to be employed in replication services to guarantee a correct coordination among the distributed replicas. Several past works address agreement and replication techniques that tolerate benign faults and Byzantine faults [15] under the asynchronous model based on state machine, such as BFT, Q/U and HQ [15][17].

Simultaneously, few projects have been done to resolve the replication protocols in primary-backup architecture based on state machine. Seneca [6] is the only famous replication protocol to be defined in terms of a pair of state machines, one for each of the primary and secondary Seneca modes. It defines a series of the possible failure and recovery events for both the primary and secondary modes, and deals with them based on state machine in face of various conditions. The state transitions of a Seneca instance in response to failure and recovery events proves that Seneca performs replication well. However, the events given by Seneca are not enough to describe the complex environment for designing an asynchronous replication protocol.

With our survey of the related approaches used in practice, we give a description of the typical asynchronous replication protocol and then define the state machine for it. The state machine is just an event-driven replication scheduler to deal with the asynchronous replication protocol.

# 3 FSM

## 3.1 Asynchronous Replication Protocol

A typical replication system is composed of Client, Primary and Backup, and its architecture is shown in *Fig. 1*-(a). It usually adopts two-step submission process to implement the asynchronous replication protocol as shown in *Fig. 1*-(b):

1) When Primary receives a request committed by Client, it writes a copy of the data of the request into the log (*Write-log*);
2) Then Primary writes the data of the request to the storage (*Write-storage*), and simultaneously sends the data to Backup (*Replicate*).

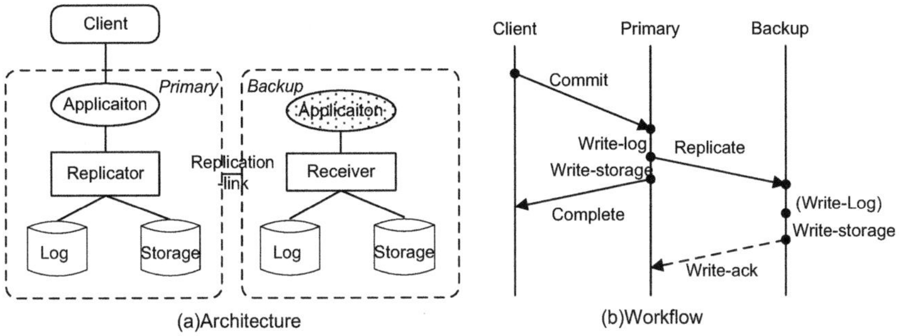


**Fig. 1.** An overview of an typical asynchronous replication protocol

After receiving the data, Backup does not need to return an acknowledgement to Primary. Moreover, Primary returns a complement signal to Client and then handle the next request directly and successively. Both of these can increase the efficiency of replication and the process is just like pipelining. But it is necessary for Primary to wait for the acknowledgement of each request completing writing from Backup (referred to as *Write-ack*) and to drain the log as quickly as possible. This can guarantee the consistency between Primary and Backup to a certain extent.

Obviously, asynchronous replication protocol achieves the replication efficiency by partly reducing the consistency between Primary and Backup: Under normal circumstances, asynchronous replication protocol is more efficient than synchronous replication protocol; but under abnormal circumstances, it is more possible for asynchronous replication protocol to suffer inconsistency than synchronous replication protocol.

### 3.2 Deterministic Finite State Machine

A typical finite state machine consists of a set of states, a start state, an input alphabet, and a transition function that maps input symbols and current states to a next state. Deterministic finite state machine (or Deterministic finite automation) is a special kind of finite state machine, and there is exactly one transition out of each state for each input symbol. A deterministic finite state machine (DFSM) $M$ is a quintuple $M = (S, \Sigma, \delta, s_0, F)$, where:

1) $S$ is a finite set of states;
2) $\Sigma$ is the finite input alphabet;
3) $s_0 \in S$ in $S$ is the initial state;
4) $F \subseteq S$ is the set of final states;
5) $\delta : S \times \Sigma \rightarrow S$ is the transition function

DFSM begins in the start state $s_0$ with an input string. When the machine is in a current state $s$ in $S$ and receives an input $i$ from $\Sigma$, it moves to the next state specified by $\delta(s,i)$.

For a replication system, its asynchronous replication protocol must change to a fixed state when being triggered by an event. Therefore, it is ideal for us to adopt a DFSM to schedule asynchronous replication protocol. The protocol will adopt an action change depending on an event. Hence, $s_0$ is the initial state where the system starts working and does not do any operation. The set of final states $F = \{s_0, s_n\}$ includes two states: $s_n$ is a normal replication state where the system keeps replicating continuously; $s_0$ is the initial state where the system is stopped by administrators after replicating for a while. In *Fig. 1*-(1), there are three main parts (i.e. storage, log and replication-link) and their states determine the states of the system jointly. If $S_s$, $S_l$ and $S_r$ are sets of storage states, log states and replication-link states respectively, the set of the system states is the Cartesian Product $S_{max} = S_s \times S_l \times S_r$. $\Sigma$ is the set of all the events which can change the states of the system. $\delta$ is the set of all the actions which can keep the states of the system transiting correctly. Both $\Sigma$ and $\delta$ are the most important in this machine, so we discuss them in the next section.

## 4 Events and Actions

### 4.1 A Taxonomy of Events

Administrators start or stop the system. Networks, log disk or data disk fail. Large-area blackouts or terrorism attacks occur. These events may affect the running of the replication system and block replicating. In order to tolerate all kinds of disasters, we must give a specific taxonomy of events. The events of the system can be divided into four categories as follows:

1) The category of the initialization events is referred to as $\sum_i$, which includes all the events to synchronize Backup with Primary and to initialize the system:

   A) Automatic initialization
   B) Static initialization
   C) Zero initialization

2) The category of the normal events is referred to as $\sum_n$, which describes all operation events of administrators:

   D) Replication starting
   E) Replication stopping
   F) Replication pausing
   G) Replication resuming

3) The category of the recoverable failure events is referred to as $\sum_r$, which may lead to inconsistency between Primary and Backup, but the system can be recovered back to the right state:

   H) Primary log overflow
   I) Primary log recovery
   J) BCB[1] starting
   K) BCB completing
   L) Atomic operation[2] starting
   M) Atomic operation completing
   N) Networks failure
   O) Networks recovery
   P) System error (e.g. OS shutdown/System poweroff/Memory error/…)
   Q) System reboot

4) The category of the unrecoverable failure events is referred as $\sum_u$, which may not only result in some data loss, inaccessibility or inconsistency, but also make the system fall into the wrong state:

   R) Primary log failure
   S) Primary storage failure
   T) Backup log failure
   U) Backup storage failure

All the events of each category will produce the same result, i.e. they will transit the system to the same state. We call all the events in the same category as equivalent

---

[1] BCB (Block change bitmap) is a bitmap table associated with the storage and used to track writes when the log overflows. It is the key method to protect the overflowing log. It is also the key method to initialize Backup automatically. Each bit in the BCB represents a region of blocks whose contents are different between Primary and Backup. We can incrementally synchronize Backup with Primary by looking up the bitmap.

[2] Atomic operation is used to transmit a set of blocks in the log to Backup as a single update. Backup only needs to return an acknowledgement to Primary after the whole set.

ones. All of the four categories make up the set of all events that the system encounters, i.e. $\Sigma = \Sigma_i \cup \Sigma_n \cup \Sigma_r \cup \Sigma_u = \{A, B, \cdots, T, U\}$.

### 4.2 Actions

For a replication system, actions corresponding to events are the key part to schedule the replication protocol correctly. According to the taxonomy of events, we design four types of the actions (i.e. transition functions): initialization actions, normal actions, actions for recoverable failure events and actions for unrecoverable failure events. In normal situations, the system runs from $s_0$ to $s_n$ by adopting initialization and normal actions; in abnormal situations, the system has to deal with failure events by adopting actions for recoverable or unrecoverable failure events.

**Initialization actions.** In Section 4.1, we define the category of the initialization events: automatic initialization, static initialization and zero initialization. By the former two initialization events, we can synchronize Backup with Primary and then achieve the consistency between them. Their processes are described as follows:

1) *Automatic initialization*: administrators can synchronize Backup over the network either when the application is active or inactive. That is, we can read the data from the storage on Primary and transfer them to Backup by adopting automatic replication while recordng the log of the applications. After automatic initialization, the replication is running base on the log on Primary.
2) *Static initialization*: administrators can use the network or the third-party medium to synchronize Backup with Primary when the application is inactive. That is, we can transfer the data to Backup by adopting on-line full copy over the network or off-line manual copy with tapes or movable disks. Because using third-party medium for initialization is easy, we just consider the static initialization over the network (i.e. on-line full copy). After completing static initialization, the applications are activated and the replication starts.

It is important for automatic initialization to keep network bandwidth abundant, but it can be implemented when applications are active. It is important for static initialization to keep the applications inactive, but its implementation is easy. Although both automatic and static initializations have their own merits, their actions (the state transitions) are the same as shown in *Fig.* 2-(1). In *Fig.* 2-(1), Zero initialization is a special event to initialize the system where Backup has the same data as Primary and synchronization is a null and skipped operation.

**Normal actions.** As defined in Section 4.1, starting, stopping, pausing and resuming are the normal events for the replication system. By the four events, we can operate the system and schedule the protocol to replicate the data from Primary to Backup after completing initialization. During normal operations, Backup is always keeping consistent with Primary and all the data are recoverable. We design a set of normal actions for those events and show the state transitions in *Fig. 2-(2).*

**Actions for recoverable failure events.** Abnormal events can be divided into two kinds: recoverable and unrecoverable. Of the two kinds, the recoverable failure events occur when there are something wrong with the log, BCB, networks, OS or the system. It is more significant to deal with this kind of events, because they maybe result in inconsistency between Backup and Primary.

It is very difficult for us to solve or repair inconsistency during replicating. Therefore, we firstly stop the replication and then adopt some actions to solve those events and their inconsistency. At last, we go on with the normal replication. For example, if the log overflows, we can use BCB to record the block changes of the storage and replicate the data of the changed blocks to Backup. If networks, OS or the system fails, the replication usually is paused, and continues after failures are removed. We design a set of normal actions for those events and show the state transitions in *Fig. 2-(3).*

**Actions for unrecoverable failure events.** The unrecoverable failure events may result in the data loss, so Backup is inconsistent with Primary, even the data on Primary are stale. We usually replace the disk or volume of the stale log or storage with a new disk or volume, and then initialize Backup according to Primary. For example, when the storage on Primary fails, we supply Primary with a new storage and run initialization again. Because the log on Backup is used during atomic replication, its failure can not result in inconsistency. We only need substitute it with a new log and continue replicating. In a word, actions for unrecoverable failure events are simple and we do not discuss them in more detail.

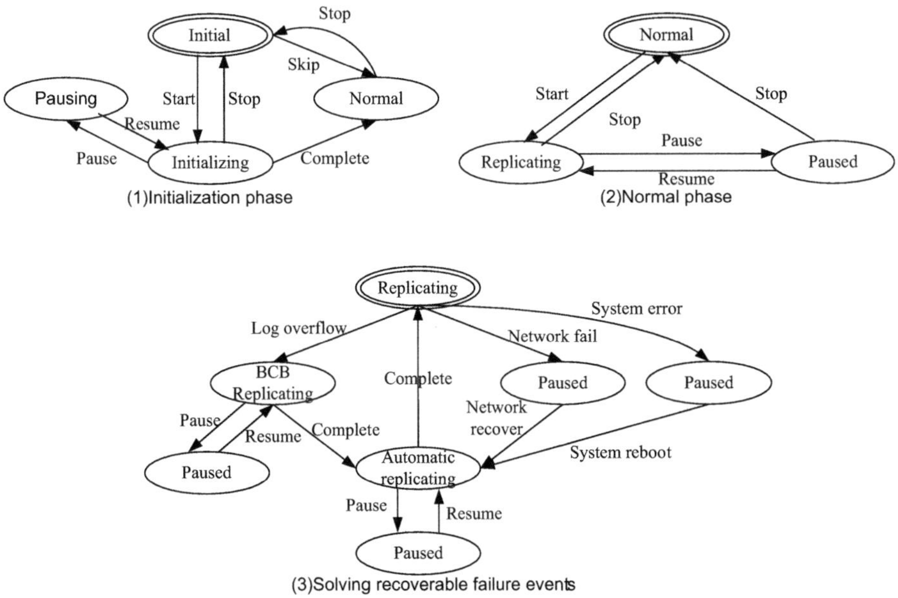


**Fig. 2.** State transitions of three main phases

In this section, we give the general profile of every type of actions, and give the state transitions of three main phases. It will guide us to implement the FSM into practice in Section 5.

# 5 Implementation and Evaluation

## 5.1 Implementation

Logical volume replicator (LVR) [18] is a remote and online replication system prototype. We design it based on logical volume manager (LVM) [19] on Linux. LVM provides a higher-level view of the disk storage on a computer system than the traditional view of disks and partitions. This gives the system administrator much more flexibility in allocating storage to applications and users.

So we can create one replicated volume group on Primary under the control of LVM, which includes a volume as the log and a set of volumes as the storage. We also can create the other replicated volume group on Backup. Both the replicated volume groups are a pair of peers to build up the replication link and implement asynchronous replication protocol based on DFSM.

DFSM for asynchronous replication protocol of LVR is also a quintuple $M = (S, \Sigma, \delta, s_0, F)$, and its state transitions is shown in *Fig. 3*:

1) $S_{max} = S_s \times S_l \times S_r$, where $S_s = \{stale, consistent, inconsistent\}$, $S_l = \{normal, BCB, atomic\}$, *and* $S_r = \{running, stopped, paused\_with\_trying, paused\_without\_trying, statpaused\_with\_trying, statpaused\_without\_trying, autopaused\_with\_trying, autopaused\_with\_trying, statinit, autoinit\}$. $S_s$ means the state of the storage volumes can be stale, consistent and inconsistent. $S_l$ means the state of the log volume can be normal-replicating, BCB-replicating or atomic-replicating. $S_r$ means the state of replication link can be running, stopped, paused and initialization, where the suffixes (i.e. *_with_trying* and *_without_trying*) mean whether the link will try to connect or not, and the prefixes (i.e. *stat-* and *auto-*) mean the system adopts static initialization or automatic initialization. Because some state combinations in $S_{max}$ do not exist in the real world, we can get the finite set of the states $S$ in LVR as shown in *Table 1*. $S$ includes 22 real states (i.e. $S = \{s_0, s_1, \cdots, s_{21}\}$) and then $S \subset S_{max}$.
2) Because it is easy to deal with the unrecoverable failure events $\Sigma_u$ by resynchronization, we define $\Sigma = \{A, B, \cdots S\}$ here. $\delta$ will be detailed and reified.
3) $s_0$ is the same as mentioned in Section 3.2, and $F = \{s_0, s_{10}\}$.

In *Fig. 3*, the graph (a) is the state transitions of the initialization phase and the graph (b) is the state transitions of the running phase. During both phases, DFSM solves all kinds of initialization, normal operations and recoverable failure events. Of all the events, $E$ is a special one and is not shown in *Fig. 3*. We define that $s_i \xrightarrow{E} s_0, (i = 1, 2, \cdots, 21)$.

**Table 1.** The set of states in LVR

| State | $S_s$ | $S_l$ | $S_r$ |
|---|---|---|---|
| $s_0$ | stale | normal | stopped |
| $s_1$ | inconsistent | BCB | autoinit |
| $s_2$ | inconsistent | BCB | autopaused_with_trying |
| $s_3$ | inconsistent | BCB | autopaused_without_trying |
| $s_4$ | inconsistent | atomic | autoinit |
| $s_5$ | inconsistent | atomic | autopaused_with_trying |
| $s_6$ | inconsistent | atomic | autopaused_without_trying |
| $s_7$ | inconsistent | normal | statinit |
| $s_8$ | inconsistent | normal | statpaused_with_trying |
| $s_9$ | inconsistent | normal | statpaused_without_trying |
| $s_{10}$ | consistent | normal | running |
| $s_{11}$ | consistent | normal | paused_without_trying |
| $s_{12}$ | consistent | BCB | paused_without_trying |
| $s_{13}$ | inconsistent | BCB | running |
| $s_{14}$ | inconsistent | BCB | paused_with_trying |
| $s_{15}$ | inconsistent | BCB | paused_without_trying |
| $s_{16}$ | inconsistent | atomic | running |
| $s_{17}$ | inconsistent | atomic | paused_with_trying |
| $s_{18}$ | inconsistent | atomic | paused_without_trying |
| $s_{19}$ | consistent | normal | paused_with_trying |
| $s_{20}$ | consistent | BCB | paused_with_trying |
| $s_{21}$ | consistent | atomic | running |

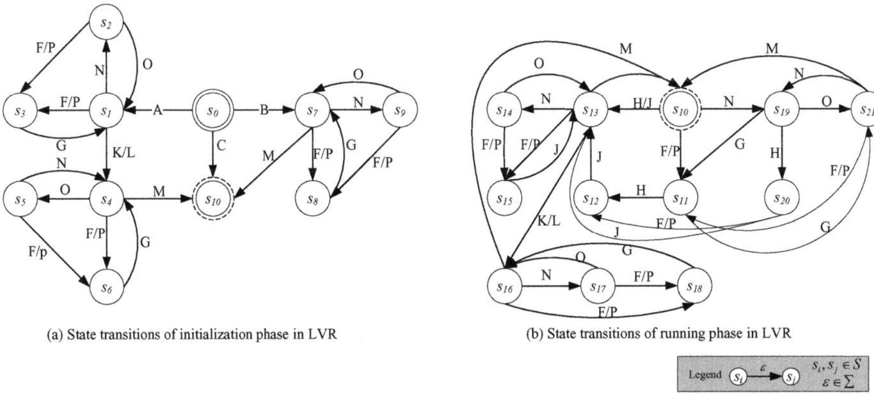


**Fig. 3.** State transitions of DFSM for asynchronous replication protocol of LVR

## 5.2 Evaluation

In order to test the feasibility of our FSM, we give a simple evaluation for asynchronous replication protocol in LVR here. The evaluation includes two aspects:

1) *Reliability and availability*: We implement asynchronous replication protocol in LVR. We use MTBF (Mean Time Between Failures) and MTTR (Mean Time To Recover) to measure the reliability and the maintainability of LVR. Because asynchronous replication protocol in LVR can deal with all kinds of events, the failure rate $\lambda \to 0$ and the maintenance rate $\mu \to 1$. So we can get that $MTBF = \frac{1}{\lambda} \to \infty$ and then $Availability = \frac{MTBF}{MTBF + MTTR} \times 100\%$ $= \frac{\mu}{\lambda + \mu} \times 100\% \to 100\%$. Then we do an experiment by running LVR for a month and simulating all kinds of events as mentioned in Section 4.1. It's proved that the availability is up to $99.999\%$.
2) *Efficiency:* We do the other experiment to evaluate the impact of our FSM on the performance of replication by using IOzone-3.263 benchmark [20]. IOzone is a benchmark tool to test file I/O performance by some standard operations (e.g. read, write, re-read, re-write, read backwards, read strided, fread, fwrite and random read/write). When running asynchronous replication protocol in LVR on Linux, we choose the most complex operation, *re-fwrite,* to reflect and analyze file system performance. The configuration of hardware and software is shown in *Table 2*, and the results are shown in *Fig. 4*. From *Fig. 4*, we can find that the file system still gains a good I/O performance. So our DFSM is very efficient and has few effections.

**Table 2.** The configuration of experimental environment

| Components | Configures Information |
|---|---|
| **Hardware** | Primary: Intel Celeron CPU 1.70GHz /Memory 512MB<br>Backup: Intel Celeron CPU 1.70GHz /Memory 512MB |
| **Network** | 1000BaseT Ethernet |
| **Software** | OS : Red Hat Linux release 9 with Kernel 2.4.20-8<br>FS: EXT3 FS<br>Logical Volume Replicator: LVR 1.0 |
| **Replication protocol** | Asynchronous |
| **Primary Device** | Hard Disk : Quantum Fireball 20G<br>Storage Volume: 2G<br>Log Volume: 1G |
| **Backup Device** | Hard Disk : Quantum Fireball 20G<br>Storage Volume: 2G<br>Log Volume: 1G |
| **Benchmark tools** | Iozone3.263 |
| **Operation** | Re-fwrite |

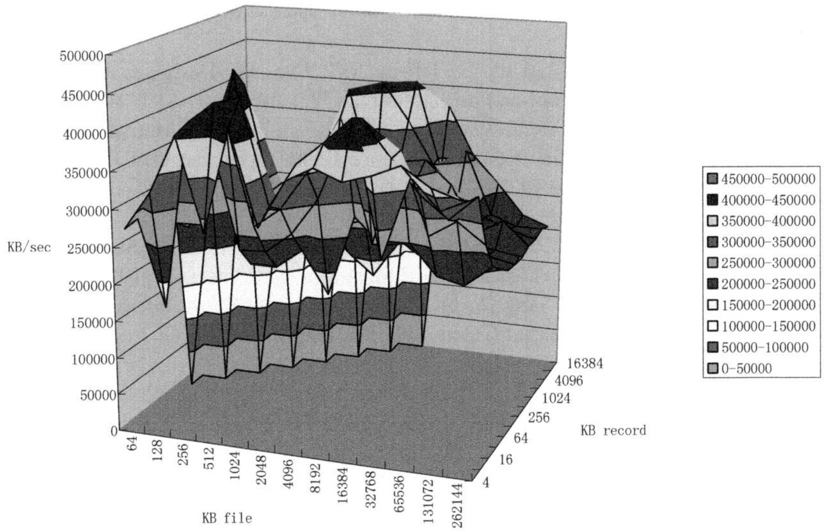


**Fig. 4.** *Re-fwrite* performance with asynchronous replication protocol of LVR

# 6 Conclusion and Future Work

With data becoming more and more important, all kinds of information systems require 24x7 availability. Replication is a rising technique for disastertolerance, and it adopts synchronous or asynchronous replication protocols to make true business continuity a realistic objective. How to deal with various events during replicating is important to design a replication protocol, especially an asynchronous replication protocol.

Finite state machine is presented to schedule and manage an asynchronous replication protocol in this paper. The five items of our FSM are introduced in detail, especially current events and triggered actions are classified. Then we implement the FSM in logical volume replicator and evaluate asynchronous replication protocol. It's proved that the FSM makes state transitions of the protocol easier and guarantees the reliability, availability and efficiency of the protocol. Therefore, the protocol can be widely deployed in replication-based disaster-tolerance systems.

FSM complies with the needs of the asynchronous replication protocol to solving all kinds of normal or abnormal events, but there still exits some future work. For example, we need to further optimizing the actions of state transitions to improve the efficiency of replication. With the optimization of FSM, asynchronous replication protocol will be more and more reliable and efficient.

**Acknowledgments.** This work is supported by the National Natural Science Foundation of China (60573096).

## References


1. Kimberly, K., Cipriano, A.S., Dirk, B., Jeffrey, S.C., John, W.: Designing for Disasters. In: Proc of the 3rd USENIX Conf on File and Storage Technologies (FAST'04) (2004) 59–72
2. David, P., Aaron, B., et al.: Recovery Oriented Computing (ROC): Motivation, Definition, Techniques, and Case Studies. Computer Science Technical Report, U.C. Berkeley (2002)
3. Gray, J., Helland, P., Patrick O., Shasha, D.: The dangers of Replication and a Solution. In: Proceedings of the SIGMOD Conf (1996)
4. Ann, C., Vivekenand, V., Zachary, K.: Protecting File Systems: A Survey of Backup Techniques. In: Proc Joint NASA and IEEE Mass Storage Conference (1998)
5. Yang, Z.H., Gong, Y.Z., Sang, W.Q., Liu, H.Y., Li, Q.Y.: A Primary-Backup Lazy Replication System for Disaster Tolerance (in Chinese). Journal Of Computer Research And Development (2003)1104–1109
6. Ji, M.W., Veitch, A., Wilkes, J.: Seneca: Remote Mirroring Done Write. In: Proc of the 2003 USENIX Technical Conference (USENIX'03). San Antonio, TX, USA (2003)
7. Azagury, A.C., Factor, M.E., Micka, W.F.: Advanced Functions for Storage Subsystems: Supporting Continuous Availability, IBM SYSTEM Journal VOL 42 (2003)
8. Using EMC SnapView and MirrorView for Remote Backup, Engineering White Paper, EMC Corporation (2002)
9. Software Solutions Guide for Enterprise Storage, Hitachi Data Systems Corporation (2000)
10. VERITAS Volume Replicator 3.5: Administrator's Guide (Solaris). Veritas Software Corp (2002)
11. Patterson, H., Manley, S., Federwisch, M., Hitz, D., Kleiman, S., Owara, S., Mirror, S.: File-System-Based Asynchronous Mirroring for Disaster Recovery. In: Proc of the First USENIX conference on File and Storage Technologies. Monterey, CA, USA (2002)
12. Yan, R., Shu, J.W., Wen, D.C.: An Implementation of Semi-Synchronous Remote Mirroring for SANs. In: Proc of the 3rd International Conference on Grid and Cooperative Computing (GCC'2004) Workshop on Storage Grid and Technologies, LNCS 3252, Springer, Berlin (2004) 229–237
13. Zhou, J., Wang, Y.J., Ruan, W., Li, S.K.: Research on Massive Data Oriented Data Consistency (in Chinese). Computer Science. (2006) 137–140
14. Lee, D., Yannakakis, M.: Principles and Methods of Testing Finite State Machines - A survey. In: Proc. of the IEEE. (1996) 1090–1123
15. Lamport, L., Shostak, R., Pease, M.: The Byzantine Generals Problem. ACM Transactions on Programming Languages and Systems. (1982) 382–401
16. Fred B.S.: Implementing Fault-tolerant Services Using the Sstate Machine Approach: a tutorial. ACM Computing Surveys. (1990) 299–319
17. Cowling, J., Myers, D., Liskov, B., Rodrigues, R., Shrira, L.: HQ Replication: A Hybrid Quorum Protocol for Byzantine Fault Tolerance. In: Proc of the Seventh USENIX Symposium on Operating Systems Design and Implementation (OSDI 2006), Seattle, WA. (2006)
18. Dong, H.Q., Li, Z.H., Lin, W., Wang, M.Z.: Design of Remote Logical Volume Replication System on Linux (in Chinese). Computer Engineering And Applications. (2004) 109–112
19. Redhat. http://sources.redhat.com/lvm2/, 2005
20. IOzone. http://www.iozone.org/, 2006

# Unbalanced Underground Distribution Systems Fault Detection and Section Estimation

Karen Rezende Caino de Oliveira, Rodrigo Hartstein Salim, André Darós Filomena, Mariana Resener, and Arturo Suman Bretas

Electric Engineering Department, Federal University of Rio Grande do Sul, Av. Osvaldo Aranha, 103, Porto Alegre, Brazil, 90035-190
{karen,salim,afilomena,mariana,abretas}@ece.ufrgs.br

**Abstract.** This paper presents a novel fault detection and section estimation method for unbalanced underground distribution systems (UDS). The method proposed is based on artificial neural networks (ANNs) and wavelet transforms (WTs). The majority of UDS are characterized by having several single/double phase laterals and non-symmetrical lines. Also, Digital Fourier Transforms (DFT), used in the majority of traditional protection relays, supplies a low level of robustness to the fault diagnosis process due to its inversely proportional time-frequency characteristic. These characteristics compromise the traditional fault diagnosis methods performance. ANNs are capable of learning and generalizing, whereas WTs are robust tools capable of evaluating a signal's frequency range that can characterize the fault phenomenon. This paper describes the proposed diagnosis method and discusses the results obtained from simulated implementation. The obtained results demonstrate the capability and robustness of the technique indicating its potential for on-line applications.

**Keywords:** Fault Detection, Fault Section Estimation, Wavelet Transforms, Artificial Neural Networks, Underground Distribution Feeders.

## 1 Introduction

Electromagnetic transients in Electric Power Systems (EPS) are common, and constantly induced by short circuits, switching operations and lightning cases. These events can be permanent or non permanent to the EPS. In both cases, however, the occurrence of such events constantly implies the protection relays operation to isolate the correct faulted equipment. After the protection scheme operation, the faulted equipment can be restored. Permanent faults, however, need to be detected and located first, in order to send maintenance crews to fix the equipment. Restoration of faulted equipments can generate the reoccurrence of the fault and lead to power system blackouts.

The great majority of today's electric power company's maintenance crews have for several years located permanent faults through visual inspection. This procedure can last days in large distribution systems and be inadequate for underground systems, where inspection boxes are the only access to the conductors, which makes impossible the complete visual inspection of the line.

D.-S. Huang, L. Heutte, and M. Loog (Eds.): ICIC 2007, LNAI 4682, pp. 1054–1065, 2007.

In recent years, several fault diagnosis methods for transmission [1], [2] and distribution systems [3], [4], [12], [13] have been proposed in order to reduce these problems. However, these methods don't consider the specific characteristics of underground distribution lines, the unbalanced nature of their operation and the systems with laterals.

In this paper it is presented a novel hybrid fault detection and section estimation method capable of correctly detecting and locating faults on UDS. The proposed method is composed of two subroutines, a fault detection subroutine and a fault location subroutine. WTs are used in the fault detection subroutine and ANNs in the fault location subroutine. In order to test the proposed method's performance, the scheme is implemented in MATLAB platform [5] and applied on two real data underground distribution systems of the Electric Energy State Company of Rio Grande do Sul (CEEE), southern Brazil, simulated with the use of ATP-EMTP [6].

In the second section of this paper it is presented a review of underground distribution systems modeling. In the third and fourth sections, a WTs review and an ANNs review are presented, respectively. In the fifth section the proposed method is demonstrated. The results and conclusions are presented in the sixth and seventh sections.

## 2 Underground Distribution Systems

Distribution lines modeling constitute in the calculation of the series parameters that represent the phase conductors. These parameters are a result of the electric and magnetic fields interaction that surrounds each cable. Such series parameters depend on the line type used, aerial or underground, which must be treated separately for a correct modeling. In this section, Carson's equations [7] and the Kron's reduction method [7] are presented and used in such modeling.

### 2.1 Carson's Equations Applied for Underground Distribution Cables

Due to unbalanced operation and line asymmetry, the UDS line modeling should not make the standard approximations of aerial lines. Thus, this paper uses Carson's technique to obtain the self and mutual impedances of underground distribution line cables. The technique is based in the use of conductor's images. Additionally, Carson considers that the earth is an infinite surface, with uniform and constant resistance, by which the neutral conductors introduced effects on nominal frequency, can be disregarded. The Carson's equation for a certain conductor $i$ self-impedance in [$\Omega$/m] and the mutual impedance between the conductors $i$ and $j$ is shown in equations 1 and 2, respectively.
Where:

$GMR_i$ – Medium Geometric radius of conductor $i$ [ft].
$D_{ij}$ – Distance between the conductors $i$ and $j$ [ft].
$z_{ii}$ – Self-Impedance of conductor $i$ [$\Omega$/m].
$z_{ji}$ – Mutual Impedance of conductors $i$ and $j$ [$\Omega$/m].

$$z_{ii} = r_{ii} + j0.12134 \cdot \left( \ln \frac{1}{GMR_i} + 7.93402 \right) \tag{1}$$

$$z_{ij} = j0.12134 \cdot \left( \ln \frac{1}{D_{ij}} + 7.93402 \right) \tag{2}$$

### 2.2 Kron's Reduction Method

In the case that the system is composed of 3 phase conductors and 1 neutral, the impedance matrix will have a *4x4* dimension which would reduce considerably a great number of applications use. In order to reduce this matrix to a *3x3* dimension, that will enable the use of these existing applications, it is used in this paper the Kron's Reduction Method. The method is well known as it uses Kirchoff's voltage law to reduce the matrix order.

The CEEE underground distribution test systems use tape shielded cables. In such case, specific values used in the Carson's equations can be obtained and are also applied in this work.

## 3 Wavelet Transform

The Wavelet Transform (WT) is a digital signal processing (DSP) technique based on translation and dilation of a window, named the mother wavelet [8]. The technique allows high frequency event's location with a greater time resolution.

### 3.1 Stationary Discrete Wavelet Transform

In the Discrete Wavelet Transform (DWT) calculation process, signal decimation occurs. The analyzed signal passes through the filters (wavelets), decreasing the maximum frequency of the sampled signal and also the wavelet coefficients total number after the signal's transformation. In the case that this decimation doesn't occur, the DWT is called stationary, or Stationary Wavelet Transform (SWT). In the SWT the signal is not decimated.

The SWT's great advantage is its time invariance characteristic, which doesn't occur with the DWT. This means that even in a periodic signal, the DWT of a translated signal *x[n]* is not necessarily equal to the translated DWT of the signal *x[n]*. For singularity detection applications, the use of SWT is highly recommended, and, therefore, used in this paper.

## 4 Artificial Neural Networks

Artificial Neural Networks are the mathematical models of biological neurons and their interconnection, aiming the reproduction of the human brain's processing properties [10]. The characteristics of ANNs can be summarized by:

- *Learning*: ANNs have the ability of learning about their ambient through an iterative adjustments process;
- *Generalization*: ANNs are able to generalize and obtain an adequate output for unforeseen inputs;
- *Nonlinearity*: An ANN is nonlinear if it is composed of nonlinear artificial neurons;
- *Adaptability*: ANNs are capable of adapting their synaptic weights due to changes in the ambient;
- *Fault tolerance*: ANNs have their knowledge distributed though the synaptic connections, in this way in the presence of a fault, a part of the connections can be inoperative without significant performance loss.

In this paper feedforward ANNs with sigmoid activation function were used, as well as the *Levenberg-Marquardt* supervised learning algorithm.

# 5 Hybrid Fault Diagnosis Method

The proposed methodology is composed of two interconnected subroutines. The first one, called here *fault detection subroutine* is responsible for the fault's detection, the fault inception point identification and the fault type classification. The second subroutine of the method, called the *fault location subroutine*, as the name suggests, is responsible for the fault section determination. The second subroutine only processes its data when a fault in the distribution system is previously detected, identified and classified. In the following it is presented the two subroutines.

## 5.1 Fault Detection Subroutine

The proposed fault detection subroutine is composed of 5 distinct processes:

- *Base characteristics extraction*: In this subroutine's part, the analyzed signals base characteristics are extracted;
- *Online characteristics extraction*: In this process, online characteristics are extracted, using the SWT;
- *Fault detection*: Utilizing data from both processes previously mentioned, it is determined the fault existence;
- *Fault inception identification*: This process determines the exact time in which the fault occurred;
- *Fault classification*: After the fault detection and the fault instant identification, this part of the subroutine classifies the fault type.

In the following, each process of the fault detection subroutine is explained.

### 5.1.1 Base Characteristics Extraction

The fault detection subroutine is based on the idea of certain current signal's frequency components energy's change [9]. During a fault, signals with frequency range up to 1 kHz will appear [11]. In the proposed method, the measurement of signal components in the frequency range of 750 Hz – 1 kHz is applied. In order to detect minimum signal energy's variation in this frequency range, it is necessary to determine an initial base energy from the pre-fault system. This is the output of the *Base Extraction Characteristics* process.

The process calculates initially the SWT of the 3 phase signal currents measured at the substation where the distribution feeder is located. The SWT is calculated on 3 cycles of the current signals in ¼ cycle steps with 1 cycle windows. The energy of each phase is calculated only on the detail referent to the interest frequency range (750 Hz – 1 kHz).

After this, medium signal energy is calculated. With these previous values, an energy base is chosen as the highest medium signal energy of the 3 phases.

### 5.1.2 Online Characteristics Extraction

After the step described previously, an online characteristic extraction process is performed. In this process, the current signals' SWT and the energy of the interest detail for the 3 phases are calculated. These values are afterwards normalized with the base values and sent to the fault detection process. The process is illustrated in Fig. 1.

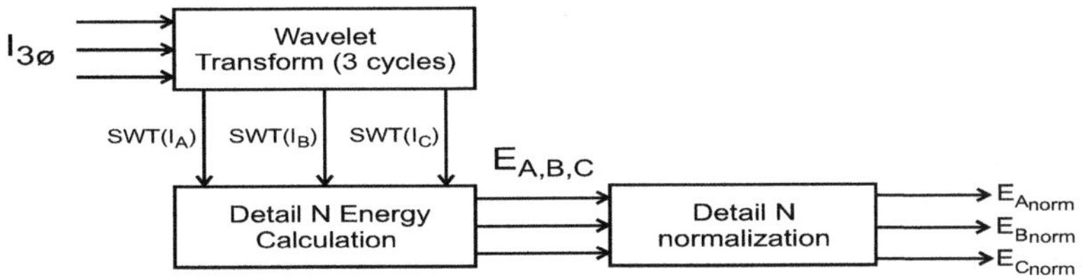


**Fig. 1.** Online characteristic extraction

### 5.1.3 Fault Detection

The fault occurrence determination is based on the characteristics extracted from the current signals. The fault detection process is illustrated in Fig. 2.

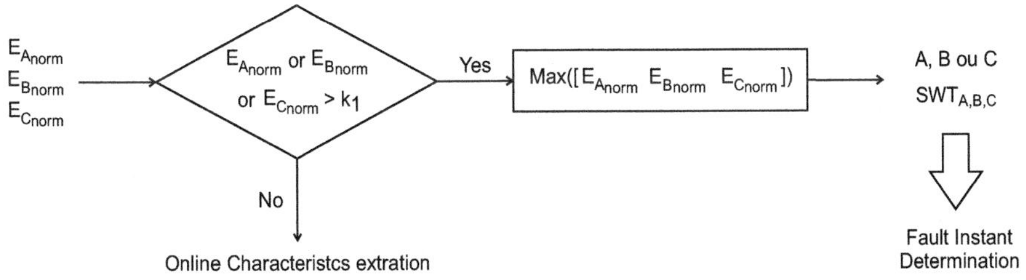


**Fig. 2.** Fault Detection

In this process the normalized energies are directly compared with a limit value. In the case that any of the phases' normalized energies is higher than a minimum value $k_1$, called here *minimum detection index*, fault detection is achieved. The value of $k_1$ is determined by the protection engineer and depends on the local distribution system characteristics.

### 5.1.4 Fault Inception Identification

This subroutine's process utilizes the values of the SWT coefficients to determine the instant of fault occurrence. Fig. 3. illustrates the process.

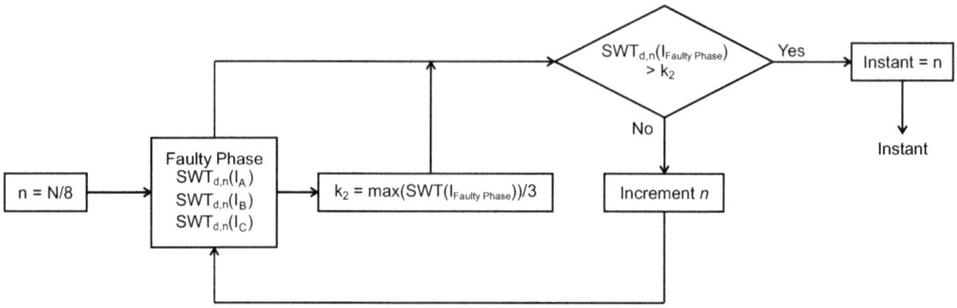


**Fig. 3.** Fault Inception Identification

Initially, the fault information is used to determine $k_2$, the *minimum instantaneous index*, which is a dependent parameter of the interest detail's energy. After, $k_2$ is compared directly with the interest details' coefficients absolute values of the SWT. It is compared only after a certain coefficient, due to high erroneous coefficient values encountered. These coefficients are encountered after the filtering process, when filters composed by a high number of coefficients in relation to the number of signal coefficients are applied. In this process, the signal is divided in 8 parts, and the first one is not considered in the analysis.

### 5.1.5 Fault Classification

In order to classify the fault type, it is used in this process the information of the interest details' energies of all system phases. The phase with the highest energy is considered as *main faulted phase*. The process uses the information of the other phases to determine the type of fault as illustrated in Fig. 4.

Firstly, it is verified if the fault is a three-phase type. Thus, the energy of the current signals' interest detail of the system phases ($E_m$ and $E_n$), except the main phase ($E_{FF}$), passes through two different comparisons. The first compares $E_m$ and $E_n$ directly with $k_3$, the *minimum classification index*. In reality, $k_3$ has the same meaning than $k_1$ with the same design projects. The second comparison is made between $E_m$ and $E_n$ with $E_{FF}$, weighed by $k_4$, the *minimum relative classification index* in order to avoid a wrong fault classification due to a small energy difference between the signals of the system.

In case one of these conditions is met for the two system's phases not classified as main phases, the fault is classified as three-phase. In a different case, the process tests the same signals to check if it is a double-phase fault. In the case only one of the phases fills in the necessary conditions, the fault is considered double-phase. If none of the phases fills in the requirements, the fault is considered single phase and the main phase is determined as the faulty phase (higher energy on the interest detail). When the fault is classified as double-phase, the process determines if the fault involves ground. This process is made through a weighed comparison of the zero sequence current before and after the fault. The comparison is weighed by $k_5$, the *minimum relationship zero sequence current index*, which represents the minimum variation of zero sequence current admissible between the steady-state operation and fault condition. This index should also be determined by the protection engineer.

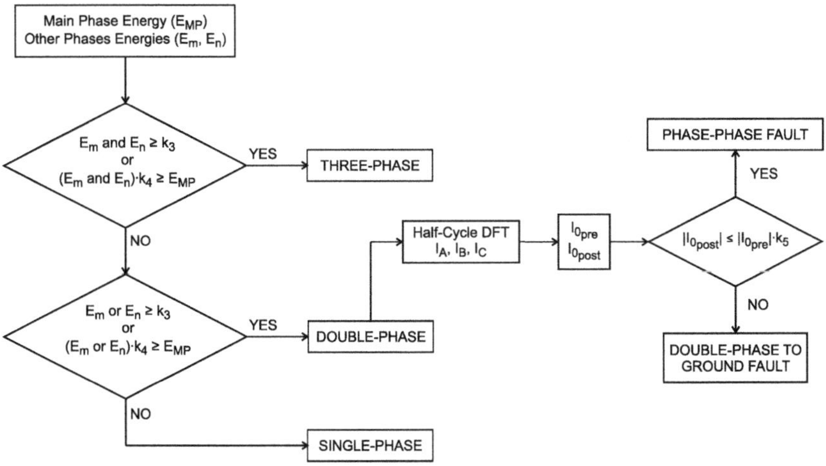


**Fig. 4.** Fault Classification

## 5.2 Fault Location Subroutine

The fault location subroutine estimates the faulted distribution line section after the fault is detected and classified by the previously described subroutine. The subroutine uses a different ANN for each phase-to-ground fault, totalizing 3 ANNs (A-g, B-g, C-g). The input vector of all ANNs is the same and is illustrated in Eq. 4.

$$\vec{x} = \left[ \left|V^{1h}\right| \quad \left|I^{1h}\right| \quad \phi^{1h} \quad \left|I\right|^{1h,3h} \quad \phi^{1h,3h} \quad R_F \right] \tag{4}$$

Where:

$\left|V^{1h}\right|$ is the fundamental voltage phasor magnitude;

$\left|I^{1h}\right|$ is the fundamental current phasor magnitude;

$\phi^{1h}$ is the angle difference between the fundamental phasors of voltage and current;

$\left|I\right|^{1h,3h}$ is the relation between the complex numbers that represent the first and third currents harmonics;

$\phi^{1h,3h}$ is the angle difference between the complex numbers that represent the phasors of the first and third harmonics;

$R_F$ is the fault resistance;

The fault resistance is the only unknown variable of the input vector. The process uses [2] in order to estimate $R_F$. Fig. 5 illustrates the subroutine:

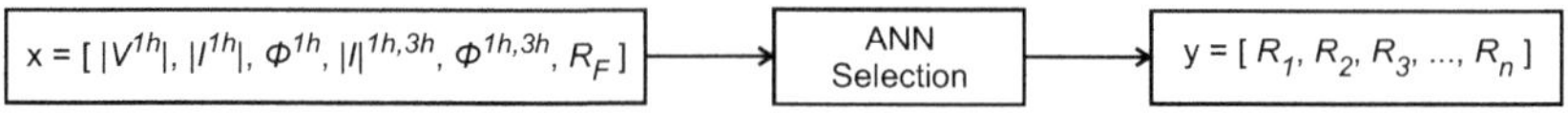


**Fig. 5.** Fault Location subroutine

As shown in Fig. 5, $x$ is the explained input vector and $y$ is the output vector, composed by binary values that represent the $n$ sections of the distribution system.

# 6 Results

In order to test the proposed technique, the method was implemented in Matlab and two underground distribution systems were simulated in ATP-EMTP. Figs. 6 and 7 show the one-line diagrams of the CEEE test systems. The feeder shown in the Fig. 6, *AL-1PL*, is an underground system with 3363 m of extension, with 3 main laterals. The system uses aluminum 750MCM and 4/0 AWG tape-shielded cables. The UDS illustrated in Fig. 7, *AL-1PW*, has 2046 m of extension, with one main derivation. *AL-1PW* uses only 750MCM tape-shielded cables. In order to correctly model the two systems, Carson equations together with the Kron's reduction method were used. The impedance matrixes of these systems were also obtained from this process.

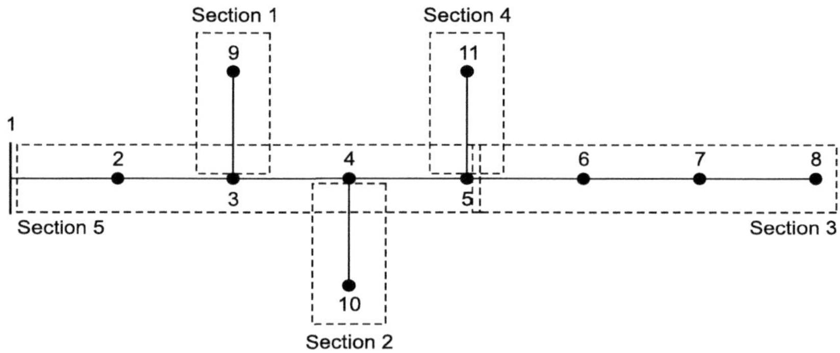


**Fig. 6.** Test Underground Distribution System(AL-1PL)

$$Z_{750MCM} = 1\times10^{-3}\cdot\begin{bmatrix} 0{,}4260+j0{,}4516 & 0{,}2998+j0{,}1550 & 0{,}2998+j0{,}1550 \\ 0{,}2998+j0{,}1550 & 0{,}4260+j0{,}4516 & 0{,}2998+j0{,}1550 \\ 0{,}2998+j0{,}1550 & 0{,}2998+j0{,}1550 & 0{,}4260+j0{,}4516 \end{bmatrix}$$

$$Z_{4/0AWG} = 1\times10^{-3}\cdot\begin{bmatrix} 0{,}6595+j0{,}5914 & 0{,}3293+j0{,}2654 & 0{,}3293+j0{,}2654 \\ 0{,}3293+j0{,}2654 & 0{,}6595+j0{,}5914 & 0{,}3293+j0{,}2654 \\ 0{,}3293+j0{,}2654 & 0{,}3293+j0{,}2654 & 0{,}6595+j0{,}5914 \end{bmatrix}$$

Both matrixes are expressed in [Ω/m].

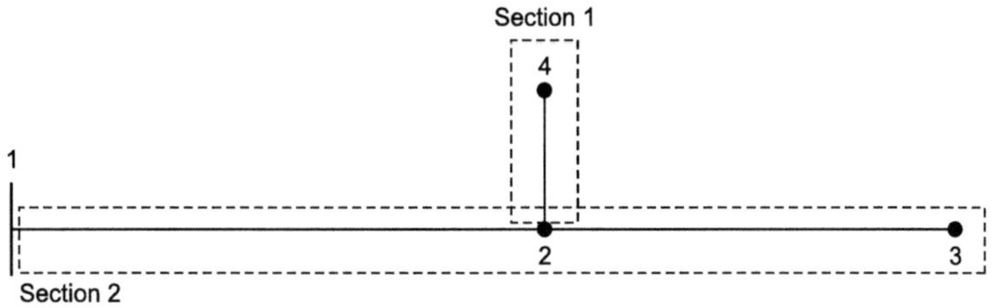


**Fig. 7.** Test Underground Distribution System(AL-1PW)

## 6.1 Detection Results

In order to determine the performance of the fault detection subroutine in the tests systems *AL-1PL* and *AL-1PW*, several fault simulations with different locations, resistances and fault types were performed.

The proposed methodology showed to be extremely robust for single phase-to-ground fault detection, inception identification and fault type classification as the results for all test cases simulated were correct, including for different fault resistance, location and type. The precision established for a correct fault instant identification was ±¼ cycle. In order to test the methodology's capacity for different fault inception angles in phase A ($\alpha$ = *0°, 30°, 90°, 150°, 180°, 210°, 270°* e *330°*), fault resistances ($R_F$ = *0* Ω, *10* Ω, *20* Ω, *50* Ω e *100* Ω ) and locations (31 different locations), other 3720 test cases were simulated and tested for the AL-1PL system. Once more, all faults were correctly detected, classified and its fault instant identified, in different fault types (AB-g, BC-g, AC-g, AB, BC, AC, ABC-g).

## 6.2 Fault Location Results

In order to determine the performance of the fault location subroutine, several simulations on both test distribution systems were performed. The test cases had different fault resistances ($R_F$ = *0* Ω, *10* Ω, *20* Ω, *50* Ω e *100* Ω) and fault types (3 different fault types). The test data was initially composed of 1440 simulated fault cases (480 for each fault type). The test systems *AL-1PL* and *AL-1PW* were divided into 5 and 2 different sections, respectively. This division made possible the application of the subroutine. Fig. 6 and Fig. 7 illustrate the division.

In the following table, the fault resistance of simulated fault tests applied in *AL-1PL* and *AL-1PW* was varied as the method was tested.

**Table 1.** Fault Resistance Effect on the Fault Section Identification (*AL-1PL*)

| $R_F$ | Phase-to-ground Faults | | |
|---|---|---|---|
| | **A-g** | **B-g** | **C-g** |
| **0** | 4,2 | 7,3 | 6,3 |
| **10** | 1,0 | 0,0 | 0,0 |
| **20** | 1,0 | 1,0 | 0,0 |
| **50** | 1,0 | 1,0 | 1,0 |
| **100** | 0,0 | 2,1 | 1,5 |
| **Total** | 1,5 | 2,1 | 1,5 |

**Table 2.** Fault Resistance Effect on the Fault Section Identification (*AL-1PW*)

| $R_F$ | Phase-to-ground Faults | | |
|---|---|---|---|
| | A-g | B-g | C-g |
| **0** | 3,5 | 4,7 | 2,3 |
| **10** | 1,2 | 1,2 | 2,3 |
| **20** | 0,0 | 4,7 | 3,5 |
| **50** | 3,5 | 1,2 | 3,3 |
| **100** | 2,3 | 2,3 | 2,3 |
| **Total** | 2,1 | 2,8 | 2,3 |

As it can be seen, the maximum error was of 2.1% for the B-g faults in *AL-1PL* system and 2.8% for *AL-1PW* system. Also, the table shows that there is a certain uniformity on the results, as the total error was limited between 1% and 3% for all test cases. In order to check the fault location effect on the fault section determination methodology, new test cases were simulated and had their results applied to the method. For these tests, several different fault locations were simulated in the two systems. In *AL-1PL*, sections 1, 2 and 3 had, respectively, 4, 8, and 4 different fault points, while sections 4 and 5 had 18 and 62 fault points, respectively. For *AL-1PW*, section 1 had 27 test fault location points and section 2, 59 points. With the variation of other fault conditions, it was generated for each section 20, 40 and 20 test cases for each single phase fault type in *AL-1PL* respectively, and 135 test cases for each fault type in *AL-1PW*. The results can be seen in Tables 3 and 4.

**Table 3.** Fault Location Effect [%] in the Fault Section Identification (*AL-1PL*)

| Section | Phase-to-ground Faults | | |
|---|---|---|---|
| | A-g | B-g | C-g |
| **1** | 5,0 | 5,0 | 5,0 |
| **2** | 0,0 | 5,0 | 2,5 |
| **3** | 10,0 | 15,0 | 10,0 |
| **4** | 2,2 | 2,2 | 3,3 |
| **5** | 0,6 | 0,6 | 0,0 |

**Table 4.** Fault Location Effect [%] in the Fault Section Identification (*AL-1PW*)

| Section | Phase-to-ground Faults | | |
|---|---|---|---|
| | A-g | B-g | C-g |
| **1** | 3,0 | 5,9 | 3,7 |
| **2** | 1,7 | 1,4 | 1,7 |

As it can be seen for all tests the subroutine had errors smaller than 6%. Only tests in section 3 from *AL-1PL* resulted in errors greater than 6%. Analyzing the results, it is possible to say (based on inference) that the highest errors encountered are a consequence of the smaller number of training patterns used on the training stage of this section, and also due to it's distance to the substation.

# 7 Conclusions

In this paper a novel fault detection and section estimation method for underground distribution systems is presented. The proposed method is divided in two major subroutines. The first subroutine, called the *Fault Detection Subroutine*, is responsible for detecting the existence of a fault, identifying the fault inception point and classifying the fault type. The second subroutine of the method, called the *Fault Location Subroutine*, is responsible for the faulty section estimation. The first subroutine is based on the use of SWT, while the second subroutine of the proposed method is based on the use of ANNs. The first subroutine uses as input data the current signals measured at the substation while the second subroutine uses, besides the same current signals, the voltage signals measured at the substation as well.

In order to adapt the method to the use in underground distribution systems, a system modelling development based on the Carlson's equations and the Kron's reduction method is presented and applied for two real underground distribution systems used by the Electric Energy Company of the State of Rio Grande do Sul (CEEE). The tests results applied on simulated data demonstrate the robustness of the proposed scheme and its potential for on-line applications.

**Acknowledgments.** The authors would like to thank the Electric Energy Company of the State of Rio Grande do Sul (CEEE) – Brazil and the Conselho Nacional de Pesquisa e Desenvolvimento (CNPq) for the financial support.

# References

1. Takagi, T. et al.: A New Algorithm of an Accurate Fault Location for EHV/UHV Transmission Lines: Part I – Fourier Transformation Method. IEEE Transactions on Power Apparatus and Systems, Vol. PAS-100, USA (1981) 1316–1323
2. Takagi, T. T. et al.: Development of a New Type Fault Locator Using the One-Terminal Voltage and Current Data. IEEE Transactions on Power Apparatus and Systems, Vol. PAS-101, USA (1982) 2892–2898
3. Lee, S. J. et al.: An Intelligent and Efficient Fault Location and Diagnosis Scheme for Radial Distribution Systems. IEEE Transactions on Power Delivery, Vol. 19, n. 2, USA (2004) 524–532
4. Xinzhou, D., Yaozhong, G., Bingyin, X.: Fault Position Relay Based on Current Traveling Waves and Wavelets. IEEE Power Engineering Society Winter Meeting, Proceedings. Power Engineering Society Winter Meeting, Singapore (2000)
5. Matlab User's Guide Version 3: Math works Inc. (2000)
6. Alternative Transient Program / Electromagnetic Transient Program (ATP/EMTP) (2000)
7. Kersting, W. H.: Distribution System Modeling and Analysis. Boca Raton: CRC Press. (2002) 314
8. Daubechies, I.: The Wavelet Transform, Time-Frequency Localization and Signal Analysis. IEEE Transaction on Information Theory, Vol. 36, New York (1990) 961–1005
9. Haykin, S., Veen, B. V.: Sinais e Sistemas. Porto Alegre, RS, Brasil: Bookman (2001)

10. Haykin, S.: Neural Networks: A Comprehensive Foundation, 2nd Edition. New Jersey, USA: Prentice-Hall Inc. (1999)
11. Elgerd, O. I.: Introdução à Teoria de Sistemas de Energia Elétrica. São Paulo, SP, Brasil: McGraw-Hill do Brasil (1977)
12. Bretas, A. S.; Pires, L. O.; Moreto, M.; Salim, R. H.: A BP Neural Network Based Technique for HIF Detection and Location on Distribution Systems with Distributed Generation. Lecture Notes in Computer Science, Vol. 4114, Berlin Heidelberg New York (2006) 608 – 613
13. Thukaram, D. et al.: Artificial Neural Network and Support Vector Machine Approach for Locating Faults in Radial Distribution Systems. IEEE Transactions on Power Delivery, Vol. 20, New York (2005) 710–721

# Stability Analysis and Synthesis of Robust Fuzzy Systems with State and Input Delays*

Xiaoguang Yang[1,2], Li Li[3], Qingling Zhang[1], Xiaodong Liu[3], and Quanying Zhu[2]

[1] Institute of Systems Science, Northeastern University, Shenyang 110004 P.R. China
[2] Department of Mathematics, Dalian Maritime University, Dalian 116026 P.R. China
xgyang57@163.com
[3] Research Center of Information and Control, Dalian University of Technology, Dalian, China, 116024
gmgmsb@163.com

**Abstract.** This paper addresses the problem of designing state feedback and output-feedback controllers for uncertain fuzzy systems with states and control input delays. By employing the Lyapunov function, some sufficient conditions for guaranteeing the robust stability of the considered systems are derived. Then the stability conditions are presented into linear matrix inequalities so that the desired controller can be easily obtained by using the Matlab linear matrix inequality (LMI) toolbox. Finally, an example is provided to illustrate the effectiveness of the proposed approach.

**Keywords:** Fuzzy Control, Linear Matrix Inequality (LMI), Robust Control, Time-Delay.

## 1 Introduction

Recently, researchers have proposed many methods of designing robust controllers for uncertain nonlinear systems with the use of the Takagi–Sugeno (T–S) fuzzy model [1], [2]. So T–S fuzzy model has been paid considerable attention and is widely used to the control design of nonlinear systems[2]-[4].

The study of uncertain or time-delay systems have received considerable attention over the past years [5]-[6]. For nonlinear fuzzy systems, the research found in the literature is only with uncertainties and state time-delay. Currently, the delay systems in state and control input have been reported [7]. But the uncertainties are not consider and only state feedback controllers are considered in previous results. However in practical applications, system states are not always measurable. Thus, state observers are required to estimate states of the

* This work is supported in part by the National Natural Science Foundation of China under Grant 60534010 60575039 and in part by the National Key Basic Research and Development Program of China under Grant 2002CB312201-06.

D.-S. Huang, L. Heutte, and M. Loog (Eds.): ICIC 2007, LNAI 4682, pp. 1066–1075, 2007.

fuzzy systems [9], which only presented state delays. It is easy to see that when state observers are included into the synthesis of controllers, the design problem becomes more complicated. Therefore, our result deal with state controller and observer controller design methods of more generalized uncertain time-delay systems.

## 2 Problem Statement

Consider a nonlinear system with time-delay that can be represented by the following fuzzy model:

Plant Rule
IF $\theta_1(t)$ is $M_{i1}$ and $\cdots$ $\theta_p(t)$ is $M_{ip}$, THEN

$$\begin{aligned} \dot{x}(t) &= (A_i + \Delta A_i)x(t) + (A_{id} + \Delta A_{id})x(t - d_1(t)) \\ &\quad +(B_i + \Delta B_i)u(t) + (B_{id} + \Delta B_{id})u(t - d_2(t)) \\ y(t) &= C_i x(t) + C_{id}x(t - d_1(t)) \\ x(t) &= \varphi(t), \quad t \in [-\sigma_0, 0], \quad i = 1, 2, \cdots, r. \end{aligned} \tag{1}$$

where $M_{ij}$ is the fuzzy set, $x(t) \in \mathbb{R}^n$ is the state vector, $u(t) \in \mathbb{R}^m$ is the control input, and $y(t) \in \mathbb{R}^q$ is the output vector. $A_i, A_{id} \in \mathbb{R}^{n\times n}$, $B_i, B_{id} \in \mathbb{R}^{n\times m}$, $2 \leq k < \infty$, and $\theta_1(t), \theta_2(t), \cdots \theta_p(t)$ are the premise variables. It is assumed that the premise variables do not depend on the input $u(t)$. The real valued functional $d_1(t)$ and $d_2(t)$ are the time-varying delays in the state and input and satisfy $0 \leq d_1(t) \leq \overline{d}_1 < \infty$, $0 \leq d_2(t) \leq \overline{d}_2 < \infty$, real positive constants as the upper bound of the time-varying delays. It is also assumed that $\dot{d}_1(t) \leq \beta_1 < 1$, $\dot{d}_2(t) \leq \beta_2 < 1$ and $\beta_1, \beta_2$ are known constants. The matrices $\Delta A_i, \Delta A_{id}, \Delta B_i$ and $\Delta B_{id}$ in (1) denote the uncertainties in the system. Suppose $\Delta A_i, \Delta A_{id}, \Delta B_i$ and $\Delta B_{id}$ take the form of

$$\left[\Delta A_i \ \Delta A_{id} \ \Delta B_i \ \Delta B_{id}\right] = MF(t)\left[N_{ia} \ N_{i1} \ N_{ib} \ N_{i2}\right]$$

where $M, N_{ia}, N_{ib}, N_{i1}$ and $N_{i2}$ are known constant matrices and $F(t)$ is an unknown function matrix and satisfies$F^T(t)F(t) \leq I$, where $I$ is the unity matrix. For the simplicity, let us introduce the following notations:

$\overline{A}_i = A_i + \Delta A_i, \overline{A}_{id} = A_{id} + \Delta A_{id}, \overline{B}_i = B_i + \Delta B_i, \overline{B}_{id} = B_{id} + \Delta B_{id}$.
Then, the (1) can be expressed by the following global model

$$\begin{aligned} \dot{x}(t) &= \sum_{i=1}^{r} h_i(\theta)[\overline{A}_i x(t) + \overline{A}_{id}x(t - d_1(t)) + \overline{B}_i u(t) + \overline{B}_{id}u(t - d_2(t))] \\ y(t) &= \sum_{i=1}^{r} h_i(\theta)[C_i x(t) + C_{id}x(t - d_1(t))] \end{aligned} \tag{2}$$

where $\theta = [\theta_1, \cdots, \theta_l]^T, \mu_i : \mathbb{R}^l \rightarrow [0, 1], i = 1, 2, \cdots, r$ is the membership function of the system with respect to the $i$th plant rule, and $h_i(\theta) = \frac{\mu_i(\theta)}{\sum_{i=1}^{r}\mu_i(\theta)}$. In this paper, we assume that $\mu_i(\theta) \geq 0$ for $i = 1, 2, \cdots, r$ and $\sum_{i=1}^{r} \mu_i(\theta) > 0$ for all $t$. Therefore, $h_i(\theta) \geq 0$ $(i = 1, 2, \cdots, r)$ and $\sum_{i=1}^{r} h_i(\theta) = 1$.

The objective of this paper is to design a state-feedback control law and an observer-based output feedback control law.

## 3 State Feedback Control Law

In this section, we present the design of state feedback controllers. Based on the parallel distributed compensation, the following fuzzy control law via state feedback is employed to deal with the problem of stabilization of the fuzzy systems (2).

**Control Rule** $i$ :

IF $\theta_1(t)$ is $M_{i1}, \cdots$, and $\theta_p(t)$ is $M_{ip}$, then

$$u(t) = \sum_{i=1}^{r} h_i(\theta) K_i x(t) \quad , i = 1, 2, \cdots, r. \tag{3}$$

where $K_i (i = 1, 2, \cdots, r)$ are the local control gains. The design of the fuzzy controller is to determine the feedback gains $K_i (i = 1, 2, \cdots, r)$ such that the resulting closed-loop system is asymptotically stable. With the control law (3), the overall closed-loop system can be written as

$$\dot{x}(t) = \sum_{i=1}^{r}\sum_{j=1}^{r} h_i(\theta) h_j(\theta) [\overline{A}_i x(t) + \overline{A}_{id} x(t - d_1(t)) + \overline{B}_i K_j x(t) + \overline{B}_{id} K_j x(t - d_2(t))]$$

$$y(t) = \sum_{i=1}^{r} h_i(\theta) [C_i x(t) + C_{id} x(t - d_1(t))] \tag{4}$$

The following lemmas will play important roles in obtaining results in this paper. We show them as follows.

**Lemma 1.** *[13] Given any $x, y \in R^{n \times n}$ and for any positive symmetric definite matrix $P \in R^{n \times n}$, Then, we have $2x^T y \leq x^T P^{-1} x + y^T P y$*

**Lemma 2.** *[14] Let $A, D, E$ and $F$ be real matrices of appropriate dimensions with $\|F\| \leq 1$. Then, for any matrix $P = P^T > 0$ and scalar $\varepsilon > 0$ such that $P - \varepsilon DD^T > 0$, we have*

$(A + DFE)^T P^{-1} (A + DFE) \leq A^T (P - \varepsilon DD^T)^{-1} A + \varepsilon^{-1} E^T E$

In the following, we present the stabilization conditions via state feedback for the closed-loop systems (4).

**Theorem 1.** *The controller (3) can stabilize the system (4), if the following matrices $X > 0, T_1 > 0, R_2 > 0, \varepsilon_1 > 0, \varepsilon_2 > 0, \varepsilon_3 > 0, \varepsilon_4 > 0$ and $G_i$ satisfy the LMIs as follows*

$$\begin{bmatrix} \delta_{11} & \delta_{12} & \delta_{13} & \delta_{14} & \delta_{15} & X & 0 & 0 \\ * & \delta_{22} & 0 & 0 & 0 & 0 & 0 & 0 \\ * & * & \delta_{33} & 0 & 0 & 0 & 0 & 0 \\ * & * & * & \delta_{44} & 0 & 0 & \delta_{47} & 0 \\ * & * & * & * & \delta_{55} & 0 & 0 & \delta_{58} \\ * & * & * & * & * & \delta_{66} & 0 & 0 \\ * & * & * & * & * & * & \delta_{77} & 0 \\ * & * & * & * & * & * & * & \delta_{88} \end{bmatrix} < 0 \tag{5}$$

$$\begin{bmatrix} \varrho_{11} & \varrho_{12} & \varrho_{13} & \varrho_{14} & \varrho_{15} & \varrho_{16} & \varrho_{17} & \varrho_{18} & \varrho_{19} & X & 0 & 0 & 0 & 0 \\ * & \varrho_{22} & 0 & 0 & 0 & 0 & 0 & 0 & 0 & 0 & 0 & 0 & 0 & 0 \\ * & * & \varrho_{33} & 0 & 0 & 0 & 0 & 0 & 0 & 0 & 0 & 0 & 0 & 0 \\ * & * & * & \varrho_{44} & 0 & 0 & 0 & 0 & 0 & 0 & 0 & 0 & 0 & 0 \\ * & * & * & * & \varrho_{55} & 0 & 0 & 0 & 0 & 0 & 0 & 0 & 0 & 0 \\ * & * & * & * & * & \varrho_{66} & 0 & 0 & 0 & 0 & \varrho_{611} & 0 & 0 & 0 \\ * & * & * & * & * & * & \varrho_{77} & 0 & 0 & 0 & 0 & \varrho_{712} & 0 & 0 \\ * & * & * & * & * & * & * & \varrho_{88} & 0 & 0 & 0 & 0 & \varrho_{813} & 0 \\ * & * & * & * & * & * & * & * & \varrho_{99} & 0 & 0 & 0 & 0 & \varrho_{914} \\ * & * & * & * & * & * & * & * & * & \varrho_{00} & 0 & 0 & 0 & 0 \\ * & * & * & * & * & * & * & * & * & * & \varrho_{01} & 0 & 0 & 0 \\ * & * & * & * & * & * & * & * & * & * & * & \varrho_{02} & 0 & 0 \\ * & * & * & * & * & * & * & * & * & * & * & * & \varrho_{03} & 0 \\ * & * & * & * & * & * & * & * & * & * & * & * & * & \varrho_{04} \end{bmatrix} \leq 0 \tag{6}$$

*where*
$\Phi(i,i) = XA_i^T + G_i^T B_i^T + A_i X + B_i G_i + (\varepsilon_1^{-1} + \varepsilon_2^{-1} + \varepsilon_3^{-1} + \varepsilon_4^{-1})MM^T + \frac{1}{1-\beta_2}R_2$; $\Phi(i,j) = XA_i^T + G_j^T B_i^T + A_i X + B_i G_j + XA_j^T + G_i^T B_j^T + A_j X + B_j G_i + 2(\varepsilon_1^{-1} + \varepsilon_2^{-1} + \varepsilon_3^{-1} + \varepsilon_4^{-1})MM^T + \frac{2}{1-\beta_2}R_2; a = \frac{(1-\beta_1)}{2}, X = P^{-1}, T_1 = R_1^{-1}, G_i = K_i X, i = 1, 2, \cdots, r; \delta_{11} = \Phi(i,i); \delta_{12} = XN_{ia}^T; \delta_{13} = G_i^T N_{ib}^T; \delta_{14} = A_{id}^T T_1; \delta_{15} = B_{id} G_i; \delta_{22} = -\varepsilon_1^{-1}; \delta_{33} = -\varepsilon_3^{-1}; \delta_{44} = -T_1; \delta_{47} = T_1 N_{i1}^T; \delta_{55} = -R_2 \delta_{58} G_i^T N_{i2}^T; \delta_{66} = -2aT_1; \delta_{77} = -\varepsilon_2^{-1}; \delta_{88} = -\varepsilon_4^{-1}; \varrho_{11} = \Phi(i,j); \varrho_{12} = XN_{ia}^T; \varrho_{13} = XN_{ja}^T; \varrho_{14} = G_j^T N_{ib}^T; \varrho_{15} = G_i^T N_{jb}^T; \varrho_{16} = A_{id}^T T_1; \varrho_{17} = A_{jd}^T T_1; \varrho_{18} = B_{id} G_j; \varrho_{19} = B_{jd} G_i; \varrho_{22} = -\varepsilon_1^{-1}; \varrho_{33} = -\varepsilon_1^{-1}; \varrho_{44} = -\varepsilon_3^{-1}; \varrho_{55} = -\varepsilon_3^{-1}; \varrho_{66} = -T_1; \varrho_{77} = -T_1; \varrho_{88} = -R_2; \varrho_{99} = -R_2; \varrho_{00} = -aT_1; \varrho_{01} = -\varepsilon_2^{-1}; \varrho_{02} = -\varepsilon_2^{-1}; \varrho_{03} = -\varepsilon_4^{-1}; \varrho_{04} = -\varepsilon_4^{-1}; \varrho_{611} = T_1 N_{i1}^T; \varrho_{712} = T_1 N_{j1}^T; \varrho_{813} = G_j^T N_{i2}^T; \varrho_{914} = G_i^T N_{j2}^T$.

*In addition, the control law (3) is obtained by* $K_i = G_i X^{-1}, i = 1, 2, \cdots, r$.

*Proof.* Choose a Lyapunov function candidate[4] for (4) as

$$V(x_t,t) = x^T(t)Px(t) + \frac{1}{1-\beta_1}\int_{t-d_1(t)}^{t} x^T(s)R_1 x(s)ds + \frac{1}{1-\beta_2}\int_{t-d_2(t)}^{t} x^T(s)PR_2Px(s)ds$$

By differentiating $V(x_t,t)$ along the trajectory of the system (4),

$\dot{V}(x_t,t) = 2\sum_{i=1}^{r}\sum_{j=1}^{r} h_i(\theta)h_j(\theta)x^T(t)P\{(A_i + B_i K_j)x(t) + (A_{id} + \Delta A_{id})x(t - d_1(t)) + (B_{id} + \Delta B_{id})K_j x(t - d_2(t)) + (\Delta A_i + \Delta B_i K_j)x(t)\} + \frac{1}{1-\beta_1}[x^T(t)R_1 x(t) -$

$(1-\dot{d}_1(t))x^T(t-d_1(t))R_1x(t-d_1(t))] + \frac{1}{1-\beta_2}[x^T(t)P \times R_2Px(t) - (1-\dot{d}_2(t))x^T(t$

$$-d_2(t))PR_2Px(t-d_2(t))] \tag{7}$$

By Lemma 1, we have

$$2P\Delta A_i \le \varepsilon_1^{-1}PMM^TP + \varepsilon_1 N_{ia}^T N_{ia} \tag{8}$$

$$2P\Delta B_i K_j \le \varepsilon_3^{-1}PMM^TP + \varepsilon_3 K_j^T N_{ib}^T N_{ib} K_j \tag{9}$$

$$\begin{aligned} &2x^T(t)P(A_{id} + \Delta A_{id})x(t-d_1(t)) \\ &\le x^T(t)P(A_{id}+\Delta A_{id})R_1^{-1}(A_{id}+\Delta A_{id})^TPx(t) + x^T(t-d_1(t))R_1x(t-d_1(t)) \end{aligned} \tag{10}$$

$$\begin{aligned} &2x^T(t)P(B_{id}+\Delta B_{id})K_jx(t-d_2(t)) \\ &\le x^T(t)P(B_{id}+\Delta B_{id})K_jP^{-1}R_2^{-1}P^{-1}K_j^T(B_{id}+ \\ &\Delta B_{id})^TPx(t) + x^T(t-d_2(t))PR_2Px(t-d_2(t)) \end{aligned} \tag{11}$$

Thus, substituting (8) (9), (10) and (11) into (7) yields

$$\begin{aligned} \dot{V}(x_t,t) \le &\sum_{i=1}^{r}\sum_{j=1}^{r} h_i(\theta)h_j(\theta)\{x^T(t)[(A_i+B_iK_j)^TP + P(A_i+B_iK_j) + (\varepsilon_1^{-1} + \\ &\varepsilon_3^{-1})PMM^TP + \varepsilon_1N_{ia}^TN_{ia} + \varepsilon_3K_j^TN_{ib}^TN_{ib}K_j + P(A_{id}+\Delta A_{id})R_1^{-1}(A_{id}+\Delta A_{id})^TP \\ &+P(B_{id}+\Delta B_{id})K_jP^{-1}R_2^{-1}P^{-1}K_j^T(B_{id}+\Delta B_{id})^TP + \frac{1}{1-\beta_1}R_1 + \frac{1}{1-\beta_2}PR_2P]x(t)\} \end{aligned} \tag{12}$$

By Lemma 2, we have

$$(A_{id}+\Delta A_{id})R_1^{-1}(A_{id}+\Delta A_{id})^T \le A_{id}(R_1-\varepsilon_2N_{i1}^TN_{i1})^{-1}A_{id}^T + \varepsilon_2^{-1}MM^T \tag{13}$$

$$\begin{aligned} &(B_{id}+\Delta B_{id})K_jP^{-1}R_2^{-1}P^{-1}K_j^T(B_{id}+\Delta B_{id})^T \\ &\le (B_{id}K_jP^{-1})(R_2-\varepsilon_4P^{-1}K_j^TN_{i2}^TN_{i2}K_jP^{-1})^{-1} \times (B_{id}K_jP^{-1})^T + \varepsilon_4^{-1}MM^T \end{aligned} \tag{14}$$

Substituting (13), (14) into (12), we have

$$\dot{V}(x_t,t) \le \sum_{i=1}^{r} h_i^2(\theta)x^T(t)\Delta_{ii}x(t) + \sum_{i=1}^{r}\sum_{j>i}^{r} h_i(\theta)h_j(\theta)x^T(t)\Delta_{ij}x(t) \tag{15}$$

$$\begin{aligned} \Delta_{ii} = &(A_i+B_iK_i)^TP + P(A_i+B_iK_i) + (\varepsilon_1^{-1}+\varepsilon_2^{-1}+\varepsilon_3^{-1}+\varepsilon_4^{-1})PMM^TP + \\ &\varepsilon_1N_{ia}^TN_{ia} + \varepsilon_3K_i^TN_{ib}^TN_{ib}K_i + PA_{id}(R_1-\varepsilon_2N_{i1}^TN_{i1})^{-1}A_{id}^TP + P(B_{id}K_iP^{-1})(R_2 \\ &-\varepsilon_4P^{-1}K_i^TN_{i2}^TN_{i2}K_iP^{-1})^{-1}(B_{id}K_iP^{-1})^TP + \frac{1}{1-\beta_1}R_1 + \frac{1}{1-\beta_2}PR_2P \end{aligned} \tag{16}$$

$$\Delta_{ij} = (A_i + B_iK_j)^TP + P(A_i + B_iK_j) + (A_j + B_jK_i)^TP + P(A_j + B_jK_i) + 2(\varepsilon_1^{-1} + \varepsilon_2^{-1} + \varepsilon_3^{-1} + \varepsilon_4^{-1})PMM^TP + \varepsilon_1N_{ia}^TN_{ia} + \varepsilon_1N_{ja}^TN_{ja} + \varepsilon_3K_j^TN_{ib}^TN_{ib}K_j + \varepsilon_3K_i^TN_{jb}^TN_{jb}K_i + PA_{id}(R_1 - \varepsilon_2N_{i1}^TN_{i1})^{-1}A_{id}^TP + P(B_{id}K_jP^{-1})(R_2 - \varepsilon_4P^{-1}K_j^T \times N_{i2}^TN_{i2}K_jP^{-1})^{-1} \times (B_{id}K_jP^{-1})^TP + PA_{jd}(R_1 - \varepsilon_2N_{i1}^TN_{i1})^{-1}A_{jd}^TP + P(B_{jd}K_iP^{-1})(R_2 - \varepsilon_4P^{-1}K_i^TN_{j2}^TN_{j2}K_iP^{-1})^{-1}(B_{jd}K_iP^{-1})^TP + \frac{2}{1-\beta_1}R_1 + \frac{2}{1-\beta_2}PR_2P \quad (17)$$

Pre- and post-multiplying conclusion (5) and (6) with $\Upsilon_1 = diag\left[P\ I\ I\ I\ I\ I\ I\ I\right]$ and $\Upsilon_2 = \mathrm{diag}\left[P\ I\ I\ I\ I\ I\ I\ I\ I\ I\ I\ I\ I\ I\right]$, respectively. And we note that $X = P^{-1}, G_i = K_iX$. Then applying the Schur complement yields (16) and (17). Therefore, it follows from (15) that $\dot{V}(x_t, t) < 0$, which implies that (4) is asymptotically stable.

This completes the proof.

## 4 Observer Feedback Design

In this part, we consider the fuzzy system with state and input delay without uncertain. The form is following

$$\begin{aligned} \dot{x}(t) &= \sum_{i=1}^{r}\sum_{j=1}^{r}h_i(\theta)h_j(\theta)[A_ix(t) + A_{id}x(t-d_1(t)) + B_iu(t) + B_{id}u(t-d_2(t))] \\ y(t) &= \sum_{i=1}^{r}h_i(\theta)[C_ix(t) + C_{id}x(t-d_1(t))], \end{aligned} \quad (18)$$

In the following, a observer-based output feedback controller design for the system (18) is considered. As in [3], the fuzzy memory observer is constructed as follows.

The overall fuzzy observer is represented as the following

$$\begin{aligned} \hat{x}(t) &= \sum_{i=1}^{r}h_i(\theta)[A_i\hat{x}(t) + A_{id}\hat{x}(t-d_1(t)) + B_iu(t) \quad (19) \\ &\quad + B_{id}u(t-d_2(t)) + L_i(y(t) - \hat{y}(t))] \\ \hat{y}(t) &= \sum_{i=1}^{r}h_i(\theta)[C_i\hat{x}(t) + C_{id}\hat{x}(t-d_1(t))], \quad (20) \end{aligned}$$

With this fuzzy observer, the observer-based control law should be

$$u(t) = \sum_{i=1}^{r}h_i(\theta)K_i\hat{x}(t) \quad , i = 1, 2, \cdots, r. \quad (21)$$

Let us denote the estimation error as

$$e(t) = x(t) - \hat{x}(t) \quad (22)$$

By differentiating (22), we get

$$\dot{e}(t) = \sum_{i=1}^{r}\sum_{j=1}^{r} h_i(\theta)h_j(\theta)[(A_i - L_iC_j)e(t) + (A_{id} - L_iC_{jd})e(t - d_1(t))] \quad (23)$$

From (18), (21), (23), the augmented systems can be written as the following form:

$$\dot{\widetilde{x}} = \sum_{i=1}^{r}\sum_{j=1}^{r} h_i(\theta)h_j(\theta)[\widetilde{A}_{ij}\widetilde{x}(t) + \widetilde{A}_{ijd}\widetilde{x}(t - d_1(t)) + \widetilde{B}_{ijd}\widetilde{x}(t - d_2(t))], \quad (24)$$

where $\widetilde{x}(t) = \begin{bmatrix} x(t) \\ e(t) \end{bmatrix}$, $\widetilde{B}_{ijd} = \begin{bmatrix} B_{id}K_j & -B_{id}K_j \\ 0 & 0 \end{bmatrix}$, $\widetilde{A}_{ij} = \begin{bmatrix} A_i + B_iK_j & -B_iK_j \\ 0 & A_i - L_iC_j \end{bmatrix}$, $\widetilde{A}_{ijd} = \begin{bmatrix} A_{1i} & 0 \\ 0 & A_{id} - L_iC_{jd} \end{bmatrix}$.

For the observer-based output feedback fuzzy control problem, we have the following results.

**Theorem 2.** *A sufficient condition for the existence of the generalized output feedback controller (21) which stabilizes the closed-loop system (24) is that there exist matrices $X > 0, Y > 0, T_{11} > 0, T_{12} > 0, R_{21} > 0, N > 0, G_i$ and $W_i$ satisfying LMIs (25)-(26)*

$$\begin{bmatrix} \Phi_1(i,i) & B_{id}G_i & A_{id}R_{11}^{-1} & P_1^{-1} \\ * & -R_{21} & 0 & 0 \\ * & * & -T_{11} & 0 \\ * & * & * & -\theta T_{11} \end{bmatrix} < 0, \quad \begin{bmatrix} \Phi_2(i,i) & 0 & Y(A_{id} - L_iC_{id}) \\ * & -N & 0 \\ * & * & -T_{12} \end{bmatrix} < 0 \quad (25)$$

$$\begin{bmatrix} \Phi_1(i,j) & \rho_{ij} & A_{1ij}T_{11} & P_1^{-1} \\ * & -2R_{21} & 0 & 0 \\ * & * & 0 & 0 \\ * & * & -2T_{11} & 0 \\ * & * & * & -0.5\theta T_{11} \end{bmatrix} \leq 0, \quad \begin{bmatrix} \Phi_2(i,j) & 0 & \begin{matrix} Y[(A_{id} - L_iC_{id}) \\ +(A_{id} - L_iC_{id})] \end{matrix} \\ * & -2N & 0 \\ * & * & -2T_{12} \end{bmatrix} \leq 0 \quad (26)$$

where $\Phi_1(i,i) = P_1^{-1}(A_i + B_iK_j)^T + (A_i + B_iK_j)P_1^{-1} + \frac{1}{1-\beta_2}R_{21}; \Phi_2(i,i) = (A_i - L_iC_j)^TY + Y(A_i - L_iC_j) + \frac{1}{1-\beta_1}T_{12} + \frac{1}{1-\beta_2}N; \Phi_1(i,j) = P_1^{-1}(A_i + B_iK_j)^T + (A_i + B_iK_j)P_1^{-1} + P_1^{-1}(A_j + B_jK_i)^T + (A_j + B_jK_i)P_1^{-1} + \frac{2}{1-\beta_2}R_{21}; \Phi_2(i,j) = (A_i - L_iC_j)^TY + Y(A_i - L_iC_j) + (A_j - L_jC_i)^TY + Y(A_j - L_jC_i) + \frac{2}{1-\beta_1}T_{12} + \frac{2}{1-\beta_2}N; N = P_2T_{22}P_2; T_{11} = R_{11}^{-1}; \theta = 1 - \beta_1; A_{1ij} = A_{1i} + A_{1j}; \rho_{ij} = B_{id}G_j + B_{jd}G_i$.

In addition, the observer-based controller (34) can be obtained by $G_i = K_iP_1^{-1}$, $W_i = YL_i$

*Proof.* Take Lyapunov function candidate as

$$V(x_t, e_t, t) = \widetilde{x}^T(t)\mathbb{Z}\widetilde{x}(t) + \frac{1}{1-\beta_1}\int_{t-d_1(t)}^{t} \widetilde{x}^T(s)\mathbb{R}_1\widetilde{x}(s)ds + \frac{1}{1-\beta_2}\int_{t-d_2(t)}^{t} \widetilde{x}^T(s)\mathbb{R}_2\widetilde{x}(s)ds$$

where $\mathbb{Z} = \begin{bmatrix} P_1 & 0 \\ 0 & P_2 \end{bmatrix}, \mathbb{R}_1 = \begin{bmatrix} R_{11} & 0 \\ 0 & R_{12} \end{bmatrix}, \mathbb{R}_2 = \begin{bmatrix} P_1 & 0 \\ 0 & P_2 \end{bmatrix} \begin{bmatrix} R_{21} & 0 \\ 0 & R_{22} \end{bmatrix} \begin{bmatrix} P_1 & 0 \\ 0 & P_2 \end{bmatrix}$, $P_1 = X^{-1}$, $P_2 = \lambda Y$, $R_{11} = T_{11}^{-1}$, $R_{12} = \lambda T_{12}$, and $R_{22} = \lambda T_{22}$, $\lambda > 0$ is a scalar to be specified later. We let $\widetilde{\Delta} = \left[ \widetilde{x}(t)\ \widetilde{x}(t-d_1(t))\ \widetilde{x}(t-d_2(t)) \right]$. Then, the derivative of along the system (24) is as followed $\dot{V} \le \sum_{i=1}^{r} h_i^2(\theta) \widetilde{\Delta}^T$

$$\times \begin{bmatrix} \Pi_1(i,i) & -P_1B_iK_i & P_1A_{1i} & 0 & P_1B_{id}K_i & -P_1B_{id}K_i \\ -(B_iK_i)^TP_1 & \Pi_2(i,i) & 0 & P_2(A_{id}-L_iC_{id}) & 0 & 0 \\ * & * & -R_{11} & 0 & 0 & 0 \\ * & * & * & -R_{12} & 0 & 0 \\ * & * & * & * & -P_1R_{21}P_1 & 0 \\ * & * & * & * & * & -P_2R_{22}P_2 \end{bmatrix}$$

$$\times \widetilde{\Delta} + \sum_{i=1}^{r} \sum_{j>i}^{r} h_i(\theta)h_j(\theta)\widetilde{\Delta}^T \begin{bmatrix} I_{11} & I_{12} & I_{13} & 0 & I_{15} & I_{16} \\ * & I_{22} & 0 & I_{24} & 0 & 0 \\ * & * & -2R_{11} & 0 & 0 & 0 \\ * & * & * & -2R_{12} & 0 & 0 \\ * & * & * & * & I_{55} & 0 \\ * & * & * & * & * & I_{66} \end{bmatrix} \widetilde{\Delta}$$

where $I_{11} = \Pi_1(i,j) + \Pi_1(j,i); I_{12} = -P_1B_iK_j - P_1B_jK_i; I_{13} = P_1(A_{1i} + A_{1j}); I_{15} = P_1(B_{id}K_j + B_{jd}K_i); I_{16} = -P_1(B_{id}K_j + B_{jd}K_i); I_{22} = \Pi_2(i,j) + \Pi_2(j,i); I_{24} = P_2(A_{id} - L_iC_{jd} + A_{jd} - L_jC_{id}); I_{55} = -2P_1R_{21}P_1; I_{66} = -2P_2R_{22}P_2$.

Now multiplying (25) on both sides by diag$\left[ P_1\ P_1\ R_{11}\ I\ I\ I \right]$ under the conditions in Theorem 2. Then applying Schur complements yields

$$\begin{bmatrix} \Pi_1(i,i) & P_1B_{id}K_i & P_1A_{id} \\ * & -P_1R_{21}P_1 & 0 \\ * & * & -R_{11} \end{bmatrix} < 0, \begin{bmatrix} \Pi_2(i,i) & 0 & Y(A_{id}-L_iC_{id}) \\ * & -P_2T_{22}P_2 & 0 \\ * & * & -T_{12} \end{bmatrix} < 0 \tag{27}$$

Similar to the proof of Th.2 in [12], it is easy to prove that (27) are equivalent to (25). By repeating the above process, we can be proved that (26) imply (28).

$$\begin{bmatrix} II_{11} & II_{12} & II_{13} & 0 & II_{15} & II_{16} \\ II_{21} & II_{22} & 0 & II_{24} & 0 & 0 \\ * & * & -R_{11} & 0 & 0 & 0 \\ * & * & * & -R_{12} & 0 & 0 \\ * & * & * & * & II_{55} & 0 \\ * & * & * & * & * & II_{66} \end{bmatrix} < 0, \begin{bmatrix} V_{11} & V_{12} & V_{13} & 0 & V_{15} & V_{16} \\ * & V_{22} & 0 & V_{24} & 0 & 0 \\ * & * & V_{33} & 0 & 0 & 0 \\ * & * & * & V_{44} & 0 & 0 \\ * & * & * & * & V_{55} & 0 \\ * & * & * & * & * & V_{66} \end{bmatrix} < 0 \tag{28}$$

where $II_{11} = \Pi_1(i,i); II_{12} = -P_1B_iK_i; II_{13} = P_1A_{1i}; II_{15} = P_1B_{id}K_i; II_{16} = -P_1B_{id}K_i; II_{21} = -(B_iK_i)^TP_1; II_{22} = \Pi_2(i,i); II_{24} = P_2(A_{id} - L_iC_{id}); II_{55} = -P_1R_{21}P_1; II_{66} = -P_2R_{22}P_2; V_{11} = \Pi_1(i,j) + \Pi_1(j,i); V_{12} = -P_1B_iK_j - P_1B_jK_i; V_{13} = P_1(A_{1i} + A_{1j}); V_{15} = P_1(B_{id}K_j + B_{jd}K_i); V_{16} = -P_1(B_{id}K_j + B_{jd}K_i); V_{22} = \Pi_2(i,j) + \Pi_2(j,i); V_{24} = P_2(A_{id} - L_iC_{jd} + A_{jd} - L_jC_{id}); V_{33} = -2R_{11}; V_{44} = -2R_{12}; V_{55} = -2P_1R_{21}P_1; V_{66} = -2P_2R_{22}P_2$.

It follows from (28) that $\dot{V}$ is reduced to $\dot{V}(x_t, e_t, t) < 0$. It implies that the closed-loop system is asymptotically stable.

# 5 Example

In this section, to illustrate the proposed results, we apply the above design technique to design state feedback controllers and observer-based controllers. Consider a uncertain fuzzy system with state and input delays, The Takagi-Sugeno model of this fuzzy system is of the following form:

$\dot{x}(t) = (A_i + \Delta A_i)x(t) + (A_{id} + \Delta A_{id})x(t - d_1(t)) + (B_i + \Delta B_i)u(t) + (B_{id} + \Delta B_{id})u(t - d_2(t))(i = 1, 2)$

Select the membership functions as

$M_{11}(x_1(t)) = \frac{1}{1+\exp(-2x_1(t))}, M_{12}(x_1(t)) = 1 - M_{11}(x_1(t))$.

The parameters are as following

$A_1 = \begin{bmatrix} 0 & -1 \\ 1 & -3 \end{bmatrix}, A_2 = \begin{bmatrix} 0 & -2 \\ 1 & -3 \end{bmatrix}, A_{1d} = \begin{bmatrix} 0.4 & 0 \\ 0 & 0.1 \end{bmatrix}, A_{2d} = \begin{bmatrix} 0.5 & 0 \\ 0 & 0.1 \end{bmatrix}, B_1 = B_2 = [-1\ 1]^T, B_{1d} = B_{2d} = [0\ 0.05]^T, M = [0.3\ 0]^T, N_{1a} = N_{2a} = [0.2\ 0], N_{1b} = N_{2b} = 0.2, N_{12} = N_{22} = 0.01, N_{11} = N_{21} = [0.1\ 0]$,

For the state feedback controllers, applying Theorem 1, the feasible solutions to (5) and (6) are given as follows:

$$P = \begin{bmatrix} 17.3302 & -5.4140 \\ -5.4140 & 4.1727 \end{bmatrix}, K_1 = [17.1463\ \text{-}5.1148]; K_2 = [17.9696\ \text{-}5.6902].$$

For the observer-based controllers, the feasible solutions to (25)–(26) are as followed by Theorem 2.

$P_1 = \begin{bmatrix} 1.0596 & -0.2139 \\ -0.2139 & 0.3648 \end{bmatrix}; Y = \begin{bmatrix} 2.6107 & -0.4971 \\ -0.4971 & 4.8338 \end{bmatrix}; K_1 = [6.8610\ -0.9638]; K_2 = [7.3855\ -1.3698]; L_1 = [10.3898\ 2.6108]^T; L_2 = [10.1781\ 1.5097]^T$. Where the time-delay $d_1(t) = 0.3\sin(t)$ and $d_2(t) = 0.2\cos(t)$, and $F(t) = \cos(t)$.

When the initial function $\varphi(t) = [1 - 1]^T$ for $t \in [-2\ 0]$ and the state feedback control law is (3), the closed-loop system is asymptotically stable. When the initial function $\widehat{\varphi}(t) = [-0.5\ 0.5]^T$ for $t \in [-2\ 0]$ with the observer-based control law (21). The simulation results are shown in Fig.1. It is clear that the closed-loop system is asymptotically stable and the estimation error asymptotically tends to zero.

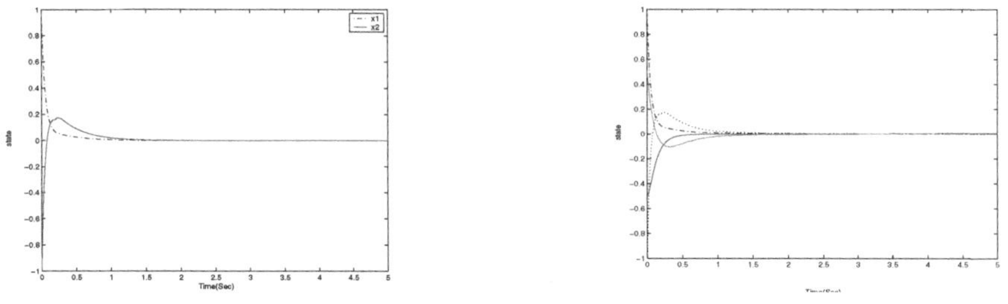


**Fig. 1.** The responses of state and the estimation states

## 6 Conclusion

In this paper, The robust stability problem of uncertain fuzzy systems with state and input delays has been studied. The linear state feedback control law and observer-based output feedback control law have been obtained by solving an LMI feasibility problem. The closed-loop system is asymptotic stability. A numerical example is also given to illustrate the design procedures and the effectiveness of the approaches developed in this paper.

## References

1. Takagi, T., Sugeno, M.: Fuzzy Identification of Systems and Its Applications to Modeling and Control. IEEE Trans. Syst., Man, Cybern, Vol. 15, No. 1 (1985) 116-132
2. Nguang, S.K., Shi, P.: Stabilization of a Class of Nonlinear Time-Delay Systems Using Fuzzy Models. in Proc. 39th IEEE Conf. Decision and Control (2000) 4415–4419
3. Cao, Y.Y., Frank, P.M.: Analysis and Synthesis of Nonlinear Timedelay System Via Fuzzy Control Approach. IEEE Trans. Fuzzy Syst., Vol. 8, No. 2 (2000) 200-211
4. Tanaka, K., Wang, H.O.: Fuzzy Control Systems Design and Analysis: A Linear Matrix Inequality Approach. New York, USA (2001)
5. Zhang, Y., Pheng, A.H.: Stability of Fuzzy Control Systems with Bounded Uncertain Delays. Fuzzy Systems, IEEE Transactions on, Vol. 10, No. 1 (2002) 92-97
6. Fridman, E., Shaked, U.: Parameter Dependent Stability and Stabilization of Uncertain Time-delay System. IEEE Trans. Automat. Control, Vol. 48, No. 5 (2003) 861-866
7. Lin, C., Wang, Q.G., Lee, T.H.: Delay-dependent LMI Conditions for Stability and Stabilization of T-S Fuzzy Systems with Bounded Time-delay. Fuzzy sets and systems, in press.
8. Chang, Y.C.,Chen, S.S.,Su, S.F.,Lee, T.T.: Static Output Feedback Stabilization for Nonlinear Interval Time-delay Systems via Fuzzy Control Approach. Fuzzy Sets and Systems, Vol. 148, No. 3 (2004) 395-410
9. Jiang, X.F.,Xu, W.L.,Han, Q.L.: Observer-based Fuzzy Control Design with Adaptation to Delay Parameter for Time-delay Systems. Fuzzy Sets and Systems, Vol. 152, No. 3 (2005) 637-649
10. Liu, X.D., Zhang, Q.L.: New Approaches to $H^\infty$ Controller Designs Based on Fuzzy Observers for T-S Fuzzy Systems via LMI. Automatica, Vol. 39 (2003) 1571-1582
11. Liu, X.D., Zhang, Q.L.: Approaches to Quadratic Stability Conditions and $H^\infty$ Control Designs for T-S Fuzzy Systems. IEEE Trans. on Fuzzy System, Vol.11, No. 6 (2003) 830-839
12. Chen, B., Liu, X.P.: Reliable Control Design of Fuzzy Dynamic Systems with Time-varying. Fuzzy Sets Syst, Vol. 146, No. 5 (2004) 349-374
13. Xie, L., Souza, C.E.D.: Robust $H^\infty$ Control for Linear Systems with Norm-bounded Time-varying Uncertainty. IEEE Trans. Autom. Control, Vol. 37, No. 8 (1992) 1188-1191
14. Wang, Y.,Xie, L., Souza, C.E.D.: Robust Control of a Class of Uncertain Nonlinear Systems. Syst. Control Lett, Vol. 19 (1992) 139-149

# Biometric User Authentication Based on 3D Face Recognition Under Ubiquitous Computing Environment

Hyeonjoon Moon[1] and Taehwa Hong[2]

[1] School of Computer Engineering, Sejong University
hmoon@sejong.ac.kr
[2] School of Electrical and Electronic Engineering, Yonsei University
riccati@yonsei.ac.kr

**Abstract.** In this paper, we developed a biometric user authentication system based on 3-dimensional (3D) face recognition under ubiquitous computing environment. Since 2D based face recognition has been shown its structural limitation, 3D model based approach for face recognition has been spotlighted as a robust solution under variant conditions of pose and illumination. Since 3D face model consists of a large number of vertices, 3D model based face recognition system is generally inefficient for real-time computation. We propose a novel 3D face representation algorithm to reduce the number of vertices and optimize its computation time while maintaining reasonable recognition performance. We evaluate the performance of proposed algorithm with the Korean face database collected using a stereo-vision based 3D face capturing device. Additionally, various decision making similarity measures were explored for recognition performance. Our experimental results indicated that our proposed algorithm is robust for biometric user authentication and is also reasonably fast for real-time processing.

**Keywords:** biometrics, 3D face recognition, 3D Model, user authentication, identification, verification, ubiquitous computing, mobile device.

## 1 Introduction

Biometrics is a rapidly evolving technology which has been widely used in variety of applications such as person identification and verification. Recently, there is a huge demand for biometric user authentication which can make an important role in ubiquitous computing environment to protect from invasion of private information. Modern devices that may serve the ubiquitous computing model which include mobile phones, personal multimedia players, and radio-frequency identification tags, etc. Biometric user authentication based on face recognition can be a valuable tool to prevent unauthorized access under ubiquitous computing related applications.

Traditional face recognition systems have primarily relied on 2D images. However, they tend to give a higher degree of recognition performance only when images are of good quality and the acquisition process can be tightly controlled. Recently, many literatures on 3D based face recognition have been published with various methods and experiments. It has several advantages over traditional 2D face recognition: First, 3D data provides absolute geometrical shape and size information of a face.

D.-S. Huang, L. Heutte, and M. Loog (Eds.): ICIC 2007, LNAI 4682, pp. 1076–1081, 2007.

Additionally, face recognition using 3D data is more robust to pose and posture changes since the model can be rotated to any arbitrary position. Also, 3D face recognition can be less sensitive to illumination since it does not solely depend on pixel intensity for calculating facial similarity. Finally, it provides face segmentation information since the background is typically not synthesized in the reconstruction process [1].

In this paper, we propose a novel 3D face representation which has flexibility to reduce the vertex number of each 3D face remarkably. All the 3D faces can be aligned with correspondence information based on the vertex number of a reference face simultaneously. This paper is organized as follows. Section 2 presents the proposed biometric user authentication process for face recognition. Section 3 describes the procedure of the 3D face recognition process under ubiquitous computing environment. Experimental results are presented in Section 4 based on Korean face database. Finally, conclusions and future work are discussed in Section 5.

## 2 User Authentication Process

We assume several constraints for this experiment.

- Ubiquitous computing device can be used by limited number of users
- Medium size and quality digital image/video camera is utilized.
- Maximum of 100 users are to be identified per mobile device.

Figure 1 shows the block diagram for user authentication process.

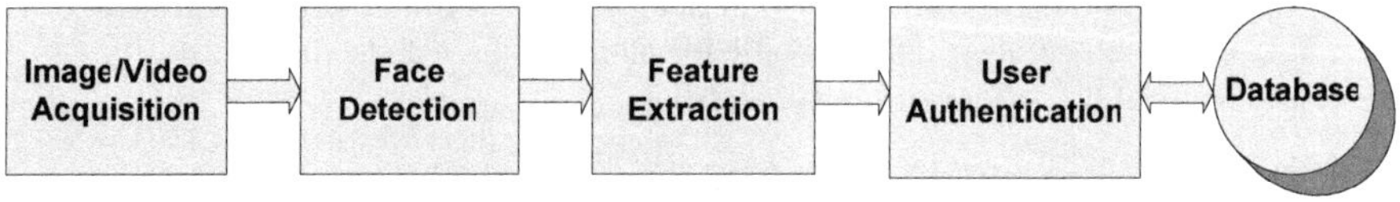


**Fig. 1.** Block diagram of the user authentication system

When the user starts using the ubiquitous computing device, i.e. mobile device, image/video frames are acquired by digital image/video camera. Since face recognition algorithm is sensitive to indoor/outdoor conditions such as amount of lights, angle of camera to the face, etc, each frame is investigated for good image quality and good position of the face. Then face detection and preprocessing techniques are applied to improve the quality of the images for further processes. Then the face recognition system processes the face image and look up the user database to authenticate the specific person. Authenticated user is informed to the device for user-dependent customization of services.

Since the face recognition algorithm is very sensitive to light conditions such as intensity, contrast, etc, we developed key frame extraction method which can adequately reflect those properties. We used object segmentation algorithm in [3] to extract object area. The algorithm is suitable for real time process and performs well especially for the object with stationary background. The extracted object contains user's face and upper body. In order to segment the face, we applied filtering followed by masking procedure. If the segmented face is tilted, geometrical transform is applied to correct the position of face. Histogram equalization and histogram specification [4] was also performed for improve the contrast of the face.

User authentication module is divided into feature extraction and similarity measure. Principal component analysis (PCA) based algorithm [5] has been widely used for many face recognition applications. However, face recognition performance is highly affected by pose and lighting variations, we used 3D based face recognition algorithm. For similarity measure, we explored L1 + Mahalanobis distance, L2 + Angle distance [6].

# 3 3D Based Face Recognition

In general, a 3D face scan consists of texture intensity in 2D frame and geometrical shape in 3D. Especially, the shape information is represented by close connections of many vertices and polygons. Since the vertex number of each of 3D face scans is different from each other, it is necessary to manipulate them to have the same number of vertices for consistent mathematical expression. We propose a novel 3D face representation algorithm as shown in Figure 2.

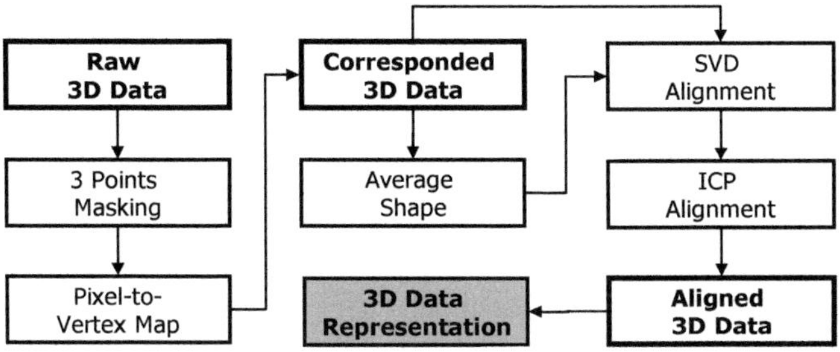


**Fig. 2.** 3D face representation process

## 3.1 Face Alignment and Model Generation

It is possible that all 3D face scans in our database are expressed with the same number of vertex points. To construct a more accurate model, it is necessary to utilize some techniques for face alignment, which is transforming the geometrical factors (scale, rotation angles and displacements) of a target face based on a reference face. Face alignment in our research is achieved by adopting singular value decomposition (SVD) [7] and Iterative Closest Points (ICP) [8][9] sequentially.

We constructed separate models from shapes and textures of 100 Korean people by applying PCA [10] independently. The separate models are generated by linear combination of the shapes and textures as given by equation below.

$$\mathbf{S} = \mathbf{S}_0 + \sum_{j=1}^{N_S} \alpha_j \mathbf{S}_j, \qquad \mathbf{T} = \mathbf{T}_0 + \sum_{j=1}^{N_T} \beta_j \mathbf{T}_j .$$

where $\boldsymbol{\alpha} = [\alpha_1\ \alpha_2 \cdots \alpha_{N_S}]$ and $\boldsymbol{\beta} = [\beta_1\ \beta_2 \cdots \beta_{N_T}]$ are the shape and texture coefficient vectors (should be estimated by a fitting procedure). Also, $\mathbf{S}_0$ and $\mathbf{S}_j$ are the shape average model and the eigenvector associated with the $j^{th}$ largest eigenvalue of the shape covariance matrix, $\mathbf{T}_0$ and $\mathbf{T}_j$ in textures likewise.

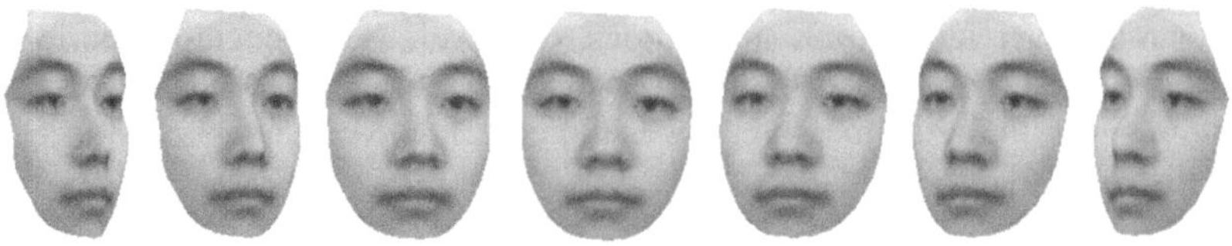

**Fig. 3.** A generative 3D model. These are rotated versions of the 3D model in Y-axis direction. Each version from the first is ranged from -45 degrees to 45 degrees at 15 degrees interval.

### 3.2 Fitting the 3D Model to a 2D Image

Shape and texture coefficients of the generative 3D model are estimated by fitting it to a given 2D input face image. This is performed iteratively as close as possible to the input face. Fitting algorithms, called stochastic Newton optimization (SNO) and inverse compositional image alignment (ICIA) were utilized in [11] and [12], respectively. We also explore the ICIA algorithm as a fitting method to guarantee the computational efficiency.

Given an input image, initial coefficients of shape and texture and projection parameters for the model are selected appropriately. Initial coefficients of shape and texture usually have zero values but projection parameters are manually decided by the registration of some important features. Then, fitting steps are iterated until convergence to a given threshold value, minimizing the texture difference between the projected model image and the input image. During the fitting process, texture coefficients are updated without an additive algorithm. In case of shape coefficients, their updated values are not acquired with ease because of the nonlinear problem of structure from motion (SFM) [13]. To solve it, we recover the shape coefficients using SVD based global approach [14] after the convergence.

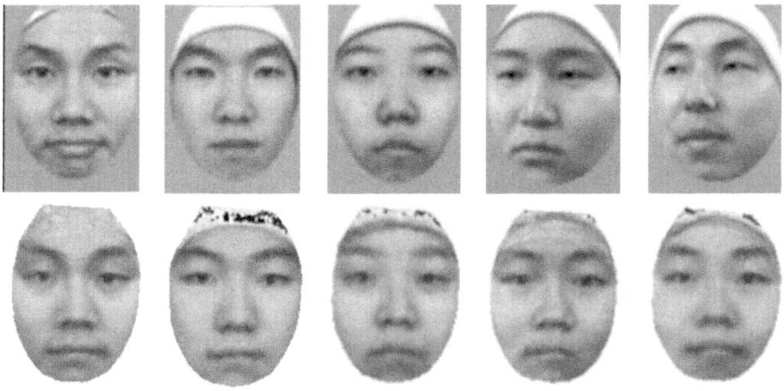

**Fig. 4.** Fitting results. The images in top row are input images and those in bottom row are the fitted versions of our 3D model. Especially, the inputs to the third column are frontal and the others are rotated 30 degrees approximately.

## 4 Experimental Results

We evaluate our face recognition system based on a 3D model generated using proposed representation algorithm. As mentioned in previous sections, 3D Korean faces are collected using a Geometrix Facevision 200 [15], which is a stereo-camera based capturing device offering a 3D face scan including approximately 30,000 ~ 40,000 vertices and corresponding 2D texture image. There are 218 3D face scans collected from 110 people during 3 sessions, which are limited with frontal views and neutral expressions, in our database. We used 100 scans in session 1 for 3D model generation and 52 scans in other sessions for the performance evaluation. Also, 7 profile face images with range from 15 to 40 degrees are acquired separately using the same device for variant pose test.

The experimental results to frontal and profile faces are shown in Table 1. In both experiments, we utilized the L2 norm and angle combined with Mahalanobis distance as a distance metric, denoted by L2+ Mahalanobis and Angle+ Mahalanobis respectively [16]. Also, we performed additional experiments on two cases, one is to use only texture coefficients and the other is to combine texture and shape coefficients. Recognition accuracy with rank 1 in both tests was 90.4% (47 out of 52 subjects) and 85.7% (6 out of 7 subjects) respectively. The average fitting time taken without the shape recovery was 3.7s on 1.73GHz Pentium-M and 1GB RAM.

**Table 1.** Recognition accuracy with rank 1 to frontal faces and pose variant faces

| | Only Texture | | Texture + Shape | |
|---|---|---|---|---|
| | L2+Mah | L2+Angle | L2+Mah | L2+Angle |
| Frontal faces | 90.4% | 86.5% | 90.4% | 88.5% |
| Pose variant faces | 71.4% | 85.7% | 71.4% | 85.7% |
| Computation time | 3.7s | | 11.2s | |

## 5 Conclusion

In this paper, we developed user authentication system based on 3D face recognition. Since the algorithm is video-based, key frame extraction was performed followed by face area detection and input data correction. For pose and illumination-invariant properties, we presented a novel 3D face representation algorithm for 3D model based face recognition system. On the basis of the presented method, an original 3D face scan including 30,000 ~ 40,000 vertices could be represented with about 5,000 vertices.

We have generated 3D model using 100 3D face images (each 3D face image composed of 4822 vertices). Then, shape and texture coefficients of the model were estimated by fitting into an input face using the ICIA algorithm. For 3D model generation and performance evaluation, we have made the Korean 3D face database from a stereo-camera based device.

Experimental results show that face recognition system using the proposed representation method is more efficient in computation time. Our experimental results indicated that the algorithm is robust and also reasonably fast for real-time processing.

Our future works will focus on developing optimized face recognition algorithm for mobile devices and real-time processing for ubiquitous computing environment.

## Acknowledgment

This research was performed for the Promotion and Support Program of Strategic Industry Innovation Cluster, one of the Collaboration Programs for Industry, Academy and Research Institute funded by City of Seoul, Korea.

## References

1. Papatheodorou, T., Rueckert, D.: Evaluation of 3D Face Recognition Using Registration and PCA. AVBPA 2005, LNCS 3546 (2005) 997-1009
2. Holger, W., Sven, C. O., Bruce G.: Real-time Video Abstraction. ACM Transactions on Graphics, ACM Press, New York 25 (2006) 1221-1226
3. Kim, C., Hwang, J. N.: Fast and Automatic Video Object Segmentation and Tracking for Content-Based Applications. IEEE Transactions on Circuits and Systems for Video Technology (CSVT), IEEE, 12 (2002) 122-129
4. Rafael, C. G., Richard E. W.: Digital Image Processing, 2nd Ed. Prentice Hall, (2002)
5. Turk, M., Pentland, A.: Eigenfaces for Recognition, J. Cognitive Neuroscience, 3 (1991) 71-86
6. Development of the Performance Evaluation Tool for the Face Recognition System, Technical Report, Korea Information Security Agency (2004)
7. Horn, B. K. P., Hilden, H. M., Negahdaripour, S.: Closed-Form Solution of Absolute Orientation Using Orthonormal Matrices. Journal of the Optical Society of America A, 5 (1988) 1127-1135
8. Besl, P. J., Mckay, N. D.: A Method for Registration of 3D Shapes. IEEE Transactions on Pattern Analysis and Machine Intelligence, 14(2) (1992) 239-256
9. Lu, X., Jain, A., Colbry, D.: Matching 2.5D Face Scans to 3D Models. IEEE Transactions on Pattern Analysis and Machine Intelligence, 28(1) (2006) 31-43
10. Vetter, T., Poggio, T.: Linear Object Classes and Image Synthesis from a Single Example Image. IEEE Transactions on Pattern Analysis and machine Intelligence, 19(7): (1997) 733-742
11. Blanz, V., Vetter, T.: Face Recognition Based on Fitting a 3D Morphable Model. IEEE Transactions on Pattern Analysis and Machine Intelligence. 25(9) (2003) 1063-1074
12. Romdhani, S., Vetter, T.: Efficient, Robust and Accurate Fitting of a 3D Morphable Model. In IEEE International Conference on Computer Vision (2003).
13. Bascle, B., Blake, A: Separability of Pose and Expression in Facial Tracking and Animation. In Sixth International Conference on Computer Vision (1998) 323-328
14. Romdhani, S., Canterakis, N., Vetter, T.: Selective vs. Global Recovery of Rigid and Non-rigid Motion. Technical report, CS Dept., Univ. of Basel (2003)
15. http://www.geometrix.com.
16. Moon, H., Phillips, P.: Computational and Performance Aspects of PCA-Based Face-Recognition Algorithms. Perception. 30 (2001) 303-321

# Score Normalization Technique for Text-Prompted Speaker Verification with Chinese Digits

Jing Li[1], Yuan Dong[1,2], Chengyu Dong[2], and Haila Wang[2]

[1] Beijing University of Posts and Telecommunications, 100876, P.R. China
Soundofsilence.li@gmail.com, yuandong@bupt.edu.cn
[2] France Telecom R&D Beijing Co, Ltd., Beijing, P.R. China
{chengyu.dong, haila.wang}@orange-ftgroup.com

**Abstract.** A text prompted speaker verification system is presented in this paper. This system is based on ten Chinese digits. Basic acoustic models are speaker dependent and content dependent phoneme HMMs which were generated by adapting speaker independent models to the utterances of specific speakers. An obvious constraint for normalization techniques used in TDSV is that the phrases with the same content should be used for competitive cohort models. So many of the score normalization techniques are either difficult to implement because of lack of data or not good for performance improvement because of poor estimation of the normalization parameters. We propose a method which combines the traditional T-Norm and Cohort Norm together to find a good tradeoff of testing utterance normalization and target speaker model normalization. The proposed method improved the system performance from the baseline equal error rate 3.42% for T-Norm and 2.72% for Cohort Norm to 2.50%.

**Keywords:** Text Dependent Speaker Verification (TDSV), Text-Prompted Passwords, Score Normalization, Cohort Norm, T-Norm.

## 1 Introduction

Text-dependent speaker verification (TDSV) system is to verify a person's claimed identity using a spoken password with specific content. Because of the specific content of voice passwords, TDSV performs better than text-independent speaker verification (TISV) system. However passwords with fixed contents suffer high risk of stolen by spiteful impostors. TDSV system with variable text passwords is to solve this problem and increase the flexibility of the systems [1]. Our objective system is a special kind of variable text TDSV system which is called text prompted speaker verification system. In this kind of system, the passwords are prompted by the system and the contents of the passwords are generated randomly based on a predefined vocabulary. Till now most of the text prompted systems are based on digital vocabularies [1] [2] [3], so the vocabulary used in our system is the set of ten Chinese digits.

In speaker verification applications, the final decision about impostors or true speakers is a potential reason for verification error because scores variability is not

D.-S. Huang, L. Heutte, and M. Loog (Eds.): ICIC 2007, LNAI 4682, pp. 1082–1089, 2007.

only from the difference of impostors and true speakers, but also from other irrelevant sources which will confuse the judgment [4]. Scores variability may be model dependent variability, testing speech dependent variability, handset dependent variability [5], and variability caused by other reasons. Some of this variability has regularity and can be compensated. Different kinds of score normalization techniques are developed to solve different problems. T-Norm is used to compensate the score variability caused by testing utterances. Cohort Normalization is a technique dedicated to the score variability coming from target models. All these techniques are already successfully used in text independent speaker recognition systems. However an obvious constraint for normalization techniques used in TDSV is that the utterances from impostors with the same content should be used for comparable and competitive impostor models because the target models are trained using specific contents. In practice the utterances with various specific contents are difficult or impossible to get from a reasonable amount of impostors. As a result in text dependent speaker verification system many of the score normalization techniques are difficult to implement because lack of data will cause poor estimation of the normalization parameters. Fortunately constraints of text [6] in normal TDSV systems are relaxed in the text prompted speaker verification system by the randomly generated passwords. The passwords are generated from a predefined the vocabulary, and this predefined vocabulary in text prompted TDSV system makes it possible to build competitive models with the same content utterances. T-Norm and Cohort Norm are combined in our text prompted speaker verification system for better performance in text dependent system. This combined method can make use of the information from both the testing utterances and the speaker models.

The remainder sections of this paper are arranged and introduced here. In section 2 our text prompted speaker verification system based on Chinese digital vocabulary is presented. In section 3 score normalization technique is stated. Experiments are presented in section 4. And the last section is conclusions.

## 2 Text-Prompted Speaker Verification with Chinese Digits

As stated above TDSV with text prompted passwords is a kind of variable text speaker verification system which means the users can speak different passwords into the system each time they log into the system. However the words of the passwords must be randomly chosen by the system from a predefined vocabulary. This vocabulary is different cross different systems. Another important factor that influences the system performance is the manner in which the systems prompt the randomly generated passwords to the users. In a telephony application the obvious way of prompting is by playing the password through the telephone with a prompting voice, which is called speech prompting (orally). An alternative approach would be prompting the passwords by text through a webpage or a document which is called text prompting (visually). Researches have shown that text-prompting work better than speech prompting from the performance point of view [3]. Our prompting manner is text prompting.

### 2.1 Enrollment Models

As stated above the vocabulary used in our system was based on ten Chinese digits and the passwords were prompted by text directly. In this way, we could make the following assumption: all speakers no matter target users or impostors would speak exactly the prompted content when they log in. In enrollment stage, the system prepared some digital sequences and let the enrollment speakers to speak them. In order to get high quality modeling of target speakers, these digital sequences should be designed elaborately. A lowest requirement for the enrollment data is to cover all the keywords in the vocabulary. Another requirement is that every keyword should occur in most variant context in order to cover the different acoustic characteristics which may occur in the prompted passwords. The occurrence time of every keyword should be balanced and should be more than a lowest level. And also the enrollment procedure should be guaranteed accurately by supervised speaking or automatic rejection of wrong speaking digital sequences.

For variable text TDSV system, phoneme level or word level models should be used because these small unit models make the system more flexible to cope with the variable text situation. Our basic acoustic model was phoneme based HMM [7]. The definition of phonemes was the same as typical “initial” and “final” form used in Chinese speech recognition. Totally 14 different phonemes were involved to present the ten Chinese digits.

In model training stage, the enrollment utterances of the speakers were used to update an existing speaker independent but phoneme dependent HMM to make it match the characteristics of the target speakers' voices [7]. This was very like a typical model adaptation procedure and MAP (Maximum a Posteriori) estimation approach was used. This method greatly decreased the amount of data needed to train speaker dependent phoneme models.

### 2.2 Verification

In the system, verification process was composed of two stages. Firstly the input utterances were mapped to the speaker independent phoneme models to do force alignment; the boundaries of phonemes were labeled, and also the speaker independent scores of phonemes were recorded as the universal background scores. Then the labeled speeches were mapped to the speaker dependent phoneme models to generate the speaker dependent scores of each phone. Our baseline system score was computed by subtracted the universal background scores from the speaker dependent scores. The scores were used to judge whether the speakers are the true speaker or impostors. If the scores were higher than the threshold, the testing speeches were decided to be generated by the true speakers, otherwise they were from impostors. The threshold was decided to balance the error rates of false rejection and false acceptance.

## 3 Score Normalization Techniques

Till now most of the popularly used score normalization techniques are based on impostor models. This kind of score normalization techniques can be divided into two approaches. The first approach is based on the use of numerous utterances collected from large amount of human voices. T-Norm and universal background model normalization belong to this category. A second approach is cohort normalization, which uses a set of cohort speakers who are close to the target speaker. As stated above, T-Norm is used to compensate the score variability caused by testing utterances. Cohort Normalization is a technique dedicated to the score variability coming from target models. Combination of the two methods can make use of the information from both the testing utterances and the speaker models. This kind of score normalization method improved the system performance compared with baseline UBM normalization or the T-Norm or Cohort-Norm only.

### 3.1 Test Normalization

T-Norm is a kind of distribution scaling technique [8] which intends to center the impostor score distribution to standard Gaussian distribution. The baseline T-Norm implementation is defined as:

$$\tilde{L}_{\lambda}^{T}(X)=\frac{L_{\lambda}(X)-\mu}{\sigma} \tag{1}$$

where the parameter $\mu$ and $\sigma$ are the mean and standard variance of the impostor scores and computed as :

$$\mu=\frac{1}{N}\sum_{i=1}^{N}L_{\bar{\lambda}_i}(X) \tag{2}$$

$$\sigma^2=\frac{1}{N}\sum_{i=1}^{N}\left(L_{\bar{\lambda}_i}(X)-\mu\right)^2 \tag{3}$$

A norm set which contain $M$ impostor models $\bar{\lambda}_i$ had been trained during the enrollment procedure. A difference of our T-Norm was that the testing utterances were compared with all the $M$ models in the norm set but not all the scores were used to calculate the norm parameters, in stead only the $N$-Best scores were selected. This idea was illuminated by AT-Norm approach [9].

### 3.2 Cohort Normalization and Phoneme Level Cohort Selection

Cohort normalization is to get a relative score [8], and researches show that the relative score performs much better than the absolute target score. The cohort normalization can be presented like:

$$\tilde{L}_{\lambda}^{C}(X)=L_{\lambda}(X)-\frac{1}{N_C}\sum_{s\in C}\log P(X\mid\lambda_s) \tag{4}$$

where $C$ is the cohort set, $N_C$ is the number of cohort speakers. $\lambda_s$ is the cohort speakers selected according to the target models.

Cohort model selection techniques have been found to be important for providing good verification performance. To define the cohort selection method, higher order statistics can be used with Bhattacharyya distance for measuring the distance between the output distributions of a pair of HMMs [10]. The form of this distance, for the one dimensional log-likelihoods, is:

$$d_B\left(\lambda_i,\lambda_j\right)=\frac{1}{4}\frac{\left(m_i-m_j\right)^2}{\sigma_i^2+\sigma_j^2}+\frac{1}{2}\log\left(\frac{\frac{1}{2}\left(\sigma_i^2+\sigma_j^2\right)}{\sigma_i\sigma_j}\right) \tag{5}$$

where $m_i$ represents the enrollment mean and $\sigma_i^2$ represents the enrollment variance.

Traditionally we say cohort speakers to present the speakers whose voices are close to the target speaker. However voices are influenced by the content of speech, so an idea is to select the cohort models not only based on a target speaker but also based on a specific content. In our system phoneme HMM was used so the cohort models could be selected on phoneme level.

### 3.3 Proposed Normalization Method

T-Norm is dedicated to remove the score variability caused by testing utterances while cohort normalization is to compensate the score variability coming from the target models. Combination of these two methods can contribute to the verification procedure.

If we define

$$\mu'=\alpha\left(X\right)\mu_C+\left(1-\alpha\left(X\right)\right)\mu_T \tag{6}$$

$$\sigma'=\alpha\left(X\right)+\left(1-\alpha\left(X\right)\right)\sigma \tag{7}$$

where $\mu_C$ and $\mu_T$ are the mean scores of Cohort Norm and T-Norm respectively. Then the final smoothed T-Norm score is defined as [6]:

$$L'\left(X\right)=\frac{L_\lambda\left(X\right)-\mu'}{\sigma'} \tag{8}$$

It can be seen from (6, 7), the smoothing function involves a scaling factor to join the cohort normalizations scores and the T-Norm scores. The form of this factor is a sigmoid function which can be proved to lie between 0 and 1. If the smoothing factor $\alpha\left(X\right)$ is a const of 0, the proposed method can be seen as a kind of T-Norm, and $\alpha\left(X\right)=1$ denotes cohort normalization respectively.

# 4 Experimental Results

Totally 103 males and 111 females were involved in the experimental corpus. The recording was completed in 20 sessions and there was about a week's duration between two closet sessions. In each session 12 prepared digital sequences were spoken by each speaker, these digital speech data is used as training data. In addition, two other digital passwords are recorded each session to be used as testing data. To be particularly, the training data and testing data in our experiments were selected from different sessions and the largest intervals between training and testing were ten weeks.

As stated above, in model training stage, the target speaker model was generated by adapt an existing speaker independent but phoneme dependent HMM with the enrollment utterances of the speakers. Speaker independent but phoneme dependent HMMs used in our system were borrowed from a speech recognition application. The HMMs for the "initials" had three HMM states, and had five HMM states for the "finals". All HMM states consisted of 16 Gaussian distributions. The speech feature used was relative spectral PLP coefficients and their first order derivatives and second order derivatives.

## 4.1 Performance Comparison Between Cohort Normalization and T-Norm

In the first experiment, 50 females and 50 males were randomly chosen as normalization set. The models of these 100 speakers were trained during training time with the same procedure of the target speakers. The testing speech was compared with the claimed model and the 100 normalization models respectively. For T-Norm N (N is from 1 to 100) best scores were averaged as normalization parameter $\mu$ and standard variance $\sigma$ was computed too. For cohort normalization, the N-Best cohort models were also selected from the 100 models.

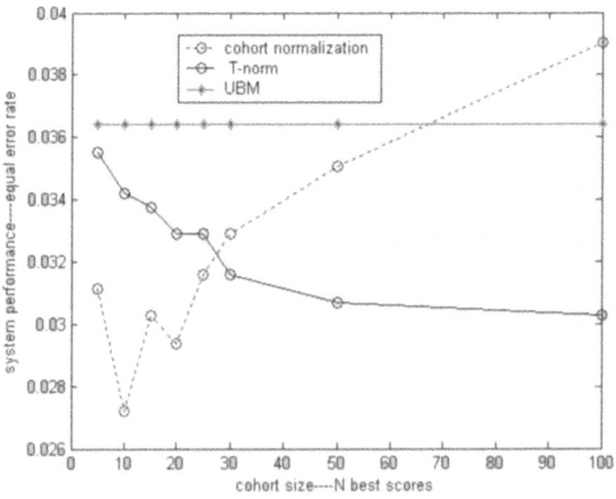


**Fig. 1.** Equal error rate transformation as the cohort size increases

It is shown in Fig.1 T-Norm performed better than Cohort Norm only when large number of normalization speakers was used. If the normalization set was small, cohort normalization performs better than T-norm. Best performance occurred when cohort normalization was applied with ten best models (Cohort normalization performs best when N is about 10%~20% of the whole norm set). The best performance of T-norm was occurred when all the 100 scores were used. In fact, within a 100 speakers' norm set, the error rate was decreasing monotonically which means more normalization speakers were needed to estimate the parameter.

### 4.2 Phoneme Level and Speaker Level Cohort Normalization

Because cohort normalization could perform well with small normalization set, in the second experiment a smaller normalization speaker set was then used for cohort normalization with only 40 impostors composed of 20 males and 20 females. Compared with baseline system which has equal error rate of 4.28%, the normalization decreased it to 3.12%, about 27% improvement (shown in Fig.2).

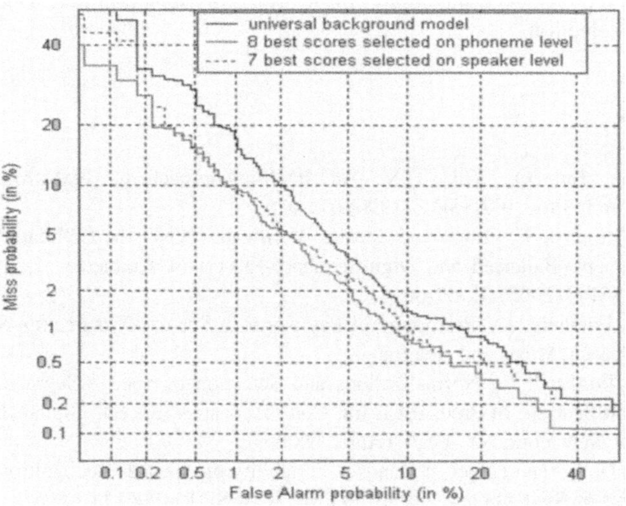


**Fig. 2.** DET curve of baseline system and cohort normalization system

If the best cohort models were chosen on the phoneme level and speaker level respectively, the experiments showed that phoneme level selection of best scores performed better than speaker level selection.

### 4.3 Proposed Normalization

As stated above T-Norm is dedicated to remove the score variability caused by testing utterances while cohort normalization is to compensate the score variability coming from the target models. Combination of these two methods can contribute to the verification procedure.

**Table 1.** The transform of EER as the changing of $\alpha$

| $\alpha$ | 0 | 0.2 | 0.5 | 0.6 | 0.7 | 0.8 | 1 |
|---|---|---|---|---|---|---|---|
| EER | 3.42% | 2.91% | 2.55% | 2.50% | 2.53% | 2.66% | 2.72% |

# 5 Conclusions

A text dependent speaker verification system with text prompted passwords is presented in this paper. Phoneme based HMM is a good choice as acoustic models to cope with the variable text situation. If impostor models or impostor scores are planned to be used for score normalization, the impostor utterances with the same content should be used for competitive normalization models or scores and the small vocabulary in our system make this feasible. Combination of T-Norm and Cohort Norm can make use of the information from both the testing utterances and the speaker models to improve the system performance. Cohort models can be selected both on the speaker level as the traditional approach and on the phoneme level as proposed in the paper. And the small unit level cohort selection performed better than speaker level selection.

# References

1. Che, C.W., Lin, Q., Yuk, D.S.: An HMM Approach to Text-prompted Speaker Verification. In Proc. ICASSP 2 (1996) 673-676
2. Kato, T., Shimizu, T., "Improved Speaker Verification Over the Cellular Phone Network Using Phoneme-Balanced and Digit-Sequence-Preserving Connected Digit Patterns", In Proc. ICASSP 2 (2003), II- 57-60
3. Melin, H., Lindberg, J.: "Prompting of Passwords in Speaker Verification System", KTH, Dept. of Speech, Music and Hearing
4. Li, K. P., Porter, J. E.: Normalizations and Selection of Speech Segments for Speaker Recognition Scoring. In Proc. IEEE Int. Conf. Acoustics, Speech, Signal Processing, vol. 1, 595-598, New York, NY, USA, (April 1988)
5. Reynolds, D. A.: The Effect of Handset Variability on Speaker Recognition Performance: Experiments on Switchboard Corpus. In Proc. ICASSP 1 (1996) 113-116
6. Hebert, M., Boies, D., Communication, N. : T-Norm for Text-Dependent Commercial Speaker Verification Applications: Effect of Lexical Mismatch. In Proc. ICASSP, 1 (2005) 729- 732
7. Matsui, T., Furui, S.: Concatenated Phoneme Models for Text Variable Speaker Recognition. In Proc. ICASSP, 2 (1993) 391-394
8. Auckenthaler, R., Carey, M., Lloyd-Thomas, H.: Score Normalization for Text-Independent Speaker Verification System. On Digital Signal Processing, 10 (2000) 42-54
9. Sturim, D.E., Reynolds, D.A.: Speaker Adaptive Cohort Selection for Tnorm in Text-Independent Speaker Verification. In Proc. ICASSP, 1 (2005) 741- 744
10. Colombi, J.M., Ruck, D.W., Anderson, T.R., Rogers, S.K., Oxley, M.: Cohort Selection and Word Grammar Effects for Speaker Recognition. In Proc. ICASSP 1 (1996) 85-88

# Identifying Modules in Complex Networks by a Graph-Theoretical Method and Its Application in Protein Interaction Networks

Rui-Sheng Wang[a], Shihua Zhang[c,d], Xiang-Sun Zhang[c,‡], and Luonan Chen[b,e,‡]

[a] School of Information, Renmin University of China, Beijing 100872, China
[b] Osaka Sangyo University, Osaka 574-8530, Japan
[c] Academy of Mathematics and Systems Science, CAS, Beijing 100080, China
[d] Graduate University of Chinese Academy of Sciences, Beijing 100049, China
[e] Institute of Systems Biology, Shanghai University, Shanghai 200444, China
[‡] zxs@amt.ac.cn, chen@eic.osaka-sandai.ac.jp

**Abstract.** Detecting community structure/modules in complex networks recently attracts increasing attention from various fields including mathematics, physics and biology. In this paper, we propose a method based on graph-theoretical clustering for identifying modularity structure in complex networks. Compared with the existing algorithms, this method, based on minimum spanning tree, has several advantages. For example, unlike many algorithms, this method is deterministic and not sensitive to the initialization. In addition, the method does not require a prior knowledge about the number of the modules. It can easily obtain the number of clusters by analyzing the edge weight distribution of minimum spanning tree. Moreover, this algorithm has computational compexity of polynomial-time with low order and can be used to deal with large-scale networks. Experimental results show that our method produces good results for real networks and can also uncover meaningful functional modules in protein interaction networks.

**Keywords:** Minimum spanning tree (MST); functional module; protein interaction network; complex network.

## 1 Introduction

Many complex systems in real world can be expressed by a network in which the agents are represented by nodes and the interactions between different agents are represented by edges [1,2,3]. Complex network has become a hot research topic in recent years. In addition to various statistical properties, such as small-world property, power-law degree distribution, network transitivity etc., community/modularity structure in complex networks has also attracted much attention. This is motivated by the fact that many complex networks such as social networks (the Internet [1], food webs [3]) and biological networks (protein-protein interaction networks [2,4,5,6], gene regulatory networks [7,8]) have modular structures. Modules, though without a consistent definition, are

D.-S. Huang, L. Heutte, and M. Loog (Eds.): ICIC 2007, LNAI 4682, pp. 1090–1101, 2007.

characterized by groups of individual nodes within which nodes are densely linked and between which the connection is sparser. Detecting modules has fundamental importance for exploiting the networks because such substructures often correspond to important functions. For example, a module in a protein interaction network may correspond to a well-known functional module or protein complex and play special biological roles in cellular system.

In recent years, many methods have been developed to detect modules in complex networks, such as betweenness-based methods [3,9], hierarchical clustering methods [9], spectral clustering methods [10], heuristic methods [11,12,13], and graph theoretic methods [14,15]. In 2004, Newman and Girvan [16] developed a modularity measure $Q$ to evaluate community structure. Generally, a bigger $Q$ corresponds to a better or more obvious modularity structure. Based on $Q$ as an objective function, another class of methods including heuristic random search algorithms aiming to maximize modularity are developed [17,18,19]. However, Fortunato and Barthélemy [20] recently found that modularity optimization may fail to identity modules smaller than a scale which depends on the size of the network and the degree of interconnectedness of the modules, even in cases where modules are unambiguously defined. In contrast to above methods, Gustafsson et al. [21] explore some standard ways of clustering data, such as $k$-means and hierarchical clustering, to cluster networks. More algorithms for detecting modularity structure in complex networks can be referred to the review papers [11,22]. A close related subject is the modular organization of biological systems. Abundant evidence suggests that biological systems are composed of interacting, separable, functional modules. Identifying these modules is essential to understand the organization of biological systems and cellular processes [5,6]. Some studies on protein interaction networks have been done in recent years but they generally focused on detecting highly connected protein modules and many peripheral proteins are omitted [5,6].

Although many module detection methods have been proposed, there are several obvious defects in them. For example, some methods are very sensitive to the initialization and may return different assignments of nodes at different initial partitions [13,19,21]. In addition, an important problem in detecting modules in complex networks is how to determine $k$, the number of modules in a network. Many methods are required to search a proper $k$ before detecting modules [10,12,21] which makes the total computation time-costly even with an efficient module-detecting subroutine. In this paper, we propose a graph-theoretical clustering approach for identifying functional modules in complex networks and protein interaction network. Compared with the existing algorithms, this method, based on minimum spanning tree, has several advantages. For example, this method is deterministic and not sensitive to the initialization. In addition, the method does not require a prior knowledge about the number of the modules. It can easily obtain the number of clusters by analyzing the edge weight distribution of minimum spanning tree. Moreover, this algorithm is of polynomial-time with low order and can be used to deal with large-scale networks. Experimental results show that our method produces good results for real networks and can also uncover meaningful functional modules in protein interaction networks.

## 2 Methods

Minimum spanning tree (MST) has been related with clustering analysis for a long time. Gower and Ross, 1969 [23] found that all the information required for the single linkage cluster of a set of data points is contained in their MST. Zahn, 1972 [24] proposed a graph-theoretical method based on MST to detect clusters from data with arbitrary shape. MST-based clustering algorithms have been extended later and applied by many authors in several fields such as pattern recognition and image processing [25,26,27,28]. In this section, we will describe how MST-based graph-theoretical clustering can be developed for identifying modules in large-scale networks.

A MST of a graph is a weighted connected spanning subgraph, where the sum of the edge weights is minimum among all of its spanning subgraphs. In a graph denoted by $G = (V, E)$, $V$ is the vertex set and $E$ is the edge set whose elements are represented by $e_{ij} = (v_i, v_j)$, where vertices $v_i, v_j \in V$. In addition to $V$ and $E$, a weight function $w$ is defined which determines a weight $w_{ij}$ for each edge $e_{ij}$. Finding a MST of $G$ is to search a connected subgraph $G' = (V, E')$ of $G$ such that $E' \subset E$ and the cost of $G'$ is minimized. The cost of $G'$ is the sum of its edge costs:

$$\sum_{e \in E'} w(e).$$

Obviously, MST is a connected acyclic graph. For a graph $G$ with $n$ vertices, its MST has exactly $n-1$ edges. Any connected component of a tree is called a subtree. A MST can be efficiently computed in $O(n^2)$ time using either Prim's or Kruskal's algorithm.

Given a complex network $G$ with adjacency matrix $A$, its MST can not provide much information for identifying clusters or modules, because such MST is obtained directly based on adjacent matrix of this network which can not characterize the similarities between nodes very well. Therefore, when MST is used in identifying modules in a network, a weight function $w$ should be defined to represent the overall dissimilarity/distances between different nodes, and acts as the objective function to be minimized. Generally, any feature matrix representing the similarities/distances between different nodes of a network can be developed into a weight function, such as shortest path matrix $P = [p_{ij}]_{n \times n}$, where $p_{ij}$ is the length of the shortest path between nodes $v_i$ and $v_j$, diffusion kernel feature matrix

$$K = \exp(\beta L) = \lim_{n \to \infty} \left(1 + \frac{\beta L}{n}\right)^n \quad (1)$$

where $\exp(L)$ represents the matrix exponential of the Laplacian matrix $L$. In this paper, a dissimilarity based on adjacent matrix is introduced which is related to the number of the common partners of two nodes:

$$H(i,j) = \frac{\mathrm{HamDis}(i,j)}{d_i + d_j}. \tag{2}$$

where $\mathrm{HamDis}(i,j)$ is the Hamming distance between vectors $a_i$ (the $i$th row of the adjacency matrix $A$) and $a_j$, and $d_i$ is the degree of the node $v_i$. We can see that the more common neighbors two nodes $v_i$ and $v_j$ have, the more similar they are in terms of this measure.

With such a weight function defined from dissimilarity matrix, the nodes of the network $V = \{v_1, v_2, \cdots, v_n\}$ can be seen as data points to be clustered and their pairwise distances are known: $d_{v_i,v_j} = w_{ij}$ denotes the distance between any $v_i$ and $v_j$. In fact, $w$ determines a complete graph $G_c$ for the network $G$. The MST of $G_c$ can be found by Prim's or Kruskal's algorithm. Since MST is a connected acyclic graph, removing an edge leads to two subtrees. Removal of a set of edges from the MST will lead to a collection of connected subgraphs of $G_c$, which can be considered as clusters. From the fact that nodes belonging to same modules tend to be similar and the edge weights between them are relatively small, we are interested in finding the inconsistent edges (with high edge weights) to be removed. Representing a set of nodes in a network as a simple tree structure seems to lose some inter-node relationships, but no essential information is lost for the purpose of clustering [27]. Xu et al. [27] proved that each cluster corresponds to one subtree, which does not overlap the representing subtree of any other cluster. Hence a clustering problem is equivalent to identifying these subtrees through solving a tree partitioning problem.

The number of inconsistent edges to be removed should be determined according to the edge weight distribution of the MST. For networks with obvious modularity structure, the inconsistent edges of their corresponding MST are also clear. For general networks, we can adopt a threshold $\theta_0$ so that the edges with cost greater than $\theta_0$ will be removed. The value of $\theta_0$ is determined according to the edge cost distribution of the MST. For example, Zahn [24] suggested a global threshold value $\delta$ which considers the distribution of the data in the feature space and is based on the average weight (distances) of the MST:

$$\delta = \lambda \frac{1}{n-1} \sum_{e \in E'} w(e)$$

where $\lambda$ is a user-defined parameter.

Unlike the situation in general data points, when MST-based clustering is applied in finding modules in a network, one thing should be considered. If a node has very low degree (an extreme case, leaf nodes) and all of its degree contributes to a module, the cost of the edges from this node to other nodes in this module will be high. If we adopt threshold to determine inconsistent edges, such edges will be wrongly removed and make the node isolated. For such isolated parts, a post-processing should be made to assign them to the most associated modules.

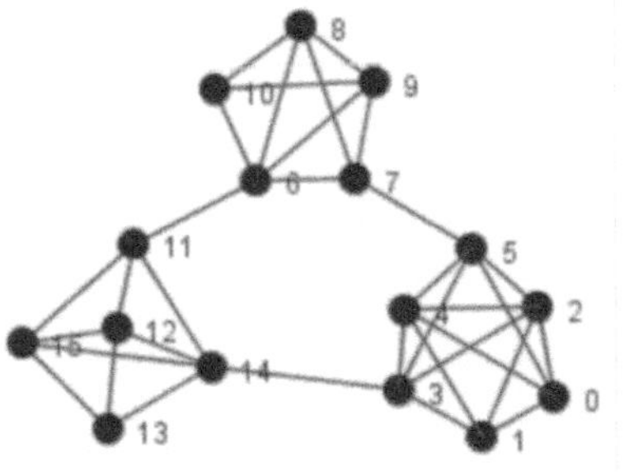


**Fig. 1.** An artificial network with three modules

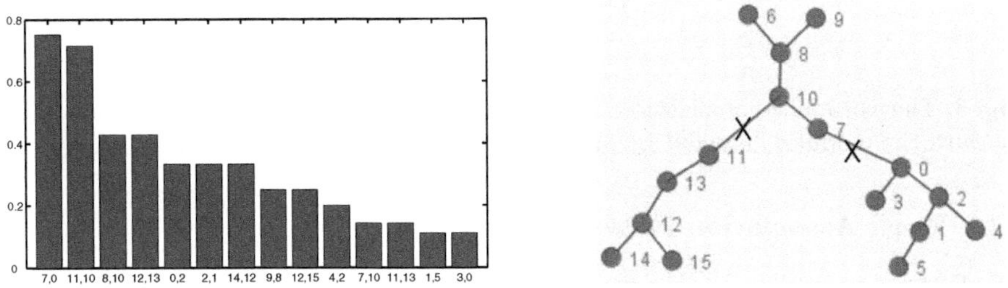


(a) (b)

**Fig. 2.** (a) The edge weight distribution of the MST (b) The MST corresponding to the artificial network

In Figure 1, we give an example to illustrate our method. The network in this example has 16 nodes with three modules. After computing the dissimilarity matrix of this network (i.e. a complete graph $G_c$ with a distance function), a MST of $G_c$ can be found which is illustrated in Figure 2(b). The weight distribution of the edges of this tree is illustrated in Figure 2(a) , from which we can find two distinct inconsistent edges: (7, 0), (11, 10). When these two edges act as the boundaries of modules, we can find three communities $\{0,1,2,3,4,5\}, \{6,7,8,9,10\}, \{11,12,13,14,15\}$ with modularity $Q = 0.5698$ which are exactly consistent with what we see from the network. If we continue to select (8, 10) and (12, 13) as the inconsistent edges, the modularity $Q$ of the five-cluster will be much lower than that of the three-cluster partition, and furthermore, the inconsistence of these edges is not prominent.

## 3 Experiment Results

In this section, we apply our method to two well studied real-world networks and a large protein-protein interaction network to test its ability of detecting meaningful modules. The algorithm is coded and implemented by Python 2.4.

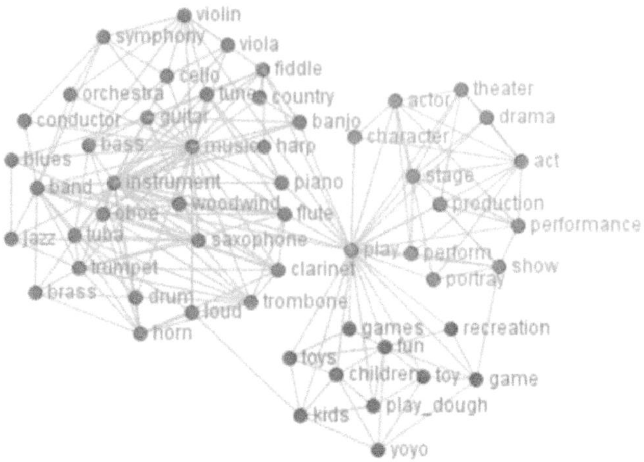


**Fig. 3.** The word 'play' association network in the South Florida Free Association norm list with the modules detected by our algorithm (indicated by different vertex colors)

### 3.1 Word Association Network

The first network that we adopt is the word association network picked from the South Florida Free Association norm list (http:// www.usf.edu/ FreeAssociation/). In the South Florida Free Association norm list, the weight of a directed link from one word to another indicates the frequency with which the people in the survey associate the end point of the link with its starting point. Here we use the word 'play' association network which has been replaced with an undirected one and tested in [14,15]. This network has 53 nodes representing different words and 197 association edges. After computing the distance function $w$ according to the dissimilarity measure (2), a complete graph $G_c$ is obtained for this network. By checking the edge weight distribution of $G_c$'s MST (not shown due to the limit of space), we find two inconsistent edges with cost 0.706 and 0.6. When removing these two edges, we can identify three modules shown in Figure 3, from which we see that words with frequent associations are in the same modules. In fact, if we set the threshold $\theta_0$ as $0.56 \sim 0.6$, we can always obtain such a partition. When $\theta_0$ is set as $0.5 \sim 0.56$, such a partition can also be obtained except that 'play' becomes an isolated node (owing to its strong association with three modules). According to the post-processing described in Methods, 'play' is placed into a determined module shown in Figure 3. The edges with cost lower than 0.5 are too many to be inconsistent edges. This result indicates that our method, though with a threshold, is not so sensitive to it. For comparison, we apply another graph-theoretical method — Clique Percolation Method (CPM) [14,15] to this network. CPM obtains almost the same partition except that 'play' has membership in all three modules which is reasonable in some

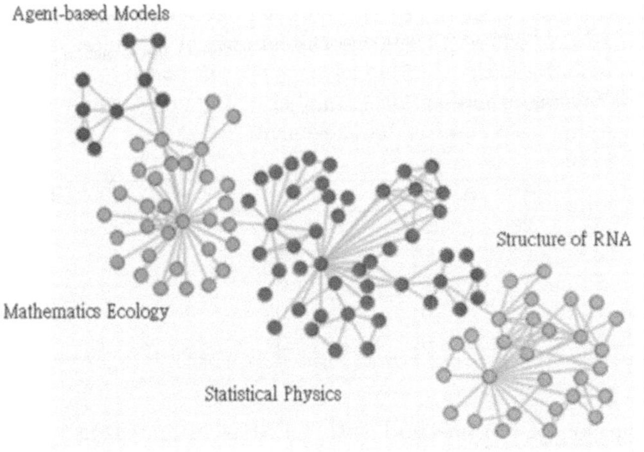


**Fig. 4.** The scientific collaboration network with the modules detected by our algorithm (indicated by different vertex colors)

manner. We point the future potential improvement of our method for fuzzy module detection in Conclusion and Discussion Section.

### 3.2 Scientific Collaboration Network

Now we test another network — the scientific collaboration network collected by Girvan and Newman [3] and also widely used to evaluate various module-detection algorithms. The modularity structure in this network is not so obvious like 'play' word association network. This network consists of 118 nodes (scientists) and the edges represent the collaboration relationships among scientists. The distance function is constructed based on (2). According to the edge weight distribution of the MST, when $\theta_0$ is set as 0.6, a rough partition with three clusters can be obtained. When $\theta_0$ is set as 0.5, the network is divided into 6 clusters which is shown by different vertex colors in Figure 4. The primary divisions of the network by our method are visually reasonable. Although the deterministic partition of this network is unknown, the detected divisions are consistent with those in [3] on the whole. The disciplines corresponding to the modules are marked in the figure. The modules denoted by agent-based models is the least well defined community and represents a group of scientists using agent-based models to study problems in economics and traffic flow [3]. The module neighboring to the agent-based model community is comprised of a group of scientists who work on mathematical models in ecology and form a strong modularity structure. The next community representing a group working primarily in statistical physics is subdivided into smaller groups. The last community is a group working primarily on the structure of RNA and has a strong modularity structure.

**Table 1.** The comparison of the results between our method and CPM on scientific collaboration network, where $|\mathcal{C}|$ denotes the number of modules, $n_c : 1 \sim 5$ denotes the number of modules with $1 \sim 5$ nodes. $n_c : 6 \sim 10$ and $n_c :> 10$ have the similar meaning. Total coverage represents the number of the nodes contained in modules. $k$ is the clique size that CPM uses as basic elements.

| Methods | $\|\mathcal{C}\|$ | $n_c : 1 \sim 5$ | $n_c : 6 \sim 10$ | $n_c :> 10$ | Total coverage |
|---|---|---|---|---|---|
| CPM, $k = 3$ | 26 | 22 | 3 | 1 | 110 |
| CPM, $k = 4$ | 12 | 11 | 1 | 0 | 56 |
| CPM, $k = 5$ | 5 | 5 | 0 | 0 | 25 |
| MST, $\theta_0 = 0.6$ | 3 | 0 | 0 | 3 | 118 |
| MST, $\theta_0 = 0.5$ | 6 | 0 | 3 | 3 | 118 |

The comparison of our method with CPM is summarized in Table 1. From Table 1, we can see that although CPM has good performance in networks with clear modularity structure, it does not do well in sparse network owing to the tight requirement of clique. CPM with $k = 3$ (3-clique percolation) almost cover all nodes, but it returns too many modules with small size which is obviously not consistent with the network structure and the prior knowledge of the discipline distribution. CPM with $k = 4$ and $k = 5$ cover very few nodes (no more than half of nodes are in modules). In contrast, the size and number of the modules detected by our method are reasonable and consistent with the prior knowledge of the discipline distribution.

### 3.3 Protein-Protein Interaction Network

Finally, we consider a larger network — a protein-protein interaction network of *S. cerevisiae*. This network has 1257 proteins as nodes and 6835 interactions as edges. The original data set [2] has 1298 proteins involved in 54406 interactions. All of these protein interactions have a confidence score. Here we only use the reliable interactions (the scores are not lower than 0.3). By applying the proposed method, we obtain a MST of the corresponding complete graph. According to the edge weight distribution of the MST, we set $\theta_0 = 0.55$ and totally find 73 modules. Among these, the largest one has 65 proteins and the smallest one has 4 proteins. Although protein functional modules are formed not only by the topology of the network, protein-protein interactions are believed to be very informative for protein function. The analysis result of the detected modules based on the function annotation table of MIPS [29] (http://mips.gsf.de/) is summarized in Figure 5. We find that most of the modules detected by our method match MIPS function categories with high consistency. We also can see that most of modules match with more than one function categories. This is owing to the fact that most proteins are involved multiple biological processes. For example, all the proteins in module 15, module 24 and module 40 shown in Figure 6 and Table 2 have multiple functions. In addition to the ability of

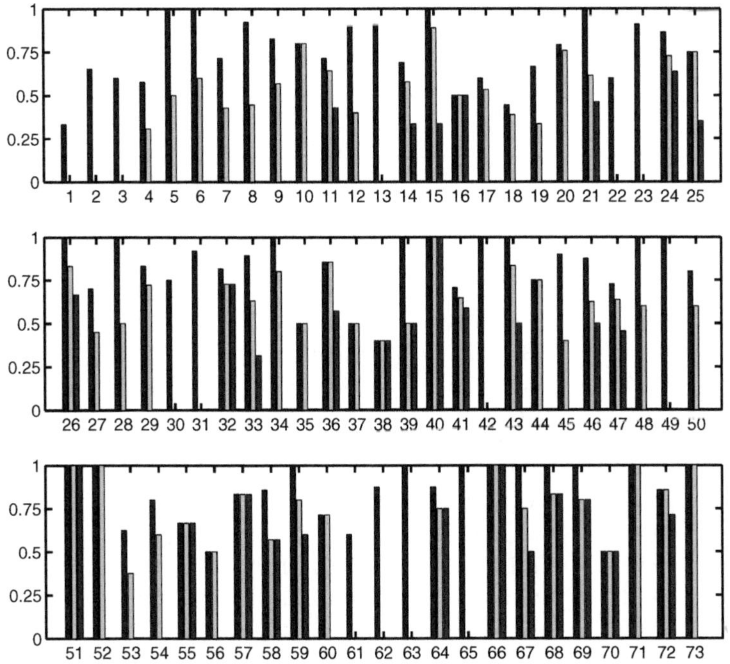


**Fig. 5.** The matching ratio of the detected modules with MIPS functional categories by analysis of constituent protein functions. Those modules with more than one matched functional categories have one ratio for each function category.

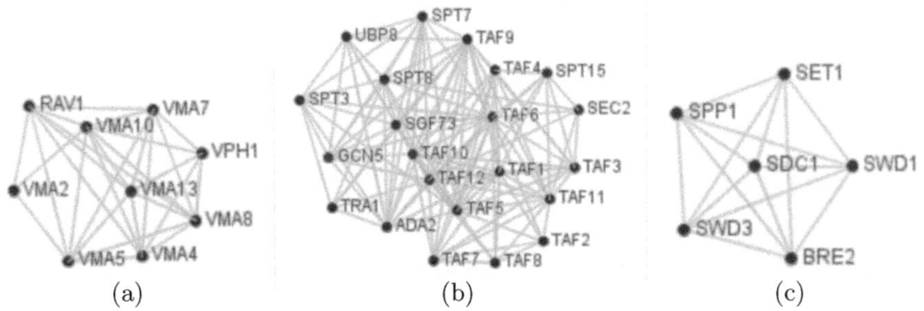


**Fig. 6.** Three examples of detected modules by the propsed method: (a) Module 15 (b) Module 40 (c) Module 24. Their functional annotations are listed in Table 2.

detecting meaningful functional modules, our method is very fast and solves this network only in several seconds on PC. Such a result indicates that it can be used to deal with large-scale biological networks.

**Table 2.** Functional annotations of three modules showed in Figure 6. The code represents functional category. The proteins involved in functions inconsistent with main functions of the detected modules are denoted in Italic type.

| Modules | Genes | ORFs | Functional categories | | |
|---|---|---|---|---|---|
| Module 15 | VPH1 | YOR270C | 34.01.01.03 | 20.01.01.01 | 14.10 |
| | VMA7 | YGR020C | 34.01.01.03 | 20.01.01.01 | |
| | VMA10 | YHR039C-A | 34.01.01.03 | 20.01.01.01 | |
| | VMA8 | YEL051W | 34.01.01.03 | 20.01.01.01 | |
| | VMA4 | YOR332W | 34.01.01.03 | 20.01.01.01 | 14.10 |
| | VMA13 | YPR036W | 34.01.01.03 | 20.01.01.01 | |
| | VMA2 | YBR127C | 34.01.01.03 | 20.01.01.01 | |
| | VMA5 | YKL080W | 34.01.01.03 | 20.01.01.01 | 14.10 |
| | RAV1 | YJR033C | 34.01.01.03 | | |
| Module 24 | TAF2 | YCR042C | 10.03.01.01.01 | 11.02.03.04 | |
| | TAF8 | YML114C | | 11.02.03.01 | |
| | TAF4 | YMR005W | | 11.02.03.01 | |
| | TAF3 | YPL011C | 10.03.01.01.01 | 11.02.03.04 | |
| | TAF11 | YML015C | 10.03.01.01.01 | 11.02.03.04 | |
| | TAF1 | YGR274C | 10.03.01.01.01 | 11.02.03.04 | 14.07.04 |
| | TAF5 | YBR198C | 10.03.01.01.01 | 11.02.03.04 | 14.07.04 |
| | TAF6 | YGL112C | 10.03.01.01.01 | 11.02.03.04 | 14.07.04 |
| | TAF10 | YDR167W | 10.03.01.01.01 | 11.02.03.04 | 14.07.04 |
| | TAF9 | YMR236W | 10.03.01.01.01 | 11.02.03.04 | 14.07.04 |
| | TAF12 | YDR145W | 10.03.01.01.01 | 11.02.03.04 | 14.07.04 |
| | *SEC2* | *YNL272C* | *18.02.03* | *20.09.07* | *20.09.16.09.03* |
| | ADA2 | YDR448W | 10.01.09.05 | 11.02.03.04 | 14.07.04 |
| | SGF73 | YGL066W | | | 14.07.04 |
| | SPT8 | YLR055C | 10.01.09.05 | 11.02.03.04 | 14.07.04 |
| | GCN5 | YGR252W | 10.01.09.05 | 11.02.03.04 | 14.07.04 |
| | SPT7 | YBR081C | 10.01.09.05 | 11.02.03.04 | 14.07.04 |
| | TRA1 | YHR099W | 10.01.09.05 | 11.02.03.04 | 14.07.04 |
| | UBP8 | YMR223W | | | 14.07.05 |
| | SPT3 | YDR392W | 10.01.09.05 | 11.02.03.04 | 14.07.04 |
| | TAF7 | YMR227C | 10.03.01.01.01 | 11.02.03.04 | |
| | SPT15 | YER148W | | 11.02.03.04 | |
| Module 40 | SWD1 | YAR003W | 10.01.09.05 | 14.07.09 | 42.10.03 |
| | BRE2 | YLR015W | 10.01.09.05 | 14.07.09 | 42.10.03 |
| | SET1 | YHR119W | 10.01.09.05 | 14.07.09 | 42.10.03 |
| | SDC1 | YDR469W | 10.01.09.05 | 14.07.09 | 42.10.03 |
| | SWD3 | YBR175W | 10.01.09.05 | 14.07.09 | 42.10.03 |
| | SPP1 | YPL138C | 10.01.09.05 | 14.07.09 | 42.10.03 |

## 4 Conclusion and Discussion

Detecting community structures in complex networks recently attracts much attention in various fields. Especially, detecting functional modules in protein interaction networks has particular significance for biologists. In this paper, we introduced a graph-theoretical clustering method to detect modules in complex networks. Our method has very good performance in real-world networks and protein interaction network. In the protein interaction network, we have detected 73 modules which have high functional consistency or even multiple functional consistencies. This method is deterministic and not sensitive to the initialization. An advantage of this method is that it does not require to find the tentative number of modules in a network. In addition, the proposed method is of polynomial-time with low order and thus can be applied to large-scale networks such as biological networks.

A potential improvement of our method is that we can further consider how to assign a node to more than one module. This is reasonable for some nodes with membership to multiple modules like 'play' in the word association network or a bridge protein between two functional modules. As a future topic, we plan to solve this point by incorporating fuzzy clustering into our framework as Vathy et al. [30] have done.

**Acknowledgement.** This research work is supported by Research Foundation of Renmin University of China "Several fundamental problems in complex networks" (No.06XNB054) and the Ministry of Science and Technology, China, under grant No.2006CB503905.

## References

1. Gibson, D., Kleinberg, J., Raghavan, P.: Inferring Web Communities from Link Topology. In Proceedings of the 9th ACM Conference on Hypertext and Hypermedia, (1998), N.Y. ACM Press
2. Gavin, A.C., Aloyand, P., Grandi, P., et al.: Proteome Survey Reveals Modularity of the Yeast Cell Machinery. Nature 440(2006) 631-636
3. Girvan, M., Newman, M.E.J.: Community Structure in Social and Biological Networks. Proc. Natl. Acad. Sci. USA, 99(2002) 7821-7826
4. Chen, L., Wu, L-Y., Wang, Y., Zhang, X-S.: Inferring Protein Interactions from Experimental Data by Association Probabilistic Method. Proteins, 62(2006) 833-837
5. Zhang, S., Ning, X.-M., Zhang, X.-S.: Identification of Functional Modules in a PPI Network by Clique Percolation Clustering. Computational Biology and Chemistry, 30(2006) 445-451
6. Zhang, S., Liu, H.-W., Ning, X.-M., Zhang, X.-S.: A Graph-Theoretic Method for Mining Functional Modules in Large Sparse Protein Interaction Networks. Proceeding of Sixth IEEE International Conference on Data Mining-Workshops (ICDMW'06), (2006) 130-135
7. Wang, R., Zhou, T., Jing, Z., Chen, L.: Modelling Periodic Oscillation of Biological Systems with Multiple Time Scale Networks. Systems Biology, 1(2004) 71-84
8. Wang, Y., Joshi,T., Xu, D., Zhang, X-S., Chen, L.: Inferring Gene Regulatory Networks from Multiple Microarray Datasets. Bioinformatics, 22(2006), 2413-2420
9. Wasserman, S., Faust, K.: Social Network Analysis: Methods and Applications. (1994), Cambridge University Press

10. White, S., Smyth, P.: A Spectral Clustering Approach to Finding Communities in Graphs. SIAM International Conference on Data Mining, (2005)
11. Danon, L., Daz-Guilera, A., Duch, J., Arenas, A.: Comparing Community Structure Identification. J. Statist. Mech.: Theory and Experiment, 09(2005), P09008
12. Angelini, L., Boccaletti, S., Marinazzo, D., et al.: Fast Identification of Network Modules by Optimization of Ratio Association. (2006), cond-mat/0610182
13. Reichardt, J., Bornholdt,S.: Detecting Fuzzy Community Structures in Somplex Networks with a Potts Model. Physical Review Letters, 93(2004), 218701
14. Pallal, G., Derenyi, I., Farkasl, I., et al.: Uncovering the Overlapping Community Structure of Complex Networks in Nature and Society. Nature, 435(2005) 814-818
15. Vicsek, T.: Phase Transitions and Overlapping Modules in Complex Networks. Physica A, 378(2007) 20-32
16. Newman,M.E.J., Girvan,M.: Finding and Evaluating Community Structure in Networks. Phys. Rev. E 69(2004) 026113
17. Duch, J., Arenas,A.: Community Identification Using Extremal Optimization. Phys. Rev. E, 72(2005) 027104
18. Newman, M.E.J.: Modularity and Community Structure in Networks. Proc. Natl Acad. Sci. USA 103(23)(2006) 8577-8582
19. Guimer, R., Amaral, L.A.N.: Functional Cartography of Complex Metabolic Networks. Nature, 438(2005) 895-900
20. Fortunato, S., Barthélemy, M.: Resolution Limit in Community Detection. Proc. Natl. Acad. Sci. USA, 104(2007) 36-41
21. Gustafsson, M., Hö rnquista,M., Lombardi, A.: Comparison and Validation of Community Structures in Complex Networks. Physica A, 367(2006) 559-576
22. Newman, M.E.J.: Detecting Community Structure in Networks. Eur. Phys. J. B 38(2004) 321-330
23. Gower, J.C., Ross, G.J.S.: Minimum Spanning Trees and Single-Linkage Cluster Analysis. Applied Statistics, 18(1969) 54-64
24. Zahn, C.T.: Graph-Theoretical Methods for Detecting and Describing gestalt clusters. IEEE Transaction on Computers, C20(1971) 68-86
25. Pävinen, N.: Clustering with a Minimum Spanning Tree of Scale-free Like Structure. Pattern Recognition Letters, 26(2005) 921-930
26. Varma, S., Simon, R.: Iterative Class Discovery and Feature Selection Using Minimal Spanning Trees. BMC Bioinformatics, 5(2004) 126-134
27. Xu, Y., Olman, V., Xu, D.: Clustering Gene Expression Data Using a Graph-Theoretic Approach: an Application of Minimum Spanning Trees. Bioinformatics, 18(2002) 536-45
28. Xu, Y., Olman, V., and Uberbacher, E.C.: A Segmentation Algorithm for Noisy Images: Design and Evaluation. Pattern Recognition Letters, 19(1998) 1213-1224
29. Mewes, H.W., Frishman, D., Guldener, U. et al.: MIPS: a Database for Genomes and Protein Sequences. Nucleic Acids Res, 30(2002) 31-34
30. Vathy-Fogarassy, A., Feil, B., Abonyi, J.: Minimal Spanning Tree Based Fuzzy Clustering. ENFORMATIKA Transactions on Engineering, Computing and Technology, 8(2005) 7-12

# Autonomous Kinematic Calibration of the Robot Manipulator with a Linear Laser-Vision Sensor

Hee-Jun Kang[1], Jeong-Woo Jeong[2], Sung-Weon Shin[2], Young-Soo Suh[1], and Young-Schick Ro[1]

[1] School of Electrical Engineering, University of Ulsan, South Korea
[2] Graduate School of Electrical Engineering, University of Ulsan, South Korea

**Abstract.** This paper presents a new autonomous kinematic calibration technique by using a laser-vision sensor called "Perceptron TriCam Contour". Because the sensor measures by capturing the image of a projected laser line on the surface of the object, we set up a long, straight line of a very fine string inside the robot workspace, and then allow the sensor mounted on a robot to measure the point intersection of the line of string and the projected laser line. The data collected by changing robot configuration and measuring the intersection points are constrained to on a single straght line such that the closed-loop calibration method can be applied. The obtained calibration method is simple and accurate and also suitable for on-site calibration in an industrial environment. The method is implemented using Hyundai VORG-35 for its effectiveness.

**Keywords:** Autonomous Kinematic Calibration, Linear Laser-Vision Sensor. Closed-Loop Constraints, Straight Line String.

## 1 Introduction

In order to extend the use of an industrial robot to more sophisticated tasks such as position correction, visual servo and inspection of 3D parts, a linear laser vision sensor is often mounted on a robot hand. Whenever a sensor is mounted on a robot, it is important to determine the kinematic relationship between the sensor and hand coordinate frames for accurate measuring and precise robot positioning. The related kinematic parameters are 3 rotation and 3 translation parameters. The problem of determining these parameters is often referred to as SCP(sensor center position) calibration. In this paper, a linear laser sensor called "Perceptron TriCam Contour Sensor" is mounted on a robot hand and the SCP calibration algorithm and entire robot kinematic calibration algorithm is presented for the most accurate use of the robot-laser vision system.

Autonomous robot calibration is defined as the automated process of determining a robot`s model by using its internal sensors[1]. In this case, a kind of constraints must be deduced from the configuration of the calibration system: task constraints utilizing laser line tracking[2], plane measurements[1,4] and plane constraint with implicit loop method[5]. The presented SCP calibration can be understood as an autonomous calibration with consideration of the linear laser-vision sensor called Perceptron. The

D.-S. Huang, L. Heutte, and M. Loog (Eds.): ICIC 2007, LNAI 4682, pp. 1102–1109, 2007.

jig suitably devised for the Perceptron allows the closed-loop constraints, only using robot joint readings and Perceptron readings. Therefore, the calibration scheme can be implemented autonomously and is suitable for on-site calibration in an industrial environment.

In this paper, autonomous kinematic calibration is presented for the most accurate use of the robot-Perceptron system. First, Section 2 gives the brief description of Perceptron sensor. With this sensor, the problem of autonomous kinematic calibration can be formulated based on a specially devised jig and explains its related parameter identification procedure in Section 3. In Section 4, the obtained algorithm is implemented to show its effectiveness, on the Hyundai VORG-35 industrial robot and the results are discussed in terms of the constraint violation quantity defined in this paper.

## 2 A Linear Vision Sensor Called Perceptron

This section simply gives the brief description of Perceptron TriCam Contour Sensor for better understanding of the following sections. Perceptron sensor is a fusion sensor of laser and vision, and gives more accurate measurements under the variation of environment than the conventional vision systems. The sensor measures by capturing the shape and position of the projected laser line as it strikes the contour of the surface being measured. The software calculates two dimensional results(Y, Z coordinate values) with respect to the sensor coordinate frame. Figure 1 shows the view of measuring the gap and flush by using the Perceptron sensor and Fig. 2 shows the Process Window on a PC which processes the image data and computes measured values, corresponding to the measurement depicted in Fig. 1.

**Fig. 1.** The View of Measuring Gap and Flush by Perceptron

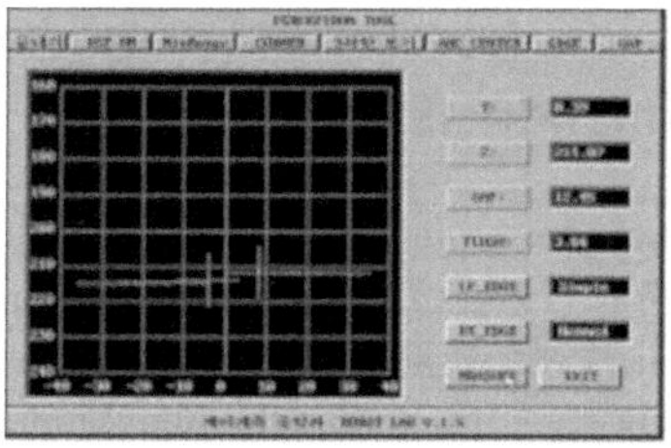

**Fig. 2.** The Process Window for the by Measurement shown in Fig. 1

## 3 Autonomous Kinematic Calibration

Unlike the robot calibration with external end-point system measurements[6,7], autonomous calibration uses only robot's internal sensors and instead, requires the existence of internal constraints deduced from the entire calibration system. And no human intervention during measurement phase is allowed so that low cost and on- site calibration be possible in an industrial environment. Fig. 3 shows the system being considered currently. The Perceptron sensor is mounted on a robot hand. A specially devised jig is very simple straight line string which made of a fine steel cable and just passes through the robot work space. Since the intersection point of laser line and fine string line will be a measurement point, the diameter of a string line must be small enough to reduce the measurement error.

The required coordinate frames for our problem are described in Fig. 4 and Fig. 5. The base coordinate frame {B} of the manipulator is set up on the string seen in Fig. 4., which comes from mutually perpendicular line to both the string line and the first joint axes line, following DH parameter convention. In other words, we set the z-axes of the base coordinate frame to be coincided with the string line such that any measurement point by Perceptron sensor is always located on the z-axes line of the base coordinate frame {B}. It leads us to have a closed loop constraints. The detailed procedure to get constraint equation is now explained.

As seen in Fig. 5, the transformation equation between {B}-frame and {1}-frame is

$$T_1^{\,B} = Trn\ (x)Rot\ (a)Trn\ (d_1)Rot\ (\theta_1) \tag{1}$$

The rest of coordinate transformation from {1}-frame to {6}-frame could be easily obtained as following DH-parameter convention and the careful consideration is required in case the manipulator has nearly parallel neighboring joints. The Modified DH parameter transformation suggested by Hayati[7] could be used. Our sample manipulator is the case.

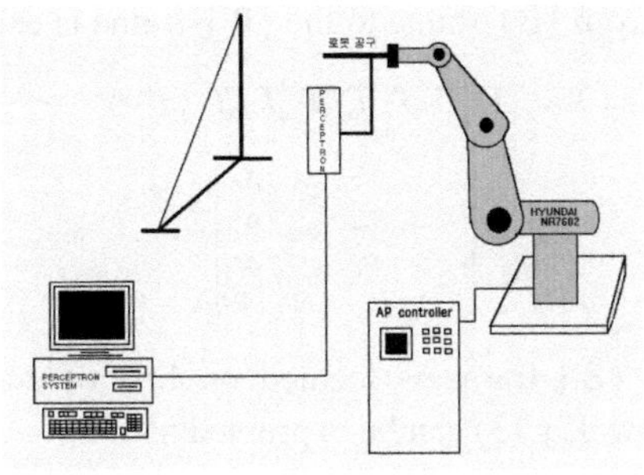


**Fig. 3.** Autonomous Kinematic Calibration System

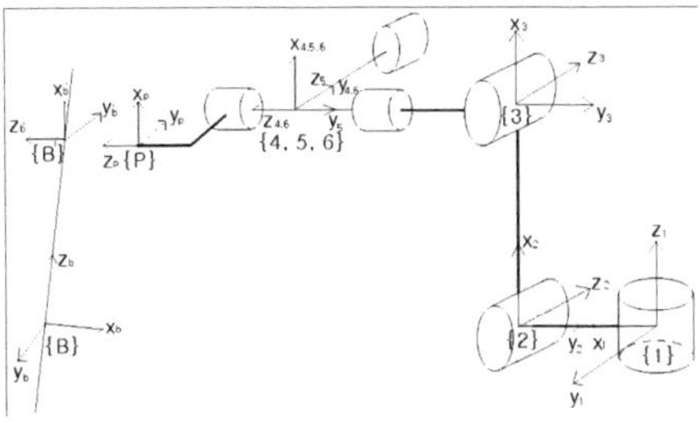

**Fig. 4.** Coordinate Assignment of Calibration System

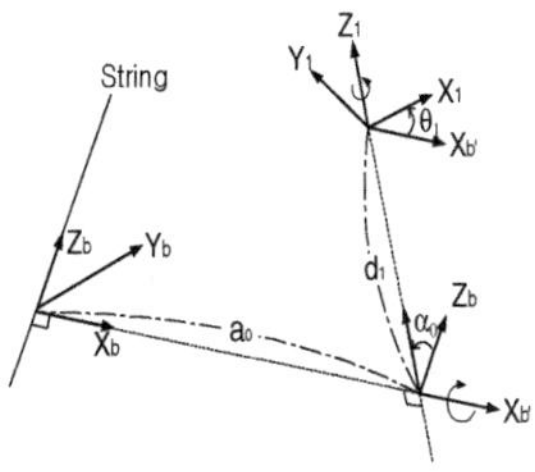


**Fig. 5.** Coordinate Transformation from to{B} to {1}

The transformation from {6}-frame to the Perceptron coordinate frame {P} is initially modeled with 3 translation and 3 rotation parameters because its convenient placement doesn't allow to use 4 DH parameters. However, the $\theta_{offset}$ of joint 6 and 3 translation parameters have the dependency on each other and causes singularity of identification procedure. Therefore, this relationship is modeled with 2 translation and 3 rotation parameters. The initial values of the 6 parameters are used as its design value, but their identification is the main purpose of the SCP(sensor center position) calibration and autonomous kinematic calibration.

As explained in Sec. 2, the Perceptron sensor system gives two dimensional results(Y, Z coordinate values) with respect to the Perceptron sensor coordinate frame {P}. The transformation from {P}-frame to {B'}-frame defined at the intersection point B' of a laser line and a fine string in Fig. 4 is described as

$$ {}^{P}_{B'}T = \begin{bmatrix} & I_{3\times3} & & 0 \\ & & & y_p \\ & & & z_p \\ 0 & 0 & 0 & 1 \end{bmatrix} \tag{2} $$

where $y_p$, $z_p$ are sensor measurement values and $x_p$ is always equal to 0, because a laser line is identical to x-axes of {P}-frame due to Sensor Characteristics.

The transformation from {B}-frame to the {B'}-frame is expressed as

$$ \begin{aligned} {}^{B}_{B'}T &= {}^{B}_{1}T\,{}^{1}_{2}T\,{}^{2}_{3}T\,{}^{3}_{4}T\,{}^{4}_{5}T\,{}^{5}_{6}T\,{}^{6}_{P}T\,{}^{P}_{B'}T \\ &= \begin{bmatrix} & {}^{B}_{B'}R_{3\times3} & & \Delta x \\ & & & \Delta y \\ & & & \Delta z \\ 0 & 0 & 0 & 1 \end{bmatrix} \end{aligned} \tag{3} $$

Since the origin of {B'}-frame is located on line of z-axes of {B}-frame, the $\Delta x$, $\Delta y$ should be 0. Now, Eq. (3) can be expressed as

$$ {}^{B}_{B'}T = \begin{bmatrix} & {}^{B}_{B'}R_{3\times3} & & 0 \\ & & & 0 \\ & & & \Delta z \\ 0 & 0 & 0 & 1 \end{bmatrix} \tag{4} $$

Likewise, two constraints $\Delta x = \Delta y = 0$ are obtained from one sensor measurement corresponding to one robot configuration such that we could derive the autonomous kinematic calibration algorithm by using conventional least square method. Its algorithm procedure is as follows.

From each sensor measurement, 2x1 position vector (r) is set as the position values x, y from {B} to {B'} and the real position vector(r*) is equal to 0 since x = y = 0. The difference between them leads the nonlinear Eq. (3) to the linearlized Eq. (5) as

$$\gamma_i - \gamma^* = \gamma_i = \Delta\gamma_i = [J_{\phi i}^{xy}]\Delta\Phi = C_i\Delta\Phi$$

$$= [[J_{\theta i}^{xy}]\vdots[J_{ai}^{xy}]\vdots[J_{\alpha i}^{xy}]\vdots[J_{di}^{xy}]] \begin{bmatrix} \Delta\theta_{offset} \\ \Delta a \\ \Delta\alpha \\ \Delta d \end{bmatrix}, \ i = 1 .... n \tag{5}$$

where $[J_i^{xy}] = [[J_i^x]\vdots[J_i^y]]^T$ and a, $\alpha$ , d are the link length, link twist, link offset respectively, and also, J is the conventional Jacobian matrix and the real joint angle is assumed to be $\theta = \theta_R + \theta_{offset}$ where $\theta_R$ is joint readings and $\theta_{offset}$ is joint error due to the non-geometric factors such as joint compliance, friction and joint clearance. Now, n equations in Eq. (5) could be augmented into a matrix form and simply expressed as

$$\Delta r = C\Delta\Phi \tag{6}$$

Based on the iterative least square method, the procedure to identify the kinematic parameters($\Phi$) is given as follows:

1) Compute $\Delta r_i$ and $C_i$ based on the current nominal values $\theta_i, a_i, d_i, \alpha_i$
2) $\Delta\Phi_i = (C_i^T C_i)^{-1} C_i^T \Delta r_i$
3) $\Phi_{i+1} = \Phi_i + \Delta\Phi_i$
4) Repeat the above steps until $\Delta\Phi$ falls within the considered tolerance value near zero.

In order to prevent the transient singularity, step 2) can be replaced by $\Delta\Phi = (C^T C + \lambda I)^{-1} C^T \Delta\gamma$ where I and λ are the identity matrix and a weight value, respectively. Selection of a weight value affects the convergence characteristics of the above algorithm and the switching technique can be considered for control of convergence speed[6].

## 4 Real Implementation and Results

In this section, the presented algorithm to validate its effectiveness has been implemented on the Hyundai VORG 35 robot and its results are discussed in terms of

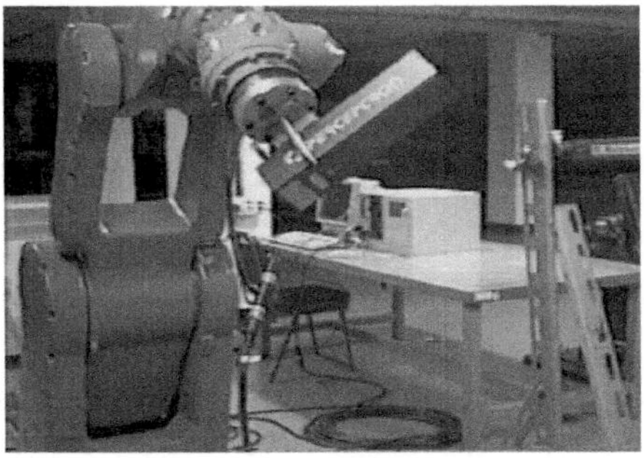

**Fig. 6.** Experimental Setup

the constraint violation quantity defined later. Fig. 6 shows the experimental set-up for data collection, which comprises of the robot, the Perceptron sensor and the fine straight string passing through the robot workspace. Table 1 shows the nominal parameters of the robot manipulator and the transformation between {6} and {P}-frame. Here, the 2 translation parameters are modeled in x-, and z-direction to prevent the singularity as explained in Section 3.

**Table 1.** Nominal Kinematic Parameters of VORG-35 Robot and Designed Transformation Parameters from {6}-frame to {P}-frame

| i | $\alpha_{i-1}$ | $a_{i-1}$ | $d_i$ | $\theta_i$ | |
|---|---|---|---|---|---|
| 1 | 0 | 0 | 0 | 0 | |
| 2 | 90 | 200 | 0 | 0 | |
| 3 | 0 | 700 | 0 | 0 | |
| 4 | 90 | 155 | 750 | 0 | |
| 5 | -90 | 0 | 140 | 0 | |
| 6 | 90 | 0 | 160 | 0 | |
| 7 | Trn_x | Trn_z | Rot_x | Rot_y | Rot_z |
| | -60 | 122 | 0 | 0 | -88 |

Translation Parameter: mm,
Rotation Parameter: deg(。)

**Table 2.** Calibration Results

| i | $\alpha_{i-1}$ | $a_{i-1}$ | $d_i$ | $\theta_i$ | |
|---|---|---|---|---|---|
| 1 | -1532.34 | 158.6061 | 43.2621 | -15.0230 | |
| 2 | 90.109 | 200.298 | -1.300 | 0.097 | |
| 3 | -0.108 | 700.451 | 0.017 | 0.166 | |
| 4 | 90.069 | 154.070 | 749.459 | 0.234 | |
| 5 | -90.103 | 0.7736 | 139.341 | 0.171 | |
| 6 | 90.069 | -0.053 | 160.000 | 88.745 | |
| 7 | Trn_x | Trn_z | Rot_x | Rot_y | Rot_z |
| | -59.593 | 121.899 | 0.395 | -0.234 | -88.562 |

Translation Parameter: mm,
Rotation Parameter: deg(。)

In this experiment, the total 28 parameters including 5 SCP parameters between {6} to {P}-frame are selected and identified. Therefore, the number of required equations is 28 and one measurement gives 2 equations. At least 14 measurements are required but 70 measurements were performed to obtain the enough constraint equations. Its calibration results are given in Table 2. In order to validate the calibration results, the constraint violation quantity about the constraint ( $\Delta x = \Delta y = 0$ ) is defined as $e = \sqrt{(\Delta x^2 + \Delta y^2)}$ . The constraint violation quantities in both the uncalibrated case with parameters in Table 1 and the calibrated case with parameters in Table 2 are shown in Fig 7. The transformation parameters between {B} to {1}-frame can not be measured in real situation such that the calibrated parameters are used for the {B} to {1} computation process in uncalibrated case. The less e means that the parameters are the closer to the corresponding real parameters. As shown in Fig 7. the presented algorithm dramatically reduces the constraint violation quantity from 7.019 mm of maximum quantity to 0.4698 mm of that.

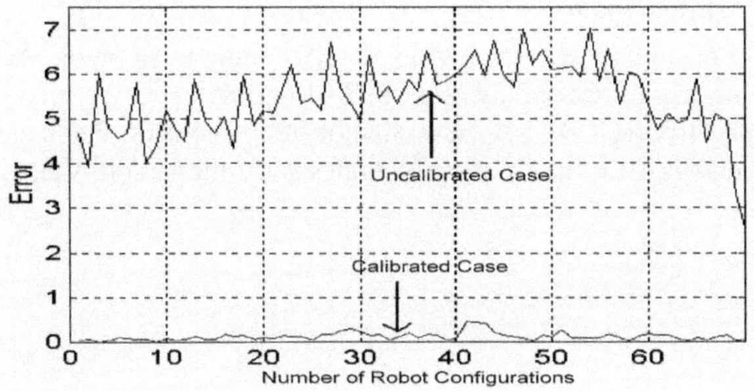


**Fig. 7.** The comparison of the error quantities between calibrated and uncalibrated case

## 5 Conclusion

An autonomous SCP and entire kinematic calibration method for the linear laser vision sensor has been developed with simple straight string. The method allows automated process of implementation and is suitable for on-site calibration in an industrial environment. The effectiveness of this method was shown through the real implementation on the Hyundai VORG 35 industrial robot such that the Perceptron's measurement accuracy & robot accuracy are improved in terms of constraint violation quantity, compared with in uncalibrated case. It was dramatically reduced from 7.019 mm of maximum quantity to 0.4698 mm of that. The proposed method is very effective to increase the robot accuracy so that it is very suitable in not only SCP calibration but also robot kinematic calibration and has very good advantage in on-site autonomous calibration with a simple straight string hung in the convenient place. It is almost jig-free robot calibration with a laser-vision sensor.

## References


1. Bennett, D.J., Geiger, D., Hollerbach, J.M.: autonomous Robot Calibration For Hand_Eye Coordination, The International Journal of Robotic Research Vol. 10, No. 5, (1991) 550-559
2. 2. Newman, W.S., Osborn, D.W.: A New Method for Kinematic Parameter Calibration via Laser Line Tracking, Proc. IEEE International Conference. on robotics and Automation, Vil.2, (1993) 160-165
3. Zhong, X.L., Lewis, J.M.: A New Method for Autonomous Robot Calibration, Proc. IEEE international Conference. on Robotics and Automation. on Robotics and Automation, (1995) 1790-1795
4. Zhuang, H., Wang, L., Roth, X.S.: Simultaneous Calibration of a Robot and a Hand-Mounted Camera, Proc.IEEE International Conference. on Robotics and Automation, Vol.2, (1993) 149-154
5. Zhuang, H., Wang, L., Roth, X.S.: Simultaneous Calibration of a Robot and a Hand-Mounted Camera, Proc.IEEE International Conference. on Robotics and Automation, Vol.2, (1993) 149-154
6. Kelmar, L., Khosla, P.K.: Automatic Generation of Kinematics for a Reconfigurable Modular Manipulator System, Proc. IEEE International Conference. on Robotics and Automation, (1988) 663-668
7. Hayati, S.A., Mirmirani, M.: Improving the Absolute Positioning Accuracy of Robot Manipulators, Journal of Robotic Systems, 397-413 (1985)
8. Newman, W.S., Osborn, D.W.: A New Method for Kinematic Parameter Calibration via Laser Line Tracking, IEEE Int. Conf. on Robotics and Automation, Vol. 2, 160-165, (1993)

# Robust Human Face Detection for Moving Pictures Based on Cascade-Typed Hybrid Classifier

Phuong-Trinh Pham-Ngoc, Tae-Ho Kim, and Kang-Hyun Jo

Graduate School of Electrical Engineering, University of Ulsan,
San 29, Mugeo-dong, Nam-ku, Ulsan, 680-749 Korea
equationsuper@gmail.com, {thkim,jkh2005}@islab.ulsan.ac.kr

**Abstract.** Face detection has been a key step in face analysis systems for decades. However, it is still a challenging task due to the variation in image background, view, pose, facial expression, etc. This paper proposes a simple and effective tool to detect human faces in moving pictures under such conditions. An improved approach aiming to reduce impacts of illumination, scale and connection of faces to receive rapidly skin homogeneous regions considered as the most potential face candidates is presented. A cascade-typed hybrid classifier, applied in retrieved face candidates, is based on template matching and appearance-based method providing a robust detection of multiply posed and viewed faces. This verification achieves advantages of the powerful discrimination of Local Binary Patterns (LBPs) and the high speed detection capability of embedded Hidden Markov Models (eHMMs). Experiments were performed out with different image databases and video sequences so that the system shows effective to detect human face for real-time uses.

**Keywords:** Face detection, skin detection, LBPs, eHMMs, video sequences.

## 1 Introduction

Human attention is essential for human-robot or -computer interaction so that a user communicates with robots or computers. Many researchers have investigated how to deal with human attention for multiple person-to-robot interaction. Tasaki et al. [16] proposed the method that enables robots to communicate with multiple people using the selection priority of the interactive partner based on the concept of proxemics [6]. Lang et al. [7] presented the attention system for mobile robot that enables the robot to shift its attention to the person of interest and to maintain attention during interaction.

For decades, the term 'face detection' has strongly related to automatic face analysis system such as video surveillance, visual communication system, multimedia applications, especially face recognition and HCI. Many face detection methods in image [22] and in video sequences [4,9,10,13,17] have been published and achieved good results. But these works still face up to difficult problems caused by variability in scale, location, facial expression, orientation, occlusion and pose. This paper presents a flexible face detector overcoming those obstacles to detect multiple faces and choose the most attentive one to apply to human-computer interaction.

D.-S. Huang, L. Heutte, and M. Loog (Eds.): ICIC 2007, LNAI 4682, pp. 1110–1119, 2007.

We propose the robust face detection by combining skin color characteristics, unsupervised face candidate localization and cascade-typed hybrid three-stage verification based on LBPs and eHMMs. This approach performs the face detection very fast keeping the detection quality for real-time uses.

## 2 Face Detection System

Our proposed face detection is robust to variations such as in illumination, human skin tone, pose, occlusion and orientation deformed by facial expressions. Moreover, when working with digital video, face detection algorithms have to deal with a significant constraint, ie. computation time. Thus the algorithm applied to an individual video frame should be able to locate face candidates and classify them as faces/non-faces in the shortest time possible. To achieve these goals, our method integrates information regarding face color, shape and texture to identify the regions which are most likely to be faces. An overview of our system is shown in Fig. 1.

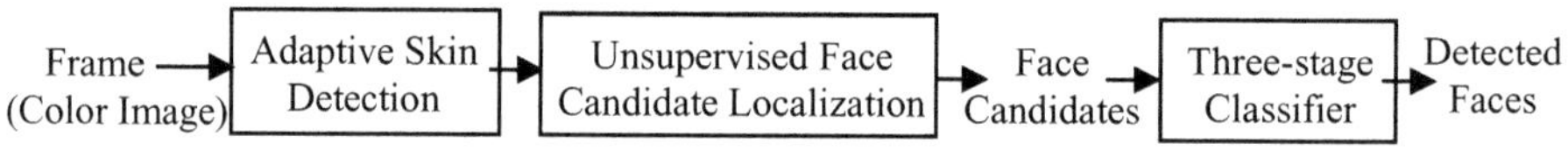


**Fig. 1.** Proposed face detection system

We use color information of human skin as the primary tool for locating face regions. In general, skin color based face detection algorithms are computationally modest, but are sensitive to noise, lighting conditions and the color content of the objects presenting in the scene. To solve these problems, we propose an adaptive selection of skin models.

### 2.1 Adaptive Skin Detection

The use of color information can simplify the task of localization in complex environment. It allows fast processing and is highly robust to geometric variations of the face pattern. Many color space with different properties have been applied such as RGB, normalized rgb, HSI, YCrCb, etc. A survey of skin color detection can be found in [18]. There are several challenges affecting skin detection work with different illumination, human races and also ambiguous similar-skin colors such as yellow, white, orange, pink and red.

Under normal illumination condition, skin color falls into a small region of the color space and some works [3,19,20,21] use this information to vote each pixel of the image as skin or non-skin. These algorithms achieved quite reasonable results; however there remain noise caused by similar-skin colors. For examples, Peer model [12] is sensitive to keep red and pink color as skin color and Lin model [8] often retains white, yellow and orange. This weak point sometimes causes a serious problem because non-skin regions are left too many to determine correct face regions. Our proposed method stems from the primary idea of Peer model to create a new skin model which can discard effectively similar-skin colors to retrieve more reasonable skin segmentation. Thus, we build a look-up table shown in Table 1 to define skin pixel.

**Table 1.** A look-up table defines skin pixel. These conditions must be satisfied simultaneously.

| R | R-G | G-B | R-B | (R-G)-(G-B) | B | G |
|---|---|---|---|---|---|---|
| [70,85] | [30,55] | [-5,35] | - | - | [20,255] | [30,255] |
| [86,100] | [30,60] | [-5,40] | - | - | [30,255] | [40,255] |
| [101,150] | [0,15] | [-5,10] | [15,75] | - | - | - |
| | [16,30] | [-255,-10] or [25,45] | - | [-15,285] | - | - |
| | [31,75] | [-5,90] | [-255,120] | [-20,285] | - | [50,255] |
| [150,200] | [15,20] | [0,40] | [20,255] | [-20,285] | - | - |
| | [21,30] | [-5,0] | [20,255] | [35,285] | - | - |
| | [31,85] | [-15,70] | [20,255] | [0,285] | [40,255] | [40,255] |
| [201,255] | [5,25] | [40,70] | - | [-30,285] | - | - |
| | [26,100] | [0,70] | - | - | - | - |

**Fig. 2.** Comparison of skin detection results: (a) Original color image, (b), (c) and (d) Skin detection results from the skin color models of Lin, Peer and proposed skin model, respectively; (e) An adaptive selection of skin color models for skin detection task

In fact, the strongest component among R, G, B decides the color. For skin color, generally R component is always the strongest one because human skin has the special expression of blood color. According to [12] and our experiences, in general, if R value is smaller than 70, the effect of R component is not enough to describe skin color; the color can be dark red, brown, green or black. Comparing to Peer model, we extend the value of R to get more dark-skin tone. The second image in Fig. 2 (d) can prove this idea. If G and B are smaller than 40 and 20, respectively, the color is strong red. Moreover, if R is larger than G or B too much, the color can be strong red, pink, yellow or orange. Also, the color is white or light yellow, if R value is too close to G and B. It means that the color is not skin color if the difference between R, G and B are too big or small; or R is smaller than 70. Approximately, our look-up table divides R components into five main ranges. Our work is adjusting reasonable differences between R, G and B in each range. Several skin detection results shown in Fig. 2 prove the advantage of our skin model comparing to others.

However, the automatic focus and white balancing of camera often cause the changing illumination. Our proposed skin model can work well in normal illumination, but it seems not good enough to work in bad illumination conditions. To overcome this problem, we use an adaptive skin model selection shown in Fig. 2 (e) to achieve skin detection flexibility. The model of Lin [8] has wide range of skin segmentation. We use it in the case of really high or low illumination condition. Our proposed skin model is fixed to use in normal illumination condition. In Fig. 2 (e), $I_P$ and $I_L$ are the same size binary images which are skin detection results from our proposed model and Lin model, respectively. For each captured color image, these two images are calculated and compared together. They are considered as different images if the ratio R given in equation (1) is larger than 10%.

$$R = \frac{\sum |I_P - I_L|}{size\ \ (I_P)} \tag{1}$$

In general, the illumination is good if two skin detection results are nearly similar. This is the case of normal illumination condition and we should use $I_P$ as the most adaptive skin detection result. Otherwise, we use $I_L$ to receive more information of skin segmentation.

## 2.2 Unsupervised Face Candidate Localization

After skin detection, we label connected skin regions and rule out the regions whose areas are smaller than 108 pixels considered as half of the smallest face to be detected. Comparing to [2], our system is improved to detect smaller faces. We call this step reducing small noise which ignores unreasonable candidates to save computation time. In some cases, skin segmentation is affected by different illumination condition, thus we lose some face regions. Different from others, our system pays attention on recovering these necessary lost regions by labeling connected non-skin regions in each skin region and change them to skin ones. This process ignores non-skin regions connecting directly to boundaries of their skin regions. This method seems to be better than using morphological filtering [2] because of not causing unexpected merging problem. It also keeps the original shape of skin regions and protects each skin region from splitting into unconnected regions which are not valid for detection by themselves. This is

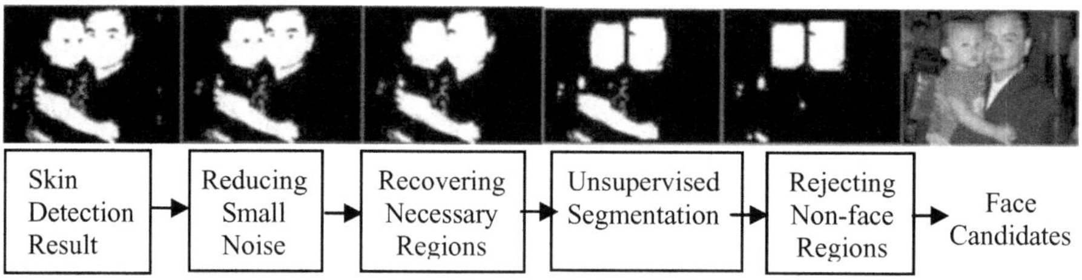


**Fig. 3.** Unsupervised face candidate localization scheme

a problem which was not solved in [2]. Our system needs to recover those lost regions because we use unsupervised segmentation to separate skin regions into smaller regions which are homogeneous in color. In some cases, human faces in pictures can be connected together or with other things such as hand, arm, shoulder, etc. This is one of challenges for face detection task. Some researchers have tried to use various ways to locate single faces such as using Hough transform to find ellipse skin region considered as face candidate [14] or the watershed segmentation algorithm [1]. Our approach is to use simple unsupervised segmentation based on intensity histogram of skin pixels. This idea is a little similar to watershed segmentation algorithm. For each skin region bordered by its rectangle, we calculate horizontal and vertical intensity histograms of skin pixels. In fact, by these histograms, there are concave regions called intersection marks at the intersection between different parts. We set all histogram values at those concave regions become zero to divide these parts according to equation (2),

$$h(i) = \begin{cases} h(i), & h(i) \geq t \\ 0 \quad , & otherwise \end{cases} \tag{2}$$

where *h(i)* is a histogram value of $i^{th}$ bin and *t* is selected threshold. To avoid over-segmentation, t is selected flexibility due to the geometry of skin regions. Before doing unsupervised segmentation process, each skin region is checked if it has properties of human faces. If so, we preserve it by covering its area with skin ellipse. After that, we reject non-face skin regions by several geometric conditions as shown in equation (3),

$$\delta(H,W,S) = \begin{cases} 1, & S \geq threshold \quad and \; \{(H \leq W \leq 3 \cdot H) \cup (W \leq H \leq 4 \cdot W)\} \\ 0, & otherwise \end{cases} \tag{3}$$

where *H* and *W* are the height and width of skin region rectangle and *S* is the number of skin pixels belonging to skin region rectangle. Therefore, we receive the most potential face candidates at this classification step.

### 2.3 Cascade-Typed Hybrid Classifier

We use a hybrid method based on template matching and appearance-based method for recognizing objects. Firstly, we create the face database for training. In our method, from face candidates extracted automatically as described above, we collect 200 frontal and profile face images to create face samples. With non-face class, we choose three main non-face objects: arm, hand and noise (50 samples for each object). All samples are normalized to 72x93 size grayscale pictures. In experiments, those samples are enough to represent two reciprocal classes.

Human face is a near-regular texture pattern generated by facial components and their configurations. Considering facial components such as eyebrow, eye, pupil, nose and face boundary, we select eight main different spatial templates shown in Fig. 4 to preserve shape information of facial components.

With only these spatial templates, we can describe all facial components; for example, eyebrow can be described by a chain of templates d, b and c. However, we combine both those spatial and local texture information to improve the capacity of describing faces. Instead of considering the central pixel $P_C$ only with its each neighborhood pixel as original LBP operator did [15], our method uses each pair of two neighborhood pixels ($P_{i1}$,$P_{i2}$) according to spatial templates to compare with the central pixel $P_C$. Eight spatial templates form eight binary digits of modified Local Binary Pattern (mLBP) number. Then mLBP produces 256 different values using equation (4).

$$mLBP = \sum_{i=0}^{7} S_i(x) \cdot 2^i \tag{4}$$

where $S_i$ is the $i^{th}$ binary digit of mLBP,

$$S_i(x) = \begin{cases} 1, & (P_C > P_{i1}) \, and \, (P_C > P_{i2}) \\ 0, & otherwise \end{cases}$$

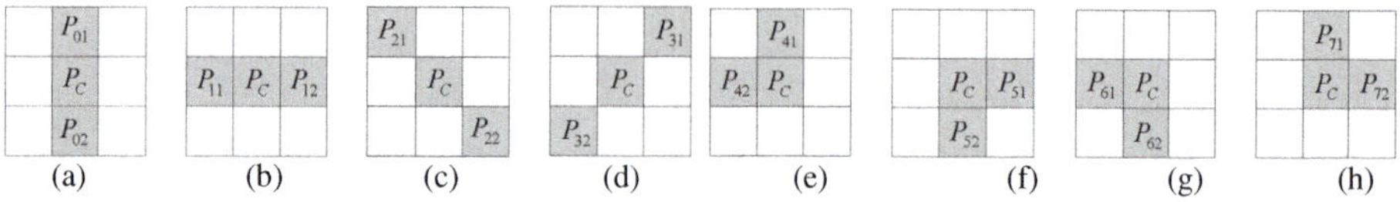


**Fig. 4.** Eight main spatial templates

In fact, mLBP gives us more information to distinguish face and non-face objects. We use the histogram of mLBP coefficients to represent a face. If we use single histogram for the whole face candidate image, occlusion will affect template matching algorithm seriously. To reduce this influence, we divide human face into two equal parts: the upper part from the nose up to forehead and the other from nose down to neck. We calculate individual histogram for each part and connect them sequentially to create one mixed 512 bin histogram representing to face candidate image. By this way, we can reduce effectively the influence of occlusion. Given an image I, one mixed mLBP histogram is denoted by $H^{mLBP}(I)$. We adopt error measurement for simple and fast computation using equation (5).

$$D(H(I_1), H(I_2)) = \sum_{i=1}^{n} \left| H_i^{mLBP}(I_1) - H_i^{mLBP}(I_2) \right| \tag{5}$$

where $H^{mLBP}(I_1)$ and $H^{mLBP}(I_2)$ are two mixed mLBP histograms, and $n$ is the number of bins.

Given a face database with m samples, for any sample P, we change it from color image to grayscale one and define its histogram-matching feature as the average distance to face training samples as follows:

$$f_{face}(P) = \frac{1}{m} \sum_{i=1}^{m} D(H(P), H(X_i)) \tag{6}$$

where $X_i$ is a face training sample. In fact, this histogram-matching feature has the discriminating ability between face and non-face patterns. With this feature $f_{face}$, we use thresholds $T_{face}$ to classify face and non-face objects. In experiment, if $f_{face}$ is smaller than 1800, face candidate can be considered as face with 99% of correct detection. If $f_{face}$ is bigger than 3500, the face candidate is almost not a human face. Only in the range [1800,3500] of $T_{face}$, it is still hard to say if the face candidate is a face. We improve face detection in this range by the following feature. With non-face database, for any sample P, we also define its histogram-matching feature as the minimum of three average distances to three non-face object samples, given by

$$f_{nonface}(P) = \min(f_{arm}(P), f_{hand}(P), f_{noise}(P)) \tag{7}$$

where $f_{arm}$, $f_{hand}$, $f_{noise}$ are calculated following to (6).
The difference between $f_{nonface}$ and $f_{face}$ also discriminates between both patterns. This difference $D_f$ is calculated as equation (8),

$$D_f(P) = f_{nonface}(P) - f_{face}(P) \tag{8}$$

We use $D_f$ to improve the face detection rate when $T_{face}$ is in [1800,3500]. We specify matching conditions for both $f_{face}$ and $D_f$ and use them jointly for the first two matching stages in three-stage classifier. With our selected matching conditions, the face detection system can reach over 80% correct detection.

To enhance the performance of our system, eHMMs [11] is used as the last step to check face candidate which are not satisfied the first two template-matching steps to give the final conclusion. It means our system is changed to four-class pattern classification problem. There are four eHMMs corresponding to face, arm, hand and noisy dataset. EHMMs algorithm performs pattern recognition for four-class problem by determining the maximum likelihood to find the most similar class for candidate object. A face candidate which was ignored by the first two stages of face detection system is checked by eHMMs. Finally, this is not human face if the result of this face candidate under this last stage is non-face.

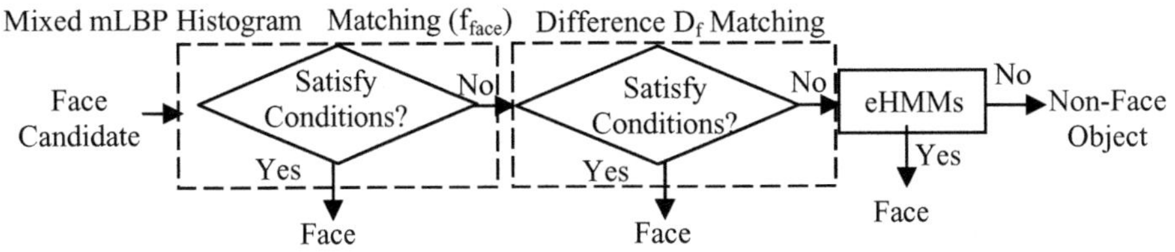


**Fig. 5.** Hybrid three-stage classifier

This hybrid classifier is advantageous for the robust detection and the speed of system. Comparing with conventional 'coarse to fine' classifier, this method is faster because one face candidate is not necessarily passed to all three stages. If it satisfies any stage in the order of stronger condition, the classifier process for the candidate stops immediately. Therefore, only in the worst case, full three stages will pass. Especially, the use of mLBP histogram and non-frontal face samples gives our system the capability of detecting multi-pose and multi-view faces.

## 3 Experimental Results

We have performed three different experiments to evaluate the system performance. In the first two experiments, we examine our face detector in NRC-IIT facial video database and our ISLAB facial database. We have included these experiments for comparison purposes regarding the computational complexity and the variation in number of persons in picture, orientation, pose, facial expression, occlusion, connection. The last one refers to video sequences taken from some movies in order to check the performance of face detector under different illumination conditions and complex environment.

NRC-IIT database consists of 23 single-face video sequences taken from 11 persons in different pose and orientation (about 15 degree). Each video clip captured with the rate 20fps is about 15 seconds long and has 160x120 size images. Comparing to the results of the other algorithms tested by NRC-IIT database [4,5], ours achieves better face detection result (correct detection: 95%) but in a bit slower (7fps). This is the result of tradeoff with using the facial geometry information in comparison of system accuracy.

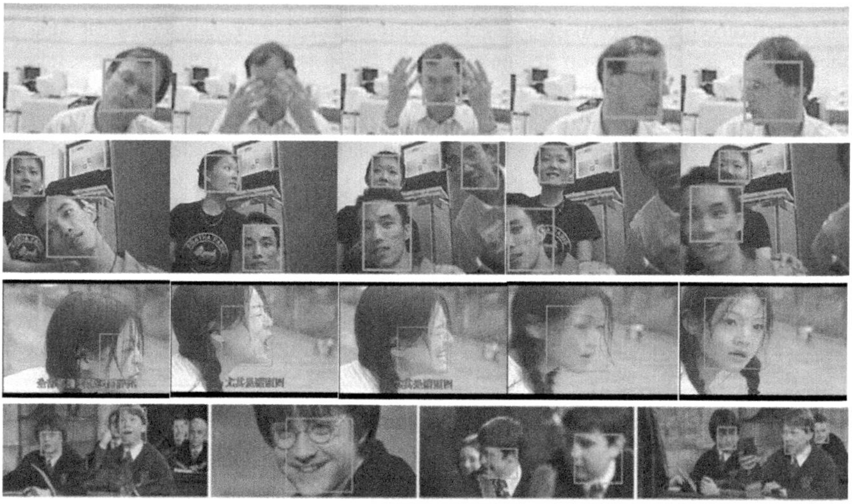

(a) (b) (c) (d)

**Fig. 6.** Several examples of face detection results: (a) Video clip in NRC-IIT database, (b) Video sequence in our ISLAB database, (c) and (d) Video sequences from movies 'Meteor Garden' and 'Harry Potter', respectively.

The proposed system was also experimented with our own ISLAB facial database. This database includes 20 video sequences with multi-face appearance from 15 different persons in different days under indoor environment. These video sequences were taken by different cameras such as Genious videocam series V4, Cosy net PC camera-PC590, USB PC camera and Canon Powershot A95. Each video sequence is about 45 seconds long and has 320x240 size picture. The result from those video sequences proved a good multiple face detection and robust detection capability in occlusion, variety of poses, orientations and facial expressions as shown in Fig. 6 (b). The speed is from 3 to 5 fps for this size of moving pictures. It also shows how well our system works under different cameras. It proves that our system can be applied in real human computer interaction. We also did further work of human attention. From detected face in each frame, our system chose the biggest face consider as most attentive one. Then we also tried to apply our system to recognize persons by using eHMMs. We can achieve good result of face detection (93%); however, the face recognition rate is still not good as we expected. Our future work is improving this face recognition rate and enhancing the speed of our system as real time processing (15fps).

Thirdly, we tried to check our system with 10 video sequences taken from several movies such as 'Harry Potter', 'The Chronicles of Narnia', 'Meteor Garden', 'Full House' etc. Each video sequence has the same size with original movie. The biggest size is 1028x240 and the smallest is 320x240. All clips are about from 15 to 45 seconds long. In those clips, there are multi-face appearance with blur motion, sudden action, complex human appearance and background as shown in Fig. 6 (c) and (d). Our face detector can work well with 83% correct detection rate in these clips with the speed 2 fps for 640x272 size image sequences. However, because color effect is often used in films, the skin detection step was influenced to cause more missing rate (17%) and false detection.

All our experiments were done with Pentium 4 CPU, 2.6GHz, 512MB of RAM in Visual C++ environment. The comparison between our method and others is given in Table 2.

**Table 2.** The comparison of proposed method and others

| | Our Proposed Algorithm | LBP Histogram Matching | eHMMs |
|---|---|---|---|
| Detection Rate | 93% | 80% | 64% |

## 4 Conclusions

The proposed technique provides with efficient mean for detecting human faces in moving pictures and can be used for variety of face analysis applications such as HCI or HRI as discussed in the above sections. Its performance is increased due to the use of adaptive skin detection, unsupervised face localization and hybrid three-stage classifier. All boosted ideas introduced in face candidate localization process gave us faster and more effective determination of face regions. A cascade-type hybrid three-stage classifier added the power of discrimination to achieve more robust face detection for multi-pose and oriented faces. Especially, we could achieve correct detection of 90 degree pose faces as shown in Fig. 6 (c) while other algorithm can detect with only 45 degree pose faces [10]. Our face detection showed more robust capability in detecting faces than using separately LBP histogram matching, eHMMs or other system [2].

**Acknowledgments.** The authors would like to thank to Ulsan Metropolitan City and MOCIE and MOE of Korean Government which partly supported this research through the NARC and post BK21 project at University of Ulsan.

## References

1. Albiot, A., Torres, L., Bournan, A.C., Delp, E.J.: A Simple and Efficient Face Detection Algorithm for Video database Applications. ICIP, Vol. 2 (2000) 239-242
2. Czirjek, C., O'Connor, N., Marlow, S., Murphy, N.: Face Detection and Clustering for Video Indexing Applications. Advanced Concepts for Intelligent Vision Systems, Belgium (2003)

3. Gacia, C., Tziritas, G.: Face Detection using Quantized Skin Color Regions, Merging and Wavelet Packet Analysis. IEEE Transaction on Multimedia, Vol. 1 (1999) 264–277
4. Gorodnichy, D.O.: Facial Recognition in Video. Proceeding of IAPR International Conference on Audio and Video-Based Biometric Person Authentication. Lecture Notes in Computer Science, Vol. 2688. UK (2003) 505–514
5. Gorodnichy, D.O.: Video-based Framework for Face Recognition in Video. Proc. Second Canadian Conference on Computer and Robot Vision, Canada (2005) 330-338
6. Hall, E. T.: Hidden Dimension. Doubleday Publishing. ISBN 0385084765 (1996)
7. Lang, S., Kleinehagenbrock, M., Hohenner, S., Fritsch, J., Fink, G.A., Sagerer, G.: Providing The Basis for Human-robot-interaction: A Multi-model Attention System for a Mobile Robot. Proceeding of the International Conference on Multimodal Interfaces, Canada (2003) 28-35
8. Lin, C., Fan, K.C.: A Color-triangle-based Approach to The Detection of Human Face. BMCV, Vol. 1811 (2000) 359-368
9. Liposcak, Z., Loncarie, S.: Prediction-and-Verification for Face Detection. Proceeding of IWISPA, Croatia (2000) 415–438
10. Mikolajczyk, K., Choudhury, R., Schimid, C.: Face Detection in Video Sequence – A Temporal Approach. Proceeding of the IEEE Computer Society Conference on CPVR, Vol.2. Hawaii (2001) 96-101
11. Nefian, A., Hayes, M.: Face Recognition using an Embedded HMM. Proceeding of IEEE Audio and Video-based Biometric Person Authentication, USA (1999) 19-44
12. Peer, P., Kovac, N., Solina, F.: Human Skin Colour Clustering for Face Detection. EUROCON, Vol. 2 (2003) 144-148
13. Séguier, R., Glaunec, A.L, Loriferne, B.: Human Faces Detection and Tracking in Video Sequence. Proceeding of the 7th Portuguese Conference on Pattern Recognition (1995)
14. Séguier, R.: A Very Fast Adaptive Face Detection System. International Conference on Visualization, Imaging and Image Processing (2004)
15. Shapiro, L.G., Stockman, G.C.: Computer Vision. Prentice Hall, New Jersey (2001)
16. Tasaki, T.,Komatani, K., Ogata, T., Okuno, H.G.: Spatially Mapping of Friendliness for Human-robot Interaction. IEEE/RSJ International Conference on Intelligent Robots and Systems (IROS 2005), Canada (2005) 1277-1282
17. Tsapatsoulis, N., Kollias, S.: Face Detection in Color Images and Video Sequences. Proceeding of the 10th Mediterranean Electrotechnical Conference MEleCon, Vol. II, Greece (2000) 498–501
18. Vezhnevets, V., Sazonov, V., Andreeva, A.: A Survey on Pixel-based Skin Color Detection Techniques. Proceeding of Graphicon, Moscow (2003) 85-92
19. Vilaplana, V., Marques, F., Salembier, P., Garrido, L.: Region-based Segmentation and Tracking of Human Faces. European Signal Processing. Rhodes (1998) 593–602
20. Wang, H., Chang, S.-F.: A Highly Efficient System for Automatic Face Region Detection in mpeg Video. IEEE Transaction on Circuit and Systems for Video Technology, Vol. 7 (1997)
21. Yang, M.H., Ahuja, N.: Detecting Human Faces in Color Image. IEEE International Conference on Image Processing. Chicago (1998) 127-130
22. Yang, M.H., Kriegman, D.J., Ahuja, N.: Detecting Faces in Images: A Survey. IEEE Trans. On PAMI, Vol. 24 (2002) 34–58

# Multimodality Image Registration by Particle Swarm Optimization of Mutual Information

Qi Li and Isao Sato

National Institute of Advanced Industrial Science and Technology (AIST),
Central 7, 1-1-1 Higashi, Tsukuba 305-8567, Japan
qi.li@aist.go.jp

**Abstract.** The goal of image registration is to align two or more images of the same scene. To fully automate the registration process by optimization of the mutual information criterion, a robust global optimizer is indispensable. This paper focuses mainly on application and evaluation of particle swarm optimization for the rigid transformation registration of multimodality images. Four different modes of particle swarm optimization are proposed to globally optimize the challenging intensity based image registration. The functional manifest of different modes is comparatively evaluated by rigid experiments. The results show that the particle swarm optimization algorithm is to be promising for optimal multimodality image registration. With consideration of the trade-off between the successful rate and the computation time, the proposed PSO Mode II gives relatively best performance during the experimental studies of multimodality image registration.

**Keywords:** image registration, mutual information, multimodality, particle swarm optimization, artificial intelligence.

## 1 Introduction

With the demands of advanced applications of multimodal imagery, registration and geometric alignment of two-dimensional and/or three-dimensional images are becoming increasingly important than before. Up to now, intensity based image registration is regarded as an effective method for multimodality. However, intensity based registration usually requires optimization of a certain similarity measure/metric between the images. Local optimization techniques frequently fail because registration functions of similarity measure or metric with respect to transformation parameters are generally nonconvex and irregular, therefore, global optimization technique is often required. Starting in 1995, with the successful implementation of mutual information as a similarity measure to the multimodality image registration [1], [2], the advance of automated image registration becomes possible. To fully automate the registration process by optimization of the mutual information criterion, a robust global optimizer is also indispensable.

Although image registration has been studied over many years and many approaches have been proposed in the literature [3], it is still a challenging topic in the research fields such as the computer vision, medical imaging and remote sensing.

D.-S. Huang, L. Heutte, and M. Loog (Eds.): ICIC 2007, LNAI 4682, pp. 1120–1130, 2007.

Among the proposed approaches in the literature, they are distinctly categorized into two types [4]: feature-based strategy and intensity-based strategy. During our research, the intensity-based method is adopted in consideration of four major reasons as follow: 1) to avoid the preprocessing of feature detection; 2) to consider the bottleneck of the feature matching; 3) to maximumly use the information in the whole image set; 4) to possibly register nearly almost kinds of image data.

In this paper, we focus mainly on development of an efficient global optimization technique, called particle swarm optimization (PSO), for the rigid transformation registration of multimodality images. The PSO is a relatively new population-based evolutionary computation technique [5], [6]. In contrast to genetic algorithms and other evolutionary strategies, which exploit the competitive characteristics of biological evolution, PSO exploits cooperative and social aspects, such as fish schooling, birds flocking, and insects swarming. Starting from a diffuse population (called a swarm), individuals (termed as particles) tend to move about the search space, eventually clustering in regions where optima are identified. Investigation into the theoretical properties of PSO is an active research field [7], [8], [9], [10], [11]. Practical applications of PSO have also appeared, including NC (numerical control) programming, classifier selection, optimizing power flow, neural network training, and image registration [12], [13]. Our work adopted PSO as a global optimization technique to register the rigid transformation for multimodal images, and assessed the performances of four different PSO modes during the optimization of the registration function using the mutual information criterion. The motivation of our studies is to look for a robust PSO based global optimizer for the determination of initial guess of multiresolution registration scheme and/or for the direct fast search of optimal parameters of the desired image pairs.

The paper is organized as follows. In Section 2, we describe the constituent parts of the automatic multimodal image registration. Section 3 details the implementation methods of different modes of the particle swarm optimization technique. In Section 4, we present and discuss the experimental results. Section 5 gives the concluding remarks of the paper.

## 2 Mathematical Model for Multimodality Image Registration

### 2.1 Optimal Registration

Image registration is to determine a mathematical mapping that is the best match of two or more images of the same scene acquired by different sensors, or taken by the same sensor at different times. To register a float image ($I_F$) to a reference image ($I_R$) can be expressed mathematically as:

$$I_R(x, y) = \zeta(I_F(T_\alpha(x, y))) , \tag{1}$$

where $T_\alpha$ is a transformation function, which maps two spatial coordinates $x$ and $y$, to the new spatial coordinates $x'$ and $y'$ by the set of parameters $\alpha$ :

$$(x', y') = T_\alpha(x, y) , \tag{2}$$

and $\zeta$ is the intensity or radiometric calibration function.

Intensity based image registration can be mapped as a typical optimization problem [3]. This can be expressed mathematically as:

$$\alpha^* = \underset{T_\alpha}{Arg\, optima}(S_{T_\alpha}(I_F, I_R)) , \tag{3}$$

where $S$ is a certain metric, and $\alpha^*$ is the optima estimated by the optimization algorithm.

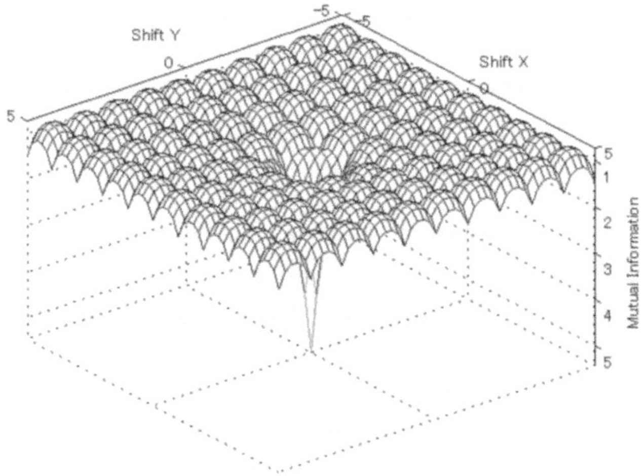


**Fig. 1.** Artifact pattern of partial volume interpolation in registration function

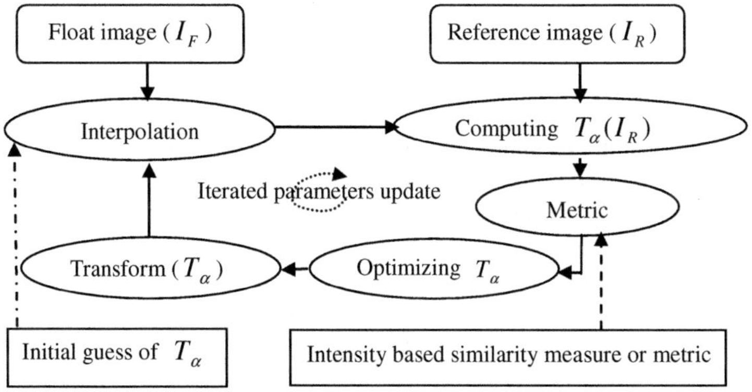


**Fig. 2.** Our proposed automated registration framework

Although image registration can be regarded as an optimization problem, the interpolation artifact complicates the registration function in Eq. (3). Furthermore, with the increase of the transformation parameters, the registration becomes into a complicated nonconvex optimization problem (see Fig. 1). Therefore, the optimization module is the engine for the whole registration process. Considering Eq. (3), our proposed iterated optimization framework for the image registration is depicted in Fig. 2.

## 2.2 Mutual Information Based Similarity Measure

Based on the information theory [14], the standard mutual information, $MI(I_F, I_R)$, of two images $I_F$ and $I_R$ can be calculated as follows:

$$MI(I_F, I_R) = H(I_F) + H(I_R) - H(I_F, I_R) , \tag{4}$$

where $H(I_F)$ and $H(I_R)$ are the marginal entropies of $I_F$ and $I_R$, and $H(I_F, I_R)$ is their joint entropy. Considering its definition, the mutual information is maximal when the two images are totally geometrically aligned by a certain transformation matrix.

In the computation of mutual information [15], the marginal entropies and joint entropy can be calculated as follows:

$$H(I_F) = \sum_{DN_F} - p_F(DN_F) \log p_F(DN_F) , \tag{5}$$

$$H(I_R) = \sum_{DN_R} - p_R(DN_R) \log p_R(DN_R) , \tag{6}$$

$$H(I_F, I_R) = \sum_{DN_F} \sum_{DN_R} - p_{F,R}(DN_F, DN_R) \log p_{F,R}(DN_F, DN_R) , \tag{7}$$

where $p_F(DN_F)$ and $p_R(DN_R)$ are the marginal probability mass functions, and $p_{F,R}(DN_F, DN_R)$ is the joint probability mass function. In practice, two methods are usually adopted to estimate these probability mass functions: Parzen window approach and histogram approach. In this study, the histogram of images is used to estimate such probability mass functions as follows:

$$p_{F,R}(DN_F, DN_R) = \frac{h(DN_F, DN_R)}{\sum_{DN_F} \sum_{DN_R} h(DN_F, DN_R)} , \tag{8}$$

$$p_F(DN_F) = \sum_{DN_R} p_{F,R}(DN_F, DN_R) , \tag{9}$$

$$p_R(DN_R) = \sum_{DN_F} p_{F,R}(DN_F, DN_R), \tag{10}$$

where $h$ is the joint histogram of the image pair ( $I_F$ , $I_R$ ). The value of $h(DN_F, DN_R)$ is the statistic numbers of corresponding pairs having intensity value $DN_F$ in the float image $I_F$ and intensity value $DN_R$ in the reference image $I_R$.

### 2.3 Rigid Transformation

In our studies, the rigid transformations are considered to register the multimodal images. The general rigid transformation matrix can be expressed mathematically as:

$$T_p^{rigid} = \begin{bmatrix} a & b & e \\ c & d & f \\ 0 & 0 & 1 \end{bmatrix}. \tag{11}$$

In our experiments, the three-parameter rigid transformation $T_p^{rigid,3}$ used can be written as:

$$T_p^{rigid,3}(x,y) = \begin{bmatrix} \cos(\theta) & \sin(\theta) & tx \\ -\sin(\theta) & \cos(\theta) & ty \\ 0 & 0 & 1 \end{bmatrix} \begin{bmatrix} x \\ y \\ 1 \end{bmatrix}, \tag{12}$$

where $\theta$ and $(tx, ty)$ denote the rotation and two translation parameters respectively.

## 3 Particle Swarm Optimization

The standard PSO (called Mode I) algorithm was originally designed and developed by Eberhart and Kennedy. In PSO, each particle (individual) adjusts its "flying" according to its own flying experience and its companions' flying experience instead of using genetic operators. Each particle is treated as a point in a *D*-dimensional space. The *i*th particle is represented as $X_I = (x_{i1}, x_{i2},..., x_{id},..., x_{iD})$. The best previous position (i.e., the position giving the best fitness value) of the *i*th particle is recorded and represented as $P_I = (p_{i1}, p_{i2},..., p_{id},..., p_{iD})$. The index of the best particle among all the particles in the population is represented by the symbol *g*. The rate of the position change (velocity) for particle $i$ is represented as $V_I = (v_{i1}, v_{i2},..., v_{id},..., v_{iD})$. For the PSO Mode I, the particles are manipulated according to the following equations:

$$v_{id} = w * v_{id} + c_1 * rand\_1 * (p_{id} - x_{id}) + c_2 * rand\_2 * (p_{gd} - x_{id}), \quad (13)$$

$$x_{id} = x_{id} + v_{id}, \quad (14)$$

where $c_1$ and $c_2$ are two positive constants, $rand\_n$, $n = 1,2,3...$ is a random function in the range [0,1], and $w$ is the inertia weight. Eq. (13) is used to calculate the particle's new velocity according to its previous velocity and the distances of its current position from its own best experience (position) and the group's best experience. Then, the particle flies toward a new position according to Eq. (14). The performance of each particle is measured according to a pre-defined fitness function, which is related to the problem to be solved. The inertia weight $w$ is employed to control the impact of the previous history of velocities on the current velocity, thus to influence the trade-off between global (wide-ranging) and local (nearby) exploration abilities of the "flying points". A larger inertia weight $w$ facilitates global exploration (searching new areas) while a smaller inertia weight tends to facilitate local exploration to fine-tune the current search area. Suitable selection of the inertia weight $w$ can provide a balance between global and local exploration abilities and thus require less iterations on average to find the optimum.

In this paper, we proposed other three PSO modes (called Mode II, III, IV) for comparative studies with Mode I during the image registration optimization. The major difference of Mode I~IV is the velocity update equation. The velocity update equations of Mode II, III, IV are mathematically expressed in Eqs. (15), (16) and (17) respectively.

$$\begin{aligned} v_{id} = &(1 - rand\_1) * rand\_2 * v_{id} + \\ &c_1 * rand\_1 * (p_{id} - x_{id}) + c_2 * rand\_2 * (p_{gd} - x_{id}), \end{aligned} \quad (15)$$

$$\begin{aligned} v_{id} = &w * v_{id} + c_1 * rand\_1 * (p_{id} - x_{id}) + \\ &c_2 * rand\_2 * (p_{gd} - x_{id}) + rand\_3 * (p_{gd} - p_{id}), \end{aligned} \quad (16)$$

$$\begin{aligned} v_{id} = &rand\_1 * rand\_2 * rand\_3 * v_{id} + \\ &c_1 * rand\_1 * (p_{id} - x_{id}) + c_2 * rand\_2 * (p_{gd} - x_{id}) + \\ &rand\_3 * (p_{gd} - p_{id}). \end{aligned} \quad (17)$$

As we have known, the performance of PSO is dependent on the settings of control parameters: inertial weights, acceleration constants, the maximum number of iterations, and the initialization of the population. The inertial weight $w$ is usually a monotonically decreasing function of the iteration. In this paper, we do not want to discuss how to best determine such control parameters. For the comparative purpose, the parameters used in four PSO modes are fixed as following: maximum iteration times: 2000; swarm size: 20; maximum velocity: 4.0; $c_1 = c_2 = 2.0$; maximum inertia weight: 0.9; minimum inertia weight: 0.2; desired accuracy: 1.0e-5. In registration experiments, the same initialization is conducted for each PSO mode.

## 4 Global Optimum Experiments

The goal of image registration is to find the optimal parameters, discussed in Section 2, which determine the relative position and orientation of the two sensed images. In this section, some experiments were conducted to register four typical multimodal image pairs. The registration processes were globally optimized by four different PSO modes with the same parameter settings. For the accurate evaluation of the performance of each PSO mode, ground truths of all the image pairs are known in advance. For each image pair, registration experiments were tried 100 times, and the successful rates of final registration results were calculated for the performance assessment. The four image pairs used in experiments are shown in Fig. 3. The successful rates of PSO registrations for each image pair are listed in Table 1-4 respectively.

From the successful rates in Table 1-4, Mode I and Mode II show the better performance on the whole than Mode III and Mode IV. Global optimizer PSO can be improved by Mode II "without considering" inertia weight during the velocity update. The merits (or robustness) of Mode II are shown through all the experimental pairs, especially for the registration cases of the image pair 1 and 4. In contrast to Mode I and Mode II, Mode III and Mode IV could not give the good successful rates for the registration cases of the image pair 1 and 4. However, each mode gives high scores for the registration cases of the image pair 2 and 3. All in all, the total performance of the four PSO modes is dependent on the practical image pair in the scope of the current studies. With consideration of the trade-off between the successful rate and the computation time (see Fig. 4), our proposed PSO Mode II gives relatively best performance during the multimodal registration studies.

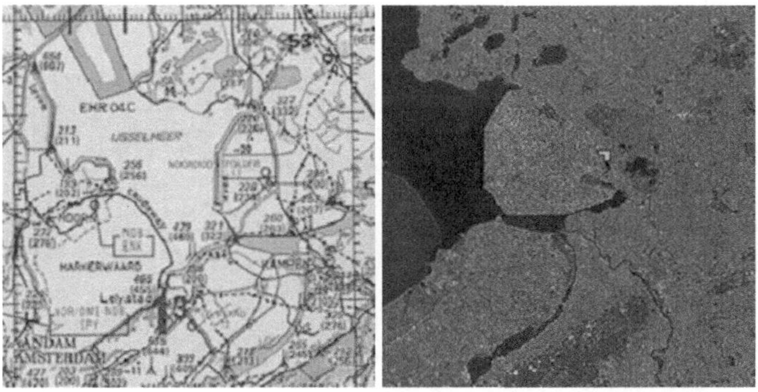

Image pair 1: Color administrative map vs. Pesudo-color image
Size: 256x256 pixels

**Fig. 3.** Data set of multimodal images

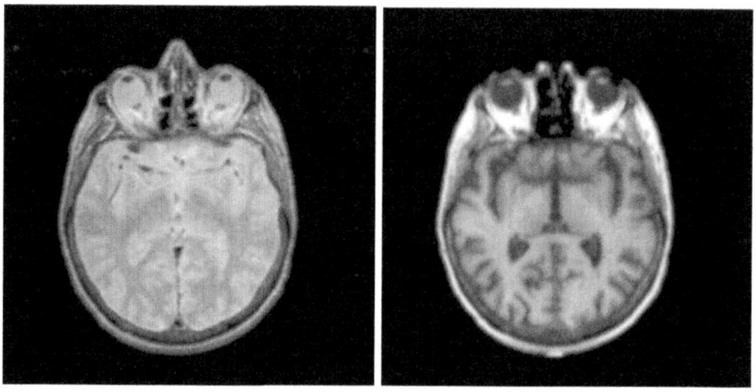

Image pair 2: PD (Proton Density) vs. MRI T1
Size: 250x250 pixels

Image pair 3: QuickBird satellite images before vs. after flooding
Size: 300x500 pixels

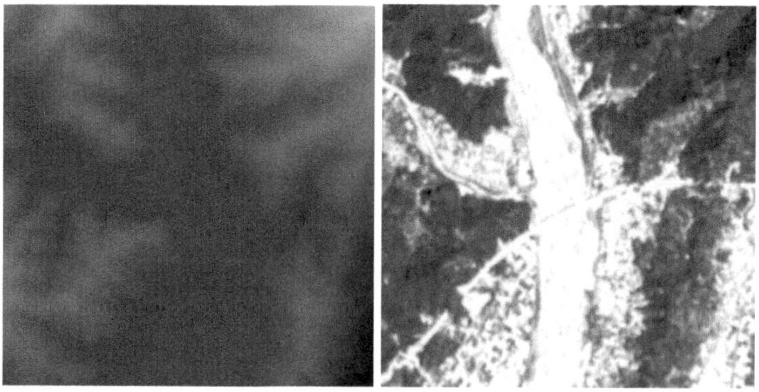

Image pair 4: DSM (Digital Surface Model) vs. Satellite image (ASTER VNIR)
Size: 256x256 pixels

**Fig. 3.** (*continued*)

**Table 1.** Successful rates of PSO registration of image pair 1

| PSO modes | Successful rates (%) |
|---|---|
| I | 59 |
| II | 67 |
| III | 36 |
| IV | 41 |

**Table 2.** Successful rates of PSO registration of image pair 2

| PSO modes | Successful rates (%) |
|---|---|
| I | 93 |
| II | 100 |
| III | 86 |
| IV | 95 |

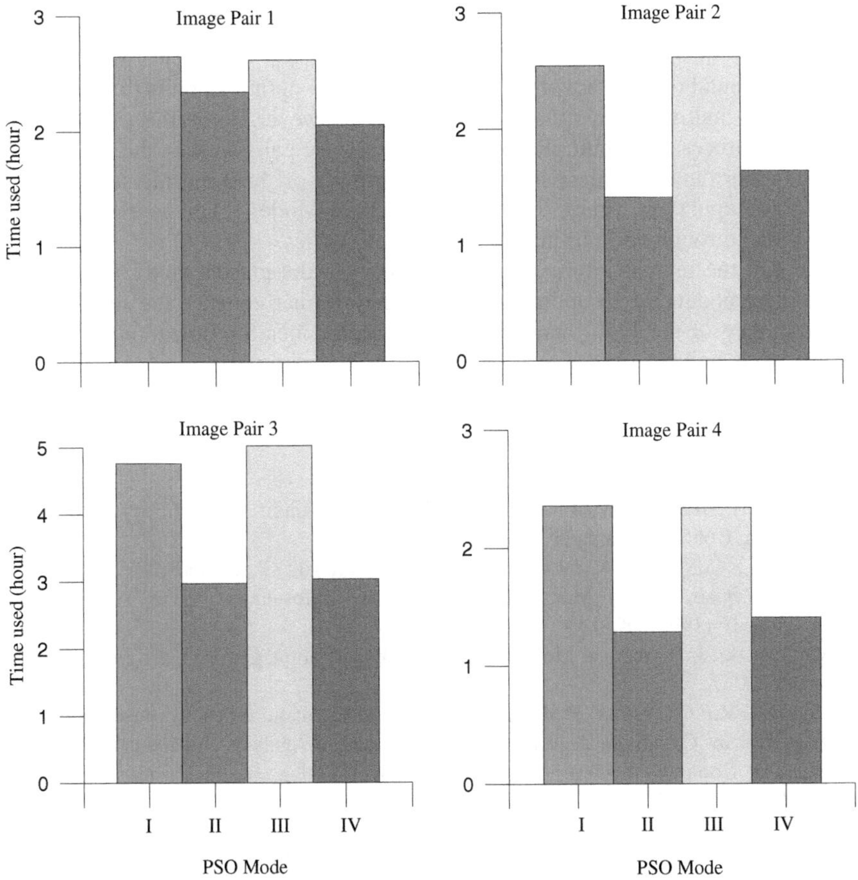


**Fig. 4.** Total computation time used by each PSO mode

**Table 3.** Successful rates of PSO registration of image pair 3

| PSO modes | Successful rates (%) |
|---|---|
| I | 99 |
| II | 88 |
| III | 80 |
| IV | 90 |

**Table 4.** Successful rates of PSO registration of image pair 4

| PSO modes | Successful rates (%) |
|---|---|
| I | 58 |
| II | 80 |
| III | 36 |
| IV | 54 |

## 5 Conclusions

The experimental results show that the particle swarm optimization is to be promising for optimal multimodality image registration. However, both the complexity of registration process and multiplicity of practical image pairs weaken the robustness of the PSO algorithms. With consideration of the trade-off between the successful rate and the computation time, the proposed PSO Mode II gives relatively best performance through our multimodal registration studies.

Although the mutual information was utilized in the present study, other metrics maybe more conducive to optimization, and may further enhance the successful rate and efficiency of the PSO (and other global and/or local optimization) approaches. These aspects will be covered to study in the future experiments.

## References

1. Viola, P.A., Wells III, W.M.: Alignment by Maximization of Mutual Information. In: Proc. of 5th Int. Conf. on Computer Vision. (1995) 20-23
2. Maes, F., Collignon, A., Vandermeulen, D., Marchal, G., Suetens, P.: Multimodality Image Registration by Maximization of Mutual Information. IEEE Trans. on Med. Imaging. 16 (1997) 187-198
3. Modersitzki, J.: Numerical Methods for Image Registration. Oxford University Press, New York (2004)
4. Guerreio, R.F.C., Aguiar, P.M.Q.: Global Motion Estimation: Feature-Based, Featureless, or Both?! In: Campilho, A., Kamel, M. (eds.): Image Analysis and Recognition. Lecture Notes in Computer Science, Vol. 4141. Springer-Verlag, Berlin Heidelberg New York (2006) 721-730
5. Shi, Y., Eberhart, R.C.: Parameter Selection in Particle Swarm Optimization. In: Proc. of 7th Int. Conf. on Evolutionary Programming VII. (1998) 591-600
6. Veldhuizen, D.A.V., Lamont, G.B.: Multiobjective Evolutionary Algorithms: Analyzing the State-Of-The-Art. Evolutionary Computation. 8 (2000) 125-147

7. Clerc, M., Kennedy, J.: The Particle Swarm - Explosion, Stability, and Convergence in A Multidimensional Complex Space. IEEE Trans. Evol. Comput. 6 (2002) 58-73
8. Zhang, L.P., Yu, H.J., Hu, S.X.: Optimal Choice of Parameters for Particle Swarm Optimization. Journal of Zhejiang University: Science. 6A (2005) 528-534
9. Cong, L., Sha, Y.H., Jiao, L.C.: Numerical Optimization Using Organizational Particle Swarm Algorithm. In: Proc. of Simulated Evolution and Learning. (2006) 150-157
10. Cui, Z.H., Zeng, J.C., Sun, G.J.: A Fast Particle Swarm Optimization. Int. J. of Innovative Computing Information and Control. 2 (2006) 1365-1380
11. Liu, D.S., Tan, K.C., Goh, C.K., Ho, W.K.: A Multiobjective Memetic Algorithm Based on Particle Swarm Optimization. IEEE Trans. on Systems Man and Cybernetics, Part B-Cybernetics. 37 (2007) 42-50
12. Phan, H.V., Lech, M., Nguyen, T.D.: Registration of 3D Range Images Using Particle Swarm Optimization. In: Proc. of Advances in Computer Science - Asian 2004. (2004) 223-235
13. Wachowiak, M.P., Smolikova, R., Zheng, Y.F., Zurada, J.M., Elmaghraby, A.S.: An Approach to Multimodal Biomedical Image Registration Utilizing Particle Swarm Optimization. IEEE Trans. Evol. Comput. 8 (2004) 289-301
14. Cover, T.M., Thomas, J.A.: Elements of Information Theory. John Wiley & Sons, New York (1991)
15. Chen, H., Varshney, P.K., Arora, M.K.: Mutual Information Based Image Registration for Remote Sensing Data. Int. J. Remote Sens. 24 (2003) 3701-3706

# Multiobjective Constriction Particle Swarm Optimization and Its Performance Evaluation

Yifeng Niu and Lincheng Shen

College of Mechatronic Engineering and Automation
National University of Defense Technology
410073, Changsha, China
{niuyifeng, lcshen}@nudt.edu.cn

**Abstract.** A novel multiobjective constriction particle swarm optimization (MOCPSO) is presented. MOCPSO not only uses mutation operator to avoid earlier convergence and uses adaptive weight to raise the search capacity, but also uses a new crowding operator to improve the distribution of nondominated solutions along the Pareto front, and uses the uniform design to obtain the optimal parameter combination. The sound evaluation criteria for multiobjective optimization algorithm are given, and some typical test functions are introduced. Experimental results show that MOCPSO has faster convergent speed and better search capacity than other multiobjective particle swarm optimization algorithms, especially when there are more than two objectives.

**Keywords:** Particle swarm optimization, multiobjective optimization, performance evaluation.

## 1 Introduction

At present, there are lots of multiobjective optimization algorithms, where multiobjective particle swarm optimization (MOPSO) has better search capacity for multiobjective optimization problems than evolutionary multiobjective optimization algorithms [1, 2, 3], MOPSO algorithm has faster convergent speed and their solutions can be uniformly distributed along the Pareto front. However, MOPSO adopts adaptive grid to record the particle searched, the number recommended to divide the object space is 30, which needs memory about $n^{30}$ ($n$ denotes the number of objective functions). When $n$ is bigger than 2, even though used integer format, it will still needs too longer computing time and may cause the failure in allocating the memory. On the other side, if the number is too small, MOPSO can't embody its superiority. On the condition that $n$ is bigger than 2, MOPSO shows its ability not equal to its ambition.

Based on the idea of MOPSO and NSGA-II [4], we present multiobjective constriction particle swarm optimization (MOCPSO). In MOCPSO, we discard the adaptive grid, design a new crowding distance to maintain the population diversity, use an adaptive mutation operator to improve the search capacities and avoid the earlier convergence, and use the uniform design to obtain the optimal parameter combination. In order to evaluate the performance of MOCPSO, we present sound evaluation

D.-S. Huang, L. Heutte, and M. Loog (Eds.): ICIC 2007, LNAI 4682, pp. 1131–1140, 2007.

criteria, including the objective distance, the inverse objective distance, the spacing and the error ratio. In contrast to MOPSO, MOCPSO has faster convergent speed and better search capacity.

The remainder of this paper is organized as follows. The algorithm of MOCPSO is introduced in section 2. The evaluation criteria of multi-objective optimization are given in section 3. The test functions are designed in section 4. The experimental results are given in section 5. Finally, a summary of our studies is given in section 6.

## 2 MOCPSO Algorithm

Kennedy and Eberhart brought forward particle swarm optimization (PSO) inspired by the choreography of a bird flock in 1995 [5]. PSO has shown a high convergence speed in single objective optimization [6], and it is also particularly suitable for multi-objective optimization [3]. In order to improve the performances of the algorithm, we present a proposal, called "multi-objective constriction particle swarm optimization" (MOCPSO), in which a new crowding operator is used to improve the distribution of nondominated solutions along the Pareto front and maintain the population diversity; an adaptive mutation operator are introduced to improve the search capacities and avoid the earlier convergence; the uniform design is used to obtain the optimal combination of the algorithm parameters.

### 2.1 MOCPSO Flow

The algorithm of MOCPSO is the following.

Step 1. Initialize the position of each particle: $pop[i]$=arbitrary, where $i=1,\dots,N_P$, $N_P$ is the particle number; initialize the velocity of each particle: $vel[i]=0$; initialize the record of each particle: $pbests[i]=pop[i]$; evaluate each of the particles in the $POP$ where $POP$ is defined as $\{pop[1], \dots, pop[N_P]\}$: $fun[i, j]$, where $j=1,\dots,N_F$, and $N_F$ is the objective number; and store the positions that represent nondominated particles in the repository of the $REP$ according to the Pareto optimality, where $REP$ is defined as $\{rep[1], \dots, rep[N_{MEM}]\}$, $MEM$ denotes the allowed maximum capacity.

Step 2. Update the velocity of each particle using the following canonical form.

$$vel[i] = \chi \cdot (vel[i] + \varphi \cdot (pop_m[i] - pop[i])) \tag{1}$$

where $\chi$ is constriction factor that causes convergence of the individual trajectory in the search space and whose value is typically approximately 0.7289 [7], $\varphi = \varphi_1 + \varphi_2$, $pop_m[i]$ denotes the temporary convergence point and can be calculated as

$$pop_m[i] = (\varphi_1 \cdot pbests[i] + \varphi_2 \cdot rep[h]) / (\varphi_1 + \varphi_2) \tag{2}$$

$\varphi_1$ and $\varphi_2$ are the acceleration coefficients with random values in the range [0, 2.05] [7], $pbests[i]$ is the best position that particle $i$ has had; $rep[h]$ is the position of particle $h$ in the repository, $h$ is the index of the solution in $REP$ with maximum crowding distance that implies the particle is located in the sparse region, as aims to maintain the population diversity; $pop[i]$ is the current position of particle $i$.

Step 3. Update the new positions of the particles adding the velocity produced from the previous step.

$$pop[i] = pop[i] + vel[i] \tag{3}$$

Step 4. Maintain the particles within the search space.

Step 5. Adaptively mutate each of the particles in the *POP* at a probability of $P_m$.

Step 6. Evaluate each of the particles in the *POP*.

Step 7. Update the contents in the *REP*, and insert all the current nondominated positions into the repository.

Step 8. Update the record of each particle. When the current position of the particle is better than the position contained in its memory, the latter is updated.

$$pbests[i] = pop[i] \tag{4}$$

Step 9. If the maximum cycle number $G_{max}$ is reached, stop the process and output the Pareto solutions in the repository; else go to Step 2.

## 2.2 Repository Control

The external repository of *REP* is used to record the nondominated particles in the primary population. At the beginning of the search, the *REP* is empty. The nondominated particles found at each iteration are compared with respect to the contents of the *REP*. If the *REP* is empty, the current solution will be accepted. If this new solution is dominated by an individual within the *REP*, such a solution will be automatically discarded. Otherwise, if none of the elements contained in the *REP* dominates the solution wishing to enter, such a solution will be stored in the *REP*. If there are solutions in the *REP* that are dominated by a new element, such solutions will be removed out of the *REP*. Finally, if the *REP* has reached its allowed maximum capacity, the new nondominated solution and the contents of the *REP* will be combined into a new population, according to the objectives, the individuals with lower crowding distances (locating the dense region) will not enter into the *REP*.

## 2.3 Adaptive Mutation

PSO is known to have a very high convergence speed. However, such a convergence speed may be harmful in the context of multi-objective optimization. An adaptive mutation operator is applied not only to the particles of the swarm, but also to the range of each design variable of the problem [1]. What this does is to cover the full range of each design variable at the beginning of the search and then we narrow the range covered over time, using a nonlinear function.

$$R = (Upper - Lower) \cdot (1 - g/(p_m G_{\max}))^2 \tag{5}$$

where *Upper* is the upper of the design variable, *Lower* is the lower, *R* is the value range of the variables.

## 2.4 New Crowding Distance

In order to improve the distribution of nondominated solutions along the Pareto front, we introduce a concept of crowding distance from NSGA-II [4] that indicates the population density. When comparing the Pareto optimality between two individuals, we find that the one with a higher crowding distance (locating the sparse region) is superior. In [4], the crowding distance has $O(mn \log n)$ computational complexity, and may need too much time because of sorting order.

Here we propose a new crowding distance that can be calculated using level sorting, and doesn't need to sort order for each objective. The crowding distance of the boundary points is set to infinity so that they can be selected into the next generation. The others can be calculated with respect to their objectives. For objective $j$, we divide its value range into a special level according to the boundary, then sort these levels in a descending order of the particle numbers, and compute the crowding distance using

$$d_{ij} = S_{ij} / N_{ij} \tag{6}$$

where $d_{ij}$ is the crowding distance of particle $i$ at objective $j$, $S_{ij}$ is the sequence number of the level where particle $i$ locates, $N_{ij}$ is the number of the particles in level $S_{ij}$. The crowding distance of particle $i$ is defined as

$$Dis[i] = \sum_j d_{ij} \tag{7}$$

The new crowding distance has less complexity than that in [4] because it doesn't need to sort order for every objective. In practice, the level number should be chosen carefully. If it is too small, many particles will have identical crowding distances. If it is too large, the computation time will be comparable to that of the original crowding distance computation algorithm in [4]. In general, the number equals 10.

## 2.5 Uniform Design

The relatively important parameters of MOCPSO include the number of particles, the number of cycles, the size of the repository, the constriction factor, and the mutation probability. In order to attain the optimal combination of these parameters, we introduce the uniform design.

Uniform design is used to convert the problem of parameter establishment into the experimental design of multi-factor and multi-level, which can reduce the work load of experiment greatly of simulation [9]. We introduce the uniform design to optimize the parameters of MOCPSO. The main objective of uniform design is to sample a small set of points from a given set of points, such that the sampled points are uniformly scattered. Let there be $n$ factors and $q$ levels per factor. When $n$ and $q$ are given, the uniform design selects $q$ combinations out of $q^n$ possible combinations, such that these $q$ combinations are scattered uniformly over the space of all possible combinations. The selected combinations are expressed in terms of a uniform array $U(n, q)=[U_{i,j}]_{q\times n}$, where $U_{i,j}$ is the level of the $j^{th}$ factor in the $i^{th}$ combination. When $q$ is prime and $q>n$, it has been proved that is given by

$$U_{i,j} = (i\sigma^{j-1} \bmod q) + 1 \tag{8}$$

where $\sigma$ is a parameter determined by the number of factors and of levels per factor.

# 3 Evaluation Criteria

In order to allow a quantitative evaluation of the performance of a multiobjective optimization algorithm, three issues are normally taken into consideration [10].

- Minimize the distance of the Pareto front produced by our algorithm with respect to the global Pareto front.
- Maximize the spread of solutions found, so that we can have a distribution of vectors as smooth and uniform as possible.
- Maximize the number of elements of the Pareto optimal set found.

Based on this notion, we adopted some criteria to evaluate the aspects previously indicated.

## 3.1 Objective Distance

The concept of objective distance (OD) was introduced by Van Veldhuizen [11] as a way of estimating how far the elements are in the set of nondominated vectors found so far from those in the Pareto optimal set. Let $S$ and $S^*$ be a non-dominated solution set and the Pareto-optimal solution set respectively, OD is defined as.

$$OD(S) = \frac{1}{|S|} \sum_{\mathbf{x} \in S} \min \|\mathbf{f}(\mathbf{x}) - \mathbf{f}(\mathbf{x}^*)\|, \ (\mathbf{x}^* \in S^*) \tag{9}$$

where $\|\mathbf{f}(\mathbf{x})\text{-}\mathbf{f}(\mathbf{x}^*)\|$ is the Euclidean distance between the two solutions $\mathbf{x}$ and $\mathbf{x}^*$ in the objective space. The objective distance is the average distance from each solution in $S$ to its nearest Pareto-optimal solution in $S^*$.

## 3.2 Inverse Objective Distance

In order to measure not only the convergence but also the diversity, Czyzak [12] used the following performance criterion that we call "inverse objective distance" (IOD) as

$$IOD(S^*) = \frac{1}{|S^*|} \sum_{\mathbf{x} \in S^*} \min \|\mathbf{f}(\mathbf{x}^*) - \mathbf{f}(\mathbf{x})\|, \ (\mathbf{x} \in S) \tag{10}$$

where the inverse objective distance measure is the average distance from each Pareto-optimal solution $\mathbf{x}^*$ in $S^*$ to its nearest solution in $S$.

## 3.3 Spread

In order to measure the spread (SP) (distribution) of vectors throughout the nondominated vectors found so far. Since the "beginning" and "end" of the current Pareto front found are known, a suitably defined metric judges how well the solutions in such front are distributed. Schott [13] proposed such a criterion measuring the distance variance of neighboring vectors in the nondominated vectors found so far. This criterion is defined as

$$SP = \sqrt{\frac{1}{n-1}\sum_{i=1}^{n}(\bar{d}-d_i)^2} \tag{11}$$

where $d_i = \min_j \| \mathbf{f}_i(\mathbf{x}) - \mathbf{f}_j(\mathbf{x}) \|$, $i, j=1\ldots,n$, $i \neq j$, $\bar{d}$ is the mean of all $d_i$, and $n$ is the number of nondominated vectors found so far. A value of zero for this metric indicates all members of the Pareto front currently available are equidistantly spaced.

### 3.4 Error Ratio

Error Ratio (ER) was proposed by Van Veldhuizen [14] to indicate the percentage of solutions (from the nondominated vectors found so far) that are not members of the true Pareto optimal set.

$$ER = \frac{\sum_{i=1}^{n} e_i}{n} \tag{12}$$

where $n$ is the number of vectors in the current set of nondominated vectors available, $e_i$=0 if vector $i$ is a member of the Pareto optimal set, and $e_i$=1 , otherwise. It should then be clear that $ER = 0$ indicates an ideal behavior, since it would mean that all the vectors generated by our algorithm belong to the Pareto optimal set of the problem.

## 4 Experiments

### 4.1 Test Functions

Several typical test functions were taken from the specialized literature to compare and analyze the performance of MOCPSO algorithm proposed. Test function 1 has two variables, two objectives, and three constraints [15].

$$\begin{gathered} \max f_1(\vec{x}) = -x_1^2 + x_2;\ \max f_2(\vec{x}) = x_1/2 + x_2 + 1 \\ x_1/6 + x_2 - 13/2 \le 0;\ x_1/2 + x_2 - 15/2 \le 0;\ 5/x_1 + x_2 - 30 \le 0;\ 0 \le x_1, x_2 \le 7 \end{gathered} \tag{13}$$

Test function 2 has three variables and two objectives [16].

$$\begin{gathered} \min f_1(\vec{x}) = \sum_{i=1}^{2}\left(-10\exp\left(-0.2\sqrt{x_i^2 + x_{i+1}^2}\right)\right);\ \min f_2(\vec{x}) = \sum_{i=1}^{3}(|x_i|^{0.8} + 5\sin(x_i^3)) \\ -5 \le x_1, x_2, x_3 \le 5 \end{gathered} \tag{14}$$

Test function 3 has two variables and two objectives [17].

$$\begin{gathered} \min f_1(x_1, x_2) = x_1;\ \min f_2(x_1, x_2) = g(x_2)/x_1 \\ where\ \ g(x_2) = 2 - \exp\left[-\left(\frac{x_2 - 0.2}{0.004}\right)^2\right] - 0.8\exp\left[-\left(\frac{x_2 - 0.6}{0.4}\right)^2\right] \\ 0.1 \le x_1 \le 1,\ 0.1 \le x_2 \le 1.0 \end{gathered} \tag{15}$$

Test function 4 (DTLZ1) has eight variables and four objectives [18].

$$\begin{aligned}
&\min f_1(\vec{x}) = \frac{1}{2}x_1x_2x_3(1+g(\vec{x})); \qquad \min f_2(\vec{x}) = \frac{1}{2}x_1x_2(1-x_3)(1+g(\vec{x})) \\
&\min f_3(\vec{x}) = \frac{1}{2}x_1(1-x_2)(1+g(\vec{x})); \ \min f_4(\vec{x}) = \frac{1}{2}(1-x_1)(1+g(\vec{x})) \\
&where \ \ g(\vec{x}) = 100\left[5+\sum_{i=4}^{n}((x_i-0.5)^2-\cos(20\pi(x_i-0.5)))\right] \\
&n=8,\ 0\le x_i\le 1,\ i=1,2,\ldots n
\end{aligned} \tag{16}$$

Test function 5 (DTLZ7) has 25 variables and six objectives [18].

$$\begin{aligned}
&\min f_1(\vec{x}) = x_1;\ \ \min f_2(\vec{x}) = x_2;\ \ \min f_3(\vec{x}) = x_3 \\
&\min f_4(\vec{x}) = x_4;\ \ \min f_5(\vec{x}) = x_5;\ \ \min\ f_6(\vec{x}) = (1+g(\vec{x}))\cdot h(f_1,f_2,\ldots f_5,g(\vec{x})) \\
&where\ g(\vec{x}) = 1+\frac{9}{20}\sum_{i=6}^{n}(x_i);\ \ h(f_1,f_2,\ldots f_5,g) = 6-\sum_{i=1}^{5}\left(\frac{f_i}{1+g}(1+\sin(3\pi f_i))\right) \\
&n=25,\ 0\le x_i\le 1,\ i=1,2,\ldots n
\end{aligned} \tag{17}$$

## 4.2 Uniform Design

In order to make MOCPSO perform well, the uniform design is firstly used to attain the optimal combination of algorithm parameters. We construct a uniform array with five factors and seven levels as follows, where $\sigma$ is equal to 3. We compute $U(5, 7)$ based on (8) and get

$$U(5,7) = \begin{bmatrix} 2 & 4 & 3 & 7 & 5 \\ 3 & 7 & 5 & 6 & 2 \\ 4 & 3 & 7 & 5 & 6 \\ 5 & 6 & 2 & 4 & 3 \\ 6 & 2 & 4 & 3 & 7 \\ 7 & 5 & 6 & 2 & 4 \\ 1 & 1 & 1 & 1 & 1 \end{bmatrix} \tag{18}$$

In the first combination of (18), the five factors have respective levels 2, 4, 3, 7, 5; in the second combination, the five factors have respective levels 3, 7, 5, 6, 2, etc. The value range of the number of particles is [20, 200]; the range of the number of cycles is [50, 350]; the range of the size of the repository is [70, 250]; the range of the constriction factor is [0.70, 0.76]; the range of the mutation probability is [0.01, 0.07].

**Table 1.** Evaluation criteria of different combinations

| Com | OD | IOD | SP | ER |
|---|---|---|---|---|
| U1 | 0.00422 | 0.00406 | 0.07634 | 0.2165 |
| U2 | 0.00965 | 0.00567 | 0.15488 | 0.3461 |
| U3 | 0.01773 | 0.01454 | 0.35632 | 0.3684 |
| U4 | 0.00302 | 0.00399 | 0.08363 | 0.2316 |
| U5 | 0.00148 | 0.00156 | 0.03542 | 0.1043 |
| U6 | 0.00587 | 0.00998 | 0.28643 | 0.2457 |
| U7 | 0.64291 | 0.06025 | 1.47912 | 0.5426 |

Take example for test function 1, the all combinations are run for a maximum value of 100 evaluations. As shown in Table 1, results indicate that the fifth combination is the optimal parameter combination.

By the uniform design, the parameters of MOCPSO are as follow: the particle number of *NP* is 170; the maximum cycle number of $G_{max}$ is 100; the allowed maximum capacity of *MEM* is 160; the range of the constriction factor is 0.72; the mutation probability of $P_m$ is 0.07. The inertia weight of $W_{max}$ is 1.2, and $W_{min}$ is 0.2; the learning factor of $c_1$ is 1, and $c_2$ is 1, the parameters of MOPSO are the same, while the inertia weight of *W* is 0.4, the grid number of $N_{div}$ is 20, for a greater number may cause the failure of program execution, e.g. 30.

### 4.3 Optimization Results

Fig. 1 shows the graphical results produced by our MOCPSO in the first three functions with two objectives chosen. The true Pareto front of the problem is shown as a continuous line in red. The solutions displayed in cyan correspond to the median result with respect to the criteria. It can be seen that MOCPSO can effectively achieve the true Pareto solutions for constraint problems and two-objective problems.

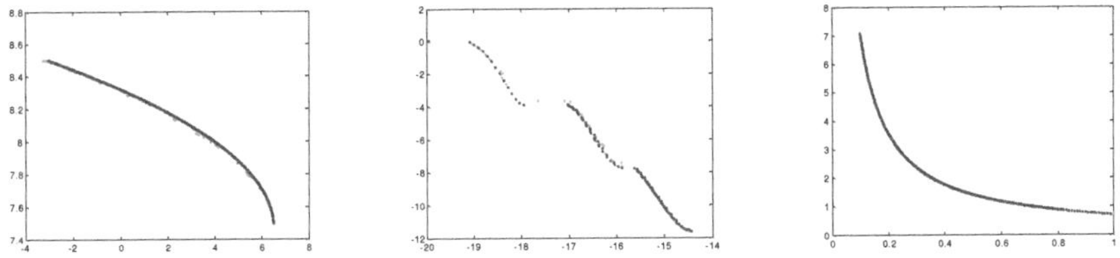

(a) Function 1 (b) Function 2 (c) Function 3

**Fig. 1.** The Pareto fronts produced by MOCPSO for three test functions

The results of optimization from MOCPSO and MOPSO are given in Table 2, which shows that MOCPSO has better search capacity and faster convergent speed than MOPSO. The results indicate that the uniform design can make optimization algorithms perform well, and the adaptive mutation operator can avoid earlier convergence and improve the search capacities.

**Table 2.** Evaluation metrics of optimization algorithms for two-objective problems

| Schemes | Algorithm | OD | IOD | SP | ER | Time(/s) |
|---|---|---|---|---|---|---|
| Fun 1 | MOPSO | 0.036535 | 0.032842 | 0.109452 | 0.1325 | 0.251 |
| | MOCPSO | 0.001689 | 0.002134 | 0.046496 | 0.1045 | 0.110 |
| Fun 2 | MOPSO | 0.008451 | 0.007242 | 0.097472 | 0.2035 | 0.179 |
| | MOCPSO | 0.007568 | 0.007729 | 0.033447 | 0.2164 | 0.158 |
| Fun 3 | MOPSO | 0.032730 | 0.031453 | 0.083584 | 0.2565 | 0.276 |
| | MOCPSO | 0.004030 | 0.002286 | 0.080783 | 0.2365 | 0.203 |

Both of Function 4 and Function 5 have multiple objectives more than two. The results of optimization form MOCPSO and MOPSO are given in Table 3. In this case, the metrics of MOPSO are entirely inferior to those of MOCPSO, which indicates that MOPSO needs too much memory and time, for the grid is worse for too many objectives, e.g. 4 or 6. In MOCPSO, the new crowding operator can increase the running speed and improve the distribution of nondominated solutions along the Pareto front, especially while there are too many objectives and too many variables.

**Table 3.** Evaluation metrics of optimization algorithms for multiple objective problems

| Schemes | Algorithm | OD | IOD | SP | ER | Time(/s) |
|---|---|---|---|---|---|---|
| Fun 4 | MOPSO | 0.039314 | 0.033206 | 0.254474 | 0.3535 | 0.363 |
| | MOCPSO | 0.036212 | 0.030961 | 0.215588 | 0.3014 | 0.253 |
| Fun 5 | MOPSO | 0.052007 | 0.045121 | 0.341241 | 0.4535 | 0.472 |
| | MOCPSO | 0.047551 | 0.037476 | 0.268331 | 0.4025 | 0.281 |

# 5 Conclusions

The proposed multiobjective constriction particle swarm optimization (MOCPSO) is an effective algorithm to solve the multi-objective problems, especially when the number of objectives is large, which can get to the Pareto front of optimization problems quickly and attain the Pareto optimal solutions. The uniform design can effectively be used to obtain the optimal parameter combination. MOCPSO can also be effectively applied to many multi-objective applications.

# References

1. Coello, C.A., Pulido, G.T., Lechuga, M.S.: Handling Multiple Objectives with Particle Swarm Optimization. IEEE Transactions on Evolutionary Computation 3 (2004) 256-279
2. Huang, V.L., Suganthan, P.N., Liang, J.J.: Comprehensive Learning Particle Swarm Optimizer for Solving Multiobjective Optimization Problems. Int. J. Intell. Syst. 2 (2006) 209-226
3. Reyes-Sierra, M., Coello, C.A.: Multi-objective Particle Swarm Optimizers: A Survey of the State-of-The-Art. Int. J. Comput. Intell. Research 3 (2006) 287-308
4. Deb, K., Pratap, A., Agarwal, S., et al.: A Fast and Elitist Multiobjective Genetic Algorithm: NSGA-II. IEEE Transactions on Evolutionary Computation 2 (2002) 182-197
5. Kennedy, J., Eberhart, R.C.: Particle Swarm Optimization. In: Proceedings of IEEE International Conference on Neural Networks (ICNN95), Perth (1995) 1942-1948
6. Kennedy, J., Eberhart, R.C.: Swarm Intelligence. Morgan Kaufmann, San Mateo (2001)
7. Kennedy, J., Mendes, R.: Neighborhood Topologies in Fully Informed and Best-of-Neighborhood Particle Swarms. IEEE Trans. Syst. Man Cybern. Pt. C: Appl. Rev. 4 (2006) 515-519
8. Shi, Y., Eberhart, R.C.: A Modified Particle Swarm Optimizer. In: Proceedings of IEEE International Conference of Evolutionary Computation, Anchorage (1998) 69-73
9. Leung, Y.W., Wang, Y.P.: Multiobjective Programming Using Uniform Design and Genetic Algorithm. IEEE Transactions on Systems, Man, and Cybernetics-Part C: Applications and Reviews 3 (2000) 293-304

10. Zitzler, E., Deb, K., Thiele, L.: Comparison of Multiobjective Evolutionary Algorithms: Empirical Results. Evol. Comput. 2 (2000) 173-195
11. Van Veldhuizen, D. A., Lamont, G. B.: Multiobjective Evolutionary Algorithm Research: A History and Analysis. Air Force Inst. Technol., Wright-Patterson AFB, OH, Tech. Rep. TR-98-03 (1998)
12. Czyzak, P., Jaszkiewicz A.: Pareto-Simulated Annealing – A Metaheuristic Technique for Multi-Objective Combinatorial Optimization. Journal of Multi-Criteria Decision Analysis 1 (1998) 34-47
13. Schott, J. R.: Fault Tolerant Design Using Single and Multicriteria Genetic Algorithm Optimization. M.S. thesis, Massachusetts Inst. Technol., Cambridge, MA (1995)
14. Van Veldhuizen, D. A.: Multiobjective Evolutionary Algorithms: Classifications, Analyzes, and New Innovations. Ph.D. Dissertation, Graduate School of Eng., Air Force Inst. Technol., Wright-Patterson AFB, OH (1999)
15. Kita, H., Yabumoto, Y., Mori, N., Nishikawa, Y.: Multi-objective Optimization by Means of the Thermodynamical Genetic Algorithm. In: Voigt H.M. et al. (Eds.): Parallel Problem Solving From Nature - PPSN IV, Lecture Notes in Computer Science, Springer-Verlag, Berlin (1996) 504-512
16. Kursawe, F.: A Variant of Evolution Strategies for Vector Optimization. In: Schwefel H. P. et al. (Eds.): Parallel Problem Solving From Nature - PPSN I, Lecture Notes in Computer Science, Vol. 496, Springer-Verlag, Berlin (1991) 193-197
17. Deb, K.: Multi-objective Genetic Algorithms: Problem Difficulties and Construction of Test Problems. Evol. Comput. 3 (1999) 205-230
18. Deb, K., Thiele, L., Laumanns, M., Zitzler, E.: Scalable Test Problems for Evolutionary Multi-Objective Optimization. Technical Report 112, Swiss Federal Institute of Technology (ETH), Zurich, Switzerland (2001)

# An Intelligent Fingerprint-Biometric Image Scrambling Scheme

Muhammad Khurram Khan[1,2] and Jiashu Zhang[1]

[1] Research Group for Biometrics & Security Engineering,
Bahria University, Dept. of Computer Science & Engineering,
13- National Stadium Road, Karachi, Pakistan
khurram.khan@scientist.com
[2] Sichuan Key Lab of Signal & Information Processing,
Southwest Jiaotong University, Chengdu 610031, China

**Abstract.** To obstruct the attacks, and to hamper with the liveness and retransmission issues of biometrics images, we have researched on the challenge/response-based biometrics scrambled image transmission. We proposed an intelligent biometrics sensor, which has computational power to receive challenges from the authentication server and generate response against the challenge with the encrypted biometric image. We utilized the FRT for biometric image encryption and used its scaling factors and random phase mask as the additional secret keys. In addition, we chaotically generated the random phase masks by a chaotic map to further improve the encryption security. Experimental and simulation results have shown that the presented system is secure, robust, and deters the risks of attacks of biometrics image transmission.

## 1 Introduction

With the widespread proliferation of network-based biometrics authentication, there is a risk that Biometrics signals can be intercepted, stored on the local machine, and obtained from applications where a user uses his biometrics. The recorded biometrics signals can then later be reused for illegal and fraudulent purposes for example to impersonate a registered user of online banking. The simplest method is that a biometrics signal is acquired once and reused several times [1]. An adversary can add simple perturbations to this previously acquired signal to demonstrate an impression that it is newly acquired by a biometrics sensor and detection of those amendments in the original biometrics signal would be difficult to determine at the server side. The financial implications of these kinds of attacks can be substantial.

Cryptography is a technique to protect the data from the illegal tampering and use. Cryptography techniques are widely adopted to protect the transmitted data over the insecure network channels. Some widely used cryptographic algorithms are RSA, DES, AES, and so on. A very critical problem with cryptographic techniques is that previously captured data, e.g. biometrics image, can be retransmitted to the authentication sever that can happily decrypt and identify the data [2]. Digital signature or hash of that data can be computed and transmitted to the authentication

D.-S. Huang, L. Heutte, and M. Loog (Eds.): ICIC 2007, LNAI 4682, pp. 1141–1151, 2007.

server that can ensure that the image was valid and captured by a registered device. However, there is no confirmation that when this image was captured that poses a challenge of replay attack in the biometrics regime.

Mostly used biometrics devices don't have ability to perform bidirectional transmission and computational power to perform hashing or any kind of digital signature computation on the sensor. This case becomes more vulnerable when the authentication server remotely validates the authenticity of the transmitted biometrics images. Most of the biometrics authentication systems also don't have ability to validate that the images were captured from the proper source and at the valid time interval.

A supplement to overcome the above mentioned problem of the cryptographic-based secure transmission of data is challenge/response authentication [1]. A challenge/response authentication system issues a challenge on the client request and verifies it in the response of the claimed identity. A very common of challenge/response system is asking the user's mother maiden name as a challenge and verifying it against the registered name in the database [1][2]. Implementing challenge/response-based authentication in the biometrics image transmission could solve the problem of replay attack or retransmission attack.

In this paper, we propose a challenge/response-based biometrics image scrambling method to tackle the attack 2 in biometric systems [2]. In contrast to the general challenge/response systems, which issues challenge to the user, our scheme is based on challenge to the biometrics sensor. We assume that biometrics sensor is intelligent enough to perform the appropriate response by the sent challenge.

Our proposed system can ensure that the transmitted biometrics images are live and newly captured by a specific sensor that is already registered in the system. The proposed system has the ability to overcome the problems of the previous state-of-the-art in the related field.

Rest of the paper is organized as follows. The Section 2 describes about the Fractional Fourier transform. The Section 3 gives the details of our proposed scheme. Experimental results are given in the Section 4. At the end, the Section 5 concludes the findings of this paper.

## 2 Mathematical Background of FRT

Recent advances in Fractional Fourier transform (FRT) have attracted much attention in the encryption field [3]. FRT is a generalization of conventional Fourier transform and can be used as a good encryption tool. FRT is a good representation tool, which expresses a signal in both spatial and frequency domain [4]. Some of the applications of the FRT are in the areas of solution of differential equations, quantum mechanics and quantum optics, optical systems and optical signal processing, swept-frequency filters, time-variant filtering and multiplexing, pattern recognition, and study of time-frequency distributions [3-6]. FRT provides more security due to its scaling factors in both $x$ and $y$ axis, random phase masks, and fractional orders which can be used as the secret keys in the encryption/decryption to protect from the attacks.

The two-dimensional fractional Fourier transform of a function $f(x, y)$ with a separable kernel can be defined as [3-6]:

$$F^{\phi_1,\phi_2}[f(x,y)](u,v) = \int\int_{-\infty}^{\infty} K_{P_1,P_2}(x,y;u,v) f(x,y)dxdy, \tag{1}$$

with the kernel,

$$K_{P_1,P_2}(x,y;u,v) = K_{P_1}(x,u), K_{P_2}(y,v) \tag{2}$$

where,

$$K_{P_1}(x,u) = \frac{\exp\{-i[\pi \operatorname{sgn}(\sin\phi_1)/4 - \phi_1/2]\}}{|\lambda f_{s1} \sin\phi_1|^{1/2}} \times \exp\left(i\pi \frac{x^2+u^2}{\lambda f_{s1} \tan\phi_1} - 2\pi i \frac{xu}{\lambda f_{s1} \sin\phi_1}\right), \tag{3}$$

$$K_{P_2}(y,v) = \frac{\exp\{-i[\pi \operatorname{sgn}(\sin\phi_2)/4 - \phi_2/2]\}}{|\lambda f_{s2} \sin\phi_2|^{1/2}} \times \exp\left(i\pi \frac{y^2+v^2}{\lambda f_{s2} \tan\phi_2} - 2\pi i \frac{yv}{\lambda f_{s2} \sin\phi_2}\right), \tag{4}$$

where $\phi_1 = P_1\pi/2$ and $\phi_2 = P_2\pi/2$, and $\phi_1$ and $\phi_2$ are transform angles, $(x, y)$ and $(u, v)$ are coordinates on input and output planes, respectively. $P_1$ and $P_2$ are fractional orders.

FRT could be reduced to the conventional Fourier transform when *P=1*. At *P=0* and *2*, the FRT is separately defined as follows [5]:

$$F^0\{g(x_0,y_0)\} = g(x_0,y_0), F^2\{g(x_0,y_0)\} = g(-x_0,-y_0) \tag{5}$$

An important FRT property is its index additiveness i.e. $F^\alpha F^\beta = F^{\alpha+\beta}$, so FRT can also be written as:

$$F^{-P}\{F^P\{g(x_0,y_0)\}\} = F_0\{g(x_0,y_0)\} = g(x_0,y_0) \tag{6}$$

where $F^{-P}$ is the inverse FRT of $F^P$.

## 3 Proposed Encryption/Decryption Scheme

Optical biometrics scanners are widely utilized to capture the biometric samples, but the main stumbling block with them is that they don't have bidirectional communication and encryption power. Mostly available biometrics devices only capture the image or video, process it, and happily send to the destination for identification process. The main drawback with this kind of traditional framework is that if an intruder intercepts the biometric image, then he can reuse, modify, and share the biometric image with others. One solution to overcome this problem is to encrypt biometric image before transmitting over the insecure network. But the main problem still exists that how to ensure authentication server that the transmitted image is live and not reuse? This question poses a very key concern in the biometrics regime.

Our proposed remedy to cope with the above mentioned concern is to design and develop an intelligent biometric sensor and technique, which can ensure that the sent biometric images are real-time and lively captured by a registered remote biometric scanner. The proposed biometric scanner has bidirectional communication capability and is intelligent enough to receive and process the secret challenges received from the authentication server, which is used as the challenge-key. This module is named as Challenge/Response Processor as shown in Fig. 1.

The second module is the 'Encryption Processor', which encrypts the captured biometric image by using the challenge-key and sends to the 'Intelligent Sensor CPU' that transmits the encrypted image to the authentication server.

The schematic diagram of the proposed encryption/decryption system is shown in Fig. 2, where Fig 2(a) depicts the proposed encryption process and Fig. 2(b) shows the decryption process.

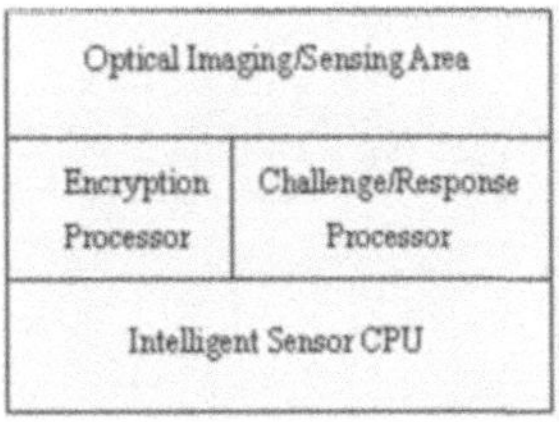


**Fig. 1.** Proposed challenge/response biometric sensor

The biometric image to be encrypted can be represented as $f_1(x_1)$. Note that, to retain simplicity, we demonstrate the mathematical proofs by using one-dimensional analysis. Image $f_1(x_1)$ is multiplied by the chaotically randomly generated phase mask $\phi_1(x_1)$, which passes through the first FRT system with fractional order $P_1$. The output of the first FRT system is multiplied by the second chaotically generated phase mask $\phi_2(x_2)$, which passes through the second FRT system with fractional order $P_2$. This process finishes at $n^{th}$ FRT system with random phase mask $\phi_n(x_n)$ and fractional order $P_n$, respectively.

The encryption process at the start can be represented as [5]:

$$g_1(x_1) = f_1(x_1)\phi_1(x_1) \tag{7}$$

The output of the first FRT system with fractional order $P_1$ is:

$$f_2(x_2) = \int g_1(x_1)K_1(x_1, x_2, \phi_1)dx_1, \tag{8}$$

where;

$$K_1(x_1, x_2, \phi_1) = \frac{\exp\{-i[\pi \operatorname{sgn}(\sin\phi_1)/4 - \phi_1/2]\}}{|\lambda f_{s1}\sin\phi_1|^{1/2}} \times \exp\left(i\pi\frac{x_1^2 + x_2^2}{\lambda f_{s1}\tan\phi_1} - 2\pi i\frac{x_1x_2}{\lambda f_{s1}\sin\phi_1}\right) \tag{9}$$

and $\phi_1 = P_1\pi/2$.

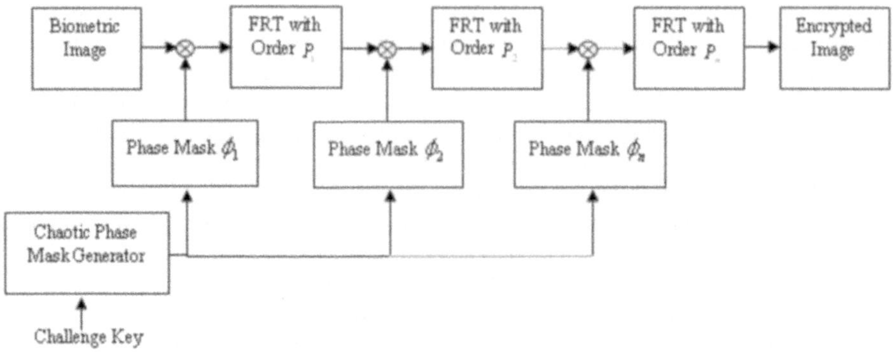


2(a) Encryption process

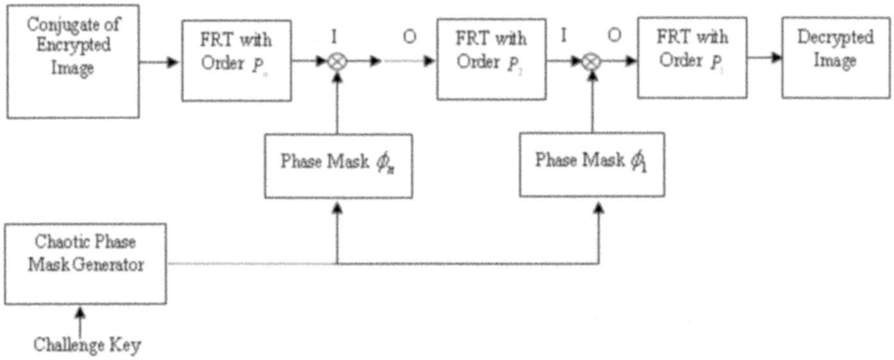


2(b) Decryption process

**Fig. 2.** Proposed biometric image encryption/decryption system

To chaotically generate the random phase masks, we use the piecewise linear chaotic map (PWLCM) [7]. The most attractive features of chaos are its extreme sensitivity to initial conditions and the outspreading of orbits over the entire space [8]. Chaos-based encryption and random generation algorithms have been widely exploited, which utilize the nonlinear dynamic characteristics of chaos. PWLCM shows superiority against the some widely used pseudorandom sequences and has controllable correlation properties. PWLCM is shown in equation (3.10):

$$C(w) = \begin{cases} w / p, & 0 \le w < p \\ (w - p)/(0.5 - p), & p \le w < 0.5 \\ (1 - w - p)/(0.5 - p), & 0.5 \le w < 1 - p \\ (1 - w)/p, & 1 - p \le w \le 1 \end{cases} \tag{10}$$

where $p \in (0,1)$, which is used as the challenge-key sent by the authentication server. The input of the second FRT system can be shown as:

$$g_2(x_2) = f_2(x_2)\phi_2(x_2) \tag{11}$$

The output of the second FRT system with fractional order $P_2$ can be generated as:

$$f_3(x_3) = \int g_2(x_2)K_2(x_2, x_3, \phi_2)dx_2, \tag{12}$$

$$= \int f_2(x_2)\phi_2(x_2)K_2(x_2, x_3, \phi_2)dx_2, \tag{13}$$

$$= \int\int g_1(x_1)K_1(x_1, x_2, \phi_1)K_2(x_2, x_3, \phi_2)\phi_2(x_2)dx_1dx_2, \tag{14}$$

$$= \int\int f_1(x_1)\phi_1(x_1)K_1(x_1, x_2, \phi_1)K_2(x_2, x_3, \phi_2)\phi_2(x_2)dx_1dx_2, \tag{15}$$

where;

$$K_1(x_2, x_3, \phi_2) = \frac{\exp\{-i[\pi \operatorname{sgn}(\sin\phi_2)/4 - \phi_2/2]\}}{|\lambda f_{s1} \sin\phi_2|^{1/2}} \times \exp\left(i\pi \frac{x_2^{\,2} + x_3^{\,2}}{\lambda f_{s1} \tan\phi_2} - 2\pi i \frac{x_2 x_3}{\lambda f_{s1} \sin\phi_2}\right) \tag{16}$$

and $\phi_2 = P_2\pi / 2$.

At the end, the final version of encrypted biometric image can be obtained at the output of $n^{th}$ FRT system with order $P_n$.

$$f_{n+1}(x_{n+1}) = \int g_n(x_n)K_n(x_n, x_{n+1}, \phi_n)dx_n, \tag{17}$$

$$= \int\int \cdots \int f_1(x_1)\phi_1(x_1)K_1(x_1, x_2, \phi_1)\cdots\phi_n(x_n)K_n(x_n, x_{n+1}, \phi_n)dx_1 \ldots dx_n, \tag{18}$$

where;

$$K_1(x_n, x_{n+1}, \phi_n) = \frac{\exp\{-i[\pi \operatorname{sgn}(\sin\phi_n)/4 - \phi_n/2]\}}{|\lambda f_{s1} \sin\phi_n|^{1/2}} \times \exp\left(i\pi \frac{x_n^{\,2} + x_{n+1}^{\,2}}{\lambda f_{s1} \tan\phi_n} - 2\pi i \frac{x_n x_{n+1}}{\lambda f_{s1} \sin\phi_n}\right) \tag{19}$$

and $\phi_n = P_n\pi / 2$.

Now the encrypted biometric image can be transmitted to the authentication server for the identification of a person. Note that, we used fractional orders as the response-secret key that can be transmitted to the server through a private channel or by using any key-exchange protocol.

The encrypted biometric image, at the authentication server end, can be decrypted with the setup as shown in figure 2(b). The conjugate of the received encrypted-biometric image is fed into the FRT system with the fractional orders of $P_n$, which are used as the response-key of the client, and we get the following output:

$$g_n^*(x_n) = \int f_{n+1}^*(x_{n+1}) K_n(x_n, x_{n+1}, \phi_n) dx_{n+1}, \tag{20}$$

$$= \left( \int \int \ldots \int f_1(x_1) \phi_1(x_1) K_1(x_1, x_2, \phi_1) \ldots \phi_n(x_n) K_n(x_n, x_{n+1}, \phi_n) dx_1 \ldots dx_{n-1} \right)^* \tag{21}$$

where '*' denotes the complex conjugate. Random phase mask $\phi(x)$ is generated by the same piecewise linear chaotic map (PWLCM) with the challenge-key that was sent by the authentication server to the client at the encryption stage, and we get the following distribution:

$$f_n^*(x_n) = \left( \int \int \ldots \int f_1(x_1) \phi_1(x_1) K_1(x_1, x_2, \phi_1) \ldots \phi_{n-1}(x_{n-1}) K_n(x_n, x_{n-1}, \phi_n) dx_1 \ldots dx_{n-1} \right)^* \tag{22}$$

Hence, the encrypted image is passed through all the FRT systems with the challenge and response secret keys and at the end, we get the decrypted version of the image. If the image is rightly decrypted with the secret keys, it ensures that it is real time and live and now, authentication server can perform identification of the person's claimed identity against the biometrics data saved into the database

# 4 Experimental Results and Discussions

An efficient and secure encryption system should be robust enough against all kind of brute force, statistical, differential, and various cryptanalysis attacks [9][10]. In this section, we perform the experiments to show the analytical view of the security of the proposed system.

Experiments are conducted on the public domain fingerprint image dataset DB3 FVC2004 [11]. FVC2004 dataset DB3 contains a total of 120 fingers and 12 impressions per finger (1440 impressions) using 30 volunteers. The size of fingerprint images is 300×480 pixels captured at a resolution of 512 dpi.

## 4.1 Key Space Analysis

Key space size is the total number of different keys that can be used in the encryption [9]. A good encryption system must be sensitive to the cipher keys and key space should be large enough to make brute force attacks infeasible.

In our system, there are many parameters used as the secret keys e.g. fractional orders, random phase masks, and challenged-chaotic key by the authentication server. The chaotic key is used as the parameter value of the piecewise linear chaotic map, which chaotically generates the random phase mask for the each iteration of the encryption process. The attractor of PWLCM is shown in Fig. 3(a). If the precision decimal value of key is $10^{-16}$, the key space size for the chaotic Tent map is over $2^{260}$, which is large enough to resist all kinds of brute force attacks. Fig. 3(b) shows the

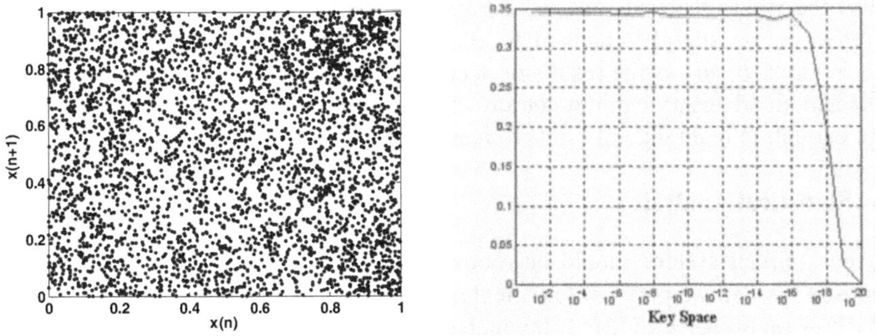


(a) Attractor of piecewise linear chaotic map (b) Key space analysis

**Fig. 3.** Simulation of PWLCM

simulation of key space analysis. In addition, we use fractional orders of FRT as the response-secret key, which is transferred to the authentication server with the encrypted biometric image.

Figures 4(a) shows the fingerprint by using secret keys $(N, P)$, (0.579143014523832 39,0.5568793109899014), respectively, where $N$ is the challenge-key sent the authentication server to generate the random phase mask and $P$ is the response-key used to generate fractional order of FRT. Fig 4(b) shows the white Gaussian noise and apparently, the original contents cannot be perceived from the encrypted image.

At the authentication server end, for the decryption of the biometric image, both secret keys i.e. *(N,P)* are needed to rightly decrypt the encrypted image. Figure 5(a)

(a) (b)

**Fig. 4.** DB3 fingerprint image of a subject (a) Plain image (b) Encrypted image

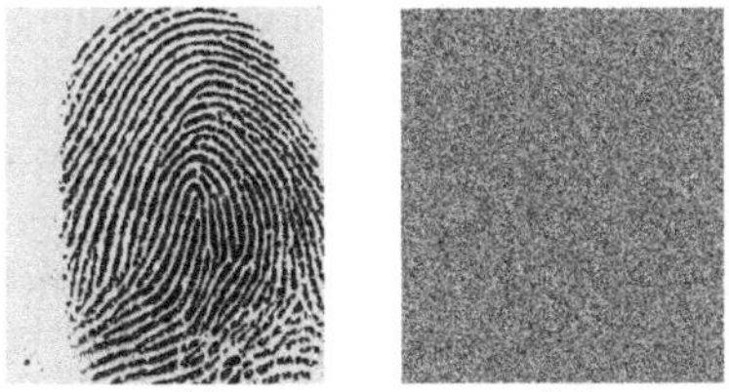

(a) (b)

**Fig. 5.** Decrypted fingerprint image (a) Decrypted image with correct secret key (b) Decrypted image with wrong key shows the right retrieved images of Figure 4(a) with the correct secret keys. If the secret keys are slightly changed to (0.57914301452383290,0.5568793109899010), then Figure 5(b) shows the results of decrypted images which also look like Gaussian noise and an adversary can not get any clue from the images. Hence, the key space is huge enough in real range of (-1,1) to resist the exhaustive key search.

## 4.2 Statistical Analysis

A secure cipher system should be robust against statistical attacks. To evaluate our proposed system, we perform statistical analysis to demonstrate its superior confusion and diffusion properties [12] [13], which strongly resist statistical attacks.

### 4.2.1 Histogram Analysis

Image histogram shows that how the pixels in an image are distributed by graphing the number of pixels at each intensity level [9]. We analyzed the histogram of the original and encrypted fingerprint images as shown in Figure 6. It can be seen from Figure 6(b) that the histogram of enciphered images are fairly uniform and is significantly different from that of the original images. Hence, there is no clue to perform any statistical attack on the encrypted biometric images.

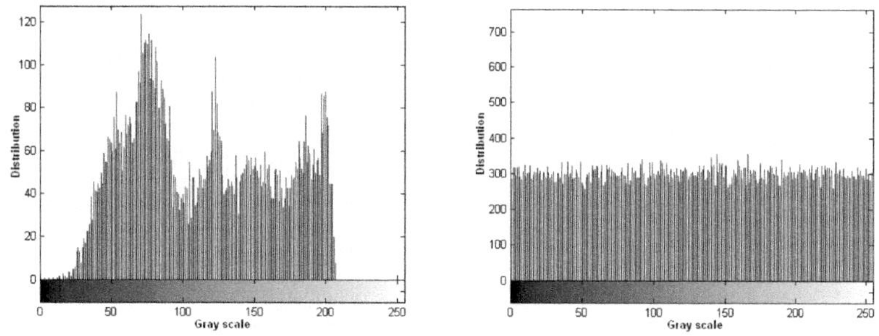


(a) Fingerprint plain image histogram (b) Fingerprint encrypted image histogram

**Fig. 6.** Histogram analysis of plain and encrypted fingerprint image

### 4.2.2 Pixels Correlation Analysis

In this subsection, we perform the pixel correlation analysis of enciphered images between two vertically adjacent pixels, two horizontally adjacent pixels, and two diagonally adjacent pixels, respectively. In experimentations, we randomly select 1000 pairs of two adjacent (vertical, horizontal, and diagonal direction) pixels from the plain and ciphered images and then, we compute the correlation coefficient of each pair by using the formula of [9]:

Figure 7 shows the correlation distribution of the two horizontally adjacent pixels in plain and ciphered fingerprint images, respectively.

Table 1 shows the correlation coefficients of horizontal, diagonal, and vertical adjacent pixels in plain and ciphered fingerprint image.

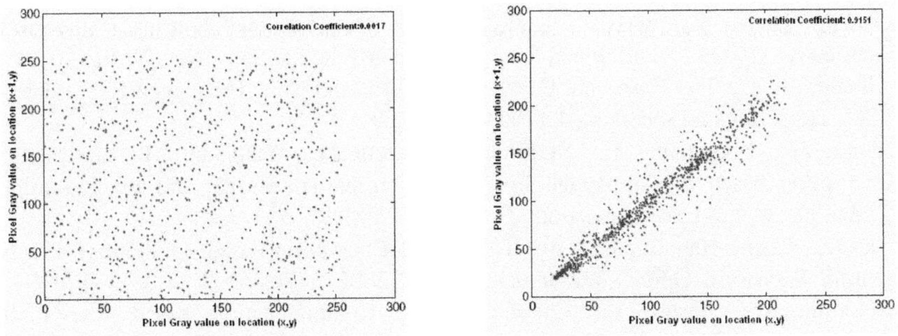


(a) Correlation analysis of plain fingerprint image (b) Correlation analysis of encrypted fingerprint image

**Fig. 7.** Correlations of two horizontally adjacent pixels in plain and encrypted fingerprint images

**Table 1.** Correlation coefficient of two adjacent pixels fingerprint image

| Attack | Plane Image | Ciphered Image |
|---|---|---|
| Horizontal | 0.9151 | 0.0017 |
| Vertical | 0.9048 | 0.0024 |
| Diagonal | 0.9189 | 0.0017 |

# 5 Conclusion

In this paper, an intelligent fingerprint image scrambling system was proposed to thwart the risks of non-liveness and retransmission of biometric image by an adversary who can play role of man-in-the-middle from client workstation to the local or remote authentication server. We proposed a novel architecture of intelligent biometrics sensor, which has computational power to receive challenges from the authentication server and generate response to the challenge with the encrypted biometric image. Proposed biometric sensor has the encryption and intelligence power in its embedded system, thus it is infeasible for an intruder to modify or interrupt the central processing power of the sensor. In addition, the encryption and decryption performance of the proposed system are encouraging to use in the real-time systems of biometrics image transmission. Experimental and simulation results have shown that the presented system is secure, robust, and deters the risks of attacks of biometrics image transmission over insecure network.

# References

1. Uludag, U., Pankanti, S., Prabhakar, S., Jain, A.K.: Biometric Cryptosystems: Issues and Challenges. Proceedings of the IEEE. 92 (2004) 948-960.
2. Ratha, N., Connell, J., Bolle, R.: Enhancing Security and Privacy in Biometrics-based Authentication Systems. IBM System Journal. 40(3) (2001) 614-634.

3. Ozaktas, H. M., Zalevsky, Z., Kutay, M,. A.: The Fractional Fourier Transform with Applications in Optics and Signal Processing. John Wiley & Sons, New York, (2000)
4. Almeida, L. B,: The Fractional Fourier Transform and Time-Frequency Representations. IEEE Trans. Signal Processing 42(1994) 3084-3091
5. Jianlin, Z., Hongqiang, L., Xiaoshan, S., Jifeng, L., Yanghua, M.: Optical Image Encryption based on Multistage Fractional Fourier Transforms and Pixel Scrambling Technique. Optics Communications 249(2005) 493-499.
6. Yan, Z., Cheng-Han, Z., Naohiro, T.: Optical Encryption based on Iterative Fractional Fourier Transform. Optics Communications 202(2002) 277-285
7. Xiaomin, W., Jiashu, Z.: Chaotic Secure Communication based on Nonlinear Autoregressive Filter with Changeable Parameters. Physics Letters A, Elsevier Science, Article in press
8. Khan, M.K., Jiashu, Z., Lei, T.: Chaotic Secure Content-based Hidden Transmission of Biometrics Templates. Chaos, Solitons, and Fractals, Elsevier Science 32(5) (2007) 1749-1759
9. Chen, G.R., Mao, Y.B., Chui, C.K.: A Symmetric Image Encryption Scheme based on 3D Chaotic Cat Maps. Chaos, Solitons & Fractals 21(2004) 749-761
10. Linhua, Z, Xiaofeng L., Xuebing, W.: An Image Encryption Approach based on Chaotic Maps. Chaos, Solitons, and Fractals 24(2005) 759-765
11. FVC DB Fingerprint online
12. Shannon, C. E.: A Mathematical Theory of Communication. Bell Systems Technical Journal 27(1948) 379-423
13. Shannon, C. E.: Communication Theory of Secrecy Systems. Bell Systems Technical Journal 28(1949) 656-715

# Reversible Data Hiding Based on Histogram

Wen-Chung Kuo[1], Dong-Jin Jiang[1], and Yu-Chih Huang[2]

[1] Department of Computer Science and Information Engineering,
National Formosa University, Taiwan, R.O.C
[2] Department of Information Management
Taiwan University of Technology, Tainan, R.O.C.
simonkuo@nfu.edu.tw

**Abstract.** Recently, the reversible data hiding technology has been discussed extensively. Its major characteristic is that allows an original image to be completely reconstructed from the stego-image after the extraction of the embedded data. 2006, the reversible data hiding based on histogram is proposed by Ni, *et al.* (NSAS). However, the embedded data can not be recovered unless the knowledge of peak point is known to the receiver in NSAS. More recently, Hwang, *et al.* (HKC) proposed a robust reversible data hiding scheme based on histogram shifting method to improve this shortcoming. At the same time, they also discuss when the element of selected minimum point is not zero and then give the solution to this problem. Unfortunately, the proposed solution will decrease the original data hiding capacity. In order to enhance the data hiding capacity and keep the goal of reversible data, we use one bit to record the change of the selected minimum point to replace the method using in [6]. According to our proposed method and experience analysis, this reversible data hiding scheme is not only to improve the original data hiding capacity but also to reach the goal of data recovering.

**Keywords:** Reversible data hide, Histogram.

## 1 Introduction

Nowadays, the rapid growth of computer and network technologies promotes the digital multimedia transmission over the Internet increasing day by day. However, these digital messages over Internet still have to face some problems, such as data security, copyright, right management, etc. Usually, there are two methods used to enhance their security, one is the well-know method, cryptography and the other is steganography. In cryptography, the only user with the private key can recover the encryption message when the encryption message has produced. Even though the attacker got an encryption message, it is also unable to find the content of message. Unfortunately, this method will be insecure when the private key is stolen or broken. Another way to solve this problem is to hide secret message behind a cover such that an unintended observer will not be aware of the existence of the hidden secret message. This data hiding technique is called steganography.

D.-S. Huang, L. Heutte, and M. Loog (Eds.): ICIC 2007, LNAI 4682, pp. 1152–1161, 2007.

So far, many techniques about data hiding have been proposed but most of them are irreversible, [1], [3], [7], [8], [9], [11], [13], [17], which means that the original cover image cannot be recovered with lossless from the stego-image. However, some applications such as military, medical, and high-energy particle physical experimental investigation, it is also desired that the original cover media can be recovered because of the required high-precision nature. The marking techniques satisfying this requirement are referred to as reversible data hiding techniques [10].

In 2001, the first reversible embedding method is carried out in the spatial domain by Honsinger, *et al.* [5]. Until now, there are many reversible data hiding techniques based on difference method have been reported in the literature [2], [4], [6], [10], [12], [14], [15], [16]. For example, Fridrich *et al.* [4] embedded the secret data into the cover image by compressing one of the least-significant bit (LSB) planes of a cover image in 2002. Then, Celik *et al.* [1] improved the embedding capacity by using arithmetic coding. Later, a reversible data hiding scheme based on the difference-expansion method is proposed by Tian [12].

In the literature, many reversible data hiding schemes based on histogram have been proposed [6], [10], [14], [15]. Recently, Ni, *et al.* [10] (NSAS) utilized the zero or the minimum points of the histogram of a cover image and then slightly modify the pixel grayscale values to allow secret data to be appended. After experimental results and performance comparison analysis, they guarantee that the computational complexity is low and the execution time is short in their scheme. However, the peak point will be changed after embedding secret data in NSAS. The embedded secret data cannot be recovered when the knowledge of peak and zero point of histogram are not transmitted to the receiver. This major problem how to transmit the additional information, peak point and zero point of histogram, from sender to receiver is not discussed in [10]. To improve the above disadvantage, Hwang, *et al.* (HKC) proposed a robust reversible data hiding scheme based on histogram shifting method. In fact, they proposed the location map to store the reversible data and give an answer to solve when the selected minimum point is not zero. However, their solution will waste the storage quantity and decrease the original data hiding capacity. In this paper, we will set up an efficient location map by using only one bit to record the change of the selected minimum point to improve the data hiding capacity.

The rest of the paper is organized as follows. The NSAS and HKC schemes will be reviewed briefly in Section 2. Then, an efficient reversible data hiding scheme is described in Section 3. Experimental results and analysis are presented in Section 4, and conclusions are drawn in Sections 5.

## 2 Review Two Reversible Data Hiding Schemes Based on Histogram

In [10], Ni *et al.* introduced a high-capacity method for reversible data embedding, which we will describe briefly in this section. For a more detailed discussion on NSAS scheme, we refer to the reader to [10].

Here, we also use the "Lena" image as an example to illustrate NSAS algorithm. For a given grayscale image, say, the Lena image 512×512×8, we first generate its histogram as shown in Fig.1(b).

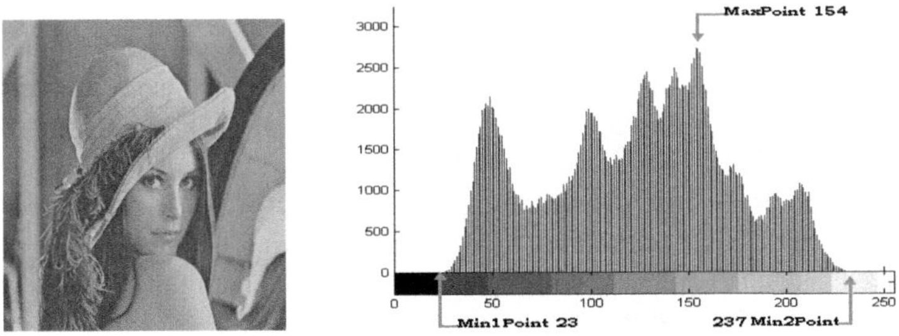


**Fig. 1.** (a) Lena, (b) The histogram of Lena image

According to the NSAS scheme [10], the following three steps complete the data embedding process.

***Step 1.*** Find out the peak point and the zero point in Fig.1(b). A peak (zero) point corresponds to the grayscale value which the maximum (no) number of pixels in the given image assumes, e.g., as shown in Fig. 1(b). For simplicity, we assume that only one zero point at 255 and one peak point at 154 are used in this scheme.

***Step 2.*** Scan the whole image in a sequential order, such as row-by-row, from top to bottom. The grayscale value of pixels between 155 (including 155) and 254 (including 254) is incremented by "1", i.e., shifting the range of the histogram, [155 254] to the right-hand side by 1 unit and leaving the grayscale value 155 empty.

***Step 3.*** Embed the secret data into the grayscale value of 154 and 155.

In summarization, the Embedding data process in NSAS scheme is shown in Fig.2.

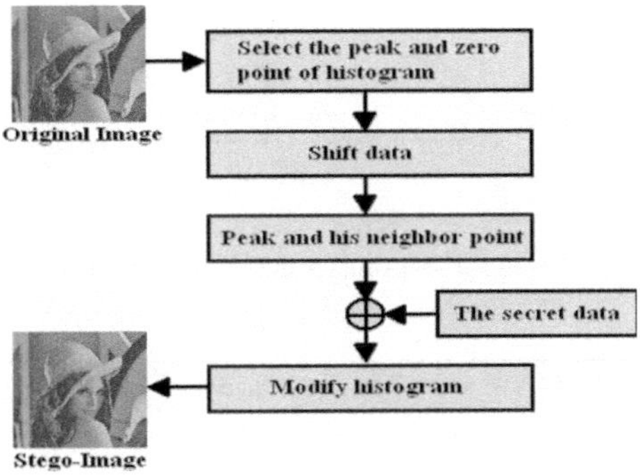


**Fig. 2.** The Embedding process in NSAS scheme

Then, the original image can be recovered to utilize reverse direction operation from stego-image in the receiver. Obviously, the data embedding capacity of NSAS scheme is equal to the grayscale value of the peak point. However, the peak point in cover-image is not the same as the peak point in stego-image because the peak point in original image will be changed after embedding secret data. In order to recover the embedded secret data, the grayscale value of peak point and zero point in the original cover-image must be transmitted to receiver. Unfortunately, this major problem how to transmit the additional information, the grayscale value of peak point and zero point, from sender to receiver is not discussed in [10].

In order to improve this shortcoming, Hwang, *et al.* (HKC) proposed a robust reversible data hiding scheme based on histogram shifting method. We also use the "Lena" image as an example. According to [6], the following four steps complete the data embedding process and show it in Fig.3.

***Step 1.*** Find out the maximum point which is location of 154 and two minimum points which are location of 237 and 23 in Fig. 1(b).

***Step 2.*** In order to recover original images, there is a location map to store location information of the pixels that have maximum point, left minimum point, and right minimum point in HKC scheme. The structure of location map is show in Fig. 4.

***Step 3.*** Generate the embedding space. These pixels that are located in histogram between left minimum point and left side of the maximum point (pixel value of 154) are shifted one pixel left.

***Step 4.*** Embed the secret data into the grayscale value of 152 and 153 or 155 and 156.

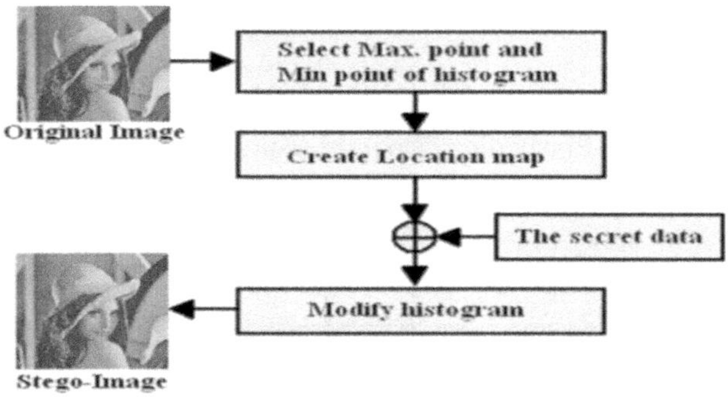


**Fig. 3.** The Embedding process in HKC scheme

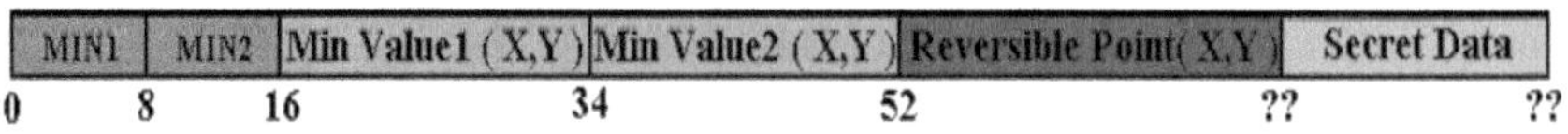


**Fig. 4.** The structure of location map

Here, we use the figure.5 to shows process of hiding data extraction and recovery of original image.

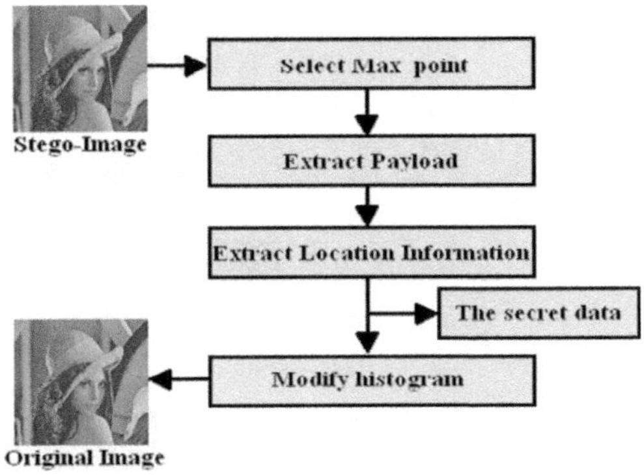


**Fig. 5.** The extraction process in HKC scheme

From the Fig.5, we just only look for the maximum point in the stego-image and then the secret data can be recovered in the HKC scheme. Obviously, the information hiding capacity is finite when we select one pair in original image. Therefore, they increase several pairs in location map to improve the capacity of embedding information into the cover-image. However, the capacity of embedding information will decrease 18 bits when it increases one minimum point in the location map each time. Furthermore, to achieve the goal of recovering the cover-image, it will waste a lot of concealing space because it is impossible for the numbers of minimum point is just only one.

## 3 Proposed a New Reversible Data Hiding Scheme

As everyone knows, the neighbor points of maximum point are regular. Therefore, the fact capacity of the embedding information is an inverse ratio to the storing the numbers of pairs in the location map. Obviously, the major disadvantage in the HKC scheme is to waste a lot of concealing space to store these coordinates of minimum points in the location map. Here, we propose another record method to replace the method using in HKC scheme to improve the above disadvantage. The following three steps complete the data embedding process in our proposed scheme.

***Step 1.*** Generate the histogram of cover-image and then find out the maximum point (154) and two minimum points (237 and 23) in Fig.6.

***Step 2.*** Set up the location map in order to recover original images. The structure of location map is show in Fig.8.

***Step 3.*** Pixels shift and embed the secret data into the grayscale value of 152 and 153 or 155 and 156. The histogram of stego-image is shown in Fig.7.

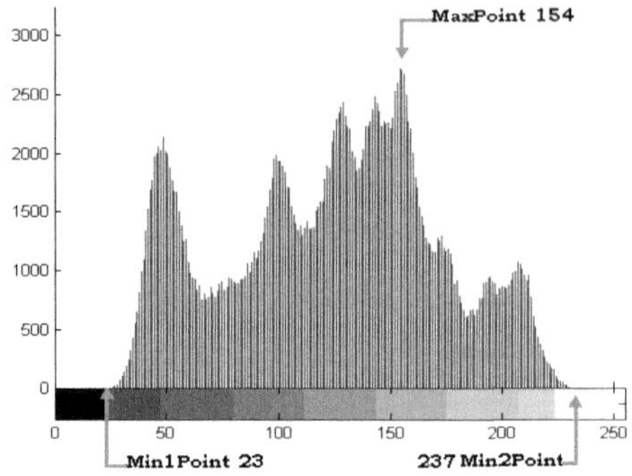


Fig. 6. The histogram of Lena

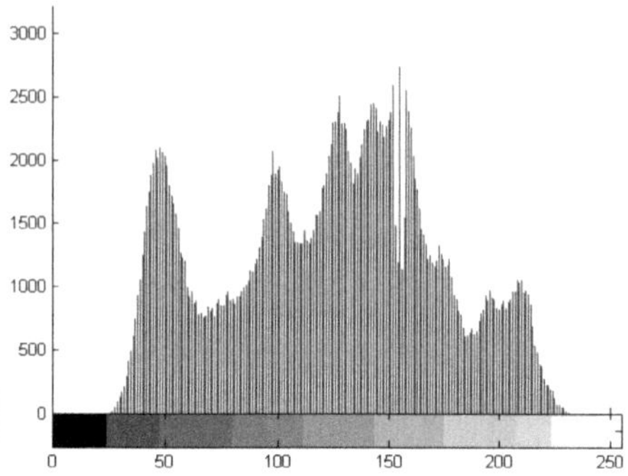

Fig. 7. The histogram of stego-image

Fig. 8. The structure of location map

In fact, the embedding process in our scheme is similar to HKC scheme but the method using to set up the location map is difference. In this proposed scheme, we utilize one bit to write down the primitive position to replace using to remember the coordinates of minimum points in the original image. Hence, it will reduce the size of storage quantity effectively and then increase the fact capacity of embedding information in our scheme.

Here, we explain the content of proposed location map shown as Fig. 9.

**MIN1：** The minimum point corresponds to the grayscale value which the minimum number of pixels in the left side of peak point in the histogram.
**MIN2：** The minimum point corresponds to the grayscale value which the minimum number of pixels in the right side of peak point in the histogram.
**Value1：** It is used to remember how many points must be recovered in left side.
**Value2：** It is used to remember how many points must be recovered in right side.
**Length1：** It is used to note these points relationship in left side.
**Length2：** It is used to note these points relationship in right side.
**Hide Data：** The fact capacity of embedding information。

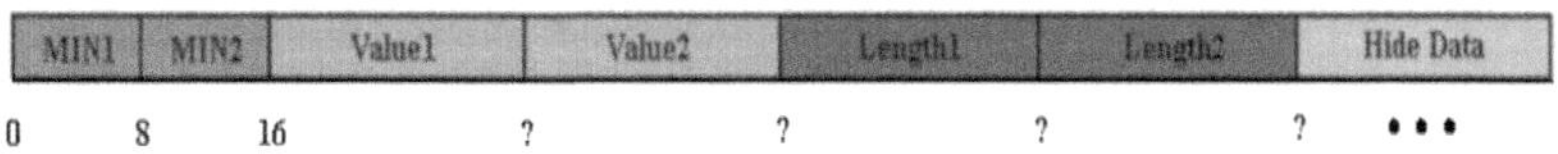


Fig. 9. The structure of location map

Here, we give an example to explain the difference of location map between our scheme and HKC scheme. There is a 3x3 picture shown as Fig.10. Then, it is used to remember the coordinate of the reversible points shown in Fig.10 (a) and just only use one bit to record the neighbor point shown in Fig.10 (b). Comparing HKC scheme and our scheme, we can find that it needs to store the reversible data space in HKC scheme more than our scheme. In particular, it will waste more storage capacities to remember the reversible points' coordinates when the minimum points are not zero.

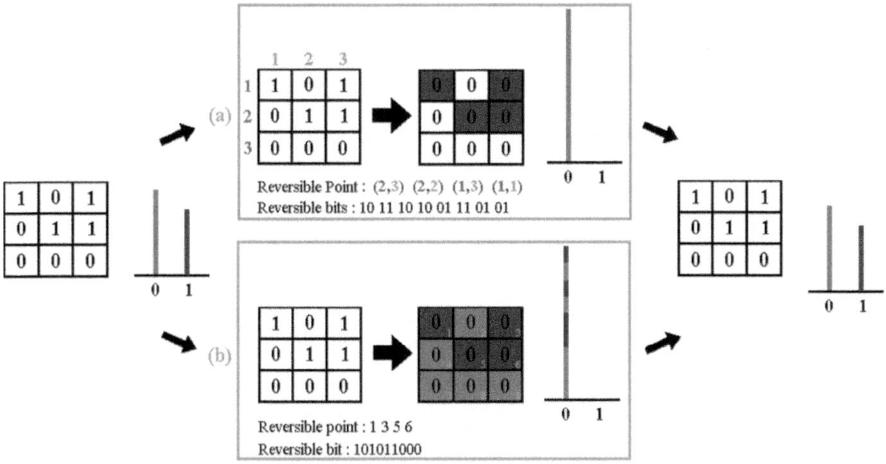


**Fig. 10.** (a) HKC scheme, (b) Our proposed scheme

## 4 Experimental Results and Analysis

These following four images of 512×512 grayscale, i.e., Lena, Baboon, boat and girl shown in Fig.11-13.(a), are used in experimenting invisibility and quantity of information hiding for the reversible data hiding algorithm suggested in the proposed scheme. From these pictures shown in Fig. 11-13 (a) and (b), we are not able to distinguish the difference between the original image and stego-images because the similar degree is too high between them.

**Fig. 11.**(a) Lena

**Fig. 11.**(b) Stego-Lena

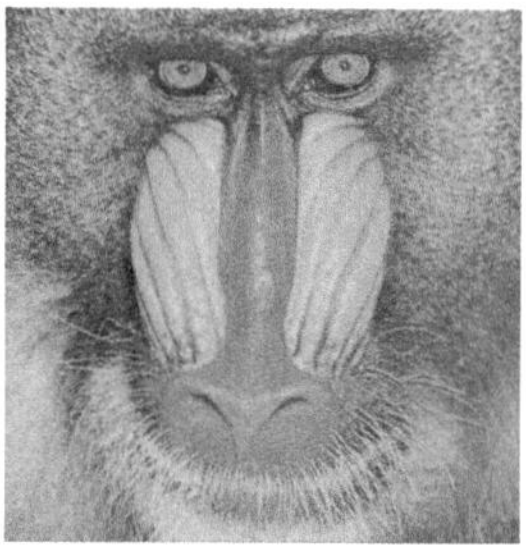

**Fig. 12.**(a) Baboon

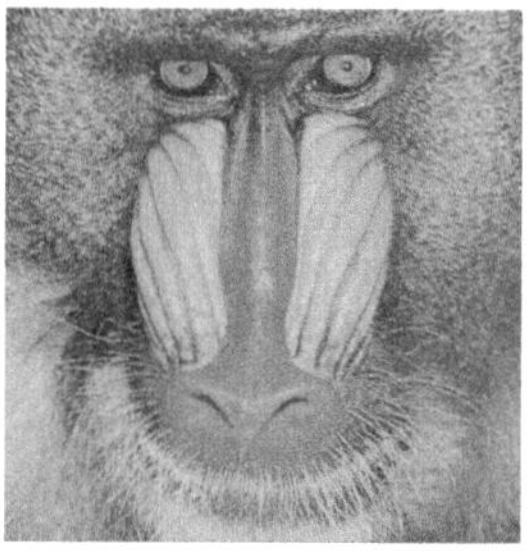

**Fig. 12.**(b) Stego- Baboon

**Fig. 13.**(a) Tiffany

**Fig. 13.**(b) Stego- Tiffany

**Table 1.** Experimental Results of the HKC scheme

| Test image | PSNR | Reversible bits | The fact capacity of hidden bits |
|---|---|---|---|
| **LENA** | 48.2db | 52 | 5,331 |
| **Baboon** | 48.2db | 213 | 5,168 |
| **Elaine** | 48.2db | 52 | 4,347 |
| **Airplane** | 48.3db | 52 | 15,294 |
| **Boat** | 48.3db | 105 | 10,036 |
| **Tiffany** | 48.3db | 52 | 4,304 |

**Table 2.** Experimental Results of the proposed scheme

| Test image | PSNR | Reversible bits | The fact capacity of hidden bits |
|---|---|---|---|
| **LENA** | 48.2db | 42 | 5,341 |
| **Baboon** | 48.2db | 69 | 5,312 |
| **Elaine** | 48.2db | 45 | 4,354 |
| **Airplane** | 48.3db | 46 | 15,300 |
| **Boat** | 48.3db | 52 | 10,089 |
| **Tiffany** | 48.3db | 1,423 | 7,047 |

We summarize the experimental results of the HKC method and the proposed method in Table 1 and 2, respectively. Even though these tables show that two algorithms results in about the same PSNR values, the needed quantity of reversible bits in our method is less than HKC method.

## 5 Conclusions

Recently, Hwang, *et al.* proposed a reversible data hiding scheme based on histogram shifting method to improve the shortcoming, the embedded data can not be recovered without the knowledge of peak point in [14]. In fact, Hwang, *et al.* stored the reversible point's data into the location map. However, this method will be effected the original data hiding capacity. In this paper, we utilize one bit to write down the primitive position to replace using to remember the coordinates of minimum points in the original image. After experiences analysis, our proposed scheme is not only to improve the original data hiding capacity but also to reach the goal of reversible data.

**Acknowledgement.** This work was supported by TWISC@NCKU, National Science Council under the Grants NSC 94-3114-P-006-001-Y.

## References

1. Celik, M. U., Sharma, G., Tekalp, A. M., Saber, E.: Reversible data hiding. In *Proc. IEEE Int. Conf. Image Processing*, Rochester, NY, (2002) 157–160
2. Chang, C. C., Tai, W.L., Lin, C.C.: A Reversible Data Hiding Scheme Based on Side Match Vector Quantization. IEEE Transactions on Circuits and Systems for Video Technology, 16(10) (2006) 1301-1308
3. Chang, C. C., Hsiao, J. Y., Chan, C. S.: Finding optimal LSB substitution in image hiding by dynamic programming strategy. *Pattern Recognit.*, vol. 36, no. 7 (2003) 1583–1595
4. Fridrich, J., Goljan, M., Du, R.: Invertible Authentication. Security and Watermarking of Multimedia Contents III (2001) 197-208
5. C. W. Honsinger, P. Jones, M. Rabbani, and J. C. Stoffel, 'Lossless recovery of an original image containing embedded data,' US Patent application, Docket No: 77102/E-D, 2001
6. J.H. Hwang, J.W. Kim, and J.U. Choi, 'A Reversible Watermarking Based on Histogram Shifting,' IWDW 2006, LNCS 4283, pp.348-361, 2006
7. S. Katzenbeisser and F. A. P. Petitcolas, *Information Hiding Techniques for Steganography and Digital Watermarking*. Norwood, MA: Artech House, 2000
8. S. H. Liu, T. H. Chen, H. X. Yao and W. Gao, 'A variable depth LSB data hiding technique in images,' Machine Learning and Cybernetics, 2004, pp.3990 – 3994, 2004
9. Z. M. Lu, J. S. Pan, and S. H. Sun, "VQ-based digital image watermarking method," *Electron. Lett.*, vol. 36, no. 14, pp. 1201–1202, 2000
10. Z. Ni, Y. Q. Shi, N. Ansari and W. Su, 'Reversible data hiding,' IEEE Transactions on Circuits and Systems for Video Technology, Vol.16, No.3, pp.354-362, 2006
11. C. I. Podilchuk and E. J. Delp, "Digital watermarking: Algorithms and applications," *IEEE Signal Process. Mag.*, vol. 18, no. 4, pp. 33–46, Jul. 2001

12. J. Tian, 'Reversible Data Embedding Using a Difference Expansion,' IEEE Transactions on Circuits and Systems for Video Technology, Vol.13, No.8, pp.890-896, 2003
13. H. C. Wu, N. I. Wu, C. S. Tsai, M. S. Hwang , 'Image Steganographic scheme based on pixel-value differencing and LSB replacement methods,' IEE Proc.-Vis. Image Signal Process., Volume 152, Issue 5, pp.611 – 615, 2005
14. G. Xuan, C. Yang, Y. Zhen , Y. Q. Shi and Z. Ni, 'Reversible data hiding based on wavelet spread spectrum,' IEEE Workshop on Multimedia Signal Processing, pp.211- 214, 2004
15. Xuan, G. Yao, Q., Yang, C., Gao, J.: Lossless Data Hiding Using Histogram Shifting Method Based on Integer Wavelets. IWDW 2006, LNCS-4283 (2006) 323-332
16. Yang, B., Schmucker, M., Funk, W., Busch, C., Sun, S.: Integer DCT-based Reversible Watermarking for Images Using Companding Technique. Steganography and Watermarking of Multimedia Contents VI, (2004) 405-415
17. Yu, Y. H., Chang, C. C., Hu, Y. C.,: Hiding Secret Data in Images via Predictive Coding. *Pattern Recognit.*, vol. 38, no. 5, (2005) 691–705

# Evolutionary Ensemble for *In Silico* Prediction of Ames Test Mutagenicity

Huanhuan Chen and Xin Yao

The Centre of Excellence for Research in Computational Intelligence and Applications
School of Computer Science, University of Birmingham, UK

**Abstract.** Driven by new regulations and animal welfare, the need to develop *in silico* models has increased recently as alternative approaches to safety assessment of chemicals without animal testing. This paper describes a novel machine learning ensemble approach to building an *in silico* model for the prediction of the Ames test mutagenicity, one of a battery of the most commonly used experimental in vitro and in vivo genotoxicity tests for safety evaluation of chemicals. Evolutionary random neural ensemble with negative correlation learning (ERNE) [1] was developed based on neural networks and evolutionary algorithms. ERNE combines the method of bootstrap sampling on training data with the method of random subspace feature selection to ensure diversity in creating individuals within an initial ensemble. Furthermore, while evolving individuals within the ensemble, it makes use of the negative correlation learning, enabling individual NNs to be trained as accurate as possible while still manage to maintain them as diverse as possible. Therefore, the resulting individuals in the final ensemble are capable of cooperating collectively to achieve better generalization of prediction. The empirical experiment suggest that ERNE is an effective ensemble approach for predicting the Ames test mutagenicity of chemicals.

**Keywords:** Ames Test Mutagenicity, *In silico* models, Evolutionary Ensemble, Negative Correlation Learning.

## 1 Introduction

The Ames test in *Salmonella typhimurium* is an in vitro biological assay to assess the mutagenic potential of chemical compounds. It is also considered as a quick assay to estimate the carcinogenic potential of a compound. Hence it serves as one of a battery of the most commonly used experimental in vitro and in vivo genotoxicity tests for safety evaluation of chemicals. Nevertheless, driven by new regulations and animal welfare, recently the needs of development of *in silico* models as alternative approaches to mutagenicity assessment of chemicals without animal testing is constantly increasing, and has attracted much attention from researchers in both fields of toxicology and computer science. Machine learning technique inevitably plays a major role in establishing relationships between chemical structural descriptors and mutagenicity for reducing, refining or replacing (3R) animal testing.

D.-S. Huang, L. Heutte, and M. Loog (Eds.): ICIC 2007, LNAI 4682, pp. 1162–1171, 2007.

The classification problem addressed in this paper is to predict whether an Ames test of a chemical is positive or negative. A positive test outcome means that the chemical tested is more likely to be a mutagen whilst a negative test outcome means that the chemical is more likely a non-mutagen. Application of different classification methods to the prediction of Aems test mutagenicity has been studied already. While most of studies use single learner based classification methods such as decision trees, $k$-nearest neighbors, neural networks [2,3], and support vector machine [4], few use ensemble methods such as bagging, boosting, and random forest [5]. However, ensemble approaches have generally been shown to be superior to their corresponding base classifiers for most of the classification problems in machine learning [6]. The preliminary analysis using a variety of classifiers also showed us similar scenarios. First, ensemble approaches of bagging, boosting and random forests using the decision tree as a base classifier outperform non-ensemble classifiers. Second, random forest approach performs best among three ensemble approaches [7]. The superiority of random forest approach is largely due to that it uniquely adopts the feature subset selection, which is particularly of value for handling a data set with a large amount of feature variables [8]. Nevertheless, randomization of both data and features taken in random forests seems to pay more attention to diversity than to accuracy of individual classifiers in an ensemble, which may degrade the performance of the generated ensemble. Both theoretical [9] and empirical studies [10] demonstrated that the generalization ability of ensemble depends crucially on both accuracy and diversity among individual classifiers in the ensemble. In this study, we focus on a novel machine learning ensemble approach based on NNs and evolutionary computation, which has the potential to pay attention to both diversity and accuracy of individuals within an ensemble. We are particularly interested in an investigation of its efficacy of the approach to building *in silico* models for the prediction of the Ames test mutagenicity of chemicals.

It is non-trivial to design an accurate yet diverse ensemble due to a trade-off between accuracy and diversity in the ensemble. In [1], we have proposed to incorporate evolutionary random neural network ensemble with negative correlation learning [11] (ERNE) to design accurate and diverse ensemble for machine learning problems. Experimental results of ERNE have shown improvement over the existing ensemble algorithms, i.e. Bagging, Adaboost and random forests. Since ERNE employs bootstrap sampling and random subspace method to generate the initialized neural ensemble and maintains the randomization in the evolving stage, diversity in the ensemble is encouraged/kept in ERNE. Evolutionary algorithm with negative correlation learning is adopted to search for a population of diverse and accurate individual NNs that collectively solve a problem. In negative correlation learning, the individual networks are trained simultaneously, rather than independently or sequentially. Evolving the ensemble with negative correlation learning emphasizes not only the accuracy of individual NNs but also the cooperation among different individual NNs and thus improve the generalization. In ERNE, as each member in the ensemble is learned from bootstrap sample of the training examples, which typically omits $1/e \approx 37\%$ of the training examples, out-of-bag (OOB) estimation, based on recording the votes of each member on those training examples omitted from its bootstrap sample and aggregating the votes for each training examples for an estimation of the generalization error, serves as another benefits of the algorithm.

The aim of this paper is to apply ERNE to the prediction of the Ames test mutagenicity of chemicals. The rest of this paper is organized as follows: The proposed algorithm is present in Section 2. Experimental results and discussion are reported in Section 3. Finally, Section 4 concludes the paper with future work.

## 2 Evolving Random Neural Ensembles with Negative Correlation Learning (ERNE)

It is widely believed that the success of ensemble algorithms depends on the accuracy and diversity among these base classifiers [12]. In general, the individual classifiers in ensemble are designed to be accurate and diverse among each other. For example, Bagging relies on bootstrap that produces different subsets of the training data; Ensemble of features employs different features instead of training data to generate diverse ensemble [8]. Random forests [13] combines Bootstrap sampling and random subspace method to generate more diverse ensembles. However, the predictions of random forests may fluctuate because of randomization of data and features simultaneously. Although the disadvantage of this could be slightly offset by including more and more decision trees in the ensemble, this of course leads to extended training times and more resources consumed.

The existing methods, random sampling of data and features, may promote the diversity but degrade the accuracy. How to improve the accuracy and simultaneously maintain the diversity for individual ensemble members to make sure that the obtained ensemble is both accurate and diverse is a key factor for ensemble algorithms. Chen et. al [1] proposed ERNE algorithm that offers a natural way to optimize accuracy and simultaneously maintain the diversity among the individuals in the ensemble. In this algorithm, randomization of both data and features has been adopted/kept to generate/maintain the diversity in the ensemble. Evolutionary ensemble with negative correlation learning provides the opportunities for these individual NNs to negatively correlated with each other and thus improve the accuracy of these individual NNs.

ERNE firstly generates an initial population of Neural Networks (NNs), each of which is trained on bootstrap of training data and random feature subspace. Then the diverse population is evolved with negative correlation learning to improve the accuracy of individual NN.

The whole process could be illustrated as following.

1. Sampling the original training set and obtain $M$ replications of training set $\{B_i\}_{i=1}^{M}$.
2. Generate an initial population of $M$ Neural Networks (NNs), the number of hidden nodes for each NN, $n_i (i = 1, ..., M)$ is specified randomly restricted

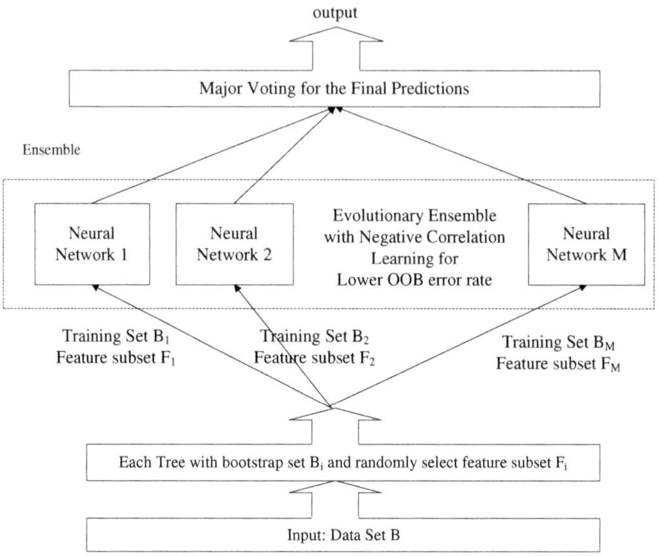


**Fig. 1.** The architecture of ERNE [1]

by the maximal number of hidden nodes. The random initial weights are distributed uniformly inside a small range.

3. Train each NN on each bootstrap set $B_i$ with randomly selected feature subset $\{F_i\}_{i=1}^{M}$ for a certain number of epochs that is proportional to the number of hidden number of neural network using negative correlation learning and calculate the out-of-bag estimation as the ensemble fitness function.
4. In each generation, randomly choose $s$ NNs to create offspring NNs[1]. For each offspring $s_i$, evolve each $s$ NNs with Gauss mutation[2], and train it with its corresponding parent's bootstrap set $B_i$ and feature subset $F_i$. $s$ is specified by the user.
5. Compare the fitness of each $s_i$ NN with their respective parents and include the better one in the population and recalculate the out-of-bag error as the fitness.
6. Go to the next step if the maximum number of generations has been reached. Otherwise, and go to Step 3.
7. Combining the population to form the ensembles.

There are four advantages of this algorithm: (1) Ensemble of different data subset and feature subset promotes the diversity among individual classifier in the ensemble. (2) Evolving the individual NN in the ensemble helps to improve

[1] Each individual, selected to be mutated with equal probability, reflects the emphasis on evolving a diverse set of individuals.

[2] Add Gauss noise to the weight vector of neural network. The parameter of Gauss noise is: $mean = 0$ and $variance = \mu$, will be specified manually.

the accuracy of these NN. (3) Negative correlation learning enables these individual NNs in the ensemble correlated with each other and improves the generalization performance. (4) It generates an internal unbiased estimate of the generalization error, OOB, as the NN ensemble building progresses.

### 2.1 Negative Correlation Learning

Negative Correlation Learning (NCL), a successful neural network ensemble learning technique developed in the evolutionary computation literature [11,14], has shown a number of empirical successes and varied applications, including regression problems [15] and classification problems [16]. It has consistently shown promising results compared with other techniques like Mixtures of Experts, Bagging, and Boosting [11,17].

NCL introduces a correlation penalty term into the error function of each individual network in the ensemble so that all the networks can be trained simultaneously and interactively on the same training data set. Liu et al. [16] implemented NCL by gradient descent method for training neural network. In fact, negative correlation learning provides a novel way to decompose the learning task of the ensemble into a number of subtasks for different individual networks.

### 2.2 Out-of-Bag Fitness Evaluation

In ERNE, out-of-bag (OOB) estimation error is taken as the objective to be optimized. As each member in the ensemble is learned from bootstrap sample of the training examples, which typically omits $1/e \approx 37\%$ of the training examples. The out-of-bag estimate is based on recording the votes of each member on those training examples omitted from its bootstrap sample and aggregating the votes for each training examples for an estimation of the generalization error. Out-of-bag estimates is proposed as an ingredient in estimates of generalization error, which has been empirically supported by [18] that the out-of-bag estimate is as accurate as using a test set of the same size as the training set. However, out-of-bag estimate requires considerably less time than the 10-fold cross-validation.

## 3 Experiments

In this section, we shall evaluate ERNE for the Ames test problem by comparing with some traditional classifiers, i.e. classification and regression tree (CART), and multilayer perceptions (MLP), as well as their corresponding ensemble learning algorithms respectively, i.e. Bagging of CART/MLP, Adaboost.M1 of CART/MLP and random forests/MLP.

### 3.1 Chemical Data Set

The chemical data used in this study were obtained from a paper of Kirkland et al. [19]. The total number of chemicals is 691, which consists of 357 chemicals with positive results from Ames tests and 334 chemicals with negative results.

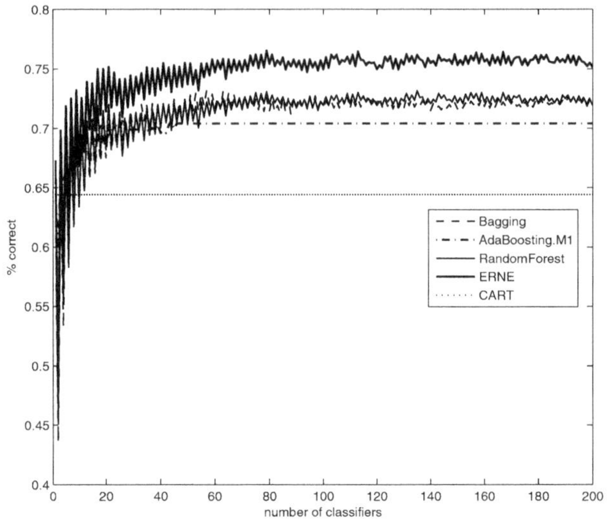


**Fig. 2.** Performance comparison of ERNE against CART and three other tree-based ensemble approaches in predicting the Ames test mutagenicity based on two-fold cross validation. The differences between ERNE and other classifiers are significant at the 5% significance level, see Table 2.

The variables that are used as features to build models are a set of 197 descriptors, which consist of atom/fragment counts, graph descriptors, topological descriptors and chemical structural descriptors [7].

### 3.2 Experimental Setup

As we know, the experimental results depend on the partitions of data set. In this paper, two-fold cross validation, allowing a sufficient test set to estimate the generalization error, is employed to evaluate these algorithms.

We prepossess all 197 features by normalizing them into values between 0.0 and 1.0. The network we used in this paper is three layer feedback NN. The number of hidden nodes will be initialized randomly but restricted in the range 3 to 8. Initial connection weights for individual NNs in an ensemble are randomly chosen. The parameter $\lambda$ is set to 0.8 and the variance of Gaussian mutation is 0.1. The parameters in use are set to: the population size $M$ (varied from 1 to 200), the number of offspring $s$ (max[20, M]), the number of generations (100). These parameters are chosen after some preliminary experiments. They are not meant to be optimal.

### 3.3 Experimental Results

The MLPs used in Bagging, Adaboosting.M1 and random MLPs are three-layer feedback NN with five hidden nodes. These MLPs are trained using scaled conjugate gradient (SCG) algorithm for 200 epoches. Figure 2 and 3 show the results

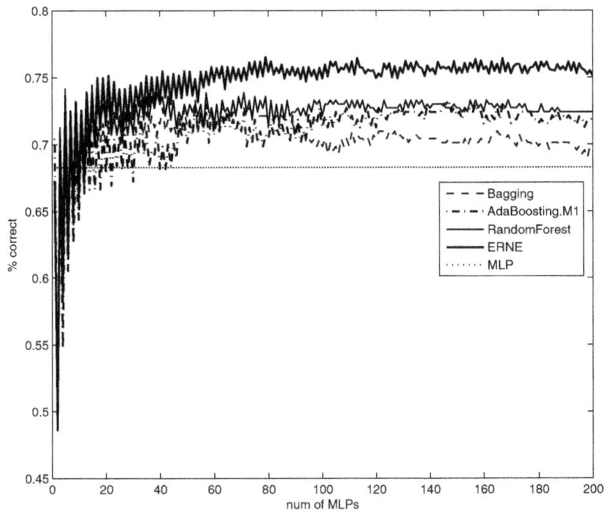


**Fig. 3.** Performance comparison of ERNE against MLP and three other MLP-based ensemble approaches in predicting the Ames test mutagenicity based on two-fold cross validation. The differences between ERNE and other classifiers are significant at the 5% significance level, see Table 2.

**Table 1.** Comparison Among ERNE with classification and regression tree (CART) and MLP based ensembles in terms of average cross validation error for Ames data set. The results are averaged on 2-fold cross-validation, respectively.

| % error | ERNR | RMLPs/RF | Bagging | Adaboost | MLP/Tree |
|---|---|---|---|---|---|
| MLP | 24.90 | 27.57 | 30.46 | 28.14 | 31.74 |
| CART | - | 28.08 | 27.57 | 29.60 | 35.59 |

of ERNE over 40 independent cross validations. In each figure, we record the performance of these algorithms with respect to the size of the ensemble, i.e. the number of classifiers in this ensemble. For comparison, Table 1 lists the the average errors of all kinds of classifiers over the 40 runs and Table 2 gives the result of two-tail pared t-test in terms of prediction error between ERNE and other classifiers.

From Figure 2 and 3, ERNE consistently outperforms other algorithms in terms of cross validation error. This is understandable since the performance of random forests is better than or similar as the other ensemble algorithms in most of the cases, and ERNE maintains good diversity by adopting bootstrap and random feature subspace, which is similar as random forests, and evolves the ensemble to optimize the accuracy and cooperation. Generally speaking, ERNE will perform no worse than random forests.

In the preliminary analysis for Ames test problem, random forest approach performs best among a variety of approaches [7], including Decision Tree (C4.5), Naive Bayesain, K-NN, Logistic Regression, support vector machine (SVM),

**Table 2.** Results (P value) of two-tail pared t-test in terms of prediction error between ERNE and one other classifier

| Methods vs ERNE | RMLPs/RF | Bagging | Adaboost | MLP/Tree |
|---|---|---|---|---|
| MLP | 0.0376 | 0.0168 | 0.0093 | 0.0001 |
| CART | 0.0336 | 0.0236 | 0.0084 | 0.0000 |

Bagging and Boosting. The state-of-the-art performance of random forest approach is largely due to that it uniquely adopts both bootstrap of data and feature subset selection, which is particularly of value for handling a data set with a large amount of feature variables [8]. ERNE not only keeps the benefits of random forests but also improves its performance by optimizing the accuracy. The superiority of ERNE over random forests can be observed in the experimental results. In our experiments, we found that Adaboost.M1 of trees sometimes will overfit when adding more and more trees in the ensemble, which can be found in Figure 2 but for neural network ensemble, Adaboost.M1 algorithm behaves better when adding more and more MLPs in the ensemble. Another interesting point to say is that MLP-based ensemble, (Bagging, Adaboosting and random MLP), is not so stable as tree based ensemble. This is because tree can be seen as a deterministic algorithm and MLP would output differently given the same input because of randomization of initialization of weight vectors.

In this experimental results, all of the NNs are ensembled to constitute the final classifier. However, from Figure 2 and 3, ensembling some effective combination of NNs would be better than ensembling all of them. This will be considered as the future work.

The following two reasons might explain why the performance of our algorithm is better than that of others.

- ERNE generates a diverse ensemble. Firstly, bootstrap sampling of data and random feature subsets generate a diverse ensemble in the initial population, which inherits the merits of random forests. Secondly, in the evolving stage, the diversity is maintained by only mutating the weight of individual NN but not changing the bootstrap of data and the feature subset used by this individual NN.
- Evolving ensembles with negative correlation learning has a potential to simultaneously optimize both the accuracy and cooperation of the existing individual NNs in the ensemble , resulting in reducing the generalization error. The potential is indicated by our empirical results of this study.

## 4 Conclusions

This paper describes a novel machine learning ensemble approach, called ERNE, to building an *in silico* model for the prediction of the Ames test mutagenicity. *In silico* models serve major roles in reducing, refining and replacing animal testing for the risk assessment of the safety of chemicals. ERNE was developed based on neural networks and evolutionary algorithms. Technically, it firstly combines the method of bootstrap sampling on training data with the method of random subspace feature selection to ensure diversity in creating individuals within an initial ensemble. Secondly, while evolving individuals within the ensemble, it makes use of the negative correlation learning, enabling individual NNs to be trained as accurate as possible while still manage to maintaining them as diverse as possible. Finally, it takes out-of-bag estimation as the fitness functions of individual NNs, which potentially enhances the generalization capabilities of individual NNs. Consequently, the resulting individuals in the final ensemble are capable of cooperating collectively well to achieve better generalization of prediction.

Empirical experiments have been carried out in this paper to evaluate ERNE on the Ames test mutagenicity prediction problem in comparison with other ensemble algorithms. ERNE has shown promising performance. The reasons to explain the superiority of ERNE are also given in this paper.

An immediate future work regarding the algorithm is to develop an ensemble subset selection method to choose a subset of individual NNs rather than all of individuals in the population to form the final ensemble, The hope is to improve the generalization performance and reduce computational complexity. We would also like to carry out some work to further understand the trade-off between accuracy and diversity of individual NNs in an ensemble in the context of improving overall generalization performance of ERNE.

For Ames test mutagenicity and other similar applications in chemoinformatics, the common characteristic of these problems is that there are plenty of features, each of which denotes a descriptor (graph descriptor, topological descriptor, chemical structural descriptor and so on). Although ensemble approaches, including ERNE, could achieve a high predict accuracy, they lack interpretability and thus make them less interesting from the viewpoint of toxicity decision making. The following work will focus on interpretability of approaches, e.g. feature selection and decision rules extraction.

## Acknowledgment

This work is partially supported by a Dorothy Hodgkin Postgraduate Scholarship to the first author.

## References

1. Chen, H., Yao, X.: Evolutionary random neural ensemble based on negative correlation learning. In: Proceedings of IEEE Congress on Evolutionary Computation (CEC'07). (2007) submitted.
2. He, L.N., Jurs, P.C., Custer, L.L., Durham, S.K., Pearl, G.M.: Predicting the genotoxicity of polycyclic aromatic compounds from molecular structure with different classifiers. Chemical Research in Toxicology **16** (2003) 1567–1580

3. Votano, J.R., Parham, M., Hall, L.H., Kier, L.B., Oloff, S., Tropsha, A., Xie, Q.A., Tong, W.: Three new consensus qsar models for the prediction of ames genotoxicity. Mutagenesis **19** (2004) 365–377
4. Mahe, P., Ueda, N., Akutsu, T., Perret, J.L., Vert, J.P.: Graph kernels for molecular structure-activity relationship analysis with support vector machines. Journal of Chemical Information and Modeling **45** (2005) 939–951
5. Zhang, Q.Y., de Sousa, J.A.: Random forest prediction of mutagenicity from empirical physicochemical descriptors. Journal of Chemical Information and Modeling **47**(1) (2007) 1–8
6. Dietterich, T.G.: An experimental comparison of three methods for constructing ensembles of decision trees: Bagging, boosting, and randomization. Machine Learning **40**(2) (2003) 139–157
7. Li, J., Dierkes, P., Gutsell, S., Stott, I.: Assessing different classifiers for in silico prediction of ames test mutagenicity. In: A poster in the 4th Joint Sheffield Conference on Chemoinformatics. (2007) submitted.
8. Ho, T.K.: The random subspace method for constructing decision forests. IEEE Transaction on Pattern Analysis and Machine Intelligence **20**(8) (1998) 832–844
9. Hansen, L.K., Salamon, P.: Neural network ensembles. IEEE Transactions on Pattern Analysis and Machine Intelligence **12**(10) (1990) 993–1001
10. Hashem, S.: Optimal linear combinations of neural networks. Neural Networks **10**(4) (1997) 599–614
11. Liu, Y., Yao, X.: Ensemble learning via negative correlation. Neural Networks **12**(10) (1999) 1399–1404
12. Brown, G., Wyatt, J., Tino, P.: Managing diversity in regression ensembles. Journal of Machine Learning Research **6** (2005) 1621–1650
13. Breiman, L.: Random forests. Machine Learning **45**(1) (2001) 5–32
14. Liu, Y., Yao, X.: Simultaneous training of negatively correlated neural networks in an ensemble. IEEE Transactions on Systems, Man, and Cybernetics, Part B: Cybernetics **29**(6) (1999) 716–725
15. Yao, X., Fischer, M., Brown, G.: Neural network ensembles and their application to traffic flow prediction in telecommunications networks. In: Proceedings of International Joint Conference on Neural Networks. (2001) 693–698
16. Liu, Y., Yao, X., Higuchi, T.: Evolutionary ensembles with negative correlation learning. IEEE Transaction on Evolutionary Computation **4**(4) (2000) 380–387
17. McKay, R., Abbass, H.: Analyzing anticorrelation in ensemble learning. In: Proceedings of 2001 Conference on Australian Artificial Neural Networks and Expert Systems. (2001) 22–27
18. Breiman, L.: Out-of-bag estimation. Technical report, Stanford University (1996)
19. Kirkland, D., Aardema, M., Henderson, L., Muller, L.: Evaluation of the ability of a battery of three in vitro genotoxicity tests to discriminate rodent carcinogens and non-carcinogens i. sensitivity, specificity and relative predictivity. Mutation Research **584** (2005) 1–256

# Parallel Filter: A Visual Classifier Based on Parallel Coordinates and Multivariate Data Analysis

Yonghong Xu, Wenxue Hong, Na Chen, Xin Li, WenYuan Liu, and Tao Zhang

Department of Biomedical Engineering, University of Yanshan,
Qinhuangdao, Hebei Province, China
xyhhit@163.com

**Abstract.** Multivariate visualization techniques are often used as assistant tools for classification tasks up to now. However, few classification systems fully utilize the capability of multivariate visualization and integrate them with multivariate analysis algorithms into a compact system. We propose an interactive visual classification model based on some multivariate graphical presentation in this paper. As an example of it, a visual classifier based on parallel coordinates plot is developed. The multivariate data is first mapped to the parallel coordinates plot, and then an optimizer based on linear discriminant analysis optimizes it into the visualization more fit for classification tasks. This optimized visualization then can be processed by decision tree algorithm and attain classification rules. It has the merit of making the invisible visible and users can steer the classification process, consequently favor the understanding and knowledge discovery of original data.

**Keywords:** multivariate visualization, multivariate data analysis, linear discriminant analysis, parallel coordinates, decision trees.

## 1 Introduction

Most classification systems nowadays are not integrated with visualization techniques and the human's pattern recognition capability and domain knowledge are rarely exploited, but there is an emerging trend of focusing more on this important feedback mechanism recently. Ankerst et al. propose an interactive decision tree construction approach based on a multidimensional visualization technique called Circle Segments [1]. They argue in another paper [2] for increased user involvement in the classification process, with three important reasons: (1) with the help from data visualization, the human capability to find useful patterns can be greatly improved. (2) The user will have more confidence in the trust that they place in the created patterns generated from this interactive process. (3) It is possible to incorporate domain knowledge to the algorithm. Note that classification visualization is not only the visualization of classification results. The key idea is to steer the pattern recognition process and discovery knowledge from the data.

Several visual classification techniques have been proposed these years. Besides Ankerst et al's Perception Based Classification [1], other techniques include the

D.-S. Huang, L. Heutte, and M. Loog (Eds.): ICIC 2007, LNAI 4682, pp. 1172–1183, 2007.

StarClass [3] and PaintingClass [4] by Soon Tee Teoh et al, and the PolyCluster [5] by Danyu Liu et al. The PBC is based on the pixel-oriented Circle Segments. The Circle Segments technique maps d-dimensional objects to a circle, which is partitioned into d segments presenting one attribute each. StarClass uses Star Coordinates for visual classification. PaintingClass allows the users to choose between Star Coordinates and parallel coordinates. Parallel Coordinates in this system is simply modified and used to visualize datasets with categorical attributes. Parallel coordinates is not fully exploited and appreciated by these authors.

In the approximately 20 years since the formal introduction of parallel coordinates into the referred literature by A. Inselberg [6] as a device for computational geometry and by Wegman [7] for hyper-dimensional data analysis, it has become famous and widely used in the statistics, data analysis and visualization domains. Parallel coordinates plot is usually used as a tool for multivariate visualization. Their potential as a classification tool is not fully recognized and explored. We think parallel coordinates plot is fit for classification task due to its projective geometry interpretation. Parallel coordinates plot is a coordinate-based visualization technique and contain the whole projection information on all directions. However, some valuable projection information is hidden in traditional parallel coordinates so some optimization is needed.

Based on parallel coordinates and multivariate data analysis algorithms, a new kind of visual classifier called parallel filter is proposed in this paper. This method unified the traditional parallel coordinates plot, the linear discriminant analysis (LDA) and the decision trees. By combining parallel coordinates, linear discriminant analysis and decision trees into a visual classifier, users can steer the classification process, consequently favor the understanding and knowledge discovery of original data.

The rest of this paper is organized as follows. In section 2, we introduce the principle of parallel coordinates and indicate that it has a good potential for visual classification due to its projective geometry interpretation. In section 3, the visual classification model based on some multivariate graphical presentation is presented. We discuss the optimization of parallel coordinates using linear discriminant analysis and classification algorithms based on decision trees in section 4. An application example is described in section 5. Section 6 summarizes this paper and outlines our plan for future research.

## 2 The Principle of Parallel Coordinates Plot

The basic principle and drawing method of parallel coordinates are described in this section. Readers are referred to literatures such as [6] or [7] for some more detail information. In order to overcome the limitation of 3-dimensional physical space, the n-dimensional coordinate axes in original data space are drawn as parallel on a 2-dimensional plane. As indicated in Fig. 1, on the plane with Cartesian coordinates, and start on the y-axis, $N$ copies of the real line, label as $x_1, x_2, \ldots, x_n$ are placed equidistant and perpendicular to the axis. A sample $x$ $(x_1, x_2, \ldots, x_n)$ is represented as a polygonal line $\bar{x}$. Its N vertices are at $((i-1)d, x_i)$ on the $x_i$ axes for $i = 1,2,\ldots,N$.

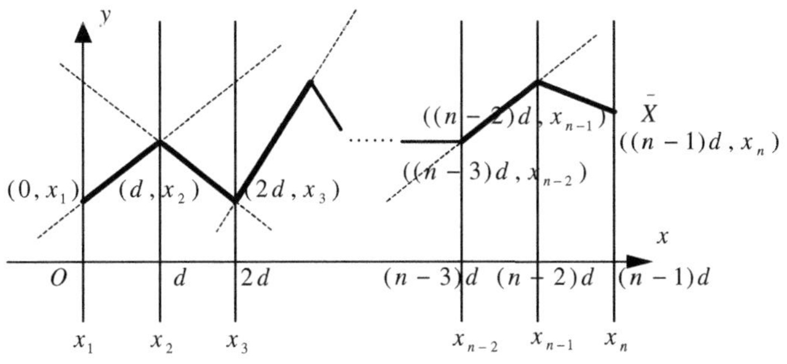


**Fig. 1.** Parallel coordinates of a sample $X(x_1, x_2,..., x_n)$

In effect, a one-to-one correspondence between points in $R^N$ and planar polygonal lines with vertices on $((i-1)d, x_i)$ is established.

The idea of parallel coordinates is so straightforward that many people stop here and ignore the mathematics underlying it. However, the attracting characteristics of parallel coordinates for visual data analysis are just resting with its projective geometry interpretation. The parallel coordinate representation enjoys some elegant duality properties with the usual Cartesian orthogonal coordinate representation. Consider a line $l$ in the Cartesian coordinate plane given by $l : x_2 = mx_1 + b$ and consider two points lying on that line, says A $(x_{1a}, x_{2a})$ and B $(x_{1b}, x_{2b})$, as shown in Fig. 2. We superimpose a Cartesian coordinate axes $x$ $y$ on the parallel coordinates plane so that the two parallel axes has the equation $x = 0$ and $x = d$ respectively. It is clear that the point $A(x_{1a}, x_{2a})$ maps into the line joining $(0, x_{1a})$ to $(d, x_{2a})$. Similarly, the point $B(x_{1b}, x_{2b})$ maps into the line joining $(0, x_{1b})$ to $(d, x_{2b})$. It is a straightforward computation to show that these two lines intersect at a point (in the $x$ $y$ plane) given by $\bar{l} : (d/(1-m), b/(1-m))$. Notice that in the parallel coordinates plot this point depends only on $m$ and $b$ the parameters of the original line in the Cartesian plot. Thus $\bar{l}$ is the dual of $l$ and we have the interesting duality result that points in Cartesian coordinates map into lines in parallel coordinates while lines in Cartesian coordinates map into points in parallel coordinates. Consider both the $x_1 x_2$ plane and the $xy$ plane to be augmented by suitable ideal points so that we may regard both as projective planes. So a parallel coordinate presentation can be looked upon as a duality transformation.

It shows that parallel coordinates plot has a potential for classification task due to its projective transformation interpretation. Nevertheless, traditional parallel coordinates plot is not very fit for classification tasks because some projection information of data objects is invisible. Furthermore, some information is valueless for pattern recognition tasks and they can be hidden to let the computer screen to reveal the most valuable information. So we consider using linear discriminant algorithms to find those valuable directions to make it more fit for classification.

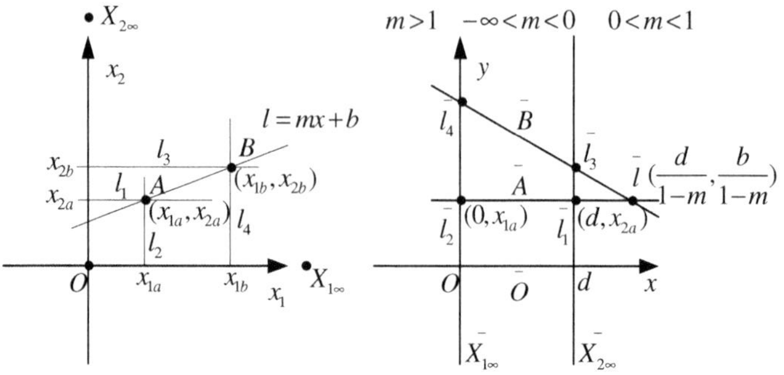

**Fig. 2.** Duality properties of the parallel coordinates plane

# 3 Visual Classification Model

Some kind of visualization serves as the basis of our approach of interactive classification. Fig.3 depicts our model for visual interactive classification. Multivariate data to be classified are first preprocessed by some algorithms before mapping to the graphical presentation. These preprocessing often include some general operation such as normalization, reducing noise etc. Then the preprocessed data is mapped to some kind of graphical presentation. This graphical presentation is optimized by class optimizer and finally turned into a visualization more fit for classification task. The information of original data remains mostly, while the most valuable information for classification are highlighted and strengthened. Class

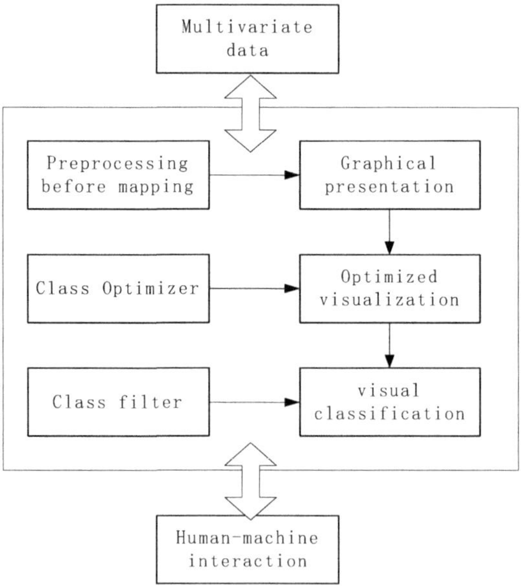


**Fig. 3.** Visual interactive classification model

optimizer is designed to be automatic operated by computer or manual operated by man. Computer algorithms function as an assistant in the background and providing some guiding when need. Optimized visualization then can be processed by class filter in order to filter off the classification irrelevant information from the visualization. So the computer can draw boundary between classes automatically or by man manually with the help of human-machine interaction. Class filter can be based on some kind of projection-based pattern recognition algorithms such as decision trees. Human-machine interactions play key roles in the visualization system. It provides a friendly interface to facilitate compact collaboration of human and machine.

As an example of our visual classification model, we will present a visual classifier based on optimized parallel coordinates in next section.

# 4 The Principle and Algorithm of Parallel Filter

Based on the model we proposed in previous section, we now present a visual classifier we call parallel filter. The parallel coordinates plot is the multivariate graphical presentation serving as a tool for visualization. Linear discriminant analysis is used as an optimizer and the decision tree algorithm are used as a class filter. The procedure of parallel filter is shown in the Table 1.

**Table 1.** Procedure of parallel filter

| Algorithms | Steps |
|---|---|
| Normalization | Processing before mapping |
| Mapping | Mapping data to parallel coordinates plot |
| Linear discriminant analysis | Optimizing parallel coordinates plot |
| Decision tree algorithm | Classification |

The procedure of the parallel filter is divided into four steps. Step 1: the original data is preprocessed by common preprocessing methods such as normalization, noise reduction and dimension reduction etc. Step 2: the data after preprocessing are mapped to parallel coordinates plot. Step 3: parallel coordinates are optimized by linear discriminant analysis, so only those directions that contains most valuable information is displayed, and step 4: use the decision tree algorithm to filter the optimized parallel coordinates. Consequently, a filtered plot is attained and can be used to classify other samples.

## 4.1 Preprocessing

*Dimension reduction or feature selection.* For some real dataset, such as the biological mass spectrum data, dimension reduction is necessary before mapping to parallel coordinates. This is also a process of feature selection or extraction.

For example, suppose there are two classes $\omega_i$ and $\omega_j$, $m_{ik}$ and $m_{jk}$ are their means, and their variances are $\sigma_{ik}$ and $\sigma_{jk}$. We can use fisher ratio law (1) to carry on the judgment.

$$G_k = (m_{ik} - m_{jk})^2 / (\sigma_{ik}^2 + \sigma_{jk}^2) \quad k=1,2,\ldots,n . \tag{1}$$

The bigger $G_k$ is, the easier it can be separated in this dimension. Arrange the data $\{G_k,\ k=1, 2, \ldots, n\}$ according to its size, and choose the first m characteristics. Good classification effect can be guaranteed.

*Normalization.* The normalization method: Use the following formula to normalize the original data.

$$X_{ij} = (X_{ij}' - X_{min}) / (X_{max} - X_{min}) . \tag{2}$$

The notation $X_{min}$, $X_{max}$, $X_{ij}'$, $X_{ij}$ represent the minimum value, the maximum value, the value without normalization (original data), and the normalized value respectively. The scope of the data obtained from this formula is between 0 and 1.

## 4.2 Mapping

The mapping step is straightforward and its task is to map data to parallel coordinates presentation, as shown in Fig. 4, where original vegetable oil data [8] (before normalization) are represented by parallel coordinates plot.

The seven classes of it are displayed in parallel coordinates using seven different colors. The fist dimension is the category, and other 7-dimension express seven variables, corresponding to contents of seven fat acids respectively. The parallel coordinates of the same dataset after normalization is indicated in Fig. 5.

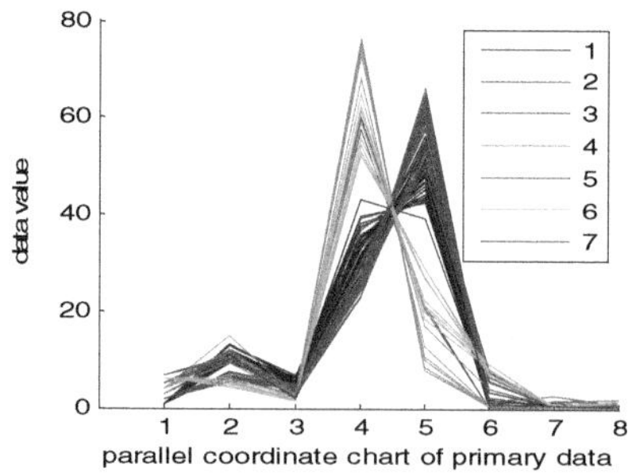


**Fig. 4.** Parallel coordinates plot of original data

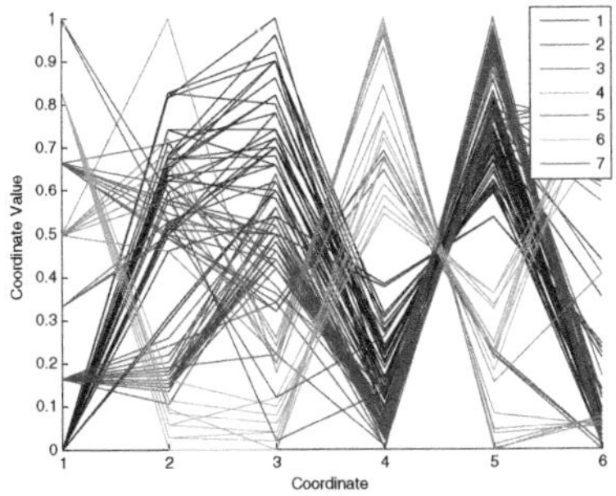


**Fig. 5.** Parallel coordinates plot of normalized oil data

We can see from Fig. 5 that: although there is some information in the original parallel coordinates. The most valuable information about the difference of one class from the others is not distinct. In other words, original parallel coordinates need to be optimized to be more fit for classification.

### 4.3 Class Optimization

Linear discriminant analysis algorithm is used as an optimizer to find the most valuable directions on which one class is well separated from others.

The basic idea of linear discriminant analysis is to find the directions on which training samples are projected and can be well separated by a point. This is usually done by the following way.

In case of a two-class classification task, the Within-class scatter matrix of the two classes is:

$$S_i = \sum_{x \in X_i} (x - m_i)(x - m_i)^T \quad . \; i = 1,2 \tag{3}$$

The whole Within-class scatter matrix is:

$$S_W = S_1 + S_2 \; . \tag{4}$$

The Between-class scatter matrix is formula (5) where $m_1$ and $m_2$ are the means of two classes. They are row vectors.

$$S_B = (m_1 - m_2)(m_1 - m_2)^T \; . \tag{5}$$

So the criterion will be:

$$J_2 = tr(S_W^{-1} S_B) \; . \tag{6}$$

We hope that the within-class scatter is small and the between-class scatter is big as far as possible. Suppose that the data $\lambda_1 \lambda_2 \ldots \lambda_D$ are the feature values of matrix $S_W^{-1} S_B$. According to the size, the order arranged is:

$$\lambda_1 \geq \lambda_2 \geq \ldots\ldots \geq \lambda_D \ . \tag{7}$$

Choose the feature vectors that correspond to the fist d feature values as W

$$W = [u_1, u_2, \ldots, u_d] \tag{8}$$

The $J_2$( W) is:

$$J_2(W) = \sum_{i=1}^{d} \lambda_i \ . \tag{9}$$

In case of a multi-class classification task, it can be regrouped to several two-class tasks and the algorithm is used recursively. Firstly the class 1 is looked as one class, and the remains are looked as another. Carry on linear discriminant analysis algorithm on the two classes. So the most valuable directions that can separate them well are found. Then using linear discriminant analysis algorithm to found the most valuable directions of class 2. Going on like this and finally the most valuable direction corresponding to each class are found and a new optimized parallel coordinates is drawn to serve for classification.

### 4.4 Class Filtering by Decision Trees Algorithm

After optimization by linear discriminant analysis algorithm, the optimized parallel coordinates can be further processed by decision trees and acquire the final classification rules.

Decision trees are a kind of classification algorithm whose simple form is a binary tree. The so-called binary tree is that the trees' each node only divides into two branches except the leaf-nodes. In other words, each node has, but also only has two sub-nodes. The classifier that has the binary tree structure may change a complex multi-category classification question into multistage two categories questions to solve. In each node, the sample collection is divided into two subsets. Each part possibly still contains many categories. It can be divided into two subsets again. Divide the subsets until each part only contains the identical category, or some kind of sample gains an advantage. Then this part is filtered off from our parallel coordinates, so as to overcome the over-plotting problem of parallel coordinates and more fit for the user to do classification task visually. This is the why we call our method as the parallel filter.

## 5 Empirical Studies

In this section, we will apply our method to a real vegetable oil data set. This dataset comes from a paper published in 2005 by Darinka Brodnjak-Voncina et al [9]. This dataset is interesting because it is a multi-class problem of 96 samples which own to seven classes of vegetable oils respectively. And there are 36 samples whose class tag is unknown (They are bought from the market, even the authors do not know to which class they own). Every sample has seven variables, which are the contents of seven fatty acids. We consider this dataset can be used as a good example of visual classification. Because this dataset has great differences in their samples numbers, the maximum is 36 and the minimum is only 2. Besides, visual classification will give us much more things other than only some class tags as many classification algorithms do.

Feature selection is first carry out by fisher ratio law criterion and five dimensions of original data are chosen, which corresponding to contents of five fat acids respectively.

Then linear discriminant analysis algorithm is performed to found the most valuable directions of the seven classes respectively. Then a new optimized parallel coordinates is drawn to serve for classification, as indicated in Fig. 6 as follows:

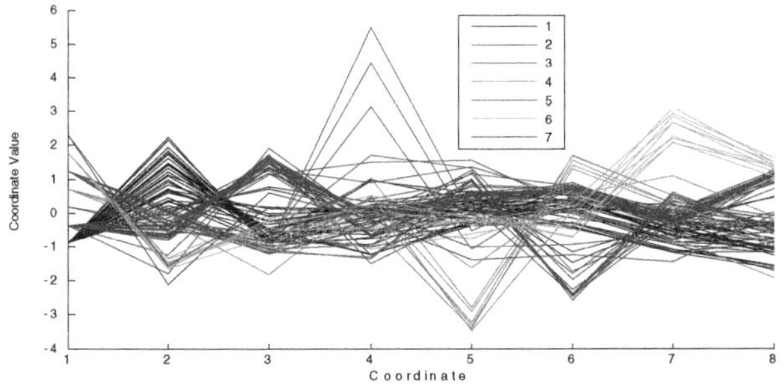


**Fig. 6.** Optimized parallel coordinates for a vegetable oil dataset

From the optimized parallel coordinates, we can see that class 2 is well separated by direction 3 and class 3 is well separated by direction 4 and so on (direction 1 expresses the category).

Using the decision tree algorithm to filter every category one by one, we will get a filtered parallel coordinates plot. Training samples are located on various intervals of various vertical lines. As indicated in Fig. 7, training samples of class 1 are located on direction 2, and training samples of class 2 are located on direction 3 etc. Because other directions contain little information for classification, so samples on those directions are filtered off to avoid the over-plotting problem. In this way, the original data and the rules of decision tree are display in the same classification optimized plot thus classification is visible and more understandable.

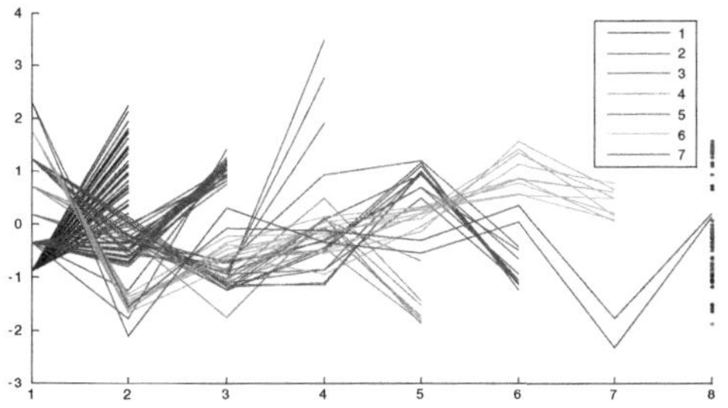


**Fig. 7.** Filtered parallel coordinates by decision tree algorithm

The difficulty of classification of this dataset lies that the categories are up to seven, they have the obvious outlier data, and some sample numbers are too few (the fewest are only two). When used the traditional classification methods like the linear discriminant analysis, the result is not ideal. When used the neural network classification, the precision are high, but its parameters are not easy to determine and it is apt to over-fit and the classified results are not easy to interpret by chemical experts. In view of this kind of problem, we use the parallel filter for visual classification and knowledge discovery.

Make the 36 samples whose class tag is unknown carry on the same transformation as above. The result is expressed in the filtered parallel coordinates plot, as shown in Fig. 8:

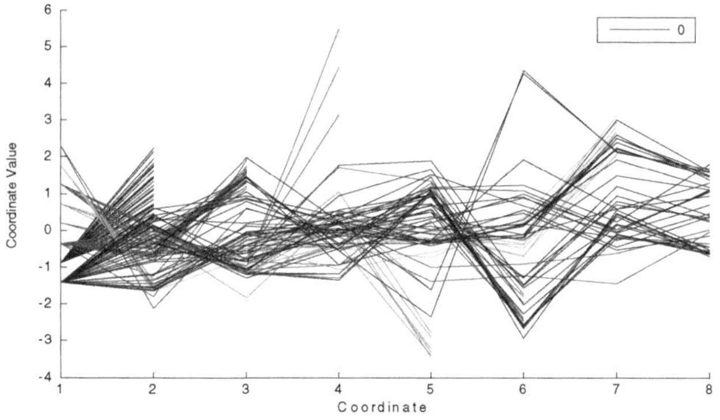


**Fig. 8.** Filtered parallel coordinates with test samples

Using classification rules obtained by parallel filter, the classification results are shown in the following Fig. 10. (Different colors demonstrate different classes):

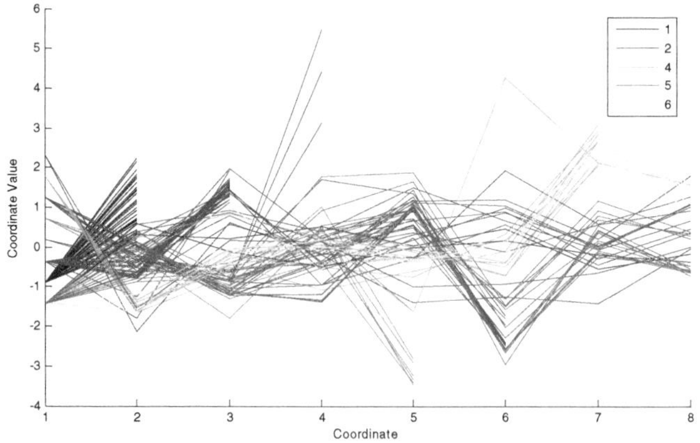


**Fig. 9.** Classification results in parallel coordinates

By comparison, we find that the result of us is only a little different from the result reported by [8]. We show the differences in the following table:

**Table 2.** Results of parallel filter comparing with [8]

| ID | Parallel filter results | Results of [8] |
|---|---|---|
| 24 | 5 | 1 |
| 87 | 2 | 1 |

The biggest merit of this method is that it can realize the cooperation of the man and machine, and the original data, classification process and final results are more transparent and understandable. We also carried on the classification with this method to another data that have 4 categories. The averaged classification accuracy is 94.85%. It is higher than 93.81% got by the linear discriminant analysis method.

## 6 Conclusions

We present a visual classifier based on parallel coordinates, linear discriminant analysis and decision trees. This is a special case of our interactive visual classification model. We use linear discriminant analysis to find some valuable projections on certain directions (This is actually a dimension reduction process). Linear discriminant analysis can be function as a optimizer for parallel coordinates here, which expose the valuable information for classification on parallel coordinates so make it more fit fore classification tasks. Classification is performed using decision trees algorithms. Parallel coordinates are well combined with machine algorithms and make the classification visible and controllable.

Compared with the traditional classification methods, this method has the merits of making the invisible visible and user can steer the classification process. Moreover, with help from visualization, the human capability to find useful patterns can be greatly improved. Consequently, the user will have more confidence in the patterns generated from this interactive process. On the other hand, classification algorithms such as linear discriminant analysis and decision trees are possible to be strengthened by incorporating domain knowledge to them.

We are now preparing to apply this method to visually analyze the pattern of serum proteomic mass spectrum data. We consider visual classification method can improve the diagnosis ways of some complicate diseases.

**Acknowledgments.** The National Science Foundation of China (No. 60605006, No.60474065, No.60671025) supports this work.

## References

1. Ankerst, M., Elsen, C., Ester, M., Kriegel, H, P.: Visual Classification: an Interactive Aapproach to Decision Tree Construction. Proc.5$^{th}$ Intl. Conf. On knowledge Discovery and Data Mining (KDD'99) (1999) 392-396

2. Ankerst, M., Ester, M., Kriegel, H. -P.: Towards an Effective Cooperation of the User and the Computer for Classification. Proc. Int'l Conf. Knowledge Discovery and Data Mining (ACM SIGKDD'00) (2000) 179-188
3. Teoh, S.T., Ma, K. -L.: StarClass: Interactive Visual Classification Using Star Coordinates. Proceedings of the 3rd SIAM International Conference on Data Mining (2003)
4. Teoh, S.T., Ma, K. -L.: PaintingClass: Interactive Construction, Visualization and Exploration of Decision Trees. Proc.9th ACM SIGKDD International Conference on Knowledge Discovery and Data mining (2003)
5. Liu, D., Sprague, A., Gray, J.: PolyCluster: an Interactive Visualization Approach to Construct Classification Rules. International Conference on Machine Learning and Applications (ICMLA), Louisville, KY, December (2004) 280-287
6. Inselberg, A.: The Plane with Parallel Coordinates. The Visual Computer 1 (1985) 69-91
7. Wegman, E.: Hyper-dimensional Data Analysis Using Parallel Coordinates. Journal of the American Statistical Association (1985) 664-675
8. Darinka, B. -V., Zdenka, C. K., Marjana, N.: Multivariate Data Analysis in Classification of Vegetable Oils Characterized by the Content of Fatty Acids. Chemometrics and Intelligent Laboratory Systems, Vol. 75. (2005) 31– 43

# Constrained Nonlinear State Estimation – A Differential Evolution Based Moving Horizon Approach

Wang Yudong, Wang Jingchun, and Liu Bo

Department of Automation, Tsinghua University, Beijing, China, 100084
wangyd04@mails.tsinghua.edu.cn

**Abstract.** A solution is proposed to estimate the states in the nonlinear discrete time system. Moving Horizon Estimation (MHE) is used to obtain the approximated states by minimizing a criterion that is the Euclidean form of the difference between the estimated outputs and the measured ones over a finite time horizon. The differential evolution (DE) algorithm is incorporated into the implementation of MHE in order to solve the optimization problem which is presented as a nonlinear programming problem due to the constraints. The effectiveness of the approach is illustrated in simulated systems that have appeared in the moving horizon estimation literature.

**Keywords:** moving horizon estimation; differential evolution; state estimation; extended Kalman filter.

## 1 Introduction

Due to the success of the model predictive control (MPC) and especially its ability to handle various constraints on-line, the moving horizon estimation (MHE) has been developed for constrained nonlinear state estimation. MHE has been suggested as a practical strategy to incorporate constraints in estimation[1]. MHE reformulates the estimate problem to an optimization problem and obtains the estimated states by solving the constrained optimization problem. In MHE only a fixed window of data is considered in order to bound the size of the optimization problem. However, when the nonlinear system is considered, the optimization problem involved in MHE becomes a nonlinear programming problem (NLP) due to the nonlinear model constraints. And NLP gives rise to nonconvexities which lead to multiple local optimal. In this case gradient optimization techniques have only been able to tackle special formulations.

Differential evolution[2],[3] developed in 90's is one of the best evolution computing methods. Unlike simple GA[4] that uses binary coding for representation, candidates in DE are represented as individuals based on floating-point members. The DE algorithm begins with randomly generating a population of *Np* parameter vectors which lie in the user defined area. At each generation, the target population is perturbed with a mutation factor and a crossover constant in order to generate a new trial population. And then the selection operation is applied to compare objective value of the two populations, namely, the target population and the trial population. The individuals who provide the smaller value of the cost function become members of the next generation. The process repeated until a predefined stopping criterion is met. Due to its simple concept, easy

D.-S. Huang, L. Heutte, and M. Loog (Eds.): ICIC 2007, LNAI 4682, pp. 1184–1192, 2007.

implementation and fast convergence, DE has been applied in a variety of unconstrained and constrained optimization problems[5].

In this paper, we incorporate the DE algorithm into the application of MHE in nonlinear system. The DE algorithm has been used to solve the NLP in the MHE. In the following part of the paper, we firstly present the formulation of the nonlinear state estimation problem. And then we give a brief introduction on MHE and DE algorithm. Next, we show how to incorporate the DE into the MHE. Finally the approach is tested in a simulated example.

## 2 Backgrounds

### 2.1 Nonlinear System Description

Let the system generating the noisy measurement sequences be described by the following discrete time model

$$x_{k+1} = f(x_k, u_k) + w_k \tag{1}$$

$$y_k = h(x_k, u_k) + v_k \tag{2}$$

where the time point $k$ takes the integer values; the $x_k \in R^{n_x}$ is the state vector; the $u_k \in R^{n_u}$ is the known input; the $y_k \in R^{n_y}$ is the measured output vector; $w_k \in R^{n_w}$ and $v_k \in R^{n_v}$ represent the system noise and the measurement noise respectively which are often assumed to be independent and Gaussian with zero mean; $f(\cdot)$ and $h(\cdot)$ represent the discretized form of the ODE or PDE. It is also assumed that the states and disturbances satisfy the following constraints:

$$x_k \in \mathrm{X}, w_k \in W, v_k \in V \tag{3}$$

Typically the constraints take the upper and lower bounds of the vector components such as $x_{i,k}^{\min} \le x_{i,k} \le x_{i,k}^{\max}$ where the subscript $i$ refer to the $i$th component of the vector $x_k$.

### 2.2 Moving Horizon Estimation

MHE at time $T$ is formulated as the solution to the following mathematical program named ***PROBLEM1***, the details about how to get the formulation can be found in [6], [7]:

**Problem1:**

$$\begin{aligned} &\left\{\hat{x}_{T-N|T}, \hat{w}_{T-N|T}, \cdots, \hat{w}_{T-1|T}\right\} \\ &= \underset{\hat{x}_{T-N|T}, \{\hat{w}_k\}_{k=T-N}^{T-1}}{\arg\min} \sum_{k=T-N}^{T} \hat{v}_{k|T}^T R^{-1} \hat{v}_{k|T} + \sum_{k=T-N}^{T-1} \hat{w}_{k|T}^T Q^{-1} \hat{w}_{k|T} + z_{T-N}(x_{T-N}) \end{aligned} \tag{4}$$

Subject to the constraints

$$\hat{x}_{k+1|T} = f(\hat{x}_{k|T}, u_k) + \hat{w}_{k|T} \tag{5}$$

$$y_k = h(\hat{x}_{k|T}, u_k) + \hat{v}_{k|T} \tag{6}$$

$$\hat{x}_{k|T} \in \mathrm{X} \tag{7}$$

$$\hat{w}_{k|T} \in W \tag{8}$$

$$\hat{v}_{k|T} \in V \tag{9}$$

where $k = T - N,...,T$, and $s_{m|n}$ denotes the estimated value of $S$ at the time point $m$, given the measurement up to time $n$. If $\hat{x}_{T-N}$ and $\{\hat{w}_k\}_{k=T-N}^{T-1}$ denote the solution to equation (4), then the optimal states estimate at time T can be obtained by solving the equation (5).

In the equation (4), the matrices $Q$ and $R$ are the tuning members in order to reconcile the model with the measurements. The matrices have the following simple interpretation: the matrix $Q$ provides a measure of confidence in the model while the matrix $R$ provides a measure of confidence in the process data. For example, if the matrix $R$ is relatively "large" to $Q$, we are more confident in the model that in the process measurement and vice-versa. In general, the matrices $Q$ and $R$ can be chosen equal to the estimated covariance matrices of system noise $\{w_k\}$ and measurement noise $\{v_k\}$ separately.

$Z_{T-N}(x_{T-N})$ is the arrival cost which is the fundamental in moving horizon estimation and it provides a mean to compress the history data. By the arrival cost the unbounded mathematical problem can be recast as an equivalent fixed-dimension mathematical program. The arrival cost summarizes the effect of the data sequence $\{y\}_0^{\tau-1}$ on the state $x_\tau$. So how to obtain or estimate the arrival cost becomes an essential part in moving horizon estimation.

However when either the constraints are considered or the system model is nonlinear, the algebraic expression for the arrival cost doesn't exist[8]. So estimation of the arrival cost is necessary. In this paper the extended Kalman filter has been used to approximate the arrival cost. The arrival cost is approximated by using a first-order Taylor series approximation of the model around the estimated trajectory $\{x_k\}_{k=0}^{T}$. So the arrival cost is taken as:

$$z_T(x) = (x_T - \hat{x}_T)^T \Pi_T^{-1} (x_T - \hat{x}_T) \tag{10}$$

And the extended Kalman filter covariance update formula is

$$\Pi_T = Q + A_{T-1}[\Pi_{T-1} - \Pi_{T-1} C_{T-1}^T (R + C_{T-1} \Pi_{T-1} C_{T-1}^T)^{-1} C_{T-1} \Pi_{T-1}] A_{T-1}^T \tag{11}$$

Where

$$A_T = \frac{\partial f(x_k, u_k)}{\partial x_k} |_{\hat{x}_T}, \quad (12)$$

$$C_T = \frac{\partial g(x_k, u_k)}{\partial x_k} |_{\hat{x}_T} \quad (13)$$

Form the above discussion we can see that when $T \le N$ MHE is equivalent to the full information estimator; and when T>N, due to the approximation of the arrival cost, MHE become a fix dimension mathematical problem.

## 2.3 Differential Evolution

In DE, there are three key factors: the population size *Np*, the mutation factor *F* and the crossover constant *CR* [2][3]. The DE algorithm starts with the random initialization of a population of *Np* variables sectors in the search space. Then a trial member will be created by mutation and crossover for each individual. The mutation operation is completed by the following formulation:

$$\mathbf{v}_i = \mathbf{x}_1 + F(\mathbf{x}_2 - \mathbf{x}_3) \quad (14)$$

where $x_1$, $x_2$, and $x_3$ are three different members in the population, and $x_i = (x_{i,1}, x_{i,2}, \ldots, x_{i,k})$, $F$ is the mutation factor which is between (0, 1), and $x_2$-$x_3$ denotes the differential item.

The crossover operation is completed by the following formulation:

$$u_{i,j} = \begin{cases} x_{i,j}, \text{if } (rand\ (j) \le CR) \text{ or } j = rnbr\ (i), \\ v_{i,j}, \text{otherwise} \end{cases} \quad i = 1,2,\cdots k \quad (15)$$

where *CR* is called crossover constant which is between (0, 1), $x_{i,j}$ denotes the old individual, and $v_{i,j}$ denotes the new individual, *rnbr*(*i*) is a random integer number in {1,2,…,*k*} that assures at least one component of *x* and *v* are the same.

Then the population is updated by the following formulation:

$$\mathbf{x}_{t+1} = \begin{cases} \mathbf{x}_t & \text{if } f(\mathbf{x}_t) \le f(\mathbf{v}_t) \\ \mathbf{v}_t & \text{if } f(\mathbf{x}_t) > f(\mathbf{v}_t) \end{cases} \quad (16)$$

Where $x_{t+1}$ denotes the updated member in the next generation, $x_t$ denotes the old member, and $v_t$ denotes the competitor.

The mutation and crossover operation are repeated until a predefined stopping criterion is met.

## 3 Differential Evolution Based Moving Horizon Estimation

The DE based MHE algorithm consists of two parts: the arrival cost estimation and state estimation. In the estimation process, the information within the horizon is utilized while the information before the horizon is summarized by the arrival cost. The MHE method is illustrated in Fig. 1.

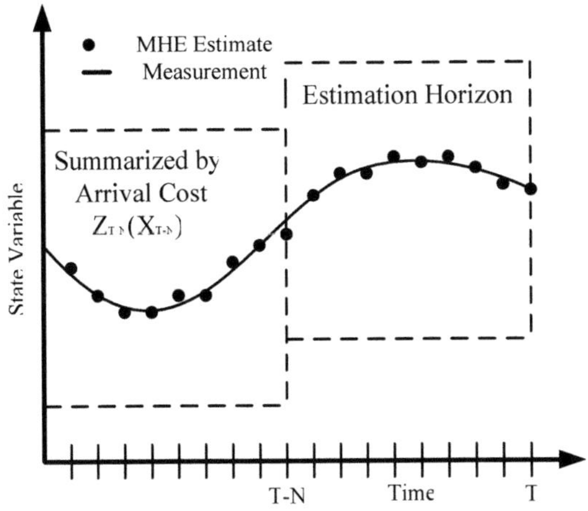


**Fig. 1.** Frame of the Moving Horizon Estimation

DE is incorporated into MHE to solve the constrained optimization problem (4)-(9). The vectors in the population of the DE algorithm are chosen as $(\hat{x}_{T-N|T}, \hat{w}_{T-N|T}, \cdots, \hat{w}_{T-1|T})$ at $T$th time. And the initial distribution of $\hat{x}_{T-N}$ is chosen as normal, whose mean and covariance are $\hat{x}_{T-N|T-1}$ and $\Pi_{T-N}$ separately. $\hat{x}_{T-N|T-1}$ is the predictive value at time $T$-1 and $\Pi_{T-N}$ is the covariance that is updated by the EKF formula at time $T$. When the searching has been finished, the current estimated state $\hat{x}_{T|T}$ can be obtained by solving equation (5).

## 4 Example

In this section, two examples are presented to test the effectiveness of the DE based MHE. First, we consider a linear example in which Kalman filter is used as a benchmark. Then a nonlinear example is considered and extended Kalman filter is taken as a benchmark. Both of the examples here are proposed in [9].

### 4.1 Linear Example

Consider the following constrained linear discrete-time system:

$$x_{k+1} = \begin{bmatrix} 0.99 & 0.2 \\ -0.1 & 0.3 \end{bmatrix} x_k + \begin{bmatrix} 0 \\ 1 \end{bmatrix} w_k \tag{17}$$

$$y_k = \begin{bmatrix} 1 & -3 \end{bmatrix} x_k + v_k \tag{18}$$

where $\{v_k\}$ is a sequence of independent, zero mean, normally distributed random variables with covariance 0.01; $w_k = |z_k|$ and $\{z_k\}$ is a sequence of independent, zero mean, normally distributed random variables with unit covariance.

We formulate constrained estimation problem with $R=1$, $Q=1$, $\hat{x}_0 = [0;0]$, $\Pi_0 = 1$. We choose the estimation horizon $N$=10 for MHE. For DE algorithm, we choose the population size is 30, the mutation factor is 0.5 and the crossover constant is 0.4. And the constraint $w_k \geq 0$ is also taken into consideration. A comparison of the Kalman filter, unconstrained MHE (U-MHE) and MHE is shown in Fig 2.

The sum square errors are used to evaluate the result

$$\sum_{k=0}^{T} (x_k^{(j)} - \hat{x}_k^{(j)}) \tag{19}$$

where $x_k^{(j)}$ denotes the $j$th entry value of the vector $x$ at time $k$ while $\hat{x}_k^{(j)}$ denotes the predictive one. Form Fig 2 we can see that the MHE is able to track the state while the EKF and U-MHE diverge. The average sum square error of 20 trials for $x(1)$ and $x(2)$ by using KF, U-MHE and MHE are list in Table 1.

**Table 1.** Sum square error of the state estimation for linear system

| | KF | U-MHE | MHE |
|---|---|---|---|
| $x(1)$ | 1666.7 | 1234.5 | 44.2884 |
| $x(2)$ | 170.7359 | 145.1248 | 47.7836 |

### 4.2 Nonlinear Example

Consider the following constrained nonlinear discrete-time system:

$$x_{k+1}(1) = 0.99x_k(1) + 0.2x_k(2) \tag{20}$$

$$x_{k+1}(2) = -0.1x_k(1) + \frac{0.5x_k(2)}{1 + x_k^2(2)} + w_k \tag{21}$$

$$y_k = x_k(1) - 3x_k(2) + v_k \tag{22}$$

where $\{v_k\}$ is a sequence of independent, zero mean, normally distributed random variables with covariance 0.01; $w_k = |z_k|$ and $\{z_k\}$ is a sequence of independent, zero mean, normally distributed random variables with unit covariance.

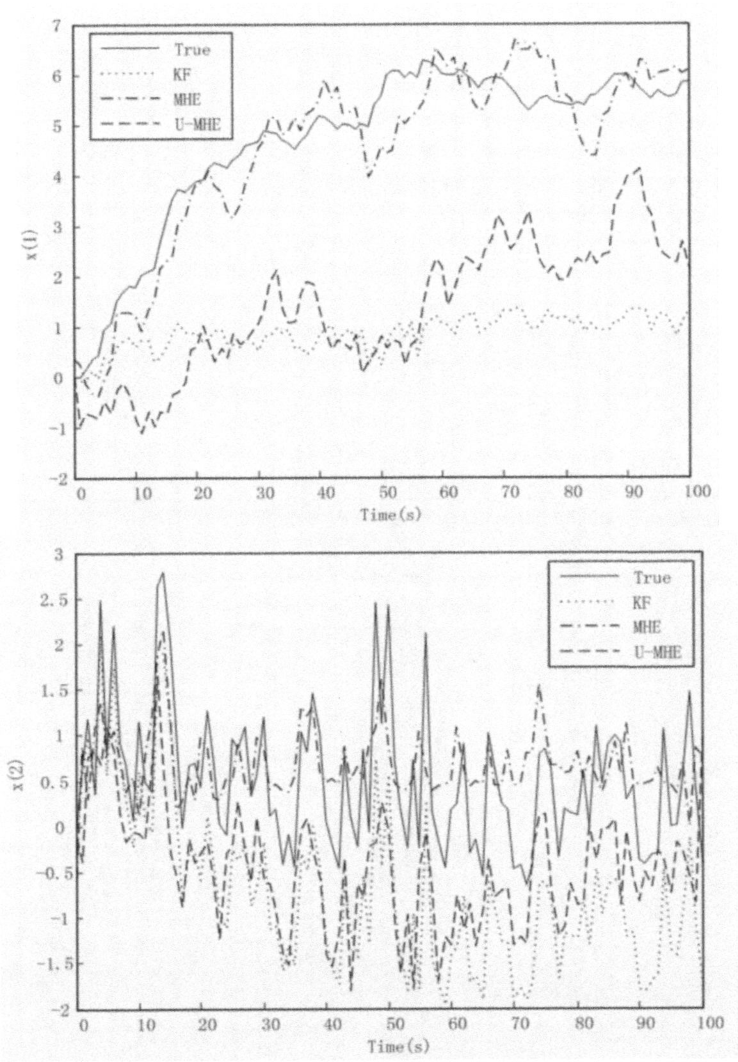


**Fig. 2.** Comparison of estimation for model (17)-(18)

From Fig. we can see that the MHE is able to track the state while the EKF and U-MHE diverge. The average sum square error of 20 trials for $x(1)$ and $x(2)$ by using EKF, U-MHE and MHE are list in Table 2.

**Table 2.** Sum square error of the state estimation for nonlinear system

| | EKF | U-MHE | MHE |
|---|---|---|---|
| $x(1)$ | 1295.3 | 1686.7 | 79.2797 |
| $x(2)$ | 133.9631 | 184.1980 | 47.1264 |

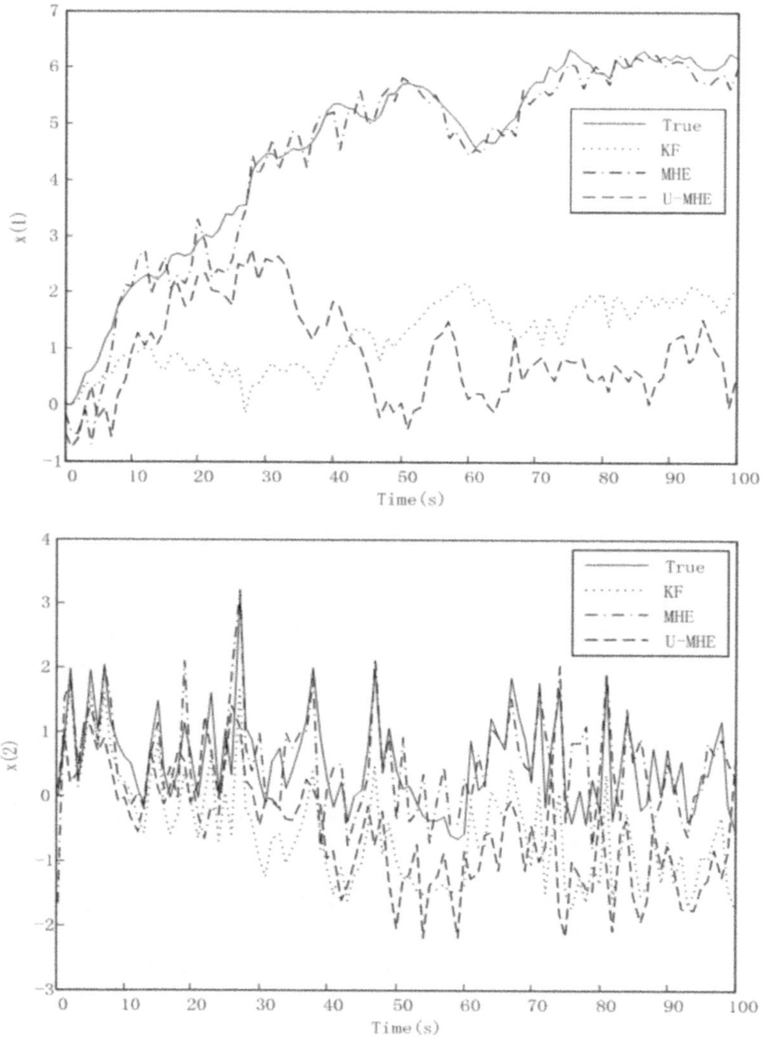


**Fig. 3.** Comparison of estimation for model (25)-(27)

For the linear and nonlinear examples, the constrained MHE can involve the constraints of the system to improve the estimation and the DE algorithm can treat these constrained optimization problems effectively. The simulated results show an excellent coherence between the simulated states and the estimated state which confirms the strength of the proposed DE based MHE strategy.

## 5 Conclusions

In this paper, we investigate the MHE as an online state estimation strategy and constraints of the system are considered in order to improve the accuracy of the estimation. The DE algorithm is incorporated into the MHE scheme to cope with the optimization problem. There are several advantages for the DE based MHE strategy. Firstly DE algorithm finds solutions by random search and only the objective function and constraints need to be considered. Secondly the parameters of DE algorithm are only three or four (for formulation of DE algorithm) which are relatively easy to handle. Thirdly it is very convenient for DE algorithm to deal with the constraints no matter they are equalities or inequalities. Fourthly the DE algorithm can work with functions (as criterion to minimize) that do not need differentiable. Thus DE based MHE strategy is easy to tune and implemented and these properties should make the DE based MHE useful to practicing engineers.

## References


1. Rao, C. V., Rawlings, J. B.: Nonlinear Moving Horizon Estimation. In A. Z. F.AllgoK wer(ed.): Nonlinear model predictive control, Progress in systems and control theory, (2000) 45-69
2. Price, K., Storn, R.: Differential Evolution - A Simple Evolution Strategy for Fast Optimization. In Dr. Dobb's Journal, vol. 22, no. 4, (1997) 18-24
3. Storn, R., Price, K.: Minimizing the Real Functions of the ICEC'96 Contest by Differential Evolution, (1996) 842-844
4. Wang(ed.), L.: Intelligent Optimization Algorithms with Application,Tsinghua University & Springer Press, (2001)
5. Huang, F.Z., Wang, L., He, Q.: An Effective Co-evolutionary Differential Evolution for Constrained Optimization. In Applied Mathematics and Computation, vol. doi:10.1016/ j.amc.(2006)07-105
6. Christopher V.R., James B.R., Jay H.L.: Constrained Linear State Estimation-a Moving Horizon Approach. In Automatica, vol. 37, (2001) 1619-1628
7. Rao, C.V.: Moving Horizon Strategies for the Constrained Monitoring and Control of Nonlinear Discrete-time Systems. In Ph.D. thesis, University of Wisconsin-Madison,(2000)
8. Rao, C. V., Rawlings, J.B.: Constrained Process Monitoring: Moving-horizon Approach. In Aiche Journal, vol. 48, no. 1, (2002) 97-109
9. Rao, C. V., Rawlings,J. B., Mayne, D.Q.: Constrained State Estimation for Nonlinear Discrete-time Systems: Stability and Moving Horizon Approximations. In Ieee Transactions on Automatic Control, vol. 48, no. 2, (2003) 246-258

# Multi-agent Optimization Design for Multi-resource Job Shop Scheduling Problems

Fan Xue and Wei Fan

College of Computer Science and Technology, Civil Aviation University of China, Tianjin 300300, P.R. China
wfan@cauc.edu.cn

**Abstract.** As a practical generalization of the job shop scheduling problem, multi-resource job shop scheduling problem (MRJSSP) is discussed in this paper. In this problem, operations may be processed by a type of resources and jobs have individual deadlines. How to design and optimize this problem with DSAFO, a novel multi-agent algorithm, is introduced in detail by a case study, including problem analysis, agent role specification, and parameter selection. Experimental results show the effectiveness and efficiency of designing and optimizing MRJSSPs with multi-agent.

## 1 Introduction

A practical generalization of the job shop scheduling problem (JSSP), which we call the multi-resource job shop scheduling problem (MRJSSP), is concerned in this paper. Informally, the problem can be stated as follows. There are a set of jobs and a set of resources. Each job consists of a lattice of operations that must be processed in a given order, and has, individually, a job ready time and a job deadline. Each operation is given an integeral processing time, and a longer resource usage time (plan time) for extra traffic (spatial distribution), preparation, and reset actions. Each operation needs one resource to process, and the processing is uninterruptible. Each resource can process only one operation simultaneously. The objective of MRJSSP is to find the best scheduling solution with minimal resource consumption, i.e. maximal resource utility.

JSSP has been studied by both academic and industrial society for decades [1], however in many practical situations, (i) an operation can be processed by any one resource (or machine) from a group; (ii) jobs have individual deadlines; (iii) the requirement of no tardiness for any jobs is more important than makespan; and (iv) to reduce consumption as much as possible in order to maximize the machine utilities. Those are, fitly, the cases of MRJSSPs.

M. Perregaard (1995) proposed multi-processor job shop scheduling problem (MPJSSP), which concerned multiple processing capacity as well, and A. Cesta, A. Oddi, and S. F. Smith (2000) developed an iterative improvement search approach for it. W. P. M. Nuijten and E. H. L. Aarts (1996) [4] presented another problem: multiple capacitated job shop scheduling problem (MCJSSP), which

D.-S. Huang, L. Heutte, and M. Loog (Eds.): ICIC 2007, LNAI 4682, pp. 1193–1204, 2007.

extended MPJSSP by allowing each operation has a size. Nevertheless, both MPJSSP and MCJSSP ignored the spatial conditions and supporting handling in real-world engineering processes. Furthermore most of scheduling works are, practically, pre-scheduled by domain experts, so what we need to optimize is, usually, to maximize the resource utility to meet a timetable, not the makespan. All of these are well presented in MRJSSP, which is detailed in Section 2.

The remainder of the paper is structured as follows: Section 2 represents a definition of the multi-resource job shop scheduling problem. Section 3 reviews the DSAFO algorithm briefly. Section 4 demonstrates the design procedure via a case study. Experimental results appear in Section 5 and a brief conclusion is given in Section 6.

## 2 The Multi-resource Job Shop Scheduling Problem

**Definition 1 (MRJSSP).** *An instance of multi-resource job shop scheduling problem is a tuple* $\langle \mathcal{F}, \mathcal{O}, \mathcal{R}, \mathcal{T}, \preceq, D, F, rt, st, ut, et, tt, \Omega, \gamma, s \rangle$ *where*

$\mathcal{J} = \{j_1, j_2, \ldots, j_n\}$ - *a set of n jobs;*

$\mathcal{O} = \{o_1, o_2, \ldots, o_p\} = \mathcal{O}_1 \cup \mathcal{O}_2 \cup \ldots \cup \mathcal{O}_m$, *where* $\forall_{m_i \neq m_j} \mathcal{O}_{m_i} \cap \mathcal{O}_{m_j} = \emptyset$ - *a set of p operations in m types (partitions);*

$\mathcal{R} = \{r_1, r_2, \ldots, r_q\} = \mathcal{R}_1 \cup \mathcal{R}_2 \cup \ldots \cup \mathcal{R}_t$, *where* $\forall_{t_i \neq t_j} \mathcal{R}_{t_i} \cap \mathcal{R}_{t_j} = \emptyset$ - *a set of q resources in t types (partitions);*

$C = \mathcal{O}_1 \times \mathcal{R}_{t_1} \cup \mathcal{O}_2 \times \mathcal{R}_{t_2} \cup \ldots \cup \mathcal{O}_m \times \mathcal{R}_{t_m}$ - *processing capabilities;*

$\preceq : \mathcal{O} \mapsto \mathcal{O}$ - *precedence or equality, decomposing* $\mathcal{O}$ *into lattices (specially, chains) of jobs;*

$D : \mathcal{J} \mapsto \mathbb{Z}_0^+$ - *deadline of a job;*

$J : \mathcal{O} \mapsto \mathcal{J}$ - *job belonging to;*

$rt : \mathcal{O} \mapsto \mathbb{Z}_0^+$ - *operation ready time;*

$st : \mathcal{O} \mapsto \mathbb{Z}^+$ - *non-zero operation service time;*

$ut : \mathcal{O} \mapsto \mathbb{Z}_0^+$ - *operation setup time;*

$et : \mathcal{O} \mapsto \mathbb{Z}_0^+$ - *operation reset time;*

$tt : \mathcal{R} \times \mathcal{O} \mapsto \mathbb{Z}_0^+$ - *resource traffic time for an operation;*

$\Omega : \mathcal{R} \times \mathbb{Z}_0^+ \mapsto \mathcal{O} \cup \{\emptyset\}$ - *which operation is in process at a certain time, returns* $\emptyset$ *when non-single operations assigned.*

*The objective is to find two functions: s and* $\gamma$*, where*

$\gamma : \mathcal{O} \mapsto \mathcal{R}$ - *assign resources;*

$s : \mathcal{O} \mapsto \mathbb{Z}_0^+$ - *assign service start time,*

$$
\textbf{s.t.} \forall_{o \in \mathcal{O}} [\, \langle o, \gamma(o) \rangle \in C
\wedge rt(o) \leq s(o)
\wedge s(o) + st(o) \leq D(F(o))
\wedge \forall_{o \prec o'} s(o) + st(o) \leq rt(o')
\wedge \forall_{s(o) - ut(o) - tt(\gamma(o), o) \leq \tau < s(o) + st(o) + et(o)} \Omega(\gamma(o), \tau) = o]
$$

$$
\wedge |ran(\gamma)| = \min_{\gamma_i \in \Gamma} |ran(\gamma_i)|,
$$

*where* $o \prec o'$ *is abbreviated from* $o \preceq o' \wedge o \neq o'$, $\Gamma$ *is the set of all valid schedules,* $|ran(\gamma)|$ *is the range size of function* $\gamma$.

MRJSSP is a special case of distributed constraint satisfaction problem (Dis-CSP), according to the definition by [6]. Remark that the decision variant of the MRJSSP is NP-complete, as (i) it is in NP, because for a given schedule $s$ all constraints can be checked in polynomial time, and (ii) the decision variant of the JSSP is an NP-complete special case of the decision variant of MRJSSP [5].

## 3 DSAFO: Overview and Strategies

DSAFO, i.e. Dynamic Scheduling Agents with Federation Organization, which had been formalized in [7], is a novel multi-agent approach to Dis-CSP, especially for MRJSSPs. DSAFO employs blackboard mechanism, federation organization, meta-level guided resource borrowing and domain knowledge guided plan backtrack. This algorithm is based on *run-and-schedule*, a dynamic distributed scheduling environment, which uninterruptedly observes real-time job data from enterprise systems. As shown in Fig. 1, DSAFO runs as follow:

1. to read real-time job data from *run-and-schedule* environment;
2. to decompose jobs into operations;
3. to divide the solution space dynamically into rational partitions with multi-agents;
4. to conquer each partition with local heuristics;
5. to optimize the solution simultaneously via coordination among partitions from global view;
6. to dispatch the solution to real world environment simultaneously.

Internally, DSAFO algorithm can be viewed as a multi-agent approach with strategies of local heuristics and global coordination.

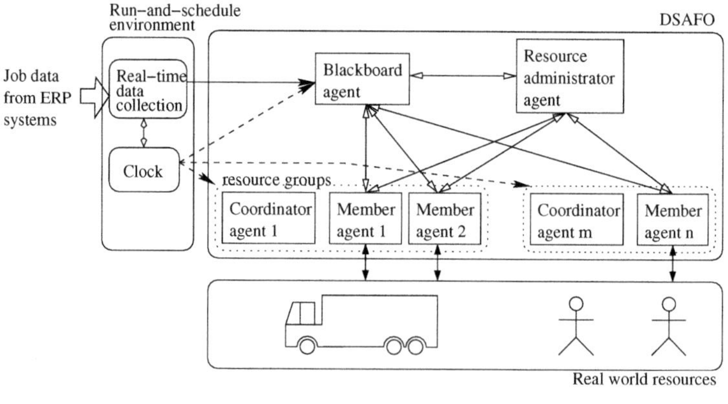


**Fig. 1.** An overview of DSAFO

There are two parts of local heuristics in DSAFO, both are EDD (Earliest Due Date first). One is to select the first operation to do according to weighted *latest finish time*, and the other is to assign an operation to a resource from some committed resources according to the earliest *committed service start time*. However, when the problem scale increases, EDD falls, generally, into local minimum. So DSAFO employs local heuristics (EDD) in dynamic cooperative agents to jump out of local minimum.

Global coordination is a mechanism that exchanges resource textures of agents and realizes a complete resource borrow procedure for agents. And there are two kinds of relationships between two *Member* agents with the same type of resources: *buddy* and *competitor*. If two *Member* agents share same type of resources and same type of operations, they are competitors; if they share same type of resources but different type of operations, they are buddies. A group of buddies always help each other, by lending its own resource friendly.

## 4 Design with DSAFO: A Case Study

Airport ground service is the service process from flight landing to takeoff, including gate assignment, baggage handling, catering, fueling, cleaning, etc. Airport ground service scheduling (AGSS) is to schedule many kinds of dynamic ground resources (baggage trucks, fuel trucks, etc.), to fulfill all constrained service operations of flights timely to meet their arrival and departure deadlines [8]. The typical operations for a job, transfer flight service, is shown as Fig. 2.

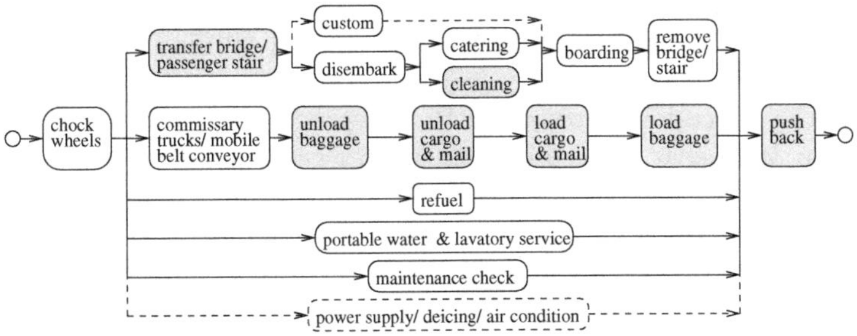


**Fig. 2.** Typical operations for a transfer flight

The AGSS problem well meets the qualifications of the MRJSSP definition. For each type of operations shown in Fig. 2, one (or a group) of certain type(s) of materials, engineers, aviation ground support facilities and equipments, and transportation equipments are required. For instance, process "*cleaning*" requires cleaners (and cleaners' bus), "*catering*" requires catering truck, etc. However, AGSS problem has one more constraint: resources, i.e. engineers and equipments, have limited job time, e.g. 8 hours for an engineer per day.

Then we demonstratively design AGSS problem optimization with DSAFO in three phases: entity and extra constraint analysis, algorithm specification, and parameter decision. Similar design procedures could be applied to other MRJSSPs.

In the analysis phase, in AGSS problem the pre-determined flight schedule and real-time operation systems (such as Flight Operation Control system, Airport Operation Data Base) provide efficient input data (ready time and due date) for jobs (flights). The jobs come, commonly, one by one, and peak not too much in busy hours. In airports, service support vehicles and engineers must travel along specific roads, load materials, and unload at job-specific points. These geographic distance and the maximal airport ground travel speed (5km/h) could provide us efficient traffic time estimation on service support actions. Some resources are always bounded together, such as a cleaners' group, they can be considered as one resource.

As to the job time limitation of resources, we should modify the resource allocation mechanism in DSAFO. Another constraint is that in minor cases several operations around a physical airplane may conflict with each other, we should conclude the cases and properly relax these operations in operation dispatching process.

In the specification phase, agent roles in DSAFO are assigned with capabilities and meanings. We design the *Blackboard* role to be in charge of decision-making of operation dispatch, the *ResourceAdmin* role to be in charge of decision-making of resource allocation, and *Member* roles to be in charge of making operation-resource match-up dynamically. The *Coordinator* role is, invariably, assigned to facilitate cooperations among *Member* agents.

The real-time job data are gathered and decomposed into operations by the *Blackboard* role, with considering the possible spatial collision among operations. the job time limitation of resources is monitored by the *ResourceAdmin* role, for it controls the allocation of resource. Proper processing plans for resources are made by the *Member* role, according to airport geographic information.

In the last phase, the parameters are designed, experimented, and analyzed one by one. In DSAFO, there are some main parameters: (*Member*) AgentNumber Blockfactor, Delayfactor, and Syncycle. We choose baggage tractor (BT) consumption and related drivers' 4-hour job optimization to test the parameters, and have spontaneous 250 runs to get the distributions of the solutions in each test.

Parameter *Member* agent number (accompanied with Reqcycle) stands for how many *Member* agents in DSAFO are maintained for a certain type of resources. And in all the parameters, this number should be notably influential. When Blockfactor=1/12, Delayfactor=1/6, and Reqcycle is from 1/6 to 2 respectively, Fig. 3 and Fig. 4 show the solution distributions with different BT *Member* agents in 3D map and contour map, as well as marginal distributions. When the number increases from 1 to 4, the solutions are going better, because the coordination strategy makes more insufficient effect when the number increases from only one (no coordination). On the contrary, when the number increases from 4

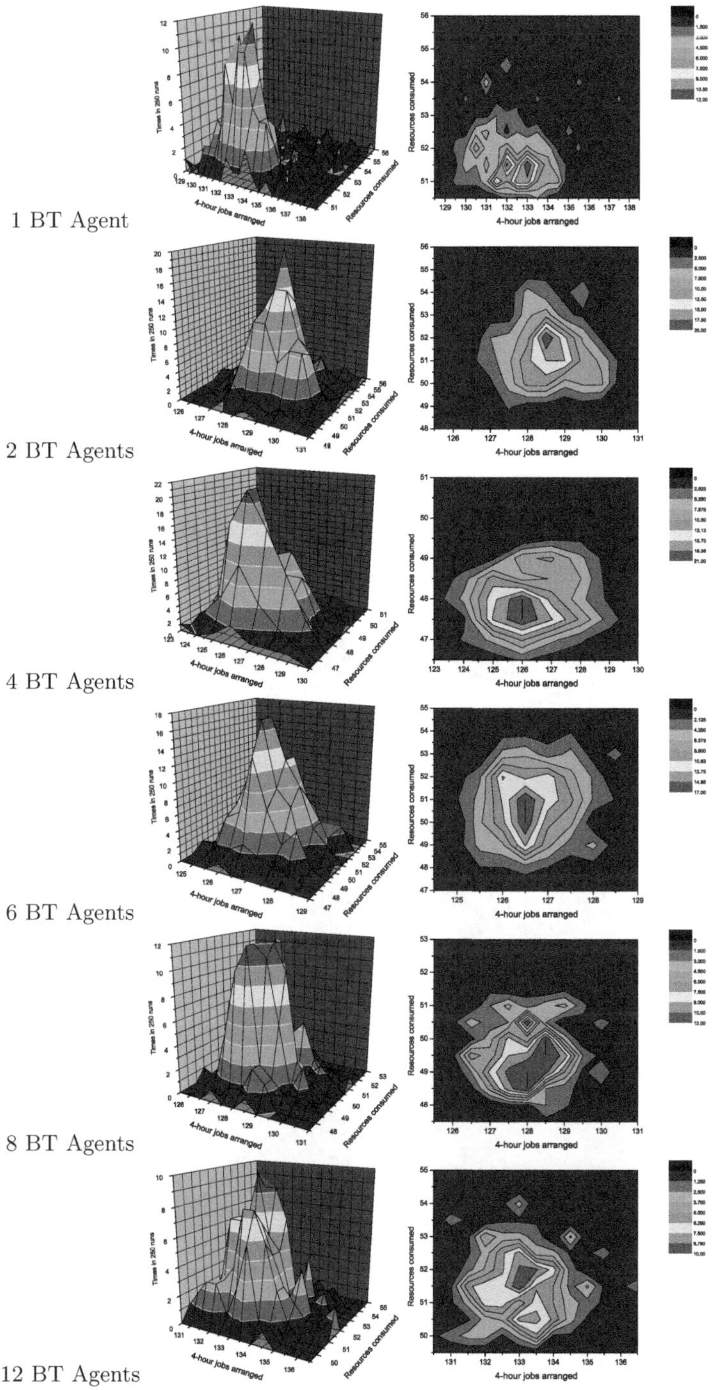


**Fig. 3.** BT solution distributions with respect to AgentNumber

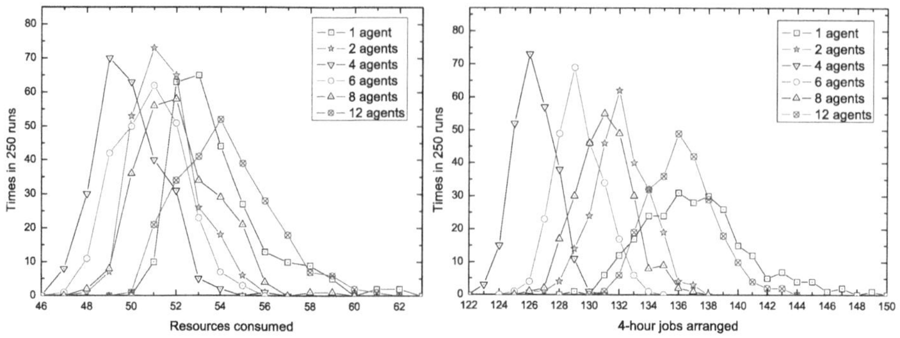


**Fig. 4.** BT solution marginal distributions with respect to AgentNumber

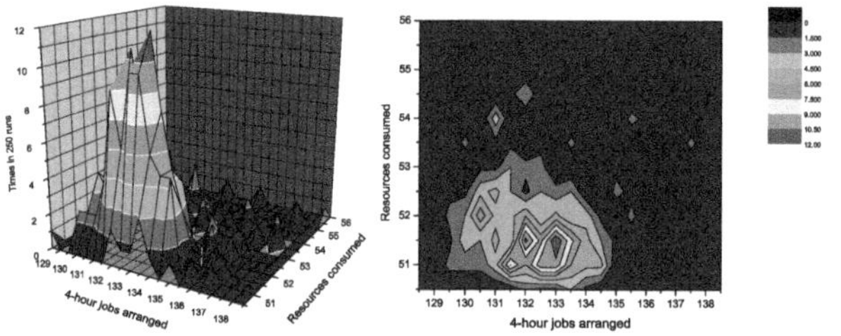


1 BT *Member* Agent with Blockfactor=1/12

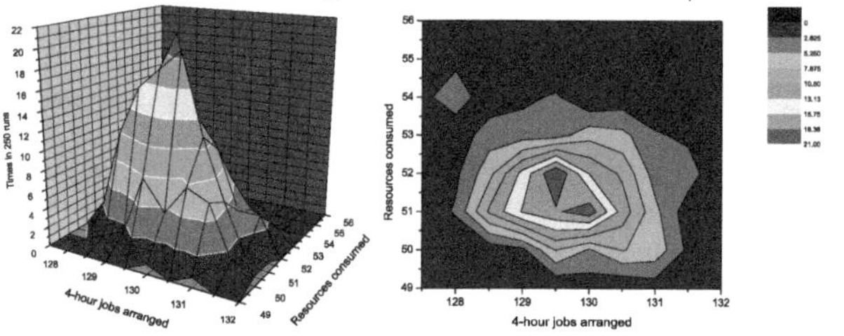


1 BT *Member* Agent with Blockfactor=1/24

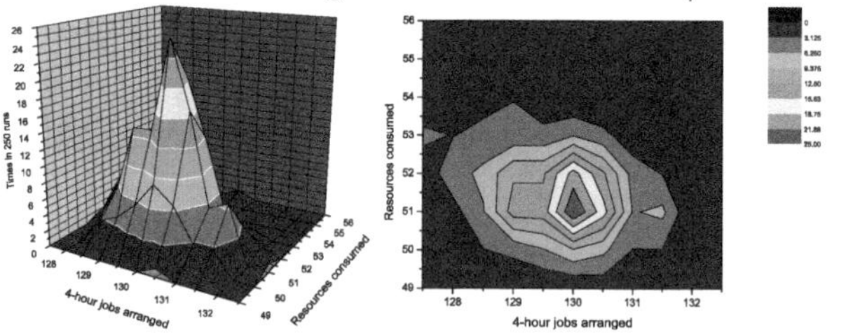


1 BT *Member* Agent with Blockfactor=1/36

**Fig. 5.** BT solution distributions with respect to Blockfactor

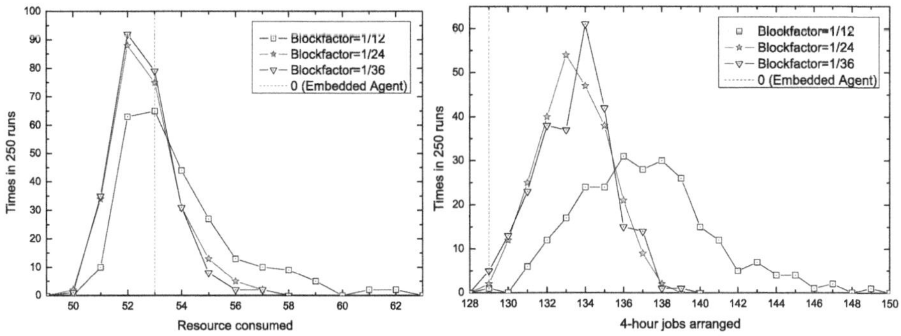


**Fig. 6.** BT solution marginal distributions with respect to Blockfactor

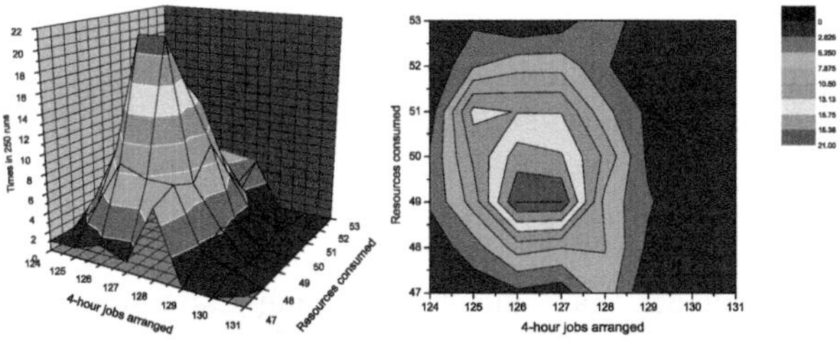


4 BT *Member* agents with Delayfactor=1/3

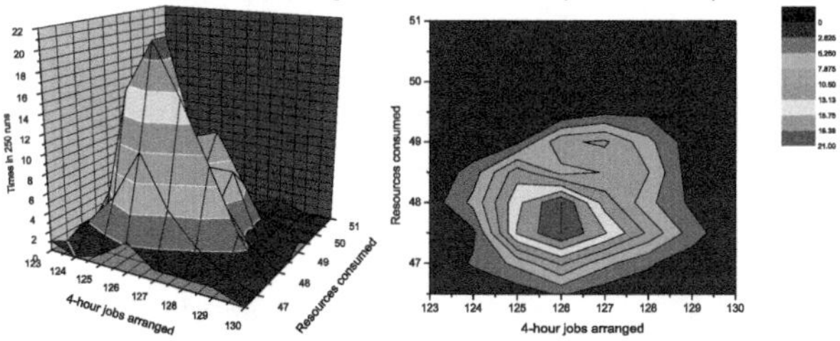


4 BT *Member* agents with Delayfactor=1/6

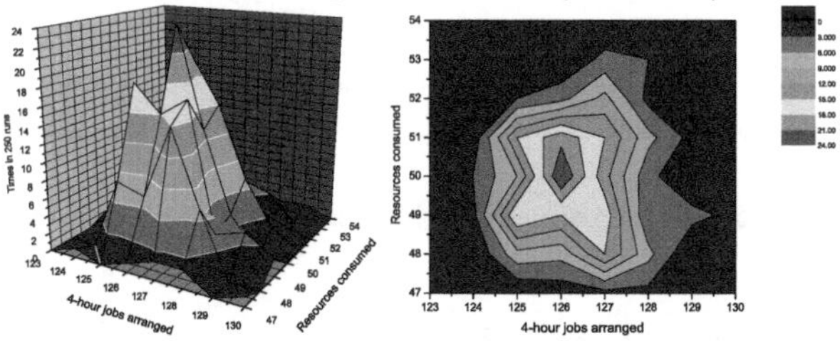


4 BT *Member* agents with Delayfactor=1/24

**Fig. 7.** BT solution distributions with respect to Delayfactor to 12, the solutions are going worse, because too many agents divide the global solution into too many fragments, so that local heuristics in these fragments lack of enough global viewpoints, and coordination cannot optimize these fragments very well.

Blockfactor is another influential parameter. We set AgentNumber=1, Delayfactor=1/6, Syncycle=5 and Reqcycle = 1/6, and then we get solution distributions with respect to Blockfactor in 3D map and contour map, and marginal distributions, as shown in Fig. 5 and Fig. 6. From the figures we can conclude that when Blockfactor decreases, resource consumption stays the same approximately and 4-hour jobs arranged decrease a bit; and when the Blockfactor goes smaller, the solutions distributes turn more centralized, and the algorithm acts more similar as a stable algorithm. Specially, we embedded one single *Member* agent in *Blackboard* agent, i.e. without any unstable factors from communication, to eliminate the interference from network message communication. The new embedded algorithm degrades to a deterministic algorithm as shown in Fig. 6 (approximately the same as EDD* in Section 5).

Delayfactor stands for how much time *Member* agent should delay extra after a successful operation commitment, and it is not a very influential parameter. We set AgentNumber=4, Blockfactor=1/12, Syncycle=5 and Reqcycle = 1/2. Then we get distributions in 3D map and contour map, and marginal distributions, as shown in Fig. 7 and Fig. 8. No significant changes are there for 4-hour jobs, however the peaks of 3D map and contour map move up-down a bit, which means resource consumption is influenced, faintly.

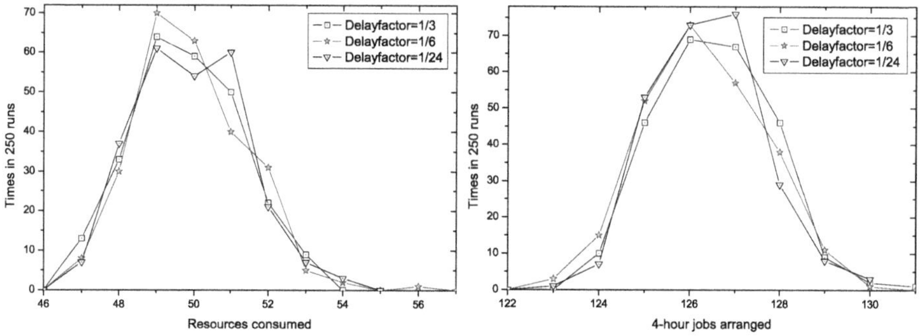


**Fig. 8.** BT solution marginal distributions with respect to Delayfactor

# 5 Experimental Comparison

Ant system have succeeded in many optimization and scheduling areas, and $\mathcal{MM}$AS ($\mathcal{MAX}$-$\mathcal{MIN}$ ant system) is an improved ant system which gives an upper and a lower bound to intensity quantity of trails in ant system [9,10]. We implemented a $\mathcal{MM}$AS for BT arrangement to compare. $m$ BT related operations in AGSS problem are transformed into $m$ nodes in trade salesman problem (TSP). A distance between two nodes (operations) $r_i$ and $r_j$ is represented as

$$d_{r_i r_j} \stackrel{\mathrm{def}}{=} \begin{cases} (DueDate_{r_j} - DueDate_{r_i})/10, & DueDate_{r_j} > DueDate_{r_i}, \\ 0.01, & DueDate_{r_j} = DueDate_{r_i}, \\ (DueDate_{r_i} - DueDate_{r_j})/30, & DueDate_{r_j} < DueDate_{r_i}. \end{cases}$$

And rest parameters of $\mathcal{MMAS}$ are: $\alpha = 1.5$, $\beta = 2$, $\rho = 0.05$, $\tau_{\text{init}} = 1$, $\tau_{\max} = 100$, $\tau_{\min} = 0.01$, $\text{N}_{\text{ant}} = \lceil n/5 \rceil$ (upper integer), $\text{NC}_{\max} = 150$. And for each successful ant run $R$ done by ant $i$, all of edges in Hamilton circle of $R$ get a positive feedback

$$\Delta\tau^i_{r_i r_j} = \begin{cases} \frac{10}{(\text{res}_R + \text{job}_R/3)^2}, & < r_i, r_j > \in \text{circle of } R; \\ 0, & otherwise. \end{cases}$$

to reinforce the whole algorithm for fewer resources and jobs.

Then DSAFO (with parameters: $\text{AgentNumber}_{\text{BT}}$=4, Blockfactor=1/12, Delayfactor=1/6, Syncycle=5), EDD* (EDD in *run-and-schedule*), ERT* (earliest ready time first in *run-and-schedule*) and $\mathcal{MMAS}$ were tested with real-world AGSS test data with 252 transfer flights. We choose BT related operations (1,008 activities in total) to test performances of these algorithms. A comparison in BT consumption and 4-hour BT jobs arrangement is shown in Table 1. Additionally we put time cost and average CPU rate in the table. The best value in each group is bolded.

**Table 1.** Optimization algorithm comparison

| Algorithm | Time | CPU | Resources | | | 4-hour jobs | | |
|---|---|---|---|---|---|---|---|---|
| | | | MIN | MAX | AVG | MIN | MAX | AVG |
| DSAFO | ≈144 sec | <1% | **47** | 56 | **49.9** | **123** | 130 | **126.3** |
| $\mathcal{MMAS}$ | ≈ 12 hours | ≈100% | **47** | — | — | 141 | — | — |
| EDD* | **≈43 sec** | <1% | 53 | 53 | 53 | 129 | **129** | 129 |
| ERT* | **≈43 sec** | <1% | 52 | **52** | 52 | 130 | 130 | 130 |

From Table 1, it can be concluded that DSAFO and $\mathcal{MMAS}$ both do well in resource consumption for MRJSSPs, and DSAFO do better in BT jobs arrangement. Furthermore, DSAFO cost not too much time (several minutes) and very low CPU rate, in fact most time is used to maintain effective message transmission. On the contrary, $\mathcal{MMAS}$ costs very much time and near 100% CPU rate.

## 6 Conclusion and Future Works

In this paper, we present a practical general model of the job shop scheduling problem, i.e. multi-resource job shop scheduling problem, and demonstrates a design and optimization process on this problem with a novel multi-agent algorithm DSAFO. We have shown that this design and optimization process is comprehensive for extra scheduling constraints and the experimental results shows its effectiveness and efficiency.

One of the future works is to put forward an easy-to-use schedule software to simplify the design process. A preliminary development environment AGSAP has been developed to apply DSAFO to aid common AGSS optimization in [11]. In future, more easy-to-use development environments should be put forward for general MRJSSPs.

## Acknowledgement

The authors acknowledge the support by National Natural Science Foundation of China (NSFC) under Grant No. 60472123.

## References

1. Jain, A.S., Meeran,S.: Deterministic Job-Shop Scheduling: Past, Present and Future. European Journal of Operational Research, **113**(2) (1999) 390-434
2. Perregaard, M.: Branch and Bound Method for the Multiprocessor Jobshop and Flowshop Scheduling Problem. Master Thesis, Departement of Computer Science, University of Copenhagen, Danemark. (1995)
3. Cesta, A., Oddi, A., and Smith, S. F: Iterative flattening: a scalable method for solving multi-capacity scheduling problems. In Proceedings of the Seventeenth National Conference on Artificial intelligence and Twelfth Conference on innovative Applications of Artificial intelligence (July 30 - August 03, 2000). AAAI Press / The MIT Press (2000) 742-747
4. Nuijten, W.P.M., Aarts, E.H.L.: A Computational Study of Constraint Satisfaction for Multiple Capacitated Job Shop Scheduling. European Journal of Operational Research, **90**(2) (1996) 269-284
5. Garey, M.R., Johnson, D.S.: Computers and Intractability - a Guide to the Theory of NP-completeness. W.H. Freeman and Company, New York. (1979)
6. Yokoo, M., Durfee, E.H., Ishida, T., Kuwabara K.: The Distributed Constraint Satisfaction Problem: Formalization and Algorithms. IEEE Transactions on Knowledge and Data Engineering **10**(5) (1998) 673-685
7. Fan, W., Xue, F.: Optimize Cooperative Agents with Organization in Distributed Scheduling System. in *Second International Conference on Intelligent Computing* (ICIC 2006), Kunming, China, 2006. D.-S. Huang, K. Li, and G.W. Irwin (Eds.): Lecture Notes in Artificial Intelligence **4114** (2006) 502-509
8. Xing, J., Liu, S., Fan, W., Ji, L.: Design of Airport Ground Service System Based on Multi-Agent. *Journal of Civil Aviation University of China.* **24**(3) (2006) 24–27 (in Chinese)
9. Stützle, T., Hoos, H.: The $\mathcal{MAX}$-$\mathcal{MIN}$ Ant System and Local Search for The Traveling Salesman Problem. In *Proceedings of the Fourth International Conference on Evolutionary Computation* (ICEC'97), IEEE Press. (1997) 308-313

10. Stützle, T., Hoos, H.: Improvements on the Ant System: Introducing $\mathcal{MAX}$–$\mathcal{MIN}$ Ant system. In *Proceedings of the International Conference on Artificial Neural Networks and Genetic Algorithms*, Springer Verlag, Wien. (1997) 245-249
11. Fan, W., Zhang, G., Xue, F.: Design and Implementation of Airline Ground Services Mas Development Platform. In *First Conference on Multi-agent Theory and Application*, Yantai, China. C. Y. Shi, Z. Z. Shi, *et al* (Eds.): Journal of Computer Research and Development **43** (s1) (2006) 414-419.

# Multi-units Unified Process Optimization Under Uncertainty Based on Differential Evolution with Hypothesis Test*

Wenxiang Lv, Bin Qian, Dexian Huang, and Yihui Jin

Department of Automation, Tsinghua University, Beijing 100084, China
lwx02@mails.thu.edu.cn, qianb04@mails.thu.edu.cn,
huangdx@tsinghua.edu.cn, jyh-dau@tsinghua.edu.cn

**Abstract.** For large-scale chemical process, which consists of lots of production units, all units have their respective optimization objects which are often conflicting with each other for a series of constraints on material and energy balance. In this paper, the total solution with two layers structure strategy made up of multi-units unified optimization and predictive control of each unit is realized. For the global optimization has high dimension, serious nonlinearity and uncertainty, the optimization algorithm based on differential evolution (DE) is performed, while a hybrid DE approach combining hypothesis test (HT) to compare the optimization objects under uncertainty is proposed. The simulation results of an application example to a 20Mt/a gas separation process show that the proposed total solution with two layers structure strategy is successful and multi-units unified optimization method based on HTDE is effective and robust for solving the optimization problem under uncertainty.

## 1 Introduction

System optimization has played an important role in raising produce efficiency and economic benefit from practice, but in large-scale chemical process, the optimization objects are conflicted with each other for there are a series of constraints on material balance and energy balance among all the units and for the local optimum is not the global optimum. Now, multi-variables predictive control with steady state optimization has more and more applied in process advanced control. The dynamics control of predictive controller can overcome disturbances to achieve tracking control and setpoint control, while the steady state optimization can search the optimum, and ensure the system more windless and efficient for its consideration of economic object. The process optimization can be performed using either predictive model [1,2,3,4] or nonlinear mechanism model [4,5]. The former is easier to realize for its model is a linear steady state gain model picked up from dynamic control model and its economic object is linear or quadratic. And the control model is achieved by the local linearization of the nonlinear process, so the optimization has enough precision

* This research is supported by National Science Foundation of China (Grant No. 60574072) as well as the National high tech. project of China(863/CIMS 2006AA04Z168).

D.-S. Huang, L. Heutte, and M. Loog (Eds.): ICIC 2007, LNAI 4682, pp. 1205–1214, 2007.

to satisfy the process with weak nonlinearity. This method has been widely used in industrial predictive control software [3] to single unit.

The global optimization has high dimension, serious nonlinearity and uncertainty to form a complex optimization problem. Recently, a novel population-based evolutionary mechanism, differential evolution (DE), has been proposed [6] for optimization complex problems over a continuous domain. DE finds the global optima by utilizing differences between contemporary population members. This allows the search behavior of each individual to self-tune, and gives an appropriate search direction and step length. Due to its simple concept, easy implementation and relatively fast convergence, DE has attracted much attention and wide applications in many real optimization problems [7,8,9]. For there is essential uncertainty and measure noise in the real production process, the normal optimization methods ignoring the uncertainty are not proper. A hybrid DE approach combining hypothesis test (HT) to compare the optimization objects under uncertainty is proposed.

The total solution with two layers structure strategy of unified optimization to multi process units based on predictive control model is discussed in this paper. The total solution with two layers structure strategy is made up of multi-units unified optimization model constructed with the predictive control models and the global optimization algorithm based on DE with hypothesis test under uncertainty and every unit's predictive control. With a 20Mt/a gas separation project, the unified optimization strategy is performed on columns which have been controlled with multi-variables model predictive controller.

The organization of the remaining content is as follows. In Section 2, the mathematical formulation of the total solution with two layers structure strategy and the unified optimization problem under uncertainty is presented. The hybrid algorithm is proposed after briefly introducing DE and hypothesis test in Section 3. Then, the practice unified optimization problem for testing is provided in Section 4 and the simulation results are shown in Section 5. Finally, we end with some conclusions.

## 2 Unified Optimization Scheme Based on Predictive Control Model

### 2.1 Process Unit Control Problems Description

For $n$ process units correlated with each other, the $i$th transfer function models are

$$\boldsymbol{Y}_i(s)=\boldsymbol{G}_i(s)\boldsymbol{U}_i(s)+\boldsymbol{G}_{Di}(s)\boldsymbol{D}_i(s), i=1,\ldots,n, \tag{1}$$

where controlled variables (CVs) $\boldsymbol{Y}_i(s)\in \boldsymbol{C}^{r_i}$, manipulated variables (MVs) $\boldsymbol{U}_i(s)\in \boldsymbol{C}^{m_i}$, disturbed variables (DVs) $\boldsymbol{D}_i(s)\in \boldsymbol{C}^{m_{Di}}$, transfer function matrices $\boldsymbol{G}_i(s)\in \boldsymbol{C}^{r_i\times m_i}$, $\boldsymbol{G}_{Di}(s)\in \boldsymbol{C}^{r_i\times m_{Di}}$.

From this model we can pick up the steady state gain models are

$$\Delta\boldsymbol{y}_i=\boldsymbol{K}_i\Delta\boldsymbol{u}_i+\boldsymbol{K}_{Di}\Delta\boldsymbol{d}_i, i=1,\ldots,n, \tag{2}$$

where $\Delta\boldsymbol{y}_i\in \boldsymbol{R}^{r_i}, \Delta\boldsymbol{u}_i\in \boldsymbol{R}^{m_i}, \Delta\boldsymbol{d}_i\in \boldsymbol{R}^{m_{Di}}$, while $\boldsymbol{K}_i\in \boldsymbol{R}^{r_i\times m_i}, \boldsymbol{K}_{Di}\in \boldsymbol{R}^{r_i\times m_{Di}}$ are the steady state gains picked up from $\boldsymbol{G}_i(s), \boldsymbol{G}_{Di}(s)$. Thus,

$$\Delta y = \begin{bmatrix} K_1 & & 0 \\ & \ddots & \\ 0 & & K_n \end{bmatrix} \Delta u + \begin{bmatrix} K_{D1} & & 0 \\ & \ddots & \\ 0 & & K_{Dn} \end{bmatrix} \Delta d = K\Delta u + K_D \Delta d \text{,} \tag{3}$$

where $\Delta y = \lfloor \Delta y_1^T, \ldots, \Delta y_n^T \rfloor, \Delta u = \lfloor \Delta u_1^T, \ldots, \Delta u_n^T \rfloor, \Delta d = \lfloor \Delta d_1^T, \ldots, \Delta d_n^T \rfloor$.

For the correlations between the process units, including cascade, parallel, and feed backward, the steady state models are

$$A\Delta y + B\Delta u + B_D \Delta d = c \,. \tag{4}$$

## 2.2 Transformation of the Optimization Problem

It is important to implement unified optimization on these process units seriously coupled. The unified optimization scheme of multi-column is to build a uniform mathematic model for the whole flow sheet and perform optimization calculation. The key is to build the global optimization model which has been researched much by many researchers [1,10]. In this paper, unified model is built by picking up steady state gains from each dynamic control model and connecting them with correlation models. At last, the optimization results are implemented through the independent predictive controllers.

In the correlation model shown in Eq. (4), there are some disturbed variables $\Delta d_c$ correlated with other variables and the other disturbed variables $\Delta d_{\bar{c}}$ independent to other variables from production practice experience. The assumption that $\Delta d_c$ has the only solution from Eq. (4) can be supposed. While $\Delta d_{\bar{c}}$ is set to zero for their unpredictability.

Set

$$\Delta \tilde{d} = \begin{bmatrix} \Delta d_c \\ \Delta d_{\bar{c}} \end{bmatrix} = S\Delta d \text{,} \tag{5}$$

where $S$ is a row transform matrix which rearranges the elements in $\Delta d$, so Eq. (4) is transformed to

$$\begin{bmatrix} A \\ 0 \end{bmatrix} \Delta y + \begin{bmatrix} B \\ 0 \end{bmatrix} \Delta u + \begin{bmatrix} \tilde{B}_D & 0 \\ 0 & I \end{bmatrix} \Delta \tilde{d} = \begin{bmatrix} c \\ 0 \end{bmatrix}, \tag{6}$$

viz.

$$\overline{A}\Delta y + \overline{B}\Delta u + \overline{B}_D \Delta \tilde{d} = \overline{c} \,, \tag{7}$$

where

$$\begin{bmatrix} \tilde{B}_D & 0 \end{bmatrix} = B_D S^{-1} \,. \tag{8}$$

Since $\overline{B}_D$ is a nonsingular matrix as assumption above, then

$$\Delta d = S^{-1} \overline{B}_D^{-1} (\overline{c} - \overline{A}\Delta y - \overline{B}\Delta u) \tag{9}$$

can be attained from Eq. (7).

Then the unified steady state model of multi process units can be attained from Eq. (3) and (5) as

$$\Delta y = K\Delta u + K_D S^{-1}\overline{B}_D^{-1}(\overline{c} - \overline{A}\Delta y - \overline{B}\Delta u) \quad (10)$$

viz.

$$(I + K_D S^{-1}\overline{B}_D^{-1}\overline{A})\Delta y = (K - K_D S^{-1}\overline{B}_D^{-1}\overline{B})\Delta u + K_D S^{-1}\overline{B}_D^{-1}\overline{c}\,, \quad (11)$$

For the controllers, there are also limit constraints for every variables.

Now, the unified optimization problem of multi process units is shown as

$$\begin{aligned} \min J &= \min f(\Delta u) \\ &= (y(t)+\Delta y - y_s)^T \Lambda_y (y(t)+\Delta y - y_s) + c_y^T(y(t)+\Delta y) \\ &\quad + (u(t)+\Delta u - u_s)^T \Lambda^T (u(t)+\Delta u - u_s) + c_u^T(u(t)+\Delta u) \\ s.t. &\begin{cases} (I + K_D S^{-1}\overline{B}_D^{-1}\overline{A})\Delta y = (K - K_D S^{-1}\overline{B}_D^{-1}\overline{B})\Delta u + K_D S^{-1}\overline{B}_D^{-1}\overline{c} \\ y_{\min} \le y(t)+\Delta y \le y_{\max} \\ u_{\min} \le u(t)+\Delta u \le u_{\max} \\ \Delta u_{\min} \le \Delta u \le \Delta u_{\max} \end{cases} \end{aligned} \quad (12)$$

where $y_s \in R^{r_i}, u_s \in R^{m_i}$ are the quadratic desired values of $y$ and $u$, the diagonal matrix $\Lambda_y \in R^{r_i \times r_i}, \Lambda_u \in R^{m_i \times m_i}$ are the quadratic coefficients, $c_y \in R^{r_i}, c_u \in R^{m_i}$ are the linear coefficients, while subscript *min* means the minimal value and *max* means the maximal.

### 2.4 The Optimization Problem Under Uncertainty

For the essential uncertainty and measure noise of the real production process, the uncertainty of the process parameters and variables, such as $K$ and $K_D$, should be considered seriously.

Based on Eq. (12), the optimization problem under uncertainty can be formulated as follows:

$$\min \hat{J} = \min Ef(\Delta u, K, K_D) = \min \frac{1}{n}\sum_{j=1}^{n} f(\Delta u, K_j, K_{D,j})$$

$$s.t. \begin{cases} f(\Delta u, K_j, K_{D,j}) = (y(t)+\Delta y_j - y_s)^T \Lambda_y (y(t)+\Delta y_j - y_s) + c_y^T(y(t)+\Delta y_j) \\ \qquad + (u(t)+\Delta u - u_s)^T \Lambda^T (u(t)+\Delta u - u_s) + c_u^T(u(t)+\Delta u) \\ (I + K_{D,j} S^{-1}\overline{B}_D^{-1}\overline{A})\Delta y_i = (K_i - K_{D,j} S^{-1}\overline{B}_D^{-1}\overline{B})\Delta u + K_{D,j} S^{-1}\overline{B}_D^{-1}\overline{c} \\ y_{\min} \le y(t)+\Delta y_j \le y_{\max}, \qquad j=1\cdots n, \\ u_{\min} \le u(t)+\Delta u \le u_{\max} \\ \Delta u_{\min} \le \Delta u \le \Delta u_{\max} \end{cases} \quad (13)$$

where $n$ is sample size or evaluation number, $j$ denotes each sample.

Then, we can apply DE to the uncertain optimization problem mentioned above. Obviously, it is worthy studying the optimization performance and robustness of DE when uncertainty with different magnitude is present and limited sampling number is allowed. Next, a hybrid DE-based approach will be proposed to for optimization under uncertainty.

## 3 Differential Evolution with Hypothesis Test

DE is a branch of metaheuristic proposed by Storn and Price [6] for optimization problems over continuous domains. Though DE's apparent is relatively simple but the key evolutionary operators of *mutation* and *crossover* are very efficient. Therefore, DE has been applied with high success in a variety of fields. In DE, it starts with the random initialization of a population of individuals in the search space and works on the cooperative behaviors of the individuals in the population. At each generation, the *mutation* and *crossover* operators are applied to individuals to generate a new population. Then, one-to-one greedy *selection* takes place and the population is updated. Currently, several variants of DE [11] have been proposed depending on the selection of the base vector to be perturbed, the number and selection of the differentiation vectors and the type of crossover operators.

Hypothesis test (HT) is an important statistical technique used to make test for predefined hypothesis using experiment data [12]. Through the search, HT can be used to reserve good individuals for new generation and to reduce repeated search, so as to maintain the quality and diversity of the new population. Thus, we propose a hybrid DE approach for optimization under uncertainty, whose optimization procedure is shown as follows:

Let $i$th individual in the $N$-dimensional search space at generation $t$ be $\boldsymbol{X}_i(t)=[x_{i,1}(t), x_{i,2}(t), \ldots, x_{i,N}(t)]$ ($i$=1, 2, …, $M$), where $M$ denotes the size of the population, and $\boldsymbol{X}_i(t)$ is a real vector.

Step1: Randomly initialize the population of individual for DE.

Step2: Sample the uncertain parameters $n_s$ times and evaluate the estimated objective values of all individuals, and determine the best individual $\boldsymbol{X}^*$ which has the best objective value.

Step3: DE's *mutation.* In order to obtain each individual's corresponding *mutant vector* $\boldsymbol{V}_i(t+1)=[v_{i,1}(t+1), v_{i,2}(t+1), \ldots, v_{i,N}(t+1)]$, *mutation* operation is performed for each individual according to the following equation:

$$V_i(t+1)=\boldsymbol{X}_{r1}(t)+F*\left(\boldsymbol{X}_{r2}(t)-\boldsymbol{X}_{r3}(t)\right), \tag{14}$$

where $r1$, $r2$, $r3\in\{1, 2, \ldots, M\}$ are randomly chosen and mutually different and also different from the current index $i$, $F\in(0,2)$ is constant called scaling factor which controls amplification of the differential variation $\boldsymbol{X}_{r2}(t)-\boldsymbol{X}_{r3}(t)$. $\boldsymbol{X}_{r1}(t)$ is the base vector to be perturbed.

Step4: DE's *crossover*. For the sake of getting each individual's *trial vector* $\boldsymbol{U}_i(t+1)=[u_{i,1}(t+1), u_{i,2}(t+1), \ldots, u_{i,N}(t+1)]$, *crossover* operation is performed between each individual and its corresponding *mutant vector* by the following equation:

$$u_{i,j}(t+1)=\begin{cases} v_{i,j}(t+1), \text{if } (rand(j)\le CR) \text{ or } j=randn(i), \\ x_{i,j}(t), \text{otherwise.} \end{cases} \quad j=1,\cdots,N, \tag{15}$$

where *rand(j)* is the *j*th evaluation of a random number uniformly distributed in the range of [0, 1], *randn(i)*is a randomly chosen index from the set {1, 2, …, *N*}, $CR \in [0, 1]$ is constant called crossover parameter that controls the diversity of the population.

Step5: Sample the uncertain parameters $n_s$ times and evaluate the estimated objective values of the *trial vectors*.

Step6: DE's *selection.* In order to generate the new individual for the next generation, *selection* operation is performed between each individual and its corresponding *trial vector* by the following selection criterion with hypothesis test:

$$\boldsymbol{X}_i(t+1) = \begin{cases} \boldsymbol{U}_i(t+1), \text{ if } \bar{J}(\boldsymbol{U}_i(t+1)) - \bar{J}(\boldsymbol{X}_i(t)) < \tau, \\ \boldsymbol{X}_i(t), \text{ otherwise.} \end{cases} \tag{16}$$

where the unbiased estimated mean value $\bar{J}_i$ comes from the following equations of hypothesis test

$$\bar{J}_i = \bar{J}(\boldsymbol{X}_i) = \frac{1}{n_i}\sum_{j=1}^{n_i} J(\boldsymbol{X}_i, \xi_i), i = 1, 2 \tag{17}$$

$$s_i^2 = \frac{1}{n_i - 1}\sum_{j=1}^{n_i} [J(\boldsymbol{X}_i, \xi_j) - \bar{J}_i]^2, i = 1, 2 \tag{18}$$

$$\tau = t_{\alpha/2}(2n_s - 2)\cdot\sqrt{\frac{2}{n_s}}\cdot\sqrt{\frac{(n_s-1)s_1^2 + (n_s-1)s_2^2}{2n_s - 2}}. \tag{19}$$

where $s_i^2$ is the estimated variance.

Step7: Determine the best individual of the current new population with the best objective value. If the objective value of the current best individual is better, then update $\boldsymbol{X}^*$ and its objective value.

Step8: If a stopping criterion is met, then output $\boldsymbol{X}^*$ and its objective value; otherwise go back to Step 3.

# 4 Analysis of Practical Case

## 4.1 Practical Process Description

A 20Mt/a gas separation plant of some refinery is chosen as the object. This plant is composed of five cascaded distillation columns, which are propane off column (T1), ethane off column (T2), propylene columns (T3 and T4), and pentane off column (T5). The raw material is the liquefied petroleum gas after desulphurization, viz. composition of alkanes or alkenes from C2 to C5. The final products include ethane, propylene, propane, light C4 mixture and heavy C4 mixture with C5, and all of them should satisfy enough purity. In this flow sheet, the primary economic object is to get more propylene satisfied polymerization purity, viz. the mass faction of propylene lager than 98.5%. The secondary object is to reduce the energy consumption, including the flows of steam and hot water.

## 4.2 Predictive Control Scheme of Each Column

Before the unified optimization, the predictive control scheme of each column should be decided according to the mechanism analysis of these columns and implemented. The core arithmetic of the controllers is DMC. The process variables selected as CVs, MVs or DVs are shown as Tab. 1.

**Table 1.** Predictive control scheme of every column

| | T1(n=1) | T2(n=2) | T3 and T4(n=3) |
|---|---|---|---|
| CVn,1 | Top C4 mass fraction | Bottom C2 mass fraction | T4 top propylene mass fraction |
| CVn,2 | Bottom C3 mass fraction | Bottom temperature | T3 bottom propylene mass fraction |
| CVn,3 | Top temperature | | T4 top temperature |
| CVn,4 | Bottom temperature | | T3 bottom temperature |
| MVn,1 | Reflux flow | Hot water power | T4 reflux flow |
| MVn,2 | Steam flow | Top pressure | T3 hot water power |
| MVn,3 | Top pressure | | T4 top pressure |
| DVn,1 | Feed flow | Feed flow | T3 feed flow |
| DVn,2 | Feed temperature | Feed temperature | T3 feed temperature |

## 4.3 Unified Optimization Scheme

The optimization scheme of this gas separation plant should consider both raising the yield of the valuable product and reducing the energy consumption within the constraints of model and operation security. The detailed optimization model is

$$\begin{aligned}\min J = & -600(CV_{1,1}+\Delta CV_{1,1})+1.5(MV_{1,1}+\Delta MV_{1,1})+50(MV_{1,2}+\Delta MV_{1,2}) \\ & -250(CV_{2,1}+\Delta CV_{2,1})+2(MV_{2,1}+\Delta MV_{2,1}) \\ & +50000(CV_{3,1}+\Delta CV_{3,1}-98.7)^2 \\ & +(MV_{3,1}+\Delta MV_{3,1})+0.5(MV_{3,2}+\Delta MV_{3,2})\end{aligned} \tag{20}$$

$$s.t.\begin{cases}\Delta CV_{n,r}=\sum_{m} a_{n,r,m}\Delta MV_{n,m}+\sum_{m_d} b_{n,r,m_d}\Delta DV_{n,m_d} \\ \Delta DV_{2,1}=-0.5\Delta MV_{1,1} \\ \Delta DV_{2,2}=0.5\Delta CV_{1,3} \\ \Delta DV_{3,2}=\Delta CV_{2,2} \\ (CV_{n,r})_{\min}\le(CV+\Delta CV)_{n,r}\le(CV_{n,r})_{\max} \\ (MV_{n,m})_{\min}\le(MV+\Delta MV)_{n,m}\le(MV_{n,m})_{\max} \\ (\Delta MV_{n,m})_{\min}\le\Delta MV_{n,m}\le(\Delta MV_{n,m})_{\max}\end{cases}$$

where the given coefficients are transferred from the production data, and can be set and changed through human machine interface (HMI); $n$ is the index number of the controller; $r$ is the index number of the CV in the controller $n$; $m$ is the index number of MV in that controller $n$; $m_d$ is the index number of DV in that controller $n$;

$a_{n,r,m}, b_{n,r,m_d}$ are the steady state gains of CVs to MVs and DVs picked up from dynamic model, which have uncertainty.

## 5 Unified Optimization Simulation

### 5.1 Structure of the Multi-units Unified Optimization Realization

Using HYSYS, known as a flow sheet simulation software, the steady state and dynamic simulation of various chemical industrial processes can be realized. The software can perform bidirectional data communication between the simulation and other programs through an interface program [13]. Fig. 1 shows the virtual produce plant developed based on the HYSYS and structure of the unified optimization.

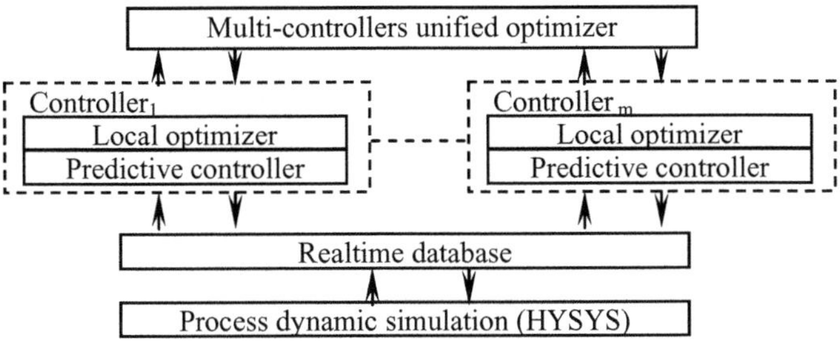


**Fig. 1.** Structure of unified optimization on virtual produce plant

### 5.2 Simulation Results

In Eq. (20), all of model gains $a_{n,r,m}, b_{n,r,m_d}$ could be uncertain. Their uncertainty is taken as 5% of their normal distribution of expectation. For DEHT and normal DE, the parameters are all set as follows: population size is 20, the maximum generation is 500, scaling factor is 1.2, and crossover parameter is 0.8. The simulation results of the unified optimization are listed in Tab. 2, Tab.3 and Fig. 3. Tab. 2 shows the results of the first period of the real-time optimization. From this table, it can be seen that DEHT is of better performance than normal DE and QP.

The unified optimization results are implemented indirectly through local optimizer. The local optimization of CVs stops and that of MVs is set to quadratic optimization with the desired values set to unified optimization results. Fig. 4 shows the results of the parameters adjustment of MVs in local optimizer of propylene columns. When unified optimizer stops, the settings of local optimizers will recover. Tab. 3 and Fig. 3 show the longtime results of the unified optimization. From the table, it can be seen that the unified optimization can balance variables of multi-units to reach the global optimum.

Fig. 2 reveals the simulation result of the unified optimization. The three curves are the result of DEHT, normal DE and QP respectively. The arrow shows the time when the unified optimizer begins to work. It is shown in the figure that the unified optimizer can optimize the three correlated process units controlled by three

**Table 2.** Optimization results of the first period under uncertain model gains

| | Initial | DEHT | DE | QP |
|---|---|---|---|---|
| T1 Top C4 mass fraction/% | 0.55 | 0.57 | 0.56 | 0.56 |
| T1 Bottom C3 mass fraction/% | 1.1 | 0.95 | 0.96 | 0.96 |
| T1 Reflux flow/( $t \cdot h^{-1}$) | 23.7 | 23.6 | 23.6 | 23.6 |
| T1 Steam flow/( $t \cdot h^{-1}$) | 6.9 | 6.9 | 6.9 | 6.9 |
| T2 Bottom C2 mass fraction/% | 0.13 | 0.14 | 0.13 | 0.13 |
| T2 Hot water power/(℃ $\cdot t \cdot h^{-1}$) | 1002 | 995 | 996 | 996 |
| T4 top propylene mass fraction/% | 98.5 | 98.5 | 98.5 | 98.5 |
| T3 bottom propylene mass fraction/% | 1.25 | 1.2 | 1.23 | 1.23 |
| T4 reflux flow/( $t \cdot h^{-1}$) | 112.8 | 112 | 112.1 | 112.2 |
| T3 hot water power/(℃ $\cdot t \cdot h^{-1}$) | 8362 | 8380 | 8378 | 8379 |
| Optimization Object | 8115.85 | 8095.4 | 8105 | 8105.6 |

**Table 3.** Unified optimization results under uncertain model gains

| | Initial | DEHT | DE | QP |
|---|---|---|---|---|
| T1 Top C4 mass fraction/% | 0.55 | 0.78 | 0.76 | 0.77 |
| T1 Bottom C3 mass fraction/% | 1.1 | 0.95 | 0.96 | 0.96 |
| T1 Reflux flow/( $t \cdot h^{-1}$) | 23.7 | 20.1 | 20.5 | 20.5 |
| T1 Steam flow/( $t \cdot h^{-1}$) | 6.9 | 7 | 7 | 7.1 |
| T2 Bottom C2 mass fraction/% | 0.13 | 0.14 | 0.13 | 0.13 |
| T2 Hot water power/(℃ $\cdot t \cdot h^{-1}$) | 1002 | 947 | 950 | 950 |
| T4 top propylene mass fraction/% | 98.5 | 98.7 | 98.7 | 98.7 |
| T3 bottom propylene mass fraction/% | 1.25 | 0.47 | 0.48 | 0.48 |
| T4 reflux flow/( $t \cdot h^{-1}$) | 112.8 | 109.9 | 110 | 110 |
| T3 hot water power/(℃ $\cdot t \cdot h^{-1}$) | 8362 | 8538 | 8540 | 8540 |
| Optimization Object | 8115.85 | 6350.05 | 6372.25 | 6371.25 |

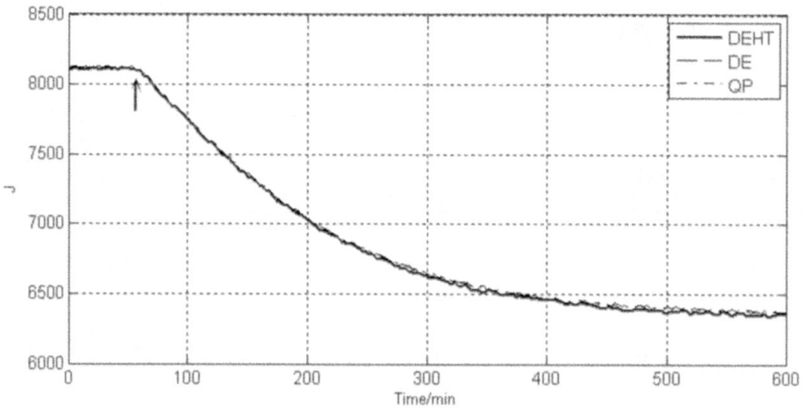


**Fig. 2.** Optimization object curves under uncertain model gains

controllers as a whole, and the trends under uncertainty account for the DEHT method is better than the other two.

## 6 Conclusion

In this paper, the total solution with two layers structure strategy to the control and optimization of large-scale chemical process is realized, which is made up of multi-units unified optimization model constructed with the predictive control models and the global optimization algorithm based on differential evolution (DE) under uncertainty and predictive control of each unit, and applied to a real production process successfully. A hybrid DE approach combining hypothesis test (HT) named as DEHT to compare the optimization objects under uncertainty is proposed. The simulation results of an application example to a 20Mt/a gas separation process show that the proposed total solution with two layers structure strategy is successful and multi-units unified optimization method based on HTDE is effective and robust for solving the optimization problem under uncertainty.

## References

1. Vargas-Villamil, F.D., Rivera, D.E.: A Model Predictive Control Approach for Real-time Optimization of Reentrant Manufacturing Lines. Computers in Industry, (2001) 45(1): 45-57
2. Huang, D.X., Wang, J.C., Jin, Y.H.: Stable MIMO Constrained Predictive Control with Steady-state Objective Optimization. Chinese Journal of Chemical Engineering, (2000) 8(4): 332-338
3. Huang, D.X.: Studies and Applications on the Methodology of Multi-variables Predictive Control. Beijing: Dep. Of Automation, Tsinghua University, (1999) (in Chinese)
4. Qin, V., Badgwell, T.A.: A Survey of Industrial Model Predictive Control Technology. Control Engineering Practice, (2003) 11: 733-764
5. Pham, Q.T.: Dynamic Optimization of Chemical Engineering Processes by an Evolutionary Method. Computers and Chemical Engineering, (1998) 22: 1089-1097
6. Storn, R., Price, K.: Differential Evolution-A Simple and Efficient Heuristic for Global Optimization over Continuous Spaces. J Global Optim II, (1997) 341-359
7. Ilonen, J., Kamarainen, J.K., Lampinen, J.: Differential Evolution Training Algorithm for Feed-forward Neural Networks. Neural Process Lett, (2003) 17(1): 93-105
8. Chang, F.P., Hwang, C.: Design of Digital PID Controllers for Continuous-time Plants with Integral Performance Criteria. J Chin Inst Chem Eng, (2004) 35(6): 683-696
9. Chang, Y.P., Wu, C.J.: Optimal Multiobjective Planning of Large-scale Passive Harmonic Filters using Hybrid Differential Evolution Method Considering Parameter and Loading Uncertainty. IEEE T Power Deliver, (2005) 20(1): 408-416
10. Zhang, Y.Y., Shao, Z.J., Zhong, W.T. et al: Centralized Optimization and Decentralized Optimization of Distilation Columns in Series. Automation in Petro-Chemical Industry, (1999) 3: 19-23(in Chinese)
11. Price, K., Storn, R.: Differential Evolution (DE) for Continuous Function Optimization. http://www.icsi.berkeley.edu/%7Estorn/code.html, (2007)
12. Y. Zhang, D. Monder, J.F. Forbes: Real-time Optimization under Parametric Uncertainty: A Probability Constrained Approach. J. Process Contr. 12 (2002) 373-389
13. Huang, D.X., Gu, J., Wang, Y.H.: On-Line Application Platform Based on Flowsheeting Simulation. Process Control Science Technology and Applications, (2002) 823-827 (in Chinese)

# An Angle-Based Crossover Tabu Search for Vehicle Routing Problem*

Ning Yang[1], Ping Li[1], and Mingsen Li[2]

[1] Faculty of Electric and Automatic Engineering, Shanghai University of Electric Power, Shanghai China 200090
yangning@shiep.edu.cn
[2] Jilin Academy of Agricultural Machinery, Changchun China 130022

**Abstract.** An improved tabu search – crossover tabu search (CTS) is presented which adopt the crossover operator of the genetic algorithm as the diversification strategy, and selecting elite solutions as the intensification strategies. To improve the performances, the angle-based idea of the sweep heuristic is used to confirm the neighborhood, and an object function with punishment. The angle-based CTS is applied for the vehicle routing problem. The simulating results which compared the tradition sweep heuristic and the standard tabu search shows the results got by angle-based CTS are better than those got by other two heuristics. The experiment shows the angle-based CTS has good performance on the vehicle routing problem.

**Keywords:** vehicle routing problem, tabu search, crossover operator, crossover tabu search.

## 1 Introduction

The basic vehicle routing problem (VRP) consists of a number of customers, each requiring a specified weight of goods to be delivered. One or more vehicles dispatched from a single or more depots must deliver the good required, and then return to the depot. As a very complicated combinatorial optimization problem, a traditional heuristics can't get good solution, and we will focus our attention on meta-heuristic algorithms that lead to reasonably good solutions. Toth and Vigo report that the use of computerized methods in distribution processes often result in savings ranging from 5% to 20% [1].

For any meta-heuristic algorithms, it is a key problem that how to improve and balance local search and global search. TS was proposed and developed by Glover and Glover et al [2,3], which had been used to solve a wide range of tough optimization problems such as job shop scheduling, the Salesman Problem (TSP) and the VRP. For TS, some intensification strategies and diversification strategies were proposed for this problem. Intensification strategies are used to enhance the efficiency

* This work was supported by the Shanghai Municipal Education Commission program under grant KZ-2006-14 and Shanghai University of Electrical Power program under K-2006-27.

D.-S. Huang, L. Heutte, and M. Loog (Eds.): ICIC 2007, LNAI 4682, pp. 1215–1222, 2007.

of local search, which are based on modifying choice rules to encourage move combinations. Diversification strategies are designed to drive the search into new regions to enhance the global search. In fact, many meta-heuristics have their intensification strategies and diversification strategies, such as genetic algorithm.

Genetic algorithms are general-purpose searching algorithms based upon the principles of evolution observed in nature. Genetic algorithm combines coding, selection, crossover, and mutation operators [4]. The core operator, selection is used as intensification strategy, and the crossover and mutation are used as diversification strategies.

As analysis above, we presented a crossover tabu search. The crossover operator is adopted as the diversification strategy of the CTS, and the selecting elite solutions is adopted as the intensification strategies. The vehicle routing problem is used to test the performances of the CTS.

The solution quality of the VRP is not only depends on good meta-heuristic, but also on appropriate neighborhood. The solution quality and resolution speed is influenced by appropriate neighborhood. In this research, the angle-based idea of the Sweep heuristic is used to confirm neighborhood.

The paper is organized as the followings. We will discuss the main idea in section 1, the angle-based idea of the sweep heuristic and the sweep based CTS will be introduced in section 2, the vehicle routing problem will be described in section 3, and the results analyses of four algorithms will be presented in section 4, followed by conclusion in the final section 5.

## 2 Angle-Based CTS

### 2.1 Standard TS

In some sense, TS originates from the local search [3], and uses the character of tabu to simulate the way of human thinking. The standard TS includes neighborhood function $\phi$, move, tabu list $T$, tabu tenure and aspiration strategy. The standard TS is described in figure 1.

***Standard Tabu Search***

*Initial:* the tabu list $T$; $s$: the initial solution; $s^*$: the best solution; $s^* = s$

*Loop /* the main cycle of TS * /*

According to the neighborhood function $\phi$, the tabu list $T$ and Aspiration criterion, find the new possible best solution $s'$, $s' \in \phi(s')$.

$s = s'$ /* replace the current solution by the new one * /

Insert the attribute of the solution $s$ into the tabu list $T$;

If $f(s)$ better than $f(s^*)$, then $s^* = s$; /* save the best so far solution * /

Update the tabu list $T$

**Fig. 1.** Program of the standard tabu search

In the Fig.1, $s$ stands for an initial solution, and $\phi(s)$ is neighborhood of current solution. The detail of the TS theory can be found in [2,3]. After the neighborhood being confirmed, 2-opt [5] algorithm is used to decide the move.

## 2.2 Intensification Strategy

In general, intensification is based on some intermediate-term memory, where one records the number of consecutive iterations in which various "solution components" have been presented in the current solution without interruption [6]. Of course, the TS itself is an intensification strategy

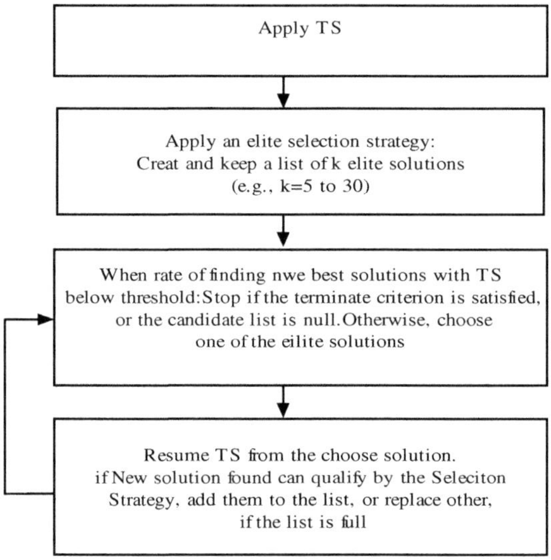


**Fig. 2.** The intensification of selecting solutions strategy

A selecting solutions strategy [3] is used in this research. Creating and keeping a candidate elite resolution list of k elite solutions, (e.g., k=3 to 50), and applying an elite selection strategy means selecting the better resolution in the current iteration of TS, and saving to the candidate elite resolution list. If the candidate list is full, the worst candidate solution will be cancelled. When the rate of finding new best solutions falls below threshold and the terminating criterion is not satisfied, one of the elite solutions from the candidate list is chosen and the TS will continue. Fig.2 described the process.

## 2.3 Diversification Strategy

Diversification strategies drive the search to new search space. Diversification strategies are based on modifying choice rules to bring attributes to the solutions that are infrequently used. Alternatively, they may introduce such attributes by partially or fully re-starting the solution process.

As the core operator of the GA [4,7], the crossover operator doesn't destroy the two father individuals completely, but strives to inherit better characteristics of the father individuals and accept some other new characteristics, and reconstructs two new son individuals, which means to exploit new search spaces. In this research, we use two-point crossover.

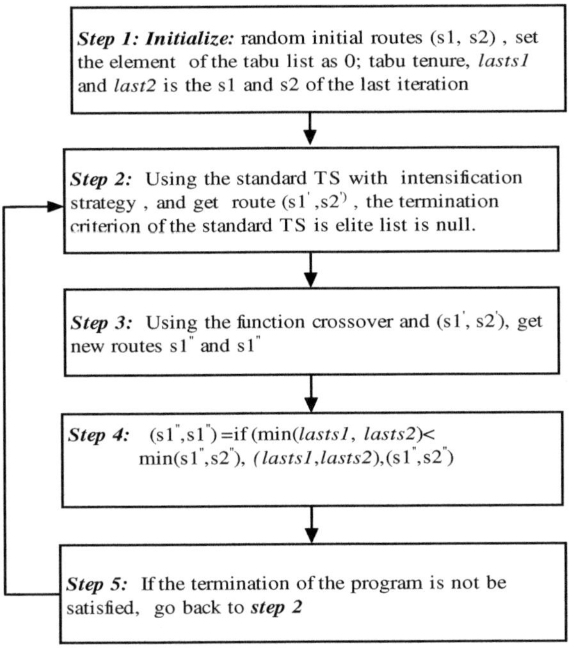


**Fig. 3.** The flow of the CTS

## 2.4 Crossover Tabu Search

According to the above analyses, Fig.3 described the program of the CTS. First, two initial routes ( $s1$ , $s2$ ) are randomly generated. Using the standard TS with intensification strategy to search until the elite list is null and the new route ( $s1'$ , $s2'$ ) is obtained. The crossover function will generate new route ( $s1'$ , $s2'$ ), and $lasts1$ and $lasts2$ will represent $s1'$ and $s2'$ of the last iteration. Speaking in an algorithmic language, the update of ( $s1$ , $s2$ ) is as $(s1, s2) = if(\min(lasts1, lasts2) < \min(s1'', s2''), (lasts1, lasts2), (s1'', s2''))$ and then take the ( $s1$ , $s2$ ) to standard TS again.

## 2.5 The Angle-Based Idea of the Sweep Heuristic

As the city number increase, it is important that how to confirm the neighborhood for the TSP. The scale of the neighborhood influences the resolution speed and quality. In this research, the angle-based idea of the sweep heuristic is used. The sweep heuristic has been applied for VRP [8].

All cities can be positioned on a plan. In the sweep heuristic, cities have geographic positions. Without loss of generality, suppose any point in the plan as zero point and is given Cartesian coordinates (0,0), and suppose each cities $i$ has polar coordinates $(r_i, \theta_i)$ relative to the that the zeros point. Here, suppose $0 \le \theta_i \le 2\pi$ is measured clockwise from the positive x-axis and the angle difference of each two cities can be calculate. So the neighborhood of the city $i$ can be confirmed by the angle difference of the city $i$ and other cities. If the angle difference of the city $j$ and the city $i$ is smaller than some angle threshold designed, the city $j$ is one member of the neighborhood of the city $i$.

## 3 Vehicle Routing Problem

The classical VRP is defined on an undirected graph $G = (V, E)$ where $V = \{v_0, v_1, \ldots, v_n\}$ is a vertex set and $E = \{(v_i, v_j) : v_i, v_j \in V, i < j\}$ is an edge set. Vertex $v_0$ represents a depot and vertex set $v_1, v_2, \ldots v_n$ represent customers and cities. A non-negative cost, distance or travel time matrix $C = (c_{ij})$ is defined on $E$. Each customer $v_i (i = 1 \ldots n)$ has a non-negative demand $q_i$ and a non-negative service time $s_i$ [9, 10]. Each route has to satisfy the follow rules:

1) Of minimum total cost
2) Starting and ending at the depot
3) Each customer is visited exactly once by exactly one vehicle
4) The total demand of any route does not exceed $Q$

The resolution of the VRP is not depend on appropriate meta-heuristics, but also depend on the neighborhood, neighbor operator, object function and the parameter adjustment.

### 3.1 Neighborhood

The neighborhood is important for the searching efficient and searching quality. Bigger neighborhood maybe includes more resolutions, but the searching time would be added. Smaller neighborhood improves the searching velocity, but the optimal solution would be omitted. So, how to confirm the neighborhood is important point for the VRP. The angle-based idea of the sweep heuristic is adopted to confirm the neighborhood in this article.

### 3.2 Neighbor Operator

For the different customers of the neighborhood operator, which are on the same rout or different routes, the different neighborhood operators are adopted. When the objects on the same route, the $\lambda$-opt is adopted, and when on different routes, the $\lambda$-interchange and $\lambda$-exchange is adopted.

### 3.3 Object Function

When the neighborhood is confirmed, an object function is needed to evaluate the neighborhood operator. For VRP, the object function need include two parts. i) Length: the length change of the total routes. ii) Weight: the load change of the route. Considering of the load change of the route, sometimes, move will cause some bad resolutions. It is appropriate to keep these resolutions with some punishment, not discard them, which can enlarge the searching space. *Bad_weight* is the load weight of the route.

```
For i=1: Bad_weight
  Sum_Weight1=Sum_Weight1+k1*i;
End
```

When a bad resolution becomes a good resolution, the praise is used. *Good_Weight* is the load weight change of the route

```
For i=1: Good_weigt
  Sum_Weight2=Sum_Weight2+k2*i;
End
Sum_Weight= Sum_Weight1- Sum_Weight2;
```

*k1,k2* are adjustable parameter, *Sum_Weight* is the parameter of the load weight of the route in the object function, *Save_Dist* is the parameter of the distance of the route in the object function. The object function is as (1), which is used to evaluate the individuals of the neighborhood.

$$F(i,j) = Save\_Dist(i,j) + Sum\_Weight(i,j) \quad . \tag{1}$$

### 3.4 Adjustable of the Parameter

During the search, how to balance the global search and local search is depend on the change frequency of the optimal resolution. When the frequency is low, the search should pay more attention on global search, on the other hand, on local search. Using the adjustable parameter can fit the change. In addition, $k1,\ k2, \phi$, the tabu length of the tabu search can be changed to change the different weightiness of the global search and local search which also depend on the change frequency of the optimal resolution.

## 4 Resolution of VRP Instances

### 4.1 Program Structure

The whole program includes a main program and four subprograms. The subprograms are described as follow:

*Initial_Route ():* Use the angle-based idea or random method to generate the initial route.

*Candidate_choice ():* Confirm the neighborhood using the idea described in 3.1 section.

*Cross_Tabu ()* : Use the crossover tabu search to get better resolution. The end condition of the subprogram is the cycle number or the candidate elite resolution number in the candidate elite resolution list is zero.

*Neighborhood ()* : According to the change frequency of the optimal solution and the customer's position, Use different neighbor operator, 2-opt, 2-interchange, 2-exchange and insert to get new solutions.

### 4.2 Resolution of VRP Instances

There are twelve VRP instances of Augerat ea al, (http:// neo.lcc.uma.es/radi-aeb/WebVRP/index.html), which were adopted to evaluate the efficiency of the CTS. The algorithms used in the experimentation are as follows:

1) Sweep algorithm
2) Tabu search
3) Angle-based CTS.

The results of the comparison are presented in Table 1. The customer number and the vehicle number are assigned in the examples. The optimal values are the shortest distance of examples, which had been proved and opened on internet. First, the result of the Sweep algorithm is the worst, which proves the traditional heuristic can not get satisfied result. Secondly, the results of the TS are better than that of the Sweep

**Table 1.** Comparison of the algorithms

| | Customer number | Vehicle number | Optimal value | Sweep | TS | Angle-based CTS |
|---|---|---|---|---|---|---|
| A-n32-k5 | 32 | 5 | 784 | 832 | 806 | 784 |
| A-n44-k6 | 44 | 6 | 937 | 1056 | 997 | 939 |
| A-n63-k10 | 63 | 10 | 1315 | 1406 | 1405 | 1366 |
| B-n34-k5 | 34 | 5 | 788 | 828 | 797 | 788 |
| B-n68-k9 | 68 | 9 | 1304 | 1374 | 1357 | 1339 |
| F-n72-k4 | 72 | 4 | 238 | 305 | 279 | 245 |
| A-n38-k5 | 38 | 5 | 730 | 827 | 799 | 731 |
| A-n53-k7 | 53 | 7 | 1010 | 1229 | 1184 | 1070 |
| A-n80-k10 | 80 | 10 | 1764 | 1880 | 1843 | 1820 |
| B-n56-k7 | 56 | 7 | 707 | 737 | 716 | 710 |
| B-n78-k10 | 78 | 10 | 1266 | 1350 | 1312 | 1280 |
| F-n135-k7 | 135 | 7 | 1165 | 1426 | 1320 | 1236 |

algorithm, which proves the meta-heuristic is a good way to resolve the optimization problems. Finally, the result of the sweep-based CTS is the best, although have litter difference with the optimal value, which proves the crossover operator improves the global search, and the selecting solutions strategy improves the local search. All the comparison results show that the sweep-based CTS can balance the global search and local search, and can resolve the VRP efficiently.

## 5 Conclusion

The vehicle routing problem has great significance for the development of the city transportation, port business and tourism, and even the overall economic development of the cities. According to the idea of intensification strategy and diversification strategy, an improved crossover tabu search is presented. The crossover operator of the genetic is adopted as intensification strategy and elite and the selecting solutions strategy as diversification strategy. With the angle-based idea of the Sweep algorithm to confirm neighborhood, and an object function with punishment, the angle-base CTS is applied to resolve the VRP. The tradition sweep heuristic and the standard tabu search also applied to resolve the same VRP. The compared results shows the results got by angle-based CTS are better than those got by other two heuristics, and close to the exist best results. The experiment shows the CTS with the angle-based idea of the sweep heuristic has good performance on the vehicle routing problem.

## References

1. Toth, P., Vigo, D.: The Vehicle Routing Problem: Monographs on Discrete Mathematics and Applications. Philadelphia: SIAM, (2001)
2. Glover, F.: Heuristic for Integer Programming using Surrogate Constraints. Decision Sciences. 8. (1977). 156-66;
3. Glover, F.: Tabu Search Fundamentals and Uses. Graduate School of Business, University of Colorado, Boulder. (1995)
4. Cheng, G.N., Zhuang Z.Q.: Theory and Application of the Genetic Algorithm. Posts & Telecom Press. June (1999)
5. Laporte, G., Gendreau,M., Potvin, J.Y., Semetf.: Classical and Modern Heuristics for the Vehicle Routing Problem. International Transactions in Operational Research, Vol.7. (2000) 285-300.
6. Gendreau, M.: An Introduction to Tabu Search. Centre de recherche sur les transports and Département d´informatique et de recherche opérationnelle Université de Montréal ase postale 6128, Succursale "Centre-ville" Montréal, Canada H3C 3J7 July (2002)
7. Wang, W.L., Yao, M.H., Wu, Y.G., Wu, Q.D.: Hybrid Flow-shop Scheduling Approach Based on Genetic Algorithm. Journal of System Simulation, Vol.14. (2002) 863
8. Renaud, J., Fayez, F., Boctor.: A Sweep-based Algorithm for the Fleet Size and Mix Vehicle Routing Problem. European Journal of Operational Research, Vol.140. (2002) 618-628
9. Gouveia, L.: A Result on Projection for the Vehicle Routing Problem. Theory and Methodology, Vol. 85. (1995) 610-624
10. Laporte, G., Gendreau, M.: Classic and Modern Heuristics for the Vehicle Routing Problem. International Transactions in Operational Research, Vol 7. (2000) 285-300

# Saturation Throughput Analysis of IEEE 802.11e EDCA

Yutae Lee, Kye-Sang Lee, and Jong Min Jang

Dongeui University, Busan 614-714, Korea
ylee@deu.ac.kr, ksl789@gmail.com, jmjang@deu.ac.kr

**Abstract.** This paper introduces a simple three-dimensional Markov chain model for the Enhanced Distributed Channel Access (EDCA) mechanism of IEEE 802.11e MAC under saturation condition. This new Markov chain model can be used to evaluate the throughput performance of IEEE 802.11e EDCA. This new model captures all of the major Quality of Service (QoS) specific features, namely Arbitration Interframe Space (AIFS), minimum contention window size, maximum contention window size and virtual collision, for the IEEE 802.11e EDCA mechanism and hence can provide a novel numerical approach to pick the set of EDCA parameter values to meet the QoS requirements. The results of our analytical model are then verified using simulations.

## 1 Introduction

With the increasing demand of WLANs, especially of IEEE 802.11, the support of differentiated QoS has become one of the recent critical issues on the success of IEEE 802.11 [3] MAC protocols for the future wireless communications. It is important to develop new medium access scheme that can support the differentiated QoS requirements over IEEE 802.11 WLANs, which is specified by the IEEE 802.11e. The IEEE 802.11e standard [4] specifies differentiated service classes in the MAC layer to support the delivery of priority packets and have drawn tremendous interest from both industry and academia. IEEE 802.11e defines HCF for access mechanism, which uses two mechanisms for the support of QoS differentiation. They are EDCA and HCF HCCA. EDCA is a contention based access mechanism and delivers traffic based on differentiated Access Categories (ACs).

Performance analysis of IEEE 802.11e EDCA mechanism has attracted lots of research efforts over the past years. It has been shown, using simulation models, that the EDCA protocol provides a significant improvement in QoS support over the 802.11 MAC. Xiao [10] modified the Markov chain in [2] to model resource sharing by different classes. Mangold et al. [6] introduced a separate one-dimensional Markov chain to be used along with Bianch's model, so that the effect of varying interframe spaces can also be investigated. Robinson and Randhawa [7] extended the Bianchi's model [2] to analyze the saturation throughput performance of the EDCA mechanism. Xiao [11] developed a model to analyze the contention window size differentiation in the EDCA mechanism. Kong et al. [5] analyzed the throughput performance of differentiated service traffic. But,

D.-S. Huang, L. Heutte, and M. Loog (Eds.): ICIC 2007, LNAI 4682, pp. 1223–1232, 2007.

they assumed that the collision probability is independent of the system situation. In fact, the collision probability under the situation where all ACs have started their backoff counter counting down is different from that under the situation where some ACs are still in their AIFS period. Tao and Panwar [9] proposed a Markov chain model for the 802.11e EDCA mode and computed the throughput when under saturation load. But, they assumed that an unsuccessful station will keep retransmitting with maximum backoff window.

This paper introduces a simple three-dimensional Markov chain model for the Enhanced Distributed Channel Access (EDCA) mechanism of IEEE 802.11e MAC under saturation condition. This model can be used to evaluate the throughput performance of IEEE 802.11e EDCA. This model captures all of the major Quality of Service (QoS) specific features, such as AIFS, minimum contention window size, maximum contention window size and and virtual collision, for the IEEE 802.11e EDCA mechanism. The results of our analytical model are then verified using simulations.

## 2 Enhanced Distributed Channel Access (EDCA)

The IEEE 802.11 DCF has no functionality to support QoS requirements. To overcome this drawback and enhance the traditional DCF, as an extension to the basic DCF mechanism of the legacy 802.11, EDCA has been proposed to provide differentiated and distributed channel access for packets with 8 different priorities in a station, which are mapped into four different ACs at the MAC layer. Individual AC contends to channel access and is differentiated by a set of EDCA parameters, namely AIFS, minimum contention window size and maximum contention window size [1]. AIFS is the time interval a packet of a given AC has to wait after the channel becomes idle before it can start the backoff process or transmit. After $i$ collisions, the backoff counter for that particular AC is selected uniformly from range $[0, 2^i CW_{min} - 1]$, until $i$ reaches the backoff stage $m$ such that $2^m CW_{min} = CW_{max}$. At that point, the packet will still be retransmitted, if a collision occurs, with the backoff counters chosen from the range $[0, CW_{max} - 1]$. When the total number of retransmissions equals the maximum number of allowable retransmissions, no further retransmission is attempted, and the packet is discarded [9]. In EDCA, each station implements a queue for each AC. Each queue has its own QoS parameters and backoff counter. Packets belonging to different ACs within a single station may collide with each other when their backoff counters decrement to zero simultaneously [8]. This phenomenon is called an virtual collision in 802.11e, and is prevented by letting the highest priority involved in the collision win the contention [9].

## 3 Saturation Throughput Analysis Without Virtual Collision

We model the backoff operation of each station with a Markov chain. Our approach is similar to that of [2]. In this section, we assume that each station only implements one queue of the four ACs in EDCA.

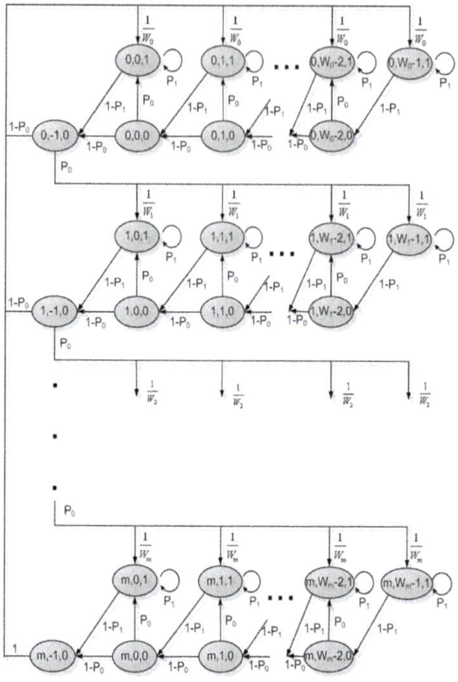


**Fig. 1.** State transition diagram of IEEE 802.11e EDCA for $AC_0$

It is assumed that AIFS[$AC_0$] =DIFS+Slot Time and AIFS[$AC_l$]=DIFS for $l = 1, 2, 3$. To capture the operation of AIFS differentiation, we use different types of Markov chains for different AIFS values. Two different types of chains are required according to the values of AIFS. The Markov chain model for stations implementing $AC_0$ is shown in Fig. 1. The states $(i, j, 1)$ and $(i, j, 0)$ correspond to the $i$th backoff stage of a station and that station has $j$ as its backoff counter. The variable $i$ ranges from 0 as the first backoff stage to $m$ the retransmission limit. The value $j$ ranges from -1 to $W_i - 1$, where $W_i$ is the backoff window of stage $i$; the backoff window $W_i$ is given by $W_i = \min(2^i CW_{min}[\text{AC}_0], CW_{max}[\text{AC}_0])$. Note that, when the backoff counter expires, the station cannot transmit immediately, but it has to wait for an extra idle backoff slot. The value $j = -1$ corresponds to the extra idle slot. $AC_0$ stations need to wait for a slot time longer than the other stations when a transmission occurs. The states $(i, j, 1)$ model this waiting. These states have loop transitions, which model the backoff counter freezing by $AC_0$ stations when transmissions by higher priority stations occur before the AIFS for $AC_0$. States $(i, j, 0)$ model the situation after the AIFS period for $AC_0$. The probabilities $p_0$ and $p_1$ in Figure 1 are the probabilities that, from the tagged $AC_0$ station's point of view, at least one of the other stations

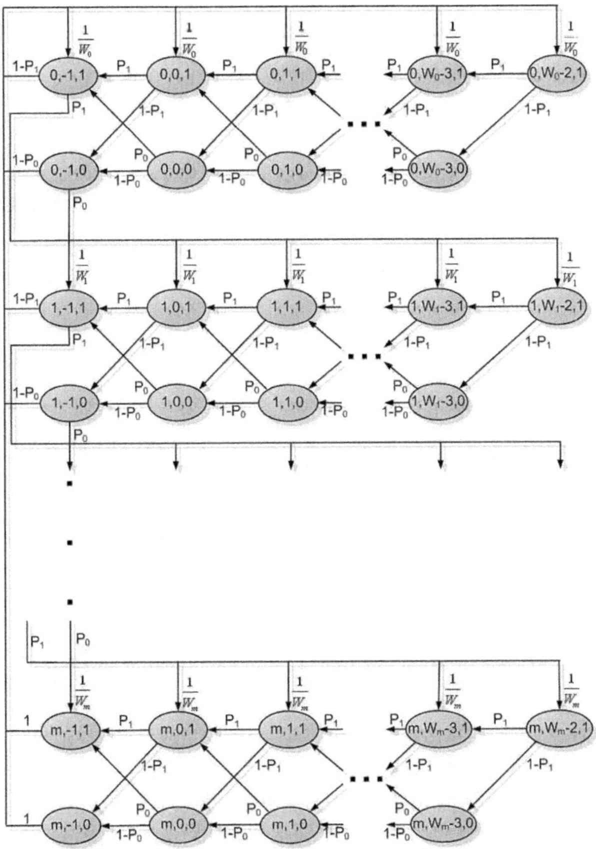

**Fig. 2.** State transition diagram of IEEE 802.11e EDCA for $\mathrm{AC}_l$, $l = 1, 2, 3$

transmit during a slot time after the AIFS period for $\mathrm{AC}_0$ and just before the AIFS period for $\mathrm{AC}_0$.

Fig. 2 shows the Markov chain model for stations implementing $\mathrm{AC}_1$, $\mathrm{AC}_2$ or $\mathrm{AC}_3$. For $\mathrm{AC}_l$, $l = 1, 2, 3$, the conditional probabilities $p_0$ and $p_1$ in Figure 2 are the probabilities that, from the tagged $\mathrm{AC}_l$ station's point of view, at least one of the other stations transmit during a slot time after the AIFS period for $\mathrm{AC}_0$ and just before the AIFS period for $\mathrm{AC}_0$.

Let $b_{i,j,k}[\mathrm{AC}_l]$ be the stationary distribution of the Markov chain for $\mathrm{AC}_l$. For $\mathrm{AC}_0$, owing to the chain regularities, we have

$$b_{i,j,0}[\mathrm{AC}_0] = \frac{W_i[\mathrm{AC}_0] - j - 1}{W_i[\mathrm{AC}_0]} \left(p_0[\mathrm{AC}_0]\right)^i b_{0,-1,0}[\mathrm{AC}_0],$$

$$i = 0, \cdots, m[\mathrm{AC}_0], \quad j = -1, 0, \cdots, W_i[\mathrm{AC}_0] - 2,$$

$$b_{i,j,1}[\mathrm{AC}_0] = \frac{1 + (W_i[\mathrm{AC}_0] - j - 1)\, p_0[\mathrm{AC}_0]}{W_i[\mathrm{AC}_0]\,(1 - p_1[\mathrm{AC}_0])} (p_0[\mathrm{AC}_0])^i\, b_{0,-1,0}[\mathrm{AC}_0],$$
$$i = 0, \cdots, m[\mathrm{AC}_0], \quad j = 0, \cdots, W_i[\mathrm{AC}_0] - 1,$$

where $b_{i,j,k}[\mathrm{AC}_l]$, $W_i[\mathrm{AC}_l]$, $p_i[\mathrm{AC}_l]$ and $m[\mathrm{AC}_l]$ corresponds to the values $b_{i,j,k}$, $W_i$, $p_i$ and $m$ respectively for $\mathrm{AC}_l$. With normalization condition, we obtain $b_{0,-1,0}[\mathrm{AC}_0]$.

Similarly for $\mathrm{AC}_l$, $l = 1, 2, 3$, owing to the chain regularities, we have

$$b_{i,W_i[\mathrm{AC}_l]-2,1}[\mathrm{AC}_l] = \frac{1}{W_i[\mathrm{AC}_l]} \left(p_0[\mathrm{AC}_l] b_{i-1,-1,0}[\mathrm{AC}_l] + p_1[\mathrm{AC}_l] b_{i-1,-1,1}[\mathrm{AC}_l]\right),$$
$$i = 1, \cdots, m[\mathrm{AC}_l],$$
$$b_{i,W_i[\mathrm{AC}_l]-3,1}[\mathrm{AC}_l] = (1 + p_1[\mathrm{AC}_l])\, b_{i,W_i[\mathrm{AC}_l]-2,1}[\mathrm{AC}_l], \quad i = 0, \cdots, m[\mathrm{AC}_l],$$
$$b_{i,W_i[\mathrm{AC}_l]-3,0}[\mathrm{AC}_l] = (1 - p_1[\mathrm{AC}_l])\, b_{i,W_i[\mathrm{AC}_l]-2,1}[\mathrm{AC}_l], \quad i = 0, \cdots, m[\mathrm{AC}_l],$$
$$b_{i,j,1}[\mathrm{AC}_l] = b_{i,W_i[\mathrm{AC}_l]-2,1}[\mathrm{AC}_l] + p_1[\mathrm{AC}_l] b_{i,j+1,1}[\mathrm{AC}_l] + p_0[\mathrm{AC}_l] b_{i,j+1,0}[\mathrm{AC}_l],$$
$$i = 0, \cdots, m[\mathrm{AC}_l], \quad j = -1, 0, \cdots, W_i[\mathrm{AC}_l] - 4,$$
$$b_{i,j,0}[\mathrm{AC}_l] = (1 - p_1[\mathrm{AC}_l])\, b_{i,j+1,1}[\mathrm{AC}_l] + (1 - p_0[\mathrm{AC}_l])\, b_{i,j+1,0}[\mathrm{AC}_l],$$
$$i = 0, \cdots, m[\mathrm{AC}_l], \quad j = -1, 0, \cdots, W_i[\mathrm{AC}_l] - 4.$$

The state stationary probabilities can be expressed in terms of $b_{0,W_0[\mathrm{AC}_l]-2,1}[\mathrm{AC}_l]$, which is obtained by imposing the normalization condition.

The probabilities $\tau_0[\mathrm{AC}_l]$ and $\tau_1[\mathrm{AC}_l]$ that a station of $\mathrm{AC}_l$ transmits in a slot during an AIFS period of $\mathrm{AC}_0$ and during other period respectively are given by

$$\tau_0[\mathrm{AC}_0] = \frac{\sum_{i=0}^{m[\mathrm{AC}_0]} b_{i,-1,0}[\mathrm{AC}_0]}{\sum_{i=0}^{m[\mathrm{AC}_0]} \sum_{j=-1}^{W_i[\mathrm{AC}_0]-2} b_{i,j,0}[\mathrm{AC}_0]},$$
$$\tau_1[\mathrm{AC}_0] = 0,$$
$$\tau_0[\mathrm{AC}_l] = \frac{\sum_{i=0}^{m[\mathrm{AC}_l]} b_{i,-1,0}[\mathrm{AC}_l]}{\sum_{i=0}^{m[\mathrm{AC}_l]} \sum_{j=-1}^{W_i[\mathrm{AC}_l]-3} b_{i,j,0}[\mathrm{AC}_l]}, \quad l = 1, 2, 3,$$
$$\tau_1[\mathrm{AC}_l] = \frac{\sum_{i=0}^{m[\mathrm{AC}_l]} b_{i,-1,1}[\mathrm{AC}_l]}{\sum_{i=0}^{m[\mathrm{AC}_l]} \sum_{j=-1}^{W_i[\mathrm{AC}_l]-2} b_{i,j,1}[\mathrm{AC}_l]}, \quad l = 1, 2, 3,$$

The probabilities $\tau_i[\mathrm{AC}_l]$ depend on the collision probability $p_i[\mathrm{AC}_l]$, which is given by

$$p_i[\mathrm{AC}_l] = 1 - (1 - \tau_i[\mathrm{AC}_l])^{n[\mathrm{AC}_l]-1} \prod_{x \neq l} (1 - \tau_i[\mathrm{AC}_x])^{n[\mathrm{AC}_x]}$$

for $i = 0, 1$ and $l = 0, 1, 2, 3$, where $n[\mathrm{AC}_l]$ is the number of stations implementing $\mathrm{AC}_l$.

The probability $P_I$ that a slot is idle is

$$P_I = P_0 \prod_{l=0}^{3} (1 - \tau_0[\mathrm{AC}_l])^{n[\mathrm{AC}_l]} + P_1 \prod_{l=1}^{3} (1 - \tau_1[\mathrm{AC}_l])^{n[\mathrm{AC}_l]},$$

where the probabilities $P_0$ and $P_1$ that an arbitrary slot is in an AIFS period of $\mathrm{AC}_0$ and in other period respectively are given by

$$P_1 = \sum_{i=0}^{m[\mathrm{AC}_0]} \sum_{j=0}^{W_i[\mathrm{AC}_0]-1} b_{i,j,1}[\mathrm{AC}_0],$$
$$P_0 = 1 - P_1.$$

The probabilities $P_{S0}[\mathrm{AC}_l]$ and $P_{S1}[\mathrm{AC}_l]$ that a slot time contains a successful transmission of $\mathrm{AC}_l$ under the condition that the slot is in an AIFS period of $\mathrm{AC}_0$ and in other period respectively are given by

$$P_{Si}[\mathrm{AC}_l] = \frac{n[\mathrm{AC}_l]\tau_i[\mathrm{AC}_l]}{1 - \tau_i[\mathrm{AC}_l]} \prod_{x=i}^{3} (1 - \tau_i[\mathrm{AC}_x])^{n[\mathrm{AC}_x]}, \quad i = 0, 1.$$

The probability $P_C$ that a slot time contains a collision is $P_C = 1 - P_I - P_S$, where

$$P_S = \sum_{l=0}^{3} \left[P_0 P_{S0}[\mathrm{AC}_l] + P_1 P_{S1}[\mathrm{AC}_l]\right].$$

The saturation throughput of $\mathrm{AC}_l$ is given by

$$S[\mathrm{AC}_l] = \frac{E[P]\left[P_0 P_{S0}[\mathrm{AC}_l] + P_1 P_{S1}[\mathrm{AC}_l]\right]}{P_I \sigma + P_S T_S + P_C T_C},$$

where $E[P]$ is the average length of frame payload, $\sigma$ is the length of a slot time, $T_S$ is the average length of successful transmission and $T_C$ is the average length of collision. The value $T_S$ and $T_C$ for the basic and the RTS-CTS access methods are given in [2].

## 4 Saturation Throughput Analysis with Virtual Collision

Consider the case where each station runs four queues where each queue corresponds to an AC. With virtual collision, when two or more queues of a station have backoff counters of zero, the highest priority queue is favored and given the chance to access the medium. The lower priority queues still see the collision, and they will increase their backoff stages and choose other backoff counters.

We use the same approach as in the previous scenario for virtual collision case with some notable differences. Firstly, as each station now has four queues, we use a Markov chain to model each queue instead of each station. Secondly, virtual collision changes the collision probability seen by each queue as higher priority queues will not see the collision with the lower priority queues of the same station. The collision probability $p_i[\mathrm{AC}_l]$ is now given by

$$p_i[\mathrm{AC}_l] = 1 - \prod_{x \leq l} (1 - \tau_i[\mathrm{AC}_x])^{n-1} \prod_{x > l} (1 - \tau_i[\mathrm{AC}_x])^n ,$$

where $n$ is the number of stations.

## 5 Numerical Results

In order to evaluate the performance of EDCA, we use the values of system parameters given in Table 1. These values conform to those recommended in the IEEE standard. We show the throughput performances with virtual collision. Fig. 3 shows the saturation throughput of the four ACs with virtual collision for basic access method. In order to validate our model, we have conducted simulations, using the ns-2 simulator. The results indicate that our analysis is closely matched with simulations. This means that our new model for the analysis of the EDCA can show faithfully the performance of the EDCA. Fig. 4 shows the analytical result of the saturation throughput for basic access and RTS/CTS method.The figure indicates that EDCA can provide rate differentiation among

**Table 1.** System Parameters

| | |
|---|---|
| Payload Size | 8000 bits |
| Phy Header (including preamble) | 192 bits |
| Mac Header (including CRC bits) | 272 bits |
| RTS Frame | Phy Header + 160 bits |
| CTS Frame | Phy Header + 112 bits |
| ACK Frame | Phy Header + 112 bits |
| Time Slot | 20e-6 sec |
| SIFS | 10e-6 sec |
| DIFS | 50e-6 sec |
| Data Rate | 1e6 bits/sec |
| $CW_{min}[\mathrm{AC}_0]; CW_{min}[\mathrm{AC}_1]; CW_{min}[\mathrm{AC}_2]; CW_{min}[\mathrm{AC}_3]$ | 32; 32; 16; 8 |
| $CW_{max}[\mathrm{AC}_0]; CW_{max}[\mathrm{AC}_1]; CW_{max}[\mathrm{AC}_2]; CW_{max}[\mathrm{AC}_3]$ | 1024; 1024; 256; 128 |
| $m[\mathrm{AC}_l]$ for all $l$ | 5 |

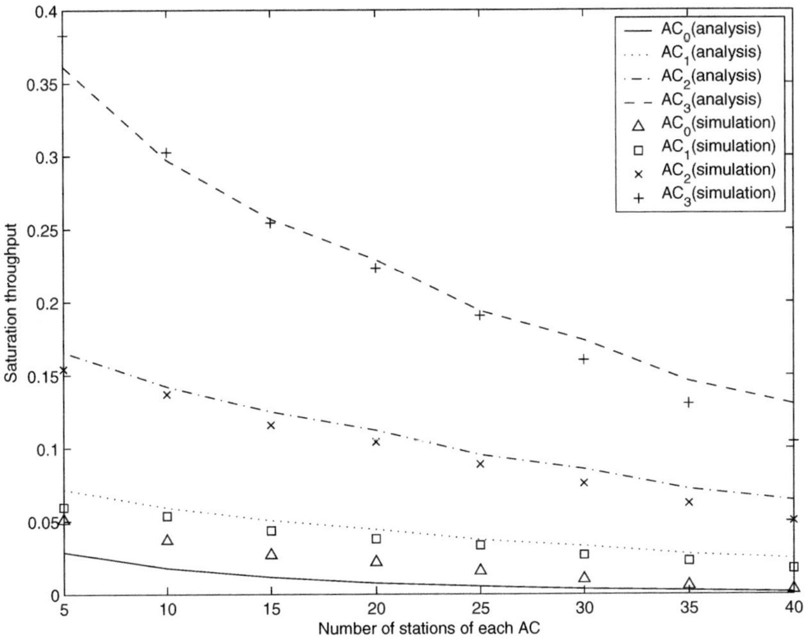


**Fig. 3.** Saturation throughput for basic access method

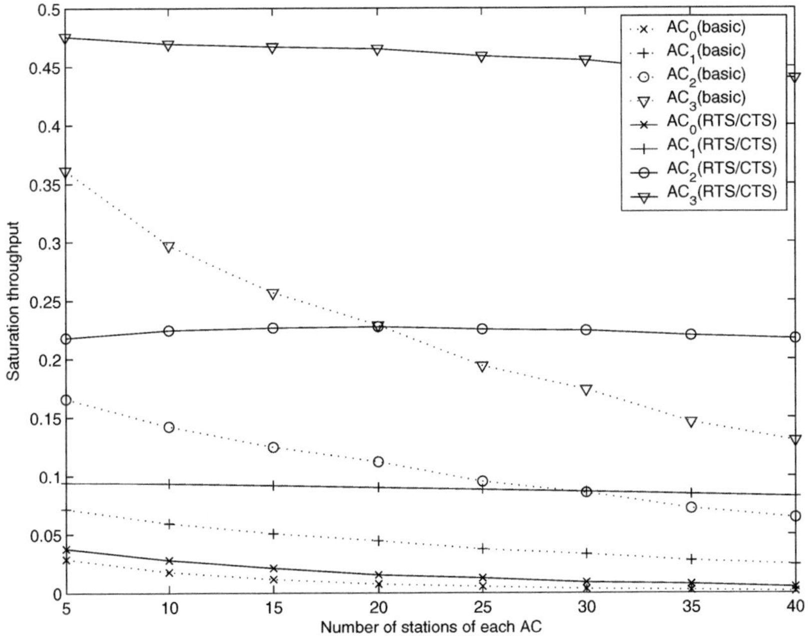


**Fig. 4.** Saturation throughput for basic access and RTS/CTS method

stations of various ACs. As expected, the performance of EDCA with RTS/CTS enabled shows better throughput over all ACs than that of EDCA for basic access method.

## 6 Conclusions

This paper introduced a simple three-dimensional Markov chain model to analyze the saturation throughput of the 802.11e EDCA mechanism. This Markov chain model captures all of the major QoS specific features, namely AIFS, minimum contention window size, maximum contention window size and virtual collision, introduced in EDCA, and hence, leads to results that match with simulations extremely well. This means that our new model for the analysis of the EDCA can show faithfully the performance of the EDCA.

**Acknowledgment.** This work was supported by Dongeui University Grant (2005AA162), and by the MIC(Ministry of Information and Communication), Korea, under the ITRC(Information Technology Research Center) support program supervised by the IITA(Institute of Information Technology Assessment). Prof. K. S. Lee's work was supported by Dongeui University during his sabbatical year(2003-2004).

## References

1. Abu-Tair, M. I., Min, G.: Performance Evaluation of an Enhanced Distributed Channel Access Protocol under Heterogeneous Traffic. Proc. IPDPS (2006)
2. Bianch, G.: Performance Analysis of the IEEE 802.11 Distributed Coordination Function. IEEE J. Select. Areas Commun. **18** (2000) 535–547
3. IEEE 802.11 Standard Part II: Wireless LAN Medium Access Control(MAC) and Physical Layer(PHY) Specifications. (1999)
4. IEEE 802.11e Standard Part II: Wireless LAN Medium Access Control(MAC) and Physical Layer(PHY) Specifications. Amendment 8: Medium Access Control (MAC) Quality of Service Enhancements. (2005)
5. Kong, Z., Tsang, D. H. K., Bensaou, B., Gao, D.: Performance Analysis of IEEE 802.11e Contention-based Channel Access. IEEE J. Select. Areas Commun. **22(10)** (2004) 2095–2106
6. Mangold, S., Hiertz, G., Walke, B.: IEEE 802.11e Wireless LAN-Resource Sharing with Contention based Medium Access. Proc. IEEE PIMRC 2003 Beijing China (2003) 2019–2026
7. Robinson, J. W., Randhawa, T. S.: Saturation throughput Analysis of IEEE 802.11e Enhanced Distributed Coordination Function. IEEE J. Select. Areas Commun. **22(5)** (2004) 917–928
8. Tantra, J. W., Foh, C. H., Mnaouer, A. B.: Throughput and Delay Analysis of the IEEE 802.11e EDCA Saturation. Proc. ICC 2005 (May 2005) Seoul Korea

9. Tao, Z., Panwar, S.: Throughput and Delay Analysis for the IEEE 802.11e Enhanced Distributed Channel Access. IEEE Transactions on Communications **54(4)** (2006) 596–603
10. Xiao, Y.: Enhanced DCF of IEEE 802.11e to Support QoS. Proc. IEEE WCNC 2003 Mar. LA USA (2003)1291–1296
11. Xiao, Y.: Performance Analysis of IEEE 802.11e EDCF under Saturation Condition. Proc. ICC 2004,Jun., Paris France (2004)

# A Wavelet Neural Network Optimal Control Model for Traffic-Flow Prediction in Intelligent Transport Systems

Da-rong Huang[1] and Xing-rong Bai[2]

[1] Institute of Information and Computational Science, Chongqing Jiaotong University, Chongqing 400074, China
hcx1978@163.com
[2] School of Civil Engineering & Architecture, Chongqing Jiaotong University Chongqing 400074, China
xingrongbai@126.com

**Abstract.** Based on wavelet transform and neural network theory, a traffic-flow prediction model, which was used in optimal control of Intelligent Traffic system, is constructed. First of all, we have extracted the scale coefficient and wavelet coefficient from the online measured raw data of traffic flow via wavelet transform; Secondly, an Artificial Neural Network model of Traffic-flow Prediction was constructed and trained using the coefficient sequences as inputs and raw data as outputs; Simultaneous, we have designed the running principium of the optimal control system of traffic-flow Forecasting model, the network topological structure and the data transmitted model; Finally, a simulated example has shown that the technique is effectively and exactly. The theoretical results indicated that the wavelet neural network prediction model and algorithms have a broad prospect for practical application.

## 1 Introduction

In recent years, with the booming development of the Intelligent Traffic System (ITS), the intelligent traffic control and the inducement system have become a pop-field in ITS. To the implement of these systems, the real-time accurate forecasting traffic flow is very vital. At the same time, it is also the necessary step of the optimal control of the urban traffic. The accuracy of the forecasting traffic flow would directly influence the effect of the traffic control and inducement.

However, because the traffic system is a time-variable system in consists of human and machines, so it has high uncertainty. And that, the uncertainty comes from the artificial factor (for example, traffic accidents, paroxysmal incidents and the mental state of the drivers, etc.) as well as the natural factor (for example, season change, etc.). Owing to these uncertain disturbances, the traffic flow forecasting, especially the short-time traffic flow forecasting is very difficult. Therefore, the problem how to appropriately forecast the traffic flow became a very vital step in ITS. To solve the problem, some scholars have analyzed and researched it from different angle. Many scholars [1-4] had promoted to use Kalman filter, fuzzy theory and chaos theory to study the prediction of the traffic flow: The filter structure was described using the state

D.-S. Huang, L. Heutte, and M. Loog (Eds.): ICIC 2007, LNAI 4682, pp. 1233–1244, 2007.

equations and the measured equations, i.e. the curve of datum function was established by the most approximate curve from some day in recent weeks to now. They had gotten the local optimal control model of forecasting traffic flow. Whereas, it is a pity that these models were constructed based the linear theory. In other words, the local optimal control project of practical traffic flow is only an approximate conclusion. To overcome these problems, in [5-7], some scholars had constructed some global optimal control model for Forecasting Traffic-flow, and that had gotten some excellent results. But these fitting models couldn't perfectly and essentially reflect the complexity and intrinsic structure of the dynamic data forecasting, as these causalities were constructed by analyzing the recursive model and the time-series model. Thus, some information was lost and the forecasting value isn't quite effective, so the accuracy and real-time of forecasting Traffic-flow need to improve. So, according to historical traffic flow detecting at the nodes of road-network, the authors have analyzed the structures and the forecasting methods of the real-time dynamic traffic system combining Wavelet transform and Artificial Neural Network. And we have constructed a Wavelet Neural Network (WNN) model for forecasting Traffic-flow.

The layout of the rest of the paper is as follow: The traffic flow forecasting model of the real-time dynamic traffic flow systems was constructed in Section 2.Section 3 describes the corresponding algorithms and designs the network topological control structure of traffic flow forecasting. Section 4 discusses the simulative results by experiments. Finally, Section 5 concludes this paper with final inferences and directions for future work.

## 2 Construct and Analysis of the Traffic-Flow Forecasting Model

Because the traffic-flow signals are Non-Stationary Stochastic Processes and that not all information are same importance to the real control of the traffic flow system. Thus, we need to extract the useful information from all detecting traffic-flow data for information coding. So the optimal control of traffic systems is implemented. Simultaneously, because the wavelet transforms developing since 80's of last century have ability analyzing irregular and nonsymmetrical signals, we decomposed the raw data using wavelet transform and deleted the disturbance. Then, the Neural Network forecasting model, which were used to optimal control of traffic flow, were constructed according to the useful data.

### 2.1 Signal Processing Model Based Wavelet Transform

The wavelet function is clearly defined: supposing $\psi(x)$ is a square integral function, if its Fourier satisfies the condition:

$$C_{\psi} = \int_{R^*} \frac{|\Psi(\omega)|^2}{|\omega|} \mathrm{d}\omega < \infty . \tag{1}$$

Where $\psi(t)$ is called a base wavelet, (1) is the tolerance condition. Carrying the expansion and contraction and the translation based on the base wavelet, supposes its expansion and contraction factor is $a$ , the translation factor is $b$ . Then has:

$$\psi_{(a,b)}(t) = \frac{1}{\sqrt{|a|}} \psi\left(\frac{t-b}{a}\right) (a > 0, b \in R) \ . \tag{2}$$

$\psi_{(a,b)}(t)$ is called the wavelet base function relies on parameter scale factor $a$ and the translation factor $b$.

So, for an arbitrary signal $f(t)$, its continuous wavelet transform is defined as following.

$$W_f(a,b) = \int_R f(t) \overline{\psi}_{(a,b)}(t) \mathrm{d}t = \frac{1}{\sqrt{|a|}} \int_R f(t) \overline{\psi}\left(\frac{t-b}{a}\right) \mathrm{d}t \ . \tag{3}$$

In practical, because most equations be treated by computer are discrete, put $a = 2^{-j}$, $b = 2^{-j}k$, $W_f(a,b)$ became a discrete binary wavelet function as following.

$$\psi_{j,k} = 2^{j/2} \psi(2^j t - k) \ , W_f(j,k) = 2^{j/2} \int_{-\infty}^{\infty} f(t) \tilde{\Psi}(2^j t - k) \, dx \ .$$

Where $j, k \in Z$.

So, an arbitrary traffic flow signal $f(t)$ can be decomposed as following wavelet series.

$$f(t) = \sum_{j,k=-\infty}^{\infty} W_f(j,k) \Psi_{j,k}(t) \ . \tag{4}$$

In this paper we have selected the orthogonal wavelet transform to decomposing the signal, the corresponding algorithms is Mallat algorithm as following.

$$\begin{cases} C_{j+1} = H * C_j \\ D_{j+1} = G * D_j \end{cases} . \tag{5}$$

Where $H$ and $G$ are respectively high-filter and low-filter.

If raw traffic flow $f(t)$ is described as $C_0$, thus Fig. 1 shows wavelet decomposing process.

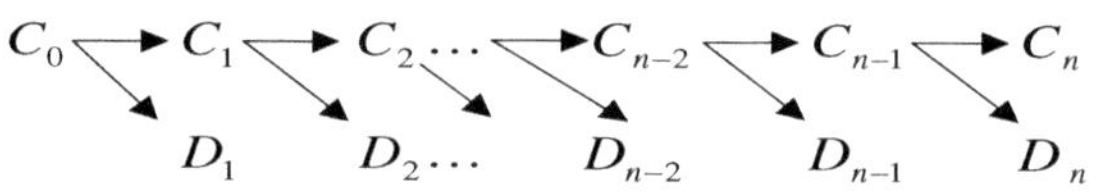


**Fig. 1.** The wavelet decomposing process of traffic signal

After the wavelet decomposing, number of the detailed and approximate signals decreased 50%, but the decrease in the number is bad for forecasting traffic flow. To overcome the disadvantage, we have reconfigured the decomposing signal to the original signal and insert the corresponding redundancy information into the reconstruction signal using the reconstruction algorithm as following.

$$C_j = \overline{H} * C_{j+1} + \overline{G} * D_{j+1} \; ( j = n-1, n-2, \cdots, 1, 0) \; . \qquad (6)$$

Here $\overline{H}$ and $\overline{G}$ is respectively the dual operator of $H$ and $G$. Fig. 2 shows the wavelet reconstruction process.

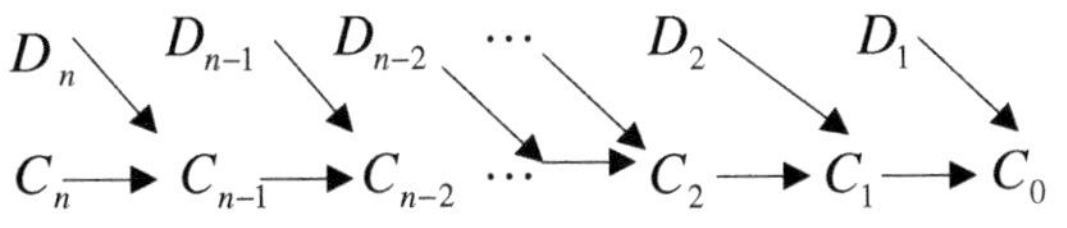


**Fig. 2.** The wavelet reconstruction process of traffic signal

In reconstruction process the wavelet reconstruction number of signals increased, so that original signal $C_0$ satisfied

$$C_0 = C_n + \sum_{i=0}^{n} D_n \; . \qquad (7)$$

Thus, the disturbance signals of real traffic flow were deleted using the wavelet decomposing and the wavelet reconstruction, and the useful traffic signals which would be used to construct the forecasting traffic flow model had be gotten.

### 2.2 Neural Network Forecasting Model and Algorithm of Traffic-Flow

To implement the optimal control of the traffic flow, first the new signal flow $\{(t, Y_t), t = 1,2,\cdots n\}$ were extracted from the historical traffic flow $\{f(t), t = 1,2,\cdots n\}$ by the wavelet transform. Because these disturbances in the historical data were deleted, so these new traffic flows $Y_t$ may be used to construct the effective forecasting model.

On the other hand, we have only gotten some data points $\{(t, Y_t), t = 1,2,\cdots n\}$ through processing Traffic-flow signals, and that the structure of the functions was unknown, as a result of that the concrete relationship between Traffic-flow data couldn't be constructed. And yet, owing to that Artificial Neural Network is a distributed computing model: the nonlinear mapping relation between inputs and outputs could be extracted and approximated through learning function of network oneself, but the knowledge, which is about dynamic characters, parameters and structures of being modeled object, need not to know. That indicated that Artificial Neural Network model have a broad practical applied prospect in function approximation theory, and the intrinsic ideas of forecasting traffic-flow just is function approximation. So using Artificial Neural Network theory we may construct a Traffic-flow forecasting model to implement optimal control in ITS. In this paper we would construct forecasting model of traffic flow using BP Neural Network. Its fundamental structure of network model is shown as Fig. 3.

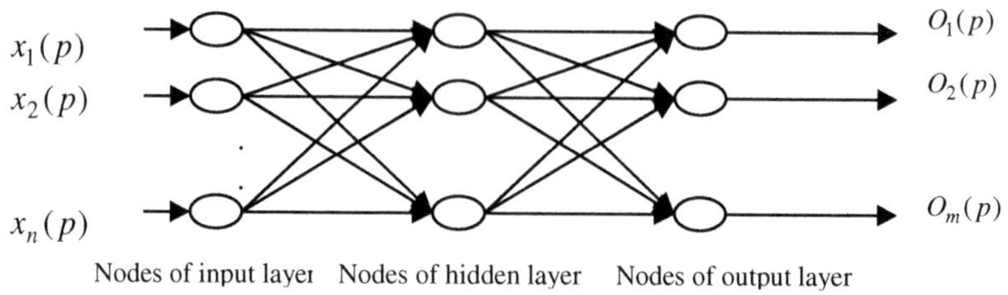


**Fig. 3.** Fundamental structure of BP Neural Network model

In Neural Network model every neuron was shown by a node, Network was consisted of input layer, hidden layer and output layer. Where the hidden layer may not only be one layer shown as Fig. 3, but also be multilayer. The nodes through proceeding layer to next layer were connected by weights.

In input layer of Neural Network model, the numbers of nodes were $n$ and the input variables $x_i(p)$ $(i=1,2,\cdots;n)$ was $\{(t,Y_t),t=1,2,\cdots n\}$ which were processed by wavelet transform; the numbers of nodes of hidden layers were $l$ and the connection weights between input layer and hidden layer was $w_{ji}$ $(i=1,2,\cdots,n\ ;j=1,2,\cdots,l)$, and we have given a threshold $\theta_j$ $(j=1,2,\cdots,l)$ for every hidden node; at the same time, the inputs variables of hidden nodes were $y_j(p)$ and the connection weights between hidden layer and output layer was $v_{kj}$ $(j=1,2,\cdots,l\ ;k=1,2,\cdots,m)$, the numbers of nodes of output layer were $m$ and the outputs of output layer were $O_k(p)$ with threshold $\varphi_k$ $(k=1,2,\cdots,m)$. To implement the arbitrary nonlinear mapping relationship between outputs and inputs , Sigmoid differentiable function were usually regarded as the transfer function of hidden layer of network structure, i.e.

$$f(x)=\frac{1}{1+e^{-Qx}}. \tag{8}$$

Where $Q$ denoted the nonlinear parameter of neurons and were called Gain or Adjust parameter. $Q$ is larger, $S$ curve is sharper; otherwise $S$ curve is smoother. In generally, let $Q=1$. The transfer function reflected the saturation of neurons and their value is from 0 to 1.

To designing an ANN forecasting model of Traffic-flow, putting $x_i$ was regarded as the output for a node $i$, and that the network input for a node $j$ was defined as following.

$$net_j=\sum_i w_{ji}x_i-\theta_j\ . \tag{9}$$

Thus, the output of node $i$ is determined by the transfer function and the net input of the node. It is given by: $y_j=f(net_j)$

And the error function was defined as following.

$$E = \frac{1}{2N} \sum_{p=1}^{N} (O_p - T_p)^2 . \tag{10}$$

Where $O_P$ was real-time output of network, $T_p$ was anticipant output of network.

Let $E = \frac{1}{2}(O(p) - T(p))^2$ , $\delta_j = \frac{\partial E}{\partial net_j}$ and $y_j = f(net_j)$ , so

$$\frac{\partial E}{\partial W_{ji}} = \frac{\partial E}{\partial net_j} . \frac{\partial net_j}{\partial W_{ji}} = \frac{\partial E}{\partial net_j} .x_i = \delta_j x_i . \tag{11}$$

Next we would discuss the problem:

(1) When $j$ is output node, $y_j = O$ , then

$$\delta_j = \frac{\partial E}{\partial net_j} = \frac{\partial E}{\partial O} . \frac{\partial O}{\partial net_j} = -(T - O) f'(net_j) . \tag{12}$$

(2) When $j$ is hidden node, then

$$\delta_j = \frac{\partial E}{\partial net_j} = \frac{\partial E}{\partial y_j} . \frac{\partial y_j}{\partial net_j} = \frac{\partial E}{\partial y_j} .f'(net_j) . \tag{13}$$

Where $y_j$ is input to next layer $(j+1)$ , $\frac{\partial E}{\partial y_j}$ was computed by $(j+1)$ layer.

Thus, for the $m$th node of $(j+1)$ layer have that.

$$\frac{\partial E}{\partial y_j} = \sum_m \frac{\partial E}{\partial net_j} . \frac{\partial net_j}{\partial y_j} = \sum_m \frac{\partial E}{\partial net_m} . \frac{\partial}{\partial y_j} \sum_j W_{jm} O_j = \sum_m \frac{\partial E}{\partial net_m} . \sum_j W_{jm} = \sum_m \delta_m W_{jm} \tag{14}$$

According to (13) and (14), we have gotten.

$$\begin{cases} \delta_j = f'(net_j) \sum\limits_m \delta_m W_{jm} \\ \dfrac{\partial E}{\partial W_{jm}} = \delta_j y_j \end{cases} . \tag{15}$$

At the same time the different function satisfied:

$$\begin{aligned} f'(net_j) &= \frac{dE}{dnet_j} [\frac{1}{1+e^{-Qnet_j}}] = Q(1+e^{-Qnet_j})^{-2} .e^{-Qnet_j} \\ &= Q.f(net_j).(1+f(net_j)) = Q.y_j.(1+y_j) \end{aligned} . \tag{16}$$

According to the pre-derivation, the training step of BP neural network was concluded as following.

(1) Selecting the initial weight $\omega_{ji}$ and the initial threshold $\theta_j$ ;

(2) Supplying randomly a set of input and expected samples for network;

(3) Computing $y_j$ , $net_j$ and $O$ for every node $j$ of each layer (positive direction diffusion process); and computing $\delta_j$ by (11) and (14) (negative direction process);

(4) Modifying the weights value by:

$$W_{ji}(t+1) = W_{ji}(t) - \eta \frac{\partial E}{\partial W_{ji}}, \eta > 0 .$$

Where $\frac{\partial E}{\partial W_{ji}} = \sum_{p=1}^{N} \frac{\partial E}{\partial W_{ji}}$ , $t$ was iterative time, $\eta$ was Step Length.

(5) Repeating above step through to the performance of network was satisfied. The implementation algorithm is shown in the flowing chart as Fig. 4.

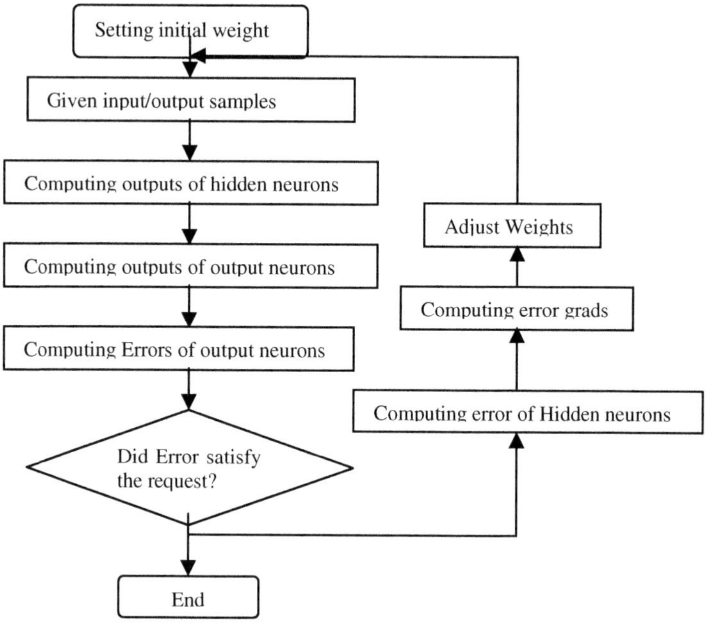


**Fig. 4.** Learning algorithms of BP Network model

So, the parameter structure of ANN model had been gotten, and then the optimal control of Real-time Traffic-flow can be implemented according to the detecting data of Road network. Simultaneous, we had selected 5 indexes that would be used to evaluate model's performance on account of the accuracy of the forecasting traffic flow would directly influence the effect of the real-time traffic optimal control, i.e.

Average Absolute Error:

$$MAE = \frac{1}{n}\sum_{t=1}^{n}\left|Y_t - \hat{Y}_t\right| . \tag{17}$$

Square Error:

$$MSE = \frac{1}{n}\sqrt{\sum_{t=1}^{n}(Y_t - \hat{Y}_t)^2} . \tag{18}$$

Average Absolute Percentage Error:

$$MAPE = \frac{1}{n}\sum_{t=1}^{n}\left|\frac{Y_t - \hat{Y}_t}{Y_t}\right| . \tag{19}$$

Square Percentage Error:

$$MSP = \frac{1}{n}\sqrt{\sum_{t=1}^{n}\left(\frac{Y_t - \hat{Y}_t}{Y_t}\right)^2} . \tag{20}$$

Equalization Coefficient:

$$EC = 1 - \frac{\sqrt{\sum_{t=1}^{n}(Y_t - \hat{Y}_t)^2}}{\sqrt{\sum_{t=1}^{n}Y^2{}_t} + \sqrt{\sum_{t=1}^{n}\hat{Y}_t{}^2}} . \tag{21}$$

## 3 Design of Traffic-Flow Forecasting Algorithm

In practical, because the real-time optimal project of traffic flow signals is given continuously, so we would design the running principles of traffic flow optimal control systems as following.

1) Initialization of system. Loading historical data $\{f(t), t = 1,2,\cdots n\}$ of traffic flow, and pre-storage in the computer.
2) Processing the historical data by (1)-(7), and output the new signals $\{(t, Y_t), t = 1,2,\cdots n\}$, and regard them as the input signal of Network model.
3) Separating the experimental data into training samples and testing samples, and establishing ANN forecasting model, going on next step.
4) Computing output error of every neuron when the training data was regarded as the input of every neuron: if the error less than given threshold, stop training; otherwise, repeating the step.
5) Forecasting future traffic flow according to the historical data and outputting the simulation curves: if the evaluating effect is excellent, return directly to step 2. Otherwise, after designing the project of optimal control, return to step 2.
6) Printed the analyzing report of traffic flow optimal control system.

The implementation algorithm is shown in the flowing chart as Fig. 5.

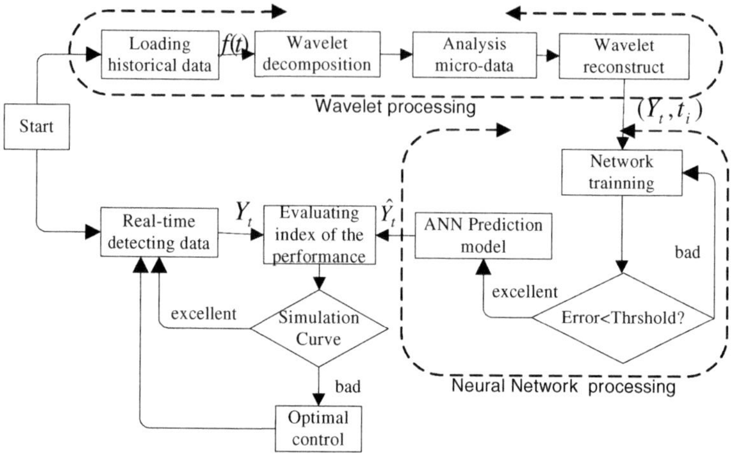


**Fig. 5.** Topological structure of Wavelet Neural Network Prediction Model for Traffic-flow

Fig. 5 shows that the network topological structure was composed of Wavelet processing module, ANN Prediction module, evaluating module, decision module and simulation module. Where the Wavelet processing module was element and the ANN Prediction module was kernel. Thus in the running process of traffic optimal control systems the control flow of transmitted data is shown in the flowing chart as Fig. 6.

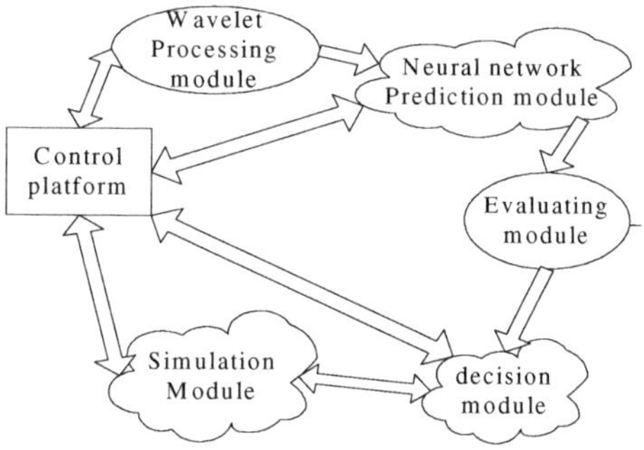


**Fig. 6.** Control flow of data for forecasting traffic flow

Obviously, all data may be returned to the control center and be real-time controlled, so the effective online detecting of optimal traffic control can be carried out.

## 4 Simulation Examples

The paper applied the traffic-flow forecasting optimal control system, which was designed based on wavelet ANN model, to real-time traffic optimal control. On 17 to

20 July 2005, at the highway detecting station we had real-time measured Traffic –flow. At the first day (16 July) the detecting data were gotten by conventional measured method, at another four days (17 to 20) the Traffic-flow data were measured by singular measured method as following: for real-time detecting traffic flows one roadway was closed at these times interval, i.e. 8:00~8:10,9:30~9:40,11:00~11:10, 12:30~12:40,14:00~14:10,15:30~15:40,17:00~17:10,18:30~18:40.

On 16 July the detecting time is from a.m7:00 to 20:025 and the time interval is 10s: the detecting traffic flow data contained 780 data. In our experiment we have separated original data into 4 parts using wavelet transform, i.e. $data=(data_1, data_2, data_3, data_4)$. Thus, we had selected anterior 100 data to process by wavelet transform. The corresponding simulation curves of original data and fitting data are shown as Fig. 7, respectively.

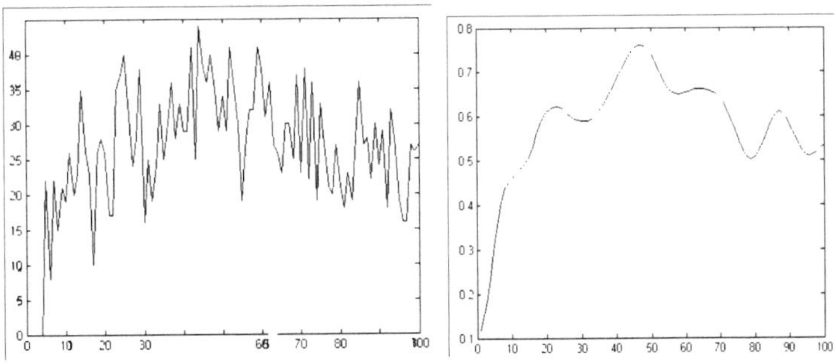

**Fig. 7.** Distribution plot of raw detecting data (1-100) and Distribution plot of new data (1-100)

We were regarded the new data that were processed by wavelet transform as the input of ANN model, then the training simulation curves were shown as Fig. 8.

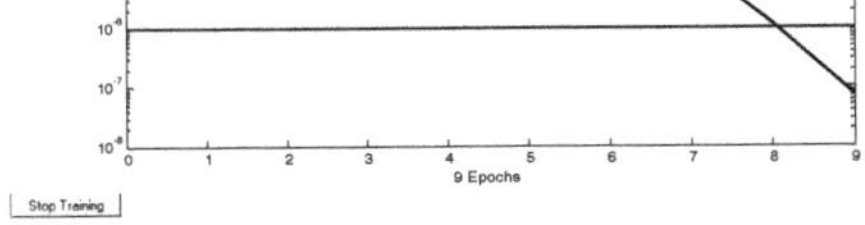


**Fig. 8.** Simulation curve of Network training

Using Network model, the fitting between forecasting values and detecting values of anterior 100 Traffic-flow data was shown as Fig. 9.

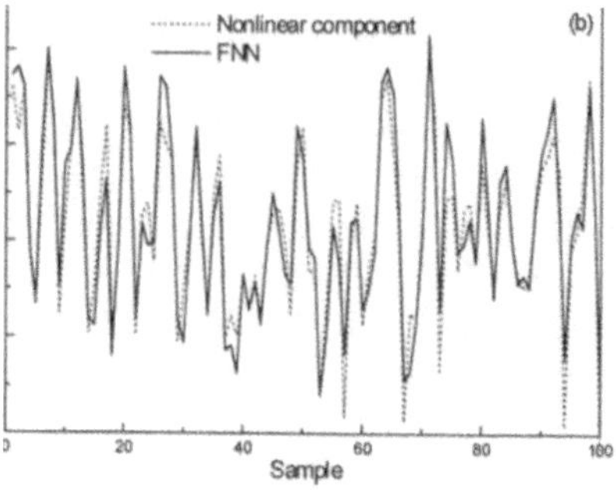


**Fig. 9.** Fitting curve of WANN model( 1-100)

Thus, on 17 July we have forecasted Traffic-flow By WANN model. The fitting curves between forecasting-value and actual detecting-value are as Fig. 10.

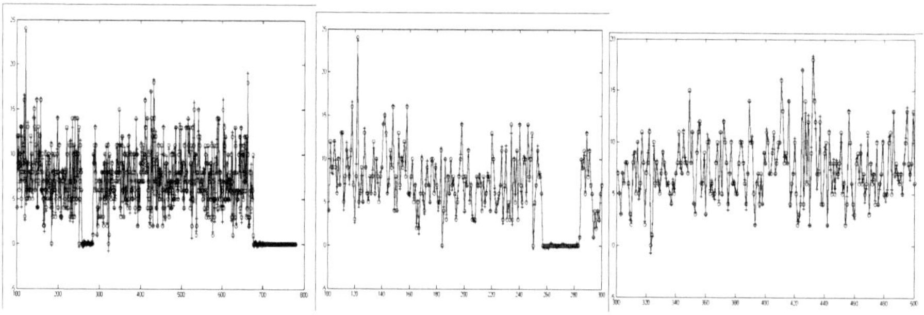

**Fig. 10.** (a) Fitting curve between forecasting and detecting data (101-780) ;(b) Fitting curve between forecasting and detecting data (101-301); (c) Fitting curve between forecasting and detecting data (301-501) ( Notes: + indicates forecasting data, ○ indicates raw data)

Fig. 10 (a) showed the whole effect between forecasting data and detecting data. Correspondingly, Fig. 10 (b) and Fig. 10 (c) describes the part fitting effect. When the detecting data is anterior 100,500 or 780, these evaluating indexes are respectively: $MAE_{100}$=3.0287, $MSE_{100}$=42.7248, $MAPE_{100}$=32.423, $MSP_{100}$=2.4325, $EC_{100}$=0.8565; $MAE_{500}$=4.8036, $MSE_{500}$=97.2343, $MAPE_{500}$=34.505, $MSP_{500}$=2.4354, $EC_{500}$=0.8658; $MAE_{780}$=6.4033, $MSE_{780}$=162.8236, $MAPE_{780}$=48.41, $MSP_{780}$=2.2297, $EC_{780}$=0.9243.

Also we have forecasted and concluded for Traffic-flow data on 18, 19 and 20 July by WANN model. All analyses indicated that the forecasting model can be used to implement different optimal traffic control, thus we had also carried out effective scheme of traffic resource.

## 5 Conclusions

In this paper a WANN Forecasting Model of traffic-flow time-series had been constructed based on wavelet transform and Neural Network. And the simulation curve showed that the fitting degree between the forecasting data and the detecting data is very high using WANN Forecasting model, so we can carry out the dynamic traffic optimal control.

And yet, owing to that Artificial Neural Network is a distributed computing model: the nonlinear mapping relation between inputs and outputs could be extracted and approximated through learning function of network oneself, but the knowledge, which is about dynamic characters, parameters and structures of being modeled object, need not to know. Thus, the number of data samples is smaller. At the same time, because the future traffic flow were forecasting according to the foregone traffic flow and the current traffic flow, And those algorithms have definite convergence. Therefore, their accuracy and real-time is very excellent, so the real-time dynamic traffic flow can be forecasted effectively.

**Acknowledgments.** The authors acknowledge the research was supported by Natural Science Foundation Project of CQ CSTC (No. 2006BB2422) and CQ CMEC (No. KJ060414).

## References

1. Yi, J.: Dynamic Prediction of Traffic Flow and Congestion at Freeway Construction Zones. Journal of Construction Education. Spring 2002,Vol. 7, No.1 (2002) 45-57
2. Heribert, K., Claire, C.: Traffic Situation Prediction Applying Pattern Matching and Fuzzy Classification. European Symposium on Intelligent Techniques, June 3-4, Orthodox Academy of Crete, Greece (1999)
3. Jianming, H., Jingyan, S., Yi, Zh.: Modeling and Analysis for Self-organization of Urban Traffic Flow. The 8th International IEEE Conference on Intelligent Transportation Systems, Vienna, Austria, September 13-16 (2005) 490-495
4. Jianming, H., Chunguang, Z., Jingyan, S.: An Applicable Short-term Traffic Flow Forecasting Method Based on Chaotic Theory. The 2003 IEEE International Conference on Intelligent Transportation Systems, Shanghai (2003)
5. Eul-Bun, L., Changmo, K., John, T.: Application of Macro- and Microscopic Simulations for the Traffic Planning of Urban Highway Reconstructions. Transportation Research 84th annual meeting, January 9-13 (2005) 1-23
6. Da-rong, H., Dacheng, W., Jun, S.: Forecasting Model of Traffic Flow Based Wavelet Transform and Fractal Theory. Proceeding of International Conference on Sensing, computing and Automation `06/5, chongqing, china, an added volume: Dynamics of Continuous Discrete and Impulsive Systems, Series B: Application and Algorithms (2006) 494-498
7. Da-rong, H., Da-cheng, W., Jun, S.: Forecasting Model of Traffic Flow Based ARMA and Wavelet Transform. Proceeding of Dynamics and Impulsive Systems `06/7, Qingdao, China, An supplementary volume: Dynamics of Continuous, Discrete and Impulsive Systems, series A: Mathematic analysis (2006) 860-869

# Conditional Density Estimation with HMM Based Support Vector Machines

Fasheng Hu[1], Zhenqiu Liu[2,⋆], Chunxin Jia[3], and Dechang Chen[4]

[1] School of Mathematics and System Science, Shandong University, Jinan, Shandong Province, P.R. China
[2] Division of Biostatistics, Greenebaum Cancer Center, University of Maryland Medicine, Baltimore, MD 21201, USA
zliu@umm.edu
[3] Department of Finance, Guanghua School of Management, Peking University, Beijing, China
[4] Division of Epidemiology and Biostatistics Department of Preventive Medicine and Biometrics, Uniformed Services University of the Health Sciences, Bethesda, MD 20814, USA

**Abstract.** Conditional density estimation is very important in financial engineer, risk management, and other engineering computing problem. However, most regression models have a latent assumption that the probability density is a Gaussian distribution, which is not necessarily true in many real life applications. In this paper, we give a framework to estimate or predict the conditional density mixture dynamically. Through combining the Input-Output HMM with SVM regression together and building a SVM model in each state of the HMM, we can estimate a conditional density mixture instead of a single gaussian. With each SVM in each node, this model can be applied for not only regression but classifications as well. We applied this model to denoise the ECG data. The proposed method has the potential to apply to other time series such as stock market return predictions.

## 1 Introduction

A general regression or pattern recognition problem is posed as follows:

- Give $n$ i.i.d. sample: $(\mathbf{x}_0, y_0), \ldots, (\mathbf{x}_n, y_n)$
  where $\mathbf{x}_i$ for $i = 0, \ldots, n$ is a column vector of length $d$ and $y_i$ is the output.
- Find the decision function, $f(\mathbf{x})$ such that $y = f(\mathbf{x})$, where $y$ is the output for data point $\mathbf{x}$.

The primal problem for the SVM regression is as follows:

$$\textbf{Min} \ \ \lambda||\mathbf{w}||^2 + \sum_{i=0}^{n} \xi_i^2 \tag{1}$$

⋆ Corresponding author.

D.-S. Huang, L. Heutte, and M. Loog (Eds.): ICIC 2007, LNAI 4682, pp. 1245–1254, 2007.

$$\textbf{Subject to: } y_i - \mathbf{w}'\Phi(\mathbf{x}_i) = \xi_i,\ i = 0, \ldots, n$$

The dual of SVM regression in vector form can be shown as: **Max** $W(\alpha) = \mathbf{y}'\alpha - \frac{1}{4\lambda}\alpha'\mathbf{K}\alpha - \frac{1}{4}\alpha'\alpha$, where $\mathbf{K}$ denotes the kernel matrix $\mathbf{K}_{ij} = K(\mathbf{x}_i, \mathbf{x}_j)$, $\lambda$ is the regular parameter, and $\alpha$ are the dual variables. We have the solution for the dual is $\alpha = 2\lambda(\mathbf{K} + \lambda\mathbf{I})^{-1}\mathbf{k}$ and the corresponding regression function

$$f(\mathbf{x}) = \mathbf{y}'(\mathbf{K} + \lambda\mathbf{I})^{-1}\mathbf{k}, \tag{2}$$

where $\mathbf{k}$ is the vector with entries $k_i = K(\mathbf{x}_i, \mathbf{x})$ for $i = 0, \ldots, n$. SVM regression problem can also be explained from Gaussian process or kernel ridge regression ( Shawe-Taylor & Cristianini 2000, Rasmussen 1996). When $\Phi(\mathbf{x}_i) = \mathbf{x}_i$ and $K(\mathbf{x}_i, \mathbf{x}_j) = \mathbf{x}'\mathbf{x}$, we have the linear support vector machine. Equation (2) can also be used for classifications, even if its theoretical explanations are not completely clear. People have claimed success in solving classification problems recently (Van Gestel etc 2004, Mangasarian & Wild 2006). Classification using equation (2) has been called either as Least Squared SVM (LS-SVM) or Lagrangian SVM.

Support vector machines (SVM) have been very successful in many regression and classification applications over the past several years. Recently applications can be seen in financial engineering, web mining, bioinformatics, and speech recognition. However in SVM regression, we assume that the conditional probability density of the output is Gaussian. In SVM classification, even we do not assume that the conditional probability is Gaussian, some researchers still do the classification used the Gaussian like assumption together with a sigmoid function. In this paper, however, we will pay more attention to the regression and conditional density mixture problems.

HMM models has been used extensively in sequential data analysis especially in speech recognition ( Sanches 2000). The parameters of HMM model are estimated using the Maximum likelihood methods and EM algorithm (Y, Bengio 1999). There are different version of HMM such as HMM with Gaussian or mixture of the Gaussian, tree based HMM (L. Rabiner 1989). We will concentrate on Input-Output HMM model. Input-Output HMMs are simply HMMS for which the emission and transition distributions are conditional on some other sequences, called the input sequences. There are several drawbacks for the Input-Output HMM. EM is a kind of unsupervised learning scheme. It is easily stick to the local minimum and the maximum likelihood estimates does not necessarily translate to the better performance in regression and classification. The HMM models also have a two-Markov assumption which is not true in general.

SVM/HMM hybrids attempt to address some of the modelling weaknesses we mentioned above. There have been several papers discussing the hybrid training of HMM/SVM from speech recognition point of view (Ganapathiraju et al. 2000, 2001, Huang et al. 2006). However, SVM and HMM are trained separately in their approach and HMM is seen more as a preprocessor of SVM. Their HMM model also have a very special structure for speech recognition. Our approach is applied to a general Input-Output HMM structure. We also integrate the HMM and SVM together and train the both with respect to a single error function. In this way, SVM can be used to capture temporally local, but possibly complex and nonlinear dependencies, while the HMM is used to handle the temporal structures. Our approach attempt to combine the strengths of supervised learning with those of unsupervised learning: it builds on the advantage of supervised learning that allows for performance evaluation, while providing the flexibility of unsupervised learning that has the advantage of discovering and interpreting hidden states. Based on the model we proposed, we can predict not only the point value but also the conditional density mixture which is very important in many applications.

This paper is organized as follows. In section 2, we will introduce the HMM based SVM algorithm. We will first introduce the likelihood function and the EM algorithm used in HMM based SVM in great detail, and then we will show how to predict the conditional density and point value. The computational results are given in Section 3. Conclusions and remarks are discussed in Section 4.

## 2 HMM and HMM Based SVM Algorithm

A hidden Markov model (HMM) is model where we generate a sequence of outputs in addition to the Markov sequence. The output sequence is also affected by input sequences $\mathbf{x}$.

$$\begin{array}{ccccccccc} \mathbf{x}_0 & & \mathbf{x}_1 & & \mathbf{x}_2 & & \mathbf{x}_3 & & \\ \downarrow & & \downarrow & & \downarrow & & \downarrow & & \\ s_0 & \rightarrow & s_1 & \rightarrow & s_2 & \rightarrow & s_3 & \rightarrow & \ldots \\ \downarrow & & \downarrow & & \downarrow & & \downarrow & & \\ y_0 & & y_1 & & y_2 & & y_3 & & \end{array}$$

A HMM is defined by the following elements: the number of states $m$, the initial stat distribution $P_0(s_0)$, the state transition model $P_1(s_{t+1}|s_t)$, and the output model $P_o(y_t|s_t, \mathbf{x}_t)$.

This is a latent variable model in the sense that we will only observe the output $\{y_0, y_1, \ldots, y_n\}$; the state sequence remain "hidden". There are several problems we have to solve: 1. how do we evaluate the probability that our model generated the observation sequence $\{y_0, y_1, \ldots, y_n\}$? 2. how do we uncover the most likely hidden state sequence corresponding to the observations? and 3. how do we adapt the parameters of the HMM to better account for the observations.

### 2.1 Probability of Observed Data

Given a HMM model, this is the first problem we have to solve. In principle computing the probability of the observed sequence involve summing over exponentially many possible hidden state sequences

$$P(y_0, \ldots, y_n) = \sum_{s_0, \ldots, s_n} P_0(s_0) P_o(y_0|s_0, \mathbf{x}_0) \ldots P_1(s_n|s_{n-1}) P_o(y_n|s_n, \mathbf{x}_n),$$

where $P_0(s_0)$ is the initial probability. We can, however, exploit the structure of the model to evaluate the probability much more efficiently. We will introduce two simple algorithms here. First we define a forward probabilities $\alpha_t(i)$:

$$\alpha_t(i) = P(y_0, \ldots, y_t, s_t = i), \tag{3}$$

where $\alpha_t(i)$ is the probability of the partial observation sequence, $y_0, \ldots, y_t$, (until time t) and state $s_i$ at time t. From this forward probability and Bayesian therm, we may get: $\frac{\alpha_t(i)}{\sum_j \alpha_t(j)} = P(s_t = i|y_o, \ldots, y_t)$. On the other hand, the backward probabilities $\beta_t(i)$ are defined as follows:

$$\beta_t(i) = P(y_{t+1}, \ldots, y_n|s_t = i), \tag{4}$$

which is the probability of the partial observation sequence from $t+1$ to the end, given state $s_i$ at time t. Based solely on $\alpha_t(i)$, we can have the following Algorithm I.

**Algorithm I: Forward Algorithm**

- Initialization: $\alpha_0(i) = P_0(s_0 = i)P_o(y_0|s_0 = i, \mathbf{x}_0)$
- Induction: $\alpha_t(i) = \left[\sum_j \alpha_{t-1}(j)P_1(s_t = i|s_{t-1} = j)\right] P_o(y_t|s_t = i, \mathbf{x}_t)$
- Termination: $P(y_0, \ldots, y_n) = \sum_{i=0}^{m} \alpha_n(i)$, where $m$ is the number of states.

Based on both $\alpha_t(i)$ and $\beta_t(i)$, we may get the following forward-backward algorithm:

**Algorithm II: Forward-Backward Algorithm**

- Initialization: $\alpha_0(i) = P_0(s_0 = i)P_o(y_0|s_0 = i, \mathbf{x}_0)$ and $\beta_n(i) = 1$
- Forward and backward recursion:

$$\begin{aligned}\alpha_t(i) &= P(y_0, \ldots, y_t, s_t = i) \\ &= [\sum_j \alpha_{t-1}(j)P_1(s_{t=i}|s_{t-1} = j)]P_o(y_t|s_t = i, \mathbf{x}_t)\end{aligned}$$

$$\begin{aligned}\beta_t(i) &= P(y_{t+1}, \ldots, y_n|s_t = i, \mathbf{x}_t) \\ &= \sum_j P_1(s_{t+1} = j|s_t = i)P_o(y_{t+1}|s_{t+1} = j, \mathbf{x}_{t+1})\beta_{t+1}(j)\end{aligned}$$

- Calculating the probability: The probability of generating the observations and going through state i at time t is $P(y_0, \ldots, y_n, s_t = i) = \alpha_t(i)\beta_t(i)$. Summing over the possible states at time t gives $P(y_0, \ldots, y_n) = \sum_j \alpha_t(j)\beta_t(j)$, for any $t = 0, \ldots, n$.

In above algorithms , $y_t$ depends on $\mathbf{x}_t$ latently through $P_o(y_t|s_t, \mathbf{x}_t)$. We do not write it explicitly in $\alpha$ and $\beta$ for notation simplicity.

### 2.2 The Optimal State Sequence Associated with the Observation Sequence

This is the second problem we need to solve. There are several ways to solve this problem, we have implemented both ways in our program. The first based on the criteria to choose the states $s_t$ which are individually most likely. This criteria maximizes the expected number of correct individual states. More specifically, we compute the posterior probability that the HMM was in a particular state $i$ at time $t$.

$$P(s_t = i|y_0, \ldots, y_n) = \frac{\alpha_t(i)\beta_t(i)}{\sum_j \alpha_t(j)\beta_t(j)} = \gamma_t(i).$$

Using $\gamma_t(i)$, we can solve for the most likely state $s_t$ at time t through maximizing $\gamma_t(i)$ with respect to $i$, as $s_t = \arg\max_i[\gamma_t(i)]$ for each $0 \leq t \leq n$.

Another way is to find the single best state sequence based on dynamic programming methods, called *Viterbi algorithm.* We wish to find the sequence of hidden states $s_0, \ldots, s_n$ that has the highest probability of generating the sequence of observations $y_0, \ldots, y_n$: $s_0^*, \ldots, s_n^* = \arg\max_{s_0,\ldots,s_n} P(y_0, \ldots, y_n, s_0, \ldots, s_n)$. To do so, let 's define the probability of generating the first t observations along the best path that ends with $s_t = i$:

$$\delta_t(i) = \max_{s_0,\ldots,s_n} P(y_0, \ldots, y_t, s_0, \ldots, s_{t-1}, s_t = i)$$

The Viterbi algorithm (dynamic programming) is shown as follows:

**Viterbi Algorithm:**

- Initialization: $\delta_0 = P_0(s_0 = i)P(y_0|s_0 = i, \mathbf{x}_0)$
- Forward recursion:

$$\delta_t(i) = \max_j\{\delta_{t-1}(j)P_1(s_t = i|s_{t-1} = j)\}P(y_t|s_t = i, \mathbf{x}_t), \text{ for each } 0 \leq t \leq n.$$

- Backward recursion: $s_n^* = \arg\max_i \delta_n(i)$ and

$$s_t^* = \arg\max_i \delta_t(i)P_1(s_{t+1} = s_{t+1}^*|s_t = i)$$

### 2.3 EM Algorithm for HMM

This is the third question we have to solve and is the most difficult problem. It is to determine a method to adjust the model parameters to maximize the probability of the observation sequence. The EM algorithm is as follows:

**E-step:**
Compute the posterior probabilities.

$$\gamma_t(i) \stackrel{def}{=} P(s_t = i|y_0, \ldots, y_n) = \frac{\alpha_t(i)\beta_t(i)}{\sum_j \alpha_t(j)\beta_t(j)}.$$

and

$$\xi_t(i,j) \stackrel{def}{=} P(s_t = i, s_{t+1} = j | y_0, \ldots, y_n)$$
$$= \frac{\alpha_t(i) P_1(s_{t+1} = j | s_t = i) P_o(y_{t+1} | s_{t+1} = j, \mathbf{x}_{t+1}) \beta_{t+1}(j)}{\sum_j \alpha_t(j) \beta_t(j)},$$

where $t = 0, \ldots, n-1$.

**M-step:**
The initial state distribution can be updated according to the expected fraction of times the sequences started from a specific state i: $\hat{P}_0(i) = \gamma_0(i)$ To update the transition probabilities, we first define the expected number of transitions from $i$ to $j$ $\hat{N}(i,j) = \sum_{t=0}^{n-1} \xi_t(i,j)$. So the maximum likelihood estimate of the transition probabilities is then a ratio of these soft counts.

$$\hat{P}_1(j|i) = \frac{\hat{N}(i,j)}{\sum_{j'} \hat{N}(i,j')}$$

We assume the output are continuous, we have to solve a weighted maximum likelihood estimation problem separately for each state $i$ we maximize: $\sum_{t=0}^{n} \gamma_t(i) \log P(y_t|i, \mathbf{x}_t)$. We have not discuss the distribution yet. If we assume that $P(y_t|s_t, \mathbf{x}_t)$ is a gaussian at each state $i$, the probability at state $i$ is

$$P(y_t|s_t = i, \mathbf{x}_t) = \frac{1}{\sqrt{2\pi\sigma_i^2}} exp(-\frac{(y_t - \hat{y}_i(\mathbf{x}_t))^2}{2\sigma_i^2}).$$

This is an error function

$$\sum_{t=0}^{n} \gamma_t(i)(y_t - \hat{y}_i(\mathbf{x}_t))^2$$

for each state $i$. This can be viewed as a weighted least squared error function. For noise data, the least squared error function always lead to over-fitting. Therefore we need to add an regular term in support vector machines. Hence, we have the following error function with a regular term for each state $i$ in SVM.

$$E = \sum_{t=0}^{n} \gamma_t(i)(y_t - \hat{y}_i(\mathbf{x}_t))^2 + \lambda ||\mathbf{w}||^2, \tag{5}$$

where $\mathbf{w}$ is are coefficients for SVM regression. Therefore

$$\hat{y}_i(t) = (\gamma_i I y)'(\gamma_i I K + \lambda I)^{-1} k, \tag{6}$$

where $I$ is the identity matrix and $\gamma_i = [\gamma_0(i), \gamma_1(i), \ldots, \gamma_n(i)]'$ is a column vector. Equation (6) is similar to equation (2) with one more coefficient vector $\gamma_i$. In this way we integrate the HMM and SVM model together and build a HMM based SVM model in each state of the HMM.

### 2.4 Predict the Values and Density Distribution

We can predict the density distribution of a state in terms of the transition probability $P(s_{t+1} = j|s_t = i) = a_{ij}$ and the joint probability $\alpha_t(i)$ and the observations through time t.

$$P(s_{t+1} = j|y_0, \ldots, y_t) = \frac{P(y_0, \ldots, y_t, s_{t+1} = j)}{P(y_0, \ldots, y_t)} = \frac{\sum_{i=1}^{m} \alpha_t(i)a_{ij}}{\sum_{j=1}^{m}\sum_{i=1}^{m} \alpha_t(i)a_{ij}} = c_j$$

The density for $y_{t+1}$ is a linear weighted mixture model of the densities of the individual support vector machine:

$$\begin{aligned} P(y_{t+1}|\mathbf{x}_{t+1}, y_0, \ldots, y_t) &= \sum_{j=1}^{m} P(y_{t+1}|\mathbf{x}_{t+1}, s_{t+1} = j)P(s_{t+1}|y_0, \ldots, y_t) \\ &= \sum_{j=1}^{m} c_j P(y_{t+1}|\mathbf{x}_{t+1}, s_{t+1} = j) \end{aligned} \quad (7)$$

For the special case of regression, Gaussian distribution are assumed for each individual SVM model. The individual densities are described by their conditional mean $\hat{y_i}(t)$ and the variances $\sigma_j^2$.

Should we be interested in the overall mean (median) of the predicted density at time $t+1$, we have the linear combination of each individual mean (median):

$$\hat{y_{t+1}} = \sum_{j=1}^{m} c_j \hat{y_j}(t+1).$$

## 3 Computational Results

Our first example is a toy problem. The data is generalized with two distinct processes: the Markov chain of hidden states, and the dynamics of each states. The transitional matrix of the hidden chain is given by

$$A = \begin{pmatrix} 0.99\ 0.01 \\ 0.03\ 0.97 \end{pmatrix},$$

and the dynamics of each individual state is:

$$y_{t+1} = \begin{cases} 0.1y_t^2 + 0.5y_t + 0.8\varepsilon \\ -0.1y_t^3 - 0.4y_t + 0.5\delta \end{cases},$$

where $\varepsilon \sim N(0,1)$ and $\delta \sim N(0,1)$. We use the simple linear kernel $(x'x)$ and second order polynomial kernel $(x'.x+1)^2$ to simulate it. The output is in Table (1).

Knowing the true model of this toy problem allows us to compare the estimated parameters with the the true parameters. We compared the diagonal

**Table 1.** Computational Results with Different Kernels

| Models | $a_{11}$ | $a_{22}$ | $\sigma_1$ | $\sigma_2$ |
|---|---|---|---|---|
| True Value | 0.99 | 0.97 | 0.8 | 0.5 |
| Linear Kernel | 0.969 | 0.962 | 0.93 | 0.44 |
| Polynomial Kernel | 0.987 | 0.969 | 0.807 | 0.492 |

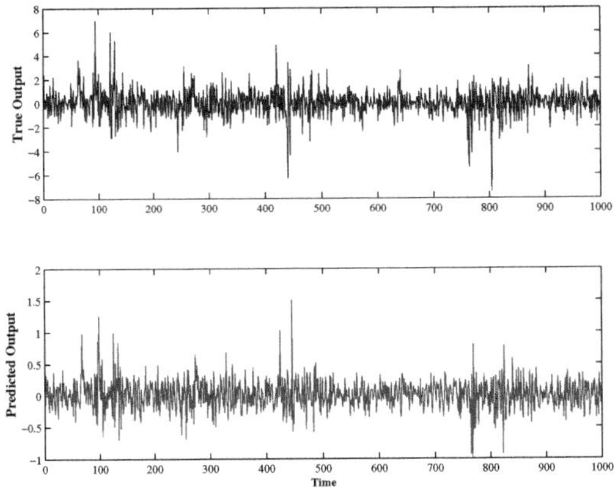


**Fig. 1.** The True and One Step Predicted Output with HMMSVM

elements of the transitional probability matrix A and the noise level $\sigma_i$. The results with second order polynomial kernel are closer to the true values. The True and predicted outputs are given in Figure (1). Obviously this model predicted well of the original data.

Our second application is for ECG time series denoising. In statistical signal and image denoising, wavelet methods are extensively used in the literature. One of assumptions about the noises is that they have a gaussian distributed (white noise), which is not held for most of real life application. In this study, we apply the HMMSVM method to ECG denoising. ECG signals have many useful information about the conditions of the patients. ECG with noise contamination have to be cleaned before the doctors can make any reference from the signal. Here we give a contaminated ECG signal and denoised with different methods. The data was obtained from physionet of MIT. We predict the corrent value based on previous 7 lagged ECG values. Our HMMSVM with four state of the following transition matrix:

$$A = \begin{pmatrix} 0.909 & 0.031 & 0.020 & 0.040 \\ 0.014 & 0.940 & 0.029 & 0.017 \\ 0.011 & 0.014 & 0.968 & 0.007 \\ 0.012 & 0.003 & 0.004 & 0.981 \end{pmatrix}.$$

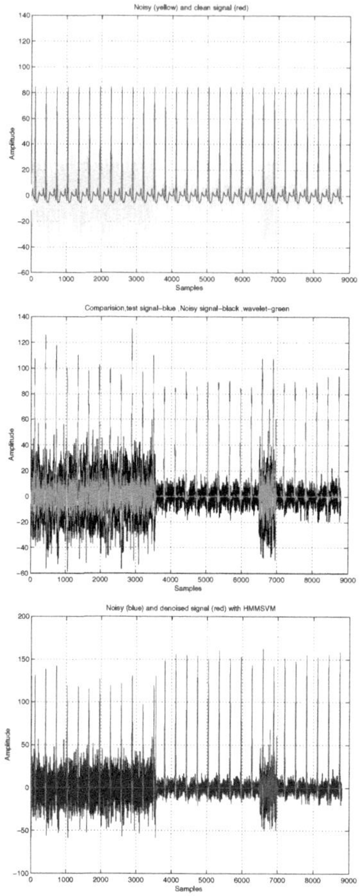


**Fig. 2.** Performance of our HMMSVM and Wavelet: top panel-True signal, middle panel denoised with standard wavelet, and the bottom panel: denoised with HMMSVM

**Table 2.** Mean and Standard Deviation of the $\sigma_i$

| HMMSVM | Mixture 1 | Mixture 2 | Mixture 3 | Mixture 4 |
|---|---|---|---|---|
| Mean $\sigma_i$ | 2.37 | 1.92 | 0.83 | 0.65 |
| STD of $\sigma_i$ | 0.05 | 0.12 | 0.04 | 0.03 |

The average noise levels $\sigma_i$ of the individual mixture for HMMSVM is given in Table (2). The denoising results of our model is given in Figure (2). With HMMSVM model, we can denoise the signal with different model for different noise level. Our model certainly outperformed the standard wavelet model.

## 4 Conclusions

In this paper, we introduce a HMM based SVM methods for non gaussian conditional probability and point value prediction. We give a toy problem to validate the model and show its applications in signal denoising. The algorithm works well and can have many potential applications. With little revision, it can be utilized for stock market return prediction and financial risk management. The proposed framework can also be used for classification as well. We will explore more applications in the near future.

## References

1. Bengio,Y.:Markovian Models for Sequential Data. Neural Computing Surveys, 2(1999) 129–162
2. Ganapathiraju,A., Hamaker,J., Picone,J.:Hybrid HMM/SVM Architectures for Speech Recognition.Speech Transcription Workshop, College Park, Maryland, USA(2000)
3. Ganapathiraju,A., Hamaker, J., Ordowski, M., Doddington,G.,Picone,J.:Syllable-Based Large Vocabulary Continuous Speech Recognition. IEEE Transactions on Speech and Audio Processing, 9(4)(2001)358-366
4. Huang, B.Q., Du, C.J., Zhang, Y.B., Kechadi, M.T.: A Hybrid HMM-SVM Method for Online Handwriting Symbol Recognition. Intelligent Systems Design and Applications, 6(2006)
5. Mangasarian, O.L.,Wild, E.W.:Multisurface Proximal Support Vector Machine Classification via Generalized Eigenvalues.IEEE Transactions on Pattern Analysis and Machine Intelligence,28(1)(2006)69-74
6. Rabiner,L.:A Tutorial on Hidden Markov Models and Selected Applications in Speech Recognition. Proc. IEEE, 77(2)(1989)257–286
7. Rasmussen, C.E.:Evaluation of Gaussian Processes and Other Methods for Non-linear Regression. PhD thesis, Department of Computer Science, University of Toronto (1996)
8. Sanches, I.: Noise-Compensated Hidden Markov Models. IEEE transactions on Speech and Audio Processing, 8(5) (2000)533
9. Shawe-Taylor, J.,Cristianini,N.: Support Vector Machines and Other Kernel-based Learning Methods. Cambridge University Press(2000)
10. Gestel,T.V., Suykens, J.A.K.,Baesens,B.,Viaene, S.,Vanthienen, J.,Dedene, G., Moor,B.D., Vandewalle, J.: Benchmarking Least Squares Support Vector Machine Classifiers. Machine Learning, 54(2004)5-32

# Estimating Selectivity for Current Query of Moving Objects Using Index-Based Histogram

Jeong Hee Chi[1] and Sang Ho Kim[2]

[1] Department of Internet&Multimedia Eng., Konkuk University, Korea
jhchi@konkuk.ac.kr
[2] Center for Rural Information&Culture, Korea Rural Economic Institute, Korea
kimsh@krei.re.kr

**Abstract.** Selectivity estimation is one of the query optimization techniques. It is difficult for the previous selectivity estimation techniques for moving objects to apply the location change of moving objects to synopsis. Therefore, they result in much error when estimating selectivity for queries, because they are based on the extended spatial synopsis which does not consider the property of the moving objects. In order to reduce the estimation error, the existing techniques should often rebuild the synopsis. Consequently problem occurs, that is, the whole database should be read frequently. In this paper, we proposed a moving object histogram method based on quadtree to develop a selectivity estimation technique for moving object queries. We then analyzed the performance of the proposed method through the implementation and evaluation of the proposed method. Our method can be used in various location management systems such as vehicle location tracking systems, location based services, telematics services, emergency rescue service, etc in which the location information of moving objects changes over time.

**Keywords:** Selectivity Estimation, Histogram, Quadtree, Moving object.

## 1 Introduction

Recently, along with the development of wireless computing and GPS technologies, which can support a real-time tracking of moving objects, many applications using the moving object data have been developed. These types of applications include vehicle tracking, digital battlefields, airplane traffic management systems, as well as location-based services. Mobile data management systems efficiently supporting such kinds of application services should be able to manage the moving objects that continuously change their locations over time.

In this paper, we propose QTH(Quadtree based Histogram) technique to estimate selectivity for current query on current location of moving objects. Selectivity estimation is one of the query optimization techniques. Until now, many works have been performed to estimate selectivity for spatial data[1],[2],[3],[4],[5]. However, the spatial and moving object selectivity estimation technique treats data with different character. That is, spatial objects are static objects which do not change their location

D.-S. Huang, L. Heutte, and M. Loog (Eds.): ICIC 2007, LNAI 4682, pp. 1255–1264, 2007.

over time, but moving objects are dynamic objects which continuously change their location over time. Therefore the existing spatial histogram just can support selectivity for query on location of moving objects at specific time point; it is difficult for previous selectivity estimation techniques to apply the real-time location data of moving objects to a synopsis. In order to reduce the estimation error according to data location change, the existing techniques should often rebuild the synopsis. Consequently problem occurs, that is, the whole database should be read frequently.

Recent several works have proposed histogram techniques for moving object changing their location over time[6],[7],[8],[9]. The previous techniques make a histogram based on spatial partition algorithm or dimension transformation method. These techniques make a histogram using spatial partition strategy based on initial spatial distribution of object; the partitioned spatial information can be broken easily according to location change of moving object. In order to solve this problem like the spatial histogram, the histogram should often rebuild. However, it occurs overhead because of rebuilding histogram frequently. Therefore, it is crucial for technique which can provide exact selectivity for moving object changing their location over time.

In this paper, we propose a moving object histogram which can reflect the location change of moving objects in the histogram. In order to build such a histogram we use quadtree and hash table. Our method can reduce the overhead which the whole index should be scanned to use the index based histogram, since the synopsis of our method maintains as hash table in memory. To reduce the spatial overhead occurring when buckets are assigned to a target node in the index, we adjust the number of buckets according to the height level of the index.

The rest of this paper is organized as follows. In the next section we summarize related work. The proposed method is presented in section 3. In section 4 we describe experimental results and evaluations for the proposed method. Finally, we draw conclusion in Section 5.

## 2 Related Works

There are several techniques addressing selectivity estimation of range query for moving objects. The problem was first investigated in [6]. They focus on managing moving objects and static future query. They constructed a spatio-temporal histogram by partitioning the spatial domain into several disjoint buckets through minskew technique[1], and then assigning MBR, velocity and additional information in each bucket. They derive formulae for one-dimensional space, and then extend it for the multi-dimensional case. This technique has the problem which frequent updates are needed in maintaining a good performance, and they may overestimate the query results in two-dimensional space.

Y. Tao proposed a comprehensive query model and histogram to estimate query results in [7]. Although the query model can deal with various kinds of query type, static/dynamic, the histogram based on minskew should be rebuilt since the uniformity in each bucket can be easily lost.

M. Hadjieleftheriou proposed two new approaches using the duality transformation in [8]. The first method converts the trajectory of each object to a point in the four dimensional space, which permits the direct employment of conventional multi-dimensional histograms. Accordingly, a query is transformed to a simplex search region in the dual space. The second approach, instead of building a separate histogram, uses the extents of the leaf nodes of an R-tree index as the histogram buckets. Although this method can support dynamic maintenance by using the index, the histogram based R-tree which permits overlap of nodes incurs large error, since several nodes overlapped with query can be counted twice.

# 3 QuadTree Based Histogram

## 3.1 Data Model

We assume that moving objects are point objects and change their positions continuously over time such as car, person, and aircraft. Location information of objects is regularly extracted through equipments like GPS, and they are stored in database when updates such as insertion, deletion of objects and change of speed occur. In other words, let O be a moving objects in n-dimensional space, $[U_{min\_i}, U_{max\_i}]$ be an spatial domain, moving object O is stored in form of $<O_{id}, S>$ when updates occur, where $O_{id}$ is identifier of moving object, $S = \{ ps_1, ps_2, .., ps_i, .., ps_n \}$is spatial position information of object where $ps_1, ps_2, .., ps_i, .., ps_n$ is coordinate of object ($U_{min\_i} \leq ps_i \leq U_{max\_i}$).

The proposed histogram is constructed based on position of object at current time t=0. Also, since objects issue regular updates, the histogram needs to be adjusted accordingly. An object update is given in form of < Oid, (S), (S')>, where (S') are a new position of updated object. Window query Q can also be defined as spatial object in n-dimensional space, with the space range $[qs_{i-}, qs_{i+}]$. Tables 1 summarize the symbols that will appear in our technique.

**Table 1.** Symbols for histogram

| Parameters | Description |
|---|---|
| $n$ | dimensionality of the data space |
| $[U_{min\_i}, U_{max\_i}]$ | space range on the i-th dimension |
| $S=\{ps_1,ps_2,...,ps_n\}$ | space information of moving object |
| $ps_i$ | coordinate of object on i-th dimensional space |
| $qs$-, $qs$+ | space range of query |

## 3.2 Construction of Quadtree Histogram

In this section, we introduce the proposed histogram based quadtree. Index-based histogram has the advantage that it does not have to be rebuilt frequently to reflect the change according to movement of objects, and can be also dynamically updated whenever update of index occurs.

QTH consist of quadtree and hash table. Quadtree[10] is index technique which does not allow overlap of nodes. It can reflect well data distribution through by non-uniformity spatial partition method compared with index based on gird. It also can be implemented and maintained simply since it supports simple partitioning and merging strategy compared with R-tree, R*-tree, TPR-tree, TPR*-tree and HTPR-tree supporting object based spatial partitioning strategy. QTH take advantage of spatial partitioning algorithm and update algorithm of quadtree, the former is for making buckets of histogram, and the latter is for reflecting update information of objects on histogram. QTH also use hash table for maintaining histogram constructed by buckets for node of quadtree. QTH do not have to read whole database to construct histogram, since it can be constructed index and histogram at once. Figure 1 shows the structure of QTH.

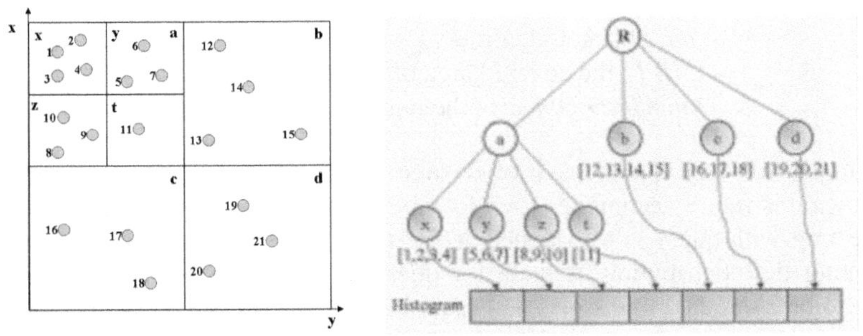

(a) dataset (b) moving object histogram

**Fig. 1.** The Structure of QTH

We usually construct a QTH through the following two phases.

**Phase 1: Build quadtree.** First step is to build quadtree over the entire space by insert all objects. This process is identical to the regular quadtree algorithm. Quadtree change dynamically the cover area of each node. When a node contains too many objects, we split it into several smaller ones and reallocation the objects in the old node. On the other hand, if several nodes do not have enough objects, we merge them into one big node and put the objects together. Though this construction process is more complex than uniformity space partition method, it does not take a long time to process because it just read objects within changed space not whole data. This process also is performed at object update time not query time, so it does not affect query selectivity estimation time. Nodes of quadtree store the following information.

***<numberOfObjects, nodeArea, cPointer, flag, objectList>***

-. *numberOfObjects* : the number of objects current inside the node
-. *nodeArea* : the covered area of node
-. *cPointer* : pointer to children if not leaf node
-. *flag* : whether or not it is a leaf node
-. *objectList* : list of object inside node.

**Phase 2: Construct buckets of histogram.** It constructed by maintaining buckets corresponding to leaf nodes using node id as hash key. QTH is able to solve the searching overhead problem according to the height of index in case of skewed dataset as maintaining histogram based on hash table. To get a good estimate for moving object changing their position over time, an update of histogram is very important. It is very simple for our histogram. Our histogram just is adjusted as necessary whenever updates of leaf node corresponding to bucket occur. To minimize storage size of histogram and selectivity estimation time, we do not assign a bucket which corresponds to leaf node which does not have object. After constructing buckets, the buckets store the following information.

***<nid, MBR, numerOfObjects>***

-. *nid* : bucket identifier
-. *MBR* : the covered area of bucket
-. *numberOfObjects* : the number of objects inside bucket

Figure 2 show the query result of R*-tree and quadtree for dataset of figure 2(a). As shown this figure, technique based R*-tree may come to overestimate, since nodes intersecting with query is overlapped. This is because that selectivity is obtained by computing the contribution of node for query. So, the area with thick dashed line comes to be computed four times. However, our histogram computes the contribution of each bucket only one time since quadtree does not permit overlapping between nodes.

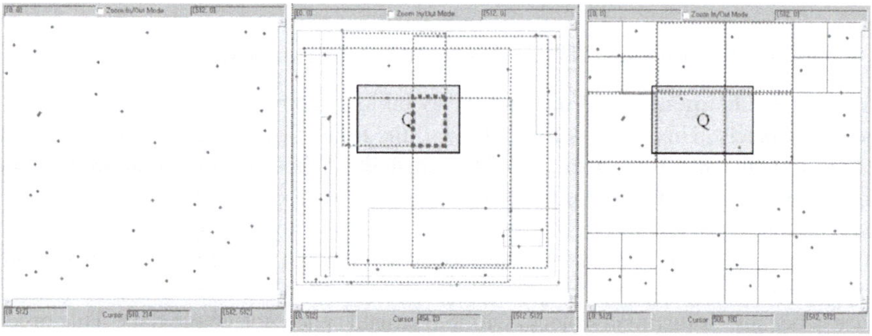


(a) dataset (b) R*-tree (c) Quadtree

**Fig. 2.** An example of query result for R*-tree and Quadtree

### 3.3 Current Query for Moving Objects

Moving objects change their position and velocity over time. Current query is a query for position of moving objects at current time. The selectivity for query is the sum of the contributions from all histogram buckets. Under the query model proposed by [7],[11], since our estimation method is based on assuming a uniform distribution within each bucket like other methods, the contribution of bucket is proportional to the area of its intersection with query and the number of objects contained in bucket. To reduce the intersection examination between bucket and query, we take advantage of dimensional transformation method of [7].

**Lemma 1.** Let query Q and bucket B be a 2-dimensional rectangle whose spatail extent is $\{qR_{imin}, qR_{imax}\}$ and $\{bR_{imin}, bR_{imax}\}$. Qiven query Q, we formulate another query Q' such that (i) its spatial extent is $q'R_{imin} = qR_{imin}$ - ($bR_{imax}$ - bRimin), and (ii) $q'R_{imax} = qR_{imax}$. If query Q' covers the left corner point of bucket B, then query Q intersect with the bucket B.

Existing intersection between Query Q and bucket B means max{$qR_{imin}$, bRimin} <= min{$qR_{imax}$, $bR_{imax}$}, and if query covers the minimum point $bR_{imin}$ of bucket, then the point satisfy query. In other words, if $qR_{imin} <= bR_{imin} <= qR_{imax}$, query intersect the point. Let $\triangle x$ be $bR_{imax}$-$bR_{imin}$, the former equation is same with $qR_{imin} <= bR_{imin} + \triangle x <= qR_{imax} + \triangle x$. This equation would be $qR_{imin} - (bR_{imax} - bR_{imin}) <= bR_{imin} <= qR_{imax}$, where $qR_{imin} - (bR_{imax} - bR_{imin})$ is same with $q'R_{imin}$, and $qR_{imax} = q'R_{imax}$, therefore $q'R_{imin} <= bR_{imin} <= q'R_{imax}$. ■

Figure 3 shows intersection examination between query and buckets. In figure 3(a), there is object A and B, given query Q, then we perform intersection examination of point and query instead of rectangle object and query.

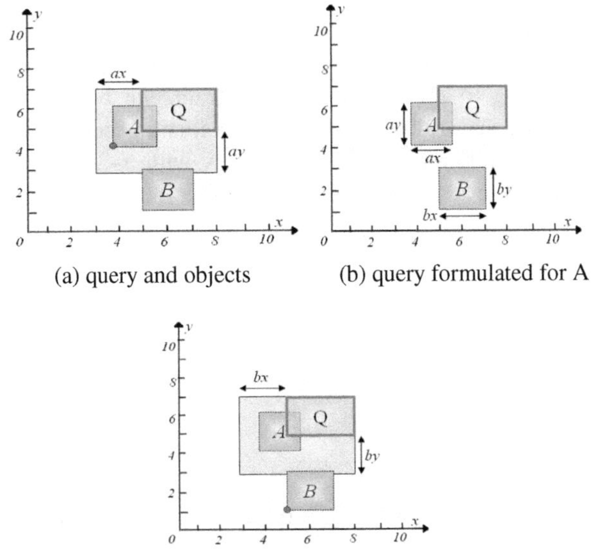


(a) query and objects (b) query formulated for A

(c) query formulated for B

**Fig. 3.** The example of query transformation

To formulate transformed query for object 'A' like figure 3(b), we transform the query using ax and ay, the length of each dimension of object 'A'. In figure 3(b), since transformed query Q' covers the left corner point of object 'A', original query Q intersect with object 'A'. In case of object 'B' in figure 3(c), since the transformed query do not cover the left corner point of object 'B', object 'B' do not satisfy original query. The bucket filtering method through by dimensional transformation can reduce the number of candidate object for which selectivity estimation is necessary.

## 4 Experiments

This section experimentally evaluates the proposed methods. We compared the proposed histogram with MinSkew-based MH and R*-Tree-based RH. We generated dataset with 100,000 point which has the skewness coefficient 0.8 using GSTD, the dataset exist within space range <(0,0),(10000,10000)>. We experimented on an accuracy of different method according to storage size and query size. In order to evaluate the effect of data update; we changed the position of 10%~40% dataset. We use the following equation to evaluate accuracy.

$$E_r = \frac{\left(\sum_{q_i=1}^{q_i=n} |S'(q_i) - S(q_i)|\right)}{\sum_{q_i=1}^{q_i=n} S(q_i)} \times 100 \quad (1)$$

where qi is i[th] query, we evaluated the proposed method with average relative error for 1,000 queries, S(qi) is actual selectivity for qi and S'(qi) is selectivity estimated by using histogram.

In order to evaluate the proposed method, we experiment on the accuracy of proposed method according to query size, storage size, and update rate. For evaluating performance according to the storage of histogram, we build a histogram which consisted of 300,600, and 1,000 buckets. For evaluating performance according to query size, we make 1000 number of range query with 5~20% size of whole space extent using GSTD. We also experiment on estimation time using bucket filtering method.

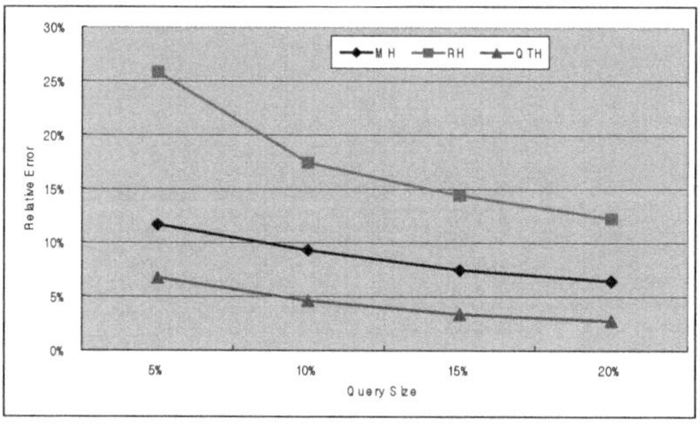


**Fig. 4.** Accuracy according to Query Size

Figure 4 shows the average relative error according to 5%~20% query size. The result shows that the error rates of MH, RH, and QTH decrease as query size increases. Especially, the GTH has better accuracy than the others. This means that in case of RH technique overlap between buckets exist, so it occur double counting problem, and has high error rates.

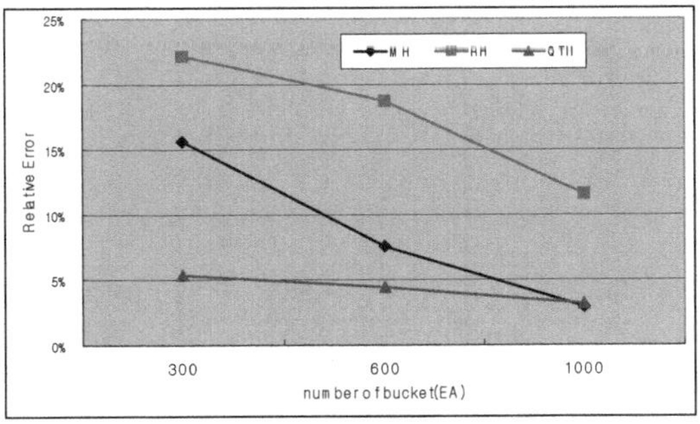


**Fig. 5.** Accuracy according to Storage Size

Figure 5 shows the relative error of different methods according to storage size. As can be expected, the result shows that the error rates of all methods decrease as the number of buckets increase. The result also shows that the proposed method, QTH has better accuracy, this is because QTH method does not assign bucket to nodes that does not have object, so we can get many buckets reflecting data distribution.

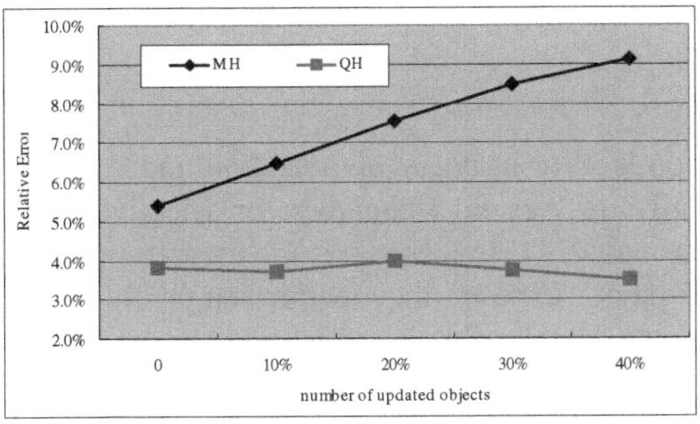


**Fig. 6.** Accuracy according to data update

Figure 6 shows the experimental result according to histogram update. In this experiment we tested the proposed method on update rates of five different update rates: 0%, 10%, 20%, 30%, and 40%, where update rates means that 10~40% of whole datasets change their position. The result shows that in case QTH, its accuracy has no big change, but the accuracy of MH method decrease sharply as update rates increase. This is because QTH update buckets of histogram according to update strategy of quadtree, but in case of MH the extent of buckets partitioned by the existing datasets before update does not reflect the updated data distribution. In order to increase the accuracy, MH should rebuild very often. However, rebuilding histogram occur overhead to read whole database.

Figure 7 shows the experimental results in terms of bucket filtering. As can be expected, the results show that the estimation time increases as the number of bucket

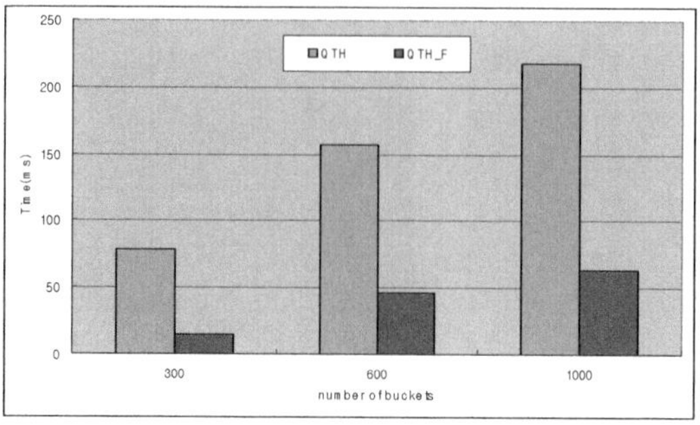


**Fig. 7.** Estimation Time according to Bucket Filtering

increases. From this experiment, we can know that QTH_F applying filtering method is faster three times than QTH which does not applying filtering method. This is because in case of rectangle objects the selectivity can be obtained by computing the proportion of intersection between rectangle objects and window query, computing the proportion of intersection takes a long time because of examining whether overlap, contain and containedby relationship between objects exist or not.

## 5 Conclusions

Selectivity estimation is used in query optimization and decision of optimal access path cardinally. Until now, several techniques for selectivity estimation on moving objects have been proposed. These techniques are focused on obtaining high accuracy for several kinds of query. However, they require frequently updating or rebuilding of histogram to capture the data distribution of moving objects with time.

To solve this problem, we propose an index-based histogram. Our histogram is based on quadtree and hash table, so it does not have to be rebuilt frequently to reflect the change according to movement of objects, and can be also dynamically updated whenever update of index occurs. The various experimental results showed that the proposed method has better performance than the existing methods since the location change of moving objects over time is reflected in the histogram.

Hence, we hope that the proposed method is able to be used in various location management systems such as vehicle location tracking systems, location based services, telematics services, emergency rescue service, etc in which the location information of moving objects changes over time.

## References

1. Acharya, S., Poosala, V., Ramaswamy, S.: Selectivity Estimation in Spatial Databases. ACM SIGMOD(1999) 13-24
2. Jin, J., An, N., Sivasubramaniam, A.: Analyzing Range Queries on Spatial Data. IEEE ICDE(2000) 525-534

3. Ning, A., Y. Zhen, Y., Sivasubramaniam, A.: Selectivity Estimation for Spatial Joins. IEEE ICDE(2001) 368-375
4. Sun, C., Agrawal, D., Abbadi, A. E.: Exploring Spatial Datasets with Histograms. IEEE ICDE(2002) 93-102
5. Chi, J. H., Kim. S. H., Ryu, K. H.: Spatial Selectivity Estimation Using Compressed Histogram Information. APWEB(2005) 489-494
6. Choi, Y. J., Chung, C. W.: Selectivity Estimation for Spatio-temporal Queries to Moving Objects. ACM SIGMOD(2002) 440-451
7. Tao, Y., Sun, J., Papadias, D.: Selectivity Estimation for Predictive Spatio-Temporal Queries. IEEE ICDE(2003) 417-428
8. Hadjieleftheriou, M., Kollios, G., Tsotras, V.: Performance Evaluation of Spatio-temporal Selectivity Estimation Techniques. IEEE SSDBM(2003) 202-211
9. Elmongui, H. G., Mokbel, M. F., Aref, W. G.: Spatio-temporal Histogram. SSTD(2005) 19-36
10. Samet, H.: The Design and Analysis of Spatial Data Structures. Addison Wesley Publishing Company, INC(1994)
11. Kollios, G., Gunopulos, D., Tsotras, V.: On indexing Mobile Object. ACM PODS(1999) 261-272

# Forecasting Approach Using Hybrid Model ASVR/NGARCH with Quantum Minimization

Bao Rong Chang[1,*] and Hsiu Fen Tsai[2]

[1] Department of Computer Science and Information Engineering
National Taitung University, Taitung, Taiwan 950
brchang@nttu.edu.tw
[2] Department of International Business
Shu-Te University, Kaohsiung, Taiwan 824
soenfen@mail.stu.edu.tw

**Abstract.** Two crucial problems of overshoot and volatility clustering are encountered in time series prediction frequently, which may give rise to big residual errors and then deteriorate the predictive accuracy. Thus this study introduces a novel scheme to overcome the preceding problems at the same time. First, adaptive support vector regression (ASVR) with fewer data is applied to tackling the overshoot results and its achievement outperforms the traditional prediction methods like ARMAX or grey model. Next, the effect of volatility clustering can be resolved by incorporating nonlinear generalized autoregressive conditional heteroscedasticity (NGARCH) into ASVR to form a hybrid model. Finally, in order to avoid the over-fit or under-fit modeling resulted from quadratic optimization or back-propagation neural network, instead a quantum-based algorithm called quantum minimization (QM) is proposed herein to tune the hybrid model of ASVR and NGARCH to overcome the problems of overshoot and volatility clustering simultaneously due to quantum parallelism finding the optimum in depth. In summary, the proposed method obtains the satisfactory results because of best-fitting the dynamics of time series and thus significantly improving the predictive accuracy.

**Keywords:** adaptive support vector regression, nonlinear generalized autoregressive conditional heteroscedasticity, quantum minimization.

## 1 Introduction

Traditional model like ARMAX [1] cannot resolve the volatility clustering effect in time series prediction as shown in Fig. 1. Another one is a grey model (GM) [2] just acquiring fewer data for modeling without any training process; however, it always encounters the overshoot problem [3] as shown in Fig. 2. Both volatility clustering and overshoot effects make a lot of big residual errors in time series prediction and hence deteriorate the predictive accuracy. This study proposes a novel scheme for dealing with the preceding both problems simultaneous. First, adaptive support vector regression (ASVR) [4] is used to overcome the problem of overshoot. This kind of

* Corresponding author. Phone: +886-89-318855 ext. 2607; Fax: +886-89-350214.

D.-S. Huang, L. Heutte, and M. Loog (Eds.): ICIC 2007, LNAI 4682, pp. 1265–1277, 2007.

regression particularly employs a fast self-tuned parameterization to fulfill the procedure of constraint optimization [5] [6] during the SVR training phase. Second, a nonlinear version of GARCH [7] called nonlinear generalized autoregressive conditional heteroscedasticity (NGARCH) [8] is incorporated into ASVR so as for resolving the volatility clustering effect. Furthermore, how to construct a hybrid model to combine ASVR and NGARCH appropriately is deeply concerned. In order to avoid the over-fit or under-fit modeling resulted from either quadratic optimization or back-propagation neural network, instead quantum computing [9] [10] is considered as a good way to optimize the hybrid model ASVR/NGARCH because of quantum parallelism finding the optimum in depth. To do so, a quantum-based algorithm called quantum minimization (QM) [11] is introduced to coordinate the above-mentioned hybrid model so as for achieving near-optimal or optimal solutions.

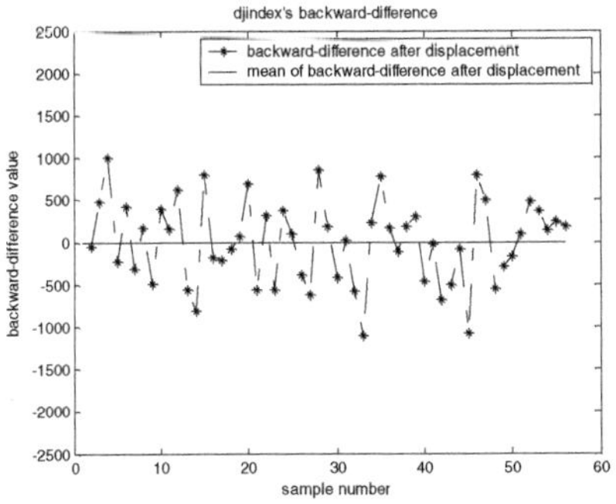


**Fig. 1.** A sequence of backward-difference from New York D.J. Industrials Index has displaced to its mean value dated from January 1999 to August 2003 for a period of 56 months. This plot showing the volatility clustering as big changes, appearing around sample number 2-4, 12-16, 27-35, and 45-47 as well as small changes, revealing around sample number 5-11,17-19,21-26, 36-44, and 48-56.

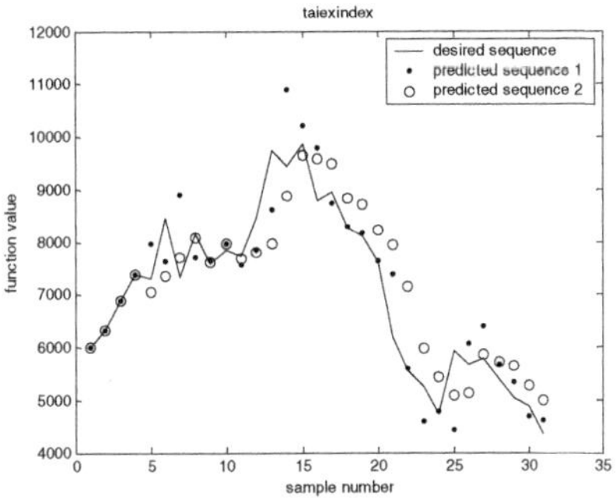


**Fig. 2.** The desired sequence represents the monthly Taiwan's stock price index for a period of 31 months from January 1999 to July 2001. The outputs of grey model are denoted by "•", revealing the overshoot effect around the turning-point region at sample number 7, 14, 25, 26, and 27. The underestimated predicted outputs, resulted from a cumulated 3-point least squared linear model, is denoted by "o" at sample number 6, 12,13, 25 and 26.

# 2 Hybrid Forecasting Model

## 2.1 Adaptive Support Vector Regression (ASVR)

Initially developed for solving classification problems, support vectors machines (SVMs) technology [6] can also be successfully applied in regression problems, i.e. functional approximation. We consider approximating functions $f(\cdot)$ solved by support vector regression (SVR) [5] with the form of

$$f(\mathbf{x},\mathbf{w}) = \sum_{i=1}^{l} w_i \phi(x_i) + b \tag{1}$$

where $\phi(\cdot)$, $w_i$, and $b$ denote a nonlinear mapping, a weighted value, and a bias, respectively. Furthermore, Vapnik introduced a general type of loss function, namely the linear loss function with $\varepsilon$-insensitivity zone [6], as

$$|y-f(\mathbf{x},\mathbf{w})|_{\varepsilon}=\begin{cases} 0 & \text{if } |y-f(\mathbf{x},\mathbf{w})|\le\varepsilon, \\ |y-f(\mathbf{x},\mathbf{w})|-\varepsilon & \text{otherwise} \end{cases} \tag{2}$$

According to the learning theory of SVMs [6], this can be expressed by maximizing dual variables Lagrangian $L_d(\boldsymbol{\alpha},\boldsymbol{\alpha}^*)$ where $l$, $\mathbf{x_i}$, $\mathbf{y_i}$, and $K(\cdot,\cdot)$ denote the number of vectors, an input vector, an output vector, and the kernel function, respectively.

$$L_d(\boldsymbol{\alpha},\boldsymbol{\alpha}^*)=-\frac{1}{2}\sum_{i,j=1}^{l}(\alpha_i-\alpha_i^*)(\alpha_j-\alpha_j^*)K(\mathbf{x_i},\mathbf{x_j})-\varepsilon\sum_{i=1}^{l}(\alpha_i+\alpha_i^*)-\sum_{i=1}^{l}(\alpha_i-\alpha_i^*)\mathbf{y_i}, \tag{3}$$

subject to the constraints

$$\sum_{i=1}^{l}\alpha_i=\sum_{i=1}^{l}\alpha_i^*, \tag{4}$$

$$0\le\alpha_i\le C,\quad i=1,\dots,l \tag{5}$$

$$0\le\alpha_i^*\le C,\quad i=1,\dots,l \tag{6}$$

After obtaining the Lagrange multipliers $\alpha_i$ and $\alpha_i^*$, we find the optimal weights of regression

$$\mathbf{w_0}=\sum_{i=1}^{l}(\alpha_i-\alpha_i^*)\phi(\mathbf{x_i}), \tag{7}$$

and an optimal bias

$$\mathbf{b_0}=\frac{1}{l}\left(\sum_{i=1}^{l}(\mathbf{y_i}-\phi(\mathbf{x_i})^T\mathbf{w_0})\right). \tag{8}$$

A fast algorithm applied to constraint optimization for support vector regression (SVR) is called adaptive support vector regression (ASVR) [4]. It is designed for exploring three free parameters C, $\varepsilon$ and $\sigma_{rbkf}$ [6] such that the computation time of quadratic programming (QP) is significantly reduced and achieved rapid convergence to the near-optimal solution. In the ASVR algorithm, two scale factors, $v$ and $\upsilon$ refer to [4], are evaluated in advance and then applied these evaluated values for calculating the free parameters $\varepsilon$ in Eq. (9) and $\sigma_{rbkf}$ in Eq. (10), respectively, where the vector $\mathbf{x}$ stands for an input vector. An order $n$ ,see [4], is also predetermined and used in computing the free parameter C in Eq. (10) and Eq. (3). In this manner, a straightforward parameter-seeking is done rather than using a heuristic method due to a long time for searching. Note that Eq. (10) and Eq. (11) are based on the modified Bessel function of second kind with the order $n$ [12] as follows:

$$\varepsilon = v \times \frac{|\max(\mathbf{X}) - \min(\mathbf{X})|}{2} \tag{9}$$

$$C = K_n(\varepsilon) = (-1)^{n+1}\{\ln(\varepsilon/2)+\gamma\}I_n(\varepsilon) + \frac{1}{2}\sum_{k=0}^{n-1}(-1)^k(n-k-1)!(\varepsilon/2)^{2k-n} + \frac{(-1)^n}{2}\sum_{k=0}^{\infty}\frac{(\varepsilon/2)^{n+2k}}{k!(n+k)!}\{\Phi(k)+\Phi(n+k)\} \tag{10}$$

$$I_n(\varepsilon) = \sum_{k=0}^{\infty}\frac{(\varepsilon/2)^{n+2k}}{k!\Gamma(n+k+1)} \tag{11}$$

where $\gamma = 0.5772156...$ is Euler's constant and $\Phi(p) = 1+\frac{1}{2}+\frac{1}{3}+...+\frac{1}{p}$, $\Phi(0)=0$ [12].

Eq. (11) determines a free parameter $\sigma_{rbkf}$ of radial basis kernel function for quadratic programming in SVR.

$$\sigma_{rbkf} = \upsilon \cdot \left( \sum_{i=1}^{l}(x_i - \bar{x})^2 \Big/ l-1 \right)^{1/2}, \quad \bar{x} = \sum_{j=1}^{l} x_j \Big/ l \tag{12}$$

In short, adaptive support vector regression firstly finds three tunable free parameters C, $\varepsilon$, and $\sigma_{rbkf}$ and subsequently utilizes these parameters in SVR optimization to search for the optimal weights and bias mentioned above.

## 2.2 NGARCH Resolving Volatility Clustering

ARMAX(r,m,Nx) [13] encompasses autoregressive (AR), moving-average (MA) and regression (X) models, in any combination, as expressed below

$$y_{armax}(t) = C^{armax} + \sum_{i=1}^{r} R_i^{armax} y(t-i) + e_{resid}(t) + \sum_{j=1}^{m} M_j^{armax} e_{resid}(t-j) + \sum_{k=1}^{N_x} \beta_k^{armax}\mathbf{X}(t,k) \tag{13}$$

where $C^{armax}$ = a constant coefficient, $R_i^{armax}$ = autoregressive coefficients, $M_j^{armax}$ = moving average coefficients, $e_{resid}(t)$ = residuals, $y_{armax}(t)$ = responses, $\beta_k^{armax}$ = regression coefficients, $\mathbf{X}$ = an explanatory regression matrix in which each column is a time series and $X(t,k)$ denotes a element at the $t$ th row and $k$ th column of input matrix.

NGARCH(p,q) [8] describes nonlinear time-varying conditional variances and Gaussian residuals $e_{resid}(t)$. Its mathematical formula is

$$\sigma_{ntvcv}^2(t) = K^{ng} + \sum_{i=1}^{p} G_i^{ng}\sigma_{ntvcv}^2(t-i) + \sum_{j=1}^{q} A_j^{ng}\sigma_{ntvcv}^2(t-j)\left[\frac{e_{resid}(t-j)}{\sqrt{\sigma_{ntvcv}^2(t-j)}} - C_j^{ng}\right]^2 \tag{14}$$

with constraints

$$\sum_{i=1}^{p} G_i^{ng} + \sum_{j=1}^{q} A_j^{ng} < 1, \; K^{ng} > 0, \; G_i^{ng} \ge 0, \; i = 1,...,p, \; A_j^{ng} \ge 0, \; j = -1,...,q$$

where $K^{ng}$ = a constant coefficient, $G_i^{ng}$ = linear-term coefficients, $A_j^{ng}$ = nonlinear-term coefficients, $C_j^{ng}$ = nonlinear-term thresholds, $\sigma_{ntvcv}^2(t)$ = a nonlinear time-varying conditional variance and $e_{resid}(t-j)$ = j-lag Gaussian distributed residual in ARMAX.

In the presence of conditional heteroscedasticity, this hybrid model can perform ARMAX and NGARCH separately over every period in a time series. For simplicity as employed in [14], it is possible to merge the outputs of ARMAX and NGARCH linearly to attain better results as

$$y_{Composite\ Model}(t) = f(y_{ARMAX}(t), y_{NGARCH}(t)) = Cf_1 \cdot y_{ARMAX}(t) + Cf_2 \cdot y_{NGARCH}(t) \quad (15)$$

where $f$ is defined as a linear function of ARMAX and NGARCH outputs, $y_{ARMAX}(t)$ and $y_{NGARCH}(t)$. $Cf_1$ and $Cf_2$ in Eq. (15) are the coefficients of a linear combination of ARMAX and NGARCH outputs. The resulting residual $y_{NGARCH}(t)$ at time $t$ is obtained from a product of $\sqrt{\sigma_{ntvcv}^2(t)}$ in Eq. (14) and a normalized random number $randn(1)$ where $0 \le randn(1) \le 1$.

### 2.3 ASVR Coordinated with NGARCH to Improve Regression

ARMAX cannot fit data sequences very well for irregular or non-periodic time series due to the lack of a dynamic learning mechanism. So, we propose an improved approach, i.e. to replace ARMAX with ASVR for the conditional mean component of hybrid model because ASVR has its own self-adaptive learning ability to fit irregular or non-periodic time series. This proposed hybrid model is rewritten as ASVR/NGARCH. Formulation of the linear combination [14] is expressed as

$$y_{Proposed\ Composite\ Model}(t) = g(y_{ANFIS}(t), y_{NGARCH}(t)) = Coef_1 \cdot y_{ANFIS}(t) + Coef_2 \cdot y_{NGARCH}(t) \quad (16)$$

where $g$ is defined as a linear function of the ANFIS and NGARCH outputs, respectively, $y_{ASVR}(t)$ and $y_{NGARCH}(t)$, while $Coef_1$ and $Coef_2$ are respectively the coefficients of the linear combination of the ASVR and NGARCH outputs.

A novel adaptation mechanism, called quantum minimization (QM) [11], is presented in the next section and will be exploited to search for optimal or near-optimal coefficients $Coef_1$ and $Coef_2$ in Eq. (16).

## 3 Quantum Adaptation

### 3.1 Quantum Exponential Searching Algorithm

As reported in [15] where the searching problem is to find the index $i$ such that $T[i] = x$, we are ready to describe the algorithm for finding a solution when the number $t$ of solutions is known. For simplicity, we assume at first that $1 \le t \le 3N/4$.

Step 1: Initialize $m=1$ and set $\lambda=6/5$. (Any value of $\lambda$ strictly between 1 and 4/3 would do.)
Step 2: Choose $j$ uniformly at random among the nonnegative integers small than $m$.
Step 3: Apply $j$ iterations of Grover's algorithm [16] starting from initial state

$$|\Psi_0\rangle=\sum_i \frac{1}{\sqrt{N}}|i\rangle .$$

Step 4: Observe the register: let $i$ be the outcome.
Step 5: If $T[i]=x$, the problem is solved: exit.
Step 6: Otherwise, set $m$ to $\min(\lambda m,\sqrt{N})$ and go back to step 2.

### 3.2 Quantum Minimum Searching Algorithm

We second give the minimum searching algorithm [11] in which the minimum searching problem is to find the index $i$ such that $T[i]$ is minimum where $T[0,...,N-1]$ is to be an unsorted table of $N$ items, each holding a value from an ordered set.

Step 1: Choose threshold index $0\le i\le N-1$ uniformly at random.
Step 2: Repeat the following stages and interrupt it when the total running time is more than $22.5\sqrt{N}+1.4\lg^2 N$. Then go to step 3.
- (a) Initialize the memory as $\sum_j \frac{1}{\sqrt{N}}|j\rangle|i\rangle$. Mark every item $j$ for which $T[j]<T[i]$.
- (b) Apply the quantum exponential searching algorithm [15].
- (c) Observe the first register: let $i'$ be the outcome. If $T[i']<T[i]$, then set threshold index $i$ to $i'$.

Step 3: Return $i$

This process is repeated until the probability that the threshold index selects the minimum is sufficiently large.

### 3.3 Forecasting Based on Signal Deviation

Single-step-look-ahead prediction, as shown in Fig. 3 and Fig. 4, can be arranged by adding the most recent predicted signal deviation $\delta\hat{o}(k+1)$ of Eq. (17) to the observed current output $o(k)$.

$$\delta\hat{o}(k+1)=h(o(k),o(k-1),...,o(k-s),\delta o(k),\delta o(k-1),...,\delta o(k-s)) \tag{17}$$

$$\hat{o}(k+1)=o(k)+\delta\hat{o}(k+1) \tag{18}$$

Based on the QM-ASVR/NGARCH structure, one can form the function $p$ of the ASVR output, $\delta\hat{o}_{asvr}(k+1)$, and the square-root of NGARCH's output, $\hat{\sigma}_{\delta o}(k+1)$, as presented below and shown in Fig. 3.

$$\delta\hat{o}_{qm-asvr/ngarch}(k+1) = p(\delta\hat{o}_{asvr}(k+1),\ \hat{\sigma}_{\delta o}(k+1)) \tag{19}$$

A weighted-average function is assumed to combine both $\delta\hat{o}_{asvr}(k+1)$ and $\hat{\sigma}_{\delta o}(k+1)$ to attain a near-optimal result $\delta\hat{o}_{qm-asvr/ngarch}(k+1)$.

$$\begin{aligned} &\delta\hat{o}_{qm-asvr/ngarch}(k+1) = w_{asvr}\cdot\delta\hat{o}_{asvr}(k+1) + w_{ngarch}\cdot\hat{\sigma}_{\delta o}(k+1) \\ &s.t. \qquad w_{asvr} + w_{ngarch} = 1 \end{aligned} \tag{20}$$

Here, the linear combination of two nonlinear functions, $\delta\hat{o}_{asvr}(k+1)$ and $\hat{\sigma}_{\delta o}(k+1)$, can also optimally approximate an unknown nonlinear target $\delta\hat{o}_{qm-asvr/ngarch}(k+1)$. Let $W_{asng} = [w_{asvr}\ w_{ngarch}]^T$ denote a weight-vector of $w_{asvr}$ and $w_{ngarch}$. A digital cost-function (DCF) [10] is defined as

$$DCF = \frac{\left\|W_{asng}\right\|^2}{2} + Const.\sum_{k=0}^{l-1}\left\|y(k+1) - y(k) - o(k) - \delta\hat{o}_{qm-asvr/ngarch}(k+1)\right\|^2 \tag{21}$$

$Const.$: a constant,
which can be used for measuring the accuracy when the respected cost is minimized. Quantum minimization mentioned above is employed for adapting the appropriate weights, $w_{asvr}$ and $w_{ngarch}$, for the forecast $\delta\hat{o}_{asvr}(k+1)$ and $\hat{\sigma}_{\delta o}(k+1)$ as per Eq. (20), respectively. Quantum minimization gives an order of computational cost as $O(\sqrt{N})$.

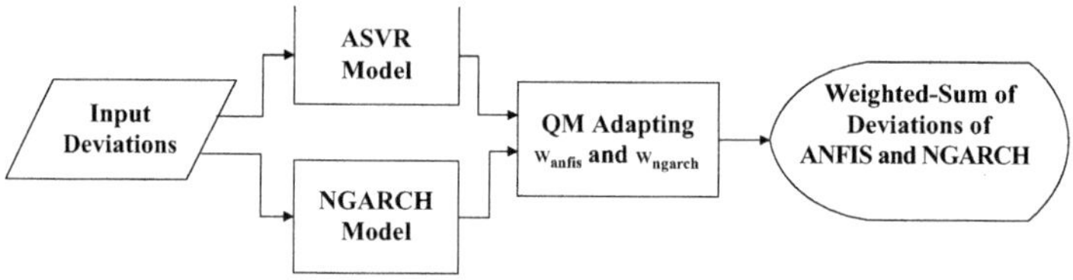


**Fig. 3.** Diagram of QM adapting ASVR/NGARCH outputs

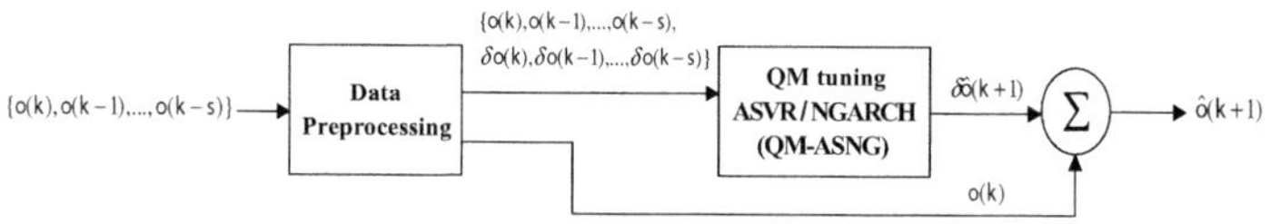


**Fig. 4.** Prediction using QM-ASVR/NGARCH system

## 4 Experimental Results and Discussions

As shown in Figs. 5 to 10, with a sliding widow size of 7 data points, the predicted sequences indicate the predicted results for the following competing methods: grey model (GM), cumulated 3-point least square polynomial (C3LSP) [3], auto-regressive moving-average (ARMA), back-propagation neural network (BPNN), adaptive neuro-fuzzy inference system (ANFIS), support vector regression (SVR), real-coded genetic algorithm [17] tuning hybrid ASVR/NGARCH model (RGA-ASVR/NGARCH), and quantum-minimized hybrid ASVR/NGARCH model (QM-ASVR/NGARCH). Single-step-look-ahead prediction methodology is employed in all experiments. In single-step-look-ahead design, a small number of the most recent observed data are collected as a sliding window (i.e. data queue) for modeling an intermediate predictor to predict the next period output. Once the next period's sampled datum is obtained, we drop a datum at the bottom of the data queue and add the most recent sampled datum into the data queue at the top position, thereby forming the new data queue used for the next prediction. This process continues until the task is terminated. To simplify comparison of the tested methods as plotted curves, only the three most representatives are shown in the figures. Thus GM, C3LSP and the proposed QM-ASVR/NGARCH are illustrated in Figs. 5 to 10, where "•" represents the sequential output of GM prediction, "∘" represents the sequential output of C3LSP prediction and "$-*-$" represents the sequential output of QM-ASVR/NGARCH prediction. All eight methods, however, are compared for goodness-of-fit in Tables 1 to 6.

In order to justify reasonable accuracy for a time series forecast, three well-known criteria [18] are commonly utilized. The terminology of these criteria is indicated as: (a) mean squared error (MSE); (b) mean absolute percent error (MAPE); (c) Theil'U inequality coefficient (Theil'U).

$$MSE = \frac{\sum_{t=1}^{l} \left(y_{t_c+t} - \hat{y}_{t_c+t}\right)^2}{l} \tag{22}$$

$$MAPE = \frac{100}{l} \sum_{t=1}^{l} \left| \frac{y_{t_c+t} - \hat{y}_{t_c+t}}{y_{t_c+t}} \right| \% \tag{23}$$

$$Theil'U = \sqrt{\frac{MSE}{MS}} = \sqrt{\frac{\sum_{t=1}^{l} \left(y_{t_c+t} - \hat{y}_{t_c+t}\right)^2 \Big/ l}{\sum_{t=1}^{l} y_{t_c+t}^2 \Big/ l}} \tag{24}$$

where $l$ = the number of periods in forecasting, $t_c$ = the current period, $y_{t_c+t}$ = a desired value at the $t_c+t$ th period and $\hat{y}_{t_c+t}$ = a predicted value at the $t_c+t$ th period.

First, the forecast of international stock price indices prediction for four markets (New York Dow Jones Industrial Index, Taipei TAIEX Index, Tokyo Nikkei Index, and London FTSE-100 Index) [19] are shown in Figs. 5 to 8. In addition, this study shows performance evaluation based on (a) mean squared error (MSE) (unit=$10^5$), (b) mean absolute percent error (MAPE), and (c) Theil'U inequality coefficient (Theil'U) between the actual sampled values and the predicted results of international stock price monthly indices over 48 months from Jan. 2002 to Dec. 2005. Forecasting performance of all eight methods is summarized in Tables 1 to 3, showing QM-ASVR/NGARCH obtains the best prediction results. The goodness-of-fit of QM-ASVR/NGARCH prediction for four stock price indices is tested by Ljung-Box Q-test [20] with p-values of 0.4371, 0.2518, 0.3076 and 0.3410, where each p-value is greater than the level of significance (0.05).

Second, Figs. 9 and 10 show the comparative forecasts of typhoon moving path of two typhoons (Nari typhoon during September 6-19, 2001 and Mindulle typhoon during June 28 – July 3, 2004) [21]. Performance evaluation is again made on the basis of MSE, MAPE, and Theil'U between the actual and predicted values. Tables 4 to 6 summarize prediction performance of our alternative methods and shows that QM-ASVR/NGARCH achieves superior results. The goodness of fit of QM-ASVR/NGARCH prediction for two typhoon moving paths is also tested by Ljung-Box Q-test with p-values of 0.0984 and 0.1051, in which each p-value is greater than level of significance (0.05).

**Table 1.** Mean Square Error (MSE) between the desired values and the predicted results for international stock price monthly indices (unit=$10^5$)

| Methods | Mean Square Error (MSE) | | | | |
|---|---|---|---|---|---|
| | N Y- D.J. Industrials Index | Taiwan TAIEX Index | Tokyo Nikkei Index | London FTSE-100 Index | Average |
| GM | 1.9582 | 1.7472 | 3.2209 | 4.0063 | 2.7332 |
| C3LSP | 1.7125 | 1.6481 | 5.1591 | 4.7420 | 3.3154 |
| ARMA | 1.8230 | 1.4737 | 2.9384 | 3.8832 | 2.5296 |
| BPNN | 1.2803 | 1.0461 | 3.0189 | 3.0656 | 2.1027 |
| ANFIS | 2.5016 | 1.4512 | 3.3282 | 3.8494 | 2.7826 |
| SVR | 1.2744 | 1.1255 | 3.1843 | 3.2062 | 2.1976 |
| RGA-ASVR/NGARCH | 1.1642 | 1.0298 | 3.0114 | 3.0165 | 2.0555 |
| QM-ASVR/NGARCH | 1.0875 | 0.9113 | 2.9029 | 2.9536 | 1.9638 |

**Table 2.** Mean Absolute Percent Error (MAPE) between the desired values and the predicted results for international stock price monthly indices

| Methods | Mean Absolute Percent Error (MAPE) | | | | |
|---|---|---|---|---|---|
| | N Y- D.J. Industrials Index | Taiwan TAIEX Index | Tokyo Nikkei Index | London FTSE-100 Index | Average |
| GM | 0.0365 | 0.0649 | 0.0449 | 0.0354 | 0.0454 |
| C3LSP | 0.0335 | 0.0599 | 0.0539 | 0.0355 | 0.0457 |
| ARMA | 0.0361 | 0.0581 | 0.0414 | 0.0353 | 0.0427 |
| BPNN | 0.0559 | 0.0505 | 0.0419 | 0.0306 | 0.0447 |
| ANFIS | 0.0403 | 0.6152 | 0.0434 | 0.0331 | 0.1830 |
| SVR | 0.0300 | 0.0532 | 0.0436 | 0.0310 | 0.0395 |
| RGA-ASVR/NGARCH | 0.0294 | 0.0486 | 0.0412 | 0.0305 | 0.0374 |
| QM-ASVR/NGARCH | 0.0283 | 0.0445 | 0.0401 | 0.0291 | 0.0355 |

**Table 3.** Theil'U Inequality Coefficient (Theil'U) between the desired values and the predicted results for international stock price monthly indices

| Methods | Theil'U Inequality Coefficient (Theil'U) | | | | |
|---|---|---|---|---|---|
| | N Y- D.J. Industrials Index | Taiwan TAIEX Index | Tokyo Nikkei Index | London FTSE-100 Index | Average |
| GM | 0.0435 | 0.0721 | 0.0501 | 0.0414 | 0.0518 |
| C3LSP | 0.0407 | 0.0700 | 0.0634 | 0.0451 | 0.0548 |
| ARMA | 0.0420 | 0.0662 | 0.0479 | 0.0408 | 0.0492 |
| BPNN | 0.0352 | 0.0558 | 0.0485 | 0.0362 | 0.0439 |
| ANFIS | 0.0491 | 0.0657 | 0.0510 | 0.0411 | 0.0517 |
| SVR | 0.0351 | 0.0579 | 0.0498 | 0.0371 | 0.0450 |
| RGA-ASVR/NGARCH | 0.0264 | 0.0529 | 0.0472 | 0.0358 | 0.0406 |
| QM-ASVR/NGARCH | 0.0217 | 0.0502 | 0.0436 | 0.0345 | 0.0375 |

**Table 4.** Mean Square Error (MSE) between the desired values and the predicted results for Nari typhoon moving path

| Methods | Mean-Square-Error (MSE) | | |
|---|---|---|---|
| | Nari Typhoon Moving Path | Mindulle Typhoon Moving Path | Average |
| GM | 0.0648 | 0.0703 | 0.0676 |
| C3LSP | 0.3698 | 0.4494 | 0.4096 |
| ARMA | 0.4497 | 0.7812 | 0.6155 |
| BPNN | 0.1535 | 0.6205 | 0.3870 |
| ANFIS | 0.0816 | 0.5372 | 0.3094 |
| SVR | 0.0609 | 0.0775 | 0.0692 |
| RGA-ASVR/NGARCH | 0.0603 | 0.0679 | 0.0641 |
| QM-ASVR/NGARCH | 0.0594 | 0.0656 | 0.0625 |

**Table 5.** Mean Absolute Percent Error (MAPE) between the desired values and the predicted results for Nari typhoon moving path

| Methods | Mean Absolute Percent Error (MAPE) | | |
|---|---|---|---|
| | Nari Typhoon Moving Path | Mindulle Typhoon Moving Path | Average |
| GM | 0.0047 | 0.0043 | 0.0045 |
| C3LSP | 0.0086 | 0.0140 | 0.0113 |
| ARMA | 0.0095 | 0.0177 | 0.0136 |
| BPNN | 0.0066 | 0.0127 | 0.0097 |
| ANFIS | 0.0050 | 0.0172 | 0.0111 |
| SVR | 0.0044 | 0.0050 | 0.0047 |
| RGA-ASVR/NGARCH | 0.0042 | 0.0046 | 0.0044 |
| QM-ASVR/NGARCH | 0.0038 | 0.0043 | 0.0041 |

**Table 6.** Theil'U Inequality Coefficient (Theil'U)between the desired values and the predicted results for Nari typhoon moving path

| Methods | Theil'U Inequality Coefficient (Theil'U) | | |
|---|---|---|---|
| | Nari Typhoon Moving Path | Mindulle Typhoon Moving Path | Average |
| GM | 0.0022 | 0.0023 | 0.0023 |
| C3LSP | 0.0060 | 0.0045 | 0.0053 |
| ARMA | 0.0048 | 0.0089 | 0.0069 |
| BPNN | 0.0036 | 0.0077 | 0.0057 |
| ANFIS | 0.0025 | 0.0056 | 0.0041 |
| SVR | 0.0021 | 0.0022 | 0.0022 |
| RGA-ASVR/NGARCH | 0.0018 | 0.0019 | 0.0019 |
| QM-ASVR/NGARCH | 0.0013 | 0.0016 | 0.0015 |

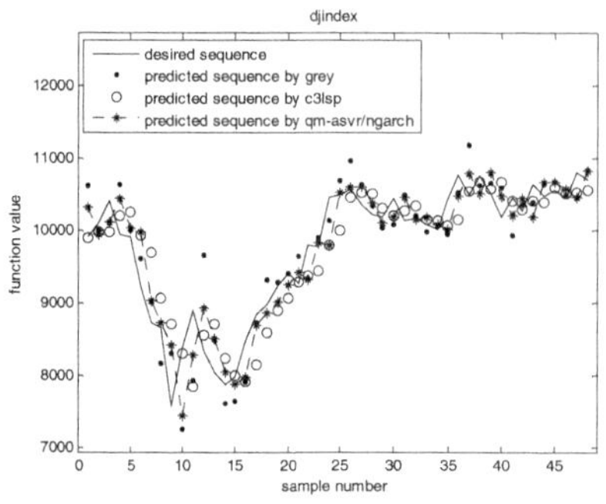


**Fig. 5.** Forecasts of N. Y. -D. J. Industrials. monthly index

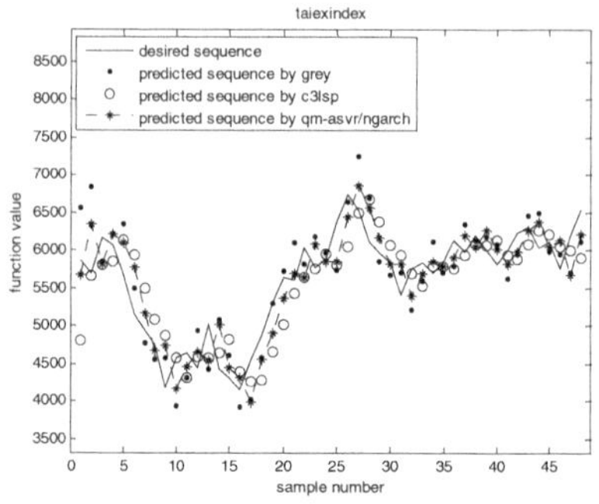


**Fig. 6.** Forecasts of Taiwan TAIEX monthly index

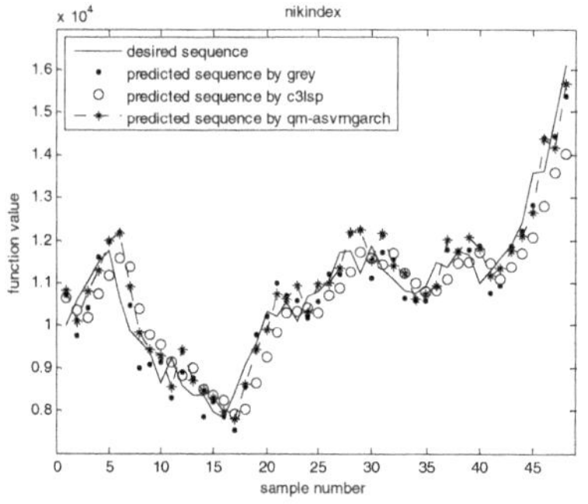


**Fig. 7.** Forecasts of Japan Nikkei monthly index

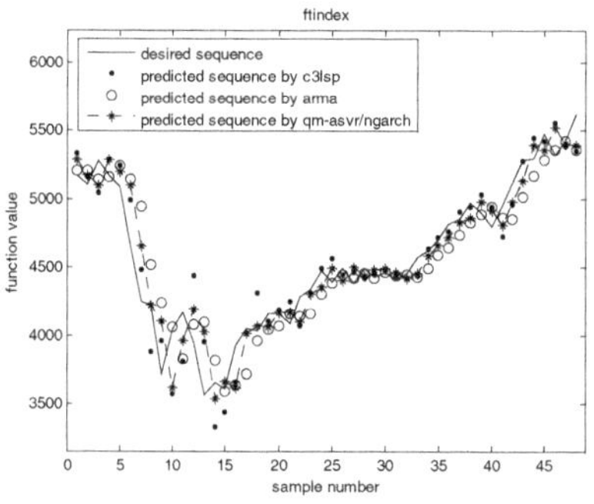


**Fig. 8.** Forecasts of London FTSE-100 monthly index

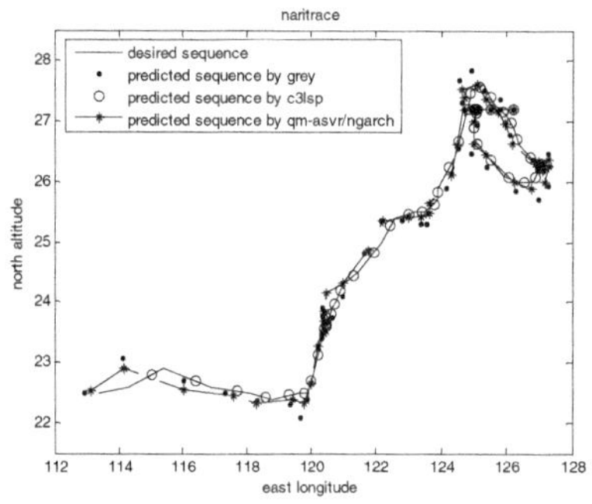


**Fig. 9.** Forecasts of Nari typhoon moving path

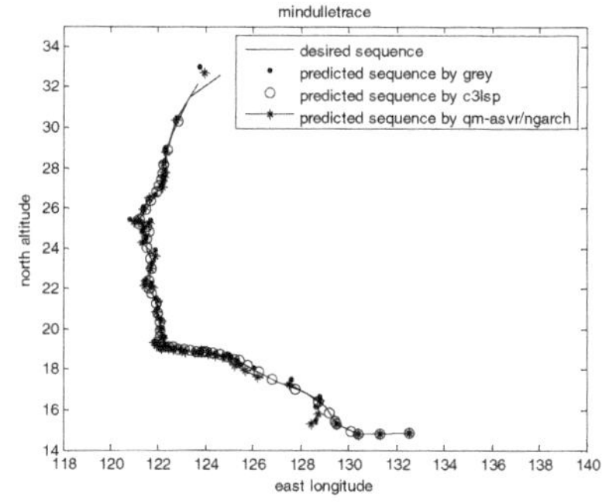


**Fig. 10.** Forecasts of Mindulle typhoon moving path

## 5 Conclusions

This study has proposed a novel scheme that incorporates a nonlinear generalized autoregressive conditional heteroscedasticity (NGARCH) into adaptive support vector regression (ASVR) to form a hybrid model. This hybrid model is tuned optimally by quantum minimization (QM) in such a way that two effects, the overshoot and the volatility clustering, are reduced dramatically at the same time. As a result, big residual errors decrease a lot and thus the predictive accuracy increase significantly. Several methods have been performed under the same two experiments in this study, and consequently the proposed one has outperformed the other methods.

**Acknowledgments.** This work is fully supported by the National Science Council, Taiwan, Republic of China, under grant number **NSC 94-2218-E-143-001**.

## References

1. Box, G. E. P., Jenkins, G. M., Reinsel, G. C.: Time Series Analysis: Forecasting & Control. Prentice-Hall, New Jersey (1994)
2. Deng, J. L.: Control Problems of Grey System. System and Control Letter. 1(5) (1982) 288-294
3. Chang, B. R.: A Study of Non-Periodic Short-Term Random Walk Forecasting Based on RBFNN, ARMA, or SVR-GM(1,1,tau) Approach. In Proc. IJCNN. (2003) 254-259
4. Chang, B. R.: Compensation and Regularization for Improving the Forecasting Accuracy by Adaptive Support Vector Regression. International Journal of Fuzzy Systems. 7(3) (2005) 110-119
5. Vapnik, V.:The Nature of Statistical Learning Theory. Springer-Verlag, New York (1995)
6. Cristianini, N., Shawe-Taylor, J.: An Introduction to Support Vector Machines (and other kernel-based learning methods). Cambridge University Press, London (2000)
7. Bellerslve, T.: Generalized Autoregressive Conditional Heteroscedasticity. Journal of Econometrics. 31 (1986) 307-327
8. Gourieroux, C.: ARCH Models and Financial Applications. Springer-Verlag, New York (1997)
9. Nielsen, M. A., Chuang, I. L.: Quantum computation and quantum information. Cambridge University Press, London (2000)
10. Anguita, D., Ridella, S., Rivieccio, F., Zunino, R.: Training Support Vector Machines: a Quantum- Computing Perspective. In Proc. IEEE IJCNN. (2003) 1587-1592
11. Durr, C., Hoyer, P.: A Quantum Algorithm for Finding the Minimum. (1996) http://arxiv.org/abs/quant-ph/9607014
12. Kreyszig, E.: Advanced Engineering Mathematics. 8th edn. Wiley, New York (1999)
13. Hamilton, J. D.: Time Series Analysis. Princeton University Press, New Jersey (1994)
14. Pshenichnyj, B. N., Wilson, S. S.: The Linearization Method for Constrained Optimization. Springer-Verlag, New York (1994)
15. Grover, L. K.: A Fast Quantum Mechanical Algorithm for Database Search. In Proc.28th Ann. ACM Symp, Theory of Comp., ACM Press, (1996) 212-219
16. Boyer, M., Brassard, G., Hoyer, P., Tapp, A.: Tight Bounds on Quantum Searching. Fortschritte Der Physik, 46 (1998) 493-505

17. Ono, I., Kobayashi, S.: A Real-coded Genetic Algorithm for Function Optimization Using Unimodal Normal Distribution Crossover. In Proc. 7th Int. Conf. on Genetic Algorithms. (1997) 246-253
18. Diebold, F. X..: Elements of Forecasting. South-Western, Cincinnati (1998)
19. FIBV FOCUS MONTHLY STATISTICS, International Stock Price Index (2005)
20. Ljung, G. M., Box, G. E. P.: On a Measure of Lack of Fit in Time Series Models. Biometrika, 65 (1978) 67-72
21. Central Weather Bureau, Typhoon Moving Trace Inquire, Taipei, Taiwan, (2005) http://rdc28.cwb.gov.tw/

# Forecasting of Market Clearing Price by Using GA Based Neural Network

Bo Yang[1,3], Yun-ping Chen[1], Zun-lian Zhao[2], and Qi-ye Han[3]

[1] School of Electrical Engineering, Wuhan University, Wuhan 430072, China
yangbo@ccpg.com.cn
[2] State Grid Corporation of China, Beijing 100031, China
zl-zhao@sgcc.com.cn
[3] Central China Grid Company Limited, Wuhan 430077, China
hanqy@ccpg.com.cn

**Abstract.** Forecasting of Market Clearing Price (MCP) is important to economic benefits of electricity market participants. To accurately forecast MCP, a novel two-stage GA-based neural network model (GA-NN) is proposed. In the first stage, GA chromosome is designed into two parts: boolean coding part for neural network topology and real coding part for connection weights. By hybrid genetic operation of selection, crossover and mutation under the criterion of error minimization between the actual output and the desired output, optimal architecture of neural network is obtained. In the second stage, gradient learning algorithm with momentum rate is imposed on neural network with optimal architecture. After learning process, optimal connection weights are obtained. The proposed model is tested on MCP forecasting in California electricity market. The test results show that GA-NN has self-adaptive ability in its topology and connection weights and can obtain more accurate MCP forecasting values than BP neural network.

## 1 Introduction

In the deregulated electricity industry, market participants, such as independent power producers, transmission and distribution service providers, energy retailers and terminal customers, increasingly care about fluctuation of Market Clearing Price (MCP) and have tried many methods to forecast MCP. MCP forecasting is important to their economic benefits. By accurately forecasting MCP, independent power producers can adjust price and quantity in energy bidding and take strategic actions to obtain additional profits; energy retailers can change price and quantity in energy offering and take effective measures to reduce costs of energy purchased; terminal customers can vary period and quantity in energy consumption according to spot price and lower their costs of energy consumed.

In many existing electricity market, MCP series have the following features [1-6]: uncertainty, probability and complexity. Until now, many methods have been proposed to forecast MCP. These methods can be divided into two main categories: time series and neural network. Time series models, such as ARIMA [7][8], GARCH [9], dynamic regression model and transfer function model [10], grey system model [11-13], have been widely used to forecast MCP, but they have limitations in

D.-S. Huang, L. Heutte, and M. Loog (Eds.): ICIC 2007, LNAI 4682, pp. 1278–1286, 2007.

forecasting accuracy because time series models are predefined mathematic model with given parameters. Neural network is an adaptable system that can learn relationships through repeated presentation of data. It does not construct predefined mathematic model, but construct mapping relationships between input data and output data. It has the potential of forecasting MCP. Artificial neural network [14][15], radial basis function network [16], cascaded neural network [17], neuro-fuzzy [18] and neural networks committee machine [19][20] have been proposed to forecast MCP. But neural network methods have inherited disadvantages. For example, the topology of neural network is hard to determine; the search for optimal weights can get caught into local optimum; network training is slow to converge or even diverges. These disadvantages mentioned above lower MCP forecasting accuracy.

To improve the accuracy of MCP forecasting, a novel two-stage GA-based neural network model (GA-NN), which can automatically determine optimal topology of neural network and train connection weights with faster speed, is proposed. The rest of the paper is arranged as follows. In Section 2, MCP forecasting formulation is given. In Section 3, data structure and training algorithm of GA-NN is provided. In Section 4, GA-NN is tested on California MCP forecasting and test results are presented. At last, some conclusions and remarks are given.

## 2 Formulation of MCP Forecasting

Neural network is an adaptive system composed of many simple processing elements operating in parallel. The function of processing element is determined by network topology, connection weights, and the processing performed at computing elements or nodes. Based on neural network model, MCP forecasting problem can be formulated as follows:

$$\text{Min } E = \sum E_p \tag{1}$$

$$E_p = \frac{1}{2}\sum_{j=1}^{m}\left(y_{pj} - o_{pj}\right)^2 \tag{2}$$

$$\boldsymbol{O}_P = \boldsymbol{F}_l\left(\cdots\left(\boldsymbol{F}_2\left(\boldsymbol{F}_1\left(\boldsymbol{X}_P \boldsymbol{W}_1\right)\boldsymbol{W}_2\right)\cdots\right)\boldsymbol{W}_l\right) \tag{3}$$

where

$$\boldsymbol{W}_l = \begin{bmatrix} w_{1,1}, \cdots, w_{1,h_l} \\ w_{2,1}, \cdots, w_{2,h_l} \\ \cdots\cdots \\ w_{h_{l-1},1}, \cdots, w_{h_{l-1},h_l} \end{bmatrix}_{h_{l-1}\times h_l} \tag{4}$$

and

$$\boldsymbol{W} = \left(\boldsymbol{W}_1, \cdots, \boldsymbol{W}_l\right) \tag{5}$$

$$\boldsymbol{X}_p = \left(x_{p1}, \cdots x_{pi}, \cdots, x_{pn}\right) \tag{6}$$

$$\boldsymbol{Y}_p = \left(y_{p1}, \cdots, y_{pj}, \cdots, y_{pm}\right) \tag{7}$$

$$\boldsymbol{O}_p = \left(o_{p1}, \cdots, o_{pj}, \cdots, o_{pm}\right) \tag{8}$$

In the proposed formulation, $n$ is the dimension of input layer, $m$ is the dimension of output layer, $l$ is the layer number of neural network, $x_{pi}$, $y_{pj}$, and $o_{pj}$ are the *i*th dimension input data, the *j*th dimension desired output data, the *j*th dimension actual output data in the *p*th neural network sample, respectively. $h_l$ is the number of processing elements in the *l*th layer. $\boldsymbol{W}_l$ and $\boldsymbol{F}_l$ are connection weight matrix and activation function matrix in the *l*th layer, respectively.

## 3 GA-NN Forecasting Model

In previous MCP forecasting methods, the topology of neural network is predetermined by user experience. Optimal architecture can not be identified automatically and generalization power of neural network is jeopardized. Besides, due to stochastic of initial weights and search complexity for optimal weights, training algorithm can easily be trapped into local optimum, or be slow to converge or even diverge. Therefore, MCP forecasting accuracy by using traditional neural network can not be improved further. Genetic Algorithm (GA) is a powerful, general-purpose optimization program which mimics the theory of biological evolution. GA starts with a randomized population of parent chromosomes representing various possible solutions to a problem. New child chromosomes are created by selection, crossover and mutation operations. All chromosomes are then evaluated according to a fitness function, with the fittest surviving into the next generation. The result is a gene pool that evolves over time to produce better and better solutions to a problem.

Inspired by GA ability in global searching, we design a novel two-stage GA-based neural network model (GA-NN) by combining powers of genetic algorithm and neural network to improve MCP forecasting accuracy in day-ahead electricity market. In the proposed model, MCP forecasting problem is converted into optimization problem of error minimization between the actual output and the desired output. GA-NN training process is divided into two stages. In the first stage, GA is used to determine optimal topology of neural network automatically and also provides initial connection weights for further training by using hybrid genetic operations of selection, crossover and mutation. In the second stage, gradient learning algorithm with momentum rate is used to train neural network. The output of GA-NN model that has been trained is regarded as MCP forecasting value.

### 3.1 Data Structure of Chromosome

Chromosome design is an artifact for GA-NN forecasting model. In our opinion, chromosome can be designed to be two matrixes: $\boldsymbol{U}$ for network topology and $\boldsymbol{V}$ for connection weights. Row number of the topology matrix represents layer number of neural network and column number represents the number of processing elements in each layer. The *i-j*th element of the topology matrix is a boolean variable which can only be set to zero or one.

$$i \le h = \max(h_1, \cdots, h_l),\ j \le l \tag{9}$$

$u_{i,j} = 0$, if node is absent at *j*th processing elements of *i*th layer

$u_{i,j} = 1$, if node exists at *j*th processing elements of *i*th layer

Connection weight matrix can be defined as follows:

$$V = [W_1, \cdots, W_l] \tag{10}$$

## 3.2 Training Algorithm

Training process of GA-NN can be divided into two stages. In the first stage, GA is the training algorithm, and selection, crossover, and mutation operators are used to evolve population. Optimal architecture of neural network is the result of first stage. We customize crossover and mutation operators according to MCP forecasting problem. Crossover occurs as a probabilistic exchange of genes between two or more chromosomes. One-point crossover of the topology matrix can be defined as follows:

$$\boldsymbol{U}_1 = \begin{bmatrix} u^1_{1,1}, \cdots, u^1_{1,l} \\ u^1_{2,1}, \cdots, u^1_{2,l} \\ \cdots, u^1_{i,j}, \cdots \\ u^1_{h,1}, \cdots, u^1_{h,l} \end{bmatrix}_{h \times l}, \boldsymbol{U}_2 = \begin{bmatrix} u^2_{1,1}, \cdots, u^2_{1,l} \\ u^2_{2,1}, \cdots, u^2_{2,l} \\ \cdots, u^2_{i,j}, \cdots \\ u^2_{h,1}, \cdots, u^2_{h,l} \end{bmatrix}_{h \times l} \tag{11}$$

$$\boldsymbol{U}_{c1} = \begin{bmatrix} u^1_{1,1}, \cdots, u^1_{1,l} \\ u^1_{2,1}, \cdots, u^1_{2,l} \\ \cdots, u^2_{i,j}, \cdots \\ u^1_{h,1}, \cdots, u^1_{h,l} \end{bmatrix}_{h \times l}, \boldsymbol{U}_{c2} = \begin{bmatrix} u^2_{1,1}, \cdots, u^2_{1,l} \\ u^2_{2,1}, \cdots, u^2_{2,l} \\ \cdots, u^1_{i,j}, \cdots \\ u^2_{h,1}, \cdots, u^2_{h,l} \end{bmatrix}_{h \times l} \tag{12}$$

$i$ and $j$ are generated randomly. Crossover of the connection weight matrix can be defined as follows:

$$\begin{aligned} \boldsymbol{V}_{c1} &= \lambda \boldsymbol{V}_1 + (1-\lambda)\boldsymbol{V}_2 \\ \boldsymbol{V}_{c2} &= \lambda \boldsymbol{V}_2 + (1-\lambda)\boldsymbol{V}_1 \end{aligned} \tag{13}$$

$\lambda$ is a random real number between 0 and 1.Mutation involves the random replacement of genes in a chromosome. Mutation of element in the topology matrix can be defined as follows:

$$u_{i,j} = 0,\ \text{if}\ u_{i,j} = 1\ ;\ u_{i,j} = 1,\ \text{if}\ u_{i,j} = 0 \tag{14}$$

Mutation of element in the connection weight matrix can be defined as follows:

$$w_{k,i} = random(0,1),\ w_{k,i} \in \boldsymbol{W}_j\ \text{if}\ u_{i,j} \ne 0 \cap u_{k,j-1} \ne 0 \tag{15}$$

In the second stage, gradient learning algorithm with momentum rate is used to train neural network. The optimal topology and initial weights of GA-NN has been determined by GA in the first stage. The update of weight in the second stage is given as follows:

$$\Delta W = -\alpha \frac{\partial E}{\partial W} + \beta \Delta W' \qquad (16)$$

$\alpha$ and $\beta$ are learning rate and momentum rate respectively. $\Delta W$ and $\Delta W'$ are weight increment matrix in current iteration and in the last iteration respectively.

## 4 Simulation Results

To test the performance of GA-NN for MCP forecasting, MCP series in California day-ahead electricity market is chosen as examples. The California electricity restructuring began from its planning in the early 1990s, and unexpectedly experienced market crisis in 2000 and 2001. The crisis began with very high prices in the summer of 2000 in the San Diego, and then widespread through out the region in early 2001 as the state experienced rolling blackouts, prices over 10 times usual levels, and bankruptcy or near-bankruptcy of market utilities [21]. Since then, regulatory institutions and market participants have realized that MCP is index of market stability and MCP forecasting is needed both in short-run and in long-run period.

Because of data acquisition limitation, history MCP series and energy demand series are regarded as determining factors of MCP forecasting. MCP series from January 16, 1999 to October 15, 1999 are listed in Figure 1 below.

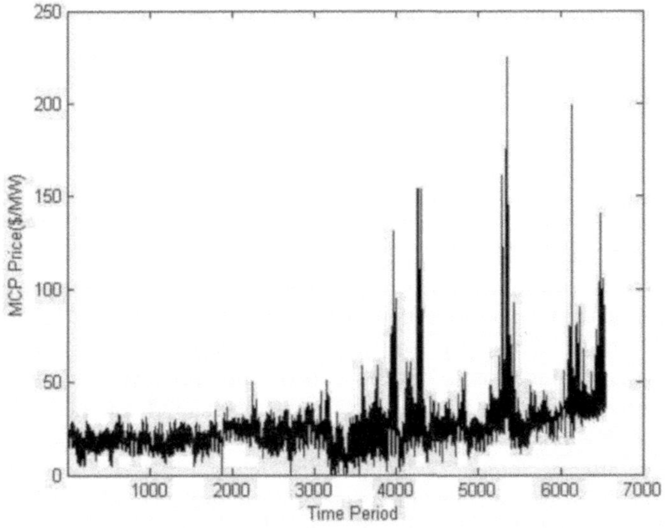


**Fig. 1.** MCP series from January 16, 1999 to October 15, 1999

Accordingly, energy demand series from January 16, 1999 to October 15, 1999 are listed in Figure 2 below.

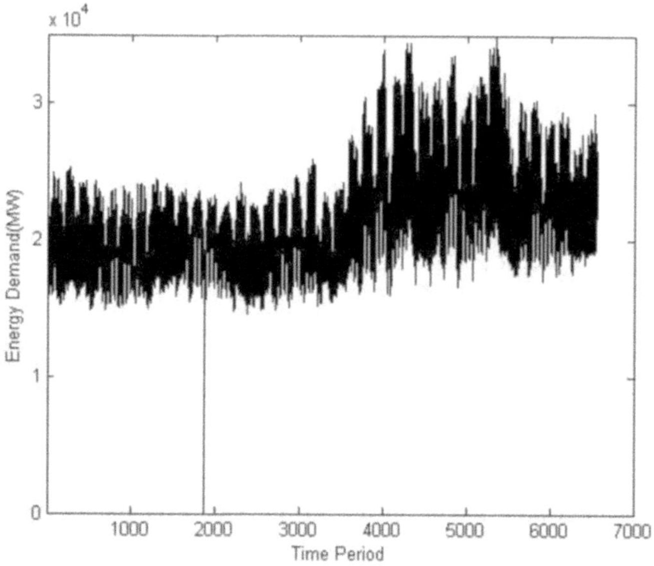


**Fig. 2.** Energy demand series from January 16, 1999 to October 15, 1999

The input data of GA-NN is given below.

$$X = (p_{d-14,t-1}, p_{d-7,t-1}, p_{d-2,t-1}, p_{d-1,t-1}, p_{d-14,t}, p_{d-7,t}, p_{d-2,t}, p_{d-1,t}, p_{d-14,t+1}, p_{d-7,t+1}, p_{d-2,t+1}, p_{d-1,t+1}, d_{d-14,t-1}, d_{d-7,t-1}, d_{d-2,t-1}, d_{d-1,t-1}, d_{d-14,t}, d_{d-7,t}, d_{d-2,t}, d_{d-1,t}, d_{d-14,t+1}, d_{d-7,t+1}, d_{d-2,t+1}, d_{d-1,t+1}) \quad (17)$$

$d-14$, $d-7$, $d-2$ ,or $d-1$ refer to the day ahead of two weeks, one week, two days, or one day compared with the forecasting day. $t-1, t$, or $t+1$ refer to the time period that previous to, same to, or next to the forecasting time period. $p$ and $d$ are MCP and energy demand respectively.

According to neural network theory, a single hidden layer multilayer perception can learn any desired continuous input-output mapping if there are sufficient numbers of axons in the hidden layer, so we choose three-layer neural network in our experiments. The number of processing elements in input layer and output layer is 24 and 1, respectively. The number of processing elements in hidden layer ranges from 4 to 30. The parameters of GA-NN model are given as follows. Population size is 50, selection operator is roulette selection, and maximum iteration number is 500. Crossover probability and mutation probability are 0.9 and 0.01 respectively. Forecasting day is October 16, 1999. Training data set ranges from January 16, 1999 to October 15, 1999. To improve generalization ability and avoid overtraining, 20% of training samples are extracted from training data set and belong to validation set that is used for cross validation. After the first-stage training, optimal topology and initial forecasting value of GA-NN are achieved and given in Table 1 below. The average error of initial MCP forecasting values is 20.36%.

**Table 1.** MCP forecasting value and error of GA-NN in the first stage

| Time period | Optimal topology | Value ($/MW) | Error (%) | Time period | Optimal topology | Value ($/MW) | Error (%) |
|---|---|---|---|---|---|---|---|
| 1 | 24-9-1 | 32.8126 | 6.93 | 13 | 24-10-1 | 55.7982 | 20.03 |
| 2 | 24-15-1 | 32.7392 | 6.43 | 14 | 24-9-1 | 61.7361 | 36.11 |
| 3 | 24-27-1 | 31.2610 | 6.67 | 15 | 24-18-1 | 79.3032 | 74.84 |
| 4 | 24-28-1 | 27.9829 | 13.2 | 16 | 24-28-1 | 63.2268 | 30.27 |
| 5 | 24-23-1 | 26.6673 | 20.38 | 17 | 24-14-1 | 71.5177 | 41.66 |
| 6 | 24-23-1 | 26.6723 | 21.52 | 18 | 24-13-1 | 68.5145 | 38.72 |
| 7 | 24-19-1 | 28.9956 | 10.09 | 19 | 24-8-1 | 69.7585 | 39.54 |
| 8 | 24-21-1 | 34.0097 | 5.46 | 20 | 24-26-1 | 51.8163 | 1.26 |
| 9 | 24-18-1 | 33.4414 | 4.44 | 21 | 24-13-1 | 53.1949 | 6.35 |
| 10 | 24-21-1 | 47.9033 | 31.21 | 22 | 24-23-1 | 41.7173 | 1.72 |
| 11 | 24-25-1 | 49.7123 | 18.37 | 23 | 24-21-1 | 37.1843 | 13.77 |
| 12 | 24-21-1 | 57.6719 | 29.60 | 24 | 24-9-1 | 31.7125 | 10.05 |

After the second-stage training, better MCP forecasting values are achieved and listed in Table 2 below. The average error of final MCP forecasting values is 11.37%.

**Table 2.** MCP forecasting value and error of GA-NN in the second stage

| Time period | Desired MCP ($/MW) | Value ($/MW) | Error (%) | Time period | Desired MCP ($/MW) | Value ($/MW) | Error (%) |
|---|---|---|---|---|---|---|---|
| 1 | 35.2551 | 32.8129 | 6.93 | 13 | 46.4887 | 55.7982 | 20.03 |
| 2 | 34.9901 | 32.7392 | 6.43 | 14 | 45.3587 | 61.7361 | 36.11 |
| 3 | 33.4933 | 31.2610 | 6.67 | 15 | 45.3583 | 44.5038 | 1.88 |
| 4 | 32.2475 | 28.3337 | 12.14 | 16 | 48.5365 | 63.2268 | 30.27 |
| 5 | 33.4919 | 28.0701 | 16.19 | 17 | 50.4844 | 46.3503 | 8.19 |
| 6 | 33.9843 | 26.6723 | 21.52 | 18 | 49.3895 | 53.3205 | 7.96 |
| 7 | 32.2493 | 28.9956 | 10.09 | 19 | 49.9904 | 40.3973 | 19.19 |
| 8 | 32.2477 | 31.5164 | 2.27 | 20 | 51.1735 | 50.8192 | 0.69 |
| 9 | 34.9935 | 33.4414 | 4.44 | 21 | 50.018 | 53.1949 | 6.35 |
| 10 | 36.5087 | 39.2587 | 7.53 | 22 | 41.0108 | 39.9886 | 2.49 |
| 11 | 41.9972 | 47.6822 | 13.54 | 23 | 43.1224 | 37.6873 | 12.60 |
| 12 | 44.5001 | 49.3477 | 10.89 | 24 | 35.256 | 32.2793 | 8.44 |

To compare GA-NN with BP in MCP forecasting, BP neural network is also used to forecast MCP on October 16, 1999. The architecture of BP neural network is 24-24-1. Results are listed in Figure 3. From the Figure below, we can know that in the time periods with low demand, for example, from 1th to 11th time period, both GA-NN and BP can achieve satisfactory forecasting results, while in the periods with peak demand, for example, from 12th to 19th time period, GA-NN can get much better forecasting results than BP. The cause for the conclusion above maybe is that in peak demand period, there leaves two much room for independent power providers to exert their market power. Apparently, whether in low demand period or in peak demand period, GA-NN is superior to BP neural network.

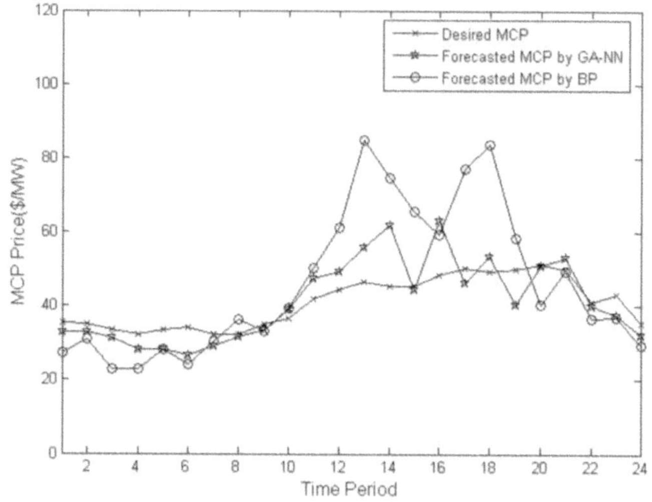


**Fig. 3.** MCP forecasting results by GA-NN and BP

## 5 Conclusions

MCP forecasting can greatly affect economic benefits of participants in electricity market. The proposed GA-NN model in the paper can be used to forecast MCP. In a electricity market without too much market power exerted by independent power producers, our GA-NN model can achieve satisfactory results. Compared with BP neural network, GA-NN model dose not need user experience to predetermine optimal topology of neural network. From our experiments, GA-NN is superior to BP neural network.

## References

1. Benini, M., Marracci, M., Pelacchi, P., Venturini, A.: Day-ahead Market Price Volatility Analysis in Deregulated Electricity Markets. Proceedings of IEEE Power Engineering Society Summer Meeting. 3 (2002) 1354-1359
2. Breipohl, A. M.: Electricity Price Forecasting Models. Proceedings of IEEE Power Engineering Society Winter Meeting. 2(2002)963-966

3. Bunn, D. W.: Forecasting Loads and Prices in Competitive Power Markets. Proceedings of IEEE. 2 (88) (2000)163-169
4. Obradovic, Z., Tomsovic, K.: Time Series Methods for Forecasting Electricity Market Pricing. Proceedings of IEEE Power Engineering Society Summer Meeting. 2(1999)1264-1265
5. Ni, E., Luh, P. B.: Forecasting Power Market Clearing Price and Its Discrete PDF Using a Bayesian-based Classification Method. Proceedings of IEEE Power Engineering Society Winter Meeting. 3(2001)1518-1523
6. Bastian, J., Zhu, J. X., Banunarayanan, V., Mukerji, R.: Forecasting Energy Prices in a Competitive Market. Computer Applications in Power. 3 (12)(1999)40-45
7. Contreras, J., Espinola, R., Nogales, F. J., Conejo, A. J.: ARIMA Models to Predict Next-day Electricity Prices. IEEE Trans on Power Systems. 3(18)(2003)1014-1020
8. Conejo, A. J., Plazas, M. A., Espinola, R., Molina, A. B.: Day-ahead Electricity Price Forecasting Using the Wavelet Transform and ARIMA Models. IEEE Trans on Power Systems. 2(20)(2005)1035-1042
9. Garcia, R. C., Contreras, J., Akkeren, M. V., Garcia, B. C.: A GARCH Forecasting Model to Predict Day-ahead Electricity Prices. IEEE Trans on Power Systems. 2(20)(2005)867-873
10. Nogales, F. J., Contreras, J., Conejo, A. J., Espinola, R.: Forecasting Next-day Electricity Prices by Time Series Models. IEEE Trans on Power Systems. 2(17)(2002)342-348
11. Du, S. H., Hou, Z. J., Jiang, C. W.: A New Short-term Grey Forecasting Procedure of Spot Price. Journal of Grey System. 4(14)(2002)351-358
12. Du, S. H., Hou, Z. J., Jiang, C. W.: Grey Forecasting Price Mutation and Its Simulation. Journal of Grey System. 1(15)(2003)43-48
13. Ma, X., Hou, Z. J., Jiang, C. W.: Grey Forecasting Electricity Forward Price. Journal of Grey System. 3 (15) (2003) 263-266
14. Szkuta, B. R., Sanabria, L. A., Dillon, T. S.: Electricity Price Short-term Forecasting Using Artificial Neural Networks. IEEE Trans on Power Systems. 3(14)(1999) 851-857
15. Yamin, H. Y., Shahidehpour, S. M., Li, Z.: Adaptive Short-term Electricity Price Forecasting Using Artificial Neural Networks in The Restructured Power Markets. Electrical power and energy systems. 8 (26) (2004) 571-581
16. Guo, J. J., Luh, P. B.: Selecting Input Factors for Clusters of Gaussian Radial Basis Function Networks to Improve Market Clearing Price Prediction. IEEE Trans Power Systems. 3(18)(2003)665-672
17. Zhang, L., Luh, P. B., Kasiviswanathan, K.: Energy Clearing Price Prediction and Confidence Interval Estimation With Cascaded Neural Networks. IEEE Trans on Power Systems. 1(18)(2003)99-105
18. Rodriguez, C. P., Anders, G. J.: Energy Price Forecasting in the Ontario Competitive Power System Market. IEEE Trans on Power Systems. 1(19)(2004)366-374
19. Guo, J. J., Luh, P. B.: Improving Market Clearing Price Prediction by Using a Committee Machine of Neural Networks. IEEE Trans on Power Systems. 4(19)(2004)1867-1876
20. Guo, J. J., Luh, P. B.: Market Clearing Price Prediction Using a Committee Machine With Adaptive Weighting Coefficients. Proceedings of IEEE Power Engineering Society Winter Meeting. 1(2002)77-82
21. Blumstein, C., Friedman, L. S., Green, R. J.: The History of Electricity Restructuring in California. Http://www.ucei.org

# A Difference Scheme for the Camassa-Holm Equation

Ahamed Adam Abdelgadir, Yang-xin Yao, Yi-ping Fu, and Ping Huang

School of Mathematical Sciences, South China University of Technology, Guangzhou 510640, P.R. China

**Abstract.** In this paper, a difference scheme satisfying two conservative laws for the periodic Camassa-Holm equation is given. Convergence is proved. The numerical simulation exhibits the time evolution and the interaction of solitary wave solutions.

## 1 Introduction

Recently, a considerable amount of research activity has focussed on the study of the Camassa-Holm equation

$$u_t - u_{xxt} + 3uu_x = 2u_x u_{xx} + uu_{xxx}, \quad t > 0, \ x \in \mathbb{R} \tag{1}$$

which models the unidirectional propagation of shallow water waves over a flat bottom, $u(x,t)$ stands for the fluid velocity at time $t$ in the spatial $x$ direction (see [1]). There are a great amount of works (see [2-9] and their references) contributed to the study of the analytic solutions for the equation, such as the existence and regularity of solutions, the solitary wave solutions and "blow-up" behaviors of the solutions. Also, there are some works contributed to the study of the numerical approximation method and the numerical computations for the solutions of Camassa-Holm equation, such as the works in [10] and [11]. The papers [10] and [11] explored numerically the different aspects of periodic travelling-wave solutions and the phenomenon of singularith formation for the Eq.(1) by using a spectral method, respectively.

In this paper, we consider the difference method for the periodic initial-boundary problem of Eq.(1). The boundary condition is

$$u(x+L,t) = u(x,t), \quad t \geq 0, \ x \in \mathbb{R}, \tag{2}$$

and the initial value condition is

$$u(x,0) = u_0(x), \quad x \in \mathbb{R}. \tag{3}$$

The problem (1)–(3) has at least the following two conserved quantities

$$I_1 = \int_0^L u, \quad I_2 = \int_0^L (u^2 + u_x^2).$$

D.-S. Huang, L. Heutte, and M. Loog (Eds.): ICIC 2007, LNAI 4682, pp. 1287–1295, 2007.

Here we give a difference scheme keeping the above two conserved quantities. It is known that satisfying conservative laws is important for a numerical scheme. In comparison to the numerical methods used in [10] and [11], the two conserved quantities are kept better in this paper.

The paper is organized as follows. In Sect.2, we give a conservative difference scheme and prove its convergence. Sec.3 is devoted to simulate the evolution of solitary wave solutions of Eq.(1).

## 2 Difference Scheme, Conservative Laws and Convergence

For any given $T > 0$, divide the domain $[0, L] \times [0, T]$ by parallel lines $x = jh(j = 0, 1, \cdots, J)$ and $t = n\tau(n = 0, 1, \cdots, N)$, where $Jh = L$, $N\tau = T$, $h$ and $\tau$ are meshsteps. For convenience, the following notations are used:

$$\Delta_+ u_j^n = \frac{u_{j+1}^n - u_j^n}{h}, \quad \Delta_- u_j^n = \frac{u_j^n - u_{j-1}^n}{h},$$

$$\Delta_0 u_j^n = \frac{1}{2}(\Delta_+ + \Delta_-)u_j^n, \quad u_j^{n+\frac{1}{2}} = \frac{1}{2}(u_j^{n+1} + u_j^n),$$

$$(u^n, v^n) = \sum_{j=1}^{J} u_j^n v_j^n h, \quad \|u^n\|^2 = (u^n, u^n), \quad \|u^n\|_\infty = \max_{1\leq j\leq J} |u_j^n|.$$

$C$ denotes a general constant which may have different values in different places.

Under the periodic boundary condition, there are the relations

$$(\Delta_+ u, v) = -(u, \Delta_- v), \quad (\Delta_0 u, v) = -(u, \Delta_0 v). \tag{4}$$

Now we construct the finite difference scheme for the system (1)–(3)

$$\begin{aligned} &\frac{u_j^{n+1} - u_j^n}{\tau} - \frac{\Delta_+\Delta_-(u_j^{n+1} - u_j^n)}{\tau} + \Delta_0(u_j^{n+\frac{1}{2}})^2 + u_j^{n+\frac{1}{2}}\Delta_0 u_j^{n+\frac{1}{2}} \\ &= 2\Delta_0 u_j^{n+\frac{1}{2}}\Delta_+\Delta_- u_j^{n+\frac{1}{2}} + \frac{1}{2}u_{j-1}^{n+\frac{1}{2}}\Delta_-\Delta_+\Delta_- u_j^{n+\frac{1}{2}} \\ &+ \frac{1}{2}u_{j+1}^{n+\frac{1}{2}}\Delta_+\Delta_+\Delta_- u_j^{n+\frac{1}{2}}, \end{aligned} \tag{5}$$

$$u_{j+J}^n = u_j^n, \tag{6}$$

$$u_j^0 = u_0(jh). \tag{7}$$

The discrete forms of two conservative laws $I_1$ and $I_2$ are

$$I_1^n = \sum_{j=1}^{J} u_j^n h, \quad I_2^n = \sum_{j=1}^{J}((u_j^n)^2 + (\Delta_+ u_j^n)^2)h.$$

**Lemma 1.** *The solution* $\{u_j^n\}$ *of the difference scheme (5)–(7) satisfies the following conservative laws*

$$I_1^n = const, \quad I_2^n = const.$$

*Proof.* Computing the inner product of the difference Eq.(5) with 1 and $u^n$ respectively, we have

$$\sum_{j=1}^{J} u_j^{n+1} h - \sum_{j=1}^{J} u_j^n h = 0,$$

$$\|u^{n+1}\|^2 - \|u^n\|^2 + \|\Delta_+ u^{n+1}\|^2 - \|\Delta_+ u^n\|^2 = 0,$$

where we have used (4). Then Lemma 1 is gotten from the above relations.

In order to prove the convergence of the scheme (5)–(7), the following two lemmas will be needed.

**Lemma 2 (Discrete Sobolev's inequality [12]).** *Suppose that* $\{u_j^n\}$ *is mesh functions. Given* $\varepsilon > 0$, *there exists a constant* $C$ *dependent on* $\varepsilon$ *such that*

$$\|u^n\|_\infty \leq \varepsilon \|u_x^n\| + C\|u^n\|.$$

**Lemma 3 (Gronwall's inequality [12]).** *Suppose that nonnegative mesh functions* $\{w(n), \rho(n), n = 1, 2, \cdots, N, N\tau = T\}$ *satisfy the inequality*

$$w(n) \leq \rho(n) + \sum_{k=1}^{n} B_k w(k)\tau,$$

*where* $B_k$ $(k = 1, 2, \cdots, N)$ *are nonnegative constant. Then for any* $0 \leq n \leq N$, *there is*

$$w(n) \leq \rho(n) \exp(n\tau \sum_{k=1}^{N} B_k).$$

**Corollary 1.** *The solution* $\{u_j^n\}$ *of the difference scheme (5)–(7) satisfies the estimation*

$$\|u^n\|_\infty \leq C.$$

*Proof.* According to Lemma 1 and Lemma 2, the corollary follows immediately.

From Taylor's expansion, it can be easily obtained that

**Lemma 4.** *Assume* $u(x,t) \in C^{(5,3)}$, *then the truncation error of the scheme (5)–(7) is of order* $O(h^2 + \tau^2)$.

**Theorem 1.** *Under the condition of Lemma 4, the numerical solution of the scheme (5)–(7) converges to the solution of the system (1)–(3) with order* $O(h^2 + \tau^2)$ *by norm* $\|\cdot\|_\infty$.

*Proof.* Suppose $u(x,t)$ is the exact solution of the system (1)–(3) and $v_j^n = u(x_j, t_n)$. Let $e_j^n = u_j^n - v_j^n$, then $e_j^n$ satisfies

$$\frac{e_j^{n+1}-e_j^n}{\tau} - \frac{\Delta_+\Delta_-(e_j^{n+1}-e_j^n)}{\tau} = -\Delta_0((v_j^{n+\frac{1}{2}}+u_j^{n+\frac{1}{2}})e_j^{n+\frac{1}{2}})$$
$$-e_j^{n+\frac{1}{2}}\Delta_0 u_j^{n+\frac{1}{2}} - v_j^{n+\frac{1}{2}}\Delta_0 e_j^{n+\frac{1}{2}} + 2\Delta_0 e_j^{n+\frac{1}{2}}\Delta_+\Delta_- u_j^{n+\frac{1}{2}}$$
$$+2\Delta_0 v_j^{n+\frac{1}{2}}\Delta_+\Delta_- e_j^{n+\frac{1}{2}} + \frac{1}{2}e_{j-1}^{n+\frac{1}{2}}\Delta_-\Delta_+\Delta_- u_j^{n+\frac{1}{2}} + \frac{1}{2}v_{j-1}^{n+\frac{1}{2}}\Delta_-\Delta_+\Delta_- e_j^{n+\frac{1}{2}}$$
$$+\frac{1}{2}e_{j+1}^{n+\frac{1}{2}}\Delta_+\Delta_+\Delta_- u_j^{n+\frac{1}{2}} + \frac{1}{2}v_{j+1}^{n+\frac{1}{2}}\Delta_+\Delta_+\Delta_- e_j^{n+\frac{1}{2}} + O(h^2+\tau^2), \qquad (8)$$

with periodic boundary condition

$$e_{j+J}^n = e_j^n,$$

and initial data

$$e_j^0 = 0.$$

Computing the inner product of (8) with $e_j^{n+\frac{1}{2}}$, we obtain

$$\frac{\|e^{n+1}\|^2-\|e^n\|^2}{2\tau} + \frac{\|\Delta_+e^{n+1}\|^2-\|\Delta_+e^n\|^2}{2\tau}$$
$$= -(\Delta_0((v^{n+\frac{1}{2}}+u^{n+\frac{1}{2}})e^{n+\frac{1}{2}}), e^{n+\frac{1}{2}}) - (e^{n+\frac{1}{2}}\Delta_0 u^{n+\frac{1}{2}} - v^{n+\frac{1}{2}}\Delta_0 e^{n+\frac{1}{2}}, e^{n+\frac{1}{2}})$$
$$+ 2(\Delta_0 e^{n+\frac{1}{2}}\Delta_+\Delta_- u^{n+\frac{1}{2}}, e^{n+\frac{1}{2}}) + 2(\Delta_0 v^{n+\frac{1}{2}}\Delta_+\Delta_- e^{n+\frac{1}{2}}, e^{n+\frac{1}{2}})$$
$$+\frac{1}{2}\sum_{j=1}^{J} e_j^{n+\frac{1}{2}} e_{j-1}^{n+\frac{1}{2}}\Delta_-\Delta_+\Delta_- u_j^{n+\frac{1}{2}} h + \frac{1}{2}\sum_{j=1}^{J} e_j^{n+\frac{1}{2}} v_{j-1}^{n+\frac{1}{2}}\Delta_-\Delta_+\Delta_- e_j^{n+\frac{1}{2}} h$$
$$+\frac{1}{2}\sum_{j=1}^{J} e_j^{n+\frac{1}{2}} e_{j+1}^{n+\frac{1}{2}}\Delta_+\Delta_+\Delta_- u_j^{n+\frac{1}{2}} h + \frac{1}{2}\sum_{j=1}^{J} e_j^{n+\frac{1}{2}} v_{j+1}^{n+\frac{1}{2}}\Delta_+\Delta_+\Delta_- e_j^{n+\frac{1}{2}} h$$
$$+ (O(h^2+\tau), e^{n+\frac{1}{2}}).$$
$$= A_1 + A_2 + A_3 + A_4 + A_5 + A_6 + A_7 + A_8 + A_9 \ . \qquad (9)$$

Now, we estimate the terms $A_1$–$A_9$ of Eq.(9). By using (4), Corollary 1, Young's inequality, and paying attention to $u(x,t) \in C^{(5,3)}([0,L]\times[0,T])$, we have

$$A_1 = (((v^{n+\frac{1}{2}}+u^{n+\frac{1}{2}})e^{n+\frac{1}{2}}), \Delta_0 e^{n+\frac{1}{2}})$$
$$\leq \|u^{n+\frac{1}{2}}+v^{n+\frac{1}{2}}\|_\infty \|e^{n+\frac{1}{2}}\|\|\Delta_0 e^{n+\frac{1}{2}}\|$$
$$\leq C(\|e^{n+1}\|^2 + \|e^n\|^2 + \|\Delta_+e^{n+1}\|^2 + \|\Delta_+e^n\|^2), \qquad (10)$$

$$A_2 = (u^{n+\frac{1}{2}}, \Delta_0(e^{n+\frac{1}{2}})^2) - (v^{n+\frac{1}{2}}\Delta_0 e^{n+\frac{1}{2}}, e^{n+\frac{1}{2}})$$
$$\leq C(\|e^{n+1}\|^2 + \|e^n\|^2 + \|\Delta_+e^{n+1}\|^2 + \|\Delta_+e^n\|^2), \qquad (11)$$

$$A_9 \leq (O(h^2+\tau^2))^2 + \|e^{n+1}\|^2 + \|e^n\|^2, \tag{12}$$

$$\begin{aligned} A_3 + A_5 + A_7 &= 2(\Delta_0(e^{n+\frac{1}{2}}\Delta_+\Delta_- u^{n+\frac{1}{2}}, e^{n+\frac{1}{2}}) \\ &\quad - \frac{1}{2}\sum_{j=1}^{J} \Delta_+(e_j^{n+\frac{1}{2}} e_{j-1}^{n+\frac{1}{2}})\Delta_+\Delta_- u_j^{n+\frac{1}{2}} h \\ &\quad - \frac{1}{2}\sum_{j=1}^{J} \Delta_-(e_j^{n+\frac{1}{2}} e_{j+1}^{n+\frac{1}{2}})\Delta_+\Delta_- u_j^{n+\frac{1}{2}} h \\ &= 2(\Delta_0(e^{n+\frac{1}{2}}\Delta_+\Delta_- u^{n+\frac{1}{2}}, e^{n+\frac{1}{2}}) \\ &\quad - \frac{1}{2}(\Delta_+\Delta_- u^{n+\frac{1}{2}}, e^{n+\frac{1}{2}}\Delta_+ e^{n+\frac{1}{2}} + e^{n+\frac{1}{2}}\Delta_- e^{n+\frac{1}{2}}) \\ &\quad - \frac{1}{2}(\Delta_+\Delta_- u^{n+\frac{1}{2}}, e^{n+\frac{1}{2}}\Delta_- e^{n+\frac{1}{2}} + e^{n+\frac{1}{2}}\Delta_+ e^{n+\frac{1}{2}}) \\ &= 0, \end{aligned} \tag{13}$$

$$\begin{aligned} A_4 + A_6 + A_8 &= 2(\Delta_0(v^{n+\frac{1}{2}}\Delta_+\Delta_- e^{n+\frac{1}{2}}, e^{n+\frac{1}{2}}) \\ &\quad - \frac{1}{2}\sum_{j=1}^{J} \Delta_+\Delta_- e_j^{n+\frac{1}{2}} \Delta_+(e_j^{n+\frac{1}{2}} v_{j-1}^{n+\frac{1}{2}}) h \\ &\quad - \frac{1}{2}\sum_{j=1}^{J} \Delta_+\Delta_- e_j^{n+\frac{1}{2}} \Delta_-(e_j^{n+\frac{1}{2}} v_{j+1}^{n+\frac{1}{2}}) h \\ &= (\Delta_+\Delta_- e^{n+\frac{1}{2}}, e^{n+\frac{1}{2}}\Delta_0 v^{n+\frac{1}{2}}) - (\Delta_+\Delta_- e^{n+\frac{1}{2}}, v^{n+\frac{1}{2}}\Delta_0 e^{n+\frac{1}{2}}) \\ &= A_{10} + A_{11}, \end{aligned} \tag{14}$$

where

$$\begin{aligned} A_{10} &= -(\Delta_+ e^{n+\frac{1}{2}}, \Delta_+(e^{n+\frac{1}{2}}\Delta_0 v^{n+\frac{1}{2}})) \\ &\leq C(\|e^{n+1}\|^2 + \|e^n\|^2 + \|\Delta_+ e^{n+1}\|^2 + \|\Delta_+ e^n\|^2), \end{aligned} \tag{15}$$

$$\begin{aligned} A_{11} &= \frac{1}{2}(\Delta_- e^{n+\frac{1}{2}}, \Delta_-(v^{n+\frac{1}{2}}\Delta_+ e^{n+\frac{1}{2}})) + \frac{1}{2}(\Delta_+ e^{n+\frac{1}{2}}, \Delta_+(v^{n+\frac{1}{2}}\Delta_- e^{n+\frac{1}{2}})) \\ &= (\Delta_0 e^{n+\frac{1}{2}}, v^{n+\frac{1}{2}}\Delta_+\Delta_- e^{n+\frac{1}{2}}) + \frac{1}{2}(\Delta_- e^{n+\frac{1}{2}}, \Delta_- v^{n+\frac{1}{2}}\Delta_- e^{n+\frac{1}{2}}) \\ &\quad + \frac{1}{2}(\Delta_+ e^{n+\frac{1}{2}}, \Delta_+ v^{n+\frac{1}{2}}\Delta_+ e^{n+\frac{1}{2}}). \end{aligned} \tag{16}$$

Equation (16) leads to

$$\begin{aligned} 2A_{11} &= \frac{1}{2}(\Delta_- e^{n+\frac{1}{2}}, \Delta_-(v^{n+\frac{1}{2}}\Delta_- e^{n+\frac{1}{2}}) + \frac{1}{2}(\Delta_+ e^{n+\frac{1}{2}}, \Delta_+(v^{n+\frac{1}{2}}\Delta_+ e^{n+\frac{1}{2}}) \\ &\leq C(\|\Delta_+ e^{n+1}\|^2 + \|\Delta_+ e^n\|^2). \end{aligned} \tag{17}$$

Inserting the inequalities (10)–(15) and (17) into (9), it follows that

$$\frac{1}{\tau}(\|e^{n+1}\|^2 - \|e^n\|^2 + \|\Delta_+ e^{n+1}\|^2 - \|\Delta_+ e^n\|^2)$$
$$\leq C(\|e^{n+1}\|^2 + \|e^n\|^2 + \|\Delta_+ e^{n+1}\|^2 + \|\Delta_+ e^n\|^2) + (O(h^2+\tau^2))^2.$$

Thus, it is easy to get that

$$\|e^{n+1}\|^2 \leq (O(h^2+\tau^2))^2 + \tau C\sum_{k=1}^{n+1}(\|e^k\|^2 + \|\Delta_+ e^k\|^2)$$

According to Lemma 3, we obtain

$$\|e^n\| + \|\Delta_+ e^n\| \leq O(h^2+\tau^2).$$

Lastly, according to Corollary 1, this implies

$$\|e^n\|_\infty \leq O(h^2+\tau^2).$$

Hence the convergence is proved.

## 3 Numerical Simulation

Camassa and Hlom [1] discovered that Eq.(1) has solitons such as $u(x,t) = ce^{-|x-ct|}$ ($c$ is wave speed). Constantin and Strauss [6] proved the orbital stability of the solitons. We know that solitons asymptotically preserve its shape and velocity upon nonlinear interaction with other solitary waves [13]. In this section, by using the difference scheme, we simulate numerically the changing patterns of solitary wave solutions.

The scheme (5)–(7) is a nonlinear algebraic system about $u_j^{n+1}$. To solve it, the following iterative process can be applied for $s = 0, 1, 2, \cdots$,

$$\frac{u_j^{n+1(s+1)} - u_j^n}{\tau} - \frac{\Delta_+\Delta_-(u_j^{n+1(s+1)} - u_j^n)}{\tau}$$
$$= -\Delta_0(u_j^{n(s)+\frac{1}{2}})^2 + u_j^{n(s)+\frac{1}{2}}\Delta_0 u_j^{n(s)+\frac{1}{2}} + 2\Delta_0 u_j^{n(s)+\frac{1}{2}}\Delta_+\Delta_- u_j^{n(s)+\frac{1}{2}}$$
$$+\frac{1}{2}u_{j-1}^{n(s)+\frac{1}{2}}\Delta_-\Delta_+\Delta_- u_j^{n(s)+\frac{1}{2}} + \frac{1}{2}u_{j+1}^{n(s)+\frac{1}{2}}\Delta_+\Delta_+\Delta_- u_j^{n(s)+\frac{1}{2}},$$

$$u_{j+J}^{n+1(s)} = u_j^{n+1(s)},$$

$$u_j^{n+1(0)} = u_j^n,$$

where $u_j^{n(s)+\frac{1}{2}} = \frac{1}{2}(u^{n+1(s)} + u_j^n)$ and $u_j^{n+1(s)}$ expresses the s-th iterative value. The catch-ran method may be used to solve the next iterative value.

First, we explore numerically the changing patterns of a solitary wave. Camassa and Holm [1] found the solitary wave solutions (peakons)

$$u(x,t) = c\varphi(x-ct) = ce^{-|x-ct|}, \quad t>0, \quad x\in \mathbb{R}, \tag{18}$$

where $c$ is wave speed. These solutions have a corner (that is, a finite jump in their derivative) at their peak of height $c$.

We set the initial data $u_0(x) = e^{-|x|}$ with wave speed $c = 1$, the periodic interval $[-10, 10]$. In the computing, we take $h = 20/2000$, $\tau = 0.0005$. Fig.1 give the time evolution for $u$ at $T = 0$ and $T = 5$ respectively, which confirms

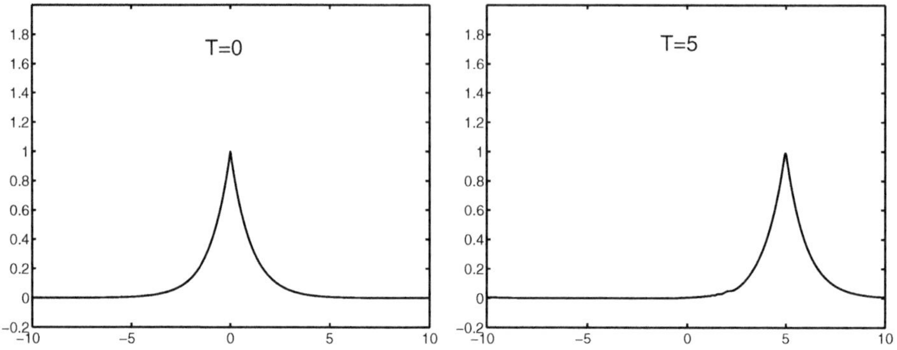


**Fig. 1.**

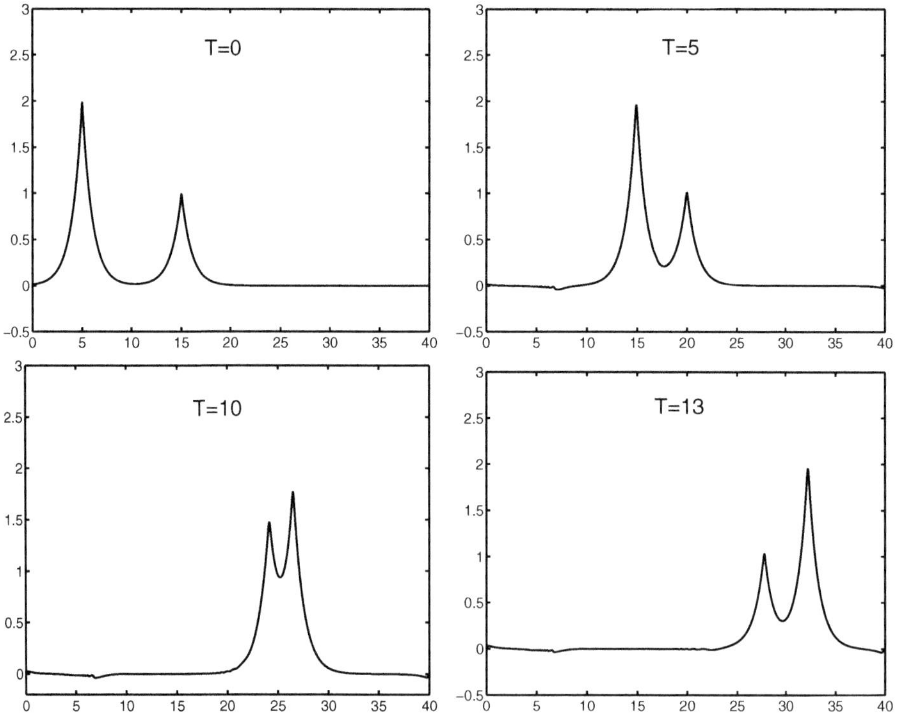


**Fig. 2.**

that Eq.(1) possess the solitary wave solution such as (18) and that the solitary wave does not change its shape for all time.

Then we investigate numerically the interaction of two solitary wave. In the algorithm, we take $u_0(x) = 2e^{-|x-5|} + e^{-|x-15|}$, $L = 40$, and $h = 40/2000, \tau = 0.0005$. Fig.2 gives the time evolution of two interacting solitary waves of different speeds, and confirms numerically that two interacting solitary waves do not change their shape and velocity for all time.

From Fig.2, we can see that two solitary waves (with different wave speeds) are initially well separated, with the larger one to the left (the waves are travelling from left to right), then the faster, taller wave overlaps the slower, smaller one, and the waves interact nonlinearly. After the interaction, the wave separate, with the large one on the right, having regained their initial amplitude and velocity and the only effect of the interaction are phase shifts, that is the centers of the waves are at different positions than where they would have been had there been no interaction. Finally, after a long time the initial profile recurs.

## 4 Conclusion

In this work we present a conservative difference scheme for the Camassa-Holm equation. We then make some convergence analysis. Numerical experiments are carried to effectively study the time evolution of a solitary wave and two solitary waves. Numerical results coincide with those based on the study of the analytic solution [1,6,8,13].

## Acknowledgement

This work is supported by Natural Science Foundation of China under grant number 10471047.

## References

1. Camassa, R., Holm, D.D.: An integrable shallow water equation with peaked solitons. Phys. Rev. Lett. **71** (1993) 1661–1664
2. Constantin, A., Escher, J.: Global weak solutions for a shallow water equation. Indiana Univ. Math. J. **47** (1998) 1527–1545
3. Constantin, A., Escher, J.: Well-posedness, global existence, and blow-up phenomena for a periodic quasi-linear hyperbolic equation. Comm. Pure Appl. Math. **51** (1998) 475–504
4. Constantin, A.: On the blow-up of solutions a periodic shallow water equation. J. Nonlinear Sci. **10** (2000) 391–399
5. Constantin, A., Molinet, L.: Global weak solutions for a shallow water equation. Comm. Math. Phys. **211** (2000) 45–61
6. Constantin, A., Strauss, W.A.: Stability of peakons. Commun. Pure Appl. Math. **LIII** (2000) 0603–0610
7. Constantin, A., Molinet, L.: Global weak solutions for a shallow water equation. Comm. Math. Phys. **211** (2000) 45–61.

8. Constantin, A., Strauss, W.A.: Stability of the Camassa-Holm solitons. J Nonlinear. Sci. **12** (2002) 415–422
9. McKean, H.P.: Breakdown of the Camassa-Holm equation. Comm. Pure Appl. Math. **LVII** (2004) 416–418
10. Kalisch, H., Lenells, J.: Numerical study of traveling-wave solutions for the Camassa-Hlom equation. Chaos, Solitons and Fractals **25** (2005) 287–298
11. Artebrant, R., Schroll, H.J.: Numerical simulation of Camassa-Holm peakons by adaptive upwinding. Appl. Numer. Math. **56** (2006) 695–711
12. Zhang, L.M., Chang, Q.S.: A conservative numerical scheme for a class of nonliner Schrodinger equation with wave operator. Appl. Math. Comput. **145** (2003) 603–612
13. Ablowitz, M.J., Clarkson, P.A.: Solitons, nonlinear evolution equations and inverse scattering. London Mathematical Society, Lecture Note Series **149**, Cambridge University Press, 1991

# Research on Design of a Planar Hybrid Actuator Based on a Hybrid Algorithm

Ke Zhang

School of Mechanical and Automation Engineering
Shanghai Institute of Technology, 200235 Shanghai, China
zkwy@hotmail.com

**Abstract.** Hybrid actuators, which are a combination of two types of motor and mechanism, have good flexibility. In this paper, the kinematics analysis for a hybrid five-bar actuator is introduced. An optimization design of hybrid actuator is performed with reference to kinematics objective function. By the use of the properties of ergodicity, stochastic property, and "regularity" of chaos, a hybrid optimization algorithm is proposed based on chaos optimization algorithm (COA) and a gradient-based search. The efficiency of COA is much higher than some stochastic algorithms such as simulated anneal algorithm (SAA) and genetic algorithm (GA) when COA is used to a kind of continuous problems. The chaos optimization algorithm can improve the efficiency of searching in the whole field by gradually shrinking the area of optimization variable. The precision of optimum dimensions obtained by using the hybrid optimization method can be improved evidently. Finally, a numerical example is carried out, and the simulation results show that the optimization method is feasible and satisfactory in the design of hybrid actuator.

## 1 Introduction

A hybrid actuator is a configuration that combines the motions of two characteristically different electric motors by means of a mechanism to produce programmable output. Where one of the motions coming from a constant speed motor provides the main power, a small servomotor introduces programmability to the resultant actuator. Such machines will introduce to users greater flexibility with programmability option, and energy utilization will be realized at maximum. Although some points are partially explored, there is still a need for kinematics analysis and optimal design studies to guide potential users for possible industrial applications with hybrid machines.

Tokuz and Jones [1] have used a hybrid machine configuration to produce a reciprocating motion. A slider crank mechanism was driven by a differential gearbox having two separate inputs; constant speed motor and a pancake servomotor to simulate a stamping press. A mathematical model was developed for the hybrid machine-motor system. The model results were compared with the experimental ones, and model validation was achieved. Conner [2] has studied on the synthesis of a hybrid mechanism,

D.-S. Huang, L. Heutte, and M. Loog (Eds.): ICIC 2007, LNAI 4682, pp. 1296–1305, 2007.

also performing kinematic analysis. Greenough et al. [3] have later presented a study on design of hybrid machines, and a Svoboda linkage is considered as a two degree of mechanism. Kinematic analysis of Svoboda linkage was presented with inverse kinematics issue. Later Kireçci and Dülger [4] have proposed a different hybrid arrangement driven by two permanent magnet DC servomotors and a constant speed motor. In the configuration, a slider crank mechanism having an adjustable crank was used together with two power screw arrangements to produce a motion in x-y plane. Dülger and Kireçci [5] had given description of a different hybrid actuator configuration consisting of a servomotor driven seven link mechanism with an adjustable crank.

The above design of hybrid mechanism employed mainly traditional optimization design methods. However, these traditional optimization methods have drawbacks in finding the global optimal solution, because it is so easy for these traditional methods to trap in local minimum points [6]. Chaos optimization algorithm (COA) is a stochastic optimization technique. By the use of the properties of ergodicity, stochastic property, and "regularity" of chaos variables, COA is prone to leap over local minimum points, and that continuity and differentiability of objective function are not required.

Optimal design is an important subject in designing a hybrid actuator. The main purpose of the paper is to present kinematics analysis and to investigate the optimal kinematics design of hybrid actuator by deriving its mathematical model. By means of these equations, kinematics optimization design for hybrid five bar actuator is taken by using a hybrid optimization method based on COA. The calculation results of the example are obtained in this system herein.

## 2 Hybrid Actuator Description

Figure 1 represents five link mechanism configuration having all revolute joints except one slider on output link. Notations shown in Fig.1 are applied throughout the study. The hybrid mechanism has an adjustable link designed to include a power screw mechanism for converting rotary motion to linear motion by means of a small slider. The slider is assumed to move on a frictionless plane.

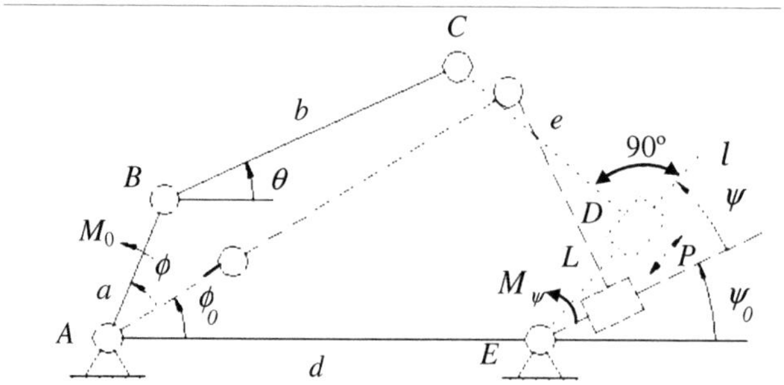


**Fig. 1.** Schematic diagram of five link mechanism

The crank is driven by a DC motor (the main motor) through a reduction gearbox; the slider is driven by a lead screw coupled the second servomotor (the assist motor). Here the main motor is applied as a constant speed motor, and the constant speed motor profile is applied. Point-to-point positioning is certainly achieved for both motors, and the system output is taken from the last link.

$a, b, d, e$ link lengths of the mechanism (m)
$\phi, \theta, \varphi, \psi$ angular displacement of the links (rad)
$\dot{\phi}, \dot{\theta}, \dot{\varphi}, \dot{\psi}$ angular velocity of the links (rad/s)
$\ddot{\phi}, \ddot{\theta}, \ddot{\varphi}, \ddot{\psi}$ angular acceleration of the links (rad/s$^2$)
$L, \dot{L}, \ddot{L}$ displacement, velocity, and acceleration of the slider on output link (m, m/s, m/s$^2$)
$P$ the assist driving force (N)
$M_0$ the main driving torque (Nm)
$M_\psi$ drag torque on output link (Nm)
$X_i, Y_i$ positions to the mass centre of the links in fixed coordinates (m)

## 3 Kinematics Analysis of Hybrid Actuator

Kinematic analysis of five bar linkage is needed before carrying out derivations for the mathematical model. The mechanism is shown with its position vectors in Fig. 1. The output of system is dependent on two separate motor inputs and the geometry of five bar mechanism. Referring to Fig. 1, the output is given by $\psi$, and the configuration represents in-line mechanism. The output motion profiles $\psi$, $\dot{\psi}$, $\ddot{\psi}$ can be designed for the system as $\overline{\psi}$, $\dot{\overline{\psi}}$, $\ddot{\overline{\psi}}$.

In general, the model of a mechanical system can simply be considered as inertial rigid system. Simplifying assumptions are required while developing the mathematical model. Friction and clearance in all joints are neglected. The mechanism operates in vertical plane and gravity effects are included.

Since the hybrid five-bar actuator has two degrees of freedom ($\phi$, $L$) in Fig. 1, equations (1) and (2) is found as below.

$$d\psi = \frac{\partial \psi}{\partial \phi} d\phi + \frac{\partial \psi}{\partial L} dL \tag{1}$$

or

$$d\psi = d\psi_1 + d\psi_2 \tag{2}$$

where $d\psi_1, d\psi_2$ are tiny displacement of output link caused solely by the main motion $d\phi$ and the assist motion $dL$ respectively.

If friction losses in all joints and the change in the kinetic energy are neglected, we can obtain

$$M_0 d\phi + PdL + M_\psi d\psi = 0 \tag{3}$$

Since both $d\phi$ and $dL$ are independent variables, $M_0$ and $P$ can be obtained from equations (1) and (3),

$$M_0 = -M_\psi \frac{\partial \psi}{\partial \phi} \tag{4}$$

$$P = -M_\psi \frac{\partial \psi}{\partial L} \tag{5}$$

By referring to Fig. 1, the loop closure equation is written as:

$$\overrightarrow{AB} + \overrightarrow{BC} + \overrightarrow{CD} = \overrightarrow{AE} + \overrightarrow{ED} \tag{6}$$

By solving vector loop equation (6), angular positions of the link $b$ and $e$ are obtained. Having found the angular displacements of each linkage as $\theta$, and $\phi$ in the five bar linkage, time derivatives can be taken to find angular velocity and accelerations. Some partial derivatives also can be found. They are definitely needed during the analysis of dynamic model. We can obtain

$$L^2 + 2 \cdot CC \cdot L + EE^2 + FF^2 - b^2 = 0 \tag{7}$$

where

$$EE = d - e \cdot \sin(\psi + \psi_0) - a \cdot \cos(\phi + \phi_0)$$

$$FF = e \cdot \cos(\psi + \psi_0) - a \cdot \sin(\phi + \phi_0)$$

$$CC = d \cos(\psi + \psi_0) - a \cos(\psi + \psi_0 - \phi - \phi_0)$$

$$A_1 = d - (a + b)\cos\phi_0$$

$$B_1 = (a + b)\sin\phi_0$$

$$\psi_0 = 2\arctan\left(\frac{A_1 + \sqrt{A_1^2 + B_1^2 - e^2}}{B_1 + e}\right)$$

Therefore,

$$L = -CC + \sqrt{CC^2 + b^2 - EE^2 - FF^2} \tag{8}$$

$\dot{L}, \ddot{L}$, $\partial\psi / \partial L$ and $\partial\psi / \partial\phi$ can be found from equation (8). Angular positions of link $b$ and $e$ are

$$\theta = \arctan\frac{FF + L \cdot \sin(\psi + \psi_0)}{EE + L \cdot \cos(\psi + \psi_0)} \tag{9}$$

$$\varphi = \tfrac{2}{3}\pi + \psi + \psi_0 \tag{10}$$

Thus, we can get $\dot{\theta},\ddot{\theta}$ , $\partial\theta/\partial\phi$, $\partial\theta/\partial L$ from equation (9), and $\dot{\varphi}=\dot{\psi}$ , $\ddot{\varphi}=\ddot{\psi}$ from equation (10).

According to the geometrical relationship of linkage in Fig. 1, the expressions of $X_i$ and $Y_i$ (the positions to the mass centre of each links in fixed coordinates) can be easily obtained as follows.

$$X_a = S_a^x \cos(\phi+\phi_0) - S_a^y \sin(\phi+\phi_0)$$

$$Y_a = S_a^x \sin(\phi+\phi_0) + S_a^y \cos(\phi+\phi_0)$$

$$X_b = a\cos(\phi+\phi_0) + S_b^x \cos\theta - S_b^y \sin\theta$$

$$Y_b = a\sin(\phi+\phi_0) + S_b^x \sin\theta + S_b^y \cos\theta$$

$$X_e = d + L\cos(\psi+\psi_0) - S_e^y \sin(\psi+\psi_0) - S_e^y \cos(\psi+\psi_0)$$

$$Y_e = L\sin(\psi+\psi_0) + S_e^y \cos(\psi+\psi_0) - S_e^y \sin(\psi+\psi_0)$$

$$X_l = d + S_l^x \cos(\psi+\psi_0) - S_l^y \sin(\psi+\psi_0)$$

$$Y_l = S_l^x \sin(\psi+\psi_0) - S_l^y \cos(\psi+\psi_0)$$

By using the above equations of $X_i$ and $Y_i$ , we can get $\ddot{X}_i$ , $\ddot{Y}_i$ , $\partial X_i/\partial\phi$ , $\partial X_i/\partial L$, $\partial Y_i/\partial\phi$ and $\partial Y_i/\partial L$ $(i=a,b,e,l)$ one by one.

# 4 Optimum Design of Hybrid Actuator

## 4.1 Design Variables

Hybrid actuator can be determined by selecting a design vector as follows:

$$x = [x_1, x_2, x_3, x_4]^T \tag{11}$$

where , $x_1 = \dfrac{a}{d}, x_2 = \dfrac{b}{d}, x_3 = \dfrac{e}{d}, x_4 = \phi_0, d = 1$.

## 4.2 Objective Functions

The problem of determining mechanism dimensions can be expressed as a constrained optimization problem.

In order to ensure the smaller adjustment for the assist motion under constraint functions, objective functions can be designed respectively as follows.

$$\min f_1 = L_{\max} - L_{\min} \tag{12}$$

$$\min f_2 = \max\left(\frac{P}{-M_\psi}\right)^2 = \max\left(\frac{\partial\psi}{\partial L}\right)^2 \tag{13}$$

$$\min f_3 = \max\left(\frac{\partial\Psi}{\partial L}\cdot\frac{dL}{d\Phi}\right)^2 = \max\left(\left(\frac{d\overline{\psi}}{d\phi} - \frac{\partial\psi}{\partial\phi}\right)^2\right) \tag{14}$$

### 4.3 Constraint Functions

These functions consist of inequality constraints with stand type according to MATLAB Optimization Toolbox. They are functions of design variables.

1) Inequality constraint related to the movable condition of hybrid actuator

To ensure existence of the hybrid five-bar actuator, the follow inequality constraints are to be satisfied.

$$\begin{aligned} &a + d - b - \sqrt{e^2 + L^2} < 0 \\ &a + \sqrt{e^2 + L^2} - b - d < 0 \\ &a + b - \sqrt{e^2 + L^2} - d < 0 \\ &a - \min(b, \sqrt{e^2 + L^2}, d) < 0 \end{aligned} \tag{15}$$

2) Inequality constraint due to the transmission angle

$$\begin{aligned} &\frac{b^2 + c^2 + d^2 - (d-a)^2}{2\cdot b\cdot\sqrt{e^2 + L^2}} - \cos[\gamma] \le 0 \\ &\frac{b^2 + c^2 + d^2 - (d+a)^2}{2\cdot b\cdot\sqrt{e^2 + L^2}} - \cos[\gamma] \le 0 \end{aligned} \tag{16}$$

where $[\gamma]$ is allowable transmission angle of mechanism.

## 5 Hybrid Optimization Algorithm

Being a global search method, the COA algorithm is expected to give its best results if it is augmented with a local search method that is responsible for fine search. The COA algorithm is hybridized with a gradient based search. The ''candidate'' solution found by the COA algorithm at each iteration step is used as an initial solution to commence a gradient-based search using ''*fmincon*'' function in MATLAB [7], which based on the interior-reflective Newton method. The ''*fmincon*'' function can solve the constraint problem here.

Integrating MATLAB Optimization Toolbox and chaos optimization algorithm, we has written a hybrid optimization design program.

We choose Logistic mapping formula [8] to generate chaos variables as follows.

$$x(k+1) = \lambda x(k)(1-x(k)) \tag{17}$$

where $\lambda$ is control parameter, $0 \le x(0) \le 1$. This map is often used to model population dynamics. The parameter $\lambda$ is the nonlinearity parameter; when $\lambda$ =3.79 the mapping is chaotic [9]. The logistic map is a one dimensional map. By the use of sensitivity of chaos motion for initial values , we can obtain $n$ chaos variables by endowing $n$ tiny difference initial values with (17).

The procedure of hybrid optimization based on COA is as follows:

Step 1: Algorithm initialization

Set $k=1$, $k'=1, \varepsilon$ (commonly, choose optimization objective function allowed error $\varepsilon = 10^{-8}$ ) , and give two large positive integer $N_1, N_2$. Generate different initial values $x_n$ in (18) by using random number, and then we can obtain chaos variables $x_{i,n+1} (i=1,2,\ldots,n)$ by the use of (17).

Step 2: Calculate the range $(x'_{\min}, x'_{\max})$ of optimization variables , and let chaos variables translate into solution space variables.

$$x'_{i,n+1} = x'_{\min} + (x'_{\max} - x'_{\min})x'_{i,n+1} \tag{18}$$

Step 3: Rough iterative search by using chaos variables

Let $x_i(k) = x'_{i,n+1}$, calculate optimization solution $f_i(k)$ by using *fmincon* function of Matlab Optimization Toolbox. Set $x_i^*(1) = x_1$, $f_i^*(1) = f_1$.

If $f_i(k) \le \varepsilon$, then save $x_1$, $k$ and $f_i(k)$ into optimization solution file. If $f_i(k) \le f_1$, then $f_1 = f_i(k)$ , $x_1 = x_i(k)$, else $f_i(k) > f_1$, then abandon $x_i(k)$.

If $k > N_1$, end rough iterative search, otherwise $k = k+1$.

Step 4: Exact iterative search by using chaos variables

Having obtained optimization solution $x_1$ and $f_1$ of rough iterative search, we perform exact iterative search, then solution space variables is

$$x^*_{i,n+1} = x_1 + \alpha x_{i,n+1} \tag{19}$$

where $\alpha$ is adjustment constant and may be small than one. It is worth noting that $x_1$ is the optimization solution of rough iterative search.

Let $x_i(k') = x'_{i,n+1}$, calculate optimization solution $f_i(k')$ by using *fmincon* function. Set $x_i^*(1) = x_2$, $f_i^*(1) = f_2$.

If $f_i(k') \le \varepsilon$ , then save $x_2$ , $k'$ and $f_i(k')$ into optimization solution file. If $f_i(k') \le f_2$, then $f_2 = f_i(k')$, $x_2 = x_i(k')$, else $f_i(k') > f_2$, then abandon $x_i(k')$.

If the stopping condition is satisfied or $k' > N_2$ then end exact iterative, output the global optimization solution and the optimal function values, otherwise $k' = k'+1$.

## 6 Numerical Examples

Initial values of design varibles can be obtained with link lengths of link and angle $\phi_0$ in Table. 1. These link lengths and angle values for hybrid five bar actuator in the studies of optimal kinematic design were obtained by Wang [10]. The output motion profile is designed a sine acceleration law.

We choose $[\gamma] = 45°$, kinematic and performance parameters obtained for hybrid five bar actuator in this studies of optimal kinematic design are shown in Table. 2.

For kinematic performance index of the actuator, we choose the actuator manipulability index $\mu$, $\mu = \sqrt{\det(\boldsymbol{J} \cdot \boldsymbol{J}^T)}$ [11], where matrix $\boldsymbol{J}$ is known as the Jacobian. Most of the measures of kinematic performance of linkage actuators are based on the Jacobian matrix. According to kinematic analysis of hybrid five bar actuator in Fig. 1, the Jacobian matrix can be found as follows

$$\boldsymbol{J} = \begin{bmatrix} \dfrac{a \cdot sin(\phi + \phi_0 - \theta)}{e \cdot sin(\varphi - \theta) + L \cdot sin(\psi + \psi_0 - \theta)} & \dfrac{cos(\psi + \psi_0 - \theta)}{e \cdot sin(\varphi - \theta) + L \cdot sin(\psi + \psi_0 - \theta)} \end{bmatrix} \quad (20)$$

Figure 2 shows the change of manipulability index $\mu$ values of the actuator. As shown in these graphs, manipulability index $\mu$ obtained by using the third objective function (equation (14)) is satisfactory, compared with the others. Hybrid actuator can get good kinematic performance. As seen in Table. 2, the results obtained by using the third optimal objective function (min $f_3$) are to be satisfied.

In order to test validity of the hybrid optimization algorithm, we perform optimization design by using different optimization methods. The manipulability index $\mu$ of optimization design from MATLAB Optimization Box and the chaos algorithm are provided respectively in Fig. 3. Here, equation (14) is chosen as objective function. As shown in Fig. 3, the comparison shows that the global optimum solution of optimization design of hybrid five bar actuator using COA proposed can be obtained.

**Table 1.** Initial value of hybrid five bar actuator

| | $a$ | $b$ | $e$ | $\phi_0$ (°) |
|---|---|---|---|---|
| initial value | 0.1319 | 0.9424 | 0.1 | 13.146 |

**Table 2.** The results of optimization design

| | $a$ | $b$ | $e$ | $\phi_0$ (°) | $\psi_0$ (°) | min $f_1$ | min $f_2$ | min $f_3$ |
|---|---|---|---|---|---|---|---|---|
| min $f_1$ | 0.1 | 0.9787 | 0.1708 | 8.5543 | -11.94 | 0.0221 | 39.740 | 0.0352 |
| min $f_2$ | 0.1615 | 0.9904 | 0.2194 | 8.5759 | -32.29 | 0.1062 | 28.967 | 0.1869 |
| min $f_3$ | 0.1 | 0.9755 | 0.1734 | 8.7751 | -11.70 | 0.0207 | 37.768 | 0.0321 |

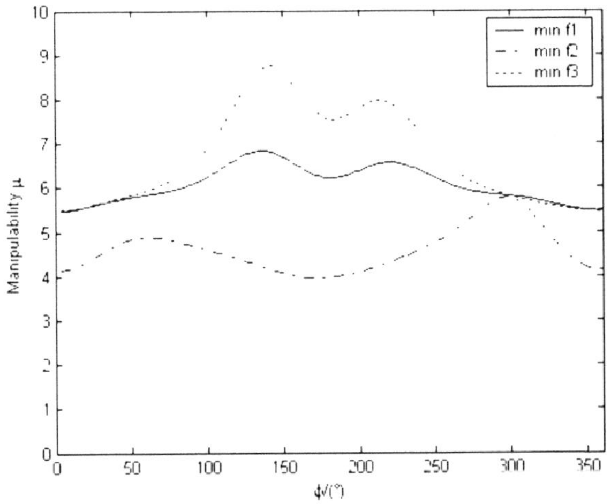


**Fig. 2.** The manipulability index of the hybrid five bar actuator for three objective funtions

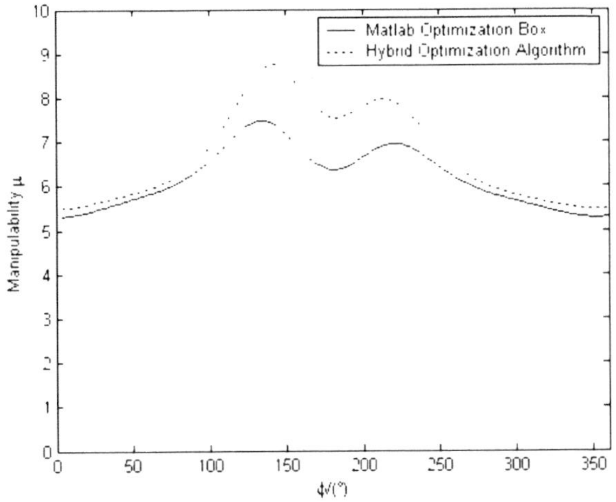


**Fig. 3.** The manipulability index of the hybrid five bar actuator by using different optimization methods

# 7 Conclusions

This paper has described theoretical investigation of a hybrid actuator with its kinematics. The method of kinematics analysis and design for hybrid actuator is proposed on the basis of geometric configuration of the hybrid actuator and chaos optimization algorithm. Making use of the method, an optimal kinematics design for a hybrid five bar actuator is given in this paper. As a result of the comparisons, better performances have been obtained in terms of kinematics objective functions. Optimal kinematics design is more propitious to the applied effects of assist motion control and the development of hybrid actuator.

**Acknowledgements.** This research is supported by Science Foundation of Shanghai Institute of Technology (Grant No. YJ200609).

## References


1. Tokuz, L.C., Jones, J.R.: Hybrid Machine Modelling and Control. PhD thesis, Liverpool Polytechnic (1992)
2. Conner, A.M.: The Synthesis of Hybrid Mechanism Using Genetic Algorithm. PhD thesis, Livepool John Moores University (1996)
3. Greenough, J. D., Bradshaw, W. K., Gilmartin, M. J.: 9th World Congress on the Theory of Machines and Mechanisms 4 (1995) 2501--2505
4. Kireçci, A., Dülger, L.C.: A Study on a Hybrid Actuator. Mechanism and Machine Theory 35 (8) (2000) 1141--1149
5. Dülger, L.C., Kireçci, A., Topalbekiroglu. M.: Modelling and Simulation of Hybrid Actuator. Mechanism and Machine Theory 38 (3) (2003) 395--407
6. Singiresu, S.R.: Engineering Optimization. John Wiley & Sons, New York (1996)
7. Grace, A.: Optimization Toolbox for Use with Matlab, User's Guide, Math Works Inc., (1994)
8. Holden, A. V.: Chaos, Manchester University Press (1986)
9. Otawara, K. and Jordan, M. I.: Controlling Chaos with an Artificial Neural Network, in Proc. IEEE Int. Joint Conf. Fuzzy Syst. 14 (1995) 1943--1948
10. Wang, S. Z., Shen, Y. J.: Dynamics Analysis and Design for a Variable Structure Genealized linkage. China Mechanical Engineering 13(2) (2002) 245--248
11. Yoshikawa, T.: Manipulability of Robotic Mechanism. International Journal of Robotic Research 4 (2) (1985) 3--9

# Network Traffic Prediction and Applications Based on Time Series Model

Jun Lv[1], Xing Li[2], and Tong Li[1]

[1] The Academy of Armored Force Engineering, Beijing ,China
[2] China Education and Research Network, Tsinghua University
lvjun@mails.tsinghua.edu.cn

**Abstract.** Network traffic prediction is a very complex and difficult issue in the network management and design. This paper shows a model with a new algorithm (MLSL), and the model parameters can be modified by the new algorithm, which improves the adaptive ability of the model and makes the model adaptive function. Simulation and actual network traffic data experiment has proved that this algorithm has the advantage of high prediction accuracy and fast convergence, and its computing complexity is lower than other related algorithms.

## 1 Introduction

Nowadays, there are many important problems, such as network congestion, network virus spasm and etc, which will cause network traffic anomaly and network paralysis. If we should predict the network abnormal change ahead of one hour or one day, we would take action to protect the net and reduce the loss.

How to enhance the prediction accuracy of the actual data? Usually, most of the models do not consider the adaptive capacity. Therefore, we have a new idea in the network models, and propose a new algorithm to update the model coefficient continuously. Our methodology relies on the adaptive filter theory. On the basis of LSL algorithm, we modify the LSL algorithm and reduce the computation cost, which will improve the convergence speed and prediction accuracy.

We monitor the backbone traffic of CERNET, collecting the original data. In total port traffic, 80 port traffic data accounts for 30%. We use two sets of data: one group of data is transmitted between CERNET and Beijing Internet exchange center, the other is netflow data, which is collected at the export of CERNET international flow, and obtained from the buffer of CISCO router.

In Section II, we propose the modified algorithm to build the traffic model and using model to forecast time series. Simulation will discuss in Section III. In Section IV, we discuss the CERNET traffic data collected and experimental result. We give a summary in Section V.

## 2 Propose a New Algorithm Based on Time Series Model

Many people have discussed the network models[1][2] to simulate the Internet. long-range dependant: fbm, fgn, FARIMA ,GARIMA [3] etc. Short memory model: (1)

D.-S. Huang, L. Heutte, and M. Loog (Eds.): ICIC 2007, LNAI 4682, pp. 1306–1315, 2007.

Markov models. (2) Regressive models: AR, ARMA, ARIMA.. For short time prediction, AR model with high precision, a small number of parameters, the advantages of simple calculation. But ARMA model can be only viewed as the approximate prediction result of the true value [7].

For stationary residual process, we start to establish AR model. AR(P) :

$$X_t = \varphi_1 X_{t-1} + \varphi_2 X_{t-2} + \ldots .. + \varphi_p X_{t-p} + \alpha_t \quad (1)$$

## 2.1 Modified LSL Algorithm [5]

After choosing the model orders, we need to estimate coefficient of AR model in Eq.(1). LS( Least-square) method is often used to calculate the model parameters. Our purpose is to make the model adaptive recursive capabilities. So we apply the modified LSL algorithm (MLSL) to keep the filter updated and modify the coefficients with the input samples updated.

Adaptive coefficient estimation algorithm can make the filter updated for each new input data. The application of this principle, the Least-square lattice (LSL) filter can produce new reflect coefficient and new prediction error corresponding to the order. It is only affected by the new input data, and updated with the input data one by one.

The LSL adaptive filter can be formed under the criterion of minimizing both the forward predictor error power and backward predictor error power at the same time. Let $f_{m,n}$ and $b_{m,n}$ be the forward and backward errors at the output of the m-th section of the lattice in correspondence to the n-th time step. They are related to the outputs of the previous section of the lattice by:[6]

$$\begin{aligned} f_{m,n} &= f_{m,n-1} + k^f_{m,n} b_{m-1,n-1} \\ b_{m,n} &= b_{m-1,n-1} + k^b_{m,n} f_{m-1,n} \end{aligned} \quad (2)$$

Input parameter: P: lattice order, $x_n$ : the data sample at time n

$k^f_{m,n}$ : forward reflection coefficient, $k^b_{m,n}$ : backward reflection coefficient

Initial: $f_{0,0} = b_{0,0} = x_0$, $E^f_{0,0} = E^b_{0,0} = x_0^{\ 2}$

For each new input data sample compute

$f_{0,n} = b_{0,n} = x_n$, $E^f_{0,n} = E^b_{0,n} = E^2_{0,n-1} + x_n^{\ 2}$

The LSL algorithm updates at the n-th time step the errors and the parameters by the following recursion to be used in connection with Eq.(2)

$$S_{m,n} = S_{m,n-1} - f_{m-1,n} b_{m-1,n-1} / \cos^2 \theta_{m,n} \quad (2a)$$

$$\begin{aligned} k^f_{m,n} &= S_{m,n} / E^f_{m-1,n} \\ k^b_{m,n} &= S_{m,n} / E^b_{m-1,n-1} \\ E^f_{m,n} &= E^f_{m-1,n} - S^2_{m,n} / E^b_{m-1,n-1} \\ E^b_{m,n} &= E^b_{m-1,n-1} - S^2_{m,n} / E^f_{m-1,n} \\ \cos^2\theta_{m,n} &= \cos^2\theta_{m-1,n} - b^2_{m-1,n-1} / E^b_{m-1,n-1} \end{aligned} \tag{2b}$$

m=1,2….min.(p, n) Eq.(2) to Eq.(2b) may be further simplified by:

(1) $k^f_{m,n} = k^b_{m,n}$

(2) Since $\theta$ is very small, so $\cos\theta$ can be approximated to one, thus the reflection coefficient simplifies to:

$$k_{m,n} = P_{m,n} / D_{m,n} = 2S_{m,n} / (E^f_{m-1,n} + E^b_{m-1,n-1}) \tag{3}$$

and the values of $E^f_{m,n}$ and $E^b_{m,n}$ are given by:

$$\begin{aligned} E^f_{m,n} &= E^f_{m,n-1} + f^2_{m,n} \\ E^b_{m,n} &= E^b_{m,n-1} + b^2_{m,n} \end{aligned}$$

Therefore, $P_{m,n}$ becomes

$$P_{m,n} = P_{m,n-1} - 2f_{m-1,n} b_{m-1,n-1} \tag{4}$$

Eq.(3) may be updated with a reduced CC as follows:

$$D_{m,n} = D_{m,n-1} + f^2_{m-1,n} + b^2_{m-1,n-1} \tag{5}$$

So by using the above formulas we can get a new reflection coefficient for each new input data sample, the CC of the new algorithm is only 5N, rather, LSL is 10N. The coefficient of prediction filter { $a_i$ } is evaluated by Levinson-Durbin recursion:

$$a_m(i) = \begin{cases} a_{m-1}(i) + k_m a_{m-1}(m-i) & i=1,2,3\ldots\ldots m\text{-}1 \\ k_m & i=m \end{cases} \tag{6}$$

## 2.2 Forecasting with New Algorithm

We may use MLSL algorithm to calculate the prediction filter coefficient { $a_i$ }, by replacing { $\varphi_i$ } of eq.(1) with {- $a_i$ }, the model will be constructed.

AR(P) linear least error recursion prediction formula:

$$\hat{X}_k = \varphi_1 \hat{X}_k(l-1) + .... + \varphi_p \hat{X}_k(l-p) \qquad (l>0)$$

$$\text{for } l \leq 0 , \qquad \hat{X}_k(l) = X_{k+1} \tag{7}$$

### 2.3 ARMA Model Forecasts

ARMA model expression:

$$X_t - \varphi_1 X_{t-1} - \varphi_2 X_{t-2} - ... - \varphi_p X_{t-p} = \alpha_t - \theta_1 \alpha_{t-1} - ...\theta_q \alpha_{t-q} \tag{8}$$

We can only get the approximate prediction [7] by using ARMA model. For the use of AR model, we have been able to obtain the precise result. Choosing ARMA (2, 1) to test and calculating the parameters: $\varphi_1$ , $\varphi_2$ and $\theta_1$.

$$X_t = \varphi_1 X_{t-1} + \varphi_2 X_{t-2} + a_t - \theta_1 a_{t-1}$$

By using model recursive method:

$$a_t = X_t - \varphi_1 X_{t-1} - \varphi_2 X_{t-2} + \theta_1 a_{t-1}$$

$$a_{t-1} = X_{t-1} - \varphi_1 X_{t-2} - \varphi_2 X_{t-3} + \theta_1 a_{t-2}$$

$$a_{t-2} = X_{t-2} - \varphi_1 X_{t-3} - \varphi_2 X_{t-4} + \theta_1 a_{t-3}$$

Let $a_{t-3} = 0$, so one step prediction values is:

$$\hat{X}_t(1) = \varphi_1 X_t + \varphi_2 X_{t-1} - \theta_1 a_t$$

Multi step prediction recursive formula:

When $l>q$ , $\hat{X}_k(l) = \sum_{j=1}^{p} \varphi_j \hat{X}_k(l-j)$

When $l \leq j$ , $\hat{X}_k(l-j) = X_{k+l-j}$

## 3 Simulation

### 3.1 Generation of the Traffic Data

In order to verify the accuracy of the algorithm, we test the algorithm by simulating. Firstly, we need to generate network traffic. Due to the characteristics of burst and complex, it is difficult to simulate the network traffic data. Usually, we generate the self-similar process to simulate the real traffic data. Article [9] has introduced that a state space representation for self-similar signals and systems based on scale stationary ARMA models.

- **Self-Similar traffic data generation**

We select the Inverse Fourier Transform algorithm to generate the Fractal Gaussian Noise (FGN) for simple and rapid property. Giving Hurst parameter and the length of the data, we may synthesise the self-similar traffic sample which has FGN power spectrum [8]. Anomaly pulse signal is the stochastic data produced by computer. Fig.1 is the self-similar data and prediction result of simulating.

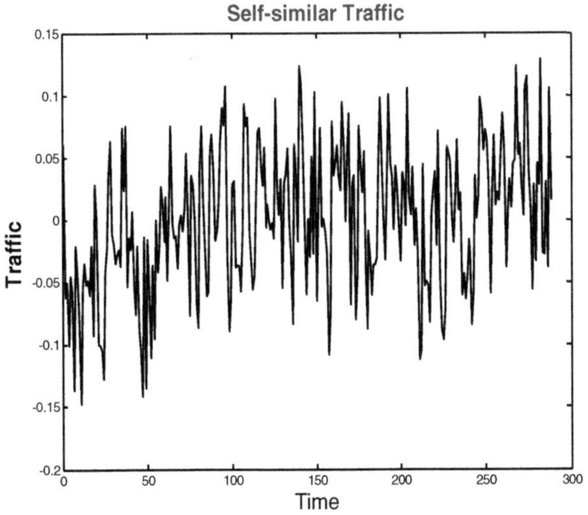


**Fig. 1.** (a) N=288, H=0.8, original data

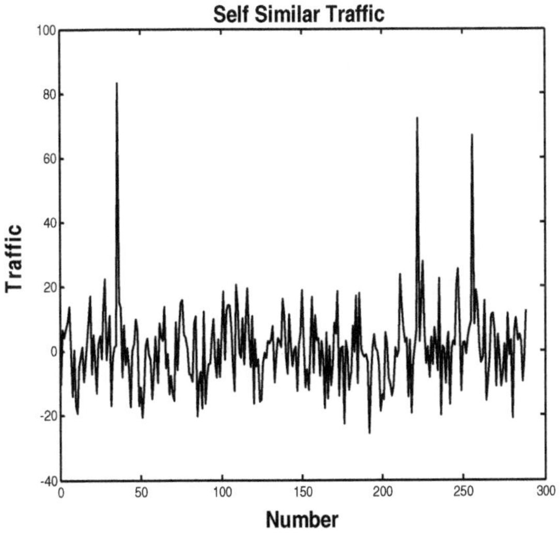


**Fig. 1.** (b) N=288, H=0.8, insert anomaly pulse. SN=9.953db

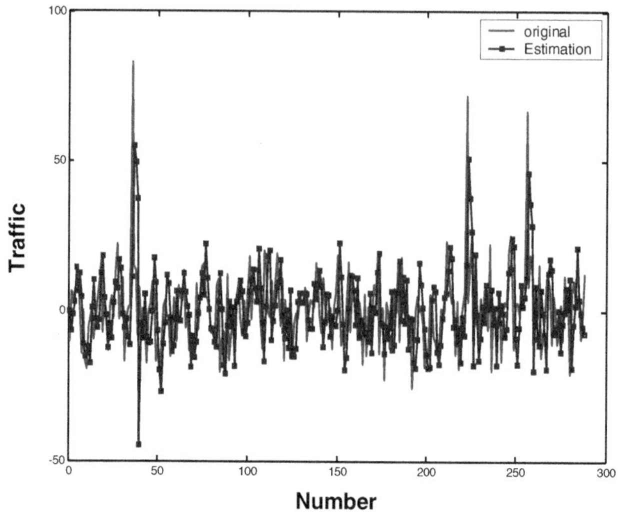


**Fig. 1.** (c) One step forecast figure. N=288, H=0.8, SN=9.953db

# 4 Real Network Traffic Experiment

## 4.1 Data Collected and Analysis

In this work, we use historical information collected continuously since 2004 on the CERNET IP backbone network. There are many factors that contribute to trends and variations in the overall traffic. We collect values for two particular MIB (Management information Base) objects, incoming and outgoing link utilization in bps, for the links between CERNET and Beijing Internet Centre, time spans from January 1st 2004 until June of 2005, time interval is 1 hour. We collect the netflow data in CERNET international gateway, time interval is 5 minutes. The date of this observation spans from June 6th 2005 to June 19th 2005. Fig.2 shows the process for data collecting. Fig.3 plots raw data(MIB) for one week.

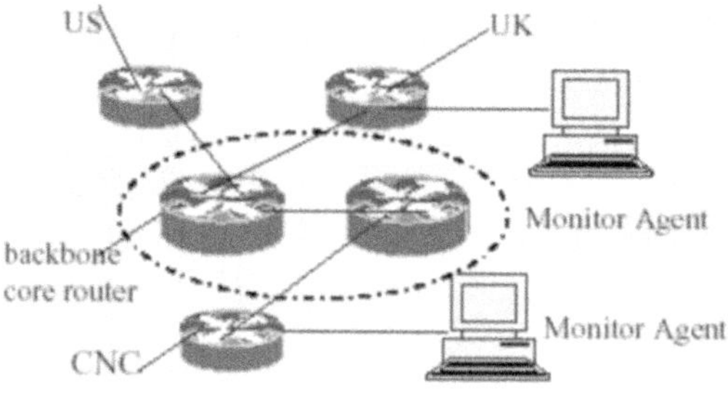


**Fig. 2.** Data collecting system

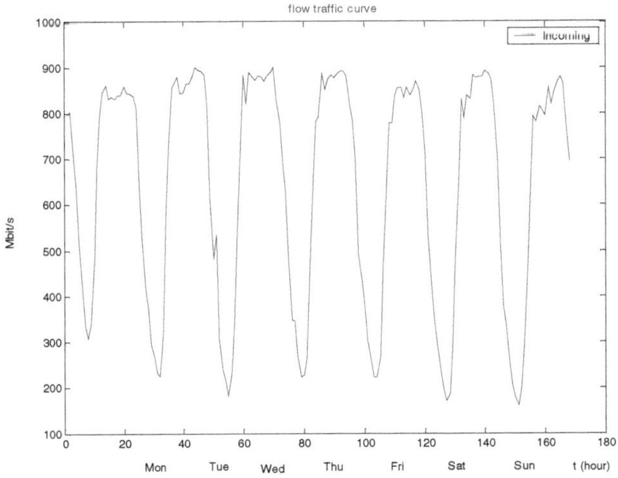


**Fig. 3.** One week traffic, the x-axis is time in hours, and the y-axis is flow traffic

The figure above shows that network traffic has non-stationary characteristic. Regular or periodic seasonal and unexpected phenomena also have shown. For non-stationary time series, we can also use ANVOA [4] variance analysis method to remove the periodic seasonal component and acquire a stationary time series to build model[10].

### 4.2 Forecast Results of the Algorithm

Fig.4(a) shows the estimated values of the original data(SNMP MIB). Fig.4(b) is the forecast figure comparing of different models and algorithms.( MLSL, LS, ARMA)

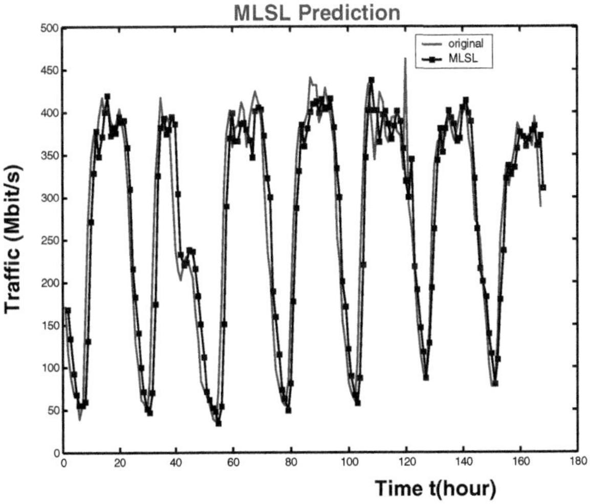


**Fig. 4.** (a) This is one step forecast graph of MLSL algorithm in the original data

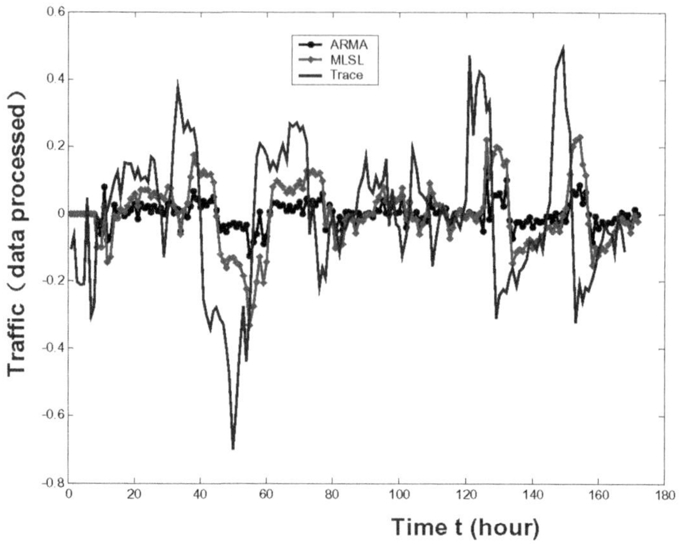


**Fig. 4.** (b) Forecast figure of different algorithms (10 steps)

### 4.3 Model Comparison

Mean squared error (MSE) of autocorrelation function is often used to compare the precision of different model.

$$MSE(m) = \frac{1}{n}\sum (ACF_m(i) - ACF_s(i))^2$$

MSE ( m ) is the autocorrelation mean squared errors of model m , $ACF_m$ [i] denotes the i-th autocorrelation of model m, and $ACF_s$ [i] is the i-th autocorrelation of the residual data serial. MSE statistical values indicate that the simulation precision of autocorrelation between model and data sample, the smaller the MSE is, the higher the precision degree shows. MSE values belongs to different models is given below. Table 1 shows the simulation results of different algorithms (N=288, H=0.8, SN=9.5db). Table 2 is the experimental result with SNMP MIB data. Table 3 is the MSE of different model with netflow traffic data.

**Table 1.** MSE values of different models (simulation)

| Model \ MSE | 30 | 50 | 100 | 200 | 400 |
|---|---|---|---|---|---|
| AR(4)(MLSL) | 0.0148 | 1.2e-004 | 4.3e-004 | 1.5e-004 | 5.4e-004 |
| AR(4)(LS) | 0.0315 | 1.8e-004 | 7.6e-004 | 1.7e-004 | 5.6e-004 |
| ARMA(2,1) | 0.0417 | 0.0026 | 0.0099 | 0.0025 | 8.39e-004 |

**Table 2.** MSE values of different models (SNMP MIB data)

| Number | 30 | 50 | 100 | 150 |
|---|---|---|---|---|
| AR(4)MLSL | 0.0105 | 0.0046 | 0.0032 | 9.26e-004 |
| AR(4)LS | 0.0116 | 0.0053 | 0.0039 | 0.0013 |
| ARMA(4) | 0.0110 | 0.0050 | 0.0048 | 0.0031 |

**Table 3.** MSE Comparison of different models (Netflow data)

| MSE / Model | 150 | 250 | 300 | 400 |
|---|---|---|---|---|
| AR(4)(MLSL) | 1.41e-004 | 1.16e-004 | 1.41e-004 | 9.15e-005 |
| AR(4) ( LS ) | 1.42e-004 | 1.243e-004 | 1.52e-004 | 9.27e-005 |
| ARMA(2,1) | 0.0309 | 4.7e-004 | 5.16e-004 | 3.61e-004 |

### 4.4 Discussion and Conclusions

A set of Experiments above, for example, simulation and actual network test, the prediction precision (MSE) in the Table 1,Tables and Table 3, which denotes that AR model is superior to ARMA for stationary short-term prediction. On the whole, we can obtain the conclusion: The short-term forecasts for stationary time series, the AR model is the first to consider the model. Certainly, the number of parameters is the least and the calculation is very simple.

Tables shown above indicate that MLSL algorithm has the features of converging faster and a high degree of accuracy. MLSL algorithm recursive continuously and update the coefficients, thereby increasing accuracy. Moreover, the CC (computation cost) of MLSL is only 5N, while LL algorithm has a CC of 10N.

## 5 Summary

This paper shows two part contribution:

(1) CERNET IP flow process is similar to the process of AR model, this conclusion may obtained by analyzing the autocorrelation and partial autocorrelation function, also through comparing the performance of different models. Certainly, the original data must be pre-processed to become stationary time series.

(2) By far, most people spend quantity time to investigate the model type, whereas, the calculation has become increasingly complicated. Based on the idea of simplifying, we propose the simplest algorithm (MLSL) to build traffic model and predict the future values of CERNET IP backbone flow. The model parameters can be updated with each new input data, so that MLSL algorithm has the characteristic of fast convergence and high accuracy. Simulative and Experimental results show that our algorithm outperforms the LS or ARMA algorithm, and its computing complexity is lower than other related algorithms.

## References

1. Paxson, V., Flod, S.: Wide-area Traffic: the Failure of Poisson Modelling, Proc of the ACM/SIGCOMM'94. (1994) 257-268
2. Legend, W.E.: On the self-sinilar Nature of Ethernet traffic. IEEE/ACM transactions on networking. 2(1) (1994) 1-15
3. Shu, Y.T., Jin, Z.G.: Prediction-based Admission Control FARIMA Models. 2000 IEEE ICC 2000, 1325-1329
4. Hellerstrin, J.L., Zhang, F., Shalhabuddin, P.: Characterizing Normal Operation of a Web Server: Application to Workload Forecasting and Problem Detection. Proceedings of the computer measurement group,1998
5. Martinelli, G.: Efficient Adaptive Linear Predictor. Electron lett. (1982) 617-618
6. Haykin, S.: Adaptive Filter Theory, 3 rd ed, 1998
7. Yang, W.Q., Gu, L.: Time Series Analysis and Dynamic Data Modelling. Press of Beijing University of Science and Engineering, 1988
8. Paxson, V.: Fast Approximate Synthesis of Fractional Gaussian Noise for Generating Self-similar Network Traffic. Computer communication review. 10 (1997) 5-18
9. Meltem, I., Birsen, Y., Banu,O.: Kalman Filtering for Self-similar Processes. Proceedings of the 11$^{th}$ IEEE Workshop on Statistical Signal processing. (2001) 82-85
10. Lv, J., Li, X., Ran, C.S.: Network Traffic Prediction and Fault Detection based on Adaptive Linear Model. Proceedings of 2004 IEEE International Conference on Industrial technology (ICIT2004). (2004) 880-886

# On Approach of Intelligent Soft Computing for Variables Estimate of Process Control System

Zaiwen Liu[1], Xiaoyi Wang[1], and Lifeng Cui[2]

[1] School of Information Engineering,
Beijing Technology and Business University
No.33,Fucheng Road, Beijing, 100037, China
[2] School of Chymistry and Environment Engineering,
Beijing Technology and Business University
No.33, Fucheng Road, Beijing, 100037, China
liuzw@th.btbu.edu.cn

**Abstract.** A new approach of intelligent soft computing based on process neural network for variables estimate of process control system was proposed. Process neural network (PNN) is a new type of artificial neural network put forward in recent years. Some algorithms of PNN were discussed, and convergence rate was comparatively low. An improved algorithm for raising training speed based on function orthogonal basis expansion in PNN for soft computing was researched. After increasing the normalizing rule on original algorithm, and introducing function momentum adjustment item and learning rate automatically adjustment method for network weight function, the training time of learning algorithm for PNN was reduced. The fact showed that the stability and training precision was improved with the learning rate automatic adjustment method, and it can also restrain the network falls into local least by introducing momentum adjustment item, and a good result of application in sewage disposal system was represented.

**Keywords:** intelligence computing, variables estimate, neural network, learning algorithm, sewage disposal, process control system.

## 1 Introduction

Some output variables in the process control system can't be measured directly by instruments, such as BOD(Biochemical Oxygen Demand) in sewage disposal process, so they can't be controlled in closed loop, and artificial neural networks were used to model the industrial production process in resent years. A new method of soft computing with mathematic model based on PNN is suitable for predicting the variables of process control system.

PNN is an extent of traditional neural network, in which the inputs and outputs may be time-variation, and PNN and BP neuron network are of the same structure. Converge operation includes not only the multi-inputs aggregation to space but also the accumulation to time. PNN decreases synchronously instantaneous restrictions

D.-S. Huang, L. Heutte, and M. Loog (Eds.): ICIC 2007, LNAI 4682, pp. 1316–1326, 2007.

which exist in the input of traditional neural network. Consequently, PNN extends traditional neuron in time domain and traditional neuron is a special example in process neuron. Now, as to the studying of PNN, He Xingui and Xu Shaohua presented algorithms which based on the expanding of weighting function orthogonal basis and discrete *Walsh* transform separately [1-2].

## 2 Model and Configuration of PNN

### 2.1 Features of Process Neuron

PNN is an artificial neuron system based on certain topological structure, which is made of some process neurons, and process neuron usually consists of three parts: weighting function, converge and energizing threshold. But the input and weighting value of process neuron can be time-varying, namely function depending on time and even multivariate function including time. So, PNN extends traditional neuron in time domain, and traditional neuron is a special example in process neuron [3]. The configuration of single process neuron is shown as Fig.1.

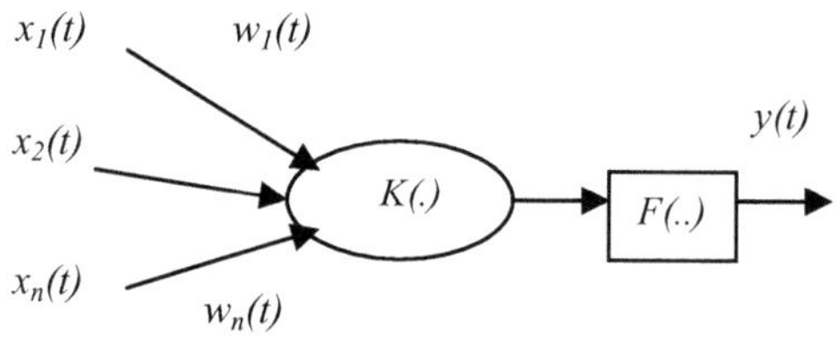


**Fig. 1.** Model of process neuron

The relation between inputs and outputs of process neuron can be written as(1):

$$Y=F((w(t)\oplus x(t)) \otimes K(t)- \theta) \quad (1)$$

Where $\oplus$ represents a space-converge operation, $\otimes$ represents a time-converge operation, and $k(.)$ is time-converged function, $F(.)$ is energizing function[4], and the output can be obtained commonly by (2).

$$Y = \int_0^T \sum_{i=1}^{n} w_i(t)x_i(t)dt \quad (2)$$

$x_1(t),x_2(t)...x_n(t)$ are process neuron inputs respectively, and $w_1(t),w_2(t)...w_n(t)$ are connection weighting. As equation (2) showed above, converge operation of PNN includes not only the multi-inputs converge to space but also the accumulation to time.

### 2.2 Configuration of PNN

PNN with the configuration of three layers in which hidden layer is process neuron and input and output are common neurons is shown as Fig.2.

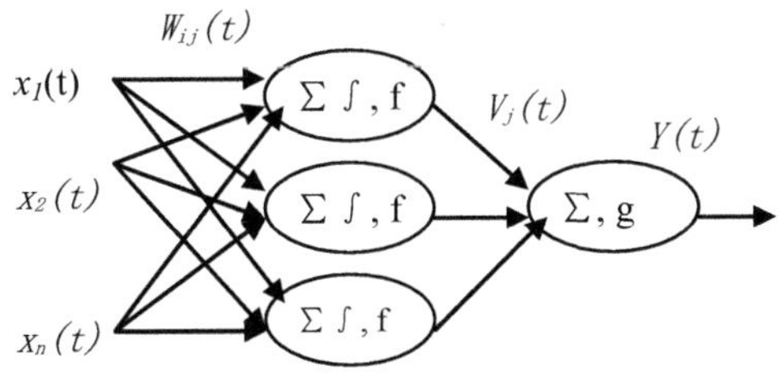


**Fig. 2.** The configuration of PNN

Suppose the number of input-layer neurons is *n*, the number of hidden layer neurons is *m*, the mapping relation between network input and output is stated as (3).

$$y = g\left\{\sum_{j=1}^{m} v_j f\left(\sum_{i=1}^{n}\left[\int_0^T (W_{ij}(t)x_i(t))dt - \theta_j^l\right]\right) - \theta\right\} \tag{3}$$

Where $W_{ij}(t)$ is the connection weighting value from input layer to hidden layer, $V_j$ is the connection weighting value form hidden layer to output layer, $\theta_j^1$ is the energizing threshold of hidden layer node *j*, [0,T] is sampling period, *f* and *g* are energizing functions, $\theta$ is the threshold of output node.

## 3 Algorithm Based on the Orthogonal Basis Function

There are complexity between learning process neural network and training traditional feed forward neural network. An aggregation operator is added in process neuron, and it makes the neuron network has the ability to deal with the information of space-time two dimensions at the same time, so the information processing enginery of biological neuron is imitated better than traditional neuron. A group of proper function orthogonal bases are brought in input space. Input functions are transformed into limited series of this group by given precision. Meanwhile, network weighting functions are expressed as the expanded forms in the same group of basis functions whose orthogonality can be used to simplify the complexity of process neuron about time-aggregation operation.

The application shows that this algorithm has not only simplified the operation of PNN but also increased the stability and aggregation in network learning. There are many kinds of methods to choose function orthogonal basis. Basis function can be chosen as circular function, multinomial function, such as *Fourier* basis function, *Wavelet* basis function, *Walsh* function etc. The algorithm based on the expanding of weighting function orthogonal basis is described as following:

### 3.1 Expand Basis Function About Input Function

If the input space of PNN is *[0 T]* and $b_1(t), b_2(t) \ldots\ldots b_k(t)$ are a group of standard orthogonal basis functions, input function [$x_1(t)$, $x_2(t)$,……$x_n(t)$]is expanded in the form of basis functions, and showed as (4).

$$X = (\sum_{l=1}^{L} a_{1l} b_l(t), \sum_{l=1}^{L} a_{2l} b_l(t), \ldots\ldots \sum_{l=1}^{L} a_{nl} b_l(t)) \tag{4}$$

### 3.2 Initialize Network Parameters and Determining Hidden Layer Number

The structure of no time-varying neuron is adopted from hidden layer to output layer, so let the initialized value of weighting $V_j$ be a tiny random number between *[0,1]*.

The neural threshold of hidden layer and output layer are initialized as the way of $V_j$. Because $W_{ij}(t)$ is functions about time, for the convenience of calculation, weighting functions are expanded in the form of basis functions in the same group:

$$W_{ij}(t) = (\sum_{l=1}^{L} w_{ij}^{(l)} b_l(t)) \tag{5}$$

$w_{ij}^{(l)}$ is the connection weighting from input node *i* to hidden layer node *j* relative to $b_l(t)$.

### 3.3 Calculate Forward Outputs and Errors

The forward outputs and errors can be calculated as the follows:

$$y = g\left\{\sum_{j=1}^{m} v_j f\left[\sum_{i=1}^{3}\sum_{l=1}^{L} a_{il}^{(k)} w_{ij}^{jl} - \theta_j^{(l)}\right] - \theta\right\} \tag{6}$$

$$E = \sum_{k=1}^{K} (y_k - d_k)^2 = \sum_{k=1}^{K} \left\{ g\left\{\sum_{j=1}^{m} v_j f\left[\sum\sum a_{il}^{(k)} w_{ij}^{jl} - \theta_j^{(l)}\right] - \theta\right\} - d_k \right\}^2 \tag{7}$$

In (6), $a_{il}^{(k)}$, which is relative to $b_l(t)$ based on expanded basis function, is the expanded coefficient of $i_{th}$ input neuron in group *k* of training sample. Equation (7) can be stated as the total error of group training. $d_k$ is the expected output of $k_{th}$ training.

### 3.4 Regulate Network Weighting and Threshold

The regulation functions of network weighting and threshold according to gradient descent algorithm can be stated as follows:

$$v_j = v_j + a\Delta v_j, \; w_{ij}^{jl} = w_{ij}^{jl} + b\Delta w_{ij}^{jl}, \; \theta_j^1 = \theta_j^1 + \Upsilon\Delta\theta_j^1 \tag{8}$$

$$\Delta v_j = -2\sum_{k=1}^{K}\left[(g(z_k) - d_k) g'(z_k) f(u_{kj})\right] \quad j=1,2\ldots\ldots\ldots m \tag{9}$$

$$\Delta w_{ij}^{(jl)} = -2\sum_{k=1}^{K}\left[(g(z_k) - d_k) g'(z_k) f'(u_{kj}) a_{il}^{(k)}\right] \tag{10}$$

$$i=1,2,3 \;, l=1,2\ldots\ldots L,, \; j=1,2\ldots m$$

$$\Delta\theta_j^1 = -2\sum_{k=1}^{K}\left[(g(z_k)-d_k)g'(z_k)f'(u_{kj})(-1)\right] \tag{11}$$

Where $u_{kj} = \sum_{i=1}^{3}\sum_{l=1}^{L} a_{il}^{(k)} w_{ij}^{(jl)} - \theta_j^1$ , $z_k = \sum_{j=1}^{m} v_j f\left(\sum_{i=1}^{3}\sum_{l=1}^{L} a_{il}^{(k)} w_{ij}^{(jl)} - \theta_j^1\right) - \theta$ (12)

### 3.5 Check Precision and Biggest Circulation Times

Checking the error $E$ and the biggest circulation times to meet the expectation of precision, if error $E<\varepsilon$ or training times $S<N_{max}$, the expectation of network has been achieved, then turn to 3.5. Otherwise, the value of circulation times is added 1, then back to 3.3.

## 4 Improving of Algorithm for Raising the Convergence Speed

A gradient descent algorithm can be used to training algorithm of PNN, but there exist some flaws, such as lower convergence speed and smallest in part etc. in practical applications. so it is very hard to obtain satisfying training results. Using ameliorated traditional BP algorithm is as follows:

### 4.1 Add Momentum Item

Adding momentum into the regulation of weighting can not only raise the convergence speed of network, but also prevent the oscillation when error curved surface is regulated [4]. The regulation equation:

$$\Delta w(t) = w(t) - w(t-1) \tag{13}$$

So (8) is transformed into following equations:

$$v_j(t+1) = v_j(t) + a\Delta v_j + \zeta\Delta v(t) \tag{14}$$

$$w_{ij}^{jl}(t+1) = w_{ij}^{jl}(t) + b\Delta w_{ij}^{jl} + \zeta\Delta w_{ij}^{jl}(t) \tag{15}$$

$$\theta_j^1(t+1) = \theta_j^1(t) + \Upsilon\Delta\theta_j^1 + \zeta\Delta\theta_j^1(t) \tag{16}$$

Where $\zeta\Delta v(t)$ , $\zeta\Delta w_{ij}^{jl}(t)$ , $\zeta\Delta\theta_j^1(t)$ are regulation values of momentum separately. $\zeta$ is regulation coefficient of momentum.

### 4.2 Using Adaptive Learning Speed

Choosing proper learning speed is of great importance to network training. If the learning speed is too big, network oscillation will be diverged. Similarly, if it is too small, oscillation will be prevented but the convergence speed will slow down.

Let $a,b,\Upsilon \leq 2/\lambda_{max}$ commonly, where $\lambda_{max}$ is the biggest eigenvalue of autocorrelation matrix of input vector $\boldsymbol{X}$. Learning speed can be regulated as the following formulas[3].

$$\eta = \eta.\Phi \quad (\zeta = \zeta \quad \Delta E < 0) \tag{17}$$

$$\eta = \eta.\delta \quad (\zeta = \zeta \quad \Delta E > 0) \tag{18}$$

Where $\Phi > 1$, $\delta < 1$. $\zeta$ is the coefficient of momentum regulation. $\eta$ is network learning speed $a,b,\Upsilon$.

# 5 Application of the Method

## 5.1 Selection of Variables for PNN for Soft Computing

The sewage disposal process is varied with time, and its output variable ($BOD_5$), which is difficult to be measured directly, can be gotten by means of a method of mathematical calculation and estimation through computing the relationship between the output variable and other process variables (such as DO, Ph, TP, COD) which can be measured in real-time.

The soft computing technology in sewage disposal process has often been studied in recent years [6]. Most of soft computing models adopt traditional artificial neural network model, whose inputs are constants independent of time. Every input is instantaneous in the form of geometrical point, however, sewage disposal process is a reaction process associated with time nearly, and the inputs of its neural network are dependent on time.

Time accumulation of inputs is neglected by traditional neural network, so there are big errors on prediction and the values of prediction can not be applied well in practical production. Based on the research of PNN, it can be brought into the predictive control of soft computing in sewage disposal process [7].The structure of PNN for soft computing in sewage disposal adopts the three-layer network, in which hidden layer is process neuron and input and output are common neurons. The input layer has three nodes: DO (Dissolved Oxygen), COD or TOC (Total Organic Carbon), MLSS (Mixed liquor suspended solid or Strength of Sludge).

According to the analysis of process in sewage disposal, the three assistant variables mentioned above, which is detected easily, can almost show the changing rules of output variables $BOD_5$ and the three variables are independent of each other, which is fit for soft sensing theory. Let process layer have $m$ nodes, so the reflection between input and output of network is defined as (2).

Where *Sigmoid* function should be adopted as the basic function for the soft computing in sewage disposal process. $\theta$ is the threshold of output node. $x_i(t)$ is the parameter of input node, $j=1,2,3...m$, $i=1,2,3$. Equation (2) is the forward-time and space accumulation values of a group of training data.

## 5.2 Acquisition and Processing of Sample Data

According to the data requirements and network structure of PNN, some groups of training data are selected to simulate network convergence and error curve, and training data are showed as table1.

**Table 1.** Groups of training data ( *mg/L* )

| TIME | *0.5* | *1.0* | *1.5* | *2.0* | *2.5* | *3.0* | *3.5* |
|---|---|---|---|---|---|---|---|
| *TOC* | 524 | 456.6 | 400 | 365 | 300 | 210 | 125 |
| *DO* | 2.1 | 3.1 | 5.1 | 6.8 | 8.5 | 10.3 | 12.3 |
| *MLSS* | 4000 | 3850 | 3000 | 2530 | 2230 | 2100 | 2000 |
| $BOD_5$ | | | | 32 | | | |
| *TOC* | 456 | 386 | 324 | 300 | 265 | 168 | 75 |
| *DO* | 1.9 | 2.5 | 3.8 | 5.1 | 6.43 | 9.5 | 10.3 |
| *MLSS* | 3800 | 3600 | 3200 | 2650 | 2300 | 2150 | 2100 |
| $BOD_5$ | | | | 33 | | | |
| *TOC* | 531 | 469 | 394 | 300 | 242 | 158 | 75 |
| *DO* | 1.5 | 2.1 | 2.6 | 3 | 3.5 | 4.9 | 6.3 |
| *MLSS* | 3600 | 3250 | 2800 | 2650 | 2400 | 2100 | 2100 |
| $BOD_5$ | | | | 32.5 | | | |
| *TOC* | 421 | 391 | 269 | 200 | 162 | 110 | 45 |
| *DO* | 1.4 | 1.4 | 1.5 | 1.9 | 2.8 | 3.5 | 4.6 |
| *MLSS* | 2800 | 2750 | 2550 | 2300 | 2250 | 2100 | 2100 |
| $BOD_5$ | | | | 34 | | | |

## 5.3 Curves of Input Variables

The curves of input variables(*TOC,DO.MISS*) of the network can be created according to the sample data of table1,and showed as Fig.3.

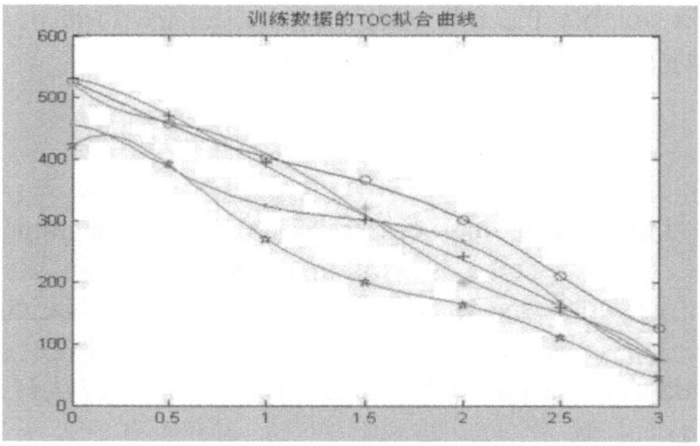


**Fig. 3.** Curves of input variables of the network

### 5.4 Data Processing

First the sampling interval of input variables should be obtained and the detailed interval can be decreased according to the increasing changing current of data. Based on the technological condition of SBR method in sewage disposal process, the circle of whole sewage disposal is settled as 8 hours and sampling interval is 1 hour. The form of one group of sampling data is:

$$X(t, DO, TOC, MLSS)=( 0,DO_1 ,TOC_1 ,MLSS_1; 1,DO_2 ,TOC_2 ,MLSS_2;...8, DO_8,TOC_8, MLSS_8) \tag{19}$$

Data like (19) is fitted into the form of fifth order multinomial in sampling circle so that a group of training sampling can be obtained as follows:

$$X = (\sum_{i=0}^{5} a_{1i}x^i, \sum_{i=0}^{5} a_{2i}x^i, \sum_{i=0}^{5} a_{3i}x^i) \tag{20}$$

Because fifth order multinomial can approach any function by fitting, this fitting can be adopted to show and gather the space geometrical points of input parameters in sampling circle. Selecting a group of orthogonal trigonometric basis functions:

$$1,sin(x),cos(x),sin(2x),cos(2x),...sin(nx) ,cos(nx). \tag{21}$$

The number of basis functions can be adjusted according to the precision requirement of approaching function and the ability of generalization after network training. Do not choose many basis functions or the calculation will be increased largely. It goes against the network training.

Based on the orthogonal basis functions selected, the fitting fifth-order multinomial of input can be transformed into Fourier series with *[0, T]*:

$$X = (\sum_{i=0}^{L}(a_{1i}\cos(ix)+b_{1i}\sin(ix)), \sum_{i=0}^{L}(a_{2i}\cos(ix)+b_{2i}\sin(ix)), \sum_{i=0}^{L}(a_{3i}\cos(ix)+b_{3i}\sin(ix))) \tag{22}$$

Where $L$ is the number of orthogonal basis, $a_{1i}$,$b_{1i}$ are expanding coefficients of orthogonal basis. There are many kinds of expanded forms of Fourier series, and the Fourier series with dissymmetrical interval is adopted in this algorithm.

There are $2*L+1$ items in the expanded coefficient which should be noticed in programming. To simplify the following formula, let the form of orthogonal basis selected be $b_1(t),b_2(t)......b_l(t)$, so (22) is transformed into (23):

$$X = (\sum_{l=1}^{L} a_{1l}b_l(t), \sum_{l=1}^{L} a_{2l}b_l(t), \sum_{l=1}^{L} a_{3l}b_l(t)) \tag{23}$$

There is big difference among orders of coefficient magnitudes in expanded orthogonal basis. To prevent the saturation output of neuron because of the big absolute value of net input and to prevent the weighting regulation to come to the even area of error curved surface, the coefficient $a_{1i}$, $a_{2i}$, $a_{3i}$ of (22) should be treated as a whole. The data coefficients are changed in the range of *[-1,1]* to complete pre-treatment of the data.

### 5.5 Realization of Algorithm

Calculating the network parameters such as the number of hidden layer neuron, learning speed of weighting from input layer to hidden layer, et al. mentioned as above.

Given the number of hidden layer nodes is $m$, the weighting matrix can be expressed as:

$$w = \begin{bmatrix} \sum_{l=1}^{L} w_{11}^{(1l)} b_l(t), \sum_{l=1}^{L} w_{12}^{(2l)} b_l(t), \ldots\ldots, \sum_{l=1}^{L} w_{1m}^{(ml)} b_l(t) \\ \sum_{l=1}^{L} w_{21}^{(1l)} b_l(t), \sum_{l=1}^{L} w_{22}^{(2l)} b_l(t), \ldots\ldots, \sum_{l=1}^{L} w_{2m}^{(ml)} b_l(t) \\ \sum_{l=1}^{L} w_{31}^{(1l)} b_l(t), \sum_{l=1}^{L} w_{32}^{(2l)} b_l(t), \ldots\ldots, \sum_{l=1}^{L} w_{3m}^{(ml)} b_l(t) \end{bmatrix} \tag{24}$$

Where $w_{1m}^{(ml)}$ is the expanded coefficient of the first input neuron and $m_{th}$ hidden layer neuron under basis function $b_l(t)$ .If there are $K$ groups of training samplings, according to the gradient descent algorithm, the output and error are obtained as (6)and (7). In (6), $a_{il}^{(k)}$, which is relative to $b_l(t)$ based on expanded basis function, is the expanded coefficient of $i_{th}$ input neuron in group $k$ of training sample. Equation (7) can be stated as the total error of group training. $d_k$ is the expected output of $k_{th}$ training. Equations (6), (7) are the simplified results of calculation based on orthogonal basis.

The equations of network weighting value and threshold regulated are as (4)~(15).Because both $f$ and $g$ use *Sigmoid* function, the equation (25) is brought into (9), (10), (11) to simplify the calculation..

$$f'(x) = f(x)(1 - f(x)) \tag{25}$$

Check the error and biggest circulation times to meet the expectation of precision ,if error $E<\varepsilon$ or training times $S<N_{max}$, the expectation of network has been achieved. Then save the network weighting, threshold and error.

### 5.6 Analysis of Simulation

MATLAB programming can be used to realize the network training algorithms. Last *M* profile function is defined as:

$$[W, V, E, EN, YCYZ, SCCYZ] = process\_neural\ (x, d, T, hidnum, funnum, N_{max}, e, s1, s2, s3\ ) \tag{26}$$

Where $W$ is weighting function, $V$ is weighting value from hidden layer to output layer, $E$ is error value of every training, *YCYZ* is threshold of hidden layer, *SCCYZ* is threshold of output layer, $x$ is sampling basis of training, $d$ is signal from tutor, $T$ is sampling cycle, *hidnum* is the number of hidden-layer neurons, *funnum* is the number of basis functions, $N_{\max}$ is the biggest time of training, *e* is training precision, *s1* is the learning speed of weighting value from input layer to hidden layer,*s2* is the learning speed of threshold in hidden layer, *s3* is the learning speed of weighting value from hidden layer to output layer.

Based on the condition of *hidnum*=12; *e*=0.01; *s1*=0.38; *s2*=0.3; *s3*=0.4; $N_{max}$=1000,the regulation of momentum item and fixed learning speed, neural network is trained to obtain the error curve as following Figs.

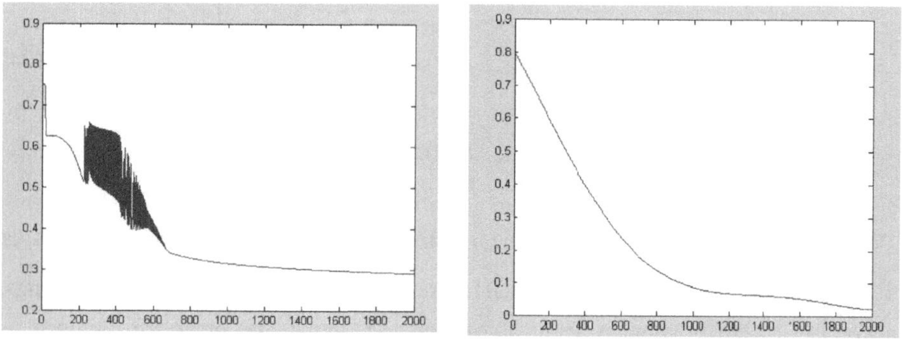


(a) (b)

**Fig. 4.** Error curves of training results (a) Error curve of convergence of unimproved algorithm. (b) Error curve of improved learning speed algorithm.

Fig.4 (a) is the training results of unimproved algorithm, and oscillation in training is evidence. Fig.4(b) is the training results of adaptive learning speed algorithm. Seen as these figures, the convergence speed of networks can be raised obviously and oscillation in training can be prevented successfully by using improved algorithm. Meanwhile, it can be seen that the adaptive algorithm of learning speed is more effective than the one of adding momentum obviously.

## 5.7 Precision of the Simulation

The training results compared with actual values are showed as table2.

**Table 2.** Compare of training results with actual values

| *Data group No.* | *1* | *2* | *3* | *4* |
|---|---|---|---|---|
| Actual value | 32 | 31 | 32.5 | 34 |
| Training results | 33.29 | 32.156 | 33.654 | 33.127 |
| Absolute error | 1.295 | 1.156 | 1.154 | 0.873 |
| Relative error | 4.1% | 3.79% | 3.56% | 2.57% |

Seen from the figures and table2 above, the convergence speed of network can be accelerated by changing learning speed. However, according to other groups of training curves, it is hard to find a balance point between the regulation value of coefficient in momentum items and learning speed. Especially, the oscillation can be brought into the beginning of network through raising learning speed. The oscillation decreases along with the increase of training times and the convergence speed of network error curve turns slowly. So disadvantages of PNN algorithms should be paid attention to, which is like traditional BP algorithm in many aspects.

## 6 Conclusion

An improved learning algorithm of BP network can be introduced to the soft computing method based on PNN. From the point of biology, PNN is fit for the operation mechanism of organic neuron furthermore. To raise the training speed of PNN obviously, an improved algorithm was brought into this network, and the algorithm can reason network weighting and threshold regulation by using traditional BP gradient descent algorithm. Error function can be optimized to accelerate the convergence of network. The training speed of network is affected obviously by the initialization of network weighting function. Consequently, the improved algorithms of BP network are not only all fit to PNN theoretically, but also raise the training speed of networks. Many process input problems, such as sewage disposal process, in relative to time can be settled by PNN.

The intelligent soft computing based on PNN may be used to fulfill measurement of the effluent BOD from sewage disposal process. A good result of soft computing was obtained by simulation for the process.

**Acknowledgements.** This study is supported by Beijing Natural Science Foundation under grant *No*.4062011, and is also supported by the Beijing Education Committee under grant *No*.200589. Those supports are gratefully acknowledged.

## References

1. Xu, S.H., He, X.G..: Learning Algorithms of Expanded PNN Based on Orthogonal Basis. Computer Transaction. (2004) 645~650
2. He, X.G., Liang, Y.Z., Xu, S.H..: Training of Process Neural Network and Its Application, China Engineering Science. ( 2001) 31~35
3. Yao, Y., Wang, J..: Improve the Research of the Training Speed of BP Network. Information Technology.1 (2002)
4. Li ,P,.: Prediction and Analysis of Market Demand Based on Process Neural Network, Information Technology. (2003) 44~45
5. He, X.G., Liang, Y.Z.: Some Academic Questions of Process Neural Network. China Engineering Science (2000) 40~44
6. Liu,Z.W., Cui, L.F., Qi, G.Q.: Soft Sensing Method Based on RBF Neural Network for Effluent BOD from SBR.China Water & Wastewater.Beijing. 20 (2004) 17-20
7. Liu, Z.W., Lian, X.Q., Wang, Z.X..: A Soft Sensing Method Based on Process Neural Network. Proceedings of 2006 IEEE International Conference on Information Acquisition. (2006) 611-615

# ICA Based on KPCA and Hierarchical RBF Network for Face Recognition

Jin Zhou[1], Haokui Tang[1], and Weidong Zhou[2]

[1] School of Information Science and Engineering,
University of Jinan, Jinan 250022, P.R. China
ise_zhouj@ujn.edu.cn
[2] School College of Information Science and Engineering,
Shandong University, Jinan 250100, P.R. China
wdzhou@sdu.edu.cn

**Abstract.** This paper proposes a new face recognition approach by using Independent Component Analysis (ICA) and Hierarchical Radial Basis Function (HRBF) network classification model. To improve the quality of the face images, a series of image pre-processing techniques, which include histogram equalization, edge detection and geometrical transformation are used. The ICA based on Kernel Principal Component Analysis (KPCA) and FastICA is employed to extract features, and the HRBF network is used to identify the faces. To accelerate the convergence of the HRBF network and improve the quality of the solutions, the Extended Compact Genetic Programming (ECGP) and Particle Swarm Optimization (PSO) are applied to optimize the HRBF network structure and parameters. The experimental results show that the proposed framework is efficient for face recognition.

**Keywords:** ICA, KPCA, HRBF, Face Recognition.

## 1 Introduction

Face recognition has become one of the most active research areas of pattern recognition since the early 1990s. In the past 20 years, significant advances have been made in the design of the successful classifier of face recognition [1,2]. But the diversity of the face patterns makes it difficult to realize the robust recognition system, and the complexity of the pattern makes it hard to implement fast. Recently, sets of approaches that use different techniques to correct perspective distortion are being proposed. These techniques are sometimes referred to as view-tolerant. For a complete review on the topic of face recognition the reader is referred to [3].

Over the last few years, ICA has aroused wide research interests and become a popular tool for blind source separation and feature extraction [4], and a number of algorithms for performing ICA have been proposed [5]. Recently, the kernel-based nonlinear feature extraction techniques have been of wide concern. And KPCA is proposed for nonlinear PCA [10]. In this paper, ICA based on KPCA and FastICA is used for feature extraction.

D.-S. Huang, L. Heutte, and M. Loog (Eds.): ICIC 2007, LNAI 4682, pp. 1327–1338, 2007.

Neural network has been widely applied in pattern recognition [6]. Recently, RBF neural network has been used in many engineering and scientific applications including face recognition [7]. Hierarchical RBF network consist of multiple RBF networks assembled in different level or cascade architecture, and has been proved effective than single RBF neural network [12,13]. In this paper, an automatic method for constructing HRBF network is proposed to identify the faces.

The paper is organized as follows: The image pre-processing schemes are discussed briefly in section 2. The feature extraction scheme, ICA based on KPCA and FastICA, is presented in section 3. The design of the HRBF network is discussed in section 4. The experimental results of the approach are given in section 5. Section 6 presents the conclusions.

## 2 Image Pre-processing

To augment the interesting information, image pre-processing is required. There are lots of techniques of image pre-processing discussed [8,9].

Our scheme is as follows: To decrease the influence of the variations of the light, histogram equalization is used. And because the person's identity tightly relates with his or her facial features and is almost independent of his or her hair and clothes, we find a sub-region, which consists of the person's main facial features to represent the face based on edge detection and projection. The Sobel edge operators are adopted and according to the projection information and the distribution of the faces are used to locate the sub-region. At last, geometrical transformations, zooming out and zooming in and rotations are used to obtain the same size and the similar pose images [8]. And the size of the image in our system is 60×60.

## 3 Feature Extraction

### 3.1 KPCA for Whitening the Data

Before applying an ICA algorithm on the data in observation space, it is usually very useful to do some preprocessing work (e.g. whitening data). The preprocessing can make the problem of ICA estimation simpler and better conditioned. Generally, a whitened PCA is used to sphere the data $x$

$$z = \mathrm{W}_{\mathrm{P}} x \tag{1}$$

and make the transformed data $z$ satisfy

$$E\{zz^{\mathrm{T}}\} = \mathrm{I} \tag{2}$$

where $\mathrm{W}_{\mathrm{P}}$ is a whitened matrix, and I is an identity matrix.

But the traditional PCA is based exclusively on the second-order statistics with smooth Gaussian distribution. It is difficult to describe the data with non-Gaussian distribution, so the Kernel-based algorithm (KPCA algorithm [10]) is proposed for nonlinear PCA. KPCA uses kernel function to obtain the arbitrary high-order correlation between input variants, and find the principal components needed through the inner production between input data (observation data).

Based on this idea, we will develop a concise ICA algorithm using KPCA to whitening the data in feature space.

Given an observation sequence $x_1, x_2, \ldots, x_M$ in $R^n$, Then, a nonlinear mapping $\Phi$ is used to map the input data space $R^n$ into the feature space F:

$$\begin{aligned} \Phi:\ & R^n \rightarrow F \\ & x \rightarrow \Phi(x) \end{aligned} \tag{3}$$

correspondingly, a pattern in the original input space $R^n$ is mapped into a potentially much higher dimensional feature vector in the feature space F.

Let us construct the covariance matrix in the feature space F:

$$C = \frac{1}{M}\sum_{j=1}^{M}(\Phi(x_j) - \overline{\Phi})(\Phi(x_j) - \overline{\Phi})^{\mathrm{T}} \tag{4}$$

where

$$\overline{\Phi} = \frac{1}{M}\sum_{j=1}^{M}\Phi(x_j) \tag{5}$$

However, it is not easy to centralize data directly in the feature space F. To avoid this difficulty, we make the assumption that

$$\sum_{j=1}^{M}\Phi(x_j) = 0 \tag{6}$$

So let us consider the following noncentralized covariance matrix:

$$\tilde{C} = \frac{1}{M}\sum_{j=1}^{M}\Phi(x_j)\Phi(x_j)^{\mathrm{T}} \tag{7}$$

Now we have to solve the Eigenvalue equation:

$$\lambda V = \tilde{C} V \tag{8}$$

for Eigenvalues $\lambda \geq 0$ and Eigenvectors $V \in F \setminus \{0\}$.

It is very computationally intensive or even impossible to calculate $\tilde{C}$ 's eigenvectors in a high-dimensional (even infinite-dimensional) feature space. KPCA can be viewed as utilizing two key techniques to solve this problem artfully. One is the SVD technique adopted in Eigenfaces [11], and the other is the so-called kernel-tricks [10]. SVD technique can be used to transform the eigenvector calculation problem of a large-size matrix to the eigenvector calculation problem of a small-size matrix and, kernel-tricks can be used to avoid the computation of dot products in the feature space by virtue of the following formula:

$$K(x_i, x_j) = (\Phi(x_i) \cdot \Phi(x_j)) \tag{9}$$

Specifically, let $Q = [\Phi(x_1),...,\Phi(x_M)]$, then $\tilde{C}$ can also be expressed by

$$\tilde{C} = \frac{1}{M} QQ^T \tag{10}$$

Let us form the matrix $\tilde{R} = Q^T Q$: By virtue of kernel-tricks, we can determine the elements of the $M \times M$ matrix $\tilde{R}$ by

$$\tilde{R}_{ij} = \Phi(x_i)^T \cdot \Phi(x_j) = (\Phi(x_i) \cdot \Phi(x_j)) = K(x_i, x_j) \tag{11}$$

Let us calculate the orthonormal eigenvectors $u_1, u_2, \ldots, u_m$ of $\tilde{R}$ corresponding to m largest eigenvalues $\lambda_1 \geq \lambda_2 \geq \ldots \geq \lambda_m$. Then, by SVD technique, the orthonormal eigenvectors $v_1, v_2, \ldots, v_m$ of $\tilde{C}$ corresponding to $m$ largest eigenvalues $\lambda_1, \lambda_2, \ldots, \lambda_m$ are

$$v_j = \frac{1}{\sqrt{\lambda_j}} Qu_j, \quad j = 1, \ldots, m. \tag{12}$$

let $U = [u_1,...,u_M]$, $\Lambda = [\lambda_1,...,\lambda_M]$, $V = [v_1,...,v_M]$, then

$$V = \Lambda^{-1/2} QU, \tag{13}$$

$$\tilde{C} = E\{\Phi(x)\Phi(x)^T\} = V\Lambda V^T, \tag{14}$$

$$V^T V = I \tag{15}$$

let whiten matrix $W_p = \Lambda^{-1/2} V^T$, and $z = W_p \Phi(x)$, we will find $z$ satisfy

$$\begin{aligned} E\{zz^T\} &= E\{ W_p \Phi(x) \Phi(x)^T W_p^T\} \\ &= E\{ \Lambda^{-1/2} V^T \Phi(x) \Phi(x)^T V \Lambda^{-1/2} \} \\ &= \Lambda^{-1/2} V^T E\{ \Phi(x) \Phi(x)^T \} V \Lambda^{-1/2} \\ &= \Lambda^{-1/2} V^T V \Lambda V^T V \Lambda^{-1/2} \\ &= I \end{aligned} \tag{16}$$

specifically,

$$\begin{aligned} z &= W_p \Phi(x) \\ &= \Lambda^{-1/2} V^T \Phi(x) \\ &= \Lambda^{-1} U^T Q^T \Phi(x) \\ &= \Lambda^{-1} U^T [k(x_1, x), k(x_2, x), \ldots, k(x_M, x)]^T \\ &= \Lambda^{-1} U^T \tilde{R}_x \end{aligned} \tag{17}$$

Now, a question is: how to get the covariance matrix C's eigenvectors? Actually, if we centralize $\tilde{R}$ by

$$\mathrm{R}_{ij} = \tilde{\mathrm{R}}_{ij} - \frac{1}{M}\sum_{m=1}^{M} 1_{im}\tilde{\mathrm{R}}_{mj} - \frac{1}{M}\sum_{n=1}^{M}\tilde{\mathrm{R}}_{in}1_{nj} + \frac{1}{M^2}\sum_{m,n=1}^{M} 1_{im}\tilde{\mathrm{R}}_{mn}1_{nj} \tag{18}$$

(where $1_{ij}=1/M$) and get R's eigenvectors $u_1, u_2, \ldots, u_m$, we can obtain C's eigenvectors $v_1, v_2, \ldots, v_m$.

After the projection of the mapped sample $\Phi(x)$ onto the eigenvector $v_j$, we can obtain the $j$-th feature

$$\begin{aligned} z &= \mathrm{W_p}\,\Phi(x) \\ &= \Lambda^{-1}\,\mathrm{U^T}\,[k(x_1, x), k(x_2, x), \ldots, k(x_M, x)]^{\mathrm{T}} \\ &= \Lambda^{-1}\,\mathrm{U^T}\,\mathrm{R}_x \end{aligned} \tag{19}$$

## 3.2 Independent Component Extraction Using FastICA

After whitening, the following task is to find a new unmixing matrix W in the KPCA-transformed space to recover the independent source $s$ from $z$, i.e.,

$$s = \mathrm{W}z \tag{20}$$

Note that the new unmixing matrix W should be orthogonal. This can be seen from

$$E\{ss^{\mathrm{T}}\} = \mathrm{W}\,E\{zz^{\mathrm{T}}\}\,\mathrm{W^T} = \mathrm{W}\,\mathrm{W^T} \tag{21}$$

By far, numerous ICA algorithms have been developed for the calculation of the unmixing matrix. For simplicity, here, we only outline the FastICA algorithm (also called fix-point algorithm) that was proposed by Hyvärinen et al. [4]. First of all, let us consider the one-unit version of FastICA. The learning rule is to find a maximally "nongaussian" projection direction, i.e., a unit vector $w$ such that the projection $w^{\mathrm{T}}z$ maximizes non-gaussianity. The non-gaussianity can be measured via a properly chosen non-quadratic function (contrast function) $G(u)$. Denote the derivative of $G(u)$ by $g(u)$. The basic form of FastICA algorithm is as follows:

*Step* 1: Whiten the observation data $z = \mathrm{W_p}\,\Phi(x)$;
*Step* 2: Choose an initial (e.g. random) weight vector $w(0)$; $n = 0$;
*Step* 3: Let $n = n + 1$;

$$w(n) = E\{zg(w^{\mathrm{T}}(n-1)z)\} - E\{g'(w^{\mathrm{T}}(n-1)z)\}\,w(n-1); \tag{22}$$

*Step* 4: Let $w(n) = w(n) / \| w(n) \|$;
*Step* 5: If not converged, go back to *Step* 3.
*Step* 6: Get one independent component $s_1 = w_1 z$.

We can estimate a set of orthogonal projection directions $w_1, \ldots, w_d$ one by one using the foregoing one-unit FastICA algorithm by virtue of a Gram–Schmidt-like decorrelation scheme. Specifically, after obtaining $i$ directions $w_1, \ldots, w_i$ ($i < d$), we run the one-unit FastICA algorithm for $w_{i+1}$, and after every iteration step we normalize $w_{i+1}$ as

$$w_{i+1} = w_{i+1} - \sum_{j=1}^{i} (w_{i+1}^{\mathrm{T}} w_j) w_j \text{ and } w_{i+1} = w_{i+1} \Big/ \sqrt{w_{i+1}^{\mathrm{T}} w_{i+1}} \tag{23}$$

The resulting unmixing matrix is W= $(w_1, \ldots, w_d)^{\mathrm{T}}$.

## 4 Design of the Classifier

To identify the person, the fine classifier is needed. Because the pattern is complex and nonlinear, hierarchical radial basis function (HRBF) classification model are proposed for face recognition [12,13].

### 4.1 Encode and Calculation

A function set $F$ and terminal instruction set $T$ used for generating a HRBF network model are described as $S = F \cup T = \{+_2, +_3, \ldots, +_k\} \cup \{x_1, \ldots, x_n\}$, where $+_i$ ($i = 2, 3, \ldots, k$) denotes non-leaf nodes' instructions and taking $i$ arguments. $x_1, x_2, \ldots, x_n$ are leaf nodes' instructions and taking no arguments. The output of a non-leaf node is calculated as a HRBF network model. In this research, Gaussian radial basis function is used and the number of radial basis functions used in hidden layer of the network is same with the number of inputs, that is, $m = n$ (see Fig.1).

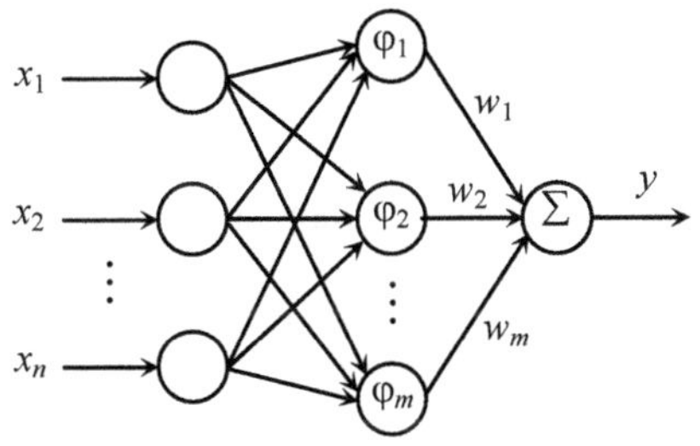


**Fig. 1.** An example of RBF neural network

In the creation process of HRBF network tree (see Fig.2), if a nonterminal instruction, i.e., $+_i$ ($i = 2, 3, 4, \ldots, k$) is selected, $i$ real values are randomly generated and used for representing the connection strength between the node $+_i$ and its children. In addition, $2 \times n^2$ adjustable parameters $a_i$ and $b_i$ are randomly created as radial basis function parameters. For developing the face classifier, the radial basis function used is as follows

$$f(a_i, b_i, x) = \exp\left( - \left\| \frac{(x - a_i)^2}{b_i^{\,2}} \right\| \right) \tag{24}$$

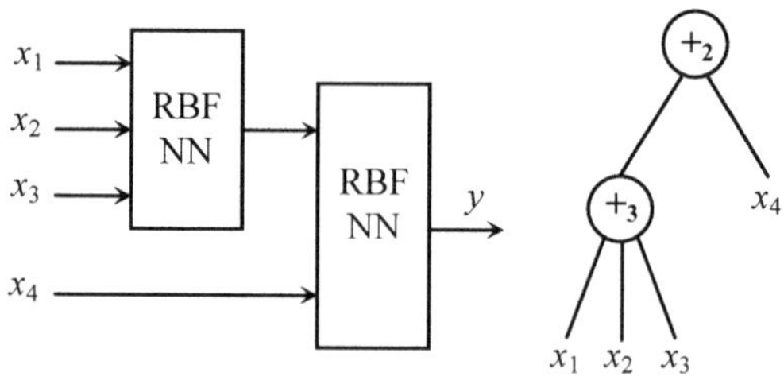


**Fig. 2.** An example of HRBF network (left), and a tree-structural representation of the HRBF network (right)

The total excitation of $+_k$ is

$$net_k = \sum_{j=1}^{k} w_j * x_j \tag{25}$$

where $x_j$ ($j$=1, 2, ..., $k$) are the inputs to node $+_k$. The output of the node $+_k$ is then calculated by

$$out_k = f(a_k, b_k, net_k) = \exp\left(-\left\|\frac{(net_k - a_k)^2}{b_k^{\ 2}}\right\|\right) \tag{26}$$

The overall output of HRBF network tree can be computed from left to right by depth-first method, recursively.

### 4.2 Tree Structure Optimization

Finding an optimal or near-optimal HRBF is formulated as a product of evolution. In this paper, the Extended Compact Genetic Programming (ECGP [14]) is employed to find an optimal or near-optimal HRBF structure. ECGP is a direct extension of Extended Compact Genetic Algorithm (ECGA) to the tree representation, which is based on the Probabilistic Incremental Program Evolution (PIPE) prototype tree. In ECGA, Marginal Product Models (MPMs) are used to model the interaction among genes, represented as random variables, given a population of Genetic Algorithm individuals. MPMs are represented as measures of marginal distributions on partitions of random variables. ECGP is based on the PIPE prototype tree, and thus each node in the prototype tree is a random variable. ECGP decomposes or partitions the prototype tree into sub-trees, and the MPM factorizes the joint probability of all nodes of the prototype tree, to a product of marginal distributions on a partition of its sub-trees. A greedy search heuristic is used to find an optimal MPM mode under the framework of minimum encoding inference. ECGP can represent the probability distribution for more than one node at a time. Thus, it extends PIPE in that the interactions among multiple nodes are considered.

### 4.3 Parameter Optimization

The Particle Swarm Optimization (PSO) [15] conducts searches using a population of particles, which correspond to individuals in Evolutionary Algorithm (EA). A population of particles is randomly generated initially. Each particle represents a potential solution and has a position represented by a position vector $x_i$. A swarm of particles moves through the problem space, with the moving velocity of each particle represented by a velocity vector $v_i$. At each time step, a function $f_i$ representing a quality measure is calculated by using $x_i$ as input. Each particle keeps track of its own best position, which is associated with the best fitness it has achieved so far in a vector $p_i$. Furthermore, the best position among all the particles obtained so far in the population is kept track of as $p_g$. In addition to this global version, another version of PSO keeps track of the best position among all the topological neighbors of a particle. At each time step $t$, by using the individual best position, $p_i(t)$, and the global best position, $p_g(t)$, a new velocity for particle $i$ is updated by

$$v_i(t+1) = v_i(t) + c_1 * rand_1 * (p_i(t) - x_i(t)) + c_2 * rand_2 * (p_g(t) - x_i(t)) \tag{27}$$

where $c_1$ and $c_2$ are positive constant and $rand_1$ and $rand_2$ are uniformly distributed random number in [0, 1]. The term $v_i$ is limited to the range of $\pm v_{\max}$. If the velocity violates this limit, it is set to its proper limit. Changing velocity this way enables the particle $i$ to search around its individual best position, $p_i$, and global best position, $p_g$. Based on the updated velocities, each particle changes its position according to the following equation

$$x_i(t+1) = x_i(t) + v_i(t+1) \tag{28}$$

### 4.4 Procedure of the General Learning Algorithm

The general learning procedure for constructing the HRBF network can be described as follows:

1) Create an initial population randomly (HRBF network trees and its corresponding parameters);
2) Structure optimization is achieved by using ECGP algorithm;
3) If a better structure is found, then go to step 4, otherwise go to step 2;
4) Parameter optimization is achieved by PSO algorithm. In this stage, the architecture of HRBF network model is fixed, and it is the best tree developed during the end of run of the structure search;
5) If the maximum number of local search is reached, or no better parameter vector is found for a significantly long time then go to step 6; otherwise go to step 4;
6) If solution is found, then the algorithm is stopped; otherwise go to step 2.

## 5 Experimental Results

### 5.1 Face Database

We performed extensive experiments on two benchmark face datasets, namely the ORL and the Yale face database.

For ORL face dataset, 40 persons with variations in facial expression. All images were taken under a dark background, and the subjects were in an upright frontal position, with tilting and rotation tolerance up to 20 degree, and tolerance of up to about 10%. Fig.3 shows 10 images of one person from the selected dataset. The Yale face database contains 165 images of 15 persons. There are 11 images per person with different facial expressions or lightings. Fig.4 shows the 11 images of one person. For each experiment, 5 images per person are generated randomly to form the training data set and the remaining were chosen as test data set. This process was repeated to 20 times for each experiment.

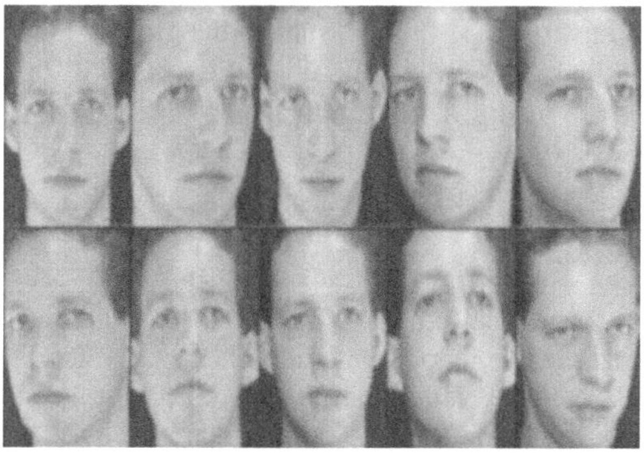

**Fig. 3.** Example in ORL face dataset

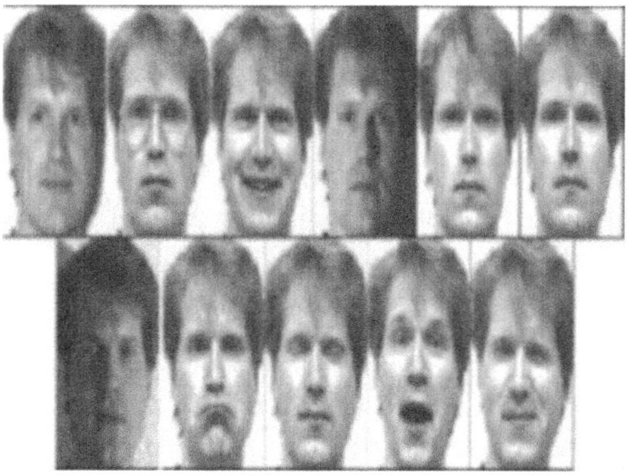

**Fig. 4.** Example in Yale face dataset

## 5.2 Image Pre-processing

The size of the faces in the database is 92×112. After image pre-processing, we use the faces whose sizes are 60×60. Some examples are given in Fig.5.

## 5.3 Parameter Selection

In ICA, 60 principal components are used in the whitening phase, the Gaussian RBF kernel function is considered in KPCA, the basic contrast function $G(u) = x\exp(-x^2/2)$ is adopted in the FastICA.

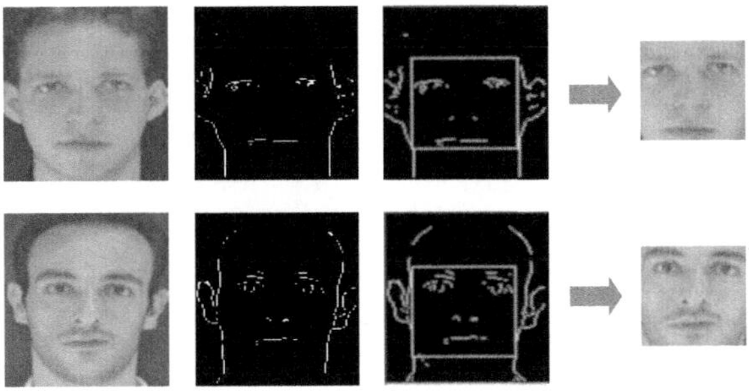

**Fig. 5.** Results After Pre-processing

We need 40 HRBFs to represent 40 different persons, and pertaining to the training error of the HRBF, we make following modifications: As the error occurs to the positive samples, i.e. the sample is corresponding to the HRBF, but its output is 0, we call this error *p_error*. On the contrary, if the output of HRBF of the negative samples is 1, this error is called *n_error*. Because in our training tasks, the positive samples are far less than the negative ones, we adapt our training error as following:

$$f_e = c * p\_error + n\_error \quad (c \geq 1) \tag{29}$$

### 5.4 Result Comparison

For this simulation, the ICA is employed to the training and testing datasets, respectively. The extracted 60 input features are used for constructing a HRBF model. A HRBF classifier was constructed using the training dataset and then the classifier was used on the testing dataset to classify the data as an face ID or not. The instruction sets used to create an optimal HRBF classifier is $S = \{+_2, +_3, \ldots, +_{10}, x_1, x_2, \ldots, x_{60}\}$. Where $x_i$ ($i = 1, 2, \ldots, 60$) denotes the 60 features extracted by ICA.

A comparison of different feature extraction methods and different face classification methods for ORL face dataset (average recognition rate for 20 independent runs) is shown in Table 1. Table 2 depicts the face recognition performance of the HRBF by using the 60 features for Yale data set. The HRBF method helps to reduce the features from 60 to 6-15. For each experiment, the true positive rate (TP), false positive rate (FP) were also computed. For save space, they're not shown here.

**Table 1.** Comparison of different approaches for ORL face recognition

| Method | Recognition rate |
|---|---|
| PCA[17] | 88.31% |
| LDA[17] | 88.87% |
| NN[17] | 94.64% |
| PCA+RBF[16] | 94.5% |
| LDA+RBF[16] | 94.0% |
| ICA(PCA)+HRBF | 96.71% |
| ICA(KPCA)+HRBF | 97.87% |

**Table 2.** Comparison of different approaches for Yale face recognition

| Method | Recognition rate |
|---|---|
| PCA[17] | 81.13% |
| LDA[17] | 98.69% |
| NN[17] | 83.51% |
| ICA(PCA)+HRBF | 97.45% |
| ICA(KPCA)+HRBF | 98.79% |

## 6 Discussion

When using the KPCA method the following two problems will be encountered: One is the storage of the training data in terms of the dot products is too expensive since the size of kernel matrix R increases quadratically with the number of training data, another is the computation of the kernel functions $k(x_i, x_j)$ is expensive. Several approaches to the problem of non-sparse kernel expansion were proposed. These methods are based on approximating the found solution [18,19]. The next step we will improve the KPCA method using in ICA to get higher computational efficiency.

## 7 Conclusions

In this paper ICA based feature extraction method and HRBF classification model are proposed for face recognition. The ORL and Yale database images are used for conducting all the experiments. Facial features are first extracted by the ICA using KPCA to whiten the data and FastICA to extract the independent component of the original face image as well as maintains the main facial features. Compared with the traditional PCA approach, the KPCA can extract features, which are more useful and principal components for classification purposes. The presented HRBF model for face recognition is focused on improving the face recognition performance by reducing the input features. Simulation results on ORL and Yale face database also show that the proposed method achieves high training and recognition speed, as well as high recognition rate. More importantly, it is insensitive to illumination variations.

**Acknowledgment.** This research was partially supported by the Natural Science Foundation of China under contract number 60573065, and the Key Subject Research Foundation of Shandong Province.

## References

1. Zhang, C. P., Su, G. D.: Human Face Recognition: A survey. Journal of Image and Graphic, 5(11) (2000) 885-894
2. Yuille, A., Hallien, P.: Feature Extraction from Faces Using Deformable Templates. International Journal of Computer Vision, 8(2) (1992) 99-111
3. Zhao, W., Chellappa, R., Rosenfeld, A. and Phillips, P.J.: Face Recognition: A Literature Survey. Technical Report CART-TR-948, University of Maryland (2002)

4. Hyvärinen, A., Karhunen, J.: Independent Component Analysis. Wiley, New York (2001)
5. Bartlett, M. S., Movellan, J. R., Sejnowski, T. J.: Face Recognition by Independent Component Analysis. IEEE Trans. Neural Networks, 13(6) (2002) 1450-1464
6. Valentin, D., Abdi, H., Toole, A. J. O., Cottrell, G. W.: Connectionist Models of Face Processing: A survey. Pattern Recognition, 27 (1994) 1209-1230
7. Er, M. J., Wu, S., Lu, J., Toh, H. L.: Face Recognition with Radial Basis Function (RBF) Neural Networks. IEEE Trans. Neural Network, 13(3) (2002) 697-710
8. Chen, X. L., Shan, S. G.: Pose-varied Face Recognition. Journal of Image and Graphics, 4(10) (1999) 818-824
9. Zhu, C. R., Wang, R. S.: Multi-pose Face Recognition Based on A Single View. Chinese Journal of Computers, 26(1) (2003) 104-109
10. Schölkopf, B., Smola, A., Muller, K. R.: Nonlinear Component Analysis as a Kernel Eigenvalue Problem. Neural Comput., 10(5) (1998) 1299-1319
11. Golub, G. H., Van Loan, C. F.: Matrix Computations, Thirded. The Johns Hopkins University Press, Baltimore (1996)
12. Mat Isa, N. A., Mashor, M. Y., Othman, N. H.: Diagnosis of Cervical Cancer using Hierarchical Radial Basis Function (HRBF) Network. Proceedings of the International Conference on Artificial Intelligence in Engineering and Technology, (2002) 458-463
13. Chen, Y. H., Peng, L. Z., Abraham, A.: Hierarchical Radial Basis Function Neural Networks for Classification Problems. Springer LNCS, 3971(2006) 873-879
14. Sastry, K., Goldberg, D. E.: Probabilistic Model Building and Competent Genetic Programming. Genetic Programming Theory and Practise, (2003) 205-220
15. Kennedy, J., Eberhart, R.: Particle Swarm Optimization. IEEE International Conference on Neural Networks, (1995) 1942-1948
16. Su, H., Feng, D., Zhao, R. C.: Face Recognition Using Multi-feature and Radial Basis Function Network. Proc. of the Pan-Sydney Area Workshop on Visual Information Processing (VIP2002), Sydney, Australia, (2002) 183-189
17. Huang, R., Pavlovic, V., Metaxas, D. N.: A Hybrid Face Recognition Method Using Markov Random Fields. ICPR04, (2004) 157-160
18. Schölkopf, B., Knirsch, P., Smola, C., Burges, A.: Fast Approximation of Support Vector Kernel Expansions, and an Interpretation of Clustering as Approximation in Feature Spaces. DAGM Symp. Mustererkennung, Springer LNCS, (1998) 124-132
19. Franc, V., Hlavac, V.: Training Set Approximation for Kernel Methods. Proceedings of the 8th Computer Vision Winter Workshop, Prague, Czech Republic, (2003) 121-126

# Long-Range Temporal Correlations in the Spontaneous in vivo Activity of Interneuron in the Mouse Hippocampus

Sheng-Bo Guo[1,2], Ying Wang[3], Xing Yan[1,2], Longnian Lin[3], Joe Tsien[3,4], and De-Shuang Huang[1,2]

[1] Intelligent Computing Lab, Hefei Institute of Intelligent Machines, Chinese Academy of Sciences, Hefei, Anhui 230031, China

[2] Department of Automation, University of Science and Technology of China, Hefei, Anhui, 230026, China

[3] Shanghai Institute of Brain Functional Genomics, Key Laboratory of Chinese Ministry of Education, and Shanghai Key Laboratory of Brain Functional Genomics, East China Normal University, Shanghai 200062, China

[4] Center for System Neurobiology, Departments of Pharmacology and Biomedical Engineering, Boston University, Boston, MA 02118, USA

dshuang@iim.ac.cn

**Abstract.** The spontaneous in vivo firings of neuron in mouse hippocampus are generally considered as neuronal noise, where there is no any correlation in the inter-spike interval (ISI) sequences. In the present study, we investigate the nature of the ISI sequences of neuron in CA1 area of mouse hippocampus. By using the detrended fluctuation analysis (DFA), we calculated the fluctuation or scaling exponent of the ISI sequences. The results indicated that there exists the long-range power-law correlation over large time scale in the ISI sequences. To further investigate the long-range correlation of ISI, we studied the long-range correlation of ISI sequences from different types of neurons in mouse hippocampus, which are four types of interneurons categorized by their firing patterns. Our results show the presence of long-range correlations in the ISI sequence of different types of neurons. Furthermore, the shuffle surrogate data achieved by randomly shuffle the original ISI sequence is used to verify our conclusion. The application of shuffle surrogate shows that the long-range correlation is destroyed by randomly shuffle, which demonstrates that there is actually the long-range correlation in the ISI sequence. Furthermore, we also compare the long-range correlations of ISI sequence when mice are in different behavioral states, slow-wave sleep (SWS) and active exploration (AE). Our results indicated that the ISI sequences exhibit different extent of long-range correlations: the long-range correlation is significantly stronger when mice are in AE than that of ISI sequence when mice are in SWS, which demonstrated that the varied long-range correlations exhibiting in ISIs of interneurons might be associated with activities of neuronal network regulating the ongoing neuronal activity of different interneurons.

**Keywords:** Hippocampus, Interneuron, Long-Range Temporal Correlations, Detrended Fluctuation Analysis, Fractal.

D.-S. Huang, L. Heutte, and M. Loog (Eds.): ICIC 2007, LNAI 4682, pp. 1339–1344, 2007.

# 1 Introduction

As an important formation in the brain, hippocampus plays an important role in learning and memory. Since 1950s, hippocampus has attracted an increasing number of researchers [4, 7, 12, 13, 19]. And some researchers have investigated the long-range correlations in the neuronal models [10]. However, there are relatively less researches in the long-range correlation of the spontaneous in vivo activity of neurons in hippocampus, especially in mouse hippocampus.

Because the inter-spike interval (ISI) sequence of the spontaneous in vivo activity of neurons is irregular, ISI sequence is frequently modeled by the stochastic point process. Recently, an emerging perspective is that there exist the long-range correlations in the spontaneous spiking of neurons in the hippocampal-amygdala [3].

In recent years, there are increasing evidence indicating that many physical and biological systems exhibit long-range power-law correlations such as economics [14, 15], DNA [18, 19], human gait [8], neural receptors in biological systems [2], ion channel kinetics [11]. Detrended Fluctuation Analysis (DFA) is a scaling analysis method used to quantify long-range power-law correlations in signals [6], and the DFA method could also help distinguish distinct statues of the same system with different scaling behavior such as the healthy and sick individuals [1] as well as for waking and sleeping states [9]. To investigate the long-range correlation in neuronal spike train in mouse hippocampus, the DFA is used to characterize the presence of long-range dependence, which indicated that the spontaneous firing patterns of most of the recorded neurons could not be well described by a renewal process or by any process with short correlation; rather they present long-range power-law correlations, representing ongoing memory effects in the ISI sequence. To verify our conclusion, the shuffle surrogate dataset is achieved by randomly shuffle the original ISI sequence, thus the ISI distribution function and the mean firing rate of the shuffle surrogate is the same as the original one. By applying DFA in the shuffle surrogate dataset, we found that the long-range correlation is destroyed by randomly shuffle.

To further study the relationship between long-range correlations and behavioral states, we recorded the spontaneous in vivo activity of the neuron when mice were in different behavioral states, slow-wave sleep (SWS) and active exploration (AE), which are two significant behavioral states for mouse. After the spontaneous in vivo activities are recorded, the DFA is applied and the results indicate that 1) the ISI sequences exhibit long-range dependence when mouse are in both of these two states, and 2) the long-range correlations of the ISI sequence in SWS is significantly weaker than that of ISI sequence during mice are in AE.

# 2 Method

Detrended fluctuation analysis (DFA) is a modified root mean square analysis, which is widely used to the analysis of physiological data [18]. The method of DFA is briefly formulated as follows. First, the ISIs (of total length N) is integrated, $y(k)=\sum_{i=1}^{k}(X-\overline{X}), k=1,\cdots,N$ , where $X(i)$ is the ith inter-spike interval (ISI) and $\overline{X}$ is the average ISI sequences. Next, the integrated time series is divided into t=int(N/n) non-overlapping boxes of length, n. In each box, a least-squares line representing the trend in that box is fit to the data, Denoted by $y_n(k)$ the y coordinate of the straight line segment time series, we detrended the integrated time series, $y_n(k)$, in each box. Subsequently, the root-mean-square fluctuation of this integrated and detrended time series is calculated by $F(n) = \left\{ \frac{1}{N} \sum_{k=1}^{N} [y(k) - y_n(k)]^2 \right\}^{\frac{1}{2}}$.

This computation is repeated over all time scales (box scales) in order to provide a relationship between F(n), the average fluctuation as a function of box size, and the box size n, which is the number of spikes in a box. The system with power-law correlation obeys $F(n) \propto n^{\alpha}$, where $\alpha$ is the slope of the line relating log F(n) and log n and is called the fluctuation or scaling exponent.

Consider an ISI sequence where the value at one ISI is completely uncorrelated from any previous values, i.e. white noise. The kinds of ISI sequence can be achieved by randomly shuffle the original ISI sequences (so-called shuffle surrogate dataset). For this type of uncorrelated data, the integrated value, y(k), corresponds to a random walk, thus $\alpha = 0.5$ [16]. For $0 < \alpha < 0.5$, ISI sequence exhibits an anti-persistent correlated process such that large and small values of the ISI sequence are more likely to alternate. And for $0.5 < \alpha \leq 1$, ISI sequences persistent long-range power-law correlations, which means a large (compared to the average) ISI is more likely to be followed by large ISI and vice versa. For $\alpha = 1$, ISI sequence corresponds to 1/f noise, which is a self-similar or scale-invariant fractal sequence. The degree of deviation of $\alpha$ from 0.5 toward 1 represents the strength of long-range correlation in the sequence.

# 3 Results

## 3.1 Detrended Fluctuation Analysis: Long-Range Correlations of ISI Sequences of Interneurons

The analysis of one hippocampus interneuron is shown in Fig. 1. Fig. 1A shows the original ISI sequence of an interneuron belonging to Interneuron Type III. This ISI sequence is recorded when mouse is in AE. From Fig. 1B, we can see a clear power-law scaling relationship between average fluctuation and the box size n, which exhibits the property of scale invariance. The slope, scaling exponent, is achieved by least-square fit. Also in Fig. 1B, the result of DFA in shuffle surrogate for the original ISI sequence is evidently deviated from the original one. Although it is still exhibit power-law relationship, the scale exponent for shuffle surrogate is 0.50 indicating the effectiveness of DFA. Fig. 1C shows one of the shuffle surrogates of original ISI sequence shown in Fig. 1B. And the result of DFA in this sequence is shown in Fig. 1D, in which we noticed that there is no significant difference for the scale exponent for the ISI sequence shown in Fig. 1C and its shuffle surrogate. From the result above, we conclude that the original ISI sequence does exhibit the long-range correlations.

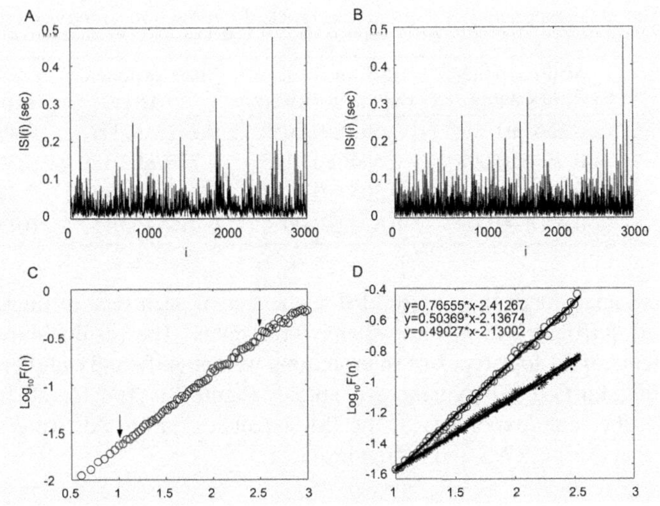


**Fig. 1.** DFA analysis on the ISI sequence and its shuffle surrogate

### 3.2 Long-Range Correlations in Different Interneurons

To further investigate the relationship between long-range correlations of ISI sequences in different firing patterns, the scale exponents of ISI sequences were estimated in two different behavioral states: SWS and AE. To maintain stable estimation, seven ISI sequences with 3000 continuous spikes were chosen from each interneuron for SWS and AE. We then statistically investigate the scale exponents of the seven sequences, and achieved the mean and standard deviation as the estimated scale exponent for each interneuron in each behavioral state. Figure 2 shows that the

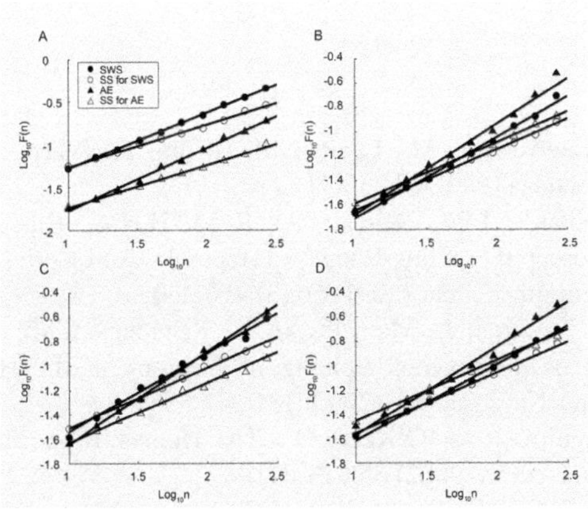


**Fig. 2.** The scale exponents were computed for four types of interneurons by performing DFA on 3000 ISIs during AE and SWS

**Table 1.** The scale exponent of recorded four types of interneurons in mouse hippocampus

| Type | Scale exponent (SWS) | SS for Scale exponent (SWS) | Scale exponent (AE) | SS for Scale exponent (AE) |
|---|---|---|---|---|
| Type Ⅰ (n=5) | 0.6567±0.0094 | 0.5066±0.0082 | 0.7373±0.0097 | 0.4986±0.0138 |
| Type Ⅱ (n=7) | 0.6516±0.0081 | 0.5046±0.0115 | 0.7339±0.0129 | 0.5069±0.0113 |
| Type Ⅲ (n=3) | 0.6506±0.0106 | 0.5054±0.0117 | 0.7538±0.0125 | 0.4961±0.0141 |
| Type Ⅳ (n=9) | 0.6301±0.0037 | 0.5002±0.0078 | 0.7142±0.0153 | 0.4981±0.0025 |

result of DFA analysis of the original ISI sequences of each type of interneurons and their shuffled surrogates in different behavioral states. The results showed that the scale exponents of all four types of interneurons were significantly higher than that of corresponding shuffled ISI sequences (Table 1, Figure 2). And for all four types of interneurons, the scale exponents of the ISI sequences during AE were significantly higher than that during SWS and more close to 1.

## 4 Discussion

To investigate the long-range correlation of the ISI sequence from these four types of interneurons, we employed the method named as DFA for the detection of long-range power-law correlations in the recorded ISI sequences. Consequently, we demonstrated that the spontaneous in vivo activity of interneuron in mouse hippocampus exhibits long-range correlations, rather than a stochastic point process. And we thus concluded that the long-range correlations of the ISI sequence indicated the presence of history effect or the memory in the firing pattern. Aside from the demonstration of the presence of long-range correlations in ISI sequences, we have further studied the relationship between the extent of long-range correlations and different behavioral states, SWS and AE.

## References


1. Ashkenazy, Y., Lewkowicz, M., Levitan, J., Havlin, S., Saermark, K., Moelgaard, H.: Thomsen PEB Fractals (1999) 4(1):85-91
2. Bahar, S., Kantelhardt, J.W., Neiman, A., Rego, H.H.A., Russell, D.F., Wilkens, L., Bunde, A., Moss, F.: Long-Range Temporal Anti-Correlations in Paddlefish Eletroreceptors. Europhys. Lett. (2001) 56 (3):454-460
3. Bhattacharya, J., Edwards, J., Mamelak, A.N., Schuman, E.M.: Long-Range Temporal Correlations in the Spontaneous Spiking of Neurons in the Hippocampal-Amygdala complex of humans. Neuroscience (2005) 131:547-555
4. Burgess, N., Maguire, E.A., O'Keefe, J.: The Human Hippocampus and Spatial and Episodic Memory. Neuron (2002) 35:625-641
5. Chen, Z., Ivanov, P.C., Hu, K., Stanley, H.E.: Effect of Nonstationarities on Detrended Fluctuation Analysis. Physical Review E 65, (2002) 04117:1-15
6. Harris, K.D., Henze, D.A., Hirase, H., Leinekugel, X., Dragoi, G., Czurko, A., Buzsaki, G.: Spike Train Dynamics Predicts Theta-related Phase Precession in Hippocampal Pyramidal Cells. Nature (2002) 417:738-741

7. Hausdorff, J.M., Peng, C.K., Ladin, Z., Wei, J.Y., Goldberger, A.L.: Is Walking a Random Walk? Evidence for Long-Range Correlations in Stride Interval of Human gait. J Appl Physiol (1995) 78:349-358
8. Ivanov, P.C., Bunde, A., Amaral, L.A.N., Havlin, S., Fritsch-Yelle, J., Baevsky, R.M., Stanley, H.E.: Sleep-Wake Differences in Scaling Behavior of the Human Heartbeat: Analysis of Terrestrial and Long-Term Space Flight Data. Europhys. Lett. 48(5):594-600
9. Jackson, B.S.: Including Long-Range Dependence in Integrated-and-Fire Models of the High Interspike-Interval Variability of Cortical Neurons. Neural Computation (2004) 16:2125-2195
10. Lan, T.H., Xu, B.Q., Yuan, H.J., Lin, J.R.: Rescaled Range Analysis Applied to the Study Delayed Rectifier Potassium Channel Kinetics. Biophys Chem (2003) 106(1):67-74
11. Lin, L.N., Osan, R., Shoham, S., Jin, W.J., Zuo, W.Q., Tsien, J.Z.: Identification of Network-Level Coding Units for Real-Time Representation of Episodic Experiences in the Hippocampus. Proc. Natl. Acad. Sci. 102, (2005) No. 17:6125-6230
12. Lin, L.N., Chen, G.F., Xie, K., Zaia, K.A., Zhang, S., Tsien, J.Z.: Large-Scale Neural Ensemble Recording in the Brains of Freely Behaving Mice. J of Neurosc Methods (2006) 155:28-38
13. Liu, Y., Cizeau, P., Meyer, M., Peng, C.K., Stanley, H.E.: Correlations in Economic Time Series. Physica A (1997) 245:437-440
14. Liu, Y., Gopikrishnan, P., Cizeau, P., Meyer, M., Peng, C.K., Stanley, H.E.: Statistical Properties of the Volatility of Price Fluctuations. Phys Rev E (1999) 60(2):1390-1400
15. Montroll, E.W., Shlesinger, M.F.: Nonequilibrium Phenomena □. From Stochastics to Hydrodynamics, edited by Lebowitz JL and Montroll EW, North-Holland, Amsterdam:1-121 (1984)
16. Moruzzi, G., Magoun, H.W.: Brainstem Reticular Formation and Activation of the EEG. Electroencephal. Clin. Neurophysiol L (1949) 455-473
17. Peng, C.K., Buldyrev, S.V., Havlin, S., Simons, M., Stanley, H.E., Goldberger, A.L.: Mosaic Organization of DNA Nucleotides. Phys Rev E Stat Phys Plasmas Fluids Relat Interdiscip Topics 49 (1994) 1685-1689
18. Peng, C.K., Havlin, S., Stanley, H.E., Goldberger, A.L.: Quantification of Scaling Exponents and Crossover Phenomena in Nonstationary Heartbeat Time Series. (1995) Chaos 5: 82-87
19. Penzel, T., Kantelhardt, J.W., Grote, L., Peter, J.H., Bunde, A. Comparison of Detrended Fluctuation Analysis and Spectral Analysis for Heart Rate Variability in Sleep and Sleep Apnea. IEEE Trans Biomed Eng 50 (2003) 1143-1151

# Implementation and Performance Analysis of Noncoherent UWB Transceiver Under LOS Residential Channel Environment

Sungsoo Choi[1], Insoo Koo[2], and Youngsun Kim[1]

[1] Korea Electrotechnology Research Institute (KERI), Republic of Korea
[2] University of Ulsan (UoU), Republic of Korea

**Abstract.** This paper focuses on two things. One is to describe implementation of a Low-Rate Ultra-Wideband (LR-UWB) transceiver for low power Wireless Personal Area Network (WPAN) applications. The other is to derive the Bit Error Rate (BER) performance of the proposed system. At first, in order to implement low-power and low-complexity transceiver, we design a non-coherent On-Off-Keying(OOK) UWB system by adopting the architecture of the simplified asynchronous transmission and the edge-triggered pulse transmission, which makes the system performance independent of the shape of the transmitted waveform, robust to multipath channels. The designed non-coherent UWB transceiver architecture has an advantage of the simple realization since any mixer, high-speed correlator, and very high-sampling A/D converter are not necessary at the cost of performance degradation of about 3dB. Further, the designed non-coherent UWB transceiver is actually implemented with the wireless CANVAS prototyping testbed in short range indoor application environment such as a lecture room. The implemented prototype testbed is proven to offer the data rate of 9.6kbps and 115kbps on the conditions of Peer-to-Peer (P-to-P) in the indoor channel within the range of about 10m. At second, we analyze the bit error rate (BER) performance of this proposed system under the line-of-sight (LOS) residential channel environment. As a result, it is necessary to choose appropriate number of Window Bank(WB)s for better BER performance.

## 1 Introduction

Recently, various low-rate alternative physical layers (PHY) have been designed and developed in order to offer short range wireless connectivity with ultra low power consumption and low cost for applications such as wireless sensor network (WSN) or wireless personal area network (WPAN)[1]. Among them, an impulse radio based Ultra-Wideband (UWB) has drawn lots of interesting as one of the most suitable PHY solutions due to the following profitable features: an ease of installation and interfacing, extremely low complexity, and a reasonable battery life while maintaining a simple and flexible protocol[2,3]. In the research area of WPAN, newly the research development and standardization of low rate UWB (LR-UWB) have been accelerated since it is believed that applications of LR-UWB will create bigger new market than those of high rate WPAN solutions

D.-S. Huang, L. Heutte, and M. Loog (Eds.): ICIC 2007, LNAI 4682, pp. 1345–1356, 2007.

[4]. On March of this year, the standardization of IEEE 802.15.4a is scheduled to release, which is based on LR-UWB physical layer adopting an impulse radio, and it is expected to outperform the performance of conventional low rate WPAN devises [1, 4, 5].

For designing the LR-UWB transceiver, numerous competing architecture issues to recover the transmitted signal from among thermal noise, multipath, and multi-user effects include coherent versus non-coherent UWB receivers, a correlation detection, a low noise amplifier, an analog front end, a linearity and dynamic range, an analog-to-digital conversion based on time interleaved architecture and frequency domain channelization, a demodulation via baseband processing or narrow band signal processing, and so on. Particularly, the design issue of coherent versus non-coherent UWB receivers is very crucial. Generally, coherent systems based on correlator in the receiver can provide fairly good performance. However, coherent systems are very sensitive to signal synchronization based on the precise timing information of transmitted and received signals for detection, and they require an additional pulse generator with specific pulse shape to achieve maximum energy of the received signal in the receiver. Therefore, these systems may increase implementation complexity, and in consequence, it results in more power consumption and more system cost. On the contrary, the structure of noncoherent system is so simple that the system consumes less power than coherent systems. Though the noncoherent system cannot perform as good as coherent systems, its cost-effective features with the relaxed performance are enough to be attractive to LR-UWB applications. For this reason, we design the LR-UWB transceiver of noncoherent On-Off-Keying (OOK) scheme, and implement it called a wireless CANVAS prototyping testbed. Particularly, instead of not considering impulse timing synchronization and sophisticated analog parts, we take the advantage of a non-coherent receiver and asynchronous transmission for our applications. In addition, we derive the bit error rate (BER) performance of the proposed system under LOS Residential Channel Environment.

The rest of the paper is organized as follows: In Section 2, we describe the implementation of a Low-Rate Ultra-Wideband (LR-UWB) transceiver, and give implementation results about CANVAS system. In Section 3, we analyze the bit error rate (BER) performance BER of the designed architecture under LOS Residential Channel Environment, and we show numerical results of BER performance. Finally, we draw a conclusion in the last Section.

## 2 Implementation of Low-Rate Ultra-wideband Transceiver

Figure 1 shows a conceptual block diagram of the wireless non-coherent LR-UWB system proposed in this paper. The transmitter consists of a Controller transforming signals to transmit an edge-triggered impulse train according to the data input through repeater, and a Filtered Impluse Generator with a level counter, a pulse train generator, an impulse generator, and an ultra wideband

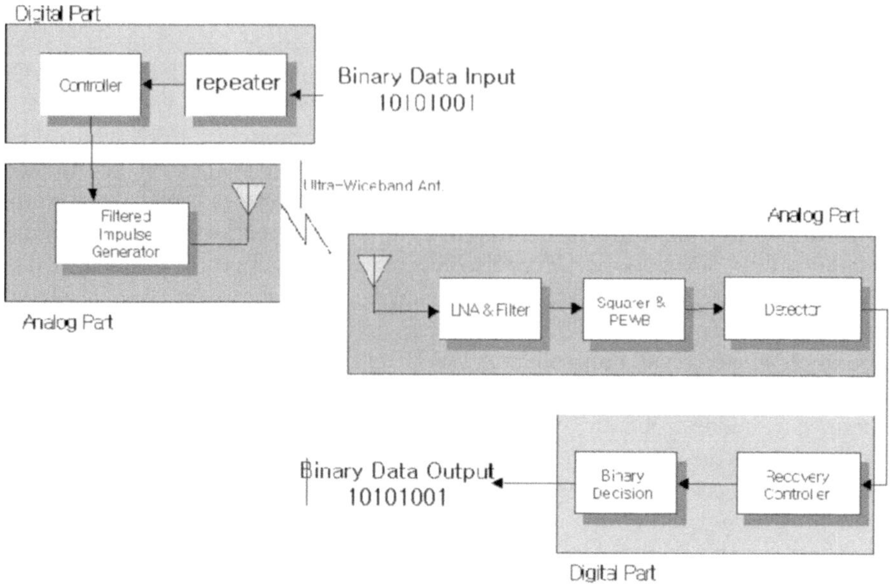


**Fig. 1.** Conceptual Block Diagram of Non-coherent LR-UWB Transceiver

antenna. The receiver architecture is largely divided into two parts of analog RF part and digital base band signal processing part. Analog RF part consists of a wideband Low Noise Amplifier (LNA), a band pass Filter, a Squarer, a Parallel Energy Window Bank (PEWB), and a signal Detector. The digital part consists of a Recovery Controller with digital logic converter and digital correlator, and a Binary Decision processor. To fully simplify Analog RF Part of system, we just consider the PEWB, which is the special module of a multiple 1-bit Analog-to-Digital Converter (ADC) with delayed phase of 1 ns instead of very high-sampling ADC on the analog part to take a digitized received signal and do not consider a fast analog correlator, mixer, and the fast OSC on the RF analog receiver part.

## 2.1 Transmitter Design

### 2.1.1 Impulse Generator

A UWB signal is defined by either a minimum instantaneous bandwidth of 500 $MHz$ or a minimum fractional bandwidth of 20 percent[1]. In that sense, the designed impulse-based LR-UWB transmitter has a pulse amplitude modulation scheme with short pulse trains, that is, very short duration and low duty cycle, when transmitting binary information. Figure 2, a block diagram of the proposed impulse generator consists of a short pulse generator part which makes a short duration of pulses and an impulse generator part which makes a pulse shape sufficient to limit the radiated frequency regulation of the lower band of UWB

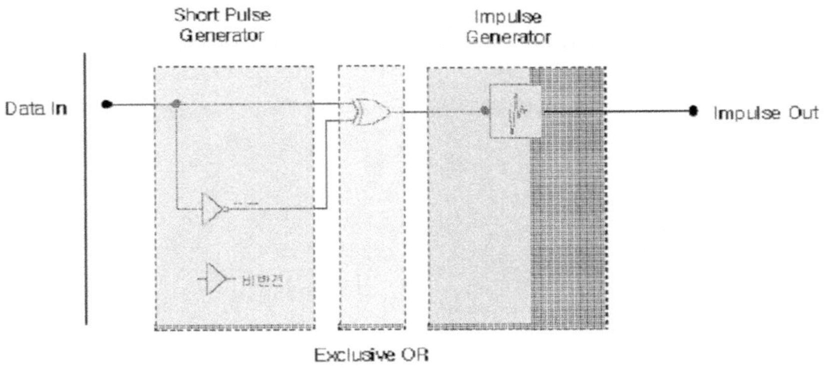


**Fig. 2.** Impulse Generator Circuit

bandwidth, $3GHz$ between $5GHz$. After the impulse generator, it transmits mono pulse like a Gaussian signal whose pulse duration is shorter than $2ns$.

### 2.1.2 Transmission Processing

Transmission controller in non-coherent LR-UWB transmitter converts signal for transmission. It is designed to have two main components. One is a pulse generator. It transforms binary data into coded signal and, further maps them into the impulse transmission. The other is asynchronous signal converter. The signal converter is coded on Xilinx CPLD (XC9536). Besides MAX232, 11.0592MHz Oscillator, DC+5V input cable, D-type 9Pin serial connector and RF connector are also constituted as main parts and other devices are used. The source of electricity is independently designed to supply DC 5V. Input serial data signal of which voltage level is +11V to -11V, is converted into transmission signals for impulse generation by programmable CPLD chip. At this moment, the level converter inverts the original signal and let it out. Then CPLD of XC9536 receives the inverted signal and when the signal is idle (no signal) it outputs Low(0). If rising edge occurs, after outputting High(1) of 5V level during 904ns, it returns to Low(0). And also when falling edge occurs, after outputting High(1), it returns to Low(0) during the same period. Namely whenever input data signal is changed, it generates a pulse with constant period of 904ns. This edge triggered signal is used as an input of impulse generator and then through wideband wireless antenna radiates the generated impulse signal.

## 2.2 Receiver Design

### 2.2.1 Detect Processing

The received RF impulsive signal, sent through two ultra wide antennas in free space channel, results in the pulse shape looking like secondary differentiated wave. The signal detector of this system is designed based on the fast step recovery diode and band pass filtered discrete type device with RC constant, in the sense of envelop detector.

**Fig. 3.** Wireless Canvas Prototype Testbed

### 2.2.2 Recovery Processing

To recover the original signal from the received one which was passed through antennas and signal detector, the controller operates the recovery processing that the detected RF impulsive signals converts the binary signals using 1-bit ADC. Signal converter consists of Xilinx (XC9536), which is a CPLD for base-band binary decision processing, MAX232, 11.0592MHz Oscillator, DC+5V input cable, D-type 9Pin serial connector and RF Connector. The source of electricity is independently designed to supply DC 5V. Signal detector on the receiver part is designed to receive signal from RF receiver. During no received impulse signal on the defined window time slot, the signal detector outputs High(1). If rising edge of the received impulse signal occurs it outputs an inverted signal after 2 Clock Time (using 11.0592Mhz about 180ns(1/Freq * 2)). The inverted signal consists of serial signals and the voltage level is changed 0V to 10V to be transmitted to PC. When receiving pulses we can confirm converting of signal data.

## 2.3 Implementation of Wireless Canvas Prototyping Testbed Using LR-UWB Transceiver

Figure 3 shows the implemented receiver part of LR-UWB transceiver which is called Canvas testbed. It first operates in receiving the canvas data written or drawn by user on PC. After them passing through the transmitting Canvas testbed with the digital part and the analogue end with ultra wideband antenna, the other distant Canvas testbed receives the transmitted pulses of the canvas

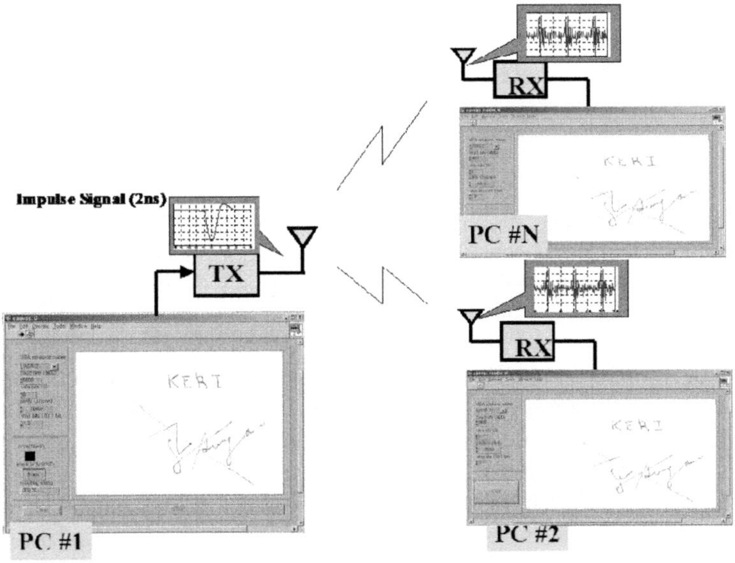


**Fig. 4.** System Model of Wireless Canvas Prototype Testbed with GUI

drawn by user. It detects the received pulse signal through the analogue part and the digital reception part. The digital information of the received signal is reliably demodulated when impulse trains are transmitted at the rate of 115kbps. Figure 4 shows a graphical user interface (GUI) environment of Canvas application system realized for the aforementioned LR-UWB transceiver. We used the LabVIEW tool of NI (National Instrument) Inc. for Canvas interfacing the LR-UWB transceiver with PC. Using this developed GUI program, we draw pictures or write letters on the transmission side, then the same picture or letters are simultaneously displayed on the canvases of the multiple receiver sides being apart from the transmitter within about $10m$.

## 3 BER Performance Analysis

The generalized impulse response of UWB channel is represented by multiple paths or rays having real positive gains, propagation delays, and associated phase shifts as follows [6], [7]:

$$h(t) = \sum_{l} \sum_{k} a_{k,l} \exp\{j\phi_{k,l}\} \delta(t - T_l - \tau_{k,l}) \quad (1)$$

where $a_{k,l}$ is the weight of the $k$-th ray of the $l$-th cluster, $T_l$ is the delay of the $l$-th cluster, and $\tau_{k,l}$ is the delay of the $k$-th ray relative to the $l$ th cluster arrival time $T_l$. The phase $\phi_{k,l}$ is uniformly distributed, i.e., the phase is an uniformly distributed random variable from the range $[0, 2\pi]$.

**Table 1.** Channel Parameters of IEEE802.15.4A Residential Environment

| Parameters | LOS | NLOS |
|---|---|---|
| Pathloss at 1m distance, $PL_0[dB]$ | 43.9 | 48.7 |
| pathloss exponent, $n$ | 1.79 | 4.58 |
| Mean number of clusters, $\bar{L}$ | 3 | 3.5 |
| Inter-cluster arrival rate, $\Lambda[1/ns]$ | 0.047 | 0.12 |
| Ray arrival rates, $\lambda_1, \lambda_2[1/ns], \beta$ | 1.54, 0.15, 0.095 | 1.77, 0.15,0.045 |
| Inter-cluster decay constant, $\gamma_0[ns]$ | 12.53 | 17.50 |
| Intra-cluster decay time constant parameters, $\Gamma[ns]$ | 22.61 | 26.27 |

IEEE802.15.4a Tasking Group (TG) categorizes UWB environments into 5 classes: residential, indoor office, outdoor, open outdoor and industrial environments. Each class is divided into two conditions which are line-of-sight (LOS) and non-line-of-sight (NLOS). The delay spread $T_d$ of this impulse response, one of the important parameters of channel model which is the longest excess delay of the impulse response, is about 180 $ns$. Under the assumption that there are no additional techniques, such as channel coding, spread spectrum, etc, pulse repetition time has to be longer than the delay spread of the impulse response to avoid the performance degradation that can be resulted from inter-symbol-interference (ISI). The channel parameters of residential environment are presented in the Table I [6].

We design the transmitted signal $s(t)$ as shown in Fig. 5. Figure 5(a) depicts the edge-triggered signal, i.e., two signal sets relative to two successive symbols '1's. The transmitted pulses, a positive $p(t)$ and a negative $p(t)$ in Fig. 5(b), are generated at rising edge and falling edge of the bit trains, respectively. There is a signal to noise ratio (SNR) improvement with this signal set. Additionally, the signal set can be repeated to get more energy in receiver with the bit repetition rate $R$ at the cost of data rate. The bit repetition time $T_b$ can be represented by

$$T_b = R \cdot T_s \tag{2}$$

where $T_s$ is the signal set duration. Therefore, the closed form of $s(t)$ is expressed as

$$s(t) = \sum_{n=-\infty}^{\infty} \sum_{i=0}^{2R-1} B_n(-1)^i p(t - (\frac{i}{2R} + n)T_s) \tag{3}$$

where $B_n$ is information bits, '0' or '1'.

### 3.1 Performance Under AWGN Channel Model

Under AWGN channel, the output of the square of the received signal under each hypothesis is given by

$$y = \int_0^{T_i} [w(t)]^2 dt \qquad under \;\; H_0(SPACE) \tag{4}$$

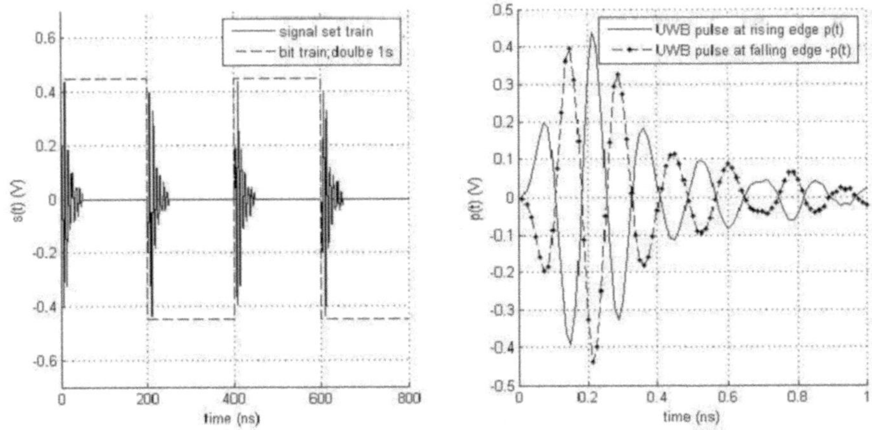


**Fig. 5.** Transmitted signal characteristics: (a) transmitted signal relative to bit train. (b) UWB pulses at rising edge and falling edge.

$$y = \int_0^{T_i} [r_{AWGN}(t)]^2 dt = \int_0^{T_i} [s(t) + w(t)]^2 dt \qquad under \;\; H_1(MARK) \tag{5}$$

where $T_i$ is integration time, $w(t)$ is AWGN with spectral density of $N_0/2$, and $r_{AWGN}(t)$ is the received signal. If there are n window banks (WB) in parallel energy window bank (PEWB)s, $n$ samplings are carried out. Therefore, the output of PEWB $Y$ is formulated by

$$Y = \sum_{i=1}^{n} X_i^2 \tag{6}$$

where $X_i$, $i$ =1, 2, $\cdots$ , $n$ are statistically independent Gaussian random variables. When a symbol '0' is sent from transmitter, $X_i$ are zero mean Gaussian with variance $\sigma^2$. On the contrary, when a symbol '1' is sent, $X_i$ are nonzero mean Gaussian with variance $\sigma^2$. In addition, $Y$ is a random variable whose distribution is central or non-central *chi-square* with degrees of freedom of $n$ [8]. For that reason, the integration time is given by the product of the time delay between two adjacent WBs and the number of WBs in PEWB which is equal to the degrees of freedom of chi-square distribution. Therefore, the probability density functions (PDFs) of Y under SPACE and MARK hypotheses are given by

$$p_{Y|H_0}(y) = \frac{1}{N_0 \Gamma(M)} (\frac{y}{N_0})^{M-1} \exp(-\frac{y}{N_0}), \;\; y \geq 0 \qquad under \;\; H_0(SPACE) \tag{7}$$

$$p_{Y|H_1}(y) = \frac{1}{N_0} (\frac{y}{E})^{(M-1)/2} \exp(-\frac{y+E}{N_0}) I_{M-1}(\frac{2\sqrt{yE}}{N_0}), \;\; y \geq 0 \; under \;\; H_1(MARK) \tag{8}$$

where $M$ is equal to the half of degrees of freedom $n$; $n = 2M$, $\Gamma(\cdot)$ is the gamma function, $I_\alpha(\cdot)$ is the $\alpha$-order modified Bessel function of the first kind, and $E$ is the non-centrality parameter which is given by

$$E = \sum_{i=1}^{2M} m_i^2 \tag{9}$$

The $m_i$'s are the means of Gaussian random variable $X_i$ with variance $\sigma$[8].

The optimum threshold $\gamma_{opt}$, which leads to minimum BER, is a solution of following equation that can be obtained by substituting $\gamma_{opt}$ for in Eqns.(7) and (8) under equal probabilities of hypotheses such that we have

$$p_{Y|H_0}(\gamma_{opt}) = p_{Y|H_1}(\gamma_{opt}) \tag{10}$$

In order to solve above equation in closed form is quite difficult. The optimum threshold can be obtained by various methods [9], [10]. In this paper, we acquire the optimum threshold with iteration process. In general, the optimum threshold $\gamma_{opt}$ approaches $\varepsilon/2 + MN_0$ for large $\varepsilon/(N_0 M^2)$ where $\varepsilon = E/2$ is the average signal energy for OOK modulation scheme [10].

For the calculated optimum threshold, the probability of deciding MARK when SPACE is sent is calculated by

$$P(1|0) = \int_{\gamma_{opt}}^{\infty} p_{Y|H_0}(y)dy = \exp\left(-\gamma_{opt}/N_0\right) \sum_{k=0}^{M-1} \frac{1}{k!} \left(\frac{\gamma_{opt}}{N_0}\right)^k \tag{11}$$

Similarly the probability of deciding SPACE when MARK is sent can be expressed in terms of the generalized Marcum's Q function of order $M$,

$$P(0|1) = \int_{0}^{\gamma_{opt}} p_{Y|H_1}(y)dy = 1 - Q_M\left(\sqrt{\frac{4\varepsilon}{N_0}}, \sqrt{\frac{2\gamma_{opt}}{N_0}}\right) \tag{12}$$

[8]. The error probability is obtained by averaging up Eqns. (11) and (12);

$$P_e = \frac{1}{2}\{P(1|0) + P(0|1)\} \tag{13}$$

### 3.2 Performance Under LOS Residential Channel Model

The difference between LOS residential channel model and Gaussian channel model is that the channel impulse response $h(t)$ is added. Because the specific channel impulse response is considered, the BER performance of the proposed system has a more practical meaning, especially for an IEEE802.15.4a environment. The output of the square of the received signal under each hypothesis is given by

$$y = \int_0^{Ti} [w(t)]^2 dt \qquad under \;\; H_0(SPACE) \tag{14}$$

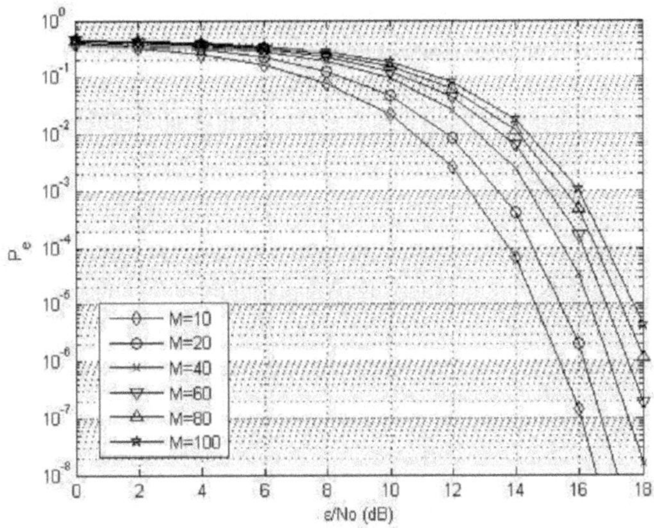


**Fig. 6.** Error probability with various M under AWGN channel

$$y = \int_0^{Ti} [r_C H(t)]^2 dt = \int_0^{Ti} [s(t) * h(t) + w(t)]^2 dt \quad under \ \ H_1(MARK) \ (15)$$

$r_{CH}(t)$ is the received signal, and $h(t)$ is the recommended channel impulse response. The delay spread of the residential environment channel equals to 180 $ns$. To avoid ISI, the half of signal set duration $T_s/2$ must be longer than the delay spread of the channel. The output of PEWB is a central or noncentral chi-square distributed random variable. Consequently, the procedure for deriving BER performance is similar to that of the Gaussian case except the noncentrality parameter $E$, because the means of $r_{CH}(t)$ are not equal to those of $r_{AWGN}(t)$ when MARK is sent.

### 3.3 Error Probability

Fig.6 shows the error probability $P_e$ versus $\frac{\varepsilon}{N_0}$ for various M under AWGN channel. According to the result, under AWGN channel, it is verified that noncoherent system with fewer number of WBs provide better BER performance than noncoherent system with larger number of WBs. Because we use the transmitted pulse of 1 $ns$ duration and AWGN channel has no delay spread characteristic, the power is concentrated on the first 1 $ns$ of the received signal. More energy collection causes the noise energy increase without signal energy increase. Consequently, it leads to the BER degradation. However, the channel impulse response of residential environment causes the delay spread effect. Therefore, considerable amount of energy collection with appropriate number of WBs is needed in order to obtain the best BER performance. Fig.7 shows the error probability for various M under LOS residential channel environment. From the figure, we know

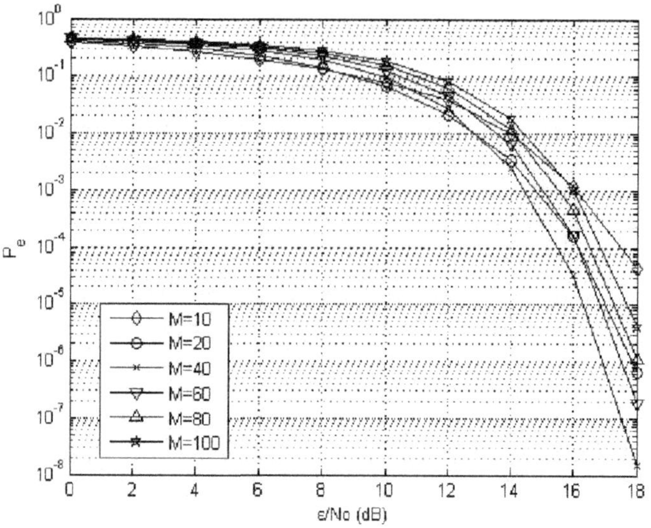


**Fig. 7.** Error probability with various M under LOS residential channel environment

that the best BER performance is achieved when 80 WBs (equal to integration time of 80 *ns*) are used. The integrated power to total received power ratios for M=10, 40, and 100 are 40%, 98%, and 100%, respectively.

## 4 Conclusion

In this paper, we implemented a wireless asynchronous non-coherent UWB transceiver with the object of reducing the power consumption and the hardware complexity. We did not employ any correlator of impulse signals and the fast A/D converter, but adopted non-coherent and asynchronous receiving method adequate. NRZ signals for binary data was utilized and it was transmitted by impulses with the pulse duration shorter than 2ns. The implemented prototype testbed is proven to offer the data rate of 9.6kbps and 115kbps on the conditions of Peer-to-Peer (P-to-P) in the indoor channel within the range of about 10m. In addition, we also analyzed the bit error rate (BER) performance of the proposed system under the line-of-sight (LOS) residential channel environment. As a result, it is necessary to choose an appropriate number of Window Bank(WB)s in order to obtain better BER performance.

## References


1. Toodor, C.: Wireless Communication Standards. Standards Information Network, IEEE Press(2004)
2. First Report and Order in the Matter of Revision of Part 15 of the Commission's Rules Regarding Ultra-Wideband Transmission Systems. FCC, released, ET Docket 98-153, FCC02-48, April 22, 2002

3. Scholtz, R.A., Win, M.Z.: Impulse Radio. In Wireless Commun., TDMA Versus CDMA, S.Glisic and P. Leppanen, Eds. London, U.K.: Kluwer(1997) 245-263
4. Bluetooth Special Interest Group.: Core Specification Version 1.1. Specification of the Bluetooth System (2001)
5. IEEE Std 802.15.4 IEEE Standard for Information Technology-Telecommunications and Information Exchange between Systems-local and Metropolitan Area Networks- Specific requirements, Part 15.4: Wireless Medium Access Control (MAC) and Physical Layer (PHY) Specifications for Low-Rate Wireless Personal Area Networks (LR-WPAN) (2003)
6. Molisch, A.F., et al.: IEEE 802.15.4a Channel Model. in Standardization Final Report of IEEE 802.15.4a Subgroup, Feb(2005)
7. Saleh, A., Valenzuela, R.A.: A Statistical Model for Indoor Multipath Propagation," IEEE J. Selected Areas Comm., vol.5 (1987) 128-137
8. Federal Communications Commission.: Part 15-Radio Frequency Devices. (2004)846-848
9. Proakis, J.G.: Digital Communications. McGraw-Hill International, 4th ed.(2001)
10. Paquelet, S., Aubert, L.M., Uguen, B.: An Impulse Radio Asynchronous Transceiver for High Data Rates," Joint UWBST & IWUWBS 2004 (2004)1-5
11. Humblet. P.A., Azizoglu, M.: On the Bit Error Rate of Lightwave Systems with Optical Amplifiers. Lightwave Technology, vol. 9 (1991) 1576-1582
12. Win, M.Z., Scholtz, R.A.: Ultra Wideband Bandwidth Time-hopping SS Impulse Radio for Wireless Multiple Access Communicatinos," IEEE Trans. Comm., vol. 48, No. 4 (2000)679-689
13. Win, M.Z., Scholtz, R.A.: On the Robustness of Ultra-Wide Bandwidth Signals in Dense Multipath Environments," IEEE Commun. Lett., vol. 2 (1998) 51-53

# MemoPA: Intelligent Personal Assistant Agents with a Case Memory Mechanism

Ke-Jia Chen and Jean-Paul Barthès

CMR UMR6599 HEUDIASYC, Computer Science Department, Université de Technologie de Compiègne, France
{chenkeji, barthes}@utc.fr

**Abstract.** A Personal Assistant (PA) agent is a software agent capable of helping people to handle tasks in their workplace. The paper proposes a memory mechanism for personal assistant agents in order to enhance agent intelligence while working with the user or with other agents. Inspired by a case memory model in the domain of Case-Based Reasoning (CBR), this paper endows PA agents with a case memory mechanism, which results in improved PA agents: MemoPAs. We present the memory mechanism of MemoPA in detail, and report a first implementation of the method. Finally, future work is outlined for improving the memory mechanism.

**Keywords:** Personal Assistant Agent, Case-Based Reasoning, Memory Model.

## 1 Introduction

Application of autonomous agents in the field of human-computer interaction gave birth to personal assistant agents (PA). Compared to the conventional direct-manipulation approach, interacting with an agent has potential advantages in productivity and creativity [1]. As a result, the PA agent has recently become one of the most compelling visions for agent research [2].

PA agents aim at reducing the workload of the user by handling tasks on her behalf, hiding the task complexity, learning and adapting to the user's preferences [3] [4] [5] as well as collaborating with other agents [6]. Two main challenges in this area are: first, how to know the user, and second, how to assist the user. Most of the previous work has contributed to the first challenge by improving the interaction between users and PA agents, learning user's preferences and goals, providing help at the right time and so on. However, only knowing the user is not enough for solving all problems. A good PA agent is supposed to have some knowledge related to the tasks to be done. Problems can be solved in an intelligent way by reasoning, making inferences and learning [7]. Currently, there is not much work along this approach.

The paper presents an attempt to improve the competence of a PA agent by making it more intelligent. Considering that the PA agent usually performs a substantial amount of repetitive tasks, previous experiences can be used to handle similar future situations. Inspired by the early work in the domain of Case-Based Reasoning (CBR), we propose to endow PA agents with a case memory mechanism, which results in MemoPA.

D.-S. Huang, L. Heutte, and M. Loog (Eds.): ICIC 2007, LNAI 4682, pp. 1357–1367, 2007.

In this paper, the memory of MemoPA is implemented using MOSS, a knowledge representation tool based on a representation model, PDM [8]. The memory mechanism of MemoPA is then designed, developed and tested using a multi-agent platform named OMAS [9]. Two examples of using MemoPA are described showing that a memory mechanism is beneficial to enhancing the intelligence of a PA agent.

The paper is organized as follows: Section 2 contains a brief introduction of CBR and case memory models. Section 3 describes in detail how to integrate CBR ideas into PA agents. Section 4 shows a first attempt for implementing MemoPA, and gives a brief discussion. Finally, the paper concludes and raises several issues for future work.

## 2 Case Memory in CBR

The main assumption of CBR is that the world is regular: events recur with regularity and similar problems have similar solutions. Generally, CBR emphasizes the use of concrete instances (cases) over abstract operators (rules or norms). Reasoning is regarded as a process of remembering one or several past cases according to a current situation. Such remembered cases are later used to suggest solutions to new problems, to warn of possible failures, or to interpret the current situation. Once the new problem is solved, a new case can be retained appropriately and integrated into the case memory.

For PA agents, CBR is appealing for several reasons. On one hand, the reasoning process is relatively simple; solutions can be quickly proposed avoiding recomputing them from scratch. On the other hand, user requests to a PA agent can be diverse and complex. Therefore, it is unfeasible to generalize rules or norms covering all possible situations and to collect enough data for training a system before using it. CBR can deal with this uncertain problem-solving world in a more efficient way.

### 2.1 Issues in CBR

In order to better use CBR in PA agents, several issues must be considered, such as case representation and organization, case retrieval and other problems like adaptation and maintenance of the case base.

**Case Representation.** A case is a contextualised piece of knowledge that represents an experience. Two main kinds of knowledge are included in a case [10]: the content characterizing the situation and the context in which the case can be used.

Rather than a simple feature vector, a case can be represented in a variety of forms like frames [11] or objects. A structured case memory is preferred to a flat case base because it can later facilitate remembering.

In this paper, individual cases (problem-solving episodes) are represented and organised in a case memory model, which will be introduced in Section 2.2.

**Case Retrieval.** In the CBR literature, past cases are retrieved by identifying descriptors of a current problem. They are often called indexes. Desired indexes should both preserve the richness of cases and simplify case access.

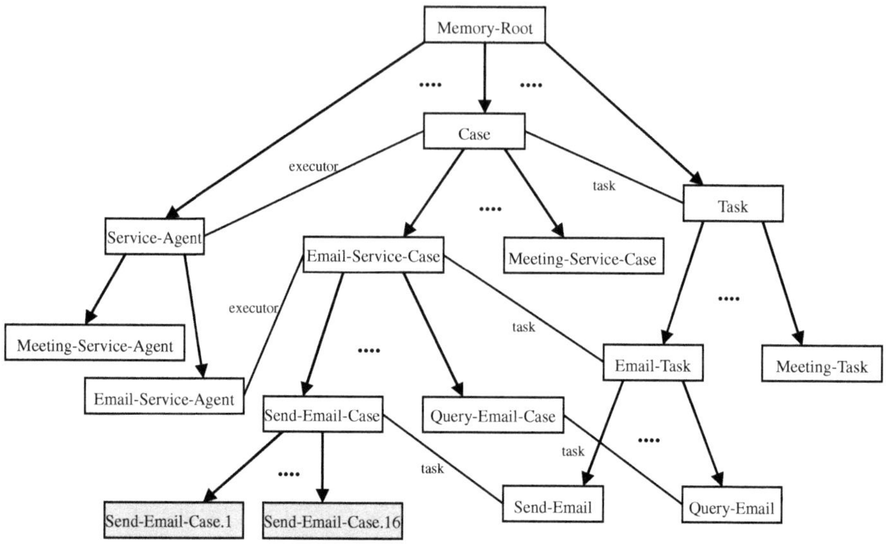


**Fig. 1.** A segment of the MOPs-based memory for a PA agent

As for similarity assessment methods, the nearest neighbor (NN) algorithm is conventionally used. The similarity depends on how well the cases match on each dimension and how important each dimension really is. But NN is usually more effective when the case base is relatively small.

**Other issues.** Case adaptation is always regarded as a challenge of CBR. But sometimes it could be a distraction from the fundamental power of CBR, namely, simplicity. Currently, no adaptation measure is considered in our work.

Eventually, once a new case is generated and added into the case memory, the memory will change dynamically. A set of cases can be generalized as more abstract knowledge and outdated cases are removed.

### 2.2 Case Memory Model

CBR integrates all the above issues with memory processes [12]. Choosing a case memory model is essential to PA agents, to provide a base for reasoning and learning.

Two influential case memory models are the dynamic memory model of Schank and Kolodner and the category-exemplar model of Porter and Bareiss [10]. In the dynamic memory model [13], the case memory is a hierarchical structure of what is called episodic memory organization packets. The memory can be dynamically re-organized by adding new cases. In the category-exemplar model, the memory is embedded in a network structure of categories, cases and index pointers. The model is mainly used to represent general domain knowledge rather than episodes or specific domain knowledge.

Considering the domain-related problem-solving characteristic of PA agents, the paper structures the case memory based on dynamic memory model. Both domain knowledge and episodes are translated into memory organization packets (MOPs) [14] with two kinds of relations: abstraction relations and packaging relations (pairs of role and filler, where the filler can also be a MOP).

Fig. 1 shows a segment of the MOPs-based memory designed for a PA agent. Arrows represent abstraction relations and lines represent packaging relations. All the episodic cases are located in the lowest level of the MOPs hierarchy.

With this organization, case retrieval is done by following the abstraction links to find the most appropriate MOP. New episode cases could be inserted into the memory (e.g. *Send-Email-Case.16* in Fig. 1), and abstraction cases can be created manually or automatically (e.g. *Email-Service-Case* and *Send-Email-Case* in Fig. 1)

Overall the MOPs-based memory model includes rich contextualized cases, demands less storage and allows a more efficient search strategy. Therefore, it was selected to be the prototype of memory mechanism in MemoPA.

# 3 MemoPA

The current section presents a task-oriented intelligent PA agent named MemoPA in details. Having a case memory mechanism, MemoPA behaves in a more intelligent way.

## 3.1 An Example

A simple example below explains how memory can enhance the intelligence of a MemoPA in an application concerning e-Government.

One day, Susan came to visit a civil servant named Tom for applying for her house allowance. Tom asked his MemoPA named SMART to handle Susan's request.

Being informed of its job, SMART searched its memory and "remembered" that a similar event had happened before. In that case, a citizen named Paul had successfully applied for his house allowance: SMART first finds a local service agent named HA who processes house allowance applications. Before processing the task, Paul's personal information is required. Unfortunately, there is no available information locally. So SMART asks other agents in the workplace to handle it. Finally, another PA agent named SMART2 replied with Paul's information because it is connected with a service agent named CD that manages citizens' documents.

Based on this experience, SMART now asks SMART2 directly for Susan's information. Once acquired, the information is sent to HA immediately since SMART now knows that HA is capable of processing the task.

Although the past similar cases cannot always be helpful because there are too many possibilities, MemoPA can at least suggest a potentially useful candidate solution.

## 3.2 Memory Mechanism in MemoPA

As mentioned before, the memory mechanism is an essential ingredient for the consciousness of MemoPA. Here two basic components are involved (Fig. 2): a case memory and a memory processor (to monitor and update the memory).

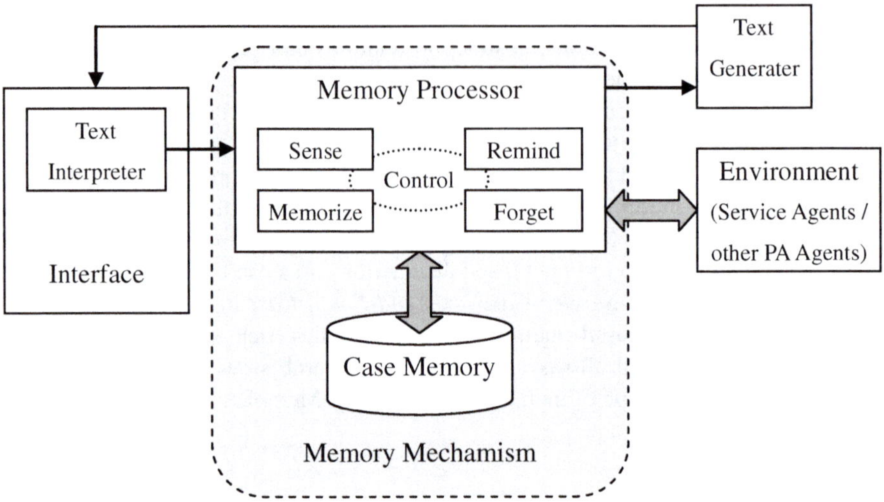


**Fig. 2.** A diagram of the memory mechanism in MemoPA

**Case Memory.** MemoPA records its task-related experiences into case memory. In the memory, domain-independent knowledge, domain-dependent knowledge, and specific episodic knowledge are uniformly represented in MOPs-like model.

Two types of information are included within an episodic case: indexed information (task type, task arguments, etc.) used for retrieval; contextual information (end time, memory owner, etc.) used for further reasoning.

In addition, we separate episodic cases into short-term cases and long-term cases. The former present a context for the task they currently process; the later store the processed tasks. In MemoPA, both kinds of cases are involved in task processing.

**Memory Processor.** The memory processor (Fig. 2) contains four cognitive functions (*Sense*, *Remind*, *Memorize* and *Forget*) and a *Control*.

The *Sense* function works when a message arrives from the user or other agents. Afterwards, either a new short-term case is built and partly filled, or a corresponding old short-term case is renewed. Results of *Sense* might trigger other cognitive functions.

The *Remind* function plays an essential role in the memory mechanism. Once a short-term case of a new task is built, MemoPA will call *Remind* to find the most similar cases in the memory. Here, a two-step indexing method is proposed. The first step finds an abstract MOP with the related features that best match the current situation, and the second step uses a weighted NN method to select the most similar episodic cases under that MOP. If no case can be found, the task is processed from scratch like before.

The *Memorize* function is triggered once a task is finished. The short-term episodic case of that task is transformed into a long-term episodic case and then saved in the memory.

The *Forget* function removes some useless experiences in order to maintain the efficiency of the case memory. In the paper, only totally repetitive task cases are to be removed.

The *Control* makes all the cognitive functions work together.

Finally, the memory mechanism should be connected with other modules such as natural language processor in the interface of PA agents. In this case, the memory mechanism can even improve the interaction by providing a current task context in form of a short-term case (e.g. a list of salient features).

# 4 Implementation

In implementation, the case memory of MemoPA is represented using a knowledge representation tool named MOSS. Both the domain ontology and the case memory are represented by MOSS.

The memory mechanism is located within a PA agent architecture (Fig. 3) proposed by Barthès and Ramos [9]. The architecture consists of several modules. The *Ontology* module specifies a vocabulary and generic knowledge. The *Tasks* module shows what tasks are currently active. The *Skills* module gives a set of services it can provide. The *World* Module contains an internal representation of other agents. The *Self* module contains the memory mechanism of the PA agent. The *Master* module reflects the user's profile that is obtained by the user interface.

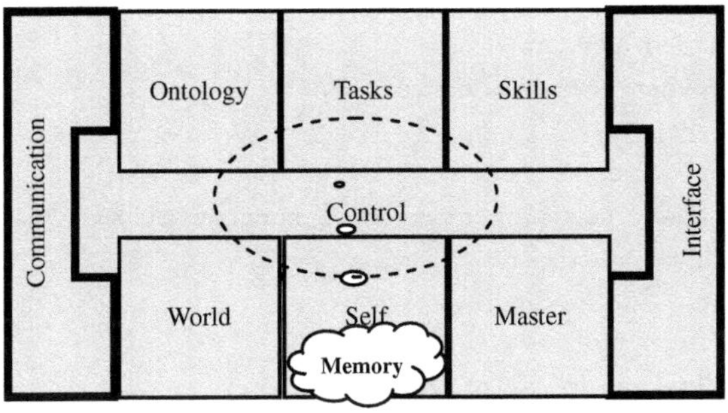


**Fig. 3.** An overall architecture of MemoPA

MemoPA works in a multi-agent platform named OMAS [9], which is programmed in the Allegro Common LISP in the Windows environment. In OMAS, PA agents and associated staff agents (a set of service agents devoted to a PA agent) are designed independently. Several PA agents can be connected in the same local *coterie* (a set of agents with same interests).

Presently, MemoPA is designed as an office assistant in order to verify the efficiency of the memory mechanism. Its ontology has basic information about the offered services. MemoPA is associated with two service agents: an Email Agent and a Meeting Agent. The Email Agent has skills for querying email addresses, sending emails and receiving emails, and the Meeting Agent has skills for arranging meetings, canceling meetings and querying meetings. Both service agents have their own domain ontology.

## 4.1 Interaction Samples

Currently, due to a lack of natural language processing module, MemoPA works in a simulated interactive environment. Here is an interaction scenario showing how the memory mechanism works. A detailed activity of MemoPA is given to make the process clearer.

User: Do I have meetings tomorrow?

Agent: Finds a similar experience: (Query-Meeting-Case.2 (:Executor "Meeting-Agent")). Creates a short-term case: (Query-Meeting-Case.6).

Agent: Sends to "Meeting-Agent" with arguments: (:date "25/05/07").

Agent: Yes. 10:00 with Tom.

Agent: (Query-Meeting-Case.6) becomes a long-term case with arguments: (:date "25/05/07") and result: (:success (meeting (:date "25/05/07" :hour "10" :attendees "Tom" :location nil :topic nil))).

User: Send an email to Tom.

Agent: No similar experience, asks other service agents.

Agent: "Email-Agent" replies with required task arguments: (:receiver :title :content). Creates a short-term case: (Send-Email-Case.1).

Agent: What's the title?

User: Meeting tomorrow.

Agent: What's the content?

User: Don't forget tomorrow's meeting.

Agent: Sends to "Email-Agent" with arguments: (:receiver "Tom" :title "Meeting tomorrow" :content "Don't forget tomorrow's meeting.").

Agent: Email Agent has sent this email successfully.

Agent: (Send-Email-Case.1) becomes a long-term case with arguments: (:receiver "Tom" :title "Meeting tomorrow" :content "Don't forget tomorrow's meeting.") and result: (:success nil).

Agent: What else can I do for you?

User: send another email to Kelly.

Agent: Finds a similar experience: (Send-Email-Case.1 (Executor: Email-Agent)). Creates a short-term case: (Send-Email-Case.2) with arguments: (:receiver "kelly" :title nil :content nil).

Agent: What's the title?

User: Hi

Agent: what's the content?

User: Can we have lunch tomorrow?

Agent: Sends to "Email-Agent" with arguments: (:receiver "Kelly" :title "Hi" :content "Can we have lunch tomorrow?").

Agent: Email agent has sent this email successfully.

Agent: (Send-Email-Case.2) becomes a long-term case with arguments: (:receiver "Kelly" :title "Hi" :content "Can we have lunch tomorrow?") and result: (:success nil).

Agent: What else can I do for you?

In the example above, when the user asked MemoPA to query meetings, MemoPA remembered a similar experience. So it sent the query task as well as its argument "time" to Meeting Agent directly. When MemoPA was first asked to "send an email to Tom", it had no "Send Email" experience in memory. So it asked all connected agents for help. A service agent named Email Agent replied with a list of task arguments. Once all the necessary information was filled, this task was sent to Email Agent again and processed. Such an experience was later filed in the memory, which is why "send another email to Kelly" was processed more easily.

Moreover, the memory mechanism of MemoPA can preview other possible situations in applications. For example, when the retrieved case is a fail experience, MemoPA will behave differently according to different failure types (no such service, no information, etc.). These experiences could avoid the similar failure in future.

The following interaction happens in another scenario that presents not only the role of long-term memory case in agent reasoning but also the role of short-term memory case in user-agent conversation.

User: Arrange a meeting for tomorrow at 10 am.

Agent: Searches in the memory and finds a similar experience: (Arrange-Meeting-Case.7 (Executor: Meeting-Agent)). Creates a short-term case (Arrange-Meeting-Case.11) with the arguments: (:date "25/05/07" :hour "10" :attendees nil :location nil :topic nil).

Agent: Checks the arguments: (:date :hour) in (Arrange-Meeting-Case.7). (:date "25/05/07" :hour "10").

Agent: Sorry, you have arranged a meeting at that time. What about another time?

User: 3 pm.

Agent: Searches in the memory and finds a similar experience. (Arrange-Meeting-Case.8 (Executor: Meeting-Agent))

Agent: Checks the arguments (:date :hour) in (Arrange-Meeting-Case.8). (:date "25/05/07" :hour "16").

Agent: OK.

Agent: Continues the short-term case: (Arrange-Meeting-Case.11) with the arguments: (:date "25/05/07" :hour "15" :attendees nil :location nil :topic nil).

Agent: Who are the attendees?

User: Tom and Mike.

Agent: Where will the meeting be held?

User: Room 220.

Agent: What's the topic?

User: About the use of ontology in agent systems.

Agent: Sends to "Meeting-Agent" with arguments: (:date "25/05/07" :hour "15" :attendees ("Tom" "Mike") :location "Room 220" :topic "About the use of ontology in agent systems").

Agent: Meeting agent has arranged this meeting successfully.

Agent: (Arrange-Meeting-Case.11) becomes a long-term case with arguments: (:date "25/05/07" :hour "15" :attendees ("Tom" "Mike") :location "Room

```
220" :topic "About the use of ontology in agent systems") and result:
(:success nil).
  Agent: What else can I do for you?
```

MemoPA in the above example was asked to arrange a meeting. From a similar experience, MemoPA recognized the task as an "Arrange Meeting" task. But a conflict of time was detected between the past case and the current one. So MemoPA had to reschedule the time with the user. Afterwards, using the context provided in the short-term case, MemoPA can take the "Arrange Meeting" task back, meaning that the agent can remember what just happened without asking the user to repeat it.

Notice that for each task, the agent needs different reasoning rules to use memory cases. In the "Arrange Meeting" task, for example, MemoPA checks the time arguments from the remembered case to avoid time conflicts. While in the "Query Meeting" task, the agent only checks the arguments and the result from the remembered case to propose the answer directly.

### 4.2 Discussion

Actually, the task like "send email" is relatively simple so the advantage of using a memory is not quite impressive. For the complicate tasks with a lot of steps or communications like the example in Section 3, MemoPA could make more favorable improvements. Besides, in the emerging field of pedagogical agents [15], the teaching agents can also benefit from a memory mechanism. On one side, the memory can provide the interpretation of learners' responses by remembering relevant teaching experiences. On the other side, the agent can adapt its teaching plan according to the differences between learners' current learning results and past ones in the memory.

In some related work [6] [16], the agent memory has been organized to predict future actions. However, the term "memory" in those papers usually refers to a sensory memory restricted to simple states and actions. They might not be appropriate for complicated tasks with many possibilities. Another related work [17] implements a case-based memory for an interface agent operating in a process plant. But this dialogue agent works in a quite specific domain without interacting with a multi-agent environment. In this paper, with a well-formed knowledge representation and a multi-agent framework, MemoPA could better solve the problems in practice.

## 5 Conclusion

The paper proposed a memory mechanism for a PA agent in order to enhance its intelligence in assisting users. After a brief review of CBR and case memory models, a memory mechanism for MemoPA was presented. The method is later implemented in a MemoPA prototype to handle office affaires. Compared to other related work, MemoPA has advantages in knowledge representation and agent cooperation.

Several main efforts are considered for future work:

First, a powerful dialogue mechanism will be developed. The memory can be used to solve the problems of conversation interruption and pronoun references. In this case, short-term cases will be further studied since they represent current situations.

The second important effort aims at better using memory cases, especially the long-term cases. For a MemoPA, the capability of reasoning is mostly decided by the manner of case utilizations.

Third, the learning ability of MemoPA will be improved. Current abstraction cases are built manually. When the sum of episodic cases becomes large, abstraction cases should be created automatically using inductive learning methods.

Last, in order to better present the memory mechanism, some evaluation methods will be studied. In task-based agent experimentation, the comparison of task processing time or result correctness could be used to evaluate the MemoPA approach.

The work will finally be applied to the data from a European project named TerreGov [18] in the domain of e-Government.

**Acknowledgments.** This research is funded by the Chinese Government Scholarship in UT-INSA program and uses data from the TerreGov Project (IST-507749).

## References

1. Lieberman, H.: Autonomous Interface Agents. In: Pemberton, S. (eds.): Proceedings of the SIGCHI Conference on Human Factors in Computing Systems. ACM Press, New York (1997) 67–73
2. Modi, P.J., Veloso, M., Smith, S.F., Oh, J.: CMRadar: A Personal Assistant Agent for Calendar Management. Agent-Oriented Information Systems. In: Bresciani, P., Giorgini, P., Henderson-Sellers, B. (eds.): Agent-Oriented Information Systems II. Lecture Notes in Computer Science, Vol. 2. Springer-Verlag, Berlin Heidelberg New York (2005) 169–181
3. Maes, P.: Agents that Reduce Work and Information Overload. Communications of the ACM, Vol. 37, No. 7. ACM Press, New York (1994) 30–40
4. Maes, P., Kozierok, R.: Learning Interface Agents. In: Fikes, R., Lehnert, W. (eds.): Proceedings of the 11th National Conference on Artificial Intelligence. AAAI Press/MIT Press, Washington DC (1993) 459–465
5. Mitchell, T., Caruana, R., Freitag, D., McDermott, J., Zabowski, D.: Experience with a Learning Personal Assistant. Communications of the ACM, Vol. 37, No. 7. ACM Press, New York (1994) 81–91
6. Lashkari, Y., Metral, M., Maes, P.: Collaborative Interface Agents. In: Hayes-Roth, B., Korf, R. (eds.): Proceedings of the 12th National Conference on Artificial Intelligence. AAAI Press, Seattle, WA (1994) 444–449
7. Wobcke, W.R., Ho, V., Nguyen, A., Krzywicki, A.: A BDI Agent Architecture for Dialogue Modeling and Coordination in a Smart Personal Assistant. In: Skowron, A., Barthès, J.-P.A., Jain, L., Sun, R., Mahoudeaux, P.M., Liu, J., Zhong, N. (eds.): Proceedings of the IEEE/WIC/ACM International Conference on Web Intelligence and Intelligent Agent Technology. IEEE press, Compiègne, France (2005) 323–329
8. Shen, W., Barthès, J.-P.A.: Description and Applications of an Object-Oriented Model PDM. Modeling Complex Data for Creating Information: Real and Virtual Objects. (1995) 15–24
9. Barthès, J.-P.A., Ramos, M.: Agents Assistants Personnels dans les Systèmes Multi-Agents Mixtes - Réalisation sur la Plate-forme OMAS. In : Briffault, X., Guessoum, Z., Occello, M. (eds.): Technique et Science Informatiques, Vol. 21, No. 4. Paris, France (2002) 473–498

10. Watson, I., Marir, F.: Case-Based Reasoning: A Review. In: McBurney, P., Parsons S. (eds.): Knowledge Engineering Review, Vol. 9, No. 4. Cambridge University Press, Cambridge, UK (1994) 355–381
11. Minsky, M.: Form and Content in Computer Science. Journal of the ACM, Vol. 17, No. 2. ACM Press, New York (1970) 197–215
12. Kolodner, J.L.: Case-Based Reasoning. Morgan Kaufmann, San Mateo, CA (1993)
13. Schank, R.C.: Dynamic Memory: A Theory of Reminding and Learning in Computers and People. Cambridge University Press, Cambridge, UK (1982)
14. Riesbeck, C., Schank, R.: Inside Case-Based Reasoning. Lawrence Erlbaum, Hillsdale, NJ (1989)
15. Johnson W.L.: Pedagogical Agent Research at CARTE. AI Magazine, Vol. 22, No. 4. AAAI Press, Menlo Park, CA (2001) 85–94
16. Lerman, K., Galstyan, A.: Agent Memory and Adaptation in Multi-Agent Systems. In: Proceedings of the 2nd International Joint Conference on Autonomous Agents and Multi-Agent Systems. ACM Press, New York (2003) 797–803
17. Koide, S., Yamauchi, S.: Interface Agent for Process Plant Operation towards the Next Generation Interface. In: Proceedings of the IEEE/ASME International Conference on Advanced Intelligent Mechatronics. IEEE Press, Atlanta, GA (1999) 197–202
18. TERREGOV: Impact of E-Government on Territorial Government Service (2004)

# Author Index

# Lecture Notes in Artificial Intelligence (LNAI)

Vol. 4682: D.-S. Huang, L. Heutte, M. Loog (Eds.), Advanced Intelligent Computing Theories and Applications. XXVII, 1373 pages. 2007.

Vol. 4660: S. Džeroski, J. Todorovski (Eds.), Computational Discovery of Scientific Knowledge. X, 327 pages. 2007.

Vol. 4651: F. Azevedo, P. Barahona, F. Fages, F. Rossi (Eds.), Recent Advances in Constraints. VIII, 185 pages. 2007.

Vol. 4632: R. Alhajj, H. Gao, X. Li, J. Li, O.R. Zaïane (Eds.), Advanced Data Mining and Applications. XV, 634 pages. 2007.

Vol. 4626: R.O. Weber, M.M. Richter (Eds.), Case-Based Reasoning Research and Development. XIII, 534 pages. 2007.

Vol. 4617: V. Torra, Y. Narukawa, Y. Yoshida (Eds.), Modeling Decisions for Artificial Intelligence. XII, 502 pages. 2007.

Vol. 4612: I. Miguel, W. Ruml (Eds.), Abstraction, Reformulation, and Approximation. XI, 418 pages. 2007.

Vol. 4604: U. Priss, S. Polovina, R. Hill (Eds.), Conceptual Structures: Knowledge Architectures for Smart Applications. XII, 514 pages. 2007.

Vol. 4603: F. Pfenning (Ed.), Automated Deduction – CADE-21. XII, 522 pages. 2007.

Vol. 4597: P. Perner (Ed.), Advances in Data Mining. XI, 353 pages. 2007.

Vol. 4594: R. Bellazzi, A. Abu-Hanna, J. Hunter (Eds.), Artificial Intelligence in Medicine. XVI, 509 pages. 2007.

Vol. 4585: M. Kryszkiewicz, J.F. Peters, H. Rybinski, A. Skowron (Eds.), Rough Sets and Intelligent Systems Paradigms. XIX, 836 pages. 2007.

Vol. 4578: F. Masulli, S. Mitra, G. Pasi (Eds.), Applications of Fuzzy Sets Theory. XVIII, 693 pages. 2007.

Vol. 4573: M. Kauers, M. Kerber, R. Miner, W. Windsteiger (Eds.), Towards Mechanized Mathematical Assistants. XIII, 407 pages. 2007.

Vol. 4571: P. Perner (Ed.), Machine Learning and Data Mining in Pattern Recognition. XIV, 913 pages. 2007.

Vol. 4570: H.G. Okuno, M. Ali (Eds.), New Trends in Applied Artificial Intelligence. XXI, 1194 pages. 2007.

Vol. 4565: D.D. Schmorrow, L.M. Reeves (Eds.), Foundations of Augmented Cognition. XIX, 450 pages. 2007.

Vol. 4562: D. Harris (Ed.), Engineering Psychology and Cognitive Ergonomics. XXIII, 879 pages. 2007.

Vol. 4548: N. Olivetti (Ed.), Automated Reasoning with Analytic Tableaux and Related Methods. X, 245 pages. 2007.

Vol. 4539: N.H. Bshouty, C. Gentile (Eds.), Learning Theory. XII, 634 pages. 2007.

Vol. 4529: P. Melin, O. Castillo, L.T. Aguilar, J. Kacprzyk, W. Pedrycz (Eds.), Foundations of Fuzzy Logic and Soft Computing. XIX, 830 pages. 2007.

Vol. 4511: C. Conati, K. McCoy, G. Paliouras (Eds.), User Modeling 2007. XVI, 487 pages. 2007.

Vol. 4509: Z. Kobti, D. Wu (Eds.), Advances in Artificial Intelligence. XII, 552 pages. 2007.

Vol. 4496: N.T. Nguyen, A. Grzech, R.J. Howlett, L.C. Jain (Eds.), Agent and Multi-Agent Systems: Technologies and Applications. XXI, 1046 pages. 2007.

Vol. 4483: C. Baral, G. Brewka, J. Schlipf (Eds.), Logic Programming and Nonmonotonic Reasoning. IX, 327 pages. 2007.

Vol. 4482: A. An, J. Stefanowski, S. Ramanna, C.J. Butz, W. Pedrycz, G. Wang (Eds.), Rough Sets, Fuzzy Sets, Data Mining and Granular Computing. XIV, 585 pages. 2007.

Vol. 4481: J. Yao, P. Lingras, W.-Z. Wu, M. Szczuka, N.J. Cercone, D. Ślęzak (Eds.), Rough Sets and Knowledge Technology. XIV, 576 pages. 2007.

Vol. 4476: V. Gorodetsky, C. Zhang, V.A. Skormin, L. Cao (Eds.), Autonomous Intelligent Systems: Multi-Agents and Data Mining. XIII, 323 pages. 2007.

Vol. 4455: S. Muggleton, R. Otero, A. Tamaddoni-Nezhad (Eds.), Inductive Logic Programming. XII, 456 pages. 2007.

Vol. 4452: M. Fasli, O. Shehory (Eds.), Agent-Mediated Electronic Commerce. VIII, 249 pages. 2007.

Vol. 4451: T.S. Huang, A. Nijholt, M. Pantic, A. Pentland (Eds.), Artifical Intelligence for Human Computing. XVI, 359 pages. 2007.

Vol. 4441: C. Müller (Ed.), Speaker Classification. X, 309 pages. 2007.

Vol. 4438: L. Maicher, A. Sigel, L.M. Garshol (Eds.), Leveraging the Semantics of Topic Maps. X, 257 pages. 2007.

Vol. 4434: G. Lakemeyer, E. Sklar, D.G. Sorrenti, T. Takahashi (Eds.), RoboCup 2006: Robot Soccer World Cup X. XIII, 566 pages. 2007.

Vol. 4429: R. Lu, J.H. Siekmann, C. Ullrich (Eds.), Cognitive Systems. X, 161 pages. 2007.

Vol. 4428: S. Edelkamp, A. Lomuscio (Eds.), Model Checking and Artificial Intelligence. IX, 185 pages. 2007.

Vol. 4426: Z.-H. Zhou, H. Li, Q. Yang (Eds.), Advances in Knowledge Discovery and Data Mining. XXV, 1161 pages. 2007.

Vol. 4411: R.H. Bordini, M. Dastani, J. Dix, A.E.F. Seghrouchni (Eds.), Programming Multi-Agent Systems. XIV, 249 pages. 2007.

Vol. 4410: A. Branco (Ed.), Anaphora: Analysis, Algorithms and Applications. X, 191 pages. 2007.

Vol. 4399: T. Kovacs, X. Llorà, K. Takadama, P.L. Lanzi, W. Stolzmann, S.W. Wilson (Eds.), Learning Classifier Systems. XII, 345 pages. 2007.

Vol. 4390: S.O. Kuznetsov, S. Schmidt (Eds.), Formal Concept Analysis. X, 329 pages. 2007.

Vol. 4389: D. Weyns, H.V.D. Parunak, F. Michel (Eds.), Environments for Multi-Agent Systems III. X, 273 pages. 2007.

Vol. 4384: T. Washio, K. Satoh, H. Takeda, A. Inokuchi (Eds.), New Frontiers in Artificial Intelligence. IX, 401 pages. 2007.

Vol. 4371: K. Inoue, K. Satoh, F. Toni (Eds.), Computational Logic in Multi-Agent Systems. X, 315 pages. 2007.

Vol. 4369: M. Umeda, A. Wolf, O. Bartenstein, U. Geske, D. Seipel, O. Takata (Eds.), Declarative Programming for Knowledge Management. X, 229 pages. 2006.

Vol. 4343: C. Müller (Ed.), Speaker Classification. X, 355 pages. 2007.

Vol. 4342: H. de Swart, E. Orłowska, G. Schmidt, M. Roubens (Eds.), Theory and Applications of Relational Structures as Knowledge Instruments II. X, 373 pages. 2006.

Vol. 4335: S.A. Brueckner, S. Hassas, M. Jelasity, D. Yamins (Eds.), Engineering Self-Organising Systems. XII, 212 pages. 2007.

Vol. 4334: B. Beckert, R. Hähnle, P.H. Schmitt (Eds.), Verification of Object-Oriented Software. XXIX, 658 pages. 2007.

Vol. 4333: U. Reimer, D. Karagiannis (Eds.), Practical Aspects of Knowledge Management. XII, 338 pages. 2006.

Vol. 4327: M. Baldoni, U. Endriss (Eds.), Declarative Agent Languages and Technologies IV. VIII, 257 pages. 2006.

Vol. 4314: C. Freksa, M. Kohlhase, K. Schill (Eds.), KI 2006: Advances in Artificial Intelligence. XII, 458 pages. 2007.

Vol. 4304: A. Sattar, B.-h. Kang (Eds.), AI 2006: Advances in Artificial Intelligence. XXVII, 1303 pages. 2006.

Vol. 4303: A. Hoffmann, B.-h. Kang, D. Richards, S. Tsumoto (Eds.), Advances in Knowledge Acquisition and Management. XI, 259 pages. 2006.

Vol. 4293: A. Gelbukh, C.A. Reyes-Garcia (Eds.), MICAI 2006: Advances in Artificial Intelligence. XXVIII, 1232 pages. 2006.

Vol. 4289: M. Ackermann, B. Berendt, M. Grobelnik, A. Hotho, D. Mladenič, G. Semeraro, M. Spiliopoulou, G. Stumme, V. Svátek, M. van Someren (Eds.), Semantics, Web and Mining. X, 197 pages. 2006.

Vol. 4285: Y. Matsumoto, R.W. Sproat, K.-F. Wong, M. Zhang (Eds.), Computer Processing of Oriental Languages. XVII, 544 pages. 2006.

Vol. 4274: Q. Huo, B. Ma, E.-S. Chng, H. Li (Eds.), Chinese Spoken Language Processing. XXIV, 805 pages. 2006.

Vol. 4265: L. Todorovski, N. Lavrač, K.P. Jantke (Eds.), Discovery Science. XIV, 384 pages. 2006.

Vol. 4264: J.L. Balcázar, P.M. Long, F. Stephan (Eds.), Algorithmic Learning Theory. XIII, 393 pages. 2006.

Vol. 4259: S. Greco, Y. Hata, S. Hirano, M. Inuiguchi, S. Miyamoto, H.S. Nguyen, R. Słowiński (Eds.), Rough Sets and Current Trends in Computing. XXII, 951 pages. 2006.

Vol. 4253: B. Gabrys, R.J. Howlett, L.C. Jain (Eds.), Knowledge-Based Intelligent Information and Engineering Systems, Part III. XXXII, 1301 pages. 2006.

Vol. 4252: B. Gabrys, R.J. Howlett, L.C. Jain (Eds.), Knowledge-Based Intelligent Information and Engineering Systems, Part II. XXXIII, 1335 pages. 2006.

Vol. 4251: B. Gabrys, R.J. Howlett, L.C. Jain (Eds.), Knowledge-Based Intelligent Information and Engineering Systems, Part I. LXVI, 1297 pages. 2006.

Vol. 4248: S. Staab, V. Svátek (Eds.), Managing Knowledge in a World of Networks. XIV, 400 pages. 2006.

Vol. 4246: M. Hermann, A. Voronkov (Eds.), Logic for Programming, Artificial Intelligence, and Reasoning. XIII, 588 pages. 2006.

Vol. 4223: L. Wang, L. Jiao, G. Shi, X. Li, J. Liu (Eds.), Fuzzy Systems and Knowledge Discovery. XXVIII, 1335 pages. 2006.

Vol. 4213: J. Fürnkranz, T. Scheffer, M. Spiliopoulou (Eds.), Knowledge Discovery in Databases: PKDD 2006. XXII, 660 pages. 2006.

Vol. 4212: J. Fürnkranz, T. Scheffer, M. Spiliopoulou (Eds.), Machine Learning: ECML 2006. XXIII, 851 pages. 2006.

Vol. 4211: P. Vogt, Y. Sugita, E. Tuci, C.L. Nehaniv (Eds.), Symbol Grounding and Beyond. VIII, 237 pages. 2006.

Vol. 4203: F. Esposito, Z.W. Raś, D. Malerba, G. Semeraro (Eds.), Foundations of Intelligent Systems. XVIII, 767 pages. 2006.

Vol. 4201: Y. Sakakibara, S. Kobayashi, K. Sato, T. Nishino, E. Tomita (Eds.), Grammatical Inference: Algorithms and Applications. XII, 359 pages. 2006.

Vol. 4200: I.F.C. Smith (Ed.), Intelligent Computing in Engineering and Architecture. XIII, 692 pages. 2006.

Vol. 4198: O. Nasraoui, O. Zaïane, M. Spiliopoulou, B. Mobasher, B. Masand, P.S. Yu (Eds.), Advances in Web Mining and Web Usage Analysis. IX, 177 pages. 2006.

Vol. 4196: K. Fischer, I.J. Timm, E. André, N. Zhong (Eds.), Multiagent System Technologies. X, 185 pages. 2006.

Vol. 4188: P. Sojka, I. Kopeček, K. Pala (Eds.), Text, Speech and Dialogue. XV, 721 pages. 2006.

Vol. 4183: J. Euzenat, J. Domingue (Eds.), Artificial Intelligence: Methodology, Systems, and Applications. XIII, 291 pages. 2006.

Vol. 4180: M. Kohlhase, OMDoc – An Open Markup Format for Mathematical Documents [version 1.2]. XIX, 428 pages. 2006.

GPSR Compliance

*The European Union's (EU) General Product Safety Regulation (GPSR) is a set of rules that requires consumer products to be safe and our obligations to ensure this.*

*If you have any concerns about our products, you can contact us on ProductSafety@springernature.com*

In case Publisher is established outside the EU, the EU authorized representative is:

Springer Nature Customer Service Center GmbH
Europaplatz 3
69115 Heidelberg, Germany

**Batch number: 10755036**

Printed by Printforce, the Netherlands